18. $\displaystyle\int \frac{u^2\,du}{\sqrt{a+bu}} = \frac{2}{15b^3}(3b^2u^2 - 4abu + 8a^2)\sqrt{a+bu} + C$

19. $\displaystyle\int \frac{u^n\,du}{\sqrt{a+bu}} = \frac{2u^n\sqrt{a+bu}}{b(2n+1)} - \frac{2an}{b(2n+1)}\int \frac{u^{n-1}\,du}{\sqrt{a+bu}}$

20. $\displaystyle\int \frac{du}{u\sqrt{a+bu}} = \begin{cases} \dfrac{1}{\sqrt{a}}\ln\left|\dfrac{\sqrt{a+bu}-\sqrt{a}}{\sqrt{a+bu}+\sqrt{a}}\right| + C & \text{if } a>0 \\[3mm] \dfrac{2}{\sqrt{-a}}\tan^{-1}\sqrt{\dfrac{a+bu}{-a}} + C & \text{if } a<0 \end{cases}$

21. $\displaystyle\int \frac{du}{u^n\sqrt{a+bu}} = -\frac{\sqrt{a+bu}}{a(n-1)u^{n-1}} - \frac{b(2n-3)}{2a(n-1)}\int \frac{du}{u^{n-1}\sqrt{a+bu}}$

22. $\displaystyle\int \frac{\sqrt{a+bu}\,du}{u} = 2\sqrt{a+bu} + a\int \frac{du}{u\sqrt{a+bu}}$

23. $\displaystyle\int \frac{\sqrt{a+bu}\,du}{u^n} = -\frac{(a+bu)^{3/2}}{a(n-1)u^{n-1}} - \frac{b(2n-5)}{2a(n-1)}\int \frac{\sqrt{a+bu}\,du}{u^{n-1}}$

Forms Containing $a^2 \pm u^2$

24. $\displaystyle\int \frac{du}{a^2+u^2} = \frac{1}{a}\tan^{-1}\frac{u}{a} + C$

25. $\displaystyle\int \frac{du}{a^2-u^2} = \frac{1}{2a}\ln\left|\frac{u+a}{u-a}\right| + C = \begin{cases} \dfrac{1}{a}\tanh^{-1}\dfrac{u}{a} + C & \text{if } |u|<a \\[3mm] \dfrac{1}{a}\coth^{-1}\dfrac{u}{a} + C & \text{if } |u|>a \end{cases}$

26. $\displaystyle\int \frac{du}{u^2-a^2} = \frac{1}{2a}\ln\left|\frac{u-a}{u+a}\right| + C = \begin{cases} -\dfrac{1}{a}\tanh^{-1}\dfrac{u}{a} + C & \text{if } |u|<a \\[3mm] -\dfrac{1}{a}\coth^{-1}\dfrac{u}{a} + C & \text{if } |u|>a \end{cases}$

Forms Containing $\sqrt{u^2 \pm a^2}$

In formulas 27 through 38, we may replace

$\ln(u + \sqrt{u^2+a^2})$ by $\sinh^{-1}\dfrac{u}{a}$

$\ln|u + \sqrt{u^2-a^2}|$ by $\cosh^{-1}\dfrac{u}{a}$

$\ln\left|\dfrac{a+\sqrt{u^2+a^2}}{u}\right|$ by $\sinh^{-1}\dfrac{a}{u}$

27. $\displaystyle\int \frac{du}{\sqrt{u^2\pm a^2}} = \ln|u + \sqrt{u^2\pm a^2}| + C$

28. $\displaystyle\int \sqrt{u^2\pm a^2}\,du = \frac{u}{2}\sqrt{u^2\pm a^2} \pm \frac{a^2}{2}\ln|u + \sqrt{u^2\pm a^2}| + C$

29. $\displaystyle\int u^2\sqrt{u^2\pm a^2}\,du = \frac{u}{8}(2u^2\pm a^2)\sqrt{u^2\pm a^2}$
$$-\frac{a^4}{8}\ln|u + \sqrt{u^2\pm a^2}| + C$$

30. $\displaystyle\int \frac{\sqrt{u^2+a^2}\,du}{u} = \sqrt{u^2+a^2} - a\ln\left|\frac{a+\sqrt{u^2+a^2}}{u}\right| + C$

31. $\displaystyle\int \frac{\sqrt{u^2-a^2}\,du}{u} = \sqrt{u^2-a^2} - a\sec^{-1}\left|\frac{u}{a}\right| + C$

32. $\displaystyle\int \frac{\sqrt{u^2\pm a^2}\,du}{u^2} = -\frac{\sqrt{u^2\pm a^2}}{u} + \ln|u + \sqrt{u^2\pm a^2}| + C$

33. $\displaystyle\int \frac{u^2\,du}{\sqrt{u^2\pm a^2}} = \frac{u}{2}\sqrt{u^2\pm a^2} \mp \frac{a^2}{2}\ln|u + \sqrt{u^2\pm a^2}| + C$

34. $\displaystyle\int \frac{du}{u\sqrt{u^2+a^2}} = -\frac{1}{a}\ln\left|\frac{a+\sqrt{u^2+a^2}}{u}\right| + C$

35. $\displaystyle\int \frac{du}{u\sqrt{u^2-a^2}} = \frac{1}{a}\sec^{-1}\left|\frac{u}{a}\right| + C$

36. $\displaystyle\int \frac{du}{u^2\sqrt{u^2\pm a^2}} = -\frac{\sqrt{u^2\pm a^2}}{\pm a^2 u} + C$

37. $\displaystyle\int (u^2\pm a^2)^{3/2}\,du = \frac{u}{8}(2u^2\pm 5a^2)\sqrt{u^2\pm a^2}$
$$+\frac{3a^4}{8}\ln|u + \sqrt{u^2\pm a^2}| + C$$

38. $\displaystyle\int \frac{du}{(u^2\pm a^2)^{3/2}} = \frac{u}{\pm a^2\sqrt{u^2\pm a^2}} + C$

Forms Containing $\sqrt{a^2 - u^2}$

39. $\displaystyle\int \frac{du}{\sqrt{a^2-u^2}} = \sin^{-1}\frac{u}{a} + C$

40. $\displaystyle\int \sqrt{a^2-u^2}\,du = \frac{u}{2}\sqrt{a^2-u^2} + \frac{a^2}{2}\sin^{-1}\frac{u}{a} + C$

41. $\displaystyle\int u^2\sqrt{a^2-u^2}\,du = \frac{u}{8}(2u^2-a^2)\sqrt{a^2-u^2} + \frac{a^4}{8}\sin^{-1}\frac{u}{a} + C$

42. $\displaystyle\int \frac{\sqrt{a^2-u^2}\,du}{u} = \sqrt{a^2-u^2} - a\ln\left|\frac{a+\sqrt{a^2-u^2}}{u}\right| + C$
$$= \sqrt{a^2-u^2} - a\cosh^{-1}\frac{a}{u} + C$$

43. $\displaystyle\int \frac{\sqrt{a^2-u^2}\,du}{u^2} = -\frac{\sqrt{a^2-u^2}}{u} - \sin^{-1}\frac{u}{a} + C$

(This table is continued on the back endpapers)

P9-DVH-854

$$V_{SPHERE} = \frac{4}{3}\pi r^3 \qquad A = 4\pi r^2$$

$$V_{CYL} = \pi r^2 h \qquad A = 2\pi r h$$

$$V_{CONE} = \frac{1}{3}\pi r^2 h \qquad A = \pi r s \qquad \text{s slant height}$$

Synthetic Division

$$3x^5 + 2x^3 - 5 = 0$$

$$
\begin{array}{r|rrrrrr}
1 & 3 & 0 & 2 & 0 & 0 & -5 \\
\text{add} & & 3 & 3 & 5 & 5 & 5 \\
\hline
 & 3 & 3 & 5 & 5 & 5 & 0
\end{array}
$$

$$(x-1)(3x^4 + 3x^3 + 5x^2 + 5x + 5)$$

$$x^3 - y^3 = (x-y)(x^2 + xy + y^2)$$

$$x^3 + y^3 = (x+y)(x^2 - xy + y^2$$

$$y = e^x$$

$$y = \ln x$$

THE CALCULUS
WITH ANALYTIC GEOMETRY

Fifth Edition

Louis Leithold
Pepperdine University

	0	30	45	60	90	180	270	36
	0	$\pi/6$	$\pi/4$	$\pi/3$	$\pi/2$	π	$3\pi/2$	2π
sin	0	1/2	$\sqrt{2}/2$	$\sqrt{3}/2$	1	0	-1	0
cos	1	$\sqrt{3}/2$	$\sqrt{2}/2$	1/2	0	-1	0	1
tan		$\frac{1}{\sqrt{3}}$	1	$\sqrt{3}$				

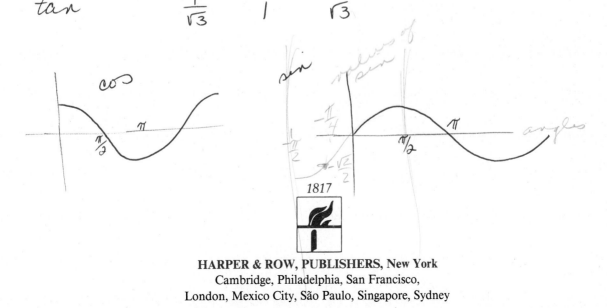

HARPER & ROW, PUBLISHERS, New York
Cambridge, Philadelphia, San Francisco,
London, Mexico City, São Paulo, Singapore, Sydney

Sponsoring Editor: Ann Trump
Project Editor: David Nickol
Text and Cover Design: Edward A. Butler
Text Art: Fineline Illustrations, Inc.
Production: Debra Forrest
Compositor: Syntax International Pte. Ltd.
Printer and Binder: Kingsport Press

THE CALCULUS WITH ANALYTIC GEOMETRY, Fifth Edition
Copyright © 1986 by Harper & Row, Publishers, Inc.

Library of Congress Cataloging-in-Publication Data
Leithold, Louis.
    The calculus with analytic geometry.
    Includes indexes.
    1. Calculus.   2. Geometry, Analytic.   I. Title.
QA303.L428   1986   515'.15   85-22037
ISBN 0-06-043926-2

HARPER INTERNATIONAL EDITION
ISBN 0-06-350413-8

85 86 87 88 9 8 7 6 5 4 3 2 1

ACKNOWLEDGMENTS

Reviewers of the fifth edition of *The Calculus with Analytic Geometry*
Linda H. Boyd, DeKalb Community College, Clarkston
Robert A. Chaffer, Central Michigan University, Mt. Pleasant
Jacqueline M. Dewar, Loyola Marymount University, Los Angeles
Nicholas J. DeLillo, Manhattan College, New York City
Larry S. Dilley, Central Missouri State University, Warrensburg
Linda L. Exley, DeKalb Community College, Clarkston
W. W. Mitchell, Jr., Phoenix College, Phoenix
Roger B. Nelsen, Lewis and Clark College, Portland
Raymond Plankinton, Central Community College, Columbus
George W. Schultz, St. Petersburg Junior College, Clearwater
Delores A. Williams, Pepperdine University, Malibu

Assistants for Answers of Exercises
Charles Stone, DeKalb Community College, Clarkston
Carol Stanton, University of California, Riverside

Cover and Chapter Opening Artist
Joyce Treiman, Pacific Palisades
Courtesy of Tortue Gallery, Santa Monica

To these people, to the staff of Harper & Row, and to all the users of the first four editions who have suggested changes, I express my deep appreciation.

L.L.

THE COVER AND CHAPTER OPENING ARTIST

Joyce Treiman had her first solo exhibition in Chicago in 1943. Since then she has become a well-known American painter with a multitude of exhibitions, prizes, and awards.

In 1985, Joyce Treiman attended some of my calculus classes at Pepperdine University where she observed me and the students, and the interaction among us. During these classes she made sketches and took photographs, and from them she created the drawings that are reproduced on the cover and the chapter opening pages.

L.L.

P R E F A C E

The fifth edition of *The Calculus with Analytic Geometry,* like the other four, is designed for prospective mathematics majors as well as for students whose primary interest is in engineering, the physical and social sciences, or nontechnical fields. As with previous editions, I have endeavored to achieve a healthy balance between the presentation of elementary calculus from a rigorous approach and that from the intuitive and computational point of view. Bearing in mind that a textbook should be written for the student, I have attempted to keep the presentation geared to a beginner's experience and maturity. I desire that the reader be aware that proofs of theorems are necessary and that these proofs be well motivated and carefully explained so that they are understandable to the student who has achieved an average mastery of the preceding sections of the book. If a theorem is stated without proof, I have generally augmented the discussion by both figures and examples, and in such cases I have always stressed that what is presented is an illustration of the content of the theorem and is not a proof.

In this edition, the structure of the text has been altered. The material is now divided into four segments:

Prelude

Part 1   Functions of a Single Variable

Part 2   Infinite Series

Part 3   Vectors and Functions of More Than One Variable

There are fifteen sections appearing at the ends of some of the chapters that are designated as *supplementary.* They are self-contained and can be included or omitted without affecting the understanding of subsequent material. These supplementary sections are of three types:

1. Additional subject matter that is not necessarily part of the traditional syllabus of a calculus course: Sections 3.12, 7.5, 8.11, 8.12, 10.9, 13.9, and 18.5.
2. More applications of the calculus: Sections 3.11, 4.10, 5.8, 5.9, and 16.6.

3. Further theoretical discussions including proofs of some theorems: Sections 1.9, 1.10, and 15.8.

"Prelude" consists of the single Chapter 0, "Topics in Precalculus." It contains basic facts about the real-number system, and this treatment is less detailed than in previous editions. The introduction to analytic geometry in this chapter includes the traditional material on straight lines as well as that on the circle. The definition of a function, operations with functions, and particular kinds of functions are discussed. The presentation of the six trigonometric functions here allows their early use in examples of differentiation and integration of nonalgebraic functions

Part 1, "Functions of a Single Variable," is comprised of Chapters 1 through 10. A thorough discussion of limits and continuity is contained in Chapter 1. The definition of limit is stated in the form "if a then b" rather than its logical equivalent "b whenever a," and this form of the definition is used throughout the text. Proofs of some theorems on limits appear in Supplementary Sections 1.9 and 1.10. The treatment of continuity has been modified, and there is an added discussion of continuity of the trigonometric functions. Some additional geometrical interpretations are included.

In Chapter 2, "The Derivative and Differentiation," before giving the formal definition of a derivative I have defined the tangent line to a curve to demonstrate in advance its geometrical application. The derivatives of all six trigonometric functions are presented here, and they are then available as examples for the initial presentation of the chain rule. Section 2.7, "The Derivative of a Composite Function," has been rewritten with more examples and illustrations. The chain rule is now stated with composite function notation and the proofs for both the special case and the general case are given with this symbolism. The discussion of notations for the derivative has been rewritten with the Leibniz notation $\dfrac{dy}{dx}$ introduced earlier than in previous editions. The section on the differential has been moved forward to this chapter.

Chapter 3 gives the traditional applications of the derivative to problems involving maxima and minima, as well as to curve sketching. There has been a reordering of topics in this chapter. Concavity and points of inflection appear before the second-derivative test, and all the graphing techniques are presented before the second section on absolute extrema. Infinite limits are introduced here because they are useful for graphing. They also can be applied when determining extreme function values. New to this edition are the discussion of oblique asymptotes in Section 3.8 and the presentation of Newton's method in Supplementary Section 3.12.

The topics of antidifferentiation and the definite integral are combined in Chapter 4. Section 4.2, "Some Techniques of Antidifferentiation," has been rewritten with more examples and illustrations, and the antidifferentiation technique of "changing the variable" is used instead of "substitution." The chain rule for antidifferentiation is now proved with composite function notation. I use the term "antidifferentiation" instead of "indefinite integration," but the standard notation $\int f(x)\, dx$ is retained. This notation will suggest that some relation must exist between definite integrals and antiderivatives, but I see no harm in this as long as the presentation gives the theoretically proper view of the definite integral as the limit of sums. In Chapter 4, there is an introduction to differential equations, and the complete discussion of area of a plane region appears here. Most of the applications of the definite integral are now in a single chapter (Chapter 5) as was the situation in the first three editions; two of them are in supplementary sections.

The coverage of inverse functions in the first two sections of Chapter 6 has been revised. Also in Chapter 6, applications of the natural exponential function include the laws of growth and decay as well as some new material on bounded growth involving the learning curve. In Chapter 7, the domains of the inverse secant and inverse cosecant functions have been redefined so that the formulas for their derivatives do not involve absolute value. The section on inverse hyperbolic functions has been designated as supplementary.

In Chapter 8, the exercise sets have been expanded to include more applications of the various techniques of integration. Section 8.6 contains new coverage of logistic growth with applications to biology and sociology. Numerical integration has been moved to this chapter. A new section, labeled supplementary, involves the use of a table of integrals.

Chapter 9, "Indeterminate Forms, Improper Integrals, and Taylor's Formula," has been moved forward to precede the one on analytic geometry topics. A discussion of the probability density function has been added to give another application of improper integrals. Chapter 10, "Polar Coordinates and the Conic Sections," contains a major rewrite of the corresponding material that appeared in the fourth edition. The cartesian equations of the conics are obtained first. The polar equations do not appear until Section 10.8, where they occur as part of a unified treatment of conic sections. New to this edition is a discussion of when the limaçon has a dent and when it hasn't.

The presentation of infinite series appears in a separate part (Part 2) to make it more apparent that it is self-contained and can be covered anytime after the completion of Part 1. Some of the material has been rewritten with more motivation added. The treatment is now in two chapters instead of one, and some of the longer sections in the fourth edition have been split into two. New to this edition is a discussion of the root test. Also new is a summary of tests for convergence of an infinite series. The exercise sets in Chapters 11 and 12 have been expanded to include more applications.

Part 3 contains the calculus of functions of more than one variable and vector calculus with a vector approach to solid analytic geometry. In Chapter 13, "Vectors in the Plane and Parametric Equations," and Chapter 14, "Vectors in Three-Dimensional Space and Solid Analytic Geometry," there has been a rewriting and reordering of some of the material. The discussions of scalar projection, vector projection, and cross product have been expanded and the triple vector product introduced. Additional applications have been incorporated and new examples and illustrations have been added.

In Chapter 15, "Differential Calculus of Functions of More Than One Variable," the treatment of functions, limits, and continuity has been revised with new examples and illustrations. The material on differentiability and the total differential has been rewritten. The proof of the theorem that gives sufficient conditions for a function of two variables to be differentiable at a point has been deferred to Supplementary Section 15.8. New topics in Chapter 16 include exact differential equations, appearing in Section 16.5, and the method of least squares, which is the subject of Supplementary Section 16.6.

Chapter 17, "Multiple Integration," remains essentially the same as in the fourth edition, although some new exercises have been incorporated. The section on Green's Theorem that appeared in the chapter on multiple integration in the fourth edition is now part of Chapter 18, "Introduction to the Calculus of Vector Fields," which has an expanded coverage of vector calculus. The discussion of line integrals has also been moved to Chapter 18. The new topics here are surface integrals, divergence and curl in three dimensions, Gauss's Divergence Theorem, and Stokes's Theorem. The approach in Chapter 18 is intuitive and the applications are to physics and engineering.

There are now over 7000 exercises that have been revised and graded to provide a wide variety of types that range from computational to applied and theoretical problems. The answers to the odd-numbered exercises are given in the back of the book, and the answers to the even-numbered ones are available in a separate booklet. Detailed step-by-step solutions for nearly half the even-numbered exercises (those having numbers divisible by four) appear in a supplement to this text, *An Outline for the Study of Calculus,* by John H. Minnick, published by Harper & Row, in three volumes.

Louis Leithold

CONTENTS

Z E R O

TOPICS IN PRECALCULUS

The **real number system** consists of a set of elements called **real numbers** and two operations called **addition** and **multiplication**. The set of real numbers is denoted by R. The operation of addition is denoted by the symbol $+$, and the operation of multiplication is denoted by the symbol $\cdot$. If a and b are elements of the set R, $a + b$ denotes the **sum** of a and b, and $a \cdot b$ (or ab) denotes their **product**. The operation of **subtraction** is defined by the equation

$$a - b = a + (-b)$$

where $-b$ denotes the **negative** of b such that $b + (-b) = 0$. The operation of **division** is defined by the equation

$$a \div b = a \cdot b^{-1}$$

where b^{-1} denotes the **reciprocal** of b such that $b \cdot b^{-1} = 1$.

The real number system can be completely described by a set of axioms (the word **axiom** is used to indicate a formal statement that is assumed to be true without proof). With these axioms we can derive the properties of the real numbers from which follow the familiar algebraic operations of addition, subtraction, multiplication, and division, as well as the algebraic concepts of solving equations, factoring, and so forth.

Properties that can be shown to be logical consequences of axioms are **theorems**. In the statement of most theorems there are two parts: the "if" part, called the **hypothesis**, and the "then" part, called the **conclusion**. The argument verifying a theorem is a **proof**. A proof consists in showing that the conclusion follows from the assumed truth of the hypothesis.

A real number is either a positive number, a negative number, or zero, and any real number can be classified as either **rational** or **irrational**. A rational number is any number that can be expressed as the ratio of two integers. That is, a rational number is a number of the form p/q, where p and q are integers and $q \neq 0$. The rational numbers consist of the following:

The **integers** (positive, negative, and zero)

$$\ldots, -5, -4, -3, -2, -1, 0, 1, 2, 3, 4, 5, \ldots$$

The positive and negative **fractions**, such as

$$\tfrac{2}{7} \qquad -\tfrac{4}{5} \qquad \tfrac{83}{5}$$

The positive and negative **terminating decimals**, such as

$$2.36 = \frac{236}{100} \qquad -0.003251 = -\frac{3{,}251}{1{,}000{,}000}$$

The positive and negative **nonterminating repeating decimals**, such as

$$0.333\ldots = \tfrac{1}{3} \qquad -0.549549549\ldots = -\tfrac{61}{111}$$

The real numbers that are not rational numbers are called **irrational numbers**. These are positive and negative **nonterminating nonrepeating** decimals, for example,

$$\sqrt{3} = 1.732\ldots \qquad \pi = 3.14159\ldots$$

We use some set notation and terminology in the discussions that follow. The idea of *set* is used extensively in mathematics and is such a basic concept that it is not given a formal definition here. We can say that a **set** is a collection of objects, and the objects in a set are called the **elements** of a set. If every element of a set S is also an element of a set T, then S is a **subset** of T. In calculus we are concerned with the set R of real numbers. Two subsets of R are the set N of natural numbers and the set Z of integers.

We use the symbol $\in$ to indicate that a specific element belongs to a set. Hence we may write $8 \in N$, which is read "8 is an element of N." The notation $a, b \in S$ indicates that both a and b are elements of S. The symbol $\notin$ is read "is not an element of." Thus we read $\frac{1}{2} \notin N$ as "$\frac{1}{2}$ is not an element of N."

A pair of braces $\{\ \ \}$ used with words or symbols can describe a set. If S is the set of natural numbers less than 6, we can write the set S as

$$\{1, 2, 3, 4, 5\}$$

We can also write the set S as

$$\{x, \text{ such that } x \text{ is a natural number less than } 6\}$$

where the symbol x is called a *variable*. A **variable** is a symbol used to represent any element of a given set. Another way of writing the above set S is to use what is called **set-builder notation**, where a vertical bar is used in place of the words "such that." Using set-builder notation to describe the set S we have

$$\{x \,|\, x \text{ is a natural number less than } 6\}$$

which is read "the set of all x such that x is a natural number less than 6."

Two sets A and B are said to be **equal**, written $A = B$, if A and B have identical elements. The **union** of two sets A and B, denoted by $A \cup B$ and read "A union B," is the set of all elements that are in A or in B or in both A and B. The **intersection** of A and B, denoted by $A \cap B$ and read "A intersection B," is the set of only those elements that are in both A and B. The set that contains no elements is called the **empty set** and is denoted by $\varnothing$.

▶ ILLUSTRATION 1   Suppose $A = \{2, 4, 6, 8, 10, 12\}$, $B = \{1, 4, 9, 16\}$,   and $C = \{2, 10\}$. Then

$$A \cup B = \{1, 2, 4, 6, 8, 9, 10, 12, 16\} \qquad A \cap B = \{4\}$$
$$B \cup C = \{1, 2, 4, 9, 10, 16\} \qquad\qquad B \cap C = \varnothing \qquad\qquad ◀$$

There is an ordering for the set R by means of a relation denoted by the symbols $<$ (read "is less than") and $>$ (read "is greater than"), which we now formally define.

0.1.1 DEFINITION   If $a, b \in R$,

(i) $a < b$ if and only if $b - a$ is positive;
(ii) $a > b$ if and only if $a - b$ is positive.

▶ ILLUSTRATION 2

$3 < 5$ because $5 - 3 = 2$, and 2 is positive

$-10 < -6$ because $-6 - (-10) = 4$, and 4 is positive

$7 > 2$ because $7 - 2 = 5$, and 5 is positive

$-2 > -7$ because $-2 - (-7) = 5$, and 5 is positive

$\frac{3}{4} > \frac{2}{3}$ because $\frac{3}{4} - \frac{2}{3} = \frac{1}{12}$, and $\frac{1}{12}$ is positive   ◀

We now define the symbols $\leq$ (read "is less than or equal to") and $\geq$ (read "is greater than or equal to").

0.1.2 DEFINITION   If $a, b \in R$,

(i) $a \leq b$ if and only if   either $a < b$ or $a = b$;
(ii) $a \geq b$ if and only if   either $a > b$ or $a = b$.

The statements $a < b$, $a > b$, $a \leq b$, and $a \geq b$ are called **inequalities**. Some examples are $2 < 7$, $-5 < 6$, $-5 < -4$, $14 > 8$, $2 > -4$, $-3 > -6$, $5 > 3$, and $-10 < -7$. In particular, $a < b$ and $a > b$ are called **strict** inequalities, whereas $a \leq b$ and $a \geq b$ are called **nonstrict** inequalities.

The following theorem follows immediately from Definition 0.1.1.

0.1.3 THEOREM   (i) $a > 0$ if and only if a is positive;
(ii) $a < 0$ if and only if a is negative.

A number x is between a and b if $a < x$ and $x < b$. We can write this as a **continued inequality** as follows:

$$a < x < b \tag{1}$$

Another continued inequality is

$$a \leq x \leq b$$

which means that both $a \leq x$ and $x \leq b$. Other continued inequalities are $a \leq x < b$ and $a < x \leq b$.

The following theorems can be proved by using axioms for the set R and 0.1.1 through 0.1.3.

0.1.4 THEOREM   (i) If $a > 0$ and $b > 0$, then $a + b > 0$.
(ii) If $a > 0$ and $b > 0$, then $ab > 0$.

Part (i) of the above theorem states that the sum of two positive numbers is positive and part (ii) states that the product of two positive numbers is positive.

0.1.5 THEOREM
Transitive Property of Order   If $a, b, c \in R$, and

if $a > b$ and $b > c$, then $a > c$

▶ ILLUSTRATION 3   If $x > 5$ and $5 > y$, then by the transitive property of order, it follows that $x > y$.   ◀

0.1.6 THEOREM   Suppose $a, b, c \in R$.

   (i) If $a > b$, then $a + c > b + c$.
   (ii) If $a > b$ and $c > 0$, then $ac > bc$.
   (iii) If $a > b$ and $c < 0$, then $ac < bc$.

▶ ILLUSTRATION 4   (a) If $x > y$, it follows from Theorem 0.1.6(i) that $x + 4 > y + 4$. For instance, $9 > 3$; thus $9 + 4 > 3 + 4$ or, equivalently, $13 > 7$.

   Furthermore, if $x > y$, then $x - 11 > y - 11$. For instance, $9 > 3$; thus $9 - 11 > 3 - 11$ or, equivalently, $-2 > -8$.

   (b) If $x > y$, it follows from Theorem 0.1.6(ii) that $7x > 7y$. For instance, because $8 > 5$, then $7 \cdot 8 > 7 \cdot 5$ or, equivalently, $56 > 35$.

   (c) Because $6 > 4$, then if $z < 0$, it follows from Theorem 0.1.6(iii) that $6z < 4z$. For instance, because $6 > 4$, then $6(-3) < 4(-3)$ or, equivalently, $-18 < -12$.   ◀

The following theorem is similar to Theorem 0.1.6 except that the direction of the inequality is reversed.

0.1.7 THEOREM   Suppose $a, b, c \in R$

   (i) If $a < b$, then $a + c < b + c$.
   (ii) If $a < b$ and $c > 0$, then $ac < bc$.
   (iii) If $a < b$ and $c < 0$, then $ac > bc$.

▶ ILLUSTRATION 5   (a) If $x < y$, from Theorem 0.1.7(i) we have $x + 5 < y + 5$. For instance, $-8 < -2$; thus $-8 + 5 < -2 + 5$ or, equivalently, $-3 < 3$.

   Furthermore, if $x < y$, then $x - 4 < y - 4$. For instance, $-8 < -2$; thus $-8 - 4 < -2 - 4$ or, equivalently, $-12 < -6$.

   (b) If $x < y$, it follows from Theorem 0.1.7(ii) that $4x < 4y$. For instance, $-5 < 3$; thus $4(-5) < 4(3)$ or, equivalently, $-20 < 12$.

   (c) Because $-5 < 3$, then if $z < 0$, it follows from Theorem 0.1.7(iii) that $-5z > 3z$. For instance, because $-5 < 3$, then $(-5)(-4) > 3(-4)$ or, equivalently, $20 > -12$.   ◀

Theorems 0.1.6 and 0.1.7 also are valid if in each theorem the symbol $>$ is replaced by $\geq$ and $<$ is replaced by $\leq$.

   Parts (ii) of Theorems 0.1.6 and 0.1.7 state that if both members of an inequality are multiplied by a positive number, the direction of the inequality remains unchanged, whereas parts (iii) state that if both members of an inequality are multiplied by a negative number, the direction of the inequality is reversed. Parts (ii) and (iii) also hold for division because dividing both members of an inequality by a number d $(d \neq 0)$ is equivalent to multiplying them by $\dfrac{1}{d}$.

0.1.8 THEOREM   If $a < b$ and $c < d$, then $a + c < b + d$.

▶ ILLUSTRATION 6   If $x < 8$ and $y < -3$, then we have from Theorem 0.1.8, $x + y < 8 + (-3)$; that is, $x + y < 5$.   ◀

We impose upon the set R a condition called the **axiom of completeness** (Axiom 11.2.5). The statement of this axiom is deferred until Section 11.2 because it requires some terminology that is best introduced and discussed later. However, we now give a geometric interpretation to the set of real numbers by associating them with the points on a line, called an **axis**. The axiom of completeness guarantees that there is a one-to-one correspondence between the set R and the set of points on an axis.

FIGURE 0.1.1

Refer to Fig. 0.1.1, where the axis is a horizontal line. A point on the axis is chosen to represent the number 0. This point is called the **origin**. A unit of distance is selected. Then each positive number x is represented by the point at a distance of x units to the right of the origin, and each negative number x is represented by the point at a distance of $-x$ units to the left of the origin (it should be noted that if x is negative, then $-x$ is positive). To each real number there corresponds a unique point on the axis, and with each point on the axis there is associated only one real number; hence there is a one-to-one correspondence between R and the points on the axis. So the points on the axis are identified with the numbers they represent, and the same symbol is used for both the number and the point representing that number on the axis. We identify R with the axis, and we call the axis the **real number line**.

We see that $a < b$ if and only if the point representing the number a is to the left of the point representing the number b. Similarly, $a > b$ if and only if the point representing a is to the right of the point representing b. For instance, the number 2 is less than the number 5 and the point 2 is to the left of the point 5. We could also write $5 > 2$ and say that the point 5 is to the right of the point 2.

The set of all numbers x satisfying the continued inequality (1) is called an **open interval** and is denoted by (a, b). Therefore

$$(a, b) = \{x \mid a < x < b\}$$

The **closed interval** from a to b is the open interval (a, b) together with the two endpoints a and b and is denoted by $[a, b]$. Thus

$$[a, b] = \{x \mid a \leq x \leq b\}$$

FIGURE 0.1.2

FIGURE 0.1.3

Figure 0.1.2 illustrates the open interval (a, b), and Fig. 0.1.3 illustrates the closed interval $[a, b]$.

The **interval half-open on the left** is the open interval (a, b) together with the right endpoint b. It is denoted by $(a, b]$; so

$$(a, b] = \{x \mid a < x \leq b\}$$

We define an **interval half-open on the right** in a similar way and denote it by $[a, b)$. Thus

$$[a, b) = \{x \mid a \leq x < b\}$$

FIGURE 0.1.4

FIGURE 0.1.5

Figure 0.1.4 illustrates the interval $(a, b]$, and Fig. 0.1.5 illustrates the interval $[a, b)$.

We shall use the symbol $+\infty$ (positive infinity) and the symbol $-\infty$ (negative infinity); however, take care not to confuse these symbols with real numbers, for they do not obey the properties of the real numbers. We have the following intervals:

$$(a, +\infty) = \{x \mid x > a\}$$

$$(-\infty, b) = \{x \mid x < b\}$$

$$[a, +\infty) = \{x \mid x \geq a\}$$

$$(-\infty, b] = \{x \mid x \leq b\}$$

$$(-\infty, +\infty) = R$$

FIGURE 0.1.6

FIGURE 0.1.7

Figure 0.1.6 illustrates the interval $(a, +\infty)$, and Fig. 0.1.7 illustrates the interval $(-\infty, b)$. Note that $(-\infty, +\infty)$ denotes the set of all real numbers.

For each of the intervals (a, b), $[a, b]$, $[a, b)$, and $(a, b]$ the numbers a and b are called the **endpoints** of the interval. The closed interval $[a, b]$ contains both its endpoints, whereas the open interval (a, b) contains neither endpoint. The interval $[a, b)$ contains its left endpoint but not its right one, and the interval $(a, b]$ contains its right endpoint but not its left one. An open interval can be thought of as one that contains none of its endpoints, and a closed interval can be regarded as one that contains all of its endpoints. Consequently, the interval $[a, +\infty)$ is considered to be a closed interval because it contains its only endpoint a. Similarly, $(-\infty, b]$ is a closed interval, whereas $(a, +\infty)$ and $(-\infty, b)$ are open. The intervals $[a, b)$ and $(a, b]$ are neither open nor closed. The interval $(-\infty, +\infty)$ has no endpoints, and it is considered both open and closed.

Intervals are used to represent *solution sets* of inequalities in one variable. The **solution set** of such an inequality is the set of all numbers that satisfy the inequality.

EXAMPLE 1
Find the solution set of the inequality $2 + 3x < 5x + 8$. Illustrate it on the real number line.

SOLUTION
If x is a number such that

$$2 + 3x < 5x + 8$$

then

$$2 + 3x - 2 < 5x + 8 - 2$$

$$3x < 5x + 6$$

$$-2x < 6$$

$$x > -3$$

What we have proved is that if $2 + 3x < 5x + 8$, then $x > -3$. Each of the steps is reversible; that is, if we start with

$$x > -3$$

then

$$-2x < 6$$

$$2 + 3x < 5x + 8$$

Therefore we can conclude that

$$2 + 3x < 5x + 8 \quad \text{if and only if} \quad x > -3$$

So the solution set of the given inequality is the interval $(-3, +\infty)$, which is illustrated in Fig. 0.1.8.

$-3 \quad 0$

FIGURE 0.1.8

EXAMPLE 2
Find the solution set of the inequality $4 < 3x - 2 \le 10$. Illustrate it on the real number line.

SOLUTION
Add 2 to each member of the inequality to obtain

$$6 < 3x \le 12$$

$$2 < x \le 4$$

Each step is reversible; so the solution set is the interval $(2, 4]$, as illustrated in Fig. 0.1.9.

$0 \qquad 2 \qquad 4$

FIGURE 0.1.9

EXAMPLE 3
Find the solution set of the inequality

$$\frac{7}{x} > 2$$

Illustrate it on the real number line.

SOLUTION
We wish to multiply both members of the inequality by x. However, the direction of the inequality that results will depend on whether x is positive or negative.

Observe that if $x < 0$, then

$$\frac{7}{x} < 0$$

which contradicts the given inequality. Therefore we need consider only $x > 0$. Multiplying on both sides of the given inequality by x we obtain

$$7 > 2x$$

$$\tfrac{7}{2} > x$$

$$x < \tfrac{7}{2}$$

FIGURE 0.1.10

Because the above steps are reversible, the solution set of the inequality is $\{x \mid x > 0\} \cap \{x \mid x < \frac{7}{2}\}$ or, equivalently, $\{x \mid 0 < x < \frac{7}{2}\}$, which is the interval $(0, \frac{7}{2})$, as illustrated in Fig. 0.1.10.

EXAMPLE 4
Find the solution set of the inequality

$$\frac{x}{x - 3} < 4$$

Illustrate it on the real number line.

SOLUTION
To multiply both members of the inequality by $x - 3$ we must consider two cases.

Case 1: $x - 3 > 0$; that is, $x > 3$.
    Multiplying on both sides of the inequality by $x - 3$ we get

$$x < 4x - 12$$

$$-3x < -12$$

$$x > 4$$

Thus the solution set of Case 1 is $\{x \mid x > 3\} \cap \{x \mid x > 4\}$ or, equivalently, $\{x \mid x > 4\}$, which is the interval $(4, +\infty)$.

Case 2: $x - 3 < 0$; that is, $x < 3$.
    Multiplying on both sides by $x - 3$ and reversing the direction of the inequality we have

$$x > 4x - 12$$

$$-3x > -12$$

$$x < 4$$

Therefore x must be less than 4 and also less than 3. Thus the solution set of Case 2 is the interval $(-\infty, 3)$.
    If the solution sets for Cases 1 and 2 are combined, we obtain $(-\infty, 3) \cup (4, +\infty)$, which is illustrated in Fig. 0.1.11.

FIGURE 0.1.11

The concept of the *absolute value* of a number is used in some important definitions in the study of calculus. Furthermore, you will need to work with inequalities involving absolute value.

0.1.9 DEFINITION   The **absolute value** of x, denoted by $|x|$, is defined by

$$|x| = \begin{cases} x & \text{if } x \geq 0 \\ -x & \text{if } x < 0 \end{cases}$$

▶ ILLUSTRATION 15

$$|3| = 3 \qquad |-5| = -(-5) \qquad |8 - 14| = |-6|$$
$$= 5 \qquad\qquad = -(-6)$$
$$= 6 \qquad\qquad ◀$$

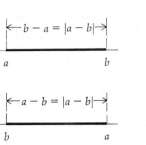

FIGURE 0.1.12

From the definition, the absolute value of a number is either a positive number or zero; that is, it is nonnegative.

In terms of geometry, the absolute value of a number x is its distance from 0, without regard to direction. In general, $|a - b|$ is the distance between a and b without regard to direction, that is, without regard to which is the larger number. Refer to Fig. 0.1.12.

We have the following properties of absolute values.

0.1.10 THEOREM $|x| < a$ if and only if $-a < x < a$, where $a > 0$.

PROOF   Because the theorem has an *if and only if* qualification, the proof requires two parts.

Part 1: Prove that $|x| < a$ if $-a < x < a$, where $a > 0$. Here, two cases must be considered: $x \geq 0$ and $x < 0$.

Case 1: $x \geq 0$.
Then $|x| = x$. Because $x < a$, we conclude that $|x| < a$.

Case 2: $x < 0$.
Then $|x| = -x$. Because $-a < x$, we apply Theorem 0.1.7(iii) and obtain $a > -x$ or, equivalently, $-x < a$. But because $-x = |x|$, we have $|x| < a$.
In both cases, then,

$$|x| < a \quad \text{if } -a < x < a \qquad \text{where } a > 0$$

Part 2: Prove that $|x| < a$ only if $-a < x < a$, where $a > 0$. Here we must show that whenever the inequality $|x| < a$ holds, the inequality $-a < x < a$ also holds. Assume that $|x| < a$ and consider the two cases $x \geq 0$ and $x < 0$.

Case 1: $x \geq 0$.
Then $|x| = x$. Because $|x| < a$, we conclude that $x < a$. Also, because $a > 0$, it follows from Theorem 0.1.6(iii) that $-a < 0$. Thus $-a < 0 \leq x < a$, or $-a < x < a$.

Case 2: $x < 0$.
Then $|x| = -x$. Because $|x| < a$, we conclude that $-x < a$. Also, because $x < 0$, it follows from Theorem 0.1.7(iii) that $0 < -x$. Therefore $-a < 0 < -x < a$, or $-a < -x < a$, which by applying Theorem 0.1.7(iii) yields $-a < x < a$.
In both cases,

$$|x| < a \quad \text{only if } -a < x < a \qquad \text{where } a > 0 \qquad ■$$

0.1.11 COROLLARY	$|x| \leq a$ if and only if $-a \leq x \leq a$, where $a > 0$.

0.1.12 THEOREM	$|x| > a$ if and only if $x > a$ or $x < -a$, where $a > 0$.

The proof of Theorem 0.1.12 is left as an exercise (see Exercise 61).

0.1.13 COROLLARY	$|x| \geq a$ if and only if $x \geq a$ or $x \leq -a$, where $a > 0$.

The following examples illustrate the solution of equations and in-equalities involving absolute values.

EXAMPLE 5
Solve each of the equations for x: (a) $|3x + 2| = 5$; (b) $|2x - 1| = |4x + 3|$; (c) $|5x + 4| = -3$.

SOLUTION
(a) $|3x + 2| = 5$

This equation will be satisfied if either

$$3x + 2 = 5 \quad \text{or} \quad -(3x + 2) = 5$$
$$x = 1 \qquad\qquad x = -\tfrac{7}{3}$$

(b) $|2x - 1| = |4x + 3|$

This equation will be satisfied if either

$$2x - 1 = 4x + 3 \quad \text{or} \quad 2x - 1 = -(4x + 3)$$
$$x = -2 \qquad\qquad x = -\tfrac{1}{3}$$

(c) $|5x + 4| = -3$

Because the absolute value of a number may never be negative, this equation has no solution.

EXAMPLE 6
Find the solution set of the inequality $|x - 5| < 4$. Illustrate it on the real number line.

SOLUTION
If x is a number such that

$$|x - 5| < 4$$

then

$$-4 < x - 5 < 4 \quad \text{(by Theorem 0.1.10)}$$
$$1 < x < 9$$

FIGURE 0.1.13

Because each step is reversible, we can conclude that

$$|x - 5| < 4 \quad \text{if and only if} \quad 1 < x < 9$$

So the solution set of the given inequality is the open interval $(1, 9)$, which is illustrated in Fig. 0.1.13.

EXAMPLE 7
Find the solution set of the inequality $|3x + 2| > 5$.

SOLUTION
By Theorem 0.1.12, the given inequality is equivalent to

$$3x + 2 > 5 \quad \text{or} \quad 3x + 2 < -5 \tag{2}$$

That is, the given inequality will be satisfied if either of the inequalities in (2) is satisfied.

Considering the first inequality we have

$$3x + 2 > 5$$

$$x > 1$$

Therefore every number in the interval $(1, +\infty)$ is a solution.

From the second inequality,

$$3x + 2 < -5$$

$$x < -\tfrac{7}{3}$$

Hence every number in the interval $(-\infty, -\tfrac{7}{3})$ is a solution.

The solution set of the given inequality is therefore $(-\infty, -\tfrac{7}{3}) \cup (1, +\infty)$.

You may recall from algebra that the symbol $\sqrt{a}$, where $a \geq 0$, is defined as the unique *nonnegative* number x such that $x^2 = a$. We read $\sqrt{a}$ as "the principal square root of a." For example,

$$\sqrt{4} = 2 \qquad \sqrt{0} = 0 \qquad \sqrt{\tfrac{9}{25}} = \tfrac{3}{5}$$

Note: $\sqrt{4} \neq -2$ even though $(-2)^2 = 4$, because $\sqrt{4}$ denotes only the *positive* square root of 4. The *negative* square root of 4 is designated by $-\sqrt{4}$.

Because we are concerned only with real numbers in this book, $\sqrt{a}$ is not defined if $a < 0$.

EXAMPLE 8
Find all values of x for which $\sqrt{x^2 + 7x + 12}$ is real.

SOLUTION

$$x^2 + 7x + 12 = (x + 3)(x + 4)$$

$$\sqrt{(x + 3)(x + 4)} \quad \text{is real when} \quad (x + 3)(x + 4) \geq 0$$

We find the solution set of this inequality. The inequality will be satisfied when both factors are nonnegative or when both factors are nonpositive, that is, if $x + 3 \geq 0$ and $x + 4 \geq 0$, or if $x + 3 \leq 0$ and $x + 4 \leq 0$. We consider two cases.

Case 1: $x + 3 \geq 0$ and $x + 4 \geq 0$. That is,

$$x \geq -3 \quad \text{and} \quad x \geq -4$$

Both inequalities hold if $x \geq -3$, which is the interval $[-3, +\infty)$.

Case 2: $x + 3 \leq 0$ and $x + 4 \leq 0$. That is,

$$x \leq -3 \quad \text{and} \quad x \leq -4$$

Both inequalities hold if $x \leq -4$, which is the interval $(-\infty, -4]$.
    If we combine the solution sets of Cases 1 and 2, then we have $(-\infty, -4] \cup [-3, +\infty)$.

From the definition of $\sqrt{a}$ it follows that

$$\sqrt{x^2} = |x|$$

▶ ILLUSTRATION 16

$$\sqrt{5^2} = |5| \qquad \sqrt{(-3)^2} = |-3|$$
$$= 5 \qquad\qquad\qquad = 3$$

◀

The following theorems about absolute value will be useful later.

0.1.14 THEOREM   If a and b are any numbers, then

$$|ab| = |a| \cdot |b|$$

PROOF

$$|ab| = \sqrt{(ab)^2}$$
$$= \sqrt{a^2 b^2}$$
$$= \sqrt{a^2} \cdot \sqrt{b^2}$$
$$= |a| \cdot |b|$$

■

0.1.15 THEOREM   If a is any number and b is any number except 0,

$$\left| \frac{a}{b} \right| = \frac{|a|}{|b|}$$

The proof of Theorem 0.1.15 is left as an exercise (see Exercise 62).

0.1.16 THEOREM
The Triangle Inequality

If a and b are any numbers, then

$$|a + b| \leq |a| + |b|$$

PROOF   By Definition 0.1.9, either $a = |a|$ or $a = -|a|$; thus

$$-|a| \leq a \leq |a| \tag{3}$$

Furthermore,

$$-|b| \leq b \leq |b| \qquad\qquad (4)$$

From inequalities (3) and (4) and Theorem 0.1.8,

$$-(|a| + |b|) \leq a + b \leq |a| + |b|$$

Hence, from Corollary 0.1.11 it follows that

$$|a + b| \leq |a| + |b| \qquad\qquad \blacksquare$$

Theorem 0.1.16 has two important corollaries, which we now state and prove.

0.1.17 COROLLARY   If a and b are any numbers, then

$$|a - b| \leq |a| + |b|$$

PROOF

$$|a - b| = |a + (-b)| \leq |a| + |(-b)| = |a| + |b| \qquad\qquad \blacksquare$$

0.1.18 COROLLARY   If a and b are any numbers, then

$$|a| - |b| \leq |a - b|$$

PROOF

$$|a| = |(a - b) + b| \leq |a - b| + |b|$$

thus, subtracting $|b|$ from both members of the inequality we have

$$|a| - |b| \leq |a - b| \qquad\qquad \blacksquare$$

$x^2 > y^2$
$|x| > |y|$

EXERCISES 0.1

In Exercises 1 through 22, find the solution set of the given inequality, and illustrate the solution set on the real number line.

1. $5x + 2 > x - 6$ **2.** $3 - x < 5 + 3x$ **3.** $\frac{2}{3}x - \frac{1}{2} \leq 0$ **4.** $3 - 2x \geq 9 + 4x$

5. $13 \geq 2x - 3 \geq 5$ **6.** $-2 < 6 - 4x \leq 8$ **7.** $2 > -3 - 3x \geq -7$ **8.** $2 \leq 5 - 3x < 11$

9. $\dfrac{4}{x} - 3 > \dfrac{2}{x} - 7$ 2 cases **10.** $\dfrac{5}{x} < \dfrac{3}{4}$ **11.** $\dfrac{1}{x + 1} < \dfrac{2}{3x - 1}$ o.d line **12.** $\dfrac{x + 1}{2 - x} < \dfrac{x}{3 + x}$

$|x| > 2$

13. $x^2 > 4$ **14.** $x^2 \leq 9$ **15.** $(x - 3)(x + 5) > 0$ **16.** $x^2 - 3x + 2 > 0$

17. $1 - x - 2x^2 \geq 0$ **18.** $x^2 + 3x + 1 > 0$ **19.** $4x^2 + 9x < 9$ **20.** $2x^2 - 6x + 3 < 0$

21. $\dfrac{1}{3x - 7} \geq \dfrac{4}{3 - 2x}$ **22.** $x^3 + 1 > x^2 + x$

In Exercises 23 through 30, solve for x.

23. $|4x + 3| = 7$ **24.** $|3x - 8| = 4$ **25.** $|5x - 3| = |3x + 5|$ **26.** $|x - 2| = |3 - 2x|$

27. $|7x| = 4 - x$ 28. $2x + 3 = |4x + 5|$ 29. $\left|\dfrac{x + 2}{x - 2}\right| = 5$ 30. $\left|\dfrac{3x + 8}{2x - 3}\right| = 4$

In Exercises 31 through 36, find all the values of x for which the number is real.

31. $\sqrt{8x - 5}$ 32. $\sqrt{x^2 - 16}$ 33. $\sqrt{x^2 - 3x - 10}$ 34. $\sqrt{2x^2 + 5x - 3}$
35. $\sqrt{x^2 - 5x + 4}$ 36. $\sqrt{x^2 + 2x - 1}$

In Exercises 37 through 52, find the solution set of the given inequality, and illustrate the solution set on the real number line.

37. $|x + 4| < 7$ 38. $|2x - 5| < 3$ 39. $|3x - 4| \leq 2$ 40. $|3x + 2| \geq 1$ $3x > 0 \quad 6$
41. $|5 - x| > 7$ 42. $|3 - x| < 5$ 43. $|7 - 4x| \leq 9$ 44. $|6 - 2x| \geq 7$ $3x < 0 \quad 6$
45. $|2x - 5| > 3$ 46. $|x + 4| \leq |2x - 6|$ 47. $|3x| > |6 - 3x|$ *Hold question of both probs*   48. $|3 + 2x| < |4 - x|$ $3x < 0 \quad 6$ $3x > 0 \quad 6$

49. $|9 - 2x| \geq |4x|$ 50. $|5 - 2x| \geq 7$ 51. $\left|\dfrac{x + 2}{2x - 3}\right| < 4$ 52. $\left|\dfrac{6 - 5x}{3 + x}\right| \leq \dfrac{1}{2}$

In Exercises 53 through 56, solve for x, and write the answer with absolute value notation.

53. $\dfrac{x - a}{x + a} > 0$ 54. $\dfrac{a - x}{a + x} \geq 0$ 55. $\dfrac{x - 2}{x - 4} > \dfrac{x + 2}{x}$ 56. $\dfrac{x + 5}{x + 3} < \dfrac{x + 1}{x - 1}$

57. Prove Theorem 0.1.5.   58. Prove Theorem 0.1.6(i).
59. Prove Theorem 0.1.6(ii) and (iii).   60. Prove that if $x < y$, then $x < \frac{1}{2}(x + y) < y$.
61. Prove Theorem 0.1.12.   62. Prove Theorem 0.1.15.

0.2
THE NUMBER PLANE AND GRAPHS OF EQUATIONS

We now consider ordered pairs of real numbers. Any set of two real numbers form a **pair**, and when the order of the pair is designated, it is called an **ordered pair of real numbers**. If x is the first real number and y is the second, we denote this ordered pair by writing them in parentheses with a comma separating them as (x, y). Note that the ordered pair $(3, 7)$ is different from the ordered pair $(7, 3)$.

0.2.1 DEFINITION   The set of all ordered pairs of real numbers is called the **number plane**, and each ordered pair (x, y) is called a **point** in the number plane. The number plane is denoted by R^2.

Just as we can identify R with points on an axis (a one-dimensional space), we can identify R^2 with points in a geometric plane (a two-dimensional space). The method used with R^2 is the one attributed to the French mathematician René Descartes (1596–1650), who is credited with originating analytic geometry in 1637.

A horizontal line is chosen in the geometric plane and is called the **x axis**. A vertical line is chosen and is called the **y axis**. The point of intersection of the x axis and the y axis is called the **origin** and is denoted by the letter O. A unit of length is chosen. Usually the unit length on each axis is the same. We establish the positive direction on the x axis to the right of the origin, and the positive direction on the y axis above the origin. See Fig. 0.2.1.

FIGURE 0.2.1

FIGURE 0.2.2

We now associate an ordered pair of real numbers (x, y) with a point P in the geometric plane. Refer to Fig. 0.2.2. The distance of P from the y axis, considered positive if P is to the right of the y axis and negative if P is to the left of the y axis, is called the **abscissa** (or **x coordinate**) of P and is denoted by x. The distance of P from the x axis, considered positive if P is above the x axis and negative if P is below the x axis, is called the **ordinate** (or **y coordinate**) of P and is denoted by y. The abscissa and the ordinate of a point are called the **rectangular cartesian coordinates** of the point. There is a one-to-one correspondence between the points in a geometric plane and R^2; that is, with each point there corresponds a unique ordered pair (x, y), and with each ordered pair (x, y) there is associated only one point. This one-to-one correspondence is called a **rectangular cartesian coordinate system**. Figure 0.2.3 illustrates a rectangular cartesian coordinate system with some points plotted.

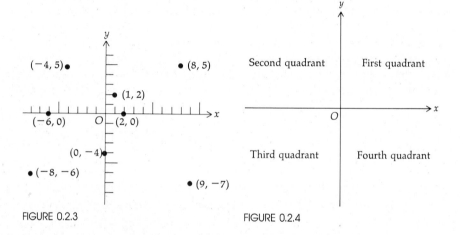

FIGURE 0.2.3   FIGURE 0.2.4

The x and y axes are called the **coordinate axes**. They divide the plane into four parts, called **quadrants**. The first quadrant is the one in which the abscissa and the ordinate are both positive, that is, the upper right quadrant. The other quadrants are numbered in the counterclockwise direction, with the fourth being the lower right quadrant. See Fig. 0.2.4.

Because of the one-to-one correspondence, we identify R^2 with the geometric plane. For this reason we call an ordered pair (x, y) a *point*. Similarly, we refer to a *line* in R^2 as the set of all points corresponding to a line in the geometric plane, and we use other geometric terms for sets of points in R^2.

Consider the equation

$$y = x^2 - 2 \tag{1}$$

where (x, y) is a point in R^2. This is an equation in R^2. A solution of this equation means an ordered pair of numbers, one for x and one for y, that satisfies the equation. For example, if x is replaced by 3 in (1), we see that $y = 7$; thus $x = 3$ and $y = 7$ constitutes a solution of this equation. If any number is substituted for x in the right side of (1), a corresponding value for

y is obtained. It is seen, then, that (1) has an unlimited number of solutions. Table 0.2.1 gives a few such solutions.

If we plot the points having as coordinates the number pairs (x, y) satisfying (1), we have a sketch of the graph of the equation. In Fig. 0.2.5 points are plotted whose coordinates are the number pairs obtained from Table 0.2.1. These points are connected by a smooth curve. Any point (x, y) on this curve has coordinates satisfying (1). Also, the coordinates of any point not on this curve do not satisfy the equation.

0.2.2 DEFINITION

The **graph of an equation** in R^2 is the set of all points (x, y) in R^2 whose coordinates are numbers satisfying the equation.

The graph of an equation is sometimes called the **locus** of the equation. The graph of an equation in R^2 is also called a **curve**. Unless otherwise stated, an equation with two unknowns, x and y, is considered an equation in R^2.

The graph in Fig. 0.2.5 is a **parabola**. The lowest point on the graph is $(0, -2)$ and is the *vertex* of the parabola. This parabola opens upward. The y axis is the *axis* of the parabola. A complete treatment of parabolas occurs in Chapter 10, where we show that the graph of an equation of the form $y = ax^2 + bx + c$, where a, b, and c are real numbers and $a \neq 0$, is a parabola having a vertical axis and opening upward if $a > 0$ and downward if $a < 0$. In the following example, the graph is a parabola whose axis is horizontal.

Table 0.2.1

x	$y = x^2 - 2$
0	-2
1	-1
2	2
3	7
4	14
-1	-1
-2	2
-3	7
-4	14

FIGURE 0.2.5

EXAMPLE 1
Draw a sketch of the graph of the equation

$$y^2 - x - 2 = 0$$

SOLUTION
Solving the given equation for y we have

$$y = \pm\sqrt{x + 2} \tag{2}$$

Equations (2) are equivalent to the two equations

$$y = \sqrt{x + 2} \tag{3}$$

$$y = -\sqrt{x + 2} \tag{4}$$

The coordinates of all points that satisfy (2) will satisfy either (3) or (4), and the coordinates of any point that satisfies either (3) or (4) will satisfy (2). Table 0.2.2 gives some of these values of x and y.

Table 0.2.2

x	0	0	1	1	2	2	3	3	-1	-1	-2
y	$\sqrt{2}$	$-\sqrt{2}$	$\sqrt{3}$	$-\sqrt{3}$	2	-2	$\sqrt{5}$	$-\sqrt{5}$	1	-1	0

Observe that for any value of $x < -2$ there is no real value for y. Also, for each value of $x > -2$ there are two values for y. A sketch of the graph of the given equation is shown in Fig. 0.2.6.

FIGURE 0.2.6

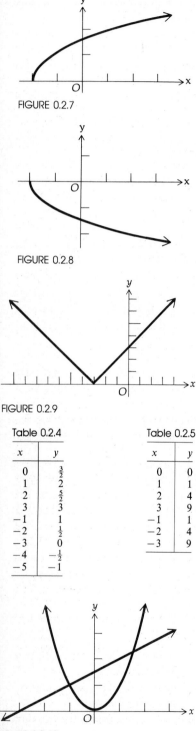

FIGURE 0.2.7

FIGURE 0.2.8

FIGURE 0.2.9

Table 0.2.4

x	y
0	$\frac{3}{2}$
1	2
2	$\frac{5}{2}$
3	3
−1	1
−2	$\frac{1}{2}$
−3	0
−4	$-\frac{1}{2}$
−5	−1

Table 0.2.5

x	y
0	0
1	1
2	4
3	9
−1	1
−2	4
−3	9

FIGURE 0.2.10

EXAMPLE 2

Draw sketches of the graphs of the equations

$$y = \sqrt{x + 2} \quad \text{and} \quad y = -\sqrt{x + 2}$$

SOLUTION

The equation $y = \sqrt{x + 2}$ is the same as (3). The value of y is nonnegative; hence the graph of the equation is the upper half of the graph of (2). A sketch of this graph is shown in Fig. 0.2.7.

Similarly, the graph of the equation $y = -\sqrt{x + 2}$, a sketch of which is shown in Fig. 0.2.8, is the lower half of the parabola of Fig. 0.2.6.

EXAMPLE 3

Draw a sketch of the graph of the equation

$$y = |x + 3|$$

SOLUTION

From the definition of the absolute value of a number it follows that

$$y = \begin{cases} x + 3 & \text{if } x + 3 \geq 0 \\ -(x + 3) & \text{if } x + 3 < 0 \end{cases}$$

or, equivalently,

$$y = \begin{cases} x + 3 & \text{if } x \geq -3 \\ -(x + 3) & \text{if } x < -3 \end{cases}$$

Table 0.2.3 gives some values of x and y satisfying the given equation.

Table 0.2.3

x	0	1	2	3	−1	−2	−3	−4	−5	−6	−7	−8	−9
y	3	4	5	6	2	1	0	1	2	3	4	5	6

A sketch of the graph of the given equation is shown in Fig. 0.2.9.

EXAMPLE 4

Draw a sketch of the graph of the equation

$$(x - 2y + 3)(y - x^2) = 0 \tag{5}$$

SOLUTION

By the property of real numbers that $ab = 0$ if and only if $a = 0$ or $b = 0$, we have from (5)

$$x - 2y + 3 = 0 \tag{6}$$

$$y - x^2 = 0 \tag{7}$$

The coordinates of all points that satisfy (5) will satisfy either (6) or (7), and the coordinates of any point that satisfies either (6) or (7) will satisfy (5). Therefore the graph of (5) will consist of the graphs of (6) and (7). Table 0.2.4 gives some values of x and y satisfying (6), and Table 0.2.5 gives some values of x and y satisfying (7). A sketch of the graph of (5) is shown in Fig. 0.2.10.

0.2.3 DEFINITION   An **equation of a graph** is an equation that is satisfied by the coordinates of those, and only those points on the graph.

▶ ILLUSTRATION 1   In R^2, $y = 4$ is an equation whose graph consists of those points having an ordinate of 4. This is a line parallel to the x axis and 4 units above it. Figure 0.2.11 shows a sketch of the graph.   ◀

In drawing a sketch of the graph of an equation it is often helpful to consider properties of *symmetry* of a graph.

0.2.4 DEFINITION   Two points P and Q are said to be **symmetric with respect to a line** if and only if the line is the perpendicular bisector of the line segment PQ. Two points P and Q are said to be **symmetric with respect to a third point** if and only if the third point is the midpoint of the line segment PQ.

FIGURE 0.2.11

▶ ILLUSTRATION 2   The points $(3, 2)$ and $(3, -2)$ are symmetric with respect to the x axis, the points $(3, 2)$ and $(-3, 2)$ are symmetric with respect to the y axis, and the points $(3, 2)$ and $(-3, -2)$ are symmetric with respect to the origin (see Fig. 0.2.12).   ◀

In general, the points (x, y) and $(x, -y)$ are symmetric with respect to the x axis, (x, y) and $(-x, y)$ are symmetric with respect to the y axis, and (x, y) and $(-x, -y)$ are symmetric with respect to the origin.

0.2.5 DEFINITION   The graph of an equation is symmetric with respect to a line l if and only if for every point P on the graph there is a point Q, also on the graph, such that P and Q are symmetric with respect to l. The graph of an equation is symmetric with respect to a point R if and only if for every point P on the graph there is a point S, also on the graph, such that P and S are symmetric with respect to R.

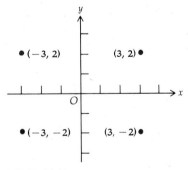

FIGURE 0.2.12

From Definition 0.2.5 it follows that if a point (x, y) is on a graph symmetric with respect to the x axis, then the point $(x, -y)$ also must be on the graph. And if both the points (x, y) and $(x, -y)$ are on the graph, then the graph is symmetric with respect to the x axis. Therefore the coordinates of the point $(x, -y)$ as well as (x, y) must satisfy an equation of the graph. Hence the graph of an equation in x and y is symmetric with respect to the x axis if and only if an equivalent equation is obtained when y is replaced by $-y$ in the equation. We have thus proved part (i) in the following theorem. The proofs of parts (ii) and (iii) are similar.

0.2.6 THEOREM

Tests for Symmetry

The graph of an equation in x and y is

 (i) symmetric with respect to the x axis if and only if an equivalent equation is obtained when y is replaced by $-y$ in the equation;
 (ii) symmetric with respect to the y axis if and only if an equivalent equation is obtained when x is replaced by $-x$ in the equation;
 (iii) symmetric with respect to the origin if and only if an equivalent equation is obtained when x is replaced by $-x$ and y is replaced by $-y$ in the equation.

The graph in Fig. 0.2.5 is symmetric with respect to the y axis, and for (1) an equivalent equation is obtained when x is replaced by $-x$. In Example 1 we have the equation $y^2 - x - 2 = 0$ for which an equivalent equation is obtained when y is replaced by $-y$, and its graph sketched in Fig. 0.2.6 is symmetric with respect to the x axis. The following example gives a graph that is symmetric with respect to the origin.

EXAMPLE 5
Draw a sketch of the graph of the equation

$$xy = 1$$

SOLUTION
We see that if in the given equation x is replaced by $-x$ and y is replaced by $-y$, an equivalent equation is obtained; hence by Theorem 0.2.6(iii) the graph is symmetric with respect to the origin. Table 0.2.6 gives some values of x and y satisfying the given equation.

Table 0.2.6

x	1	2	3	4	$\frac{1}{2}$	$\frac{1}{3}$	$\frac{1}{4}$	-1	-2	-3	-4	$-\frac{1}{2}$	$-\frac{1}{3}$	$-\frac{1}{4}$
y	1	$\frac{1}{2}$	$\frac{1}{3}$	$\frac{1}{4}$	2	3	4	-1	$-\frac{1}{2}$	$-\frac{1}{3}$	$-\frac{1}{4}$	-2	-3	-4

From the given equation, $y = 1/x$. We see that as x increases through positive values, y decreases through positive values and gets closer and closer to zero. As x decreases through positive values, y increases through positive values and gets larger and larger. As x increases through negative values (i.e., x takes on the values -4, -3, -2, -1, $-\frac{1}{2}$, etc.), y takes on negative values having larger and larger absolute values. A sketch of the graph is shown in Fig. 0.2.13.

FIGURE 0.2.13

EXERCISES 0.2

In Exercises 1 through 6, plot the given point P and such of the following points as may apply:
(a) The point Q such that the line through Q and P is perpendicular to the x axis and bisected by it. Give the coordinates of Q.
(b) The point R such that the line through P and R is perpendicular to and bisected by the y axis. Give the coordinates of R.
(c) The point S such that the line through P and S is bisected by the origin. Give the coordinates of S.
(d) The point T such that the line through P and T is perpendicular to and bisected by the $45°$ line through the origin bisecting the first and third quadrants. Give the coordinates of T.

1. $P(1, -2)$ **2.** $P(-2, 2)$ **3.** $P(2, 2)$ **4.** $P(-2, -2)$ **5.** $P(-1, -3)$ **6.** $P(0, -3)$

In Exercises 7 through 36, draw a sketch of the graph of the equation.

7. $y = 2x + 5$ **8.** $y = 4x - 3$ **9.** $y = \sqrt{x + 5}$ **10.** $y = \sqrt{x - 3}$ **11.** $y = -\sqrt{x + 5}$

12. $y = -\sqrt{x - 3}$ **13.** $y^2 = x - 3$ **14.** $y = 5$ **15.** $x = -3$ **16.** $x = y^2 + 1$

17. $y = |x - 5|$ **18.** $y = -|x + 2|$ **19.** $y = |x| - 5$ **20.** $y = -|x| + 2$ **21.** $y = 5 - |x + 1|$

22. $y = 3 - x^2$ **23.** $4x^2 + 9y^2 = 36$ **24.** $4x^2 - 9y^2 = 36$ **25.** $y = 4 + x^2$ **26.** $2y = -x^3$

27. $y = 4x^3$ **28.** $y^2 = 4x^3$ **29.** $4x^2 - y^2 = 0$ **30.** $y = \sqrt[3]{x}$ **31.** $4x^2 + y^2 = 0$

32. $3x^2 - 13xy - 10y^2 = 0$ **33.** $x^4 - 5x^2y + 4y^2 = 0$ **34.** $(2x + y - 1)(4y + x^2) = 0$

35. $2y = \sqrt[3]{x - 8}$ **36.** $(y^2 - x + 2)(y + \sqrt{x - 4}) = 0$

In Exercises 37 through 42, draw a sketch of the graph of each of the equations.

37. (a) $x + 3y = 0$; (b) $x - 3y = 0$; (c) $x^2 - 9y^2 = 0$ **38.** (a) $2x - 5y = 0$; (b) $2x + 5y = 0$; (c) $4x^2 - 25y^2 = 0$

39. (a) $y = \sqrt{2x}$; (b) $y = -\sqrt{2x}$; (c) $y^2 = 2x$ **40.** (a) $y = \sqrt{-2x}$; (b) $y = -\sqrt{-2x}$; (c) $y^2 = -2x$

41. (a) $y = \sqrt{4 - x^2}$; (b) $y = -\sqrt{4 - x^2}$; (c) $x^2 + y^2 = 4$ **42.** (a) $y = \sqrt{x^2 - 4}$; (b) $y = -\sqrt{x^2 - 4}$; (c) $x^2 - y^2 = 4$

43. (a) Write an equation whose graph is the x axis. (b) Write an equation whose graph is the y axis. (c) Write an equation whose graph is the set of all points on either the x axis or the y axis.

44. (a) Write an equation whose graph consists of all points having an abscissa of 4. (b) Write an equation whose graph consists of all points having an ordinate of -3.

45. Prove that a graph that is symmetric with respect to both coordinate axes is also symmetric with respect to the origin.

46. The graph of an equation in x and y is symmetric with respect to the line with equation $y = x$ if and only if an equivalent equation is obtained when x is replaced by y and y is replaced by x. Show that the graph of the equation $ax^2 + by^2 = c$ is symmetric with respect to this line if and only if $a = b$.

47. Prove that a graph that is symmetric with respect to any two perpendicular lines is also symmetric with respect to their point of intersection.

0.3
THE DISTANCE FORMULA, THE CIRCLE, AND THE MIDPOINT FORMULAS

We now discuss the problem of finding the distance between two points in R^2. If A is the point (x_1, y_1) and B is the point (x_2, y_1) (i.e., A and B have the same ordinate but different abscissas), then the **directed distance** from A to B is denoted by $\overline{AB}$, and we define

$$\overline{AB} = x_2 - x_1$$

▶ ILLUSTRATION 1   Refer to Fig. 0.3.1(a)–(c). If A is the point $(3, 4)$ and B is the point $(9, 4)$, then $\overline{AB} = 9 - 3$; that is, $\overline{AB} = 6$. If A is the point $(-8, 0)$ and B is the point $(6, 0)$, then $\overline{AB} = 6 - (-8)$; that is, $\overline{AB} = 14$. If A is the point $(4, 2)$ and B is the point $(1, 2)$, then $\overline{AB} = 1 - 4$; that is, $\overline{AB} = -3$. We see that $\overline{AB}$ is positive if B is to the right of A, and $\overline{AB}$ is negative if B is to the left of A. ◀

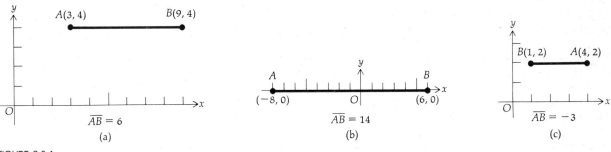

FIGURE 0.3.1

If C is the point (x_1, y_1) and D is the point (x_1, y_2), then the directed distance from C to D, denoted by $\overline{CD}$, is defined by

$$\overline{CD} = y_2 - y_1$$

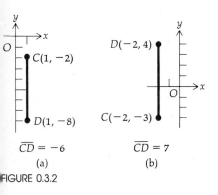

$\overline{CD} = -6$ $\overline{CD} = 7$

(a)   (b)

FIGURE 0.3.2

▶ ILLUSTRATION 2   Refer to Fig. 0.3.2(a) and (b). If C is the point $(1, -2)$ and D is the point $(1, -8)$, then $\overline{CD} = -8 - (-2)$; that is, $\overline{CD} = -6$. If C is the point $(-2, -3)$ and D is the point $(-2, 4)$, then $\overline{CD} = 4 - (-3)$; that is, $\overline{CD} = 7$. The number $\overline{CD}$ is positive if D is above C, and $\overline{CD}$ is negative if D is below C.
◀

Observe that the terminology *directed distance* indicates both a distance and a direction (positive or negative). If we are concerned only with the length of the line segment between two points P_1 and P_2 (i.e., the distance between the points P_1 and P_2 without regard to direction), then we use the terminology *undirected distance*. We denote the **undirected distance** from P_1 to P_2 by $|\overline{P_1P_2}|$, which is a nonnegative number.

We now wish to obtain a formula for computing $|\overline{P_1P_2}|$ if $P_1(x_1, y_1)$ and $P_2(x_2, y_2)$ are any two points in the plane. We use the Pythagorean theorem from plane geometry, which is as follows:

> *In a right triangle, the sum of the squares of the lengths of the perpendicular sides is equal to the square of the length of the hypotenuse.*

Figure 0.3.3 shows P_1 and P_2 in the first quadrant and the point $M(x_2, y_1)$. Note that $|\overline{P_1P_2}|$ is the length of the hypotenuse of right triangle P_1MP_2. Using the Pythagorean theorem we have

$$|\overline{P_1P_2}|^2 = |\overline{P_1M}|^2 + |\overline{MP_2}|^2$$
$$|\overline{P_1P_2}| = \sqrt{|\overline{P_1M}|^2 + |\overline{MP_2}|^2}$$
$$|\overline{P_1P_2}| = \sqrt{(x_2 - x_1)^2 + (y_2 - y_1)^2} \tag{1}$$

Notice in formula (1) we do not have a $\pm$ symbol in front of the radical in the right member because $|\overline{P_1P_2}|$ is a nonnegative number. Formula (1) holds for all possible positions of P_1 and P_2 in all four quadrants. The length of the hypotenuse is always $|\overline{P_1P_2}|$, and the lengths of the legs are always $|\overline{P_1M}|$ and $|\overline{MP_2}|$ (see Exercises 33 and 34). The result is stated as a theorem.

FIGURE 0.3.3

0.3.1 THEOREM   The undirected distance between two points $P_1(x_1, y_1)$ and $P_2(x_2, y_2)$ is given by

$$|\overline{P_1P_2}| = \sqrt{(x_2 - x_1)^2 + (y_2 - y_1)^2}$$

Observe that if P_1 and P_2 are on the same horizontal line, then $y_1 = y_2$, and

$$|\overline{P_1P_2}| = \sqrt{(x_2 - x_1)^2 + 0^2}$$

or, equivalently, because $\sqrt{a^2} = |a|$,

$$|\overline{P_1P_2}| = |x_2 - x_1|$$

Furthermore, if P_1 and P_2 are on the same vertical line, then $x_1 = x_2$, and

$$|\overline{P_1P_2}| = \sqrt{0^2 + (y_2 - y_1)^2}$$
$$|\overline{P_1P_2}| = |y_2 - y_1|$$

EXAMPLE 1

By showing that two sides have the same length, prove that the triangle with vertices at $A(-2, 4)$, $B(-5, 1)$, and $C(-6, 5)$ is isosceles.

SOLUTION

The triangle is shown in Fig. 0.3.4.

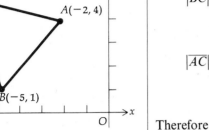

$$|\overline{BC}| = \sqrt{(-6+5)^2 + (5-1)^2} \qquad |\overline{BA}| = \sqrt{(-2+5)^2 + (4-1)^2}$$
$$= \sqrt{1+16} \qquad\qquad\qquad = \sqrt{9+9}$$
$$= \sqrt{17} \qquad\qquad\qquad\quad = 3\sqrt{2}$$

$$|\overline{AC}| = \sqrt{(-6+2)^2 + (5-4)^2}$$
$$= \sqrt{16+1}$$
$$= \sqrt{17}$$

Therefore

$$|\overline{BC}| = |\overline{AC}|$$

Hence triangle ABC is isosceles.

FIGURE 0.3.4

EXAMPLE 2

Prove analytically that the lengths of the diagonals of a rectangle are equal.

SOLUTION

Draw a general rectangle. Because the coordinate axes can be chosen anywhere in the plane, and because the choice of the position of the axes does not affect the truth of the theorem, we take the origin at one vertex, the x axis along one side, and the y axis along another side. This procedure simplifies the coordinates of the vertices on the two axes. Refer to Fig. 0.3.5.

Now the hypothesis and the conclusion of the theorem can be stated. *Hypothesis: OABC* is a rectangle with diagonals *OB* and *AC*. *Conclusion:* $|\overline{OB}| = |\overline{AC}|$.

PROOF

$$|\overline{OB}| = \sqrt{(a-0)^2 + (b-0)^2} \qquad |\overline{AC}| = \sqrt{(0-a)^2 + (b-0)^2}$$
$$= \sqrt{a^2 + b^2} \qquad\qquad\qquad = \sqrt{a^2 + b^2}$$

Therefore

$$|\overline{OB}| = |\overline{AC}|$$ ∎

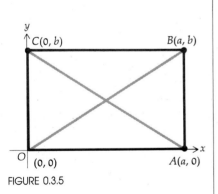

FIGURE 0.3.5

One of the simplest curves that is the graph of a second-degree equation in two variables is the *circle*, which is defined now.

0.3.2 DEFINITION   A **circle** is the set of all points in a plane equidistant from a fixed point. The fixed point is called the **center** of the circle, and the constant equal distance is called the **radius** of the circle.

The formula in the following theorem for obtaining an equation of a circle is derived by applying the distance formula.

0.3.3 THEOREM The circle with center at the point $C(h, k)$ and radius r has as an equation

$$(x - h)^2 + (y - k)^2 = r^2 \qquad (2)$$

PROOF The point $P(x, y)$ lies on the circle if and only if

$$|\overline{PC}| = r$$

that is, if and only if

$$\sqrt{(x - h)^2 + (y - k)^2} = r$$

This is true if and only if

$$(x - h)^2 + (y - k)^2 = r^2$$

which is (2). Equation (2) is satisfied by the coordinates of those and only those points that lie on the given circle. Hence (2) is an equation of the circle.   ■

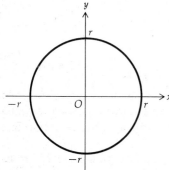

FIGURE 0.3.6

Figure 0.3.6 shows the circle with center at (h, k) and radius r. If the center of the circle is at the origin, then $h = 0$ and $k = 0$; therefore its equation is

$$x^2 + y^2 = r^2$$

Such a circle is shown in Fig. 0.3.7.

If the center and radius of a circle are known, the circle can be drawn by using a compass.

FIGURE 0.3.7

▶ ILLUSTRATION 3 The equation

$$x^2 + y^2 = 9$$

is an equation of a circle having its center at the origin and radius 3.   ◀

EXAMPLE 3
Find an equation of the circle with center at $(4, -2)$ and radius equal to 5. Draw the circle.

SOLUTION
The circle is shown in Fig. 0.3.8. An equation of the circle is of the form of (2), where $h = 4$, $k = -2$, and $r = 5$. Substituting into (2) we have

$$(x - 4)^2 + [y - (-2)]^2 = 5^2$$

$$(x - 4)^2 + (y + 2)^2 = 25$$

$$x^2 - 8x + 16 + y^2 + 4y + 4 = 25$$

$$x^2 + y^2 - 8x + 4y - 5 = 0$$

FIGURE 0.3.8

EXAMPLE 4

Show that the graph of the equation

$$x^2 + y^2 + 6x - 2y - 15 = 0$$

is a circle, and find its center and radius.

SOLUTION

The given equation may be written as

$$(x^2 + 6x) + (y^2 - 2y) = 15$$

Completing the squares of the terms in parentheses by adding 9 and 1 on both sides of the equation we have

$$(x^2 + 6x + 9) + (y^2 - 2y + 1) = 15 + 9 + 1$$

$$(x + 3)^2 + (y - 1)^2 = 25$$

Comparing this equation with (2) we see that this is an equation of a circle with its center at $(-3, 1)$ and a radius of 5.

In (2), removing parentheses and combining terms gives

$$x^2 + y^2 - 2hx - 2ky + (h^2 + k^2 - r^2) = 0 \qquad (3)$$

Equation (3) is of the form

$$x^2 + y^2 + Dx + Ey + F = 0 \qquad (4)$$

where $D = -2h$, $E = -2k$, and $F = h^2 + k^2 - r^2$.

Equation (4) is called the **general form** of an equation of a circle, whereas (2) is called the **center-radius form**. Because every circle has a center and a radius, its equation can be put in the center-radius form, and hence into the general form.

Now consider the question of whether or not the graph of every equation of the form

$$x^2 + y^2 + Dx + Ey + F = 0$$

is a circle. To determine this we attempt to write this equation in the center-radius form. The equation is rewritten as

$$x^2 + y^2 + Dx + Ey = -F$$

We complete the squares of the terms in parentheses by adding $\frac{1}{4}D^2$ and $\frac{1}{4}E^2$ on both sides, thus giving

$$(x^2 + Dx + \tfrac{1}{4}D^2) + (y^2 + Ey + \tfrac{1}{4}E^2) = -F + \tfrac{1}{4}D^2 + \tfrac{1}{4}E^2$$

$$(x + \tfrac{1}{2}D)^2 + (y + \tfrac{1}{2}E)^2 = \tfrac{1}{4}(D^2 + E^2 - 4F) \qquad (5)$$

Equation (5) is in the form of (2) if and only if

$$\tfrac{1}{4}(D^2 + E^2 - 4F) = r^2$$

We now consider three cases, namely, $D^2 + E^2 - 4F$ as positive, zero, and negative.

Case 1: $D^2 + E^2 - 4F > 0$.

Then $r^2 = \frac{1}{4}(D^2 + E^2 - 4F)$, and so (5) is an equation of a circle having a radius equal to $\frac{1}{2}\sqrt{D^2 + E^2 - 4F}$ and its center at $(-\frac{1}{2}D, -\frac{1}{2}E)$.

Case 2: $D^2 + E^2 - 4F = 0$.

Equation (5) is then of the form

$$(x + \tfrac{1}{2}D)^2 + (y + \tfrac{1}{2}E)^2 = 0 \tag{6}$$

Because the only real values of x and y satisfying (6) are $x = -\frac{1}{2}D$ and $y = -\frac{1}{2}E$, the graph is the point $(-\frac{1}{2}D, -\frac{1}{2}E)$. Comparing (6) with (2) we see that $h = -\frac{1}{2}D$, $k = -\frac{1}{2}E$, and $r = 0$. Thus this point can be called a **point-circle**.

Case 3: $D^2 + E^2 - 4F < 0$.

Then (5) has a negative number on the right side and the sum of the squares of two real numbers on the left side. There are no real values of x and y that satisfy such an equation; consequently the graph is the empty set.

Before the results of these three cases are stated as a theorem, observe that an equation of the form

$$Ax^2 + Ay^2 + Dx + Ey + F = 0 \qquad \text{where } A \neq 0 \tag{7}$$

can be written in the form of (4) by dividing by A, thereby obtaining

$$x^2 + y^2 + \frac{D}{A}x + \frac{E}{A}y + \frac{F}{A} = 0$$

Equation (7) is a special case of the general equation of the second degree,

$$Ax^2 + Bxy + Cy^2 + Dx + Ey + F = 0$$

in which the coefficients of x^2 and y^2 are equal and which has no xy term. The following theorem is a result of this discussion.

0.3.4 THEOREM   The graph of any second-degree equation in R^2 in x and y, in which the coefficients of x^2 and y^2 are equal and in which there is no xy term, is either a circle, a point-circle, or the empty set.

▶ ILLUSTRATION 4   The equation

$$2x^2 + 2y^2 + 12x - 8y + 31 = 0$$

is of the form (7), and therefore its graph is either a circle, a point-circle, or the empty set. The equation is put in the form of (2):

$$x^2 + y^2 + 6x - 4y + \tfrac{31}{2} = 0$$

$$(x^2 + 6x) + (y^2 - 4y) = -\tfrac{31}{2}$$

$$(x^2 + 6x + 9) + (y^2 - 4y + 4) = -\tfrac{31}{2} + 9 + 4$$

$$(x + 3)^2 + (y - 2)^2 = -\tfrac{5}{2}$$

Therefore the graph is the empty set.   ◀

EXAMPLE 5

Find an equation of the circle through the three points $A(4, 5)$, $B(3, -2)$, and $C(1, -4)$.

SOLUTION

The general form of an equation of the circle is

$$x^2 + y^2 + Dx + Ey + F = 0$$

Because the three points A, B, and C must lie on the circle, the coordinates of these points must satisfy the equation. Thus

$$16 + 25 + 4D + 5E + F = 0$$

$$9 + 4 + 3D - 2E + F = 0$$

$$1 + 16 + D - 4E + F = 0$$

or, equivalently,

$$4D + 5E + F = -41$$

$$3D - 2E + F = -13$$

$$D - 4E + F = -17$$

Solving these three equations simultaneously we get

$$D = 7 \qquad E = -5 \qquad F = -44$$

Thus an equation of the circle is

$$x^2 + y^2 + 7x - 5y - 44 = 0$$

Let $P_1(x_1, y_1)$ and $P_2(x_2, y_2)$ be the endpoints of a line segment. We denote this line segment by $P_1 P_2$. This is not to be confused with the notation $\overline{P_1 P_2}$, which denotes the directed distance from P_1 to P_2. That is, $\overline{P_1 P_2}$ denotes a number, whereas $P_1 P_2$ is a line segment. Let $P(x, y)$ be the midpoint of the line segment $P_1 P_2$. Refer to Fig. 0.3.9.

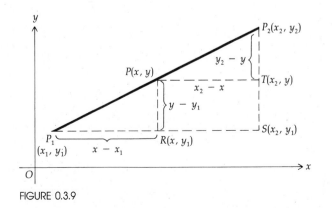

FIGURE 0.3.9

In Fig. 0.3.9, triangles P_1RP and PTP_2 are congruent. Therefore $|P_1R| = |PT|$, and so $x - x_1 = x_2 - x$, giving

$$x = \frac{x_1 + x_2}{2} \tag{8}$$

Similarly, $|RP| = |TP_2|$. Then $y - y_1 = y_2 - y$, and therefore

$$y = \frac{y_1 + y_2}{2} \tag{9}$$

Hence the coordinates of the midpoint of a line segment are, respectively, the average of the abscissas and the average of the ordinates of the end-points of the line segment.

In the derivation of formulas (8) and (9) it was assumed that $x_2 > x_1$ and $y_2 > y_1$. The same formulas are obtained by using any orderings of these numbers (see Exercises 35 and 36).

EXAMPLE 6
Find an equation of the circle having a diameter with extremities at $A(-2, 3)$ and $B(4, 5)$.

SOLUTION
Because the extremities of a diameter are at the points A and B, the mid-point of the line segment AB is the center of the circle. See Fig. 0.3.10. If $C(h, k)$ is the center of the circle, then

$$h = \frac{-2 + 4}{2} \qquad k = \frac{3 + 5}{2}$$

$$= 1 \qquad\qquad = 4$$

The center is at $C(1, 4)$. The radius of the circle can be computed as either $|CA|$ or $|CB|$. If $r = |CA|$, then

$$r = \sqrt{(1 + 2)^2 + (4 - 3)^2}$$
$$= \sqrt{10}$$

An equation of the circle is therefore

$$(x - 1)^2 + (y - 4)^2 = 10$$
$$x^2 + y^2 - 2x - 8y + 7 = 0$$

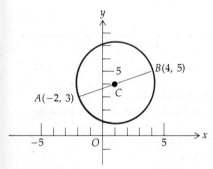

FIGURE 0.3.10

EXAMPLE 7
Prove analytically that the line segments joining the midpoints of the oppo-site sides of any quadrilateral bisect each other.

SOLUTION
Draw a general quadrilateral. Take the origin at one vertex and the x axis along one side. This method simplifies the coordinates of the two vertices on the x axis. See Fig. 0.3.11.

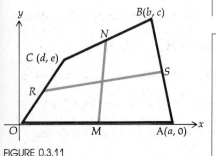

FIGURE 0.3.11

Hypothesis: OABC is a quadrilateral. *M* is the midpoint of *OA*, *N* is the midpoint of *CB*, *R* is the midpoint of *OC*, and *S* is the midpoint of *AB*.
Conclusion: MN and *RS* bisect each other.

PROOF   To prove that two line segments bisect each other, we show that they have the same midpoint. From formulas (8) and (9) we obtain the coordinates of *M*, *N*, *R*, and *S*. *M* is the point $(\frac{1}{2}a, 0)$, *N* is the point $(\frac{1}{2}(b + d), \frac{1}{2}(c + e))$, *R* is the point $(\frac{1}{2}d, \frac{1}{2}e)$, and *S* is the point $(\frac{1}{2}(a + b), \frac{1}{2}c)$.

The abscissa of the midpoint of *MN* is $\frac{1}{2}[\frac{1}{2}a + \frac{1}{2}(b + d)] = \frac{1}{4}(a + b + d)$.

The ordinate of the midpoint of *MN* is $\frac{1}{2}[0 + \frac{1}{2}(c + e)] = \frac{1}{4}(c + e)$.

Therefore, the midpoint of *MN* is the point $(\frac{1}{4}(a + b + d), \frac{1}{4}(c + e))$.

The abscissa of the midpoint of *RS* is $\frac{1}{2}[\frac{1}{2}d + \frac{1}{2}(a + b)] = \frac{1}{4}(a + b + d)$.

The ordinate of the midpoint of *RS* is $\frac{1}{2}[\frac{1}{2}e + \frac{1}{2}c] = \frac{1}{4}(c + e)$.

Therefore, the midpoint of *RS* is the point $(\frac{1}{4}(a + b + d), \frac{1}{4}(c + e))$.

Thus, the midpoint of *MN* is the same point as the midpoint of *RS*.

Therefore, *MN* and *RS* bisect each other.   ∎

EXERCISES 0.3

1. A median of a triangle is a line segment from a vertex to the midpoint of the opposite side. Find the length of the medians of the triangle having vertices $A(2, 3)$, $B(3, -3)$, and $C(-1, -1)$.

2. Find the length of the medians of the triangle having vertices $A(-3, 5)$, $B(2, 4)$, and $C(-1, -4)$.

3. Prove that the triangle with vertices $A(3, -6)$, $B(8, -2)$, and $C(-1, -1)$ is a right triangle. Find the area of the triangle. (*Hint:* Use the converse of the Pythagorean theorem.)

4. Find the midpoints of the diagonals of the quadrilateral whose vertices are $(0, 0)$, $(0, 4)$, $(3, 5)$, and $(3, 1)$.

5. Prove that the points $A(-7, 2)$, $B(3, -4)$, and $C(1, 4)$ are the vertices of an isosceles triangle.

6. Prove that the points $A(-4, -1)$, $B(-2, -3)$, $C(4, 3)$, and $D(2, 5)$ are the vertices of a rectangle.

7. By using distance formula (1), prove that the points $(-3, 2)$, $(1, -2)$, and $(9, -10)$ lie on a line.

8. Determine whether the points $(14, 7)$, $(2, 2)$, and $(-4, -1)$ lie on a line by using distance formula (1).

9. Prove that the points $A(6, -13)$, $B(-2, 2)$, $C(13, 10)$, and $D(21, -5)$ are the vertices of a square. Find the length of a diagonal.

10. If one end of a line segment is the point $(-4, 2)$ and the midpoint is $(3, -1)$, find the coordinates of the other end of the line segment. $(10, -4)$

11. If one end of a line segment is the point $(6, -2)$ and the midpoint is $(-1, 5)$, find the coordinates of the other end of the line segment.

12. The abscissa of a point is -6, and its distance from the point $(1, 3)$ is $\sqrt{74}$. Find the ordinate of the point.

13. If two vertices of an equilateral triangle are $(-4, 3)$ and $(0, 0)$, find the third vertex.

14. Find an equation that must be satisfied by the coordinates of any point that is equidistant from the two points $(-3, 2)$ and $(4, 6)$.

15. Find an equation that must be satisfied by the coordinates of any point whose distance from the point $(5, 3)$ is always two units greater than its distance from the point $(-4, -2)$.

16. Given the two points $A(-3, 4)$ and $B(2, 5)$, find the coordinates of a point *P* on the line through *A* and *B* such that *P* is (a) twice as far from *A* as from *B*, and (b) twice as far from *B* as from *A*.

In Exercises 17 through 20, find an equation of the circle with center at C and radius r. Write the equation in both the center-radius form and the general form.

17. $C(4, -3), r = 5$ **18.** $C(0, 0), r = 8$ **19.** $C(-5, -12), r = 3$ **20.** $C(-1, 1), r = 2$

In Exercises 21 and 22, find an equation of the circle satisfying the given conditions.

21. Center is at $(1, 2)$ and through the point $(3, -1)$.   **22.** Through the three points $(2, 8)$, $(7, 3)$, and $(-2, 0)$.

In Exercises 23 through 26, find the center and radius of each circle, and draw a sketch of the graph.

23. $x^2 + y^2 - 6x - 8y + 9 = 0$ **24.** $2x^2 + 2y^2 - 2x + 2y + 7 = 0$

25. $3x^2 + 3y^2 + 4y - 7 = 0$ **26.** $x^2 + y^2 - 10x - 10y + 25 = 0$

In Exercises 27 through 32, determine whether the graph is a circle, a point-circle, or the empty set.

27. $x^2 + y^2 - 2x + 10y + 19 = 0$ **28.** $4x^2 + 4y^2 + 24x - 4y + 1 = 0$

29. $x^2 + y^2 - 10x + 6y + 36 = 0$ **30.** $x^2 + y^2 + 2x - 4y + 5 = 0$

31. $36x^2 + 36y^2 - 48x + 36y - 119 = 0$ **32.** $9x^2 + 9y^2 + 6x - 6y + 5 = 0$

33. Derive distance formula (1) if P_1 is in the third quadrant and P_2 is in the second quadrant. Draw a figure.

34. Derive distance formula (1) if P_1 is in the second quadrant and P_2 is in the fourth quadrant. Draw a figure.

35. Derive midpoint formulas (8) and (9) if P_1 is in the first quadrant and P_2 is in the third quadrant.

36. Derive midpoint formulas (8) and (9) if $P_1(x_1, y_1)$ and $P_2(x_2, y_2)$ are both in the second quadrant and $x_2 > x_1$ and $y_1 > y_2$.

37. Find the coordinates of the three points that divide the line segment from $A(-5, 3)$ to $B(6, 8)$ into four equal parts.

38. If the point $P(x, y)$ is on the line through $P_1(x_1, y_1)$ and $P_2(x_2, y_2)$ such that $|\overline{P_1 P}|/|\overline{P_1 P_2}| = k$, prove that the coordinates of P are given by

$$x = x_1 + k(x_2 - x_1) \quad \text{and} \quad y = y_1 + k(y_2 - y_1)$$

where

$0 < k < 1$ if P is between P_1 and P_2

$k > 1$ if P is not between P_1 and P_2 and closer to P_2

$k < 0$ if P is not between P_1 and P_2 and closer to P_1

In Exercises 39 through 43, use the formulas of Exercise 38 to find the coordinates of point P.

39. The point P is on the line segment between points $P_1(1, 3)$ and $P_2(6, 2)$ and is three times as far from P_2 as it is from P_1.

40. The point P is on the line segment between points $P_1(1, 3)$ and $P_2(6, 2)$ and is three times as far from P_1 as it is from P_2.

41. The point P is on the line through P_1 and P_2 and is three times as far from $P_2(6, 2)$ as it is from $P_1(1, 3)$ but is not between P_1 and P_2.

42. The point P is on the line through P_1 and P_2 and is three times as far from $P_1(1, 3)$ as it is from $P_2(6, 2)$ but is not between P_1 and P_2.

43. The point P is on the line through $P_1(-3, 5)$ and $P_2(-1, 2)$ so that $\overline{PP_1} = 4 \cdot \overline{P_1 P_2}$.

44. From the origin, chords of the circle $x^2 + y^2 + 4x = 0$ are drawn. Prove that the set of midpoints of these chords is a circle.

45. Find an equation whose graph consists of all points equidistant from the points $(-1, 2)$ and $(3, 4)$. Draw a sketch of the graph of this equation.

46. Find an equation that must be satisfied by the coordinates of any point the sum of whose distances from $(0, -3)$ and $(0, 3)$ is 9. Draw a sketch of the graph of this equation.

47. Prove analytically that the sum of the squares of the distances of any point from two opposite vertices of any rectangle is equal to the sum of the squares of its distances from the other two vertices.

48. Prove analytically that the midpoint of the hypotenuse of any right triangle is equidistant from each of the three vertices.

49. Prove analytically that the line segment joining the midpoints of two opposite sides of any quadrilateral and the line segment joining the midpoints of the diagonals of the quadrilateral bisect each other.

50. Prove analytically that if the lengths of two of the medians of a triangle are equal, the triangle is isosceles.

0.4
EQUATIONS OF A LINE

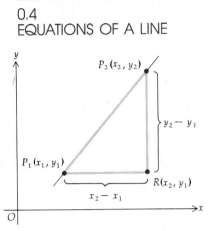

FIGURE 0.4.1

Let l be a nonvertical line and $P_1(x_1, y_1)$ and $P_2(x_2, y_2)$ be any two distinct points on l. Figure 0.4.1 shows such a line. In the figure, R is the point (x_2, y_1), and the points P_1, P_2, and R are vertices of a right triangle; furthermore, $\overline{P_1R} = x_2 - x_1$ and $\overline{RP_2} = y_2 - y_1$. The number $y_2 - y_1$ gives the measure of the change in the ordinate from P_1 to P_2, and it may be positive, negative, or zero. The number $x_2 - x_1$ gives the measure of the change in the abscissa from P_1 to P_2, and it may be positive or negative. Because the line l is not vertical, $x_2 \neq x_1$, and therefore $x_2 - x_1$ is not zero. Let

$$m = \frac{y_2 - y_1}{x_2 - x_1} \tag{1}$$

The value of m computed from this equation is independent of the choice of the two points P_1 and P_2 on l. To show this, suppose we choose two different points $\bar{P}_1(\bar{x}_1, \bar{y}_1)$ and $\bar{P}_2(\bar{x}_2, \bar{y}_2)$ and compute a number $\bar{m}$ from (1).

$$\bar{m} = \frac{\bar{y}_2 - \bar{y}_1}{\bar{x}_2 - \bar{x}_1}$$

We shall show that $\bar{m} = m$. Refer to Fig. 0.4.2. Triangles $\bar{P}_1\bar{R}\bar{P}_2$ and P_1RP_2 are similar; so the lengths of corresponding sides are proportional. Therefore

$$\frac{\bar{y}_2 - \bar{y}_1}{\bar{x}_2 - \bar{x}_1} = \frac{y_2 - y_1}{x_2 - x_1}$$

or

$$\bar{m} = m$$

Thus the value of m computed from (1) is the same number no matter what two points on l are selected. This number m is called the *slope* of the line. Following is the formal definition.

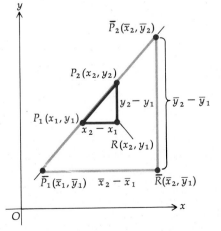

FIGURE 0.4.2

0.4.1 DEFINITION   If $P_1(x_1, y_1)$ and $P_2(x_2, y_2)$ are any two distinct points on line l, which is not parallel to the y axis, then the **slope** of l, denoted by m, is given by

$$m = \frac{y_2 - y_1}{x_2 - x_1} \tag{2}$$

If we multiply on both sides of (2) by $x_2 - x_1$, we obtain

$$y_2 - y_1 = m(x_2 - x_1)$$

It follows from this equation that if we consider a particle moving along a line, the change in the ordinate of the particle is equal to the product of the slope and the change in the abscissa.

▶ ILLUSTRATION 1   If l is the line through the points $P_1(2, 3)$ and $P_2(4, 7)$, and m is the slope of l, then by Definition 0.4.1,

$$m = \frac{7 - 3}{4 - 2}$$

$$= 2$$

FIGURE 0.4.3

FIGURE 0.4.4

FIGURE 0.4.5

Refer to Fig. 0.4.3. If a particle is moving along the line l, the change in the ordinate is two times the change in the abscissa. That is, if the particle is at $P_2(4, 7)$ and the abscissa is increased by one unit, then the ordinate is increased by two units, and the particle is at the point $P_3(5, 9)$. Similarly, if the particle is at $P_1(2, 3)$ and the abscissa is decreased by three units, then the ordinate is decreased by six units, and the particle is at $P_4(-1, -3)$. ◄

If the slope of a line is positive, then as the abscissa of a point on the line increases, the ordinate increases. Such a line is shown in Fig. 0.4.4. In Fig. 0.4.5 there is a line whose slope is negative. For this line, as the abscissa of a point on the line increases, the ordinate decreases.

If a line is parallel to the x axis, then $y_2 = y_1$, and so the slope of the line is zero. If a line is parallel to the y axis, $x_2 = x_1$; thus (1) is meaningless, because we cannot divide by zero. This is the reason that lines parallel to the y axis, or vertical lines, are excluded in Definition 0.4.1. Thus a vertical line does not have a slope.

Since two points $P_1(x_1, y_1)$ and $P_2(x_2, y_2)$ determine a unique line, we should be able to obtain an equation of the line through these two points. Consider $P(x, y)$ any point on the line. We want an equation that is satisfied by x and y if and only if $P(x, y)$ is on the line through $P_1(x_1, y_1)$ and $P_2(x_2, y_2)$. There are two cases.

Case 1: $x_2 = x_1$.

In this case the line through P_1 and P_2 is parallel to the y axis, and all points on this line have the same abscissa. So $P(x, y)$ is any point on the line if and only if

$$x = x_1 \tag{3}$$

Equation (3) is an equation of a line parallel to the y axis. Note that this equation is independent of y; that is, the ordinate may have any value whatsoever, and the point $P(x, y)$ is on the line whenever the abscissa is x_1.

Case 2: $x_2 \neq x_1$.

The slope of the line through P_1 and P_2 is given by

$$m = \frac{y_2 - y_1}{x_2 - x_1} \tag{4}$$

If $P(x, y)$ is any point on the line except (x_1, y_1), the slope is also given by

$$m = \frac{y - y_1}{x - x_1} \tag{5}$$

The point P will be on the line through P_1 and P_2 if and only if the value of m from (4) is the same as the value of m from (5), that is, if and only if

$$\frac{y - y_1}{x - x_1} = \frac{y_2 - y_1}{x_2 - x_1}$$

Multiplying on both sides of this equation by $x - x_1$ we obtain

$$y - y_1 = \frac{y_2 - y_1}{x_2 - x_1}(x - x_1) \tag{6}$$

Equation (6) is satisfied by the coordinates of P_1 as well as by the coordinates of any other point on the line through P_1 and P_2.

Equation (6) is called the **two-point** form of an equation of the line. It gives an equation of the line if two points on the line are known.

If in (6) we replace $\frac{y_2 - y_1}{x_2 - x_1}$ by m, we get

$$y - y_1 = m(x - x_1) \tag{7}$$

Equation (7) is called the **point-slope** form of an equation of the line. It gives an equation of the line if a point $P_1(x_1, y_1)$ and the slope m of the line are known. It is recommended that you use the point-slope form even when two points are given, as shown in the following illustration.

▶ ILLUSTRATION 2   To find an equation of the line through the two points $A(6, -3)$ and $B(-2, 3)$ we first compute m.

$$m = \frac{3 - (-3)}{-2 - 6}$$

$$= \frac{6}{-8}$$

$$= -\tfrac{3}{4}$$

Using the point-slope form of an equation of the line with A as P_1 we have

$$y - (-3) = -\tfrac{3}{4}(x - 6)$$

$$4y + 12 = -3x + 18$$

$$3x + 4y = 6$$

Of course, the point B may be taken as P_1; in which case we have

$$y - 3 = -\tfrac{3}{4}(x + 2)$$

$$4y - 12 = -3x - 6$$

$$3x + 4y = 6 \qquad\qquad ◀$$

If we choose the particular point $(0, b)$ (i.e., the point where the line intersects the y axis) for the point (x_1, y_1) in (7), we have

$$y - b = m(x - 0)$$

$$y = mx + b \tag{8}$$

The number b, which is the ordinate of the point where the line intersects the y axis, is called the **y intercept** of the line. Consequently, (8) is called the **slope-intercept** form of an equation of the line. This form is especially important because it enables us to find the slope of a line from its equation. It is also important because it expresses the y coordinate explicitly in terms of the x coordinate.

EXAMPLE 1

Given the line having the equation

$$6x + 5y - 7 = 0$$

find the slope of the line.

SOLUTION

The equation is solved for y.

$$5y = -6x + 7$$

$$y = -\tfrac{6}{5}x + \tfrac{7}{5}$$

Comparing this equation with (8) we see that $m = -\tfrac{6}{5}$ and $b = \tfrac{7}{5}$. Therefore the slope is $-\tfrac{6}{5}$.

Another form of an equation of a line is the one involving the intercepts of a line. The **x intercept** of a line is defined as the abscissa of the point at which the line intersects the x axis. The x intercept is denoted by a. If the x intercept a and the y intercept b are given, two points $(a, 0)$ and $(0, b)$ on the line are known. Applying (6), the two-point form, we have

$$y - 0 = \frac{b - 0}{0 - a}(x - a)$$

$$-ay = bx - ab$$

$$bx + ay = ab$$

Dividing by ab, if $a \neq 0$ and $b \neq 0$, we obtain

$$\frac{x}{a} + \frac{y}{b} = 1 \tag{9}$$

Equation (9) is called the **intercept** form of an equation of the line. Obviously it does not apply to a line through the origin, because for such a line both a and b are zero.

EXAMPLE 2

The point $(2, 3)$ bisects the portion of a line that is cut off by the coordinate axes. Find an equation of the line.

SOLUTION

Refer to Fig. 0.4.6. If a is the x intercept of the line and b is the y intercept of the line, then the point $(2, 3)$ is the midpoint of the line segment joining

FIGURE 0.4.6

$(a, 0)$ and $(0, b)$. By the midpoint formulas,

$$2 = \frac{a + 0}{2} \qquad 3 = \frac{0 + b}{2}$$

$$a = 4 \qquad\qquad b = 6$$

The intercept form, (9), gives

$$\frac{x}{4} + \frac{y}{6} = 1$$

$$3x + 2y = 12$$

0.4.2 THEOREM   The graph of the equation

$$Ax + By + C = 0 \qquad\qquad (10)$$

where A, B, and C are constants and where not both A and B are zero, is a straight line.

PROOF   Consider the two cases $B \neq 0$ and $B = 0$.

Case 1: $B \neq 0$.

Because $B \neq 0$, we divide on both sides of (10) by B and obtain

$$y = -\frac{A}{B} x - \frac{C}{B} \qquad\qquad (11)$$

Equation (11) is an equation of a straight line because it is in the slope-intercept form, where $m = -A/B$ and $b = -C/B$.

Case 2: $B = 0$.

Because $B = 0$, we may conclude that $A \neq 0$ and thus have

$$Ax + C = 0$$

$$x = -\frac{C}{A} \qquad\qquad (12)$$

Equation (12) is in the form of (3); thus the graph is a straight line parallel to the y axis. This completes the proof.   ■

Because the graph of (10) is a straight line, it is called a **linear equation**. Equation (10) is the general equation of the first degree in x and y.

Since two points determine a line, to draw a sketch of the graph of a straight line from its equation we need only determine the coordinates of two points on the line, plot the two points, and then draw the line. Any two points will suffice, but it is usually convenient to plot the two points where the line intersects the two axes. These points are given by the intercepts.

EXAMPLE 3

Given line l_1, having the equation $2x - 3y = 12$, and line l_2, having the equation $4x + 3y = 6$, draw a sketch of each of the lines. Then find the co-ordinates of the point of intersection of l_1 and l_2.

FIGURE 0.4.7

SOLUTION
To draw a sketch of the graph of l_1 we find the intercepts a and b. In the equation of l_1 substitute 0 for x and get $b = -4$. In the equation of l_1 substitute 0 for y and get $a = 6$. Similarly, we obtain the intercepts a and b for l_2, and for l_2 $a = \frac{3}{2}$ and $b = 2$. The two lines are plotted in Fig. 0.4.7.

To find the coordinates of the point of intersection of l_1 and l_2 we solve the two equations simultaneously. Because the point must lie on both lines, its coordinates must satisfy both equations. Both equations are put in the slope-intercept form.

$$y = \tfrac{2}{3}x - 4 \quad \text{and} \quad y = -\tfrac{4}{3}x + 2$$

Eliminating y gives

$$\tfrac{2}{3}x - 4 = -\tfrac{4}{3}x + 2$$

$$2x - 12 = -4x + 6$$

$$x = 3$$

Thus

$$y = \tfrac{2}{3}(3) - 4$$

$$= -2$$

Therefore the point of intersection is $(3, -2)$.

0.4.3 THEOREM   If l_1 and l_2 are two distinct nonvertical lines having slopes m_1 and m_2, respectively, then l_1 and l_2 are parallel if and only if $m_1 = m_2$.

PROOF   Let an equation of l_1 be $y = m_1 x + b_1$, and let an equation of l_2 be $y = m_2 x + b_2$. Because there is an "if and only if" qualification, the proof consists of two parts.

Part 1: Prove that l_1 and l_2 are parallel if $m_1 = m_2$.
Assume that l_1 and l_2 are not parallel. Let us show that this assumption leads to a contradiction. If l_1 and l_2 are not parallel, then they intersect. Call this point of intersection $P(x_0, y_0)$. The coordinates of P must satisfy the equations of l_1 and l_2, and so

$$y_0 = m_1 x_0 + b_1 \quad \text{and} \quad y_0 = m_2 x_0 + b_2$$

But $m_1 = m_2$, which gives

$$y_0 = m_1 x_0 + b_1 \quad \text{and} \quad y_0 = m_1 x_0 + b_2$$

from which it follows that $b_1 = b_2$. Thus, because $m_1 = m_2$ and $b_1 = b_2$, both lines l_1 and l_2 have the same equation, $y = m_1 x + b_1$, and so the lines are the same. But this contradicts the hypothesis that l_1 and l_2 are distinct lines. Therefore our assumption is false. So we conclude that l_1 and l_2 are parallel.

Part 2: Prove that l_1 and l_2 are parallel only if $m_1 = m_2$.
Here we must show that if l_1 and l_2 are parallel, then $m_1 = m_2$. Assume that $m_1 \neq m_2$. Solving the equations for l_1 and l_2 simultaneously we get,

upon eliminating y,

$$m_1 x + b_1 = m_2 x + b_2$$

$$(m_1 - m_2)x = b_2 - b_1$$

Because we have assumed that $m_1 \neq m_2$, this gives

$$x = \frac{b_2 - b_1}{m_1 - m_2}$$

Hence l_1 and l_2 have a point of intersection, which contradicts the hypothesis that l_1 and l_2 are parallel. So our assumption is false, and therefore $m_1 = m_2$. ■

▶ ILLUSTRATION 3 Let l_1 be the line through the points $A(1,2)$ and $B(3, -6)$ and m_1 be the slope of l_1; and let l_2 be the line through the points $C(2, -5)$ and $D(-1, 7)$ and m_2 be the slope of l_2. Then

$$m_1 = \frac{-6 - 2}{3 - 1} \qquad m_2 = \frac{7 - (-5)}{-1 - 2}$$

$$= \frac{-8}{2} \qquad\qquad = \frac{12}{-3}$$

$$= -4 \qquad\qquad = -4$$

Because $m_1 = m_2$, it follows that l_1 and l_2 are parallel. See Fig. 0.4.8. ◀

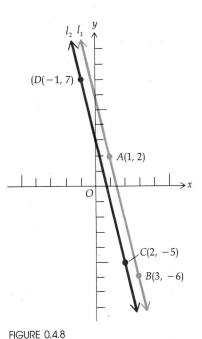

FIGURE 0.4.8

Any two distinct points determine a line. Three distinct points may or may not lie on the same line. If three or more points lie on the same line, they are said to be **collinear**. Hence three points A, B, and C are collinear if and only if the line through the points A and B is the same as the line through the points B and C. Because the line through A and B and the line through B and C both contain the point B, they are the same line if and only if their slopes are equal.

EXAMPLE 4
Determine by means of slopes if the points $A(-3, -4)$, $B(2, -1)$, and $C(7, 2)$ are collinear.

SOLUTION
If m_1 is the slope of the line through A and B, and m_2 is the slope of the line through B and C, then

$$m_1 = \frac{-1 - (-4)}{2 - (-3)} \qquad m_2 = \frac{2 - (-1)}{7 - 2}$$

$$= \frac{3}{5} \qquad\qquad = \frac{3}{5}$$

Hence $m_1 = m_2$. Therefore the line through A and B and the line through B and C have the same slope and contain the common point B. Thus they are the same line, and therefore A, B, and C are collinear.

In Fig. 0.4.9 the two lines l_1 and l_2 are perpendicular. We state and prove the following theorem about the slopes of two perpendicular lines.

0.4.4 THEOREM

If neither line l_1 nor line l_2 is vertical, then l_1 and l_2 are perpendicular if and only if the product of their slopes is -1. That is, if m_1 is the slope of l_1 and m_2 is the slope of l_2, then l_1 and l_2 are perpendicular if and only if $m_1 m_2 = -1$.

PROOF

Part 1: Prove that l_1 and l_2 are perpendicular only if $m_1 m_2 = -1$.

Let L_1 be the line through the origin parallel to l_1, and let L_2 be the line through the origin parallel to l_2. See Fig. 0.4.9. Therefore, by Theorem 0.4.3, the slope of line L_1 is m_1 and the slope of line L_2 is m_2. Because neither L_1 nor L_2 is vertical, these two lines intersect the line $x = 1$ at points P_1 and P_2, respectively. The abscissa of both P_1 and P_2 is 1. Let $\bar{y}$ be the ordinate of P_1. Since L_1 contains the points $(0, 0)$ and $(1, \bar{y})$ and its slope is m_1, it follows from Definition 0.4.1 that

$$m_1 = \frac{\bar{y} - 0}{1 - 0}$$

Thus $\bar{y} = m_1$. Similarly, the ordinate of P_2 is shown to be m_2.

Applying the Pythagorean theorem to right triangle $P_1 O P_2$ we get

$$|\overline{OP_1}|^2 + |\overline{OP_2}|^2 = |\overline{P_1 P_2}|^2 \tag{13}$$

From the distance formula,

$$|\overline{OP_1}|^2 = (1 - 0)^2 + (m_1 - 0)^2 \qquad |\overline{OP_2}|^2 = (1 - 0)^2 + (m_2 - 0)^2$$
$$= 1 + m_1{}^2 \qquad\qquad\qquad = 1 + m_2{}^2$$
$$|\overline{P_1 P_2}|^2 = (1 - 1)^2 + (m_2 - m_1)^2$$
$$= m_2{}^2 - 2m_1 m_2 + m_1{}^2$$

Substituting into (13) gives

$$(1 + m_1{}^2) + (1 + m_2{}^2) = m_2{}^2 - 2m_1 m_2 + m_1{}^2$$
$$2 = -2m_1 m_2$$
$$m_1 m_2 = -1$$

Part 2: Prove that l_1 and l_2 are perpendicular if $m_1 m_2 = -1$.

Starting with $m_1 m_2 = -1$ we can reverse the steps of the proof of Part 1 in order to obtain

$$|\overline{OP_1}|^2 + |\overline{OP_2}|^2 = |\overline{P_1 P_2}|^2$$

from which it follows, from the converse of the Pythagorean theorem, that L_1 and L_2 are perpendicular; hence l_1 and l_2 are perpendicular. ∎

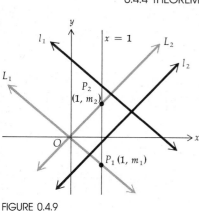

FIGURE 0.4.9

Theorem 0.4.4 states that if two lines are perpendicular and neither one is vertical, the slope of one of the lines is the negative reciprocal of the slope of the other line.

▶ ILLUSTRATION 4   If line l_1 is perpendicular to line l_2 and the slope of l_1 is $\frac{2}{3}$, then the slope of l_2 must be $-\frac{3}{2}$. ◀

EXAMPLE 5
Prove by means of slopes that the four points $A(6, 2)$, $B(8, 6)$, $C(4, 8)$, and $D(2, 4)$ are the vertices of a rectangle.

SOLUTION
See Fig. 0.4.10. Let

$m_1 =$ the slope of AB

$$= \frac{6 - 2}{8 - 6}$$

$$= 2$$

$m_2 =$ the slope of BC

$$= \frac{8 - 6}{4 - 8}$$

$$= -\frac{1}{2}$$

$m_3 =$ the slope of DC

$$= \frac{8 - 4}{4 - 2}$$

$$= 2$$

$m_4 =$ the slope of AD

$$= \frac{4 - 2}{2 - 6}$$

$$= -\frac{1}{2}$$

Because

$$m_1 = m_3, \ AB \parallel DC$$

$$m_2 = m_4, \ BC \parallel AD$$

$$m_1 m_2 = -1, \ AB \perp BC$$

Therefore quadrilateral $ABCD$ has opposite sides parallel and two adjacent sides perpendicular; thus $ABCD$ is a rectangle.

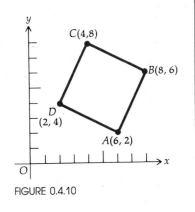

FIGURE 0.4.10

EXAMPLE 6
Given the line l having the equation $2x + 3y - 5 = 0$, find an equation of the line perpendicular to line l and passing through the point $A(-1, 3)$.

SOLUTION
Because the required line is perpendicular to line l, its slope must be the negative reciprocal of the slope of l. We find the slope of l by putting its equation into the slope-intercept form. Solving the given equation for y we obtain

$$y = -\tfrac{2}{3}x + \tfrac{5}{3}$$

Therefore the slope of l is $-\frac{2}{3}$, and the slope of the required line is $\frac{3}{2}$. Because it is also known that the required line contains the point $(-1, 3)$, we use the point-slope form, which gives

$$y - 3 = \tfrac{3}{2}(x + 1)$$

$$2y - 6 = 3x + 3$$

$$3x - 2y + 9 = 0$$

In the following example a line is tangent to a circle. The definition of the tangent line to a general curve at a specific point is given in Sec. 2.1. However, for a circle we use the definition from plane geometry, which states that a tangent line at a point P on the circle is the line intersecting the circle at only the point P.

EXAMPLE 7

Find an equation of the circle with its center at the point $C(1, 6)$ and tangent to the line l having the equation $x - y - 1 = 0$.

SOLUTION

See Fig. 0.4.11. Given that $h = 1$ and $k = 6$, if we find r, we can obtain an equation of the circle by using the center-radius form.

Let l_1 be the line through C and the point P, which is the point of tangency of line l with the circle.

$$r = |\overline{PC}|$$

Hence we must find the coordinates of P. This is done by finding an equation of l_1 and then finding the point of intersection of l_1 with l. Since l_1 is along a diameter of the circle, and l is tangent to the circle, l_1 is perpendicular to l. Because the slope of l is 1, the slope of l_1 is -1. Therefore, using the point-slope form of an equation of a line we obtain as an equation of l_1

$$y - 6 = -1(x - 1)$$

$$x + y - 7 = 0$$

Solving this equation simultaneously with the given equation of l, namely,

$$x - y - 1 = 0$$

we get $x = 4$ and $y = 3$. Thus P is the point $(4, 3)$. Because $r = |\overline{PC}|$,

$$r = \sqrt{(4 - 1)^2 + (3 - 6)^2}$$
$$= \sqrt{18}$$

So an equation of the circle is

$$(x - 1)^2 + (y - 6)^2 = (\sqrt{18})^2$$

$$x^2 + y^2 - 2x - 12y + 19 = 0$$

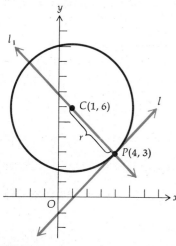

FIGURE 0.4.11

EXERCISES 0.4

In Exercises 1 through 4, find the slope of the line through the given points.

1. $(2, -3), (-4, 3)$ **2.** $(5, 2), (-2, -3)$ **3.** $(\frac{1}{3}, \frac{1}{2}), (-\frac{5}{6}, \frac{2}{3})$ **4.** $(-2.1, 0.3), (2.3, 1.4)$

In Exercises 5 through 19, find an equation of the line satisfying the given conditions.

5. The slope is 4 and through the point $(2, -3)$. **6.** The slope is 3 and through the point $(-4, -1)$.

7. The slope is -2 and through the point $(-3, 5)$. **8.** Through the points $(-2, 7)$ and $(6, 0)$.

9. Through the points $(4, 6)$ and $(0, -7)$. **10.** Through the points $(3, 1)$ and $(-5, 4)$.

11. The x intercept is -3 and the y intercept is 4.

12. Through the point $(1, 4)$ and parallel to the line whose equation is $2x - 5y + 7 = 0$.

13. Through the point $(-2, 3)$ and perpendicular to the line whose equation is $2x - y - 2 = 0$.

14. Through the point $(-3, -4)$ and parallel to the y axis.

15. Through the point $(1, -7)$ and parallel to the x axis.

16. The slope is -2 and the x intercept is 4.

17. Through the point $(-2, -5)$ and having a slope of $\sqrt{3}$.

18. Through the origin and bisecting the angle between the axes in the first and third quadrants.

19. Through the origin and bisecting the angle between the axes in the second and fourth quadrants.

In Exercises 20 through 23, find the slope of the given line.

20. $x + 3y = 7$ **21.** $4x - 6y = 5$ **22.** $3x - 5 = 0$ **23.** $2y + 9 = 0$

24. Find an equation of the line through the points $(3, -5)$ and $(1, -2)$, and write the equation in slope-intercept form.

25. Find an equation of the line through the points $(1, 3)$ and $(2, -2)$, and write the equation in intercept form.

26. Show that the lines having the equations $3x + 5y + 7 = 0$ and $6x + 10y - 5 = 0$ are parallel.

27. Show that the lines having the equations $3x + 5y + 7 = 0$ and $5x - 3y - 2 = 0$ are perpendicular.

28. Find the value of k such that the lines whose equations are $3kx + 8y = 5$ and $6y - 4kx = -1$ are perpendicular.

In Exercises 29 through 31, for each set of points determine by means of slopes if the points are collinear.

29. (a) $(2, 3), (-4, -7), (5, 8)$; (b) $(2, -1), (1, 1), (3, 4)$ **30.** (a) $(4, 6), (1, 2), (-5, -4)$; (b) $(-3, 6), (3, 2), (9, -2)$

31. (a) $(2, 5), (-1, 4), (3, -2)$; (b) $(0, 2), (-3, -1), (4, 6)$

32. Three consecutive vertices of a parallelogram are $(-4, 1)$, $(2, 3)$, and $(8, 9)$. Find the coordinates of the fourth vertex.

33. Show by means of slopes that the points $(-4, -1)$, $(3, \frac{8}{3})$, $(8, -4)$, and $(2, -9)$ are the vertices of a trapezoid.

34. Prove by means of slopes that the three points $A(3, 1)$, $B(6, 0)$, and $C(4, 4)$ are the vertices of a right triangle, and find the area of the triangle.

35. Given the line l having the equation $2y - 3x = 4$ and the point $P(1, -3)$, find (a) an equation of the line through P and perpendicular to l; (b) the shortest distance from P to line l.

36. Find an equation of the line that has equal intercepts and that passes through the point $(8, -6)$.

37. Given the line l having the equation $Ax + By + C = 0$, $B \neq 0$, find (a) the slope; (b) the y intercept; (c) the x intercept; (d) an equation of the line through the origin perpendicular to l.

38. If A, B, C, and D are constants, show that (a) the lines $Ax + By + C = 0$ and $Ax + By + D = 0$ are parallel and (b) the lines $Ax + By + C = 0$ and $Bx - Ay + D = 0$ are perpendicular.

39. Find equations of the three medians of the triangle having vertices $A(3, -2)$, $B(3, 4)$, and $C(-1, 1)$, and prove that they meet in a point.

40. Find equations of the perpendicular bisectors of the sides of the triangle having vertices $A(-1, -3)$, $B(5, -3)$, and $C(5, 5)$, and prove that they meet in a point.

41. Find an equation of each of the lines through the point $(3, 2)$ that forms with the coordinate axes a triangle of area 12.

42. Let l_1 be the line $A_1x + B_1y + C_1 = 0$, and let l_2 be the line $A_2x + B_2y + C_2 = 0$. If l_1 is not parallel to l_2 and if k is any constant, the equation

$$A_1x + B_1y + C_1 + k(A_2x + B_2y + C_2) = 0$$

represents an unlimited number of lines. Prove that each of these lines contains the point of intersection of l_1 and l_2.

43. Given that an equation of l_1 is $2x + 3y - 5 = 0$ and that an equation of l_2 is $3x + 5y - 8 = 0$, by using Exercise 42 and without finding the coordinates of the point of intersection of l_1 and l_2, find an equation of the line through this point of intersection and (a) passing through the point $(1, 3)$; (b) parallel to the x axis; (c) parallel to the y axis; (d) having slope -2; (e) perpendicular to the line having the equation $2x + y = 7$; (f) forming an isosceles triangle with the coordinate axes.

44. Find an equation of each straight line that is perpendicular to the line having the equation $5x - y = 1$ and that forms with the coordinate axes a triangle having an area of measure 5.

In Exercises 45 through 48, find an equation of the circle satisfying the given conditions.

45. Center is at $(-3, -5)$ and tangent to the line $12x + 5y = 4$.

46. Center is at $(-2, 5)$ and tangent to the line $x = 7$.

47. Tangent to the line $3x + y + 2 = 0$ at $(-1, 1)$ and through the point $(3, 5)$.

48. Tangent to the line $3x + 4y - 16 = 0$ at $(4, 1)$ and with a radius of 5 (two possible circles).

49. Find an equation of the line that is tangent to the circle $x^2 + y^2 - 4x + 6y - 12 = 0$ at the point $(5, 1)$.

50. Find an equation of each of the two lines having slope $-\frac{4}{3}$ that are tangent to the circle $x^2 + y^2 + 2x - 8y - 8 = 0$.

51. Prove analytically that the diagonals of a rhombus (an equilateral parallelogram) are perpendicular.

52. Prove analytically that the line segments joining consecutive midpoints of the sides of any quadrilateral form a parallelogram.

53. Prove analytically that the diagonals of a parallelogram bisect each other.

54. Prove analytically that if the diagonals of a quadrilateral bisect each other, then the quadrilateral is a parallelogram.

55. Given the line $y = mx + b$ tangent to the circle $x^2 + y^2 = r^2$, find an equation involving m, b, and r.

56. Prove analytically that a line from the center of any circle bisecting any chord is perpendicular to the chord.

0.5
FUNCTIONS

Many instances occur in practice where the value of one quantity depends on the value of another. In particular, a person's salary may depend on the number of hours worked, the distance traveled by an object may depend on the time elapsed since it left a specific point, the volume of the space occupied by a gas having a constant pressure depends on the temperature of the gas, the resistance of an electrical cable of fixed length depends on its diameter, and so forth. A relationship between such quantities is often described by means of a *function*. In this section we discuss the concept of a function, function notation, and types of functions.

We intuitively consider y to be a function of x if there is some rule by which a unique value is assigned to y by a corresponding value of x. Familiar examples of such relationships are given by equations such as

$$y = 2x^2 + 5 \tag{1}$$

and

$$y = \sqrt{x^2 - 9} \tag{2}$$

The formal definition makes the concept of a function precise.

0.5.1 DEFINITION   A **function** is a set of ordered pairs of numbers (x, y) in which no two distinct ordered pairs have the same first number. The set of all admissible values of x is called the **domain** of the function, and the set of all resulting values of y is called the **range** of the function.

In Definition 0.5.1 the restriction that no two distinct ordered pairs can have the same first number assures that y is unique for a specific value of x.

Equation (1) defines a function. Call this function f. The equation gives the rule by which a unique value of y can be determined whenever x is given; that is, multiply the number x by itself, then multiply that product by 2, and add 5. The function f is the set of all ordered pairs (x, y) such that x and y satisfy (1), that is,

$$f = \{(x, y) \mid y = 2x^2 + 5\}$$

When $x = 1$, $y = 7$. Therefore the function f contains the ordered pair $(1, 7)$. When $x = 2$, $y = 13$; hence f contains the ordered pair $(2, 13)$. Some other ordered pairs contained in the function f are $(0, 5)$, $(-1, 7)$, $(-2, 13)$, $(\frac{1}{2}, \frac{11}{2})$, and $(10, 205)$. Of course, f contains an unlimited number of ordered pairs, because for every real number assigned to x there is a resulting value of y.

The numbers x and y are **variables**. Because for the function f values are assigned to x, and because the value of y is dependent on the choice of x, x is the **independent variable** and y is the **dependent variable**. The domain of the function is the set of all admissible values of the independent variable, and the range of the function is the set of all resulting values of the dependent variable. For the function f under consideration the domain is the set of all real numbers, and it can be denoted with interval notation as $(-\infty, +\infty)$. The smallest value that y can assume is 5 (when $x = 0$). The range of f is then the set of all positive numbers greater than or equal to 5, which is $[5, +\infty)$.

▶ ILLUSTRATION 1   Let f be the function that is the set of all ordered pairs (x, y) such that

$$f = \{(x, y) \mid y = \sqrt{x - 2}\}$$

Because the numbers are confined to real numbers, y is a function of x only for $x - 2 \geq 0$ because for any x satisfying this inequality, a unique value of y is determined. However, if $x < 2$, a square root of a negative number is obtained, and hence no real number y exists. Therefore we must restrict x so that $x \geq 2$. Thus the domain of f is the interval $[2, +\infty)$, and the range of f is $[0, +\infty)$.   ◀

▶ ILLUSTRATION 2   Let g be the function that is the set of all ordered pairs (x, y) defined by (2); that is,

$$g = \{(x, y) \mid y = \sqrt{x^2 - 9}\}$$

We observe that y is a function of x only for $x \geq 3$ or $x \leq -3$ (or simply $|x| \geq 3$), because for any x satisfying either of these inequalities a unique value of y is determined. No real value of y is determined if x is in the open interval $(-3, 3)$, because for these values of x a square root of a negative number is obtained. Hence the domain of g is $(-\infty, -3] \cup [3, +\infty)$, and the range is $[0, +\infty)$.   ◀

0.5.2 DEFINITION   If f is a function, then the **graph** of f is the set of points (x, y) in R^2 for which (x, y) is an ordered pair in f.

FIGURE 0.5.1

FIGURE 0.5.2

FIGURE 0.5.3

▶ ILLUSTRATION 3   (a) A sketch of the graph of the function f of Illustration 1 appears in Fig. 0.5.1. It is the top half of a parabola.

(b) Figure 0.5.2 shows a sketch of the graph of the function g of Illustration 2.   ◀

Recall that to have a function there must be *exactly one value* of the dependent variable y for a value of the independent variable x in the domain of the function. Thus the graph of a function can be intersected by a vertical line in at most one point. Observe that this is the situation for the graphs of the functions in Figs. 0.5.1 and 0.5.2.

▶ ILLUSTRATION 4   Consider the set

$$\{(x, y) \mid x^2 + y^2 = 25\}$$

A sketch of the graph of this set is shown in Fig. 0.5.3. This set of ordered pairs is not a function, because for any x in the interval $(-5, 5)$ there are two ordered pairs having that number as a first element. For example, both $(3, 4)$ and $(3, -4)$ are ordered pairs in the given set. Furthermore, observe that the graph of the given set is a circle with center at the origin and radius 5, and a vertical line having the equation $x = a$ (where $-5 < a < 5$) intersects the circle in two points. Refer to the figure.   ◀

Section 1.1 deals with graphs of functions, and more examples appear there.

To introduce the notation for a *function value*, let f be the function having as its domain variable x and as its range variable y. Then the symbol $f(x)$ (read "f of x") denotes the particular value of y corresponding to a value of x.

▶ ILLUSTRATION 5   In Illustration 1, $f = \{(x, y) \mid y = \sqrt{x - 2}\}$. Thus

$$f(x) = \sqrt{x - 2}$$

Because when $x = 3$, $\sqrt{x - 2} = 1$, we have $f(3) = 1$. Similarly, $f(5) = \sqrt{3}$, $f(6) = 2$, $f(9) = \sqrt{7}$, and so on.   ◀

Often an equation giving the function value is used to define a function. However, the domain of the function must be given either explicitly or implicitly. For instance, if f is defined by

$$f(x) = 3x^2 - 5x + 2$$

it is implied that x can be any real number. However, if f is defined by

$$f(x) = 3x^2 - 5x + 2 \qquad 1 \le x \le 10$$

then the domain of f consists of all real numbers between and including 1 and 10.

Similarly, if g is defined by the equation

$$g(x) = \frac{5x - 2}{x + 4}$$

it is implied that $x \neq -4$, because the quotient is undefined for $x = -4$; hence, the domain of g is the set of all real numbers except -4.

If

$$h(x) = \sqrt{9 - x^2}$$

it is implied that x is in the closed interval $[-3, 3]$ because $\sqrt{9 - x^2}$ is not a real number for $x > 3$ or $x < -3$. Thus the domain of h is $[-3, 3]$ and the range is $[0, 3]$.

EXAMPLE 1

Given that f is the function defined by

$$f(x) = x^2 + 3x - 4$$

find: (a) $f(0)$; (b) $f(2)$; (c) $f(h)$; (d) $f(2h)$; (e) $f(2x)$; (f) $f(x + h)$; (g) $f(x) + f(h)$.

SOLUTION

(a) $f(0) = 0^2 + 3 \cdot 0 - 4$ (b) $f(2) = 2^2 + 3 \cdot 2 - 4$

        $= 4$ $= 6$

(c) $f(h) = h^2 + 3h - 4$ (d) $f(2h) = (2h)^2 + 3(2h) - 4$

                                          $= 4h^2 + 6h - 4$

(e) $f(2x) = (2x)^2 + 3(2x) - 4$

            $= 4x^2 + 6x - 4$

(f) $f(x + h) = (x + h)^2 + 3(x + h) - 4$

                 $= x^2 + 2hx + h^2 + 3x + 3h - 4$

                 $= x^2 + (2h + 3)x + (h^2 + 3h - 4)$

(g) $f(x) + f(h) = (x^2 + 3x - 4) + (h^2 + 3h - 4)$

                  $= x^2 + 3x + (h^2 + 3h - 8)$

Compare the computations in parts (f) and (g) of Example 1. In part (f) the computation is for $f(x + h)$, which is the function value at the sum of x and h. In part (g), where $f(x) + f(h)$ is computed, we obtain the sum of the two function values $f(x)$ and $f(h)$.

In Chapter 2 we need to compute quotients of the form

$$\frac{f(x + h) - f(x)}{h} \qquad h \neq 0$$

This quotient arises as the slope of the line through the points $(x, f(x))$ and $(x + h, f(x + h))$ on the graph of the function defined by $y = f(x)$. See Fig. 0.5.4.

FIGURE 0.5.4

EXAMPLE 2
Find

$$\frac{f(x + h) - f(x)}{h}$$

where $h \neq 0$, if (a) $f(x) = 4x^2 - 5x + 7$; (b) $f(x) = \sqrt{x}$.

SOLUTION

(a) $\dfrac{f(x + h) - f(x)}{h} = \dfrac{4(x + h)^2 - 5(x + h) + 7 - (4x^2 - 5x + 7)}{h}$

$$= \frac{4x^2 + 8hx + 4h^2 - 5x - 5h + 7 - 4x^2 + 5x - 7}{h}$$

$$= \frac{8hx - 5h + 4h^2}{h}$$

$$= 8x - 5 + 4h$$

(b) $\dfrac{f(x + h) - f(x)}{h} = \dfrac{\sqrt{x + h} - \sqrt{x}}{h}$

$$= \frac{(\sqrt{x + h} - \sqrt{x})(\sqrt{x + h} + \sqrt{x})}{h(\sqrt{x + h} + \sqrt{x})}$$

$$= \frac{(x + h) - x}{h(\sqrt{x + h} + \sqrt{x})}$$

$$= \frac{h}{h(\sqrt{x + h} + \sqrt{x})}$$

$$= \frac{1}{\sqrt{x + h} + \sqrt{x}}$$

In the second step of this solution the numerator and denominator are multiplied by the conjugate of the numerator in order to rationalize the numerator, and this gives a common factor of h in the numerator and denominator.

We now define some operations on functions. In the definition new functions are formed from given functions by adding, subtracting, multiplying, and dividing function values. Accordingly, these new functions are known as the *sum*, *difference*, *product*, and *quotient* of the original functions.

0.5.3 DEFINITION   Given the two functions f and g:

(i) their **sum**, denoted by $f + g$, is the function defined by

$$(f + g)(x) = f(x) + g(x)$$

(ii) their **difference**, denoted by $f - g$, is the function defined by

$$(f - g)(x) = f(x) - g(x)$$

(iii) their **product**, denoted by $f \cdot g$, is the function defined by

$$(f \cdot g)(x) = f(x) \cdot g(x)$$

(iv) their **quotient**, denoted by f/g, is the function defined by

$$(f/g)(x) = f(x)/g(x)$$

In each case the *domain* of the resulting function consists of those values of x common to the domains of f and g, with the additional requirement in case (iv) that the values of x for which $g(x) = 0$ are excluded.

EXAMPLE 3

Given that f and g are the functions defined by

$$f(x) = \sqrt{x + 1} \quad \text{and} \quad g(x) = \sqrt{x - 4}$$

find: (a) $(f + g)(x)$; (b) $(f - g)(x)$; (c) $(f \cdot g)(x)$; (d) $(f/g)(x)$.

SOLUTION

(a) $(f + g)(x) = \sqrt{x + 1} + \sqrt{x - 4}$ (b) $(f - g)(x) = \sqrt{x + 1} - \sqrt{x - 4}$

(c) $(f \cdot g)(x) = \sqrt{x + 1} \cdot \sqrt{x - 4}$ (d) $(f/g)(x) = \dfrac{\sqrt{x + 1}}{\sqrt{x - 4}}$

The domain of f is $[-1, +\infty)$, and the domain of g is $[4, +\infty)$. So in parts (a), (b), and (c) the domain of the resulting function is $[4, +\infty)$. In part (d) the denominator is zero when $x = 4$; thus 4 is excluded from the domain, and the domain is therefore $(4, +\infty)$.

In addition to combining two functions by the operations given in Definition 0.5.3 we consider the *composite function* of two given functions.

0.5.4 DEFINITION

Given the two functions f and g, the **composite function**, denoted by $f \circ g$, is defined by

$$(f \circ g)(x) = f(g(x))$$

and the domain of $f \circ g$ is the set of all numbers x in the domain of g such that $g(x)$ is in the domain of f.

EXAMPLE 4

Given that f and g are defined by

$$f(x) = \sqrt{x} \quad \text{and} \quad g(x) = 2x - 3$$

find $F(x)$ if $F = f \circ g$, and determine the domain of F.

SOLUTION

$$\begin{aligned}
F(x) &= (f \circ g)(x) \\
&= f(g(x)) \\
&= f(2x - 3) \\
&= \sqrt{2x - 3}
\end{aligned}$$

The domain of g is $(-\infty, +\infty)$, and the domain of f is $[0, +\infty)$. So the domain of F is the set of real numbers for which $2x - 3 \geq 0$ or, equivalently, $[\frac{3}{2}, +\infty)$.

EXAMPLE 5

Given that f and g are defined by

$$f(x) = \sqrt{x} \quad \text{and} \quad g(x) = x^2 - 1$$

find: (a) $f \circ f$; (b) $g \circ g$; (c) $f \circ g$; (d) $g \circ f$. Also determine the domain of the composite function in each part.

SOLUTION

(a) $(f \circ f)(x) = f(f(x))$
$$= f(\sqrt{x})$$
$$= \sqrt{\sqrt{x}}$$
$$= \sqrt[4]{x}$$

The domain of $f \circ f$ is $[0, +\infty)$.

(b) $(g \circ g)(x) = g(g(x))$
$$= g(x^2 - 1)$$
$$= (x^2 - 1)^2 - 1$$
$$= x^4 - 2x^2$$

The domain of $g \circ g$ is $(-\infty, +\infty)$.

(c) $(f \circ g)(x) = f(g(x))$
$$= f(x^2 - 1)$$
$$= \sqrt{x^2 - 1}$$

(d) $(g \circ f)(x) = g(f(x))$
$$= g(\sqrt{x})$$
$$= (\sqrt{x})^2 - 1$$
$$= x - 1$$

(c) The domain of $f \circ g$ is $(-\infty, -1] \cup [1, +\infty)$ or, equivalently, all x not in $(-1, 1)$.

(d) The domain of $g \circ f$ is $[0, +\infty)$. Note that even though $x - 1$ is defined for all values of x, the domain of $g \circ f$, by Definition 0.5.4, is the set of all numbers x in the domain of f (which is $\{x \mid x \geq 0\}$) such that $f(x)$ is in the domain of g.

0.5.5 DEFINITION

(i) A function f is said to be an **even** function if for every x in the domain of f, $f(-x) = f(x)$.

(ii) A function f is said to be an **odd** function if for every x in the domain of f, $f(-x) = -f(x)$.

In both parts (i) and (ii) it is understood that $-x$ is in the domain of f whenever x is.

▶ ILLUSTRATION 6   (a) If $f(x) = 3x^4 - 2x^2 + 7$, then

$$f(-x) = 3(-x)^4 - 2(-x)^2 + 7$$
$$= 3x^4 - 2x^2 + 7$$
$$= f(x)$$

Therefore f is an even function.

(b) If $g(x) = 3x^5 - 4x^3 - 9x$, then

$$g(-x) = 3(-x)^5 - 4(-x)^3 - 9(-x)$$
$$= -3x^5 + 4x^3 + 9x$$
$$= -(3x^5 - 4x^3 - 9x)$$
$$= -g(x)$$

Therefore g is an odd function.

(c) If $h(x) = 2x^4 + 7x^3 - x^2 + 9$, then

$$h(-x) = 2(-x)^4 + 7(-x)^3 - (-x)^2 + 9$$
$$= 2x^4 - 7x^3 - x^2 + 9$$

We see that the function h is neither even nor odd.   ◀

From the definition of an even function and Theorem 0.2.6(ii) it follows that the graph of an even function is symmetric with respect to the y axis. From Theorem 0.2.6(iii) and the definition of an odd function, the graph of an odd function is symmetric with respect to the origin.

▶ ILLUSTRATION 7   (a) If $f(x) = x^2$, f is an even function, and its graph is a parabola that is symmetric with respect to the y axis. See Fig. 0.5.5.

(b) If $g(x) = x^3$, g is an odd function. The graph of g, shown in Fig. 0.5.6, is symmetric with respect to the origin.   ◀

If the range of a function f consists of only one number, then f is called a **constant function**. So if $f(x) = c$, and if c is any real number, then f is a constant function and its graph is a straight line parallel to the x axis at a directed distance of c units from the x axis.

If a function f is defined by

$$f(x) = a_n x^n + a_{n-1} x^{n-1} + a_{n-2} x^{n-2} + \ldots + a_1 x + a_0$$

where $a_0, a_1, \ldots, a_n$ are real numbers ($a_n \neq 0$) and n is a nonnegative integer, then f is called a **polynomial function** of degree n. Thus the function defined by

$$f(x) = 3x^5 - x^2 + 7x - 1$$

is a polynomial function of degree 5.

If the degree of a polynomial function is 1, then the function is called a **linear function**; if the degree is 2, the function is called a **quadratic function**; and if the degree is 3, the function is called a **cubic function**. The particular linear function defined by $f(x) = x$ is called the **identity function**.

FIGURE 0.5.5

FIGURE 0.5.6

▶ ILLUSTRATION 8

The function f defined by $f(x) = 3x + 4$ is a linear function.

The function g defined by $g(x) = 5x^2 - 8x + 1$ is a quadratic function.

The function h defined by $h(x) = 8x^3 - x + 4$ is a cubic function.   ◀

If a function can be expressed as the quotient of two polynomial functions, it is called a **rational function**.

An **algebraic function** is a function formed by a finite number of algebraic operations on the identity function and the constant function. These algebraic operations include addition, subtraction, multiplication, division, raising to powers, and extracting roots. An example of an algebraic function is the function f defined by

$$f(x) = \frac{(x^2 - 3x + 1)^3}{x^4 + 1}$$

In addition to algebraic functions, **transcendental functions** are considered in calculus. Examples of transcendental functions are the trigonometric functions discussed in the next section, and the logarithmic and exponential functions introduced in Chapter 6.

EXERCISES 0.5

In Exercises 1 through 4, determine if the given set is a function. If it is a function, what is its domain?

1. (a) $\{(x, y) \mid y = \sqrt{x - 4}\}$; (b) $\{(x, y) \mid y = \sqrt{x^2 - 4}\}$; (c) $\{(x, y) \mid y = \sqrt{4 - x^2}\}$; (d) $\{(x, y) \mid x^2 + y^2 = 4\}$.

2. (a) $\{(x, y) \mid y = \sqrt{x + 1}\}$; (b) $\{(x, y) \mid y = \sqrt{x^2 - 1}\}$; (c) $\{(x, y) \mid y = \sqrt{1 - x^2}\}$; (d) $\{(x, y) \mid x^2 + y^2 = 1\}$.

3. (a) $\{(x, y) \mid y = x^2\}$; (b) $\{(x, y) \mid x = y^2\}$; (c) $\{(x, y) \mid y = x^3\}$; (d) $\{(x, y) \mid x = y^3\}$.

4. (a) $\{(x, y) \mid y = (x - 1)^2 + 2\}$; (b) $\{(x, y) \mid x = (y - 2)^2 + 1\}$; (c) $\{(x, y) \mid y = (x + 2)^3 - 1\}$; (d) $\{(x, y) \mid x = (y + 1)^3 - 2\}$.

5. Given $f(x) = 2x - 1$, find (a) $f(3)$; (b) $f(-2)$; (c) $f(0)$; (d) $f(a + 1)$; (e) $f(x + 1)$; (f) $f(2x)$; (g) $2f(x)$; (h) $f(x + h)$; (i) $f(x) + f(h)$; (j) $\dfrac{f(x + h) - f(x)}{h}$, $h \neq 0$.

6. Given $f(x) = \dfrac{3}{x}$, find (a) $f(1)$; (b) $f(-3)$; (c) $f(6)$; (d) $f(\frac{1}{3})$; (e) $f\left(\dfrac{3}{a}\right)$; (f) $f\left(\dfrac{3}{x}\right)$; (g) $\dfrac{f(3)}{f(x)}$; (h) $f(x - 3)$; (i) $f(x) - f(3)$; (j) $\dfrac{f(x + h) - f(x)}{h}$, $h \neq 0$.

7. Given $f(x) = 2x^2 + 5x - 3$, find (a) $f(-2)$; (b) $f(-1)$; (c) $f(0)$; (d) $f(3)$; (e) $f(h + 1)$; (f) $f(2x^2)$; (g) $f(x^2 - 3)$; (h) $f(x + h)$; (i) $f(x) + f(h)$; (j) $\dfrac{f(x + h) - f(x)}{h}$, $h \neq 0$.

8. Given $g(x) = 3x^2 - 4$, find (a) $g(-4)$; (b) $g(\frac{1}{2})$; (c) $g(x^2)$; (d) $g(3x^2 - 4)$; (e) $g(x - h)$; (f) $g(x) - g(h)$; (g) $\dfrac{g(x + h) - g(x)}{h}$, $h \neq 0$.

9. Given $F(x) = \sqrt{2x + 3}$, find (a) $F(-1)$; (b) $F(4)$; (c) $F(\frac{1}{2})$; (d) $F(11)$; (e) $F(2x + 3)$; (f) $\dfrac{F(x + h) - F(x)}{h}$, $h \neq 0$.

10. Given $G(x) = \sqrt{2x^2 + 1}$, find (a) $G(-2)$; (b) $G(0)$; (c) $G(1)$; (d) $G(\frac{4}{7})$; (e) $G(2x^2 - 1)$; (f) $\dfrac{G(x + h) - G(x)}{h}$; $h \neq 0$.

In Exercises 11 through 20, the functions f and g are defined. In each exercise, define the following functions and determine the domain of the resulting function: (a) $f + g$; (b) $f - g$; (c) $f \cdot g$; (d) f/g; (e) g/f.

11. $f(x) = x - 5$; $g(x) = x^2 - 1$

12. $f(x) = \sqrt{x}$; $g(x) = x^2 + 1$

13. $f(x) = \dfrac{x + 1}{x - 1}$; $g(x) = \dfrac{1}{x}$

14. $f(x) = \sqrt{x}$; $g(x) = 4 - x^2$

15. $f(x) = \sqrt{x}$; $g(x) = x^2 - 1$

16. $f(x) = |x|$; $g(x) = |x - 3|$

17. $f(x) = x^2 + 1$; $g(x) = 3x - 2$

18. $f(x) = \sqrt{x + 4}$; $g(x) = x^2 - 4$

19. $f(x) = \dfrac{1}{x + 1}$; $g(x) = \dfrac{x}{x - 2}$

20. $f(x) = x^2$; $g(x) = \dfrac{1}{\sqrt{x}}$

In Exercises 21 through 30, the functions f and g are defined. In each exercise, define the following functions and determine the domain of the composite function: (a) $f \circ g$; (b) $g \circ f$; (c) $f \circ f$; (d) $g \circ g$.

21. $f(x) = x - 2$; $g(x) = x + 7$

22. $f(x) = 3 - 2x$; $g(x) = 6 - 3x$

23. The functions of Exercise 11.

24. The functions of Exercise 12.

25. $f(x) = \sqrt{x - 2}$; $g(x) = x^2 - 2$

26. $f(x) = x^2 - 1$; $g(x) = \dfrac{1}{x}$

27. $f(x) = \dfrac{1}{x}$; $g(x) = \sqrt{x}$

28. $f(x) = \sqrt{x}$; $g(x) = -\dfrac{1}{x}$

29. $f(x) = |x|$; $g(x) = |x + 2|$

30. $f(x) = \sqrt{x^2 - 1}$; $g(x) = \sqrt{x - 1}$

In Exercises 31 and 32, a function f is defined. In each exercise, define the following functions, and determine the domain of the resulting function: (a) $f(x^2)$; (b) $[f(x)]^2$; (c) $(f \circ f)(x)$.

31. $f(x) = 2x - 3$

32. $f(x) = \dfrac{2}{x - 1}$

33. Given $G(x) = |x - 2| - |x| + 2$, express $G(x)$ without absolute-value bars if x is in the given interval: (a) $[2, +\infty)$; (b) $(-\infty, 0)$; (c) $[0, 2)$.

34. Given $f(t) = \dfrac{|3 + t| - |t| - 3}{t}$, express $f(t)$ without absolute-value bars if t is in the given interval: (a) $(0, +\infty)$; (b) $[-3, 0)$; (c) $(-\infty, -3)$.

35. Given

$$f(x) = \begin{cases} \dfrac{|x|}{x} & \text{if } x \neq 0 \\ 1 & \text{if } x = 0 \end{cases}$$

find (a) $f(1)$; (b) $f(-1)$; (c) $f(4)$; (d) $f(-4)$; (e) $f(-x)$; (f) $f(x + 1)$; (g) $f(x^2)$; (h) $f(-x^2)$.

In each part of Exercises 36 and 37, determine whether the given function is even, odd, or neither.

36. (a) $f(x) = 2x^4 - 3x^2 + 1$
(b) $f(x) = 5x^3 - 7x$
(c) $f(s) = s^2 + 2s + 2$
(d) $g(x) = x^6 - 1$

(e) $h(t) = 5t^7 + 1$
(f) $f(x) = |x|$
(g) $f(y) = \dfrac{y^3 - y}{y^2 + 1}$
(h) $g(z) = \dfrac{z - 1}{z + 1}$

37. (a) $g(x) = 5x^2 - 4$ E
(b) $f(x) = x^3 + 1$ N
(c) $f(t) = 4t^5 + 3t^3 - 2t$ O
(d) $g(r) = \dfrac{r^2 - 1}{r^2 + 1}$ E

(e) $f(x) = \begin{cases} 1 & \text{if } x \geq 0 \\ -1 & \text{if } x < 0 \end{cases}$ O
(f) $h(x) = \dfrac{4x^2 - 5}{2x^3 + x}$ O
(g) $f(z) = (z - 1)^2$ N
(h) $g(x) = \dfrac{|x|}{x^2 + 1}$ E

(i) $g(x) = \sqrt{x^2 - 1}$ E
(j) $f(x) = \sqrt[3]{x}$ O

38. There is one function that is both even and odd. What is it? $f(x) = 0$

39. Determine whether the composite function $f \circ g$ is odd or even in each of the following cases: (a) f and g are both even; (b) f and g are both odd; (c) f is even and g is odd; (d) f is odd and g is even.

If f and g are two functions such that $(f \circ g)(x) = x$ and $(g \circ f)(x) = x$, then f and g are *inverse functions*. In Exercises 40 through 42, show that f and g are inverse functions.

40. $f(x) = 2x - 3$ and $g(x) = \dfrac{x + 3}{2}$ 　　**41.** $f(x) = \dfrac{1}{x + 1}$ and $g(x) = \dfrac{1 - x}{x}$ 　　**42.** $f(x) = x^2$, $x \geq 0$ and $g(x) = \sqrt{x}$

0.6
THE TRIGONOMETRIC FUNCTIONS

It is assumed that you have studied trigonometry in a previous course. However, because of the importance of the trigonometric functions in calculus, a brief review of them, together with a development of the identities you will need, is presented here.

　　In geometry an **angle** is defined as the union of two rays called the **sides**, having a common endpoint called the **vertex**. Any angle is congruent to some angle having its vertex at the origin and one side, called the **initial side**, lying on the positive side of the x axis. Such an angle is said to be in **standard position**. Figure 0.6.1 shows an angle AOB in standard position with OA as the initial side. The other side, OB, is called the **terminal side**. The angle AOB can be formed by rotating the side OA to the side OB, and under such a rotation the point A moves along the circumference of a circle having its center at O and radius $|\overline{OA}|$ to the point B.

　　In dealing with problems involving angles of triangles, the measurement of an angle is usually given in degrees. However, in the calculus we are concerned with trigonometric functions of real numbers, and these functions are defined in terms of *radian measure*.

　　The length of an arc of a circle is used to define the radian measure of an angle.

FIGURE 0.6.1

0.6.1 DEFINITION 　　Let AOB be an angle in standard position and $|\overline{OA}| = 1$. If s units is the length of the arc of the circle traveled by point A as the initial side OA is rotated to the terminal side OB, the **radian measure**, t, of angle AOB is given by

$$t = s \quad \text{if the rotation is counterclockwise}$$

and

$$t = -s \quad \text{if the rotation is clockwise}$$

▶ **ILLUSTRATION 1** 　From the fact that the measure of the length of the unit circle's circumference is 2π, the radian measures of the angles in Fig. 0.6.2(a)–(f) are determined. They are $\frac{1}{2}\pi$, $\frac{1}{4}\pi$, $-\frac{1}{2}\pi$, $\frac{3}{2}\pi$, $-\frac{3}{4}\pi$, and $\frac{7}{4}\pi$, respectively. ◀

　　In Definition 0.6.1 it is possible that there may be more than one complete revolution in the rotation of OA.

FIGURE 0.6.2

(a)

(b)

FIGURE 0.6.3

▶ ILLUSTRATION 2   Figure 0.6.3(a) shows an angle whose radian measure is $\frac{5}{2}\pi$, and Fig. 0.6.3(b) shows one whose radian measure is $-\frac{13}{4}\pi$.   ◀

An angle formed by one complete revolution so that OA is coincident with OB has degree measure of 360 and radian measure of 2π. Hence there is the following correspondence between degree measure and radian measure (where the symbol $\sim$ indicates that the given measurements are for the same or congruent angles):

$$360° \sim 2\pi \text{ rad} \qquad 180° \sim \pi \text{ rad}$$

From this it follows that

$$1° \sim \tfrac{1}{180}\pi \text{ rad} \qquad 1 \text{ rad} \sim \frac{180°}{\pi}$$

$$\approx 57°18'$$

Note that the symbol $\approx$ before 57°18′ indicates that 1 rad and approximately 57°18′ are measurements for the same or congruent angles.

From this correspondence the measurement of an angle can be converted from one system of units to the other.

▶ ILLUSTRATION 3

$$162° \sim 162 \cdot \tfrac{1}{180}\pi \text{ rad} \qquad \tfrac{5}{12}\pi \text{ rad} \sim \tfrac{5}{12}\pi \cdot \frac{180°}{\pi}$$

$$162° \sim \tfrac{9}{10}\pi \text{ rad} \qquad\qquad \tfrac{5}{12}\pi \text{ rad} \sim 75°$$

◀

Table 0.6.1 gives the corresponding degree and radian measures of certain angles.

We now define the sine and cosine functions of any real number.

0.6.2 DEFINITION   Suppose that t is a real number. Place an angle, having radian measure t, in standard position and let point P be at the intersection of the terminal side of the angle with the unit circle having its center at the origin. If P is the point (x, y), then the **cosine function** is defined by

$$\cos t = x$$

and the **sine function** is defined by

$$\sin t = y$$

From the above definition it is seen that $\sin t$ and $\cos t$ are defined for any value of t. Hence the domain of the sine and cosine functions is the set of all real numbers. The largest value either function may have is 1, and the smallest value is -1. It will be shown later that the sine and cosine functions take on all values between -1 and 1, and from this it follows that the range of the two functions is $[-1, 1]$.

For certain values of t, the cosine and sine are easily obtained from a figure. From Fig. 0.6.4 we see that $\cos 0 = 1$ and $\sin 0 = 0$, $\cos \frac{1}{4}\pi = \frac{1}{2}\sqrt{2}$ and $\sin \frac{1}{4}\pi = \frac{1}{2}\sqrt{2}$, $\cos \frac{1}{2}\pi = 0$ and $\sin \frac{1}{2}\pi = 1$, $\cos \pi = -1$ and $\sin \pi = 0$, and $\cos \frac{3}{2}\pi = 0$ and $\sin \frac{3}{2}\pi = -1$. Table 0.6.2 gives these values and some others that are frequently used.

An equation of the unit circle having its center at the origin is $x^2 + y^2 = 1$. Because $x = \cos t$ and $y = \sin t$, it follows that

$$\cos^2 t + \sin^2 t = 1 \tag{1}$$

Note that $\cos^2 t$ and $\sin^2 t$ stand for $(\cos t)^2$ and $(\sin t)^2$. Equation (1) is called an **identity** because it is valid for any real number t.

Figures 0.6.5 and 0.6.6 show angles having a negative radian measure of $-t$ and corresponding angles having a positive radian measure of t. From these figures observe that

$$\cos(-t) = \cos t \quad \text{and} \quad \sin(-t) = -\sin t \tag{2}$$

Table 0.6.1

Degree Measure	Radian Measure
30	$\frac{1}{6}\pi$
45	$\frac{1}{4}\pi$
60	$\frac{1}{3}\pi$
90	$\frac{1}{2}\pi$
120	$\frac{2}{3}\pi$
135	$\frac{3}{4}\pi$
150	$\frac{5}{6}\pi$
180	π
270	$\frac{3}{2}\pi$
360	2π

FIGURE 0.6.4

Table 0.6.2

x	$\sin x$	$\cos x$
0	0	1
$\frac{1}{6}\pi$	$\frac{1}{2}$	$\frac{1}{2}\sqrt{3}$
$\frac{1}{4}\pi$	$\frac{1}{2}\sqrt{2}$	$\frac{1}{2}\sqrt{2}$
$\frac{1}{3}\pi$	$\frac{1}{2}\sqrt{3}$	$\frac{1}{2}$
$\frac{1}{2}\pi$	1	0
$\frac{2}{3}\pi$	$\frac{1}{2}\sqrt{3}$	$-\frac{1}{2}$
$\frac{3}{4}\pi$	$\frac{1}{2}\sqrt{2}$	$-\frac{1}{2}\sqrt{2}$
$\frac{5}{6}\pi$	$\frac{1}{2}$	$-\frac{1}{2}\sqrt{3}$
π	0	-1
$\frac{3}{2}\pi$	-1	0
2π	0	1

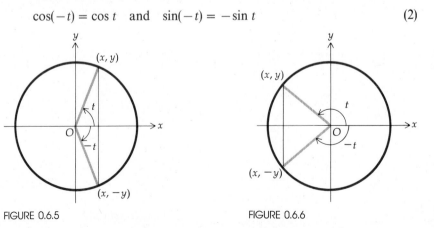

FIGURE 0.6.5   FIGURE 0.6.6

These equations hold for any real number t because the points where the terminal sides of the angles (having radian measures t and $-t$) intersect the unit circle have equal abscissas and ordinates that differ only in sign. Hence Eqs. (2) are identities. From these equations it follows that the cosine is an even function and the sine is an odd function.

From Definition 0.6.2 the following identities are obtained:

$$\cos(t + 2\pi) = \cos t \quad \text{and} \quad \sin(t + 2\pi) = \sin t \tag{3}$$

The property of cosine and sine stated by Eqs. (3) is called *periodicity*, which is now defined.

0.6.3 DEFINITION   A function f is said to be **periodic** if there exists a real number $p \neq 0$ such that whenever x is in the domain of f, then $x + p$ is also in the domain of f and

$$f(x + p) = f(x)$$

The number p is called a **period** of f. The least positive period is called the **fundamental period**.

From the above definition and Eqs. (3) it is seen that the sine and cosine functions are periodic with period 2π; that is, whenever the value of the independent variable t is increased by 2π, the value of each of the functions is repeated. It is because of the periodicity of the sine and cosine that these functions have important applications in physics and engineering in connection with periodically repetitive phenomena, such as wave motion and vibrations.

We now proceed to derive a formula that is useful because other important identities can be obtained from it. The formula is:

$$\cos(a + b) = \cos a \cos b - \sin a \sin b \tag{4}$$

where a and b are any real numbers.

Refer to Fig. 0.6.7 showing a unit circle with the points $Q(1, 0)$, $P_1(\cos a, \sin a)$, $P_2(\cos b, \sin b)$, $P_3(\cos(a + b), \sin(a + b))$, and $P_4(\cos a, -\sin a)$. Using the notation $\overset{\frown}{RS}$ to denote the measure of the length of arc from R to S we have $\overset{\frown}{QP_1} = a$, $\overset{\frown}{QP_2} = b$, $\overset{\frown}{P_2P_3} = a$, and $\overset{\frown}{P_4Q} = |-a| = a$. Because

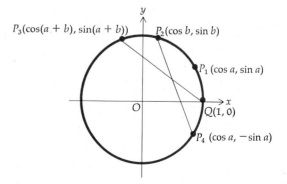

FIGURE 0.6.7

$$\widehat{QP_3} = \widehat{QP_2} + \widehat{P_2P_3}, \text{ then}$$

$$\widehat{QP_3} = b + a \tag{5}$$

and because $\widehat{P_4P_2} = \widehat{P_4Q} + \widehat{QP_2}$, it follows that

$$\widehat{P_4P_2} = a + b \tag{6}$$

From (5) and (6), $\widehat{QP_3} = \widehat{P_4P_2}$; therefore the length of the chord joining the points Q and P_3 is the same as the length of the chord joining the points P_4 and P_2. By squaring the measures of these lengths,

$$|\overline{QP_3}|^2 = |\overline{P_4P_2}|^2 \tag{7}$$

From the distance formula,

$$|\overline{QP_3}|^2 = [\cos(a + b) - 1]^2 + [\sin(a + b) - 0]^2$$
$$= \cos^2(a + b) - 2\cos(a + b) + 1 + \sin^2(a + b)$$

and because $\cos^2(a + b) + \sin^2(a + b) = 1$ the above may be written as

$$|\overline{QP_3}|^2 = 2 - 2\cos(a + b) \tag{8}$$

Again from the distance formula,

$$|\overline{P_4P_2}|^2 = (\cos b - \cos a)^2 + (\sin b + \sin a)^2$$
$$= \cos^2 b - 2\cos a \cos b + \cos^2 a + \sin^2 b + 2\sin a \sin b + \sin^2 a$$

and because $\cos^2 b + \sin^2 b = 1$ and $\cos^2 a + \sin^2 a = 1$ the above becomes

$$|\overline{P_4P_2}|^2 = 2 - 2\cos a \cos b + 2\sin a \sin b \tag{9}$$

Substituting from (8) and (9) into (7) we obtain

$$2 - 2\cos(a + b) = 2 - 2\cos a \cos b + 2\sin a \sin b$$

from which follows

$$\cos(a + b) = \cos a \cos b - \sin a \sin b$$

which is formula (4).

A formula for $\cos(a - b)$ is obtained from formula (4) by substituting $-b$ for b:

$$\cos(a + (-b)) = \cos a \cos(-b) - \sin a \sin(-b)$$

Because $\cos(-b) = \cos b$ and $\sin(-b) = -\sin b$, then

$$\cos(a - b) = \cos a \cos b + \sin a \sin b \tag{10}$$

for all real numbers a and b.

With $a = \frac{1}{2}\pi$ in (10) we have $\cos(\frac{1}{2}\pi - b) = \cos \frac{1}{2}\pi \cos b + \sin \frac{1}{2}\pi \sin b$; and because $\cos \frac{1}{2}\pi = 0$ and $\sin \frac{1}{2}\pi = 1$ it becomes $\cos(\frac{1}{2}\pi - b) = \sin b$. Now if in this equation we let $\frac{1}{2}\pi - b = c$, then $b = \frac{1}{2}\pi - c$ and we obtain $\cos c = \sin(\frac{1}{2}\pi - c)$. We have therefore proved the following two identities:

$$\cos(\tfrac{1}{2}\pi - t) = \sin t \quad \text{and} \quad \sin(\tfrac{1}{2}\pi - t) = \cos t \tag{11}$$

Formulas similar to (4) and (10) for the sine of the sum and difference of two numbers follow easily. A formula for the sine of the sum of two numbers is obtained from (10) and (11). From (11),

$$\sin(a + b) = \cos(\tfrac{1}{2}\pi - (a + b))$$
$$= \cos((\tfrac{1}{2}\pi - a) - b)$$

With (10) it follows that

$$\sin(a + b) = \cos(\tfrac{1}{2}\pi - a)\cos b + \sin(\tfrac{1}{2}\pi - a)\sin b$$

and from (11),

$$\sin(a + b) = \sin a \cos b + \cos a \sin b \tag{12}$$

for all real numbers a and b.

To obtain a formula for the sine of the difference of two numbers, write $a - b$ as $a + (-b)$ and apply formula (12). Thus

$$\sin(a + (-b)) = \sin a \cos(-b) + \cos a \sin(-b)$$

and because $\cos(-b) = \cos b$ and $\sin(-b) = -\sin b$,

$$\sin(a - b) = \sin a \cos b - \cos a \sin b \tag{13}$$

which is valid for all real numbers a and b.

By letting $a = t$ and $b = t$ in formulas (12) and (4), respectively, the following formulas are obtained:

$$\sin 2t = 2 \sin t \cos t \tag{14}$$

$$\cos 2t = \cos^2 t - \sin^2 t \tag{15}$$

From the identity $\sin^2 t + \cos^2 t = 1$ we may rewrite formula (15) as

$$\cos 2t = 2 \cos^2 t - 1 \tag{16}$$

or as

$$\cos 2t = 1 - 2 \sin^2 t \tag{17}$$

Replacing t by $\tfrac{1}{2}t$ in formulas (16) and (17) and performing some algebraic manipulations we get, respectively,

$$\cos^2 \frac{1}{2} t = \frac{1 + \cos t}{2} \tag{18}$$

and

$$\sin^2 \frac{1}{2} t = \frac{1 - \cos t}{2} \tag{19}$$

Formulas (14) through (19) are identities because they hold for all real numbers t.

By subtracting the terms of formula (4) from the corresponding terms of formula (10) the following is obtained:

$$\sin a \sin b = \tfrac{1}{2}\left[-\cos(a + b) + \cos(a - b)\right] \tag{20}$$

By adding corresponding terms of formulas (4) and (10),

$$\cos a \cos b = \tfrac{1}{2}[\cos(a + b) + \cos(a - b)] \tag{21}$$

and by adding corresponding terms of formulas (12) and (13),

$$\sin a \cos b = \tfrac{1}{2}[\sin(a + b) + \sin(a - b)] \tag{22}$$

By letting $c = a + b$ and $d = a - b$ in formulas (22), (21), and (20) we obtain, respectively,

$$\sin c + \sin d = 2 \sin \frac{c + d}{2} \cos \frac{c - d}{2} \tag{23}$$

$$\cos c + \cos d = 2 \cos \frac{c + d}{2} \cos \frac{c - d}{2} \tag{24}$$

and

$$\cos c - \cos d = -2 \sin \frac{c + d}{2} \sin \frac{c - d}{2} \tag{25}$$

If the terms of formula (13) are subtracted from the corresponding terms of formula (12) and $a + b$ and $a - b$ are replaced by c and d, respectively, then

$$\sin c - \sin d = 2 \cos \frac{c + d}{2} \sin \frac{c - d}{2} \tag{26}$$

You are asked to perform the indicated operations in the derivation of some of the above formulas in Exercises 17 to 22.

We now deal with the other four trigonometric functions. They are defined in terms of the sine and cosine.

0.6.4 DEFINITION   The **tangent** and **secant** functions are defined by

$$\tan t = \frac{\sin t}{\cos t} \qquad \sec t = \frac{1}{\cos t}$$

for all real numbers t for which $\cos t \neq 0$.

The **cotangent** and **cosecant** functions are defined by

$$\cos t = \frac{\cos t}{\sin t} \qquad \csc t = \frac{1}{\sin t}$$

for all real numbers t for which $\sin t \neq 0$.

The tangent and secant functions are not defined when $\cos t = 0$, which occurs when t is $\tfrac{1}{2}\pi, \tfrac{3}{2}\pi$, or $\tfrac{1}{2}\pi + k\pi$, where k is any positive or negative integer or zero. Therefore the domain of the tangent and secant functions is the set of all real numbers except numbers of the form $\tfrac{1}{2}\pi + k\pi$, where k is any integer. Similarly, because $\cot t$ and $\csc t$ are not defined when $\sin t = 0$, the domain of the cotangent and cosecant functions is the set of all real numbers except numbers of the form $k\pi$, where k is any integer.

By using identity (1) and Definition 0.6.4 we obtain two other important identities. If on both sides of (1) we divide by $\cos^2 t$ when $\cos t \neq 0$, we get

$$\frac{\cos^2 t}{\cos^2 t} + \frac{\sin^2 t}{\cos^2 t} = \frac{1}{\cos^2 t}$$

and because $\sin t / \cos t = \tan t$ and $1/\cos t = \sec t$ we have the identity

$$1 + \tan^2 t = \sec^2 t \tag{27}$$

By dividing on both sides of (1) by $\sin^2 t$ when $\sin t \neq 0$ we obtain in a similar way the identity

$$\cot^2 t + 1 = \csc^2 t \tag{28}$$

Three other important formulas that follow immediately from Definition 0.6.4 are

$$\sin t \csc t = 1 \tag{29}$$

$$\cos t \sec t = 1 \tag{30}$$

and

$$\tan t \cot t = 1 \tag{31}$$

Formulas (29), (30), and (31) are valid for all values of t for which the functions are defined, and therefore they are identities.

Another formula that we need is one that expresses the tangent of the difference of two numbers in terms of the tangents of the two numbers. By using formulas (13) and (10) we have

$$\tan(a - b) = \frac{\sin(a - b)}{\cos(a - b)}$$

$$= \frac{\sin a \cos b - \cos a \sin b}{\cos a \cos b + \sin a \sin b}$$

Dividing both the numerator and denominator by $\cos a \cos b$ we get

$$\tan(a - b) = \frac{\dfrac{\sin a \cos b}{\cos a \cos b} - \dfrac{\cos a \sin b}{\cos a \cos b}}{\dfrac{\cos a \cos b}{\cos a \cos b} + \dfrac{\sin a \sin b}{\cos a \cos b}}$$

$$= \frac{\dfrac{\sin a}{\cos a} - \dfrac{\sin b}{\cos b}}{1 + \dfrac{\sin a}{\cos a} \cdot \dfrac{\sin b}{\cos b}}$$

Therefore we have the identity

$$\tan(a - b) = \frac{\tan a - \tan b}{1 + \tan a \tan b} \tag{32}$$

Taking $a = 0$ and $b = t$ in formula (32) gives

$$\tan(0 - t) = \frac{\tan 0 - \tan t}{1 + \tan 0 \tan t}$$

Because $\tan 0 = 0$, we have from the above equation

$$\tan(-t) = -\tan t \tag{33}$$

Therefore the tangent is an odd function.

Taking $b = -b$ in formula (32) gives

$$\tan(a - (-b)) = \frac{\tan a - \tan(-b)}{1 + \tan a \tan(-b)}$$

and replacing $\tan(-b)$ by $-\tan b$ we get the identity

$$\tan(a + b) = \frac{\tan a + \tan b}{1 - \tan a \tan b} \tag{34}$$

In Exercises 25 and 26 you are asked to derive the formulas

$$\tan \tfrac{1}{2}t = \frac{1 - \cos t}{\sin t} \tag{35}$$

and

$$\tan \tfrac{1}{2}t = \frac{\sin t}{1 + \cos t} \tag{36}$$

We have defined the trigonometric functions with real-number domains. There are important uses of trigonometric functions for which the domains are sets of angles. Thus we wish to consider $\sin \theta$, $\cos \theta$, $\tan \theta$, $\cot \theta$, $\sec \theta$, and $\csc \theta$ when θ represents an angle. For this purpose a trigonometric function of θ is the corresponding function of the real number t, where t is the radian measure of θ.

0.6.5 DEFINITION   If θ is an angle having radian measure t, then

$$\sin \theta = \sin t \qquad \cos \theta = \cos t \qquad \tan \theta = \tan t$$

$$\cot \theta = \cot t \qquad \sec \theta = \sec t \qquad \csc \theta = \csc t$$

When considering a trigonometric function of an angle θ, often the measurement of the angle is used in place of θ. For instance, if the degree measure of an angle θ is 60 (or, equivalently, the radian measure of θ is $\tfrac{1}{3}\pi$), then in place of $\sin \theta$ we could write $\sin 60°$ or $\sin \tfrac{1}{3}\pi$. Notice that when the measurement of an angle is in degrees, the degree symbol is written. However, when there is no symbol attached, the measurement of the angle is in radians. For instance, $\cos 2°$ means the cosine of an angle having degree measure 2, while $\cos 2$ means the cosine of an angle having radian measure 2. This is consistent with the fact that the cosine of an angle having radian measure 2 is equal to the cosine of the real number 2.

The graphs of the trigonometric functions are discussed in Section 2.6, where we can obtain more information about these graphs than here.

The exercises of this section are intended as a review of some of the fundamental concepts of trigonometry. As you do them, you may find it necessary to consult a text in trigonometry.

EXERCISES 0.6

In Exercises 1 and 2, find the equivalent radian measurement.

1. (a) $60°$; (b) $135°$; (c) $210°$; (d) $-150°$; (e) $20°$; (f) $450°$; (g) $-75°$; (h) $100°$

2. (a) $45°$; (b) $120°$; (c) $240°$; (d) $-225°$; (e) $15°$; (f) $540°$; (g) $-48°$; (h) $2°$

In Exercises 3 and 4, find the equivalent degree measurement.

3. (a) $\frac{1}{4}\pi$ rad; (b) $\frac{2}{3}\pi$ rad; (c) $\frac{11}{6}\pi$ rad; (d) $-\frac{1}{2}\pi$ rad; (e) $\frac{1}{2}$ rad; (f) 3π rad; (g) -2 rad; (h) $\frac{1}{12}\pi$ rad

4. (a) $\frac{1}{6}\pi$ rad; (b) $\frac{4}{3}\pi$ rad; (c) $\frac{3}{4}\pi$ rad; (d) -5π rad; (e) $\frac{1}{3}$ rad; (f) -5 rad; (g) $\frac{11}{12}\pi$ rad; (h) 0.2 rad

In Exercises 5 through 12, determine the indicated function value without using tables or a calculator.

5. (a) $\sin \frac{1}{6}\pi$; (b) $\cos \frac{1}{4}\pi$; (c) $\sin(-\frac{3}{2}\pi)$; (d) $\cos(-\frac{2}{3}\pi)$

6. (a) $\cos \frac{1}{3}\pi$; (b) $\sin \frac{1}{4}\pi$; (c) $\cos(-\frac{1}{2}\pi)$; (d) $\sin(-2\pi)$

7. (a) $\cos \frac{5}{6}\pi$; (b) $\sin(-\frac{5}{4}\pi)$; (c) $\cos 3\pi$; (d) $\sin(-5\pi)$

8. (a) $\sin \frac{4}{3}\pi$; (b) $\cos(-\frac{1}{6}\pi)$; (c) $\sin 7\pi$; (d) $\cos(-\frac{5}{2}\pi)$

9. (a) $\tan \frac{1}{3}\pi$; (b) $\cot \frac{1}{4}\pi$; (c) $\sec(-\pi)$; (d) $\csc \frac{1}{2}\pi$

10. (a) $\cot \frac{1}{6}\pi$; (b) $\tan \frac{1}{4}\pi$; (c) $\csc(-\frac{3}{2}\pi)$; (d) $\sec \pi$

11. (a) $\sec(-\frac{1}{6}\pi)$; (b) $\csc \frac{3}{4}\pi$; (c) $\tan \frac{5}{6}\pi$; (d) $\cot(-\frac{3}{4}\pi)$

12. (a) $\csc(-\frac{1}{3}\pi)$; (b) $\sec \frac{5}{6}\pi$; (c) $\tan(-\frac{5}{4}\pi)$; (d) $\cot \frac{3}{2}\pi$

In Exercises 13 and 14, find the indicated value by using formula (4) or (12).

13. (a) $\sin \frac{11}{12}\pi$ (Hint: $\frac{11}{12}\pi = \frac{2}{3}\pi + \frac{1}{4}\pi$); (b) $\cos \frac{7}{12}\pi$

14. (a) $\cos \frac{11}{12}\pi$; (b) $\sin \frac{7}{12}\pi$

In Exercises 15 and 16, find the indicated value by using formula (18) or (19).

15. (a) $\cos \frac{1}{12}\pi$; (b) $\sin \frac{1}{8}\pi$

16. (a) $\sin \frac{1}{12}\pi$; (b) $\cos \frac{1}{8}\pi$

17. Derive formula (20) by subtracting the terms of formula (4) from the corresponding terms of formula (10).

18. Derive formula (21) by a method similar to that suggested in Exercise 17.

19. Derive formula (22) by a method similar to that suggested in Exercise 17.

20. Derive formula (23) by using formula (22).

21. Derive formula (24) by using formula (21).

22. Derive formula (25) by using formula (20).

23. Derive a formula for $\sin 3t$ in terms of $\sin t$ by using formulas (12), (14), (15), and (1).

24. Derive a formula for $\cos 3t$ in terms of $\cos t$. Use a method similar to that suggested in Exercise 23.

25. Derive formula (35) from the definition of the tangent function, and formulas (18) and (19).

26. Derive formula (36) by a method similar to that suggested in Exercise 25.

In Exercises 27 through 30, find the indicated function value without using tables or a calculator. (Hint: $\frac{1}{12}\pi = \frac{1}{2} \cdot \frac{1}{6}\pi$; $\frac{1}{8}\pi = \frac{1}{2} \cdot \frac{1}{4}\pi$; $\frac{5}{12}\pi = \frac{1}{4}\pi + \frac{1}{6}\pi$; and so on.)

27. (a) $\sin \frac{1}{12}\pi$: (b) $\cos \frac{3}{8}\pi$; (c) $\tan \frac{5}{8}\pi$; (d) $\cot \frac{11}{12}\pi$

28. (a) $\sin \frac{1}{8}\pi$; (b) $\cos \frac{5}{12}\pi$; (c) $\tan \frac{13}{12}\pi$; (d) $\cot \frac{7}{8}\pi$

29. (a) $\cos \frac{19}{12}\pi$; (b) $\sin(-\frac{5}{8}\pi)$; (c) $\tan \frac{7}{12}\pi$; (d) $\cot \frac{11}{8}\pi$

30. (a) $\sin \frac{19}{12}\pi$; (b) $\cos(-\frac{5}{8}\pi)$; (c) $\cot \frac{7}{12}\pi$; (d) $\tan \frac{11}{8}\pi$

In Exercises 31 through 36, find all values of t in the interval $[0, 2\pi)$ for which the given equation is satisfied.

31. (a) $\sin t = 1$; (b) $\cos t = -1$; (c) $\tan t = 1$; (d) $\sec t = 1$

32. (a) $\sin t = -1$; (b) $\cos t = 1$; (c) $\tan t = -1$; (d) $\csc t = 1$

33. (a) $\sin t = 0$; (b) $\cos t = 0$; (c) $\tan t = 0$; (d) $\cot t = 0$

34. (a) $\sin t = \frac{1}{2}$; (b) $\cos t = -\frac{1}{2}$; (c) $\cot t = 1$; (d) $\sec t = 2$

35.   (a) $\sin t = -\frac{1}{2}$; (b) $\cos t = \frac{1}{2}$; (c) $\cot t = -1$; (d) $\csc t = 2$

36.   (a) $\sin t = -\frac{1}{2}\sqrt{2}$; (b) $\cos t = \frac{1}{2}\sqrt{2}$; (c) $\tan t = -\frac{1}{3}\sqrt{3}$; (d) $\cot t = \frac{1}{3}\sqrt{3}$

37.   For what values of t in $[0, 2\pi)$ is (a) $\tan t$ undefined and (b) $\csc t$ undefined?

38.   For what values of t in $[0, \pi)$ is (a) $\cot t$ undefined and (b) $\sec t$ undefined?

39.   Suppose f is a periodic function for which the fundamental period is π, and the domain of f is the set of all real numbers. Prove that f is also periodic with period 2π.

40.   Suppose f is a periodic function for which the fundamental period is 2π, and the domain of f is the set of all real numbers. Prove that f is also periodic with period -2π.

41.   Prove that the function of Exercise 39 is periodic with period $n\pi$, for every integer n. (*Hint:* Use mathematical induction.)

42.   Prove that the function of Exercise 40 is periodic with period $2\pi n$ for every integer n. (*Hint:* Use mathematical induction.)

0.7
AN APPLICATION OF THE TANGENT FUNCTION TO THE SLOPE OF A LINE

The tangent function can be used in connection with the slope of a straight line. We first define the *angle of inclination* of a line.

0.7.1 DEFINITION   The **angle of inclination** of a line not parallel to the x axis is the smallest angle measured counterclockwise from the positive direction of the x axis to the line. The angle of inclination of a line parallel to the x axis is defined to have measure zero.

If α is the angle of inclination of a line, then $0° \le \alpha < 180°$. Figure 0.7.1 shows a line L for which $0° < \alpha < 90°$, and Fig. 0.7.2 shows one for which $90° < \alpha < 180°$.

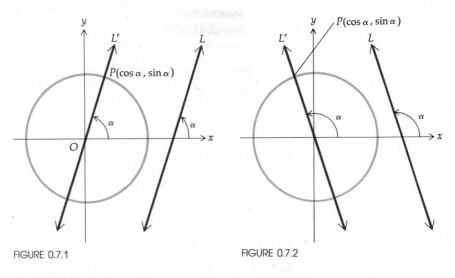

FIGURE 0.7.1   FIGURE 0.7.2

0.7.2 THEOREM   If α is the angle of inclination of line L, not parallel to the y axis, then the slope m of L is given by

$$m = \tan \alpha$$

FIGURE 0.7.3

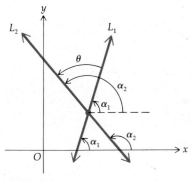

FIGURE 0.7.4

PROOF   Refer to Figs. 0.7.1 and 0.7.2, which show the given line L whose angle of inclination is α and whose slope is m. The line L' that passes through the origin and is parallel to L also has slope m and an angle of inclination α. The point $P(\cos\alpha, \sin\alpha)$ at the intersection of L and the unit circle U lies on L'. And because the point $(0,0)$ also lies on L', it follows from Definition 0.4.1 that

$$m = \frac{\sin\alpha - 0}{\cos\alpha - 0}$$

$$= \frac{\sin\alpha}{\cos\alpha}$$

$$= \tan\alpha \qquad \blacksquare$$

If line L is parallel to the y axis, the angle of inclination of L is $90°$ and $\tan 90°$ is not defined. This is consistent with the fact that a vertical line has no slope.

Theorem 0.7.2 can be used to obtain a formula for finding the angle between two nonvertical intersecting lines. If two lines intersect, two supplementary angles are formed at their point of intersection. To distinguish these two angles let L_2 be the line with the greater angle of inclination α_2 and let L_1 be the other line whose angle of inclination is α_1. If θ is the angle between the two lines, then we define

$$\theta = \alpha_2 - \alpha_1 \qquad (1)$$

If L_1 and L_2 are parallel, then $\alpha_1 = \alpha_2$ and the angle between the two lines is $0°$. Thus if L_1 and L_2 are two distinct lines, then $0° \leq \theta < 180°$. Refer to Figs. 0.7.3 and 0.7.4. The following theorem enables us to find θ when the slopes of L_1 and L_2 are known.

0.7.3 THEOREM   Let L_1 and L_2 be two nonvertical lines that intersect and are not perpendicular, and let L_2 be the line having the greater angle of inclination. Then if m_1 is the slope of L_1, m_2 is the slope of L_2, and θ is the angle between L_1 and L_2,

$$\tan\theta = \frac{m_2 - m_1}{1 + m_1 m_2}$$

PROOF   If α_1 and α_2 are the angles of inclination of L_1 and L_2, respectively, from (4),

$$\theta = \alpha_2 - \alpha_1$$

Therefore

$$\tan\theta = \tan(\alpha_2 - \alpha_1)$$

Applying formula (32) of Sec. 0.6 to the right side of the above we get

$$\tan\theta = \frac{\tan\alpha_2 - \tan\alpha_1}{1 + \tan\alpha_2 \tan\alpha_1}$$

$$= \frac{m_2 - m_1}{1 + m_1 m_2} \qquad \blacksquare$$

FIGURE 0.7.5

▶ ILLUSTRATION 1 (a) Refer to Fig. 0.7.5, where θ is the angle between the lines whose slopes are $\frac{3}{4}$ and $\frac{2}{5}$. Because L_2 is the slope of the line having the greater inclination, in the formula of Theorem 0.7.3 we take $m_2 = \frac{3}{4}$ and $m_1 = \frac{2}{5}$, and we have

$$\tan \theta = \frac{m_2 - m_1}{1 + m_1 m_2}$$

$$= \frac{\frac{3}{4} - \frac{2}{5}}{1 + \frac{3}{4} \cdot \frac{2}{5}}$$

$$= \frac{15 - 8}{20 + 6}$$

$$= \frac{7}{26}$$

Therefore, to the nearest degree,

$$\theta = 15°$$

FIGURE 0.7.6

(b) See Fig. 0.7.6, where θ is the angle between the lines having slopes 2 and $-\frac{2}{5}$. In the formula of Theorem 0.7.3 we take $m_2 = -\frac{2}{5}$ and $m_1 = 2$, and we have

$$\tan \theta = \frac{m_2 - m_1}{1 + m_1 m_2}$$

$$= \frac{-\frac{2}{5} - 2}{1 + (-\frac{2}{5})(2)}$$

$$= \frac{-2 - 10}{5 - 4}$$

$$= -12$$

Thus, to the nearest degree,

$$\theta = 95°$$ ◀

EXAMPLE 1
Find to the nearest degree the measurements of the interior angles of the triangle having vertices $B(-2, 1)$, $C(2, 2)$, $D(-3, 4)$.

SOLUTION
Let β, γ, and δ be the interior angles of the triangle at the vertices B, C, and D, respectively. Let u denote the line through B and C, let v denote the line through C and D, let w denote the line through B and D, and let m_u, m_v, and m_w be their respective slopes. Refer to Fig. 0.7.7.

Using the formula for the slope of a line through two given points we get

$$m_u = \frac{1}{4} \qquad m_v = -\frac{2}{5} \qquad m_w = -3$$

FIGURE 0.7.7

To determine β we observe that line w has a greater angle of inclination than line u; so in the formula of Theorem 0.7.3, $m_2 = m_w$; therefore $m_2 = -3$. Also $m_1 = m_u$; hence $m_1 = \frac{1}{4}$. The angle between lines u and v is β. Thus, from Theorem 0.7.3,

$$\tan \beta = \frac{-3 - \frac{1}{4}}{1 + \frac{1}{4}(-3)}$$

$$= -13$$

Therefore

$$\beta = 94°$$

To find γ, because line v has a greater angle of inclination than line u, $m_2 = m_v$; then $m_2 = -\frac{2}{5}$. Furthermore, $m_1 = m_u$; hence $m_1 = \frac{1}{4}$. The angle between lines u and v is by definition $180° - \gamma$. From Theorem 0.7.3,

$$\tan(180° - \gamma) = \frac{-\frac{2}{5} - \frac{1}{4}}{1 + \frac{1}{4}(-\frac{2}{5})}$$

$$= -\frac{13}{18}$$

Hence

$$180° - \gamma = 144°$$

$$\gamma = 36°$$

To determine δ, because line v has a greater angle of inclination than line w, $m_2 = m_v$ and $m_1 = m_w$. Thus $m_2 = -\frac{2}{5}$ and $m_1 = -3$. The angle between lines v and w is δ. From Theorem 0.7.3,

$$\tan \delta = \frac{-\frac{2}{5} - (-3)}{1 + (-\frac{2}{5})(-3)}$$

$$= \frac{13}{11}$$

Therefore

$$\delta = 50°$$

The sum of the degree measures of the angles of a triangle must be 180. We use this to check the results, and we have

$$\beta + \gamma + \delta = 94° + 36° + 50°$$

$$= 180°$$

EXERCISES 0.7

In Exercises 1 through 4, find $\tan \theta$ if θ is the angle between the lines having the given slopes.

1. (a) 1 and $\frac{1}{4}$; (b) 4 and $-\frac{5}{3}$

2. (a) $\frac{1}{2}$ and $-\frac{3}{4}$; (b) $\frac{2}{7}$ and $\frac{7}{2}$

3. (a) $-\frac{1}{2}$ and $\frac{2}{3}$; (b) $-\frac{5}{4}$ and $-\frac{7}{5}$

4. (a) $-\frac{3}{5}$ and 2; (b) $-\frac{1}{3}$ and $-\frac{1}{10}$

In Exercises 5 and 6, find to the nearest degree the measurement of the angle between the lines having the given slopes.

5. (a) 5 and $-\frac{7}{9}$; (b) $-\frac{3}{2}$ and $-\frac{3}{4}$ **6.** (a) -3 and 2; (b) $\frac{3}{2}$ and $\frac{1}{4}$

7. Find to the nearest degree the measurements of the interior angles of the triangle formed by the lines that have equations $2x + y - 6 = 0$, $3x - y - 4 = 0$, and $3x + 4y + 8 = 0$.

8. Find to the nearest degree the measurements of the interior angles of the triangle having vertices at $(1, 0)$, $(-3, 2)$, and $(2, 3)$.

9. Find an equation of a line through the point $(-1, 4)$ making an angle of radian measure $\frac{1}{4}\pi$ with the line having equation $2x + y - 5 = 0$ (two solutions).

10. Find an equation of a line through the point $(-3, -2)$ making an angle of radian measure $\frac{1}{3}\pi$ with the line having equation $3x - 2y - 7 = 0$ (two solutions).

11. Use Theorem 0.7.3 to prove that the triangle having vertices at $(2, 3)$, $(6, 2)$, and $(3, -1)$ is isosceles.

12. Use Theorem 0.7.3 to prove that the triangle having vertices at $(-3, 0)$, $(-1, 0)$, and $(-2, \sqrt{3})$ is equilateral.

13. Find the slope of the bisector of the angle at A in the triangle having the vertices $A(4, 1)$, $B(6, 5)$, and $C(-1, 8)$.

14. Find the slope of the bisector of the angle at A in the triangle having the vertices $A(-3, 5)$, $B(2, -4)$, and $C(-1, 7)$.

15. Let A, B, and C be the vertices of an oblique triangle (no right angle), and let α, β, and γ be the interior angles at vertices A, B, and C, respectively. Prove that $\tan \alpha + \tan \beta + \tan \gamma = \tan \alpha \tan \beta \tan \gamma$. (*Hint:* First show that $\tan(\alpha + \beta) = -\tan \gamma$.)

16. Find the tangents of the interior angles of the triangle having vertices at $(-3, -2)$, $(-6, 3)$, and $(5, 1)$, and check by applying the equality of Exercise 15.

17. Follow the instructions of Exercise 16 for the triangle having vertices at $(5, 7)$, $(-1, 4)$, and $(2, -6)$.

18. Prove Theorem 0.4.3 by using Theorem 0.7.2.

19. Prove Theorem 0.4.4 by using Theorem 0.7.2.

REVIEW EXERCISES FOR CHAPTER 0

In Exercises 1 through 12, find the solution set of the given inequality, and illustrate the solution on the real number line.

1. $3x - 7 \le 5x - 17$ **2.** $8 < 5x + 4 \le 10$ **3.** $\dfrac{x}{x - 1} > \dfrac{1}{4}$ *2 cases* *a subset*   **4.** $\dfrac{3}{x + 4} < \dfrac{2}{x - 5}$

5. $2x^2 + x < 3$ **6.** $|x - 2| < 4$ **7.** $|3 + 5x| \le 9$ **8.** $|5 - 2x| \le 3$

9. $|4 - 3x| < 8$ **10.** $|2x - 5| > 7$ **11.** $|2x + 7| > 5$ **12.** $\left|\dfrac{2 - 3x}{3 + x}\right| \ge \dfrac{1}{4}$

In Exercises 13 through 15, solve for x.

13. $|3 - 4x| = 15$ **14.** $\left|\dfrac{2x - 1}{x + 3}\right| = 4$ **15.** $|3x - 4| = |6 - 2x|$

16. Find all the values of x for which $\sqrt{x^2 + 2x - 15}$ is real.

In Exercises 17 through 22, draw a sketch of the graph of the equation.

17. $y^2 = x - 4$ **18.** $y = |x - 4|$ **19.** $y = x^2 - 2$

20. $x^2 + 4y^2 = 16$ **21.** $9x^2 - 25y^2 = 0$ **22.** $4x^2 - y^2 = 16$

23. Find the abscissas of the points having ordinate 4 that are at a distance of $\sqrt{117}$ from the point $(5, -2)$.

24. Find an equation that must be satisfied by the coordinates of any point whose distance from $(-1, 2)$ is twice its distance from $(3, -4)$.

25. Find an equation that must be satisfied by the coordinates of any point whose distance from the point $(-3, 4)$ is $\sqrt{10}$.

26. Define the following sets of points by either an equation or an inequality: (a) the point-circle $(3, -5)$; (b) the set of all points whose distance from the point $(3, -5)$ is less than 4; (c) the set of all points whose distance from the point $(3, -5)$ is at least 5.

27. Prove that the quadrilateral having vertices at $(1,2)$, $(5,-1)$, $(11,7)$, and $(7,10)$ is a rectangle.
28. Prove that the triangle with vertices at $(-8,1)$, $(-1,-6)$, and $(2,4)$ is isosceles, and find its area.
29. Prove that the points $(2,4)$, $(1,-4)$, and $(5,-2)$ are the vertices of a right triangle, and find the area of the triangle.
30. Prove that the points $(1,-1)$, $(3,2)$, and $(7,8)$ are collinear in two ways: (a) by using the distance formula; (b) by using slopes.
31. Two vertices of a parallelogram are at $(-3,4)$ and $(2,3)$, and its center is at $(0,-1)$. Find the other two vertices.
32. Two opposite vertices of a square are at $(3,-4)$ and $(9,-4)$. Find the other two vertices.
33. Prove that the points $(2,13)$, $(-2,5)$, $(3,-1)$, and $(7,7)$ are the vertices of a parallelogram.
34. Find an equation of the circle having its center at $(-2,4)$ and radius $\sqrt{5}$. Write the equation in the general form.
35. Find an equation of the circle having the points $(-3,2)$ and $(5,6)$ as endpoints of a diameter.
36. Find an equation of the circle through the points $(3,-1)$, $(2,2)$, and $(-4,5)$.
37. Find the center and radius of the circle having the equation $4x^2 + 4y^2 - 12x + 8y + 9 = 0$.
38. Find an equation of the line through the points $(2,-4)$ and $(7,3)$, and write the equation in intercept form.
39. Find an equation of the line through the point $(-1,6)$ and having slope 3.
40. Find an equation of the line that is the perpendicular bisector of the line segment from $(-1,5)$ to $(3,2)$.
41. Find an equation of the line through the point $(5,-3)$ and perpendicular to the line whose equation is $2x - 5y = 1$.
42. Find an equation of the circle having as its diameter the common chord of the two circles $x^2 + y^2 + 2x - 2y - 14 = 0$ and $x^2 + y^2 - 4x + 4y - 2 = 0$.
43. Find an equation of the circle circumscribed about the triangle having sides on the lines $x - 3y + 2 = 0$, $3x - 2y + 6 = 0$, and $2x + y - 3 = 0$.
44. Find the shortest distance from the point $(2,-5)$ to the line having the equation $3x + y = 2$.
45. Find an equation of the line through the point of intersection of the lines $5x + 6y - 4 = 0$ and $x - 3y + 2 = 0$ and perpendicular to the line $x - 4y - 20 = 0$ without finding the point of intersection of the two lines. (*Hint:* See Exercise 42 in Exercises 0.4.)
46. The sides of a parallelogram are on the line $x + 2y = 10$, $3x - y = -20$, $x + 2y = 15$, and $3x - y = -10$. Find equations of the diagonals without finding the vertices of the parallelogram. (*Hint:* See Exercise 42 in Exercises 0.4.)
47 Determine all values of k for which the graphs of the two equations $x^2 + y^2 = k$ and $x + y = k$ intersect.
48. Determine the values of k and h if $3x + ky + 2 = 0$ and $5x - y + h = 0$ are equations of the same line.
49. Given $f(x) = 3x^2 - x + 5$, find (a) $f(-3)$; (b) $f(-x^2)$; (c) $\dfrac{f(x+h) - f(x)}{h}$, $h \neq 0$.
50. Given $g(x) = \sqrt{x - 1}$, find $\dfrac{g(x+h) - g(x)}{h}$, $h \neq 0$.
51. Determine if the function is even, odd, or neither: (a) $f(x) = 2x^3 - 3x$; (b) $g(x) = 5x^4 + 2x^2 - 1$; (c) $h(x) = 3x^5 - 2x^3 + x^2 - x$; (d) $F(x) = \dfrac{x^2 + 1}{x^3 - x}$.

In Exercises 52 through 56, for the given functions f and g define the following functions and determine the domain of the resulting function: (a) $f + g$; (b) $f - g$; (c) $f \cdot g$; (d) f/g; (e) g/f; (f) $f \circ g$; (g) $g \circ f$.
52. $f(x) = \sqrt{x + 2}$ and $g(x) = x^2 + 4$
53. $f(x) = x^2 - 4$ and $g(x) = 4x - 3$
54. $f(x) = x^2 - 9$ and $g(x) = \sqrt{x + 5}$
55. $f(x) = \dfrac{1}{x - 3}$ and $g(x) = \dfrac{x}{x + 1}$
56. $f(x) = \sqrt{x}$ and $g(x) = \dfrac{1}{x^2}$

In Exercises 57 and 58, determine the indicated function value without using tables or a calculator.
57. (a) $\sin \frac{2}{3}\pi$; (b) $\cos \frac{5}{4}\pi$; (c) $\tan(-\frac{5}{6}\pi)$; (d) $\cot \frac{13}{12}\pi$; (e) $\sec \pi$; (f) $\csc(-\frac{1}{8}\pi)$
58. (a) $\cos \frac{7}{6}\pi$; (b) $\sin(-\frac{7}{4}\pi)$; (c) $\cot(-\frac{1}{2}\pi)$; (d) $\tan \frac{5}{8}\pi$; (e) $\sec(-\frac{1}{12}\pi)$; (f) $\csc \frac{5}{3}\pi$

In Exercises 59 and 60, find all values of t in the interval $[0, 2\pi)$ for which the given equation is satisfied.

59. (a) $\sin t = \frac{1}{2}$; (b) $\cos t = 1$; (c) $\tan t = -1$; (d) $\cot t = \sqrt{3}$; (e) $\sec t = -2$; (f) $\csc t = \sqrt{2}$

60. (a) $\sin t = \frac{1}{2}\sqrt{3}$; (b) $\cos t = -1$; (c) $\tan t = -\sqrt{3}$; (d) $\cot t = 1$; (e) $\sec t = -\sqrt{2}$; (f) $\csc t = 2$

61. Find to the nearest degree the measurements of the four interior angles of the quadrilateral having vertices at $(5, 6)$, $(-2, 4)$, $(-2, 1)$, and $(3, 1)$, and verify that the sum is $360°$.

62. Find an equation of a line through the point $(2, 5)$ making an angle of radian measure $\frac{1}{4}\pi$ with the line having equation $3x - 4y + 7 = 0$ (two solutions).

63. Find equations of the lines through the origin that are tangent to the circle having its center at $(2, 1)$ and radius 2.

64. Prove that the two lines $A_1x + B_1y + C_1 = 0$ and $A_2x + B_2y + C_2 = 0$ are parallel if and only if $A_1B_2 - A_2B_1 = 0$.

65. Prove analytically that if the diagonals of a rectangle are perpendicular, then the rectangle is a square.

66. Prove analytically that the line segment joining the midpoints of any two sides of a triangle is parallel to the third side and that its length is one-half the length of the third side.

67. In a triangle the point of intersection of the medians, the point of intersection of the altitudes, and the center of the circumscribed circle are collinear. Find these three points and prove that they are collinear for the triangle having vertices at $(2, 8)$, $(5, -1)$, and $(6, 6)$.

68. Prove analytically that the set of points equidistant from two given points is the perpendicular bisector of the line segment joining the two points.

69. Prove that if x is any real number, $|x| < x^2 + 1$.

70. Prove analytically that the three medians of any triangle meet in a point.

FUNCTIONS OF A SINGLE VARIABLE

Learning calculus can be your most stimulating and exciting educational experience because it is the basis for much of mathematics and for many of the greatest accomplishments of the modern world.

Some of the ideas of calculus can be found in the works of the ancient Greek mathematicians at the time of Archimedes (287–212 B.C.) and in works of the early seventeenth century by René Descartes (1596–1650), Pierre de Fermat (1601–1665), John Wallis (1616–1703), and Isaac Barrow (1630–1677). However, the invention of calculus is often attributed to Sir Isaac Newton (1642–1727) and Gottfried Wilhelm Leibniz (1646–1716) because they began the generalization and unification of the subject. There were other mathematicians of the seventeenth and eighteenth century who joined in the development of the calculus; some of them were Jakob Bernoulli (1654–1705), Johann Bernoulli (1667–1748), Leonhard Euler (1707–1783), and Joseph L. Lagrange (1736–1813). However, it wasn't until the nineteenth century that the processes of calculus were given a sound foundation by such mathematicians as Bernhard Bolzano (1781–1848), Augustin L. Cauchy (1789–1857), Karl Weierstrass (1815–1897), and Richard Dedekind (1831–1916).

The two fundamental mathematical operations in calculus are *differentiation* and *integration*. These operations involve the computation of the *derivative* and the *definite integral*, each of which is based on the notion of *limit*. The treatment of limits, along with that of *continuity*, appears in Chapter 1. The idea of a limit of a function is first given a step-by-step motivation, which brings the discussion from computing the value of a function near a number, through an intuitive treatment of the limiting process, up to a rigorous epsilon-delta definition.

The derivative is introduced in Chapter 2. The theory and applications of differentiation are presented there and in Chapter 3. The interpretation of the derivative as a rate of change leads to its importance in many fields. For instance, in physics, velocity in rectilinear motion is defined in terms of a derivative because it is the measure of the rate of change of distance traveled with respect to time. The rate of growth of bacteria gives an application of the derivative in biology. The rate of change of a chemical reaction is of interest to a chemist. Economists are concerned with marginal concepts such as marginal revenue, marginal cost, and marginal profit, which are all rates of change. The derivative is used to define the tangent line to a curve. This application gives information about the behavior of functions; thus it is an aid in sketching their graphs. Furthermore, the derivative enables us to compute extreme function values. For example, we can determine the strongest rectangular beam that can be cut from a given cylindrical log or the dimensions of a box requiring the least amount of material for a specific volume.

The *antiderivative* is treated in the first two sections of Chapter 4. The introduction of the definite integral follows the definition of the measure of the area under a curve as a limit of sums. Exercises involving the evaluation of definite integrals by finding limits of sums are given to stress that this is how they are calculated. Elementary properties of the definite integral are derived and used to prove the two fundamental theorems of the calculus. These theorems are important because they provide us with an alternative to computing limits of sums.

In Chapter 5 there are numerous applications of the definite integral including the calculation of volumes, lengths of arcs of graphs of functions, centroids, work, and liquid pressure. The presentation highlights not only the manipulative techniques but also the fundamental principles involved.

The first two sections of Chapter 6 pertain to *inverse functions* and their properties. The treatment of logarithmic and exponential functions in the remaining sections of the chapter is the modern approach. The *natural logarithm* is defined as an integral, and the *natural exponential function* is defined as the inverse of the natural logarithmic function. With this background, we are able to define an irrational power of a real number. The *inverse trigonometric functions*, and the derivatives and integrals of them, appear in Chapter 7. Also in Chapter 7, the *hyperbolic functions* and their inverses are discussed.

Chapter 8, on techniques of integration, involves one of the most important computational aspects of the calculus. The theoretical backgrounds of each different method are explained after an introductory motivation. The mastery of integration techniques depends upon the examples, and problems that you will certainly meet in practice are used as illustrations. These problems require patience and persistence to solve.

Methods for computing certain limits involving *indeterminate forms* are treated in the first two sections of Chapter 9. The next two sections involve *improper integrals* with infinite limits of integration and those with discontinuous integrands. *Taylor's formula*, also discussed in Chapter 9, gives a procedure for approximating functions by means of polynomials.

Chapter 10, the final one in Part 1, is concerned with topics in analytic geometry. The first three sections are devoted to a discussion of *polar coordinates* and some of their applications. Then each of the *conic sections* is introduced by indicating how it is formed by intersecting a plane and a cone. The analytic definition of each conic is given, and its equation in rectangular coordinates is obtained. There is also a unified treatment of conic sections that utilizes their polar equations.

LIMITS AND CONTINUITY

1.1
GRAPHS OF FUNCTIONS

Section 0.5 of the prelude is devoted to the subject of functions. In that section a function is defined, function notation is introduced, operations on functions are given, and particular types of functions are presented. In this chapter we are concerned with limits and continuity of functions. In the treatment of these topics, graphs of functions are used extensively, and this section deals with a discussion of such graphs. The interval and set notation introduced in Sec. 0.1 are used in the development.

In Definition 0.5.1 a function is defined as follows:

A **function** is a set of ordered pairs of real numbers (x, y) in which no two distinct ordered pairs have the same first number. The set of all admissible values of x is called the **domain** of the function, and the set of all resulting values of y is called the **range** of the function.

From Definition 0.5.2, the **graph** of a function f is the set of all points (x, y) in the plane R^2 for which (x, y) is an ordered pair in f.

FIGURE 1.1.1

EXAMPLE 1
The **absolute value function** is defined by

$$f(x) = |x|$$

Determine the domain and range of the absolute value function, and draw a sketch of its graph.

SOLUTION
The domain of f is $(-\infty, +\infty)$, and the range is $[0, +\infty)$. A sketch of the graph appears in Fig. 1.1.1.

In the following illustration, a function is defined by more than one equation. Such a definition is permissible provided that for each number x in the domain there is a unique function value in the range.

▶ ILLUSTRATION 1   Let f be the function defined by

$$f(x) = \begin{cases} -3 & \text{if } x \leq -1 \\ 1 & \text{if } -1 < x \leq 2 \\ 4 & \text{if } 2 < x \end{cases}$$

The domain of f is $(-\infty, +\infty)$, while the range of f consists of the three numbers -3, 1, and 4. A sketch of the graph is shown in Fig. 1.1.2.   ◀

FIGURE 1.1.2

EXAMPLE 2
Let g be the function defined by

$$g(x) = \begin{cases} 3x - 2 & \text{if } x < 1 \\ x^2 & \text{if } 1 \leq x \end{cases}$$

Determine the domain and range of g, and draw a sketch of its graph.

SOLUTION
The domain of g is $(-\infty, +\infty)$, and the range is also $(-\infty, +\infty)$. The graph consists of the portion of the straight line $y = 3x - 2$ for which $x < 1$ and

FIGURE 1.1.3

FIGURE 1.1.4

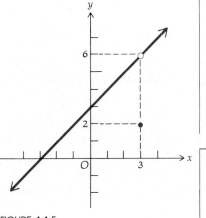

FIGURE 1.1.5

the portion of the parabola $y = x^2$ for which $1 \leq x$. A sketch of the graph is shown in Fig. 1.1.3.

EXAMPLE 3
The function h is defined by

$$h(x) = \frac{x^2 - 9}{x - 3}$$

Determine the domain and range of h, and draw a sketch of its graph.

SOLUTION
Because $h(x)$ is defined for all x except 3, the domain of h consists of all real numbers except 3. When $x = 3$, both the numerator and denominator are zero, and $0/0$ is undefined.

Factoring the numerator into $(x - 3)(x + 3)$ we obtain

$$h(x) = \frac{(x - 3)(x + 3)}{x - 3}$$

or $h(x) = x + 3$, provided that $x \neq 3$. In other words, the function h can be defined by

$$h(x) = x + 3 \qquad \text{if } x \neq 3$$

The range of h consists of all real numbers except 6. The graph consists of all points on the line $y = x + 3$ except the point $(3, 6)$, and it appears in Fig. 1.1.4.

EXAMPLE 4
Let H be the function defined by

$$H(x) = \begin{cases} x + 3 & \text{if } x \neq 3 \\ 2 & \text{if } x = 3 \end{cases}$$

Determine the domain and range of H, and draw a sketch of its graph.

SOLUTION
A sketch of the graph of H is shown in Fig. 1.1.5. The graph consists of the point $(3, 2)$ and all points on the line $y = x + 3$ except the point $(3, 6)$. Function H is defined for all values of x, and therefore its domain is $(-\infty, +\infty)$. The range of H consists of all real numbers except 6.

EXAMPLE 5
Let ϕ be the function defined by

$$\phi(x) = \frac{(x^2 + 3x - 4)(x^2 - 9)}{(x^2 + x - 12)(x + 3)}$$

Determine the domain and range of ϕ, and draw a sketch of its graph.

SOLUTION
Factoring the numerator and denominator we obtain

$$\phi(x) = \frac{(x + 4)(x - 1)(x - 3)(x + 3)}{(x + 4)(x - 3)(x + 3)}$$

FIGURE 1.1.6

FIGURE 1.1.7

FIGURE 1.1.8

The denominator is zero when x is -4, -3, and 3; therefore ϕ is undefined for these three values of x. For values of x other than -4, -3, or 3 we may divide the numerator and denominator by the common factors and obtain

$$\phi(x) = x - 1 \qquad \text{if } x \neq -4,\, x \neq -3,\, x \neq 3$$

Thus the domain of ϕ is the set of all real numbers except -4, -3, and 3, and the range of ϕ is the set of all real numbers except those values of $x - 1$ obtained by replacing x by -4, -3, or 3, that is, all real numbers except -5, -4, and 2. The graph of this function, shown in Fig. 1.1.6, is the straight line $y = x - 1$, with the points $(-4, -5)$, $(-3, -4)$, and $(3, 2)$ deleted.

EXAMPLE 6

The function f is defined by

$$f(x) = \begin{cases} x^2 & \text{if } x \neq 2 \\ 7 & \text{if } x = 2 \end{cases}$$

Determine the domain and range of f, and draw a sketch of its graph.

SOLUTION

A sketch of the graph of f appears in Fig. 1.1.7. The graph consists of the point $(2, 7)$ and all points on the parabola $y = x^2$ except the point $(2, 4)$. Because f is defined for all values of x, the domain is $(-\infty, +\infty)$. The range of f is $[0, +\infty)$.

EXAMPLE 7

Let F be the function defined by

$$F(x) = \begin{cases} x - 1 & \text{if } x < 3 \\ 5 & \text{if } x = 3 \\ 2x + 1 & \text{if } 3 < x \end{cases}$$

Determine the domain and range of F, and draw a sketch of its graph.

SOLUTION

Figure 1.1.8 shows a sketch of the graph of F; it consists of the portion of the line $y = x - 1$ for which $x < 3$, the point $(3, 5)$, and the portion of the line $y = 2x + 1$ for which $3 < x$. The domain of F is $(-\infty, +\infty)$. The function values are either numbers less than 2, the number 5, or numbers greater than 7. Therefore the range of F is the number 5 and those numbers in $(-\infty, 2) \cup (7, +\infty)$.

EXAMPLE 8

The function g is defined by

$$g(x) = \sqrt{x(x - 2)}$$

Determine the domain and range of g, and draw a sketch of its graph.

SOLUTION

Because $\sqrt{x(x-2)}$ is not a real number when $x(x-2) < 0$, the domain of g consists of the values of x for which $x(x-2) \geq 0$. This inequality will be satisfied when one of the following two cases holds: $x \geq 0$ and $x - 2 \geq 0$; or $x \leq 0$ and $x - 2 \leq 0$.

Case 1: $x \geq 0$ and $x - 2 \geq 0$. That is,

$$x \geq 0 \quad \text{and} \quad x \geq 2$$

Both inequalities hold if $x \geq 2$, which is the interval $[2, +\infty)$.

Case 2: $x \leq 0$ and $x - 2 \leq 0$. That is,

$$x \leq 0 \quad \text{and} \quad x \leq 2$$

Both inequalities hold if $x \leq 0$, which is the interval $(-\infty, 0]$.

The solutions for the two cases are combined to obtain the domain of g. It is $(-\infty, 0] \cup [2, +\infty)$. The range of g is the interval $[0, +\infty)$. Figure 1.1.9 shows a sketch of the graph of g.

FIGURE 1.1.9

In the next illustration, the symbol $[\![x]\!]$ is used. $[\![x]\!]$ is defined as the greatest integer less than or equal to x; that is,

$$[\![x]\!] = n \quad \text{if } n \leq x < n + 1, \qquad \text{where } n \text{ is an integer}$$

This function is called the **greatest integer function**. It follows that $[\![1]\!] = 1$, $[\![1.3]\!] = 1$, $[\![\tfrac{1}{2}]\!] = 0$, $[\![-4.2]\!] = -5$, $[\![-8]\!] = -8$, $[\![9.8]\!] = 9$, and so on.

▶ ILLUSTRATION 2   Let I be the greatest integer function; that is,

$$I(x) = [\![x]\!]$$

A sketch of the graph of I is shown in Fig. 1.1.10. The following values are used to draw the sketch.

FIGURE 1.1.10

$$
\begin{array}{ll}
-5 \leq x < -4 & [\![x]\!] = -5 \\
-4 \leq x < -3 & [\![x]\!] = -4 \\
-3 \leq x < -2 & [\![x]\!] = -3 \\
-2 \leq x < -1 & [\![x]\!] = -2 \\
-1 \leq x < 0 & [\![x]\!] = -1 \\
0 \leq x < 1 & [\![x]\!] = 0 \\
1 \leq x < 2 & [\![x]\!] = 1 \\
2 \leq x < 3 & [\![x]\!] = 2 \\
3 \leq x < 4 & [\![x]\!] = 3 \\
4 \leq x < 5 & [\![x]\!] = 4
\end{array}
$$

The domain of I is the set of all real numbers, and the range is the set of all the integers.   ◀

EXAMPLE 9

Let the function G be defined by

$$G(x) = [\![x]\!] - x$$

Draw a sketch of the graph of G. State its domain and range.

SOLUTION

If $0 \leq x < 1, [\![x]\!] =$ 0; so $G(x) = -x$.

If $1 \leq x < 2, [\![x]\!] =$ 1; so $G(x) = 1 - x$.

If $2 \leq x < 3, [\![x]\!] =$ 2; so $G(x) = 2 - x$.

If $-1 \leq x < 0, [\![x]\!] = -1$; so $G(x) = -1 - x$.

and so on.

A sketch of the graph is shown in Fig. 1.1.11. The domain of G is the set of all real numbers, and the range of G is $(-1, 0]$.

FIGURE 1.1.11

EXERCISES 1.1

In each exercise, determine the domain and range of the given function, and draw a sketch of its graph.

1. $f(x) = 3x - 1$

2. $g(x) = x^2 + 2$

3. $F(x) = 3x^2 - 6$

4. $G(x) = 5 - x^2$

5. $G(x) = \sqrt{x + 1}$

6. $f(x) = \sqrt{5 - 3x}$

7. $h(x) = \sqrt{3x - 4}$

8. $f(x) = \sqrt{4 - x^2}$

9. $g(x) = \sqrt{x^2 - 4}$

10. $H(x) = |x - 3|$

11. $\phi(x) = |3x + 2|$

12. $F(x) = \dfrac{4x^2 - 1}{2x + 1}$

13. $f(x) = \dfrac{x^2 - 4x + 3}{x - 1}$

14. $g(x) = \dfrac{x^3 - 3x^2 - 4x + 12}{x^2 - x - 6}$

15. $G(x) = \begin{cases} -2 & \text{if } x \leq 3 \\ 2 & \text{if } 3 < x \end{cases}$

16. $h(x) = \begin{cases} -4 & \text{if } x < -2 \\ -1 & \text{if } -2 \leq x \leq 2 \\ 3 & \text{if } 2 < x \end{cases}$

17. $f(x) = \begin{cases} 2x - 1 & \text{if } x \neq 2 \\ 0 & \text{if } x = 2 \end{cases}$

18. $f(x) = \begin{cases} x^2 - 4 & \text{if } x \neq -3 \\ -2 & \text{if } x = -3 \end{cases}$

19. $H(x) = \begin{cases} x^2 - 4 & \text{if } x < 3 \\ 2x - 1 & \text{if } 3 \leq x \end{cases}$

20. $\phi(x) = \begin{cases} x + 5 & \text{if } x < -5 \\ \sqrt{25 - x^2} & \text{if } -5 \leq x \leq 5 \\ x - 5 & \text{if } 5 < x \end{cases}$

21. $f(x) = \begin{cases} x + 6 & \text{if } x \leq -4 \\ \sqrt{16 - x^2} & \text{if } -4 < x < 4 \\ 6 - x & \text{if } 4 \leq x \end{cases}$

22. $g(x) = \begin{cases} 6x + 7 & \text{if } x < -2 \\ 3 & \text{if } x = -2 \\ 4 - x & \text{if } -2 < x \end{cases}$

23. $F(x) = \begin{cases} x - 2 & \text{if } x < 0 \\ 0 & \text{if } x = 0 \\ x^2 + 1 & \text{if } 0 < x \end{cases}$

24. $G(x) = \dfrac{(x^2 + 3x - 4)(x^2 - 5x + 6)}{(x^2 - 3x + 2)(x - 3)}$

25. $F(x) = \dfrac{(x + 1)(x^2 + 3x - 10)}{x^2 + 6x + 5}$

26. $h(x) = \sqrt{x^2 - 5x + 6}$

27. $f(x) = \sqrt{x^2 - 3x - 4}$

28. $f(x) = \dfrac{x^3 + 3x^2 + x + 3}{x + 3}$

29. $g(x) = \dfrac{x^3 - 2x^2}{x - 2}$

30. $F(x) = \dfrac{x^4 + x^3 - 9x^2 - 3x + 18}{x^2 + x - 6}$

31. $h(x) = \dfrac{x^3 + 5x^2 - 6x - 30}{x + 5}$ **32.** $g(x) = |x| \cdot |x - 1|$ **33.** $f(x) = |x| + |x - 1|$

34. $F(x) = [\![x + 2]\!]$ **35.** $g(x) = [\![x - 4]\!]$ **36.** $H(x) = |x| + [\![x]\!]$ **37.** $G(x) = x - [\![x]\!]$

38. $h(x) = [\![x^2]\!]$ **39.** $g(x) = \dfrac{[\![x]\!]}{|x|}$ **40.** $h(x) = \dfrac{|x|}{[\![x]\!]}$

1.2
THE LIMIT OF A FUNCTION

The concept of the limit of a function is basic to the study of calculus. In this section the notion of the limit of a function is first given a step-by-step motivation, which brings the discussion from computing the value of a function near a number, through an intuitive treatment of the limiting process, up to a rigorous definition.

As a particular example we shall consider the function f defined by

$$f(x) = \frac{2x^2 + x - 3}{x - 1} \tag{1}$$

Observe that $f(x)$ exists for all x except $x = 1$. We shall investigate the function values when x is close to 1 but not equal to 1. The following illustration shows how the function defined by (1) could arise and why we would wish to consider such function values.

▶ ILLUSTRATION 1   The point $P(1, 2)$ is on the curve having the equation

$$y = 2x^2 + x - 1 \tag{2}$$

FIGURE 1.2.1

Let $Q(x, 2x^2 + x - 1)$ be another point on this curve, distinct from P. See Figs. 1.2.1 and 1.2.2, each showing a portion of the graph of (2) and the secant line through Q and P where Q is near P. In Fig. 1.2.1 the x coordinate of Q is less than 1, and in Fig. 1.2.2 it is greater than 1. Suppose $f(x)$ is the slope of the line PQ. Then

$$f(x) = \frac{(2x^2 + x - 1) - 2}{x - 1}$$

$$f(x) = \frac{2x^2 + x - 3}{x - 1}$$

FIGURE 1.2.2

which is Eq. (1). Furthermore, $x \neq 1$ because P and Q are distinct points. As x gets closer and closer to 1, the values of $f(x)$ get closer and closer to the number that we shall define in Sec. 2.1 as the slope of the tangent line to the curve at the point P.   ◀

For the function f defined by (1) let x take on the values 0, 0.25, 0.50, 0.75, 0.9, 0.99, 0.999, 0.9999, and so on. We are taking values of x closer and closer to 1 but less than 1; in other words, the variable x is approaching 1

Table 1.2.1

x	$f(x) = \dfrac{2x^2 + x - 3}{x - 1}$
0	3
0.25	3.5
0.5	4
0.75	4.5
0.9	4.8
0.99	4.98
0.999	4.998
0.9999	4.9998
0.99999	4.99998

Table 1.2.2

x	$f(x) = \dfrac{2x^2 + x - 3}{x - 1}$
2	7
1.75	6.5
1.5	6.0
1.25	5.5
1.1	5.2
1.01	5.02
1.001	5.002
1.0001	5.0002
1.00001	5.00002

through values that are less than 1. This is illustrated in Table 1.2.1. Now let the variable x approach 1 through values that are greater than 1; that is, let x take on the values 2, 1.75, 1.5, 1.25, 1.1, 1.01, 1.001, 1.0001, 1.00001, and so on. Refer to Table 1.2.2.

Observe from both tables that as x gets closer and closer to 1, $f(x)$ gets closer and closer to 5; and the closer x is to 1, the closer $f(x)$ is to 5. For instance, from Table 1.2.1, when $x = 0.9$, $f(x) = 4.8$; that is, when x is 0.1 less than 1, $f(x)$ is 0.2 less than 5. When $x = 0.999$, $f(x) = 4.998$; that is, when x is 0.001 less than 1, $f(x)$ is 0.002 less than 5. Furthermore, when $x = 0.9999$, $f(x) = 0.49998$; that is, when x is 0.0001 less than 1, $f(x)$ is 0.0002 less than 5.

Table 1.2.2 shows that when $x = 1.1$, $f(x) = 5.2$; that is, when x is 0.1 greater than 1, $f(x)$ is 0.2 greater than 5. When $x = 1.001$, $f(x) = 5.002$; that is, when x is 0.001 greater than 1, $f(x)$ is 0.002 greater than 5. When $x = 1.0001$, $f(x) = 5.0002$; that is, when x is 0.0001 greater than 1, $f(x)$ is 0.0002 greater than 5.

Therefore, from the two tables we see that when x differs from 1 by ± 0.001 (i.e., $x = 0.999$ or $x = 1.001$), $f(x)$ differs from 5 by ± 0.002 (i.e., $f(x) = 4.998$ or $f(x) = 5.002$). And when x differs from 1 by ± 0.0001, $f(x)$ differs from 5 by ± 0.0002.

Now, looking at the situation another way, we consider the values of $f(x)$ first. We see that we can make the value of $f(x)$ as close to 5 as we please by taking x close enough to 1. Another way of saying this is that we can make the absolute value of the difference between $f(x)$ and 5 as small as we please by making the absolute value of the difference between x and 1 small enough. That is, $|f(x) - 5|$ can be made as small as we please by making $|x - 1|$ small enough. But bear in mind that $f(x)$ never takes on the value 5.

A more precise way of noting this is by using two symbols for these small differences. The symbols usually used are the Greek letters ϵ (epsilon) and δ (delta). So we state that for any given positive number ϵ there is an appropriately chosen positive number δ such that, if $|x - 1|$ is less than δ and $|x - 1| \neq 0$ (i.e., $x \neq 1$), then $|f(x) - 5|$ will be less than ϵ. It is important to realize that the size of δ depends on the size of ϵ. Still another way of phrasing this is: Given any positive number ϵ, we can make $|f(x) - 5| < \epsilon$ by taking $|x - 1|$ small enough; that is, there is some sufficiently small positive number δ such that

$$\text{if} \quad 0 < |x - 1| < \delta \quad \text{then} \quad |f(x) - 5| < \epsilon \tag{3}$$

Observe that the numerator of the fraction in (1) can be factored so that

$$f(x) = \frac{(2x + 3)(x - 1)}{x - 1}$$

If $x \neq 1$, the numerator and denominator can be divided by $x - 1$ to obtain

$$f(x) = 2x + 3 \qquad x \neq 1 \tag{4}$$

Equation (4) with the stipulation that $x \neq 1$ is just as suitable as (1) for a definition of $f(x)$. From (4) and the two tables notice that if $|x - 1| < 0.1$, then $|f(x) - 5| < 0.2$. Thus given $\epsilon = 0.2$, we take $\delta = 0.1$ and state that

$$\text{if} \quad 0 < |x - 1| < 0.1 \quad \text{then} \quad |f(x) - 5| < 0.2$$

This is statement (3) with $\epsilon = 0.2$ and $\delta = 0.1$.

Also, if $|x - 1| < 0.001$, then $|f(x) - 5| < 0.002$. Hence if $\epsilon = 0.002$, we take $\delta = 0.001$ and state that

$$\text{if} \quad 0 < |x - 1| < 0.001 \quad \text{then} \quad |f(x) - 5| < 0.002$$

This is statement (3) with $\epsilon = 0.002$ and $\delta = 0.001$.

Similarly, if $\epsilon = 0.0002$, we take $\delta = 0.0001$ and state that

$$\text{if} \quad 0 < |x - 1| < 0.0001 \quad \text{then} \quad |f(x) - 5| < 0.0002$$

This is statement (3) with $\epsilon = 0.0002$ and $\delta = 0.0001$.

We could go on and give ϵ any small positive value and find a suitable value for δ such that if $|x - 1|$ is less than δ and $x \neq 1$ (or $|x - 1| > 0$), then $|f(x) - 5|$ will be less than ϵ. Now, because for any $\epsilon > 0$ we can find a $\delta > 0$ such that if $0 < |x - 1| < \delta$ then $|f(x) - 5| < \epsilon$, we state that the limit of $f(x)$ as x approaches 1 is equal to 5, or, expressed with symbols,

$$\lim_{x \to 1} f(x) = 5 \tag{5}$$

Observe that in (5) we have a new use of the "equals" symbol. Here for no value of x does $f(x)$ have the value 5. The "equals" symbol is appropriate in (5) because the left side is written as $\lim_{x \to 1} f(x)$.

From (4) it is apparent that $f(x)$ can be made as close to 5 as we please by taking x sufficiently close to 1, and this property of the function f does not depend on f being defined when $x = 1$. This fact gives the distinction between $\lim_{x \to 1} f(x)$ and the function value at 1; that is, $\lim_{x \to 1} f(x) = 5$, but $f(1)$ does not exist. Consequently, in statement (3), $0 < |x - 1|$ because we are concerned only with values of $f(x)$ for x close to 1, but not for $x = 1$.

Let us see what all this means geometrically for the particular function defined by either (1) or (4). Figure 1.2.3 illustrates the geometric significance of ϵ and δ. Observe that if x on the horizontal axis lies between $1 - \delta$ and $1 + \delta$, then $f(x)$ on the vertical axis will lie between $5 - \epsilon$ and $5 + \epsilon$; or

$$\text{if} \quad 0 < |x - 1| < \delta \quad \text{then} \quad |f(x) - 5| < \epsilon$$

Another way of stating this is that $f(x)$ on the vertical axis can be restricted to lie between $5 - \epsilon$ and $5 + \epsilon$ by restricting x on the horizontal axis to lie between $1 - \delta$ and $1 + \delta$.

Note that the values of ϵ are chosen arbitrarily and can be as small as desired, and that the value of δ is dependent on the ϵ chosen. It should also be pointed out that the smaller the value of ϵ, the smaller will be the corresponding value of δ.

Summing up for this example, we state that $\lim_{x \to 1} f(x) = 5$ because for any $\epsilon > 0$, however small, there exists a $\delta > 0$ such that

$$\text{if} \quad 0 < |x - 1| < \delta \quad \text{then} \quad |f(x) - 5| < \epsilon$$

The limit of a function in general is now defined.

FIGURE 1.2.3

1.2.1 DEFINITION

Let f be a function that is defined at every number in some open interval containing a, except possibly at the number a itself. The **limit of $f(x)$ as x approaches a is L**, written as

$$\lim_{x \to a} f(x) = L$$

if the following statement is true:

Given any $\epsilon > 0$, however small, there exists a $\delta > 0$ such that

$$\text{if} \quad 0 < |x - a| < \delta \quad \text{then} \quad |f(x) - L| < \epsilon \tag{6}$$

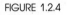

FIGURE 1.2.4

In words, Definition 1.2.1 states that the function values $f(x)$ approach a limit L as x approaches a number a if the absolute value of the difference between $f(x)$ and L can be made as small as we please by taking x sufficiently near a but not equal to a.

It is important to realize that in the above definition nothing is mentioned about the value of the function when $x = a$. That is, it is not necessary that the function be defined for $x = a$ in order for $\lim_{x \to a} f(x)$ to exist.

A geometric interpretation of Definition 1.2.1 for a function f is shown in Fig. 1.2.4. A portion of the graph of f near the point where $x = a$ is shown in the figure. Because f is not necessarily defined at a, there need be no point on the graph with abscissa a. Observe that if x on the horizontal axis lies between $a - \delta_1$ and $a + \delta_1$, then $f(x)$ on the vertical axis will lie between $L - \epsilon_1$ and $L + \epsilon_1$. Stated another way, by restricting x on the horizontal axis to lie between $a - \delta_1$ and $a + \delta_1$, then $f(x)$ on the vertical axis can be restricted to lie between $L - \epsilon_1$ and $L + \epsilon_1$. Thus

$$\text{if} \quad 0 < |x - a| < \delta_1 \quad \text{then} \quad |f(x) - L| < \epsilon_1$$

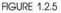

FIGURE 1.2.5

Figure 1.2.5 shows how a smaller value of ϵ can require a different choice for δ. In the figure it is seen that for $\epsilon_2 < \epsilon_1$, the δ_1 value is too large; that is, there are values of x for which $0 < |x - a| < \delta_1$, but $|f(x) - L|$ is not less than ϵ_2. For instance, $0 < |\bar{x} - a| < \delta_1$ but $|f(\bar{x}) - L| > \epsilon_2$. So we must choose a smaller value δ_2 shown in Fig. 1.2.6 such that

$$\text{if} \quad 0 < |x - a| < \delta_2 \quad \text{then} \quad |f(x) - L| < \epsilon_2$$

However, for any choice of $\epsilon > 0$, no matter how small, there exists a $\delta > 0$ such that statement (6) holds. Therefore $\lim_{x \to a} f(x) = L$.

In the examples of this section and from time to time throughout the text we use the following two symbols that are helpful in our discussions:

The arrow $\Rightarrow$ means *implies*.

The double arrow $\Leftrightarrow$ means the statement (or equation) preceding it and the statement (or equation) following it are *equivalent*.

FIGURE 1.2.6

EXAMPLE 1

Let the function f be defined by the equation

$$f(x) = 4x - 7$$

$f(x) = 4x - 7$

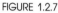

FIGURE 1.2.7

and assume

$$\lim_{x \to 3} f(x) = 5$$

(a) By using a figure similar to Fig. 1.2.3, for $\epsilon = 0.01$, determine a $\delta > 0$ such that

$$\text{if} \quad 0 < |x - 3| < \delta \quad \text{then} \quad |f(x) - 5| < 0.01 \qquad (7)$$

(b) By using properties of inequalities, determine a $\delta > 0$ such that statement (7) holds.

SOLUTION

(a) Refer to Fig. 1.2.7 and observe that the function values increase as the values of x increase. Thus the figure indicates that we need a value of x_1 such that $f(x_1) = 4.99$ and a value of x_2 such that $f(x_2) = 5.01$; that is, we need an x_1 and an x_2 such that

$$4x_1 - 7 = 4.99 \qquad 4x_2 - 7 = 5.01$$

$$x_1 = \frac{11.99}{4} \qquad x_2 = \frac{12.01}{4}$$

$$x_1 = 2.9975 \qquad x_2 = 3.0025$$

Because $3 - 2.9975 = 0.0025$ and $3.0025 - 3 = 0.0025$, we choose $\delta = 0.0025$ so that we have the statement

$$\text{if} \quad 0 < |x - 3| < 0.0025 \quad \text{then} \quad |f(x) - 5| < 0.01$$

(b) Because $f(x) = 4x - 7$,

$$
\begin{aligned}
|f(x) - 5| &= |(4x - 7) - 5| \\
&= |4x - 12| \\
&= 4|x - 3|
\end{aligned}
$$

We wish to determine a $\delta > 0$ such that

$$\text{if} \quad 0 < |x - 3| < \delta \quad \text{then} \quad |f(x) - 5| < 0.01$$

$$\Leftrightarrow \quad \text{if} \quad 0 < |x - 3| < \delta \quad \text{then} \quad 4|x - 3| < 0.01$$

$$\Leftrightarrow \quad \text{if} \quad 0 < |x - 3| < \delta \quad \text{then} \quad |x - 3| < 0.0025 \qquad (8)$$

Statement (8) indicates that a suitable choice for δ is 0.0025. Then we have the following argument:

$$0 < |x - 3| < 0.0025$$

$$\Rightarrow \quad 4|x - 3| < 4(0.0025)$$

$$\Rightarrow \quad |4x - 12| < 0.01$$

$$\Rightarrow \quad |(4x - 7) - 5| < 0.01$$

$$\Rightarrow \quad |f(x) - 5| < 0.01$$

We have shown that

$$\text{if}\quad 0 < |x - 3| < 0.0025 \quad \text{then}\quad |f(x) - 5| < 0.01 \tag{9}$$

In this example any positive number less than 0.0025 can be used in place of 0.0025 as the required δ. Observe this fact in Fig. 1.2.7. Furthermore, if $0 < \gamma < 0.0025$ and statement (9) holds, we have

$$\text{if}\quad 0 < |x - 3| < \gamma \quad \text{then}\quad |f(x) - 5| < 0.01$$

because every number x satisfying the inequality $0 < |x - 3| < \gamma$ also satisfies the inequality $0 < |x - 3| < 0.0025$.

The solution of Example 1 consisted of finding a δ for a specific ϵ. If for any $\epsilon > 0$ we can find a $\delta > 0$ such that

$$\text{if}\quad 0 < |x - 3| < \delta \quad \text{then}\quad |f(x) - 5| < \epsilon$$

we shall have established that the limit is 5. This is done in the next example.

EXAMPLE 2

Use Definition 1.2.1 to prove that

$$\lim_{x \to 3} (4x - 7) = 5$$

SOLUTION

The first requirement of Definition 1.2.1 is that $4x - 7$ be defined at every number in some open interval containing 3 except possibly at 3. Because $4x - 7$ is defined for all real numbers, any open interval containing 3 will satisfy this requirement. Now we must show that for any $\epsilon > 0$ there exists a $\delta > 0$ such that

$$\text{if}\quad 0 < |x - 3| < \delta \quad \text{then}\quad |(4x - 7) - 5| < \epsilon \tag{10}$$

In Example 1 we showed that $|(4x - 7) - 5| = 4|x - 3|$. Therefore (10) is equivalent to the statement

$$\text{if}\quad 0 < |x - 3| < \delta \quad \text{then}\quad 4|x - 3| < \epsilon$$

$$\Leftrightarrow \quad \text{if}\quad 0 < |x - 3| < \delta \quad \text{then}\quad |x - 3| < \tfrac{1}{4}\epsilon \tag{11}$$

Statement (11) indicates that $\tfrac{1}{4}\epsilon$ is a satisfactory δ. With this choice of δ we have the following argument:

$$0 < |x - 3| < \delta$$
$$\Rightarrow \quad 4|x - 3| < 4\delta$$
$$\Rightarrow \quad |4x - 12| < 4\delta$$
$$\Rightarrow \quad |(4x - 7) - 5| < 4\delta$$
$$\Rightarrow \quad |(4x - 7) - 5| < \epsilon \quad (\text{because } \delta = \tfrac{1}{4}\epsilon)$$

Thus we have established that if $\delta = \tfrac{1}{4}\epsilon$, statement (10) holds. This proves that $\lim_{x \to 3} (4x - 7) = 5$.

In particular, if $\epsilon = 0.01$, then we take $\delta = \frac{1}{4}(0.01)$, that is, $\delta = 0.0025$. This value of δ corresponds to the one found in Example 1.

Any positive number less than $\frac{1}{4}\epsilon$ can be used in place of $\frac{1}{4}\epsilon$ as the required δ.

FIGURE 1.2.8

EXAMPLE 3

Let the function f be defined by the equation

$$f(x) = x^2$$

and assume

$$\lim_{x \to 2} f(x) = 4$$

By using a figure for $\epsilon = 0.001$, determine a $\delta > 0$ such that

$$\text{if}\quad 0 < |x - 2| < \delta \quad \text{then} \quad |f(x) - 4| < 0.001$$

SOLUTION

Refer to Fig. 1.2.8 and observe that if $x > 0$, the function values increase as the values of x increase. Therefore the figure indicates that we need a positive value of x_1 such that $f(x_1) = 3.999$ and a positive value of x_2 such that $f(x_2) = 4.001$; that is, we need an x_1 and an x_2 such that

$$x_1{}^2 = 3.999 \qquad x_2{}^2 = 4.001$$

Each of these equations has two solutions. In each case we reject the negative square root because x_1 and x_2 are positive. Thus

$$x_1 = \sqrt{3.999} \qquad x_2 = \sqrt{4.001}$$
$$x_1 \approx 1.9997 \qquad x_2 \approx 2.0002$$

Then $2 - 1.9997 = 0.0003$ and $2.0002 - 2 = 0.0002$. Because $0.0002 < 0.0003$, we choose $\delta = 0.0002$ so that we have the statement

$$\text{if}\quad 0 < |x - 2| < 0.0002 \quad \text{then} \quad |f(x) - 4| < 0.001$$

Any positive number less than 0.0002 can be selected as the required δ.

EXAMPLE 4

Use Definition 1.2.1 to prove that

$$\lim_{x \to 2} x^2 = 4$$

SOLUTION

Because x^2 is defined for all real numbers, any open interval containing 2 will satisfy the first requirement of Definition 1.2.1. We must show that for any $\epsilon > 0$ there exists a $\delta > 0$ such that

$$\text{if}\quad 0 < |x - 2| < \delta \quad \text{then} \quad |x^2 - 4| < \epsilon$$
$$\Leftrightarrow \quad \text{if}\quad 0 < |x - 2| < \delta \quad \text{then} \quad |x - 2||x + 2| < \epsilon \tag{12}$$

Observe that on the right end of (12), in addition to the factor $|x - 2|$ we have the factor $|x + 2|$. Thus to prove (12) we wish to place a restriction on δ that will give us an inequality involving $|x + 2|$. Such a restriction is to choose the open interval required by Definition 1.2.1 to be the interval $(1, 3)$, and this implies that $\delta \leq 1$. Then

$$0 < |x - 2| < \delta \quad \text{and} \quad \delta \leq 1$$

$$\Rightarrow \qquad |x - 2| < 1$$

$$\Rightarrow \qquad -1 < x - 2 < 1$$

$$\Rightarrow \qquad 3 < x + 2 < 5$$

$$\Rightarrow \qquad |x + 2| < 5 \tag{13}$$

Now

$$0 < |x - 2| < \delta \quad \text{and} \quad |x + 2| < 5$$

$$\Rightarrow \quad |x - 2||x + 2| < \delta \cdot 5 \tag{14}$$

Remember that our goal is to have $|x - 2||x + 2| < \epsilon$. Statement (14) indicates that we should require $\delta \cdot 5 \leq \epsilon$, that is, $\delta \leq \epsilon/5$. This means that we have put two restrictions on δ: $\delta \leq 1$ and $\delta \leq \epsilon/5$. So that both restrictions hold we take δ as the smaller of the two numbers 1 and $\epsilon/5$; with symbols we write this as $\delta = \min(1, \epsilon/5)$. Using this δ we have the following argument:

$$0 < |x - 2| < \delta$$

$$\Rightarrow \qquad |x - 2||x + 2| < \delta|x + 2|$$

$$\Rightarrow \quad |(x - 2)(x + 2)| < \delta|x + 2|$$

$$\Rightarrow \qquad |x^2 - 4| < \delta|x + 2| \tag{15}$$

However, we showed in (13) that if $\delta \leq 1$ and $0 < |x - 2| < \delta$, then $|x + 2| < 5$; that is, $\delta|x + 2| < \delta \cdot 5$. Continuing on where we left off at (15) we have

$$|x^2 - 4| < \delta|x + 2| \quad \text{and} \quad \delta|x + 2| < \delta \cdot 5$$

$$\Rightarrow \quad |x^2 - 4| < \delta \cdot 5$$

$$\Rightarrow \quad |x^2 - 4| < \frac{\epsilon}{5} \cdot 5 \quad \text{(because } \delta \leq \epsilon/5\text{)}$$

$$\Rightarrow \quad |x^2 - 4| < \epsilon$$

We have demonstrated that for any $\epsilon > 0$ choose $\delta = \min(1, \epsilon/5)$ and the following statement is true:

$$\text{if} \quad 0 < |x - 2| < \delta \quad \text{then} \quad |x^2 - 4| < \epsilon$$

This proves that $\lim_{x \to 2} x^2 = 4$.

EXAMPLE 5
Use Definition 1.2.1 to prove that

$$\lim_{t \to 7} \frac{8}{t - 3} = 2$$

SOLUTION

Because $8/(t - 3)$ is defined for all real numbers except 3, the open interval required by Definition 1.2.1 can be any open interval containing 7 but not containing 3. We wish to show that for any $\epsilon > 0$ there exists a $\delta > 0$ such that

$$\text{if} \quad 0 < |t - 7| < \delta \quad \text{then} \quad \left| \frac{8}{t - 3} - 2 \right| < \epsilon \tag{16}$$

Because

$$\left| \frac{8}{t - 3} - 2 \right| = \left| \frac{8 - 2(t - 3)}{t - 3} \right|$$

$$= \left| \frac{14 - 2t}{t - 3} \right|$$

$$= \frac{2 \cdot |7 - t|}{|t - 3|}$$

$$= |t - 7| \cdot \frac{2}{|t - 3|}$$

statement (16) is equivalent to

$$\text{if} \quad 0 < |t - 7| < \delta \quad \text{then} \quad |t - 7| \cdot \frac{2}{|t - 3|} < \epsilon \tag{17}$$

Notice that on the right end of (17), in addition to the factor $|t - 7|$, we have as another factor the quotient $\dfrac{2}{|t - 3|}$. Therefore to prove (17) we need to restrict δ so that we will have an inequality involving $\dfrac{2}{|t - 3|}$. By choosing the open interval stipulated in Definition 1.2.1 to be the interval $(6, 8)$, which contains 7 but not 3, we are requiring that $\delta \leq 1$. Then

$$0 < |t - 7| < \delta \quad \text{and} \quad \delta \leq 1$$

$$\Rightarrow \qquad |t - 7| < 1$$

$$\Rightarrow \quad -1 < t - 7 < 1$$

$$\Rightarrow \qquad 3 < t - 3 < 5$$

$$\Rightarrow \qquad 3 < |t - 3| < 5$$

$$\Rightarrow \quad \frac{1}{5} < \frac{1}{|t - 3|} < \frac{1}{3}$$

$$\Rightarrow \quad \frac{2}{5} < \frac{2}{|t - 3|} < \frac{2}{3} \tag{18}$$

Now

$$0 < |t - 7| < \delta \quad \text{and} \quad \frac{2}{|t - 3|} < \frac{2}{3}$$

$$\Rightarrow \quad |t - 7| \cdot \frac{2}{|t - 3|} < \delta \cdot \frac{2}{3} \tag{19}$$

Because our goal is to have $|t - 7| \cdot \dfrac{2}{|t - 3|} < \epsilon$, statement (19) indicates that we should require $\delta \cdot \frac{2}{3} \leq \epsilon$, that is, $\delta \leq \frac{3}{2}\epsilon$. Thus we have put two restrictions on δ: $\delta \leq 1$ and $\delta \leq \frac{3}{2}\epsilon$. Hence we choose $\delta = \min(1, \frac{3}{2}\epsilon)$. With this δ we have the following argument:

$$0 < |t - 7| < \delta$$

$$\Rightarrow \quad |t - 7| \cdot \frac{2}{|t - 3|} < \delta \cdot \frac{2}{|t - 3|}$$

$$\Rightarrow \quad \left|\frac{14 - 2t}{t - 3}\right| < \delta \cdot \frac{2}{|t - 3|}$$

$$\Rightarrow \quad \left|\frac{8 - 2(t - 3)}{t - 3}\right| < \delta \cdot \frac{2}{|t - 3|}$$

$$\Rightarrow \quad \left|\frac{8}{t - 3} - 2\right| < \delta \cdot \frac{2}{|t - 3|} \tag{20}$$

We showed in (18) that if $\delta \leq 1$ and $0 < |t - 7| < \delta$, then $\dfrac{2}{|t - 3|} < \dfrac{2}{3}$, that is,

$\delta \cdot \dfrac{2}{|t - 3|} < \delta \cdot \dfrac{2}{3}$. Continuing on from (20) we have

$$\left|\frac{8}{t - 3} - 2\right| < \delta \cdot \frac{2}{|t - 3|} \quad \text{and} \quad \delta \cdot \frac{2}{|t - 3|} < \delta \cdot \frac{2}{3}$$

$$\Rightarrow \quad \left|\frac{8}{t - 3} - 2\right| < \delta \cdot \frac{2}{3}$$

$$\Rightarrow \quad \left|\frac{8}{t - 3} - 2\right| < \frac{3}{2}\epsilon \cdot \frac{2}{3} \quad \text{(because } \delta \leq \tfrac{3}{2}\epsilon\text{)}$$

$$\Rightarrow \quad \left|\frac{8}{t - 3} - 2\right| < \epsilon$$

We have shown that for any $\epsilon > 0$, if $\delta = \min(1, \frac{3}{2}\epsilon)$, the following statement is true:

$$\text{if} \quad 0 < |t - 7| < \delta \quad \text{then} \quad \left|\frac{8}{t - 3} - 2\right| < \epsilon$$

Thus we have proved that $\lim\limits_{t \to 7} \dfrac{8}{t - 3} = 2$.

The following theorem states that a function cannot approach two different limits at the same time. It is called a *uniqueness theorem* because it guarantees that if the limit of a function exists, it is unique. We state the theorem here but defer its proof until Supplementary Section 1.9.

1.2.2 THEOREM If $\lim\limits_{x \to a} f(x) = L_1$ and $\lim\limits_{x \to a} f(x) = L_2$, then $L_1 = L_2$.

Because of Theorem 1.2.2, we can state that if a function f has a limit L at the number a, then L is *the* limit of f at a.

EXERCISES 1.2

In Exercises 1 through 14, you are given $f(x)$, a, and L, as well as $\lim\limits_{x \to a} f(x) = L$. (a) By using a figure and arguments similar to those in Examples 1 and 3, determine a $\delta > 0$ for the given ϵ such that

$$\text{if} \quad 0 < |x - a| < \delta \quad \text{then} \quad |f(x) - L| < \epsilon \tag{21}$$

(b) By using properties of inequalities, determine a $\delta > 0$ such that statement (21) holds for the given value of ϵ.

1. $\lim\limits_{x \to 3} (2x + 4) = 10$; $\epsilon = 0.01$

2. $\lim\limits_{x \to 2} (4x - 5) = 3$; $\epsilon = 0.001$

3. $\lim\limits_{x \to -1} (3 - 4x) = 7$; $\epsilon = 0.02$

4. $\lim\limits_{x \to -2} (2 + 5x) = -8$; $\epsilon = 0.002$

5. $\lim\limits_{x \to 3} x^2 = 9$; $\epsilon = 0.005$

6. $\lim\limits_{x \to -4} x^2 = 16$; $\epsilon = 0.03$

7. $\lim\limits_{x \to 2} (x^2 - 2x + 1) = 1$; $\epsilon = 0.001$

8. $\lim\limits_{x \to 1} (x^2 - 5) = -4$; $\epsilon = 0.01$

9. $\lim\limits_{x \to 1} \sqrt{x} = 1$; $\epsilon = 0.002$

10. $\lim\limits_{x \to 4} \sqrt{x} = 2$; $\epsilon = 0.005$

11. $\lim\limits_{x \to -2} \dfrac{x^2 - 4}{x + 2} = -4$; $\epsilon = 0.01$

12. $\lim\limits_{x \to 1/3} \dfrac{9x^2 - 1}{3x - 1} = 2$; $\epsilon = 0.01$

13. $\lim\limits_{x \to 2} \dfrac{1}{x} = \dfrac{1}{2}$; $\epsilon = 0.003$

14. $\lim\limits_{x \to 2} \dfrac{1}{x + 1} = \dfrac{1}{3}$; $\epsilon = 0.02$

In Exercises 15 through 46, prove the limit by applying Definition 1.2.1.

15. $\lim\limits_{x \to 2} 7 = 7$

16. $\lim\limits_{x \to 5} (-4) = -4$

17. $\lim\limits_{x \to 4} (2x + 1) = 9$

18. $\lim\limits_{x \to 1} (4x + 3) = 7$

19. $\lim\limits_{x \to -1} (5x + 8) = 3$

20. $\lim\limits_{x \to 3} (3x - 5) = 4$

21. $\lim\limits_{x \to 3} (7 - 3x) = -2$

22. $\lim\limits_{x \to -4} (2x + 7) = -1$

23. $\lim\limits_{x \to -2} (1 + 3x) = -5$

24. $\lim\limits_{x \to -2} (7 - 2x) = 11$

25. $\lim\limits_{x \to -1} \dfrac{x^2 - 1}{x + 1} = -2$

26. $\lim\limits_{x \to 3} \dfrac{x^2 - 9}{x - 3} = 6$

27. $\lim\limits_{x \to 1} x^2 = 1$

28. $\lim\limits_{x \to -3} x^2 = 9$

29. $\lim\limits_{x \to 5} (x^2 - 3x) = 10$

30. $\lim\limits_{x \to 2} (x^2 + 2x - 1) = 7$

31. $\lim\limits_{x \to -3} (5 - x - x^2) = -1$

32. $\lim\limits_{x \to -1} (3 + 2x - x^2) = 0$

33. $\lim\limits_{x \to -3} \dfrac{1}{x} = -\dfrac{1}{3}$

34. $\lim\limits_{x \to 4} \dfrac{2}{x + 2} = \dfrac{1}{3}$

35. $\lim\limits_{x \to 3} \dfrac{4}{x - 1} = 2$

36. $\lim\limits_{x \to -4} \dfrac{1}{x + 3} = -1$

37. $\lim\limits_{x \to 5} \dfrac{2}{x - 4} = 2$

38. $\lim\limits_{x \to 6} \dfrac{x}{x - 3} = 2$

39. $\lim\limits_{x \to -2} \dfrac{3x}{3x + 2} = \dfrac{3}{2}$

40. $\lim\limits_{x \to 1/2} \dfrac{3 + 2x}{5 - x} = \dfrac{8}{9}$

41. $\lim\limits_{x \to 4} \sqrt{x + 5} = 3$ **42.** $\lim\limits_{x \to -1} \sqrt{1 - 3x} = 2$ **43.** $\lim\limits_{x \to 1} \dfrac{\sqrt{x} - 1}{x - 1} = \dfrac{1}{2}$ **44.** $\lim\limits_{x \to 1} \dfrac{1}{\sqrt{5 - x}} = \dfrac{1}{2}$

45. $\lim\limits_{x \to -3} \dfrac{1}{\sqrt{3 - 2x}} = \dfrac{1}{3}$ **46.** $\lim\limits_{x \to 4} \dfrac{\sqrt{x} - 2}{x - 4} = \dfrac{1}{4}$

47. Prove that $\lim\limits_{x \to a} x^2 = a^2$ if a is any positive number.   **48.** Prove that $\lim\limits_{x \to a} x^2 = a^2$ if a is any negative number.

49. Prove that $\lim\limits_{x \to a} \sqrt{x} = \sqrt{a}$ if a is any positive number.   **50.** Prove that $\lim\limits_{x \to a} x^3 = a^3$ if a is any positive number.

51. Prove that $\lim\limits_{x \to a} x^3 = a^3$ if a is any negative number.

52. Prove that $\lim\limits_{x \to a} \sqrt[3]{x} = \sqrt[3]{a}$. (*Hint:* $a^3 - b^3 = (a - b)(a^2 + ab + b^2)$.)

1.3
THEOREMS ON LIMITS
OF FUNCTIONS

In Sec. 1.2 we proved that the limit of a function was a particular number by applying Definition 1.2.1. To compute limits of functions by easier methods we use theorems whose proofs are based on Definition 1.2.1. These theorems, as well as others on limits of functions that appear in later sections of this chapter, are labeled "limit theorems" and are so designated as they are presented.

1.3.1 LIMIT THEOREM 1 If m and b are any constants,

$$\lim_{x \to a} (mx + b) = ma + b$$

PROOF   To prove this theorem we use Definition 1.2.1. For any $\epsilon > 0$ we must prove that there exists a $\delta > 0$ such that

$$\text{if } \quad 0 < |x - a| < \delta \quad \text{then} \quad |(mx + b) - (ma + b)| < \epsilon \tag{1}$$

Case 1: $m \neq 0$.

Because $|(mx + b) - (ma + b)| = |m| \cdot |x - a|$, we want to find a $\delta > 0$ for any $\epsilon > 0$ such that

$$\text{if } \quad 0 < |x - a| < \delta \quad \text{then} \quad |m| \cdot |x - a| < \epsilon$$

or, because $m \neq 0$,

$$\text{if } \quad 0 < |x - a| < \delta \quad \text{then} \quad |x - a| < \frac{\epsilon}{|m|} \tag{2}$$

Statement (2) will hold if $\delta = \epsilon/|m|$; so we conclude that

$$\text{if } \quad 0 < |x - a| < \delta \text{ and } \delta = \frac{\epsilon}{|m|} \quad \text{then} \quad |(mx + b) - (ma + b)| < \epsilon$$

This proves the theorem for Case 1.

Case 2: $m = 0$.

If $m = 0$, then $|(mx + b) - (ma + b)| = 0$ for all values of x. So take δ to be any positive number, and statement (1) holds. This proves the theorem for Case 2.

▶ ILLUSTRATION 1   From Limit Theorem 1 it follows that

$$\lim_{x \to 2} (3x + 5) = 3 \cdot 2 + 5$$
$$= 11$$

◀

1.3.2 LIMIT THEOREM 2   If c is a constant, then for any number a,

$$\lim_{x \to a} c = c$$

PROOF   This follows immediately from Limit Theorem 1 by taking $m = 0$ and $b = c$.   ■

1.3.3 LIMIT THEOREM 3

$$\lim_{x \to a} x = a$$

PROOF   This also follows immediately from Limit Theorem 1 by taking $m = 1$ and $b = 0$.   ■

▶ ILLUSTRATION 2   From Limit Theorem 2,

$$\lim_{x \to 5} 7 = 7$$

and from Limit Theorem 3,

$$\lim_{x \to -6} x = -6$$

◀

1.3.4 LIMIT THEOREM 4   If $\lim_{x \to a} f(x) = L$ and $\lim_{x \to a} g(x) = M$, then

$$\lim_{x \to a} [f(x) \pm g(x)] = L \pm M$$

$$\Leftrightarrow \quad \lim_{x \to a} [f(x) \pm g(x)] = \lim_{x \to a} f(x) \pm \lim_{x \to a} g(x)$$

PROOF   We shall prove this theorem with the plus sign. Given

$$\lim_{x \to a} f(x) = L \tag{3}$$

and

$$\lim_{x \to a} g(x) = M \tag{4}$$

we wish to prove that

$$\lim_{x \to a} [f(x) + g(x)] = L + M \tag{5}$$

To prove (5), use Definition 1.2.1; that is, for any $\epsilon > 0$ we must prove that there exists a $\delta > 0$ such that

$$\text{if} \quad 0 < |x - a| < \delta \quad \text{then} \quad |[f(x) + g(x)] - (L + M)| < \epsilon \tag{6}$$

Because (3) is given, it follows from the definition of a limit that for $\frac{1}{2}\epsilon > 0$ there exists a $\delta_1 > 0$ such that

$$\text{if}\quad 0 < |x - a| < \delta_1 \quad \text{then}\quad |f(x) - L| < \tfrac{1}{2}\epsilon$$

Similarly, from (4), for $\frac{1}{2}\epsilon > 0$ there exists a $\delta_2 > 0$ such that

$$\text{if}\quad 0 < |x - a| < \delta_2 \quad \text{then}\quad |g(x) - M| < \tfrac{1}{2}\epsilon$$

Now let δ be the smaller of the two numbers δ_1 and δ_2. Therefore $\delta \le \delta_1$ and $\delta \le \delta_2$. So

$$\text{if}\quad 0 < |x - a| < \delta \quad \text{then}\quad |f(x) - L| < \tfrac{1}{2}\epsilon$$

and

$$\text{if}\quad 0 < |x - a| < \delta \quad \text{then}\quad |g(x) - M| < \tfrac{1}{2}\epsilon$$

Hence if $0 < |x - a| < \delta$, then

$$\begin{aligned}
\big|[f(x) + g(x)] - (L + M)\big| &= \big|(f(x) - L) + (g(x) - M)\big| \\
&\le \big|f(x) - L\big| + \big|g(x) - M\big| \\
&< \tfrac{1}{2}\epsilon + \tfrac{1}{2}\epsilon \\
&= \epsilon
\end{aligned}$$

In this way we have obtained statement (6), thereby proving that

$$\lim_{x \to a} [f(x) + g(x)] = L + M$$

The proof of Limit Theorem 4 using the minus sign is left as an exercise (see Exercise 42).   ■

Limit Theorem 4 can be extended to any finite number of functions.

1.3.5 LIMIT THEOREM 5   If $\lim_{x \to a} f_1(x) = L_1$, $\lim_{x \to a} f_2(x) = L_2, \ldots,$ and $\lim_{x \to a} f_n(x) = L_n$, then

$$\lim_{x \to a} [f_1(x) \pm f_2(x) \pm \ldots \pm f_n(x)] = L_1 \pm L_2 \pm \ldots \pm L_n$$

$\Leftrightarrow\quad \lim_{x \to a} [f_1(x) \pm f_2(x) \pm \ldots \pm f_n(x)] = \lim_{x \to a} f_1(x) \pm \lim_{x \to a} f_2(x) \pm \ldots \pm \lim_{x \to a} f_n(x)$

This theorem may be proved by applying Limit Theorem 4 and mathematical induction (see Exercise 43).

1.3.6 LIMIT THEOREM 6   If $\lim_{x \to a} f(x) = L$ and $\lim_{x \to a} g(x) = M$, then

$$\lim_{x \to a} [f(x) \cdot g(x)] = L \cdot M$$

$\Leftrightarrow\quad \lim_{x \to a} [f(x) \cdot g(x)] = \lim_{x \to a} f(x) \cdot \lim_{x \to a} g(x)$

The proof of this theorem is more sophisticated than those of the preceding theorems. The steps of the proof are indicated in Exercises 45 and 46.

▶ ILLUSTRATION 3   From Limit Theorem 3, $\lim_{x \to 3} x = 3$, and from Limit Theorem 1, $\lim_{x \to 3} (2x + 1) = 7$. Thus from Limit Theorem 6,

$$\lim_{x \to 3} [x(2x + 1)] = \lim_{x \to 3} x \cdot \lim_{x \to 3} (2x + 1)$$

$$= 3 \cdot 7$$

$$= 21 \qquad \blacktriangleleft$$

Limit Theorem 6 also can be extended to any finite number of functions by applying mathematical induction.

1.3.7 LIMIT THEOREM 7   If $\lim_{x \to a} f_1(x) = L_1$, $\lim_{x \to a} f_2(x) = L_2, \ldots$, and $\lim_{x \to a} f_n(x) = L_n$, then

$$\lim_{x \to a} [f_1(x)f_2(x) \ldots f_n(x)] = L_1 L_2 \ldots L_n$$

$\Leftrightarrow \quad \lim_{x \to a} [f_1(x)f_2(x) \ldots f_n(x)] = \lim_{x \to a} f_1(x) \cdot \lim_{x \to a} f_2(x) \cdot \ldots \cdot \lim_{x \to a} f_n(x)$

The proof is left as an exercise (see Exercise 44).

1.3.8 LIMIT THEOREM 8   If $\lim_{x \to a} f(x) = L$ and n is any positive integer, then

$$\lim_{x \to a} [f(x)]^n = L^n$$

$\Leftrightarrow \quad \lim_{x \to a} [f(x)]^n = \left[\lim_{x \to a} f(x) \right]^n$

The proof follows immediately from Limit Theorem 7 by taking $f_1(x)$, $f_2(x), \ldots, f_n(x)$ all equal to $f(x)$ and $L_1, L_2, \ldots, L_n$ all equal to L.

▶ ILLUSTRATION 4   From Limit Theorem 1, $\lim_{x \to -2} (5x + 7) = -3$. Therefore, from Limit Theorem 8 it follows that

$$\lim_{x \to -2} (5x + 7)^4 = \left[\lim_{x \to -2} (5x + 7) \right]^4$$

$$= (-3)^4$$

$$= 81 \qquad \blacktriangleleft$$

1.3.9 LIMIT THEOREM 9   If $\lim_{x \to a} f(x) = L$ and $\lim_{x \to a} g(x) = M$, then

$$\lim_{x \to a} \frac{f(x)}{g(x)} = \frac{L}{M} \qquad \text{if } M \neq 0$$

$\Leftrightarrow \quad \lim_{x \to a} \frac{f(x)}{g(x)} = \dfrac{\lim_{x \to a} f(x)}{\lim_{x \to a} g(x)} \quad \text{if } \lim_{x \to a} g(x) \neq 0$

The proof, based on Definition 1.2.1, is given in Supplementary Section 1.9.

▶ ILLUSTRATION 5   From Limit Theorem 3, $\lim\limits_{x \to 4} x = 4$, and from Limit Theorem 1, $\lim\limits_{x \to 4} (-7x + 1) = -27$. Therefore from Limit Theorem 9,

$$\lim_{x \to 4} \frac{x}{-7x + 1} = \frac{\lim\limits_{x \to 4} x}{\lim\limits_{x \to 4} (-7x + 1)}$$

$$= \frac{4}{-27}$$

$$= -\frac{4}{27}$$ ◀

1.3.10 LIMIT THEOREM 10   If n is a positive integer and $\lim\limits_{x \to a} f(x) = L$, then

$$\lim_{x \to a} \sqrt[n]{f(x)} = \sqrt[n]{L}$$

$$\Leftrightarrow \quad \lim_{x \to a} \sqrt[n]{f(x)} = \sqrt[n]{\lim_{x \to a} f(x)}$$

with the restriction that if n is even, $L > 0$.

The proof of this theorem is also given in Supplementary Section 1.9.

▶ ILLUSTRATION 6   From the result of Illustration 5 and Limit Theorem 10 it follows that

$$\lim_{x \to 4} \sqrt[3]{\frac{x}{-7x + 1}} = \sqrt[3]{\lim_{x \to 4} \frac{x}{-7x + 1}}$$

$$= \sqrt[3]{-\frac{4}{27}}$$

$$= -\frac{\sqrt[3]{4}}{3}$$ ◀

Following are some examples illustrating the application of the above theorems. To indicate the limit theorem being used we use the abbreviation "L.T." followed by the theorem number; for example, "L.T. 2" refers to Limit Theorem 2.

EXAMPLE 1
Find $\lim\limits_{x \to 3} (x^2 + 7x - 5)$, and, when applicable, indicate the limit theorems being used.

SOLUTION

$$\lim_{x \to 3} (x^2 + 7x - 5) = \lim_{x \to 3} x^2 + \lim_{x \to 3} 7x - \lim_{x \to 3} 5 \qquad \text{(L.T. 5)}$$

$$= \lim_{x \to 3} x \cdot \lim_{x \to 3} x + \lim_{x \to 3} 7 \cdot \lim_{x \to 3} x - \lim_{x \to 3} 5 \qquad \text{(L.T. 6)}$$

$$= 3 \cdot 3 + 7 \cdot 3 - 5 \qquad \text{(L.T. 3 and L.T. 2)}$$

$$= 9 + 21 - 5$$

$$= 25$$

It is important at this point to realize that the limit in Example 1 was evaluated by direct application of the theorems on limits. For the function f defined by $f(x) = x^2 + 7x - 5$, note that $f(3) = 25$, which is the same as $\lim_{x \to 3} (x^2 + 7x - 5)$. It is not always true that $\lim_{x \to a} f(x) = f(a)$ (see Example 4).

EXAMPLE 2
Find

$$\lim_{x \to 2} \sqrt{\frac{x^3 + 2x + 3}{x^2 + 5}}$$

and, when applicable, indicate the limit theorems being used.

SOLUTION

$$\lim_{x \to 2} \sqrt{\frac{x^3 + 2x + 3}{x^2 + 5}} = \sqrt{\lim_{x \to 2} \frac{x^3 + 2x + 3}{x^2 + 5}} \qquad \text{(L.T. 10)}$$

$$= \sqrt{\frac{\lim_{x \to 2} (x^3 + 2x + 3)}{\lim_{x \to 2} (x^2 + 5)}} \qquad \text{(L.T. 9)}$$

$$= \sqrt{\frac{\lim_{x \to 2} x^3 + \lim_{x \to 2} 2x + \lim_{x \to 2} 3}{\lim_{x \to 2} x^2 + \lim_{x \to 2} 5}} \qquad \text{(L.T. 5)}$$

$$= \sqrt{\frac{(\lim_{x \to 2} x)^3 + \lim_{x \to 2} 2 \cdot \lim_{x \to 2} x + \lim_{x \to 2} 3}{(\lim_{x \to 2} x)^2 + \lim_{x \to 2} 5}}$$

$$\text{(L.T. 6 and L.T. 8)}$$

$$= \sqrt{\frac{2^3 + 2 \cdot 2 + 3}{2^2 + 5}} \qquad \text{(L.T. 3 and L.T. 2)}$$

$$= \sqrt{\frac{8 + 4 + 3}{9}}$$

$$= \frac{\sqrt{15}}{3}$$

EXAMPLE 3
Find

$$\lim_{x \to 3} \frac{x^3 - 27}{x - 3}$$

and, when applicable, indicate the limit theorems being used.

SOLUTION
Here we have a situation different than in the preceding examples. Limit Theorem 9 cannot be applied to the quotient $(x^3 - 27)/(x - 3)$ because $\lim_{x \to 3} (x - 3) = 0$. However, factoring the numerator we obtain

$$\frac{x^3 - 27}{x - 3} = \frac{(x - 3)(x^2 + 3x + 9)}{x - 3}$$

This quotient is $x^2 + 3x + 9$ if $x \neq 3$ (since if $x \neq 3$, the numerator and denominator can be divided by $x - 3$).

When evaluating $\lim_{x \to 3} [(x^3 - 27)/(x - 3)]$, we are considering values of x close to 3 but not equal to 3. Therefore it is possible to divide the numerator and denominator by $x - 3$. The solution takes the following form:

$$\lim_{x \to 3} \frac{x^3 - 27}{x - 3} = \lim_{x \to 3} \frac{(x - 3)(x^2 + 3x + 9)}{x - 3}$$

$$= \lim_{x \to 3} (x^2 + 3x + 9)$$

$$= \lim_{x \to 3} x^2 + \lim_{x \to 3} (3x + 9) \qquad \text{(L.T. 4)}$$

$$= (\lim_{x \to 3} x)^2 + 18 \qquad \text{(L.T. 8 and L.T. 1)}$$

$$= 3^2 + 18 \qquad \text{(L.T. 3)}$$

$$= 27$$

Note that in Example 3, $(x^3 - 27)/(x - 3)$ is not defined when $x = 3$, but $\lim_{x \to 3} [(x^3 - 27)/(x - 3)]$ exists and is equal to 27.

EXAMPLE 4
Given that f is the function defined by

$$f(x) = \begin{cases} x - 3 & \text{if } x \neq 4 \\ 5 & \text{if } x = 4 \end{cases}$$

find $\lim_{x \to 4} f(x)$.

FIGURE 1.3.1

SOLUTION

When evaluating $\lim\limits_{x \to 4} f(x)$, we are considering values of x close to 4 but not equal to 4. Thus

$$\lim_{x \to 4} f(x) = \lim_{x \to 4} (x - 3)$$

$$= 1 \tag{L.T. 1}$$

In Example 4, $\lim\limits_{x \to 4} f(x) = 1$ but $f(4) = 5$; therefore $\lim\limits_{x \to 4} f(x) \neq f(4)$. In terms of geometry, there is a break in the graph of the function at the point where $x = 4$ (see Fig. 1.3.1). The graph of the function consists of the isolated point $(4, 5)$ and the straight line whose equation is $y = x - 3$, with the point $(4, 1)$ deleted.

EXAMPLE 5
Find

$$\lim_{x \to 4} \frac{\sqrt{x} - 2}{x - 4}$$

and, when applicable, indicate the limit theorems being used.

SOLUTION

As in Example 3, Limit Theorem 9 cannot be applied to the quotient $(\sqrt{x} - 2)/(x - 4)$ because $\lim\limits_{x \to 4} (x - 4) = 0$. To simplify the quotient we rationalize the numerator by multiplying the numerator and denominator by $\sqrt{x} + 2$.

$$\frac{\sqrt{x} - 2}{x - 4} = \frac{(\sqrt{x} - 2)(\sqrt{x} + 2)}{(x - 4)(\sqrt{x} + 2)}$$

$$= \frac{x - 4}{(x - 4)(\sqrt{x} + 2)}$$

Because we are evaluating the limit as x approaches 4, we are considering values of x close to 4 but not equal to 4. Hence we can divide the numerator and denominator by $x - 4$. Therefore

$$\frac{\sqrt{x} - 2}{x - 4} = \frac{1}{\sqrt{x} + 2} \qquad \text{if } x \neq 4$$

The solution is as follows:

$$\lim_{x \to 4} \frac{\sqrt{x} - 2}{x - 4} = \lim_{x \to 4} \frac{(\sqrt{x} - 2)(\sqrt{x} + 2)}{(x - 4)(\sqrt{x} + 2)}$$

$$= \lim_{x \to 4} \frac{x - 4}{(x - 4)(\sqrt{x} + 2)}$$

$$= \lim_{x \to 4} \frac{1}{\sqrt{x} + 2}$$

$$= \frac{\displaystyle\lim_{x \to 4} 1}{\displaystyle\lim_{x \to 4} (\sqrt{x} + 2)} \qquad \text{(L.T. 9)}$$

$$= \frac{1}{\displaystyle\lim_{x \to 4} \sqrt{x} + \lim_{x \to 4} 2} \qquad \text{(L.T. 2 and L.T. 4)}$$

$$= \frac{1}{\sqrt{\displaystyle\lim_{x \to 4} x} + 2} \qquad \text{(L.T. 10 and L.T. 2)}$$

$$= \frac{1}{\sqrt{4} + 2} \qquad \text{(L.T. 3)}$$

$$= \tfrac{1}{4}$$

From time to time we will need two other equations that are equivalent to the equation

$$\lim_{x \to a} f(x) = L$$

They are given in the following two theorems.

1.3.11 THEOREM $\displaystyle\lim_{x \to a} f(x) = L$ if and only if $\displaystyle\lim_{x \to a} [f(x) - L] = 0$

PROOF   Because the theorem has an *if and only if* qualification, the proof requires two parts.

Part 1: Prove that $\displaystyle\lim_{x \to a} f(x) = L$ if $\displaystyle\lim_{x \to a} [f(x) - L] = 0$.

We start with $\displaystyle\lim_{x \to a} f(x)$, replace $f(x)$ by $[f(x) - L] + L$, and then apply Limit Theorem 4.

$$\lim_{x \to a} f(x) = \lim_{x \to a} ([f(x) - L] + L)$$

$$= \lim_{x \to a} [f(x) - L] + \lim_{x \to a} L$$

$$= 0 + L$$

$$= L$$

Part 2: Prove that $\lim_{x \to a} f(x) = L$ only if $\lim_{x \to a} [f(x) - L] = 0$.

Here we must show that if $\lim_{x \to a} f(x) = L$, then $\lim_{x \to a} [f(x) - L] = 0$. We apply Limit Theorem 4 to $\lim_{x \to a} [f(x) - L]$.

$$\lim_{x \to a} [f(x) - L] = \lim_{x \to a} f(x) - \lim_{x \to a} L$$

$$= L - L$$

$$= 0 \qquad \blacksquare$$

| 1.3.12 THEOREM | $\lim_{x \to a} f(x) = L$ if and only if $\lim_{t \to 0} f(t + a) = L$ |

PROOF   Let $t + a = x$; then $x - a = t$. There are two parts to the proof.

Part 1: Prove that $\lim_{x \to a} f(x) = L$ if $\lim_{t \to 0} f(t + a) = L$.

If $\lim_{t \to 0} f(t + a) = L$, from Definition 1.2.1 it follows that for any $\epsilon > 0$ there exists a $\delta > 0$ such that

$$\text{if} \quad 0 < |t| < \delta \quad \text{then} \quad |f(t + a) - L| < \epsilon \tag{7}$$

or, equivalently, by replacing $t + a$ by x and t by $x - a$,

$$\text{if} \quad 0 < |x - a| < \delta \quad \text{then} \quad |f(x) - L| < \epsilon \tag{8}$$

From Definition 1.2.1, statement (8) implies that

$$\lim_{x \to a} f(x) = L$$

Part 2: Prove that $\lim_{x \to a} f(x) = L$ only if $\lim_{t \to 0} f(t + a) = L$.

If $\lim_{x \to a} f(x) = L$, then by Definition 1.2.1, for any $\epsilon > 0$ there exists a $\delta > 0$ such that statement (8) holds. By replacing x by $t + a$ and $x - a$ by t we have the equivalent statement (7). Thus from Definition 1.2.1 we can conclude that

$$\lim_{t \to 0} f(t + a) = L \qquad \blacksquare$$

EXERCISES 1.3

In Exercises 1 through 35, find the value of the limit, and, when applicable, indicate the limit theorems being used.

1. $\lim_{x \to 5} (3x - 7)$

2. $\lim_{x \to -4} (5x + 2)$

3. $\lim_{x \to 2} (x^2 + 2x - 1)$

4. $\lim_{x \to 3} (2x^2 - 4x + 5)$

5. $\lim_{z \to -2} (z^3 + 8)$

6. $\lim_{y \to -1} (y^3 - 2y^2 + 3y - 4)$

7. $\lim_{x \to 3} \dfrac{4x - 5}{5x - 1}$

8. $\lim_{x \to 2} \dfrac{3x + 4}{8x - 1}$

9. $\lim_{t \to 2} \dfrac{t^2 - 5}{2t^3 + 6}$

10. $\lim_{x \to -1} \dfrac{2x + 1}{x^2 - 3x + 4}$

11. $\lim_{x \to 2} \dfrac{x - 2}{x^2 - 4}$

12. $\lim_{z \to -5} \dfrac{z^2 - 25}{z + 5}$

13. $\lim_{x \to -3/2} \dfrac{4x^2 - 9}{2x + 3}$

14. $\lim_{x \to 1/3} \dfrac{3x - 1}{9x^2 - 1}$

15. $\lim\limits_{x \to -3} \dfrac{x^2 + 5x + 6}{x^2 - x - 12}$

16. $\lim\limits_{t \to -2} \dfrac{2 - 3t - 2t^2}{16 + 6t - t^2}$

17. $\lim\limits_{s \to 4} \dfrac{3s^2 - 8s - 16}{2s^2 - 9s + 4}$

18. $\lim\limits_{x \to 4} \dfrac{3x^2 - 17x + 20}{4x^2 - 25x + 36}$

19. $\lim\limits_{y \to -2} \dfrac{y^3 + 8}{y + 2}$

20. $\lim\limits_{s \to 1} \dfrac{s^3 - 1}{s - 1}$

21. $\lim\limits_{r \to 1} \sqrt{\dfrac{8r + 1}{r + 3}}$

22. $\lim\limits_{x \to -3} \sqrt[3]{\dfrac{5 + 2x}{5 - x}}$

23. $\lim\limits_{x \to 4} \sqrt[3]{\dfrac{x^2 - 3x + 4}{2x^2 - x - 1}}$

24. $\lim\limits_{x \to 2} \sqrt{\dfrac{x^2 + 3x + 4}{x^3 + 1}}$

25. $\lim\limits_{y \to -3} \sqrt{\dfrac{y^2 - 9}{2y^2 + 7y + 3}}$

26. $\lim\limits_{t \to 3/2} \sqrt{\dfrac{8t^3 - 27}{4t^2 - 9}}$

27. $\lim\limits_{x \to 9} \dfrac{3 - \sqrt{x}}{9 - x}$

28. $\lim\limits_{t \to 0} \dfrac{2 - \sqrt{4 - t}}{t}$

29. $\lim\limits_{x \to 0} \dfrac{\sqrt{x + 2} - \sqrt{2}}{x}$

30. $\lim\limits_{x \to 1} \dfrac{\sqrt[3]{x} - 1}{x - 1}$

31. $\lim\limits_{h \to 0} \dfrac{\sqrt[3]{h + 1} - 1}{h}$

32. $\lim\limits_{x \to -2} \dfrac{x^3 - x^2 - x + 10}{x^2 + 3x + 2}$

33. $\lim\limits_{x \to -1} \dfrac{2x^2 - x - 3}{x^3 + 2x^2 + 6x + 5}$

34. $\lim\limits_{y \to 4} \dfrac{2y^3 - 11y^2 + 10y + 8}{3y^3 - 17y^2 + 16y + 16}$

35. $\lim\limits_{x \to 3} \dfrac{2x^3 - 5x^2 - 2x - 3}{4x^3 - 13x^2 + 4x - 3}$

36. If $f(x) = x^2 + 5x - 3$, show that $\lim\limits_{x \to 2} f(x) = f(2)$.

37. If $F(x) = 2x^3 + 7x - 1$, show that $\lim\limits_{x \to -1} F(x) = F(-1)$.

38. If $g(x) = \dfrac{x^2 - 16}{x - 4}$, show that $\lim\limits_{x \to 4} g(x) = 8$ but that $g(4)$ is not defined.

39. If $h(x) = \dfrac{\sqrt{x + 9} - 3}{x}$, show that $\lim\limits_{x \to 0} h(x) = \tfrac{1}{6}$, but that $h(0)$ is not defined.

40. Given that f is the function defined by

$$f(x) = \begin{cases} 2x - 1 & \text{if } x \neq 2 \\ 1 & \text{if } x = 2 \end{cases}$$

(a) Find $\lim\limits_{x \to 2} f(x)$, and show that $\lim\limits_{x \to 2} f(x) \neq f(2)$. (b) Draw a sketch of the graph of f.

41. Given that f is the function defined by

$$f(x) = \begin{cases} x^2 - 9 & \text{if } x \neq -3 \\ 4 & \text{if } x = -3 \end{cases}$$

(a) Find $\lim\limits_{x \to -3} f(x)$, and show that $\lim\limits_{x \to -3} f(x) \neq f(-3)$. (b) Draw a sketch of the graph of f.

42. Use Definition 1.2.1 to prove that if

$$\lim\limits_{x \to a} f(x) = L \quad \text{and} \quad \lim\limits_{x \to a} g(x) = M$$

then

$$\lim\limits_{x \to a} [f(x) - g(x)] = L - M$$

43. Prove Limit Theorem 5 by applying Limit Theorem 4 and mathematical induction.

44. Prove Limit Theorem 7 by applying Limit Theorem 6 and mathematical induction.

45. Use Definition 1.2.1 to prove that if

$$\lim\limits_{x \to a} f(x) = L \quad \text{and} \quad \lim\limits_{x \to a} g(x) = 0$$

then

$$\lim\limits_{x \to a} [f(x) \cdot g(x)] = 0$$

(*Hint:* To prove that $\lim_{x \to a} [f(x) \cdot g(x)] = 0$ we must show that for any $\epsilon > 0$ there exists a $\delta > 0$ such that if $0 < |x - a| < \delta$, then $|f(x) \cdot g(x)| < \epsilon$. First show that there is a $\delta_1 > 0$ such that if $0 < |x - a| < \delta_1$, then $|f(x)| < 1 + |L|$, by applying Definition 1.2.1 to $\lim_{x \to a} f(x) = L$ with $\epsilon = 1$ and $\delta = \delta_1$, and then use the triangle inequality. Then show that there is a $\delta_2 > 0$ such that if $0 < |x - a| < \delta_2$, then $|g(x)| < \epsilon/(1 + |L|)$ by applying Definition 1.2.1 to $\lim_{x \to a} g(x) = 0$. By taking δ as the smaller of the two numbers δ_1 and δ_2, the theorem is proved.)

46. Prove Limit Theorem 6: If $\lim_{x \to a} f(x) = L$ and $\lim_{x \to a} g(x) = M$, then

$$\lim_{x \to a} [f(x) \cdot g(x)] = L \cdot M$$

(*Hint:* Write $f(x) \cdot g(x) = [f(x) - L]g(x) + L[g(x) - M] + L \cdot M$. Apply Limit Theorem 5, Theorem 1.3.11, and the result of Exercise 45.)

1.4
ONE-SIDED LIMITS

When considering $\lim_{x \to a} f(x)$, we are concerned with values of x in an open interval containing a but not at a itself, that is, at values of x close to a and either greater than a or less than a. However, suppose that we have the function f for which, say, $f(x) = \sqrt{x - 4}$. Because $f(x)$ does not exist if $x < 4$, f is not defined on any open interval containing 4. Hence $\lim_{x \to 4} \sqrt{x - 4}$ has no meaning. However, if x is restricted to values greater than 4, the value of $\sqrt{x - 4}$ can be made as close to 0 as we please by taking x sufficiently close to 4 but greater than 4. In such a case we let x approach 4 from the right and consider the **one-sided limit from the right**, or the **right-hand limit**, which is now defined.

1.4.1 DEFINITION   Let f be a function that is defined at every number in some open interval (a, c). Then the **limit of $f(x)$, as x approaches a from the right, is L**, written

$$\lim_{x \to a^+} f(x) = L$$

if for any $\epsilon > 0$, however small, there exists a $\delta > 0$ such that

$$\text{if} \quad 0 < x - a < \delta \quad \text{then} \quad |f(x) - L| < \epsilon \tag{1}$$

Note that in statement (1) there are no absolute-value bars around $x - a$ since $x - a > 0$, because $x > a$.

It follows from Definition 1.4.1 that

$$\lim_{x \to 4^+} \sqrt{x - 4} = 0$$

If when considering the limit of a function the independent variable x is restricted to values less than a number a, we say that x approaches a from the left; the limit is called the **one-sided limit from the left**, or the **left-hand limit**.

1.4.2 DEFINITION   Let f be a function that is defined at every number in some open interval (d, a). Then the **limit of $f(x)$, as x approaches a from the left, is L**, written

$$\lim_{x \to a^-} f(x) = L$$

if for any $\epsilon > 0$, however small, there exists a $\delta > 0$ such that

$$\text{if } \quad 0 < a - x < \delta \quad \text{ then } \quad |f(x) - L| < \epsilon$$

We can refer to $\lim\limits_{x \to a} f(x)$ as the **two-sided limit** to distinguish it from the one-sided limits.

Limit Theorems 1–10 given in Sec. 1.3 remain unchanged when "$x \to a$" is replaced by "$x \to a^{+}$" or "$x \to a^{-}$."

EXAMPLE 1
The **signum function** is defined by

$$\text{sgn } x = \begin{cases} -1 & \text{if } x < 0 \\ 0 & \text{if } x = 0 \\ 1 & \text{if } 0 < x \end{cases}$$

FIGURE 1.4.1

(a) Draw a sketch of the graph of this function. (b) Determine $\lim\limits_{x \to 0^{-}} \text{sgn } x$ and $\lim\limits_{x \to 0^{+}} \text{sgn } x$, if they exist.

SOLUTION
(a) A sketch of the graph appears in Fig. 1.4.1.
(b) Because sgn $x = -1$ if $x < 0$ and sgn $x = 1$ if $0 < x$, we have

$$\lim_{x \to 0^{-}} \text{sgn } x = \lim_{x \to 0^{-}} (-1) \qquad \lim_{x \to 0^{+}} \text{sgn } x = \lim_{x \to 0^{+}} 1$$

$$= -1 \qquad\qquad\qquad = 1$$

In Example 1, $\lim\limits_{x \to 0^{-}} \text{sgn } x \neq \lim\limits_{x \to 0^{+}} \text{sgn } x$. Because the left-hand limit and the right-hand limit are not equal, the two-sided limit $\lim\limits_{x \to 0} \text{sgn } x$ does not exist. The concept of the two-sided limit failing to exist because the two one-sided limits are unequal is stated in the following theorem.

1.4.3 THEOREM $\lim\limits_{x \to a} f(x)$ exists and is equal to L if and only if $\lim\limits_{x \to a^{-}} f(x)$ and $\lim\limits_{x \to a^{+}} f(x)$ both exist and both are equal to L.

The proof of Theorem 1.4.3 is left as an exercise (See Exercise 34).

FIGURE 1.4.2

▶ ILLUSTRATION 1   A wholesaler sells a product by the pound (or fraction of a pound); if not more than 10 pounds are ordered, the wholesaler charges \$1 per pound. However, to invite large orders the wholesaler charges only 90 cents per pound if more than 10 pounds are purchased. Thus if x pounds of the product are purchased and $C(x)$ dollars is the total cost of the order, then

$$C(x) = \begin{cases} x & \text{if } 0 \leq x \leq 10 \\ 0.9x & \text{if } 10 < x \end{cases} \tag{2}$$

A sketch of the graph of C is shown in Fig. 1.4.2. Observe that $C(x)$ is obtained from the equation $C(x) = x$ when $0 \leq x \leq 10$ and from the equation $C(x) = 0.9x$ when $10 < x$. Because of this situation, when considering the limit of $C(x)$ as x approaches 10, we must distinguish between the left-hand limit at 10 and the right-hand limit at 10. For the function C defined by (2) we have

$$\lim_{x \to 10^-} C(x) = \lim_{x \to 10^-} x \qquad \lim_{x \to 10^+} C(x) = \lim_{x \to 10^+} 0.9x$$

$$= 10 \qquad\qquad\qquad = 9$$

Because $\lim\limits_{x \to 10^-} C(x) \neq \lim\limits_{x \to 10^+} C(x)$, we conclude from Theorem 1.4.3 that $\lim\limits_{x \to 10} C(x)$ does not exist. Observe in Fig. 1.4.2 that at $x = 10$ there is a break in the graph of the function C. We return to this function in Sec. 1.6.   ◄

FIGURE 1.4.3

EXAMPLE 2

Let g be defined by

$$g(x) = \begin{cases} |x| & \text{if } x \neq 0 \\ 2 & \text{if } x = 0 \end{cases}$$

(a) Draw a sketch of the graph of g. (b) Find $\lim\limits_{x \to 0} g(x)$ if it exists.

SOLUTION

(a) A sketch of the graph is shown in Fig. 1.4.3.

(b) $\lim\limits_{x \to 0^-} g(x) = \lim\limits_{x \to 0^-} (-x) \qquad \lim\limits_{x \to 0^+} g(x) = \lim\limits_{x \to 0^+} x$

$$= 0 \qquad\qquad\qquad = 0$$

Because $\lim\limits_{x \to 0^-} g(x) = \lim\limits_{x \to 0^+} g(x)$, it follows from Theorem 1.4.3 that $\lim\limits_{x \to 0} g(x)$ exists and is equal to 0.

Observe in Example 2 that $g(0) = 2$, which has no effect on $\lim\limits_{x \to 0} g(x)$. Also observe that there is a break in the graph of g at $x = 0$.

FIGURE 1.4.4

EXAMPLE 3

Let h be defined by

$$h(x) = \begin{cases} 4 - x^2 & \text{if } x \leq 1 \\ 2 + x^2 & \text{if } 1 < x \end{cases}$$

(a) Draw a sketch of the graph of h. (b) Find each of the following limits if they exist: $\lim\limits_{x \to 1^-} h(x)$, $\lim\limits_{x \to 1^+} h(x)$, $\lim\limits_{x \to 1} h(x)$.

SOLUTION

(a) A sketch of the graph appears in Fig. 1.4.4.

(b) $\lim\limits_{x \to 1^-} h(x) = \lim\limits_{x \to 1^-} (4 - x^2) \qquad \lim\limits_{x \to 1^+} h(x) = \lim\limits_{x \to 1^+} (2 + x^2)$

$$= 3 \qquad\qquad\qquad\qquad = 3$$

Because $\lim\limits_{x \to 1^-} h(x) = \lim\limits_{x \to 1^+} h(x)$ and both are equal to 3, it follows from Theorem 1.4.3 that $\lim\limits_{x \to 1} h(x) = 3$.

FIGURE 1.4.5

EXAMPLE 4

Let f be defined by

$$f(x) = \begin{cases} x + 5 & \text{if } x < -3 \\ \sqrt{9 - x^2} & \text{if } -3 \le x \le 3 \\ 5 - x & \text{if } 3 < x \end{cases}$$

(a) Draw a sketch of the graph of f. (b) Find, if they exist, each of the following limits: $\lim\limits_{x \to -3^-} f(x)$, $\lim\limits_{x \to -3^+} f(x)$, $\lim\limits_{x \to -3} f(x)$, $\lim\limits_{x \to 3^-} f(x)$, $\lim\limits_{x \to 3^+} f(x)$, $\lim\limits_{x \to 3} f(x)$.

SOLUTION

(a) A sketch of the graph of f is shown in Fig. 1.4.5.

(b)
$$\lim_{x \to -3^-} f(x) = \lim_{x \to -3^-} (x + 5) \qquad \lim_{x \to -3^+} f(x) = \lim_{x \to -3^+} \sqrt{9 - x^2}$$
$$= 2 \qquad\qquad\qquad = 0$$

Because $\lim\limits_{x \to -3^-} f(x) \ne \lim\limits_{x \to -3^+} f(x)$, then $\lim\limits_{x \to -3} f(x)$ does not exist.

$$\lim_{x \to 3^-} f(x) = \lim_{x \to 3^-} \sqrt{9 - x^2} \qquad \lim_{x \to 3^+} f(x) = \lim_{x \to 3^+} (5 - x)$$
$$= 0 \qquad\qquad\qquad = 2$$

Because $\lim\limits_{x \to 3^-} f(x) \ne \lim\limits_{x \to 3^+} f(x)$, then $\lim\limits_{x \to 3} f(x)$ does not exist.

EXERCISES 1.4

In Exercises 1 through 22, draw a sketch of the graph, and find the indicated limit if it exists; if the limit does not exist, state the reason.

1. $f(x) = \begin{cases} 2 & \text{if } x < 1 \\ -1 & \text{if } x = 1 \\ -3 & \text{if } 1 < x \end{cases}$

(a) $\lim\limits_{x \to 1^+} f(x)$; (b) $\lim\limits_{x \to 1^-} f(x)$; (c) $\lim\limits_{x \to 1} f(x)$

2. $f(x) = \begin{cases} -2 & \text{if } x < 0 \\ 2 & \text{if } 0 \le x \end{cases}$

(a) $\lim\limits_{x \to 0^+} f(x)$; (b) $\lim\limits_{x \to 0^-} f(x)$; (c) $\lim\limits_{x \to 0} f(x)$

3. $f(t) = \begin{cases} t + 4 & \text{if } t \le -4 \\ 4 - t & \text{if } -4 < t \end{cases}$

(a) $\lim\limits_{t \to -4^+} f(t)$; (b) $\lim\limits_{t \to -4^-} f(t)$; (c) $\lim\limits_{t \to -4} f(t)$

4. $g(s) = \begin{cases} s + 3 & \text{if } s \le -2 \\ 3 - s & \text{if } -2 < s \end{cases}$

(a) $\lim\limits_{s \to -2^+} g(s)$; (b) $\lim\limits_{s \to -2^-} g(s)$; (c) $\lim\limits_{s \to -2} g(s)$

5. $F(x) = \begin{cases} x^2 & \text{if } x \le 2 \\ 8 - 2x & \text{if } 2 < x \end{cases}$

(a) $\lim\limits_{x \to 2^+} F(x)$; (b) $\lim\limits_{x \to 2^-} F(x)$; (c) $\lim\limits_{x \to 2} F(x)$

6. $h(x) = \begin{cases} 2x + 1 & \text{if } x < 3 \\ 10 - x & \text{if } 3 \le x \end{cases}$

(a) $\lim\limits_{x \to 3^+} h(x)$; (b) $\lim\limits_{x \to 3^-} h(x)$; (c) $\lim\limits_{x \to 3} h(x)$

7. $g(r) = \begin{cases} 2r + 3 & \text{if } r < 1 \\ 2 & \text{if } r = 1 \\ 7 - 2r & \text{if } 1 < r \end{cases}$

(a) $\lim\limits_{r \to 1^+} g(r)$; (b) $\lim\limits_{r \to 1^-} g(r)$; (c) $\lim\limits_{r \to 1} g(r)$

8. $g(t) = \begin{cases} 3 + t^2 & \text{if } t < -2 \\ 0 & \text{if } t = -2 \\ 11 - t^2 & \text{if } -2 < t \end{cases}$

(a) $\lim\limits_{t \to -2^+} g(t)$; (b) $\lim\limits_{t \to -2^-} g(t)$; (c) $\lim\limits_{t \to -2} g(t)$

9. $f(x) = \begin{cases} x^2 - 4 & \text{if } x < 2 \\ 4 & \text{if } x = 2 \\ 4 - x^2 & \text{if } 2 < x \end{cases}$

(a) $\lim_{x \to 2^+} f(x)$; (b) $\lim_{x \to 2^-} f(x)$; (c) $\lim_{x \to 2} f(x)$

10. $f(x) = \begin{cases} 2x + 3 & \text{if } x < 1 \\ 4 & \text{if } x = 1 \\ x^2 + 2 & \text{if } 1 < x \end{cases}$

(a) $\lim_{x \to 1^+} f(x)$; (b) $\lim_{x \to 1^-} f(x)$; (c) $\lim_{x \to 1} f(x)$

11. $F(x) = |x - 5|$

(a) $\lim_{x \to 5^+} F(x)$; (b) $\lim_{x \to 5^-} F(x)$; (c) $\lim_{x \to 5} F(x)$

12. $f(x) = 3 + |2x - 4|$

(a) $\lim_{x \to 2^+} f(x)$; (b) $\lim_{x \to 2^-} f(x)$; (c) $\lim_{x \to 2} f(x)$

13. $G(x) = |2x - 3| - 4$

(a) $\lim_{x \to 3/2^+} G(x)$; (b) $\lim_{x \to 3/2^-} G(x)$; (c) $\lim_{x \to 3/2} G(x)$

14. $F(x) = \begin{cases} |x - 1| & \text{if } x < -1 \\ 0 & \text{if } x = -1 \\ |1 - x| & \text{if } -1 < x \end{cases}$

(a) $\lim_{x \to -1^+} F(x)$; (b) $\lim_{x \to -1^-} F(x)$; (c) $\lim_{x \to -1} F(x)$

15. $f(x) = \dfrac{|x|}{x}$

(a) $\lim_{x \to 0^+} f(x)$; (b) $\lim_{x \to 0^-} f(x)$; (c) $\lim_{x \to 0} f(x)$

16. $S(x) = |\text{sgn } x|$ (sgn x is defined in Example 1)

(a) $\lim_{x \to 0^+} S(x)$; (b) $\lim_{x \to 0^-} S(x)$; (c) $\lim_{x \to 0} S(x)$

17. $f(x) = \begin{cases} 2 & \text{if } x < -2 \\ \sqrt{4 - x^2} & \text{if } -2 \le x \le 2 \\ -2 & \text{if } 2 < x \end{cases}$

(a) $\lim_{x \to -2^-} f(x)$; (b) $\lim_{x \to -2^+} f(x)$; (c) $\lim_{x \to -2} f(x)$; (d) $\lim_{x \to 2^-} f(x)$; (e) $\lim_{x \to 2^+} f(x)$; (f) $\lim_{x \to 2} f(x)$

18. $f(x) = \begin{cases} x + 1 & \text{if } x < -1 \\ x^2 & \text{if } -1 \le x \le 1 \\ 1 - x & \text{if } 1 < x \end{cases}$

(a) $\lim_{x \to -1^-} f(x)$; (b) $\lim_{x \to -1^+} f(x)$; (c) $\lim_{x \to -1} f(x)$; (d) $\lim_{x \to 1^-} f(x)$; (e) $\lim_{x \to 1^+} f(x)$; (f) $\lim_{x \to 1} f(x)$

19. $f(t) = \begin{cases} \sqrt[3]{t} & \text{if } t < 0 \\ \sqrt{t} & \text{if } 0 \le t \end{cases}$

(a) $\lim_{t \to 0^+} f(t)$; (b) $\lim_{t \to 0^-} f(t)$; (c) $\lim_{t \to 0} f(t)$

20. $g(x) = \begin{cases} \sqrt[3]{-x} & \text{if } x \le 0 \\ \sqrt[3]{x} & \text{if } 0 < x \end{cases}$

(a) $\lim_{x \to 0^+} g(x)$; (b) $\lim_{x \to 0^-} g(x)$; (c) $\lim_{x \to 0} g(x)$

21. $F(x) = \begin{cases} \sqrt{x^2 - 9} & \text{if } x \le -3 \\ \sqrt{9 - x^2} & \text{if } -3 < x < 3 \\ \sqrt{x^2 - 9} & \text{if } 3 \le x \end{cases}$

(a) $\lim_{x \to -3^-} F(x)$; (b) $\lim_{x \to -3^+} F(x)$; (c) $\lim_{x \to -3} F(x)$; (d) $\lim_{x \to 3^-} F(x)$; (e) $\lim_{x \to 3^+} F(x)$; (f) $\lim_{x \to 3} F(x)$

22. $G(t) = \begin{cases} \sqrt[3]{t + 1} & \text{if } t \le -1 \\ \sqrt{1 - t^2} & \text{if } -1 < t < 1 \\ \sqrt[3]{t - 1} & \text{if } 1 \le t \end{cases}$

(a) $\lim_{t \to -1^-} G(t)$; (b) $\lim_{t \to -1^+} G(t)$; (c) $\lim_{t \to -1} G(t)$; (d) $\lim_{t \to 1^-} G(t)$; (e) $\lim_{t \to 1^+} G(t)$; (f) $\lim_{t \to 1} G(t)$

23. $F(x) = x - 2 \text{ sgn } x$, where sgn x is defined in Example 1. Find, if they exist: (a) $\lim_{x \to 0^+} F(x)$; (b) $\lim_{x \to 0^-} F(x)$; (c) $\lim_{x \to 0} F(x)$.

24. $h(x) = \text{sgn } x - U(x)$, where sgn x is defined in Example 1, and U is the unit step function defined by

$$U(x) = \begin{cases} 0 & \text{if } x < 0 \\ 1 & \text{if } 0 \le x \end{cases}$$

Find, if they exist: (a) $\lim_{x \to 0^+} h(x)$; (b) $\lim_{x \to 0^-} h(x)$; (c) $\lim_{x \to 0} h(x)$.

25. Find, if they exist: (a) $\lim_{x \to 2^+} [\![x]\!]$; (b) $\lim_{x \to 2^-} [\![x]\!]$; (c) $\lim_{x \to 2} [\![x]\!]$.

26. Find, if they exist: (a) $\lim_{x \to 4^+} [\![x - 3]\!]$; (b) $\lim_{x \to 4^-} [\![x - 3]\!]$; (c) $\lim_{x \to 4} [\![x - 3]\!]$.

27. Let $h(x) = (x - 1) \text{ sgn } x$. Draw a sketch of the graph of h. Find, if they exist: (a) $\lim_{x \to 0^+} h(x)$; (b) $\lim_{x \to 0^-} h(x)$; (c) $\lim_{x \to 0} h(x)$.

28. Let $G(x) = [\![x]\!] + [\![4 - x]\!]$. Draw a sketch of the graph of G. Find, if they exist: (a) $\lim_{x \to 3^+} G(x)$; (b) $\lim_{x \to 3^-} G(x)$; (c) $\lim_{x \to 3} G(x)$.

29. Given $f(x) = \begin{cases} 3x + 2 & \text{if } x < 4 \\ 5x + k & \text{if } 4 \le x \end{cases}$. Find the value of k such that $\lim_{x \to 4} f(x)$ exists.

30. Given $f(x) = \begin{cases} kx - 3 & \text{if } x \le -1 \\ x^2 + k & \text{if } -1 < x \end{cases}$. Find the value of k such that $\lim_{x \to -1} f(x)$ exists.

31. Given $f(x) = \begin{cases} x^2 & \text{if } x \le -2 \\ ax + b & \text{if } -2 < x < 2 \\ 2x - 6 & \text{if } 2 \le x \end{cases}$. Find the values of a and b such that $\lim_{x \to -2} f(x)$ and $\lim_{x \to 2} f(x)$ both exist.

32. Given $f(x) = \begin{cases} 2x - a & \text{if } x < -3 \\ ax + 2b & \text{if } -3 \le x \le 3 \\ b - 5x & \text{if } 3 < x \end{cases}$. Find the values of a and b such that $\lim_{x \to -3} f(x)$ and $\lim_{x \to 3} f(x)$ both exist.

33. Let $f(x) = \begin{cases} -1 & \text{if } x < 0 \\ 1 & \text{if } 0 < x \end{cases}$. Show that $\lim_{x \to 0} f(x)$ does not exist but that $\lim_{x \to 0} |f(x)|$ does exist.

34. Prove Theorem 1.4.3.

35. Shipping charges are often based on a formula that offers a lower charge per pound as the size of the shipment is increased. Suppose x pounds is the weight of a shipment, $C(x)$ dollars is the total cost of the shipment, and

$$C(x) = \begin{cases} 0.80x & \text{if } 0 < x \le 50 \\ 0.70x & \text{if } 50 < x \le 200 \\ 0.65x & \text{if } 200 < x \end{cases}$$

(a) Draw a sketch of the graph of C. Find each of the following limits: (b) $\lim_{x \to 50^-} C(x)$; (c) $\lim_{x \to 50^+} C(x)$; (d) $\lim_{x \to 200^-} C(x)$; (e) $\lim_{x \to 200^+} C(x)$.

36. Given $f(x) = \begin{cases} x^2 + 3 & \text{if } x \le 1 \\ x + 1 & \text{if } 1 < x \end{cases}$ and $g(x) = \begin{cases} x^2 & \text{if } x \le 1 \\ 2 & \text{if } 1 < x \end{cases}$

(a) Show that $\lim_{x \to 1^-} f(x)$ and $\lim_{x \to 1^+} f(x)$ both exist but are not equal, and hence $\lim_{x \to 1} f(x)$ does not exist.

(b) Show that $\lim_{x \to 1^-} g(x)$ and $\lim_{x \to 1^+} g(x)$ both exist but are not equal, and hence $\lim_{x \to 1} g(x)$ does not exist.

(c) Find formulas for $f(x) \cdot g(x)$.

(d) Prove that $\lim_{x \to 1} [f(x) \cdot g(x)]$ exists by showing that $\lim_{x \to 1^-} [f(x) \cdot g(x)] = \lim_{x \to 1^+} [f(x) \cdot g(x)]$.

1.5
INFINITE LIMITS

FIGURE 1.5.1

Table 1.5.1

x	$f(x) = \dfrac{3}{(x-2)^2}$
3	3
$\frac{5}{2}$	12
$\frac{7}{3}$	27
$\frac{9}{4}$	48
$\frac{21}{10}$	300
$\frac{201}{100}$	30,000
$\frac{2001}{1000}$	3,000,000

Table 1.5.2

x	$f(x) = \dfrac{3}{(x-2)^2}$
1	3
$\frac{3}{2}$	12
$\frac{5}{3}$	27
$\frac{7}{4}$	48
$\frac{19}{10}$	300
$\frac{199}{100}$	30,000
$\frac{1999}{1000}$	3,000,000

Let f be the function defined by

$$f(x) = \frac{3}{(x-2)^2}$$

A sketch of the graph of this function is in Fig. 1.5.1.

We investigate the function values of f when x is close to 2. Let x approach 2 from the right, and consider the values of $f(x)$ given in Table 1.5.1. From this table you see intuitively that as x gets closer and closer to 2 through values greater than 2, $f(x)$ increases without bound. In other words, we can make $f(x)$ greater than any preassigned positive number (i.e., $f(x)$ can be made as large as we please) for all values of x close enough to 2 and x greater than 2.

To indicate that $f(x)$ increases without bound as x approaches 2 through values greater than 2 we write

$$\lim_{x \to 2^+} \frac{3}{(x-2)^2} = +\infty$$

Now let x approach 2 from the left, and consider the values of $f(x)$ given in Table 1.5.2. You see intuitively from this table that as x gets closer and closer to 2 through values less than 2, $f(x)$ increases without bound; so we write

$$\lim_{x \to 2^-} \frac{3}{(x-2)^2} = +\infty$$

Therefore, as x approaches 2 from either the right or the left, $f(x)$ increases without bound, and we write

$$\lim_{x \to 2} \frac{3}{(x-2)^2} = +\infty$$

1.5.1 DEFINITION   Let f be a function that is defined at every number in some open interval I containing a, except possibly at the number a itself. **As x approaches a, $f(x)$ increases without bound**, which is written

$$\lim_{x \to a} f(x) = +\infty \qquad (1)$$

if for any number $N > 0$ there exists a $\delta > 0$ such that

$$\text{if} \quad 0 < |x - a| < \delta \quad \text{then} \quad f(x) > N$$

Another way of stating Definition 1.5.1 is as follows: "The function values $f(x)$ increase without bound as x approaches a number a if $f(x)$ can be made as large as we please (i.e., greater than any positive number N) for all values of x sufficiently close to a but not equal to a."

Note: It should be stressed again that $+\infty$ is not a symbol for a real number; hence, when we write $\lim\limits_{x \to a} f(x) = +\infty$, it does not have the same meaning as $\lim\limits_{x \to a} f(x) = L$, where L is a real number. Equation (1) can be read

FIGURE 1.5.2.

as "the limit of $f(x)$ as x approaches a is positive infinity." In such a case the limit does not exist, but the symbol $+\infty$ indicates the behavior of the function values $f(x)$ as x gets closer and closer to a.

In an analogous manner we can indicate the behavior of a function whose function values decrease without bound. To lead up to this, consider the function g defined by the equation

$$g(x) = \frac{-3}{(x-2)^2}$$

A sketch of the graph of this function is in Fig. 1.5.2.

The function values given by $g(x) = -3/(x-2)^2$ are the negatives of the function values given by $f(x) = 3/(x-2)^2$. So for the function g, as x approaches 2, either from the right or the left, $g(x)$ decreases without bound, and we write

$$\lim_{x \to 2} \frac{-3}{(x-2)^2} = -\infty$$

1.5.2 DEFINITION Let f be a function that is defined at every number in some open interval I containing a, except possibly at the number a itself. **As x approaches a, $f(x)$ decreases without bound**, which is written

$$\lim_{x \to a} f(x) = -\infty \tag{2}$$

if for any number $N < 0$ there exists a $\delta > 0$ such that

if $\quad 0 < |x - a| < \delta \quad$ then $\quad f(x) < N$

Note: Equation (2) can be read as "the limit of $f(x)$ as x approaches a is negative infinity," observing again that the limit does not exist and that the symbol $-\infty$ indicates only the behavior of the function values as x approaches a.

We can consider one-sided limits that are "infinite." In particular, $\lim\limits_{x \to a^+} f(x) = +\infty$ if f is defined at every number in some open interval (a, c) and if for any number $N > 0$ there exists a $\delta > 0$ such that

if $\quad 0 < x - a < \delta, \quad$ then $\quad f(x) > N$

Similar definitions can be given if $\lim\limits_{x \to a^-} f(x) = +\infty$, $\lim\limits_{x \to a^+} f(x) = -\infty$, and $\lim\limits_{x \to a^-} f(x) = -\infty$.

Now suppose that h is the function defined by the equation

$$h(x) = \frac{2x}{x-1} \tag{3}$$

FIGURE 1.5.3

A sketch of the graph of this function is in Fig. 1.5.3. By referring to Figs. 1.5.1, 1.5.2, and 1.5.3, note the difference in the behavior of the function whose graph is sketched in Fig. 1.5.3 from the functions of the other two figures. Observe that

$$\lim_{x \to 1^-} \frac{2x}{x-1} = -\infty \tag{4}$$

$$\lim_{x \to 1^+} \frac{2x}{x-1} = +\infty \tag{5}$$

That is, for the function defined by (3), as x approaches 1 through values less than 1, the function values decrease without bound, and as x approaches 1 through values greater than 1, the function values increase without bound.

Before giving some examples, we need two limit theorems involving "infinite" limits.

1.5.3 LIMIT THEOREM 11   If r is any positive integer, then

(i) $\lim\limits_{x \to 0^+} \dfrac{1}{x^r} = +\infty$;

(ii) $\lim\limits_{x \to 0^-} \dfrac{1}{x^r} = \begin{cases} -\infty & \text{if } r \text{ is odd} \\ +\infty & \text{if } r \text{ is even} \end{cases}$.

PROOF   We prove part (i). The proof of part (ii) is analogous and is left as an exercise (see Exercise 49). We must show that for any $N > 0$ there exists a $\delta > 0$ such that

$$\text{if} \quad 0 < x < \delta \quad \text{then} \quad \frac{1}{x^r} > N$$

or, equivalently, because $x > 0$ and $N > 0$,

$$\text{if} \quad 0 < x < \delta \quad \text{then} \quad x^r < \frac{1}{N}$$

or, equivalently, because $r > 0$,

$$\text{if} \quad 0 < x < \delta \quad \text{then} \quad x < \left(\frac{1}{N}\right)^{1/r}$$

The above statement holds if $\delta = \left(\dfrac{1}{N}\right)^{1/r}$. Therefore when $\delta = \left(\dfrac{1}{N}\right)^{1/r}$

$$\text{if} \quad 0 < x < \delta \quad \text{then} \quad \frac{1}{x^r} > N \qquad\blacksquare$$

▶ ILLUSTRATION 1   From Limit Theorem 11(i) it follows that

$$\lim_{x \to 0^+} \frac{1}{x^3} = +\infty \quad \text{and} \quad \lim_{x \to 0^+} \frac{1}{x^4} = +\infty$$

From Limit Theorem 11(ii)

$$\lim_{x \to 0^-} \frac{1}{x^3} = -\infty \quad \text{and} \quad \lim_{x \to 0^-} \frac{1}{x^4} = +\infty \qquad\qquad\blacktriangleleft$$

Limit Theorem 12, which follows, involves the limit of a rational function for which the limit of the denominator is zero and the limit of the numerator is a nonzero constant. Such a situation occurs in (4) and (5).

1.5.4 LIMIT THEOREM 12   If a is any real number, and if $\lim\limits_{x \to a} f(x) = 0$ and $\lim\limits_{x \to a} g(x) = c$, where c is a constant not equal to 0, then

(i) if $c > 0$ and if $f(x) \to 0$ through positive values of $f(x)$,

$$\lim_{x \to a} \frac{g(x)}{f(x)} = +\infty \qquad \frac{+}{+}$$

(ii) if $c > 0$ and if $f(x) \to 0$ through negative values of $f(x)$,

$$\lim_{x \to a} \frac{g(x)}{f(x)} = -\infty \qquad \frac{+}{-}$$

(iii) if $c < 0$ and if $f(x) \to 0$ through positive values of $f(x)$,

$$\lim_{x \to a} \frac{g(x)}{f(x)} = -\infty \qquad \frac{-}{+}$$

(iv) if $c < 0$ and if $f(x) \to 0$ through negative values of $f(x)$,

$$\lim_{x \to a} \frac{g(x)}{f(x)} = +\infty \qquad \frac{-}{-}$$

The theorem is also valid if "$x \to a$" is replaced by "$x \to a^{+}$" or "$x \to a^{-}$."

We shall prove part (i) of Limit Theorem 12 and leave the proofs of the other parts as exercises (see Exercises 50 through 52). However, before giving the proof, we demonstrate the content of the four parts of the theorem in the following illustration.

▶ ILLUSTRATION 2   (a) To evaluate

$$\lim_{x \to 4^{+}} \frac{2x - 1}{x - 4}$$

observe that $\lim\limits_{x \to 4^{+}} (2x - 1) = 7 > 0$ and $\lim\limits_{x \to 4^{+}} (x - 4) = 0$, where $x - 4$ is approaching 0 through positive values. Therefore we apply Limit Theorem 12(i) and have

$$\lim_{x \to 4^{+}} \frac{2x - 1}{x - 4} = +\infty$$

(b) To evaluate

$$\lim_{x \to 4^{-}} \frac{2x - 1}{x - 4}$$

note that $\lim\limits_{x \to 4^{-}} (2x - 1) = 7 > 0$ and $\lim\limits_{x \to 4^{-}} (x - 4) = 0$, where $x - 4$ is approaching 0 through negative values. Thus from Limit Theorem 12(ii),

$$\lim_{x \to 4^-} \frac{2x - 1}{x - 4} = -\infty$$

(c) To evaluate

$$\lim_{x \to 4^+} \frac{1 - 2x}{x - 4}$$

observe that $\lim_{x \to 4^+} (1 - 2x) = -7 < 0$ and $\lim_{x \to 4^+} (x - 4) = 0$, where $x - 4$ is approaching 0 through positive values. Hence from Limit Theorem 12(iii),

$$\lim_{x \to 4^+} \frac{1 - 2x}{x - 4} = -\infty$$

(d) To evaluate

$$\lim_{x \to 4^-} \frac{1 - 2x}{x - 4}$$

note that $\lim_{x \to 4^-} (1 - 2x) = -7 < 0$ and $\lim_{x \to 4^-} (x - 4) = 0$, where $x - 4$ is approaching 0 through negative values. Therefore, from Limit Theorem 12(iv) it follows that

$$\lim_{x \to 4^-} \frac{1 - 2x}{x - 4} = +\infty$$

◀

PROOF OF PART (i) OF LIMIT THEOREM 12   To prove that

$$\lim_{x \to a} \frac{g(x)}{f(x)} = +\infty$$

we must show that for any $N > 0$ there exists a $\delta > 0$ such that

$$\text{if} \quad 0 < |x - a| < \delta \quad \text{then} \quad \frac{g(x)}{f(x)} > N \tag{6}$$

Since $\lim_{x \to a} g(x) = c > 0$, by taking $\epsilon = \frac{1}{2}c$ in Definition 1.2.1 it follows that there exists a $\delta_1 > 0$ such that

$$\text{if} \quad 0 < |x - a| < \delta_1 \quad \text{then} \quad |g(x) - c| < \frac{1}{2}c$$

By applying Theorem 0.1.10 to the above inequality it follows that there exists a $\delta_1 > 0$ such that

$$\text{if} \quad 0 < |x - a| < \delta_1 \quad \text{then} \quad -\frac{1}{2}c < g(x) - c < \frac{1}{2}c$$

$$\Leftrightarrow \quad \text{if} \quad 0 < |x - a| < \delta_1 \quad \text{then} \quad \frac{1}{2}c < g(x) < \frac{3}{2}c$$

So there exists a $\delta_1 > 0$ such that

$$\text{if} \quad 0 < |x - a| < \delta_1 \quad \text{then} \quad g(x) > \frac{1}{2}c \tag{7}$$

Now $\lim\limits_{x \to a} f(x) = 0$. Thus for any $\epsilon > 0$ there exists a $\delta_2 > 0$ such that

$$\text{if}\quad 0 < |x - a| < \delta_2 \quad \text{then}\quad |f(x)| < \epsilon$$

Since $f(x)$ is approaching zero through positive values of $f(x)$, the absolute value bars around $f(x)$ can be removed; hence for any $\epsilon > 0$ there exists a $\delta_2 > 0$ such that

$$\text{if}\quad 0 < |x - a| < \delta_2 \quad \text{then}\quad 0 < f(x) < \epsilon \tag{8}$$

From statements (7) and (8) we can conclude that for any $\epsilon > 0$ there exist a $\delta_1 > 0$ and a $\delta_2 > 0$ such that

$$\text{if}\quad 0 < |x - a| < \delta_1 \quad \text{and}\quad 0 < |x - a| < \delta_2 \quad \text{then}\quad \frac{g(x)}{f(x)} > \frac{\frac{1}{2}c}{\epsilon}$$

Hence, if $\epsilon = c/(2N)$ and $\delta = \min(\delta_1, \delta_2)$, then

$$\text{if}\quad 0 < |x - a| < \delta \quad \text{then}\quad \frac{g(x)}{f(x)} > \frac{\frac{1}{2}c}{c/(2N)} = N$$

which is statement (6). Hence part (i) is proved.   ■

When Limit Theorem 12 is applied, we can often get an indication of whether the result is $+\infty$ or $-\infty$ by taking a *suitable value* of x near a to ascertain if the quotient is positive or negative, as shown in the following illustration.

▶ ILLUSTRATION 3   In (4) we have

$$\lim_{x \to 1^-} \frac{2x}{x - 1}$$

Limit Theorem 12 is applicable because $\lim\limits_{x \to 1^-} 2x = 2$ and $\lim\limits_{x \to 1^-} (x - 1) = 0$. We wish to determine if we have $+\infty$ or $-\infty$. Because $x \to 1^-$, take a value of x near 1 and less than 1; for instance, take $x = 0.9$. Then

$$\frac{2x}{x - 1} = \frac{2(0.9)}{0.9 - 1} \quad \Leftrightarrow \quad \frac{2x}{x - 1} = -18$$

The negative quotient leads us to suspect that

$$\lim_{x \to 1^-} \frac{2x}{x - 1} = -\infty$$

This result follows from part (ii) of Limit Theorem 12, because when $x \to 1^-$, $x - 1$ is approaching 0 through negative values.

For the limit in (5), because $x \to 1^+$, take $x = 1.1$. Then

$$\frac{2x}{x - 1} = \frac{2(1.1)}{1.1 - 1} \quad \Leftrightarrow \quad \frac{2x}{x - 1} = 22$$

Because the quotient is positive we suspect that

$$\lim_{x \to 1^+} \frac{2x}{x - 1} = +\infty$$

This result follows from part (i) of Limit Theorem 12, because when $x \to 1^+$, $x - 1$ is approaching 0 through positive values. ◀

When using the procedure described in Illustration 3, be careful that the value of x selected is close enough to a to indicate the true behavior of the quotient. For instance, when computing $\lim\limits_{x \to 1^-} \dfrac{2x}{x - 1}$, the value of x selected must not only be less than 1 but also greater than 0.

EXAMPLE 1
Find

(a) $\lim\limits_{x \to 3^+} \dfrac{x^2 + x + 2}{x^2 - 2x - 3}$ (b) $\lim\limits_{x \to 3^-} \dfrac{x^2 + x + 2}{x^2 - 2x - 3}$

SOLUTION

(a) $\lim\limits_{x \to 3^+} \dfrac{x^2 + x + 2}{x^2 - 2x - 3} = \lim\limits_{x \to 3^+} \dfrac{x^2 + x + 2}{(x - 3)(x + 1)}$

The limit of the numerator is 14, which can be verified easily.

$$\lim_{x \to 3^+} (x - 3)(x + 1) = \lim_{x \to 3^+} (x - 3) \cdot \lim_{x \to 3^+} (x + 1)$$
$$= 0 \cdot 4$$
$$= 0$$

The limit of the denominator is 0, and the denominator is approaching 0 through positive values. Then from Limit Theorem 12(i),

$$\lim_{x \to 3^+} \frac{x^2 + x + 2}{x^2 - 2x - 3} = +\infty$$

(b) $\lim\limits_{x \to 3^-} \dfrac{x^2 + x + 2}{x^2 - 2x - 3} = \lim\limits_{x \to 3^-} \dfrac{x^2 + x + 2}{(x - 3)(x + 1)}$

As in part (a), the limit of the numerator is 14.

$$\lim_{x \to 3^-} (x - 3)(x + 1) = \lim_{x \to 3^-} (x - 3) \cdot \lim_{x \to 3^-} (x + 1)$$
$$= 0 \cdot 4$$
$$= 0$$

In this case, the limit of the denominator is zero, but the denominator is approaching zero through negative values. From Limit Theorem 12(ii),

$$\lim_{x \to 3^-} \frac{x^2 + x + 2}{x^2 - 2x - 3} = -\infty$$

EXAMPLE 2
Find

(a) $\lim\limits_{x \to 2^+} \dfrac{\sqrt{x^2 - 4}}{x - 2}$ (b) $\lim\limits_{x \to 2^-} \dfrac{\sqrt{4 - x^2}}{x - 2}$

SOLUTION

(a) Because $x \to 2^+$, $x - 2 > 0$; so $x - 2 = \sqrt{(x-2)^2}$. Thus

$$\lim_{x \to 2^+} \frac{\sqrt{x^2-4}}{x-2} = \lim_{x \to 2^+} \frac{\sqrt{(x-2)(x+2)}}{\sqrt{(x-2)^2}}$$

$$= \lim_{x \to 2^+} \frac{\sqrt{x-2}\sqrt{x+2}}{\sqrt{x-2}\sqrt{x-2}}$$

$$= \lim_{x \to 2^+} \frac{\sqrt{x+2}}{\sqrt{x-2}}$$

The limit of the numerator is 2. The limit of the denominator is 0, and the denominator is approaching 0 through positive values. Therefore, by Limit Theorem 12(i) it follows that

$$\lim_{x \to 2^+} \frac{\sqrt{x^2-4}}{x-2} = +\infty$$

(b) Because $x \to 2^-$, $x - 2 < 0$; so $x - 2 = -\sqrt{(2-x)^2}$. Therefore

$$\lim_{x \to 2^-} \frac{\sqrt{4-x^2}}{x-2} = \lim_{x \to 2^-} \frac{\sqrt{2-x}\sqrt{2+x}}{-\sqrt{2-x}\sqrt{2-x}}$$

$$= \lim_{x \to 2^-} \frac{\sqrt{2+x}}{-\sqrt{2-x}}$$

The limit of the numerator is 2. The limit of the denominator is 0, and the denominator is approaching 0 through negative values. Hence by Limit Theorem 12(ii),

$$\lim_{x \to 2^-} \frac{\sqrt{4-x^2}}{x-2} = -\infty$$

EXAMPLE 3

Find

$$\lim_{x \to 4^-} \frac{[\![x]\!] - 4}{x - 4}$$

SOLUTION

$\lim\limits_{x \to 4^-} [\![x]\!] = 3$. Therefore $\lim\limits_{x \to 4^-} ([\![x]\!] - 4) = -1$. Furthermore, $\lim\limits_{x \to 4^-} (x - 4) = 0$, and $x - 4$ is approaching 0 through negative values. Hence from Limit Theorem 12(iv),

$$\lim_{x \to 4^-} \frac{[\![x]\!] - 4}{x - 4} = +\infty$$

Remember that because $+\infty$ and $-\infty$ are not symbols for numbers, the Limit Theorems 1–10 of Sec. 1.3 do not hold for "infinite" limits. However, there are the following properties regarding such limits. The proofs are left as exercises (see Exercises 53–55).

1.5.5 THEOREM   **(i)** If $\lim\limits_{x \to a} f(x) = +\infty$, and $\lim\limits_{x \to a} g(x) = c$, where c is any constant, then

$$\lim_{x \to a} [f(x) + g(x)] = +\infty$$

(ii) If $\lim\limits_{x \to a} f(x) = -\infty$, and $\lim\limits_{x \to a} g(x) = c$, where c is any constant, then

$$\lim_{x \to a} [f(x) + g(x)] = -\infty$$

The theorem is valid if "$x \to a$" is replaced by "$x \to a^+$" or "$x \to a^-$."

▶ ILLUSTRATION 3   Because $\lim\limits_{x \to 2^+} \dfrac{1}{x - 2} = +\infty$ and $\lim\limits_{x \to 2^+} \dfrac{1}{x + 2} = \dfrac{1}{4}$, it follows from Theorem 1.5.5(i) that

$$\lim_{x \to 2^+} \left[\frac{1}{x - 2} + \frac{1}{x + 2} \right] = +\infty$$ ◀

1.5.6 THEOREM   If $\lim\limits_{x \to a} f(x) = +\infty$ and $\lim\limits_{x \to a} g(x) = c$, where c is any constant except 0, then

(i) if $c > 0$, $\lim\limits_{x \to a} f(x) \cdot g(x) = +\infty$;

(ii) if $c < 0$, $\lim\limits_{x \to a} f(x) \cdot g(x) = -\infty$.

The theorem is valid if "$x \to a$" is replaced by "$x \to a^+$" or "$x \to a^-$."

▶ ILLUSTRATION 4

$$\lim_{x \to 3} \frac{5}{(x - 3)^2} = +\infty \quad \text{and} \quad \lim_{x \to 3} \frac{x + 4}{x - 4} = -7$$

Therefore, from Theorem 1.5.6(ii),

$$\lim_{x \to 3} \left[\frac{5}{(x - 3)^2} \cdot \frac{x + 4}{x - 4} \right] = -\infty$$ ◀

1.5.7 THEOREM   If $\lim\limits_{x \to a} f(x) = -\infty$ and $\lim\limits_{x \to a} g(x) = c$, where c is any constant except 0, then

(i) if $c > 0$, $\lim\limits_{x \to a} f(x) \cdot g(x) = -\infty$;

(ii) if $c < 0$, $\lim\limits_{x \to a} f(x) \cdot g(x) = +\infty$.

The theorem is valid if "$x \to a$" is replaced by "$x \to a^+$" or "$x \to a^-$."

▶ ILLUSTRATION 5   In Example 2(b) we showed

$$\lim_{x \to 2^-} \frac{\sqrt{4 - x^2}}{x - 2} = -\infty$$

Furthermore,

$$\lim_{x \to 2^-} \frac{x - 3}{x + 2} = -\frac{1}{4}$$

Thus, from Theorem 1.5.7(ii) it follows that

$$\lim_{x \to 2^-} \left[\frac{\sqrt{4 - x^2}}{x - 2} \cdot \frac{x - 3}{x + 2} \right] = +\infty$$

FIGURE 1.5.4

An aid in drawing the sketch of the graph of a function is to find, if there are any, the *asymptotes* of the graph. In this section we discuss *vertical asymptotes*. Let the function f be defined by

$$f(x) = \frac{1}{(x - a)^2} \tag{9}$$

A sketch of the graph of f is in Fig. 1.5.4. Any line parallel to and above the x axis will intersect this graph in two points: one point to the left of the line $x = a$ and one point to the right of this line. Thus for any $k > 0$, no matter how large, the line $y = k$ will intersect the graph of f in two points; the distance of these two points from the line $x = a$ gets smaller and smaller as k gets larger and larger. The line $x = a$ is called a *vertical asymptote* of the graph of f.

1.5.8 DEFINITION

The line $x = a$ is said to be a **vertical asymptote** of the graph of the function f if at least one of the following statements is true:

(i) $\lim\limits_{x \to a^+} f(x) = +\infty$

(ii) $\lim\limits_{x \to a^+} f(x) = -\infty$

(iii) $\lim\limits_{x \to a^-} f(x) = +\infty$

(iv) $\lim\limits_{x \to a^-} f(x) = -\infty$

▶ ILLUSTRATION 6   Each of the Figs. 1.5.5 through 1.5.8 shows a portion of the graph of a function for which the line $x = a$ is a vertical asymptote. In Fig.

FIGURE 1.5.5   FIGURE 1.5.6   FIGURE 1.5.7   FIGURE 1.5.8

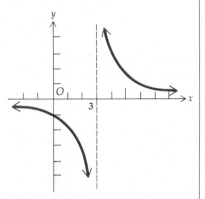

FIGURE 1.5.9

FIGURE 1.5.10

1.5.5, part (i) of Definition 1.5.8 applies; in Fig. 1.5.6, part (ii) applies; and in Figs. 1.5.7 and 1.5.8, parts (iii) and (iv), respectively, apply.   ◀

For the function defined by (9), both parts (i) and (iii) of the above definition are true. See Fig. 1.5.4. If g is the function defined by

$$g(x) = -\frac{1}{(x-a)^2}$$

then both parts (ii) and (iv) are true, and the line $x = a$ is a vertical asymptote of the graph of g. This is shown in Fig. 1.5.9.

EXAMPLE 4
Find the vertical asymptote of the graph of the function defined by the equation

$$f(x) = \frac{3}{x-3}$$

and draw a sketch of the graph of f.

SOLUTION

$$\lim_{x \to 3^+} \frac{3}{x-3} = +\infty \qquad \lim_{x \to 3^-} \frac{3}{x-3} = -\infty$$

It follows from Definition 1.5.8 that the line $x = 3$ is a vertical asymptote of the graph of f. A sketch of the graph of f is shown in Fig. 1.5.10.

EXERCISES 1.5
In Exercises 1 through 32, evaluate the limit.

1. $\displaystyle\lim_{x \to 5^+} \frac{1}{x-5}$

2. $\displaystyle\lim_{x \to 5^-} \frac{1}{x-5}$

3. $\displaystyle\lim_{x \to 5} \frac{1}{(x-5)^2}$

4. $\displaystyle\lim_{x \to 1^-} \frac{x+2}{1-x}$

5. $\displaystyle\lim_{x \to 1^+} \frac{x+2}{1-x}$

6. $\displaystyle\lim_{x \to 1} \frac{x+2}{(x-1)^2}$

7. $\displaystyle\lim_{x \to -1^+} \frac{x-2}{x+1}$

8. $\displaystyle\lim_{x \to -1^-} \frac{x-2}{x+1}$

9. $\displaystyle\lim_{x \to -4^-} \frac{x}{x+4}$

10. $\displaystyle\lim_{x \to 4^+} \frac{x}{x-4}$

11. $\displaystyle\lim_{x \to -3^-} \frac{4x}{9-x^2}$

12. $\displaystyle\lim_{x \to 3^+} \frac{4x^2}{9-x^2}$

13. $\displaystyle\lim_{t \to 2^+} \frac{t+2}{t^2-4}$

14. $\displaystyle\lim_{t \to 2^-} \frac{-t+2}{(t-2)^2}$

15. $\displaystyle\lim_{t \to 2^-} \frac{t+2}{t^2-4}$

16. $\displaystyle\lim_{x \to 0^+} \frac{\sqrt{3+x^2}}{x}$

17. $\displaystyle\lim_{x \to 0^-} \frac{\sqrt{3+x^2}}{x}$

18. $\displaystyle\lim_{x \to 0} \frac{\sqrt{3+x^2}}{x^2}$

19. $\displaystyle\lim_{x \to 3^+} \frac{\sqrt{x^2-9}}{x-3}$

20. $\displaystyle\lim_{x \to 4^-} \frac{\sqrt{16-x^2}}{x-4}$

21. $\displaystyle\lim_{x \to 0^+} \left(\frac{1}{x} - \frac{1}{x^2}\right)$

22. $\displaystyle\lim_{x \to 0^+} \frac{x^2-3}{x^3+x^2}$

23. $\displaystyle\lim_{x \to 0^-} \frac{2-4x^3}{5x^2+3x^3}$

24. $\displaystyle\lim_{s \to 2^-} \left(\frac{1}{s-2} - \frac{3}{s^2-4}\right)$

25. $\lim\limits_{t \to -4^-} \left(\dfrac{2}{t^2 + 3t - 4} - \dfrac{3}{t + 4} \right)$

26. $\lim\limits_{x \to 1^-} \dfrac{2x^3 - 5x^2}{x^2 - 1}$

27. $\lim\limits_{x \to 3^-} \dfrac{[\![x]\!] - x}{3 - x}$

28. $\lim\limits_{x \to 1^-} \dfrac{[\![x^2]\!] - 1}{x^2 - 1}$

29. $\lim\limits_{x \to 3^-} \dfrac{x^3 + 9x^2 + 20x}{x^2 + x - 12}$

30. $\lim\limits_{x \to -2^+} \dfrac{6x^2 + x - 2}{2x^2 + 3x - 2}$

31. $\lim\limits_{x \to 1^+} \dfrac{x - 1}{\sqrt{2x - x^2 - 1}}$

32. $\lim\limits_{x \to 2^-} \dfrac{x - 2}{2 - \sqrt{4x - x^2}}$

33. For each of the following functions, find the vertical asymptote of the graph of the function, and draw a sketch of the graph: (a) $f(x) = \dfrac{1}{x}$; (b) $g(x) = \dfrac{1}{x^2}$; (c) $h(x) = \dfrac{1}{x^3}$; (d) $\phi(x) = \dfrac{1}{x^4}$.

34. For each of the following functions, find the vertical asymptote of the graph of the function, and draw a sketch of the graph: (a) $f(x) = -\dfrac{1}{x}$; (b) $g(x) = -\dfrac{1}{x^2}$; (c) $h(x) = -\dfrac{1}{x^3}$; (d) $\phi(x) = -\dfrac{1}{x^4}$.

In Exercises 35 through 45, find the vertical asymptote(s) of the graph of the function, and draw a sketch of the graph.

35. $f(x) = \dfrac{2}{x - 4}$

36. $f(x) = \dfrac{3}{x + 1}$

37. $f(x) = \dfrac{-2}{x + 3}$

38. $f(x) = \dfrac{-4}{x - 5}$

39. $f(x) = \dfrac{-2}{(x + 3)^2}$

40. $f(x) = \dfrac{4}{(x - 5)^2}$

41. $f(x) = \dfrac{5}{x^2 + 8x + 15}$

42. $f(x) = \dfrac{1}{x^2 + 5x - 6}$

43. $f(x) = \dfrac{2}{4x^2 - 4x - 35}$

44. $f(x) = \dfrac{x^2 + x - 2}{x^2 - x - 6}$

45. $f(x) = \dfrac{x^2 - x - 12}{x^2 - 6x + 8}$

46. Prove that $\lim\limits_{x \to 4} \dfrac{-2}{(x - 4)^2} = -\infty$ by using Definition 1.5.2.

47. Prove that $\lim\limits_{x \to 2} \dfrac{3}{(x - 2)^2} = +\infty$ by using Definition 1.5.1.

48. Use Definition 1.5.1 to prove that $\lim\limits_{x \to -3} \left| \dfrac{5 - x}{3 + x} \right| = +\infty$.

49. Prove Theorem 1.5.3(ii).
50. Prove Theorem 1.5.4(ii).
51. Prove Theorem 1.5.4(iii).
52. Prove Theorem 1.5.4(iv).
53. Prove Theorem 1.5.5.
54. Prove Theorem 1.5.6.
55. Prove Theorem 1.5.7.

1.6
CONTINUITY OF A FUNCTION AT A NUMBER

In Illustration 1 of Sec. 1.4 we discussed the function C defined by

$$C(x) = \begin{cases} x & \text{if } 0 \le x \le 10 \\ 0.9x & \text{if } 10 < x \end{cases} \tag{1}$$

where $C(x)$ dollars is the total cost of x pounds of a product. We showed that $\lim\limits_{x \to 10} C(x)$ does not exist because $\lim\limits_{x \to 10^-} C(x) \neq \lim\limits_{x \to 10^+} C(x)$. A sketch of

FIGURE 1.6.6

Because $f(2)$ is not defined, condition (i) is not satisfied. Hence f is discontinuous at 2. ◀

▶ ILLUSTRATION 5   Let g be defined by

$$g(x) = \begin{cases} \dfrac{1}{x-2} & \text{if } x \neq 2 \\ 3 & \text{if } x = 2 \end{cases}$$

A sketch of the graph of g appears in Fig. 1.6.6. The three conditions of Definition 1.6.1 are investigated at 2.

Since $g(2) = 3$, condition (i) is satisfied.

$$\lim_{x \to 2^-} g(x) = \lim_{x \to 2^-} \frac{1}{x-2} \qquad \lim_{x \to 2^+} g(x) = \lim_{x \to 2^+} \frac{1}{x-2}$$

$$= -\infty \qquad\qquad = +\infty$$

Because $\lim_{x \to 2} g(x)$ does not exist, condition (ii) is not satisfied. Therefore g is discontinuous at 2. ◀

FIGURE 1.6.7

▶ ILLUSTRATION 6   Let h be defined by

$$h(x) = \begin{cases} 3 + x & \text{if } x \leq 1 \\ 3 - x & \text{if } 1 < x \end{cases}$$

A sketch of the graph of h is shown in Fig. 1.6.7. Because there is a break in the graph at the point where $x = 1$, we investigate the conditions of Definition 1.6.1 at 1.

Because $h(1) = 4$, condition (i) is satisfied.

$$\lim_{x \to 1^-} h(x) = \lim_{x \to 1^-} (3 + x) \qquad \lim_{x \to 1^+} h(x) = \lim_{x \to 1^+} (3 - x)$$

$$= 4 \qquad\qquad = 2$$

Because $\lim_{x \to 1^-} h(x) \neq \lim_{x \to 1^+} h(x)$, then $\lim_{x \to 1} h(x)$ does not exist; therefore condition (ii) fails to hold at 1. Hence h is discontinuous at 1. ◀

FIGURE 1.6.8

▶ ILLUSTRATION 7   Let F be defined by

$$F(x) = \begin{cases} |x - 3| & \text{if } x \neq 3 \\ 2 & \text{if } x = 3 \end{cases}$$

A sketch of the graph of F appears in Fig. 1.6.8. We investigate the three conditions of Definition 1.6.1 at the point where $x = 3$.

Since $F(3) = 2$, condition (i) is satisfied.

$$\lim_{x \to 3^-} F(x) = \lim_{x \to 3^-} (3 - x) \qquad \lim_{x \to 3^+} F(x) = \lim_{x \to 3^+} (x - 3)$$

$$= 0 \qquad\qquad = 0$$

Therefore, $\lim\limits_{x \to 3} F(x) = 0$, and so condition (ii) is satisfied. $\lim\limits_{x \to 3} F(x) = 0$, but $F(3) = 2$; therefore condition (iii) is not satisfied. Thus F is discontinuous at 3. ◀

It should be apparent that the geometric notion of a break in the graph at a certain point is synonymous with the concept of a function being discontinuous at a certain value of the independent variable.

After Illustration 3 it was mentioned that if $f(1)$ had been defined to be 5, then f would be continuous at 1. This illustrates the concept of a *removable discontinuity*. In general, suppose that f is a function that is discontinuous at the number a, but for which $\lim\limits_{x \to a} f(x)$ exists. Then either $f(a) \neq \lim\limits_{x \to a} f(x)$ or else $f(a)$ does not exist. Such a discontinuity is called a **removable discontinuity** because if f is redefined at a such that $f(a)$ is equal to $\lim\limits_{x \to a} f(x)$, the new function becomes continuous at a. If the discontinuity is not removable, it is called an **essential discontinuity**.

EXAMPLE 1

In each of the Illustrations 3 through 7, determine if the discontinuity is removable or essential.

SOLUTION

REMOV. In Illustration 3 the function is discontinuous at 1, but $\lim\limits_{x \to 1} f(x) = 5$. By redefining $f(1) = 5$, then $\lim\limits_{x \to 1} f(x) = f(1)$; so the discontinuity is removable.

ESSENTIAL In Illustration 4 the function f is discontinuous at 2. $\lim\limits_{x \to 2} f(x)$ does not exist; hence the discontinuity is essential.

ESSENTIAL In Illustration 5, $\lim\limits_{x \to 2} g(x)$ does not exist; so the discontinuity is essential.

ESSENTIAL In Illustration 6 the function h is discontinuous because $\lim\limits_{x \to 1} h(x)$ does not exist; so again the discontinuity is essential.

REMOV. In Illustration 7, $\lim\limits_{x \to 3} F(x) = 0$, but $F(3) = 2$; so F is discontinuous at 3. However, if $F(3)$ is redefined to be 0, then the function is continuous at 3; so the discontinuity is removable.

EXAMPLE 2

The function defined by

$$f(x) = \frac{\sqrt{x} - 2}{x - 4}$$

is discontinuous at 4. Show that the discontinuity is removable and redefine $f(4)$ so that the discontinuity is removed.

SOLUTION

The function f is discontinuous at 4 because $f(4)$ does not exist. If $\lim\limits_{x \to 4} f(x)$ exists, the discontinuity can be removed by redefining $f(4) = \lim\limits_{x \to 4} f(x)$. We compute the limit.

$$\lim_{x \to 4} f(x) = \lim_{x \to 4} \frac{\sqrt{x} - 2}{x - 4}$$

$$= \lim_{x \to 4} \frac{(\sqrt{x} - 2)(\sqrt{x} + 2)}{(x - 4)(\sqrt{x} + 2)}$$

$$= \lim_{x \to 4} \frac{x - 4}{(x - 4)(\sqrt{x} + 2)}$$

$$= \lim_{x \to 4} \frac{1}{\sqrt{x} + 2}$$

$$= \tfrac{1}{4}$$

Therefore we let $f(4) = \tfrac{1}{4}$, and we have the new function defined by

$$f(x) = \begin{cases} \dfrac{\sqrt{x} - 2}{x - 4} & \text{if } x \neq 4 \\[2mm] \tfrac{1}{4} & \text{if } x = 4 \end{cases}$$

and this function is continuous at 4.

The following theorem about functions that are continuous at a number is obtained by applying Definition 1.6.1 and limit theorems.

1.6.2 THEOREM   If f and g are two functions that are continuous at the number a, then

(i) $f + g$ is continuous at a;
(ii) $f - g$ is continuous at a;
(iii) $f \cdot g$ is continuous at a;
(iv) f/g is continuous at a, provided that $g(a) \neq 0$.

To illustrate the kind of proof required for each part of this theorem, we prove part (i).

Because f and g are continuous at a, from Definition 1.6.1 it follows that

$$\lim_{x \to a} f(x) = f(a) \qquad\qquad (3)$$

$$\lim_{x \to a} g(x) = g(a) \qquad\qquad (4)$$

Therefore from (3) and (4) and Limit Theorem 4,

$$\lim_{x \to a} [f(x) + g(x)] = f(a) + g(a) \qquad\qquad (5)$$

Equation (5) is the condition that $f + g$ is continuous at a, which furnishes the proof of (i).

The proofs of parts (ii), (iii), and (iv) are similar.

Consider the polynomial function f defined by

$$f(x) = b_0 x^n + b_1 x^{n-1} + b_2 x^{n-2} + \ldots + b_{n-1} x + b_n \qquad b_0 \neq 0$$

where n is a nonnegative integer and $b_0, b_1, \ldots, b_n$ are real numbers. By successive applications of limit theorems we can show that if a is any number,

$$\lim_{x \to a} f(x) = b_0 a^n + b_1 a^{n-1} + b_2 a^{n-2} + \ldots + b_{n-1} a + b_n$$

$$\Rightarrow \quad \lim_{x \to a} f(x) = f(a)$$

thus establishing the following theorem.

1.6.3 THEOREM   A polynomial function is continuous at every number.

▶ ILLUSTRATION 8   If $f(x) = x^3 - 2x^2 + 5x + 1$, then f is a polynomial function and therefore, by Theorem 1.6.3, is continuous at every number. In particular, because f is continuous at 3, then $\lim\limits_{x \to 3} f(x) = f(3)$. Thus

$$\lim_{x \to 3} (x^3 - 2x^2 + 5x + 1) = 3^3 - 2(3)^2 + 5(3) + 1$$
$$= 27 - 18 + 15 + 1$$
$$= 25 \qquad ◀$$

1.6.4 THEOREM   A rational function is continuous at every number in its domain.

PROOF   If f is a rational function, it can be expressed as the quotient of two polynomial functions. So f can be defined by

$$f(x) = \frac{g(x)}{h(x)}$$

where g and h are two polynomial functions, and the domain of f consists of all numbers except those for which $h(x) = 0$.

If a is any number in the domain of f, then $h(a) \neq 0$; so by Limit Theorem 9,

$$\lim_{x \to a} f(x) = \frac{\lim\limits_{x \to a} g(x)}{\lim\limits_{x \to a} h(x)} \qquad (6)$$

Because g and h are polynomial functions, by Theorem 1.6.3 they are continuous at a; so $\lim\limits_{x \to a} g(x) = g(a)$ and $\lim\limits_{x \to a} h(x) = h(a)$. Consequently, from (6),

$$\lim_{x \to a} f(x) = \frac{g(a)}{h(a)}$$

Thus f is continuous at every number in its domain.   ∎

EXAMPLE 3
Given

$$f(x) = \frac{x^3 + 1}{x^2 - 9}$$

Determine the numbers at which f is continuous.

SOLUTION
The domain of f is the set R of real numbers except those for which $x^2 - 9 = 0$. Because $x^2 - 9 = 0$ when $x = \pm 3$, it follows that the domain of f is the set of all real numbers except 3 and -3.

Because f is a rational function, it follows from Theorem 1.6.4 that f is continuous at all real numbers except 3 and -3.

EXAMPLE 4
Let f be defined by

$$f(x) = \begin{cases} 2x - 3 & \text{if } x \le 1 \\ x^2 & \text{if } 1 < x \end{cases}$$

Determine the numbers at which f is continuous.

SOLUTION
The functions having values $2x - 3$ and x^2 are polynomial functions and are therefore continuous everywhere. Thus the only number at which continuity is questionable is 1. We check the three conditions for continuity at 1.

(i) $f(1) = -1$. Thus condition (i) holds.

(ii) $\lim\limits_{x \to 1^-} f(x) = \lim\limits_{x \to 1^-} (2x - 3)$ $\lim\limits_{x \to 1^+} f(x) = \lim\limits_{x \to 1^+} x^2$

$$= -1 \qquad\qquad\qquad = 1$$

Because $\lim\limits_{x \to 1^-} f(x) \ne \lim\limits_{x \to 1^+} f(x)$, the two-sided limit $\lim\limits_{x \to 1} f(x)$ does not exist. Therefore f is discontinuous at 1. However, f is continuous at every real number except 1.

In Supplementary Section 1.9 we prove the following theorem about the continuity of the function defined by $f(x) = \sqrt[n]{x}$.

1.6.5 THEOREM   If n is a positive integer and

$$f(x) = \sqrt[n]{x}$$

then

(i) if n is odd, f is continuous at every number;
(ii) if n is even, f is continuous at every positive number.

FIGURE 1.6.9

FIGURE 1.6.10

▶ **ILLUSTRATION 9** (a) If $f(x) = \sqrt[3]{x}$, it follows from Theorem 1.6.5(i) that f is continuous at every real number. A sketch of the graph of f appears in Fig. 1.6.9.

(b) If $g(x) = \sqrt{x}$, then from Theorem 1.6.5(ii) it follows that g is continuous at every positive number. In Fig. 1.6.10 there is a sketch of the graph of g. ◀

In Sec. 1.7 we apply a definition of continuity of a function that uses ϵ and δ notation. To obtain this alternate definition we start with Definition 1.6.1, which states that the function f is continuous at the number a if $f(a)$ exists, if $\lim_{x \to a} f(x)$ exists, and if

$$\lim_{x \to a} f(x) = f(a) \tag{7}$$

Applying Definition 1.2.1 where L is $f(a)$, it follows that (7) will hold if for any $\epsilon > 0$ there exists a $\delta > 0$ such that

$$\text{if } 0 < |x - a| < \delta \quad \text{then} \quad |f(x) - f(a)| < \epsilon \tag{8}$$

If f is to be continuous at a, $f(a)$ must exist; thus in statement (8) the condition that $|x - a| > 0$ is not necessary, because when $x = a$, $|f(x) - f(a)|$ will be 0 and thus less than ϵ. We have, then, the following theorem, which serves as our desired alternate definition of continuity.

1.6.6 THEOREM The function f is said to be continuous at the number a if f is defined on some open interval containing a and if for any $\epsilon > 0$ there exists a $\delta > 0$ such that

$$\text{if } |x - a| < \delta \quad \text{then} \quad |f(x) - f(a)| < \epsilon$$

EXERCISES 1.6

In Exercises 1 through 22, draw a sketch of the graph of the function; then, by observing where there are breaks in the graph, determine the values of the independent variable at which the function is discontinuous, and show why Definition 1.6.1 is not satisfied at each discontinuity.

1. $f(x) = \dfrac{x^2 + x - 6}{x + 3}$

2. $F(x) = \dfrac{x^2 - 3x - 4}{x - 4}$

3. $g(x) = \begin{cases} \dfrac{x^2 + x - 6}{x + 3} & \text{if } x \neq -3 \\ 1 & \text{if } x = -3 \end{cases}$

4. $G(x) = \begin{cases} \dfrac{x^2 - 3x - 4}{x - 4} & \text{if } x \neq 4 \\ 2 & \text{if } x = 4 \end{cases}$

5. $h(x) = \dfrac{5}{x - 4}$

6. $H(x) = \dfrac{1}{x + 2}$

7. $f(x) = \begin{cases} \dfrac{5}{x - 4} & \text{if } x \neq 4 \\ 2 & \text{if } x = 4 \end{cases}$

8. $g(x) = \begin{cases} \dfrac{1}{x + 2} & \text{if } x \neq -2 \\ 0 & \text{if } x = -2 \end{cases}$

9. $F(x) = \dfrac{x^4 - 16}{x^2 - 4}$

10. $h(x) = \dfrac{(x - 1)(x^2 - x - 12)}{x^2 - 5x + 4}$

11. $G(x) = \dfrac{x^2 - 4}{x^4 - 16}$

12. $H(x) = \dfrac{x^2 - 5x + 4}{(x - 1)(x^2 - x - 12)}$

13. $f(x) = \begin{cases} -1 & \text{if } x < 0 \\ 0 & \text{if } x = 0 \\ \sqrt{x} & \text{if } 0 < x \end{cases}$

14. $f(x) = \begin{cases} x - 1 & \text{if } x < 1 \\ 1 & \text{if } x = 1 \\ 1 - x & \text{if } 1 < x \end{cases}$

15. $g(t) = \begin{cases} t^2 - 4 & \text{if } t < 2 \\ 4 & \text{if } t = 2 \\ 4 - t^2 & \text{if } 2 < t \end{cases}$

16. $H(x) = \begin{cases} 1 + x & \text{if } x \le -2 \\ 2 - x & \text{if } -2 < x \le 2 \\ 2x - 1 & \text{if } 2 < x \end{cases}$

17. $g(x) = \begin{cases} \sqrt{-x} & \text{if } x < 0 \\ \sqrt[3]{x + 1} & \text{if } 0 \le x \end{cases}$

18. $f(x) = \begin{cases} |x + 2| & \text{if } x \ne -2 \\ 3 & \text{if } x = -2 \end{cases}$

19. $f(x) = \dfrac{|x|}{x}$

20. $g(x) = \begin{cases} \dfrac{|x|}{x} & \text{if } x \ne 0 \\ 1 & \text{if } x = 0 \end{cases}$

21. The greatest integer function (see Illustration 2 in Sec. 1.1).
22. The signum function (see Example 1 in Sec. 1.4).

In Exercises 23 through 32, prove that the function is discontinuous at the number a. Then determine if the discontinuity is removable or essential. If the discontinuity is removable, redefine $f(a)$ so that the discontinuity is removed.

23. $f(x) = \dfrac{9x^2 - 4}{3x - 2}; \; a = \dfrac{2}{3}$

24. $f(s) = \begin{cases} \dfrac{1}{s + 5} & \text{if } s \ne -5 \\ 0 & \text{if } s = -5 \end{cases}; \; a = -5$

25. $f(t) = \begin{cases} 9 - t^2 & \text{if } t \le 2 \\ 3t + 2 & \text{if } 2 < t \end{cases}; \; a = 2$

26. $f(x) = \dfrac{x^2 - x - 12}{x^2 + 2x - 3}; \; a = -3$

27. $f(x) = \begin{cases} |x - 3| & \text{if } x \ne 3 \\ 2 & \text{if } x = 3 \end{cases}; \; a = 3$

28. $f(x) = \begin{cases} \dfrac{x^2 - 4x + 3}{x - 3} & \text{if } x \ne 3 \\ 5 & \text{if } x = 3 \end{cases}; \; a = 3$

29. $f(t) = \begin{cases} t^2 - 4 & \text{if } t \le 2 \\ t & \text{if } 2 < t \end{cases}; \; a = 2$

30. $f(y) = \dfrac{\sqrt{y + 5} - \sqrt{5}}{y}; \; a = 0$

31. $f(x) = \dfrac{3 - \sqrt{x + 9}}{x}; \; a = 0$

32. $f(x) = \dfrac{\sqrt[3]{x + 1} - 1}{x}; \; a = 0$

In Exercises 33 through 42, determine the numbers at which the given function is continuous.

33. $f(x) = x^2(x + 3)^2$

34. $f(x) = (x - 5)^3(x^2 + 4)^5$

35. $g(x) = \dfrac{x}{x - 3}$

36. $h(x) = \dfrac{x + 1}{2x + 5}$

37. $F(x) = \dfrac{x^3 + 7}{x^2 - 4}$

38. $G(x) = \dfrac{x - 2}{x^2 + 2x - 8}$

39. $f(x) = \begin{cases} 3x - 1 & \text{if } x < 2 \\ 4 - x^2 & \text{if } 2 \le x \end{cases}$

40. $f(x) = \begin{cases} (x + 2)^2 & \text{if } x \le 0 \\ x^2 + 2 & \text{if } 0 < x \end{cases}$

41. $f(x) = \begin{cases} \dfrac{1}{x - 2} & \text{if } x \le 1 \\ \dfrac{1}{x} & \text{if } 1 < x \end{cases}$

42. $f(x) = \begin{cases} \dfrac{1}{x + 1} & \text{if } x < 4 \\ \sqrt{x - 4} & \text{if } 4 \le x \end{cases}$

43. The function C of Exercise 35 in Exercises 1.4 is defined by

$$C(x) = \begin{cases} 0.80x & \text{if } 0 < x \le 50 \\ 0.70x & \text{if } 50 < x \le 200 \\ 0.65x & \text{if } 200 < x \end{cases}$$

(a) Draw a sketch of the graph of C. (b) At what numbers is C discontinuous? (c) Show why Definition 1.6.1 is not satisfied at each discontinuity in part (b).

44. Suppose the postage of a letter is computed as follows: 22 cents for the first ounce or less, then 17 cents for each ounce (or fractional part of an ounce) for the next 11 oz. If x ounces is the weight of the letter and $0 < x \leq 12$, express the number of cents in the postage as a function of x. (a) Draw a sketch of the graph of this function. (b) At what numbers in the open interval $(0, 12)$ is the function discontinuous? (c) Show why Definition 1.6.1 is not satisfied at each discontinuity in part (b).

45. Let f be the function defined by

$$f(x) = \begin{cases} |x - [\![x]\!]| & \text{if } [\![x]\!] \text{ is even} \\ |x - [\![x + 1]\!]| & \text{if } [\![x]\!] \text{ is odd} \end{cases}$$

Draw a sketch of the graph of f. At what numbers is f discontinuous?

46. The function f defined by

$$f(x) = \frac{\sqrt{2 + \sqrt[3]{x}} - 2}{x - 8}$$

is discontinuous at 8. Show that the discontinuity is removable, and redefine $f(8)$ so that the discontinuity is removed.

47. The function g defined by

$$g(x) = \frac{\sqrt[3]{x + a^3} - a}{x}$$

is discontinuous at 0. Show that the discontinuity is removable, and redefine $g(0)$ so that the discontinuity is removed.

48. The function f is defined by

$$f(x) = \lim_{n \to 0} \frac{2nx}{n^2 - nx}$$

Draw a sketch of the graph of f. At what values of x is f discontinuous?

49. If

$$f(x) = \begin{cases} -x & \text{if } x < 0 \\ 1 & \text{if } 0 \leq x \end{cases} \quad \text{and} \quad g(x) = \begin{cases} 1 & \text{if } x < 0 \\ x & \text{if } 0 \leq x \end{cases}$$

prove that f and g are both discontinuous at 0 but that the product $f \cdot g$ is continuous at 0.

50. Give an example to show that the product of two functions f and g may be continuous at a number a, where f is continuous at a but g is discontinuous at a.

51. Give an example of two functions that are both discontinuous at a number a but whose sum is continuous at a.

52. Prove that if f is continuous at a and g is discontinuous at a, then $f + g$ is discontinuous at a.

1.7

CONTINUITY OF A COMPOSITE FUNCTION AND CONTINUITY ON AN INTERVAL

In Sec. 0.5 a composite function was defined as follows:

Given the two functions f and g, the composite function, denoted by $f \circ g$, is defined by

$$(f \circ g)(x) = f(g(x))$$

and the domain of $f \circ g$ is the set of all numbers in the domain of g such that $g(x)$ is in the domain of f.

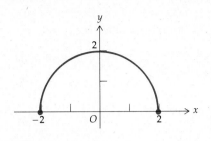

FIGURE 1.7.1

▶ ILLUSTRATION 1   If $f(x) = \sqrt{x}$ and $g(x) = 4 - x^2$, and if h is the composite function $f \circ g$, then

$$h(x) = f(g(x))$$
$$= f(4 - x^2)$$
$$= \sqrt{4 - x^2}$$

Because the domain of g is the set of all real numbers and the domain of f is the set of all nonnegative numbers, the domain of h is the set of all real numbers such that $4 - x^2 \geq 0$, that is, all numbers in the closed interval $[-2, 2]$. A sketch of the graph of h is shown in Fig. 1.7.1.   ◀

From Fig. 1.7.1 it appears that h is continuous at every number in the open interval $(-2, 2)$. We show later in Example 1 how this fact can be proved by theorems on continuity. But first we need an important theorem regarding the limit of a composite function. The proof of this theorem makes use of Theorem 1.6.6, which is the definition of continuity involving ϵ and δ notation.

1.7.1 THEOREM   If $\lim\limits_{x \to a} g(x) = b$ and if the function f is continuous at b,

$$\lim_{x \to a} (f \circ g)(x) = f(b)$$

$$\Leftrightarrow \quad \lim_{x \to a} f(g(x)) = f(\lim_{x \to a} g(x))$$

PROOF   Because f is continuous at b, we have the following statement from Theorem 1.6.6. For any $\epsilon_1 > 0$ there exists a $\delta_1 > 0$ such that

$$\text{if} \quad |y - b| < \delta_1 \quad \text{then} \quad |f(y) - f(b)| < \epsilon_1 \tag{1}$$

Because $\lim\limits_{x \to a} g(x) = b$, for any $\delta_1 > 0$ there exists a $\delta_2 > 0$ such that

$$\text{if} \quad 0 < |x - a| < \delta_2 \quad \text{then} \quad |g(x) - b| < \delta_1 \tag{2}$$

If $0 < |x - a| < \delta_2$, we replace y in statement (1) by $g(x)$ and obtain the following: For any $\epsilon_1 > 0$ there exists a $\delta_1 > 0$ such that

$$\text{if} \quad |g(x) - b| < \delta_1 \quad \text{then} \quad |f(g(x)) - f(b)| < \epsilon_1 \tag{3}$$

From statements (3) and (2) we conclude that for any $\epsilon_1 > 0$ there exists a $\delta_2 > 0$ such that

$$\text{if} \quad 0 < |x - a| < \delta_2 \quad \text{then} \quad |f(g(x)) - f(b)| < \epsilon_1$$

from which it follows that

$$\lim_{x \to a} f(g(x)) = f(b)$$

$$\Leftrightarrow \quad \lim_{x \to a} f(g(x)) = f(\lim_{x \to a} g(x)) \qquad \blacksquare$$

Theorem 1.7.1 is applied in Supplementary Section 1.9 to prove Limit Theorems 9 and 10. Another application of Theorem 1.7.1 is in proving the following theorem about the continuity of a composite function.

1.7.2 THEOREM   If the function g is continuous at a and the function f is continuous at $g(a)$, then the composite function $f \circ g$ is continuous at a.

PROOF   Because g is continuous at a,

$$\lim_{x \to a} g(x) = g(a) \qquad\qquad (4)$$

Now f is continuous at $g(a)$; thus we can apply Theorem 1.7.1 to the composite function $f \circ g$, thereby giving

$$\lim_{x \to a} (f \circ g)(x) = \lim_{x \to a} f(g(x))$$
$$= f(\lim_{x \to a} g(x))$$
$$= f(g(a)) \quad \text{(by Eq. (4))}$$
$$= (f \circ g)(a)$$

which proves that $f \circ g$ is continuous at a.   ∎

Theorem 1.7.2 states that *a continuous function of a continuous function is continuous*. The following example shows how it is used to determine the numbers for which a particular function is continuous.

EXAMPLE 1
If

$$h(x) = \sqrt{4 - x^2}$$

determine the values of x for which h is continuous.

SOLUTION
The function h is the one obtained in Illustration 1 as the composite function $f \circ g$, where $f(x) = \sqrt{x}$ and $g(x) = 4 - x^2$. Because g is a polynomial function, it is continuous everywhere. Furthermore, f is continuous at every positive number by Theorem 1.6.5(ii). Therefore, by Theorem 1.7.2, h is continuous at every number x for which $g(x) > 0$, that is, when $4 - x^2 > 0$. Hence h is continuous at every number in the open interval $(-2, 2)$.

Because the function h of Example 1 is continuous at every number in the open interval $(-2, 2)$, we say that h is *continuous on the open interval* $(-2, 2)$.

1.7.3 DEFINITION   A function is said to be **continuous on an open interval** if and only if it is continuous at every number in the open interval.

We refer again to the function h of Example 1. Because h is not defined on any open interval containing either -2 or 2, we cannot consider $\lim_{x \to -2} h(x)$

or $\lim\limits_{x \to 2} h(x)$. Hence, to discuss the question of the continuity of h on the closed interval $[-2, 2]$, we must extend the concept of continuity to include continuity at an endpoint of a closed interval. This is done by first defining *right-hand continuity* and *left-hand continuity*.

1.7.4 DEFINITION   The function f is said to be **continuous from the right at the number a** if and only if the following three conditions are satisfied.

(i) $f(a)$ exists;

(ii) $\lim\limits_{x \to a^+} f(x)$ exists;

(iii) $\lim\limits_{x \to a^+} f(x) = f(a)$.

1.7.5 DEFINITION   The function f is said to be **continuous from the left at the number a** if and only if the following three conditions are satisfied:

(i) $f(a)$ exists;

(ii) $\lim\limits_{x \to a^-} f(x)$ exists;

(iii) $\lim\limits_{x \to a^-} f(x) = f(a)$.

1.7.6 DEFINITION   A function whose domain includes the closed interval $[a, b]$ is said to be **continuous on $[a, b]$** if and only if it is continuous on the open interval (a, b), as well as continuous from the right at a and continuous from the left at b.

EXAMPLE 2
Prove that the function h of Example 1 is continuous on the closed interval $[-2, 2]$.

SOLUTION
The function h is defined by

$$h(x) = \sqrt{4 - x^2}$$

and in Example 1 we showed that h is continuous on the open interval $(-2, 2)$. By applying Theorem 1.7.1 we compute $\lim\limits_{x \to -2^+} h(x)$ and $\lim\limits_{x \to 2^-} h(x)$.

$$\lim_{x \to -2^+} h(x) = \lim_{x \to -2^+} \sqrt{4 - x^2} \qquad \lim_{x \to 2^-} h(x) = \lim_{x \to 2^-} \sqrt{4 - x^2}$$
$$= 0 \qquad\qquad\qquad\qquad = 0$$
$$= h(-2) \qquad\qquad\qquad = h(2)$$

Thus h is continuous from the right at -2 and continuous from the left at 2. Hence by Definition 1.7.6, h is continuous on the closed interval $[-2, 2]$.

A sketch of the graph of h appears in Fig. 1.7.1.

1.7.7 DEFINITION   **(i)** A function whose domain includes the interval half-open on the right $[a, b)$ is said to be **continuous on $[a, b)$** if and only if it is continuous on the open interval (a, b) and continuous from the right at a.

(ii) A function whose domain includes the interval half-open on the left $(a, b]$ is said to be **continuous on $(a, b]$** if and only if it is continuous on the open interval (a, b) and continuous from the left at b.

Definitions similar to those in Definition 1.7.7 can be given for continuity on the intervals $[a, +\infty)$ and $(-\infty, b]$.

EXAMPLE 3
Given

$$f(x) = \sqrt{\frac{2 - x}{3 + x}}$$

(a) Determine the domain of f. (b) Determine whether f is continuous or discontinuous on each of the following intervals: $(-3, 2)$, $[-3, 2]$, $[-3, 2)$, $(-3, 2]$.

SOLUTION
(a) Let g and h be the functions defined by

$$g(x) = \sqrt{x} \quad \text{and} \quad h(x) = \frac{2 - x}{3 + x}$$

The domain of g is the set of nonnegative numbers, and the domain of h is the set of all real numbers except -3.

$$(g \circ h)(x) = g(h(x))$$

$$= g\left(\frac{2 - x}{3 + x}\right)$$

$$= \sqrt{\frac{2 - x}{3 + x}}$$

Therefore

$$f(x) = (g \circ h)(x)$$

We wish to determine the domain of f. The domain of f is the set of all numbers for which $h(x)$ is nonnegative, that is, all numbers for which

$$\frac{2 - x}{3 + x} \geq 0 \tag{5}$$

To determine the values of x that satisfy (5), refer to Table 1.7.1. In the table we consider the intervals for which $x < -3$, $-3 < x < 2$, and $2 < x$, because the numerator changes sign when $x = 2$ and the denominator changes sign when $x = -3$.

Table 1.7.1

	$2 - x$	$3 + x$	$\dfrac{2 - x}{3 + x}$	$f(x) = \sqrt{\dfrac{2 - x}{3 + x}}$
$x < -3$	+	−	−	does not exist
$x = -3$	5	0	undefined	does not exist
$-3 < x < 2$	+	+	+	+
$x = 2$	0	5	0	0
$2 < x$	−	+	−	does not exist

From the table observe that $f(x)$ exists if x is in the interval half-open on the left $(-3, 2]$; so this interval is the domain of f.

(b) Because h is a rational function, it follows from Theorem 1.6.4 that h is continuous at every number in its domain, that is, at all numbers except -3. From Theorem 1.6.5(ii), the function g is continuous at all positive numbers. Thus from Theorem 1.7.2 it follows that f is continuous at all values of x for which

$$\frac{2 - x}{3 + x} > 0$$

From Table 1.7.1 these are the numbers in the open interval $(-3, 2)$, and so f is continuous on that interval.

Because -3 is not in the domain of f, the function is discontinuous on the closed interval $[-3, 2]$ and the interval $[-3, 2)$. Because

$$\lim_{x \to 2^-} \sqrt{\frac{2 - x}{3 + x}} = 0$$

$$= f(2)$$

it follows that f is continuous from the left at 2. Therefore f is continuous on $(-3, 2]$.

FIGURE 1.7.2

FIGURE 1.7.3

EXAMPLE 4

A cardboard box manufacturer wishes to make open boxes from square pieces of cardboard of side 12 in. by cutting equal squares from the four corners and turning up the sides. (a) If x in. is the length of the side of the square to be cut out, express the number of cubic inches in the volume of the box as a function of x. (b) What is the domain of the function? (c) Prove that the function is continuous on its domain.

SOLUTION

(a) Figure 1.7.2 represents a given piece of cardboard, and Fig. 1.7.3 represents the box obtained from the cardboard. The numbers of inches in the dimensions of the box are then x, $12 - 2x$, and $12 - 2x$. The volume of the box is the product of the three dimensions. Therefore, if $V(x)$ cubic

inches is the volume of the box,

$$V(x) = x(12 - 2x)(12 - 2x)$$
$$V(x) = 144x - 48x^2 + 4x^3 \tag{6}$$

(b) From (6) we note that $V(0) = 0$ and $V(6) = 0$. From the conditions of the problem we see that x cannot be negative and x cannot be greater than 6. Thus the domain of V is the closed interval $[0, 6]$.

(c) Because V is a polynomial function, V is continuous everywhere. Thus V is continuous on the closed interval $[0, 6]$.

The function of Example 4 is discussed again in Secs. 2.1 and 3.2 where we use the fact that V is continuous on the closed interval $[0, 6]$ to determine the value of x that will give a box having the largest possible volume.

We now discuss an important theorem about a function that is continuous on a closed interval. It is called the **intermediate-value theorem**, and it is needed in later sections.

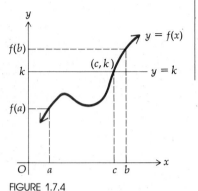

FIGURE 1.7.4

1.7.8 THEOREM

Intermediate-Value Theorem

If the function f is continuous on the closed interval $[a, b]$, and if $f(a) \neq f(b)$, then for any number k between $f(a)$ and $f(b)$ there exists a number c between a and b such that $f(c) = k$.

The proof of this theorem is beyond the scope of this book; it can be found in an advanced calculus text. However, we discuss the geometric interpretation of the theorem. In Fig. 1.7.4, $(0, k)$ is any point on the y axis between the points $(0, f(a))$ and $(0, f(b))$. Theorem 1.7.8 states that the line $y = k$ must intersect the curve whose equation is $y = f(x)$ at the point (c, k), where c lies between a and b. Figure 1.7.4 shows this intersection.

Note that for some values of k there may be more than one possible value for c. The theorem states that there is always at least one value of c but that it is not necessarily unique. Figure 1.7.5 shows three possible values of c (c_1, c_2, and c_3) for a particular k. Theorem 1.7.8 states that if the function f is continuous on a closed interval $[a, b]$, then f assumes every value between $f(a)$ and $f(b)$ as x assumes all values between a and b. The importance of the continuity of f on $[a, b]$ is demonstrated in the following illustration.

FIGURE 1.7.5

▶ ILLUSTRATION 2   Consider the function f defined by

$$f(x) = \begin{cases} x - 1 & \text{if } 0 \leq x \leq 2 \\ x^2 & \text{if } 2 < x \leq 3 \end{cases}$$

A sketch of the graph of this function is shown in Fig. 1.7.6.

The function f is discontinuous at 2, which is in the closed interval $[0, 3]$; $f(0) = -1$ and $f(3) = 9$. If k is any number between 1 and 4, there is no value of c such that $f(c) = k$ because there are no function values between 1 and 4.   ◀

FIGURE 1.7.6

Another function for which Theorem 1.7.8 does not hold is given in the next illustration.

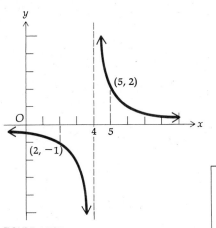

FIGURE 1.7.7

▶ ILLUSTRATION 3   Let the function g be defined by

$$g(x) = \frac{2}{x - 4}$$

A sketch of the graph of this function is shown in Fig. 1.7.7.

The function g is discontinuous at 4, which is in the closed interval $[2, 5]$; $g(2) = -1$ and $g(5) = 2$. If k is any number between -1 and 2, there is no value of c between 2 and 5 such that $g(c) = k$. In particular, if $k = 1$, then $g(6) = 1$, but 6 is not in the interval $[2, 5]$.   ◀

EXAMPLE 5

Given the function f defined by

$$f(x) = 4 + 3x - x^2 \qquad 2 \le x \le 5$$

(a) Verify the intermediate-value theorem if $k = 1$; that is, find a number c in the interval $[2, 5]$ such that $f(c) = 1$. (b) Draw a sketch of the graph of f on $[2, 5]$, and show the point $(c, 1)$.

SOLUTION

(a) Because f is a polynomial function, it is continuous everywhere, and thus continuous on $[2, 5]$. Since $f(2) = 6$ and $f(5) = -6$, the intermediate-value theorem guarantees that there is a number c between 2 and 5 such that $f(c) = 1$; that is,

$$4 + 3c - c^2 = 1$$

$$c^2 - 3c - 3 = 0$$

$$c = \frac{3 \pm \sqrt{9 + 12}}{2}$$

$$c = \frac{3 \pm \sqrt{21}}{2}$$

We reject $\frac{1}{2}(3 - \sqrt{21})$ because this number is outside the interval $[2, 5]$. The number $\frac{1}{2}(3 + \sqrt{21})$ is in the interval $[2, 5]$, and

$$f\left(\frac{3 + \sqrt{21}}{2}\right) = 1$$

(b) The required sketch appears in Fig. 1.7.8.

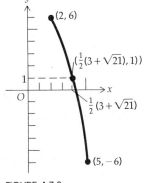

FIGURE 1.7.8

EXERCISES 1.7

In Exercises 1 through 10, functions f and g are defined. In each exercise define $f \circ g$ and determine the numbers at which $f \circ g$ is continuous.

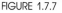

1. $f(x) = \sqrt{x}$; $g(x) = x^2 - 16$

4. $f(x) = \sqrt{x}$; $g(x) = x^2 + 4$

2. $f(x) = \sqrt{x}$; $g(x) = 16 - x^2$

5. $f(x) = x^3$; $g(x) = \sqrt{x}$

3. $f(x) = \sqrt[4]{x}$; $g(x) = x^2 + 1$

6. $f(x) = x^2$; $g(x) = x^2 - 3$

7. $f(x) = \sqrt{x}$; $g(x) = \dfrac{1}{x-2}$ **8.** $f(x) = \sqrt{x+1}$; $g(x) = \sqrt[3]{x}$ **9.** $f(x) = \dfrac{1}{x-2}$; $g(x) = \sqrt{x}$

10. $f(x) = \sqrt[3]{x}$; $g(x) = \sqrt{x+1}$

In Exercises 11 through 20, find the domain of the given function, and then determine whether the function is continuous or discontinuous on each of the indicated intervals.

11. $f(x) = \dfrac{2}{x+5}$; $(3,7)$, $[-6,4]$, $(-\infty,0)$, $(-5,+\infty)$, $[-5,+\infty)$, $[-10,-5)$

12. $g(x) = \dfrac{x}{x-2}$; $(-\infty,0]$, $[0,+\infty)$, $(0,2)$, $(0,2]$, $[2,+\infty)$, $(2,+\infty)$

13. $f(t) = \dfrac{t}{t^2-1}$; $(0,1)$, $(-1,1)$, $[0,1]$, $(-1,0]$, $(-\infty,-1]$, $(1,+\infty)$

14. $f(r) = \dfrac{r+3}{r^2-4}$; $(0,4]$, $(-2,2)$, $(-\infty,-2]$, $(2,+\infty)$, $[-4,4]$, $(-2,2]$

15. $g(x) = \sqrt{x^2-9}$; $(-\infty,-3)$, $(-\infty,-3]$, $(3,+\infty)$, $[3,+\infty)$, $(-3,3)$

16. $f(x) = [\![x]\!]$; $(-\frac{1}{2},\frac{1}{2})$, $(\frac{1}{4},\frac{1}{2})$, $(1,2)$, $[1,2)$, $(1,2]$

17. $f(t) = \dfrac{|t-1|}{t-1}$; $(-\infty,1)$, $(-\infty,1]$, $[-1,1]$, $(-1,+\infty)$, $(1,+\infty)$

18. $h(x) = \begin{cases} 2x-3 & \text{if } x < -2 \\ x-5 & \text{if } -2 \le x \le 1 \\ 3-x & \text{if } 1 < x \end{cases}$; $(-\infty,1)$, $(-2,+\infty)$, $(-2,1)$, $[-2,1)$, $[-2,1]$

19. $f(x) = \sqrt{4-x^2}$; $(-2,2)$, $[-2,2]$, $[-2,2)$, $(-2,2]$, $(-\infty,-2]$, $(2,+\infty)$

20. $F(y) = \dfrac{1}{3+2y-y^2}$; $(-1,3)$, $[-1,3]$, $[-1,3)$, $(-1,3]$

In Exercises 21 through 26, the given function is continuous on an unlimited number of intervals. For each function determine the interval (or union of intervals) having all of these intervals as subsets.

21. (a) $f(x) = \sqrt{9-x^2}$; (b) $g(x) = \sqrt{\dfrac{x+4}{x-4}}$ **22.** (a) $g(x) = \sqrt{x^2-9}$; (b) $f(x) = \sqrt{\dfrac{4-x}{4+x}}$

23. (a) $h(x) = \sqrt{\dfrac{2-x}{5+x}}$; (b) $f(x) = |x-5|$ **24.** (a) $F(x) = \sqrt{6x^2-x-2}$; (b) $f(x) = |9-x^2|$

25. (a) $g(x) = \dfrac{|x|}{x}$; (b) $f(x) = x^2(x^{-2} + x^{-1/2})^3$ **26.** (a) $h(x) = \dfrac{x-1}{|x-1|}$; (b) $G(x) = \left(\dfrac{x^2}{x^2-4} - \dfrac{1}{x}\right)^{1/3}$

In Exercises 27 through 30, draw a sketch of the graph of a function f that satisfies the given conditions.

27. f is continuous on $(-\infty,2]$ and $(2,+\infty)$; $\lim\limits_{x \to 0} f(x) = 4$; $\lim\limits_{x \to 2^-} f(x) = -3$; $\lim\limits_{x \to 2^+} f(x) = +\infty$; $\lim\limits_{x \to 5} f(x) = 0$

28. f is continuous on $(-\infty,0)$ and $[0,+\infty)$; $\lim\limits_{x \to -4} f(x) = 0$; $\lim\limits_{x \to 0^-} f(x) = 3$; $\lim\limits_{x \to 0^+} f(x) = -3$; $\lim\limits_{x \to 4} f(x) = 2$

29. f is continuous on $(-\infty,-3]$, $(-3,3)$, and $[3,+\infty)$; $\lim\limits_{x \to -5} f(x) = 2$; $\lim\limits_{x \to -3^-} f(x) = 0$; $\lim\limits_{x \to -3^+} f(x) = 4$; $\lim\limits_{x \to 0} f(x) = 1$; $\lim\limits_{x \to 3^-} f(x) = 0$; $\lim\limits_{x \to 3^+} f(x) = -5$; $\lim\limits_{x \to 4} f(x) = 0$

30. f is continuous on $(-\infty, -2), [-2,4],$ and $(4, +\infty);$ $\lim\limits_{x \to -5} f(x) = 0;$ $\lim\limits_{x \to -2^-} f(x) = -\infty;$ $\lim\limits_{x \to -2^+} f(x) = -3;$ $\lim\limits_{x \to 0} f(x) = -1;$
$\lim\limits_{x \to 4^-} f(x) = 2;$ $\lim\limits_{x \to 4^+} f(x) = 5;$ $\lim\limits_{x \to 6} f(x) = 0$

31. Prove that the function of Exercise 17 in Exercises 1.4 is continuous on the closed interval $[-2,2]$.

32. Prove that the function of Example 4 in Sec. 1.4 is continuous on the closed interval $[-3,3]$.

33. A manufacturer of open tin boxes wishes to make use of pieces of tin with dimensions 8 in. by 15 in. by cutting equal squares from the four corners and turning up the sides. (a) If x in. is the length of the side of the square cut out, express the number of cubic inches in the volume of the box as a function of x. (b) What is the domain of the function? (c) Prove that the function is continuous on its domain.

34. Suppose that the manufacturer of Exercise 33 makes the open boxes from square pieces of tin that measure k cm on a side. (a) If x cm is the length of the side of the square cut out, express the number of cubic centimeters in the volume of the box as a function of x. (b) What is the domain of the function? (c) Prove that the function is continuous on its domain.

35. A rectangular field is to be enclosed with 240 m of fence. (a) If x meters is the length of the field, express the number of square meters in the area of the field as a function of x. (b) What is the domain of the function? (c) Prove that the function is continuous on its domain.

36. A rectangular garden is to be placed so that a side of a house serves as a boundary and 100 ft of fencing material is to be used for the other three sides. (a) If x ft is the length of the side of the garden that is parallel to the house, express the number of square feet in the area of the garden as a function of x. (b) What is the domain of the function? (c) Prove that the function is continuous on its domain.

In Exercises 37 through 40, find the values of the constants c and k that make the function continuous on $(-\infty, +\infty)$, and draw a sketch of the graph of the resulting function.

37. $f(x) = \begin{cases} 3x + 7 & \text{if } x \le 4 \\ kx - 1 & \text{if } 4 < x \end{cases}$

38. $f(x) = \begin{cases} kx - 1 & \text{if } x < 2 \\ kx^2 & \text{if } 2 \le x \end{cases}$

39. $f(x) = \begin{cases} x & \text{if } x \le 1 \\ cx + k & \text{if } 1 < x < 4 \\ -2x & \text{if } 4 \le x \end{cases}$

40. $f(x) = \begin{cases} x + 2c & \text{if } x < -2 \\ 3cx + k & \text{if } -2 \le x \le 1 \\ 3x - 2k & \text{if } 1 < x \end{cases}$

In Exercises 41 through 48, a function f and a closed interval $[a,b]$ are given. Determine if the intermediate-value theorem holds for the given value of k. If the theorem holds, find a number c such that $f(c) = k$. If the theorem does not hold, give the reason. Draw a sketch of the curve and the line $y = k$.

41. $f(x) = 2 + x - x^2;$ $[a,b] = [0,3];$ $k = 1$

42. $f(x) = x^2 + 5x - 6;$ $[a,b] = [-1,2];$ $k = 4$

43. $f(x) = \sqrt{25 - x^2};$ $[a,b] = [-4.5, 3];$ $k = 3$

44. $f(x) = -\sqrt{100 - x^2};$ $[a,b] = [0,8];$ $k = -8$

45. $f(x) = \dfrac{4}{x + 2};$ $[a,b] = [-3, 1];$ $k = \frac{1}{2}$

46. $f(x) = \begin{cases} 1 + x & \text{if } -4 \le x \le -2 \\ 2 - x & \text{if } -2 < x \le 1 \end{cases};$ $[a,b] = [-4, 1];$ $k = \frac{1}{2}$

47. $f(x) = \begin{cases} x^2 - 4 & \text{if } -2 \le x < 1 \\ x^2 - 1 & \text{if } 1 \le x \le 3 \end{cases};$ $[a,b] = [-2, 3];$ $k = -1$

48. $f(x) = \dfrac{5}{2x - 1};$ $[a,b] = [0, 1];$ $k = 2$

49. Given that f is defined by

$$f(x) = \begin{cases} g(x) & \text{if } a \le x < b \\ h(x) & \text{if } b \le x \le c \end{cases}$$

If g is continuous on $[a, b)$, and h is continuous on $[b, c]$, can we conclude that f is continuous on $[a, c]$? If your answer is yes, prove it. If your answer is no, what additional condition or conditions would assure continuity of f on $[a, c]$?

50. If f is the function defined by $f(x) = \sqrt[n]{x^m}$, where m and n are positive integers, prove that f is continuous on the interval $[0, +\infty)$.

51. Find the largest value of k for which the function defined by $f(x) = [\![x^2 - 2]\!]$ is continuous on the interval $[3, 3 + k)$.

52. Suppose that f is a function for which $0 \leq f(x) \leq 1$ if $0 \leq x \leq 1$. Prove that if f is continuous on $[0, 1]$, there is at least one number c in $[0, 1]$ such that $f(c) = c$. (*Hint:* If neither 0 nor 1 qualifies as c, then $f(0) > 0$ and $f(1) < 1$. Consider the function g for which $g(x) = f(x) - x$, and apply the intermediate-value theorem to g on $[0, 1]$.)

53. Show that the intermediate-value theorem guarantees that the equation $x^3 - 4x^2 + x + 3 = 0$ has a root between 1 and 2.

54. Show that the intermediate-value theorem guarantees that the equation $x^3 + x + 3 = 0$ has a root between -2 and -1.

55. Prove that if the function f is continuous at a, then $\lim\limits_{t \to 0} f(a - t) = f(a)$.

1.8
CONTINUITY OF THE TRIGONOMETRIC FUNCTIONS AND THE SQUEEZE THEOREM

In our discussion of the continuity of the trigonometric functions we make use of the following limit:

$$\lim_{t \to 0} \frac{\sin t}{t} \tag{1}$$

It is apparent that the function given by $\dfrac{\sin t}{t}$ is not defined when $t = 0$. To get an intuitive idea about the existence of the limit in (1), we consider values of $\dfrac{\sin t}{t}$ when t is close to 0. In Table 1.8.1 there are values of the function when t is 1.0, 0.9, 0.8, ..., 0.1, and 0.01, and in Table 1.8.2 there are values of the function when t is -1.0, -0.9, -0.8, ..., -0.1, and -0.01. These values are easily computed on a calculator.

From the two tables it appears that if the limit in (1) exists, it may be equal to 1. That the limit does exist and is equal to 1 is proved in Theorem 1.8.2, but in the proof of that theorem we need to use the following theorem, sometimes referred to as the **squeeze theorem**. The squeeze theorem is not only important in the proof of Theorem 1.8.2 but is used to prove some major theorems in later sections.

Table 1.8.1

t	$\dfrac{\sin t}{t}$
1.0	0.84147
0.9	0.87036
0.8	0.89670
0.7	0.92031
0.6	0.94107
0.5	0.95885
0.4	0.97355
0.3	0.98507
0.2	0.99335
0.1	0.99833
0.01	0.99998

Table 1.8.2

t	$\dfrac{\sin t}{t}$
-1.0	0.84147
-0.9	0.87036
-0.8	0.89670
-0.7	0.92031
-0.6	0.94107
-0.5	0.95885
-0.4	0.97355
-0.3	0.98507
-0.2	0.99335
-0.1	0.99833
-0.01	0.99998

1.8.1 THEOREM
The Squeeze Theorem

Suppose that the functions f, g, and h are defined on some open interval I containing a except possibly at a itself, and that $f(x) \leq g(x) \leq h(x)$ for all x in I for which $x \neq a$. Also suppose that $\lim\limits_{x \to a} f(x)$ and $\lim\limits_{x \to a} h(x)$ both exist and are equal to L. Then $\lim\limits_{x \to a} g(x)$ exists and is equal to L.

Before Theorem 1.8.1 is proved, consider the following illustration, which interprets the theorem geometrically.

▶ ILLUSTRATION 1   Let the functions f, g, and h be defined by

$$f(x) = -4(x - 2)^2 + 3$$

$$g(x) = \frac{(x - 2)(x^2 - 4x + 7)}{x - 2}$$

$$h(x) = 4(x - 2)^2 + 3$$

FIGURE 1.8.1

The graphs of f and h are parabolas having their vertex at $(2, 3)$. The graph of g is a parabola with its vertex $(2, 3)$ deleted. Sketches of these graphs are shown in Fig. 1.8.1. The function g is not defined when $x = 2$; however, for all $x \neq 2$, $f(x) \leq g(x) \leq h(x)$. Furthermore, $\lim_{x \to 2} f(x) = 3$ and $\lim_{x \to 2} h(x) = 3$. The hypothesis of Theorem 1.8.1 is therefore satisfied, and it follows that $\lim_{x \to 2} g(x) = 3$.   ◀

PROOF OF THEOREM 1.8.1   To prove that $\lim_{x \to a} g(x) = L$ we must show that for any $\epsilon > 0$ there is a $\delta > 0$ such that

$$\text{if} \quad 0 < |x - a| < \delta \quad \text{then} \quad |g(x) - L| < \epsilon \quad \quad (2)$$

It is given that

$$\lim_{x \to a} f(x) = L \quad \text{and} \quad \lim_{x \to a} h(x) = L$$

and so for any $\epsilon > 0$ there is a $\delta_1 > 0$ such that

$$\text{if} \quad 0 < |x - a| < \delta_1 \quad \text{then} \quad |f(x) - L| < \epsilon$$
$$\Leftrightarrow \quad \text{if} \quad 0 < |x - a| < \delta_1 \quad \text{then} \quad L - \epsilon < f(x) < L + \epsilon \quad \quad (3)$$

and a $\delta_2 > 0$ such that

$$\text{if} \quad 0 < |x - a| < \delta_2 \quad \text{then} \quad |h(x) - L| < \epsilon$$
$$\Leftrightarrow \quad \text{if} \quad 0 < |x - a| < \delta_2 \quad \text{then} \quad L - \epsilon < h(x) < L + \epsilon \quad \quad (4)$$

Let $\delta = \min(\delta_1, \delta_2)$, and so $\delta \leq \delta_1$ and $\delta \leq \delta_2$. Therefore it follows from statement (3) that

$$\text{if} \quad 0 < |x - a| < \delta \quad \text{then} \quad L - \epsilon < f(x) \quad \quad (5)$$

and from statement (4) that

$$\text{if} \quad 0 < |x - a| < \delta \quad \text{then} \quad h(x) < L + \epsilon \quad \quad (6)$$

We are given that

$$f(x) \leq g(x) \leq h(x) \quad \quad (7)$$

From statements (5), (6), and (7),

$$\text{if} \quad 0 < |x - a| < \delta \quad \text{then} \quad L - \epsilon < f(x) \leq g(x) \leq h(x) < L + \epsilon$$

Therefore

$$\text{if} \quad 0 < |x - a| < \delta \quad \text{then} \quad L - \epsilon < g(x) < L + \epsilon$$
$$\Leftrightarrow \quad \text{if} \quad 0 < |x - a| < \delta \quad \text{then} \quad |g(x) - L| < \epsilon$$

which is statement (2). Hence

$$\lim_{x \to a} g(x) = L \quad \quad \blacksquare$$

EXAMPLE 1

Given $|g(x) - 2| \leq 3(x - 1)^2$ for all x. Use the squeeze theorem to find $\lim\limits_{x \to 1} g(x)$.

SOLUTION

Because $|g(x) - 2| \leq 3(x - 1)^2$ for all x, it follows from Theorem 0.1.10 that

$$-3(x - 1)^2 \leq g(x) - 2 \leq 3(x - 1)^2 \qquad \text{for all } x$$

$$\Leftrightarrow \quad -3(x - 1)^2 + 2 \leq g(x) \leq 3(x - 1)^2 + 2 \qquad \text{for all } x$$

Let $f(x) = -3(x - 1)^2 + 2$ and $h(x) = 3(x - 1)^2 + 2$. Then

$$\lim_{x \to 1} f(x) = 2 \quad \text{and} \quad \lim_{x \to 1} h(x) = 2 \tag{8}$$

Furthermore, for all x,

$$f(x) \leq g(x) \leq h(x) \tag{9}$$

It follows from (8), (9), and the squeeze theorem that

$$\lim_{x \to 1} g(x) = 2$$

EXAMPLE 2

Use the squeeze theorem to prove that

$$\lim_{x \to 0} \left| x \sin \frac{1}{x} \right| = 0$$

SOLUTION

Because $-1 \leq \sin t \leq 1$ for all t, then

$$0 \leq \left| \sin \frac{1}{x} \right| \leq 1 \qquad \text{if } x \neq 0$$

Therefore, if $x \neq 0$,

$$\left| x \sin \frac{1}{x} \right| = |x| \left| \sin \frac{1}{x} \right|$$

$$\leq |x|$$

Hence

$$0 \leq \left| x \sin \frac{1}{x} \right| \leq |x| \qquad \text{if } x \neq 0 \tag{10}$$

Because $\lim\limits_{x \to 0} 0 = 0$ and $\lim\limits_{x \to 0} |x| = 0$, it follows from inequality (10) and the squeeze theorem that

$$\lim_{x \to 0} \left| x \sin \frac{1}{x} \right| = 0$$

1.8.2 THEOREM

$$\lim_{t \to 0} \frac{\sin t}{t} = 1$$

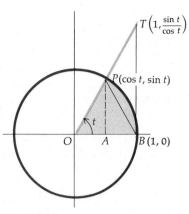

FIGURE 1.8.2

PROOF   First assume that $0 < t < \frac{1}{2}\pi$. Refer to Fig. 1.8.2, which shows the unit circle $x^2 + y^2 = 1$ and the shaded sector BOP, where B is the point $(1, 0)$ and P is the point $(\cos t, \sin t)$. The area of a circular sector of radius r and central angle of radian measure t is determined by $\frac{1}{2}r^2 t$; so if S square units is the area of sector BOP,

$$S = \tfrac{1}{2}t \tag{11}$$

Consider now the triangle BOP, and let K_1 square units be the area of this triangle. Because $K_1 = \frac{1}{2}|\overline{AP}| \cdot |\overline{OB}|$, $|\overline{AP}| = \sin t$ and $|\overline{OB}| = 1$, we have

$$K_1 = \tfrac{1}{2}\sin t \tag{12}$$

The line through the points $O(0, 0)$ and $P(\cos t, \sin t)$ has slope $\dfrac{\sin t}{\cos t}$; therefore its equation is

$$y = \frac{\sin t}{\cos t}x$$

This line intersects the line $x = 1$ at the point $(1, \sin t/\cos t)$, which is the point T in Fig. 1.8.2. If K_2 square units is the area of right triangle BOT, then $K_2 = \frac{1}{2}|\overline{BT}| \cdot |\overline{OB}|$. Because $|\overline{BT}| = \sin t/\cos t$ and $|\overline{OB}| = 1$, we have

$$K_2 = \frac{1}{2} \cdot \frac{\sin t}{\cos t} \tag{13}$$

From Fig. 1.8.2 observe that

$$K_1 < S < K_2 \tag{14}$$

Substituting from (11), (12), and (13) into inequality (14),

$$\frac{1}{2}\sin t < \frac{1}{2}t < \frac{1}{2} \cdot \frac{\sin t}{\cos t}$$

Multiplying each member of the above inequality by $2/\sin t$, which is positive because $0 < t < \frac{1}{2}\pi$,

$$1 < \frac{t}{\sin t} < \frac{1}{\cos t}$$

Taking the reciprocal of each member of the above inequality and reversing the direction of the inequality signs,

$$\cos t < \frac{\sin t}{t} < 1 \tag{15}$$

From the right-hand inequality in the above,

$$\sin t < t \tag{16}$$

and from formula (19) in Sec. 0.6,

$$\frac{1 - \cos t}{2} = \sin^2 \frac{1}{2} t \tag{17}$$

Replacing t by $\frac{1}{2}t$ in inequality (16) and squaring,

$$\sin^2 \tfrac{1}{2}t < \tfrac{1}{4}t^2 \tag{18}$$

Thus from (17) and (18) it follows that

$$\frac{1 - \cos t}{2} < \frac{t^2}{4}$$

$$\Leftrightarrow \qquad 1 - \tfrac{1}{2}t^2 < \cos t \tag{19}$$

From (15) and (19) and because $0 < t < \frac{1}{2}\pi$,

$$1 - \frac{1}{2}t^2 < \frac{\sin t}{t} < 1 \qquad \text{if } 0 < t < \tfrac{1}{2}\pi \tag{20}$$

If $-\frac{1}{2}\pi < t < 0$, then $0 < -t < \frac{1}{2}\pi$; and so from (20),

$$1 - \frac{1}{2}(-t)^2 < \frac{\sin(-t)}{-t} < 1 \qquad \text{if } -\tfrac{1}{2}\pi < t < 0$$

But $\sin(-t) = -\sin t$; thus the above can be written as

$$1 - \frac{1}{2}t^2 < \frac{\sin t}{t} < 1 \qquad \text{if } -\tfrac{1}{2}\pi < t < 0 \tag{21}$$

From (20) and (21) we conclude that

$$1 - \frac{1}{2}t^2 < \frac{\sin t}{t} < 1 \qquad \text{if } -\tfrac{1}{2}\pi < t < \tfrac{1}{2}\pi \text{ and } t \neq 0 \tag{22}$$

Because $\lim\limits_{t \to 0} (1 - \frac{1}{2}t^2) = 1$ and $\lim\limits_{t \to 0} 1 = 1$, it follows from (22) and the squeeze theorem that

$$\lim_{t \to 0} \frac{\sin t}{t} = 1 \qquad \blacksquare$$

EXAMPLE 3
Find

$$\lim_{x \to 0} \frac{\sin 3x}{\sin 5x}$$

if it exists.

SOLUTION

We wish to write the quotient $\sin 3x/\sin 5x$ in such a way that Theorem 1.8.2 can be applied. If $x \neq 0$,

$$\frac{\sin 3x}{\sin 5x} = \frac{\dfrac{\sin 3x}{x}}{\dfrac{\sin 5x}{x}} \quad \Leftrightarrow \quad \frac{\sin 3x}{\sin 5x} = \frac{3\left(\dfrac{\sin 3x}{3x}\right)}{5\left(\dfrac{\sin 5x}{5x}\right)}$$

As x approaches zero, so do $3x$ and $5x$. Hence

$$\lim_{x \to 0} \frac{\sin 3x}{3x} = \lim_{3x \to 0} \frac{\sin 3x}{3x} \qquad \lim_{x \to 0} \frac{\sin 5x}{5x} = \lim_{5x \to 0} \frac{\sin 5x}{5x}$$

$$= 1 \qquad\qquad\qquad\qquad = 1$$

Therefore

$$\lim_{x \to 0} \frac{\sin 3x}{\sin 5x} = \frac{3 \lim_{x \to 0} \left(\dfrac{\sin 3x}{3x}\right)}{5 \lim_{x \to 0} \left(\dfrac{\sin 5x}{5x}\right)}$$

$$= \frac{3 \cdot 1}{5 \cdot 1}$$

$$= \frac{3}{5}$$

From Theorem 1.8.2 we can prove that the sine function and the cosine function are continuous at 0.

1.8.3 THEOREM   The sine function is continuous at 0.

PROOF   The three conditions necessary for continuity at a number are shown to be satisfied.

(i) $\sin 0 = 0$

(ii) $\lim_{t \to 0} \sin t = \lim_{t \to 0} \dfrac{\sin t}{t} \cdot t$

$$= \lim_{t \to 0} \frac{\sin t}{t} \cdot \lim_{t \to 0} t$$

$$= 1 \cdot 0$$

$$= 0$$

(iii) $\lim_{t \to 0} \sin t = \sin 0$

Therefore the sine function is continuous at 0.   ∎

1.8.4 THEOREM   The cosine function is continuous at 0.

PROOF   We check the three conditions necessary for continuity at a number. In checking condition (ii) we use the fact that the sine function is continuous at 0, and we replace $\cos t$ by $\sqrt{1 - \sin^2 t}$ because $\cos t > 0$ when $0 < t < \frac{1}{2}\pi$ and when $-\frac{1}{2}\pi < t < 0$.

(i) $\cos 0 = 1$

(ii) $\displaystyle\lim_{t \to 0} \cos t = \lim_{t \to 0} \sqrt{1 - \sin^2 t}$

$$= \sqrt{\lim_{t \to 0} (1 - \sin^2 t)}$$

$$= \sqrt{1 - 0}$$

$$= 1$$

(iii) $\displaystyle\lim_{t \to 0} \cos t = \cos 0$

Thus the cosine function is continuous at 0.   ∎

We need to use the limit in the following theorem later. It is obtained from the previous three theorems and limit theorems.

1.8.5 THEOREM $$\lim_{t \to 0} \frac{1 - \cos t}{t} = 0$$

PROOF

$$\lim_{t \to 0} \frac{1 - \cos t}{t} = \lim_{t \to 0} \frac{(1 - \cos t)(1 + \cos t)}{t(1 + \cos t)}$$

$$= \lim_{t \to 0} \frac{(1 - \cos^2 t)}{t(1 + \cos t)}$$

$$= \lim_{t \to 0} \frac{\sin^2 t}{t(1 + \cos t)}$$

$$= \lim_{t \to 0} \frac{\sin t}{t} \cdot \lim_{t \to 0} \frac{\sin t}{1 + \cos t}$$

By Theorem 1.8.2,

$$\lim_{t \to 0} \frac{\sin t}{t} = 1$$

and because the sine and cosine functions are continuous at 0 it follows that

$$\lim_{t \to 0} \frac{\sin t}{1 + \cos t} = \frac{0}{1 + 1}$$

$$= 0$$

Therefore

$$\lim_{t \to 0} \frac{1 - \cos t}{t} = 1 \cdot 0$$

$$= 0 \qquad \blacksquare$$

EXAMPLE 4
Find

$$\lim_{x \to 0} \frac{1 - \cos x}{\sin x}$$

if it exists.

SOLUTION
Because $\lim\limits_{x \to 0} (1 - \cos x) = 0$ and $\lim\limits_{x \to 0} \sin x = 0$, the limit theorems cannot be applied to the quotient $(1 - \cos x)/\sin x$. However, if the numerator and denominator are divided by x, which is permissible because $x \neq 0$, we are able to apply Theorems 1.8.2 and 1.8.5. Thus

$$\lim_{x \to 0} \frac{1 - \cos x}{\sin x} = \lim_{x \to 0} \frac{\dfrac{1 - \cos x}{x}}{\dfrac{\sin x}{x}}$$

$$= \frac{\lim\limits_{x \to 0} \dfrac{1 - \cos x}{x}}{\lim\limits_{x \to 0} \dfrac{\sin x}{x}}$$

$$= \frac{0}{1}$$

$$= 0$$

EXAMPLE 5
Find

$$\lim_{x \to 0} \frac{2 \tan^2 x}{x^2}$$

if it exists.

SOLUTION
We use the trigonometric identity

$$\tan x = \frac{\sin x}{\cos x}$$

and we have

$$\lim_{x \to 0} \frac{2 \tan^2 x}{x^2} = 2 \lim_{x \to 0} \frac{\sin^2 x}{x^2 \cdot \cos^2 x}$$

$$= 2 \lim_{x \to 0} \frac{\sin x}{x} \cdot \lim_{x \to 0} \frac{\sin x}{x} \cdot \lim_{x \to 0} \frac{1}{\cos^2 x}$$

$$= 2 \cdot 1 \cdot 1 \cdot 1$$

$$= 2$$

By using Theorem 1.3.12 and the facts that the sine and cosine functions are continuous at 0 we can prove the following theorem.

1.8.6 THEOREM   The sine and cosine functions are continuous at every real number.

PROOF   The set of all real numbers is the domain of both the sine and cosine functions. Therefore we must show that if a is any real number,

$$\lim_{x \to a} \sin x = \sin a \quad \text{and} \quad \lim_{x \to a} \cos x = \cos a$$

or, equivalently, from Theorem 1.3.12,

$$\lim_{t \to 0} \sin(t + a) = \sin a \quad \text{and} \quad \lim_{t \to 0} \cos(t + a) = \cos a \qquad (23)$$

We use the identities

$$\sin(t + a) = \sin t \cos a + \cos t \sin a \qquad (24)$$

$$\cos(t + a) = \cos t \cos a - \sin t \sin a \qquad (25)$$

From (24),

$$\lim_{t \to 0} \sin(t + a) = \lim_{t \to 0} (\sin t \cos a + \cos t \sin a)$$

$$= \lim_{t \to 0} \sin t \cdot \lim_{t \to 0} \cos a + \lim_{t \to 0} \cos t \cdot \lim_{t \to 0} \sin a$$

$$= 0 \cdot \cos a + 1 \cdot \sin a$$

$$= \sin a$$

Therefore the first equation in (23) holds; so the sine function is continuous at every real number. From (25),

$$\lim_{t \to 0} \cos(t + a) = \lim_{t \to 0} (\cos t \cos a - \sin t \sin a)$$

$$= \lim_{t \to 0} \cos t \cdot \lim_{t \to 0} \cos a - \lim_{t \to 0} \sin t \cdot \lim_{t \to 0} \sin a$$

$$= 1 \cdot \cos a - 0 \cdot \sin a$$

$$= \cos a$$

Thus the second equation in (23) holds; so the cosine function is continuous at every real number.   ∎

By using trigonometric identities, Theorem 1.6.4 about the continuity of a rational function, and Theorem 1.8.6 we can prove that the other four trigonometric functions are continuous on their domains.

1.8.7 THEOREM   The tangent, cotangent, secant, and cosecant functions are continuous on their domains.

The proof of Theorem 1.8.7 is left as exercises (see Exercises 35 through 38).

EXERCISES 1.8

In Exercises 1 through 26, evaluate the limit if it exists.

1. $\displaystyle\lim_{x\to0}\frac{\sin 4x}{x}$

2. $\displaystyle\lim_{x\to0}\frac{2x}{\sin 3x}$

3. $\displaystyle\lim_{x\to0}\frac{\sin 9x}{\sin 7x}$

4. $\displaystyle\lim_{t\to0}\frac{\sin 3t}{\sin 6t}$

5. $\displaystyle\lim_{y\to0}\frac{3y}{\sin 5y}$

6. $\displaystyle\lim_{x\to0}\frac{\sin^3 x}{x^2}$

7. $\displaystyle\lim_{x\to0}\frac{x^2}{\sin^2 3x}$

8. $\displaystyle\lim_{x\to0}\frac{\sin^5 2x}{4x^5}$

9. $\displaystyle\lim_{x\to0}\frac{x}{\cos x}$

10. $\displaystyle\lim_{x\to0}\frac{1-\cos x}{1+\sin x}$

11. $\displaystyle\lim_{x\to0}\frac{1-\cos 4x}{x}$

12. $\displaystyle\lim_{z\to0}\frac{1-\cos 2z}{4z}$

13. $\displaystyle\lim_{x\to0}\frac{3x^2}{1-\cos^2 \frac{1}{2}x}$

14. $\displaystyle\lim_{x\to0}\frac{1-\cos^2 x}{2x^2}$

15. $\displaystyle\lim_{x\to0}\frac{\tan x}{2x}$

16. $\displaystyle\lim_{x\to0}\frac{\tan^4 2x}{4x^4}$

17. $\displaystyle\lim_{t\to0^+}\frac{\sin t}{t^2}$

18. $\displaystyle\lim_{x\to0}\frac{1-\cos x}{x^2}$

19. $\displaystyle\lim_{x\to0}\frac{1-\cos 2x}{\sin 3x}$

20. $\displaystyle\lim_{y\to0^+}\frac{\sin 4y}{\cos 3y-1}$

21. $\displaystyle\lim_{x\to\pi/2}\frac{1-\sin x}{\frac{1}{2}\pi-x}$ (*Hint:* Let $t=\frac{1}{2}\pi-x$.)

22. $\displaystyle\lim_{x\to\pi/2}\frac{\frac{1}{2}\pi-x}{\cos x}$ (*Hint:* Let $t=\frac{1}{2}\pi-x$.)

23. $\displaystyle\lim_{x\to\pi^+}\frac{\sin x}{x-\pi}$ (*Hint:* Let $t=x-\pi$.)

24. $\displaystyle\lim_{x\to\pi^+}\frac{\tan x}{x-\pi}$ (*Hint:* Let $t=x-\pi$.)

25. $\displaystyle\lim_{x\to0}\frac{x^2+3x}{\sin x}$

26. $\displaystyle\lim_{x\to0}\frac{\sin x}{3x^2+2x}$

In Exercises 27 through 30, use the squeeze theorem to find the limit.

27. $\displaystyle\lim_{x\to0}x\cos\frac{1}{x}$

28. $\displaystyle\lim_{x\to0}x^2\sin\frac{1}{\sqrt[3]{x}}$

29. $\displaystyle\lim_{x\to3}g(x)$, if $|g(x)+4|<2(3-x)^4$ for all x

30. $\displaystyle\lim_{x\to-2}g(x)$, if $|g(x)-3|<5(x+2)^2$ for all x

In Exercises 31 and 32, find the limit if it exists.

31. $\displaystyle\lim_{x\to0}\frac{\sin(\sin x)}{x}$

32. $\displaystyle\lim_{x\to0}\sin x\sin\frac{1}{x}$

33. Given: $1-\cos^2 x\le f(x)\le x^2$ for all x in the open interval $(-\frac{1}{2}\pi,\frac{1}{2}\pi)$. Find $\displaystyle\lim_{x\to0}f(x)$.

34. Given: $-\sin x\le f(x)\le 2+\sin x$ for all x in the open interval $(-\pi,0)$. Find $\displaystyle\lim_{x\to-\pi/2}f(x)$.

In Exercises 35 through 38, prove that the function is continuous on its domain.

35. the tangent function

36. the cotangent function

37. the secant function

38. the cosecant function

39. If $|f(x)| \le M$ for all x and M is a constant, use the squeeze theorem to prove that $\lim\limits_{x \to 0} x^2 f(x) = 0$.

40. Suppose that $|f(x)| \le M$ for all x, where M is a constant. Furthermore, suppose that $\lim\limits_{x \to a} |g(x)| = 0$. Use the squeeze theorem to prove that $\lim\limits_{x \to a} f(x)g(x) = 0$.

41. If $|f(x)| \le k|x - a|$ for all $x \ne a$, where k is a constant, prove that $\lim\limits_{x \to a} f(x) = 0$.

1.9 (Supplementary)
PROOFS OF SOME THEOREMS ON LIMITS OF FUNCTIONS

In Sec. 1.2 there is one theorem (Theorem 1.2.2) and in Sec. 1.3 there are two theorems (Limit Theorems 9 and 10) whose proofs were deferred to this section.

As stated in Sec. 1.2, Theorem 1.2.2 is a uniqueness theorem that guarantees that if the limit of a function exists, it is unique. We now restate it and give its proof.

1.2.2 THEOREM   If $\lim\limits_{x \to a} f(x) = L_1$ and $\lim\limits_{x \to a} f(x) = L_2$, then $L_1 = L_2$.

PROOF   We shall assume that $L_1 \ne L_2$ and show that this assumption leads to a contradiction. Because $\lim\limits_{x \to a} f(x) = L_1$, it follows from Definition 1.2.1 that for any $\epsilon > 0$ there exists a $\delta_1 > 0$ such that

$$\text{if}\quad 0 < |x - a| < \delta_1 \quad \text{then} \quad |f(x) - L_1| < \epsilon \tag{1}$$

Also, because $\lim\limits_{x \to a} f(x) = L_2$, there exists a $\delta_2 > 0$ such that

$$\text{if}\quad 0 < |x - a| < \delta_2 \quad \text{then} \quad |f(x) - L_2| < \epsilon \tag{2}$$

Now, writing $L_1 - L_2$ as $L_1 - f(x) + f(x) - L_2$ and applying the triangle inequality we have

$$
\begin{aligned}
|L_1 - L_2| &= |[L_1 - f(x)] + [f(x) - L_2]| \\
&\le |L_1 - f(x)| + |f(x) - L_2|
\end{aligned}
\tag{3}
$$

So from (1), (2), and (3) we may conclude that for any $\epsilon > 0$ there exists a $\delta_1 > 0$ and a $\delta_2 > 0$ such that

$$\text{if}\quad 0 < |x - a| < \delta_1 \text{ and } 0 < |x - a| < \delta_2 \quad \text{then} \quad |L_1 - L_2| < \epsilon + \epsilon \tag{4}$$

If δ is the smaller of δ_1 and δ_2, then $\delta \le \delta_1$ and $\delta \le \delta_2$, and (4) states that for any $\epsilon > 0$ there exists a $\delta > 0$ such that

$$\text{if}\quad 0 < |x - a| < \delta \quad \text{then} \quad |L_1 - L_2| < 2\epsilon \tag{5}$$

However, if $\epsilon = \frac{1}{2}|L_1 - L_2|$, then (5) states that there exists a $\delta > 0$ such that

$$\text{if}\quad 0 < |x - a| < \delta \quad \text{then} \quad |L_1 - L_2| < |L_1 - L_2| \tag{6}$$

Because (6) is a contradiction, our assumption is false. So $L_1 = L_2$, and the theorem is proved.   ∎

To prove Limit Theorem 9 (the limit of the quotient of two functions) and Limit Theorem 10 (the limit of the nth root of a function) we apply Theorem 1.7.1 regarding the limit of a composite function, which we now restate.

1.7.1 THEOREM   If $\lim\limits_{x \to a} g(x) = b$, and if the function f is continuous at b,

$$\lim_{x \to a} (f \circ g)(x) = f(b)$$

$\Leftrightarrow \qquad \lim\limits_{x \to a} f(g(x)) = f(\lim\limits_{x \to a} g(x))$

Theorem 1.7.1 was proved in Sec. 1.7. Before this theorem can be applied to prove Limit Theorem 9, we must prove a theorem about the continuity of the function defined by $f(x) = \dfrac{1}{x}$.

1.9.1 THEOREM   If a is any number except zero, and

$$f(x) = \frac{1}{x}$$

then f is continuous at a.

PROOF   The domain of f is the set of all real numbers except zero. Therefore a is in the domain. The proof then will be complete if we can show that

$$\lim_{x \to a} \frac{1}{x} = \frac{1}{a} \tag{7}$$

To prove (7), two cases need to be considered: $a > 0$ and $a < 0$. We prove the case when $a > 0$ and leave the proof for $a < 0$ as an exercise (see Exercise 18).

Because $1/x$ is defined for every x except zero, the open interval required by Definition 1.2.1 can be any open interval containing a but not containing 0.

Considering $a > 0$, we must show that for any $\epsilon > 0$ there exists a $\delta > 0$ such that

$$\text{if} \quad 0 < |x - a| < \delta \quad \text{then} \quad \left| \frac{1}{x} - \frac{1}{a} \right| < \epsilon \tag{8}$$

Because

$$\left| \frac{1}{x} - \frac{1}{a} \right| = \left| \frac{a - x}{ax} \right|$$

$$= \frac{|x - a|}{|a|\,|x|}$$

$$= |x - a| \cdot \frac{1}{a|x|} \quad \text{(because } a > 0\text{)}$$

statement (8) is equivalent to

$$\text{if} \quad 0 < |x - a| < \delta \quad \text{then} \quad |x - a| \cdot \frac{1}{a|x|} < \epsilon \tag{9}$$

On the right end of (9), in addition to the factor $|x - a|$ we have as another factor the quotient $\dfrac{1}{a|x|}$. Therefore, to prove (9) we need to restrict δ so that we will have an inequality involving $\dfrac{1}{a|x|}$. By choosing the open interval required by Definition 1.2.1 to be the interval $(\tfrac{1}{2}a, \tfrac{3}{2}a)$, which contains a but not 0, we are implying that $\delta \le \tfrac{1}{2}a$. Then

$$0 < |x - a| < \delta \quad \text{and} \quad \delta \le \tfrac{1}{2}a$$

$$\Rightarrow \qquad |x - a| < \tfrac{1}{2}a$$

$$\Rightarrow \quad -\tfrac{1}{2}a < x - a < \tfrac{1}{2}a$$

$$\Rightarrow \qquad \tfrac{1}{2}a < x < \tfrac{3}{2}a$$

$$\Rightarrow \qquad \tfrac{1}{2}a < |x| < \tfrac{3}{2}a \quad \text{(because } a > 0)$$

$$\Rightarrow \qquad \frac{2}{3a} < \frac{1}{|x|} < \frac{2}{a}$$

$$\Rightarrow \qquad \frac{2}{3a^2} < \frac{1}{a|x|} < \frac{2}{a^2} \tag{10}$$

Now

$$0 < |x - a| < \delta \quad \text{and} \quad \frac{1}{a|x|} < \frac{2}{a^2}$$

$$\Rightarrow \quad |x - a| \cdot \frac{1}{a|x|} < \delta \cdot \frac{2}{a^2} \tag{11}$$

Because our goal is to have $|x - a| \cdot \dfrac{1}{a|x|} < \epsilon$, statement (11) indicates that we should require $\delta \cdot \dfrac{2}{a^2} \le \epsilon$, that is, $\delta \le \tfrac{1}{2}a^2\epsilon$. Thus with the two restrictions on δ we choose $\delta = \min(\tfrac{1}{2}a, \tfrac{1}{2}a^2\epsilon)$. With this δ we have the following argument:

$$0 < |x - a| < \delta$$

$$\Rightarrow \quad |x - a| \cdot \frac{1}{a|x|} < \delta \cdot \frac{1}{a|x|}$$

$$\Rightarrow \qquad \frac{|x - a|}{|a|\,|x|} < \delta \cdot \frac{1}{a|x|} \quad \text{(because } a > 0)$$

$$\Rightarrow \qquad \left| \frac{a - x}{ax} \right| < \delta \cdot \frac{1}{a|x|}$$

$$\Rightarrow \qquad \left| \frac{1}{x} - \frac{1}{a} \right| < \delta \cdot \frac{1}{a|x|} \tag{12}$$

We showed in (10) that if $\delta \leq \frac{1}{2}a$ and $0 < |x - a| < \delta$, then $\dfrac{1}{a|x|} < \dfrac{2}{a^2}$, that is, $\delta \cdot \dfrac{1}{a|x|} < \delta \cdot \dfrac{2}{a^2}$. Continuing on from (12) we have

$$\left| \frac{1}{x} - \frac{1}{a} \right| < \delta \cdot \frac{1}{a|x|} \quad \text{and} \quad \delta \cdot \frac{1}{a|x|} < \delta \cdot \frac{2}{a^2}$$

$$\Rightarrow \quad \left| \frac{1}{x} - \frac{1}{a} \right| < \delta \cdot \frac{2}{a^2}$$

$$\Rightarrow \quad \left| \frac{1}{x} - \frac{1}{a} \right| < \frac{1}{2}a^2\epsilon \cdot \frac{2}{a^2} \quad \text{(because } \delta \leq \tfrac{1}{2}a^2\epsilon\text{)}$$

$$\Rightarrow \quad \left| \frac{1}{x} - \frac{1}{a} \right| < \epsilon$$

Thus we have shown that for any $\epsilon > 0$, if $\delta = \min(\frac{1}{2}a, \frac{1}{2}a^2\epsilon)$, then the following statement is true:

$$\text{if} \quad 0 < |x - a| < \delta \quad \text{then} \quad \left| \frac{1}{x} - \frac{1}{a} \right| < \epsilon$$

This proves that $\lim\limits_{x \to a} \dfrac{1}{x} = \dfrac{1}{a}$, if $a > 0$. Therefore f is continuous at a if $a > 0$. ∎

Before restating Limit Theorem 9 and giving its proof using Theorems 1.7.1 and 1.9.1, we give an example showing how Theorems 1.7.1 and 1.9.1 can be used to find the limit of a particular quotient.

EXAMPLE 1
By using Theorems 1.7.1 and 1.9.1, and not Limit Theorem 9 (the limit of a quotient), find

$$\lim_{x \to 2} \frac{4x - 3}{x^2 + 2x + 5}$$

SOLUTION
Let the functions f and g be defined by

$$f(x) = 4x - 3 \quad \text{and} \quad g(x) = x^2 + 2x + 5$$

Consider $f(x)/g(x)$ as the product of $f(x)$ and $1/g(x)$ and use Limit Theorem 6 (limit of a product). First, though, we must find $\lim\limits_{x \to 2} 1/g(x)$, which is done by considering $1/g(x)$ as a composite function value.
    If h is the function defined by $h(x) = 1/x$, then the composite function $h \circ g$ is the function defined by $h(g(x)) = 1/g(x)$. Now

$$\lim_{x \to 2} g(x) = \lim_{x \to 2} (x^2 + 2x + 5)$$

$$= 13$$

From Theorem 1.9.1, h is continuous at 13; thus we can use Theorem 1.7.1, and we have

$$\lim_{x \to 2} \frac{1}{x^2 + 2x + 5} = \lim_{x \to 2} \frac{1}{g(x)}$$

$$= \lim_{x \to 2} h(g(x))$$

$$= h(\lim_{x \to 2} g(x)) \quad \text{(by Theorem 1.7.1)}$$

$$= h(13)$$

$$= \tfrac{1}{13}$$

Therefore, by using Limit Theorem 6,

$$\lim_{x \to 2} \frac{4x - 3}{x^2 + 2x + 5} = \lim_{x \to 2}(4x - 3) \cdot \lim_{x \to 2} \frac{1}{x^2 + 2x + 5}$$

$$= 5 \cdot \tfrac{1}{13}$$

$$= \tfrac{5}{13}$$

LIMIT THEOREM 9   If $\lim\limits_{x \to a} f(x) = L$ and if $\lim\limits_{x \to a} g(x) = M$, then

$$\lim_{x \to a} \frac{f(x)}{g(x)} = \frac{L}{M} \qquad \text{if } M \neq 0$$

PROOF   Let h be the function defined by $h(x) = 1/x$. Then the composite function $h \circ g$ is defined by $h(g(x)) = 1/g(x)$. The function h is continuous everywhere except at 0, which follows from Theorem 1.9.1. Hence

$$\lim_{x \to a} \frac{1}{g(x)} = \lim_{x \to a} h(g(x))$$

$$= h(\lim_{x \to a} g(x)) \quad \text{(by Theorem 1.7.1)}$$

$$= h(M)$$

$$= \frac{1}{M}$$

From Limit Theorem 6 and the above result,

$$\lim_{x \to a} \frac{f(x)}{g(x)} = \lim_{x \to a} f(x) \cdot \lim_{x \to a} \frac{1}{g(x)}$$

$$= L \cdot \frac{1}{M}$$

$$= \frac{L}{M} \qquad\qquad\qquad \blacksquare$$

To prove Limit Theorem 10 by using Theorem 1.7.1 we must also use the theorem about the continuity of the function f defined by $f(x) = \sqrt[n]{x}$. This theorem was stated as Theorem 1.6.5 but not proved in Sec. 1.6. We now give its proof, which makes use of the following formula, where n is any positive integer:

$$a^n - b^n = (a - b)(a^{n-1} + a^{n-2}b + a^{n-3}b^2 + \ldots + ab^{n-2} + b^{n-1}) \tag{13}$$

Formula (13) follows from

$$a(a^{n-1} + a^{n-2}b + \ldots + ab^{n-2} + b^{n-1}) = a^n + a^{n-1}b + \ldots + ab^{n-1} \tag{14}$$

and

$$b(a^{n-1} + a^{n-2}b + \ldots + ab^{n-2} + b^{n-1}) = a^{n-1}b + \ldots + ab^{n-1} + b^n \tag{15}$$

by subtracting terms of (15) from those of (14).

1.6.5 THEOREM   If n is a positive integer and

$$f(x) = \sqrt[n]{x}$$

then

   (i) if n is odd, f is continuous at every real number a;
   (ii) if n is even, f is continuous at every positive number a.

PROOF   We prove the theorem if a is a positive number, whether n is odd or even. The case when a is negative or zero and n is odd is left as an exercise (see Exercise 19).

We wish to prove that if $a > 0$, and $f(x) = \sqrt[n]{x}$, then f is continuous at a. Because $\sqrt[n]{a}$ exists when $a > 0$, the proof will be complete if we show that

$$\lim_{x \to a} \sqrt[n]{x} = \sqrt[n]{a}$$

Because $\sqrt[n]{x}$ is defined for every nonnegative number, the open interval required by Definition 1.2.1 can be any open interval containing a and having a nonnegative number as its left endpoint. We must show that for any $\epsilon > 0$ there exists a $\delta > 0$ such that

$$\text{if} \quad 0 < |x - a| < \delta \quad \text{then} \quad |\sqrt[n]{x} - \sqrt[n]{a}| < \epsilon \tag{16}$$

To express $|\sqrt[n]{x} - \sqrt[n]{a}|$ in terms of $|x - a|$ we use (13).

$$|\sqrt[n]{x} - \sqrt[n]{a}|$$
$$= \left| \frac{(x^{1/n} - a^{1/n})[(x^{1/n})^{n-1} + (x^{1/n})^{n-2}a^{1/n} + \ldots + x^{1/n}(a^{1/n})^{n-2} + (a^{1/n})^{n-1}]}{(x^{1/n})^{n-1} + (x^{1/n})^{n-2}a^{1/n} + \ldots + x^{1/n}(a^{1/n})^{n-2} + (a^{1/n})^{n-1}} \right|$$

If (13) is applied to the numerator,

$$\left|\sqrt[n]{x} - \sqrt[n]{a}\right| = |x - a| \cdot \frac{1}{\left|x^{(n-1)/n} + x^{(n-2)/n}a^{1/n} + \ldots + x^{1/n}a^{(n-2)/n} + a^{(n-1)/n}\right|} \tag{17}$$

In this discussion we shall let

$$|\phi(x)| = \left|x^{(n-1)/n} + x^{(n-2)/n}a^{1/n} + \ldots + x^{1/n}a^{(n-2)/n} + a^{(n-1)/n}\right|$$

Then Eq. (17) is

$$\left|\sqrt[n]{x} - \sqrt[n]{a}\right| = |x - a| \cdot \frac{1}{|\phi(x)|} \tag{18}$$

Thus statement (16) is equivalent to

$$\text{if} \quad 0 < |x - a| < \delta \quad \text{then} \quad |x - a| \cdot \frac{1}{|\phi(x)|} < \epsilon \tag{19}$$

On the right end of (19), in addition to the factor $|x - a|$ we have the fraction $\frac{1}{|\phi(x)|}$. Hence to prove (19) we need to restrict δ so that we will have an inequality involving this fraction. If we choose the open interval stipulated in Definition 1.2.1 as the interval $(0, 2a)$, we are requiring that $\delta \leq a$. Then

$$0 < |x - a| < \delta \quad \text{and} \quad \delta \leq a$$
$$\Rightarrow \qquad\qquad |x - a| < a$$
$$\Rightarrow \qquad -a < x - a < a$$
$$\Rightarrow \qquad\qquad 0 < x < 2a$$
$$\Rightarrow \qquad a^{(n-1)/n} < |\phi(x)| \quad \text{(because } x > 0\text{)}$$
$$\Rightarrow \qquad\qquad \frac{1}{|\phi(x)|} < \frac{1}{a^{(n-1)/n}} \tag{20}$$

Now

$$0 < |x - a| < \delta \quad \text{and} \quad \frac{1}{|\phi(x)|} < \frac{1}{a^{(n-1)/n}}$$
$$\Rightarrow \quad |x - a| \cdot \frac{1}{|\phi(x)|} < \delta \cdot \frac{1}{a^{(n-1)/n}} \tag{21}$$

Our goal is to have $|x - a| \cdot \frac{1}{|\phi(x)|} < \epsilon$. Thus statement (21) tells us that we should require $\delta \cdot \frac{1}{a^{(n-1)/n}} \leq \epsilon$, that is, $\delta \leq a^{(n-1)/n}\epsilon$. Thus we choose

$\delta = \min(a, a^{(n-1)/n}\epsilon)$. With this δ we have the following argument:

$$0 < |x - a| < \delta$$

$$\Rightarrow \quad |x - a| \cdot \frac{1}{|\phi(x)|} < \delta \cdot \frac{1}{|\phi(x)|}$$

$$\Rightarrow \quad |\sqrt[n]{x} - \sqrt[n]{a}| < \delta \cdot \frac{1}{|\phi(x)|} \quad \text{(from Eq. (18))}$$

We showed in (20) that if $\delta \le a$ and $0 < |x - a| < \delta$, then $\dfrac{1}{|\phi(x)|} < \dfrac{1}{a^{(n-1)/n}}$,

that is, $\delta \cdot \dfrac{1}{|\phi(x)|} < \delta \cdot \dfrac{1}{a^{(n-1)/n}}$. Continuing on we have

$$|\sqrt[n]{x} - \sqrt[n]{a}| < \delta \cdot \frac{1}{|\phi(x)|} \quad \text{and} \quad \delta \cdot \frac{1}{|\phi(x)|} < \delta \cdot \frac{1}{a^{(n-1)/n}}$$

$$\Rightarrow \quad |\sqrt[n]{x} - \sqrt[n]{a}| < \delta \cdot \frac{1}{a^{(n-1)/n}}$$

$$\Rightarrow \quad |\sqrt[n]{x} - \sqrt[n]{a}| < a^{(n-1)/n}\epsilon \cdot \frac{1}{a^{(n-1)/n}} \quad \text{(because } \delta \le a^{(n-1)/n}\epsilon)$$

$$\Rightarrow \quad |\sqrt[n]{x} - \sqrt[n]{a}| < \epsilon$$

We have demonstrated that for any $\epsilon > 0$, if $\delta = \min(a, a^{(n-1)/n}\epsilon)$, then the following statement is true:

$$\text{if} \quad 0 < |x - a| < \delta \quad \text{then} \quad |\sqrt[n]{x} - \sqrt[n]{a}| < \epsilon$$

Hence we have proved that $\lim\limits_{x \to a} \sqrt[n]{x} = \sqrt[n]{a}$, if a is a positive number. Therefore f is continuous at a if $a > 0$. ∎

As we did with Limit Theorem 9, before restating Limit Theorem 10 and giving its proof using Theorems 1.7.1 and 1.6.5, we give an example showing how Theorems 1.7.1 and 1.6.5 can be used to find the limit of the nth root of a particular function.

EXAMPLE 2

By using Theorems 1.7.1 and 1.6.5, and not Limit Theorem 10 (the limit of the nth root of a function), find

$$\lim_{x \to 7} \sqrt[4]{3x - 5}$$

SOLUTION

Let the functions h and f be defined by

$$h(x) = \sqrt[4]{x} \quad \text{and} \quad f(x) = 3x - 5$$

The composite function $h \circ f$ is defined by $h(f(x)) = \sqrt[4]{f(x)}$. Now

$$\lim_{x \to 7} f(x) = \lim_{x \to 7} (3x - 5)$$

$$= 16$$

By Theorem 1.6.5, the function h is continuous at 16. Therefore we can use Theorem 1.7.1, and we have

$$\lim_{x \to 7} \sqrt[4]{3x - 5} = \lim_{x \to 7} \sqrt[4]{f(x)}$$

$$= \lim_{x \to 7} h(f(x))$$

$$= h(\lim_{x \to 7} f(x)) \quad \text{(by Theorem 1.7.1)}$$

$$= h(16)$$

$$= 2$$

LIMIT THEOREM 10   If n is a positive integer and $\lim_{x \to a} f(x) = L$, then

$$\lim_{x \to a} \sqrt[n]{f(x)} = \sqrt[n]{L}$$

with the restriction that if n is even, $L > 0$.

PROOF   Let h be the function defined by $h(x) = \sqrt[n]{x}$. Then the composite function $h \circ f$ is defined by $h(f(x)) = \sqrt[n]{f(x)}$. From Theorem 1.6.5 it follows that h is continuous at L if n is odd, or if n is even and $L > 0$. Therefore

$$\lim_{x \to a} \sqrt[n]{f(x)} = \lim_{x \to a} h(f(x))$$

$$= h(\lim_{x \to a} f(x)) \quad \text{(by Theorem 1.7.1)}$$

$$= h(L)$$

$$= \sqrt[n]{L} \qquad\qquad\qquad\qquad\qquad \blacksquare$$

EXERCISES 1.9

In Exercises 1 through 6, use Theorems 1.7.1 and 1.9.1, but not Limit Theorem 9, to find the limit.

1. $\lim\limits_{x \to 3} \dfrac{4}{x + 1}$

2. $\lim\limits_{x \to -2} \dfrac{8}{3x + 2}$

3. $\lim\limits_{t \to -1} \dfrac{t + 4}{t - 2}$

4. $\lim\limits_{y \to 3} \dfrac{y - 1}{y + 5}$

5. $\lim\limits_{x \to 2} \dfrac{3x}{2x^2 - 3x + 4}$

6. $\lim\limits_{t \to 4} \dfrac{2t}{t^2 - 4}$

In Exercises 7 through 12, use Theorems 1.7.1 and 1.6.5, but not Limit Theorem 10, to find the limit.

7. $\lim\limits_{x \to 3} \sqrt{x^2 - 5}$

8. $\lim\limits_{x \to 2} \sqrt[3]{x - 1}$

9. $\lim\limits_{y \to -4} \sqrt[3]{3y + 4}$

10. $\displaystyle\lim_{x \to -1/2} \sqrt{8t^3 + 6}$

11. $\displaystyle\lim_{t \to 1} \sqrt[4]{t^2 + 5t + 3}$

12. $\displaystyle\lim_{y \to 0} \sqrt[3]{y^2 - 1}$

13. If the function g is continuous at a number a and the function f is discontinuous at a, is it possible for the quotient of the two functions, f/g, to be continuous at a? Prove your answer.

14. If $f(x)$ is nonnegative for all x in its domain, and $\displaystyle\lim_{x \to a} [f(x)]^2$ exists and is positive, prove that $\displaystyle\lim_{x \to a} f(x) = \sqrt{\lim_{x \to a} [f(x)]^2}$.

15. Prove that if $\displaystyle\lim_{x \to a} f(x)$ exists and is L, then $\displaystyle\lim_{x \to a} |f(x)|$ exists and is $|L|$.

16. Prove that if $f(x) = g(x)$ for all values of x except $x = a$, then $\displaystyle\lim_{x \to a} f(x) = \lim_{x \to a} g(x)$ if the limits exist.

17. Prove that if $f(x) = g(x)$ for all values of x except $x = a$, then if $\displaystyle\lim_{x \to a} g(x)$ does not exist, $\displaystyle\lim_{x \to a} f(x)$ does not exist. (*Hint:* Show that the assumption that $\displaystyle\lim_{x \to a} f(x)$ does exist leads to a contradiction.)

18. Prove Theorem 1.9.1 if $a < 0$.

19. Prove Theorem 1.6.5 for the case when a is negative or zero and n is odd.

1.10 (Supplementary)
ADDITIONAL THEOREMS ON LIMITS OF FUNCTIONS

We now discuss four theorems that are needed to prove some important theorems in later sections. A graphical illustration is given after the statement of each theorem.

1.10.1 THEOREM

If $\displaystyle\lim_{x \to c} f(x)$ exists and is positive, then there is an open interval containing c such that $f(x) > 0$ for every $x \neq c$ in the interval.

▶ ILLUSTRATION 1   Consider the function f defined by

$$f(x) = \frac{5}{2x - 1}$$

A sketch of the graph of f is in Fig. 1.10.1. Because $\displaystyle\lim_{x \to 3} f(x) = 1$, and $1 > 0$, according to Theorem 1.10.1 there is an open interval containing 3 such that $f(x) > 0$ for every $x \neq 3$ in the interval. Such an interval is $(2, 4)$. Actually, any open interval (a, b) for which $\frac{1}{2} \leq a < 3$ and $b > 3$ will do. ◀

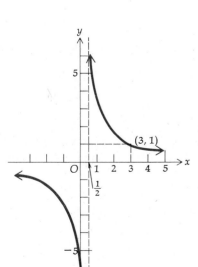

PROOF OF THEOREM 1.10.1   Let $L = \displaystyle\lim_{x \to c} f(x)$. By hypothesis, $L > 0$. Applying Definition 1.2.1 and taking $\epsilon = \frac{1}{2}L$, there is a $\delta > 0$ such that

$$\text{if} \quad 0 < |x - c| < \delta \quad \text{then} \quad |f(x) - L| < \tfrac{1}{2}L \tag{1}$$

Also, $|f(x) - L| < \frac{1}{2}L$ is equivalent to $-\frac{1}{2}L < f(x) - L < \frac{1}{2}L$ (refer to Theorem 0.1.10), which in turn is equivalent to

$$\tfrac{1}{2}L < f(x) < \tfrac{3}{2}L \tag{2}$$

Also, $0 < |x - c| < \delta$ is equivalent to $-\delta < x - c < \delta$ but $x \neq c$, which in turn is equivalent to

$$c - \delta < x < c + \delta \quad \text{but} \quad x \neq c$$

$$\Leftrightarrow \quad x \text{ is in the open interval } (c - \delta, c + \delta) \quad \text{but} \quad x \neq c \tag{3}$$

FIGURE 1.10.1

From statements (2) and (3) we can replace (1) by the statement

if x is in the open interval $(c - \delta, c + \delta)$ but $x \neq c$ then $\frac{1}{2}L < f(x) < \frac{2}{3}L$

Since $L > 0$, it follows that $f(x) > 0$ for every $x \neq c$ in the open interval $(c - \delta, c + \delta)$.   ■

EXAMPLE 1
Given

$$f(x) = \frac{-3x}{2x - 7}$$

(a) Show that $\lim\limits_{x \to 1} f(x) > 0$. (b) Verify Theorem 1.10.1 for this function by finding an open interval containing 1 such that $f(x) > 0$ for every $x \neq 1$ in the interval.

SOLUTION

(a) $\lim\limits_{x \to 1} \dfrac{-3x}{2x - 7} = \dfrac{3}{5}$ and $\dfrac{3}{5} > 0$

(b) The values of x in the required open interval must be such that

$$\frac{-3x}{2x - 7} > 0 \tag{4}$$

Inequality (4) will be satisfied when both the numerator and denominator are positive or when both are negative. Two cases are considered.

Case 1: $-3x > 0$ and $2x - 7 > 0$.
That is, $x < 0$ and $x > \frac{7}{2}$. There is no value of x for which both of these inequalities hold. Therefore the solution set of Case 1 is the empty set.

Case 2: $-3x < 0$ and $2x - 7 < 0$.
That is, $x > 0$ and $x < \frac{7}{2}$. The solution set of Case 2 is the interval $(0, \frac{7}{2})$.

Thus inequality (4) is satisfied if x is in the open interval $(0, \frac{7}{2})$.
We conclude, then, that any open interval (a, b) for which $0 \leq a < 1$ and $1 < b \leq \frac{7}{2}$ will be such that it contains 1 and $f(x) > 0$ for every $x \neq 1$ in the interval. In particular, the interval $(\frac{1}{2}, \frac{3}{2})$ will do.

1.10.2 THEOREM   If $\lim\limits_{x \to c} f(x)$ exists and is negative, there is an open interval containing c such that $f(x) < 0$ for every $x \neq c$ in the interval.

The proof of this theorem is similar to the proof of Theorem 1.10.1 and is left as an exercise (see Exercise 17).

▶ ILLUSTRATION 2   Let

$$g(x) = \frac{6 - x}{3 - 2x}$$

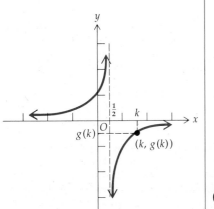

FIGURE 1.10.2

FIGURE 1.10.3

Figure 1.10.2 shows a sketch of the graph of g. $\lim\limits_{x \to 2} g(x) = -4$ and $-4 < 0$; hence, by Theorem 1.10.2 there is an open interval containing 2 such that $g(x) < 0$ for every $x \neq 2$ in the interval. Such an interval is $(\frac{3}{2}, 3)$. Any open interval (a, b) for which $\frac{3}{2} \leq a < 2$ and $2 < b \leq 6$ will suffice.   ◀

The following example also illustrates Theorem 1.10.2.

EXAMPLE 2
Given

$$g(x) = \frac{1}{1 - 2x}$$

(a) Determine the values of k such that $\lim\limits_{x \to k} g(x) < 0$. (b) Find all open intervals (a, b) containing a particular value of k satisfying part (a) such that $g(x) < 0$ for all $x \neq k$ in (a, b). (c) Draw a sketch of the graph of g, and show on the graph the geometric interpretation of the results of parts (a) and (b).

SOLUTION
(a) If $k \neq \frac{1}{2}$,

$$\lim_{x \to k} \frac{1}{1 - 2x} = \frac{1}{1 - 2k}$$

We wish to find the values of k for which

$$\frac{1}{1 - 2k} < 0 \tag{5}$$

Inequality (5) will be satisfied if and only if the denominator is negative, that is, if and only if

$$1 - 2k < 0$$
$$-2k < -1$$
$$k > \tfrac{1}{2}$$

Therefore $\lim\limits_{x \to k} g(x) < 0$ whenever k is in the interval $(\frac{1}{2}, +\infty)$.
(b) Because $g(x) < 0$ if and only if $x > \frac{1}{2}$, the required open intervals (a, b) containing k such that $g(x) < 0$ are those for which $\frac{1}{2} \leq a < k$ and $b > k$.
(c) A sketch of the graph of g is shown in Fig. 1.10.3. In the figure a value of $k > \frac{1}{2}$ is selected. Note that $\lim\limits_{x \to k} g(x) = g(k)$ and so g is continuous at k. The point $(k, g(k))$ is shown on the graph, and $g(k) < 0$. Observe that for all values of x in any open interval (a, b) where $\frac{1}{2} \leq a < k$ and $b > k$, the graph is in the fourth quadrant, and so $g(x) < 0$.

1.10.3 THEOREM   Suppose that the function f is defined on some open interval I containing c, except possibly at c. Also suppose that there is some number M for which there is a $\delta > 0$ such that if $0 < |x - c| < \delta$ then $f(x) \leq M$. Then if $\lim\limits_{x \to c} f(x)$ exists and is equal to L, $L < M$.

FIGURE 1.10.4

▶ ILLUSTRATION 3   Figure 1.10.4 shows a sketch of the graph of a function f satisfying the hypothesis of Theorem 1.10.3. From the figure note that $f(1)$ is not defined, but f is defined on the open interval $(\frac{1}{2}, \frac{3}{2})$ except at 1. Furthermore, if $0 < |x - 1| < \frac{1}{2}$, then $f(x) \leq \frac{9}{4}$. Thus it follows from Theorem 1.10.3 that if $\lim_{x \to 1} f(x)$ exists and is L, then $L \leq \frac{9}{4}$. From the figure, observe that there is an L and it is 2.   ◀

PROOF OF THEOREM 1.10.3   We assume that $M < L$ and show that this assumption leads to a contradiction. If $M < L$, there is some $\epsilon > 0$ such that $M + \epsilon = L$. Because $\lim_{x \to c} f(x) = L$, there exists a $\delta_1 > 0$ such that

$$\text{if}\quad 0 < |x - c| < \delta_1 \quad \text{then}\quad |f(x) - L| < \epsilon$$

$$\Leftrightarrow\quad \text{if}\quad 0 < |x - c| < \delta_1 \quad \text{then}\quad L - \epsilon < f(x) < L + \epsilon$$

Replace L by $M + \epsilon$; it follows that there exists a $\delta_1 > 0$ such that

$$\text{if}\quad 0 < |x - c| < \delta_1 \quad \text{then}\quad (M + \epsilon) - \epsilon < f(x)$$

$$\Leftrightarrow\quad \text{if}\quad 0 < |x - c| < \delta_1 \quad \text{then}\quad M < f(x) \tag{6}$$

But, by hypothesis, there is a δ such that

$$\text{if}\quad 0 < |x - c| < \delta \quad \text{then}\quad f(x) \leq M \tag{7}$$

Statements (6) and (7) contradict each other. Hence our assumption that $M < L$ is false. Therefore $L \leq M$.   ■

1.10.4 THEOREM   Suppose that the function f is defined on some open interval I containing c, except possibly at c. Also suppose that there is some number M for which there is a $\delta > 0$ such that if $0 < |x - c| < \delta$, then $f(x) \geq M$. Then if $\lim_{x \to c} f(x)$ exists and is equal to L, $L \geq M$.

The proof is left as an exercise (see Exercise 18).

▶ ILLUSTRATION 4   Figure 1.10.4 also illustrates Theorem 1.10.4. From the figure observe that if $0 < |x - 1| < \frac{1}{2}$, then $f(x) \geq \frac{3}{2}$; and, because, as previously stated, f is defined on the open interval $(\frac{1}{2}, \frac{3}{2})$ except at 1, Theorem 1.10.4 states that if $\lim_{x \to 1} f(x)$ exists and is L, then $L \geq \frac{3}{2}$.   ◀

EXERCISES 1.10

In Exercises 1 through 4, a function f and a number c are given. (a) Show that $\lim_{x \to c} f(x) > 0$; (b) verify Theorem 1.10.1 for the function f by finding an open interval containing c such that $f(x) > 0$ for every $x \neq c$ in the interval.

1. $f(x) = \dfrac{5}{2x + 4}$; $c = 3$ 2. $f(x) = \dfrac{7}{2x - 1}$; $c = 4$ 3. $f(x) = \dfrac{\sqrt{x} - 1}{x - 1}$; $c = 1$ 4. $f(x) = \dfrac{x^4 - 1}{x^2 - 1}$; $c = -1$

In Exercises 5 through 8, a function g and a number c are given. (a) Show that $\lim\limits_{x \to c} g(x) < 0$; (b) verify Theorem 1.10.2 for the function g by finding an open interval containing c such that $g(x) < 0$ for every $x \neq c$ in the interval.

5. $g(x) = \dfrac{x^2 - 9}{x + 3}; c = -3$

6. $g(x) = \dfrac{2 - 7x + 3x^2}{2 - x}; c = 2$

7. $g(x) = \dfrac{2 - \sqrt{4 + x}}{x}; c = 0$

8. $g(x) = \dfrac{\sqrt{x + 1} - 2}{3 - x}; c = 3$

In Exercises 9 and 10, a function f is given. (a) Determine the values of k such that $\lim\limits_{x \to k} f(x) > 0$. (b) Find all open intervals (a, b) containing a particular value of k satisfying part (a) such that $f(x) > 0$ for all $x \neq k$ in (a, b). (c) Draw a sketch of the graph of f, and show on the graph the geometric interpretation of the results of parts (a) and (b).

9. $f(x) = \dfrac{3x}{x + 1}$

10. $f(x) = \dfrac{x^2 - 25}{x^2 - 4x - 5}$

In Exercises 11 and 12, a function g is given. (a) Determine the values of k such that $\lim\limits_{x \to k} g(x) < 0$. (b) Find all open intervals (a, b) containing a particular value of k satisfying part (a) such that $g(x) < 0$ for all $x \neq k$ in (a, b). (c) Draw a sketch of the graph of g, and show on the graph the geometric interpretation of the results of parts (a) and (b).

11. $g(x) = \dfrac{2x^2 + x - 6}{2x^2 + 11x + 14}$

12. $g(x) = 2 - \sqrt{x - 1}$

13. Draw a sketch of the graph of a function f satisfying the hypothesis of Theorem 1.10.3 for which $f(2)$ is not defined but f is defined on the open interval $(1, 3)$ except at 2. Furthermore, let $f(x) \leq 5$ if $0 < |x - 2| < 1$, and let $\lim\limits_{x \to 2} f(x)$ be the number L. Show on the figure that $L \leq 5$.

14. Draw a sketch of the graph of a function f satisfying the hypothesis of Theorem 1.10.3 for which $f(0)$ is not defined but f is defined on the open interval $(-2, 2)$ except at 0. Furthermore, let $f(x) \leq 3$ if $0 < |x| < 2$, and let $\lim\limits_{x \to 0} f(x)$ be the number L. Show on the figure that $L \leq 3$.

15. Draw a sketch of the graph of a function satisfying the hypothesis of Theorem 1.10.4 for which $f(0)$ is not defined but f is defined on the open interval $(-\frac{1}{2}, \frac{1}{2})$ except at 0. Furthermore, let $f(x) \geq 2$ if $0 < |x| < \frac{1}{2}$, and let $\lim\limits_{x \to 0} f(x)$ be the number L. Show on the figure that $L \geq 2$.

16. Draw a sketch of the graph of a function f satisfying the hypothesis of Theorem 1.10.4 for which $f(1)$ is not defined but f is defined on the open interval $(\frac{3}{4}, \frac{5}{4})$ except at 1. Furthermore, let $f(x) \geq 0$ if $0 < |x - 1| < \frac{1}{4}$, and let $\lim\limits_{x \to 1} f(x)$ be the number L. Show on the figure that $L \geq 0$.

17. Prove Theorem 1.10.2.

18. Prove Theorem 1.10.4.

REVIEW EXERCISES FOR CHAPTER 1

In Exercises 1 through 8, evaluate the limit, and, when applicable, indicate the limit theorems used.

1. $\lim\limits_{x \to 2} (3x^2 - 4x + 5)$

2. $\lim\limits_{h \to 1} \dfrac{h^2 - 4}{3h^3 + 6}$

3. $\lim\limits_{z \to -3} \dfrac{z^2 - 9}{z + 3}$

4. $\lim\limits_{x \to -2} \dfrac{x^2 - x - 6}{x^2 - 5x - 14}$

5. $\lim\limits_{x \to 1/2} \sqrt[3]{\dfrac{4x^2 + 4x - 3}{4x^2 - 1}}$

6. $\lim\limits_{y \to -4} \sqrt{\dfrac{5y + 4}{y - 5}}$

7. $\lim\limits_{t \to 0} \dfrac{\sqrt{9 - t} - 3}{t}$

8. $\lim\limits_{t \to 0} \dfrac{1 - \sqrt[3]{1 + t}}{t}$

In Exercises 9 through 18, establish the limit by using Definition 1.2.1; that is, for any $\epsilon > 0$ find a $\delta > 0$ such that if $0 < |x - a| < \delta$, then $|f(x) - L| < \epsilon$.

9. $\lim\limits_{x \to 3} (2x + 1) = 7$ **10.** $\lim\limits_{x \to -2} (8 - 3x) = 14$ **11.** $\lim\limits_{x \to -4} x^2 = 16$ **12.** $\lim\limits_{x \to 3} (2x^2 - x - 6) = 9$

13. $\lim\limits_{r \to 5} \dfrac{2}{r - 3} = 1$ **14.** $\lim\limits_{y \to 5} \dfrac{3}{2y - 1} = \dfrac{1}{3}$ **15.** $\lim\limits_{t \to 4} \sqrt{t - 3} = 1$ **16.** $\lim\limits_{x \to -2} x^3 = -8$

17. $\lim\limits_{x \to 3} \dfrac{5 + 3x}{2x + 1} = 2$ **18.** $\lim\limits_{t \to 4} \sqrt{2t + 1} = 3$

In Exercises 19 through 22, draw a sketch of the graph of the function; then by observing where there are breaks in the graph, determine the values of the independent variable at which the function is discontinuous, and show why Definition 1.6.1 is not satisfied at each discontinuity.

19. $f(x) = \dfrac{x + 2}{x^2 + x - 2}$ **20.** $g(x) = \dfrac{x^4 - 1}{x^2 - 1}$

21. $g(x) = \begin{cases} 2x + 1 & \text{if } x \le -2 \\ x - 2 & \text{if } -2 < x \le 2 \\ 2 - x & \text{if } 2 < x \end{cases}$ **22.** $F(x) = \begin{cases} |4 - x| & \text{if } x \ne 4 \\ -2 & \text{if } x = 4 \end{cases}$

In Exercises 23 through 26, (a) determine the domain and range of the given function, (b) draw a sketch of its graph, and (c) determine the numbers at which the function is continuous.

23. $f(x) = \sqrt{x^2 - 4}$ **24.** $F(x) = \dfrac{|x^2 - 4|}{x + 2}$

25. $h(x) = \dfrac{(x - 2)(2x^2 + 11x + 15)}{2x^2 + x - 10}$ **26.** $f(x) = \begin{cases} 3 + x & \text{if } x < -3 \\ \sqrt{9 - x^2} & \text{if } -3 \le x \le 3 \\ 3 - x & \text{if } 3 < x \end{cases}$

In Exercises 27 through 48, evaluate the limit if it exists.

27. $\lim\limits_{x \to -1} \dfrac{2x^2 - x - 3}{3x^2 + 8x + 5}$ **28.** $\lim\limits_{y \to 3} \sqrt{\dfrac{y - 3}{y^3 - 27}}$ **29.** $\lim\limits_{x \to 3} f(x)$ if $f(x) = \begin{cases} x^2 - 1 & \text{if } x \le 3 \\ x + 5 & \text{if } 3 < x \end{cases}$

30. $\lim\limits_{x \to 1/3} (|3x - 1| - 5)$ **31.** $\lim\limits_{x \to 9} \dfrac{2\sqrt{x} - 6}{x - 9}$ **32.** $\lim\limits_{y \to 5^-} \dfrac{\sqrt{25 - y^2}}{y - 5}$

33. $\lim\limits_{x \to -4^+} \dfrac{2x}{16 - x^2}$ **34.** $\lim\limits_{s \to 7} \dfrac{5 - \sqrt{4 + 3s}}{7 - s}$ **35.** $\lim\limits_{t \to 5} \dfrac{\sqrt{t - 4}}{t^2 - 10t + 25}$

36. $\lim\limits_{x \to 0^-} \dfrac{x^2 - 5}{2x^3 - 3x^2}$ **37.** $\lim\limits_{x \to 2^+} \dfrac{[\![x]\!] - 1}{[\![x]\!] - x}$ **38.** $\lim\limits_{x \to 5} \dfrac{\sqrt{x - 1} - 2}{x - 5}$

39. $\lim\limits_{t \to 0} \dfrac{\sin 5t}{\sin 2t}$ **40.** $\lim\limits_{x \to 0} \dfrac{1 - \cos 3x}{\sin 3x}$ **41.** $\lim\limits_{x \to 0} \dfrac{1 - \cos^2 x}{x}$ **42.** $\lim\limits_{t \to 0} \dfrac{4t}{\tan t}$

43. $\lim\limits_{\theta \to 0} \dfrac{\csc 3\theta}{\cot \theta}$ **44.** $\lim\limits_{x \to 8} \dfrac{\sqrt{7 + \sqrt[3]{x}} - 3}{x - 8}$ **45.** $\lim\limits_{t \to 0} \dfrac{\sqrt[3]{(t + a)^2} - \sqrt[3]{a^2}}{t}$ **46.** $\lim\limits_{x \to 0} \dfrac{2x^2 - 3x}{2 \sin x}$

47. $\lim\limits_{x \to 0} \dfrac{\sqrt[4]{x^4 + 1} - \sqrt{x^2 + 1}}{x^2}$ $\left(\text{Hint: Write } \dfrac{\sqrt[4]{x^4 + 1} - \sqrt{x^2 + 1}}{x^2} = \dfrac{1 - \sqrt{x^2 + 1}}{x^2} + \dfrac{\sqrt[4]{x^4 + 1} - 1}{x^2}. \right)$

48. $\lim\limits_{x \to 1^+} \dfrac{[\![x^2]\!] - [\![x]\!]^2}{x^2 - 1}$

49. Draw a sketch of the graph of f if $f(x) = [\![1 - x^2]\!]$ and $-2 \leq x \leq 2$. (a) Does $\lim_{x \to 0} f(x)$ exist? (b) Is f continuous at 0?

50. Draw a sketch of the graph of g if $g(x) = (x - 1) [\![x]\!]$, and $0 \leq x \leq 2$. (a) Does $\lim_{x \to 1} g(x)$ exist? (b) Is g continuous at 1?

In Exercises 51 through 54, prove that the function is discontinuous at the number a. Then determine if the discontinuity is removable or essential. If the discontinuity is removable, redefine $f(a)$ such that the discontinuity is removed.

51. $f(x) = \dfrac{x^2 + 2x - 8}{x^2 + 3x - 4}; a = -4$

52. $f(x) = \begin{cases} 4 - x^2 & \text{if } x < 1 \\ 2x + 3 & \text{if } 1 \leq x \end{cases}; a = 1$

53. $f(x) = \dfrac{|2x - 1|}{2x - 1}; a = \dfrac{1}{2}$

54. $f(x) = \dfrac{2 - \sqrt{x + 4}}{x}; a = 0$

In Exercises 55 and 56, find the vertical asymptotes of the graph of the function, and draw a sketch of the graph.

55. $f(x) = \dfrac{5x^2}{x^2 - 4}$

56. $f(x) = \dfrac{-2}{x^2 - x - 6}$

In Exercises 57 through 60, functions f and g are defined. In each exercise define $f \circ g$, and determine all values of x for which $f \circ g$ is continuous.

57. $f(x) = \sqrt{x - 3}; g(x) = x + 2$ $\sqrt{x-1}$ $(1, \infty)$

58. $f(x) = \operatorname{sgn} x; g(x) = x^2 - x$

59. $f(x) = \operatorname{sgn} x; g(x) = x^2 - 1$

60. $f(x) = \dfrac{4}{3x - 5}; g(x) = \sqrt{x}$

In Exercises 61 and 62, a function f and a closed interval $[a, b]$ are given. Verify that the intermediate-value theorem holds for the given value of k, and find a number c such that $f(c) = k$. Draw a sketch of the curve and the line $y = k$.

61. $f(x) = x^2 + 3; [a, b] = [2, 4]; k = 10$

62. $f(x) = -\sqrt{16 - x^2}; [a, b] = [0, 4]; k = -1$

In Exercises 63 and 64, use the squeeze theorem to find the limit.

63. $\lim_{x \to 1} \left[(x - 1)^2 \sin \dfrac{1}{\sqrt[3]{x - 1}} \right]$

64. $\lim_{x \to 4} g(x)$ if $|g(x) + 5| < 3(4 - x)^2$ for all x

In Exercises 65 and 66, find the values of the constants a and b that make the function f continuous on $(-\infty, +\infty)$, and draw a sketch of the graph of f.

65. $f(x) = \begin{cases} 2x + 1 & \text{if } x \leq 3 \\ ax + b & \text{if } 3 < x < 5 \\ x^2 + 2 & \text{if } 5 \leq x \end{cases}$

66. $f(x) = \begin{cases} 3x + 6a & \text{if } x < -3 \\ 3ax - 7b & \text{if } -3 \leq x \leq 3 \\ x - 12b & \text{if } 3 < x \end{cases}$

67. Prove that $\lim_{t \to 2^+} \sqrt{t^2 - 4} = 0$ by showing that for any $\epsilon > 0$ there exists a $\delta > 0$ such that if $0 < t - 2 < \delta$, then $\sqrt{t^2 - 4} < \epsilon$.

68. Prove that $\lim_{x \to 2} \left| \dfrac{x + 3}{x - 2} \right| = +\infty$ by showing that for any $N > 0$ there exists a $\delta > 0$ such that if $0 < |x - 2| < \delta$, then $\left| \dfrac{x + 3}{x - 2} \right| > N$.

69. Give an example of a function for which $\lim_{x \to 0} |f(x)|$ exists but $\lim_{x \to 0} f(x)$ does not exist.

70. If the function g is continuous at a and f is continuous at $g(a)$, is the composite function $f \circ g$ continuous at a? Why?

71. (a) Prove that if $\lim\limits_{h \to 0} f(x + h) = f(x)$, then $\lim\limits_{h \to 0} f(x + h) = \lim\limits_{h \to 0} f(x - h)$. (b) Show that the converse of the theorem in (a) is not true by giving an example of a function for which $\lim\limits_{h \to 0} f(x + h) = \lim\limits_{h \to 0} f(x - h)$ but $\lim\limits_{h \to 0} f(x + h) \neq f(x)$.

72. Give an example of a function f that is discontinuous at 1 for which (a) $\lim\limits_{x \to 1} f(x)$ exists but $f(1)$ does not exist; (b) $f(1)$ exists but $\lim\limits_{x \to 1} f(x)$ does not exist; (c) $\lim\limits_{x \to 1} f(x)$ and $f(1)$ both exist but are not equal.

73. If the domain of f is the set of all real numbers and f is continuous at 0, prove that if $f(a + b) = f(a) + f(b)$ for all a and b, then f is continuous at every number.

74. If the domain of f is the set of all real numbers and f is continuous at 0, prove that if $f(a + b) = f(a) \cdot f(b)$ for all a and b, then f is continuous at every number.

75. Let f be the function defined by

$$f(x) = \begin{cases} 1 & \text{if } x \text{ is an integer} \\ 0 & \text{if } x \text{ is not an integer} \end{cases}$$

(a) Draw a sketch of the graph of f. (b) For what values of a does $\lim\limits_{x \to a} f(x)$ exist? (c) At what numbers is f continuous?

76. Draw a sketch of the graph of a function f that satisfies the following conditions: f is continuous on $(-\infty, -2)$, $[-2, 1)$, $[1, 3]$, and $(3, +\infty)$; $\lim\limits_{x \to -4} f(x) = 0$; $\lim\limits_{x \to -2^-} f(x) = +\infty$; $\lim\limits_{x \to -2^+} f(x) = 0$; $\lim\limits_{x \to 0} f(x) = -3$; $\lim\limits_{x \to 1^-} f(x) = -\infty$; $\lim\limits_{x \to 1^+} f(x) = 2$; $\lim\limits_{x \to 3^-} f(x) = 4$; $\lim\limits_{x \to 3^+} f(x) = -1$; $\lim\limits_{x \to 5} f(x) = 0$.

Exercises 77 and 78 pertain to Supplementary Section 1.10.

77. Given $f(x) = \dfrac{4x^2 - 9}{2x^2 + 7x + 6}$. (a) Show that $\lim\limits_{x \to -3/2} f(x) < 0$. (b) Verify Theorem 1.10.2 for the function f by finding an open interval containing $-\frac{3}{2}$ such that $f(x) < 0$ for every $x \neq -\frac{3}{2}$ in the interval.

78. Given $f(x) = \dfrac{x - 2}{x + 5}$. (a) Determine the values of k such that $\lim\limits_{x \to k} f(x) > 0$. (b) Find all open intervals (a, b) containing a particular value of k satisfying part (a) such that $f(x) > 0$ for all $x \neq k$ in (a, b). (c) Draw a sketch of the graph of f, and show on the graph the geometric interpretation of the results of parts (a) and (b).

THE DERIVATIVE AND DIFFERENTIATION

2.1
THE TANGENT LINE

FIGURE 2.1.1

FIGURE 2.1.2

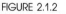

FIGURE 2.1.3

Many of the important problems in calculus depend on determining the tangent line to a given curve at a specific point on the curve. For a circle, we know from plane geometry that the tangent line at a point on the circle is the line intersecting the circle at only that point. This definition does not suffice for a curve in general. For example, in Fig. 2.1.1 the line that we wish to be the tangent line to the curve at the point P intersects the curve at another point Q. To arrive at a suitable definition of the tangent line to the graph of a function at a point on the graph, we proceed by considering how to define the slope of the tangent line at the point. Then the tangent line is determined by its slope and the point of tangency.

Consider the function f continuous at x_1. We wish to define the slope of the tangent line to the graph of f at $P(x_1, f(x_1))$. Let I be the open interval that contains x_1 and on which f is defined. Let $Q(x_2, f(x_2))$ be another point on the graph of f such that x_2 is also in I. Draw a line through P and Q. Any line through two points on a curve is called a **secant line**; therefore the line through P and Q is a secant line. In Fig. 2.1.2 the secant line is shown for various values of x_2. Figure 2.1.3 shows one particular secant line. In this figure Q is to the right of P. However, Q may be on either the right or the left side of P, as seen in Fig. 2.1.2.

Denote the difference of the abscissas of Q and P by Δx (read "delta x") so that

$$\Delta x = x_2 - x_1$$

Observe that Δx denotes a change in the value of x as x changes from x_1 to x_2 and may be either positive or negative. This change is called an *increment of x*. Be careful to note that the symbol Δx for an increment of x does not mean "delta multiplied by x."

Refer back to the secant line PQ in Fig. 2.1.3; its slope is given by

$$m_{PQ} = \frac{f(x_2) - f(x_1)}{\Delta x}$$

provided that line PQ is not vertical. Because $x_2 = x_1 + \Delta x$, the above equation can be written as

$$m_{PQ} = \frac{f(x_1 + \Delta x) - f(x_1)}{\Delta x}$$

Now think of point P as being fixed, and move point Q along the curve toward P; that is, Q approaches P. This is equivalent to stating that Δx approaches zero. As this occurs, the secant line turns about the fixed point P. If this secant line has a limiting position, it is this limiting position that we wish to be the tangent line to the graph at P. So we want the slope of the tangent line to the graph at P to be the limit of m_{PQ} as Δx approaches zero, if this limit exists. If $\lim_{\Delta x \to 0} m_{PQ}$ is $+\infty$ or $-\infty$, then as Δx approaches zero, the line PQ approaches the line through P that is parallel to the y axis. In this case we would want the tangent line to the graph at P to be the line $x = x_1$. The preceding discussion leads to the following definition.

2.1.1 DEFINITION   Suppose the function f is continuous at x_1. The **tangent line** to the graph of f at the point $P(x_1, f(x_1))$ is

(i) the line through P having slope $m(x_1)$, given by

$$m(x_1) = \lim_{\Delta x \to 0} \frac{f(x_1 + \Delta x) - f(x_1)}{\Delta x} \qquad (1)$$

   if this limit exists;

(ii) the line $x = x_1$ if

$$\lim_{\Delta x \to 0} \frac{f(x_1 + \Delta x) - f(x_1)}{\Delta x} \text{ is } +\infty \text{ or } -\infty$$

If neither (i) nor (ii) of Definition 2.1.1 holds, then there is no tangent line to the graph of f at the point $P(x_1, f(x_1))$.

EXAMPLE 1

Given the parabola $y = x^2$. In parts (a) through (c) find the slope of the secant line through the two points: (a) $(2, 4)$, $(3, 9)$; (b) $(2, 4)$, $(2.1, 4.41)$; (c) $(2, 4)$, $(2.01, 4.0401)$. (d) Find the slope of the tangent line to the parabola at the point $(2, 4)$. (e) Draw a sketch of the graph, and show a segment of the tangent line at $(2, 4)$.

SOLUTION

Let m_a, m_b, and m_c be the slopes of the secant lines for (a), (b), and (c), respectively.

(a) $m_a = \dfrac{9 - 4}{3 - 2}$ (b) $m_b = \dfrac{4.41 - 4}{2.1 - 2}$ (c) $m_c = \dfrac{4.0401 - 4}{2.01 - 2}$

$\qquad\quad = 5$ $= \dfrac{0.41}{0.1}$ $= \dfrac{0.0401}{0.01}$

$\qquad\qquad\qquad\qquad\qquad\quad = 4.1$ $= 4.01$

(d) $f(x) = x^2$. From (1) we have

$$m(2) = \lim_{\Delta x \to 0} \frac{f(2 + \Delta x) - f(2)}{\Delta x}$$

$$= \lim_{\Delta x \to 0} \frac{(2 + \Delta x)^2 - 4}{\Delta x}$$

$$= \lim_{\Delta x \to 0} \frac{4 + 4\,\Delta x + (\Delta x)^2 - 4}{\Delta x}$$

$$= \lim_{\Delta x \to 0} \frac{4\,\Delta x + (\Delta x)^2}{\Delta x}$$

$$= \lim_{\Delta x \to 0} (4 + \Delta x)$$

$$= 4$$

FIGURE 2.1.4

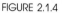

(e) Figure 2.1.4 shows a sketch of the graph and a segment of the tangent line at $(2, 4)$.

EXAMPLE 2

Find the slope of the tangent line to the curve

$$y = x^3 - 3x + 4$$

at the point (x_1, y_1).

SOLUTION

Let $f(x) = x^3 - 3x + 4$. Then

$$f(x_1) = x_1^3 - 3x_1 + 4$$
$$f(x_1 + \Delta x) = (x_1 + \Delta x)^3 - 3(x_1 + \Delta x) + 4$$

From (1),

$$m(x_1) = \lim_{\Delta x \to 0} \frac{f(x_1 + \Delta x) - f(x_1)}{\Delta x}$$

$$= \lim_{\Delta x \to 0} \frac{(x_1 + \Delta x)^3 - 3(x_1 + \Delta x) + 4 - (x_1^3 - 3x_1 + 4)}{\Delta x}$$

$$= \lim_{\Delta x \to 0} \frac{x_1^3 + 3x_1^2 \Delta x + 3x_1(\Delta x)^2 + (\Delta x)^3 - 3x_1 - 3\Delta x + 4 - x_1^3 + 3x_1 - 4}{\Delta x}$$

$$= \lim_{\Delta x \to 0} \frac{3x_1^2 \Delta x + 3x_1(\Delta x)^2 + (\Delta x)^3 - 3\Delta x}{\Delta x}$$

Because $\Delta x \neq 0$, the numerator and denominator can be divided by Δx to obtain

$$m(x_1) = \lim_{\Delta x \to 0} \left[3x_1^2 + 3x_1 \Delta x + (\Delta x)^2 - 3 \right]$$

$$m(x_1) = 3x_1^2 - 3 \tag{2}$$

Table 2.1.1

x	y	m
0	4	-3
1	2	0
2	6	9
-1	6	0
-2	2	9

To draw a sketch of the graph of the equation in Example 2 we plot some points and a segment of the tangent line at some points. Values of x are taken arbitrarily, and the corresponding value of y is computed from the given equation; the value of m is found from (2). The results are given in Table 2.1.1, and a sketch of the graph appears in Fig. 2.1.5. It is important to determine the points where the graph has a horizontal tangent. Because a horizontal line has a slope of zero, these points are found by setting $m(x_1) = 0$ and solving for x_1. Doing this calculation for this example we have $3x_1^2 - 3 = 0$, which gives $x_1 = \pm 1$. Therefore at the points having abscissas of -1 and 1 the tangent line is parallel to the x axis.

FIGURE 2.1.5

EXAMPLE 3

Find an equation of the tangent line to the curve of Example 2 at the point $(2, 6)$.

SOLUTION

Because the slope of the tangent line at any point (x_1, y_1) is given by

$$m(x_1) = 3x_1{}^2 - 3$$

the slope of the tangent line at the point $(2, 6)$ is $m(2) = 9$. Therefore an equation of the desired line in the point-slope form is

$$y - 6 = 9(x - 2)$$

$$9x - y = 12$$

EXAMPLE 4

Find an equation of the tangent line to the graph of $y = \sqrt{x - 3}$ that is perpendicular to the line $6x + 3y - 4 = 0$.

SOLUTION

To determine the slope of the tangent line to the given curve at any point (x_1, y_1) we apply Definition 2.1.1 with $f(x) = \sqrt{x - 3}$.

$$m(x_1) = \lim_{\Delta x \to 0} \frac{\sqrt{x_1 + \Delta x - 3} - \sqrt{x_1 - 3}}{\Delta x}$$

To evaluate this limit we rationalize the numerator.

$$m(x_1) = \lim_{\Delta x \to 0} \frac{(\sqrt{x_1 + \Delta x - 3} - \sqrt{x_1 - 3})(\sqrt{x_1 + \Delta x - 3} + \sqrt{x_1 - 3})}{\Delta x(\sqrt{x_1 + \Delta x - 3} + \sqrt{x_1 - 3})}$$

$$= \lim_{\Delta x \to 0} \frac{(x_1 + \Delta x - 3) - (x_1 - 3)}{\Delta x(\sqrt{x_1 + \Delta x - 3} + \sqrt{x_1 - 3})}$$

$$= \lim_{\Delta x \to 0} \frac{\Delta x}{\Delta x(\sqrt{x_1 + \Delta x - 3} + \sqrt{x_1 - 3})}$$

Numerator and denominator are divided by Δx (since $\Delta x \neq 0$) to obtain

$$m(x_1) = \lim_{\Delta x \to 0} \frac{1}{\sqrt{x_1 + \Delta x - 3} + \sqrt{x_1 - 3}}$$

$$m(x_1) = \frac{1}{2\sqrt{x_1 - 3}} \tag{3}$$

Let l be the given line. To find the slope of l, write its equation in the slope-intercept form, which is

$$y = -2x + \tfrac{4}{3}$$

Therefore the slope of l is -2, the coefficient of x. Because the product of the slopes of two perpendicular lines is -1, the slope of the desired tangent line is $\frac{1}{2}$. Equating the value of $m(x_1)$ given in (3) to $\frac{1}{2}$ we have

$$\frac{1}{2\sqrt{x_1 - 3}} = \frac{1}{2}$$
$$\sqrt{x_1 - 3} = 1$$
$$x_1 - 3 = 1$$
$$x_1 = 4$$

Thus the desired tangent line is the one with slope $\frac{1}{2}$ and through the point $(4, 1)$ on the curve. The point-slope form of an equation of a line gives

$$y - 1 = \tfrac{1}{2}(x - 4)$$
$$2y - 2 = x - 4$$
$$x - 2y - 2 = 0$$

Refer to Fig. 2.1.6, which shows a sketch of the curve together with the line l and the tangent line at $(4, 1)$.

FIGURE 2.1.6

The following illustration shows how the slope of the tangent line to the graph of a function can be used to determine a *maximum function value*.

▶ ILLUSTRATION 1   In Example 4 of Sec. 1.7 we had the following problem: "A cardboard box manufacturer wishes to make open boxes from square pieces of cardboard of side 12 in. by cutting equal squares from the four corners and turning up the sides; x in. is the length of the side of the square to be cut out." In the solution of this example we found that if $V(x)$ cubic inches is the volume of the box, then

$$V(x) = x(12 - 2x)(12 - 2x) \tag{4}$$

and the domain of V is the closed interval $[0, 6]$. Observe from (4) that $V(0) = 0$, $V(6) = 0$, and $V(x) > 0$ when x is in the open interval $(0, 6)$. A sketch of the graph of V appears in Fig. 2.1.7.

Suppose we wish to determine the length of the side of the square to be cut out to obtain a box of maximum volume. From the graph in Fig. 2.1.7 it is apparent that if $V(x)$ is to have a maximum value, it must occur for some number in the open interval $(0, 6)$. In Chap. 3 we learn that if a polynomial function has a maximum value in an open interval, then it must occur at a point where the tangent line is horizontal, that is, at a point where the slope of the tangent line is zero. Figure 2.1.7 shows a segment of the horizontal tangent line. From (4),

$$V(x) = 4(36x - 12x^2 + x^3)$$

FIGURE 2.1.7

We apply Definition 2.1.1 to find $m(x_1)$.

$$m(x_1) = \lim_{\Delta x \to 0} \frac{V(x_1 + \Delta x) - V(x_1)}{\Delta x}$$

$$= 4 \lim_{\Delta x \to 0} \frac{36(x_1 + \Delta x) - 12(x_1 + \Delta x)^2 + (x_1 + \Delta x)^3 - (36x_1 - 12x_1{}^2 + x_1{}^3)}{\Delta x}$$

$$= 4 \lim_{\Delta x \to 0} \frac{36x_1 + 36\,\Delta x - 12x_1{}^2 - 24x_1\,\Delta x - 12(\Delta x)^2 + x_1{}^3 + 3x_1{}^2\,\Delta x + 3x_1(\Delta x)^2 + (\Delta x)^3 - 36x_1 + 12x_1{}^2 - x_1{}^3}{\Delta x}$$

$$= 4 \lim_{\Delta x \to 0} \frac{36\,\Delta x - 24x_1\,\Delta x + 3x_1{}^2\,\Delta x - 12(\Delta x)^2 + 3x_1(\Delta x)^2 + (\Delta x)^3}{\Delta x}$$

$$= 4 \lim_{\Delta x \to 0} \left[36 - 24x_1 + 3x_1{}^2 - 12\,\Delta x + 3x_1\,\Delta x + (\Delta x)^2\right]$$

$$= 4(36 - 24x_1 + 3x_1{}^2)$$

Because we are searching for the value of x_1 in the open interval $(0, 6)$ for which the slope of the tangent line is zero, we set $m(x_1) = 0$ and obtain

$$12 - 8x_1 + x_1{}^2 = 0$$

$$(6 - x_1)(2 - x_1) = 0$$

$$x_1 = 6 \qquad x_1 = 2$$

We reject the number 6 because it is not in the open interval $(0, 6)$. We therefore conclude that if there is a maximum volume, it will occur when the length of the side of the square cut out is 2 in. Furthermore, because $V(2) = 128$, the maximum volume will then be 128 in.^{3}.   ◀

EXERCISES 2.1

In Exercises 1 through 6, find the slope of the tangent line to the graph at the point (x_1, y_1). Make a table of values of x, y, and m over the closed interval $[a, b]$, and include in the table all points where the graph has a horizontal tangent. Draw a sketch of the graph, and show a segment of the tangent line at each of the plotted points.

1. $y = 9 - x^2$; $[a, b] = [-3, 3]$ 2. $y = x^2 + 4$; $[a, b] = [-2, 2]$ 3. $y = -2x^2 + 4x$; $[a, b] = [-1, 3]$

4. $y = x^2 - 6x + 9$; $[a, b] = [1, 5]$ 5. $y = x^3 + 1$; $[a, b] = [-2, 2]$ 6. $y = 1 - x^3$; $[a, b] = [-2, 2]$

In Exercises 7 through 12, find the slope of the tangent line to the graph at the point (x_1, y_1). Make a table of values of x, y, and m at various points on the graph, and include in the table all points where the graph has a horizontal tangent. Draw a sketch of the graph.

7. $y = 3x^2 - 12x + 8$ 8. $y = 7 - 6x - x^2$ 9. $y = \sqrt{4 - x}$

10. $y = \sqrt{x + 1}$ 11. $y = x^3 - 6x^2 + 9x - 2$ 12. $y = x^3 - x^2 - x + 10$

In Exercises 13 through 24, find an equation of the tangent line to the given curve at the indicated point. Draw a sketch of the curve together with the tangent line.

13. $y = x^2 - 4x - 5$; $(-2, 7)$ 14. $y = x^2 - x + 2$; $(2, 4)$ 15. $y = \frac{1}{8}x^3$; $(4, 8)$

16. $y = x^2 + 2x + 1$; $(1, 4)$ 17. $y = \sqrt{9 - 4x}$; $(-4, 5)$ 18. $y = \sqrt{4x - 3}$; $(3, 3)$

19. $y = 6/x;\ (3, 2)$ **20.** $y = -8/\sqrt{x};\ (4, -4)$ **21.** $y = x^4 - 4x;\ (0, 0)$

22. $y = 2x - x^3;\ (-2, 4)$ **23.** $y = \sqrt[3]{x};\ (8, 2)$ **24.** $y = \sqrt[3]{5 - x};\ (-3, 2)$

25. Find an equation of the tangent line to the curve $y = 2x^2 + 3$ that is parallel to the line $8x - y + 3 = 0$.

26. Find an equation of the tangent line to the curve $y = 3x^2 - 4$ that is parallel to the line $3x + y = 4$.

27. Find an equation of the tangent line to the curve $y = 2 - \frac{1}{3}x^2$ that is perpendicular to the line $x - y = 0$.

28. Find an equation of each tangent line to the curve $y = x^3 - 3x$ that is perpendicular to the line $2x + 18y - 9 = 0$.

29. Refer to Exercise 33 in Exercises 1.7. A manufacturer of open tin boxes wishes to make use of pieces of tin with dimensions 8 in. by 15 in. by cutting equal squares from the four corners and turning up the sides. (a) If $V(x)$ in.^{3} is the volume of the box when x in. is the length of the side of the square cut out, write an equation defining $V(x)$. (b) Use the method of Illustration 1 to determine the length of the side of the square cut out for the volume of the box to be a maximum. Assume that there is a maximum volume.

30. Refer to Exercise 34 in Exercises 1.7. Suppose the manufacturer of Exercise 29 makes the open boxes from square pieces of tin that measure k cm on a side. (a) If $V(x)$ cm^3 is the volume of the box when x cm is the length of the side of the square cut out, write an equation defining $V(x)$. (b) Use the method of Illustration 1 to determine the length of the side of the square cut out for the volume of the box to be a maximum. Assume that there is a maximum volume.

31. Refer to Exercise 35 in Exercises 1.7. A rectangular field is to be enclosed with 240 m of fence. (a) If $A(x)$ m^2 is the area of the field when x meters is the length of the field, write an equation defining $A(x)$. (b) Under the assumption that there is a maximum area, use the method of Illustration 1 to determine what the length of the field should be for the area to be a maximum.

32. Refer to Exercise 36 in Exercises 1.7. A rectangular garden is to be placed so that a side of a house serves as a boundary, and 100 ft of fencing material is to be used for the other three sides. (a) If $A(x)$ ft^2 is the area of the garden when x ft is the length of the side of the garden that is parallel to the house, write an equation defining $A(x)$. (b) Under the assumption that there is a maximum area, use the method of Illustration 1 to determine the value of x for $A(x)$ to be a maximum.

33. A carpenter can construct bookcases at a cost of \$40 each. If the carpenter sells the bookcases for x dollars each, it is estimated that $300 - 2x$ bookcases will be sold per month. (a) If $P(x)$ dollars is the carpenter's monthly profit, write an equation defining $P(x)$. (b) Use the method of Illustration 1 to determine the selling price of each bookcase for the carpenter's monthly profit to be a maximum. Assume that there is a maximum profit.

34. A toy manufacturer can produce a particular toy at a cost of \$10 per toy. It is estimated that if the selling price of the toy is x dollars, then the number of toys sold each day is $45 - x$. (a) If $P(x)$ dollars is the manufacturer's daily profit, write an equation defining $P(x)$. (b) Under the assumption that there is a maximum daily profit, use the method of Illustration 1 to determine what the selling price of each toy should be for the manufacturer to realize the maximum daily profit.

35. If $y = 4x^2$, then at any point (x_1, y_1) on the curve, $m(x_1) = 8x_1$. Prove that there is no line through the point $(1, 5)$ that is tangent to the curve $y = 4x^2$.

36. If $y = 4 - x^2$, then at any point (x_1, y_1) on the curve, $m(x_1) = -2x_1$. Prove that there is no line through the point $(1, 2)$ that is tangent to the curve $y = 4 - x^2$.

37. The *normal line* to a curve at a given point is the line perpendicular to the tangent line at that point. Find an equation of the normal line to the curve $y = x^2 - 2x + 3$ at the point $(1, 2)$.

38. Find an equation of each of the normal lines to the curve $y = 1/x$ that is perpendicular to the line $x + 4y = 0$ (see Exercise 37).

2.2
THE DERIVATIVE

In Sec. 2.1 the slope of the tangent line to the graph of $y = f(x)$ at the point $(x_1, f(x_1))$ is defined by

$$m(x_1) = \lim_{\Delta x \to 0} \frac{f(x_1 + \Delta x) - f(x_1)}{\Delta x} \tag{1}$$

if this limit exists.

This type of limit occurs in other problems too and has a specific name.

2.2.1 DEFINITION   The **derivative** of the function f is that function, denoted by f', such that its value at any number x in the domain of f is given by

$$f'(x) = \lim_{\Delta x \to 0} \frac{f(x + \Delta x) - f(x)}{\Delta x} \qquad (2)$$

if this limit exists.

If x_1 is a particular number in the domain of f, then

$$f'(x_1) = \lim_{\Delta x \to 0} \frac{f(x_1 + \Delta x) - f(x_1)}{\Delta x} \qquad (3)$$

if this limit exists. Comparing formulas (1) and (3), note that the slope of the tangent line to the graph of $y = f(x)$ at the point $(x_1, f(x_1))$ is precisely the derivative of f evaluated at x_1.

EXAMPLE 1
Given $f(x) = 3x^2 + 12$, find the derivative of f.

SOLUTION
If x is any number in the domain of f, then from (2),

$$f'(x) = \lim_{\Delta x \to 0} \frac{f(x + \Delta x) - f(x)}{\Delta x}$$

$$= \lim_{\Delta x \to 0} \frac{[3(x + \Delta x)^2 + 12] - (3x^2 + 12)}{\Delta x}$$

$$= \lim_{\Delta x \to 0} \frac{3x^2 + 6x\,\Delta x + 3(\Delta x)^2 + 12 - 3x^2 - 12}{\Delta x}$$

$$= \lim_{\Delta x \to 0} \frac{6x\,\Delta x + 3(\Delta x)^2}{\Delta x}$$

$$= \lim_{\Delta x \to 0} (6x + 3\,\Delta x)$$

$$= 6x$$

Therefore the derivative of f is the function f' defined by $f'(x) = 6x$. The domain of f' is the set of all real numbers, which is the same as the domain of f.

Consider now formula (3), which is

$$f'(x_1) = \lim_{\Delta x \to 0} \frac{f(x_1 + \Delta x) - f(x_1)}{\Delta x}$$

In this formula let

$$x_1 + \Delta x = x \qquad (4)$$

Then

$$\text{``}\Delta x \to 0\text{''} \quad \text{is equivalent to} \quad \text{``}x \to x_1\text{''} \qquad (5)$$

From (3), (4), and (5) we obtain the following formula for $f'(x_1)$:

$$f'(x_1) = \lim_{x \to x_1} \frac{f(x) - f(x_1)}{x - x_1} \tag{6}$$

if this limit exists. Formula (6) is an alternative formula to (3) for computing $f'(x_1)$.

EXAMPLE 2

For the function f of Example 1 find the derivative of f at 2 in three ways: (a) Apply formula (3); (b) apply formula (6); (c) substitute 2 for x in the expression for $f'(x)$ in Example 1.

SOLUTION

(a) $f(x) = 3x^2 + 12$. From formula (3),

$$f'(2) = \lim_{\Delta x \to 0} \frac{f(2 + \Delta x) - f(2)}{\Delta x}$$

$$= \lim_{\Delta x \to 0} \frac{[3(2 + \Delta x)^2 + 12] - [3(2)^2 + 12]}{\Delta x}$$

$$= \lim_{\Delta x \to 0} \frac{12 + 12\,\Delta x + 3(\Delta x)^2 + 12 - 12 - 12}{\Delta x}$$

$$= \lim_{\Delta x \to 0} \frac{12\,\Delta x + 3(\Delta x)^2}{\Delta x}$$

$$= \lim_{\Delta x \to 0} (12 + 3\,\Delta x)$$

$$= 12$$

(b) From formula (6),

$$f'(2) = \lim_{x \to 2} \frac{f(x) - f(2)}{x - 2}$$

$$= \lim_{x \to 2} \frac{(3x^2 + 12) - 24}{x - 2}$$

$$= \lim_{x \to 2} \frac{3x^2 - 12}{x - 2}$$

$$= 3 \lim_{x \to 2} \frac{(x - 2)(x + 2)}{x - 2}$$

$$= 3 \lim_{x \to 2} (x + 2)$$

$$= 12$$

(c) Because, from Example 1, $f'(x) = 6x$, then $f'(2) = 12$.

The use of the symbol f' for the derivative of the function f was introduced by the French mathematician Joseph Louis Lagrange (1736–1813) in the eighteenth century. This notation emphasizes that the function f' is derived from the function f and its value at x is $f'(x)$.

If (x, y) is a point on the graph of f, then $y = f(x)$, and y' is also used as a notation for the derivative of $f(x)$. With the function f defined by the equation $y = f(x)$, we can let

$$\Delta y = f(x + \Delta x) - f(x) \tag{7}$$

where Δy is called an *increment* of y and denotes a change in the function value as x changes by Δx. By using (7) and writing $\dfrac{dy}{dx}$ in place of $f'(x)$, formula (2) becomes

$$\frac{dy}{dx} = \lim_{\Delta x \to 0} \frac{\Delta y}{\Delta x}$$

The symbol $\dfrac{dy}{dx}$ as a notation for the derivative was first used by the German mathematician Gottfried Wilhelm Leibniz (1646–1716). In the seventeenth century Leibniz and Sir Isaac Newton (1642–1727), working independently, introduced almost simultaneously the derivative. Leibniz probably thought of dx and dy as small changes in the variables x and y and of the derivative of y with respect to x as the ratio of dy to dx as dy and dx become small. The concept of a limit as we know it today was not known to Leibniz.

In the Lagrange notation the value of the derivative at $x = x_1$ is indicated by $f'(x_1)$. With the Leibniz notation we would write

$$\frac{dy}{dx}\bigg]_{x = x_1}$$

You must remember that when $\dfrac{dy}{dx}$ is used as a notation for a derivative, dy and dx have so far in this book not been given independent meaning, although later they will be defined separately. So at this time $\dfrac{dy}{dx}$ is a symbol for a derivative and should not be thought of as a ratio. As a matter of fact, $\dfrac{d}{dx}$ can be considered as an operator (a symbol for the operation of computing the derivative), and when we write $\dfrac{dy}{dx}$, it means $\dfrac{d}{dx}(y)$, that is, the derivative of y with respect to x.

EXAMPLE 3

Given $y = \dfrac{2 + x}{3 - x}$, find $\dfrac{dy}{dx}$.

SOLUTION

We are given $y = f(x)$, where $f(x) = \dfrac{2 + x}{3 - x}$.

$$\frac{dy}{dx} = \lim_{\Delta x \to 0} \frac{\Delta y}{\Delta x}$$

$$= \lim_{\Delta x \to 0} \frac{f(x + \Delta x) - f(x)}{\Delta x}$$

$$= \lim_{\Delta x \to 0} \frac{\dfrac{2 + x + \Delta x}{3 - x - \Delta x} - \dfrac{2 + x}{3 - x}}{\Delta x}$$

$$= \lim_{\Delta x \to 0} \frac{(3 - x)(2 + x + \Delta x) - (2 + x)(3 - x - \Delta x)}{\Delta x (3 - x - \Delta x)(3 - x)}$$

$$= \lim_{\Delta x \to 0} \frac{(6 + x - x^2 + 3\,\Delta x - x\,\Delta x) - (6 + x - x^2 - 2\,\Delta x - x\,\Delta x)}{\Delta x (3 - x - \Delta x)(3 - x)}$$

$$= \lim_{\Delta x \to 0} \frac{5\,\Delta x}{\Delta x (3 - x - \Delta x)(3 - x)}$$

$$= \lim_{\Delta x \to 0} \frac{5}{(3 - x - \Delta x)(3 - x)}$$

$$= \frac{5}{(3 - x)^2}$$

Two other notations for the derivative of a function f are

$$\frac{d}{dx}[f(x)] \quad \text{and} \quad D_x[f(x)]$$

Each of these notations allows us to indicate the original function in the expression for the derivative. For instance, we can write the result of Example 3 as

$$\frac{d}{dx}\left(\frac{2 + x}{3 - x}\right) = \frac{5}{(3 - x)^2}$$

or as

$$D_x\left(\frac{2 + x}{3 - x}\right) = \frac{5}{(3 - x)^2}$$

EXAMPLE 4
Compute

$$\frac{d}{dx}\left(\frac{1}{x^2}\right)$$

SOLUTION
We wish to find the derivative of $f(x)$ where $f(x) = 1/x^2$. Thus

$$\frac{d}{dx}\left(\frac{1}{x^2}\right) = \lim_{\Delta x \to 0} \frac{f(x + \Delta x) - f(x)}{\Delta x}$$

$$= \lim_{\Delta x \to 0} \frac{\dfrac{1}{(x + \Delta x)^2} - \dfrac{1}{x^2}}{\Delta x}$$

$$= \lim_{\Delta x \to 0} \frac{x^2 - (x + \Delta x)^2}{x^2(x + \Delta x)^2 \, \Delta x}$$

$$= \lim_{\Delta x \to 0} \frac{x^2 - x^2 - 2x \, \Delta x - (\Delta x)^2}{x^2(x + \Delta x)^2 \, \Delta x}$$

$$= \lim_{\Delta x \to 0} \frac{\Delta x(-2x - \Delta x)}{x^2(x + \Delta x)^2 \, \Delta x}$$

$$= \lim_{\Delta x \to 0} \frac{-2x - \Delta x}{x^2(x + \Delta x)^2}$$

$$= \frac{-2x}{x^2(x^2)}$$

$$= -\frac{2}{x^3}$$

Of course, if the function and the variables are denoted by letters other than f, x, and y, the notations for the derivative incorporate those letters. For instance, if the function g is defined by the equation $s = g(t)$, then the derivative of g can be indicated in each of the following ways:

$$g'(t) \qquad \frac{ds}{dt} \qquad \frac{d}{dt}[g(t)] \qquad D_t[g(t)]$$

EXAMPLE 5
Given $f(x) = x^{1/3}$. (a) Find $f'(x)$. (b) Show that $f'(0)$ does not exist even though f is continuous at 0. (c) Draw a sketch of the graph of f.

SOLUTION
(a) From Definition 2.2.1,

$$f'(x) = \lim_{\Delta x \to 0} \frac{(x + \Delta x)^{1/3} - x^{1/3}}{\Delta x} \tag{8}$$

We rationalize the numerator to obtain a common factor of Δx in the numerator and the denominator; this yields

$$f'(x) = \lim_{\Delta x \to 0} \frac{[(x+\Delta x)^{1/3} - x^{1/3}][(x+\Delta x)^{2/3} + (x+\Delta x)^{1/3}x^{1/3} + x^{2/3}]}{\Delta x[(x+\Delta x)^{2/3} + (x+\Delta x)^{1/3}x^{1/3} + x^{2/3}]}$$

$$= \lim_{\Delta x \to 0} \frac{(x+\Delta x) - x}{\Delta x[(x+\Delta x)^{2/3} + (x+\Delta x)^{1/3}x^{1/3} + x^{2/3}]}$$

$$= \lim_{\Delta x \to 0} \frac{1}{(x+\Delta x)^{2/3} + (x+\Delta x)^{1/3}x^{1/3} + x^{2/3}}$$

$$= \frac{1}{x^{2/3} + x^{1/3}x^{1/3} + x^{2/3}}$$

$$= \frac{1}{3x^{2/3}}$$

(b) Observe that $\dfrac{1}{3x^{2/3}}$ is not defined when $x = 0$. If (8) is used to compute $f'(0)$, we have

$$\lim_{\Delta x \to 0} \frac{(0 + \Delta x)^{1/3} - 0^{1/3}}{\Delta x} = \lim_{\Delta x \to 0} \frac{1}{(\Delta x)^{2/3}}$$

and this limit does not exist. However, the function f is continuous at 0 because

$$\lim_{x \to 0} f(x) = \lim_{x \to 0} x^{1/3}$$

$$= 0$$

$$= f(0)$$

(c) A sketch of the graph of f appears in Fig. 2.2.1.

FIGURE 2.2.1

▶ ILLUSTRATION 1   For the function f of Example 5 consider

$$\lim_{\Delta x \to 0} \frac{f(x_1 + \Delta x) - f(x_1)}{\Delta x} \qquad \text{at } x_1 = 0$$

We have

$$\lim_{\Delta x \to 0} \frac{f(0 + \Delta x) - f(0)}{\Delta x} = \lim_{\Delta x \to 0} \frac{(\Delta x)^{1/3} - 0}{\Delta x}$$

$$= \lim_{\Delta x \to 0} \frac{1}{(\Delta x)^{2/3}}$$

$$= +\infty$$

From Definition 2.1.1(ii) it follows that the line $x = 0$ is the tangent line to the graph of f at the origin.   ◀

We conclude this section by defining some terminology that is used in connection with derivatives.

The process of computing the derivative is called **differentiation**. Thus differentiation is the operation of deriving a function f' from a function f.

If a function has a derivative at x_1, the function is said to be **differentiable** at x_1. That is, the function f is differentiable at x_1 if $f'(x_1)$ exists. A function is **differentiable on an open interval** if it is differentiable at every number in the open interval.

▶ ILLUSTRATION 2   The function f of Example 5 and Illustration 1 is differentiable at every number except 0. Furthermore, this function is differentiable on every open interval not containing 0.   ◀

If a function is differentiable at every number in its domain, it is said to be a **differentiable function**.

▶ ILLUSTRATION 3   In Example 1, $f(x) = 3x^2 + 12$, and the domain of f is the set of all real numbers. Because $f'(x) = 6x$ and $6x$ exists for all real numbers, f is a differentiable function.   ◀

EXERCISES 2.2

In Exercises 1 through 8, find $f'(x)$ for the given function by applying formula (2) of this section.

1. $f(x) = 7x + 3$ **2.** $f(x) = 8 - 5x$ **3.** $f(x) = -4$ **4.** $f(x) = 3x^2 + 4$
5. $f(x) = 4 - 2x^2$ **6.** $f(x) = 3x^2 - 2x + 1$ **7.** $f(x) = 4x^2 + 5x + 3$ **8.** $f(x) = 6 - 3x - x^2$

In Exercises 9 through 18, find the indicated derivative.

9. $\dfrac{d}{dx}(8 - x^3)$ **10.** $\dfrac{d}{dx}(x^3)$ **11.** $\dfrac{d}{dx}(\sqrt{x})$ **12.** $\dfrac{d}{dx}(\sqrt{3x + 5})$ **13.** $D_x\left(\dfrac{1}{x + 1}\right)$

14. $D_x\left(\dfrac{2x + 3}{3x - 2}\right)$ **15.** $D_x\left(\dfrac{1}{x^2} - x\right)$ **16.** $D_x\left(\dfrac{3}{1 + x^2}\right)$ **17.** $\dfrac{d}{dx}\left(\dfrac{1}{\sqrt{x + 1}}\right)$ **18.** $\dfrac{d}{dx}\left(\dfrac{1}{\sqrt{3x}}\right)$

In Exercises 19 through 24, find $f'(a)$ for the given value of a by applying formula (3) of this section.

19. $f(x) = 1 - x^2; a = 3$ **20.** $f(x) = \dfrac{4}{5x}; a = 2$ **21.** $f(x) = \dfrac{2}{x^3}; a = 4$

22. $f(x) = \dfrac{2}{\sqrt{x}} - 1; a = 4$ **23.** $f(x) = \sqrt{10 - x}; a = 6$ **24.** $f(x) = \dfrac{1}{x} + x + x^2; a = -3$

In Exercises 25 through 30, find $f'(a)$ for the given value of a by applying formula (6) of this section.

25. $f(x) = 3x + 2; a = -3$ **26.** $f(x) = x^2 - x + 4; a = 4$ **27.** $f(x) = 2 - x^3; a = -2$

28. $f(x) = \sqrt{1 + 9x}; a = 7$ **29.** $f(x) = \dfrac{1}{\sqrt{2x + 3}}; a = 3$ **30.** $f(x) = \dfrac{1}{\sqrt[3]{x}} - x; a = -8$

In Exercises 31 through 36, find $\dfrac{dy}{dx}$.

31. $y = \dfrac{4}{x^2} + 3x$ **32.** $y = \dfrac{4}{2x - 5}$ **33.** $y = \sqrt{2 - 7x}$ **34.** $y = \dfrac{1}{\sqrt{x - 1}}$ **35.** $y = \sqrt[3]{1 - 2x}$ **36.** $y = \sqrt{4 - x^2}$

37. If g is continuous at a and $f(x) = (x - a)g(x)$, find $f'(a)$. (*Hint:* Use formula (6).)

38. If g is continuous at a and $f(x) = (x^2 - a^2)g(x)$, find $f'(a)$. (*Hint:* Use formula (6).)

39. Given $f(x) = \sqrt[3]{x} - 1$, find $f'(x)$. Is f differentiable at 1? Draw a sketch of the graph of f.

40. Given $f(x) = x^{2/3}$ (a) Find $f'(x)$. (b) Is f differentiable at 0? (c) Draw a sketch of the graph of f.

41. Given $f(x) = \sqrt[3]{1 - 3x}$, find $f'(x)$. For what value or values of x is f not differentiable?

42. If

$$f''(x) = \lim_{\Delta x \to 0} \frac{f'(x + \Delta x) - f'(x)}{\Delta x}$$

find $f''(x)$ if $f(x) = ax^2 + bx$.

43. Use the formula of Exercise 42 to find $f''(x)$ if $f(x) = a/x$.

44. Let f be a function whose domain is the set of all real numbers and $f(a + b) = f(a) \cdot f(b)$ for all a and b. Furthermore, suppose that $f(0) = 1$ and $f'(0)$ exists. Prove that $f'(x)$ exists for all x and that $f'(x) = f'(0) \cdot f(x)$.

45. If f is differentiable at a, prove that

$$f'(a) = \lim_{\Delta x \to 0} \frac{f(a + \Delta x) - f(a - \Delta x)}{2 \Delta x}$$

(*Hint:* $f(a + \Delta x) - f(a - \Delta x) = f(a + \Delta x) - f(a) + f(a) - f(a - \Delta x)$.)

2.3
DIFFERENTIABILITY AND
CONTINUITY

The function defined by $f(x) = x^{1/3}$ is continuous at the number zero but is not differentiable there. However, the graph of this function has a vertical tangent line at the origin, as shown in Illustration 1 of Sec. 2.2. In the following illustration there is a function that is continuous but not differentiable at zero and whose graph does not have a tangent line at the origin.

▶ ILLUSTRATION 1   Let f be the absolute value function defined by

$$f(x) = |x|$$

A sketch of the graph of this function is shown in Fig. 2.3.1. From formula (3) of Sec. 2.2,

$$f'(0) = \lim_{\Delta x \to 0} \frac{f(0 + \Delta x) - f(0)}{\Delta x}$$

FIGURE 2.3.1

if this limit exists. Because $f(0 + \Delta x) = |\Delta x|$ and $f(0) = 0$,

$$\lim_{\Delta x \to 0} \frac{f(0 + \Delta x) - f(0)}{\Delta x} = \lim_{\Delta x \to 0} \frac{|\Delta x|}{\Delta x}$$

Because $|\Delta x| = \Delta x$ if $\Delta x > 0$ and $|\Delta x| = -\Delta x$ if $\Delta x < 0$, one-sided limits at 0 must be considered.

$$\lim_{\Delta x \to 0^+} \frac{|\Delta x|}{\Delta x} = \lim_{\Delta x \to 0^+} \frac{\Delta x}{\Delta x} \qquad \lim_{\Delta x \to 0^-} \frac{|\Delta x|}{\Delta x} = \lim_{\Delta x \to 0^-} \frac{-\Delta x}{\Delta x}$$

$$= \lim_{\Delta x \to 0^+} 1 \qquad\qquad\qquad = \lim_{\Delta x \to 0^-} (-1)$$

$$= 1 \qquad\qquad\qquad\qquad = -1$$

Because $\lim\limits_{\Delta x \to 0^+} \dfrac{|\Delta x|}{\Delta x} \neq \lim\limits_{\Delta x \to 0^-} \dfrac{|\Delta x|}{\Delta x}$, it follows that the two-sided limit $\lim\limits_{\Delta x \to 0} \dfrac{|\Delta x|}{\Delta x}$ does not exist. Therefore $f'(0)$ does not exist, and so f is not differentiable at 0.

Because Definition 2.1.1 is not satisfied when $x = 0$, there is no tangent line at the origin for the graph of the absolute-value function.   ◄

Because the functions of the above illustration and Example 5 of Sec. 2.2 are continuous at a number but not differentiable there, we may conclude that continuity of a function at a number does not imply differentiability of the function at that number. However, differentiability *does* imply continuity, which is given by the next theorem.

2.3.1 THEOREM   If a function f is differentiable at x_1, then f is continuous at x_1.

PROOF   To prove that f is continuous at x_1 we must show that the three conditions of Definition 1.6.1 hold there. That is, we must show that (i) $f(x_1)$ exists; (ii) $\lim\limits_{x \to x_1} f(x)$ exists; and (iii) $\lim\limits_{x \to x_1} f(x) = f(x_1)$.

By hypothesis, f is differentiable at x_1. Therefore $f'(x_1)$ exists. Because by formula (6) of Sec. 2.2

$$f'(x_1) = \lim_{x \to x_1} \frac{f(x) - f(x_1)}{x - x_1} \tag{1}$$

$f(x_1)$ must exist; otherwise the above limit has no meaning. Therefore condition (i) holds at x_1. Now consider

$$\lim_{x \to x_1} [f(x) - f(x_1)]$$

We can write

$$\lim_{x \to x_1} [f(x) - f(x_1)] = \lim_{x \to x_1} \left[(x - x_1) \cdot \frac{f(x) - f(x_1)}{x - x_1} \right] \tag{2}$$

Because

$$\lim_{x \to x_1} (x - x_1) = 0 \quad \text{and} \quad \lim_{x \to x_1} \frac{f(x) - f(x_1)}{x - x_1} = f'(x_1)$$

we apply the theorem on the limit of a product (Theorem 1.3.6) to the right side of (2) and obtain

$$\lim_{x \to x_1} [f(x) - f(x_1)] = \lim_{x \to x_1} (x - x_1) \cdot \lim_{x \to x_1} \frac{f(x) - f(x_1)}{x - x_1}$$
$$= 0 \cdot f'(x_1)$$

so that

$$\lim_{x \to x_1} [f(x) - f(x_1)] = 0$$

FIGURE 2.3.2

FIGURE 2.3.3

FIGURE 2.3.4

Then

$$\lim_{x \to x_1} f(x) = \lim_{x \to x_1} \left[f(x) - f(x_1) + f(x_1) \right]$$

$$= \lim_{x \to x_1} \left[f(x) - f(x_1) \right] + \lim_{x \to x_1} f(x_1)$$

$$= 0 + f(x_1)$$

which gives

$$\lim_{x \to x_1} f(x) = f(x_1) \tag{3}$$

From (3) it follows that conditions (ii) and (iii) for continuity of f at x_1 hold. Therefore the theorem is proved.   ■

A function f can fail to be differentiable at a number c for one of the following reasons:

1. The function f is discontinuous at c. This follows from Theorem 2.3.1. Refer to Fig. 2.3.2 for a sketch of the graph of such a function.
2. The function f is continuous at c, and the graph of f has a vertical tangent line at the point where $x = c$. See Fig. 2.3.3 for a sketch of the graph of a function having this property. This situation also occurs in Example 5 of Sec. 2.2.
3. The function f is continuous at c, and the graph of f does not have a tangent line at the point where $x = c$. In Fig. 2.3.4 there is a sketch of the graph of a function satisfying this condition. Observe that there is a "sharp turn" in the graph at $x = c$. In Illustration 1 there is another such function.

Before giving an additional example of a function that is continuous at a number but not differentiable there, the concept of a **one-sided derivative** is introduced.

2.3.2 DEFINITION   If the function f is defined at x_1, then the **derivative from the right** of f at x_1, denoted by $f'_+(x_1)$, is defined by

$$f'_+(x_1) = \lim_{\Delta x \to 0^+} \frac{f(x_1 + \Delta x) - f(x_1)}{\Delta x} \tag{4}$$

$$\Leftrightarrow \quad f'_+(x_1) = \lim_{x \to x_1^+} \frac{f(x) - f(x_1)}{x - x_1} \tag{5}$$

if the limit exists.

2.3.3 DEFINITION   If the function f is defined at x_1, then the **derivative from the left** of f at x_1, denoted by $f'_-(x_1)$, is defined by

$$f'_-(x_1) = \lim_{\Delta x \to 0^-} \frac{f(x_1 + \Delta x) - f(x_1)}{\Delta x} \tag{6}$$

$$\Leftrightarrow \quad f'_-(x_1) = \lim_{x \to x_1^-} \frac{f(x) - f(x_1)}{x - x_1} \tag{7}$$

if the limit exists.

From Definitions 2.3.2 and 2.3.3 and Theorem 1.4.3 it follows that a function f defined on an open interval containing x_1 is differentiable at x_1 if and only if $f'_+(x_1)$ and $f'_-(x_1)$ both exist and are equal. Of course, then $f'(x_1)$, $f'_+(x_1)$, and $f'_-(x_1)$ are all equal.

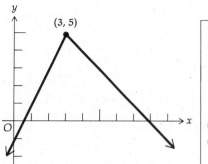

FIGURE 2.3.5

EXAMPLE 1

Let f be the function defined by

$$f(x) = \begin{cases} 2x - 1 & \text{if } x < 3 \\ 8 - x & \text{if } 3 \le x \end{cases}$$

(a) Draw a sketch of the graph of f. (b) Prove that f is continuous at 3. (c) Find $f'_-(3)$ and $f'_+(3)$. (d) Is f differentiable at 3?

SOLUTION

(a) A sketch of the graph is shown in Fig. 2.3.5.

(b) To prove that f is continuous at 3, we verify the three conditions for continuity at a number.

(i) $f(3) = 5$

(ii) $\lim\limits_{x \to 3^-} f(x) = \lim\limits_{x \to 3^-} (2x - 1)$ $\qquad$ $\lim\limits_{x \to 3^+} f(x) = \lim\limits_{x \to 3^+} (8 - x)$

$\qquad\qquad\qquad = 5$ $\qquad\qquad\qquad\qquad\qquad = 5$

Thus $\lim\limits_{x \to 3} f(x) = 5$.

(iii) $\lim\limits_{x \to 3} f(x) = f(3)$

Because conditions (i), (ii), and (iii) all hold at 3, f is continuous at 3.

(c) $f'_-(3) = \lim\limits_{\Delta x \to 0^-} \dfrac{f(3 + \Delta x) - f(3)}{\Delta x}$ $\qquad$ $f'_+(3) = \lim\limits_{\Delta x \to 0^+} \dfrac{f(3 + \Delta x) - f(3)}{\Delta x}$

$\qquad = \lim\limits_{\Delta x \to 0^-} \dfrac{[2(3 + \Delta x) - 1] - 5}{\Delta x}$ $\qquad = \lim\limits_{\Delta x \to 0^+} \dfrac{[8 - (3 + \Delta x)] - 5}{\Delta x}$

$\qquad = \lim\limits_{\Delta x \to 0^-} \dfrac{6 + 2\Delta x - 6}{\Delta x}$ $\qquad\qquad = \lim\limits_{\Delta x \to 0^+} \dfrac{8 - 3 - \Delta x - 5}{\Delta x}$

$\qquad = \lim\limits_{\Delta x \to 0^-} \dfrac{2\Delta x}{\Delta x}$ $\qquad\qquad\quad = \lim\limits_{\Delta x \to 0^+} \dfrac{-\Delta x}{\Delta x}$

$\qquad = \lim\limits_{\Delta x \to 0^-} 2$ $\qquad\qquad\qquad = \lim\limits_{\Delta x \to 0^+} (-1)$

$\qquad = 2$ $\qquad\qquad\qquad\qquad\quad = -1$

(d) Because

$$\lim\limits_{\Delta x \to 0^-} \dfrac{f(3 + \Delta x) - f(3)}{\Delta x} \ne \lim\limits_{\Delta x \to 0^+} \dfrac{f(3 + \Delta x) - f(3)}{\Delta x}$$

we conclude that

$$\lim_{\Delta x \to 0} \frac{f(3 + \Delta x) - f(3)}{\Delta x}$$

does not exist. Hence f is not differentiable at 3. However, the derivative from the left and the derivative from the right both exist at 3.

EXAMPLE 2
Let f be defined by

$$f(x) = |1 - x^2|$$

(a) Draw a sketch of the graph of f. (b) Prove that f is continuous at 1. (c) Determine if f is differentiable at 1.

SOLUTION
By definition of absolute value, if $x < -1$ or $x > 1$, then $f(x) = -(1 - x^2)$, and if $-1 \le x \le 1$, $f(x) = 1 - x^2$. Therefore f can be defined as follows:

$$f(x) = \begin{cases} x^2 - 1 & \text{if } x < -1 \\ 1 - x^2 & \text{if } -1 \le x \le 1 \\ x^2 - 1 & \text{if } 1 < x \end{cases}$$

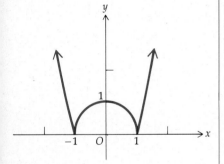

FIGURE 2.3.6

(a) A sketch of the graph of f is shown in Fig. 2.3.6.
(b) To prove that f is continuous at 1 we verify the three conditions for continuity.

   (i) $f(1) = 0$

   (ii) $\displaystyle\lim_{x \to 1^-} f(x) = \lim_{x \to 1^-} (1 - x^2)$ $\displaystyle\lim_{x \to 1^+} f(x) = \lim_{x \to 1^+} (x^2 - 1)$

                $= 0$ $= 0$

     Thus $\displaystyle\lim_{x \to 1} f(x) = 0$.

   (iii) $\displaystyle\lim_{x \to 1} f(x) = f(1)$

   Because conditions (i)–(iii) all hold at 1, f is continuous at 1.

(c) $\displaystyle f'_-(1) = \lim_{x \to 1^-} \frac{f(x) - f(1)}{x - 1}$ $\displaystyle f'_+(1) = \lim_{x \to 1^+} \frac{f(x) - f(1)}{x - 1}$

          $\displaystyle = \lim_{x \to 1^-} \frac{(1 - x^2) - 0}{x - 1}$ $\displaystyle = \lim_{x \to 1^+} \frac{(x^2 - 1) - 0}{x - 1}$

          $\displaystyle = \lim_{x \to 1^-} \frac{(1 - x)(1 + x)}{x - 1}$ $\displaystyle = \lim_{x \to 1^+} \frac{(x - 1)(x + 1)}{x - 1}$

          $\displaystyle = \lim_{x \to 1^-} [-(1 + x)]$ $\displaystyle = \lim_{x \to 1^+} (x + 1)$

          $= -2$ $= 2$

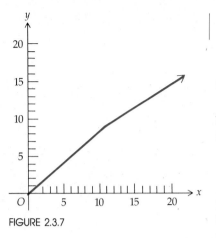

FIGURE 2.3.7

Because $f'_-(1) \neq f'_+(1)$, it follows that $f'(1)$ does not exist, and so f is not differentiable at 1.

The function of Example 2 is also not differentiable at $x = -1$, as can be shown by a method similar to that used for $x = 1$ (see Exercise 26).

▶ ILLUSTRATION 2   In Illustration 2 of Sec. 1.6 we had the function C defined by

$$C(x) = \begin{cases} x & \text{if } 0 \le x \le 10 \\ 0.7x + 3 & \text{if } 10 < x \end{cases}$$

where $C(x)$ dollars is the total cost of x pounds of a product. Figure 2.3.7 shows a sketch of the graph of C. In Sec. 1.6 we showed that C is continuous at 10.

$$C'_-(10) = \lim_{x \to 10^-} \frac{C(x) - C(10)}{x - 10} \qquad C'_+(10) = \lim_{x \to 10^+} \frac{C(x) - C(10)}{x - 10}$$

$$= \lim_{x \to 10^-} \frac{x - 10}{x - 10} \qquad\qquad = \lim_{x \to 10^+} \frac{(0.7x + 3) - 10}{x - 10}$$

$$= \lim_{x \to 10^-} 1 \qquad\qquad = \lim_{x \to 10^+} \frac{0.7(x - 10)}{x - 10}$$

$$= 1 \qquad\qquad = \lim_{x \to 10^+} 0.7$$

$$= 0.7$$

Because $C'_-(10) \neq C'_+(10)$, C is not differentiable at 10.   ◀

EXAMPLE 3
Given

$$f(x) = \begin{cases} \dfrac{1}{x} & \text{if } 0 < x < b \\ 1 - \tfrac{1}{4}x & \text{if } b \le x \end{cases}$$

(a) Determine a value of b so that f is continuous at b. (b) Is f differentiable at the value of b found in part (a)?

SOLUTION
(a) The function f will be continuous at b if $\lim\limits_{x \to b^-} f(x) = f(b)$ and $\lim\limits_{x \to b^+} f(x) = f(b)$.

$$\lim_{x \to b^-} f(x) = \lim_{x \to b^-} \frac{1}{x} \qquad \lim_{x \to b^+} f(x) = \lim_{x \to b^+} (1 - \tfrac{1}{4}x)$$

$$= \frac{1}{b} \qquad\qquad = 1 - \tfrac{1}{4}b$$

$f(b) = 1 - \frac{1}{4}b$; therefore f will be continuous at b if

$$\frac{1}{b} = 1 - \frac{1}{4}b$$

$$4 = 4b - b^2$$

$$b^2 - 4b + 4 = 0$$

$$(b - 2)^2 = 0$$

$$b = 2$$

Thus

$$f(x) = \begin{cases} \dfrac{1}{x} & \text{if } 0 < x < 2 \\ 1 - \frac{1}{4}x & \text{if } 2 \le x \end{cases}$$

and f is continuous at 2.

(b) To determine if f is differentiable at 2 we compute $f'_-(2)$ and $f'_+(2)$.

$$f'_-(2) = \lim_{x \to 2^-} \frac{f(x) - f(2)}{x - 2} \qquad f'_+(2) = \lim_{x \to 2^+} \frac{f(x) - f(2)}{x - 2}$$

$$= \lim_{x \to 2^-} \frac{\dfrac{1}{x} - \dfrac{1}{2}}{x - 2} \qquad\qquad = \lim_{x \to 2^+} \frac{(1 - \frac{1}{4}x) - \frac{1}{2}}{x - 2}$$

$$= \lim_{x \to 2^-} \frac{2 - x}{2x(x - 2)} \qquad\qquad = \lim_{x \to 2^+} \frac{\frac{1}{2} - \frac{1}{4}x}{x - 2}$$

$$= \lim_{x \to 2^-} \frac{-1}{2x} \qquad\qquad\quad = \lim_{x \to 2^+} \frac{2 - x}{4(x - 2)}$$

$$= -\tfrac{1}{4} \qquad\qquad\qquad\quad = \lim_{x \to 2^+} \frac{-1}{4}$$

$$\qquad\qquad\qquad\qquad\qquad = -\tfrac{1}{4}$$

Because $f'_-(2) = f'_+(2)$, it follows that $f'(2)$ exists, and hence f is differentiable at 2.

EXERCISES 2.3

In Exercises 1 through 20, do each of the following: (a) Draw a sketch of the graph of the function; (b) determine if f is continuous at x_1; (c) find $f'_-(x_1)$ and $f'_+(x_1)$ if they exist; (d) determine if f is differentiable at x_1.

1. $f(x) = \begin{cases} x + 2 & \text{if } x \le -4 \\ -x - 6 & \text{if } -4 < x \end{cases}$ $\qquad$ 2. $f(x) = \begin{cases} 3 - 2x & \text{if } x < 2 \\ 3x - 7 & \text{if } 2 \le x \end{cases}$ $\qquad$ 3. $f(x) = |x - 3|; \ x_1 = 3$

$\qquad x_1 = -4 \qquad\qquad\qquad\qquad\qquad x_1 = 2$

4. $f(x) = 1 + |x + 2|$; $x_1 = -2$

5. $f(x) = \begin{cases} -1 & \text{if } x < 0 \\ x - 1 & \text{if } 0 \leq x \end{cases}$

$x_1 = 0$

6. $f(x) = \begin{cases} x & \text{if } x \leq 0 \\ x^2 & \text{if } 0 < x \end{cases}$

$x_1 = 0$

7. $f(x) = \begin{cases} x^2 & \text{if } x \leq 0 \\ -x^2 & \text{if } 0 < x \end{cases}$

$x_1 = 0$

8. $f(x) = \begin{cases} x^2 - 4 & \text{if } x < 2 \\ \sqrt{x - 2} & \text{if } 2 \leq x \end{cases}$

$x_1 = 2$

9. $f(x) = \begin{cases} \sqrt{1 - x} & \text{if } x < 1 \\ (1 - x)^2 & \text{if } 1 \leq x \end{cases}$

$x_1 = 1$

10. $f(x) = \begin{cases} x^2 & \text{if } x < -1 \\ -1 - 2x & \text{if } -1 \leq x \end{cases}$

$x_1 = -1$

11. $f(x) = \begin{cases} 2x^2 - 3 & \text{if } x \leq 2 \\ 8x - 11 & \text{if } 2 < x \end{cases}$

$x_1 = 2$

12. $f(x) = \begin{cases} x^2 - 9 & \text{if } x < 3 \\ 6x - 18 & \text{if } 3 \leq x \end{cases}$

$x_1 = 3$

13. $f(x) = \sqrt[3]{x + 1}$; $x_1 = -1$

14. $f(x) = (x - 2)^{-2}$; $x_1 = 2$

15. $f(x) = \begin{cases} 5 - 6x & \text{if } x \leq 3 \\ -4 - x^2 & \text{if } 3 < x \end{cases}$

$x_1 = 3$

16. $f(x) = \begin{cases} -x^{2/3} & \text{if } x \leq 0 \\ x^{2/3} & \text{if } 0 < x \end{cases}$

$x_1 = 0$

17. $f(x) = \begin{cases} x - 2 & \text{if } x < 0 \\ x^2 & \text{if } 0 \leq x \end{cases}$

$x_1 = 0$

18. $f(x) = \begin{cases} x^3 & \text{if } x \leq 1 \\ x + 1 & \text{if } 1 < x \end{cases}$

$x_1 = 1$

19. $f(x) = \begin{cases} 3x^2 & \text{if } x \leq 2 \\ x^3 & \text{if } 2 < x \end{cases}$

$x_1 = 2$

20. $f(x) = \begin{cases} x^2 + 1 & \text{if } x < -1 \\ 1 - x^2 & \text{if } -1 \leq x \end{cases}$

$x_1 = -1$

21. Given $f(x) = \sqrt{x - 4}$. (a) Prove that f is continuous from the right at 4. (b) Prove that $f'_+(4)$ does not exist. (c) Draw a sketch of the graph of f.

22. Given $f(x) = \sqrt{4 - x^2}$. (a) Prove that f is continuous on the closed interval $[-2, 2]$. (b) Prove that neither $f'_+(-2)$ nor $f'_-(2)$ exists. (c) Draw a sketch of the graph of f.

23. Given $f(x) = \sqrt{x^2 - 9}$. (a) Prove that f is continuous on $(-\infty, -3]$ and $[3, +\infty)$. (b) Prove that neither $f'_-(-3)$ nor $f'_+(3)$ exists. (c) Draw a sketch of the graph of f.

24. Given $f(x) = \sqrt{8 - x^3}$. (a) Prove that f is continuous from the left at 2. (b) Prove that $f'_-(2)$ does not exist. (c) Draw a sketch of the graph of f.

25. Given $f(x) = \operatorname{sgn} x$. (a) Prove that $f'_+(0)$ and $f'_-(0)$ do not exist. (b) Prove that $\lim\limits_{x \to 0^+} f'(x) = 0$ and $\lim\limits_{x \to 0^-} f'(x) = 0$. (c) Draw a sketch of the graph of f.

26. Prove that the function of Example 2 is continuous at -1 but not differentiable there.

27. Given $f(x) = x^{3/2}$. (a) Prove that f is continuous from the right at 0. (b) Prove that $f'_+(0)$ exists, and find its value. (c) Draw a sketch of the graph of f.

28. Given $f(x) = (1 - x^2)^{3/2}$. (a) Prove that f is continuous on the closed interval $[-1, 1]$. (b) Prove that f is differentiable on the open interval $(-1, 1)$ and that both $f'_+(-1)$ and $f'_-(1)$ exist. (c) Draw a sketch of the graph of f.

29. Given

$$f(x) = \begin{cases} x^2 - 7 & \text{if } 0 < x \leq b \\ \dfrac{6}{x} & \text{if } b < x \end{cases}$$

(a) Determine a value of b for which f is continuous at b. (b) Is f differentiable at the value of b found in part (a)?

30. Find the values of a and b such that f is differentiable at 1 if

$$f(x) = \begin{cases} x^2 & \text{if } x < 1 \\ ax + b & \text{if } 1 \leq x \end{cases}$$

31. Find the values of a and b such that f is differentiable at 2 if

$$f(x) = \begin{cases} ax + b & \text{if } x < 2 \\ 2x^2 - 1 & \text{if } 2 \le x \end{cases}$$

32. Given $f(x) = [\![x]\!]$, find $f'(x_1)$ if x_1 is not an integer. Prove by applying Theorem 2.3.1 that $f'(x_1)$ does not exist if x_1 is an integer. If x_1 is an integer, what can you say about $f'_-(x_1)$ and $f'_+(x_1)$?

33. Given $f(x) = (x - 1)[\![x]\!]$. Draw a sketch of the graph of f for x in $[0, 2]$. Find, if they exist: (a) $f'_-(1)$; (b) $f'_+(1)$; (c) $f'(1)$.

34. Given $f(x) = (5 - x)[\![x]\!]$. Draw a sketch of the graph of f for x in $[4, 6]$. Find, if they exist: (a) $f'_-(5)$; (b) $f'_+(5)$; (c) $f'(5)$.

35. Given $f(x) = (x - a)[\![x]\!]$, show that $f'_-(a) + 1 = f'_+(a)$.

36. Let the function f be defined by

$$f(x) = \begin{cases} \dfrac{g(x) - g(a)}{x - a} & \text{if } x \neq a \\ g'(a) & \text{if } x = a \end{cases}$$

Prove that if $g'(a)$ exists, f is continuous at a.

37. In Illustration 1 we have the absolute-value function, defined by $f(x) = |x|$, and this function is not differentiable at 0. Prove that $f'(x) = |x|/x$ for all $x \neq 0$. (*Hint:* Let $|x| = \sqrt{x^2}$.)

38. If $f(x) = |x|$ and $g(x) = -|x|$, (a) find a formula for $(f + g)(x)$. (b) Prove that $f + g$ is differentiable at 0 while neither f nor g is differentiable there.

39. A school-sponsored trip that can accommodate up to 250 students will cost each student \$15 if not more than 150 students make the trip; however, the cost per student will be reduced \$0.05 for each student in excess of 150 until the cost reaches \$10 per student. (a) If x students make the trip, express the number of dollars in the gross income as a function of x. (b) Because x is the number of students, x is a nonnegative integer. To have the continuity requirements to apply the calculus, x must be a nonnegative real number. With this assumption show that the function in part (a) is continuous at 150 but not differentiable there.

40. A manufacturer can make a profit of \$20 on each item if not more than 800 items are produced each week. The profit decreases \$0.02 per item over 800. (a) If x items are produced each week, express the number of dollars in the manufacturer's weekly profit as a function of x. (b) Because x is the number of items, x is a nonnegative integer. To have the continuity requirements to apply the calculus, x must be a nonnegative real number. With this assumption show that the function in part (a) is continuous at 800 but not differentiable there.

41. Given

$$f(x) = \begin{cases} 0 & \text{if } x \le 0 \\ x^n & \text{if } 0 < x \end{cases}$$

where n is a positive integer. (a) For what values of n is f differentiable for all values of x? (b) For what values of n is f' continuous for all values of x?

2.4
THEOREMS ON
DIFFERENTIATION OF
ALGEBRAIC FUNCTIONS

Recall from Sec. 2.2 that the operation of finding the derivative of a function is called differentiation, which can be performed by applying Definition 2.2.1. However, because this process is usually rather lengthy, we state and prove some theorems that enable us to find the derivative of certain functions more easily. These theorems are proved by applying Definition 2.2.1. Following the proof of each theorem we state the corresponding formula of differentiation.

2.4.1 THEOREM   If c is a constant and if $f(x) = c$ for all x, then

$$f'(x) = 0$$

PROOF

$$f'(x) = \lim_{\Delta x \to 0} \frac{f(x + \Delta x) - f(x)}{\Delta x}$$

$$= \lim_{\Delta x \to 0} \frac{c - c}{\Delta x}$$

$$= \lim_{\Delta x \to 0} 0$$

$$= 0 \qquad \blacksquare$$

$$D_x(c) = 0 \qquad\qquad (1)$$

The derivative of a constant is zero.

▶ ILLUSTRATION 1   If $f(x) = 5$, then

$$f'(x) = 0$$ ◀

2.4.2 THEOREM   If n is a positive integer and if $f(x) = x^n$, then

$$f'(x) = nx^{n-1}$$

PROOF

$$f'(x) = \lim_{\Delta x \to 0} \frac{f(x + \Delta x) - f(x)}{\Delta x}$$

$$= \lim_{\Delta x \to 0} \frac{(x + \Delta x)^n - x^n}{\Delta x}$$

Applying the binomial theorem to $(x + \Delta x)^n$ we have

$$f'(x) = \lim_{\Delta x \to 0} \frac{\left[x^n + nx^{n-1}\Delta x + \frac{n(n-1)}{2!} x^{n-2}(\Delta x)^2 + \ldots + nx(\Delta x)^{n-1} + (\Delta x)^n \right] - x^n}{\Delta x}$$

$$= \lim_{\Delta x \to 0} \frac{nx^{n-1}\Delta x + \frac{n(n-1)}{2!} x^{n-2}(\Delta x)^2 + \ldots + nx(\Delta x)^{n-1} + (\Delta x)^n}{\Delta x}$$

The numerator and denominator are divided by Δx to obtain

$$f'(x) = \lim_{\Delta x \to 0} \left[nx^{n-1} + \frac{n(n-1)}{2!} x^{n-2}(\Delta x) + \ldots + nx(\Delta x)^{n-2} + (\Delta x)^{n-1} \right]$$

Every term except the first has a factor of Δx; therefore every term except the first approaches zero as Δx approaches zero. Thus

$$f'(x) = nx^{n-1} \qquad \blacksquare$$

$$D_x(x^n) = nx^{n-1} \qquad\qquad (2)$$

▶ ILLUSTRATION 2   (a) If $f(x) = x^8$, then

$$f'(x) = 8x^7$$

(b) If $f(x) = x$, then

$$f'(x) = 1 \cdot x^0$$
$$= 1 \cdot 1$$
$$= 1$$ ◀

2.4.3 THEOREM   If f is a function, c is a constant, and g is the function defined by

$$g(x) = c \cdot f(x)$$

then if $f'(x)$ exists,

$$g'(x) = c \cdot f'(x)$$

PROOF

$$g'(x) = \lim_{\Delta x \to 0} \frac{g(x + \Delta x) - g(x)}{\Delta x}$$

$$= \lim_{\Delta x \to 0} \frac{cf(x + \Delta x) - cf(x)}{\Delta x}$$

$$= \lim_{\Delta x \to 0} c \cdot \left[\frac{f(x + \Delta x) - f(x)}{\Delta x} \right]$$

$$= c \cdot \lim_{\Delta x \to 0} \frac{f(x + \Delta x) - f(x)}{\Delta x}$$

$$= cf'(x)$$ ■

$$D_x[c \cdot f(x)] = c \cdot D_x f(x) \tag{3}$$

The derivative of a constant times a function is the constant times the derivative of the function if this derivative exists.

By combining Theorems 2.4.2 and 2.4.3 we obtain the following result: If $f(x) = cx^n$, where n is a positive integer and c is a constant,

$$f'(x) = cnx^{n-1}$$
$$D_x(cx^n) = cnx^{n-1} \tag{4}$$

▶ ILLUSTRATION 3   If $f(x) = 5x^7$, then

$$f'(x) = 5 \cdot 7x^6$$
$$= 35x^6$$ ◀

2.4.4 THEOREM   If f and g are functions and if h is the function defined by

$$h(x) = f(x) + g(x)$$

then if $f'(x)$ and $g'(x)$ exist,

$$h'(x) = f'(x) + g'(x)$$

PROOF

$$
\begin{aligned}
h'(x) &= \lim_{\Delta x \to 0} \frac{h(x + \Delta x) - h(x)}{\Delta x} \\
&= \lim_{\Delta x \to 0} \frac{[f(x + \Delta x) + g(x + \Delta x)] - [f(x) + g(x)]}{\Delta x} \\
&= \lim_{\Delta x \to 0} \left[\frac{f(x + \Delta x) - f(x)}{\Delta x} + \frac{g(x + \Delta x) - g(x)}{\Delta x} \right] \\
&= \lim_{\Delta x \to 0} \frac{f(x + \Delta x) - f(x)}{\Delta x} + \lim_{\Delta x \to 0} \frac{g(x + \Delta x) - g(x)}{\Delta x} \\
&= f'(x) + g'(x)
\end{aligned}
$$

∎

$$D_x[f(x) + g(x)] = D_x f(x) + D_x g(x) \tag{5}$$

The derivative of the sum of two functions is the sum of their derivatives if these derivatives exist.

The result of the preceding theorem can be extended to any finite number of functions, by mathematical induction, and this is stated as another theorem.

2.4.5 THEOREM   The derivative of the sum of a finite number of functions is equal to the sum of their derivatives if these derivatives exist.

From the preceding theorems the derivative of any polynomial function can be found easily.

EXAMPLE 1
Given

$$f(x) = 7x^4 - 2x^3 + 8x + 5$$

find $f'(x)$.

SOLUTION

$$
\begin{aligned}
f'(x) &= D_x(7x^4 - 2x^3 + 8x + 5) \\
&= D_x(7x^4) + D_x(-2x^3) + D_x(8x) + D_x(5) \\
&= 28x^3 - 6x^2 + 8
\end{aligned}
$$

2.4.6 THEOREM   If f and g are functions and if h is the function defined by

$$h(x) = f(x)g(x)$$

then if $f'(x)$ and $g'(x)$ exist,

$$h'(x) = f(x)g'(x) + g(x)f'(x)$$

PROOF

$$h'(x) = \lim_{\Delta x \to 0} \frac{h(x + \Delta x) - h(x)}{\Delta x}$$

$$= \lim_{\Delta x \to 0} \frac{f(x + \Delta x) \cdot g(x + \Delta x) - f(x) \cdot g(x)}{\Delta x}$$

If $f(x + \Delta x) \cdot g(x)$ is subtracted and added in the numerator, then

$$h'(x) = \lim_{\Delta x \to 0} \frac{f(x + \Delta x) \cdot g(x + \Delta x) - f(x + \Delta x) \cdot g(x) + f(x + \Delta x) \cdot g(x) - f(x) \cdot g(x)}{\Delta x}$$

$$= \lim_{\Delta x \to 0} \left[f(x + \Delta x) \cdot \frac{g(x + \Delta x) - g(x)}{\Delta x} + g(x) \cdot \frac{f(x + \Delta x) - f(x)}{\Delta x} \right]$$

$$= \lim_{\Delta x \to 0} \left[f(x + \Delta x) \cdot \frac{g(x + \Delta x) - g(x)}{\Delta x} \right] + \lim_{\Delta x \to 0} \left[g(x) \cdot \frac{f(x + \Delta x) - f(x)}{\Delta x} \right]$$

$$= \lim_{\Delta x \to 0} f(x + \Delta x) \cdot \lim_{\Delta x \to 0} \frac{g(x + \Delta x) - g(x)}{\Delta x} + \lim_{\Delta x \to 0} g(x) \cdot \lim_{\Delta x \to 0} \frac{f(x + \Delta x) - f(x)}{\Delta x}$$

Because f is differentiable at x, by Theorem 2.3.1 f is continuous at x; therefore $\lim_{\Delta x \to 0} f(x + \Delta x) = f(x)$. Also,

$$\lim_{\Delta x \to 0} \frac{g(x + \Delta x) - g(x)}{\Delta x} = g'(x) \qquad \lim_{\Delta x \to 0} \frac{f(x + \Delta x) - f(x)}{\Delta x} = f'(x)$$

and

$$\lim_{\Delta x \to 0} g(x) = g(x)$$

thus giving

$$h'(x) = f(x)g'(x) + g(x)f'(x) \qquad \blacksquare$$

$$D_x[f(x)g(x)] = f(x) \cdot D_x g(x) + g(x) \cdot D_x f(x) \tag{6}$$

The derivative of the product of two functions is the first function times the derivative of the second function plus the second function times the derivative of the first function if these derivatives exist.

EXAMPLE 2
Given

$$h(x) = (2x^3 - 4x^2)(3x^5 + x^2)$$

find $h'(x)$.

SOLUTION

$$h'(x) = (2x^3 - 4x^2)(15x^4 + 2x) + (3x^5 + x^2)(6x^2 - 8x)$$
$$= (30x^7 - 60x^6 + 4x^4 - 8x^3) + (18x^7 - 24x^6 + 6x^4 - 8x^3)$$
$$= 48x^7 - 84x^6 + 10x^4 - 16x^3$$

In Example 2, note that if multiplication is performed before the differentiation, the same result is obtained. Doing this we have

$$h(x) = 6x^8 - 12x^7 + 2x^5 - 4x^4$$
$$h'(x) = 48x^7 - 84x^6 + 10x^4 - 16x^3$$

2.4.7 THEOREM   If f and g are functions and if h is the function defined by

$$h(x) = \frac{f(x)}{g(x)}, \qquad \text{where } g(x) \neq 0$$

then if $f'(x)$ and $g'(x)$ exist,

$$h'(x) = \frac{g(x)f'(x) - f(x)g'(x)}{[g(x)]^2}$$

PROOF

$$h'(x) = \lim_{\Delta x \to 0} \frac{h(x + \Delta x) - h(x)}{\Delta x}$$

$$= \lim_{\Delta x \to 0} \frac{\dfrac{f(x + \Delta x)}{g(x + \Delta x)} - \dfrac{f(x)}{g(x)}}{\Delta x}$$

$$= \lim_{\Delta x \to 0} \frac{f(x + \Delta x) \cdot g(x) - f(x) \cdot g(x + \Delta x)}{\Delta x \cdot g(x) \cdot g(x + \Delta x)}$$

If $f(x) \cdot g(x)$ is subtracted and added in the numerator, then

$$h'(x) = \lim_{\Delta x \to 0} \frac{f(x + \Delta x) \cdot g(x) - f(x) \cdot g(x) - f(x) \cdot g(x + \Delta x) + f(x) \cdot g(x)}{\Delta x \cdot g(x) \cdot g(x + \Delta x)}$$

$$= \lim_{\Delta x \to 0} \frac{\left[g(x) \cdot \dfrac{f(x + \Delta x) - f(x)}{\Delta x} \right] - \left[f(x) \cdot \dfrac{g(x + \Delta x) - g(x)}{\Delta x} \right]}{g(x) \cdot g(x + \Delta x)}$$

$$= \frac{\displaystyle\lim_{\Delta x \to 0} g(x) \cdot \lim_{\Delta x \to 0} \frac{f(x + \Delta x) - f(x)}{\Delta x} - \lim_{\Delta x \to 0} f(x) \cdot \lim_{\Delta x \to 0} \frac{g(x + \Delta x) - g(x)}{\Delta x}}{\displaystyle\lim_{\Delta x \to 0} g(x) \cdot \lim_{\Delta x \to 0} g(x + \Delta x)}$$

Because g is differentiable at x, then g is continuous at x; thus we have $\lim_{\Delta x \to 0} g(x + \Delta x) = g(x)$. Furthermore, $\lim_{\Delta x \to 0} g(x) = g(x)$ and $\lim_{\Delta x \to 0} f(x) = f(x)$.

With these results and the definitions of $f'(x)$ and $g'(x)$ we get

$$h'(x) = \frac{g(x) \cdot f'(x) - f(x) \cdot g'(x)}{g(x) \cdot g(x)}$$

$$= \frac{g(x)f'(x) - f(x)g'(x)}{[g(x)]^2}$$ ∎

$$D_x\left[\frac{f(x)}{g(x)}\right] = \frac{g(x)\,D_x f(x) - f(x)\,D_x g(x)}{[g(x)]^2}\qquad(7)$$

The derivative of the quotient of two functions is the fraction having as its denominator the square of the original denominator, and as its numerator the denominator times the derivative of the numerator minus the numerator times the derivative of the denominator if these derivatives exist.

EXAMPLE 3
Find

$$D_x\left(\frac{2x^3 + 4}{x^2 - 4x + 1}\right)$$

SOLUTION

$$D_x\left(\frac{2x^3 + 4}{x^2 - 4x + 1}\right) = \frac{(x^2 - 4x + 1)(6x^2) - (2x^3 + 4)(2x - 4)}{(x^2 - 4x + 1)^2}$$

$$= \frac{6x^4 - 24x^3 + 6x^2 - 4x^4 + 8x^3 - 8x + 16}{(x^2 - 4x + 1)^2}$$

$$= \frac{2x^4 - 16x^3 + 6x^2 - 8x + 16}{(x^2 - 4x + 1)^2}$$

2.4.8 THEOREM   If $f(x) = x^{-n}$, where $-n$ is a negative integer and $x \neq 0$, then

$$f'(x) = -nx^{-n-1}$$

PROOF   If $-n$ is a negative integer, then n is a positive integer. We write

$$f(x) = \frac{1}{x^n}$$

From Theorem 2.4.7,

$$f'(x) = \frac{x^n \cdot 0 - 1 \cdot nx^{n-1}}{(x^n)^2}$$

$$= \frac{-nx^{n-1}}{x^{2n}}$$

$$= -nx^{n-1-2n}$$

$$= -nx^{-n-1}$$ ∎

EXAMPLE 4
Find

$$\frac{d}{dx}\left(\frac{3}{x^5}\right)$$

SOLUTION

$$\frac{d}{dx}\left(\frac{3}{x^5}\right) = \frac{d}{dx}(3x^{-5})$$

$$= 3(-5x^{-6})$$

$$= -\frac{15}{x^6}$$

If r is any positive or negative integer, it follows from Theorems 2.4.2 and 2.4.8 that

$$D_x(x^r) = rx^{r-1} \tag{8}$$

and from Theorems 2.4.2, 2.4.3, and 2.4.8 we obtain

$$D_x(cx^r) = crx^{r-1} \tag{9}$$

EXERCISES 2.4

In Exercises 1 through 24, differentiate the given function by applying the theorems of this section.

1. $f(x) = 7x - 5$
2. $g(x) = 8 - 3x$
3. $g(x) = 1 - 2x - x^2$
4. $f(x) = 4x^2 + 4x + 1$
5. $f(x) = x^3 - 3x^2 + 5x - 2$
6. $f(x) = 3x^4 - 5x^2 + 1$
7. $f(x) = \frac{1}{8}x^8 - x^4$
8. $g(x) = x^7 - 2x^5 + 5x^3 - 7x$
9. $F(t) = \frac{1}{4}t^4 - \frac{1}{2}t^2$
10. $H(x) = \frac{1}{3}x^3 - x + 2$
11. $v(r) = \frac{4}{3}\pi r^3$
12. $G(y) = y^{10} + 7y^5 - y^3 + 1$
13. $F(x) = x^2 + 3x + \dfrac{1}{x^2}$
14. $f(x) = \dfrac{x^3}{3} + \dfrac{3}{x^3}$
15. $g(x) = 4x^4 - \dfrac{1}{4x^4}$
16. $f(x) = x^4 - 5 + x^{-2} + 4x^{-4}$
17. $g(x) = \dfrac{3}{x^2} + \dfrac{5}{x^4}$
18. $H(x) = \dfrac{5}{6x^5}$
19. $f(s) = \sqrt{3}(s^3 - s^2)$
20. $g(x) = (2x^2 + 5)(4x - 1)$
21. $f(x) = (2x^4 - 1)(5x^3 + 6x)$
22. $f(x) = (4x^2 + 3)^2$
23. $G(y) = (7 - 3y^3)^2$
24. $F(t) = (t^3 - 2t + 1)(2t^2 + 3t)$

In Exercises 25 through 36, compute the indicated derivative by applying the theorems of this section.

25. $D_x[(x^2 - 3x + 2)(2x^3 + 1)]$
26. $D_x\left(\dfrac{2x}{x + 3}\right)$
27. $D_x\left(\dfrac{x}{x - 1}\right)$
28. $D_y\left(\dfrac{2y + 1}{3y + 4}\right)$
29. $\dfrac{d}{dx}\left(\dfrac{x^2 + 2x + 1}{x^2 - 2x + 1}\right)$
30. $\dfrac{d}{dx}\left(\dfrac{4 - 3x - x^2}{x - 2}\right)$
31. $\dfrac{d}{dt}\left(\dfrac{5t}{1 + 2t^2}\right)$
32. $\dfrac{d}{dx}\left(\dfrac{x^4 - 2x^2 + 5x + 1}{x^4}\right)$
33. $\dfrac{d}{dy}\left(\dfrac{y^3 - 8}{y^3 + 8}\right)$

34. $\dfrac{d}{ds}\left(\dfrac{s^2 - a^2}{s^2 + a^2}\right)$ **35.** $D_x\left[\dfrac{2x + 1}{x + 5}(3x - 1)\right]$ **36.** $D_x\left[\dfrac{x^3 + 1}{x^2 + 3}(x^2 - 2x^{-1} + 1)\right]$

37. Find an equation of the tangent line to the curve $y = x^3 - 4$ at the point $(2, 4)$.
38. Find an equation of the normal line to the curve $y = 4x^2 - 8x$ at the point $(1, -4)$ (see Exercise 37 in Exercises 2.1).
39. Find an equation of the normal line to the curve $y = 10/(14 - x^2)$ at the point $(4, -5)$ (see Exercise 37 in Exercises 2.1).
40. Find an equation of the tangent line to the curve $y = 8/(x^2 + 4)$ at the point $(2, 1)$.
41. Find an equation of the line tangent to the curve $y = 3x^2 - 4x$ and parallel to the line $2x - y + 3 = 0$.
42. Find an equation of the tangent line to the curve $y = x^4 - 6x$ that is perpendicular to the line $x - 2y + 6 = 0$.
43. Find an equation of each of the normal lines to the curve $y = x^3 - 4x$ that is parallel to the line $x + 8y - 8 = 0$ (see Exercise 37 in Exercises 2.1).
44. Find an equation of each of the tangent lines to the curve $3y = x^3 - 3x^2 + 6x + 4$ that is parallel to the line $2x - y + 3 = 0$.
45. Find an equation of each of the lines through the point $(4, 13)$ that is tangent to the curve $y = 2x^2 - 1$.
46. Given $f(x) = \frac{1}{3}x^3 + 2x^2 + 5x + 5$. Show that $f'(x) \geq 0$ for all values of x.
47. If f, g, and h are functions and $\phi(x) = f(x) \cdot g(x) \cdot h(x)$, prove that if $f'(x)$, $g'(x)$, and $h'(x)$ exist,

$$\phi'(x) = f(x) \cdot g(x) \cdot h'(x) + f(x) \cdot g'(x) \cdot h(x) + f'(x) \cdot g(x) \cdot h(x)$$

(*Hint:* Apply Theorem 2.4.6 twice.)

Use the result of Exercise 47 to differentiate the functions in Exercises 48 through 51.
48. $f(x) = (x^2 + 3)(2x - 5)(3x + 2)$ **49.** $h(x) = (3x + 2)^2(x^2 - 1)$
50. $g(x) = (3x^3 + x^{-3})(x + 3)(x^2 - 5)$ **51.** $\phi(x) = (2x^2 + x + 1)^3$

**2.5
RECTILINEAR MOTION AND
THE DERIVATIVE AS A
RATE OF CHANGE**

Consider a particle moving along a straight line. Such a motion is called **rectilinear motion**. One direction is chosen arbitrarily as positive, and the opposite direction is negative. For simplicity assume that the motion of the particle is along a horizontal line, with distance to the right as positive and distance to the left as negative. Select some point on the line and denote it by the letter O. Let f be the function determining the directed distance of the particle from O at any particular time.

To be more specific, let s centimeters (cm) be the directed distance of the particle from O at t seconds (sec) of time. Then f is the function defined by the equation

$$s = f(t) \tag{1}$$

which gives the directed distance from the point O to the particle at a particular instant of time.

▶ ILLUSTRATION 1   Let

$$s = t^2 + 2t - 3$$

Then when $t = 0$, $s = -3$; therefore the particle is 3 cm to the left of point O when $t = 0$. When $t = 1$, $s = 0$; so the particle is at point O at 1 sec. When $t = 2$, $s = 5$; so the particle is 5 cm to the right of point O at 2 sec. When $t = 3$, $s = 12$; so the particle is 12 cm to the right of point O at 3 sec.

Figure 2.5.1 illustrates the various positions of the particle for specific values of t.

FIGURE 2.5.1

Between the time for $t = 1$ and $t = 3$, the particle moves from the point where $s = 0$ to the point where $s = 12$; thus in the 2-sec interval the change in the directed distance from O is 12 cm. The average velocity of the particle is the ratio of the change in the directed distance from a fixed point to the change in the time. So the number of centimeters per second in the average velocity of the particle from $t = 1$ to $t = 3$ is $\frac{12}{2} = 6$. From $t = 0$ to $t = 2$, the change in the directed distance from O of the particle is 8 cm, and so the number of centimeters per second in the average velocity of the particle in this 2-sec interval is $\frac{8}{2} = 4$.   ◀

In Illustration 1 the average velocity of the particle is obviously not constant; and the average velocity supplies no specific information about the motion of the particle at any particular instant. For example, if a car travels a distance of 100 km in the same direction and it takes 2 hr, we say that the average velocity in traveling that distance is 50 km/hr. However, from this information we cannot determine the speedometer reading of the car at any particular time in the 2-hr period. The speedometer reading at a specific time is referred to as the *instantaneous velocity*. The following discussion enables us to arrive at a definition of what is meant by "instantaneous velocity."

Let (1) define s (the number of centimeters in the directed distance of the particle from point O) as a function of t (the number of seconds in the time). When $t = t_1$, $s = s_1$. The change in the directed distance from O is $(s - s_1)$ cm over the interval of time $(t - t_1)$ sec, and the number of centimeters per second in the average velocity of the particle over this interval of time is given by

$$\frac{s - s_1}{t - t_1}$$

or, because $s = f(t)$ and $s_1 = f(t_1)$, the average velocity is found from

$$\frac{f(t) - f(t_1)}{t - t_1} \tag{2}$$

Now the shorter the interval is from t_1 to t, the closer the average velocity will be to what we would intuitively think of as the instantaneous velocity at t_1.

For example, if the speedometer reading of a car as it passes a point P_1 is 80 km/hr and if a point P is, for instance, 10 m from P_1, then the average velocity of the car as it travels this 10 m will very likely be close to 80 km/hr because the variation of the velocity of the car along this short stretch is probably slight. Now if the distance from P_1 to P were shortened to 5 m, the average velocity of the car in this interval would be even closer to the speedometer reading of the car as it passes P_1. We can continue this process, and the speedometer reading at P_1 can be represented as the limit of the average velocity between P_1 and P as P approaches P_1. That is, the *instantaneous velocity* can be defined as the limit of quotient (2) as t approaches t_1, provided the limit exists. This limit is the derivative of the function f at t_1. We have, then, the following definition.

2.5.1 DEFINITION   If f is a function given by the equation

$$s = f(t)$$

and a particle is moving along a straight line such that s is the number of units in the directed distance of the particle from a fixed point on the line at t units of time, then the **instantaneous velocity** of the particle at t units of time is v units of velocity, where

$$v = f'(t) \quad \Leftrightarrow \quad v = \frac{ds}{dt}$$

if it exists.

The instantaneous velocity may be either positive or negative, depending on whether the particle is moving along the line in the positive or the negative direction. When the instantaneous velocity is zero, the particle is at rest.

The **speed** of a particle at any time is defined as the absolute value of the instantaneous velocity. Hence the speed is a nonnegative number. The terms "speed" and "instantaneous velocity" are often confused. It should be noted that the speed indicates only how fast the particle is moving, whereas the instantaneous velocity also tells the direction of motion.

EXAMPLE 1

A particle is moving along a horizontal line according to the equation

$$s = 2t^3 - 4t^2 + 2t - 1$$

Determine the intervals of time when the particle is moving to the right and when it is moving to the left. Also determine the instant when the particle reverses its direction.

SOLUTION

$$v = \frac{ds}{dt}$$

$$= 6t^2 - 8t + 2$$
$$= 2(3t^2 - 4t + 1)$$
$$= 2(3t - 1)(t - 1)$$

The instantaneous velocity is zero when $t = \frac{1}{3}$ and $t = 1$. Therefore the particle is at rest at these two times. The particle is moving to the right when v is positive, and it is moving to the left when v is negative. We determine the sign of v for various intervals of t, and the results are given in Table 2.5.1.

Table 2.5.1

	$3t - 1$	$t - 1$	Conclusion
$t < \frac{1}{3}$	$-$	$-$	v is positive, and the particle is moving to the right
$t = \frac{1}{3}$	0	$-$	v is zero, and the particle is changing direction from right to left
$\frac{1}{3} < t < 1$	$+$	$-$	v is negative, and the particle is moving to the left
$t = 1$	$+$	0	v is zero, and the particle is changing direction from left to right
$1 < t$	$+$	$+$	v is positive, and the particle is moving to the right

The motion of the particle, indicated in Fig. 2.5.2, is along the horizontal line; however, the behavior of the motion is indicated above the line. Table 2.5.2 gives values of s and v for specific values of t.

Table 2.5.2

t	s	v
-1	-9	16
0	-1	2
$\frac{1}{3}$	$-\frac{19}{27}$	0
1	-1	0
2	3	10

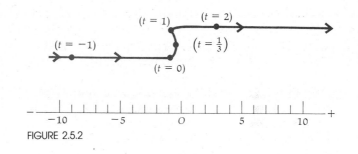

FIGURE 2.5.2

EXAMPLE 2

A ball is thrown vertically upward from the ground with an initial velocity of 64 ft/sec. If the positive direction of the distance from the starting point is up, the equation of motion is

$$s = -16t^2 + 64t$$

Let t be the number of seconds in the time that has elapsed since the ball was thrown and s be the number of feet in the distance of the ball from the starting point at t sec. (a) Find the instantaneous velocity of the ball at the end of 1 sec. Is the ball rising or falling at the end of 1 sec? (b) Find the instantaneous velocity of the ball at the end of 3 sec. Is the ball rising or falling at the end of 3 sec? (c) How many seconds does it take the ball to reach its highest point? (d) How high will the ball go? (e) Find the speed of the ball at the end of 1 sec and at the end of 3 sec. (f) How many seconds does it take the ball to reach the ground? (g) Find the instantaneous velocity of the ball when it reaches the ground.

SOLUTION

Let $v(t)$ be the number of feet per second in the instantaneous velocity of the ball at t sec. Then

$$v(t) = \frac{ds}{dt}$$

$$= -32t + 64$$

(a) $v(1) = -32(1) + 64$; that is, $v(1) = 32$; so at the end of 1 sec the ball is rising with an instantaneous velocity of 32 ft/sec.
(b) $v(3) = -32(3) + 64$; that is, $v(3) = -32$; so at the end of 3 sec the ball is falling with an instantaneous velocity of -32 ft/sec.
(c) The ball reaches its highest point when the direction of motion changes, that is, when $v(t) = 0$. Setting $v(t) = 0$ we obtain $-32t + 64 = 0$. Thus $t = 2$.
(d) When $t = 2$, $s = 64$; therefore the ball reaches a highest point of 64 ft above the starting point.
(e) $|v(t)|$ is the number of feet per second in the speed of the ball at t sec.

$$|v(1)| = 32 \quad \text{and} \quad |v(3)| = 32$$

(f) The ball will reach the ground for $s = 0$. Setting $s = 0$ we have $-16t^2 + 64t = 0$, from which we obtain $t = 0$ and $t = 4$. Therefore the ball will reach the ground in 4 sec.
(g) $v(4) = -64$; when the ball reaches the ground, its instantaneous velocity is -64 ft/sec.

Table 2.5.3 gives values of s and v for some specific values of t. The motion of the ball is indicated in Fig. 2.5.3. The motion is assumed to be in a straight vertical line, and the behavior of the motion is indicated to the left of the line.

The concept of velocity in rectilinear motion corresponds to the more general concept of instantaneous rate of change. For example, if a particle is moving along a straight line according to the equation $s = f(t)$, the velocity of the particle at t units of time is determined by the derivative of s with respect to t. Because velocity can be interpreted as a rate of change of distance per unit change in time, the derivative of s with respect to t is the rate of change of s per unit change in t.

In a similar way, if a quantity y is a function of a quantity x, we may express the rate of change of y per unit change in x. The discussion is analogous to the discussions of the slope of a tangent line to a graph and the instantaneous velocity of a particle moving along a straight line.

If the functional relationship between y and x is given by

$$y = f(x)$$

and if x changes from the value x_1 to $x_1 + \Delta x$, then y changes from $f(x_1)$ to $f(x_1 + \Delta x)$. So the change in y, which we denote by Δy, is $f(x_1 + \Delta x) - f(x_1)$ when the change in x is Δx. The average rate of change of y per unit change

Table 2.5.3

t	s	v
0	0	64
$\frac{1}{2}$	28	48
1	48	32
2	64	0
3	48	-32
$\frac{7}{2}$	28	-48
4	0	-64

FIGURE 2.5.3

in x, as x changes from x_1 to $x_1 + \Delta x$, is then

$$\frac{f(x_1 + \Delta x) - f(x_1)}{\Delta x} = \frac{\Delta y}{\Delta x} \tag{3}$$

If the limit of this quotient exists as $\Delta x \to 0$, this limit is what we intuitively think of as the instantaneous rate of change of y per unit change in x at x_1. Accordingly, we have the following definition.

2.5.2 DEFINITION   If $y = f(x)$, the **instantaneous rate of change of y per unit change in x at x_1** is $f'(x_1)$ or, equivalently, the derivative of y with respect to x at x_1, if it exists there.

To illustrate geometrically, let $f'(x_1)$ be the instantaneous rate of change of y per unit change in x at x_1. Then if $f'(x_1)$ is multiplied by Δx (the change in x), we have the change that would occur in y if the point (x, y) were to move along the tangent line at (x_1, y_1) of the graph of $y = f(x)$. See Fig. 2.5.4. The average rate of change of y per unit change in x is given by the fraction in (3), and if this fraction is multiplied by Δx, we get Δy, which is the actual change in y caused by a change of Δx in x when the point (x, y) moves along the graph.

FIGURE 2.5.4

EXAMPLE 3
Let e cm be the length of an edge of a cube. Find the average rate of change of the volume per centimeter change in the length of the edge as e changes from (a) 3.00 to 3.20; (b) 3.00 to 3.10; (c) 3.00 to 3.01. (d) What is the instantaneous rate of change of the volume per centimeter change in the length of the edge when $e = 3$?

SOLUTION
Let $V(e)$ cm^3 be the volume of the cube. Then

$$V(e) = e^3$$

Then the average rate of change of $V(e)$ per unit change in e as e changes from e_1 to $e_1 + \Delta e$ is

$$\frac{V(e_1 + \Delta e) - V(e_1)}{\Delta e}$$

(a) $e_1 = 3$, $\Delta e = 0.2$

$$\frac{V(3.2) - V(3)}{0.2} = \frac{(3.2)^3 - 3^3}{0.2}$$

$$\approx \frac{5.77}{0.2}$$

$$\approx 28.8$$

(b) $e_1 = 3$, $\Delta e = 0.1$

$$\frac{V(3.1) - V(3)}{0.1} = \frac{(3.1)^3 - 3^3}{0.1}$$

$$\approx \frac{2.79}{0.1}$$

$$= 27.9$$

(c) $e_1 = 3$, $\Delta e = 0.01$

$$\frac{V(3.01) - V(3)}{0.01} = \frac{(3.01)^3 - 3^3}{0.01}$$

$$\approx \frac{0.271}{0.01}$$

$$= 27.1$$

In part (a) we see that as the length of the edge of the cube changes from 3.00 cm to 3.20 cm, the change in the volume is 5.77 cm^3 and the average rate of change of the volume is 28.8 cm^3 per centimeter change in the length of the edge. There are similar interpretations of parts (b) and (c).

(d) The instantaneous rate of change of $V(e)$ per unit change in e at 3 is $V'(3)$.

$$V'(e) = 3e^2 \qquad V'(3) = 27$$

Therefore, when the length of the edge of the cube is 3 cm, the instantaneous rate of change of the volume is 27 cm^3 per centimeter change in the length of the edge.

EXAMPLE 4

The annual gross earnings of a particular corporation t years from January 1, 1984, is p millions of dollars, and

$$p = \tfrac{2}{5}t^2 + 2t + 10$$

Find (a) the rate at which the gross earnings were growing on January 1, 1986; (b) the rate at which the gross earnings should be growing on January 1, 1990.

SOLUTION

(a) On January 1, 1986, $t = 2$. Thus we compute $\dfrac{dp}{dt}$ when $t = 2$.

$$\frac{dp}{dt} = \frac{4}{5}t + 2 \qquad \frac{dp}{dt}\bigg]_{t=2} = \frac{8}{5} + 2$$

$$= 3.6$$

Thus on January 1, 1986, the gross earnings were growing at the rate of $3.6 million per year.

(b) On January 1, 1990, $t = 6$, and

$$\frac{dp}{dt}\bigg]_{t=6} = \frac{24}{5} + 2$$

$$= 6.8$$

Therefore on January 1, 1990, the gross earnings should be growing at the rate of $6.8 million per year.

The results of Example 4 are meaningful only if they are compared to the actual earnings of the corporation. For example, if on January 1, 1985, it was found that the earnings of the corporation for the year 1984 had been $3 million, then the rate of growth on January 1, 1986, of $3.6 million annually would have been excellent. However, if the earnings in 1984 had been $300 million, then the growth rate on January 1, 1986, would have been very poor. The measure used to compare the rate of change with the amount of the quantity that is being changed is called the *relative rate*.

2.5.3 DEFINITION   If $y = f(x)$, the **relative rate of change of y with respect to x** at x_1 is given by $\dfrac{f'(x_1)}{f(x_1)}$ or, equivalently, $\dfrac{dy/dx}{y}$ evaluated at $x = x_1$.

If the relative rate is multiplied by 100, the percent rate of change is obtained.

EXAMPLE 5
Find to the nearest 0.1 percent the relative rate of growth of the gross earnings on (a) January 1, 1986, and (b) January 1, 1990, for the corporation of Example 4.

SOLUTION
(a) When $t = 2$,

$$p = \tfrac{2}{5}(4) + 2(2) + 10$$
$$= 15.6$$

Hence on January 1, 1986, the relative rate of growth of the corporation's annual gross earnings was

$$\left.\frac{dp/dt}{p}\right]_{p=2} = \frac{3.6}{15.6}$$
$$\approx 0.231$$
$$= 23.1 \text{ percent}$$

(b) When $t = 6$,

$$p = \tfrac{2}{5}(36) + 2(6) + 10$$
$$= 36.4$$

Therefore on January 1, 1990, the relative rate of growth of the corporation's annual gross earnings should be

$$\left.\frac{dp/dt}{p}\right]_{t=6} = \frac{6.8}{36.4}$$
$$\approx 0.187$$
$$= 18.7 \text{ percent}$$

Observe that the growth rate of $6.8 million for January 1, 1990, is greater than the $3.6 million for January 1, 1986; however, the relative

growth rate of 18.7 percent for January 1, 1990, is less than the relative
growth rate of 23.1 percent for January 1, 1986.

EXERCISES 2.5

In Exercises 1 through 8, a particle is moving along a horizontal line according to the given equation, where s cm is the directed distance of the particle from a point O at t sec. Find the instantaneous velocity $v(t)$ cm/sec at t sec; and then find $v(t_1)$ for the particular value of t_1 given.

1. $s = 3t^2 + 1; t_1 = 3$

2. $s = 8 - t^2; t_1 = 5$

3. $s = \dfrac{1}{4t}; t_1 = \frac{1}{2}$

4. $s = \dfrac{3}{t^2}; t_1 = -2$

5. $s = 2t^3 - t^2 + 5; t_1 = -1$

6. $s = 4t^3 + 2t - 1; t_1 = \frac{1}{2}$

7. $s = \dfrac{2t}{4 + t}; t_1 = 0$

8. $s = \dfrac{1}{t} + \dfrac{3}{t^2}; t_1 = 2$

In Exercises 9 through 14, the motion of a particle is along a horizontal line according to the given equation, where s cm is the directed distance of the particle from a point O at t sec. The positive direction is to the right. Determine the intervals of time when the particle is moving to the right and when it is moving to the left. Also determine when the particle reverses its direction. Show the behavior of the motion by a figure similar to Fig. 2.5.2, and choose values of t at random but include the values of t when the particle reverses its direction.

9. $s = t^3 + 3t^2 - 9t + 4$

10. $s = 2t^3 - 3t^2 - 12t + 8$

11. $s = \frac{2}{3}t^3 + \frac{3}{2}t^2 - 2t + 4$

12. $s = \dfrac{t}{1 + t^2}$

13. $s = \dfrac{t}{9 + t^2}$

14. $s = \dfrac{t + 1}{t^2 + 4}$

15. An object falls from rest, and $s = -16t^2$ where s ft is the distance of the object from the starting point at t sec, and the positive direction is upward. If a stone is dropped from a building 256 ft high, find (a) the instantaneous velocity of the stone 1 sec after it is dropped; (b) the instantaneous velocity of the stone 2 sec after it is dropped; (c) how long it takes the stone to reach the ground; (d) the instantaneous velocity of the stone when it reaches the ground.

16. A stone is dropped from a height of 64 ft. If s ft is the height of the stone t sec after being dropped, then $s = -16t^2 + 64$. (a) How long does it take the stone to reach the ground? (b) Find the instantaneous velocity of the stone when it reaches the ground.

17. A billiard ball is hit and travels in a straight line. If s cm is the distance of the ball from its initial position at t sec, then $s = 100t^2 + 100t$. If the ball hits a cushion that is 39 cm from its initial position, at what velocity does it hit the cushion?

18. If a stone is thrown vertically upward from the ground with an initial velocity of 32 ft/sec, then $s = -16t^2 + 32t$, where s ft is the distance of the stone from the starting point at t sec, and the positive direction is upward. Find (a) the average velocity of the stone during the time interval $\frac{3}{4} \le t \le \frac{5}{4}$; (b) the instantaneous velocity of the stone at $\frac{3}{4}$ sec and at $\frac{5}{4}$ sec; (c) the speed of the stone at $\frac{3}{4}$ sec and at $\frac{5}{4}$ sec; (d) the average velocity of the stone during the time interval $\frac{1}{2} \le t \le \frac{3}{4}$; (e) how many seconds it will take the stone to reach the highest point; (f) how high the stone will go; (g) how many seconds it will take the stone to reach the ground; (h) the instantaneous velocity of the stone when it reaches the ground. Show the behavior of the motion by a figure similar to Fig. 2.5.3.

19. If a ball is given a push so that it has an initial velocity of 24 ft/sec down a certain inclined plane, then $s = 24t + 10t^2$, where s ft is the distance of the ball from the starting point at t sec and the positive direction is down the inclined plane. (a) What is the instantaneous velocity of the ball at t_1 sec? (b) How long does it take for the velocity to increase to 48 ft/sec?

20. A rocket is fired vertically upward, and it is s ft above the ground t sec after being fired, where $s = 560t - 16t^2$ and the positive direction is upward. Find (a) the velocity of the rocket 2 sec after being fired, and (b) how long it takes for the rocket to reach its maximum height.

21. If A cm^2 is the area of a square and s cm is the length of a side of the square, find the average rate of change of A with respect to s as s changes from (a) 4.00 to 4.60; (b) 4.00 to 4.30; (c) 4.00 to 4.10 (d) What is the instantaneous rate of change of A with respect to s when s is 4.00? $\quad A = s^2 \quad \frac{dA}{ds} = 2s = 8$

22. The length of a rectangle is 4 in. more than its width, and the 4-in. difference is maintained as the rectangle increases in size. If A in.^{2} is the area of the rectangle and w in. is the width of the rectangle, find the average rate of change of A with respect to w as w changes from (a) 3.00 to 3.20; (b) 3.00 to 3.10; (c) 3.00 to 3.01. (d) What is the instantaneous rate of change of A with respect to w when w is 3.00?

23. A particular company started doing business on April 1, 1983. The annual gross earnings of the company after t years of operation are p dollars, where $p = 50,000 + 18,000t + 600t^2$. (a) Find the rate at which the gross earnings were growing on April 1, 1985. (b) Find the relative rate of growth of the gross earnings on April 1, 1985. $\quad \frac{dp}{p}$

24. For the company of Exercise 23, find (a) the rate at which the gross earnings should be growing on April 1, 1993, and (b) the relative rate of growth of the gross earnings on April 1, 1993.

25. If water is being drained from a swimming pool and V liters is the volume of water in the pool t min after the draining starts, where $V = 250(1600 - 80t + t^2)$, find (a) the average rate at which the water leaves the pool during the first 5 min, and (b) how fast the water is flowing out of the pool 5 min after the draining starts.

26. Boyle's law for the expansion of a gas is $PV = C$, where P is the number of pounds per square unit of pressure, V is the number of cubic units in the volume of the gas, and C is a constant. Find the instantaneous rate of change of V with respect to P when $P = 4$ and $V = 8$.

27. In an electric circuit, if E volts is the electromotive force, R ohms is the resistance, and I amperes is the current, Ohm's law states that $IR = E$. Assuming that E is constant, show that R decreases at a rate proportional to the inverse square of I. $\quad R = \frac{E}{I} \qquad \frac{dR}{dI} = -\frac{E}{I^2}$

28. A balloon maintains the shape of a sphere as it is being inflated. Find the rate of change of the surface area with respect to the radius at the instant when the radius is 2 m.

29. A cold front approaches the college campus. The temperature is T degrees t hours after midnight, and

$$T = 0.1(400 - 40t + t^2) \qquad 0 \le t \le 12$$

(a) Find the average rate of change of T with respect to t between 5 A.M. and 6 A.M. (b) Find the instantaneous rate of change of T with respect to t at 5 A.M.

30. It is estimated that a worker in a shop that makes picture frames can paint y frames x hours after starting work at 8 A.M., and

$$y = 3x + 8x^2 - x^3 \qquad 0 \le x \le 4$$

(a) Find the rate at which the worker is painting at 10 A.M. (b) Find the number of frames that the worker paints between 10 A.M. and 11 A.M.

31. Suppose that the number of people in the population of a particular city t years after January 1, 1986, will be $40t^2 + 200t + 10,000$. (a) Find the rate at which the population will be growing on January 1, 1995. (b) Find the rate at which the population will be growing on January 1, 2001. (c) Find the relative rate of growth of the population on January 1, 1995. (d) Find the relative rate of growth of the population on January 1, 2001.

32. Two particles, A and B, move to the right along a horizontal line. They start at a point O, s ft is the directed distance of the particle from O at t sec, and the equations of motion are

$$s = 4t^2 + 5t \quad \text{(for particle } A\text{)}$$

$$s = 7t^2 + 3t \quad \text{(for particle } B\text{)}$$

If $t = 0$ at the start, for what values of t will the velocity of particle A exceed the velocity of particle B?

33. The profit of a retail store is $100y$ dollars when x dollars are spent daily on advertising and $y = 2500 + 36x - 0.2x^2$. Use the derivative to determine if it would be profitable for the daily advertising budget to be increased if the daily advertising budget is (a) $60 and (b) $300. (c) What is the maximum value for x below which it is profitable to increase the advertising budget?

34. Show that for any linear function f the average rate of change of $f(x)$ as x changes from x_1 to $x_1 + k$ is the same as the instantaneous rate of change of $f(x)$ at x_1.

**2.6
DERIVATIVES OF THE
TRIGONOMETRIC
FUNCTIONS**

To show that the sine function has a derivative we apply the trigonometric identity

$$\sin(a + b) = \sin a \cos b + \cos a \sin b \tag{1}$$

as well as Theorems 1.8.2 and 1.8.5.

Let f be the sine function, so that

$$f(x) = \sin x$$

From the definition of a derivative,

$$f'(x) = \lim_{\Delta x \to 0} \frac{f(x + \Delta x) - f(x)}{\Delta x}$$

$$= \lim_{\Delta x \to 0} \frac{\sin(x + \Delta x) - \sin x}{\Delta x}$$

Formula (1) for $\sin(x + \Delta x)$ is used to obtain

$$f'(x) = \lim_{\Delta x \to 0} \frac{\sin x \cos(\Delta x) + \cos x \sin(\Delta x) - \sin x}{\Delta x}$$

$$= \lim_{\Delta x \to 0} \frac{\sin x [\cos(\Delta x) - 1]}{\Delta x} + \lim_{\Delta x \to 0} \frac{\cos x \sin(\Delta x)}{\Delta x}$$

$$= -\lim_{\Delta x \to 0} \frac{1 - \cos(\Delta x)}{\Delta x} \left(\lim_{\Delta x \to 0} \sin x \right) + \left(\lim_{\Delta x \to 0} \cos x \right) \lim_{\Delta x \to 0} \frac{\sin(\Delta x)}{\Delta x} \tag{2}$$

From Theorem 1.8.5,

$$\lim_{\Delta x \to 0} \frac{1 - \cos(\Delta x)}{\Delta x} = 0 \tag{3}$$

and from Theorem 1.8.2,

$$\lim_{\Delta x \to 0} \frac{\sin(\Delta x)}{\Delta x} = 1 \tag{4}$$

Substituting from (3) and (4) into (2) we get

$$f'(x) = -0 \cdot \sin x + \cos x \cdot 1$$

$$= \cos x$$

We have proved the following theorem.

2.6.1 THEOREM $D_x(\sin x) = \cos x$

EXAMPLE 1
Given

$$f(x) = x^2 \sin x$$

find $f'(x)$.

SOLUTION

We find the derivative of a product of two functions by applying Theorem 2.4.6.

$$f'(x) = x^2 D_x(\sin x) + D_x(x^2)\sin x$$
$$= x^2 \cos x + 2x \sin x$$

To find the derivative of the cosine function we proceed as with the sine function. Here we apply the identity

$$\cos(a + b) = \cos a \cos b - \sin a \sin b \tag{5}$$

If g is the cosine function, then

$$g(x) = \cos x$$

$$g'(x) = \lim_{\Delta x \to 0} \frac{g(x + \Delta x) - g(x)}{\Delta x}$$

$$= \lim_{\Delta x \to 0} \frac{\cos(x + \Delta x) - \cos x}{\Delta x}$$

Formula (5) for $\cos(x + \Delta x)$ is used to obtain

$$g'(x) = \lim_{\Delta x \to 0} \frac{\cos x \cos(\Delta x) - \sin x \sin(\Delta x) - \cos x}{\Delta x}$$

$$= \lim_{\Delta x \to 0} \frac{\cos x[\cos(\Delta x) - 1]}{\Delta x} - \lim_{\Delta x \to 0} \frac{\sin x \sin(\Delta x)}{\Delta x}$$

$$= -\lim_{\Delta x \to 0} \frac{1 - \cos(\Delta x)}{\Delta x} \left(\lim_{\Delta x \to 0} \cos x \right) - \left(\lim_{\Delta x \to 0} \sin x \right) \lim_{\Delta x \to 0} \frac{\sin(\Delta x)}{\Delta x} \tag{6}$$

We substitute from (3) and (4) into (6) and obtain

$$g'(x) = -0 \cdot \cos x - \sin x \cdot 1$$
$$= -\sin x$$

The following theorem has been proved.

2.6.2 THEOREM $D_x(\cos x) = -\sin x$

EXAMPLE 2
Given

$$y = \frac{\sin x}{1 - 2\cos x}$$

find $\dfrac{dy}{dx}$.

SOLUTION

We apply Theorem 2.4.7 (the derivative of a quotient).

$$\frac{dy}{dx} = \frac{(1 - 2\cos x)\,D_x(\sin x) - \sin x \cdot D_x(1 - 2\cos x)}{(1 - 2\cos x)^2}$$

$$= \frac{(1 - 2\cos x)(\cos x) - \sin x(2\sin x)}{(1 - 2\cos x)^2}$$

$$= \frac{\cos x - 2(\cos^2 x + \sin^2 x)}{(1 - 2\cos x)^2}$$

$$= \frac{\cos x - 2}{(1 - 2\cos x)^2}$$

The derivatives of the tangent, cotangent, secant, and cosecant functions are obtained from trigonometric identities involving the sine and cosine as well as the derivatives of the sine and cosine and theorems on differentiation. For the derivative of the tangent we apply the identities

$$\tan x = \frac{\sin x}{\cos x} \qquad \sec x = \frac{1}{\cos x} \qquad \sin^2 x + \cos^2 x = 1$$

2.6.3 THEOREM $D_x(\tan x) = \sec^2 x$

PROOF

$$D_x(\tan x) = D_x\left(\frac{\sin x}{\cos x}\right)$$

$$= \frac{\cos x \cdot D_x(\sin x) - \sin x \cdot D_x(\cos x)}{\cos^2 x}$$

$$= \frac{(\cos x)(\cos x) - (\sin x)(-\sin x)}{\cos^2 x}$$

$$= \frac{\cos^2 x + \sin^2 x}{\cos^2 x}$$

$$= \frac{1}{\cos^2 x}$$

$$= \sec^2 x \qquad \blacksquare$$

2.6.4 THEOREM $D_x(\cot x) = -\csc^2 x$

The proof of this theorem is left as an exercise (see Exercise 1). It is analogous to the proof of Theorem 2.6.3. The following identities are used:

$$\cot x = \frac{\cos x}{\sin x} \qquad \csc x = \frac{1}{\sin x}$$

2.6.5 THEOREM $D_x(\sec x) = \sec x \tan x$

PROOF

$$D_x(\sec x) = D_x\left(\frac{1}{\cos x}\right)$$

$$= \frac{\cos x \cdot D_x(1) - 1 \cdot D_x(\cos x)}{\cos^2 x}$$

$$= \frac{\cos x \cdot 0 - 1 \cdot (-\sin x)}{\cos^2 x}$$

$$= \frac{\sin x}{\cos^2 x}$$

$$= \frac{1}{\cos x} \cdot \frac{\sin x}{\cos x}$$

$$= \sec x \tan x \qquad \blacksquare$$

EXAMPLE 3
Compute

$$\frac{d}{dx}(\tan x \sec x)$$

SOLUTION

$$\frac{d}{dx}(\tan x \sec x) = \tan x \cdot \frac{d}{dx}(\sec x) + \frac{d}{dx}(\tan x) \cdot \sec x$$

$$= \tan x(\sec x \tan x) + \sec^2 x(\sec x)$$

$$= \sec x \tan^2 x + \sec^3 x$$

2.6.6 THEOREM $$D_x(\csc x) = -\csc x \cot x$$

The proof of this theorem is also left as an exercise (see Exercise 2).

In a trigonometry course the graphs of the functions are sketched by intuitive considerations. Now we can obtain these graphs in a formal manner by using the derivatives of the trigonometric functions. We first discuss the graphs of the sine and cosine. For each of these functions the domain is the set of all real numbers, and the range is $[-1, 1]$. Let

$$f(x) = \sin x \qquad f'(x) = \cos x$$

To determine where the graph has a horizontal tangent, set $f'(x) = 0$ and get $x = \frac{1}{2}\pi + k\pi$, where k is any integer. At these values of x, $\sin x$ is either $+1$ or -1, and these are the largest and smallest values that $\sin x$ assumes. The graph intersects the x axis at the points where $\sin x = 0$, that is, at the points where $x = k\pi$ and k is any integer. Furthermore, when k is an even integer, $f'(k\pi) = 1$, and when k is an odd integer, $f'(k\pi) = -1$. Thus at the points of intersection of the graph with the x axis the slope of the tangent line is either

1 or -1. From this information we draw the sketch of the graph of the sine function shown in Fig. 2.6.1.

$$f(x) = \sin x$$

FIGURE 2.6.1

For the graph of the cosine function we use the identity

$$\cos x = \sin(\tfrac{1}{2}\pi + x)$$

Thus the graph of the cosine is obtained from the graph of the sine by translating the y axis $\tfrac{1}{2}\pi$ units to the right. See Fig. 2.6.2.

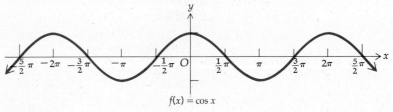

$$f(x) = \cos x$$

FIGURE 2.6.2

EXAMPLE 4

Find an equation of the tangent line to the graph of the cosine function at the point $(\tfrac{3}{2}\pi, 0)$.

SOLUTION

If $f(x) = \cos x$, $f'(x) = -\sin x$. Thus $f'(\tfrac{3}{2}\pi) = -\sin \tfrac{3}{2}\pi$. Because $\sin \tfrac{3}{2}\pi = -1$, $f'(\tfrac{3}{2}\pi) = 1$. From the point-slope form of an equation of the tangent line having slope 1 and containing the point $(\tfrac{3}{2}\pi, 0)$, we have

$$y - 0 = 1(x - \tfrac{3}{2}\pi)$$

$$y = x - \tfrac{3}{2}\pi$$

The graph of the tangent function is now considered. Because

$$\tan(-x) = -\tan x$$

the graph is symmetric with respect to the origin. Furthermore,

$$\tan(x + \pi) = \tan x$$

and so by Definition 0.6.3, the tangent function is periodic with fundamental period π. The tangent function is continuous at all numbers in its domain, which is the set of all real numbers except those of the form $\tfrac{1}{2}\pi + k\pi$, where

k is any integer. The range is the set of all real numbers. If k is any integer, $\tan k\pi = 0$. Therefore the graph intersects the x axis at the points $(k\pi, 0)$. Let

$$f(x) = \tan x \qquad f'(x) = \sec^2 x$$

Because $f'(k\pi) = \sec^2 k\pi$ and $\sec^2 k\pi = 1$ for k any integer, it follows that where the graph intersects the x axis, the slope of the tangent line is 1. Setting $f'(x) = 0$ gives $\sec^2 x = 0$. Because $\sec^2 x \geq 1$ for all x, we conclude that there are no horizontal tangent lines.

Consider the interval $[0, \frac{1}{2}\pi)$ on which the tangent function is defined everywhere.

$$\lim_{x \to \pi/2^-} \tan x = \lim_{x \to \pi/2^-} \frac{\sin x}{\cos x}$$

Because $\lim\limits_{x \to \pi/2^-} \sin x = 1$ and $\lim\limits_{x \to \pi/2^-} \cos x = 0$, where $\cos x$ is approaching zero through positive values,

$$\lim_{x \to \pi/2^-} \tan x = +\infty$$

Therefore the line $x = \frac{1}{2}\pi$ is a vertical asymptote of the graph. In Table 2.6.1 there are some values of x in the interval $[0, \frac{1}{2}\pi)$ and the corresponding values of $\tan x$. By plotting the points having as coordinates the number pairs $(x, \tan x)$ we get the portion of the graph for x in $[0, \frac{1}{2}\pi)$. Because of symmetry with respect to the origin, the portion of the graph for x in $(-\frac{1}{2}\pi, 0]$ is obtained. Since the period is π, we complete the sketch of the graph shown in Fig. 2.6.3.

Table 2.6.1

x	$\tan x$
0	0
$\frac{1}{6}\pi$	$\dfrac{1}{\sqrt{3}} \approx 0.58$
$\frac{1}{4}\pi$	1
$\frac{1}{3}\pi$	$\sqrt{3} \approx 1.73$

$$f(x) = \tan x$$

FIGURE 2.6.3

We can get the graph of the cotangent function from that of the tangent function by using the identity

$$\cot x = -\tan(\tfrac{1}{2}\pi + x)$$

From this identity it follows that the graph of the cotangent is obtained from the graph of the tangent by translating the y axis $\frac{1}{2}\pi$ units to the right and then taking a reflection of the graph with respect to the x axis. A sketch of the graph of the cotangent function appears in Fig. 2.6.4.

FIGURE 2.6.4

Because

$$\sec(x + 2\pi) = \sec x$$

the secant function is periodic with fundamental period 2π. The domain of the secant function is the set of all real numbers except those of the form $\frac{1}{2}\pi + k\pi$, where k is any integer. The range is $(-\infty, -1] \cup [1, +\infty)$. The function is continuous at all numbers in its domain. There is no intersection of the graph with the x axis because $\sec x$ is never zero.

We use the derivative to determine if the graph has any horizontal tangent lines. Let

$$f(x) = \sec x \qquad f'(x) = \sec x \tan x$$

Setting $f'(x) = 0$ gives $\sec x \tan x = 0$. Because $\sec x \neq 0$, $f'(x) = 0$ when $\tan x = 0$, which is when $x = k\pi$, where k is any integer.

We first consider the graph for x in $(-\frac{1}{2}\pi, \frac{1}{2}\pi) \cup (\frac{1}{2}\pi, \frac{3}{2}\pi)$. There are horizontal tangent lines at $x = 0$ and $x = \pi$.

$$\lim_{x \to -\pi/2^+} \sec x = \lim_{x \to -\pi/2^+} \frac{1}{\cos x} \qquad \lim_{x \to \pi/2^-} \sec x = \lim_{x \to \pi/2^-} \frac{1}{\cos x}$$

$$= +\infty \qquad\qquad = +\infty$$

$$\lim_{x \to \pi/2^+} \sec x = \lim_{x \to \pi/2^+} \frac{1}{\cos x} \qquad \lim_{x \to 3\pi/2^-} \sec x = \lim_{x \to 3\pi/2^-} \frac{1}{\cos x}$$

$$= -\infty \qquad\qquad = -\infty$$

Therefore the lines $x = -\frac{1}{2}\pi$, $x = \frac{1}{2}\pi$, and $x = \frac{3}{2}\pi$ are vertical asymptotes of the graph.

With the above information and plotting a few points we get a sketch of the graph of the secant function for x in $(-\frac{1}{2}\pi, \frac{1}{2}\pi) \cup (\frac{1}{2}\pi, \frac{3}{2}\pi)$. Because the period is 2π, the sketch of the graph shown in Fig. 2.6.5 is obtained.

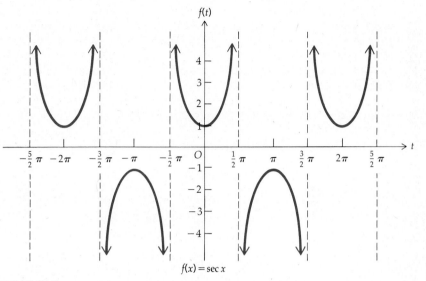

$f(x) = \sec x$

FIGURE 2.6.5

From the identity

$$\csc x = -\sec(x + \tfrac{1}{2}\pi)$$

we get the graph of the cosecant function from that of the secant by translating the y axis $\frac{1}{2}\pi$ units to the right and taking a reflection of the graph with respect to the x axis. A sketch of the graph of the cosecant function appears in Fig. 2.6.6.

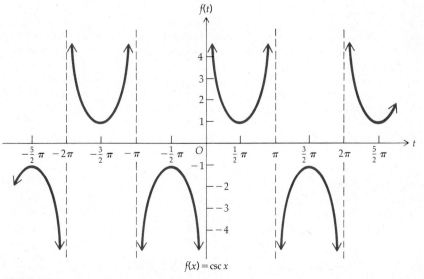

$f(x) = \csc x$

FIGURE 2.6.6

EXERCISES 2.6

1. Prove: $D_x(\cot x) = -\csc^2 x$.

2. Prove: $D_x(\csc x) = -\csc x \cot x$.

In Exercises 3 through 16, find the derivative of the given function.

3. $f(x) = 3 \sin x$

4. $g(x) = \sin x + \cos x$

5. $g(x) = \tan x + \cot x$

6. $f(x) = 4 \sec x - 2 \csc x$

7. $f(t) = 2t \cos t$

8. $f(x) = 4x^2 \cos x$

9. $g(x) = x \sin x + \cos x$

10. $g(y) = 3 \sin y - y \cos y$

11. $h(x) = 4 \sin x \cos x$

12. $f(x) = x^2 \sin x + 2x \cos x$

13. $f(x) = x^2 \cos x - 2x \sin x - 2 \cos x$

14. $h(y) = y^3 - y^2 \cos y + 2y \sin y + 2 \cos y$

15. $f(x) = 3 \sec x \tan x$

16. $f(t) = \sin t \tan t$

In Exercises 17 through 30, compute the indicated derivative.

17. $D_y(\cot y \csc y)$

18. $D_x(\cos x \cot x)$

19. $D_z\left(\dfrac{2 \cos z}{z + 1}\right)$

20. $D_t\left(\dfrac{\sin t}{t}\right)$

21. $\dfrac{d}{dx}\left(\dfrac{\sin x}{1 - \cos x}\right)$

22. $\dfrac{d}{dx}\left(\dfrac{x + 4}{\cos x}\right)$

23. $\dfrac{d}{dt}\left(\dfrac{\tan t}{\cos t - 4}\right)$

24. $\dfrac{d}{dy}\left(\dfrac{\cot y}{1 - \sin y}\right)$

25. $\dfrac{d}{dy}\left(\dfrac{1 + \sin y}{1 - \sin y}\right)$

26. $\dfrac{d}{dx}\left(\dfrac{\sin x - 1}{\cos x + 1}\right)$

27. $D_x[(x - \sin x)(x + \cos x)]$

28. $D_z[(z^2 + \cos z)(2z - \sin z)]$

29. $D_t\left(\dfrac{2 \csc t - 1}{\csc t + 2}\right)$

30. $D_y\left(\dfrac{\tan y + 1}{\tan y - 1}\right)$

In Exercises 31 through 42, find $f'(a)$ for the given value of a.

31. $f(x) = x \cos x$; $a = 0$

32. $f(x) = x \sin x$; $a = \frac{3}{2}\pi$

33. $f(x) = \dfrac{\cos x}{x}$; $a = \frac{1}{2}\pi$

34. $f(x) = \dfrac{\sec x}{x^2}$; $a = \pi$

35. $f(x) = x^2 \tan x$; $a = \pi$

36. $f(x) = x^2 \cos x - \sin x$; $a = 0$

37. $f(x) = \sin x(\cos x - 1)$; $a = \pi$

38. $f(x) = (\cos x + 1)(x \sin x - 1)$; $a = \frac{1}{2}\pi$

39. $f(x) = x \cos x + x \sin x$; $a = \frac{1}{4}\pi$

40. $f(x) = \tan x + \sec x$; $a = \frac{1}{6}\pi$

41. $f(x) = 2 \cot x - \csc x$; $a = \frac{2}{3}\pi$

42. $f(x) = \dfrac{1}{\cot x - 1}$; $a = \frac{3}{4}\pi$

43. Find an equation of the tangent line to the graph of the sine function at the point where (a) $x = 0$; (b) $x = \frac{1}{3}\pi$; (c) $x = \pi$.

44. Find an equation of the tangent line to the graph of the cosine function at the point where (a) $x = \frac{1}{2}\pi$; (b) $x = -\frac{1}{2}\pi$; (c) $x = \frac{1}{6}\pi$.

45. Find an equation of the tangent line to the graph of the tangent function at the point where (a) $x = 0$; (b) $x = \frac{1}{4}\pi$; (c) $x = -\frac{1}{4}\pi$.

46. Find an equation of the tangent line to the graph of the secant function at the point where (a) $x = \frac{1}{4}\pi$; (b) $x = -\frac{1}{4}\pi$; (c) $x = \frac{3}{4}\pi$.

In Exercises 47 through 50, a particle is moving along a straight line according to the given equation, where s cm is the directed distance of the particle from the origin at t sec. (a) What is the instantaneous velocity of the particle at t sec? (b) Find the instantaneous velocity of the particle at t_1 sec for each of the given values of t_1.

47. $s = 4 \sin t$; t_1 is $0, \frac{1}{3}\pi, \frac{1}{2}\pi, \frac{2}{3}\pi$, and π

48. $s = 6 \cos t$; t_1 is $0, \frac{1}{6}\pi, \frac{1}{2}\pi, \frac{5}{6}\pi$, and π

49. $s = -3 \cos t$; t_1 is $0, \frac{1}{6}\pi, \frac{1}{3}\pi, \frac{1}{2}\pi, \frac{2}{3}\pi, \frac{5}{6}\pi$, and π

50. $s = -\frac{1}{2} \sin t$; t_1 is $0, \frac{1}{6}\pi, \frac{1}{3}\pi, \frac{1}{2}\pi, \frac{2}{3}\pi, \frac{5}{6}\pi$, and π

51. If a body of weight W lb is dragged along a horizontal floor by means of a force of magnitude F lb and directed at an angle of θ radians with the plane of the floor, then F is given by the equation

$$F = \frac{kW}{k \sin \theta + \cos \theta}$$

where k is a constant and is called the coefficient of friction. If $k = 0.5$, find the instantaneous rate of change of F with respect to θ when (a) $\theta = \frac{1}{4}\pi$; (b) $\theta = \frac{1}{2}\pi$.

52. A projectile is shot from a gun at an angle of elevation having radian measure $\frac{1}{2}\alpha$ and an initial velocity of v_0 ft/sec. If R ft is the range of the projectile, then

$$R = \frac{v_0^2}{g} \sin \alpha \qquad 0 \leq \alpha \leq \pi$$

where g ft/sec^2 is the acceleration due to gravity. (a) If $v_0 = 480$, find the rate of change of R with respect to α when $\alpha = \frac{1}{2}\pi$ (i.e., the angle of elevation has radian measure $\frac{1}{4}\pi$). Take $g = 32$. (b) Find the values of α for which $D_\alpha R > 0$.

2.7
THE DERIVATIVE OF A COMPOSITE FUNCTION

To find the derivative of a composite function we use one of the important theorems in calculus called the **chain rule**. Before stating this theorem we give three illustrations showing how previous theorems can be used to determine the derivatives of some particular composite functions. In each illustration the final expression for the derivative is written in a form that we can associate later with the chain rule.

▶ ILLUSTRATION 1   If

$$F(x) = (4x^2 + 1)^3$$

we can obtain $F'(x)$ by twice applying Theorem 2.4.6 (the derivative of a product). The computation is as follows:

$$F(x) = (4x^2 + 1)^2(4x^2 + 1)$$

$$F'(x) = (4x^2 + 1)^2 \cdot D_x(4x^2 + 1) + (4x^2 + 1) \cdot D_x[(4x^2 + 1)(4x^2 + 1)]$$

$$= (4x^2 + 1)^2(8x) + (4x^2 + 1)[(4x^2 + 1)(8x) + (4x^2 + 1)(8x)]$$

$$= (4x^2 + 1)^2(8x) + (4x^2 + 1)[2(4x^2 + 1)(8x)]$$

$$= (4x^2 + 1)^2(8x) + 2[(4x^2 + 1)^2(8x)]$$

Thus

$$F'(x) = [3(4x^2 + 1)^2](8x) \qquad (1)$$

◀

▶ ILLUSTRATION 2   If

$$G(x) = \sin 2x$$

then to find $G'(x)$ we can use the trigonometric identities

$$\sin 2x = 2 \sin x \cos x \qquad \cos 2x = \cos^2 x - \sin^2 x$$

and Theorem 2.4.6. We have

$$G(x) = 2 \sin x \cos x$$

$$G'(x) = (2 \sin x) D_x(\cos x) + (2 \cos x) D_x(\sin x)$$

$$= (2 \sin x)(-\sin x) + (2 \cos x)(\cos x)$$

$$= 2(\cos^2 x - \sin^2 x)$$

Therefore

$$G'(x) = (\cos 2x)(2) \qquad (2)$$

◀

▶ ILLUSTRATION 3   If

$$H(x) = (\cos x)^{-1}$$

we can compute $H'(x)$ by first using the identity $(\cos x)^{-1} = \sec x$.

$$H(x) = \sec x$$

$$H'(x) = \sec x \tan x$$

$$= \frac{1}{\cos x} \cdot \frac{\sin x}{\cos x}$$

$$= (-1) \frac{1}{\cos^2 x} (-\sin x)$$

Hence

$$H'(x) = [-1(\cos x)^{-2}](-\sin x) \qquad (3)$$

◀

In Illustration 1, if $f(x) = x^3$ and $g(x) = 4x^2 + 1$, then F is the composite function $f \circ g$; that is,

$$F(x) = f(g(x))$$
$$= f(4x^2 + 1)$$
$$= (4x^2 + 1)^3$$

Because $f'(x) = 3x^2$ and $g'(x) = 8x$, we can write (1) in the form

$$F'(x) = f'(g(x))g'(x) \qquad (4)$$

In Illustration 2, if $f(x) = \sin x$ and $g(x) = 2x$, then G is the composite function $f \circ g$; that is,

$$G(x) = f(g(x))$$
$$= f(2x)$$
$$= \sin 2x$$

Because $f'(x) = \cos x$ and $g'(x) = 2$, we can write (2) in the form

$$G'(x) = f'(g(x))g'(x) \qquad (5)$$

In Illustration 3, if $f(x) = x^{-1}$ and $g(x) = \cos x$, then H is the composite function $f \circ g$; that is,

$$H(x) = f(g(x))$$
$$= f(\cos x)$$
$$= (\cos x)^{-1}$$

Since $f'(x) = -1 \cdot x^{-2}$ and $g'(x) = -\sin x$, we can write (3) in the form

$$H'(x) = f'(g(x))g'(x) \tag{6}$$

Observe that the right-hand sides of (4), (5), and (6) are all $f'(g(x))g'(x)$, which is the right-hand side of the chain rule, stated as the following theorem.

2.7.1 THEOREM   If the function g is differentiable at x and the function f is differentiable at

The Chain Rule $g(x)$, then the composite function $f \circ g$ is differentiable at x, and

$$(f \circ g)'(x) = f'(g(x))g'(x) \tag{7}$$

Before discussing a proof of the chain rule we give two more illustrations showing its application.

▶ ILLUSTRATION 4   Let

$$f(x) = x^{10} \qquad g(x) = 2x^3 - 5x^2 + 4$$

Then the composite function $f \circ g$ is defined by

$$(f \circ g)(x) = f(g(x))$$
$$= (2x^3 - 5x^2 + 4)^{10}$$

To apply (7) we need to compute $f'(g(x))$ and $g'(x)$. Because $f(x) = x^{10}$, $f'(x) = 10x^9$; thus

$$f'(g(x)) = 10[g(x)]^9$$
$$f'(g(x)) = 10(2x^3 - 5x^2 + 4)^9 \tag{8}$$

Furthermore, because $g(x) = 2x^3 - 5x^2 + 4$,

$$g'(x) = 6x^2 - 10x \tag{9}$$

Therefore, from (7), (8), and (9) we have

$$(f \circ g)'(x) = f'(g(x))g'(x)$$
$$= 10(2x^3 - 5x^2 + 4)^9(6x^2 - 10x) \qquad \blacktriangleleft$$

▶ ILLUSTRATION 5   Let

$$f(x) = \sin x \qquad g(x) = x^2 + 3$$

Then the composite function $f \circ g$ is defined by

$$(f \circ g)(x) = f(g(x))$$
$$= \sin(x^2 + 3)$$

We compute $f'(g(x))$ and $g'(x)$. Because $f(x) = \sin x$, $f'(x) = \cos x$. Hence

$$f'(g(x)) = \cos[g(x)]$$
$$f'(g(x)) = \cos(x^2 + 3) \tag{10}$$

Because $g(x) = x^2 + 3$,

$$g'(x) = 2x \tag{11}$$

Thus from (7), (10), and (11) we obtain

$$\begin{aligned}
(f \circ g)'(x) &= f'(g(x))g'(x) \\
&= [\cos(x^2 + 3)](2x) \\
&= 2x \cos(x^2 + 3)
\end{aligned}$$ ◀

The proof of the chain rule for all functions f and g satisfying the hypothesis is sophisticated, and we defer it to the end of this section. However, at this time we give a simpler proof that is valid for many functions. The requirement of this simpler proof is that if g is differentiable at x_1 and f is differentiable at $g(x_1)$, then for all $x \neq x_1$ in some open interval containing x_1, $g(x) - g(x_1) \neq 0$; that is, there is an open interval I containing x_1 such that

$$\text{if} \quad x \neq x_1 \quad \text{then} \quad g(x) - g(x_1) \neq 0 \qquad \text{for all } x \text{ in } I \tag{12}$$

The proof then is as follows:

Let h be the composite function $f \circ g$. Applying formula (6) from Sec. 2.2,

$$h'(x_1) = \lim_{x \to x_1} \frac{h(x) - h(x_1)}{x - x_1} \tag{13}$$

We wish to show that this limit exists. Because $h(x) = f(g(x))$, (13) can be written as

$$h'(x_1) = \lim_{x \to x_1} \frac{f(g(x)) - f(g(x_1))}{x - x_1} \tag{14}$$

Applying requirement (12) we can multiply the numerator and denominator of the fraction in (14) by $g(x) - g(x_1)$ and obtain

$$h'(x_1) = \lim_{x \to x_1} \frac{f(g(x)) - f(g(x_1))}{x - x_1} \cdot \frac{g(x) - g(x_1)}{g(x) - g(x_1)}$$

and if the limits exist,

$$h'(x_1) = \lim_{x \to x_1} \frac{f(g(x)) - f(g(x_1))}{g(x) - g(x_1)} \cdot \lim_{x \to x_1} \frac{g(x) - g(x_1)}{x - x_1} \tag{15}$$

Because g is differentiable at x_1, the second limit on the right-hand side of (15) exists, and

$$\lim_{x \to x_1} \frac{g(x) - g(x_1)}{x - x_1} = g'(x_1) \tag{16}$$

Because g is differentiable at x_1, g is continuous at x_1. Therefore

$$\lim_{x \to x_1} g(x) = g(x_1)$$

that is, as $x \to x_1$, $g(x) \to g(x_1)$. Therefore in the first limit on the right-hand side of (15) we can replace $x \to x_1$ by $g(x) \to g(x_1)$, and we have

$$\lim_{x \to x_1} \frac{f(g(x)) - f(g(x_1))}{g(x) - g(x_1)} = \lim_{g(x) \to g(x_1)} \frac{f(g(x)) - f(g(x_1))}{g(x) - g(x_1)} \qquad (17)$$

Compare the limit on the right-hand side of (17) with (6) in Sec. 2.2 and you will observe that it is $f'(g(x_1))$, which exists because $f \circ g$ is differentiable at $g(x_1)$. Therefore we have

$$\lim_{g(x) \to g(x_1)} \frac{f(g(x)) - f(g(x_1))}{g(x) - g(x_1)} = f'(g(x_1)) \qquad (18)$$

Substituting from (18) and (16) in (15) and replacing $h'(x_1)$ by its equivalent $(f \circ g)'(x_1)$, we have

$$(f \circ g)'(x_1) = f'(g(x_1)) \cdot g'(x_1)$$

which is (7) with x_1 in place of x. This completes the proof of the chain rule with the additional requirement (12).

▶ ILLUSTRATION 6   We apply the chain rule to find $h'(x)$ if

$$h(x) = \cos 5x$$

Let

$$f(x) = \cos x \qquad g(x) = 5x$$

Then

$$f'(x) = -\sin x \qquad g'(x) = 5$$

Because $h(x) = f(g(x))$, from the chain rule we have

$$h'(x) = f'(g(x)) \cdot g'(x)$$
$$= (-\sin 5x)(5)$$
$$= -5 \sin 5x$$

◀

▶ ILLUSTRATION 7   Suppose

$$h(x) = \left(\frac{2}{x - 1} \right)^5$$

To determine $h'(x)$, let

$$f(x) = x^5 \qquad g(x) = \frac{2}{x - 1}$$

Then

$$f'(x) = 5x^4 \qquad g'(x) = \frac{-2}{(x - 1)^2}$$

Because $h(x) = f(g(x))$, we have from the chain rule

$$h'(x) = f'(g(x)) \cdot g'(x)$$

$$= 5\left(\frac{2}{x-1}\right)^4 \cdot \frac{-2}{(x-1)^2}$$

$$= \frac{-160}{(x-1)^6}$$

◀

When computing derivatives by the chain rule we don't actually write the functions f and g as we did in Illustrations 4 through 7, but we bear them in mind. For instance, for the computation in Illustration 6 we would write

$$h(x) = \cos 5x$$

$$h'(x) = (-\sin 5x) \cdot D_x(5x)$$

$$= (-\sin 5x)(5)$$

$$= -5 \sin 5x$$

and for the computation in Illustration 7 we would write

$$h(x) = \left(\frac{2}{x-1}\right)^5$$

$$h'(x) = 5\left(\frac{2}{x-1}\right)^4 \cdot D_x\left(\frac{2}{x-1}\right)$$

$$= 5\left(\frac{2}{x-1}\right)^4 \cdot \frac{-2}{(x-1)^2}$$

$$= \frac{-160}{(x-1)^6}$$

After some practice you can also eliminate some of the intermediate steps.

EXAMPLE 1
Given

$$f(x) = \frac{1}{4x^3 + 5x^2 - 7x + 8}$$

find $f'(x)$ by the chain rule.

SOLUTION
Write $f(x) = (4x^3 + 5x^2 - 7x + 8)^{-1}$, and apply the chain rule to obtain

$$f'(x) = -1(4x^3 + 5x^2 - 7x + 8)^{-2} \cdot D_x(4x^3 + 5x^2 - 7x + 8)$$

$$= -1(4x^3 + 5x^2 - 7x + 8)^{-2}(12x^2 + 10x - 7)$$

$$= \frac{-12x^2 - 10x + 7}{(4x^3 + 5x^2 - 7x + 8)^2}$$

EXAMPLE 2
Compute

$$\frac{d}{dx}\left[\left(\frac{2x+1}{3x-1}\right)^4\right]$$

SOLUTION
From the chain rule,

$$\frac{d}{dx}\left[\left(\frac{2x+1}{3x-1}\right)^4\right] = 4\left(\frac{2x+1}{3x-1}\right)^3 \cdot \frac{d}{dx}\left(\frac{2x+1}{3x-1}\right)$$

$$= 4\left(\frac{2x+1}{3x-1}\right)^3 \frac{(3x-1)(2)-(2x+1)(3)}{(3x-1)^2}$$

$$= \frac{4(2x+1)^3(-5)}{(3x-1)^5}$$

$$= -\frac{20(2x+1)^3}{(3x-1)^5}$$

If the Leibniz notation is used for the derivative, the chain rule can be stated as follows:

If y is a function of u, defined by $y = f(u)$ and $\dfrac{dy}{du}$ exists, and if u is a function of x, defined by $u = g(x)$ and $\dfrac{du}{dx}$ exists, then y is a function of x and $\dfrac{dy}{dx}$ exists and is given by

$$\frac{dy}{dx} = \frac{dy}{du} \cdot \frac{du}{dx} \tag{19}$$

Observe from (19) the convenient form for remembering the chain rule. The formal statement suggests a symbolic "division" of du in the numerator and denominator of the right-hand side. However, remember from Sec. 2.2 that when we introduced the Leibniz notation $\dfrac{dy}{dx}$, it was emphasized that neither dy nor dx has been given independent meaning. Therefore you should consider (19) as an equation involving formal differentiation notation.

Another way of writing the chain rule is to let $u = g(x)$. Then

$$(f \circ g)(x) = f(u) \qquad (f \circ g)'(x) = D_x f(u) \qquad f'(g(x)) = f'(u) \qquad g'(x) = D_x u$$

With these substitutions (7) becomes

$$D_x[f(u)] = f'(u)\,D_x u \tag{20}$$

We shall use this form of the chain rule to state important differentiation formulas. If u is a differentiable function of x, we have from (20) and

Theorems 2.6.1–2.6.6 the following formulas involving the derivatives of the trigonometric functions:

$$D_x(\sin u) = \cos u\, D_x u \qquad\qquad D_x(\cos u) = -\sin u\, D_x u$$

$$D_x(\tan u) = \sec^2 u\, D_x u \qquad\qquad D_x(\cot u) = -\csc^2 u\, D_x u$$

$$D_x(\sec u) = \sec u \tan u\, D_x u \qquad\qquad D_x(\csc u) = -\csc u \cot u\, D_x u$$

EXAMPLE 3
Given

$$F(t) = \tan(3t^2 + 2t)$$

find $F'(t)$.

SOLUTION
We use the chain rule and obtain

$$\begin{aligned} F'(t) &= \sec^2(3t^2 + 2t) \cdot D_t(3t^2 + 2t) \\ &= \sec^2(3t^2 + 2t) \cdot (6t + 2) \\ &= 2(3t + 1)\sec^2(3t^2 + 2t) \end{aligned}$$

EXAMPLE 4
Given

$$y = \sin(\cos x)$$

find $\dfrac{dy}{dx}$.

SOLUTION
We apply the chain rule.

$$\begin{aligned} \frac{dy}{dx} &= \cos(\cos x)[D_x(\cos x)] \\ &= \cos(\cos x)[-\sin x] \\ &= -\sin x\,[\cos(\cos x)] \end{aligned}$$

EXAMPLE 5
Given

$$f(x) = (3x^2 + 2)^2(x^2 - 5x)^3$$

find $f'(x)$.

SOLUTION
Consider f as the product of the two functions g and h, where

$$g(x) = (3x^2 + 2)^2 \qquad h(x) = (x^2 - 5x)^3$$

From Theorem 2.4.6 for the derivative of the product of two functions,

$$f'(x) = g(x)h'(x) + h(x)g'(x)$$

We find $h'(x)$ and $g'(x)$ by the chain rule.

$$
\begin{aligned}
f'(x) &= (3x^2 + 2)^2[3(x^2 - 5x)^2(2x - 5)] + (x^2 - 5x)^3[2(3x^2 + 2)(6x)] \\
&= 3(3x^2 + 2)(x^2 - 5x)^2[(3x^2 + 2)(2x - 5) + 4x(x^2 - 5x)] \\
&= 3(3x^2 + 2)(x^2 - 5x)^2[6x^3 - 15x^2 + 4x - 10 + 4x^3 - 20x^2] \\
&= 3(3x^2 + 2)(x^2 - 5x)^2(10x^3 - 35x^2 + 4x - 10)
\end{aligned}
$$

EXAMPLE 6
Compute

$$D_x(\sec^4 2x^2)$$

SOLUTION
We use the chain rule twice.

$$
\begin{aligned}
D_x(\sec^4 2x^2) &= 4\sec^3 2x^2[D_x(\sec 2x^2)] \\
&= 4\sec^3 2x^2[(\sec 2x^2 \tan 2x^2)\,D_x(2x^2)] \\
&= (4\sec^4 2x^2 \tan 2x^2)(4x) \\
&= 16x\sec^4 2x^2 \tan 2x^2
\end{aligned}
$$

FIGURE 2.7.1

We now return to a discussion of the proof of the chain rule. We gave the simplified proof that is valid under the additional requirement (12) that is satisfied by many functions. For example, let

$$f(x) = x^2 \qquad g(x) = x^2 + 1$$

Then

$$(f \circ g)(x) = (x^2 + 1)^2$$

Both g and f are differentiable everywhere. A sketch of the graph of g appears in Fig. 2.7.1. For any x_1 there is an open interval I containing x_1 so that requirement (12) is satisfied. However, suppose

FIGURE 2.7.2

$$
f(x) = x^2 \qquad g(x) = \begin{cases} 0 & \text{if } x < 0 \\ 1 & \text{if } 0 \le x \end{cases}
$$

See Fig. 2.7.2 for a sketch of the graph of g. The function f is differentiable everywhere, and g is differentiable at every number except 0. But if x_1 is any of these numbers, there is no open interval I containing x_1 that satisfies requirement (12).

The following proof of the chain rule does not depend on requirement (12). An important part of the discussion consists of introducing the new function F that has properties useful in the proof. This device of "making up" a function is a common one for mathematicians.

PROOF OF THE CHAIN RULE   Let x_1 be any number in the domain of g such that g is differentiable at x_1 and f is differentiable at $g(x_1)$. Form the function

F defined by

$$F(t) = \begin{cases} \dfrac{f(t) - f(g(x_1))}{t - g(x_1)} & \text{if } t \neq g(x_1) \\ f'(g(x_1)) & \text{if } t = g(x_1) \end{cases} \tag{21}$$

Then

$$\lim_{t \to g(x_1)} F(t) = \lim_{t \to g(x_1)} \frac{f(t) - f(g(x_1))}{t - g(x_1)}$$

From (6) in Sec. 2.2, the function on the right-hand side of the above equation is $f'(g(x_1))$. Therefore

$$\lim_{t \to g(x_1)} F(t) = f'(g(x_1)) \tag{22}$$

But from (21),

$$f'(g(x_1)) = F(g(x_1)) \tag{23}$$

Substituting from (23) into (22) we get

$$\lim_{t \to g(x_1)} F(t) = F(g(x_1))$$

Therefore F is continuous at $g(x_1)$. Furthermore, from (21),

$$F(t) = \frac{f(t) - f(g(x_1))}{t - g(x_1)} \qquad \text{if } t \neq g(x_1)$$

Multiplying on both sides of this equation by $t - g(x_1)$ gives

$$f(t) - f(g(x_1)) = F(t)[t - g(x_1)] \qquad \text{if } t \neq g(x_1) \tag{24}$$

Observe that Eq. (24) holds even if $t = g(x_1)$ because the left-hand side is

$$f(g(x_1)) - f(g(x_1)) = 0$$

and the right-hand side is

$$F(g(x_1))[g(x_1) - g(x_1)] = 0$$

Therefore the stipulation in (24) that $t \neq g(x_1)$ is not necessary, and we write

$$f(t) - f(g(x_1)) = F(t)[t - g(x_1)] \tag{25}$$

Now let h be the composite function $f \circ g$, so that

$$h(x) = f(g(x)) \tag{26}$$

Then from (6) in Sec. 2.2,

$$h'(x_1) = \lim_{x \to x_1} \frac{h(x) - h(x_1)}{x - x_1} \tag{27}$$

if the limit exists. Substituting from (26) in the right-hand side of (27) we get

$$h'(x_1) = \lim_{x \to x_1} \frac{f(g(x)) - f(g(x_1))}{x - x_1} \tag{28}$$

if the limit exists. Now by letting $t = g(x)$ in (25) it follows that for every x in the domain of g such that $g(x)$ is in the domain of f,

$$f(g(x)) - f(g(x_1)) = F(g(x))[g(x) - g(x_1)] \tag{29}$$

Substituting from (29) into (28) we have

$$h'(x_1) = \lim_{x \to x_1} \frac{F(g(x))[g(x) - g(x_1)]}{x - x_1}$$

Thus if the limits exist,

$$h'(x_1) = \lim_{x \to x_1} F(g(x)) \cdot \lim_{x \to x_1} \frac{g(x) - g(x_1)}{x - x_1} \tag{30}$$

Because F is continuous at $g(x_1)$,

$$\lim_{x \to x_1} F(g(x)) = F(g(x_1)) \tag{31}$$

But from (21),

$$F(g(x_1)) = f'(g(x_1)) \tag{32}$$

Substituting from (32) into (31) we get

$$\lim_{x \to x_1} F(g(x)) = f'(g(x_1)) \tag{33}$$

Furthermore, because g is differentiable at x_1,

$$\lim_{x \to x_1} \frac{g(x) - g(x_1)}{x - x_1} = g'(x_1) \tag{34}$$

Substituting from (33) and (34) into (30) and replacing $h'(x_1)$ by $(f \circ g)'(x_1)$ we have

$$(f \circ g)'(x_1) = f'(g(x_1)) \cdot g'(x_1)$$

which is (7) with x replaced by x_1. Thus we have proved the chain rule. ∎

EXERCISES 2.7

In Exercises 1 through 12, find the derivative of the given function.

1. $f(x) = (2x + 1)^3$
2. $f(x) = (10 - 5x)^4$
3. $F(x) = (x^2 + 4x - 5)^4$
4. $g(r) = (2r^4 + 8r^2 + 1)^5$
5. $f(t) = (2t^4 - 7t^3 + 2t - 1)^2$
6. $H(z) = (z^3 - 3z^2 + 1)^{-3}$
7. $f(x) = (x^2 + 4)^{-2}$
8. $g(x) = \sin x^2$
9. $f(x) = 4 \cos 3x - 3 \sin 4x$
10. $G(x) = \sec^2 x$
11. $h(t) = \frac{1}{3} \sec^3 2x - \sec 2x$
12. $f(x) = \cos(3x^2 + 1)$

In Exercises 13 through 24, compute the indicated derivative.

13. $\dfrac{d}{dx}(\sec^2 x \tan^2 x)$
14. $\dfrac{d}{dt}(2 \sin^3 t \cos^2 t)$
15. $\dfrac{d}{dt}(\cot^4 t - \csc^4 t)$

16. $\dfrac{d}{dx}[(4x^2 + 7)^2(2x^3 + 1)^4]$
17. $D_u[(3u^2 + 5)^3(3u - 1)^2]$
18. $D_x[(x^2 - 4x^{-2})^2(x^2 + 1)^{-1}]$

19. $D_x[(2x - 5)^{-1}(4x + 3)^{-2}]$

20. $D_x[(2x - 9)^2(x^3 + 4x - 5)^3]$

21. $D_r[(r^2 + 1)^3(2r^2 + 5r - 3)^2]$

22. $D_y[(y + 3)^3(5y + 1)^2(3y^2 - 4)]$

23. $\dfrac{d}{dy}\left[\left(\dfrac{y - 7}{y + 2}\right)^2\right]$

24. $\dfrac{d}{dt}\left[\left(\dfrac{2t^2 + 1}{3t^3 + 1}\right)^2\right]$

In Exercises 25 through 36, find the derivative of the given function.

25. $f(x) = \left(\dfrac{2x - 1}{3x^2 + x - 2}\right)^3$

26. $F(x) = \dfrac{(x^2 + 3)^3}{(5x - 8)^2}$

27. $f(z) = \dfrac{(z^2 - 5)^3}{(z^2 + 4)^2}$

28. $G(x) = \dfrac{(4x - 1)^3(x^2 + 2)^4}{(3x^2 + 5)^2}$

29. $g(t) = \sin^2(3t^2 - 1)$

30. $f(x) = \tan^2 x^2$

31. $f(x) = (\tan^2 x - x^2)^3$

32. $G(x) = (2 \sin x - 3 \cos x)^3$

33. $f(y) = \dfrac{3 \sin 2y}{\cos^2 2y + 1}$

34. $g(x) = \dfrac{\cot^2 2x}{1 + x^2}$

35. $F(x) = 4 \cos(\sin 3x)$

36. $f(x) = \sin^2(\cos 2x)$

37. Find an equation of the tangent line to the curve $y = (x^2 - 1)^2$ at each of the following points: $(-2, 9), (-1, 0), (0, 1), (1, 0),$ and $(2, 9)$. Draw a sketch of the graph and segments of the tangent lines at the given points.

38. Find an equation of the tangent line to the curve $y = 4 \tan 2x$ at the point where $x = \frac{1}{8}\pi$.

In Exercises 39 through 42, a particle is moving along a straight line according to the given equation, where s cm is the directed distance of the particle from an origin at t sec. (a) What is the instantaneous velocity of the particle at t sec? (b) Find the instantaneous velocity of the particle at t_1 sec for each of the given values of t_1.

39. $s = \dfrac{(t^2 - 1)^2}{(t^2 + 1)^2}, t \geq 0; t_1$ is 1, 2

40. $s = \left(\dfrac{3t}{2t + 1}\right)^4; t \geq 0; t_1$ is $\frac{1}{2}$, 1

41. $s = 5 \sin \pi t + 3 \cos \pi t; t_1$ is $\frac{1}{4}, \frac{3}{2}$

42. $s = 2 \cos \pi(t + 1); t_1$ is $\frac{1}{4}, \frac{3}{2}$

43. The electromotive force for an electric circuit with a simplified generator is $E(t)$ volts at t sec, where $E(t) = 50 \sin 120\pi t$. Find the instantaneous rate of change of $E(t)$ with respect to t at (a) 0.02 sec and (b) 0.2 sec.

44. A wave produced by a simple sound has the equation $P(t) = 0.003 \sin 1800\pi t$, where $P(t)$ dynes per square centimeter is the difference between the atmospheric pressure and the air pressure at the eardrum at t sec. Find the instantaneous rate of change of $P(t)$ with respect to t at (a) $\frac{1}{9}$ sec; (b) $\frac{1}{8}$ sec; (c) $\frac{1}{7}$ sec.

45. When a pendulum of length 10 cm has swung so that θ is the radian measure of the angle formed by the pendulum and a vertical line, then if $h(\theta)$ cm is the vertical height of the end of the pendulum above its lowest position, $h(\theta) = 20 \sin^2 \frac{1}{2}\theta$. Find the instantaneous rate of change of $h(\theta)$ with respect to θ when (a) $\theta = \frac{1}{3}\pi$; (b) $\theta = \frac{1}{2}\pi$.

46. If K square units is the area of a right triangle, 10 units is the length of the hypotenuse, and α is the radian measure of an acute angle, then $K = 25 \sin 2\alpha$. Find the instantaneous rate of change of K with respect to α when (a) $\alpha = \frac{1}{6}\pi$; (b) $\alpha = \frac{1}{4}\pi$; (c) $\alpha = \pi$.

47. An airplane is flying parallel to the ground at an altitude of 2 km and at a speed of $4\frac{1}{2}$ km/min. If the plane flies directly over the Statue of Liberty, at what rate is the line-of-sight distance between the plane and the statue changing 20 sec later?

48. At 8 A.M. a ship sailing due north at 24 knots (nautical miles per hour) is at a point P. At 10 A.M. a second ship sailing due east at 32 knots is at P. At what rate is the distance between the two ships changing at (a) 9 A.M. and (b) 11 A.M.?

In Exercises 49 through 52, show that the simplified proof of the chain rule with the additional requirement (12) is not valid for the given functions f and g. In each exercise draw a sketch of the graph of g.

49. $f(x) = x^4; g(x) = [\![x]\!]$

50. $f(x) = \tan x; g(x) = \begin{cases} -1 & \text{if } x < 0 \\ 0 & \text{if } x = 0 \\ 1 & \text{if } 0 < x \end{cases}$

51. $f(x) = x^2; g(x) = |x| + |x - 1|$

52. $g(x) = x^2 + 1; g(x) = |x - 2| + |x + 2|$

53. Given $f(x) = x^3$ and $g(x) = f(x^2)$. Find (a) $f'(x^2)$; (b) $g'(x)$.

54. Given $f(u) = u^2 + 5u + 5$ and $g(x) = (x + 1)/(x - 1)$. Find the derivative of $f \circ g$ in two ways: (a) by first finding $(f \circ g)(x)$ and then finding $(f \circ g)'(x)$; (b) by using the chain rule.

SOLUTION

$f(r) = (4 \sin^2 r + 9 \cos^2 r)^{1/2}$. We apply Theorem 2.8.2.

$$f'(r) = \tfrac{1}{2}(4 \sin^2 r + 9 \cos^2 r)^{-1/2} \cdot D_r(4 \sin^2 r + 9 \cos^2 r)$$

$$= \frac{8 \sin r \cdot D_r(\sin r) + 18 \cos r \cdot D_r(\cos r)}{2\sqrt{4 \sin^2 r + 9 \cos^2 r}}$$

$$= \frac{8 \sin r \cos r + 18 \cos r(-\sin r)}{2\sqrt{4 \sin^2 r + 9 \cos^2 r}}$$

$$= \frac{-10 \sin r \cos r}{2\sqrt{4 \sin^2 r + 9 \cos^2 r}}$$

$$= -\frac{5 \sin r \cos r}{\sqrt{4 \sin^2 r + 9 \cos^2 r}}$$

EXERCISES 2.8

In Exercises 1 through 24, find the derivative of the given function.

1. $f(x) = 4x^{1/2} + 5x^{-1/2}$ **2.** $f(x) = 3x^{2/3} - 6x^{1/3} + x^{-1/3}$ **3.** $g(x) = \sqrt{1 + 4x^2}$

4. $f(s) = \sqrt{2 - 3s^2}$ **5.** $f(x) = (5 - 3x)^{2/3}$ **6.** $g(x) = \sqrt[3]{4x^2 - 1}$ **7.** $g(y) = \dfrac{1}{\sqrt{25 - y^2}}$

8. $f(x) = (5 - 2x^2)^{-1/3}$ **9.** $h(t) = 2 \cos \sqrt{t}$ **10.** $f(x) = 4 \sec \sqrt{x}$ **11.** $g(r) = \cot \sqrt{3r}$

12. $g(x) = \sqrt{3 \sin x}$ **13.** $f(x) = (\sin 3x)^{-1/2}$ **14.** $f(y) = \sqrt{1 + \csc^2 y}$ **15.** $f(x) = \tan \sqrt{x^2 + 1}$

16. $f(y) = 3 \cos \sqrt[3]{2y^2}$ **17.** $g(x) = \sqrt{\dfrac{2x - 5}{3x + 1}}$ **18.** $h(t) = \dfrac{\sqrt{t - 1}}{\sqrt{t + 1}}$

19. $F(x) = \sqrt[3]{2x^3 - 5x^2 + x}$ **20.** $G(t) = \sqrt{\dfrac{5t + 6}{5t - 4}}$ **21.** $g(t) = \sqrt{2t} + \sqrt{\dfrac{2}{t}}$

22. $g(x) = \sqrt[3]{(3x^2 + 5x - 1)^2}$ **23.** $f(x) = (5 - x^2)^{1/2}(x^3 + 1)^{1/4}$ **24.** $g(y) = (y^2 + 3)^{1/3}(y^3 - 1)^{1/2}$

In Exercises 25 through 36, compute the indicated derivative.

25. $\dfrac{d}{dx}\left(\dfrac{\sqrt{x^2 - 1}}{x}\right)$ **26.** $\dfrac{d}{dx}(\sqrt{x^2 - 5}\sqrt[3]{x^2 + 3})$ **27.** $\dfrac{d}{dt}\left(\sqrt{\dfrac{\sin t + 1}{1 - \sin t}}\right)$ **28.** $\dfrac{d}{dz}(\sin \sqrt[3]{z} \cos \sqrt[3]{z})$

29. $\dfrac{d}{dy}(\tan \sqrt{y} \sec \sqrt{y})$ **30.** $\dfrac{d}{dx}\left(\sqrt{\dfrac{\cos x - 1}{\sin x}}\right)$ **31.** $D_x\left(\dfrac{\sqrt{x - 1}}{\sqrt[3]{x + 1}}\right)$ **32.** $D_x\left(\dfrac{4x + 6}{\sqrt{x^2 + 3x + 4}}\right)$

33. $D_x(\sqrt{9 + \sqrt{9 - x}})$ **34.** $D_y\left(\sqrt[4]{\dfrac{y^3 + 1}{y^3 - 1}}\right)$ **35.** $D_z\left(\dfrac{1}{\sqrt{1 + \cos^2 2z}}\right)$ **36.** $D_x\left(\sqrt{x} \tan \sqrt{\dfrac{1}{x}}\right)$

In Exercises 37 and 38, find an equation of the tangent line to the curve at each of the given points. Draw a sketch of the graph and segments of the tangent lines at the given points.

37. $y = (2x - 2)^{2/3}$; $(-3, 4)$, $(0, \sqrt[3]{4})$, $(1, 0)$, $(2, \sqrt[3]{4})$, $(5, 4)$ **38.** $y = (6 - 2x)^{1/3}$; $(-1, 2)$, $(1, \sqrt[3]{4})$, $(3, 0)$, $(5, -\sqrt[3]{4})$, $(7, -2)$

39. Find an equation of the tangent line to the curve $y = \sqrt{x^2 + 9}$ at the point $(4, 5)$.

40. Find an equation of the tangent line to the curve $y = (7x - 6)^{-1/3}$ that is perpendicular to the line $12x - 7y + 2 = 0$.

41. Find an equation of the normal line to the curve $y = x\sqrt{16 + x^2}$ at the origin.

42. Find an equation of the tangent line to the curve $y = \sqrt{\sin x + \cos x}$ at the point where $x = \frac{1}{4}\pi$.

43. An object is moving along a straight line according to the equation of motion $s = \sqrt{4t^2 + 3}$, with $t \geq 0$. Find the value of t for which the measure of the instantaneous velocity is (a) 0; (b) 1; (c) 2.

44. An object is moving along a straight line according to the equation of motion $s = \sqrt{5 + t^2}$, with $t \geq 0$. Find the value of t for which the measure of the instantaneous velocity is (a) 0; (b) 1.

45. A property development company rents each apartment at p dollars per month when x apartments are rented, and $p = 30\sqrt{300 - 2x}$. If R dollars is the revenue received from the rental of x apartments, then $R = px$. How many apartments must be rented before the rate of change of R with respect to x (the marginal revenue) is zero? (*Note:* Because x is the number of apartments rented, x is a nonnegative integer. However, to apply the calculus, assume that x is a nonnegative real number and then round off any noninteger values of x to the nearest integer.)

46. The daily production at a particular factory is $f(x)$ units when the capital investment is x thousands of dollars, and $f(x) = 200\sqrt{2x + 1}$. If the current capitalization is \$760,000, use the derivative to estimate the change in the daily production if the capital investment is increased by \$1000.

In Exercises 47 through 50, find the derivative of the given function. (*Hint:* $|a| = \sqrt{a^2}$.)

47. $f(x) = |x^2 - 4|$

48. $g(x) = x|x|$

49. $g(x) = |x|^3$

50. $h(x) = \sqrt[3]{|x| + x}$

51. Given $f(u) = 1/u^2$ and $g(x) = \sqrt{x}/\sqrt{2x^3 - 6x + 1}$, find the derivative of $f \circ g$ in two ways: (a) by first finding $(f \circ g)(x)$ and then finding $(f \circ g)'(x)$; (b) by using the chain rule.

52. Suppose that $g(x) = \sqrt{9 - x^2}$ and $h(x) = f(g(x))$, where f is differentiable at 3. Prove that $h'(0) = 0$.

53. Suppose $g(x) = |f(x)|$. Prove that if $f'(x)$ and $g'(x)$ exist, then $|g'(x)| = |f'(x)|$.

2.9
IMPLICIT DIFFERENTIATION

If $f = \{(x, y) | y = 3x^2 + 5x + 1\}$, then the equation

$$y = 3x^2 + 5x + 1 \tag{1}$$

defines the function f explicitly. However, not all functions are defined by such an equation. For example, if we have the equation

$$x^6 - 2x = 3y^6 + y^5 - y^2 \tag{2}$$

we cannot solve for y in terms of x; however, there may exist one or more functions f such that if $y = f(x)$, Eq. (2) is satisfied, that is, such that the equation

$$x^6 - 2x = 3[f(x)]^6 + [f(x)]^5 - [f(x)]^2$$

holds for all values of x in the domain of f. In this case the function f is defined *implicitly* by the given equation.

With the assumption that (2) defines y as at least one differentiable function of x, the derivative of y with respect to x can be found by *implicit differentiation*.

Equation (2) is a special type of equation involving x and y because it can be written so that all the terms involving x are on one side of the equation and all the terms involving y are on the other side. It serves as a first example to illustrate the process of implicit differentiation.

The left side of (2) is a function of x, and the right side is a function of y. Let F be the function defined by the left side of (2), and let G be the function defined by the right side of (2). Thus

$$F(x) = x^6 - 2x \qquad G(y) = 3y^6 + y^5 - y^2$$

where y is a function of x, say $y = f(x)$. So (2) can be written as

$$F(x) = G(f(x)) \tag{3}$$

Equation (3) is satisfied by all values of x in the domain of f for which $G(f(x))$ exists.

Then for all values of x for which f is differentiable,

$$D_x(x^6 - 2x) = D_x(3y^6 + y^5 - y^2) \tag{4}$$

The derivative on the left side of (4) is easily found, and

$$D_x(x^6 - 2x) = 6x^5 - 2 \tag{5}$$

We find the derivative on the right side of (4) by the chain rule.

$$D_x(3y^6 + y^5 - y^2) = 18y^5 \cdot \frac{dy}{dx} + 5y^4 \cdot \frac{dy}{dx} - 2y \cdot \frac{dy}{dx} \tag{6}$$

Substituting the values from (5) and (6) into (4) we obtain

$$6x^5 - 2 = (18y^5 + 5y^4 - 2y)\frac{dy}{dx}$$

$$\frac{dy}{dx} = \frac{6x^5 - 2}{18y^5 + 5y^4 - 2y}$$

Observe that by using implicit differentiation we have obtained an expression for $\dfrac{dy}{dx}$ that involves both variables x and y.

In the following illustration the method of implicit differentiation is used to find $\dfrac{dy}{dx}$ from a more general type of equation.

▶ ILLUSTRATION 1   Consider the equation

$$3x^4y^2 - 7xy^3 = 4 - 8y \tag{7}$$

and assume that there exists at least one differentiable function f such that if $y = f(x)$, Eq. (7) is satisfied. Differentiating on both sides of (7) (bearing in mind that y is a differentiable function of x) and applying the theorems for the derivative of a product, the derivative of a power, and the chain rule, we obtain

$$12x^3y^2 + 3x^4\left(2y\frac{dy}{dx}\right) - 7y^3 - 7x\left(3y^2\frac{dy}{dx}\right) = 0 - 8\frac{dy}{dx}$$

$$\frac{dy}{dx}(6x^4y - 21xy^2 + 8) = 7y^3 - 12x^3y^2$$

$$\frac{dy}{dx} = \frac{7y^3 - 12x^3y^2}{6x^4y - 21xy^2 + 8} \qquad ◀$$

Remember that we assumed that both (2) and (7) define y as at least one differentiable function of x. It may be that an equation in x and y does not imply the existence of any real-valued function, as is the case for the equation

$$x^2 + y^2 + 4 = 0$$

which is not satisfied by any real values of x and y. Furthermore, it is possible that an equation in x and y may be satisfied by many different functions, some of which are differentiable and some of which are not. A general discussion is beyond the scope of this book but can be found in an advanced calculus text. In subsequent discussions when we state that an equation in x and y defines y implicitly as a function of x, it is assumed that one or more of these functions is differentiable. Example 4, which follows, illustrates the fact that implicit differentiation gives the derivative of two differentiable functions defined by the given equation.

EXAMPLE 1

Given $(x + y)^2 - (x - y)^2 = x^4 + y^4$, find $\dfrac{dy}{dx}$.

SOLUTION

Differentiating implicitly with respect to x we have

$$2(x + y)\left(1 + \frac{dy}{dx}\right) - 2(x - y)\left(1 - \frac{dy}{dx}\right) = 4x^3 + 4y^3 \frac{dy}{dx}$$

$$2x + 2y + (2x + 2y)\frac{dy}{dx} - 2x + 2y + (2x - 2y)\frac{dy}{dx} = 4x^3 + 4y^3 \frac{dy}{dx}$$

$$\frac{dy}{dx}(4x - 4y^3) = 4x^3 - 4y$$

$$\frac{dy}{dx} = \frac{x^3 - y}{x - y^3}$$

EXAMPLE 2

Find an equation of the tangent line to the curve $x^3 + y^3 = 9$ at the point $(1, 2)$.

SOLUTION

We differentiate implicitly with respect to x.

$$3x^2 + 3y^2 \frac{dy}{dx} = 0$$

$$\frac{dy}{dx} = -\frac{x^2}{y^2}$$

Therefore, at the point $(1, 2)$, $\dfrac{dy}{dx} = -\dfrac{1}{4}$. An equation of the tangent line is then

$$y - 2 = -\tfrac{1}{4}(x - 1)$$

$$x + 4y - 9 = 0$$

EXAMPLE 3

Given $x \cos y + y \cos x = 1$, find $\dfrac{dy}{dx}$.

SOLUTION

Differentiating implicitly with respect to x we get

$$1 \cdot \cos y + x(-\sin y)\frac{dy}{dx} + \frac{dy}{dx}(\cos x) + y(-\sin x) = 0$$

$$\frac{dy}{dx}(\cos x - x \sin y) = y \sin x - \cos y$$

$$\frac{dy}{dx} = \frac{y \sin x - \cos y}{\cos x - x \sin y}$$

EXAMPLE 4

Given the equation $x^2 + y^2 = 9$, find (a) $\dfrac{dy}{dx}$ by implicit differentiation; (b) two functions defined by the equation; (c) the derivative of each of the functions obtained in part (b) by explicit differentiation. (d) Verify that the result obtained in part (a) agrees with the results obtained in part (c).

SOLUTION

(a) We differentiate implicitly.

$$2x + 2y\frac{dy}{dx} = 0$$

$$\frac{dy}{dx} = -\frac{x}{y}$$

(b) If the given equation is solved for y,

$$y = \sqrt{9 - x^2} \quad \text{and} \quad y = -\sqrt{9 - x^2}$$

Let f_1 and f_2 be the two functions for which

$$f_1(x) = \sqrt{9 - x^2} \quad \text{and} \quad f_2(x) = -\sqrt{9 - x^2}$$

(c) Because $f_1(x) = (9 - x^2)^{1/2}$ and $f_2(x) = -(9 - x^2)^{1/2}$ from the chain rule we obtain

$$f_1'(x) = \tfrac{1}{2}(9 - x^2)^{-1/2}(-2x) \qquad f_2'(x) = -\tfrac{1}{2}(9 - x^2)^{-1/2}(-2x)$$

$$= -\frac{x}{\sqrt{9 - x^2}} \qquad\qquad\qquad = \frac{x}{\sqrt{9 - x^2}}$$

(d) For $y = f_1(x)$, where $f_1(x) = \sqrt{9 - x^2}$, it follows from part (c) that

$$f_1{}'(x) = -\frac{x}{\sqrt{9 - x^2}}$$

$$= -\frac{x}{y}$$

which agrees with the answer in part (a).

For $y = f_2(x)$, where $f_2(x) = -\sqrt{9 - x^2}$, we have from part (c)

$$f_2{}'(x) = \frac{x}{\sqrt{9 - x^2}}$$

$$= -\frac{x}{-\sqrt{9 - x^2}}$$

$$= -\frac{x}{y}$$

which also agrees with the answer in part (a).

EXERCISES 2.9

In Exercises 1 through 28, find $\dfrac{dy}{dx}$ by implicit differentiation.

1. $x^2 + y^2 = 16$ **2.** $4x^2 - 9y^2 = 1$ **3.** $x^3 + y^3 = 8xy$ **4.** $x^2 + y^2 = 7xy$

5. $\dfrac{1}{x} + \dfrac{1}{y} = 1$ **6.** $\dfrac{3}{x} - \dfrac{3}{y} = 2x$ **7.** $\sqrt{x} + \sqrt{y} = 4$ **8.** $2x^3 y + 3xy^3 = 5$

9. $x^2 y^2 = x^2 + y^2$ **10.** $(2x + 3)^4 = 3y^4$ **11.** $x^2 = \dfrac{x + 2y}{x - 2y}$ *mult first*   **12.** $\dfrac{x}{\sqrt{y}} - 4y = x$

13. $\sqrt[3]{x} + \sqrt[3]{xy} = 4y^2$ **14.** $\sqrt{y} + \sqrt[3]{y} + \sqrt[4]{y} = x$ **15.** $\sqrt{xy} + 2x = \sqrt{y}$ **16.** $y + \sqrt{xy} = 3x^3$

17. $\dfrac{y}{\sqrt{x} - y} = 2 + x^2$ **18.** $x^2 y^3 = x^4 - y^4$ **19.** $y = \cos(x - y)$ **20.** $x = \sin(x + y)$

21. $\sec^2 x + \csc^2 y = 4$ **22.** $\cot xy + xy = 0$ **23.** $x \sin y + y \cos x = 1$ **24.** $\cos(x + y) = y \sin x$

25. $\sec^2 y + \cot(x - y) = \tan^2 x$ **26.** $\csc(x - y) + \sec(x + y) = x$

27. $(x + y)^2 - (x - y)^2 = x^3 + y^3$ **28.** $y\sqrt{2 + 3x} + x\sqrt{1 + y} = x$

In Exercises 29 through 32, consider y as the independent variable and find $\dfrac{dx}{dy}$.

29. $x^4 + y^4 = 12x^2 y$ **30.** $y = 2x^3 - 5x$ **31.** $x^3 y + 2y^4 - x^4 = 0$ **32.** $y\sqrt{x} - x\sqrt{y} = 9$

33. Find an equation of the tangent line to the curve $16x^4 + y^4 = 32$ at the point $(1, 2)$.

34. Find an equation of the normal line to the curve $9x^3 - y^3 = 1$ at the point $(1, 2)$.

35. Find an equation of the normal line to the curve $x^2 + xy + y^2 - 3y = 10$ at the point $(2, 3)$.

36. Find an equation of the tangent line to the curve $\sqrt[3]{xy} = 14x + y$ at the point $(2, -32)$.

37. Find the rate of change of y with respect to x at the point $(3, 2)$ if $7y^2 - xy^3 = 4$.

38. For the circle $x^2 + y^2 = r^2$, show that the tangent line at any point (x_1, y_1) on the curve is perpendicular to the line through (x_1, y_1) and the center of the circle.

39. At what point of the curve $x + \sqrt{xy} + y = 1$ is the tangent line parallel to the x axis?

40. There are two lines through the point $(-1, 3)$ that are tangent to the curve $x^2 + 4y^2 - 4x - 8y + 3 = 0$. Find an equation of each of these lines.

In Exercises 41 through 46, an equation is given. Do the following in each of these problems: (a) Find two functions defined by the equation, and state their domains. (b) Draw a sketch of the graph of each of the functions obtained in part (a). (c) Draw a sketch of the graph of the equation. (d) Find the derivative of each of the functions obtained in part (a), and state the domains of the derivatives. (e) Find $\dfrac{dy}{dx}$ by implicit differentiation from the given equation, and verify that the result so obtained agrees with the results in part (d). (f) Find an equation of each tangent line at the given value of x_1.

41. $y^2 = 4x - 8$; $x_1 = 3$ **42.** $x^2 + y^2 = 25$; $x_1 = 4$ **43.** $x^2 - y^2 = 9$; $x_1 = -5$ **44.** $y^2 - x^2 = 16$; $x_1 = -3$
45. $x^2 + y^2 - 2x - 4y - 4 = 0$; $x_1 = 1$ **46.** $x^2 + 4y^2 + 6x - 40y + 93 = 0$; $x_1 = -2$

47. Prove that the sum of the x and y intercepts of any tangent line to the curve $x^{1/2} + y^{1/2} = k^{1/2}$ is constant and equal to k.

48. If $x^n y^m = (x + y)^{n+m}$, prove that $x \cdot \dfrac{dy}{dx} = y$.

49. Let f be the power function defined by $f(x) = x^r$, where r is any rational number. Under the assumption that f is differentiable, use implicit differentiation to show that $f'(x) = rx^{r-1}$. $\left(Hint: \text{Let } r = \dfrac{p}{q}, \text{ where } p \text{ and } q \text{ are integers and } q > 0. \right.$
Then replace $f(x)$ by y and write the equation as $y^q = x^p$. Use implicit differentiation to find $\dfrac{dy}{dx}.\bigg)$

2.10
RELATED RATES

There are many problems concerned with the rate of change of two or more related variables with respect to time in which it is not necessary to express each of these variables directly as functions of time. For example, suppose we have an equation involving the variables x and y, and both x and y are functions of a third variable t, where t units denotes time. Then, because the rate of change of x with respect to t and the rate of change of y with respect to t are given by $\dfrac{dx}{dt}$ and $\dfrac{dy}{dt}$, respectively, we differentiate on both sides of the given equation with respect to t by applying the chain rule and proceed as in the following example.

EXAMPLE 1
In (a) and (b), x and y are functions of a third variable t. (a) If $5x + 3xy = 4$ and $\dfrac{dy}{dt} = 3$, find $\dfrac{dx}{dt}$ when $x = 2$. (b) If $x \cos y = 5$ and $\dfrac{dx}{dt} = -4$, find $\dfrac{dy}{dt}$ when $y = \frac{1}{3}\pi$.

SOLUTION

(a) The given equation is

$$5x + 3xy = 4 \tag{1}$$

We differentiate on both sides of the equation with respect to t.

$$5\frac{dx}{dt} + 3y\frac{dx}{dt} + 3x\frac{dy}{dt} = 0$$

$$\frac{dx}{dt} = -\frac{3x}{3y + 5} \cdot \frac{dy}{dt} \tag{2}$$

To find the value of y when $x = 2$ we replace x by 2 in (1) and obtain $y = -1$. From (2), with $x = 2$, $y = -1$, and $\frac{dy}{dt} = 3$,

$$\frac{dx}{dt}\bigg]_{x=2} = -\tfrac{6}{2}(3)$$

$$= -9$$

(b) On both sides of the equation

$$x \cos y = 5 \tag{3}$$

we differentiate with respect to t.

$$(\cos y)\frac{dx}{dt} - (x \sin y)\frac{dy}{dt} = 0$$

$$\frac{dy}{dt} = \frac{\cos y}{x \sin y} \cdot \frac{dx}{dt} \tag{4}$$

From (3), when $y = \tfrac{1}{3}\pi$, $x = 10$. From (4), with $y = \tfrac{1}{3}\pi$, $x = 10$, and $\frac{dx}{dt} = -4$,

$$\frac{dy}{dt}\bigg]_{y=\pi/3} = \frac{\tfrac{1}{2}}{10(\tfrac{1}{2}\sqrt{3})}(-4)$$

$$= -\tfrac{2}{15}\sqrt{3}$$

EXAMPLE 2

A ladder 25 ft long is leaning against a vertical wall. If the bottom of the ladder is pulled horizontally away from the wall at 3 ft/sec, how fast is the top of the ladder sliding down the wall when the bottom is 15 ft from the wall?

SOLUTION

Let t be the number of seconds in the time that has elapsed since the ladder started to slide down the wall, y be the number of feet in the distance from the ground to the top of the ladder at t sec, and x be the number of feet in the distance from the bottom of the ladder to the wall at t sec. See Fig. 2.10.1.

25 ft

y ft

x ft

FIGURE 2.10.1

Because the bottom of the ladder is pulled horizontally away from the wall at 3 ft/sec, $\dfrac{dx}{dt} = 3$. We wish to find $\dfrac{dy}{dt}$ when $x = 15$. From the Pythagorean theorem,

$$y^2 = 625 - x^2 \tag{5}$$

Because x and y are functions of t, we differentiate on both sides of (5) with respect to t and obtain

$$2y \frac{dy}{dt} = -2x \frac{dx}{dt}$$

$$\frac{dy}{dt} = -\frac{x}{y} \frac{dx}{dt} \tag{6}$$

When $x = 15$, it follows from (5) that $y = 20$. Because $\dfrac{dx}{dt} = 3$, we get from (6)

$$\frac{dy}{dt}\Bigg]_{y=20} = -\tfrac{15}{20} \cdot 3$$

$$= -\tfrac{9}{4}$$

Therefore the top of the ladder is sliding down the wall at the rate of $2\tfrac{1}{4}$ ft/sec when the bottom is 15 ft from the wall. The significance of the minus sign is that y is decreasing as t is increasing.

Example 2 involves a problem with rates of change of related variables. It is called a problem in related rates. In such problems the variables have a specific relationship for values of t, where t is a measure of time. This relationship is usually expressed in the form of an equation, as in Example 2 with Eq. (5). Values of the variables and rates of change of the variables with respect to t are often given at a particular instant. In Example 2, at the instant when $x = 15$, then $y = 20$ and $\dfrac{dx}{dt} = 3$, and we wish to find $\dfrac{dy}{dt}$.

The following steps represent a possible procedure for solving a problem involving related rates.

1. Draw a figure if it is feasible to do so.
2. Define the variables. Generally define t first, because the other variables usually depend on t.
3. Write down any numerical facts known about the variables and their derivatives with respect to t.
4. Obtain an equation involving the variables that depend on t.
5. Differentiate with respect to t on both sides of the equation found in step 4.
6. Substitute values of known quantities in the equation of step 5, and solve for the desired quantity.

FIGURE 2.10.2

EXAMPLE 3

A tank is in the form of an inverted cone having an altitude of 16 m and a base radius of 4 m. Water is flowing into the tank at the rate of 2 m³/min. How fast is the water level rising when the water is 5 m deep?

SOLUTION

Let t be the number of minutes in the time that has elapsed since water started to flow into the tank, h be the number of meters in the height of the water level at t min, r be the number of meters in the radius of the surface of the water at t min, and V be the number of cubic meters in the volume of water in the tank at t min.

At any time, the volume of water in the tank may be expressed in terms of the volume of a cone (see Fig. 2.10.2).

$$V = \tfrac{1}{3}\pi r^2 h \tag{7}$$

$V, r,$ and h are all functions of t. Because water is flowing into the tank at the rate of 2 m³/min, $\dfrac{dV}{dt} = 2$. We wish to find $\dfrac{dh}{dt}$ when $h = 5$. To express r in terms of h we have from similar triangles

$$\frac{r}{h} = \frac{4}{16} \quad \Leftrightarrow \quad r = \tfrac{1}{4}h$$

Substituting this value of r into (7) we obtain

$$V = \tfrac{1}{3}\pi(\tfrac{1}{4}h)^2(h) \quad \Leftrightarrow \quad V = \tfrac{1}{48}\pi h^3 \tag{8}$$

By differentiating on both sides of (8) with respect to t,

$$\frac{dV}{dt} = \frac{1}{16}\pi h^2 \frac{dh}{dt}$$

Substituting 2 for $\dfrac{dV}{dt}$ and solving for $\dfrac{dh}{dt}$ we get

$$\frac{dh}{dt} = \frac{32}{\pi h^2}$$

Therefore

$$\frac{dh}{dt}\bigg]_{h=5} = \frac{32}{25\pi}$$

Therefore the water level is rising at the rate of $\dfrac{32}{25\pi}$ m/min when the water is 5 m deep.

EXAMPLE 4

Two cars, one going due east at the rate of 90 km/hr and the other going due south at the rate of 60 km/hr, are traveling toward an intersection of the two roads. At what rate are the two cars approaching each other at the instant when the first car is 200 m and the second car is 150 m from the intersection?

FIGURE 2.10.3

SOLUTION

Refer to Fig. 2.10.3, where the point P is the intersection of the two roads.

Let t sec be the time elapsed since the cars started to approach P, x meters be the distance of the first car from P at t sec, y meters be the distance of the second car from P at t sec, and z meters be the distance between the two cars at t sec.

The first car is approaching P at the rate of 90 km/hr, which is 90,000/3600 m/sec, or 25 m/sec. Because x is decreasing as t is increasing, it follows that $\dfrac{dx}{dt} = -25$. Similarly, because 60 km/hr is 60,000/3600 m/sec, which is $\frac{50}{3}$ m/sec, $\dfrac{dy}{dt} = -\frac{50}{3}$. We wish to find $\dfrac{dz}{dt}$ when $x = 200$ and $y = 150$.

From the Pythagorean theorem,

$$z^2 = x^2 + y^2 \tag{9}$$

Differentiating on both sides of (9) with respect to t we obtain

$$2z\frac{dz}{dt} = 2x\frac{dx}{dt} + 2y\frac{dy}{dt}$$

$$\frac{dz}{dt} = \frac{x\dfrac{dx}{dt} + y\dfrac{dy}{dt}}{z} \tag{10}$$

When $x = 200$ and $y = 150$, it follows from (9) that $z = 250$. In (10) let $\dfrac{dx}{dt} = -25$, $\dfrac{dy}{dt} = -\frac{50}{3}$, $x = 200$, $y = 150$, and $z = 250$ to get

$$\frac{dz}{dt}\bigg]_{z=250} = \frac{(200)(-25) + (150)(-\frac{50}{3})}{250}$$

$$= -30$$

Therefore, at the instant in question the cars are approaching each other at the rate of 30 m/sec.

EXAMPLE 5

Suppose in a certain market that x thousands of crates of oranges are supplied daily when p dollars is the price per crate, and the supply equation is

$$px - 20p - 3x + 105 = 0$$

If the daily supply is decreasing at the rate of 250 crates per day, at what rate is the price changing when the daily supply is 5000 crates?

SOLUTION

Let t days be the time that has elapsed since the daily supply of oranges started to decrease. Then p and x are both functions of t. Because the daily supply is decreasing at the rate of 250 crates per day, $\dfrac{dx}{dt} = -\dfrac{250}{1000}$; that is, $\dfrac{dx}{dt} = -\dfrac{1}{4}$.

We wish to find $\dfrac{dp}{dt}$ when $x = 5$. From the given supply equation we differentiate implicitly with respect to t and obtain

$$p\frac{dx}{dt} + x\frac{dp}{dt} - 20\frac{dp}{dt} - 3\frac{dx}{dt} = 0$$

$$\frac{dp}{dt} = \frac{3 - p}{x - 20} \cdot \frac{dx}{dt} \qquad (11)$$

When $x = 5$, it follows from the supply equation that $p = 6$. Because $\dfrac{dx}{dt} = -\dfrac{1}{4}$, we have from (11)

$$\frac{dp}{dt}\bigg]_{p=6} = \frac{3 - 6}{5 - 20}\left(-\frac{1}{4}\right)$$

$$= -\frac{1}{20}$$

Thus the price of a crate of oranges is decreasing at the rate of $0.05 per day when the daily supply is 5000 crates.

EXAMPLE 6

An airplane is flying west at 500 ft/sec at an altitude of 4000 ft. The airplane is in a vertical plane with a searchlight on the ground. If the light is to be kept on the plane, how fast is the searchlight revolving when the airplane is due east of the searchlight at an airline distance of 2000 ft?

SOLUTION

Refer to Fig. 2.10.4. The searchlight is at point L, and at a particular instant of time the airplane is at point P.

Let t be the number of seconds in the time, x be the number of feet due east in the airline distance of the airplane from the searchlight at time t sec, and θ be the number of radians in the angle of elevation of the airplane at the searchlight at time t sec.

We are given $\dfrac{dx}{dt} = -500$, and we wish to find $\dfrac{d\theta}{dt}$ when $x = 2000$.

$$\tan \theta = \frac{4000}{x} \qquad (12)$$

Differentiating on both sides of (12) with respect to t we obtain

$$\sec^2 \theta \frac{d\theta}{dt} = -\frac{4000}{x^2}\frac{dx}{dt}$$

P

4000 ft

θ

L x ft

FIGURE 2.10.4

Substituting $\dfrac{dx}{dt} = -500$ in the above and dividing by $\sec^2 \theta$ gives

$$\frac{d\theta}{dt} = \frac{2,000,000}{x^2 \sec^2 \theta} \tag{13}$$

When $x = 2000$, $\tan \theta = 2$. Because $\sec^2 \theta = 1 + \tan^2 \theta$, $\sec^2 \theta = 5$. Substituting these values into (13) we have, when $x = 2000$,

$$\frac{d\theta}{dt} = \frac{2,000,000}{4,000,000(5)}$$

$$= \tfrac{1}{10}$$

We conclude that at the given instant the measurement of the angle is increasing at the rate of $\tfrac{1}{10}$ rad/sec, and this is how fast the searchlight is revolving.

EXERCISES 2.10

In Exercises 1 through 8, x and y are functions of a third variable t.

1. If $2x + 3y = 8$ and $\dfrac{dy}{dt} = 2$, find $\dfrac{dx}{dt}$.

2. If $\dfrac{x}{y} = 10$ and $\dfrac{dx}{dt} = -5$, find $\dfrac{dy}{dt}$.

3. If $xy = 20$ and $\dfrac{dy}{dt} = 10$, find $\dfrac{dx}{dt}$ when $x = 2$.

4. If $2 \sin x + 4 \cos y = 3$ and $\dfrac{dy}{dt} = 3$, find $\dfrac{dx}{dt}$ at $(\tfrac{1}{6}\pi, \tfrac{1}{3}\pi)$.

5. If $\sin^2 x + \cos^2 y = \tfrac{5}{4}$ and $\dfrac{dx}{dt} = -1$, find $\dfrac{dy}{dt}$ at $(\tfrac{2}{3}\pi, \tfrac{3}{4}\pi)$.

6. If $x^2 + y^2 = 25$ and $\dfrac{dx}{dt} = 5$, find $\dfrac{dy}{dt}$ when $y = 4$.

7. If $\sqrt{x} + \sqrt{y} = 5$ and $\dfrac{dy}{dt} = 3$, find $\dfrac{dx}{dt}$ when $x = 1$.

8. If $y(\tan x + 1) = 4$ and $\dfrac{dy}{dt} = -4$, find $\dfrac{dx}{dt}$ when $x = \pi$.

9. A kite is flying at a height of 40 ft. A child is flying it so that it is moving horizontally at a rate of 3 ft/sec. If the string is taut, at what rate is the string being paid out when the length of the string released is 50 ft?

10. A spherical balloon is being inflated so that its volume is increasing at the rate of 5 m³/min. At what rate is the diameter increasing when the diameter is 12 m?

11. A spherical snowball is being made so that its volume is increasing at the rate of 8 ft³/min. Find the rate at which the radius is increasing when the snowball is 4 ft in diameter.

12. Suppose that when the diameter is 6 ft, the snowball in Exercise 11 stopped growing and started to melt at the rate of $\tfrac{1}{4}$ ft³/min. Find the rate at which the radius is changing when the radius is 2 ft.

13. Sand is being dropped at the rate of 10 m³/min onto a conical pile. If the height of the pile is always twice the base radius, at what rate is the height increasing when the pile is 8 m high?

14. A light is hung 15 ft above a straight horizontal path. If a man 6 ft tall is walking away from the light at the rate of 5 ft/sec, how fast is his shadow lengthening?

15. In Exercise 14, at what rate is the tip of the man's shadow moving?

16. A man 6 ft tall is walking toward a building at the rate of 5 ft/sec. If there is a light on the ground 50 ft from the building, how fast is the man's shadow on the building growing shorter when he is 30 ft from the building?

17. A water tank in the form of an inverted cone is being emptied at the rate of 6 m³/min. The altitude of the cone is 24 m, and the base radius is 12 m. Find how fast the water level is lowering when the water is 10 m deep.

18. A trough is 12 ft long and its ends are in the form of inverted isosceles triangles having an altitude of 3 ft and a base of 3 ft. Water is flowing into the trough at the rate of 2 ft³/min. How fast is the water level rising when the water is 1 ft deep?

19. Boyle's law for the expansion of gas is $PV = C$, where P is the number of pounds per square unit of pressure, V is the number of cubic units of volume of the gas, and C is a constant. At a certain instant the pressure is 3000 lb/ft², the volume is 5 ft³, and the volume is increasing at the rate of 3 ft³/min. Find the rate of change of the pressure at this instant.

20. The adiabatic law (no gain or loss of heat) for the expansion of air is $PV^{1.4} = C$, where P is the number of pounds per square unit of pressure, V is the number of cubic units of volume, and C is a constant. At a specific instant the pressure is 40 lb/in.² and is increasing at the rate of 8 lb/in.² each second. What is the rate of change of volume at this instant?

21. A stone is dropped into a still pond. Concentric circular ripples spread out, and the radius of the disturbed region increases at the rate of 16 cm/sec. At what rate does the area of the disturbed region increase when its radius is 4 cm?

22. Oil is running into an inverted conical tank at the rate of 3π m³ per minute. If the tank has a radius of 2.5 m at the top and a depth of 10 m, how fast is the depth of the oil changing when it is 8 m?

23. An automobile traveling at a rate of 30 ft/sec is approaching an intersection. When the automobile is 120 ft from the intersection, a truck traveling at the rate of 40 ft/sec crosses the intersection. The automobile and the truck are on roads that are at right angles to each other. How fast are the automobile and the truck separating 2 sec after the truck leaves the intersection?

24. A rope is attached to a boat at water level, and a woman on a dock is pulling on the rope at the rate of 50 ft/min. If the woman's hands are 16 ft above the water level, how fast is the boat approaching the dock when the amount of rope out is 20 ft?

25. This week a factory is producing 50 units of a particular commodity, and the amount being produced is increasing at the rate of 2 units per week. If $C(x)$ dollars is the total cost of producing x units and $C(x) = 0.08x^3 - x^2 + 10x + 48$, find the current rate at which the production cost is increasing.

26. The demand for a particular kind of breakfast cereal is given by the demand equation $px + 50p = 16,000$, where x thousands of boxes are demanded when p cents is the price per box. If the current price of the cereal is $1.60 per box and the price per box is increasing at the rate of 0.4 cent each week, find the rate of change in the demand.

27. The supply equation for a certain commodity is $x = 1000\sqrt{3p^2 + 20p}$, where x units are supplied per month when p dollars is the price per unit. Find the rate of change in the supply if the current price is $20 per unit and the price is increasing at the rate of $0.50 per month.

28. Suppose that y is the number of workers in the labor force needed to produce x units of a certain commodity, and $x = 4y^2$. If the production of the commodity this year is 250,000 units and the production is increasing at the rate of 18,000 units per year, what is the current rate at which the labor force should be increased?

29. The demand equation for a particular shirt is $2px + 65p - 4950 = 0$, where x hundreds of shirts are demanded per week when p dollars is the price of a shirt. If the shirt is selling this week at $30 and the price is increasing at the rate of $0.20 per week, find the rate of change in the demand.

30. The measure of one of the acute angles of a right triangle is decreasing at the rate of $\frac{1}{36}\pi$ rad/sec. If the length of the hypotenuse is constant and 40 cm, find how fast the area is changing when the measure of the acute angle is $\frac{1}{6}\pi$.

31. Two trucks, one traveling west and the other traveling south, are approaching an intersection. If both trucks are traveling at the rate of k km/hr, show that they are approaching each other at the rate of $k\sqrt{2}$ km/hr when they are each m km from the intersection.

32. A horizontal trough is 16 m long, and its ends are isosceles trapezoids with an altitude of 4 m, a lower base of 4 m, and an upper base of 6 m. Water is being poured into the trough at the rate of 10 m³/min. How fast is the water level rising when the water is 2 m deep?

33. In Exercise 32, if the water level is decreasing at the rate of 25 cm/min when the water is 3 m deep, at what rate is water being drawn from the trough?

34. A ladder 7 m long is leaning against a wall. If the bottom of the ladder is pushed horizontally toward the wall at 1.5 m/sec, how fast is the top of the ladder sliding up the wall when the bottom is 2 m from the wall?

35. A ladder 20 ft long is leaning against an embankment inclined 60° to the horizontal. If the bottom of the ladder is being moved horizontally toward the embankment at 1 ft/sec, how fast is the top of the ladder moving when the bottom is 4 ft from the embankment?

36. The volume of a balloon is decreasing at a rate proportional to its surface area. Show that the radius of the balloon shrinks at a constant rate.

37. An airplane is flying at a constant speed at an altitude of 10,000 ft on a line that will take it directly over an observer on the ground. At a given instant the observer notes that the angle of elevation of the airplane is $\frac{1}{3}\pi$ radians and is increasing at the rate of $\frac{1}{60}$ rad/sec. Find the speed of the airplane.

38. A radar antenna is located on a ship 4 mi from a straight shore, and it is rotating at 32 rpm. How fast is the radar beam moving along the shoreline when the beam makes an angle of 45° with the shore?

39. If a ladder of length 30 ft that is leaning against a wall has its upper end sliding down the wall at the rate of $\frac{1}{2}$ ft/sec, what is the rate of change of the measure of the acute angle made by the ladder with the ground when the upper end is 18 ft above the ground?

40. Water is being poured at the rate of 8 ft^3/min into a tank in the form of a cone. The cone is 20 ft deep and 10 ft in diameter at the top. If there is a leak in the bottom and the water level is rising at the rate of 1 in./min, when the water is 16 ft deep, how fast is the water leaking?

2.11
DERIVATIVES OF HIGHER ORDER

If f' is the derivative of the function f, then f' is also a function, and it is the **first derivative** of f. It is sometimes referred to as the **first derived function**. If the derivative of f' exists, it is called the **second derivative** of f, or the second derived function, and can be denoted by f'' (read as "f double prime"). Similarly, the **third derivative** of f, or the third derived function, is defined as the first derivative of f'' if it exists. The third derivative of f is denoted by f''' (read as "f triple prime").

The **nth derivative** of the function f, where n is a positive integer greater than 1, is the first derivative of the $(n-1)$st derivative of f. We denote the nth derivative of f by $f^{(n)}$. Thus if $f^{(n)}$ is the nth derived function, we can write the function f itself as $f^{(0)}$.

EXAMPLE 1

Find all the derivatives of the function f defined by

$$f(x) = 8x^4 + 5x^3 - x^2 + 7$$

SOLUTION

$$f'(x) = 32x^3 + 15x^2 - 2x$$
$$f''(x) = 96x^2 + 30x - 2$$
$$f'''(x) = 192x + 30$$
$$f^{(4)}(x) = 192$$
$$f^{(5)}(x) = 0$$
$$f^{(n)}(x) = 0 \qquad n \geq 5$$

The Leibniz notation for the first derivative is $\dfrac{dy}{dx}$. For the second derivative of y with respect to x the Leibniz notation is $\dfrac{d^2y}{dx^2}$, because it represents

$\dfrac{d}{dx}\left[\dfrac{d}{dx}(y)\right]$. The symbol $\dfrac{d^n y}{dx^n}$ is a notation for the nth derivative of y with respect to x.

Other symbols for the nth derivative of f are

$$\frac{d^n}{dx^n}[f(x)] \qquad D_x{}^n[f(x)]$$

EXAMPLE 2

Compute

$$\frac{d^3}{dx^3}(2 \sin x + 3 \cos x - x^3)$$

SOLUTION

$$\frac{d}{dx}(2 \sin x + 3 \cos x - x^3) = 2 \cos x - 3 \sin x - 3x^2$$

$$\frac{d^2}{dx^2}(2 \sin x + 3 \cos x - x^3) = -2 \sin x - 3 \cos x - 6x$$

$$\frac{d^3}{dx^3}(2 \sin x + 3 \cos x - x^3) = -2 \cos x + 3 \sin x - 6$$

Because $f'(x)$ gives the instantaneous rate of change of $f(x)$ with respect to x, $f''(x)$, being the derivative of $f'(x)$, gives the instantaneous rate of change of $f'(x)$ with respect to x. Furthermore, if (x, y) is any point on the graph of $y = f(x)$, then $\dfrac{dy}{dx}$ gives the slope of the tangent line to the graph at the point (x, y). Thus $\dfrac{d^2 y}{dx^2}$ is the instantaneous rate of change of the slope of the tangent line with respect to x at the point (x, y).

EXAMPLE 3

Let $m(x)$ be the slope of the tangent line to the curve

$$y = x^3 - 2x^2 + x$$

at the point (x, y). Find the instantaneous rate of change of $m(x)$ with respect to x at the point $(2, 2)$.

SOLUTION

$$m(x) = \frac{dy}{dx}$$

$$= 3x^2 - 4x + 1$$

The instantaneous rate of change of $m(x)$ with respect to x is given by $m'(x)$ or, equivalently, $\dfrac{d^2y}{dx^2}$.

$$m'(x) = \frac{d^2y}{dx^2}$$

$$= 6x - 4$$

At the point $(2, 2)$, $\dfrac{d^2y}{dx^2} = 8$.

The second derivative $f''(x)$ is expressed in units of $f'(x)$ per unit of x, which is units of $f(x)$ per unit of x, per unit of x. For example, in straight-line motion, if $f(t)$ centimeters is the distance of a particle from the origin at t sec, then $f'(t)$ centimeters per second is the velocity of the particle at t sec, and $f''(t)$ centimeters per second per second is the instantaneous rate of change of the velocity at t sec. In physics, the instantaneous rate of change of the velocity is called the **instantaneous acceleration**. Therefore, if a particle is moving along a straight line according to the equation of motion $s = f(t)$, where the instantaneous velocity at t sec is given by v cm/sec, and the instantaneous acceleration at t sec is given by a cm/sec^2, then a is the first derivative of v with respect to t or, equivalently, the second derivative of s with respect to t; that is,

$$v = \frac{ds}{dt}$$

$$a = \frac{dv}{dt} \quad \Leftrightarrow \quad a = \frac{d^2s}{dt^2}$$

When $a > 0$, v is increasing, and when $a < 0$, v is decreasing. When $a = 0$, then v is not changing. Because the speed of the particle at t sec is $|v|$ cm/sec, we have the following results:

(i) If $v \geq 0$ and $a > 0$, the speed is increasing.
(ii) If $v \geq 0$ and $a < 0$, the speed is decreasing.
(iii) If $v \leq 0$ and $a > 0$, the speed is decreasing.
(iv) If $v \leq 0$ and $a < 0$, the speed is increasing.

▶ ILLUSTRATION 1   A particle is moving along a horizontal line according to the equation

$$s = 3t^2 - t^3 \qquad t \geq 0 \tag{1}$$

where s cm is the distance of the particle from the origin at t sec. If v cm/sec is the instantaneous velocity at t sec, then $v = \dfrac{ds}{dt}$. Therefore

$$v = 6t - 3t^2 \tag{2}$$

If a cm/sec^2 is the instantaneous acceleration at t sec, then $a = \dfrac{dv}{dt}$. Thus

$$a = 6 - 6t \tag{3}$$

Let us determine the values of t when any one of the quantities s, v, or a is 0. From (1),

$$s = 0 \quad \text{when} \quad t = 0 \text{ or } t = 3$$

From (2),

$$v = 0 \quad \text{when} \quad t = 0 \text{ or } t = 2$$

From (3),

$$a = 0 \quad \text{when} \quad t = 1$$

In Table 2.11.1 there are values of s, v, and a for these values of t: 0, 1, 2, and 3. Also indicated is the sign of the quantities s, v, and a in the intervals of t excluding 0, 1, 2, and 3. A conclusion is formed regarding the position and motion of the particle for the various values of t.

Table 2.11.1

	s	v	a	Conclusion
$t = 0$	0	0	6	Particle is at the origin. The velocity is 0 and is increasing. The speed is increasing.
$0 < t < 1$	$+$	$+$	$+$	Particle is at the right of the origin, and it is moving to the right. The velocity is increasing. The speed is increasing.
$t = 1$	2	3	0	Particle is 2 cm to the right of the origin, and it is moving to the right at 3 cm/sec. The velocity is not changing; so the speed is not changing.
$1 < t < 2$	$+$	$+$	$-$	Particle is at the right of the origin, and it is moving to the right. The velocity is decreasing. The speed is decreasing.
$t = 2$	4	0	-6	Particle is 4 cm to the right of the origin, and it is changing its direction of motion from right to left. The velocity is decreasing. The speed is increasing.
$2 < t < 3$	$+$	$-$	$-$	Particle is at the right of the origin, and it is moving to the left. The velocity is decreasing. The speed is increasing.
$t = 3$	0	-9	-12	Particle is at the origin, and it is moving to the left at 9 cm/sec. The velocity is decreasing. The speed is increasing.
$3 < t$	$-$	$-$	$-$	Particle is at the left of the origin, and it is moving to the left. The velocity is decreasing. The speed is increasing.

In Fig. 2.11.1 the motion of the particle is along the horizontal line, and the behavior of the motion is indicated above the line.

FIGURE 2.11.1

EXAMPLE 4

A particle is moving along a straight line according to the equation of motion

$$s = \frac{1}{2}t^2 + \frac{4t}{t+1}$$

where s cm is the directed distance of the particle from the origin at t sec. If v cm/sec is the instantaneous velocity at t sec and a cm/sec^2 is the instantaneous acceleration at t sec, find t, s, and v when $a = 0$.

SOLUTION

$$v = \frac{ds}{dt} \qquad\qquad a = \frac{dv}{dt}$$

$$= t + \frac{4}{(t+1)^2} \qquad = 1 - \frac{8}{(t+1)^3}$$

Setting $a = 0$ we have

$$\frac{(t+1)^3 - 8}{(t+1)^3} = 0$$

$$(t+1)^3 = 8$$

from which the only real value of t is obtained from the principal cube root of 8, so that $t + 1 = 2$; that is, $t = 1$. When $t = 1$,

$$s = \frac{1}{2}(1)^2 + \frac{4 \cdot 1}{1+1} \qquad v = 1 + \frac{4}{(1+1)^2}$$

$$= \tfrac{5}{2} \qquad\qquad\qquad = 2$$

Therefore the acceleration is 0 at 1 sec when the particle is $\frac{5}{2}$ cm from the origin and moving to the right at a velocity of 2 cm/sec.

A particle moving in a straight line is said to have **simple harmonic motion** if the measure of its acceleration is always proportional to the measure of its displacement from a fixed point on the line and its acceleration and displacement are oppositely directed.

EXAMPLE 5

Show that if a particle is moving along a straight line according to the equation of motion

$$s = b \sin(kt + \theta) \tag{4}$$

where b, k, and θ are constants and s cm is the directed distance of the particle from the origin at t sec, then the motion is simple harmonic.

SOLUTION

We wish to show that if a cm/sec^2 is the acceleration of the particle at t sec, then a is proportional to s, and a and s have opposite signs. To determine a we first find v, where v cm/sec is the velocity of the particle at t sec.

$$v = \frac{ds}{dt}$$

$$= b[\cos(kt + \theta)](k)$$
$$= bk \cos(kt + \theta)$$

$$a = \frac{dv}{dt}$$

$$= bk[-\sin(kt + \theta)](k)$$
$$= -bk^2 \sin(kt + \theta) \tag{5}$$

Substituting from (4) into (5) we have

$$a = -k^2 s$$

Because $-k^2$ is a constant, a is proportional to s. Furthermore, because $-k^2$ is negative, a and s are oppositely directed. Therefore the motion is simple harmonic.

Further applications of the second derivative are its uses in the sketch of the graph of a function (Sec. 3.5) and the second-derivative test for relative extrema (Sec. 3.6). An important application of other higher ordered derivatives is to determine infinite series, as shown in Chap. 12.

The following example illustrates how the second derivative is found for functions defined implicitly.

EXAMPLE 6
Given

$$4x^2 + 9y^2 = 36$$

find $\dfrac{d^2y}{dx^2}$ by implicit differentiation.

SOLUTION
Differentiating implicitly with respect to x we have

$$8x + 18y\frac{dy}{dx} = 0$$

$$\frac{dy}{dx} = \frac{-4x}{9y} \tag{6}$$

To find $\dfrac{d^2y}{dx^2}$ we compute the derivative of a quotient and keep in mind that y is a function of x. Thus

$$\frac{d^2y}{dx^2} = \frac{9y(-4) - (-4x)\left(9 \cdot \dfrac{dy}{dx}\right)}{81y^2} \tag{7}$$

Substituting the value of $\dfrac{dy}{dx}$ from (6) into (7) we get

$$\frac{d^2y}{dx^2} = \frac{-36y + (36x)\dfrac{-4x}{9y}}{81y^2}$$

$$= \frac{-36y^2 - 16x^2}{81y^3}$$

$$= \frac{-4(9y^2 + 4x^2)}{81y^3} \tag{8}$$

Because any values of x and y satisfying (8) must also satisfy the original equation, we can replace $9y^2 + 4x^2$ by 36 and obtain

$$\frac{d^2y}{dx^2} = \frac{-4(36)}{81y^3}$$

$$= -\frac{16}{9y^3}$$

EXERCISES 2.11

In Exercises 1 through 16, find the first and second derivative of the function defined by the given equation.

1. $f(x) = x^5 - 2x^3 + x$ **2.** $F(x) = 7x^3 - 8x^2$ **3.** $g(s) = 2s^4 - 4s^3 + 7s - 1$ **4.** $G(t) = t^3 - t^2 + t$

5. $F(x) = x^2\sqrt{x} - 5x$ **6.** $g(r) = \sqrt{r} + \dfrac{1}{\sqrt{r}}$ **7.** $f(x) = \sqrt{x^2 + 1}$ **8.** $h(y) = \sqrt[3]{2y^3 + 5}$

9. $f(t) = 4\cos t^2$ **10.** $g(t) = 2\sin^3 t$ **11.** $G(x) = \cot^2 x$ **12.** $f(x) = \dfrac{2 - \sqrt{x}}{2 + \sqrt{x}}$

13. $g(x) = \dfrac{x^2}{x^2 + 4}$ **14.** $g(x) = (2x - 3)^2(x + 4)^3$ **15.** $f(x) = \sqrt{\sin x + 1}$ **16.** $f(x) = \sec 2x + \tan 2x$

17. Find $D_x{}^3(x^4 - 2x^2 + x - 5)$.   **18.** Find $D_t{}^3(\sqrt{4t + 1})$.   **19.** Find $\dfrac{d^4}{dx^4}\left(\dfrac{3}{2x - 1}\right)$.

20. Find $f^{(4)}(x)$ if $f(x) = \dfrac{2}{x - 1}$.   **21.** Find $D_x{}^3(2\tan 3x)$.   **22.** Find $\dfrac{d^4}{dt^4}(3\sin^2 2t)$.

23. Find $f^{(5)}(x)$ if $f(x) = \cos 2x - \sin 2x$.   **24.** Find $\dfrac{d^3u}{dv^3}$ if $u = v\sqrt{v - 2}$.

25. Given $x^2 + y^2 = 1$, show that $\dfrac{d^2y}{dx^2} = -\dfrac{1}{y^3}$.   **26.** Given $x^2 + 25y^2 = 100$, show that $\dfrac{d^2y}{dx^2} = -\dfrac{4}{25y^3}$.

27. Given $x^3 + y^3 = 1$, show that $\dfrac{d^2y}{dx^2} = \dfrac{-2x}{y^5}$.   **28.** Given $x^{1/2} + y^{1/2} = 2$, show that $\dfrac{d^2y}{dx^2} = \dfrac{1}{x^{3/2}}$.

29. Given $x^4 + y^4 = a^4$ (a is a constant), find $\dfrac{d^2y}{dx^2}$ in simplest form.

30. Given $b^2x^2 - a^2y^2 = a^2b^2$ (a and b are constants), find $\dfrac{d^2y}{dx^2}$ in simplest form.

31. Find the slope of the tangent line at each point of the graph of $y = x^4 + x^3 - 3x^2$ where the rate of change of the slope is zero.

32. Find the instantaneous rate of change of the slope of the tangent line to the graph of $y = 2x^3 - 6x^2 - x + 1$ at the point $(3, -2)$.

In Exercises 33 and 34, a particle is moving along a horizontal line according to the given equation, where s cm is the directed distance of the particle from the origin at t sec, v cm/sec is the velocity of the particle at t sec, and a cm/sec^2 is the acceleration of the particle at t sec. Find v and a in terms of t. Make a table similar to Table 2.11.1 that gives a description of the position and motion of the particle. Include in the table the intervals of time when the particle is moving to the left, when it is moving to the right, when the velocity is increasing, when the velocity is decreasing, when the speed is increasing, when the speed is decreasing, and the position of the particle with respect to the origin during these intervals of time. Show the behavior of the motion by a figure similar to Fig. 2.11.1.

33. $s = t^3 - 9t^2 + 15t, t \geq 0$ **34.** $s = \frac{1}{6}t^3 - 2t^2 + 6t - 2, t \geq 0$

In Exercises 35 through 39, a particle is moving along a straight line according to the given equation, where s ft is the directed distance of the particle from the origin at t sec. Find the time when the instantaneous acceleration is zero, and then find the directed distance of the particle from the origin and the instantaneous velocity at this time.

35. $s = \frac{1}{3}t^3 - \frac{3}{2}t^2 + 2t + 1, t \geq 0$ **36.** $s = 2t^3 - 6t^2 + 3t - 4, t \geq 0$ **37.** $s = \dfrac{125}{16t + 32} - \dfrac{2}{5}t^5, t \geq 0$

38. $s = 9t^2 + 2\sqrt{2t + 1}, t \geq 0$ **39.** $s = \frac{4}{9}t^{3/2} + 2t^{1/2}, t \geq 0$

In Exercises 40 through 45, a particle is moving along a straight line according to the given equation of motion, where s cm is the directed distance of the particle from the origin at t sec. Show that the motion is simple harmonic.

40. $s = A \sin 2\pi kt + B \cos 2\pi kt$, where A, B, and k are constants
41. $s = b \cos(kt + \theta)$, where b, k, and θ are constants
42. $s = 6 \sin(t + \frac{1}{3}\pi) + 4 \sin(t - \frac{1}{6}\pi)$ **43.** $s = \sin(6t - \frac{1}{3}\pi) + \sin(6t + \frac{1}{6}\pi)$
44. $s = 8 \cos^2 6t - 4$ **45.** $s = 5 - 10 \sin^2 2t$

In Exercises 46 through 49, find formulas for $f'(x)$ and $f''(x)$, and state the domains of f' and f''.

46. $f(x) = \begin{cases} \dfrac{x^2}{|x|} & \text{if } x \neq 0 \\ 0 & \text{if } x = 0 \end{cases}$ **47.** $f(x) = \begin{cases} -x^2 & \text{if } x < 0 \\ x^2 & \text{if } 0 \leq x \end{cases}$ **48.** $f(x) = |x|^3$ **49.** $f(x) = \begin{cases} \dfrac{x^5}{|x|} & \text{if } x \neq 0 \\ 0 & \text{if } x = 0 \end{cases}$

50. For the function of Exercise 48, find $f'''(x)$ when it exists.
51. For the function of Exercise 49, find $f'''(x)$ when it exists.

52. Show that if $xy = 1$, then $\dfrac{d^2y}{dx^2} \cdot \dfrac{d^2x}{dy^2} = 4$.

53. If f', g', f'', and g'' exist and if $h = f \circ g$, express $h''(x)$ in terms of the derivatives of f and g.
54. If f and g are two functions such that their first and second derivatives exist and h is the function defined by the equation $h(x) = f(x) \cdot g(x)$, prove that

$$h''(x) = f(x) \cdot g''(x) + 2f'(x) \cdot g'(x) + f''(x) \cdot g(x)$$

55. If $y = x^n$, where n is any positive integer, prove by mathematical induction that $\dfrac{d^n y}{dx^n} = n!$

56. If

$$y = \frac{1}{1 - 2x}$$

prove by mathematical induction that

$$\frac{d^n y}{dx^n} = \frac{2^n n!}{(1 - 2x)^{n+1}}$$

57. If k is any positive integer, prove by mathematical induction that

$$D_x^n(\sin x) = \begin{cases} \sin x & \text{if } n = 4k \\ \cos x & \text{if } n = 4k + 1 \\ -\sin x & \text{if } n = 4k + 2 \\ -\cos x & \text{if } n = 4k + 3 \end{cases}$$

58. Obtain a formula similar to that in Exercise 57 for $D_x^n(\cos x)$.

2.12
THE DIFFERENTIAL

Suppose the function f is defined by the equation

$$y = f(x)$$

Then, when $f'(x)$ exists

$$f'(x) = \lim_{\Delta x \to 0} \frac{\Delta y}{\Delta x} \qquad (1)$$

where

$$\Delta y = f(x + \Delta x) - f(x) \qquad (2)$$

From (1) it follows that for any $\epsilon > 0$ there exists a $\delta > 0$ such that

$$\text{if } \quad 0 < |\Delta x| < \delta \quad \text{then} \quad \left| \frac{\Delta y}{\Delta x} - f'(x) \right| < \epsilon$$

$$\Leftrightarrow \quad \text{if } \quad 0 < |\Delta x| < \delta \quad \text{then} \quad \frac{|\Delta y - f'(x)\Delta x|}{|\Delta x|} < \epsilon$$

This means that $|\Delta y - f'(x)\Delta x|$ is small compared to $|\Delta x|$. That is, for a sufficiently small $|\Delta x|$, $f'(x)\Delta x$ is a good approximation to the value of Δy, and we write

$$\Delta y \approx f'(x)\Delta x \qquad (3)$$

if $|\Delta x|$ is sufficiently small.

For a graphical interpretation of statement (3), refer to Fig. 2.12.1. In the figure, an equation of the curve is $y = f(x)$. The line PT is tangent to the curve at $P(x, f(x))$, Q is the point $(x + \Delta x, f(x + \Delta x))$, and the directed distance MQ is $\Delta y = f(x + \Delta x) - f(x)$. In the figure, Δx and Δy are both positive; however, they could be negative. For a small value of Δx, the slope of the secant line

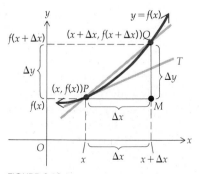

FIGURE 2.12.1

PQ and the slope of the tangent line at P are approximately equal; that is,

$$\frac{\Delta y}{\Delta x} \approx f'(x)$$

$$\Delta y \approx f'(x)\,\Delta x$$

which is statement (3).

The right side of statement (3) is defined to be the *differential* of y.

2.12.1 DEFINITION   If the function f is defined by $y = f(x)$, then the **differential of y**, denoted by dy, is given by

$$dy = f'(x)\,\Delta x \tag{4}$$

where x is in the domain of f' and Δx is an arbitrary increment of x.

Refer now to Fig. 2.12.2, which is the same as Fig. 2.12.1 except the vertical line segment MR is shown, where the directed distance $\overline{MR} = dy$. Observe that dy represents the change in y along the tangent line to the graph of the equation $y = f(x)$ at the point $P(x, f(x))$, when x is changed by Δx.

This concept of the differential involves a special type of function of two variables and a detailed study of such functions appears in Chap. 15. The symbol df may be used to represent this function. The variable x can be any number in the domain of f', and Δx can be any number whatsoever. To state that df is a function of the two independent variables x and Δx means that to each ordered pair $(x, \Delta x)$ in the domain of df there corresponds one and only one number in the range of df, and this number can be represented by $df(x, \Delta x)$ so that

FIGURE 2.12.2

$$df(x, \Delta x) = f'(x)\,\Delta x \tag{5}$$

Comparing (4) and (5) we see that when $y = f(x)$, dy and $df(x, \Delta x)$ are two different notations for $f'(x)\,\Delta x$. The dy symbolism is used in subsequent discussions.

▶ ILLUSTRATION 1   If $y = 3x^2 - x$, then $f(x) = 3x^2 - x$; so $f'(x) = 6x - 1$. Therefore, from Definition 2.12.1

$$dy = (6x - 1)\,\Delta x$$

In particular, if $x = 2$, then $dy = 11\,\Delta x$.   ◀

When $y = f(x)$, Definition 2.12.1 indicates what is meant by dy, the differential of the dependent variable. We also wish to define the differential of the independent variable, or dx. To arrive at a suitable definition for dx that is consistent with the definition of dy we consider the identity function, which is the function defined by $f(x) = x$. Then $f'(x) = 1$ and $y = x$; thus from (4), $dy = 1 \cdot \Delta x$; that is,

$$\text{if} \quad y = x \quad \text{then} \quad dy = \Delta x \tag{6}$$

For the identity function we would want dx to be equal to dy; that is, because of statement (6) we would want dx to be equal to Δx. It is this reasoning that leads to the following definition.

2.12.2 DEFINITION   If the function f is defined by $y = f(x)$, then the **differential of** x, denoted by dx, is given by

$$dx = \Delta x \tag{7}$$

where Δx is an arbitrary increment of x, and x is any number in the domain of f'.

From (4) and (7),

$$dy = f'(x)\,dx \tag{8}$$

Dividing on both sides of (8) by dx we have

$$\frac{dy}{dx} = f'(x) \qquad \text{if } dx \neq 0 \tag{9}$$

Equation (9) expresses the derivative as the quotient of two differentials. Recall that when the notation $\dfrac{dy}{dx}$ for a derivative was introduced in Sec. 2.2, dy and dx had not been given independent meaning.

EXAMPLE 1
Given $y = 4x^2 - 3x + 1$, find Δy, dy, and $\Delta y - dy$ for (a) any x and Δx; (b) $x = 2$, $\Delta x = 0.1$; (c) $x = 2$, $\Delta x = 0.01$; (d) $x = 2$, $\Delta x = 0.001$.

SOLUTION
(a) Because $y = 4x^2 - 3x + 1$, let

$$f(x) = 4x^2 - 3x + 1$$

Then from (2),

$$\begin{aligned}
\Delta y &= f(x + \Delta x) - f(x) \\
&= 4(x + \Delta x)^2 - 3(x + \Delta x) + 1 - (4x^2 - 3x + 1) \\
&= 4x^2 + 8x\,\Delta x + 4(\Delta x)^2 - 3x - 3\,\Delta x + 1 - 4x^2 + 3x - 1 \\
&= (8x - 3)\,\Delta x + 4(\Delta x)^2
\end{aligned}$$

From (8),

$$\begin{aligned}
dy &= f'(x)\,dx \\
&= (8x - 3)\,dx \\
&= (8x - 3)\,\Delta x
\end{aligned}$$

Thus

$$\begin{aligned}
\Delta y - dy &= (8x - 3)\,\Delta x + 4(\Delta x)^2 - (8x - 3)\,\Delta x \\
&= 4(\Delta x)^2
\end{aligned}$$

The results for parts (b), (c), and (d) are given in Table 2.12.1, where $\Delta y = (8x - 3)\Delta x + 4(\Delta x)^2$ and $dy = (8x - 3)\Delta x$.

Table 2.12.1

	x	Δx	Δy	dy	$\Delta y - dy$
(b)	2	0.1	1.34	1.3	0.04
(c)	2	0.01	0.1304	0.13	0.0004
(d)	2	0.001	0.013004	0.013	0.000004

Note from Table 2.12.1 that the closer Δx is to zero, the smaller is the difference between Δy and dy. Furthermore, observe that for each value of Δx, the corresponding value of $\Delta y - dy$ is smaller than the value of Δx. More generally, dy is an approximation of Δy when Δx is small, and the approximation is of better accuracy than the size of Δx.

For a fixed value of x, say x_0,

$$dy = f'(x_0)\,dx \tag{10}$$

That is, dy is a linear function of dx; consequently dy is usually easier to compute than Δy, as was seen in Example 1. Because

$$f(x_0 + \Delta x) - f(x_0) = \Delta y$$

then

$$f(x_0 + \Delta x) = f(x_0) + \Delta y$$

Thus

$$f(x_0 + \Delta x) \approx f(x_0) + dy \tag{11}$$

Our results are illustrated in Fig. 2.12.3. The equation of the curve in the figure is $y = f(x)$, and the graph is concave upward. The line PT is

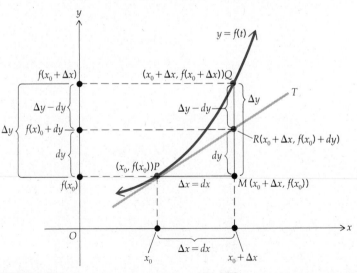

FIGURE 2.12.3

tangent to the curve at $P(x_0, f(x_0))$; Δx and dx are equal and are represented by the directed distance $\overline{PM}$, where M is the point $(x_0 + \Delta x, f(x_0))$. We let Q be the point $(x_0 + \Delta x, f(x_0 + \Delta x))$, and the directed distance $\overline{MQ}$ is Δy or, equivalently, $f(x_0 + \Delta x) - f(x_0)$. The slope of PT is $f'(x) = dy/dx$. Also, the slope of PT is $\overline{MR}/\overline{PM}$, and because $\overline{PM} = dx$, we have $dy = \overline{MR}$ and $\overline{RQ} = \Delta y - dy$. Note that the smaller the value of dx (i.e., the closer the point Q is to the point P), then the smaller will be the value of $\Delta y - dy$ (i.e., the smaller will be the length of the line segment RQ). An equation of the tangent line PT is

$$y = f(x_0) + f'(x_0)(x - x_0)$$

Thus, if $\bar{y}$ is the ordinate of R, then

$$\bar{y} = f(x_0) + f'(x_0)[(x_0 + \Delta x) - x_0]$$
$$\bar{y} = f(x_0) + f'(x_0)\Delta x$$
$$\bar{y} = f(x_0) + dy \tag{12}$$

By comparing Eqs. (11) and (12), observe that when using $f(x_0) + dy$ to approximate the value of $f(x_0 + \Delta x)$, we are approximating the ordinate of the point $Q(x_0 + \Delta x, f(x_0 + \Delta x))$ on the curve by the ordinate of the point $R(x_0 + \Delta x, f(x_0) + dy)$ on the line that is tangent to the curve at $P(x_0, f(x_0))$.

In Fig. 2.12.3, $\Delta y > dy$, so that $\Delta y - dy > 0$. Refer now to Fig. 2.12.4, where $\Delta y < dy$, so that $\Delta y - dy < 0$. Again observe that the smaller the value of dx, then the smaller will be the values of $|\Delta y - dy|$; that is, the closer the point Q is to the point P, the smaller will be the length of the line segment QR.

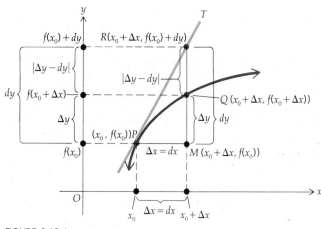

FIGURE 2.12.4

The calculations in the following examples are intended to illustrate the concept of the differential.

EXAMPLE 2
Find an approximate value for cos 61° by using differentials.

SOLUTION
We first write the equivalent radian measure for 61°.

$$61° \sim \tfrac{61}{180}\pi$$

Consider the function defined by $y = \cos x$. Then

$$dy = -\sin x \, dx$$

Because $\tfrac{61}{180}\pi = \tfrac{1}{3}\pi + \tfrac{1}{180}\pi$, and $\cos \tfrac{1}{3}\pi = 0.5$, we compute dy, where x is $\tfrac{1}{3}\pi$, and dx and Δx are $\tfrac{1}{180}\pi$.

$$dy = (-\sin \tfrac{1}{3}\pi)(\tfrac{1}{180}\pi)$$

$$= -\frac{\sqrt{3}}{2}\left(\frac{\pi}{180}\right)$$

$$\approx -0.0151$$

We apply formula (11) with $x_0 = \tfrac{1}{3}\pi$, $\Delta x = \tfrac{1}{180}\pi$, and $dy = -0.0151$, and we have

$$\cos(\tfrac{1}{3}\pi + \tfrac{1}{180}\pi) \approx \cos(\tfrac{1}{3}\pi) + (-0.0151)$$

$$\cos \tfrac{61}{180}\pi \approx \cos \tfrac{1}{3}\pi - 0.0151$$

$$\cos \tfrac{61}{180}\pi \approx 0.5000 - 0.0151$$

Therefore $\cos 61° \approx 0.4849$.

From Table 5 in the back of the book, $\cos 61° = 0.4848$. Therefore the approximation is accurate to three decimal places.

EXAMPLE 3
Find the approximate volume of a spherical shell whose inner radius is 4 in. and whose thickness is $\tfrac{1}{16}$ in.

SOLUTION
We consider the volume of the spherical shell as an increment of the volume of a sphere. Let r be the number of inches in the radius of a sphere, V be the number of cubic inches in the volume of a sphere, and ΔV be the number of cubic inches in the volume of a spherical shell.

$$V = \tfrac{4}{3}\pi r^3 \qquad dV = 4\pi r^2 \, dr$$

Substituting $r = 4$ and $dr = \tfrac{1}{16}$ into the above we obtain

$$dV = 4\pi(4)^2 \tfrac{1}{16}$$

$$= 4\pi$$

Therefore $\Delta V \approx 4\pi$, and we conclude that the volume of the spherical shell is approximately 4π in.[3].

Suppose that y is a function x and that x, in turn, is a function of a third variable t; that is,

$$y = f(x) \quad \text{and} \quad x = g(t) \tag{13}$$

The two equations in (13) together define y as a function of t. For example, suppose that $y = x^3$ and $x = 2t^2 - 1$. Combining these two equations we get $y = (2t^2 - 1)^3$. In general, if the two equations in (13) are combined, we obtain

$$y = f(g(t)) \tag{14}$$

The derivative of y with respect to t can be found by the chain rule, which yields

$$\frac{dy}{dt} = \left(\frac{dy}{dx}\right)\left(\frac{dx}{dt}\right) \tag{15}$$

Equation (15) expresses $\dfrac{dy}{dt}$ as a function of x and t because $\dfrac{dy}{dx}$ is a function of x and $\dfrac{dx}{dt}$ is a function of t.

▶ ILLUSTRATION 2   If $y = x^3$ and $x = 2t^2 - 1$, then

$$\frac{dy}{dt} = \left(\frac{dy}{dx}\right)\left(\frac{dx}{dt}\right)$$

$$= 3x^2(4t)$$

$$= 12x^2 t \qquad\qquad ◀$$

Because (14) defines y as a function of the independent variable t, the differential of y is obtained from (8):

$$dy = \left(\frac{dy}{dt}\right) dt \tag{16}$$

Equation (16) expresses dy as a function of t and dt. Substituting from (15) into (16) we get

$$dy = \left(\frac{dy}{dx}\right)\left(\frac{dx}{dt}\right) dt \tag{17}$$

Now, because x is a function of the independent variable t, (8) can be applied to obtain the differential of x, and we have

$$dx = \left(\frac{dx}{dt}\right) dt \tag{18}$$

Equation (18) expresses dx as a function of t and dt. From (17) and (18), by replacing $\left(\dfrac{dx}{dt}\right) dt$ in (17) by dx, we get

$$dy = \left(\frac{dy}{dx}\right) dx \tag{19}$$

You should bear in mind that in (19), dy is a function of t and dt, and that dx is a function of t and dt. If $\dfrac{dy}{dx}$ in (19) is replaced by $f'(x)$, we have

$$dy = f'(x)\,dx \qquad (20)$$

Equation (20) resembles (8). However, in (8), x is the independent variable, and dy is expressed in terms of x and dx, whereas in (20), t is the independent variable, and both dy and dx are expressed in terms of t and dt. Thus we have the following theorem.

2.12.3 THEOREM   If $y = f(x)$, then when $f'(x)$ exists,

$$dy = f'(x)\,dx$$

whether or not x is an independent variable.

If on both sides of (20) we divide by dx (provided $dx \neq 0$), we obtain

$$f'(x) = \frac{dy}{dx} \qquad dx \neq 0 \qquad (21)$$

Equation (21) states that if $y = f(x)$, then $f'(x)$ is the quotient of the two differentials dy and dx, even though x may not be an independent variable.

▶ ILLUSTRATION 3   Suppose $x = \cos t$, $y = \sin t$, and $0 < t < \pi$. Then

$$dx = -\sin t\,dt \quad \text{and} \quad dy = \cos t\,dt$$

Therefore

$$\frac{dy}{dx} = \frac{\cos t\,dt}{-\sin t\,dt}$$

$$\frac{dy}{dx} = -\frac{\cos t}{\sin t}$$

$$\frac{dy}{dx} = -\frac{x}{y} \qquad (22)$$

Because $x^2 + y^2 = \cos^2 t + \sin^2 t$, then $x^2 + y^2 = 1$. Thus

$$y^2 = 1 - x^2$$

Furthermore $y > 0$ because $y = \sin t$ and $0 < t < \pi$. Thus from the above equation,

$$y = \sqrt{1 - x^2} \qquad (23)$$

Substituting the value of y from (23) into (22) we have

$$\frac{dy}{dx} = -\frac{x}{\sqrt{1 - x^2}}$$

This result can also be obtained by computing $\dfrac{dy}{dx}$ from (23).   ◀

In Sec. 2.4 we proved theorems for finding derivatives of algebraic functions. The formulas from these theorems are now stated with the Leibniz notation. Along with the formula for the derivative there is a corresponding formula for the differential. In these formulas, u and v are functions of x, and it is understood that the formulas hold providing $\dfrac{du}{dx}$ and $\dfrac{dv}{dx}$ exist. When c appears, it is a constant.

I $\dfrac{d(c)}{dx} = 0$ I′ $d(c) = 0$

II $\dfrac{d(x^n)}{dx} = nx^{n-1}$ II′ $d(x^n) = nx^{n-1}\,dx$

III $\dfrac{d(cu)}{dx} = c\dfrac{du}{dx}$ III′ $d(cu) = c\,du$

IV $\dfrac{d(u+v)}{dx} = \dfrac{du}{dx} + \dfrac{dv}{dx}$ IV′ $d(u+v) = du + dv$

V $\dfrac{d(uv)}{dx} = u\dfrac{dv}{dx} + v\dfrac{du}{dx}$ V′ $d(uv) = u\,dv + v\,du$

VI $\dfrac{d\left(\dfrac{u}{v}\right)}{dx} = \dfrac{v\dfrac{du}{dx} - u\dfrac{dv}{dx}}{v^2}$ VI′ $d\left(\dfrac{u}{v}\right) = \dfrac{v\,du - u\,dv}{v^2}$

VII $\dfrac{d(u^n)}{dx} = nu^{n-1}\dfrac{du}{dx}$ VII′ $d(u^n) = nu^{n-1}\,du$

The operation of differentiation is extended to include the process of finding the differential as well as finding the derivative. If $y = f(x)$, dy can be found either by applying formulas I′–VII′ or by finding $f'(x)$ and multiplying it by dx.

EXAMPLE 4
Given
$$y = \frac{\sqrt{x^2 + 1}}{2x + 1}$$
find dy.

SOLUTION
From formula VI′,
$$dy = \frac{(2x + 1)\,d(\sqrt{x^2 + 1}) - \sqrt{x^2 + 1}\,d(2x + 1)}{(2x + 1)^2} \tag{24}$$

From formula VII′,
$$d(\sqrt{x^2 + 1}) = \tfrac{1}{2}(x^2 + 1)^{-1/2}2x\,dx$$
$$= x(x^2 + 1)^{-1/2}\,dx \tag{25}$$

and

$$d(2x + 1) = 2\,dx \tag{26}$$

Substituting values from (25) and (26) into (24) we get

$$dy = \frac{x(2x + 1)(x^2 + 1)^{-1/2}\,dx - 2(x^2 + 1)^{1/2}\,dx}{(2x + 1)^2}$$

$$= \frac{(2x^2 + x)\,dx - 2(x^2 + 1)\,dx}{(2x + 1)^2(x^2 + 1)^{1/2}}$$

$$= \frac{x - 2}{(2x + 1)^2\sqrt{x^2 + 1}}\,dx$$

EXAMPLE 5

Given

$$2x^2y^2 - 3x^3 + 5y^3 + 6xy^2 = 5$$

where x and y are functions of a third variable, find $\dfrac{dy}{dx}$ by computing the differential term by term.

SOLUTION

This is a problem in implicit differentiation. Taking the differential term by term we get

$$4xy^2\,dx + 4x^2y\,dy - 9x^2\,dx + 15y^2\,dy + 6y^2\,dx + 12xy\,dy = 0$$

Dividing by dx, if $dx \neq 0$, we have

$$(4x^2y + 15y^2 + 12xy)\frac{dy}{dx} = -4xy^2 + 9x^2 - 6y^2$$

$$\frac{dy}{dx} = \frac{9x^2 - 6y^2 - 4xy^2}{4x^2y + 15y^2 + 12xy}$$

EXERCISES 2.12

In Exercises 1 through 4, (a) find dy and Δy for the indicated values of x and Δx. (b) Draw a sketch of the graph, and indicate the line segments whose lengths are dy and Δy.

1. $y = x^2$; $x = 2$ and $\Delta x = 1$

2. $y = x^3$; $x = 2$ and $\Delta x = 1$

3. $y = \sqrt[3]{x}$; $x = 8$ and $\Delta x = 2$

4. $y = \sqrt{x}$; $x = 4$ and $\Delta x = 3$

In Exercises 5 through 10, find (a) Δy; (b) dy; (c) $\Delta y - dy$.

5. $y = 6 - 3x - 2x^2$ **6.** $y = 3x^2 - x$ **7.** $y = x^3 - x^2$ **8.** $y = \dfrac{1}{x^2 + 1}$ **9.** $y = \dfrac{2}{x - 1}$ **10.** $y = \dfrac{1}{x} - x^3$

In Exercises 11 through 16, find, for the given values, (a) Δy; (b) dy; (c) $\Delta y - dy$.

11. $y = x^2 - 3x;\ x = 2;\ \Delta x = 0.03$ **12.** $y = x^2 - 3x;\ x = -1;\ \Delta x = 0.02$ **13.** $y = \dfrac{1}{x};\ x = -2;\ \Delta x = -0.1$

14. $y = \dfrac{1}{x};\ x = 3;\ \Delta x = -0.2$ **15.** $y = x^3 + 1;\ x = 1;\ \Delta x = -0.5$ **16.** $y = x^3 + 1;\ x = -1;\ \Delta x = 0.1$

In Exercises 17 through 24, find dy.

17. $y = (3x^2 - 2x + 1)^3$ **18.** $y = \sqrt{4 - x^2}$ **19.** $y = x^2\sqrt[3]{2x + 3}$ **20.** $y = \dfrac{3x}{x^2 + 2}$

21. $y = \dfrac{2 + \cos x}{2 - \sin x}$ **22.** $y = \cot 2x \csc 2x$ **23.** $y = \tan^2 x \sec^2 x$ **24.** $y = x^2 \sin\dfrac{1}{x} - x \cos\dfrac{1}{x}$

In Exercises 25 through 30, x and y are functions of a third variable. Find $\dfrac{dy}{dx}$ by finding the differential term by term (see Example 5).

25. $3x^2 + 4y^2 = 48$ **26.** $8x^2 - y^2 = 32$ **27.** $\sqrt{x} + \sqrt{y} = 4$
28. $2x^2y - 3xy^3 + 6y^2 = 1$ **29.** $\sin x \cos y - \cos x \sin y = 1$ **30.** $3 \tan^2 x + 4 \sec^2 y = 1$

In Exercises 31 through 34, do the following: (a) Find dx and dy in terms of t and dt; (b) use the results of part (a) to find $\dfrac{dy}{dx}$; (c) express y in terms of x by eliminating t between the pair of equations, and then find $\dfrac{dy}{dx}$ by using theorems of differentiation.

31. $x = 2t^2,\ y = 3t$ **32.** $x = 1 + t,\ y = 1 - t^2$
33. $x = 2 \cos t,\ y = 3 \sin t,\ 0 < t < \pi$ **34.** $x = 1 - \cos t,\ y = 2 + \sin t,\ 0 < t < \pi$
(*Hint for Exercises 33 and 34:* Eliminate t by using the identity $\sin^2 t + \cos^2 t = 1$.)

In Exercises 35 through 38, use differentials to approximate the given trigonometric function value after converting the degree measurement to radian measurement. Express the answer to three significant digits.

35. $\sin 46°$ **36.** $\cos 148°$ **37.** $\tan 298°$ **38.** $\cot 224°$

39. The measurement of an edge of a cube is found to be 15 cm with a possible error of 0.01 cm. Use differentials to find the approximate error in computing from this measurement (a) the volume; (b) the area of one of the faces.

40. A metal box in the form of a cube is to have an interior volume of 1000 cm³. The six sides are to be made of metal $\frac{1}{2}$ cm thick. If the cost of the metal to be used is \$0.20 per cubic centimeter, use differentials to find the approximate cost of the metal to be used in the manufacture of the box.

41. An open cylindrical tank is to have an outside coating of thickness 2 cm. If the inner radius is 6 m and the altitude is 10 m, find by differentials the approximate amount of coating material to be used.

42. The stem of a particular mushroom is cylindrical in shape, and a stem of height 2 cm and radius r cm has a volume of V cm³, where $V = 2\pi r^2$. Use the differential to find the approximate increase in the volume of the stem when the radius increases from 0.4 cm to 0.5 cm.

43. A burn on a person's skin is in the shape of a circle such that if r cm is the radius and A cm² is the area of the burn, then $A = \pi r^2$. Use the differential to find the approximate decrease in the area of the burn when the radius decreases from 1 cm to 0.8 cm.

44. A certain bacterial cell is spherical in shape such that if r micrometers (μm) is its radius and V μm³ is its volume, then $V = \frac{4}{3}\pi r^3$. Use the differential to find the approximate increase in the volume of the cell when the radius increases from 2.2 μm to 2.3 μm.

45. A tumor in a person's body is spherical in shape such that if r cm is the radius and V cm^3 is the volume of the tumor, then $V = \frac{4}{3}\pi r^3$. Use the differential to find the approximate increase in the volume of the tumor when the radius increases from 1.5 cm to 1.6 cm.

46. If t sec is the time for one complete swing of a simple pendulum of length l ft, then $4\pi^2 l = gt^2$, where $g = 32.2$. A clock having a pendulum of length 1 ft gains 5 min each day. Find the approximate amount by which the pendulum should be lengthened to correct the inaccuracy.

47. The measure of the electrical resistance of a wire is proportional to the measure of its length and inversely proportional to the square of the measure of its diameter. Suppose the resistance of a wire of given length is computed from a measurement of the diameter with a possible 2 percent error. Find the possible percent error in the computed value of the resistance.

48. A contractor agrees to paint both sides of 1000 circular signs each of radius 3 m. Upon receiving the signs it is discovered that the radius of each sign is 1 cm too large. Use differentials to find the approximate percent increase of paint that will be needed.

49. If the possible error in the measurement of the volume of a gas is 0.1 ft^3 and the allowable error in the pressure is $0.001C$ lb/ft^2, find the size of the smallest container for which Boyle's law (Exercise 26 in Exercises 2.5) holds.

50. For the adiabatic law for the expansion of air (Exercise 20 in Exercises 2.10), prove that $\dfrac{dP}{P} = -1.4\dfrac{dV}{V}$.

REVIEW EXERCISES FOR CHAPTER 2

In Exercises 1 through 12 find the derivative of the given function.

1. $f(x) = \dfrac{4}{x^2} - \dfrac{3}{x^4}$

2. $g(x) = 5(x^4 + 3x^7)$

3. $G(t) = (3t^2 - 4)(4t^3 + t - 1)$

4. $f(x) = (x^4 - 2x)(4x^2 + 2x + 5)$

5. $g(x) = \dfrac{x^3 + 1}{x^3 - 1}$

6. $h(y) = \dfrac{y^2}{y^3 + 8}$

7. $f(s) = (2s^3 - 3s + 7)^4$

8. $F(x) = (4x^4 - 4x^2 + 1)^{-1/3}$

9. $f(x) = \sqrt[3]{\dfrac{x}{x^3 + 1}}$

10. $g(x) = \left(\dfrac{3x^2 + 4}{x^7 + 1}\right)^{10}$

11. $F(x) = (x^2 - 1)^{3/2}(x^2 - 4)^{1/2}$

12. $g(x) = (x^4 - x)^{-3}(5 - x^2)^{-1}$

In Exercises 13 through 20, compute the indicated derivative.

13. $D_x[(x + 1)\sin x - x \cos x]$

14. $D_r(\tan^2 3r)$

15. $\dfrac{d}{dx}\left(x\tan\dfrac{1}{x}\right)$

16. $D_t(\sin^2 3t\sqrt{\cos 2t})$

17. $\dfrac{d}{dt}\left(\dfrac{\sec^2 t}{1 + t^2}\right)$

18. $\dfrac{d}{dx}\left(\dfrac{1 + \sin x}{x \cos x}\right)$

19. $D_w[\sin(\cos 3w) - 3\cos^2 2w]$

20. $D_x(\tan^3 x \cdot \sec x)$

In Exercises 21 through 30, find $\dfrac{dy}{dx}$.

21. $4x^2 + 4y^2 - y^3 = 0$

22. $y = \sqrt{1 + x} + \sqrt{1 - x}$

23. $y = \dfrac{1}{x - \sqrt{x^2 - 1}}$

24. $xy^2 + 2y^3 = x - 2y$

25. $\sin(x + y) + \sin(x - y) = 1$

26. $x^{2/3} + y^{2/3} = a^{2/3}$

27. $y = x^2 + [x^3 + (x^4 + x)^2]^3$ **28.** $y = \dfrac{x\sqrt{3 + 2x}}{4x - 1}$ **29.** $\tan x + \tan y = xy$

30. $\sec(x + y) - \sec(x - y) = 1$

31. Find equations of the tangent lines to the curve $y = 2x^3 + 4x^2 - x$ that have the slope $\frac{1}{2}$.

32. Find an equation of the normal line to the curve $x - y = \sqrt{x + y}$ at the point $(3, 1)$.

33. Find equations of the tangent and normal lines to the curve $2x^3 + 2y^3 - 9xy = 0$ at the point $(2, 1)$.

34. Find equations of the tangent and normal lines to the curve $y = 8 \sin^3 2x$ at the point $(\frac{1}{12}\pi, 1)$.

35. Prove that the line tangent to the curve $y = -x^4 + 2x^2 + x$ at the point $(1, 2)$ is also tangent to the curve at another point, and find this point.

36. Prove that the tangent lines to the curves $4y^3 - x^2y - x + 5y = 0$ and $x^4 - 4y^3 + 5x + y = 0$ at the origin are perpendicular.

37. Find $\dfrac{d^3 y}{dx^3}$ if $y = \sqrt{3 - 2x}$.

38. Given $\dfrac{dy}{dx} = y^k$, where k is a constant and y is a function of x. Express $\dfrac{d^3 y}{dx^3}$ in terms of y and k.

39. Given $f(x) = \frac{1}{12}x^4 + \frac{2}{3}x^3 + \frac{3}{2}x^2 + 8x + 2$. For what values of x is $f''(x) > 0$?

40. Show that if $xy = k$, where k is a nonzero constant, then $\dfrac{d^2 y}{dx^2} \cdot \dfrac{d^2 x}{dy^2} = \dfrac{4}{k}$.

41. A particle is moving in a horizontal line according to the equation $s = t^3 - 11t^2 + 24t + 100$, where s cm is the directed distance of the particle from the origin at t sec. (a) The particle is at the starting point when $t = 0$. For what other values of t is the particle at the starting point? (b) Determine the velocity of the particle at each instant that it is at the starting point, and interpret the sign of the velocity in each case.

42. A particle is moving in a horizontal line according to the equation $s = t^3 - 3t^2 - 9t + 2$, where s cm is the directed distance of the particle from a point O at t sec. The positive direction is to the right. Determine the intervals of time when the particle is moving to the right and when it is moving to the left. Also determine when the particle reverses its direction. Show the behavior of the motion by a figure similar to Fig. 2.5.2, and choose values of t at random, but include the values of t when the particle reverses its direction.

43. A particle is moving along a straight line according to the equation $s = 5 - 2 \cos^2 t$, where s cm is the directed distance of the particle from the origin at t sec. If v cm/sec and a cm/sec^2 are, respectively, the velocity and acceleration of the particle at t sec, find v and a in terms of s.

44. An object is sliding down an inclined plane according to the equation $s = 12t^2 + 6t$, where s meters is the directed distance of the object from the top t sec after starting. (a) Find the velocity 3 sec after the start. (b) Find the initial velocity.

45. A ball is thrown vertically upward from the top of a house 112 ft high. Its equation of motion is $s = -16t^2 + 96t$, where s ft is the directed distance of the ball from the starting point at t sec. Find (a) the instantaneous velocity of the ball at 2 sec; (b) how high the ball will go; (c) how long it takes for the ball to reach the ground; (d) the instantaneous velocity of the ball when it reaches the ground.

46. Stefan's law states that a body emits radiant energy according to the formula $R = kT^4$, where R is the measure of the rate of emission of the radiant energy per square unit of area, T is the measure of the Kelvin temperature of the surface, and k is a constant. Find (a) the average rate of change of R with respect to T as T increases from 200 to 300; (b) the instantaneous rate of change of R with respect to T when T is 200.

47. If A square units is the area of an isosceles right triangle for which each leg has a length of x units, find (a) the average rate of change of A with respect to x as x changes from 8.00 to 8.01; (b) the instantaneous rate of change of A with respect to x when x is 8.00.

48. If $y = x^{2/3}$, find the relative rate of change of y with respect to x when (a) $x = 8$, and (b) $x = c$, where c is a constant.

49. The supply equation for a calculator is $y = m^2 + \sqrt{m}$, where $100y$ calculators are supplied when m dollars is the price per calculator. Find (a) the average rate of change of the supply with respect to the price when the price is increased from $16 to $17; (b) the instantaneous (or marginal) rate of change of the supply with respect to the price when the price is $16.

50. Use the definition of a derivative to find $f'(x)$ if $f(x) = 3x^2 - 5x + 1$.

51. Use the definition of a derivative to find $f'(-5)$ if $f(x) = \dfrac{3}{x+2}$.

52. Use the definition of a derivative to find $f'(5)$ if $f(x) = \sqrt{3x+1}$.

53. Use the definition of a derivative to find $f'(x)$ if $f(x) = \sqrt{4x-3}$.

54. Find $f''(x)$ if $f(x) = 3\sin^2 x - 4\cos^2 x$.

55. Find $f''(\pi)$ if $f(x) = \sqrt{2 + \cos x}$.

56. Find $f'(-3)$ if $f(x) = (|x| - x)\sqrt[3]{9x}$.

57. Find $f'(x)$ if $f(x) = (|x+1| - |x|)^2$.

58. Given

$$f(x) = \begin{cases} x^2 - 16 & \text{if } x < 4 \\ 8x - 32 & \text{if } 4 \le x \end{cases}$$

(a) Draw a sketch of the graph of f. (b) Determine if f is continuous at 4. (c) Determine if f is differentiable at 4.

59. Given

$$f(x) = \begin{cases} x^2 + 2 & \text{if } x \le 3 \\ 20 - x^2 & \text{if } 3 < x \end{cases}$$

(a) Draw a sketch of the graph of f. (b) Determine if f is continuous at 3. (c) Determine if f is differentiable at 3.

60. The remainder theorem of elementary algebra states that if $P(x)$ is a polynomial in x and r is any real number, then there is a polynomial $Q(x)$ such that $P(x) = Q(x)(x - r) + P(r)$. What is $\lim\limits_{x \to r} Q(x)$?

61. Given $f(x) = |x|^3$. (a) Draw a sketch of the graph of f. (b) Find $\lim\limits_{x \to 0} f(x)$ if it exists. (c) Find $f'(0)$ if it exists.

62. Given $f(x) = x^2 \operatorname{sgn} x$. (a) Discuss the differentiability of f. (b) Is f' continuous on its domain?

63. Given

$$f(x) = \begin{cases} ax^2 + b & \text{if } x \le 1 \\ \dfrac{1}{|x|} & \text{if } 1 < x \end{cases}$$

Find the values of a and b such that $f'(1)$ exists.

64. Suppose

$$f(x) = \begin{cases} x^3 & \text{if } x < 1 \\ ax^2 + bx + c & \text{if } 1 \le x \end{cases}$$

Find the values of a, b, and c such that $f''(1)$ exists.

65. A particle is moving along a straight line, and

$$s = \sin(4t + \tfrac{1}{3}\pi) + \sin(4t + \tfrac{1}{6}\pi)$$

where s ft is the directed distance of the particle from the origin at t sec. Prove that the motion is simple harmonic.

66. If an equation of motion is $s = \cos 2t + 2\sin 2t$, prove that the motion is simple harmonic.

67. If a particle is moving along a straight line and $s = \cos 2t + \cos t$, prove that the motion is not simple harmonic.

68. A particle is moving in a straight line according to the equation $s = \sqrt{a + bt^2}$, where a and b are positive constants. Prove that the measure of the acceleration of the particle is inversely proportional to s^3 for any t.

69. A ship leaves a port at noon and travels due west at 20 knots. At noon the next day a second ship leaves the same port and travels northwest at 15 knots. How fast are the two ships separating when the second ship has traveled 90 nautical miles?

70. A reservoir is 80 m long and its cross section is an isosceles trapezoid having equal sides of 10 m, an upper base of 17 m, and a lower base of 5 m. At the instant when the water is 5 m deep, find the rate at which the water is leaking out if the water level is falling at the rate of 0.1 m/hr.

71. A funnel in the form of a cone is 10 in. across the top and 8 in. deep. Water is flowing into the funnel at the rate of 12 in.^{3}/sec and out at the rate of 4 in.^{3}/sec. How fast is the surface of the water rising when it is 5 in. deep?

72. As the last car of a train passes under a bridge, an automobile crosses the bridge on a roadway perpendicular to the track and 30 ft above it. The train is traveling at the rate of 80 ft/sec and the automobile is traveling at the rate of 40 ft/sec. How fast are the train and the automobile separating after 2 sec?

73. A man 6 ft tall is walking toward a building at the rate of 4 ft/sec. If there is a light on the ground 40 ft from the building, how fast is the man's shadow on the building growing shorter when he is 30 ft from the building?

74. Use differentials to find an approximate value of sin 121°.

75. If $y = 80x - 16x^2$, find the difference $\Delta y - dy$ if (a) $x = 2$ and $\Delta x = 0.1$; (b) $x = 4$ and $\Delta x = -0.2$.

76. Find $\dfrac{dy}{dt}$ if $y = x^3 - 2x + 1$ and $x^3 + t^3 - 3t = 0$.

77. Find $\dfrac{dy}{dt}$ if $8x^3 + 27y^3 - 4xy = 0$ and $t^3 + t^2x - x^3 = 4$.

78. If $x^3 + y^3 - 3xy^2 + 1 = 0$, find dy at the point $(1, 1)$ if $dx = 0.1$.

79. Use differentials to approximate the volume of material needed to make a rubber ball if the radius of the hollow inner core is 2 in. and the thickness of the rubber is $\frac{1}{8}$ in.

80. A container in the form of a cube having a volume of 1000 in.^{3} is to be made by using 6 equal squares of material costing 20 cents per square inch. How accurately must the side of each square be made so that the total cost of the material shall be correct to within $5.00?

81. If t sec is the time for one complete swing of a simple pendulum of length l ft, then $4\pi^2l = gt^2$, where $g = 32.2$. What is the effect upon the time if an error of 0.01 ft is made in measuring the length of the pendulum?

82. The measure of the radius of a right-circular cone is $\frac{4}{3}$ times the measure of the altitude. How accurately must the altitude be measured if the error in the computed volume is not to exceed 3 percent?

83. Suppose $f(x) = 3x + |x|$ and $g(x) = \frac{3}{4}x - \frac{1}{4}|x|$. Prove that neither $f'(0)$ nor $g'(0)$ exists but that $(f \circ g)'(0)$ does exist.

84. Give an example of two functions f and g for which f is differentiable at $g(0)$, g is not differentiable at 0, and $f \circ g$ is differentiable at 0.

85. Give an example of two functions f and g for which f is not differentiable at $g(0)$, g is differentiable at 0, and $f \circ g$ is differentiable at 0.

86. In Exercise 45 of Exercises 2.2, you are to prove that if f is differentiable at a, then

$$f'(a) = \lim_{\Delta x \to 0} \frac{f(a + \Delta x) - f(a - \Delta x)}{2 \, \Delta x}$$

Show by using the absolute-value function that it is possible for the limit in the above equation to exist even though $f'(a)$ does not exist.

87. If $f'(x_1)$ exists, prove that

$$\lim_{x \to x_1} \frac{xf(x_1) - x_1f(x)}{x - x_1} = f(x_1) - x_1f'(x_1)$$

88. Let f and g be two functions whose domains are the set of all real numbers. Furthermore, suppose that (i) $g(x) = xf(x) + 1$; (ii) $g(a + b) = g(a) \cdot g(b)$ for all a and b; (iii) $\lim_{x \to 0} f(x) = 1$. Prove that $g'(x) = g(x)$.

89. If the two functions f and g are differentiable at the number x_1, is the composite function $f \circ g$ necessarily differentiable at x_1? If your answer is yes, prove it. If your answer is no, give a counterexample.

90. Suppose $g(x) = |f(x)|$. If $f^{(n)}(x)$ exists and $f(x) \neq 0$, prove that

$$g^{(n)}(x) = \frac{f(x)}{|f(x)|} f^{(n)}(x)$$

91. Prove that $D_x{}^n(\sin x) = \sin(x + \frac{1}{2}n\pi)$. (*Hint:* Use mathematical induction and the formulas $\sin(x + \frac{1}{2}\pi) = \cos x$ or $\cos(x + \frac{1}{2}\pi) = -\sin x$ after each differentiation.)

92. Suppose the function f is defined on the open interval $(0, 1)$ and

$$f(x) = \frac{\sin \pi x}{x(x - 1)}$$

Define f at 0 and 1 so that f is continuous on the closed interval $[0, 1]$.

T H R E E

EXTREME FUNCTION VALUES AND TECHNIQUES OF GRAPHING

3.1
MAXIMUM AND MINIMUM
FUNCTION VALUES

We have seen that the geometrical interpretation of the derivative of a function is the slope of the tangent line to the graph of the function at a point. This fact enables us to apply derivatives as an aid in sketching graphs. For example, the derivative may be used to determine at what points the tangent line is horizontal; these are the points where the derivative is zero. Also, the derivative may be used to find the intervals for which the graph of a function lies above the tangent line and the intervals for which the graph lies below the tangent line. Before applying the derivative to draw sketches of graphs, we need some definitions and theorems.

3.1.1 DEFINITION The function f is said to have a **relative maximum value** at c if there exists an open interval containing c, on which f is defined, such that $f(c) \geq f(x)$ for all x in this interval.

Figures 3.1.1 and 3.1.2 each show a sketch of a portion of the graph of a function having a relative maximum value at c.

3.1.2 DEFINITION The function f is said to have a **relative minimum value** at c if there exists an open interval containing c, on which f is defined, such that $f(c) \leq f(x)$ for all x in this interval.

Figures 3.1.3 and 3.1.4 each show a sketch of a portion of the graph of a function having a relative minimum value at c.

If the function f has either a relative maximum or a relative minimum value at c, then f is said to have a **relative extremum** at c.

The following theorem is used to locate the possible values of c for which there is a relative extremum.

3.1.3 THEOREM If $f(x)$ exists for all values of x in the open interval (a, b), and if f has a relative extremum at c, where $a < c < b$, then if $f'(c)$ exists, $f'(c) = 0$.

The geometrical interpretation of Theorem 3.1.3 is that if f has a relative extremum at c, and if $f'(c)$ exists, then the graph of f must have a horizontal tangent line at the point where $x = c$.

The proof of Theorem 3.1.3 makes use of Theorems 1.10.3 and 1.10.4. You may wish to refer to those theorems at this time and also refer to Illustrations 3 and 4 in Supplementary Section 1.10, which give geometrical interpretations of the theorems.

PROOF OF THEOREM 3.1.3   The proof will be given for the case when f has a relative minimum value at c.

If $f'(c)$ exists, then

$$f'(c) = \lim_{x \to c} \frac{f(x) - f(c)}{x - c} \tag{1}$$

Because f has a relative minimum value at c, by Definition 3.1.2 there exists a $\delta > 0$ such that

$$\text{if} \quad 0 < |x - c| < \delta \quad \text{then} \quad f(x) - f(c) \geq 0$$

FIGURE 3.1.1

FIGURE 3.1.2

FIGURE 3.1.3

FIGURE 3.1.4

FIGURE 3.1.5

If x is approaching c from the right, $x - c > 0$, and therefore

$$\text{if} \quad 0 < x - c < \delta \quad \text{then} \quad \frac{f(x) - f(c)}{x - c} \geq 0$$

By Theorem 1.10.4, if the limit exists,

$$\lim_{x \to c^+} \frac{f(x) - f(c)}{x - c} \geq 0 \tag{2}$$

Similarly, if x is approaching c from the left, $x - c < 0$, and therefore

$$\text{if} \quad -\delta < x - c < 0 \quad \text{then} \quad \frac{f(x) - f(c)}{x - c} \leq 0$$

so that by Theorem 1.10.3, if the limit exists,

$$\lim_{x \to c^-} \frac{f(x) - f(c)}{x - c} \leq 0 \tag{3}$$

Because $f'(c)$ exists, the limits in inequalities (2) and (3) must be equal, and both must be equal to $f'(c)$. So from (2),

$$f'(c) \geq 0 \tag{4}$$

and from (3),

$$f'(c) \leq 0 \tag{5}$$

Because both (4) and (5) are taken to be true, we conclude that

$$f'(c) = 0$$

which was to be proved.

The proof for the case when f has a relative maximum value at c is similar and is left as an exercise (see Exercise 59). ■

If f is a differentiable function, then the only possible values of x for which f can have a relative extremum are those for which $f'(x) = 0$. However, $f'(x)$ can be equal to zero for a specific value of x, and yet f may not have a relative extremum there, as shown in the following illustration.

▶ ILLUSTRATION 1   Consider the function f defined by

$$f(x) = (x - 1)^3$$

A sketch of the graph of this function is shown in Fig. 3.1.5. $f'(x) = 3(x - 1)^2$, and so $f'(1) = 0$. However, $f(x) < 0$ if $x < 1$, and $f(x) > 0$ if $x > 1$. So f does not have a relative extremum at 1. ◀

A function f may have a relative extremum at a number and f' may fail to exist there. This situation is shown in Illustration 2.

FIGURE 3.1.6

▶ ILLUSTRATION 2   Let the function f be defined as follows:

$$f(x) = \begin{cases} 2x - 1 & \text{if } x \leq 3 \\ 8 - x & \text{if } 3 < x \end{cases}$$

A sketch of the graph of this function is shown in Fig. 3.1.6. The function f has a relative maximum value at 3. The derivative from the left at 3 is given by $f'_-(3) = 2$, and the derivative from the right at 3 is given by $f'_+(3) = -1$. Therefore we conclude that $f'(3)$ does not exist.   ◀

Illustration 2 demonstrates why the condition "$f'(c)$ exists" must be included in the hypothesis of Theorem 3.1.3.

It is possible that a function f can be defined at a number c where $f'(c)$ does not exist and yet f may not have a relative extremum there. The following illustration gives such a function.

▶ ILLUSTRATION 3   Let the function f be defined by

$$f(x) = x^{1/3}$$

The domain of f is the set of all real numbers.

FIGURE 3.1.7

$$f'(x) = \frac{1}{3x^{2/3}} \qquad \text{if } x \neq 0$$

Furthermore, $f'(0)$ does not exist. Figure 3.1.7 shows a sketch of the graph of f. The function has no relative extrema.   ◀

In summary, then, if a function f is defined at a number c, a necessary condition for f to have a relative extremum there is that either $f'(c) = 0$ or $f'(c)$ does not exist. But this condition is not sufficient.

3.1.4 DEFINITION   If c is a number in the domain of the function f, and if either $f'(c) = 0$ or $f'(c)$ does not exist, then c is called a **critical number** of f.

Because of this definition and the previous discussion, a necessary (but not sufficient) condition for a function to have a relative extremum at c is for c to be a critical number.

EXAMPLE 1
Find the critical numbers of the function f defined by

$$f(x) = x^{4/3} + 4x^{1/3}$$

SOLUTION

$$f'(x) = \tfrac{4}{3}x^{1/3} + \tfrac{4}{3}x^{-2/3}$$
$$= \tfrac{4}{3}x^{-2/3}(x + 1)$$
$$= \frac{4(x + 1)}{3x^{2/3}}$$

When $x = -1$, $f'(x) = 0$, and when $x = 0$, $f'(x)$ does not exist. Both -1 and 0 are in the domain of f; therefore the critical numbers of f are -1 and 0.

EXAMPLE 2
Find the critical numbers of the function g defined by

$$g(x) = \sin x \cos x$$

SOLUTION
Because $\sin 2x = 2 \sin x \cos x$,

$$g(x) = \tfrac{1}{2} \sin 2x$$

$$g'(x) = \tfrac{1}{2}(\cos 2x)2$$

$$= \cos 2x$$

Since $g'(x)$ exists for all x, the only critical numbers are those for which $g'(x) = 0$. Because $\cos 2x = 0$ when

$$2x = \tfrac{1}{2}\pi + k\pi \qquad \text{where } k \text{ is any integer}$$

the critical numbers of g are $\tfrac{1}{4}\pi + \tfrac{1}{2}k\pi$, where k is any integer.

We are frequently concerned with a function defined on a given interval, and we wish to find the largest or smallest function value on the interval. These intervals can be either closed, open, or closed at one end and open at the other. The greatest function value on an interval is called the *absolute maximum value*, and the smallest function value on an interval is called the *absolute minimum value*. Following are the precise definitions.

3.1.5 DEFINITION   The function f is said to have an **absolute maximum value on an interval** if there is some number c in the interval such that $f(c) \geq f(x)$ for all x in the interval. In such a case, $f(c)$ is the absolute maximum value of f on the interval.

3.1.6 DEFINITION   The function f is said to have an **absolute minimum value on an interval** if there is some number c in the interval such that $f(c) \leq f(x)$ for all x in the interval. In such a case, $f(c)$ is the absolute minimum value of f on the interval.

An **absolute extremum** of a function on an interval is either an absolute maximum value or an absolute minimum value of the function on the interval. A function may or may not have an absolute extremum on a given interval. In each of the following illustrations, a function and an interval are given, and we determine the absolute extrema of the function on the interval if there are any.

▶ ILLUSTRATION 4   Suppose f is the function defined by

$$f(x) = 2x$$

A sketch of the graph of f on $[1, 4]$ is shown in Fig. 3.1.8. The function f

FIGURE 3.1.8

FIGURE 3.1.9

FIGURE 3.1.10

has an absolute minimum value of 2 on $[1, 4)$. There is no absolute maximum value of f on $[1, 4)$ because $\lim\limits_{x \to 4^-} f(x) = 8$, but $f(x)$ is always less than 8 on the given interval. ◀

▶ ILLUSTRATION 5   Consider the function f defined by

$$f(x) = -x^2$$

A sketch of the graph of f on $(-3, 2]$ is shown in Fig. 3.1.9. The function f has an absolute maximum value of 0 on $(-3, 2]$. There is no absolute minimum value of f on $(-3, 2]$ because $\lim\limits_{x \to -3^+} f(x) = -9$, but $f(x)$ is always greater than -9 on the given interval. ◀

▶ ILLUSTRATION 6   The function f defined by

$$f(x) = \frac{x}{1 - x^2}$$

has neither an absolute maximum value nor an absolute minimum value on $(-1, 1)$. A sketch of the graph of f on $(-1, 1)$ is shown in Fig. 3.1.10. Observe that

$$\lim_{x \to -1^+} f(x) = -\infty \qquad \lim_{x \to 1^-} f(x) = +\infty \qquad ◀$$

▶ ILLUSTRATION 7   Let f be the function defined by

$$f(x) = \begin{cases} x + 1 & \text{if } x < 1 \\ x^2 - 6x + 7 & \text{if } 1 \leq x \end{cases}$$

A sketch of the graph of f on $[-5, 4]$ is shown in Fig. 3.1.11. The absolute maximum value of f on $[-5, 4]$ occurs at 1, and $f(1) = 2$; the absolute minimum value of f on $[-5, 4]$ occurs at -5, and $f(-5) = -4$. Note that f has a relative maximum value at 1 and a relative minimum value at 3. Also observe that 1 is a critical number of f because f' does not exist at 1, and 3 is a critical number of f because $f'(3) = 0$. ◀

FIGURE 3.1.11

FIGURE 3.1.12

▶ ILLUSTRATION 8   The function f defined by

$$f(x) = \frac{1}{x - 3}$$

has neither an absolute maximum value nor an absolute minimum value on $[1, 5]$. See Fig. 3.1.12 for a sketch of the graph of f. $\lim\limits_{x \to 3^-} f(x) = -\infty$; so $f(x)$ can be made less than any negative number by taking $3 - x > 0$ and less than a suitable positive δ. Also, $\lim\limits_{x \to 3^+} f(x) = +\infty$; so $f(x)$ can be made greater than any positive number by taking $x - 3 > 0$ and less than a suitable positive δ.   ◀

We may speak of an absolute extremum of a function when no interval is specified. In such a case we are referring to an absolute extremum of the function on the entire domain of the function.

3.1.7 DEFINITION	$f(c)$ is said to be the **absolute maximum value** of the function f if c is in the domain of f and if $f(c) \geq f(x)$ for all values of x in the domain of f.

3.1.8 DEFINITION	$f(c)$ is said to be the **absolute minimum value** of the function f if c is in the domain of f and if $f(c) \leq f(x)$ for all values of x in the domain of f.

▶ ILLUSTRATION 9   The graph of the function f defined by

$$f(x) = x^2 - 4x + 8$$

is a parabola, and a sketch is shown in Fig. 3.1.13. The lowest point of the parabola is at $(2, 4)$, and the parabola opens upward. The function has an absolute minimum value of 4 at 2. There is no absolute maximum value of f.   ◀

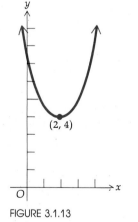

(2, 4)

FIGURE 3.1.13

Referring back to Illustrations 4–9, we see that the only case in which there is both an absolute maximum function value and an absolute minimum function value is in Illustration 7, where the function is continuous on the closed interval $[-5, 4]$. In the other illustrations, either we do not have a closed interval or we do not have a continuous function. If a function is continuous on a closed interval, there is a theorem, called the *extreme-value theorem*, which assures that the function has both an absolute maximum value and an absolute minimum value on the interval. The proof of this theorem is beyond the scope of this book. You are referred to an advanced calculus text for the proof.

3.1.9 THEOREM Extreme-Value Theorem	If the function f is continuous on the closed interval $[a, b]$, then f has an absolute maximum value and an absolute minimum value on $[a, b]$.

Theorem 3.1.9 states that continuity of a function on a closed interval is a sufficient condition to guarantee that the function has both an absolute maximum value and an absolute minimum value on the interval. However,

it is not a necessary condition. For example, the function whose graph appears in Fig. 3.1.14 has an absolute maximum value at $x = c$ and an absolute minimum value at $x = d$, even though the function is discontinuous on the open interval $(-1, 1)$

An absolute extremum of a function continuous on a closed interval must be either a relative extremum or a function value at an endpoint of the interval. Because a necessary condition for a function to have a relative extremum at a number c is for c to be a critical number, the absolute maximum value and the absolute minimum value of a continuous function f on a closed interval $[a, b]$ can be determined by the following procedure:

1. Find the function values at the critical numbers of f on (a, b).
2. Find the values of $f(a)$ and $f(b)$.
3. The largest of the values from steps 1 and 2 is the absolute maximum value, and the smallest of the values from steps 1 and 2 is the absolute minimum value.

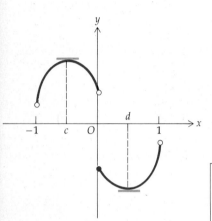

FIGURE 3.1.14

Table 3.1.1

x	-2	-1	$\frac{1}{3}$	$\frac{1}{2}$
$f(x)$	-1	2	$\frac{22}{27}$	$\frac{7}{8}$

FIGURE 3.1.15

EXAMPLE 3
Given

$$f(x) = x^3 + x^2 - x + 1$$

find the absolute extrema of f on $[-2, \frac{1}{2}]$.

SOLUTION
Because f is continuous on $[-2, \frac{1}{2}]$, the extreme-value theorem applies. To find the critical numbers of f, first find f':

$$f'(x) = 3x^2 + 2x - 1$$

Because $f'(x)$ exists for all real numbers, the only critical numbers of f will be the values of x for which $f'(x) = 0$. Set $f'(x) = 0$.

$$(3x - 1)(x + 1) = 0$$

$$x = \tfrac{1}{3} \qquad x = -1$$

The critical numbers of f are -1 and $\frac{1}{3}$, and each of these numbers is in the given closed interval $[-2, \frac{1}{2}]$. The function values at the critical numbers and at the endpoints of the interval are given in Table 3.1.1.

The absolute maximum value of f on $[-2, \frac{1}{2}]$ is therefore 2, which occurs at -1, and the absolute minimum value of f on $[-2, \frac{1}{2}]$ is -1, which occurs at the left endpoint -2. A sketch of the graph of this function on $[-2, \frac{1}{2}]$ is shown in Fig. 3.1.15.

EXAMPLE 4
Given

$$f(x) = (x - 2)^{2/3}$$

find the absolute extrema of f on $[1, 5]$.

SOLUTION
Because f is continuous on $[1, 5]$, the extreme-value theorem applies.

$$f'(x) = \frac{2}{3(x - 2)^{1/3}}$$

There is no value of x for which $f'(x) = 0$. However, because $f'(x)$ does not exist at 2, we conclude that 2 is a critical number of f; so the absolute extrema occur either at 2 or at one of the endpoints of the interval. The function values at these numbers are given in Table 3.1.2.

From the table we conclude that the absolute minimum value of f on $[1, 5]$ is 0, occurring at 2, and the absolute maximum value of f on $[1, 5]$ is $\sqrt[3]{9}$, occurring at 5. A sketch of the graph of this function on $[1, 5]$ is shown in Fig. 3.1.16.

Table 3.1.2

x	1	2	5
$f(x)$	1	0	$\sqrt[3]{9}$

FIGURE 3.1.16

EXERCISES 3.1

In Exercises 1 through 20, find the critical numbers of the given function.

1. $f(x) = x^3 + 7x^2 - 5x$

2. $g(x) = 2x^3 - 2x^2 - 16x + 1$

3. $f(x) = x^4 + 4x^3 - 2x^2 - 12x$

4. $f(x) = x^{7/3} + x^{4/3} - 3x^{1/3}$

5. $g(x) = x^{6/5} - 12x^{1/5}$

6. $f(x) = x^4 + 11x^3 + 34x^2 + 15x - 2$

7. $f(t) = (t^2 - 4)^{2/3}$

8. $f(x) = (x^3 - 3x^2 + 4)^{1/3}$

9. $h(x) = \dfrac{x - 3}{x + 7}$

10. $f(t) = t^{5/3} - 3t^{2/3}$

11. $f(x) = \dfrac{x}{x^2 - 9}$

12. $f(x) = \dfrac{x + 1}{x^2 - 5x + 4}$

13. $f(x) = \sin^2 3x$

14. $f(z) = \cos^2 4z$

15. $g(t) = \sin 2t \cos 2t$

16. $f(x) = \sin 2x + \cos 2x$

17. $f(x) = \tan^2 4x$

18. $g(x) = \sec^2 3x$

19. $G(x) = (x - 2)^3(x + 1)^2$

20. $F(x) = (5 + x)^3(2 - x)^2$

In Exercises 21 through 40, find the absolute extrema of the given function on the indicated interval, if there are any, and determine the values of x at which the absolute extrema occur. Draw a sketch of the graph of the function on the interval.

21. $f(x) = 4 - 3x$; $(-1, 2]$

22. $f(x) = x^2 - 2x + 4$; $(-\infty, +\infty)$

23. $g(x) = \dfrac{1}{x}$; $[-2, 3]$

24. $f(x) = \dfrac{1}{x}$; $[2, 3)$

25. $f(x) = 2 \cos x$; $[-\frac{2}{3}\pi, \frac{1}{3}\pi)$

26. $G(x) = -3 \sin x$; $[0, \frac{3}{4}\pi)$

27. $f(x) = \sqrt{3 + x}$; $[-3, +\infty)$

28. $f(x) = \sqrt{4 - x^2}$; $(-2, 2)$

29. $h(x) = \dfrac{4}{(x - 3)^2}$; $[2, 5]$

30. $g(x) = \dfrac{3x}{9 - x^2}$; $(-3, 2)$

31. $F(x) = |x - 4| + 1$; $(0, 6)$

32. $f(x) = |4 - x^2|$; $(-\infty, +\infty)$

33. $g(x) = \sqrt{4 + 7x}$; $[0, 3)$

34. $F(x) = U(x) - U(x - 1)$ where $U(x) = \begin{cases} 0 & \text{if } x < 0 \\ 1 & \text{if } 0 \le x \end{cases}$; $(-1, 1)$

35. $f(x) = \begin{cases} \dfrac{2}{x - 5} & \text{if } x \ne 5 \\ 2 & \text{if } x = 5 \end{cases}$; $[3, 5]$

36. $f(x) = \begin{cases} |x + 1| & \text{if } x \ne -1 \\ 3 & \text{if } x = -1 \end{cases}$; $[-2, 1]$

37. $f(x) = x - [\![x]\!]$; $(1, 3)$

38. $h(x) = 2x + [\![2x - 1]\!]$; $(1, 2]$

39. $g(x) = \sec 3x$; $[-\frac{1}{6}\pi, \frac{1}{6}\pi]$

40. $f(x) = \tan 2x$; $[-\frac{1}{4}\pi, \frac{1}{6}\pi]$

In Exercises 41 through 58, find the absolute maximum value and the absolute minimum value of the given function on the indicated interval by the method used in Examples 3 and 4 of this section. Draw a sketch of the graph of the function on the interval.

41. $g(x) = x^3 + 5x - 4$; $[-3, -1]$ **42.** $f(x) = x^3 + 3x^2 - 9x$; $[-4, 4]$ **43.** $f(x) = x^4 - 8x^2 + 16$; $[-4, 0]$

44. $f(x) = x^4 - 8x^2 + 16$; $[-3, 2]$ **45.** $f(x) = x^4 - 8x^2 + 16$; $[0, 3]$ **46.** $g(x) = x^4 - 8x^2 + 16$; $[-1, 4]$

47. $f(t) = 2 \sin t$; $[-\pi, \pi]$ **48.** $f(w) = 3 \cos 2w$; $[\frac{1}{6}\pi, \frac{3}{4}\pi]$ **49.** $f(x) = \frac{1}{2} \csc 2x$; $[-\frac{1}{4}\pi, \frac{1}{6}\pi]$

50. $h(x) = 2 \sec \frac{1}{2}x$; $[-\frac{1}{3}\pi, \frac{1}{2}\pi]$ **51.** $f(x) = \dfrac{x}{x+2}$; $[-1, 2]$ **52.** $f(x) = \dfrac{x+5}{x-3}$; $[-5, 2]$

53. $f(x) = \dfrac{x+1}{2x-3}$; $[0, 1]$ **54.** $f(x) = \begin{cases} 2x - 7 & \text{if } -1 \le x \le 2 \\ 1 - x^2 & \text{if } 2 < x \le 4 \end{cases}$; $[-1, 4]$

55. $F(x) = \begin{cases} 3x - 4 & \text{if } -3 \le x < 1 \\ x^2 - 2 & \text{if } 1 \le x \le 3 \end{cases}$; $[-3, 3]$ **56.** $G(x) = \begin{cases} 4 - (x+5)^2 & \text{if } -6 \le x \le -4 \\ 12 - (x+1)^2 & \text{if } -4 < x \le 0 \end{cases}$; $[-6, 0]$

57. $f(x) = (x+1)^{2/3}$; $[-2, 1]$ **58.** $g(x) = 1 - (x-3)^{2/3}$; $[-5, 4]$

59. Prove Theorem 3.1.3 for the case when f has a relative maximum value at c.

3.2 APPLICATIONS INVOLVING AN ABSOLUTE EXTREMUM ON A CLOSED INTERVAL

We consider some problems in which the solution is an absolute extremum of a function on a closed interval. Use is made of the extreme-value theorem, which assures us that both an absolute maximum value and an absolute minimum value of a function exist on a closed interval if the function is continuous on that closed interval. In the following illustration we demonstrate the procedure by considering the problem discussed in Example 4 of Section 1.7 and then again in Illustration 1 of Section 2.1.

▶ ILLUSTRATION 1   A cardboard box manufacturer wishes to make open boxes from pieces of cardboard 12 in. square by cutting equal squares from the four corners and turning up the sides. We wish to find the length of the side of the square to be cut out to obtain a box of the largest possible volume. Figure 3.2.1 represents a given piece of cardboard, and Fig. 3.2.2 represents the box. We showed in Example 4 of Section 1.7 that if x in. is the length of the side of the square to be cut out and $V(x)$ in.^{3} is the volume of the box, then

$$V(x) = 144x - 48x^2 + 4x^3 \tag{1}$$

and the domain of V is the closed interval $[0, 6]$. Because V is continuous on $[0, 6]$, it follows from the extreme-value theorem that V has an absolute maximum value on this interval. We also know that this absolute maximum value of V must occur at either a critical number or at an endpoint of the interval. To find the critical numbers of V we find $V'(x)$ and then find the values of x for which either $V'(x) = 0$ or $V'(x)$ does not exist. From (1),

$$V'(x) = 144 - 96x + 12x^2$$

$V'(x)$ exists for all values of x. If $V'(x) = 0$,

$$12(x^2 - 8x + 12) = 0$$

$$x = 6 \qquad x = 2$$

FIGURE 3.2.1

FIGURE 3.2.2

The critical numbers of V are 2 and 6, both of which are in the closed interval $[0,6]$. The absolute maximum value of V on $[0,6]$ must occur at either a critical number or at an endpoint of the interval. Because $V(0) = 0$ and $V(6) = 0$, while $V(2) = 128$, the absolute maximum value of V on $[0,6]$ is 128, occurring when $x = 2$.

Therefore the largest possible volume is 128 in.^{3}, and this is obtained when the length of the side of the square cut out is 2 in. This result agrees with that found in Illustration 1 of Section 2.1.   ◀

EXAMPLE 1

Points A and B are opposite each other on shores of a straight river 3 km wide. Point C is on the same shore as B but 2 km down the river from B. A telephone company wishes to lay a cable from A to C. If the cost per kilometer of the cable is 25 percent more under the water than it is on land, what line of cable would be least expensive for the company?

SOLUTION

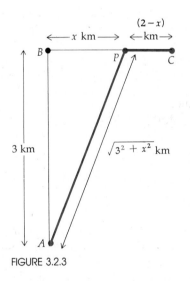

FIGURE 3.2.3

Refer to Fig. 3.2.3. Let P be a point on the same shore as B and C and between B and C so that the cable will run from A to P to C. Let x km be the distance from B to P. Then $(2 - x)$ km is the distance from P to C, and $x \in [0, 2]$. Let k dollars be the cost per kilometer on land and $\frac{5}{4}k$ dollars be the cost per kilometer under the water (k is a constant). If $C(x)$ dollars is the total cost of running the cable from A to P and from P to C, then

$$C(x) = \tfrac{5}{4}k\sqrt{3^2 + x^2} + k(2 - x)$$

Because C is continuous on $[0,2]$, the extreme-value theorem applies; thus C has both an absolute maximum value and an absolute minimum value on $[0,2]$. We wish to find the absolute minimum value.

$$C'(x) = \frac{5kx}{4\sqrt{9 + x^2}} - k$$

$C'(x)$ exists for all values of x. Setting $C'(x) = 0$ and solving for x we have

$$\frac{5kx}{4\sqrt{9 + x^2}} - k = 0$$

$$5x = 4\sqrt{9 + x^2} \qquad (2)$$

$$25x^2 = 16(9 + x^2)$$

$$9x^2 = 16 \cdot 9$$

$$x^2 = 16$$

$$x = \pm 4$$

The number -4 is an extraneous root of (2), and 4 is not in the interval $[0,2]$. Therefore there are no critical numbers of C in $[0,2]$. The absolute minimum value of C on $[0,2]$ must therefore occur at an endpoint of the interval. Computing $C(0)$ and $C(2)$ we get

$$C(0) = \tfrac{23}{4}k \quad \text{and} \quad C(2) = \tfrac{5}{4}k\sqrt{13}$$

Because $\tfrac{5}{4}k\sqrt{13} < \tfrac{23}{4}k$, the absolute minimum value of C on $[0,2]$ is $\tfrac{5}{4}k\sqrt{13}$, occurring when $x = 2$. Therefore, for the cost of the cable to be the least, the cable should go directly from A to C under the water.

EXAMPLE 2

A rectangular field is to be fenced off along the bank of a river; no fence is required along the river. If the material for the fence costs \$8 per running foot for the two ends and \$12 per running foot for the side parallel to the river, find the dimensions of the field of largest possible area that can be enclosed with \$3600 worth of fence.

SOLUTION

Let x ft be the length of an end of the field, y ft be the length of the side parallel to the river, and A ft^2 be the area of the field. See Fig. 3.2.4. Hence

$$A = xy \tag{3}$$

Because the cost of the material for each end is \$8 per running foot and the length of an end is x ft, the total cost for the fence for each end is $8x$ dollars. Similarly, the total cost of the fence for the third side is $12y$ dollars. Then

$$8x + 8x + 12y = 3600 \tag{4}$$

To express A in terms of a single variable we solve (4) for y in terms of x and substitute this value into (3), yielding A as a function of x, and

$$A(x) = x(300 - \tfrac{4}{3}x) \tag{5}$$

From (4), if $y = 0$, $x = 225$, and if $x = 0$, $y = 300$. Because both x and y must be nonnegative, the value of x that will make A an absolute maximum is in the closed interval $[0, 225]$. Because A is continuous on the closed interval $[0, 225]$, from the extreme-value theorem A has an absolute maximum value on this interval. From (5),

$$A(x) = 300x - \tfrac{4}{3}x^2$$

$$A'(x) = 300 - \tfrac{8}{3}x$$

Because $A'(x)$ exists for all x, the critical numbers of A are found by setting $A'(x) = 0$, which gives

$$x = 112\tfrac{1}{2}$$

The only critical number of A is $112\tfrac{1}{2}$, which is in the closed interval $[0, 225]$. Thus the absolute maximum value of A must occur at either 0, $112\tfrac{1}{2}$, or 225. Because $A(0) = 0$ and $A(225) = 0$, while $A(112\tfrac{1}{2}) = 16{,}875$, the absolute maximum value of A on $[0, 225]$ is 16,875, occurring when $x = 112\tfrac{1}{2}$ and $y = 150$ (obtained from (4) by substituting $112\tfrac{1}{2}$ for x).

river

x ft x ft

y ft

FIGURE 3.2.4

Therefore the largest possible area that can be enclosed for $3600 is 16,875 square feet, and this is obtained when the side parallel to the river is 150 ft long and the ends are each $112\frac{1}{2}$ ft long.

EXAMPLE 3

In the planning of a coffee shop it is estimated that if there are places for 40 to 80 people, the daily gross profit will be $8 per place. However, if the seating capacity is above 80 places, the daily gross profit on each place will be decreased by $0.04 times the number of places above 80. What should be the seating capacity to yield the greatest daily gross profit?

SOLUTION

Let x be the number of places in the seating capacity and $P(x)$ dollars be the daily gross profit. $P(x)$ is obtained by multiplying x by the number of dollars in the profit per place. When $40 \leq x \leq 80$, $8 is the profit per place; so $P(x) = 8x$. However, when $x > 80$, the number of dollars in the profit per place is $[8 - 0.04(x - 80)]$, thus giving $P(x) = x[8 - 0.04(x - 80)]$; that is, $P(x) = 11.20x - 0.04x^2$. So

$$P(x) = \begin{cases} 8x & \text{if } 40 \leq x \leq 80 \\ 11.20x - 0.04x^2 & \text{if } 80 < x \leq 280 \end{cases}$$

The upper bound of 280 for x is obtained by noting that $11.20x - 0.04x^2 = 0$ when $x = 280$; and when $x > 280$, $11.20x - 0.04x^2$ is negative.

Even though x, by definition, is an integer, to have a continuous function we let x take on all real values in the interval $[40, 280]$. There is continuity at 80 because $P(80) = 640$ and

$$\lim_{x \to 80^-} P(x) = \lim_{x \to 80^-} 8x \qquad \lim_{x \to 80^+} P(x) = \lim_{x \to 80^+} (11.20x - 0.04x^2)$$
$$= 640 \qquad\qquad\qquad = 640$$

Thus P is continuous on the closed interval $[40, 280]$, and the extreme-value theorem guarantees an absolute maximum value of P on this interval.

When $40 < x < 80$, $P'(x) = 8$; when $80 < x < 280$, $P'(x) = 11.20 - 0.08x$. $P'(80)$ does not exist since $P'_-(80) = 8$ and $P'_+(80) = 4.80$. Set $P'(x) = 0$.

$$11.20 - 0.08x = 0$$

$$x = 140$$

The critical numbers of P are then 80 and 140. We evaluate $P(x)$ at the endpoints of the interval $[40, 280]$ and at the critical numbers.

$$P(40) = 320 \qquad P(80) = 640 \qquad P(140) \doteq 784 \qquad P(280) = 0$$

The absolute maximum value of P, then, is 784, occurring when $x = 140$.

The seating capacity should be 140 places, which gives a daily gross profit of $784.

EXAMPLE 4

Find the dimensions of the right-circular cylinder of greatest volume that can be inscribed in a right-circular cone with a radius of 5 cm and a height of 12 cm.

SOLUTION

Let r cm be the radius of the cylinder, h cm be the height of the cylinder, and V cm^3 be the volume of the cylinder.

Figure 3.2.5 illustrates the cylinder inscribed in the cone, and Fig. 3.2.6 illustrates a plane section through the axis of the cone.

If $r = 0$ and $h = 12$, we have a degenerate cylinder, which is the axis of the cone. If $r = 5$ and $h = 0$, we also have a degenerate cylinder, which is a diameter of the base of the cone. The number r is in the closed interval $[0, 5]$ and h is in the closed interval $[0, 12]$.

The following formula expresses V in terms of r and h:

$$V = \pi r^2 h \tag{6}$$

To express V in terms of a single variable we need another equation involving r and h. From Fig. 3.2.6, and by similar triangles,

$$\frac{12 - h}{r} = \frac{12}{5}$$

$$h = \frac{60 - 12r}{5} \tag{7}$$

Substituting from (7) into formula (6) we obtain V as a function of r and write

$$V(r) = \tfrac{12}{5}\pi(5r^2 - r^3) \qquad \text{with } r \text{ in } [0, 5] \tag{8}$$

Because V is continuous on the closed interval $[0, 5]$, it follows from the extreme-value theorem that V has an absolute maximum value on this interval. The values of r and h that give this absolute maximum value for V are the numbers to find.

$$V'(r) = \tfrac{12}{5}\pi(10r - 3r^2)$$

To find the critical numbers of V, set $V'(r) = 0$ and solve for r:

$$r(10 - 3r) = 0$$

$$r = 0 \qquad r = \tfrac{10}{3}$$

Because $V'(r)$ exists for all values of r, the only critical numbers of V are 0 and $\tfrac{10}{3}$, both of which are in the closed interval $[0, 5]$. The absolute maximum value of V on $[0, 5]$ must occur at either 0, $\tfrac{10}{3}$, or 5. From (8) we obtain

$$V(0) = 0 \qquad V(\tfrac{10}{3}) = \tfrac{400}{9}\pi \qquad V(5) = 0$$

Therefore the absolute maximum value of V is $\tfrac{400}{9}\pi$, and this occurs when $r = \tfrac{10}{3}$. When $r = \tfrac{10}{3}$, we find from (7) that $h = 4$.

Thus the greatest volume of an inscribed cylinder in the given cone is $\tfrac{400}{9}\pi$ cm^3, which occurs when the radius is $\tfrac{10}{3}$ cm and the height is 4 cm.

FIGURE 3.2.5

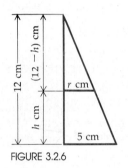

FIGURE 3.2.6

EXERCISES 3.2

In some of the exercises, the independent variable, by definition, may represent a nonnegative integer. For instance, in Exercise 17, if x represents the number of students, then x must be a nonnegative integer. In such exercises, to have the continuity requirements necessary to apply the calculus, allow the independent variable to represent a nonnegative real number. Then if the number found is not an integer compute the function values at the two integers nearest to the number and choose the larger (smaller) as the maximum (minimum) value.

1. Refer to Exercise 33 in Exercises 1.7 and Exercise 29 in Exercises 2.1. A manufacturer of open tin boxes wishes to make use of pieces of tin with dimensions 8 in. by 15 in. by cutting equal squares from the four corners and turning up the sides. Use the method of this section to find the length of the side of the square to be cut out if an open box having the largest possible volume is to be obtained from each piece of tin.

2. Refer to Exercise 34 in Exercises 1.7 and Exercise 30 in Exercises 2.1. Suppose the manufacturer of Exercise 1 makes the open boxes from square pieces of tin that measure k cm on a side. Use the method of this section to determine the length of the side of the square cut out for the volume of the box to be a maximum.

3. Refer to Exercise 35 in Exercises 1.7 and Exercise 31 in Exercises 2.1. Find the dimensions of the largest rectangular field that can be enclosed with 240 m of fence.

4. Find the dimensions of the largest rectangular garden that can be fenced off with 100 ft of fencing material.

5. If one side of a rectangular field is to have a river as a natural boundary, find the dimensions of the largest rectangular field that can be enclosed by using 240 m of fence for the other three sides.

6. Refer to Exercise 36 in Exercises 1.7 and Exercise 32 in Exercises 2.1. Find the dimensions of the largest rectangular garden that can be placed so that a side of a house serves as a boundary and 100 ft of fencing material is to be used for the other three sides.

7. Find the number in the interval $[0, 1]$ such that the difference between the number and its square is a maximum.

8. Find the number in the interval $[\frac{1}{3}, 2]$ such that the sum of the number and its reciprocal is a maximum.

9. Find the area of the largest rectangle having two vertices on the x axis and two vertices on or above the x axis and on the parabola $y = 9 - x^2$.

10. Find the area of the largest rectangle that can be inscribed in a given circle of radius r.

11. An island is at point A, 6 km offshore from the nearest point B on a straight beach. A woman on an island wishes to go to a point C, 9 km down the beach from B. The woman can rent a boat for \$15 per kilometer and travel by water to a point P between B and C, and then she can hire a car with a driver at a cost of \$12 per kilometer and travel a straight road from P to C. Find the least expensive route from point A to point C.

12. Solve Exercise 11 if point C is only 7 km down the beach from B.

13. Solve Example 1 of this section if point C is 6 km down the river from B.

14. Example 1 and Exercises 11, 12, and 13 are special cases of the following more general problem. Let

$$f(x) = u\sqrt{a^2 + x^2} + v(b - x)$$

where x is in $[0, b]$ and $u > v > 0$. Show that for the absolute minimum value of f to occur at a number in the open interval $(0, b)$, the following inequality must be satisfied: $av < b\sqrt{u^2 - v^2}$.

15. If R ft is the range of a projectile, then

$$R = \frac{v_0^2 \sin 2\theta}{g} \qquad 0 \le \theta \le \tfrac{1}{2}\pi$$

where v_0 ft/sec is the initial velocity, g ft/sec^2 is the acceleration due to gravity, and θ is the radian measure of the angle that the gun makes with the horizontal. Find the value of θ that makes the range a maximum.

16. If a body of weight W lb is dragged along a horizontal floor by means of a force of magnitude F lb and directed at an angle of θ radians with the plane of the floor, then F is given by the equation

$$F = \frac{kW}{k \sin \theta + \cos \theta}$$

where k is a constant called the coefficient of friction and $0 < k < 1$. If $0 \le \theta \le \tfrac{1}{2}\pi$, find $\cos \theta$ when F is least.

17. Refer to Exercise 39 in Exercises 2.3. A school-sponsored trip that can accommodate up to 250 students will cost each student $15 if not more than 150 students make the trip; however, the cost per student will be reduced $0.05 for each student in excess of 150 until the cost reaches $10 per student. How many students should make the trip for the school to receive the largest gross income?

18. Solve Exercise 17 if the reduction per student in excess of 150 is $0.07.

19. A private club charges annual membership dues of $100 per member, less $0.50 for each member over 600 and plus $0.50 for each member less than 600. How many members will give the club the most revenue from annual dues?

20. Refer to Exercise 40 in Exercises 2.3. A manufacturer can make a profit of $20 on each item if not more than 800 items are produced each week. The profit decreases $0.02 per item over 800. How many items should the manufacturer produce each week to have the greatest profit?

21. Orange trees grown in California produce 600 oranges per year if no more than 20 trees are planted per acre. For each additional tree planted per acre, the yield per tree decreases by 15 oranges. How many trees per acre should be planted to obtain the greatest number of oranges?

22. Show that the largest rectangle having a given perimeter of p units is a square.

23. Find the dimensions of the right-circular cylinder of greatest lateral surface area that can be inscribed in a sphere with a radius of 6 in.

24. Find the dimensions of the right-circular cylinder of greatest volume that can be inscribed in a sphere with a radius of 6 in.

25. For a package to be accepted by a particular mailing service, the sum of the length and girth (the perimeter of a cross section) must not be greater than 100 in. If a package is to be in the shape of a rectangular box with a square cross section, find the dimensions of the package having the greatest possible volume that can be mailed by the service.

26. Two products A and B are manufactured at a particular factory. If C dollars is the total cost of production for an 8-hour day, then $C = 3x^2 + 42y$, where x machines are used to produce product A and y machines are used to produce product B. If during an 8-hour day there are 15 machines working, determine how many of these machines should be used to produce A and how many should be used to produce B for the total cost to be least.

27. Given the circle having the equation $x^2 + y^2 = 9$, find (a) the shortest distance from the point $(4, 5)$ to a point on the circle; (b) the longest distance from the point $(4, 5)$ to a point on the circle.

28. Suppose that a weight is to be held 10 ft below a horizontal line AB by a wire in the shape of a Y. If the points A and B are 8 ft apart, what is the shortest total length of wire that can be used?

29. Assume that the decrease in a person's blood pressure depends on the amount of a particular drug taken by the person. Thus if x mg of the drug is taken, the decrease in blood pressure is a function of x. Suppose that $f(x)$ defines this function and

$$f(x) = \tfrac{1}{2}x^2(k - x)$$

and x is in $[0, k]$, where k is a positive constant. Determine the value of x that causes the greatest decrease in blood pressure.

30. During a cough there is a decrease in the radius of a person's trachea (or windpipe). Suppose that the normal radius of the trachea is R cm and the radius of the trachea during a cough is r cm, where R is a constant and r is a variable. The velocity of air through the trachea can be shown to be a function of r, and if $V(r)$ cm/sec is this velocity, then

$$V(r) = kr^2(R - r)$$

where k is a positive constant and r is in $[\tfrac{1}{2}R, R]$. Determine the radius of the trachea during a cough for which the velocity of air through the trachea is greatest.

31. In a particular small town the rate at which a rumor spreads is jointly proportional to the number of people who have heard the rumor and the number of people who have not heard it. Show that the rumor is being spread at the greatest rate when half the population of the town knows the rumor.

32. A particular lake can support up to 14,000 fish, and the rate of growth of the fish population is jointly proportional to the number of fish present and the difference between 14,000 and the number present. What should be the size of the fish population for the growth rate to be a maximum?

33. The strength of a rectangular beam is jointly proportional to its breadth and the square of its depth. Find the dimensions of the strongest beam that can be cut from a log in the shape of a right-circular cylinder of radius 72 cm.

34. The stiffness of a rectangular beam is jointly proportional to the breadth and the cube of the depth. Find the dimensions of the stiffest beam that can be cut from a log in the shape of a right-circular cylinder of radius a cm.

35. A piece of wire 10 ft long is cut into two pieces. One piece is bent into the shape of a circle and the other into the shape of a square. How should the wire be cut so that (a) the combined area of the two figures is as small as possible; (b) the combined area of the two figures is as large as possible?

36. Solve Exercise 35 if one piece of wire is bent into the shape of an equilateral triangle and the other piece is bent into the shape of a square.

3.3
ROLLE'S THEOREM AND THE MEAN-VALUE THEOREM

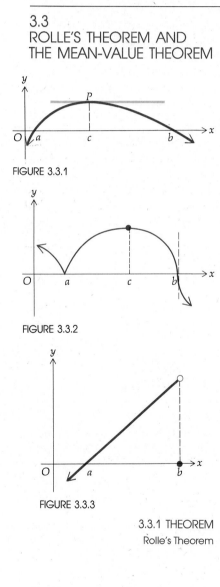

FIGURE 3.3.1

FIGURE 3.3.2

FIGURE 3.3.3

One of the most important theorems in calculus is the *mean-value theorem*. It is used to prove many theorems of both differential and integral calculus, as well as of other subjects, such as numerical analysis. The proof of the mean-value theorem is based on a special case of it known as *Rolle's theorem*, which we discuss first.

Let f be a function that is continuous on the closed interval $[a, b]$, differentiable on the open interval (a, b), and such that $f(a) = 0$ and $f(b) = 0$. The French mathematician Michel Rolle (1652–1719) proved that if a function f satisfies these conditions, there is at least one number c between a and b for which $f'(c) = 0$.

Let us see what this means geometrically. Figure 3.3.1 shows a sketch of the graph of a function f that satisfies the conditions in the preceding paragraph. We see intuitively that there is at least one point on the curve between the points $(a, 0)$ and $(b, 0)$ where the tangent line is parallel to the x axis; that is, the slope of the tangent line is zero. This situation is illustrated in Fig. 3.3.1 at the point P. So the abscissa of P is the c such that $f'(c) = 0$.

The function, whose graph is sketched in Fig. 3.3.1, not only is differentiable on the open interval (a, b) but also is differentiable at the endpoints of the interval. However, the condition that f be differentiable at the endpoints is not necessary for the graph to have a horizontal tangent line at some point in the interval; Fig. 3.3.2 illustrates this. We see in Fig. 3.3.2 that the function is not differentiable at a and b; there is, however, a horizontal tangent line at the point where $x = c$, and c is between a and b.

It is necessary, however, that the function be continuous at the endpoints of the interval to guarantee a horizontal tangent line at an interior point. Figure 3.3.3 shows a sketch of the graph of a function that is continuous on the interval $[a, b)$ but discontinuous at b; the function is differentiable on the open interval (a, b), and the function values are zero at both a and b. However, there is no point at which the graph has a horizontal tangent line.

We now state and prove Rolle's theorem.

3.3.1 THEOREM
Rolle's Theorem

Let f be a function such that

(i) it is continuous on the closed interval $[a, b]$;
(ii) it is differentiable on the open interval (a, b);
(iii) $f(a) = 0$ and $f(b) = 0$.

Then there is a number c in the open interval (a, b) such that

$$f'(c) = 0$$

PROOF   Consider two cases.

Case 1: $f(x) = 0$ for all x in $[a, b]$.

Then $f'(x) = 0$ for all x in (a, b); therefore any number between a and b can be taken for c.

Case 2: $f(x)$ is not zero for some value of x in the open interval (a, b).

Because f is continuous on the closed interval $[a, b]$, from the extreme-value theorem, f has an absolute maximum value on $[a, b]$ and an absolute minimum value on $[a, b]$. From (iii), $f(a) = 0$ and $f(b) = 0$. Furthermore, $f(x)$ is not zero for some x in (a, b). Hence f will have either a positive absolute maximum value at some c_1 in (a, b) or a negative absolute minimum value at some c_2 in (a, b), or both. Thus for $c = c_1$, or $c = c_2$ as the case may be, there is an absolute extremum at an interior point of the interval $[a, b]$. Therefore the absolute extremum $f(c)$ is also a relative extremum, and because $f'(c)$ exists by hypothesis, it follows from Theorem 3.1.3 that $f'(c) = 0$ This proves the theorem.   ∎

There may be more than one number in the open interval (a, b) for which the derivative of f is zero. This is illustrated geometrically in Fig. 3.3.4, where there is a horizontal tangent line at the point where $x = c_1$ and also at the point where $x = c_2$, so that both $f'(c_1) = 0$ and $f'(c_2) = 0$.

The converse of Rolle's theorem is not true. That is, we cannot conclude that if a function f is such that $f'(c) = 0$, with $a < c < b$, then the conditions (i), (ii), and (iii) must hold. (See Exercise 32.)

FIGURE 3.3.4

EXAMPLE 1

Given

$$f(x) = 4x^3 - 9x$$

verify that conditions (i), (ii), and (iii) of the hypothesis of Rolle's theorem are satisfied for each of the following intervals: $[-\frac{3}{2}, 0]$, $[0, \frac{3}{2}]$, and $[-\frac{3}{2}, \frac{3}{2}]$. Then find a suitable value for c in each of these intervals for which $f'(c) = 0$.

SOLUTION

$$f'(x) = 12x^2 - 9$$

Because $f'(x)$ exists for all values of x, f is differentiable on $(-\infty, +\infty)$ and therefore continuous on $(-\infty, +\infty)$. Conditions (i) and (ii) of Rolle's theorem thus hold on any interval. To determine on which intervals condition (iii) holds, we find the values of x for which $f(x) = 0$. If $f(x) = 0$,

$$4x(x^2 - \tfrac{9}{4}) = 0$$

$$x = -\tfrac{3}{2} \qquad x = 0 \qquad x = \tfrac{3}{2}$$

With $a = -\frac{3}{2}$ and $b = 0$, Rolle's theorem holds on $[-\frac{3}{2}, 0]$. Similarly, Rolle's theorem holds on $[0, \frac{3}{2}]$ and $[-\frac{3}{2}, \frac{3}{2}]$.

To find the suitable values for c, set $f'(x) = 0$ and get

$$12x^2 - 9 = 0$$

$$x = -\tfrac{1}{2}\sqrt{3} \qquad x = \tfrac{1}{2}\sqrt{3}$$

Therefore, in the interval $[-\frac{3}{2}, 0]$ a suitable choice for c is $-\frac{1}{2}\sqrt{3}$. In the interval $[0, \frac{3}{2}]$, take $c = \frac{1}{2}\sqrt{3}$. In the interval $[-\frac{3}{2}, \frac{3}{2}]$ there are two possibilities for c: either $-\frac{1}{2}\sqrt{3}$ or $\frac{1}{2}\sqrt{3}$.

We now apply Rolle's theorem to prove the mean-value theorem. You should become thoroughly familiar with the content of this theorem.

3.3.2 THEOREM

Mean-Value Theorem

Let f be a function such that

(i) it is continuous on the closed interval $[a, b]$;
(ii) it is differentiable on the open interval (a, b).

Then there is a number c in the open interval (a, b) such that

$$f'(c) = \frac{f(b) - f(a)}{b - a}$$

Before proving this theorem, we interpret it geometrically. In a sketch of the graph of the function f, $[f(b) - f(a)]/(b - a)$ is the slope of the line segment joining the points $A(a, f(a))$ and $B(b, f(b))$. The mean-value theorem states that there is some point on the curve between A and B where the tangent line is parallel to the secant line through A and B; that is, there is some number c in (a, b) such that

$$f'(c) = \frac{f(b) - f(a)}{b - a}$$

FIGURE 3.3.5

Refer to Fig. 3.3.5.

Take the x axis along the line segment AB and observe that the mean-value theorem is a generalization of Rolle's theorem, which is used in its proof.

PROOF   An equation of the line through A and B in Fig. 3.3.5 is

$$y - f(a) = \frac{f(b) - f(a)}{b - a}(x - a)$$

$$\Leftrightarrow \qquad y = \frac{f(b) - f(a)}{b - a}(x - a) + f(a)$$

Now if $F(x)$ measures the vertical distance between a point $(x, f(x))$ on the graph of the function f and the corresponding point on the secant line through A and B, then

$$F(x) = f(x) - \frac{f(b) - f(a)}{b - a}(x - a) - f(a) \qquad (1)$$

We show that this function F satisfies the three conditions of the hypothesis of Rolle's theorem.

The function F is continuous on the closed interval $[a, b]$ because it is the sum of f and a linear polynomial function, both of which are continuous there. Therefore condition (i) is satisfied by F. Condition (ii) is satisfied by F because f is differentiable on (a, b). From (1) it follows that $F(a) = 0$ and $F(b) = 0$. Therefore condition (iii) of Rolle's theorem is satisfied by F.

The conclusion of Rolle's theorem states that there is a c in the open interval (a, b) such that $F'(c) = 0$. But

$$F'(x) = f'(x) - \frac{f(b) - f(a)}{b - a}$$

Thus

$$F'(c) = f'(c) - \frac{f(b) - f(a)}{b - a}$$

Therefore there is a number c in (a, b) such that

$$0 = f'(c) - \frac{f(b) - f(a)}{b - a}$$

$$\Leftrightarrow \quad f'(c) = \frac{f(b) - f(a)}{b - a}$$

which was to be proved.   ∎

EXAMPLE 2
Given

$$f(x) = x^3 - 5x^2 - 3x$$

verify that the hypothesis of the mean-value theorem is satisfied for $a = 1$ and $b = 3$. Then find all numbers c in the open interval $(1, 3)$ such that

$$f'(c) = \frac{f(3) - f(1)}{3 - 1}$$

SOLUTION
Because f is a polynomial function, f is continuous and differentiable for all values of x. Therefore the hypothesis of the mean-value theorem is satisfied for any a and b.

$$f'(x) = 3x^2 - 10x - 3$$

$$f(1) = -7 \quad \text{and} \quad f(3) = -27$$

Hence

$$\frac{f(3) - f(1)}{3 - 1} = \frac{-27 - (-7)}{2}$$

$$= -10$$

Set $f'(c) = -10$ to obtain

$$3c^2 - 10c - 3 = -10$$

$$3c^2 - 10c + 7 = 0$$

$$(3c - 7)(c - 1) = 0$$

$$c = \tfrac{7}{3} \qquad c = 1$$

Because 1 is not in the open interval $(1, 3)$, the only possible value for c is $\tfrac{7}{3}$.

EXAMPLE 3
Given

$$f(x) = x^{2/3}$$

draw a sketch of the graph of f. Show that there is no number c in the open interval $(-2, 2)$ such that

$$f'(c) = \frac{f(2) - f(-2)}{2 - (-2)}$$

Which condition of the hypothesis of the mean-value theorem fails to hold for f when $a = -2$ and $b = 2$?

SOLUTION
A sketch of the graph of f is shown in Fig. 3.3.6.

$$f'(x) = \tfrac{2}{3}x^{-1/3}$$

So

$$f'(c) = \frac{2}{3c^{1/3}}$$

$$\frac{f(2) - f(-2)}{2 - (-2)} = \frac{4^{1/3} - 4^{1/3}}{4}$$

$$= 0$$

There is no number c for which $\dfrac{2}{3c^{1/3}} = 0$.

The function f is continuous on the closed interval $[-2, 2]$; however, f is not differentiable on the open interval $(-2, 2)$ because $f'(0)$ does not exist. Therefore condition (ii) of the hypothesis of the mean-value theorem fails to hold for f when $a = -2$ and $b = 2$.

Prior to the statement of the mean-value theorem, we indicated that it is one of the most important theorems in the calculus because it is used to prove many other theorems. In such cases it is not necessary to find the value of the number c guaranteed by the theorem. The crucial fact of the theorem is that such a number c exists. To indicate the power of the mean-value theorem we show its use in the proof of the following theorem, which is needed in Chapter 4.

FIGURE 3.3.6

3.3.3 THEOREM   If f is a function such that $f'(x) = 0$ for all values of x in an interval I, then f is constant on I.

PROOF   Assume that f is not constant on the interval I. Then there exist two distinct numbers x_1 and x_2 in I, where $x_1 < x_2$, such that $f(x_1) \neq f(x_2)$. Because, by hypothesis, $f'(x) = 0$ for all x in I, then $f'(x) = 0$ for all x in the closed interval $[x_1, x_2]$. Hence f is differentiable at all x in $[x_1, x_2]$ and f is continuous on $[x_1, x_2]$. Therefore the hypothesis of the mean-value theorem is satisfied, and so there is a number c, with $x_1 < c < x_2$, such that

$$f'(c) = \frac{f(x_1) - f(x_2)}{x_1 - x_2} \tag{2}$$

But because $f'(x) = 0$ for all x in the interval $[x_1, x_2]$, then $f'(c) = 0$, and from (2) it follows that $f(x_1) = f(x_2)$. Yet we assumed that $f(x_1) \neq f(x_2)$. Hence there is a contradiction, and so f is constant on I.   ■

Another significant theorem that the mean-value theorem is used to prove is Theorem 3.4.3, given in the next section.

EXERCISES 3.3

In Exercises 1 through 4, verify that conditions (i), (ii), and (iii) of the hypothesis of Rolle's theorem are satisfied by the given function on the indicated interval. Then find a suitable value for c that satisfies the conclusion of Rolle's theorem.

1. $f(x) = x^2 - 4x + 3$; $[1, 3]$
2. $f(x) = x^3 - 2x^2 - x + 2$; $[1, 2]$
3. $f(x) = \sin 2x$; $[0, \frac{1}{2}\pi]$
4. $f(x) = 3\cos^2 x$; $[\frac{1}{2}\pi, \frac{3}{2}\pi]$

In Exercises 5 through 10, verify that the hypothesis of the mean-value theorem is satisfied for the given function on the indicated interval. Then find a suitable value for c that satisfies the conclusion of the mean-value theorem.

5. $f(x) = x^2 + 2x - 1$; $[0, 1]$
6. $f(x) = x^3 + x^2 - x$; $[-2, 1]$
7. $f(x) = x^{2/3}$; $[0, 1]$

8. $f(x) = \sqrt{1 - \sin x}$; $[0, \frac{1}{2}\pi]$
9. $f(x) = \sqrt{1 + \cos x}$; $[-\frac{1}{2}\pi, \frac{1}{2}\pi]$
10. $f(x) = \dfrac{x^2 + 4x}{x - 7}$; $[2, 6]$

In Exercises 11 through 16, (a) draw a sketch of the graph of the given function on the indicated interval; (b) test the three conditions (i), (ii), and (iii) of the hypothesis of Rolle's theorem, and determine which conditions are satisfied and which, if any, are not satisfied; and (c) if the three conditions in part (b) are satisfied, determine a point at which there is a horizontal tangent line.

11. $f(x) = x^{4/3} - 3x^{1/3}$; $[0, 3]$
12. $f(x) = x^{3/4} - 2x^{1/4}$; $[0, 4]$
13. $f(x) = \dfrac{x^2 - x - 12}{x - 3}$; $[-3, 4]$

14. $f(x) = \begin{cases} 3x + 6 & \text{if } x < 1 \\ x - 4 & \text{if } 1 \le x \end{cases}$; $[-2, 4]$
15. $f(x) = \begin{cases} x^2 - 4 & \text{if } x < 1 \\ 5x - 8 & \text{if } 1 \le x \end{cases}$; $[-2, \frac{8}{5}]$
16. $f(x) = 1 - |x|$; $[-1, 1]$

The geometric interpretation of the mean-value theorem is that for a suitable c in the open interval (a, b), the tangent line to the curve $y = f(x)$ at the point $(c, f(c))$ is parallel to the secant line through the points $(a, f(a))$ and $(b, f(b))$. In Exercises 17 through 20, find a value of c satisfying the conclusion of the mean-value theorem, draw a sketch of the graph on the closed interval $[a, b]$, and show the tangent line and secant line.

17. $f(x) = x^2$; $a = 3$, $b = 5$
18. $f(x) = x^2$; $a = 2$, $b = 4$
19. $f(x) = \sin x$; $a = 0$, $b = \frac{1}{2}\pi$
20. $f(x) = 2\cos x$; $a = \frac{1}{3}\pi$, $b = \frac{2}{3}\pi$

For each of the functions in Exercises 21 through 24, there is no number c in the open interval (a, b) that satisfies the conclusion of the mean-value theorem. In each exercise, determine which part of the hypothesis of the mean-value theorem fails to hold. Draw a sketch of the graph of $y = f(x)$ and the line through the points $(a, f(a))$ and $(b, f(b))$.

33. $f'(x) > 0$ if $x < c$; $f'(x) < 0$ if $x > c$; $f''(x) < 0$ if $x < c$; $f''(x) < 0$ if $x > c$
34. $f'(x) > 0$ if $x < c$; $f'(x) > 0$ if $x > c$; $f''(x) > 0$ if $x < c$; $f''(x) < 0$ if $x > c$
35. $f'(x) > 0$ if $x < c$; $f'(x) < 0$ if $x > c$; $f''(x) > 0$ if $x < c$; $f''(x) > 0$ if $x > c$
36. $f'(x) < 0$ if $x < c$; $f'(x) > 0$ if $x > c$; $f''(x) > 0$ if $x < c$; $f''(x) < 0$ if $x > c$
37. $f''(c) = 0$; $f'(c) = 0$; $f''(x) > 0$ if $x < c$; $f''(x) < 0$ if $x > c$
38. $f'(c) = 0$; $f'(x) > 0$ if $x < c$; $f''(x) > 0$ if $x > c$
39. $f''(c) = 0$; $f'(c) = 0$; $f''(x) > 0$ if $x < c$; $f''(x) > 0$ if $x > c$
40. $f'(c) = 0$; $f'(x) < 0$ if $x < c$; $f''(x) > 0$ if $x > c$
41. $f''(c) = 0$; $f'(c) = -1$; $f''(x) < 0$ if $x < c$; $f''(x) > 0$ if $x > c$
42. $f''(c) = 0$; $f'(c) = \frac{1}{2}$; $f''(x) > 0$ if $x < c$; $f''(x) < 0$ if $x > c$
43. $f'(c)$ does not exist; $f''(x) > 0$ if $x < c$; $f''(x) > 0$ if $x > c$
44. $f'(c)$ does not exist; $f''(c)$ does not exist; $f''(x) < 0$ if $x < c$; $f''(x) > 0$ if $x > c$

45. $\lim\limits_{x \to c^-} f'(x) = +\infty$; $\lim\limits_{x \to c^+} f'(x) = 0$; $f''(x) > 0$ if $x < c$; $f''(x) < 0$ if $x > c$

46. $\lim\limits_{x \to c^-} f'(x) = +\infty$; $\lim\limits_{x \to c^+} f'(x) = -\infty$; $f''(x) > 0$ if $x < c$; $f''(x) > 0$ if $x > c$

47. Draw a sketch of the graph of a function f for which $f(x)$, $f'(x)$, and $f''(x)$ exist and are positive for all x.
48. Draw a sketch of the graph of a function f for which $f(x)$, $f'(x)$, and $f''(x)$ exist and are negative for all x.
49. If $f(x) = x^5$, show that 0 is a critical number of f but that $f(0)$ is not a relative extremum. Is the origin a point of inflection of the graph of f? Prove your answer.
50. If $f(x) = 3x^2 + x|x|$, prove that $f''(0)$ does not exist but the graph of f is concave upward everywhere.
51. It is estimated that a worker in a shop that makes picture frames can paint y frames x hours after starting work at 8 A.M., and

$$y = 3x + 8x^2 - x^3 \qquad 0 \le x \le 4$$

Find at what time the worker is working most efficiently (that is, at what time does the worker reach the point of diminishing returns?). (*Hint:* See Illustration 4.)

3.6
THE SECOND-DERIVATIVE
TEST FOR RELATIVE EXTREMA

(c, f(c))

a   c   b

FIGURE 3.6.1

In Section 3.4 you learned how to determine whether a function f has a relative maximum value or a relative minimum value at a critical number c by checking the algebraic sign of f' at numbers in intervals to the left and right of c. Another test for relative extrema is one that involves only the critical number c. Before stating the test in the form of a theorem, we give an informal geometric discussion that should appeal to your intuition.

Suppose that f is a function such that f' and f'' exist on some open interval (a, b) containing c and that $f'(c) = 0$. Also suppose that f'' is negative on (a, b). From Theorem 3.4.3(ii), because $f''(x) < 0$ on (a, b), then f' is decreasing on $[a, b]$. Since the value of f' at a point on the graph of f gives the slope of the tangent line at the point, it follows that the slope of the tangent line is decreasing on $[a, b]$. In Fig. 3.6.1 there is a sketch of the graph of a function f having these properties. From Theorem 3.5.3(ii), the graph of f is concave downward at all points in the figure, and a segment of the tangent line is shown at some of these points. Observe that the slope of the tangent line is decreasing on $[a, b]$. Note that f has a relative maximum value at c, where $f'(c) = 0$ and $f''(c) < 0$.

Now suppose that f is a function having the properties of the function in the previous paragraph except that f'' is positive on (a, b). Then from Theorem 3.4.3(i), because $f''(x) > 0$ on (a, b), it follows that f' is increasing on

$[a, b]$. Thus the slope of the tangent line is increasing on $[a, b]$. Figure 3.6.2 shows a sketch of the graph of a function f having these properties. From Theorem 3.5.3(i), the graph of f is concave upward at all points in the figure, and a segment of the tangent line is shown at some of these points. The slopes of these tangent lines are increasing on $[a, b]$. The function f has a relative minimum value at c, where $f'(c) = 0$ and $f''(c) > 0$.

The facts in the preceding two paragraphs are given in the **second-derivative test for relative extrema**, which is now stated.

3.6.1 THEOREM
Second-Derivative Test for Relative Extrema

Let c be a critical number of a function f at which $f'(c) = 0$, and let f' exist for all values of x in some open interval containing c. If $f''(c)$ exists and

(i) if $f''(c) < 0$, then f has a relative maximum value at c;
(ii) if $f''(c) > 0$, then f has a relative minimum value at c.

The proof of part (i) makes use of Theorem 1.10.2 and that of part (ii) utilizes Theorem 1.10.1. You may wish to refer to those theorems now and also refer to Illustrations 1 and 2 in Supplementary Section 1.10, which give geometrical interpretations of the theorems.

PROOF OF (i)   By hypothesis, $f''(c)$ exists and is negative; so

$$f''(c) = \lim_{x \to c} \frac{f'(x) - f'(c)}{x - c} < 0$$

Therefore, by Theorem 1.10.2 there is an open interval I containing c such that

$$\frac{f'(x) - f'(c)}{x - c} < 0 \tag{1}$$

for every $x \neq c$ in the interval.

Let I_1 be the open interval containing all values of x in I for which $x < c$; therefore c is the right endpoint of the open interval I_1. Let I_2 be the open interval containing all values of x in I for which $x > c$; so c is the left endpoint of the open interval I_2.

Then if x is in I_1, $x - c < 0$, and it follows from inequality (1) that $f'(x) - f'(c) > 0$ or, equivalently, $f'(x) > f'(c)$. If x is in I_2, $x - c > 0$, and it follows from (1) that $f'(x) - f'(c) < 0$ or, equivalently, $f'(x) < f'(c)$.

But because $f'(c) = 0$, we conclude that if x is in I_1, $f'(x) > 0$, and if x is in I_2, $f'(x) < 0$. Therefore $f'(x)$ changes algebraic sign from positive to negative as x increases through c, and so, by Theorem 3.4.4, f has a relative maximum value at c.

The proof of part (ii) is similar and is left as an exercise (see Exercise 46). ∎

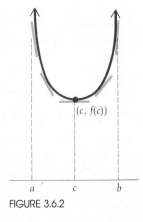

$(c, f(c))$

$a \quad c \quad b$

FIGURE 3.6.2

EXAMPLE 1
Given

$$f(x) = x^4 + \tfrac{4}{3}x^3 - 4x^2$$

find the relative maxima and minima of f by applying the second-derivative test. Draw a sketch of the graph of f.

SOLUTION
We compute the first and second derivatives of f.

$$f'(x) = 4x^3 + 4x^2 - 8x \qquad f''(x) = 12x^2 + 8x - 8$$

Set $f'(x) = 0$.

$$4x(x + 2)(x - 1) = 0$$
$$x = 0 \qquad x = -2 \qquad x = 1$$

Thus the critical numbers of f are -2, 0, and 1. We determine whether or not there is a relative extremum at any of these critical numbers by finding the sign of the second derivative there. The results are summarized in Table 3.6.1.

Table 3.6.1

	$f(x)$	$f'(x)$	$f''(x)$	Conclusion
$x = -2$	$-\frac{32}{3}$	0	$+$	f has a relative minimum value
$x = 0$	0	0	$-$	f has a relative maximum value
$x = 1$	$-\frac{5}{3}$	0	$+$	f has a relative minimum value

From the facts in the table and by plotting a few more points we obtain the sketch of the graph of f shown in Fig. 3.6.3.

FIGURE 3.6.3

EXAMPLE 2
Find the relative extrema of the sine function by applying the second-derivative test.

SOLUTION
Let

$$f(x) = \sin x$$

Then

$$f'(x) = \cos x \qquad f''(x) = -\sin x$$

$f'(x)$ exists for all x. The critical numbers are obtained by setting $f'(x) = 0$.

$$\cos x = 0$$
$$x = \tfrac{1}{2}\pi + k\pi \qquad k \text{ is any integer}$$

We determine whether or not there is a relative extremum at any of these critical numbers by finding the sign of the second derivative there. If k is an even integer, then

FIGURE 3.6.4

FIGURE 3.6.5

FIGURE 3.6.6

$$f''(\tfrac{1}{2}\pi + k\pi) = -\sin(\tfrac{1}{2}\pi + k\pi) \qquad k \text{ is an even integer}$$
$$= -\sin \tfrac{1}{2}\pi$$
$$= -1$$

If k is an odd integer, then

$$f''(\tfrac{1}{2}\pi + k\pi) = -\sin(\tfrac{1}{2}\pi + k\pi) \qquad k \text{ is an odd integer}$$
$$= -\sin(\tfrac{3}{2}\pi + (k-1)\pi) \qquad (k-1) \text{ is an even integer}$$
$$= -\sin \tfrac{3}{2}\pi$$
$$= 1$$

The results of applying the second-derivative test are summarized in Table 3.6.2.

Table 3.6.2

	$f(x)$	$f'(x)$	$f''(x)$	Conclusion
$x = \tfrac{1}{2}\pi + k\pi$ (k is an even integer)	1	0	−	f has a relative maximum value
$x = \tfrac{1}{2}\pi + k\pi$ (k is an odd integer)	−1	0	+	f has a relative minimum value

If $f''(c) = 0$, as well as $f'(c) = 0$, nothing can be concluded regarding a relative extremum of f at c. The following three illustrations justify this statement.

▶ ILLUSTRATION 1   If $f(x) = x^4$, then $f'(x) = 4x^3$ and $f''(x) = 12x^2$. Thus $f(0)$, $f'(0)$, and $f''(0)$ all have the value zero. By applying the first-derivative test we see that f has a relative minimum value at 0. A sketch of the graph of f is shown in Fig. 3.6.4.   ◀

▶ ILLUSTRATION 2   If $g(x) = -x^4$, then $g'(x) = -4x^3$ and $g''(x) = -12x^2$. Hence $g(0)$, $g'(0)$, and $g''(0)$ are all zero. In this case g has a relative maximum value at 0, as can be seen by applying the first-derivative test. A sketch of the graph of g is shown in Fig. 3.6.5.   ◀

▶ ILLUSTRATION 3   If $h(x) = x^3$, then $h'(x) = 3x^2$ and $h''(x) = 6x$; so $h(0)$, $h'(0)$ and $h''(0)$ are all zero. The function h does not have a relative extremum at 0 because if $x < 0$, $h(x) < h(0)$; and if $x > 0$, $h(x) > h(0)$. A sketch of the graph of h is shown in Fig. 3.6.6.   ◀

In Illustrations 1, 2, and 3 we have examples of three functions, each of which has zero for its second derivative at a number for which its first derivative is zero; yet one function has a relative minimum value at that number, another function has a relative maximum value at that number, and the third function has neither a relative maximum value nor a relative minimum value at that number.

EXAMPLE 3

Given

$$f(x) = x^{2/3} - 2x^{1/3}$$

find the relative extrema of f by applying the second-derivative test when possible. Use the second derivative to find any points of inflection of the graph of f, and determine where the graph is concave upward and where it is concave downward. Draw a sketch of the graph.

SOLUTION

$$f'(x) = \tfrac{2}{3}x^{-1/3} - \tfrac{2}{3}x^{-2/3} \qquad f''(x) = -\tfrac{2}{9}x^{-4/3} + \tfrac{4}{9}x^{-5/3}$$

Because $f'(0)$ does not exist, 0 is a critical number of f. Other critical numbers are found by setting $f'(x) = 0$.

$$\frac{2}{3x^{1/3}} - \frac{2}{3x^{2/3}} = 0$$

$$2x^{1/3} - 2 = 0$$

$$x^{1/3} = 1$$

$$x = 1$$

Thus 1 is also a critical number. We can determine if there is a relative extremum at 1 by applying the second-derivative test. We cannot use the second-derivative test at the critical number 0 because $f'(0)$ does not exist. We apply the first-derivative test at $x = 0$. Table 3.6.3 shows the results of these tests.

Because $f''(0)$ does not exist, $(0, 0)$ is a possible point of inflection. To find other possible points of inflection we set $f''(x) = 0$.

$$-\frac{2}{9x^{4/3}} + \frac{4}{9x^{5/3}} = 0$$

$$-2x^{1/3} + 4 = 0$$

$$x^{1/3} = 2$$

$$x = 8$$

To determine if there are points of inflection where x is 0 and 8, we check to see if $f''(x)$ changes sign; at the same time we learn about the concavity of the graph in the respective intervals. At a point of inflection it is necessary that the graph have a tangent line there. At the origin there is a vertical tangent line because

$$\lim_{x \to 0} f'(x) = \lim_{x \to 0} \frac{2x^{1/3} - 2}{3x^{2/3}}$$

$$= -\infty$$

37. $f'(c) = 0$; $\lim\limits_{x \to d^-} f'(x) = +\infty$; $\lim\limits_{x \to d^+} f'(x) = +\infty$; $f'(e) = 0$; $f''(x) > 0$ if $x < d$; $f''(x) < 0$ if $x > d$

38. $f'(c) = 0$; $\lim\limits_{x \to d^-} f'(x) = -\infty$; $\lim\limits_{x \to d^+} f'(x) = -\infty$; $f'(e) = 0$; $f''(x) < 0$ if $x < d$; $f''(x) > 0$ if $x > d$

39. $f'(c)$ does not exist; $f'(d) = -1$; $f''(d) = 0$; $f'(e) = 0$; $f''(x) > 0$ if $x < c$; $f''(x) < 0$ if $c < x < d$; $f''(x) > 0$ if $x > d$

40. $f'(c) = 0$; $f'(d) = -1$; $f''(d) = 0$; $f'(e)$ does not exist; $f''(x) < 0$ if $x < d$; $f''(x) > 0$ if $d < x < e$; $f''(x) < 0$ if $x > e$

41. $f'(c) = 0$; $f'(d)$ does not exist; $f'(e) = 0$; $f''(e) = 0$; $f''(x) < 0$ if $x < d$; $f''(x) < 0$ if $d < x < e$; $f''(x) > 0$ if $x > e$

42. $f'(c) = 0$; $f''(c) = 0$; $f'(d)$ does not exist; $f'(e) = 0$; $f''(x) < 0$ if $x < c$; $f''(x) > 0$ if $c < x < d$; $f''(x) > 0$ if $x > d$

43. Suppose that $\frac{1}{2}\sqrt{2}$ and $-\frac{1}{2}\sqrt{3}$ are critical numbers of a function f and that $f''(x) = x[\![\frac{1}{2}x^2 + 1]\!]$. At each of these numbers, determine if f has a relative extremum, and if so, whether it is a relative maximum or a relative minimum.

44. Given $f(x) = x^r - rx + k$, where $r > 0$ and $r \neq 1$, prove that (a) if $0 < r < 1$, f has a relative maximum value at 1; (b) if $r > 1$, f has a relative minimum value at 1.

45. Given $f(x) = x^3 + 3rx + 5$, prove that (a) if $r > 0$, f has no relative extrema; (b) if $r < 0$, f has both a relative maximum value and a relative minimum value.

46. Prove Theorem 3.6.1(ii).

47. Given $f(x) = x^2 + rx^{-1}$, prove that regardless of the value of r, f has a relative minimum value and no relative maximum value.

48. If $f(x) = ax^2 + bx + c$, use the second-derivative test to show that f has a relative maximum value if $a < 0$. Find the number at which the relative maximum value occurs.

49. Suppose that f is a function for which $f''(x)$ exists for all values of x in some open interval I and that at a number c in I, $f''(c) = 0$ and $f'''(c)$ exists and is not zero. Prove that the point $(c, f(c))$ is a point of inflection of the graph of f. (*Hint:* The proof is similar to the proof of the second-derivative test.)

3.7
LIMITS AT INFINITY

Section 1.5 was devoted to infinite limits where function values either increased or decreased without bound as the independent variable approached a fixed number. We now consider limits of functions when the independent variable either increases or decreases without bound. We postponed this topic until now because such limits are especially useful when discussing techniques of graphing. Furthermore, these limits can be applied when determining extreme function values on open intervals.

We begin with a particular function. Let f be defined by

$$f(x) = \frac{2x^2}{x^2 + 1}$$

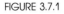

FIGURE 3.7.1

A sketch of the graph of this function is shown in Fig. 3.7.1. Let x take on the values 0, 1, 2, 3, 4, 5, 10, 100, 1000, and so on, allowing x to increase without bound. The corresponding function values are given in Table 3.7.1. Observe from the table that as x increases through positive values, the function values $f(x)$ get closer and closer to 2.

Table 3.7.1

x	0	1	2	3	4	5	10	100	1000
$f(x) = \dfrac{2x^2}{x^2 + 1}$	0	1	$\dfrac{8}{5}$	$\dfrac{18}{10}$	$\dfrac{32}{17}$	$\dfrac{50}{26}$	$\dfrac{200}{101}$	$\dfrac{20{,}000}{10{,}001}$	$\dfrac{2{,}000{,}000}{1{,}000{,}001}$

In particular, when $x = 4$,

$$2 - \frac{2x^2}{x^2 + 1} = 2 - \frac{32}{17}$$

$$= \tfrac{2}{17}$$

Therefore the difference between 2 and $f(x)$ is $\tfrac{2}{17}$ when $x = 4$. When $x = 100$,

$$2 - \frac{2x^2}{x^2 + 1} = 2 - \frac{20,000}{10,001}$$

$$= \frac{2}{10,001}$$

Table 3.7.2

x	$f(x) = \dfrac{2x^2}{x^2 + 1}$
0	0
-1	1
-2	$\dfrac{8}{5}$
-3	$\dfrac{18}{10}$
-4	$\dfrac{32}{17}$
-5	$\dfrac{50}{26}$
-10	$\dfrac{200}{101}$
-100	$\dfrac{20,000}{10,001}$
-1000	$\dfrac{2,000,000}{1,000,001}$

Hence the difference between 2 and $f(x)$ is $2/10,001$ when $x = 100$.

Continuing on, we see intuitively that the value of $f(x)$ can be made as close to 2 as we please by taking x large enough. In other words, the difference between 2 and $f(x)$ can be made as small as we please by taking x any number greater than some sufficiently large positive number. Or, going a step further, for any $\epsilon > 0$, however small, we can find a number $N > 0$ such that if $x > N$, then $|f(x) - 2| < \epsilon$.

When an independent variable x is increasing without bound through positive values, we write "$x \to +\infty$." From the illustrative example above, then, we can say that

$$\lim_{x \to +\infty} \frac{2x^2}{x^2 + 1} = 2$$

3.7.1 DEFINITION   Let f be a function that is defined at every number in some interval $(a, +\infty)$. The **limit of $f(x)$, as x increases without bound, is L**, written

$$\lim_{x \to +\infty} f(x) = L \tag{1}$$

if for any $\epsilon > 0$, however small, there exists a number $N > 0$ such that

$$\text{if} \quad x > N \quad \text{then} \quad |f(x) - L| < \epsilon$$

Note: When $x \to +\infty$ is written, it does not have the same meaning as, for instance, $x \to 1000$. The symbol $x \to +\infty$ indicates the behavior of the variable x.

Now consider the same function, and let x take on the values $0, -1, -2, -3, -4, -5, -10, -100, -1000$, and so on, allowing x to decrease through negative values without bound. Table 3.7.2 gives the corresponding function values of $f(x)$.

Observe that the function values are the same for the negative numbers as for the corresponding positive numbers. So we see intuitively that as x

decreases without bound, $f(x)$ approaches 2; that is, $|f(x) - 2|$ can be made as small as we please by taking x any number less than some negative number having a sufficiently large absolute value. Formally we say that for any $\epsilon > 0$, however small, we can find a number $N < 0$ such that if $x < N$, then $|f(x) - 2| < \epsilon$. Using the symbol $x \to -\infty$ to denote that the variable x is decreasing without bound we write

$$\lim_{x \to -\infty} \frac{2x^2}{x^2 + 1} = 2$$

3.7.2 DEFINITION   Let f be a function that is defined at every number in some interval $(-\infty, a)$. The **limit of $f(x)$, as x decreases without bound, is L**, written

$$\lim_{x \to -\infty} f(x) = L \tag{2}$$

if for any $\epsilon > 0$, however small, there exists a number $N < 0$ such that

$$\text{if} \quad x < N \quad \text{then} \quad |f(x) - L| < \epsilon$$

Note: As in the note following Definition 3.7.1, the symbol $x \to -\infty$ indicates only the behavior of the variable x.

Limit Theorems 2, 4, 5, 6, 7, 8, 9, and 10 in Section 1.3 and Limit Theorems 11 and 12 in Section 1.5 remain unchanged when "$x \to a$" is replaced by "$x \to +\infty$" or "$x \to -\infty$." We have the following additional limit theorem.

3.7.3 LIMIT THEOREM 13   If r is any positive integer, then

(i) $\displaystyle\lim_{x \to +\infty} \frac{1}{x^r} = 0$

(ii) $\displaystyle\lim_{x \to -\infty} \frac{1}{x^r} = 0$

PROOF OF (i)   To prove part (i) we must show that Definition 3.7.1 holds for $f(x) = 1/x^r$ and $L = 0$; that is, we must show that for any $\epsilon > 0$ there exists a number $N > 0$ such that

$$\text{if} \quad x > N \quad \text{then} \quad \left|\frac{1}{x^r} - 0\right| < \epsilon$$

$$\Leftrightarrow \quad \text{if} \quad x > N \quad \text{then} \quad |x|^r > \frac{1}{\epsilon}$$

or, equivalently, since $r > 0$,

$$\text{if} \quad x > N \quad \text{then} \quad |x| > \left(\frac{1}{\epsilon}\right)^{1/r}$$

For the above to hold, take $N = (1/\epsilon)^{1/r}$. Thus

$$\text{if} \quad N = \left(\frac{1}{\epsilon}\right)^{1/r} \quad \text{and} \quad x > N \quad \text{then} \quad \left|\frac{1}{x^r} - 0\right| < \epsilon$$

This proves part (i).   ∎

The proof of part (ii) is analogous and is left as an exercise (see Exercise 42).

EXAMPLE 1
Find

$$\lim_{x \to +\infty} \frac{4x - 3}{2x + 5}$$

SOLUTION
To use Limit Theorem 13, divide the numerator and the denominator by x, thus giving

$$\lim_{x \to +\infty} \frac{4x - 3}{2x + 5} = \lim_{x \to +\infty} \frac{4 - \dfrac{3}{x}}{2 + \dfrac{5}{x}}$$

$$= \frac{\displaystyle\lim_{x \to +\infty} 4 - \lim_{x \to +\infty} 3 \cdot \lim_{x \to +\infty} \frac{1}{x}}{\displaystyle\lim_{x \to +\infty} 2 + \lim_{x \to +\infty} 5 \cdot \lim_{x \to +\infty} \frac{1}{x}}$$

$$= \frac{4 - 3 \cdot 0}{2 + 5 \cdot 0}$$

$$= 2$$

EXAMPLE 2
Find

$$\lim_{x \to -\infty} \frac{2x^2 - x + 5}{4x^3 - 1}$$

SOLUTION
To use Limit Theorem 13, divide the numerator and the denominator by the highest power of x occurring in either the numerator or denominator, which in this case is x^3.

$$\lim_{x \to -\infty} \frac{2x^2 - x + 5}{4x^3 - 1} = \lim_{x \to -\infty} \frac{\dfrac{2}{x} - \dfrac{1}{x^2} + \dfrac{5}{x^3}}{4 - \dfrac{1}{x^3}}$$

$$= \frac{\displaystyle\lim_{x \to -\infty} 2 \cdot \lim_{x \to -\infty} \frac{1}{x} - \lim_{x \to -\infty} \frac{1}{x^2} + \lim_{x \to -\infty} 5 \cdot \lim_{x \to -\infty} \frac{1}{x^3}}{\displaystyle\lim_{x \to -\infty} 4 - \lim_{x \to -\infty} \frac{1}{x^3}}$$

$$= \frac{2 \cdot 0 - 0 + 5 \cdot 0}{4 - 0}$$

$$= 0$$

EXAMPLE 3

Find

$$\lim_{x \to +\infty} \frac{3x + 4}{\sqrt{2x^2 - 5}}$$

SOLUTION

Because the highest power of x is 2 and it appears under the radical sign, we divide numerator and denominator by $\sqrt{x^2}$, which is $|x|$. We have, then,

$$\lim_{x \to +\infty} \frac{3x + 4}{\sqrt{2x^2 - 5}} = \lim_{x \to +\infty} \frac{\dfrac{3x}{\sqrt{x^2}} + \dfrac{4}{\sqrt{x^2}}}{\dfrac{\sqrt{2x^2 - 5}}{\sqrt{x^2}}}$$

$$= \lim_{x \to +\infty} \frac{\dfrac{3x}{|x|} + \dfrac{4}{|x|}}{\sqrt{2 - \dfrac{5}{x^2}}}$$

Because $x \to +\infty$, $x > 0$; therefore $|x| = x$. Thus we have

$$\lim_{x \to +\infty} \frac{3x + 4}{\sqrt{2x^2 - 5}} = \lim_{x \to +\infty} \frac{\dfrac{3x}{x} + \dfrac{4}{x}}{\sqrt{2 - \dfrac{5}{x^2}}}$$

$$= \frac{\displaystyle\lim_{x \to +\infty} 3 + \lim_{x \to +\infty} 4 \cdot \lim_{x \to +\infty} \left(\frac{1}{x}\right)}{\sqrt{\displaystyle\lim_{x \to +\infty} 2 - \lim_{x \to +\infty} 5 \cdot \lim_{x \to +\infty} \left(\frac{1}{x^2}\right)}}$$

$$= \frac{3 + 4 \cdot 0}{\sqrt{2 - 5 \cdot 0}}$$

$$= \frac{3}{\sqrt{2}}$$

EXAMPLE 4
Find

$$\lim_{x \to -\infty} \frac{3x + 4}{\sqrt{2x^2 - 5}}$$

SOLUTION

The function is the same as the one in Example 3. Again we begin by dividing numerator and denominator by $\sqrt{x^2}$ or, equivalently, $|x|$.

$$\lim_{x \to -\infty} \frac{3x + 4}{\sqrt{2x^2 - 5}} = \lim_{x \to -\infty} \frac{\dfrac{3x}{|x|} + \dfrac{4}{|x|}}{\sqrt{2 - \dfrac{5}{x^2}}}$$

Because $x \to -\infty$, $x < 0$; therefore $|x| = -x$. We have, then,

$$\lim_{x \to -\infty} \frac{3x + 4}{\sqrt{2x^2 - 5}} = \lim_{x \to -\infty} \frac{\dfrac{3x}{-x} + \dfrac{4}{-x}}{\sqrt{2 - \dfrac{5}{x^2}}}$$

$$= \frac{\lim\limits_{x \to -\infty} (-3) - \lim\limits_{x \to -\infty} 4 \cdot \lim\limits_{x \to -\infty} \dfrac{1}{x}}{\sqrt{\lim\limits_{x \to -\infty} 2 - \lim\limits_{x \to -\infty} 5 \cdot \lim\limits_{x \to -\infty} \dfrac{1}{x^2}}}$$

$$= \frac{-3 - 4 \cdot 0}{\sqrt{2 - 5 \cdot 0}}$$

$$= -\frac{3}{\sqrt{2}}$$

"Infinite" limits at infinity can be considered. There are formal definitions for each of the following.

$$\lim_{x \to +\infty} f(x) = +\infty \qquad \lim_{x \to -\infty} f(x) = +\infty$$

$$\lim_{x \to +\infty} f(x) = -\infty \qquad \lim_{x \to -\infty} f(x) = -\infty$$

For example, $\lim\limits_{x \to +\infty} f(x) = +\infty$ if the function f is defined on some interval $(a, +\infty)$ and if for any number $N > 0$ there exists an $M > 0$ such that if $x > M$, then $f(x) > N$. The other definitions are left as an exercise (see Exercise 39).

EXAMPLE 5
Find

$$\lim_{x \to +\infty} \frac{x^2}{x + 1}$$

SOLUTION

Divide the numerator and the denominator by x^2.

$$\lim_{x \to +\infty} \frac{x^2}{x + 1} = \lim_{x \to +\infty} \frac{1}{\dfrac{1}{x} + \dfrac{1}{x^2}}$$

Evaluating the limit of the denominator, we have

$$\lim_{x \to +\infty} \left(\frac{1}{x} + \frac{1}{x^2} \right) = \lim_{x \to +\infty} \frac{1}{x} + \lim_{x \to +\infty} \frac{1}{x^2}$$

$$= 0 + 0$$

$$= 0$$

Therefore the limit of the denominator is 0, and the denominator is approaching 0 through positive values.

The limit of the numerator is 1, and so by Limit Theorem 12(i) (1.5.4) it follows that

$$\lim_{x \to +\infty} \frac{x^2}{x + 1} = +\infty$$

EXAMPLE 6

Find

$$\lim_{x \to +\infty} \frac{2x - x^2}{3x + 5}$$

SOLUTION

$$\lim_{x \to +\infty} \frac{2x - x^2}{3x + 5} = \lim_{x \to +\infty} \frac{\dfrac{2}{x} - 1}{\dfrac{3}{x} + \dfrac{5}{x^2}}$$

The limits of the numerator and the denominator are considered separately.

$$\lim_{x \to +\infty} \left(\frac{2}{x} - 1 \right) = \lim_{x \to +\infty} \frac{2}{x} - \lim_{x \to +\infty} 1 \qquad \lim_{x \to +\infty} \left(\frac{3}{x} + \frac{5}{x^2} \right) = \lim_{x \to +\infty} \frac{3}{x} + \lim_{x \to +\infty} \frac{5}{x^2}$$

$$= 0 - 1 \qquad\qquad\qquad\qquad = 0 + 0$$

$$= -1 \qquad\qquad\qquad\qquad\quad = 0$$

Therefore we have the limit of a quotient in which the limit of the numerator is -1 and the limit of the denominator is 0, where the denominator is approaching 0 through positive values. By Limit Theorem 12(iii) it follows that

$$\lim_{x \to +\infty} \frac{2x - x^2}{3x + 5} = -\infty$$

EXAMPLE 7

A closed box with a square base is to have a volume of 2000 in.^{3}. The material for the top and bottom of the box is to cost $3 per square inch, and the material for the sides is to cost $1.50 per square inch. Let x in. be the length of a side of the square base and $C(x)$ dollars be the total cost of the material. (a) Write an equation defining $C(x)$, and state the domain of the function C. (b) Find $\lim\limits_{x \to 0^+} C(x)$ and $\lim\limits_{x \to +\infty} C(x)$, and explain these results in terms of the problem. (c) Draw a sketch of the graph of C.

SOLUTION

(a) Because x in. is the length of a side of the square base, x^2 in.^{2} is the area of the base. Let y in. be the depth of the box. See Fig. 3.7.2. Because the volume of the box is the product of the area of the base and the depth

$$x^2 y = 2000$$

$$y = \frac{2000}{x^2} \tag{3}$$

y in.

x in.

x in.

FIGURE 3.7.2

the total number of square inches in the combined area of the top and bottom is $2x^2$, and for the sides it is $4xy$. Therefore the number of dollars in the total cost of the material is

$$3(2x^2) + \tfrac{3}{2}(4xy)$$

Replacing y by its equal from (3) we have

$$C(x) = 6x^2 + 6x\left(\frac{2000}{x^2}\right)$$

$$C(x) = 6x^2 + \frac{12,000}{x} \tag{4}$$

The domain of C is $(0, +\infty)$.

(b) When computing $\lim\limits_{x \to 0^+} C(x)$, we observe that

$$\lim_{x \to 0^+} 6x^2 = 0 \quad \text{and} \quad \lim_{x \to 0^+} \frac{12,000}{x} = +\infty$$

and when computing $\lim\limits_{x \to +\infty} C(x)$, we notice that

$$\lim_{x \to +\infty} 6x^2 = +\infty \quad \text{and} \quad \lim_{x \to +\infty} \frac{12,000}{x} = 0$$

Therefore

$$\lim_{x \to 0^+} C(x) = +\infty \quad \text{and} \quad \lim_{x \to +\infty} C(x) = +\infty$$

These results indicate that the total cost has no absolute maximum value.

(c) To draw a sketch of the graph of C we use the results of part (b) and also determine where the graph has a horizontal tangent line by computing

FIGURE 3.7.3

$C'(x)$ and setting it equal to zero. From (4),

$$C'(x) = 12x - \frac{12,000}{x^2}$$

$C'(x) = 0$ when

$$12x^3 - 12,000 = 0$$

$$x^3 = 1000$$

$$x = 10$$

Thus the graph has a horizontal tangent line at the point $(10, 1800)$. A sketch of the graph of C appears in Fig. 3.7.3.

From Fig. 3.7.3 we can conclude that if there is a minimum total cost, it will occur when the length of a side of the square base in 10 in. In Illustration 1 of Section 3.10 we prove that C has an absolute minimum value when $x = 10$.

EXERCISES 3.7

In Exercises 1 through 22, find the limit.

1. $\displaystyle\lim_{x \to +\infty} \frac{1}{x^3}$

2. $\displaystyle\lim_{x \to -\infty} \frac{3}{x^4}$

3. $\displaystyle\lim_{t \to +\infty} \frac{2t + 1}{5t - 2}$

4. $\displaystyle\lim_{x \to -\infty} \frac{6x - 4}{3x + 1}$

5. $\displaystyle\lim_{x \to -\infty} \frac{2x + 7}{4 - 5x}$

6. $\displaystyle\lim_{x \to +\infty} \frac{1 + 5x}{2 - 3x}$

7. $\displaystyle\lim_{x \to +\infty} \frac{7x^2 - 2x + 1}{3x^2 + 8x + 5}$

8. $\displaystyle\lim_{s \to -\infty} \frac{4s^2 + 3}{2s^2 - 1}$

9. $\displaystyle\lim_{x \to +\infty} \frac{x + 4}{3x^2 - 5}$ 0

10. $\displaystyle\lim_{x \to +\infty} \frac{x^2 + 5}{x^3}$

11. $\displaystyle\lim_{y \to +\infty} \frac{2y^2 - 3y}{y + 1}$ $+\infty$

12. $\displaystyle\lim_{x \to +\infty} \frac{x^2 - 2x + 5}{7x^3 + x + 1}$

13. $\displaystyle\lim_{x \to -\infty} \frac{4x^3 + 2x^2 - 5}{8x^3 + x + 2}$

14. $\displaystyle\lim_{x \to +\infty} \frac{3x^4 - 7x^2 + 2}{2x^4 + 1}$

15. $\displaystyle\lim_{y \to +\infty} \frac{2y^3 - 4}{5y + 3}$ $+\infty$

16. $\displaystyle\lim_{x \to -\infty} \frac{5x^3 - 12x + 7}{4x^2 - 1}$

17. $\displaystyle\lim_{x \to -\infty} \left(3x + \frac{1}{x^2}\right)$ $-\infty$

18. $\displaystyle\lim_{t \to +\infty} \left(\frac{2}{t^2} - 4t\right)$

19. $\displaystyle\lim_{x \to +\infty} \frac{\sqrt{x^2 + 4}}{x + 4}$ 1

20. $\displaystyle\lim_{x \to -\infty} \frac{\sqrt{x^2 + 4}}{x + 4}$

21. $\displaystyle\lim_{w \to -\infty} \frac{\sqrt{w^2 - 2w + 3}}{w + 5}$

22. $\displaystyle\lim_{y \to -\infty} \frac{\sqrt{y^4 + 1}}{2y^2 - 3}$

23. A rectangular field having an area of 2700 m² is to be enclosed by a fence, and an additional fence is to be used to divide the field down the middle. The cost of the fence down the middle is $12 per running meter, and the fence along the sides costs $18 per running meter. Let x meters be the length of the dividing fence and $C(x)$ dollars be the cost of the fencing. (a) Write an equation defining $C(x)$, and state the domain of the function C. (b) Find $\displaystyle\lim_{x \to 0^+} C(x)$ and $\displaystyle\lim_{x \to +\infty} C(x)$, and explain these results in terms of the problem. (c) Draw a sketch of the graph of C.

24. A rectangular open tank is to have a square base, and its volume is to be 125 m³. The cost per square meter for the bottom is $24 and for the sides is $12. Let x meters be the length of a side of the square base and $C(x)$ dollars be the cost of the material. (a) Write an equation defining $C(x)$, and state the domain of the function C. (b) Find $\displaystyle\lim_{x \to 0^+} C(x)$ and $\displaystyle\lim_{x \to +\infty} C(x)$, and explain these results in terms of the problem. (c) Draw a sketch of the graph of C.

25. A page of print is to contain 24 in.^{2} of printed region, a margin of $1\frac{1}{2}$ in. at the top and bottom, and a margin of 1 in. at the sides. Let x in. be the width of the printed portion of the page and $A(x)$ in.^{2} be the total area of the page. (a) Write an equation defining $A(x)$, and state the domain of the function A. (b) Find $\lim\limits_{x \to 0^+} A(x)$ and $\lim\limits_{x \to +\infty} A(x)$, and explain these results in terms of the problem. (c) Draw a sketch of the graph of A.

26. A one-story building having a rectangular floor space of 13,200 ft^2 is to be constructed where a walkway 22 ft wide is required in the front and back and a walkway 15 ft wide is required on each side. Let x ft be the length of the front and back of the building and $A(x)$ ft^2 be the total area of the lot on which the building and walkways will be located. (a) Write an equation defining $A(x)$, and state the domain of the function A. (b) Find $\lim\limits_{x \to 0^+} A(x)$ and $\lim\limits_{x \to +\infty} A(x)$, and explain these results in terms of the problem. (c) Draw a sketch of the graph of A.

In Exercises 27 through 31, find the limit. *Hint:* First obtain a fraction with a rational numerator.

27. $\lim\limits_{x \to +\infty} (\sqrt{x^2 + 1} - x)$

28. $\lim\limits_{x \to +\infty} (\sqrt{x^2 + x} - x)$

29. $\lim\limits_{r \to +\infty} (\sqrt{3r^2 + r} - 2r)$ $-\infty$

$\left(r\sqrt{3 + \frac{1}{r}} - 2\right)$ $r(\sqrt{3}-2)$

30. $\lim\limits_{x \to +\infty} (\sqrt[3]{x^3 + 1} - x)$

31. $\lim\limits_{x \to -\infty} (\sqrt[3]{x^3 + x} - \sqrt[3]{x^3 + 1})$

32. Find $\lim\limits_{t \to +\infty} \dfrac{\sqrt{t + \sqrt{t + \sqrt{t}}}}{\sqrt{t + 1}}$.

$\lim\limits_{x \to +\infty} \sqrt{x^2+1} - x \left(\dfrac{\sqrt{x^2+1} + x}{\sqrt{x^2+1} + x}\right) = \lim\limits_{x \to +\infty} \dfrac{1}{\sqrt{x^2+1} + x} = 0$

In Exercises 33 through 36, prove that $\lim\limits_{x \to +\infty} f(x) = 1$ by applying Definition 3.7.1; that is, for any $\epsilon > 0$ show that there exists a number $N > 0$ such that if $x > N$, then $|f(x) - 1| < \epsilon$.

33. $f(x) = \dfrac{x}{x - 1}$

34. $f(x) = \dfrac{2x}{2x + 3}$

35. $f(x) = \dfrac{x^2 - 1}{x^2 + 1}$

36. $f(x) = \dfrac{x^2 + 2x}{x^2 - 1}$

37. Prove that $\lim\limits_{x \to -\infty} \dfrac{8x + 3}{2x - 1} = 4$ by showing that for any $\epsilon > 0$ there exists a number $N < 0$ such that if $x < N$ then $\left| \dfrac{8x + 3}{2x - 1} - 4 \right| < \epsilon$.

38. Prove part (i) of Limit Theorem 12 (1.5.4) if "$x \to a$" is replaced by "$x \to +\infty$."

39. Give a definition for each of the following: (a) $\lim\limits_{x \to +\infty} f(x) = -\infty$; (b) $\lim\limits_{x \to -\infty} f(x) = +\infty$; (c) $\lim\limits_{x \to -\infty} f(x) = -\infty$.

40. Prove that $\lim\limits_{x \to +\infty} (x^2 - 4) = +\infty$ by showing that for any $N > 0$ there exists an $M > 0$ such that if $x > M$ then $x^2 - 4 > N$.

41. Prove that $\lim\limits_{x \to +\infty} (6 - x - x^2) = -\infty$ by applying the definition in Exercise 39(a).

42. Prove part (ii) of Limit Theorem 13 (3.7.3).

3.8
ASYMPTOTES OF A GRAPH

In Section 1.5 we indicated that an aid in drawing the sketch of the graph of a function is to find, if there are any, the asymptotes of the graph. In that section the definition (1.5.8) of a vertical asymptote is given as follows:

The line $x = a$ is said to be a **vertical asymptote** of the graph of the function f if at least one of the following statements is true:

(i) $\lim\limits_{x \to a^+} f(x) = +\infty$

(ii) $\lim\limits_{x \to a^+} f(x) = -\infty$

$$\lim_{x \to +\infty} f(x) = b$$

FIGURE 3.8.1

(iii) $\lim_{x \to a^-} f(x) = +\infty$

(iv) $\lim_{x \to a^-} f(x) = -\infty$

A *horizontal asymptote* of a graph is a line parallel to the x axis, and the following is the formal definition.

3.8.1 DEFINITION

$$\lim_{x \to +\infty} f(x) = b$$

FIGURE 3.8.2

The line $y = b$ is said to be a **horizontal asymptote** of the graph of the function f if at least one of the following statements is true:

(i) $\lim_{x \to +\infty} f(x) = b$, and for some number N, if $x > N$, then $f(x) \neq b$;

(ii) $\lim_{x \to -\infty} f(x) = b$, and for some number N, if $x < N$, then $f(x) \neq b$.

$$\lim_{x \to -\infty} f(x) = b$$

FIGURE 3.8.3

▶ ILLUSTRATION 1   In each of Figs. 3.8.1 through 3.8.4 is shown a portion of the graph of a function for which the line $y = b$ is a horizontal asymptote. In Figs. 3.8.1 and 3.8.2, part (i) of Definition 3.8.1 applies, and in Figs. 3.8.3 and 3.8.4, part (ii) is true. Both parts (i) and (ii) hold for the function whose graph appears in Fig. 3.8.5.   ◀

$$\lim_{x \to -\infty} f(x) = b$$

FIGURE 3.8.4

In Fig. 3.8.6 there is a sketch of the graph of a function for which $\lim_{x \to +\infty} f(x) = b$, but there is no number N such that if $x > N$, then $f(x) \neq b$. Consequently the line $y = b$ is not a horizontal asymptote of the graph of f. An example of such a function is given in Exercise 59 of Exercises 6.5.

$$\lim_{x \to +\infty} f(x) = b$$

and

$$\lim_{x \to -\infty} f(x) = b$$

FIGURE 3.8.5

EXAMPLE 1
Find the horizontal asymptotes of the graph of the function f defined by

$$f(x) = \frac{x}{\sqrt{x^2 + 1}}$$

and draw a sketch of the graph.

SOLUTION
First consider $\lim_{x \to +\infty} f(x)$.

$$\lim_{x \to +\infty} f(x) = \lim_{x \to +\infty} \frac{x}{\sqrt{x^2 + 1}}$$

We divide the numerator and denominator by $\sqrt{x^2}$ or, equivalently, $|x|$, and we have

FIGURE 3.8.6

$$\lim_{x \to +\infty} \frac{x}{\sqrt{x^2+1}} = \lim_{x \to +\infty} \frac{\dfrac{x}{\sqrt{x^2}}}{\sqrt{\dfrac{x^2}{x^2}+\dfrac{1}{x^2}}}$$

$$= \lim_{x \to +\infty} \frac{\dfrac{x}{|x|}}{\sqrt{1+\dfrac{1}{x^2}}}$$

Because $x \to +\infty$, $x > 0$; therefore $|x| = x$. Thus

$$\lim_{x \to +\infty} \frac{x}{\sqrt{x^2+1}} = \lim_{x \to +\infty} \frac{\dfrac{x}{x}}{\sqrt{1+\dfrac{1}{x^2}}}$$

$$= \frac{\displaystyle\lim_{x \to +\infty} 1}{\sqrt{\displaystyle\lim_{x \to +\infty} 1 + \lim_{x \to +\infty} \frac{1}{x^2}}}$$

$$= \frac{1}{\sqrt{1+0}}$$

$$= 1$$

Therefore, by Definition 3.8.1(i), the line $y = 1$ is a horizontal asymptote.

Now consider $\lim\limits_{x \to -\infty} f(x)$. Again we divide numerator and denominator by $\sqrt{x^2}$, which is $|x|$; because $x \to -\infty$, $x < 0$ and so $|x| = -x$. We have

$$\lim_{x \to -\infty} f(x) = \lim_{x \to -\infty} \frac{\dfrac{x}{-x}}{\sqrt{1+\dfrac{1}{x^2}}}$$

$$= \frac{\displaystyle\lim_{x \to -\infty} (-1)}{\sqrt{\displaystyle\lim_{x \to -\infty} 1 + \lim_{x \to -\infty} \frac{1}{x^2}}}$$

$$= \frac{-1}{\sqrt{1+0}}$$

$$= -1$$

Accordingly, by Definition 3.8.1 (ii), the line $y = -1$ is a horizontal asymptote. A sketch of the graph appears in Fig. 3.8.7.

FIGURE 3.8.7

EXAMPLE 2

Find the vertical and horizontal asymptotes of the graph of the equation

$$xy^2 - 2y^2 - 4x = 0$$

and draw a sketch of the graph.

SOLUTION

Solve the given equation for y and obtain

$$y = \pm 2 \sqrt{\frac{x}{x-2}} \tag{1}$$

Equation (1) defines two functions:

$$y = f_1(x) \quad \text{where } f_1 \text{ is defined by} \quad f_1(x) = +2 \sqrt{\frac{x}{x-2}}$$

and

$$y = f_2(x) \quad \text{where } f_2 \text{ is defined by} \quad f_2(x) = -2 \sqrt{\frac{x}{x-2}}$$

The graph of the given equation is composed of the graphs of the two functions f_1 and f_2. The domains of the two functions consist of those values of x for which $x/(x-2) \geq 0$. By using the result of Example 8 in Section 1.1 and excluding $x = 2$ we see that the domain of f_1 and f_2 is $(-\infty, 0] \cup (2, +\infty)$.

Now consider f_1. Because

$$\lim_{x \to 2^+} f_1(x) = \lim_{x \to 2^+} 2 \sqrt{\frac{x}{x-2}}$$

$$= +\infty$$

by Definition 1.5.8(i) the line $x = 2$ is a vertical asymptote of the graph of f_1.

$$\lim_{x \to +\infty} f_1(x) = \lim_{x \to +\infty} 2 \sqrt{\frac{1}{1 - \dfrac{2}{x}}}$$

$$= 2$$

Thus by Definition 3.8.1(i) the line $y = 2$ is a horizontal asymptote of the graph of f_1.

Similarly, $\displaystyle \lim_{x \to -\infty} f_1(x) = 2$. A sketch of the graph of f_1 is shown in Fig. 3.8.8.

$$\lim_{x \to 2^+} f_2(x) = \lim_{x \to 2^+} \left[-2 \sqrt{\frac{x}{x-2}} \right]$$

$$= -\infty$$

Hence, by Definition 1.5.8(ii) the line $x = 2$ is a vertical asymptote of the graph of f_2.

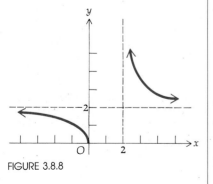

y

O 2 x

FIGURE 3.8.8

FIGURE 3.8.9

FIGURE 3.8.10

$$\lim_{x \to +\infty} f_2(x) = \lim_{x \to +\infty} \left[-2 \sqrt{\dfrac{1}{1 - \dfrac{2}{x}}} \right]$$

$$= -2$$

Thus, by Definition 3.8.1(i) the line $y = -2$ is a horizontal asymptote of the graph of f_2.

Also, $\lim\limits_{x \to -\infty} f_2(x) = -2$. A sketch of the graph of f_2 is shown in Fig. 3.8.9.

The graph of the given equation is the union of the graphs of f_1 and f_2, and a sketch is shown in Fig. 3.8.10.

EXAMPLE 3
Find the horizontal and vertical asymptotes of the graph of the function f if

$$f(x) = \frac{8x - 2x^2}{x^2 - 9}$$

Draw a sketch of the graph of f.

SOLUTION

$$\lim_{x \to +\infty} f(x) = \lim_{x \to +\infty} \frac{\dfrac{8}{x} - 2}{1 - \dfrac{9}{x^2}}$$

$$= \frac{0 - 2}{1 - 0}$$

Thus

$$\lim_{x \to +\infty} f(x) = -2 \qquad (2)$$

Also,

$$\lim_{x \to -\infty} f(x) = -2 \qquad (3)$$

Thus, from Definition 3.8.1 the line $y = -2$ is a horizontal asymptote.

If the numerator and denominator are factored,

$$f(x) = \frac{2x(4 - x)}{(x - 3)(x + 3)}$$

Consider the four limits $\lim\limits_{x \to 3^+} f(x)$, $\lim\limits_{x \to 3^-} f(x)$, $\lim\limits_{x \to -3^+} f(x)$, and $\lim\limits_{x \to -3^-} f(x)$. To find $\lim\limits_{x \to 3^+} f(x)$, observe that $\lim\limits_{x \to 3^+} 2x(4 - x) = 6$ and $\lim\limits_{x \to 3^+} (x - 3)(x + 3) = 0$, where $(x - 3)(x + 3)$ is approaching 0 through positive values. Therefore, by Limit Theorem 12(i) (1.5.4),

$$\lim_{x \to 3^+} f(x) = +\infty \qquad (4)$$

Since $\lim\limits_{x \to 3^-} 2x(4 - x) = 6$ and $\lim\limits_{x \to 3^-} (x - 3)(x + 3) = 0$, where $(x - 3)(x + 3)$ is approaching 0 through negative values, it follows from Limit Theorem 12(ii) that

$$\lim_{x \to 3^-} f(x) = -\infty \tag{5}$$

From (4) and (5) and Definition 1.5.8(i) and (iv), the line $x = 3$ is a vertical asymptote.

Limit Theorem 12(iv) is used for $\lim\limits_{x \to -3^+} f(x)$; $\lim\limits_{x \to -3^+} 2x(4 - x) = -42$ and $\lim\limits_{x \to -3^+} (x - 3)(x + 3) = 0$, where $(x - 3)(x + 3)$ is approaching 0 through negative values. Therefore

$$\lim_{x \to -3^+} f(x) = +\infty \tag{6}$$

Because $\lim\limits_{x \to -3^-} 2x(4 - x) = -42$ and $\lim\limits_{x \to -3^-} (x - 3)(x + 3) = 0$, where $(x - 3)(x + 3)$ is approaching 0 through positive values, it follows from Limit Theorem 12(iii) that

$$\lim_{x \to -3^-} f(x) = -\infty \tag{7}$$

From (6) and (7) and Definition 1.5.8(i) and (iv), the line $x = -3$ is a vertical asymptote.

From the given equation it follows that the graph of f contains the points $(0, 0)$ and $(4, 0)$. We plot a few other points and use the information given in (2) through (7) to obtain the sketch of the graph appearing in Fig. 3.8.11.

FIGURE 3.8.11

An asymptote of a graph that is neither horizontal nor vertical is called an **oblique asymptote**. In Section 10.6 we give the formal definition (10.6.3) of an oblique asymptote. At this time we show that the graph of a rational function of the form $f(x)/g(x)$, where the degree of $f(x)$ is one more than the degree of $g(x)$, has the line $y = mx + b$ as an oblique asymptote by proving that

$$\lim_{x \to +\infty} \left| \frac{f(x)}{g(x)} - (mx + b) \right| = 0 \tag{8}$$

If we divide the polynomial $f(x)$ in the numerator by the polynomial $g(x)$ in the denominator, we obtain the sum of a linear function and a rational function; that is,

$$\frac{f(x)}{g(x)} = mx + b + \frac{h(x)}{g(x)}$$

where the degree of the polynomial $h(x)$ is less than the degree of $g(x)$. Then

$$\lim_{x \to +\infty} \left| \frac{f(x)}{g(x)} - (mx + b) \right| = \lim_{x \to +\infty} \left| \frac{h(x)}{g(x)} \right|$$

When the numerator and denominator of $h(x)/g(x)$ are divided by the highest power of x appearing in $g(x)$, there will be a constant term in the denominator and all other terms in the denominator and every term in the numerator will be of the form k/x^r where k is a constant. Therefore, as $x \to +\infty$, the limit of the numerator will be 0 and the limit of the denominator will be a constant. Thus

$$\lim_{x \to +\infty} \left| \frac{h(x)}{g(x)} \right| = 0$$

and (8) is established.

EXAMPLE 4

Find the asymptotes of the graph of the function h defined by

$$h(x) = \frac{x^2 + 3}{x - 1} \tag{9}$$

and draw a sketch of the graph.

SOLUTION
Because

$$\lim_{x \to 1^-} h(x) = -\infty \quad \text{and} \quad \lim_{x \to 1^+} h(x) = +\infty$$

the line $x = 1$ is a vertical asymptote. There are no horizontal asymptotes, because if the numerator and denominator of $h(x)$ are divided by x^2, we obtain

$$\frac{1 + \dfrac{3}{x^2}}{\dfrac{1}{x} - \dfrac{1}{x^2}}$$

and as $x \to +\infty$ or $x \to -\infty$, the limit of the numerator is 1 and the limit of the denominator is 0. However, the degree of the numerator of $h(x)$ is one more than the degree of the denominator, and when the numerator is divided by the denominator, we obtain

$$h(x) = x + 1 + \frac{4}{x - 1}$$

Therefore the line $y = x + 1$ is an oblique asymptote.

    To draw a sketch of the graph of h we determine if there are any horizontal tangent lines. From (9),

FIGURE 3.8.12

$$h'(x) = \frac{2x(x - 1) - (x^2 + 3)}{(x - 1)^2}$$

$$= \frac{x^2 - 2x - 3}{(x - 1)^2}$$

Setting $h'(x) = 0$ we get

$$x^2 - 2x - 3 = 0$$

$$(x + 1)(x - 3) = 0$$

$$x = -1 \qquad x = 3$$

Thus there are horizontal tangent lines at the points $(-1, -2)$ and $(3, 6)$.

By indicating the asymptotes and the horizontal tangent lines and plotting a few points we obtain the sketch of the graph of h shown in Fig. 3.8.12.

EXERCISES 3.8

In Exercises 1 through 12, find the horizontal and vertical asymptotes of the graph of the function defined by the given equation, and draw a sketch of the graph.

1. $f(x) = \dfrac{2x + 1}{x - 3}$　　**2.** $f(x) = \dfrac{4 - 3x}{x + 1}$　　**3.** $g(x) = 1 - \dfrac{1}{x}$　　**4.** $h(x) = 1 + \dfrac{1}{x^2}$

5. $f(x) = \dfrac{2}{\sqrt{x^2 - 4}}$　　**6.** $F(x) = \dfrac{-3x}{\sqrt{x^2 + 3}}$　　**7.** $G(x) = \dfrac{4x^2}{x^2 - 9}$　　**8.** $g(x) = \dfrac{x^2}{4 - x^2}$

9. $h(x) = \dfrac{2x}{6x^2 + 11x - 10}$　　**10.** $f(x) = \dfrac{-1}{\sqrt{x^2 + 5x + 6}}$　　**11.** $f(x) = \dfrac{4x^2}{\sqrt{x^2 - 2}}$ $\pm\sqrt{2}$　　**12.** $h(x) = \dfrac{x}{\sqrt{x^2 - 9}}$

In Exercises 13 through 20, find the horizontal and vertical asymptotes of the graph of the given equation, and draw a sketch of the graph.

13. $3xy - 2x - 4y - 3 = 0$　　**14.** $2xy + 4x - 3y + 6 = 0$　　**15.** $x^2y^2 - x^2 + 4y^2 = 0$
16. $xy^2 + 3y^2 - 9x = 0$　　**17.** $(y^2 - 1)(x - 3) = 6$　　**18.** $2xy^2 + 4y^2 - 3x = 0$
19. $x^2y - 2x^2 - y - 2 = 0$　　**20.** $x^2y + 4xy - x^2 + x + 4y - 6 = 0$

In Exercises 21 through 32, find the vertical and oblique asymptotes of the graph of the function defined by the given equation, and draw a sketch of the graph.

21. $f(x) = \dfrac{x^2}{x - 1}$　　**22.** $f(x) = \dfrac{x^2 - 3}{x - 2}$　　**23.** $f(x) = \dfrac{x^2 - 8}{x - 3}$　　**24.** $f(x) = \dfrac{x^2 + 4}{x}$

25. $f(x) = \dfrac{2x^2 + 2}{x}$　　**26.** $f(x) = \dfrac{x^2 - 5}{x + 3}$　　**27.** $f(x) = \dfrac{x^2 - 4}{x}$　　**28.** $f(x) = \dfrac{x^2 - 3x + 2}{x + 4}$

29. $f(x) = \dfrac{x^2 - 4x - 5}{x + 2}$　　**30.** $f(x) = \dfrac{x^3 - 4}{x^2}$　　**31.** $f(x) = \dfrac{x^3 + 2x^2 + 4}{x^2}$　　**32.** $f(x) = \dfrac{(x + 1)^3}{(x - 1)^2}$

3.9
APPLICATIONS TO DRAWING A SKETCH OF THE GRAPH OF A FUNCTION

When applying the properties discussed in previous sections to obtain a sketch of the graph of a function f, you should proceed as follows:

1. Determine the domain of f.
2. Find any y intercepts of the graph. Locate any x intercepts if the resulting equation is easy to solve.
3. Test for symmetry with respect to the y axis and origin.
4. Compute $f'(x)$ and $f''(x)$.
5. Determine the critical numbers of f. These are the values of x in the domain of f for which either $f'(x)$ does not exist or $f'(x) = 0$.
6. Apply either the first-derivative test (Theorem 3.4.4) or the second-derivative test (Theorem 3.6.1) to determine whether at a critical number there is a relative maximum value, a relative minimum value, or neither.
7. Determine the intervals on which f is increasing by finding the values of x for which $f'(x)$ is positive; determine the intervals on which f is decreasing by finding the values of x for which $f'(x)$ is negative. In locating the intervals on which f is monotonic, also check the critical numbers at which f does not have a relative extremum.
8. To obtain possible points of inflection, find the critical numbers of f', that is, the values of x for which $f''(x)$ does not exist or $f''(x) = 0$. At each of these values of x check to see if $f''(x)$ changes sign and if the graph has a tangent line there to determine if there actually is a point of inflection.
9. Check for concavity of the graph. Find the values of x for which $f''(x)$ is positive to obtain the points at which the graph is concave upward; to obtain the points at which the graph is concave downward find the values of x for which $f''(x)$ is negative.
10. It is helpful to find the slope of each inflectional tangent.
11. Check for any possible horizontal, vertical, or oblique asymptotes.

It is suggested that all the information obtained be incorporated into a table, as in the following examples.

EXAMPLE 1
Given

$$f(x) = x^3 - 3x^2 + 3$$

Draw a sketch of the graph of f by first finding the following: the relative extrema of f; the points of inflection of the graph of f; the intervals on which f is increasing; the intervals on which f is decreasing; where the graph is concave upward; where the graph is concave downward; and the slope of any inflectional tangent.

SOLUTION
The domain of f is the set of all real numbers. The y intercept is 3.

$$f'(x) = 3x^2 - 6x \qquad f''(x) = 6x - 6$$

Set $f'(x) = 0$ to obtain $x = 0$ and $x = 2$. From $f''(x) = 0$ we get $x = 1$. In making the table, consider the points at which $x = 0$, $x = 1$, and $x = 2$, and the intervals excluding these values of x:

$$x < 0 \qquad 0 < x < 1 \qquad 1 < x < 2 \qquad 2 < x$$

From the information in Table 3.9.1 and by plotting a few points, we obtain the sketch of the graph shown in Fig. 3.9.1.

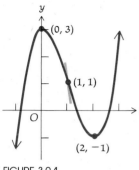

(0, 3)

(1, 1)

(2, −1)

FIGURE 3.9.1

Table 3.9.1

	$f(x)$	$f'(x)$	$f''(x)$	Conclusion
$x < 0$		+	−	f is increasing; graph is concave downward
$x = 0$	3	0	−	f has a relative maximum value; graph is concave downward
$0 < x < 1$		−	−	f is decreasing; graph is concave downward
$x = 1$	1	−3	0	f is decreasing; graph has a point of inflection
$1 < x < 2$		−	+	f is decreasing; graph is concave upward
$x = 2$	−1	0	+	f has a relative minimum value; graph is concave upward
$2 < x$		+	+	f is increasing; graph is concave upward

EXAMPLE 2

Given

$$f(x) = \frac{x^2}{x^2 - 4}$$

Draw a sketch of the graph of f by following the instructions of Example 1. Also find the horizontal and vertical asymptotes.

SOLUTION

The domain of f is the set of all real numbers except ± 2. The only intersection of the graph with an axis is at the origin. Because $f(-x) = f(x)$, the graph is symmetric with respect to the y axis.

$$f'(x) = \frac{2x(x^2 - 4) - 2x(x^2)}{(x^2 - 4)^2} \qquad f''(x) = \frac{-8(x^2 - 4)^2 + 8x[2(x^2 - 4)(2x)]}{(x^2 - 4)^4}$$

$$= \frac{-8x}{(x^2 - 4)^2} \qquad\qquad = \frac{24x^2 + 32}{(x^2 - 4)^3}$$

Set $f'(x) = 0$ to obtain $x = 0$; $f''(x)$ is never zero. For Table 3.9.2 consider the points at which $x = 0$ and $x = \pm 2$, because 2 and −2 are not in the domain of f. Also for the table consider the intervals excluding these values of x:

$$x < -2 \qquad -2 < x < 0 \qquad 0 < x < 2 \qquad 2 < x$$

Because 2 and −2 are excluded from the domain of f, we compute the following limits:

$$\lim_{x \to 2^+} \frac{x^2}{x^2 - 4} = +\infty$$

$$\lim_{x \to 2^-} \frac{x^2}{x^2 - 4} = -\infty$$

$$\lim_{x \to -2^+} \frac{x^2}{x^2 - 4} = -\infty$$

$$\lim_{x \to -2^-} \frac{x^2}{x^2 - 4} = +\infty$$

Therefore $x = 2$ and $x = -2$ are vertical asymptotes of the graph.

$$\lim_{x \to +\infty} \frac{x^2}{x^2 - 4} = \lim_{x \to +\infty} \frac{1}{1 - \dfrac{4}{x^2}} \qquad \lim_{x \to -\infty} \frac{x^2}{x^2 - 4} = \lim_{x \to -\infty} \frac{1}{1 - \dfrac{4}{x^2}}$$

$$= 1 \qquad\qquad\qquad = 1$$

Hence $y = 1$ is a horizontal asymptote of the graph.

With the facts from Table 3.9.2, the asymptotes as guides, the plotting of a few points, and the symmetry property, we obtain the sketch of the graph of f shown in Fig. 3.9.2.

FIGURE 3.9.2

Table 3.9.2

	$f(x)$	$f'(x)$	$f''(x)$	Conclusion
$x < -2$		+	+	f is increasing; graph is concave upward
$x = -2$	does not exist	does not exist	does not exist	
$-2 < x < 0$		+	−	f is increasing; graph is concave downward
$x = 0$	0	0	−	f has a relative maximum value
$0 < x < 2$		−	−	f is decreasing; graph is concave downward
$x = 2$	does not exist	does not exist	does not exist	
$2 < x$		−	+	f is decreasing; graph is concave upward

Observe in Example 2 that because the graph is symmetric with respect to the y axis, it can be obtained by doing the calculus computations for x in $[0, +\infty)$ and then applying the symmetry property.

EXAMPLE 3
Given

$$f(x) = 5x^{2/3} - x^{5/3}$$

Draw a sketch of the graph of f by following the instructions of Example 1.

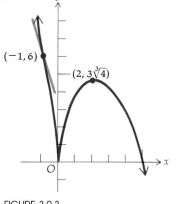

$(-1, 6)$

$(2, 3\sqrt[3]{4})$

FIGURE 3.9.3

SOLUTION

The domain of f is the set of all real numbers. The y intercept is 0. Setting $f(x) = 0$ we get

$$x^{2/3}(5 - x) = 0$$

$$x = 0 \qquad x = 5$$

Therefore the x intercepts are 0 and 5.

$$f'(x) = \tfrac{10}{3}x^{-1/3} - \tfrac{5}{3}x^{2/3} \qquad f''(x) = -\tfrac{10}{9}x^{-4/3} - \tfrac{10}{9}x^{-1/3}$$

$$= \tfrac{5}{3}x^{-1/3}(2 - x) \qquad\qquad = -\tfrac{10}{9}x^{-4/3}(1 + x)$$

When $x = 0$, neither $f'(x)$ nor $f''(x)$ exists. Set $f'(x) = 0$ to obtain $x = 2$. Therefore the critical numbers of f are 0 and 2. From $f''(x) = 0$ we obtain $x = -1$. In making the table, consider the points at which x is -1, 0, and 2, and the following intervals:

$$x < -1 \qquad -1 < x < 0 \qquad 0 < x < 2 \qquad 2 < x$$

A sketch of the graph, drawn from the information in Table 3.9.3 and by plotting a few points, appears in Fig. 3.9.3.

Table 3.9.3

	$f(x)$	$f'(x)$	$f''(x)$	Conclusion
$x < -1$		$-$	$+$	f is decreasing; graph is concave upward
$x = -1$	6	-5	0	f is decreasing; graph has a point of inflection
$-1 < x < 0$		$-$	$-$	f is decreasing; graph is concave downward
$x = 0$	0	does not exist	does not exist	f has a relative minimum value
$0 < x < 2$		$+$	$-$	f is increasing; graph is concave downward
$x = 2$	$3\sqrt[3]{4} \approx 4.8$	0	$-$	f has a relative maximum value; graph is concave downward
$2 < x$		$-$	$-$	f is decreasing; graph is concave downward

EXERCISES 3.9

For each function f, draw a sketch of its graph by first finding the following: the relative extrema of f; the points of inflection of the graph of f; the intervals on which f is increasing; the intervals on which f is decreasing; where the graph is concave upward; where the graph is concave downward; the slope of any inflectional tangent; and the horizontal, vertical, and oblique asymptotes, if there are any.

1. $f(x) = 2x^3 - 6x + 1$
2. $f(x) = x^3 + x^2 - 5x$
3. $f(x) = x^4 - 2x^3$
4. $f(x) = 3x^4 + 2x^3$
5. $f(x) = x^3 + 5x^2 + 3x - 4$
6. $f(x) = 2x^3 - \tfrac{1}{2}x^2 - 12x + 1$
7. $f(x) = x^4 - 3x^3 + 3x^2 + 1$
8. $f(x) = x^4 - 4x^3 + 16x$
9. $f(x) = \tfrac{1}{4}x^4 - \tfrac{1}{3}x^3 - x^2 + 1$
10. $f(x) = \tfrac{1}{4}x^4 - x^3$
11. $f(x) = \tfrac{1}{2}x^4 - 2x^3 + 3x^2 + 2$
12. $f(x) = 3x^4 + 4x^3 + 6x^2 - 4$

13. $f(x) = \begin{cases} x^2 & \text{if } x < 0 \\ 2x^2 & \text{if } 0 \le x \end{cases}$

14. $f(x) = \begin{cases} -x^3 & \text{if } x < 0 \\ x^3 & \text{if } 0 \le x \end{cases}$

15. $f(x) = \begin{cases} -x^4 & \text{if } x < 0 \\ x^4 & \text{if } 0 \le x \end{cases}$

16. $f(x) = \begin{cases} 2(x-1)^3 & \text{if } x < 1 \\ (x-1)^4 & \text{if } 1 \le x \end{cases}$

17. $f(x) = \begin{cases} 3(x-2)^2 & \text{if } x \le 2 \\ (2-x)^3 & \text{if } 2 < x \end{cases}$

18. $f(x) = x^2(x+4)^3$

19. $f(x) = (x+1)^3(x-2)^2$

20. $f(x) = 3x^5 + 5x^3$

21. $f(x) = 3x^5 + 5x^4$

22. $f(x) = |4 - x^2|$

23. $f(x) = 3\cos 2x;\ x \in [-\pi, \pi]$

24. $f(x) = 2\sin 3x;\ x \in [-\pi, \pi]$

25. $f(x) = \begin{cases} \sin x & \text{if } 0 \le x < \frac{1}{2}\pi \\ \sin(x - \frac{1}{2}\pi) & \text{if } \frac{1}{2}\pi \le x \le \pi \end{cases}$

26. $f(x) = \begin{cases} \cos x & \text{if } -\pi \le x \le 0 \\ \cos(\pi - x) & \text{if } 0 < x \le \pi \end{cases}$

27. $f(x) = 2\tan \frac{1}{2}x;\ x \in (-\pi, \pi)$

28. $f(x) = 3\sec \frac{1}{4}x;\ x \in (-2\pi, 2\pi)$

29. $f(x) = \sin x + \cos x;\ x \in [-2\pi, 2\pi]$

30. $f(x) = |\sin x|;\ x \in [-2\pi, 2\pi]$

31. $f(x) = \dfrac{x^2}{x-1}$

32. $f(x) = \dfrac{x}{x^2 - 4}$

33. $f(x) = \dfrac{x^2 + 1}{x^2 - 1}$

34. $f(x) = \dfrac{x^2 + 1}{x - 3}$

35. $f(x) = \dfrac{2x}{x^2 + 1}$

36. $f(x) = \dfrac{x^2 - 4}{x^2 - 9}$

37. $f(x) = 3x^{2/3} - 2x$

38. $f(x) = x^{1/3} + 2x^{4/3}$

39. $f(x) = 3x^{4/3} - 4x$

40. $f(x) = 3x^{1/3} - x$

41. $f(x) = 2 + (x-3)^{1/3}$

42. $f(x) = 2 + (x-3)^{4/3}$

43. $f(x) = 2 + (x-3)^{5/3}$

44. $f(x) = 2 + (x-3)^{2/3}$

45. $f(x) = x^2\sqrt{4 - x}$

46. $f(x) = x\sqrt{9 - x^2}$

47. $f(x) = (x+2)\sqrt{-x}$

48. $f(x) = \dfrac{9x}{x^2 + 9}$

49. $f(x) = (x+1)^{2/3}(x-2)^{1/3}$

50. $f(x) = \dfrac{x^3}{x^2 - 1}$

3.10
FURTHER TREATMENT OF ABSOLUTE EXTREMA AND APPLICATIONS

The extreme-value theorem (3.1.9) guarantees an absolute maximum value and an absolute minimum value for a function that is continuous on a closed interval. We now consider some functions defined on intervals for which the extreme-value theorem does not apply and which may or may not have absolute extrema.

EXAMPLE 1
Given

$$f(x) = \frac{x^2 - 27}{x - 6}$$

find the absolute extrema of f on the interval $[0, 6)$ if there are any.

SOLUTION
The function f is continuous on the interval $[0, 6)$ because the only discontinuity of f is at 6, which is not in the interval.

$$f'(x) = \frac{2x(x-6) - (x^2 - 27)}{(x-6)^2}$$

$$= \frac{x^2 - 12x + 27}{(x-6)^2}$$

$$= \frac{(x-3)(x-9)}{(x-6)^2}$$

Observe that $f'(x)$ exists for all values of x in $[0, 6)$ and $f'(x) = 0$ when x is 3 or 9; so the only critical number of f in the interval $[0, 6)$ is 3. The first-derivative test is applied to determine if f has a relative extremum at 3, and the results are summarized in Table 3.10.1.

Table 3.10.1

	$f(x)$	$f'(x)$	Conclusion
$0 \le x < 3$		$+$	f is increasing
$x = 3$	6	0	f has a relative maximum value
$3 < x < 6$		$-$	f is decreasing

Because f has a relative maximum value at 3, and f is increasing on the interval $[0, 3)$ and decreasing on the interval $(3, 6)$, then on $[0, 6)$ f has an absolute maximum value at 3, and it is $f(3)$, which is 6. Note that $\lim\limits_{x \to 6^-} f(x) = -\infty$, and conclude that there is no absolute minimum value of f on $[0, 6)$.

EXAMPLE 2
Given

$$f(x) = \frac{-x}{(x^2 + 6)^2}$$

find the absolute extrema of f on $(0, +\infty)$ if there are any.

SOLUTION
The function f is continuous for all values of x.

$$f'(x) = \frac{-1(x^2 + 6)^2 + 4x^2(x^2 + 6)}{(x^2 + 6)^4}$$

$$= \frac{-(x^2 + 6) + 4x^2}{(x^2 + 6)^3}$$

$$= \frac{3x^2 - 6}{(x^2 + 6)^3}$$

$f'(x)$ exists for all values of x. Setting $f'(x) = 0$ we obtain $x = \pm\sqrt{2}$; so $\sqrt{2}$ is the only critical number of f in $(0, +\infty)$. The first-derivative test is applied at $\sqrt{2}$, and the results are summarized in Table 3.10.2.

Table 3.10.2

	$f(x)$	$f'(x)$	Conclusion
$0 < x < \sqrt{2}$		$-$	f is decreasing
$x = \sqrt{2}$	$-\frac{1}{64}\sqrt{2}$	0	f has a relative minimum value
$\sqrt{2} < x$		$+$	f is increasing

Because f has a relative minimum value at $\sqrt{2}$ and because f is decreasing on $(0, \sqrt{2})$ and increasing on $(\sqrt{2}, +\infty)$, f has an absolute minimum value at $\sqrt{2}$ on $(0, +\infty)$. The absolute minimum value is $-\frac{1}{64}\sqrt{2}$. There is no absolute maximum value of f on $(0, +\infty)$.

For certain functions the following theorem is useful to determine if a relative extremum is an absolute extremum.

3.10.1 THEOREM   Let the function f be continuous on the interval I containing the number c. If $f(c)$ is a relative extremum of f on I and c is the only number in I for which f has a relative extremum, then $f(c)$ is an absolute extremum of f on I. Furthermore,

(i) if $f(c)$ is a relative maximum value of f on I, then $f(c)$ is an absolute maximum value of f on I;
(ii) if $f(c)$ is a relative minimum value of f on I, then $f(c)$ is an absolute minimum value of f on I.

PROOF   We prove part (i). The proof of part (ii) is similar.

Because $f(c)$ is a relative maximum value of f on I, then by Definition 3.1.1 there is an open interval J, where $J \subset I$, and where J contains c, such that

$$f(c) \geq f(x) \qquad \text{for all } x \in J$$

Because c is the only number in I for which f has a relative maximum value, it follows that

$$f(c) > f(k) \qquad \text{if } k \in J \text{ and } k \neq c \qquad (1)$$

To show that $f(c)$ is an absolute maximum value of f on I we show that if d is any number other than c in I, then $f(c) > f(d)$. We assume that

$$f(c) \leq f(d) \qquad (2)$$

and show that this assumption leads to a contradiction. Because $d \neq c$, then either $c < d$ or $d < c$. We consider the case that $c < d$ (the proof is similar if $d < c$).

Because f is continuous on I, then f is continuous on the closed interval $[c, d]$. Therefore, by the extreme-value theorem, f has an absolute minimum value on $[c, d]$. Assume this absolute minimum value occurs at e, where $c \leq e \leq d$. From inequality (1) it follows that $e \neq c$, and from inequalities (1) and (2) it follows that $e \neq d$. Consequently $c < e < d$, and hence f has a relative minimum value at e. But this statement contradicts the hypothesis that c is the only number in I for which f has a relative extremum. Thus our assumption that $f(c) \leq f(d)$ is false. Therefore $f(c) > f(d)$ if $d \in I$ and $d \neq c$, and consequently $f(c)$ is an absolute maximum value of f on I.   ■

EXAMPLE 3
Given

$$f(x) = 3x^4 - 8x^3 + 12x^2 - 12x + 3$$

find the absolute extrema of f on $(-\infty, +\infty)$, if there are any.

SOLUTION

$$f'(x) = 12x^3 - 24x^2 + 24x - 12 \qquad f''(x) = 36x^2 - 48x + 24$$

$f'(x)$ exists for all values of x. Set $f'(x) = 0$ and obtain

$$12(x^3 - 2x^2 + 2x - 1) = 0 \tag{3}$$

We use synthetic division to determine that $x - 1$ is a factor of the left member of (3) and have

$$12(x - 1)(x^2 - x + 1) = 0 \tag{4}$$

Because the equation $x^2 - x + 1 = 0$ has only imaginary roots, the only real solution to (4) is 1. Therefore $f'(1) = 0$. To determine if $f(1)$ is a relative extremum we apply the second-derivative test, the results of which are summarized in Table 3.10.3.

Table 3.10.3

	$f(x)$	$f'(x)$	$f''(x)$	Conclusion
$x = 1$	-2	0	$+$	f has a relative minimum value

The function f is continuous on $(-\infty, +\infty)$, and the one and only relative extremum of f on $(-\infty, +\infty)$ is at $x = 1$. Therefore it follows from Theorem 3.10.1(ii) that -2, the relative minimum value of f, is the absolute minimum value of f.

In Section 3.2 the applications involved finding absolute extrema of functions continuous on a closed interval, and the extreme-value theorem was used in the solutions of the problems. We now deal with applications involving absolute extrema for which the extreme-value theorem cannot be applied. The procedure is demonstrated in the following illustration by considering the problem presented in Example 7 of Section 3.7.

▶ ILLUSTRATION 1   A closed box with a square base is to have a volume of 2000 in.^{3}. The material for the top and bottom of the box is to cost \$3 per square inch, and the material for the sides is to cost \$1.50 per square inch. We wish to find the dimensions of the box so that the total cost of the material is least. We showed in Example 7 of Section 3.7 that if x in. is the length of the side of the square base and $C(x)$ dollars is the total cost of the material, then

$$C(x) = 6x^2 + \frac{12{,}000}{x} \tag{5}$$

and the domain of C is $(0, +\infty)$; furthermore C is continuous on its domain. From (5),

$$C'(x) = 12x - \frac{12{,}000}{x^2} \tag{6}$$

Observe that $C'(x)$ does not exist when $x = 0$, but 0 is not in the domain of C. Therefore the only critical numbers will be those obtained by setting $C'(x) = 0$, which gives

$$12x - \frac{12{,}000}{x^2} = 0$$

$$x^3 = 1000$$

The only real solution is 10. Thus 10 is the only critical number. To determine if $x = 10$ makes C a relative minimum we apply the second-derivative test. From (6),

$$C''(x) = 12 + \frac{24{,}000}{x^3}$$

The results of the second-derivative test are summarized in Table 3.10.4.

Table 3.10.4

	$C'(x)$	$C''(x)$	Conclusion
$x = 10$	0	+	C has a relative minimum value

Because C is continuous on its domain $(0, +\infty)$ and the one and only relative extremum of C on $(0, +\infty)$ is at $x = 10$, it follows from Theorem 3.10.1(ii) that this relative minimum value of C is the absolute minimum value. Thus the total cost of the material will be least when the side of the square base is 10 in. The depth then will be 20 in., because the area of the base will be 100 in.² and the volume is 2000 in.³. ◀

EXAMPLE 4
Find the shortest distance from the point $A(2, \frac{1}{2})$ to a point on the parabola $y = x^2$, and find the point on the parabola that is closest to A.

SOLUTION
Let z be the number of units in the distance from the point $A(2, \frac{1}{2})$ to a point $P(x, y)$ on the parabola. Refer to Fig. 3.10.1.

$$z = \sqrt{(x - 2)^2 + (y - \tfrac{1}{2})^2} \tag{7}$$

Because $P(x, y)$ is on the parabola, its coordinates satisfy the equation $y = x^2$. Substituting this value of y into (7) we obtain z as a function of x and write

$$z(x) = \sqrt{(x - 2)^2 + (x^2 - \tfrac{1}{2})^2}$$

$$z(x) = \sqrt{x^4 - 4x + \tfrac{17}{4}} \tag{8}$$

Because P can be any point on the parabola, x is any real number. We wish to find the absolute minimum value of z on $(-\infty, +\infty)$.

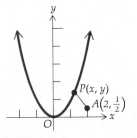

FIGURE 3.10.1

$$z'(x) = \tfrac{1}{2}(4x^3 - 4)(x^4 - 4x + \tfrac{17}{4})^{-1/2}$$

$$= \frac{2(x - 1)(x^2 + x + 1)}{\sqrt{x^4 - 4x + \tfrac{17}{4}}}$$

$z'(x)$ exists for all values of x because the denominator is $z(x)$, which is never zero. Therefore the critical numbers are obtained by setting $z'(x) = 0$. The only real solution is 1; thus 1 is the only critical number of z. Rather than computing the second derivative, which involves a fraction and radicals, it is easier to apply the first-derivative test in this problem. Table 3.10.5 summarizes the results of this test.

Table 3.10.5

	$z(x)$	$z'(x)$	Conclusion
$x < 1$		$-$	z is decreasing
$x = 1$	$\tfrac{1}{2}\sqrt{5}$	0	z has a relative minimum value
$1 < x$		$+$	z is increasing

Because z has a relative minimum value at 1, and because z is decreasing on $(-\infty, 1)$ and increasing on $(1, +\infty)$, z has an absolute minimum value of $\tfrac{1}{2}\sqrt{5}$ at 1. So the point on the parabola closest to A is $(1, 1)$.

We can show that the point $A(2, \tfrac{1}{2})$ lies on the normal line at $(1, 1)$ of the graph of the parabola. Because the slope of the tangent line at any point (x, y) of the parabola is $2x$, the slope of the tangent line at $(1, 1)$ is 2. Therefore the slope of the normal line at $(1, 1)$ is $-\tfrac{1}{2}$, which is the same as the slope of the line through $A(2, \tfrac{1}{2})$ and $(1, 1)$.

EXAMPLE 5

If a closed tin can of volume 16π in.^{3} is to be in the form of a right-circular cylinder, find the height and radius if the least amount of material is to be used in its manufacture.

SOLUTION

Let r in. be the base radius of the cylinder, h in. be the height of the cylinder, and S in.^{2} be the total surface area of the cylinder. See Fig. 3.10.2. The lateral surface area is $2\pi rh$ in.^{2}, the area of the top is πr^2 in.^{2}, and the area of the bottom is πr^2 in.^{2}. Therefore

$$S = 2\pi rh + 2\pi r^2 \qquad (9)$$

If V in.^{3} is the volume of a right-circular cylinder, then $V = \pi r^2 h$. Thus

$$16\pi = \pi r^2 h \qquad (10)$$

Solving (10) for h and substituting into (9) we obtain S as a function of r:

$$S(r) = 2\pi r\left(\frac{16}{r^2}\right) + 2\pi r^2$$

$$S(r) = \frac{32\pi}{r} + 2\pi r^2 \qquad (11)$$

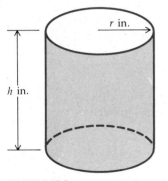

FIGURE 3.10.2

The domain of S is $(0, +\infty)$, and S is continuous on its domain. From (11),

$$S'(r) = -\frac{32\pi}{r^2} + 4\pi r \qquad (12)$$

$S'(r)$ does not exist when $r = 0$, but 0 is not in the domain of S. The only critical numbers are those obtained by setting $S'(r) = 0$, from which we have

$$4\pi r^3 = 32\pi$$

$$r^3 = 8$$

$$r = 2$$

The only critical number of S is 2. From (12),

$$S''(r) = \frac{64\pi}{r^3} + 4\pi$$

The results of applying the second-derivative test are summarized in Table 3.10.6.

Table 3.10.6

	$S'(r)$	$S''(r)$	Conclusion
$r = 2$	0	+	S has a relative minimum value

Because S is continuous on its domain $(0, +\infty)$ and the one and only relative extremum of S on $(0, +\infty)$ is at $r = 2$, it follows from Theorem 3.10.1(ii) that this relative minimum value of S is the absolute minimum value of S. When $r = 2$, we have, from (10), $h = 4$. Therefore the least amount of material will be used in the manufacture of the can when the radius is 2 in. and the height is 4 in.

In the preceding examples and in the exercises of Section 3.2, the variable for which we wished to find an absolute extremum was expressed as a function of only one variable. Sometimes this procedure is either too difficult or too laborious, or occasionally even impossible. Often the given information enables us to obtain two equations involving three variables. Instead of eliminating one of the variables, it may be more advantageous to differentiate implicitly. The following example illustrates this method. The problem is similar to the one in Example 5, but in this example the volume of the required can is not specified.

EXAMPLE 6
If a closed tin can of fixed volume is to be in the form of a right-circular cylinder, find the ratio of the height to the base radius if the least amount of material is to be used in its manufacture.

SOLUTION
We wish to find a relationship between the height and the base radius of the

right-circular cylinder in order for the total surface area to be an absolute minimum for a fixed volume. Therefore we consider the volume of the cylinder a constant.

Let V cubic units be the volume of a cylinder (a constant).

We now define the variables.

Let r units be the base radius of the cylinder; $r > 0$. Let h units be the height of the cylinder; $h > 0$. Let S square units be the total surface area of the cylinder.

We have the following equations:

$$S = 2\pi r^2 + 2\pi rh \tag{13}$$

$$V = \pi r^2 h \tag{14}$$

Because V is a constant, we could solve (14) for either r or h in terms of the other and substitute into (13), which will give S as a function of one variable. The alternative method is to consider S as a function of two variables r and h; however, r and h are not independent of each other. That is, if we choose r as the independent variable, then S depends on r; also, h depends on r.

Differentiating S and V with respect to r and bearing in mind that h is a function of r, we have

$$\frac{dS}{dr} = 4\pi r + 2\pi h + 2\pi r \frac{dh}{dr} \tag{15}$$

$$\frac{dV}{dr} = 2\pi rh + \pi r^2 \frac{dh}{dr} \tag{16}$$

Because V is a constant, $\dfrac{dV}{dr} = 0$; therefore, from (16),

$$2\pi rh + \pi r^2 \frac{dh}{dr} = 0$$

with $r \neq 0$. Divide by r and solve for $\dfrac{dh}{dr}$ to obtain

$$\frac{dh}{dr} = -\frac{2h}{r} \tag{17}$$

Substituting from (17) into (15) we obtain

$$\frac{dS}{dr} = 2\pi \left[2r + h + r\left(-\frac{2h}{r} \right) \right]$$

$$\frac{dS}{dr} = 2\pi(2r - h) \tag{18}$$

To find when S has a relative minimum value, set $\dfrac{dS}{dr} = 0$ and obtain $2r - h = 0$, which gives

$$r = \tfrac{1}{2}h$$

To determine if this relationship between r and h makes S a relative minimum we apply the second-derivative test. Then from (18),

$$\frac{d^2S}{dr^2} = 2\pi\left(2 - \frac{dh}{dr}\right) \qquad (19)$$

Substituting from (17) into (19) we get

$$\frac{d^2S}{dr^2} = 2\pi\left[2 - \left(\frac{-2h}{r}\right)\right]$$

$$= 2\pi\left(2 + \frac{2h}{r}\right)$$

The results of the second-derivative test are summarized in Table 3.10.7.

Table 3.10.7

	$\dfrac{dS}{dr}$	$\dfrac{d^2S}{dr^2}$	Conclusion
$r = \frac{1}{2}h$	0	+	S has a relative minimum value

From (13) and (14), S is a continuous function of r on $(0, +\infty)$. Because the one and only relative extremum of S on $(0, +\infty)$ is at $r = \frac{1}{2}h$, we conclude from Theorem 3.10.1(ii) that S has an absolute minimum value at $r = \frac{1}{2}h$. Therefore the total surface area of the tin can will be least for a specific volume when the ratio of the height to the base radius is 2.

Geometric problems involving absolute extrema occasionally are more easily solved by using trigonometric functions. The following example illustrates this fact.

EXAMPLE 7
A right-circular cylinder is to be inscribed in a sphere of given radius. Find the ratio of the altitude to the base radius of the cylinder having the largest lateral surface area.

SOLUTION
Refer to Fig. 3.10.3. The measure of the constant radius of the sphere is taken as a.

Let θ radians be the angle at the center of the sphere subtended by the radius of the cylinder, r units be the radius of the cylinder, h units be the altitude of the cylinder, and S square units be the lateral surface area of the cylinder. From Fig. 3.10.3,

$$r = a \sin\theta \quad \text{and} \quad h = 2a \cos\theta$$

Because $S = 2\pi rh$,

$$S = 2\pi(a\sin\theta)(2a\cos\theta)$$

$$= 2\pi a^2(2\sin\theta\cos\theta)$$

$$= 2\pi a^2 \sin 2\theta$$

FIGURE 3.10.3

Thus S is a function of θ and its domain is $(0, \frac{1}{2}\pi)$.

$$\frac{dS}{d\theta} = 4\pi a^2 \cos 2\theta \quad \text{and} \quad \frac{d^2S}{d\theta^2} = -8\pi a^2 \sin 2\theta$$

Set $\dfrac{dS}{d\theta} = 0$.

$$\cos 2\theta = 0$$

Because $0 < \theta < \frac{1}{2}\pi$,

$$\theta = \tfrac{1}{4}\pi$$

We apply the second-derivative test for relative extrema and summarize the results in Table 3.10.8.

Table 3.10.8

	$\dfrac{dS}{d\theta}$	$\dfrac{d^2S}{d\theta^2}$	Conclusion
$\theta = \frac{1}{4}\pi$	0		S has a relative maximum value

S is continuous on its domain, and because there is only one relative extremum, it follows from Theorem 3.10.1(i) that the relative maximum value of S is the absolute maximum value of S.

When $\theta = \frac{1}{4}\pi$,

$$r = a \sin \tfrac{1}{4}\pi \qquad h = 2a \cos \tfrac{1}{4}\pi$$
$$= \tfrac{1}{2}\sqrt{2}a \qquad\quad = \sqrt{2}a$$

So for the cylinder having the largest lateral surface area, $h/r = 2$.

EXERCISES 3.10

In Exercises 1 through 16, find the absolute extrema of the given function on the indicated interval if there are any.

1. $f(x) = x^2; (-3, 2]$

2. $g(x) = x^3 + 2x^2 - 4x + 1; (-3, 2)$

3. $F(x) = \dfrac{x+2}{x-2}; [-4, 4]$

4. $f(x) = \dfrac{x^2}{x+3}; [-4, -1]$

5. $g(x) = 4x^2 - 2x + 1; (-\infty, +\infty)$

6. $f(x) = x^3 - 3x + 5; (-\infty, 0)$

7. $f(x) = (x-1)^{1/3}; (-\infty, +\infty)$

8. $G(x) = (x-5)^{2/3}; (-\infty, +\infty)$

9. $f(x) = x^4 + x^3 - \frac{9}{2}x^2 + 2x + 1; (-\infty, +\infty)$

10. $h(x) = 3x^4 - 2x^3 - 12x^2 - 6x - 1; (-\infty, +\infty)$

11. $g(x) = \dfrac{x^2+7}{x-3}; [-3, 3)$

12. $f(x) = \dfrac{x^2-30}{x-4}; (-\infty, 4)$

13. $f(x) = \dfrac{x}{(x^2+4)^{3/2}}; [0, +\infty)$

14. $g(x) = \sqrt{x^4 - 4x + 7}; (-\infty, +\infty)$

15. $f(x) = \tan x + 2 \sec x; (-\frac{1}{2}\pi, \frac{1}{2}\pi)$

16. $h(x) = x + \cot x; [\frac{1}{6}\pi, \frac{2}{3}\pi]$

17. Refer to Exercise 23 in Exercises 3.7. A rectangular field having an area of 2700 m² is to be enclosed by a fence, and an additional fence is to be used to divide the field down the middle. The cost of the fence down the middle is $12 per running meter, and the fence along the sides costs $18 per running meter. Find the dimensions of the field such that the cost of the fencing will be the least.

18. Refer to Exercise 24 in Exercises 3.7. A rectangular open tank is to have a square base, and its volume is to be 125 m³. The cost per square meter for the bottom is $24 and for the sides is $12. Find the dimensions of the tank for the cost of the material to be the least.

19. Refer to Exercise 25 in Exercises 3.7. A page of print is to contain 24 in.² of printed region, a margin of $1\frac{1}{2}$ in. at the top and bottom, and a margin of 1 in. at the sides. What are the dimensions of the smallest page that will fill these requirements?

20. Refer to Exercise 26 in Exercises 3.7. A one-story building having a rectangular floor space of 13,200 ft² is to be constructed where a walkway 22 ft wide is required in the front and back and a walkway 15 ft wide is required on each side. Find the dimensions of the lot having the least area on which this building can be located.

21. A box manufacturer is to produce a closed box of volume 288 in.³, where the base is a rectangle having a length three times its width. Find the dimensions of the box constructed from the least amount of material.

22. Solve Exercise 21 if the box is to have an open top.

23. Find an equation of the tangent line to the curve $y = x^3 - 3x^2 + 5x$ that has the least slope.

24. Find the point on the curve $y^2 - x^2 = 1$ that is closest to the point $(2, 0)$.

25. Find the point on the line $3x + y = 6$ that is closest to the origin.

26. A cardboard poster containing 32 in.² of printed region is to have a margin of 2 in. at the top and bottom and $\frac{4}{3}$ in. at the sides. Determine the dimensions of the smallest piece of cardboard that can be used to make the poster.

27. For the tin can of Example 5, suppose that the cost of material for the top and bottom is twice as much as it is for the sides. Find the height and base radius for the cost of the material to be the least.

28. Solve Example 5 if the tin can is open instead of closed.

29. In a particular community, a certain epidemic spreads in such a way that x months after the start of the epidemic, P percent of the population is infected, where

$$P = \frac{30x^2}{(1 + x^2)^2}$$

In how many months will the most people be infected, and what percent of the population is this?

30. A direct current generator has an electromotive force of E volts and an internal resistance of r ohms, where E and r are constants. If R ohms is the external resistance, the total resistance is $(r + R)$ ohms, and if P watts is the power, then

$$P = \frac{E^2 R}{(r + R)^2}$$

Show that the most power is consumed when the external resistance is equal to the internal resistance.

31. A Norman window consists of a rectangle surmounted by a semicircle. If the perimeter of a Norman window is to be 32 ft, determine what should be the radius of the semicircle and the height of the rectangle such that the window will admit the most light.

32. Solve Exercise 31 if the window is to be such that the semicircle transmits only half as much light per square foot of area as the rectangle.

33. It is determined that if salaries are excluded, the number of dollars in the cost per kilometer for operating a truck is $8 + \frac{1}{300}x$, where x km/hr is the speed of the truck. If the combined salary of the driver and the driver's assistant is $27 per hour, what should be the average speed of the truck for the cost per kilometer to be the least?

34. The number of dollars in the cost per hour of fuel for a cargo ship is $\frac{1}{50}v^3$, where v knots (nautical miles per hour) is the speed of the ship. If there are additional costs of $400 per hour, what should be the average speed of the ship for the cost per nautical mile to be the least?

35. An automobile traveling at the rate of 30 ft/sec is approaching an intersection. When the automobile is 120 ft from the intersection, a truck traveling at the rate of 40 ft/sec crosses the intersection. The automobile and the truck are on roads that are at right angles to each other. How long after the truck leaves the intersection are the two vehicles closest?

36. Two airplanes A and B are flying horizontally at the same altitude. The position of plane B is southwest of plane A and 20 km to the west and 20 km to the south of A. If plane A is flying due west at 16 km/min and plane B is flying due north at $\frac{64}{3}$ km/min, (a) in how many seconds will they be closest, and (b) what will be their closest distance?

37. A steel girder 27 ft long is moved horizontally along a passageway 8 ft wide and into a corridor at right angles to the passageway. How wide must the corridor be for the girder to go around the corner? Neglect the horizontal width of the girder.

38. If two corridors at right angles to each other are 10 ft and 15 ft wide, respectively, what is the length of the longest steel girder that can be moved horizontally around the corner? Neglect the horizontal width of the girder.

39. A funnel of specific volume is to be in the shape of a right-circular cone. Find the ratio of the height to the base radius if the least amount of material is to be used in its manufacture.

40. A right-circular cone is to be inscribed in a sphere of given radius. Find the ratio of the altitude to the base radius of the cone of largest possible volume.

41. A right-circular cone is to be circumscribed about a sphere of given radius. Find the ratio of the altitude to the base radius of the cone of least possible volume.

42. Prove by the method of this section that the shortest distance from the point $P_1(x_1, y_1)$ to the line l having the equation $Ax + By + C = 0$ is $|Ax_1 + By_1 + C|/\sqrt{A^2 + B^2}$. (*Hint:* If s is the number of units from P_1 to a point $P(x, y)$ on l, then s will be an absolute minimum when s^2 is an absolute minimum.)

43. Find the altitude of the right-circular cone of largest possible volume that can be inscribed in a sphere of radius a units. Let 2θ be the radian measure of the vertical angle of the cone.

44. The cross section of a trough has the shape of an inverted isosceles triangle. If the lengths of the equal sides are 15 in., find the size of the vertex angle that will give maximum capacity for the trough.

3.11 (*Supplementary*)
APPLICATIONS OF DIFFERENTIATION IN ECONOMICS AND BUSINESS

In economics the variation of one quantity with respect to another may be described by either an *average* concept or a *marginal* concept. The average concept expresses the variation of one quantity over a specified range of values of a second quantity, whereas the marginal concept is the instantaneous change in the first quantity that results from a very small unit change in the second quantity. We begin our examples in economics with the definitions of average cost and marginal cost. To define a marginal concept precisely we use the notion of a limit, and this leads to the derivative.

Suppose $C(x)$ dollars is the total cost of producing x units of a commodity. The function C is called a **total cost function**. In normal circumstances x and $C(x)$ are positive. Since x represents the number of units of a commodity, x is usually a nonnegative integer. However, to apply the calculus we assume that x is a nonnegative real number to give the continuity requirements for the function C.

The **average cost** of producing each unit of a commodity is obtained by dividing the total cost by the number of units produced. If $Q(x)$ dollars is the average cost,

$$Q(x) = \frac{C(x)}{x}$$

and Q is called an **average cost function**.

Now suppose that the number of units in a particular output is x_1, and this is changed by Δx. Then the change in the total cost is given by $C(x_1 + \Delta x) - C(x_1)$, and the average change in the total cost with respect to the change in the number of units produced is given by

$$\frac{C(x_1 + \Delta x) - C(x_1)}{\Delta x} \qquad (1)$$

Economists use the term *marginal cost* for the limit of the quotient (1) as Δx approaches zero, provided the limit exists. This limit, being the derivative of C at x_1, gives the following definition.

3.11.1 DEFINITION   If $C(x)$ dollars is the total cost of producing x units of a commodity, then the **marginal cost**, when $x = x_1$, is given by $C'(x_1)$, if it exists. The function C' is called the **marginal cost function**.

In Definition 3.11.1, $C'(x_1)$ may be interpreted as the rate of change of the total cost when x_1 units are produced.

▶ ILLUSTRATION 1   Suppose that $C(x)$ dollars is the total cost of manufacturing x toys, and

$$C(x) = 110 + 4x + 0.02x^2$$

(a) The marginal cost function is C', and

$$C'(x) = 4 + 0.04x$$

(b) The marginal cost when $x = 50$ is $C'(50)$, and

$$C'(50) = 4 + 0.04(50)$$
$$= 6$$

Therefore the rate of change of the total cost, when 50 toys are manufactured, is $6 per toy.

(c) The number of dollars in the actual cost of manufacturing the fifty-first toy is $C(51) - C(50)$, and

$$C(51) - C(50) = [110 + 4(51) + 0.02(51)^2] - [110 + 4(50) + 0.02(50)^2]$$
$$= 366.02 - 360$$
$$= 6.02$$

Note that the answers in (b) and (c) differ by 0.02. This discrepancy occurs because the marginal cost is the instantaneous rate of change of $C(x)$ with respect to a unit change in x. Hence $C'(50)$ is the approximate number of dollars in the cost of producing the fifty-first toy.   ◀

Observe that the computation of $C'(50)$ in Illustration 1 is simpler than computing $C(51) - C(50)$. Economists frequently approximate the cost of producing one additional unit by using the marginal cost function. Specifically, $C'(k)$ dollars is the approximate cost of the $(k + 1)$st unit after the first k units have been produced.

Certain economic restrictions are imposed on the total cost function C in normal situations. They are as follows:

1. $C(0) \geq 0$. That is, when nothing is produced, the total cost must be positive or zero. $C(0)$ is called the **overhead cost** of production.
2. $C'(x) > 0$ for all x. That is, the total cost must increase as the number of units produced increases.
3. $C''(x) \geq 0$ for x greater than some positive number N. When the number of units of the commodity produced is large, the marginal cost will eventually be increasing or zero. So unless $C''(x) = 0$, the graph of the total cost function is concave upward for $x > N$. However, the marginal cost

may decrease for some values of x; hence, for these values of x, $C''(x) < 0$, and therefore the graph of the total cost function will be concave downward for these values of x (see Example 2 for this situation).

▶ ILLUSTRATION 2   Consider a linear total cost function

$$C(x) = mx + b$$

Observe that b represents the overhead cost. The marginal cost is given by $C'(x) = m$. If Q is the average cost function,

$$Q(x) = m + \frac{b}{x}$$

$$Q'(x) = -\frac{b}{x^2}$$

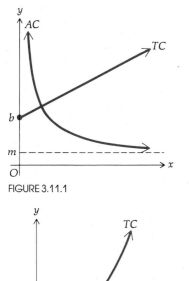

FIGURE 3.11.1

Refer to Fig. 3.11.1 for sketches of the total cost curve (labeled TC) and the average cost curve (labeled AC). The total cost curve is a segment of a straight line in the first quadrant having slope m and y intercept b. The average cost curve is a branch of an equilateral hyperbola in the first quadrant that has the line $y = m$ as horizontal asymptote. Because $Q'(x)$ is always negative, the average cost function is always decreasing, and as x increases, the value of $Q(x)$ gets closer and closer to m. The concept of the average cost approaching a constant frequently occurs in large manufacturing situations where the number of units produced is very great.   ◀

▶ ILLUSTRATION 3   Consider a quadratic total cost function

$$C(x) = ax^2 + bx + c$$

where a and c are positive. Here c dollars is the overhead cost. The total cost curve is a parabola opening upward. Because $C'(x) = 2ax + b$, a critical number of C is $-b/(2a)$, which gives the vertex of the parabola. We distinguish two cases: $b \geq 0$ and $b < 0$.

Case 1: $b \geq 0$. $-b/(2a)$ is then either negative or zero, and the vertex of the parabola is either to the left of the y axis or on the y axis. Hence the domain of C is the set of all nonnegative numbers. A sketch of TC for which $b > 0$ is shown in Fig. 3.11.2.

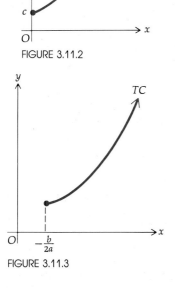

FIGURE 3.11.2

FIGURE 3.11.3

Case 2: $b < 0$. $-b/(2a)$ is positive; so the vertex of the parabola is to the right of the y axis, and the domain of C is restricted to numbers in the interval $[-b/(2a), +\infty)$. A sketch of TC for which $b < 0$ is shown in Fig. 3.11.3.   ◀

EXAMPLE 1

Suppose that $100C(x)$ dollars is the total cost of producing $100x$ units of a commodity, and

$$C(x) = \tfrac{1}{2}x^2 - 2x + 5$$

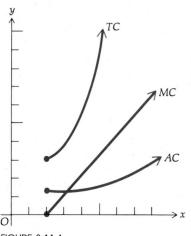

FIGURE 3.11.4

Find the function giving (a) the average cost and (b) the marginal cost. (c) Find the absolute minimum average unit cost. (d) Draw sketches of the total cost curve, average cost curve, and marginal cost curve on the same set of axes.

SOLUTION

(a) If Q is the average cost function,

$$Q(x) = \frac{C(x)}{x}$$

$$= \frac{1}{2}x - 2 + \frac{5}{x}$$

(b) The marginal cost function is C', and

$$C'(x) = x - 2$$

(c) We compute $Q'(x)$ and $Q''(x)$.

$$Q'(x) = \frac{1}{2} - \frac{5}{x^2} \qquad Q''(x) = \frac{10}{x^3}$$

From $Q'(x) = 0$, $\sqrt{10}$ is a critical number of Q, and $Q(\sqrt{10}) = \sqrt{10} - 2$; thus $Q(\sqrt{10}) \approx 1.16$. Because $Q''(\sqrt{10}) > 0$, Q has a relative minimum value of approximately 1.16 at $x = \sqrt{10}$. From the equation defining $Q(x)$ we see that Q is continuous on $(0, +\infty)$. Because the only relative extremum of Q on $(0, +\infty)$ is at $x = \sqrt{10}$, it follows from Theorem 3.7.2(ii) that Q has an absolute minimum value there. When $x = \sqrt{10}$, $100x \approx 316$; so we conclude that the absolute minimum average unit cost is \$1.16 when 316 units are produced.

The sketches of the curves, TC, AC, and MC (the marginal cost curve) are shown in Fig. 3.11.4.

In Fig. 3.11.4, observe that the lowest point on curve AC occurs at the point of intersection of the curves AC and MC, which is the point where the average cost and the marginal cost are equal. This is true in general, and it follows from the fact that $Q'(x) = 0$ here. Because $Q(x) = C(x)/x$, then

$$Q'(x) = \frac{xC'(x) - C(x)}{x^2}$$

Thus $Q'(x) = 0$ when $xC'(x) - C(x) = 0$ or, equivalently, when

$$C'(x) = \frac{C(x)}{x}$$

You should note the economic significance that when the marginal cost and the average cost are equal, the commodity is being produced at the very lowest average unit cost.

The following example involving a cubic cost function illustrates the case in which the concavity of the graph of the total cost function changes.

EXAMPLE 2

Draw a sketch of the graph of the total cost function C for which

$$C(x) = x^3 - 6x^2 + 13x + 1$$

Determine where the graph is concave upward and where it is concave downward. Find any points of inflection and an equation of any inflectional tangent. Draw a segment of the inflectional tangent.

SOLUTION

$$C'(x) = 3x^2 - 12x + 13 \qquad C''(x) = 6x - 12$$

$C'(x)$ can be written as $3(x - 2)^2 + 1$. Hence $C'(x) > 0$ for all x. $C''(x) = 0$ when $x = 2$. To determine the concavity of the graph for the intervals $(0, 2)$ and $(2, +\infty)$ and if the graph has a point of inflection at $x = 2$, we use the results summarized in Table 3.11.1.

Table 3.11.1

	$C(x)$	$C'(x)$	$C''(x)$	Conclusion
$0 < x < 2$		$+$	$-$	C is increasing; graph is concave downward
$x = 2$	11	1	0	graph has a point of inflection
$2 < x$		$+$	$+$	C is increasing; graph is concave upward

An equation of the inflectional tangent is

$$y - 11 = 1(x - 2)$$

$$x - y + 9 = 0$$

A sketch of the graph of the total cost function together with a segment of the inflectional tangent is shown in Fig. 3.11.5.

FIGURE 3.11.5

Consider circumstances affecting a manufacturer in which the only variables are the price and the quantity of the commodity demanded. Let x units of the commodity be demanded when p dollars is the price of one unit.

Upon reflection, it should seem reasonable that the amount of the commodity demanded in the marketplace by consumers will depend on the price of the commodity. As the price falls, consumers generally demand more of the commodity. Should the price rise, the opposite will occur. Consumers will demand less.

An equation giving the relationship between the amount, given by x, of a commodity demanded and the price, given by p, is called a **demand equation**. Such an equation is arrived at by the application of statistical methods to economic data. If the equation is solved for p, we have

$$p = f(x) \tag{2}$$

The function f in this equation is called the **price function**, and $f(x)$ dollars is the price of one unit of the commodity when x units are demanded. In normal economic situations, the domain of the price function consists, as you would expect, of nonnegative numbers.

The graph of the demand equation is called the **demand curve**. When drawing a sketch of the demand curve, it is customary in economics to use the vertical axis to represent the price and the horizontal axis to represent the demand. Because the given demand equation may apply for only certain values of x and p, it is often necessary to restrict them to closed intervals; that is, $x \in [0, a]$ and $p \in [0, b]$. Even though in actual practice quantities and prices usually take on rational values only, we allow x and p to be any real numbers within these closed intervals.

▶ ILLUSTRATION 4   Consider the demand equation

$$p^2 + 2x - 16 = 0 \qquad (3)$$

Because in normal economic situations the variables x and p will be nonnegative, when (3) is solved for p we reject the negative values of p and get

$$p = \sqrt{16 - 2x} \qquad (4)$$

FIGURE 3.11.6

which is of the form of (2). Hence the price function for the demand equation (3) is the function f for which $f(x) = \sqrt{16 - 2x}$. A sketch of the demand curve is shown in Fig. 3.11.6. The graph is restricted to the first quadrant because of the requirement that x and p be nonnegative. From (4) we see that $p \le 4$ and $16 - 2x \ge 0$ or, equivalently, $x \le 8$. Therefore $x \in [0, 8]$ and $p \in [0, 4]$.

◀

In addition to the restriction that x and p be nonnegative under normal circumstances, we impose the condition that as the price per unit decreases, the demand for a commodity increases, and as the price per unit increases, the demand for a commodity decreases; that is, if p_1 dollars is the price of x_1 units of a commodity and p_2 dollars is the price per unit of x_2 units, then $x_2 > x_1$ if and only if $p_2 < p_1$. This condition, in turn, reflects nothing more than "common economic sense" and is illustrated in Fig. 3.11.7.

Another function important in economics is the **total revenue function**, denoted by R, and

$$R(x) = px$$

FIGURE 3.11.7

Because p and x are nonnegative under normal circumstances, so is $R(x)$. When $x \ne 0$, from the above equation

$$\frac{R(x)}{x} = p$$

which shows that the revenue per unit (the average revenue) and the price per unit are equal.

3.11.2 DEFINITION   If $R(x)$ dollars is the total revenue obtained when x units of a commodity are demanded, then the **marginal revenue**, when $x = x_1$, is given by $R'(x_1)$, if it exists. The function R' is called the **marginal revenue function**.

$R'(x_1)$ may be positive, negative, or zero, and it may be interpreted as the rate of change of the total revenue when x_1 units are demanded. Just as

$C'(k)$ dollars is the approximate cost of the $(k + 1)$st unit after the first k units have been produced, $R'(k)$ dollars is the approximate revenue from the sale of the $(k + 1)$st unit after the first k units have been sold.

▶ ILLUSTRATION 5   Suppose that $R(x)$ dollars is the total revenue received from the sale of x tables, and

$$R(x) = 300x - \frac{x^2}{2}$$

(a) The marginal revenue function is R', and

$$R'(x) = 300 - x$$

(b) The marginal revenue when $x = 40$ is given by $R'(40)$, and

$$R'(40) = 300 - 40$$
$$= 260$$

Thus the rate of change of the total revenue when 40 tables are sold is $260 per table.

(c) The number of dollars in the actual revenue from the sale of the forty-first table is $R(41) - R(40)$, and

$$R(41) - R(40) = \left[300(41) - \frac{(41)^2}{2} \right] - \left[300(40) - \frac{(40)^2}{2} \right]$$
$$= [12,300 - 840.50] - [12,000 - 800]$$
$$= 11,459.50 - 11,200$$
$$= 259.50$$

Hence the actual revenue from the sale of the forty-first table is $259.50. In part (b) we obtained $R'(40) = 260$, and $260 is an approximation of the revenue received from the sale of the forty-first table.   ◀

The graphs of the total revenue function and the marginal revenue function are called the **total revenue curve** (labeled TR) and the **marginal revenue curve** (labeled MR), respectively.

EXAMPLE 3
The demand equation for a particular commodity is

$$p^2 + x - 12 = 0$$

Find the total revenue function and the marginal revenue function. Draw sketches of the demand curve, total revenue curve, and marginal revenue curve on the same set of axes.

SOLUTION
If the demand equation is solved for p, we find $p = \pm\sqrt{12 - x}$. Because $R(x) = px$ and $p \geq 0$,

Table 3.11.2

x	p	$R(x)$	$R'(x)$
0	$\sqrt{12}$	0	$\sqrt{12}$
3	3	9	$\frac{5}{2}$
8	2	16	0
11	1	11	$-\frac{9}{2}$
12	0	0	does not exist

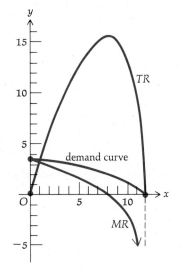

FIGURE 3.11.8

$$R(x) = x\sqrt{12 - x}$$

and

$$R'(x) = \sqrt{12 - x} - \frac{x}{2\sqrt{12 - x}}$$

$$= \frac{24 - 3x}{2\sqrt{12 - x}}$$

Setting $R'(x) = 0$ we obtain $x = 8$. From the information in Table 3.11.2, the required sketches are drawn as shown in Fig. 3.11.8.

The total profit earned by a business is defined to be the difference between the total revenue and the total cost. That is, if $P(x)$ dollars is the total profit obtained by producing and selling x units of a commodity, then

$$P(x) = R(x) - C(x) \tag{5}$$

where $R(x)$ dollars is the total revenue and $C(x)$ dollars is the total cost. The function P is called the **total profit function**. From (5) we obtain

$$P'(x) = R'(x) - C'(x) \tag{6}$$

The function P' is called the **marginal profit function**, and $P'(k)$ is the approximate profit realized from the $(k + 1)$st unit after k units have been produced and sold. From (6), the marginal profit is the marginal revenue minus the marginal cost.

In Fig. 3.11.9(a) there are sketches of the graphs of a total cost function and a total revenue function of a particular company, and in Fig. 3.11.9(b) there is the sketch of the graph of the company's corresponding total profit function (denoted by TP). Observe in Fig. 3.11.9(a) that when curve TC is above curve TR (when the cost is greater than the revenue), then in Fig. 3.11.9(b) curve TP is below the x axis (the profit is negative; i.e., the company shows a loss); this occurs when $0 \le x < x_2$ and when $x > x_3$. When curve TC is below curve TR (when the cost is less than the revenue), curve TP is above the x axis (the profit is positive; i.e., the company shows a profit); this occurs when $x_2 < x < x_3$. At the points of intersection of curves TC and TR (when the cost equals the revenue), curve TP intersects the x axis (the profit is zero; i.e., the company is breaking even); this occurs at x_2 and x_3.

The company is showing a profit when its output is between x_2 and x_3. Let us determine what level of production is necessary to obtain the maximum profit. In Fig. 3.11.9(a) the vertical distance between curves TR and TC for a particular value of x is $P(x)$, which gives the total profit corresponding to that value of x. When this vertical distance is largest, $P(x)$ has its maximum value. Observe from Fig. 3.11.9(a) that the distance AB is the largest vertical distance between the two curves in the interval $[x_2, x_3]$, and this occurs at the critical number x_1 of the function P. From (6) we see that $P'(x) = 0$ if and only if $R'(x) = C'(x)$. Thus *for maximum profit the marginal revenue must equal the marginal cost*. Note in Fig. 3.11.9(a) that at points B and A on the two curves the tangent lines are parallel, and hence $R'(x_1) = C'(x_1)$.

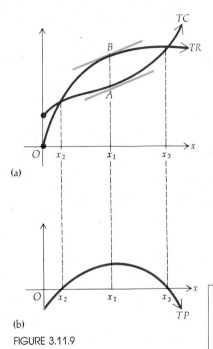

(a)

(b)

FIGURE 3.11.9

From (6),

$$P''(x) = R''(x) - C''(x) \qquad (7)$$

Because P will have a relative maximum function value at a number x for which $P'(x) = 0$ and $P''(x) < 0$, we can conclude from (6) and (7) that this will occur for a value of x for which $R'(x) = C'(x)$ and $R''(x) < C''(x)$. Therefore the profit function will have a relative maximum function value when the marginal revenue equals the marginal cost and the slope of the graph of the marginal revenue function is less than the slope of the graph of the marginal cost function. These concepts are illustrated in the following example, in which we consider *perfect competition*. When a company is operating under **perfect competition**, there is a great number of small firms and so any one firm cannot affect price by increasing production. Therefore, under perfect competition the price of a commodity is constant, and the company can sell as much as it wishes to sell at this constant price. The demand curve in such a case is horizontal; it is a line parallel to the x axis.

EXAMPLE 4

Under perfect competition a firm can sell at a price of $100 per unit all of a particular commodity it produces. Let $C(x)$ dollars be the total cost of each day's production when x units are produced, and

$$C(x) = x^2 + 20x + 700 \qquad (8)$$

Find the number of units that should be produced daily for the firm to have the greatest daily total profit.

SOLUTION

If $R(x)$ dollars is the total revenue from the sale of x units, then

$$R(x) = 100x \qquad (9)$$

If P is the total profit function, then

$$P(x) = R(x) - C(x)$$
$$P(x) = 100x - (x^2 + 20x + 700)$$
$$P(x) = -x^2 + 80x - 700 \qquad (10)$$

Observe that x is in $[0, +\infty)$.

From (8) and (9),

$$C'(x) = 2x + 20 \quad \text{and} \quad C''(x) = 2$$
$$R'(x) = 100 \quad \text{and} \quad R''(x) = 0$$

Equating $C'(x)$ and $R'(x)$ we get

$$2x + 20 = 100$$
$$x = 40$$

Because $R''(x) < C''(x)$ for all x, $P''(x) = R''(x) - C''(x) < 0$. Therefore $P(40)$

is a relative maximum value. It is also an absolute maximum value, because P is continuous on $[0, +\infty)$ and $P(40)$ is the only extremum of P for x in $[0, +\infty)$.

Therefore the daily total profit is greatest when 40 units are produced for the day.

An alternate method of solution is to find $P'(x)$ and $P''(x)$ from (10) and proceed as customary in extremum problems.

For a given demand equation, the amount demanded by the consumer depends only upon the price of the commodity. Under a **monopoly**, which means that there is only one producer of a certain commodity, price and hence demand can be controlled by regulating the quantity of the commodity produced. Generally the price decreases as the producer increases the output. The producer under a monopoly is called a **monopolist**. The monopolist wishes to control the quantity produced and hence the price per unit (determined from the demand equation) so that the profit will be as large as possible.

EXAMPLE 5

Suppose that under a monopoly the demand equation for a certain commodity is

$$p = 4 - 0.0002x$$

where x units are produced each day and p dollars is the price of each unit. The number of dollars in the total cost of producing x units is $600 + 3x$. If the daily profit is to be as large as possible, find the number of units that the monopolist should produce each day, the price of each unit, and the daily profit.

SOLUTION

Because the demand equation is $p = 4 - 0.0002x$ and x and p must be nonnegative, it follows that x is in the closed interval $[0, 20{,}000]$. Because $R(x) = xp$, we have

$$R(x) = 4x - 0.0002x^2 \qquad x \in [0, 20{,}000] \tag{11}$$

The total cost function is given by

$$C(x) = 600 + 3x \tag{12}$$

If $P(x)$ dollars is the total profit, $P(x) = R(x) - C(x)$, and so

$$P(x) = x - 0.0002x^2 - 600 \qquad x \in [0, 20{,}000]$$

From (11) and (12),

$$R'(x) = 4 - 0.0004x \quad \text{and} \quad R''(x) = -0.0004$$

$$C'(x) = 3 \quad \text{and} \quad C''(x) = 0$$

Equating $R'(x)$ and $C'(x)$ we get

$$4 - 0.0004x = 3$$

$$x = 2500$$

Because $R''(x)$ is always less than $C''(x)$, $P(2500)$ is an absolute maximum value (observe that $P(x)$ is negative for $x = 0$ and $x = 20{,}000$). $P(2500) = 650$, and when $x = 2500$, $p = 3.50$.

Therefore, to have the greatest daily profit, 2500 units should be produced each day to be sold at $3.50 each for a total profit of $650.

EXAMPLE 6

Solve Example 5 if a tax of $0.20 is levied by the government on the monopolist for each unit produced.

SOLUTION

With the added tax, the total cost function is now given by

$$C(x) = (600 + 3x) + 0.20x$$

and so the marginal cost function is given by

$$C'(x) = 3.20$$

Equating $R'(x)$ and $C'(x)$ we get

$$4 - 0.0004x = 3.20$$

$$x = 2000$$

As in Example 5, it follows that when $x = 2000$, P has an absolute maximum value on $[0, 20{,}000]$. From $P(x) = 0.8x - 0.0002x^2 - 600$, $P(2000) = 200$. From the demand equation $p = 4 - 0.0002x$ we obtain $p = 3.60$ when $x = 2000$.

Therefore, if the tax of $0.20 per unit is levied, the monopolist should produce only 2000 units each day, and they should be sold at $3.60 per unit to attain a maximum daily total profit of $200.

It is interesting to note that in comparing the results of Examples 5 and 6, the entire $0.20 increase should not be passed on to the consumer to achieve the greatest daily profit. That is, it is most profitable to raise the unit price by only $0.10. The economic significance of this result is that consumers are sensitive to price changes, which prohibits the monopolist from passing on the tax completely to the consumer.

A further note of interest in Examples 5 and 6 is how the overhead cost of a company does not affect the determination of the number of units to be produced or the unit price such that maximum profit is obtained. Regardless of whether anything is produced, the overhead cost must be met. In Examples 5 and 6, because $C(x)$ is $600 + 3x$ and $600 + 3.2x$, respectively, the overhead cost is $600. If the 600 in the expression for $C(x)$ is replaced by any constant k, $C'(x)$ is not affected; hence the value of x for which the marginal cost equals the marginal revenue is not affected by any such change. Of course, a change in overhead cost affects the unit cost and hence the actual profit; however, if a company is to have the greatest profit possible, a change in its overhead cost will not affect the number of units to be produced nor the price per unit.

An important business application of finding absolute extrema is **inventory control**, where a company is concerned with minimizing the costs of obtaining and storing inventory.

In addition to the costs of producing each item, a manufacturing company has two other costs, called the *setup cost* and the *carrying cost*, which is often a cost of storage. The **setup cost** is the expense of preparing a production run, and the **carrying cost** is the amount incurred when the produced items are stored until they are sold. We make two assumptions:

1. Production is instantaneous, in which case the inventory will become zero before there is a new production run.
2. No shortages are permitted, which prohibits a negative inventory.

We assume that if there is a uniform demand when a production run consists of x units, the average inventory is $\frac{1}{2}x$ units. This assumption should seem reasonable, and it can be proved using the definite integral, as shown in Section 4.10 (see Exercise 34 in Exercises 4.10). Suppose, for example, that there is a uniform monthly demand of 600 units of a particular commodity. There could be one production run of 600 units at the beginning of the month, and these items would be stored and sold during the month. Because of uniform demand, the average inventory during the month would be 300 units. If there were two production runs—say, 300 units produced at the beginning of the month and another 300 units produced after 15 days—then the average monthly inventory would be reduced to 150 units, and if 200 units were produced each 10 days, this would be further reduced to 100 units. As the number of production runs increases, the average monthly inventory decreases, and so as the total setup cost increases, the total carrying cost decreases. We wish to minimize the sum of these two costs.

EXAMPLE 7

A company has a uniform demand of 6000 items per month. The setup cost for each production run is $60, and the monthly storage cost for each item is 50 cents. If the production is instantaneous and shortages are not permitted, how many items should be produced in each run to minimize the total monthly cost of obtaining and storing inventory?

SOLUTION

Let x be the number of items produced in each run; $0 < x \leq 6000$. Let $C(x)$ be the number of dollars in the total monthly cost of obtaining and storing inventory.

Because 6000 items are to be produced each month, the number of runs per month is $6000/x$. The cost of preparing each run is $60, and hence the number of dollars in the total setup cost is

$$60\left(\frac{6000}{x}\right)$$

Because we are assuming a uniform demand, and since production is instantaneous, the average number of items on inventory is $x/2$. The monthly storage cost is 50 cents per item, and therefore the number of dollars in the total carrying cost is

$$\frac{1}{2} \cdot \frac{x}{2}$$

Hence

$$C(x) = 60 \left(\frac{6000}{x} \right) + \frac{1}{2} \cdot \frac{x}{2}$$

By the definition of x, it is a positive integer. However, we consider the function C to have as its domain the set of all real numbers in the interval $(0, 6000]$, and if necessary we shall adjust the solution to the nearest positive integer. Hence C is continuous on $(0, 6000]$.

$$C'(x) = -\frac{360{,}000}{x^2} + \frac{1}{4} \qquad C''(x) = \frac{720{,}000}{x^3}$$

$C'(x)$ exists for all x in the interval $(0, 6000]$. Hence the only critical numbers of C are obtained by setting $C'(x) = 0$. We get

$$x^2 = 1{,}440{,}000$$

from which we obtain the critical number 1200 (-1200 is rejected since it is not in the interval $(0, 6000]$). The results of applying the second-derivative test at 1200 are summarized in Table 3.11.3.

Table 3.11.3

	$C'(x)$	$C''(x)$	Conclusion
$x = 1200$	0	$+$	C has a relative minimum value

The only extremum of C on the interval $(0, 6000]$ occurs at 1200, and therefore the relative minimum value of C is an absolute minimum value. We conclude, then, that to minimize the total monthly cost of obtaining and storing inventory, 1200 items should be produced in each run, and the number of runs per month should be $6000 \div 1200 = 5$.

EXERCISES 3.11

1. The number of dollars in the total cost of manufacturing x watches in a certain plant is given by $C(x) = 1500 + 3x + x^2$. Find (a) the marginal cost function; (b) the marginal cost when $x = 40$; (c) the actual cost of manufacturing the forty-first watch.

2. If $C(x)$ dollars is the total cost of manufacturing x paperweights and

$$C(x) = 200 + \frac{50}{x} + \frac{x^2}{5}$$

find (a) the marginal cost function; (b) the marginal cost when $x = 10$; (c) the actual cost of manufacturing the eleventh paperweight.

3. Suppose that a liquid is produced by a certain chemical process and that the total cost function C is given by $C(x) = 6 + 4\sqrt{x}$, where $C(x)$ dollars is the total cost of producing x liters of the liquid. Find (a) the marginal cost when 16 liters are produced and (b) the number of liters produced when the marginal cost is $0.40 per liter.

4. The number of dollars in the total cost of producing x units of a certain commodity is given by $C(x) = 40 + 3x + 9\sqrt{2x}$. Find (a) the marginal cost when 50 units are produced and (b) the number of units produced when the marginal cost is \$4.50.

5. Let $C(x)$ dollars be the total cost of producing x units of a commodity, where $C(x) = x^2 + 4x + 8$. Find (a) the average cost function and (b) the marginal cost function. (c) Find the absolute minimum average unit cost. (d) Draw sketches of the total cost, average cost, and marginal cost curves on the same set of axes. Observe that the average cost and marginal cost are equal when the average cost has its least value.

6. Follow the instructions of Exercise 5 if $C(x) = 3x^2 + x + 3$.

7. If $C(x)$ dollars is the total cost of producing x units of a commodity and $C(x) = 3x^2 - 6x + 4$, find (a) the average cost function and (b) the marginal cost function. (c) What is the range of C? (d) Find the absolute minimum average unit cost. (e) Draw sketches of the total cost, average cost, and marginal cost curves on the same set of axes. Observe that the average cost and marginal cost are equal when the average cost has its least value.

8. If $C(x)$ dollars is the total cost of producing x units of a commodity and $C(x) = 2x^2 - 8x + 18$, find (a) the domain and range of C, (b) the average cost function, (c) the absolute minimum average unit cost, and (d) the marginal cost function. (e) Draw sketches of the total cost, average cost, and marginal cost curves on the same set of axes.

9. The total cost function C is given by $C(x) = \frac{1}{3}x^3 - 2x^2 + 5x + 2$. (a) Determine the range of C. (b) Find the marginal cost function. (c) Find the interval on which the marginal cost function is decreasing and the interval on which it is increasing. (d) Draw a sketch of the graph of the total cost function; determine where the graph is concave upward and where it is concave downward, and find the points of inflection and an equation of any inflectional tangent.

10. The number of hundreds of dollars in the total cost of producing $100x$ radios per day in a certain factory is given by $C(x) = 4x + 5$. Find (a) the average cost function and (b) the marginal cost function. (c) Show that there is no absolute minimum average unit cost. (d) What is the smallest number of radios that the factory must produce in a day so that the average cost per radio is less than \$7? (e) Draw sketches of the total cost, average cost, and marginal cost curves on the same set of axes.

11. The overhead cost of a manufacturer of children's toys is \$400 per week, and other costs amount to \$3 for each toy produced. Find (a) the total cost function, (b) the average cost function, and (c) the marginal cost function. (d) Show that there is no absolute minimum average unit cost. (e) What is the smallest number of toys that must be produced so that the average cost per toy is less than \$3.42? (f) Draw sketches of the graphs of the functions in (a), (b), and (c) on the same set of axes.

12. The total revenue received from the sale of x desks is $R(x)$ dollars, and $R(x) = 200x - \frac{1}{3}x^2$. Find (a) the marginal revenue function; (b) the marginal revenue when $x = 30$; (c) the actual revenue from the sale of the thirty-first desk.

13. If $R(x)$ dollars is the total revenue received from the sale of x television sets and $R(x) = 600x - \frac{1}{20}x^3$, find (a) the marginal revenue function; (b) the marginal revenue when $x = 20$; (c) the actual revenue from the sale of the twenty-first television set.

14. If the demand equation for a particular commodity is $3x + 4p = 12$, find (a) the price function, (b) the total revenue function, and (c) the marginal revenue function. Draw sketches of the demand, total revenue, and marginal revenue curves on the same set of axes. Verify that the marginal revenue curve intersects the x axis at the point whose abscissa is the value of x for which the total revenue is greatest and that the demand curve intersects the x axis at the point whose abscissa is twice that.

15. The demand equation for a particular commodity is $x^2 + p^2 = 36$, where $100x$ units are demanded when p dollars is the price per unit. Find (a) the total revenue function and (b) the marginal revenue function. (c) Find the absolute maximum total revenue.

16. Follow the instructions of Exercise 15 if the demand equation is $px^2 + 9p - 18 = 0$.

17. A company that builds and sells desks is operating under perfect competition and can sell at a price of \$200 per desk all the desks it produces. If x desks are produced and sold each week and $C(x)$ dollars is the total cost of the week's production, then $C(x) = x^2 + 40x + 3000$. Determine how many desks should be built each week for the manufacturer to have the greatest weekly total profit. What is the greatest weekly total profit?

18. A firm operating under perfect competition manufactures and sells portable radios. The firm can sell at a price of \$75 per radio all the radios it produces. If x radios are manufactured each day and $C(x)$ dollars is the daily total cost of production, then $C(x) = x^2 + 25x + 100$. How many radios should be produced each day for the firm to have the greatest daily total profit? What is the greatest daily total profit?

19. Under a monopoly the demand equation for a certain commodity is $x + p = 140$, where x is the number of units produced daily and p is the number of dollars in the price of each unit. The number of dollars in the total cost of producing x units

is given by $C(x) = x^2 + 20x + 300$, and x is in the closed interval $[0, 140]$. (a) Find the total profit function, and draw a sketch of its graph. (b) On a set of axes different from that in (a) draw sketches of the total revenue and total cost curves, and show the geometrical interpretation of the total profit function. (c) Find the marginal revenue and marginal cost functions. (d) Find the maximum daily profit. (e) Draw sketches of the graphs of the marginal revenue function and the marginal cost function on the same set of axes, and show that they intersect at the point for which the value of x makes the total profit a maximum.

20. Follow the instructions of Exercise 19 if the demand equation is $x^2 + p = 320$, $C(x) = 20x$, and x is in the closed interval $[0, 8\sqrt{5}]$.

21. Suppose that under a monopoly the demand equation of a particular article of merchandise is $p = 6 - \frac{1}{5}\sqrt{x - 100}$, where p dollars is the price per article when x articles are demanded and $x \in [100, 1000]$. If $C(x)$ dollars is the total cost of producing x articles, then $C(x) = 2x + 100$. (a) Find the marginal revenue and marginal cost functions. (b) Find the value of x that yields the maximum profit.

22. The demand equation for a certain commodity is $p = (8 - \frac{1}{100}x)^2$, where p dollars is the price per unit when x units are demanded and $x \in [0, 800]$. The total cost function is given by $C(x) = 18x - \frac{1}{100}x^2$, where $C(x)$ dollars is the total cost of producing x units. (a) Find the marginal revenue and marginal cost functions. (b) Find the value of x that yields the maximum total profit.

23. A monopolist determines that if $C(x)$ cents is the total cost of producing x units of a certain commodity, then $C(x) = 20,000 + 25x$. The demand equation is $p = 100 - 0.02x$, where p cents is the price per unit when x units are demanded. If the total profit is to be maximized, find (a) the number of units that should be produced, (b) the price of each unit, and (c) the total profit.

24. Solve Exercise 23 if the government levies a tax on the monopolist of 10 cents per unit produced.

25. Solve Exercise 23 if the government imposes a 10 percent tax based on the consumer's price.

26. For the monopolist of Exercise 23, determine the amount of tax that should be levied by the government on each unit produced for the tax revenue received by the government to be maximized.

27. Under a monopoly the demand equation for a certain commodity is $x + 2p = 24$, where x units are demanded when p dollars is the price per unit. If $C(x)$ dollars is the total cost of producing x units, then $C(x) = \frac{1}{100}x^3 + \frac{3}{10}x^2$. (a) Find the price per unit if the total profit is to be maximized. (b) Solve part (a) if the government levies on the monopolist a tax of $2 for each unit produced.

28. Find the maximum tax revenue that can be received by the government if an additive tax for each unit produced is levied on a monopolist for which the demand equation is $x + 3p = 75$, where p dollars is the price of one unit when x units are demanded; and $C(x) = 3x + 100$, where $C(x)$ dollars is the total cost of producing x units.

29. The demand equation for a certain commodity produced by a monopolist is $p = a - bx$, where x units are demanded when p dollars is the price per unit, and the total cost, $C(x)$ dollars, of producing x units is determined by $C(x) = c + dx$. The numbers $a, b, c,$ and d are positive constants. If the government levies a tax on the monopolist of t dollars per unit produced, show that to maximize profits the monopolist should pass on to the consumer only one-half of the tax; that is, the monopolist should increase the unit price by $\frac{1}{2}t$ dollars. Note: The demand equation and the total cost function of Example 6 and Exercise 24 are special cases of this more general situation.

30. The demand equation for a monopolist is $(100 - x)^2 = 100p$, where x units are demanded when p dollars is the price per unit. The total cost function is given by $C(x) = 55x - \frac{4}{5}x^2$, where $C(x)$ dollars is the total cost of producing x units. (a) Find the price the monopolist should charge to make the greatest profit. (b) If the government imposes on the monopolist a tax of $9 for each unit produced, determine the price that should be charged for the greatest total profit. (c) Show that the increase in price from part (a) to part (b) is greater than the tax levied by the government.

31. A table manufacturer has a contract to supply 3000 tables per year at a uniform rate. The cost of starting a production run is $96, and the annual storage cost is $40 per table. If the excess tables are stored until delivery, how many tables should be produced in each run if the total cost of obtaining and storing inventory is to be minimized? Assume that production is instantaneous and shortages are not permitted.

32. A firm has a contract to supply 5 units of a certain commodity each day. The cost of starting a production run is $250, and the storage cost of each unit is $1 per day. Assuming that production is instantaneous and shortages are not permitted, determine the number of units that should be produced in each run to minimize the total cost of obtaining and storing inventory.

33. A distributor of a certain nonperishable food product has a monthly demand of 12,000 cases, and they are being bought at a constant rate. The cost of placing an order, which includes such expenses as clerical costs, delivery charges, and han-

dling costs, is $25 regardless of the size of the order. The monthly storage cost is 60 cents per case. Assuming that deliveries are instantaneous and that shortages are not allowed, determine how many orders should be placed each month and the size of each order such that the cost of maintaining an inventory is minimized.

34. An appliance store sells 2500 television sets per year, and they are being sold at a constant rate. The cost of placing an order is $20 regardless of the size of the order, and the annual storage cost is $10 per set. Assuming that deliveries are instantaneous and that shortages are not allowed, determine how many orders should be placed each year and the size of each order such that the cost of maintaining an inventory is minimized.

35. Solve Exercise 33 if the monthly demand is R cases, the cost of placing an order is K dollars, and the monthly storage cost is S dollars per case.

3.12 (Supplementary) NUMERICAL SOLUTIONS OF EQUATIONS BY NEWTON'S METHOD

The solutions of an equation of the form

$$f(x) = 0 \qquad (1)$$

are called the **roots** of the equation or the **zeros** of the function f. If f is a polynomial function of degree less than five, there are formulas for obtaining the zeros. You are familiar with the formulas if f has degree one (a linear function) or two (a quadratic function). For a polynomial function of degree three or four, the general method for obtaining the zeros is complicated. Furthermore, for the zeros of a polynomial function of degree five or more there is a theorem, credited to Niels Abel (1802–1829), that states that there can be no general formula in terms of a finite number of operations on the coefficients. There are, however, numerical processes for approximating solutions of equations, and they are more important now than ever before because of the widespread use of computers and programmable calculators. One of the processes involves an application of the derivative, and it was devised by Sir Isaac Newton in the seventeenth century. It is known as Newton's method and is the subject of this section.

Consider Eq. (1), where f is a differentiable function. Newton's method gives a procedure for approximating a root of this equation or, equivalently, a zero of f, that is, a number r such that $f(r) = 0$. We first give a geometric interpretation of the concepts involved. Refer to Fig. 3.12.1, that shows a sketch of the graph of $y = f(x)$. The number r is an x intercept of the graph. To obtain an approximation of r we first select a number x_1. The choice of x_1 can be made by referring to a sketch of the graph, and it should be reasonably close to the number r. We then consider the tangent line to the graph of f at the point $(x_1, f(x_1))$. The tangent line, denoted by T_1, appears in Fig. 3.12.1, and T_1 has an x intercept x_2. The number x_2 now serves as a second approximation of r. We then repeat the process with the tangent line T_2 at the point $(x_2, f(x_2))$. The x intercept of T_2 is x_3. We continue the process until we have the required degree of accuracy. For this graph it appears that the numbers x_1, x_2, x_3, and so on are getting closer and closer to the number r. This situation occurs for many functions.

To obtain the successive approximations $x_2, x_3, \ldots$ from the first approximation x_1 we use the equations of the tangent lines. The tangent line T_1 at the point $(x_1, f(x_1))$ has a slope of $f'(x_1)$. Thus an equation of T_1 is

$$y - f(x_1) = f'(x_1)(x - x_1) \qquad (2)$$

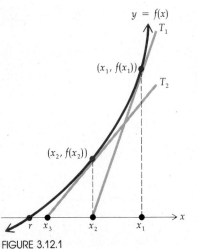

FIGURE 3.12.1

The x intercept of T_1 is x_2, and we determine x_2 by letting $x = x_2$ and $y = 0$ in (2). We get

$$0 - f(x_1) = f'(x_1)(x_2 - x_1)$$

$$x_2 = x_1 - \frac{f(x_1)}{f'(x_1)} \qquad \text{if } f'(x_1) \neq 0$$

With this value of x_2 an equation of T_2 is

$$y - f(x_2) = f'(x_2)(x - x_2) \tag{3}$$

Then in (3) we let $x = x_3$ and $y = 0$, and we have

$$0 - f(x_2) = f'(x_2)(x_3 - x_2)$$

$$x_3 = x_2 - \frac{f(x_2)}{f'(x_2)} \qquad \text{if } f'(x_2) \neq 0$$

Continuing on in this manner we get the general formula for the approximation x_{n+1} in terms of the preceding approximation x_n:

$$x_{n+1} = x_n - \frac{f(x_n)}{f'(x_n)} \qquad \text{if } f'(x_n) \neq 0 \tag{4}$$

Formula (4) is easily adapted for use on a computer or a programmable calculator.

From formula (4) we can obtain the $(n + 1)$st approximation from the nth approximation, provided $f'(x_n) \neq 0$. When $f'(x_n) = 0$, the tangent line is horizontal, and in such a case, unless the tangent line is the x axis itself, it has no x intercept. Figure 3.12.2 shows this happening when $f'(x_2) = 0$. So Newton's method is not applicable if $f'(x_n) = 0$ for some x_n. You should also be aware that the value of x_{n+1} obtained from (4) is not necessarily a better approximation of r than x_n. If, for instance, x_1 is not reasonably close to r, then $|f'(x_1)|$ may be small so that the tangent line T_1 is nearly horizontal. Then x_2, the x intercept of T_1, could be farther away from r than x_1. See Fig. 3.12.3, where this situation occurs.

In the following illustration we show how Newton's method is applied to an equation for which we know the answer.

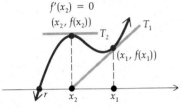

FIGURE 3.12.2

► ILLUSTRATION 1   Let us use Newton's method to obtain the positive root of the equation $x^2 = 9$ by starting with a first approximation of 4. We write the equation as $x^2 - 9 = 0$ and let

$$f(x) = x^2 - 9$$

$$f'(x) = 2x$$

FIGURE 3.12.3

From (4) we obtain

$$x_{n+1} = x_n - \frac{f(x_n)}{f'(x_n)}$$

$$x_{n+1} = x_n - \frac{x_n^2 - 9}{2x_n} \tag{5}$$

We first apply (5) with $n = 1$ and $x_1 = 4$ and we get

$$x_2 = x_1 - \frac{x_1^2 - 9}{2x_1}$$

$$= 4 - \frac{16 - 9}{8}$$

$$= 4 - 0.875$$

$$= 3.125$$

From (5) with $n = 2$ and $x_2 = 3.125$, we have

$$x_3 = x_2 - \frac{x_2^2 - 9}{2x_2}$$

$$= 3.125 - \frac{(3.125)^2 - 9}{2(3.125)}$$

$$= 3.125 - 0.1225$$

$$= 3.0025$$

From (5) with $n = 3$ and $x_3 = 3.0025$, we obtain

$$x_4 = x_3 - \frac{x_3^2 - 9}{2x_3}$$

$$= 3.0025 - \frac{(3.0025)^2 - 9}{2(3.0025)}$$

$$= 3.0025 - 0.002499$$

$$= 3.0000$$

We apply (5) again; this time we take $n = 4$ and $x_4 = 3.0000$ and get

$$x_5 = x_4 - \frac{x_4^2 - 9}{2x_4}$$

$$= 3.0000 - \frac{(3.0000)^2 - 9}{2(3.0000)}$$

$$= 3.0000$$

Certainly all successive approximations obtained from (5) will be 3.0000. Thus the positive root of the equation $x^2 - 9 = 0$ is 3.0000 to four decimal places.

◄

Observe that when x_n is a solution of (1), $f(x_n) = 0$. Thus from (4),

$$x_{n+1} = x_n - \frac{f(x_n)}{f'(x_n)}$$

$$= x_n - 0$$

$$= x_n$$

Consequently all subsequent approximations are equal to x_n. Note that this situation occurs in Illustration 1, where all approximations after and including x_4 have the same value to four decimal places.

Also observe from (4) that $x_{n+1} = x_n$ implies that $f(x_n) = 0$. Therefore we can conclude that when two successive approximations are equal, we have an approximation for a zero of f.

It is possible, however, that for certain functions, if your initial choice of x_1 is not close enough to the desired zero, you may obtain approximations for a different zero. See Fig. 3.12.4 for a sketch of the graph of a function where this situation could happen. Note that the indicated choice of x_1 near the desired zero r gives successive approximations $x_2, x_3, x_4, \ldots$ near another zero s. Thus when applying Newton's method you should first draw a rough sketch of the graph of the function to obtain your initial approximation. Refer to the graph as you proceed to make sure you are getting successive approximations to the zero you are seeking.

FIGURE 3.12.4

In summary, then, when using Newton's method to solve an equation of the form (1), do the following:

1. Make a *good guess* for the first approximation x_1. A rough sketch of the graph of f will help to obtain a reasonable choice.
2. With the value of x_1 in formula (4), get a second approximation x_2. Then use x_2 in (4) to get a third approximation x_3, and so on until $x_{n+1} = x_n$ to the required degree of accuracy.

EXAMPLE 1

Use Newton's method to find the real root of the equation

$$x^3 - 2x - 2 = 0$$

to four decimal places.

SOLUTION

Let $f(x) = x^3 - 2x - 2$; thus $f'(x) = 3x^2 - 2$. Then from (4) we have

$$x_{n+1} = x_n - \frac{x_n^3 - 2x_n - 2}{3x_n^2 - 2} \qquad (6)$$

To obtain a sketch of the graph of f we plot the points $(-2, -6)$, $(-1, -1)$, $(0, -2)$, $(1, -3)$, and $(2, 2)$. There are relative extrema of f when $f'(x) = 0$, that is, when $x = \pm\frac{1}{3}\sqrt{6}$. The graph appears as in Fig. 3.12.5. Because the graph intersects the x axis at only one point, there is one real root of the given equation. Because $f(1) = -3$ and $f(2) = 2$, this root lies between 1 and 2. A suitable choice for our first approximation is $x_1 = 1.5$. Table 3.12.1

FIGURE 3.12.5

Table 3.12.1

n	x_n	$x_n^3 - 2x_n - 2$	$3x_n^2 - 2$	$\dfrac{x_n^3 - 2x_n - 2}{3x_n^2 - 2}$	x_{n+1}
1	1.50000	-1.62500	4.75000	-0.34211	1.84211
2	1.84211	0.56674	8.18011	0.06928	1.77283
3	1.77283	0.02621	7.42878	0.00353	1.76930
4	1.76930	0.00006	7.39127	0.00001	1.76929
5	1.76929	-0.00002	7.39116	0.00000	1.76929

shows the successive approximations computed from (6) with this x_1. We wish the root to be accurate to four decimal places; thus we use five places in the computations. Because x_5 and x_6 are equal (to five decimal places), we round off that number to four places and obtain 1.7693 as the required root.

EXAMPLE 2

Use Newton's method to find to three decimal places the x coordinate of the point of intersection in the first quadrant of the line $y = \frac{1}{3}x$ and the curve $y = \sin x$.

SOLUTION

Figure 3.12.6 shows the line and the curve. We wish to find the positive value of x for which

$$\sin x = \tfrac{1}{3}x$$

$$\Leftrightarrow \quad 3 \sin x - x = 0$$

Let

$$f(x) = 3 \sin x - x$$

$$f'(x) = 3 \cos x - 1$$

From formula (4),

$$x_{n+1} = x_n - \frac{f(x_n)}{f'(x_n)}$$

$$x_{n+1} = x_n - \frac{3 \sin x_n - x_n}{3 \cos x_n - 1} \tag{7}$$

From Fig. 3.12.6 it appears that a reasonable choice of x_1 is 2. We compute the successive approximations from formula (7); these appear in Table 3.12.2. The calculations are carried out to four decimal places. Observe that to four decimal places x_4 and x_5 are both equal to 2.2789. Thus to three decimal places the positive value of x for which $\sin x = \frac{1}{3}x$ is 2.279.

$y = \frac{1}{3}x$

$y = \sin x$

FIGURE 3.12.6

Table 3.12.2

n	x_n	$3 \sin x_n - x_n$	$3 \cos x_n - 1$	$\dfrac{3 \sin x_n - x_n}{3 \cos x_n - 1}$	x_{n+1}
1	2.0000	0.7279	-2.2484	-0.3237	2.3237
2	2.3237	-0.1346	-3.0513	0.0441	2.2796
3	2.2796	-0.0022	-2.9528	0.0007	2.2789
4	2.2789	-0.0001	-2.9512	0.0000	2.2789

There are theorems that state conditions for which Newton's method is applicable, as well as theorems relating to its accuracy. Such theorems can be found in texts on numerical analysis.

EXERCISES 3.12

In Exercises 1 through 4, use Newton's method to find the real root of the given equation to four decimal places.

1. $x^3 - 4x^2 - 2 = 0$ 2. $6x^3 + 9x + 1 = 0$ 3. $x^5 - x + 1 = 0$ 4. $x^5 + x - 1 = 0$

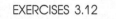

In Exercises 5 through 10, use Newton's method to find, to the nearest thousandth, the approximate value of the indicated root.

5. $x^3 - 4x - 8 = 0$; the positive root

6. $x^3 - 2x + 7 = 0$; the negative root

7. $x^4 - 10x + 5 = 0$; the smallest positive root

8. $x^4 - 10x + 5 = 0$; the largest positive root

9. $2x^4 - 2x^3 + x^2 + 3x - 4 = 0$; the negative root

10. $x^4 + x^3 - 3x^2 - x - 4 = 0$; the positive root

In Exercises 11 through 14, use Newton's method to find the value of the given radical to five decimal places.

11. $\sqrt{3}$ by solving the equation $x^2 - 3 = 0$

12. $\sqrt{10}$ by solving the equation $x^2 - 10 = 0$

13. $\sqrt[3]{6}$ by solving the equation $x^3 - 6 = 0$

14. $\sqrt[3]{7}$ by solving the equation $x^3 - 7 = 0$

In Exercises 15 through 18, use Newton's method to find to four decimal places the x coordinate of the point of intersection in the first quadrant of the graphs of the two equations.

15. $y = x$; $y = \cos x$

16. $y = \frac{1}{2}x$; $y = \sin x$

17. $y = x^2$; $y = \sin x$

18. $y = x^2$; $y = \cos x$

19. Equations of the form $\tan x + ax = 0$, where a is an integer, arise in heat conduction problems. The positive roots of the equation in increasing order are $\alpha_1, \alpha_2, \alpha_3, \ldots$. If $a = 1$, find α_1 and α_2 to four decimal places.

20. Follow the instructions of Exercise 19 if $a = -2$.

In Exercises 21 and 22, obtain an approximation for π to five decimal places by using Newton's method to solve the given equation.

21. $\tan x = 0$

22. $\cos x + 1 = 0$

REVIEW EXERCISES FOR CHAPTER 3

In Exercises 1 through 12, find the absolute extrema of the given function on the indicated interval, if there are any, and find the values of x at which the absolute extrema occur. Draw a sketch of the graph of the function on the interval.

1. $f(x) = \sqrt{5 + x}$; $[-5, +\infty)$

2. $f(x) = \sqrt{9 - x^2}$; $(-3, 3)$

3. $f(x) = \frac{5}{2}x^6 - 3x^5$; $[-1, 2]$

4. $f(x) = x^4 - 12x^2 + 36$; $[-2, 6]$

5. $f(x) = |9 - x^2|$; $[-2, 3]$

6. $f(x) = \dfrac{3}{x - 2}$; $[0, 4]$

7. $f(x) = x^4 - 12x^2 + 36$; $[0, 5]$

8. $f(x) = \begin{cases} 2x + 3 & \text{if } -2 \le x < 1 \\ x^2 + 4 & \text{if } 1 \le x \le 2 \end{cases}$; $[-2, 2]$

9. $f(x) = 2 \sin 3x$; $[-\frac{1}{3}\pi, \frac{1}{3}\pi]$

10. $f(x) = 4 \cos^2 2x$; $[0, \frac{3}{4}\pi]$

11. $f(x) = \tan 4x$; $[-\frac{1}{8}\pi, \frac{1}{12}\pi]$

12. $f(x) = \csc 3x$; $[0, \frac{1}{3}\pi]$

In Exercises 13 through 32, draw a sketch of the graph of the given function f by first finding the following: the relative extrema of f; the points of inflection of the graph of f; the intervals on which f is increasing; the intervals on which f is decreasing; where the graph is concave upward; where the graph is concave downward; the slope of any inflectional tangent; the horizontal, vertical, and oblique asymptotes, if there are any.

13. $f(x) = x^3 + 3x^2 - 4$

14. $f(x) = x^3 + 2x^2 + x - 5$

15. $f(x) = (x - 4)^2(x + 2)^3$

16. $f(x) = (x - 1)^3(x - 3)$

17. $f(x) = (x - 4)^{1/3} - 3$

18. $f(x) = (x + 2)^3 + 2$

19. $f(x) = \dfrac{4x^2}{3x^2 + 1}$

20. $f(x) = \dfrac{x^2}{x^2 - 9}$

21. $f(x) = \dfrac{5x^2}{x^2 - 4}$

22. $f(x) = \dfrac{x}{x^2 - 1}$

23. $f(x) = \dfrac{x^2}{x - 3}$

24. $f(x) = \dfrac{x^2 + 9}{x}$

25. $f(x) = \begin{cases} (1 - x)^3 & \text{if } x \le 1 \\ (x - 1)^3 & \text{if } 1 < x \end{cases}$

26. $f(x) = \begin{cases} x^3 - 3x & \text{if } x < 2 \\ 6 - x^2 & \text{if } 2 \le x \end{cases}$

27. $f(x) = (x - 3)^{5/3} + 1$

28. $f(x) = (x + 2)^{4/3}$

29. $f(x) = (x + 1)^{2/3}(x - 3)^2$

30. $f(x) = x\sqrt{25 - x^2}$

31. $f(x) = \sin 2x - \cos 2x$; $x \in [-\frac{3}{8}\pi, \frac{5}{8}\pi]$

32. $f(x) = x - \tan x$; $x \in (-\frac{1}{2}\pi, \frac{1}{2}\pi)$

In Exercises 33 through 36, evaluate the limit if it exists.

33. $\displaystyle\lim_{x \to -\infty} \frac{x^2 + 5}{2x - 4}$

34. $\displaystyle\lim_{x \to +\infty} \left(\frac{8x^3 + 7x - 2}{7x^3 + 3x^2 + 5x}\right)^2$

35. $\displaystyle\lim_{x \to +\infty} (\sqrt{x + 1} - \sqrt{x})$

36. $\displaystyle\lim_{t \to +\infty} (\sqrt{t^2 + t} - \sqrt{t^2 + 4})$

In Exercises 37 and 38, verify that the three conditions of the hypothesis of Rolle's theorem are satisfied by the given function on the indicated interval. Then find a suitable value for c that satisfies the conclusion of Rolle's theorem. Draw a sketch of the graph of f on the interval, and use the sketch to give the geometric interpretation of Rolle's theorem.

37. $f(x) = x^3 - x^2 - 4x + 4$; $[-2, 1]$ **38.** $f(x) = 2 \sin 3x$; $[0, \frac{1}{3}\pi]$

In Exercises 39 and 40, verify that the hypothesis of the mean-value theorem is satisfied for the given function on the indicated interval. Then find a suitable value for c that satisfies the conclusion of the mean-value theorem. Draw a sketch of the graph of f on the interval, and use the sketch to give the geometric interpretation of the mean-value theorem.

39. $f(x) = \sqrt{3 - x}$; $[-6, -1]$ **40.** $f(x) = x^3$; $[-2, 2]$

41. (a) If f is a polynomial function and $f(a)$, $f(b)$, $f'(a)$, and $f'(b)$ are zero, prove that there are at least two numbers in the open interval (a, b) that are roots of the equation $f''(x) = 0$. (b) Show that the function defined by the equation $f(x) = (x^2 - 4)^2$ satisfies part (a) for the open interval $(-2, 2)$.

42. If f is the function defined by $f(x) = |2x - 4| - 6$, then $f(-1) = 0$ and $f(5) = 0$. However, f' never has the value 0. Show why Rolle's theorem does not hold.

43. Suppose that f and g are two functions that satisfy the hypothesis of the mean-value theorem on $[a, b]$. Furthermore, suppose that $f'(x) = g'(x)$ for all x in the open interval (a, b). Prove that $f(x) - g(x) = f(a) - g(a)$ for all x in the closed interval $[a, b]$.

44. If f is a polynomial function, show that between any two consecutive roots of the equation $f'(x) = 0$ there is at most one root of the equation $f(x) = 0$.

45. Find the absolute maximum value attained by the function f if $f(x) = A \sin kx + B \cos kx$, where A, B, and k are positive constants.

46. Find the point (or points) of inflection of the graph of the function defined by $f(x) = x^{1/5}$, and determine where the graph is concave upward and where it is concave downward.

47. Find the point (or points) of inflection of the graph of the function defined by

$$f(x) = \begin{cases} 2 - x^2 & \text{if } x < 0 \\ x^2 + 2 & \text{if } 0 \le x \end{cases}$$

and determine where the graph is concave upward and where it is concave downward.

48. If $f(x) = ax^3 + bx^2$, determine a and b so that the graph of f will have a point of inflection at $(2, 16)$.

49. If $f(x) = ax^3 + bx^2 + cx$, determine a, b, and c so that the graph of f will have a point of inflection at $(1, -1)$ and so that the slope of the inflectional tangent there will be -3.

50. If $f(x) = \dfrac{x + 1}{x^2 + 1}$, prove that the graph of f has three points of inflection that are collinear. Draw a sketch of the graph.

51. If $f(x) = x|x|$, show that the graph of f has a point of inflection at the origin.

52. Let $f(x) = x^n$, where n is a positive integer. Prove that the graph of f has a point of inflection at the origin if and only if n is an odd integer and $n > 1$. Furthermore show that if n is even, f has a relative minimum value at 0.

53. If $f(x) = (x^2 + a^2)^p$, where p is a rational number and $p \ne 0$, prove that if $p < \frac{1}{2}$, the graph of f has two points of inflection, and if $p \ge \frac{1}{2}$, the graph of f has no points of inflection.

54. Suppose that the graph of a function has a point of inflection at $x = c$. What can you conclude, if anything, about (a) the continuity of f at c; (b) the continuity of f' at c; (c) the continuity of f'' at c?

In Exercises 55 and 56, find the horizontal and vertical asymptotes of the graph of the given equation, and draw a sketch of the graph.

55. $4xy - 5x + 7y = 0$ **56.** $xy - 5x - 3y + 2 = 0$

57. Find two nonnegative numbers whose sum is 12 such that the sum of their squares is an absolute minimum.

58. Find two nonnegative numbers whose sum is 12 such that their product is an absolute maximum.

59. Show that among all the rectangles having a perimeter of 36 in., the square of side 9 in. has the greatest area.

60. Show that among all the rectangles having an area of 81 in.^{2}, the square of side 9 in. has the least perimeter.

61. Find the shortest distance from the point $(\frac{9}{2}, 0)$ to the curve $y = \sqrt{x}$.

62. Find the shortest distance from the point $A(0, 4)$ to a point on the hyperbola $x^2 - y^2 = 16$, and find the point on the hyperbola that is closest to A.

63. An open box having a square base is to have a volume of 4000 in.^{3}. Find the dimensions of the box that can be constructed with the least amount of material.

64. Solve Exercise 63 if the box is to be closed.

65. Solve Exercise 63 if the open box is to have a volume of k in.^{3}.

66. In a town of population 11,000 the rate of growth of an epidemic is jointly proportional to the number of people infected and the number of people not infected. Determine the number of people infected when the epidemic is growing at a maximum rate.

67. Because of various restrictions, the size of a particular community is limited to 3000 inhabitants, and the rate of increase of the population is jointly proportional to its size and the difference between 3000 and its size. Determine the size of the population for which the rate of growth of the population is a maximum.

68. A manufacturer offers to deliver to a dealer 300 chairs at $90 per chair and to reduce the price per chair on the entire order by $0.25 for each additional chair over 300. Find the dollar total involved in the largest possible transaction between the manufacturer and the dealer under these circumstances.

69. Two towns A and B are to get their water supply from the same pumping station to be located on the bank of a straight river that is 15 km from town A and 10 km from town B. If the points on the river nearest to A and B are 20 km apart and A and B are on the same side of the river, where should the pumping station be located so that the least amount of piping is required?

70. One of the acute angles of a triangle is to have a radian measure of $\frac{1}{6}\pi$, and the side opposite this angle is to have a length of 10 in. Prove that of all the triangles satisfying these requirements, the one having the maximum area is isosceles. (*Hint:* Express the measure of the area of the triangle in terms of trigonometric functions of one of the other acute angles.)

71. In a warehouse, goods weighing 1000 pounds are transported along a level floor by securing a heavy rope under a low mobile platform and pulling it with a motorized vehicle. If the rope is directed at an angle of θ radians with the plane of the floor, then the force of magnitude F lb along the rope is given by

$$F = \frac{1000k}{k \sin \theta + \cos \theta}$$

where k is the constant coefficient of friction and $0 < k < 1$. If $0 \le \theta \le \frac{1}{2}\pi$, show that F is least when $\tan \theta = k$.

72. A closed tin can having a volume of 27 in.^{3} is to be in the form of a right-circular cylinder. If the circular top and bottom are cut from square pieces of tin, find the radius and height of the can if the least amount of tin is to be used in its manufacture. Include the tin that is wasted when obtaining the top and bottom.

73. Two particles start their motion at the same time. One particle is moving along a horizontal line and its equation of motion is $x = t^2 - 2t$, where x cm is the directed distance of the particle from the origin at t sec. The other particle is moving along a vertical line that intersects the horizontal line at the origin, and its equation of motion is $y = t^2 - 2$, where y cm is the directed distance of the particle from the origin at t sec. Find when the directed distance between the two particles is least, and their velocities at that time.

74. A ladder is to reach over a fence $\frac{27}{8}$ m high to a wall 8 m behind the fence. Find the length of the shortest ladder that may be used.

75. Solve Exercise 74 if the fence is h meters high and the wall is w meters behind the fence.

76. A tent is to be in the shape of a cone. Find the ratio of the measure of the radius to the measure of the altitude for a tent of given volume to require the least material.

77. A large sign is to contain 50 m^2 of printed material. If margins of 4 m at the top and bottom and 2 m on each side are required, find the dimensions of the smallest sign that will meet these specifications.

78. Find the volume of the largest right-circular cylinder that can be inscribed in a right-circular cone having a radius of 4 in. and a height of 8 in.

79. Find the dimensions of the right-circular cone of least volume that can be circumscribed about a right-circular cylinder of radius r cm and height h cm.

80. A piece of wire 80 cm long is bent to form a rectangle. Find the dimensions of the rectangle so that its area is as large as possible.

81. A piece of wire 20 cm long is cut into two pieces, and each piece is bent into the shape of a square. How should the wire be cut so that the total area of the two squares is as small as possible?

82. If $f(x) = x + \sin x$, prove that f has no relative extrema but that the graph of f has horizontal tangent lines. Draw a sketch of the graph for $x \in [-2\pi, 2\pi]$.

83. (a) If $f(x) = 3|x| + 4|x - 1|$, prove that f has an absolute minimum value of 3. (b) If $g(x) = 4|x| + 3|x - 1|$, prove that g has an absolute minimum value of 3. (c) If $h(x) = a|x| + b|x - 1|$, where $a > 0$ and $b > 0$, prove that h has an absolute minimum value that is the smaller of the two numbers a and b.

84. If $f(x) = |x|^a \cdot |x - 1|^b$, where a and b are positive rational numbers, prove that f has a relative maximum value of $a^a b^b / (a + b)^{a+b}$.

85. Let f and g be two functions that are differentiable at every number in the closed interval $[a, b]$. Suppose further that $f(a) = g(a)$ and $f(b) = g(b)$. Prove that there exists a number c in the open interval (a, b) such that $f'(c) = g'(c)$. (*Hint:* Consider the function $f - g$.)

86. Draw a sketch of the graph of a function f on the interval I in each case: (a) I is the open interval $(0, 2)$ and f is continuous on I. At 1, f has a relative maximum value but $f'(1)$ does not exist. (b) I is the closed interval $[0, 2]$. The function f has a relative minimum value at 1, but the absolute minimum value of f is at 0. (c) I is the open interval $(0, 2)$, and f' has a relative maximum value at 1.

Exercises 87 through 92 pertain to Supplementary Section 3.11.

87. The demand equation for a particular commodity is $(p + 4)(x + 3) = 48$, where $100x$ units are demanded when p dollars is the price per unit. Find (a) the price function, (b) the total revenue function, (c) the marginal revenue function. (d) Find the absolute maximum total revenue. (e) Draw sketches of the demand, total revenue, and marginal revenue curves on the same set of axes.

88. The demand equation for a certain commodity is $p + 2\sqrt{x - 1} = 6$ and the total cost function is given by $C(x) = 2x - 1$. (a) Determine the permissible values of x. (b) Find the marginal revenue and marginal cost functions. (c) Find the value of x that yields the maximum profit.

89. The demand equation for a certain commodity is $10^6 px = 10^9 - 2 \cdot 10^6 x + 18 \cdot 10^3 x^2 - 6x^3$, where x units are produced weekly and p dollars is the price of each unit; and $x \geq 100$. The number of dollars in the average cost of producing each unit is given by $Q(x) = \frac{1}{50}x - 24 + 11 \cdot 10^3 x^{-1}$. Find the number of units that should be produced each week and the price of each unit for the weekly profit to be maximized.

90. When $1000x$ boxes of a certain kind of material are produced, the number of dollars in the total cost of production is given by $C(x) = 135x^{1/3} + 450$. Find (a) the marginal cost when 8000 boxes are produced and (b) the number of boxes produced when the marginal cost (per thousand boxes) is $20.

91. A company that manufactures lamps has a contract to supply 3000 lamps per year at a uniform rate. The cost of starting a production run is $120. The excess lamps are stored until delivery, and the annual storage cost is $50 per lamp. How many lamps should be produced in each run if the total cost of obtaining and storing inventory is to be a minimum? Assume that production is instantaneous and that shortages are not permitted.

92. A manufacturer has to supply 10,000 desks per year at a uniform rate. The cost of starting a production run is $180, the excess desks are stored until delivery, and the annual storage cost is $40 per desk. Assuming that deliveries are instantaneous and that shortages are not allowed, how many desks should be produced in each run if the total cost of obtaining and storing inventory is to be minimized?

Exercises 93 through 96 pertain to Supplementary Section 3.12.

93. Use Newton's method to find to three decimal places the positive root of the equation $4x^4 - 3x^3 + 2x - 5 = 0$.

94. Use Newton's method to find to three decimal places the negative root of the equation $3x^4 - 4x^3 + 36x^2 + 2x - 8 = 0$.

95. Find to four decimal places by Newton's method the x coordinate of the point of intersection of the curve $y = \sin x$ and the line $y = 2x - 3$.

96. Find to four decimal places the value of x in the interval $(\frac{1}{2}\pi, \frac{3}{2}\pi)$ for which $\tan x = x$ by applying Newton's method.

But, by hypothesis, $f'(x) = g'(x)$ for all x in I. Therefore

$$h'(x) = 0 \qquad \text{for all } x \text{ in } I$$

Thus Theorem 3.3.3 applies to the function h, and there is a constant K such that

$$h(x) = K \qquad \text{for all } x \text{ in } I$$

Replacing $h(x)$ by $f(x) - g(x)$ we have

$$f(x) = g(x) + K \qquad \text{for all } x \text{ in } I$$

and the theorem is proved.   ∎

The next theorem follows immediately from Theorem 4.1.2.

4.1.3 THEOREM   If F is a particular antiderivative of f on an interval I, then every antiderivative of f on I is given by

$$F(x) + C \tag{2}$$

where C is an arbitrary constant, and all antiderivatives of f on I can be obtained from (2) by assigning particular values to C.

PROOF   Let G represent any antiderivative of f on I; then

$$G'(x) = f(x) \qquad \text{for all } x \text{ in } I \tag{3}$$

Because F is a particular antiderivative of f on I,

$$F'(x) = f(x) \qquad \text{for all } x \text{ in } I \tag{4}$$

From (3) and (4) it follows that

$$G'(x) = F'(x) \qquad \text{for all } x \text{ in } I$$

Therefore, from Theorem 4.1.2 there is a constant K such that

$$G(x) = F(x) + K \qquad \text{for all } x \text{ in } I$$

Because G represents any antiderivative of f on I, it follows that all antiderivatives of f can be obtained from $F(x) + C$, where C is an arbitrary constant. Hence the theorem is proved.   ∎

Antidifferentiation is the process of finding the set of all antiderivatives of a given function. The symbol $\int$ denotes the operation of antidifferentiation, and we write

$$\int f(x)\,dx = F(x) + C \tag{5}$$

where

$$F'(x) = f(x)$$

and

$$d(F(x)) = f(x)\,dx \tag{6}$$

Leibniz introduced the convention of writing the differential of a function after the antidifferentiation symbol. The advantage of using the differential in this manner will be apparent to you when we compute antiderivatives by changing the variable in Section 4.2. From (5) and (6) we can write

$$\int d(F(x)) = F(x) + C \tag{7}$$

This equation will be used to obtain formulas for antidifferentiation in the sections that follow. Equation (7) states that when we antidifferentiate the differential of a function, we obtain that function plus an arbitrary constant. So we can think of the $\int$ symbol for antidifferentiation as meaning that operation which is the inverse of the operation denoted by d for computing a differential.

If $\{F(x) + C\}$ is the set of all functions whose differentials are $f(x)\,dx$, it is also the set of all functions whose derivatives are $f(x)$. Therefore antidifferentiation is considered as the operation of finding the set of all functions having a given derivative.

Because antidifferentiation is the inverse operation of differentiation, antidifferentiation theorems can be obtained from differentiation theorems. Thus the following theorems can be proved from the corresponding ones for differentiation.

4.1.4 THEOREM

$$\int dx = x + C$$

4.1.5 THEOREM

$$\int af(x)\,dx = a\int f(x)\,dx$$

where a is a constant.

Theorem 4.1.5 states that to find an antiderivative of a constant times a function, first find an antiderivative of the function and then multiply it by the constant.

4.1.6 THEOREM

If f_1 and f_2 are defined on the same interval, then

$$\int [f_1(x) + f_2(x)]\,dx = \int f_1(x)\,dx + \int f_2(x)\,dx$$

Theorem 4.1.6 states that to find an antiderivative of the sum of two functions, find an antiderivative of each of the functions separately and then add the results, it being understood that both functions are defined on the same interval. Theorem 4.1.6 can be extended to any finite number of functions. Combining Theorem 4.1.6 with Theorem 4.1.5 we have the following one.

4.1.7 THEOREM

If $f_1, f_2, \ldots, f_n$ are defined on the same interval,

$$\int [c_1 f_1(x) + c_2 f_2(x) + \ldots + c_n f_n(x)]\,dx$$

$$= c_1 \int f_1(x)\,dx + c_2 \int f_2(x)\,dx + \ldots + c_n \int f_n(x)\,dx$$

where $c_1, c_2, \ldots, c_n$ are constants.

4.1.8 THEOREM   If n is a rational number,

$$\int x^n \, dx = \frac{x^{n+1}}{n+1} + C \qquad n \neq -1$$

PROOF

$$D_x\left(\frac{x^{n+1}}{n+1}\right) = \frac{(n+1)x^n}{n+1}$$

$$= x^n \qquad \blacksquare$$

▶ ILLUSTRATION 2   Applying Theorem 4.1.8 for particular values of n we have

$$\int x^2 \, dx = \frac{x^3}{3} + C \qquad\qquad \int x^3 \, dx = \frac{x^4}{4} + C$$

$$\int \frac{1}{x^2} \, dx = \int x^{-2} \, dx \qquad \int \sqrt[3]{x} \, dx = \int x^{1/3} \, dx$$

$$= \frac{x^{-2+1}}{-2+1} + C \qquad\qquad = \frac{x^{1/3+1}}{\frac{1}{3}+1} + C$$

$$= \frac{x^{-1}}{-1} + C \qquad\qquad = \frac{x^{4/3}}{\frac{4}{3}} + C$$

$$= -\frac{1}{x} + C \qquad\qquad = \tfrac{3}{4}x^{4/3} + C$$

◀

The next illustration shows how Theorems 4.1.4 through 4.1.8 are used to antidifferentiate.

▶ ILLUSTRATION 3

$$\int (3x + 5)\,dx = \int 3x\,dx + \int 5\,dx \quad \text{(by Theorem 4.1.6)}$$

$$= 3\int x\,dx + 5\int dx \quad \text{(by Theorem 4.1.5)}$$

$$= 3\left(\frac{x^2}{2} + C_1\right) + 5(x + C_2) \quad \text{(by Theorems 4.1.8 and 4.1.4)}$$

$$= \tfrac{3}{2}x^2 + 5x + (3C_1 + 5C_2)$$

Because $3C_1 + 5C_2$ is an arbitrary constant, it may be denoted by C; so the result can be written as

$$\tfrac{3}{2}x^2 + 5x + C$$

The answer can be checked by finding its derivative.

$$D_x(\tfrac{3}{2}x^2 + 5x + C) = 3x + 5 \qquad\qquad ◀$$

EXAMPLE 1

Evaluate $\int (5x^4 - 8x^3 + 9x^2 - 2x + 7)\, dx$.

SOLUTION

$$\int (5x^4 - 8x^3 + 9x^2 - 2x + 7)\, dx$$

$$= 5\int x^4\, dx - 8\int x^3\, dx + 9\int x^2\, dx - 2\int x\, dx + 7\int dx$$

$$= 5 \cdot \frac{x^5}{5} - 8 \cdot \frac{x^4}{4} + 9 \cdot \frac{x^3}{3} - 2 \cdot \frac{x^2}{2} + 7x + C$$

$$= x^5 - 2x^4 + 3x^3 - x^2 + 7x + C$$

EXAMPLE 2

Evaluate $\int \sqrt{x}\left(x + \dfrac{1}{x}\right) dx$.

SOLUTION

$$\int \sqrt{x}\left(x + \frac{1}{x}\right) dx = \int x^{1/2}(x + x^{-1})\, dx$$

$$= \int (x^{3/2} + x^{-1/2})\, dx$$

$$= \frac{x^{5/2}}{\frac{5}{2}} + \frac{x^{1/2}}{\frac{1}{2}} + C$$

$$= \tfrac{2}{5}x^{5/2} + 2x^{1/2} + C$$

EXAMPLE 3

Evaluate $\int \dfrac{5t^2 + 7}{t^{4/3}}\, dt$.

SOLUTION

$$\int \frac{5t^2 + 7}{t^{4/3}}\, dt = 5\int \frac{t^2}{t^{4/3}}\, dt + 7\int \frac{1}{t^{4/3}}\, dt$$

$$= 5\int t^{2/3}\, dt + 7\int t^{-4/3}\, dt$$

$$= 5\left(\frac{t^{5/3}}{\frac{5}{3}}\right) + 7\left(\frac{t^{-1/3}}{-\frac{1}{3}}\right) + C$$

$$= 5(\tfrac{3}{5}t^{5/3}) + 7(-3t^{-1/3}) + C$$

$$= 3t^{5/3} - \frac{21}{t^{1/3}} + C$$

The theorems for the antiderivative of the sine and cosine functions follow immediately from the corresponding ones for differentiation.

4.1.9 THEOREM

$$\int \sin x \, dx = -\cos x + C$$

PROOF $D_x(-\cos x) = -(-\sin x)$
$$= \sin x \qquad \blacksquare$$

4.1.10 THEOREM

$$\int \cos x \, dx = \sin x + C$$

PROOF $D_x(\sin x) = \cos x \qquad \blacksquare$

The following theorems are consequences of the theorems for the derivatives of the tangent, cotangent, secant, and cosecant functions. The proofs are again immediate by finding the derivatives of the right-hand side of the equations.

4.1.11 THEOREM

$$\int \sec^2 x \, dx = \tan x + C$$

4.1.12 THEOREM

$$\int \csc^2 x \, dx = -\cot x + C$$

4.1.13 THEOREM

$$\int \sec x \tan x \, dx = \sec x + C$$

4.1.14 THEOREM

$$\int \csc x \cot x \, dx = -\csc x + C$$

EXAMPLE 4
Evaluate $\int (3 \sec x \tan x - 5 \csc^2 x) \, dx$.

SOLUTION
We apply Theorems 4.1.13 and 4.1.12.

$$\int (3 \sec x \tan x - 5 \csc^2 x) \, dx = 3 \int \sec x \tan x \, dx - 5 \int \csc^2 x \, dx$$
$$= 3 \sec x - 5(-\cot x) + C$$
$$= 3 \sec x + 5 \cot x + C$$

Trigonometric identities are often used when computing antiderivatives involving trigonometric functions. The following eight fundamental identities are crucial:

$$\sin x \csc x = 1 \qquad \cos x \sec x = 1 \qquad \tan x \cot x = 1$$

$$\tan x = \frac{\sin x}{\cos x} \qquad \cot x = \frac{\cos x}{\sin x}$$

$$\sin^2 x + \cos^2 x = 1 \qquad \tan^2 x + 1 = \sec^2 x \qquad \cot^2 x + 1 = \csc^2 x$$

EXAMPLE 5

Evaluate $\displaystyle\int \frac{2 \cot x - 3 \sin^2 x}{\sin x}\, dx$.

SOLUTION

$$\int \frac{2 \cot x - 3 \sin^2 x}{\sin x}\, dx$$

$$= 2 \int \frac{1}{\sin x} \cdot \cot x \, dx - 3 \int \frac{\sin^2 x}{\sin x}\, dx$$

$$= 2 \int \csc x \cot x \, dx - 3 \int \sin x \, dx$$

$$= 2(-\csc x) - 3(-\cos x) + C \quad \text{(from Theorems 4.1.14 and 4.1.9)}$$

$$= -2 \csc x + 3 \cos x + C$$

EXAMPLE 6

Evaluate $\int (\tan^2 x + \cot^2 x + 4)\, dx$.

SOLUTION

$$\int (\tan^2 x + \cot^2 x + 4)\, dx$$

$$= \int \left[(\sec^2 x - 1) + (\csc^2 x - 1) + 4 \right] dx$$

$$= \int \sec^2 x \, dx + \int \csc^2 x \, dx + 2 \int dx$$

$$= \tan x - \cot x + 2x + C \quad \text{(from Theorems 4.1.11 and 4.1.12)}$$

Often in applications involving antidifferentiation it is desired to find a particular antiderivative that satisfies certain conditions called **initial** or **boundary conditions**, depending on whether they occur at one or more than one point. For example, if an equation involving $\dfrac{dy}{dx}$ is given as well as the initial condition that $y = y_1$ when $x = x_1$, then after the set of all antiderivatives is found, if x and y are replaced by x_1 and y_1, a particular value of the arbitrary constant C is determined. With this value of C a particular antiderivative is obtained.

▶ ILLUSTRATION 4   Suppose we wish to find the particular antiderivative satisfying the equation

$$\frac{dy}{dx} = 2x \tag{8}$$

and the initial condition that $y = 6$ when $x = 2$. From (8),

$$y = \int 2x \, dx$$

$$y = x^2 + C \tag{9}$$

In (9) we substitute 2 for x and 6 for y and get

$$6 = 4 + C$$

$$C = 2$$

When this value of C is substituted back in (9), we obtain

$$y = x^2 + 2$$

which gives the particular antiderivative desired.   ◄

EXAMPLE 7

At any point (x, y) on a particular curve the tangent line has a slope equal to $4x - 5$. If the curve contains the point $(3, 7)$, find its equation.

SOLUTION

Because the slope of the tangent line to a curve at any point (x, y) is the value of the derivative at that point, we have

$$\frac{dy}{dx} = 4x - 5$$

$$y = \int (4x - 5)\, dx$$

$$y = 4\left(\frac{x^2}{2}\right) - 5x + C$$

$$y = 2x^2 - 5x + C \tag{10}$$

Equation (10) represents a *family* of curves. Because we wish to determine the particular curve of this family that contains the point $(3, 7)$, we substitute 3 for x and 7 for y in (10) and get

$$7 = 2(9) - 5(3) + C$$

$$7 = 18 - 15 + C$$

$$C = 4$$

Replacing C by 4 in (10) we get the required equation, which is

$$y = 2x^2 - 5x + 4$$

EXAMPLE 8

An art collector purchased a painting for $1000 by an artist, the value of whose works are currently increasing with respect to time according to the formula

$$\frac{dV}{dt} = 5t^{3/2} + 10t + 50$$

where V dollars is the anticipated value of a painting t years after its purchase. If this formula were valid for the next 6 years, what would be the anticipated value of the painting 4 years from now?

SOLUTION

From the given equation,

$$V = \int (5t^{3/2} + 10t + 50) \, dt$$

$$= 5 \cdot \left(\frac{t^{5/2}}{\frac{5}{2}}\right) + 10\left(\frac{t^2}{2}\right) + 50t + C$$

$$= 2t^{5/2} + 5t^2 + 50t + C$$

Because $V = 1000$ when $t = 0$, we get $C = 1000$. Hence

$$V = 2t^{5/2} + 5t^2 + 50t + 1000$$

Letting V_4 be the value of V when $t = 4$ we have

$$V_4 = 2(4)^{5/2} + 5(4)^2 + 50(4) + 1000$$

$$= 64 + 80 + 200 + 1000$$

$$= 1344$$

Therefore, 4 years from now the anticipated value of the painting would be $1344.

EXERCISES 4.1

In Exercises 1 through 36, perform the antidifferentiation. In Exercises 1 through 8, 15 through 18, and 31 through 34, check by finding the derivative of your answer.

1. $\displaystyle\int 3x^4 \, dx$ **2.** $\displaystyle\int 2x^7 \, dx$ **3.** $\displaystyle\int \frac{1}{x^3} \, dx$ **4.** $\displaystyle\int \frac{3}{t^5} \, dt$ **5.** $\displaystyle\int 5u^{3/2} \, du$

6. $\displaystyle\int 10\sqrt[3]{x^2} \, dx$ **7.** $\displaystyle\int \frac{2}{\sqrt[3]{x}} \, dx$ **8.** $\displaystyle\int \frac{3}{\sqrt{y}} \, dy$ **9.** $\displaystyle\int 6t^2 \sqrt[3]{t} \, dt$ **10.** $\displaystyle\int 7x^3 \sqrt{x} \, dx$

11. $\displaystyle\int (4x^3 + x^2) \, dx$ **12.** $\displaystyle\int (3u^5 - 2u^3) \, du$ **13.** $\displaystyle\int y^3(2y^2 - 3) \, dy$ **14.** $\displaystyle\int x^4(5 - x^2) \, dx$ **15.** $\displaystyle\int (3 - 2t + t^2) \, dt$

16. $\displaystyle\int (4x^3 - 3x^2 + 6x - 1) \, dx$ **17.** $\displaystyle\int (8x^4 + 4x^3 - 6x^2 - 4x + 5) \, dx$ **18.** $\displaystyle\int (2 + 3x^2 - 8x^3) \, dx$

19. $\displaystyle\int \sqrt{x}(x + 1) \, dx$ **20.** $\displaystyle\int (ax^2 + bx + c) \, dx$ **21.** $\displaystyle\int (x^{3/2} - x) \, dx$ **22.** $\displaystyle\int \left(\sqrt{x} - \frac{1}{\sqrt{x}}\right) dx$

23. $\displaystyle\int \left(\frac{2}{x^3} + \frac{3}{x^2} + 5\right) dx$ **24.** $\displaystyle\int \left(3 - \frac{1}{x^4} + \frac{1}{x^2}\right) dx$ **25.** $\displaystyle\int \frac{x^2 + 4x - 4}{\sqrt{x}} \, dx$ **26.** $\displaystyle\int \frac{y^4 + 2y^2 - 1}{\sqrt{y}} \, dy$

27. $\displaystyle\int \left(\sqrt[3]{x} + \frac{1}{\sqrt[3]{x}}\right) dx$ **28.** $\displaystyle\int \frac{27t^3 - 1}{\sqrt[3]{t}} \, dt$ **29.** $\displaystyle\int (3 \sin t - 2 \cos t) \, dt$ **30.** $\displaystyle\int (5 \cos x - 4 \sin x) \, dx$

31. $\displaystyle\int \frac{\sin x}{\cos^2 x} \, dx$ **32.** $\displaystyle\int \frac{\cos x}{\sin^2 x} \, dx$ **33.** $\displaystyle\int (4 \csc x \cot x + 2 \sec^2 x) \, dx$

34. $\displaystyle\int (3 \csc^2 t - 5 \sec t \tan t) \, dt$ **35.** $\displaystyle\int (2 \cot^2 \theta - 3 \tan^2 \theta) \, d\theta$ **36.** $\displaystyle\int \frac{3 \tan \theta - 4 \cos^2 \theta}{\cos \theta} \, d\theta$

37. The point $(3, 2)$ is on a curve, and at any point (x, y) on the curve the tangent line has a slope equal to $2x - 3$. Find an equation of the curve.

38. The slope of the tangent line at any point (x, y) on a curve is $3\sqrt{x}$. If the point $(9, 4)$ is on the curve, find an equation of the curve.

39. The points $(-1, 3)$ and $(0, 2)$ are on a curve, and at any point (x, y) on the curve $\dfrac{d^2y}{dx^2} = 2 - 4x$. Find an equation of the curve. $\left(Hint: \text{ Let } \dfrac{d^2y}{dx^2} = \dfrac{dy'}{dx}, \text{ and obtain an equation involving } y', x, \text{ and an arbitrary constant } C_1. \text{ From this equation obtain another equation involving } y, x, C_1, \text{ and } C_2. \text{ Compute } C_1 \text{ and } C_2 \text{ from the conditions.} \right)$

40. An equation of the tangent line to a curve at the point $(1, 3)$ is $y = x + 2$. If at any point (x, y) on the curve, $\dfrac{d^2y}{dx^2} = 6x$, find an equation of the curve. See the hint for Exercise 39.

41. At any point (x, y) on a curve, $\dfrac{d^2y}{dx^2} = 1 - x^2$, and an equation of the tangent line to the curve at the point $(1, 1)$ is $y = 2 - x$. Find an equation of the curve. See the hint for Exercise 39.

42. At any point (x, y) on a curve, $\dfrac{d^3y}{dx^3} = 2$, and $(1, 3)$ is a point of inflection at which the slope of the inflectional tangent is -2. Find an equation of the curve.

43. The volume of water in a tank is V cubic meters when the depth of the water is h meters. If the rate of change of V with respect to h is $\pi(4h^2 + 12h + 9)$, find the volume of water in the tank when the depth is 3 m.

44. The efficiency of a factory worker is expressed as a percent. For instance, if the worker's efficiency at a particular time is given as 70 percent, then the worker is performing at 70 percent of her full potential. Suppose that E percent is a factory worker's efficiency t hours after beginning work, and the rate at which E is changing is $(35 - 8t)$ percent per hour. If the worker's efficiency is 81 percent after working 3 hr, find her efficiency after working (a) 4 hr and (b) 8 hr.

45. Let $f(x) = 1$ for all x in $(-1, 1)$, and let

$$g(x) = \begin{cases} -1 & \text{if } -1 < x \le 0 \\ 1 & \text{if } 0 < x < 1 \end{cases}$$

Then $f'(x) = 0$ for all x in $(-1, 1)$ and $g'(x) = 0$ whenever g' exists in $(-1, 1)$. However, $f(x) \ne g(x) + K$ for x in $(-1, 1)$. Why doesn't Theorem 4.1.2 hold?

46. Let

$$f(x) = \begin{cases} -1 & \text{if } x < 0 \\ 0 & \text{if } x = 0 \\ 1 & \text{if } 0 < x \end{cases}$$

and $F(x) = |x|$. Show that $F'(x) = f(x)$ if $x \ne 0$. Is F an antiderivative of f on $(-\infty, +\infty)$? Explain.

47. Let $f(x) = |x|$ and F be defined by

$$F(x) = \begin{cases} -\frac{1}{2}x^2 & \text{if } x < 0 \\ \frac{1}{2}x^2 & \text{if } 0 \le x \end{cases}$$

Show that F is an antiderivative of f on $(-\infty, +\infty)$.

48. Let

$$U(x) = \begin{cases} 0 & \text{if } x < 0 \\ 1 & \text{if } 0 \le x \end{cases}$$

Show that U does not have an antiderivative on $(-\infty, +\infty)$. (*Hint:* Assume that U has an antiderivative F on $(-\infty, +\infty)$, and a contradiction is obtained by showing that it follows from the mean-value theorem that there exists a number k such that $F(x) = x + k$ if $x > 0$, and $F(x) = k$ if $x < 0$.)

4.2
SOME TECHNIQUES OF ANTIDIFFERENTIATION

Many antiderivatives cannot be found directly by applying the theorems of Section 4.1. It is therefore necessary to learn certain techniques that can be used in the computation of such antiderivatives. In this section we discuss techniques requiring the *chain rule for antidifferentiation* and those involving a change of variable.

▶ ILLUSTRATION 1   To differentiate $\frac{1}{10}(1 + x^2)^{10}$ we apply the chain rule for differentiation and obtain

$$D_x\left[\tfrac{1}{10}(1 + x^2)^{10}\right] = (1 + x^2)^9(2x)$$

Suppose we wish to antidifferentiate $(1 + x^2)^9(2x)$. Then we need to compute

$$\int (1 + x^2)^9(2x\,dx) \tag{1}$$

To lead up to a procedure that can be used in such a situation, let

$$g(x) = 1 + x^2 \qquad g'(x)\,dx = 2x\,dx \tag{2}$$

Then (1) can be written as

$$\int [g(x)]^9[g'(x)\,dx] \tag{3}$$

From Theorem 4.1.8,

$$\int u^9\,du = \tfrac{1}{10}u^{10} + C \tag{4}$$

Observe that (3) is of the same form as the left-hand side of (4). Thus

$$\int [g(x)]^9[g'(x)\,dx] = \tfrac{1}{10}[g(x)]^{10} + C$$

and with $g(x)$ and $g'(x)\,dx$ given in (2) we have

$$\int (1 + x^2)^9(2x\,dx) = \tfrac{1}{10}(1 + x^2)^{10} + C \tag{5}$$

◀

Justification of the procedure used to obtain (5) is provided by the following theorem, which is analogous to the chain rule for differentiation and is called the *chain rule for antidifferentiation*.

4.2.1 THEOREM
Chain Rule for Antidifferentiation

Let g be a differentiable function, and let the range of g be an interval I. Suppose that f is a function defined on I and that F is an antiderivative of f on I. Then

$$\int f(g(x))[g'(x)\,dx] = F(g(x)) + C \tag{6}$$

PROOF   By hypothesis,

$$F'(g(x)) = f(g(x)) \tag{7}$$

By the chain rule for differentiation,

$$D_x[F(g(x))] = F'(g(x))[g'(x)] \tag{8}$$

Substituting from (7) in (8) we get

$$D_x[F(g(x))] = f(g(x))[g'(x)] \tag{9}$$

From (9) it follows that

$$\int f(g(x))[g'(x)\,dx] = F(g(x)) + C$$

which is what we wished to prove.   ■

As a particular case of Theorem 4.2.1, from Theorem 4.1.8, we have the generalized power formula for antiderivatives, which we now state.

4.2.2 THEOREM   If g is a differentiable function, and n is a rational number,

$$\int [g(x)]^n[g'(x)\,dx] = \frac{[g(x)]^{n+1}}{n+1} + C \qquad n \neq -1 \tag{10}$$

EXAMPLE 1
Find $\int \sqrt{3x + 4}\,dx$.

SOLUTION
To apply Theorem 4.2.2 we first write

$$\int \sqrt{3x + 4}\,dx = \int (3x + 4)^{1/2}\,dx$$

We observe that if

$$g(x) = 3x + 4 \quad \text{then} \quad g'(x)\,dx = 3\,dx \tag{11}$$

Therefore we need a factor of 3 to go with dx to give $g'(x)\,dx$. Hence we write

$$\int (3x + 4)^{1/2}\,dx = \int (3x + 4)^{1/2}\tfrac{1}{3}(3\,dx)$$

$$= \tfrac{1}{3}\int (3x + 4)^{1/2}(3\,dx)$$

Thus from Theorem 4.2.2, with $g(x)$ and $g'(x)\,dx$ given in (11), we have

$$\frac{1}{3}\int (3x + 4)^{1/2}(3\,dx) = \frac{1}{3}\cdot\frac{(3x + 4)^{3/2}}{\frac{3}{2}} + C$$

$$= \tfrac{2}{9}(3x + 4)^{3/2} + C$$

EXAMPLE 2

Find $\int x^2(5 + 2x^3)^8 \, dx$.

SOLUTION

Observe that if

$$g(x) = 5 + 2x^3 \quad \text{then} \quad g'(x) \, dx = 6x^2 \, dx \tag{12}$$

Because

$$\int x^2(5 + 2x^3)^8 \, dx = \int (5 + 2x^3)^8(x^2 \, dx)$$

we need a factor of 6 to go with $x^2 \, dx$ to give $g'(x) \, dx$. Therefore we write

$$\int x^2(5 + 2x^3)^8 \, dx = \tfrac{1}{6} \int (5 + 2x^3)^8(6x^2 \, dx)$$

Applying Theorem 4.2.2 with $g(x)$ and $g'(x) \, dx$ given in (12) we get

$$\frac{1}{6} \int (5 + 2x^3)^8(6x^2 \, dx) = \frac{1}{6} \cdot \frac{(5 + 2x^3)^9}{9} + C$$

$$= \tfrac{1}{54}(5 + 2x^3)^9 + C$$

Equation (6) giving the chain rule for antidifferentiation is

$$\int f(g(x))[g'(x) \, dx] = F(g(x)) + C$$

where F is an antiderivative of f. If in this formula f is the cosine function, then F is the sine function and we have

$$\int \cos(g(x))[g'(x) \, dx] = \sin(g(x)) + C \tag{13}$$

We use this formula in the next example.

EXAMPLE 3

Find $\int x \cos x^2 \, dx$.

SOLUTION

If

$$g(x) = x^2 \quad \text{then} \quad g'(x) \, dx = 2x \, dx \tag{14}$$

Because

$$\int x \cos x^2 \, dx = \int (\cos x^2)(x \, dx)$$

we need a factor of 2 to go with $x \, dx$ to give $g'(x) \, dx$. Thus we write

$$\int x \cos x^2 \, dx = \tfrac{1}{2} \int (\cos x^2)(2x \, dx)$$

We apply (13) with $g(x)$ and $g'(x) \, dx$ given in (14), and we obtain

$$\tfrac{1}{2} \int (\cos x^2)(2x \, dx) = \tfrac{1}{2} \sin x^2 + C$$

The details of the solutions of Examples 1, 2, and 3 can be shortened by not specifically stating $g(x)$ and $g'(x) \, dx$. The solution of Example 1 then takes the following form:

$$\int \sqrt{3x + 4} \, dx = \frac{1}{3} \int (3x + 4)^{1/2}(3 \, dx)$$

$$= \frac{1}{3} \cdot \frac{(3x + 4)^{3/2}}{\frac{3}{2}} + C$$

$$= \tfrac{2}{9}(3x + 4)^{3/2} + C$$

The solution of Example 2 can be written as

$$\int x^2(5 + 2x^3)^8 \, dx = \frac{1}{6} \int (5 + 2x^3)^8(6x^2 \, dx)$$

$$= \frac{1}{6} \cdot \frac{(5 + 2x^3)^9}{9} + C$$

$$= \tfrac{1}{54}(5 + 2x^3)^9 + C$$

and the solution of Example 3 can be shortened as follows:

$$\int x \cos x^2 \, dx = \tfrac{1}{2} \int (\cos x^2)(2x \, dx)$$

$$= \tfrac{1}{2} \sin x^2 + C$$

EXAMPLE 4

Find $\displaystyle\int \frac{4x^2 \, dx}{(1 - 8x^3)^4}$.

SOLUTION
Because $d(1 - 8x^3) = -24x^2 \, dx$, we write

$$\int \frac{4x^2 \, dx}{(1 - 8x^3)^4} = 4 \int (1 - 8x^3)^{-4}(x^2 \, dx)$$

$$= 4\left(-\frac{1}{24}\right) \int (1 - 8x^3)^{-4}(-24x^2 \, dx)$$

$$= -\frac{1}{6} \cdot \frac{(1 - 8x^3)^{-3}}{-3} + C$$

$$= \frac{1}{18(1 - 8x^3)^3} + C$$

The results of each of the above examples can be checked by finding the derivative of the answer.

▶ ILLUSTRATION 2   In Example 2 we have

$$\int x^2(5 + 2x^3)^8 \, dx = \tfrac{1}{54}(5 + 2x^3)^9 + C$$

Checking by differentiation gives

$$D_x\left[\tfrac{1}{54}(5 + 2x^3)^9\right] = \tfrac{1}{54} \cdot 9(5 + 2x^3)^8(6x^2)$$
$$= x^2(5 + 2x^3)^8 \qquad \blacktriangleleft$$

Sometimes it is possible to compute an antiderivative after a change of variable, as shown in the following example.

EXAMPLE 5

Find $\int x^2\sqrt{1 + x}\,dx$.

SOLUTION
Let

$$u = 1 + x \qquad du = dx \qquad x = u - 1$$

We have

$$\int x^2\sqrt{1 + x}\,dx = \int (u - 1)^2 u^{1/2}\,du$$

$$= \int (u^2 - 2u + 1)u^{1/2}\,du$$

$$= \int u^{5/2}\,du - 2\int u^{3/2}\,du + \int u^{1/2}\,du$$

$$= \frac{u^{7/2}}{\frac{7}{2}} - 2 \cdot \frac{u^{5/2}}{\frac{5}{2}} + \frac{u^{3/2}}{\frac{3}{2}} + C$$

$$= \tfrac{2}{7}(1 + x)^{7/2} - \tfrac{4}{5}(1 + x)^{5/2} + \tfrac{2}{3}(1 + x)^{3/2} + C$$

▶ ILLUSTRATION 3   An alternate method for the solution of Example 5 is to let

$$v = \sqrt{1 + x} \qquad v^2 = 1 + x$$
$$x = v^2 - 1 \qquad dx = 2v\,dv$$

The computation then takes the following form:

$$\int x^2\sqrt{1 + x}\,dx = \int (v^2 - 1)^2 \cdot v \cdot (2v\,dv)$$

$$= 2\int v^6\,dv - 4\int v^4\,dv + 2\int v^2\,dv$$

$$= \tfrac{2}{7}v^7 - \tfrac{4}{5}v^5 + \tfrac{2}{3}v^3 + C$$

$$= \tfrac{2}{7}(1 + x)^{7/2} - \tfrac{4}{5}(1 + x)^{5/2} + \tfrac{2}{3}(1 + x)^{3/2} + C$$

Checking by differentiation gives

$$D_x\left[\tfrac{2}{7}(1 + x)^{7/2} - \tfrac{4}{5}(1 + x)^{5/2} + \tfrac{2}{3}(1 + x)^{3/2}\right]$$

$$= (1 + x)^{5/2} - 2(1 + x)^{3/2} + (1 + x)^{1/2}$$
$$= (1 + x)^{1/2}[(1 + x)^2 - 2(1 + x) + 1]$$
$$= (1 + x)^{1/2}[1 + 2x + x^2 - 2 - 2x + 1]$$
$$= x^2\sqrt{1 + x} \qquad \blacktriangleleft$$

EXAMPLE 6

Find $\displaystyle\int \frac{\sin \sqrt{x}}{\sqrt{x}}\, dx$

SOLUTION

Let

$$u = \sqrt{x} \qquad du = \frac{1}{2\sqrt{x}}\, dx$$

Therefore

$$\int \frac{\sin \sqrt{x}}{\sqrt{x}} = 2 \int \sin \sqrt{x}\left(\frac{1}{2\sqrt{x}}\, dx\right)$$

$$= 2 \int \sin u \, du$$

$$= -2 \cos u + C$$

$$= -2 \cos \sqrt{x} + C$$

EXAMPLE 7

Find $\int \sin x \sqrt{1 - \cos x}\, dx$.

SOLUTION

Let

$$u = 1 - \cos x \qquad du = \sin x\, dx$$

Thus

$$\int \sin x \sqrt{1 - \cos x}\, dx = \int u^{1/2}\, du$$

$$= \tfrac{2}{3} u^{3/2} + C$$

$$= \tfrac{2}{3}(1 - \cos x)^{3/2} + C$$

EXAMPLE 8

Find $\int \tan x \sec^2 x\, dx$ by two methods: (a) Let $u = \tan x$; (b) Let $v = \sec x$. (c) Explain the difference in appearance of the answers in (a) and (b).

SOLUTION

(a) If

$$u = \tan x \quad \text{then} \quad du = \sec^2 x\, dx$$

We have

$$\int \tan x \sec^2 x\, dx = \int u\, du$$

$$= \frac{u^2}{2} + C$$

$$= \tfrac{1}{2} \tan^2 x + C$$

(b) If

$$v = \sec x \quad \text{then} \quad dv = \sec x \tan x \, dx$$

Thus

$$\int \tan x \sec^2 x \, dx = \int \sec x (\sec x \tan x \, dx)$$

$$= \int v \, dv$$

$$= \frac{v^2}{2} + C$$

$$= \tfrac{1}{2} \sec^2 x + C$$

(c) Because $\sec^2 x = 1 + \tan^2 x$, the functions defined by $\frac{1}{2} \tan^2 x$ and $\frac{1}{2} \sec^2 x$ differ by a constant; so each serves as an antiderivative of $\tan x \sec^2 x$. Furthermore we can write

$$\tfrac{1}{2} \sec^2 x + C = \tfrac{1}{2}(\tan^2 x + 1) + C$$

$$= \tfrac{1}{2} \tan^2 x + \tfrac{1}{2} + C$$

$$= \tfrac{1}{2} \tan^2 x + K \quad \text{where } K = \tfrac{1}{2} + C$$

EXAMPLE 9

A wound is healing in such a way that t days since Monday the area of the wound has been decreasing at a rate of $-3(t + 2)^{-2}$ cm^2 per day. If on Tuesday the area of the wound was 2 cm^2, (a) what was the area of the wound on Monday, and (b) what is the anticipated area of the wound on Friday if it continues to heal at the same rate?

SOLUTION

Let A cm^2 be the area of the wound t days since Monday. Then

$$\frac{dA}{dt} = -3(t + 2)^{-2}$$

$$A = -3 \int (t + 2)^{-2} \, dt$$

Because $d(t + 2) = dt$ we obtain

$$A = -3 \cdot \frac{(t + 2)^{-1}}{-1} + C$$

$$A = \frac{3}{t + 2} + C \tag{15}$$

Because on Tuesday the area of the wound was 2 cm^2, we know that $A = 2$ when $t = 1$. Substituting these values in (15) we obtain

$$2 = 1 + C$$

$$C = 1$$

Therefore from (15),

$$A = \frac{3}{t+2} + 1 \tag{16}$$

(a) On Monday, $t = 0$. Let A_0 be the value of A when $t = 0$. From (16),

$$A_0 = \tfrac{3}{2} + 1$$
$$= \tfrac{5}{2}$$

Thus on Monday the area of the wound was 2.5 cm².

(b) On Friday, $t = 4$. Let A_4 be the value of A when $t = 4$. From (16),

$$A_4 = \tfrac{3}{6} + 1$$
$$= \tfrac{3}{2}$$

Hence on Friday the anticipated area of the wound is 1.5 cm².

EXERCISES 4.2

In Exercises 1 through 52, perform the antidifferentiation.

1. $\int \sqrt{1 - 4y}\, dy$

2. $\int \sqrt[3]{3x - 4}\, dx$

3. $\int \sqrt[3]{6 - 2x}\, dx$

4. $\int \sqrt{5r + 1}\, dr$

5. $\int x\sqrt{x^2 - 9}\, dx$

6. $\int 3x\sqrt{4 - x^2}\, dx$

7. $\int x^2(x^3 - 1)^{10}\, dx$

8. $\int x(2x^2 + 1)^6\, dx$

9. $\int 5x\sqrt[3]{(9 - 4x^2)^2}\, dx$

10. $\int \dfrac{x\, dx}{(x^2 + 1)^3}$

11. $\int \dfrac{y^3\, dy}{(1 - 2y^4)^5}$

12. $\int \dfrac{s\, ds}{\sqrt{3s^2 + 1}}$

13. $\int (x^2 - 4x + 4)^{4/3}\, dx$

14. $\int x^4\sqrt{3x^5 - 5}\, dx$

15. $\int x\sqrt{x + 2}\, dx$

16. $\int \dfrac{t\, dt}{\sqrt{t + 3}}$

17. $\int \dfrac{2r\, dr}{(1 - r)^7}$

18. $\int x^3(2 - x^2)^{12}\, dx$

19. $\int \sqrt{3 - 2x}\, x^2\, dx$

20. $\int (x^3 + 3)^{1/4}x^5\, dx$

21. $\int \cos 4\theta\, d\theta$

22. $\int \sin \tfrac{1}{3}x\, dx$

23. $\int 6x^2 \sin x^3\, dx$

24. $\int \tfrac{1}{2}t \cos 4t^2\, dt$

25. $\int \sec^2 5x\, dx$

26. $\int \csc^2 2\theta\, d\theta$

27. $\int y \csc 3y^2 \cot 3y^2\, dy$

28. $\int r^2 \sec^2 r^3\, dr$

29. $\int \cos x(2 + \sin x)^5\, dx$

30. $\int \dfrac{4 \sin x\, dx}{(1 + \cos x)^2}$

31. $\int \sqrt{1 + \dfrac{1}{3x}} \dfrac{dx}{x^2}$

32. $\int \sqrt{\dfrac{1}{t} - 1} \dfrac{dt}{t^2}$

33. $\int 2 \sin x \sqrt[3]{1 + \cos x}\, dx$

34. $\int \sin 2x\sqrt{2 - \cos 2x}\, dx$

35. $\int \cos^2 t \sin t\, dt$

36. $\int \sin^3 \theta \cos \theta\, d\theta$

37. $\int (\tan 2x + \cot 2x)^2\, dx$

38. $\int \dfrac{\tfrac{1}{2} \cos \tfrac{1}{4}x}{\sqrt{\sin \tfrac{1}{4}x}}\, dx$

39. $\displaystyle\int \frac{\cos 3x}{\sqrt{1 - 2 \sin 3x}}\,dx$

40. $\displaystyle\int \frac{\sec^2 3\sqrt{t}}{\sqrt{t}}\,dt$

41. $\displaystyle\int \frac{(x^2 + 2x)\,dx}{\sqrt{x^3 + 3x^2 + 1}}$

42. $\displaystyle\int x(x^2 + 1)\sqrt{4 - 2x^2 - x^4}\,dx$

43. $\displaystyle\int \frac{x(3x^2 + 1)\,dx}{(3x^4 + 2x^2 + 1)^2}$

44. $\displaystyle\int \sqrt{3 + s}(s + 1)^2\,ds$

45. $\displaystyle\int \frac{(y + 3)\,dy}{(3 - y)^{2/3}}$

46. $\displaystyle\int (2t^2 + 1)^{1/3}t^3\,dt$

47. $\displaystyle\int \frac{(r^{1/3} + 2)^4\,dr}{\sqrt[3]{r^2}}$

48. $\displaystyle\int \left(t + \frac{1}{t}\right)^{3/2}\left(\frac{t^2 - 1}{t^2}\right)\,dt$

49. $\displaystyle\int \frac{x^3\,dx}{(x^2 + 4)^{3/2}}$

50. $\displaystyle\int \frac{x^3\,dx}{\sqrt{1 - 2x^2}}$

51. $\displaystyle\int \sin x \sin(\cos x)\,dx$

52. $\displaystyle\int \sec x \tan x \cos(\sec x)\,dx$

53. Evaluate $\int (2x + 1)^3\,dx$ by two methods: (a) Expand $(2x + 1)^3$ by the binomial theorem; (b) let $u = 2x + 1$. (c) Explain the difference in appearance of the answers obtained in (a) and (b).

54. Evaluate $\int x(x^2 + 2)^2\,dx$ by two methods: (a) Expand $(x^2 + 2)^2$ and multiply the result by x; (b) let $u = x^2 + 2$. (c) Explain the difference in appearance of the answers obtained in (a) and (b).

55. Evaluate $\int \frac{(\sqrt{x} - 1)^2}{\sqrt{x}}\,dx$ by two methods: (a) Expand $(\sqrt{x} - 1)^2$ and multiply the result by $x^{-1/2}$; (b) let $u = \sqrt{x} - 1$. (c) Explain the difference in appearance of the answers obtained in (a) and (b).

56. Evaluate $\int \sqrt{x - 1}\,x^2\,dx$ by two methods: (a) Let $u = x - 1$; (b) let $v = \sqrt{x - 1}$.

57. Evaluate $\int 2 \sin x \cos x\,dx$ by three methods: (a) Let $u = \sin x$; (b) let $u = \cos x$; (c) use the identity $2 \sin x \cos x = \sin 2x$. (d) Explain the difference in appearance of the answers obtained in (a), (b), and (c).

58. Evaluate $\int \csc^2 x \cot x\,dx$ by two methods: (a) Let $u = \cot x$; (b) let $v = \csc x$. (c) Explain the difference in appearance of the answers obtained in (a) and (b).

59. If q coulombs is the charge of electricity received by a condenser from an electric circuit of i amperes at t sec, then $i = \dfrac{dq}{dt}$. If $i = 5 \sin 60t$ and $q = 0$ when $t = \frac{1}{2}\pi$, find the greatest positive charge on the condenser.

60. Do Exercise 59 if $i = 4 \cos 120t$ and $q = 0$ when $t = 0$.

61. The cost of a certain piece of machinery is \$700, and its value is depreciating with time according to the formula $\dfrac{dV}{dt} = -500(t + 1)^{-2}$, where V dollars is its value t years after its purchase. What is its value three years after its purchase?

62. The volume of a balloon is increasing according to the formula $\dfrac{dV}{dt} = \sqrt{t + 1} + \frac{2}{3}t$, where V cm^3 is the volume of the balloon at t sec. If $V = 33$ when $t = 3$, find (a) a formula for V in terms of t; (b) the volume of the balloon at 8 sec.

63. For the first 10 days in December a plant cell grew in such a way that t days after December 1 the volume of the cell was increasing at a rate of $(12 - t)^{-2}$ cubic micrometers per day. If on December 3 the volume of the cell was 3 μm^3, what was the volume on December 8?

64. The volume of water in a tank is V m^3 when the depth of the water is h meters. If the rate of change of V with respect to h is given by $\dfrac{dV}{dh} = \pi(2h + 3)^2$, find the volume of water in the tank when the depth is 3 m.

Let $s = \bar{s}$ when $t = \frac{1}{3}$. Then

$$\bar{s} = \frac{1}{2\pi} \sin \frac{2}{3} \pi + 5$$

$$= \frac{1}{2\pi} \frac{\sqrt{3}}{2} + 5$$

$$\approx 5.14$$

Thus the particle is 5.14 cm to the right of the origin $\frac{1}{3}$ sec after the start of the motion.

If an object is moving freely in a vertical line and is being pulled toward the earth by a force of gravity, the acceleration due to *gravity*, denoted by g ft/sec^2, varies with the distance of the object from the center of the earth. However, for small changes of distances the accleration due to gravity is almost constant, and an approximate value of g, if the object is near sea level, is 32.

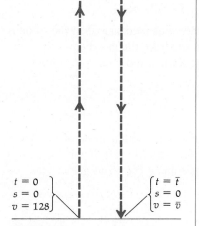

$$\begin{cases} v = 0 \\ s = \bar{s} \end{cases}$$

$$\begin{aligned} t &= 0 \\ s &= 0 \\ v &= 128 \end{aligned}$$

$$\begin{cases} t = \bar{t} \\ s = 0 \\ v = \bar{v} \end{cases}$$

FIGURE 4.3.2

Table 4.3.1

t	s	v
0	0	128
	$\bar{s}$	0
$\bar{t}$	0	$\bar{v}$

EXAMPLE 6

A stone is thrown vertically upward from the ground with an initial velocity of 128 ft/sec. If the only force considered is that attributed to the acceleration due to gravity, find (a) how long it will take for the stone to strike the ground, (b) the speed with which it will strike the ground, and (c) how high the stone will rise.

SOLUTION

The motion of the stone is illustrated in Fig. 4.3.2. The positive direction is upward.

Let t sec be the time that has elapsed since the stone was thrown, s ft be the distance of the stone from the ground at t sec, v ft/sec be the velocity of the stone at t sec, and $|v|$ ft/sec be the speed of the stone at t sec.

When the stone strikes the ground, $s = 0$. Let $\bar{t}$ and $\bar{v}$ be the particular values of t and v when $s = 0$ and $t \neq 0$. The stone will be at its highest point when the velocity is zero. Let $\bar{s}$ be the particular value of s when $v = 0$. Table 4.3.1 shows the boundary conditions.

Because the only acceleration is due to gravity, which is the downward direction, the accleration has a constant value of -32 ft/sec^2. Because the acceleration is given by $\dfrac{dv}{dt}$ we have

$$\frac{dv}{dt} = -32$$

$$dv = -32 \, dt$$

$$\int dv = -32 \int dt$$

$$v = -32t + C_1$$

Since $v = 128$ when $t = 0$, we substitute these values in the above equation and get $C_1 = 128$. Therefore

$$v = -32t + 128 \tag{16}$$

Because $v = \dfrac{ds}{dt}$,

$$\frac{ds}{dt} = -32t + 128$$

$$ds = (-32t + 128)\,dt$$

$$\int ds = \int (-32t + 128)\,dt$$

$$s = -16t^2 + 128t + C_2$$

Since $s = 0$ when $t = 0$, then $C_2 = 0$, and substituting 0 for C_2 in the above equation gives

$$s = -16t^2 + 128t \tag{17}$$

(a) In (17) we substitute $\bar{t}$ for t and 0 for s and get

$$0 = -16\bar{t}(\bar{t} - 8)$$

from which $\bar{t} = 0$ or $\bar{t} = 8$. However, the value 0 occurs when the stone is thrown; so it takes 8 sec for the stone to strike the ground.

(b) To obtain $\bar{v}$ we use (16) and substitute 8 for t and $\bar{v}$ for v to obtain

$$\bar{v} = -32(8) + 128$$

$$= -128$$

Thus $|\bar{v}| = 128$; so the stone strikes the ground with a speed of 128 ft/sec.

(c) To find $\bar{s}$ we first find the value of t for which $v = 0$. From (16), $t = 4$ when $v = 0$. In (17) we substitute 4 for t and $\bar{s}$ for s and get

$$\bar{s} = -16(16) + 128(4)$$

$$= 256$$

Therefore the stone will rise 256 ft.

EXERCISES 4.3

In Exercises 1 through 14, find the complete solution of the differential equation.

1. $\dfrac{dy}{dx} = 4x - 5$

2. $\dfrac{dy}{dx} = 6 - 3x^2$

3. $\dfrac{dy}{dx} = 3x^2 + 2x - 7$

4. $\dfrac{ds}{dt} = 5\sqrt{s}$

5. $\dfrac{dy}{dx} = 3xy^2$

6. $\dfrac{dy}{dx} = \dfrac{\sqrt{x} + x}{\sqrt{y} - y}$

7. $\dfrac{du}{dv} = \dfrac{3v\sqrt{1 + u^2}}{u}$

8. $\dfrac{dy}{dx} = \dfrac{x^2\sqrt{x^3 - 3}}{y^2}$

9. $\dfrac{dy}{dx} = \dfrac{\sec^2 x}{\tan^2 y}$

10. $\dfrac{du}{dv} = \dfrac{\cos 2v}{\sin 3u}$

11. $\dfrac{d^2 y}{dx^2} = 5x^2 + 1$

12. $\dfrac{d^2 y}{dx^2} = \sqrt{2x - 3}$

13. $\dfrac{d^2 s}{dt^2} = \sin 3t + \cos 3t$

14. $\dfrac{d^2 u}{dv^2} = \tan v \sec^2 v$

In Exercises 15 through 20, find the particular solution of the given differential equation determined by the initial conditions.

15. $\dfrac{dy}{dx} = x^2 - 2x - 4$; $y = -6$ when $x = 3$

16. $\dfrac{dy}{dx} = (x + 1)(x + 2)$; $y = -\frac{3}{2}$ when $x = -3$

17. $\dfrac{dy}{dx} = \dfrac{\cos 3x}{\sin 2y}$; $y = \frac{1}{3}\pi$ when $x = \frac{1}{2}\pi$

18. $\dfrac{ds}{dt} = \cos \frac{1}{2}t$; $s = 3$ when $t = \frac{1}{3}\pi$

19. $\dfrac{d^2 u}{dv^2} = 4(1 + 3v)^2$; $u = -1$ and $\dfrac{du}{dv} = -2$ when $v = -1$

20. $\dfrac{d^2 y}{dx^2} = -\dfrac{3}{x^4}$; $y = \frac{1}{2}$ and $\dfrac{dy}{dx} = -1$ when $x = 1$

In Exercises 21 through 32, a particle is moving on a straight line, s ft is the directed distance of the particle from the origin at t sec, v ft/sec is the velocity of the particle at t sec, and a ft/sec^2 is the acceleration of the particle at t sec.
$\left(\text{Hint for Exercises 29 through 32: } a = \dfrac{dv}{dt} = \dfrac{dv}{ds}\dfrac{ds}{dt} = v\dfrac{dv}{ds}.\right)$

21. $v = \sqrt{2t + 4}$; $s = 0$ when $t = 0$. Express s in terms of t.
22. $v = 4 - t$; $s = 0$ when $t = 2$. Express s in terms of t.
23. $a = 5 - 2t$; $v = 2$ and $s = 0$ when $t = 0$. Express v and s in terms of t.
24. $a = 17$; $v = 0$ and $s = 0$ when $t = 0$. Express v and s in terms of t.
25. $a = t^2 + 2t$; $s = 1$ when $t = 0$ and $s = -3$ when $t = 2$. Express v and s in terms of t.
26. $a = 3t - t^2$; $v = \frac{7}{6}$ and $s = 1$ when $t = 1$. Express v and s in terms of t.
27. $a = -4\sqrt{2}\cos(2t - \frac{1}{4}\pi)$; $v = 2$ and $s = 1$ when $t = 0$. Express v and s in terms of t.
28. $a = 18 \sin 3t$; $v = -6$ and $s = 4$ when $t = 0$. Express v and s in terms of t.
29. $a = 800$; $v = 20$ when $s = 1$. Find an equation involving v and s.
30. $a = 500$; $v = 10$ when $s = 5$. Find an equation involving v and s.
31. $a = 5s + 2$; $v = 4$ when $s = 2$. Find an equation involving v and s.
32. $a = 2s + 1$; $v = 2$ when $s = 1$. Find an equation involving v and s.

In Exercises 33 through 40, the only force considered is that due to the acceleration of gravity, which we take as 32 ft/sec^2 in the downward direction.

33. A stone is thrown vertically upward from the ground with an initial velocity of 20 ft/sec. (a) How long will it take the stone to strike the ground? (b) With what speed will the stone strike the ground? (c) How long will the stone be going upward? (d) How high will the stone go?
34. A stone is thrown vertically upward from the ground with an initial velocity of 24 ft/sec. (a) How long will it take the stone to strike the ground, and (b) with what speed will the stone strike the ground?
35. A ball is dropped from the top of the Washington Monument, which is 555 ft high. (a) How long will it take the ball to reach the ground, and (b) with what speed will it strike the ground?
36. A stone is thrown vertically upward from the ground with an initial velocity of 16 ft/sec. (a) How long will the stone be going upward, and (b) how high will the stone go?
37. A woman in a balloon drops her binoculars when the balloon is 150 ft above the ground and rising at the rate of 10 ft/sec. (a) How long will it take the binoculars to strike the ground, and (b) what is their speed on impact?
38. A ball is thrown downward from a window that is 80 ft above the ground with an initial velocity of -64 ft/sec. (a) When does the ball strike the ground, and (b) with what speed will it strike the ground?

39. A ball is thrown vertically upward with an initial velocity of 40 ft/sec from a point 20 ft above the ground. (a) If v ft/sec is the velocity of the ball when it is s ft from the starting point, express v in terms of s. (b) What is the velocity of the ball when it is 36 ft from the ground and rising? (*Hint:* See hint for Exercises 29 through 32.)

40. A stone is thrown vertically upward from the top of a house 60 ft above the ground with an initial velocity of 40 ft/sec. (a) At what time will the stone reach its greatest height, and (b) what is its greatest height? (c) How long will it take the stone to pass the top of the house on its way down, and (d) what is its velocity at that instant? (e) How long will it take the stone to strike the ground, and (f) with what velocity will it strike the ground?

41. A particle is moving along a straight line in such a way that if v cm/sec is the velocity of the particle at t sec, then $v = \sin \pi t$, where the positive direction is to the right of the origin. If the particle is at the origin at the start of the motion, find its position $\frac{2}{3}$ sec later.

42. If a ball is rolled across level ground with an initial velocity of 20 ft/sec, and if the speed of the ball is decreasing at the rate of 6 ft/sec^2 due to friction, how far will the ball roll?

43. If the driver of an automobile wishes to increase the speed from 40 km/hr to 100 km/hr while traveling a distance of 200 m, what constant acceleration should be maintained?

44. What constant negative acceleration will enable a driver to decrease the speed from 120 km/hr to 60 km/hr while traveling a distance of 100 m?

45. If the brakes are applied on a car traveling 100 km/hr and the brakes can give the car a constant negative acceleration of 8 m/sec^2, (a) how long will it take the car to come to a stop, and (b) how far will the car travel before stopping?

46. A ball started upward from the bottom of an inclined plane with an initial velocity of 6 ft/sec. If there was a downward acceleration of 4 ft/sec^2, how far up the plane did the ball go before rolling down?

47. If the brakes on a car can give the car a constant negative acceleration of 8 m/sec^2, what is the greatest speed it may be going if it is necessary to be able to stop the car within 25 m after the brake is applied?

48. A block of ice slides down a chute with a constant acceleration of 3 m/sec^2. The chute is 36 m long and it takes 4 sec for the ice to reach the bottom. (a) What is the initial velocity of the ice? (b) What is the speed of the ice after it has traveled 12 m? (c) How long does it take the ice to go the 12 m?

49. The equation $x^2 = 4ay$ represents a one-parameter family of parabolas. Find an equation of another one-parameter family of curves such that at any point (x, y) there is a curve of each family through it and the tangent lines to the two curves at this point are perpendicular. (*Hint:* First show that the slope of the tangent line at any point (x, y), not on the y axis, of the parabola of the given family through that point is $2y/x$.)

50. Solve Exercise 49 if the given one-parameter family of curves has the equation $x^3 + y^3 = a^3$.

4.4
AREA

FIGURE 4.4.1

You probably have an intuitive idea of what is meant by the *area* of certain geometrical figures. It is a measurement that in some way gives the size of the region enclosed by the figure. The area of a rectangle is the product of its length and width, and the area of a triangle is half the product of the lengths of the base and the altitude.

The area of a polygon can be defined as the sum of the areas of triangles into which it is decomposed, and it can be proved that the area thus obtained is independent of how the polygon is decomposed into triangles (see Fig. 4.4.1). However, how do we define the area of a region in a plane if the region is bounded by a curve? In this section we give the definition of the area of such a region, and in Section 4.5 we use this definition to motivate the definition of the *definite integral*.

In our treatment of area we shall be concerned with the sums of many terms, and so a notation called the **sigma notation** is introduced to facilitate writing these sums. This notation involves the use of the symbol $\sum$, the capital sigma of the Greek alphabet. Some examples of the sigma notation are given in the following illustration.

▶ ILLUSTRATION 1

$$\sum_{i=1}^{5} i^2 = 1^2 + 2^2 + 3^2 + 4^2 + 5^2$$

$$\sum_{i=-2}^{2} (3i + 2) = [3(-2) + 2] + [3(-1) + 2] + [3 \cdot 0 + 2]$$
$$+ [3 \cdot 1 + 2] + [3 \cdot 2 + 2]$$
$$= (-4) + (-1) + 2 + 5 + 8$$

$$\sum_{j=1}^{n} j^3 = 1^3 + 2^3 + 3^3 + \ldots + n^3$$

$$\sum_{k=3}^{8} \frac{1}{k} = \frac{1}{3} + \frac{1}{4} + \frac{1}{5} + \frac{1}{6} + \frac{1}{7} + \frac{1}{8}$$ ◀

We have the formal definition of the sigma notation.

4.4.1 DEFINITION

$$\sum_{i=m}^{n} F(i) = F(m) + F(m + 1) + F(m + 2) + \ldots + F(n - 1) + F(n) \quad (1)$$

where m and n are integers, and $m \leq n$.

The right side of (1) consists of the sum of $(n - m + 1)$ terms, the first of which is obtained by replacing i by m in $F(i)$, the second by replacing i by $m + 1$ in $F(i)$, and so on, until the last term is obtained by replacing i by n in $F(i)$.

The number m is called the **lower limit** of the sum, and n is called the **upper limit**. The symbol i is called the **index of summation**. It is a "dummy" symbol because any other letter can be used for this purpose. For example,

$$\sum_{k=3}^{5} k^2 = 3^2 + 4^2 + 5^2$$

is equivalent to

$$\sum_{i=3}^{5} i^2 = 3^2 + 4^2 + 5^2$$

▶ ILLUSTRATION 2   From Definition 4.4.1,

$$\sum_{i=3}^{6} \frac{i^2}{i + 1} = \frac{3^2}{3 + 1} + \frac{4^2}{4 + 1} + \frac{5^2}{5 + 1} + \frac{6^2}{6 + 1}$$ ◀

Sometimes the terms of a sum involve subscripts, as shown in the next illustration.

▶ ILLUSTRATION 3

$$\sum_{i=1}^{n} A_i = A_1 + A_2 + \ldots + A_n$$

$$\sum_{k=4}^{9} k b_k = 4b_4 + 5b_5 + 6b_6 + 7b_7 + 8b_8 + 9b_9$$

$$\sum_{i=1}^{5} f(x_i)\Delta x = f(x_1)\Delta x + f(x_2)\Delta x + f(x_3)\Delta x + f(x_4)\Delta x + f(x_5)\Delta x$$ ◀

The following theorems involving the sigma notation are useful for computation and are easily proved.

4.4.2 THEOREM

$$\sum_{i=1}^{n} c = cn, \text{ where } c \text{ is any constant}$$

PROOF

$$\sum_{i=1}^{n} c = c + c + \ldots + c \quad (n \text{ terms})$$

$$= cn$$ ∎

4.4.3 THEOREM

$$\sum_{i=1}^{n} c \cdot F(i) = c \sum_{i=1}^{n} F(i), \text{ where } c \text{ is any constant}$$

PROOF

$$\sum_{i=1}^{n} c \cdot F(i) = c \cdot F(1) + c \cdot F(2) + c \cdot F(3) + \ldots + c \cdot F(n)$$

$$= c[F(1) + F(2) + F(3) + \ldots + F(n)]$$

$$= c \sum_{i=1}^{n} F(i)$$ ∎

4.4.4 THEOREM

$$\sum_{i=1}^{n} [F(i) + G(i)] = \sum_{i=1}^{n} F(i) + \sum_{i=1}^{n} G(i)$$

The proof is left as an exercise (see Exercise 51). Theorem 4.4.4 can be extended to the sum of any number of functions.

4.4.5 THEOREM

$$\sum_{i=a}^{b} F(i) = \sum_{i=a+c}^{b+c} F(i - c) \tag{2}$$

and

$$\sum_{i=a}^{b} F(i) = \sum_{i=a-c}^{b-c} F(i + c) \tag{3}$$

The proof of Theorem 4.4.5 is left as an exercise (see Exercise 52). The following illustration shows the application of this theorem.

▶ ILLUSTRATION 4   From Eq. (2) of Theorem 4.4.5,

$$\sum_{i=3}^{10} F(i) = \sum_{i=5}^{12} F(i-2) \quad \text{and} \quad \sum_{i=6}^{11} i^2 = \sum_{i=7}^{12} (i-1)^2$$

From Eq. (3) of Theorem 4.4.5,

$$\sum_{i=3}^{10} F(i) = \sum_{i=1}^{8} F(i+2) \quad \text{and} \quad \sum_{i=6}^{11} i^2 = \sum_{i=1}^{6} (i+5)^2 \qquad \blacktriangleleft$$

4.4.6 THEOREM

$$\sum_{i=1}^{n} [F(i) - F(i-1)] = F(n) - F(0)$$

PROOF

$$\sum_{i=1}^{n} [F(i) - F(i-1)] = \sum_{i=1}^{n} F(i) - \sum_{i=1}^{n} F(i-1)$$

On the right side of this equation we write the first summation in another form and apply Eq. (3) of Theorem 4.4.5 with $c = 1$ to the second summation. Then

$$\sum_{i=1}^{n} [F(i) - F(i-1)] = \left(\sum_{i=1}^{n-1} F(i) + F(n) \right) - \sum_{i=1-1}^{n-1} F[(i+1) - 1]$$

$$= \sum_{i=1}^{n-1} F(i) + F(n) - \sum_{i=0}^{n-1} F(i)$$

$$= \sum_{i=1}^{n-1} F(i) + F(n) - \left(F(0) + \sum_{i=1}^{n-1} F(i) \right)$$

$$= F(n) - F(0) \qquad \blacksquare$$

EXAMPLE 1

Evaluate $\displaystyle\sum_{i=1}^{n} (4^i - 4^{i-1})$.

SOLUTION

From Theorem 4.4.6, where $F(i) = 4^i$, it follows that

$$\sum_{i=1}^{n} (4^i - 4^{i-1}) = 4^n - 4^0$$
$$= 4^n - 1$$

In the following theorem there are four formulas that are useful for computation with sigma notation. They are numbered for future reference.

4.4.7 THEOREM   If n is a positive integer, then

$$\sum_{i=1}^{n} i = \frac{n(n+1)}{2} \qquad \text{(Formula 1)}$$

$$\sum_{i=1}^{n} i^2 = \frac{n(n+1)(2n+1)}{6} \qquad \text{(Formula 2)}$$

$$\sum_{i=1}^{n} i^3 = \frac{n^2(n+1)^2}{4} \qquad \text{(Formula 3)}$$

$$\sum_{i=1}^{n} i^4 = \frac{n(n+1)(6n^3+9n^2+n-1)}{30} \qquad \text{(Formula 4)}$$

PROOF   We prove Formula 1 first.

$$\sum_{i=1}^{n} i = 1 + 2 + 3 + \ldots + (n-1) + n$$

and

$$\sum_{i=1}^{n} i = n + (n-1) + (n-2) + \ldots + 2 + 1$$

If the left sides of these two equations are added, the result is

$$2 \sum_{i=1}^{n} i$$

and if the right sides are added term by term, we obtain n terms, each having the value $(n+1)$. Hence

$$2 \sum_{i=1}^{n} i = (n+1) + (n+1) + (n+1) + \ldots + (n+1) \quad n \text{ terms}$$
$$= n(n+1)$$

Therefore

$$\sum_{i=1}^{n} i = \frac{n(n+1)}{2}$$

which is Formula 1. Formula 1, as well as the other three formulas, can be proved by mathematical induction. We now give the proof of Formula 2 by mathematical induction. The proof consists of two parts and a conclusion.

Part 1:   First Formula 2 is verified for $n = 1$. When $n = 1$, the left side of Formula 2 is

$$\sum_{i=1}^{1} i^2 = 1$$

and the right side is

$$\frac{1(1+1)(2+1)}{6} = \frac{1 \cdot 2 \cdot 3}{6}$$

$$= 1$$

Hence Formula 2 is true when $n = 1$.

Part 2: We assume that Formula 2 is true when $n = k$, where k is any positive integer; that is, we assume

ASSUME $$\sum_{i=1}^{k} i^2 = \frac{k(k+1)(2k+1)}{6} \qquad (4)$$

With this assumption we wish to prove that the formula is also true when $n = k + 1$; that is, we wish to prove

PROVE $$\sum_{i=1}^{k+1} i^2 = \frac{(k+1)[(k+1)+1][2(k+1)+1]}{6} \qquad (5)$$

When $n = k + 1$, we have

$$\sum_{i=1}^{k+1} i^2 = 1^2 + 2^2 + 3^2 + \ldots + k^2 + (k+1)^2$$

$$= \sum_{i=1}^{k} i^2 + (k+1)^2$$

$$= \frac{k(k+1)(2k+1)}{6} + (k+1)^2 \quad \text{(by applying (4))}$$

$$= \frac{k(k+1)(2k+1) + 6(k+1)^2}{6} \qquad \textit{common denom}$$

$$= \frac{(k+1)[k(2k+1) + 6(k+1)]}{6} \qquad \textit{factor} \qquad {}^{2k^2+k\ +6k+6}$$

$$= \frac{(k+1)(2k^2 + 7k + 6)}{6} \qquad \textit{mult commutive}$$

$$= \frac{(k+1)(k+2)(2k+3)}{6} \qquad \textit{factor}$$

$$= \frac{(k+1)[(k+1)+1][2(k+1)+1]}{6}$$

which is (5).

Conclusion: From part 1 we know that Formula 2 is true for $n = 1$. Because it is true for $n = 1$, it follows from part 2 that the formula is true for $n = 1 + 1$, or 2; because it is true for $n = 2$, it is true for $n = 2 + 1$, or 3; and so on. By the principle of mathematical induction the formula is true for all positive-integer values of n. ∎

The proofs of Formulas 3 and 4 are left as exercises (see Exercises 45 through 50).

EXAMPLE 2

Evaluate $\displaystyle\sum_{i=1}^{n} i(3i - 2)$.

SOLUTION

$$\sum_{i=1}^{n} i(3i - 2) = \sum_{i=1}^{n} (3i^2 - 2i)$$

$$= \sum_{i=1}^{n} (3i^2) + \sum_{i=1}^{n} (-2i) \quad \text{(by Theorem 4.4.4)}$$

$$= 3 \sum_{i=1}^{n} i^2 - 2 \sum_{i=1}^{n} i \quad \text{(by Theorem 4.4.3)}$$

$$= 3 \cdot \frac{n(n + 1)(2n + 1)}{6} - 2 \cdot \frac{n(n + 1)}{2} \quad \text{(by Formulas 2 and 1)}$$

$$= \frac{2n^3 + 3n^2 + n - 2n^2 - 2n}{2}$$

$$= \frac{2n^3 + n^2 - n}{2}$$

Before discussing the area of a plane region, we wish to indicate why the terminology "*measure* of the area" is used. The word *measure* refers to a number (no units are included). For example, if the area of a triangle is 20 cm², we say that the square-centimeter measure of the area of the triangle is 20. When the word *measurement* is applied, the units are included. Thus the measurement of the area of the triangle is 20 cm².

Consider now a region R in the plane as shown in Fig. 4.4.2. The region R is bounded by the x axis, the lines $x = a$ and $x = b$, and the curve having the equation $y = f(x)$, where f is a function continuous on the closed interval $[a, b]$. For simplicity, take $f(x) \geq 0$ for all x in $[a, b]$. We wish to assign a number A to be the measure of the area of R, and we use a limiting process similar to the one used in defining the area of a circle: The area of a circle is defined as the limit of the areas of inscribed regular polygons as the number of sides increases without bound. We realize intuitively that, whatever number is chosen to represent A, that number must be at least as great as the measure of the area of any polygonal region contained in R, and it must be no greater than the measure of the area of any polygonal region containing R.

We first define a polygonal region contained in R. Divide the closed interval $[a, b]$ into n subintervals. For simplicity, we now take each of these subintervals as being of equal length, for instance, Δx. Therefore $\Delta x = (b - a)/n$. Denote the endpoints of these subintervals by $x_0, x_1, x_2, \ldots, x_{n-1}, x_n$, where $x_0 = a$, $x_1 = a + \Delta x, \ldots, x_i = a + i\,\Delta x, \ldots, x_{n-1} = a + (n - 1)\Delta x$, $x_n = b$. Let the ith subinterval be denoted by $[x_{i-1}, x_i]$. Because f is continuous on the closed interval $[a, b]$, it is continuous on each closed subinterval.

$y = f(x)$

R

FIGURE 4.4.2

FIGURE 4.4.3

By the extreme-value theorem there is a number in each subinterval for which f has an absolute minimum value. In the ith subinterval let this number be c_i, so that $f(c_i)$ is the absolute minimum value of f on the subinterval $[x_{i-1}, x_i]$. Consider n rectangles, each having a width Δx units and an altitude $f(c_i)$ units (see Fig. 4.4.3). Let the sum of the areas of these n rectangles be given by S_n square units; then

$$S_n = f(c_1)\Delta x + f(c_2)\Delta x + \ldots + f(c_i)\Delta x + \ldots + f(c_n)\Delta x$$

or, with the sigma notation,

$$S_n = \sum_{i=1}^{n} f(c_i)\Delta x \qquad (6)$$

The summation on the right side of (6) gives the sum of the measures of the areas of n inscribed rectangles. Thus, however we define A, it must be such that

$$A \geq S_n$$

In Fig. 4.4.3 the shaded region has an area of S_n square units. Now let n increase. Specifically, multiply n by 2; then the number of rectangles is doubled, and the width of each rectangle is halved. This is illustrated in Fig. 4.4.4, showing twice as many rectangles as Fig. 4.4.3. By comparing the two figures, notice that the shaded region in Fig. 4.4.4 appears to approximate the region R more nearly than that of Fig. 4.4.3. So the sum of the measures of the areas of the rectangles in Fig. 4.4.4 is closer to the number we wish to represent the measure of the area of R.

As n increases, the values of S_n found from Eq. (6) increase, and successive values of S_n differ from each other by amounts that become arbitrarily

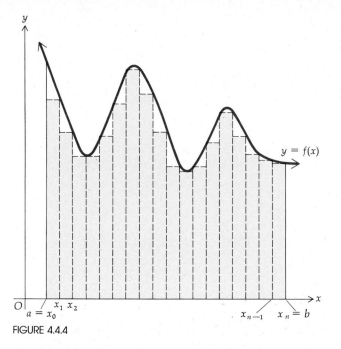

FIGURE 4.4.4

small. This is proved in advanced calculus by a theorem that states that if f is continuous on $[a,b]$, then as n increases without bound, the value of S_n given by (6) approaches a limit. It is this limit that we take as the definition of the measure of the area of region R.

4.4.8 DEFINITION   Suppose that the function f is continuous on the closed interval $[a,b]$, with $f(x) \geq 0$ for all x in $[a,b]$, and that R is the region bounded by the curve $y = f(x)$, the x axis, and the lines $x = a$ and $x = b$. Divide the interval $[a,b]$ into n subintervals, each of length $\Delta x = (b - a)/n$, and denote the ith subinterval by $[x_{i-1}, x_i]$. Then if $f(c_i)$ is the absolute minimum function value on the ith subinterval, the measure of the area of region R is given by

$$A = \lim_{n \to +\infty} \sum_{i=1}^{n} f(c_i) \Delta x \tag{7}$$

Equation (7) means that for any $\epsilon > 0$ there is a number $N > 0$ such that if n is a positive integer and

$$\text{if} \quad n > N \quad \text{then} \quad \left| \sum_{i=1}^{n} f(c_i) \Delta x - A \right| < \epsilon$$

We could take circumscribed rectangles instead of inscribed rectangles. In this case we take as the measures of the altitudes of the rectangles the absolute maximum value of f on each subinterval. The existence of an absolute maximum value of f on each subinterval is guaranteed by the extreme-value theorem. The corresponding sums of the measures of the areas of the circumscribed rectangles are at least as great as the measure of the area of the region R, and it can be shown that the limit of these sums as n increases without bound is exactly the same as the limit of the sum of the measures of the areas of the inscribed rectangles. This is also proved in advanced calculus. Thus we could define the measure of the area of the region R by

$$A = \lim_{n \to +\infty} \sum_{i=1}^{n} f(d_i)\,\Delta x \qquad (8)$$

where $f(d_i)$ is the absolute maximum value of f in $[x_{i-1}, x_i]$.

The measure of the altitude of the rectangle in the ith subinterval actually can be taken as the function value of any number in that subinterval, and the limit of the sum of the measures of the areas of the rectangles is the same no matter what numbers are selected. In Section 4.5 we extend the definition of the measure of the area of a region to be the limit of such a sum.

EXAMPLE 3

Find the area of the region bounded by the curve $y = x^2$, the x axis, and the line $x = 3$ by taking inscribed rectangles.

SOLUTION

Figure 4.4.5 shows the region and the ith inscribed rectangle. We apply Definition 4.4.8. Divide the closed interval $[0, 3]$ into n subintervals, each of length Δx: $x_0 = 0$, $x_1 = \Delta x$, $x_2 = 2\,\Delta x, \ldots, x_i = i\,\Delta x, \ldots, x_{n-1} = (n-1)\,\Delta x$, $x_n = 3$.

$$\Delta x = \frac{3-0}{n} \qquad f(x) = x^2$$

$$= \frac{3}{n}$$

Because f is increasing on $[0, 3]$, the absolute minimum value of f on the ith subinterval $[x_{i-1}, x_i]$ is $f(x_{i-1})$. Therefore, from (7),

$$A = \lim_{n \to +\infty} \sum_{i=1}^{n} f(x_{i-1})\,\Delta x \qquad (9)$$

Because $x_{i-1} = (i-1)\,\Delta x$ and $f(x) = x^2$,

$$f(x_{i-1}) = [(i-1)\,\Delta x]^2$$

Therefore

$$\sum_{i=1}^{n} f(x_{i-1})\,\Delta x = \sum_{i=1}^{n} (i-1)^2 (\Delta x)^3$$

But $\Delta x = 3/n$; so

$$\sum_{i=1}^{n} f(x_{i-1})\,\Delta x = \sum_{i=1}^{n} (i-1)^2 \frac{27}{n^3}$$

$$= \frac{27}{n^3} \sum_{i=1}^{n} (i-1)^2$$

$$= \frac{27}{n^3} \left[\sum_{i=1}^{n} i^2 - 2 \sum_{i=1}^{n} i + \sum_{i=1}^{n} 1 \right]$$

FIGURE 4.4.5

and using Formulas 2 and 1 and Theorem 4.4.2 we get

$$\sum_{i=1}^{n} f(x_{i-1})\Delta x = \frac{27}{n^3}\left[\frac{n(n+1)(2n+1)}{6} - 2\cdot\frac{n(n+1)}{2} + n\right]$$

$$= \frac{27}{n^3}\cdot\frac{2n^3 + 3n^2 + n - 6n^2 - 6n + 6n}{6}$$

$$= \frac{9}{2}\cdot\frac{2n^2 - 3n + 1}{n^2}$$

Then, from (9),

$$A = \lim_{n\to+\infty}\left[\frac{9}{2}\cdot\frac{2n^2 - 3n + 1}{n^2}\right]$$

$$= \frac{9}{2}\cdot\lim_{n\to+\infty}\left(2 - \frac{3}{n} + \frac{1}{n^2}\right)$$

$$= \tfrac{9}{2}(2 - 0 + 0)$$

$$= 9$$

Therefore the area of the region is 9 square units.

EXAMPLE 4
Find the area of the region in Example 3 by taking circumscribed rectangles.

SOLUTION
With circumscribed rectangles the measure of the altitude of the ith rectangle is the absolute maximum value of f on the ith subinterval $[x_{i-1}, x_i]$, which is $f(x_i)$. From (8),

$$A = \lim_{n\to+\infty}\sum_{i=1}^{n} f(x_i)\Delta x \tag{10}$$

Because $x_i = i\,\Delta x$, then $f(x_i) = (i\,\Delta x)^2$, and so

$$\sum_{i=1}^{n} f(x_i)\Delta x = \sum_{i=1}^{n} i^2(\Delta x)^3$$

$$= \frac{27}{n^3}\sum_{i=1}^{n} i^2$$

$$= \frac{27}{n^3}\left[\frac{n(n+1)(2n+1)}{6}\right]$$

$$= \frac{9}{2}\cdot\frac{2n^2 + 3n + 1}{n^2}$$

Therefore, from Eq. (10),

$$A = \lim_{n\to+\infty}\frac{9}{2}\cdot\left(2 + \frac{3}{n} + \frac{1}{n^2}\right)$$

$$= 9 \quad \text{(as in Example 3)}$$

FIGURE 4.4.6

EXAMPLE 5
Find the area of the trapezoid that is the region bounded by the lines $x = 1$ and $x = 3$, the x-axis, and the line $2x + y = 8$. Take inscribed rectangles.

SOLUTION
The region and the ith inscribed rectangle are shown in Fig. 4.4.6. The closed interval $[1, 3]$ is divided into n subintervals, each of length Δx: $x_0 = 1$, $x_1 = 1 + \Delta x$, $x_2 = 1 + 2\Delta x, \ldots, x_i = 1 + i\Delta x, \ldots, x_{n-1} = 1 + (n-1)\Delta x$, $x_n = 3$.

$$\Delta x = \frac{3 - 1}{n}$$

$$= \frac{2}{n}$$

Solving the equation of the line for y we obtain $y = -2x + 8$. Therefore $f(x) = -2x + 8$, and because f is decreasing on $[1, 3]$, the absolute minimum value of f on the ith subinterval $[x_{i-1}, x_i]$ is $f(x_i)$. Because $x_i = 1 + i\Delta x$ and $f(x) = -2x + 8$, then $f(x_i) = -2(1 + i\Delta x) + 8$, that is, $f(x_i) = 6 - 2i\Delta x$. From (7),

$$A = \lim_{n \to +\infty} \sum_{i=1}^{n} f(x_i)\Delta x$$

$$= \lim_{n \to +\infty} \sum_{i=1}^{n} (6 - 2i\Delta x)\Delta x$$

$$= \lim_{n \to +\infty} \sum_{i=1}^{n} \left[6\Delta x - 2i(\Delta x)^2\right]$$

$$= \lim_{n \to +\infty} \sum_{i=1}^{n} \left[6\left(\frac{2}{n}\right) - 2i\left(\frac{2}{n}\right)^2\right]$$

$$= \lim_{n \to +\infty} \left[\frac{12}{n} \sum_{i=1}^{n} 1 - \frac{8}{n^2} \sum_{i=1}^{n} i\right]$$

From Theorem 4.4.2 and Formula 1,

$$A = \lim_{n \to +\infty} \left[\frac{12}{n} \cdot n - \frac{8}{n^2} \cdot \frac{n(n+1)}{2}\right]$$

$$= \lim_{n \to +\infty} \left(8 - \frac{4}{n}\right)$$

$$= 8$$

Therefore the area is 8 square units. Using the formula from plane geometry for the area of a trapezoid, $A = \frac{1}{2}h(b_1 + b_2)$, where h, b_1, and b_2 are, respectively, the number of units in the lengths of the altitude and the two bases, we get $A = \frac{1}{2}(2)(6 + 2)$; that is, $A = 8$, which agrees with the result.

EXERCISES 4.4

In Exercises 1 through 12, find the given sum.

1. $\displaystyle\sum_{i=1}^{6} (3i - 2)$ **2.** $\displaystyle\sum_{i=1}^{20} (5i + 4)$ **3.** $\displaystyle\sum_{i=1}^{7} (i^2 + 1)$ **4.** $\displaystyle\sum_{i=1}^{7} (i + 1)^2$ **5.** $\displaystyle\sum_{i=1}^{10} (i - 1)^3$ **6.** $\displaystyle\sum_{i=1}^{10} (i^3 - 1)$

7. $\displaystyle\sum_{i=2}^{5} \frac{i}{i-1}$ **8.** $\displaystyle\sum_{j=3}^{6} \frac{2}{j(j-2)}$ **9.** $\displaystyle\sum_{i=-2}^{3} 2^i$ **10.** $\displaystyle\sum_{i=0}^{3} \frac{1}{1+i^2}$ **11.** $\displaystyle\sum_{k=1}^{4} \frac{(-1)^{k+1}}{k}$ **12.** $\displaystyle\sum_{k=-2}^{3} \frac{k}{k+3}$

In Exercises 13 through 20, evaluate the indicated sum by using Theorems 4.4.2 through 4.4.7.

13. $\displaystyle\sum_{i=1}^{25} 2i(i - 1)$ **14.** $\displaystyle\sum_{i=1}^{20} 3i(i^2 + 2)$ **15.** $\displaystyle\sum_{k=1}^{n} (2^k - 2^{k-1})$ **16.** $\displaystyle\sum_{i=1}^{n} (10^{i+1} - 10^i)$ **17.** $\displaystyle\sum_{k=1}^{100} \left[\frac{1}{k} - \frac{1}{k+1}\right]$

18. $\displaystyle\sum_{i=1}^{n} 2i(1 + i^2)$ **19.** $\displaystyle\sum_{i=1}^{n} 4i^2(i - 2)$ **20.** $\displaystyle\sum_{k=1}^{n} [(3^{-k} - 3^k)^2 - (3^{k-1} + 3^{-k-1})^2]$

In Exercises 21 through 36, use the method of this section to find the area of the given region; use inscribed or circumscribed rectangles as indicated. For each exercise draw a figure showing the region and the ith rectangle.

21. The region bounded by $y = x^2$, the x axis, and the line $x = 2$; inscribed rectangles.

22. The region of Exercise 21; circumscribed rectangles.

23. The region bounded by $y = 2x$, the x axis, and the lines $x = 1$ and $x = 4$; circumscribed rectangles.

24. The region of Exercise 23; inscribed rectangles.

25. The region above the x axis and to the right of the line $x = 1$ bounded by the x axis, the line $x = 1$, and the curve $y = 4 - x^2$; inscribed rectangles.

26. The region of Exercise 25; circumscribed rectangles.

27. The region lying to the left of the line $x = 1$ bounded by the curve and lines of Exercise 25; circumscribed rectangles.

28. The region of Exercise 27; inscribed rectangles.

29. The region bounded by $y = 3x^4$, the x axis, and the line $x = 1$; inscribed rectangles.

30. The region of Exercise 29; circumscribed rectangles.

31. The region bounded by $y = x^3$, the x axis, and the lines $x = -1$ and $x = 2$; inscribed rectangles.

32. The region of Exercise 31; circumscribed rectangles.

33. The region bounded by $y = x^3 + x$, the x axis, and the lines $x = -2$ and $x = 1$; circumscribed rectangles.

34. The region of Exercise 33; inscribed rectangles.

35. The region bounded by $y = mx$, with $m > 0$, the x axis, and the lines $x = a$ and $x = b$, with $b > a > 0$; circumscribed rectangles.

36. The region of Exercise 35; inscribed rectangles.

37. Use the method of this section to find the area of an isosceles trapezoid whose bases have measures b_1 and b_2 and whose altitude has measure h.

38. The graph of $y = 4 - |x|$ and the x axis from $x = -4$ to $x = 4$ form a triangle. Use the method of this section to find the area of this triangle.

In Exercises 39 through 44, find the area of the region by taking as the measure of the altitude of the ith rectangle $f(m_i)$, where m_i is the midpoint of the ith subinterval. (*Hint:* $m_i = \frac{1}{2}(x_i + x_{i-1})$.)

39. The region of Example 3.   **40.** The region of Exercise 21.   **41.** The region of Exercise 23.

42. The region of Exercise 25.   **43.** The region of Exercise 27.   **44.** The region of Exercise 29.

45. Prove Formula 2 of Theorem 4.4.7 without mathematical induction. (*Hint:* $i^3 - (i - 1)^3 = 3i^2 - 3i + 1$; so $\displaystyle\sum_{i=1}^{n} [i^3 - (i - 1)^3] = \sum_{i=1}^{n} (3i^2 - 3i + 1)$. On the left side of this equation use Theorem 4.4.6; on the right side use Theorems 4.4.2, 4.4.3, and 4.4.4 and Formula 1.)

46. Prove Formula 3 of Theorem 4.4.7. (*Hint:* $i^4 - (i - 1)^4 = 4i^3 - 6i^2 + 4i - 1$, and use a method similar to the one for Exercise 45.)

47. Prove Formula 4 of Theorem 4.4.7. (See the hints for Exercises 45 and 46.)
48. Prove Formula 1 of Theorem 4.4.7 by mathematical induction.
49. Prove Formula 3 of Theorem 4.4.7 by mathematical induction.
50. Prove Formula 4 of Theorem 4.4.7 by mathematical induction.
51. Prove Theorem 4.4.4.   52. Prove Theorem 4.4.5.

4.5
THE DEFINITE INTEGRAL

In Section 4.4 the measure of the area of a region was defined as the following limit:

$$\lim_{n \to +\infty} \sum_{i=1}^{n} f(c_i)\,\Delta x \tag{1}$$

To lead up to this definition we divided the closed interval $[a, b]$ into subintervals of equal length and then took c_i as the point in the ith subinterval for which f has an absolute minimum value. We also restricted the function values $f(x)$ to be nonnegative on $[a, b]$ and further required f to be continuous on $[a, b]$.

The limit in (1) is a special case of a new kind of limiting process that leads to the definition of the *definite integral*. We now discuss this "new kind of limit."

Let f be a function defined on the closed interval $[a, b]$. Divide this interval into n subintervals by choosing *any* $n - 1$ intermediate points between a and b. Let $x_0 = a$ and $x_n = b$, and let $x_1, x_2, \ldots, x_{n-1}$ be the intermediate points so that

$$x_0 < x_1 < x_2 < \ldots < x_{n-1} < x_n$$

The points $x_0, x_1, x_2, \ldots, x_{n-1}, x_n$ are not necessarily equidistant. Let $\Delta_1 x$ be the length of the first subinterval so that $\Delta_1 x = x_1 - x_0$; let $\Delta_2 x$ be the length of the second subinterval so that $\Delta_2 x = x_2 - x_1$; and so forth so that the length of the ith subinterval is $\Delta_i x$, and

$$\Delta_i x = x_i - x_{i-1}$$

A set of all such subintervals of the interval $[a, b]$ is called a **partition** of the interval $[a, b]$. Let Δ be such a partition. Figure 4.5.1 illustrates one such partition Δ of $[a, b]$.

FIGURE 4.5.1

The partition Δ contains n subintervals. One of these subintervals is longest; however, there may be more than one such subinterval. The length of the longest subinterval of the partition Δ, called the **norm** of the partition, is denoted by $\|\Delta\|$.

Choose a point in each subinterval of the partition Δ: Let ξ_1 be the point chosen in $[x_0, x_1]$ so that $x_0 \leq \xi_1 \leq x_1$. Let ξ_2 be the point chosen in $[x_1, x_2]$, so that $x_1 \leq \xi_2 \leq x_2$, and so forth, so that ξ_i is the point chosen in $[x_{i-1}, x_i]$, and $x_{i-1} \le \xi_i \le x_i$. Form the sum

$$f(\xi_1)\,\Delta_1 x + f(\xi_2)\,\Delta_2 x + \ldots + f(\xi_i)\,\Delta_i x + \ldots + f(\xi_n)\,\Delta_n x$$

or

$$\sum_{i=1}^{n} f(\xi_i)\,\Delta_i x$$

Such a sum is called a **Riemann sum**, named for the mathematician Georg Friedrich Bernhard Riemann (1826–1866).

▶ ILLUSTRATION 1   Suppose that $f(x) = 10 - x^2$, with $\frac{1}{4} \le x \le 3$. We will find the Riemann sum for the function f on $\left[\frac{1}{4}, 3\right]$ for the partition Δ: $x_0 = \frac{1}{4}$, $x_1 = 1$, $x_2 = 1\frac{1}{2}$, $x_3 = 1\frac{3}{4}$, $x_4 = 2\frac{1}{4}$, $x_5 = 3$, and $\xi_1 = \frac{1}{2}$, $\xi_2 = 1\frac{1}{4}$, $\xi_3 = 1\frac{3}{4}$, $\xi_4 = 2$, $\xi_5 = 2\frac{3}{4}$.

Figure 4.5.2 shows a sketch of the graph of f on $\left[\frac{1}{4}, 3\right]$ and the five rectangles, the measures of whose areas are the terms of the Riemann sum.

$$\begin{aligned}
\sum_{i=1}^{5} f(\xi_i)\,\Delta_i x &= f(\xi_1)\,\Delta_1 x + f(\xi_2)\,\Delta_2 x + f(\xi_3)\,\Delta_3 x + f(\xi_4)\,\Delta_4 x + f(\xi_5)\,\Delta_5 x \\
&= f(\tfrac{1}{2})(1 - \tfrac{1}{4}) + f(\tfrac{5}{4})(1\tfrac{1}{2} - 1) + f(\tfrac{7}{4})(1\tfrac{3}{4} - 1\tfrac{1}{2}) \\
&\qquad + f(2)(2\tfrac{1}{4} - 1\tfrac{3}{4}) + f(\tfrac{11}{4})(3 - 2\tfrac{1}{4}) \\
&= (9\tfrac{3}{4})(\tfrac{3}{4}) + (8\tfrac{7}{16})(\tfrac{1}{2}) + (6\tfrac{15}{16})(\tfrac{1}{4}) + (6)(\tfrac{1}{2}) + (2\tfrac{7}{16})(\tfrac{3}{4}) \\
&= 18\tfrac{3}{32}
\end{aligned}$$

The norm of Δ is the length of the longest subinterval. Hence $\|\Delta\| = \frac{3}{4}$. ◀

Because the function values $f(x)$ are not restricted to nonnegative values, some of the $f(\xi_i)$ could be negative. In such a case the geometric interpretation of the Riemann sum would be the sum of the measures of the areas of the rectangles lying above the x axis and the negatives of the measures

FIGURE 4.5.2

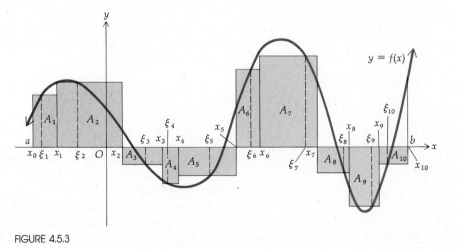

FIGURE 4.5.3

of the areas of the rectangles lying below the x axis. This situation is illustrated in Fig. 4.5.3. Here

$$\sum_{i=1}^{10} f(\xi_i)\Delta_i x = A_1 + A_2 - A_3 - A_4 - A_5 + A_6 + A_7 - A_8 - A_9 - A_{10}$$

because $f(\xi_3)$, $f(\xi_4)$, $f(\xi_5)$, $f(\xi_8)$, $f(\xi_9)$, and $f(\xi_{10})$ are negative numbers.

Let f be a function whose domain includes the closed interval $[a,b]$. Suppose there is a number L such that $\left| \sum_{i=1}^{n} f(\xi_i)\Delta_i x - L \right|$ can be made as small as we please for all partitions Δ whose norms are sufficiently small, and for any ξ_i in the closed interval $[x_{i-1}, x_i]$, $i = 1, 2, \ldots, n$. In such a case f is said to be *integrable* on $[a,b]$. Following is the formal definition of a function being integrable on a closed interval; it involves ϵ and δ notation.

4.5.1 DEFINITION   Let f be a function whose domain includes the closed interval $[a,b]$. Then f is said to be **integrable** on $[a,b]$ if there is a number L satisfying the condition that, for any $\epsilon > 0$, there exists a $\delta > 0$ such that for every partition Δ for which $\|\Delta\| < \delta$, and for any ξ_i in the closed interval $[x_{i-1}, x_i]$, $i = 1, 2, \ldots, n$, then

$$\left| \sum_{i=1}^{n} f(\xi_i)\Delta_i x - L \right| < \epsilon \tag{2}$$

For such a situation we write

$$\lim_{\|\Delta\| \to 0} \sum_{i=1}^{n} f(\xi_i)\Delta_i x = L \tag{3}$$

Definition 4.5.1 states that, for a given function f defined on the closed interval $[a,b]$, we can make the values of the Riemann sums as close to L as we please by taking the norms $\|\Delta\|$ of all partitions Δ of $[a,b]$ sufficiently small for all possible choices of the numbers ξ_i for which $x_{i-1} \leq \xi_i \leq x_i$, $i = 1, 2, \ldots, n$.

The limiting process given by (3) is different from that discussed in Chapter 1. From Definition 4.5.1 the number L in (3) exists if for any $\epsilon > 0$ there exists a $\delta > 0$ such that for every partition Δ for which $\|\Delta\| < \delta$, and for any ξ_i in the closed interval $[x_{i-1}, x_i]$, $i = 1, 2, \ldots, n$, then inequality (2) holds.

In Definition 1.2.1 we had

$$\lim_{x \to a} f(x) = L \tag{4}$$

if for any $\epsilon > 0$ there exists a $\delta > 0$ such that

$$\text{if} \quad 0 < |x - a| < \delta \quad \text{then} \quad |f(x) - L| < \epsilon$$

In limiting process (3), for a particular $\delta > 0$ there are infinitely many partitions Δ having norm $\|\Delta\| < \delta$. This is analogous to the fact that in limiting process (4), for a given $\delta > 0$ there are infinitely many values of x for which $0 < |x - a| < \delta$. However, in limiting process (3), for each partition Δ there are infinitely many choices of ξ_i. It is in this respect that the two limiting processes differ.

In Chapter 1 (Theorem 1.2.2 proved in Supplementary Section 1.9) we showed that if the number L in limiting process (4) exists, it is unique. In a similar manner we can show that if there is a number L satisfying Definition 4.5.1, then it is unique. Now we can define the *definite integral*.

4.5.2 DEFINITION   If f is a function defined on the closed interval $[a, b]$, then the **definite integral** of f from a to b, denoted by $\int_a^b f(x)\,dx$, is given by

$$\int_a^b f(x)\,dx = \lim_{||\Delta|| \to 0} \sum_{i=1}^n f(\xi_i)\,\Delta_i x \tag{5}$$

if the limit exists.

Note that the statement "the function f is integrable on the closed interval $[a, b]$" is synonymous with the statement "the definite integral of f from a to b exists."

In the notation for the definite integral $\int_a^b f(x)\,dx$, $f(x)$ is called the **integrand**, a is called the **lower limit**, and b is called the **upper limit**. The symbol $\int$ is called an **integral sign**. The integral sign resembles a capital S, which is appropriate because the definite integral is the limit of a sum. It is the same symbol we have been using to indicate the operation of antidifferentiation. The reason for the common symbol is that a theorem (4.8.2), called the second fundamental theorem of the calculus, enables us to evaluate a definite integral by finding an antiderivative (also called an **indefinite integral**).

The following question now arises: Under what conditions does a number L satisfying Definition 4.5.1 exist; that is, under what conditions is a function integrable? An answer to this question is given by the next theorem.

4.5.3 THEOREM   If a function is continuous on the closed interval $[a, b]$, then it is integrable on $[a, b]$.

The proof of this theorem is beyond the scope of this book and is given in advanced calculus texts. The condition that f is continuous on $[a, b]$, while being sufficient to guarantee that f is integrable on $[a, b]$, is not a necessary condition for the existence of $\int_a^b f(x)\,dx$. That is, if f is continuous on $[a, b]$, then Theorem 4.5.3 assures us that $\int_a^b f(x)\,dx$ exists; however, it is possible for the definite integral to exist even if the function is discontinuous at some numbers in $[a, b]$. The following example gives a function that is discontinuous and yet integrable on a closed interval.

EXAMPLE 1
Let f be the function defined by

$$f(x) = \begin{cases} 0 & \text{if } x \neq 0 \\ 1 & \text{if } x = 0 \end{cases}$$

Let $[a, b]$ be any interval for which $a < 0 < b$. Show that f is discontinuous on $[a, b]$, yet integrable on $[a, b]$.

SOLUTION

Because $\lim\limits_{x \to 0} f(x) = 0$, but $f(0) = 1$, f is discontinuous at 0 and hence discontinuous on $[a, b]$.

To prove that f is integrable on $[a, b]$ we show that Definition 4.5.1 is satisfied. Consider the Riemann sum

$$\sum_{i=1}^{n} f(\xi_i) \Delta_i x$$

If none of the numbers $\xi_1, \xi_2, \ldots, \xi_n$ is zero, then the Riemann sum is zero. Suppose that $\xi_j = 0$. Then

$$\sum_{i=1}^{n} f(\xi_i) \Delta_i x = 1 \cdot \Delta_j x$$

In either case,

$$\left| \sum_{i=1}^{n} f(\xi_i) \Delta_i x \right| \leq \|\Delta\|$$

Hence

$$\text{if} \quad \|\Delta\| < \epsilon \quad \text{then} \quad \left| \sum_{i=1}^{n} f(\xi_i) \Delta_i x - 0 \right| < \epsilon$$

Comparing the above with Definition 4.5.1 where $\delta = \epsilon$ and $L = 0$, we see that f is integrable on $[a, b]$.

Observe that the function f of Example 1 has a finite number (only 1) of discontinuities on $[a, b]$ and $|f(x)| \leq 1$ for all x in $[a, b]$. This function f is a member of a special set of functions that are integrable on a closed interval $[a, b]$. Three other functions in this set appear in Exercises 32–34.

At the beginning of this section we stated that the limit used in Definition 4.4.8 to define the measure of the area of a region is a special case of the limit used in Definition 4.5.2 to define the definite integral. In the discussion of area, the interval $[a, b]$ was divided into n subintervals of equal length. Such a partition of the interval $[a, b]$ is called a **regular partition**. If Δx is the length of each subinterval in a regular partition, then each $\Delta_i x = \Delta x$, and the norm of the partition is Δx. Making these substitutions in (5) we have

$$\int_a^b f(x)\, dx = \lim_{\Delta x \to 0} \sum_{i=1}^{n} f(\xi_i) \Delta x \tag{6}$$

Furthermore,

$$\Delta x = \frac{b - a}{n} \quad \text{and} \quad n = \frac{b - a}{\Delta x}$$

Thus

$$\lim_{n \to +\infty} \Delta x = 0 \quad \text{and} \quad \lim_{\Delta x \to 0} n = +\infty \tag{7}$$

The reason that $\lim\limits_{\Delta x \to 0} n = +\infty$ is that $b > a$ and Δx approaches zero through positive values (because $\Delta x > 0$). From the limits in (7) we conclude that

$$\Delta x \to 0 \quad \text{is equivalent to} \quad n \to +\infty \tag{8}$$

Thus we have from Eq. (6) and statement (8),

$$\int_a^b f(x)\,dx = \lim_{n \to +\infty} \sum_{i=1}^n f(\xi_i)\Delta x \tag{9}$$

It should be remembered that ξ_i can be any point in the ith subinterval $[x_{i-1}, x_i]$.

Comparing the limit used in Definition 4.4.8, which gives the measure of the area of a region, with the limit on the right side of (9), we have in the first case

$$\lim_{n \to +\infty} \sum_{i=1}^n f(c_i)\Delta x \tag{10}$$

where $f(c_i)$ is the absolute minimum function value on $[x_{i-1}, x_i]$. In the second case we have

$$\lim_{n \to +\infty} \sum_{i=1}^n f(\xi_i)\Delta x \tag{11}$$

where ξ_i is any number in $[x_{i-1}, x_i]$.

Because the function f is continuous on $[a,b]$, by Theorem 4.5.3, $\int_a^b f(x)\,dx$ exists; therefore this definite integral is the limit of all Riemann sums of f on $[a, b]$ including those in (10) and (11). Because of this, we redefine the area of a region in a more general way.

4.5.4 DEFINITION   Let the function f be continuous on $[a, b]$ and $f(x) \geq 0$ for all x in $[a, b]$. Let R be the region bounded by the curve $y = f(x)$, the x axis, and the lines $x = a$ and $x = b$. Then the measure A of the area of region R is given by

$$A = \lim_{\|\Delta\| \to 0} \sum_{i=1}^n f(\xi_i)\Delta_i x$$

$$\Leftrightarrow \quad A = \int_a^b f(x)\,dx$$

The above definition states that if $f(x) \geq 0$ for all x in $[a, b]$, the definite integral $\int_a^b f(x)\,dx$ can be interpreted geometrically as the measure of the area of the region R shown in Fig. 4.5.4.

Equation (9) can be used to find the exact value of a definite integral, as illustrated in the following example.

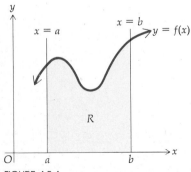

FIGURE 4.5.4

EXAMPLE 2

Find the exact value of the definite integral $\int_1^3 x^2\,dx$. Interpret the result geometrically.

SOLUTION

Consider a regular partition of the closed interval $[1, 3]$ into n subintervals. Then $\Delta x = 2/n$.

If we choose ξ_i as the right endpoint of each subinterval, we have

$$\xi_1 = 1 + \frac{2}{n}, \xi_2 = 1 + 2\left(\frac{2}{n}\right), \xi_3 = 1 + 3\left(\frac{2}{n}\right), \ldots,$$

$$\xi_i = 1 + i\left(\frac{2}{n}\right), \ldots, \xi_n = 1 + n\left(\frac{2}{n}\right)$$

Because $f(x) = x^2$,

$$f(\xi_i) = \left(1 + \frac{2i}{n}\right)^2$$

$$= \left(\frac{n + 2i}{n}\right)^2$$

Therefore, by using (9) and applying theorems from Section 4.4, we get

$$\int_1^3 x^2\,dx = \lim_{n \to +\infty} \sum_{i=1}^n \left(\frac{n + 2i}{n}\right)^2 \frac{2}{n}$$

$$= \lim_{n \to +\infty} \frac{2}{n^3} \sum_{i=1}^n (n^2 + 4ni + 4i^2)$$

$$= \lim_{n \to +\infty} \frac{2}{n^3}\left[n^2 \sum_{i=1}^n 1 + 4n \sum_{i=1}^n i + 4 \sum_{i=1}^n i^2 \right]$$

$$= \lim_{n \to +\infty} \frac{2}{n^3}\left[n^2 n + 4n \cdot \frac{n(n+1)}{2} + \frac{4n(n+1)(2n+1)}{6} \right]$$

$$= \lim_{n \to +\infty} \frac{2}{n^3}\left[n^3 + 2n^3 + 2n^2 + \frac{2n(2n^2 + 3n + 1)}{3} \right]$$

$$= \lim_{n \to +\infty}\left[6 + \frac{4}{n} + \frac{8n^2 + 12n + 4}{3n^2} \right]$$

$$= \lim_{n \to +\infty}\left[6 + \frac{4}{n} + \frac{8}{3} + \frac{4}{n} + \frac{4}{3n^2} \right]$$

$$= 6 + 0 + \tfrac{8}{3} + 0 + 0$$

$$= \tfrac{26}{3}$$

We interpret the result geometrically. Because $x^2 \geq 0$ for all x in $[1, 3]$, the region bounded by the curve $y = x^2$, the x axis, and the lines $x = 1$ and $x = 3$ has an area of $\frac{26}{3}$ square units. The region is shown in Fig. 4.5.5.

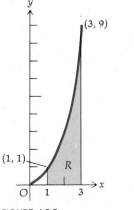

FIGURE 4.5.5

In Definition 4.5.2 the closed interval $[a, b]$ is given, and so we assume that $a < b$. To consider the definite integral of a function f from a to b when $a > b$, or when $a = b$, we have the following definitions.

4.5.5 DEFINITION   If $a > b$, then

$$\int_a^b f(x)\,dx = -\int_b^a f(x)\,dx$$

if $\int_b^a f(x)\,dx$ exists.

▶ ILLUSTRATION 2   In Example 2 we showed that $\int_1^3 x^2\,dx = \frac{26}{3}$. Therefore, from Definition 4.5.5,

$$\int_3^1 x^2\,dx = -\int_1^3 x^2\,dx$$

$$= -\frac{26}{3} \qquad\qquad ◀$$

4.5.6 DEFINITION   If $f(a)$ exists, then

$$\int_a^a f(x)\,dx = 0$$

▶ ILLUSTRATION 3   From Definition 4.5.6,

$$\int_1^1 x^2\,dx = 0 \qquad\qquad ◀$$

EXERCISES 4.5

In Exercises 1 through 9, find the Riemann sum for the function on the interval, using the given partition Δ and the given values of ξ_i. Draw a sketch of the graph of the function on the given interval, and show the rectangles the measure of whose areas are the terms of the Riemann sum. (See Illustration 1 and Fig. 4.5.2.)

1. $f(x) = x^2$, $0 \le x \le 3$; for Δ: $x_0 = 0$, $x_1 = \frac{1}{2}$, $x_2 = 1\frac{1}{4}$, $x_3 = 2\frac{1}{4}$, $x_4 = 3$; $\xi_1 = \frac{1}{4}$, $\xi_2 = 1$, $\xi_3 = 1\frac{1}{2}$, $\xi_4 = 2\frac{1}{2}$

2. $f(x) = x^2$, $0 \le x \le 3$; for Δ: $x_0 = 0$, $x_1 = \frac{3}{4}$, $x_2 = 1\frac{1}{4}$, $x_3 = 2$, $x_4 = 2\frac{3}{4}$, $x_5 = 3$; $\xi_1 = \frac{1}{2}$, $\xi_2 = 1$, $\xi_3 = 1\frac{3}{4}$, $\xi_4 = 2\frac{1}{4}$, $\xi_5 = 2\frac{3}{4}$

3. $f(x) = 1/x$, $1 \le x \le 3$; for Δ: $x_0 = 1$, $x_1 = 1\frac{2}{3}$, $x_2 = 2\frac{1}{4}$, $x_3 = 2\frac{2}{3}$, $x_4 = 3$; $\xi_1 = 1\frac{1}{4}$, $\xi_2 = 2$, $\xi_3 = 2\frac{1}{2}$, $\xi_4 = 2\frac{3}{4}$

4. $f(x) = x^3$, $-1 \le x \le 2$; for Δ: $x_0 = -1$, $x_1 = -\frac{1}{3}$, $x_2 = \frac{1}{2}$, $x_3 = 1$, $x_4 = 1\frac{1}{4}$, $x_5 = 2$; $\xi_1 = -\frac{1}{2}$, $\xi_2 = 0$, $\xi_3 = \frac{2}{3}$, $\xi_4 = 1$, $\xi_5 = 1\frac{1}{2}$

5. $f(x) = x^2 - x + 1$, $0 \le x \le 1$; for Δ: $x_0 = 0$, $x_1 = 0.2$, $x_2 = 0.5$, $x_3 = 0.7$, $x_4 = 1$; $\xi_1 = 0.1$, $\xi_2 = 0.4$, $\xi_3 = 0.6$, $\xi_4 = 0.9$

6. $f(x) = 1/(x + 2)$, $-1 \le x \le 3$; for Δ: $x_0 = -1$, $x_1 = -\frac{1}{4}$, $x_2 = 0$, $x_3 = \frac{1}{2}$, $x_4 = 1\frac{1}{4}$, $x_5 = 2$, $x_6 = 2\frac{1}{4}$, $x_7 = 2\frac{3}{4}$, $x_8 = 3$; $\xi_1 = -\frac{3}{4}$, $\xi_2 = 0$, $\xi_3 = \frac{1}{4}$, $\xi_4 = 1$, $\xi_5 = 1\frac{1}{2}$, $\xi_6 = 2$, $\xi_7 = 2\frac{1}{2}$, $\xi_8 = 3$

7. $f(x) = \sin x$, $0 \le x \le \pi$; for Δ: $x_0 = 0$, $x_1 = \frac{1}{4}\pi$, $x_2 = \frac{1}{2}\pi$, $x_3 = \frac{2}{3}\pi$, $x_4 = \frac{3}{4}\pi$, $x_5 = \pi$; $\xi_1 = \frac{1}{6}\pi$, $\xi_2 = \frac{1}{3}\pi$, $\xi_3 = \frac{1}{2}\pi$, $\xi_4 = \frac{3}{4}\pi$, $\xi_5 = \frac{5}{6}\pi$.

8. $f(x) = 3\cos\frac{1}{2}x$, $-\pi \le x \le \pi$; for Δ: $x_0 = -\pi$, $x_1 = -\frac{1}{2}\pi$, $x_2 = -\frac{1}{3}\pi$, $x_3 = \frac{1}{3}\pi$, $x_4 = \frac{7}{12}\pi$, $x_5 = \pi$; $\xi_1 = -\frac{2}{3}\pi$, $\xi_2 = -\frac{1}{3}\pi$, $\xi_3 = 0$, $\xi_4 = \frac{1}{2}\pi$, $\xi_5 = \frac{2}{3}\pi$

9. $f(x) = [\![x]\!] + 2$, $-3 \le x \le 3$; for Δ: $x_0 = -3$, $x_1 = -1$, $x_2 = 0$, $x_3 = 2$, $x_4 = 3$; $\xi_1 = -2.5$, $\xi_2 = -0.5$, $\xi_3 = 1$, $\xi_4 = 2.5$

In Exercises 10 through 18, find the exact value of the definite integral. Use the method of Example 2 of this section.

10. $\int_2^7 3x\,dx$ **11.** $\int_0^2 x^2\,dx$ **12.** $\int_2^4 x^2\,dx$ **13.** $\int_1^2 x^3\,dx$ **14.** $\int_0^5 (x^3 - 1)\,dx$

15. $\int_1^4 (x^2 + 4x + 5)\,dx$ **16.** $\int_0^4 (x^2 + x - 6)\,dx$ **17.** $\int_{-2}^2 (x^3 + 1)\,dx$ **18.** $\int_{-2}^1 x^4\,dx$

In Exercises 19 through 28, find the exact area of the region in the following way: (a) Express the measure of the area as the limit of a Riemann sum with regular partitions; (b) express this limit with definite integral notation; (c) evaluate the definite integral by the method of this section and a suitable choice of ξ_i. Draw a figure showing the region.

19. Bounded by the line $y = 2x - 1$, the x axis, and the lines $x = 1$ and $x = 5$.
20. Bounded by the line $y = 2x - 6$, the x axis, and the lines $x = 4$ and $x = 7$.
21. Bounded by the line $y = -3x + 2$, the x axis, and the lines $x = -5$ and $x = -1$.
22. Bounded by the line $y = 3x + 2$, the x axis, and the lines $x = -4$ and $x = -1$.
23. Bounded by the curve $y = 4 - x^2$, the x axis, and the lines $x = 1$ and $x = 2$.
24. Bounded by the curve $y = (x + 3)^2$, the x axis, and the lines $x = -3$ and $x = 0$.
25. Bounded by the curve $y = 12 - x - x^2$, the x axis, and the lines $x = -3$ and $x = 2$.
26. Bounded by the curve $y = 10 + x - x^2$, the x axis, and the lines $x = -2$ and $x = 3$.
27. Bounded by the curve $y = x^3 - 4$, the x axis, and the lines $x = -2$ and $x = -1$.
28. Bounded by the curve $y = 6x + x^2 - x^3$, the x axis, and the lines $x = -1$ and $x = 3$.

29. Express as a definite integral: $\lim\limits_{n \to +\infty} \sum\limits_{i=1}^{n} \dfrac{8i^2}{n^3}$. (*Hint:* Consider the function f for which $f(x) = x^2$.)

30. Express as a definite integral: $\lim\limits_{n \to +\infty} \sum\limits_{i=1}^{n} \dfrac{1}{n+i}$. $\left(\textit{Hint:} \text{ Consider the function } f \text{ for which } f(x) = \dfrac{1}{x} \text{ on } [1,2].\right)$

31. Express as a definite integral: $\lim\limits_{n \to +\infty} \sum\limits_{i=1}^{n} \dfrac{n}{i^2}$. $\left(\textit{Hint:} \text{ Consider the function } f \text{ for which } f(x) = \dfrac{1}{x^2}.\right)$

32. Let $[a, b]$ be any interval such that $a < 0 < b$. Prove that even though the unit step function (Exercise 24 in Exercises 1.4) is discontinuous on $[a, b]$, it is integrable on $[a, b]$ and $\int_a^b U(x)\,dx = b$.

33. Prove that the signum function is discontinuous on $[-1, 1]$ and yet integrable on $[-1, 1]$. Furthermore show that $\int_{-1}^1 \operatorname{sgn} x\,dx = 0$.

34. Prove that the greatest integer function is discontinuous on $[0, \frac{3}{2}]$ and yet integrable on $[0, \frac{3}{2}]$. Furthermore show that $\int_0^{3/2} [\![x]\!]\,dx = \frac{1}{2}$.

4.6
PROPERTIES OF THE
DEFINITE INTEGRAL

Evaluating a definite integral from the definition, by actually finding the limit of a sum as we did in Section 4.5, is usually quite tedious and frequently almost impossible. To establish a much simpler method we first need to develop some properties of the definite integral. First the following two theorems about Riemann sums are needed.

4.6.1 THEOREM   If Δ is any partition of the closed interval $[a, b]$, then

$$\lim_{\|\Delta\| \to 0} \sum_{i=1}^{n} \Delta_i x = b - a$$

PROOF

$$\sum_{i=1}^{n} \Delta_i x - (b - a) = (b - a) - (b - a)$$
$$= 0$$

Hence, for any $\epsilon > 0$, any choice of $\delta > 0$ guarantees that

$$\text{if}\quad \|\Delta\| < \delta \quad \text{then} \quad \left| \sum_{i=1}^{n} \Delta_i x - (b - a) \right| < \epsilon$$

Thus by Definition 4.5.1,

$$\lim_{\|\Delta\| \to 0} \sum_{i=1}^{n} \Delta_i x = b - a \qquad \blacksquare$$

4.6.2 THEOREM   If f is defined on the closed interval $[a, b]$, and if

$$\lim_{\|\Delta\| \to 0} \sum_{i=1}^{n} f(\xi_i) \Delta_i x$$

exists, where Δ is any partition of $[a, b]$, then if k is any constant,

$$\lim_{\|\Delta\| \to 0} \sum_{i=1}^{n} kf(\xi_i) \Delta_i x = k \lim_{\|\Delta\| \to 0} \sum_{i=1}^{n} f(\xi_i) \Delta_i x$$

The proof of this theorem is left as an exercise (see Exercise 43).

▶ ILLUSTRATION 1   Refer to Fig. 4.6.1. If $k > 0$, the definite integral $\int_a^b k\, dx$ gives the measure of the area of the shaded region, which is a rectangle whose dimensions are k units and $(b - a)$ units. This fact is a geometric interpretation of the following theorem when $k > 0$.   ◀

FIGURE 4.6.1

4.6.3 THEOREM   If k is any constant, then

$$\int_a^b k\, dx = k(b - a)$$

PROOF   From Definition 4.5.2,

$$\int_a^b f(x)\, dx = \lim_{\|\Delta\| \to 0} \sum_{i=1}^{n} f(\xi_i) \Delta_i x$$

If $f(x) = k$ for all x in $[a, b]$, we have from this equation

$$\int_a^b k\, dx = \lim_{\|\Delta\| \to 0} \sum_{i=1}^{n} k \Delta_i x$$

$$= k \lim_{\|\Delta\| \to 0} \sum_{i=1}^{n} \Delta_i x \quad \text{(by Theorem 4.6.2)}$$

$$= k(b - a) \quad \text{(by Theorem 4.6.1)} \qquad \blacksquare$$

EXAMPLE 1

Evaluate $\int_{-3}^{5} 4\,dx$.

SOLUTION

We apply Theorem 4.6.3.

$$\int_{-3}^{5} 4\,dx = 4[5 - (-3)]$$
$$= 4(8)$$
$$= 32$$

4.6.4 THEOREM   If the function f is integrable on the closed interval $[a, b]$, and if k is any constant, then

$$\int_{a}^{b} kf(x)\,dx = k \int_{a}^{b} f(x)\,dx$$

PROOF   Because f is integrable on $[a, b]$, $\lim\limits_{||\Delta|| \to 0} \sum\limits_{i=1}^{n} f(\xi_i)\Delta_i x$ exists; so by Theorem 4.6.2,

$$\lim_{||\Delta|| \to 0} \sum_{i=1}^{n} kf(\xi_i)\Delta_i x = k \lim_{||\Delta|| \to 0} \sum_{i=1}^{n} f(\xi_i)\Delta_i x$$

Therefore

$$\int_{a}^{b} kf(x)\,dx = k \int_{a}^{b} f(x)\,dx \qquad \blacksquare$$

Observe the similarity of the following theorem to Limit Theorem 4 (1.3.4) about the limit of the sum of two functions. Also notice that the proofs of the two theorems are alike.

4.6.5 THEOREM   If the functions f and g are integrable on $[a, b]$, then $f + g$ is integrable on $[a, b]$ and

$$\int_{a}^{b} [f(x) + g(x)]\,dx = \int_{a}^{b} f(x)\,dx + \int_{a}^{b} g(x)\,dx$$

PROOF   The functions f and g are integrable on $[a, b]$; thus let

$$\int_{a}^{b} f(x)\,dx = M \quad \text{and} \quad \int_{a}^{b} g(x)\,dx = N$$

To prove that on $[a, b]$ $f + g$ is integrable and $\int_{a}^{b} [f(x) + g(x)]\,dx = M + N$, we must show that for any $\epsilon > 0$ there exists a $\delta > 0$ such that for all partitions Δ and for any ξ_i in $[x_{i-1}, x_i]$, if $||\Delta|| < \delta$, then

$$\left| \sum_{i=1}^{n} [f(\xi_i) + g(\xi_i)]\Delta_i x - (M + N) \right| < \epsilon$$

Because

$$M = \lim_{||\Delta|| \to 0} \sum_{i=1}^{n} f(\xi_i)\Delta_i x \quad \text{and} \quad N = \lim_{||\Delta|| \to 0} \sum_{i=1}^{n} g(\xi_i)\Delta_i x$$

it follows that for any $\epsilon > 0$ there exist a $\delta_1 > 0$ and a $\delta_2 > 0$ such that for all partitions Δ and for any ξ_i in $[x_{i-1}, x_i]$, if $\|\Delta\| < \delta_1$ and $\|\Delta\| < \delta_2$, then

$$\left| \sum_{i=1}^{n} f(\xi_i) \Delta_i x - M \right| < \frac{\epsilon}{2} \quad \text{and} \quad \left| \sum_{i=1}^{n} g(\xi_i) \Delta_i x - N \right| < \frac{\epsilon}{2}$$

Therefore, if $\delta = \min(\delta_1, \delta_2)$, then for any $\epsilon > 0$, for all partitions Δ and for any ξ_i in $[x_{i-1}, x_i]$, if $\|\Delta\| < \delta$, then

$$\left| \sum_{i=1}^{n} f(\xi_i) \Delta_i x - M \right| + \left| \sum_{i=1}^{n} g(\xi_i) \Delta_i x - N \right| < \frac{\epsilon}{2} + \frac{\epsilon}{2} = \epsilon \qquad (1)$$

By the triangle inequality we have

$$\left| \left(\sum_{i=1}^{n} f(\xi_i) \Delta_i x - M \right) + \left(\sum_{i=1}^{n} g(\xi_i) \Delta_i x - N \right) \right|$$

$$\leq \left| \sum_{i=1}^{n} f(\xi_i) \Delta_i x - M \right| + \left| \sum_{i=1}^{n} g(\xi_i) \Delta_i x - N \right| \qquad (2)$$

From inequalities (1) and (2) we have

$$\left| \left(\sum_{i=1}^{n} f(\xi_i) \Delta_i x + \sum_{i=1}^{n} g(\xi_i) \Delta_i x \right) - (M + N) \right| < \epsilon \qquad (3)$$

From Theorem 4.4.4,

$$\sum_{i=1}^{n} f(\xi_i) \Delta_i x + \sum_{i=1}^{n} g(\xi_i) \Delta_i x = \sum_{i=1}^{n} [f(\xi_i) + g(\xi_i)] \Delta_i x \qquad (4)$$

So by substituting from (4) into (3) we are able to conclude that for any $\epsilon > 0$, for all partitions Δ and for any ξ_i in $[x_{i-1}, x_i]$, if $\|\Delta\| < \delta$ where $\delta = \min(\delta_1, \delta_2)$, then

$$\left| \sum_{i=1}^{n} [f(\xi_i) + g(\xi_i)] \Delta_i x - (M + N) \right| < \epsilon$$

This proves that $f + g$ is integrable on $[a, b]$ and that

$$\int_a^b [f(x) + g(x)] \, dx = \int_a^b f(x) \, dx + \int_a^b g(x) \, dx \qquad \blacksquare$$

The plus sign in the statement of Theorem 4.6.5 can be replaced by a minus sign as a result of applying Theorem 4.6.4, where $k = -1$.

Theorem 4.6.5 can be extended to any number of functions. That is, if the functions $f_1, f_2, \ldots, f_n$ are all integrable on $[a, b]$, then $(f_1 \pm f_2 \pm \cdots \pm f_n)$ is integrable on $[a, b]$ and

$$\int_a^b [f_1(x) \pm f_2(x) \pm \ldots \pm f_n(x)] \, dx$$

$$= \int_a^b f_1(x) \, dx \pm \int_a^b f_2(x) \, dx \pm \ldots \pm \int_a^b f_n(x) \, dx$$

EXAMPLE 2

Use the result of Example 2, Section 4.5, and the fact that $\int_1^3 x \, dx = 4$ to evaluate $\int_1^3 (3x^2 - 5x + 2) \, dx$.

SOLUTION

In Example 2, Section 4.5, we had the result

$$\int_1^3 x^2 \, dx = \tfrac{26}{3}$$

From properties of the definite integral,

$$\int_1^3 (3x^2 - 5x + 2) \, dx = \int_1^3 3x^2 \, dx - \int_1^3 5x \, dx + \int_1^3 2 \, dx$$

$$= 3 \int_1^3 x^2 \, dx - 5 \int_1^3 x \, dx + 2(3 - 1)$$

$$= 3(\tfrac{26}{3}) - 5(4) + 4$$

$$= 26 - 20 + 4$$

$$= 10$$

FIGURE 4.6.2

▶ ILLUSTRATION 2   A geometric interpretation of Theorem 4.6.6, which fol-lows, is shown in Fig. 4.6.2 where $f(x) \geq 0$. For all x in $[a, b]$, the measure of the area of the region bounded by the curve $y = f(x)$ and the x axis from a to b is equal to the sum of the measures of the areas of the regions from a to c and from c to b.   ◀

4.6.6 THEOREM   If the function f is integrable on the closed intervals $[a, b]$, $[a, c]$, and $[c, b]$,

$$\int_a^b f(x) \, dx = \int_a^c f(x) \, dx + \int_c^b f(x) \, dx$$

where $a < c < b$.

PROOF   Let Δ be a partition of $[a, b]$. Form the partition Δ' of $[a, b]$ in the following way. If c is one of the partitioning points of Δ (i.e., $c = x_i$ for some i), then Δ' is exactly the same as Δ. If c is not one of the partitioning points of Δ but is contained in the subinterval $[x_{i-1}, x_i]$, then the partition Δ' has as its partitioning points all the partitioning points of Δ and, in addition, the point c. Therefore the subintervals of the partition Δ' are the same as the subintervals of Δ, with the exception that the subinterval $[x_{i-1}, x_i]$ of Δ is divided into the two subintervals $[x_{i-1}, c]$ and $[c, x_i]$.

If $\|\Delta'\|$ is the norm of Δ' and if $\|\Delta\|$ is the norm of Δ, then

$$\|\Delta'\| \leq \|\Delta\|$$

If in the partition Δ' the interval $[a, c]$ is divided into r subintervals and the interval $[c, b]$ is divided into $(n - r)$ subintervals, then the part of the parti-tion Δ' from a to c gives a Riemann sum of the form

$$\sum_{i=1}^r f(\xi_i) \Delta_i x$$

and the other part of the partition Δ', from c to b, gives a Riemann sum of the form

$$\sum_{i=r+1}^{n} f(\xi_i)\Delta_i x$$

Using the definition of the definite integral and properties of the sigma notation we have

$$\int_a^b f(x)\,dx = \lim_{\|\Delta\| \to 0} \sum_{i=1}^{n} f(\xi_i)\Delta_i x$$

$$= \lim_{\|\Delta\| \to 0} \left[\sum_{i=1}^{r} f(\xi_i)\Delta_i x + \sum_{i=r+1}^{n} f(\xi_i)\Delta_i x \right]$$

$$= \lim_{\|\Delta\| \to 0} \sum_{i=1}^{r} f(\xi_i)\Delta_i x + \lim_{\|\Delta\| \to 0} \sum_{i=r+1}^{n} f(\xi_i)\Delta_i x$$

Because $0 < \|\Delta'\| \leq \|\Delta\|$, we can replace $\|\Delta\| \to 0$ by $\|\Delta'\| \to 0$, giving

$$\int_a^b f(x)\,dx = \lim_{\|\Delta'\| \to 0} \sum_{i=1}^{r} f(\xi_i)\Delta_i x + \lim_{\|\Delta'\| \to 0} \sum_{i=r+1}^{n} f(\xi_i)\Delta_i x$$

Applying the definition of the definite integral to the right side of the above equation we have

$$\int_a^b f(x)\,dx = \int_a^c f(x)\,dx + \int_c^b f(x)\,dx \qquad \blacksquare$$

The result of Theorem 4.6.6 is true for any ordering of the numbers a, b, and c. This is stated as another theorem.

4.6.7 THEOREM   If f is integrable on a closed interval containing the three numbers a, b, and c, then

$$\int_a^b f(x)\,dx = \int_a^c f(x)\,dx + \int_c^b f(x)\,dx \qquad (5)$$

regardless of the order of a, b, and c.

PROOF   If a, b, and c are distinct, there are six possible orderings of these three numbers: $a < b < c$, $a < c < b$, $b < a < c$, $b < c < a$, $c < a < b$, and $c < b < a$. The second ordering, $a < c < b$, is Theorem 4.6.6. We make use of Theorem 4.6.6 in proving that Eq. (5) holds for the other orderings.

Suppose that $a < b < c$; then from Theorem 4.6.6 we have

$$\int_a^b f(x)\,dx + \int_b^c f(x)\,dx = \int_a^c f(x)\,dx \qquad (6)$$

From Definition 4.5.5,

$$\int_b^c f(x)\,dx = -\int_c^b f(x)\,dx \qquad (7)$$

Substituting from (7) into (6) we obtain

$$\int_a^b f(x)\,dx - \int_c^b f(x)\,dx = \int_a^c f(x)\,dx$$

Thus

$$\int_a^b f(x)\,dx = \int_a^c f(x)\,dx + \int_c^b f(x)\,dx$$

which is the desired result.

The proofs for the other four orderings are similar and left as exercises (see Exercises 44 through 47).

There is also the possibility that two of the three numbers are equal; for example, $a = c < b$. Then

$$\int_a^c f(x)\,dx = \int_a^a f(x)\,dx$$

$$= 0 \quad \text{(by Definition 4.5.6)}$$

Also, because $a = c$,

$$\int_c^b f(x)\,dx = \int_a^b f(x)\,dx$$

Therefore

$$\int_a^c f(x)\,dx + \int_c^b f(x)\,dx = 0 + \int_a^b f(x)\,dx$$

which is the desired result.   ■

FIGURE 4.6.3

▶ ILLUSTRATION 3   In Fig. 4.6.3, $f(x) \geq g(x) \geq 0$ for all x in $[a, b]$. The definite integral $\int_a^b f(x)\,dx$ gives the measure of the area of the region bounded by the curve $y = f(x)$, the x axis, and the lines $x = a$ and $x = b$; $\int_a^b g(x)\,dx$ gives the measure of the area of the region bounded by the curve $y = g(x)$, the x axis, and the lines $x = a$ and $x = b$. In the figure we see that the first area is greater than the second area. We have given a geometric interpretation of the following theorem when $f(x)$ and $g(x)$ are nonnegative on $[a, b]$. ◀

4.6.8 THEOREM   If the functions f and g are integrable on the closed interval $[a, b]$, and if $f(x) \geq g(x)$ for all x in $[a, b]$, then

$$\int_a^b f(x)\,dx \geq \int_a^b g(x)\,dx$$

PROOF   Because f and g are integrable on $[a, b]$, $\int_a^b f(x)\,dx$ and $\int_a^b g(x)\,dx$ both exist. Therefore,

$$\int_a^b f(x)\,dx - \int_a^b g(x)\,dx = \int_a^b f(x)\,dx + \int_a^b [-g(x)]\,dx \quad \text{(by Theorem 4.6.4)}$$

$$= \int_a^b [f(x) - g(x)]\,dx \quad \text{(by Theorem 4.6.5)}$$

Let h be the function defined by

$$h(x) = f(x) - g(x)$$

Then $h(x) \geq 0$ for all x in $[a, b]$ because $f(x) \geq g(x)$ for all x in $[a, b]$. We wish to prove that $\int_a^b h(x) \geq 0$. Because

$$\int_a^b h(x)\, dx = \lim_{\|\Delta\| \to 0} \sum_{i=1}^n h(\xi_i) \Delta_i x$$

we assume that

$$\lim_{\|\Delta\| \to 0} \sum_{i=1}^n h(\xi_i) \Delta_i x = L < 0 \tag{8}$$

Then by Definition 4.5.1, with $\epsilon = -L$, there exists a $\delta > 0$ such that

$$\text{if} \quad \|\Delta\| < \delta \quad \text{then} \quad \left| \sum_{i=1}^n h(\xi_i) \Delta_i x - L \right| < -L \tag{9}$$

But because

$$\sum_{i=1}^n h(\xi_i) \Delta_i x - L \leq \left| \sum_{i=1}^n h(\xi_i) \Delta_i x - L \right|$$

from (9) we have

$$\text{if} \quad \|\Delta\| < \delta \quad \text{then} \quad \sum_{i=1}^n h(\xi_i) \Delta_i x - L < -L$$

$$\Leftrightarrow \quad \text{if} \quad \|\Delta\| < \delta \quad \text{then} \quad \sum_{i=1}^n h(\xi_i) \Delta_i x < 0 \tag{10}$$

But statement (10) is impossible, because every $h(\xi_i)$ is nonnegative and every $\Delta_i x > 0$; thus we have a contradiction to our assumption (8). Therefore (8) is false, and

$$\lim_{\|\Delta\| \to 0} \sum_{i=1}^n h(\xi_i) \Delta_i x \geq 0$$

$$\int_a^b h(x) \geq 0$$

Because $h(x) = f(x) - g(x)$, we have

$$\int_a^b [f(x) - g(x)]\, dx \geq 0$$

$$\int_a^b f(x)\, dx - \int_a^b g(x)\, dx \geq 0$$

$$\int_a^b f(x)\, dx \geq \int_a^b g(x)\, dx \qquad\blacksquare$$

FIGURE 4.6.4

▶ ILLUSTRATION 4   In Fig. 4.6.4, $f(x) \geq 0$ for all x in $[a, b]$, and m and M are, respectively, the absolute minimum and absolute maximum function values of f on $[a, b]$. The integral $\int_a^b f(x)\,dx$ gives the measure of the area of the region bounded by the curve $y = f(x)$, the x axis, and the lines $x = a$ and $x = b$. This area is greater than that of the rectangle whose dimensions are m and $b - a$ and less than that of the rectangle whose dimensions are M and $b - a$. Thus we have a geometric interpretation of the next theorem if $f(x) \geq 0$ for all x in $[a, b]$. ◀

4.6.9 THEOREM   Suppose that the function f is continuous on the closed interval $[a, b]$. If m and M are, respectively, the absolute minimum and absolute maximum function values of f on $[a, b]$ so that

$$m \leq f(x) \leq M \qquad \text{for } a \leq x \leq b$$

then

$$m(b - a) \leq \int_a^b f(x)\,dx \leq M(b - a)$$

PROOF   Because f is continuous on $[a, b]$, the extreme-value theorem guarantees the existence of m and M.

By Theorem 4.6.3,

$$\int_a^b m\,dx = m(b - a) \tag{11}$$

and

$$\int_a^b M\,dx = M(b - a) \tag{12}$$

Because f is continuous on $[a, b]$, it follows from Theorem 4.5.3 that f is integrable on $[a, b]$. Then because $f(x) \geq m$ for all x in $[a, b]$, we have from Theorem 4.6.8

$$\int_a^b f(x)\,dx \geq \int_a^b m\,dx$$

which from (11) gives

$$\int_a^b f(x)\,dx \geq m(b - a) \tag{13}$$

Similarly, because $M \geq f(x)$ for all x in $[a, b]$, it follows from Theorem 4.6.8 that

$$\int_a^b M\,dx \geq \int_a^b f(x)\,dx$$

which from (12) gives

$$M(b - a) \geq \int_a^b f(x)\,dx \tag{14}$$

Combining inequalities (13) and (14) we have

$$m(b - a) \leq \int_a^b f(x)\,dx \leq M(b - a)$$ ■

EXAMPLE 3

Apply Theorem 4.6.9 to find a closed interval containing the value of $\int_{1/2}^{4} (x^3 - 6x^2 + 9x + 1)\,dx$. Use the results of Example 1, Section 3.4.

SOLUTION

If

$$f(x) = x^3 - 6x^2 + 9x + 1$$

from Example 1 in Section 3.4, f has a relative minimum value of 1 at $x = 3$ and a relative maximum value of 5 at $x = 1$. $f(\tfrac{1}{2}) = \tfrac{33}{8}$ and $f(4) = 5$. Hence the absolute minimum value of f on $[\tfrac{1}{2}, 4]$ is 1, and the absolute maximum value is 5. Taking $m = 1$ and $M = 5$ in Theorem 4.6.9 we have

$$1(4 - \tfrac{1}{2}) \le \int_{1/2}^{4} (x^3 - 6x^2 + 9x + 1)\,dx \le 5(4 - \tfrac{1}{2})$$

$$\tfrac{7}{2} \le \int_{1/2}^{4} (x^3 - 6x^2 + 9x + 1)\,dx \le \tfrac{35}{2}$$

Therefore the closed interval $[\tfrac{7}{2}, \tfrac{35}{2}]$ contains the value of the definite integral. In Example 2, Section 4.8, the exact value of the definite integral is shown to be $\tfrac{679}{64}$.

EXAMPLE 4

Apply Theorem 4.6.9 to find a closed interval containing the value of $\int_{\pi/4}^{3\pi/4} \sqrt{\sin x}\,dx$.

SOLUTION

If $f(x) = \sqrt{\sin x}$, then

$$f'(x) = \frac{\cos x}{2\sqrt{\sin x}}$$

For x in $[\tfrac{1}{4}\pi, \tfrac{3}{4}\pi]$, $f'(x) = 0$ when $x = \tfrac{1}{2}\pi$. Since $f'(x) > 0$ when $\tfrac{1}{4}\pi < x < \tfrac{1}{2}\pi$, and $f'(x) < 0$ when $\tfrac{1}{2}\pi < x < \tfrac{3}{4}\pi$, it follows that f has a relative maximum value at $\tfrac{1}{2}\pi$; and $f(\tfrac{1}{2}\pi) = 1$. Furthermore, $f(\tfrac{1}{4}\pi) = \sqrt[4]{2}/\sqrt{2} \approx 0.841$, and $f(\tfrac{3}{4}\pi) \approx 0.841$. Thus, on $[\tfrac{1}{4}\pi, \tfrac{3}{4}\pi]$ the absolute minimum value of f is 0.841 and the absolute maximum value is 1. So, with $m = 0.841$ and $M = 1$ in Theorem 4.6.9,

$$0.841[\tfrac{3}{4}\pi - \tfrac{1}{4}\pi] \le \int_{\pi/4}^{3\pi/4} \sqrt{\sin x}\,dx \le 1[\tfrac{3}{4}\pi - \tfrac{1}{4}\pi]$$

$$0.420\pi \le \int_{\pi/4}^{3\pi/4} \sqrt{\sin x}\,dx \le 0.5\pi$$

$$1.32 \le \int_{\pi/4}^{3\pi/4} \sqrt{\sin x}\,dx \le 1.57$$

The value of the definite integral is therefore in the closed interval $[1.32, 1.57]$.

EXERCISES 4.6

In Exercises 1 through 6, evaluate the definite integral.

1. $\int_2^5 4\,dx$ 2. $\int_{-3}^4 7\,dx$ 3. $\int_{-2}^2 \sqrt{5}\,dx$ 4. $\int_5^{-1} 6\,dx$ 5. $\int_{-5}^{-10} dx$ 6. $\int_3^3 dx$

In Exercises 7 through 18, evaluate the given definite integral by using the following results:

$\int_{-1}^2 x^2\,dx = 3$ $\int_{-1}^2 x\,dx = \frac{3}{2}$ $\int_0^\pi \sin x\,dx = 2$ $\int_0^\pi \cos x\,dx = 0$ $\int_0^\pi \sin^2 x\,dx = \frac{1}{2}\pi$

7. $\int_{-1}^2 (2x^2 - 4x + 5)\,dx$ 8. $\int_{-1}^2 (8 - x^2)\,dx$ 9. $\int_{-1}^2 (2 - 5x + \frac{1}{2}x^2)\,dx$

10. $\int_{-1}^2 (3x^2 - 4x - 1)\,dx$ 11. $\int_2^{-1} (2x + 1)^2\,dx$ 12. $\int_{-1}^2 (5x^2 + \frac{1}{3}x - \frac{1}{2})\,dx$

13. $\int_{-1}^2 (x - 1)(2x + 3)\,dx$ 14. $\int_2^{-1} 3x(x - 4)\,dx$ 15. $\int_0^\pi (2\sin x + 3\cos x + 1)\,dx$

16. $\int_0^\pi 3\cos^2 x\,dx$ 17. $\int_0^\pi (\cos x + 4)^2\,dx$ 18. $\int_\pi^0 (\sin x - 2)^2\,dx$

In Exercises 19 through 34, apply Theorem 4.6.9 to find a closed interval containing the value of the given definite integral.

19. $\int_3^7 2x\,dx$ 20. $\int_2^5 3x\,dx$ 21. $\int_0^4 x^2\,dx$ 22. $\int_{-2}^1 x^3\,dx$ 23. $\int_{-3}^6 \sqrt{3 + x}\,dx$

24. $\int_{-2}^1 (x + 1)^{2/3}\,dx$ 25. $\int_{-4}^0 (x^4 - 8x^2 + 16)\,dx$ 26. $\int_{-1}^4 (x^4 - 8x^2 + 16)\,dx$ 27. $\int_{\pi/6}^{\pi/3} \sin x\,dx$

28. $\int_{-\pi/3}^{2\pi/3} \cos x\,dx$ 29. $\int_1^4 |x - 2|\,dx$ 30. $\int_{-1}^2 \sqrt{x^2 + 5}\,dx$ 31. $\int_{-1}^2 \frac{x}{x + 2}\,dx$

32. $\int_{-5}^2 \frac{x + 5}{x - 3}\,dx$ 33. $\int_{-\pi/2}^{\pi/2} (4\cos^3 x - 9\cos x)\,dx$ 34. $\int_{-\pi/2}^{\pi/2} 3\sin^3 x\,dx$

In Exercises 35 through 38, use Theorem 4.6.8 to determine which of the symbols $\geq$ or $\leq$ should be inserted in the blank to make the inequality correct.

35. $\int_{-1}^3 (2x^2 - 4)\,dx$ _____ $\int_{-1}^3 (x^2 - 6)\,dx$ 36. $\int_4^5 \sqrt{6 - x}\,dx$ _____ $\int_4^5 \sqrt{x - 2}\,dx$

37. $\int_{3\pi/4}^{5\pi/4} \sin^2 x\,dx$ _____ $\int_{3\pi/4}^{5\pi/4} \cos^2 x\,dx$ 38. $\int_0^{\pi/4} \cos x\,dx$ _____ $\int_0^{\pi/4} \sin x\,dx$

39. Show that if f is continuous on $[-3, 4]$, then

$$\int_3^{-1} f(x)\,dx + \int_4^3 f(x)\,dx + \int_{-3}^4 f(x)\,dx + \int_{-1}^{-3} f(x)\,dx = 0$$

40. Show that if f is continuous on $[-1, 2]$, then

$$\int_{-1}^2 f(x)\,dx + \int_2^0 f(x)\,dx + \int_0^1 f(x)\,dx + \int_1^{-1} f(x)\,dx = 0$$

41. Show that $\int_0^1 x\,dx \geq \int_0^1 x^2\,dx$ but $\int_1^2 x\,dx \leq \int_1^2 x^2\,dx$. Do not evaluate the definite integrals.

42. If f is continuous on $[a, b]$, prove that $\left| \int_a^b f(x)\,dx \right| \leq \int_a^b |f(x)|\,dx$. (*Hint:* $-|f(x)| \leq f(x) \leq |f(x)|$.)

43. Prove Theorem 4.6.2.

In Exercises 44 through 49, prove that Theorem 4.6.7 is valid for the given ordering of a, b, and c; in each case use Theorem 4.6.6.

44. $b < a < c$ 45. $b < c < a$ 46. $c < a < b$ 47. $c < b < a$ 48. $a < c = b$ 49. $a = b < c$

4.7
THE MEAN-VALUE THEOREM
FOR INTEGRALS

We are now ready to state and prove the *mean-value theorem for integrals*. This theorem is used in the proof of many important theorems, the most prominent of which is Theorem 4.8.1, given in the next section. Before the statement of the mean-value theorem for integrals, we give an illustration that interprets the theorem geometrically.

▶ ILLUSTRATION 1   Consider $f(x) \geq 0$ for all values of x in $[a, b]$. Then $\int_a^b f(x)\,dx$ gives the measure of the area of the region bounded by the curve whose equation is $y = f(x)$, the x axis, and the lines $x = a$ and $x = b$. See Fig. 4.7.1. The mean-value theorem for integrals states that there is a number χ in $[a, b]$ such that the area of the rectangle $AEFB$ of height $f(\chi)$ units and width $(b - a)$ units is equal to the area of the region $ADCB$. ◀

4.7.1 THEOREM
Mean-Value Theorem for Integrals

If the function f is continuous on the closed interval $[a, b]$, there exists a number χ in $[a, b]$ such that

$$\int_a^b f(x)\,dx = f(\chi)(b - a)$$

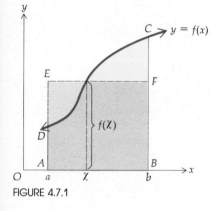

FIGURE 4.7.1

PROOF Because f is continuous on $[a, b]$, from the extreme-value theorem f has an absolute maximum value and an absolute minimum value on $[a, b]$.

Let m be the absolute minimum value occurring at $x = x_m$. Thus

$$f(x_m) = m \qquad a \leq x_m \leq b \tag{1}$$

Let M be the absolute maximum value, occurring at $x = x_M$. Thus

$$f(x_M) = M \qquad a \leq x_M \leq b \tag{2}$$

We have, then,

$$m \leq f(x) \leq M \qquad \text{for all } x \text{ in } [a, b]$$

From Theorem 4.6.9 it follows that

$$m(b - a) \leq \int_a^b f(x)\,dx \leq M(b - a)$$

Dividing by $b - a$ and noting that $b - a$ is positive because $b > a$ we get

$$m \leq \frac{\int_a^b f(x)\,dx}{b - a} \leq M$$

But from (1) and (2), $m = f(x_m)$ and $M = f(x_M)$; so we have

$$f(x_m) \leq \frac{\int_a^b f(x)\,dx}{b - a} \leq f(x_M) \tag{3}$$

From inequalities (3) and the intermediate-value theorem there is some number χ in a closed interval containing x_m and x_M such that

$$f(\chi) = \frac{\int_a^b f(x)\,dx}{b - a}$$

$$\Leftrightarrow \quad \int_a^b f(x)\,dx = f(\chi)(b - a) \qquad a \leq \chi \leq b \qquad \blacksquare$$

The value of χ in Theorem 4.7.1 is not necessarily unique. The theorem does not provide a method for finding χ, but it states that a value of χ exists, and this fact is used to prove other theorems. In some particular cases we can find the value of χ guaranteed by the theorem, as we illustrate in the following example.

EXAMPLE 1

Find the value of χ such that $\int_1^3 f(x)\,dx = f(\chi)(3 - 1)$ if $f(x) = x^2$. Use the result of Example 2 in Section 4.5.

SOLUTION

In Example 2 of Section 4.5 we obtained

$$\int_1^3 x^2\,dx = \frac{26}{3}$$

Therefore we wish to find χ such that

$$f(\chi) \cdot (2) = \frac{26}{3}$$

that is,

$$\chi^2 = \frac{13}{3}$$

$$\chi = \pm\tfrac{1}{3}\sqrt{39}$$

We reject $-\tfrac{1}{3}\sqrt{39}$ because it is not in the interval $[1, 3]$, and we have

$$\int_1^3 f(x)\,dx = f(\tfrac{1}{3}\sqrt{39})(3 - 1)$$

The value $f(\chi)$ given by Theorem 4.7.1 is called the *mean value* (or *average value*) of f on the interval $[a, b]$. It is a generalization of the arithmetic mean of a finite set of numbers. That is, if $\{f(x_1), f(x_2), \ldots, f(x_n)\}$ is a set of n numbers, then the arithmetic mean is given by

$$\frac{\sum\limits_{i=1}^{n} f(x_i)}{n}$$

To generalize this definition, consider a regular partition of the closed interval $[a, b]$, which is divided into n subintervals of equal length $\Delta x = (b - a)/n$.

Let ξ_i be any point in the ith subinterval. Form the sum:

$$\frac{\sum_{i=1}^{n} f(\xi_i)}{n} \tag{4}$$

The quotient (4) corresponds to the arithmetic mean of n numbers. Because $\Delta x = (b - a)/n$ we have

$$n = \frac{b - a}{\Delta x} \tag{5}$$

Substituting from (5) into (4) we obtain

$$\frac{\sum_{i=1}^{n} f(\xi_i)}{\dfrac{b-a}{\Delta x}} = \frac{\sum_{i=1}^{n} f(\xi_i)\,\Delta x}{b - a}$$

Taking the limit as $n \to +\infty$ (or $\Delta x \to 0$) we have, if the limit exists,

$$\lim_{n \to +\infty} \frac{\sum_{i=1}^{n} f(\xi_i)\,\Delta x}{b - a} = \frac{\int_a^b f(x)\,dx}{b - a}$$

This leads to the following definition.

4.7.2 DEFINITION   If the function f is integrable on the closed interval $[a, b]$, the **average value** of f on $[a, b]$ is

$$\frac{\int_a^b f(x)\,dx}{b - a}$$

FIGURE 4.7.2

EXAMPLE 2

(a) If $f(x) = x^2$, find the average value of f on the interval $[1, 3]$.

(b) Interpret the result of part (a) geometrically.

SOLUTION

(a) In Example 2, Section 4.5, we obtained

$$\int_1^3 x^2\,dx = \tfrac{26}{3}$$

So if A.V. is the average value of f on $[1, 3]$, we have

$$\text{A.V.} = \frac{\tfrac{26}{3}}{3 - 1}$$

$$= \tfrac{13}{3}$$

(b) In Example 1 we found for this function

$$f(\tfrac{1}{3}\sqrt{39}) = \tfrac{13}{3}$$

Therefore the average value of f occurs at $\tfrac{1}{3}\sqrt{39}$. In Fig. 4.7.2 there is a sketch of the graph of f on $[1, 3]$ and the line segment from the point

$E(\frac{1}{3}\sqrt{39}, 0)$ on the x axis to the point $F(\frac{1}{3}\sqrt{39}, \frac{13}{3})$ on the graph of f. The area of rectangle $AGHB$ having height $\frac{13}{3}$ and width 2 is equal to the area of the region $ACDB$. Consequently, the area of the shaded region CGF is equal to the area of the shaded region FDH.

An application of the average value of a function occurs in physics and engineering in connection with the concept of center of mass. This topic is discussed in Chapter 5.

EXERCISES 4.7

In Exercises 1 through 8, find the value of χ satisfying the mean-value theorem for integrals. For the value of the definite integral use the result of the indicated exercise in Exercises 4.5. Draw a figure illustrating the application of the theorem.

1. $\int_0^2 x^2 \, dx$; Exercise 11

2. $\int_2^4 x^2 \, dx$; Exercise 12

3. $\int_1^2 x^3 \, dx$; Exercise 13

4. $\int_0^5 (x^3 - 1) \, dx$; Exercise 14

5. $\int_1^4 (x^2 + 4x + 5) \, dx$; Exercise 15

6. $\int_0^4 (x^2 + x - 6) \, dx$; Exercise 16

7. $\int_{-2}^2 (x^3 + 1) \, dx$; Exercise 17

8. $\int_{-2}^1 x^4 \, dx$; Exercise 18

In Exercises 9 through 12, use the mean-value theorem for integrals to prove the inequality.

9. $\int_0^2 \frac{1}{x^2 + 4} \, dx \le \frac{1}{2}$

10. $\int_{-3}^3 \frac{1}{x^2 + 6} \, dx \le 1$

11. $\int_{-\pi/6}^{\pi/6} \cos x^2 \, dx \le \frac{\pi}{3}$

12. $\int_0^\pi \sin \sqrt{x} \, dx \le \pi$

In Exercises 13 through 16, use the mean-value theorem for integrals and Theorem 4.6.8 to prove the inequality.

13. $0 \le \int_2^5 \frac{1}{x^3 + 1} \, dx \le \frac{1}{3}$

14. $0 \le \int_5^9 \frac{1}{\sqrt{x-1}} \, dx \le 2$

15. $0 \le \int_0^2 \sin \frac{1}{2}\pi x \, dx \le 2$

16. $0 \le \int_{-1/2}^{1/2} \cos \pi x \le 1$

17. Given that $\int_{-1}^2 x \, dx = \frac{3}{2}$, find the average value of the identity function on the interval $[-1, 2]$. Also find the value of x at which the average value occurs. Give a geometric interpretation of the results.

18. Find the average value of the function f defined by $f(x) = x^2$ on the interval $[-1, 2]$ given that $\int_{-1}^2 x^2 \, dx = 3$. Also find the value of x at which the average value occurs. Give a geometric interpretation of the results.

19. Given that $\int_0^\pi \sin x \, dx = 2$, find the average value of the sine function on the interval $[0, \pi]$. Also find the smallest value of x at which the average value occurs. Give a geometric interpretation of the results.

20. Find the average value of the function f defined by $f(x) = \sec^2 x$ on the interval $[0, \frac{1}{4}\pi]$ given that $\int_0^{\pi/4} \sec^2 x \, dx = 1$. Also find the value of x at which the average value occurs. Give a geometric interpretation of the results.

21. Suppose a ball is dropped from rest and after t sec its velocity is v ft/sec. Neglecting air resistance, express v in terms of t as $v = f(t)$, and find the average value of f on $[0, 2]$. (*Hint:* Find the value of the definite integral by interpreting it as the measure of the area of a region enclosed by a triangle.)

22. Find the average value of the function f defined by $f(x) = \sqrt{49 - x^2}$ on the interval $[0, 7]$. Draw a figure. (*Hint:* Find the value of the definite integral by interpreting it as the measure of the area of a region enclosed by a quarter-circle.)

23. Find the average value of the function f defined by $f(x) = \sqrt{16 - x^2}$ on the interval $[-4, 4]$. Draw a figure. (*Hint:* Find the value of the definite integral by interpreting it as the measure of the area of a region enclosed by a semicircle.)

24. Suppose that f is integrable on $[-4, 7]$. If the average value of f on the interval $[-4, 7]$ is $\frac{17}{4}$, find $\int_{-4}^{7} f(x)\, dx$.

25. If f is continuous on $[a, b]$ and $\int_{a}^{b} f(x)\, dx = 0$, prove that there is at least one number c in $[a, b]$ such that $f(c) = 0$.

26. The following theorem is a generalization of Theorem 4.7.1: If f and g are two functions that are continuous on the closed interval $[a, b]$ and $g(x) > 0$ for all x in the open interval (a, b), then there exists a number χ in $[a, b]$ such that

$$\int_{a}^{b} f(x)g(x)\, dx = f(\chi) \int_{a}^{b} g(x)\, dx$$

Prove this theorem by a method similar to that for Theorem 4.7.1: Obtain the inequality $m \leq f(x) \leq M$ and then conclude that $mg(x) \leq f(x)g(x) \leq Mg(x)$; apply Theorem 4.6.8 and proceed as in the proof of Theorem 4.7.1.

27. Show that when $g(x) = 1$, the theorem of Exercise 26 becomes Theorem 4.7.1.

In Exercises 28 through 32, use the theorem of Exercise 26 to prove the inequality.

28. $\displaystyle\int_{0}^{4} \frac{x\, dx}{x^3 + 2} < \int_{0}^{4} x\, dx$

29. $\displaystyle\int_{-1}^{1} \frac{x^2\, dx}{\sqrt{x^2 + 4}} < \int_{-1}^{1} x^2\, dx$

30. $\displaystyle\int_{0}^{\pi} x \sin x\, dx \leq \int_{0}^{\pi} x\, dx$

31. $\displaystyle\int_{-1/2}^{1/2} \sin^2 \pi x \cos \pi x\, dx \leq \int_{-1/2}^{1/2} \cos \pi x\, dx$

32. $\displaystyle\int_{0}^{1} \frac{x \cos x}{x^2 + 1}\, dx \leq \int_{0}^{1} x\, dx$

4.8
THE FUNDAMENTAL THEOREMS OF THE CALCULUS

Historically, the basic concepts of the definite integral were used by the ancient Greeks, principally Archimedes (287–212 B.C.), more than 2000 years ago, which was many years before the differential calculus was discovered.

In the seventeenth century, almost simultaneously but working independently, Newton and Leibniz showed how the calculus could be used to find the area of a region bounded by a curve or a set of curves, by evaluating a definite integral by antidifferentiation. The procedure involves what are known as the *fundamental theorems of the calculus*. Before we state and prove these important theorems, we discuss definite integrals having a variable upper limit.

Let f be a function continuous on the closed interval $[a, b]$. Then the value of the definite integral $\int_{a}^{b} f(x)\, dx$ depends only on the function f and the numbers a and b, and not on the symbol x, which is used here as the independent variable. In Example 2, Section 4.5, we found the value of $\int_{1}^{3} x^2\, dx$ to be $\frac{26}{3}$. Any other symbol instead of x could have been used; for example,

$$\int_{1}^{3} t^2\, dt = \frac{26}{3} \qquad \int_{1}^{3} u^2\, du = \frac{26}{3} \qquad \int_{1}^{3} r^2\, dr = \frac{26}{3}$$

If f is continuous on the closed interval $[a, b]$, then by Theorem 4.5.3 the definite integral $\int_{a}^{b} f(t)\, dt$ exists. We previously stated that if the definite integral exists, it is a unique number. If x is a number in $[a, b]$, then f is continuous on $[a, x]$ because it is continuous on $[a, b]$. Consequently, $\int_{a}^{x} f(t)\, dt$ exists and is a unique number whose value depends on x. Therefore $\int_{a}^{x} f(t)\, dt$

defines a function F having as its domain all numbers in the closed interval $[a, b]$ and whose function value at any number x in $[a, b]$ is given by

$$F(x) = \int_a^x f(t)\,dt \tag{1}$$

As a notational observation, if the limits of the definite integral are variables, different symbols are used for these limits and for the independent variable in the integrand. Hence, in (1), because x is the upper limit, we use the letter t as the independent variable in the integrand.

If, in Eq. (1), $f(t) \geq 0$ for all values of t in $[a, b]$, then the function value $F(x)$ can be interpreted geometrically as the measure of the area of the region bounded by the curve whose equation is $y = f(t)$, the t axis, and the lines $t = a$ and $t = x$. (See Fig. 4.8.1.) Note that $F(a) = \int_a^a f(t)\,dt$, which by Definition 4.5.6 equals 0.

We now state and prove an important theorem giving the derivative of a function F defined as a definite integral having a variable upper limit. This theorem is called *the first fundamental theorem of the calculus.*

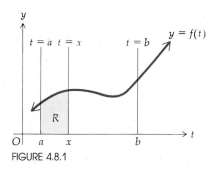

FIGURE 4.8.1

4.8.1 THEOREM

First Fundamental Theorem of the Calculus

Let the function f be continuous on the closed interval $[a, b]$ and let x be any number in $[a, b]$. If F is the function defined by

$$F(x) = \int_a^x f(t)\,dt$$

then

$$F'(x) = f(x) \tag{2}$$

(If $x = a$, the derivative in (2) may be a derivative from the right, and if $x = b$, the derivative in (2) may be a derivative from the left.)

PROOF   Consider two numbers x_1 and $x_1 + \Delta x$ in $[a, b]$. Then

$$F(x_1) = \int_a^{x_1} f(t)\,dt$$

and

$$F(x_1 + \Delta x) = \int_a^{x_1 + \Delta x} f(t)\,dt$$

so that

$$F(x_1 + \Delta x) - F(x_1) = \int_a^{x_1 + \Delta x} f(t)\,dt - \int_a^{x_1} f(t)\,dt \tag{3}$$

By Theorem 4.6.7,

$$\int_a^{x_1} f(t)\,dt + \int_{x_1}^{x_1 + \Delta x} f(t)\,dt = \int_a^{x_1 + \Delta x} f(t)\,dt$$

$$\int_a^{x_1 + \Delta x} f(t)\,dt - \int_a^{x_1} f(t)\,dt = \int_{x_1}^{x_1 + \Delta x} f(t)\,dt \tag{4}$$

Substituting from (4) into (3) we get

$$F(x_1 + \Delta x) - F(x_1) = \int_{x_1}^{x_1 + \Delta x} f(t)\,dt \tag{5}$$

By the mean-value theorem for integrals (4.7.1) there is some number χ in the closed interval bounded by x_1 and $x_1 + \Delta x$ such that

$$\int_{x_1}^{x_1 + \Delta x} f(t)\, dt = f(\chi)\, \Delta x \tag{6}$$

From (5) and (6) we obtain

$$F(x_1 + \Delta x) - F(x_1) = f(\chi)\, \Delta x$$

$$\frac{F(x_1 + \Delta x) - F(x_1)}{\Delta x} = f(\chi)$$

Taking the limit as Δx approaches zero we have

$$\lim_{\Delta x \to 0} \frac{F(x_1 + \Delta x) - F(x_1)}{\Delta x} = \lim_{\Delta x \to 0} f(\chi) \tag{7}$$

The left side of (7) is $F'(x_1)$. To determine $\lim\limits_{\Delta x \to 0} f(\chi)$, recall that χ is in the closed interval bounded by x_1 and $x_1 + \Delta x$, and because

$$\lim_{\Delta x \to 0} x_1 = x_1 \quad \text{and} \quad \lim_{\Delta x \to 0} (x_1 + \Delta x) = x_1$$

it follows from the squeeze theorem (1.8.1) that $\lim\limits_{\Delta x \to 0} \chi = x_1$. Thus we have $\lim\limits_{\Delta x \to 0} f(\chi) = \lim\limits_{\chi \to x_1} f(\chi)$. Because f is continuous at x_1, $\lim\limits_{\chi \to x_1} f(\chi) = f(x_1)$; thus $\lim\limits_{\Delta x \to 0} f(\chi) = f(x_1)$, and from (7) we get

$$F'(x_1) = f(x_1) \tag{8}$$

If the function f is not defined for values of x less than a but is continuous from the right at a, then in the above argument, if $x_1 = a$ in (7), Δx must approach 0 from the right. Hence the left side of (8) will be $F'_+(x_1)$. Similarly, if f is not defined for values of x greater than b but is continuous from the left at b, then if $x_1 = b$ in (7), Δx must approach 0 from the left. Hence we have $F'_-(x_1)$ on the left side of (8).

Because x_1 is any number in $[a, b]$, Eq. (8) states what we wished to prove. ■

Theorem 4.8.1 states that the definite integral $\int_a^x f(t)\, dt$, with variable upper limit x, is an antiderivative of f.

EXAMPLE 1

Apply Theorem 4.8.1 to find $D_x \int_1^x \sqrt{2 + t}\, dt$.

SOLUTION

With the notation of the statement of Theorem 4.8.1,

$$F(x) = \int_1^x \sqrt{2 + t}\, dt \qquad f(t) = \sqrt{2 + t}$$

and $a = 1$. Then from the theorem,

$$F'(x) = f(x)$$

Thus

$$D_x \int_1^x \sqrt{2 + t}\, dt = \sqrt{2 + x}$$

We now use Theorem 4.8.1 to prove the *second fundamental theorem of the calculus.*

4.8.2 THEOREM
Second Fundamental Theorem of the Calculus

Let the function f be continuous on the closed interval $[a, b]$ and let g be a function such that

$$g'(x) = f(x) \tag{9}$$

for all x in $[a, b]$. Then

$$\int_a^b f(t)\, dt = g(b) - g(a)$$

(If $x = a$, the derivative in (9) may be a derivative from the right, and if $x = b$, the derivative in (9) may be a derivative from the left.)

PROOF   If f is continuous at all numbers in $[a, b]$, we know from Theorem 4.8.1 that the definite integral $\int_a^x f(t)\, dt$, with variable upper limit x, defines a function F whose derivative on $[a, b]$ is f. Because by hypothesis $g'(x) = f(x)$, it follows from Theorem 4.1.2 that

$$g(x) = \int_a^x f(t)\, dt + k \tag{10}$$

where k is some constant.

Letting $x = b$ and $x = a$, successively, in (10) we get

$$g(b) = \int_a^b f(t)\, dt + k \tag{11}$$

and

$$g(a) = \int_a^a f(t)\, dt + k \tag{12}$$

From (11) and (12),

$$g(b) - g(a) = \int_a^b f(t)\, dt - \int_a^a f(t)\, dt$$

But, by Definition 4.5.6, $\int_a^a f(t)\, dt = 0$; so

$$g(b) - g(a) = \int_a^b f(t)\, dt$$

which is what we wished to prove.

If f is not defined for values of x greater than b but is continuous from the left at b, the derivative in (9) is a derivative from the left, and we have

$g'_-(b) = F'_-(b)$, from which (11) follows. Similarly, if f is not defined for values of x less than a but is continuous from the right at a, then the derivative in (9) is a derivative from the right, and we have $g'_+(a) = F'_+(a)$, from which (12) follows.   ∎

We are now in a position to find the exact value of a definite integral by applying Theorem 4.8.2. In applying the theorem we denote

$$[g(b) - g(a)] \quad \text{by} \quad g(x)\Big]_a^b$$

▶ ILLUSTRATION 1   We apply the second fundamental theorem of the calculus to evaluate

$$\int_1^3 x^2 \, dx$$

Here $f(x) = x^2$. An antiderivative of x^2 is $\frac{1}{3}x^3$. From this we choose

$$g(x) = \frac{x^3}{3}$$

Therefore, from Theorem 4.8.2,

$$\int_1^3 x^2 \, dx = \frac{x^3}{3}\Big]_1^3$$
$$= 9 - \tfrac{1}{3}$$
$$= \tfrac{26}{3}$$

Compare this result with that of Example 2, Section 4.5.   ◀

Because of the connection between definite integrals and antiderivatives, we used the integral sign $\int$ for the notation $\int f(x) \, dx$ for an antiderivative. We now dispense with the terminology of antiderivatives and antidifferentiation and begin to call $\int f(x) \, dx$ an **indefinite integral**. The process of evaluating an indefinite integral or a definite integral is called **integration**.

The difference between an indefinite integral and a definite integral should be emphasized. The indefinite integral $\int f(x) \, dx$ is defined as a function g such that its derivative $D_x[g(x)] = f(x)$. However, the definite integral $\int_a^b f(x) \, dx$ is a number whose value depends on the function f and the numbers a and b, and it is defined as the limit of a Riemann sum. The definition of the definite integral makes no reference to differentiation.

The indefinite integral involves an arbitrary constant; for instance,

$$\int x^2 \, dx = \frac{x^3}{3} + C$$

This arbitrary constant C is called a **constant of integration**. In applying the second fundamental theorem to evaluate a definite integral, we do not need to include the arbitrary constant C in the expression for $g(x)$ because the theorem permits us to select *any* antiderivative, including the one for which $C = 0$.

EXAMPLE 2

Evaluate $\displaystyle\int_{1/2}^{4} (x^3 - 6x^2 + 9x + 1)\,dx.$

SOLUTION

$$\int_{1/2}^{4} (x^3 - 6x^2 + 9x + 1)\,dx = \int_{1/2}^{4} x^3\,dx - 6 \int_{1/2}^{4} x^2\,dx$$

$$+ 9 \int_{1/2}^{4} x\,dx + \int_{1/2}^{4} dx$$

$$= \frac{x^4}{4} - 6 \cdot \frac{x^3}{3} + 9 \cdot \frac{x^2}{2} + x \Big]_{1/2}^{4}$$

$$= (64 - 128 + 72 + 4) - (\tfrac{1}{64} - \tfrac{1}{4} + \tfrac{9}{8} + \tfrac{1}{2})$$

$$= \tfrac{679}{64}$$

EXAMPLE 3

Evaluate $\displaystyle\int_{-1}^{1} (x^{4/3} + 4x^{1/3})\,dx.$

SOLUTION

$$\int_{-1}^{1} (x^{4/3} + 4x^{1/3})\,dx = \tfrac{3}{7}x^{7/3} + 4 \cdot \tfrac{3}{4}x^{4/3} \Big]_{-1}^{1}$$

$$= \tfrac{3}{7} + 3 - (-\tfrac{3}{7} + 3)$$

$$= \tfrac{6}{7}$$

EXAMPLE 4

Evaluate $\displaystyle\int_{0}^{2} 2x^2 \sqrt{x^3 + 1}\,dx.$

SOLUTION

$$\int_{0}^{2} 2x^2 \sqrt{x^3 + 1}\,dx = \frac{2}{3} \int_{0}^{2} \sqrt{x^3 + 1}\,(3x^2\,dx)$$

$$= \frac{2}{3} \cdot \frac{(x^3 + 1)^{3/2}}{\frac{3}{2}} \Big]_{0}^{2}$$

$$= \tfrac{4}{9}(8 + 1)^{3/2} - \tfrac{4}{9}(0 + 1)^{3/2}$$

$$= \tfrac{4}{9}(27 - 1)$$

$$= \tfrac{104}{9}$$

EXAMPLE 5
Evaluate $\int_0^3 x\sqrt{1 + x}\, dx$.

SOLUTION

To evaluate the indefinite integral $\int x\sqrt{1 + x}\, dx$ we let

$$u = \sqrt{1 + x} \qquad u^2 = 1 + x \qquad x = u^2 - 1 \qquad dx = 2u\, du$$

Substituting we have

$$\int x\sqrt{1 + x}\, dx = \int (u^2 - 1)u(2u\, du)$$

$$= 2\int (u^4 - u^2)\, du$$

$$= \tfrac{2}{5}u^5 - \tfrac{2}{3}u^3 + C$$

$$= \tfrac{2}{5}(1 + x)^{5/2} - \tfrac{2}{3}(1 + x)^{3/2} + C$$

Therefore the definite integral

$$\int_0^3 x\sqrt{1 + x}\, dx = \tfrac{2}{5}(1 + x)^{5/2} - \tfrac{2}{3}(1 + x)^{3/2}\Big]_0^3$$

$$= \tfrac{2}{5}(4)^{5/2} - \tfrac{2}{3}(4)^{3/2} - \tfrac{2}{3}(1)^{5/2} + \tfrac{2}{3}(1)^{3/2}$$

$$= \tfrac{64}{5} - \tfrac{16}{3} - \tfrac{2}{5} + \tfrac{2}{3}$$

$$= \tfrac{116}{15}$$

Another method for evaluating the definite integral in Example 5 is provided by a formula that follows from Theorems 4.8.2 and 4.2.1 (the chain rule for antidifferentiation). From these theorems, if F is an antiderivative of f,

$$\int_a^b f(g(x))g'(x)\, dx = F(g(x))\Big]_a^b$$

$$\Leftrightarrow \quad \int_a^b f(g(x))g'(x)\, dx = F(g(b)) - F(g(a))$$

Thus

$$\int_a^b f(g(x))g'(x)\, dx = F(u)\Big]_{g(a)}^{g(b)}$$

$$\Leftrightarrow \quad \int_a^b f(g(x))g'(x)\, dx = \int_{g(a)}^{g(b)} f(u)\, du \tag{13}$$

To apply (13), change the variables in the given integral by letting $u = g(x)$. Then $du = g'(x)\, dx$. Then change the x-limits of integration a and b to u-limits, which are $g(a)$ and $g(b)$.

▶ ILLUSTRATION 2   To evaluate the integral of Example 5, let $u = \sqrt{1 + x}$, $x = u^2 - 1$, and $dx = 2u\, du$. Furthermore, when $x = 0$, $u = 1$, and when

$x = 3$, $u = 2$. Thus, from (13) we have

$$\int_0^3 x\sqrt{1+x}\,dx = 2\int_1^2 (u^4 - u^2)\,du$$

$$= \tfrac{2}{5}u^5 - \tfrac{2}{3}u^3\Big]_1^2$$

$$= \tfrac{64}{5} - \tfrac{16}{3} - \tfrac{2}{5} + \tfrac{2}{3}$$

$$= \tfrac{116}{15}$$

◄

EXAMPLE 6
Evaluate $\displaystyle\int_0^{\pi/2} \sin^3 x \cos x\,dx$.

SOLUTION
Let

$$u = \sin x \qquad du = \cos x\,dx$$

When $x = 0$, $u = 0$; when $x = \tfrac{1}{2}\pi$, $u = 1$. Therefore

$$\int_0^{\pi/2} \sin^3 x \cos x\,dx = \int_0^1 u^3\,du$$

$$= \frac{u^4}{4}\Big]_0^1$$

$$= \tfrac{1}{4}$$

EXAMPLE 7
Evaluate $\displaystyle\int_{-3}^4 |x + 2|\,dx$.

SOLUTION
If $f(x) = |x + 2|$, instead of finding an antiderivative of f directly, write $f(x)$ as

$$f(x) = \begin{cases} -x - 2 & \text{if } x \le -2 \\ x + 2 & \text{if } -2 \le x \end{cases}$$

From Theorem 4.6.7,

$$\int_{-3}^4 |x + 2|\,dx = \int_{-3}^{-2} (-x - 2)\,dx + \int_{-2}^4 (x + 2)\,dx$$

$$= \left[-\frac{x^2}{2} - 2x\right]_{-3}^{-2} + \left[\frac{x^2}{2} + 2x\right]_{-2}^4$$

$$= \left[(-2 + 4) - (-\tfrac{9}{2} + 6)\right] + \left[(8 + 8) - (2 - 4)\right]$$

$$= \tfrac{1}{2} + 18$$

$$= \tfrac{37}{2}$$

EXAMPLE 8
Given $f(x) = \sec^2 x$. Find the average value of f on the interval $\left[-\tfrac{1}{4}\pi, \tfrac{1}{4}\pi\right]$.

SOLUTION
Let A.V. be the average value of f on $\left[-\tfrac{1}{4}\pi, \tfrac{1}{4}\pi\right]$. From Definition 4.7.2,

$$\text{A.V.} = \frac{1}{(\tfrac{1}{4}\pi) - (-\tfrac{1}{4}\pi)} \int_{-\pi/4}^{\pi/4} \sec^2 x \, dx$$

$$= \frac{2}{\pi} \Big[\tan x\Big]_{-\pi/4}^{\pi/4}$$

$$= \frac{2}{\pi} \big[1 - (-1)\big]$$

$$= \frac{4}{\pi}$$

EXERCISES 4.8

In Exercises 1 through 36, evaluate the definite integral.

1. $\int_0^3 (3x^2 - 4x + 1)\,dx$

2. $\int_0^4 (x^3 - x^2 + 1)\,dx$

3. $\int_3^6 (x^2 - 2x)\,dx$

4. $\int_{-1}^3 (3x^2 + 5x - 1)\,dx$

5. $\int_1^2 \frac{x^2 + 1}{x^2}\,dx$

6. $\int_{-3}^5 (y^3 - 4y)\,dy$

7. $\int_0^1 \frac{z}{(z^2 + 1)^3}\,dz$

8. $\int_1^4 \sqrt{x}(2 + x)\,dx$

9. $\int_1^{10} \sqrt{5x - 1}\,dx$

10. $\int_0^{\sqrt{5}} t\sqrt{t^2 + 1}\,dt$

11. $\int_{-2}^0 3w\sqrt{4 - w^2}\,dw$

12. $\int_{-1}^3 \frac{dy}{(y + 2)^3}$

13. $\int_0^{\pi/2} \sin 2x \, dx$

14. $\int_0^\pi \cos\tfrac{1}{2}x \, dx$

15. $\int_1^2 t^2\sqrt{t^3 + 1}\,dt$

16. $\int_1^3 \frac{x\,dx}{(3x^2 - 1)^3}$

17. $\int_0^1 \frac{(y^2 + 2y)\,dy}{\sqrt[3]{y^3 + 3y^2 + 4}}$

18. $\int_2^4 \frac{w^4 - w}{w^3}\,dw$

19. $\int_0^{15} \frac{w\,dw}{(1 + w)^{3/4}}$

20. $\int_4^5 x^2\sqrt{x - 4}\,dx$

21. $\int_{-2}^5 |x - 3|\,dx$

22. $\int_{-4}^4 |x - 2|\,dx$

23. $\int_{-1}^1 \sqrt{|x| - x}\,dx$

24. $\int_{-3}^3 \sqrt{3 + |x|}\,dx$

25. $\int_0^3 (x + 2)\sqrt{x + 1}\,dx$

26. $\int_{-2}^1 (x + 1)\sqrt{x + 3}\,dx$

27. $\int_0^1 \frac{x^3 + 1}{x + 1}\,dx$ (*Hint:* Divide the numerator by the denominator.)

28. $\int_1^4 \frac{x^5 - x}{3x^3}\,dx$

29. $\int_1^{64} \left(\sqrt{t} - \frac{1}{\sqrt{t}} + \sqrt[3]{t}\right)dt$

30. $\int_0^1 \sqrt{x}\sqrt{1 + x\sqrt{x}}\,dx$

31. $\int_1^2 \frac{x^3 + 2x^2 + x + 2}{(x + 1)^2}\,dx$

32. $\int_{-3}^2 \frac{3x^3 - 24x^2 + 48x + 5}{x^2 - 8x + 16}\,dx$

(*Hint* for Exercises 31 and 32: Divide the numerator by the denominator.)

33. $\int_0^1 \sin \pi x \cos \pi x \, dx$

34. $\int_0^{1/2} \sec^2 \tfrac{1}{2}\pi t \tan \tfrac{1}{2}\pi t \, dt$

35. $\int_{\pi/8}^{\pi/4} 3\csc^2 2x \, dx$

36. $\int_0^{\pi/6} (\sin 2x + \cos 3x)\,dx$

At top right (handwritten):
$$u = x^2$$
$$\frac{du}{dx} = 2x \left[\frac{d}{du} \int_1^u \sqrt{1+t^4}\, dt \right] \frac{du}{dx} = \sqrt{1+u^4}\,(2x)$$
$$= \sqrt{1+x^8}\,(2x)$$

In Exercises 37 through 42, use Theorem 4.8.1 to find the indicated derivative.

37. $D_x \int_0^x \sqrt{4 + t^6}\, dt$ $\sqrt{4+x^6}$

38. $D_x \int_2^x \frac{dt}{t^4 + 4}$ $\frac{1}{x^4+4}$

39. $D_x \int_x^3 \sqrt{\sin t}\, dt$ $-\sqrt{\sin x}$

40. $D_x \int_x^5 \sqrt{1 + t^4}\, dt$ $-\sqrt{1+x^4}$

41. $D_x \int_{-x}^x \frac{dt}{3 + t^2}$ (handwritten) $\int_{-x}^0 + \int_0^x$ $= -\int_0^{-x} \frac{dt}{3+t^2} + \int_0^x \frac{dt}{3+t^2}$ $u=-x,\ du=-dx$ $-\frac{d}{du}\int_0^u \frac{dt}{3+t^2} \frac{du}{dx} + \cdots = -\frac{1}{3+u^2}(-1) + \frac{1}{3+x^2} = \frac{2}{3+x^2}$

42. $D_x \int_1^{x^2} \sqrt{1 + t^4}\, dt$

In Exercises 43 through 46, find the average value of the function f on the given interval $[a, b]$. In Exercises 43 and 44, find the value of x at which the average value of f occurs, and make a sketch.

43. $f(x) = 9 - x^2;\ [a, b] = [0, 3]$

44. $f(x) = 8x - x^2;\ [a, b] = [0, 4]$

45. $f(x) = 3x\sqrt{x^2 - 16};\ [a, b] = [4, 5]$

46. $f(x) = x^2\sqrt{x - 3};\ [a, b] = [7, 12]$

47. In an electric circuit, suppose that E volts is the electromotive force at t sec and that $E = 2 \sin 3t$. Find the average value of E from $t = 0$ to $t = \frac{1}{3}\pi$.

48. For the electric circuit of Exercise 47, find the square root of the average value of E^2 from $t = 0$ to $t = \frac{1}{3}\pi$. (*Hint:* Use the identity $\sin^2 x = \frac{1}{2}(1 - \cos 2x)$.)

49. A ball is dropped from rest, and after t sec its velocity is v ft/sec. Neglecting air resistance, show that the average velocity during the first $\frac{1}{2}T$ sec is one-third of the average velocity during the next $\frac{1}{2}T$ sec.

50. A stone is thrown downward with an initial velocity of v_0 ft/sec. Neglect air resistance. (a) Show that if v ft/sec is the velocity of the stone after falling s ft, then $v = \sqrt{v_0^2 + 2gs}$. (b) Find the average velocity during the first 100 ft of fall if the initial velocity is 60 ft/sec. (Take $g = 32$ and downward as the positive direction.)

51. Find $\int_4^{16} \left[D_x \int_5^x (2\sqrt{t} - 1)\, dt \right] dx$.

52. Let f be a function whose derivative f' is continuous on $[a, b]$. Find the average value of the slope of the tangent line of the graph of f on $[a, b]$, and give a geometric interpretation of the result.

4.9
AREA OF A REGION IN A PLANE

From Definition 4.5.4, if f is a function continuous on the closed interval $[a, b]$, and if $f(x) \geq 0$ for all x in $[a, b]$, then the number of square units in the area of the region bounded by the curve $y = f(x)$, the x axis, and the lines $x = a$ and $x = b$ is

$$\lim_{\|\Delta\| \to 0} \sum_{i=1}^n f(\xi_i)\, \Delta_i x$$

which is the definite integral $\int_a^b f(x)\, dx$.

In the examples that follow we first express the measure of the required area as the limit of a Riemann sum to reinforce the procedure for setting up such sums for future applications.

EXAMPLE 1
Find the area of the region in the first quadrant bounded by the curve

$$y = x\sqrt{x^2 + 5}$$

the x axis, and the line $x = 2$. Make a sketch.

$y = x\sqrt{x^2 + 5}$

$(\xi_i, \xi_i\sqrt{\xi_i^2 + 5})$

$\Delta_i x$

FIGURE 4.9.1

SOLUTION

See Fig. 4.9.1. The region is shown together with one of the rectangular elements of area.

We take a partition of the interval $[0, 2]$. The width of the ith rectangle is $\Delta_i x$, and the altitude is $\xi_i\sqrt{\xi_i^2 + 5}$ units, where ξ_i is any number in the ith subinterval. Therefore the measure of the area of the rectangular element is $\xi_i\sqrt{\xi_i^2 + 5}\,\Delta_i x$. The sum of the measures of the areas of n such rectangles is

$$\sum_{i=1}^{n} \xi_i\sqrt{\xi_i^2 + 5}\,\Delta_i x$$

which is a Riemann sum. The limit of this sum as $\|\Delta\|$ approaches 0 gives the measure of the desired area. The limit of the Riemann sum is a definite integral that we evaluate by the second fundamental theorem of the calculus. Let A square units be the area of the region. Then

$$A = \lim_{\|\Delta\|\to 0} \sum_{i=1}^{n} \xi_i\sqrt{\xi_i^2 + 5}\,\Delta_i x$$

$$= \int_0^2 x\sqrt{x^2 + 5}\,dx$$

$$= \tfrac{1}{2}\int_0^2 \sqrt{x^2 + 5}\,(2x\,dx)$$

$$= \tfrac{1}{2}\cdot\tfrac{2}{3}(x^2 + 5)^{3/2}\Big]_0^2$$

$$= \tfrac{1}{3}\big[(9)^{3/2} - (5)^{3/2}\big]$$

$$= \tfrac{1}{3}(27 - 5\sqrt{5})$$

$$\approx 5.27$$

Hence the area is $\tfrac{1}{3}(27 - 5\sqrt{5})$ square units, or approximately 5.27 square units.

So far we have considered the area of a region for which the function values are nonnegative on $[a, b]$. Suppose now that $f(x) < 0$ for all x in $[a, b]$. Then each $f(\xi_i)$ is a negative number; so we define the number of square units in the area of the region bounded by $y = f(x)$, the x axis, and the lines $x = a$ and $x = b$ to be

$$\lim_{\|\Delta\|\to 0} \sum_{i=1}^{n} \big[-f(\xi_i)\big]\Delta_i x$$

which equals

$$-\int_a^b f(x)\,dx$$

EXAMPLE 2

Find the area of the region bounded by the curve

$$y = x^2 - 4x$$

the x axis, and the lines $x = 1$ and $x = 3$.

$y = x^2 - 4x$

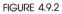

$(\xi_i, \xi_i{}^2 - 4\xi_i)$

FIGURE 4.9.2

SOLUTION

The region, together with a rectangular element of area, is shown in Fig. 4.9.2.

We take a partition of the interval $[1, 3]$; the width of the ith rectangle is $\Delta_i x$. Because $x^2 - 4x < 0$ on $[1, 3]$, the altitude of the ith rectangle is $-(\xi_i{}^2 - 4\xi_i) = 4\xi_i - \xi_i{}^2$. Hence the sum of the measures of the areas of n rectangles is given by

$$\sum_{i=1}^{n} (4\xi_i - \xi_i{}^2) \Delta_i x$$

The measure of the desired area is given by the limit of this sum as $\|\Delta\|$ approaches 0; so if A square units is the area of the region,

$$
\begin{aligned}
A &= \lim_{\|\Delta\| \to 0} \sum_{i=1}^{n} (4\xi_i - \xi_i{}^2) \Delta_i x \\
&= \int_1^3 (4x - x^2) \, dx \\
&= 2x^2 - \tfrac{1}{3}x^3 \Big]_1^3 \\
&= \tfrac{22}{3}
\end{aligned}
$$

Thus the area of the region is $\tfrac{22}{3}$ square units.

EXAMPLE 3

Find the area of the region bounded by the curve

$$y = x^3 - 2x^2 - 5x + 6$$

the x axis, and the lines $x = -1$ and $x = 2$.

SOLUTION

The region is shown in Fig. 4.9.3. Let

$$f(x) = x^3 - 2x^2 - 5x + 6$$

Because $f(x) \geq 0$ when x is in the closed interval $[-1, 1]$ and $f(x) \leq 0$ when x is in the closed interval $[1, 2]$, we separate the region into two parts. Let A_1 be the number of square units in the area of the region when x is in $[-1, 1]$, and let A_2 be the number of square units in the area of the region when x is in $[1, 2]$. Then

$$
\begin{aligned}
A_1 &= \lim_{\|\Delta\| \to 0} \sum_{i=1}^{n} f(\xi_i) \Delta_i x \\
&= \int_{-1}^{1} f(x) \, dx \\
&= \int_{-1}^{1} (x^3 - 2x^2 - 5x + 6) \, dx
\end{aligned}
$$

and

$$
\begin{aligned}
A_2 &= \lim_{\|\Delta\| \to 0} \sum_{i=1}^{n} [-f(\xi_i)] \Delta_i x \\
&= \int_{1}^{2} -(x^3 - 2x^2 - 5x + 6) \, dx
\end{aligned}
$$

$(\xi_i, f(\xi_i))$

$y = f(x)$

$(\xi_i, f(\xi_i))$

FIGURE 4.9.3

If A square units is the area of the entire region, then

$$A = A_1 + A_2$$
$$= \int_{-1}^{1} (x^3 - 2x^2 - 5x + 6)\,dx - \int_{1}^{2} (x^3 - 2x^2 - 5x + 6)\,dx$$
$$= \left[\tfrac{1}{4}x^4 - \tfrac{2}{3}x^3 - \tfrac{5}{2}x^2 + 6x\right]_{-1}^{1} - \left[\tfrac{1}{4}x^4 - \tfrac{2}{3}x^3 - \tfrac{5}{2}x^2 + 6x\right]_{1}^{2}$$
$$= \left[(\tfrac{1}{4} - \tfrac{2}{3} - \tfrac{5}{2} + 6) - (\tfrac{1}{4} + \tfrac{2}{3} - \tfrac{5}{2} - 6)\right]$$
$$\qquad\qquad - \left[(4 - \tfrac{16}{3} - 10 + 12) - (\tfrac{1}{4} - \tfrac{2}{3} - \tfrac{5}{2} + 6)\right]$$
$$= \tfrac{32}{3} - (-\tfrac{29}{12})$$
$$= \tfrac{157}{12}$$

The area of the region is therefore $\tfrac{157}{12}$ square units.

FIGURE 4.9.4

Now consider two functions f and g continuous on the closed interval $[a, b]$ and such that $f(x) \geq g(x)$ for all x in $[a, b]$. We wish to find the area of the region bounded by the two curves $y = f(x)$ and $y = g(x)$ and the two lines $x = a$ and $x = b$. Such a situation is shown in Fig. 4.9.4.

Take a partition of the interval $[a, b]$, with the ith subinterval having a length of $\Delta_i x$. In each subinterval choose a point ξ_i. Consider the rectangle having altitude $[f(\xi_i) - g(\xi_i)]$ units and width $\Delta_i x$ units. A rectangle is shown in Fig. 4.9.4. There are n such rectangles, one associated with each subinterval. The sum of the measures of the areas of these n rectangles is given by the following Riemann sum:

$$\sum_{i=1}^{n} [f(\xi_i) - g(\xi_i)]\,\Delta_i x$$

This Riemann sum is an approximation to what we intuitively think of as the number representing the "measure of the area" of the region. The smaller the value of $\|\Delta\|$, the better is this approximation. If A square units is the area of the region, we define

$$A = \lim_{\|\Delta\| \to 0} \sum_{i=1}^{n} [f(\xi_i) - g(\xi_i)]\,\Delta_i x \qquad (1)$$

Because f and g are continuous on $[a, b]$, so too is $f - g$; therefore the limit in (1) exists and is equal to the definite integral

$$\int_{a}^{b} [f(x) - g(x)]\,dx$$

EXAMPLE 4
Find the area of the region bounded by the curves $y = x^2$ and $y = -x^2 + 4x$.

SOLUTION
To find the points of intersection of the two curves we solve the equations simultaneously and obtain the points $(0, 0)$ and $(2, 4)$. The region is shown in Fig. 4.9.5.
Let

$$f(x) = -x^2 + 4x \qquad g(x) = x^2$$

FIGURE 4.9.5

Therefore, in the interval $[0, 2]$ the curve $y = f(x)$ is above the curve $y = g(x)$. We draw a vertical rectangular element of area, having altitude $[f(\xi_i) - g(\xi_i)]$ units and width $\Delta_i x$ units. The measure of the area of this rectangle then is given by $[f(\xi_i) - g(\xi_i)] \Delta_i x$. The sum of the measures of the areas of n such rectangles is given by the Riemann sum

$$\sum_{i=1}^{n} [f(\xi_i) - g(\xi_i)] \Delta_i x$$

If A square units is the area of the region, then

$$A = \lim_{||\Delta|| \to 0} \sum_{i=1}^{n} [f(\xi_i) - g(\xi_i)] \Delta_i x$$

and the limit of the Riemann sum is a definite integral. Hence

$$A = \int_0^2 [f(x) - g(x)] \, dx$$
$$= \int_0^2 [(-x^2 + 4x) - x^2] \, dx$$
$$= \int_0^2 (-2x^2 + 4x) \, dx$$
$$= \left[-\tfrac{2}{3}x^3 + 2x^2 \right]_0^2$$
$$= -\tfrac{16}{3} + 8 - 0$$
$$= \tfrac{8}{3}$$

The area of the region is $\tfrac{8}{3}$ square units.

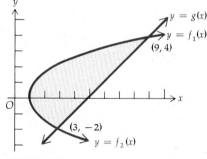

$y = g(x)$

$y = f_1(x)$

$(9, 4)$

$(3, -2)$

$y = f_2(x)$

FIGURE 4.9.6

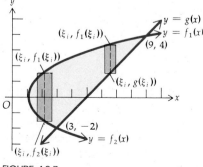

$(\xi_i, f_1(\xi_i))$

$y = g(x)$

$y = f_1(x)$

$(9, 4)$

$(\xi_i, f_1(\xi_i))$

$(\xi_i, g(\xi_i))$

$(3, -2)$

$y = f_2(x)$

$(\xi_i, f_2(\xi_i))$

FIGURE 4.9.7

EXAMPLE 5

Find the area of the region bounded by the parabola $y^2 = 2x - 2$ and the line $y = x - 5$.

SOLUTION

The two curves intersect at the points $(3, -2)$ and $(9, 4)$. The region is shown in Fig. 4.9.6.

The equation $y^2 = 2x - 2$ is equivalent to the two equations

$$y = \sqrt{2x - 2} \quad \text{and} \quad y = -\sqrt{2x - 2}$$

with the first equation giving the upper half of the parabola and the second equation giving the bottom half. If we let $f_1(x) = \sqrt{2x - 2}$ and let $f_2(x) = -\sqrt{2x - 2}$, the equation of the top half of the parabola is $y = f_1(x)$, and the equation of the bottom half is $y = f_2(x)$. If we let $g(x) = x - 5$, the equation of the line is $y = g(x)$.

In Fig. 4.9.7 we see two vertical rectangular elements of area. Each rectangle has the upper base on the curve $y = f_1(x)$. Because the lower base of the first rectangle is on the curve $y = f_2(x)$, the altitude is $[f_1(\xi_i) - f_2(\xi_i)]$ units. Because the lower base of the second rectangle lies on the curve $y = g(x)$, its altitude is $[f_1(\xi_i) - g(\xi_i)]$ units. If we wish to solve this problem by using vertical rectangular elements of area, we must divide the region into

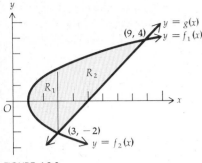

FIGURE 4.9.8

two separate regions, for instance R_1 and R_2, where R_1 is the region bounded by the curves $y = f_1(x)$ and $y = f_2(x)$ and the line $x = 3$, and where R_2 is the region bounded by the curves $y = f_1(x)$ and $y = g(x)$ and the line $x = 3$ (see Fig. 4.9.8).

If A_1 is the number of square units in the area of region R_1, we have

$$A_1 = \lim_{||\Delta|| \to 0} \sum_{i=1}^{n} [f_1(\xi_i) - f_2(\xi_i)] \Delta_i x$$

$$= \int_1^3 [f_1(x) - f_2(x)] \, dx$$

$$= \int_1^3 [\sqrt{2x - 2} + \sqrt{2x - 2}] \, dx$$

$$= 2 \int_1^3 \sqrt{2x - 2} \, dx$$

$$= \tfrac{2}{3}(2x - 2)^{3/2} \Big]_1^3$$

$$= \tfrac{16}{3}$$

If A_2 is the number of square units in the area of region R_2, we have

$$A_2 = \lim_{||\Delta|| \to 0} \sum_{i=1}^{n} [f_1(\xi_i) - g(\xi_i)] \Delta_i x$$

$$= \int_3^9 [f_1(x) - g(x)] \, dx$$

$$= \int_3^9 [\sqrt{2x - 2} - (x - 5)] \, dx$$

$$= \left[\tfrac{1}{3}(2x - 2)^{3/2} - \tfrac{1}{2}x^2 + 5x \right]_3^9$$

$$= [\tfrac{64}{3} - \tfrac{81}{2} + 45] - [\tfrac{8}{3} - \tfrac{9}{2} + 15]$$

$$= \tfrac{38}{3}$$

Hence $A_1 + A_2 = \tfrac{16}{3} + \tfrac{38}{3}$. Therefore the area of the entire region is 18 square units.

EXAMPLE 6
Find the area of the region in Example 5 by taking horizontal rectangular elements of area.

SOLUTION
Figure 4.9.9 illustrates the region with a horizontal rectangular element of area.

If in the equations of the parabola and the line we solve for x,

$$x = \tfrac{1}{2}(y^2 + 2) \qquad x = y + 5$$

Letting $\phi(y) = \tfrac{1}{2}(y^2 + 2)$ and $\lambda(y) = y + 5$, the equation of the parabola may be written as $x = \phi(y)$ and the equation of the line as $x = \lambda(y)$. Consider the closed interval $[-2, 4]$ on the y axis, and take a partition of this interval. The ith subinterval will have a length of $\Delta_i y$. In the ith subinterval $[y_{i-1}, y_i]$

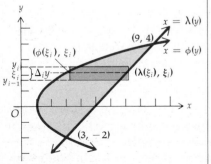

FIGURE 4.9.9

choose a point ξ_i. Then the length of the ith rectangular element is $[\lambda(\xi_i) - \phi(\xi_i)]$ units and the width is $\Delta_i y$ units. The measure of the area of the region can be approximated by the Riemann sum

$$\sum_{i=1}^{n} [\lambda(\xi_i) - \phi(\xi_i)] \Delta_i y$$

If A square units is the area of the region, then

$$A = \lim_{\|\Delta\| \to 0} \sum_{i=1}^{n} [\lambda(\xi_i) - \phi(\xi_i)] \Delta_i y$$

Because λ and ϕ are continuous on $[-2, 4]$, so too is $\lambda - \phi$, and the limit of the Riemann sum is a definite integral:

$$A = \int_{-2}^{4} [\lambda(y) - \phi(y)] \, dy$$

$$= \int_{-2}^{4} [(y + 5) - \tfrac{1}{2}(y^2 + 2)] \, dy$$

$$= \tfrac{1}{2} \int_{-2}^{4} (-y^2 + 2y + 8) \, dy$$

$$= \tfrac{1}{2} \left[-\tfrac{1}{3} y^3 + y^2 + 8y \right]_{-2}^{4}$$

$$= \tfrac{1}{2} [(-\tfrac{64}{3} + 16 + 32) - (\tfrac{8}{3} + 4 - 16)]$$

$$= 18$$

Comparing the solutions in Examples 5 and 6 we see that in the first case there are two definite integrals to evaluate, whereas in the second case there is only one. In general, if possible, the rectangular elements of area should be constructed so that a single definite integral is obtained. The following example illustrates a situation where two definite integrals are necessary.

EXAMPLE 7

Find the area of the region bounded by the two curves $y = x^3 - 6x^2 + 8x$ and $y = x^2 - 4x$.

SOLUTION

The points of intersection of the two curves are $(0, 0)$, $(3, -3)$, and $(4, 0)$. The region is shown in Fig. 4.9.10.

Let

$$f(x) = x^3 - 6x^2 + 8x \qquad g(x) = x^2 - 4x$$

In the interval $[0, 3]$ the curve $y = f(x)$ is above the curve $y = g(x)$, and in the interval $[3, 4]$ the curve $y = g(x)$ is above the curve $y = f(x)$. So the region must be divided into two separate regions R_1 and R_2, where R_1 is the region bounded by the two curves in the interval $[0, 3]$ and R_2 is the region bounded by the two curves in the interval $[3, 4]$. If A_1 square units is the

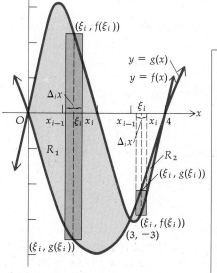

FIGURE 4.9.10

area of R_1 and A_2 square units is the area of R_2,

$$A_1 = \lim_{||\Delta|| \to 0} \sum_{i=1}^{n} [f(\xi_i) - g(\xi_i)] \Delta_i x \qquad A_2 = \lim_{||\Delta|| \to 0} \sum_{i=1}^{n} [g(\xi_i) - f(\xi_i)] \Delta_i x$$

so that

$$A_1 + A_2 = \int_0^3 [(x^3 - 6x^2 + 8x) - (x^2 - 4x)]\, dx$$

$$+ \int_3^4 [(x^2 - 4x) - (x^3 - 6x^2 + 8x)]\, dx$$

$$= \int_0^3 (x^3 - 7x^2 + 12x)\, dx + \int_3^4 (-x^3 + 7x^2 - 12x)\, dx$$

$$= \left[\tfrac{1}{4}x^4 - \tfrac{7}{3}x^3 + 6x^2 \right]_0^3 + \left[-\tfrac{1}{4}x^4 + \tfrac{7}{3}x^3 - 6x^2 \right]_3^4$$

$$= \tfrac{45}{4} + \tfrac{7}{12}$$

$$= \tfrac{71}{6}$$

Therefore the required area is $\tfrac{71}{6}$ square units.

EXERCISES 4.9

In Exercises 1 through 38, find the area of the region bounded by the given curves. In each problem do the following: (a) Draw a figure showing the region and a rectangular element of area; (b) express the area of the region as the limit of a Riemann sum; (c) find the limit in part (b) by evaluating a definite integral by the second fundamental theorem of the calculus.

1. $y = 4 - x^2$; x axis

2. $y = x^2 - 2x + 3$; x axis; $x = -2$; $x = 1$

3. $y = 4x - x^2$; x axis; $x = 1$; $x = 3$

4. $y = 6 - x - x^2$; x axis

5. $y = \sqrt{x + 1}$; x axis; y axis; $x = 8$

6. $y = \dfrac{1}{x^2} - x$; x axis; $x = 2$; $x = 3$

7. $y = x^2 + x - 12$; x axis

8. $y = x^2 - 6x + 5$; x axis

9. $y = \sin x$; x axis; $x = \tfrac{1}{3}\pi$; $x = \tfrac{2}{3}\pi$

10. $y = \cos x$; x axis; y axis; $x = \tfrac{1}{6}\pi$

11. $y = \sec^2 x$; x axis; y axis; $x = \tfrac{1}{4}\pi$

12. $y = \csc^2 x$; x axis; $x = \tfrac{1}{4}\pi$; $x = \tfrac{1}{3}\pi$

13. $x^2 = -y$; $y = -4$. Take the elements of area perpendicular to the y axis.

14. $y^2 = -x$; $x = -2$; $x = -4$

15. $x^2 + y + 4 = 0$; $y = -8$. Take the elements of area perpendicular to the y axis.

16. The same region as in Exercise 15. Take the elements of area parallel to the y axis.

17. $x^2 - y + 1 = 0$; $x - y + 1 = 0$. Take the elements of area perpendicular to the x axis.

18. The same region as in Exercise 17. Take the elements of area parallel to the x axis.

19. $x^3 = 2y^2$; $x = 0$; $y = -2$

20. $y^3 = 4x$; $x = 0$; $y = -2$

21. $y = 2 - x^2$; $y = -x$

22. $y = x^2$; $y = x^4$

23. $y^2 = x - 1$; $x = 3$

24. $y = x^2$; $x^2 = 18 - y$

25. $y = \sqrt{x}$; $y = x^3$

26. $x = 4 - y^2$; $x = 4 - 4y$

27. $y^3 = x^2$; $x - 3y + 4 = 0$

28. $xy^2 = y^2 - 1$; $x = 1$; $y = 1$; $y = 4$

29. $x = y^2 - 2$; $x = 6 - y^2$

30. $x = y^2 - y$; $x = y - y^2$

31. $y = 2x^3 - 3x^2 - 9x$; $y = x^3 - 2x^2 - 3x$

32. $3y = x^3 - 2x^2 - 15x$; $y = x^3 - 4x^2 - 11x + 30$

33. $y = x^3 + 3x^2 + 2x$; $y = 2x^2 + 4x$

34. $y = |x - 1| + 3$; $y = 0$; $x = -2$; $x = 4$

35. $y = \cos x - \sin x$; $x = 0$; $y = 0$

36. $y = \sin x$; $y = -\sin x$; $x = -\tfrac{1}{2}\pi$; $x = \tfrac{1}{2}\pi$

37. $y = |x|$; $y = x^2 - 1$; $x = -1$; $x = 1$

38. $y = |x + 1| + |x|$; $y = 0$; $x = -2$; $x = 3$

39. Find by integration the area of the triangle having vertices at $(5, 1)$, $(1, 3)$, and $(-1, -2)$.

40. Find by integration the area of the triangle having vertices at $(3, 4)$, $(2, 0)$, and $(0, 1)$.

41. Find the area of the region bounded by the curve $x^3 - x^2 + 2xy - y^2 = 0$ and the line $x = 4$. (*Hint*: Solve the cubic equation for y in terms of x, and express y as two functions of x.)

42. Find the area of the region bounded by the three curves $y = x^2$, $x = y^3$, and $x + y = 2$.

43. Find the area of the region bounded by the three curves $y = x^2$, $y = 8 - x^2$, and $4x - y + 12 = 0$.

44. Find by integration the area of the trapezoid having vertices at $(-1, -1)$, $(2, 2)$, $(6, 2)$, and $(7, -1)$.

45. Find the area of the region bounded by the curve $y = \sin x$, the line $y = 1$, and the y axis to the right of the y axis.

46. Find the area of the region bounded by the two curves $y = \sin x$ and $y = \cos x$ between two consecutive points of intersection.

47. Find the area of the region bounded by the curve $y = \tan^2 x$, the x axis, and the line $x = \frac{1}{4}\pi$.

48. Find the area of the region above the parabola $x^2 = 4py$ and inside the triangle formed by the x axis and the lines $y = x + 8p$ and $y = -x + 8p$.

49. Find the area of the region bounded by the two parabolas $y^2 = 4px$ and $x^2 = 4py$.

50. Find the rate of change of the measure of the area of Exercise 48 with respect to p when $p = \frac{3}{8}$.

51. Find the rate of change of the measure of the area of Exercise 49 with respect to p when $p = 3$.

52. Determine m so that the region above the line $y = mx$ and below the parabola $y = 2x - x^2$ has an area of 36 square units.

53. Determine m so that the region above the curve $y = mx^2$ $(m > 0)$, to the right of the y axis, and below the line $y = m$ has an area of K square units, where $K > 0$.

54. If A square units is the area of the region bounded by the parabola $y^2 = 4x$ and the line $y = mx$ $(m > 0)$, find the rate of change of A with respect to m.

4.10 (*Supplementary*) APPLICATIONS OF INTEGRATION IN ECONOMICS AND BUSINESS

In Supplementary Section 3.11 we discussed the marginal cost and marginal revenue functions. They are the first derivatives, C' and R' of the total cost function C and the total revenue function R, respectively. Thus C and R can be obtained from C' and R' by indefinite integration (antidifferentiation). When finding the function C from C', the arbitrary constant can be evaluated if we know either the overhead cost (i.e., the cost when no units are produced) or the cost of production of a specific number of units of the commodity. Because it is generally true that the total revenue is zero when the number of units produced is zero, this fact may be used to evaluate the arbitrary constant when finding the function R from R'.

EXAMPLE 1
The marginal cost function C' is given by

$$C'(x) = 4x - 8$$

where $C(x)$ dollars is the total cost of producing x units. If the cost of producing 5 units is $20, find the total cost function. Draw sketches of the total cost curve and the marginal cost curve on the same set of axes.

SOLUTION
The marginal cost must be nonnegative. Hence $4x - 8 \geq 0$, and therefore the permissible values of x are $x \geq 2$. Because $C'(x) = 4x - 8$,

$$C(x) = \int (4x - 8)\, dx$$

$$= 2x^2 - 8x + k$$

Because $C(5) = 20$ we obtain $k = 10$. Hence

$$C(x) = 2x^2 - 8x + 10 \qquad x \geq 2$$

The required sketches are shown in Fig. 4.10.1.

FIGURE 4.10.1

EXAMPLE 2

If the marginal revenue is given by

$$R'(x) = 27 - 12x + x^2$$

find the total revenue function and the demand equation. Draw sketches of the demand curve, the total revenue curve, and the marginal revenue curve on the same set of axes.

SOLUTION

If R is the total revenue function, then from the given equation

$$R(x) = \int (27 - 12x + x^2)\, dx$$

$$= 27x - 6x^2 + \tfrac{1}{3}x^3 + C$$

Because $R(0) = 0$ we get $C = 0$. Hence

$$R(x) = 27x - 6x^2 + \tfrac{1}{3}x^3$$

If f is the price function, $R(x) = xf(x)$; so

$$f(x) = 27 - 6x + \tfrac{1}{3}x^2$$

If p dollars is the price of one unit of the commodity when x units are demanded, then because $p = f(x)$, the demand equation is

$$3p = 81 - 18x + x^2$$

To determine the permissible values of x we use the facts that $x \geq 0$, $p \geq 0$, and f is a decreasing function. Because

$$f'(x) = -6 + \tfrac{2}{3}x$$

f is decreasing when $x < 9$ (i.e., when $f'(x) < 0$). Also, when $x = 9$, $p = 0$; so the permissible values of x are the numbers in the closed interval $[0, 9]$. The required sketches are shown in Fig. 4.10.2.

FIGURE 4.10.2

EXAMPLE 3

After experimentation, a certain manufacturer determined that if x units of a certain article of merchandise are produced per day, the marginal cost is given by

$$C'(x) = 0.3x - 11$$

where $C(x)$ dollars is the total cost of producing x units. If the selling price of the article is fixed at \$19 per unit and the overhead cost is \$100 per day, find the maximum daily total profit that can be obtained.

SOLUTION

Let $R(x)$ dollars be the total revenue obtained by selling x units and $P(x)$ dollars be the total profit obtained by selling x units. Because x units are sold at \$19 per unit,

$$R(x) = 19x \qquad R'(x) = 19$$

We are given that

$$C'(x) = 0.3x - 11$$

The total profit will be a maximum when the marginal revenue equals the marginal cost (see Section 3.11). Equating $R'(x)$ and $C'(x)$ we get $x = 100$. So 100 units should be produced each day for maximum total profit.

$$C(x) = \int (0.3x - 11)\,dx$$
$$= 0.15x^2 - 11x + k$$

Because the overhead cost is $100, $C(0) = 100$, and hence $k = 100$. Therefore

$$C(x) = 0.15x^2 - 11x + 100$$

Because $P(x) = R(x) - C(x)$,

$$P(x) = 19x - (0.15x^2 - 11x + 100)$$
$$= -0.15x^2 + 30x - 100$$

Hence $P(100) = 1400$. Therefore the maximum daily total profit is $1400, which is obtained if 100 units are produced daily.

For an application of integration in business, suppose that a producer of a modern electronic calculator determines that during the first 3 years of production, if x years have elapsed since the calculator was first introduced, then $f(x)$ units per year should be produced, where

$$f(x) = 60 + 288x^2 \qquad 0 \leq x \leq 3$$

How should this equation be interpreted? Because

$$f(1) = 60 + 288(1)^2$$
$$= 348$$

one might conclude that 348 calculators are produced 1 year after the commodity first came on the market. However, this interpretation is valid only if the rate of production is constant on a yearly basis. That is, 348 units would be produced during the second year only if the annual rate of production during the second year is a constant 348 units, which is the level at which it was at the end of the first year. This situation does not occur. For instance,

$$f(1\tfrac{1}{2}) = 60 + 288(\tfrac{3}{2})^2 \qquad f(2) = 60 + 288(2)^2$$
$$= 708 \qquad\qquad\qquad = 1212$$

Since when $x = 1$ the production is 348 units, and when $x = 2$ the production is 1212 units, it follows that the number of calculators produced during the second year is between 348 and 1212. A better approximation arises from the argument that during the first half of the second year the number of units produced should be at least $\tfrac{1}{2}(348) = 174$ and during the second half it should be at least $\tfrac{1}{2}(708) = 345$; thus during the second year the number of calculators produced should be at least $174 + 354 = 528$. A geometric interpretation of this reasoning is shown in Fig. 4.10.3. The figure

FIGURE 4.10.3

shows a sketch of the graph of

$$f(x) = 60 + 288x^2 \qquad 0 \le x \le 3$$

and two shaded rectangles. The number of square units in the area of the first rectangle is $\frac{1}{2}(348) = 174$, which is the number of calculators produced during the first half of the second year if production during that time is at a constant yearly rate of $f(1) = 348$. The number of square units in the area of the second rectangle is $\frac{1}{2}(708) = 354$, which is the number of calculators produced during the second half of the year if production during that time is at a constant yearly rate of $f(1\frac{1}{2}) = 708$.

For an even better approximation of the second-year production, we assume that during each month the production rate is the constant amount at which it is at the first of the month. With this assumption as well as that each month is $\frac{1}{12}$ of a year, the number of units produced during the first month of the second year is $\frac{1}{12}f(1) = \frac{1}{12}(348)$ or 29; during the second month the number of units produced is $\frac{1}{12}f(1\frac{1}{12}) = \frac{1}{12}[60 + 288(\frac{13}{12})^2]$ or $33\frac{1}{6}$; similarly during the third month, the number of units produced is $\frac{1}{12}f(1\frac{2}{12}) = \frac{1}{12}[60 + 288(\frac{7}{6})^2]$ or $37\frac{2}{3}$; and so on; during the twelfth month, the number of units produced is $\frac{1}{12}f(1\frac{11}{12}) = \frac{1}{12}[60 + 288(\frac{23}{12})^2]$ or $93\frac{1}{6}$. In a similar way we compute the number of units produced during the other eight months; and the total number of units produced during the second year is then

$$\frac{1}{12}[f(1) + f(1\tfrac{1}{12}) + f(1\tfrac{2}{12}) + f(1\tfrac{3}{12}) + f(1\tfrac{4}{12}) + f(1\tfrac{5}{12}) + f(1\tfrac{6}{12})$$
$$+ f(1\tfrac{7}{12}) + f(1\tfrac{8}{12}) + f(1\tfrac{9}{12}) + f(1\tfrac{10}{12}) + f(1\tfrac{11}{12})]$$

$$= 29 + 33\tfrac{1}{6} + 37\tfrac{2}{3} + 42\tfrac{1}{2} + 47\tfrac{2}{3} + 53\tfrac{1}{6} + 59$$
$$+ 65\tfrac{1}{6} + 71\tfrac{2}{3} + 78\tfrac{1}{2} + 85\tfrac{2}{3} + 93\tfrac{1}{6}$$

$$= 696\tfrac{1}{3} \qquad (1)$$

We can conclude from this approximation that the number of calculators produced during the second year is at least 696.

Summation (1) represents the sum of the areas of the 12 shaded rectangles in Fig. 4.10.4. The first rectangle has a height of $f(1)$ units and a width of $\frac{1}{12}$ unit; thus the number of square units in its area is $\frac{1}{12}f(1) = \frac{1}{12}(348)$, that is, 29. For the second rectangle the height is $f(1\frac{1}{12})$ units and its width is $\frac{1}{12}$ unit, and so the number of square units in its area is $\frac{1}{12}f(1\frac{1}{12}) = \frac{1}{12}(398)$, that is $33\frac{1}{6}$; and so on. The twelfth rectangle has a height of $f(1\frac{11}{12})$ units and a width of $\frac{1}{12}$ unit; therefore the number of square units in its area is $\frac{1}{12}f(1\frac{11}{12}) = \frac{1}{12}(1118)$, that is, $93\frac{1}{6}$.

We can improve the approximation of the second-year production by keeping the production rate constant during a length of time shorter than a month. For instance, suppose that we assume that during each week the production rate is the constant amount at which it is at the beginning of the week. Then the number of calculators produced during the second year would be given by the number of square units in the sum of the areas of 52 rectangles. We can improve upon the approximation even further by assuming that during each day the production rate is the constant amount at

FIGURE 4.10.4

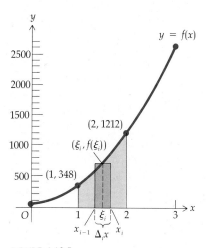

FIGURE 4.10.5

which it is at the start of the day. Then the number of square units in the sum of the areas of 365 rectangles will be the approximation of the number of calculators produced during the second year. If we take smaller and smaller lengths of time over which the production rate is assumed to be constant, the number of rectangles increases without bound and the total number of calculators produced will get closer and closer to the number of square units in the area of the shaded region shown in Fig. 4.10.5.

Therefore we divide the interval $[1, 2]$ into n subintervals, where $\Delta_i x$ is the length of the ith subinterval $[x_{i-1}, x_i]$. Let ξ_i be any number in the ith subinterval. If N calculators are produced during the second year, then

$$N = \lim_{||\Delta|| \to 0} \sum_{i=1}^{n} f(\xi_i) \, \Delta_i x$$

$$= \int_1^2 f(x) \, dx$$

$$= \int_1^2 (60 + 288x^2) \, dx$$

$$= 60x + 288 \cdot \frac{x^3}{3} \Big]_1^2$$

$$= 60x + 96x^3 \Big]_1^2$$

$$= 120 + 768 - (60 + 96)$$

$$= 732$$

We now consider some more applications of the definite integral in business. The quantity to be found is first expressed as the limit of a Riemann sum, which is a definite integral evaluated by using the second fundamental theorem of the calculus.

EXAMPLE 4
Suppose that during the first five years for which a new commodity has been on the market, y units per year are sold when x years have elapsed since the product was first introduced, and

$$y = 3000\sqrt{x} + 1000 \qquad 0 \le x \le 5 \tag{2}$$

Find the total sales during the first 4 years.

SOLUTION
A sketch of the graph of (2) is shown in Fig. 4.10.6. The number of units sold during the first 4 years is the measure of the area of the shaded region shown in the figure. Let

$$f(x) = 3000\sqrt{x} + 1000$$

The interval $[0, 4]$ is divided into n subintervals, where $\Delta_i x$ is the width of the ith subinterval $[x_{i-1}, x_i]$. Let ξ_i be any number in the ith subinterval. If S

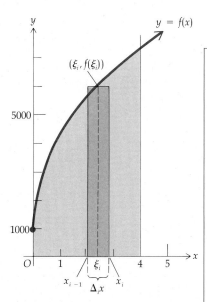

FIGURE 4.10.6

units are sold during the first 4 years,

$$S = \lim_{\|\Delta\| \to 0} \sum_{i=1}^{n} f(\xi_i)\,\Delta_i x$$

$$= \int_0^4 f(x)\,dx$$

$$= \int_0^4 (3000x^{1/2} + 1000)\,dx$$

$$= 3000 \cdot \frac{x^{3/2}}{\frac{3}{2}} + 1000x \bigg]_0^4$$

$$= 2000(4)^{3/2} + 1000 \cdot 4$$

$$= 2000(8) + 4000$$

$$= 20{,}000$$

Therefore 20,000 units are sold during the first 4 years.

EXAMPLE 5

A company manager estimates that the purchase of a particular piece of equipment will result in a savings of operating costs to the company. The rate of operating-cost savings is $f(x)$ dollars per year when the equipment has been in use for x years, and

$$f(x) = 4000x + 1000 \qquad 0 \le x \le 10$$

(a) What is the savings in operating costs for the first 5 years? (b) If the purchase price is $36,000, how many years of use is required for the equipment to pay for itself?

SOLUTION

(a) A sketch of the graph of f on $[0, 5]$ is shown in Fig. 4.10.7. The number of dollars in the savings in operating costs for the first 5 years is the measure of the area of the shaded region in the figure. If S dollars is this number, then

$$S = \lim_{\|\Delta\| \to 0} \sum_{i=1}^{n} f(\xi_i)\,\Delta_i x$$

$$= \int_0^5 f(x)\,dx$$

$$= \int_0^5 (4000x + 1000)\,dx$$

$$= 2000x^2 + 1000x \bigg]_0^5$$

$$= 2000(25) + 1000(5)$$

$$= 55{,}000$$

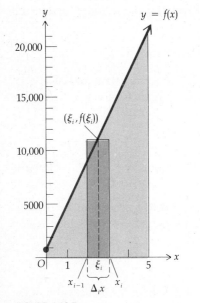

FIGURE 4.10.7

Therefore, the savings in operating costs for the first 5 years is $55,000.

(b) Because the purchase price is $36,000, the number of years of use required for the equipment to pay for itself is n, where

$$\int_0^n f(x)\,dx = 36{,}000$$

$$\int_0^n (4000x + 1000)\,dx = 36{,}000$$

$$2000x^2 + 1000x \Big]_0^n = 36{,}000$$

$$2000n^2 + 1000n = 36{,}000$$

$$2n^2 + n - 36 = 0$$

$$(n - 4)(2n + 9) = 0$$

$$n = 4 \qquad n = -\tfrac{9}{2}$$

Thus 4 years of use are required for the equipment to pay for itself.

The next example shows how the definite integral is used to compute the cost of maintaining an inventory.

EXAMPLE 6

The concession manager of a theater chain receives a shipment of a certain foodstuff every Monday. Because attendance is low at the beginning of the week and high on the weekend, the demand increases as the week progresses, so that after x days the inventory is y units, where

$$y = 49{,}000 - 1000x^2 \qquad 0 \le x \le 7 \tag{3}$$

If the daily storage cost is 0.03 cent per unit, find the total cost of maintaining inventory for 7 days.

SOLUTION

Figure 4.10.8 shows a sketch of the graph of (3). If A square units is the area of the shaded region in the figure and C cents is the total cost of maintaining inventory for 7 days, then

$$C = 0.03A$$

To compute A, the interval $[0, 7]$ is divided into n subintervals, where $\Delta_i x$ is the length of the ith subinterval $[x_{i-1}, x_i]$. If ξ_i is any number in the ith subinterval and

$$f(x) = 49{,}000 - 1000x^2$$

$$A = \lim_{\|\Delta\| \to 0} \sum_{i=1}^{n} f(\xi_i)\,\Delta_i x$$

$$= \int_0^7 f(x)\,dx$$

$$= \int_0^7 (49{,}000 - 1000x^2)\,dx$$

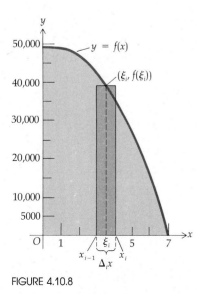

FIGURE 4.10.8

Therefore

$$C = 0.03 \int_0^7 (49{,}000 - 1000x^2)\, dx$$

$$= 0.03 \left[49{,}000x - 1000 \cdot \frac{x^3}{3} \right]_0^7$$

$$= 0.03 \left[49{,}000 \cdot 7 - 1000 \cdot \frac{7^3}{3} \right]$$

$$= 0.03(7)^3 [1000 - \tfrac{1000}{3}]$$

$$= 0.01(343)(2000)$$

$$= 6860$$

Thus the total cost of maintaining inventory for 7 days is $68.60.

EXAMPLE 7
For the theater chain of Example 6 find the average inventory of the particular foodstuff over the 7-day period from Monday to Sunday.

SOLUTION
The inventory of the particular foodstuff after x days is $f(x)$ units, and

$$f(x) = 49{,}000 - 1000x^2$$

From Definition 4.7.2, if S units is the average inventory over the 7-day period from Monday to Sunday, then

$$S = \frac{\int_0^7 f(x)\, dx}{7 - 0}$$

$$= \frac{1}{7} \int_0^7 (49{,}000 - 1000x^2)\, dx$$

$$= \frac{1}{7} \left[49{,}000x - 1000 \cdot \frac{x^3}{3} \right]_0^7$$

$$= \frac{1}{7} \left[49{,}000 \cdot 7 - 1000 \cdot \frac{7^3}{3} \right]$$

$$= \frac{7^2}{3}(3000 - 1000)$$

$$= 32{,}667$$

EXERCISES 4.10

1. The marginal revenue function for a certain commodity is given by $R'(x) = 12 - 3x$. If x units are demanded when p dollars is the price per unit, find (a) the total revenue function and (b) the demand equation. (c) Draw sketches of the total revenue curve and the demand curve on the same set of axes.

2. For a particular article of merchandise the marginal revenue function is given by $R'(x) = 15 - 4x$. Find (a) the total revenue function and (b) the demand equation. (c) Draw sketches of the total revenue curve and the demand curve on the same set of axes.

3. The marginal cost function for a particular article of merchandise is given by $C'(x) = 3(5x + 4)^{-1/2}$. If the overhead cost is $10, find the total cost function.

4. For a certain commodity the marginal cost function is given by $C'(x) = 3\sqrt{2x + 4}$. If the overhead cost is zero, find the total cost function.

5. Find the demand equation of a commodity for which the marginal revenue function is given by $R'(x) = 4 + 10(x + 5)^{-2}$.

6. Find the price function for a commodity for which the marginal revenue function is given by $R'(x) = 14 - 9x + x^2$.

7. The marginal cost function for a particular commodity is given by $C'(x) = 6x - 17$. If the cost of producing 2 units is $25, find the total cost function.

8. If the marginal revenue function is given by $R'(x) = 3x^2 - 12x + 10$, find (a) the total revenue function and (b) the price function.

9. The marginal revenue function is given by $R'(x) = 16 - 3x^2$. Find (a) the total revenue function and (b) the demand equation. (c) Draw sketches of the demand curve, the total revenue curve, and the marginal revenue curve on the same set of axes.

10. Find (a) the total revenue function and (b) the demand equation if the marginal revenue function is given by $R'(x) = \frac{3}{4}x^2 - 10x + 12$. (c) Draw sketches of the demand curve, the total revenue curve, and the marginal revenue curve on the same set of axes.

11. The marginal cost function is given by $C'(x) = 3x^2 + 8x + 4$, and the overhead cost is $6. If $C(x)$ dollars is the total cost of x units, (a) find the total cost function, and (b) draw sketches of the total cost curve and the marginal cost curve on the same set of axes.

12. The marginal cost function is defined by $C'(x) = 6x$, where $C(x)$ is the number of hundreds of dollars in the total cost of x hundred units of a certain commodity. If the cost of 200 units is $2000, find (a) the total cost function and (b) the overhead cost. (c) Draw sketches of the total cost curve and the marginal cost curve on the same set of axes.

13. A company has determined that the marginal cost function for the production of a particular commodity is given by $C'(x) = 125 + 10x + \frac{1}{9}x^2$, where $C(x)$ dollars is the total cost of producing x units of the commodity. If the overhead cost is $250, what is the cost of producing 15 units?

14. The marginal revenue function is given by $R'(x) = ab(x + b)^{-2} - c$. Find (a) the total revenue function and (b) the demand equation.

15. During the first 4 months, the sales of a new product is expected to be $f(x)$ units per month x months after the product was first put on the market, where $f(x) = 100 + 150x^2$ and $0 \le x \le 4$. (a) Interpret the expected second-month sales as the measure of the area of a plane region, and draw a figure showing the plane region. (b) Compute the expected second-month sales by finding the area of the plane region in part (a).

16. For the product of Exercise 15, do the following: (a) Interpret the expected first 3 months' sales as the measure of the area of a plane region, and draw a figure showing the plane region. (b) Compute the expected first 3 months' sales by finding the area of the region in part (a).

17. Suppose that for a 5-minute period a secretary can type at the rate of $f(x)$ words per minute x min from the starting time, where $f(x) = 75 + 10x - 3x^2$ and $0 \le x \le 5$. How many words does the secretary type during the 5 min?

18. How many words does the secretary in Exercise 17 type during the third minute?

19. The purchase of some new machinery is expected to result in a savings of operating costs so that when the machinery is x years old, the operating cost savings is $f(x)$ dollars per year, where $f(x) = 1000 + 5000x$. (a) How much is saved in operating costs during the first 6 years the machinery is in use? (b) If the machinery was purchased at a price of $67,500, how long will it take for the machinery to pay for itself?

20. It is estimated that x years from now the population of a certain city will be increasing at the rate of $f(x)$ people per year, where $f(x) = 400 + 100x^{3/2}$ and $0 \le x \le 5$. Determine the increase in population for the next 4 years.

21. The book value of a certain piece of equipment is changing at the rate of $f(x)$ dollars per year when the equipment is x years old, and $f(x) = 4000x - 60,000$. Determine by how much the equipment depreciates during the third year.

22. If the equipment in Exercise 21 was purchased for $450,000, how long will it take for it to be fully depreciated?

23. The manager of an amusement park expects that during the first month of operation the number of daily admissions will increase so that $f(t)$ admissions per day will occur at t days from the opening, where $f(t) = 9800 + 40t$. On what day is the 100,000th visitor expected?

24. For the amusement park of Exercise 23, how many admissions are expected (a) the fifth day; (b) during the first 5 days?

25. A merchant receives an annual shipment of 7200 Christmas ornaments on October 23. The sales pattern is essentially the same each year: Over a 60-day period the inventory moves slowly at the beginning, but as Christmas approaches, the demand increases so that x days after October 23 the inventory is y ornaments, where $y = 7200 - 2x^2$ and $0 \le x \le 60$. If the daily storage cost is 0.02 cent per ornament, find the total cost of maintaining inventory for 60 days.

26. A candy dealer stocks 9000 Valentine gift boxes at the end of January and plans to sell them over a 15-day period. The sales are low at the beginning of the period, but the demand increases as the date gets closer to February 14, so that x days after January 30 the inventory is y boxes, where $y = 9000 - 40x^2$ and $0 \le x \le 15$. If the storage cost per box of candy is 0.04 cent per day, find the total cost of maintaining inventory for 15 days.

27. A grocer receives a shipment of 3600 cans of soup that are expected to be sold at a constant rate of 120 cans per day for a 30-day period. (a) If the daily storage cost is 0.01 cent per can, what is the grocer's total cost of maintaining inventory for 30 days? (b) Show that the result of part (a) is the same as the cost of storing 1800 (half of 3600) cans of soup for the entire 30 days.

28. A wholesale distributor has a standing order for 25,000 boxes of detergent that arrive every 20 weeks. These boxes are shipped out by the distributor at a constant rate of 1250 boxes per week. (a) If storage costs are 0.3 cent per box per week, what is the total cost of maintaining inventory for 20 weeks? (b) Show that the result of part (a) is the same as the cost of storing 12,500 (half of 25,000) boxes for the entire 20 weeks.

29. For the merchant of Exercise 25, determine the average number of ornaments on hand for the 60-day period.

30. For the candy dealer of Exercise 26, determine the average number of Valentine gift boxes on hand for the 15-day period.

31. For the secretary of Exercise 17, find the average typing speed during the first 4 min of the 5-min period.

32. The demand equation for a certain commodity is $x^2 + 100p^2 = 10,000$, where x units are demanded when p dollars is the price per unit. Find the average total revenue when the number of units demanded takes on all values from 60 to 80.

33. For the x months from June 1 the price of a 1-pound can of coffee was $f(x)$ dollars, where $f(x) = 3.40 - 0.40x + 0.06x^2$ and $0 \le x \le 5$. Determine the average price for the 3 months from July 1 to October 1.

34. A manufacturing company produces a monthly supply of n items of a certain commodity, and they are stored until they are sold. The items are sold at a constant rate until the inventory becomes zero at the end of the month. If the storage costs are p cents per month per unit, show that the monthly storage costs are $p \cdot \dfrac{n}{2}$, which is the same as the cost of storing $\frac{1}{2}n$ units for the entire month.

35. A manufacturer of children's toys has a new toy coming on the market and wishes to determine a selling price for the toy such that the total profit will be a maximum. From analyzing the price and demand of another similar toy, it is anticipated that if x toys are demanded when p dollars is the price per toy, then $\dfrac{dp}{dx} = -\dfrac{p^2}{30,000}$, and the demand should be 1800 when the price is \$10. If $C(x)$ dollars is the total cost of producing x toys, then $C(x) = x + 7500$. Find the price that should be charged for the manufacturer's total profit to be a maximum.

REVIEW EXERCISES FOR CHAPTER 4

In Exercises 1 through 24, perform the antidifferentiation; that is, evaluate the indefinite integral.

1. $\displaystyle \int (2x^3 - x^2 + 3)\, dx$

2. $\displaystyle \int (5x^4 + 3x - 1)\, dx$

3. $\displaystyle \int (4y + 6\sqrt{y})\, dy$

4. $\displaystyle \int 3z^{-2}\, dz$

5. $\displaystyle \int \sin 3t\, dt$

6. $\displaystyle \int \frac{\sin \sqrt{w}}{\sqrt{w}}\, dw$

7. $\displaystyle \int \cos^2 x \sin x\, dx$

8. $\displaystyle \int \sqrt{x}(1 + x^2)\, dx$

9. $\displaystyle \int \left(\frac{2}{x^4} - \frac{5}{x^2} \right) dx$

10. $\displaystyle \int \left(\sqrt[3]{t} - \frac{1}{\sqrt[3]{t}} \right) dt$

11. $\displaystyle \int 5x(2 + 3x^2)^8\, dx$

12. $\displaystyle \int x^4 \sqrt{x^5 - 1}\, dx$

13. $\displaystyle \int \left(\sqrt{3x} + \frac{1}{\sqrt{5x}} \right) dx$

14. $\displaystyle \int \sqrt[3]{7w + 3}\, dw$

15. $\displaystyle \int \frac{x^3 + x}{(x^4 + 2x^2)^7}\, dx$

16. $\displaystyle \int (x^3 + x)\sqrt{x^2 + 3}\, dx$

17. $\displaystyle\int \frac{s}{\sqrt{2s+3}}\,ds$

18. $\displaystyle\int t \sec^2 t^2\,dt$

19. $\displaystyle\int \tan^2 3\theta\,d\theta$

20. $\displaystyle\int (3\cot^2 2x - 2\csc^2 3x)\,dx$

21. $\displaystyle\int \frac{5\cos^2 x - 3\tan x}{\cos x}\,dx$

22. $\displaystyle\int \sin^3 2\theta \cot 2\theta\,d\theta$

23. $\displaystyle\int \sqrt{4x+3}\,(x^2+1)\,dx$

24. $\displaystyle\int \left(\frac{x^3+2}{x^3}\right)\sqrt{x-\frac{1}{x^2}}\,dx$

In Exercises 25 through 28, find the exact value of the definite integral by using the definition of the definite integral; do not use the fundamental theorems of the calculus.

25. $\displaystyle\int_0^2 (3x^2-1)\,dx$

26. $\displaystyle\int_0^1 (2x^2+4x+1)\,dx$

27. $\displaystyle\int_{-2}^1 (x^3+2x)\,dx$

28. $\displaystyle\int_{-1}^3 (x^2-1)^2\,dx$

In Exercises 29 through 38, evaluate the definite integral by using the second fundamental theorem of the calculus.

29. $\displaystyle\int_{-2}^2 (t^3-3t)\,dt$

30. $\displaystyle\int_1^5 \frac{dy}{\sqrt{2y-1}}$

31. $\displaystyle\int_2^3 \frac{12x\,dx}{(x^2-1)^2}$

32. $\displaystyle\int_{-5}^5 2x\sqrt[3]{x^2+2}\,dx$

33. $\displaystyle\int_0^{\pi/6} \frac{\sin 2\theta\,d\theta}{\cos^2 2\theta}$

34. $\displaystyle\int_{\pi/3}^{\pi} \sin^2 \tfrac{1}{2}t \cos \tfrac{1}{2}t\,dt$

35. $\displaystyle\int_{-1}^7 \frac{x^2\,dx}{\sqrt{x+2}}$

36. $\displaystyle\int_1^2 \frac{y\,dy}{\sqrt{5-y}}$

37. $\displaystyle\int_0^{\pi/2} (\tan^2 \tfrac{1}{2}x + \sec^2 \tfrac{1}{2}x)\,dx$

38. $\displaystyle\int_{\pi/6}^{\pi/3} (1-\cos\theta)\csc^2\theta\,d\theta$

In Exercises 39 through 42, find the complete solution of the given differential equation.

39. $x^2 y\dfrac{dy}{dx} = (y^2-1)^2$

40. $\dfrac{d^2y}{dx^2} = 12x^2 - 30x$

41. $\dfrac{d^2y}{dx^2} = \sqrt{2x-1}$

42. $\dfrac{dy}{dx} = \dfrac{x\sqrt{1-y^2}}{y\sqrt{2x^2+1}}$

43. Evaluate $\int (x^3+1)^2 x^2\,dx$ by two methods: (a) Let $u = x^3+1$; (b) first expand $(x^3+1)^2$. Compare the results, and explain the difference in the appearance of the answers obtained in (a) and (b).

44. Evaluate $\int (x^4)^6 4x^3\,dx$ as $\int u^6\,du$ and as $\int 4x^{27}\,dx$, and compare the results.

45. The slope of the tangent line at any point (x, y) on a curve is $10 - 4x$, and the point $(1, -1)$ is on the curve. Find an equation of the curve.

46. Suppose that a particular company estimates its growth income from sales by the formula $\dfrac{dS}{dt} = 2(t-1)^{2/3}$, where S millions of dollars is the gross income from sales t years hence. If the gross income from the current year's sales is \$8 million, what should be the expected gross income from sales 2 years from now?

47. It is July 31 and a tumor has been growing inside a person's body in such a way that t days since July 1 the volume of the tumor has been increasing at a rate of $\frac{1}{100}(t+6)^{1/2}$ cm^3 per day. If the volume of the tumor on July 4 was 0.20 cm^3, what is the volume today?

48. The enrollment at a certain college has been increasing at the rate of $1000(t+1)^{-1/2}$ students per year since 1981. If the enrollment in 1984 was 10,000, (a) what was the enrollment in 1981, and (b) what is the anticipated enrollment in 1989 if it is expected to be increasing at the same rate?

49. The volume of a balloon is increasing according to the formula $\dfrac{dV}{dt} = \sqrt{t+1} + \tfrac{2}{3}t$, where V cm^3 is the volume of the balloon at t sec. If $V = 33$ when $t = 3$, find (a) a formula for V in terms of t; (b) the volume of the balloon at 8 sec.

In Exercises 50 and 51, a particle is moving on a straight line, s ft is the directed distance of the particle from the origin at t sec, v ft/sec is the velocity of the particle at t sec, and a ft/sec^2 is the acceleration of the particle at t sec.

50. $a = 3t + 4$; $v = 5$ and $s = 0$, when $t = 0$. Express v and s in terms of t.

51. $a = 6 \cos 2t$; $v = 3$ and $s = 4$ when $t = \frac{1}{2}\pi$. Express v and s in terms of t.

52. Neglecting air resistance, if an object is dropped from an airplane at a height of 30,000 ft above the ocean, how long will it take the object to strike the water?

53. Suppose a bullet is fired directly downward from the airplane in Exercise 52 with a muzzle velocity of 2500 ft/sec. If air resistance is neglected, how long will it take the bullet to reach the ocean?

54. A ball is thrown vertically upward from the top of a house 64 ft above the ground, and the initial velocity is 48 ft/sec. (a) How long will it take the ball to reach its greatest height, and (b) what is its greatest height? (c) How long will it take the ball to strike the ground, and (d) with what velocity will it strike the ground?

55. Suppose the ball in Exercise 54 is thrown downward with an initial velocity of 48 ft/sec. (a) How long will it take the ball to strike the ground, and (b) with what velocity will it strike the ground?

56. Suppose the ball in Exercise 54 is dropped from the top of the house. (a) How long will it take for the ball to strike the ground, and (b) with what velocity will it strike the ground?

In Exercises 57 and 58, find the sum.

57. $\displaystyle\sum_{i=1}^{100} 2i(i^3 - 1)$

58. $\displaystyle\sum_{i=1}^{41} (\sqrt[3]{3i - 1} - \sqrt[3]{3i + 2})$

59. Prove that $\displaystyle\sum_{i=1}^{n} i^3 = \left(\sum_{i=1}^{n} i\right)^2$, and verify the formula for $n = 1, 2$, and 3.

60. Express as a definite integral and evaluate the definite integral: $\displaystyle\lim_{n \to +\infty} \sum_{i=1}^{n} (8\sqrt{i}/n^{3/2})$ (*Hint:* Consider the function f for which $f(x) = \sqrt{x}$.)

61. Show that each of the following inequalities holds: (a) $\displaystyle\int_{-2}^{-1} \frac{dx}{x - 3} \geq \int_{-2}^{-1} \frac{dx}{x}$; (b) $\displaystyle\int_{1}^{2} \frac{dx}{x} \geq \int_{1}^{2} \frac{dx}{x - 3}$; (c) $\displaystyle\int_{4}^{5} \frac{dx}{x - 3} \geq \int_{4}^{5} \frac{dx}{x}$.

62. If

$$f(x) = \begin{cases} 0 & \text{if } a \leq x < c \\ k & \text{if } x = c \\ 1 & \text{if } c < x \leq b \end{cases}$$

prove that f is integrable on $[a, b]$ and $\displaystyle\int_{a}^{b} f(x)\,dx = b - c$ regardless of the value of k.

In Exercises 63 and 64, apply Theorem 4.6.9 to find a closed interval containing the value of the given definite integral.

63. $\displaystyle\int_{-\pi/2}^{\pi/2} \sqrt{\cos t}\,dt$

64. $\displaystyle\int_{-2}^{1} \sqrt[3]{2x^3 - 3x^2 + 1}\,dx$

In Exercises 65 and 66, evaluate the definite integral.

65. $\displaystyle\int_{-3}^{3} |x - 2|^3\,dx$

66. $\displaystyle\int_{-2}^{2} x|x - 3|\,dx$

67. Find $D_x \displaystyle\int_{x}^{4} (3t^2 - 4)^{3/2}\,dt$.

68. Find $D_x \displaystyle\int_{-x}^{x} \frac{4}{1 + t^2}\,dt$.

In Exercises 69 and 70, use the second fundamental theorem of the calculus to evaluate the definite integral. Then find the value of χ satisfying the mean-value theorem for integrals.

69. $\displaystyle\int_{0}^{3} (x^2 + 1)\,dx$

70. $\displaystyle\int_{1}^{4} \sqrt{x}\,dx$

71. Find the average value of the cosine function on the closed interval $[a, a + 2\pi]$.

72. Interpret the mean-value theorem for integrals (4.7.1) in terms of an average function value.

73. If $f(x) = x^2\sqrt{x - 3}$, find the average value of f on $[7, 12]$.

74. (a) Find the average value of the function f defined by $f(x) = 1/x^2$ on the interval $[1, r]$. (b) If A is the average value found in part (a), find $\lim\limits_{r \to +\infty} A$.

75. A body falls from rest and travels a distance of s ft before striking the ground. If the only force acting is that of gravity, which gives the body an acceleration of g ft/sec^2 toward the ground, show that the average value of the velocity, expressed as a function of distance, while traveling this distance is $\frac{2}{3}\sqrt{2gs}$ ft/sec, and that this average velocity is two-thirds of the final velocity.

76. Suppose a ball is dropped from rest and after t sec its directed distance from the starting point is s ft and its velocity is v ft/sec. Neglect air resistance. When $t = t_1$, $s = s_1$, and $v = v_1$. (a) Express v as a function of t as $v = f(t)$, and find the average value of f on $[0, t_1]$. (b) Express v as a function of s as $v = h(s)$, and find the average value of h on $[0, s_1]$. (c) Write the results of parts (a) and (b) in terms of t_1, and determine which average velocity is larger.

In Exercises 77 through 84, find the area of the region bounded by the given curve and lines. Draw a figure showing the region and a rectangular element of area. Express the measure of the area as the limit of a Riemann sum, and then with definite integral notation. Evaluate the definite integral by the second fundamental theorem of the calculus.

77. $y = 9 - x^2$; x axis; y axis; $x = 3$

78. $y = 3\cos\frac{1}{2}x$; x axis; $x = -\frac{1}{2}\pi$; $x = \frac{1}{2}\pi$

79. $y = 2\sqrt{x - 1}$; x axis; $x = 5$; $x = 17$

80. $y = 16 - x^2$; x axis

81. $y = x\sqrt{x + 5}$; x axis; $x = -1$; $x = 4$

82. $y = \dfrac{4}{x^2} - x$; x axis; $x = -2$; $x = -1$

83. $x^2 + y - 5 = 0$; $y = -4$

84. $y = x^2 - 7x$; x axis; $x = 2$; $x = 4$

85. Find the area of the region bounded by the curves $x = y^2$ and $x = y^3$.

86. Find the area of the region bounded by the curves $y = \sin 2x$ and $y = \sin x$ from $x = 0$ to $x = \frac{1}{3}\pi$.

87. Find the area of the region bounded by the curves $y = \cos x$ and $y = \sin x$ from $x = \frac{1}{4}\pi$ to $x = \frac{5}{4}\pi$.

88. Find the area of the region bounded by the loop of the curve $y^2 = x^2(4 - x)$.

89. Find the area of the region in the first quadrant bounded by the y axis and the curves $y = \sec^2 x$ and $y = 2\tan^2 x$.

90. An automobile traveling at a constant speed of 60 mi/hr along a straight highway fails to stop at a stop sign. If 3 sec later a highway patrol car starts from rest from the stop sign and maintains a constant acceleration of 8 ft/sec^2, how long will it take the patrol car to overtake the automobile, and how far from the stop sign will this occur? Also determine the speed of the patrol car when it overtakes the automobile.

91. Suppose that on a particular day in a certain city the Fahrenheit temperature is $f(t)$ degrees t hours since midnight, where

$$f(t) = 60 - 15\sin\tfrac{1}{12}\pi(8 - t) \qquad 0 \le t \le 24$$

(a) Draw a sketch of the graph of f. Find the temperature at (b) 12 midnight; (c) 8 A.M.; (d) 12 noon; (e) 2 P.M.; and (f) 6 P.M. (g) Find the average temperature between 8 A.M. and 6 P.M.

92. If n is any positive integer, prove that $\int_0^\pi \sin^2 nx\, dx = \frac{1}{2}\pi$.

93. Evaluate $\int_0^\pi |\cos x + \frac{1}{2}|\, dx$.

94. Given $F(x) = \displaystyle\int_x^{2x} \frac{1}{t}\, dt$ and $x > 0$. Prove that F is a constant function by showing that $F'(x) = 0$. (*Hint:* Use the first fundamental theorem of the calculus (4.9.1) after writing the given integral as the difference of two integrals.)

95. Let f be continuous on $[a, b]$ and $\int_a^b f(t)\, dt \ne 0$. Show that for any number k in $(0, 1)$ there is a number c in (a, b) such that $\int_a^c f(t)\, dt = k\int_a^b f(t)\, dt$. $\left(\text{*Hint:* Consider the function } F \text{ for which } F(x) = \int_a^x f(t)\, dt \middle/ \int_a^b f(t)\, dt, \text{ and apply the intermediate-value theorem.}\right)$

96. Make up an example of a discontinuous function for which the mean-value theorem for integrals (a) does not hold, and (b) does hold.

97. If $f(x) = x + |x - 1|$, and

$$F(x) = \begin{cases} x & \text{if } x < 1 \\ x^2 - x + 1 & \text{if } 1 \le x \end{cases}$$

show that F is an antiderivative of f on $(-\infty, +\infty)$.

98. Let f and g be two functions such that for all x in $(-\infty, +\infty)$, $f'(x) = g(x)$ and $g'(x) = -f(x)$. Further suppose that $f(0) = 0$ and $g(0) = 1$. Prove that $[f(x)]^2 + [g(x)]^2 = 1$. (*Hint:* Consider the functions F and G where $F(x) = [f(x)]^2$ and $G(x) = -[g(x)]^2$, and show that $F'(x) = G'(x)$ for all x.)

Exercises 99 through 106 pertain to Supplementary Section 4.10.

99. If x units of a certain article of merchandise are produced per day, the marginal cost is given by $C'(x) = 0.5x - 10$, where $C(x)$ dollars is the total cost of producing x units and the overhead cost is \$50 per day. If the selling price of the article is \$20 per unit, find the maximum daily profit that can be obtained.

100. The demand equation of a certain commodity is $p^2 + x = 10$, where $100x$ units of the commodity are demanded when p dollars is the price per unit. Find the average total revenue when the number of units demanded takes on all values from 600 to 800.

101. For the first 5 months that a product was on the market the sales were $f(x)$ units per month x months after the product was first introduced, and $f(x) = 50 + 180x^2$, where $0 \le x \le 5$. (a) Interpret the third-month sales as the measure of the area of a plane region, and draw a figure showing the plane region. (b) Find the third-month sales by computing the area of the plane region in part (a).

102. The marginal revenue for a particular commodity is given by $R'(x) = x^2 - 16x + 55$, where $R(x)$ dollars is the total revenue received from the sale of x units of the commodity. Find (a) the total revenue function and (b) the demand equation. (c) Draw sketches of the demand curve, the total revenue curve, and the marginal revenue curve on the same set of axes.

103. A company that sells men's overcoats starts its fiscal year on July 1. For the three fiscal years beginning on July 1, 1983, the profit from sales was given approximately by

$$P(t) = 20{,}000(1 - \cos \tfrac{1}{6}\pi t) \qquad 0 \le t \le 36$$

where $P(t)$ dollars per month is the profit t months since July 1, 1983. (a) Draw a sketch of the graph of P. Find the profit per month on (b) October 1, 1983; (c) January 1, 1984; (d) April 1, 1984; and (e) July 1, 1984. (f) Determine the total profit for the three fiscal years beginning on July 1, 1983, by finding the area of the region bounded by the graph of P and the t axis.

104. When some particular machinery is x years old, its book value is changing at the rate of $f(x)$ dollars per year, where $f(x) = 3000x - 18{,}000$. By how much does the machinery depreciate the first 4 years?

105. A college bookstore receives a shipment of 600 calculus textbooks on August 10 prior to the beginning of classes on September 9. Over the 30-day period the sales are low at first, and the demand increases as the opening of school approaches so that x days after August 10 the inventory is y books, where $y = 600 - \tfrac{2}{3}x^2$ and $0 \le x \le 30$. If the daily storage cost is 0.15 cent per book, find the total cost of maintaining inventory for 30 days.

106. A particular machine generates total revenue at the rate of $R(x)$ dollars per year x years since its purchase, and $R(x) = 2500 - 5x^2$. Furthermore, the total cost of operating and maintaining the machine increases with time so that in x years the total cost is $C(x)$ dollars per year, where $C(x) = 880 + 15x^2$. (a) Draw a figure showing sketches of the graphs of the functions R and C on the same set of axes. (b) How many years is it profitable to operate the machine? (c) What is the total profit realized from the operation of the machine for the number of years determined in part (b)? (d) Show on the figure obtained in part (a) the region whose area is N square units, where N is the number found in part (c).

APPLICATIONS OF THE DEFINITE INTEGRAL

5.1
VOLUME OF A SOLID
OF REVOLUTION:
CIRCULAR-DISK AND
CIRCULAR-RING METHODS

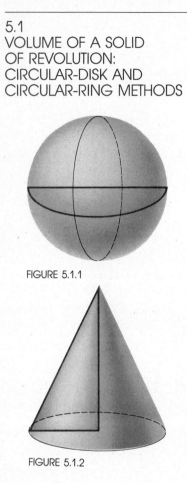

FIGURE 5.1.1

FIGURE 5.1.2

We have developed a method for finding the area of a plane region. Now this process is extended to find the volume of a **solid of revolution**, which is a solid obtained by revolving a region in a plane about a line in the plane, called the **axis of revolution**, which either touches the boundary of the region or does not intersect the region at all. For example, if the region bounded by a semicircle and its diameter is revolved about the diameter, a sphere is generated (see Fig. 5.1.1). A right-circular cone is generated if the region bounded by a right triangle is revolved about one of its legs (see Fig. 5.1.2).

We must first define what is meant by the *volume* of a solid of revolution; that is, we wish to assign a number, for example V, to what we intuitively think of as the measure of the volume of such a solid. We assume that the volume of a cylinder is the product of the area of the base and the altitude. Thus the measure of the volume of a right-circular cylinder is $\pi r^2 h$, where r and h are, respectively, the number of units in the base radius and altitude.

Consider first the case where the axis of revolution is a boundary of the region that is revolved. Let the function f be continuous on the closed interval $[a, b]$, and assume that $f(x) \geq 0$ for all x in $[a, b]$. Let R be the region bounded by the curve $y = f(x)$, the x axis, and the lines $x = a$ and $x = b$. Let S be the solid of revolution obtained by revolving the region R about the x axis. We proceed to find a suitable definition for the number V that gives the measure of the volume of S.

Let Δ be a partition of the closed interval $[a, b]$ given by

$$a = x_0 < x_1 < x_2 < \ldots < x_{n-1} < x_n = b$$

Then there are n subintervals of the form $[x_{i-1}, x_i]$, where $i = 1, 2, \ldots, n$, with the length of the ith subinterval being $\Delta_i x = x_i - x_{i-1}$. Choose any point ξ_i, with $x_{i-1} \leq \xi_i \leq x_i$, in each subinterval, and draw the rectangles having widths $\Delta_i x$ units and altitudes $f(\xi_i)$ units. Figure 5.1.3 shows the region R together with the ith rectangle.

When the ith rectangle is revolved about the x axis, we obtain a circular disk in the form of a right-circular cylinder whose base radius is given by $f(\xi_i)$ units and whose altitude is given by $\Delta_i x$ units, as shown in Fig. 5.1.4. The measure of the volume of this circular disk, which we denote by $\Delta_i V$, is

$$\Delta_i V = \pi [f(\xi_i)]^2 \Delta_i x \tag{1}$$

FIGURE 5.1.3

FIGURE 5.1.4

FIGURE 5.1.5

Because there are n rectangles, n circular disks are obtained in this way, and the sum of the measures of the volumes of these n circular disks is

$$\sum_{i=1}^{n} \Delta_i V = \sum_{i=1}^{n} \pi [f(\xi_i)]^2 \Delta_i x \qquad (2)$$

which is a Riemann sum.

The Riemann sum in (2) is an approximation of what we intuitively think of as the number of cubic units in the volume of the solid of revolution. The smaller we take the norm $\|\Delta\|$ of the partition, the larger will be n, and the closer this approximation will be to the number V we wish to assign to the measure of the volume. We therefore define V to be the limit of the Riemann sum in (2) as $\|\Delta\|$ approaches zero. This limit exists because f^2 is continuous on $[a, b]$, which is true because f is continuous there. We have, then, the following definition.

5.1.1 DEFINITION   Let the function f be continuous on the closed interval $[a, b]$, and assume that $f(x) \geq 0$ for all x in $[a, b]$. If S is the solid of revolution obtained by revolving about the x axis the region bounded by the curve $y = f(x)$, the x axis, and the lines $x = a$ and $x = b$, and if V is the number of cubic units in the volume of S, then

$$V = \lim_{\|\Delta\| \to 0} \sum_{i=1}^{n} \pi [f(\xi_i)]^2 \Delta_i x$$

$$= \pi \int_a^b [f(x)]^2 dx \qquad (3)$$

▶ ILLUSTRATION 1   We find the volume of the solid of revolution generated when the region bounded by the curve $y = x^2$, the x axis, and the lines $x = 1$ and $x = 2$ is revolved about the x axis. Refer to Fig. 5.1.5, showing the region and a rectangular element of area. Figure 5.1.6 shows an element of volume and the solid of revolution. The measure of the volume of the circular disk is given by

$$\Delta_i V = \pi (\xi_i^2)^2 \Delta_i x$$

$$= \pi \xi_i^4 \Delta_i x$$

Then

$$V = \lim_{\|\Delta\| \to 0} \sum_{i=1}^{n} \pi \xi_i^4 \Delta_i x$$

$$= \pi \int_1^2 x^4 dx$$

$$= \pi (\tfrac{1}{5} x^5) \Big]_1^2$$

$$= \tfrac{31}{5} \pi$$

FIGURE 5.1.6

Therefore the volume of the solid of revolution is $\tfrac{31}{5}\pi$ cubic units.   ◀

A definition similar to Definition 5.1.1 applies when both the axis of revolution and a boundary of the revolved region is the y axis or any line parallel to either the x axis or the y axis.

EXAMPLE 1

Find the volume of the solid generated by revolving about the line $x = 1$ the region bounded by the curve

$$(x - 1)^2 = 20 - 4y$$

and the lines $x = 1$, $y = 1$, and $y = 3$ and to the right of $x = 1$.

SOLUTION

The region and a rectangular element of area are shown in Fig. 5.1.7. An element of volume and the solid of revolution appear in Fig. 5.1.8.

In the equation of the curve we solve for x and have

$$x = \sqrt{20 - 4y} + 1$$

Let $g(y) = \sqrt{20 - 4y} + 1$. We take a partition of the interval $[1, 3]$ on the y axis. Then if $\Delta_i V$ cubic units is the volume of the ith circular disk,

$$\Delta_i V = \pi[g(\xi_i) - 1]^2 \, \Delta_i y$$
$$= \pi[(\sqrt{20 - 4\xi_1} + 1) - 1]^2 \, \Delta_i y$$
$$= \pi(20 - 4\xi_i) \, \Delta_i y$$

If V cubic units is the volume of the solid of revolution,

$$V = \lim_{\|\Delta\| \to 0} \sum_{i=1}^{n} \pi(20 - 4\xi_i) \, \Delta_i y$$
$$= \pi \int_{1}^{3} (20 - 4y) \, dy$$
$$= \pi \left[20y - 2y^2 \right]_{1}^{3}$$
$$= \pi[(60 - 18) - (20 - 2)]$$
$$= 24\pi$$

The volume of the solid of revolution is therefore 24π cubic units.

Now suppose that the axis of revolution is not a boundary of the region being revolved. Let f and g be two continuous functions on the closed interval $[a, b]$, and assume that $f(x) \geq g(x) \geq 0$ for all x in $[a, b]$. Let R be the region bounded by the curves $y = f(x)$ and $y = g(x)$ and the lines $x = a$ and $x = b$. Let S be the solid of revolution generated by revolving the region R about the x axis. If V cubic units is the volume of S, we wish to find a value for V.

Let Δ be a partition of the interval $[a, b]$, given by

$$a = x_0 < x_1 < x_2 < \ldots < x_{n-1} < x_n = b$$

FIGURE 5.1.7

FIGURE 5.1.8

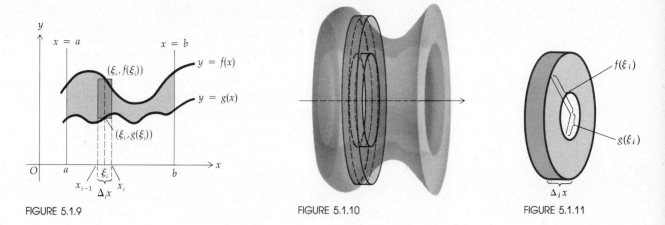

FIGURE 5.1.9   FIGURE 5.1.10   FIGURE 5.1.11

and the ith subinterval $[x_{i-1}, x_i]$ has length $\Delta_i x = x_i - x_{i-1}$. In the ith sub-interval choose any ξ_i, with $x_{i-1} \le \xi_i \le x_i$. Consider the n rectangular elements of area for the region R. See Fig. 5.1.9 illustrating the region and the ith rectangle, and Fig. 5.1.10 showing the solid of revolution.

When the ith rectangle is revolved about the x axis, a circular ring (or "washer"), as shown in Fig. 5.1.11, is obtained. The number giving the difference of the measures of the areas of the two circular regions is $(\pi[f(\xi_i)]^2 - \pi[g(\xi_i)]^2)$, and the thickness is $\Delta_i x$ units. Therefore the measure of the volume of the circular ring is given by

$$\Delta_i V = \pi([f(\xi_i)]^2 - [g(\xi_i)]^2)\Delta_i x$$

The sum of the measures of the volumes of the n circular rings formed by revolving the n rectangular elements of area about the x axis is

$$\sum_{i=1}^{n} \Delta_i V = \sum_{i=1}^{n} \pi([f(\xi_i)]^2 - [g(\xi_i)]^2)\Delta_i x \tag{4}$$

The number of cubic units in the volume, then, is defined to be the limit of the Riemann sum in (4) as $\|\Delta\|$ approaches zero. The limit exists since $f^2 - g^2$ is continuous on $[a, b]$ because f and g are continuous there.

5.1.2 DEFINITION   Let the functions f and g be continuous on the closed interval $[a, b]$, and assume that $f(x) \ge g(x) \ge 0$ for all x in $[a, b]$. Then if V cubic units is the volume of the solid of revolution generated by revolving about the x axis the region bounded by the curves $y = f(x)$ and $y = g(x)$ and the lines $x = a$ and $x = b$,

$$V = \lim_{\|\Delta\| \to 0} \sum_{i=1}^{n} \pi([f(\xi_i)]^2 - [g(\xi_i)]^2)\Delta_i x$$

$$= \pi \int_a^b ([f(x)]^2 - [g(x)]^2)\, dx \tag{5}$$

As before, a similar definition applies when the axis of revolution is the y axis or any line parallel to either the x axis or the y axis.

FIGURE 5.1.12

FIGURE 5.1.13

EXAMPLE 2

Find the volume of the solid generated by revolving about the x axis the region bounded by the parabola $y = x^2 + 1$ and the line $y = x + 3$.

SOLUTION

The points of intersection are $(-1, 2)$ and $(2, 5)$. Figure 5.1.12 shows the region and a rectangular element of area. An element of volume and the solid of revolution are shown in Fig. 5.1.13.

If $f(x) = x + 3$ and $g(x) = x^2 + 1$, the measure of the volume of the circular ring is

$$\Delta_i V = \pi([f(\xi_i)]^2 - [g(\xi_i)]^2) \Delta_i x$$

If V cubic units is the volume of the solid, then

$$V = \lim_{\|\Delta\| \to 0} \sum_{i=1}^{n} \pi([f(\xi_i)]^2 - [g(\xi_i)]^2) \Delta_i x$$

$$= \pi \int_{-1}^{2} ([f(x)]^2 - [g(x)]^2) \, dx$$

$$= \pi \int_{-1}^{2} [(x + 3)^2 - (x^2 + 1)^2] \, dx$$

$$= \pi \int_{-1}^{2} [-x^4 - x^2 + 6x + 8] \, dx$$

$$= \pi \left[-\tfrac{1}{5}x^5 - \tfrac{1}{3}x^3 + 3x^2 + 8x \right]_{-1}^{2}$$

$$= \pi[(-\tfrac{32}{5} - \tfrac{8}{3} + 12 + 16) - (\tfrac{1}{5} + \tfrac{1}{3} + 3 - 8)]$$

$$= \tfrac{117}{5}\pi$$

Therefore the volume of the solid of revolution is $\tfrac{117}{5}\pi$ cubic units.

EXAMPLE 3

Find the volume of the solid generated by revolving about the line $x = -4$ the region bounded by the two parabolas $x = y - y^2$ and $x = y^2 - 3$.

SOLUTION

The curves intersect at the points $(-2, -1)$ and $(-\tfrac{3}{4}, \tfrac{3}{2})$. The region and a rectangular element of area are shown in Fig. 5.1.14. Figure 5.1.15 shows the solid of revolution as well as an element of volume, which is a circular ring.

FIGURE 5.1.14

FIGURE 5.1.15

Let $F(y) = y - y^2$ and $G(y) = y^2 - 3$. The number of cubic units in the volume of the circular ring is

$$\Delta_i V = \pi([4 + F(\xi_i)]^2 - [4 + G(\xi_i)]^2) \Delta_i y$$

Thus

$$V = \lim_{||\Delta|| \to 0} \sum_{i=1}^{n} \pi([4 + F(\xi_i)]^2 - [4 + G(\xi_i)]^2) \Delta_i y$$

$$= \pi \int_{-1}^{3/2} [(4 + y - y^2)^2 - (4 + y^2 - 3)^2] \, dy$$

$$= \pi \int_{-1}^{3/2} (-2y^3 - 9y^2 + 8y + 15) \, dy$$

$$= \pi \left[-\tfrac{1}{2}y^4 - 3y^3 + 4y^2 + 15y \right]_{-1}^{3/2}$$

$$= \tfrac{875}{32} \pi$$

The volume of the solid of revolution is then $\tfrac{875}{32}\pi$ cubic units.

EXERCISES 5.1

In Exercises 1 through 8, find the volume of the solid of revolution generated when the given region of Fig. 5.1.16 is revolved about the indicated line. An equation of the curve in the figure is $y^2 = x^3$.

1. OAC about the x axis ✓
2. OAC about the line AC
3. OAC about the line BC
4. OAC about the y axis ✓
5. OBC about the y axis ✓
6. OBC about the line BC
7. OBC about the line AC
8. OBC about the x axis ✓

FIGURE 5.1.16

In Exercises 9 through 12, find the volume of the solid of revolution generated by revolving about the indicated line the region bounded by the curve $y = \sqrt{x}$, the x axis, and the line $x = 4$.

9. the line $x = 4$ 10. the x axis ✓   11. the y axis ✓   12. the line $y = 2$

13. Find the volume of the solid generated by revolving about the x axis the region bounded by the curve $y = x^3$, the x axis, and the line $x = 2$.

14. Find the volume of the solid generated by revolving the region in Exercise 13 about the y axis.

15. Find the volume of the sphere generated by revolving about a diameter the region enclosed by the circle $x^2 + y^2 = r^2$.

16. Find by integration the volume of a right-circular cone of altitude h units and base radius a units.

17. The region bounded by the curve $y = \sec x$, the x axis, the y axis, and the line $x = \tfrac{1}{4}\pi$ is revolved about the x axis. Find the volume of the solid generated.

18. Find the volume of the solid generated when the region bounded by the curve $y = \csc x$, the x axis, and the lines $x = \tfrac{1}{6}\pi$ and $x = \tfrac{1}{3}\pi$ is revolved about the x axis.

19. Find the volume of the solid of revolution generated if the region bounded by one arch of the sine curve is revolved about the x axis. (*Hint:* Use the identity $\sin^2 x = \frac{1}{2}(1 - \cos 2x)$.)

20. The region bounded by the y axis and the curves $y = \sin x$ and $y = \cos x$ for $0 \le x \le \frac{1}{4}\pi$ is revolved about the x axis. Find the volume of the solid of revolution generated. (*Hint:* Use the following identities: $\sin^2 x = \frac{1}{2}(1 - \cos 2x)$ and $\cos^2 x = \frac{1}{2}(1 + \cos 2x)$.)

21. Find the volume of the solid generated if the region of Exercise 19 is revolved about the line $y = 1$.

22. Find the volume of the solid generated if the region of Exercise 20 is revolved about the line $y = 1$.

23. The region bounded by the curve $y = \cot x$, the line $x = \frac{1}{6}\pi$, and the x axis is revolved about the x axis. Find the volume of the solid generated.

24. The region bounded by the curve $y = \tan x$, the line $x = \frac{1}{3}\pi$, and the x axis is revolved about the x axis. Find the volume of the solid generated.

25. Find the volume of the solid generated by revolving about the line $x = -4$ the region bounded by that line and the parabola $x = 4 + 6y - 2y^2$.

26. Find the volume of the solid generated by revolving about the x axis the region bounded by the parabola $y^2 = 4x$ and the line $y = x$.

27. Find the volume of the solid generated by revolving the region of Exercise 26 about the line $x = 4$.

28. Find the volume of the solid generated by revolving about the y axis the region bounded by the line through $(1, 3)$ and $(3, 7)$, and the lines $y = 3$, $y = 7$, and $x = 0$.

29. Find the volume of the solid generated by revolving about the line $y = -3$ the region bounded by the two parabolas $y = x^2$ and $y = 1 + x - x^2$.

30. Find the volume of the solid generated by revolving about the x axis the region bounded by the graph of $y = |x - 3|$, and the lines $x = 1$, $x = 4$, and $y = 0$.

31. Find the volume of the solid generated by revolving about the x axis the region bounded by the lines $y = [\![\sqrt{x + 1}]\!]$, $x = 3$, $x = 7$ and $y = 0$.

32. An oil tank in the shape of a sphere has a diameter of 60 ft. How much oil does the tank contain if the depth of the oil is 25 ft?

33. A paraboloid of revolution is obtained by revolving the parabola $y^2 = 4px$ about the x axis. Find the volume bounded by a paraboloid of revolution and a plane perpendicular to its axis if the plane is 10 cm from the vertex, and if the plane section of intersection is a circle having a radius of 6 cm.

34. Find the volume of the solid generated by revolving about the x axis the region bounded by the loop of the curve whose equation is $2y^2 = x(x^2 - 4)$.

35. Find the volume of the solid generated when the region bounded by one loop of the curve, having the equation $x^2y^2 = (x^2 - 9)(1 - x^2)$, is revolved about the x axis.

36. The region bounded by a pentagon having vertices at $(-4, 4)$, $(-2, 0)$, $(0, 8)$, $(2, 0)$, and $(4, 4)$ is revolved about the x axis. Find the volume of the solid generated.

37. A solid of revolution is formed by revolving about the x axis the region bounded by the curve $y = \sqrt{2x + 4}$, the x axis, the y axis, and the line $x = c$ $(c > 0)$. For what value of c will the volume be 12π cubic units?

38. The region in the first quadrant bounded by the curve $y = \sec x$, the y axis, and the line $y = 2$ is revolved about the x axis. Find the volume of the solid generated.

39. The region bounded by the curve $y = \csc x$ and the lines $y = 2$, $x = \frac{1}{6}\pi$, and $x = \frac{5}{6}\pi$ is revolved about the x axis. Find the volume of the solid generated.

40. The region in the first quadrant bounded by the coordinate axes, the line $y = 1$, and the curve $y = \cot x$ is revolved about the x axis. Find the volume of the solid generated.

5.2
VOLUME OF A SOLID
OF REVOLUTION:
CYLINDRICAL-SHELL
METHOD

In the preceding section we found the volume of a solid of revolution by taking the rectangular elements of area perpendicular to the axis of revolution, and the element of volume was either a circular disk or a circular ring. If a rectangular element of area is parallel to the axis of revolution, then when this element of area is revolved about the axis of revolution, a *cylindrical*

FIGURE 5.2.1

FIGURE 5.2.2

FIGURE 5.2.3

shell is obtained. A **cylindrical shell** is a solid contained between two cylinders having the same center and axis. Such a cylindrical shell is shown in Fig. 5.2.1.

If the cylindrical shell has an inner radius r_1 units, outer radius r_2 units, and altitude h units, then its volume V cubic units is given by

$$V = \pi r_2{}^2 h - \pi r_1{}^2 h \tag{1}$$

Let R be the region bounded by the curve $y = f(x)$, the x axis, and the lines $x = a$ and $x = b$, where f is continuous on the closed interval $[a, b]$ and $f(x) \geq 0$ for all x in $[a, b]$; further assume that $a \geq 0$. Such a region is shown in Fig. 5.2.2. If R is revolved about the y axis, a solid of revolution S is generated. Such a solid is shown in Fig. 5.2.3. To find the volume of S when the rectangular elements of area are taken parallel to the y axis we proceed in the following manner.

Let Δ be a partition of the closed interval $[a, b]$ given by

$$a = x_0 < x_1 < x_2 < \ldots < x_{n-1} < x_n = b$$

Let m_i be the midpoint of the ith subinterval $[x_{i-1}, x_i]$. Then we have $m_i = \frac{1}{2}(x_{i-1} + x_i)$. Consider the rectangle having altitude $f(m_i)$ units and width $\Delta_i x$ units. If this rectangle is revolved about the y axis, a cylindrical shell is obtained. Figure 5.2.3 shows the cylindrical shell generated by the rectangular element of area.

If $\Delta_i V$ gives the measure of the volume of this cylindrical shell, we have, from formula (1), where $r_1 = x_{i-1}$, $r_2 = x_i$, and $h = f(m_i)$,

$$\Delta_i V = \pi x_i{}^2 f(m_i) - \pi x_{i-1}{}^2 f(m_i)$$

$$\Delta_i V = \pi(x_i{}^2 - x_{i-1}{}^2) f(m_i)$$

$$\Delta_i V = \pi(x_i - x_{i-1})(x_i + x_{i-1}) f(m_i) \tag{2}$$

Because $x_i - x_{i-1} = \Delta_i x$, and because $x_i + x_{i-1} = 2m_i$, then from (2),

$$\Delta_i V = 2\pi m_i f(m_i) \Delta_i x \tag{3}$$

If n rectangular elements of area are revolved about the y axis, n cylindrical shells are obtained. The sum of the measures of their volumes is

$$\sum_{i=1}^{n} \Delta_i V = \sum_{i=1}^{n} 2\pi m_i f(m_i) \Delta_i x \tag{4}$$

which is a Riemann sum because any particular number (such as m_i) in $[x_{i-1}, x_i]$ can replace ξ_i.

Then we define the measure of the volume of the solid of revolution to be the limit of the Riemann sum in (4) as $\|\Delta\|$ approaches zero. The limit exists because if f is continuous on $[a, b]$, so is the function having function values $2\pi x f(x)$.

5.2.1 DEFINITION   Let the function f be continuous on the closed interval $[a, b]$, where $a \geq 0$. Assume that $f(x) \geq 0$ for all x in $[a, b]$. If R is the region bounded by the curve $y = f(x)$, the x axis, and the lines $x = a$ and $x = b$, if S is the solid of revolution obtained by revolving R about the y axis, and if V cubic units is

the volume of S, then

$$V = \lim_{||\Delta|| \to 0} \sum_{i=1}^{n} 2\pi m_i f(m_i) \Delta_i x$$

$$= 2\pi \int_a^b x f(x)\,dx \qquad (5)$$

Definition 5.2.1 is consistent with Definitions 5.1.1 and 5.1.2 for the measure of the volume of a solid of revolution for which the definitions apply.

The formula for the measure of the volume of the shell is easily remembered by noticing that $2\pi m_i$, $f(m_i)$, and $\Delta_i x$ are, respectively, the numbers that are the measures of the circumference of the circle having as radius the mean of the inner and outer radii of the shell, the altitude of the shell, and the thickness.

FIGURE 5.2.4

FIGURE 5.2.5

EXAMPLE 1

The region bounded by the curve $y = x^2$, the x axis, and the line $x = 2$ is revolved about the y axis. Find the volume of the solid generated. Take the elements of area parallel to the axis of revolution.

SOLUTION
Figure 5.2.4 shows the region and a rectangular element of area. Figure 5.2.5 shows the solid of revolution and the cylindrical shell obtained by revolving the rectangular element of area about the y axis.

The element of volume is a cylindrical shell the measure of whose volume is

$$\Delta_i V = 2\pi m_i (m_i^2)\,\Delta_i x$$
$$= 2\pi m_i^3\,\Delta_i x$$

Thus

$$V = \lim_{||\Delta|| \to 0} \sum_{i=1}^{n} 2\pi m_i^3\,\Delta_i x$$

$$= 2\pi \int_0^2 x^3\,dx$$

$$= 2\pi (\tfrac{1}{4}x^4) \Big]_0^2$$

$$= 8\pi$$

Therefore the volume of the solid of revolution is 8π cubic units.

EXAMPLE 2

The region bounded by the curve $y = x^2$ and the lines $y = 1$ and $x = 2$ is revolved about the line $y = -3$. Find the volume of the solid generated by taking the rectangular elements of area parallel to the axis of revolution.

SOLUTION
The region and a rectangular element of area are shown in Fig. 5.2.6.

The equation of the curve is $y = x^2$. Solving for x we obtain $x = \pm\sqrt{y}$. Because $x > 0$ for the given region, $x = \sqrt{y}$.

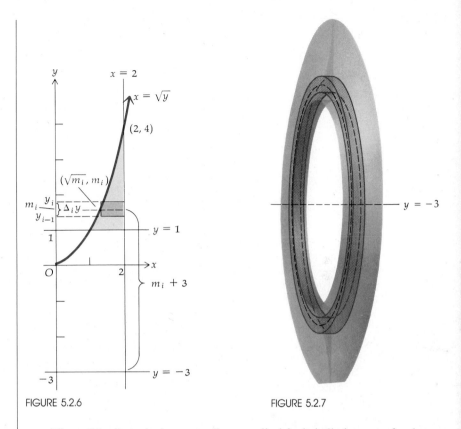

FIGURE 5.2.6   FIGURE 5.2.7

The solid of revolution as well as a cylindrical shell element of volume is shown in Fig. 5.2.7. The outer radius of the cylindrical shell is $(y_i + 3)$ units and the inner radius is $(y_{i-1} + 3)$ units. Hence the mean of the inner and outer radii is $(m_i + 3)$ units. Because the altitude and thickness of the cylindrical shell are, respectively, $(2 - \sqrt{m_i})$ units and $\Delta_i y$ units,

$$\Delta_i V = 2\pi(m_i + 3)(2 - \sqrt{m_i})\,\Delta_i y$$

Hence, if V cubic units is the volume of the solid of revolution,

$$\begin{aligned}
V &= \lim_{\|\Delta\| \to 0} \sum_{i=1}^{n} 2\pi(m_i + 3)(2 - \sqrt{m_i})\,\Delta_i y \\
&= \int_1^4 2\pi(y + 3)(2 - \sqrt{y})\,dy \\
&= 2\pi \int_1^4 (-y^{3/2} + 2y - 3y^{1/2} + 6)\,dy \\
&= 2\pi\left[-\tfrac{2}{5}y^{5/2} + y^2 - 2y^{3/2} + 6y \right]_1^4 \\
&= \tfrac{66}{5}\pi
\end{aligned}$$

Therefore the volume is $\frac{66}{5}\pi$ cubic units.

EXERCISES 5.2

1–12. Solve Exercises 1 through 12 in Section 5.1 by the cylindrical-shell method.

In Fig. 5.2.8 the region bounded by the x axis, the line $x = 1$, and the curve $y = x^2$ is denoted by R_1; the region bounded by the two curves $y = x^2$ and $y^2 = x$ is denoted by R_2; the region bounded by the y axis, the line $y = 1$, and the curve $y^2 = x$ is denoted by R_3. In Exercises 13 through 20, find the volume of the solid generated when the indicated region is revolved about the given line.

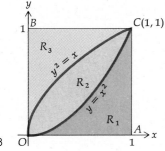

FIGURE 5.2.8

13. R_1 is revolved about the y axis; the rectangular elements are parallel to the axis of revolution.
14. Same as Exercise 13, but the rectangular elements are perpendicular to the axis of revolution.
15. R_2 is revolved about the x axis; the rectangular elements are parallel to the axis of revolution.
16. Same as Exercise 15, but the rectangular elements are perpendicular to the axis of revolution.
17. R_3 is revolved about the line $y = 2$; the rectangular elements are parallel to the axis of revolution.
18. Same as Exercise 17, but the rectangular elements are perpendicular to the axis of revolution.
19. R_2 is revolved about the line $x = -2$; the rectangular elements are parallel to the axis of revolution.
20. Same as Exercise 19, but the rectangular elements are perpendicular to the axis of revolution.

In Exercises 21 through 24, the region bounded by the curves $x = y^2 - 2$ and $x = 6 - y^2$ is revolved about the indicated axis. Find the volume of the solid generated.

21. the x axis   **22.** the y axis   **23.** the line $x = 2$ **24.** the line $y = 2$
25. Find the volume of the solid generated if the region bounded by the parabola $y^2 = 4ax$ $(a > 0)$ and the line $x = a$ is revolved about $x = a$.
26. Find the volume of the solid generated by revolving about the x axis the region bounded by the curves $y = x^3$ and $x = y^3$. Take the rectangular elements of area parallel to the axis of revolution.
27. Find the volume of the solid generated by revolving about the line $y = 1$ the region bounded by that line and the parabola $x^2 = 4y$. Take the rectangular elements of area parallel to the axis of revolution.
28. Find the volume of the solid generated by revolving about the y axis the region bounded by the curve $x^{2/3} + y^{2/3} = a^{2/3}$.
29. Find the volume of the solid generated by revolving about the y axis the region bounded by the curve $y = \sin x^2$, the x axis, and the lines $x = \frac{1}{2}\sqrt{\pi}$ and $x = \sqrt{\pi}$.
30. Find the volume of the solid generated by revolving about the y axis the region in the first quadrant bounded by the curve $y = \cos x^2$ and the coordinate axes.
31. The region in the first quadrant bounded by the curve $x = \cos y^2$, the y axis, and the x axis, where $0 \le x \le 1$, is revolved about the x axis. Find the volume of the solid of revolution generated.
32. A hole of radius 2 cm is bored through a spherical shaped solid of radius 6 cm, and the axis of the hole is a diameter of the sphere. Find the volume of the part of the solid that remains.
33. A hole of radius $2\sqrt{3}$ in. is bored through the center of a spherical shaped solid of radius 4 in. Find the volume of the portion of the solid cut out.
34. Find the volume of the solid generated if the region bounded by the parabola $y^2 = 4px$ and the line $x = p$ is revolved about the y axis.
35. Find the volume of the solid generated by revolving about the y axis the region bounded by the graph of $y = |x - 3|$, and the lines $x = 1$, $x = 5$, and $y = 0$. Take the rectangular elements of area parallel to the axis of revolution.
36. Find the volume of the solid generated by revolving about the y axis the region outside the curve $y = x^2$ and between the lines $y = 2x - 1$ and $y = x + 2$.

37. A solid of revolution is formed by revolving about the y axis the region bounded by the curve $y = \sqrt[3]{x}$, the x axis, and the line $x = c$ $(c > 0)$. Take the rectangular elements of area parallel to the axis of revolution to determine the value of c that will give a volume of 12π cubic units.

38. Find the volume of the solid generated by revolving about the x axis the region bounded by the x axis and the lines $y = [\![\sqrt{2x + 1}\,]\!]$, $x = 4$, and $x = 7$. Take the rectangular elements of area parallel to the axis of revolution.

5.3
VOLUME OF A SOLID
HAVING KNOWN PARALLEL
PLANE SECTIONS

Let S be a solid. By a **plane section** of S is meant a plane region formed by the intersection of a plane with S. In Section 5.1 we learned how to find the volume of a solid of revolution for which all plane sections perpendicular to the axis of revolution are circular. We now generalize this method to find the volume of a solid for which it is possible to express the area of any plane section perpendicular to a fixed line in terms of the perpendicular distance of the plane section from a fixed point. We first define a *cylinder*.

5.3.1 DEFINITION   A solid is a **right cylinder** if it is bounded by two congruent plane regions R_1 and R_2 lying in parallel planes and by a lateral surface generated by a line segment, having its endpoints on the boundaries of R_1 and R_2, which moves so that it is always perpendicular to the planes of R_1 and R_2.

Figure 5.3.1 illustrates a right cylinder. The **height** of the cylinder is the perpendicular distance between the planes of R_1 and R_2, and the base is either R_1 or R_2. If the base of the right cylinder is a region enclosed by a circle, we have a **right-circular cylinder** (see Fig. 5.3.2); if the base is a region enclosed by a rectangle, we have a **rectangular parallelepiped** (see Fig. 5.3.3).

FIGURE 5.3.1   FIGURE 5.3.2   FIGURE 5.3.3.

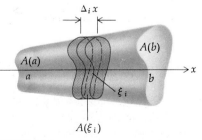

FIGURE 5.3.4

If the area of the base of a right cylinder is A square units and the height is h units, then by definition the volume of the right cylinder is measured by the product of A and h. As stated above, we are considering solids for which the area of any plane section that is perpendicular to a fixed line is a function of the perpendicular distance of the plane section from a fixed point. The solid S of Fig. 5.3.4 is such a solid, and it lies between the planes perpendicular to the x axis at a and b. We represent the number of square units in the area of the plane section of S in the plane perpendicular to the x axis at x by $A(x)$, where A is continuous on $[a, b]$.

Let Δ be a partition of the closed interval $[a, b]$ given by

$$a = x_0 < x_1 < x_2 < \ldots < x_n = b$$

There are, then, n subintervals of the form $[x_{i-1}, x_i]$, where $i = 1, 2, \ldots, n$, with the length of the ith subinterval being $\Delta_i x = x_i - x_{i-1}$. Choose any number ξ_i, with $x_{i-1} \leq \xi_i \leq x_i$, in each subinterval, and construct the right cylinders of heights $\Delta_i x$ units and plane section areas $A(\xi_i)$ square units. The volume of the ith cylinder is $A(\xi_i) \Delta_i x$ cubic units. We obtain n right cylinders, and the sum of the measures of the volumes of these n cylinders is given by the Riemann sum

$$\sum_{i=1}^{n} A(\xi_i) \Delta_i x \tag{1}$$

This Riemann sum is an approximation of what we intuitively think of as the measure of the volume of S, and the smaller we take the norm $\|\Delta\|$ of the partition Δ, the larger n will be, and the closer this approximation will be to the number we wish to assign to the measure of the volume. We have, then, the following definition.

5.3.2 DEFINITION   Let S be a solid such that S lies between planes drawn perpendicular to the x axis at a and b. If the measure of the area of the plane section of S drawn perpendicular to the x axis at x is given by $A(x)$, where A is continuous on $[a, b]$, then the measure of the volume of S is given by

$$V = \lim_{\|\Delta\| \to 0} \sum_{i=1}^{n} A(\xi_i) \Delta_i x$$

$$= \int_a^b A(x)\, dx \tag{2}$$

Definitions 5.1.1 and 5.1.2 are special cases of Definition 5.3.2. If in (2) we take $A(x) = \pi[f(x)]^2$, we have Eq. (3) of Section 5.1. Furthermore, if $A(x) = \pi([f(x)]^2 - [g(x)]^2)$, we have Eq. (5) of Section 5.1.

EXAMPLE 1
If the base of a solid is the region enclosed by a circle with a radius of r units, and if all plane sections perpendicular to a fixed diameter of the base are squares, find the volume of the solid.

SOLUTION
Take the circle in the xy plane, the center at the origin, and the fixed diameter along the x axis. Therefore an equation of the circle is $x^2 + y^2 = r^2$. Figure 5.3.5 shows the solid and an element of volume, which is a right cylinder of altitude $\Delta_i x$ units and with an area of the base given by $[2f(\xi_i)]^2$ square units, where $f(x)$ is obtained by solving the equation of the circle for y and setting $y = f(x)$. This computation gives $f(x) = \sqrt{r^2 - x^2}$. Therefore, if

FIGURE 5.3.5

V cubic units is the volume of the solid,

$$V = \lim_{||\Delta|| \to 0} \sum_{i=1}^{n} [2f(\xi_i)]^2 \, \Delta_i x$$

$$= 4 \int_{-r}^{r} (r^2 - x^2) \, dx$$

$$= 4\left[r^2 x - \tfrac{1}{3}x^3 \right]_{-r}^{r}$$

$$= \tfrac{16}{3}r^3$$

EXAMPLE 2

A wedge is cut from a right-circular cylinder with a radius of r cm by two planes, one perpendicular to the axis of the cylinder and the other intersecting the first along a diameter of the circular plane section at an angle of measurement 60°. Find the volume of the wedge.

SOLUTION

The wedge is shown in Fig. 5.3.6. The xy plane is taken as the plane perpendicular to the axis of the cylinder, and the origin is at the point of perpendicularity. An equation of the circular plane section is then $x^2 + y^2 = r^2$. Every plane section of the wedge perpendicular to the x axis is a right triangle. An element of volume is a right cylinder having altitude $\Delta_i x$ cm, and area of the base given by $\frac{1}{2}\sqrt{3}[f(\xi_i)]^2$ cm^2, where $f(x)$ is obtained by solving the equation of the circle for y and setting $y = f(x)$. Therefore, we have $f(x) = \sqrt{r^2 - x^2}$. Thus, if V cm^3 is the volume of the wedge,

$$V = \lim_{||\Delta|| \to 0} \sum_{i=1}^{n} \tfrac{1}{2}\sqrt{3}(r^2 - \xi_i^2) \, \Delta_i x$$

$$= \tfrac{1}{2}\sqrt{3} \int_{-r}^{r} (r^2 - x^2) \, dx$$

$$= \tfrac{1}{2}\sqrt{3} \left[r^2 x - \tfrac{1}{3}x^3 \right]_{-r}^{r}$$

$$= \tfrac{2}{3}\sqrt{3}r^3$$

Hence the volume of the wedge is $\tfrac{2}{3}\sqrt{3}r^3$ cm^3.

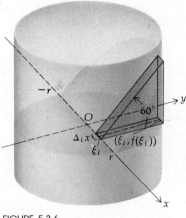

FIGURE 5.3.6

EXERCISES 5.3

1. The base of a solid is the region enclosed by an ellipse having the equation $3x^2 + y^2 = 6$. Find the volume of the solid if all plane sections perpendicular to the x axis are squares.

2. The base of a solid is the region enclosed by the hyperbola $25x^2 - 4y^2 = 100$ and the line $x = 4$. Find the volume of the solid if all plane sections perpendicular to the x axis are squares.

3. The base of a solid is the region enclosed by a circle having a radius of 7 cm. Find the volume of the solid if all plane sections perpendicular to a fixed diameter of the base are equilateral triangles.

4. The base of a solid is the region of Exercise 2. Find the volume of the solid if all plane sections perpendicular to the x axis are equilateral triangles.

5. The base of a solid is the region of Exercise 3. Find the volume of the solid if all plane sections perpendicular to a fixed diameter of the base are isosceles triangles of height equal to the distance of the plane section from the center of the circle. The side of the triangle lying in the base of the solid is not one of the two sides of equal length.

6. The base of a solid is the region enclosed by a circle with a radius of r units, and all plane sections perpendicular to a fixed diameter of the base are isosceles right triangles having the hypotenuse in the plane of the base. Find the volume of the solid.

7. Solve Exercise 6 if the isosceles right triangles have one leg in the plane of the base.

8. Find the volume of a right pyramid having a height of h units and a square base of side a units.

9. Find the volume of a tetrahedron having 3 mutually perpendicular faces and 3 mutually perpendicular edges whose lengths are 3 in., 4 in., and 7 in.

10. Solve Exercise 9 if the edges of the tetrahedron have lengths of measures a, b, and c.

11. The base of a solid is the region enclosed by an ellipse having the equation $2x^2 + y^2 = 8$. Find the volume of the solid if all plane sections perpendicular to the x axis are isosceles triangles whose height and length of base are equal.

12. The base of a solid is the region enclosed by a circle with a radius of 4 in., and each plane section perpendicular to a fixed diameter of the base is an isosceles triangle having an altitude of 10 in. and a chord of the circle as a base. Find the volume of the solid.

13. A layer through the center of a solid is the region enclosed by a circle with a radius of 9 cm, and each plane section perpendicular to a fixed diameter of the layer is a square having a chord of the circle as a diagonal. Find the volume of the solid.

14. The base of a solid is the region enclosed by a circle having a radius of r units. Find the volume of the solid if all plane sections perpendicular to a fixed diameter of the base are rectangles having a height equal to one-half the length of the chord of the circle forming the base of the rectangle.

15. The base of a solid is the region enclosed by the curve $x = 2\sqrt{y}$ and the lines $x + y = 0$ and $y = 9$. Find the volume of the solid if all plane sections perpendicular to the y axis are squares having a diagonal with one endpoint on the line $x + y = 0$ and the other endpoint on the curve $x = 2\sqrt{y}$.

16. Two right-circular cylinders, each having a radius of r units, have axes that intersect at right angles. Find the volume of the solid common to the two cylinders.

17. A wedge is cut from a solid in the shape of a right-circular cylinder with a radius of r cm by a plane through a diameter of the base and inclined to the plane of the base at an angle of measurement 45°. Find the volume of the wedge.

18. A wedge is cut from a solid in the shape of a right-circular cone having a base radius of 5 ft and an altitude of 20 ft by two half planes containing the axis of the cone. The angle between the two planes has a measurement of 30°. Find the volume of the wedge cut out.

5.4
WORK

The *work* done by a force acting on an object is defined in physics as "force times displacement." For example, suppose that an object is moving to the right along the x axis from a point a to a point b, and a constant force of F lb is acting on the object in the direction of motion. Then if the displacement is measured in feet, $b - a$ is the number of feet in the displacement. And if W is the number of foot-pounds of work done by the force, W is defined by

$$W = F(b - a) \tag{1}$$

▶ ILLUSTRATION 1   If W foot-pounds is the work necessary to lift a 70-lb weight to a height of 3 ft, then

$$W = 70 \cdot 3$$
$$= 210$$ ◀

The unit of measurement for work depends on the units of force and distance. In the English system, where the force is measured in pounds and the distance is measured in feet, work is measured in foot-pounds. In the metric system, if the unit of mass is the kilogram and distance is measured in meters, then the unit of force is the *newton*, and the unit of work is a newton-meter, which is called a *joule*. If in the metric system the unit of mass is the gram and distance is measured in centimeters, then the unit of force is the *dyne*, and the unit of work is a dyne-centimeter, which is called an *erg*. For conversion purposes, 1 newton is 10^5 dynes, and 1 joule is 10^7 ergs.

The following illustration shows a computation for work using units in the metric system.

▶ ILLUSTRATION 2   We wish to find the work done in lifting a rock of mass 8 kg a distance of 4 m. We use the formula $F = Ma$, where F newtons (nt) is the force required to give a mass of M kilograms an acceleration of a meters per second per second. The force in this case is the force of gravity and the acceleration is that due to gravity, which is 9.81 m/sec^2. The mass is 8 kg. Therefore $M = 8$, $a = 9.81$, and

$$F = 8(9.81)$$
$$= 78.48$$

Thus we wish to find the work done by a force of 78.48 nt and a displacement of 4 m. If W joules is the work,

$$W = (78.48)(4)$$
$$= 313.92$$

Hence the work done is 313.92 joules.   ◀

We now consider the work done by a variable force, which is a function of the position of the object on which the force is acting. We wish to define what is meant by the term "work" in such a case.

Suppose that $f(x)$, where f is continuous on $[a, b]$, is the number of units in the force acting in the direction of motion on an object as it moves to the right along the x axis from point a to point b. Let Δ be a partition of the closed interval $[a, b]$:

$$a = x_0 < x_1 < x_2 < \ldots < x_{n-1} < x_n = b$$

The ith subinterval is $[x_{i-1}, x_i]$; and if x_{i-1} is close to x_i, the force is almost constant in this subinterval. If we assume that the force is constant in the ith subinterval and if ξ_i is any point such that $x_{i-1} \leq \xi_i \leq x_i$, then if $\Delta_i W$ is the number of units of work done on the object as it moves from the point x_{i-1} to the point x_i, from formula (1) we have

$$\Delta_i W = f(\xi_i)(x_i - x_{i-1})$$

Replacing $x_i - x_{i-1}$ by $\Delta_i x$ we have

$$\Delta_i W = f(\xi_i) \Delta_i x$$

$$\sum_{i=1}^{n} \Delta_i W = \sum_{i=1}^{n} f(\xi_i) \Delta_i x \tag{2}$$

The smaller we take the norm of the partition Δ, the larger n will be and the closer the Riemann sum in (2) will be to what we intuitively think of as the measure of the total work done. We therefore define the measure of the total work as the limit of the Riemann sum in (2).

5.4.1 DEFINITION   Let the function f be continuous on the closed interval $[a, b]$ and $f(x)$ be the number of units in the force acting on an object at the point x on the x axis. Then if W units is the **work** done by the force as the object moves from a to b, W is given by

$$W = \lim_{\|\Delta\| \to 0} \sum_{i=1}^{n} f(\xi_i) \Delta_i x$$

$$= \int_a^b f(x) \, dx \tag{3}$$

EXAMPLE 1

A particle is moving along the x axis under the action of a force of $f(x)$ lb when the particle is x ft from the origin. If $f(x) = x^2 + 4$, find the work done as the particle moves from the point where $x = 2$ to the point where $x = 4$.

SOLUTION

We take a partition of the closed interval $[2, 4]$. If W ft-lb is the work done as the particle moves from the point where $x = 2$ to the point where $x = 4$, then from Definition 5.4.1,

$$W = \lim_{\|\Delta\| \to 0} \sum_{i=1}^{n} f(\xi_i) \Delta_i x$$

$$= \int_2^4 f(x) \, dx$$

$$= \int_2^4 (x^2 + 4) \, dx$$

$$= \frac{x^3}{3} + 4x \Big]_2^4$$

$$= \tfrac{64}{3} + 16 - (\tfrac{8}{3} + 8)$$

$$= 26\tfrac{2}{3}$$

Therefore the work done is $26\tfrac{2}{3}$ ft-lb.

In the following example we use **Hooke's law**, which states that if a spring is stretched x units beyond its natural length, it is pulled back with a force equal to kx units, where k is a constant depending on the material and size of the spring.

EXAMPLE 2

A spring has a natural length of 14 cm. If a force of 50 dynes is required to keep the spring stretched 2 cm, how much work is done in stretching the spring from its natural length to a length of 18 cm?

SOLUTION

Place the spring along the x axis with the origin at the point where the stretching starts. See Fig. 5.4.1.

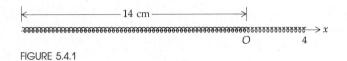

FIGURE 5.4.1

Let x be the number of centimeters the spring is stretched and $f(x)$ be the number of dynes in the force acting on the spring x cm beyond its natural length. Then, by Hooke's law,

$$f(x) = kx$$

Because $f(2) = 50$ we have

$$50 = k \cdot 2$$

$$k = 25$$

Therefore

$$f(x) = 25x$$

Consider a partition of the interval $[0, 4]$ on the x axis, let $\Delta_i x$ be the length of the ith subinterval $[x_{i-1}, x_i]$, and let ξ_i be any point in the ith subinterval. If W ergs (dyne-centimeters) is the work done in stretching the spring from its natural length of 14 cm to a length of 18 cm,

$$W = \lim_{\|\Delta\| \to 0} \sum_{i=1}^{n} f(\xi_i) \Delta_i x$$

$$= \int_0^4 f(x)\, dx$$

$$= \int_0^4 25x\, dx$$

$$= \tfrac{25}{2} x^2 \Big]_0^4$$

$$= 200$$

Therefore the work done in stretching the spring is 200 ergs.

EXAMPLE 3

A water tank in the form of an inverted right-circular cone is 20 ft across the top and 15 ft deep. If the surface of the water is 5 ft below the top of the tank, find the work done in pumping the water to the top of the tank.

SOLUTION

Refer to Fig. 5.4.2. The positive x axis is chosen in the downward direction because the motion is vertical. Take the origin at the top of the tank. We consider a partition of the closed interval $[5, 15]$ on the x axis and let ξ_i be

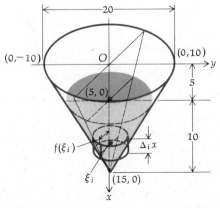

FIGURE 5.4.2

any point in the ith subinterval $[x_{i-1}, x_i]$. An element of volume is a circular disk having thickness $\Delta_i x$ ft and radius $f(\xi_i)$ ft, where the function f is determined by an equation of the line through the points $(0, 10)$ and $(15, 0)$ in the form $y = f(x)$. The number of cubic feet in the volume of this element is given by $\Delta_i V = \pi[f(\xi_i)]^2 \Delta_i x$. If w is the number of pounds in the weight of 1 ft^3 of water, then the number of pounds in the weight of this element is $w\pi[f(\xi_i)]^2 \Delta_i x$, which is the force acting on the element. If x_{i-1} is close to x_i, then the distance through which this element moves is approximately ξ_i ft. Thus if $\Delta_i W$ ft-lb is the work done in pumping the element to the top of the tank, $\Delta_i W$ is approximately $(w\pi[f(\xi_i)]^2 \Delta_i x) \cdot \xi_i$; so if W is the number of foot-pounds of work done,

$$W = \lim_{\|\Delta\| \to 0} \sum_{i=1}^{n} w\pi[f(\xi_i)]^2 \cdot \xi_i \Delta_i x$$

$$= w\pi \int_{5}^{15} [f(x)]^2 x \, dx$$

To determine $f(x)$ we find an equation of the line through the points $(15, 0)$ and $(0, 10)$ by using the intercept form:

$$\frac{x}{15} + \frac{y}{10} = 1 \quad \Leftrightarrow \quad y = -\tfrac{2}{3}x + 10$$

Therefore $f(x) = -\tfrac{2}{3}x + 10$, and

$$W = w\pi \int_{5}^{15} (-\tfrac{2}{3}x + 10)^2 x \, dx$$

$$= w\pi \int_{5}^{15} (\tfrac{4}{9}x^3 - \tfrac{40}{3}x^2 + 100x) \, dx$$

$$= w\pi \left[\tfrac{1}{9}x^4 - \tfrac{40}{9}x^3 + 50x^2 \right]_{5}^{15}$$

$$= \frac{10{,}000\pi w}{9}$$

Therefore the work done is $10{,}000\pi w/9$ ft-lb.

EXAMPLE 4

As a water tank is being raised, water spills out at a constant rate of 2 ft^3 per foot of rise. If the weight of the tank is 200 lb and it originally contains 1000 ft^3 of water, find the work done in raising the tank 20 ft.

SOLUTION

Refer to Fig. 5.4.3. Here we are taking the origin at the starting point of the bottom of the tank and the positive x axis is in the upward direction because the motion is vertically upward from O. We consider a partition of the closed interval $[0, 20]$ on the x axis. Let ξ_i be any point in the ith subinterval $[x_{i-1}, x_i]$. When the bottom of the tank is at ξ_i, the number of cubic feet of water in the tank is $1000 - 2\xi_i$. If w lb is the weight of 1 ft^3 of water, then the number of pounds in the weight of the tank and its contents when

FIGURE 5.4.3

it is at ξ_i is $[200 + w(1000 - 2\xi_i)]$, which is the force acting on the tank. If $\Delta_i W$ ft-lb is the work done in raising the tank through the ith subinterval, then $\Delta_i W$ is approximately $[200 + w(1000 - 2\xi_i)]\Delta_i x$. We use the terminology "approximately" because we are assuming that the amount of water in the tank is constant throughout this subinterval. If W ft-lb is the work done in raising the tank 20 ft, we have

$$W = \lim_{\|\Delta\| \to 0} \sum_{i=1}^{n} [200 + w(1000 - 2\xi_i)] \Delta_i x$$

$$= \int_0^{20} [200 + w(1000 - 2x)]\, dx$$

$$= \Big[200x + w(1000x - x^2)\Big]_0^{20}$$

$$= 4000 + 19,600w$$

Therefore the work done is $(4000 + 19,600w)$ ft-lb.

EXERCISES 5.4

In Exercises 1 and 2, a particle is moving along the x axis under the action of a force of $f(x)$ lb when the particle is x ft from the origin. Find the work done as the particle moves from the point where $x = a$ to the point where $x = b$.

1. $f(x) = (2x + 1)^2$; $a = 1$; $b = 3$ $\int_1^3 (2x+1)^2 dx$

2. $f(x) = x^2\sqrt{x^3 + 1}$; $a = 0$; $b = 2$

In Exercises 3 and 4, a particle is moving along the x axis under the action of a force of $f(x)$ newtons when the particle is x meters from the origin. Find the work done as the particle moves from the point where $x = a$ to the point where $x = b$.

3. $f(x) = x\sqrt{x + 1}$; $a = 3$; $b = 8$ $\int_3^8 x\sqrt{x+1}\, dx$

4. $f(x) = (4x - 1)^2$; $a = 1$; $b = 4$

5. An object is moving along the x axis under the action of a force of $f(x)$ dynes when the object is x cm from the origin. If 96 ergs is the work done in moving the object from the origin to the point where $x = K$, and $f(x) = 2x - 3$, find K if $K > 0$. $96 = \int_0^K (2x-3)\, dx$

6. Solve Exercise 5 if 90 ergs is the work done, and $f(x) = 4x - 3$.

7. A spring has a natural length of 8 in. If a force of 20 lb stretches the spring $\frac{1}{2}$ in., find the work done in stretching the spring from 8 in. to 11 in. $F = kx$ $k = 40$ $20 = k\frac{1}{2}$ $F = 40x$ $\int_0^3 40x\, dx$

8. A spring has a natural length of 10 in., and a 30-lb force stretches it to $11\frac{1}{2}$ in. (a) Find the work done in stretching the spring from 10 in. to 12 in. (b) Find the work done in stretching the spring from 12 in. to 14 in.

9. A force of 8 nt stretches a spring of natural length 4 m to an additional 50 cm. Find the work done in stretching the spring from its natural length to 5 m. $F = kx$ $k = 16$ $8 = k\frac{1}{2}$ $F = 16x$ $\int_0^1 16x\, dx$

10. A force of 500 dynes stretches a spring from its natural length of 20 cm to a length of 24 cm. Find the work done in stretching the spring from its natural length to a length of 28 cm.

11. A spring has a natural length of 12 cm. A force of 600 dynes compresses the spring to 10 cm. Find the work done in compressing the spring from 12 cm to 9 cm. Hooke's law holds for compression as well as for extension. $F = kx$ $F = 300x$ $600 = k2$ $\frac{600}{2} = k$ $\int 300x$

12. A spring has a natural length of 6 in. A 1200-lb force compresses it to $5\frac{1}{2}$ in. Find the work done in compressing it from 6 in. to $4\frac{1}{2}$ in.

13. A tank full of water is in the form of a rectangular parallelepiped 5 ft deep, 15 ft wide, and 25 ft long. Find the work required to pump the water in the tank up to a level 1 ft above the surface of the tank.

14. A trough full of water is 10 ft long, and its cross section is in the shape of an isosceles triangle 2 ft wide across the top and 2 ft high. How much work is done in pumping all the water out of the trough over the top?

15. A hemispherical tank, placed so that the top is a circular region of radius 6 ft, is filled with water to a depth of 4 ft. Find the work done in pumping the water to the top of the tank.

16. A right-circular cylindrical tank with a depth of 12 ft and a radius of 4 ft is half full of oil weighing 60 lb/ft^3. Find the work done in pumping the oil to a height 6 ft above the tank.

17. A cable 200 ft long and weighing 4 lb/ft is hanging vertically down a well. If a weight of 100 lb is suspended from the lower end of the cable, find the work done in pulling the cable and weight to the top of the well.

18. A bucket weighing 20 lb containing 60 lb of sand is attached to the lower end of a chain 100 ft long and weighing 10 lb that is hanging in a deep well. Find the work done in raising the bucket to the top of the well.

19. Solve Exercise 18 if the sand is leaking out of the bucket at a constant rate and has all leaked out just as soon as the bucket is at the top of the well.

20. As a flour sack is being raised a distance of 9 ft, flour leaks out at such a rate that the number of pounds lost is directly proportional to the square root of the distance traveled. If the sack originally contained 60 lb of flour and it loses a total of 12 lb while being raised the 9 ft, find the work done in raising the sack.

21. A tank in the form of an inverted right-circular cone is 8 m across the top and 10 m deep. The tank is filled to a height of 8 m with oil weighing 950 kg/m^3. Find the work done in pumping the oil to the top of the tank. (*Hint:* The number of newtons of force necessary to lift an element is the product of the number of kilograms of mass (the same as the number of kilograms of weight) and 9.81, the number of m/sec^2 in the acceleration due to gravity.)

22. If the tank of Exercise 21 is filled to a height of 9 m with water, find the work done in pumping the water to the top of the tank. The weight of 1 m^3 of water at 0°C is 1000 kg. (See hint for Exercise 21.)

23. A right-circular cylindrical tank with a depth of 10 m and a radius of 5 m is half filled with water. Find the work necessary to pump the water to the top of the tank. The weight of 1 m^3 of water at 0°C is 1000 kg. (See hint for Exercise 21.)

24. If in Exercise 22 only half of the water is to be pumped to the top of the tank, find the work.

25. A 1-horsepower motor can do 550 ft-lb of work per second. If a 0.1 hp motor is used to pump water from a full tank in the shape of a rectangular parallelepiped 2 ft deep, 2 ft wide, and 6 ft long to a point 5 ft above the top of the tank, how long will it take?

26. A meteorite is a miles from the center of the earth and falls to the surface of the earth. The force of gravity is inversely proportional to the square of the distance of a body from the center of the earth. Find the work done by gravity if the weight of the meteorite is w lb at the surface of the earth. Let R miles be the radius of the earth.

27. A tank in the form of a rectangular parallelepiped 6 ft deep, 4 ft wide, and 12 ft long is full of oil weighing 50 lb/ft^3. When one-third of the work necessary to pump the oil to the top of the tank has been done, find by how much the surface of the oil is lowered.

28. A cylindrical tank 10 ft high and 5 ft in radius is standing on a platform 50 ft high. Find the depth of the water when one-half of the work required to fill the tank from the ground level through a pipe in the bottom has been done.

5.5
LENGTH OF ARC OF THE
GRAPH OF A FUNCTION

In our discussion of length of arc we use the word "length" in place of the words "measure of the length." It is understood then that the length of an arc is a number without any units of measurement attached to it.

Let the function f be continuous on the closed interval $[a, b]$. Consider the graph of this function defined by the equation $y = f(x)$, of which a sketch is shown in Fig. 5.5.1. The portion of the curve from the point $A(a, f(a))$ to the point $B(b, f(b))$ is called an *arc*. We wish to assign a number to what we intuitively think of as the length of such an arc. If the arc is a line segment from the point (x_1, y_1) to the point (x_2, y_2), we know from the formula for the distance between two points that its length is given by $\sqrt{(x_1 - x_2)^2 + (y_1 - y_2)^2}$. We use this formula for defining the length of an arc in general. Recall from geometry that the circumference of a circle is defined as the limit of the perimeters of regular polygons inscribed in the circle. For other curves we proceed in a similar way.

Let Δ be a partition of the closed interval $[a, b]$ formed by dividing the interval into n subintervals by choosing any $n - 1$ intermediate numbers

FIGURE 5.5.1

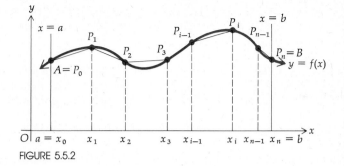

FIGURE 5.5.2

between a and b. Let $x_0 = a$ and $x_n = b$, and let $x_1, x_2, x_3, \ldots, x_{n-1}$ be the intermediate numbers so that $x_0 < x_1 < x_2 < \ldots < x_{n-1} < x_n$. Then the ith subinterval is $[x_{i-1}, x_i]$; its length, denoted by $\Delta_i x$, is $x_i - x_{i-1}$, where $i = 1, 2, 3, \ldots, n$. Then if $\|\Delta\|$ is the norm of the partition Δ, each $\Delta_i x \leq \|\Delta\|$.

Associated with each point $(x_i, 0)$ on the x axis is a point $P_i(x_i, f(x_i))$ on the curve. Draw a line segment from each point P_{i-1} to the next point P_i, as shown in Fig. 5.5.2. The length of the line segment from P_{i-1} to P_i is denoted by $\overline{|P_{i-1}P_i|}$ and is given by the distance formula

$$\overline{|P_{i-1}P_i|} = \sqrt{(x_i - x_{i-1})^2 + (y_i - y_{i-1})^2} \tag{1}$$

The sum of the lengths of the line segments is

$$\overline{|P_0P_1|} + \overline{|P_1P_2|} + \overline{|P_2P_3|} + \ldots + \overline{|P_{i-1}P_i|} + \ldots + \overline{|P_{n-1}P_n|}$$

which can be written with sigma notation as

$$\sum_{i=1}^{n} \overline{|P_{i-1}P_i|} \tag{2}$$

It seems plausible that if n is sufficiently large, the sum in (2) will be "close to" what we would intuitively think of as the length of the arc AB. So we define the length of the arc as the limit of the sum in (2) as the norm of Δ approaches zero, in which case n increases without bound. We have, then, the following definition.

5.5.1 DEFINITION   Suppose the function f is continuous on the closed interval $[a, b]$. Further suppose that there exists a number L having the following property:

For any $\epsilon > 0$ there is a $\delta > 0$ such that for every partition Δ of the interval $[a, b]$ it is true that

$$\text{if} \quad \|\Delta\| < \delta \quad \text{then} \quad \left| \sum_{i=1}^{n} \overline{|P_{i-1}P_i|} - L \right| < \epsilon$$

Then we write

$$L = \lim_{\|\Delta\| \to 0} \sum_{i=1}^{n} \overline{|P_{i-1}P_i|} \tag{3}$$

and L is called the **length of the arc** of the curve $y = f(x)$ from the point $A(a, f(a))$ to the point $B(b, f(b))$.

If the limit in (3) exists, the arc is said to be **rectifiable**.

We derive a formula for finding the length L of an arc that is rectifiable. The derivation requires that the derivative of f be continuous on $[a, b]$; such a function is said to be *smooth* on $[a, b]$.

5.5.2 DEFINITION   A function f is said to be **smooth** on an interval I if f' is continuous on I.

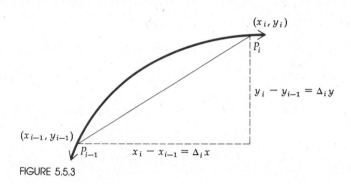

FIGURE 5.5.3

Refer now to Fig. 5.5.3. If P_{i-1} has coordinates (x_{i-1}, y_{i-1}) and P_i has coordinates (x_i, y_i), then the length of the chord $P_{i-1}P_i$ is given by formula (1). Letting $x_i - x_{i-1} = \Delta_i x$ and $y_i - y_{i-1} = \Delta_i y$ we have

$$\left|\overline{P_{i-1}P_i}\right| = \sqrt{(\Delta_i x)^2 + (\Delta_i y)^2} \tag{4}$$

or, equivalently, because $\Delta_i x \neq 0$,

$$\left|\overline{P_{i-1}P_i}\right| = \sqrt{1 + \left(\frac{\Delta_i y}{\Delta_i x}\right)^2} \, (\Delta_i x) \tag{5}$$

Because we required that f' be continuous on $[a, b]$, the hypothesis of the mean-value theorem (3.3.2) is satisfied by f; so there is a number z_i in the open interval (x_{i-1}, x_i) such that

$$f(x_i) - f(x_{i-1}) = f'(z_i)(x_i - x_{i-1}) \tag{6}$$

Because $\Delta_i y = f(x_i) - f(x_{i-1})$ and $\Delta_i x = x_i - x_{i-1}$, from (6) we have

$$\frac{\Delta_i y}{\Delta_i x} = f'(z_i) \tag{7}$$

Substituting from (7) into (5) we get

$$\left|\overline{P_{i-1}P_i}\right| = \sqrt{1 + [f'(z_i)]^2} \, \Delta_i x \tag{8}$$

For each i from 1 to n there is an equation of the form (8).

$$\sum_{i=1}^{n} \left|\overline{P_{i-1}P_i}\right| = \sum_{i=1}^{n} \sqrt{1 + [f'(z_i)]^2} \, \Delta_i x \tag{9}$$

Taking the limit on both sides of (9) as $\|\Delta\|$ approaches zero we obtain

$$\lim_{\|\Delta\|\to 0} \sum_{i=1}^{n} |\overline{P_{i-1}P_i}| = \lim_{\|\Delta\|\to 0} \sum_{i=1}^{n} \sqrt{1 + [f'(z_i)]^2}\, \Delta_i x \tag{10}$$

if this limit exists.

To show that the limit on the right side of (10) exists, let g be the function defined by

$$g(x) = \sqrt{1 + [f'(x)]^2}$$

Because f' is continuous on $[a,b]$, g is continuous on $[a,b]$. Therefore, because $x_{i-1} < z_i < x_i$, for $i = 1, 2, \ldots, n$,

$$\lim_{\|\Delta\|\to 0} \sum_{i=1}^{n} \sqrt{1 + [f'(z_i)]^2}\, \Delta_i x = \lim_{\|\Delta\|\to 0} \sum_{i=1}^{n} g(z_i)\, \Delta_i x$$

$$\Leftrightarrow \quad \lim_{\|\Delta\|\to 0} \sum_{i=1}^{n} \sqrt{1 + [f'(z_i)]^2}\, \Delta_i x = \int_a^b g(x)\, dx \tag{11}$$

Because $g(x) = \sqrt{1 + [f'(x)]^2}$, from (11) we have

$$\lim_{\|\Delta\|\to 0} \sum_{i=1}^{n} \sqrt{1 + [f'(z_i)]^2}\, \Delta_i x = \int_a^b \sqrt{1 + [f'(x)]^2}\, dx \tag{12}$$

Substituting from (12) into (10) we get

$$\lim_{\|\Delta\|\to 0} \sum_{i=1}^{n} |\overline{P_{i-1}P_i}| = \int_a^b \sqrt{1 + [f'(x)]^2}\, dx \tag{13}$$

Then from (3) and (13) we obtain

$$L = \int_a^b \sqrt{1 + [f'(x)]^2}\, dx \tag{14}$$

In this way we have proved the following theorem.

5.5.3 THEOREM   If the function f and its derivative f' are continuous on the closed interval $[a,b]$, then the length of arc of the curve $y = f(x)$ from the point $(a, f(a))$ to the point $(b, f(b))$ is given by

$$L = \int_a^b \sqrt{1 + [f'(x)]^2}\, dx$$

We also have the following theorem, which gives the length of the arc of a curve when x is expressed as a function of y.

5.5.4 THEOREM   If the function F and its derivative F' are continuous on the closed interval $[c,d]$, then the length of arc of the curve $x = F(y)$ from the point $(F(c), c)$ to the point $(F(d), d)$ is given by

$$L = \int_c^d \sqrt{1 + [F'(y)]^2}\, dy$$

The proof of Theorem 5.5.4 is identical to that of Theorem 5.5.3; here we interchange x and y as well as the functions f and F.

The definite integral obtained when applying Theorem 5.5.3 or Theorem 5.5.4 is often difficult to evaluate. Because our techniques of integration have so far been limited to integration of powers and some trigonometric functions, we are further restricted in finding equations of curves for which we can evaluate the resulting definite integrals to find the length of an arc.

FIGURE 5.5.4

EXAMPLE 1

Find the length of the arc of the curve $y = x^{2/3}$ from the point $(1, 1)$ to $(8, 4)$ by using Theorem 5.5.3.

SOLUTION

See Fig. 5.5.4. Because $f(x) = x^{2/3}$, $f'(x) = \frac{2}{3}x^{-1/3}$. From Theorem 5.5.3,

$$L = \int_1^8 \sqrt{1 + \frac{4}{9x^{2/3}}}\, dx$$

$$= \frac{1}{3}\int_1^8 \frac{\sqrt{9x^{2/3} + 4}}{x^{1/3}}\, dx$$

To evaluate this definite integral let $u = 9x^{2/3} + 4$; then $du = 6x^{-1/3}\, dx$. When $x = 1$, $u = 13$; when $x = 8$, $u = 40$. Therefore

$$L = \tfrac{1}{18}\int_{13}^{40} u^{1/2}\, du$$

$$= \tfrac{1}{18}\left[\tfrac{2}{3}u^{3/2}\right]_{13}^{40}$$

$$= \tfrac{1}{27}(40^{3/2} - 13^{3/2})$$

$$\approx 7.6$$

EXAMPLE 2

Find the length of the arc in Example 1 by using Theorem 5.5.4.

SOLUTION

Because $y = x^{2/3}$ and $x > 0$, we solve for x and obtain $x = y^{3/2}$. Letting $F(y) = y^{3/2}$ we have

$$F'(y) = \tfrac{3}{2}y^{1/2}$$

Then, from Theorem 5.5.4,

$$L = \int_1^4 \sqrt{1 + \tfrac{9}{4}y}\, dy$$

$$= \tfrac{1}{2}\int_1^4 \sqrt{4 + 9y}\, dy$$

$$= \tfrac{1}{18}\left[\tfrac{2}{3}(4 + 9y)^{3/2}\right]_1^4$$

$$= \tfrac{1}{27}(40^{3/2} - 13^{3/2})$$

$$\approx 7.6$$

If f' is continuous on $[a,b]$, the definite integral $\int_a^x \sqrt{1 + [f'(t)]^2}\, dt$ is a function of x; and it gives the length of the arc of the curve $y = f(x)$ from the point $(a, f(a))$ to the point $(x, f(x))$, where x is any number in the closed interval $[a, b]$. Let s denote the length of this arc; thus s is a function of x, and

$$s(x) = \int_a^x \sqrt{1 + [f'(t)]^2}\, dt$$

From Theorem 4.8.1,

$$s'(x) = \sqrt{1 + [f'(x)]^2}$$

or, because $s'(x) = ds/dx$ and $f'(x) = dy/dx$,

$$\frac{ds}{dx} = \sqrt{1 + \left(\frac{dy}{dx}\right)^2}$$

Multiplying by dx we obtain

$$ds = \sqrt{1 + \left(\frac{dy}{dx}\right)^2}\, dx \tag{15}$$

Similarly, for the length of arc of the curve $x = g(y)$ from $(g(c), c)$ to $(g(y), y)$,

$$ds = \sqrt{\left(\frac{dx}{dy}\right)^2 + 1}\, dy \tag{16}$$

Squaring on both sides of either (15) or (16) gives

$$(ds)^2 = (dx)^2 + (dy)^2 \tag{17}$$

From (17) we get the geometric interpretation of ds, which is shown in Fig. 5.5.5. In the figure, line T is tangent to the curve $y = f(x)$ at the point P. $|\overline{PM}| = \Delta x = dx$; $|\overline{MQ}| = \Delta y$; $|\overline{MR}| = dy$; $|\overline{PR}| = ds$; the length of arc PQ is Δs.

FIGURE 5.5.5

EXERCISES 5.5

1. Compute the length of the segment of the line $y = 3x$ from the point $(1, 3)$ to the point $(2, 6)$ by three methods: (a) Use the distance formula; (b) use Theorem 5.3.3; (c) use Theorem 5.5.4.

2. Compute the length of the segment of the line $x + 3y = 4$ from the point $(-2, 2)$ to the point $(4, 0)$ by three methods: (a) Use the distance formula; (b) use Theorem 5.3.3; (c) use Theorem 5.5.4.

3. Compute the length of the segment of the line $4x + 9y = 36$ between its x and y intercepts by three methods: (a) Use the Pythagorean theorem; (b) use Theorem 5.3.3; (c) use Theorem 5.5.4.

$$L = \int_1^2 \sqrt{1 + \left(\frac{y^3}{2} - \frac{y^{-3}}{2}\right)^2} = \int_1^2 \sqrt{1 + \frac{x^6}{4} - \frac{1}{2} + \frac{x^{-6}}{4}}\, dx \cdots = \frac{1}{2}\int_1^2 y^3 + x^{-3}\, dy \qquad \frac{33}{16}$$

4. Follow the instructions of Exercise 3 for the line $5x - 2y = 10$.

5. Find the length of the arc of the curve $9y^2 = 4x^3$ from the origin to the point $(3, 2\sqrt{3})$. $y = \frac{4}{9} N^{3/2}$ $f'(x) = x^{1/2}$

6. Find the length of the arc of the curve $x^2 = (2y + 3)^3$ from $(1, -1)$ to $(7\sqrt{7}, 2)$. $L = \int_c^3 \sqrt{1+x}$ $\frac{14}{3}$

7. Find the length of the arc of the curve $8y = x^4 + 2x^{-2}$ from the point where $x = 1$ to the point where $x = 2$.

8. Use Theorem 5.5.3 to find the length of the arc of the curve $y^3 = 8x^2$ from the point $(1, 2)$ to the point $(27, 18)$. $y = \frac{2x^{2/3}}{-1/3}$

9. Solve Exercise 8 by using Theorem 5.5.4. $\frac{dy}{dx} = \frac{4}{3}x^{-1/3}$

10. Find the length of the arc of the curve $y = \frac{2}{3}(x - 5)^{3/2}$ from the point where $x = 6$ to the point where $x = 8$.

11. Find the length of the arc of the curve $y = \frac{1}{3}(x^2 + 2)^{3/2}$ from the point where $x = 0$ to the point where $x = 3$.

12. Find the length of the arc of the curve $6xy = y^4 + 3$ from the point where $y = 1$ to the point where $y = 2$.

13. Find the length of the arc of the curve $y = \frac{1}{3}\sqrt{x}(3x - 1)$ from the point where $x = 1$ to the point where $x = 4$.

14. Find the length of the arc of the curve $y = \frac{1}{6}x^3 + \frac{1}{2}x^{-1}$ from the point $(2, \frac{19}{12})$ to the point $(5, \frac{314}{15})$.

15. Find the length of the arc of the curve $x^{2/3} + y^{2/3} = 1$ in the first quadrant from the point where $x = \frac{1}{8}$ to the point where $x = 1$.

16. Find the length of the arc of the curve $x^{2/3} + y^{2/3} = a^{2/3}$ (a is a constant, $a > 1$) in the first quadrant from the point where $x = 1$ to the point where $x = a$.

17. Find the length of the curve $(x/a)^{2/3} + (y/b)^{2/3} = 1$ in the first quadrant from the point where $x = \frac{1}{8}a$ to the point where $x = a$.

18. Find the length of the curve $9y^2 = x^2(2x + 3)$ in the second quadrant from the point where $x = -1$ to the point where $x = 0$.

19. Find the length of the curve $9y^2 = x(x - 3)^2$ in the first quadrant from the point where $x = 1$ to the point where $x = 3$.

20. Find the length of the curve $9y^2 = 4(1 + x^2)^3$ in the first quadrant from the point where $x = 0$ to the point where $x = 2\sqrt{2}$.

21. If $f(x) = \int_0^x \sqrt{\cos t}\, dt$, find the length of the arc of the graph of f from the point where $x = 0$ to the point where $x = \frac{1}{2}\pi$. (*Hint:* Use Theorem 4.8.1 and the identity $\cos^2 \frac{1}{2}x = \frac{1}{2}(1 + \cos x)$.)

22. If $f(x) = \int_0^x \sqrt{\sin t}\, dt$, find the length of the arc of the graph of f from the point where $x = 0$ to the point where $x = \frac{1}{2}\pi$. (*Hint:* Use the hint for Exercise 21 and the identity $\sin x = \cos(\frac{1}{2}\pi - x)$.)

5.6
CENTER OF MASS OF
A ROD

In Section 4.7 you learned that if a function f is continuous on the closed interval $[a, b]$, the average value of f on $[a, b]$ is given by

$$\frac{\int_a^b f(x)\, dx}{b - a}$$

An important application of the average value of a function occurs in physics in connection with the concept of *center of mass*.

    To arrive at a definition of *mass*, consider a particle that is set into motion along an axis by a force exerted on the particle. Assume that all velocities are small compared with the velocity of light. So long as the force is acting on the particle, the velocity of the particle is increasing; that is, the particle is accelerating. The ratio of the force to the acceleration is constant regardless of the magnitude of the force, and this constant ratio is called the **mass** of the particle. Therefore, if the force is F units, the acceleration is a units, and the mass is M units, then

$$M = \frac{F}{a}$$

We shall be measuring force, mass, and acceleration in units of both the British engineering system and the metric system. We now discuss these units.

In the British engineering system we first have a unit of force that is 1 lb and a unit of acceleration that is 1 ft/sec². The unit of mass is then defined as the mass of a particle whose acceleration is 1 ft/sec² when the particle is subjected to a force of 1 lb. This unit of mass is called 1 *slug*. Hence

$$1 \text{ slug} = \frac{1 \text{ lb}}{1 \text{ ft/sec}^2}$$

▶ ILLUSTRATION 1   The acceleration of a certain particle moving on a horizontal line is 10 ft/sec² when the force is 30 lb. Therefore the mass of the particle is

$$\frac{30 \text{ lb}}{10 \text{ ft/sec}^2} = \frac{3 \text{ lb}}{1 \text{ ft/sec}^2}$$

Thus for every 1 ft/sec² of acceleration, a force of 3 lb must be exerted on the particle. The mass of the particle is then 3 slugs.   ◀

In the meter-kilogram-second (mks) system we have the standard unit of mass, the kilogram. The unit of acceleration in the mks system is 1 m/sec². Therefore the unit of force is the force that provides a mass of 1 kg an acceleration of 1 m/sec². This force is called 1 *newton*, written as 1 nt.

▶ ILLUSTRATION 2   A particle of mass 6 kg is subjected to a constant horizontal force of 3 nt. The acceleration of the particle is obtained by dividing the force by the mass, and so it is

$$\frac{3 \text{ nt}}{6 \text{ kg}} = 0.5 \text{ m/sec}^2$$

With symbols, if M kg is the mass, F nt is the force, and a m/sec² is the acceleration, then because $a = F/M$, $F = 3$, and $M = 6$, $a = 0.5$.   ◀

In the mks system, the acceleration due to gravity near the surface of the earth is 9.81 m/sec². If M kg is the mass of an object, and if F nt is the force on the object due to gravity near the surface of the earth, then

$$F = 9.81M$$

In the centimeter-gram-second (cgs) system, the standard unit of mass is the gram, and 1 g = 0.001 kg. The unit of acceleration in the cgs system is 1 cm/sec². Therefore the unit of force is the force that gives a particle of mass 1 g an acceleration of 1 cm/sec². This force is called 1 *dyne*. Because

$$1 \text{ kg} = 10^3 \text{ g} \quad \text{and} \quad 1 \text{ m/sec}^2 = 10^2 \text{ cm/sec}^2$$

then

$$1 \text{ nt} = 10^5 \text{ dynes}$$

We summarize the discussion of units in the British engineering system and the metric system in Table 5.6.1. This table gives the units of force, mass, and acceleration in each system.

Table 5.6.1

System of Units	Force	Mass	Acceleration
British engineering	pound	slug	ft/sec^2
mks	newton	kilogram	m/sec^2
cgs	dyne	gram	cm/sec^2

FIGURE 5.6.1

Consider now a horizontal rod, of negligible weight and thickness, placed on the x axis. On the rod is a system of n particles located at points $x_1, x_2, \ldots, x_n$. The ith particle ($i = 1, 2, \ldots, n$) is at a directed distance x_i m from the origin and its mass is m_i kg. See Fig. 5.6.1. The number of kilograms in the total mass of the system is $\sum_{i=1}^{n} m_i$. We define the number of kilogram-meters in the *moment of mass* of the ith particle with respect to the origin as $m_i x_i$. The **moment of mass** for the system is defined as the sum of the moments of mass of all the particles. Hence, if M_0 kg-m is the moment of mass of the system with respect to the origin, then

$$M_0 = \sum_{i=1}^{n} m_i x_i$$

If the measurement of the distance is in feet and of the mass in slugs, then the moment of mass is measured in slug-ft.

We wish to find a point $\bar{x}$ such that if the total mass of the system were concentrated there, its moment of mass with respect to the origin would be equal to the moment of mass of the system with respect to the origin. Then $\bar{x}$ must satisfy the equation

$$\bar{x} \sum_{i=1}^{n} m_i = \sum_{i=1}^{n} m_i x_i$$

$$\bar{x} = \frac{\sum_{i=1}^{n} m_i x_i}{\sum_{i=1}^{n} m_i} \tag{1}$$

The point $\bar{x}$ is called the **center of mass** of the system, and it is the point where the system will balance. The position of the center of mass is independent of the position of the origin; that is, the location of the center of mass relative to the positions of the particles does not change when the origin is changed. The center of mass is important because the behavior of an entire system of particles can be described by the behavior of the center of mass of the system.

EXAMPLE 1

Given four particles of masses 2, 3, 1, and 5 kg located on the x axis at the points having coordinates 5, 2, -3, and -4, respectively, where distance measurement is in meters, find the center of mass of this system.

SOLUTION

If $\bar{x}$ is the coordinate of the center of mass, we have from formula (1)

$$\bar{x} = \frac{2(5) + 3(2) + (1)(-3) + 5(-4)}{2 + 3 + 1 + 5}$$

$$= -\frac{7}{11}$$

Thus the center of mass is $\frac{7}{11}$ m to the left of the origin.

The preceding discussion is now extended to a rigid horizontal rod having a continuously distributed mass. The rod is said to be **homogeneous** if it has constant linear density, that is, if its mass is directly proportional to its length. In other words, if the segment of the rod whose length is $\Delta_i x$ meters has a mass of $\Delta_i m$ kg, and $\Delta_i m = k\,\Delta_i x$, then the rod is homogeneous. The number k is a constant, and k kg/m is called the **linear density** of the rod.

FIGURE 5.6.2

Suppose that we have a nonhomogeneous rod, in which case the linear density varies along the rod. Let L meters be the length of the rod, and place the rod on the x axis so the left endpoint of the rod is at the origin and the right endpoint is at L. See Fig. 5.6.2. The linear density at any point x on the rod is $\rho(x)$ kg/m, where ρ is continuous on $[0, L]$. To find the total mass of the rod we consider a partition Δ of the closed interval $[0, L]$ into n subintervals. The ith subinterval is $[x_{i-1}, x_i]$, and its length is $\Delta_i x$ meters. If ξ_i is any point in $[x_{i-1}, x_i]$, an approximation to the mass of the part of the rod contained in the ith subinterval is $\Delta_i m$ kg, where

$$\Delta_i m = \rho(\xi_i)\,\Delta_i x$$

The number of kilograms in the total mass of the rod is approximated by

$$\sum_{i=1}^{n} \Delta_i m = \sum_{i=1}^{n} \rho(\xi_i)\,\Delta_i x \tag{2}$$

The smaller we take the norm of the partition Δ, the closer the Riemann sum in (2) will be to what we intuitively think of as the measure of the mass of the rod, and so we define the measure of the mass as the limit of the Riemann sum in (2).

5.6.1 DEFINITION   A rod of length L meters has its left endpoint at the origin. If the number of kilograms per meter in the linear density at a point x meters from the origin is $\rho(x)$, where ρ is continuous on $[0, L]$, then the total **mass** of the rod is M kg, where

$$M = \lim_{\|\Delta\| \to 0} \sum_{i=1}^{n} \rho(\xi_i) \Delta_i x$$

$$= \int_0^L \rho(x)\,dx \tag{3}$$

In Definition 5.6.1, if the distance is measured in feet and the mass is measured in slugs, then the density is measured in slugs per foot.

EXAMPLE 2
The density at any point of a rod 4 m long varies directly as the distance from the point to an external point in the line of the rod and 2 m from an end, where the density is 5 kg/m. Find the total mass of the rod.

SOLUTION
Figure 5.6.3 shows the rod placed on the x axis. If $\rho(x)$ is the number of kilograms per meter in the density of the rod at the point x meters from the end having the greater density, then

$$\rho(x) = c(6 - x)$$

where c is the constant of proportionality. Because $\rho(4) = 5$, then $5 = 2c$ or $c = \frac{5}{2}$. Hence $\rho(x) = \frac{5}{2}(6 - x)$. Therefore, if M kg is the total mass of the rod, we have from Definition 5.6.1

FIGURE 5.6.3

$$M = \lim_{\|\Delta\| \to 0} \sum_{i=1}^{n} \tfrac{5}{2}(6 - \xi_i) \Delta_i x$$

$$= \int_0^4 \tfrac{5}{2}(6 - x)\,dx$$

$$= \tfrac{5}{2}\left[6x - \tfrac{1}{2}x^2 \right]_0^4$$

$$= 40$$

The total mass of the rod is therefore 40 kg.

FIGURE 5.6.4

We now proceed to define the center of mass of the rod of Definition 5.6.1. However, first we must define the moment of mass of the rod with respect to the origin.

As before, place the rod on the x axis with the left endpoint at the origin and the right endpoint at L. See Fig. 5.6.4. Let Δ be a partition of $[0, L]$ into n subintervals, with the ith subinterval $[x_{i-1}, x_i]$ having length $\Delta_i x$ meters. If ξ_i is any point in $[x_{i-1}, x_i]$, an approximation to the moment of mass with respect to the origin of the part of the rod contained in the ith subinterval is $\xi_i \Delta_i m$ kg-m, where $\Delta_i m = \rho(\xi_i) \Delta_i x$. The number of kilogram-meters in the moment of mass of the entire rod is approximated by

$$\sum_{i=1}^{n} \xi_i \Delta_i m = \sum_{i=1}^{n} \xi_i \rho(\xi_i) \Delta_i x \tag{4}$$

The smaller we take the norm of the partition Δ, the closer the Riemann sum in (4) will be to what we intuitively think of as the measure of the moment of mass of the rod with respect to the origin. We have, then, the following definition.

5.6.2 DEFINITION A rod of length L meters has its left endpoint at the origin and the number of kilograms per meter in the linear density at a point x meters from the origin is $\rho(x)$, where ρ is continuous on $[0, L]$. The **moment of mass** of the rod with respect to the origin is M_0 kg-m, where

$$M_0 = \lim_{\|\Delta\| \to 0} \sum_{i=1}^{n} \xi_i \rho(\xi_i) \Delta_i x$$

$$= \int_0^L x\rho(x)\, dx \tag{5}$$

The center of mass of the rod is at the point $\bar{x}$ such that if M kg is the total mass of the rod, $\bar{x}M = M_0$. Thus, from (3) and (5),

$$\bar{x} = \frac{\displaystyle\int_0^L x\rho(x)\, dx}{\displaystyle\int_0^L \rho(x)\, dx} \tag{6}$$

EXAMPLE 3
Find the center of mass of the rod in Example 2.

SOLUTION
In Example 2, $M = 40$. From (6) with $\rho(x) = \frac{5}{2}(6 - x)$,

$$\bar{x} = \frac{\displaystyle\int_0^4 \frac{5}{2}x(6 - x)\, dx}{40}$$

$$= \tfrac{1}{16}\left[3x^2 - \tfrac{1}{3}x^3\right]_0^4$$

$$= \tfrac{5}{3}$$

Therefore the center of mass is at $\tfrac{5}{3}$ m from the end having the greater density.

▶ ILLUSTRATION 3   If a rod is of uniform density k kg/m, where k is a constant, then from formula (6),

$$\bar{x} = \frac{\displaystyle\int_0^L xk\,dx}{\displaystyle\int_0^L k\,dx}$$

$$= \frac{\dfrac{kx^2}{2}\bigg]_0^L}{kx\bigg]_0^L}$$

$$= \frac{\dfrac{kL^2}{2}}{kL}$$

$$= \frac{L}{2}$$

Thus the center of mass is at the center of the rod, as is to be expected. ◀

EXERCISES 5.6

In Exercises 1 through 4, a particle is moving on a horizontal line. Find the force exerted on the particle if it has the given mass and acceleration.

1. Mass is 50 slugs; acceleration is 5 ft/sec^2.

2. Mass is 10 kg; acceleration is 6 m/sec^2.

3. Mass is 80 g; acceleration is 50 cm/sec^2.

4. Mass is 22 slugs; acceleration is 4 ft/sec^2.

In Exercises 5 through 8, a particle is subjected to the given horizontal force, and either the mass or acceleration of the particle is given. Find the other quantity.

5. Force is 6 nt; mass is 4 kg.

6. Force is 32 lb; mass is 8 slugs.

7. Force is 24 lb; acceleration is 9 ft/sec^2.

8. Force is 700 dynes; acceleration is 80 cm/sec^2.

In Exercises 9 through 12, a system of particles is located on the x axis. The number of kilograms in the mass of each particle and the coordinate of its position are given. Distance is measured in meters. Find the center of mass of each system.

9. $m_1 = 5$ at 2; $m_2 = 6$ at 3; $m_3 = 4$ at 5; $m_4 = 3$ at 8

10. $m_1 = 2$ at -4; $m_2 = 8$ at -1; $m_3 = 4$ at 2; $m_4 = 2$ at 3

11. $m_1 = 2$ at -3; $m_2 = 4$ at -2; $m_3 = 20$ at 4; $m_4 = 10$ at 6; $m_5 = 30$ at 9

12. $m_1 = 5$ at -7; $m_2 = 3$ at -2; $m_3 = 5$ at 0; $m_4 = 1$ at 2; $m_5 = 8$ at 10

In Exercises 13 through 21, find the total mass of the given rod and the center of mass.

13. The length of a rod is 6 m and the linear density of the rod at a point x m from one end is $(2x + 3)$ kg/m.

14. The length of a rod is 20 cm and the linear density of the rod at a point x cm from one end is $(3x + 2)$ g/cm.

15. The length of a rod is 9 in. and the linear density of the rod at a point x in. from one end is $(4x + 1)$ slugs/in.

16. The length of a rod is 3 ft, and the linear density of the rod at a point x ft from one end is $(5 + 2x)$ slugs/ft.

17. The length of a rod is 12 cm, and the measure of the linear density at a point is a linear function of the measure of the distance from the left end of the rod. The linear density at the left end is 3 g/cm and at the right end is 4 g/cm.

18. The length of a rod is 10 m and the measure of the linear density at a point is a linear function of the measure of the distance of the point from the left end of the rod. The linear density at the left end is 2 kg/m and at the right end is 3 kg/m.

19. The measure of the linear density at any point of a rod 6 m long varies directly as the distance from the point to an external point in the line of the rod and 4 m from an end, where the density is 3 kg/m.

20. A rod is 10 ft long, and the measure of the linear density at a point is a linear function of the measure of the distance from the center of the rod. The linear density at each end of the rod is 5 slugs/ft and at the center the linear density is $3\frac{1}{2}$ slugs/ft.

21. The measure of the linear density at a point of a rod varies directly as the third power of the measure of the distance of the point from one end. The length of the rod is 4 ft and the linear density is 2 slugs/ft at the center.

22. The linear density at any point of a rod 5 m long varies directly as the distance from the point to an external point in the line of the rod and 2 m from an end, where the density is K kg/m. Find K if the total mass of the rod is 135 kg.

23. The linear density at any point of a rod 3 m long varies directly as the distance from a point to an external point in the line of the rod and 1 m from an end where the density is 2 kg/m. If the total mass of the rod is 15 kg, find the center of mass of the rod.

24. The measure of the linear density at a point on a rod varies directly as the fourth power of the measure of the distance of the point from one end. The length of the rod is 2 m and the linear density is 2 kg/m at the center. If the total mass of the rod is $\frac{64}{5}$ kg, find the center of mass of the rod.

25. A rod is L ft long and the center of mass of the rod is at the point $\frac{3}{4}L$ ft from the left end. If the measure of the linear density at a point is proportional to a power of the measure of the distance of the point from the left end and the linear density at the right end is 20 slugs/ft, find the linear density at a point x ft from the left end. The mass is in slugs.

26. The total mass of a rod of length L ft is M slugs, and the measure of the linear density at a point x ft from the left end is proportional to the measure of the distance of the point from the right end. Show that the linear density at a point on the rod x ft from the left end is $2M(L - x)/L^2$ slugs/ft.

27. A rod is 6 m long and its mass is 24 kg. If the measure of the linear density at any point of the rod varies directly as the square of the distance of the point from one end, find the largest value of the linear density.

5.7
CENTROID OF A PLANE REGION

FIGURE 5.7.1

Let the masses of n particles located at the points $(x_1, y_1), (x_2, y_2), \ldots, (x_n, y_n)$ in the xy plane be measured by $m_1, m_2, \ldots, m_n$, and consider the problem of finding the center of mass of this system. We may imagine the particles being supported by a sheet of negligible weight and negligible thickness and may assume that each particle has its position at exactly one point. The center of mass is the point where the sheet will balance. Refer to Fig. 5.7.1, which shows eight particles placed on the sheet. The label of the ith particle in the figure is m_i, which is the measure of its mass. The sheet will balance on a fulcrum located at the center of mass denoted by $(\bar{x}, \bar{y})$. To determine the center of mass we first define the moment of mass of a system of particles with respect to an axis.

Suppose a particle at a distance of d meters from an axis has a mass of m kg. If M_1 kg-m is the moment of mass of the particle with respect to the axis, then

$$M_1 = md \tag{1}$$

If the ith particle, having mass m_i kg, is located at the point (x_i, y_i), its distance from the y axis is x_i meters; thus, from formula (1), the moment of mass of this particle with respect to the y axis is $m_i x_i$ kg-m. Similarly, the moment of mass of the particle with respect to the x axis is $m_i y_i$ kg-m. The moment of the system of n particles with respect to the y axis is M_y kg-m, where

$$M_y = \sum_{i=1}^{n} m_i x_i \tag{2}$$

and the moment of the system with respect to the x axis is M_x kg-m, where

$$M_x = \sum_{i=1}^{n} m_i y_i \tag{3}$$

The total mass of the system is M kg, where

$$M = \sum_{i=1}^{n} m_i \tag{4}$$

The center of mass of the system is at the point $(\bar{x}, \bar{y})$, where

$$\bar{x} = \frac{M_y}{M} \tag{5}$$

and

$$\bar{y} = \frac{M_x}{M} \tag{6}$$

The point $(\bar{x}, \bar{y})$ can be interpreted as the point such that, if the total mass M kg of the system were concentrated there, its moment of mass with respect to the y axis, M_y kg-m, would be determined by $M_y = M\bar{x}$, and its moment of mass with respect to the x axis, M_x kg-m, would be determined by $M_x = M\bar{y}$.

EXAMPLE 1
Find the center of mass of the four particles having masses 2, 6, 4, and 1 kg located at the points $(5, -2)$, $(-2, 1)$, $(0, 3)$, and $(4, -1)$, respectively.

SOLUTION

$$M_y = \sum_{i=1}^{4} m_i x_i \qquad\qquad M_x = \sum_{i=1}^{4} m_i y_i$$
$$= 2(5) + 6(-2) + 4(0) + 1(4) \qquad = 2(-2) + 6(1) + 4(3) + 1(-1)$$
$$= 2 \qquad\qquad\qquad\qquad\qquad = 13$$

$$M = \sum_{i=1}^{4} m_i$$
$$= 2 + 6 + 4 + 1$$
$$= 13$$

Therefore

$$\bar{x} = \frac{M_y}{M} \qquad \bar{y} = \frac{M_x}{M}$$

$$= \tfrac{2}{13} \qquad\qquad = \tfrac{13}{13}$$

$$\qquad\qquad\qquad = 1$$

The center of mass is at $(\tfrac{2}{13}, 1)$.

Consider now a thin sheet of continuously distributed mass, for example, a piece of paper or a flat strip of tin. We regard such sheets as being two-dimensional and call such a plane region a *lamina*. In this section we confine our discussion to homogeneous laminae, that is, laminae having constant area density. Laminae of variable area density are considered in Chapter 17 in connection with applications of multiple integrals.

Let a homogeneous lamina of area A m^2 have a mass of M kg. Then if the constant area density is k kg/m^2, $M = kA$. If the homogeneous lamina is a rectangle, its center of mass is defined to be at the center of the rectangle. We use this definition to define the center of mass of a more general homogeneous lamina.

FIGURE 5.7.2

Let L be the homogeneous lamina whose constant area density is k kg/m^2 and which is bounded by the curve $y = f(x)$, the x axis, and the lines $x = a$ and $x = b$. The function f is continuous on the closed interval $[a, b]$, and $f(x) \geq 0$ for all x in $[a, b]$. See Fig. 5.7.2. Let Δ be a partition of the interval $[a, b]$ into n subintervals. The ith subinterval is $[x_{i-1}, x_i]$, and $\Delta_i x = x_i - x_{i-1}$. The midpoint of $[x_{i-1}, x_i]$ is γ_i. Associated with each subinterval is a rectangular lamina whose width, altitude, and area density are given by $\Delta_i x$ meters, $f(\gamma_i)$ meters, and k kg/m^2, respectively, and whose center of mass is at the point $(\gamma_i, \tfrac{1}{2} f(\gamma_i))$. The area of the rectangular lamina is $f(\gamma_i) \Delta_i x$ m^2; hence $k f(\gamma_i) \Delta_i x$ kg is its mass. Consequently, if $\Delta_i M_y$ kg-m is the moment of mass of this rectangular element with respect to the y axis,

$$\Delta_i M_y = \gamma_i k f(\gamma_i) \Delta_i x$$

The sum of the measures of the moments of mass of n such rectangular laminae with respect to the y axis is given by the Riemann sum

$$\sum_{i=1}^{n} k \gamma_i f(\gamma_i) \Delta_i x$$

We define the *moment of mass of L with respect to the y axis* as the limit of this Riemann sum as $\|\Delta\| \to 0$. We state this formally in Definition 5.7.1.

Similarly, if $\Delta_i M_x$ kg-m is the moment of mass of the ith rectangular lamina with respect to the x axis,

$$\Delta_i M_x = \tfrac{1}{2} f(\gamma_i) k f(\gamma_i) \Delta_i x$$

and the sum of the measures of the moments of mass of n such rectangular laminae with respect to the x axis is given by the Riemann sum

$$\sum_{i=1}^{n} \tfrac{1}{2} k [f(\gamma_i)]^2 \Delta_i x$$

The limit of this Riemann sum as $\|\Delta\| \to 0$ is defined in Definition 5.7.1 as the *moment of mass of L with respect to the x axis.*

The mass of the ith rectangular lamina is $kf(\gamma_i)\Delta_i x$ kg; so the sum of the measures of the masses of n rectangular laminae is given by

$$\sum_{i=1}^{n} kf(\gamma_i)\Delta_i x$$

In Definition 5.7.1, the limit of this Riemann sum is defined as the *total mass of L.*

5.7.1 DEFINITION   Let L be the homogeneous lamina whose constant area density is k kg/m² and which is bounded by the curve $y = f(x)$, the x axis, and the lines $x = a$ and $x = b$. The function f is continuous on $[a,b]$ and $f(x) \geq 0$ for all x in $[a,b]$. If M_y kg-m is the **moment of mass of the lamina L with respect to the y axis**, then

$$M_y = \lim_{\|\Delta\| \to 0} \sum_{i=1}^{n} k\gamma_i f(\gamma_i)\Delta_i x$$

$$= k \int_a^b xf(x)\,dx \tag{7}$$

If M_x kg-m is the **moment of mass of the lamina L with respect to the x axis**, then

$$M_x = \lim_{\|\Delta\| \to 0} \sum_{i=1}^{n} \tfrac{1}{2}k[f(\gamma_i)]^2 \Delta_i x$$

$$= \tfrac{1}{2}k \int_a^b [f(x)]^2\,dx \tag{8}$$

If M kg is the **total mass of the lamina L**, then

$$M = \lim_{\|\Delta\| \to 0} \sum_{i=1}^{n} kf(\gamma_i)\Delta_i x$$

$$= k \int_a^b f(x)\,dx \tag{9}$$

If $(\bar{x}, \bar{y})$ is the **center of mass of the lamina L**, then

$$\bar{x} = \frac{M_y}{M} \qquad \bar{y} = \frac{M_x}{M} \tag{10}$$

Substituting from (7), (8), and (9) into (10) we have

$$\bar{x} = \frac{k \int_a^b xf(x)\,dx}{k \int_a^b f(x)\,dx} \qquad \bar{y} = \frac{\tfrac{1}{2}k \int_a^b [f(x)]^2\,dx}{k \int_a^b f(x)\,dx}$$

Dividing both the numerator and denominator by k we get

$$\bar{x} = \frac{\int_a^b xf(x)\,dx}{\int_a^b f(x)\,dx} \qquad \bar{y} = \frac{\tfrac{1}{2} \int_a^b [f(x)]^2\,dx}{\int_a^b f(x)\,dx} \quad \text{area} \tag{11}$$

In formulas (11) the denominator is the number of square units in the area of the region; so we have expressed a physical problem in terms of a geometric one. That is, $\bar{x}$ and $\bar{y}$ can be considered as the average abscissa and the average ordinate, respectively, of a geometric region. In such a case $\bar{x}$ and $\bar{y}$ depend only on the region, not on the mass of the lamina. So we refer to the center of mass of a plane region instead of to the center of mass of a homogeneous lamina. In such a case the center of mass is called the *centroid* of the region. Instead of moments of mass we consider moments of the region.

5.7.2 DEFINITION

Let R be the region bounded by the curve $y = f(x)$, the x axis, and the lines $x = a$ and $x = b$. The function f is continuous on $[a, b]$ and $f(x) \geq 0$ for all x in $[a, b]$. If M_x denotes the **moment of R with respect to the x axis** and M_y denotes the **moment of R with respect to the y axis**, then

$$M_x = \lim_{||\Delta|| \to 0} \sum_{i=1}^{n} \tfrac{1}{2}[f(\gamma_i)]^2 \Delta_i x \qquad M_y = \lim_{||\Delta|| \to 0} \sum_{i=1}^{n} \gamma_i f(\gamma_i) \Delta_i x$$

$$= \tfrac{1}{2} \int_a^b [f(x)]^2 \, dx \qquad\qquad = \int_a^b x f(x) \, dx$$

If $(\bar{x}, \bar{y})$ is the **centroid** of the plane region R whose area is A square units, and M_x and M_y are defined as above,

$$\bar{x} = \frac{M_y}{A} \qquad \bar{y} = \frac{M_x}{A}$$

EXAMPLE 2

Find the centroid of the first quadrant region bounded by the curve $y^2 = 4x$, the x axis, and the lines $x = 1$ and $x = 4$.

SOLUTION

Let $f(x) = 2x^{1/2}$. The equation of the curve is then $y = f(x)$. In Fig. 5.7.3 the region is shown together with the ith rectangular element. The centroid of the rectangle is at $(\gamma_i, \tfrac{1}{2} f(\gamma_i))$. The area A square units of the region is given by

$$A = \lim_{||\Delta|| \to 0} \sum_{i=1}^{n} f(\gamma_i) \Delta_i x$$

$$= \int_1^4 f(x) \, dx$$

$$= \int_1^4 2x^{1/2} \, dx$$

$$= \tfrac{4}{3} x^{3/2} \Big]_1^4$$

$$= \tfrac{28}{3}$$

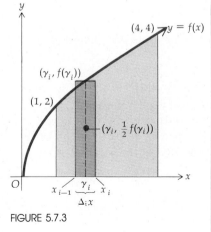

FIGURE 5.7.3

We now compute M_y and M_x.

$$M_y = \lim_{||\Delta|| \to 0} \sum_{i=1}^{n} \gamma_i f(\gamma_i) \Delta_i x \qquad M_x = \lim_{||\Delta|| \to 0} \sum_{i=1}^{n} \tfrac{1}{2} f(\gamma_i) \cdot f(\gamma_i) \Delta_i x$$

$$M_y = \int_1^4 xf(x)\,dx \qquad\qquad M_x = \tfrac{1}{2}\int_1^4 [f(x)]^2\,dx$$

$$= \int_1^4 x(2x^{1/2})\,dx \qquad\qquad = \tfrac{1}{2}\int_1^4 4x\,dx$$

$$= 2\int_1^4 x^{3/2}\,dx \qquad\qquad = x^2\Big]_1^4$$

$$= \tfrac{4}{5}x^{5/2}\Big]_1^4 \qquad\qquad = 15$$

$$= \tfrac{124}{5}$$

Hence

$$\bar{x} = \frac{M_y}{A} \qquad \bar{y} = \frac{M_x}{A}$$

$$= \frac{\frac{124}{5}}{\frac{28}{3}} \qquad = \frac{15}{\frac{28}{3}}$$

$$= \tfrac{93}{35} \qquad = \tfrac{45}{28}$$

Therefore the centroid is at the point $(\tfrac{93}{35}, \tfrac{45}{28})$.

In the following example the region is bounded by two curves instead of one curve and a coordinate axis. The method for finding the centroid is the same as before, but the equations for M_x and M_y are now dependent on the equations defining the two curves. The procedure to be followed is shown in the example.

FIGURE 5.7.4

EXAMPLE 3

Find the centroid of the region bounded by the curves $y = x^2$ and $y = 2x + 3$.

SOLUTION

The points of intersection of the two curves are $(-1, 1)$ and $(3, 9)$. The region is shown in Fig. 5.7.4, together with the ith rectangular element.

Let $f(x) = x^2$ and $g(x) = 2x + 3$. The centroid of the ith rectangular element is at the point $(\gamma_i, \tfrac{1}{2}[f(\gamma_i) + g(\gamma_i)])$, where γ_i is the midpoint of the ith subinterval $[x_{i-1}, x_i]$. The measure of the area of the region is given by

$$A = \lim_{\|\Delta\|\to 0} \sum_{i=1}^{n} [g(\gamma_i) - f(\gamma_i)]\,\Delta_i x$$

$$= \int_{-1}^{3} [g(x) - f(x)]\,dx$$

$$= \int_{-1}^{3} [2x + 3 - x^2]\,dx$$

$$= \tfrac{32}{3}$$

We now compute M_y and M_x.

$$M_y = \lim_{\|\Delta\| \to 0} \sum_{i=1}^{n} \gamma_i[g(\gamma_i) - f(\gamma_i)]\Delta_i x$$

$$= \int_{-1}^{3} x[g(x) - f(x)]\, dx$$

$$= \int_{-1}^{3} x[2x + 3 - x^2]\, dx$$

$$= \tfrac{32}{3}$$

$$M_x = \lim_{\|\Delta\| \to 0} \sum_{i=1}^{n} \tfrac{1}{2}[g(\gamma_i) + f(\gamma_i)][g(\gamma_i) - f(\gamma_i)]\Delta_i x$$

$$= \tfrac{1}{2}\int_{-1}^{3} [g(x) + f(x)][g(x) - f(x)]\, dx$$

$$= \tfrac{1}{2}\int_{-1}^{3} [2x + 3 + x^2][2x + 3 - x^2]\, dx$$

$$= \tfrac{1}{2}\int_{-1}^{3} [4x^2 + 12x + 9 - x^4]\, dx$$

$$= \tfrac{544}{15}$$

Therefore

$$\bar{x} = \frac{M_y}{A} \qquad \bar{y} = \frac{M_x}{A}$$

$$= \frac{\frac{32}{3}}{\frac{32}{3}} \qquad\qquad = \frac{\frac{544}{15}}{\frac{32}{3}}$$

$$= 1 \qquad\qquad = \tfrac{17}{5}$$

Hence the centroid is at the point $(1, \tfrac{17}{5})$.

If a plane region has an axis of symmetry, the centroid of the region lies on the axis of symmetry. This is now stated and proved as a theorem.

5.7.3 THEOREM   If the plane region R has the line L as an axis of symmetry, the centroid of R lies on L.

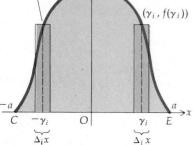

FIGURE 5.7.5

PROOF   Choose the coordinate axes so that L is on the y axis and the origin is in the region R. Figure 5.7.5 illustrates an example of this situation. In the figure, R is the region CDE, C is the point $(-a, 0)$, E is the point $(a, 0)$, and an equation of the curve CDE is $y = f(x)$.

Consider a partition of the interval $[0, a]$. Let γ_i be the midpoint of the ith subinterval $[x_{i-1}, x_i]$. The moment with respect to the y axis of the rectangular element having an altitude $f(\gamma_i)$ and a width $\Delta_i x$ is $\gamma_i[f(\gamma_i)\Delta_i x]$. Because of symmetry, for a similar partition of the interval $[-a, 0]$ there is a corresponding element having as its moment with respect to the y axis $-\gamma_i f(\gamma_i)\Delta_i x$. The sum of these two moments is 0; therefore $M_y = 0$. Because $\bar{x} = M_y/A$, we conclude that $\bar{x} = 0$. Thus the centroid of the region R lies on the y axis, which is what was to be proved.   ∎

By applying the preceding theorem we can simplify the problem of finding the centroid of a plane region that can be divided into regions having axes of symmetry.

EXAMPLE 4

Find the centroid of the region bounded by the semicircle $y = \sqrt{4 - x^2}$ and the x axis.

SOLUTION

The region is shown in Fig. 5.7.6.

Because the y axis is an axis of symmetry, the centroid lies on the y axis; so $\bar{x} = 0$.

The moment of the region with respect to the x axis is given by

$$M_x = \lim_{\|\Delta\| \to 0} \sum_{i=1}^{n} \tfrac{1}{2}[\sqrt{4 - \gamma_i^2}]^2 \, \Delta_i x$$

$$= 2 \cdot \tfrac{1}{2} \int_0^2 (4 - x^2) \, dx$$

$$= 4x - \tfrac{1}{3}x^3 \Big]_0^2$$

$$= \tfrac{16}{3}$$

The area of the region is 2π square units; so

$$\bar{y} = \frac{\frac{16}{3}}{2\pi}$$

$$= \frac{8}{3\pi}$$

The centroid is at the point $\left(0, \dfrac{8}{3\pi}\right)$.

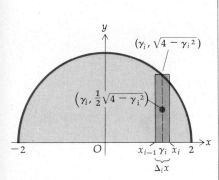

FIGURE 5.7.6

EXERCISES 5.7

1. Find the center of mass of the three particles having masses of 1, 2, and 3 kg and located at the points $(-1, 3)$, $(2, 1)$, and $(3, -1)$, respectively.
2. Find the center of mass of the four particles having masses of 2, 3, 3, and 4 kg and located at the points $(-1, -2)$, $(1, 3)$, $(0, 5)$, and $(2, 1)$, respectively.
3. The y coordinate of the center of mass of four particles is 5. The particles have masses 2, 5, 4, and m slugs and are located at the points $(3, 2)$, $(-1, 0)$, $(0, 20)$, and $(2, -2)$, respectively. Find m.
4. Find the center of mass of the three particles having masses 3, 7, and 2 slugs located at the points $(2, 3)$, $(-1, 4)$, and $(0, 2)$, respectively.
5. Find the center of mass of three particles of equal mass located at the points $(4, -2)$, $(-3, 0)$, and $(1, 5)$.
6. Prove that the center of mass of three particles of equal mass in a plane lies at the point of intersection of the medians of the triangle having as vertices the points at which the particles are located.

In Exercises 7 through 14, find the centroid of the region with the indicated boundaries.

7. The parabola $y = 4 - x^2$ and the x axis.
8. The parabola $x = 2y - y^2$ and the y axis.
9. The parabola $y = x^2$ and the line $y = 4$.
10. The parabola $y^2 = 4x$, the y axis, and the line $y = 4$.

11. The curves $y = x^3$ and $y = 4x$ in the first quadrant.

12. The lines $y = 2x + 1$, $x + y = 7$, and $x = 8$.

13. The curves $y = x^2 - 4$ and $y = 2x - x^2$.

14. The curves $y = x^2$ and $y = x^3$.

15. Find the center of mass of the lamina bounded by the parabola $2y^2 = 18 - 3x$ and the y axis if the area density at any point (x, y) is $\sqrt{6 - x}$ kg/m².

16. Solve Exercise 15 if the area density at any point (x, y) is x kg/m².

17. If the centroid of the region bounded by the parabola $y^2 = 4px$ and the line $x = a$ is to be at the point $(p, 0)$, find the value of a.

18. Prove that the distance from the centroid of a triangle to any side of the triangle is equal to one-third of the length of the altitude to that side.

19. Let R be the region bounded by the curves $y = f_1(x)$ and $y = f_2(x)$ (see Fig. 5.7.7). If A is the measure of the area of R and if $\bar{y}$ is the ordinate of the centroid of R, prove that the measure of the volume V of the solid of revolution obtained by revolving R about the x axis is given by $V = 2\pi\bar{y}A$. Stating this equation in words we have:

> *If a plane region is revolved about a line in its plane that does not cut the region, then the measure of the volume of the solid of revolution generated is equal to the product of the measure of the area of the region and the measure of the distance traveled by the centroid of the region.*

The above statement is known as the *theorem of Pappus* for volumes of solids of revolution.

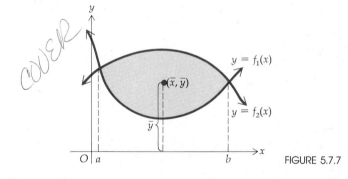

FIGURE 5.7.7

20. Use the theorem of Pappus to find the volume of the torus (doughnut shape) generated by revolving a circle with a radius of r units about a line in its plane at a distance of b units from its center, where $b > r$.

21. Use the theorem of Pappus to find the centroid of the region bounded by a semicircle and its diameter.

22. Use the theorem of Pappus to find the volume of a sphere with a radius of r units.

23. Use the theorem of Pappus to find the volume of a right-circular cone with base radius r units and height h units.

24. Let R be the region bounded by the semicircle $y = \sqrt{r^2 - x^2}$ and the x axis. Use the theorem of Pappus to find the moment of R with respect to the line $y = -r$.

25. If R is the region of Exercise 24, use the theorem of Pappus to find the volume of the solid of revolution generated by revolving R about the line $x - y = r$. (*Hint:* Use the result of Exercise 42 in Exercises 3.10.)

26. Give an example to show that the centroid of a plane region is not necessarily a point within the region.

5.8 (Supplementary) CENTROID OF A SOLID OF REVOLUTION

To find the center of mass of a solid, in general we must make use of multiple integration. This procedure is taken up in Chapter 17 as an application of multiple integrals. However, if the shape of the solid is that of a solid of revolution and its volume density is constant, we find the center of mass by a method similar to the one used to obtain the center of mass of a homogeneous lamina. Following is the procedure for finding the center of mass of a homogeneous solid of revolution, with the assumption that the center of mass is on the axis of revolution.

We first set up a three-dimensional coordinate system. The x and y axes are taken as in two dimensions, and the third axis, the z axis, is taken perpendicular to them at the origin. A point in three dimensions is then given by (x, y, z). The plane containing the x and y axes is called the xy plane, and the xz plane and the yz plane are defined similarly.

Suppose that the x axis is the axis of revolution. Then under the assumption that the center of mass lies on the axis of revolution, the y and z coordinates of the center of mass are each zero; so it is necessary only to find the x coordinate, which we call $\bar{x}$. To find $\bar{x}$ we make use of the moment of the solid of revolution with respect to the yz plane.

Let f be a function that is continuous on the closed interval $[a, b]$, and assume that $f(x) \geq 0$ for all x in $[a, b]$. R is the region bounded by the curve $y = f(x)$, the x axis, and the lines $x = a$ and $x = b$; S is the homogeneous solid of revolution whose volume density is k kg/m³, where k is a constant, and which is generated by revolving the region R about the x axis. Take a partition Δ of the closed interval $[a, b]$, and denote the ith subinterval by $[x_{i-1}, x_i]$ (with $i = 1, 2, \ldots, n$). Let γ_i be the midpoint of $[x_{i-1}, x_i]$. Form n rectangles having altitudes of $f(\gamma_i)$ meters and bases whose widths are $\Delta_i x$ meters. Refer to Fig. 5.8.1, showing the region R and the ith rectangle. If each of the n rectangles is revolved about the x axis, n circular disks are generated. The ith rectangle generates a circular disk having a radius of $f(\gamma_i)$ meters and a thickness of $\Delta_i x$ meters; its volume is $\pi[f(\gamma_i)]^2 \Delta_i x$ m³, and its mass is $k\pi[f(\gamma_i)]^2 \Delta_i x$ kg. Figure 5.8.2 shows the solid of revolution S and the ith circular disk.

The center of mass of the circular disk lies on the axis of revolution at the center of the disk: $(\gamma_i, 0, 0)$. The moment of mass of the disk with respect to the yz plane is then $\Delta_i M_{yz}$ kg-m, where

$$\Delta_i M_{yz} = \gamma_i (k\pi[f(\gamma_i)]^2 \Delta_i x)$$

The sum of the measures of the moments of mass of the n circular disks with respect to the yz plane is given by the Riemann sum

$$\sum_{i=1}^{n} \gamma_i k\pi[f(\gamma_i)]^2 \Delta_i x$$

The limit of this sum as $\|\Delta\|$ approaches zero is defined to be the *moment of mass of S with respect to the yz plane*. We state this formally in Definition 5.8.1.

The volume, V m³, of S was defined in Section 5.1 by

$$V = \lim_{\|\Delta\| \to 0} \sum_{i=1}^{n} \pi[f(\gamma_i)]^2 \Delta_i x$$

$$= \pi \int_a^b [f(x)]^2 \, dx \tag{1}$$

The *mass of S* is defined as kV.

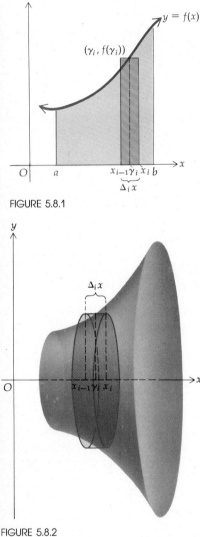

FIGURE 5.8.1

FIGURE 5.8.2

5.8.1 DEFINITION   The region R is bounded by the curve $y = f(x)$, the x axis, and the lines $x = a$ and $x = b$, where the function f is continuous on the closed interval $[a, b]$, and $f(x) \geq 0$ for all x in $[a, b]$. The homogeneous solid of revolution S, whose volume density is k kg/m^3, where k is a constant, is generated by revolving the region R about the x axis. The number of kilogram-meters in the **moment of mass of S with respect to the yz plane** is denoted by M_{yz}, and

$$
\begin{aligned}
M_{yz} &= \lim_{||\Delta|| \to 0} \sum_{i=1}^{n} \gamma_i k \pi [f(\gamma_i)]^2 \Delta_i x \\
&= k\pi \int_a^b x[f(x)]^2 \, dx
\end{aligned}
\tag{2}
$$

The **mass**, M kg, of solid S is defined by

$$
\begin{aligned}
M &= \lim_{||\Delta|| \to 0} \sum_{i=1}^{n} k\pi [f(\gamma_i)]^2 \Delta_i x \, . \\
&= k\pi \int_a^b [f(x)]^2 \, dx
\end{aligned}
\tag{3}
$$

The **center of mass** of S is the point $(\bar{x}, 0, 0)$ such that

$$
\bar{x} = \frac{M_{yz}}{M}
\tag{4}
$$

If we substitute from (2) and (3) into (4), we get

$$
\bar{x} = \frac{k\pi \int_a^b x[f(x)]^2 \, dx}{k\pi \int_a^b [f(x)]^2 \, dx}
$$

$$
\bar{x} = \frac{\int_a^b x[f(x)]^2 \, dx}{\int_a^b [f(x)]^2 \, dx}
\tag{5}
$$

From (5) we observe that the center of mass of a homogeneous solid of revolution depends only on the shape of the solid, not on its substance. Therefore, as we did for a homogeneous lamina, we refer to the center of mass as the *centroid* of the solid of revolution. When we have a homogeneous solid of revolution, instead of the moment of mass we consider the *moment of the solid*.

5.8.2 DEFINITION   If S is the solid of revolution of Definition 5.8.1, then the **moment of the solid S with respect to the yz plane** is given by

$$
\begin{aligned}
M_{yz} &= \lim_{||\Delta|| \to 0} \sum_{i=1}^{n} \gamma_i \pi [f(\gamma_i)]^2 \Delta_i x \\
&= \pi \int_a^b x[f(x)]^2 \, dx
\end{aligned}
\tag{6}
$$

FIGURE 5.8.3

FIGURE 5.8.4

The **centroid** of S, whose volume is V m^3, is $(\bar{x}, 0, 0)$, where

$$\bar{x} = \frac{M_{yz}}{V} \qquad (7)$$

Equation (7) is obtained from (1), (5), and (6).

EXAMPLE 1

Find the centroid of the solid of revolution generated by revolving about the x axis the region bounded by the curve $y = x^2$, the x axis, and the line $x = 3$.

SOLUTION

The region and a rectangular element are shown in Fig. 5.8.3. The solid of revolution and an element of volume are shown in Fig. 5.8.4.

$$M_{yz} = \lim_{||\Delta|| \to 0} \sum_{i=1}^{n} \gamma_i \pi (\gamma_i{}^2)^2 \Delta_i x \qquad V = \lim_{||\Delta|| \to 0} \sum_{i=1}^{n} \pi (\gamma_i{}^2)^2 \Delta_i x$$

$$= \pi \int_0^3 x^5 \, dx \qquad\qquad = \pi \int_0^3 x^4 \, dx$$

$$= \frac{243}{2} \pi \qquad\qquad\qquad = \frac{243}{5} \pi$$

Therefore

$$\bar{x} = \frac{M_{yz}}{V}$$

$$= \frac{\frac{243}{2} \pi}{\frac{243}{5} \pi}$$

$$= \frac{5}{2}$$

Thus the centroid is at the point $(\frac{5}{2}, 0, 0)$.

In the preceding discussion and example the element of volume is a circular disk. When the element of volume is a circular ring, the concepts involved are the same, but the equations used to define the moment of the solid and the volume of the solid are obtained by considering the Riemann sum of volumes of circular rings. We proceed to derive these equations. The functions f and g are continuous on the closed interval $[a, b]$, where $f(x) \geq g(x) \geq 0$ for all x in $[a, b]$. Let R be the region bounded above by the curve $y = f(x)$, below by the curve $y = g(x)$, and on the sides by the lines $x = a$ and $x = b$. The rectangular elements of area are taken perpendicular to the x axis; so the element of volume is a circular ring. Figure 5.8.5 shows the region R and a rectangular element of area. The ith rectangle has a width of $\Delta_i x = x_i - x_{i-1}$, and γ_i is the midpoint of the interval $[x_{i-1}, x_i]$. The centroid of the circular ring obtained by revolving the rectangular element of area about the x axis is at the center of the ring, which is the point $(\gamma_i, 0, 0)$. Let S be the solid of revolution obtained by revolving R about the x axis.

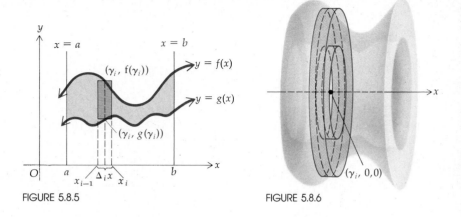

FIGURE 5.8.5

FIGURE 5.8.6

Figure 5.8.6 shows S and the ith circular ring. The moment M_{yz} of S with respect to the yz plane is given by

$$M_{yz} = \lim_{||\Delta|| \to 0} \sum_{i=1}^{n} \gamma_i \pi ([f(\gamma_i)]^2 - [g(\gamma_i)]^2) \Delta_i x$$

$$= \pi \int_a^b x([f(x)]^2 - [g(x)]^2) \, dx \qquad (8)$$

If V cubic units is the volume of S, then

$$V = \lim_{||\Delta|| \to 0} \sum_{i=1}^{n} \pi ([f(\gamma_i)]^2 - [g(\gamma_i)]^2) \Delta_i x$$

$$= \pi \int_a^b ([f(x)]^2 - [g(x)]^2) \, dx \qquad (9)$$

If the centroid of S is at the point $(\bar{x}, 0, 0)$, then

$$\bar{x} = \frac{M_{yz}}{V}$$

FIGURE 5.8.7

FIGURE 5.8.8

EXAMPLE 2

Find the centroid of the solid of revolution generated by revolving about the x axis the region bounded by the curve $y = x^2 + 1$ and the lines $x = 1$ and $y = 1$.

SOLUTION

Figure 5.8.7 shows the region and a rectangular element of area, and Fig. 5.8.8 shows the solid of revolution and an element of volume, which is a circular ring. From (8),

$$M_{yz} = \lim_{||\Delta|| \to 0} \sum_{i=1}^{n} \gamma_i \pi [(\gamma_i^2 + 1)^2 - 1^2] \Delta_i x$$

$$= \pi \int_0^1 x[(x^2 + 1)^2 - 1] \, dx$$

$$= \pi \int_0^1 (x^5 + 2x^3)\, dx$$

$$= \pi \left[\tfrac{1}{6}x^6 + \tfrac{1}{2}x^4\right]_0^1$$

$$= \pi\left[\tfrac{1}{6} + \tfrac{1}{2}\right]$$

$$= \tfrac{2}{3}\pi$$

$$V = \lim_{\|\Delta\| \to 0} \sum_{i=1}^n \pi\left[(\gamma_i{}^2 + 1)^2 - 1^2\right]\Delta_i x$$

$$= \pi \int_0^1 \left[(x^2 + 1)^2 - 1\right] dx$$

$$= \pi \int_0^1 (x^4 + 2x^2)\, dx$$

$$= \pi \left[\tfrac{1}{5}x^5 + \tfrac{2}{3}x^3\right]_0^1$$

$$= \pi\left[\tfrac{1}{5} + \tfrac{2}{3}\right]$$

$$= \tfrac{13}{15}\pi$$

Therefore

$$\bar{x} = \frac{M_{yz}}{V}$$

$$= \frac{\tfrac{2}{3}\pi}{\tfrac{13}{15}\pi}$$

$$= \frac{10}{13}$$

The centroid is at the point $(\tfrac{10}{13}, 0, 0)$.

FIGURE 5.8.9

FIGURE 5.8.10

The centroid of a solid of revolution also can be found by the cylindrical-shell method. Let R be the region bounded by the curve $y = f(x)$, where f is continuous and $f(x) \geq 0$ on $[a, b]$, the x axis, and the lines $x = a$ and $x = b$. Let S be the solid of revolution generated by revolving R about the y axis. If the rectangular elements are taken parallel to the y axis, then the element of volume is a cylindrical shell. Let the ith rectangle have a width of $\Delta_i x = x_i - x_{i-1}$, and let γ_i be the midpoint of the interval $[x_{i-1}, x_i]$. The centroid of the cylindrical shell obtained by revolving this rectangle about the y axis is at the center of the cylindrical shell, which is the point $(0, \tfrac{1}{2}f(\gamma_i), 0)$. Figure 5.8.9 shows the region R and a rectangular element of area, and Fig. 5.8.10 shows the cylindrical shell. The moment, M_{xz}, of S with respect to the xz plane is given by

$$M_{xz} = \lim_{\|\Delta\| \to 0} \sum_{i=1}^n \tfrac{1}{2}f(\gamma_i)2\pi\gamma_i f(\gamma_i)\,\Delta_i x$$

$$= \pi \int_a^b x[f(x)]^2\, dx$$

If V cubic units is the volume of S,

$$V = \lim_{||\Delta|| \to 0} \sum_{i=1}^{n} 2\pi \gamma_i f(\gamma_i) \Delta_i x$$

$$= 2\pi \int_a^b x f(x)\, dx$$

Then the centroid of S is at the point $(0, \bar{y}, 0)$, where

$$\bar{y} = \frac{M_{xz}}{V}$$

$y = x^2$

$(3, 9)$

(γ_i, γ_i^2)

$(\gamma_i, \frac{1}{2}\gamma_i^2)$

$x_{i-1}\; \gamma_i\; x_i$

$\Delta_i x$

FIGURE 5.8.11

FIGURE 5.8.12

EXAMPLE 3
Use the cylindrical-shell method to find the centroid of the solid of revolution generated by revolving the region in Example 1 about the y axis.

SOLUTION
Figure 5.8.11 shows the region and a rectangular element of area. The solid of revolution and a cylindrical-shell element of volume are shown in Fig. 5.8.12.

$$M_{xz} = \lim_{||\Delta|| \to 0} \sum_{i=1}^{n} \tfrac{1}{2}\gamma_i^2 \, 2\pi \gamma_i \gamma_i^2 \, \Delta_i x \qquad V = \lim_{||\Delta|| \to 0} \sum_{i=1}^{n} 2\pi \gamma_i \gamma_i^2 \, \Delta_i x$$

$$= \pi \int_0^3 x^5\, dx \qquad\qquad = 2\pi \int_0^3 x^3\, dx$$

$$= \tfrac{243}{2}\pi \qquad\qquad\qquad = \tfrac{81}{2}\pi$$

Therefore

$$\bar{y} = \frac{M_{xz}}{V}$$

$$= \frac{\frac{243}{2}\pi}{\frac{81}{2}\pi}$$

$$= 3$$

Hence the centroid is at the point $(0, 3, 0)$.

EXAMPLE 4
Solve Example 1 by the cylindrical-shell method.

SOLUTION
Figure 5.8.13 shows the region and a rectangular element of area, and Fig. 5.8.14 shows the cylindrical-shell element of volume obtained by revolving the rectangle about the x axis. The centroid of the cylindrical shell is at its center, which is the point $(\tfrac{1}{2}(3 + \sqrt{\gamma_i}), 0, 0)$. The centroid of the solid of revolution is at $(\bar{x}, 0, 0)$.

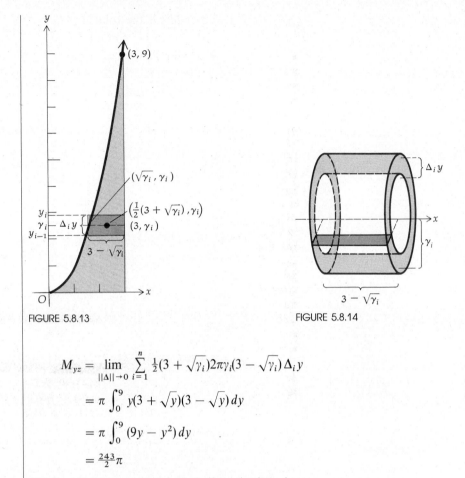

FIGURE 5.8.13   FIGURE 5.8.14

$$M_{yz} = \lim_{||\Delta|| \to 0} \sum_{i=1}^{n} \tfrac{1}{2}(3 + \sqrt{\gamma_i})2\pi\gamma_i(3 - \sqrt{\gamma_i})\Delta_i y$$

$$= \pi \int_0^9 y(3 + \sqrt{y})(3 - \sqrt{y})\,dy$$

$$= \pi \int_0^9 (9y - y^2)\,dy$$

$$= \tfrac{243}{2}\pi$$

Finding V by the cylindrical-shell method we have

$$V = \lim_{||\Delta|| \to 0} \sum_{i=1}^{n} 2\pi\gamma_i(3 - \sqrt{\gamma_i})\Delta_i y$$

$$= 2\pi \int_0^9 y(3 - y^{1/2})\,dy$$

$$= \tfrac{243}{5}\pi$$

Thus because $\bar{x} = M_{yz}/V$, we get $\bar{x} = \tfrac{5}{2}$ and the centroid is at $(\tfrac{5}{2}, 0, 0)$. The result agrees with that of Example 1.

EXERCISES 5.8

In Exercises 1 through 16, find the centroid of the solid of revolution generated by revolving the given region about the indicated line.

1. The region bounded by $y = 4x - x^2$ and the x axis, about the y axis. Take the rectangular elements parallel to the axis of revolution.

2. Same as Exercise 1, but take the rectangular elements perpendicular to the axis of revolution.
3. The region bounded by $x + 2y = 2$, the x axis, and the y axis, about the x axis. Take the rectangular elements perpendicular to the axis of revolution.
4. Same as Exercise 3, but take the rectangular elements parallel to the axis of revolution.
5. The region bounded by $y^2 = x^3$ and $x = 4$, about the x axis. Take the rectangular elements parallel to the axis of revolution.
6. Same as Exercise 5, but take the rectangular elements perpendicular to the axis of revolution.
7. The region bounded by $y = x^3$, $x = 2$, and the x axis, about the line $x = 2$. Take the rectangular elements perpendicular to the axis of revolution.
8. Same as Exercise 7, but take the rectangular elements parallel to the axis of revolution.
9. The region bounded by $x^4y = 1$, $y = 1$, and $y = 4$, about the y axis.
10. The region in Exercise 9, about the x axis.
11. The region bounded by the lines $y = x$, $y = 2x$, and $x + y = 6$, about the y axis.
12. The region bounded by the portion of the circle $x^2 + y^2 = 4$ in the first quadrant, the portion of the line $2x - y = 4$ in the fourth quadrant, and the y axis, about the y axis.
13. The region bounded by $y = x^2$ and $y = x + 2$, about the line $y = 4$.
14. The region bounded by $y^2 = 4x$ and $y^2 = 16 - 4x$, about the x axis.
15. The region bounded by $y = \sqrt{4px}$, the x axis, and the line $x = p$, about the line $x = p$.
16. The region of Exercise 15, about the line $y = 2p$.
17. Find the centroid of the right-circular cone of altitude h units and base radius r units.
18. The region in the first quadrant bounded by a line through the origin with positive slope, the x axis, and the line $x = c$ $(c > 0)$ is revolved about the x axis. Show that the centroid of the solid of revolution is at the point $(\frac{3}{4}c, 0, 0)$.
19. Find the center of mass of the solid of revolution of Exercise 3 if the measure of the volume density of the solid at any point is equal to a constant k times the measure of the distance of the point from the yz plane.
20. Find the center of mass of the solid of revolution of Exercise 5 if the measure of the volume density of the solid at any point is equal to a constant k times the measure of the distance of the point from the yz plane.
21. The region bounded by $y = x^{5/2}$, the x axis, and the line $x = 1$ is revolved about the x axis. Find the center of mass of the solid of revolution if the measure of the volume density of the solid at any point is equal to a constant k times the square of the measure of the distance of the point from the yz plane.
22. The region bounded by the parabola $x^2 = 4y$, the x axis, and the line $x = 4$ is revolved about the y axis. Find the center of mass of the solid of revolution if the measure of the volume density of the solid at any point is equal to the measure of the distance of the point from the xz plane.
23. The region bounded by the curve $y = x^2$, the x axis, and the line $x = b$ $(b > 0)$ is revolved about the x axis. Its moment of mass with respect to the yz plane is $\frac{32}{3}k\pi$ kg-m, where the volume density of the solid is k kg/m^3. Find the value of b.
24. Suppose that a cylindrical hole with a radius of r units is bored through a solid wooden hemisphere of radius $2r$ units such that the axis of the cylinder is the same as the axis of the hemisphere. Find the centroid of the solid remaining.

5.9 *(Supplementary)*
LIQUID PRESSURE

Another application of the definite integral in physics is to find the force caused by liquid pressure on a plate submerged in the liquid or on a side of a container holding the liquid. First suppose that a flat plate is inserted horizontally into a liquid in a container. The weight of the liquid exerts a force on the plate. The force per square unit of area exerted by the liquid on the plate is called the **pressure** of the liquid.

Let ρ lb/ft^3 be the weight density of the liquid and h ft be the depth of a point below the surface of the liquid. If P lb/ft^2 is the pressure exerted by the liquid at the point, then

$$P = \rho h \tag{1}$$

If A ft^2 is the area of a flat plate that is submerged horizontally in the liquid, and F lb is the force caused by liquid pressure acting on the upper face of the plate, then

$$F = PA \tag{2}$$

Substituting from (1) into (2) gives

$$F = \rho hA \tag{3}$$

From (1) it follows that the size of the container is immaterial so far as liquid pressure is concerned. For example, at a depth of 5 ft in a swimming pool filled with salt water the pressure is the same as at a depth of 5 ft in the Pacific Ocean, assuming the density of the water is the same.

▶ ILLUSTRATION 1   A rectangular sheet of tin 8 ft × 12 ft is submerged in a tank of water at a depth of 10 ft. If P lb/ft^2 is the pressure exerted by the water at a point on the upper face of the sheet of tin,

$$P = 10\rho \tag{4}$$

The area of the piece of tin is 96 ft^2. So if F lb is the force caused by liquid pressure acting on the upper face of the sheet of tin,

$$F = 96P \tag{5}$$

Substituting from (4) into (5) we get

$$F = 960\rho$$

Letting $\rho = 62.5$ we have $F = 60{,}000$. Therefore the force due to water pressure on the upper face of the sheet of tin is 60,000 lb.   ◀

Now suppose that the plate is submerged vertically in the liquid. Then at points on the plate at different depths the pressure, computed from (1), will be different and will be greater at the bottom of the plate than at the top. We now proceed to define the force caused by liquid pressure when the plate is submerged vertically in the liquid. We use Pascal's principle: At any point in a liquid, the pressure is the same in all directions.

In Fig. 5.9.1 let $ABCD$ be the region bounded by the x axis, the lines $x = a$ and $x = b$, and the curve $y = f(x)$, where the function f is continuous and $f(x) \geq 0$ on the closed interval $[a, b]$. Choose the coordinate axes so the y axis lies along the line of the surface of the liquid. Take the x axis vertical with the positive direction downward. The length of the plate at a depth x ft is given by $f(x)$ ft.

Let Δ be a partition of the closed interval $[a, b]$ that divides the interval into n subintervals. Choose a point ξ_i in the ith subinterval, with $x_{i-1} \leq \xi_i \leq x_i$. Draw n horizontal rectangles. The ith rectangle has a length of $f(\xi_i)$ ft and a width of $\Delta_i x$ ft (see Fig. 5.9.1).

If we rotate each rectangular element through an angle of 90°, each element becomes a plate submerged in the liquid at a depth of ξ_i ft below

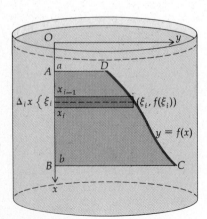

FIGURE 5.9.1

the surface of the liquid and perpendicular to the region $ABCD$. Then the force on the ith rectangular element is given by $\rho \xi_i f(\xi_i) \Delta_i x$ lb. An approximation to F, the number of pounds in the total force on the vertical plate, is given by

$$\sum_{i=1}^{n} \rho \xi_i f(\xi_i) \Delta_i x \tag{6}$$

which is a Riemann sum.

The smaller we take $\|\Delta\|$, the larger n will be and the closer the approximation given by (6) will be to what we wish to be the measure of the total force. We have, then, the following definition.

5.9.1 DEFINITION   Suppose that a flat plate is submerged vertically in a liquid of weight ρ pounds per cubic unit. The length of the plate at a depth of x units below the surface of the liquid is $f(x)$ units, where f is continuous on the closed interval $[a, b]$ and $f(x) \geq 0$ on $[a, b]$. Then F, the number of pounds of **force caused by liquid pressure** on the plate, is given by

$$F = \lim_{\|\Delta\| \to 0} \sum_{i=1}^{n} \rho \xi_i f(\xi_i) \Delta_i x$$

$$= \int_a^b \rho x f(x)\, dx \tag{7}$$

EXAMPLE 1

A trough having a trapezoidal cross section is full of water. If the trapezoid is 3 ft wide at the top, 2 ft wide at the bottom, and 2 ft deep, find the total force owing to water pressure on one end of the trough.

SOLUTION

Figure 5.9.2 illustrates one end of the trough together with a rectangular element of area. An equation of line AB is $y = \frac{3}{2} - \frac{1}{4}x$. Let $f(x) = \frac{3}{2} - \frac{1}{4}x$. If we rotate the rectangular element through 90°, the force on the element is given by $2\rho \xi_i f(\xi_i) \Delta_i x$ lb. If F is the number of pounds in the total force on the side of the trough,

$$F = \lim_{\|\Delta\| \to 0} \sum_{i=1}^{n} 2\rho \xi_i f(\xi_i) \Delta_i x$$

$$= 2\rho \int_0^2 x f(x)\, dx$$

$$= 2\rho \int_0^2 x(\tfrac{3}{2} - \tfrac{1}{4}x)\, dx$$

$$= 2\rho \left[\tfrac{3}{4}x^2 - \tfrac{1}{12}x^3 \right]_0^2$$

$$= \tfrac{14}{3}\rho$$

Taking $\rho = 62.5$ we find that the total force is 291.2 lb.

$(0, -\frac{3}{2})$ O $B(0, \frac{3}{2})$ y

$(\xi_i, f(\xi_i))$

$\Delta_i x$

$(2, -1)$ $(2, 0)$ $A(2, 1)$

x

FIGURE 5.9.2

EXAMPLE 2

The ends of a trough are semicircular regions, each with a radius of 2 ft. Find the force caused by liquid pressure on one end if the trough is full of water.

SOLUTION

Figure 5.9.3 shows one end of the trough together with a rectangular element of area. An equation of the semicircle is $x^2 + y^2 = 4$. Solving for y gives $y = \sqrt{4 - x^2}$. The force on the rectangular element is given by $2\rho\xi_i\sqrt{4 - \xi_i^2}\,\Delta_i x$.

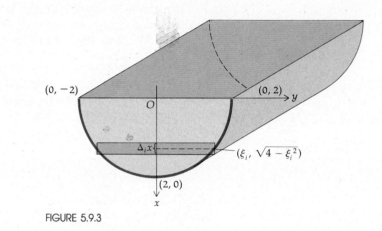

FIGURE 5.9.3

So if F lb is the total force on the side of the trough,

$$F = \lim_{\|\Delta\| \to 0} \sum_{i=1}^{n} 2\rho\xi_i\sqrt{4 - \xi_i^2}\,\Delta_i x$$

$$= 2\rho \int_0^2 x\sqrt{4 - x^2}\,dx$$

$$= -\tfrac{2}{3}\rho(4 - x^2)^{3/2}\Big]_0^2$$

$$= \tfrac{16}{3}\rho$$

Therefore the total force is $\tfrac{16}{3}\rho$ lb.

There is a useful relation between the force caused by liquid pressure on a plane region and the location of the centroid of the region. Refer to Fig. 5.9.1, where $ABCD$ is the region bounded by the x axis, the lines $x = a$ and $x = b$, and the curve $y = f(x)$, where f is continuous and $f(x) \ge 0$ on the closed interval $[a, b]$. We consider the region $ABCD$ as a vertical plate immersed in a liquid having weight density ρ pounds per cubic unit of the liquid. If F lb is the force owing to liquid pressure on the vertical plate, from (7),

$$F = \rho \int_a^b xf(x)\,dx \qquad (8)$$

If $\bar{x}$ is the abscissa of the centroid of the region $ABCD$, then $\bar{x} = M_y/A$.

Because $M_y = \int_a^b xf(x)\,dx,$

$$\bar{x} = \frac{\int_a^b xf(x)\,dx}{A}$$

$$\int_a^b xf(x)\,dx = \bar{x}A \tag{9}$$

Substituting from (9) into (8) we obtain

$$F = \rho \bar{x} A \tag{10}$$

From (10) it follows that the total force owing to liquid pressure against a vertical plane region is the same as it would be if the region were horizontal at a depth $\bar{x}$ units below the surface of a liquid.

▶ ILLUSTRATION 2   Consider a trough full of water having as ends semicircular regions each with a radius of 2 ft. Using the result of Example 4 in Section 5.7 we find that the centroid of the region is at a depth of $8/(3\pi)$ ft. Therefore, using (10) we see that if F lb is the force on one end of the trough,

$$F = \rho \cdot \frac{8}{3\pi} \cdot 2\pi$$

$$= \frac{16}{3}\rho$$

This agrees with the result found in Example 2.   ◀

For various simple plane regions the centroid may be found in a table. When both the area of the region and the centroid of the region may be obtained directly, (10) is easy to apply and is used in such cases by engineers to find the force caused by liquid pressure.

If mass is measured in kilograms, then pressure is measured in newtons per square meter and force is measured in newtons. If P nt/m^2 is the pressure, h meters is the depth of a point below the surface of the liquid, and ρ kg/m^3 is the density of the liquid, then instead of Eq. (1) we have

$$P = \rho h(9.81) \tag{11}$$

where 9.81 m/sec^2 is the acceleration due to gravity. From (2), if A m^2 is the area of a flat plate that is submerged horizontally in the liquid and F nt is the force caused by liquid pressure acting on the upper face of the plate, then

$$F = PA \tag{12}$$

Substituting from (11) into (12) gives

$$F = 9.81\rho h A \tag{13}$$

Thus in the mks system, Eq. (7) of Definition 5.9.1 becomes

$$F = \lim_{\|\Delta\| \to 0} \sum_{i=1}^{n} 9.81\rho \xi_i f(\xi_i)\,\Delta_i x$$

$$= \int_a^b 9.81\rho x f(x)\,dx$$

where the length of the plate x meters below the surface of the liquid is $f(x)$ meters. The density of water is approximately 1000 kg/m^3. So if the liquid causing the pressure is water, we take $\rho = 1000$.

EXAMPLE 3

A container in the shape of a right-circular cylinder having a base of radius 3 m is on its side at the bottom of a tank full of water. The depth of the tank is 13 m. Find the total force due to water pressure on one end of the container.

SOLUTION

Figure 5.9.4 shows one end of the container in the tank and a rectangular element of area. The coordinate system has been chosen so that the origin is at the center of the circle. An equation of the circle is $x^2 + y^2 = 9$. Solving for x gives $x = \sqrt{9 - y^2}$. The force on the rectangular element is given by

$$9.81\rho(10 - \xi_i)[2\sqrt{9 - \xi_i^2}]\Delta_i y$$

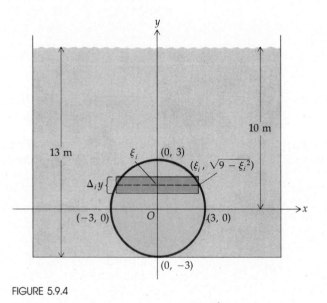

FIGURE 5.9.4

So if F nt is the total force on the end of the container,

$$F = \lim_{||\Delta|| \to 0} \sum_{i=1}^{n} 9.81\rho(10 - \xi_i)[2\sqrt{9 - \xi_i^2}]\Delta_i y$$

$$= 19.62\rho \int_{-3}^{3} (10 - y)\sqrt{9 - y^2}\, dy$$

Because $\rho = 1000$ we have

$$F = 196{,}200 \int_{-3}^{3} \sqrt{9 - y^2}\, dy - 19{,}620 \int_{-3}^{3} y\sqrt{9 - y^2}\, dy \qquad (14)$$

To evaluate $\int_{-3}^{3} \sqrt{9 - y^2}\, dy$ requires a technique of integration you will learn in Section 8.5. At present we determine its value by considering it as the measure of the area of the region enclosed by a semicircle of radius 3. Therefore

$$\int_{-3}^{3} \sqrt{9 - y^2}\, dy = \tfrac{9}{2}\pi$$

Substituting this value into (14) and evaluating the second integral we have

$$F = 196,200(\tfrac{9}{2}\pi) + 19,620\left[\tfrac{1}{3}(9 - y^2)^{3/2}\right]_{-3}^{3}$$

$$= 882,900\pi$$

Thus the total force is $882,900\pi$ nt.

EXERCISES 5.9

1. A plate in the shape of a rectangle is submerged vertically in a tank of water, with the upper edge lying in the surface. If the width of the plate is 10 ft and the depth is 8 ft, find the force due to liquid pressure on one side of the plate.

2. A square plate of side 4 ft is submerged vertically in a tank of water and its center is 2 ft below the surface. Find the force due to liquid pressure on one side of the plate.

3. Solve Exercise 2 if the center of the plate is 4 ft below the surface.

4. A plate in the shape of an isosceles right triangle is submerged vertically in a tank of water, with one leg lying in the surface. The legs are each 6 ft long. Find the force due to liquid pressure on one side of the plate.

5. A rectangular tank full of water is 2 ft wide and 18 in. deep. Find the force due to liquid pressure on one end of the tank.

6. The ends of a trough are equilateral triangles having sides with lengths of 2 ft. If the water in the trough is 1 ft deep, find the force due to liquid pressure on one end.

7. The face of the gate of a dam is in the shape of an isosceles triangle 4 m wide at the top and 3 m high. If the upper edge of the face of the gate is 15 m below the surface of the water, find the total force due to water pressure on the gate.

8. The face of a gate of a dam is vertical and in the shape of an isosceles trapezoid 3 m wide at the top, 4 m wide at the bottom, and 3 m high. If the upper base is 20 m below the surface of the water, find the total force due to water pressure on the gate.

9. The face of a dam adjacent to the water is vertical, and its shape is in the form of an isosceles triangle 250 m wide across the top and 100 m high in the center. If the water is 10 m deep in the center, find the total force on the dam due to water pressure.

10. An oil tank is in the shape of a right-circular cylinder 4 m in diameter, and its axis is horizontal. If the tank is half full of oil having density 750 kg/m^3, find the total force on one end due to liquid pressure.

11. An oil tank is in the shape of a right-circular cylinder with a radius of r meters, and its axis is horizontal. If the tank is half full of oil having density 750 kg/m^3, find r if the total force on one end of the tank due to liquid pressure is 80,000 nt.

12. Solve Exercise 4 by using Eq. (10).   13. Solve Exercise 5 by using Eq. (10).

14. Solve Exercise 6 by using Eq. (10).

15. The face of a dam adjacent to the water is vertical and is in the shape of an isosceles trapezoid 90 ft wide at the top, 60 ft wide at the bottom, and 20 ft high. Use Eq. (10) to find the total force due to water pressure on the face of the dam.

16. A semicircular plate with a radius of 3 ft is submerged vertically in a tank of water, with its diameter lying in the surface. Use Eq. (10) to find the total force due to water pressure on one side of the plate.

17. Find the moment about the lower base of the trapezoid of the force in Exercise 15.

18. A plate in the shape of a region bounded by the parabola $x^2 = 6y$ and the line $2y = 3$ is placed in a water tank with

its vertex downward and the line in the surface of the water. Find the total force due to water pressure on one side of the plate if distance is measured in meters.

19. A cylindrical tank is half full of gasoline having weight density 42 lb/ft^3. If the axis is horizontal and the diameter is 6 ft, find the force on an end due to liquid pressure.

20. If the end of a water tank is in the shape of a rectangle and the tank is full, show that the measure of the force due to water pressure on the end is the product of the measure of the area of the end and the measure of the force at the geometrical center.

21. The bottom of a swimming pool is an inclined plane. The pool is 2 ft deep at one end and 8 ft deep at the other. If the width of the pool is 25 ft and the length is 40 ft, find the total force due to water pressure on the bottom.

22. The face of a dam adjacent to the water is inclined at an angle of 45° from the vertical. The face is a rectangle of width 80 ft and slant height 40 ft. If the dam is full of water, find the total force due to water pressure on the face.

23. The face of a dam adjacent to the water is inclined at an angle of 30° from the vertical. The shape of the face is a rectangle of width 50 ft and slant height 30 ft. If the dam is full of water, find the total force due to water pressure on the face.

24. Solve Exercise 23 if the face of the dam is an isosceles trapezoid 120 ft wide at the top, 80 ft wide at the bottom, and with a slant height of 40 ft.

REVIEW EXERCISES FOR CHAPTER 5

1. Find the volume of the solid generated by revolving about the x axis the region bounded by the curve $y = x^4$, the line $x = 1$, and the x axis.

2. Find the volume of the solid generated if the region of Exercise 1 is revolved about the y axis.

3. The region bounded by the curve $y = \sqrt{\sin x}$, the x axis, and the line $x = \frac{1}{2}\pi$ is revolved about the x axis. Find the volume of the solid generated.

4. The region bounded by the curve $x = \sqrt{\cos y}$, the line $y = \frac{1}{6}\pi$, and the y axis, where $\frac{1}{6}\pi \leq y \leq \frac{1}{2}\pi$, is revolved about the y axis. Find the volume of the solid generated.

5. The region bounded by the curve $y = \csc x$, the x axis, and the lines $x = \frac{1}{4}\pi$ and $x = \frac{1}{2}\pi$ is revolved about the x axis. Find the volume of the solid generated.

6. The region in the first quadrant bounded by the curves $x = y^2$ and $x = y^4$ is revolved about the y axis. Find the volume of the solid generated.

7. Find the volume of the solid generated by revolving about the y axis the region bounded by the parabola $x = y^2 + 2$ and the line $x = y + 8$.

8. Find the volume of the solid of revolution generated when the region bounded by the parabola $y^2 = x$, the x axis, and the line $x = 4$ is revolved about the line $x = 4$. Take the elements of area parallel to the axis of revolution.

9. The base of a solid is the region bounded by the parabola $y^2 = 8x$ and the line $x = 8$. Find the volume of the solid if every plane section perpendicular to the axis of the base is a square.

10. The base of a solid is the region enclosed by a circle having a radius of r units, and every plane section perpendicular to a fixed diameter of the base is a square for which a chord of the circle is a diagonal. Find the volume of the solid.

11. Find the volume of the solid generated by revolving the region bounded by the curve $y = |x - 2|$, the x axis, and the lines $x = 1$ and $x = 4$ about the x axis.

12. Use integration to find the volume of a segment of a sphere if the sphere has a radius of r units and the altitude of the segment is h units.

13. Find the volume of the solid generated by revolving about the line $y = -1$ the region above the x axis bounded by the line $2y = x + 3$ and the curves $y^2 + x = 0$ and $y^2 - 4x = 0$ from $x = -1$ to $x = 1$.

14. Find the length of the arc of the curve $ay^2 = x^3$ from the origin to $(4a, 8a)$.

15. Find the length of the arc of the curve $6y^2 = x(x - 2)^2$ from $(2, 0)$ to $(8, 4\sqrt{3})$.

16. A force of 500 lb is required to compress a spring whose natural length is 10 in. to a length of 9 in. Find the work done to compress the spring to a length of 8 in.

17. A force of 600 dynes stretches a spring from its natural length of 30 cm to a length of 35 cm. Find the work done in stretching the spring from its natural length to a length of 40 cm.

18. A trough full of water is 6 ft long, and its cross section is in the shape of a semicircle with a diameter of 2 ft at the top. How much work is required to pump the water out over the top?

19. A cable 20 ft long and weighing 2 lb/ft is hanging vertically from the top of a pole. Find the work done in raising the entire cable to the top of the pole.

20. The work necessary to stretch a spring from 9 in. to 10 in. is $\frac{3}{2}$ times the work necessary to stretch it from 8 in. to 9 in. What is the natural length of the spring?

21. A tank full of water is in the form of a rectangular parallelepiped 4 m deep, 15 m wide, and 30 m long. Find the work required to pump the water in the tank up to a level 50 cm above the top of the tank. The weight of 1 m^3 of water at $0°C$ is 1000 kg. (*Hint:* See hint for Exercise 21 in Exercises 5.4.)

22. A hemispherical tank having a diameter of 10 m is filled with water to a depth of 3 m. Find the work done in pumping the water to the top of the tank. The weight of 1 m^3 of water at $0°C$ is 1000 kg. (*Hint:* See hint for Exercise 21 in Exercises 5.4.)

23. A water tank is in the shape of a hemisphere surmounted by a right-circular cylinder. The radius of both the hemisphere and the cylinder is 4 ft, and the altitude of the cylinder is 8 ft. If the tank is full of water, find the work necessary to empty the tank by pumping it through an outlet at the top of the tank.

24. A container has the same shape and dimensions as a solid of revolution formed by revolving about the y axis the region in the first quadrant bounded by the parabola $x^2 = 4py$, the y axis, and the line $y = p$. If the container is full of water, find the work done in pumping all the water up to a point $3p$ ft above the top of the container.

25. The surface of a tank is the same as that of a paraboloid of revolution obtained by revolving the parabola $y = x^2$ about the y axis. The vertex of the parabola is at the bottom of the tank, and the tank is 36 ft high. If the tank is filled with water to a depth of 20 ft, find the work done in pumping all of the water out over the top.

26. A wedge is cut from a right-circular cylinder with a radius of r units by two planes, one perpendicular to the axis of the cylinder and the other intersecting the first along a diameter of the circular plane section at an angle of measurement $30°$. Find the volume of the wedge.

27. A church steeple is 30 ft high, and every horizontal plane section is a square having sides of length one-tenth of the distance of the plane section from the top of the steeple. Find the volume of the steeple.

28. A sphere of radius 10 cm is intersected by two parallel planes on the same side of the center of the sphere. The distance from the center of the sphere to one of the planes is 1 cm, and the distance between the two planes is 6 cm. Find the volume of the solid portion of the sphere between the two planes.

29. Solve Exercise 28 if the two planes lie on opposite sides of the center of the sphere but the other facts are the same.

30. A solid is formed by revolving about the y axis the region bounded by the curve $y^3 = x$, the x axis, and the line $x = c$, where $c > 0$. For what value of c will the volume of the solid be 12π cubic units?

31. Find the length of the arc of the curve $9x^{2/3} + 4y^{2/3} = 36$ in the second quadrant from the point where $x = -1$ to the point where $x = -\frac{1}{8}$.

32. Find the length of the arc of the curve $3y = (x^2 - 2)^{3/2}$ from the point where $x = 3$ to the point where $x = 6$.

33. Three particles of masses 4, 2, and 7 kg are located on the x axis at the points having coordinates -5, 4, and 2, respectively, where the distance is measured in meters. Find the center of mass of the system.

34. Three particles having masses 5, 2, and 8 slugs are located, respectively, at the points $(-1,3)$, $(2, -1)$, and $(5,2)$. Find the center of mass, where distance is measured in feet.

35. Find the coordinates of the center of mass of the four particles having equal masses located at the points $(3,0)$, $(2,2)$, $(2,4)$, and $(-1,2)$.

36. Three particles, each having the same mass, are located on the x axis at the points having coordinates -4, 1, and 5, where the distance is measured in meters. Find the coordinates of the center of mass of the system.

37. The length of a rod is 8 in. and the linear density of the rod at a point x in. from the left end is $2\sqrt{x + 1}$ slugs/in. Find the total mass of the rod and the center of mass.

38. The length of a rod is 4 m and the linear density of the rod at a point x meters from the left end is $(3x + 1)$ kg/m. Find the total mass of the rod and the center of mass.

39. Find the centroid of the region in the first quadrant bounded by the coordinate axes and the parabola $y = 9 - x^2$.

40. Find the centroid of the region bounded by the parabola $y^2 = x$ and the line $y = x - 2$.

41. Find the centroid of the region bounded by the curves $y = \sqrt{x}$ and $y = x^2$.

42. Find the centroid of the region bounded above by the parabola $4x^2 = 36 - 9y$ and below by the x axis.

43. Use the theorem of Pappus to find the volume of a sphere of radius 4 m.

44. Use the theorem of Pappus to find the volume of a right-circular cone with base radius 2 m and height 3 m.

45. The region from $x = 0$ to $x = \frac{1}{2}\pi$ bounded by the curve $y = \sin x$, the line $y = 1$, and the y axis is revolved about the x axis. Find the volume of the solid generated. (*Hint:* Use the identity $\sin^2 x = \frac{1}{2}(1 - \cos 2x)$.)

46. The region bounded by the curves $y = \tan x$ and $y = \cot x$ and the x axis, where $0 \le x \le \frac{1}{2}\pi$, is revolved about the x axis. Find the volume of the solid generated.

47. If $f(x) = \int_0^x \sqrt{\cos t}\, dt$, find the length of the arc of the graph of f from the point where $x = \frac{1}{3}\pi$ to the point where $x = \frac{1}{2}\pi$. (*Hint:* Use Theorem 4.8.1 and the identity $\cos^2 \frac{1}{2}x = \frac{1}{2}(1 + \cos x)$.)

Exercises 48 through 51 pertain to Supplementary Section 5.8.

48. The region bounded by the parabola $x^2 = 4y$, the x axis, and the line $x = 4$ is revolved about the y axis. Find the centroid of the solid of revolution formed.

49. The region bounded by the curve $y = x^3$, the line $x = 1$, and the x axis is revolved about the x axis. Find the centroid of the solid of revolution formed.

50. Find the center of mass of the solid of revolution of Exercise 48 if the measure of the volume density of the solid at any point is equal to the measure of the distance of the point from the xz plane.

51. The region bounded by the curve $x = \sqrt{y}$, the y axis, and the line $y = 4$ is revolved about the x axis. Find the centroid of the solid of revolution formed.

Exercises 52 through 55 pertain to Supplementary Section 5.9.

52. A semicircular plate with a radius of 4 ft is submerged vertically in a tank of water, with its diameter lying in the surface. Use Eq. (10) of Section 5.9 to find the force due to water pressure on one side of the plate.

53. A plate in the shape of a region bounded by the parabola $x^2 = 6y$ and the line $2y = 3$ is placed in a water tank with its vertex downward and the line in the surface of the water. Find the force due to water pressure on one side of the plate if distance is measured in feet.

54. The face of a dam adjacent to the water is inclined at an angle of 45° from the vertical. The face is a rectangle of width 80 ft and slant height 40 ft. If the dam is full of water, find the total force due to water pressure on the face.

55. A cylindrical tank is half full of gasoline having weight density 40 lb/ft³. If the axis is horizontal and the diameter is 8 ft, find the force on an end due to liquid pressure.

S I X

INVERSE FUNCTIONS, LOGARITHMIC FUNCTIONS, AND EXPONENTIAL FUNCTIONS

6.1
INVERSE FUNCTIONS

At the beginning of Section 4.1, before discussing antidifferentiation (the inverse operation of differentiation), we indicated that you were familiar with the following facts: Addition and subtraction are inverse operations; multiplication and division are inverse operations; raising to powers and extracting roots are inverse operations. One of a pair of inverse operations essentially "undoes" the other operation. For instance, if 4 is added to x, the sum is $x + 4$; if then 4 is subtracted from this sum the difference is x. If x is multiplied by 2, the product is $2x$; if this product is then divided by 2, the quotient is x. The third power of x is x^3; if we take the cube root of x^3, we obtain x. In the following illustration we use pairs of functions associated with these pairs of inverse operations. You will learn subsequently (in Illustration 5) that each pair is a set of *inverse functions*.

▶ ILLUSTRATION 1   (a) Let $f(x) = x + 4$ and $g(x) = x - 4$. Then

$$f(g(x)) = f(x - 4) \qquad g(f(x)) = g(x + 4)$$
$$= (x - 4) + 4 \qquad\qquad = (x + 4) - 4$$
$$= x \qquad\qquad\qquad = x$$

(b) Let $f(x) = 2x$ and $g(x) = \dfrac{x}{2}$. Then

$$f(g(x)) = f\left(\frac{x}{2}\right) \qquad g(f(x)) = g(2x)$$

$$= 2\left(\frac{x}{2}\right) \qquad\qquad = \frac{2x}{2}$$

$$= x \qquad\qquad\qquad = x$$

(c) Let $f(x) = x^3$ and $g(x) = \sqrt[3]{x}$. Then

$$f(g(x)) = f(\sqrt[3]{x}) \qquad g(f(x)) = g(x^3)$$
$$= (\sqrt[3]{x})^3 \qquad\qquad = \sqrt[3]{x^3}$$
$$= x \qquad\qquad\qquad = x \qquad\qquad ◀$$

Each pair of functions f and g in Illustration 1 satisfies the following two statements:

$$f(g(x)) = x \qquad \text{for } x \text{ in the domain of } g \qquad\qquad (1)$$

and

$$g(f(x)) = x \qquad \text{for } x \text{ in the domain of } f \qquad\qquad (2)$$

Theorem 6.1.4, which comes later, concludes with the fact that statements (1) and (2) are satisfied by a pair of inverse functions.

We lead up to the formal definition of the *inverse of a function* by considering some more particular functions. Let the function f be defined by

$$f(x) = x^2$$

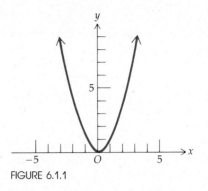

FIGURE 6.1.1

A sketch of the graph of f is shown in Fig. 6.1.1. The domain of f is the set of real numbers, and the range of f is the interval $[0, +\infty)$. To each value of x in the domain there corresponds one and only one number in the range. For instance, because $f(2) = 4$, the number in the range that corresponds to the number 2 in the domain is 4. However, because $f(-2) = 4$, the number corresponding to the number -2 in the domain is also the number 4 in the range. So 4 is the function value of two distinct numbers in the domain. Furthermore, every number except 0 in the range of this function is the function value of two distinct numbers in the domain. In particular, $\frac{25}{4}$ is the function value of both $\frac{5}{2}$ and $-\frac{5}{2}$, 1 is the function value of both 1 and -1, and 9 is the function value of both 3 and -3.

A different situation occurs with the function g defined by

$$g(x) = x^3 \qquad -2 \le x \le 2$$

The domain of g is the closed interval $[-2, 2]$, and the range is $[-8, 8]$. A sketch of the graph of g is shown in Fig. 6.1.2. This function is one for which a number in its range is the function value of one and only one number in the domain. Such a function is called *one-to-one*.

6.1.1 DEFINITION   A function f is said to be **one-to-one** if and only if whenever x_1 and x_2 are any two distinct numbers in the domain of f, then $f(x_1) \ne f(x_2)$.

▶ ILLUSTRATION 2   As discussed above, for the function f defined by

$$f(x) = x^2$$

every number except 0 in the range of f is the function value of two distinct numbers in the domain. Therefore, by Definition 6.1.1, this function is not one-to-one. ◀

FIGURE 6.1.2

We have seen that a vertical line can intersect the graph of a function in only one point. Observe that this is the situation in Fig. 6.1.1 for any vertical line and the function defined by $f(x) = x^2$, and in Fig. 6.1.2 for a line $x = a$ with $-2 \le a \le 2$, and the function defined by $g(x) = x^3$, where $-2 \le x \le 2$. For a one-to-one function it is also true that a horizontal line can intersect the graph in only one point. You can verify this fact for the one-to-one function whose graph appears in Fig. 6.1.2. Furthermore, note that for the function of Fig. 6.1.1 a line $y = b$, with $b > 0$, intersects the graph in two points, and this function is not one-to-one.

EXAMPLE 1
(a) If $f(x) = 4x - 3$, prove that f is one-to-one. Draw a sketch of the graph of f. (b) If $h(x) = 4 - x^2$, prove that h is not one-to-one. Draw a sketch of the graph of h.

FIGURE 6.1.3

SOLUTION
(a) We are given $f(x) = 4x - 3$. The domain of f is the set of real numbers. If x_1 and x_2 are two real numbers and $x_1 \neq x_2$, then

$$4x_1 \neq 4x_2$$
$$4x_1 - 3 \neq 4x_2 - 3$$
$$f(x_1) \neq f(x_2)$$

Therefore, by Definition 6.1.1, f is one-to-one. A sketch of the graph of f is shown in Fig. 6.1.3. Observe that a horizontal line intersects the graph in only one point.

(b) We are given $h(x) = 4 - x^2$. The domain of h is the set of real numbers. The numbers 1 and -1 are in the domain of h, and

$$h(1) = 3 \qquad h(-1) = 3$$

Because $1 \neq -1$ and $h(1) = h(-1)$, it follows from Definition 6.1.1 that h is not one-to-one. A sketch of the graph of h appears in Fig. 6.1.4. Note that a horizontal line $y = b$, with $b < 4$, intersects the graph in two points.

It is not always a simple matter to use Definition 6.1.1 to prove that a function is one-to-one. The following theorem gives a test that can sometimes be used.

6.1.2 THEOREM	A function that is either increasing on an interval or decreasing on an interval is one-to-one on the interval.

FIGURE 6.1.4

PROOF   Assume that the function f is increasing on an interval. If x_1 and x_2 are two numbers in the interval and $x_1 \neq x_2$, then either $x_1 < x_2$ or $x_2 < x_1$. If $x_1 < x_2$, it follows from the definition of an increasing function (Defini-292 tion 3.4.1) that $f(x_1) < f(x_2)$; so $f(x_1) \neq f(x_2)$. If $x_2 < x_1$, then $f(x_2) < f(x_1)$, and so again $f(x_1) \neq f(x_2)$. Therefore it follows from Definition 6.1.1 that f is one-to-one on the interval. The proof is similar if f is decreasing on an interval. ∎

To apply Theorem 6.1.2 it is necessary first to determine if the function is monotonic (increasing or decreasing) on an interval. Theorem 3.4.3 can usually be used for this purpose.   292

EXAMPLE 2
If

$$f(x) = \frac{2x + 3}{x - 1}$$

prove that f is one-to-one on each of the intervals $(-\infty, 1)$ and $(1, +\infty)$. Draw a sketch of the graph of f to show that f is one-to-one on its entire domain.

SOLUTION

The domain of f is the set of all real numbers except 1, or, equivalently, the set $(-\infty, 1) \cup (1, +\infty)$. We compute $f'(x)$.

$$f'(x) = \frac{2(x-1) - (2x+3)}{(x-1)^2}$$

$$= -\frac{5}{(x-1)^2}$$

Because $f'(x) < 0$ for all $x \neq 1$, we can conclude from Theorem 3.4.3 that f is decreasing on each of the intervals $(-\infty, 1)$ and $(1, +\infty)$. Therefore, from Theorem 6.1.2, f is one-to-one on each of these intervals.

Furthermore,

$$\lim_{x \to -\infty} f(x) = \lim_{x \to -\infty} \frac{2 + \dfrac{3}{x}}{1 - \dfrac{1}{x}} \qquad \lim_{x \to +\infty} f(x) = \lim_{x \to +\infty} \frac{2 + \dfrac{3}{x}}{1 - \dfrac{1}{x}}$$

$$= 2 \qquad\qquad\qquad\qquad = 2$$

$$\lim_{x \to 1^-} f(x) = \lim_{x \to 1^-} \frac{2x + 3}{x - 1} \qquad \lim_{x \to 1^+} f(x) = \lim_{x \to 1^+} \frac{2x + 3}{x - 1}$$

$$= -\infty \qquad\qquad\qquad\qquad = +\infty$$

With this information we obtain the sketch of the graph of f shown in Fig. 6.1.5. From the graph we observe that f is one-to-one on its domain.

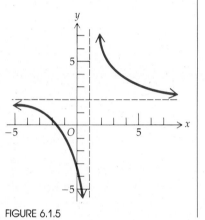

FIGURE 6.1.5

▶ ILLUSTRATION 3   Consider the equation

$$y = x^3 \qquad -2 \leq x \leq 2 \tag{3}$$

This equation defines the one-to-one function g discussed prior to Definition 6.1.1, where

$$g(x) = x^3 \qquad -2 \leq x \leq 2 \tag{4}$$

Function g is the set of ordered pairs (x, y) satisfying (3). If we solve (3) for x, we obtain

$$x = \sqrt[3]{y} \qquad -8 \leq y \leq 8 \tag{5}$$

which defines a function G where

$$G(y) = \sqrt[3]{y} \qquad -8 \leq y \leq 8 \tag{6}$$

The function G is the set of ordered pairs (y, x) satisfying (5).   ◀

The function G of Illustration 3 is called the *inverse* of the function g. The notation g^{-1} denotes the inverse of g; it is read "g inverse." Note that the symbol -1 used to denote the inverse of a function should not be confused with the exponent -1. We have the following formal definition of the inverse of a function.

6.1.3 DEFINITION   If f is a one-to-one function that is the set of ordered pairs (x, y), then there is a function f^{-1}, called the **inverse** of f, where f^{-1} is the set of ordered pairs (y, x) defined by

$$x = f^{-1}(y) \quad \text{if and only if} \quad y = f(x) \tag{7}$$

The domain of f^{-1} is the range of f and the range of f^{-1} is the domain of f.

We eliminate y between the equations in (7) by writing the equation

$$f^{-1}(y) = x$$

and replacing y by $f(x)$ to obtain

$$f^{-1}(f(x)) = x \tag{8}$$

where x is in the domain of f.

We eliminate x between the same pair of equations by writing the equation

$$f(x) = y$$

and replacing x by $f^{-1}(y)$. We get

$$f(f^{-1}(y)) = y \tag{9}$$

where y is in the domain of f^{-1}. Because the symbol used for the independent variable is arbitrary, we can replace y by x in (9) to obtain

$$f(f^{-1}(x)) = x \tag{10}$$

where x is in the domain of f^{-1}.

From (8) and (10) we see that if the inverse of the function f is the function f^{-1}, then the inverse of f^{-1} is f. We state this fact as well as (8) and (10) formally as the following theorem.

6.1.4 THEOREM   If f is a one-to-one function having f^{-1} as its inverse, then f^{-1} is a one-to-one function having f as its inverse. Furthermore,

$$f^{-1}(f(x)) = x \qquad \text{for } x \text{ in the domain of } f$$

and

$$f(f^{-1}(x)) = x \qquad \text{for } x \text{ in the domain of } f^{-1}$$

We use the terminology *inverse functions* when referring to a function and its inverse.

▶ ILLUSTRATION 4   Because the functions g and G defined by (4) and (6) in Illustration 3 are inverse functions, g^{-1} can be written in place of G. We have, from (4),

$$g(x) = x^3 \qquad -2 \le x \le 2$$

and from (6),

$$g^{-1}(y) = \sqrt[3]{y} \qquad -8 \le y \le 8$$

or, equivalently, if we replace y by x,

$$g^{-1}(x) = \sqrt[3]{x} \qquad -8 \le x \le 8$$

Observe that the domain of g is $[-2, 2]$, which is the range of g^{-1}; also the range of g is $[-8, 8]$, which is the domain of g^{-1}.   ◀

If a function f is defined by the equation

$$y = f(x)$$

and if f has an inverse, then f^{-1} can be found by solving the equation for x in terms of y and obtaining

$$x = f^{-1}(y)$$

Then if we interchange x and y in this equation, we get

$$y = f^{-1}(x)$$

▶ ILLUSTRATION 5   Each of the functions f in parts (a), (b), and (c) of Illustration 1 is one-to-one. Therefore $f^{-1}(x)$ exists. For each function f we compute $f^{-1}(x)$ from the definition of $f(x)$ by substituting y for $f(x)$ and solving the resulting equation for x. This procedure gives the equation $x = f^{-1}(y)$. We then have the definition of $f^{-1}(y)$ from which we obtain $f^{-1}(x)$.

(a) $f(x) = x + 4$ (b) $f(x) = 2x$ (c) $f(x) = x^3$

$\qquad y = x + 4$ $\qquad\qquad y = 2x$ $\qquad\qquad y = x^3$

$\qquad x = y - 4$ $\qquad\qquad\qquad x = \dfrac{2}{y}$ $\qquad\qquad x = \sqrt[3]{y}$

$\qquad f^{-1}(y) = y - 4$ $\qquad\qquad\qquad\qquad f^{-1}(y) = \sqrt[3]{y}$

$\qquad f^{-1}(x) = x - 4$ $\qquad\quad f^{-1}(y) = \dfrac{2}{y}$ $\qquad f^{-1}(x) = \sqrt[3]{x}$

$$f^{-1}(x) = \frac{2}{x}$$

Observe that the function f^{-1} in each part is the function g in the corresponding part of Illustration 1.   ◀

EXAMPLE 3
Find $f^{-1}(x)$ for the function f of Example 1(a). Verify the equations of Theorem 6.1.4 for f and f^{-1}. Draw a sketch of the graph of f^{-1}.

SOLUTION
In Example 1(a) we showed that the function f defined by

$$f(x) = 4x - 3$$

is one-to-one. Therefore f^{-1} exists. To find $f^{-1}(x)$ write the equation

$$y = 4x - 3$$

and solve for x. We obtain

$$x = \frac{y + 3}{4}$$

Therefore

$$f^{-1}(y) = \frac{y + 3}{4} \quad \Leftrightarrow \quad f^{-1}(x) = \frac{x + 3}{4}$$

We verify the equations of Theorem 6.1.4.

$$f^{-1}(f(x)) = f^{-1}(4x - 3) \qquad f(f^{-1}(x)) = f\left(\frac{x + 3}{4}\right)$$
$$= \frac{(4x - 3) + 3}{4} \qquad\qquad = 4\left(\frac{x + 3}{4}\right) - 3$$
$$= \frac{4x}{4} \qquad\qquad\qquad = (x + 3) - 3$$
$$= x \qquad\qquad\qquad\quad = x$$

FIGURE 6.1.6

A sketch of the graph of f^{-1} is shown in Fig. 6.1.6.

In Fig. 6.1.3 there is a sketch of the graph of the function f of Example 3, and in Fig. 6.1.6 there is a sketch of the graph of f^{-1}. In Fig. 6.1.7 the graphs of f and f^{-1} are shown on the same set of axes. In this figure it appears intuitively true that if the point $Q(u, v)$ is on the graph of f, then the point $R(v, u)$ is on the graph of f^{-1}. That this situation actually occurs is demonstrated in the following paragraph.

In general, if Q is the point (u, v) and R is the point (v, u), the line segment QR is perpendicular to the line $y = x$ and is bisected by it. We state that the point Q is a *reflection of the point R* with respect to the line $y = x$, and the point R is a *reflection of the point Q* with respect to the line $y = x$. If x and y are interchanged in the equation $y = f(x)$, we obtain the equation $x = f(y)$, and the graph of the equation $x = f(y)$ is said to be a *reflection of the graph* of the equation $y = f(x)$ with respect to the line $y = x$. Because the equation $x = f(y)$ is equivalent to the equation $y = f^{-1}(x)$, we conclude that the graph of the equation $y = f^{-1}(x)$ is a reflection of the graph of the equation $y = f(x)$ with respect to the line $y = x$. Therefore, if a function has an inverse, the graphs of the functions are reflections of each other with respect to the line $y = x$.

FIGURE 6.1.7

▶ ILLUSTRATION 6   Functions g and g^{-1} of Illustration 4 are defined by

$$g(x) = x^3 \qquad -2 \le x \le 2$$

and

$$g^{-1}(x) = \sqrt[3]{x} \qquad -8 \le x \le 8$$

FIGURE 6.1.8

Sketches of the graphs of g and g^{-1} are shown on the same set of axes in Fig. 6.1.8. Observe that these graphs are reflections of each other with respect to the line $y = x$. ◄

EXAMPLE 4
Find $f^{-1}(x)$ for the function f of Example 2. Verify the equations of Theorem 6.1.4 for f and f^{-1}. Draw sketches of the graphs of f and f^{-1} on the same set of axes.

SOLUTION
The function f of Example 2 is defined by

$$f(x) = \frac{2x + 3}{x - 1}$$

In Example 2 we showed that f is one-to-one. Therefore f has an inverse f^{-1}. To find $f^{-1}(x)$, let $y = f(x)$ and solve for x, thus giving $x = f^{-1}(y)$. So we have

$$y = \frac{2x + 3}{x - 1}$$

$$xy - y = 2x + 3$$

$$x(y - 2) = y + 3$$

$$x = \frac{y + 3}{y - 2}$$

Therefore

$$f^{-1}(y) = \frac{y + 3}{y - 2} \quad \Leftrightarrow \quad f^{-1}(x) = \frac{x + 3}{x - 2}$$

FIGURE 6.1.9

The domain of f^{-1} is the set of all real numbers except 2. We verify the equations of Theorem 6.1.4.

$$f^{-1}(f(x)) = f^{-1}\left(\frac{2x+3}{x-1}\right) \qquad f(f^{-1}(x)) = f\left(\frac{x+3}{x-2}\right)$$

$$= \frac{\left(\dfrac{2x+3}{x-1}\right)+3}{\left(\dfrac{2x+3}{x-1}\right)-2} \qquad\qquad = \frac{2\left(\dfrac{x+3}{x-2}\right)+3}{\left(\dfrac{x+3}{x-2}\right)-1}$$

$$= \frac{(2x+3)+3(x-1)}{(2x+3)-2(x-1)} \qquad = \frac{2(x+3)+3(x-2)}{(x+3)-(x-2)}$$

$$= \frac{5x}{5} \qquad\qquad\qquad = \frac{5x}{5}$$

$$= x \qquad\qquad\qquad\qquad = x$$

Sketches of the graphs of f and f^{-1} on the same set of axes are shown in Fig. 6.1.9.

We have the following theorem about the inverse of a function that is continuous and monotonic on a closed interval.

6.1.5 THEOREM   Suppose the function f has the closed interval $[a, b]$ as its domain. Then

(i) if f is continuous and increasing on $[a, b]$, f has an inverse f^{-1} that is defined on $[f(a), f(b)]$;

(ii) if f is continuous and decreasing on $[a, b]$, f has an inverse f^{-1} that is defined on $[f(b), f(a)]$.

PROOF   For part (i), if f is continuous on $[a, b]$ and if k is any number such that $f(a) < k < f(b)$, then by the intermediate-value theorem (1.7.8) there exists a number c in (a, b) such that $f(c) = k$. Therefore the range of f is the closed interval $[f(a), f(b)]$. Because f is increasing on $[a, b]$, f is one-to-one and so f has an inverse f^{-1}. Because the domain of f^{-1} is the range of f, f^{-1} is defined on $[f(a), f(b)]$.

The proof of part (ii) is similar. However, because f is decreasing on $[a, b]$, $f(a) > f(b)$; so the range of f is $[f(b), f(a)]$. Thus f^{-1} is defined on $[f(b), f(a)]$.   ■

▶ ILLUSTRATION 7   Let f be the function defined by

$$f(x) = x^2 - 1$$

A sketch of the graph of f is shown in Fig. 6.1.10. Because f is not one-to-one, f does not have an inverse.

However, let us restrict the domain to the closed interval $[0, c]$ and consider the function f_1 defined by

$$f_1(x) = x^2 - 1 \qquad x \in [0, c] \tag{11}$$

FIGURE 6.1.10

Also consider the function f_2 for which the domain is restricted to the closed interval $[-c, 0]$; that is

$$f_2(x) = x^2 - 1 \qquad x \in [-c, 0] \tag{12}$$

The functions f_1 and f_2 defined by (11) and (12) are distinct functions because their domains are different. We find the derivatives of f_1 and f_2 and obtain

$$f_1'(x) = 2x \qquad x \in [0, c]$$
$$f_2'(x) = 2x \qquad x \in [-c, 0]$$

Because f_1 is continuous on $[0, c]$ and $f_1'(x) > 0$ for all x in $(0, c)$, by Theorem 3.4.3(i) f_1 is increasing on $[0, c]$. Therefore, by Theorem 6.1.5(i), f_1 has an inverse f_1^{-1} that is defined on $[f_1(0), f_1(c)]$ which is $[-1, c^2 - 1]$. We obtain $f_1^{-1}(x)$ by replacing $f_1(x)$ by y in (11) and solving for x, where $x \geq 0$. We get

$$y = x^2 - 1$$
$$x^2 = y + 1$$
$$x = \sqrt{y + 1}$$

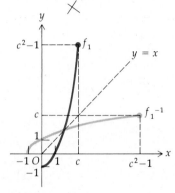

Therefore $f_1^{-1}(y) = \sqrt{y + 1}$, and if we replace y by x,

$$f_1^{-1}(x) = \sqrt{x + 1} \qquad x \in [-1, c^2 - 1] \tag{13}$$

FIGURE 6.1.11

Because f_2 is continuous on $[-c, 0]$ and because $f_2'(x) < 0$ for all x in $(-c, 0)$, by Theorem 3.4.3(ii) f_2 is decreasing on $[-c, 0]$. Thus, by Theorem 6.1.5(ii), f_2 has an inverse f_2^{-1} that is defined on $[f_2(0), f_2(-c)]$ which is $[-1, c^2 - 1]$. If in (12) we replace $f_2(x)$ by y and solve for x, where $x \leq 0$, we obtain

$$x = -\sqrt{y + 1}$$

Therefore $f_2^{-1}(y) = -\sqrt{y + 1}$, and

$$f_2^{-1}(x) = -\sqrt{x + 1} \qquad x \in [-1, c^2 - 1] \tag{14}$$

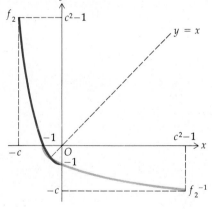

In Fig. 6.1.11 there are sketches of the graphs of f_1 and its inverse f_1^{-1}, as defined in (11) and (13), plotted on the same set of axes. Sketches of the graphs of f_2 and its inverse f_2^{-1}, as defined in (12) and (14), are plotted on the same set of axes in Fig. 6.1.12. In both figures observe that the graphs of the function and its inverse are reflections of each other with respect to the line $y = x$. ◄

FIGURE 6.1.12

There are functions that have an inverse but for which we cannot obtain an equation defining the inverse function explicitly. For example, let

$$f(x) = x^5 + 5x^3 + 2x - 4 \tag{15}$$
$$f'(x) = 5x^4 + 15x^2 + 2$$

Because $f'(x) > 0$ for all x, f is an increasing function, and therefore f has an inverse. However, if we replace $f(x)$ by y in (15), we obtain a fifth-degree equation in x. Equations of the fifth degree and higher in general do not have solutions that can be expressed by a formula. Nevertheless we can determine some properties of the inverse function. These properties are discussed in Section 6.2.

EXERCISES 6.1

In Exercises 1 through 18, determine if the given function is one-to-one. Draw a sketch of the graph of the function.

1. $f(x) = 2x + 3$
2. $g(x) = 8 - 4x$
3. $f(x) = \frac{1}{2}x^2 - 2$
4. $f(x) = 3 - x^2$
5. $g(x) = 4 - x^3$
6. $h(x) = \frac{1}{2}x^3 + 1$
7. $f(x) = \sqrt{x + 3}$
8. $g(x) = \sqrt{1 - x^2}$
9. $h(x) = 2 \sin x, -\frac{1}{2}\pi \le x \le \frac{1}{2}\pi$
10. $f(x) = 1 - \cos x, 0 \le x \le \pi$
11. $f(x) = \frac{1}{2}\tan x, -\frac{1}{2}\pi < x < \frac{1}{2}\pi$
12. $F(x) = \cot \frac{1}{2}x, 0 < x < 2\pi$
13. $G(x) = \sec x, x \in [0, \frac{1}{2}\pi) \cup [\pi, \frac{3}{2}\pi)$
14. $g(x) = \csc x, x \in (0, \frac{1}{2}\pi] \cup (\pi, \frac{3}{2}\pi]$
15. $f(x) = \dfrac{2}{x + 3}$
16. $g(x) = 5$
17. $g(x) = |x - 2|$
18. $f(x) = \dfrac{1}{2x - 4}$

In Exercises 19 through 40, determine if the given function has an inverse. If the inverse exists, do the following: (a) Find it and state its domain and range; (b) draw sketches of the graphs of the function and its inverse on the same set of axes. If the function does not have an inverse, show that a horizontal line intersects the graph of the function in more than one point.

19. $f(x) = 5x - 7$
20. $f(x) = 3x + 6$
21. $g(x) = 1 - x^2$
22. $g(x) = x^5$
23. $f(x) = (4 - x)^3$
24. $F(x) = 3(x^2 + 1)$
25. $h(x) = \sqrt{2x - 6}$
26. $g(x) = \sqrt{1 - x^2}$
27. $F(x) = \sqrt[3]{x + 1}$
28. $f(x) = |x| + x$
29. $f(x) = (x + 2)^4$
30. $g(x) = 3\sqrt[3]{x} + 1$
31. $f(x) = 2\sqrt[5]{x}$
32. $f(x) = \dfrac{2x - 1}{x}$
33. $f(x) = \dfrac{x - 3}{x + 1}$
34. $g(x) = \dfrac{8}{x^3 + 1}$
35. $g(x) = x^2 + 5, x \ge 0$
36. $f(x) = (2x - 1)^2, x \le \frac{1}{2}$
37. $f(x) = (2x + 1)^3, -\frac{1}{2} \le x \le \frac{1}{2}$
38. $f(x) = \frac{1}{8}x^3, -1 \le x \le 1$
39. $F(x) = \sqrt{9 - x^2}, 0 \le x \le 3$
40. $G(x) = \sqrt{4x^2 - 9}, x \ge \frac{3}{2}$

In Exercises 41 through 46, (a) prove that the given function f has an inverse, (b) find $f^{-1}(x)$, and (c) verify the equations of Theorem 6.1.4 for f and f^{-1}.

41. $f(x) = 4x - 3$
42. $f(x) = 5x + 2$
43. $f(x) = x^3 + 2$
44. $f(x) = (x + 2)^3$
45. $f(x) = \dfrac{3x + 1}{2x + 4}$
46. $f(x) = \dfrac{x - 3}{3x - 6}$

In Exercises 47 through 50, do the following: (a) Show that the given function f is not one-to-one and hence does not have an inverse; (b) restrict the domain and obtain two one-to-one functions, f_1 and f_2, having the same range as f; (c) find $f_1^{-1}(x)$ and $f_2^{-1}(x)$ and state the domains of f_1^{-1} and f_2^{-1}; (d) draw sketches of the graphs of f_1 and f_1^{-1} on the same set of axes; (e) draw sketches of the graphs of f_2 and f_2^{-1} on the same set of axes.

47. $f(x) = x^2 + 4$
48. $f(x) = 2x^2 - 6$
49. $f(x) = \sqrt{9 - x^2}$
50. $f(x) = \frac{1}{2}\sqrt{x^2 - 16}$

51. Given

$$f(x) = \begin{cases} x & \text{if } x < 1 \\ x^2 & \text{if } 1 \le x \le 9 \\ 27\sqrt{x} & \text{if } 9 < x \end{cases}$$

Prove that f has an inverse function and find $f^{-1}(x)$.

52. Determine the value of the constant k so that the function defined by

$$f(x) = \frac{x + 5}{x + k}$$

will be its own inverse.

6.2
INVERSE FUNCTION THEOREM AND THE DERIVATIVE OF THE INVERSE OF A FUNCTION

At the conclusion of Section 6.1 we stated that a function f can have an inverse f^{-1} that is not defined explicitly by an equation. However, information about the continuity and differentiability of f^{-1} can be acquired from properties of f.

▶ ILLUSTRATION 1   In Example 3 of Section 6.1 we had the function f and its inverse f^{-1} defined by

$$f(x) = 4x - 3 \qquad f^{-1}(x) = \frac{x + 3}{4}$$

In Fig. 6.1.7 the graphs of f and f^{-1} are shown on the same set of axes. We see that both f and f^{-1} are continuous and increasing on their domains.   ◀

▶ ILLUSTRATION 2   In Illustration 6 of Section 6.1 the function g and its inverse g^{-1} are defined by

$$g(x) = x^3 \qquad -2 \le x \le 2$$
$$g^{-1}(x) = \sqrt[3]{x} \qquad -8 \le x \le 8$$

and the graphs of g and g^{-1} are shown in Fig. 6.1.8. Each of these functions is continuous and increasing on its domain.   ◀

In Illustrations 1 and 2 there are examples of a function and its inverse that satisfy the conditions in the following theorem, known as the *inverse function theorem* for increasing functions. The existence of the inverse function in the statement of the theorem is guaranteed by Theorem 6.1.5.

6.2.1 THEOREM
Inverse Function Theorem

Suppose that the function f is continuous and increasing on the closed interval $[a, b]$. Then if f^{-1} is its inverse, which is defined on $[f(a), f(b)]$,

(i) f^{-1} is increasing on $[f(a), f(b)]$.
(ii) f^{-1} is continuous on $[f(a), f(b)]$.

PROOF OF (I)   To prove that f^{-1} is increasing on $[f(a), f(b)]$ we must show that if y_1 and y_2 are two numbers in $[f(a), f(b)]$ such that $y_1 < y_2$, then $f^{-1}(y_1) < f^{-1}(y_2)$. Because f^{-1} is defined on $[f(a), f(b)]$, there exist numbers x_1 and x_2 in $[a, b]$ such that $y_1 = f(x_1)$ and $y_2 = f(x_2)$. Therefore

$$f^{-1}(y_1) = f^{-1}(f(x_1)) \qquad f^{-1}(y_2) = f^{-1}(f(x_2))$$
$$= x_1 \qquad\qquad\qquad\qquad = x_2 \qquad\qquad (1)$$

If $x_2 < x_1$, then because f is increasing on $[a, b]$, $f(x_2) < f(x_1)$ or, equivalently, $y_2 < y_1$. But $y_1 < y_2$; therefore x_2 cannot be less than x_1.

If $x_2 = x_1$, then because f is a function, $f(x_1) = f(x_2)$ or, equivalently, $y_1 = y_2$, but this also contradicts the fact that $y_1 < y_2$. Therefore $x_2 \neq x_1$.

So if x_2 is not less than x_1 and $x_2 \neq x_1$, it follows that $x_1 < x_2$; hence, from the two equations in (1), $f^{-1}(y_1) < f^{-1}(y_2)$. Thus we have proved that f^{-1} is increasing on $[f(a), f(b)]$.

PROOF OF (ii)   To prove that f^{-1} is continuous on the closed interval $[f(a), f(b)]$ we must show that if r is any number in the open interval $(f(a), f(b))$, then f^{-1} is continuous at r, and f^{-1} is continuous from the right at $f(a)$, and f^{-1} is continuous from the left at $f(b)$.

We prove that f^{-1} is continuous at any r in the open interval $(f(a), f(b))$ by showing that Theorem 1.6.6 holds at r. We wish to show that, for any $\epsilon > 0$ small enough so that $f^{-1}(r) - \epsilon$ and $f^{-1}(r) + \epsilon$ are both in $[a, b]$, there exists a $\delta > 0$ such that

$$\text{if} \quad |y - r| < \delta \quad \text{then} \quad |f^{-1}(y) - f^{-1}(r)| < \epsilon$$

Let $f^{-1}(r) = s$. Then $f(s) = r$. Because, from (i), f^{-1} is increasing on $[f(a), f(b)]$, we conclude that $a < s < b$. Therefore

$$a \leq s - \epsilon < s < s + \epsilon \leq b$$

Because f is increasing on $[a, b]$,

$$f(a) \leq f(s - \epsilon) < r < f(s + \epsilon) \leq f(b) \tag{2}$$

Let δ be the smaller of the two numbers $r - f(s - \epsilon)$ and $f(s + \epsilon) - r$; so $\delta \leq r - f(s - \epsilon)$ and $\delta \leq f(s + \epsilon) - r$ or, equivalently,

$$f(s - \epsilon) \leq r - \delta \tag{3}$$

and

$$r + \delta \leq f(s + \epsilon) \tag{4}$$

If $|y - r| < \delta$, then $-\delta < y - r < \delta$ or, equivalently,

$$r - \delta < y < r + \delta \tag{5}$$

From (2), (3), (4), and (5) we have, if $|y - r| < \delta$, then

$$f(a) \leq f(s - \epsilon) < y < f(s + \epsilon) \leq f(b)$$

Because f^{-1} is increasing on $[f(a), f(b)]$, it follows from the above that

$$\text{if} \quad |y - r| < \delta \quad \text{then} \quad f^{-1}(f(s - \epsilon)) < f^{-1}(y) < f^{-1}(f(s + \epsilon))$$

$\Leftrightarrow \quad$ if $\quad |y - r| < \delta \quad$ then $\quad s - \epsilon < f^{-1}(y) < s + \epsilon$

$\Leftrightarrow \quad$ if $\quad |y - r| < \delta \quad$ then $\quad -\epsilon < f^{-1}(y) - s < \epsilon$

$\Leftrightarrow \quad$ if $\quad |y - r| < \delta \quad$ then $\quad |f^{-1}(y) - f^{-1}(r)| < \epsilon$

So f^{-1} is continuous on the open interval $(f(a), f(b))$.

The proofs that f^{-1} is continuous from the right at $f(a)$ and continuous from the left at $f(b)$ are left as an exercise (see Exercise 41).   ∎

The inverse function theorem for decreasing functions is now stated. The proof is similar to that for Theorem 6.2.1 and is left as an exercise (see Exercises 42 and 43).

6.2.2 THEOREM
(Inverse Function Theorem)

Suppose that the function f is continuous and decreasing on the closed interval $[a, b]$. Then if f^{-1} is its inverse, which is defined on $[f(b), f(a)]$,

(i) f^{-1} is decreasing on $[f(b), f(a)]$;
(ii) f^{-1} is continuous on $[f(b), f(a)]$.

The inverse function theorem is used to prove the following theorem, which expresses a relationship between the derivative of a function and the derivative of its inverse if the function has an inverse. In the statement and proof of the theorem we use the Leibniz notation for a derivative. Notice how this notation used in Eq. (6) makes the equation easy to remember.

6.2.3 THEOREM

Suppose that the function f is continuous and monotonic on the closed interval $[a, b]$, and let $y = f(x)$. If f is differentiable on $[a, b]$ and $f'(x) \neq 0$ for any x in $[a, b]$, then the derivative of the inverse function f^{-1}, defined by $x = f^{-1}(y)$, is given by

$$\frac{dx}{dy} = \frac{1}{\dfrac{dy}{dx}} \tag{6}$$

PROOF   Because f is continuous and monotonic on $[a, b]$, then by Theorems 6.2.1 and 6.2.2, f has an inverse that is continuous and monotonic on $[f(a), f(b)]$ (or $[f(b), f(a)]$ if $f(b) < f(a)$).

If x is a number in $[a, b]$, let Δx be an increment of x, $\Delta x \neq 0$, such that $x + \Delta x$ is also in $[a, b]$. Then the corresponding increment of y is given by

$$\Delta y = f(x + \Delta x) - f(x) \tag{7}$$

$\Delta y \neq 0$ since $\Delta x \neq 0$ and f is monotonic on $[a, b]$; that is, either

$$f(x + \Delta x) < f(x) \quad \text{or} \quad f(x + \Delta x) > f(x) \qquad \text{on } [a, b]$$

If x is in $[a, b]$ and $y = f(x)$, then y is in $[f(a), f(b)]$ (or $[f(b), f(a)]$). Also, if $x + \Delta x$ is in $[a, b]$, then $y + \Delta y$ is in $[f(a), f(b)]$ (or $[f(b), f(a)]$) because $y + \Delta y = f(x + \Delta x)$ by (7). So

$$x = f^{-1}(y) \tag{8}$$

and

$$x + \Delta x = f^{-1}(y + \Delta y) \tag{9}$$

It follows from (8) and (9) that

$$\Delta x = f^{-1}(y + \Delta y) - f^{-1}(y) \tag{10}$$

From the definition of a derivative,

$$\frac{dx}{dy} = \lim_{\Delta y \to 0} \frac{f^{-1}(y + \Delta y) - f^{-1}(y)}{\Delta y}$$

Substituting from (7) and (10) into the above equation we get

$$\frac{dx}{dy} = \lim_{\Delta y \to 0} \frac{\Delta x}{f(x + \Delta x) - f(x)}$$

and because $\Delta x \neq 0$,

$$\frac{dx}{dy} = \lim_{\Delta y \to 0} \frac{1}{\dfrac{f(x + \Delta x) - f(x)}{\Delta x}} \tag{11}$$

Before we find the limit in (11) we show that under the hypothesis of this theorem $\Delta x \to 0$ is equivalent to $\Delta y \to 0$. First we show that $\lim\limits_{\Delta y \to 0} \Delta x = 0$. From (10),

$$\lim_{\Delta y \to 0} \Delta x = \lim_{\Delta y \to 0} \left[f^{-1}(y + \Delta y) - f^{-1}(y) \right]$$

Because f^{-1} is continuous on $[f(a), f(b)]$ (or $[f(b), f(a)]$), the limit on the right side of the above equation is zero. So

$$\lim_{\Delta y \to 0} \Delta x = 0 \tag{12}$$

Now we demonstrate that $\lim\limits_{\Delta x \to 0} \Delta y = 0$. From (7),

$$\lim_{\Delta x \to 0} \Delta y = \lim_{\Delta x \to 0} \left[f(x + \Delta x) - f(x) \right]$$

Because f is continuous on $[a, b]$, the limit on the right side of the above equation is zero, and therefore

$$\lim_{\Delta x \to 0} \Delta y = 0 \tag{13}$$

From (12) and (13) it follows that

$$\Delta x \to 0 \quad \text{if and only if} \quad \Delta y \to 0 \tag{14}$$

Thus, applying the limit theorem (regarding the limit of a quotient) to (11) and using (14), we have

$$\frac{dx}{dy} = \frac{1}{\lim\limits_{\Delta x \to 0} \dfrac{f(x + \Delta x) - f(x)}{\Delta x}}$$

Because f is differentiable on $[a, b]$, the limit in the denominator of the above is $f'(x)$ or, equivalently, $\dfrac{dy}{dx}$. Thus

$$\frac{dx}{dy} = \frac{1}{\dfrac{dy}{dx}} \qquad \blacksquare$$

▶ ILLUSTRATION 3   We verify Theorem 6.2.3 for the function f defined by $f(x) = \sqrt{x}$. If we let $y = f(x)$, we have the equation

$$y = \sqrt{x} \qquad x \geq 0, \, y \geq 0 \tag{15}$$

Because f is one-to-one, f^{-1} exists and is defined by $f^{-1}(y) = y^2$. If we let $x = f^{-1}(y)$, we have the equation

$$x = y^2 \qquad y \geq 0, \, x \geq 0 \tag{16}$$

From (15),

$$\frac{dy}{dx} = \frac{1}{2\sqrt{x}}$$

and from (16),

$$\frac{dx}{dy} = 2y \tag{17}$$

Substituting from (15) into (17) we have

$$\frac{dx}{dy} = 2\sqrt{x}$$

$$= \frac{1}{\dfrac{1}{2\sqrt{x}}}$$

$$= \frac{1}{\dfrac{dy}{dx}} \tag{18}$$

When $x = 0$, $\dfrac{dy}{dx}$ does not exist; thus (18) is not satisfied for this value of x. Because the domain of f is the closed interval $[0, +\infty)$, the theorem is valid for any x in the open interval $(0, +\infty)$.   ◀

EXAMPLE 1
Show that Theorem 6.2.3 holds for the function f of Examples 2 and 4 in Section 6.1.

SOLUTION
The function f is defined by $(2x + 3)/(x - 1)$. If we let $y = f(x)$, we have

$$y = \frac{2x + 3}{x - 1} \tag{19}$$

$$\frac{dy}{dx} = -\frac{5}{(x - 1)^2}$$

In the solution of Example 4 in Section 6.1 we showed that

$$x = \frac{y + 3}{y - 2}$$

We compute $\dfrac{dx}{dy}$ from this equation and get

$$\frac{dx}{dy} = -\frac{5}{(y - 2)^2}$$

In the above equation we substitute the value of y from (19) and obtain

$$\frac{dx}{dy} = -\frac{5}{\left(\dfrac{2x + 3}{x - 1} - 2\right)^2}$$

$$= -\frac{5(x - 1)^2}{(2x + 3 - 2x + 2)^2}$$

$$= -\tfrac{5}{25}(x - 1)^2$$

$$= -\tfrac{1}{5}(x - 1)^2$$

$$= \frac{1}{\dfrac{dy}{dx}}$$

EXAMPLE 2

Given that f is the function defined by

$$f(x) = x^3 + x$$

determine if f has an inverse. If it does, find the derivative of the inverse function.

SOLUTION

$$f'(x) = 3x^2 + 1$$

Therefore $f'(x) > 0$ for all real numbers; so f is increasing on its domain. Thus f is a one-to-one function, and hence it has an inverse f^{-1}. Let $y = f(x)$, and then $x = f^{-1}(y)$. By Theorem 6.2.3,

$$\frac{dx}{dy} = \frac{1}{\dfrac{dy}{dx}}$$

$$= \frac{1}{3x^2 + 1}$$

When using Theorem 6.2.3 to compute the value of the derivative of the inverse of a function at a particular number, it is more convenient to have the statement of the theorem with f' and $(f^{-1})'$ notations for the derivatives. With this notation we restate the theorem as Theorem 6.2.4.

6.2.4 THEOREM   Suppose the function f is continuous and monotonic on a closed interval $[a, b]$ containing the number c, and let $f(c) = d$. If $f'(c)$ exists and $f'(c) \neq 0$, then $(f^{-1})'(d)$ exists and

$$(f^{-1})'(d) = \frac{1}{f'(c)}$$

▶ ILLUSTRATION 4   We show that Theorem 6.2.4 holds for a particular function and particular values of c and d. If f is the function of Illustration 3,

$$f(x) = \sqrt{x} \qquad f'(x) = \frac{1}{2\sqrt{x}}$$

$$f^{-1}(x) = x^2 \qquad (f^{-1})'(x) = 2x \qquad x \geq 0$$

The function f is continuous and monotonic on any closed interval $[a, b]$ for which $0 \leq a < b$. Let $c = 9$, and then $d = f(9)$; that is, $d = 3$. Theorem 6.2.4 states that

$$(f^{-1})'(3) = \frac{1}{f'(9)} \tag{20}$$

Because $(f^{-1})'(3) = 6$ and $f'(9) = \frac{1}{6}$, Eq. (20) is verified.   ◀

▶ ILLUSTRATION 5   The function f mentioned at the conclusion of Section 6.1 is defined by

$$f(x) = x^5 + 5x^3 + 2x - 4 \tag{21}$$

and as stated there, f has an inverse f^{-1}. But we do not have an equation defining explicitly the function value of f^{-1}. Nevertheless it is possible to compute the derivative of f^{-1} at a particular point on the graph of $y = f^{-1}(x)$. For instance, because $(1, 4)$ is on the graph of $y = f(x)$, the point $(4, 1)$ is on the graph of $y = f^{-1}(x)$. We compute the value of $(f^{-1})'(4)$ by using Theorem 6.2.4, which states that

$$(f^{-1})'(4) = \frac{1}{f'(1)}$$

We first find $f'(x)$ from (21).

$$f'(x) = 5x^4 + 15x^2 + 2$$

Therefore

$$f'(1) = 22$$

Thus

$$(f^{-1})'(4) = \tfrac{1}{22}$$ ◀

EXAMPLE 3
(a) On the same set of axes draw sketches of the graphs of the functions f and f^{-1} of Example 2. (b) Find the slope of the tangent line to the graph of f at the point $(1, 2)$. (c) Find the slope of the tangent line to the graph of f^{-1} at the point $(2, 1)$.

SOLUTION
(a) For the function f,

$$f(x) = x^3 + x$$

$$f'(x) = 3x^2 + 1$$

$$f''(x) = 6x$$

The function is increasing on its domain. Because $f''(0) = 0$ and $f''(x)$ changes sign at $x = 0$, the graph of f has a point of inflection at the origin. The slope of the inflectional tangent at the origin is $f'(0) = 1$. We use this information to draw the sketch of the graph of f shown in Fig. 6.2.1. This figure also shows a sketch of the graph of f^{-1}, which is obtained from the fact that it is the reflection of the graph of f with respect to the line $y = x$. Observe that to find an equation defining $f^{-1}(x)$ explicitly it is necessary to solve the third degree equation $y = x^3 + x$.
(b) The slope of the tangent line to the graph of f at the point $(1, 2)$ is $f'(1) = 4$.
(c) The slope of the tangent line to the graph of f^{-1} at the point $(2, 1)$ is $(f^{-1})'(2)$. From Theorem 6.2.4,

$$(f^{-1})'(2) = \frac{1}{f'(1)}$$

$$= \tfrac{1}{4}$$

FIGURE 6.2.1

EXERCISES 6.2

In Exercises 1 through 10, let $y = f(x)$ and $x = f^{-1}(y)$, and verify that $\dfrac{dx}{dy} = \dfrac{1}{\frac{dy}{dx}}$.

1. $f(x) = 4x - 3$
2. $f(x) = 7 - 2x$
3. $f(x) = \sqrt{x + 1}$
4. $f(x) = 8x^3$
5. $f(x) = \tfrac{1}{5}x^5$
6. $f(x) = \sqrt{4 - 3x}$
7. $f(x) = \sqrt[3]{x - 8}$
8. $f(x) = \sqrt[5]{x}$
9. $f(x) = \dfrac{2x - 3}{x + 2}$
10. $f(x) = \dfrac{3x + 4}{2x + 6}$

In Exercises 11 through 28, find $(f^{-1})'(d)$ for the given f and d.
11. $f(x) = \sqrt{3x + 1};\ d = 1$
12. $f(x) = x^5 + 2;\ d = 1$
13. $f(x) = x^2 - 16,\ x \geq 0;\ d = 9$
14. $f(x) = \sqrt{4 - x};\ d = 3$
15. $f(x) = x^3 + 5;\ d = -3$
16. $f(x) = 4x^3 + 2x;\ d = 6$
17. $f(x) = 3x^5 + 2x^3;\ d = 5$
18. $f(x) = \sin x;\ -\tfrac{1}{2}\pi \leq x \leq \tfrac{1}{2}\pi;\ d = \tfrac{1}{2}$
19. $f(x) = \tfrac{1}{2}\cos^2 x;\ 0 \leq x \leq \tfrac{1}{2}\pi;\ d = \tfrac{1}{4}$
20. $f(x) = \tan 2x;\ -\tfrac{1}{4}\pi < x < \tfrac{1}{4}\pi;\ d = 1$
21. $f(x) = 2 \cot x;\ 0 < x < \pi;\ d = 2$
22. $f(x) = \sec \tfrac{1}{2}x;\ 0 \leq x < \pi;\ d = 2$

23. $f(x) = \frac{1}{2} \csc x; \ 0 < x < \frac{1}{2}\pi; \ d = 1$
24. $f(x) = x^2 - 6x + 7, \ x \leq 3; \ d = 0$
25. $f(x) = 2x^2 + 8x + 7, \ x \leq -2; \ d = 1$
26. $f(x) = 2x^3 + x + 20; \ d = 2$
27. $f(x) = \int_{-3}^{x} \sqrt{t + 3} \, dt, \ x > -3; \ d = 18$ $\frac{2}{3}(t+3)^{3/2}\Big|_{-3} = 18$
28. $f(x) = \int_{x}^{2} t \, dt, \ x < 0; \ d = -6$

29. Given $f(x) = \int_{1}^{x} \sqrt{16 - t^4} \, dt, \ -2 \leq x \leq 2$. Prove that f has an inverse f^{-1}, and compute $(f^{-1})'(0)$.

30. Given $f(x) = \int_{2}^{x} \sqrt{9 + t^4} \, dt$. Prove that f has an inverse f^{-1}, and compute $(f^{-1})'(0)$.

31. Given $f(x) = x^3 + 3x - 1$. (a) On the same set of axes draw sketches of the graphs of the functions f and f^{-1}. (b) Find the slope of the tangent line to the graph of f at the point $(1, 3)$. (c) Find the slope of the tangent line to the graph of f^{-1} at the point $(3, 1)$.

32. Given $f(x) = 6 - x - x^3$. (a) On the same set of axes draw sketches of the graphs of the functions f and f^{-1}. (b) Find the slope of the tangent line to the graph of f at the point $(2, -4)$. (c) Find the slope of the tangent line to the graph of f^{-1} at the point $(-4, 2)$.

In Exercises 33 through 38, do the following: (a) Solve the equation for y in terms of x, and express y as one or more functions of x; (b) for each of the functions obtained in (a) determine if the function has an inverse, and, if it does, determine the domain of the inverse function; (c) use implicit differentiation to find $\dfrac{dy}{dx}$ and $\dfrac{dx}{dy}$, and determine the values of x and y for which $\dfrac{dy}{dx}$ and $\dfrac{dx}{dy}$ are reciprocals.

33. $x^2 + y^2 = 9$ ⟶ not 1-1
34. $x^2 - 4y^2 = 16$
35. $xy = 4$
36. $9y^2 - 8x^3 = 0$
37. $2x^2 - 3xy + 1 = 0$
38. $2x^2 + 2y + 1 = 0$

39. Show that the formula of Theorem 6.2.3 can be written as

$$(f^{-1})'(x) = \frac{1}{f'(f^{-1}(x))}$$

40. Use the formula of Exercise 39 to show that

$$(f^{-1})''(x) = -\frac{f''(f^{-1}(x))}{[f'(f^{-1}(x))]^3}$$

41. Given that the function f is continuous and increasing on the closed interval $[a, b]$, by assuming Theorem 6.2.1(i) prove that f^{-1} is continuous from the right at $f(a)$ and continuous from the left at $f(b)$.

42. Prove Theorem 6.2.2(i).

43. Prove Theorem 6.2.2(ii).

6.3

THE NATURAL
LOGARITHMIC FUNCTION

The definition of the logarithmic function that you encountered in algebra is based on exponents. The properties of logarithms are then proved from corresponding properties of exponents. One such property of exponents is

$$a^x \cdot a^y = a^{x+y} \tag{1}$$

If the exponents x and y are positive integers and if a is any real number, (1) follows from the definition of a positive integer exponent and mathematical induction. If the exponents are allowed to be any integers, either positive, negative, or zero, and $a \neq 0$, then (1) will hold if a zero exponent and a negative integer exponent are defined by

$$a^0 = 1$$

and

$$a^{-n} = \frac{1}{a^n} \qquad n > 0$$

If the exponents are rational numbers and $a \geq 0$, then (1) holds when $a^{m/n}$ is defined by

$$a^{m/n} = (\sqrt[n]{a})^m$$

It is not quite so simple to define a^x when x is an irrational number. For example, what is meant by $2^{\sqrt{3}}$? The definition of the logarithmic function, as given in elementary algebra, is based on the assumption that a^x exists if a is any positive number and x is any real number.

This definition states that the equation

$$a^x = N$$

where a is any positive number except 1 and N is any positive number, can be solved for x, and x is uniquely determined by

$$x = \log_a N$$

The following properties of logarithms are proved from those of exponents:

$$\log_a 1 = 0 \tag{2}$$

$$\log_a MN = \log_a M + \log_a N \tag{3}$$

$$\log_a \frac{M}{N} = \log_a M - \log_a N \tag{4}$$

$$\log_a M^n = n \log_a M \tag{5}$$

$$\log_a a = 1 \tag{6}$$

In this chapter we define the logarithmic function by using calculus and prove the properties of logarithms by means of this definition. Then the exponential function is defined in terms of the logarithmic function. This definition enables us to define a^x when x is any real number and $a > 0$. The properties of exponents will then be proved if the exponent is any real number.

Recall the formula

$$\int t^n \, dt = \frac{t^{n+1}}{n+1} + C \qquad n \neq -1$$

This formula does not hold when $n = -1$.

To evaluate $\int t^n \, dt$ for $n = -1$ we need a function whose derivative is $\frac{1}{x}$. The first fundamental theorem of the calculus (4.8.1) gives us one; it is

$$\int_a^x \frac{1}{t} \, dt$$

FIGURE 6.3.1

FIGURE 6.3.2

where a can be any real number having the same sign as x. To interpret such a function let R_1 be the region bounded by the curve $y = 1/t$, by the t axis, on the left by the line $t = 1$, and on the right by the line $t = x$, where $x > 1$. This region R_1 is shown in Fig. 6.3.1. The measure of the area of R_1 is a function of x; call it $A(x)$ and define it as a definite integral by

$$A(x) = \int_1^x \frac{1}{t}\, dt \qquad (7)$$

Now consider the integral in (7) if $0 < x < 1$. From Definition 4.5.5,

$$\int_1^x \frac{1}{t}\, dt = -\int_x^1 \frac{1}{t}\, dt$$

Then the integral $\int_x^1 (1/t)\, dt$ represents the measure of the area of the region R_2 bounded by the curve $y = 1/t$, by the t axis, on the left by the line $t = x$, and on the right by the line $t = 1$. So the integral $\int_1^x (1/t)\, dt$ is then the negative of the measure of the area of the region R_2 shown in Fig. 6.3.2.

If $x = 1$, the integral $\int_1^x (1/t)\, dt$ becomes $\int_1^1 (1/t)\, dt$, which equals 0 by Definition 4.5.6. In this case the left and right boundaries of the region are the same and the measure of the area is 0.

Thus the integral $\int_1^x (1/t)\, dt$ for $x > 0$ can be interpreted in terms of the measure of the area of a region. Its value depends on x and is therefore a function of x. This integral is used to define the *natural logarithmic function*.

6.3.1 DEFINITION   The **natural logarithmic function** is the function defined by

$$\ln x = \int_1^x \frac{1}{t}\, dt \qquad x > 0$$

The domain of the natural logarithmic function is the set of all positive numbers. We read $\ln x$ as "the natural logarithm of x."

The natural logarithmic function is differentiable because by applying the first fundamental theorem of the calculus (4.8.1)

$$D_x(\ln x) = D_x\left(\int_1^x \frac{1}{t}\, dt \right)$$

$$= \frac{1}{x}$$

From this result and the chain rule we have the following theorem.

6.3.2 THEOREM   If u is a differentiable function of x and $u(x) > 0$, then

$$D_x(\ln u) = \frac{1}{u} \cdot D_x u$$

EXAMPLE 1
Given

$$f(x) = \ln(3x^2 - 6x + 8)$$

find $f'(x)$.

SOLUTION
From Theorem 6.3.2,

$$f'(x) = \frac{1}{3x^2 - 6x + 8}(6x - 6)$$

$$= \frac{6x - 6}{3x^2 - 6x + 8}$$

EXAMPLE 2
Given

$$y = \ln[(4x^2 + 3)(2x - 1)]$$

find $\dfrac{dy}{dx}$.

SOLUTION
Applying Theorem 6.3.2 we get

$$\frac{dy}{dx} = \frac{1}{(4x^2 + 3)(2x - 1)} \cdot [8x(2x - 1) + 2(4x^2 + 3)]$$

$$= \frac{24x^2 - 8x + 6}{(4x^2 + 3)(2x - 1)} \tag{8}$$

EXAMPLE 3
Given

$$y = \ln\left(\frac{x}{x + 1}\right)$$

find $\dfrac{dy}{dx}$.

SOLUTION
From Theorem 6.3.2,

$$\frac{dy}{dx} = \frac{1}{\dfrac{x}{x + 1}} \cdot \frac{(x + 1) - x}{(x + 1)^2}$$

$$= \frac{x + 1}{x} \cdot \frac{1}{(x + 1)^2}$$

$$= \frac{1}{x(x + 1)}$$

It should be emphasized that when using Theorem 6.3.2, $u(x)$ must be positive; that is, a number in the domain of the derivative must be in the domain of the given function, ln u.

▶ ILLUSTRATION 1   In Example 1 the domain of the given function is the set of all real numbers, because $3x^2 - 6x + 8 > 0$ for all x. This can be seen from the fact that the parabola having equation $y = 3x^2 - 6x + 8$ has its vertex at $(1, 5)$ and opens upward. Hence $(6x - 6)/(3x^2 - 6x + 8)$ is the derivative for all values of x.

In Example 2, because $(4x^2 + 3)(2x - 1) > 0$ only when $x > \frac{1}{2}$, the domain of the given function is the interval $(\frac{1}{2}, +\infty)$. Therefore it is understood that fraction (8) is the derivative only if $x > \frac{1}{2}$.

Because $x/(x + 1) > 0$ when either $x < -1$ or $x > 0$, the domain of the function in Example 3 is $(-\infty, -1) \cup (0, +\infty)$; so $1/[x(x + 1)]$ is the derivative if either $x < -1$ or $x > 0$.   ◀

We show that the natural logarithmic function obeys the properties of logarithms given earlier.

6.3.3 THEOREM	$\ln 1 = 0$

PROOF   If $x = 1$ in Definition 6.3.1,

$$\ln 1 = \int_1^1 \frac{1}{t}\, dt$$

The right side of the above is zero by Definition 4.5.6. Thus

$$\ln 1 = 0$$ ■

Theorem 6.3.3 corresponds to the property of logarithms given by (2). The next three theorems correspond to the properties of logarithms given by (3), (4), and (5). The discussion of property (6) is postponed until Section 6.5.

6.3.4 THEOREM	If a and b are any positive numbers, then
	$\ln(ab) = \ln a + \ln b$

PROOF   We apply Definition 6.3.1 and then Theorem 4.6.7 to get

$$\ln(ab) = \int_1^{ab} \frac{1}{t}\, dt$$

$$\ln(ab) = \int_1^a \frac{1}{t}\, dt + \int_a^{ab} \frac{1}{t}\, dt \tag{9}$$

In the second integral on the right side of (9) let $t = au$. Then $dt = a\, du$. When $t = a$, $u = 1$; and when $t = ab$, $u = b$. Thus

$$\int_a^{ab} \frac{1}{t}\, dt = \int_1^b \frac{1}{au} (a\, du)$$

$$\int_a^{ab} \frac{1}{t}\, dt = \int_1^b \frac{1}{u}\, du \tag{10}$$

Substituting from (10) in (9) we have

$$\ln(ab) = \int_1^a \frac{1}{t}\, dt + \int_1^b \frac{1}{u}\, du$$

$$= \ln a + \ln b \qquad \blacksquare$$

6.3.5 THEOREM   If a and b are any positive numbers, then

$$\ln \frac{a}{b} = \ln a - \ln b$$

PROOF   Because $a = (a/b) \cdot b$,

$$\ln a = \ln\left(\frac{a}{b} \cdot b\right)$$

Applying Theorem 6.3.4 to the right side of the above equation we get

$$\ln a = \ln \frac{a}{b} + \ln b$$

$$\ln \frac{a}{b} = \ln a - \ln b \qquad \blacksquare$$

6.3.6 THEOREM   If a is any positive number and r is any rational number, then

$$\ln a^r = r \ln a$$

PROOF   From Theorem 6.3.2, if r is any rational number and $x > 0$,

$$D_x(\ln x^r) = \frac{1}{x^r} \cdot r x^{r-1}$$

$$= \frac{r}{x}$$

and

$$D_x(r \ln x) = \frac{r}{x}$$

Therefore

$$D_x(\ln x^r) = D_x(r \ln x)$$

From the above equation the derivatives of $\ln x^r$ and $r \ln x$ are equal; so it follows from Theorem 4.1.2 that there is a constant K such that

$$\ln x^r = r \ln x + K \qquad \text{for all } x > 0 \tag{11}$$

To determine K, substitute 1 for x in (11) and get

$$\ln 1^r = r \ln 1 + K$$

But $\ln 1 = 0$; hence $K = 0$. Replacing K by 0 in (11) gives

$$\ln x^r = r \ln x \qquad \text{for all } x > 0$$

from which it follows that if $x = a$, where a is any positive number, then

$$\ln a^r = r \ln a \qquad\qquad\qquad\qquad\qquad \blacksquare$$

▶ ILLUSTRATION 2   In Example 2, if Theorem 6.3.4 is applied before finding the derivative, we have

$$y = \ln(4x^2 + 3) + \ln(2x - 1) \tag{12}$$

The domain of the function defined by the above equation is the interval $(\frac{1}{2}, +\infty)$, which is the same as the domain of the given function. From (12),

$$\frac{dy}{dx} = \frac{8x}{4x^2 + 3} + \frac{2}{2x - 1}$$

and combining the fractions gives

$$\frac{dy}{dx} = \frac{8x(2x - 1) + 2(4x^2 + 3)}{(4x^2 + 3)(2x - 1)}$$

which is equivalent to the first line of the solution of Example 2.   ◀

▶ ILLUSTRATION 3   If we apply Theorem 6.3.5 before finding the derivative in Example 3, we have

$$y = \ln x - \ln(x + 1) \tag{13}$$

Because $\ln x$ is defined only when $x > 0$, and $\ln(x + 1)$ is defined only when $x > -1$, the domain of the function defined by (13) is the interval $(0, +\infty)$. But the domain of the function given in Example 3 consists of the two intervals $(-\infty, -1)$ and $(0, +\infty)$. Computing the derivative from (13) we have

$$\frac{dy}{dx} = \frac{1}{x} - \frac{1}{x + 1}$$

$$= \frac{1}{x(x + 1)}$$

but remember here that x must be greater than 0, whereas in the solution of Example 3 values of x less than -1 are also included.   ◀

Illustration 3 shows the care that must be taken when applying Theorems 6.3.4, 6.3.5, and 6.3.6 to functions involving the natural logarithm.

EXAMPLE 4
Given

$$f(x) = \ln(2x - 1)^3$$

find $f'(x)$.

SOLUTION
From Theorem 6.3.6,

$$f(x) = 3 \ln(2x - 1)$$

Observe that $\ln(2x - 1)^3$ and $3 \ln(2x - 1)$ both have the same domain: $x > \frac{1}{2}$. Applying Theorem 6.3.2 gives

$$f'(x) = 3 \cdot \frac{1}{2x - 1} \cdot 2$$

$$= \frac{6}{2x - 1}$$

To draw a sketch of the graph of the natural logarithmic function we must first consider some properties of this function.

Let f be the function defined by $f(x) = \ln x$; that is,

$$f(x) = \int_1^x \frac{1}{t}\,dt \qquad x > 0$$

The domain of f is the set of all positive numbers. The function f is differentiable for all x in its domain, and

$$f'(x) = \frac{1}{x} \tag{14}$$

Because f is differentiable for all $x > 0$, f is continuous for all $x > 0$. From (14), $f'(x) > 0$ for all $x > 0$, and therefore f is an increasing function.

$$f''(x) = -\frac{1}{x^2} \tag{15}$$

From (15), $f''(x) < 0$ when $x > 0$. Therefore the graph of $y = f(x)$ is concave downward at every point. Because $f(1) = \ln 1$ and $\ln 1 = 0$, the x intercept of the graph is 1.

We can determine an approximate numerical value for $\ln 2$ by using the equation

$$\ln 2 = \int_1^2 \frac{1}{t}\,dt$$

and interpreting the definite integral as the number of square units in the area of the region appearing in Fig. 6.3.3. From this figure we observe that $\ln 2$ is between the measures of the areas of the rectangles, each having a base

FIGURE 6.3.3

of length 1 unit and altitudes of lengths $\frac{1}{2}$ unit and 1 unit; that is,

$$0.5 < \ln 2 < 1$$

An inequality can be obtained analytically, using Theorem 4.6.8, by proceeding as follows. Let $f(t) = 1/t$ and $g(t) = \frac{1}{2}$. Then $f(t) \geq g(t)$ for all t in $[1, 2]$. Because f and g are continuous on $[1, 2]$, they are integrable on $[1, 2]$, and from Theorem 4.6.8,

$$\int_1^2 \frac{1}{t} \, dt \geq \int_1^2 \frac{1}{2} \, dt$$

$$\ln 2 \geq \tfrac{1}{2} \tag{16}$$

Similarly, if $f(t) = 1/t$ and $h(t) = 1$, then $h(t) \geq f(t)$ for all t in $[1, 2]$. Because h and f are continuous on $[1, 2]$, they are integrable there; and again using Theorem 4.6.8 we obtain

$$\int_1^2 1 \, dt \geq \int_1^2 \frac{1}{t} \, dt$$

$$1 \geq \ln 2 \tag{17}$$

Combining (16) and (17) we get

$$0.5 \leq \ln 2 \leq 1 \tag{18}$$

The number 0.5 is a lower bound of $\ln 2$ and 1 is an upper bound of $\ln 2$. In a similar manner we can obtain a lower and upper bound for the natural logarithm of any positive real number. Later we learn, by applying infinite series, how to compute the natural logarithm of any positive real number to any desired number of decimal places.

The value of $\ln 2$ to five decimal places is given by

$$\ln 2 \approx 0.69315$$

Using this value of $\ln 2$ and applying Theorem 6.3.6 we can find an approximate value for the natural logarithm of any power of 2. In particular,

$\ln 4 = \ln 2^2$	$\ln 8 = \ln 2^3$	$\ln \frac{1}{2} = \ln 2^{-1}$	$\ln \frac{1}{4} = \ln 2^{-2}$
$= 2 \ln 2$	$= 3 \ln 2$	$= -1 \cdot \ln 2$	$= -2 \ln 2$
≈ 1.3863	≈ 2.0795	≈ -0.69315	≈ -1.3863

Natural logarithmic function values to four decimal places are given in Table 4 in the back of the book. A calculator with an $\boxed{\ln}$ key can also be used to obtain these values.

We now determine the behavior of the natural logarithmic function for large values of x by considering $\lim\limits_{x \to +\infty} \ln x$.

Because the natural logarithmic function is an increasing function, if we take p as any positive number, we have

$$\text{if} \quad x > 2^p \quad \text{then} \quad \ln x > \ln 2^p \tag{19}$$

From Theorem 6.3.6,

$$\ln 2^p = p \ln 2 \tag{20}$$

Substituting from (20) into (19) we get

if $x > 2^p$ then $\ln x > p \ln 2$

Because from (18) $\ln 2 \geq \frac{1}{2}$, we have from the above

if $x > 2^p$ then $\ln x > \frac{1}{2}p$

Letting $p = 2n$, where $n > 0$, we have

if $x > 2^{2n}$ then $\ln x > n \tag{21}$

It follows from (21), by taking $N = 2^{2n}$, that for any $n > 0$

if $x > N$ then $\ln x > n$

So we may conclude that

$$\lim_{x \to +\infty} \ln x = +\infty \tag{22}$$

To determine the behavior of the natural logarithmic function for positive values of x near zero we investigate $\lim_{x \to 0^+} \ln x$. Because $\ln x = \ln(x^{-1})^{-1}$,

$$\ln x = -\ln \frac{1}{x} \tag{23}$$

The expression "$x \to 0^+$" is equivalent to "$1/x \to +\infty$"; so from (23) we write

$$\lim_{x \to 0^+} \ln x = -\lim_{1/x \to +\infty} \ln \frac{1}{x} \tag{24}$$

From (22) we have

$$\lim_{1/x \to +\infty} \ln \frac{1}{x} = +\infty \tag{25}$$

Therefore, from (24) and (25) we get

$$\lim_{x \to 0^+} \ln x = -\infty \tag{26}$$

From (26) and (22) and the intermediate-value theorem (1.7.8) we conclude that the range of the natural logarithmic function is the set of all real numbers. From (26) we conclude that the graph of the natural logarithmic function is asymptotic to the negative part of the y axis through the fourth quadrant. The properties of the natural logarithmic function are summarized as follows

  (i) The domain is the set of all positive numbers.
 (ii) The range is the set of all real numbers.
(iii) The function is increasing on its entire domain.
(iv) The function is continuous at all numbers in its domain.

(v) The graph of the function is concave downward at all points.
(vi) The graph of the function is asymptotic to the negative part of the y axis through the fourth quadrant.

By using these properties and plotting a few points with a segment of the tangent line at the points, we can draw a sketch of the graph of the natural logarithmic function. In Fig. 6.3.4 we have plotted the points having abscissas of $\frac{1}{4}$, $\frac{1}{2}$, 1, 2, 4, and 8. The slope of the tangent line is found from the formula $D_x(\ln x) = 1/x$.

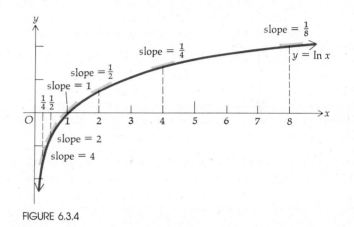

FIGURE 6.3.4

EXERCISES 6.3

In Exercises 1 through 26, differentiate the given function and simplify the result.

1. $f(x) = \ln(4 + 5x)$
2. $g(x) = \ln(1 + 4x^2)$
3. $h(x) = \ln\sqrt{4 + 5x}$
4. $f(x) = \ln(8 - 2x)$

5. $f(t) = \ln(3t + 1)^2$
6. $h(x) = \ln(8 - 2x)^5$
7. $g(t) = \ln^2(3t + 1)$
8. $G(x) = \ln\sqrt{1 + 4x^2}$

9. $f(x) = \ln\sqrt[3]{4 - x^2}$
10. $g(y) = \ln(\ln y)$
11. $F(y) = \ln(\sin 5y)$
12. $f(x) = x \ln x$

13. $f(x) = \cos(\ln x)$
14. $g(x) = \ln \cos \sqrt{x}$
15. $G(x) = \ln(\sec 2x + \tan 2x)$

16. $h(y) = \csc(\ln y)$
17. $f(x) = \ln \sqrt{\tan x}$
18. $f(t) = \ln \sqrt[4]{\dfrac{t^2 - 1}{t^2 + 1}}$

19. $f(w) = \ln \sqrt{\dfrac{3w + 1}{2w - 5}}$
20. $f(x) = \ln[(5x - 3)^4(2x^2 + 7)^3]$
21. $h(x) = \dfrac{x}{\ln x}$

22. $g(x) = \ln(\cos 2x + \sin 2x)$
23. $g(x) = \ln \sqrt[3]{\dfrac{x + 1}{x^2 + 1}}$
24. $f(x) = \sqrt[3]{\ln x^3}$

25. $F(x) = \sqrt{x + 1} - \ln(1 + \sqrt{x + 1})$
26. $G(x) = x \ln(x + \sqrt{1 + x^2}) - \sqrt{1 + x^2}$

In Exercises 27 through 32, find $\dfrac{dy}{dx}$ by implicit differentiation.

27. $\ln xy + x + y = 2$
28. $\ln \dfrac{y}{x} + xy = 1$
29. $x = \ln(x + y + 1)$

30. $\ln(x + y) - \ln(x - y) = 4$ **31.** $x + \ln x^2y + 3y^2 = 2x^2 - 1$ **32.** $x \ln y + y \ln x = xy$

33. Draw a sketch of the graph of $y = \ln x$ by plotting the points having the abscissas $\frac{1}{9}, \frac{1}{3}, 1, 3,$ and 9, and use $\ln 3 \approx 1.1$. At each of the five points find the slope of the tangent line and draw a segment of the tangent line.

In Exercises 34 through 41, draw a sketch of the graph of the curve having the given equation.

34. $x = \ln y$ **35.** $y = \ln(-x)$ **36.** $y = \ln(x + 1)$ **37.** $y = \ln|x|$

38. $y = \ln \dfrac{1}{x - 1}$ **39.** $y = x - \ln x$ **40.** $y = x \ln x$ **41.** $y = \ln \sin x$

42. Find an equation of the tangent line to the curve $y = \ln x$ at the point whose abscissa is 2.

43. Find an equation of the normal line to the curve $y = \ln x$ that is parallel to the line $x + 2y - 1 = 0$.

44. Find an equation of the normal line to the graph of $y = x \ln x$ that is perpendicular to the line having the equation $x - y + 7 = 0$.

45. Find an equation of the tangent line to the graph of $y = \ln(4x^2 - 3)^5$ at the point whose abscissa is 1.

46. A particle is moving on a straight line according to the equation of motion $s = (t + 1)^2 \ln(t + 1)$, where s is the directed distance of the particle from the starting point at t sec. Find the velocity and acceleration when $t = 3$.

47. In a telegraph cable, the measure of the speed of the signal is proportional to $x^2 \ln(1/x)$, where x is the ratio of the measure of the radius of the core of the cable to the measure of the thickness of the cable's winding. Find the value of $\ln x$ for which the speed of the signal is greatest.

48. A manufacturer of electric generators began operations on January 1, 1976. During the first year there were no sales because the company concentrated on product development and research. After the first year the sales increased steadily according to the equation $y = x \ln x$, where x is the number of years during which the company has been operating and y is the number of millions of dollars in the sales volume. (a) Draw a sketch of the graph of the equation. Determine the rate at which the sales were increasing on (b) January 1, 1981, and (c) January 1, 1986.

49. A particular company has determined that when its weekly advertising expense is x dollars, then if S dollars is its total weekly income from sales, $S = 4000 \ln x$. (a) Determine the rate of change of sales income with respect to advertising expense when $800 is the weekly advertising budget. (b) If the weekly advertising budget is increased to $950, what is the approximate increase in the total weekly income from sales?

50. The linear density of a rod at a point x cm from one end is $2/(1 + x)$ g/cm. If the rod is 15 cm long, find the mass and center of mass of the rod.

51. Prove that $x - 1 - \ln x > 0$ and $1 - \ln x - 1/x < 0$ for all $x > 0$ and $x \neq 1$, thus establishing the inequality

$$1 - \frac{1}{x} < \ln x < x - 1$$

for all $x > 0$ and $x \neq 1$. (*Hint:* Let $f(x) = x - 1 - \ln x$ and $g(x) = 1 - \ln x - 1/x$, and determine the signs of $f'(x)$ and $g'(x)$ on the intervals $(0, 1)$ and $(1, +\infty)$.)

52. Show the geometric interpretation of the inequality in Exercise 51 by drawing on the same pair of axes sketches of the graphs of the functions f, g, and h defined by the equations

$$f(x) = 1 - \frac{1}{x} \qquad g(x) = \ln x \qquad h(x) = x - 1$$

53. Use the result of Exercise 51 to prove that

$$\lim_{x \to 0} \frac{\ln(1 + x)}{x} = 1$$

54. Establish the limit of Exercise 53 by using the definition of the derivative to find $F'(0)$ for the function F for which $F(x) = \ln(1 + x)$.

6.4
LOGARITHMIC DIFFERENTIATION AND INTEGRALS YIELDING THE NATURAL LOGARITHMIC FUNCTION

For the discussion of both of the topics, logarithmic differentiation and integrals yielding the natural logarithmic function, we need a formula for $D_x(\ln|x|)$. To find such a formula by using Theorem 6.3.2 we first substitute $\sqrt{x^2}$ for $|x|$. Thus

$$D_x(\ln|x|) = D_x(\ln \sqrt{x^2})$$

$$= \frac{1}{\sqrt{x^2}} \cdot D_x(\sqrt{x^2})$$

$$= \frac{1}{\sqrt{x^2}} \cdot \frac{x}{\sqrt{x^2}}$$

$$= \frac{x}{x^2}$$

Hence

$$D_x(\ln|x|) = \frac{1}{x} \tag{1}$$

FIGURE 6.4.1

In Exercise 37 of Exercises 6.3 you were asked to draw a sketch of the graph of $y = \ln|x|$. This graph appears in Fig. 6.4.1. The slope of the tangent line at each point (x, y) of the graph is $\frac{1}{x}$.

From formula (1) and the chain rule we obtain the following theorem.

6.4.1 THEOREM   If u is a differentiable function of x,

$$D_x(\ln|u|) = \frac{1}{u} \cdot D_x u$$

The following example illustrates how the properties of the natural logarithmic function, given in Theorems 6.3.4, 6.3.5, and 6.3.6, can simplify the work involved in differentiating complicated expressions involving products, quotients, and powers.

EXAMPLE 1
Given

$$y = \frac{\sqrt[3]{x + 1}}{(x + 2)\sqrt{x + 3}}$$

find $\dfrac{dy}{dx}$.

SOLUTION
From the given equation,

$$|y| = \left| \frac{\sqrt[3]{x+1}}{(x+2)\sqrt{x+3}} \right|$$

$$= \frac{|\sqrt[3]{x+1}|}{|x+2||\sqrt{x+3}|}$$

Taking the natural logarithm and applying the properties of logarithms we obtain

$$\ln|y| = \tfrac{1}{3}\ln|x+1| - \ln|x+2| - \tfrac{1}{2}\ln|x+3|$$

Differentiating on both sides implicitly with respect to x and applying Theorem 6.4.1 we get

$$\frac{1}{y}\frac{dy}{dx} = \frac{1}{3(x+1)} - \frac{1}{x+2} - \frac{1}{2(x+3)}$$

Multiplying on both sides by y we obtain

$$\frac{dy}{dx} = y \cdot \frac{2(x+2)(x+3) - 6(x+1)(x+3) - 3(x+1)(x+2)}{6(x+1)(x+2)(x+3)}$$

Replacing y by its given value we obtain

$$\frac{dy}{dx} = \frac{(x+1)^{1/3}}{(x+2)(x+3)^{1/2}} \cdot \frac{2x^2 + 10x + 12 - 6x^2 - 24x - 18 - 3x^2 - 9x - 6}{6(x+1)(x+2)(x+3)}$$

$$= \frac{-7x^2 - 23x - 12}{6(x+1)^{2/3}(x+2)^2(x+3)^{3/2}}$$

The process illustrated in Example 1 is called **logarithmic differentiation**, which was developed in 1697 by Johann Bernoulli (1667–1748).

From Theorem 6.4.1 we obtain the following one for indefinite integration.

6.4.2 THEOREM

$$\int \frac{1}{u}\,du = \ln|u| + C$$

From Theorems 6.4.2 and 4.1.8, for n any real number,

$$\int u^n\,du = \begin{cases} \dfrac{u^{n+1}}{n+1} + C & \text{if } n \neq -1 \\[2mm] \ln|u| + C & \text{if } n = -1 \end{cases}$$

(2)

EXAMPLE 2

Evaluate

$$\int \frac{x^2\,dx}{x^3 + 1}$$

SOLUTION

$$\int \frac{x^2\,dx}{x^3 + 1} = \frac{1}{3}\int \frac{3x^2\,dx}{x^3 + 1}$$

$$= \tfrac{1}{3}\ln|x^3 + 1| + C$$

EXAMPLE 3

Evaluate

$$\int_0^2 \frac{x^2 + 2}{x + 1}\,dx$$

SOLUTION

Because $(x^2 + 2)/(x + 1)$ is an improper fraction, we divide the numerator by the denominator and obtain

$$\frac{x^2 + 2}{x + 1} = x - 1 + \frac{3}{x + 1}$$

Therefore

$$\int_0^2 \frac{x^2 + 2}{x + 1}\,dx = \int_0^2 \left(x - 1 + \frac{3}{x + 1} \right) dx$$

$$= \tfrac{1}{2}x^2 - x + 3\ln|x + 1|\Big]_0^2$$

$$= 2 - 2 + 3\ln 3 - 3\ln 1$$

$$= 3\ln 3 - 3\cdot 0$$

$$= 3\ln 3$$

Because $3\ln 3 = \ln 3^3$, the answer can be written as $\ln 27$.

EXAMPLE 4

Evaluate

$$\int \frac{\ln x}{x}\,dx$$

SOLUTION

Let

$$u = \ln x \qquad du = \frac{1}{x}\,dx$$

Therefore

$$\int \frac{\ln x}{x}\, dx = \int u\, du$$

$$= \frac{u^2}{2} + C$$

$$= \tfrac{1}{2}(\ln x)^2 + C$$

We have delayed obtaining the formulas for the indefinite integrals of the tangent, cotangent, secant, and cosecant functions until now because the natural logarithmic function is needed.

A formula for the indefinite integral of the tangent function is derived as follows: Because

$$\int \tan u\, du = \int \frac{\sin u}{\cos u}\, du$$

we let

$$v = \cos u \qquad dv = -\sin u\, du$$

and obtain

$$\int \tan u\, du = -\int \frac{dv}{v}$$

$$= -\ln|v| + C$$
$$= -\ln|\cos u| + C$$
$$= \ln|(\cos u)^{-1}| + C$$
$$= \ln|\sec u| + C$$

The result is stated formally as a theorem.

6.4.3 THEOREM

$$\int \tan u\, du = \ln|\sec u| + C$$

▶ ILLUSTRATION 1

$$\int \tan 3x\, dx = \tfrac{1}{3} \int \tan 3x\, (3\, dx)$$

$$= \tfrac{1}{3} \ln|\sec 3x| + C$$ ◀

The theorem giving the indefinite integral of the cotangent function is proved in a way similar to the proof of Theorem 6.4.3. See Exercise 45.

6.4.4 THEOREM

$$\int \cot u\, du = \ln|\sin u| + C$$

To obtain the formula for $\int \sec u \, du$ we multiply the numerator and denominator of the integrand by $\sec u + \tan u$, and we have

$$\int \sec u \, du = \int \frac{\sec u(\sec u + \tan u)}{\sec u + \tan u} \, du$$

$$= \int \frac{(\sec^2 u + \sec u \tan u)}{\sec u + \tan u} \, du$$

Let

$$v = \sec u + \tan u \qquad dv = (\sec u \tan u + \sec^2 u) \, du$$

Therefore we have

$$\int \sec u \, du = \int \frac{dv}{v}$$

$$= \ln|v| + C$$

$$= \ln|\sec u + \tan u| + C$$

We have proved the following theorem.

6.4.5 THEOREM $\int \sec u \, du = \ln|\sec u + \tan u| + C$

In Section 8.8 we obtain a formula for the integral of the secant function by a method that does not depend on the "trick" of multiplying the numerator and denominator by $\sec u + \tan u$ used to prove Theorem 6.4.5.

A formula for $\int \csc u \, du$ can be derived by multiplying the numerator and denominator of the integrand by $\csc u - \cot u$ and proceeding as we did for Theorem 6.4.5. Another procedure is to let $\int \csc u \, du = \int \sec(u - \frac{1}{2}\pi) \, du$ and use Theorem 6.4.5 and trigonometric identities. You are asked to provide these derivations in Exercise 45. The formula obtained is given in the next theorem.

6.4.6 THEOREM $\int \csc u \, du = \ln|\csc u - \cot u| + C$

▶ ILLUSTRATION 2

$$\int \frac{dx}{\sin 2x} = \int \csc 2x \, dx$$

$$= \tfrac{1}{2} \int \csc 2x(2 \, dx)$$

$$= \tfrac{1}{2} \ln|\csc 2x - \cot 2x| + C$$ ◀

EXAMPLE 5

Evaluate

$$\int_{\pi/8}^{\pi/6} (\csc 4x - \cot 4x)\, dx$$

SOLUTION

$$\int_{\pi/8}^{\pi/6} (\csc 4x - \cot 4x)\, dx$$

$$= \tfrac{1}{4} \int_{\pi/8}^{\pi/6} (\csc 4x - \cot 4x)(4\, dx)$$

$$= \tfrac{1}{4} \Big[\ln|\csc 4x - \cot 4x| - \ln|\sin 4x| \Big]_{\pi/8}^{\pi/6}$$

$$= \tfrac{1}{4} \big[(\ln|\csc \tfrac{2}{3}\pi - \cot \tfrac{2}{3}\pi| - \ln|\sin \tfrac{2}{3}\pi|) - (\ln|\csc \tfrac{1}{2}\pi - \cot \tfrac{1}{2}\pi| - \ln|\sin \tfrac{1}{2}\pi|) \big]$$

$$= \frac{1}{4} \left[\left(\ln \left| \frac{2}{\sqrt{3}} - \left(-\frac{1}{\sqrt{3}} \right) \right| - \ln \left| \frac{\sqrt{3}}{2} \right| \right) - (\ln|1 - 0| - \ln|1|) \right]$$

$$= \frac{1}{4} \left[\left(\ln \frac{3}{\sqrt{3}} - \ln \frac{\sqrt{3}}{2} \right) - (0 - 0) \right]$$

$$= \frac{1}{4} \left[\ln \sqrt{3} - \ln \frac{\sqrt{3}}{2} \right]$$

$$= \frac{1}{4} \left[\ln \frac{\sqrt{3}}{\frac{\sqrt{3}}{2}} \right] \quad \text{(by Theorem 6.3.5)}$$

$$= \tfrac{1}{4} \ln 2$$

EXERCISES 6.4

In Exercises 1 through 8, use Theorem 6.4.1 to find $\dfrac{dy}{dx}$.

1. $y = \ln|x^3 + 1|$　　　　**2.** $y = \ln|x^2 - 1|$　　　　**3.** $y = \ln|\cos 3x|$　　　　**4.** $y = \ln|\sec 2x|$

5. $y = \ln|\tan 4x + \sec 4x|$　　**6.** $y = \ln|\cot 3x - \csc 3x|$　　**7.** $y = \ln \left| \dfrac{3x}{x^2 + 4} \right|$　　　**8.** $y = \sin(\ln|2x + 1|)$

In Exercises 9 through 14, find $\dfrac{dy}{dx}$ by logarithmic differentiation.

9. $y = x^2(x^2 - 1)^3(x + 1)^4$　　　**10.** $y = (5x - 4)(x^2 + 3)(3x^3 - 5)$　　　**11.** $y = \dfrac{x(x - 1)(x + 2)}{(x - 4)^3}$

12. $y = \dfrac{x^5(x+2)}{x-3}$

13. $y = \dfrac{x^3 + 2x}{\sqrt[5]{x^7 + 1}}$

14. $y = \dfrac{\sqrt{1-x^2}}{(x+1)^{2/3}}$

In Exercises 15 through 36, evaluate the indefinite integral.

15. $\displaystyle\int \frac{dx}{3-2x}$

16. $\displaystyle\int \frac{dx}{7x+10}$

17. $\displaystyle\int \frac{3x}{x^2+4}\,dx$

18. $\displaystyle\int \frac{x}{2-x^2}\,dx$

19. $\displaystyle\int \frac{3x^2}{5x^3-1}\,dx$

20. $\displaystyle\int \frac{2x-1}{x(x-1)}\,dx$

21. $\displaystyle\int \frac{\cos t}{1+2\sin t}\,dt$

22. $\displaystyle\int \frac{\sin 3t}{\cos 3t - 1}\,dt$

23. $\displaystyle\int (\cot 5x + \csc 5x)\,dx$

24. $\displaystyle\int (\tan 2x - \sec 2x)\,dx$

25. $\displaystyle\int \frac{2 - 3\sin 2x}{\cos 2x}\,dx$

26. $\displaystyle\int \frac{\cos 3x + 3}{\sin 3x}\,dx$

27. $\displaystyle\int \frac{2x^3}{x^2-4}\,dx$

28. $\displaystyle\int \frac{5-4y^2}{3+2y}\,dy$

29. $\displaystyle\int \frac{dx}{x\ln x}$

30. $\displaystyle\int \frac{dx}{\sqrt{x}(1+\sqrt{x})}$

31. $\displaystyle\int \frac{\ln^2 3x}{x}\,dx$

32. $\displaystyle\int \frac{(2 + \ln^2 x)}{x(1 - \ln x)}\,dx$

33. $\displaystyle\int \frac{2\ln x + 1}{x[(\ln x)^2 + \ln x]}\,dx$

34. $\displaystyle\int \frac{3x^5 - 2x^3 + 5x^2 - 2}{x^3 + 1}\,dx$

35. $\displaystyle\int \frac{\tan(\ln x)}{x}\,dx$

36. $\displaystyle\int \frac{\cot \sqrt{t}}{\sqrt{t}}\,dt$

In Exercises 37 through 44, evaluate the definite integral.

37. $\displaystyle\int_3^5 \frac{2x}{x^2-5}\,dx$

38. $\displaystyle\int_4^5 \frac{x}{4-x^2}\,dx$

39. $\displaystyle\int_1^3 \frac{2t+3}{t+1}\,dt$

40. $\displaystyle\int_1^5 \frac{4z^3-1}{2z-1}\,dz$

41. $\displaystyle\int_0^{\pi/6} (\tan 2x + \sec 2x)\,dx$

42. $\displaystyle\int_{\pi/12}^{\pi/6} (\cot 3x + \csc 3x)\,dx$

43. $\displaystyle\int_2^4 \frac{dx}{x\ln^2 x}$

44. $\displaystyle\int_2^4 \frac{\ln x}{x}\,dx$

45. (a) Prove Theorem 6.4.4. (b) Prove Theorem 6.4.6 by multiplying the numerator and denominator of the integrand by $\csc u - \cot u$. (c) Prove Theorem 6.4.6 by letting $\int \csc u\,du = \int \sec(u - \tfrac{1}{2}\pi)\,du$ and using Theorem 6.4.5 and trigonometric identities.

46. Prove that $\int \csc u\,du = -\ln|\csc u + \cot u| + C$ in two ways: (a) Use Theorem 6.4.6; (b) multiply the numerator and denominator of the integrand by $\csc u + \cot u$.

In Exercises 47 through 55, give the exact value of the number to be found, and then obtain an approximation of this number to three decimal places by using either a calculator or the table of natural logarithms in the back of the book.

47. If $f(x) = 1/x$, find the average value of f on the interval $[1, 5]$.

48. If $f(x) = (x + 2)/(x - 3)$, find the average value of f on the interval $[4, 6]$.

49. Use Boyle's law for the expansion of a gas (see Exercise 26 in Exercises 2.5) to find the average pressure with respect to the volume as the volume increases from 4 ft³ to 8 ft³ and the pressure is 2000 lb/ft² when the volume is 4 ft³.

50. Find the area of the region bounded by the curve $y = x/(2x^2 + 4)$, the x axis, the y axis, and the line $x = 4$.

51. Find the area of the region bounded by the curve $y = 2/(x - 3)$, the x axis, and the lines $x = 4$ and $x = 5$.

52. Find the volume of the solid of revolution generated when the region bounded by the curve $y = 1 - 3/x$, the x axis, and the line $x = 1$ is revolved about the x axis.

53. Find the volume of the solid of revolution generated when the region bounded by the curve $y = 1 + 2/\sqrt{x}$, the x axis, and the lines $x = 1$ and $x = 4$ is revolved about the x axis.

54. Find the length of the arc of the curve $y = \ln \sec x$ from $x = 0$ to $x = \frac{1}{4}\pi$.

55. Find the length of the arc of the curve $y = \ln \csc x$ from $x = \frac{1}{6}\pi$ to $x = \frac{1}{2}\pi$.

56. Prove: $\displaystyle\lim_{x \to +\infty} \frac{\ln x}{x} = 0$. $\left(\textit{Hint:}\ \text{First prove that } \displaystyle\int_1^x \frac{1}{\sqrt{t}}\,dt \geq \int_1^x \frac{1}{t}\,dt \text{ by using Theorem 4.6.8.}\right)$

6.5
THE NATURAL EXPONENTIAL FUNCTION

Because the natural logarithmic function is increasing on its entire domain, then by Theorems 6.1.5 and 6.2.1 it has an inverse that is also an increasing function. The inverse of the natural logarithmic function is called the *natural exponential function*, which we now define.

6.5.1 DEFINITION The **natural exponential function** is the inverse of the natural logarithmic function; thus it is defined by

$$\exp(x) = y \quad \text{if and only if} \quad x = \ln y \tag{1}$$

The notation $\exp(x)$ is read as "the value of the natural exponential function at x."

Because the range of the natural logarithmic function is the set of all real numbers, the domain of the natural exponential function is the set of all real numbers. The range of the natural exponential function is the set of positive numbers, because this is the domain of the natural logarithmic function.

Because the natural logarithmic function and the natural exponential function are inverses of each other, it follows from Theorem 6.1.4 that

$$\ln(\exp(x)) = x \tag{2}$$

and

$$\exp(\ln x) = x \tag{3}$$

We now wish to define what is meant by a^x, where a is a positive number and x is an irrational number. To arrive at a reasonable definition consider the case a^r, where $a > 0$ and r is a rational number. We have from (3), where x is a^r,

$$a^r = \exp[\ln(a^r)] \tag{4}$$

But by Theorem 6.3.6, $\ln a^r = r \ln a$; so from (4),

$$a^r = \exp(r \ln a) \tag{5}$$

Because the right side of (5) has a meaning if r is any real number, we use it for our definition.

6.5.2 DEFINITION If a is any positive number and x is any real number, we define

$$a^x = \exp(x \ln a) \tag{6}$$

Theorem 6.3.6 states that if a is any positive number and r is any rational number, then $\ln a^r = r \ln a$. Because of Definition 6.5.2, this equation is also valid if r is any real number, and we now state this fact as a theorem.

6.5.3 THEOREM   If a is any positive number and x is any real number,

$$\ln a^x = x \ln a$$

PROOF   From Definition 6.5.2,

$$a^x = \exp(x \ln a)$$

Hence, from Definition 6.5.1,

$$\ln a^x = x \ln a \qquad \blacksquare$$

Following is the definition of one of the most important numbers in mathematics.

6.5.4 DEFINITION   The number e is defined by the formula

$$e = \exp 1$$

The letter e was chosen because of the Swiss mathematician and physicist Leonhard Euler (1707–1783).

The number e is a *transcendental* number; that is, it cannot be expressed as the root of any polynomial with integer coefficients. The number π is another example of a transcendental number. The proof that e is transcendental was first given in 1873 by Charles Hermite, and its value can be expressed to any required degree of accuracy. In Chapter 12 we learn a method for doing this. The value of e to seven decimal places is 2.7182818. Thus

$$e \approx 2.7182818$$

The importance of the number e will become apparent as you proceed through this chapter.

6.5.5 THEOREM $\ln e = 1$

PROOF   By Definition 6.5.4,

$$e = \exp 1$$

Therefore

$$\ln e = \ln(\exp 1)$$

Because the natural logarithmic function and the natural exponential function are inverses, it follows that the right side of this equation is 1. Thus

$$\ln e = 1 \qquad \blacksquare$$

Observe that Theorem 6.5.5 corresponds to property (6) in Section 6.3. We have now completed showing that the function $\ln$ satisfies properties (2)–(6) of logarithms, given in Section 6.3.

6.5.6 THEOREM   For all values of x,

$$\exp(x) = e^x$$

PROOF   By Definition 6.5.2, with $a = e$,

$$e^x = \exp(x \ln e) \tag{7}$$

But by Theorem 6.5.5, $\ln e = 1$, and substituting in (7) we obtain

$$e^x = \exp(x) \qquad\blacksquare$$

From now on we write e^x in place of $\exp(x)$, and so from Definition 6.5.1,

$$e^x = y \quad \text{if and only if} \quad x = \ln y \tag{8}$$

With e^x in place of $\exp(x)$, (2) and (3) become

$$\ln e^x = x \tag{9}$$

and

$$e^{\ln x} = x \tag{10}$$

If we replace $\exp(x \ln a)$ by $e^{x \ln a}$ in (6) of Definition 6.5.2, we have

$$a^x = e^{x \ln a} \qquad \text{for every } a > 0 \tag{11}$$

Equation (11) can be used to compute a^x if x is any real number. For values of powers of e you can use Table 3 in the back of the book. Such values can also be obtained from a calculator with an $\boxed{e^x}$ key. If your calculator has no $\boxed{e^x}$ key, powers of e can be obtained by pressing the $\boxed{\text{INV}}$ key followed by the $\boxed{\ln}$ key. The reason that these two successive operations give a value for e^x is that the natural exponential function and the natural logarithmic function are inverses of each other.

EXAMPLE 1
Compute the value of $2^{\sqrt{3}}$ to two decimal places.

SOLUTION
From (11),

$$2^{\sqrt{3}} = e^{\sqrt{3} \ln 2}$$
$$= e^{1.732(0.6931)}$$
$$= e^{1.200}$$
$$\approx 3.32$$

Because $0 = \ln 1$, we have from statement (8)

$$e^0 = 1$$

We now state some properties of the natural exponential function as theorems.

6.5.7 THEOREM   If a and b are any real numbers, then

$$e^a \cdot e^b = e^{a+b}$$

PROOF   Let $A = e^a$ and $B = e^b$. Then from statement (8),

$$\ln A = a \qquad (12)$$

and

$$\ln B = b \qquad (13)$$

From Theorem 6.3.4,

$$\ln AB = \ln A + \ln B \qquad (14)$$

Substituting from (12) and (13) into (14) we obtain

$$\ln AB = a + b$$

Thus

$$e^{\ln AB} = e^{a+b} \qquad (15)$$

From (10) it follows that the left side of (15) is AB; so

$$AB = e^{a+b}$$

Replacing A and B by their values we get

$$e^a \cdot e^b = e^{a+b} \qquad \blacksquare$$

6.5.8 THEOREM   If a and b are any real numbers, then

$$e^a \div e^b = e^{a-b}$$

The proof is analogous to the proof of Theorem 6.5.7, where Theorem 6.3.4 is replaced by Theorem 6.3.5.

6.5.9 THEOREM   If a and b are any real numbers, then

$$(e^a)^b = e^{ab}$$

PROOF   If in (10) $x = (e^a)^b$, we have

$$(e^a)^b = e^{\ln(e^a)^b}$$

Applying Theorem 6.5.3 to the exponent in the right side of this equation we obtain

$$(e^a)^b = e^{b \ln e^a}$$

But from (9), $\ln e^a = a$, and therefore

$$(e^a)^b = e^{ab} \qquad \blacksquare$$

Because the natural exponential function is the inverse of the natural logarithmic function, it follows from Theorem 6.2.3 that it is differentiable.

We obtain the theorem for the derivative of the natural exponential function by implicit differentiation. Let

$$y = e^x$$

Then from statement (8),

$$x = \ln y$$

On both sides of this equation we differentiate implicitly with respect to x to get

$$1 = \frac{1}{y} \cdot \frac{dy}{dx}$$

$$\frac{dy}{dx} = y$$

Replacing y by e^x we obtain

$$\frac{d}{dx}(e^x) = e^x \tag{16}$$

The next theorem follows from (16) and the chain rule.

6.5.10 THEOREM   If u is a differentiable function of x,

$$D_x(e^u) = e^u D_x u$$

Observe that the derivative of the function defined by $f(x) = ke^x$, where k is a constant, is itself. The only other function we have previously encountered that has this property is the constant function zero; actually, this is the special case of $f(x) = ke^x$ when $k = 0$. It can be proved that the most general function that is its own derivative is given by $f(x) = ke^x$ (see Exercise 60).

EXAMPLE 2
Given

$$y = e^{1/x^2}$$

find $\dfrac{dy}{dx}$.

SOLUTION
From Theorem 6.5.10,

$$\frac{dy}{dx} = e^{1/x^2}\left(-\frac{2}{x^3}\right)$$

$$= -\frac{2e^{1/x^2}}{x^3}$$

EXAMPLE 3
Given

$$y = e^{2x + \ln x}$$

find $\dfrac{dy}{dx}$.

SOLUTION
Because $e^{2x + \ln x} = e^{2x}e^{\ln x}$ and $e^{\ln x} = x$, then

$$y = xe^{2x}$$

Therefore

$$\frac{dy}{dx} = e^{2x} + 2xe^{2x}$$

The indefinite integration formula given in the following theorem is a consequence of Theorem 6.5.10.

6.5.11 THEOREM

$$\int e^u \, du = e^u + C$$

EXAMPLE 4
Find $\displaystyle\int \frac{e^{\sqrt{x}}}{\sqrt{x}} \, dx$.

SOLUTION
Let

$$u = \sqrt{x} \qquad du = \frac{1}{2\sqrt{x}} \, dx$$

Therefore

$$\int \frac{e^{\sqrt{x}}}{\sqrt{x}} \, dx = 2 \int e^u \, du$$

$$= 2e^u + C$$

$$= 2e^{\sqrt{x}} + C$$

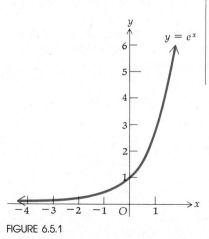

FIGURE 6.5.1

Because from (8) $e^x = y$ if and only if $x = \ln y$, the graph of $y = e^x$ is identical to the graph of $x = \ln y$. So we can obtain the graph of $y = e^x$, shown in Fig. 6.5.1, by interchanging the x and y axes in Fig. 6.3.4.

The graph of $y = e^x$ can be obtained without referring to the graph of the natural logarithmic function. Because the range of the exponential function is the set of all positive numbers, it follows that $e^x > 0$ for all values of x. So the graph lies entirely above the x axis. $\dfrac{dy}{dx} = e^x > 0$ for all x; so the function is increasing for all x. $\dfrac{d^2y}{dx^2} = e^x > 0$ for all x; so the graph is concave upward at all points.

We have the following two limits:

$$\lim_{x \to +\infty} e^x = +\infty \tag{17}$$

$$\lim_{x \to -\infty} e^x = 0 \tag{18}$$

The proofs of (17) and (18) are left as exercises (see Exercises 57 and 58). To plot some specific points use Table 3 in the back of the book or a calculator for powers of e.

EXAMPLE 5
Find the area of the region bounded by the curve $y = e^x$, the coordinate axes, and the line $x = 2$.

SOLUTION
The region is shown in Fig. 6.5.2.

$$A = \lim_{\|\Delta\| \to 0} \sum_{i=1}^{n} e^{\xi_i} \Delta_i x$$

$$= \int_0^2 e^x \, dx$$

$$= e^x \Big]_0^2$$

$$= e^2 - e^0$$

$$= e^2 - 1$$

The value of e^2 to two decimal places is 7.39. Therefore the area of the region is approximately 6.39 square units.

Functions having values of the form Ce^{kx}, where C and k are constants, occur frequently in applications in various fields. Some of these applications are discussed in Section 6.6.

In Definition 6.5.4 the number e was defined as the value of the natural exponential function at 1; that is, $e = \exp 1$. To arrive at another way of defining e we consider the natural logarithmic function

$$f(x) = \ln x$$

We know that the derivative of f is given by $f'(x) = 1/x$; hence $f'(1) = 1$. However, let us apply the definition of the derivative to find $f'(1)$. We have

$$f'(1) = \lim_{\Delta x \to 0} \frac{f(1 + \Delta x) - f(1)}{\Delta x}$$

$$= \lim_{\Delta x \to 0} \frac{\ln(1 + \Delta x) - \ln 1}{\Delta x}$$

$$= \lim_{\Delta x \to 0} \frac{1}{\Delta x} \ln(1 + \Delta x)$$

FIGURE 6.5.2

Therefore

$$\lim_{\Delta x \to 0} \frac{1}{\Delta x} \ln(1 + \Delta x) = 1$$

Replacing Δx by h we have from the above equation and Theorem 6.5.3

$$\lim_{h \to 0} \ln(1 + h)^{1/h} = 1 \tag{19}$$

Now, because the natural exponential function and the natural logarithmic function are inverse functions, we have

$$\lim_{h \to 0} (1 + h)^{1/h} = \lim_{h \to 0} \exp[\ln(1 + h)^{1/h}] \tag{20}$$

Because the natural exponential function is continuous and $\lim_{h \to 0} \ln(1 + h)^{1/h}$ exists and equals 1, as shown in Eq. (19), we can apply Theorem 1.7.1 to the right side of (20) and get

$$\lim_{h \to 0} (1 + h)^{1/h} = \exp\left[\lim_{h \to 0} \ln(1 + h)^{1/h} \right]$$

$$= \exp 1$$

Hence

$$\lim_{h \to 0} (1 + h)^{1/h} = e \tag{21}$$

Equation (21) is sometimes used to define e; however, to use this as a definition it is necessary to prove that the limit exists.

Let us consider the function F defined by

$$F(h) = (1 + h)^{1/h} \tag{22}$$

and determine the function values for some values of h close to zero. These values are obtained from a calculator. When h is positive, the values appear in Table 6.5.1, and when h is negative they appear in Table 6.5.2.

The two tables lead us to suspect that $\lim_{h \to 0} (1 + h)^{1/h}$ is probably a number that lies between 2.7181 and 2.7184. As previously mentioned, in Chapter 12 we learn a method for finding the value of e to any desired number of decimal places.

Table 6.5.1

h	$F(h) = (1 + h)^{1/h}$
1	2
0.5	2.25
0.05	2.6533
0.01	2.7048
0.001	2.7169
0.0001	2.7181

Table 6.5.2

h	$F(h) = (1 + h)^{1/h}$
-0.5	4
-0.05	2.7895
-0.01	2.7320
-0.001	2.7196
-0.0001	2.7184

EXERCISES 6.5

In Exercises 1 through 16, find $\dfrac{dy}{dx}$.

1. $y = e^{5x}$ **2.** $y = e^{-7x}$ **3.** $y = e^{-3x^2}$ **4.** $y = e^{x^2 - 3}$ **5.** $y = e^{\cos x}$ **6.** $y = e^{2 \sin 3x}$

7. $y = e^x \sin e^x$ **8.** $y = \dfrac{e^x}{x}$ **9.** $y = \tan e^{\sqrt{x}}$ **10.** $y = e^{e^x}$ **11.** $y = \dfrac{e^x - e^{-x}}{e^x + e^{-x}}$ **12.** $y = \ln \dfrac{e^{4x} - 1}{e^{4x} + 1}$

13. $y = x^5 e^{-3 \ln x}$ **14.** $y = \ln(e^x + e^{-x})$ **15.** $y = \sec e^{2x} + e^{2 \sec x}$ **16.** $y = \tan e^{3x} + e^{\tan 3x}$

In Exercises 17 through 20, find $\dfrac{dy}{dx}$ by implicit differentiation.

17. $e^x + e^y = e^{x+y}$ 18. $e^y = \ln(x^3 + 3y)$ 19. $y^2 e^{2x} + xy^3 = 1$ 20. $ye^{2x} + xe^{2y} = 1$

In Exercises 21 through 28, evaluate the indefinite integral.

21. $\displaystyle\int e^{2-5x}\, dx$ 22. $\displaystyle\int e^{2x+1}\, dx$ 23. $\displaystyle\int \frac{1+e^{2x}}{e^x}\, dx$ 24. $\displaystyle\int e^{3x} e^{2x}\, dx$

25. $\displaystyle\int \frac{e^{3x}}{(1-2e^{3x})^2}\, dx$ 26. $\displaystyle\int x^2 e^{2x^3}\, dx$ 27. $\displaystyle\int \frac{e^{2x}}{e^x+3}\, dx$ 28. $\displaystyle\int \frac{dx}{1+e^x}$

In Exercises 29 through 36, evaluate the definite integral.

29. $\displaystyle\int_0^1 e^2\, dx$ 30. $\displaystyle\int_1^{e^2} \frac{dx}{x}$ 31. $\displaystyle\int_e^{e^3} \frac{dx}{x}$ 32. $\displaystyle\int_1^e \frac{\ln x}{x}\, dx$

33. $\displaystyle\int_e^{e^2} \frac{dx}{x(\ln x)^2}$ 34. $\displaystyle\int_0^3 \frac{e^x + e^{-x}}{2}\, dx$ 35. $\displaystyle\int_0^2 xe^{4-x^2}\, dx$ 36. $\displaystyle\int_1^2 \frac{e^x}{e^x + e}\, dx$

37. Draw a sketch of the graph of $y = e^{-x}$.   38. Draw a sketch of the graph of $y = e^{|x|}$.

In Exercises 39 and 40, find the relative extrema of f, the points of inflection of the graph of f, the intervals on which f is increasing, the intervals on which f is decreasing, where the graph is concave upward, where the graph is concave downward, and the slope of any inflectional tangent. Draw a sketch of the graph of f.

39. $f(x) = xe^{-x}$ 40. $f(x) = e^{-x^2}$

In Exercises 41 through 44, compute the value to two decimal places.

41. $2^{\sqrt{2}}$ 42. $(\sqrt{2})^{\sqrt{2}}$ 43. $(\sqrt{2})^e$ 44. e^e

45. Find an equation of the tangent line to the curve $y = e^{-x}$ that is perpendicular to the line $2x - y = 5$.

46. Find the area of the region bounded by the curve $y = e^x$ and the line through the points $(0, 1)$ and $(1, e)$.

47. A body is moving along a straight line and at t sec the velocity is v ft/sec, where $v = e^3 - e^{2t}$. Find the distance traveled by the particle while $v > 0$ after $t = 0$.

48. A particle is moving along a straight line, where s ft is the directed distance of the particle from the origin, v ft/sec is the velocity of the particle, and a ft/sec^2 is the acceleration of the particle at t sec. If $a = e^t + e^{-t}$ and $v = 1$ and $s = 2$ when $t = 0$, find v and s in terms of t.

49. If p lb/ft^2 is the atmospheric pressure at a height of h ft above sea level, then $p = 2116e^{-0.0000318h}$. Find the time rate of change of the atmospheric pressure outside an airplane that is 5000 ft high and rising at the rate of 160 ft/sec.

50. At a certain height the gauge on an airplane indicates that the atmospheric pressure is 1500 lb/ft^2. Applying the formula of Exercise 49, approximate by differentials how much higher the airplane must rise so that the pressure will be 1480 lb/ft^2.

51. If l ft is the length of an iron rod when t degrees is its temperature, then $l = 60e^{0.00001t}$. Use differentials to find the approximate increase in l when t increases from 0 to 10.

52. A simple electric circuit containing no condensers, a resistance of R ohms, and an inductance of L henrys has the electromotive force cut off when the current is I_0 amperes. The current dies down so that at t sec the current is i amperes, and $i = I_0 e^{-(R/L)t}$. Show that the rate of change of the current is proportional to the current.

53. A tank is in the shape of the solid of revolution formed by rotating about the x axis (with the positive x axis downward) the region bounded by the curve $y^2 x = e^{-2x}$ and the lines $x = 1$ and $x = 4$. If the tank is full of water, find the work done in pumping all the water to a point 1 ft above the top of the tank. Distance is measured in feet.

54. An advertising agency determined statistically that if a breakfast food manufacturer increases its budget for television commercials by x thousand dollars, there will be an increase in the total profit of $25x^2 e^{-0.2x}$ hundred dollars. What

should be the advertising budget increase in order for the manufacturer to have the greatest profit? What will be the corresponding increase in the company's profit?

55. Draw a sketch of the graph of the function defined by

$$f(x) = \begin{cases} (1 + x)^{1/x} & \text{if } x \neq 0 \\ e & \text{if } x = 0 \end{cases}$$

on the interval $[-0.5, 0.5]$ and plotting the points for which x has the values -0.5, -0.1, -0.15, -0.01, 0.01, 0.05, 0.1, and 0.5. Use a calculator for the function values, and take different scales on the two axes. The y coordinate of the point at which the graph intersects the y axis is the number e.

56. An important function in statistics is the probability density function defined by

$$f(x) = \frac{1}{\sigma\sqrt{2\pi}} \exp\left[-\frac{(x - \mu)^2}{2\sigma^2} \right]$$

where σ and μ are constants such that $\sigma > 0$ and μ is any real number. Find (a) the relative extrema of f; (b) the points of inflection of the graph of f; (c) $\lim\limits_{x \to +\infty} f(x)$; (d) $\lim\limits_{x \to -\infty} f(x)$. (e) Draw a sketch of the graph of f.

57. Prove that $\lim\limits_{x \to +\infty} e^x = +\infty$ by showing that for any $N > 0$ there exists an $M > 0$ such that if $x > M$, then $e^x > N$.

58. Prove that $\lim\limits_{x \to -\infty} e^x = 0$ by showing that for any $\epsilon > 0$ there exists an $N < 0$ such that if $x < N$, then $e^x < \epsilon$.

59. Draw a sketch of the graph of the function f for which $f(x) = e^{-x} \sin x$. Show that $\lim\limits_{x \to +\infty} f(x) = 0$ but that the line $y = 0$ is not a horizontal asymptote of the graph of f. Why does Definition 3.8.1(i) not hold for this function?

60. Prove that the most general function that is equal to its derivative is given by $f(x) = ke^x$. $\left(\textit{Hint:} \text{ Let } y = f(x), \text{ and solve} \right.$ the differential equation $\dfrac{dy}{dx} = y.\left.\right)$

**6.6
APPLICATIONS OF THE
NATURAL EXPONENTIAL
FUNCTION**

Mathematical models involving powers of e occur in many fields, such as chemistry, physics, biology, psychology, sociology, business, and economics. We begin by discussing models involving the *laws of growth and decay*. These laws arise when the rate of change of the amount of a quantity with respect to time is proportional to the amount of the quantity present at a given instant. For example, it may be that the rate of growth of the population of a community is proportional to the actual population at any given instant. In biology, under certain circumstances, the rate of growth of a culture of bacteria is proportional to the amount of bacteria present at any specific time. In a chemical reaction it is often the case that the rate of the reaction is proportional to the quantity of the substance present; for instance, it is known from experiments that the rate of decay of radium is proportional to the amount of radium present at a given moment. An application in business occurs when interest is compounded continuously.

In such cases, if the time is represented by t units, and if y units represents the amount of the quantity present at any time, then

$$\frac{dy}{dt} = ky$$

where k is a constant and $y > 0$ for all $t \geq 0$. If y increases as t increases, then $k > 0$, and we have the **law of natural growth**. If y decreases as t increases, then $k < 0$, and we have the **law of natural decay**.

If by definition y is a positive integer (for instance, if y is the population of a certain community), we assume that y can be any positive real number in order for y to be a continuous function of t.

Suppose that we have a mathematical model involving the law of natural growth or decay and the initial condition that $y = y_0$ when $t = 0$. The differential equation is

$$\frac{dy}{dt} = ky$$

Separating the variables we obtain

$$\frac{dy}{y} = k \, dt$$

Integrating we have

$$\int \frac{dy}{y} = k \int dt$$

$$\ln|y| = kt + \bar{c}$$

$$|y| = e^{kt + \bar{c}}$$

$$|y| = e^{\bar{c}} \cdot e^{kt}$$

Letting $e^{\bar{c}} = C$ we have $|y| = Ce^{kt}$, and because y is positive we can omit the absolute-value bars, thereby giving

$$y = Ce^{kt}$$

Because $y = y_0$ when $t = 0$, we obtain $C = y_0$. Thus

$$y = y_0 e^{kt}$$

We have proved the following theorem.

6.6.1 THEOREM   Suppose that y is a continuous function of t with $y > 0$ for all $t \geq 0$. Furthermore,

$$\frac{dy}{dt} = ky$$

where k is a constant and $y = y_0$ when $t = 0$. Then

$$y = y_0 e^{kt}$$

EXAMPLE 1

In a certain culture the rate of growth of bacteria is proportional to the amount present. If 1000 bacteria are present initially and the amount doubles in 12 min, how long will it take before there will be 1,000,000 bacteria present?

Table 6.6.1

t	0	12	T
y	1000	2000	1,000,000

SOLUTION

Let t min be the time from now and let y bacteria be present at t min. Table 6.6.1 gives the boundary conditions, where T min is the time it takes until there are 1,000,000 bacteria present.

The differential equation is

$$\frac{dy}{dt} = ky$$

where k is a constant and $y = 1000$ when $t = 0$. We have the law of natural growth, and from Theorem 6.6.1,

$$y = 1000e^{kt} \tag{1}$$

Because $y = 2000$ when $t = 12$ we obtain from (1)

$$e^{12k} = 2$$

Thus

$$12k = \ln 2$$

$$k = \tfrac{1}{12} \ln 2$$

$$k = 0.05776$$

Hence from (1) with this value of k we get

$$y = 1000e^{0.05776t}$$

Replacing t by T and y by 1,000,000 we have

$$1,000,000 = 1000e^{0.05776T}$$

$$e^{0.05776T} = 1000$$

$$0.05776T = \ln 1000$$

$$T = \frac{\ln 1000}{0.05776}$$

$$T = 119.6$$

Therefore there will be 1,000,000 bacteria present in 119.6 min, which is 1 hr, 59 min, 36 sec.

EXAMPLE 2

The rate of increase of the population of a certain city is proportional to the population. If the population in 1950 was 50,000 and in 1980 it was 75,000, what is the expected population in 2010?

SOLUTION

Let t years be the time since 1950. Let y be the population in t years. We have the boundary conditions given in Table 6.6.2, where y_{60} is the expected population in 2010.

Table 6.6.2

t	0	30	60
y	50,000	75,000	y_{60}

The differential equation is

$$\frac{dy}{dt} = ky$$

where k is a constant and $y = 50{,}000$ when $t = 0$. We have the law of natural growth, and from Theorem 6.6.1,

$$y = 50{,}000e^{kt} \tag{2}$$

Because $y = 75{,}000$ when $t = 30$, we get from (2)

$$75{,}000 = 50{,}000e^{30k}$$

$$e^{30k} = \tfrac{3}{2} \tag{3}$$

When $t = 60$, $y = y_{60}$. Therefore from (2),

$$y_{60} = 50{,}000e^{60k}$$

$$y_{60} = 50{,}000(e^{30k})^2 \tag{4}$$

Substituting from (3) into (4) we obtain

$$y_{60} = 50{,}000(\tfrac{3}{2})^2$$
$$= 112{,}500$$

Therefore the expected population in 2010 is 112,500.

In the above example, because the population is increasing with time we have a case of the law of natural growth. If a population decreases with time, which can occur if the death rate is greater than the birth rate, then we have a case of the law of natural decay (see Exercise 3). In the next example there is another situation involving the law of natural decay. In problems involving the law of natural decay, the **half-life** of a substance is the time for half of it to decay.

EXAMPLE 3
The rate of decay of radium is proportional to the amount present at any time. If 60 mg of radium are present now and its half-life is 1690 years, how much radium will be present 100 years from now?

SOLUTION
Let t be the number of years in the time from now. Let y mg of radium be present at t years. We have the boundary conditions given in Table 6.6.3, where y_{100} mg of radium will be present 100 years from now.
    The differential equation is

$$\frac{dy}{dt} = ky$$

where k is a constant and $y = 60$ when $t = 0$. We have the law of natural decay, and from Theorem 6.6.1,

$$y = 60e^{kt} \tag{5}$$

TABLE 6.6.3

t	0	1690	100
y	60	30	y_{100}

Because $y = 30$ when $t = 1690$, from (5) we get $30 = 60e^{1690k}$; that is,

$$e^{1690k} = 0.5$$

$$1690k = \ln 0.5$$

$$k = \frac{\ln 0.5}{1690}$$

$$k = -0.000410$$

Substituting this value of k in (5) we obtain

$$y = 60e^{-0.000410t} \tag{6}$$

When $t = 100$, $y = y_{100}$; thus from (6),

$$y_{100} = 60e^{-0.0410}$$

$$= 57.6$$

Therefore 57.6 mg of radium will be present 100 years from now.

If the function of Theorem 6.6.1 is denoted by f, then $y = f(t)$. Furthermore, if $f(0) = B$ $(B > 0)$, then from the theorem, when

$$f'(t) = kt \qquad t \geq 0 \tag{7}$$

we have

$$f(t) = Be^{kt} \qquad t \geq 0 \tag{8}$$

If $k > 0$, then (7) is the law of natural growth, and (8) defines a function that is said to have **exponential growth**. Because when $k > 0$,

$$\lim_{t \to +\infty} f(t) = B \lim_{t \to +\infty} e^{kt}$$

$$= +\infty$$

then $f(t)$ increases without bound as t increases without bound. A sketch of the graph of (8) when $k > 0$ appears in Fig. 6.6.1.

If $k < 0$, then (7) is the law of natural decay and (8) defines a function that is said to have **exponential decay**. Observe from (8) that if $k < 0$,

$$\lim_{t \to +\infty} f(t) = B \lim_{t \to +\infty} e^{kt}$$

$$= 0$$

and $f(t)$ is approaching 0 through positive values. A sketch of the graph of (8) when $k < 0$ is shown in Fig. 6.6.2.

Suppose now that a quantity increases at a rate proportional to the difference between a fixed positive number A and its size. Then if time is represented by t units and y units is the amount of the quantity present at any time,

$$\frac{dy}{dt} = k(A - y) \tag{9}$$

FIGURE 6.6.1

$f(t) = Be^{kt}$; $k > 0$

FIGURE 6.6.2

$f(t) = Be^{kt}$; $k < 0$

where k is a positive constant, and $y < A$ for all $t \geq 0$. Separating the variables in (9) we obtain

$$\frac{dy}{A - y} = kt$$

Integrating we have

$$\int \frac{dy}{A - y} = k \int dt$$

$$-\ln|A - y| = kt + C$$

$$\ln|A - y| = -kt - C$$

$$|A - y| = e^{-C}e^{-kt}$$

Let $e^{-C} = B$. Because $A - y$ is positive, we can omit the absolute-value bars. Thus we have

$$A - y = Be^{-kt}$$

$$y = A - Be^{-kt} \tag{10}$$

If in this equation we let $y = f(t)$, it becomes

$$f(t) = A - Be^{-kt} \tag{11}$$

where A, B, and k are positive constants. Equation (11) describes **bounded growth**. From (11),

$$\lim_{t \to +\infty} f(t) = \lim_{t \to +\infty} (A - Be^{-kt})$$

$$= A - B \lim_{t \to +\infty} e^{-kt}$$

$$= A - B \cdot 0$$

$$= A$$

and $f(t)$ is approaching A through values less than A. Therefore, from Definition 3.8.1(i) the graph of f has as a horizontal asymptote the line A units above the t axis. Also note that

$$f(0) = A - Be^{-k \cdot 0}$$

$$= A - B$$

From this information we obtain the sketch of the graph of f in Fig. 6.6.3. This graph is sometimes called a **learning curve**. The name appears appropriate when $f(t)$ represents the competence at which a person performs a job. As a person's experience increases, the competence increases rapidly at first and then slows down as additional experience has little effect on the skill at which the task is performed.

FIGURE 6.6.3

EXAMPLE 4

A new employee is performing a job more efficiently each day in such a way that if y units are produced per day after t days on the job,

$$\frac{dy}{dt} = k(80 - y)$$

where k is a positive constant and $y < 80$ for all $t \geq 0$. The employee produces 20 units the first day of work and 50 units per day after being on the job 10 days. (a) How many units per day does the employee produce after being on the job 30 days? (b) Show that after being on the job 60 days the employee is producing just 1 unit less than full potential.

SOLUTION

The given differential equation is the same as (9) with $A = 80$. Therefore the general solution is of the form of (10) with $A = 80$; thus it is

$$y = 80 - Be^{-kt} \qquad (12)$$

Table 6.6.4 shows the boundary conditions, where y_{30} units are produced per day after being on the job 30 days and y_{60} units are produced per day after being on the job 60 days. Because $y = 20$ when $t = 0$, we obtain from (12)

$$20 = 80 - Be^0$$
$$20 = 80 - B$$
$$B = 60$$

Replacing B by 60 in (12) we get

$$y = 80 - 60e^{-kt} \qquad (13)$$

From (13), with $y = 50$ when $t = 10$, we have

$$50 = 80 - 60e^{-10k}$$
$$e^{-10k} = 0.5$$

Substituting this value of e^{-10k} into (13) we have

$$y = 80 - 60(e^{-10k})^{t/10}$$
$$= 80 - 60(0.5)^{t/10} \qquad (14)$$

(a) Because $y = y_{30}$ when $t = 30$, we get from (14)

$$y_{30} = 80 - 60(0.5)^3$$
$$= 80 - 60(0.125)$$
$$= 80 - 7.5$$
$$= 72.5$$

Therefore the employee is producing 72 units per day after being on the job 30 days.

Table 6.6.4

t	0	10	30	60
y	20	50	y_{30}	y_{60}

(b) From (14), with $y = y_{60}$ when $t = 60$, we obtain

$$y_{60} = 80 - 60(0.5)^6$$
$$= 80 - 60(0.0156)$$
$$= 80 - 0.938$$
$$= 79.062$$

After being on the job 60 days, the employee is producing 79 units per day. Because

$$\lim_{t \to +\infty} [80 - 60(0.5)^{t/10}] = 80 - 60 \lim_{t \to +\infty} \frac{1}{2^{t/10}}$$
$$= 80 - 60 \cdot 0$$
$$= 80$$

full potential is 80 units per day. Thus after 60 days the employee is producing just 1 unit less than full potential.

In Section 8.6 we consider another type of bounded growth, one that gives a model of population growth that takes into account environmental factors.

An application of the natural exponential function in physics is afforded by Newton's law of cooling, given in the next example.

EXAMPLE 5

Newton's law of cooling states that the rate at which a body changes temperature is proportional to the difference between its temperature and that of the surrounding medium. If a body is in air of temperature 35° and the body cools from 120° to 60° in 40 min, find the temperature of the body after 100 min.

SOLUTION

Let t min be the time that has elapsed since the body started to cool. Let y degrees be the temperature of the body at t min. Table 6.6.5 gives the boundary conditions, where y_{100} degrees is the temperature of the body after 100 min.

From Newton's law of cooling we have

$$\frac{dy}{dt} = k(y - 35)$$

where k is a constant and $y > 35$ for all $t \geq 0$. Separating the variables we obtain

$$\frac{dy}{y - 35} = k \, dt$$

Thus

$$\int \frac{dy}{y - 35} = k \int dt$$
$$\ln|y - 35| = kt + \bar{c}$$

Table 6.6.5

t	0	40	100
y	120	60	y_{100}

We can delete the absolute-value bars because $y > 35$. Therefore we have

$$y - 35 = e^{kt + \bar{c}}$$

$$y = e^{\bar{c}}e^{kt} + 35$$

$$y = Ce^{kt} + 35$$

where $C = e^{\bar{c}}$. When $t = 0$, $y = 120$; so $C = 85$. Hence

$$y = 85e^{kt} + 35 \qquad (15)$$

When $t = 40$, $y = 60$, and we obtain

$$60 = 85e^{40k} + 35$$

$$e^{40k} = \tfrac{5}{17}$$

We substitute this value of e^{40k} in (15), and we have

$$y = 85(e^{40k})^{t/40} + 35$$

$$= 85(\tfrac{5}{17})^{t/40} + 35 \qquad (16)$$

Because $y = y_{100}$ when $t = 100$, we have from (16)

$$y_{100} = 85(\tfrac{5}{17})^{5/2} + 35$$

$$= 39$$

Therefore after 100 min the temperature of the body is $39°$.

EXAMPLE 6

There are 100 million liters of fluoridated water in the reservoir containing a city's water supply, and the water contains 700 kg of fluoride. To decrease the fluoride content, fresh water runs into the reservoir at the rate of 3 million liters per day, and the mixture of water and fluoride, kept uniform, runs out of the reservoir at the same rate. How many kilograms of fluoride are in the reservoir 60 days after the pure water started to flow into the reservoir?

SOLUTION

Let t days be the time elapsed since the pure water started to flow into the reservoir. Let x kg be the weight of the fluoride in the reservoir at t days.

Because 100 million liters of fluoridated water are in the tank at all times, at t days the weight of fluoride per million liters is $x/100$ kg. Three million liters of the fluoridated water run out of the reservoir each day; so the reservoir loses $3(x/100)$ kg of fluoride per day. Because $\dfrac{dx}{dt}$ is the rate of change of x with respect to t, and x is decreasing as t increases, we have the differential equation

$$\frac{dx}{dt} = -\frac{3x}{100}$$

TABLE 6.6.6

t	0	60
x	700	x_{60}

where $x > 0$ for all $t > 0$. We also have the initial conditions given in Table 6.6.6, where x_{60} kg is the weight of fluoride in the reservoir after 60 days. Separating the variables and integrating we get

$$\frac{dx}{x} = -0.03\, dt$$

$$\int \frac{dx}{x} = -0.03 \int dt$$

$$\ln|x| = -0.03t + \bar{c}$$

$$|x| = e^{\bar{c}} e^{-0.03t}$$

$$x = C e^{-0.03t}$$

where $C = e^{\bar{c}}$ and the absolute-value bars are omitted because $x > 0$. When $t = 0$, $x = 700$; thus $C = 700$. Letting $t = 60$ and $x = x_{60}$ we have

$$x_{60} = 700 e^{-1.8}$$
$$= 700(0.1653)$$
$$= 115.7$$

Hence there are 115.7 kg of fluoride in the reservoir 60 days after the pure water started to flow into the reservoir.

The calculus is often very useful to the economist for evaluating certain business decisions. However, to use the calculus we must deal with continuous functions. Consider, for example, the following formula, which gives A, the number of dollars in the amount after t years, if P dollars is invested at an annual rate of $100i$ percent, compounded m times per year:

$$A = P\left(1 + \frac{i}{m}\right)^{mt} \tag{17}$$

Let us conceive of a situation in which the interest is continuously compounding; that is, consider formula (17), where we let the number of interest periods per year increase without bound. Then going to the limit in formula (17) we have

$$A = P \lim_{m \to +\infty} \left(1 + \frac{i}{m}\right)^{mt}$$

which can be written as

$$A = P \lim_{m \to +\infty} \left[\left(1 + \frac{i}{m}\right)^{m/i}\right]^{it} \tag{18}$$

To compute this limit by Theorem 1.7.1 we must first determine if

$$\lim_{m \to +\infty} \left(1 + \frac{i}{m}\right)^{m/i}$$

exists. Letting $h = i/m$ we have $m/i = 1/h$; and because $m \to +\infty$ is equivalent to $h \to 0^+$ we have

$$\lim_{m \to +\infty} \left(1 + \frac{i}{m}\right)^{m/i} = \lim_{h \to 0^+} (1 + h)^{1/h}$$

$$= e$$

Hence, using Theorem 1.7.1 we have

$$\lim_{m \to +\infty} \left[\left(1 + \frac{i}{m}\right)^{m/i}\right]^{it} = \left[\lim_{m \to +\infty} \left(1 + \frac{i}{m}\right)^{m/i}\right]^{it}$$

$$= e^{it}$$

and so (18) becomes

$$A = Pe^{it} \tag{19}$$

By letting t vary through the set of nonnegative real numbers we see that (19) expresses A as a continuous function of t.

Another way of looking at the same situation is to consider an investment of P dollars, which increases at a rate proportional to its size. This is the law of natural growth. Then if A dollars is the amount at t years, we have

$$\frac{dA}{dt} = kA$$

$$\int \frac{dA}{A} = k \int dt$$

$$\ln|A| = kt + \bar{c}$$

$$|A| = e^{\bar{c}}e^{kt}$$

$$A = Ce^{kt}$$

where $C = e^{\bar{c}}$ and the absolute-value bars are deleted because $A > 0$. When $t = 0$, $A = P$; so $C = P$. Therefore we have

$$A = Pe^{kt} \tag{20}$$

Comparing (20) with (19) we see that they are the same if $k = i$. So if an investment increases at a rate proportional to its size, we say that the interest is **compounded continuously**, and the annual interest rate is the constant of proportionality.

▶ ILLUSTRATION 1   If P dollars is invested at a rate of 8% per year compounded continuously, and A dollars is the amount of the investment at t years,

$$\frac{dA}{dt} = 0.08A$$

and

$$A = Pe^{0.08t}$$

◀

If in (19) we take $P = 1$, $i = 1$, and $t = 1$, we get $A = e$, which gives a justification for the economist's interpretation of the number e as the yield on an investment of \$1 for a year at an interest rate of 100% compounded continuously.

In the following example we use the terminology *effective rate* of interest, which is the rate that gives the same amount of interest compounded once a year.

EXAMPLE 7

If \$5000 is borrowed at an interest rate of 12% per year, compounded continuously, and the loan is to be repaid in one payment at the end of a year, how much must the borrower repay? Also, find the effective annual rate of interest.

SOLUTION

Let A dollars be the amount to be repaid, and because $P = 5000$, $i = 0.12$, and $t = 1$, we have from (19)

$$A = 5000e^{0.12}$$
$$= 5000(1.1275)$$
$$= 5637.50$$

Hence the borrower must repay \$5637.50. Letting j be the effective annual rate of interest we have

$$5000(1 + j) = 5000e^{0.12}$$
$$1 + j = e^{0.12}$$
$$j = 1.1275 - 1$$
$$j = 0.1275$$
$$j = 12.75\%$$

The effective annual interest rate is 12.75%.

EXERCISES 6.6

1. The rate of natural increase of the population of a certain city is proportional to the population. If the population increases from 40,000 to 60,000 in 40 years, when will the population be 80,000?

2. The population of a particular city doubled in 60 years from 1890 to 1950. If the rate of natural increase of the population at any time is proportional to the population at the time, and the population in 1950 was 60,000, estimate the population in the year 2000.

3. The population of a town is decreasing at a rate proportional to its size. In 1975 the population was 50,000 and in 1985 it was 44,000. What is the expected population in 1995?

4. After the pre-opening and opening day publicity of a certain exploitation film stopped, the attendance decreased at a rate proportional to its size. If the opening day's attendance at a specific theater was 5000 and the attendance on the third day was 2000, what is the expected attendance on the sixth day?

5. After an automobile is 1 year old, its rate of depreciation at any time is proportional to its value at that time. If an automobile was purchased on June 1, 1984, and its values on June 1, 1985, and June 1, 1986, were, respectively, $7000 and $5800, what is its expected value on June 1, 1990?

6. Suppose that the value of a certain antique collection increases with age and its rate of appreciation at any time is proportional to its value at that time. If the value of the collection was $25,000 10 years ago and its present value is $35,000, in how many years is its value expected to be $50,000?

7. Bacteria grown in a certain culture increase at a rate proportional to the number present. If 1000 bacteria are present initially and the number doubles in 30 min, how many bacteria will there be in 2 hours?

8. In a certain bacterial culture where the rate of growth of bacteria is proportional to the number present, the number triples in 1 hour. If at the end of 4 hours there were 10 million bacteria, how many bacteria were present initially?

9. If the half-life of radium is 1690 years, what percent of the amount present now will be remaining after (a) 100 years and (b) 1000 years.

10. Thirty percent of a radioactive substance disappears in 15 years. Find the half-life of the substance.

11. The winter mortality rate of a certain species of wildlife in a particular geographical region is proportional to the number of the species present at any time. There were 2400 of the species present in the region on December 21, the first day of winter; 30 days later there were 2000. How many of the species were expected to survive the winter? That is, how many will be living 90 days after December 21?

12. When a simple electric circuit, containing no capacitors but having inductance and resistance, has the electromotive force removed, the rate of decrease of the current is proportional to the current. The current is i amperes t sec after the cutoff, and $i = 40$ when $t = 0$. If the current dies down to 15 amperes in 0.01 sec, find i in terms of t.

13. (a) Find the interest earned on an investment of $500 at the end of one year if the annual interest rate is 10% compounded continuously. (b) What is the effective annual rate of interest?

14. A loan of $1000 is repaid in one payment at the end of a year. If the interest rate is 8% compounded continuously, determine (a) the total amount repaid and (b) the effective rate of interest.

15. If an amount of money invested doubles itself in 10 years at interest compounded continuously, how long will it take for the original amount to triple itself?

16. If the purchasing power of a dollar is decreasing at the rate of 10% annually, compounded continuously, how long will it take for the purchasing power to be $0.50?

17. How long will it take for $500 to accumulate to $1000 if money is invested at 8 percent per year compounded continuously?

18. A new worker on an assembly line can do a particular task in such a way that if y units are completed per day after t days on the assembly line, then

$$\frac{dy}{dt} = k(90 - y)$$

where k is a positive constant and $y < 90$ for all $t \geq 0$. On the day the worker starts, 60 units are completed, and after being on the job 5 days, the worker does 75 units per day. (a) How many units per day are completed after the worker is on the job 9 days? (a) Show that the worker is producing at almost full potential after 30 days.

19. Suppose that a student has 3 hours to cram for an examination and during this time wishes to memorize a set of 60 facts. According to psychologists, the rate at which a person can memorize a set of facts is proportional to the number of facts remaining to be memorized. Thus if the student memorizes y facts in t min,

$$\frac{dy}{dt} = k(60 - y)$$

where k is a positive constant and $y < 60$ for all $t \geq 0$. It is assumed that initially zero facts are memorized. If the student memorizes 15 facts in the first 20 min, how many facts will the student memorize in (a) 1 hour and (b) 3 hours?

20. Under the conditions of Example 5, after how many minutes will the temperature of the body be 45°?

21. If a body in air at a temperature of 0° cools from 200° to 100° in 40 min, how many more minutes will it take for the body to cool to 50°? Use Newton's law of cooling given in Example 5.

22. If a thermometer is taken from a room in which the temperature is $75°$ into the open, where the temperature is $35°$, and the reading of the thermometer is $65°$ after 30 sec, (a) how long after the removal will the reading be $50°$? (b) What is the thermometer reading 3 min after the removal? Use Newton's law of cooling given in Example 5.

23. In a certain chemical reaction the rate of conversion of a substance is proportional to the amount of the substance still unreacted at that time. After 10 min one-third of the original amount of the substance has been reacted and 20 g has been reacted after 15 min. What was the original amount of the substance?

24. Sugar decomposes in water at a rate proportional to the amount still unchanged. If 50 kg of sugar were present initially and at the end of 5 hr this is reduced to 20 kg, how long will it take until 90 percent of the sugar is decomposed?

25. There are 100 liters of brine in a tank, and the brine contains 70 kg of dissolved salt. Fresh water runs into the tank at the rate of 3 liters/min, and the mixture, kept uniform by stirring, runs out at the same rate. How many kilograms of salt are there in the tank at the end of 1 hour?

26. A tank contains 200 liters of brine in which there are 3 kg of salt per liter. It is desired to dilute this solution by adding brine containing 1 kg of salt per liter, which flows into the tank at the rate of 4 liters/min and runs out at the same rate. When will the tank contain $1\frac{1}{2}$ kg of salt per liter?

27. Professor Willard Libby of the University of California at Los Angeles was awarded the Nobel prize in chemistry for discovering a method of determining the date of death of a once-living object. Professor Libby made use of the fact that the tissue of a living organism is composed of two kinds of carbons, a radioactive carbon-14 (commonly written ^{14}C) and a stable carbon-12 (^{12}C), in which the ratio of the amount of ^{14}C to the amount of ^{12}C is approximately constant. When the organism dies, the law of natural decay applies to ^{14}C. If it is determined that the amount of ^{14}C in a piece of charcoal is only 15 percent of its original amount and the half-life of ^{14}C is 5600 years, when did the tree from which the charcoal came die?

28. Refer to Exercise 27. Suppose that after finding a fossil an archaeologist determines that the amount of ^{14}C present in the fossil is 25 percent of its original amount. Using the fact that the half-life of ^{14}C is 5600 years, what is the age of the fossil?

29. In biology an equation sometimes used to describe the restricted growth of a population is the Gompertz growth equation,

$$\frac{dy}{dt} = ky \ln \frac{a}{y}$$

where a and k are positive constants. Find the general solution of this differential equation.

6.7
OTHER EXPONENTIAL AND LOGARITHMIC FUNCTIONS

From Definition 6.5.2, if $a > 0$,

$$a^x = e^{x \ln a} \tag{1}$$

The expression on the left side of (1) is called the *exponential function to the base a.*

6.7.1 DEFINITION   If a is any positive number and x is any real number, then the function f defined by

$$f(x) = a^x$$

is called the **exponential function to the base *a*.**

The exponential function to the base a satisfies the same properties as the natural exponential function.

▶ ILLUSTRATION 1   If x and y are any real numbers and a is positive, then from (1),

$$a^x a^y = e^{x \ln a} e^{y \ln a}$$
$$= e^{x \ln a + y \ln a}$$
$$= e^{(x+y)\ln a}$$
$$= a^{x+y} \qquad \qquad ◀$$

From Illustration 1 we have the property

$$a^x a^y = a^{x+y} \qquad \qquad (2)$$

We also have the following properties:

$$a^x \div a^y = a^{x-y} \qquad \qquad (3)$$
$$(a^x)^y = a^{xy} \qquad \qquad (4)$$
$$(ab)^x = a^x b^x \qquad \qquad (5)$$
$$a^0 = 1 \qquad \qquad (6)$$

The proofs of (3) through (6) are left as exercises (see Exercises 37 to 40). To find the derivative of the exponential function to the base a we set $a^x = e^{x \ln a}$ and apply the chain rule.

$$a^x = e^{x \ln a}$$
$$D_x(a^x) = e^{x \ln a} D_x(x \ln a)$$
$$= e^{x \ln a}(\ln a)$$
$$= a^x \ln a$$

Therefore we have the following theorem.

6.7.2 THEOREM   If a is any positive number and u is a differentiable function of x,

$$D_x(a^u) = a^u \ln a \, D_x u$$

▶ ILLUSTRATION 2   If $y = 3^{x^2}$, then from Theorem 6.7.2,

$$\frac{dy}{dx} = 3^{x^2}(\ln 3)(2x)$$
$$= 2(\ln 3)x 3^{x^2} \qquad \qquad ◀$$

The next theorem, giving the indefinite integration formula for the exponential function to the base a, follows from Theorem 6.7.2.

6.7.3 THEOREM   If a is any positive number other than 1,

$$\int a^u \, du = \frac{a^u}{\ln a} + C$$

EXAMPLE 1

Find $\int \sqrt{10^{3x}}\, dx$.

SOLUTION

Because $\sqrt{10^{3x}} = 10^{3x/2}$, we apply Theorem 6.7.3 with $u = \frac{3}{2}x$. We have, then,

$$\int \sqrt{10^{3x}}\, dx = \int 10^{3x/2}\, dx$$

$$= \tfrac{2}{3} \int 10^{3x/2}(\tfrac{3}{2}\, dx)$$

$$= \frac{2}{3} \cdot \frac{10^{3x/2}}{\ln 10} + C$$

$$= \frac{2\sqrt{10^{3x}}}{3 \ln 10} + C$$

EXAMPLE 2

Draw sketches of the graphs of $y = 2^x$ and $y = 2^{-x}$ on the same set of axes. Find the area of the region bounded by these two graphs and the line $x = 2$.

SOLUTION

The required sketches are shown in Fig. 6.7.1. The region is shaded in the figure. If A square units is the desired area,

$$A = \lim_{||\Delta|| \to 0} \sum_{i=1}^{n} \left[2^{\xi_i} - 2^{-\xi_i} \right] \Delta_i x$$

$$= \int_0^2 (2^x - 2^{-x})\, dx$$

$$= \frac{2^x}{\ln 2} + \frac{2^{-x}}{\ln 2} \Bigg]_0^2$$

$$= \frac{4}{\ln 2} + \frac{\frac{1}{4}}{\ln 2} - \frac{1}{\ln 2} - \frac{1}{\ln 2}$$

$$= \frac{9}{4 \ln 2}$$

$$\approx 3.25$$

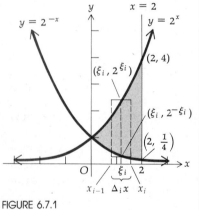

FIGURE 6.7.1

We can now define the *logarithmic function to the base a* if a is any positive number other than 1.

6.7.4 DEFINITION   If a is any positive number except 1, the **logarithmic function to the base a** is the inverse of the exponential function to the base a; we write

$$y = \log_a x \qquad \text{if and only if} \quad a^y = x \tag{7}$$

The above is the definition of the logarithmic function to the base a usually given in elementary algebra; however, (7) has meaning for y any real number because a^y has been precisely defined. If $a = e$, we have the logarithmic function to the base e, which is the natural logarithmic function.

$\log_a x$ is read as "the logarithm of x to the base a."

The logarithmic function to the base a obeys the same laws as the natural logarithmic function. We list them:

$$\log_a(xy) = \log_a x + \log_a y \tag{8}$$

$$\log_a(x \div y) = \log_a x - \log_a y \tag{9}$$

$$\log_a 1 = 0 \tag{10}$$

$$\log_a x^y = y \log_a x \tag{11}$$

The proofs of (8) through (11) are left as exercises (see Exercises 41 to 44).

A relationship between logarithms to the base a and natural logarithms follows easily. Let

$$y = \log_a x$$

Then

$$a^y = x$$

$$\ln a^y = \ln x$$

$$y \ln a = \ln x$$

$$y = \frac{\ln x}{\ln a}$$

Replacing y by $\log_a x$ we obtain

$$\log_a x = \frac{\ln x}{\ln a} \tag{12}$$

Most calculators do not have a $\boxed{\log_a x}$ key. Thus (12) is a convenient formula to apply to compute values of $\log_a x$ on a calculator.

Equation (12) sometimes is used as the definition of the logarithmic function to the base a. Because the natural logarithmic function is continuous at all $x > 0$, it follows from (12) that the logarithmic function to the base a is continuous at all $x > 0$.

If in (12) $x = e$, we have

$$\log_a e = \frac{\ln e}{\ln a}$$

$$\Leftrightarrow \quad \log_a e = \frac{1}{\ln a} \tag{13}$$

We now find the derivative of the logarithmic function to the base a. We differentiate on both sides of (12) with respect to x, and we obtain

$$D_x(\log_a x) = \frac{1}{\ln a} D_x(\ln x)$$

$$D_x(\log_a x) = \frac{1}{\ln a} \cdot \frac{1}{x} \tag{14}$$

Substituting from (13) into (14) we get

$$D_x(\log_a x) = \frac{\log_a e}{x} \qquad (15)$$

From (15) and (14) and the chain rule we have the following theorem.

6.7.5 THEOREM   If u is a differentiable function of x,

$$D_x(\log_a u) = \frac{\log_a e}{u} \cdot D_x u$$

$$\Leftrightarrow \quad D_x(\log_a u) = \frac{1}{(\ln a)u} \cdot D_x u$$

If in Theorem 6.7.5 $a = e$, then we have

$$D_x(\log_e u) = \frac{\log_e e}{u} D_x u$$

$$\Leftrightarrow \quad D_x(\ln u) = \frac{1}{u} D_x u$$

which is Theorem 6.3.2 for the derivative of the natural logarithmic function.

EXAMPLE 3
Given

$$y = \log_{10} \frac{x+1}{x^2+1}$$

find dy/dx.

SOLUTION
Using (9) we write

$$y = \log_{10}(x+1) - \log_{10}(x^2+1)$$

From Theorem 6.7.5,

$$\frac{dy}{dx} = \frac{\log_{10} e}{x+1} - \frac{\log_{10} e}{x^2+1} \cdot 2x$$

$$= \log_{10} e\left(\frac{1}{x+1} - \frac{2x}{x^2+1}\right)$$

$$= \frac{\log_{10} e(1 - 2x - x^2)}{(x+1)(x^2+1)}$$

Because x^n has been defined for any real number n, we can now prove the theorem for the derivative of the power function if the exponent is any real number.

6.7.6 THEOREM   If n is any real number and the function f is defined by

$$f(x) = x^n \qquad \text{where } x > 0$$

then

$$f'(x) = nx^{n-1}$$

PROOF   From Definition 6.5.2,

$$f(x) = e^{n \ln x}$$

Thus

$$f'(x) = e^{n \ln x} D_x(n \ln x)$$

$$= e^{n \ln x}\left(\frac{n}{x}\right)$$

$$= x^n \cdot \frac{n}{x}$$

$$= nx^{n-1} \qquad \blacksquare$$

Theorem 6.7.6 enables us to find the derivative of a variable to a constant power. Previously in this section we learned how to differentiate a constant to a variable power. We now consider the derivative of a function whose function value is a variable to a variable power.

EXAMPLE 4

If $y = x^x$, where $x > 0$, find $\dfrac{dy}{dx}$.

SOLUTION
From Definition 6.5.2, if $x > 0$, $x^x = e^{x \ln x}$. Therefore

$$y = e^{x \ln x}$$

$$\frac{dy}{dx} = e^{x \ln x} D_x(x \ln x)$$

$$= e^{x \ln x}\left(x \cdot \frac{1}{x} + \ln x\right)$$

$$= x^x(1 + \ln x)$$

The method of logarithmic differentiation can also be used to find the derivative of a function whose function value is a variable to a variable power, as shown in the following example.

EXAMPLE 5

Find $\dfrac{dy}{dx}$ in Example 4 by using logarithmic differentiation.

SOLUTION

We are given $y = x^x$ with $x > 0$. We take the natural logarithm on both sides of the equation and obtain

$$\ln y = \ln x^x$$

$$\ln y = x \ln x$$

Differentiating on both sides of the above equation with respect to x gives

$$\frac{1}{y} \cdot \frac{dy}{dx} = x \cdot \frac{1}{x} + \ln x$$

$$\frac{dy}{dx} = y(1 + \ln x)$$

$$= x^x(1 + \ln x)$$

EXERCISES 6.7

In Exercises 1 through 24, find the derivative of the given function.

1. $f(x) = 3^{5x}$ **2.** $f(x) = 6^{-3x}$ **3.** $f(t) = 4^{3t^2}$ **4.** $g(x) = 10^{x^2 - 2x}$

5. $f(x) = 4^{\sin 2x}$ **6.** $f(z) = 2^{\csc 3z}$ **7.** $g(x) = 2^{5x}3^{4x^2}$ **8.** $f(x) = (x^3 + 3)2^{-7x}$

9. $h(x) = \dfrac{\log_{10} x}{x}$ **10.** $f(t) = \log_{10} \dfrac{t}{t + 1}$ **11.** $f(x) = \sqrt{\log_a x}$ **12.** $f(x) = \log_a[\log_a(\log_a x)]$

13. $f(x) = \log_{10}[\log_{10}(x + 1)]$ **14.** $g(w) = \tan 2^{3w}$ **15.** $f(t) = \sec 3^{t^2}$

16. $f(x) = x^{x^2}; \; x > 0$ **17.** $f(x) = x^{\sqrt{x}}; \; x > 0$ **18.** $f(x) = x^{\ln x}; \; x > 0$

19. $g(z) = z^{\cos z}; \; z > 0$ **20.** $g(t) = (\cos t)^t; \; \cos t > 0$ **21.** $h(x) = (\sin x)^{\tan x}; \; \sin x > 0$

22. $f(x) = (x)^{x^x}; \; x > 0$ **23.** $f(x) = x^{e^x}; \; x > 0$ **24.** $f(x) = (\ln x)^{\ln x}; \; x > 1$

In Exercises 25 through 32, evaluate the indefinite integral.

25. $\int 3^{2x}\, dx$ **26.** $\int a^{nx}\, dx$ **27.** $\int a^t e^t\, dt$ **28.** $\int 5^{x^4 + 2x}(2x^3 + 1)\, dx$

29. $\int x^2 10^{x^3}\, dx$ **30.** $\int a^{z \ln z}(\ln z + 1)\, dz$ **31.** $\int e^y 2^{e^y} 3^{e^y}\, dy$ **32.** $\int \dfrac{4^{\ln(1/x)}}{x}\, dx$

In Exercises 33 and 34, compute the value of the given logarithm.

33. $\log_5 e$ **34.** $\log_2 10$

In Exercises 35 and 36, use differentials to find an approximate value of the given logarithm, and express the answer to three decimal places.

35. $\log_{10} 997$ **36.** $\log_{10} 1.015$

In Exercises 37 through 40, prove the given property if a and b are any positive numbers and x and y are any real numbers.

37. $a^x \div a^y = a^{x - y}$ **38.** $(a^x)^y = a^{xy}$ **39.** $(ab)^x = a^x b^x$ **40.** $a^0 = 1$

In Exercises 41 through 44, prove the given property if a is any positive number except 1, and x and y are any positive numbers.

41. $\log_a(xy) = \log_a x + \log_a y$
42. $\log_a(x \div y) = \log_a x - \log_a y$
43. $\log_a 1 = 0$
44. $\log_a x^y = y \log_a x$

45. A company has learned that when it initiates a new sales campaign, the number of sales per day increases. However, the number of extra daily sales per day decreases as the impact of the campaign wears off. For a specific campaign the company has determined that if $S(t)$ is the number of extra daily sales as a result of the campaign and t is the number of days that have elapsed since the campaign ended, then $S(t) = 1000(3^{-t/2})$. Find the rate at which the extra daily sales are decreasing when (a) $t = 4$ and (b) $t = 10$.   *find* $S'(t) = 1000 \; 3^{-t/2}(-\tfrac{1}{2})\ln 3 = -500 \ln 3 \; 3^{-t/2}$

46. A company estimates that in t years the number of its employees will be $N(t)$, where $N(t) = 1000(0.8)^{t/2}$. (a) How many employees does the company expect to have in 4 years? (b) At what rate is the number of employees expected to be changing in 4 years?

47. A particle is moving along a straight line according to the equation of motion $s = A \cdot 2^{kt} + B \cdot 2^{-kt}$, where A, B, and k are constants and s ft is the directed distance of the particle from the origin at t sec. Prove that the motion is simple harmonic by showing that if a ft/sec^2 is the acceleration at t sec, then a is proportional to s.

48. A particle moves along a straight line according to the equation of motion $s = t^{1/t}$, where s ft is the directed distance of the particle from the origin at t sec. Find the velocity and acceleration at 2 sec.

49. An historically important abstract painting was purchased in 1920 for \$200, and its value has doubled every 10 years since its purchase. If y dollars is the value of the painting t years after its purchase, (a) define y in terms of t. (b) What was the value of the painting in 1980? (c) Find the rate at which the value of the painting was increasing in 1980.

50. Draw sketches of the graphs of $y = \log_{10} x$ and $y = \ln x$ on the same set of axes.
51. Draw sketches of the graphs of $y = e^x$ and $y = 2^x$ on the same set of axes. Find the area of the region bounded by these two graphs and the line $x = 1$.
52. Find the area of the region bounded by the graph of $y = 3^x$ and the lines $x = 1$ and $y = 1$.
53. Find the volume of the solid generated if the region of Exercise 52 is revolved about the x axis.
54. Given $f(x) = \tfrac{1}{2}(a^x + a^{-x})$. Prove that $f(b + c) + f(b - c) = 2f(b)f(c)$.

$$\pi \int_0^1 (3^{2x} - 1)\,dx \qquad\qquad \int_0^1 (3^x - 1)\,dx$$

REVIEW EXERCISES FOR CHAPTER 6

In Exercises 1 through 6, determine if the given function has an inverse. If the inverse exists, do the following: (a) Define it and state its domain and range; (b) draw sketches of the graphs of the function and its inverse on the same set of axes. If the function does not have an inverse, show that a horizontal line intersects the graph of the function in more than one point.

1. $f(x) = x^3 - 4$
2. $f(x) = 2\sqrt[3]{x} - 1$
3. $f(x) = 9 - x^2$

4. $f(x) = \sqrt{4 - x^2}$
5. $f(x) = \dfrac{3x - 4}{x}$
6. $f(x) = |2x - 3|$

In Exercises 7 and 8, (a) prove that the given function f has an inverse, (b) find $f^{-1}(x)$, and (c) verify the equations of Theorem 6.1.4 for f and f^{-1}.

7. $f(x) = \sqrt[3]{x + 1}$
8. $f(x) = \dfrac{2x - 1}{2x + 1}$

$c^2 - 6c + 8 = 3$
$c = 5$
2^{N-6}

In Exercises 9 through 12, find $(f^{-1})'(d)$ for the given f and d.
9. $f(x) = x^2 - 6x + 8, \; x \geq 3; \; d = 3$ $(f^{-1})'(3) = \dfrac{1}{f'(5)} = \dfrac{1}{4}$
10. $f(x) = \sqrt{3x + 4}; \; d = 5$
11. $f(x) = 8x^3 + 6x; \; d = 4$
12. $f(x) = x^5 + x - 22; \; d = 12$

In Exercises 13 through 24, differentiate the given function.

13. $f(x) = (\ln x^2)^2$
14. $f(t) = 3\sin(e^{4t})$
15. $g(x) = 2^{\cos 4x}$
16. $f(x) = \dfrac{e^x}{(e^x + e^{-x})^2}$

17. $f(x) = e^{x/(4 + x^2)}$

18. $f(x) = 10^{-5x^2}$

19. $f(x) = \sqrt{\log_{10} \dfrac{1 + x}{1 - x}}$

20. $g(x) = \ln \sqrt{\dfrac{2x + 1}{x - 3}}$

21. $f(t) = t^{3/\ln t}$

22. $F(x) = \log_{10}(x^2 2^{x^2} + 1)$

23. $f(x) = (x)^{xe^x}; \; x > 0$

24. $f(x) = 3^{x^{x^2}}$

In Exercises 25 through 32, evaluate the indefinite integral.

25. $\displaystyle\int \frac{3e^{2x}}{1 + e^{2x}} \, dx$

26. $\displaystyle\int e^{2x^2 - 4x}(x - 1) \, dx$

27. $\displaystyle\int (e^{3x} + 2^{3x}) \, dx$

28. $\displaystyle\int \frac{10^{\ln x^2}}{x} \, dx$

29. $\displaystyle\int \frac{xe^{6x^2}}{\sqrt{1 + e^{6x^2}}} \, dx$

30. $\displaystyle\int (x + 1)e^x 7^{xe^x} \, dx$

31. $\displaystyle\int \frac{2^x \, dx}{\sqrt{3 \cdot 2^x + 4}}$

32. $\displaystyle\int \frac{10^x + 1}{10^x - 1} \, dx$

In Exercises 33 through 38, evaluate the definite integral.

33. $\displaystyle\int_0^2 x^2 e^{x^3} \, dx$

34. $\displaystyle\int_0^1 (e^{2x} + 1)^2 \, dx$

35. $\displaystyle\int_1^8 \frac{x^{1/3}}{x^{4/3} + 4} \, dx$

36. $\displaystyle\int_{1/3}^{1/2} \frac{4x^{-3} + 2}{x^{-2} - x} \, dx$

37. $\displaystyle\int_0^{\ln 2} \frac{e^{2x}}{e^x - 5} \, dx$

38. $\displaystyle\int_e^{e^2} \frac{dx}{x(\ln x)}$

39. Find $\dfrac{dy}{dx}$ if $ye^x + xe^y + x + y = 0$.

40. If $f(x) = \log_{(e^x)}(x + 1)$, find $f'(x)$.

41. Use differentials to find an approximate value to three decimal places of $\log_{10} 100937$. Use the fact that $\log_{10} e = 0.434$, with accuracy to three decimal places.

42. Find an equation of the tangent line to the curve $y = x^{x-1}$ at $(2, 2)$.

43. A particle is moving on a straight line, where s ft is the directed distance of the particle from the origin, v ft/sec is the velocity of the particle, and a ft/sec^2 is the acceleration of the particle at t sec. If $a = e^t + e^{-t}$ and $v = 1$ and $s = 2$ when $t = 0$, find v and s in terms of t.

44. The area of the region bounded by the curve $y = e^{-x}$, the coordinate axes, and the line $x = b \, (b > 0)$ is a function of b. If f is this function, find $f(b)$. Also find $\displaystyle\lim_{b \to +\infty} f(b)$.

45. The volume of the solid of revolution obtained by revolving the region in Exercise 44 about the x axis is a function of b. If g is this function, find $g(b)$. Also find $\displaystyle\lim_{b \to +\infty} g(b)$.

46. The rate of natural increase of the population of a certain city is proportional to the population. If the population doubles in 60 years and if the population in 1950 was 60,000, estimate the population in the year 2000.

47. The rate of decay of a radioactive substance is proportional to the amount present. If half of a given deposit of the substance disappears in 1900 years, how long will it take for 95% of the deposit to disappear?

48. Prove that if a rectangle is to have its base on the x axis and two of its vertices on the curve $y = e^{-x^2}$, then the rectangle will have the largest possible area if the two vertices are at the points of inflection of the graph.

49. Given $f(x) = \ln|x|$ and $x < 0$. Show that f has an inverse function. If g is the inverse function, find $g(x)$ and the domain of g.

50. Prove that if $x < 1$, $\ln x < x$. (*Hint:* Let $f(x) = x - \ln x$, and show that f is decreasing on $(0, 1)$ and find $f(1)$.)

51. When a gas undergoes an adiabatic (no gain or loss of heat) expansion or compression, then the rate of change of the pressure with respect to the volume varies directly as the pressure and inversely as the volume. If the pressure is p lb/in.^{2} when the volume is v in.^{3}, and the initial pressure and volume are p_0 lb/in.^{2} and v_0 in.^{3}, show that $pv^k = p_0 v_0^k$.

52. If W in.-lb is the work done by a gas expanding against a piston in a cylinder and P lb/in.^{2} is the pressure of the gas when the volume of the gas is V in.^{3}, show that if V_1 in.^{3} and V_2 in.^{3} are the initial and final volumes, respectively, then

$$W = \int_{V_1}^{V_2} P \, dV$$

53. Suppose that a piston compresses a gas in a cylinder from an initial volume of 60 in.^{3} to a volume of 40 in.^{3}. If Boyle's law (Exercise 26 in Exercises 2.5) holds, and the initial pressure is 50 lb/in.^{2}, find the work done by the piston. (Use the result of Exercise 52.)

54. The charge of electricity on a spherical surface leaks off at a rate proportional to the charge. Initially, the charge of electricity was 8 coulombs and one-fourth leaks off in 15 min. When will there be only 2 coulombs remaining?

55. How long will it take for an investment to double itself if interest is paid at the rate of 8% per year compounded continuously?

56. Interest on a savings account is computed at 10% per year compounded continuously. If one wishes to have $1000 in the account at the end of a year by making a single deposit now, what should be the amount of the deposit?

57. The rate of bacterial growth in a certain culture is proportional to the number present, and the number doubles in 20 min. If at the end of 1 hour there were 1,500,000 bacteria, how many bacteria were present initially?

58. Refer to Exercise 27 in Exercises 6.6. A paleontologist discovered an insect preserved inside a transparent amber, which is hardened tree pitch, and the amount of ^{14}C present in the insect was determined to be 2 percent of its original amount. Use the fact that the half-life of ^{14}C is 5600 years to determine the age of the insect at the time of discovery.

59. A student studying a foreign language has 50 verbs to memorize. The rate at which the student can memorize these verbs is proportional to the number of verbs remaining to be memorized; that is, if the student memorizes y verbs in t min,

$$\frac{dy}{dt} = k(50 - y)$$

Assume that initially no verbs are memorized, and suppose that 20 verbs are memorized in the first 30 minutes. How many verbs will the student memorize in (a) 1 hour and (b) 2 hours? (c) After how many hours will the student have only one verb left to memorize?

60. A tank contains 100 liters of fresh water, and brine containing 2 kg of salt per liter flows into the tank at the rate of 3 liters/min. If the mixture, kept uniform by stirring, flows out at the same rate, how many kilograms of salt are there in the tank at the end of 30 min?

61. A tank contains 60 gal of salt water with 120 lb of dissolved salt. Salt water with 3 lb of salt per gallon flows into the tank at the rate of 2 gal/min, and the mixture, kept uniform by stirring, flows out at the same rate. How long will it be before there are 135 lb of salt in the tank?

62. Use Newton's law of cooling, given in Example 5 of Section 6.6, to determine the current temperature of a body in air of temperature 40° if 30 min ago the body's temperature was 150° and 10 min ago it was 90°.

63. Find the point on the graph of $f(x) = e^x$ for which the tangent line to the graph there passes through the origin.

64. Find the volume of the solid of revolution generated if the region bounded by the curve $y = 2^{-x}$ and the lines $x = 1$ and $x = 4$ is revolved about the x axis.

65. Prove that

$$\lim_{x \to 0} \frac{\log_a(1 + x)}{x} = \log_a e$$

(*Note:* Compare with Exercise 53 in Exercises 6.3.)

66. Prove that

$$\lim_{x \to 0} \frac{a^x - 1}{x} = \ln a$$

(*Hint:* Let $y = a^x - 1$ and express $(a^x - 1)/x$ as a function of y, say $g(y)$. Then show that $y \to 0$ as $x \to 0$, and find $\lim_{y \to 0} g(y)$.)

67. Use the results of Exercises 65 and 66 to prove that

$$\lim_{x \to 0} \frac{x^b - 1}{x - 1} = b$$

(*Hint:* Write

$$\frac{x^b - 1}{x - 1} = \frac{e^{b \ln x} - 1}{b \ln x} \cdot \frac{b \ln x}{x - 1}$$

Then let $s = b \ln x$ and $t = x - 1$.)

68. Prove that

$$\lim_{x \to 0} \frac{e^{ax} - 1}{x} = a$$

(*Hint:* Let $f(x) = e^{ax}$, and find $f'(0)$ by two methods.)

69. If the domain of f is the set of all real numbers and $f'(x) = cf(x)$ for all x where c is a constant, prove that there is a constant k for which $f(x) = ke^{cx}$ for all x. (*Hint:* Consider the function g for which $g(x) = f(x)e^{-cx}$, and find $g'(x)$.)

70. Prove that

$$D_x^{\,n}(\ln x) = (-1)^{n-1} \frac{(n-1)!}{x^n}$$

(*Hint:* Use mathematical induction.)

71. Find $\displaystyle\int_0^t e^{-|x|}\, dx$ if t is any real number.

72. Prove that if $x > 0$, and $\displaystyle\int_1^x t^{h-1}\, dt = 1$, then $\displaystyle\lim_{h \to 0} x = \lim_{h \to 0} (1 + h)^{1/h}$.

73. Do Exercise 61 in the Review Exercises for Chapter 4 by evaluating each integral.

S E V E N

INVERSE TRIGONOMETRIC FUNCTIONS
AND HYPERBOLIC FUNCTIONS

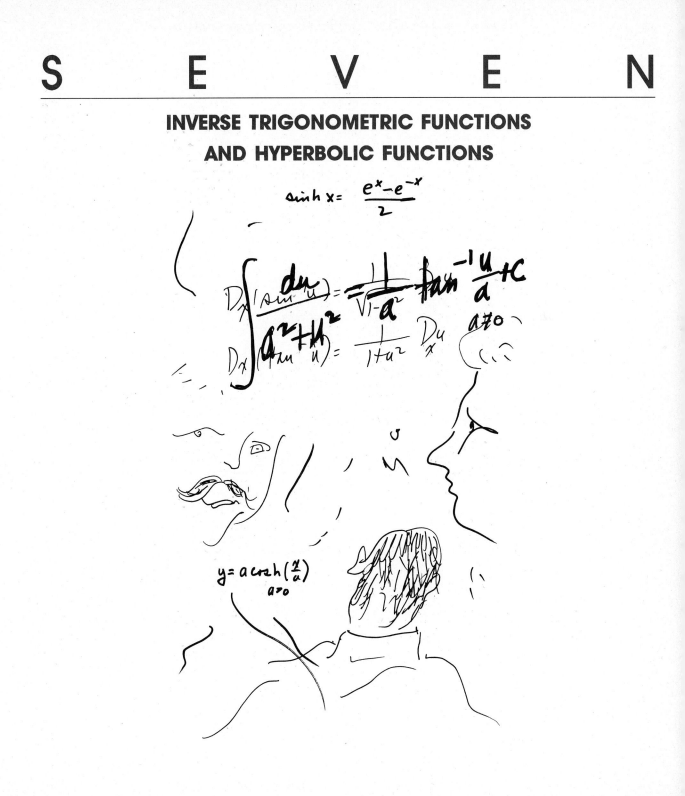

**7.1
THE INVERSE
TRIGONOMETRIC
FUNCTIONS**

Figure 7.1.1 shows a sketch of the graph of the sine function. This function is not one-to-one, because every number in its range is the function value of more than one number in its domain. Therefore the sine function does not have an inverse. However, observe from Fig. 7.1.1 that the sine is increasing

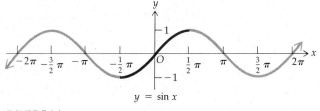

$y = \sin x$

FIGURE 7.1.1

on the interval $[-\frac{1}{2}\pi, \frac{1}{2}\pi]$. This fact follows from Theorem 3.4.3, because if $f(x) = \sin x$, then $f'(x) = \cos x > 0$ for all x in $(-\frac{1}{2}\pi, \frac{1}{2}\pi)$. Therefore, even though the sine function does not have an inverse, it follows from Theorem 6.1.5 that the function F for which

$$F(x) = \sin x \quad \text{and} \quad -\tfrac{1}{2}\pi \le x \le \tfrac{1}{2}\pi \tag{1}$$

does have an inverse. The domain of F is $[-\frac{1}{2}\pi, \frac{1}{2}\pi]$, and its range is $[-1, 1]$. A sketch of the graph of F is shown in Fig. 7.1.2. The inverse of the function defined by (1) is called the *inverse sine function* and is denoted by the symbol $\sin^{-1}$. Following is the formal definition.

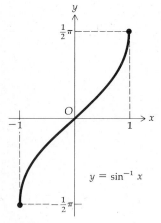

FIGURE 7.1.2

7.1.1 DEFINITION   The **inverse sine function**, denoted by $\sin^{-1}$, is defined as follows:

$$y = \sin^{-1} x \quad \text{if and only if} \quad x = \sin y \text{ and } -\tfrac{1}{2}\pi \le y \le \tfrac{1}{2}\pi$$

The domain of $\sin^{-1}$ is the closed interval $[-1, 1]$, and the range is the closed interval $[-\frac{1}{2}\pi, \frac{1}{2}\pi]$. A sketch of the graph is shown in Fig. 7.1.3.

The use of the symbol -1 to represent the inverse sine function makes it necessary to denote the reciprocal of $\sin x$ by $(\sin x)^{-1}$ to avoid confusion. A similar convention is applied when using any negative exponent with a trigonometric function. For instance,

$$\frac{1}{\sin x} = (\sin x)^{-1} \qquad \frac{1}{\sin^2 x} = (\sin x)^{-2} \qquad \frac{1}{\cos^3 x} = (\cos x)^{-3}$$

and so on.

$y = \sin^{-1} x$

FIGURE 7.1.3

▶ ILLUSTRATION 1

(a) $\sin^{-1} \dfrac{1}{\sqrt{2}} = \dfrac{1}{4}\pi$ (b) $\sin^{-1}\left(-\dfrac{1}{\sqrt{2}}\right) = -\dfrac{1}{4}\pi$ ◀

▶ ILLUSTRATION 2   (a) To find an approximate value of $\sin^{-1}(0.4695)$ from Table 5 in the appendix, look for 0.4695 in the sine column of the table, and observe that $\sin 0.4887 \approx 0.4695$. Therefore

$$\sin^{-1}(0.4695) \approx 0.4887$$

(b) If a calculator is used to determine $\sin^{-1}(0.4695)$, first set the calculator in the radian mode. Obtain 0.4695 in the display, and then press the $\boxed{\sin^{-1}}$ key (or the $\boxed{\text{INV}}$ key followed by the $\boxed{\sin}$ key) and read 0.4887.

◀

It follows from Definition 7.1.1 that

$$\sin(\sin^{-1} x) = x \qquad \text{for } x \text{ in } [-1, 1] \tag{2}$$

$$\sin^{-1}(\sin y) = y \qquad \text{for } y \text{ in } [-\tfrac{1}{2}\pi, \tfrac{1}{2}\pi] \tag{3}$$

In (1) the domain of F is restricted to the closed interval $[-\tfrac{1}{2}\pi, \tfrac{1}{2}\pi]$ so that the function is monotonic on its domain and therefore has an inverse function. However, the sine function has period 2π and is increasing on other intervals, for instance, $[-\tfrac{5}{2}\pi, -\tfrac{3}{2}\pi]$ and $[\tfrac{3}{2}\pi, \tfrac{5}{2}\pi]$. Also, the function is decreasing on certain closed intervals, in particular the intervals $[-\tfrac{3}{2}\pi, -\tfrac{1}{2}\pi]$ and $[\tfrac{1}{2}\pi, \tfrac{3}{2}\pi]$. Any one of these intervals could just as well be chosen for the domain of the function F of Eq. (1). The choice of the interval $[-\tfrac{1}{2}\pi, \tfrac{1}{2}\pi]$, however, is customary because it is the largest interval containing the number 0 on which the function is monotonic.

EXAMPLE 1
Find $\sin^{-1}(\cos \tfrac{5}{6}\pi)$.

SOLUTION
Let $t = \sin^{-1}(\cos \tfrac{5}{6}\pi)$. Then

$$\sin t = \cos \tfrac{5}{6}\pi \quad \text{and} \quad -\tfrac{1}{2}\pi \le t \le \tfrac{1}{2}\pi$$

Because $\cos \tfrac{5}{6}\pi = -\dfrac{\sqrt{3}}{2}$,

$$\sin t = -\dfrac{\sqrt{3}}{2} \quad \text{and} \quad -\tfrac{1}{2}\pi \le t \le \tfrac{1}{2}\pi$$

Therefore $t = -\tfrac{1}{3}\pi$, and we write

$$\sin^{-1}(\cos \tfrac{5}{6}\pi) = -\tfrac{1}{3}\pi$$

A sketch of the graph of the cosine function appears in Fig. 7.1.4. The cosine function does not have an inverse because it also is not one-to-one. To define the inverse cosine function we restrict the cosine to an interval on which the function is monotonic. We choose the interval $[0, \pi]$ on which

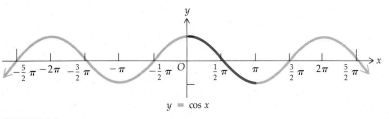

$$y = \cos x$$

FIGURE 7.1.4

the cosine is decreasing, as shown in Fig. 7.1.4. So consider the function G defined by

$$G(x) = \cos x \quad \text{and} \quad 0 \leq x \leq \pi \tag{4}$$

The range of this function G is the closed interval $[-1, 1]$, and its domain is the closed interval $[0, \pi]$. A sketch of the graph of G is shown in Fig. 7.1.5. Because this function is continuous and decreasing on its domain, it has an inverse function called the *inverse cosine function*, which is denoted by $\cos^{-1}$.

7.1.2 DEFINITION   The **inverse cosine function**, denoted by $\cos^{-1}$, is defined as follows:

$$y = \cos^{-1} x \quad \text{if and only if} \quad x = \cos y \text{ and } 0 \leq y \leq \pi$$

The domain of $\cos^{-1}$ is the closed interval $[-1, 1]$, and the range is the closed interval $[0, \pi]$. A sketch of its graph is shown in Fig. 7.1.6.

▶ ILLUSTRATION 3

(a) $\cos^{-1} \dfrac{1}{\sqrt{2}} = \dfrac{1}{4}\pi$ (b) $\cos^{-1}\left(-\dfrac{1}{\sqrt{2}}\right) = \dfrac{3}{4}\pi$ ◀

From Definition 7.1.2 it follows that

$$\cos(\cos^{-1} x) = x \qquad \text{for } x \text{ in } [-1, 1] \tag{5}$$

$$\cos^{-1}(\cos y) = y \qquad \text{for } y \text{ in } [0, \pi] \tag{6}$$

FIGURE 7.1.5

FIGURE 7.1.6

EXAMPLE 2

Find the exact value of $\sin[2 \cos^{-1}(-\tfrac{3}{5})]$.

SOLUTION

Let $t = \cos^{-1}(-\tfrac{3}{5})$. Then

$$\cos t = -\tfrac{3}{5} \quad \text{and} \quad \tfrac{1}{2}\pi < t < \pi$$

We wish to find the exact value of $\sin 2t$. We use the identity

$$\sin 2t = 2 \sin t \cos t \tag{7}$$

From the identity $\sin^2 t + \cos^2 t = 1$, and because $\sin t > 0$ since t is in $(\tfrac{1}{2}\pi, \pi)$, $\sin t = \sqrt{1 - \cos^2 t}$. Thus

$$\sin t = \sqrt{1 - (-\tfrac{3}{5})^2}$$

$$= \tfrac{4}{5}$$

Therefore, from (7),

$$\sin 2t = 2(\tfrac{4}{5})(-\tfrac{3}{5})$$

$$= -\tfrac{24}{25}$$

from which we conclude that

$$\sin[2 \cos^{-1}(-\tfrac{3}{5})] = -\tfrac{24}{25}$$

EXAMPLE 3
Prove

$$\cos^{-1} x = \tfrac{1}{2}\pi - \sin^{-1} x \qquad \text{for } |x| \le 1 \tag{8}$$

SOLUTION
Let x be in $[-1, 1]$, and let

$$t = \cos(\tfrac{1}{2}\pi - \sin^{-1} x) \tag{9}$$

Applying the reduction formula $\cos(\tfrac{1}{2}\pi - \theta) = \sin \theta$ with $\theta = \sin^{-1} x$ on the right side of (9) we get

$$t = \sin(\sin^{-1} x)$$

and using (2) we have

$$t = x$$

Replacing t by x in (9) gives

$$x = \cos(\tfrac{1}{2}\pi - \sin^{-1} x) \tag{10}$$

Because $-\tfrac{1}{2}\pi \le \sin^{-1} x \le \tfrac{1}{2}\pi$, by adding $-\tfrac{1}{2}\pi$ to each member we have

$$-\pi \le -\tfrac{1}{2}\pi + \sin^{-1} x \le 0$$

Multiplying each member of the above inequality by -1 and reversing the direction of the inequality signs gives

$$0 \le \tfrac{1}{2}\pi - \sin^{-1} x \le \pi \tag{11}$$

From (10), (11), and Definition 7.1.2 it follows that

$$\cos^{-1} x = \tfrac{1}{2}\pi - \sin^{-1} x \qquad \text{for } |x| \le 1$$

which is (8).

Observe in the above solution that identity (8) depends on choosing the range of the inverse cosine function to be $[0, \pi]$.

To obtain the inverse tangent function we first consider the graph of the tangent function shown in Fig. 7.1.7. The function is continuous and in-

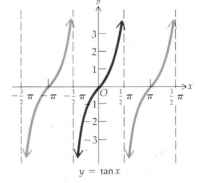

$y = \tan x$

FIGURE 7.1.7

creasing on the open interval $(-\frac{1}{2}\pi, \frac{1}{2}\pi)$, as indicated in the figure. We restrict the tangent function to this interval and let H be the function defined by

$$H(x) = \tan x \quad \text{and} \quad -\tfrac{1}{2}\pi < x < \tfrac{1}{2}\pi \tag{12}$$

The domain of H is the open interval $(-\frac{1}{2}\pi, \frac{1}{2}\pi)$, and the range is the set of all real numbers. A sketch of its graph is shown in Fig. 7.1.8. Because the function H is continuous and increasing on its domain, it has an inverse function, called the *inverse tangent function*, which is denoted by $\tan^{-1}$.

7.1.3 DEFINITION   The **inverse tangent function**, denoted by $\tan^{-1}$, is defined as follows:

$$y = \tan^{-1} x \quad \text{if and only if} \quad x = \tan y \text{ and } -\tfrac{1}{2}\pi < y < \tfrac{1}{2}\pi$$

The domain of $\tan^{-1}$ is the set of all real numbers, and the range is the open interval $(-\frac{1}{2}\pi, \frac{1}{2}\pi)$. A sketch of its graph is shown in Fig. 7.1.9.

▶ ILLUSTRATION 4

(a) $\tan^{-1} \sqrt{3} = \tfrac{1}{3}\pi$ (b) $\tan^{-1}\left(-\dfrac{1}{\sqrt{3}}\right) = -\dfrac{1}{6}\pi$ (c) $\tan^{-1} 0 = 0$ ◀

From Definition 7.1.3 we have

$$\tan(\tan^{-1} x) = x \qquad \text{for } x \text{ in } (-\infty, +\infty) \tag{13}$$

$$\tan^{-1}(\tan y) = y \qquad \text{for } y \text{ in } (-\tfrac{1}{2}\pi, \tfrac{1}{2}\pi) \tag{14}$$

FIGURE 7.1.8

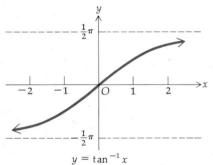

$y = \tan^{-1} x$

FIGURE 7.1.9

EXAMPLE 4
Find the exact value of $\sec[\tan^{-1}(-3)]$.

SOLUTION
Let $\theta = \tan^{-1}(-3)$. Then $\tan \theta = -3$. Because the range of the inverse tangent function is $(-\frac{1}{2}\pi, \frac{1}{2}\pi)$, and because $\tan \theta$ is negative, it follows that $-\frac{1}{2}\pi < \theta < 0$. Figure 7.1.10 shows the angle having radian measure θ that satisfies these requirements. From the figure we observe that $\sec \theta = \sqrt{10}$. Therefore

$$\sec[\tan^{-1}(-3)] = \sqrt{10}$$

Before defining the inverse cotangent function we refer back to Example 3, in which we proved identity (8) involving $\cos^{-1}$ and $\sin^{-1}$. This identity can be used to define the inverse cosine function, and then it can be proved that the range of $\cos^{-1}$ is $[0, \pi]$. We use this kind of procedure in discussing the inverse cotangent function.

7.1.4 DEFINITION   The **inverse cotangent function**, denoted by $\cot^{-1}$, is defined by

$$\cot^{-1} x = \tfrac{1}{2}\pi - \tan^{-1} x \qquad \text{where } x \text{ is any real number} \tag{15}$$

FIGURE 7.1.10

FIGURE 7.1.11

FIGURE 7.1.12

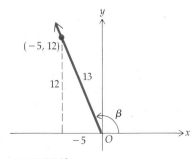

FIGURE 7.1.13

It follows from the definition that the domain of $\cot^{-1}$ is the set of all real numbers. To obtain the range we write (15) as

$$\tan^{-1} x = \tfrac{1}{2}\pi - \cot^{-1} x \tag{16}$$

Because

$$-\tfrac{1}{2}\pi < \tan^{-1} x < \tfrac{1}{2}\pi$$

by substituting from (16) into this inequality we get

$$-\tfrac{1}{2}\pi < \tfrac{1}{2}\pi - \cot^{-1} x < \tfrac{1}{2}\pi$$

Subtracting $\tfrac{1}{2}\pi$ from each member we get

$$-\pi < -\cot^{-1} x < 0$$

and now multiplying each member by -1 and reversing the direction of the inequality signs we obtain

$$0 < \cot^{-1} x < \pi$$

Therefore the range of the inverse cotangent function is the open interval $(0, \pi)$. A sketch of its graph is in Fig. 7.1.11.

▶ ILLUSTRATION 5

(a) $\tan^{-1} 1 = \tfrac{1}{4}\pi$

(b) $\tan^{-1}(-1) = -\tfrac{1}{4}\pi$

(c) $\cot^{-1} 1 = \tfrac{1}{2}\pi - \tan^{-1} 1$
$= \tfrac{1}{2}\pi - \tfrac{1}{4}\pi$
$= \tfrac{1}{4}\pi$

(d) $\cot^{-1}(-1) = \tfrac{1}{2}\pi - \tan^{-1}(-1)$
$= \tfrac{1}{2}\pi - (-\tfrac{1}{4}\pi)$
$= \tfrac{3}{4}\pi$ ◀

EXAMPLE 5
Find the exact value of $\cos[\cot^{-1} \tfrac{3}{4} + \cot^{-1}(-\tfrac{5}{12})]$.

SOLUTION
Let $\alpha = \cot^{-1} \tfrac{3}{4}$ and $\beta = \cot^{-1}(-\tfrac{5}{12})$. Then

$$\cot \alpha = \tfrac{3}{4} \quad \text{and} \quad 0 < \alpha < \tfrac{1}{2}\pi$$

$$\cot \beta = -\tfrac{5}{12} \quad \text{and} \quad \tfrac{1}{2}\pi < \beta < \pi$$

We wish to find $\cos(\alpha + \beta)$. From the cosine sum identity,

$$\cos(\alpha + \beta) = \cos \alpha \cos \beta - \sin \alpha \sin \beta \tag{17}$$

To determine $\cos \alpha$ and $\sin \alpha$, refer to Fig. 7.1.12, which shows a first-quadrant angle α for which $\cot \alpha = \tfrac{3}{4}$. From the figure,

$$\sin \alpha = \tfrac{4}{5} \qquad \cos \alpha = \tfrac{3}{5} \tag{18}$$

Figure 7.1.13 shows a second-quadrant angle β for which $\cot \beta = -\tfrac{5}{12}$. From the figure,

$$\sin \beta = \tfrac{12}{13} \qquad \cos \beta = -\tfrac{5}{13} \tag{19}$$

FIGURE 7.1.14

Substituting from (18) and (19) into (17) we have

$$\cos(\alpha + \beta) = \tfrac{3}{5}(-\tfrac{5}{13}) - \tfrac{4}{5}(\tfrac{12}{13})$$
$$= -\tfrac{63}{65}$$

To lead up to the definition of the *inverse secant function* we first consider a sketch of the graph of the secant function shown in Fig. 7.1.14. Observe that the secant function is increasing on the interval $[0, \tfrac{1}{2}\pi)$ and decreasing on the interval $[\pi, \tfrac{3}{2}\pi)$. Furthermore, if $x \in [0, \tfrac{1}{2}\pi) \cup [\pi, \tfrac{3}{2}\pi)$, then $\sec x \in (-\infty, -1] \cup [1, +\infty)$. We choose the interval $[\pi, \tfrac{3}{2}\pi)$ because differentiation of the inverse secant function, as well as some other computations with it (in Section 8.5, for instance), is simplified if the inverse secant function is defined so that when $x < 0$, then $\pi \leq \sec^{-1} x < \tfrac{3}{2}\pi$. Thus we make the following definition.

7.1.5. DEFINITION

The **inverse secant function**, denoted by $\sec^{-1}$, is defined as follows:

$$y = \sec^{-1} x \quad \text{if and only if} \quad x = \sec y \text{ and } \begin{cases} 0 \leq y < \tfrac{1}{2}\pi & \text{if } x \geq 1 \\ \pi \leq y < \tfrac{3}{2}\pi & \text{if } x \leq -1 \end{cases}$$

The domain of $\sec^{-1}$ is $(-\infty, -1) \cup [1, +\infty)$. The range of $\sec^{-1}$ is $[0, \tfrac{1}{2}\pi) \cup [\pi, \tfrac{3}{2}\pi)$. A sketch of the graph of $\sec^{-1}$ appears in Fig. 7.1.15.

$y = \sec^{-1} x$

FIGURE 7.1.15

From Definition 7.1.5 we have

$$\sec(\sec^{-1} x) = x \qquad \text{for } x \text{ in } (-\infty, -1] \cup [1, +\infty) \tag{20}$$

$$\sec^{-1}(\sec y) = y \qquad \text{for } y \text{ in } [0, \tfrac{1}{2}\pi) \cup [\pi, \tfrac{3}{2}\pi) \tag{21}$$

We now define the *inverse cosecant function* in terms of the inverse secant function.

7.1.6 DEFINITION

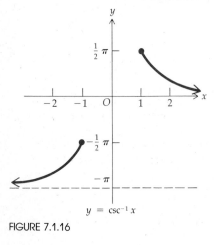

The **inverse cosecant function,** denoted by $\csc^{-1}$, is defined by

$$\csc^{-1} x = \tfrac{1}{2}\pi - \sec^{-1} x \qquad \text{for } |x| \geq 1$$

From the definition the domain of $\csc^{-1}$ is $(-\infty, -1] \cup [1, +\infty)$. The range of $\csc^{-1}$ can be found in a manner similar to that used to determine the range of $\cot^{-1}$. The range of $\csc^{-1}$ is $(-\pi, -\tfrac{1}{2}\pi] \cup (0, \tfrac{1}{2}\pi]$, and you are asked to show this in Exercise 41. A sketch of the graph of $\csc^{-1}$ appears in Fig. 7.1.16.

$y = \csc^{-1} x$

FIGURE 7.1.16

▶ ILLUSTRATION 6

(a) $\sec^{-1} 2 = \tfrac{1}{3}\pi$

(b) $\sec^{-1}(-2) = \tfrac{4}{3}\pi$

(c) $\csc^{-1} 2 = \tfrac{1}{2}\pi - \sec^{-1} 2$
$\qquad = \tfrac{1}{2}\pi - \tfrac{1}{3}\pi$
$\qquad = \tfrac{1}{6}\pi$

(d) $\csc^{-1}(-2) = \tfrac{1}{2}\pi - \sec^{-1}(-2)$
$\qquad = \tfrac{1}{2}\pi - \tfrac{4}{3}\pi$
$\qquad = -\tfrac{5}{6}\pi$ ◀

EXERCISES 7.1

In Exercises 1 through 6, determine the exact function value.

1. (a) $\sin^{-1} \frac{1}{2}$; (b) $\sin^{-1}(-\frac{1}{2})$; (c) $\cos^{-1} \frac{1}{2}$; (d) $\cos^{-1}(-\frac{1}{2})$

2. (a) $\sin^{-1} \frac{\sqrt{3}}{2}$; (b) $\sin^{-1}\left(-\frac{\sqrt{3}}{2}\right)$; (c) $\cos^{-1} \frac{\sqrt{3}}{2}$; (d) $\cos^{-1}\left(-\frac{\sqrt{3}}{2}\right)$

3. (a) $\tan^{-1} \frac{1}{\sqrt{3}}$; (b) $\tan^{-1}(-\sqrt{3})$; (c) $\sec^{-1} \frac{2}{\sqrt{3}}$; (d) $\sec^{-1}\left(-\frac{2}{\sqrt{3}}\right)$

4. (a) $\cot^{-1} \frac{1}{\sqrt{3}}$; (b) $\cot^{-1}(-\sqrt{3})$; (c) $\csc^{-1} \frac{2}{\sqrt{3}}$; (d) $\csc^{-1}\left(-\frac{2}{\sqrt{3}}\right)$

5. (a) $\sin^{-1} 1$; (b) $\sin^{-1}(-1)$; (c) $\csc^{-1} 1$; (d) $\csc^{-1}(-1)$; (e) $\sin^{-1} 0$

6. (a) $\cos^{-1} 1$; (b) $\cos^{-1}(-1)$; (c) $\sec^{-1} 1$; (d) $\sec^{-1}(-1)$; (e) $\cos^{-1} 0$

7. Given $x = \sin^{-1} \frac{1}{3}$, find the exact value of each of the following: (a) $\cos x$; (b) $\tan x$; (c) $\cot x$; (d) $\sec x$; (e) $\csc x$.

8. Given $x = \cos^{-1} \frac{2}{3}$, find the exact value of each of the following: (a) $\sin x$; (b) $\tan x$; (c) $\cot x$; (d) $\sec x$; (e) $\csc x$.

9. Do Exercise 7 if $x = \sin^{-1}(-\frac{1}{3})$.   **10.** Do Exercise 8 if $x = \cos^{-1}(-\frac{2}{3})$.

11. Given $y = \tan^{-1}(-2)$, find the exact value of each of the following: (a) $\sin y$; (b) $\cos y$; (c) $\cot y$; (d) $\sec y$; (e) $\csc y$.

12. Given $y = \cot^{-1}(-\frac{1}{2})$, find the exact value of each of the following: (a) $\sin y$; (b) $\cos y$; (c) $\tan y$; (d) $\sec y$; (e) $\csc y$.

13. Given $t = \csc^{-1}(-\frac{3}{2})$, find the exact value of each of the following: (a) $\sin t$; (b) $\cos t$; (c) $\tan t$; (d) $\cot t$; (e) $\sec t$.

14. Given $t = \sec^{-1}(-3)$, find the exact value of each of the following: (a) $\sin t$; (b) $\cos t$; (c) $\tan t$; (d) $\cot t$; (e) $\csc t$.

In Exercises 15 through 28, find the exact value of the given quantity.

15. (a) $\tan[\sin^{-1} \frac{1}{2}\sqrt{3}]$; (b) $\sin[\tan^{-1} \frac{1}{2}\sqrt{3}]$

16. (a) $\cos[\tan^{-1}(-3)]$; (b) $\tan[\sec^{-1}(-3)]$

17. (a) $\cos[\sin^{-1}(-\frac{1}{2})]$; (b) $\sin[\cos^{-1}(-\frac{1}{2})]$

18. (a) $\tan[\cot^{-1}(-1)]$; (b) $\cot[\tan^{-1}(-1)]$

19. $\cos[2 \sin^{-1}(-\frac{5}{13})]$

20. $\tan[2 \sec^{-1}(-\frac{5}{4})]$

21. $\sin(\sin^{-1} \frac{2}{3} + \cos^{-1} \frac{1}{3})$

22. $\cos[\sin^{-1}(-\frac{1}{2}) + \sin^{-1} \frac{1}{4}]$

23. $\cos[\sin^{-1} \frac{2}{3} + 2 \sin^{-1}(-\frac{1}{3})]$

24. $\tan[\tan^{-1}(-\frac{2}{5}) - \cos^{-1}(-\frac{1}{2}\sqrt{2})]$

25. $\tan(\tan^{-1} \frac{3}{4} - \sin^{-1} \frac{1}{2})$

26. $\tan[\sec^{-1} \frac{5}{3} + \csc^{-1}(-\frac{13}{12})]$

27. $\cos(\sin^{-1} \frac{1}{3} - \tan^{-1} \frac{1}{2})$

28. $\sin[\cos^{-1}(-\frac{2}{3}) + 2 \sin^{-1}(-\frac{1}{3})]$

29. Prove: $\cos^{-1} \frac{3}{\sqrt{10}} + \cos^{-1} \frac{2}{\sqrt{5}} = \frac{1}{4} \pi$.

30. Prove: $2 \tan^{-1} \frac{1}{3} - \tan^{-1}(-\frac{1}{7}) = \frac{1}{4} \pi$.

In Exercises 31 through 38, draw a sketch of the graph of the given equation.

31. $y = \frac{1}{2} \sin^{-1} x$ **32.** $y = \sin^{-1} \frac{1}{2}x$ **33.** $y = \tan^{-1} 2x$ **34.** $y = 2 \tan^{-1} x$

35. $y = \cos^{-1} 3x$ **36.** $y = \frac{1}{2} \sec^{-1} 2x$ **37.** $y = 2 \cot^{-1} \frac{1}{2}x$ **38.** $y = \frac{1}{2} \csc^{-1} \frac{1}{2}x$

39. A weight is suspended from a spring and vibrating vertically according to the equation $y = 2 \sin 4\pi(t + \frac{1}{8})$, where y cm is the directed distance of the weight from its central position t sec after the start of the motion and the positive direction is upward. (a) Solve the equation for t. (b) Use the equation in part (a) to determine the smallest three positive values of t for which the weight is 1 cm above its central position.

40. A 60-cycle alternating current is described by the equation $x = 20 \sin 120\pi(t - \frac{11}{720})$, where x amperes is the current at t sec. (a) Solve the equation for t. (b) Use the equation in part (a) to determine the smallest three positive values of t for which the current is 10 amperes.

41. Prove that the range of $\csc^{-1}$ is $(-\pi, -\frac{1}{2}\pi] \cup (0, \frac{1}{2}\pi]$.

42. Prove that $\tan^{-1} x + \tan^{-1}\left(\frac{1}{x}\right) = \frac{1}{2} \pi$, if $x > 0$. (*Hint:* Use the identity $\cot A = \tan(\frac{1}{2}\pi - A)$).

7.2
DERIVATIVES OF THE
INVERSE TRIGONOMETRIC
FUNCTIONS

Because the inverse sine function is continuous and increasing on its domain, it follows from Theorem 6.2.3 that it has a derivative. To derive the formula for this derivative let

$$y = \sin^{-1} x$$

which is equivalent to

$$x = \sin y \quad \text{and} \quad y \text{ is in } \left[-\tfrac{1}{2}\pi, \tfrac{1}{2}\pi\right] \tag{1}$$

Differentiating on both sides of (1) with respect to y we obtain

$$\frac{dx}{dy} = \cos y \quad \text{and} \quad y \text{ is in } \left[-\tfrac{1}{2}\pi, \tfrac{1}{2}\pi\right] \tag{2}$$

From the identity $\sin^2 y + \cos^2 y = 1$, and replacing $\sin y$ by x, we obtain

$$\cos^2 y = 1 - x^2 \tag{3}$$

If y is in $\left[-\tfrac{1}{2}\pi, \tfrac{1}{2}\pi\right]$, $\cos y$ is nonnegative; so from (3) it follows that

$$\cos y = \sqrt{1 - x^2} \quad \text{if } y \text{ is in } \left[-\tfrac{1}{2}\pi, \tfrac{1}{2}\pi\right] \tag{4}$$

Substituting from (4) into (2) we get

$$\frac{dx}{dy} = \sqrt{1 - x^2}$$

From Theorem 6.2.3, $\dfrac{dy}{dx}$ is the reciprocal of $\dfrac{dx}{dy}$; hence

$$D_x(\sin^{-1} x) = \frac{1}{\sqrt{1 - x^2}} \tag{5}$$

The domain of the derivative of the inverse sine function is the open interval $(-1, 1)$.

From (5) and the chain rule we have the following theorem.

7.2.1 THEOREM   If u is a differentiable function of x,

$$D_x(\sin^{-1} u) = \frac{1}{\sqrt{1 - u^2}} D_x u$$

▶ ILLUSTRATION 1   If $y = \sin^{-1} x^2$, then

$$\frac{dy}{dx} = \frac{1}{\sqrt{1 - (x^2)^2}} (2x)$$

$$= \frac{2x}{\sqrt{1 - x^4}} \qquad\qquad ◀$$

To derive the formula for the derivative of the inverse cosine function we use (8) of Section 7.1, which is

$$\cos^{-1} x = \tfrac{1}{2}\pi - \sin^{-1} x$$

Differentiating with respect to x we have

$$D_x(\cos^{-1} x) = D_x(\tfrac{1}{2}\pi - \sin^{-1} x)$$

Thus

$$D_x(\cos^{-1} x) = -\frac{1}{\sqrt{1-x^2}} \tag{6}$$

where x is in $(-1, 1)$.

    The next theorem follows from (6) and the chain rule.

7.2.2 THEOREM   If u is a differentiable function of x,

$$D_x(\cos^{-1} u) = -\frac{1}{\sqrt{1-u^2}} D_x u$$

    We now derive the formula for the derivative of the inverse tangent function. If

$$y = \tan^{-1} x$$

then

$$x = \tan y \quad \text{and} \quad y \text{ is in } (-\tfrac{1}{2}\pi, \tfrac{1}{2}\pi)$$

Differentiating on both sides of the above equation with respect to y we obtain

$$\frac{dx}{dy} = \sec^2 y \quad \text{and} \quad y \text{ is in } (-\tfrac{1}{2}\pi, \tfrac{1}{2}\pi) \tag{7}$$

From the identity $\sec^2 y = 1 + \tan^2 y$, and replacing $\tan y$ by x, we have

$$\sec^2 y = 1 + x^2 \tag{8}$$

Substituting from (8) into (7) we get

$$\frac{dx}{dy} = 1 + x^2$$

So from Theorem 6.2.3,

$$D_x(\tan^{-1} x) = \frac{1}{1+x^2} \tag{9}$$

    The domain of the derivative of the inverse tangent function is the set of all real numbers.

    From (9) and the chain rule we obtain the following theorem.

7.2.3 THEOREM   If u is a differentiable function of x,

$$D_x(\tan^{-1} u) = \frac{1}{1+u^2} D_x u$$

▶ ILLUSTRATION 2   If $f(x) = \tan^{-1} \dfrac{1}{x+1}$, then

$$f'(x) = \frac{1}{1 + \dfrac{1}{(x+1)^2}} \cdot \frac{-1}{(x+1)^2}$$

$$= \frac{-1}{(x+1)^2 + 1}$$

$$= \frac{-1}{x^2 + 2x + 2}$$ ◀

From Definition 7.1.4,

$$\cot^{-1} x = \tfrac{1}{2}\pi - \tan^{-1} x$$

Differentiating with respect to x we obtain the formula

$$D_x(\cot^{-1} x) = -\frac{1}{1+x^2}$$

From this formula and the chain rule we have the following theorem.

7.2.4 THEOREM   If u is a differentiable function of x,

$$D_x(\cot^{-1} u) = -\frac{1}{1+u^2} D_x u$$

EXAMPLE 1

Given $y = x^3 \cot^{-1} \tfrac{1}{3}x$, find $\dfrac{dy}{dx}$.

SOLUTION

$$\frac{dy}{dx} = 3x^2 \cot^{-1} \tfrac{1}{3}x + x^3 \cdot \frac{-1}{1 + \tfrac{1}{9}x^2} \cdot \frac{1}{3}$$

$$= 3x^2 \cot^{-1} \tfrac{1}{3}x - \frac{3x^3}{9 + x^2}$$

EXAMPLE 2

Given $\ln(x + y) = \tan^{-1}\left(\dfrac{x}{y}\right)$, find $\dfrac{dy}{dx}$.

SOLUTION

Differentiating implicitly on both sides of the given equation with respect to x we get

$$\frac{1}{x+y}\left(1+\frac{dy}{dx}\right)=\frac{1}{1+\dfrac{x^2}{y^2}}\cdot\frac{y-x\dfrac{dy}{dx}}{y^2}$$

$$\frac{1+\dfrac{dy}{dx}}{x+y}=\frac{y-x\dfrac{dy}{dx}}{y^2+x^2}$$

$$y^2+x^2+(y^2+x^2)\frac{dy}{dx}=xy+y^2-(x^2+xy)\frac{dy}{dx}$$

$$\frac{dy}{dx}=\frac{xy-x^2}{2x^2+xy+y^2}$$

To obtain the formula for the derivative of the inverse secant function let

$$y=\sec^{-1}x\qquad\text{and }|x|\geq 1$$

Then

$$x=\sec y\qquad\text{and } y\text{ is in }[0,\tfrac{1}{2}\pi)\cup[\pi,\tfrac{3}{2}\pi)\qquad(10)$$

Differentiating on both sides of (10) with respect to y we get

$$\frac{dx}{dy}=\sec y\tan y\qquad\text{and } y\text{ is in }[0,\tfrac{1}{2}\pi)\cup[\pi,\tfrac{3}{2}\pi)\qquad(11)$$

From the identity $\tan^2 y=\sec^2 y-1$, with $\sec y=x$, we get

$$\tan^2 y=x^2-1\qquad(12)$$

Because y is in $[0,\tfrac{1}{2}\pi)\cup[\pi,\tfrac{3}{2}\pi)$, $\tan y$ is nonnegative. Thus from (12),

$$\tan y=\sqrt{x^2-1}\qquad\text{if } y\text{ is in }[0,\tfrac{1}{2}\pi)\cup[\pi,\tfrac{3}{2}\pi)\qquad(13)$$

Substituting from (10) and (13) in (11) we have

$$\frac{dx}{dy}=x\sqrt{x^2-1}$$

Thus from Theorem 6.2.3,

$$D_x(\sec^{-1}x)=\frac{1}{x\sqrt{x^2-1}}\qquad(14)$$

where $|x|>1$. From (14) and the chain rule the next theorem follows.

7.2.5 THEOREM   If u is a differentiable function of x,

$$D_x(\sec^{-1}u)=\frac{1}{u\sqrt{u^2-1}}D_xu$$

▶ ILLUSTRATION 3   If $f(x)=\sec^{-1}(3e^x)$, then

$$f'(x)=\frac{1}{3e^x\sqrt{(3e^x)^2-1}}(3e^x)$$

$$= \frac{1}{\sqrt{9e^{2x} - 1}}$$ ◀

From Definition 7.1.6,

$$\csc^{-1} x = \tfrac{1}{2}\pi - \sec^{-1} x \qquad \text{for } |x| \geq 1$$

Differentiating with respect to x we obtain

$$D_x(\csc^{-1} x) = -\frac{1}{x\sqrt{x^2 - 1}} \tag{15}$$

where $|x| > 1$. From (15) and the chain rule we have the following theorem.

7.2.6 THEOREM   If u is a differentiable function of x,

$$D_x(\csc^{-1} u) = -\frac{1}{u\sqrt{u^2 - 1}} D_x u$$

EXAMPLE 3
Find $f'(x)$ if

$$f(x) = x \csc^{-1} \frac{1}{x}$$

SOLUTION

$$f'(x) = \csc^{-1}\frac{1}{x} + x\left[-\frac{1}{\frac{1}{x}\sqrt{\left(\frac{1}{x}\right)^2 - 1}}\left(-\frac{1}{x^2}\right) \right]$$

$$= \csc^{-1}\frac{1}{x} + \left[\frac{-x^2}{\frac{\sqrt{1 - x^2}}{\sqrt{x^2}}}\left(-\frac{1}{x^2}\right) \right]$$

$$= \csc^{-1}\frac{1}{x} + \frac{1}{\frac{\sqrt{1 - x^2}}{|x|}}$$

$$= \csc^{-1}\frac{1}{x} + \frac{|x|}{\sqrt{1 - x^2}}$$

In the following example, an observer is looking at a picture placed high on a wall. See Fig. 7.2.1. When the observer is far away from the wall, the angle subtended at the observer's eye by the picture is small. As the observer gets closer to the wall, that angle increases until it reaches a maximum value. Then as the observer gets even closer to the wall, the angle gets smaller. When the angle is a maximum, we say that the observer has the "best view" of the picture.

FIGURE 7.2.1

EXAMPLE 4
A picture 7 ft high is placed on a wall with its base 9 ft above the eye level of an observer. How far from the wall should the observer stand to get the best view of the picture, that is, so that the angle subtended at the observer's eye by the picture is a maximum?

SOLUTION
Let x ft be the distance of the observer from the wall, θ be the radian measure of the angle subtended at the observer's eye by the picture, α be the radian measure of the angle subtended at the observer's eye by the portion of the wall above eye level and below the picture, and $\beta = \alpha + \theta$. Refer to Fig. 7.2.1.
    We wish to find the value of x that will make θ an absolute maximum. Because x is in the interval $(0, +\infty)$, the absolute maximum value of θ will be a relative maximum value. We see from the figure that

$$\cot \beta = \frac{x}{16} \quad \text{and} \quad \cot \alpha = \frac{x}{9}$$

Because $0 < \beta < \tfrac{1}{2}\pi$ and $0 < \alpha < \tfrac{1}{2}\pi$,

$$\beta = \cot^{-1} \frac{x}{16} \quad \text{and} \quad \alpha = \cot^{-1} \frac{x}{9}$$

Substituting these values of β and α in the equation $\theta = \beta - \alpha$ we get

$$\theta = \cot^{-1} \frac{x}{16} - \cot^{-1} \frac{x}{9}$$

Differentiating with respect to x gives

$$\frac{d\theta}{dx} = -\frac{\frac{1}{16}}{1 + \left(\frac{x}{16}\right)^2} + \frac{\frac{1}{9}}{1 + \left(\frac{x}{9}\right)^2}$$

$$= -\frac{16}{16^2 + x^2} + \frac{9}{9^2 + x^2}$$

Setting $\dfrac{d\theta}{dx} = 0$ we obtain

$$9(16^2 + x^2) - 16(9^2 + x^2) = 0$$

$$-7x^2 + 9 \cdot 16(16 - 9) = 0$$

$$x^2 = 9 \cdot 16$$

$$x = 12$$

The -12 is rejected because it is not in the interval $(0, +\infty)$. The results of the first-derivative test are shown in Table 7.2.1. Because the relative maximum value of θ is an absolute maximum value, we conclude that the observer should stand 12 ft from the wall.

Table 7.2.1

	$\dfrac{d\theta}{dx}$	Conclusion
$0 < x < 12$	$+$	
$x = 12$	0	θ has a relative maximum value
$12 < x < +\infty$	$-$	

EXERCISES 7.2

In Exercises 1 through 26, find the derivative of the given function.

1. $f(x) = \sin^{-1} \frac{1}{2}x$

2. $f(x) = \cos^{-1} 3x$

3. $g(x) = \tan^{-1} 2x$

4. $f(x) = \csc^{-1} 2x$

5. $F(x) = 2 \cos^{-1} \sqrt{x}$

6. $g(x) = \frac{1}{2} \sin^{-1} x^2$

7. $g(t) = \sec^{-1} 5t + \csc^{-1} 5t$

8. $f(y) = \cot^{-1} e^y$

9. $f(x) = \sin^{-1} \sqrt{1 - x^2}$

10. $f(w) = 2 \tan^{-1} \dfrac{1}{w}$

11. $F(x) = \cot^{-1} \dfrac{2}{x} + \tan^{-1} \dfrac{x}{2}$

12. $f(x) = \ln(\tan^{-1} 3x)$

13. $h(y) = y \sin^{-1} 2y$

14. $f(t) = t^2 \cos^{-1} t$

15. $g(x) = x^2 \sec^{-1} \dfrac{1}{x}$

16. $g(s) = \cos^{-1} s + \dfrac{s}{1 - s^2}$

17. $f(x) = \cos^{-1}(\sin x)$

18. $h(x) = \tan^{-1} \dfrac{2x}{1 - x^2}$

19. $F(x) = \ln(\tan^{-1} x^2)$

20. $F(x) = \sec^{-1} \sqrt{x^2 + 4}$

21. $f(x) = 4 \sin^{-1} \frac{1}{2}x + x\sqrt{4 - x^2}$

22. $f(t) = a \sin^{-1} \dfrac{t}{a} + \sqrt{a^2 - t^2}$

23. $h(x) = \csc^{-1}(2e^{3x})$

24. $f(x) = \sin^{-1} x + \cos^{-1} x$

25. $G(x) = x \cot^{-1} x + \ln \sqrt{1 + x^2}$

26. $f(t) = \sin^{-1} \dfrac{t - 1}{t + 1}$

In Exercises 27 through 30, find $\dfrac{dy}{dx}$ by implicit differentiation.

27. $e^x + y = \cos^{-1} x$

28. $\ln(\sin^2 3x) = e^x + \cot^{-1} y$

29. $x \sin y + x^3 = \tan^{-1} y$

30. $\sin^{-1}(xy) = \cos^{-1}(x + y)$

31. Find equations of the tangent line and normal line to the graph of the equation $y = \sec^{-1}(2x + 1)$ at the point $(\frac{1}{2}, \frac{1}{3}\pi)$.

32. Use differentials to find an approximate value of $\sin^{-1} 0.52$ to three decimal places.

33. A sign 3 ft high is placed on a wall with its base 2 ft above the eye level of a woman attempting to read it. How far from the wall should she stand to get the best view of the sign, that is, so that the angle subtended at the woman's eye by the picture is a maximum?

34. Example 5 and Exercise 33 are particular cases of the following more general situation: An object (for instance, a picture or a sign) a ft high is placed on a wall with its base b ft above the eye level of an observer. Show that the observer gets the best view of the object when the distance of the observer from the wall is $\sqrt{b(a + b)}$ ft.

35. A picture 40 cm high is placed on a wall with its base 30 cm above the level of the eye of an observer. If the observer is approaching the wall at the rate of 40 cm/sec, how fast is the measure of the angle subtended at the observer's eye by the picture changing when the observer is 1 m from the wall?

36. A ladder 25 ft long is leaning against a vertical wall. If the bottom of the ladder is pulled horizontally away from the wall so that the top is sliding down at 3 ft/sec, how fast is the measure of the angle between the ladder and the ground changing when the bottom of the ladder is 15 ft from the wall?

37. A light is 3 km from a straight beach. If the light revolves and makes 2 rpm, find the speed of light along the beach when the spot is 2 km from the point on the beach nearest the light.

38. A man on a dock is pulling in at the rate of 2 ft/sec a rowboat by means of a rope. The man's hands are 20 ft above the level of the point where the rope is attached to the boat. How fast is the measure of the angle of depression of the rope changing when there are 52 ft of rope out?

39. A woman is walking at the rate of 5 ft/sec along the diameter of a circular courtyard. A light at one end of a diameter perpendicular to her path casts a shadow on the circular wall. How fast is the shadow moving along the wall when the distance from the woman to the center of the courtyard is $\frac{1}{2}r$, where r ft is the radius of the courtyard?

40. In Exercise 39, how far is the woman from the center of the courtyard when the speed of her shadow along the wall is 9 ft/sec?

41. A rope is attached to a weight and it passes over a hook that is 8 ft above the ground. The rope is pulled over the hook at the rate of $\frac{3}{4}$ ft/sec and drags the weight along level ground. If the length of the rope between the weight and the hook is x ft when the radian measure of the angle between the rope and the floor is θ, find the time rate of change of θ in terms of x.

$$\frac{dx}{dt} = -\frac{3}{4} \qquad \csc\theta = \frac{x}{8} \qquad \theta = \csc^{-1}\frac{x}{8} \qquad \frac{d\theta}{dt} = \qquad \frac{dx}{dt} : \frac{6}{x\sqrt{x^2-64}}$$

42. Do Exercise 42 of Exercises 7.1 by showing that $\tan^{-1} x$ and $-\tan^{-1}\left(\dfrac{1}{x}\right)$ differ by a constant and then evaluating the constant.

43. Do Example 3 of Section 7.1 by showing that $\sin^{-1} x$ and $-\cos^{-1} x$ differ by a constant and then evaluating the constant.

44. Given: $f(x) = \tan^{-1}(1/x) - \cot^{-1} x$. (a) Show that $f'(x) = 0$ for all x in the domain of f. (b) Prove that there is no constant C for which $f(x) = C$ for all x in its domain. (c) Why doesn't the statement in part (b) contradict Theorem 4.1.2?

**7.3
INTEGRALS YIELDING
INVERSE TRIGONOMETRIC
FUNCTIONS**

From the theorems for the derivatives of the inverse trigonometric functions we obtain the following one giving some indefinite integral formulas.

7.3.1 THEOREM

$$\int \frac{du}{\sqrt{1-u^2}} = \sin^{-1} u + C \tag{1}$$

$$\int \frac{du}{1+u^2} = \tan^{-1} u + C \tag{2}$$

$$\int \frac{du}{u\sqrt{u^2-1}} = \sec^{-1} u + C \tag{3}$$

The proof of each formula is immediate by taking the derivative of the right-hand side. We also have the formulas given in the next theorem.

7.3.2 THEOREM

$$\int \frac{du}{\sqrt{a^2-u^2}} = \sin^{-1}\frac{u}{a} + C \qquad \text{where } a > 0 \tag{4}$$

$$\int \frac{du}{a^2+u^2} = \frac{1}{a}\tan^{-1}\frac{u}{a} + C \qquad \text{where } a \neq 0 \tag{5}$$

$$\int \frac{du}{u\sqrt{u^2-a^2}} = \frac{1}{a}\sec^{-1}\frac{u}{a} + C \qquad \text{where } a > 0 \tag{6}$$

PROOF   These formulas can be proved by finding the derivative of the right side and obtaining the integrand. We prove formula (4).

$$D_u\left(\sin^{-1}\frac{u}{a}\right) = \frac{1}{\sqrt{1-\left(\dfrac{u}{a}\right)^2}} \, D_u\left(\frac{u}{a}\right)$$

$$= \frac{\sqrt{a^2}}{\sqrt{a^2-u^2}} \cdot \frac{1}{a}$$

$$= \frac{a}{\sqrt{a^2-u^2}} \cdot \frac{1}{a} \qquad \text{if } a > 0$$

$$= \frac{1}{\sqrt{a^2-u^2}} \qquad \text{if } a > 0$$

The proofs of (5) and (6) are left as exercises (see Exercises 38 and 39). ∎

EXAMPLE 1
Evaluate

$$\int \frac{dx}{\sqrt{4-9x^2}}$$

SOLUTION

$$\int \frac{dx}{\sqrt{4-9x^2}} = \frac{1}{3}\int \frac{d(3x)}{\sqrt{4-(3x)^2}}$$

$$= \frac{1}{3}\sin^{-1}\frac{3x}{2} + C$$

EXAMPLE 2
Evaluate

$$\int \frac{dx}{3x^2-2x+5}$$

SOLUTION

$$\int \frac{dx}{3x^2-2x+5} = \int \frac{dx}{3(x^2-\frac{2}{3}x)+5}$$

To complete the square of $x^2 - \frac{2}{3}x$ we add $\frac{1}{9}$, and because $\frac{1}{9}$ is multiplied by 3, we actually add $\frac{1}{3}$ to the denominator, and so we also subtract $\frac{1}{3}$ from the denominator. Therefore we have

$$\int \frac{dx}{3x^2-2x+5} = \int \frac{dx}{3(x^2-\frac{2}{3}x+\frac{1}{9})+5-\frac{1}{3}}$$

$$= \int \frac{dx}{3(x - \frac{1}{3})^2 + \frac{14}{3}}$$

$$= \frac{1}{3} \int \frac{dx}{(x - \frac{1}{3})^2 + \frac{14}{9}}$$

$$= \frac{1}{3} \cdot \frac{3}{\sqrt{14}} \tan^{-1} \left(\frac{x - \frac{1}{3}}{\frac{1}{3}\sqrt{14}} \right) + C$$

$$= \frac{1}{\sqrt{14}} \tan^{-1} \left(\frac{3x - 1}{\sqrt{14}} \right) + C$$

EXAMPLE 3

Evaluate

$$\int \frac{(2x + 7)\,dx}{x^2 + 2x + 5}$$

SOLUTION

Because $d(x^2 + 2x + 5) = (2x + 2)\,dx$,   we   write   the   numerator   as $(2x + 2)\,dx + 5\,dx$ and express the original integral as the sum of two integrals.

$$\int \frac{(2x + 7)\,dx}{x^2 + 2x + 5} = \int \frac{(2x + 2)\,dx}{x^2 + 2x + 5} + 5 \int \frac{dx}{x^2 + 2x + 5}$$

$$= \ln|x^2 + 2x + 5| + 5 \int \frac{dx}{(x + 1)^2 + 4}$$

$$= \ln(x^2 + 2x + 5) + \frac{5}{2} \tan^{-1} \frac{x + 1}{2} + C$$

Note: Because $x^2 + 2x + 5 > 0$, for all x, $|x^2 + 2x + 5| = x^2 + 2x + 5$.

EXAMPLE 4

Evaluate

$$\int \frac{3\,dx}{(x + 2)\sqrt{x^2 + 4x + 3}}$$

SOLUTION

$$\int \frac{3\,dx}{(x + 2)\sqrt{x^2 + 4x + 3}} = \int \frac{3\,dx}{(x + 2)\sqrt{(x^2 + 4x + 4) - 1}}$$

$$= 3 \int \frac{d(x + 2)}{(x + 2)\sqrt{(x + 2)^2 - 1}}$$

$$= 3 \sec^{-1}(x + 2) + C$$

EXAMPLE 5

Find the area of the region in the first quadrant bounded by the curve

$$y = \frac{1}{1 + x^2}$$

the x axis, the y axis, and the line $x = 1$.

SOLUTION

Figure 7.3.1 shows the region and a rectangular element of area. If A square units is the area of the region,

$$A = \lim_{\|\Delta\| \to 0} \sum_{i=1}^{n} \frac{1}{1 + \xi_i^2} \Delta_i x$$

$$= \int_0^1 \frac{dx}{1 + x^2}$$

$$= \tan^{-1} x \Big]_0^1$$

$$= \tan^{-1} 1 - \tan^{-1} 0$$

$$= \tfrac{1}{4}\pi - 0$$

$$= \tfrac{1}{4}\pi$$

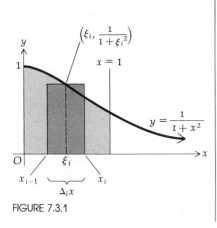

FIGURE 7.3.1

EXERCISES 7.3

In Exercises 1 through 25, evaluate the indefinite integral.

1. $\displaystyle\int \frac{dx}{\sqrt{1 - 4x^2}}$

2. $\displaystyle\int \frac{dx}{x^2 + 25}$

3. $\displaystyle\int \frac{dx}{9x^2 + 16}$

4. $\displaystyle\int \frac{dt}{\sqrt{1 - 16t^2}}$

5. $\displaystyle\int \frac{dx}{4 + (x - 1)^2}$

6. $\displaystyle\int \frac{dx}{9 + (3 - x)^2}$

7. $\displaystyle\int \frac{dx}{4x\sqrt{x^2 - 16}}$

8. $\displaystyle\int \frac{x\,dx}{x^4 + 16}$

9. $\displaystyle\int \frac{dx}{\sqrt{2 - 5x^2}}$

10. $\displaystyle\int \frac{3\,dx}{x\sqrt{x^2 - 9}}$

11. $\displaystyle\int \frac{r\,dr}{\sqrt{16 - 9r^4}}$

12. $\displaystyle\int \frac{du}{u\sqrt{16u^2 - 9}}$

13. $\displaystyle\int \frac{e^x\,dx}{7 + e^{2x}}$

14. $\displaystyle\int \frac{\sin x\,dx}{\sqrt{2 - \cos^2 x}}$

15. $\displaystyle\int \frac{dx}{(1 + x)\sqrt{x}}$

16. $\displaystyle\int \frac{ds}{\sqrt{2s - s^2}}$

17. $\displaystyle\int \frac{dx}{x^2 - x + 2}$

18. $\displaystyle\int \frac{dx}{\sqrt{3x - x^2 - 2}}$

19. $\displaystyle\int \frac{dx}{\sqrt{15 + 2x - x^2}}$

20. $\displaystyle\int \frac{dx}{2x^2 + 2x + 3}$

21. $\displaystyle\int \frac{x\,dx}{\sqrt{3 - 2x - x^2}}$

22. $\displaystyle\int \frac{x\,dx}{x^2 + x + 1}$

23. $\displaystyle\int \frac{(2+x)\,dx}{\sqrt{4-2x-x^2}}$

24. $\displaystyle\int \frac{2\,dt}{(t-3)\sqrt{t^2-6t+5}}$

25. $\displaystyle\int \frac{2x^3\,dx}{2x^2-4x+3}$

In Exercises 26 through 33, evaluate the definite integral.

26. $\displaystyle\int_0^{1/2} \frac{dx}{\sqrt{1-x^2}}$

27. $\displaystyle\int_0^1 \frac{1+x}{1+x^2}\,dx$

28. $\displaystyle\int_0^{\sqrt{3}} \frac{x\,dx}{\sqrt{1+x^2}}$

29. $\displaystyle\int_{-4}^{-2} \frac{dt}{\sqrt{-t^2-6t-5}}$

30. $\displaystyle\int_2^5 \frac{dx}{x^2-4x+13}$

31. $\displaystyle\int_0^1 \frac{dx}{e^x+e^{-x}}$

32. $\displaystyle\int_{1/\sqrt{2}}^1 \frac{dx}{x\sqrt{4x^2-1}}$

33. $\displaystyle\int_1^e \frac{dx}{x[1+(\ln x)^2]}$

34. Find the area of the region bounded by the curve $y = 8/(x^2+4)$, the x axis, the y axis, and the line $x = 2$.

35. Find the area of the region bounded by the curve $y = 1/\sqrt{5-4x-x^2}$, the x axis, and the lines $x = -\frac{7}{2}$ and $x = -\frac{1}{2}$.

36. Find the area of the region bounded by the curves $x^2 = 4ay$ and $y = 8a^3/(x^2+4a^2)$.

37. Find the circumference of the circle $x^2 + y^2 = r^2$ by integration.

In Exercises 38 through 42, prove the given formula by showing that the derivative of the right side is equal to the integrand.

38. $\displaystyle\int \frac{du}{a^2+u^2} = \frac{1}{a}\tan^{-1}\frac{u}{a} + C$

39. $\displaystyle\int \frac{du}{u\sqrt{u^2-a^2}} = \frac{1}{a}\sec^{-1}\frac{u}{a} + C$ if $a > 0$

40. $\displaystyle\int \frac{du}{\sqrt{1-u^2}} = -\cos^{-1}u + C$

Is this formula equivalent to formula (1)? Why?

41. $\displaystyle\int \frac{du}{1+u^2} = -\cot^{-1}u + C$

Is this formula equivalent to formula (2)? Why?

42. $\displaystyle\int \frac{du}{u\sqrt{u^2-1}} = -\csc^{-1}u + C$

Is this formula equivalent to formula (3)? Why?

43. In Section 2.11 we stated that a particle moving in a straight line is said to have *simple harmonic motion* if the measure of its acceleration is always proportional to the measure of its displacement from a fixed point on the line and its acceleration and displacement are oppositely directed. Therefore, if at t sec s cm is the directed distance of the particle from the origin and v cm/sec is the velocity of the particle, then a differential equation for simple harmonic motion is

$$\frac{dv}{dt} = -k^2 s \tag{7}$$

where k^2 is the constant of proportionality and the minus sign indicates that the acceleration is opposite in direction from the displacement. Because $\dfrac{dv}{dt} = \dfrac{dv}{ds}\cdot\dfrac{ds}{dt}$, it follows that $\dfrac{dv}{dt} = v\dfrac{dv}{ds}$. Thus (7) may be written as

$$v\frac{dv}{ds} = -k^2 s \tag{8}$$

(a) Solve (8) for v to get $v = \pm k\sqrt{a^2-s^2}$. *Note:* Take a^2k^2 as the arbitrary constant of integration, and justify this choice.

(b) Letting $v = ds/dt$ in the solution of part (a) we obtain the differential equation

$$\frac{ds}{dt} = \pm k\sqrt{a^2-s^2} \tag{9}$$

Taking $t = 0$ at the instant when $v = 0$ (and hence $s = a$), solve (9) to obtain

$$s = a\cos kt \tag{10}$$

(c) Show that the largest value for $|s|$ is a. The number a is called the *amplitude* of the motion. (d) The particle will oscillate between the points where $s = a$ and $s = -a$. If T sec is the time for the particle to go from a to $-a$ and return, show that $T = 2\pi/k$. The number T is called the *period* of the motion.

44. A particle is moving in a straight line according to the equation of motion $s = 5 - 10 \sin^2 2t$, where s cm is the directed distance of the particle from the origin at t sec. Use the result of part (b) of Exercise 43 to show that the motion is simple harmonic. Find the amplitude and period of this motion.

45. Show that the motion of Exercise 44 is simple harmonic by showing that differential equation (7) is satisfied.

7.4 THE HYPERBOLIC FUNCTIONS	Certain combinations of e^x and e^{-x} appear so frequently in applications of mathematics that they are given special names. Two of these functions are the *hyperbolic sine function* and the *hyperbolic cosine function*. At the end of this section we show that the function values are related to the coordinates of the points on an equilateral hyperbola in a manner similar to that in which the values of the corresponding trigonometric functions are related to the coordinates of points on a circle. Following are the definitions of the hyperbolic sine function and the hyperbolic cosine function.

7.4.1 DEFINITION   The **hyperbolic sine function** is defined by

$$\sinh x = \frac{e^x - e^{-x}}{2}$$

The domain and range are the set of all real numbers.

7.4.2 DEFINITION   The **hyperbolic cosine function** is defined by

$$\cosh x = \frac{e^x + e^{-x}}{2}$$

The domain is the set of all real numbers and the range is the set of all numbers in the interval $[1, +\infty)$.

Because

$$\sinh(-x) = \frac{e^{-x} - e^x}{2} \qquad\qquad \cosh(-x) = \frac{e^{-x} + e^x}{2}$$

$$= -\frac{e^x - e^{-x}}{2} \qquad\qquad\qquad = \cosh x$$

$$= -\sinh x$$

the hyperbolic sine is an odd function and the hyperbolic cosine is an even function.

The formulas for the derivatives of the hyperbolic sine and hyperbolic cosine functions are obtained by applying Definitions 7.4.1 and 7.4.2 and differentiating the resulting expressions involving exponential functions. Thus

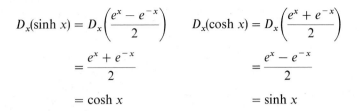

$$D_x(\sinh x) = D_x\left(\frac{e^x - e^{-x}}{2}\right) \qquad D_x(\cosh x) = D_x\left(\frac{e^x + e^{-x}}{2}\right)$$

$$= \frac{e^x + e^{-x}}{2} \qquad\qquad = \frac{e^x - e^{-x}}{2}$$

$$= \cosh x \qquad\qquad\qquad = \sinh x$$

From these formulas and the chain rule we have the following theorem.

7.4.3 THEOREM   If u is a differentiable function of x,

$$D_x(\sinh u) = \cosh u \, D_x u$$

$$D_x(\cosh u) = \sinh u \, D_x u$$

Because $D_x(\sinh x) > 0$ for all x, the hyperbolic sine function is increasing on its entire domain. With this information, the knowledge that it is an odd function, and some values obtained from the table of hyperbolic functions in the appendix, we have the graph of the hyperbolic sine function shown in Fig. 7.4.1.

The hyperbolic cosine function is decreasing on the interval $(-\infty, 0]$ because $D_x(\cosh x) < 0$ if $x < 0$, and it is increasing on the interval $[0, +\infty)$ because $D_x(\cosh x) > 0$ if $x > 0$. Furthermore, the hyperbolic cosine is an even function. With these facts and some values of $\cosh x$ from the table in the appendix, we obtain the graph of the hyperbolic cosine function shown in Fig. 7.4.2.

The remaining four hyperbolic functions are defined in terms of the hyperbolic sine and hyperbolic cosine functions. Observe that each satisfies an identity analogous to one satisfied by corresponding trigonometric functions.

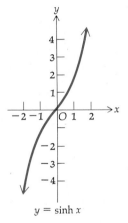

$y = \sinh x$

FIGURE 7.4.1

7.4.4 DEFINITION   The **hyperbolic tangent function**, **hyperbolic cotangent function**, **hyperbolic secant function**, and **hyperbolic cosecant function** are defined, respectively, as follows:

$$\tanh x = \frac{\sinh x}{\cosh x} \tag{1}$$

$$\coth x = \frac{\cosh x}{\sinh x} \tag{2}$$

$$\operatorname{sech} x = \frac{1}{\cosh x} \tag{3}$$

$$\operatorname{csch} x = \frac{1}{\sinh x} \tag{4}$$

$y = \cosh x$

FIGURE 7.4.2

The hyperbolic functions in Definition 7.4.4 can be expressed in terms of exponential functions by using Definitions 7.4.1 and 7.4.2. We have

$$\tanh x = \frac{e^x - e^{-x}}{e^x + e^{-x}} \qquad \coth x = \frac{e^x + e^{-x}}{e^x - e^{-x}}$$

$$\text{sech } x = \frac{2}{e^x + e^{-x}} \qquad \text{csch } x = \frac{2}{e^x - e^{-x}}$$

A sketch of the graph of the hyperbolic tangent function is shown in Fig. 7.4.3. Notice from Figs. 7.4.1, 7.4.2, and 7.4.3 that none of these functions is periodic, while the corresponding trigonometric functions are periodic.

There are identities that are satisfied by the hyperbolic functions which are similar to those satisfied by the trigonometric functions. Four of the fundamental identities are given in Definition 7.4.4. The other four fundamental identities are as follows:

$y = \tanh x$

FIGURE 7.4.3

$$\tanh x = \frac{1}{\coth x} \tag{5}$$

$$\cosh^2 x - \sinh^2 x = 1 \tag{6}$$

$$1 - \tanh^2 x = \text{sech}^2 x \tag{7}$$

$$1 - \coth^2 x = -\text{csch}^2 x \tag{8}$$

Equation (5) follows immediately from (1) and (2). Following is the proof of (6).

$$\cosh^2 x - \sinh^2 x = \left(\frac{e^x + e^{-x}}{2}\right)^2 - \left(\frac{e^x - e^{-x}}{2}\right)^2$$

$$= \tfrac{1}{4}(e^{2x} + 2e^0 + e^{-2x} - e^{2x} + 2e^0 - e^{-2x})$$

$$= 1$$

Equation (7) can be proved by using the formulas for $\tanh x$ and $\text{sech } x$ in terms of e^x and e^{-x} as in the proof above, or an alternate proof is obtained by using other identities as follows:

$$1 - \tanh^2 x = 1 - \frac{\sinh^2 x}{\cosh^2 x}$$

$$= \frac{\cosh^2 x - \sinh^2 x}{\cosh^2 x}$$

$$= \frac{1}{\cosh^2 x}$$

$$= \text{sech}^2 x$$

The proof of (8) is left as an exercise (see Exercise 1). From the eight fundamental identities it is possible to prove others. Some of these appear in the exercises. Other identities, such as the hyperbolic functions of the sum or difference of two numbers and the hyperbolic functions of twice a number or one-half a number, are similar to the corresponding trigonometric identities. Sometimes it is helpful to make use of the following two relations that follow from Definitions 7.4.1 and 7.4.2.

$$\cosh x + \sinh x = e^x \tag{9}$$

$$\cosh x - \sinh x = e^{-x} \tag{10}$$

We use them in proving the following identity:

$$\sinh(x + y) = \sinh x \cosh y + \cosh x \sinh y \tag{11}$$

From Definition 7.4.1,

$$\sinh(x + y) = \frac{e^{x+y} - e^{-(x+y)}}{2}$$

$$= \frac{e^x e^y - e^{-x} e^{-y}}{2}$$

Applying (9) and (10) to the right side of this equation we obtain

$$\sinh(x + y) = \tfrac{1}{2}[(\cosh x + \sinh x)(\cosh y + \sinh y)$$
$$-(\cosh x - \sinh x)(\cosh y - \sinh y)]$$

Expanding the right side of the above equation and combining terms we have formula (11).

In a similar manner we can prove

$$\cosh(x + y) = \cosh x \cosh y + \sinh x \sinh y \tag{12}$$

If in (11) and (12) y is replaced by x, the following two formulas are obtained:

$$\sinh 2x = 2 \sinh x \cosh x \tag{13}$$

$$\cosh 2x = \cosh^2 x + \sinh^2 x \tag{14}$$

Formula (14) combined with the identity (6) gives two alternate formulas for $\cosh 2x$, which are

$$\cosh 2x = 2 \sinh^2 x + 1 \tag{15}$$

$$\cosh 2x = 2 \cosh^2 x - 1 \tag{16}$$

Solving (15) and (16) for $\sinh x$ and $\cosh x$, respectively, and replacing x by $\tfrac{1}{2}x$ we get

$$\sinh \frac{x}{2} = \begin{cases} \sqrt{\dfrac{\cosh x - 1}{2}} & \text{if } x \geq 0 \\[2ex] -\sqrt{\dfrac{\cosh x - 1}{2}} & \text{if } x < 0 \end{cases} \tag{17}$$

$$\cosh \frac{x}{2} = \sqrt{\frac{\cosh x + 1}{2}} \tag{18}$$

We do not have a $\pm$ symbol on the right side of (18) because the range of the hyperbolic cosine function is $[1, +\infty)$. The details of the proofs of (12) through (18) are left as exercises (see Exercises 2, 4, and 5).

To find the derivative of the hyperbolic tangent function we use some of the identities.

$$D_x(\tanh x) = D_x \frac{\sinh x}{\cosh x}$$

$$= \frac{\cosh^2 x - \sinh^2 x}{\cosh^2 x}$$

$$= \frac{1}{\cosh^2 x}$$

$$= \operatorname{sech}^2 x$$

The formulas for the derivatives of the remaining hyperbolic functions are as follows: $D_x(\coth x) = -\operatorname{csch}^2 x$; $D_x(\operatorname{sech} x) = -\operatorname{sech} x \tanh x$; $D_x(\operatorname{csch} x) = -\operatorname{csch} x \coth x$. The proofs of these formulas are left as exercises (see Exercises 15 and 16).

From these formulas and the chain rule we have the following theorem.

THEOREM 7.4.5   If u is a differentiable function of x,

$$D_x(\tanh u) = \operatorname{sech}^2 u \, D_x u$$

$$D_x(\coth u) = -\operatorname{csch}^2 u \, D_x u$$

$$D_x(\operatorname{sech} u) = -\operatorname{sech} u \tanh u \, D_x u$$

$$D_x(\operatorname{csch} u) = -\operatorname{csch} u \coth u \, D_x u$$

Observe that the formulas for the derivatives of the hyperbolic sine, cosine, and tangent all have a plus sign, whereas those for the derivatives of the hyperbolic cotangent, secant, and cosecant all have a minus sign. Otherwise the formulas are similar to the corresponding ones for the derivatives of the trigonometric functions.

EXAMPLE 1

Find $\dfrac{dy}{dx}$ if

$$y = \tanh(1 - x^2)$$

SOLUTION

$$\frac{dy}{dx} = \operatorname{sech}^2(1 - x^2) \cdot D_x(1 - x^2)$$

$$= -2x \operatorname{sech}^2(1 - x^2)$$

EXAMPLE 2

Find $f'(x)$ if

$$f(x) = \ln \sinh x$$

SOLUTION

$$f'(x) = \frac{1}{\sinh x} \cdot \cosh x$$

$$= \frac{\cosh x}{\sinh x}$$

$$= \coth x$$

The indefinite integration formulas in the next theorem follow from the differentiation formulas in Theorems 7.4.3 and 7.4.5.

7.4.6 THEOREM

$$\int \sinh u \, du = \cosh u + C$$

$$\int \cosh u \, du = \sinh u + C$$

$$\int \operatorname{sech}^2 u \, du = \tanh u + C$$

$$\int \operatorname{csch}^2 u \, du = -\coth u + C$$

$$\int \operatorname{sech} u \tanh u \, du = -\operatorname{sech} u + C$$

$$\int \operatorname{csch} u \coth u \, du = -\operatorname{csch} u + C$$

The methods used to integrate the hyperbolic functions are similar to those used for the trigonometric functions. The following two examples illustrate the method.

EXAMPLE 3

Evaluate $\int \sinh x \cosh^2 x \, dx$.

SOLUTION

$$\int \sinh x \cosh^2 x \, dx = \int \cosh^2 x \, (\sinh x \, dx)$$

$$= \tfrac{1}{3} \cosh^3 x + C$$

EXAMPLE 4

Evaluate $\int \tanh^2 x \, dx$.

SOLUTION

$$\int \tanh^2 x \, dx = \int (1 - \operatorname{sech}^2 x) \, dx$$

$$= \int dx - \int \operatorname{sech}^2 x \, dx$$

$$= x - \tanh x + C$$

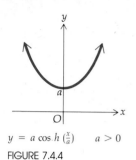

$y = a \cos.h \left(\frac{x}{a}\right)$ $a > 0$

FIGURE 7.4.4

A **catenary** is the curve formed by a flexible cable of uniform density hanging from two points under its own weight. Some cables of suspension bridges and some attached to telephone poles hang in this shape. If the lowest point of the catenary is at $(0, a)$, it can be shown that an equation of it is

$$y = a \cosh\left(\frac{x}{a}\right) \qquad a > 0 \tag{19}$$

A sketch of the graph of (19) appears in Fig. 7.4.4.

EXAMPLE 5
Find the length of the arc of the catenary defined by (19) from the point $(0, a)$ to the point (x_1, y_1), where $x_1 > 0$.

SOLUTION
If

$$f(x) = a \cosh\left(\frac{x}{a}\right) \quad \text{then} \quad f'(x) = a \sinh\left(\frac{x}{a}\right) \cdot \left(\frac{1}{a}\right)$$

$$= \sinh\left(\frac{x}{a}\right)$$

From Theorem 5.5.3, if L units is the length of the given arc,

$$L = \int_0^{x_1} \sqrt{1 + [f'(x)]^2}\, dx$$

$$= \int_0^{x_1} \sqrt{1 + \sinh^2\left(\frac{x}{a}\right)}\, dx$$

$$= \int_0^{x_1} \sqrt{\cosh^2\left(\frac{x}{a}\right)}\, dx \quad \text{(from (6))}$$

$$= \int_0^{x_1} \cosh\left(\frac{x}{a}\right) dx \quad \left(\text{because } \cosh\left(\frac{x}{a}\right) \geq 1\right)$$

$$= a \sinh\left(\frac{x}{a}\right)\Big]_0^{x_1}$$

$$= a \sinh\left(\frac{x_1}{a}\right) - a \sinh 0$$

$$= a \sinh\left(\frac{x_1}{a}\right)$$

In Section 0.6 the cosine and sine functions were defined as coordinates of a point on the unit circle

$$x^2 + y^2 = 1 \tag{20}$$

FIGURE 7.4.5

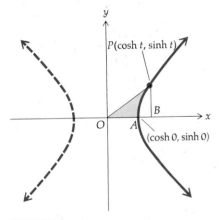

FIGURE 7.4.6

Recall from Definition 0.6.2 that if t is the radian measure of the angle between the x axis and a line from the origin to the point $P(x, y)$ on the unit circle then

$$\cos t = x \qquad \sin t = y \tag{21}$$

See Fig. 7.4.5. By substituting the values of x and y from (21) in (20) we obtain the identity

$$\cos^2 t + \sin^2 t = 1$$

In Section 10.6 you will learn that the curve

$$x^2 - y^2 = 1$$

is an equilateral hyperbola, called the unit hyperbola. Refer to Fig. 7.4.6. If t is any real number, then the point $(\cosh t, \sinh t)$ is on this hyperbola because

$$\cosh^2 t - \sinh^2 t = 1$$

Observe that because $\cosh t$ is never less than 1, all points $(\cosh t, \sinh t)$ are on the right branch of the hyperbola.

We now show how the areas of the shaded regions in Figs. 7.4.5 and 7.4.6 are related. Because the area of a circular sector of radius r units and a central angle of radian measure t is given by $\frac{1}{2}r^2 t$ square units, the area of the circular sector in Fig. 7.4.5 is $\frac{1}{2}t$ square units, since $r = 1$. The sector AOP in Fig. 7.4.6 is the region bounded by the x axis, the line OP, and the arc AP of the unit hyperbola. If A_1 square units is the area of sector AOP, A_2 square units is the area of triangle OBP, and A_3 square units is the area of region ABP,

$$A_1 = A_2 - A_3 \tag{22}$$

From the formula for determining the area of a triangle,

$$A_2 = \tfrac{1}{2}\cosh t \sinh t \tag{23}$$

We find A_3 by integration:

$$A_3 = \int_0^t \sinh u \, d(\cosh u)$$

$$= \int_0^t \sinh^2 u \, du$$

$$= \tfrac{1}{2}\int_0^t (\cosh 2u - 1) \, du$$

$$= \tfrac{1}{4}\sinh 2u - \tfrac{1}{2}u \Big]_0^t$$

Therefore

$$A_3 = \tfrac{1}{2}\cosh t \sinh t - \tfrac{1}{2}t \tag{24}$$

Substituting from (23) and (24) into (22) we have

$$A_1 = \tfrac{1}{2}\cosh t \sinh t - (\tfrac{1}{2}\cosh t \sinh t - \tfrac{1}{2}t)$$

$$= \tfrac{1}{2}t$$

Thus the number of square units in the area of circular sector AOP of Fig. 7.4.5 and the number of square units in the area of sector AOP of Fig. 7.4.6 is, in each case, one-half the value of the parameter associated with the point P. For the unit circle, the parameter t is the radian measure of the angle AOP. The parameter t for the unit hyperbola is not interpreted as the measure of an angle; however, sometimes the term *hyperbolic radian* is used in connection with t.

We have therefore shown that the function values of sinh and cosh have the same relationship to the unit hyperbola as the function values of sine and cosine have to the unit circle. So just as the sine and cosine are called *circular functions*, sinh and cosh are called hyperbolic functions.

EXERCISES 7.4

In Exercises 1 through 10, prove the identity.

1. $1 - \coth^2 x = -\operatorname{csch}^2 x$

2. $\cosh(x + y) = \cosh x \cosh y + \sinh x \sinh y$

3. $\tanh(x + y) = \dfrac{\tanh x + \tanh y}{1 + \tanh x \tanh y}$

4. (a) $\sinh 2x = 2 \sinh x \cosh x$; (b) $\cosh 2x = \cosh^2 x + \sinh^2 x$; (c) $\cosh 2x = 2 \sinh^2 x + 1$; (d) $\cosh 2x = 2 \cosh^2 x - 1$

5. (a) $\cosh \tfrac{1}{2}x = \sqrt{\dfrac{\cosh x + 1}{2}}$; (b) $\sinh \tfrac{1}{2}x = \pm\sqrt{\dfrac{\cosh x - 1}{2}}$

6. $\operatorname{csch} 2x = \tfrac{1}{2} \operatorname{sech} x \operatorname{csch} x$

7. (a) $\tanh(-x) = -\tanh x$; (b) $\operatorname{sech}(-x) = \operatorname{sech} x$

8. $\sinh 3x = 3 \sinh x + 4 \sinh^3 x$

9. $\cosh 3x = 4 \cosh^3 x - 3 \cosh x$

10. $\sinh^2 x - \sinh^2 y = \sinh(x + y) \sinh(x - y)$

11. Prove: $(\sinh x + \cosh x)^n = \cosh nx + \sinh nx$, if n is any positive integer. (*Hint:* Use formula (9).)

12. Prove that the hyperbolic tangent function is an odd function and the hyperbolic secant function is an even function.

13. Prove: $\tanh(\ln x) = \dfrac{x^2 - 1}{x^2 + 1}$

14. Prove: $\dfrac{1 + \tanh x}{1 - \tanh x} = e^{2x}$

15. Prove: (a) $D_x(\sinh x) = \cosh x$; (b) $D_x(\coth x) = -\operatorname{csch}^2 x$.

16. Prove: (a) $D_x(\operatorname{sech} x) = -\operatorname{sech} x \tanh x$; (b) $D_x(\operatorname{csch} x) = -\operatorname{csch} x \coth x$.

In Exercises 17 through 30, find the derivative of the given function.

17. $f(x) = \tanh \dfrac{4x + 1}{5}$

18. $g(t) = \cosh t^3$

19. $f(y) = \sinh e^{2y}$

20. $f(x) = \coth(\ln x)$

21. $f(w) = \operatorname{sech}^2 4w$

22. $g(x) = \ln(\sinh x^3)$

23. $f(x) = e^x \cosh x$

24. $h(x) = \coth \dfrac{1}{x}$

25. $h(t) = \ln(\tanh t)$

26. $F(x) = \tan^{-1}(\sinh x^2)$

27. $G(x) = \sin^{-1}(\tanh x^2)$

28. $f(x) = \ln(\coth 3x - \operatorname{csch} 3x)$

29. $f(x) = x^{\sinh x}$; $x > 0$

30. $g(x) = (\cosh x)^x$

31. Prove: (a) $\displaystyle\int \tanh u \, du = \ln|\cosh u| + C$; (b) $\displaystyle\int \coth u \, du = \ln|\sinh u| + C$.

32. Prove: (a) $\displaystyle\int \operatorname{sech} u \, du = 2 \tan^{-1} e^u + C$.

33. Prove: $\displaystyle\int \operatorname{csch} u \, du = \ln|\tanh \tfrac{1}{2}u| + C$.

In Exercises 34 through 43, evaluate the indefinite integral.

34. $\int \sinh^4 x \cosh x \, dx$ **35.** $\int e^t \cosh(e^t) \sinh(e^t) \, dt$ **36.** $\int e^x \operatorname{csch}^2 e^x \, dx$
$-\coth e^x + C$ **37.** $\int \dfrac{\sinh \sqrt{x}}{\sqrt{x}} \, dx$

38. $\int x \operatorname{sech}^2 x^2 \, dx$ **39.** $\int \coth^2 3x \, dx$ **40.** $\int \operatorname{sech}^3 x \tanh x \, dx$ **41.** $\int \operatorname{sech}^2 x \tanh^5 x \, dx$

42. $\int \dfrac{e^{\cosh t}}{\operatorname{csch} t} \, dt$ **43.** $\int \tanh x \ln(\cosh x) \, dx$

In Exercises 44 through 47, evaluate the definite integral.

44. $\int_0^{\ln 2} \tanh z \, dz$ **45.** $\int_0^{\ln 3} \operatorname{sech}^2 t \, dt$

46. $\int_{-1}^1 (\cosh x - \sinh x) \, dx$ **47.** $\int_0^2 \sinh^3 x \cosh x \, dx$

48. Draw a sketch of the graph of the hyperbolic cotangent function. Find equations of the asymptotes.
49. Draw a sketch of the graph of the hyperbolic secant function. Find the relative extrema of the function, the points of inflection of the graph, the intervals on which the graph is concave upward, and the intervals on which the graph is concave downward.
50. Prove that a catenary is concave upward at each point.
51. Find the area of the region bounded by the catenary of Example 5, the y axis, the x axis, and the line $x = x_1$, where $x_1 > 0$.
52. Find the volume of the solid of revolution if the region of Exercise 51 is revolved about the x axis.
53. A particle is moving along a straight line according to the equation of motion $s = e^{-ct/2}(A \sinh t + B \cosh t)$, where s cm is the directed distance of the particle from the origin at t sec. If v cm/sec and a cm/sec^2 are the velocity and acceleration, respectively, of the particle at t sec, find v and a. Also show that a is the sum of two numbers, one of which is proportional to s and the other proportional to v.
54. Prove that the hyperbolic sine function is continuous and increasing on its entire domain.
55. Prove that the hyperbolic tangent function is continuous and increasing on its entire domain.
56. Prove that the hyperbolic cosine function is continuous on its entire domain but is not monotonic on its entire domain. Find the intervals on which the function is increasing and the intervals on which it is decreasing.

7.5 *(Supplementary)*
THE INVERSE
HYPERBOLIC FUNCTIONS

From the graph of the hyperbolic sine function (Fig. 7.4.1) we see that a horizontal line intersects the graph in one and only one point; therefore the hyperbolic sine function is one-to-one. Furthermore, the hyperbolic sine function is continuous and increasing on its domain (see Exercise 54 in Exercises 7.4). Therefore, by Theorem 6.1.5, the function has an inverse.

7.5.1 DEFINITION   The **inverse hyperbolic sine function**, denoted by $\sinh^{-1}$, is defined as follows:

$$y = \sinh^{-1} x \quad \text{if and only if} \quad x = \sinh y$$

A sketch of the graph of the inverse hyperbolic sine function is shown in Fig. 7.5.1. Its domain is the set R of all real numbers, and its range is also the set R.

It follows from Definition 7.5.1 that

$$\sinh(\sinh^{-1} x) = x \tag{1}$$

and

$$\sinh^{-1}(\sinh y) = y \tag{2}$$

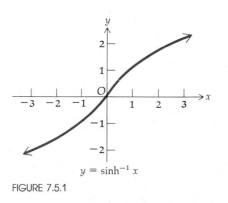

$y = \sinh^{-1} x$

FIGURE 7.5.1

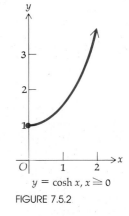

$y = \cosh x,\, x \geq 0$

FIGURE 7.5.2

From Fig. 7.4.2 we notice that a horizontal line, $y = k$, where $k > 1$, intersects the graph of the hyperbolic cosine function in two points. Therefore, for each number greater than 1 in the range of this function there correspond two numbers in the domain. So the hyperbolic cosine function is not one-to-one and does not have an inverse. However, we define a function F as follows:

$$F(x) = \cosh x \qquad \text{for } x \geq 0 \tag{3}$$

The domain of F is the interval $[0, +\infty)$, and the range is the interval $[1, +\infty)$. A sketch of the graph of F is shown in Fig. 7.5.2. The function F is continuous and increasing on its domain (see Exercise 1), and so by Theorem 6.1.5 F has an inverse, which is called the *inverse hyperbolic cosine function*.

$y = \cosh^{-1} x$

FIGURE 7.5.3

7.5.2 DEFINITION   The **inverse hyperbolic cosine function**, denoted by $\cosh^{-1}$, is defined as follows:

$$y = \cosh^{-1} x \quad \text{if and only if} \quad x = \cosh y \text{ and } y \geq 0$$

A sketch of the graph of the inverse hyperbolic cosine function is shown in Fig. 7.5.3. The domain of this function is the interval $[1, +\infty)$, and the range is the interval $[0, +\infty)$. From Definition 7.5.2,

$$\cosh(\cosh^{-1} x) = x \qquad \text{if } x \geq 1 \tag{4}$$

and

$$\cosh^{-1}(\cosh y) = y \qquad \text{if } y \geq 0 \tag{5}$$

As with the hyperbolic sine function, a horizontal line intersects each of the graphs of the hyperbolic tangent function, the hyperbolic cotangent function, and the hyperbolic cosecant function at one and only one point. For the hyperbolic tangent function this may be seen in Fig. 7.4.3. Each of these three functions then is one-to-one; furthermore, each is continuous and monotonic on its domain. Thus each has an inverse function.

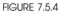

$y = \tanh^{-1} x$

FIGURE 7.5.4

7.5.3 DEFINITION The **inverse hyperbolic tangent function**, the **inverse hyperbolic cotangent function**, and the **inverse hyperbolic cosecant function**, denoted respectively by $\tanh^{-1}$, $\coth^{-1}$, and csch^{-1}, are defined as follows:

$$y = \tanh^{-1} x \quad \text{if and only if} \quad x = \tanh y$$
$$y = \coth^{-1} x \quad \text{if and only if} \quad x = \coth y$$
$$y = \operatorname{csch}^{-1} x \quad \text{if and only if} \quad x = \operatorname{csch} y$$

A sketch of the graph of the inverse hyperbolic tangent function is in Fig. 7.5.4.

Because $\operatorname{sech} x = 1/\cosh x$ and the hyperbolic cosine function is not one-to-one, it follows that the hyperbolic secant function is not one-to-one. Therefore it does not have an inverse. However, we proceed as we did with the hyperbolic cosine function and define a new function that has an inverse function. Let the function G be defined by

$$G(x) = \operatorname{sech} x \qquad \text{for } x \geq 0 \tag{6}$$

It can be shown that G is continuous and monotonic on its domain (see Exercise 2); therefore G has an inverse, which is called the *inverse hyperbolic secant function*.

7.5.4 DEFINITION The **inverse hyperbolic secant function**, denoted by sech^{-1}, is defined as follows:

$$y = \operatorname{sech}^{-1} x \quad \text{if and only if} \quad x = \operatorname{sech} y \text{ and } y \geq 0 \tag{7}$$

The inverse hyperbolic functions can be expressed in terms of natural logarithms. This should not be surprising, because the hyperbolic functions were defined in terms of the exponential function, and the natural logarithmic function is the inverse of the exponential function.

Following are these expressions for $\sinh^{-1}$, $\cosh^{-1}$, $\tanh^{-1}$, and $\coth^{-1}$.

$$\sinh^{-1} x = \ln(x + \sqrt{x^2 + 1}) \qquad x \text{ any real number} \tag{8}$$

$$\cosh^{-1} x = \ln(x + \sqrt{x^2 - 1}) \qquad x \geq 1 \tag{9}$$

$$\tanh^{-1} x = \frac{1}{2} \ln \frac{1 + x}{1 - x} \qquad |x| < 1 \tag{10}$$

$$\coth^{-1} x = \frac{1}{2} \ln \frac{x+1}{x-1} \qquad |x| > 1 \tag{11}$$

We prove (9) and leave the proofs of the other three formulas as exercises (see Exercises 3 through 5).

To prove (9), let $y = \cosh^{-1} x$, $x \geq 1$. Then from Definition 7.5.2, we have $x = \cosh y$, $y \geq 0$. Applying Definition 7.4.2 to $\cosh y$ we get

$$x = \frac{e^y + e^{-y}}{2} \qquad y \geq 0$$

from which we obtain

$$e^{2y} - 2xe^y + 1 = 0 \qquad y \geq 0$$

Solving this equation for e^y by the quadratic formula we obtain

$$e^y = \frac{2x \pm \sqrt{4x^2 - 4}}{2}$$

$$= x \pm \sqrt{x^2 - 1} \qquad y \geq 0 \tag{12}$$

We know that $y \geq 0$ and $x \geq 1$. Therefore $e^y \geq 1$. When $x = 1$, $x + \sqrt{x^2 - 1} = 1$ and $x - \sqrt{x^2 - 1} = 1$. Furthermore, when $x > 1$, we have $0 < x - 1 < x + 1$; so $\sqrt{x-1} < \sqrt{x+1}$. Therefore

$$\sqrt{x-1}\sqrt{x-1} < \sqrt{x-1}\sqrt{x+1}$$

giving $x - 1 < \sqrt{x^2 - 1}$. Hence when $x > 1$, $x - \sqrt{x^2 - 1} < 1$. Consequently, we can safely reject the minus sign in (12). And since $y = \cosh^{-1} x$, we have

$$\cosh^{-1} x = \ln(x + \sqrt{x^2 - 1}) \qquad x \geq 1$$

which is (9).

EXAMPLE 1

Express each of the following in terms of a natural logarithm: (a) $\sinh^{-1} 2$; (b) $\tanh^{-1}(-\frac{4}{5})$.

SOLUTION

(a) From (8),

$$\sinh^{-1} 2 = \ln(2 + \sqrt{5})$$

(b) From (10),

$$\tanh^{-1}(-\tfrac{4}{5}) = \frac{1}{2} \ln \frac{\frac{1}{5}}{\frac{9}{5}}$$

$$= \tfrac{1}{2} \ln 3^{-2}$$

$$= -\ln 3$$

To obtain a formula for the derivative of the inverse hyperbolic sine function let $y = \sinh^{-1} x$, and then $x = \sinh y$. Thus, because $\dfrac{dx}{dy} = \cosh y$ and $\dfrac{dy}{dx}$ is the reciprocal of $\dfrac{dx}{dy}$,

$$\frac{dy}{dx} = \frac{1}{\cosh y} \qquad (13)$$

From the identity $\cosh^2 y - \sinh^2 y = 1$ we get $\cosh y = \sqrt{\sinh^2 y + 1}$, where when we take the square root, the minus sign is rejected because $\cosh y \geq 1$. Therefore, because $x = \sinh y$, $\cosh y = \sqrt{x^2 + 1}$. Substituting this value of $\cosh y$ in (13) we obtain

$$D_x(\sinh^{-1} x) = \frac{1}{\sqrt{x^2 + 1}} \qquad (14)$$

Formula (14) can also be derived by using (8), as follows:

$$D_x(\sinh^{-1} x) = D_x \ln(x + \sqrt{x^2 + 1})$$

$$= \frac{1 + \dfrac{x}{\sqrt{x^2 + 1}}}{x + \sqrt{x^2 + 1}}$$

$$= \frac{\sqrt{x^2 + 1} + x}{\sqrt{x^2 + 1}(x + \sqrt{x^2 + 1})}$$

$$= \frac{1}{\sqrt{x^2 + 1}}$$

The formulas for the derivatives of the other five inverse hyperbolic functions are obtained in a way analogous to that for the derivative of the inverse hyperbolic sine function. Their derivations are left as exercises (see Exercises 6 through 10). From these formulas and the chain rule the next theorem follows.

7.5.5 THEOREM   If u is a differentiable function of x,

$$D_x(\sinh^{-1} u) = \frac{1}{\sqrt{u^2 + 1}} D_x u \qquad (15)$$

$$D_x(\cosh^{-1} u) = \frac{1}{\sqrt{u^2 - 1}} D_x u \qquad u > 1 \qquad (16)$$

$$D_x(\tanh^{-1} u) = \frac{1}{1 - u^2} D_x u \qquad |u| < 1 \qquad (17)$$

$$D_x(\coth^{-1} u) = \frac{1}{1 - u^2} D_x u \qquad |u| > 1 \qquad (18)$$

$$D_x(\text{sech}^{-1} u) = -\frac{1}{u\sqrt{1 - u^2}} D_x u \qquad 0 < u < 1 \tag{19}$$

$$D_x(\text{csch}^{-1} u) = -\frac{1}{|u|\sqrt{1 + u^2}} D_x u \qquad u \neq 0 \tag{20}$$

EXAMPLE 2
Find $f'(x)$ if

$$f(x) = \tanh^{-1}(\cos 2x)$$

SOLUTION
From (17),

$$f'(x) = \frac{1}{1 - \cos^2 2x}(-2 \sin 2x)$$

$$= \frac{-2 \sin 2x}{\sin^2 2x}$$

$$= \frac{-2}{\sin 2x}$$

$$= -2 \csc 2x$$

EXAMPLE 3
Verify that $D_x(x \cosh^{-1} x - \sqrt{x^2 - 1}) = \cosh^{-1} x$.

SOLUTION

$$D_x(x \cosh^{-1} x - \sqrt{x^2 - 1}) = \cosh^{-1} x + x \cdot \frac{1}{\sqrt{x^2 - 1}} - \frac{x}{\sqrt{x^2 - 1}}$$

$$= \cosh^{-1} x$$

The main application of the inverse hyperbolic functions is in connection with integration. This topic is the subject of Supplementary Section 8.11.

EXERCISES 7.5

1. If $F(x) = \cosh x$ and $x \geq 0$, prove that F is continuous and increasing on its domain.
2. If $G(x) = \text{sech } x$ and $x \geq 0$, prove that G is continuous and monotonic on its domain.

In Exercises 3 through 10, prove the indicated formula of this section.

3. Formula (8)	4. Formula (10)	5. Formula (11)	6. Formula (16)
7. Formula (17)	8. Formula (18)	9. Formula (19)	10. Formula (20)

In Exercises 11 through 14, express the given quantity in terms of a natural logarithm.
11. $\sinh^{-1}\frac{1}{4}$ **12.** $\cosh^{-1}3$ **13.** $\tanh^{-1}\frac{1}{2}$ **14.** $\coth^{-1}(-2)$

In Exercises 15 through 32, find the derivative of the given function.
15. $f(x) = \sinh^{-1}x^2$ **16.** $G(x) = \cosh^{-1}\frac{1}{3}x$ **17.** $F(x) = \tanh^{-1}4x$ **18.** $h(w) = \tanh^{-1}w^3$
19. $g(x) = \coth^{-1}(3x+1)$ **20.** $f(r) = \operatorname{csch}^{-1}\frac{1}{2}r^2$ **21.** $f(x) = x^2\cosh^{-1}x^2$ **22.** $h(x) = (\operatorname{sech}^{-1}x)^2$
23. $f(x) = \sinh^{-1}(\tan x)$ **24.** $g(x) = \tanh^{-1}(\sin 3x)$ **25.** $h(x) = \cosh^{-1}(\csc x)$ **26.** $F(x) = \coth^{-1}(\cosh x)$
27. $f(z) = (\coth^{-1}z^2)^3$ **28.** $f(t) = \sinh^{-1}e^{2t}$ **29.** $g(x) = \tanh^{-1}(\cos e^x)$
30. $h(x) = \cosh^{-1}(\ln x)$ **31.** $G(x) = x\sinh^{-1}x - \sqrt{1+x^2}$ **32.** $H(x) = \ln\sqrt{x^2+1} - x\tanh^{-1}x$

In Exercises 33 through 36, draw a sketch of the graph of the indicated function.
33. The inverse hyperbolic tangent function.   **34.** The inverse hyperbolic cotangent function.
35. The inverse hyperbolic secant function.   **36.** The inverse hyperbolic cosecant function.

REVIEW EXERCISES FOR CHAPTER 7

1. Given $x = \sin^{-1}\frac{3}{5}$ and $y = \cos^{-1}(-\frac{4}{5})$, find the exact value of each of the following: (a) $\cos x$; (b) $\sin y$; (c) $\tan x$; (d) $\tan y$.
2. Given $x = \cos^{-1}\frac{7}{25}$ and $y = \tan^{-1}(-\frac{7}{24})$, find the exact value of each of the following: (a) $\sin x$; (b) $\sin y$; (c) $\cos y$; (d) $\tan x$.

In Exercises 3 and 4, find the exact value of the given quantity.
3. (a) $\sin[2\cos^{-1}(-\frac{12}{13})]$; (b) $\tan[\cos^{-1}\frac{3}{5} + \sin^{-1}(-\frac{7}{25})]$ **4.** (a) $\tan[2\sin^{-1}(-\frac{24}{25})]$; (b) $\cos[\tan^{-1}\frac{4}{3} - \sin^{-1}(-\frac{5}{13})]$

In Exercises 5 through 16, find the derivative of the given function.
5. $f(x) = \tan^{-1}2^x$ **6.** $g(x) = x^2\sin^{-1}2x$ **7.** $F(x) = \sinh^3 2x$ **8.** $h(x) = e^x(\cosh x + \sinh x)$

9. $f(x) = \sec^{-1}\sqrt{x^2+1}$ **10.** $G(w) = \tan^{-1}\left(\frac{w-3}{3w+1}\right)$ **11.** $g(x) = \sin^{-1}(\tanh\sqrt{x})$ **12.** $f(x) = \operatorname{sech}(\tan 3x)$

13. $h(x) = \operatorname{csch}^2 e^{2x}$ **14.** $g(x) = \cos^{-1}(\tanh 2x)$ **15.** $f(x) = (\cosh x)^{1/x}$ **16.** $f(x) = \sec^{-1}(\cosh x)$

In Exercises 17 and 18, draw a sketch of the graph of the given function.
17. $f(x) = 2\tanh\frac{1}{2}x$ **18.** $f(x) = \frac{1}{2}\sin^{-1}3x$

In Exercises 19 and 20, find $\dfrac{dy}{dx}$ by implicit differentiation.

19. $\cot^{-1}\dfrac{x^2}{3y} + xy^2 = 0$ **20.** $\tan^{-1}(x+3y) = \ln y$

In Exercises 21 through 28, evaluate the indefinite integral.

21. $\displaystyle\int \frac{3x\,dx}{\sqrt{1-x^4}}$ **22.** $\displaystyle\int x\coth\frac{1}{2}x^2\,dx$ **23.** $\displaystyle\int \tanh^2 3x\,dx$ **24.** $\displaystyle\int \frac{dx}{\sqrt{7+5x-2x^2}}$

25. $\displaystyle\int \frac{dt}{2t^2+3t+5}$ **26.** $\displaystyle\int \frac{dy}{9e^y + e^{-y}}$ **27.** $\displaystyle\int \frac{dx}{\sqrt{e^{2x}-8}}$ **28.** $\displaystyle\int \frac{\cosh t\,dt}{\sqrt{\sinh t}}$

In Exercises 29 through 32, evaluate the definite integral.

29. $\displaystyle\int_1^2 \frac{(t+2)\,dt}{\sqrt{4t-t^2}}$ **30.** $\displaystyle\int_0^2 \operatorname{sech}^2\frac{1}{2}x\,dx$ **31.** $\displaystyle\int_0^1 \sqrt{1+\cosh y}\,dy$ **32.** $\displaystyle\int_{-1}^1 \frac{(2x+6)\,dx}{x^2+2x+5}$

33. In an electric circuit the electromotive force is x volts at t sec, where $x = 20 \cos 120\pi t$. (a) Solve the equation for t. Use the equation in part (a) to find the smallest positive value of t for which the electromotive force is (b) 10 volts; (c) 5 volts; (d) -10 volts; (e) -5 volts.

34. A weight is suspended from a spring and vibrating vertically according to the equation $y = 4 \sin 2\pi(t + \frac{1}{6})$, where y cm is the directed distance of the weight from its central position t sec after the start of the motion, and the positive direction is upward. (a) Solve the equation for t. Use the equation in part (a) to determine the smallest positive value of t for which the displacement of the weight above its central position is (b) 2 cm and (c) 3 cm.

35. Show that $\cosh(\ln x) = \dfrac{x^2 + 1}{2x}$.

36. Prove: (a) $\lim\limits_{x \to +\infty} \coth x = 1$; (b) $\lim\limits_{x \to +\infty} \operatorname{csch} x = 0$.

37. Find the area of the region bounded by the curve $y = 9/\sqrt{9 - x^2}$, the two coordinate axes, and the line $x = 2\sqrt{2}$.

38. Find the length of the catenary $y = \cosh x$ from $(\ln 2, \frac{5}{4})$ to $(\ln 3, \frac{5}{3})$.

39. Find the volume of the solid of revolution generated when the region bounded by the curve $y = \sqrt{\sinh x}$, the x axis, and the lines $x = 0$ and $x = \ln 2$ is revolved about the x axis.

40. The *gudermannian*, named after the German mathematician Christoph Gudermann (1798–1852), is the function defined by $\operatorname{gd} x = \tan^{-1}(\sinh x)$. Show that $D_x(\operatorname{gd} x) = \operatorname{sech} x$.

41. A searchlight is $\frac{1}{2}$ km from a straight road and it keeps a light trained on an automobile that is traveling at a constant speed of 60 km/hr. Find the rate at which the light beam is changing direction (a) when the car is at the point on the road nearest the searchlight and (b) when the car is $\frac{1}{2}$ km down the road from this point.

42. A picture 5 ft high is placed on a wall with its base 7 ft above the level of the eye of an observer. If the observer is approaching the wall at the rate of 3 ft/sec, how fast is the measure of the angle subtended at her eye by the picture changing when the observer is 10 ft from the wall?

43. An airplane is flying at a speed of 300 mi/hr at an altitude of 4 mi. If an observer is on the ground, find the time rate of change of the measure of the observer's angle of elevation of the airplane when the airplane is directly over a point on the ground 2 mi from the observer.

44. A helicopter leaves the ground at a point 800 ft from an observer and rises vertically at 25 ft/sec. Find the time rate of change of the measure of the observer's angle of elevation of the helicopter when the helicopter is 600 ft above the ground.

45. Two points A and B are diametrically opposite each other on the shores of a circular lake 1 km in diameter. A man desires to go from point A to point B. He can row at the rate of $1\frac{1}{2}$ km/hr and walk at the rate of 5 km/hr. Find the least amount of time it can take him to get from point A to point B.

46. Solve Exercise 45 if the rates of rowing and walking are, respectively, 2 km/hr and 4 km/hr.

Exercises 47 through 53 pertain to Supplementary Section 7.5. In Exercises 47 and 48, express the given quantity in terms of a natural logarithm.

47. $\cosh^{-1} 2$

48. $\tanh^{-1} \frac{1}{4}$

In Exercises 49 through 52, find the derivative of the given function.

49. $h(w) = w^2 \sinh^{-1} 2w$ 50. $f(x) = \tanh^{-1}(\sin \sqrt{x})$ 51. $f(x) = \operatorname{sech}^{-1} e^{-2x}$ 52. $g(t) = \cosh^{-1} \sqrt{t^2 + 1}$

53. The graph of the equation $x = a \sinh^{-1} \sqrt{\dfrac{a^2}{y^2} - 1} - \sqrt{a^2 - y^2}$ is called a *tractrix*. Prove that the slope of the curve at any point (x, y) is $-y/\sqrt{a^2 - y^2}$.

E I G H T

TECHNIQUES OF INTEGRATION

8.1
A SYNOPSIS OF
INTEGRATION FORMULAS

In Definition 4.5.2 the definite integral of a function f from a to b was defined as the limit of a Riemann sum as follows:

$$\int_a^b f(x)\,dx = \lim_{||\Delta|| \to 0} \sum_{i=1}^n f(\xi_i)\,\Delta_i x \tag{1}$$

if the limit exists. We stated a theorem, proved in advanced calculus, that the limit on the right side of (1) exists if f is continuous on the closed interval $[a, b]$.

The exact value of a definite integral may be calculated by the second fundamental theorem of the calculus, provided that an antiderivative (or indefinite integral) of the integrand can be found. In practice, it is not always possible to find the indefinite integral. That is, we may have a definite integral that exists, but the integrand has no antiderivative that can be expressed in terms of elementary functions. An example of such a definite integral is

$$\int_0^{1/2} e^{-t^2}\,dt$$

A method for computing the value of this definite integral to any required degree of accuracy by using infinite series is given in Example 2 of Section 12.3.

Numerous definite integrals that occur in practice can be evaluated by finding an antiderivative of the integrand. Some methods for doing this have already been given, and additional ones are presented in this chapter. Often you may find it desirable to resort to a table of integrals instead of performing a complicated integration. A short table of integrals can be found on the front and back endpapers, and Supplementary Section 8.12 provides a discussion of the use of such a table. Sometimes it may be necessary to employ some of the techniques of integration to express the integrand in a form that is found in a table. Therefore you should acquire proficiency in recognizing which technique to apply to a given integral. Furthermore, development of computational skills is important in all branches of mathematics, and the exercises in this chapter provide a good training ground. For these reasons you are advised to use a table of integrals only after you have mastered integration.

Standard indefinite integration formulas that you learned in previous chapters and that are used frequently are listed below and numbered for reference.

1. $\int du = u + C$ **2.** $\int a\,du = au + C$ where a is any constant

3. $\int [f(u) + g(u)]\,du = \int f(u)\,du + \int g(u)\,du$

4. $\int u^n\,du = \dfrac{u^{n+1}}{n+1} + C$ $n \neq -1$ **5.** $\int \dfrac{du}{u} = \ln|u| + C$

6. $\int a^u\,du = \dfrac{a^u}{\ln a} + C$ where $a > 0$ and $a \neq 1$ **7.** $\int e^u\,du = e^u + C$

8. $\int \sin u\,du = -\cos u + C$ **9.** $\int \cos u\,du = \sin u + C$

10. $\int \sec^2 u\,du = \tan u + C$ **11.** $\int \csc^2 u\,du = -\cot u + C$

12. $\int \sec u\,\tan u\,du = \sec u + C$ **13.** $\int \csc u\,\cot u\,du = -\csc u + C$

14. $\int \tan u\,du = \ln|\sec u| + C$ **15.** $\int \cot u\,du = \ln|\sin u| + C$

16. $\int \sec u\,du = \ln|\sec u + \tan u| + C$ **17.** $\int \csc u\,du = \ln|\csc u - \cot u| + C$

18. $\displaystyle\int \frac{du}{\sqrt{a^2 - u^2}} = \sin^{-1}\frac{u}{a} + C$ where $a > 0$

19. $\displaystyle\int \frac{du}{a^2 + u^2} = \frac{1}{a}\tan^{-1}\frac{u}{a} + C$ where $a \neq 0$

20. $\displaystyle\int \frac{du}{u\sqrt{u^2 - a^2}} = \frac{1}{a}\sec^{-1}\frac{u}{a} + C$ where $a > 0$

21. $\int \sinh u\,du = \cosh u + C$ **22.** $\int \cosh u\,du = \sinh u + C$

23. $\int \operatorname{sech}^2 u\,du = \tanh u + C$ **24.** $\int \operatorname{csch}^2 u\,du = -\coth u + C$

25. $\int \operatorname{sech} u\,\tanh u\,du = -\operatorname{sech} u + C$

26. $\int \operatorname{csch} u\,\coth u\,du = -\operatorname{csch} u + C$

8.2
INTEGRATION BY PARTS

From the formula for the derivative of the product of two functions we obtain a very useful method of integration called *integration by parts*. If f and g are differentiable functions, then

$$D_x[f(x)g(x)] = f(x)g'(x) + g(x)f'(x)$$

$$\Leftrightarrow \qquad f(x)g'(x) = D_x[f(x)g(x)] - g(x)f'(x)$$

Integrating on each side of this equation we obtain

$$\int f(x)g'(x)\,dx = \int D_x[f(x)g(x)]\,dx - \int g(x)f'(x)\,dx$$

$$\int f(x)g'(x)\,dx = f(x)g(x) - \int g(x)f'(x)\,dx \qquad (1)$$

We call (1) the **formula for integration by parts**. For computational purposes a more convenient way of writing this formula is obtained by letting

$$u = f(x) \qquad \text{and} \qquad v = g(x)$$

Then

$$du = f'(x)\,dx \quad \text{and} \quad dv = g'(x)\,dx$$

so that (1) becomes

$$\int u\,dv = uv - \int v\,du \tag{2}$$

This formula expresses the integral $\int u\,dv$ in terms of another integral, $\int v\,du$. By a suitable choice of u and dv, it may be easier to evaluate the second integral than the first. When choosing the substitutions for u and dv, we usually want dv to be the most complicated factor of the integrand that can be integrated directly and u to be a function whose derivative is a simpler function. The method is shown by the following illustrations and examples.

▶ ILLUSTRATION 1   We wish to evaluate

$$\int x^3 e^{x^2}\,dx$$

To determine the substitutions for u and dv, bear in mind that to find v we must be able to integrate dv. This suggests letting $dv = xe^{x^2}\,dx$ and $u = x^2$. Then

$$v = \tfrac{1}{2}e^{x^2} + C_1 \quad \text{and} \quad du = 2x\,dx$$

From formula (2),

$$\int x^3 e^{x^2}\,dx = x^2(\tfrac{1}{2}e^{x^2} + C_1) - \int (\tfrac{1}{2}e^{x^2} + C_1)2x\,dx$$

$$= \tfrac{1}{2}x^2 e^{x^2} + C_1 x^2 - \int xe^{x^2}\,dx - 2C_1 \int x\,dx$$

$$= \tfrac{1}{2}x^2 e^{x^2} + C_1 x^2 - \tfrac{1}{2}e^{x^2} - C_1 x^2 + C_2$$

$$= \tfrac{1}{2}e^{x^2}(x^2 - 1) + C_2 \qquad\qquad ◀$$

In Illustration 1 observe that the first constant of integration C_1 does not appear in the final answer. The C_1 is used only to show that all choices for v of the form $\tfrac{1}{2}e^{x^2} + C_1$ produce the same result for $\int x^3 e^{x^2}\,dx$. This situation is true in general, and we prove it as follows: By writing $v + C_1$ in formula (2) we have

$$\int u\,dv = u(v + C_1) - \int (v + C_1)\,du$$

$$= uv + C_1 u - \int v\,du - C_1 \int du$$

$$= uv + C_1 u - \int v\,du - C_1 u$$

$$= uv - \int v\,du$$

Therefore it is not necessary to write C_1 when finding v from dv.

EXAMPLE 1

Find $\int x \ln x \, dx$.

SOLUTION

Let $u = \ln x$ and $dv = x \, dx$. Then

$$du = \frac{dx}{x} \quad \text{and} \quad v = \frac{x^2}{2}$$

So

$$\int x \ln x \, dx = \frac{x^2}{2} \ln x - \int \frac{x^2}{2} \cdot \frac{dx}{x}$$

$$= \frac{x^2}{2} \ln x - \frac{1}{2} \int x \, dx$$

$$= \tfrac{1}{2}x^2 \ln x - \tfrac{1}{4}x^2 + C$$

▶ ILLUSTRATION 2   The answer in Example 1 can be written as

$$\tfrac{1}{2}x^2(\ln x - \tfrac{1}{2}) + C$$

We check this result by computing the derivative of a product.

$$D_x\left[\frac{1}{2}x^2(\ln x - 1)\right] = \frac{1}{2}x^2\left(\frac{1}{x}\right) + x\left(\ln x - \frac{1}{2}\right)$$

$$= \tfrac{1}{2}x + x \ln x - \tfrac{1}{2}x$$

$$= x \ln x \qquad\qquad ◀$$

EXAMPLE 2

Find $\int x \cos x \, dx$.

SOLUTION

Let $u = x$ and $dv = \cos x \, dx$. Then

$$du = dx \quad \text{and} \quad v = \sin x$$

Therefore

$$\int x \cos x \, dx = x \sin x - \int \sin x \, dx$$

$$= x \sin x + \cos x + C$$

▶ ILLUSTRATION 3   In Example 2, if instead of our choices of u and dv as above we let

$$u = \cos x \quad \text{and} \quad dv = x \, dx$$

then

$$du = -\sin x \, dx \quad \text{and} \quad v = \tfrac{1}{2}x^2$$

Thus

$$\int x \cos x\, dx = \frac{x^2}{2} \cos x + \frac{1}{2} \int x^2 \sin x\, dx$$

The integral on the right is more complicated than the one with which we started, thereby indicating that these are not desirable choices for u and dv.

◀

It may happen that a particular integral requires repeated applications of integration by parts. This is illustrated in the following example.

EXAMPLE 3

Find $\int x^2 e^x\, dx$.

SOLUTION

Let $u = x^2$ and $dv = e^x\, dx$. Then

$$du = 2x\, dx \quad \text{and} \quad v = e^x$$

We have, then,

$$\int x^2 e^x\, dx = x^2 e^x - 2 \int x e^x\, dx$$

We now apply integration by parts to the integral on the right. Let

$$\bar{u} = x \quad \text{and} \quad d\bar{v} = e^x\, dx$$

Then

$$d\bar{u} = dx \quad \text{and} \quad \bar{v} = e^x$$

So we obtain

$$\int x e^x\, dx = x e^x - \int e^x\, dx$$
$$= x e^x - e^x + \bar{C}$$

Therefore

$$\int x^2 e^x\, dx = x^2 e^x - 2(x e^x - e^x + \bar{C})$$
$$= x^2 e^x - 2x e^x + 2e^x + C \qquad \text{where } C = -2\bar{C}$$

Integration by parts often is used when the integrand involves logarithms, inverse trigonometric functions, and products of functions.

EXAMPLE 4

Find $\int \tan^{-1} x\, dx$.

SOLUTION

Let $u = \tan^{-1} x$ and $dv = dx$. Then

$$du = \frac{dx}{1 + x^2} \quad \text{and} \quad v = x$$

Thus

$$\int \tan^{-1} x \, dx = x \tan^{-1} x - \int \frac{x \, dx}{1 + x^2}$$

$$= x \tan^{-1} x - \tfrac{1}{2} \ln(1 + x^2) + C$$

A situation that sometimes occurs when using integration by parts is shown in Example 5.

EXAMPLE 5

Find $\int e^x \sin x \, dx$.

SOLUTION

Let $u = e^x$ and $dv = \sin x \, dx$. Then

$$du = e^x \, dx \quad \text{and} \quad v = -\cos x$$

Therefore

$$\int e^x \sin x \, dx = -e^x \cos x + \int e^x \cos x \, dx$$

The integral on the right is similar to the first integral except that it has $\cos x$ in place of $\sin x$. We apply integration by parts again by letting

$$\bar{u} = e^x \quad \text{and} \quad d\bar{v} = \cos x \, dx$$

So

$$d\bar{u} = e^x \, dx \quad \text{and} \quad \bar{v} = \sin x$$

Thus

$$\int e^x \sin x \, dx = -e^x \cos x + \left(e^x \sin x - \int e^x \sin x \, dx \right)$$

Now on the right there is the same integral as on the left. So we add $\int e^x \sin x \, dx$ to both sides of the equation and obtain

$$2 \int e^x \sin x \, dx = -e^x \cos x + e^x \sin x + 2C$$

Observe that the right side of the above equation contains an arbitrary constant because on the left side we have an indefinite integral. This arbitrary constant is written as $2C$ so that when we divide on both sides of the equation by 2, the arbitrary constant in the answer becomes C. Thus we have

$$\int e^x \sin x \, dx = \tfrac{1}{2} e^x(\sin x - \cos x) + C$$

In applying integration by parts to a specific integral, one pair of choices for u and dv may work while another pair may not. We saw this in Illustration 3, and another case occurs in Illustration 4.

▶ ILLUSTRATION 4   In Example 5, in the step where we have

$$\int e^x \sin x \, dx = -e^x \cos x + \int e^x \cos x \, dx$$

if we evaluate the integral on the right by letting

$$\bar{u} = \cos x \qquad \text{and} \quad d\bar{v} = e^x \, dx$$

$$d\bar{u} = -\sin x \, dx \quad \text{and} \qquad \bar{v} = e^x$$

we get

$$\int e^x \sin x \, dx = -e^x \cos x + \left(e^x \cos x + \int e^x \sin x \, dx \right)$$

$$= \int e^x \sin x \, dx \qquad\qquad \blacktriangleleft$$

EXAMPLE 6
Find the area of the region bounded by the graph of the curve $y = \ln x$, the x axis, and the line $x = e$.

SOLUTION
The region appears in Fig. 8.2.1. If A square units is the area of this region,

$$A = \lim_{\|\Delta\| \to 0} \sum_{i=1}^{n} \ln \xi_i \, \Delta_i x$$

$$= \int_1^e \ln x \, dx$$

To evaluate the integral we use integration by parts with $u = \ln x$ and $dv = dx$. Then $du = \dfrac{dx}{x}$ and $v = x$. Thus

$$A = x \ln x \Big]_1^e - \int_1^e dx$$

$$= x \ln x - x \Big]_1^e$$

$$= (e \ln e - e) - (1 \cdot \ln 1 - 1)$$

$$= e \cdot 1 - e - (1 \cdot 0 - 1)$$

$$= 1$$

Therefore the area is 1 square unit.

FIGURE 8.2.1

EXERCISES 8.2

In Exercises 1 through 24, evaluate the indefinite integral.

1. $\displaystyle\int x e^{3x} \, dx$ **2.** $\displaystyle\int x \cos 2x \, dx$ **3.** $\displaystyle\int x \sec x \tan x \, dx$ **4.** $\displaystyle\int x \, 3^x \, dx$ **5.** $\displaystyle\int \ln(2x) \, dx$

6. $\displaystyle\int \sin^{-1} w \, dw$ **7.** $\displaystyle\int (\ln x)^2 \, dx$ **8.** $\displaystyle\int x \sec^2 x \, dx$ **9.** $\displaystyle\int x \tan^{-1} x \, dx$ **10.** $\displaystyle\int x^2 \ln x \, dx$

11. $\displaystyle\int \frac{x e^x}{(x + 1)^2} \, dx$ **12.** $\displaystyle\int x^2 \sin 3x \, dx$ **13.** $\displaystyle\int \sin x \, \ln(\cos x) \, dx$ **14.** $\displaystyle\int \sin(\ln x) \, dx$

15. $\int e^x \cos x \, dx$

16. $\int x^3 \, e^{x^2} \, dx$ ($u = N^2$)

17. $\int \frac{x^3 \, dx}{\sqrt{1 - x^2}}$ ($u = N^2$)

18. $\int \frac{\sin 2x}{e^x} \, dx$

19. $\int x^2 \sinh x \, dx$

20. $\int \frac{e^{2x}}{\sqrt{1 - e^x}} \, dx$

21. $\int \frac{\cot^{-1} \sqrt{z}}{\sqrt{z}} \, dz$ ($t = \sqrt{z}$, $u = \cot^{-1} u$)

22. $\int \cos^{-1} 2x \, dx$

23. $\int \cos \sqrt{x} \, dx$

24. $\int \tan^{-1} \sqrt{x} \, dx$

In Exercises 25 through 34, evaluate the definite integral.

25. $\int_0^2 x^2 3^x \, dx$

26. $\int_{-1}^2 \ln(x + 2) \, dx$

27. $\int_0^{\pi/3} \sin 3x \cos x \, dx$

28. $\int_0^{\pi^2/2} \cos \sqrt{2x} \, dx$

29. $\int_0^2 x e^{2x} \, dx$

30. $\int_{-\pi}^{\pi} z^2 \cos 2z \, dz$

31. $\int_0^{\pi/4} e^{3x} \sin 4x \, dx$

32. $\int_0^1 x \sin^{-1} x \, dx$

33. $\int_1^4 \sec^{-1} \sqrt{t} \, dt$

34. $\int_{\pi/4}^{3\pi/4} x \cot x \csc x \, dx$

35. Find the area of the region bounded by the curve $y = \ln x$, the x axis, and the line $x = e^2$.

36. Find the volume of the solid generated by revolving the region in Exercise 35 about the x axis.

37. Find the volume of the solid generated by revolving the region in Exercise 35 about the y axis.

38. Find the area of the region bounded by the curve $y = x \csc^2 x$, the x axis, and the lines $x = \frac{1}{6}\pi$ and $x = \frac{1}{4}\pi$.

39. Find the area of the region bounded by the curve $y = 2x e^{-x/2}$, the x axis, and the line $x = 4$.

40. Find the volume of the solid of revolution generated by revolving about the x axis the region of Exercise 39.

41. The linear density of a rod at a point x meters from one end is $2e^{-x}$ kg/m. If the rod is 6 m long, find the mass and center of mass of the rod.

42. Find the centroid of the region bounded by the curve $y = e^x$, the coordinate axes, and the line $x = 3$.

43. Find the centroid of the region in the first quadrant bounded by the curves $y = \sin x$ and $y = \cos x$, and the y axis.

44. The region in the first quadrant bounded by the curve $y = \cos x$ and the lines $y = 1$ and $x = \frac{1}{2}\pi$ is revolved about the line $x = \frac{1}{2}\pi$. Find the volume of the solid generated.

45. A water tank full of water is in the shape of the solid of revolution formed by rotating about the x axis the region bounded by the curve $y = e^{-x}$, the coordinate axes, and the line $x = 4$. Find the work done in pumping all the water to the top of the tank. Distance is measured in feet. Take the positive x axis vertically downward.

46. A particle is moving along a straight line, and s ft is the directed distance of the particle from the origin at t sec. If v ft/sec is the velocity at t sec, $s = 0$ when $t = 0$, and $v \cdot s = t \sin t$, find s in terms of t and also s when $t = \frac{1}{2}\pi$.

47. (a) Show that if $m \geq 2$ and m is an integer,

$$\int \sec^m x \, dx = \frac{\sec^{m-2} x \tan x}{m - 1} + \frac{m - 2}{m - 1} \int \sec^{m-2} x \, dx$$

(b) Use the formula in part (a) to find $\int \sec^6 x \, dx$.

48. (a) Derive the following formula, where r is any real number:

$$\int x^r e^x \, dx = x^r e^x - r \int x^{r-1} e^x \, dx$$

(b) Use the formula derived in (a) to find $\int x^4 e^x \, dx$.

49. (a) Derive the following formula, where r is any real number:

$$\int x^r \ln x \, dx = \begin{cases} \dfrac{x^{r+1}}{r + 1} \ln x - \dfrac{x^{r+1}}{(r + 1)^2} + C & \text{if } r \neq -1 \\ \frac{1}{2}(\ln x)^2 + C & \text{if } r = -1 \end{cases}$$

(b) Use the formula in part (a) to find $\int x^3 \ln x \, dx$.

50. (a) Derive the following formula, where r and q are any real numbers:

$$\int x^r (\ln x)^q \, dx = \begin{cases} \dfrac{x^{r+1}(\ln x)^q}{r+1} - \dfrac{q}{r+1} \displaystyle\int x^r (\ln x)^{q-1} \, dx & \text{if } r \neq -1 \\[4mm] \dfrac{(\ln x)^{q+1}}{q+1} + C & \text{if } r = -1 \text{ and } q \neq -1 \end{cases}$$

(b) Use the formula derived in (a) to find $\int x^4 (\ln x)^2 \, dx$.

Exercises 51 through 56 pertain to the indicated Supplementary Section.

51. *Section 4.10.* The marginal cost function is C' and $C'(x) = \ln x$, where $x > 1$. Find the total cost function if $C(x)$ dollars is the total cost of producing x units and $C(1) = 5$.

52. *Section 4.10.* A manufacturer has discovered that if $100x$ units of a particular commodity are produced per week, the marginal cost is determined by $x2^{x/2}$ and the marginal revenue is determined by $8 \cdot 2^{-x/2}$, where both the production cost and the revenue are in thousands of dollars. If the weekly fixed costs amount to \$2000, find the maximum weekly profit that can be obtained.

53. *Section 5.8.* Find the centroid of the solid of revolution generated by revolving about the x axis the region bounded by the curve $y = e^x$ and the lines $x = 1$ and $y = 1$.

54. *Section 5.8.* Find the centroid of the solid of revolution generated if the region bounded by $y = e^{-x}$, the coordinate axes, and the line $x = 1$ is revolved about the x axis.

55. *Section 5.9.* A board is in the shape of a region bounded by a straight line and one arch of the sine curve. If the board is submerged vertically in water so that the straight line is the lower boundary 2 ft below the surface of the water, find the force on the board due to water pressure.

56. *Section 5.8.* Find the centroid of the solid of revolution generated if the region of Exercise 55 is revolved about the y axis.

8.3
INTEGRATION OF POWERS OF SINE AND COSINE

We consider four cases of indefinite integrals involving powers of sine and cosine. The method used in each case is shown by an illustration.

Case 1: $\int \sin^n u \, du$ or $\int \cos^n u \, du$, where n is an odd integer.

▶ ILLUSTRATION 1

$$\int \cos^3 x \, dx = \int \cos^2 x (\cos x \, dx)$$

$$= \int (1 - \sin^2 x)(\cos x \, dx)$$

So

$$\int \cos^3 x \, dx = \int \cos x \, dx - \int \sin^2 x \cos x \, dx \qquad (1)$$

For the second integral on the right side of (1) observe that because $d(\sin x) = \cos x \, dx$, we have

$$\int \sin^2 x (\cos x \, dx) = \tfrac{1}{3} \sin^3 x + C_1$$

Because the first integral on the right side of (1) is $\sin x + C_2$,

$$\int \cos^3 x \, dx = \sin x - \tfrac{1}{3} \sin^3 x + C \qquad \blacktriangleleft$$

EXAMPLE 1

Find $\int \sin^5 x \, dx$.

SOLUTION

$$\int \sin^5 x \, dx = \int (\sin^2 x)^2 \sin x \, dx$$

$$= \int (1 - \cos^2 x)^2 \sin x \, dx$$

$$= \int (1 - 2 \cos^2 x + \cos^4 x) \sin x \, dx$$

$$= \int \sin x \, dx - 2 \int \cos^2 x \sin x \, dx + \int \cos^4 x \sin x \, dx$$

$$= -\cos x + 2 \int \cos^2 x (-\sin x \, dx) - \int \cos^4 x (-\sin x \, dx)$$

$$= -\cos x + \tfrac{2}{3} \cos^3 x - \tfrac{1}{5} \cos^5 x + C$$

Case 2: $\int \sin^n u \, du$ and $\int \cos^n u \, du$, where n is an even integer.

The method used in Case 1 does not work for this case. We use the following identities from trigonometry:

$$\sin^2 x = \frac{1 - \cos 2x}{2} \qquad \cos^2 x = \frac{1 + \cos 2x}{2}$$

▶ ILLUSTRATION 2

$$\int \sin^2 x \, dx = \int \frac{1 - \cos 2x}{2} \, dx$$

$$= \tfrac{1}{2} x - \tfrac{1}{4} \sin 2x + C \qquad\qquad ◀$$

EXAMPLE 2

Find $\int \cos^4 x \, dx$.

SOLUTION

$$\int \cos^4 x \, dx = \int (\cos^2 x)^2 \, dx$$

$$= \int \left(\frac{1 + \cos 2x}{2} \right)^2 dx$$

$$= \frac{1}{4} \int dx + \frac{1}{2} \int \cos 2x \, dx + \frac{1}{4} \int \cos^2 2x \, dx$$

$$= \frac{1}{4} x + \frac{1}{4} \sin 2x + \frac{1}{4} \int \frac{1 + \cos 4}{2} x \, dx$$

$$= \tfrac{1}{4} x + \tfrac{1}{4} \sin 2x + \tfrac{1}{8} x + \tfrac{1}{32} \sin 4x + C$$

$$= \tfrac{3}{8} x + \tfrac{1}{4} \sin 2x + \tfrac{1}{32} \sin 4x + C$$

Case 3: $\int \sin^n x \cos^m x \, dx$, where at least one of the exponents is odd.

The solution of this case is similar to the method used for Case 1.

▶ ILLUSTRATION 3

$$\int \sin^3 x \cos^4 x \, dx = \int \sin^2 x \cos^4 x (\sin x \, dx)$$

$$= \int (1 - \cos^2 x) \cos^4 x (\sin x \, dx)$$

$$= \int \cos^4 x \sin x \, dx - \int \cos^6 x \sin x \, dx$$

$$= -\tfrac{1}{5} \cos^5 x + \tfrac{1}{7} \cos^7 x + C$$ ◀

Case 4: $\int \sin^n x \cos^m x \, dx$, where both m and n are even.

The solution of this case is similar to the method used for Case 2.

▶ ILLUSTRATION 4

$$\int \sin^2 x \cos^4 x \, dx$$

$$= \int \left(\frac{1 - \cos 2x}{2} \right) \left(\frac{1 + \cos 2x}{2} \right)^2 dx$$

$$= \frac{1}{8} \int dx + \frac{1}{8} \int \cos 2x \, dx - \frac{1}{8} \int \cos^2 2x \, dx - \frac{1}{8} \int \cos^3 2x \, dx$$

$$= \frac{1}{8} x + \frac{1}{16} \sin 2x - \frac{1}{8} \int \frac{1 + \cos 4x}{2} \, dx - \frac{1}{8} \int (1 - \sin^2 2x) \cos 2x \, dx$$

$$= \frac{x}{8} + \frac{\sin 2x}{16} - \frac{x}{16} - \frac{\sin 4x}{64} - \frac{1}{8} \int \cos 2x \, dx + \frac{1}{8} \int \sin^2 2x \cos 2x \, dx$$

$$= \frac{x}{16} + \frac{\sin 2x}{16} - \frac{\sin 4x}{64} - \frac{\sin 2x}{16} + \frac{\sin^3 2x}{48} + C$$

$$= \frac{x}{16} + \frac{\sin^3 2x}{48} - \frac{\sin 4x}{64} + C$$ ◀

EXAMPLE 3
Find $\int \sin^4 x \cos^4 x \, dx$.

SOLUTION
If we use the identity

$$\sin x \cos x = \tfrac{1}{2} \sin 2x$$

we have

$\left(\sin x \cos x \right)^4 = \left(\tfrac{1}{2} \sin 2x \right)^4$

$= \tfrac{1}{16} \sin^4 2x$

$$\int \sin^4 x \cos^4 x \, dx = \frac{1}{16} \int \sin^4 2x \, dx$$

$$= \frac{1}{16} \int \left(\frac{1 - \cos 4x}{2}\right)^2 dx$$

$$= \frac{1}{64} \int dx - \frac{1}{32} \int \cos 4x \, dx + \frac{1}{64} \int \cos^2 4x \, dx$$

$$= \frac{x}{64} - \frac{\sin 4x}{128} + \frac{1}{64} \int \frac{1 + \cos 8x}{2} \, dx$$

$$= \frac{x}{64} - \frac{\sin 4x}{128} + \frac{x}{128} + \frac{\sin 8x}{1024} + C$$

$$= \frac{3x}{128} - \frac{\sin 4x}{128} + \frac{\sin 8x}{1024} + C$$

EXAMPLE 4

Find the centroid of the region in the first quadrant to the left of the line $x = \frac{1}{2}\pi$ and bounded by the curve $y = \sin x$, the x axis, and the line $x = \frac{1}{2}\pi$.

SOLUTION

We use the symbols A, M_x, M_y, $\bar{x}$, and $\bar{y}$, as defined in Section 5.7. The region and the ith rectangular element are shown in Fig. 8.3.1. We first compute the area of the region.

$$A = \lim_{||\Delta|| \to 0} \sum_{i=1}^{n} \sin \gamma_i \Delta_i x$$

$$= \int_0^{\pi/2} \sin x \, dx$$

$$= -\cos x \Big]_0^{\pi/2}$$

$$= 1$$

We apply Definition 5.7.2 to compute M_x and M_y. To evaluate the integral for M_y we use integration by parts with $u = x$ and $dv = \sin x \, dx$.

$$M_x = \lim_{||\Delta|| \to 0} \sum_{i=1}^{n} \tfrac{1}{2}[\sin \gamma_i]^2 \Delta_i x \qquad M_y = \lim_{||\Delta|| \to 0} \sum_{i=1}^{n} \gamma_i \sin \gamma_i \Delta_i x$$

$$= \frac{1}{2} \int_0^{\pi/2} \sin^2 x \, dx \qquad\qquad = \int_0^{\pi/2} x \sin x \, dx$$

$$= \frac{1}{2} \int_0^{\pi/2} \frac{1 - \cos 2x}{2} \, dx \qquad = -x \cos x \Big]_0^{\pi/2} + \int_0^{\pi/2} \cos x \, dx$$

$$= \tfrac{1}{4}\left[x - \tfrac{1}{2}\sin 2x\right]_0^{\pi/2} \qquad = -x \cos x + \sin x \Big]_0^{\pi/2}$$

$$= \tfrac{1}{4}\left[\tfrac{1}{2}\pi\right] \qquad\qquad\qquad = -\tfrac{1}{2}\pi \cos \tfrac{1}{2}\pi + \sin \tfrac{1}{2}\pi$$

$$= \tfrac{1}{8}\pi \qquad\qquad\qquad\qquad = 1$$

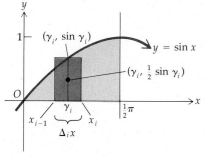

FIGURE 8.3.1

Therefore

$$\bar{x} = \frac{M_y}{A} \qquad \bar{y} = \frac{M_x}{A}$$

$$= \frac{1}{1} \qquad = \frac{\frac{1}{8}\pi}{1}$$

$$= 1 \qquad = \frac{1}{8}\pi$$

Thus the centroid is at the point $(1, \frac{1}{8}\pi)$.

The next example involves another type of integral containing a product of a sine and a cosine.

EXAMPLE 5

Find $\int \sin 3x \cos 2x \, dx$.

SOLUTION
We use the following identity from trigonometry:

$$\sin mx \cos nx = \tfrac{1}{2}\sin(m - n)x + \tfrac{1}{2}\sin(m + n)x$$

So

$$\int \sin 3x \cos 2x \, dx = \int (\tfrac{1}{2}\sin x + \tfrac{1}{2}\sin 5x)\, dx$$

$$= \tfrac{1}{2}\int \sin x \, dx + \tfrac{1}{2}\int \sin 5x \, dx$$

$$= -\tfrac{1}{2}\cos x - \tfrac{1}{10}\cos 5x + C$$

EXERCISES 8.3

In Exercises 1 through 24, evaluate the indefinite integral.

1. $\int \sin^4 x \cos x \, dx$

2. $\int \sin^5 x \cos x \, dx$

3. $\int \cos^3 4x \sin 4x \, dx$

4. $\int \cos^6 \tfrac{1}{2}x \sin \tfrac{1}{2}x \, dx$

5. $\int \sin^3 x \, dx$

6. $\int \sin^2 3x \, dx$

7. $\int \sin^4 z \, dz$

8. $\int \cos^5 x \, dx$

9. $\int \cos^2 \tfrac{1}{2}x \, dx$

10. $\int \sin^3 x \cos^3 x \, dx$

11. $\int \sin^2 x \cos^3 x \, dx$

12. $\int \cos^6 x \, dx$

13. $\int \sin^5 x \cos^2 x \, dx$

14. $\int \sin^2 2t \cos^4 2t \, dt$

15. $\int \sin^2 3t \cos^2 3t \, dt$

16. $\int \sqrt{\cos z} \sin^3 z \, dz$

17. $\int \dfrac{\cos^3 3x}{\sqrt[3]{\sin 3x}} \, dx$

18. $\int \sin^3 \tfrac{1}{2}y \cos^2 \tfrac{1}{2}y \, dy$

19. $\int \cos 4x \cos 3x \, dx$

20. $\int \sin 2x \cos 4x \, dx$

21. $\int \sin 3y \cos 5y \, dy$

22. $\int \cos t \cos 3t \, dt$

23. $\int (\sin 3t - \sin 2t)^2 \, dt$

24. $\int \sin x \sin 3x \sin 5x \, dx$

In Exercises 25 through 32, evaluate the definite integral.

25. $\int_0^{\pi/2} \cos^3 x \, dx$

26. $\int_0^1 \sin^3 \tfrac{1}{2}\pi t \, dt$

27. $\int_0^1 \sin^4 \tfrac{1}{2}\pi x \, dx$

28. $\int_0^{\pi/3} \sin^3 t \cos^2 t \, dt$

$\left(\text{and } \frac{\partial \pi t}{2}\right)^2$

29. $\int_0^1 \sin^2 \pi t \cos^2 \pi t \, dt$ 30. $\int_0^{\pi/6} \sin 2x \cos 4x \, dx$ 31. $\int_0^{\pi/8} \sin 3x \cos 5x \, dx$ 32. $\int_0^{\pi/2} \sin^2 \frac{1}{2}x \cos^2 \frac{1}{2}x \, dx$

33. Evaluate $\int 2 \sin x \cos x \, dx$ by three methods: (a) Make the substitution $u = \sin x$; (b) make the substitution $u = \cos x$; (c) use the identity $2 \sin x \cos x = \sin 2x$. Explain the difference in appearance of the answers obtained in (a), (b), and (c).

34. If n is any positive integer, prove that $\int_0^\pi \sin^2 nx \, dx = \frac{1}{2}\pi$.

35. If n is a positive odd integer, prove that $\int_0^\pi \cos^n x \, dx = 0$.

In Exercises 36 through 38, m and n are any positive integers; show that the given formula is true.

36. $\int_{-1}^1 \cos n\pi x \cos m\pi x \, dx = \begin{cases} 0 & \text{if } m \neq n \\ 1 & \text{if } m = n \end{cases}$ 37. $\int_{-1}^1 \cos n\pi x \sin m\pi x \, dx = 0$

38. $\int_{-1}^1 \sin n\pi x \sin m\pi x \, dx = \begin{cases} 0 & \text{if } m \neq n \\ 1 & \text{if } m = n \end{cases}$

39. Find the area of the region bounded by the curve $y = \sin^2 x$ and the x axis from $x = 0$ to $x = \pi$.
40. Find the volume of the solid of revolution generated by revolving one arch of the sine curve about the x axis.
41. Find the volume of the solid of revolution generated if the region of Exercise 39 is revolved about the x axis.
42. The region in the first quadrant bounded by the curve $y = \cos x$ and the lines $y = 1$ and $x = \frac{1}{2}\pi$ is revolved about the x axis. Find the volume of the solid generated.
43. Find the volume of the solid of revolution generated if the region of Exercise 39 is revolved about the line $y = 1$.
44. The region bounded by the y axis and the curves $y = \sin x$ and $y = \cos x$ for $0 \leq x \leq \frac{1}{4}\pi$ is revolved about the x axis. Find the volume of the solid of revolution generated.
45. Find the centroid of the region from $x = 1$ to $x = \frac{1}{2}\pi$ bounded by the curve $y = \cos x$ and the coordinate axes.
46. Find the centroid of the region described in Exercise 44.

Exercises 47 and 48 pertain to the indicated Supplementary Section.
47. *Section 5.8.* Find the centroid of the solid of revolution generated if the region bounded by $y = \sin x$, the x axis, and the line $x = \frac{1}{2}\pi$ is revolved about the x axis.
48. *Section 5.9.* The face of a dam is in the shape of one arch of the curve $y = \frac{1}{100} \cos \frac{1}{200}\pi x$, and the surface of the water is at the top of the dam. Find the force due to water pressure on the face of the dam if distance is measured in feet.

8.4
INTEGRATION OF POWERS
OF TANGENT, COTANGENT,
SECANT, AND COSECANT

Recall the following integration formulas involving tangent, cotangent, secant, and cosecant:

$$\int \tan u \, du = \ln|\sec u| + C \qquad \int \cot u \, du = \ln|\sin u| + C$$

$$\int \sec u \, du = \ln|\sec u + \tan u| + C \qquad \int \csc u \, du = \ln|\csc u - \cot u| + C$$

$$\int \sec^2 u \, du = \tan u + C \qquad \int \csc^2 u \, du = -\cot u + C$$

$$\int \sec u \tan u \, du = \sec u + C \qquad \int \csc u \cot u \, du = -\csc u + C$$

With these formulas and the trigonometric identities

$$1 + \tan^2 u = \sec^2 u \qquad 1 + \cot^2 u = \csc^2 u$$

we can evaluate integrals of the form

$$\int \tan^m u \sec^n u \, du \quad \text{and} \quad \int \cot^m u \csc^n u \, du \tag{1}$$

where m and n are nonnegative integers.

▶ ILLUSTRATION 1

(a) $\displaystyle\int \tan^2 x \, dx = \int (\sec^2 x - 1) \, dx$ (b) $\displaystyle\int \cot^2 x \, dx = \int (\csc^2 x - 1) \, dx$

$$= \int \sec^2 x \, dx - \int dx \qquad\qquad = \int \csc^2 x \, dx - \int dx$$

$$= \tan x - x + C \qquad\qquad\qquad = -\cot x - x + C$$

◀

We now distinguish various cases of integrals of the form (1).

Case 1: $\displaystyle\int \tan^n u \, du$ or $\displaystyle\int \cot^n u \, du$, where n is a positive integer.
We write

$$\tan^n u = \tan^{n-2} u \tan^2 u \qquad\qquad \cot^n u = \cot^{n-2} u \cot^2 u$$

$$= \tan^{n-2} u(\sec^2 u - 1) \qquad\qquad = \cot^{n-2} u(\csc^2 u - 1)$$

EXAMPLE 1

Find $\displaystyle\int \tan^3 x \, dx$.

SOLUTION

$$\int \tan^3 x \, dx = \int \tan x(\sec^2 x - 1) \, dx$$

$$= \int \tan x \sec^2 x \, dx - \int \tan x \, dx$$

$$= \tfrac{1}{2} \tan^2 x + \ln|\cos x| + C$$

EXAMPLE 2

Find $\displaystyle\int \cot^4 3x \, dx$.

SOLUTION

$$\int \cot^4 3x \, dx = \int \cot^2 3x(\csc^2 3x - 1) \, dx$$

$$= \int \cot^2 3x \csc^2 3x \, dx - \int \cot^2 3x \, dx$$

$$= \tfrac{1}{9}(-\cot^3 3x) - \int (\csc^2 3x - 1) \, dx$$

$$= -\tfrac{1}{9} \cot^3 3x + \tfrac{1}{3} \cot 3x + x + C$$

Case 2: $\int \sec^n u \, du$ or $\int \csc^n u \, du$, where n is a positive even integer.

We write

$$\sec^n u = \sec^{n-2} u \sec^2 u \qquad\qquad \csc^n u = \csc^{n-2} u \csc^2 u$$
$$= (\tan^2 u + 1)^{(n-2)/2} \sec^2 u \qquad = (\cot^2 u + 1)^{(n-2)/2} \csc^2 u$$

EXAMPLE 3

Find $\int \csc^6 x \, dx$.

SOLUTION

$$\int \csc^6 x \, dx = \int (\cot^2 x + 1)^2 \csc^2 x \, dx$$
$$= \int \cot^4 x \csc^2 x \, dx + 2 \int \cot^2 x \csc^2 x \, dx + \int \csc^2 x \, dx$$
$$= -\tfrac{1}{5} \cot^5 x - \tfrac{2}{3} \cot^3 x - \cot x + C$$

Case 3: $\int \sec^n u \, du$ or $\int \csc^n u \, du$, where n is a positive odd integer.

To integrate odd powers of the secant and cosecant we use integration by parts. The process is illustrated in the following example.

EXAMPLE 4

Find $\int \sec^3 x \, dx$.

SOLUTION

Let $u = \sec x$ and $dv = \sec^2 x \, dx$. Then

$$du = \sec x \tan x \, dx \quad \text{and} \quad v = \tan x$$

Therefore

$$\int \sec^3 x \, dx = \sec x \tan x - \int \sec x \tan^2 x \, dx$$
$$\int \sec^3 x \, dx = \sec x \tan x - \int \sec x (\sec^2 x - 1) \, dx$$
$$\int \sec^3 x \, dx = \sec x \tan x - \int \sec^3 x \, dx + \int \sec x \, dx$$

Adding $\int \sec^3 x \, dx$ to both sides we get

$$2 \int \sec^3 x \, dx = \sec x \tan x + \ln|\sec x + \tan x| + 2C$$
$$\int \sec^3 x \, dx = \tfrac{1}{2} \sec x \tan x + \tfrac{1}{2} \ln|\sec x + \tan x| + C$$

Case 4: $\int \tan^m u \sec^n u \, du$ or $\int \cot^m u \csc^n u \, du$, where n is a positive even integer.

This case is illustrated by the following example.

EXAMPLE 5

Find $\int \tan^5 x \sec^4 x \, dx$.

SOLUTION

$$\int \tan^5 x \sec^4 x \, dx = \int \tan^5 x (\tan^2 x + 1) \sec^2 x \, dx$$

$$= \int \tan^7 x \sec^2 x \, dx + \int \tan^5 x \sec^2 x \, dx$$

$$= \tfrac{1}{8} \tan^8 x + \tfrac{1}{6} \tan^6 x + C$$

Case 5: $\int \tan^m u \sec^n u \, du$ or $\int \cot^m u \csc^n u \, du$, where n is a positive odd integer.

The next example illustrates this case.

EXAMPLE 6

Find $\int \tan^5 x \sec^7 x \, dx$.

SOLUTION

$$\int \tan^5 x \sec^7 x \, dx = \int \tan^4 x \sec^6 x \sec x \tan x \, dx$$

$$= \int (\sec^2 x - 1)^2 \sec^6 x (\sec x \tan x \, dx)$$

$$= \int \sec^{10} x (\sec x \tan x \, dx) - 2 \int \sec^8 x (\sec x \tan x \, dx)$$

$$+ \int \sec^6 x (\sec x \tan x \, dx)$$

$$= \tfrac{1}{11} \sec^{11} x - \tfrac{2}{9} \sec^9 x + \tfrac{1}{7} \sec^7 x + C$$

Case 6: $\int \tan^m u \sec^n u \, du$ or $\int \cot^m u \csc^n u \, du$, where m is a positive even integer and n is a positive odd integer.

The integrand can be expressed in terms of odd powers of secant or cosecant. For example,

$$\int \tan^2 x \sec^3 x \, dx = \int (\sec^2 x - 1) \sec^3 x \, dx$$

$$= \int \sec^5 x \, dx - \int \sec^3 x \, dx$$

To evaluate each of these integrals we use integration by parts, as indicated in case 3.

EXERCISES 8.4

In Exercises 1 through 30, evaluate the indefinite integral.

1. $\int \tan^2 5x \, dx$ **2.** $\int \cot^2 4t \, dt$ **3.** $\int x \cot^2 2x^2 \, dx$ **4.** $\int e^x \tan^2(e^x) \, dx$ **5.** $\int \cot^3 t \, dt$

6. $\int \tan^4 x \, dx$ **7.** $\int \tan^6 3x \, dx$ **8.** $\int \cot^5 2x \, dx$ **9.** $\int \sec^4 x \, dx$ **10.** $\int \csc^4 x \, dx$

11. $\int \csc^3 x \, dx$ **12.** $\int \sec^5 x \, dx$ **13.** $\int e^x \tan^4(e^x) \, dx$ **14.** $\int \dfrac{\sec^4(\ln x)}{x} \, dx$ **15.** $\int \tan^6 x \sec^4 x \, dx$

16. $\int \tan^5 x \sec^3 x \, dx$ **17.** $\int \cot^2 3x \csc^4 3x \, dx$ **18.** $\int (\sec 5x + \csc 5x)^2 \, dx$ **19.** $\int (\tan 2x + \cot 2x)^2 \, dx$

20. $\int \dfrac{dx}{1 + \cos x}$ **21.** $\int \dfrac{2 \sin w - 1}{\cos^2 w} \, dw$ **22.** $\int \dfrac{\tan^3 \sqrt{x}}{\sqrt{x}} \, dx$ **23.** $\int \tan^5 3x \, dx$

24. $\int \dfrac{\tan^4 y}{\sec^5 y} \, dy$ **25.** $\int \dfrac{du}{1 + \sec \frac{1}{2}u}$ **26.** $\int \dfrac{\csc^4 x}{\cot^2 x} \, dx$ **27.** $\int \dfrac{\sec^3 x}{\tan^4 x} \, dx$

28. $\int \dfrac{\sin^2 \pi x}{\cos^6 \pi x} \, dx$ **29.** $\int \dfrac{\tan^3(\ln x) \sec^6(\ln x)}{x} \, dx$ **30.** $\int \dfrac{\sec^4 w}{\sqrt{\tan w}} \, dw$

In Exercises 31 through 36, evaluate the definite integral.

31. $\int_{\pi/16}^{\pi/12} \tan^3 4x \, dx$ **32.** $\int_{\pi/8}^{\pi/6} 3 \sec^4 2t \, dt$ **33.** $\int_{-\pi/4}^{\pi/4} \sec^6 x \, dx$

34. $\int_0^{\pi/3} \dfrac{\tan^3 x}{\sec x} \, dx$ **35.** $\int_{\pi/4}^{\pi/2} \dfrac{\cos^4 t}{\sin^6 t} \, dt$ **36.** $\int_{\pi/6}^{\pi/4} \cot^3 w \, dw$

37. Find the area of the region bounded by the curve $y = \tan^2 x$, the x axis, and the line $x = \frac{1}{4}\pi$.

38. Find the volume of the solid of revolution generated if the region bounded by the curve $y = 3 \csc^3 x$, the x axis, and the lines $x = \frac{1}{6}\pi$ and $x = \frac{1}{2}\pi$ is revolved about the x axis.

39. Find the volume of the solid of revolution generated if the region bounded by the curve $y = \sec^2 x$, the axes, and the line $x = \frac{1}{4}\pi$ is revolved about the x axis.

40. Prove: $\int \cot x \csc^n x \, dx = -\dfrac{\csc^n x}{n} + C$ if $n \neq 0$.

41. Prove: $\int \tan^n x \, dx = \dfrac{\tan^{n-1} x}{n - 1} - \int \tan^{n-2} x \, dx$ if n is a positive integer greater than 1.

42. Derive a formula similar to that in Exercise 40 for $\int \tan x \sec^n x \, dx$ if $n \neq 0$.

43. Derive a formula similar to that in Exercise 41 for $\int \cot^n x \, dx$, if n is a positive integer greater than 1.

8.5
INTEGRATION BY TRIGONOMETRIC SUBSTITUTION

If the integrand contains an expression of the form $\sqrt{a^2 - u^2}$, $\sqrt{a^2 + u^2}$, or $\sqrt{u^2 - a^2}$, where $a > 0$, it is often possible to perform the integration by making a trigonometric substitution that results in an integral involving trigonometric functions. We consider each form as a separate case.

Case 1: The integrand contains an expression of the form $\sqrt{a^2 - u^2}$, where $a > 0$.

We introduce a new variable θ by letting $u = a \sin \theta$, where

$$0 \leq \theta \leq \tfrac{1}{2}\pi \quad \text{if } u \geq 0 \quad \text{and} \quad -\tfrac{1}{2}\pi \leq \theta < 0 \quad \text{if } u < 0$$

Then $du = a \cos \theta \, d\theta$, and

$$\begin{aligned} \sqrt{a^2 - u^2} &= \sqrt{a^2 - a^2 \sin^2 \theta} \\ &= \sqrt{a^2(1 - \sin^2 \theta)} \\ &= a\sqrt{\cos^2 \theta} \end{aligned}$$

Because $-\tfrac{1}{2}\pi \leq \theta \leq \tfrac{1}{2}\pi$, $\cos \theta \geq 0$. Hence $\sqrt{\cos^2 \theta} = \cos \theta$, and

$$\sqrt{a^2 - u^2} = a \cos \theta$$

Because $\sin \theta = u/a$ and $-\tfrac{1}{2}\pi \leq \theta \leq \tfrac{1}{2}\pi$,

$$\theta = \sin^{-1} \frac{u}{a}$$

EXAMPLE 1
Evaluate

$$\int \frac{\sqrt{9 - x^2}}{x^2} \, dx$$

SOLUTION
Let $x = 3 \sin \theta$, where $0 < \theta \leq \tfrac{1}{2}\pi$ if $x > 0$ and $-\tfrac{1}{2}\pi \leq \theta < 0$ if $x < 0$. Then $dx = 3 \cos \theta \, d\theta$ and

$$\begin{aligned} \sqrt{9 - x^2} &= \sqrt{9 - 9 \sin^2 \theta} \\ &= 3\sqrt{\cos^2 \theta} \\ &= 3 \cos \theta \end{aligned}$$

Therefore

$$\int \frac{\sqrt{9 - x^2}}{x^2} \, dx = \int \frac{3 \cos \theta}{9 \sin^2 \theta} (3 \cos \theta \, d\theta)$$

$$= \int \cot^2 \theta \, d\theta$$

$$= \int (\csc^2 \theta - 1) \, d\theta$$

$$= -\cot \theta - \theta + C$$

Because $\sin \theta = \tfrac{1}{3}x$ and $-\tfrac{1}{2}\pi \leq \theta \leq \tfrac{1}{2}\pi$, $\theta = \sin^{-1} \tfrac{1}{3}x$. To find $\cot \theta$, refer to Figs. 8.5.1 (for $x > 0$) and 8.5.2 (for $x < 0$). Observe that in either case $\cot \theta = \sqrt{9 - x^2}/x$. Therefore

$$\int \frac{\sqrt{9 - x^2}}{x^2} \, dx = -\frac{\sqrt{9 - x^2}}{x} - \sin^{-1} \frac{x}{3} + C$$

Case 2: The integrand contains an expression of the form $\sqrt{a^2 + u^2}$, where $a > 0$.

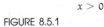

$x > 0$

FIGURE 8.5.1

$x < 0$

FIGURE 8.5.2

We introduce a new variable θ by letting $u = a \tan \theta$, where

$$0 \le \theta < \tfrac{1}{2}\pi \quad \text{if } u \ge 0 \quad \text{and} \quad -\tfrac{1}{2}\pi < \theta < 0 \quad \text{if } u < 0$$

Then $du = a \sec^2 \theta \, d\theta$, and

$$\begin{aligned}
\sqrt{a^2 + u^2} &= \sqrt{a^2 + a^2 \tan^2 \theta} \\
&= a\sqrt{1 + \tan^2 \theta} \\
&= a\sqrt{\sec^2 \theta}
\end{aligned}$$

Because $-\tfrac{1}{2}\pi < \theta < \tfrac{1}{2}\pi$, $\sec \theta \ge 1$. Thus $\sqrt{\sec^2 \theta} = \sec \theta$, and

$$\sqrt{a^2 + u^2} = a \sec \theta$$

Because $\tan \theta = u/a$ and $-\tfrac{1}{2}\pi < \theta < \tfrac{1}{2}\pi$,

$$\theta = \tan^{-1} \frac{u}{a}$$

EXAMPLE 2
Evaluate $\displaystyle\int \sqrt{x^2 + 5} \, dx$.

SOLUTION
Substitute $x = \sqrt{5} \tan \theta$, where $0 \le \theta < \tfrac{1}{2}\pi$ if $x \ge 0$ and $-\tfrac{1}{2}\pi < \theta < 0$ if $x < 0$. Then $dx = \sqrt{5} \sec^2 \theta \, d\theta$ and

$$\begin{aligned}
\sqrt{x^2 + 5} &= \sqrt{5 \tan^2 \theta + 5} \\
&= \sqrt{5}\sqrt{\sec^2 \theta} \\
&= \sqrt{5} \sec \theta
\end{aligned}$$

Therefore

$$\begin{aligned}
\int \sqrt{x^2 + 5} \, dx &= \int \sqrt{5} \sec \theta (\sqrt{5} \sec^2 \theta \, d\theta) \\
&= 5 \int \sec^3 \theta \, d\theta
\end{aligned}$$

Using the result of Example 4 of Section 8.4 we have

$$\int \sqrt{x^2 + 5} \, dx = \tfrac{5}{2} \sec \theta \tan \theta + \tfrac{5}{2} \ln|\sec \theta + \tan \theta| + C$$

We determine $\sec \theta$ from Figs. 8.5.3 (for $x \ge 0$) and 8.5.4 (for $x < 0$), where $\tan \theta = x/\sqrt{5}$. We see in either case that $\sec \theta = \sqrt{x^2 + 5}/\sqrt{5}$. Hence

$$\begin{aligned}
\int \sqrt{x^2 + 5} \, dx &= \frac{5}{2} \cdot \frac{\sqrt{x^2 + 5}}{\sqrt{5}} \cdot \frac{x}{\sqrt{5}} + \frac{5}{2} \ln\left|\frac{\sqrt{x^2 + 5}}{\sqrt{5}} + \frac{x}{\sqrt{5}}\right| + C \\
&= \tfrac{1}{2}x\sqrt{x^2 + 5} + \tfrac{5}{2} \ln|\sqrt{x^2 + 5} + x| - \tfrac{5}{2} \ln\sqrt{5} + C \\
&= \tfrac{1}{2}x\sqrt{x^2 + 5} + \tfrac{5}{2} \ln(\sqrt{x^2 + 5} + x) + C_1
\end{aligned}$$

Observe that we replaced $-\tfrac{5}{2} \ln \sqrt{5} + C$ by the arbitrary constant C_1. Furthermore, because $\sqrt{x^2 + 5} + x > 0$, we drop the absolute-value bars.

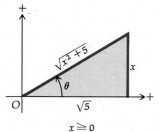

$x \ge 0$

FIGURE 8.5.3

$x < 0$

FIGURE 8.5.4

Case 3: The integrand contains an expression of the form $\sqrt{u^2 - a^2}$, where $a > 0$.

We introduce a new variable by letting $u = a \sec \theta$, where

$$0 \le \theta < \tfrac{1}{2}\pi \quad \text{if } u \ge a \quad \text{and} \quad \pi \le \theta < \tfrac{3}{2}\pi \quad \text{if } u \le -a$$

Then $du = a \sec \theta \tan \theta \, d\theta$ and

$$\sqrt{u^2 - a^2} = \sqrt{a^2 \sec^2 \theta - a^2}$$
$$= \sqrt{a^2(\sec^2 \theta - 1)}$$
$$= a\sqrt{\tan^2 \theta}$$

Because either $0 \le \theta < \tfrac{1}{2}\pi$ or $\pi \le \theta < \tfrac{3}{2}\pi$, $\tan \theta \ge 0$. Thus $\sqrt{\tan^2 \theta} = \tan \theta$, and we have

$$\sqrt{u^2 - a^2} = a \tan \theta$$

Because $\sec \theta = u/a$ and θ is in $[0, \tfrac{1}{2}\pi) \cup [\pi, \tfrac{3}{2}\pi)$,

$$\theta = \sec^{-1} \frac{u}{a}$$

EXAMPLE 3
Evaluate

$$\int \frac{dx}{x^3 \sqrt{x^2 - 9}}$$

SOLUTION
Let $x = 3 \sec \theta$, where $0 < \theta < \tfrac{1}{2}\pi$ if $x > 3$ and $\pi < \theta < \tfrac{3}{2}\pi$ if $x < -3$. Then $dx = 3 \sec \theta \tan \theta \, d\theta$ and

$$\sqrt{x^2 - 9} = \sqrt{9 \sec^2 \theta - 9}$$
$$= 3\sqrt{\tan^2 \theta}$$
$$= 3 \tan \theta$$

Hence

$$\int \frac{dx}{x^3 \sqrt{x^2 - 9}} = \int \frac{3 \sec \theta \tan \theta \, d\theta}{27 \sec^3 \theta \cdot 3 \tan \theta}$$

$$= \tfrac{1}{27} \int \cos^2 \theta \, d\theta$$

$$= \tfrac{1}{54} \int (1 + \cos 2\theta) \, d\theta$$

$$= \tfrac{1}{54} (\theta + \tfrac{1}{2} \sin 2\theta) + C$$

$$= \tfrac{1}{54} (\theta + \sin \theta \cos \theta) + C$$

Because $\sec \theta = \tfrac{1}{3}x$ and θ is in $(0, \tfrac{1}{2}\pi) \cup (\pi, \tfrac{3}{2}\pi)$, $\theta = \sec^{-1} \tfrac{1}{3}x$. When $x > 3$, $0 < \theta < \tfrac{1}{2}\pi$, and we obtain $\sin \theta$ and $\cos \theta$ from Fig. 8.5.5. When $x < -3$, $\pi < \theta < \tfrac{3}{2}\pi$, and we obtain $\sin \theta$ and $\cos \theta$ from Fig. 8.5.6. In either case

$x > 3$

FIGURE 8.5.5

$x < -3$

FIGURE 8.5.6

$\sin \theta = \sqrt{x^2 - 9}/x$ and $\cos \theta = 3/x$. Therefore

$$\int \frac{dx}{x^3\sqrt{x^2 - 9}} = \frac{1}{54}\left(\sec^{-1}\frac{x}{3} + \frac{\sqrt{x^2 - 9}}{x} \cdot \frac{3}{x}\right) + C$$

$$= \frac{1}{54}\sec^{-1}\frac{x}{3} + \frac{\sqrt{x^2 - 9}}{18x^2} + C$$

EXAMPLE 4

Evaluate

$$\int \frac{dx}{\sqrt{x^2 - 25}}$$

SOLUTION

Let $x = 5\sec\theta$, where $0 < \theta < \frac{1}{2}\pi$ if $x > 5$ and $\pi < \theta < \frac{3}{2}\pi$ if $x < -5$. Then $dx = 5\sec\theta\tan\theta\, d\theta$ and

$$\sqrt{x^2 - 25} = \sqrt{25\sec^2\theta - 25}$$
$$= 5\sqrt{\tan^2\theta}$$
$$= 5\tan\theta$$

Therefore

$$\int \frac{dx}{\sqrt{x^2 - 25}} = \int \frac{5\sec\theta\tan\theta\, d\theta}{5\tan\theta}$$

$$= \int \sec\theta\, d\theta$$

$$= \ln|\sec\theta + \tan\theta| + C$$

To find $\tan\theta$ refer to Fig. 8.5.7 (for $x > 5$) and Fig. 8.5.8 (for $x < -5$). In either case, $\sec\theta = \frac{1}{5}x$ and $\tan\theta = \frac{1}{5}\sqrt{x^2 - 25}$. We have, then,

$$\int \frac{dx}{\sqrt{x^2 - 25}} = \ln\left|\frac{x}{5} + \frac{\sqrt{x^2 - 25}}{5}\right| + C$$

$$= \ln|x + \sqrt{x^2 - 25}| - \ln 5 + C$$
$$= \ln|x + \sqrt{x^2 - 25}| + C_1$$

$x > 5$

FIGURE 8.5.7

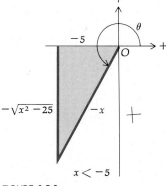

$x < -5$

FIGURE 8.5.8

EXAMPLE 5

Evaluate

$$\int_1^2 \frac{dx}{(6 - x^2)^{3/2}}$$

SOLUTION

To evaluate the indefinite integral $\int dx/(6 - x^2)^{3/2}$ we make the substitution $x = \sqrt{6}\sin\theta$. In this case we can restrict θ to the interval $0 < \theta < \frac{1}{2}\pi$ because we are evaluating a definite integral for which $x > 0$ since x is in $[1, 2]$. So

$x = \sqrt{6} \sin\theta$, $0 < \theta < \frac{1}{2}\pi$, and $dx = \sqrt{6} \cos\theta \, d\theta$. Furthermore,

$$(6 - x^2)^{3/2} = (6 - 6\sin^2\theta)^{3/2}$$
$$= 6\sqrt{6}(1 - \sin^2\theta)^{3/2}$$
$$= 6\sqrt{6}(\cos^2\theta)^{3/2}$$
$$= 6\sqrt{6}\cos^3\theta$$

Hence

$$\int \frac{dx}{(6 - x^2)^{3/2}} = \int \frac{\sqrt{6}\cos\theta \, d\theta}{6\sqrt{6}\cos^3\theta}$$

$$= \frac{1}{6}\int \frac{d\theta}{\cos^2\theta}$$

$$= \frac{1}{6}\int \sec^2\theta \, d\theta$$

$$= \frac{1}{6}\tan\theta + C$$

We find $\tan\theta$ from Fig. 8.5.9, in which $\sin\theta = x/\sqrt{6}$ and $0 < \theta < \frac{1}{2}\pi$. Thus $\tan\theta = x/\sqrt{6 - x^2}$, and so

$$\int \frac{dx}{(6 - x^2)^{3/2}} = \frac{x}{6\sqrt{6 - x^2}} + C$$

Therefore

$$\int_1^2 \frac{dx}{(6 - x^2)^{3/2}} = \frac{x}{6\sqrt{6 - x^2}}\Bigg]_1^2$$

$$= \frac{1}{3\sqrt{2}} - \frac{1}{6\sqrt{5}}$$

$$= \frac{\sqrt{2}}{6} - \frac{\sqrt{5}}{30}$$

$$= \frac{5\sqrt{2} - \sqrt{5}}{30}$$

FIGURE 8.5.9

EXERCISES 8.5

In Exercises 1 through 24, evaluate the indefinite integral.

1. $\displaystyle\int \frac{dx}{x^2\sqrt{4 - x^2}}$

2. $\displaystyle\int \frac{\sqrt{4 - x^2}}{x^2}\,dx$

3. $\displaystyle\int \frac{dx}{x\sqrt{x^2 + 4}}$

4. $\displaystyle\int \frac{x^2\,dx}{\sqrt{x^2 + 6}}$

5. $\displaystyle\int \frac{dx}{x\sqrt{25 - x^2}}$

6. $\displaystyle\int \sqrt{1 - u^2}\,du$

7. $\displaystyle\int \frac{dx}{\sqrt{x^2 - a^2}}$

8. $\displaystyle\int \frac{dw}{w^2\sqrt{w^2 - 7}}$

9. $\displaystyle\int \frac{x^2\,dx}{(x^2 + 4)^2}$

10. $\displaystyle\int \frac{dx}{(4 + x^2)^{3/2}}$

11. $\displaystyle\int \frac{dx}{(4x^2 - 9)^{3/2}}$

12. $\displaystyle\int \frac{dx}{x^4\sqrt{16 + x^2}}$

13. $\displaystyle\int \frac{2\,dt}{t\sqrt{t^4 + 25}}$

14. $\displaystyle\int \frac{x^3\,dx}{(25 - x^2)^2}$

15. $\displaystyle\int \frac{dx}{\sqrt{4x + x^2}}$

16. $\displaystyle\int \frac{dx}{\sqrt{4x - x^2}}$

17. $\displaystyle\int \frac{dx}{(5 - 4x - x^2)^{3/2}}$

18. $\displaystyle\int \frac{dx}{x\sqrt{x^4 - 4}}$

19. $\displaystyle\int \frac{\sec^2 x\,dx}{(4 - \tan^2 x)^{3/2}}$

20. $\displaystyle\int \frac{e^{-x}\,dx}{(9e^{-2x} + 1)^{3/2}}$

21. $\displaystyle\int \frac{\ln^3 w\,dw}{w\sqrt{\ln^2 w - 4}}$

22. $\displaystyle\int \frac{dz}{(z^2 - 6z + 18)^{3/2}}$

23. $\displaystyle\int \frac{e^t\,dt}{(e^{2t} + 8e^t + 7)^{3/2}}$

24. $\displaystyle\int \frac{\sqrt{16 - e^{2x}}}{e^x}\,dx$

In Exercises 25 through 32, evaluate the definite integral.

25. $\displaystyle\int_0^2 \frac{x^3\,dx}{\sqrt{16 - x^2}}$

26. $\displaystyle\int_0^4 \frac{dx}{(16 + x^2)^{3/2}}$

27. $\displaystyle\int_{\sqrt3}^{3\sqrt3} \frac{dx}{x^2\sqrt{x^2 + 9}}$

28. $\displaystyle\int_0^1 \frac{x^2\,dx}{\sqrt{4 - x^2}}$

29. $\displaystyle\int_4^6 \frac{dx}{x\sqrt{x^2 - 4}}$

30. $\displaystyle\int_1^3 \frac{dx}{x^4\sqrt{x^2 + 3}}$

31. $\displaystyle\int_0^5 x^2\sqrt{25 - x^2}\,dx$

32. $\displaystyle\int_4^8 \frac{dw}{(w^2 - 4)^{3/2}}$

33. Use methods previous to this section (i.e., without a trigonometric substitution) to evaluate the integrals:

(a) $\displaystyle\int \frac{3\,dx}{x\sqrt{4x^2 - 9}}$ (b) $\displaystyle\int \frac{5x\,dx}{\sqrt{3 - 2x^2}}$

34. Find the area of the region bounded by the curve $y = \sqrt{x^2 - 9}/x^2$, the x axis, and the line $x = 5$.

35. Find the length of the arc of the curve $y = \ln x$ from $x = 1$ to $x = 3$.

36. Find the volume of the solid of revolution generated by revolving the region of Exercise 34 about the y axis.

37. Find the volume of the solid of revolution generated when the region to the right of the y axis bounded by the curve $y = x\sqrt[4]{9 - x^2}$ and the x axis is revolved about the x axis.

38. Find the length of the arc of the parabola $y = x^2$ from $(0,0)$ to $(1, 1)$.

39. Find the center of mass of a rod 8 cm long if the linear density at a point x cm from the left end is $\rho(x)$ g/cm, where $\rho(x) = \sqrt{x^2 + 36}$.

40. The linear density of a rod at a point x meters from one end is $\sqrt{9 + x^2}$ kg/m. Find the mass and center of mass of the rod if it is 3 m long.

41. Find the centroid of the region bounded by the curve $yx^2 = \sqrt{x^2 - 9}$, the x axis, and the line $x = 5$.

42. Use integration to obtain πr^2 square units as the area of the region enclosed by a circle of radius r units.

Exercises 43 through 45 pertain to Supplementary Section 5.9.

43. A horizontal cylindrical pipe has a 4-ft inner diameter and is closed at one end by a circular gate that just fits over the pipe. If the pipe contains water at a depth of 3 ft, find the force on the gate due to water pressure.

44. A gate in an irrigation ditch is in the shape of a segment of a circle of radius 4 ft. The top of the gate is horizontal and 3 ft above the lowest point on the gate. If the water level is 2 ft above the top of the gate, find the force on the gate due to the water pressure.

45. An automobile's gasoline tank is in the shape of a right-circular cylinder of radius 8 in. with a horizontal axis. Find the total force on one end when the gasoline is 12 in. deep and ρ oz/in.^{3} is the weight density of gasoline.

8.6
INTEGRATION OF RATIONAL
FUNCTIONS BY PARTIAL
FRACTIONS WHEN THE
DENOMINATOR HAS ONLY
LINEAR FACTORS

In Section 0.5 a rational function was defined as one that can be expressed as the quotient of two polynomial functions. That is, the function H is a rational function if $H(x) = P(x)/Q(x)$, where $P(x)$ and $Q(x)$ are polynomials. We saw previously that if the degree of the numerator is not less than the degree of the denominator, we have an improper fraction, and in that case we divide the numerator by the denominator until we obtain a proper fraction, one in which the degree of the numerator is less than the degree of the denominator.

For example,

$$\frac{x^4 - 10x^2 + 3x + 1}{x^2 - 4} = x^2 - 6 + \frac{3x - 23}{x^2 - 4}$$

So if we wish to integrate

$$\int \frac{x^4 - 10x^2 + 3x + 1}{x^2 - 4} \, dx$$

the problem is reduced to integrating

$$\int (x^2 - 6) \, dx + \int \frac{3x - 23}{x^2 - 4} \, dx$$

In general, then, we are concerned with the integration of expressions of the form

$$\int \frac{P(x)}{Q(x)} \, dx$$

where the degree of $P(x)$ is less than the degree of $Q(x)$.

To do this it is often necessary to write $P(x)/Q(x)$ as the sum of *partial fractions*. The denominators of the partial fractions are obtained by factoring $Q(x)$ into a product of linear and quadratic factors where the quadratic factors have no real zeros. Sometimes it may be difficult to find these factors of $Q(x)$; however, a theorem from advanced algebra states that theoretically this can always be done. We state this theorem without proof.

8.6.1 THEOREM   Any polynomial with real coefficients can be expressed as a product of linear and quadratic factors in such a way that each of the factors has real coefficients.

After $Q(x)$ has been factored into products of linear and quadratic factors, the method of determining the partial fractions depends on the nature of these factors. We consider various cases separately. The results of advanced algebra, which are not proved here, provide us with the form of the partial fractions in each case.

Case 1: The factors of $Q(x)$ are all linear, and none is repeated. That is,

$$Q(x) = (a_1 x + b_1)(a_2 x + b_2) \dots (a_n x + b_n)$$

where no two of the factors are identical. In this case we write

$$\frac{P(x)}{Q(x)} \equiv \frac{A_1}{a_1 x + b_1} + \frac{A_2}{a_2 x + b_2} + \dots + \frac{A_n}{a_n x + b_n} \tag{1}$$

where $A_1, A_2, \dots, A_n$ are constants to be determined.

Note that we used $\equiv$ (read as "identically equal") instead of $=$ in (1) because (1) is an identity.

The following illustration shows how the values of A_i are found.

▶ ILLUSTRATION 1   To evaluate

$$\int \frac{(x - 1)\,dx}{x^3 - x^2 - 2x}$$

we factor the denominator and have

$$\frac{x - 1}{x^3 - x^2 - 2x} \equiv \frac{x - 1}{x(x - 2)(x + 1)}$$

So

$$\frac{x - 1}{x(x - 2)(x + 1)} \equiv \frac{A}{x} + \frac{B}{x - 2} + \frac{C}{x + 1} \qquad (2)$$

Because (2) is an identity it must hold for all x except 0, 2, and -1. From (2),

$$x - 1 \equiv A(x - 2)(x + 1) + Bx(x + 1) + Cx(x - 2) \qquad (3)$$

Equation (3) is an identity that is true for all values of x including 0, 2, and -1. We wish to find the constants A, B, and C. Substituting 0 for x in (3) we obtain

$$-1 = -2A \quad \Leftrightarrow \quad A = \tfrac{1}{2}$$

Substituting 2 for x in (3) we get

$$1 = 6B \quad \Leftrightarrow \quad B = \tfrac{1}{6}$$

Substituting -1 for x in (3) we obtain

$$-2 = 3C \quad \Leftrightarrow \quad C = -\tfrac{2}{3}$$

There is another method for finding the values of A, B, and C. If on the right side of (3) we combine terms,

$$x - 1 \equiv (A + B + C)x^2 + (-A + B - 2C)x - 2A \qquad (4)$$

For (4) to be an identity the coefficients on the left must equal the corresponding coefficients on the right. Hence

$$A + B + C = 0$$
$$-A + B - 2C = 1$$
$$-2A = -1$$

Solving these equations simultaneously we get $A = \tfrac{1}{2}$, $B = \tfrac{1}{6}$, and $C = -\tfrac{2}{3}$. Substituting these values in (2) we get

$$\frac{x - 1}{x(x - 2)(x + 1)} \equiv \frac{\tfrac{1}{2}}{x} + \frac{\tfrac{1}{6}}{x - 2} + \frac{-\tfrac{2}{3}}{x + 1}$$

So the given integral can be expressed as follows:

$$\int \frac{x-1}{x^3-x^2-2x}\,dx = \frac{1}{2}\int \frac{dx}{x} + \frac{1}{6}\int \frac{dx}{x-2} - \frac{2}{3}\int \frac{dx}{x+1}$$

$$= \tfrac{1}{2}\ln|x| + \tfrac{1}{6}\ln|x-2| - \tfrac{2}{3}\ln|x+1| + \tfrac{1}{6}\ln C$$

$$= \tfrac{1}{6}(3\ln|x| + \ln|x-2| - 4\ln|x+1| + \ln C)$$

$$= \frac{1}{6}\ln\left|\frac{Cx^3(x-2)}{(x+1)^4}\right| \qquad \blacktriangleleft$$

Case 2: The factors of $Q(x)$ are all linear, and some are repeated.

Suppose that $(a_i x + b_i)$ is a p-fold factor. Then, corresponding to this factor there will be the sum of p partial fractions

$$\frac{A_1}{(a_i x + b_i)^p} + \frac{A_2}{(a_i x + b_i)^{p-1}} + \cdots + \frac{A_{p-1}}{(a_i x + b_i)^2} + \frac{A_p}{a_i x + b_i}$$

where $A_1, A_2, \ldots, A_p$ are constants to be determined.

Example 1 following illustrates this case and the method of determining each A_i.

EXAMPLE 1
Find

$$\int \frac{(x^3-1)\,dx}{x^2(x-2)^3}$$

SOLUTION
The fraction in the integrand is written as a sum of partial fractions as follows:

$$\frac{x^3-1}{x^2(x-2)^3} \equiv \frac{A}{x^2} + \frac{B}{x} + \frac{C}{(x-2)^3} + \frac{D}{(x-2)^2} + \frac{E}{x-2} \tag{5}$$

Multiplying on both sides of (5) by the lowest common denominator we get

$$x^3 - 1 \equiv A(x-2)^3 + Bx(x-2)^3 + Cx^2 + Dx^2(x-2) + Ex^2(x-2)^2 \tag{6}$$

We substitute 2 for x in (6) and obtain

$$7 = 4C \quad \Leftrightarrow \quad C = \tfrac{7}{4}$$

Substituting 0 for x in (6) we get

$$-1 = -8A \quad \Leftrightarrow \quad A = \tfrac{1}{8}$$

We substitute these values for A and C in (6) and expand the powers of the binomials, and we have

$$x^3 - 1 \equiv \tfrac{1}{8}(x^3 - 6x^2 + 12x - 8) + Bx(x^3 - 6x^2 + 12x - 8) + \tfrac{7}{4}x^2$$
$$+ Dx^3 - 2Dx^2 + Ex^2(x^2 - 4x + 4)$$

$$x^3 - 1 \equiv (B + E)x^4 + (\tfrac{1}{8} - 6B + D - 4E)x^3$$
$$+ (-\tfrac{3}{4} + 12B + \tfrac{7}{4} - 2D + 4E)x^2 + (\tfrac{3}{2} - 8B)x - 1$$

Equating the coefficients of like powers of x we obtain

$$B + E = 0$$

$$\tfrac{1}{8} - 6B + D - 4E = 1$$

$$-\tfrac{3}{4} + 12B + \tfrac{7}{4} - 2D + 4E = 0$$

$$\tfrac{3}{2} - 8B = 0$$

Solving, we get

$$B = \tfrac{3}{16} \qquad D = \tfrac{5}{4} \qquad E = -\tfrac{3}{16}$$

Therefore, from (5),

$$\frac{x^3 - 1}{x^2(x - 2)^3} \equiv \frac{\tfrac{1}{8}}{x^2} + \frac{\tfrac{3}{16}}{x} + \frac{\tfrac{7}{4}}{(x - 2)^3} + \frac{\tfrac{5}{4}}{(x - 2)^2} + \frac{-\tfrac{3}{16}}{x - 2}$$

Thus

$$\int \frac{x^3 - 1}{x^2(x - 2)^3}\, dx$$

$$= \frac{1}{8} \int \frac{dx}{x^2} + \frac{3}{16} \int \frac{dx}{x} + \frac{7}{4} \int \frac{dx}{(x - 2)^3} + \frac{5}{4} \int \frac{dx}{(x - 2)^2} - \frac{3}{16} \int \frac{dx}{x - 2}$$

$$= -\frac{1}{8x} + \frac{3}{16} \ln|x| - \frac{7}{8(x - 2)^2} - \frac{5}{4(x - 2)} - \frac{3}{16} \ln|x - 2| + C$$

$$= \frac{-11x^2 + 17x - 4}{8x(x - 2)^2} + \frac{3}{16} \ln\left|\frac{x}{x - 2}\right| + C$$

EXAMPLE 2

Find

$$\int \frac{du}{u^2 - a^2}$$

SOLUTION

$$\frac{1}{u^2 - a^2} \equiv \frac{A}{u - a} + \frac{B}{u + a}$$

Multiplying by $(u - a)(u + a)$ we get

$$1 \equiv A(u + a) + B(u - a)$$

$$1 \equiv (A + B)u + Aa - Ba$$

Equating coefficients we have

$$A + B = 0$$

$$Aa - Ba = 1$$

Solving simultaneously we get

$$A = \frac{1}{2a} \qquad B = -\frac{1}{2a}$$

Therefore

$$\int \frac{du}{u^2 - a^2} = \frac{1}{2a} \int \frac{du}{u - a} - \frac{1}{2a} \int \frac{du}{u + a}$$

$$= \frac{1}{2a} \ln|u - a| - \frac{1}{2a} \ln|u + a| + C$$

$$= \frac{1}{2a} \ln\left|\frac{u - a}{u + a}\right| + C$$

The type of integral of the above example occurs frequently enough for it to be listed as a formula. It is not necessary to memorize it because an integration by partial fractions is fairly simple.

$$\int \frac{du}{u^2 - a^2} = \frac{1}{2a} \ln\left|\frac{u - a}{u + a}\right| + C \tag{7}$$

If we have $\int du/(a^2 - u^2)$, we write

$$\int \frac{du}{a^2 - u^2} = -\int \frac{du}{u^2 - a^2}$$

$$= -\frac{1}{2a} \ln\left|\frac{u - a}{u + a}\right| + C$$

$$= \frac{1}{2a} \ln\left|\frac{u + a}{u - a}\right| + C$$

This is also listed as a formula.

$$\int \frac{du}{a^2 - u^2} = \frac{1}{2a} \ln\left|\frac{u + a}{u - a}\right| + C \tag{8}$$

In Section 6.6 we discussed exponential growth that occurs when the rate of increase of the amount of a quantity is proportional to the amount present at a given instant. It has the mathematical model

$$f(t) = Be^{kt} \tag{9}$$

where k is a positive constant, B units is the amount present initially, and $f(t)$ units is the amount present at t units of time, where $f(t) \geq B$ for $t \geq 0$. We also discussed in Section 6.6 bounded growth that happens when a quantity increases at a rate proportional to the difference between a fixed positive number A and its size. A mathematical model for this bounded growth is

$$f(t) = A - Be^{-kt} \tag{10}$$

FIGURE 8.6.1

FIGURE 8.6.2

where B and k are positive constants and $A - B \leq f(t) < A$ for $t \geq 0$. Sketches of the graphs of (9) and (10) appear in Figs. 8.6.1 and 8.6.2, respectively.

Consider now the growth of a population that is affected by the environment imposing an upper bound on its size. For instance, space or reproduction may be factors limited by the environment. In such cases a mathematical model of the form (9) does not apply because the population does not increase beyond a certain point. A model that takes into account environmental factors is obtained when a quantity is increasing at a rate that is jointly proportional to its size and the difference between a fixed positive number A and its size. Thus if y units is the amount of the quantity present at t units of time,

$$\frac{dy}{dt} = ky(A - y) \tag{11}$$

where k is a positive constant, and $0 < y < A$ for $t \geq 0$.

To solve (11) we first separate the variables and obtain

$$\frac{dy}{y(A - y)} = k\,dt$$

$$\int \frac{dy}{y(A - y)} = k \int dt \tag{12}$$

Writing the integrand on the left as the sum of partial fractions gives

$$\frac{1}{y(A - y)} = \frac{1}{A}\left(\frac{1}{y} + \frac{1}{A - y}\right)$$

Thus

$$\int \frac{dy}{y(A - y)} = \frac{1}{A} \int \left(\frac{1}{y} + \frac{1}{A - y}\right) dy$$

$$= \frac{1}{A}\left(\ln|y| - \ln|A - y|\right) + C_1$$

Therefore, from (12), we have

$$\frac{1}{A}\left(\ln|y| - \ln|A - y|\right) = kt + C_2$$

$$\ln|A - y| - \ln|y| = -Akt - AC_2$$

$$\ln\left|\frac{A - y}{y}\right| = -Akt - AC_2$$

$$\left|\frac{A - y}{y}\right| = e^{-Akt}e^{-AC_2} \tag{13}$$

Because $0 < y < A$, $(A - y)/y > 0$. Therefore we can omit the absolute-value bars. With $B = e^{-AC_2}$ we have, from (13),

$$A - y = Bye^{-Akt}$$

$$y(1 + Be^{-Akt}) = A$$

$$y = \frac{A}{1 + Be^{-Akt}} \tag{14}$$

If we let $y = f(t)$, then (14) is written as

$$f(t) = \frac{A}{1 + Be^{-Akt}} \qquad t \geq 0 \tag{15}$$

where A, B, and k are positive constants. To draw a sketch of the graph of f, first consider $\lim\limits_{t \to +\infty} f(t)$. From (15),

$$\lim_{t \to +\infty} f(t) = \frac{A}{1 + B \lim\limits_{t \to +\infty} e^{-Akt}}$$

$$= \frac{A}{1 + B \cdot 0}$$

$$= A$$

and $f(t)$ is approaching A through values less than A. Therefore, from Definition 3.8.1 the line A units above the t axis is a horizontal asymptote of the graph of f. Because $f(0) = \dfrac{A}{1 + B}$, the graph intersects the $f(t)$ axis at $\dfrac{A}{1 + B}$. In Exercise 38 you are asked to show that the graph of f has a point of inflection at $t = \dfrac{1}{Ak} \ln B$. With this information we draw a sketch of the graph as shown in Fig. 8.6.3. It is called a curve of *logistic growth*. Observe that when t is small, the graph is similar to the one for exponential growth in Fig. 8.6.1, and as t increases the curve is analogous to that shown in Fig. 8.6.2 for bounded growth.

An application of logistic growth in economics is the distribution of information about a particular product. Logistic growth is used by biologists to describe the spread of a disease and by sociologists to describe the spread of a rumor or joke.

FIGURE 8.6.3

$f(t) = \dfrac{A}{1 + Be^{Akt}}$

EXAMPLE 3

In a community of 45,000 people, the rate of growth of a flu epidemic is jointly proportional to the number of people who have contracted the flu and the number of people who have not contracted it. (a) If 200 people had the flu at the outbreak of the epidemic and 2800 people have it now after 3 weeks, find a mathematical model describing the epidemic. How many people are expected to have the flu (b) after 5 weeks, and (c) after 10 weeks? (d) If the epidemic continues indefinitely, how many people will contract the flu?

SOLUTION

(a) If t weeks have elapsed since the outbreak of the epidemic and y people have the flu after t weeks, then

$$\frac{dy}{dt} = ky(45{,}000 - y) \tag{16}$$

where k is a constant and $0 < y < 45{,}000$ for $t \geq 0$. We have the boundary conditions given in Table 8.6.1, where y_5 and y_{10} are the number of people having the flu after 5 weeks and 10 weeks, respectively.

Differential equation (16) is of the form of (11), and its general solution is of the form of (14). Therefore the general solution of (16) is

$$y = \frac{45{,}000}{1 + Be^{-45{,}000kt}} \tag{17}$$

Because $y = 200$ when $t = 0$, from (17) we get

$$200 = \frac{45{,}000}{1 + Be^0}$$

$$1 + B = 225$$

$$B = 224$$

Substituting this value of B in (17) we obtain

$$y = \frac{45{,}000}{1 + 224e^{-45{,}000kt}} \tag{18}$$

When $t = 3$, $y = 2800$. Thus from (18) we get

$$2800 = \frac{45{,}000}{1 + 224e^{-135{,}000k}}$$

$$1 + 224e^{-135{,}000k} = \frac{45{,}000}{2800}$$

$$1 + 224e^{-135{,}000k} = 16.0714$$

$$e^{-135{,}000k} = \frac{15.0714}{224}$$

$$-135{,}000k = \ln 0.0672830$$

$$k = -\frac{\ln 0.0672830}{135{,}000}$$

$$k = 0.0000199915$$

Substituting this value of k in (18) we obtain

$$y = \frac{45{,}000}{1 + 224e^{-0.899616t}} \tag{19}$$

which is the mathematical model desired.

Table 8.6.1

t	0	3	5	10
y	200	2800	y_5	y_{10}

(b) Because $y = y_5$ when $t = 5$,

$$y_5 = \frac{45{,}000}{1 + 224e^{-4.49808}}$$

$$= \frac{45{,}000}{1 + 224(0.0111303)}$$

$$= 12{,}882.2$$

(c) Because $y = y_{10}$ when $t = 10$,

$$y_{10} = \frac{45{,}000}{1 + 224e^{-8.99616}}$$

$$= \frac{45{,}000}{1 + 224(0.000123885)}$$

$$= 43{,}785.0$$

Therefore 12,882 people are expected to have the flu after 5 weeks and 43,785 are expected to have it after 10 weeks.

(d) Because

$$\lim_{t \to +\infty} \frac{45{,}000}{1 + 224e^{-0.899616t}} = \frac{45{,}000}{1 + 224 \cdot 0}$$

$$= 45{,}000$$

the entire community of 45,000 people will contract the flu if the epidemic continues indefinitely.

In chemistry, the *law of mass action* affords an application of integration that leads to the use of partial fractions. Under certain conditions it is found that a substance A reacts with a substance B to form a third substance C in such a way that the rate of change of the amount of C is proportional to the product of the amounts of A and B remaining at any given time.

Suppose that initially there are α grams of A and β grams of B and that r grams of A combine with s grams of B to form $(r + s)$ grams of C. If x grams of substance C is present at t units of time, then C contains $rx/(r + s)$ grams of A and $sx/(r + s)$ grams of B. The number of grams of substance A remaining is then $\alpha - rx/(r + s)$, and the number of grams of substance B remaining is $\beta - sx/(r + s)$. Therefore the law of mass action gives

$$\frac{dx}{dt} = K\left(\alpha - \frac{rx}{r + s}\right)\left(\beta - \frac{sx}{r + s}\right)$$

where K is the constant of proportionality. This equation can be written as

$$\frac{dx}{dt} = \frac{Krs}{(r + s)^2}\left(\frac{r + s}{r}\alpha - x\right)\left(\frac{r + s}{s}\beta - x\right) \tag{20}$$

Letting

$$k = \frac{Krs}{(r + s)^2} \qquad a = \frac{r + s}{r}\alpha \qquad b = \frac{r + s}{s}\beta$$

Eq. (20) becomes

$$\frac{dx}{dt} = k(a - x)(b - x) \tag{21}$$

We can separate the variables in (21) and get

$$\frac{dx}{(a - x)(b - x)} = k \, dt$$

If $a = b$, then the left side of the above equation can be integrated by the power formula. If $a \neq b$, partial fractions can be used for the integration.

EXAMPLE 4

A chemical reaction causes a substance A to combine with a substance B to form a substance C so that the law of mass action is obeyed. If in Eq. (21) $a = 8$ and $b = 6$, and 2 g of substance C are formed in 10 min, how many grams of C are formed in 15 min?

SOLUTION

If x grams of substance C is present at t min, we have the boundary conditions shown in Table 8.6.2, where x_{15} grams of substance C is present at 15 min. Equation (21) becomes

Table 8.6.2

t	0	10	15
x	0	2	x_{15}

$$\frac{dx}{dt} = k(8 - x)(6 - x)$$

Separating the variables we have

$$\int \frac{dx}{(8 - x)(6 - x)} = k \int dt \qquad (22)$$

Writing the integrand as the sum of partial fractions gives

$$\frac{1}{(8 - x)(6 - x)} \equiv \frac{A}{8 - x} + \frac{B}{6 - x}$$

$$1 \equiv A(6 - x) + B(8 - x)$$

Substituting 6 for x gives $B = \frac{1}{2}$, and substituting 8 for x gives $A = -\frac{1}{2}$. Hence (22) is written as

$$-\frac{1}{2} \int \frac{dx}{8 - x} + \frac{1}{2} \int \frac{dx}{6 - x} = k \int dt$$

Integrating we have

$$\tfrac{1}{2} \ln|8 - x| - \tfrac{1}{2} \ln|6 - x| + \tfrac{1}{2} \ln|C| = kt$$

$$\ln \left| \frac{6 - x}{C(8 - x)} \right| = -2kt$$

$$\frac{6 - x}{8 - x} = Ce^{-2kt} \qquad (23)$$

Substituting $x = 0$, $t = 0$ in (23) gives $C = \frac{3}{4}$. Hence

$$\frac{6 - x}{8 - x} = \frac{3}{4} e^{-2kt} \qquad (24)$$

Substituting $x = 2$, $t = 10$ in (24) we have

$$\tfrac{4}{6} = \tfrac{3}{4}e^{-20k}$$

$$e^{-20k} = \tfrac{8}{9}$$

Substituting $x = x_{15}$, $t = 15$ into (24) we get

$$\frac{6 - x_{15}}{8 - x_{15}} = \frac{3}{4}e^{-30k}$$

$$4(6 - x_{15}) = 3(e^{-20k})^{3/2}(8 - x_{15})$$

$$24 - 4x_{15} = 3(\tfrac{8}{9})^{3/2}(8 - x_{15})$$

$$24 - 4x_{15} = \frac{16\sqrt{2}}{9}(8 - x_{15})$$

$$x_{15} = \frac{54 - 32\sqrt{2}}{9 - 4\sqrt{2}}$$

$$x_{15} \approx 2.6$$

Therefore 2.6 g of substance C will be formed in 15 min.

EXERCISES 8.6

In Exercises 1 through 20, evaluate the indefinite integral.

1. $\displaystyle\int \frac{dx}{x^2 - 4}$

2. $\displaystyle\int \frac{x^2\,dx}{x^2 + x - 6}$

3. $\displaystyle\int \frac{5x - 2}{x^2 - 4}\,dx$

4. $\displaystyle\int \frac{(4x - 2)\,dx}{x^3 - x^2 - 2x}$

5. $\displaystyle\int \frac{4w - 11}{2w^2 + 7w - 4}\,dw$

6. $\displaystyle\int \frac{9t^2 - 26t - 5}{3t^2 - 5t - 2}\,dt$

7. $\displaystyle\int \frac{6x^2 - 2x - 1}{4x^3 - x}\,dx$

8. $\displaystyle\int \frac{x^2 + x + 2}{x^2 - 1}\,dx$

9. $\displaystyle\int \frac{dx}{x^3 + 3x^2}$

10. $\displaystyle\int \frac{x^2 + 4x - 1}{x^3 - x}\,dx$

11. $\displaystyle\int \frac{dx}{x^2(x + 1)^2}$

12. $\displaystyle\int \frac{3x^2 - x + 1}{x^3 - x^2}\,dx$

13. $\displaystyle\int \frac{x^2 - 3x - 7}{(2x + 3)(x + 1)^2}\,dx$

14. $\displaystyle\int \frac{dt}{(t + 2)^2(t + 1)}$

15. $\displaystyle\int \frac{3z + 1}{(z^2 - 4)^2}\,dz$

16. $\displaystyle\int \frac{(5x^2 - 11x + 5)\,dx}{x^3 - 4x^2 + 5x - 2}$

17. $\displaystyle\int \frac{x^4 + 3x^3 - 5x^2 - 4x + 17}{x^3 + x^2 - 5x + 3}\,dx$

18. $\displaystyle\int \frac{2x^4 - 2x + 1}{2x^5 - x^4}\,dx$

19. $\displaystyle\int \frac{-24x^3 + 30x^2 + 52x + 17}{9x^4 - 6x^3 - 11x^2 + 4x + 4}\,dx$

20. $\displaystyle\int \frac{dx}{16x^4 - 8x^2 + 1}$

In Exercises 21 through 28, evaluate the definite integral.

21. $\displaystyle\int_1^2 \frac{x - 3}{x^3 + x^2}\,dx$

22. $\displaystyle\int_0^4 \frac{(x - 2)\,dx}{2x^2 + 7x + 3}$

23. $\displaystyle\int_1^3 \frac{x^2 - 4x + 3}{x(x + 1)^2}\,dx$

24. $\displaystyle\int_1^4 \frac{(2x^2 + 13x + 18)\,dx}{x^3 + 6x^2 + 9x}$

25. $\displaystyle\int_1^2 \frac{5x^2 - 3x + 18}{9x - x^3}\,dx$

26. $\displaystyle\int_0^1 \frac{(3x^2 + 7x)\,dx}{x^3 + 6x^2 + 11x + 6}$

27. $\displaystyle\int_0^5 \frac{(x^2 - 3)\,dx}{x^3 + 4x^2 + 5x + 2}$

28. $\displaystyle\int_0^4 \frac{x^2\,dx}{2x^3 + 9x^2 + 12x + 4}$

29. Find the area of the region bounded by the curve $y = (x - 1)/(x^2 - 5x + 6)$, the x axis, and the lines $x = 4$ and $x = 6$.

30. Find the area of the region in the first quadrant bounded by the curve $(x + 2)^2 y = 4 - x$.

31. Find the volume of the solid of revolution generated by revolving the region in Exercise 29 about the y axis.

32. Find the volume of the solid of revolution generated if the region in Exercise 30 is revolved about the x axis.

33. Find the centroid of the region bounded by the curve $y = (x - 1)/(x^2 - 5x + 6)$, the x axis, and the lines $x = 4$ and $x = 6$.

34. Find the centroid of the region in the first quadrant bounded by the curve $(x + 2)^2 y = 4 - x$.

35. One day on a college campus, when there were 5000 people in attendance, a particular student heard that a certain controversial speaker was going to make an unscheduled appearance. This information was told to friends who in turn related it to others, and the rate of growth of the spread of this information was jointly proportional to the number of people who had heard it and the number of people who had not heard it. (a) If after 10 min 144 people had heard the rumor, find a mathematical model describing the spread of information. How many people had heard the rumor (b) after 15 min and (c) after 20 min? (d) How many people will eventually hear the rumor?

36. In a particular town of population A, 20 percent of the residents heard a radio announcement about a local political scandal. The rate of growth of the spread of information about the scandal was jointly proportional to the number of people who had heard it and the number of people who had not heard it. If 50 percent of the population heard about the scandal after 1 hour, how long was it until 80 percent of the population heard it?

37. In a community in which A people are susceptible to a particular virus, the rate of growth of the spread of the virus was jointly proportional to the number of people who had caught the virus and the number of susceptible people who had not caught it. If 10 percent of those susceptible had the virus initially and 25 percent had been infected after 3 weeks, what percent of those susceptible had been infected after 6 weeks?

38. Show that the graph of the function f defined by (15) has a point of inflection at $t = \dfrac{1}{Ak} \ln B$.

39. A manufacturer who began operations four years ago has determined that income from sales has increased steadily at the rate of $\dfrac{t^3 + 3t^2 + 6t + 7}{t^2 + 3t + 2}$ millions of dollars per year, where t is the number of years that the company has been operating. It is estimated that the total income from sales will increase at the same rate for the next 2 years. If the total income from sales for the year just ended was $6 million, what is the total income from sales expected for the period ending 1 year from now? Give the answer to the nearest $100.

40. A particle is moving along a straight line so that if v ft/sec is the velocity of the particle at t sec, then

$$v = \frac{t + 3}{t^2 + 3t + 2}$$

Find the distance traveled by the particle from the time when $t = 0$ to the time when $t = 2$.

41. Suppose in Example 4 that $a = 5$ and $b = 4$ and 1 g of substance C is formed in 5 min. How many grams of C are formed in 10 min?

42. Suppose in Example 4 that $a = 6$ and $b = 3$ and 1 g of substance C is formed in 4 min. How long will it take 2 g of substance C to be formed?

43. At any instant the rate at which a substance dissolves is proportional to the product of the amount of the substance present at that instant and the difference between the concentration of the substance in solution at that instant and the concentration of the substance in a saturated solution. A quantity of insoluble material is mixed with 10 lb of salt initially, and the salt is dissolving in a tank containing 20 gal of water. If 5 lb of salt dissolves in 10 min and the concentration of salt in a saturated solution is 3 lb/gal, how much salt will dissolve in 20 min?

**8.7
INTEGRATION OF RATIONAL FUNCTIONS BY PARTIAL FRACTIONS WHEN THE DENOMINATOR CONTAINS QUADRATIC FACTORS**

The discussion of integration of rational functions by partial fractions continues with the two cases where the denominator contains quadratic factors.

Case 3: The factors of $Q(x)$ are linear and quadratic, and none of the quadratic factors is repeated.

Corresponding to the quadratic factor $ax^2 + bx + c$ in the denominator is the partial fraction of the form

$$\frac{Ax + B}{ax^2 + bx + C}$$

EXAMPLE 1
Find

$$\int \frac{(x^2 - 2x - 3)\,dx}{(x - 1)(x^2 + 2x + 2)}$$

SOLUTION
The fraction in the integrand is written as a sum of partial fractions as follows:

$$\frac{x^2 - 2x - 3}{(x - 1)(x^2 + 2x + 2)} \equiv \frac{Ax + B}{x^2 + 2x + 2} + \frac{C}{x - 1} \tag{1}$$

Multiplying on both sides of (1) by the lowest common denominator we have

$$x^2 - 2x - 3 \equiv (Ax + B)(x - 1) + C(x^2 + 2x + 2) \tag{2}$$

We can compute C by substituting 1 for x in (2), and we have

$$-4 = 5C \quad \Leftrightarrow \quad C = -\tfrac{4}{5}$$

We replace C by $-\tfrac{4}{5}$ in (2) and multiply on the right side to obtain

$$x^2 - 2x - 3 \equiv (A - \tfrac{4}{5})x^2 + (B - A - \tfrac{8}{5})x + (-\tfrac{8}{5} - B)$$

Equating coefficients of like powers of x gives

$$A - \tfrac{4}{5} = 1$$
$$B - A - \tfrac{8}{5} = -2$$
$$-\tfrac{8}{5} - B = -3$$

Therefore

$$A = \tfrac{9}{5} \qquad B = \tfrac{7}{5}$$

Substituting the values of A, B, and C in (1) we obtain

$$\frac{x^2 - 2x - 3}{(x - 1)(x^2 + 2x + 2)} \equiv \frac{\tfrac{9}{5}x + \tfrac{7}{5}}{x^2 + 2x + 2} + \frac{-\tfrac{4}{5}}{x - 1}$$

So

$$\int \frac{x^2 - 2x - 3}{(x - 1)(x^2 + 2x + 2)}\,dx$$

$$= \frac{9}{5}\int \frac{x\,dx}{x^2 + 2x + 2} + \frac{7}{5}\int \frac{dx}{x^2 + 2x + 2} - \frac{4}{5}\int \frac{dx}{x - 1} \tag{3}$$

To integrate $\int (x\,dx)/(x^2 + 2x + 2)$ we see that the differential of the denominator is $2(x + 1)\,dx$; so we add and subtract 1 in the numerator, thereby giving

$$\frac{9}{5} \int \frac{x\,dx}{x^2 + 2x + 2} = \frac{9}{5} \int \frac{(x + 1)\,dx}{x^2 + 2x + 2} - \frac{9}{5} \int \frac{dx}{x^2 + 2x + 2} \qquad (4)$$

Substituting from (4) into (3) and combining terms we get

$$\int \frac{x^2 - 2x - 3}{(x - 1)(x^2 + 2x + 2)}\,dx$$

$$= \frac{9}{5} \cdot \frac{1}{2} \int \frac{2(x + 1)\,dx}{x^2 + 2x + 2} - \frac{2}{5} \int \frac{dx}{x^2 + 2x + 2} - \frac{4}{5} \int \frac{dx}{x - 1} \qquad (5)$$

$$= \frac{9}{10} \ln|x^2 + 2x + 2| - \frac{2}{5} \int \frac{dx}{(x + 1)^2 + 1} - \frac{4}{5} \ln|x - 1|$$

$$= \tfrac{9}{10} \ln|x^2 + 2x + 2| - \tfrac{2}{5} \tan^{-1}(x + 1) - \tfrac{8}{10} \ln|x - 1| + \tfrac{1}{10} \ln C$$

$$= \frac{1}{10} \ln \left| \frac{C(x^2 + 2x + 2)^9}{(x - 1)^8} \right| - \frac{2}{5} \tan^{-1}(x + 1)$$

▶ ILLUSTRATION 1   In Example 1 some steps are saved if instead of (1) the original fraction is expressed as

$$\frac{x^2 - 2x - 3}{(x - 1)(x^2 + 2x + 2)} \equiv \frac{D(2x + 2) + E}{x^2 + 2x + 2} + \frac{F}{x - 1}$$

Note: We write $D(2x + 2) + E$ instead of $Ax + B$ because

$$2x + 2 = D_x(x^2 + 2x + 2)$$

Then, solving for D, E, and F we obtain

$$D = \tfrac{9}{10} \qquad E = -\tfrac{2}{5} \qquad F = -\tfrac{4}{5}$$

giving (5) directly.   ◀

Case 4:   The factors of $Q(x)$ are linear and quadratic, and some of the quadratic factors are repeated.

If $ax^2 + bx + c$ is a p-fold quadratic factor of $Q(x)$, then corresponding to this factor $(ax^2 + bx + c)^p$ we have the sum of the following p partial fractions:

$$\frac{A_1 x + B_1}{(ax^2 + bx + c)^p} + \frac{A_2 x + B_2}{(ax^2 + bx + c)^{p-1}} + \cdots + \frac{A_p x + B_p}{ax^2 + bx + c}$$

▶ ILLUSTRATION 2   If the denominator contains the factor $(x^2 - 5x + 2)^3$, we have, corresponding to this factor,

$$\frac{Ax + B}{(x^2 - 5x + 2)^3} + \frac{Cx + D}{(x^2 - 5x + 2)^2} + \frac{Ex + F}{x^2 - 5x + 2}$$

or, more conveniently,

$$\frac{A(2x-5)+B}{(x^2-5x+2)^3} + \frac{C(2x-5)+D}{(x^2-5x+2)^2} + \frac{E(2x-5)+F}{x^2-5x+2}$$ ◀

EXAMPLE 2
Find

$$\int \frac{(x-2)\,dx}{x(x^2-4x+5)^2}$$

SOLUTION

$$\frac{x-2}{x(x^2-4x+5)^2} \equiv \frac{A}{x} + \frac{B(2x-4)+C}{(x^2-4x+5)^2} + \frac{D(2x-4)+E}{x^2-4x+5}$$ (6)

Multiplying on both sides of (6) by the lowest common denominator we have

$$x-2 \equiv A(x^2-4x+5)^2 + x(2Bx-4B+C)$$
$$+ x(x^2-4x+5)(2Dx-4D+E)$$ (7)

$$x-2 \equiv Ax^4 + 16Ax^2 + 25A - 8Ax^3 + 10Ax^2 - 40Ax + 2Bx^2 - 4Bx$$
$$+ Cx + 2Dx^4 - 12Dx^3 + Ex^3 + 26Dx^2 - 4Ex^2 - 20Dx + 5Ex$$

$$x-2 \equiv (A+2D)x^4 + (-8A-12D+E)x^3 + (26A+2B+26D-4E)x^2$$
$$+ (-40A-4B+C-20D+5E)x + 25A$$ (8)

The value of A can be computed from (7) by substituting 0 for x. If we equate coefficients in (8) and solve the resulting equations simultaneously, we obtain

$$A = -\tfrac{2}{25} \qquad B = \tfrac{1}{5} \qquad C = \tfrac{1}{5} \qquad D = \tfrac{1}{25} \qquad E = -\tfrac{4}{25}$$

Therefore

$$\int \frac{(x-2)\,dx}{x(x^2-4x+5)^2}$$

$$= -\frac{2}{25}\int \frac{dx}{x} + \frac{1}{5}\int \frac{(2x-4)\,dx}{(x^2-4x+5)^2} + \frac{1}{5}\int \frac{dx}{(x^2-4x+5)^2}$$

$$+ \frac{1}{25}\int \frac{(2x-4)\,dx}{x^2-4x+5} - \frac{4}{25}\int \frac{dx}{x^2-4x+5}$$

$$= -\frac{2}{25}\ln|x| - \frac{1}{5(x^2-4x+5)} + \frac{1}{5}\int \frac{dx}{[(x^2-4x+4)+1]^2}$$

$$+ \frac{1}{25}\ln|x^2-4x+5| - \frac{4}{25}\int \frac{dx}{(x^2-4x+4)+1}$$ (9)

We evaluate separately the integrals in the third and fifth terms on the right side of (9).

$$\int \frac{dx}{[(x^2 - 4x + 4) + 1]^2} = \int \frac{dx}{[(x - 2)^2 + 1]^2}$$

Let $x - 2 = \tan \theta$, where $0 \le \theta < \frac{1}{2}\pi$ if $x \ge 2$, and $-\frac{1}{2}\pi < \theta < 0$ if $x < 2$. Then $dx = \sec^2 \theta \, d\theta$ and $(x - 2)^2 + 1 = \tan^2 \theta + 1$. Hence

$$\int \frac{dx}{[(x - 2)^2 + 1]^2} = \int \frac{\sec^2 \theta \, d\theta}{(\tan^2 \theta + 1)^2}$$

$$= \int \frac{\sec^2 \theta \, d\theta}{\sec^4 \theta}$$

$$= \int \frac{d\theta}{\sec^2 \theta}$$

$$= \int \cos^2 \theta \, d\theta$$

$$= \int \frac{1 + \cos 2\theta}{2} \, d\theta$$

$$= \frac{\theta}{2} + \frac{1}{4} \sin 2\theta + C_1$$

$$= \frac{\theta}{2} + \frac{1}{2} \sin \theta \cos \theta + C_1$$

Because $\tan \theta = x - 2$ and $-\frac{1}{2}\pi < \theta < \frac{1}{2}\pi$, $\theta = \tan^{-1}(x - 2)$. We find $\sin \theta$ and $\cos \theta$ from Figs. 8.7.1 (if $x \ge 2$) and 8.7.2 (if $x < 2$). In either case

$$\sin \theta = \frac{x - 2}{\sqrt{x^2 - 4x + 5}} \qquad \cos \theta = \frac{1}{\sqrt{x^2 - 4x + 5}}$$

Thus

$$\int \frac{dx}{[(x - 2)^2 + 1]^2} = \frac{1}{2} \tan^{-1}(x - 2) + \frac{1}{2} \cdot \frac{x - 2}{\sqrt{x^2 - 4x + 5}} \cdot \frac{1}{\sqrt{x^2 - 4x + 5}} + C_1$$

$$\int \frac{dx}{[(x - 2)^2 + 1]^2} = \frac{1}{2} \tan^{-1}(x - 2) + \frac{x - 2}{2(x^2 - 4x + 5)} + C_1 \qquad (10)$$

Now, considering the other integral on the right side of (9), we have

$$\int \frac{dx}{(x^2 - 4x + 4) + 1} = \int \frac{dx}{(x - 2)^2 + 1}$$

$$\int \frac{dx}{(x^2 - 4x + 4) + 1} = \tan^{-1}(x - 2) + C_2 \qquad (11)$$

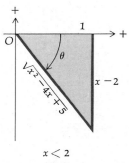

FIGURE 8.7.1

$x \ge 2$

FIGURE 8.7.2

$x < 2$

Substituting from (10) and (11) into (9) we get

$$\int \frac{(x-2)\,dx}{x(x^2-4x+5)^2}$$

$$= -\frac{2}{25}\ln|x| - \frac{1}{5(x^2-4x+5)} + \frac{1}{10}\tan^{-1}(x-2) + \frac{x-2}{10(x^2-4x+5)}$$

$$+ \frac{1}{25}\ln|x^2-4x+5| - \frac{4}{25}\tan^{-1}(x-2) + C$$

$$= \frac{1}{25}\ln\left|\frac{x^2-4x+5}{x^2}\right| - \frac{3}{50}\tan^{-1}(x-2) + \frac{x-4}{10(x^2-4x+5)} + C$$

EXERCISE 8.7

In Exercises 1 through 20, evaluate the indefinite integral.

1. $\displaystyle\int \frac{dx}{2x^3 + x}$

2. $\displaystyle\int \frac{(x+4)\,dx}{x(x^2+4)}$

3. $\displaystyle\int \frac{dx}{16x^4 - 1}$

4. $\displaystyle\int \frac{(x^2 - 4x - 4)\,dx}{x^3 - 2x^2 + 4x - 8}$

5. $\displaystyle\int \frac{(t^2 + t + 1)\,dt}{(2t+1)(t^2+1)}$

6. $\displaystyle\int \frac{3w^3 + 13w + 4}{w^3 + 4w}\,dw$

7. $\displaystyle\int \frac{(x^2 + x)\,dx}{x^3 - x^2 + x - 1}$

8. $\displaystyle\int \frac{dx}{9x^4 + x^2}$

9. $\displaystyle\int \frac{dx}{x^3 + x^2 + x}$

10. $\displaystyle\int \frac{(x+3)\,dx}{4x^4 + 4x^3 + x^2}$

11. $\displaystyle\int \frac{(2x^2 - x + 2)\,dx}{x^5 + 2x^3 + x}$

12. $\displaystyle\int \frac{(2x^3 + 9x)\,dx}{(x^2+3)(x^2-2x+3)}$

13. $\displaystyle\int \frac{(5z^3 - z^2 + 15z - 10)\,dz}{(z^2 - 2z + 5)^2}$

14. $\displaystyle\int \frac{dt}{(t^2 + 1)^3}$

15. $\displaystyle\int \frac{(x^2 + 2x - 1)\,dx}{27x^3 - 1}$

16. $\displaystyle\int \frac{e^{5x}\,dx}{(e^{2x} + 1)^2}$

17. $\displaystyle\int \frac{18\,dx}{(4x^2 + 9)^2}$

18. $\displaystyle\int \frac{(2x^2 + 3x + 2)\,dx}{x^3 + 4x^2 + 6x + 4}$

19. $\displaystyle\int \frac{(\sec^2 x + 1)\sec^2 x\,dx}{1 + \tan^3 x}$

20. $\displaystyle\int \frac{(6w^4 + 4w^3 + 9w^2 + 24w + 32)\,dw}{(w^3 + 8)(w^2 + 3)}$

In Exercises 21 through 29, evaluate the definite integral.

21. $\displaystyle\int_1^4 \frac{(4 + 5x^2)\,dx}{x^3 + 4x}$

22. $\displaystyle\int_0^1 \frac{x\,dx}{x^3 + 2x^2 + x + 2}$

23. $\displaystyle\int_3^4 \frac{(5x^3 - 4x)\,dx}{x^4 - 16}$

24. $\displaystyle\int_0^1 \frac{9\,dx}{8x^3 + 1}$

25. $\displaystyle\int_{-1}^0 \frac{x^2\,dx}{(2x^2 + 2x + 1)^2}$

26. $\displaystyle\int_0^{1/2} \frac{(x+1)\,dx}{x^3 - 1}$

27. $\displaystyle\int_0^1 \frac{(x^2 + 3x + 3)\,dx}{x^3 + x^2 + x + 1}$

28. $\displaystyle\int_{\pi/6}^{\pi/2} \frac{\cos x\,dx}{\sin x + \sin^3 x}$

29. $\displaystyle\int_{\ln 2}^{\ln 5} \frac{12\,dt}{e^{2t} + 16}$

30. Use methods previous to those in this section (i.e., without partial fractions) to evaluate the integrals:

(a) $\displaystyle\int \frac{(x^2 - 4x + 6)\,dx}{x^3 - 6x^2 + 18x}$ (b) $\displaystyle\int \frac{3x + 1}{(x+2)^4}\,dx$

31. Find the area of the region bounded by the curve $y(x^2 + 1)^3 = x^3$, the x axis, the y axis, and the line $x = 1$.

32. Find the area of the region bounded by the curve $y(x^3 + 8) = 4$, the x axis, the y axis, and the line $x = 1$.

33. Find the volume of the solid of revolution generated by revolving the region of Exercise 32 about the y axis.

34. Find the abscissa of the centroid of the region of Exercise 32.

35. A particle is moving along a straight line so that if v cm/sec is the velocity of the particle at t sec, then

$$v = \frac{t^2 - t + 1}{(t + 2)^2(t^2 + 1)}$$

Find a formula for the distance traveled by the particle from the time when $t = 0$ to the time when $t = t_1$.

8.8
INTEGRATION OF
RATIONAL FUNCTIONS OF
SINE AND COSINE

If an integrand is a rational function of $\sin x$ and $\cos x$, it can be reduced to a rational function of z by the substitution

$$z = \tan \tfrac{1}{2}x \tag{1}$$

as will be shown by an example. To obtain the formulas for $\sin x$ and $\cos x$ in terms of z we use the following identities: $\sin 2y = 2 \sin y \cos y$ and $\cos 2y = 2 \cos^2 y - 1$ with $y = \tfrac{1}{2}x$. We have, then,

$$\sin x = 2 \sin \tfrac{1}{2}x \cos \tfrac{1}{2}x \qquad\qquad \cos x = 2 \cos^2 \tfrac{1}{2}x - 1$$

$$= 2 \cdot \frac{\sin \tfrac{1}{2}x \cos^2 \tfrac{1}{2}x}{\cos \tfrac{1}{2}x} \qquad\qquad = \frac{2}{\sec^2 \tfrac{1}{2}x} - 1$$

$$= 2 \tan \tfrac{1}{2}x \cdot \frac{1}{\sec^2 \tfrac{1}{2}x} \qquad\qquad = \frac{2}{1 + \tan^2 \tfrac{1}{2}x} - 1$$

$$= \frac{2 \tan \tfrac{1}{2}x}{1 + \tan^2 \tfrac{1}{2}x} \qquad\qquad = \frac{2}{1 + z^2} - 1$$

$$= \frac{2z}{1 + z^2} \qquad\qquad = \frac{1 - z^2}{1 + z^2}$$

Because $z = \tan \tfrac{1}{2}x$,

$$dz = \tfrac{1}{2} \sec^2 \tfrac{1}{2}x \, dx$$
$$= \tfrac{1}{2}(1 + \tan^2 \tfrac{1}{2}x) \, dx$$

Thus

$$dx = \frac{2 \, dz}{1 + z^2}$$

We state these results as a theorem.

8.8.1 THEOREM   If $z = \tan \tfrac{1}{2}x$, then

$$\sin x = \frac{2z}{1 + z^2} \qquad \cos x = \frac{1 - z^2}{1 + z^2} \qquad dx = \frac{2 \, dz}{1 + z^2}$$

EXAMPLE 1

Find

$$\int \frac{dx}{1 - \sin x + \cos x}$$

SOLUTION

Let $z = \tan \frac{1}{2}x$. Then from the formulas of Theorem 8.8.1 we have

$$\int \frac{dx}{1 - \sin x + \cos x} = \int \frac{\dfrac{2\,dz}{1 + z^2}}{1 - \dfrac{2z}{1 + z^2} + \dfrac{1 - z^2}{1 + z^2}}$$

$$= 2 \int \frac{dz}{(1 + z^2) - 2z + (1 - z^2)}$$

$$= 2 \int \frac{dz}{2 - 2z}$$

$$= \int \frac{dz}{1 - z}$$

$$= -\ln|1 - z| + C$$

$$= -\ln|1 - \tan \tfrac{1}{2}x| + C$$

EXAMPLE 2

Find $\int \sec x\,dx$ by letting $z = \tan \frac{1}{2}x$.

SOLUTION

With $z = \tan \frac{1}{2}x$ and the formulas of Theorem 8.8.1 we have

$$\int \sec x\,dx = \int \frac{dx}{\cos x}$$

$$= \int \frac{2\,dz}{1 + z^2} \cdot \frac{1 + z^2}{1 - z^2}$$

$$= 2 \int \frac{dz}{1 - z^2}$$

$$= \ln \left| \frac{1 + z}{1 - z} \right| + C \quad \text{(from (8) in Section 8.6)}$$

$$= \ln \left| \frac{1 + \tan \tfrac{1}{2}x}{1 - \tan \tfrac{1}{2}x} \right| + C$$

The value of $\int \sec x \, dx$ from Example 2 can be written in another form by letting $1 = \tan \frac{1}{4}\pi$ and using the trigonometric identity

$$\tan(a + b) = \frac{\tan a + \tan b}{1 - \tan a \tan b}$$

Thus

$$\int \sec x \, dx = \ln \left| \frac{\tan \frac{1}{4}\pi + \tan \frac{1}{2}x}{1 - \tan \frac{1}{4}\pi \cdot \tan \frac{1}{2}x} \right| + C$$

$$\int \sec x \, dx = \ln \left| \tan(\tfrac{1}{4}\pi + \tfrac{1}{2}x) \right| + C \tag{2}$$

In Theorem 6.4.5 we have the formula

$$\int \sec x \, dx = \ln |\sec x + \tan x| + C$$

which is obtained by the trick of multiplying the numerator and denominator of the integrand by $\sec x + \tan x$. Still another form for $\int \sec x \, dx$ is obtained as follows:

$$\int \sec x \, dx = \int \frac{dx}{\cos x}$$

$$= \int \frac{\cos x \, dx}{\cos^2 x}$$

$$= \int \frac{\cos x \, dx}{1 - \sin^2 x}$$

$$= \int \frac{du}{1 - u^2} \quad \text{(by letting } u = \sin x \text{ and } du = \cos x \, dx)$$

$$= \frac{1}{2} \ln \left| \frac{1 + u}{1 - u} \right| + C$$

$$= \ln \left| \frac{1 + \sin x}{1 - \sin x} \right|^{1/2} + C$$

Because $-1 \leq \sin x \leq 1$ for all x, $1 + \sin x$ and $1 - \sin x$ are nonnegative. Hence the absolute-value bars can be removed, and we have

$$\int \sec x \, dx = \ln \sqrt{\frac{1 + \sin x}{1 - \sin x}} + C \tag{3}$$

EXERCISES 8.8

In Exercises 1 through 18, evaluate the indefinite integral.

1. $\displaystyle\int \frac{3\,dx}{8 + 7\cos x}$

2. $\displaystyle\int \frac{dx}{1 + \sin x}$

3. $\displaystyle\int \frac{3\,dx}{7 + 8\cos x}$

4. $\displaystyle\int \frac{dx}{\sin x - \cos x + 2}$

5. $\displaystyle\int \frac{dx}{\sin x + \tan x}$

6. $\displaystyle\int \frac{dx}{5 + 4\cos x}$

7. $\displaystyle\int \frac{dx}{3 - 5\sin x}$

8. $\displaystyle\int \frac{dx}{\tan x - 1}$

9. $\displaystyle\int \frac{dx}{4\sin x - 3\cos x}$

10. $\displaystyle\int \frac{dx}{3\cos x - 2\sin x + 3}$

11. $\displaystyle\int \frac{8\,dx}{3\cos 2x + 1}$

12. $\displaystyle\int \frac{\cos x\,dx}{3\cos x - 5}$

13. $\displaystyle\int \frac{\cos x\,dx}{1 + 2\cos x}$

14. $\displaystyle\int \frac{dx}{\sin x - \tan x}$

15. $\displaystyle\int \frac{dx}{\cot x(6 + 7\cos 2x)}$

16. $\displaystyle\int \frac{dx}{\cot 2x(1 - \cos 2x)}$

17. $\displaystyle\int \frac{dx}{2\sin x + 2\cos x + 3}$

18. $\displaystyle\int \frac{5\,dx}{6 + 4\sec x}$

In Exercises 19 through 24, evaluate the definite integral.

19. $\displaystyle\int_0^{\pi/2} \frac{dx}{5\sin x + 3}$

20. $\displaystyle\int_0^{\pi/4} \frac{8\,dx}{\tan x + 1}$

21. $\displaystyle\int_{-\pi/3}^{\pi/2} \frac{3\,dx}{2\cos x + 1}$

22. $\displaystyle\int_0^{\pi/2} \frac{dx}{3 + \cos 2x}$

23. $\displaystyle\int_{\pi/6}^{\pi/3} \frac{3\,dx}{2\sin 2x + 1}$

24. $\displaystyle\int_0^{\pi/2} \frac{\sin 2x\,dx}{2 + \cos x}$

25. Show that formula (3) of this section is equivalent to the formula $\int \sec x\,dx = \ln|\sec x + \tan x| + C$. (*Hint:* Multiply the numerator and denominator under the radical sign by $(1 + \sin x)$.)

26. By using the substitution $z = \tan \frac{1}{2}x$, prove that

$$\int \csc x\,dx = \ln\sqrt{\frac{1 - \cos x}{1 + \cos x}} + C$$

27. Show that the result in Exercise 26 is equivalent to the formula $\int \csc x\,dx = \ln|\csc x - \cot x| + C$. (*Hint:* Use a method similar to that suggested in the hint for Exercise 25.)

28. Evaluate the integral

$$\int \frac{\tan \frac{1}{2}x}{\sin x}\,dx$$

by two methods: (a) Let $z = \tan \frac{1}{2}x$; (b) let $u = \frac{1}{2}x$ and obtain an integral involving trigonometric functions of u.

29. Use the substitution of this section, $z = \tan \frac{1}{2}x$, to show that $\int \sin x\,dx = -\cos x + C$.

8.9
MISCELLANEOUS
SUBSTITUTIONS

If an integrand involves fractional powers of a variable x, the integrand can be simplified by the substitution

$$x = z^n$$

where n is the lowest common denominator of the denominators of the exponents. This is illustrated by the following example.

EXAMPLE 1
Find

$$\int \frac{\sqrt{x}\,dx}{1 + \sqrt[3]{x}}$$

SOLUTION
We let $x = z^6$; then $dx = 6z^5\,dz$. So

$$\int \frac{x^{1/2}\,dx}{1 + x^{1/3}} = \int \frac{z^3(6z^5\,dz)}{1 + z^2}$$

$$= 6 \int \frac{z^8}{z^2 + 1}\,dz$$

Dividing the numerator by the denominator we have

$$\int \frac{x^{1/2}\,dx}{1 + x^{1/3}} = 6 \int \left(z^6 - z^4 + z^2 - 1 + \frac{1}{z^2 + 1} \right) dz$$

$$= 6 \left(\tfrac{1}{7}z^7 - \tfrac{1}{5}z^5 + \tfrac{1}{3}z^3 - z + \tan^{-1} z \right) + C$$

$$= \tfrac{6}{7}x^{7/6} - \tfrac{6}{5}x^{5/6} + 2x^{1/2} - 6x^{1/6} + 6\tan^{-1} x^{1/6} + C$$

No general rule can be given to determine a substitution that will result in a simpler integrand. Sometimes a substitution that does not rationalize the given integrand may still result in a simpler integrand.

EXAMPLE 2
Evaluate $\int x^5 \sqrt{x^2 + 4}\,dx$.

SOLUTION
Let $z = \sqrt{x^2 + 4}$. Then $z^2 = x^2 + 4$, and $2z\,dz = 2x\,dx$. So

$$\int x^5 \sqrt{x^2 + 4}\,dx = \int (x^2)^2 \sqrt{x^2 + 4}\; x\,dx$$

$$= \int (z^2 - 4)^2 z(z\,dz)$$

$$= \int (z^6 - 8z^4 + 16z^2)\,dz$$

$$= \tfrac{1}{7}z^7 - \tfrac{8}{5}z^5 + \tfrac{16}{3}z^3 + C$$

$$= \tfrac{1}{105}z^3 [15z^4 - 168z^2 + 560] + C$$

$$= \tfrac{1}{105}(x^2 + 4)^{3/2}[15(x^2 + 4)^2 - 168(x^2 + 4) + 560] + C$$

$$= \tfrac{1}{105}(x^2 + 4)^{3/2}(15x^4 - 48x^2 + 128) + C$$

EXAMPLE 3
Evaluate

$$\int \frac{dx}{x^2 \sqrt{27x^2 + 6x - 1}}$$

by using the reciprocal substitution $x = 1/z$.

SOLUTION
If $x = 1/z$, then $dx = -dz/z^2$. Therefore

$$\int \frac{dx}{x^2 \sqrt{27x^2 + 6x - 1}} = \int \frac{\dfrac{-dz}{z^2}}{\left(\dfrac{1}{z^2}\right) \sqrt{\dfrac{27}{z^2} + \dfrac{6}{z} - 1}}$$

$$= -\int \frac{\sqrt{z^2}\, dz}{\sqrt{27 + 6z - z^2}}$$

$$= -\int \frac{|z|\, dz}{\sqrt{27 + 6z - z^2}}$$

$$= \begin{cases} -\displaystyle\int \frac{z\, dz}{\sqrt{27 + 6z - z^2}} & \text{if } z > 0 \\[2ex] \displaystyle\int \frac{z\, dz}{\sqrt{27 + 6z - z^2}} & \text{if } z < 0 \end{cases} \tag{1}$$

We evaluate $\int z\, dz/\sqrt{27 + 6z - z^2}$.

$$\int \frac{z\, dz}{\sqrt{27 + 6z - z^2}} = -\frac{1}{2} \int \frac{(-2z + 6)\, dz}{\sqrt{27 + 6z - z^2}} + 3 \int \frac{dz}{\sqrt{27 + 6z - z^2}}$$

$$= -\frac{1}{2} \cdot 2\sqrt{27 + 6z - z^2} + 3 \int \frac{dz}{\sqrt{27 + 9 - (z^2 - 6z + 9)}}$$

$$= -\sqrt{27 + 6z - z^2} + 3 \int \frac{dz}{\sqrt{36 - (z - 3)^2}}$$

$$= -\sqrt{27 + 6z - z^2} + 3 \sin^{-1} \frac{z - 3}{6} + C$$

$$= -\sqrt{27 + \frac{6}{x} - \frac{1}{x^2}} + 3 \sin^{-1} \frac{\dfrac{1}{x} - 3}{6} + C$$

$$= -\frac{\sqrt{27x^2 + 6x - 1}}{\sqrt{x^2}} + 3 \sin^{-1} \frac{1 - 3x}{6x} + C$$

$$= -\frac{\sqrt{27x^2 + 6x - 1}}{|x|} + 3 \sin^{-1} \frac{1 - 3x}{6x} + C \tag{2}$$

Substituting from (2) into (1) we get

$$\int \frac{dx}{x^2 \sqrt{27x^2 + 6x - 1}}$$

$$= \begin{cases} \dfrac{\sqrt{27x^2 + 6x - 1}}{|x|} - 3 \sin^{-1} \dfrac{1 - 3x}{6x} + C & \text{if } x > 0 \\[4mm] -\dfrac{\sqrt{27x^2 + 6x - 1}}{|x|} + 3 \sin^{-1} \dfrac{1 - 3x}{6x} + C & \text{if } x < 0 \end{cases}$$

$$= \begin{cases} \dfrac{\sqrt{27x^2 + 6x - 1}}{x} - 3 \sin^{-1} \dfrac{1 - 3x}{6x} + C & \text{if } x > 0 \\[4mm] \dfrac{\sqrt{27x^2 + 6x - 1}}{x} + 3 \sin^{-1} \dfrac{1 - 3x}{6x} + C & \text{if } x < 0 \end{cases}$$

EXERCISES 8.9

In Exercises 1 through 19, evaluate the indefinite integral. In Exercises 14 through 19 use the reciprocal substitution $x = 1/z$.

1. $\displaystyle \int \frac{x\,dx}{3 + \sqrt{x}}$

2. $\displaystyle \int \frac{dx}{\sqrt[3]{x} - x}$

3. $\displaystyle \int \frac{dx}{x\sqrt{1 + 4x}}$

4. $\displaystyle \int x(1 + x)^{2/3}\,dx$

5. $\displaystyle \int \frac{\sqrt{1 + x}}{1 - x}\,dx$

6. $\displaystyle \int \frac{dx}{3 + \sqrt{x + 2}}$

7. $\displaystyle \int \frac{dx}{1 + \sqrt[3]{x - 2}}$

8. $\displaystyle \int \frac{dx}{2\sqrt[3]{x} + \sqrt{x}}$

9. $\displaystyle \int \frac{dx}{\sqrt{2x} - \sqrt{x + 4}}$

10. $\displaystyle \int \frac{x^3\,dx}{\sqrt{5x^2 + 4}}$

11. $\displaystyle \int \frac{(2x^5 + 3x^2)\,dx}{\sqrt{1 + 2x^3}}$

12. $\displaystyle \int \frac{dx}{\sqrt{\sqrt{x} + 1}}$

13. $\displaystyle \int \frac{dx}{\sqrt{x}\sqrt[3]{x}(1 + \sqrt[3]{x})^2}$

14. $\displaystyle \int \frac{dx}{x\sqrt{x^2 + 4x - 4}}$

15. $\displaystyle \int \frac{dx}{x\sqrt{1 + x + x^2}}$

16. $\displaystyle \int \frac{dx}{x^2\sqrt{1 + 2x + 3x^2}}$

17. $\displaystyle \int \frac{dx}{x\sqrt{9x^2 + 4x - 1}}$

18. $\displaystyle \int \frac{dx}{x\sqrt{x^2 + 2x - 1}}$

19. $\displaystyle \int \frac{dx}{x^2\sqrt{5x^2 + 8x - 1}}$

20. Evaluate the integral in Exercise 18 by using the substitution $\sqrt{x^2 + 2x - 1} = z - x$.

In Exercises 21 through 28, evaluate the definite integral.

21. $\displaystyle \int_0^4 \frac{dx}{1 + \sqrt{x}}$

22. $\displaystyle \int_0^1 \frac{x^{3/2}}{x + 1}\,dx$

23. $\displaystyle \int_{1/2}^2 \frac{dx}{\sqrt{2x}(\sqrt{2x} + 9)}$

24. $\displaystyle \int_{16}^{18} \frac{dx}{\sqrt{x} - \sqrt[4]{x^3}}$

25. $\displaystyle \int_{1/3}^3 \frac{\sqrt[3]{x - x^3}}{x^4}\,dx$

26. $\displaystyle \int_1^2 \frac{dx}{x\sqrt{x^2 + 4x - 4}}$

27. $\displaystyle \int_0^1 \frac{\sqrt{x}}{1 + \sqrt[3]{x}}\,dx$

28. $\displaystyle \int_2^{11} \frac{x^3\,dx}{\sqrt[3]{x^2 + 4}}$

29. Evaluate $\displaystyle \int \frac{dx}{x - \sqrt{x}}$ by two methods: (a) Let $x = z^2$; (b) write $x - \sqrt{x} = \sqrt{x}(\sqrt{x} - 1)$ and let $u = \sqrt{x} - 1$.

You have seen how many different kinds of problems can be solved by evaluating a definite integral. When applying the second fundamental theorem of the calculus for this purpose it is necessary to find an indefinite integral (antiderivative), and in this chapter you have learned various techniques for such computations. However, there still remain many functions for which a definite integral cannot be evaluated exactly by using elementary functions. Examples of such definite integrals are

$$\int_0^1 \sqrt{1 + x^4}\, dx \qquad \int_0^{\pi/2} \sin x^2\, dx$$

Other examples appear in the exercises. In this section we learn two methods for computing an approximate value of a definite integral of a function that is continuous on a closed interval. These devices often give fairly good accuracy, and variations of them are used for evaluating a definite integral by computers. The first method is known as the *trapezoidal rule*.

Let f be a function continuous on a closed interval $[a, b]$. The definite integral $\int_a^b f(x)\, dx$ is the limit of a Riemann sum; that is,

$$\int_a^b f(x)\, dx = \lim_{\|\Delta\| \to 0} \sum_{i=1}^n f(\xi_i)\, \Delta_i x$$

The geometric interpretation of the Riemann sum

$$\sum_{i=1}^n f(\xi_i)\, \Delta_i x$$

is that it is equal to the sum of the measures of the areas of the rectangles lying above the x axis plus the negative of the measures of the areas of the rectangles lying below the x axis (see Fig. 4.5.3).

To approximate the measure of the area of a region we shall use trapezoids instead of rectangles. Let us also use regular partitions and function values at equally spaced points.

Thus for the definite integral $\int_a^b f(x)\, dx$ we divide the interval $[a, b]$ into n subintervals, each of length $\Delta x = (b - a)/n$. This gives the following $n + 1$ points: $x_0 = a$, $x_1 = a + \Delta x$, $x_2 = a + 2\Delta x, \ldots, x_i = a + i\Delta x, \ldots,$ $x_{n-1} = a + (n - 1)\Delta x$, $x_n = b$. Then the definite integral $\int_a^b f(x)\, dx$ may be expressed as the sum of n definite integrals as follows:

$$\int_a^b f(x)\, dx = \int_a^{x_1} f(x)\, dx + \int_{x_1}^{x_2} f(x)\, dx + \ldots$$
$$+ \int_{x_{i-1}}^{x_i} f(x)\, dx + \ldots + \int_{x_{n-1}}^b f(x)\, dx \qquad (1)$$

To interpret (1) geometrically, refer to Fig. 8.10.1, in which $f(x) \geq 0$ for all x in $[a, b]$; however, (1) holds for any function that is continuous on $[a, b]$.

Then the integral $\int_a^{x_1} f(x)\, dx$ is the measure of the area of the region bounded by the x axis, the lines $x = a$ and $x = x_1$, and the portion of the curve from P_0 to P_1. This integral may be approximated by the measure

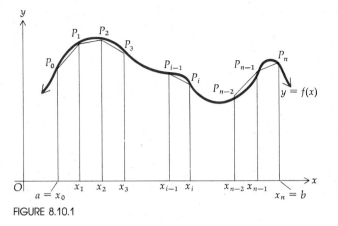

FIGURE 8.10.1

of the area of the trapezoid formed by the lines $x = a$, $x = x_1$, P_0P_1, and the x axis. By a formula from geometry the measure of the area of this trapezoid is

$$\tfrac{1}{2}[f(x_0) + f(x_1)]\,\Delta x$$

Similarly, the other integrals on the right side of (1) may be approximated by the measure of the area of a trapezoid. For the ith integral,

$$\int_{x_{i-1}}^{x_i} f(x)\,dx \approx \tfrac{1}{2}[f(x_{i-1}) + f(x_i)]\,\Delta x \tag{2}$$

So using (2) for each of the integrals on the right side of (1) we have

$$\int_a^b f(x)\,dx \approx \tfrac{1}{2}[f(x_0) + f(x_1)]\,\Delta x + \tfrac{1}{2}[f(x_1) + f(x_2)]\,\Delta x + \dots$$
$$+ \tfrac{1}{2}[f(x_{n-2}) + f(x_{n-1})]\,\Delta x + \tfrac{1}{2}[f(x_{n-1}) + f(x_n)]\,\Delta x$$

Thus

$$\int_a^b f(x)\,dx \approx \tfrac{1}{2}\Delta x[f(x_0) + 2f(x_1) + 2f(x_2) + \dots + 2f(x_{n-1}) + f(x_n)] \tag{3}$$

Formula (3) is known as the *trapezoidal rule* and is stated formally in the following theorem.

8.10.1 THEOREM   If the function f is continuous on the closed interval $[a, b]$ and the num-
The Trapezoidal Rule   bers $a = x_0, x_1, x_2, \dots, x_n = b$ form a regular partition of $[a, b]$, then

$$\int_a^b f(x)\,dx \approx \frac{b - a}{2n}\,[f(x_0) + 2f(x_1) + 2f(x_2) + \dots + 2f(x_{n-1}) + f(x_n)]$$

EXAMPLE 1
Find an approximation for

$$\int_0^3 \frac{dx}{16 + x^2}$$

by using the trapezoidal rule with $n = 6$. Express the result to three decimal places. Check by finding the exact value of the definite integral.

SOLUTION

Because $[a, b] = [0, 3]$ and $n = 6$,

$$\Delta x = \frac{b - a}{n} \qquad \frac{b - a}{2n} = \frac{3 - 0}{12}$$

$$= \frac{3 - 0}{6} \qquad\qquad = 0.25$$

$$= 0.5$$

Therefore

$$\int_0^3 \frac{dx}{16 + x^2} \approx 0.25[f(x_0) + 2f(x_1) + 2f(x_2) + 2f(x_3) + 2f(x_4) + 2f(x_5) + f(x_6)]$$

where $f(x) = 1/(16 + x^2)$. The computation of the sum in brackets in the above is shown in Table 8.10.1, where the entries can be obtained by using a hand-held calculator. Thus

$$\int_0^3 \frac{dx}{16 + x^2} = 0.25(0.6427)$$

$$\approx 0.1607$$

$$\approx 0.161$$

The exact value of the definite integral is obtained from the inverse tangent function. We have

$$\int_0^3 \frac{dx}{16 + x^2} = \frac{1}{4} \tan^{-1} \frac{x}{4} \Big]_0^3$$

$$= \tfrac{1}{4} \tan^{-1} \tfrac{3}{4}$$

Because $\tan^{-1} \tfrac{3}{4} \approx 0.6435$, we get from this result

$$\int_0^3 \frac{dx}{16 + x^2} \approx 0.161$$

which agrees with our approximation by the trapezoidal rule.

To consider the accuracy of the approximation of a definite integral by the trapezoidal rule, we refer to two kinds of errors that are introduced. One is the error due to the approximation of the graph of the function by segments of straight lines. The term **truncation error** refers to this kind of error. The other kind of error, which is unavoidable, is called the **round-off error**. It arises because numbers having a finite number of digits are used to approximate actual numbers. As the value of n (the number of subintervals) increases, the accuracy of the approximation of the area of the region by areas of trapezoids improves; thus the truncation error is reduced. However, as n increases, more computations are necessary; hence there is an increase in the round-off error. By methods discussed in *numerical analysis*, it is possible in a particular problem to determine the value of n that minimizes the combined errors. Obviously the round-off error is affected by how the calculations are performed, whether on a hand-held calculator or a computer,

Table 8.10.1

i	x_i	$f(x_i)$	k_i	$k_i \cdot f(x_i)$
0	0	0.0625	1	0.0625
1	0.5	0.0615	2	0.1230
2	1	0.0588	2	0.1176
3	1.5	0.0548	2	0.1096
4	2	0.0500	2	0.1000
5	2.5	0.0450	2	0.0900
6	3	0.0400	1	0.0400

$$\sum_{i=0}^{6} k_i f(x_i) = 0.6427$$

or whether no such mechanical devices are used. The truncation error can be estimated by a theorem. We prove first that as Δx approaches zero and n increases without bound, the limit of the approximation by the trapezoidal rule is the exact value of the definite integral. Let

$$T = \tfrac{1}{2}\Delta x[f(x_0) + 2f(x_1) + \ldots + 2f(x_{n-1}) + f(x_n)]$$

Then

$$T = [f(x_1) + f(x_2) + \ldots + f(x_n)]\Delta x + \tfrac{1}{2}[f(x_0) - f(x_n)]\Delta x$$

$$\Leftrightarrow \quad T = \sum_{i=1}^{n} f(x_i)\Delta x + \tfrac{1}{2}[f(a) - f(b)]\Delta x$$

Therefore, if $n \to +\infty$ and $\Delta x \to 0$,

$$\lim_{\Delta x \to 0} T = \lim_{\Delta x \to 0} \sum_{i=1}^{n} f(x_i)\Delta x + \lim_{\Delta x \to 0} \tfrac{1}{2}[f(a) - f(b)]\Delta x$$

$$= \int_{a}^{b} f(x)\,dx + 0$$

Thus we can make the difference between T and the value of the definite integral as small as we please by taking n sufficiently large (and consequently Δx sufficiently small).

The following theorem, which is proved in numerical analysis, gives a method for estimating the truncation error obtained when using the trapezoidal rule. The truncation error is denoted by ϵ_T.

8.10.2 THEOREM   Let the function f be continuous on the closed interval $[a, b]$, and f' and f'' both exist on $[a, b]$. If

$$\epsilon_T = \int_{a}^{b} f(x)\,dx - T$$

where T is the approximate value of $\int_{a}^{b} f(x)\,dx$ found by the trapezoidal rule, then there is some number η in $[a, b]$ such that

$$\epsilon_T = -\tfrac{1}{12}(b - a)f''(\eta)(\Delta x)^2 \tag{4}$$

EXAMPLE 2
Find the bounds for the truncation error in the result of Example 1.

SOLUTION
We first find the absolute minimum and absolute maximum values of $f''(x)$ on $[0, 3]$.

$$f(x) = (16 + x^2)^{-1}$$

$$f'(x) = -2x(16 + x^2)^{-2}$$

$$f''(x) = 8x^2(16 + x^2)^{-3} - 2(16 + x^2)^{-2}$$

$$= (6x^2 - 32)(16 + x^2)^{-3}$$

$$f'''(x) = -6x(6x^2 - 32)(16 + x^2)^{-4} + 12x(16 + x^2)^{-3}$$

$$= 24x(16 - x^2)(16 + x^2)^{-4}$$

Because $f'''(x) > 0$ for all x in the open interval $(0, 3)$, then f'' is increasing on the open interval $(0, 3)$. Therefore the absolute minimum value of f'' on $[0, 3]$ is $f''(0)$, and the absolute maximum value of f'' on $[0, 3]$ is $f''(3)$.

$$f''(0) = -\frac{1}{128} \qquad f''(3) = \frac{22}{15,625}$$

Taking $\eta = 0$ on the right side of (4) we get

$$-\frac{3}{12}\left(-\frac{1}{128}\right)\frac{1}{4} = \frac{1}{2048}$$

Taking $\eta = 3$ on the right side of (4) we have

$$-\frac{3}{12}\left(\frac{22}{15,625}\right)\frac{1}{4} = -\frac{11}{125,000}$$

Therefore if ϵ_T is the truncation error in the result of Example 1,

$$-\frac{11}{125,000} \leq \epsilon_T \leq \frac{1}{2048}$$

$$-0.0001 \leq \epsilon_T \leq 0.0005$$

If in Theorem 8.10.2 $f(x) = mx + b$, then $f''(x) = 0$ for all x. Therefore $\epsilon_T = 0$; so the trapezoidal rule gives the exact value of the definite integral of a linear function.

Another method for approximating the value of a definite integral is provided by *Simpson's rule* (sometimes referred to as the *parabolic rule*), named after the British mathematician Thomas Simpson (1710–1761). For a given partition of the closed interval $[a, b]$, Simpson's rule usually gives a better approximation than the trapezoidal rule. In the trapezoidal rule, successive points on the graph of $y = f(x)$ are connected by segments of straight lines, whereas in Simpson's rule the points are connected by segments of parabolas. Before Simpson's rule is developed, we state and prove a theorem that will be needed.

8.10.3 THEOREM   If $P_0(x_0, y_0)$, $P_1(x_1, y_1)$, and $P_2(x_2, y_2)$ are three noncollinear points on the parabola having the equation $y = Ax^2 + Bx + C$, where $y_0 \geq 0$, $y_1 \geq 0$, $y_2 \geq 0$, $x_1 = x_0 + h$, and $x_2 = x_0 + 2h$, then the measure of the area of the region bounded by the parabola, the x axis, and the lines $x = x_0$ and $x = x_2$ is given by

$$\tfrac{1}{3}h(y_0 + 4y_1 + y_2)$$

PROOF   The parabola whose equation is $y = Ax^2 + Bx + C$ has a vertical axis. Refer to Fig. 8.10.2, which shows the region bounded by the parabola, the x axis, and the lines $x = x_0$ and $x = x_2$.

Because P_0, P_1, and P_2 are points on the parabola, their coordinates satisfy the equation of the parabola. So when we replace x_1 by $x_0 + h$, and

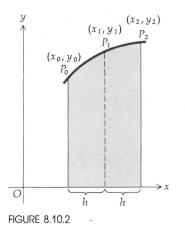

FIGURE 8.10.2

x_2 by $x_0 + 2h$, we have

$$y_0 = Ax_0^2 + Bx_0 + C$$

$$y_1 = A(x_0 + h)^2 + B(x_0 + h) + C$$
$$= A(x_0^2 + 2hx_0 + h^2) + B(x_0 + h) + C$$

$$y_2 = A(x_0 + 2h)^2 + B(x_0 + 2h) + C$$
$$= A(x_0^2 + 4hx_0 + 4h^2) + B(x_0 + 2h) + C$$

Therefore

$$y_0 + 4y_1 + y_2 = A(6x_0^2 + 12hx_0 + 8h^2) + B(6x_0 + 6h) + 6C \qquad (5)$$

Now if K square units is the area of the region, then K can be computed by the limit of a Riemann sum, and we have

$$K = \lim_{||\Delta|| \to 0} \sum_{i=1}^{n} (A\xi_i^2 + B\xi_i + C)\Delta x$$

$$= \int_{x_0}^{x_0 + 2h} (Ax^2 + Bx + C)\, dx$$

$$= \tfrac{1}{3}Ax^3 + \tfrac{1}{2}Bx^2 + Cx \Big]_{x_0}^{x_0 + 2h}$$

$$= \tfrac{1}{3}A(x_0 + 2h)^3 + \tfrac{1}{2}B(x_0 + 2h)^2 + C(x_0 + 2h) - (\tfrac{1}{3}Ax_0^3 + \tfrac{1}{2}Bx_0^2 + Cx_0)$$

$$= \tfrac{1}{3}h[A(6x_0^2 + 12hx_0 + 8h^2) + B(6x_0 + 6h) + 6C] \qquad (6)$$

Substituting from (5) in (6) we get

$$K = \tfrac{1}{3}h(y_0 + 4y_1 + y_2) \qquad \blacksquare$$

Let the function f be continuous on the closed interval $[a, b]$. Consider a regular partition of the interval $[a, b]$ of n subintervals, where n is even. The length of each subinterval is given by $\Delta x = (b - a)/n$. Let the points on the curve $y = f(x)$ having these partitioning points as abscissas be denoted by $P_0(x_0, y_0)$, $P_1(x_1, y_1)$, ..., $P_n(x_n, y_n)$; see Fig. 8.10.3, where $f(x) \geq 0$ for all x in $[a, b]$.

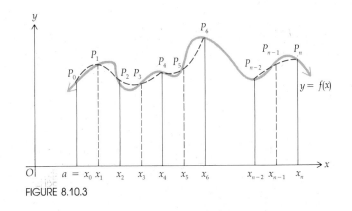

FIGURE 8.10.3

We approximate the segment of the curve $y = f(x)$ from P_0 to P_2 by the segment of the parabola with a vertical axis and through P_0, P_1, and P_2. Then by Theorem 8.10.3 the measure of the area of the region bounded by this parabola, the x axis, and the lines $x = x_0$ and $x = x_2$, with $h = \Delta x$, is given by

$$\tfrac{1}{3}\Delta x(y_0 + 4y_1 + y_2) \quad \text{or} \quad \tfrac{1}{3}\Delta x[f(x_0) + 4f(x_1) + f(x_2)]$$

In a similar manner we approximate the segment of the curve $y = f(x)$ from P_2 to P_4 by the segment of the parabola with a vertical axis and through P_2, P_3, and P_4. The measure of the area of the region bounded by this parabola, the x axis, and the lines $x = x_2$ and $x = x_4$ is given by

$$\tfrac{1}{3}\Delta x(y_2 + 4y_3 + y_4) \quad \text{or} \quad \tfrac{1}{3}\Delta x[f(x_2) + 4f(x_3) + f(x_4)]$$

This process is continued until there are $\tfrac{1}{2}n$ such regions, and the measure of the area of the last region is given by

$$\tfrac{1}{3}\Delta x(y_{n-2} + 4y_{n-1} + y_n) \quad \text{or} \quad \tfrac{1}{3}\Delta x[f(x_{n-2}) + 4f(x_{n-1}) + f(x_n)]$$

The sum of the measures of the areas of these regions approximates the measure of the area of the region bounded by the curve whose equation is $y = f(x)$, the x axis, and the lines $x = a$ and $x = b$. The measure of the area of this region is given by the definite integral $\int_a^b f(x)\,dx$. So we have as an approximation to the definite integral

$$\tfrac{1}{3}\Delta x[f(x_0) + 4f(x_1) + f(x_2)] + \tfrac{1}{3}\Delta x[f(x_2) + 4f(x_3) + f(x_4)] + \cdots$$
$$+ \tfrac{1}{3}\Delta x[f(x_{n-4}) + 4f(x_{n-3}) + f(x_{n-2})] + \tfrac{1}{3}\Delta x[f(x_{n-2}) + 4f(x_{n-1}) + f(x_n)]$$

Thus

$$\int_a^b f(x)\,dx \approx \tfrac{1}{3}\Delta x[f(x_0) + 4f(x_1) + 2f(x_2) + 4f(x_3) + 2f(x_4) + \cdots$$
$$+ 2f(x_{n-2}) + 4f(x_{n-1}) + f(x_n)] \qquad (7)$$

where $\Delta x = (b - a)/n$.

Formula (7) is known as *Simpson's rule*. It is stated formally in the next theorem.

8.10.4 THEOREM
Simpson's Rule

If the function f is continuous on the closed interval $[a, b]$, n is an even integer, and the numbers $a = x_0, x_1, x_2, \ldots, x_{n-1}, x_n = b$ form a regular partition of $[a, b]$, then

$$\int_a^b f(x)\,dx \approx \frac{b - a}{3n}[f(x_0) + 4f(x_1) + 2f(x_2) + 4f(x_3) + 2f(x_4) + \cdots$$

$$+ 2f(x_{n-2}) + 4f(x_{n-1}) + f(x_n)]$$

EXAMPLE 3
Use Simpson's rule with $n = 4$ to approximate the value of

$$\int_0^2 \frac{dx}{x + 1}$$

Express the result to four decimal places. Check by finding the exact value of the definite integral.

SOLUTION
Applying Simpson's rule with $n = 4$ we have

$$x = \frac{b - a}{n} \qquad \frac{b - a}{3n} = \frac{1 - 0}{3(4)}$$

$$= \frac{1 - 0}{4} \qquad\qquad = \tfrac{1}{12}$$

$$= \tfrac{1}{4}$$

Therefore, if $f(x) = 1/(x + 1)$,

$$\int_0^1 \frac{dx}{x + 1} \approx \tfrac{1}{12}[f(x_0) + 4f(x_1) + 2f(x_2) + 4f(x_3) + f(x_4)] \tag{8}$$

The computation of the expression in brackets on the right side of (8) is shown in Table 8.10.2.

Substituting the sum from Table 8.10.2 in (8) we get

$$\int_0^1 \frac{dx}{x + 1} \approx \frac{1}{12}(8.31906)$$

$$\approx 0.69325^+$$

Rounding off the result to four decimal places gives

$$\int_0^1 \frac{dx}{x + 1} \approx 0.6933$$

The exact value of the definite integral is found as follows:

$$\int_0^1 \frac{dx}{x + 1} = \ln|x + 1|\Big]_0^1$$

$$= \ln 2 - \ln 1$$

$$= \ln 2$$

An approximation of $\ln 2$ to four decimal places is 0.6931, which agrees with the approximation by Simpson's rule in the first three places. The error in the approximation by Simpson's rule is -0.0002.

In applying Simpson's rule, the larger we take the value of n, the smaller will be the value of Δx, and so geometrically it seems evident that the smaller will be the truncation error of the approximation, because a parabola passing through three points of a curve that are close to each other will be close to the curve throughout the subinterval of width $2\,\Delta x$.

The following theorem that is proved in numerical analysis gives a method for determining the truncation error in applying Simpson's rule. This error is denoted by ϵ_S.

Table 8.10.2

i	x_i	$f(x_i)$	k_i	$k_i \cdot f(x_i)$
0	0	1.00000	1	1.00000
1	0.25	0.80000	4	3.20000
2	0.5	0.66667	2	1.33334
3	0.75	0.57143	4	2.28572
4	1	0.50000	1	0.50000

$$\sum_{i=0}^4 k_i f(x_i) = 8.31906$$

8.10.5 THEOREM   Let the function f be continuous on the closed interval $[a, b]$, and f', f'', f''', and $f^{(iv)}$ all exist on $[a, b]$. If

$$\epsilon_S = \int_a^b f(x)\, dx - S$$

where S is the approximate value of $\int_a^b f(x)\, dx$ found by Simpson's rule, then there is some number η in $[a, b]$ such that

$$\epsilon_S = -\tfrac{1}{180}(b - a) f^{(iv)}(\eta)(\Delta x)^4 \tag{9}$$

EXAMPLE 4
Find the bounds for the truncation error in Example 3.

SOLUTION

$$f(x) = (x + 1)^{-1}$$
$$f'(x) = -1(x + 1)^{-2}$$
$$f''(x) = 2(x + 1)^{-3}$$
$$f'''(x) = -6(x + 1)^{-4}$$
$$f^{(iv)}(x) = 24(x + 1)^{-5}$$
$$f^{(v)}(x) = -120(x + 1)^{-6}$$

Because $f^{(v)}(x) < 0$ for all x in $[0, 1]$, $f^{(iv)}$ is decreasing on $[0, 1]$. Thus the absolute minimum value of $f^{(iv)}$ is at the right endpoint 1, and the absolute maximum value of $f^{(iv)}$ on $[0, 1]$ is at the left endpoint 0.

$$f^{(iv)}(0) = 24 \quad \text{and} \quad f^{(iv)}(1) = \tfrac{3}{4}$$

Substituting 0 for η in the right side of (9) we get

$$-\tfrac{1}{180}(b - a) f^{(iv)}(0)(\Delta x)^4 = -\tfrac{1}{180}(24)(\tfrac{1}{4})^4$$
$$\approx -0.00052$$

Substituting 1 for η in the right side of (9) we have

$$-\frac{1}{180}(b - a) f^{(iv)}(1)(\Delta x)^4 = -\frac{1}{180} \cdot \frac{3}{4}\left(\frac{1}{4}\right)^4$$
$$\approx -0.00002$$

So

$$-0.00052 \leq \epsilon_S \leq -0.00002 \tag{10}$$

The inequality (10) agrees with the discussion in Example 3 regarding the error in the approximation of $\int_0^1 dx/(x + 1)$ by Simpson's rule because $-0.00052 < -0.0002 < -0.00002$.

If $f(x)$ is a polynomial of degree three or less, then $f^{(iv)}(x) \equiv 0$ and therefore $\epsilon_S = 0$. In other words, Simpson's rule gives an exact result for a poly-

FIGURE 8.10.4

nomial of the third degree or lower. This statement is geometrically obvious if $f(x)$ is of the second or first degree because in the first case the graph of $y = f(x)$ is a parabola, and in the second case the graph is a straight line.

There is an important function in statistics called the **standardized normal probability density function** that is defined by

$$N(x) = \frac{1}{\sqrt{2\pi}} e^{-x^2/2}$$

A sketch of the graph of N is shown in Fig. 8.10.4. It is the bell-shaped curve well known to statisticians. The curve is symmetric with respect to the y axis. Table 8.10.3 gives values of $N(x)$ for some values of x.

Observe that as either x increases without bound or decreases without bound, then $N(x)$ rapidly approaches 0. For instance, $N(3) = 0.004$ and $N(4) = 0.0001$. The probability that a random choice of x will be in the closed interval $[a, b]$ is denoted by $P([a, b])$, and

$$P([a, b]) = \frac{1}{\sqrt{2\pi}} \int_a^b e^{-x^2/2}\, dx \tag{11}$$

The definite integral in (11) is the measure of the area of the region bounded above by the curve $y = N(x)$, below by the x axis, and on the sides by the lines $x = a$ and $x = b$. This definite integral cannot be evaluated by the second fundamental theorem of the calculus because we cannot find an antiderivative of the integrand by using elementary functions. However, we can compute an approximate value of such a definite integral by either the trapezoidal rule or Simpson's rule, as shown in the following example.

Table 8.10.3

x	$N(x)$
0	0.40
± 0.5	0.35
± 1.0	0.24
± 1.5	0.13
± 2.0	0.05
± 2.5	0.02
± 3.0	0.004
± 3.5	0.001
± 4.0	0.0001

EXAMPLE 5

For the standardized normal probability density function, determine the probability that a random choice of x will be in the interval $[0, 2]$. Approximate the value of the definite integral by (a) the trapezoidal rule with $n = 4$ and (b) Simpson's rule with $n = 4$.

SOLUTION

The probability that a random choice of x will be in the interval $[0, 2]$ is $P([0, 2])$, and from (11),

$$P([0, 2]) = \frac{1}{\sqrt{2\pi}} \int_0^2 e^{-x^2/2}\, dx \tag{12}$$

(a) We approximate the integral in (12) by the trapezoidal rule with $n = 4$. Because $[a, b] = [0, 2]$, $\Delta x = \frac{1}{2}$. Therefore, with $f(x) = e^{-x^2/2}$,

$$\int_0^2 e^{-x^2/2}\, dx \approx \tfrac{1}{4}\left[f(0) + 2f(\tfrac{1}{2}) + 2f(1) + 2f(\tfrac{3}{2}) + f(2)\right]$$
$$= \tfrac{1}{4}\left[e^0 + 2e^{-1/8} + 2e^{-1/2} + 2e^{-9/8} + e^{-2}\right]$$
$$\approx \tfrac{1}{4}\left[1 + 2(0.8825) + 2(0.6065) + 2(0.3246) + 0.1353\right]$$
$$= \tfrac{1}{4}(4.7625)$$
$$\approx 1.191$$

Thus

$$P([0,2]) \approx \frac{1}{\sqrt{2\pi}} (1.191)$$

$$\approx 0.475$$

(b) If Simpson's rule with $n = 4$ is used to approximate the integral in (12), we have

$$\int_0^2 e^{-x^2/2} \, dx \approx \tfrac{1}{6}[f(0) + 4f(\tfrac{1}{2}) + 2f(1) + 4f(\tfrac{3}{2}) + f(2)]$$

$$= \tfrac{1}{6}[e^0 + 4e^{-1/8} + 2e^{-1/2} + 4e^{-9/8} + e^{-2}]$$

$$\approx \tfrac{1}{6}[1 + 4(0.8825) + 2(0.6065) + 4(0.3246) + 0.1353]$$

$$= \tfrac{1}{6}(7.1767)$$

$$\approx 1.196$$

Therefore

$$P([0,2]) \approx \frac{1}{\sqrt{2\pi}} (1.196)$$

$$\approx 0.477$$

FIGURE 8.10.5

FIGURE 8.10.6

In Fig. 8.10.5 the shaded region is in the first quadrant and is bounded by the graph of the standardized normal probability density function, the coordinate axes, and the line $x = t$, where $t > 0$. If $A(t)$ square units is the area of this region, it can be shown, although it is difficult to do so, that $\lim_{t \to +\infty} A(t) = 0.5$. Thus the exact value of $P([0,2])$ is less than 0.5, because it is the measure of the area of the shaded region in Fig. 8.10.6. This fact agrees with the results in Example 5.

EXERCISES 8.10

In Exercises 1 through 12, compute the approximate value of the given definite integral by the trapezoidal rule for the indicated value of n. Express the result to three decimal places. In Exercises 1 through 6, find the exact value of the definite integral, and compare the result with the approximation.

1. $\int_0^2 x^3 \, dx; \; n = 4$ 2. $\int_0^2 x\sqrt{4 - x^2} \, dx; \; n = 8$ 3. $\int_1^2 \frac{dx}{x}; \; n = 5$ 4. $\int_2^{10} \frac{dx}{1 + x}; \; n = 8$

5. $\int_0^\pi \cos x \, dx; \; n = 4$ 6. $\int_0^\pi \sin x \, dx; \; n = 6$ 7. $\int_0^1 \frac{dx}{\sqrt{1 + x^2}}; \; n = 5$ 8. $\int_2^3 \sqrt{1 + x^2} \, dx; \; n = 6$

9. $\int_{\pi/2}^{3\pi/2} \frac{\sin x}{x} \, dx; \; n = 6$ 10. $\int_0^1 \sqrt{1 + x^3} \, dx; \; n = 4$ 11. $\int_0^2 \sqrt{1 + x^4} \, dx; \; n = 6$ 12. $\int_0^\pi \frac{\sin x}{1 + x} \, dx; \; n = 6$

In Exercises 13 through 18, find the bounds for the truncation error in the approximation of the indicated exercise.
13. Exercise 1   14. Exercise 5   15. Exercise 3   16. Exercise 4   17. Exercise 6   18. Exercise 8

19. Approximate $\int_0^2 x^3 \, dx$ to three decimal places by Simpson's rule with $n = 4$. Compare the result with those obtained in Exercise 1, and observe that Simpson's rule gives better accuracy than the trapezoidal rule with the same number of subintervals.

20. Approximate $\int_0^\pi \sin x \, dx$ to three decimal places by Simpson's rule with $n = 6$. Compare the result with those obtained in Exercise 6, and observe that Simpson's rule gives better accuracy than the trapezoidal rule with the same number of subintervals.

In Exercises 21 through 24, approximate the definite integral by Simpson's rule with the indicated value of n. Express the answer to three decimal places.

21. $\int_{-1}^0 \dfrac{dx}{1-x}$; $n = 4$ **22.** $\int_{-1/2}^0 \dfrac{dx}{\sqrt{1-x^2}}$; $n = 4$ **23.** $\int_0^1 \dfrac{dx}{x^2+x+1}$; $n = 4$ **24.** $\int_1^2 \dfrac{dx}{x+1}$; $n = 8$

In Exercises 25 through 28, find bounds for the truncation error in the approximation of the indicated exercise.

25. Exercise 19   **26.** Exercise 20   **27.** Exercise 21   **28.** Exercise 24

Each of the definite integrals in Exercises 29 through 34 cannot be evaluated exactly in terms of elementary functions. Use Simpson's rule, with the indicated value of n, to find an approximate value of the given definite integral. Express the result to three decimal places.

29. $\int_{\pi/2}^{3\pi/2} \dfrac{\sin x}{x} \, dx$; $n = 6$ **30.** $\int_0^2 \sqrt{1+x^4} \, dx$; $n = 6$ **31.** $\int_1^{1.8} \sqrt{1+x^3} \, dx$; $n = 4$

32. $\int_0^1 \sqrt[3]{1-x^2} \, dx$; $n = 4$ **33.** $\int_0^2 \dfrac{dx}{\sqrt{1+x^3}}$; $n = 8$ **34.** $\int_0^{\pi/2} \sqrt{\sin x} \, dx$; $n = 6$

35. Show that the exact value of the integral $\int_0^2 \sqrt{4-x^2} \, dx$ is π by interpreting it as the measure of the area of a region. Approximate the definite integral by the trapezoidal rule with $n = 8$. Give the result to three decimal places, and compare the value so obtained with the exact value.

36. Show that the exact value of the integral $\int_0^1 4\sqrt{1-x^2} \, dx$ is π by interpreting it as the measure of the area of a region. Use Simpson's rule with $n = 6$ to get an approximate value of the definite integral to three decimal places. Compare the results.

37. For the standardized normal probability density function determine the probability that a random choice of x will be in the interval $[0, 1]$. Approximate the value of the definite integral by (a) the trapezoidal rule with $n = 4$ and (b) Simpson's rule with $n = 4$.

38. For the standardized normal probability density function, determine the probability that a random choice of x will be in the interval $[-3, 3]$. Approximate the value of the definite integral by (a) the trapezoidal rule with $n = 6$ and (b) Simpson's rule with $n = 6$.

39. Express the length of one arch of the sine curve as a definite integral. Use Simpson's rule with $n = 4$ to approximate the value of the definite integral to three decimal places.

40. Express as a definite integral the length of the arc of the curve $y = \frac{1}{3}x^3$ from the origin to the point $(1, \frac{1}{3})$. Use Simpson's rule with $n = 4$ to approximate the value of the definite integral to three decimal places.

41. Suppose that f is a function continuous on the closed interval $[a, b]$. Let T be the approximate value of $\int_a^b f(x) \, dx$ by using the trapezoidal rule, and let S be the approximate value of $\int_a^b f(x) \, dx$ by using Simpson's rule, where the same partition of the interval $[a, b]$ is used for both approximations. Show that

$$S = \tfrac{2}{3}[T + \Delta x(f(x_1) + f(x_3) + f(x_5) + \ldots + f(x_{n-1}))]$$

where n is even.

42. Find the area of the region enclosed by the loop of the curve whose equation is $y^2 = 8x^2 - x^5$. Evaluate the definite integral by Simpson's rule with $n = 8$. Express the result to three decimal places.

43. The error function, denoted by erf, is defined by

$$\text{erf}(x) = \frac{2}{\sqrt{\pi}} \int_0^x e^{-t^2} \, dt$$

Find an approximate value of erf(1) to four decimal places by using Simpson's rule with $n = 10$.

44. For the error function defined in Exercise 43, find an approximate value of erf(10) to four decimal places by using Simpson's rule with $n = 10$.

8.11 *(Supplementary)*
INTEGRALS YIELDING INVERSE HYPERBOLIC FUNCTIONS

The inverse hyperbolic functions can be applied in integration, and sometimes their use shortens the computation considerably. However, no new types of integrals are evaluated by this procedure. Only new forms of the results are obtained.

From formula (15) in Section 7.5,

$$D_x(\sinh^{-1} u) = \frac{1}{\sqrt{u^2 + 1}} \, D_x u$$

from which we obtain the integration formula

$$\int \frac{du}{\sqrt{u^2 + 1}} = \sinh^{-1} u + C$$

If formula (8) of Section 7.5 is used to express $\sinh^{-1} u$ as a natural logarithm, the following theorem is obtained.

8.11.1 THEOREM

$$\int \frac{du}{\sqrt{u^2 + 1}} = \sinh^{-1} u + C$$

$$= \ln(u + \sqrt{u^2 + 1}) + C$$

Formula (16) in Section 7.5 is

$$D_x(\cosh^{-1} u) = \frac{1}{\sqrt{u^2 - 1}} \, D_x u \qquad \text{where } u > 1$$

from which it follows that

$$\int \frac{du}{\sqrt{u^2 - 1}} = \cosh^{-1} u + C \qquad \text{if } u > 1$$

By combining this with formula (9) in Section 7.5, we obtain the following theorem.

8.11.2 THEOREM

$$\int \frac{du}{\sqrt{u^2 - 1}} = \cosh^{-1} u + C$$

$$= \ln(u + \sqrt{u^2 - 1}) + C \qquad \text{if } u > 1$$

Formulas (17) and (18) of Section 7.5 are, respectively,

$$D_x(\tanh^{-1} u) = \frac{1}{1 - u^2} D_x u \qquad \text{where } |u| < 1$$

$$D_x(\coth^{-1} u) = \frac{1}{1 - u^2} D_x u \qquad \text{where } |u| > 1$$

From the above two formulas we get

$$\int \frac{du}{1 - u^2} = \begin{cases} \tanh^{-1} u + C & \text{if } |u| < 1 \\ \coth^{-1} u + C & \text{if } |u| > 1 \end{cases}$$

With this formula and (10) and (11) of Section 7.5 we have the next theorem.

8.11.3 THEOREM

$$\int \frac{du}{1 - u^2} = \begin{cases} \tanh^{-1} u + C & \text{if } |u| < 1 \\ \coth^{-1} u + C & \text{if } |u| > 1 \end{cases}$$

$$= \frac{1}{2} \ln \left| \frac{1 + u}{1 - u} \right| + C \qquad \text{if } u \neq 1$$

We also have the three formulas given in the following theorem.

8.11.4 THEOREM

$$\int \frac{du}{\sqrt{u^2 + a^2}} = \sinh^{-1} \frac{u}{a} + C$$

$$= \ln(u + \sqrt{u^2 + a^2}) + C \qquad \text{if } a > 0 \tag{1}$$

$$\int \frac{du}{\sqrt{u^2 - a^2}} = \cosh^{-1} \frac{u}{a} + C$$

$$= \ln(u + \sqrt{u^2 - a^2}) + C \qquad \text{if } u > a > 0 \tag{2}$$

$$\int \frac{du}{a^2 - u^2} = \begin{cases} \dfrac{1}{a} \tanh^{-1} \dfrac{u}{a} + C & \text{if } |u| < a \\ \dfrac{1}{a} \coth^{-1} \dfrac{u}{a} + C & \text{if } |u| > a \end{cases}$$

$$= \frac{1}{2a} \ln \left| \frac{a + u}{a - u} \right| + C \qquad \text{if } u \neq a \text{ and } a \neq 0 \tag{3}$$

PROOF   The formulas can be proved by finding the derivative of the right side and obtaining the integrand, or else more directly by using a hyperbolic function substitution. The proof of (1) by differentiating the right side is as follows:

$$D_u\left(\sinh^{-1} \frac{u}{a}\right) = \frac{1}{\sqrt{\left(\dfrac{u}{a}\right)^2 + 1}} \cdot \frac{1}{a}$$

$$= \frac{\sqrt{a^2}}{\sqrt{u^2 + a^2}} \cdot \frac{1}{a}$$

and because $a > 0$, $\sqrt{a^2} = a$; thus

$$D_u\left(\sinh^{-1}\frac{u}{a}\right) = \frac{1}{\sqrt{u^2 + a^2}}$$

To obtain the natural logarithm representation we use formula (8) of Section 7.5, and we have

$$\sinh^{-1}\frac{u}{a} = \ln\left(\frac{u}{a} + \sqrt{\left(\frac{u}{a}\right)^2 + 1}\right)$$

$$= \ln\left(\frac{u}{a} + \frac{\sqrt{u^2 + a^2}}{a}\right)$$

$$= \ln(u + \sqrt{u^2 + a^2}) - \ln a$$

Therefore

$$\sinh^{-1}\frac{u}{a} + C = \ln(u + \sqrt{u^2 + a^2}) - \ln a + C$$

$$= \ln(u + \sqrt{u^2 + a^2}) + C_1$$

where $C_1 = C - \ln a$.

The proof of (2) by a hyperbolic function substitution makes use of the identity $\cosh^2 x - \sinh^2 x = 1$. Let $u = a \cosh x$, where $x > 0$; then $du = a \sinh x\, dx$, and $x = \cosh^{-1}(u/a)$. Substituting in the integrand we get

$$\int \frac{du}{\sqrt{u^2 - a^2}} = \int \frac{a \sinh x\, dx}{\sqrt{a^2 \cosh^2 x - a^2}}$$

$$= \int \frac{a \sinh x\, dx}{\sqrt{a^2}\sqrt{\cosh^2 x - 1}}$$

$$= \int \frac{a \sinh x\, dx}{\sqrt{a^2}\sqrt{\sinh^2 x}}$$

Because $a > 0$, $\sqrt{a^2} = a$. Furthermore, because $x > 0$, $\sinh x > 0$; therefore, $\sqrt{\sinh^2 x} = \sinh x$. Then

$$\int \frac{du}{\sqrt{u^2 - a^2}} = \int \frac{a \sinh x\, dx}{a \sinh x}$$

$$= \int dx$$

$$= x + C$$

$$= \cosh^{-1}\frac{u}{a} + C$$

The natural logarithm representation may be obtained in a way similar to the way we obtained the one for formula (1). The proof of formula (3) is left as an exercise (see Exercise 23).   ■

The formulas of Theorems 8.11.1 through 8.11.4 give alternate representations of the integral in question. In evaluating a definite integral in which one of these forms occurs, the inverse hyperbolic function representation may be easier to use. Also the inverse hyperbolic function representation is a less cumbersome one to write for the indefinite integral. Observe that the natural logarithm form of formula (3) was obtained in Section 8.6 by using partial fractions.

EXAMPLE 1

Evaluate

$$\int \frac{dx}{\sqrt{x^2 - 6x + 13}}$$

SOLUTION

We apply formula (1) after rewriting the denominator.

$$\int \frac{dx}{\sqrt{x^2 - 6x + 13}} = \int \frac{dx}{\sqrt{(x^2 - 6x + 9) + 4}}$$

$$= \int \frac{dx}{\sqrt{(x - 3)^2 + 4}}$$

$$= \sinh^{-1}\left(\frac{x - 3}{2}\right) + C$$

$$= \ln(x - 3 + \sqrt{x^2 - 6x + 13}) + C$$

EXAMPLE 2

Evaluate

$$\int_6^{10} \frac{dx}{\sqrt{x^2 - 5}}$$

SOLUTION

From (2),

$$\int_6^{10} \frac{dx}{\sqrt{x^2 - 25}} = \cosh^{-1}\frac{x}{5}\bigg]_6^{10}$$

$$= \cosh^{-1} 2 - \cosh^{-1} 1.2$$

From a calculator or a table such as Table 4 in the appendix, we obtain $\cosh^{-1} 2 \approx 1.32$ and $\cosh^{-1} 1.2 \approx 0.62$. Thus

$$\int_6^{10} \frac{dx}{\sqrt{x^2 - 25}} \approx 0.70$$

Instead of applying the formulas, integrals of the forms of those in Theorems 8.11.1 through 8.11.4 can be obtained by using a hyperbolic function substitution and proceeding in a manner similar to that using a trigonometric substitution.

EXAMPLE 3

Evaluate the integral of Example 1 without using a formula but by using a hyperbolic function substitution.

SOLUTION

In the solution of Example 1 the given integral was rewritten as

$$\int \frac{dx}{\sqrt{(x-3)^2 + 4}}$$

If we let $x - 3 = 2 \sinh u$, then $dx = 2 \cosh u\,du$ and $u = \sinh^{-1} \frac{1}{2}(x - 3)$. Therefore

$$\int \frac{dx}{\sqrt{(x-3)^2 + 4}} = \int \frac{2 \cosh u\,du}{\sqrt{4 \sinh^2 u + 4}}$$

$$= \int \frac{2 \cosh u\,du}{2\sqrt{\sinh^2 u + 1}}$$

$$= \int \frac{\cosh u\,du}{\sqrt{\cosh^2 u}}$$

$$= \int \frac{\cosh u\,du}{\cosh u}$$

$$= \int du$$

$$= u + C$$

$$= \sinh^{-1}\left(\frac{x-3}{2}\right) + C$$

which agrees with the result of Example 1.

EXERCISES 8.11

In Exercises 1 through 16, express the indefinite integral in terms of an inverse hyperbolic function and as a natural logarithm.

1. $\displaystyle\int \frac{dx}{\sqrt{4 + x^2}}$

2. $\displaystyle\int \frac{dx}{\sqrt{4x^2 - 9}}$

3. $\displaystyle\int \frac{x\,dx}{\sqrt{x^4 - 1}}$

4. $\displaystyle\int \frac{dx}{25 - x^2}$

5. $\displaystyle\int \frac{dx}{9x^2 - 16}$

6. $\displaystyle\int \frac{dx}{4e^x - e^{-x}}$

7. $\displaystyle\int \frac{\cos x\,dx}{\sqrt{4 - \cos^2 x}}$

8. $\displaystyle\int \frac{dx}{\sqrt{x^2 - 4x + 1}}$

9. $\displaystyle\int \frac{dt}{\sqrt{5 - e^{-2t}}}$

10. $\displaystyle\int \frac{dw}{4w - w^2 - 3}$

11. $\displaystyle\int \frac{dx}{2 - 4x - x^2}$

12. $\displaystyle\int \frac{dx}{\sqrt{25x^2 + 9}}$

13. $\displaystyle\int \frac{dx}{x^2 + 10x + 24}$

14. $\displaystyle\int \frac{dz}{\sqrt{9z^2 - 6z - 8}}$

15. $\displaystyle\int \frac{3\,dw}{w\sqrt{4 \ln^2 w + 9}}$

16. $\displaystyle\int \frac{3x\,dx}{\sqrt{x^4 + 6x^2 + 5}}$

In Exercises 17 through 22, evaluate the definite integral and express the answer in terms of a natural logarithm.

17. $\displaystyle\int_3^5 \frac{dx}{\sqrt{x^2 - 4}}$

18. $\displaystyle\int_{-4}^{-3} \frac{dx}{1 - x^2}$

19. $\displaystyle\int_{-1/2}^{1/2} \frac{dx}{1 - x^2}$

20. $\displaystyle\int_1^2 \frac{dx}{\sqrt{x^2 + 2x}}$

21. $\displaystyle\int_2^3 \frac{dx}{\sqrt{9x^2 - 12x - 5}}$

22. $\displaystyle\int_{-2}^2 \frac{dx}{\sqrt{16 + x^2}}$

23. Prove formula (3) by using a hyperbolic function substitution.

24. A curve goes through the point $(0, a)$, $a > 0$, and the slope at any point is $\sqrt{y^2/a^2 - 1}$. Prove that the curve is a catenary.

25. A man wearing a parachute falls out of an airplane, and when the parachute opens his velocity is 200 ft/sec. If v ft/sec is his velocity t sec after the parachute opens,

$$\frac{324}{g} \cdot \frac{dv}{dt} = 324 - v^2$$

Solve this differential equation to obtain

$$t = \frac{18}{g}\left(\coth^{-1}\frac{v}{18} - \coth^{-1}\frac{100}{9}\right)$$

26. Show that $\sinh^{-1} u = \ln(u + \sqrt{u^2 + 1})$ by using a trigonometric substitution to evaluate $\displaystyle\int \frac{du}{\sqrt{u^2 + 1}}$.

27. The region bounded by the curve $y = (16 - x^2)^{-1/2}$, the x axis, and the lines $x = -2$ and $x = 3$ is revolved about the x axis. Show that the volume of the solid generated is $\frac{1}{4}\pi[\tanh^{-1}\frac{3}{4} - \tanh^{-1}(-\frac{1}{2})]$.

8.12 (Supplementary) USE OF A TABLE OF INTEGRALS

We have presented various techniques of integration, and you have seen how they are useful for evaluating many integrals. However, there may be occasions when these procedures are either not sufficient or else lead to a complicated integration. In such cases you may wish to use a *table of integrals*. Fairly complete tables of integrals appear in mathematics handbooks, and shorter tables are found in most calculus textbooks. You should be cautioned not to rely too heavily on tables when evaluating integrals. A mastery of integration skills is essential because, as mentioned in Section 8.1, it may be necessary to employ some of the techniques to express the integrand in a form that is found in a table.

A short table of integrals appears on the endpapers of this book. The formulas used in the examples and exercises of this section appear in this table. Observe that in the table there are various headings indicating the form of the integrand. The first heading is *Some Elementary Forms*, and the five formulas listed here are included in those given in Section 8.1. The second heading is *Rational Forms Containing a + bu*. The first example utilizes one of these formulas.

EXAMPLE 1
Find

$$\int \frac{x\,dx}{(4 - x)^3}$$

SOLUTION

Formula 10 in the table of integrals is

$$\int \frac{u \, du}{(a + bu)^3} = \frac{1}{b^2} \left[\frac{a}{2(a + bu)^2} - \frac{1}{a + bu} \right] + C$$

Using this formula with $u = x$, $a = 4$, and $b = -1$ we have

$$\int \frac{x \, dx}{(4 - x)^3} = \frac{1}{(-1)^2} \left[\frac{4}{2(4 - x)^2} - \frac{1}{4 - x} \right] + C$$

$$= \frac{2}{(4 - x)^2} - \frac{1}{4 - x} + C$$

EXAMPLE 2

Find

$$\int \frac{dx}{6 - 2x^2}$$

SOLUTION

Formula 25 in the table is

$$\int \frac{du}{a^2 - u^2} = \frac{1}{2a} \ln \left| \frac{u + a}{u - a} \right| + C \tag{1}$$

Observe that this formula can be used if the coefficient of x^2 in the given integral is 1 instead of 2. Thus we write

$$\int \frac{dx}{6 - 2x^2} = \frac{1}{2} \int \frac{dx}{3 - x^2} \tag{2}$$

On the right side of (2) we apply (1) with $u = x$ and $a = \sqrt{3}$, and we have

$$\int \frac{dx}{6 - 2x^2} = \frac{1}{2} \cdot \frac{1}{2\sqrt{3}} \ln \left| \frac{x + \sqrt{3}}{x - \sqrt{3}} \right| + C$$

$$= \frac{\sqrt{3}}{12} \ln \left| \frac{x + \sqrt{3}}{x - \sqrt{3}} \right| + C$$

EXAMPLE 3

Find

$$\int \frac{dx}{8x^2 + 4x}$$

SOLUTION

$$\int \frac{dx}{8x^2 + 4x} = \frac{1}{4} \int \frac{dx}{x(2x + 1)} \tag{3}$$

The integral on the right side of (3) is of the form

$$\int \frac{du}{u(a + bu)}$$

where $u = x$, $a = 1$, and $b = 2$. Formula (11) in the table is

$$\int \frac{du}{u(a + bu)} = \frac{1}{a} \ln \left| \frac{u}{a + bu} \right| + C$$

Using this formula we have

$$\frac{1}{4} \int \frac{dx}{x(2x + 1)} = \frac{1}{4} \cdot \frac{1}{1} \ln \left| \frac{x}{1 + 2x} \right| + C$$

$$= \frac{1}{4} \ln \left| \frac{x}{2x + 1} \right| + C$$

EXAMPLE 4
Find

$$\int \frac{dx}{\sqrt{x^2 + 2x - 3}}$$

SOLUTION
Formula 27 in the table is

$$\int \frac{du}{\sqrt{u^2 \pm a^2}} = \ln \left| u + \sqrt{u^2 \pm a^2} \right| + C \tag{4}$$

We may be able to apply this formula to the given integral if by completing the square under the radical sign we obtain an expression of the form $u^2 \pm a^2$. To complete the square of $x^2 + 2x$ we add 1, and so we also subtract 1. Thus we write

$$x^2 + 2x - 3 = (x^2 + 2x + 1) - 1 - 3$$
$$= (x + 1)^2 - 4$$

Therefore

$$\int \frac{dx}{\sqrt{x^2 + 2x - 3}} = \int \frac{dx}{\sqrt{(x + 1)^2 - 4}}$$

The integral is of the form

$$\int \frac{du}{\sqrt{u^2 - a^2}}$$

where $u = x + 1$ and $a = 2$. Hence, from (4),

$$\int \frac{dx}{\sqrt{(x + 1)^2 - 4}} = \ln \left| (x + 1) + \sqrt{(x + 1)^2 - 4} \right| + C$$

$$= \ln \left| x + 1 + \sqrt{x^2 + 2x - 3} \right| + C$$

EXAMPLE 5
Find

$$\int x^2 \sqrt{4x^2 + 1} \, dx$$

SOLUTION

Formula 29 in the table, with the plus sign, is

$$\int u^2 \sqrt{u^2 + a^2}\, du = \frac{u}{8}(2u^2 + a^2)\sqrt{u^2 + a^2} - \frac{a^4}{8}\ln\left|u + \sqrt{u^2 + a^2}\right| + C \qquad (5)$$

We can apply this formula by writing the given integral as follows:

$$\int x^2\sqrt{4x^2 + 1}\, dx = \int x^2\, \sqrt{4\left(x^2 + \frac{1}{4}\right)}\, dx$$

$$= 2\int x^2\, \sqrt{x^2 + \frac{1}{4}}\, dx$$

From (5), with $u = x$ and $a = \frac{1}{2}$, we have

$$2\int x^2\, \sqrt{x^2 + \frac{1}{4}}\, dx = 2\left[\frac{x}{8}\left(2x^2 + \frac{1}{4}\right)\sqrt{x^2 + \frac{1}{4}} - \frac{\frac{1}{16}}{8}\ln\left|x + \sqrt{x^2 + \frac{1}{4}}\right|\right] + C$$

$$= \frac{x}{16}(8x^2 + 1)\sqrt{x^2 + \frac{1}{4}} - \frac{1}{64}\ln\left|x + \sqrt{x^2 + \frac{1}{4}}\right| + C$$

Formulas 16, 19, 21, 73, 77, 86, and 98, among others, express one integral in terms of a simpler integral of the same form. Such formulas are called **reduction formulas**. The next example shows how they are applied.

EXAMPLE 6

Find $\int \sec^5 x\, dx$.

SOLUTION

Formula 77 is

$$\int \sec^n u\, du = \frac{1}{n-1}\sec^{n-2} u \tan u + \frac{n-2}{n-1}\int \sec^{n-2} u\, du \qquad (6)$$

We apply this formula with $u = x$ and $n = 5$, and we have

$$\int \sec^5 x\, dx = \frac{1}{4}\sec^3 x \tan x + \frac{3}{4}\int \sec^3 x\, dx \qquad (7)$$

We now use (6) to the integral on the right side of (7) with $n = 3$, and we obtain

$$\int \sec^5 x\, dx = \frac{1}{4}\sec^3 x \tan x + \frac{3}{4}\left(\frac{1}{2}\sec x \tan x + \frac{1}{2}\int \sec x\, dx\right)$$

$$= \frac{1}{4}\sec^3 x \tan x + \frac{3}{8}\sec x \tan x + \frac{3}{8}\ln|\sec x + \tan x| + C$$

Because the integral in Example 6 is an odd power of the secant, it can be evaluated by using integration by parts, as explained in Section 8.4. As a matter of fact, this integral appears as Exercise 12 in Exercises 8.4.

EXERCISES 8.12

In Exercises 1 through 36, use the table of integrals on the endpapers to evaluate the integral. In Exercises 1 through 4, the integrand is a rational form containing $a + bu$. Use one of the formulas 6 through 13.

1. $\displaystyle\int \frac{x\,dx}{2 + 3x}$

2. $\displaystyle\int \frac{x\,dx}{(5 - 2x)^3}$

3. $\displaystyle\int \frac{x^2\,dx}{(6 - x)^2}$

4. $\displaystyle\int \frac{dx}{x(7 + 3x)}$

In Exercises 5 through 8, the integrand is a form containing $\sqrt{a + bu}$. Use one of the formulas 14 through 23.

5. $\displaystyle\int x\sqrt{1 + 2x}\,dx$

6. $\displaystyle\int x^2\sqrt{1 + 2x}\,dx$

7. $\displaystyle\int \frac{\sqrt{1 + 2x}}{x}\,dx$

8. $\displaystyle\int \frac{dx}{x^2\sqrt{1 + 2x}}$

In Exercises 9 and 10, the integrand is a form containing $a^2 \pm u^2$. Use one of the formulas 24 through 26.

9. $\displaystyle\int \frac{dx}{4 - x^2}$

10. $\displaystyle\int \frac{dx}{x^2 - 25}$

In Exercises 11 through 14, the integrand is a form containing $\sqrt{u^2 \pm a^2}$. Use one of the formulas 27 through 38.

11. $\displaystyle\int \frac{dx}{\sqrt{x^2 + 6x}}$

12. $\displaystyle\int \sqrt{4x^2 + 1}\,dx$

13. $\displaystyle\int \frac{\sqrt{9x^2 + 4}}{x}\,dx$

14. $\displaystyle\int \frac{dx}{(x - 1)^2\sqrt{x^2 - 2x - 3}}$

In Exercises 15 and 16, the integrand is a form containing $\sqrt{a^2 - u^2}$. Use one of the formulas 39 through 48.

15. $\displaystyle\int \frac{\sqrt{9 - 4x^2}}{x}\,dx$

16. $\displaystyle\int \frac{dx}{x^2\sqrt{25 - 9x^2}}$

In Exercises 17 and 18, the integrand is a form containing $2au - u^2$. Use one of the formulas 49 through 58.

17. $\displaystyle\int x\sqrt{4x - x^2}\,dx$

18. $\displaystyle\int \frac{x^2\,dx}{\sqrt{4x - x^2}}$

In Exercises 19 through 24, the integrand is a form containing trigonometric functions. Use one of the formulas 59 through 88.

19. $\displaystyle\int \sin^5 x\,dx$

20. $\displaystyle\int \cos^8 x\,dx$

21. $\displaystyle\int \csc^7 x\,dx$

22. $\displaystyle\int \sin 3x \cos 5x\,dx$

23. $\displaystyle\int t^4 \cos t\,dt$

24. $\displaystyle\int \sin^3 x \cos^5 x\,dx$

In Exercises 25 and 26, the integrand is a form containing an inverse trigonometric function. Use one of the formulas 89 through 94.

25. $\displaystyle\int \sec^{-1} 3x\,dx$

26. $\displaystyle\int \tan^{-1} 4t\,dt$

In Exercises 27 through 34, the integrand is a form containing an exponential or logarithmic function. Use one of the formulas 95 through 106.

27. $\displaystyle\int x^4 e^x\,dx$

28. $\displaystyle\int x^3 2^x\,dx$

29. $\displaystyle\int x^2 e^{4x}\,dx$

30. $\displaystyle\int x^2 \ln x\,dx$

31. $\displaystyle\int x^3 \ln(3x)\,dx$

32. $\displaystyle\int 5x^2 e^{-2x}\,dx$

33. $\displaystyle\int e^{2x} \sin 5x\,dx$

34. $\displaystyle\int e^{3t} \cos 4t\,dt$

In Exercises 35 and 36, the integrand is a form containing a hyperbolic function. Use one of the formulas 107 through 124.

35. $\displaystyle\int 3y \sinh 5y\,dy$

36. $\displaystyle\int e^x \operatorname{sech} e^x\,dx$

In Exercises 37 through 48, use the table of integrals on the endpapers to evaluate the definite integral.

37. $\displaystyle\int_1^2 \frac{dx}{x(5-x^2)}$

38. $\displaystyle\int_0^3 \frac{x\,dx}{(1+x)^2}$

39. $\displaystyle\int_0^3 \frac{x^2\,dx}{\sqrt{x^2+16}}$

40. $\displaystyle\int_0^2 \frac{dx}{(9+4x^2)^{3/2}}$

41. $\displaystyle\int_1^2 x^4 \ln x\,dx$

42. $\displaystyle\int_0^1 x^2 e^{-x}\,dx$

43. $\displaystyle\int_3^4 \sqrt{x^2+2x-15}\,dx$

44. $\displaystyle\int_3^5 x^2\sqrt{x^2-9}\,dx$

45. $\displaystyle\int_1^2 \sqrt{4w-w^2}\,dw$

46. $\displaystyle\int_0^{\pi/3} \sec^5 x\,dx$

47. $\displaystyle\int_{\pi/8}^{\pi/4} \sin 3t\,\sin 5t\,dt$

48. $\displaystyle\int_0^{\pi/4} \tan^6 \theta\,d\theta$

REVIEW EXERCISES FOR CHAPTER 8 AND REVIEW OF INTEGRATION

In Exercises 1 through 62, evaluate the indefinite integral.

1. $\displaystyle\int \tan^2 4x \cos^4 4x\,dx$

2. $\displaystyle\int \frac{5x^2-3}{x^3-x}\,dx$

3. $\displaystyle\int \frac{e^x\,dx}{\sqrt{4-e^x}}$

4. $\displaystyle\int \frac{dx}{x^2\sqrt{a^2+x^2}}$

5. $\displaystyle\int \tan^{-1}\sqrt{x}\,dx$

6. $\displaystyle\int \frac{dt}{t^4+1}$

7. $\displaystyle\int \cos^2 \tfrac{1}{3}x\,dx$

8. $\displaystyle\int \frac{\sqrt{x+1}+1}{\sqrt{x+1}-1}\,dx$

9. $\displaystyle\int \frac{x^2+1}{(x-1)^3}\,dx$

10. $\displaystyle\int \frac{dy}{\sqrt{y}+1}$

11. $\displaystyle\int \sin x \sin 3x\,dx$

12. $\displaystyle\int \cos\theta \cos 2\theta\,d\theta$

13. $\displaystyle\int \frac{dx}{x+x^{4/3}}$

14. $\displaystyle\int t\sqrt{2t-t^2}\,dt$

15. $\displaystyle\int (\sec 3x + \csc 3x)^2\,dx$

16. $\displaystyle\int \frac{dx}{\sqrt{e^x-1}}$

17. $\displaystyle\int \frac{2t^3+11t+8}{t^3+4t^2+4t}\,dt$

18. $\displaystyle\int x^3 e^{3x}\,dx$

19. $\displaystyle\int \frac{x^4+1}{x^4-1}\,dx$

20. $\displaystyle\int \frac{\sqrt{x^2-4}}{x^2}\,dx$

21. $\displaystyle\int \sin^4 3x \cos^2 3x\,dx$

22. $\displaystyle\int t\sin^2 2t\,dt$

23. $\displaystyle\int \frac{dr}{\sqrt{3-4r-r^2}}$

24. $\displaystyle\int \frac{4x^2+x-2}{x^3-5x^2+8x-4}\,dx$

25. $\displaystyle\int x^3 \cos x^2\,dx$

26. $\displaystyle\int \frac{y\,dy}{9+16y^4}$

27. $\displaystyle\int e^{t/2}\cos 2t\,dt$

28. $\displaystyle\int \frac{du}{u^{5/8}-u^{1/8}}$

29. $\displaystyle\int \frac{\sin x \cos x}{4+\sin^4 x}\,dx$

30. $\displaystyle\int \frac{\sqrt{w-a}}{w}\,dw,\ a>0$

31. $\displaystyle\int \sin^5 nx\,dx$

32. $\displaystyle\int \frac{dx}{x\ln x(\ln x-1)}$

33. $\displaystyle\int \csc^5 x\,dx$

34. $\displaystyle\int \frac{dx}{5+4\sec x}$

35. $\displaystyle\int \frac{2y^2+1}{y^3-6y^2+12y-8}\,dy$

36. $\displaystyle\int \frac{x^5\,dx}{(x^2-a^2)^3}$

37. $\displaystyle\int \frac{\sin x\,dx}{1+\cos^2 x}$

38. $\displaystyle\int \frac{dx}{x\sqrt{x^2+x+1}}$

39. $\displaystyle\int \sqrt{4t-t^2}\,dt$

40. $\displaystyle\int \frac{dx}{\sqrt{1-x+3x^2}}$

41. $\displaystyle\int \frac{dx}{x^4-x}$

42. $\displaystyle\int \frac{\sqrt{t}-1}{\sqrt{t}+1}\,dt$

43. $\displaystyle\int \frac{e^x\,dx}{\sqrt{4-9e^{2x}}}$

44. $\displaystyle\int \frac{dx}{5+4\cos 2x}$

45. $\displaystyle\int \cot^2 3x \csc^4 3x\,dx$

46. $\displaystyle\int \frac{\cot x\,dx}{3+2\sin x}$

47. $\displaystyle\int x^2 \sin^{-1} x\,dx$

48. $\displaystyle\int \frac{dx}{x\sqrt{5x-6-x^2}}$

49. $\displaystyle\int \frac{dx}{\sin x - 2\csc x}$

50. $\displaystyle\int \cos x \ln(\sin x)\,dx$

51. $\displaystyle\int \frac{\cos 3t\,dt}{\sin 3t\sqrt{\sin^2 3t-\tfrac{1}{4}}}$

52. $\displaystyle\int \frac{dx}{(x^2 + 6x + 34)^2}$

53. $\displaystyle\int \frac{\sqrt{x^2 + a^2}}{x^4}\, dx$

54. $\displaystyle\int \tan x \sin x\, dx$

55. $\displaystyle\int \frac{\sin^{-1}\sqrt{2t}}{\sqrt{1 - 2t}}\, dt$

56. $\displaystyle\int \ln(x^2 + 1)\, dx$

57. $\displaystyle\int \frac{dx}{\sqrt{2 + \sqrt{x - 1}}}$

58. $\displaystyle\int \frac{dx}{2 + 2\sin x + \cos x}$

59. $\displaystyle\int \sqrt{\tan x}\, dx$

60. $\displaystyle\int \frac{dx}{\sqrt{1 + \sqrt[3]{x}}}$

61. $\displaystyle\int x^n \ln x\, dx$

62. $\displaystyle\int \tan^n x \sec^4 x\, dx,\ n > 0$

In Exercises 63 through 94, evaluate the definite integral.

63. $\displaystyle\int_0^\pi \sqrt{2 + 2\cos x}\, dx$

64. $\displaystyle\int_{1/2}^1 \sqrt{\frac{1 - x}{x}}\, dx$

65. $\displaystyle\int_1^2 \frac{2x^2 + x + 4}{x^3 + 4x^2}\, dx$

66. $\displaystyle\int_0^1 \frac{dx}{e^x + e^{-x}}$

67. $\displaystyle\int_0^2 \frac{t^3\, dt}{\sqrt{4 + t^2}}$

68. $\displaystyle\int_0^{\pi/2} \sin^3 t \cos^3 t\, dt$

69. $\displaystyle\int_{-2}^{2\sqrt3} \frac{x^2\, dx}{(16 - x^2)^{3/2}}$

70. $\displaystyle\int_0^1 \frac{xe^x\, dx}{(1 + x)^2}$

71. $\displaystyle\int_0^{\pi/4} \sec^4 x\, dx$

72. $\displaystyle\int_0^2 \frac{(1 - x)\, dx}{x^2 + 3x + 2}$

73. $\displaystyle\int_{\pi/12}^{\pi/8} \cot^3 2y\, dy$

74. $\displaystyle\int_0^2 (2^x + x^2)\, dx$

75. $\displaystyle\int_a^{a/2} \frac{\sqrt{(a^2 - x^2)^3}}{x^2}\, dx$

76. $\displaystyle\int_1^2 (\ln x)^2\, dx$

77. $\displaystyle\int_{\sqrt3/3}^1 \frac{(2x^2 - 2x + 1)\, dx}{x^3 + x}$

78. $\displaystyle\int_{\sqrt2/2}^1 \frac{x^3\, dx}{\sqrt{2 - x^2}}$

79. $\displaystyle\int_1^{10} \log_{10} \sqrt{ex}\, dx$

80. $\displaystyle\int_0^{2\pi} |\sin x - \cos x|\, dx$

81. $\displaystyle\int_1^2 \frac{x + 2}{(x + 1)^2}\, dx$

82. $\displaystyle\int_0^{\sqrt{\pi/2}} xe^{x^2} \cos x^2\, dx$

83. $\displaystyle\int_0^\pi |\cos^3 x|\, dx$

84. $\displaystyle\int_{-\pi/4}^{\pi/4} |\tan^5 x|\, dx$

85. $\displaystyle\int_0^{1/2} \frac{2x\, dx}{x^3 - x^2 - x + 1}$

86. $\displaystyle\int_0^1 x^3\sqrt{1 + x^2}\, dx$

87. $\displaystyle\int_0^{1/2} \frac{x\, dx}{\sqrt{1 - 4x^4}}$

88. $\displaystyle\int_0^{\pi/12} \frac{dx}{\cos^4 3x}$

89. $\displaystyle\int_0^1 \sqrt{2y + y^2}\, dy$

90. $\displaystyle\int_0^4 \frac{x^2\, dx}{x^3 + 4x^2 + 5x + 2}$

91. $\displaystyle\int_0^{\pi/2} \frac{dt}{12 + 13\cos t}$

92. $\displaystyle\int_0^3 \frac{dr}{(r + 2)\sqrt{r + 1}}$

93. $\displaystyle\int_0^{16} \sqrt{4 - \sqrt{x}}\, dx$

94. $\displaystyle\int_{2\pi/3}^\pi \frac{\sin\frac12 t}{1 + \cos\frac12 t}\, dt$

95. The linear density of a rod 3 m long at a point x meters from one end is ke^{-3x} kg/m. Find the mass and center of mass of the rod.

96. Find the center of mass of a rod 4 m long if the linear density at the point x meters from the left end is $\sqrt{9 + x^2}$ kg/m.

97. Find the length of the arc of the parabola $y^2 = 6x$ from $x = 6$ to $x = 12$.

98. Find the area of the region bounded by the curve $y = \sin^{-1} 2x$, the line $x = \frac14\sqrt3$, and the x axis.

99. Find the area of the region enclosed by one loop of the curve $x^2 = y^4(1 - y^2)$.

100. Find the length of the arc of the curve $y = \ln x$ from $x = 1$ to $x = e$.

101. Find the volume of the solid of revolution generated by revolving about the y axis the region bounded by the curve $y = \ln 2x$, the x axis, and the line $x = e$.

102. The region in the first quadrant bounded by the curve $y = \dfrac{5 - x}{(x + 1)^2}$, the x axis, and the y axis is revolved about the x axis. Find the volume of the solid generated.

103. Two chemicals A and B react to form a chemical C, and the rate of change of the amount of C is proportional to the product of the amounts of A and B remaining at any given time. Initially there are 60 lb of chemical A and 60 lb of chemical B, and to form 5 lb of C, 3 lb of A and 2 lb of B are required. After 1 hour, 15 lb of C are formed. (a) If x lb of C are formed at t hours, find an expression for x in terms of t. (b) Find the amount of C after 3 hours.

104. A tank is in the shape of the solid of revolution formed by rotating about the x axis the region bounded by the curve $y = \ln x$, the x axis, and the lines $x = e$ and $x = e^2$. If the tank is full of water, find the work done in pumping all the water to the top of the tank. Distance is measured in feet. Take the positive x axis vertically downward.

105. Find the centroid of the region of Exercise 99.

106. Find the centroid of the region enclosed by the loop of the curve $y^2 = x^2 - x^3$.

107. Find the centroid of the region bounded by the y axis and the curves $y = \sin x - \cos x$ and $y = \sin x + \cos x$ from $x = 0$ to $x = \frac{1}{2}\pi$.

108. Find the centroid of the region in the first quadrant bounded by the coordinate axes and the curve $y = \cos x$.

109. A pond can support a maximum of 10,000 fish so that the rate of growth of the fish population is jointly proportional to the number of fish present and the difference between 10,000 and the number present. The pond initially contained 400 fish, and 6 weeks later there were 3000 fish. (a) How many fish did the pond contain after 8 weeks? (b) When was the growth rate greatest? That is, after how many weeks did the pond contain 5000 fish?

110. In a town of population 12,000 the growth rate of a flu epidemic is jointly proportional to the number of people who have the flu and the number of people who do not have it. Five days ago 400 people in the town had the flu, and today 1000 people have it. How many people are expected to have the flu tomorrow? (b) In how many days from now will the epidemic be spreading the fastest? That is, when will half the population have the flu?

In Exercises 111 through 113, find an approximate value for the given integral (a) by using the trapezoidal rule with $n = 4$, and (b) by using Simpson's rule with $n = 4$. Express the results to three decimal places.

111. $\int_0^2 \sqrt{1 + x^2}\, dx$

112. $\int_1^{9/5} \sqrt{1 + x^3}\, dx$

113. $\int_{1/10}^{1/2} \frac{\cos x}{x}\, dx$

114. Given the integral $\int_d^{d+6} (x^2 + bx + c)\, dx$, where b, c, and d are constants. Suppose this integral is approximated by using the trapezoidal rule with $n = k$. (a) Show that the error in the approximation is exactly $-36/k^2$. (b) What is the smallest value of k such that the approximation is accurate to one decimal place?

Exercises 115 and 116 pertain to Supplementary Section 8.11. Obtain the result by a hyperbolic function substitution.

115. $\int \dfrac{dx}{x^2\sqrt{a^2 + x^2}} = -\dfrac{1}{a^2} \coth\left(\sinh^{-1}\dfrac{x}{a}\right) + C, a > 0$

116. $\int \dfrac{dx}{(a^2 - x^2)^{3/2}} = \dfrac{1}{a^2} \sinh\left(\tanh^{-1}\dfrac{x}{a}\right) + C, a > 0$

Exercises 117 through 122 pertain to Supplementary Section 8.12. Use the table of integrals in the endpapers to evaluate the indefinite or definite integral.

117. $\int \sqrt{9x^2 + 1}\, dx$

118. $\int \dfrac{dt}{\sqrt{4t^2 - 9}}$

119. $\int \sin^3 2x \cos^3 2x\, dx$

120. $\int_4^5 w^2\sqrt{w^2 - 16}\, dw$

121. $\int_0^1 x^3 e^{2x}\, dx$

122. $\int_0^{\pi/6} e^{2t} \sin 3t\, dt$

Exercises 123 through 126 pertain to the indicated Supplementary Section.

123. *Section 5.9.* The vertical end of a water trough is 3 ft wide at the top and 2 ft deep, and it has the form of the region bounded by the x axis and one arch of the curve $y = 2 \sin \frac{1}{3}\pi x$. If the trough is full of water, find the force due to water pressure on the end.

124. *Section 5.8.* Find the centroid of the solid of revolution generated by revolving about the y axis the region bounded by the curve $y = e^x$, the coordinate axes, and the line $x = 1$.

125. *Section 5.8.* Find the centroid of the solid of revolution generated by revolving about the x axis the region in the first quadrant bounded by the curve $y = \cos x$ and the coordinate axes.

126. *Section 5.8.* If the region in Exercise 125 is revolved about the y axis, find the centroid of the resulting solid of revolution.

INDETERMINATE FORMS, IMPROPER INTEGRALS, AND TAYLOR'S FORMULA

9.1
THE INDETERMINATE
FORM 0/0

Limit Theorem 9 (1.3.9) states that if $\lim\limits_{x \to a} f(x)$ and $\lim\limits_{x \to a} g(x)$ both exist, then

$$\lim_{x \to a} \frac{f(x)}{g(x)} = \frac{\lim\limits_{x \to a} f(x)}{\lim\limits_{x \to a} g(x)}$$

provided that $\lim\limits_{x \to a} g(x) \neq 0$.

There are various situations for which this theorem cannot be used. In particular, if $\lim\limits_{x \to a} g(x) = 0$ and $\lim\limits_{x \to a} f(x) = k$, where k is a constant not equal to 0, then Limit Theorem 12 (1.5.4) can be applied. Consider the case when both $\lim\limits_{x \to a} f(x) = 0$ and $\lim\limits_{x \to a} g(x) = 0$. Some limits of this type have previously been discussed.

▶ ILLUSTRATION 1   We wish to find

$$\lim_{x \to 4} \frac{x^2 - x - 12}{x^2 - 3x - 4}$$

Here, $\lim\limits_{x \to 4} (x^2 - x - 12) = 0$ and $\lim\limits_{x \to 4} (x^2 - 3x - 4) = 0$. However, the numerator and denominator can be factored, which gives

$$\lim_{x \to 4} \frac{x^2 - x - 12}{x^2 - 3x - 4} = \lim_{x \to 4} \frac{(x - 4)(x + 3)}{(x - 4)(x + 1)}$$

$$= \lim_{x \to 4} \frac{x + 3}{x + 1}$$

$$= \tfrac{7}{5} \qquad\qquad ◀$$

▶ ILLUSTRATION 2 $\lim\limits_{x \to 0} \sin x = 0$ and $\lim\limits_{x \to 0} x = 0$; and by Theorem 1.8.2,

$$\lim_{x \to 0} \frac{\sin x}{x} = 1 \qquad\qquad ◀$$

9.1.1 DEFINITION   If f and g are two functions such that $\lim\limits_{x \to a} f(x) = 0$ and $\lim\limits_{x \to a} g(x) = 0$, then the function f/g has the **indeterminate form 0/0 at a**.

▶ ILLUSTRATION 3   From Definition 9.1.1, $(x^2 - x - 12)/(x^2 - 3x - 4)$ has the indeterminate form 0/0 at 4; however, we saw in Illustration 1 that

$$\lim_{x \to 4} \frac{x^2 - x - 12}{x^2 - 3x - 4} = \frac{7}{5}$$

Also, $\sin x/x$ has the indeterminate form 0/0 at 0, while $\lim\limits_{x \to 0} (\sin x/x) = 1$, as indicated in Illustration 2.   ◀

We now consider a more general method for finding the limit, if it exists, of a function at a number where it has the indeterminate form 0/0. This method is attributed to the French mathematician Guillaume François de L'Hôpital (1661–1707), who wrote the first calculus textbook, published in 1696. It is known as *L'Hôpital's rule.*

9.1.2 THEOREM

L'Hôpital's Rule

Let f and g be functions that are differentiable on an open interval I, except possibly at the number a in I. Suppose that for all $x \neq a$ in I, $g'(x) \neq 0$. Then if $\lim_{x \to a} f(x) = 0$ and $\lim_{x \to a} g(x) = 0$, and if

$$\lim_{x \to a} \frac{f'(x)}{g'(x)} = L$$

it follows that

$$\lim_{x \to a} \frac{f(x)}{g(x)} = L$$

The theorem is valid if all the limits are right-hand limits or all the limits are left-hand limits.

Before giving the proof of Theorem 9.1.2, we show the use of the theorem by an illustration and examples.

▶ ILLUSTRATION 4   We use L'Hôpital's rule to evaluate the limits in Illustrations 1 and 2. In Illustration 1, because $\lim_{x \to 4} (x^2 - x - 12) = 0$ and $\lim_{x \to 4} (x^2 - 3x - 4) = 0$, we can apply L'Hôpital's rule and obtain

$$\lim_{x \to 4} \frac{x^2 - x - 12}{x^2 - 3x - 4} = \lim_{x \to 4} \frac{2x - 1}{2x - 3}$$
$$= \tfrac{7}{5}$$

L'Hôpital's rule can be applied in Illustration 2 because $\lim_{x \to 0} \sin x = 0$ and $\lim_{x \to 0} x = 0$. Then

$$\lim_{x \to 0} \frac{\sin x}{x} = \lim_{x \to 0} \frac{\cos x}{1}$$
$$= 1$$ ◀

EXAMPLE 1
Find

$$\lim_{x \to 0} \frac{x}{1 - e^x}$$

if it exists.

SOLUTION
Because $\lim\limits_{x \to 0} x = 0$ and $\lim\limits_{x \to 0} (1 - e^x) = 0$, L'Hôpital's rule can be applied. Thus

$$\lim_{x \to 0} \frac{x}{1 - e^x} = \lim_{x \to 0} \frac{1}{-e^x}$$

$$= \frac{1}{-1}$$

$$= -1$$

EXAMPLE 2
Find

$$\lim_{x \to 1} \frac{1 - x + \ln x}{x^3 - 3x + 2}$$

if it exists.

SOLUTION

$$\lim_{x \to 1} (1 - x + \ln x) = 1 - 1 + 0 \qquad \lim_{x \to 1} (x^3 - 3x + 2) = 1 - 3 + 2$$
$$= 0 \qquad\qquad\qquad\qquad = 0$$

Therefore, applying L'Hôpital's rule we have

$$\lim_{x \to 1} \frac{1 - x + \ln x}{x^3 - 3x + 2} = \lim_{x \to 1} \frac{-1 + \dfrac{1}{x}}{3x^2 - 3}$$

Now, because $\lim\limits_{x \to 1} (-1 + 1/x) = 0$ and $\lim\limits_{x \to 1} (3x^2 - 3) = 0$, we apply L'Hôpital's rule again, giving

$$\lim_{x \to 1} \frac{-1 + \dfrac{1}{x}}{3x^2 - 3} = \lim_{x \to 1} \frac{-\dfrac{1}{x^2}}{6x}$$

$$= -\tfrac{1}{6}$$

Therefore

$$\lim_{x \to 1} \frac{1 - x + \ln x}{x^3 - 3x + 2} = -\frac{1}{6}$$

To prove Theorem 9.1.2 we need to use the theorem known as *Cauchy's mean-value theorem*, which extends to two functions the mean-value theorem (3.3.2) for a single function. This theorem is attributed to the French mathematician Augustin L. Cauchy (1789–1857).

9.1.3 THEOREM   If f and g are two functions such that
Cauchy's Mean-Value Theorem

(i) f and g are continuous on the closed interval $[a, b]$;
(ii) f and g are differentiable on the open interval (a, b);
(iii) for all x in the open interval (a, b), $g'(x) \neq 0$

then there exists a number z in the open interval (a, b) such that

$$\frac{f(b) - f(a)}{g(b) - g(a)} = \frac{f'(z)}{g'(z)}$$

PROOF   We first show that $g(b) \neq g(a)$. Assume $g(b) = g(a)$. Because g satisfies the two conditions in the hypothesis of the mean-value theorem (3.3.2), there is some number c in (a, b) such that $g'(c) = [g(b) - g(a)]/(b - a)$. But if $g(b) = g(a)$, then there is some number c in (a, b) such that $g'(c) = 0$. But condition (iii) of the hypothesis of this theorem states that for all x in (a, b), $g'(x) \neq 0$. Therefore there is a contradiction. Hence the assumption that $g(b) = g(a)$ is false. So $g(b) \neq g(a)$, and consequently $g(b) - g(a) \neq 0$.

Now consider the function h defined by

$$h(x) = f(x) - f(a) - \left[\frac{f(b) - f(a)}{g(b) - g(a)} \right] [g(x) - g(a)]$$

Then

$$h'(x) = f'(x) - \left[\frac{f(b) - f(a)}{g(b) - g(a)} \right] g'(x) \tag{1}$$

Therefore h is differentiable on (a, b) because f and g are differentiable there, and h is continuous on $[a, b]$ because f and g are continuous there.

$$h(a) = f(a) - f(a) - \left[\frac{f(b) - f(a)}{g(b) - g(a)} \right] [g(a) - g(a)]$$

$$= 0$$

$$h(b) = f(b) - f(a) - \left[\frac{f(b) - f(a)}{g(b) - g(a)} \right] [g(b) - g(a)]$$

$$= 0$$

Hence the three conditions of the hypothesis of Rolle's theorem are satisfied by the function h. So there exists a number z in the open interval (a, b) such that $h'(z) = 0$. Thus, from (1),

$$f'(z) - \frac{f(b) - f(a)}{g(b) - g(a)} g'(z) = 0 \tag{2}$$

Because $g'(z) \neq 0$ on (a, b), we have from (2)

$$\frac{f(b) - f(a)}{g(b) - g(a)} = \frac{f'(z)}{g'(z)}$$

where z is some number in (a, b). This proves the theorem.   ■

If g is the function such that $g(x) = x$, then the conclusion of Cauchy's mean-value theorem becomes the conclusion of the former mean-value theorem because then $g'(z) = 1$. So the former mean-value theorem is a special case of Cauchy's mean-value theorem.

A geometric interpretation of Cauchy's mean-value theorem is given in Section 13.3. It is postponed until then because parametric equations are needed.

▶ ILLUSTRATION 5   Suppose that $f(x) = 3x^2 + 3x - 1$ and $g(x) = x^3 - 4x + 2$. We find a number z in $(0, 1)$, predicted by Theorem 9.1.3.

$$f'(x) = 6x + 3 \qquad g'(x) = 3x^2 - 4$$

Thus f and g are differentiable and continuous everywhere, and for all x in $(0, 1)$, $g'(x) \neq 0$. Hence, by Theorem 9.1.3 there exists a z in $(0, 1)$ such that

$$\frac{f(1) - f(0)}{g(1) - g(0)} = \frac{6z + 3}{3z^2 - 4}$$

Substituting $f(1) = 5$, $g(1) = -1$, $f(0) = -1$, and $g(0) = 2$ and solving for z we have

$$\frac{5 - (-1)}{-1 - 2} = \frac{6z + 3}{3z^2 - 4}$$

$$6z^2 + 6z - 5 = 0$$

$$z = \frac{-6 \pm \sqrt{36 + 120}}{12}$$

$$= \frac{-6 \pm 2\sqrt{39}}{12}$$

$$= \frac{-3 \pm \sqrt{39}}{6}$$

Only one of these numbers is in $(0, 1)$, namely $z = \frac{1}{6}(-3 + \sqrt{39})$.   ◀

We are now in a position to prove Theorem 9.1.2. We distinguish three cases: (i) $x \to a^+$; (ii) $x \to a^-$; (iii) $x \to a$.

PROOF OF THEOREM 9.1.2(i)   Because in the hypothesis it is not assumed that f and g are defined at a, we consider two new functions F and G for which

$$F(x) = f(x) \quad \text{if } x \neq a \text{ and } F(a) = 0$$

$$G(x) = g(x) \quad \text{if } x \neq a \text{ and } G(a) = 0$$

(3)

Let b be the right endpoint of the open interval I given in the hypothesis. Because f and g are both differentiable on I, except possibly at a, we conclude that F and G are both differentiable on the interval $(a, x]$, where $a < x < b$. Therefore F and G are both continuous on $(a, x]$. The functions F and G

are also both continuous from the right at a because $\lim\limits_{x \to a^+} F(x) = \lim\limits_{x \to a^+} f(x)$ and $\lim\limits_{x \to a^+} f(x) = 0$, which is $F(a)$; similarly, $\lim\limits_{x \to a^+} G(x) = G(a)$. Therefore F and G are continuous on the closed interval $[a, x]$. So F and G satisfy the three conditions of the hypothesis of Cauchy's mean-value theorem (Theorem 9.1.3) on the interval $[a, x]$. Hence

$$\frac{F(x) - F(a)}{G(x) - G(a)} = \frac{F'(z)}{G'(z)} \tag{4}$$

where z is some number such that $a < z < x$. From (3) and (4) we have

$$\frac{f(x)}{g(x)} = \frac{f'(z)}{g'(z)} \tag{5}$$

Because $a < z < x$, it follows that as $x \to a^+$, $z \to a^+$; therefore

$$\lim_{x \to a^+} \frac{f(x)}{g(x)} = \lim_{x \to a^+} \frac{f'(z)}{g'(z)}$$

$$= \lim_{z \to a^+} \frac{f'(z)}{g'(z)} \tag{6}$$

But by hypothesis, the limit on the right side of (6) is L. Therefore

$$\lim_{x \to a^+} \frac{f(x)}{g(x)} = L \qquad \blacksquare$$

The proof of case (ii) is similar to the proof of case (i) and is left as an exercise (see Exercise 34). The proof of case (iii) is based on the results of cases (i) and (ii) and is also left as an exercise (see Exercise 35).

L'Hôpital's rule also holds if either x increases without bound or x decreases without bound, as given in the next theorem.

9.1.4 THEOREM
L'Hôpital's Rule

Let f and g be functions that are differentiable for all $x > N$, where N is a positive constant, and suppose that for all $x > N$, $g'(x) \neq 0$. Then if $\lim\limits_{x \to +\infty} f(x) = 0$ and $\lim\limits_{x \to +\infty} g(x) = 0$, and if

$$\lim_{x \to +\infty} \frac{f'(x)}{g'(x)} = L$$

it follows that

$$\lim_{x \to +\infty} \frac{f(x)}{g(x)} = L$$

The theorem is also valid if $x \to +\infty$ is replaced by $x \to -\infty$.

PROOF   We prove the theorem for $x \to +\infty$. The proof for $x \to -\infty$ is left as an exercise (see Exercise 38).

For all $x > N$, let $x = 1/t$; then $t = 1/x$. Let F and G be the functions defined by $F(t) = f(1/t)$ and $G(t) = g(1/t)$, if $t \neq 0$. Then $f(x) = F(t)$ and

$g(x) = G(t)$, where $x > N$ and $0 < t < 1/N$. From Definitions 3.7.1 and 1.4.1 it may be shown that the statements

$$\lim_{x \to +\infty} f(x) = M \quad \text{and} \quad \lim_{t \to 0^+} F(t) = M$$

have the same meaning. It is left as an exercise to prove this (see Exercise 36). Because by hypothesis $\lim_{x \to +\infty} f(x) = 0$ and $\lim_{x \to +\infty} g(x) = 0$, we can conclude that

$$\lim_{t \to 0^+} F(t) = 0 \quad \text{and} \quad \lim_{t \to 0^+} G(t) = 0 \tag{7}$$

Considering the quotient $F'(t)/G'(t)$, we have, using the chain rule,

$$\frac{F'(t)}{G'(t)} = \frac{-\dfrac{1}{t^2} f'\left(\dfrac{1}{t}\right)}{-\dfrac{1}{t^2} g'\left(\dfrac{1}{t}\right)}$$

$$= \frac{f'\left(\dfrac{1}{t}\right)}{g'\left(\dfrac{1}{t}\right)}$$

$$= \frac{f'(x)}{g'(x)}$$

Because by hypothesis $\lim_{x \to +\infty} f'(x)/g'(x) = L$, it follows from the above that

$$\lim_{t \to 0^+} \frac{F'(t)}{G'(t)} = L \tag{8}$$

Because for all $x > N$, $g'(x) \neq 0$,

$$G'(t) \neq 0 \qquad \text{for all } 0 < t < \frac{1}{N} \tag{9}$$

From (7), (8), and (9) it follows from Theorem 9.1.2 that

$$\lim_{t \to 0^+} \frac{F(t)}{G(t)} = L$$

But because $F(t)/G(t) = f(x)/g(x)$ for all $x > N$ and $t \neq 0$, then

$$\lim_{x \to +\infty} \frac{f(x)}{g(x)} = L$$

and so the theorem is proved.   ■

Theorems 9.1.2 and 9.1.4 also hold if L is replaced by $+\infty$ or $-\infty$. The proofs for these cases are omitted.

EXAMPLE 3
Find

$$\lim_{x \to +\infty} \frac{\sin \dfrac{1}{x}}{\tan^{-1}\left(\dfrac{1}{x}\right)}$$

if it exists.

SOLUTION

$\lim\limits_{x \to +\infty} \sin(1/x) = 0$ and $\lim\limits_{x \to +\infty} \tan^{-1}(1/x) = 0$. So from L'Hôpital's rule,

$$\lim_{x \to +\infty} \frac{\sin \dfrac{1}{x}}{\tan^{-1}\left(\dfrac{1}{x}\right)} = \lim_{x \to +\infty} \frac{\cos \dfrac{1}{x} \cdot \left(-\dfrac{1}{x^2}\right)}{\dfrac{1}{1 + \dfrac{1}{x^2}} \cdot \left(-\dfrac{1}{x^2}\right)}$$

$$= \lim_{x \to +\infty} \frac{\cos \dfrac{1}{x}}{\dfrac{x^2}{x^2 + 1}}$$

Because

$$\lim_{x \to +\infty} \cos \frac{1}{x} = 1 \quad \text{and} \quad \lim_{x \to +\infty} \frac{x^2}{x^2 + 1} = \lim_{x \to +\infty} \frac{1}{1 + \dfrac{1}{x^2}}$$

$$= 1$$

it follows that

$$\lim_{x \to +\infty} \frac{\cos \dfrac{1}{x}}{\dfrac{x^2}{x^2 + 1}} = 1$$

Therefore the given limit is 1.

EXAMPLE 4
Given

$$f(x) = \begin{cases} \dfrac{e^x - 1}{x} & \text{if } x \neq 0 \\ 1 & \text{if } x = 0 \end{cases}$$

(a) Prove that f is continuous at 0 by using Definition 1.6.1. (b) Prove that f is differentiable at 0 by computing $f'(0)$.

SOLUTION

(a) We check the three conditions for continuity.

(i) $f(0) = 1$

(ii) $\lim\limits_{x \to 0} f(x) = \lim\limits_{x \to 0} \dfrac{e^x - 1}{x}$

To compute the limit we apply L'Hôpital's rule because $\lim\limits_{x \to 0} (e^x - 1) = 0$ and $\lim\limits_{x \to 0} x = 0$. We have

$$\lim_{x \to 0} f(x) = \lim_{x \to 0} \frac{e^x}{1}$$

$$= 1$$

(iii) $\lim\limits_{x \to 0} f(x) = f(0)$

Therefore f is continuous at 0.

(b) $f'(0) = \lim\limits_{x \to 0} \dfrac{f(x) - f(0)}{x - 0}$

$$= \lim_{x \to 0} \frac{\dfrac{e^x - 1}{x} - 1}{x}$$

$$= \lim_{x \to 0} \frac{e^x - 1 - x}{x^2}$$

Because $\lim\limits_{x \to 0} (e^x - 1 - x) = 0$ and $\lim\limits_{x \to 0} x^2 = 0$, we apply L'Hôpital's rule and get

$$f'(0) = \lim_{x \to 0} \frac{e^x - 1}{2x}$$

Because $\lim\limits_{x \to 0} (e^x - 1) = 0$ and $\lim\limits_{x \to 0} 2x = 0$, we apply L'Hôpital's rule again and obtain

$$f'(0) = \lim_{x \to 0} \frac{e^x}{2}$$

$$= \tfrac{1}{2}$$

EXERCISES 9.1

In Exercises 1 through 5, find all values of z in the given interval (a, b) satisfying the conclusion of Theorem 9.1.3 for the given pair of functions.

1. $f(x) = x^3$, $g(x) = x^2$; $(a, b) = (0, 2)$

2. $f(x) = \dfrac{2x}{1 + x^2}$, $g(x) = \dfrac{1 - x^2}{1 + x^2}$; $(a, b) = (0, 2)$

3. $f(x) = \sin x$, $g(x) = \cos x$; $(a, b) = (0, \pi)$

4. $f(x) = \cos 2x$, $g(x) = \sin x$; $(a, b) = (0, \frac{1}{2}\pi)$

5. $f(x) = \sqrt{x + 5}$, $g(x) = x + 3$; $(a, b) = (-4, -1)$

In Exercises 6 through 30, evaluate the limit, if it exists.

6. $\lim\limits_{x \to \pi/2} \dfrac{3 \cos x}{2x - \pi}$

7. $\lim\limits_{x \to 0} \dfrac{x}{\tan x}$

8. $\lim\limits_{x \to 0} \dfrac{\tan x - x}{x - \sin x}$

9. $\lim\limits_{x \to +\infty} \dfrac{\sin \dfrac{2}{x}}{\dfrac{1}{x}}$

10. $\lim\limits_{x \to 2} \dfrac{\sin \pi x}{2 - x}$

11. $\lim\limits_{x \to -\infty} \dfrac{2^x}{e^x}$

12. $\lim\limits_{x \to 0} \dfrac{e^x - \cos x}{x \sin x}$

13. $\lim\limits_{x \to \pi/2} \dfrac{\ln(\sin x)}{(\pi - 2x)^2}$

14. $\lim\limits_{x \to 0} \dfrac{\sin^{-1} x}{x}$

15. $\lim\limits_{x \to 0} \dfrac{2^x - 3^x}{x}$

16. $\lim\limits_{t \to 2} \dfrac{t^n - 2^n}{t - 2}$

17. $\lim\limits_{\theta \to 0} \dfrac{\theta - \sin \theta}{\tan^3 \theta}$

18. $\lim\limits_{x \to 0} \dfrac{e^{2x^2} - 1}{\sin^2 x}$

19. $\lim\limits_{z \to +\infty} \dfrac{1 - e^{1/z}}{-\dfrac{3}{z}}$

20. $\lim\limits_{y \to 0} \dfrac{y^2}{1 - \cosh y}$

21. $\lim\limits_{t \to 0} \dfrac{\sin t}{\ln(2e^t - 1)}$

22. $\lim\limits_{z \to 0} \dfrac{5z}{5^z - e^z}$

23. $\lim\limits_{t \to 0} \dfrac{\sin^2 t}{\sin t^2}$

24. $\lim\limits_{x \to 0} \dfrac{\tanh 2x}{\tanh x}$

25. $\lim\limits_{x \to 0} \dfrac{(1 + x)^{1/5} - (1 - x)^{1/5}}{(1 + x)^{1/3} - (1 - x)^{1/3}}$

26. $\lim\limits_{x \to 0} \dfrac{e^x - 10^x}{x}$

27. $\lim\limits_{x \to \pi} \dfrac{1 + \cos 2x}{1 - \sin x}$

28. $\lim\limits_{x \to +\infty} \dfrac{\dfrac{1}{x^2} - 2 \tan^{-1} \dfrac{1}{x}}{\dfrac{1}{x}}$

29. $\lim\limits_{x \to 0} \dfrac{\cos x - \cosh x}{x^2}$

30. $\lim\limits_{x \to 0} \dfrac{\sinh x - \sin x}{\sin^3 x}$

31. An electrical circuit has a resistance of R ohms, an inductance of L henrys, and an electromotive force of E volts, where R, L, and E are positive. If I amperes is the current flowing in the circuit t sec after a switch is turned on, then

$$I = \frac{E}{R}(1 - e^{-Rt/L})$$

If t, E, and L are constants, find $\lim\limits_{R \to 0^+} I$.

32. In a geometric progression, if a is the first term, r is the common ratio of two successive terms, and S is the sum of the first n terms, then if $r \neq 1$,

$$S = \frac{a(r^n - 1)}{r - 1}$$

Find $\lim\limits_{r \to 1} S$. Is the result consistent with the sum of the first n terms if $r = 1$?

33. Find values for a and b such that

$$\lim\limits_{x \to 0} \frac{\sin 3x + ax + bx^3}{x^3} = 0$$

34. Prove Theorem 9.1.2(ii).

35. Prove Theorem 9.1.2(iii).

36. Suppose that f is a function defined for all $x > N$, where N is a positive constant. If $t = 1/x$ and $F(t) = f(1/t)$, where $t \neq 0$, prove that the statements $\lim\limits_{x \to +\infty} f(x) = M$ and $\lim\limits_{t \to 0^+} F(t) = M$ have the same meaning.

37. Given:

$$f(x) = \begin{cases} \dfrac{\cos x - 1}{x} & \text{if } x \neq 0 \\ 0 & \text{if } x = 0 \end{cases}$$

(a) Prove that f is continuous at 0 by using Definition 1.6.1. (b) Prove that f is differentiable at 0 by computing $f'(0)$.

38. Prove Theorem 9.1.4 for $x \to -\infty$.

9.2
OTHER INDETERMINATE FORMS

Suppose that we wish to determine whether

$$\lim_{x \to 0^+} \frac{\ln x}{\dfrac{1}{x}}$$

exists. We cannot apply the theorem involving the limit of a quotient because $\lim_{x \to 0^+} \ln x = -\infty$ and $\lim_{x \to 0^+} (1/x) = +\infty$. In this case we say that the function defined by $\ln x / (1/x)$ has the indeterminate form $(-\infty)/(+\infty)$ at $x = 0$. L'Hôpital's rule also applies to an indeterminate form of this type as well as to $(+\infty)/(+\infty)$, $(-\infty)/(-\infty)$, and $(+\infty)/(-\infty)$. This is given by the following theorems, for which the proofs are omitted because they are beyond the scope of this book.

9.2.1 THEOREM
L'Hôpital's Rule

Let f and g be functions that are differentiable on an open interval I, except possibly at the number a in I, and suppose that for all $x \neq a$ in I, $g'(x) \neq 0$. Then if $\lim_{x \to a} f(x) = +\infty$ or $-\infty$, and $\lim_{x \to a} g(x) = +\infty$ or $-\infty$, and if

$$\lim_{x \to a} \frac{f'(x)}{g'(x)} = L$$

it follows that

$$\lim_{x \to a} \frac{f(x)}{g(x)} = L$$

The theorem is valid if all the limits are right-hand limits or if all the limits are left-hand limits.

EXAMPLE 1
Find

$$\lim_{x \to 0^+} \frac{\ln x}{\dfrac{1}{x}}$$

if it exists.

SOLUTION
Because $\lim\limits_{x \to 0^+} \ln x = -\infty$ and $\lim\limits_{x \to 0^+} (1/x) = +\infty$, we apply L'Hôpital's rule
and get

$$\lim_{x \to 0^+} \frac{\ln x}{\dfrac{1}{x}} = \lim_{x \to 0^+} \frac{\dfrac{1}{x}}{-\dfrac{1}{x^2}}$$

$$= \lim_{x \to 0^+} (-x)$$

$$= 0$$

9.2.2 THEOREM

L'Hôpital's Rule

Let f and g be functions that are differentiable for all $x > N$, where N is a positive constant, and suppose that for all $x > N$, $g'(x) \neq 0$. Then if $\lim\limits_{x \to +\infty} f(x) = +\infty$ or $-\infty$, and $\lim\limits_{x \to +\infty} g(x) = +\infty$ or $-\infty$, and if

$$\lim_{x \to +\infty} \frac{f'(x)}{g'(x)} = L$$

it follows that

$$\lim_{x \to +\infty} \frac{f(x)}{g(x)} = L$$

The theorem is also valid if $x \to +\infty$ is replaced by $x \to -\infty$.

Theorems 9.2.1 and 9.2.2 also hold if L is replaced by $+\infty$ or $-\infty$, and the proofs for these cases are also omitted.

EXAMPLE 2
Find

$$\lim_{x \to +\infty} \frac{\ln(2 + e^x)}{3x}$$

if it exists.

SOLUTION
Because $\lim\limits_{x \to +\infty} \ln(2 + e^x) = +\infty$ and $\lim\limits_{x \to +\infty} 3x = +\infty$, by applying
L'Hôpital's rule we obtain

$$\lim_{x \to +\infty} \frac{\ln(2 + e^x)}{3x} = \lim_{x \to +\infty} \frac{\dfrac{1}{2 + e^x} \cdot e^x}{3}$$

$$= \lim_{x \to +\infty} \frac{e^x}{6 + 3e^x}$$

Now because $\lim\limits_{x \to +\infty} e^x = +\infty$ and $\lim\limits_{x \to +\infty} (6 + 3e^x) = +\infty$, we apply

L'Hôpital's rule again and get

$$\lim_{x \to +\infty} \frac{\ln(2 + e^x)}{3x} = \lim_{x \to +\infty} \frac{e^x}{3e^x}$$

$$= \lim_{x \to +\infty} \frac{1}{3}$$

$$= \tfrac{1}{3}$$

EXAMPLE 3
Find

$$\lim_{x \to \pi/2^-} \frac{\sec^2 x}{\sec^2 3x}$$

if it exists.

SOLUTION

$\lim\limits_{x \to \pi/2^-} \sec^2 x = +\infty$ and $\lim\limits_{x \to \pi/2^-} \sec^2 3x = +\infty$. So from L'Hôpital's rule,

$$\lim_{x \to \pi/2^-} \frac{\sec^2 x}{\sec^2 3x} = \lim_{x \to \pi/2^-} \frac{2 \sec^2 x \tan x}{6 \sec^2 3x \tan 3x}$$

$$\lim_{x \to \pi/2^-} 2 \sec^2 x \tan x = +\infty \quad \text{and} \quad \lim_{x \to \pi/2^-} 6 \sec^2 3x \tan 3x = +\infty$$

Observe that further applications of L'Hôpital's rule will not help us. However, the original quotient may be rewritten, and we have

$$\lim_{x \to \pi/2^-} \frac{\sec^2 x}{\sec^2 3x} = \lim_{x \to \pi/2^-} \frac{\cos^2 3x}{\cos^2 x}$$

Now, because $\lim\limits_{x \to \pi/2^-} \cos^2 3x = 0$ and $\lim\limits_{x \to \pi/2^-} \cos^2 x = 0$, we may apply L'Hôpital's rule, giving

$$\lim_{x \to \pi/2^-} \frac{\cos^2 3x}{\cos^2 x} = \lim_{x \to \pi/2^-} \frac{-6 \cos 3x \sin 3x}{-2 \cos x \sin x}$$

$$= \lim_{x \to \pi/2^-} \frac{3(2 \cos 3x \sin 3x)}{(2 \cos x \sin x)}$$

$$= \lim_{x \to \pi/2^-} \frac{3 \sin 6x}{\sin 2x}$$

Because $\lim\limits_{x \to \pi/2^-} 3 \sin 6x = 0$ and $\lim\limits_{x \to \pi/2^-} \sin 2x = 0$, we use L'Hôpital's rule again and have

$$\lim_{x \to \pi/2^-} \frac{3 \sin 6x}{\sin 2x} = \lim_{x \to \pi/2^-} \frac{18 \cos 6x}{2 \cos 2x}$$

$$= \frac{18(-1)}{2(-1)}$$

$$= 9$$

Therefore

$$\lim_{x \to \pi/2^-} \frac{\sec^2 x}{\sec^2 3x} = 9$$

The limit in Example 3 can be evaluated without L'Hôpital's rule by using Theorem 1.8.2. This is left as an exercise (see Exercise 38).

In addition to $0/0$ and $\pm\infty/\pm\infty$, other indeterminate forms are $0 \cdot (+\infty)$, $+\infty - (+\infty)$, 0^0, $(\pm\infty)^0$, and $1^{\pm\infty}$. These indeterminate forms are defined analogously to the other two. For instance, if $\lim\limits_{x \to a} f(x) = +\infty$ and $\lim\limits_{x \to a} g(x) = 0$, then the function defined by $f(x)^{g(x)}$ has the indeterminate form $(+\infty)^0$ at a. To find the limit of a function having one of these indeterminate forms, it must be changed to either the form $0/0$ or $\pm\infty/\pm\infty$ before L'Hôpital's rule can be applied. The following examples illustrate the method.

EXAMPLE 4
Find $\lim\limits_{x \to 0^+} \sin^{-1} x \csc x$ if it exists.

SOLUTION
Because $\lim\limits_{x \to 0^+} \sin^{-1} x = 0$ and $\lim\limits_{x \to 0^+} \csc x = +\infty$, the function defined by $\sin^{-1} x \csc x$ has the indeterminate form $0 \cdot (+\infty)$ at 0. Before we can apply L'Hôpital's rule we rewrite $\sin^{-1} x \csc x$ as $\sin^{-1} x/\sin x$, and consider $\lim\limits_{x \to 0^+} (\sin^{-1} x/\sin x)$. Now $\lim\limits_{x \to 0^+} \sin^{-1} x = 0$ and $\lim\limits_{x \to 0^+} \sin x = 0$; so we have the indeterminate form $0/0$. Therefore, from L'Hôpital's rule we obtain

$$\lim_{x \to 0^+} \frac{\sin^{-1} x}{\sin x} = \lim_{x \to 0^+} \frac{\dfrac{1}{\sqrt{1 - x^2}}}{\cos x}$$

$$= \frac{1}{1}$$

$$= 1$$

EXAMPLE 5
Find

$$\lim_{x \to 0} \left(\frac{1}{x^2} - \frac{1}{x^2 \sec x} \right)$$

if it exists.

SOLUTION
Because

$$\lim_{x \to 0} \frac{1}{x^2} = +\infty \quad \text{and} \quad \lim_{x \to 0} \frac{1}{x^2 \sec x} = +\infty$$

we have the indeterminate form $+\infty - (+\infty)$. Rewriting the expression we have

$$\lim_{x \to 0} \left(\frac{1}{x^2} - \frac{1}{x^2 \sec x} \right) = \lim_{x \to 0} \frac{\sec x - 1}{x^2 \sec x}$$

$\lim_{x \to 0} (\sec x - 1) = 0$ and $\lim_{x \to 0} (x^2 \sec x) = 0$; so we apply L'Hôpital's rule and obtain

$$\lim_{x \to 0} \frac{\sec x - 1}{x^2 \sec x} = \lim_{x \to 0} \frac{\sec x \tan x}{2x \sec x + x^2 \sec x \tan x}$$

$$= \lim_{x \to 0} \frac{\tan x}{2x + x^2 \tan x}$$

$\lim_{x \to 0} \tan x = 0$ and $\lim_{x \to 0} (2x + x^2 \tan x) = 0$

Thus we apply the rule again and obtain

$$\lim_{x \to 0} \frac{\tan x}{2x + x^2 \tan x} = \lim_{x \to 0} \frac{\sec^2 x}{2 + 2x \tan x + x^2 \sec^2 x}$$

$$= \tfrac{1}{2}$$

Therefore

$$\lim_{x \to 0} \left(\frac{1}{x^2} - \frac{1}{x^2 \sec x} \right) = \frac{1}{2}$$

For any one of the indeterminate forms 0^0, $(\pm\infty)^0$, $1^{\pm\infty}$, the procedure for evaluating the limit is illustrated in Example 6.

EXAMPLE 6
Find $\lim_{x \to 0^+} (x + 1)^{\cot x}$ if it exists.

SOLUTION
Because $\lim_{x \to 0^+} (x + 1) = 1$ and $\lim_{x \to 0^+} \cot x = +\infty$, we have the indeterminate form $1^{+\infty}$. Let

$$y = (x + 1)^{\cot x} \tag{1}$$

Then

$$\ln y = \cot x \ln(x + 1)$$

$$= \frac{\ln(x + 1)}{\tan x}$$

So

$$\lim_{x \to 0^+} \ln y = \lim_{x \to 0^+} \frac{\ln(x + 1)}{\tan x} \tag{2}$$

Because $\lim\limits_{x\to0^+} \ln(x+1) = 0$ and $\lim\limits_{x\to0^+} \tan x = 0$,   we   may   apply L'Hôpital's rule to the right side of (2) and obtain

$$\lim_{x\to0^+} \frac{\ln(x+1)}{\tan x} = \lim_{x\to0^+} \frac{\dfrac{1}{x+1}}{\sec^2 x}$$

$$= 1$$

Therefore, substituting 1 on the right side of (2) we have

$$\lim_{x\to0^+} \ln y = 1 \qquad\qquad (3)$$

Because the exponential function is continuous on its entire domain, which is the set of all real numbers, we may apply Theorem 1.7.1, and so

$$\lim_{x\to0^+} \exp(\ln y) = \exp\left(\lim_{x\to0^+} \ln y\right) \qquad\qquad (4)$$

Therefore it follows from (3) and (4) that

$$\lim_{x\to0^+} y = e^1$$

But from (1), $y = (x+1)^{\cot x}$, and therefore

$$\lim_{x\to0^+} (x+1)^{\cot x} = e$$

EXERCISES 9.2

In Exercises 1 through 34, evaluate the limit, if it exists.

1. $\lim\limits_{x\to+\infty} \dfrac{x^2}{e^x}$

2. $\lim\limits_{x\to\pi/2^-} \dfrac{\ln(\cos x)}{\ln(\tan x)}$

3. $\lim\limits_{x\to1/2^-} \dfrac{\ln(1-2x)}{\tan \pi x}$

4. $\lim\limits_{x\to+\infty} \dfrac{(\ln x)^3}{x}$

5. $\lim\limits_{x\to0^+} x \csc x$

6. $\lim\limits_{x\to1/2^+} (2x-1) \tan \pi x$

7. $\lim\limits_{x\to+\infty} \dfrac{\ln x}{\sqrt{x}}$

8. $\lim\limits_{x\to0^+} \tan^{-1} x \cot x$

9. $\lim\limits_{x\to0^+} \tan x (\ln x)$

10. $\lim\limits_{x\to0^+} \left(\dfrac{1}{\sin x} - \dfrac{1}{x}\right)$

11. $\lim\limits_{x\to1} \left(\dfrac{1}{\ln x} - \dfrac{1}{x-1}\right)$

12. $\lim\limits_{x\to0^+} (\sinh x)^{\tan x}$

13. $\lim\limits_{x\to0^+} x^{\sin x}$

14. $\lim\limits_{x\to0} (x + e^{2x})^{1/x}$

15. $\lim\limits_{x\to+\infty} (x^2 - \sqrt{x^4 - x^2 + 2})$

16. $\lim\limits_{x\to2} \left(\dfrac{5}{x^2+x-6} - \dfrac{1}{x-2}\right)$

17. $\lim\limits_{x\to+\infty} x^{1/x}$

18. $\lim\limits_{x\to0^+} (1+x)^{\ln x}$

19. $\lim\limits_{x\to+\infty} (e^x + x)^{2/x}$

20. $\lim\limits_{x\to0^+} x^{1/\ln x}$

21. $\lim\limits_{x\to0} (1 + ax)^{1/x};\ a \neq 0$

22. $\lim\limits_{x\to+\infty} \dfrac{x^2 + 2x}{e^{3x} - 1}$

23. $\lim\limits_{x\to0^+} (\sin x)^{x^2}$

24. $\lim\limits_{x\to+\infty} \left(1 + \dfrac{1}{2x}\right)^{x^2}$

25. $\lim\limits_{x\to0} (1 + \sinh x)^{2/x}$

26. $\lim\limits_{x\to2} (x-2) \tan \frac{1}{4}\pi x$

27. $\lim\limits_{x\to0} [(\cos x)e^{x^2/2}]^{4/x^4}$

28. $\lim\limits_{x \to 0} (\cos x)^{1/x^2}$

29. $\lim\limits_{x \to +\infty} [(x^6 + 3x^5 + 4)^{1/6} - x]$

30. $\lim\limits_{x \to +\infty} \dfrac{\ln(x + e^x)}{3x}$

31. $\lim\limits_{x \to 0^+} \dfrac{e^{-1/x}}{x}$

32. $\lim\limits_{x \to 0^+} x^{x^x}$

33. $\lim\limits_{x \to +\infty} \dfrac{x}{\sqrt{1 + x^2}}$

34. $\lim\limits_{x \to +\infty} (x - \sqrt{x^2 + x})$

35. If $f(x) = \begin{cases} (1 - e^{4x})^x & \text{if } x < 0 \\ k & \text{if } 0 \le x \end{cases}$, find k so that f is continuous at $x = 0$.

36. If $f(x) = \begin{cases} (x + 1)^{(\ln k)/x} & \text{if } x \ne 0 \\ 5 & \text{if } x = 0 \end{cases}$, find k so that f is continuous at $x = 0$.

37. If $\lim\limits_{x \to +\infty} \left(\dfrac{nx + 1}{nx - 1}\right)^x = 9$, find n.

38. Evaluate the limit in Example 3 without using L'Hôpital's rule but by using Theorem 1.8.2 and the identities $\cos(\frac{1}{2}\pi - t) = \sin t$ and $\sin(\frac{1}{2}\pi - t) = \cos t$.

39. (a) Prove that $\lim\limits_{x \to 0} (e^{-1/x^2}/x^n) = 0$ for any positive integer n. (b) If $f(x) = e^{-1/x^2}$, use the result of (a) to prove that the limit of f and all of its derivatives, as x approaches 0, is 0.

40. Suppose $f(x) = \displaystyle\int_1^x e^{3t}\sqrt{9t^4 + 1}\, dt$ and $g(x) = x^n e^{3x}$. If $\lim\limits_{x \to +\infty} \left[\dfrac{f'(x)}{g'(x)}\right] = 1$, find n.

In Exercises 41 and 42, draw a sketch of the graph of f by first finding the relative extrema of f and the horizontal asymptotes of the graph, if there are any.

41. $f(x) = x^{1/x}, \ x > 0$

42. $f(x) = x^x, \ x > 0$

43. Determine each of the following limits: (a) $\lim\limits_{x \to 0^+} \sin x^{\csc x}$; (b) $\lim\limits_{x \to +\infty} \left(\sin \dfrac{1}{x}\right)^x$. Is $0^{+\infty}$ an indeterminate form?

9.3
IMPROPER INTEGRALS
WITH INFINITE LIMITS
OF INTEGRATION

In defining the definite integral $\int_a^b f(x)\, dx$, the function f was assumed to be defined on the closed interval $[a, b]$. We now extend the definition of the definite integral to consider an infinite interval of integration and call such an integral an **improper integral**. In Section 9.4 we discuss another kind of improper integral.

▶ ILLUSTRATION 1   Consider the problem of finding the area of the region bounded by the curve $y = e^{-x}$, the x axis, the y axis, and the line $x = b$, where $b > 0$. This region is shown in Fig. 9.3.1. If A square units is the area of the region,

$$A = \lim_{\|\Delta\| \to 0} \sum_{i=1}^{n} e^{-\xi_i} \Delta_i x$$

$$= \int_0^b e^{-x}\, dx$$

$$= -e^{-x}\Big]_0^b$$

$$= 1 - e^{-b}$$

FIGURE 9.3.1

If we let b increase without bound, then

$$\lim_{b \to +\infty} \int_0^b e^{-x} dx = \lim_{b \to +\infty} (1 - e^{-b})$$

$$\lim_{b \to +\infty} \int_0^b e^{-x} dx = 1 \tag{1}$$

From (1) it follows that no matter how large a value we take for b, the area of the region shown in Fig. 9.3.1 will always be less than 1 square unit.

◄

Equation (1) states that if $b > 0$, for any $\epsilon > 0$ there exists an $N > 0$ such that

$$\text{if} \quad b > N \quad \text{then} \quad \left| \int_0^b e^{-x} dx - 1 \right| < \epsilon$$

In place of (1) we write

$$\int_0^{+\infty} e^{-x} dx = 1$$

In general, we have the following definition.

9.3.1 DEFINITION   If f is continuous for all $x \geq a$, then

$$\int_a^{+\infty} f(x) dx = \lim_{b \to +\infty} \int_a^b f(x) dx$$

if this limit exists.

If the lower limit of integration is infinite, we have the following definition.

9.3.2 DEFINITION   If f is continuous for all $x \leq b$, then

$$\int_{-\infty}^b f(x) dx = \lim_{a \to -\infty} \int_a^b f(x) dx$$

if this limit exists.

Finally, we have the case when both limits of integration are infinite.

9.3.3 DEFINITION   If f is continuous for all values of x, and c is any real number, then

$$\int_{-\infty}^{+\infty} f(x) dx = \lim_{a \to -\infty} \int_a^c f(x) dx + \lim_{b \to +\infty} \int_c^b f(x) dx \tag{2}$$

if these limits exist.

In Exercise 40 you are asked to prove that when the limits exist, the right side of (2) is independent of the choice of c. When Definition 9.3.3 is applied, c is usually taken as 0.

In Definitions 9.3.1, 9.3.2, and 9.3.3, if the limits exist, we say that the improper integral is **convergent**. If the limits do not exist, we say that the improper integral is **divergent**.

EXAMPLE 1

Evaluate

$$\int_{-\infty}^{2} \frac{dx}{(4-x)^2}$$

if it converges.

SOLUTION

$$\int_{-\infty}^{2} \frac{dx}{(4-x)^2} = \lim_{a \to -\infty} \int_{a}^{2} \frac{dx}{(4-x)^2}$$

$$= \lim_{a \to -\infty} \left[\frac{1}{4-x} \right]_{a}^{2}$$

$$= \lim_{a \to -\infty} \left(\frac{1}{2} - \frac{1}{4-a} \right)$$

$$= \tfrac{1}{2} - 0$$

$$= \tfrac{1}{2}$$

EXAMPLE 2

Evaluate

$$\int_{0}^{+\infty} xe^{-x}\,dx$$

if it converges.

SOLUTION

$$\int_{0}^{+\infty} xe^{-x}\,dx = \lim_{b \to +\infty} \int_{0}^{b} xe^{-x}\,dx$$

To evaluate the integral we use integration by parts with $u = x$, $dv = e^{-x}\,dx$, $du = dx$, and $v = -e^{-x}$. Thus

$$\int_{0}^{+\infty} xe^{-x}\,dx = \lim_{b \to +\infty} \left[-xe^{-x} - e^{-x} \right]_{0}^{b}$$

$$= \lim_{b \to +\infty} (-be^{-b} - e^{-b} + 1)$$

$$= -\lim_{b \to +\infty} \frac{b}{e^b} - 0 + 1 \qquad (3)$$

To evaluate $\displaystyle\lim_{b \to +\infty} \frac{b}{e^b}$, we apply L'Hôpital's rule because $\displaystyle\lim_{b \to +\infty} b = +\infty$

and $\lim\limits_{b \to +\infty} e^b = +\infty$. We have

$$\lim_{b \to +\infty} \frac{b}{e^b} = \lim_{b \to +\infty} \frac{1}{e^b}$$

$$= 0$$

Therefore, from (3),

$$\int_0^{+\infty} x e^{-x}\, dx = 1$$

EXAMPLE 3
Evaluate if they exist:

(a) $\displaystyle\int_{-\infty}^{+\infty} x\, dx$ (b) $\displaystyle\lim_{r \to +\infty} \int_{-r}^{r} x\, dx$

SOLUTION
(a) From Definition 9.3.3 with $c = 0$ we have

$$\int_{-\infty}^{+\infty} x\, dx = \lim_{a \to -\infty} \int_a^0 x\, dx + \lim_{b \to +\infty} \int_0^b x\, dx$$

$$= \lim_{a \to -\infty} \left[\tfrac{1}{2}x^2\right]_a^0 + \lim_{b \to +\infty} \left[\tfrac{1}{2}x^2\right]_0^b$$

$$= \lim_{a \to -\infty} (-\tfrac{1}{2}a^2) + \lim_{b \to +\infty} \tfrac{1}{2}b^2$$

Because neither of these two limits exists, the improper integral diverges.

(b) $\displaystyle\lim_{r \to +\infty} \int_{-r}^{r} x\, dx = \lim_{r \to +\infty} \left[\tfrac{1}{2}x^2\right]_{-r}^r$

$$= \lim_{r \to +\infty} (\tfrac{1}{2}r^2 - \tfrac{1}{2}r^2)$$

$$= \lim_{r \to +\infty} 0$$

$$= 0$$

Example 3 illustrates why we do not use the limit in (b) to determine the convergence of an improper integral where both limits of integration are infinite. That is, the improper integral in (a) is divergent, but the limit in (b) exists and is zero.

EXAMPLE 4
Evaluate

$$\int_{-\infty}^{+\infty} \frac{dx}{x^2 + 6x + 12}$$

if it converges.

SOLUTION

$$\int_{-\infty}^{+\infty} \frac{dx}{x^2 + 6x + 12} = \lim_{a \to -\infty} \int_a^0 \frac{dx}{(x+3)^2 + 3} + \lim_{b \to +\infty} \int_0^b \frac{dx}{(x+3)^2 + 3}$$

$$= \lim_{a \to -\infty} \left[\frac{1}{\sqrt{3}} \tan^{-1} \frac{x+3}{\sqrt{3}} \right]_a^0 + \lim_{b \to +\infty} \left[\frac{1}{\sqrt{3}} \tan^{-1} \frac{x+3}{\sqrt{3}} \right]_0^b$$

$$= \lim_{a \to -\infty} \left(\frac{1}{\sqrt{3}} \tan^{-1} \sqrt{3} - \frac{1}{\sqrt{3}} \tan^{-1} \frac{a+3}{\sqrt{3}} \right)$$

$$+ \lim_{b \to +\infty} \left(\frac{1}{\sqrt{3}} \tan^{-1} \frac{b+3}{\sqrt{3}} - \frac{1}{\sqrt{3}} \tan^{-1} \sqrt{3} \right)$$

$$= \frac{1}{\sqrt{3}} \left[\lim_{a \to -\infty} \left(-\tan^{-1} \frac{a+3}{\sqrt{3}} \right) + \lim_{b \to +\infty} \tan^{-1} \frac{b+3}{\sqrt{3}} \right]$$

$$= \frac{1}{\sqrt{3}} \left[-\left(-\frac{\pi}{2} \right) + \frac{\pi}{2} \right]$$

$$= \frac{\pi}{\sqrt{3}}$$

EXAMPLE 5

Is it possible to assign a finite number to represent the measure of the area of the region to the right of the line $x = 1$, below the graph of $y = 1/x$, and above the x axis?

SOLUTION

The region is shown in Fig. 9.3.2. Let L be the number we wish to assign to the measure of the area, if possible. Let A be the measure of the area of the region bounded by the graphs of the equations $y = 1/x$, $y = 0$, $x = 1$, and $x = b$, where $b > 1$. Then

$$A = \lim_{||\Delta|| \to 0} \sum_{i=1}^{n} \frac{1}{\xi_i} \Delta_i x$$

$$= \int_1^b \frac{1}{x} dx$$

So we shall let $L = \lim_{b \to +\infty} A$ if this limit exists. But

$$\lim_{b \to +\infty} A = \lim_{b \to +\infty} \int_1^b \frac{1}{x} dx$$

$$= \lim_{b \to +\infty} \left[\ln b - \ln 1 \right]$$

$$= +\infty$$

Therefore it is not possible to assign a finite number to represent the measure of the area of the region.

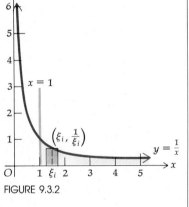

FIGURE 9.3.2

EXAMPLE 6

Is it possible to assign a finite number to represent the measure of the volume of the solid formed by revolving the region in Example 5 about the x axis?

SOLUTION

The element of volume is a circular disk having a thickness of $\Delta_i x$ and a base radius of $1/\xi_i$. Let L be the number we wish to assign to the measure of the volume, and let V be the measure of the volume of the solid formed by revolving about the x axis the region bounded by the graphs of the equations $y = 1/x$, $y = 0$, $x = 1$, and $x = b$, where $b > 1$. Then

$$V = \lim_{\|\Delta\| \to 0} \sum_{i=1}^{n} \pi \left(\frac{1}{\xi_i}\right)^2 \Delta_i x$$

$$= \pi \int_{1}^{b} \frac{1}{x^2} \, dx$$

We shall let $L = \lim\limits_{b \to +\infty} V$, if the limit exists.

$$\lim_{b \to +\infty} V = \lim_{b \to +\infty} \pi \int_{1}^{b} \frac{dx}{x^2}$$

$$= \pi \lim_{b \to +\infty} \left[-\frac{1}{x} \right]_{1}^{b}$$

$$= \pi \lim_{b \to +\infty} \left(-\frac{1}{b} + 1 \right)$$

$$= \pi$$

Therefore we assign π to represent the measure of the volume of the solid.

EXAMPLE 7

Determine if $\int_{0}^{+\infty} \sin x \, dx$ is convergent or divergent.

SOLUTION

$$\int_{0}^{+\infty} \sin x \, dx = \lim_{b \to +\infty} \int_{0}^{b} \sin x \, dx$$

$$= \lim_{b \to +\infty} \left[-\cos x \right]_{0}^{b}$$

$$= \lim_{b \to +\infty} (-\cos b + 1)$$

For any integer n, as b takes on all values from $n\pi$ to $2n\pi$, $\cos b$ takes on all values from -1 to 1. Hence $\lim\limits_{b \to +\infty} \cos b$ does not exist. Therefore the improper integral is divergent.

Example 7 illustrates the case for which an improper integral is divergent where the limit is not infinite.

An application of an improper integral with an infinite limit of integration involves probability. The probability of a particular event occurring is a number in the closed interval $[0, 1]$. If an event is certain to occur, then the probability of its happening is 1; if the event will never occur, then the probability is 0. The surer it is that an event will occur, the closer its probability is to 1.

Suppose that the set of all possible outcomes of a particular situation is the set of all numbers x in some interval I. For instance, x may be the number of minutes in the waiting time for a table at a particular restaurant, the number of hours in the life of a tube for a television set, or the number of inches in a person's height. It is sometimes necessary to determine the probability of x being in some subinterval of I. For example, one may wish to find the probability that a person will have to wait between 20 and 30 minutes for a table at a restaurant, or the probability of a television tube lasting more than 2000 hours, or the probability that someone chosen at random will have a height between 66 and 72 inches. Such problems involve evaluating an integral of a function called a *probability density function*. Probability density functions are obtained from statistical experiments. We give a brief informal discussion of them here to show how improper integrals arise.

A **probability density function** is a function f having as its domain the set R of real numbers and that satisfies the following two conditions:

1. $f(x) \geq 0$ for all x in R.

2. $\displaystyle\int_{-\infty}^{+\infty} f(x)\, dx = 1$

We shall consider here the *exponential density function* defined by

$$f(x) = \begin{cases} ke^{-kx} & \text{if } x \geq 0 \\ 0 & \text{if } x < 0 \end{cases} \tag{4}$$

where $k > 0$. To verify that this function qualifies as a probability density function we show that the two properties hold.

1. If $x < 0$, $f(x) = 0$; if $x \geq 0$, $f(x) = ke^{-kx}$, and because $k > 0$, $ke^{-kx} > 0$.

2. $\displaystyle\int_{-\infty}^{+\infty} f(x)\, dx = \int_{-\infty}^{0} f(x)\, dx + \int_{0}^{+\infty} f(x)\, dx$

$\displaystyle\qquad\qquad\quad = \int_{-\infty}^{0} 0\, dx + \int_{0}^{+\infty} ke^{-kx}\, dx$

$\displaystyle\qquad\qquad\quad = 0 + \lim_{b \to +\infty} \left[-\int_{0}^{b} e^{-kx}(-k\, dx) \right]$

$\displaystyle\qquad\qquad\quad = \lim_{b \to +\infty} \left[-e^{-kx} \right]_{0}^{b}$

$\displaystyle\qquad\qquad\quad = \lim_{b \to +\infty} (-e^{-kb} + 1)$

$\displaystyle\qquad\qquad\quad = 1$

FIGURE 9.3.3

If f is a probability density function for a particular event occurring, then the **probability that the event will occur over the closed interval $[a, b]$** is denoted by $P([a, b])$ and

$$P([a, b]) = \int_a^b f(x)\, dx$$

Figure 9.3.3 shows a sketch of the graph of the exponential density function. Because $\int_0^{+\infty} ke^{-kx}\, dx = 1$, the measure of the area of the region bounded by $y = ke^{-kx}$, the x axis, and the y axis is 1. The measure of the area of the shaded region in the figure is $P([a, b])$.

EXAMPLE 8
For a particular kind of battery the probability density function for x hours to be the life of a battery selected at random is given by

$$f(x) = \begin{cases} \frac{1}{60}e^{-x/60} & \text{if } x \geq 0 \\ 0 & \text{if } x < 0 \end{cases} \tag{5}$$

Find the probability that the life of a battery selected at random will be (a) between 15 and 25 hours and (b) at least 50 hours.

SOLUTION
The function defined by (5) is of the form of (4) with $k = \frac{1}{60}$. (a) The probability that the life of a battery selected at random will be between 15 and 25 hours is $P([15, 25])$, and (b) the probability that it will be at least 50 hours is $P([50, +\infty))$.

(a) $P([15, 25])$

$$= \int_{15}^{25} \tfrac{1}{60} e^{-x/60}\, dx$$

$$= -\int_{15}^{25} e^{-x/60}(-\tfrac{1}{60}\, dx)$$

$$= -e^{-x/60}\Big]_{15}^{25}$$

$$= -e^{-25/60} + e^{-15/60}$$

$$= -e^{-0.417} + e^{-0.250}$$

$$= -0.659 + 0.779$$

$$= 0.120$$

(b) $P([50, +\infty))$

$$= \lim_{b \to +\infty} \int_{50}^{b} \tfrac{1}{60} e^{-x/60}\, dx$$

$$= \lim_{b \to +\infty} \left[-e^{-x/60} \right]_{50}^{b}$$

$$= \lim_{b \to +\infty} (-e^{-b/60} + e^{-50/60})$$

$$= 0 + e^{-0.833}$$

$$= 0.435$$

EXERCISES 9.3
In Exercises 1 through 18, determine whether the improper integral is convergent or divergent. If it is convergent, evaluate it.

1. $\int_0^{+\infty} e^{-x}\, dx$

2. $\int_{-\infty}^{1} e^x\, dx$

3. $\int_{-\infty}^{0} x5^{-x^2}\, dx$

4. $\int_1^{+\infty} 2^{-x}\, dx$

5. $\int_0^{+\infty} x2^{-x}\, dx$

6. $\int_5^{+\infty} \dfrac{dx}{\sqrt{x-1}}$

7. $\int_{-\infty}^{+\infty} x \cosh x\, dx$

8. $\int_{-\infty}^{0} x^2 e^x\, dx$

9. $\int_5^{+\infty} \dfrac{x\, dx}{\sqrt[3]{9 - x^2}}$

10. $\int_{-\infty}^{+\infty} \dfrac{3x\, dx}{(3x^2 + 2)^3}$

11. $\displaystyle\int_{\sqrt{3}}^{+\infty} \frac{3\,dx}{x^2 + 9}$ **12.** $\displaystyle\int_{e}^{+\infty} \frac{dx}{x \ln x}$ **13.** $\displaystyle\int_{-\infty}^{+\infty} e^{-|x|}\,dx$ **14.** $\displaystyle\int_{-\infty}^{+\infty} x e^{-x^2}\,dx$ **15.** $\displaystyle\int_{e}^{+\infty} \frac{dx}{x(\ln x)^2}$

16. $\displaystyle\int_{-\infty}^{+\infty} \frac{dx}{16 + x^2}$ **17.** $\displaystyle\int_{1}^{+\infty} \ln x\,dx$ **18.** $\displaystyle\int_{0}^{+\infty} e^{-x}\cos x\,dx$

19. Evaluate if they exist: (a) $\displaystyle\int_{-\infty}^{+\infty} \sin x\,dx$; (b) $\displaystyle\lim_{r \to +\infty} \int_{-r}^{r} \sin x\,dx$.

20. Prove that if $\displaystyle\int_{-\infty}^{b} f(x)\,dx$ is convergent, then $\displaystyle\int_{-b}^{+\infty} f(-x)\,dx$ is also convergent and has the same value.

21. Show that the improper integral $\displaystyle\int_{-\infty}^{+\infty} x(1 + x^2)^{-2}\,dx$ is convergent and the improper integral $\displaystyle\int_{-\infty}^{+\infty} x(1 + x^2)^{-1}\,dx$ is divergent.

22. Prove that the improper integral $\displaystyle\int_{1}^{+\infty} \frac{dx}{x^n}$ is convergent if and only if $n > 1$.

23. Determine if it is possible to assign a finite number to represent the measure of the area of the region bounded by the curve whose equation is $y = 1/(e^x + e^{-x})$ and the x axis. If a finite number can be assigned, find it.

24. Determine if it is possible to assign a finite number to represent the measure of the area of the region bounded by the x axis, the line $x = 2$, and the curve whose equation is $y = 1/(x^2 - 1)$. If a finite number can be assigned, find it.

25. Determine if it is possible to assign a finite number to represent the measure of the volume of the solid formed by revolving about the x axis the region to the right of the line $x = 1$ and bounded by the curve whose equation is $y = 1/x^{3/2}$ and the x axis. If a finite number can be assigned, find it.

26. Determine if it is possible to assign a finite number to represent the measure of the volume of the solid formed by revolving about the x axis the region bounded by the x axis, the y axis, and the curve whose equation is $y = e^{-2x}$. If a finite number can be assigned, find it.

27. For the battery of Example 8, find the probability that the life of a battery selected at random will be (a) not more than 50 hours and (b) at least 75 hours.

28. For a certain type of light bulb, the probability density function that x hours will be the life of a bulb selected at random is given by

$$f(x) = \begin{cases} \frac{1}{40}e^{-x/40} & \text{if } x \geq 0 \\ 0 & \text{if } x < 0 \end{cases}$$

Find the probability that the life of a bulb selected at random will be (a) between 40 and 60 hours and (b) at least 60 hours.

29. In a certain city, the probability density function for x min to be the length of a telephone call selected at random is given by

$$f(x) = \begin{cases} \frac{1}{3}e^{-x/3} & \text{if } x \geq 0 \\ 0 & \text{if } x < 0 \end{cases}$$

Find the probability that a telephone call selected at random will last (a) between 1 min and 2 min, and (b) at least 5 min.

30. For a particular appliance, the probability density function that it will need servicing x months after it is purchased is given by

$$f(x) = \begin{cases} 0.02e^{-0.02x} & \text{if } x \geq 0 \\ 0 & \text{if } x < 0 \end{cases}$$

If the appliance is guaranteed for a year, what is the probability that a customer selected at random will not need servicing during the 1-year warranty period?

31. If f is a probability density function, then the *mean* (or *average value*) of the probabilities is given by $\int_{-\infty}^{+\infty} x f(x)\,dx$. Find the mean of the probabilities obtained from the exponential density function (4).

32. A *uniform* probability density function is defined by

$$f(x) = \begin{cases} 0 & \text{if } x < c \\ \dfrac{1}{d-c} & \text{if } c \le x \le d \\ 0 & \text{if } d < x \end{cases}$$

Show that this function satisfies the conditions necessary for it to qualify as a probability density function.

Exercises 33 through 36 show an application of improper integrals in the field of economics. Suppose there is a continuous flow of income for which interest is compounded continuously at the annual rate of $100i$ percent and $f(t)$ dollars is the income per year at any time t years. If the income continues indefinitely, the present value, V dollars, of all future income is defined by

$$V = \int_0^{+\infty} f(t)e^{-it}\,dt$$

33. A continuous flow of income is decreasing with time, and at t years the number of dollars in the annual income is $1000 \cdot 2^{-t}$. Find the present value of this income if it continues indefinitely using an interest rate of 8 percent compounded continuously.

34. Suppose that the owner of a piece of business property holds a permanent lease on the property so that the rent is paid perpetually. If the annual rent is \$12,000 and money is worth 10 percent compounded continuously, find the present value of all future rent payments.

35. The British Consol is a bond with no maturity (i.e., it never comes due), and it affords the holder an annual lump-sum payment. By finding the present value of a flow of payments of R dollars annually and using the current interest rate $100i$ percent, compounded continuously, show that the fair selling price of a British Consol is R/i dollars.

36. The continuous flow of profit for a company is increasing with time, and at t years the number of dollars in the profit per year is proportional to t. Show that the present value of the company is inversely proportional to i^2, where $100i$ percent is the interest rate compounded continuously.

37. Determine the values of n for which the following improper integral is convergent: $\displaystyle\int_e^{+\infty} \frac{dx}{x(\ln x)^n}$.

38. Determine a value of n for which the improper integral $\displaystyle\int_1^{+\infty}\left(\frac{n}{x+1} - \frac{3x}{2x^2+n}\right)dx$ is convergent, and evaluate the integral for this value of n.

39. Determine a value of n for which the improper integral $\displaystyle\int_1^{+\infty}\left(\frac{nx^2}{x^3+1} - \frac{1}{13x+1}\right)dx$ is convergent, and evaluate the integral for this value of n.

40. Suppose f is continuous for all values of x. Prove that if

$$\lim_{a \to -\infty} \int_a^c f(x)\,dx = L \quad \text{and} \quad \lim_{b \to +\infty} \int_c^b f(x)\,dx = M$$

then if d is any real number,

$$\lim_{a \to -\infty} \int_a^d f(x)\,dx + \lim_{b \to +\infty} \int_d^b f(x)\,dx = L + M$$

Hint: $\displaystyle\int_a^d f(x)\,dx = \int_a^c f(x)\,dx + \int_c^d f(x)\,dx$ and $\displaystyle\int_d^b f(x)\,dx = \int_d^c f(x)\,dx + \int_c^b f(x)\,dx$.

9.4
OTHER IMPROPER
INTEGRALS

FIGURE 9.4.1

Figure 9.4.1 shows the region bounded by the curve whose equation is $y = 1/\sqrt{x}$, the x axis, the y axis, and the line $x = 4$. If it is possible to assign a finite number to represent the measure of the area of this region, it would be given by

$$\lim_{\|\Delta\| \to 0} \sum_{i=1}^{n} \frac{1}{\sqrt{\xi_i}} \Delta_i x$$

If this limit exists, it is the definite integral denoted by

$$\int_0^4 \frac{dx}{\sqrt{x}} \tag{1}$$

However, the integrand is discontinuous at the lower limit zero. Furthermore, $\lim_{x \to 0^+} 1/\sqrt{x} = +\infty$, and so we state that the integrand has an infinite discontinuity at the lower limit. Such an integral is improper, and its existence can be determined from the following definition.

9.4.1 DEFINITION   If f is continuous at all x in the interval half open on the left $(a, b]$, and if $\lim_{x \to a^+} f(x) = \pm\infty$, then

$$\int_a^b f(x)\,dx = \lim_{t \to a^+} \int_t^b f(x)\,dx$$

if this limit exists.

▶ ILLUSTRATION 1   We determine if it is possible to assign a finite number to represent the measure of the area of the region shown in Fig. 9.4.1. From the discussion preceding Definition 9.4.1, the measure of the area of the given region will be the improper integral (1) if it exists. By Definition 9.4.1,

$$\int_0^4 \frac{dx}{\sqrt{x}} = \lim_{t \to 0^+} \int_t^4 \frac{dx}{\sqrt{x}}$$

$$= \lim_{t \to 0^+} 2x^{1/2} \Big]_t^4$$

$$= \lim_{t \to 0^+} (4 - 2\sqrt{t})$$

$$= 4 - 0$$

$$= 4$$

Therefore we assign 4 to the measure of the area of the given region.   ◀

If the integrand has an infinite discontinuity at the upper limit of integration, we use the following definition to determine the existence of the improper integral.

9.4.2 DEFINITION   If f is continuous at all x in the interval half open on the right $[a, b)$, and if $\lim\limits_{x \to b^-} f(x) = \pm\infty$, then

$$\int_a^b f(x)\, dx = \lim_{t \to b^-} \int_a^t f(x)\, dx$$

if this limit exists.

If there is an infinite discontinuity at an interior point of the interval of integration, the existence of the improper integral is determined from the following definition.

9.4.3 DEFINITION   If f is continuous at all x in the interval $[a, b]$ except c, where $a < c < b$, and if $\lim\limits_{x \to c} |f(x)| = +\infty$, then

$$\int_a^b f(x)\, dx = \lim_{t \to c^-} \int_a^t f(x)\, dx + \lim_{s \to c^+} \int_s^b f(x)\, dx$$

if these limits exist.

If $\int_a^b f(x)\, dx$ is an improper integral, it is convergent if the corresponding limit exists; otherwise it is divergent.

EXAMPLE 1
Evaluate

$$\int_0^2 \frac{dx}{(x - 1)^2}$$

if it is convergent.

SOLUTION
The integrand has an infinite discontinuity at 1. Applying Definition 9.4.3 we have

$$\int_0^2 \frac{dx}{(x - 1)^2} = \lim_{t \to 1^-} \int_0^t \frac{dx}{(x - 1)^2} + \lim_{s \to 1^+} \int_s^2 \frac{dx}{(x - 1)^2}$$

$$= \lim_{t \to 1^-} \left[-\frac{1}{x - 1} \right]_0^t + \lim_{s \to 1^+} \left[-\frac{1}{x - 1} \right]_s^2$$

$$= \lim_{t \to 1^-} \left[-\frac{1}{t - 1} - 1 \right] + \lim_{s \to 1^+} \left[-1 + \frac{1}{s - 1} \right]$$

Because neither of these limits exist, the improper integral is divergent.

▶ ILLUSTRATION 2   Suppose that in evaluating the integral in Example 1 we had failed to note the infinite discontinuity of the integrand at 1. We would have obtained

$$-\frac{1}{x-1}\Big]_0^2 = -\frac{1}{1} + \frac{1}{-1}$$

$$= -2$$

This is obviously an incorrect result. Because $1/(x-1)^2$ is never negative, the integral from 0 to 2 could not possibly be a negative number.   ◄

EXAMPLE 2

Evaluate $\int_0^1 x \ln x\, dx$ if it is convergent.

SOLUTION

The integrand has a discontinuity at the lower limit. From Definition 9.4.1,

$$\int_0^1 x \ln x\, dx = \lim_{t \to 0^+} \int_t^1 x \ln x\, dx$$

$$= \lim_{t \to 0^+} \left[\tfrac{1}{2}x^2 \ln x - \tfrac{1}{4}x^2 \right]_t^1$$

$$= \lim_{t \to 0^+} \left[\tfrac{1}{2}\ln 1 - \tfrac{1}{4} - \tfrac{1}{2}t^2 \ln t + \tfrac{1}{4}t^2 \right]$$

Hence

$$\int_0^1 x \ln x\, dx = 0 - \tfrac{1}{4} - \tfrac{1}{2} \lim_{t \to 0^+} t^2 \ln t + 0 \tag{2}$$

To evaluate

$$\lim_{t \to 0^+} t^2 \ln t = \lim_{t \to 0^+} \frac{\ln t}{\dfrac{1}{t^2}}$$

we apply L'Hôpital's rule, because $\lim_{t \to 0^+} \ln t = -\infty$ and $\lim_{t \to 0^+} 1/t^2 = +\infty$. We have

$$\lim_{t \to 0^+} \frac{\ln t}{\dfrac{1}{t^2}} = \lim_{t \to 0^+} \frac{\dfrac{1}{t}}{-\dfrac{2}{t^3}}$$

$$= \lim_{t \to 0^+} \left[-\frac{t^2}{2} \right]$$

$$= 0$$

Therefore, from (2),

$$\int_0^1 x \ln x\, dx = -\tfrac{1}{4}$$

EXAMPLE 3

Evaluate

$$\int_1^{+\infty} \frac{dx}{x\sqrt{x^2 - 1}}$$

if it is convergent.

SOLUTION

For this integral there is both an infinite upper limit and an infinite discontinuity of the integrand at the lower limit. We proceed as follows.

$$\int_1^{+\infty} \frac{dx}{x\sqrt{x^2 - 1}} = \lim_{t \to 1^+} \int_t^2 \frac{dx}{x\sqrt{x^2 - 1}} + \lim_{b \to +\infty} \int_2^b \frac{dx}{x\sqrt{x^2 - 1}}$$

$$= \lim_{t \to 1^+} \left[\sec^{-1} x \right]_t^2 + \lim_{b \to +\infty} \left[\sec^{-1} x \right]_2^b$$

$$= \lim_{t \to 1^+} (\sec^{-1} 2 - \sec^{-1} t) + \lim_{b \to +\infty} (\sec^{-1} b - \sec^{-1} 2)$$

$$= \tfrac{1}{3}\pi - \lim_{t \to 1^+} \sec^{-1} t + \lim_{b \to +\infty} \sec^{-1} b - \tfrac{1}{3}\pi$$

$$= -0 + \tfrac{1}{2}\pi$$

$$= \tfrac{1}{2}\pi$$

The integral in Example 3 is called an *improper integral of mixed type*.

EXERCISES 9.4

In Exercises 1 through 25, determine whether the improper integral is convergent or divergent. If it is convergent, evaluate it.

1. $\int_0^1 \frac{dx}{\sqrt{1 - x}}$ **2.** $\int_0^{16} \frac{dx}{x^{3/4}}$ **3.** $\int_{-5}^{-3} \frac{x\,dx}{\sqrt{x^2 - 9}}$ **4.** $\int_0^4 \frac{x\,dx}{\sqrt{16 - x^2}}$ **5.** $\int_2^4 \frac{dt}{\sqrt{16 - t^2}}$

6. $\int_{-4}^1 \frac{dz}{(z + 3)^3}$ **7.** $\int_{\pi/4}^{\pi/2} \sec \theta\, d\theta$ **8.** $\int_{-2}^0 \frac{dx}{\sqrt{4 - x^2}}$ **9.** $\int_0^{+\infty} \frac{dx}{x^3}$ **10.** $\int_0^{\pi/2} \tan \theta\, d\theta$

11. $\int_0^{\pi/2} \frac{dy}{1 - \sin y}$ **12.** $\int_0^2 \frac{dx}{(x - 1)^{2/3}}$ **13.** $\int_0^4 \frac{dx}{x^2 - 2x - 3}$ **14.** $\int_2^{+\infty} \frac{dx}{x\sqrt{x^2 - 4}}$ **15.** $\int_0^{+\infty} \ln x\, dx$

16. $\int_0^2 \frac{dx}{\sqrt{2x - x^2}}$ **17.** $\int_{-2}^0 \frac{dw}{(w + 1)^{1/3}}$ **18.** $\int_{-1}^1 \frac{dx}{x^2}$ **19.** $\int_{-2}^2 \frac{dx}{x^3}$ **20.** $\int_0^{+\infty} \frac{e^{-\sqrt{x}}}{\sqrt{x}}\, dx$

21. $\int_{1/2}^2 \frac{dz}{z(\ln z)^{1/5}}$ **22.** $\int_0^2 \frac{x\,dx}{1 - x}$ **23.** $\int_1^2 \frac{dx}{x\sqrt{x^2 - 1}}$ **24.** $\int_0^1 \frac{dx}{x\sqrt{4 - x^2}}$ **25.** $\int_1^3 \frac{dy}{\sqrt[3]{y - 2}}$

26. Evaluate, if they exist: (a) $\int_{-1}^1 \frac{dx}{x}$; (b) $\lim_{r \to 0^+} \left[\int_{-1}^{-r} \frac{dx}{x} + \int_r^1 \frac{dx}{x} \right]$.

In Exercises 27 through 29, find the values of n for which the improper integral converges, and evaluate the integral for these values of n.

27. $\displaystyle\int_0^1 x^n \, dx$

28. $\displaystyle\int_0^1 x^n \ln x \, dx$

29. $\displaystyle\int_0^1 x^n \ln^2 x \, dx$

30. Show that it is possible to assign a finite number to represent the measure of the area of the region bounded by the curve whose equation is $y = 1/\sqrt{x}$, the line $x = 1$, and the x and y axes, but that it is not possible to assign a finite number to represent the measure of the volume of the solid of revolution generated if this region is revolved about the x axis.

31. Determine if it is possible to assign a finite number to represent the measure of the volume of the solid formed by revolving about the x axis the region bounded by the curve whose equation is $y = x^{-1/3}$, the line $x = 8$, and the x and y axes.

32. Given the improper integral $\displaystyle\int_a^b \frac{dx}{(x-a)^n}$ where $b > a$. Determine if the integral is convergent or divergent in each case:

(a) $0 < n < 1$; (b) $n = 1$; (c) $n > 1$. If the integral is convergent, evaluate it.

33. Use integration to prove that the circumference of the circle $x^2 + y^2 = a^2$ is $2\pi a$.

9.5

TAYLOR'S FORMULA

Values of polynomial functions can be found by performing a finite number of additions and multiplications. However, there are other functions, such as the logarithmic, exponential, and trigonometric functions, that cannot be evaluated as easily. We show in this section that many functions can be approximated by polynomials and that the polynomial, instead of the original function, can be used for computations when the difference between the actual function value and the polynomial approximation is sufficiently small.

There are various methods of approximating a given function by polynomials. One of the most widely used is that involving **Taylor's formula**, named in honor of the English mathematician Brook Taylor (1685–1731). The following theorem, which can be considered as a generalization of the mean-value theorem (3.3.2), gives Taylor's formula.

9.5.1 THEOREM   Let f be a function such that f and its first n derivatives are continuous on the closed interval $[a, b]$. Furthermore, let $f^{(n+1)}(x)$ exist for all x in the open interval (a, b). Then there is a number ξ in the open interval (a, b) such that

$$f(b) = f(a) + \frac{f'(a)}{1!}(b-a) + \frac{f''(a)}{2!}(b-a)^2 + \ldots$$

$$+ \frac{f^{(n)}(a)}{n!}(b-a)^n + \frac{f^{(n+1)}(\xi)}{(n+1)!}(b-a)^{n+1} \qquad (1)$$

Equation (1) also holds if $b < a$; in such a case $[a, b]$ is replaced by $[b, a]$, and (a, b) is replaced by (b, a).

Before proving this theorem, note that when $n = 0$, (1) becomes

$$f(b) = f(a) + f'(\xi)(b-a)$$

where ξ is between a and b. This is the mean-value theorem.

There are several known proofs of Theorem 9.5.1, although none is very well motivated. The one following makes use of Cauchy's mean-value theorem (9.1.3).

PROOF OF THEOREM 9.5.1   Let F and G be two functions defined by

$$F(x) = f(b) - f(x) - f'(x)(b - x) - \frac{f''(x)}{2!}(b - x)^2 - \cdots$$

$$- \frac{f^{(n-1)}(x)}{(n-1)!}(b - x)^{n-1} - \frac{f^{(n)}(x)}{n!}(b - x)^n \qquad (2)$$

and

$$G(x) = \frac{(b - x)^{n+1}}{(n + 1)!} \qquad (3)$$

It follows that $F(b) = 0$ and $G(b) = 0$. Differentiating in (2) we get

$$F'(x) = -f'(x) + f'(x) - f''(x)(b - x) + \frac{2f''(x)(b - x)}{2!}$$

$$- \frac{f'''(x)(b - x)^2}{2!} + \frac{3f'''(x)(b - x)^2}{3!} - \frac{f^{(iv)}(x)(b - x)^3}{3!} + \cdots$$

$$+ \frac{(n - 1)f^{(n-1)}(x)(b - x)^{n-2}}{(n - 1)!} - \frac{f^{(n)}(x)(b - x)^{n-1}}{(n - 1)!}$$

$$+ \frac{nf^{(n)}(x)(b - x)^{n-1}}{n!} - \frac{f^{(n+1)}(x)(b - x)^n}{n!}$$

Combining terms we see that the sum of every odd-numbered term with the following even-numbered term is zero; so only the last term remains. Therefore

$$F'(x) = -\frac{f^{(n+1)}(x)}{n!}(b - x)^n \qquad (4)$$

Differentiating in (3) we obtain

$$G'(x) = -\frac{1}{n!}(b - x)^n \qquad (5)$$

Checking the hypothesis of Theorem 9.1.3 we see that

(i) F and G are continuous on $[a, b]$;
(ii) F and G are differentiable on (a, b);
(iii) for all x in (a, b), $G'(x) \neq 0$.

So by the conclusion of Theorem 9.1.3,

$$\frac{F(b) - F(a)}{G(b) - G(a)} = \frac{F'(\xi)}{G'(\xi)}$$

where ξ is in (a, b). But $F(b) = 0$ and $G(b) = 0$. So

$$F(a) = \frac{F'(\xi)}{G'(\xi)} G(a) \tag{6}$$

for some ξ in (a, b).

Letting $x = a$ in (3), $x = \xi$ in (4), and $x = \xi$ in (5) and substituting into (6) we obtain

$$F(a) = -\frac{f^{(n+1)}(\xi)}{n!} (b - \xi)^n \left[-\frac{n!}{(b - \xi)^n} \right] \frac{(b - a)^{n+1}}{(n + 1)!}$$

$$F(a) = \frac{f^{(n+1)}(\xi)}{(n + 1)!} (b - a)^{n+1} \tag{7}$$

If $x = a$ in (2), we obtain

$$F(a) = f(b) - f(a) - f'(a)(b - a) - \frac{f''(a)}{2!} (b - a)^2 - \dots$$

$$- \frac{f^{(n-1)}(a)}{(n - 1)!} (b - a)^{n-1} - \frac{f^{(n)}(a)}{n!} (b - a)^n \tag{8}$$

Substituting from (7) into (8) we get

$$f(b) = f(a) + f'(a)(b - a) + \frac{f''(a)}{2!} (b - a)^2 + \dots$$

$$+ \frac{f^{(n)}(a)}{n!} (b - a)^n + \frac{f^{(n+1)}(\xi)}{(n + 1)!} (b - a)^{n+1}$$

which is the desired result. The theorem holds if $b < a$ because the conclusion of Theorem 9.1.3 is unaffected if a and b are interchanged.   ■

If in (1) b is replaced by x, Taylor's formula is obtained. It is

$$f(x) = f(a) + \frac{f'(a)}{1!} (x - a) + \frac{f''(a)}{2!} (x - a)^2 + \dots$$

$$+ \frac{f^{(n)}(a)}{n!} (x - a)^n + \frac{f^{(n+1)}(\xi)}{(n + 1)!} (x - a)^{n+1} \tag{9}$$

where ξ is between a and x.

The condition under which (9) holds is that f and its first n derivatives must be continuous on a closed interval containing a and x, and the $(n + 1)$st derivative of f must exist at all points of the corresponding open interval. Formula (9) may be written as

$$f(x) = P_n(x) + R_n(x) \tag{10}$$

where

$$P_n(x) = f(a) + \frac{f'(a)}{1!}(x - a) + \frac{f''(a)}{2!}(x - a)^2 + \ldots$$

$$+ \frac{f^{(n)}(a)}{n!}(x - a)^n \qquad (11)$$

and

$$R_n(x) = \frac{f^{(n+1)}(\xi)}{(n + 1)!}(x - a)^{n+1} \qquad \text{where } \xi \text{ is between } a \text{ and } x \qquad (12)$$

$P_n(x)$ is called the nth-degree **Taylor polynomial** of the function f at the number a, and $R_n(x)$ is called the *remainder*. The term $R_n(x)$ as given in (12) is called the *Lagrange form* of the remainder, named in honor of the French mathematican Joseph L. Lagrange (1736–1813).

The special case of Taylor's formula obtained by taking $a = 0$ in (9) is

$$f(x) = f(0) + \frac{f'(0)}{1!}x + \frac{f''(0)}{2!}x^2 + \ldots + \frac{f^{(n)}(0)}{n!}x^n + \frac{f^{(n+1)}(\xi)}{(n + 1)!}x^{n+1} \quad (13)$$

where ξ is between 0 and x. Formula (13) is called **Maclaurin's formula**, named in honor of the Scottish mathematician Colin Maclaurin (1698–1746). However, the formula was obtained earlier by Taylor and by another British mathematician, James Stirling (1692–1770). The nth degree **Maclaurin polynomial** for a function f, obtained from (11) with $a = 0$, is

$$P_n(x) = f(0) + \frac{f'(0)}{1!}x + \frac{f''(0)}{2!}x^2 + \ldots + \frac{f^{(n)}(0)}{n!}x^n \qquad (14)$$

We may approximate a function by means of a Taylor polynomial at a number a or by a Maclaurin polynomial.

▶ ILLUSTRATION 1   We compute the nth degree Maclaurin polynomial for the natural exponential function. If $f(x) = e^x$, all the derivatives of f at x are e^x and the derivatives at zero are 1. Therefore, from (14),

$$P_n(x) = 1 + x + \frac{x^2}{2!} + \frac{x^3}{3!} + \ldots + \frac{x^n}{n!} \qquad (15)$$

Thus the first four Maclaurin polynomials of the natural exponential function are

$$P_0(x) = 1$$

$$P_1(x) = 1 + x$$

$$P_2(x) = 1 + x + \tfrac{1}{2}x^2$$

$$P_3(x) = 1 + x + \tfrac{1}{2}x^2 + \tfrac{1}{6}x^3$$

FIGURE 9.5.1

FIGURE 9.5.2

FIGURE 9.5.3

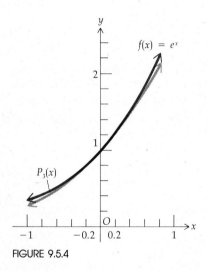

FIGURE 9.5.4

In Figs. 9.5.1 through 9.5.4 the graph of $f(x) = e^x$ appears along with the graphs of $P_0(x)$, $P_1(x)$, $P_2(x)$, and $P_3(x)$, respectively. In Fig. 9.5.5 the graphs of the four Maclaurin polynomials and the graph of $f(x) = e^x$ are shown on one coordinate system. Observe how the polynomials approximate e^x for values of x near zero, and notice that as n increases, the approximation improves. Tables 9.5.1 and 9.5.2 give values of e^x, $P_n(x)$ (when n is 0, 1, 2, and 3) and $e^x - P_n(x)$ for $x = 0.4$ and $x = 0.2$, respectively. Observe that from these two values of x it appears that the closer x is to zero, the better is the approximation for a specific $P_n(x)$. ◀

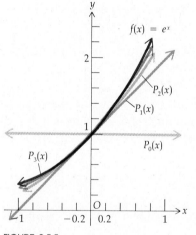

FIGURE 9.5.5

Table 9.5.1

n	$e^{0.4}$	$P_n(0.4)$	$e^{0.4} - P_n(0.4)$
0	1.4918	1	0.4918
1	1.4918	1.4	0.0918
2	1.4918	1.48	0.0118
3	1.4918	1.4907	0.0011

Table 9.5.2

n	$e^{0.2}$	$P_n(0.2)$	$e^{0.2} - P_n(0.2)$
0	1.2214	1	0.2214
1	1.2214	1.2	0.0214
2	1.2214	1.22	0.0014
3	1.2214	1.2213	0.0001

▶ ILLUSTRATION 2   We now determine the nth degree Maclaurin polynomial for the sine function. If $f(x) = \sin x$, then

$$f'(x) = \cos x \qquad f''(x) = -\sin x \qquad f'''(x) = -\cos x$$
$$f^{(iv)}(x) = \sin x \qquad f^{(v)}(x) = \cos x \qquad f^{(vi)}(x) = -\sin x$$

and so on. Thus $f(0) = 0$,

$$f'(0) = 1 \qquad f''(0) = 0 \qquad f'''(0) = -1 \qquad f^{(iv)}(0) = 0 \qquad f^{(v)}(0) = 1$$

and so on. From (14),

$$P_n(x) = x - \frac{x^3}{3!} + \frac{x^5}{5!} - \frac{x^7}{7!} + \dots + (-1)^{n-1} \frac{x^{2n-1}}{(2n-1)!}$$

Thus $P_0(x) = 0$,

$$P_1(x) = x \qquad\qquad\qquad\qquad P_2(x) = x$$

$$P_3(x) = x - \frac{x^3}{6} \qquad\qquad\qquad P_4(x) = x - \frac{x^3}{6}$$

$$P_5(x) = x - \frac{x^3}{6} + \frac{x^5}{120} \qquad\qquad P_6(x) = x - \frac{x^3}{6} + \frac{x^5}{120}$$

$$P_7(x) = x - \frac{x^3}{6} + \frac{x^5}{120} - \frac{x^7}{5040} \qquad P_8(x) = x - \frac{x^3}{6} + \frac{x^5}{120} - \frac{x^7}{5040}$$

and so on.

Figure 9.5.6 shows the graph of the sine function along with the graphs of its Maclaurin polynomials of degrees 1, 3, 5, and 7. Notice that the polynomial approximations improve as n increases. ◀

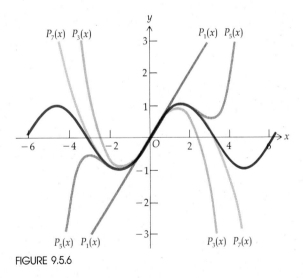

FIGURE 9.5.6

EXAMPLE 1
Find the third-degree Taylor polynomial of the cosine function at $\frac{1}{4}\pi$ and the Lagrange form of the remainder.

SOLUTION

Let $f(x) = \cos x$. Then from (11),

$$P_3(x) = f\left(\frac{\pi}{4}\right) + f'\left(\frac{\pi}{4}\right)\left(x - \frac{\pi}{4}\right) + \frac{f''(\frac{1}{4}\pi)}{2!}\left(x - \frac{\pi}{4}\right)^2 + \frac{f'''(\frac{1}{4}\pi)}{3!}\left(x - \frac{\pi}{4}\right)^3$$

Because $f(x) = \cos x$, $f(\frac{1}{4}\pi) = \frac{1}{2}\sqrt{2}$; $f'(x) = -\sin x$, $f'(\frac{1}{4}\pi) = -\frac{1}{2}\sqrt{2}$; $f''(x) = -\cos x$, $f''(\frac{1}{4}\pi) = -\frac{1}{2}\sqrt{2}$; $f'''(x) = \sin x$, $f'''(\frac{1}{4}\pi) = \frac{1}{2}\sqrt{2}$. Therefore

$$P_3(x) = \tfrac{1}{2}\sqrt{2} - \tfrac{1}{2}\sqrt{2}(x - \tfrac{1}{4}\pi) - \tfrac{1}{4}\sqrt{2}(x - \tfrac{1}{4}\pi)^2 + \tfrac{1}{12}\sqrt{2}(x - \tfrac{1}{4}\pi)^3$$

Because $f^{(iv)}(x) = \cos x$, we obtain from (12)

$$R_3(x) = \tfrac{1}{24}(\cos \xi)(x - \tfrac{1}{4}\pi)^4 \qquad \text{where } \xi \text{ is between } \tfrac{1}{4}\pi \text{ and } x$$

Because $|\cos \xi| \leq 1$, we may conclude that for all x, $|R_3(x)| \leq \tfrac{1}{24}(x - \tfrac{1}{4}\pi)^4$.

From (10),

$$|R_n(x)| = |f(x) - P_n(x)| \tag{16}$$

If $P_n(x)$ is used to approximate $f(x)$, we can obtain an upper bound for the error of this approximation if we can find a number $E > 0$ such that $|R_n(x)| \leq E$ or, because of (16), such that $|f(x) - P_n(x)| \leq E$ or, equivalently,

$$P_n(x) - E \leq f(x) \leq P_n(x) + E$$

EXAMPLE 2

Use the result of Example 1 to compute an approximate value of $\cos 47°$, and determine the accuracy of the result.

SOLUTION

$47° \sim \frac{47}{180}\pi$ radians. Thus in the solution of Example 1, take $x = \frac{47}{180}\pi$ and $x - \tfrac{1}{4}\pi = \frac{1}{90}\pi$, and

$$\cos 47° = \tfrac{1}{2}\sqrt{2}\left[1 - \tfrac{1}{90}\pi - \tfrac{1}{2}(\tfrac{1}{90}\pi)^2 + \tfrac{1}{6}(\tfrac{1}{90}\pi)^3\right] + R_3(\tfrac{47}{180}\pi)$$

where

$$R_3(\tfrac{47}{180}\pi) = \tfrac{1}{24}\cos \xi(\tfrac{1}{90}\pi)^4 \qquad \text{with } \tfrac{1}{4}\pi < \xi < \tfrac{47}{180}\pi$$

Because $0 < \cos \xi < 1$,

$$0 < R_3(\tfrac{47}{180}\pi) < \tfrac{1}{24}(\tfrac{1}{90}\pi)^4 < 0.00000007$$

Taking $\tfrac{1}{90}\pi \approx 0.0349066$ we obtain

$$\cos 47° \approx 0.681998$$

which is accurate to six decimal places.

EXAMPLE 3

Use a Maclaurin polynomial to find the value of $\sqrt{e}$ accurate to four decimal places.

SOLUTION

If $f(x) = e^x$, then from (15), the nth degree Maclaurin polynomial of f is

$$P_n(x) = 1 + x + \frac{x^2}{2!} + \frac{x^3}{3!} + \ldots + \frac{x^n}{n!}$$

and from (12),

$$R_n(x) = \frac{e^\xi}{(n+1)!} x^{n+1} \qquad \text{where } \xi \text{ is between 0 and } x$$

We want $|R_n(\frac{1}{2})|$ to be less than 0.00005. If $x = \frac{1}{2}$ in the above, and because $e^{1/2} < 2$,

$$|R_n(\tfrac{1}{2})| < \frac{e^{1/2}}{2^{n+1}(n+1)!} < \frac{2}{2^{n+1}(n+1)!} = \frac{1}{2^n(n+1)!}$$

$|R_n(\frac{1}{2})|$ will be less than 0.00005 if $1/2^n(n+1)! < 0.00005$. When $n = 5$,

$$\frac{1}{2^n(n+1)!} = \frac{1}{(32)(720)}$$

$$= 0.00004$$

Because $0.00004 < 0.00005$, we take $P_5(\frac{1}{2})$ as the approximation of $\sqrt{e}$ accurate to four decimal places. Because

$$P_5(\tfrac{1}{2}) = 1 + \tfrac{1}{2} + \tfrac{1}{8} + \tfrac{1}{48} + \tfrac{1}{384} + \tfrac{1}{3840}$$

we obtain $\sqrt{e} \approx 1.6487$.

There are other forms of the remainder in Taylor's formula. Depending on the function, one form of the remainder may be more desirable to use than another. The following theorem expresses the remainder as an integral, and it is known as *Taylor's formula with integral form of the remainder.*

9.5.2 THEOREM   If f is a function whose first $n + 1$ derivatives are continuous on a closed interval containing a and x, then $f(x) = P_n(x) + R_n(x)$, where $P_n(x)$ is the nth-degree Taylor polynomial of f at a and $R_n(x)$ is the remainder given by

$$R_n(x) = \frac{1}{n!} \int_a^x (x - t)^n f^{(n+1)}(t)\, dt \tag{17}$$

The proof of this theorem is left as an exercise (see Exercise 30).

EXERCISES 9.5

In Exercises 1 through 14, find the Taylor polynomial of degree n with the Lagrange form of the remainder at the number a for the function defined by the given equation.

1. $f(x) = \dfrac{1}{x-2}$; $a = 1$; $n = 3$

2. $f(x) = e^{-x}$; $a = 0$; $n = 4$

3. $f(x) = x^{3/2}$; $a = 4$; $n = 3$

4. $f(x) = \tan x$; $a = 0$; $n = 3$

5. $f(x) = \sin x$; $a = \frac{1}{6}\pi$; $n = 3$

6. $f(x) = \cosh x$; $a = 0$; $n = 4$

7. $f(x) = \sinh x$; $a = 0$; $n = 4$

8. $f(x) = \sqrt{x}$; $a = 4$; $n = 4$

9. $f(x) = \ln x$; $a = 1$; $n = 3$

10. $f(x) = \ln(x + 2)$; $a = -1$; $n = 3$

11. $f(x) = \ln \cos x$; $a = \frac{1}{3}\pi$; $n = 3$

12. $f(x) = e^{-x^2}$; $a = 0$; $n = 3$

13. $f(x) = (1 + x)^{3/2}$; $a = 0$; $n = 3$

14. $f(x) = (1 - x)^{-1/2}$; $a = 0$; $n = 3$

15. Compute the value of e correct to five decimal places, and prove that your answer has the required accuracy.

16. Use the Taylor polynomial in Exercise 8 to compute $\sqrt{5}$ accurate to as many places as is justified when R_4 is neglected.

17. Estimate the error that results when $\cos x$ is replaced by $1 - \frac{1}{2}x^2$ if $|x| < 0.1$.

18. Estimate the error that results when $\sqrt{1 + x}$ is replaced by $1 + \frac{1}{2}x$ if $0 < x < 0.01$.

19. Compute $\sin 31°$ accurate to three decimal places by using the Taylor polynomial in Exercise 5 at $\frac{1}{6}\pi$. (Use the approximation $\frac{1}{180}\pi \approx 0.0175$.)

20. Use the Maclaurin polynomial for the function defined by $f(x) = \ln(1 + x)$ to compute the value of $\ln 1.2$ accurate to four decimal places.

21. Use the Maclaurin polynomial for the function defined by

$$f(x) = \ln \frac{1 + x}{1 - x}$$

to compute the value of $\ln 1.2$ accurate to four decimal places. Compare the computation with that of Exercise 20.

22. Show that if $0 \le x \le \frac{1}{2}$,

$$\sin x = x - \frac{x^3}{3!} + R(x)$$

where $|R(x)| < \frac{1}{3840}$.

23. Use the result of Exercise 22 to find an approximate value of $\int_0^{1/\sqrt{2}} \sin x^2 \, dx$, and estimate the error.

24. Show that the formula $(1 + x)^{3/2} \approx 1 + \frac{3}{2}x$ is accurate to three decimal places if $-0.03 \le x \le 0$.

25. Show that the formula $(1 + x)^{-1/2} \approx 1 - \frac{1}{2}x$ is accurate to two decimal places if $-0.1 \le x \le 0$.

26. Draw sketches of the graphs of $y = \sin x$ and $y = mx$ on the same set of axes. Note that if m is positive and close to zero, then the graphs intersect at a point whose abscissa is close to π. By finding the second-degree Taylor polynomial at π for the function f defined by $f(x) = \sin x - mx$, show that an approximate solution of the equation $\sin x = mx$, when m is positive and close to zero, is given by $x \approx \pi/(1 + m)$.

27. Use the method described in Exercise 26 to find an approximate solution of the equation $\cot x = mx$ when m is positive and close to zero.

28. (a) Use the first-degree Maclaurin polynomial to approximate e^k if $0 < k < 0.01$. (b) Estimate the error in terms of k.

29. Apply Taylor's formula to express the polynomial $P(x) = x^4 - x^3 + 2x^2 - 3x + 1$ as a polynomial in powers of $x - 1$.

30. Prove Theorem 9.5.2. $\left(\textit{Hint: Let } \int_a^x f'(t) \, dt = f(x) - f(a). \text{ Solve for } f(x) \text{ and integrate } \int_a^x f'(t) \, dt \text{ by parts by letting } u = f'(t) \right.$

and $dv = dt$. Repeat this process, and the desired result follows by mathematical induction. $\Big)$

REVIEW EXERCISES FOR CHAPTER 9

In Exercises 1 through 18, evaluate the limit, if it exists.

1. $\displaystyle \lim_{x \to 0^+} \frac{\tanh 2x}{\sinh^2 x}$

2. $\displaystyle \lim_{x \to \pi/2} \left(\frac{1}{1 - \sin x} - \frac{2}{\cos^2 x} \right)$

3. $\displaystyle \lim_{x \to 0} (\csc^2 x - x^{-2})$

4. $\displaystyle \lim_{x \to 0} \frac{e - (1 + x)^{1/x}}{x}$

5. $\displaystyle \lim_{t \to +\infty} \frac{\ln(1 + e^{2t}/t)}{t^{1/2}}$

6. $\displaystyle \lim_{x \to +\infty} x \ln \frac{x + 1}{x - 1}$

7. $\lim_{x \to \pi/2} (\sin^2 x)^{\tan x}$

8. $\lim_{t \to +\infty} \dfrac{t^{100}}{e^t}$

9. $\lim_{y \to 0} \dfrac{\sin 2y}{y - \sin 5y}$

10. $\lim_{z \to \pi/2^-} \dfrac{z \cos 3z}{\cos^2 z}$

11. $\lim_{t \to 0^-} (1 + 4t)^{3/t}$

12. $\lim_{y \to +\infty} (1 + e^{2y})^{-2/y}$

13. $\lim_{x \to +\infty} \dfrac{\ln(\ln x)}{\ln(x - \ln x)}$

14. $\lim_{x \to 0} \dfrac{x - \tan^{-1} x}{4x^3}$

15. $\lim_{\theta \to \pi/2} \dfrac{\tan \theta + 3}{\sec \theta - 1}$

16. $\lim_{x \to 0} \left(\dfrac{\sin x}{x} \right)^{1/x}$

17. $\lim_{x \to +\infty} (e^x - x)^{1/x}$

18. $\lim_{x \to 0^+} \left(\dfrac{1}{x} \right)^{\tan x}$

In Exercises 19 through 32, determine whether the improper integral is convergent or divergent. If it is convergent, evaluate it.

19. $\displaystyle\int_{-2}^{0} \dfrac{dx}{2x + 3}$

20. $\displaystyle\int_{0}^{+\infty} \dfrac{dx}{\sqrt{e^x}}$

21. $\displaystyle\int_{-\infty}^{0} \dfrac{dx}{(x - 2)^2}$

22. $\displaystyle\int_{2}^{4} \dfrac{x \, dx}{\sqrt{x - 2}}$

23. $\displaystyle\int_{0}^{\pi/4} \cot^2 \theta \, d\theta$

24. $\displaystyle\int_{1}^{+\infty} \dfrac{dt}{t^4 + t^2}$

25. $\displaystyle\int_{-\infty}^{3} 4^x \, dx$

26. $\displaystyle\int_{-\infty}^{0} xe^x \, dx$

27. $\displaystyle\int_{0}^{1} \dfrac{(\ln x)^2}{x} \, dx$

28. $\displaystyle\int_{0}^{+\infty} \dfrac{3 - \sqrt{x}}{\sqrt{x}} \, dx$

29. $\displaystyle\int_{-\infty}^{+\infty} \dfrac{dx}{4x^2 + 4x + 5}$

30. $\displaystyle\int_{0}^{1} \dfrac{\ln x}{x} \, dx$

31. $\displaystyle\int_{0}^{1} \dfrac{dx}{x + x^3}$

32. $\displaystyle\int_{-3}^{0} \dfrac{dx}{\sqrt{3 - 2x - x^2}}$

33. Given:

$$f(x) = \begin{cases} \dfrac{1 - e^{2x}}{2x} & \text{if } x \neq 0 \\ -1 & \text{if } x = 0 \end{cases}$$

(a) Prove that f is continuous at 0 by using Definition 1.6.1. (b) Prove that f is differentiable at 0 by computing $f'(0)$.

34. For the function of Exercise 33, find, if they exist: (a) $\lim_{x \to +\infty} f(x)$; (b) $\lim_{x \to -\infty} f(x)$.

35. Given:

$$f(x) = \begin{cases} \dfrac{e^x - e^{-x}}{e^{2x} - e^{-2x}} & \text{if } x \neq 0 \\ 1 & \text{if } x = 0 \end{cases}$$

(a) Is f continuous at 0? (b) Find $\lim_{x \to +\infty} f(x)$, if it exists.

36. If the normal line to the curve $y = \ln x$ at the point $(x_1, \ln x_1)$ intersects the x axis at the point having an abscissa of a, prove that $\lim_{x_1 \to +\infty} (a - x_1) = 0$.

37. Evaluate if they exist: (a) $\displaystyle\int_{-\infty}^{+\infty} \sinh x \, dx$; (b) $\lim_{r \to +\infty} \displaystyle\int_{-r}^{r} \sinh x \, dx$.

38. (a) Prove that $\lim_{x \to +\infty} (x^n/e^x) = 0$ for n any positive integer. (b) Find $\lim_{x \to 0} (e^{-1/x}/x^n)$, where $x > 0$ and n is any positive integer, by letting $x = 1/t$ and using the result of part (a).

In Exercises 39 through 41, find the Taylor polynomial of degree n with the Lagrange form of the remainder at the number a for the function defined by the given equation.

39. $f(x) = \cos x$; $a = 0$; $n = 6$　　　　**40.** $f(x) = (1 + x^2)^{-1}$; $a = 1$; $n = 3$　　**41.** $f(x) = x^{-1/2}$; $a = 9$; $n = 4$

42. Express both e^{x^2} and $\cos x$ as a Maclaurin polynomial of degree 4 and use them to evaluate

$$\lim_{x \to 0} \dfrac{e^{x^2} - \cos x}{x^4}$$

43. Evaluate the limit in Exercise 42 by L'Hôpital's rule.

44. Apply Taylor's formula to express the polynomial $P(x) = 4x^3 + 5x^2 - 2x + 1$ as a polynomial in powers of $x + 2$.

45. Determine if it is possible to assign a finite number to represent the measure of the area of the region to the right of the y axis bounded by the curve $4y^2 - xy^2 - x^2 = 0$ and its asymptote. If a finite number can be assigned, find it.

46. Determine if it is possible to assign a finite number to represent the measure of the area of the region bounded by the curve whose equation is $2xy - y = 1$, the x axis, and the line $x = 1$. If a finite number can be assigned, find it.

47. Determine if it is possible to assign a finite number to represent the measure of the area of the region in the first quadrant and below the curve having the equation $y = e^{-x}$. If a finite number can be assigned, find it.

48. Find the values of n for which the improper integral $\displaystyle\int_1^{+\infty} \frac{\ln x}{x^n}\, dx$ converges, and evaluate the integral for those values of n.

49. On a certain college campus the probability density function for the length of a telephone call to be x min is given by

$$f(x) = \begin{cases} 0.4e^{-0.4x} & \text{if } x \geq 0 \\ 0 & \text{if } x < 0 \end{cases}$$

What is the probability that a telephone call selected at random will last (a) between 2 and 3 min; (b) at most 3 min; (c) at least 10 min?

50. If the first four terms of the Maclaurin polynomial are used to approximate the value of $\sqrt{e}$, how accurate is the result?

51. Assume a continuous flow of income for a particular business and that at t years from now the number of dollars in the income per year is given by $1000t - 300$. What is the present value of all future income if 8% is the interest rate compounded continuously? (*Hint:* See the paragraph preceding Exercise 33 in Exercises 9.3.)

POLAR COORDINATES AND THE CONIC SECTIONS

10.1
THE POLAR COORDINATE SYSTEM

Until now we have located a point in a plane by its rectangular cartesian coordinates. There are other coordinate systems that can be used. Probably the next in importance to the cartesian coordinate system is the **polar coordinate system**. In the cartesian coordinate system the coordinates are numbers called the abscissa and the ordinate, and these numbers are directed distances from two fixed lines. In the polar coordinate system the coordinates consist of a directed distance and the measure of an angle relative to a fixed point and a fixed ray (or half line).

The fixed point is called the **pole** (or origin), and it is designated by the letter O. The fixed ray is called the **polar axis** (or polar line), which we label OA. The ray OA is usually drawn horizontally and to the right, and it extends indefinitely (see Fig. 10.1.1).

Let P be any point in the plane distinct from O. Let θ be the radian measure of the directed angle AOP, positive when measured counterclockwise and negative when measured clockwise, having as its initial side the ray OA and as its terminal side the ray OP. Then if r is the undirected distance from O to P (i.e., $r = |\overline{OP}|$), one set of polar coordinates of P is given by r and θ, and we write these coordinates as (r, θ).

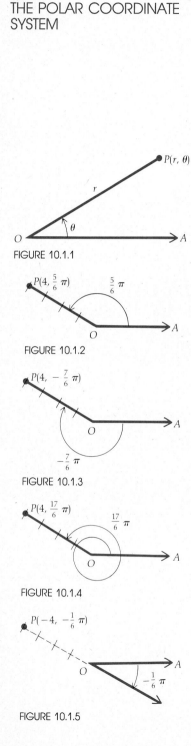

FIGURE 10.1.1

FIGURE 10.1.2

FIGURE 10.1.3

FIGURE 10.1.4

FIGURE 10.1.5

▶ ILLUSTRATION 1   The point $P(4, \frac{5}{6}\pi)$ is determined by first drawing the angle with radian measure $\frac{5}{6}\pi$, having its vertex at the pole and its initial side along the polar axis. Then the point on the terminal side that is four units from the pole is the point P (see Fig. 10.1.2). Another set of polar coordinates for this point is $(4, -\frac{7}{6}\pi)$; see Fig. 10.1.3. Furthermore, the polar coordinates $(4, \frac{17}{6}\pi)$ also yield the same point, as shown in Fig. 10.1.4.   ◀

Actually the coordinates $(4, \frac{5}{6}\pi + 2n\pi)$, where n is any integer, give the same point as $(4, \frac{5}{6}\pi)$. So a given point has an unlimited number of sets of polar coordinates. This is unlike the rectangular cartesian coordinate system because there is a one-to-one correspondence between the rectangular cartesian coordinates and the position of points in the plane, whereas there is no such one-to-one correspondence between the polar coordinates and the position of points in the plane. A further example is obtained by considering sets of polar coordinates for the pole. If $r = 0$ and θ is any real number, we have the pole, which is designated by $(0, \theta)$.

We consider polar coordinates for which r is negative. In this case, instead of the point being on the terminal side of the angle, it is on the extension of the terminal side, which is the ray from the pole extending in the direction opposite to the terminal side. So if P is on the extension of the terminal side of the angle of radian measure θ, a set of polar coordinates of P is (r, θ), where $r = -|\overline{OP}|$.

▶ ILLUSTRATION 2   The point $(-4, -\frac{1}{6}\pi)$ shown in Fig. 10.1.5 is the same point as $(4, \frac{5}{6}\pi)$, $(4, -\frac{7}{6}\pi)$, and $(4, \frac{17}{6}\pi)$ in Illustration 1. Still another set of polar coordinates for this point is $(-4, \frac{11}{6}\pi)$; see Fig. 10.1.6.   ◀

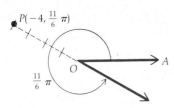

$P(-4, \frac{11}{6}\pi)$

$\frac{11}{6}\pi$

FIGURE 10.1.6

The angle is usually measured in radians; thus a set of polar coordinates of a point is an ordered pair of real numbers. For each ordered pair of real numbers there is a unique point having this set of polar coordinates. However, we have seen that a particular point can be given by an unlimited number of ordered pairs of real numbers. If the point P is not the pole, and r and θ are restricted so that $r > 0$ and $0 \le \theta < 2\pi$, then there is a unique set of polar coordinates for P.

EXAMPLE 1

(a) Plot the point having polar coordinates $(3, -\frac{2}{3}\pi)$. Find another set of polar coordinates of this point for which (b) $r < 0$ and $0 < \theta < 2\pi$; (c) $r > 0$ and $0 < \theta < 2\pi$; (d) $r < 0$ and $-2\pi < \theta < 0$.

SOLUTION

(a) The point is plotted by drawing the angle of radian measure $-\frac{2}{3}\pi$ in a clockwise direction from the polar axis. Because $r > 0$, P is on the terminal side of the angle, three units from the pole; see Fig. 10.1.7(a).

The answers to (b), (c), and (d) are, respectively, $(-3, \frac{1}{3}\pi)$, $(3, \frac{4}{3}\pi)$, and $(-3, -\frac{5}{3}\pi)$. They are illustrated in Fig. 10.1.7(b), (c), and (d).

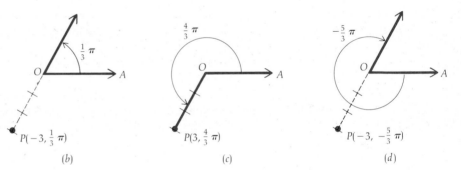

(a) $P(3, -\frac{2}{3}\pi)$ (b) $P(-3, \frac{1}{3}\pi)$ (c) $P(3, \frac{4}{3}\pi)$ (d) $P(-3, -\frac{5}{3}\pi)$

FIGURE 10.1.7

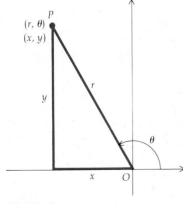

(r, θ)
(x, y)

FIGURE 10.1.8

Often we wish to refer to both the rectangular cartesian coordinates and the polar coordinates of a point. To do this we take the origin of the first system and the pole of the second system coincident, the polar axis as the positive side of the x axis, and the ray for which $\theta = \frac{1}{2}\pi$ as the positive side of the y axis.

Suppose that P is a point whose representation in the rectangular cartesian coordinate system is (x, y) and (r, θ) is a polar coordinate representation of P. We distinguish two cases: $r > 0$ and $r < 0$. In the first case, if $r > 0$, then the point P is on the terminal side of the angle of radian measure θ, and $r = |\overline{OP}|$. Such a case is shown in Fig. 10.1.8. Then

$$\cos \theta = \frac{x}{|\overline{OP}|} \qquad \sin \theta = \frac{y}{|\overline{OP}|}$$

$$= \frac{x}{r} \qquad = \frac{y}{r}$$

Thus

$$x = r \cos \theta \quad \text{and} \quad y = r \sin \theta \qquad (1)$$

In the second case, if $r < 0$, then the point P is on the extension of the terminal side and $r = -|\overline{OP}|$. See Fig. 10.1.9. Then if Q is the point $(-x, -y)$,

$$\cos \theta = \frac{-x}{|\overline{OQ}|} \qquad \sin \theta = \frac{-y}{|\overline{OQ}|}$$

$$= \frac{-x}{|\overline{OP}|} \qquad = \frac{-y}{|\overline{OP}|}$$

$$= \frac{-x}{-r} \qquad = \frac{-y}{-r}$$

$$= \frac{x}{r} \qquad = \frac{y}{r}$$

Hence

$$x = r \cos \theta \quad \text{and} \quad y = r \sin \theta \qquad (2)$$

Equations (2) and (1) are the same; thus Eqs. (1) hold in all cases.

From Eqs. (1) we can obtain the rectangular cartesian coordinates of a point when its polar coordinates are known. Also, from the equations we can obtain a polar equation of a curve if a rectangular cartesian equation is known.

To obtain equations that give a set of polar coordinates of a point when its rectangular cartesian coordinates are known, we square on both sides of each equation in (1) and obtain

$$x^2 = r^2 \cos^2 \theta \quad \text{and} \quad y^2 = r^2 \sin^2 \theta$$

Equating the sum of the left members of the above to the sum of the right members we have

$$x^2 + y^2 = r^2 \cos^2 \theta + r^2 \sin^2 \theta$$
$$x^2 + y^2 = r^2(\sin^2 \theta + \cos^2 \theta)$$
$$x^2 + y^2 = r^2$$

and so

$$r = \pm \sqrt{x^2 + y^2} \qquad (3)$$

From the equations in (1) and dividing we have

$$\frac{r \sin \theta}{r \cos \theta} = \frac{y}{x}$$

$$\Leftrightarrow \qquad \tan \theta = \frac{y}{x} \qquad (4)$$

FIGURE 10.1.9

$(-6, \frac{7}{4}\pi)$

$\frac{7}{4}\pi$

FIGURE 10.1.10

▶ ILLUSTRATION 3   The point whose polar coordinates are $(-6, \frac{7}{4}\pi)$ is plotted in Fig. 10.1.10. We find its rectangular cartesian coordinates. From (1),

$$x = r \cos \theta \qquad\qquad y = r \sin \theta$$
$$= -6 \cos \tfrac{7}{4}\pi \qquad\quad = -6 \sin \tfrac{7}{4}\pi$$
$$= -6 \cdot \frac{\sqrt{2}}{2} \qquad\quad = -6\left(-\frac{\sqrt{2}}{2}\right)$$
$$= -3\sqrt{2} \qquad\qquad = 3\sqrt{2}$$

So the point is $(-3\sqrt{2}, 3\sqrt{2})$.   ◀

The graph of an equation in polar coordinates r and θ consists of all those points and only those points P having at least one pair of coordinates that satisfy the equation. If an equation of a graph is given in polar coordinates, it is called a *polar equation* to distinguish it from a *cartesian equation*, which is the term used when an equation is given in rectangular cartesian coordinates.

EXAMPLE 2
Given that a polar equation of a graph is $r^2 = 4 \sin 2\theta$, find a cartesian equation.

SOLUTION
Because $\sin 2\theta = 2 \sin \theta \cos \theta$ we have $\sin 2\theta = 2(y/r)(x/r)$. With this substitution and $r^2 = x^2 + y^2$, we obtain from the given polar equation

$$x^2 + y^2 = 4(2)\frac{y}{r} \cdot \frac{x}{r}$$

$$x^2 + y^2 = \frac{8xy}{r^2}$$

$$x^2 + y^2 = \frac{8xy}{x^2 + y^2}$$

$$(x^2 + y^2)^2 = 8xy$$

EXAMPLE 3
Find (r, θ) if $r > 0$ and $0 \le \theta < 2\pi$ for the point whose rectangular cartesian coordinate representation is $(-\sqrt{3}, -1)$.

SOLUTION
The point $(-\sqrt{3}, -1)$ is plotted in Fig. 10.1.11. From (3), because $r > 0$,

$$r = \sqrt{3 + 1}$$
$$= 2$$

From (4), $\tan \theta = -1/(-\sqrt{3})$, and since $\pi < \theta < \tfrac{3}{2}\pi$,

$$\theta = \tfrac{7}{6}\pi$$

So the point is $(2, \tfrac{7}{6}\pi)$.

$-\sqrt{3}$

-1

r

FIGURE 10.1.11

EXAMPLE 4

Find a polar equation of a graph whose cartesian equation is

$$x^2 + y^2 - 4x = 0$$

SOLUTION

Substituting $x = r \cos \theta$ and $y = r \sin \theta$ in

$$x^2 + y^2 - 4x = 0$$

we have

$$r^2 \cos^2 \theta + r^2 \sin^2 \theta - 4r \cos \theta = 0$$

$$r^2 - 4r \cos \theta = 0$$

$$r(r - 4 \cos \theta) = 0$$

Therefore

$$r = 0 \quad \text{or} \quad r - 4 \cos \theta = 0$$

The graph of $r = 0$ is the pole. However, the pole is a point on the graph of $r - 4 \cos \theta = 0$ because $r = 0$ when $\theta = \frac{1}{2}\pi$. Therefore a polar equation of the graph is

$$r = 4 \cos \theta$$

The graph of $x^2 + y^2 - 4x = 0$ is a circle. The equation may be written in the form

$$(x - 2)^2 + y^2 = 4$$

which is an equation of the circle with center at $(2, 0)$ and radius 2.

EXERCISES 10.1

In Exercises 1 through 6, plot the point having the given set of polar coordinates; then find another set of polar coordinates for the same point for which (a) $r < 0$ and $0 \leq \theta < 2\pi$; (b) $r > 0$ and $-2\pi < \theta \leq 0$; (c) $r < 0$ and $-2\pi < \theta \leq 0$.

1. $(4, \frac{1}{4}\pi)$ **2.** $(3, \frac{5}{6}\pi)$ **3.** $(2, \frac{1}{2}\pi)$ **4.** $(3, \frac{3}{2}\pi)$ **5.** $(\sqrt{2}, \frac{7}{4}\pi)$ **6.** $(2, \frac{4}{3}\pi)$

7. Plot the point having the polar coordinates $(2, -\frac{1}{4}\pi)$. Find another set of polar coordinates for this point for which (a) $r < 0$ and $0 \leq \theta < 2\pi$; (b) $r < 0$ and $-2\pi < \theta \leq 0$; (c) $r > 0$ and $2\pi \leq \theta < 4\pi$.

8. Plot the point having the polar coordinates $(-3, -\frac{2}{3}\pi)$. Find another set of polar coordinates for this point for which (a) $r > 0$ and $0 \leq \theta < 2\pi$; (b) $r > 0$ and $-2\pi < \theta \leq 0$; (c) $r < 0$ and $2\pi \leq \theta < 4\pi$.

In Exercises 9 through 16, plot the point having the given set of polar coordinates; then give two other sets of polar coordinates of the same point, one with the same value of r and one with an r having opposite sign.

9. $(3, -\frac{2}{3}\pi)$ **10.** $(\sqrt{2}, -\frac{1}{4}\pi)$ **11.** $(-4, \frac{5}{6}\pi)$ **12.** $(-2, \frac{4}{3}\pi)$

13. $(-2, -\frac{5}{4}\pi)$ **14.** $(-3, -\pi)$ **15.** $(2, 6)$ **16.** $(5, \frac{1}{6}\pi)$

In Exercises 17 and 18, find the rectangular cartesian coordinates of the points whose polar coordinates are given.

17. (a) $(3, \pi)$; (b) $(\sqrt{2}, -\frac{3}{4}\pi)$; (c) $(-4, \frac{2}{3}\pi)$; (d) $(-1, -\frac{1}{6}\pi)$ **18.** (a) $(-2, -\frac{1}{2}\pi)$; (b) $(-1, \frac{1}{4}\pi)$; (c) $(2, -\frac{1}{6}\pi)$; (d) $(2, \frac{7}{4}\pi)$

In Exercises 19 and 20, find a set of polar coordinates of the points whose rectangular cartesian coordinates are given. Take $r > 0$ and $0 \le \theta < 2\pi$.

19. (a) $(1, -1)$; (b) $(-\sqrt{3}, 1)$; (c) $(2, 2)$; (d) $(-5, 0)$ **20.** (a) $(3, -3)$; (b) $(-1, \sqrt{3})$; (c) $(0, -2)$; (d) $(-2, -2\sqrt{3})$

In Exercises 21 through 30, find a polar equation of the graph having the given cartesian equation.

21. $x^2 + y^2 = a^2$ **22.** $x + y = 1$ **23.** $y^2 = 4(x + 1)$ **24.** $x^3 = 4y^2$ **25.** $x^2 = 6y - y^2$
26. $x^2 - y^2 = 16$ **27.** $(x^2 + y^2)^2 = 4(x^2 - y^2)$ **28.** $2xy = a^2$

29. $x^3 + y^3 - 3axy = 0$ **30.** $y = \dfrac{2x}{x^2 + 1}$

In Exercises 31 through 40, find a cartesian equation of the graph having the given polar equation.

31. $r^2 = 2 \sin 2\theta$ **32.** $r^2 \cos 2\theta = 10$ **33.** $r^2 = \cos \theta$ **34.** $r^2 = 4 \cos 2\theta$ **35.** $r^2 = \theta$

36. $r = 2 \sin 3\theta$ **37.** $r \cos \theta = -1$ **38.** $r^6 = r^2 \cos^2 \theta$ **39.** $r = \dfrac{6}{2 - 3 \sin \theta}$ **40.** $r = \dfrac{4}{3 - 2 \cos \theta}$

10.2
GRAPHS OF EQUATIONS IN POLAR COORDINATES

The equation

$$\theta = C \tag{1}$$

where C is a constant, is satisfied by all points having polar coordinates (r, C) whatever is the value of r. Therefore the graph of (1) is a straight line through the pole, and it makes an angle of radian measure C with the polar axis. See Fig. 10.2.1. The same line is given by the equation

$$\theta = C \pm n\pi$$

where n is any integer.

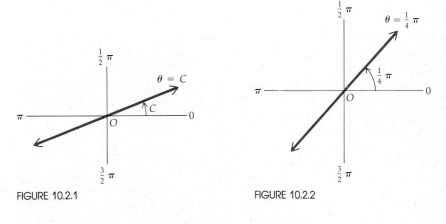

FIGURE 10.2.1   FIGURE 10.2.2

▶ ILLUSTRATION 1   (a) The graph of the equation

$$\theta = \tfrac{1}{4}\pi$$

appears in Fig. 10.2.2. It is the line passing through the pole and making an angle of radian measure $\frac{1}{4}\pi$ with the polar axis. The same line is given by

FIGURE 10.2.3

the equations

$$\theta = \tfrac{5}{4}\pi \qquad \theta = \tfrac{9}{4}\pi \qquad \theta = -\tfrac{3}{4}\pi \qquad \theta = -\tfrac{7}{4}\pi$$

and so on.

(b) The graph of the equation

$$\theta = \tfrac{2}{3}\pi$$

is shown in Fig. 10.2.3. It is the line passing through the pole and making an angle of radian measure $\tfrac{2}{3}\pi$ with the polar axis. Other equations of this line are

$$\theta = \tfrac{5}{3}\pi \qquad \theta = \tfrac{8}{3}\pi \qquad \theta = -\tfrac{1}{3}\pi \qquad \theta = -\tfrac{4}{3}\pi$$

and so on.   ◀

In general, the polar form of an equation of a line is not as simple as the cartesian form. However, if the line is parallel to either the polar axis or the $\tfrac{1}{2}\pi$ axis, the equation is fairly simple.

If a line is parallel to the polar axis and contains the point B whose cartesian coordinates are $(0, b)$ and polar coordinates are $(b, \tfrac{1}{2}\pi)$, then a cartesian equation is $y = b$. If we replace y by $r \sin \theta$, we have

$$r \sin \theta = b$$

which is a polar equation of any line parallel to the polar axis. If b is positive, the line is above the polar axis. If b is negative, it is below the polar axis.

FIGURE 10.2.4

▶ ILLUSTRATION 2   In Fig. 10.2.4 we have a sketch of the graph of the equation

$$r \sin \theta = 3$$

and in Fig. 10.2.5 we have a sketch of the graph of the equation

$$r \sin \theta = -3$$ ◀

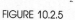

FIGURE 10.2.5

If a line is parallel to the $\tfrac{1}{2}\pi$ axis or, equivalently, perpendicular to the polar axis, and goes through the point A whose cartesian coordinates are $(a, 0)$ and polar coordinates are $(a, 0)$, a cartesian equation is $x = a$. Replacing x by $r \cos \theta$ we obtain

$$r \cos \theta = a$$

which is an equation of any line perpendicular to the polar axis. If a is positive, the line is to the right of the $\tfrac{1}{2}\pi$ axis. If a is negative, the line is to the left of the $\tfrac{1}{2}\pi$ axis.

FIGURE 10.2.6

FIGURE 10.2.7

FIGURE 10.2.8

▶ ILLUSTRATION 3   Figure 10.2.6 shows a sketch of the graph of the equation

$$r \cos \theta = 3$$

and Fig. 10.2.7 shows a sketch of the graph of the equation

$$r \cos \theta = -3$$ ◀

The graph of the equation

$$r = C$$

where C is any constant, is a circle whose center is at the pole and radius is $|C|$. The same circle is given by the equation

$$r = -C$$

▶ ILLUSTRATION 4   In Fig. 10.2.8 there is a sketch of the graph of the equation

$$r = 4$$

It is a circle with center at the pole and radius 4. The same circle is given by the equation

$$r = -4$$

although the use of such an equation is uncommon.   ◀

As was the case with the straight line, the general polar equation of a circle is not as simple as the cartesian form. However, there are special cases of an equation of a circle that are worth considering in polar form.

If a circle contains the origin (the pole) and has its center at the point having cartesian coordinates (a, b), then a cartesian equation of the circle is

$$x^2 + y^2 - 2ax - 2by = 0$$

A polar equation of this circle is

$$(r \cos \theta)^2 + (r \sin \theta)^2 - 2a(r \cos \theta) - 2b(r \sin \theta) = 0$$
$$r^2(\cos^2 \theta + \sin^2 \theta) - 2ar \cos \theta - 2br \sin \theta = 0$$
$$r^2 - 2ar \cos \theta - 2br \sin \theta = 0$$
$$r(r - 2a \cos \theta - 2b \sin \theta) = 0$$
$$r = 0 \qquad r - 2a \cos \theta - 2b \sin \theta = 0$$

Because the graph of $r = 0$ is the pole and the pole ($r = 0$ when $\theta = \tan^{-1}(-\frac{a}{b})$) is on the graph of $r - 2a \cos \theta - 2b \sin \theta = 0$, a polar equation of the circle is

$$r = 2a \cos \theta + 2b \sin \theta \qquad (2)$$

FIGURE 10.2.9

FIGURE 10.2.10

FIGURE 10.2.11

When $b = 0$ in (2), the equation becomes

$$r = 2a \cos \theta \qquad (3)$$

which is a polar equation of the circle, of radius $|a|$ units, with its center on the polar axis or its extension, and tangent to the $\frac{1}{2}\pi$ axis. If $a > 0$, the circle is to the right of the pole (see Fig. 10.2.9), and if $a < 0$, the circle is to the left of the pole.

If in (2), $a = 0$, the equation becomes

$$r = 2b \sin \theta \qquad (4)$$

which is a polar equation of the circle, of radius $|b|$ units, with its center on the $\frac{1}{2}\pi$ axis or its extension, and tangent to the polar axis. If $b > 0$, the circle is above the pole, and if $b < 0$, the circle is below the pole.

EXAMPLE 1
Draw a sketch of the graph of each of the following equations: (a) $r = 5 \cos \theta$; (b) $r = -6 \sin \theta$.

SOLUTION
(a) The equation

$$r = 5 \cos \theta$$

is of the form (3) with $a = \frac{5}{2}$. Thus the graph is a circle with center at the point having polar coordinates $(\frac{5}{2}, 0)$ and tangent to the $\frac{1}{2}\pi$ axis. A sketch of the graph appears in Fig. 10.2.10.
(b) The equation

$$r = -6 \sin \theta$$

is of the form (4) with $b = -3$. The graph is the circle with center at the point having polar coordinates $(3, \frac{3}{2}\pi)$ and tangent to the polar axis. Figure 10.2.11 shows a sketch of the graph.

To draw a sketch of the graph of a polar equation it is helpful to consider properties of symmetry of the graph. In Section 0.2 (Definition 0.2.4) we stated that two points P and Q are symmetric with respect to a line if and only if the line is the perpendicular bisector of the line segment PQ, and that two points P and Q are symmetric with respect to a third point if and only if the third point is the midpoint of the line segment PQ. Therefore the points $(2, \frac{1}{3}\pi)$ and $(2, \frac{2}{3}\pi)$ are symmetric with respect to the $\frac{1}{2}\pi$ axis, and the points $(2, \frac{1}{3}\pi)$ and $(2, -\frac{2}{3}\pi)$ are symmetric with respect to the pole. We also stated (Definition 0.2.5) that the graph of an equation is symmetric with respect to a line l if and only if for every point P on the graph there is a point Q, also on the graph, such that P and Q are symmetric with respect to l. Similarly, the graph of an equation is symmetric with respect to a point R if and only if for every point P on the graph there is a point S, also on the graph, such that P and S are symmetric with respect to R. We have three theorems giving tests for symmetry of graphs of polar equations.

10.2.1 THEOREM   If for an equation in polar coordinates an equivalent equation is obtained when (r, θ) is replaced by either $(r, -\theta + 2n\pi)$ or $(-r, \pi - \theta + 2n\pi)$, where n is any integer, the graph of the equation is symmetric with respect to the polar axis.

PROOF   If the point $P(r, \theta)$ is a point on the graph of an equation, then the graph is symmetric with respect to the polar axis if there is a point $P_1(r_1, \theta_1)$ on the graph such that the polar axis is the perpendicular bisector of the line segment $P_1 P$ (see Fig. 10.2.12). So if $r_1 = r$, then θ_1 must equal $-\theta + 2n\pi$, where n is an integer. And if $r_1 = -r$, then θ_1 must be $\pi - \theta + 2n\pi$.   ■

10.2.2 THEOREM   If for an equation in polar coordinates an equivalent equation is obtained when (r, θ) is replaced by either $(r, \pi - \theta + 2n\pi)$ or $(-r, -\theta + 2n\pi)$, where n is any integer, the graph of the equation is symmetric with respect to the $\frac{1}{2}\pi$ axis.

10.2.3 THEOREM   If for an equation in polar coordinates an equivalent equation is obtained when (r, θ) is replaced by either $(-r, \theta + 2n\pi)$ or $(r, \pi + \theta + 2n\pi)$, where n is any integer, the graph of the equation is symmetric with respect to the pole.

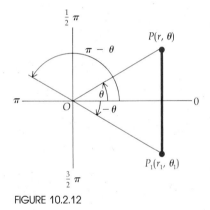

FIGURE 10.2.12

The proofs of Theorems 10.2.2 and 10.2.3 are similar to the proof of Theorem 10.2.1 and are left as exercises (see Exercises 53 and 54).

▶ ILLUSTRATION 5   For the graph of the equation

$$r = 4 \cos 2\theta$$

we test for symmetry with respect to the polar axis, the $\frac{1}{2}\pi$ axis, and the pole.

     Using Theorem 10.2.1 to test for symmetry with respect to the polar axis, we replace (r, θ) by $(r, -\theta)$ and obtain $r = 4 \cos(-2\theta)$, which is equivalent to $r = 4 \cos 2\theta$. So the graph is symmetric with respect to the polar axis.

     Using Theorem 10.2.2 to test for symmetry with respect to the $\frac{1}{2}\pi$ axis, we replace (r, θ) by $(r, \pi - \theta)$ and get $r = 4 \cos(2(\pi - \theta))$ or, equivalently, $r = 4 \cos(2\pi - 2\theta)$, which is equivalent to the equation $r = 4 \cos 2\theta$. Therefore the graph is symmetric with respect to the $\frac{1}{2}\pi$ axis.

     To test for symmetry with respect to the pole, we replace (r, θ) by $(-r, \theta)$ and obtain the equation $-r = 4 \cos 2\theta$, which is not equivalent to the given equation. But we must also determine if the other set of coordinates works. We replace (r, θ) by $(r, \pi + \theta)$ and obtain $r = 4 \cos 2(\pi + \theta)$ or, equivalently, $r = 4 \cos(2\pi + 2\theta)$, which is equivalent to the equation $r = 4 \cos 2\theta$. Therefore the graph is symmetric with respect to the pole.   ◀

     When drawing a sketch of a graph, it is desirable to determine if the pole is on the graph. This is done by substituting 0 for r and solving for θ. Also, it is advantageous to plot the points for which r has a relative maximum or relative minimum value. As a further aid in plotting, if a curve contains the pole, it is sometimes desirable to consider the tangent lines there. When helpful we use the fact shown in Supplementary Section 10.9 that if θ_1 is a value of θ that satisfies a polar equation of the curve when $r = 0$, then the line $\theta = \theta_1$ is tangent to the curve at the pole.

Table 10.2.1

θ	r
0	-1
$\frac{1}{6}\pi$	$1-\sqrt{3}$
$\frac{1}{3}\pi$	0
$\frac{1}{2}\pi$	1
$\frac{2}{3}\pi$	2
$\frac{5}{6}\pi$	$1+\sqrt{3}$
π	3

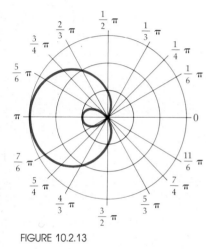

FIGURE 10.2.13

EXAMPLE 2
Draw a sketch of the graph of the equation

$$r = 1 - 2\cos\theta$$

SOLUTION
Replacing (r, θ) by $(r, -\theta)$, we obtain an equivalent equation. Therefore the graph is symmetric with respect to the polar axis.

Table 10.2.1 gives the coordinates of some points on the graph. From these points we draw half of the graph; the remainder is drawn from its symmetry with respect to the polar axis.

If $r = 0$, we obtain $\cos\theta = \frac{1}{2}$, and if $0 \leq \theta \leq \pi$, then $\theta = \frac{1}{3}\pi$. Thus the point $(0, \frac{1}{3}\pi)$ is on the graph. Furthermore, an equation of the tangent line at the pole is $\theta = \frac{1}{3}\pi$. A sketch of the graph appears in Fig. 10.2.13.

The curve in Example 2 is called a *limaçon*. The graph of an equation of the form

$$r = a \pm b\cos\theta \quad \text{or} \quad r = a \pm b\sin\theta$$

is a **limaçon**. There are four types of limaçon, and the particular type depends on the ratio a/b, where a and b are positive. We show these four types obtained from the equation

$$r = a + b\cos\theta \qquad a > 0 \text{ and } b > 0 \tag{5}$$

1. $0 < \dfrac{a}{b} < 1$ **Limaçon with a loop.** See Fig. 10.2.14(a).

2. $\dfrac{a}{b} = 1$ **Cardioid** (heart-shaped). See Fig. 10.2.14(b).

3. $1 < \dfrac{a}{b} < 2$ **Limaçon with a dent.** See Fig. 10.2.14(c).

4. $2 \leq \dfrac{a}{b}$ **Convex limaçon** (no dent). See Fig. 10.2.14(d).

If you study Supplementary Section 10.9, where horizontal and vertical tangent lines of polar curves are discussed, the distinction between limaçons of type 3 (with a dent) and those of type 4 (no dent) will be apparent.

The limaçons obtained from the equation

$$r = a + b\sin\theta \qquad a > 0 \text{ and } b > 0 \tag{6}$$

have the $\frac{1}{2}\pi$ axis as the axis of symmetry. If a limaçon has the equation

$$r = a - b\cos\theta \qquad a > 0 \text{ and } b > 0 \tag{7}$$

the limaçon points in the direction of π, and if it has the equation

$$r = a - b\sin\theta \qquad a > 0 \text{ and } b > 0 \tag{8}$$

it points in the direction of $\frac{3}{2}\pi$.

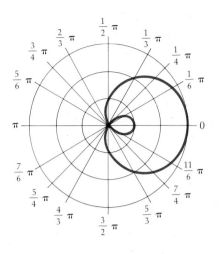

$0 < \frac{a}{b} < 1$

Limaçon with a loop

(a)

$\frac{a}{b} = 1$

Cardioid

(b)

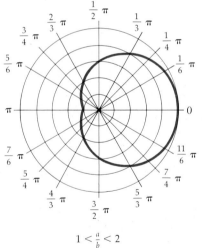

$1 < \frac{a}{b} < 2$

Limaçon with a dent

(c)

$2 \leq \frac{a}{b}$

Convex Limaçon

(d)

FIGURE 10.2.14

Table 10.2.2

θ	r
0	3
$\frac{1}{6}\pi$	4
$\frac{1}{3}\pi$	$3 + \sqrt{3}$
$\frac{1}{2}\pi$	5
π	3
$\frac{7}{6}\pi$	2
$\frac{4}{3}\pi$	$3 - \sqrt{3}$
$\frac{3}{2}\pi$	1

EXAMPLE 3

Draw a sketch of the graph of each of the following limaçons:

(a) $r = 3 + 2 \sin \theta$ (b) $r = 2 + 2 \cos \theta$ (c) $r = 2 - \sin \theta$

SOLUTION

(a) The equation $r = 3 + 2 \sin \theta$ is of the form of (6) with $a = 3$ and $b = 2$. Because $a/b = 3/2$, and $1 < \frac{3}{2} < 2$, the graph is a limaçon with a dent. It is symmetric with respect to the $\frac{1}{2}\pi$ axis. Table 10.2.2 gives the coordinates of some of the points on the graph. A sketch of the graph shown in

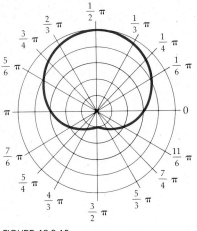

FIGURE 10.2.15

Table 10.2.3

θ	r
0	4
$\frac{1}{6}\pi$	$2 + \sqrt{3}$
$\frac{1}{3}\pi$	3
$\frac{1}{2}\pi$	2
$\frac{2}{3}\pi$	1
$\frac{5}{6}\pi$	$2 - \sqrt{3}$
π	0

Table 10.2.4

θ	r
0	2
$\frac{1}{6}\pi$	$\frac{3}{2}$
$\frac{1}{3}\pi$	$2 - \frac{1}{2}\sqrt{3}$
$\frac{1}{2}\pi$	1
π	2
$\frac{7}{6}\pi$	$\frac{5}{2}$
$\frac{4}{3}\pi$	$2 + \frac{1}{2}\sqrt{3}$
$\frac{3}{2}\pi$	3

Fig. 10.2.15 is drawn by plotting the points whose coordinates are given in Table 10.2.2 and using the symmetry property.

(b) The equation $r = 2 + 2 \cos \theta$ is of the form of (5) with $a = 2$ and $b = 2$. Because $a/b = 1$, the graph is a cardioid. It is symmetric with respect to the polar axis. The coordinates of some of the points on the graph are given in Table 10.2.3, and the sketch of the graph appearing in Fig. 10.2.16 is drawn by plotting these points and using the symmetry property.

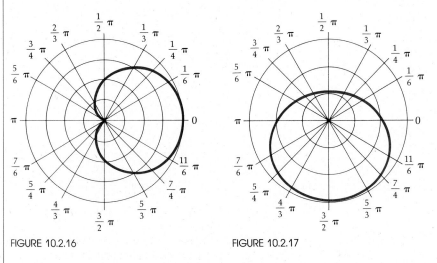

FIGURE 10.2.16   FIGURE 10.2.17

(c) The equation $r = 2 - \sin \theta$ is of the form of (8) with $a = 2$ and $b = 1$. Because $a/b = 2$, the graph is a convex limaçon. It is symmetric with respect to the $\frac{1}{2}\pi$ axis and points in the direction of $\frac{3}{2}\pi$. Figure 10.2.17 shows a sketch of the graph obtained by plotting the points whose coordinates are given in Table 10.2.4 and using the symmetry property.

The graph of an equation of the form

$$r = a \cos n\theta \quad \text{or} \quad r = a \sin n\theta \tag{9}$$

is a **rose**, having n leaves if n is odd and $2n$ leaves if n is even.

EXAMPLE 4
Draw a sketch of the four-leafed rose

$$r = 4 \cos 2\theta$$

SOLUTION
In Illustration 5 we proved that the graph is symmetric with respect to the polar axis, the $\frac{1}{2}\pi$ axis, and the pole. Substituting 0 for r in the given equation we get

$$\cos 2\theta = 0$$

Table 10.2.5

θ	r
0	4
$\frac{1}{12}\pi$	$2\sqrt{3}$
$\frac{1}{6}\pi$	2
$\frac{1}{4}\pi$	0
$\frac{1}{3}\pi$	-2
$\frac{5}{12}\pi$	$-2\sqrt{3}$
$\frac{1}{2}\pi$	-4

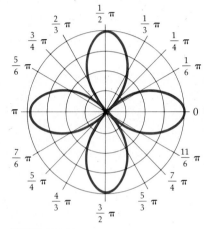

FIGURE 10.2.18

from which we obtain, for $0 \le \theta < 2\pi$,

$$\theta = \tfrac{1}{4}\pi \qquad \theta = \tfrac{3}{4}\pi \qquad \theta = \tfrac{5}{4}\pi \qquad \theta = \tfrac{7}{4}\pi$$

The lines having these equations are tangent to the graph at the pole.

Table 10.2.5 gives values of r for some values of θ from 0 to $\frac{1}{2}\pi$. From these values and the symmetry properties we draw a sketch of the graph shown in Fig. 10.2.18.

Observe that if in Eqs. (9) we take $n = 1$, we get

$$r = a \cos \theta \quad \text{or} \quad r = a \sin \theta$$

which are Eqs. (3) and (4) for a circle. Thus a circle can be considered as a one-leafed rose.

Other polar curves that occur frequently are *lemniscates* (see Exercises 29 through 32) and *spirals* (see Exercises 25 through 28). The curve in the next example is called a *spiral of Archimedes*.

EXAMPLE 5
Draw a sketch of the graph of

$$r = \theta \qquad \theta \ge 0$$

SOLUTION
When $\theta = n\pi$, where n is any integer, the graph intersects the polar axis or its extension, and when $\theta = \frac{1}{2}n\pi$, where n is any odd integer, the graph intersects the $\frac{1}{2}\pi$ axis or its extension. When $r = 0$, $\theta = 0$; thus the tangent line to the curve at the pole is the polar axis. A sketch of the graph appears in Fig. 10.2.19.

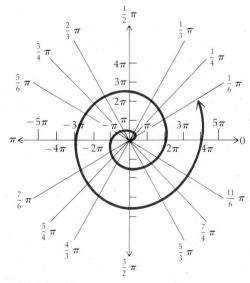

FIGURE 10.2.19

To find the points of intersection of two curves whose equations are in cartesian coordinates, we solve the two equations simultaneously. The common solutions give all the points of intersection. However, because a point has an unlimited number of sets of polar coordinates, it is possible to have as the intersection of two curves a point for which no single pair of polar coordinates satisfies both equations. This is illustrated in the following example.

EXAMPLE 6

Draw sketches of the graphs of

$$r = 2 \sin 2\theta \quad \text{and} \quad r = 1$$

with the same pole and polar axis, and find the points of intersection.

SOLUTION

The graph of $r = 2 \sin 2\theta$ is a four-leafed rose, and the graph of $r = 1$ is the circle with its center at the pole and radius 1. Sketches of the graphs appear in Fig. 10.2.20. Solving the two equations simultaneously we have

$$2 \sin 2\theta = 1$$

$$\sin 2\theta = \tfrac{1}{2}$$

Therefore

$$2\theta = \tfrac{1}{6}\pi \qquad 2\theta = \tfrac{5}{6}\pi \qquad 2\theta = \tfrac{13}{6}\pi \qquad 2\theta = \tfrac{17}{6}\pi$$

$$\theta = \tfrac{1}{12}\pi \qquad \theta = \tfrac{5}{12}\pi \qquad \theta = \tfrac{13}{12}\pi \qquad \theta = \tfrac{17}{12}\pi$$

Hence we obtain the points of intersection $(1, \tfrac{1}{12}\pi), (1, \tfrac{5}{12}\pi), (1, \tfrac{13}{12}\pi)$, and $(1, \tfrac{17}{12}\pi)$. We notice in Fig. 10.2.20 that eight points of intersection are shown. The other four points are obtained if we take another form of the equation of the circle $r = 1$; that is, consider the equation $r = -1$, which is the same circle. Solving this equation simultaneously with the equation of the four-leafed rose we have

$$\sin 2\theta = -\tfrac{1}{2}$$

Then we get

$$2\theta = \tfrac{7}{6}\pi \qquad 2\theta = \tfrac{11}{6}\pi \qquad 2\theta = \tfrac{19}{6}\pi \qquad 2\theta = \tfrac{23}{6}\pi$$

$$\theta = \tfrac{7}{12}\pi \qquad \theta = \tfrac{11}{12}\pi \qquad \theta = \tfrac{19}{12}\pi \qquad \theta = \tfrac{23}{12}\pi$$

Thus we have the four points $(-1, \tfrac{7}{12}\pi), (-1, \tfrac{11}{12}\pi), (-1, \tfrac{19}{12}\pi)$, and $(-1, \tfrac{23}{12}\pi)$. Incidentally, $(-1, \tfrac{7}{12}\pi)$ also can be written as $(1, \tfrac{19}{12}\pi), (-1, \tfrac{11}{12}\pi)$ can be written as $(1, \tfrac{23}{12}\pi), (-1, \tfrac{19}{12}\pi)$ can be written as $(1, \tfrac{7}{12}\pi)$, and $(-1, \tfrac{23}{12}\pi)$ can be written as $(1, \tfrac{11}{12}\pi)$.

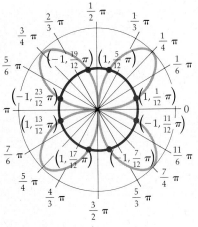

FIGURE 10.2.20

Because $(0, \theta)$ represents the pole for any θ, we determine if the pole is a point of intersection by setting $r = 0$ in each equation and solving for θ.

Often the coordinates of the points of intersection of two curves can be found directly from their graphs. However, the following is a general method.

If an equation of a curve in polar coordinates is given by $r = f(\theta)$, then the same curve is given by

$$(-1)^n r = f(\theta + n\pi) \tag{10}$$

where n is any integer.

▶ ILLUSTRATION 6   Consider the curves of Example 6. The graph of the equation $r = 2 \sin 2\theta$ also has the equation (by taking $n = 1$ in (10))

$$(-1)r = 2 \sin 2(\theta + \pi) \quad \Leftrightarrow \quad -r = 2 \sin 2\theta$$

If we take $n = 2$ in (10), the graph of $r = 2 \sin 2\theta$ also has the equation

$$(-1)^2 r = 2 \sin 2(\theta + 2\pi) \quad \Leftrightarrow \quad r = 2 \sin 2\theta$$

which is the same as the original equation. Taking n any other integer, we get either $r = 2 \sin 2\theta$ or $r = -2 \sin 2\theta$. The graph of the equation $r = 1$ also has the equation (by taking $n = 1$ in (10))

$$r = -1$$

Other integer values of n in (10) applied to the equation $r = 1$ give either $r = 1$ or $r = -1$.   ◀

If we are given the two equations $r = f(\theta)$ and $r = g(\theta)$, we obtain all the points of intersection of the graphs of the equations by doing the following:

(a) Use (10) to determine all the distinct equations of the two curves:

$$r = f_1(\theta), r = f_2(\theta), r = f_3(\theta), \ldots \tag{11}$$

$$r = g_1(\theta), r = g_2(\theta), r = g_3(\theta), \ldots \tag{12}$$

(b) Solve each equation in (11) simultaneously with each equation in (12).

(c) Check to see if the pole is a point of intersection by setting $r = 0$ in each equation, thereby giving

$$f(\theta) = 0 \quad \text{and} \quad g(\theta) = 0 \tag{13}$$

If Eqs. (13) each have a solution for θ, not necessarily the same, then the pole lies on both curves.

EXAMPLE 7

Find the points of intersection of the two curves

$$r = 2 - 2\cos\theta \quad \text{and} \quad r = 2\cos\theta$$

Draw sketches of their graphs.

SOLUTION

To find other equations of the curve represented by

$$r = 2 - 2\cos\theta$$

we have

$$(-1)r = 2 - 2\cos(\theta + \pi)$$

$$-r = 2 + 2\cos\theta$$

and

$$(-1)^2 r = 2 - 2\cos(\theta + 2\pi)$$

$$r = 2 - 2\cos\theta$$

which is the same as the original equation.

In a similar manner we find other equations of the curve given by $r = 2\cos\theta$:

$$(-1)r = 2\cos(\theta + \pi)$$

$$-r = -2\cos\theta$$

$$r = 2\cos\theta$$

which is the same as the original equation.

So there are two possible equations for the first curve, $r = 2 - 2\cos\theta$ and $-r = 2 + 2\cos\theta$, and one equation for the second curve, $r = 2\cos\theta$. Solving simultaneously $r = 2 - 2\cos\theta$ and $r = 2\cos\theta$ gives

$$2\cos\theta = 2 - 2\cos\theta$$

$$4\cos\theta = 2$$

$$\cos\theta = \tfrac{1}{2}$$

Thus $\theta = \tfrac{1}{3}\pi$ and $\tfrac{5}{3}\pi$, giving the points $(1, \tfrac{1}{3}\pi)$ and $(1, \tfrac{5}{3}\pi)$. Solving simultaneously $-r = 2 + 2\cos\theta$ and $r = 2\cos\theta$ yields

$$2 + 2\cos\theta = -2\cos\theta$$

$$4\cos\theta = -2$$

$$\cos\theta = -\tfrac{1}{2}$$

Hence $\theta = \tfrac{2}{3}\pi$ and $\theta = \tfrac{4}{3}\pi$, giving the points $(-1, \tfrac{2}{3}\pi)$ and $(-1, \tfrac{4}{3}\pi)$. However, $(-1, \tfrac{2}{3}\pi)$ is the same point as $(1, \tfrac{5}{3}\pi)$, and $(-1, \tfrac{4}{3}\pi)$ is the same point as $(1, \tfrac{1}{3}\pi)$.

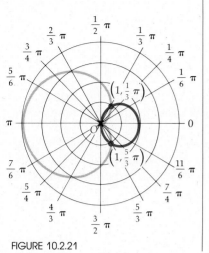

FIGURE 10.2.21

Checking to see if the pole is on the first curve by substituting $r = 0$ in the equation $r = 2 - 2 \cos \theta$, we have

$$0 = 2 - 2 \cos \theta$$

$$\cos \theta = 1$$

$$\theta = 0$$

Therefore the pole lies on the first curve. In a similar fashion, by substituting $r = 0$ in $r = 2 \cos \theta$ we get

$$0 = 2 \cos \theta$$

$$\cos \theta = 0$$

$$\theta = \tfrac{1}{2}\pi \quad \theta = \tfrac{3}{2}\pi$$

So the pole lies on the second curve.

Therefore the points of intersection of the two curves are $(1, \tfrac{1}{3}\pi)$, $(1, \tfrac{5}{3}\pi)$, and the pole. Sketches of the two curves are shown in Fig. 10.2.21.

EXERCISES 10.2

In Exercises 1 through 36, draw a sketch of the graph of the given equation.

1. (a) $\theta = \tfrac{1}{3}\pi$; (b) $r = \tfrac{1}{3}\pi$ **2.** (a) $\theta = \tfrac{3}{4}\pi$; (b) $r = \tfrac{3}{4}\pi$ **3.** (a) $\theta = 2$; (b) $r = 2$ **4.** (a) $\theta = -3$; (b) $r = -3$

5. (a) $r \cos \theta = 4$; (b) $r = 4 \cos \theta$ **6.** (a) $r \sin \theta = 2$; (b) $r = 2 \sin \theta$

7. (a) $r \sin \theta = -4$; (b) $r = -4 \sin \theta$ **8.** (a) $r \cos \theta = -5$; (b) $r = -5 \cos \theta$

9. $r = 4 - 4 \cos \theta$ **10.** $r = 3 - 3 \sin \theta$ **11.** $r = 2 + 2 \sin \theta$ **12.** $r = 3 + 3 \cos \theta$

13. $r = 2 - 3 \sin \theta$ **14.** $r = 4 - 3 \sin \theta$ **15.** $r = 3 - 2 \cos \theta$ **16.** $r = 3 - 4 \cos \theta$

17. $r = 4 + 2 \sin \theta$ **18.** $r = 6 + 2 \cos \theta$ **19.** $r = 2 \sin 3\theta$ **20.** $r = 4 \sin 5\theta$

21. $r = 2 \cos 4\theta$ **22.** $r = 3 \cos 2\theta$ **23.** $r = 4 \sin 2\theta$ **24.** $r = 3 \cos 3\theta$

25. $r = e^{\theta}$ (logarithmic spiral)   **26.** $r = e^{\theta/3}$ (logarithmic spiral)   **27.** $r = \dfrac{1}{\theta}$ (reciprocal spiral)

28. $r = 2\theta$ (spiral of Archimedes)   **29.** $r^2 = 9 \sin 2\theta$ (lemniscate)   **30.** $r^2 = 16 \cos 2\theta$ (lemniscate)

31. $r^2 = -25 \cos 2\theta$ (lemniscate)   **32.** $r^2 = -4 \sin 2\theta$ (lemniscate)   **33.** $r = 2 \sin \theta \tan \theta$ (cissoid)

34. $r = 2 \sec \theta - 1$ (conchoid of Nicomedes)   **35.** $r = |\sin 2\theta|$ **36.** $r = 2|\cos \theta|$

In Exercises 37 through 52, find the points of intersection of the graphs of the given pair of equations. Draw a sketch of each pair of graphs with the same pole and polar axis.

37. $\begin{cases} 2r = 3 \\ r = 3 \sin \theta \end{cases}$ **38.** $\begin{cases} 2r = 3 \\ r = 1 + \cos \theta \end{cases}$ **39.** $\begin{cases} r = 2 \cos \theta \\ r = 2 \sin \theta \end{cases}$ **40.** $\begin{cases} r = 2 \cos 2\theta \\ r = 2 \sin \theta \end{cases}$

41. $\begin{cases} r = 4\theta \\ r = \tfrac{1}{2}\pi \end{cases}$ **42.** $\begin{cases} r \sin \theta = 4 \\ r \cos \theta = 4 \end{cases}$ **43.** $\begin{cases} r = \tan \theta \\ r = 4 \sin \theta \end{cases}$ **44.** $\begin{cases} r = 2 \cos \theta \\ r = 2\sqrt{3} \sin \theta \end{cases}$

45. $\begin{cases} r = 3 \\ r = 2(1 + \cos \theta) \end{cases}$ **46.** $\begin{cases} r = \sin \theta \\ r = \sin 2\theta \end{cases}$ **47.** $\begin{cases} r^2 \sin 2\theta = 8 \\ r \cos \theta = 2 \end{cases}$ **48.** $\begin{cases} r = 4(1 + \sin \theta) \\ r(1 - \sin \theta) = 3 \end{cases}$

49. $\begin{cases} r = \cos \theta - 1 \\ r = \cos 2\theta \end{cases}$ **50.** $\begin{cases} r = 1 - \sin \theta \\ r = \cos 2\theta \end{cases}$ **51.** $\begin{cases} r = \sin 2\theta \\ r = \cos 2\theta \end{cases}$ **52.** $\begin{cases} r = 4 \tan \theta \sin \theta \\ r = 4 \cos \theta \end{cases}$

53. Prove Theorem 10.2.2.

54. Prove Theorem 10.2.3.

In Exercises 55 and 56, the graph of the given equation intersects itself. Find the points at which this occurs.

55. $r = \sin \frac{3}{2}\theta$

56. $r = 1 + 2\cos 2\theta$

10.3
AREA OF A REGION IN POLAR COORDINATES

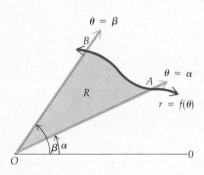

$\theta = \beta$

B

R

$\theta = \alpha$

A

$r = f(\theta)$

β α

O

0

FIGURE 10.3.1

$\theta = \beta$

$\theta = \theta_i$

$\theta = \xi_i$

$\Delta_i\theta$ $\theta = \theta_{i-1}$

$\theta = \alpha$

$r = f(\theta)$

β α

O

0

FIGURE 10.3.2

We now develop a method for finding the area of a region bounded by two lines through the pole and a curve whose equation is given in polar coordinates.

Let the function f be continuous and nonnegative on the closed interval $[\alpha, \beta]$. Let R be the region bounded by the curve whose equation is $r = f(\theta)$ and by the lines $\theta = \alpha$ and $\theta = \beta$. Then the region R is the region AOB shown in Fig. 10.3.1.

Consider a partition Δ of $[\alpha, \beta]$ defined by

$$\alpha = \theta_0 < \theta_1 < \theta_2 < \ldots < \theta_{i-1} < \theta_i < \ldots < \theta_{n-1} < \theta_n = \beta$$

Therefore we have n subintervals of the form $[\theta_{i-1}, \theta_i]$, where $i = 1, 2, \ldots, n$. Let ξ_i be a value of θ in the ith subinterval $[\theta_{i-1}, \theta_i]$. See Fig. 10.3.2, where the ith subinterval is shown together with $\theta = \xi_i$. The radian measure of the angle between the lines $\theta = \theta_{i-1}$ and $\theta = \theta_i$ is denoted by $\Delta_i\theta$. The number of square units in the area of the circular sector of radius $f(\xi_i)$ units and central angle of radian measure $\Delta_i\theta$ is given by

$$\tfrac{1}{2}[f(\xi_i)]^2 \, \Delta_i\theta$$

There is such a circular sector for each of the n subintervals. The sum of the measures of the areas of these n circular sectors is

$$\tfrac{1}{2}[f(\xi_1)]^2 \, \Delta_1\theta + \tfrac{1}{2}[f(\xi_2)]^2 \, \Delta_2\theta + \ldots + \tfrac{1}{2}[f(\xi_i)]^2 \, \Delta_i\theta + \ldots + \tfrac{1}{2}[f(\xi_n)]^2 \, \Delta_n\theta$$

which can be written, using sigma notation, as

$$\sum_{i=1}^{n} \tfrac{1}{2}[f(\xi_i)]^2 \, \Delta_i\theta \tag{1}$$

Let $\|\Delta\|$ be the norm of the partition Δ; that is, $\|\Delta\|$ is the measure of the largest $\Delta_i\theta$. Then if we let A square units be the area of the region R, we define A as the limit of Riemann sum (1) as $\|\Delta\|$ approaches 0, which is a definite integral. This definition is stated formally.

10.3.1 DEFINITION

Let R be the region bounded by the lines $\theta = \alpha$ and $\theta = \beta$ and the curve whose equation is $r = f(\theta)$, where f is continuous and nonnegative on the closed interval $[\alpha, \beta]$. Then if A square units is the area of region R,

$$A = \lim_{\|\Delta\| \to 0} \sum_{i=1}^{n} \tfrac{1}{2}[f(\xi_i)]^2 \, \Delta_i\theta$$

$$= \frac{1}{2} \int_{\alpha}^{\beta} [f(\theta)]^2 \, d\theta$$

EXAMPLE 1
Find the area of the region bounded by the graph of $r = 2 + 2 \cos \theta$.

SOLUTION
The region together with an element of area is shown in Fig. 10.3.3. Because the curve is symmetric with respect to the polar axis, we take the θ limits from 0 to π that determine the area of the region bounded by the curve above the polar axis. Then the area of the entire region is determined by multiplying that area by 2. Thus, if A square units is the required area,

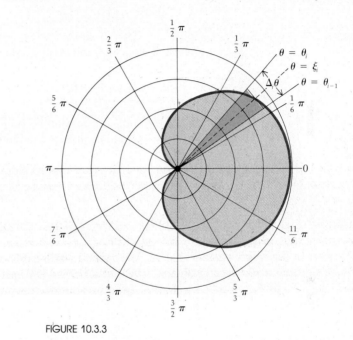

FIGURE 10.3.3

$$A = 2 \lim_{||\Delta|| \to 0} \sum_{i=1}^{n} \tfrac{1}{2}(2 + 2 \cos \xi_i)^2 \, \Delta_i \theta$$

$$= 2 \int_0^\pi \tfrac{1}{2}(2 + 2 \cos \theta)^2 \, d\theta$$

$$= 4 \int_0^\pi (1 + 2 \cos \theta + \cos^2 \theta) \, d\theta$$

$$= 4 \left[\theta + 2 \sin \theta + \tfrac{1}{2}\theta + \tfrac{1}{4} \sin 2\theta \right]_0^\pi$$

$$= 4(\pi + 0 + \tfrac{1}{2}\pi + 0 - 0)$$

$$= 6\pi$$

Therefore the area is 6π square units.

Consider now the region bounded by the lines $\theta = \alpha$ and $\theta = \beta$ and the two curves whose equations are $r = f(\theta)$ and $r = g(\theta)$, where f and g are

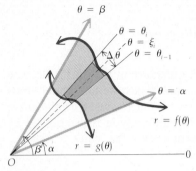

FIGURE 10.3.4

continuous on the closed interval $[\alpha, \beta]$ and $f(\theta) \geq g(\theta)$ on $[\alpha, \beta]$. See Fig. 10.3.4. We wish to find the area of this region. We take a partition of the interval $[\alpha, \beta]$ with ξ_i a value of θ in the ith subinterval $[\theta_{i-1}, \theta_i]$. The measure of the area of an element of area is the difference of the measures of the areas of two circular sectors:

$$\tfrac{1}{2}[f(\xi_i)]^2 \, \Delta_i\theta - \tfrac{1}{2}[g(\xi_i)]^2 \, \Delta_i\theta = \tfrac{1}{2}([f(\xi_i)]^2 - [g(\xi_i)]^2) \, \Delta_i\theta$$

The sum of the measures of the areas of n such elements is given by

$$\sum_{i=1}^{n} \tfrac{1}{2}([f(\xi_i)]^2 - [g(\xi_i)]^2) \, \Delta_i\theta$$

Hence, if A square units is the area of the region desired, we have

$$A = \lim_{||\Delta|| \to 0} \sum_{i=1}^{n} \tfrac{1}{2}([f(\xi_i)]^2 - [g(\xi_i)]^2) \, \Delta_i\theta \qquad (2)$$

Because f and g are continuous on $[\alpha, \beta]$, so also is $f - g$; therefore the limit in (2) exists and is equal to a definite integral. Thus

$$A = \frac{1}{2} \int_{\alpha}^{\beta} ([f(\theta)]^2 - [g(\theta)]^2) \, d\theta$$

EXAMPLE 2
Find the area of the region inside the circle $r = 3 \sin \theta$ and outside the limaçon $r = 2 - \sin \theta$.

SOLUTION
To find the points of intersection we set

$$3 \sin \theta = 2 - \sin \theta$$

$$\sin \theta = \tfrac{1}{2}$$

$$\theta = \tfrac{1}{6}\pi \quad \theta = \tfrac{5}{6}\pi$$

The curves are sketched and the region is shown together with an element of area in Fig. 10.3.5.

If we let $f(\theta) = 3 \sin \theta$ and $g(\theta) = 2 - \sin \theta$, then the equation of the circle is $r = f(\theta)$, and the equation of the limaçon is $r = g(\theta)$.

Instead of taking the limits $\tfrac{1}{6}\pi$ to $\tfrac{5}{6}\pi$ we use the property of symmetry with respect to the $\tfrac{1}{2}\pi$ axis and take the limits from $\tfrac{1}{6}\pi$ to $\tfrac{1}{2}\pi$ and multiply by 2. Then, if A square units is the area of the given region,

$$A = 2 \lim_{||\Delta|| \to 0} \sum_{i=1}^{n} \tfrac{1}{2}([f(\xi_i)]^2 - [g(\xi_i)]^2) \, \Delta_i\theta$$

$$= 2 \cdot \frac{1}{2} \int_{\pi/6}^{\pi/2} ([f(\theta)]^2 - [g(\theta)]^2) \, d\theta$$

$$= \int_{\pi/6}^{\pi/2} [9 \sin^2 \theta - (2 - \sin \theta)^2] \, d\theta$$

$$= 8 \int_{\pi/6}^{\pi/2} \sin^2 \theta \, d\theta + 4 \int_{\pi/6}^{\pi/2} \sin \theta \, d\theta - 4 \int_{\pi/6}^{\pi/2} d\theta$$

$$= 4 \int_{\pi/6}^{\pi/2} (1 - \cos 2\theta) \, d\theta + \Big[-4 \cos \theta - 4\theta \Big]_{\pi/6}^{\pi/2}$$

$$= 4\theta - 2 \sin 2\theta - 4 \cos \theta - 4\theta \Big]_{\pi/6}^{\pi/2}$$

$$= -2 \sin 2\theta - 4 \cos \theta \Big]_{\pi/6}^{\pi/2}$$

$$= (-2 \sin \pi - 4 \cos \tfrac{1}{2}\pi) - (-2 \sin \tfrac{1}{3}\pi - 4 \cos \tfrac{1}{6}\pi)$$

$$= 2 \cdot \tfrac{1}{2}\sqrt{3} + 4 \cdot \tfrac{1}{2}\sqrt{3}$$

$$= 3\sqrt{3}$$

Therefore the area is $3\sqrt{3}$ square units.

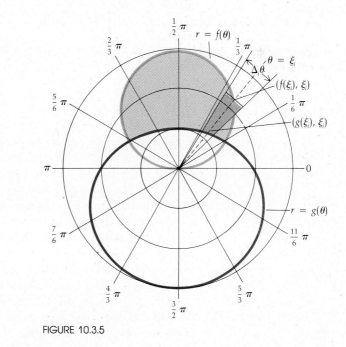

FIGURE 10.3.5

EXERCISES 10.3

In Exercises 1 through 6, find the area of the region enclosed by the graph of the given equation.

1. $r = 3 \cos \theta$ 2. $r = 2 - \sin \theta$ 3. $r = 4 \cos 3\theta$

4. $r = 4 \sin^2 \tfrac{1}{2}\theta$ 5. $r^2 = 4 \sin 2\theta$ 6. $r = 4 \sin^2 \theta \cos \theta$

7. Find the area of the region enclosed by the graph of the equation $r = \theta$ from $\theta = 0$ to $\theta = \tfrac{3}{2}\pi$.

8. Find the area of the region enclosed by the graph of $r = e^\theta$ and the lines $\theta = 0$ and $\theta = 1$.

In Exercises 9 through 12, find the area of the region enclosed by one loop of the graph of the given equation.

9. $r = 3 \cos 2\theta$ 10. $r = a(1 - 2 \cos \theta)$ 11. $r = 1 + 3 \sin \theta$ 12. $r = a \sin 3\theta$

In Exercises 13 through 16, find the area of the intersection of the regions enclosed by the graphs of the two given equations.

13. $\begin{cases} r = 2 \\ r = 3 - 2\cos\theta \end{cases}$
14. $\begin{cases} r = 4\sin\theta \\ r = 4\cos\theta \end{cases}$
15. $\begin{cases} r = 3\sin 2\theta \\ r = 3\cos 2\theta \end{cases}$
16. $\begin{cases} r^2 = 2\cos 2\theta \\ r = 1 \end{cases}$

In Exercises 15 through 21, find the area of the region that is inside the graph of the first equation and outside the graph of the second equation.

17. $\begin{cases} r = a \\ r = a(1 - \cos\theta) \end{cases}$
18. $\begin{cases} r^2 = 4\sin 2\theta \\ r = \sqrt{2} \end{cases}$
19. $\begin{cases} r = 2\sin\theta \\ r = \sin\theta + \cos\theta \end{cases}$

20. $\begin{cases} r = 2a\sin\theta \\ r = a \end{cases}$
21. $\begin{cases} r = a(1 + \cos\theta) \\ r = 2a\cos\theta \end{cases}$

22. Determine the value of a for which the area of the region enclosed by the cardioid $r = a(1 - \cos\theta)$ is 9π square units.

23. The face of a bow tie is the region enclosed by the graph of the equation $r^2 = 4\cos 2\theta$. How much material is necessary to cover the face of the tie?

24. Find the area of the region swept out by the radius vector of the spiral $r = a\theta$ during its second revolution that was not swept out during its first revolution.

25. Find the area of the region swept out by the radius vector of the curve of Exercise 24 during its third revolution that was not swept out during its second revolution.

10.4
THE PARABOLA AND
TRANSLATION OF AXES

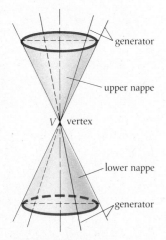

FIGURE 10.4.1

A **conic section** (or **conic**) is a curve of intersection of a plane with a right-circular cone. There are three types of curves that occur in this way: the *parabola*, the *ellipse* (including the circle as a special case), and the *hyperbola*. The curve obtained depends on the inclination of the axis of the cone to the cutting plane. The Greek mathematician Apollonius studied conic sections about 225 B.C., in terms of geometry, by using this concept.

In this section we discuss the parabola, and the following two sections are devoted to the ellipse and the hyperbola. In our treatment of each of these curves we first show how the cone and the cutting plane are taken to obtain the particular conic section. We then define the curve as a set of points in a plane. It can be proved that such a definition is a consequence of the definition of the curve as a conic section. In Section 10.5 we present this proof for an ellipse; variations of it can be given for a parabola and a hyperbola.

When considering the geometry of conic sections, a cone is regarded as having two nappes, extending indefinitely far in both directions. A portion of a right-circular cone of two nappes is shown in Fig. 10.4.1. A **generator** (or **element**) of the cone is a line lying in the cone, and all the generators of a cone contain the point V, called the **vertex**. In Fig. 10.4.2 we have a cone and a cutting plane that is parallel to one and only one generator. The conic is a *parabola*.

10.4.1 DEFINITION   A **parabola** is the set of points in a plane equidistant from a fixed point and a fixed line. The fixed point is called the **focus** and the fixed line is called the **directrix**.

We now derive an equation of a parabola from the definition. For this equation to be as simple as possible, we choose the x axis as perpendicular to the directrix and containing the focus. The origin is taken as the point on the x axis midway between the focus and the directrix. It should be stressed that we are choosing the axes (*not* the parabola) in a special way. See Fig. 10.4.3.

Let p be the directed distance $\overline{OF}$. The focus is the point $F(p, 0)$, and the directrix is the line having the equation $x = -p$. A point $P(x, y)$ is on the parabola if and only if P is equidistant from F and the directrix. That is, if $Q(-p, y)$ is the foot of the perpendicular line from P to the directrix, then P is on the parabola if and only if

$$|\overline{FP}| = |\overline{QP}|$$

Because

$$|\overline{FP}| = \sqrt{(x - p)^2 + y^2}$$

and

$$|\overline{QP}| = \sqrt{(x + p)^2 + (y - y)^2}$$

P is on the parabola if and only if

$$\sqrt{(x - p)^2 + y^2} = \sqrt{(x + p)^2}$$

By squaring on both sides of the equation we obtain

$$x^2 - 2px + p^2 + y^2 = x^2 + 2px + p^2$$
$$y^2 = 4px$$

This result is stated as a theorem.

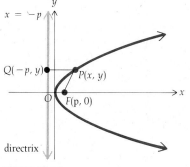

x = -p

$Q(-p, y)$ $P(x, y)$

O $F(p, 0)$

directrix

FIGURE 10.4.3

10.4.2 THEOREM An equation of the parabola having its focus at $(p, 0)$ and as its directrix the line $x = -p$ is

$$y^2 = 4px \qquad\qquad\qquad (1)$$

In Fig. 10.4.3, p is positive; p may be negative, however, because it is the directed distance $\overline{OF}$. Figure 10.4.4 shows a parabola for $p < 0$.

From Figs. 10.4.3 and 10.4.4 we see that for the equation $y^2 = 4px$ the parabola opens to the right if $p > 0$ and to the left if $p < 0$. The point midway between the focus and the directrix on the parabola is called the **vertex**. The vertex of the parabolas in Figs. 10.4.3 and 10.4.4 is the origin. The line through the vertex and the focus is called the **axis** of the parabola. The axis of the parabolas in Figs. 10.4.3 and 10.4.4 is the x axis.

In the above derivation, if the x axis and the y axis are interchanged, then the focus is at the point $F(0, p)$, and the directrix is the line having the equation $y = -p$. An equation of this parabola is $x^2 = 4py$, and the result is stated as a theorem.

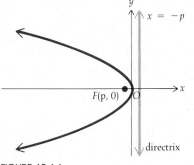

y

x = -p

$F(p, 0)$ O

directrix

FIGURE 10.4.4

10.4.3 THEOREM

An equation of the parabola having its focus at $(0, p)$ and as its directrix the line $y = -p$ is

$$x^2 = 4py \qquad (2)$$

FIGURE 10.4.5

If $p > 0$, the parabola opens upward, as shown in Fig. 10.4.5; if $p < 0$, the parabola opens downward, as shown in Fig. 10.4.6. In each case the vertex is at the origin, and the y axis is the axis of the parabola.

When drawing a sketch of the graph of a parabola, it is helpful to consider the chord through the focus, perpendicular to the axis of the parabola, because the endpoints of this chord give two points on the parabola. Such a chord is called the **latus rectum** of the parabola. The length of the latus rectum is $|4p|$. (See Exercise 20.)

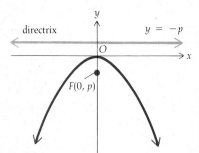

FIGURE 10.4.6

EXAMPLE 1

Find an equation of the parabola having its focus at $(0, -3)$ and as its directrix the line $y = 3$. Draw a sketch of the graph.

SOLUTION

Because the focus is on the y axis and is also below the directrix, the parabola opens downward, and $p = -3$. Hence an equation of the parabola is

$$x^2 = -12y$$

The length of the latus rectum is

$$|4(-3)| = 12$$

A sketch of the graph is shown in Fig. 10.4.7.

Any point on the parabola is equidistant from the focus and the directrix. In Fig. 10.4.7 three such points (P_1, P_2, and P_3) are shown, and we have

$$|\overline{FP_1}| = |\overline{P_1Q_1}| \qquad |\overline{FP_2}| = |\overline{P_2Q_2}| \qquad |\overline{FP_3}| = |\overline{P_3Q_3}|$$

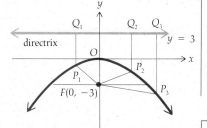

FIGURE 10.4.7

EXAMPLE 2

Given the parabola having the equation $y^2 = 7x$, find the coordinates of the focus, an equation of the directrix, and the length of the latus rectum. Draw a sketch of the graph.

SOLUTION

The given equation is of the form of (1); so

$$4p = 7$$

$$p = \tfrac{7}{4}$$

FIGURE 10.4.8

Because $p > 0$, the parabola opens to the right. The focus is at the point $F(\tfrac{7}{4}, 0)$. An equation of the directrix is $x = -\tfrac{7}{4}$. The length of the latus rectum is 7. A sketch of the graph is shown in Fig. 10.4.8.

To find the general equation of a parabola having its vertex at a point other than the origin and its directrix parallel to a coordinate axis, we first consider the concept of *translation of axis.*

Note that the shape of a curve is not affected by the position of the coordinate axes but that an equation of the curve is affected.

▶ ILLUSTRATION 1   If a circle with a radius of 3 has its center at the point $(4, -1)$, then an equation of this circle is

$$(x - 4)^2 + (y + 1)^2 = 9$$

$$x^2 + y^2 - 8x + 2y + 8 = 0$$

However, if the origin is at the center, the circle has a simpler equation, namely,

$$x^2 + y^2 = 9$$ ◀

If the coordinate axes may be taken as we please, they are generally chosen in such a way that the equations will be as simple as possible. If the axes are given, however, we often wish to find a simpler equation of a given curve referred to another set of axes.

In general, if in the plane with given x and y axes new coordinate axes are chosen parallel to the given ones, we say that there has been a **translation of axes** in the plane.

In particular, let the given x and y axes be translated to the x' and y' axes, having origin (h, k) with respect to the given axes. Also, assume that the positive numbers are on the same side of the origin on the x' and y' axes as they are on the x and y axes. See Fig. 10.4.9.

FIGURE 10.4.9

A point P in the plane, having coordinates (x, y) with respect to the given coordinate axes, will have coordinates (x', y') with respect to the new axes. To obtain relationships between these two sets of coordinates, draw a line through P parallel to the y axis and the y' axis, and also a line through P parallel to the x axis and the x' axis. Let the first line intersect the x axis at the point A and the x' axis at the point A', and the second line intersect the y axis at the point B and the y' axis at the point B'.

With respect to the x and y axes, the coordinates of P are (x, y), the coordinates of A are $(x, 0)$, and the coordinates of A' are (x, k). Because $\overline{A'P} = \overline{AP} - \overline{AA'}$,

$$y' = y - k$$

$$y = y' + k$$

With respect to the x and y axes, the coordinates of B are $(0, y)$, and the coordinates of B' are (h, y). Because $\overline{B'P} = \overline{BP} - \overline{BB'}$,

$$x' = x - h$$

$$x = x' + h$$

These results are stated as a theorem.

10.4.4 THEOREM   If (x, y) represents a point P with respect to a given set of axes, and (x', y') is a representation of P after the axes are translated to a new origin having coordinates (h, k) with respect to the given axes, then

$$x = x' + h \quad \text{and} \quad y = y' + k \tag{3}$$

$$\Leftrightarrow \quad x' = x - h \quad \text{and} \quad y' = y - k \tag{4}$$

Equations (3) or (4) are called the *equations of translating the axes*.

If an equation of a curve is given in x and y then an equation in x' and y' is obtained by replacing x by $x' + h$ and y by $y' + k$. The graph of the equation in x and y, with respect to the x and y axes, is exactly the same set of points as the graph of the corresponding equation in x' and y' with respect to the x' and y' axes.

EXAMPLE 3

Given the equation

$$x^2 + 10x + 6y + 19 = 0$$

find an equation of the graph with respect to the x' and y' axes after a translation of axes to the new origin $(-5, 1)$. Draw a sketch of the graph and show both sets of axes.

SOLUTION

A point P, represented by (x, y) with respect to the old axes, has the representation (x', y') with respect to the new axes. Then by (3), with $h = -5$ and $k = 1$,

$$x = x' - 5 \quad \text{and} \quad y = y' + 1$$

Substituting these values of x and y into the given equation we obtain

$$(x' - 5)^2 + 10(x' - 5) + 6(y' + 1) + 19 = 0$$

$$x'^2 - 10x' + 25 + 10x' - 50 + 6y' + 6 + 19 = 0$$

$$x'^2 = -6y'$$

The graph of this equation with respect to the x' and y' axes is a parabola with its vertex at the origin, opening downward, and with $4p = -6$. The graph with respect to the x and y axes is, then, a parabola having its vertex at $(-5, 1)$, its focus at $(-5, -\frac{1}{2})$, and as its directrix the line $y = \frac{5}{2}$. A sketch of the graph and both sets of axes appear in Fig. 10.4.10.

FIGURE 10.4.10

The above example illustrates how an equation can be reduced to a simpler form by a suitable translation of axes.

We now apply the translation of axes to finding the general equation of a parabola having its directrix parallel to a coordinate axis and its vertex at the point (h, k). In particular, let the directrix be parallel to the y axis. If the vertex is at the point $V(h, k)$, then the directrix has the equation

FIGURE 10.4.11

$x = h - p$, and the focus is at the point $F(h + p, k)$. Let the x' and y' axes be such that the origin O' is at $V(h, k)$. See Fig. 10.4.11.

An equation of the parabola in Fig. 10.4.11 with respect to the x' and y' axes is

$$y'^2 = 4px'$$

To obtain an equation of this parabola with respect to the x and y axes, we replace x' by $x - h$ and y' by $y - k$ from (4), which gives

$$(y - k)^2 = 4p(x - h)$$

The axis of this parabola is parallel to the x axis.

Similarly, if the directrix of a parabola is parallel to the x axis and the vertex is at $V(h, k)$, then its focus is at $F(h, k + p)$ and the directrix has the equation $y = k - p$, and an equation of the parabola with respect to the x and y axes is

$$(x - h)^2 = 4p(y - k)$$

The axis of this parabola is parallel to the y axis. We have proved, then, the following theorem.

10.4.5 THEOREM   If p is the directed distance from the vertex to the focus, an equation of the parabola with its vertex at (h, k) and with its axis parallel to the x axis is

$$(y - k)^2 = 4p(x - h) \qquad (5)$$

A parabola with the same vertex and with its axis parallel to the y axis has for an equation

$$(x - h)^2 = 4p(y - k) \qquad (6)$$

EXAMPLE 4
Find an equation of the parabola having as its directrix the line $y = 1$ and as its focus the point $F(-3, 7)$. Draw a sketch of the graph of this parabola.

SOLUTION
Because the directrix is parallel to the x axis, the axis will be parallel to the y axis, and the equation will have the form (6).

Because the vertex V is halfway between the directrix and the focus, V has coordinates $(-3, 4)$. The directed distance from the vertex to the focus is p; so

$$p = 7 - 4 \quad \Leftrightarrow \quad p = 3$$

Therefore an equation is

$$(x + 3)^2 = 12(y - 4)$$

Squaring and simplifying we have

$$x^2 + 6x - 12y + 57 = 0$$

A sketch of the graph of the parabola is shown in Fig. 10.4.12.

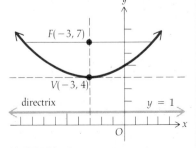

FIGURE 10.4.12

EXAMPLE 5

Given the parabola having the equation

$$y^2 + 6x + 8y + 1 = 0$$

find the vertex, the focus, an equation of the directrix, an equation of the axis, and the length of the latus rectum. Draw a sketch of the graph.

SOLUTION

Rewrite the given equation as

$$y^2 + 8y = -6x - 1$$

Completing the square of the terms involving y on the left side of this equation by adding 16 on both sides we obtain

$$y^2 + 8y + 16 = -6x + 15$$

$$(y + 4)^2 = -6(x - \tfrac{5}{2})$$

Comparing this equation with (5) we let

$$k = -4 \qquad h = \tfrac{5}{2}$$

and

$$4p = -6 \quad \Leftrightarrow \quad p = -\tfrac{3}{2}$$

Therefore the vertex is at $(\tfrac{5}{2}, -4)$, an equation of the axis is $y = -4$, the focus is at $(1, -4)$, an equation of the directrix is $x = 4$, and the length of the latus rectum is 6. A sketch of the graph is shown in Fig. 10.4.13.

$F(1, -4)$ $V(\tfrac{5}{2}, -4)$

$x = 4$

Directrix

FIGURE 10.4.13

In Section 0.3 there is a discussion of the general equation of the second degree in two variables:

$$Ax^2 + Bxy + Cy^2 + Dx + Ey + F = 0 \tag{7}$$

where $B = 0$ and $A = C$. Then the graph of (7) is either a circle, a point-circle, or the empty set. When the graph is a point-circle or the empty set, we say it is a *degenerate circle*. We now consider (7) where $B = 0$ and $AC = 0$. In such a case either $A = 0$ or $C = 0$, but not both are zero because if the three numbers A, B, and C are zero, (7) is not a second-degree equation. Suppose in (7) $B = 0$, $A = 0$, and $C \neq 0$; then (7) becomes

$$Cy^2 + Dx + Ey + F = 0 \tag{8}$$

If $D \neq 0$, this is an equation of a parabola because it can be obtained from (5) by squaring and combining terms. If in (8) $D = 0$, then the equation is

$$Cy^2 + Ey + F = 0$$

The graph of this equation may be two parallel lines, one line, or the empty set; any one of these graphs is then referred to as a *degenerate parabola*.

▶ ILLUSTRATION 2   The graph of the equation $4y^2 - 9 = 0$ is two parallel lines; $9y^2 + 6y + 1 = 0$ is an equation of one line; and $2y^2 + y + 1 = 0$ is satisfied by no real values of y.   ◀

A similar discussion holds if in (7) $B = 0$, $C = 0$, and $A \neq 0$. The results are summarized in the following theorem.

10.4.6 THEOREM   If in the general second-degree equation (7), $B = 0$, and either $A = 0$ and $C \neq 0$ or $C = 0$ and $A \neq 0$, then the graph is one of the following: a parabola, two parallel lines, one line, or the empty set.

At the beginning of this section we indicated that a parabola is obtained as a conic section when the cutting plane is parallel to one and only one generator of the cone. If the cutting plane contains the vertex of the cone and only one generator, as in Fig. 10.4.14, then a straight line is obtained, and this is a degenerate parabola. The degenerate parabola consisting of two parallel lines cannot be obtained as a plane section of a cone unless we consider a circular cylinder as a degenerate cone (one with its vertex at infinity). Then a plane parallel to the elements of the cylinder and cutting two distinct elements produces the degenerate parabola consisting of two parallel lines.

line

FIGURE 10.4.14

There is an interesting property of parabolas that has applications to the construction of searchlights, automobile headlights, and telescopes. In Fig. 10.4.15 the line PT is the tangent line at point P to the graph of a parabola. The point F is the focus of the parabola, and α is the measure of the angle between the line segment FP and the tangent line PT. Line PR is parallel to the axis of the parabola, and β is the measure of the angle between PR and PT. In Exercise 48 you are asked to prove that $\alpha = \beta$. Because of this equality, for a parabolic mirror in a searchlight, rays of light from a source at the focus are reflected along a line parallel to the axis. A similar principle involves parabolic reflectors in automobile headlights. For a parabolic mirror in a reflecting telescope, a converse situation occurs, where rays of light from an object in the sky, which go toward the mirror and are parallel to the axis, are reflected by the mirror and go through the focus.

FIGURE 10.4.15

EXAMPLE 6
A parabolic mirror has a depth of 12 cm at the center, and the distance across the top of the mirror is 32 cm. Find the distance from the vertex to the focus.

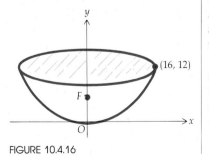

FIGURE 10.4.16

SOLUTION
See Fig. 10.4.16. The coordinate axes are chosen so that the parabola has its vertex at the origin and its axis along the y axis, and it opens upward. Therefore an equation of the parabola is of the form

$$x^2 = 4py$$

where p cm is the distance from the vertex to the focus. Because the point $(16, 12)$ is on the parabola, its coordinates satisfy the equation, and we have

$$16^2 = 4p(12)$$

$$p = \tfrac{16}{3}$$

Therefore the distance from the vertex to the focus is $5\tfrac{1}{3}$ cm.

There are other practical applications of parabolas. The path of a projectile is a parabola if motion is considered to be in a plane and air resistance is neglected. Arches are sometimes parabolic in shape, and the cable of a suspension bridge could hang in the shape of a parabola. Disk antennas for receiving satellite television signals are also parabolic in shape.

EXERCISES 10.4

For each of the parabolas in Exercises 1 through 8, find the coordinates of the focus, an equation of the directrix, and the length of the latus rectum. Draw a sketch of the curve.

1. $x^2 = 4y$
2. $y^2 = 6x$
3. $y^2 = -8x$
4. $x^2 = -16y$
5. $x^2 + y = 0$
6. $y^2 + 5x = 0$
7. $2y^2 - 9x = 0$
8. $3x^2 + 4y = 0$

In Exercises 9 through 17, find an equation of the parabola having the given properties.

9. Focus $(5, 0)$; directrix $x = -5$.
10. Focus $(0, 4)$; directrix $y = -4$.
11. Focus $(0, -2)$; directrix $y - 2 = 0$.
12. Focus $(-\tfrac{5}{3}, 0)$; directrix $5 - 3x = 0$.
13. Focus $(\tfrac{1}{2}, 0)$; directrix $2x + 1 = 0$.
14. Focus $(0, \tfrac{2}{3})$; directrix $3y + 2 = 0$.
15. Vertex $(0, 0)$; opens to the left; length of latus rectum is 6.
16. Vertex $(0, 0)$; opens upward; length of latus rectum is 3.
17. Vertex $(0, 0)$; directrix $2x = -5$.
18. Find an equation of the parabola having its vertex at the origin, the x axis as its axis, and passing through the point $(2, -4)$.
19. Find an equation of the parabola having its vertex at the origin, the y axis as its axis, and passing through the point $(-2, -4)$. $x^2 = -y$
20. Prove that the length of the latus rectum of a parabola is $|4p|$.

In Exercises 21 through 26, find a new equation of the graph of the given equation after a translation of axes to the new origin as indicated. Draw the original and the new axes and a sketch of the graph.

21. $x^2 + y^2 + 6x + 4y = 0$; $(-3, -2)$
22. $x^2 + 4y^2 + 4x + 8y + 4 = 0$; $(-2, -1)$
23. $y^2 - 6x + 9 = 0$; $(\tfrac{3}{2}, 0)$
24. $y^2 + 3x - 2y + 7 = 0$; $(-2, 1)$
25. $y - 4 = 2(x - 1)^3$; $(1, 4)$
26. $(y + 1)^2 = 4(x - 2)^3$; $(2, -1)$

In Exercises 27 through 32, find the vertex, the focus, an equation of the axis, and an equation of the directrix of the given parabola. Draw a sketch of the graph.

27. $x^2 + 6x + 4y + 8 = 0$
28. $4x^2 - 8x + 3y - 2 = 0$
29. $y^2 + 6x + 10y + 19 = 0$
30. $3y^2 - 8x - 12y - 4 = 0$
31. $2y^2 = 4y - 3x$
32. $y = 3x^2 - 3x + 3$

In Exercises 33 through 40, find an equation of the parabola having the given properties. Draw a sketch of the graph.

33. Vertex at $(2, 4)$; focus at $(-3, 4)$.   **34.** Vertex at $(1, -3)$; directrix $y = 1$.

35. Focus at $(-1, 7)$; directrix $y = 3$.   **36.** Focus at $(-\frac{3}{4}, 4)$; directrix $x = -\frac{5}{4}$.

37. Vertex at $(3, -2)$; axis $x = 3$; length of the latus rectum is 6.

38. Directrix $x = -2$; axis $y = 4$; length of the latus rectum is 8.

39. Vertex at $(-4, 2)$; axis $y = 2$; through the point $(0, 6)$.

40. Endpoints of the latus rectum are $(1, 3)$ and $(7, 3)$.

41. The endpoints of the latus rectum of a parabola are $(5, k)$ and $(-5, k)$. If the vertex of the parabola is at the origin and the parabola opens downward, find (a) the value of k; (b) an equation of the parabola.

42. Assume that water issuing from the end of a horizontal pipe 25 ft above the ground describes a parabolic curve, the vertex of the parabola being at the end of the pipe. If at a point 8 ft below the line of the pipe the flow of water has curved outward 10 ft beyond a vertical line through the end of the pipe, how far beyond this vertical line will the water strike the ground?

43. The cable of a suspension bridge hangs in the form of a parabola when the load is uniformly distributed horizontally. The distance between two towers is 150 m, the points of support of the cable on the towers are 22 m above the roadway, and the lowest point on the cable is 7 m above the roadway. Find the vertical distance to the cable from a point in the roadway 15 m from the foot of a tower.

44. A parabolic arch has a height of 20 m and a width of 36 m at the base. If the vertex of the parabola is at the top of the arch, at which height above the base is it 18 m wide?

45. Prove that on a parabola, the point closest to the focus is the vertex.

46. Suppose the orbit of a particular comet is a parabola having the sun at the focus. When the comet is 100 million km from the sun, the angle between the axis of the parabola and the line from the sun to the comet is $45°$. Use the result of Exercise 45 to determine the shortest distance from the comet to the sun.

47. A reflecting telescope has a parabolic mirror for which the distance from the vertex to the focus is 30 ft. If the distance across the top of the mirror is 64 in., how deep is the mirror at the center?

48. In Fig. 10.4.15, prove that $\alpha = \beta$. (*Hint*: Choose the coordinate axes so that the parabola has its vertex at the origin and its axis along the y axis and opens upward. Let Q be the point of intersection of the tangent line PT with the y axis. Prove that $\alpha = \beta$ by showing that $\triangle QPF$ is isosceles.)

49. An equation of the directrix of a parabola is $x + y = 0$ and its focus is at the point $(1, 1)$. Find (a) an equation of the axis of the parabola, (b) the coordinates of the vertex, and (c) the length of the latus rectum.

50. If a parabola has its focus at the origin and the x axis is its axis, prove that it must have an equation of the form $y^2 = 4kx + 4k^2$, $k \neq 0$.

51. If the axes are translated to a new origin having coordinates (h, k), show that after the translation the equation $y = \sin x$ becomes $y' = A \sin x' + B \cos x' + C$, where A, B, and C are constants. Find A, B, and C in terms of h and k. Show that $A^2 + B^2 = 1$.

52. Show that after a translation of axes to the new origin $(-\frac{1}{4}\pi, 1)$ the equation $y = \dfrac{1}{\sqrt{2}} (\sin x + \cos x) + 1$ becomes $y' = \sin x'$.

53. Refer to the results of Exercises 51 and 52, and determine a translation of axes so that the equation

$$y = \frac{1}{\sqrt{2}} (\sin x - \cos x) - 3$$

becomes $y' = \sin x'$.

54. Refer to the results of Exercises 51 and 52, and determine a translation of axes so that the equation

$$y = \tfrac{1}{2}(\sqrt{3} \sin x + \cos x) + 2$$

becomes $y' = \sin x'$.

10.5
THE ELLIPSE

An *ellipse* is obtained as a conic section if the cutting plane is parallel to no generator, in which case the cutting plane intersects each generator, as in Fig. 10.5.1. A special case of the ellipse is a circle, as shown in Fig. 10.5.2, and it is formed if the cutting plane that intersects each generator is also perpendicular to the axis of the cone. We now define an ellipse as a set of points in a plane. At the end of this section we prove that this definition is a consequence of the definition of an ellipse as a section of a cone.

10.5.1 DEFINITION   An **ellipse** is the set of points in a plane, the sum of whose distances from two fixed points is a constant. Each fixed point is called a **focus**.

Let the undirected distance between the foci (the plural of focus) be $2c$, where $c > 0$. To obtain an equation of an ellipse we select the x axis as the line through the foci F and F' and we choose the origin as the midpoint of the segment FF'. See Fig. 10.5.3. The foci F and F' have coordinates $(c, 0)$ and $(-c, 0)$, respectively. Let the constant sum referred to in Definition 10.5.1 be $2a$. Then $a > c$ and the point $P(x, y)$ in Fig. 10.5.3 is any point on the ellipse if and only if

$$|\overline{FP}| + |\overline{F'P}| = 2a \tag{1}$$

Because

$$|\overline{FP}| = \sqrt{(x - c)^2 + y^2} \quad \text{and} \quad |\overline{F'P}| = \sqrt{(x + c)^2 + y^2}$$

P is on the ellipse if and only if

$$\sqrt{(x - c)^2 + y^2} + \sqrt{(x + c)^2 + y^2} = 2a$$

We simplify this equation by first writing it so that one radical is on the left side and the other is on the right and then squaring on both sides of the equation. Thus we have

$$\sqrt{(x - c)^2 + y^2} = 2a - \sqrt{(x + c)^2 + y^2} \tag{2}$$

$$(x - c)^2 + y^2 = 4a^2 - 4a\sqrt{(x + c)^2 + y^2} + (x + c)^2 + y^2 \tag{3}$$

$$x^2 - 2cx + c^2 + y^2 = 4a^2 - 4a\sqrt{(x + c)^2 + y^2} + x^2 + 2cx + c^2 + y^2$$

$$4a\sqrt{(x + c)^2 + y^2} = 4a^2 + 4cx$$

$$\sqrt{(x + c)^2 + y^2} = a + \frac{c}{a}x \tag{4}$$

$$x^2 + 2cx + c^2 + y^2 = a^2 + 2cx + \frac{c^2}{a^2}x^2 \tag{5}$$

$$x^2\left(1 - \frac{c^2}{a^2}\right) + y^2 = a^2 - c^2$$

$$(a^2 - c^2)x^2 + a^2y^2 = a^2(a^2 - c^2)$$

$$\frac{x^2}{a^2} + \frac{y^2}{a^2 - c^2} = 1 \tag{6}$$

ellipse
FIGURE 10.5.1

circle
FIGURE 10.5.2

FIGURE 10.5.3

Because $a > c$, $a^2 - c^2 > 0$, and we can let

$$b^2 = a^2 - c^2 \tag{7}$$

Substituting from (7) in (6) we get

$$\frac{x^2}{a^2} + \frac{y^2}{b^2} = 1 \tag{8}$$

We have shown that the coordinates (x, y) of any point P on the ellipse satisfies Eq. (8). To prove that (8) is an equation of the ellipse we must also show that any point P whose coordinates (x, y) satisfy (8) is on the ellipse. To do this we start with (8) and reverse the steps to obtain (1). When reversing the steps to go from (5) to (4) we must verify that

$$a + \frac{c}{a} x \geq 0 \tag{9}$$

and to go from (3) to (2) we must verify that

$$2a - \sqrt{(x + c)^2 + y^2} \geq 0 \tag{10}$$

You are asked to do this in Exercise 42. An alternate method for obtaining (1) from (8) is outlined in Exercise 43. As a result of this discussion, we have the following theorem.

10.5.2 THEOREM   If $2a$ is the constant referred to in Definition 10.5.1 and an ellipse has its foci at $(c, 0)$ and $(-c, 0)$, then if $b^2 = a^2 - c^2$, an equation of the ellipse is

$$\frac{x^2}{a^2} + \frac{y^2}{b^2} = 1 \tag{11}$$

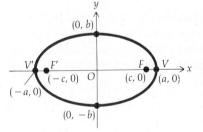

FIGURE 10.5.4

To obtain a sketch of the graph of the ellipse having Eq. (11) first observe from the equation that the graph is symmetric with respect to both the x and y axes. Furthermore, the graph intersects the x axis at the points $(a, 0)$ and $(-a, 0)$ and it intersects the y axis at the points $(0, b)$ and $(0, -b)$. See Fig. 10.5.4 and refer to it as you read the next paragraph.

The line through the foci is called the **principal axis** of the ellipse. For the ellipse of Theorem 10.5.2 the x axis is the principal axis. The points of intersection of the ellipse and its principal axis are called the **vertices**. Thus for this ellipse the vertices are at $V(a, 0)$ and $V'(-a, 0)$. The point on the principal axis that lies halfway between the two vertices is called the **center** of the ellipse. The origin is the center of this ellipse. The segment of the principal axis between the two vertices is called the **major axis** of the ellipse and its length is $2a$ units. Then we state that a is the number of units in the length of the semimajor axis. For this ellipse the segment of the y axis between the points $(0, b)$ and $(0, -b)$ is called the **minor axis** of the ellipse. Its length is $2b$ units. Hence b is the number of units in the length of the semiminor axis. Observe from (7) that $a > b$.

The ellipse is called a **central conic** in contrast to the parabola, which has no center because it has only one vertex.

EXAMPLE 1

Given the ellipse having the equation

$$\frac{x^2}{25} + \frac{y^2}{16} = 1$$

find the vertices, foci, and extremities of the minor axis. Draw a sketch of the ellipse, and show the foci.

SOLUTION

From the equation of the ellipse, $a^2 = 25$ and $b^2 = 16$; thus $a = 5$ and $b = 4$. Therefore the vertices are at the points $V(5,0)$ and $V'(-5,0)$ and the extremities of the minor axis are at the points $B(0,4)$ and $B'(0,-4)$. From (7),

$$c^2 = a^2 - b^2$$

Thus

$$c^2 = 25 - 16$$
$$= 9$$

FIGURE 10.5.5

Therefore $c = 3$ and so the foci are at $F(3,0)$ and $F'(-3,0)$. A sketch of the ellipse and its foci appear in Fig. 10.5.5.

From the definition of an ellipse (10.5.1) it follows that if P is any point on the ellipse $|\overline{FP}| + |\overline{F'P}| = 10$.

EXAMPLE 2

An arch is in the form of a semiellipse. It is 48 ft wide at the base and has a height of 20 ft. How wide is the arch at a height of 10 ft above the base?

SOLUTION

Figure 10.5.6 shows a sketch of the arch and the coordinate axes chosen so that the x axis is along the base and the origin is at the midpoint of the base. Then the ellipse has its principal axis on the x axis, its center at the origin, $a = 24$, and $b = 20$. An equation of the ellipse is of the form of (11):

FIGURE 10.5.6

$$\frac{x^2}{576} + \frac{y^2}{400} = 1$$

Let $2\bar{x}$ be the number of feet in the width of the arch at a height of 10 ft above the base. Therefore the point $(\bar{x}, 10)$ lies on the ellipse. Thus

$$\frac{\bar{x}^2}{576} + \frac{100}{400} = 1$$

and so

$$\bar{x}^2 = 432$$

$$\bar{x} = 12\sqrt{3}$$

Hence, at a height of 10 ft above the base the width of the arch is $24\sqrt{3}$ ft.

If an ellipse has its center at the origin and its principal axis on the y axis, then an equation of the ellipse is of the form

$$\frac{y^2}{a^2} + \frac{x^2}{b^2} = 1 \qquad (12)$$

which is obtained from (11) by interchanging x and y.

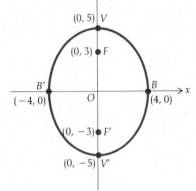

FIGURE 10.5.7

▶ ILLUSTRATION 1   Because for an ellipse $a > b$, it follows that the ellipse having the equation

$$\frac{x^2}{16} + \frac{y^2}{25} = 1$$

has its foci on the y axis. This ellipse has the same shape as the ellipse of Example 1. The vertices are at $(0, 5)$ and $(0, -5)$ and the foci are at $(0, 3)$ and $(0, -3)$. A sketch of the graph of this ellipse appears in Fig. 10.5.7.   ◀

If the center of an ellipse is at the point (h, k) rather than at the origin, and if the principal axis is parallel to one of the coordinate axes, then by a translation of axes so that the point (h, k) is the new origin, an equation of the ellipse is $\bar{x}^2/a^2 + \bar{y}^2/b^2 = 1$ if the principal axis is horizontal, and $\bar{y}^2/a^2 + \bar{x}^2/b^2 = 1$ if the principal axis is vertical. Because $\bar{x} = x - h$ and $\bar{y} = y - k$, these equations become the following in x and y:

$$\frac{(x - h)^2}{a^2} + \frac{(y - k)^2}{b^2} = 1 \qquad (13)$$

if the principal axis is horizontal, and

$$\frac{(y - k)^2}{a^2} + \frac{(x - h)^2}{b^2} = 1 \qquad (14)$$

if the principal axis is vertical.

Recall that for the general equation of the second degree in two variables

$$Ax^2 + Bxy + Cy^2 + Dx + Ey + F = 0 \qquad (15)$$

when $B = 0$ and $A = C$, the graph is either a circle or a degenerate case of a circle, which is either a point-circle or the empty set. We now discuss (15) when $B = 0$ and A and C are not necessarily equal but $AC > 0$.

If we eliminate the fractions and combine terms in (13) and (14), we obtain an equation of the form

$$Ax^2 + Cy^2 + Dx + Ey + F = 0 \tag{16}$$

where $A \neq C$ if $a \neq b$ and $AC > 0$. It can be shown by completing the squares in x and y that an equation of the form (16) can be put in the form

$$\frac{(x-h)^2}{\dfrac{1}{A}} + \frac{(y-k)^2}{\dfrac{1}{C}} = G \tag{17}$$

If $AC > 0$, then A and C have the same sign. If G has the same sign as A and C, then (17) can be written in the form of (13) or (14). Thus the graph of (16) is an ellipse.

▶ ILLUSTRATION 2   Suppose we have the equation

$$6x^2 + 9y^2 - 24x - 54y + 51 = 0$$

which can be written as

$$6(x^2 - 4x) + 9(y^2 - 6y) = -51$$

Completing the squares in x and y we get

$$6(x^2 - 4x + 4) + 9(y^2 - 6y + 9) = -51 + 24 + 81$$

$$6(x-2)^2 + 9(y-3)^2 = 54$$

$$\frac{(x-2)^2}{\frac{1}{6}} + \frac{(y-3)^2}{\frac{1}{9}} = 54$$

This is an equation of the form of (17). By dividing on both sides by 54 we have

$$\frac{(x-2)^2}{9} + \frac{(y-3)^2}{6} = 1$$

which has the form of (13).   ◀

If in (17) G has a sign opposite to that of A and C, then (17) is not satisfied by any real values of x and y. Hence the graph of (16) is the empty set.

▶ ILLUSTRATION 3   Suppose that (16) is

$$6x^2 + 9y^2 - 24x - 54y + 115 = 0$$

Then, upon completing the squares in x and y, we get

$$6(x-2)^2 + 9(y-3)^2 = -115 + 24 + 81$$

$$\frac{(x-2)^2}{\frac{1}{6}} + \frac{(y-3)^2}{\frac{1}{9}} = -10 \tag{18}$$

This is of the form of (17), where $G = -10, A = 6$, and $C = 9$. For all values of x and y the left side of (18) is nonnegative; hence the graph of (18) is the empty set.   ◀

If $G = 0$ in (17), then the equation is satisfied by only the point (h, k). Therefore the graph of (16) is a single point, which is called a **point-ellipse**.

▶ ILLUSTRATION 4   Because the equation

$$6x^2 + 9y^2 - 24x - 54y + 105 = 0$$

can be written as

$$\frac{(x - 2)^2}{\frac{1}{6}} + \frac{(y - 3)^2}{\frac{1}{9}} = 0$$

its graph is the point $(2, 3)$.   ◀

If the graph of (16) is a point-ellipse or the empty set, the graph is said to be degenerate.

If $A = C$ in (16), we have either a circle or a degenerate case of a circle, as mentioned above. A circle is a limiting form of an ellipse. This fact can be shown by considering the equation relating a, b, and c for an ellipse:

$$b^2 = a^2 - c^2$$

From this equation it is seen that as c approaches zero, b^2 approaches a^2. If $b^2 = a^2$, (13) and (14) become

$$(x - h)^2 + (y - k)^2 = a^2$$

which is an equation of a circle having its center at (h, k) and radius a. The results of Section 0.3 for a circle are the same as those obtained for (16) applied to an ellipse.

The results of the above discussion are summarized in the following theorem.

point

FIGURE 10.5.8

10.5.3 THEOREM

If in the general second-degree equation (15), $B = 0$ and $AC > 0$, then the graph is either an ellipse, a point-ellipse, or the empty set. In addition, if $A = C$, then the graph is either a circle, a point-circle, or the empty set.

The degenerate case of an ellipse, a point, is obtained as a conic section if the cutting plane contains the vertex of the cone but does not contain a generator. See Fig. 10.5.8.

EXAMPLE 3
Determine the graph of the equation

$$25x^2 + 16y^2 + 150x - 128y - 1119 = 0$$

SOLUTION

From Theorem 10.5.3, because $B = 0$ and $AC > 0$, the graph is either an ellipse or is degenerate. Completing the squares in x and y we have

$$25(x^2 + 6x + 9) + 16(y^2 - 8y + 16) = 1119 + 225 + 256$$

$$25(x + 3)^2 + 16(y - 4)^2 = 1600$$

$$\frac{(x + 3)^2}{64} + \frac{(y - 4)^2}{100} = 1 \qquad (19)$$

Equation (19) is of the form of (14); so the graph is an ellipse having its principal axis parallel to the y axis and its center at $(-3, 4)$.

EXAMPLE 4

For the ellipse of Example 3, find the vertices, foci, and extremities of the minor axis. Draw a sketch of the ellipse, and show the foci.

SOLUTION

From (19) it follows that $a = 10$ and $b = 8$. Because the center of the ellipse is at $(-3, 4)$ and the principal axis is vertical, the vertices are at the points $V(-3, 14)$ and $V'(-3, -6)$. The extremities of the minor axis are at the points $B(5, 4)$ and $B'(-11, 4)$. Because $b^2 = a^2 - c^2$,

$$64 = 100 - c^2$$

$$c^2 = 36$$

$$c = 6$$

Thus the distance from the center to a focus is 6, and so the foci are at the points $F(-3, 10)$ and $F'(-3, -2)$. A sketch of the ellipse and the foci appears in Fig. 10.5.9.

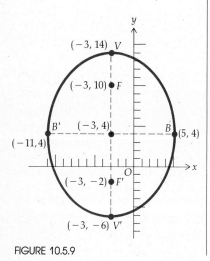

FIGURE 10.5.9

EXAMPLE 5

Find an equation of the ellipse having foci at $(-8, 2)$ and $(4, 2)$ and for which the constant referred to in Definition 10.5.1 is 18. Draw a sketch of the ellipse.

SOLUTION

The center of the ellipse is halfway between the foci and is the point $(-2, 2)$. The distance between the foci of an ellipse is $2c$ and the distance between $(-8, 2)$ and $(4, 2)$ is 12. Therefore $c = 6$. The constant referred to in Definition 10.5.1 is $2a$; thus $2a = 18$ and $a = 9$. Because $b^2 = a^2 - c^2$,

$$b^2 = 81 - 36$$

$$b^2 = 45$$

$$b = 3\sqrt{5}$$

The principal axis is parallel to the x axis; hence an equation of the ellipse is of the form of (13). Because (h, k) is the point $(-2, 2)$, $a = 9$ and $b = 3\sqrt{5}$,

FIGURE 10.5.10

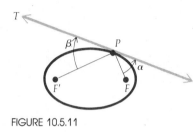

FIGURE 10.5.11

the required equation is

$$\frac{(x + 2)^2}{81} + \frac{(y - 2)^2}{45} = 1$$

A sketch of this ellipse is shown in Fig. 10.5.10.

There are applications of ellipses in astronomy because orbits of planets and satellites are ellipses. Ellipses are used in making machine gears. Arches of bridges are sometimes elliptical in shape.

There is a reflective property of the ellipse that is analogous to the one shown for the parabola in Fig. 10.4.15. For the ellipse refer to Fig. 10.5.11, where the line PT is the tangent line at P to the graph of the ellipse having foci at F and F'. The measure of the angle between the line segment FP and the tangent line PT is α, and the measure of the angle between the line segment $F'P$ and the tangent line PT is β. In Exercise 31 you are asked to prove that $\alpha = \beta$. Therefore a light ray from a source at one focus of an elliptical mirror that hits the mirror is reflected along a line through the other focus. This property of ellipses is used in so-called whispering galleries, where the ceilings have cross sections that are arcs of ellipses with common foci. A person located at one focus F can hear another person whispering at the other focus F' because the sound waves originating from the whisperer at F' hit the ceiling and are reflected by the ceiling to the listener at F. A famous example of a whispering gallery is under the dome of the Capitol in Washington, D.C. Another is at the Mormon Tabernacle in Salt Lake City.

We now prove that Definition 10.5.1 follows from the definition of an ellipse as a conic section. This proof, sometimes referred to as the "ice cream cone proof," was presented in 1822 by the Belgian mathematician G. P. Dandelin (1794–1847). Refer to Fig. 10.5.12, which shows one nappe of a cone having vertex at O and a cutting plane intersecting the cone in an ellipse. Two spheres S_1 and S_2 are inscribed in the cone. Sphere S_1 is tangent to the cone along the circle C_1 and tangent to the cutting plane at point F_1. Sphere S_2 is tangent to the cone along the circle C_2 and tangent to the cutting plane at point F_2. The planes of the circles C_1 and C_2 are parallel. We shall prove that F_1 and F_2 are the foci of the ellipse by showing that if P is any point on the ellipse, $|\overline{PF_1}| + |\overline{PF_2}|$ is a constant. To demonstrate this, we draw the line through points O and P on the surface of the cone. Points Q_1 and Q_2 are the intersections of this line with the circles C_1 and C_2, respectively. Because PF_1 and PQ_1 are two tangent lines to sphere S_1 from the point P, it follows that

$$|\overline{PF_1}| = |\overline{PQ_1}| \qquad (20)$$

Also PF_2 and PQ_2 are two tangent lines to sphere S_2 from the point P. Thus

$$|\overline{PF_2}| = |\overline{PQ_2}| \qquad (21)$$

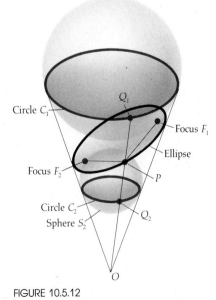

FIGURE 10.5.12

From (20) and (21),

$$|\overline{PF_1}| + |\overline{PF_2}| = |\overline{PQ_1}| + |\overline{PQ_2}|$$

Observe that $|\overline{PQ_1}| + |\overline{PQ_2}| = |\overline{Q_1Q_2}|$, which is the distance measured along the surface of the cone between the parallel planes of the circles C_1 and C_2. This distance will be the same for any choice of point P on the ellipse. Therefore $|\overline{PF_1}| + |\overline{PF_2}|$ is a constant, and F_1 and F_2 are the foci of the ellipse.

To prove that Definition 10.4.1 follows from the definition of a parabola as a conic section, you need only one sphere tangent to the cutting plane at the focus and tangent to the cone along a circle. The intersection of the plane of the circle with the cutting plane is the directrix of the parabola.

EXERCISES 10.5

In Exercises 1 through 16, find the center, vertices, foci, and extremities of the minor axis of the given ellipse. Draw a sketch of the curve, and show the foci.

1. $4x^2 + 9y^2 = 36$
2. $4x^2 + 9y^2 = 4$
3. $25x^2 + 4y^2 = 100$
4. $16x^2 + 9y^2 = 144$
5. $2x^2 + 3y^2 = 18$
6. $64x^2 + y^2 = 16$
7. $16x^2 + 4y^2 = 1$
8. $3x^2 + 4y^2 = 9$
9. $6x^2 + 9y^2 - 24x - 54y + 51 = 0$
10. $9x^2 + 4y^2 - 18x + 16y - 11 = 0$
11. $5x^2 + 3y^2 - 3y - 12 = 0$
12. $2x^2 + 2y^2 - 2x + 18y + 33 = 0$
13. $4x^2 + 4y^2 + 20x - 32y + 89 = 0$
14. $3x^2 + 4y^2 - 30x + 16y + 100 = 0$
15. $3x^2 + 5y^2 - 6x - 12 = 0$
16. $2x^2 + 3y^2 - 4x + 12y + 2 = 0$

In Exercises 17 and 18, determine whether the graph of the given equation is an ellipse, a point-ellipse, or the empty set.

17. $4x^2 + y^2 - 8x + 2y + 5 = 0$
18. $2x^2 + 3y^2 + 8x - 6y + 20 = 0$

In Exercises 19 through 28, find an equation of the ellipse having the given properties, and draw a sketch of the graph.

19. Vertices at $(-\frac{5}{2}, 0)$ and $(\frac{5}{2}, 0)$ and one focus at $(\frac{3}{2}, 0)$.
20. Foci at $(-5, 0)$ and $(5, 0)$ and for which the constant referred to in Definition 10.5.1 is 20.
21. Foci at $(0, 3)$ and $(0, -3)$ and for which the constant referred to in Definition 10.5.1 is $6\sqrt{3}$.
22. Center at the origin, its foci on the x axis, the length of the major axis equal to three times the length of the minor axis, and passing through the point $(3, 3)$.
23. Vertices at $(2, 0)$ and $(-2, 0)$ and through the point $(-1, \frac{1}{2}\sqrt{3})$.
24. Vertices at $(0, 5)$ and $(0, -5)$ and through the point $(2, -\frac{5}{3}\sqrt{5})$.
25. Center at $(4, -2)$, a vertex at $(9, -2)$, and one focus at $(0, -2)$.
26. A focus at $(2, -3)$, a vertex at $(2, 4)$, and center on the x axis.
27. Foci at $(-1, -1)$ and $(-1, 7)$ and the semimajor axis of length 8 units.
28. Foci at $(2, 3)$ and $(2, -7)$ and the length of the semiminor axis is two-thirds of the length of the semimajor axis.
29. Find an equation of the tangent line to the ellipse $4x^2 + 9y^2 = 72$ at the point $(3, 2)$.
30. Show that an equation of the tangent line to the ellipse $x^2/a^2 + y^2/b^2 = 1$ at the point (x_0, y_0) on the ellipse is $x_0x/a^2 + y_0y/b^2 = 1$.
31. In Fig. 10.5.11 prove that $\alpha = \beta$. (*Hint:* Choose the coordinate axes so that the center of the ellipse is at the origin and the axes of the ellipse are along the coordinate axes. Then use Theorem 0.7.3.)
32. The orbit of the earth around the sun is elliptical in shape with the sun at one focus and a semimajor axis of length 92.9 million miles. If the distance between the foci is 3.16 million miles, find (a) how close the earth gets to the sun and (b) the greatest possible distance between the earth and the sun.
33. The ceiling in a hallway 10 m wide is in the shape of a semiellipse and is 9 m high in the center and 6 m high at the side walls. Find the height of the ceiling 2 m from either wall.

34. The arch of a bridge is in the shape of a semiellipse having a horizontal span of 40 m and a height of 16 m at its center. How high is the arch 9 m to the right or left of the center?

35. Suppose that the orbit of a planet is in the shape of an ellipse with a major axis whose length is 500 million km. If the distance between the foci is 400 million km, find an equation of the orbit.

36. A football is 12 in. long, and a plane section containing a seam is an ellipse, of which the length of the minor axis is 7 in. Find the volume of the football if the leather is so stiff that every cross section is a square.

37. Solve Exercise 36 if every cross section is a circle.

38. Definition 10.5.1 gives a procedure for drawing the graph of an ellipse. To apply the method to the ellipse $4x^2 + 9y^2 = 36$, first determine the points of intersection with the coordinate axes. Obtain the foci on the x axis by using a compass with its center at one of the points of intersection with the y axis and with radius of 3. Then fasten thumbtacks at each focus. Take a piece of string of length 6, which is $2a$, and attach an end at one thumbtack and an end at the other thumbtack. Place a pencil against the string and make it tight. Move the pencil against the string and trace a curve. This curve is an ellipse because the pencil traces a set of points the sum of whose distances from the two tacks is the constant 6.

39. For the ellipse whose equation is

$$\frac{(x-h)^2}{a^2} + \frac{(y-k)^2}{b^2} = 1$$

where $a > b > 0$, find the coordinates of the foci in terms of h, k, a, and b.

Exercises 40 and 41 pertain to Supplementary Section 5.9.

40. A plate is in the shape of the region bounded by the ellipse having a semimajor axis of length 3 ft and a semiminor axis of length 2 ft. If the plate is lowered vertically in a tank of water until the minor axis lies in the surface of the water, find the force due to water pressure on one side of the submerged portion of the plate.

41. If the plate of Exercise 40 is lowered until the center is 3 ft below the surface of the water, find the force due to water pressure on one side of the plate. The minor axis is still horizontal.

42. By reversing the steps used to obtain Eq. (8) from Eq. (1), prove that if $P(x, y)$ is any point whose coordinates satisfy (8), then

$$|\overline{FP}| + |\overline{F'P}| = 2a$$

where F and F' are the foci of the ellipse in Fig. 10.5.3. (*Hint:* To show inequality (9) use the fact that $a > c > 0$ and that for every point (x, y) satisfying (8), $-a \le x \le a$. To show inequality (10) first show that $\sqrt{(x+c)^2 + y^2} = a + \dfrac{c}{a} x$ by replacing y^2 by $b^2 \left(1 - \dfrac{x^2}{a^2}\right)$ and using (7).)

43. Instead of reversing the steps used to obtain Eq. (8) from Eq. (1), an alternate method for obtaining (1) from (8) is as follows:

$$|\overline{FP}| + |\overline{F'P}| = \sqrt{(x-c)^2 + y^2} + \sqrt{(x+c)^2 + y^2}$$

$$= \frac{1}{a} \left(\sqrt{a^2(x-c)^2 + a^2 y^2} + \sqrt{a^2(x+c)^2 + a^2 y^2} \right)$$

Replace $a^2 y^2$ by $a^2 b^2 - b^2 x^2$ and use (7) to obtain

$$|\overline{FP}| + |\overline{F'P}| = \left| a - \frac{c}{a} x \right| + \left| a + \frac{c}{a} x \right|$$

Show that the right-hand side of the above equation is $2a$ by using the fact that $a > c > 0$ and that for every point (x, y) satisfying (8), $-a \le x \le a$.

10.6
THE HYPERBOLA

When a cutting plane of a cone is parallel to two generators, it intersects both nappes of the cone, and the conic section obtained is a *hyperbola*, shown in Fig. 10.6.1. The following definition of a hyperbola as a set of points in a plane can be proved from its definition as a conic section in a manner similar to that used for an ellipse in Section 10.5. However, for a hyperbola the proof involves one sphere in each nappe of the cone.

10.6.1 DEFINITION

A **hyperbola** is the set of points in a plane, the absolute value of the difference of whose distances from two fixed points is a constant. The two fixed points are called the **foci**.

To obtain an equation of a hyperbola we begin as we did with the ellipse by letting the undirected distance between the foci be $2c$, where $c > 0$. Then we choose the x axis as the line through the foci F and F', and we take the origin as the midpoint of the segment FF'. Refer to Fig. 10.6.2. The points $(c, 0)$ and $(-c, 0)$ are the foci F and F', respectively. Let $2a$ be the constant referred to in Definition 10.6.1. In Fig. 10.6.2 the point $P(x, y)$ represents any point on the hyperbola. Then from Definition 10.6.1,

$$\left| |\overline{FP}| - |\overline{F'P}| \right| = 2a \tag{1}$$

To determine the relationship between a and c, we use the fact that the sum of the lengths of any two sides of a triangle is greater than the length of the third side and write the two inequalities

$$|\overline{F'F}| + |\overline{FP}| > |\overline{F'P}| \qquad\qquad |\overline{F'F}| + |\overline{F'P}| > |\overline{FP}|$$

$$|\overline{F'F}| > |\overline{F'P}| - |\overline{FP}| \qquad\qquad |\overline{F'F}| > |\overline{FP}| - |\overline{F'P}| \tag{2}$$

By using absolute value bars the two inequalities in (2) can be written as the inequality

$$|\overline{F'F}| > \left| |\overline{FP}| - |\overline{F'P}| \right|$$

Because $|\overline{F'F}| = 2c$ and $\left| |\overline{FP}| - |\overline{F'P}| \right| = 2a$, we have

$$2c > 2a$$

$$c > a \tag{3}$$

Because

$$|\overline{FP}| = \sqrt{(x - c)^2 + y^2} \quad \text{and} \quad |\overline{F'P}| = \sqrt{(x + c)^2 + y^2}$$

then from (1), P is on the hyperbola if and only if

$$\left| \sqrt{(x - c)^2 + y^2} - \sqrt{(x + c)^2 + y^2} \right| = 2a$$

or, equivalently, without absolute value bars,

$$\sqrt{(x - c)^2 + y^2} - \sqrt{(x + c)^2 + y^2} = \pm 2a$$

$$\sqrt{(x - c)^2 + y^2} = \pm 2a + \sqrt{(x + c)^2 + y^2}$$

$$x^2 - 2cx + c^2 + y^2 = 4a^2 \pm 4a\sqrt{(x + c)^2 + y^2} + x^2 + 2cx + c^2 + y^2$$

hyperbola

FIGURE 10.6.1

FIGURE 10.6.2

$$\pm 4a\sqrt{(x + c)^2 + y^2} = 4a^2 + 4cx$$

$$\pm \sqrt{(x + c)^2 + y^2} = a + \frac{c}{a}x$$

$$x^2 + 2cx + c^2 + y^2 = a^2 + 2cx + \frac{c^2}{a^2}x^2$$

$$x^2\left(\frac{c^2}{a^2} - 1\right) - y^2 = c^2 - a^2$$

$$(c^2 - a^2)x^2 - a^2y^2 = a^2(c^2 - a^2)$$

$$\frac{x^2}{a^2} - \frac{y^2}{c^2 - a^2} = 1 \tag{4}$$

Because, from (3), $c > a$, we can let

$$b^2 = c^2 - a^2 \tag{5}$$

Substituting from (5) in (4) we get

$$\frac{x^2}{a^2} - \frac{y^2}{b^2} = 1 \tag{6}$$

We have proved that the coordinates (x, y) of any point P on the hyperbola satisfies Eq. (6). To prove that (6) is an equation of the hyperbola we must also show that any point P whose coordinates (x, y) satisfy (6) is on the hyperbola. You are asked to do this in Exercise 41. The procedure is similar to what we indicated in Section 10.5 for the ellipse as outlined in Exercises 42 and 43 of Exercises 10.5. This discussion gives us the following theorem.

10.6.2 THEOREM   If $2a$ is the constant referred to in Definition 10.6.1 and a hyperbola has its foci at $(c, 0)$ and $(-c, 0)$, then if $b^2 = c^2 - a^2$, an equation of the hyperbola is

$$\frac{x^2}{a^2} - \frac{y^2}{b^2} = 1 \tag{7}$$

A sketch of the graph of the hyperbola of Theorem 10.6.2 is shown in Fig. 10.6.3. We now show how this graph is obtained. From the equation we observe that the graph is symmetric with respect to both the x and y axes. As with the ellipse, the line through the foci is called the **principal axis** of the hyperbola. Thus for this hyperbola the x axis is the principal axis. The points where the graph intersects the principal axis are called the **vertices**, and the point that is halfway between the vertices is called the **center** of the hyperbola. For this hyperbola the vertices are at $V(a, 0)$ and $V'(-a, 0)$ and the center is at the origin. The segment $V'V$ of the principal axis is called the **transverse axis** of the hyperbola, and its length is $2a$ units; thus a units is the length of the semitransverse axis.

Substituting 0 for x in (7) we obtain $y^2 = -b^2$, which has no real solutions. Consequently the hyperbola does not intersect the y axis. However, the line segment having extremities at the points $(0, -b)$ and $(0, b)$ is called the

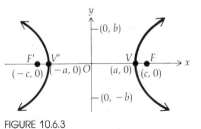

FIGURE 10.6.3

conjugate axis of the hyperbola and its length is $2b$ units. Thus b is the number of units in the length of the semiconjugate axis.

Solving (7) for y in terms of x we obtain

$$y = \pm \frac{b}{a}\sqrt{x^2 - a^2} \qquad (8)$$

We conclude from (8) that if $|x| < a$, there is no real value of y. Therefore there are no points (x, y) on the hyperbola for which $-a < x < a$. We also see from (8) that if $|x| > a$, then y has two real values. Thus the hyperbola has two *branches*. One branch contains the vertex $V(a, 0)$ and extends indefinitely far to the right of V. The other branch contains the vertex $V'(-a, 0)$ and extends indefinitely far to the left of V'.

As was the case with an ellipse, because the hyperbola has a center it is called a **central conic**.

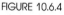

FIGURE 10.6.4

EXAMPLE 1

Given the hyperbola

$$\frac{x^2}{9} - \frac{y^2}{16} = 1$$

find the vertices, foci, and lengths of the transverse and conjugate axes. Draw a sketch of the hyperbola, and show the foci.

SOLUTION
The given equation is of the form of (7); thus $a = 3$ and $b = 4$. The vertices are therefore the points $V(3, 0)$ and $V'(-3, 0)$. The number of units in the length of the transverse axis is $2a$ or 6, and the number of units in the length of the conjugate axis is $2b$ or 8. Because from (5), $b^2 = c^2 - a^2$ we have $16 = c^2 - 9$; so $c = 5$. Hence the foci are at $F(5, 0)$ and $F'(-5, 0)$. A sketch of the hyperbola and its foci are in Fig. 10.6.4.

From Definition 10.6.1 it follows that if P is any point on this hyperbola,
$$\left| |\overline{FP}| - |\overline{F'P}| \right| = 6.$$

EXAMPLE 2

Find an equation of the hyperbola having a focus at $(5, 0)$ and the ends of its conjugate axis at $(0, 2)$ and $(0, -2)$.

SOLUTION
Because the ends of the conjugate axis are at $(0, 2)$ and $(0, -2)$, then $b = 2$, the principal axis is on the x axis, and the center is at the origin. Hence an equation is of the form

$$\frac{x^2}{a^2} - \frac{y^2}{b^2} = 1$$

Because a focus is at $(5, 0)$, $c = 5$, and because $b^2 = c^2 - a^2$, $a^2 = 25 - 4$. Thus $a = \sqrt{21}$, and an equation of the hyperbola is

$$\frac{x^2}{21} - \frac{y^2}{4} = 1$$

If in Eq. (7), x and y are interchanged we obtain

$$\frac{y^2}{a^2} - \frac{x^2}{b^2} = 1 \tag{9}$$

which is an equation of a hyperbola having its center at the origin and its principal axis on the y axis.

FIGURE 10.6.5

▶ ILLUSTRATION 1   The hyperbola with the equation

$$\frac{y^2}{9} - \frac{x^2}{16} = 1$$

has its foci and vertices on the y axis because the equation is of the form (9). A sketch of the graph of this equation appears in Fig. 10.6.5.   ◀

There is no general inequality involving a and b corresponding to the inequality $a > b$ for an ellipse. That is, for a hyperbola it is possible to have $a < b$, as in Illustration 1, where $a = 3$ and $b = 4$; or it is possible to have $a > b$, as for the hyperbola of Example 2, where $a = \sqrt{21}$ and $b = 2$. If, for a hyperbola, $a = b$, then the hyperbola is said to be **equilateral**.

We will now prove that a hyperbola has asymptotes and show how to obtain equations of these asymptotes. In Sections 1.5 and 3.8 vertical and horizontal asymptotes of the graph of a function were defined. In Section 3.8 we also discussed oblique asymptotes of a rational function. What follows is a more general definition, of which the definitions in Sections 1.5 and 3.8 are special cases.

10.6.3 DEFINITION   The graph of the equation $y = f(x)$ has the line $y = mx + b$ as an **asymptote** if either of the following statements is true:

(i) $\lim\limits_{x \to +\infty} [f(x) - (mx + b)] = 0$, and for some number $M > 0, f(x) \neq mx + b$
whenever $x > M$.

(ii) $\lim\limits_{x \to -\infty} [f(x) - (mx + b)] = 0$, and for some number $M < 0, f(x) \neq mx + b$
whenever $x < M$.

Statement (i) indicates that for any $\epsilon > 0$ there exists a number $N > 0$ such that

if $x > N$ then $0 < |f(x) - (mx + b)| < \epsilon$

that is, we can make the function value $f(x)$ as close to the value of $mx + b$ as we please by taking x large enough. This is consistent with our intuitive notion of an asymptote of a graph. A similar statement may be made for part (ii) of Definition 10.6.3.

For the hyperbola $x^2/a^2 - y^2/b^2 = 1$, upon solving for y we get

$$y = \pm \frac{b}{a} \sqrt{x^2 - a^2}$$

So if

$$f(x) = \frac{b}{a}\sqrt{x^2 - a^2}$$

then

$$\lim_{x \to +\infty} \left[f(x) - \frac{b}{a} x \right] = \lim_{x \to +\infty} \left[\frac{b}{a}\sqrt{x^2 - a^2} - \frac{b}{a} x \right]$$

$$= \frac{b}{a} \lim_{x \to +\infty} \frac{(\sqrt{x^2 - a^2} - x)(\sqrt{x^2 - a^2} + x)}{\sqrt{x^2 - a^2} + x}$$

$$= \frac{b}{a} \lim_{x \to +\infty} \frac{-a^2}{\sqrt{x^2 - a^2} + x}$$

$$= 0$$

Therefore, by Definition 10.6.3, the line $y = \dfrac{b}{a} x$ is an asymptote of the graph of $y = \dfrac{b}{a}\sqrt{x^2 - a^2}$. Similarly, it can be shown that the line $y = \dfrac{b}{a} x$ is an asymptote of the graph of $y = -\dfrac{b}{a}\sqrt{x^2 - a^2}$. Consequently, the line $y = \dfrac{b}{a} x$ is an asymptote of the hyperbola $\dfrac{x^2}{a^2} - \dfrac{y^2}{b^2} = 1$. In an analogous manner we can demonstrate that the line $y = -\dfrac{b}{a} x$ is an asymptote of this same hyperbola. We have, then, the following theorem.

10.6.4 THEOREM   The lines

$$y = \frac{b}{a} x \quad \text{and} \quad y = -\frac{b}{a} x$$

are asymptotes of the hyperbola

$$\frac{x^2}{a^2} - \frac{y^2}{b^2} = 1$$

Figure 10.6.6 shows a sketch of the hyperbola of Theorem 10.6.4 together with its asymptotes. In the figure note that the diagonals of the rectangle having vertices at (a, b), $(a, -b)$, $(-a, b)$, and $(-a, -b)$ are on the asymptotes of the hyperbola. This rectangle is called the **auxiliary rectangle** of the hyperbola. The vertices of the hyperbola are the points of intersection of the principal axis and the auxiliary rectangle. A fairly good sketch of a hyperbola can be made by first drawing the auxiliary rectangle and then drawing the branch of the hyperbola through each vertex tangent to the side of the auxiliary rectangle there and approaching asymptotically the lines on which the diagonals of the rectangle lie. Observe that because $a^2 + b^2 = c^2$, the circle having its center at the origin and passing through the vertices of the auxiliary rectangle also passes through the foci of the hyperbola.

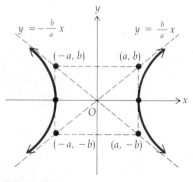

FIGURE 10.6.6

There is a mnemonic device for obtaining equations of the asymptotes of a hyperbola. For example, for the hyperbola having the equation $x^2/a^2 - y^2/b^2 = 1$, if the right side is replaced by zero, we obtain

$$\frac{x^2}{a^2} - \frac{y^2}{b^2} = 0$$

Upon factoring, this equation becomes

$$\left(\frac{x}{a} - \frac{y}{b}\right)\left(\frac{x}{a} + \frac{y}{b}\right) = 0$$

which is equivalent to the two equations

$$\frac{x}{a} - \frac{y}{b} = 0 \quad \text{and} \quad \frac{x}{a} + \frac{y}{b} = 0$$

which, by Theorem 10.6.4, are equations of the asymptotes of the given hyperbola. Using this device for the hyperbola having Eq. (9) we see that the asymptotes are the lines having equations

$$\frac{y}{a} - \frac{x}{b} = 0 \quad \text{and} \quad \frac{y}{a} + \frac{x}{b} = 0$$

which are the same lines as the asymptotes of the hyperbola with the equation $x^2/b^2 - y^2/a^2 = 1$. This hyperbola and the one having Eq. (9) are called **conjugate hyperbolas**.

The asymptotes of an equilateral hyperbola ($a = b$) are perpendicular to each other. The auxiliary rectangle for such a hyperbola is a square, and the transverse and conjugate axes have equal lengths.

If the center of a hyperbola is at (h, k) and its principal axis is parallel to the x axis, then if the axes are translated so that the point (h, k) is the new origin, an equation of the hyperbola relative to this new coordinate system is $\bar{x}^2/a^2 - \bar{y}^2/b^2 = 1$. If we replace $\bar{x}$ by $x - h$ and $\bar{y}$ by $y - k$, this equation becomes

$$\frac{(x - h)^2}{a^2} - \frac{(y - k)^2}{b^2} = 1 \tag{10}$$

Similarly, an equation of a hyperbola having its center at (h, k) and its principal axis parallel to the y axis is

$$\frac{(y - k)^2}{a^2} - \frac{(x - h)^2}{b^2} = 1 \tag{11}$$

EXAMPLE 3
The vertices of a hyperbola are at $(-5, -3)$ and $(-5, -1)$ and the extremities of the conjugate axis are at $(-7, -2)$ and $(-3, -2)$. Find an equation of the hyperbola and equations of the asymptotes. Draw a sketch of the hyperbola and the asymptotes.

SOLUTION
The distance between the vertices is $2a$; so $2a = 2$ and $a = 1$. The length of the conjugate axis is $2b$; thus $2b = 4$ and $b = 2$. Because the principal axis is parallel to the y axis, an equation of the hyperbola is of the form (11). The

center (h, k) is halfway between the vertices and is therefore at the point $(-5, -2)$. Hence an equation of the hyperbola is

$$\frac{(y + 2)^2}{1} - \frac{(x + 5)^2}{4} = 1$$

Using the mnemonic device to obtain equations of the asymptotes we have

$$\left(\frac{y + 2}{1} - \frac{x + 5}{2}\right)\left(\frac{y + 2}{1} + \frac{x + 5}{2}\right) = 0$$

which gives

$$y + 2 = \tfrac{1}{2}(x + 5) \quad \text{and} \quad y + 2 = -\tfrac{1}{2}(x + 5)$$

A sketch of the hyperbola and the asymptotes are in Fig. 10.6.7.

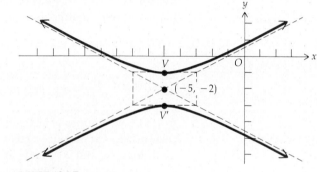

FIGURE 10.6.7

If in (10) and (11) we eliminate fractions and combine terms, the resulting equations are of the form

$$Ax^2 + Cy^2 + Dx + Ey + F = 0 \tag{12}$$

where A and C have different signs; that is, $AC < 0$. We now wish to show that the graph of an equation of the form (12), where $AC < 0$, is either a hyperbola or it degenerates. Completing the squares in x and y in (12), where $AC < 0$, the resulting equation has the form

$$\alpha^2(x - h)^2 - \beta^2(y - k)^2 = H \tag{13}$$

If $H > 0$, (13) can be written as

$$\frac{(x - h)^2}{\dfrac{H}{\alpha^2}} - \frac{(y - k)^2}{\dfrac{H}{\beta^2}} = 1$$

which has the form of (10).

▶ ILLUSTRATION 2   The equation

$$4x^2 - 12y^2 + 24x + 96y - 181 = 0$$

can be written as

$$4(x^2 + 6x) - 12(y^2 - 8y) = 181$$

and upon completing the squares in x and y we have

$$4(x^2 + 6x + 9) - 12(y^2 - 8y + 16) = 181 + 36 - 192$$

$$4(x + 3)^2 - 12(y - 4)^2 = 25$$

This has the form of (13), where $H = 25 > 0$. It may be written as

$$\frac{(x + 3)^2}{\frac{25}{4}} - \frac{(y - 4)^2}{\frac{25}{12}} = 1$$

which has the form of (10).   ◄

If in (13), $H < 0$, then (13) may be written as

$$\frac{(y - k)^2}{\frac{|H|}{\alpha^2}} - \frac{(x - h)^2}{\frac{|H|}{\beta^2}} = 1$$

which has the form of (11).

► ILLUSTRATION 3   Suppose that (12) is

$$4x^2 - 12y^2 + 24x + 96y - 131 = 0$$

Upon completing the squares in x and y we get

$$4(x + 3)^2 - 12(y - 4)^2 = -25$$

This has the form of (13), where $H = -25 < 0$, and it may be written as

$$\frac{(y - 4)^2}{\frac{25}{12}} - \frac{(x + 3)^2}{\frac{25}{4}} = 1$$

which has the form of (11).   ◄

If $H = 0$ in (13), then (13) is equivalent to the two equations

$$\alpha(x - h) - \beta(y - k) = 0 \quad \text{and} \quad \alpha(x - h) + \beta(y - k) = 0$$

which are equations of two straight lines through the point (h, k). This is the degenerate case of the hyperbola.

The following theorem summarizes the results of the above discussion.

two intersecting lines

FIGURE 10.6.8

10.6.5 THEOREM   If in the general second-degree equation

$$Ax^2 + Bxy + Cy^2 + Dx + Ey + F = 0$$

$B = 0$ and $AC < 0$, then the graph is either a hyperbola or two intersecting straight lines.

The degenerate case of a hyperbola, two intersecting lines, is obtained as a conic section if the cutting plane contains the vertex of the cone and two generators, as shown in Fig. 10.6.8.

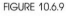

FIGURE 10.6.9

EXAMPLE 4

Determine the graph of the equation

$$9x^2 - 4y^2 - 18x - 16y + 29 = 0$$

SOLUTION

From Theorem 10.6.5, because $B = 0$ and $AC = -36 < 0$, the graph is either a hyperbola or two intersecting straight lines.

Completing the squares in x and y we get

$$9(x^2 - 2x + 1) - 4(y^2 + 4y + 4) = -29 + 9 - 16$$

$$9(x - 1)^2 - 4(y + 2)^2 = -36$$

$$\frac{(y + 2)^2}{9} - \frac{(x - 1)^2}{4} = 1 \qquad (14)$$

Equation (14) has the form of (11); so the graph is a hyperbola whose principal axis is parallel to the y axis and whose center is at $(1, -2)$.

EXAMPLE 5

Find the vertices and foci of the hyperbola of Example 4. Draw a sketch showing the hyperbola, its asymptotes, and the foci.

SOLUTION

From (14) we observe that $a = 3$ and $b = 2$. Because the principal axis is vertical, the center is at $(1, -2)$, and $a = 3$, it follows that the vertices are at the points $V(1, 1)$ and $V'(1, -5)$. For a hyperbola, $c^2 = a^2 + b^2$; therefore $c^2 = 9 + 4$ and $c = \sqrt{13}$. Therefore the foci are at $F(1, -2 + \sqrt{13})$ and $F'(1, -2 - \sqrt{13})$. Figure 10.6.9 shows the hyperbola, its asymptotes, and the foci.

The property of the hyperbola given in Definition 10.6.1 forms the basis of several important navigational systems. These systems involve a network of pairs of radio transmitters at fixed positions at a known distance from one another. The transmitters send out radio signals that are received by a navigator. The difference in arrival time of the two signals determines the difference $2a$ of the distances from the navigator. Thus the navigator's position is known to be somewhere along one arc of a hyperbola having foci at the locations of the two transmitters. One arc, rather than both, is determined because of the signal delay between the two transmitters that is built into the system. The procedure is then repeated for a different pair of radio transmitters, and another arc of a hyperbola that contains the navigator's position is determined. The point of intersection of the two hyperbolic arcs is the actual position. For example, in Fig. 10.6.10 suppose a pair of transmitters is located at points T_1 and S_1 and the signals from this pair determine the hyperbolic arc A_1. Another pair of transmitters is located at points T_2 and S_2 and hyperbolic arc A_2 is determined from their signals. Then the intersection of A_1 and A_2 is the position of the navigator.

FIGURE 10.6.10

FIGURE 10.6.11

The hyperbola has a reflective property that is used in the design of certain telescopes. In Fig. 10.6.11 the line PT is the tangent line at P to the graph of the hyperbola having foci at F and F'. The measure of the angle between the line segment FP and the tangent line PT is α and the measure of the angle between the line segment $F'P$ and the tangent line PT is β. You are asked to prove that $\alpha = \beta$ in Exercise 39. From this equality it follows that a ray of light from a source at one focus of a hyperbolic mirror (one with hyperbolic cross sections) is reflected along the line through the other focus.

Hyperbolas are also used in combat in *sound ranging* to locate the position of enemy guns by the sound of the firing of those guns. Some comets move in hyperbolic orbits. If a quantity varies inversely as another quantity, such as pressure and volume in Boyle's law for a perfect gas ($PV = k$), the graph is a hyperbola, as you will learn in Section 10.7.

From Theorems 10.4.6, 10.5.3, and 10.6.5 we may conclude that the graph of the general quadratic equation in two unknowns when $B = 0$ is either a conic or a degenerate conic. The type of conic can be determined from the product of A and C. We have the following theorem.

10.6.6 THEOREM   The graph of the equation $Ax^2 + Cy^2 + Dx + Ey + F = 0$, where A and C are not both zero, is either a conic or a degenerate conic; if it is a conic, then the graph is

(i) a *parabola* if either $A = 0$ or $C = 0$, that is, if $AC = 0$;
(ii) an *ellipse* if A and C have the same sign, that is, if $AC > 0$;
(iii) a *hyperbola* if A and C have opposite signs, that is, if $AC < 0$.

A discussion of the graph of the general quadratic equation, where $B \neq 0$, is given in Section 10.7.

EXERCISES 10.6

In Exercises 1 through 16, find the center, vertices, foci, and equations of the asymptotes of the given hyperbola. Draw a sketch of the curve and its asymptotes, and show the foci.

1. $9x^2 - 4y^2 = 36$
2. $x^2 - 9y^2 = 9$
3. $4x^2 - 25y^2 = 100$
4. $16y^2 - 9x^2 = 144$
5. $4y^2 - x^2 = 16$
6. $4y^2 - 7x^2 = 56$
7. $9y^2 - 16x^2 = 1$
8. $25x^2 - 25y^2 = 1$
9. $x^2 - y^2 + 8x - 2y - 21 = 0$
10. $4x^2 - y^2 - 8x - 12 = 0$
11. $9x^2 - 18y^2 + 54x - 36y + 79 = 0$
12. $x^2 - y^2 + 6x + 10y - 4 = 0$
13. $3y^2 - 4x^2 - 8x - 24y - 40 = 0$
14. $4x^2 - y^2 + 56x + 2y + 195 = 0$
15. $4y^2 - 9x^2 + 16y + 18x = 29$
16. $y^2 - x^2 + 2y - 2x - 1 = 0$

In Exercises 17 through 26, find an equation of the hyperbola satisfying the given conditions, and draw a sketch of the graph.

17. Vertices at $(-2, 0)$ and $(2, 0)$ and a conjugate axis of length 6.
18. Foci at $(0, 5)$ and $(0, -5)$ and a vertex at $(0, 4)$.
19. Center at the origin, its foci on the y axis, and passing through the points $(-2, 4)$ and $(-6, 7)$.
20. Extremities of its conjugate axis at $(0, -3)$ and $(0, 3)$ and one focus at $(5, 0)$.
21. One focus at $(26, 0)$ and asymptotes the lines $12y = \pm 5x$.
22. Center at $(3, -5)$, a vertex at $(7, -5)$, and a focus at $(8, -5)$.
23. Center at $(-2, -1)$, a vertex at $(-2, 11)$, and a focus at $(-2, 14)$.

24. Foci at $(3, 6)$ and $(3, 0)$ and passing through the point $(5, 3 + \frac{6}{5}\sqrt{5})$.

25. Foci at $(-1, 4)$ and $(7, 4)$ and length of the transverse axis is $\frac{8}{3}$.

26. One focus at $(-3 - 3\sqrt{13}, 1)$, asymptotes intersecting at $(-3, 1)$, and one asymptote passing through the point $(1, 7)$.

27. The vertices of a hyperbola are at $(-3, -1)$ and $(-1, -1)$ and the distance between the foci is $2\sqrt{5}$. Find (a) an equation of the hyperbola and (b) equations of the asymptotes.

28. The foci of a hyperbola are at $(2, 7)$ and $(2, -7)$ and the distance between the vertices is $8\sqrt{3}$. Find (a) an equation of the hyperbola and (b) equations of the asymptotes.

29. Find an equation of the tangent line to the hyperbola $4x^2 - y^2 = -1$ at the point $(\frac{1}{2}, \sqrt{2})$.

30. Find an equation of the normal line to the hyperbola $4x^2 - 3y^2 = 24$ at the point $(3, 2)$.

31. Find an equation of the hyperbola whose foci are the vertices of the ellipse $7x^2 + 11y^2 = 77$ and whose vertices are the foci of this ellipse.

32. Find an equation of the ellipse whose foci are the vertices of the hyperbola $11x^2 - 7y^2 = 77$ and whose vertices are the foci of this hyperbola.

33. The cost of production of a commodity is $12 less per unit at a point A than it is at a point B, and the distance between A and B is 100 km. Assuming that the route of delivery of the commodity is along a straight line, and that the delivery cost is 20 cents per unit per kilometer, find the curve at any point of which the commodity can be supplied from either A or B at the same total cost. (*Hint:* Take points A and B at $(-50, 0)$ and $(50, 0)$, respectively.)

34. Prove that there is no tangent line to the hyperbola $x^2 - y^2 = 1$ that passes through the origin.

35. Find the volume of the solid of revolution generated by revolving about the x axis the region bounded by the hyperbola $x^2/a^2 - y^2/b^2 = 1$ and the line $x = 2a$.

36. Find the centroid of the solid of revolution of Exercise 35.

37. Three listening posts are located at the points $A(0, 0)$, $B(0, \frac{21}{4})$, and $C(\frac{25}{3}, 0)$, the unit being 1 km. Microphones located at these points show that a gun is $\frac{5}{3}$ km closer to A than to C and $\frac{7}{4}$ km closer to B than to A. Determine the position of the gun by use of Definition 10.6.1.

38. Two LORAN (short for Long-Range Navigation) stations A and B lie on a line running east and west, and A is 80 mi due east of B. An airplane is traveling east on a straight line course that is 60 mi north of the line through A and B. Signals are sent at the same time from A and B, and the signal from A reaches the plane 350 μsec before the one from B. If the signals travel at the rate of 0.2 mi per microsecond, locate the position of the plane by use of Definition 10.6.1.

39. In Fig. 10.6.11 prove that $\alpha = \beta$. (*Hint:* Choose the coordinate axes so that the center of the hyperbola is at the origin and the axes of the hyperbola are along the coordinate axes. Then prove that the tangent line PT bisects the angle formed by the line segments FP and $F'P$.)

40. For an equilateral hyperbola having its center at (h, k) and principal axis parallel to the x axis, prove that the asymptotes are perpendicular.

41. Prove that if $P(x, y)$ is any point whose coordinates satisfy Eq. (6), then

$$\left| |\overline{FP}| - |\overline{F'P}| \right| = 2a$$

where F and F' are the foci of the hyperbola in Fig. 10.6.2. (*Hint:* Use a method similar to that suggested for the ellipse in either Exercise 42 or 43 of Exercises 10.5.)

10.7
ROTATION OF AXES

We have shown how a translation of coordinate axes can simplify the form of certain equations. A translation of axes gives a new coordinate system whose axes are parallel to the original axes. We now consider a rotation of coordinate axes that enables us to transform a second-degree equation having an xy term into one having no such term.

Suppose that we have two rectangular cartesian coordinate systems with the same origin. Let one system be the xy system and the other the $\bar{x}\bar{y}$ system. Suppose further that the $\bar{x}$ axis makes an angle of radian measure α with the x axis. Then of course the $\bar{y}$ axis makes an angle of radian measure α with the y axis. In such a case we state that the xy system of coordinates is *rotated* through an angle of radian measure α to form the $\bar{x}\bar{y}$ system of coordinates. A point P having coordinates (x, y) with respect to the original coordinate system will have coordinates $(\bar{x}, \bar{y})$ with respect to the new one. We now obtain relationships between these two sets of coordinates. To do this we introduce two polar coordinate systems, each system having the pole at the origin. In the first polar coordinate system the positive side of the x axis is taken as the polar axis; in the second polar coordinate system the positive side of the $\bar{x}$ axis is taken as the polar axis. See Fig. 10.7.1. Point P has two sets of polar coordinates, (r, θ) and $(\bar{r}, \bar{\theta})$, where

$$\bar{r} = r \quad \text{and} \quad \bar{\theta} = \theta - \alpha \tag{1}$$

The following equations hold:

$$\bar{x} = \bar{r} \cos \bar{\theta} \quad \text{and} \quad \bar{y} = \bar{r} \sin \bar{\theta} \tag{2}$$

Substituting from (1) into (2) we get

$$\bar{x} = r \cos(\theta - \alpha) \quad \text{and} \quad \bar{y} = r \sin(\theta - \alpha) \tag{3}$$

FIGURE 10.7.1

Using the trigonometric identities for the sine and cosine of the difference of two numbers, Eqs. (3) become

$$\bar{x} = r \cos \theta \cos \alpha + r \sin \theta \sin \alpha$$

and

$$\bar{y} = r \sin \theta \cos \alpha - r \cos \theta \sin \alpha$$

Because $r \cos \theta = x$ and $r \sin \theta = y$, we obtain from the above two equations

$$\bar{x} = x \cos \alpha + y \sin \alpha \tag{4}$$

and

$$\bar{y} = -x \sin \alpha + y \cos \alpha \tag{5}$$

Solving (4) and (5) simultaneously for x and y in terms of $\bar{x}$ and $\bar{y}$ we obtain

$$x = \bar{x} \cos \alpha - \bar{y} \sin \alpha \tag{6}$$

and

$$y = \bar{x} \sin \alpha + \bar{y} \cos \alpha \tag{7}$$

It is left as an exercise to fill in the steps going from (4) and (5) to (6) and (7). (See Exercise 26.) The results are stated as a theorem.

10.7.1 THEOREM   If (x, y) represents a point P with respect to a given set of axes and $(\bar{x}, \bar{y})$ is a representation of P after the axes have been rotated through an angle α, then

$$x = \bar{x} \cos \alpha - \bar{y} \sin \alpha \quad \text{and} \quad y = \bar{x} \sin \alpha + \bar{y} \cos \alpha$$

FIGURE 10.7.2

EXAMPLE 1

Given the equation

$$xy = 1$$

(a) Find an equation of the graph with respect to the $\bar{x}$ and $\bar{y}$ axes after a rotation of axes through an angle of radian measure $\frac{1}{4}\pi$, and (b) draw a sketch of the graph and show both sets of axes.

SOLUTION

(a) With $\alpha = \frac{1}{4}\pi$ in Theorem 10.7.1 we obtain

$$x = \frac{1}{\sqrt{2}}\bar{x} - \frac{1}{\sqrt{2}}\bar{y} \quad \text{and} \quad y = \frac{1}{\sqrt{2}}\bar{x} + \frac{1}{\sqrt{2}}\bar{y}$$

Substituting these expressions for x and y in the equation $xy = 1$ we get

$$\left(\frac{1}{\sqrt{2}}\bar{x} - \frac{1}{\sqrt{2}}\bar{y}\right)\left(\frac{1}{\sqrt{2}}\bar{x} + \frac{1}{\sqrt{2}}\bar{y}\right) = 1$$

$$\frac{\bar{x}^2}{2} - \frac{\bar{y}^2}{2} = 1$$

(b) This is an equation of an equilateral hyperbola whose asymptotes are the bisectors of the quadrants in the $\bar{x}\bar{y}$ system. Thus the graph of the equation $xy = 1$ is an equilateral hyperbola lying in the first and third quadrants and the asymptotes are the x and y axes. See Fig. 10.7.2 for the required graph.

In Section 10.6 we showed that when $B = 0$ and A and C are not both zero, the graph of the general second-degree equation in two unknowns,

$$Ax^2 + Bxy + Cy^2 + Dx + Ey + F = 0 \tag{8}$$

is either a conic or a degenerate conic. We now show that if $B \neq 0$, then any equation of the form (8) can be transformed by a suitable rotation of axes into an equation of the form

$$\bar{A}\bar{x}^2 + \bar{C}\bar{y}^2 + \bar{D}\bar{x} + \bar{E}\bar{y} + \bar{F} = 0 \tag{9}$$

where $\bar{A}$ and $\bar{C}$ are not both zero.

If the xy system is rotated through an angle α, then to obtain an equation of the graph of (8) with respect to the $\bar{x}\bar{y}$ system we replace x by $\bar{x}\cos\alpha - \bar{y}\sin\alpha$ and y by $\bar{x}\sin\alpha + \bar{y}\cos\alpha$. We get

$$\bar{A}\bar{x}^2 + \bar{B}\bar{x}\bar{y} + \bar{C}\bar{y}^2 + \bar{D}\bar{x} + \bar{E}\bar{y} + \bar{F} = 0 \tag{10}$$

where

$$\bar{A} = A\cos^2\alpha + B\sin\alpha\cos\alpha + C\sin^2\alpha$$

$$\bar{B} = -2A\sin\alpha\cos\alpha + B(\cos^2\alpha - \sin^2\alpha) + 2C\sin\alpha\cos\alpha \tag{11}$$

$$\bar{C} = A\sin^2\alpha - B\sin\alpha\cos\alpha + C\cos^2\alpha$$

We wish to find an α so that the rotation transforms (8) into an equation of the form (9). Setting $\bar{B}$ from (11) equal to zero we have

$$B(\cos^2 \alpha - \sin^2 \alpha) + (C - A)(2 \sin \alpha \cos \alpha) = 0$$

or, equivalently, with trigonometric identities,

$$B \cos 2\alpha + (C - A) \sin 2\alpha = 0$$

Because $B \neq 0$, this gives

$$\cot 2\alpha = \frac{A - C}{B} \tag{12}$$

We have shown, then, that an equation of the form (8), where $B \neq 0$, can be transformed to an equation of the form (9) by a rotation of axes through an angle α satisfying (12). We wish to show that $\bar{A}$ and $\bar{C}$ in (9) are not both zero. To prove this, notice that (10) is obtained from (8) by rotating the axes through the angle α. Also, (8) can be obtained from (10) by rotating the axes back through the angle $-\alpha$. If $\bar{A}$ and $\bar{C}$ in (10) are both zero, then the substitutions

$$\bar{x} = x \cos \alpha + y \sin \alpha \quad \text{and} \quad \bar{y} = -x \sin \alpha + y \cos \alpha$$

in (10) would result in the equation

$$\bar{D}(x \cos \alpha + y \sin \alpha) + \bar{E}(-x \sin \alpha + y \cos \alpha) + \bar{F} = 0$$

which is an equation of the first degree and hence different from (8) because we have assumed that at least $B \neq 0$. The following theorem has, therefore, been proved.

10.7.2 THEOREM   If $B \neq 0$, the equation $Ax^2 + Bxy + Cy^2 + Dx + Ey + F = 0$ can be transformed into the equation $\bar{A}\bar{x}^2 + \bar{C}\bar{y}^2 + \bar{D}\bar{x} + \bar{E}\bar{y} + \bar{F} = 0$, where $\bar{A}$ and $\bar{C}$ are not both zero, by a rotation of axes through an angle α for which $\cot 2\alpha = (A - C)/B$.

By Theorems 10.7.2 and 10.6.6 it follows that the graph of an equation of the form (8) is either a conic or a degenerate conic. To determine which type of conic is the graph of a particular equation, we use the fact that A, B, and C of (8) and $\bar{A}$, $\bar{B}$, and $\bar{C}$ of (10) satisfy the relation

$$B^2 - 4AC = \bar{B}^2 - 4\bar{A}\bar{C} \tag{13}$$

which can be proved by substituting the expressions for $\bar{A}$, $\bar{B}$, and $\bar{C}$ given in Eqs. (11) in the right side of (13). This is left as an exercise (see Exercise 25).

The expression $B^2 - 4AC$ is called the **discriminant** of (8). Equation (13) states that the discriminant of the general quadratic equation in two variables is *invariant* under a rotation of axes. If the angle of rotation is chosen so that $\bar{B} = 0$, then (13) becomes

$$B^2 - 4AC = -4\bar{A}\bar{C} \tag{14}$$

From Theorem 10.6.6 it follows that if the graph of (9) is not degenerate, then it is a parabola if $\bar{A}\bar{C} = 0$, an ellipse if $\bar{A}\bar{C} > 0$, and a hyperbola if $\bar{A}\bar{C} < 0$. So we conclude that the graph of (9) is a parabola, an ellipse, or a hyperbola depending on whether $-4\bar{A}\bar{C}$ is zero, negative, or positive. Because the graph of (8) is the same as the graph of (9), it follows from (14) that if the graph of (8) is not degenerate, then it is a parabola, an ellipse, or a hyperbola depending on whether the discriminant $B^2 - 4AC$ is zero, negative, or positive. We have proved the following theorem.

10.7.3 THEOREM   The graph of the equation

$$Ax^2 + Bxy + Cy^2 + Dx + Ey + F = 0$$

is either a conic or a degenerate conic. If it is a conic, then it is

(i) a *parabola* if $B^2 - 4AC = 0$;
(ii) an *ellipse* if $B^2 - 4AC < 0$;
(iii) a *hyperbola* if $B^2 - 4AC > 0$.

EXAMPLE 2
Simplify the equation

$$17x^2 - 12xy + 8y^2 - 80 = 0$$

by a rotation of axes. Draw a sketch of the graph of the equation and show both sets of axes.

SOLUTION

$$B^2 - 4AC = (-12)^2 - 4(17)(8)$$
$$= -400$$

Because $B^2 - 4AC < 0$, by Theorem 10.7.3 the graph is an ellipse or else it is degenerate. To eliminate the xy term by a rotation of axes we must choose an α such that

$$\cot 2\alpha = \frac{A - C}{B}$$

$$= \frac{17 - 8}{-12}$$

$$= -\tfrac{3}{4}$$

There is a 2α in the interval $(0, \pi)$ for which $\cot 2\alpha = -\tfrac{3}{4}$. Therefore α is in the interval $(0, \tfrac{1}{2}\pi)$. To apply Theorem 10.7.1 it is not necessary to find α so long as we find $\cos \alpha$ and $\sin \alpha$. These functions can be found from the value of $\cot 2\alpha$ by the trigonometric identities

$$\cos \alpha = \sqrt{\frac{1 + \cos 2\alpha}{2}} \quad \text{and} \quad \sin \alpha = \sqrt{\frac{1 - \cos 2\alpha}{2}}$$

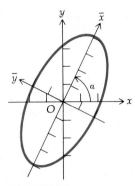

FIGURE 10.7.3

Because $\cot 2\alpha = -\frac{3}{4}$ and $0 < \alpha < \frac{1}{2}\pi$, it follows that $\cos 2\alpha = -\frac{3}{5}$. So

$$\cos \alpha = \sqrt{\frac{1 - \frac{3}{5}}{2}} \quad \text{and} \quad \sin \alpha = \sqrt{\frac{1 + \frac{3}{5}}{2}}$$

$$= \frac{1}{\sqrt{5}} \qquad\qquad = \frac{2}{\sqrt{5}}$$

Substituting $x = \bar{x}/\sqrt{5} - 2\bar{y}/\sqrt{5}$ and $y = 2\bar{x}/\sqrt{5} + \bar{y}/\sqrt{5}$ in the given equation we obtain

$$17\left(\frac{\bar{x}^2 - 4\bar{x}\bar{y} + 4\bar{y}^2}{5}\right) - 12\left(\frac{2\bar{x}^2 - 3\bar{x}\bar{y} - 2\bar{y}^2}{5}\right) + 8\left(\frac{4\bar{x}^2 + 4\bar{x}\bar{y} + \bar{y}^2}{5}\right) - 80 = 0$$

Upon simplification this equation becomes

$$\bar{x}^2 + 4\bar{y}^2 = 16$$

$$\frac{\bar{x}^2}{16} + \frac{\bar{y}^2}{4} = 1$$

So the graph is an ellipse whose major axis is 8 units long and whose minor axis is 4 units long. A sketch of the ellipse with both sets of axes is shown in Fig. 10.7.3.

EXERCISES 10.7

In Exercises 1 through 4, for the given equation, (a) find an equation of the graph with respect to the $\bar{x}$ and $\bar{y}$ axes after a rotation of axes through an angle of radian measure $\frac{1}{4}\pi$, and (b) draw a sketch of the graph and show both sets of axes.

1. $xy = 8$ 2. $xy = -4$ 3. $x^2 - y^2 = 8$ 4. $y^2 - x^2 = 16$

In Exercises 5 through 12, remove the xy term from the given equation by a rotation of axes. Draw a sketch of the graph and show both sets of axes.

5. $24xy - 7y^2 + 36 = 0$ 6. $4xy + 3x^2 = 4$ 7. $x^2 + 2xy + y^2 - 8x + 8y = 0$
8. $x^2 + xy + y^2 = 3$ 9. $xy + 16 = 0$ 10. $5x^2 + 6xy + 5y^2 = 9$
11. $31x^2 + 10\sqrt{3}xy + 21y^2 = 144$ 12. $6x^2 + 20\sqrt{3}xy + 26y^2 = 324$

In Exercises 13 through 22, simplify the given equation by a rotation and translation of axes. Draw a sketch of the graph and show the three sets of axes.

13. $x^2 + xy + y^2 - 3y - 6 = 0$ 14. $19x^2 + 6xy + 11y^2 - 26x + 38y + 31 = 0$
15. $17x^2 - 12xy + 8y^2 - 68x + 24y - 12 = 0$ 16. $x^2 - 10xy + y^2 + x + y + 1 = 0$
17. $x^2 + 2xy + y^2 + x - y - 4 = 0$ 18. $16x^2 - 24xy + 9y^2 - 60x - 80y + 400 = 0$
19. $11x^2 - 24xy + 4y^2 + 30x + 40y - 45 = 0$ 20. $3x^2 - 4xy + 8x - 1 = 0$
21. $4x^2 + 4xy + y^2 - 6x + 12y = 0$ 22. $x^2 + 2xy + y^2 - x - 3y = 0$
23. Show that the graph of $\sqrt{x} + \sqrt{y} = 1$ is a parabola by rotating the axes through an angle of radian measure $\frac{1}{4}\pi$. *Hint:* Eliminate the radicals in the equation before applying Theorem 10.7.1.
24. Given the equation $(a^2 + b^2)xy = 1$, where $a > 0$ and $b > 0$, find an equation of the graph with respect to the $\bar{x}$ and $\bar{y}$ axes after a rotation of the axes through an angle of radian measure $\tan^{-1}(b/a)$.
25. Show that for the general second-degree equation in two variables, the discriminant $B^2 - 4AC$ is invariant under a rotation of axes.

26. Derive (6) and (7) of this section by solving (4) and (5) for x and y in terms of $\bar{x}$ and $\bar{y}$. *Hint:* To solve for x, multiply each member of (4) by $\cos \alpha$ and each member of (5) by $\sin \alpha$, and then subtract corresponding members of the resulting equations. Use a similar procedure to solve for y.

10.8
A UNIFIED TREATMENT OF CONIC SECTIONS AND POLAR EQUATIONS OF CONICS

In Sections 10.4 through 10.6 we defined each of the three types of conic sections separately. An alternate approach is to start with a definition that gives a common property of conics and then introduce each of the conics as a special case of the general definition. We state this definition in the following theorem. The positive constant e in the statement of the theorem is called the **eccentricity** of the conic.

10.8.1 THEOREM   A conic section can be defined as the set of all points P in a plane such that the ratio of the undirected distance of P from a fixed point to the undirected distance of P from a fixed line that does not contain the fixed point is a positive constant e. Furthermore, if $e = 1$, the conic is a parabola; if $0 < e < 1$, it is an ellipse; and if $e > 1$, it is a hyperbola.

FIGURE 10.8.1

PROOF   If $e = 1$, we see by comparing Definition 10.4.1 and the statement of the theorem that the set is a parabola having the fixed point as its focus and the fixed line as its directrix.

Suppose now that $e \neq 1$. We first obtain a polar equation of the set of points described. Let F denote the fixed point and l denote the fixed line. We take the pole at F and the polar axis and its extension perpendicular to l. We first consider the situation when the line l is to the left of the point F. Let D be the point of intersection of l with the extension of the polar axis, and let d denote the undirected distance from F to l. Refer to Fig. 10.8.1. Let $P(r, \theta)$ be any point in the set to the right of l and on the terminal side of the angle of measure θ. Draw perpendiculars PQ and PR to the polar axis and line l, respectively. The point P is in the set described if and only if

$$|\overline{FP}| = e|\overline{RP}| \tag{1}$$

Because P is to the right of l, $\overline{RP} > 0$; thus $|\overline{RP}| = \overline{RP}$. Furthermore, $|\overline{FP}| = r$ because $r > 0$. Thus from (1),

$$r = e(\overline{RP}) \tag{2}$$

However, $\overline{RP} = \overline{DQ}$, and because $\overline{DQ} = \overline{DF} + \overline{FQ}$, we have

$$\overline{RP} = d + r \cos \theta$$

Substituting this expression for $\overline{RP}$ in (2) we get

$$r = e(d + r \cos \theta)$$

from which we obtain

$$r = \frac{ed}{1 - e \cos \theta} \tag{3}$$

A cartesian representation of (3) can be obtained by first replacing $\cos \theta$ by x/r. Then (3) becomes

$$r = \frac{ed}{1 - \dfrac{ex}{r}}$$

$$r = \frac{edr}{r - ex}$$

$$r - ex = ed$$

$$r = e(x + d)$$

We now replace r by $\pm\sqrt{x^2 + y^2}$ and get

$$\pm\sqrt{x^2 + y^2} = e(x + d)$$

Squaring on both sides of this equation gives

$$x^2 + y^2 = e^2 x^2 + 2e^2 dx + e^2 d^2$$

$$y^2 + x^2(1 - e^2) = 2e^2 dx + e^2 d^2$$

Because $e \neq 1$, we can divide on both sides of this equation by $1 - e^2$ and obtain

$$x^2 - \frac{2e^2 d}{1 - e^2} x + \frac{1}{1 - e^2} y^2 = \frac{e^2 d^2}{1 - e^2}$$

Completing the square for the terms involving x by adding $e^4 d^2/(1 - e^2)^2$ on both sides of the above equation we get

$$\left(x - \frac{e^2 d}{1 - e^2}\right)^2 + \frac{1}{1 - e^2} y^2 = \frac{e^2 d^2}{(1 - e^2)^2} \tag{4}$$

If the origin is translated to the point $(e^2 d/(1 - e^2), 0)$, (4) becomes

$$\bar{x}^2 + \frac{1}{1 - e^2} \bar{y}^2 = \frac{e^2 d^2}{(1 - e^2)^2}$$

$$\frac{\bar{x}^2}{\dfrac{e^2 d^2}{(1 - e^2)^2}} + \frac{\bar{y}^2}{\dfrac{e^2 d^2}{1 - e^2}} = 1 \tag{5}$$

Now let

$$\frac{e^2 d^2}{(1 - e^2)^2} = a^2 \qquad \text{where } a > 0 \tag{6}$$

Then (5) can be written as

$$\frac{\bar{x}^2}{a^2} + \frac{\bar{y}^2}{a^2(1 - e^2)} = 1$$

Replacing $\bar{x}$ and $\bar{y}$ by x and y we get

$$\frac{x^2}{a^2} + \frac{y^2}{a^2(1-e^2)} = 1 \tag{7}$$

If $0 < e < 1$, then $a^2(1-e^2) > 0$ and we can let

$$b^2 = a^2(1-e^2) \qquad \text{where } 0 < e < 1 \tag{8}$$

Substituting from (8) in (7) we get

$$\frac{x^2}{a^2} + \frac{y^2}{b^2} = 1 \tag{9}$$

Equation (9) is a standard form of a cartesian equation of an ellipse having its principal axis on the x axis and its center at the origin.

If $e > 1$, then $a^2(e^2 - 1) > 0$ and we can let

$$b^2 = a^2(e^2 - 1) \qquad \text{where } e > 1 \tag{10}$$

Substituting from (10) in (7) we obtain

$$\frac{x^2}{a^2} - \frac{y^2}{b^2} = 1 \tag{11}$$

Equation (11) is a standard form of a cartesian equation of a hyperbola having its principal axis on the x axis and its center at the origin.

In a similar manner we can derive an equation of a central conic (an ellipse or hyperbola) from (1) when $e \neq 1$ if the line l is to the right of the point F at the pole. In this case instead of Eq. (3) we have

$$r = \frac{ed}{1 + e \cos \theta} \tag{12}$$

The derivation of (12) is left as an exercise (see Exercise 33).

We can also derive an equation of a central conic from (1) when $e \neq 1$ if the line l is parallel to the polar axis and the point F is at the pole. In this case instead of Eq. (3) we obtain

$$r = \frac{ed}{1 \pm e \sin \theta} \tag{13}$$

where e and d are, respectively, the eccentricity and undirected distance between F and l. The plus sign is taken when l is above F, and the minus sign is taken when it is below F. The derivations of (13) are left as exercises (see Exercises 34 and 35).

We can reverse the steps in going from (1) to (7). Thus if P is any point on a central conic, Eq. (1) is satisfied.

Therefore we conclude that a conic can be defined by the described set of points.   ∎

In the above proof we showed that when the conic is a parabola, the fixed point F mentioned in the theorem is the focus of the parabola and the fixed line is the directrix. We now show that the point F is one of the foci when we have a central conic. Because (4) is an equation of a central conic for which the fixed point F is at the origin (pole) and (7) is obtained from (4) by translating the origin to the point $(e^2d/(1 - e^2), 0)$, it follows that for the conic having Eq. (7) the point F is at $(-e^2d/(1 - e^2), 0)$. We wish to show that this point is a focus of the conic. From (6),

$$a = \begin{cases} \dfrac{ed}{1 - e^2} & \text{if } 0 < e < 1 \\[2mm] \dfrac{ed}{e^2 - 1} & \text{if } e > 1 \end{cases}$$

$$\Leftrightarrow \quad \frac{-e^2d}{1 - e^2} = \begin{cases} -ae & \text{if } 0 < e < 1 \\ ae & \text{if } e > 1 \end{cases} \tag{14}$$

If (7) is an equation of an ellipse $(0 < e < 1)$, we know that the foci are at $(-c, 0)$ and $(c, 0)$, where

$$c^2 = a^2 - b^2 \qquad c > 0 \tag{15}$$

Substituting from (8) in (15) we get

$$c^2 = a^2 - a^2(1 - e^2)$$

$$c^2 = a^2e^2$$

$$c = ae \tag{16}$$

By comparing (16) and (14) we conclude that if (7) is an equation of an ellipse, the point F is the left-hand focus.

If (7) is an equation of a hyperbola $(e > 1)$, again the foci are at $(-c, 0)$ and $(c, 0)$, but for a hyperbola,

$$c^2 = a^2 + b^2 \qquad c > 0 \tag{17}$$

Substituting from (10) in (17) we have

$$c^2 = a^2 + a^2(e^2 - 1)$$

$$c^2 = a^2e^2$$

$$c = ae \tag{18}$$

By comparing (18) and (14) it follows that if (7) is an equation of a hyperbola, the point F is the right-hand focus.

The fixed line l mentioned in Theorem 10.8.1 is called the **directrix** corresponding to the focus at F. When the graph of (7) is an ellipse, the directrix corresponding to the focus at $(-c, 0)$ or, equivalently, $(-ae, 0)$ has the equation

$$x = -ae - d$$

Because when $0 < e < 1$, $d = a(1 - e^2)/e$, this equation becomes

$$x = -ae - \frac{a(1 - e^2)}{e}$$

$$x = -\frac{a}{e}$$

Similarly, if the graph of (7) is a hyperbola, the directrix corresponding to the focus at $(c, 0)$ or, equivalently, $(ae, 0)$ has the equation

$$x = ae - d$$

When $e > 1$, $d = a(e^2 - 1)/e$; thus the above equation of the directrix can be written as

$$x = \frac{a}{e}$$

Hence we have shown that if (7) is an equation of an ellipse, a focus and its corresponding directrix are $(-ae, 0)$ and $x = -a/e$; and if (7) is an equation of a hyperbola, a focus and its corresponding directrix are $(ae, 0)$ and $x = a/e$.

Because (7) contains only even powers of x and y, its graph is symmetric with respect to both the x and y axes. Therefore, if there is a focus at $(-ae, 0)$ having a corresponding directrix of $x = -a/e$, by symmetry there is also a focus at $(ae, 0)$ having a corresponding directrix of $x = a/e$. Similarly, for a focus at $(ae, 0)$ and a corresponding directrix of $x = a/e$, there is also a focus at $(-ae, 0)$ and a corresponding directrix of $x = -a/e$. These results are summarized in the following theorem.

10.8.2 THEOREM   The central conic having the equation

$$\frac{x^2}{a^2} + \frac{y^2}{a^2(1 - e^2)} = 1 \tag{19}$$

where $a > 0$, has a focus at $(-ae, 0)$, whose corresponding directrix is $x = -a/e$, and a focus at $(ae, 0)$, whose corresponding directrix is $x = a/e$.

Figures 10.8.2 and 10.8.3 show sketches of the graph of (19) together with the foci and directrices in the respective cases of an ellipse and a hyperbola.

From both Eqs. (16) and (18)

$$e = \frac{c}{a} \tag{20}$$

Because, also $e = 2c/(2a)$, the eccentricity of both the ellipse and the hyperbola is the ratio of the undirected distance between the foci to the undirected distance between the vertices. Thus the eccentricity gives an indication of the shape of a central conic.

For an ellipse the eccentricity takes on values between 0 and 1. When the foci are close together, e is close to zero, and the shape of the ellipse is close to that of a circle. See Fig. 10.8.4(a) showing an ellipse for which $e = 0.3$. If a remains fixed, then as e increases the "flatness" of the ellipse

FIGURE 10.8.2

FIGURE 10.8.3

increases. Figures 10.8.4(b) and (c) show ellipses with eccentricities of 0.7 and 0.95, respectively, each with the same value of a as in Fig. 10.8.4(a). The limiting forms of the ellipse are a circle of diameter $2a$ and a line segment of length $2a$.

For a hyperbola, the eccentricity is greater than 1. The eccentricity of an equilateral hyperbola is $\sqrt{2}$, which is obtained from (10) by taking $a = b$. See Fig. 10.8.5(b). If e approaches 1 and a remains fixed, then c approaches a, b approaches 0, and the shape of the hyperbola becomes "thin" around its principal axis. In Fig. 10.8.5(a) there is a hyperbola with $e = 1.05$ and the

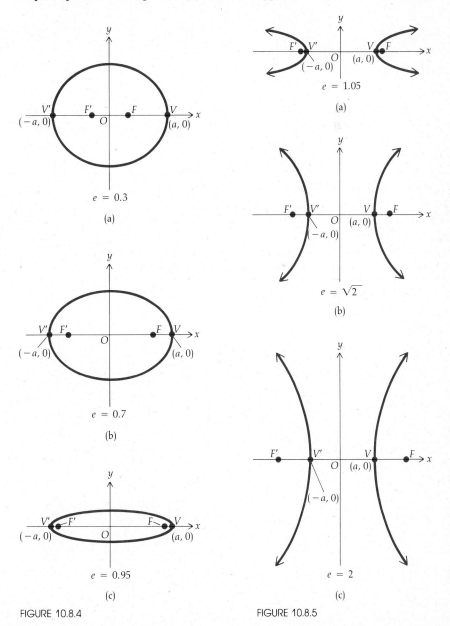

$e = 0.3$

(a)

$e = 0.7$

(b)

$e = 0.95$

(c)

FIGURE 10.8.4

$e = 1.05$

(a)

$e = \sqrt{2}$

(b)

$e = 2$

(c)

FIGURE 10.8.5

same value of a as in Fig. 10.8.5(b). If e increases as a remains fixed, then c increases and b increases, and the hyperbola becomes "fat" around its principal axis. See Fig. 10.8.5(c) for a hyperbola with $e = 2$ and the same value of a as in Figs. 10.8.5(a) and (b).

EXAMPLE 1

The ellipse of Example 1 in Section 10.5 has the equation

$$\frac{x^2}{25} + \frac{y^2}{16} = 1$$

(a) Find the eccentricity and directrices of this ellipse. (b) Draw a sketch showing the ellipse, the directrices, and the foci. Also choose any three points P on the ellipse and draw the line segments whose lengths are the undirected distances from P to a focus and its corresponding directrix. Observe that the ratio of these distances is e.

SOLUTION

(a) From the equation of the ellipse, $a = 5$ and $b = 4$. We showed in Example 1 of Section 10.5 that $c = 3$. Therefore, from (20), $e = \frac{3}{5}$. Because $a/e = 25/3$, it follows from Theorem 10.8.2 that the directrix corresponding to the focus at $(3,0)$ has the equation $x = \frac{25}{3}$, and the directrix corresponding to the focus at $(-3,0)$ has the equation $x = -\frac{25}{3}$.

(b) Figure 10.8.6 shows the ellipse, the directrices, and the foci, as well as three points P_1, P_2, and P_3 on the ellipse. For each of these points,

$$\frac{|\overline{FP}|}{|\overline{RP}|} = \frac{3}{5}$$

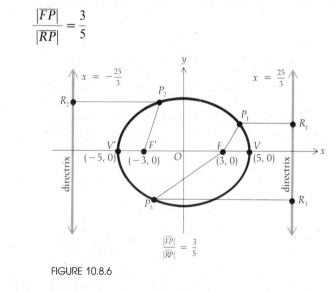

FIGURE 10.8.6

EXAMPLE 2

The hyperbola of Example 1 in Section 10.6 has the equation

$$\frac{x^2}{9} - \frac{y^2}{16} = 1$$

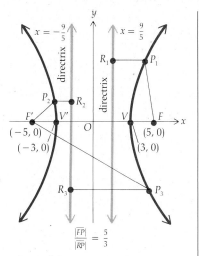

$x = -\frac{9}{5}$

$x = \frac{9}{5}$

FIGURE 10.8.7

$\dfrac{|\overline{FP}|}{|\overline{RP}|} = \dfrac{5}{3}$

(a) Find the eccentricity and directrices of this hyperbola. (b) Draw a sketch showing the hyperbola, the directrices, and the foci. Also choose any three points P on the hyperbola and draw the line segments whose lengths are the undirected distances from P to a focus and its corresponding directrix. Observe that the ratio of these distances is e.

SOLUTION

(a) From the equation of the hyperbola, $a = 3$, and $b = 4$. In Example 1 of Section 10.6 we showed that $c = 5$. From (20), $e = \frac{5}{3}$. Because $a/e = 9/5$, we conclude from Theorem 10.8.2 that the directrix corresponding to the focus at $(5, 0)$ has the equation $x = \frac{9}{5}$, and the directrix corresponding to the focus at $(-5, 0)$ has the equation $x = -\frac{9}{5}$.

(b) Figure 10.8.7 shows the hyperbola, the directrices, and the foci, as well as three points P_1, P_2, and P_3 on the hyperbola. For each of these points

$$\frac{|\overline{FP}|}{|\overline{RP}|} = \frac{5}{3}$$

In the proof of Theorem 10.8.1 we learned that all three types of conics have polar equations of the same form. When a focus is at the pole and the corresponding directrix is either perpendicular or parallel to the polar axis, an equation of the conic has the form of (3), (12), or (13). Thus we have the following theorem.

10.8.3 THEOREM   Suppose we have a conic for which e and d are, respectively, the eccentricity and the undirected distance between the focus and the corresponding directrix.

(i) If a focus of the conic is at the pole and the corresponding directrix is perpendicular to the polar axis, then an equation of the conic is

$$r = \frac{ed}{1 \pm e \cos \theta} \tag{21}$$

where the plus sign is taken when the directrix corresponding to the focus at the pole is to the right of the focus and the minus sign is taken when it is to the left of the focus.

(ii) If a focus of the conic is at the pole and the corresponding directrix is parallel to the polar axis, then an equation of the conic is

$$r = \frac{ed}{1 \pm e \sin \theta} \tag{22}$$

where the plus sign is taken when the directrix corresponding to the focus at the pole is above the focus, and the minus sign is taken when it is below the focus.

EXAMPLE 3

A parabola has its focus at the pole and its vertex at $(4, \pi)$. Find an equation of the parabola and an equation of the directrix. Draw a sketch of the parabola and the directrix.

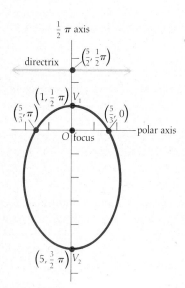

FIGURE 10.8.8

FIGURE 10.8.9

SOLUTION

Because the focus is at the pole and the vertex is at $(4, \pi)$, the polar axis and its extension are along the axis of the parabola. Furthermore, the vertex is to the left of the focus; so the directrix is also to the left of the focus. Hence an equation of the parabola is of the form of (21) with the minus sign. Because the vertex is at $(4, \pi)$, $\frac{1}{2}d = 4$; thus $d = 8$. The eccentricity $e = 1$, and therefore we obtain the equation

$$r = \frac{8}{1 - \cos \theta}$$

An equation of the directrix is given by $r \cos \theta = -d$, and because $d = 8$, then $r \cos \theta = -8$. Figure 10.8.8 shows a sketch of the parabola and the directrix.

EXAMPLE 4
An equation of a conic is

$$r = \frac{5}{3 + 2 \sin \theta}$$

Find the eccentricity, identify the conic, write an equation of the directrix corresponding to the focus at the pole, find the vertices, and draw a sketch of the curve.

SOLUTION
Dividing the numerator and denominator of the fraction in the given equation by 3 we obtain

$$r = \frac{\frac{5}{3}}{1 + \frac{2}{3} \sin \theta}$$

which is of the form of (22) with the plus sign. The eccentricity $e = \frac{2}{3}$. Because $e < 1$, the conic is an ellipse. Because $ed = \frac{5}{3}$, $d = \frac{5}{3} \div \frac{2}{3}$; thus $d = \frac{5}{2}$. The $\frac{1}{2}\pi$ axis and its extension are along the principal axis. The directrix corresponding to the focus at the pole is above the focus, and an equation of it is $r \sin \theta = \frac{5}{2}$. When $\theta = \frac{1}{2}\pi$, $r = 1$; and when $\theta = \frac{3}{2}\pi$, $r = 5$. The vertices are therefore at $(1, \frac{1}{2}\pi)$ and $(5, \frac{3}{2}\pi)$. A sketch of the ellipse appears in Fig. 10.8.9.

EXAMPLE 5
The polar axis and its extension are along the principal axis of a hyperbola having a focus at the pole. The corresponding directrix is to the left of the focus. If the hyperbola contains the point $(1, \frac{2}{3}\pi)$ and $e = 2$, find (a) an equation of the hyperbola; (b) the vertices; (c) the center; (d) an equation of the directrix corresponding to the focus at the pole. (e) Draw a sketch of the hyperbola.

SOLUTION

An equation of the hyperbola is of the form of (21) with the minus sign, where $e = 2$. We have, then,

$$r = \frac{2d}{1 - 2\cos\theta}$$

(a) Because the point $(1, \frac{2}{3}\pi)$ lies on the hyperbola, its coordinates satisfy the equation. Therefore

$$1 = \frac{2d}{1 - 2(-\frac{1}{2})}$$

from which we obtain $d = 1$. Hence an equation of the hyperbola is

$$r = \frac{2}{1 - 2\cos\theta} \tag{23}$$

(b) The vertices are the points on the hyperbola for which $\theta = 0$ and $\theta = \pi$. From (23), when $\theta = 0$, $r = -2$; and when $\theta = \pi$, $r = \frac{2}{3}$. Consequently, the left vertex V_1 is at the point $(-2, 0)$, and the right vertex V_2 is at the point $(\frac{2}{3}, \pi)$.

(c) The center of the hyperbola is the point on the principal axis halfway between the two vertices. This is the point $(\frac{4}{3}, \pi)$.

(d) An equation of the directrix corresponding to the focus at the pole is given by $r \cos\theta = -d$. Because $d = 1$, this equation is $r \cos\theta = -1$.

(e) As an aid in drawing a sketch of the hyperbola, we first draw the two asymptotes. These are lines through the center of the hyperbola that are parallel to the lines $\theta = \theta_1$ and $\theta = \theta_2$, where θ_1 and θ_2 are the values of θ in the interval $[0, 2\pi)$ for which r is not defined. From (23), r is not defined when $1 - 2\cos\theta = 0$. Therefore $\theta_1 = \frac{1}{3}\pi$ and $\theta_2 = \frac{5}{3}\pi$. Figure 10.8.10 shows a sketch of the hyperbola, as well as the two asymptotes and the directrix corresponding to the focus at the pole.

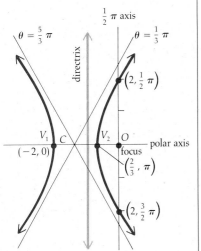

FIGURE 10.8.10

EXERCISES 10.8

In Exercises 1 through 8, (a) find the eccentricity, foci, and directrices of the given central conic. (b) Draw a sketch showing the conic, the foci, and the directrices. Also choose any four points P (one in each quadrant) on the conic and draw the line segments whose lengths are the undirected distances from P to a focus and its corresponding directrix. Observe that the ratio of these distances is e.

1. $4x^2 + 9y^2 = 36$ 2. $4x^2 + 9y^2 = 4$ 3. $25x^2 + 4y^2 = 100$ 4. $16x^2 + 9y^2 = 144$
5. $4x^2 - 25y^2 = 100$ 6. $x^2 - 9y^2 = 9$ 7. $16x^2 - 9y^2 = 144$ 8. $4y^2 - x^2 = 16$

In Exercises 9 and 10, each polar equation represents a conic having a focus at the pole. Identify the conic.

9. (a) $r = \dfrac{3}{1 - \cos\theta}$; (b) $r = \dfrac{6}{4 + 5\sin\theta}$; (c) $r = \dfrac{5}{4 - \cos\theta}$; (d) $r = \dfrac{4}{1 + \sin\pi}$

10. (a) $r = \dfrac{1}{1 - \sin\theta}$; (b) $r = \dfrac{2}{3 + \sin\theta}$; (c) $r = \dfrac{3}{2 + 4\cos\theta}$; (d) $r = \dfrac{5}{1 - \cos\pi}$

In Exercises 11 through 22, the equation is that of a conic having a focus at the pole. In each exercise, (a) find the eccentricity, (b) identify the conic, (c) write an equation of the directrix that corresponds to the focus at the pole, (d) draw a sketch of the curve.

11. $r = \dfrac{2}{1 - \cos \theta}$
 12. $r = \dfrac{4}{1 + \cos \theta}$
 13. $r = \dfrac{5}{2 + \sin \theta}$
 14. $r = \dfrac{4}{1 - 3 \cos \theta}$

15. $r = \dfrac{6}{3 - 2 \cos \theta}$
 16. $r = \dfrac{1}{2 + \sin \theta}$
 17. $r = \dfrac{9}{5 - 6 \sin \theta}$
 18. $r = \dfrac{1}{1 - 2 \sin \theta}$

19. $r = \dfrac{10}{7 - 2 \sin \theta}$
 20. $r = \dfrac{7}{3 + 4 \cos \theta}$
 21. $r = \dfrac{10}{4 + 5 \cos \theta}$
 22. $r = \dfrac{1}{5 - 3 \sin \theta}$

In Exercises 23 through 28, find a polar equation of the conic having a focus at the pole and satisfying the given conditions.

23. Parabola; vertex at $(4, \frac{3}{2}\pi)$.

24. Ellipse; $e = \frac{1}{2}$; a vertex at $(4, \pi)$.

25. Hyperbola; $e = \frac{4}{3}$; $r \cos \theta = 9$ is the directrix corresponding to the focus at the pole.

26. Hyperbola; vertices at $(1, \frac{1}{2}\pi)$ and $(3, \frac{1}{2}\pi)$.

27. Ellipse; vertices at $(3, 0)$ and $(1, \pi)$.

28. Parabola; vertex at $(6, \frac{1}{2}\pi)$.

29. (a) Find a polar equation of the hyperbola having a focus at the pole and the corresponding directrix to the left of the focus if the point $(2, \frac{4}{3}\pi)$ is on the hyperbola and $e = 3$. (b) Write an equation of the directrix that corresponds to the focus at the pole.

30. (a) Find a polar equation of the hyperbola for which $e = 3$ and which has the line $r \sin \theta = 3$ as the directrix corresponding to a focus at the pole. (b) Find the polar equations of the two lines through the pole that are parallel to the asymptotes of the hyperbola.

31. Find the area of the region inside the ellipse $r = 6/(2 - \sin \theta)$ and above the parabola $r = 3/(1 + \sin \theta)$.

32. For the ellipse and parabola of Exercise 31, find the area of the region inside the ellipse and below the parabola.

33. Show that an equation of a conic having its principal axis along the polar axis and its extension, a focus at the pole, and the corresponding directrix to the right of the focus is $r = ed/(1 + e \cos \theta)$.

34. Show that an equation of a conic having its principal axis along the $\frac{1}{2}\pi$ axis and its extension, a focus at the pole, and the corresponding directrix above the focus is $r = ed/(1 + e \sin \theta)$.

35. Show that an equation of a conic having its principal axis along the $\frac{1}{2}\pi$ axis and its extension, a focus at the pole, and the corresponding directrix below the focus is $r = ed/(1 - e \sin \theta)$.

36. Show that the equation $r = k \csc^2 \frac{1}{2}\theta$, where k is a constant, is a polar equation of a parabola.

37. A comet is moving in a parabolic orbit around the sun at the focus of the parabola. When the comet is 80 million miles from the sun, the line segment from the sun to the comet makes an angle of $\frac{1}{3}\pi$ radians with the axis of the orbit. (a) Find an equation of the comet's orbit. (b) How close does the comet come to the sun?

38. The orbit of a planet is in the form of an ellipse having the equation $r = p/(1 + e \cos \theta)$ where the pole is at the sun. Find the average measure of the distance of the planet from the sun with respect to θ.

39. Use Theorem 10.8.1 to find a polar equation of a central conic for which the center is at the pole, the principal axis is along the polar axis and its extension, and the distance from the pole to a directrix is a/e.

40. Show that the tangent lines at the points of intersection of the parabolas $r = a/(1 + \cos \theta)$ and $r = b/(1 - \cos \theta)$ are perpendicular.

10.9 (Supplementary) TANGENT LINES OF POLAR CURVES

Let $r = f(\theta)$ be a polar equation of a curve. We wish to derive a formula for finding the slope of a tangent line to a polar curve at a point (r, θ) on the curve. Consider a rectangular cartesian coordinate system and a polar coordinate system in the same plane and having the positive side of the x axis coincident with the polar axis. In Section 10.1 we learned that the two sets of coordinates are related by the equations

$$x = r \cos \theta \quad \text{and} \quad y = r \sin \theta$$

(a)

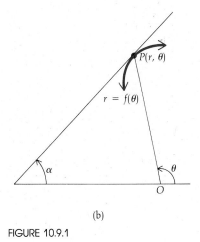

(b)

FIGURE 10.9.1

The variables x and y can be considered as functions of θ because $r = f(\theta)$. If we differentiate with respect to θ on both sides of these equations, we get, by applying the chain rule,

$$\frac{dx}{d\theta} = \cos\theta \frac{dr}{d\theta} - r\sin\theta \tag{1}$$

and

$$\frac{dy}{d\theta} = \sin\theta \frac{dr}{d\theta} + r\cos\theta \tag{2}$$

Let α be the radian measure of the inclination of the tangent line to the curve at (r, θ). See Fig. 10.9.1(a) and (b) where in (a) $\alpha \geq \theta$ and in (b) $\alpha < \theta$. In either case

$$\tan\alpha = \frac{dy}{dx}$$

and if $\dfrac{dx}{d\theta} \neq 0$,

$$\frac{dy}{dx} = \frac{\dfrac{dy}{d\theta}}{\dfrac{dx}{d\theta}} \tag{3}$$

Substituting from (1) and (2) into (3) we obtain

$$\frac{dy}{dx} = \frac{\sin\theta \dfrac{dr}{d\theta} + r\cos\theta}{\cos\theta \dfrac{dr}{d\theta} - r\sin\theta} \tag{4}$$

If $\cos\theta \neq 0$, we divide the numerator and the denominator of the right side of (4) by $\cos\theta$ and replace $\dfrac{dy}{dx}$ by $\tan\alpha$, thereby giving

$$\tan\alpha = \frac{\tan\theta \dfrac{dr}{d\theta} + r}{\dfrac{dr}{d\theta} - r\tan\theta} \tag{5}$$

If in (4), $\cos\theta = 0$, we obtain

$$\tan\alpha = -\frac{1}{r}\frac{dr}{d\theta} \tag{6}$$

which holds for points on the $\frac{1}{2}\pi$ axis or its extension.

Observe from (5) that if $r = 0$ and $\dfrac{dr}{d\theta} \neq 0$,

$$\tan\alpha = \tan\theta$$

From this equation we conclude that the values of θ that satisfy a polar equation of the curve when $r = 0$ are the inclinations of the tangent lines to the curve at the pole. Thus if $\theta_1, \theta_2, \ldots, \theta_k$ are these values of θ, then equations of the tangent lines to the curve at the pole are

$$\theta = \theta_1, \theta = \theta_2, \ldots, \theta = \theta_k$$

In Section 10.2 we used this information as an aid in drawing sketches of graphs of polar curves.

EXAMPLE 1

Find the slope of the tangent line to the curve in Example 3(c) of Section 10.2 at the point where (a) $\theta = 0$ and (b) $\theta = \frac{5}{6}\pi$.

SOLUTION

An equation of the curve is

$$r = 2 - \sin \theta \tag{7}$$

Thus

$$\frac{dr}{d\theta} = -\cos \theta$$

If α is the radian measure of the angle of inclination of the tangent line at (r, θ), then from (5),

$$\tan \alpha = \frac{\dfrac{\sin \theta}{\cos \theta}(-\cos \theta) + (2 - \sin \theta)}{-\cos \theta - (2 - \sin \theta)\dfrac{\sin \theta}{\cos \theta}}$$

$$= \frac{-\sin \theta \cos \theta + 2\cos \theta - \sin \theta \cos \theta}{-\cos^2 \theta - 2\sin \theta + \sin^2 \theta}$$

$$= \frac{-2\sin \theta \cos \theta + 2\cos \theta}{-1 + \sin^2 \theta - 2\sin \theta + \sin^2 \theta}$$

$$= \frac{2\cos \theta(1 - \sin \theta)}{2\sin^2 \theta - 2\sin \theta - 1} \tag{8}$$

(a) At $\theta = 0$, $r = 2$, and from (8),

$$\tan \alpha = \frac{2(1)(1 - 0)}{2(0) - 2(0) - 1}$$

$$= -2$$

(b) At $\theta = \frac{5}{6}\pi$, $r = \frac{3}{2}$, and

$$\tan \alpha = \frac{2\left(-\dfrac{\sqrt{3}}{2}\right)\left(1 - \dfrac{1}{2}\right)}{2(\frac{1}{4}) - 2(\frac{1}{2}) - 1}$$

$$= \frac{-\sqrt{3}(\frac{1}{2})}{\frac{1}{2} - 1 - 1}$$

$$= \frac{\sqrt{3}}{3}$$

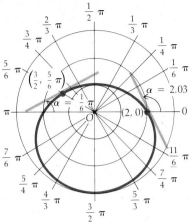

FIGURE 10.9.2

Figure 10.9.2 shows a sketch of the graph of (7) and the tangent lines at $\theta = 0$ and $\theta = \frac{5}{6}\pi$. At $\theta = 0$, because $\tan \alpha = -2$, $\alpha \approx 2.03$; at $\theta = \frac{5}{6}\pi$, because $\tan \alpha = \frac{1}{3}\sqrt{3}$, $\alpha = \frac{1}{6}\pi$.

Formula (5) can be used to determine where a polar curve has horizontal and vertical tangent lines. This information is helpful when sketching graphs of polar equations. The procedure is shown in the following illustration and next example applied to two of the limaçons in Example 3 of Section 10.2.

▶ ILLUSTRATION 1   The curve in Example 3(c) of Section 10.2 and Example 1 of this section has the equation

$$r = 2 - \sin \theta$$

The horizontal tangent lines of this curve occur when $\tan \alpha = 0$. Thus we equate the numerator of (8) to zero and solve for θ.

$$2 \cos \theta (1 - \sin \theta) = 0$$

$$\cos \theta = 0 \qquad\qquad 1 - \sin \theta = 0$$

$$\theta = \tfrac{1}{2}\pi \qquad \theta = \tfrac{3}{2}\pi \qquad\qquad \sin \theta = 1$$

$$\theta = \tfrac{1}{2}\pi$$

Therefore the curve has a horizontal tangent line at the points $(1, \frac{1}{2}\pi)$ and $(3, \frac{3}{2}\pi)$. The vertical tangent lines occur when the denominator of (8) is zero and the numerator is not zero. We solve the resulting equation.

$$2 \sin^2 \theta - 2 \sin \theta - 1 = 0$$

$$\sin \theta = \frac{-b \pm \sqrt{b^2 - 4ac}}{2a}$$

$$= \frac{2 \pm \sqrt{4 + 8}}{4}$$

$$= \frac{2 \pm 2\sqrt{3}}{4}$$

$$= \frac{1 \pm \sqrt{3}}{2}$$

$$\sin \theta = 1.3660 \qquad\qquad \sin \theta = -0.3660$$

$$\text{no solution} \qquad \theta \approx 3.51 \qquad \theta \approx 5.91$$

Hence the curve has a vertical tangent line at the points having polar coordinates $(2.37, 3.51)$ and $(2.37, 5.91)$. Figure 10.9.3 shows a sketch of the graph of the curve and the horizontal and vertical tangent lines. ◀

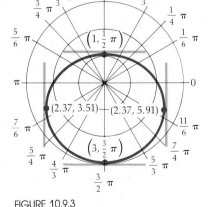

FIGURE 10.9.3

EXAMPLE 2

Determine the points at which the curve in Example 3(a) of Section 10.2 has horizontal and vertical tangent lines. Draw a sketch of the graph and show these tangent lines.

SOLUTION

The curve has the equation

$$r = 3 + 2 \sin \theta$$

Therefore

$$\frac{dr}{d\theta} = 2 \cos \theta$$

If α is the radian measure of the angle of inclination of the tangent line at (r, θ), we have, from (5),

$$\tan \alpha = \frac{\dfrac{\sin \theta}{\cos \theta}(2 \cos \theta) + (3 + 2 \sin \theta)}{2 \cos \theta - (3 + 2 \sin \theta)\dfrac{\sin \theta}{\cos \theta}}$$

$$= \frac{2 \sin \theta \cos \theta + 3 \cos \theta + 2 \sin \theta \cos \theta}{2 \cos^2 \theta - 3 \sin \theta - 2 \sin^2 \theta}$$

$$= \frac{4 \sin \theta \cos \theta + 3 \cos \theta}{2 - 2 \sin^2 \theta - 3 \sin \theta - 2 \sin^2 \theta}$$

$$= -\frac{\cos \theta(4 \sin \theta + 3)}{4 \sin^2 \theta + 3 \sin \theta - 2} \qquad (9)$$

To determine the horizontal tangent lines we equate the numerator of (9) to zero.

$$\cos \theta(4 \sin \theta + 3) = 0$$

$$\cos \theta = 0 \qquad\qquad 4 \sin \theta + 3 = 0$$

$$\theta = \tfrac{1}{2}\pi \qquad \theta = \tfrac{3}{2}\pi \qquad\qquad \sin \theta = -\tfrac{3}{4}$$

$$\theta \approx 3.99 \qquad \theta \approx 5.44$$

Thus the curve has a horizontal tangent line at the points $(5, \tfrac{1}{2}\pi)$, $(1, \tfrac{3}{2}\pi)$, $(\tfrac{3}{2}, 3.99)$ and $(\tfrac{3}{2}, 5.44)$.

The vertical tangent lines are found by equating the denominator of (9) to zero.

$$4 \sin^2 \theta + 3 \sin \theta - 2 = 0$$

$$\sin \theta = \frac{-b \pm \sqrt{b^2 - 4ac}}{2a}$$

$$= \frac{-3 \pm \sqrt{41}}{8}$$

$$\sin \theta = 0.4254 \qquad\qquad \sin \theta = -1.1754$$

$$\theta \approx 0.44 \qquad \theta \approx 2.70 \qquad \text{no solution}$$

FIGURE 10.9.4

FIGURE 10.9.5

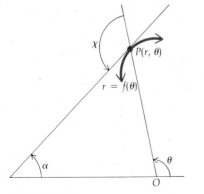

FIGURE 10.9.6

Therefore the curve has a vertical tangent line at the points with polar coordinates $(3.85, 0.44)$ and $(3.85, 2.70)$.

Refer to Fig. 10.9.4, which shows a sketch of the graph of the curve and the horizontal and vertical tangent lines.

Observe from Fig. 10.9.4 that the limaçon has a dent. This fact is apparent because of the four horizontal tangent lines. The limaçon of Fig. 10.9.3 has no dent; it has only two horizontal tangent lines.

Formula (5) is generally complicated to apply. A simpler formula is obtained by considering the angle between the line OP and the tangent line. This angle will have radian measure χ and is measured from the line OP counterclockwise to the tangent line, $0 \leq \chi < \pi$.

There are two possible cases: $\alpha \geq \theta$ and $\alpha < \theta$. These two cases are illustrated in Figs. 10.9.5 and 10.9.6. In Fig. 10.9.5, $\alpha > \theta$ and $\chi = \alpha - \theta$. In Fig. 10.9.6, $\alpha < \theta$ and $\chi = \pi - (\theta - \alpha)$. In each case

$$\tan \chi = \tan(\alpha - \theta)$$

$$\tan \chi = \frac{\tan \alpha - \tan \theta}{1 + \tan \alpha \tan \theta} \tag{10}$$

Substituting the value of $\tan \alpha$ from (5) into (10) we get

$$\tan \chi = \frac{\dfrac{\tan \theta \dfrac{dr}{d\theta} + r}{\dfrac{dr}{d\theta} - r \tan \theta} - \tan \theta}{1 + \left(\dfrac{\tan \theta \dfrac{dr}{d\theta} + r}{\dfrac{dr}{d\theta} - r \tan \theta}\right) \tan \theta}$$

$$= \frac{\tan \theta \dfrac{dr}{d\theta} + r - \tan \theta \dfrac{dr}{d\theta} + r \tan^2 \theta}{\dfrac{dr}{d\theta} - r \tan \theta + \tan^2 \theta \dfrac{dr}{d\theta} + r \tan \theta}$$

$$= \frac{r(1 + \tan^2 \theta)}{(1 + \tan^2 \theta)\dfrac{dr}{d\theta}}$$

$$= \frac{r}{\dfrac{dr}{d\theta}}$$

We have proved the following theorem.

10.9.1 THEOREM Let χ be the radian measure of the angle between the line OP and the tangent line to the graph of $r = f(\theta)$ at the point $P(r, \theta)$, where P is not on the $\frac{1}{2}\pi$ axis or its extension. The angle χ is measured from OP counterclockwise to the tangent line, and $0 \leq \chi < \pi$. Then

$$\tan \chi = \frac{r}{\dfrac{dr}{d\theta}} \tag{11}$$

Comparing (11) with (5) you can see why it is more desirable to consider χ instead of α when working with polar coordinates. Formula (11) cannot be used if P is on the $\frac{1}{2}\pi$ axis or its extension. In that case use (6) to compute $\tan \alpha$.

EXAMPLE 3
Given the cardioid having the equation

$$r = 2 + 2 \sin \theta \tag{12}$$

Find χ and α at the points where (a) $\theta = \frac{1}{6}\pi$, (b) $\theta = \frac{1}{3}\pi$, and (c) $\theta = \frac{2}{3}\pi$. (d) Draw a sketch of the cardioid, and show χ and α at each of these points.

SOLUTION
From (12),

$$\frac{dr}{d\theta} = 2 \cos \theta$$

From Theorem 10.9.1,

$$\tan \chi = \frac{2 + 2 \sin \theta}{2 \cos \theta}$$

$$= \frac{1 + \sin \theta}{\cos \theta} \tag{13}$$

(a) When $\theta = \frac{1}{6}\pi$, from (13) we get

$$\tan \chi = \frac{1 + \frac{1}{2}}{\dfrac{\sqrt{3}}{2}}$$

$$= \sqrt{3}$$

Thus $\chi = \frac{1}{3}\pi$. Because $\alpha > 0$, $\chi = \alpha - \theta$. Therefore

$$\alpha = \frac{1}{6}\pi + \frac{1}{3}\pi$$

$$= \frac{1}{2}\pi$$

(b) When $\theta = \frac{1}{3}\pi$, from (13)

$$\tan \chi = \frac{1 + \dfrac{\sqrt{3}}{2}}{\frac{1}{2}}$$

$$= 2 + \sqrt{3}$$

Therefore $\chi \approx 1.31$. Because $\alpha > 0$, $\chi = \alpha - \theta$. Thus

$$\alpha \approx \tfrac{1}{3}\pi + 1.31$$
$$\approx 2.36$$

(c) When $\theta = \tfrac{2}{3}\pi$, from (13)

$$\tan \chi = \frac{1 + \dfrac{\sqrt{3}}{2}}{-\dfrac{1}{2}}$$
$$= -2 - \sqrt{3}$$

Hence $\chi \approx 1.83$. Because $\alpha < 0$, $\chi = \pi - (\theta - \alpha)$. Therefore

$$\alpha \approx (\tfrac{2}{3}\pi + 1.83) - \pi$$
$$= 1.83 - \tfrac{1}{3}\pi$$
$$\approx 0.78$$

(d) Figure 10.9.7 shows a sketch of the cardioid and χ and α at each of the points $(3, \tfrac{1}{6}\pi)$, $(2 + \sqrt{3}, \tfrac{1}{3}\pi)$ and $(2 + \sqrt{3}, \tfrac{2}{3}\pi)$.

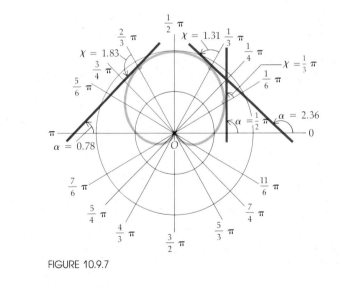

FIGURE 10.9.7

EXAMPLE 4

Find to the nearest 10 min the measurement of the smaller angle between the tangent lines to the curves

$$r = 3 \cos 2\theta \quad \text{and} \quad r = 3 \sin 2\theta$$

at the point $P(\tfrac{3}{2}\sqrt{2}, \tfrac{1}{8}\pi)$.

SOLUTION

The graph of each equation is a four-leafed rose. Refer to Fig. 10.9.8, which shows a portion of each of the curves at the point P; χ_1 is the radian measure of the angle between the line OP and the tangent line to the curve $r = 3 \cos 2\theta$, and χ_2 is the radian measure of the angle between the line OP and the tangent line to the curve $r = 3 \sin 2\theta$. If β is the angle between the tangent lines at P, then

$$\beta = \chi_1 - \chi_2$$

$$\tan \beta = \tan(\chi_1 - \chi_2)$$

$$\tan \beta = \frac{\tan \chi_1 - \tan \chi_2}{1 + \tan \chi_1 \tan \chi_2} \tag{14}$$

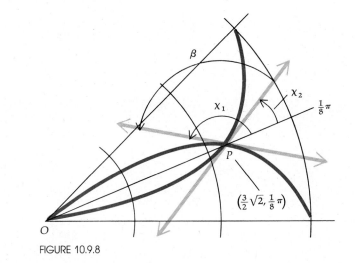

FIGURE 10.9.8

The values of $\tan \chi_1$ and $\tan \chi_2$ are found from Theorem 10.9.1. When computing $\tan \chi_1$, $r = 3 \cos 2\theta$ and $dr/d\theta = -6 \sin 2\theta$. For $\tan \chi_2$ we have $r = 3 \sin 2\theta$ and $dr/d\theta = 6 \cos 2\theta$. Thus

$$\tan \chi_1 = \frac{3 \cos 2\theta}{-6 \sin 2\theta} \qquad \tan \chi_2 = \frac{3 \sin 2\theta}{6 \cos 2\theta}$$

$$= -\tfrac{1}{2} \cot 2\theta \qquad = \tfrac{1}{2} \tan 2\theta$$

When $\theta = \tfrac{1}{8}\pi$,

$$\tan \chi_1 = -\tfrac{1}{2} \cot \tfrac{1}{4}\pi \qquad \tan \chi_2 = \tfrac{1}{2} \tan \tfrac{1}{4}\pi$$

$$= -\tfrac{1}{2} \qquad\qquad = \tfrac{1}{2}$$

Substituting these values in (14) we obtain

$$\tan \beta = \frac{-\tfrac{1}{2} - \tfrac{1}{2}}{1 + (-\tfrac{1}{2})(\tfrac{1}{2})}$$

$$= -\tfrac{4}{3}$$

The angle β has a measurement of $126°50'$ to the nearest 10 min. So, to the nearest 10 min, the measurement of the smaller angle between the two tangent lines is $180° - 126°50' = 53°10'$.

EXERCISES 10.9

In Exercises 1 through 8, find the slope of the tangent line to the given curve at the point having the indicated value of θ. Also draw a sketch of the graph and the tangent line.

1. $r = 2 \sin \theta; \theta = \frac{1}{6}\pi$ **2.** $r = 4 \cos \theta; \theta = \frac{1}{3}\pi$ **3.** $r = 1 - \cos \theta; \theta = \frac{1}{4}\pi$ **4.** $r = 2 + 2 \sin \theta; \theta = \frac{1}{4}\pi$
5. $r = 6 + 2 \sin \theta; \theta = \frac{5}{6}\pi$ **6.** $r = 3 - 2 \cos \theta; \theta = \frac{2}{3}\pi$ **7.** $r = 3 \cos 2\theta; \theta = \frac{2}{3}\pi$ **8.** $r = 2 \sin 3\theta; \theta = \frac{5}{6}\pi$

In Exercises 9 through 20, determine the points at which the given curve has horizontal and vertical tangent lines. Draw a sketch of the graph and show these tangent lines.

9. $r = 4 + 3 \sin \theta$ **10.** $r = 2 + \cos \theta$ **11.** $r = 4 - 2 \cos \theta$ **12.** $r = 3 - 2 \sin \theta$
13. $r = 1 - \sin \theta$ **14.** $r = 3 - 3 \cos \theta$ **15.** $r = 2 + 3 \cos \theta$ **16.** $r = 1 - 2 \sin \theta$
17. $r = \cos 2\theta$ **18.** $r = 2 \sin 3\theta$ **19.** $r^2 = 4 \sin 2\theta$ **20.** $r^2 = 9 \cos 2\theta$

In Exercises 21 through 28, for the given curve find χ and α at the point having the indicated value of θ. Draw a sketch of the curve and show χ and α.

21. $r\theta = 4; \theta = 1$ **22.** $r = 2\theta; \theta = \frac{5}{2}\pi$ **23.** $r = \sec 2\theta; \theta = -\frac{1}{8}\pi$ **24.** $r = 4 \sin \frac{1}{2}\theta; \theta = \frac{1}{3}\pi$
25. $r = \theta^2; \theta = \frac{1}{2}\pi$ **26.** $r = \cos 2\theta; \theta = \frac{1}{12}\pi$ **27.** $r = 3\sqrt{\cos 2\theta}; \theta = \frac{1}{6}\pi$ **28.** $r = 2(1 - \sin \theta); \theta = \pi$

In Exercises 29 through 32, find the measurement, in degrees, of the smaller angle between the tangent lines of the given pairs of curves at the indicated point of intersection.

29. $\begin{cases} r = 2 \cos \theta \\ r = 2 \sin \theta \end{cases}; (\sqrt{2}, \frac{1}{4}\pi)$

30. $\begin{cases} r = 3 \\ r = 6 \sin \theta \end{cases}; (3, \frac{1}{6}\pi)$

31. $\begin{cases} r = 4 \cos \theta \\ r = 4 \cos^2 \theta - 3 \end{cases}; (-2, \frac{2}{3}\pi)$

32. $\begin{cases} r = -4 \sin \theta \\ r = 4 \cos 2\theta \end{cases}$; the pole

In Exercises 33 through 36, find the measurement, in degrees, of the angle between the tangent lines of the given pair of curves at all points of intersection.

33. $\begin{cases} r = 1 - \sin \theta \\ r = 1 + \sin \theta \end{cases}$ **34.** $\begin{cases} r = 2 \sec \theta \\ r = \csc^2 \frac{1}{2}\theta \end{cases}$ **35.** $\begin{cases} r = \cos \theta \\ r = \sin 2\theta \end{cases}$ **36.** $\begin{cases} r = 3 \cos \theta \\ r = 1 + \cos \theta \end{cases}$

37. Given the curve $r = k$, where k is a nonzero constant and α is the radian measure of the inclination of the tangent line to the curve at the point (r, θ). Show that $\tan \alpha \tan \theta = -1$ for all θ for which $\tan \theta \neq 0$.

38. Prove that at each point of the logarithmic spiral $r = be^{a\theta}$, χ is the same.

39. Prove that $\tan \chi = \tan \frac{1}{2}\theta$ at all points of the cardioid $r = 2(1 - \cos \theta)$.

40. For the curve $r \cos \theta = 4$ find $\lim_{\theta \to \pi/2} \chi$.

41. Prove that at the points of intersection of the cardioids $r = a(1 + \sin \theta)$ and $r = b(1 - \sin \theta)$ their tangent lines are perpendicular for all values of a and b.

42. Prove that at the points of intersection of the two curves $r = a \sec^2 \frac{1}{2}\theta$ and $r = b \csc^2 \frac{1}{2}\theta$ their tangent lines are perpendicular.

REVIEW EXERCISES FOR CHAPTER 10

In Exercises 1 through 4, find a polar equation of the graph having the given cartesian equation.

1. $4x^2 - 9y^2 = 36$
2. $2xy = 1$
3. $x^2 + y^2 - 9x + 8y = 0$
4. $y^4 = x^2(a^2 - y^2)$

In Exercises 5 through 8, find a cartesian equation of the graph having the given polar equation.

5. $r = 9 \sin^2 \frac{1}{2}\theta$
6. $r(1 - \cos \theta) = 2$
7. $r = 4 \cos 3\theta$
8. $r = a \tan^2 \theta$

9. Draw a sketch of the graph of (a) $r\theta = 3$ (reciprocal spiral) and (b) $3r = \theta$ (spiral of Archimedes).

10. Show that the equations $r = 1 + \sin \theta$ and $r = \sin \theta - 1$ have the same graph.

11. Draw a sketch of the graph of $r = \sqrt{|\cos \theta|}$.
12. Draw a sketch of the graph of $r = \sqrt{|\cos 2\theta|}$.

13. Draw a sketch of the graph of $r = 2(1 + \cos \theta)$
14. Draw a sketch of the graph of $r^2 = 16 \cos \theta$.

In Exercises 15 and 16, find all the points of intersection of the graphs of the two given equations.

15. $\begin{cases} r = 2(1 + \cos \theta) \\ r^2 = 2 \cos \theta \end{cases}$

16. $\begin{cases} r \cos \theta = 1 \\ r = 1 + 2 \cos \theta \end{cases}$

17. Find the area of the region enclosed by (a) the loop of the limaçon $r = 4(1 + 2 \cos \theta)$ and (b) the outer part of the limaçon.

18. Find the area of one leaf of the rose $r = 2 \sin 3\theta$.

19. Find the area of the region inside the graph of $r = 2a \sin \theta$ and outside the graph of $r = a$.

20. Find the area of the region inside the graph of the lemniscate $r^2 = 2 \sin 2\theta$ and outside the graph of the circle $r = 1$.

21. Find the area of the region swept out by the radius vector of the logarithmic spiral $r = e^{k\theta}$ $(k > 0)$ as θ varies from 0 to 2π.

22. Find the area of the intersection of the regions enclosed by the graphs of the two equations $r = a \cos \theta$ and $r = a(1 - \cos \theta)$, $a > 0$.

23. Prove that the distance between the two points $P_1(r_1, \theta_1)$ and $P_2(r_2, \theta_2)$ is $\sqrt{r_1^2 + r_2^2 - 2r_1 r_2 \cos(\theta_2 - \theta_1)}$.

24. Find the points of intersection of the graphs of the equations $r = \tan \theta$ and $r = \cot \theta$.

25. Find a polar equation of the circle having its center at (r_0, θ_0) and a radius of a units. (*Hint:* Apply the law of cosines to the triangle having vertices at the pole, (r_0, θ_0) and (r, θ).)

26. Find the area enclosed by one loop of the curve $r = a \sin n\theta$, where n is a positive integer.

In Exercises 27 and 28, (a) find an equation of the parabola having the given focus and directrix, (b) find the length of the latus rectum, and (c) draw a sketch of the curve.

27. Focus at $(0, -3)$; directrix, $y = 3$.
28. Focus at $(6, 0)$; directrix, $x = -6$.

29. Draw a sketch of the parabola having the equation $x^2 - 6y = 0$, and find (a) the focus and (b) an equation of the directix.

30. Draw a sketch of the parabola having the equation $y = x^2 - 8x$. Find (a) the vertex, (b) an equation of the axis, (c) the focus, and (d) an equation of the directrix.

31. Given the parabola having the equation $y^2 + 8x - 6y - 7 = 0$, find (a) the vertex, (b) an equation of the axis, (c) the focus, and (d) an equation of the directrix. Draw a sketch of the curve.

32. Find an equation of the parabola having its vertex at $(-3, 5)$ and its focus at $(-3, -1)$.

33. Find an equation of the parabola having vertex at $(5, 1)$, axis parallel to the y axis, and through the point $(9, 3)$.

34. Show that any equation of the form $xy + ax + by + c = 0$ can always be written in the form $x'y' = k$ by a translation of the axes, and determine the value of k.

35. Any section of a parabolic mirror made by passing a plane through the axis of the mirror is a segment of a parabola. The altitude of the segment is 12 cm and the length of the base is 18 cm. A section of the mirror made by a plane perpendicular to its axis is a circle. Find the circumference of the circular plane section if the plane perpendicular to the axis is 3 cm from the vertex.

36. The directrix of the parabola $y^2 = 4px$ is tangent to a circle having the focus of the parabola as its center. Find an equation of the circle and the points of intersection of the two curves.

In Exercises 37 through 44, the graph of the given equation is either an ellipse or a hyperbola. Find the center, vertices, and foci of the conic. If the graph is an ellipse, also find the extremities of the minor axis. If the graph is a hyperbola, also find equations of the asymptotes. Draw a sketch of the curve and show the foci. If the curve is a hyperbola, also draw the asymptotes.

37. $4x^2 + 25y^2 = 100$ **38.** $25x^2 - 4y^2 = 100$ **39.** $16x^2 - 9y^2 = 144$ **40.** $9x^2 + 16y^2 = 144$

41. $4x^2 + y^2 + 24x - 16y + 84 = 0$ **42.** $4x^2 + 9y^2 + 32x - 18y + 37 = 0$

43. $25x^2 - y^2 + 50x + 6y - 9 = 0$ **44.** $3x^2 - 2y^2 + 6x - 8y + 11 = 0$

In Exercises 45 and 46, find an equation of the ellipse having the given properties and draw a sketch of the graph.

45. Vertices at $(0, 8)$ and $(0, -4)$ and one focus at $(0, 6)$.

46. Foci at $(-3, -2)$ and $(-3, 6)$ and the length of the semiminor axis is three-fourths the length of the semimajor axis.

In Exercises 47 and 48, find an equation of the hyperbola having the given properties, and draw a sketch of the graph.

47. Foci at $(-5, 1)$ and $(1, 1)$ and one vertex at $(-4, 1)$.

48. Center at $(-5, 2)$, a vertex at $(-1, 2)$, and a focus at $(0, 2)$.

In Exercises 49 and 50, (a) find the eccentricity, foci, and directrices of the given central conic. (b) Draw a sketch showing the conic, the foci, and the directrices.

49. $25x^2 - 9y^2 = 225$ **50.** $4x^2 + y^2 = 64$

In Exercises 51 through 54, the equation is that of a conic having a focus at the pole. In each exercise, (a) find the eccentricity; (b) identify the conic; (c) write an equation of the directrix that corresponds to the focus at the pole; (d) draw a sketch of the curve.

51. $r = \dfrac{2}{2 - \sin \theta}$ **52.** $r = \dfrac{5}{3 + 3 \sin \theta}$ **53.** $r = \dfrac{4}{2 + 3 \cos \theta}$ **54.** $r = \dfrac{4}{3 - 2 \cos \theta}$

In Exercises 55 through 58, find a polar equation of the conic satisfying the given conditions, and draw a sketch of the graph.

55. A focus at the pole; vertices at $(2, \pi)$ and $(4, \pi)$.   **56.** A focus at the pole; a vertex at $(6, \frac{1}{2}\pi)$; $e = \frac{3}{4}$.

57. A focus at the pole; a vertex at $(3, \frac{3}{2}\pi)$; $e = 1$.

58. The line $r \sin \theta = 6$ is the directrix corresponding to the focus at the pole and $e = \frac{5}{3}$.

In Exercises 59 and 60, simplify the given equation by a rotation and translation of axes. Draw a sketch of the graph and show the three sets of axes.

59. $3x^2 - 3xy - y^2 - 6y = 0$ **60.** $4x^2 + 3xy + y^2 - 6x + 12y = 0$

61. Find a polar equation of the parabola containing the point $(2, \frac{1}{3}\pi)$, whose focus is at the pole and whose vertex is on the extension of the polar axis.

62. If the distance between the two directrices of an ellipse is three times the distance between the foci, find the eccentricity.

63. Find the volume of the solid of revolution generated if the region bounded by the hyperbola $x^2/a^2 - y^2/b^2 = 1$ and the line $x = 2a$ is revolved about the y axis.

64. Show that the hyperbola $x^2 - y^2 = 4$ has the same foci as the ellipse $x^2 + 9y^2 = 9$.

65. A satellite is traveling around the earth in an elliptical orbit having the earth at one focus and an eccentricity of $\frac{1}{3}$. The closest distance that the satellite gets to the earth is 300 mi. Find the farthest distance that the satellite gets from the earth.

66. The orbit of the planet Mercury around the sun is elliptical in shape with the sun at one focus, a semimajor axis of length 36 million miles, and an eccentricity of 0.206. Find (a) how close Mercury gets to the sun and (b) the greatest possible distance between Mercury and the sun.

67. A comet is moving in a parabolic orbit around the sun at the focus F of the parabola. An observation of the comet is made when it is at point P_1, 15 million miles from the sun, and a second observation is made when it is at point P_2, 5 million miles from the sun. The line segments FP_1 and FP_2 are perpendicular. With this information there are two possible orbits for the comet. Find how close the comet comes to the sun for each orbit.

68. The arch of a bridge is in the shape of a semiellipse having a horizontal span of 40 m and a height of 16 m at its center. How high is the arch 9 m to the right or left of the center?

69. Points A and B are 1000 m apart, and it is determined from the sound of an explosion heard at these points at different times that the location of the explosion is 600 m closer to A than to B. Show that the location of the explosion is restricted to a particular curve, and find an equation of it.

70. Find the area of the region bounded by the two parabolas $r = 2/(1 - \cos \theta)$ and $r = 2/(1 + \cos \theta)$.

71. Prove that the line $y = 2x$ is an asymptote of the graph of the function f for which $f(x) = 2\sqrt{x^2 - 1}$.

72. A focal chord of a conic is divided into two segments by the focus. Prove that the sum of the reciprocals of the measures of the lengths of the two segments is the same, regardless of what chord is taken. (*Hint:* Use polar coordinates.)

73. A focal chord of a conic is a line segment passing through a focus and having its endpoints on the conic. Prove that if two focal chords of a parabola are perpendicular, the sum of the reciprocals of the measures of their lengths is a constant. (*Hint:* Use polar coordinates.)

74. Prove that the midpoints of all chords parallel to a fixed chord of a parabola lie on a line that is parallel to the axis of the parabola.

75. Show that $\sqrt{x} \pm \sqrt{y} = \pm\sqrt{a}$ represents a one-parameter family of conics, and determine the type of conic. Draw sketches of the conics for $a = 1$, $a = 2$, and $a = 4$.

76. The graph of the equation $(1 - e^2)x^2 + y^2 - 2px + p^2 = 0$ is a central conic having eccentricity e and a focus at $(p, 0)$. Simplify the equation by a translation of axes. Find the new origin with respect to the x and y axes, and determine the type of conic.

Exercises 77 through 84 pertain to Supplementary Section 10.9.

77. Find the slope of the tangent line to the graph of $r = \frac{1}{2} + \sin \theta$ at the point $(1, \frac{1}{6}\pi)$. Draw a sketch of the graph and the tangent line.

78. Find the slope of the tangent line to the graph of $r = 6 \cos \theta - 2$ at the point $(1, \frac{5}{3}\pi)$. Draw a sketch of the graph and the tangent line.

79. Determine the points at which the curve $r = 3 - 3 \sin \theta$ has horizontal and vertical tangent lines. Draw a sketch of the graph and show these tangent lines.

80. Determine the points at which the curve $r = 4 + 3 \cos \theta$ has horizontal and vertical tangent lines. Draw a sketch of the graph and show these tangent lines.

81. For the curve $r = 2\sqrt{\cos \frac{1}{2}\theta}$, find χ and α at the point $(1, \frac{2}{3}\pi)$. Draw a sketch of the curve and show χ and α.

82. For the curve $r = 2(1 + \sin \theta)$, find χ and α at the point $(3, \frac{1}{6}\pi)$. Draw a sketch of the curve and show χ and α.

83. Find the radian measure of the angle between the tangent lines at each point of intersection of the graphs of the equations $r = -6 \cos \theta$ and $r = 2(1 - \cos \theta)$.

84. Prove that the graphs of $r = a\theta$ and $r\theta = a$ have an unlimited number of points of intersection. Also prove that the tangent lines are perpendicular at only two of these points of intersection, and find these points.

P A R T 2

INFINITE SERIES

The discussion of *infinite series* utilizes concepts of calculus from Part 1. However, no subject matter in Part 3 has a knowledge of infinite series as a prerequisite. Consequently, if so desired, Part 3 can be studied prior to Part 2.

In Part 2 we study another kind of function having a set of real numbers as its domain and range. A *sequence* is defined as a function whose domain is a set of integers, and an infinite series is defined as a particular kind of sequence. You will learn how to express as an infinite series many differentiable functions such as polynomial, trigonometric, exponential, and logarithmic functions. Such representations will enable you to find approximations of irrational numbers such as $\sqrt{2}$, π, e, ln 5, and sin 0.3. Another application of infinite series is to approximate definite integrals for which the integrand has no antiderivative that can be expressed in terms of elementary functions. For example, you will learn how to use infinite series to compute values of integrals such as $\int_0^{1/2} e^{-t^2}\, dt$, $\int_0^1 \cos \sqrt{x}\, dx$, and $\int_0^{0.1} \ln(1 + \sin x)\, dx$ to any required accuracy. Furthermore, solutions of many differential equations can be expressed as infinite series.

There are two chapters in Part 2. Chapter 11 is devoted to sequences and infinite series of constant terms, while Chapter 12 is concerned with infinite series of variable terms called *power series*. The treatment of infinite series here is as complete as is feasible in an elementary calculus text. In addition to the customary computational material, you will find the proof of the equivalence of convergence and boundedness of monotonic sequences (Theorems 11.2.6 and 11.2.9) based on the completeness property of the real numbers. The proofs (Theorems 12.2.3 and 12.3.1) of the computational processes involving differentiation and integration of power series are also included.

E L E V E N

SEQUENCES AND INFINITE SERIES OF CONSTANT TERMS

11.1
SEQUENCES

Sequences of numbers are often encountered in mathematics. For instance, the numbers

$$2, 4, 6, 8, 10$$

form a sequence. This sequence is said to be **finite** because there is a last number. If the set of numbers that forms a sequence does not have a last number, the sequence is said to be **infinite**. For instance, the sequence

$$\tfrac{1}{2}, \tfrac{2}{5}, \tfrac{3}{7}, \tfrac{4}{9}, \ldots \tag{1}$$

is infinite because the three dots with no number following indicate that there is no last number. We are concerned here with infinite sequences, and when the word "sequence" is used, it is understood that we are referring to an infinite sequence. We now give a formal definition of a sequence; it is a particular kind of function.

11.1.1 DEFINITION   A **sequence** is a function whose domain is the set

$$\{1, 2, 3, \ldots, n, \ldots\}$$

of all positive integers.

The numbers in the range of a sequence are called the **elements** of the sequence.

If the nth element is given by $f(n)$, then the sequence is the set of ordered pairs of the form $(n, f(n))$, where n is a positive integer.

▶ ILLUSTRATION 1   If $f(n) = n/(2n + 1)$, then

$$f(1) = \tfrac{1}{3} \qquad f(2) = \tfrac{2}{5} \qquad f(3) = \tfrac{3}{7} \qquad f(4) = \tfrac{4}{9}$$

and so on. The range of f consists of the elements of sequence (1). Some of the ordered pairs in the sequence f are $(1, \tfrac{1}{3})$, $(2, \tfrac{2}{5})$, $(3, \tfrac{3}{7})$, $(4, \tfrac{4}{9})$, and $(5, \tfrac{5}{11})$. A sketch of the graph of this sequence is shown in Fig. 11.1.1.   ◀

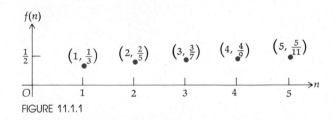

FIGURE 11.1.1

Usually the nth element $f(n)$ of the sequence is stated when the elements are listed in order. Thus the elements of sequence (1) can be written as

$$\frac{1}{3}, \frac{2}{5}, \frac{3}{7}, \frac{4}{9}, \ldots, \frac{n}{2n + 1}, \ldots$$

Because the domain of every sequence is the same, the notation $\{f(n)\}$ may be used to denote a sequence. So the sequence (1) can be denoted by $\{n/(2n + 1)\}$. The subscript notation $\{a_n\}$ is also used to denote the sequence for which $f(n) = a_n$.

A sequence

$$a_1, a_2, a_3, \ldots, a_n, \ldots$$

is said to be **equal** to a sequence

$$b_1, b_2, b_3, \ldots, b_n, \ldots$$

if and only if $a_i = b_i$ for every positive integer i. Remember that a sequence consists of an ordering of elements. Therefore it is possible for two sequences to have the same elements and be unequal. This situation is shown in the following illustration.

▶ ILLUSTRATION 2   The sequence $\{1/n\}$ has as its elements the reciprocals of the positive integers

$$1, \frac{1}{2}, \frac{1}{3}, \frac{1}{4}, \ldots, \frac{1}{n}, \ldots \tag{2}$$

The sequence for which

$$f(n) = \begin{cases} 1 & \text{if } n \text{ is odd} \\ \dfrac{2}{n+2} & \text{if } n \text{ is even} \end{cases}$$

has as its elements

$$1, \tfrac{1}{2}, 1, \tfrac{1}{3}, 1, \tfrac{1}{4}, \ldots \tag{3}$$

The elements of sequences (2) and (3) are the same; however, the sequences are different. Sketches of the graphs of sequences (2) and (3) are shown in Figs. 11.1.2 and 11.1.3, respectively.   ◀

FIGURE 11.1.2

FIGURE 11.1.3

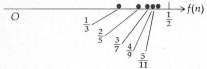

FIGURE 11.1.4

We now plot on a horizontal axis the points corresponding to successive elements of a sequence. This is done in Fig. 11.1.4 for sequence (1), which is $\{n/(2n + 1)\}$. Observe that the successive elements of the sequence get closer and closer to $\frac{1}{2}$, even though no element in the sequence has the value $\frac{1}{2}$. Intuitively we see that the element will be as close to $\frac{1}{2}$ as we please by taking the number of the element sufficiently large. Or stating this another way, $|n/(2n + 1) - \frac{1}{2}|$ can be made less than any given positive ϵ by taking n large enough. Because of this we state that the limit of the sequence $\{n/(2n + 1)\}$ is $\frac{1}{2}$.

In general, if there is a number L such that $|a_n - L|$ is arbitrarily small for n sufficiently large, the sequence $\{a_n\}$ is said to have the limit L. Following is the precise definition of the limit of a sequence.

11.1.2 DEFINITION   A sequence $\{a_n\}$ has the limit L if for any $\epsilon > 0$ there exists a number $N > 0$ such that if n is an integer and

$$\text{if } n > N \quad \text{then} \quad |a_n - L| < \epsilon$$

and we write

$$\lim_{n \to +\infty} a_n = L$$

EXAMPLE 1
Use Definition 11.1.2 to prove that the sequence

$$\left\{ \frac{n}{2n + 1} \right\}$$

has the limit $\frac{1}{2}$.

SOLUTION
We must show that for any $\epsilon > 0$ there exists a number $N > 0$ such that if n is an integer and

$$\text{if } n > N \quad \text{then} \quad \left| \frac{n}{2n + 1} - \frac{1}{2} \right| < \epsilon$$

$$\Leftrightarrow \quad \text{if } n > N \quad \text{then} \quad \left| \frac{2n - 2n - 1}{2(2n + 1)} \right| < \epsilon$$

$$\Leftrightarrow \quad \text{if } n > N \quad \text{then} \quad \left| \frac{-1}{2(2n + 1)} \right| < \epsilon$$

$$\Leftrightarrow \quad \text{if } n > N \quad \text{then} \quad \frac{1}{2(2n + 1)} < \epsilon$$

$$\Leftrightarrow \quad \text{if } n > N \quad \text{then} \quad 2n + 1 > \frac{1}{2\epsilon}$$

$$\Leftrightarrow \quad \text{if } n > N \quad \text{then} \quad n > \frac{1 - 2\epsilon}{4\epsilon} \tag{4}$$

For statement (4) to hold take $N = (1 - 2\epsilon)/(4\epsilon)$, and we have if n is an integer and

$$\text{if } n > \frac{1 - 2\epsilon}{4\epsilon} \quad \text{then} \quad \left| \frac{n}{2n + 1} - \frac{1}{2} \right| < \epsilon \tag{5}$$

Observe that, if in particular $\epsilon = \frac{1}{8}$, then $N = \frac{3}{2}$ and (5) becomes

$$\text{if } n > \tfrac{3}{2} \quad \text{then} \quad \left| \frac{n}{2n + 1} - \frac{1}{2} \right| < \frac{1}{8}$$

For instance, if $n = 4$,

$$\left| \frac{n}{2n + 1} - \frac{1}{2} \right| = \left| \frac{4}{9} - \frac{1}{2} \right|$$

$$= \frac{1}{18}$$

and $\frac{1}{18} < \frac{1}{8}$. Statement (5) proves that the given sequence has the limit $\frac{1}{2}$.

▶ ILLUSTRATION 3   Consider the sequence $\{(-1)^{n+1}/n\}$. Note that the nth element of this sequence is $(-1)^{n+1}/n$, and $(-1)^{n+1}$ is equal to $+1$ when n is odd and to -1 when n is even. Hence the elements of the sequence can be written

$$1, -\frac{1}{2}, \frac{1}{3}, -\frac{1}{4}, \frac{1}{5}, \dots, \frac{(-1)^{n+1}}{n}, \dots$$

In Fig. 11.1.5 are plotted points corresponding to successive elements of this sequence. In the figure, $a_1 = 1$, $a_2 = -\frac{1}{2}$, $a_3 = \frac{1}{3}$, $a_4 = -\frac{1}{4}$, $a_5 = \frac{1}{5}$, $a_6 = -\frac{1}{6}$, $a_7 = \frac{1}{7}$, $a_8 = -\frac{1}{8}$, $a_9 = \frac{1}{9}$, $a_{10} = -\frac{1}{10}$. The limit of the sequence is 0, and the elements oscillate about 0.   ◀

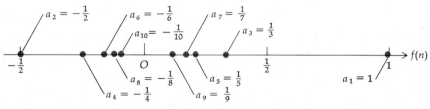

FIGURE 11.1.5

Compare Definition 11.1.2 with Definition 3.7.1 of the limit of $f(x)$ as x increases without bound. The two definitions are almost identical; however, when we state that $\lim\limits_{x \to +\infty} f(x) = L$, the function f is defined for all real numbers greater than some real number r, while when we consider $\lim\limits_{n \to +\infty} a_n$, n is restricted to positive integers. There is, however, the following theorem that follows immediately from Definition 3.7.1.

11.1.3 THEOREM   If $\lim\limits_{x \to +\infty} f(x) = L$, and f is defined for every positive integer, then also $\lim\limits_{n \to +\infty} f(n) = L$ when n is any positive integer.

The proof is left as an exercise (see Exercise 26).

▶ ILLUSTRATION 4   We verify Theorem 11.1.3 for the sequence of Example 1. For that sequence $f(n) = n/(2n + 1)$. Hence $f(x) = x/(2x + 1)$ and

$$\lim_{x \to +\infty} \frac{x}{2x + 1} = \lim_{x \to +\infty} \frac{1}{2 + \dfrac{1}{x}}$$

$$= \tfrac{1}{2}$$

It follows, then, from Theorem 11.1.3 that $\lim\limits_{n \to +\infty} f(n) = \tfrac{1}{2}$ when n is any positive integer. This agrees with the solution of Example 1.   ◀

11.1.4 DEFINITION   If a sequence $\{a_n\}$ has a limit, the sequence is said to be **convergent**, and a_n **converges** to that limit. If the sequence is not convergent, it is **divergent**.

EXAMPLE 2
Determine if the sequence

$$\left\{ \frac{4n^2}{2n^2 + 1} \right\}$$

is convergent or divergent.

SOLUTION
We wish to determine if $\lim\limits_{n \to +\infty} 4n^2/(2n^2 + 1)$ exists. Let $f(x) = 4x^2/(2x^2 + 1)$, and investigate $\lim\limits_{x \to +\infty} f(x)$.

$$\lim_{x \to +\infty} \frac{4x^2}{2x^2 + 1} = \lim_{x \to +\infty} \frac{4}{2 + \dfrac{1}{x^2}}$$

$$= 2$$

Therefore, by Theorem 11.1.3, $\lim\limits_{n \to +\infty} f(n) = 2$. Thus the given sequence is convergent and $4n^2/(2n^2 + 1)$ converges to 2.

EXAMPLE 3
Prove that if $|r| < 1$, the sequence $\{r^n\}$ is convergent and r^n converges to zero.

SOLUTION
First we consider $r = 0$. Then the sequence is $\{0\}$ and $\lim\limits_{n \to +\infty} 0 = 0$. Thus the sequence is convergent and the nth element converges to zero.

If $0 < |r| < 1$, we wish to show that Definition 11.1.2 holds with $L = 0$. Thus we wish to show that for any $\epsilon > 0$ there exists a number $N > 0$ such that if n is an integer and

$$\text{if } n > N \quad \text{then} \quad |r^n - 0| < \epsilon \tag{6}$$

$$\Leftrightarrow \quad \text{if } n > N \quad \text{then} \quad |r|^n < \epsilon$$

$$\Leftrightarrow \quad \text{if } n > N \quad \text{then} \quad \ln|r|^n < \ln \epsilon$$

$$\Leftrightarrow \quad \text{if } n > N \quad \text{then} \quad n \ln|r| < \ln \epsilon \tag{7}$$

Because $0 < |r| < 1$, $\ln|r| < 0$. Thus (7) is equivalent to

$$\text{if } n > N \quad \text{then} \quad n > \frac{\ln \epsilon}{\ln|r|}$$

Therefore, if $N = \ln \epsilon / \ln|r|$, we may conclude (6). Consequently, $\lim\limits_{n \to +\infty} r^n = 0$. Hence, by Definitions 11.1.2 and 11.1.4, $\{r^n\}$ is convergent and r^n converges to zero.

EXAMPLE 4
Determine if the sequence $\{(-1)^n + 1\}$ is convergent or divergent.

SOLUTION
The elements of this sequence are $0, 2, 0, 2, 0, 2, \ldots, (-1)^n + 1, \ldots$. Because $a_n = 0$ if n is odd and $a_n = 2$ if n is even, it appears that the sequence is divergent. To prove this, let us assume that the sequence is convergent and show that this assumption leads to a contradiction. If the sequence has the limit L, then by Definition 11.1.2 for any $\epsilon > 0$ there exists a number $N > 0$ such that if n is an integer and

$$\text{if } n > N \quad \text{then} \quad |a_n - L| < \epsilon$$

In particular, when $\epsilon = \frac{1}{2}$, there exists a number $N > 0$ such that if n is an integer and

$$\text{if } n > N \quad \text{then} \quad |a_n - L| < \tfrac{1}{2}$$

$$\Leftrightarrow \quad \text{if } n > N \quad \text{then} \quad -\tfrac{1}{2} < a_n - L < \tfrac{1}{2} \tag{8}$$

Because $a_n = 0$ if n is odd and $a_n = 2$ if n is even, it follows from (8) that

$$-\tfrac{1}{2} < -L < \tfrac{1}{2} \quad \text{and} \quad -\tfrac{1}{2} < 2 - L < \tfrac{1}{2}$$

But if $-L > -\frac{1}{2}$, then $2 - L > \frac{3}{2}$; hence $2 - L$ cannot be less than $\frac{1}{2}$. So there is a contradiction, and therefore the given sequence is divergent.

EXAMPLE 5
Determine if the sequence

$$\left\{ n \sin \frac{\pi}{n} \right\}$$

is convergent or divergent.

SOLUTION

We wish to determine if $\lim\limits_{n \to +\infty} n \sin(\pi/n)$ exists. Let $f(x) = x \sin(\pi/x)$ and investigate $\lim\limits_{x \to +\infty} f(x)$. Because $f(x)$ can be written as $[\sin(\pi/x)]/(1/x)$ and $\lim\limits_{x \to +\infty} \sin(\pi/x) = 0$ and $\lim\limits_{x \to +\infty} (1/x) = 0$, L'Hôpital's rule can be applied to obtain

$$\lim_{x \to +\infty} f(x) = \lim_{x \to +\infty} \frac{-\dfrac{\pi}{x^2} \cos \dfrac{\pi}{x}}{-\dfrac{1}{x^2}}$$

$$= \lim_{x \to +\infty} \pi \cos \frac{\pi}{x}$$

$$= \pi$$

Therefore $\lim\limits_{n \to +\infty} f(n) = \pi$ when n is a positive integer. So the given sequence is convergent and $n \sin(\pi/n)$ converges to π.

There are limit theorems for sequences that are analogous to limit theorems for functions given in Chapter 1. The statement of these theorems uses the terminology of sequences. The proofs are omitted because they are similar to the proofs of the corresponding theorems given in Chapter 1.

11.1.5 THEOREM   If $\{a_n\}$ and $\{b_n\}$ are convergent sequences and c is a constant, then

(i) the constant sequence $\{c\}$ has c as its limit;

(ii) $\lim\limits_{n \to +\infty} ca_n = c \lim\limits_{n \to +\infty} a_n$;

(iii) $\lim\limits_{n \to +\infty} (a_n \pm b_n) = \lim\limits_{n \to +\infty} a_n \pm \lim\limits_{n \to +\infty} b_n$;

(iv) $\lim\limits_{n \to +\infty} a_n b_n = \left(\lim\limits_{n \to +\infty} a_n \right)\left(\lim\limits_{n \to +\infty} b_n \right)$;

(v) $\lim\limits_{n \to +\infty} \dfrac{a_n}{b_n} = \dfrac{\lim\limits_{n \to +\infty} a_n}{\lim\limits_{n \to +\infty} b_n}$ if $\lim\limits_{n \to +\infty} b_n \neq 0$, and all $b_n \neq 0$.

EXAMPLE 6

Use Theorem 11.1.5 to prove that the sequence

$$\left\{ \frac{n^2}{2n + 1} \sin \frac{\pi}{n} \right\}$$

is convergent, and find the limit of the sequence.

SOLUTION

$$\frac{n^2}{2n+1} \sin \frac{\pi}{n} = \frac{n}{2n+1} \cdot n \sin \frac{\pi}{n}$$

In Example 1 the sequence $\{n/(2n+1)\}$ was shown to be convergent and $\lim_{n \to +\infty} [n/(2n+1)] = \frac{1}{2}$. In Example 5 we showed that the sequence $\{n \sin(\pi/n)\}$ is convergent and $\lim_{n \to +\infty} [n \sin(\pi/n)] = \pi$. Hence, by Theorem 11.1.5(iv),

$$\lim_{n \to +\infty} \left[\frac{n}{2n+1} \cdot n \sin \frac{\pi}{n} \right] = \lim_{n \to +\infty} \frac{n}{2n+1} \cdot \lim_{n \to +\infty} n \sin \frac{\pi}{n}$$

$$= \tfrac{1}{2} \cdot \pi$$

Thus the given sequence is convergent, and its limit is $\frac{1}{2}\pi$.

EXERCISES 11.1

In Exercises 1 through 19, write the first four elements of the sequence and determine if the sequence is convergent or divergent. If the sequence converges, find its limit.

1. $\left\{ \dfrac{n+1}{2n-1} \right\}$

2. $\left\{ \dfrac{2n^2+1}{3n^2-n} \right\}$

3. $\left\{ \dfrac{n^2+1}{n} \right\}$

4. $\left\{ \dfrac{3n^3+1}{2n^2+n} \right\}$

5. $\left\{ \dfrac{3-2n^2}{n^2-1} \right\}$

6. $\left\{ \dfrac{e^n}{n} \right\}$

7. $\left\{ \dfrac{\ln n}{n^2} \right\}$

8. $\left\{ \dfrac{\log_b n}{n} \right\}, b > 1$

9. $\{\tanh n\}$

10. $\{\sinh n\}$

11. $\left\{ \dfrac{n}{n+1} \sin \dfrac{n\pi}{2} \right\}$

12. $\left\{ \dfrac{\sinh n}{\sin n} \right\}$

13. $\left\{ \dfrac{1}{\sqrt{n^2+1}-n} \right\}$

14. $\{\sqrt{n+1}-\sqrt{n}\}$

15. $\left\{ \left(1+\dfrac{1}{3n}\right)^n \right\}$ (*Hint:* Use $\lim_{x \to 0} (1+x)^{1/x} = e$.)

16. $\left\{ \left(1+\dfrac{2}{n}\right)^n \right\}$ See Hint for Exercise 15.

17. $\{r^{1/n}\}$ and $r > 0$. (*Hint:* Consider two cases: $r \le 1$ and $r > 1$.)

18. $\{\cos n\pi\}$

19. $\left\{ \dfrac{n}{c^n} \right\}, c > 1$

In Exercises 20 through 25, use Definition 11.1.2 to prove that the given sequence has the limit L.

20. $\left\{ \dfrac{4}{2n-1} \right\}; L = 0$

21. $\left\{ \dfrac{3}{n-1} \right\}; L = 0$

22. $\left\{ \dfrac{1}{\sqrt{n}} \right\}; L = 0$

23. $\left\{ \dfrac{8n}{2n+3} \right\}; L = 4$

24. $\left\{ \dfrac{5-n}{2+3n} \right\}; L = -\tfrac{1}{3}$

25. $\left\{ \dfrac{2n^2}{5n^2+1} \right\}; L = \tfrac{2}{5}$

26. Prove Theorem 11.1.3.

27. Show that the sequences $\left\{\dfrac{n^2}{n-3}\right\}$ and $\left\{\dfrac{n^2}{n+4}\right\}$ are both divergent, but that the sequence $\left\{\dfrac{n^2}{n-3}-\dfrac{n^2}{n+4}\right\}$ is convergent.

28. Given the sequence

$$\left\{\frac{1-\left(1-\dfrac{1}{n}\right)^a}{1-\left(1-\dfrac{1}{n}\right)^b}\right\}, \; a \text{ and } b \text{ are constants and } b \neq 0$$

Determine if the sequence is convergent or divergent. If the sequence converges, find its limit.

29. Prove that if $|r| < 1$, the sequence $\{nr^n\}$ is convergent and nr^n converges to zero.

30. Prove that if the sequence $\{a_n\}$ converges, then $\lim\limits_{n \to +\infty} a_n$ is unique (*Hint:* Assume that $\lim\limits_{n \to +\infty} a_n$ has two different values, L and M, and show that this is impossible by taking $\epsilon = \frac{1}{2}|L - M|$ in Definition 11.1.2.)

31. Prove that if the sequence $\{a_n\}$ is convergent and $\lim\limits_{n \to +\infty} a_n = L$, then the sequence $\{a_n^2\}$ is also convergent and $\lim\limits_{n \to +\infty} a_n^2 = L^2$.

32. Prove that if the sequence $\{a_n\}$ is convergent and $\lim\limits_{n \to +\infty} a_n = L$, then the sequence $\{|a_n|\}$ is also convergent and $\lim\limits_{n \to +\infty} |a_n| = |L|$.

**11.2
MONOTONIC AND
BOUNDED SEQUENCES**

Certain kinds of sequences are given special names.

11.2.1 DEFINITION   A sequence $\{a_n\}$ is said to be

    (i) increasing if $a_n \leq a_{n+1}$ for all n;
    (ii) decreasing if $a_n \geq a_{n+1}$ for all n.

If a sequence is increasing or if it is decreasing, it is called **monotonic.**

    If $a_n < a_{n+1}$ (a special case of $a_n \leq a_{n+1}$), the sequence is **strictly increasing**; if $a_n > a_{n+1}$, the sequence is **strictly decreasing.**

EXAMPLE 1
For each of the following sequences, determine if it is increasing, decreasing, or not monotonic: (a) $\{n/(2n + 1)\}$; (b) $\{1/n\}$; (c) $\{(-1)^{n+1}/n\}$.

SOLUTION
(a) The elements of the sequence can be written

$$\frac{1}{3}, \frac{2}{5}, \frac{3}{7}, \frac{4}{9}, \dots, \frac{n}{2n+1}, \frac{n+1}{2n+3}, \dots$$

Note that a_{n+1} is obtained from a_n by replacing n by $n + 1$. Therefore, because $a_n = n/(2n + 1)$,

$$a_{n+1} = \frac{n + 1}{2(n + 1) + 1}$$

$$= \frac{n + 1}{2n + 3}$$

Look at the first four elements of the sequence and observe that the elements increase as n increases. Thus we suspect in general that

$$\frac{n}{2n + 1} \leq \frac{n + 1}{2n + 3} \tag{1}$$

Inequality (1) can be verified if an equivalent inequality can be found that we know is valid. Multiplying each member of inequality (1) by $(2n + 1)(2n + 3)$ we have

$$n(2n + 3) \leq (n + 1)(2n + 1) \tag{2}$$

Inequality (2) is equivalent to inequality (1) because (1) may be obtained from (2) by dividing each member by $(2n + 1)(2n + 3)$. Inequality (2) is equivalent to

$$2n^2 + 3n \leq 2n^2 + 3n + 1 \tag{3}$$

Inequality (3) obviously holds because the right member is 1 greater than the left member. Therefore inequality (1) holds; so the given sequence is increasing.

(b) The elements of the sequence can be written

$$1, \frac{1}{2}, \frac{1}{3}, \frac{1}{4}, \ldots, \frac{1}{n}, \frac{1}{n + 1}, \ldots$$

Because

$$\frac{1}{n} > \frac{1}{n + 1}$$

for all n, the sequence is decreasing.

(c) The elements of the sequence can be written

$$1, -\frac{1}{2}, \frac{1}{3}, -\frac{1}{4}, \ldots, \frac{(-1)^{n+1}}{n}, \frac{(-1)^{n+2}}{n + 1}, \ldots$$

Because $a_1 = 1$ and $a_2 = -\frac{1}{2}, a_1 > a_2$. But $a_3 = \frac{1}{3}$; thus $a_2 < a_3$. In a more general sense, consider three consecutive elements

$$a_n = \frac{(-1)^{n+1}}{n} \qquad a_{n+1} = \frac{(-1)^{n+2}}{n + 1} \qquad a_{n+2} = \frac{(-1)^{n+3}}{n + 2}$$

If n is odd, $a_n > a_{n+1}$ and $a_{n+1} < a_{n+2}$; for instance, $a_1 > a_2$ and $a_2 < a_3$. If n is even, $a_n < a_{n+1}$ and $a_{n+1} > a_{n+2}$; for instance, $a_2 < a_3$ and $a_3 > a_4$. Hence the sequence is neither increasing nor decreasing; thus it is not monotonic.

11.2.2 DEFINITION   The number C is called a **lower bound** of the sequence $\{a_n\}$ if $C \leq a_n$ for all positive integers n, and the number D is called an **upper bound** of the sequence $\{a_n\}$ if $a_n \leq D$ for all positive integers n.

▶ ILLUSTRATION 1   The number zero is a lower bound of the sequence $\{n/(2n + 1)\}$ whose elements are

$$\frac{1}{3}, \frac{2}{5}, \frac{3}{7}, \frac{4}{9}, \ldots, \frac{n}{2n + 1}, \ldots$$

Another lower bound of this sequence is $\frac{1}{3}$. Actually any number that is less than or equal to $\frac{1}{3}$ is a lower bound of this sequence.   ◀

▶ ILLUSTRATION 2   For the sequence $\{1/n\}$ whose elements are

$$1, \frac{1}{2}, \frac{1}{3}, \frac{1}{4}, \ldots, \frac{1}{n}, \ldots$$

1 is an upper bound; 26 is also an upper bound. Any number that is greater than or equal to 1 is an upper bound of this sequence, and any nonpositive number will serve as a lower bound.   ◀

From Illustrations 1 and 2 it is seen that a sequence may have many upper and lower bounds.

11.2.3 DEFINITION   If A is a lower bound of a sequence $\{a_n\}$ and if A has the property that for every lower bound C of $\{a_n\}$, $C \leq A$, then A is called the **greatest lower bound** of the sequence. Similarly, if B is an upper bound of a sequence $\{a_n\}$ and if B has the property that for every upper bound D of $\{a_n\}$, $B \leq D$, then B is called the **least upper bound** of the sequence.

▶ ILLUSTRATION 3   For the sequence $\{n/(2n + 1)\}$ of Illustration 1, the greatest lower bound is $\frac{1}{3}$ because every lower bound of the sequence is less than or equal to $\frac{1}{3}$. Furthermore,

$$\frac{n}{2n + 1} = \frac{1}{2 + \dfrac{1}{n}} < \frac{1}{2}$$

for all n, and $\frac{1}{2}$ is the least upper bound of the sequence.

In Illustration 2 there is the sequence $\{1/n\}$ whose least upper bound is 1 because every upper bound of the sequence is greater than or equal to 1. The greatest lower bound of this sequence is 0.   ◀

11.2.4 DEFINITION   A sequence $\{a_n\}$ is said to be **bounded** if and only if it has an upper bound and a lower bound.

Because the sequence $\{1/n\}$ has an upper bound and a lower bound, it is bounded. This sequence is also a decreasing sequence and hence is a bounded monotonic sequence. There is a theorem (11.2.6) that guarantees that a bounded monotonic sequence is convergent. In particular, the sequence $\{1/n\}$ is convergent because $\lim\limits_{n \to +\infty} (1/n) = 0$. The sequence $\{n\}$ whose elements are

$$1, 2, 3, \ldots, n, \ldots$$

is monotonic (because it is increasing) but is not bounded (because there is no upper bound). It is not convergent because $\lim\limits_{n \to +\infty} n = +\infty$.

For the proof of Theorem 11.2.6 we need a very important property of the real-number system that is now stated.

11.2.5 THE AXIOM OF COMPLETENESS Every nonempty set of real numbers that has a lower bound has a greatest lower bound. Also, every nonempty set of real numbers that has an upper bound has a least upper bound.

The second sentence in the statement of the axiom of completeness is unnecessary because it can be proved from the first sentence. It is included in the axiom here to expedite the discussion.

Suppose $\{a_n\}$ is an increasing sequence that is bounded. Let D be an upper bound of the sequence. Then if the points (n, a_n) are plotted on a rectangular cartesian coordinate system, these points will all lie below the line $y = D$. Furthermore, because the sequence is increasing, then as n increases, the points will get closer and closer to the line $y = D$. See Fig. 11.2.1. Hence, as n increases, the elements a_n increase toward D. Intuitively, it appears that the sequence $\{a_n\}$ has a limit that is either D or some number less than D. This is indeed the case and is proved in the following theorem.

11.2.6 THEOREM A bounded monotonic sequence is convergent.

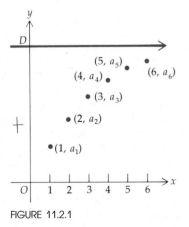

FIGURE 11.2.1

PROOF We prove the theorem for the case when the monotonic sequence is increasing. Let the sequence be $\{a_n\}$.

Because $\{a_n\}$ is bounded, there is an upper bound for the sequence. By the axiom of completeness (11.2.5), $\{a_n\}$ has a least upper bound that we call B. Then if ϵ is a positive number, $B - \epsilon$ cannot be an upper bound of the sequence because $B - \epsilon < B$ and B is the least upper bound of the sequence. So for some positive integer N,

$$B - \epsilon < a_N \tag{4}$$

Because B is the least upper bound of $\{a_n\}$, by Definition 11.2.2 it follows that

$$a_n \le B \qquad \text{for every positive integer } n \tag{5}$$

Because $\{a_n\}$ is an increasing sequence, we have from Definition 11.2.1(i)

$$a_n \le a_{n+1} \qquad \text{for every positive integer } n$$

and so

$$\text{if } n \ge N \quad \text{then} \quad a_N \le a_n \tag{6}$$

From (4), (5), and (6) it follows that

if $n \geq N$ then $B - \epsilon < a_N \leq a_n \leq B < B + \epsilon$

from which we get

if $n \geq N$ then $B - \epsilon < a_n < B + \epsilon$

$\Leftrightarrow$ if $n \geq N$ then $-\epsilon < a_n - B < \epsilon$

$\Leftrightarrow$ if $n \geq N$ then $|a_n - B| < \epsilon$ (7)

But by Definition 11.1.2, (7) is the condition that $\lim\limits_{n \to +\infty} a_n = B$. Therefore the sequence $\{a_n\}$ is convergent.

To prove the theorem when $\{a_n\}$ is a decreasing sequence, consider the sequence $\{-a_n\}$, which will be increasing, and apply the above results. We leave it as an exercise to fill in the steps (see Exercise 17).   ■

Theorem 11.2.6 states that if $\{a_n\}$ is a bounded monotonic sequence, there exists a number L such that $\lim\limits_{n \to +\infty} a_n = L$, but it does not state how to find L. For this reason Theorem 11.2.6 is called an *existence theorem*. Many important concepts in mathematics are based on existence theorems. In particular, there are many sequences for which the limit cannot be found by direct use of the definition or by use of limit theorems, but the knowledge that such a limit exists can be of value to a mathematician.

In the proof of Theorem 11.2.6 the limit of the bounded increasing sequence is the least upper bound B of the sequence. Hence, if D is an upper bound of the sequence, $\lim\limits_{n \to +\infty} a_n = B \leq D$. We have, then, the following theorem.

11.2.7 THEOREM   Let $\{a_n\}$ be an increasing sequence, and suppose that D is an upper bound of this sequence. Then $\{a_n\}$ is convergent, and

$$\lim_{n \to +\infty} a_n \leq D$$

In proving Theorem 11.2.6 for the case when the bounded monotonic sequence is decreasing, the limit of the sequence is the greatest lower bound. The next theorem follows in a way similar to that of Theorem 11.2.7.

11.2.8 THEOREM   Let $\{a_n\}$ be a decreasing sequence, and suppose that C is a lower bound of this sequence. Then $\{a_n\}$ is convergent, and

$$\lim_{n \to +\infty} a_n \geq C$$

EXAMPLE 2

Use Theorem 11.2.6 to prove that the sequence

$$\left\{ \frac{2^n}{n!} \right\}$$

is convergent.

SOLUTION

The elements of the given sequence are

$$\frac{2^1}{1!}, \frac{2^2}{2!}, \frac{2^3}{3!}, \frac{2^4}{4!}, \ldots, \frac{2^n}{n!}, \frac{2^{n+1}}{(n+1)!}, \ldots$$

$1! = 1$, $2! = 2$, $3! = 6$, $4! = 24$. Hence the elements of the sequence can be written as

$$2, 2, \frac{4}{3}, \frac{2}{3}, \ldots, \frac{2^n}{n!}, \frac{2^{n+1}}{(n+1)!}, \ldots$$

Then $a_1 = a_2 > a_3 > a_4$; so the given sequence may be decreasing. We must check to see if $a_n \geq a_{n+1}$; that is, it must be determined if

$$\frac{2^n}{n!} \geq \frac{2^{n+1}}{(n+1)!} \tag{8}$$

$$\Leftrightarrow \quad 2^n(n+1)! \geq 2^{n+1}n!$$

$$\Leftrightarrow \quad 2^n n!(n+1) \geq 2 \cdot 2^n n!$$

$$\Leftrightarrow \quad n+1 \geq 2 \tag{9}$$

When $n = 1$, inequality (9) becomes $2 = 2$, and (9) obviously holds when $n > 2$. Because inequality (8) is equivalent to (9), it follows that the given sequence is decreasing and hence monotonic. An upper bound for the given sequence is 2, and a lower bound is 0. Therefore the sequence is bounded.

The sequence $\{2^n/n!\}$ is therefore a bounded monotonic sequence, and by Theorem 11.2.6 it is convergent.

Theorem 11.2.6 states that a sufficient condition for a monotonic sequence to be convergent is that it be bounded. This is also a necessary condition and is given in the following theorem.

11.2.9 THEOREM   A convergent monotonic sequence is bounded.

PROOF   We prove the theorem for the case when the monotonic sequence is increasing. Let the sequence be $\{a_n\}$.

To prove that $\{a_n\}$ is bounded, it must be shown that it has a lower bound and an upper bound. Because $\{a_n\}$ is an increasing sequence, its first element serves as a lower bound. We must now find an upper bound.

Because $\{a_n\}$ is convergent, the sequence has a limit; call this limit L. Therefore $\lim\limits_{n \to +\infty} a_n = L$, and so by Definition 11.1.2, for any $\epsilon > 0$ there exists a number $N > 0$ such that if n is an integer and

$$\text{if } n > N \quad \text{then} \quad |a_n - L| < \epsilon$$

$$\Leftrightarrow \quad \text{if } n > N \quad \text{then} \quad -\epsilon < a_n - L < \epsilon$$

$$\Leftrightarrow \quad \text{if } n > N \quad \text{then} \quad L - \epsilon < a_n < L + \epsilon \tag{10}$$

Because $\{a_n\}$ is increasing, it follows from (10) that

$$a_n < L + \epsilon \qquad \text{for all positive integers } n$$

Therefore $L + \epsilon$ will serve as an upper bound of the sequence $\{a_n\}$.

To prove the theorem when $\{a_n\}$ is a decreasing sequence, do as suggested in the proof of Theorem 11.2.6: Consider the sequence $\{-a_n\}$, which will be increasing, and apply the above results. You are asked to do this proof in Exercise 18.   ∎

EXERCISES 11.2

In Exercises 1 through 16, determine if the given sequence is increasing, decreasing, or not monotonic.

1. $\left\{\dfrac{3n-1}{4n+5}\right\}$ 2. $\left\{\dfrac{2n-1}{4n-1}\right\}$ 3. $\left\{\dfrac{1-2n^2}{n^2}\right\}$ 4. $\{\sin n\pi\}$ 5. $\{\cos \frac{1}{3}n\pi\}$ 6. $\left\{\dfrac{n^3-1}{n}\right\}$

7. $\left\{\dfrac{1}{n+\sin n^2}\right\}$ 8. $\left\{\dfrac{2^n}{1+2^n}\right\}$ 9. $\left\{\dfrac{5^n}{1+5^{2n}}\right\}$ 10. $\left\{\dfrac{(2n)!}{5^n}\right\}$ 11. $\left\{\dfrac{n!}{3^n}\right\}$ 12. $\left\{\dfrac{n}{2^n}\right\}$

13. $\left\{\dfrac{n^n}{n!}\right\}$ 14. $\{n^2+(-1)^n n\}$ 15. $\left\{\dfrac{n!}{1\cdot3\cdot5\cdot\,\ldots\,\cdot(2n-1)}\right\}$ 16. $\left\{\dfrac{1\cdot3\cdot5\cdot\,\ldots\,\cdot(2n-1)}{2^n\cdot n!}\right\}$

17. Use the fact that Theorem 11.2.6 holds for an increasing sequence to prove that the theorem holds when $\{a_n\}$ is a decreasing sequence. (*Hint:* Consider the sequence $\{-a_n\}$.)

18. Prove Theorem 11.2.9 when $\{a_n\}$ is a decreasing sequence by a method similar to that used in Exercise 17.

In Exercises 19 and 20, determine if the given sequence is bounded.

19. $\left\{\dfrac{n^2+3}{n+1}\right\}$ 20. $\{3-(-1)^{n-1}\}$

In Exercises 21 through 30, prove that the given sequence is convergent by using Theorem 11.2.6.

21. The sequence of Exercise 1.   22. $\left\{\dfrac{n}{3^{n+1}}\right\}$ 23. $\left\{\dfrac{1\cdot3\cdot5\cdot\,\ldots\,\cdot(2n-1)}{2\cdot4\cdot6\cdot\,\ldots\,\cdot(2n)}\right\}$

24. The sequence of Exercise 8.   25. The sequence of Exercise 9.   26. The sequence of Exercise 12.

27. The sequence of Exercise 15.   28. The sequence of Exercise 16.   29. $\left\{\dfrac{n^2}{2^n}\right\}$

30. $\{k^{1/n}\}, k>1$
31. Give an example of a sequence that is bounded and convergent but not monotonic.
32. Given the sequence $\{a_n\}$ where $a_n>0$ for all n, and $a_{n+1}<ka_n$ with $0<k<1$. Prove that $\{a_n\}$ is convergent.

11.3
INFINITE SERIES OF
CONSTANT TERMS

An important part of the study of calculus involves representing functions as "infinite sums." To do this requires extending the familiar operation of addition of a finite set of numbers to addition of infinitely many numbers. To carry this out we deal with a limiting process by considering sequences. Let us associate with the sequence

$$u_1, u_2, u_3, \ldots, u_n, \ldots$$

an "infinite sum" denoted by

$$u_1 + u_2 + u_3 + \ldots + u_n + \ldots$$

But what is the meaning of such an expression? That is, what do we mean by the "sum" of an infinite number of terms, and under what circumstances does such a sum exist? To get an intuitive idea of the concept of such a sum, suppose a piece of string of length 2 ft is cut in half. One of these halves of length 1 ft is set aside and the other piece is cut in half again. One of the resulting pieces of length $\frac{1}{2}$ ft is set aside and the other piece is cut in half so that two pieces, each of length $\frac{1}{4}$ ft, are obtained. One of the pieces of length $\frac{1}{4}$ ft is set aside and then the other piece is cut in half; so two pieces, each of length $\frac{1}{8}$ ft, are obtained. Again one of the pieces is set aside and the other is cut in half. If this procedure is continued indefinitely, the number of feet in the sum of the lengths of the pieces set aside can be considered as the infinite sum

$$1 + \frac{1}{2} + \frac{1}{4} + \frac{1}{8} + \frac{1}{16} + \ldots + \frac{1}{2^{n-1}} + \ldots \tag{1}$$

Because we started with a piece of string 2 ft in length, our intuition indicates that the infinite sum (1) should be 2. We demonstrate that this is indeed the case in Illustration 2. However, we first need some preliminary definitions.

From the sequence

$$u_1, u_2, u_3, \ldots, u_n, \ldots$$

we form a new sequence $\{s_n\}$ by adding successive elements of $\{u_n\}$:

$$s_1 = u_1$$

$$s_2 = u_1 + u_2$$

$$s_3 = u_1 + u_2 + u_3$$

$$s_4 = u_1 + u_2 + u_3 + u_4$$

$$\vdots$$

$$s_n = u_1 + u_2 + u_3 + u_4 + \ldots + u_n$$

The sequence $\{s_n\}$ obtained in this manner from the sequence $\{u_n\}$ is called an *infinite series*. Following is the formal definition.

11.3.1 DEFINITION   If $\{u_n\}$ is a sequence and

$$s_n = u_1 + u_2 + u_3 + \ldots + u_n$$

then the sequence $\{s_n\}$ is called an **infinite series**. This infinite series is denoted by

$$\sum_{n=1}^{+\infty} u_n = u_1 + u_2 + u_3 + \ldots + u_n + \ldots \tag{2}$$

The numbers $u_1, u_2, u_3, \ldots, u_n, \ldots$ are called the **terms** of the infinite series. The numbers $s_1, s_2, s_3, \ldots, s_n, \ldots$ are called the **partial sums** of the infinite series.

Observe that Definition 11.3.1 states that an infinite series is a sequence of partial sums.

▶ ILLUSTRATION 1   Consider the sequence $\{u_n\}$ where $u_n = \dfrac{1}{2^{n-1}}$. Thus the sequence is

$$1, \frac{1}{2}, \frac{1}{4}, \frac{1}{8}, \frac{1}{16}, \ldots, \frac{1}{2^{n-1}}, \ldots$$

From this sequence let us form a sequence of partial sums:

$$s_1 = 1 \qquad\qquad\qquad\qquad s_1 = 1$$

$$s_2 = 1 + \frac{1}{2} \qquad\qquad\qquad \Leftrightarrow\quad s_2 = \frac{3}{2}$$

$$s_3 = 1 + \frac{1}{2} + \frac{1}{4} \qquad\qquad \Leftrightarrow\quad s_3 = \frac{7}{4}$$

$$s_4 = 1 + \frac{1}{2} + \frac{1}{4} + \frac{1}{8} \qquad \Leftrightarrow\quad s_4 = \frac{15}{8}$$

$$s_5 = 1 + \frac{1}{2} + \frac{1}{4} + \frac{1}{8} + \frac{1}{16} \quad \Leftrightarrow\quad s_5 = \frac{31}{16}$$

$$\vdots$$

$$s_n = 1 + \frac{1}{2} + \frac{1}{4} + \frac{1}{8} + \frac{1}{16} + \ldots + \frac{1}{2^{n-1}}$$

This sequence of partial sums $\{s_n\}$ is the infinite series denoted by

$$\sum_{n=1}^{+\infty} \frac{1}{2^{n-1}} = 1 + \frac{1}{2} + \frac{1}{4} + \frac{1}{8} + \frac{1}{16} + \ldots + \frac{1}{2^{n-1}} + \ldots$$

Observe that this is the infinite sum (1) obtained at the beginning of this section in the discussion of repeatedly cutting the string of length 2 ft. It is an example of a *geometric series* discussed later in this section.   ◀

When $\{s_n\}$ is a sequence of partial sums,

$$s_{n-1} = u_1 + u_2 + u_3 + \ldots + u_{n-1}$$

Thus

$$s_n = s_{n-1} + u_n \qquad\qquad\qquad (3)$$

We use formula (3) in the following example.

EXAMPLE 1
Given the infinite series

$$\sum_{n=1}^{+\infty} u_n = \sum_{n=1}^{+\infty} \frac{1}{n(n+1)}$$

[Handwritten marginal notes:]

There are two sequences involved in every series

1) sequence of terms
$u_1, u_2 \ldots u_n$

2) sequence of partial sums
$s_1, s_2 \ldots s_n$

The sum of the series (when it exists) is not limit of u_n but the limit of s_n.

(a) find the first four elements of the sequence of partial sums $\{s_n\}$, and
(b) find a formula for s_n in terms of n.

SOLUTION

(a) From formula (3),

$$s_1 = u_1 \qquad\qquad s_2 = s_1 + u_2$$

$$= \frac{1}{1 \cdot 2} \qquad\qquad = \frac{1}{2} + \frac{1}{2 \cdot 3}$$

$$= \tfrac{1}{2} \qquad\qquad\qquad = \tfrac{2}{3}$$

$$s_3 = s_2 + u_3 \qquad\qquad s_4 = s_3 + u_4$$

$$= \frac{2}{3} + \frac{1}{3 \cdot 4} \qquad\qquad = \frac{3}{4} + \frac{1}{4 \cdot 5}$$

$$= \tfrac{3}{4} \qquad\qquad\qquad = \tfrac{4}{5}$$

(b) Because $u_k = \dfrac{1}{k(k + 1)}$ we have, by partial fractions,

$$u_k = \frac{1}{k} - \frac{1}{k + 1}$$

Therefore,

$$u_1 = 1 - \tfrac{1}{2} \qquad u_2 = \tfrac{1}{2} - \tfrac{1}{3} \qquad u_3 = \tfrac{1}{3} - \tfrac{1}{4}$$

$$\cdots$$

$$u_{n-1} = \frac{1}{n - 1} - \frac{1}{n} \qquad u_n = \frac{1}{n} - \frac{1}{n + 1}$$

Thus, because $s_n = u_1 + u_2 + \ldots + u_{n-1} + u_n$,

$$s_n = \left(1 - \frac{1}{2}\right) + \left(\frac{1}{2} - \frac{1}{3}\right) + \left(\frac{1}{3} - \frac{1}{4}\right) + \ldots + \left(\frac{1}{n - 1} - \frac{1}{n}\right) + \left(\frac{1}{n} - \frac{1}{n + 1}\right)$$

Upon removing parentheses and combining terms we obtain

$$s_n = 1 - \frac{1}{n + 1}$$

$$= \frac{n}{n + 1}$$

By taking n as 1, 2, 3, and 4, we see that the previous results agree.

The method of solution of the above example applies only to a special case. In general, it is not possible to obtain such an expression for s_n.

We now define the *sum* of an infinite series.

11.3.2 DEFINITION   Let $\sum\limits_{n=1}^{+\infty} u_n$ be a given infinite series, and let $\{s_n\}$ be the sequence of partial sums defining this infinite series. Then if $\lim\limits_{n \to +\infty} s_n$ exists and is equal to S, the given series is said to be **convergent** and S is the **sum** of the given infinite series. If $\lim\limits_{n \to +\infty} s_n$ does not exist, the series is **divergent** and the series does not have a sum.

Essentially Definition 11.3.2 states that an infinite series is convergent if and only if the corresponding sequence of partial sums is convergent.

If an infinite series has a sum S, we also say that the series converges to S.

Observe that the sum of a convergent series is the limit of a sequence of partial sums and is not obtained by ordinary addition. For a convergent series the symbolism

$$\sum_{n=1}^{+\infty} u_n$$

is used to denote both the series and the sum of the series. The use of the same symbol should not be confusing because the correct interpretation is apparent from the context in which it is employed.

▶ ILLUSTRATION 2   The infinite series of Illustration 1 is

$$\sum_{n=1}^{+\infty} \frac{1}{2^{n-1}} = 1 + \frac{1}{2} + \frac{1}{4} + \frac{1}{8} + \frac{1}{16} + \ldots + \frac{1}{2^{n-1}} + \ldots \tag{4}$$

and the sequence of partial sums is $\{s_n\}$, where

$$s_n = 1 + \frac{1}{2} + \frac{1}{4} + \frac{1}{8} + \ldots + \frac{1}{2^{n-1}} \tag{5}$$

To determine if infinite series (4) has a sum we must compute $\lim\limits_{n \to +\infty} s_n$. First we find a formula for s_n. An identity from algebra is

$$a^n - b^n = (a - b)(a^{n-1} + a^{n-2}b + a^{n-3}b^2 + \ldots + ab^{n-2} + b^{n-1})$$

Applying this formula with $a = 1$ and $b = \frac{1}{2}$ we have

$$1 - \frac{1}{2^n} = \left(1 - \frac{1}{2}\right)\left(1 + \frac{1}{2} + \frac{1}{2^2} + \frac{1}{2^3} + \ldots + \frac{1}{2^{n-1}}\right)$$

$$\Leftrightarrow \quad 1 + \frac{1}{2} + \frac{1}{4} + \frac{1}{8} + \ldots + \frac{1}{2^{n-1}} = \frac{1 - \dfrac{1}{2^n}}{\frac{1}{2}} \tag{6}$$

From (5) and (6) we have

$$s_n = 2\left(1 - \frac{1}{2^n}\right)$$

Because $\lim\limits_{n \to +\infty} \dfrac{1}{2^n} = 0$ we have

$$\lim_{n \to +\infty} s_n = 2$$

Therefore infinite series (4) has the sum 2.   ◀

EXAMPLE 2
Determine if the infinite series of Example 1 has a sum.

SOLUTION
In the solution of Example 1 it was shown that the sequence of partial sums for the given series is $\{s_n\} = \{n/(n + 1)\}$. Therefore

$$\lim_{n \to +\infty} s_n = \lim_{n \to +\infty} \frac{n}{n + 1}$$

$$= \lim_{n \to +\infty} \frac{1}{1 + \dfrac{1}{n}}$$

$$= 1$$

So the infinite series has a sum equal to 1, and we write

$$\sum_{n=1}^{+\infty} \frac{1}{n(n + 1)} = \frac{1}{2} + \frac{1}{6} + \frac{1}{12} + \frac{1}{20} + \ldots + \frac{1}{n(n + 1)} + \ldots$$

$$= 1$$

EXAMPLE 3
Find the infinite series that has the following sequence of partial sums:

$$\{s_n\} = \left\{\frac{1}{2^n}\right\}$$

Also determine if the infinite series is convergent or divergent; if it is convergent, find its sum.

SOLUTION
Because $s_1 = \frac{1}{2}$, then $u_1 = \frac{1}{2}$. If $n > 1$,

$$u_n = s_n - s_{n-1}$$

$$= \frac{1}{2^n} - \frac{1}{2^{n-1}}$$

$$= -\frac{1}{2^n}$$

Therefore the infinite series is

$$\frac{1}{2} - \sum_{n=2}^{+\infty} \frac{1}{2^n}$$

Because

$$\lim_{n \to +\infty} s_n = \lim_{n \to +\infty} \frac{1}{2^n}$$

$$= 0$$

the series is convergent and its sum is 0.

As mentioned above, in most cases it is not possible to obtain an expression for s_n in terms of n; so we must have other methods for determining whether or not a given infinite series has a sum or, equivalently, whether a given infinite series is convergent or divergent.

11.3.3 THEOREM   If the infinite series $\sum\limits_{n=1}^{+\infty} u_n$ is convergent, then $\lim\limits_{n \to +\infty} u_n = 0$.

no proof will not lead to understanding

PROOF   Let $\{s_n\}$ be the sequence of partial sums for the given series, and denote the sum of the series by S. From Definition 11.3.2, $\lim\limits_{n \to +\infty} s_n = S$. Thus for any $\epsilon > 0$ there exists a number $N > 0$ such that

if $n > N$ then $|S - s_n| < \tfrac{1}{2}\epsilon$

Also

if $n > N$ then $|S - s_{n+1}| < \tfrac{1}{2}\epsilon$

Then

$$|u_{n+1}| = |s_{n+1} - s_n|$$
$$= |S - s_n + s_{n+1} - S|$$
$$\leq |S - s_n| + |s_{n+1} - S|$$

Therefore

if $n > N$ then $|u_{n+1}| < \tfrac{1}{2}\epsilon + \tfrac{1}{2}\epsilon = \epsilon$

Hence

$$\lim_{n \to +\infty} u_n = 0$$

■

CONTRAPOSITIVE OF 11.3.3

TEST FOR DIVERGENCE

If $\lim u_n \neq 0$ then $\sum u_n$ is divergent.

Theorem 11.3.3 provides a simple test for divergence because if $\lim\limits_{n \to +\infty} u_n \neq 0$, we can conclude that $\sum\limits_{n=1}^{+\infty} u_n$ is divergent.

EXAMPLE 4
Prove that the following two series are divergent.

(a) $\sum\limits_{n=1}^{+\infty} \dfrac{n^2 + 1}{n^2} = 2 + \dfrac{5}{4} + \dfrac{10}{9} + \dfrac{17}{16} + \cdots$

(b) $\sum\limits_{n=1}^{+\infty} (-1)^{n+1} 3 = 3 - 3 + 3 - 3 + \cdots$

SOLUTION

(a) $\displaystyle\lim_{n\to+\infty} u_n = \lim_{n\to+\infty} \frac{n^2+1}{n^2}$

$\displaystyle\qquad = \lim_{n\to+\infty} \frac{1+\dfrac{1}{n^2}}{1}$

$\displaystyle\qquad = 1$

$\displaystyle\qquad \neq 0$

Therefore, by Theorem 11.3.3, the series is divergent.

(b) $\displaystyle\lim_{n\to+\infty} u_n = \lim_{n\to+\infty} (-1)^{n+1}3$, which does not exist. Therefore, by Theorem 11.3.3, the series is divergent.

The converse of Theorem 11.3.3 is false. That is, if $\lim_{n\to+\infty} u_n = 0$, it does not follow that the series is necessarily convergent. In other words, it is possible to have a divergent series for which $\lim_{n\to+\infty} u_n = 0$. An example of such a series is the one known as the **harmonic series**, which is

$$\sum_{n=1}^{+\infty} \frac{1}{n} = 1 + \frac{1}{2} + \frac{1}{3} + \frac{1}{4} + \ldots + \frac{1}{n} + \ldots \tag{7}$$

Clearly, $\lim_{n\to+\infty} 1/n = 0$. In Illustration 3 the harmonic series is proved to diverge and we use the following theorem, which states that the difference between two partial sums s_R and s_T of a convergent series can be made as small as we please by taking R and T sufficiently large.

11.3.4 THEOREM Let $\{s_n\}$ be the sequence of partial sums for a given convergent series $\displaystyle\sum_{n=1}^{+\infty} u_n$. Then for any $\epsilon > 0$ there exists a number N such that

$$\text{if } R > N \text{ and } T > N \quad \text{then} \quad |s_R - s_T| < \epsilon$$

PROOF Because the series $\displaystyle\sum_{n=1}^{+\infty} u_n$ is convergent, call its sum S. Then for any $\epsilon > 0$ there exists an $N > 0$ such that if $n > N$, then $|S - s_n| < \frac{1}{2}\epsilon$. Therefore, if $R > N$ and $T > N$,

$$|s_R - s_T| = |s_R - S + S - s_T| \le |s_R - S| + |S - s_T| < \tfrac{1}{2}\epsilon + \tfrac{1}{2}\epsilon$$

So

$$\text{if } R > N \text{ and } T > N \quad \text{then} \quad |s_R - s_T| < \epsilon \qquad \blacksquare$$

▶ ILLUSTRATION 3 We prove that the harmonic series (7) is divergent. For this series,

$$s_n = 1 + \frac{1}{2} + \ldots + \frac{1}{n}$$

and

$$s_{2n} = 1 + \frac{1}{2} + \ldots + \frac{1}{n} + \frac{1}{n+1} + \ldots + \frac{1}{2n}$$

So

$$s_{2n} - s_n = \frac{1}{n+1} + \frac{1}{n+2} + \frac{1}{n+3} + \ldots + \frac{1}{2n} \tag{8}$$

If $n > 1$,

$$\frac{1}{n+1} + \frac{1}{n+2} + \frac{1}{n+3} + \ldots + \frac{1}{2n} > \frac{1}{2n} + \frac{1}{2n} + \frac{1}{2n} + \ldots + \frac{1}{2n} \tag{9}$$

There are n terms on each side of the inequality sign in (9); so the right side is $n(1/2n) = \frac{1}{2}$. Therefore, from (8) and (9),

$$\text{if } n > 1 \quad \text{then} \quad s_{2n} - s_n > \tfrac{1}{2} \tag{10}$$

But Theorem 11.3.4 states that if the given series is convergent, then $s_{2n} - s_n$ may be made as small as we please by taking n large enough; that is, if $\epsilon = \frac{1}{2}$, there exists an N such that

$$\text{if } 2n > N \text{ and } n > N \quad \text{then} \quad s_{2n} - s_n < \tfrac{1}{2}$$

But this statement contradicts (10). Therefore the harmonic series is divergent even though $\lim\limits_{n \to +\infty} 1/n = 0$.   ◄

A **geometric series** is a series of the form

$$\sum_{n=1}^{+\infty} ar^{n-1} = a + ar + ar^2 + \ldots + ar^{n-1} + \ldots \tag{11}$$

Infinite series (4) discussed in Illustrations 1 and 2 is a geometric series with $a = 1$ and $r = \frac{1}{2}$. The nth partial sum of geometric series (11) is given by

$$s_n = a(1 + r + r^2 + \ldots + r^{n-1}) \tag{12}$$

From the identity

$$1 - r^n = (1 - r)(1 + r + r^2 + \ldots + r^{n-1})$$

(12) can be written as

$$s_n = \frac{a(1 - r^n)}{1 - r} \quad \text{if } r \neq 1 \tag{13}$$

11.3.5 THEOREM   The geometric series converges to the sum $a/(1 - r)$ if $|r| < 1$, and the geometric series diverges if $|r| \geq 1$.

PROOF   In Example 3, Section 11.1, we showed that $\lim\limits_{n \to +\infty} r^n = 0$ if $|r| < 1$. Therefore, from (13) we can conclude that if $|r| < 1$,

$$\lim_{n \to +\infty} s_n = \frac{a}{1 - r}$$

So if $|r| < 1$, the geometric series converges and its sum is $a/(1 - r)$.

If $r = 1$, $s_n = na$. Then $\lim\limits_{n \to +\infty} s_n = +\infty$ if $a > 0$, and $\lim\limits_{n \to +\infty} s_n = -\infty$ if $a < 0$.

If $r = -1$, then the geometric series becomes

$$a - a + a - \ldots + (-1)^{n-1}a + \ldots$$

Thus $s_n = 0$ if n is even, and $s_n = a$ if n is odd. Therefore $\lim\limits_{n \to +\infty} s_n$ does not exist. Hence the geometric series diverges when $|r| = 1$.

If $|r| > 1$, $\lim\limits_{n \to +\infty} ar^{n-1} = a \lim\limits_{n \to +\infty} r^{n-1}$. Clearly, $\lim\limits_{n \to +\infty} r^{n-1} \neq 0$ because $|r^{n-1}|$ can be made as large as we please by taking n large enough. Therefore, by Theorem 11.3.3, the series is divergent. This completes the proof. ∎

The following example illustrates how Theorem 11.3.5 can be used to express a nonterminating repeating decimal as a common fraction.

EXAMPLE 5

Express the decimal $0.3333\ldots$ as a common fraction.

SOLUTION

$$0.3333\ldots = \frac{3}{10} + \frac{3}{100} + \frac{3}{1000} + \frac{3}{10,000} + \ldots + \frac{3}{10^n} + \ldots$$

This is a geometric series in which $a = \frac{3}{10}$ and $r = \frac{1}{10}$. Because $|r| < 1$, it follows from Theorem 11.3.5 that the series converges and its sum is $a/(1-r)$. Therefore

$$0.3333\ldots = \frac{\frac{3}{10}}{1 - \frac{1}{10}}$$

$$= \tfrac{1}{3}$$

EXERCISES 11.3

In Exercises 1 through 8, find the first four elements of the sequence of partial sums $\{s_n\}$, and find a formula for s_n in terms of n. Also determine if the infinite series is convergent or divergent; if it is convergent, find its sum.

1. $\displaystyle\sum_{n=1}^{+\infty} \frac{1}{(2n-1)(2n+1)}$

2. $\displaystyle\sum_{n=1}^{+\infty} n$

3. $\displaystyle\sum_{n=1}^{+\infty} \frac{5}{(3n+1)(3n-2)}$

4. $\displaystyle\sum_{n=1}^{+\infty} \frac{2}{(4n-3)(4n+1)}$

5. $\displaystyle\sum_{n=1}^{+\infty} \ln\frac{n}{n+1}$

6. $\displaystyle\sum_{n=1}^{+\infty} \frac{2n+1}{n^2(n+1)^2}$

7. $\displaystyle\sum_{n=1}^{+\infty} \frac{2}{5^{n-1}}$

8. $\displaystyle\sum_{n=1}^{+\infty} \frac{2^{n-1}}{3^n}$

In Exercises 9 through 13, find the infinite series that is the given sequence of partial sums. Also determine if the infinite series is convergent or divergent; if it is convergent, find its sum.

9. $\{s_n\} = \left\{\dfrac{2n}{3n+1}\right\}$

10. $\{s_n\} = \left\{\dfrac{n^2}{n+1}\right\}$

11. $\{s_n\} = \left\{\dfrac{1}{3^n}\right\}$

12. $\{s_n\} = \{3^n\}$

13. $\{s_n\} = \{\ln(2n+1)\}$

In Exercises 14 through 26, write the first four terms of the given infinite series and determine if the series is convergent or divergent. If the series is convergent, find its sum.

14. $\displaystyle\sum_{n=1}^{+\infty} \frac{n}{n+1}$

15. $\displaystyle\sum_{n=1}^{+\infty} \frac{2n+1}{3n+2}$

16. $\displaystyle\sum_{n=1}^{+\infty} [1 + (-1)^n]$

17. $\displaystyle\sum_{n=1}^{+\infty} \left(\frac{2}{3}\right)^n$

18. $\displaystyle\sum_{n=1}^{+\infty} \frac{3n^2}{n^2+1}$

19. $\displaystyle\sum_{n=1}^{+\infty} \ln\frac{1}{n}$

20. $\displaystyle\sum_{n=1}^{+\infty} \frac{2}{3^{n-1}}$

21. $\displaystyle\sum_{n=1}^{+\infty} (-1)^{n+1} \frac{3}{2^n}$

22. $\displaystyle\sum_{n=1}^{+\infty} \tan^n \frac{\pi}{6}$

23. $\displaystyle\sum_{n=1}^{+\infty} e^{-n}$

24. $\displaystyle\sum_{n=1}^{+\infty} \frac{\sinh n}{n}$

25. $\displaystyle\sum_{n=1}^{+\infty} \cos \pi n$

26. $\displaystyle\sum_{n=1}^{+\infty} \sin \pi n$

In Exercises 27 through 30, express the given nonterminating repeating decimal as a common fraction.

27. $0.27\ 27\ 27\ldots$ **28.** $2.045\ 45\ 45\ldots$ **29.** $1.234\ 234\ 234\ldots$ **30.** $0.4653\ 4653\ 4653\ldots$

31. The path of each swing, after the first, of a pendulum bob is 0.93 as long as the path of the previous swing (from one side to the other side). If the path of the first swing is 56 cm long, and air resistance eventually brings the pendulum to rest, how far does the bob travel before it comes to rest?

32. A ball is dropped from a height of 12 m. Each time it strikes the ground, it bounces back to a height of three-fourths the distance from which it fell. Find the total distance traveled by the ball before it comes to rest.

33. What is the total distance traveled by a tennis ball before coming to rest if it is dropped from a height of 100 m and if, after each fall, it rebounds eleven-twentieths of the distance from which it fell?

34. An equilateral triangle has sides of length 4 units; therefore its perimeter is 12 units. Another equilateral triangle is constructed by drawing line segments through the midpoints of the sides of the first triangle. This triangle has sides of length 2 units, and its perimeter is 6 units. If this procedure can be repeated an unlimited number of times, what is the total perimeter of all the triangles that are formed?

35. After a woman riding a bicycle removes her feet from the pedals, the front wheel rotates 300 times during the first 10 sec. Then in each succeeding 10-sec time period the wheel rotates four-fifths as many times as it did the previous period. Determine the number of rotations of the wheel before the bicycle stops.

36. Find an infinite geometric series whose sum is 6 and such that each term is four times the sum of all the terms that follow it.

11.4

FOUR THEOREMS ABOUT INFINITE SERIES

The first theorem of this section states that the convergence or divergence of an infinite series is not affected by changing a finite number of terms.

11.4.1 THEOREM   If $\displaystyle\sum_{n=1}^{+\infty} a_n$ and $\displaystyle\sum_{n=1}^{+\infty} b_n$ are two infinite series, differing only in their first m terms (i.e., $a_k = b_k$ if $k > m$), then either both series converge or both series diverge.

PROOF   Let $\{s_n\}$ and $\{t_n\}$ be the sequences of partial sums of the series $\displaystyle\sum_{n=1}^{+\infty} a_n$ and $\displaystyle\sum_{n=1}^{+\infty} b_n$, respectively. Then

$$s_n = a_1 + a_2 + \ldots + a_m + a_{m+1} + a_{m+2} + \ldots + a_n$$

and

$$t_n = b_1 + b_2 + \ldots + b_m + b_{m+1} + b_{m+2} + \ldots + b_n$$

Because $a_k = b_k$ if $k > m$, then if $n \geq m$,

$$s_n - t_n = (a_1 + a_2 + \ldots + a_m) - (b_1 + b_2 + \ldots + b_m)$$

So

$$\text{if } n \geq m \quad \text{then} \quad s_n - t_n = s_m - t_m \tag{1}$$

We wish to show that either both $\lim\limits_{n \to +\infty} s_n$ and $\lim\limits_{n \to +\infty} t_n$ exist or do not exist.

Suppose that $\lim\limits_{n \to +\infty} t_n$ exists. Then from (1),

$$\text{if } n \geq m \quad \text{then} \quad s_n = t_n + (s_m - t_m)$$

and so

$$\lim_{n \to +\infty} s_n = \lim_{n \to +\infty} t_n + (s_m - t_m)$$

Hence, when $\lim\limits_{n \to +\infty} t_n$ exists, $\lim\limits_{n \to +\infty} s_n$ also exists and both series converge.

Now suppose that $\lim\limits_{n \to +\infty} t_n$ does not exist and $\lim\limits_{n \to +\infty} s_n$ exists. From (1),

$$\text{if } n \geq m \quad \text{then} \quad t_n = s_n + (t_m - s_m)$$

Because $\lim\limits_{n \to +\infty} s_n$ exists, it follows that

$$\lim_{n \to +\infty} t_n = \lim_{n \to +\infty} s_n + (t_m - s_m)$$

and so $\lim\limits_{n \to +\infty} t_n$ has to exist, which is a contradiction. Hence, if $\lim\limits_{n \to +\infty} t_n$ does not exist, then $\lim\limits_{n \to +\infty} s_n$ does not exist, and both series diverge. ∎

EXAMPLE 1

Determine whether the infinite series

$$\sum_{n=1}^{+\infty} \frac{1}{n+4}$$

is convergent or divergent.

SOLUTION

The given series is

$$\frac{1}{5} + \frac{1}{6} + \frac{1}{7} + \ldots + \frac{1}{n+4} + \ldots$$

which can be written as

$$0 + 0 + 0 + 0 + \frac{1}{5} + \frac{1}{6} + \frac{1}{7} + \ldots + \frac{1}{n} + \ldots \tag{2}$$

Now the harmonic series, which is known to be divergent, is

$$1 + \frac{1}{2} + \frac{1}{3} + \frac{1}{4} + \frac{1}{5} + \frac{1}{6} + \frac{1}{7} + \ldots + \frac{1}{n} + \ldots \tag{3}$$

Series (2) differs from series (3) only in the first four terms. Hence, by Theorem 11.4.1, series (2) is also divergent.

EXAMPLE 2
Determine whether the following infinite series is convergent or divergent:

$$\sum_{n=1}^{+\infty} \frac{\left[\!\left[\cos \dfrac{3}{n}\pi + 2 \right]\!\right]}{3^n}$$

SOLUTION
The given series can be written as

$$\frac{[\![\cos 3\pi + 2]\!]}{3} + \frac{[\![\cos \frac{3}{2}\pi + 2]\!]}{3^2} + \frac{[\![\cos \pi + 2]\!]}{3^3} + \frac{[\![\cos \frac{3}{4}\pi + 2]\!]}{3^4}$$

$$+ \frac{[\![\cos \frac{3}{5}\pi + 2]\!]}{3^5} + \frac{[\![\cos \frac{1}{2}\pi + 2]\!]}{3^6} + \frac{[\![\cos \frac{3}{7}\pi + 2]\!]}{3^7} + \ldots$$

$$= \frac{1}{3} + \frac{2}{3^2} + \frac{1}{3^3} + \frac{1}{3^4} + \frac{1}{3^5} + \frac{2}{3^6} + \frac{2}{3^7} + \frac{2}{3^8} + \ldots \tag{4}$$

Consider the geometric series with $a = \frac{2}{3}$ and $r = \frac{1}{3}$:

$$\frac{2}{3} + \frac{2}{3^2} + \frac{2}{3^3} + \frac{2}{3^4} + \frac{2}{3^5} + \frac{2}{3^6} + \frac{2}{3^7} + \frac{2}{3^8} + \ldots \tag{5}$$

This series is convergent by Theorem 11.3.5. Because series (4) differs from series (5) only in the first five terms, it follows from Theorem 11.4.1 that series (4) is also convergent.

As a consequence of Theorem 11.4.1, for a given infinite series a finite number of terms can be added or subtracted without affecting its convergence or divergence. For instance, in Example 1 the given series may be thought of as being obtained from the harmonic series by subtracting the first four terms. And because the harmonic series is divergent, the given series is divergent. In Example 2 we could consider the convergent geometric series

$$\frac{2}{3^6} + \frac{2}{3^7} + \frac{2}{3^8} + \ldots \tag{6}$$

and obtain the given series (4) by adding five terms. Because series (6) is convergent, it follows that series (4) is convergent.

The following theorem states that if an infinite series is multiplied term by term by a nonzero constant, its convergence or divergence is not affected.

② ⇒ 11.4.2 THEOREM   Let c be any nonzero constant.

(i) If the series $\sum\limits_{n=1}^{+\infty} u_n$ is convergent and its sum is S, then the series $\sum\limits_{n=1}^{+\infty} cu_n$ is also convergent and its sum is $c \cdot S$.

(ii) If the series $\sum\limits_{n=1}^{+\infty} u_n$ is divergent, then the series $\sum\limits_{n=1}^{+\infty} cu_n$ is also divergent.

PROOF   Let the nth partial sum of the series $\sum\limits_{n=1}^{+\infty} u_n$ be s_n. Therefore $s_n = u_1 + u_2 + \ldots + u_n$. The nth partial sum of the series $\sum\limits_{n=1}^{+\infty} cu_n$ is $c(u_1 + u_2 + \ldots + u_n) = cs_n$.

PROOF OF (i)   If the series $\sum\limits_{n=1}^{+\infty} u_n$ is convergent, then $\lim\limits_{n \to +\infty} s_n$ exists and is S. Therefore

$$\lim_{n \to +\infty} cs_n = c \lim_{n \to +\infty} s_n$$

$$= c \cdot S$$

Hence the series $\sum\limits_{n=1}^{+\infty} cu_n$ is convergent and its sum is $c \cdot S$.

PROOF OF (ii)   If the series $\sum\limits_{n=1}^{+\infty} u_n$ is divergent, then $\lim\limits_{n \to +\infty} s_n$ does not exist.
Now suppose that the series $\sum\limits_{n=1}^{+\infty} cu_n$ is convergent. Then $\lim\limits_{n \to +\infty} cs_n$ exists. But $s_n = cs_n/c$; so

$$\lim_{n \to +\infty} s_n = \lim_{n \to +\infty} \frac{1}{c}(cs_n)$$

$$= \frac{1}{c} \lim_{n \to +\infty} cs_n$$

Thus $\lim\limits_{n \to +\infty} s_n$ must exist, which is a contradiction. Therefore the series $\sum\limits_{n=1}^{+\infty} cu_n$ is divergent.   ∎

EXAMPLE 3
Determine whether the infinite series

$$\sum_{n=1}^{+\infty} \frac{1}{4n}$$

is convergent or divergent.

SOLUTION

$$\sum_{n=1}^{+\infty} \frac{1}{4n} = \frac{1}{4} + \frac{1}{8} + \frac{1}{12} + \frac{1}{16} + \cdots + \frac{1}{4n} + \cdots$$

Because $\sum_{n=1}^{+\infty} \frac{1}{n}$ is the harmonic series that is divergent, then by Theorem 11.4.2(ii) with $c = \frac{1}{4}$, the given series is divergent.

Theorem 11.4.2(i) is an extension to convergent infinite series of the following property of finite sums:

$$\sum_{k=1}^{n} ca_k = c \sum_{k=1}^{n} a_k$$

Another property of finite sums is

$$\sum_{k=1}^{n} (a_k \pm b_k) = \sum_{k=1}^{n} a_k \pm \sum_{k=1}^{n} b_k$$

and its extension to convergent infinite series is given by the following theorem.

11.4.3 THEOREM   If $\sum_{n=1}^{+\infty} a_n$ and $\sum_{n=1}^{+\infty} b_n$ are convergent infinite series whose sums are S and R, respectively, then

(i) $\sum_{n=1}^{+\infty} (a_n + b_n)$ is a convergent series and its sum is $S + R$;

(ii) $\sum_{n=1}^{+\infty} (a_n - b_n)$ is a convergent series and its sum is $S - R$.

The proof of this theorem is left as an exercise (see Exercise 24).

The next theorem is a corollary of the above theorem and is sometimes used to prove that a series is divergent.

11.4.4 THEOREM   If the series $\sum_{n=1}^{+\infty} a_n$ is convergent and the series $\sum_{n=1}^{+\infty} b_n$ is divergent, then the series $\sum_{n=1}^{+\infty} (a_n + b_n)$ is divergent.

PROOF   Assume that $\sum_{n=1}^{+\infty} (a_n + b_n)$ is convergent and its sum is S. Let the sum of the series $\sum_{n=1}^{+\infty} a_n$ be R. Then because

$$\sum_{n=1}^{+\infty} b_n = \sum_{n=1}^{+\infty} [(a_n + b_n) - a_n]$$

it follows from Theorem 11.4.3(ii) that $\sum_{n=1}^{+\infty} b_n$ is convergent and its sum is $S - R$. But this is a contradiction to the hypothesis that $\sum_{n=1}^{+\infty} b_n$ is divergent. Hence $\sum_{n=1}^{+\infty} (a_n + b_n)$ is divergent.   ∎

EXAMPLE 4
Determine whether the infinite series

$$\sum_{n=1}^{+\infty} \left(\frac{1}{4n} + \frac{1}{4^n} \right)$$

is convergent or divergent.

SOLUTION

In Example 3 the series $\sum_{n=1}^{+\infty} \frac{1}{4n}$ was proved to be divergent. Because the series

$\sum_{n=1}^{+\infty} \frac{1}{4^n}$ is a geometric series with $|r| = \frac{1}{4} < 1$, it is convergent. Hence, by Theorem 11.4.4, the given series is divergent.

If both series $\sum_{n=1}^{+\infty} a_n$ and $\sum_{n=1}^{+\infty} b_n$ are divergent, the series $\sum_{n=1}^{+\infty} (a_n + b_n)$

may or may not be convergent. For example, if $a_n = \frac{1}{n}$ and $b_n = \frac{1}{n}$, then

$a_n + b_n = \frac{2}{n}$ and $\sum_{n=1}^{+\infty} \frac{2}{n}$ is divergent. But if $a_n = \frac{1}{n}$ and $b = -\frac{1}{n}$, then

$a_n + b_n = 0$ and $\sum_{n=1}^{+\infty} 0$ is convergent.

EXERCISES 11.4

In Exercises 1 through 22, determine if the series is convergent or divergent. If the series is convergent, find its sum.

1. $\sum_{n=1}^{+\infty} \frac{1}{n + 2}$ 2. $\sum_{n=3}^{+\infty} \frac{1}{n - 1}$ 3. $\sum_{n=1}^{+\infty} \frac{3}{2n}$ 4. $\sum_{n=1}^{+\infty} \frac{2}{3n}$ 5. $\sum_{n=1}^{+\infty} \frac{3}{2^n}$

6. $\sum_{n=1}^{+\infty} \frac{2}{3^n}$ 7. $\sum_{n=1}^{+\infty} \frac{4}{3} \left(\frac{5}{7} \right)^n$ 8. $\sum_{n=1}^{+\infty} \frac{7}{5} \left(\frac{3}{4} \right)^n$ 9. $\sum_{n=1}^{+\infty} \frac{\left[\sin \frac{4}{n} \pi + 3 \right]}{4^n}$ 10. $\sum_{n=1}^{+\infty} \frac{\left[\cos \frac{1}{n} \pi + 1 \right]}{2^n}$

11. $\sum_{n=1}^{+\infty} \left(\frac{1}{2n} + \frac{1}{2^n} \right)$ 12. $\sum_{n=1}^{+\infty} \left(\frac{1}{3^n} + \frac{1}{3n} \right)$ 13. $\sum_{n=1}^{+\infty} \left(\frac{1}{2^n} + \frac{1}{3^n} \right)$ 14. $\sum_{n=1}^{+\infty} \left(\frac{1}{3^n} - \frac{1}{4^n} \right)$ 15. $\sum_{n=1}^{+\infty} (e^{-n} + e^n)$

16. $\sum_{n=1}^{+\infty} (2^{-n} + 3^n)$ 17. $\sum_{n=1}^{+\infty} \left(\frac{1}{2n} - \frac{1}{3n} \right)$ 18. $\sum_{n=1}^{+\infty} \left(\frac{3}{2n} - \frac{2}{3n} \right)$ 19. $\sum_{n=1}^{+\infty} \left(\frac{3}{2^n} - \frac{2}{3^n} \right)$ 20. $\sum_{n=1}^{+\infty} \left(\frac{5}{4^n} + \frac{4}{5^n} \right)$

21. $\sum_{n=1}^{+\infty} \left(\frac{1}{n^2} + 2 \right)$ 22. $\sum_{n=1}^{+\infty} \frac{n!}{5^n}$

23. Give an example to show that even if each of the series $\sum_{n=1}^{+\infty} a_n$ and $\sum_{n=1}^{+\infty} b_n$ is divergent, it is possible for the series

$\sum_{n=1}^{+\infty} a_n b_n$ to be convergent.

24. Prove Theorem 11.4.3.

11.5
INFINITE SERIES OF
POSITIVE TERMS

If all the terms of an infinite series are positive, the sequence of partial sums is increasing. Thus the following theorem follows immediately from Theorems 11.2.6 and 11.2.9.

11.5.1 THEOREM   An infinite series of positive terms is convergent if and only if its sequence of partial sums has an upper bound.

PROOF   For an infinite series of positive terms, the sequence of partial sums has a lower bound of 0. If the sequence of partial sums also has an upper bound, it is bounded. Furthermore, the sequence of partial sums of an infinite series of positive terms is increasing. It follows, then, from Theorem 11.2.6 that the sequence of partial sums is convergent, and therefore the infinite series is convergent.

Suppose now that an infinite series of positive terms is convergent. Then the sequence of partial sums is also convergent. It follows from Theorem 11.2.9 that the sequence of partial sums is bounded, and so it has an upper bound.
∎

EXAMPLE 1
Prove that the series

$$\sum_{n=1}^{+\infty} \frac{1}{n!}$$

is convergent by using Theorem 11.5.1.

SOLUTION
We must find an upper bound for the sequence of partial sums of the series $\sum_{n=1}^{+\infty} \frac{1}{n!}$.

$$s_1 = 1, \, s_2 = 1 + \frac{1}{1 \cdot 2}, \, s_3 = 1 + \frac{1}{1 \cdot 2} + \frac{1}{1 \cdot 2 \cdot 3},$$

$$s_4 = 1 + \frac{1}{1 \cdot 2} + \frac{1}{1 \cdot 2 \cdot 3} + \frac{1}{1 \cdot 2 \cdot 3 \cdot 4}, \ldots,$$

$$s_n = 1 + \frac{1}{1 \cdot 2} + \frac{1}{1 \cdot 2 \cdot 3} + \frac{1}{1 \cdot 2 \cdot 3 \cdot 4} + \ldots + \frac{1}{1 \cdot 2 \cdot 3 \cdot \ldots \cdot n} \qquad (1)$$

Now consider the first n terms of the geometric series with $a = 1$ and $r = \frac{1}{2}$:

$$\sum_{k=1}^{n} \frac{1}{2^{k-1}} = 1 + \frac{1}{2} + \frac{1}{2^2} + \frac{1}{2^3} + \ldots + \frac{1}{2^{n-1}} \qquad (2)$$

By Theorem 11.3.5 the geometric series with $a = 1$ and $r = \frac{1}{2}$ has the sum $a/(1 - r) = 2$. Hence summation (2) is less than 2. Observe that each term of summation (1) is less than or equal to the corresponding term of summation (2); that is,

$$\frac{1}{k!} \le \frac{1}{2^{k-1}}$$

This is true because $k! = 1 \cdot 2 \cdot 3 \cdot \ldots \cdot k$, which in addition to the factor 1 contains $k - 1$ factors each greater than or equal to 2. Hence

$$s_n = \sum_{k=1}^{n} \frac{1}{k!} \le \sum_{k=1}^{n} \frac{1}{2^{k-1}} < 2$$

From the above, s_n has an upper bound of 2. Therefore, by Theorem 11.5.1, the given series is convergent.

In the above example the terms of the given series were compared with those of a known convergent series. This is a particular case of the following theorem known as the *comparison test*.

11.5.2 THEOREM

Comparison Test

Let the series $\sum_{n=1}^{+\infty} u_n$ be a series of positive terms.

(i) If $\sum_{n=1}^{+\infty} v_n$ is a series of positive terms that is known to be convergent, and $u_n \le v_n$ for all positive integers n, then $\sum_{n=1}^{+\infty} u_n$ is convergent.

(ii) If $\sum_{n=1}^{+\infty} w_n$ is a series of positive terms that is known to be divergent, and $u_n \ge w_n$ for all positive integers n, then $\sum_{n=1}^{+\infty} u_n$ is divergent.

PROOF OF (i)   Let $\{s_n\}$ be the sequence of partial sums for the series $\sum_{n=1}^{+\infty} u_n$ and $\{t_n\}$ be the sequence of partial sums for the series $\sum_{n=1}^{+\infty} v_n$. Because $\sum_{n=1}^{+\infty} v_n$ is a series of positive terms that is convergent, it follows from Theorem 11.5.1 that the sequence $\{t_n\}$ has an upper bound; call it B. Because $u_n \le v_n$ for all positive integers n, we can conclude that $s_n \le t_n \le B$ for all positive integers n. Therefore B is an upper bound of the sequence $\{s_n\}$. And because the terms of the series $\sum_{n=1}^{+\infty} u_n$ are all positive, it follows from Theorem 11.5.1 that $\sum_{n=1}^{+\infty} u_n$ is convergent.

PROOF OF (ii)   Assume that $\sum_{n=1}^{+\infty} u_n$ is convergent. Then because both $\sum_{n=1}^{+\infty} u_n$ and $\sum_{n=1}^{+\infty} w_n$ are infinite series of positive terms and $w_n \le u_n$ for all positive integers n, it follows from part (i) that $\sum_{n=1}^{+\infty} w_n$ is convergent. However, this contradicts the hypothesis; so our assumption is false. Therefore $\sum\limits_{n=1}^{+\infty} u_n$ is divergent. ∎

As stated in Section 11.4, as a result of Theorem 11.4.1, the convergence or divergence of an infinite series is not affected by discarding a finite number of terms. Therefore, when applying the comparison test, if $u_i \leq v_i$ or $u_i \geq w_i$ when $i > m$, the test is valid regardless of how the first m terms of the two series compare.

EXAMPLE 2
Determine whether the infinite series

$$\sum_{n=1}^{+\infty} \frac{4}{3^n + 1}$$

is convergent or divergent.

SOLUTION
The given series is

$$\frac{4}{4} + \frac{4}{10} + \frac{4}{28} + \frac{4}{82} + \ldots + \frac{4}{3^n + 1} + \ldots$$

Comparing the nth term of this series with the nth term of the convergent geometric series

$$\frac{4}{3} + \frac{4}{9} + \frac{4}{27} + \frac{4}{81} + \ldots + \frac{4}{3^n} + \ldots \qquad r = \tfrac{1}{3} < 1$$

we have

$$\frac{4}{3^n + 1} < \frac{4}{3^n}$$

for every positive integer n. Therefore, by the comparison test, Theorem 11.5.2(i), the given series is convergent.

EXAMPLE 3
Determine whether the infinite series $\sum\limits_{n=1}^{+\infty} \frac{1}{\sqrt{n}}$ is convergent or divergent.

SOLUTION
The given series is

$$\sum_{n=1}^{+\infty} \frac{1}{\sqrt{n}} = \frac{1}{\sqrt{1}} + \frac{1}{\sqrt{2}} + \frac{1}{\sqrt{3}} + \ldots + \frac{1}{\sqrt{n}} + \ldots$$

Comparing the nth term of this series with the nth term of the divergent harmonic series we have

$$\frac{1}{\sqrt{n}} \geq \frac{1}{n} \qquad \text{for every positive integer } n$$

So by Theorem 11.5.2(ii) the given series $\displaystyle\sum_{n=1}^{+\infty} \frac{1}{\sqrt{n}}$ is divergent.

The following theorem, known as the *limit comparison test*, is a consequence of Theorem 11.5.2 and is often easier to apply.

11.5.3 THEOREM Limit Comparison Test	Let $\displaystyle\sum_{n=1}^{+\infty} u_n$ and $\displaystyle\sum_{n=1}^{+\infty} v_n$ be two series of positive terms.

(i) If $\displaystyle\lim_{n \to +\infty} \frac{u_n}{v_n} = c > 0$, then the two series either both converge or both diverge.

(ii) If $\displaystyle\lim_{n \to +\infty} \frac{u_n}{v_n} = 0$, and if $\displaystyle\sum_{n=1}^{+\infty} v_n$ converges, then $\displaystyle\sum_{n=1}^{+\infty} u_n$ converges.

(iii) If $\displaystyle\lim_{n \to +\infty} \frac{u_n}{v_n} = +\infty$, and if $\displaystyle\sum_{n=1}^{+\infty} v_n$ diverges, then $\displaystyle\sum_{n=1}^{+\infty} u_n$ diverges.

PROOF of (i)   Because $\displaystyle\lim_{n \to +\infty} (u_n/v_n) = c$, it follows that there exists an $N > 0$ such that

$$\text{if } n > N \quad \text{then} \quad \left| \frac{u_n}{v_n} - c \right| < \frac{c}{2}$$

$$\Leftrightarrow \quad \text{if } n > N \quad \text{then} \quad -\frac{c}{2} < \frac{u_n}{v_n} - c < \frac{c}{2}$$

$$\Leftrightarrow \quad \text{if } n > N \quad \text{then} \quad \frac{c}{2} < \frac{u_n}{v_n} < \frac{3c}{2} \tag{3}$$

From the right-hand inequality (3),

$$u_n < \tfrac{3}{2}cv_n \tag{4}$$

If $\displaystyle\sum_{n=1}^{+\infty} v_n$ is convergent, so is $\displaystyle\sum_{n=1}^{+\infty} \tfrac{3}{2}cv_n$. It follows from inequality (4) and the comparison test that $\displaystyle\sum_{n=1}^{+\infty} u_n$ is convergent.

From the left-hand inequality (3),

$$v_n < \frac{2}{c} u_n \tag{5}$$

If $\sum_{n=1}^{+\infty} u_n$ is convergent, so is $\sum_{n=1}^{+\infty} \frac{2}{c} u_n$. From inequality (5) and the comparison test it follows that $\sum_{n=1}^{+\infty} v_n$ is convergent.

If $\sum_{n=1}^{+\infty} v_n$ is divergent, $\sum_{n=1}^{+\infty} u_n$ can be shown to be divergent by assuming that $\sum_{n=1}^{+\infty} u_n$ is convergent and getting a contradiction by applying inequality (5) and the comparison test.

In a similar manner, if $\sum_{n=1}^{+\infty} u_n$ is divergent, it follows that $\sum_{n=1}^{+\infty} v_n$ is divergent because a contradiction is obtained from inequality (4) and the comparison test if $\sum_{n=1}^{+\infty} v_n$ is assumed to be convergent.

We have therefore proved part (i). The proofs of parts (ii) and (iii) are left as exercises (see Exercises 30 and 31). ∎

A word of caution is in order regarding part (ii) of Theorem 11.5.3. Note that when $\lim\limits_{n \to +\infty} \dfrac{u_n}{v_n} = 0$, the divergence of the series $\sum_{n=1}^{+\infty} v_n$ does *not* imply that the series $\sum_{n=1}^{+\infty} u_n$ diverges.

EXAMPLE 4
Solve Example 2 by using the limit comparison test.

SOLUTION

Let u_n be the nth term of the given series $\sum_{n=1}^{+\infty} \dfrac{4}{3^n + 1}$ and v_n be the nth term of the convergent geometric series $\sum_{n=1}^{+\infty} \dfrac{4}{3^n}$. Therefore

$$\lim_{n \to +\infty} \frac{u_n}{v_n} = \lim_{n \to +\infty} \frac{\dfrac{4}{3^n + 1}}{\dfrac{4}{3^n}}$$

$$= \lim_{n \to +\infty} \frac{3^n}{3^n + 1}$$

$$= \lim_{n \to +\infty} \frac{1}{1 + 3^{-n}}$$

$$= 1$$

Hence, by part (i) of the limit comparison test it follows that the given series is convergent.

EXAMPLE 5
Solve Example 3 by using the limit comparison test.

SOLUTION

Let u_n be the nth term of the given series $\sum\limits_{n=1}^{+\infty} \dfrac{1}{\sqrt{n}}$ and v_n be the nth term of the divergent harmonic series. Then

$$\lim_{n \to +\infty} \frac{u_n}{v_n} = \lim_{n \to +\infty} \frac{\dfrac{1}{\sqrt{n}}}{\dfrac{1}{n}}$$

$$= \lim_{n \to +\infty} \sqrt{n}$$

$$= +\infty$$

Therefore, by part (iii) of the limit comparison test the given series is divergent.

EXAMPLE 6

Determine whether the series $\sum\limits_{n=1}^{+\infty} \dfrac{n^3}{n!}$ is convergent or divergent.

SOLUTION

In Example 1 we proved that the series $\sum\limits_{n=1}^{+\infty} \dfrac{1}{n!}$ is convergent. By the limit comparison test with $u_n = \dfrac{n^3}{n!}$ and $v_n = \dfrac{1}{n!}$,

$$\lim_{n \to +\infty} \frac{u_n}{v_n} = \lim_{n \to +\infty} \frac{\dfrac{n^3}{n!}}{\dfrac{1}{n!}}$$

$$= \lim_{n \to +\infty} n^3$$

$$= +\infty$$

Part (iii) of the limit comparison test is not applicable because $\sum\limits_{n=1}^{+\infty} v_n$ converges. However, there is a way that the limit comparison test can be used. The given series can be written as

$$\frac{1^3}{1!} + \frac{2^3}{2!} + \frac{3^3}{3!} + \frac{4^3}{4!} + \frac{5^3}{5!} + \ldots + \frac{n^3}{n!} + \ldots$$

Because Theorem 11.4.1 allows us to subtract a finite number of terms without affecting the behavior (convergence or divergence) of a series, we discard the first three terms and obtain

$$\frac{4^3}{4!} + \frac{5^3}{5!} + \frac{6^3}{6!} + \ldots + \frac{(n+3)^3}{(n+3)!} + \ldots$$

Now let $u_n = \dfrac{(n+3)^3}{(n+3)!}$ and, as before, let $v_n = \dfrac{1}{n!}$. Then

$$\lim_{n \to +\infty} \frac{u_n}{v_n} = \lim_{n \to +\infty} \frac{\dfrac{(n+3)^3}{(n+3)!}}{\dfrac{1}{n!}}$$

$$= \lim_{n \to +\infty} \frac{(n+3)^3 n!}{(n+3)!}$$

$$= \lim_{n \to +\infty} \frac{(n+3)^3 n!}{n!(n+1)(n+2)(n+3)}$$

$$= \lim_{n \to +\infty} \frac{(n+3)^2}{(n+1)(n+2)}$$

$$= \lim_{n \to +\infty} \frac{n^2 + 6n + 9}{n^2 + 3n + 2}$$

$$= \lim_{n \to +\infty} \frac{1 + \dfrac{6}{n} + \dfrac{9}{n^2}}{1 + \dfrac{3}{n} + \dfrac{2}{n^2}}$$

$$= 1$$

It follows from part (i) of the Limit Comparison Test that the given series is convergent.

▶ ILLUSTRATION 1   Consider the geometric series

$$1 + \frac{1}{2} + \frac{1}{4} + \frac{1}{8} + \frac{1}{16} + \frac{1}{32} + \ldots + \frac{1}{2^{n-1}} + \ldots \tag{6}$$

which converges to 2 as shown in Illustration 2 of Section 11.3. Regroup the terms of this series to obtain

$$\left(1 + \frac{1}{2}\right) + \left(\frac{1}{4} + \frac{1}{8}\right) + \left(\frac{1}{16} + \frac{1}{32}\right) + \ldots + \left(\frac{1}{4^{n-1}} + \frac{1}{2 \cdot 4^{n-1}}\right) + \ldots$$

which is the series

$$\frac{3}{2} + \frac{3}{8} + \frac{3}{32} + \ldots + \frac{3}{2 \cdot 4^{n-1}} + \ldots \tag{7}$$

Series (7) is the geometric series with $a = \frac{3}{2}$ and $r = \frac{1}{4}$. Thus by Theorem 11.3.5 it is convergent, and its sum is

$$\frac{a}{1-r} = \frac{\frac{3}{2}}{1 - \frac{1}{4}}$$

$$= 2$$

Therefore series (7), which is obtained from the convergent series (6) by regrouping the terms, is also convergent. Its sum is the same as that of series (6). ◀

Illustration 1 gives a particular case of the following theorem.

11.5.4 THEOREM   If $\sum\limits_{n=1}^{+\infty} u_n$ is a given convergent series of positive terms, its terms can be grouped in any manner, and the resulting series also will be convergent and will have the same sum as the given series.

PROOF   Let $\{s_n\}$ be the sequence of partial sums for the given convergent series of positive terms. Then $\lim\limits_{n \to +\infty} s_n$ exists; let this limit be S. Consider a series $\sum\limits_{n=1}^{+\infty} v_n$ whose terms are obtained by grouping the terms of $\sum\limits_{n=1}^{+\infty} u_n$ in some manner. For example, $\sum\limits_{n=1}^{+\infty} v_n$ may be the series

$$u_1 + (u_2 + u_3) + (u_4 + u_5 + u_6) + (u_7 + u_8 + u_9 + u_{10}) + \ldots$$

or it may be the series

$$(u_1 + u_2) + (u_3 + u_4) + (u_5 + u_6) + (u_7 + u_8) + \ldots$$

and so forth. Let $\{t_m\}$ be the sequence of partial sums for the series $\sum\limits_{n=1}^{+\infty} v_n$. Each partial sum of the sequence $\{t_m\}$ is also a partial sum of the sequence $\{s_n\}$. Therefore, as m increases without bound, so does n. Because $\lim\limits_{n \to +\infty} s_n = S$, we conclude that $\lim\limits_{m \to +\infty} t_m = S$. This proves the theorem.   ■

Theorem 11.5.4 and the next theorem state properties of the sum of a convergent series of positive terms that are similar to properties that hold for the sum of a finite number of terms.

11.5.5 THEOREM   If $\sum\limits_{n=1}^{+\infty} u_n$ is a given convergent series of positive terms, the order of the terms can be rearranged, and the resulting series also will be convergent and will have the same sum as the given series.

PROOF   Let $\{s_n\}$ be the sequence of partial sums for the given convergent series of positive terms, and let $\lim\limits_{n \to +\infty} s_n = S$. Let $\sum\limits_{n=1}^{+\infty} v_n$ be a series formed by rearranging the order of the terms of $\sum\limits_{n=1}^{+\infty} u_n$. For example, $\sum\limits_{n=1}^{+\infty} v_n$ may be the series

$$u_4 + u_3 + u_7 + u_1 + u_9 + u_5 + \ldots$$

Let $\{t_n\}$ be the sequence of partial sums for the series $\sum\limits_{n=1}^{+\infty} v_n$. Each partial sum of the sequence $\{t_n\}$ will be less than S because it is the sum of n terms of the infinite series $\sum\limits_{n=1}^{+\infty} u_n$. Therefore S is an upper bound of the sequence $\{t_n\}$. Furthermore, because all the terms of the series $\sum\limits_{n=1}^{+\infty} v_n$ are positive, $\{t_n\}$ is a monotonic increasing sequence. Hence, by Theorem 11.2.7 the sequence $\{t_n\}$ is convergent, and $\lim\limits_{n\to+\infty} t_n = T \le S$. Now because the given series $\sum\limits_{n=1}^{+\infty} u_n$ can be obtained from the series $\sum\limits_{n=1}^{+\infty} v_n$ by rearranging the order of the terms, we can use the same argument and conclude that $S \le T$. If both inequalities, $T \le S$ and $S \le T$, must hold, it follows that $S = T$. This proves the theorem. ∎

A series that is often used in the comparison test is the one known as the *p* **series**, or the **hyperharmonic series**. It is

$$\frac{1}{1^p} + \frac{1}{2^p} + \frac{1}{3^p} + \ldots + \frac{1}{n^p} + \ldots \qquad \text{where } p \text{ is a constant} \qquad (8)$$

In the following illustration we prove that the *p* series diverges if $p \le 1$ and converges if $p > 1$.

▶ ILLUSTRATION 2   If $p = 1$, the *p* series is the harmonic series, which diverges. If $p < 1$, then $n^p \le n$; so

$$\frac{1}{n^p} \ge \frac{1}{n} \qquad \text{for every positive integer } n$$

Hence, by Theorem 11.5.2(ii) the *p* series is divergent if $p < 1$.
If $p > 1$, group the terms as follows:

$$\frac{1}{1^p} + \left(\frac{1}{2^p} + \frac{1}{3^p}\right) + \left(\frac{1}{4^p} + \frac{1}{5^p} + \frac{1}{6^p} + \frac{1}{7^p}\right) + \left(\frac{1}{8^p} + \frac{1}{9^p} + \ldots + \frac{1}{15^p}\right) + \ldots \qquad (9)$$

Consider the series

$$\frac{1}{1^p} + \frac{2}{2^p} + \frac{4}{4^p} + \frac{8}{8^p} + \ldots + \frac{2^{n-1}}{(2^{n-1})^p} + \ldots \qquad (10)$$

This is a geometric series whose ratio is $2/2^p = 1/2^{p-1}$, which is a positive number less than 1. Hence series (10) is convergent. Rewrite the terms of series

(10) to get

$$\frac{1}{1^p} + \left(\frac{1}{2^p} + \frac{1}{2^p}\right) + \left(\frac{1}{4^p} + \frac{1}{4^p} + \frac{1}{4^p} + \frac{1}{4^p}\right) + \left(\frac{1}{8^p} + \frac{1}{8^p} + \ldots + \frac{1}{8^p}\right) + \ldots \quad (11)$$

By comparing series (9) and series (11) it is seen that the group of terms in each set of parentheses after the first group is less in sum for (9) than it is for (11). Therefore, by the comparison test, series (9) is convergent. Because (9) is merely a regrouping of the terms of the p series when $p > 1$, it follows from Theorem 11.5.4 that the p series is convergent if $p > 1$.   ◀

Note that the series in Example 3 is the p series where $p = \frac{1}{2} < 1$; therefore it is divergent.

EXAMPLE 7

Determine whether the infinite series

$$\sum_{n=1}^{+\infty} \frac{1}{(n^2 + 2)^{1/3}}$$

is convergent or divergent.

SOLUTION

Because for large values of n the number $n^2 + 2$ is close to the number n^2, so is the number $1/(n^2 + 2)^{1/3}$ close to the number $1/n^{2/3}$. The series $\sum_{n=1}^{+\infty} \frac{1}{n^{2/3}}$ is divergent because it is the p series with $p = \frac{2}{3} < 1$. From the limit comparison test with $u_n = \dfrac{1}{(n^2 + 2)^{1/3}}$ and $v_n = \dfrac{1}{n^{2/3}}$,

$$\lim_{n \to +\infty} \frac{u_n}{v_n} = \lim_{n \to +\infty} \frac{\dfrac{1}{(n^2 + 2)^{1/3}}}{\dfrac{1}{n^{2/3}}}$$

$$= \lim_{n \to +\infty} \frac{n^{2/3}}{(n^2 + 2)^{1/3}}$$

$$= \lim_{n \to +\infty} \left(\frac{n^2}{n^2 + 2}\right)^{1/3}$$

$$= \lim_{n \to +\infty} \left(\frac{1}{1 + \dfrac{2}{n^2}}\right)^{1/3}$$

$$= 1$$

Therefore the given series is divergent.

EXERCISES 11.5

In Exercises 1 through 26, determine if the given series is convergent or divergent.

1. $\displaystyle\sum_{n=1}^{+\infty} \frac{1}{n2^n}$

2. $\displaystyle\sum_{n=1}^{+\infty} \frac{1}{\sqrt{2n+1}}$

3. $\displaystyle\sum_{n=1}^{+\infty} \frac{1}{n^n}$

4. $\displaystyle\sum_{n=1}^{+\infty} \frac{n^2}{4n^3+1}$

5. $\displaystyle\sum_{n=1}^{+\infty} \frac{3n+1}{2n^2+5}$

6. $\displaystyle\sum_{n=1}^{+\infty} \frac{3}{\sqrt{n^3+n}}$

7. $\displaystyle\sum_{n=1}^{+\infty} \frac{\cos^2 n}{3^n}$

8. $\displaystyle\sum_{n=1}^{+\infty} \frac{1}{\ln(n+1)}$

9. $\displaystyle\sum_{n=1}^{+\infty} \frac{1}{\sqrt{n^2+4n}}$

10. $\displaystyle\sum_{n=1}^{+\infty} \frac{|\sin n|}{n^2}$

11. $\displaystyle\sum_{n=1}^{+\infty} \frac{n!}{(n+2)!}$

12. $\displaystyle\sum_{n=1}^{+\infty} \frac{1}{\sqrt{n^3+1}}$

13. $\displaystyle\sum_{n=1}^{+\infty} \frac{n}{5n^2+3}$

14. $\displaystyle\sum_{n=1}^{+\infty} \frac{(n-1)!}{(n+1)!}$

15. $\displaystyle\sum_{n=1}^{+\infty} \frac{n!}{(2n)!}$

16. $\displaystyle\sum_{n=1}^{+\infty} \sin\frac{1}{n}$

17. $\displaystyle\sum_{n=1}^{+\infty} \frac{|\csc n|}{n}$

18. $\displaystyle\sum_{n=1}^{+\infty} \frac{1}{n+\sqrt{n}}$

19. $\displaystyle\sum_{n=2}^{+\infty} \frac{1}{n\sqrt{n^2-1}}$

20. $\displaystyle\sum_{n=1}^{+\infty} \frac{2^n}{n!}$

21. $\displaystyle\sum_{n=1}^{+\infty} \frac{3}{2n-\sqrt{n}}$

22. $\displaystyle\sum_{n=1}^{+\infty} \frac{\sqrt{n}}{n^2+1}$

23. $\displaystyle\sum_{n=1}^{+\infty} \frac{\ln n}{n^2+2}$

24. $\displaystyle\sum_{n=1}^{+\infty} \frac{1}{3^n-\cos n}$

25. $\displaystyle\sum_{n=1}^{+\infty} \frac{(n+1)^2}{(n+2)!}$

26. $\displaystyle\sum_{n=1}^{+\infty} \frac{1}{(n+2)(n+4)}$

27. Use the series $\displaystyle\sum_{n=1}^{+\infty} (-1)^{n+1}$ to show that Theorem 11.5.1 does not apply to an infinite series of both positive and negative terms; that is, show that the sequence of partial sums has an upper bound but the series is not convergent.

28. Suppose that f is a function such that $f(n) > 0$ for n any positive integer. Furthermore suppose that if p is any positive number, $\displaystyle\lim_{n \to +\infty} n^p f(n)$ exists and is positive. Prove that the series $\displaystyle\sum_{n=1}^{+\infty} f(n)$ is convergent if $p > 1$ and divergent if $0 < p \le 1$.

29. If $\displaystyle\sum_{n=1}^{+\infty} a_n$ and $\displaystyle\sum_{n=1}^{+\infty} b_n$ are two convergent series of positive terms, use the Limit Comparison Test to prove that the series $\displaystyle\sum_{n=1}^{+\infty} a_n b_n$ is also convergent.

30. Prove Theorem 11.5.3(ii).

31. Prove Theorem 11.5.3(iii).

11.6
THE INTEGRAL TEST

The theorem known as the *integral test* makes use of the theory of improper integrals to test an infinite series of positive terms for convergence.

11.6.1 THEOREM
Integral Test

Let f be a function that is continuous, decreasing, and positive-valued for all $x \ge 1$. Then the infinite series

$$\sum_{n=1}^{+\infty} f(n) = f(1) + f(2) + f(3) + \ldots + f(n) + \ldots$$

is convergent if the improper integral

$$\int_1^{+\infty} f(x)\,dx$$

exists, and it is divergent if $\displaystyle\lim_{b \to +\infty} \int_1^b f(x)\,dx = +\infty$.

PROOF   If i is a positive integer and $i \geq 2$, then by the mean-value theorem for integrals (4.7.1) there exists a number X such that $i - 1 \leq X \leq i$ and

$$\int_{i-1}^{i} f(x)\, dx = f(X) \cdot 1 \tag{1}$$

Because f is a decreasing function,

$$f(i - 1) \geq f(X) \geq f(i)$$

and so from (1),

$$f(i - 1) \geq \int_{i-1}^{i} f(x)\, dx \geq f(i)$$

Therefore, if n is a positive integer and $n \geq 2$,

$$\sum_{i=2}^{n} f(i - 1) \geq \sum_{i=2}^{n} \int_{i-1}^{i} f(x)\, dx \geq \sum_{i=2}^{n} f(i)$$

$$\Leftrightarrow \quad \sum_{i=1}^{n-1} f(i) \geq \int_{1}^{n} f(x)\, dx \geq \sum_{i=1}^{n} f(i) - f(1) \tag{2}$$

Figures 11.6.1 and 11.6.2 show the geometric interpretation of the above discussion for $n = 6$. In Fig. 11.6.1 there is a sketch of the graph of a function f satisfying the hypothesis. The sum of the measures of the areas of the shaded rectangles is $f(1) + f(2) + f(3) + f(4) + f(5)$, which is the left member of inequality (2) when $n = 6$. Clearly, the sum of the measures of the areas of these rectangles is greater than the measure of the area given by the definite integral when $n = 6$. In Fig. 11.6.2 the sum of the measures of the areas of the shaded rectangles is $f(2) + f(3) + f(4) + f(5) + f(6)$, which is the right member of the inequality (2) when $n = 6$. This sum is less than the value of the definite integral when $n = 6$.

If the given improper integral exists, let L be its value. Then

$$\int_{1}^{n} f(x)\, dx \leq L \tag{3}$$

From the second and third members of the inequality (2) and from (3),

$$\sum_{i=1}^{n} f(i) \leq f(1) + \int_{1}^{n} f(x)\, dx \leq f(1) + L \tag{4}$$

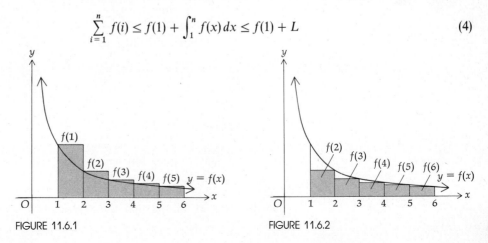

FIGURE 11.6.1   FIGURE 11.6.2

Consider now the infinite series $\sum\limits_{n=1}^{+\infty} f(n)$. Let the sequence of partial sums of this series be $\{s_n\}$, where $s_n = \sum\limits_{i=1}^{n} f(i)$. From (4), $\{s_n\}$ has an upper bound of $f(1) + L$. Hence, by Theorem 11.5.1, $\sum\limits_{n=1}^{+\infty} f(n)$ is convergent.

Suppose that $\lim\limits_{b \to +\infty} \int_1^b f(x)\,dx = +\infty$. From (2)

$$\sum_{i=1}^{n-1} f(i) \geq \int_1^n f(x)\,dx$$

for all positive integers n. Therefore

$$\lim_{n \to +\infty} s_n = \lim_{n \to +\infty} \sum_{i=1}^{n} f(i)$$

$$= +\infty$$

Hence $\sum\limits_{n=1}^{+\infty} f(n)$ is divergent. ∎

EXAMPLE 1
Use the integral test to show that the p series diverges if $p \leq 1$ and converges if $p > 1$.

SOLUTION

The p series is $\sum\limits_{n=1}^{+\infty} \dfrac{1}{n^p}$. If $f(x) = \dfrac{1}{x^p}$, then f is continuous and positive valued for all $x \geq 1$. Furthermore, if $1 \leq x_1 < x_2$, then $\dfrac{1}{x_1^{\,p}} > \dfrac{1}{x_2^{\,p}}$, and so f is decreasing for all $x \geq 1$. Therefore the hypothesis of Theorem 11.6.1 is satisfied by the function f. We consider the improper integral and have

$$\int_1^{+\infty} \frac{dx}{x^p} = \lim_{b \to +\infty} \int_1^b \frac{dx}{x^p}$$

If $p = 1$, the above integral gives

$$\lim_{b \to +\infty} \ln x \Big]_1^b = \lim_{b \to +\infty} \ln b$$

$$= +\infty$$

If $p \neq 1$, the integral gives

$$\lim_{b \to +\infty} \frac{x^{1-p}}{1-p} \Big]_1^b = \lim_{b \to +\infty} \frac{b^{1-p} - 1}{1-p}$$

This limit is $+\infty$ when $p < 1$; it is $-1/(1-p)$ if $p > 1$. Therefore, by the integral test it follows that the p series converges for $p > 1$ and diverges for $p \leq 1$.

EXAMPLE 2

Determine if the infinite series $\displaystyle\sum_{n=1}^{+\infty} ne^{-n}$ is convergent or divergent.

SOLUTION

Let $f(x) = xe^{-x}$. Then

$$f'(x) = e^{-x} - xe^{-x}$$
$$= e^{-x}(1 - x)$$

Because $f'(x) < 0$ if $x > 1$, it follows from Theorem 3.4.3 that f is decreasing if $x \geq 1$. Furthermore, f is continuous and positive valued for all $x \geq 1$. Thus the hypothesis of the integral test is satisfied. By applying integration by parts,

$$\int xe^{-x}\,dx = -e^{-x}(x + 1) + C$$

Hence

$$\int_{1}^{+\infty} xe^{-x}\,dx = \lim_{b \to +\infty} \left[-e^{-x}(x + 1) \right]_{1}^{b}$$

$$= \lim_{b \to +\infty} \left[-\frac{b + 1}{e^{b}} + \frac{2}{e} \right]$$

Because $\displaystyle\lim_{b \to +\infty}(b + 1) = +\infty$ and $\displaystyle\lim_{b \to +\infty} e^{b} = +\infty$, L'Hôpital's rule can be used to obtain

$$\lim_{b \to +\infty} \frac{b + 1}{e^{b}} = \lim_{b \to +\infty} \frac{1}{e^{b}}$$

$$= 0$$

Therefore

$$\int_{1}^{+\infty} xe^{-x}\,dx = \frac{2}{e}$$

Thus the given series is convergent.

If for an infinite series the summation index starts with $n = k$ rather than $n = 1$, we have the following modification of the integral test:

If f is a function that is continuous, decreasing, and positive-valued for all $x \geq k$, then the infinite series $\displaystyle\sum_{n=k}^{+\infty} f(n)$ is convergent if the improper integral

$$\int_{k}^{+\infty} f(x)\,dx$$

exists and is divergent if $\displaystyle\lim_{b \to +\infty} \int_{k}^{b} f(x)\,dx = +\infty$.

The proof is identical to that of Theorem 11.6.1.

EXAMPLE 3
Determine if the series

$$\sum_{n=2}^{+\infty} \frac{1}{n\sqrt{\ln n}}$$

is convergent or divergent.

SOLUTION
The function f defined by

$$f(x) = \frac{1}{x\sqrt{\ln x}}$$

is continuous and positive-valued for all $x \geq 2$. Also, if $2 \leq x_1 < x_2$, then $f(x_1) > f(x_2)$; so f is decreasing for all $x \geq 2$. Therefore the integral test can be applied.

$$\int_2^{+\infty} \frac{dx}{x\sqrt{\ln x}} = \lim_{b \to +\infty} \int_2^b (\ln x)^{-1/2} \frac{dx}{x}$$

$$= \lim_{b \to +\infty} \left[2\sqrt{\ln x}\right]_2^b$$

$$= \lim_{b \to +\infty} \left[2\sqrt{\ln b} - 2\sqrt{\ln 2}\right]$$

$$= +\infty$$

Thus the given series is divergent.

EXERCISES 11.6

In Exercises 1 through 8, use the integral test to determine if the given series is convergent or divergent.

1. $\sum_{n=1}^{+\infty} \frac{1}{2n + 1}$

2. $\sum_{n=1}^{+\infty} \frac{2}{(3n + 5)^2}$

3. $\sum_{n=1}^{+\infty} \frac{1}{(n + 2)^{3/2}}$

4. $\sum_{n=2}^{+\infty} \frac{n}{n^2 - 2}$

5. $\sum_{n=3}^{+\infty} \frac{4}{n^2 - 4}$

6. $\sum_{n=1}^{+\infty} \frac{2n + 3}{(n^2 + 3n)^2}$

7. $\sum_{n=1}^{+\infty} e^{-5n}$

8. $\sum_{n=1}^{+\infty} \frac{2n}{n^4 + 1}$

In Exercises 9 through 22, determine if the given series is convergent or divergent.

9. $\sum_{n=1}^{+\infty} \frac{\ln n}{n}$

10. $\sum_{n=2}^{+\infty} \frac{1}{n \ln n}$

11. $\sum_{n=1}^{+\infty} \frac{\tan^{-1} n}{n^2 + 1}$

12. $\sum_{n=1}^{+\infty} ne^{-n^2}$

13. $\sum_{n=1}^{+\infty} n^2 e^{-n}$

14. $\sum_{n=1}^{+\infty} ne^{-n}$

15. $\sum_{n=2}^{+\infty} \frac{\ln n}{n^3}$

16. $\sum_{n=1}^{+\infty} \cot^{-1} n$

17. $\sum_{n=1}^{+\infty} \text{csch } n$

18. $\sum_{n=1}^{+\infty} \frac{e^{\tan^{-1} n}}{n^2 + 1}$

19. $\sum_{n=1}^{+\infty} \frac{e^{1/n}}{n^2}$

20. $\sum_{n=1}^{+\infty} \text{sech}^2 n$

21. $\sum_{n=1}^{+\infty} \ln\left(\frac{n + 3}{n}\right)$

22. $\sum_{n=2}^{+\infty} \frac{1}{n(\ln n)^3}$

23. Prove that the series $\displaystyle\sum_{n=2}^{+\infty} \frac{1}{n(\ln n)^p}$ is convergent if and only if $p > 1$.

24. Prove that the series $\displaystyle\sum_{n=3}^{+\infty} \frac{1}{n(\ln n)\left[\ln(\ln n)\right]^p}$ is convergent if and only if $p > 1$.

25. Prove that the series $\displaystyle\sum_{n=1}^{+\infty} \frac{\ln n}{n^p}$ is convergent if and only if $p > 1$.

26. If s_k is the kth partial sum of the harmonic series, prove that $\ln(k + 1) < s_k < 1 + \ln k$. (*Hint:* $\dfrac{1}{m + 1} \leq \dfrac{1}{x} \leq \dfrac{1}{m}$ if $0 < m \leq x \leq m + 1$. Integrate each member of the inequality from m to $m + 1$; let m take on successively the values $1, 2, \ldots, n - 1$, and add the results.)

27. Use the result of Exercise 26 to estimate the sum $\displaystyle\sum_{m=50}^{100} \frac{1}{m} = \frac{1}{50} + \frac{1}{51} + \ldots + \frac{1}{100}$.

**11.7
ALTERNATING SERIES**

In this section and the next, infinite series having both positive and negative terms are considered. The first type of such a series that we discuss is one whose terms are alternately positive and negative—an *alternating series*.

11.7.1 DEFINITION   If $a_n > 0$ for all positive integers n, then the series

$$\sum_{n=1}^{+\infty} (-1)^{n+1} a_n = a_1 - a_2 + a_3 - a_4 + \ldots + (-1)^{n+1} a_n + \ldots \tag{1}$$

and the series

$$\sum_{n=1}^{+\infty} (-1)^n a_n = -a_1 + a_2 - a_3 + a_4 - \ldots + (-1)^n a_n + \ldots \tag{2}$$

are called **alternating series**.

▶ ILLUSTRATION 1   An example of an alternating series of the form (1), where the first term is positive, is

$$\sum_{n=1}^{+\infty} (-1)^{n+1} \frac{1}{n} = 1 - \frac{1}{2} + \frac{1}{3} - \frac{1}{4} + \ldots + (-1)^{n+1} \frac{1}{n} + \ldots$$

An alternating series of the form (2), where the first term is negative, is

$$\sum_{n=1}^{+\infty} (-1)^n \frac{1}{n!} = -1 + \frac{1}{2!} - \frac{1}{3!} + \frac{1}{4!} - \ldots + (-1)^n \frac{1}{n!} + \ldots \qquad \blacktriangleleft$$

The following theorem gives a test for the convergence of an alternating series. It is called the alternating-series test; it is also known as Leibniz's test for alternating series because Leibniz formulated it in 1705.

11.7.2 THEOREM
Alternating-Series Test

Suppose we have the alternating series $\sum\limits_{n=1}^{+\infty} (-1)^{n+1} a_n \left[\text{or} \sum\limits_{n=1}^{+\infty} (-1)^n a_n \right]$, where $a_n > 0$ and $a_{n+1} < a_n$ for all positive integers n. If $\lim\limits_{n \to +\infty} a_n = 0$, the alternating series is convergent.

PROOF   Assume that the first term of the alternating series is positive. This assumption is not a loss of generality because if this is not the case, then we discard the first term, which does not affect the convergence of the series. Thus we have the alternating series $\sum\limits_{n=1}^{+\infty} (-1)^{n+1} a_n$. Consider the partial sum

$$s_{2n} = (a_1 - a_2) + (a_3 - a_4) + \ldots + (a_{2n-1} - a_{2n}) \tag{3}$$

Because by hypothesis $a_{n+1} < a_n$, each quantity in parentheses in (3) is positive. Therefore

$$0 < s_2 < s_4 < s_6 < \ldots < s_{2n} < \ldots \tag{4}$$

We can also write s_{2n} as

$$s_{2n} = a_1 - (a_2 - a_3) - (a_4 - a_5) - \ldots - (a_{2n-2} - a_{2n-1}) - a_{2n} \tag{5}$$

Because $a_{n+1} < a_n$, each quantity in parentheses in (5) is positive. Therefore

$$s_{2n} < a_1 \qquad \text{for every positive integer } n \tag{6}$$

From (4) and (6),

$$0 < s_{2n} < a_1 \qquad \text{for every positive integer } n$$

Thus the sequence $\{s_{2n}\}$ is bounded. Furthermore, from (4), the sequence $\{s_{2n}\}$ is increasing. Therefore, by Theorem 11.2.6 the sequence $\{s_{2n}\}$ is convergent. Let $\lim\limits_{n \to +\infty} s_{2n} = S$, and from Theorem 11.2.7, $S \leq a_1$. Because $s_{2n+1} = s_{2n} + a_{2n+1}$,

$$\lim_{n \to +\infty} s_{2n+1} = \lim_{n \to +\infty} s_{2n} + \lim_{n \to +\infty} a_{2n+1}$$

But, by hypothesis, $\lim\limits_{n \to +\infty} a_{2n+1} = 0$; so $\lim\limits_{n \to +\infty} s_{2n+1} = \lim\limits_{n \to +\infty} s_{2n}$. Therefore the sequence of partial sums of the even-numbered terms and the sequence of partial sums of the odd-numbered terms have the same limit S.

We now show that $\lim\limits_{n \to +\infty} s_n = S$. Because $\lim\limits_{n \to +\infty} s_{2n} = S$, then for any $\epsilon > 0$ there exists an integer $N_1 > 0$ such that

$$\text{if } 2n \geq N_1 \quad \text{then} \quad |s_{2n} - S| < \epsilon$$

And because $\lim\limits_{n \to +\infty} s_{2n+1} = S$, there exists an integer $N_2 > 0$ such that

$$\text{if } 2n + 1 \geq N_2 \quad \text{then} \quad |s_{2n+1} - S| < \epsilon$$

If N is the larger of the two integers N_1 and N_2, it follows that if n is any integer, either odd or even, and

if $n \geq N$ then $|s_n - S| < \epsilon$

Therefore $\lim\limits_{n \to +\infty} s_n = S$; so the alternating series is convergent. ∎

EXAMPLE 1
Prove that the alternating series

$$\sum_{n=1}^{+\infty} (-1)^{n+1} \frac{1}{n}$$

is convergent.

SOLUTION
The given series is

$$1 - \frac{1}{2} + \frac{1}{3} - \frac{1}{4} + \ldots + (-1)^{n+1} \frac{1}{n} + (-1)^{n+2} \frac{1}{n+1} + \ldots$$

Because $\dfrac{1}{n+1} < \dfrac{1}{n}$ for all positive integers n, and $\lim\limits_{n \to +\infty} \dfrac{1}{n} = 0$, it follows from Theorem 11.7.2 that the given alternating series is convergent.

EXAMPLE 2
Determine if the series

$$\sum_{n=1}^{+\infty} (-1)^n \frac{n+2}{n(n+1)}$$

is convergent or divergent.

SOLUTION
The given series is an alternating series.

$$\lim_{n \to +\infty} a_n = \lim_{n \to +\infty} \frac{n+2}{n(n+1)}$$

$$= \lim_{n \to +\infty} \frac{\dfrac{1}{n} + \dfrac{2}{n^2}}{1 + \dfrac{1}{n}}$$

$$= 0$$

Before the alternating-series test can be applied, we must also show that $a_{n+1} < a_n$ or, equivalently, $\dfrac{a_{n+1}}{a_n} < 1$.

$$\frac{a_{n+1}}{a_n} = \frac{\dfrac{n+3}{(n+1)(n+2)}}{\dfrac{n+2}{n(n+1)}}$$

$$= \frac{n(n+3)}{(n+2)^2}$$

$$= \frac{n^2 + 3n}{n^2 + 4n + 4}$$

$$< 1$$

Then it follows from Theorem 11.7.2 that the given series is convergent.

11.7.3 DEFINITION   If an infinite series is convergent and its sum is S, then the **remainder** obtained by approximating the sum of the series by the kth partial sum s_k is denoted by R_k, and

$$R_k = S - s_k$$

11.7.4 THEOREM   Suppose we have the alternating series $\displaystyle\sum_{n=1}^{+\infty} (-1)^{n+1} a_n \left[\text{or} \sum_{n=1}^{+\infty} (-1)^n a_n \right]$, where $a_n > 0$ and $a_{n+1} < a_n$ for all positive integers n, and $\displaystyle\lim_{n \to +\infty} a_n = 0$. Then if R_k is the remainder obtained by approximating the sum of the series by the sum of the first k terms, $|R_k| < a_{k+1}$.

PROOF   The given series converges by the alternating-series test (Theorem 11.7.2). Assume that the odd-numbered terms of the given series are positive and the even-numbered terms are negative. Then from (4) in the proof of Theorem 11.7.2, the sequence $\{s_{2n}\}$ is increasing. So if S is the sum of the given series,

$$s_{2k} < s_{2k+2} < S \qquad \text{for all } k \geq 1 \tag{7}$$

To show that the sequence $\{s_{2n-1}\}$ is decreasing we write

$$s_{2n-1} = a_1 - (a_2 - a_3) - (a_4 - a_5) - \ldots - (a_{2n-2} - a_{2n-1}) \tag{8}$$

Because $a_{n+1} < a_n$, it follows that each quantity in parentheses in (8) is positive. Therefore, because $a_1 > 0$

$$s_1 > s_3 > s_5 > \ldots > s_{2n-1} > \ldots$$

Hence the sequence $\{s_{2n-1}\}$ is decreasing. Thus

$$S < s_{2k+1} < s_{2k-1} \qquad \text{for all } k \geq 1 \tag{9}$$

From (9), $S < s_{2k+1}$. Thus

$$S - s_{2k} < s_{2k+1} - s_{2k} = a_{2k+1} \qquad \text{for all } k \geq 1 \tag{10}$$

From (7), $s_{2k} < S$. Hence

$$0 < S - s_{2k} \qquad \text{for all } k \geq 1 \tag{11}$$

Therefore, from (10) and (11),

$$0 < S - s_{2k} < a_{2k+1} \qquad \text{for all } k \geq 1 \tag{12}$$

From (7), $-S < -s_{2k}$. Hence

$$s_{2k-1} - S < s_{2k-1} - s_{2k} = a_{2k} \qquad \text{for all } k \geq 1 \tag{13}$$

From (9),

$$0 < s_{2k-1} - S \qquad \text{for all } k \geq 1 \tag{14}$$

Then from (13) and (14),

$$0 < s_{2k-1} - S < a_{2k} \qquad \text{for all } k \geq 1 \tag{15}$$

Because from Definition 11.7.3, $R_k = S - s_k$, then (12) can be written as

$$0 < R_{2k} < a_{2k+1} \qquad \text{for all } k \geq 1 \tag{16}$$

and (15) can be written as

$$0 < -R_{2k-1} < a_{2k} \qquad \text{for all } k \geq 1 \tag{17}$$

Combining (16) and (17) we have

$$|R_k| < a_{k+1} \qquad \text{for all } k \geq 1$$

and the theorem is proved. ∎

EXAMPLE 3
A series for computing $\ln(1 + x)$ if x is in the open interval $(-1, 1)$ is

$$\ln(1 + x) = \sum_{n=1}^{+\infty} (-1)^{n+1} \frac{x^n}{n}$$

Find an upper bound for the error when the first three terms of this series are used to approximate the value of $\ln 1.1$.

SOLUTION
We use the given series with $x = 0.1$ to obtain

$$\ln 1.1 = 0.1 - \frac{(0.1)^2}{2} + \frac{(0.1)^3}{3} - \frac{(0.1)^4}{4} + \cdots$$

This series satisfies the conditions of Theorem 11.7.4; so if R_3 is the difference between the actual value of $\ln 1.1$ and the sum of the first three terms, then

$$|R_3| < 0.000025$$

Thus the sum of the first three terms will yield a value of $\ln 1.1$ accurate to at least four decimal places. Using the first three terms we get

$$\ln 1.1 \approx 0.0953$$

EXERCISES 11.7

In Exercises 1 through 14, determine if the given alternating series is convergent or divergent.

1. $\displaystyle\sum_{n=1}^{+\infty}(-1)^{n+1}\frac{1}{2n}$ **2.** $\displaystyle\sum_{n=1}^{+\infty}(-1)^{n}\frac{1}{n^2}$ **3.** $\displaystyle\sum_{n=1}^{+\infty}(-1)^{n}\frac{3}{n^2+1}$ **4.** $\displaystyle\sum_{n=1}^{+\infty}(-1)^{n+1}\frac{4}{3n-2}$

5. $\displaystyle\sum_{n=2}^{+\infty}(-1)^{n}\frac{1}{\ln n}$ **6.** $\displaystyle\sum_{n=1}^{+\infty}(-1)^{n+1}\sin\frac{\pi}{n}$ **7.** $\displaystyle\sum_{n=1}^{+\infty}(-1)^{n+1}\frac{n^2}{n^3+2}$ **8.** $\displaystyle\sum_{n=1}^{+\infty}(-1)^{n+1}\frac{\ln n}{n}$

9. $\displaystyle\sum_{n=1}^{+\infty}(-1)^{n+1}\frac{\ln n}{n^2}$ **10.** $\displaystyle\sum_{n=1}^{+\infty}(-1)^{n}\frac{e^n}{n}$ **11.** $\displaystyle\sum_{n=1}^{+\infty}(-1)^{n}\frac{3^n}{n^2}$ DIV **12.** $\displaystyle\sum_{n=1}^{+\infty}(-1)^{n}\frac{\sqrt{n}}{3n-1}$

13. $\displaystyle\sum_{n=1}^{+\infty}(-1)^{n}\frac{n}{2^n}$ **14.** $\displaystyle\sum_{n=1}^{+\infty}(-1)^{n+1}\frac{3^n}{1+3^{2n}}$

In Exercises 15 through 22, find an upper bound for the error if the sum of the first four terms is used as an approximation to the sum of the given infinite series.

15. $\displaystyle\sum_{n=1}^{+\infty}(-1)^{n+1}\frac{1}{n}$ **16.** $\displaystyle\sum_{n=1}^{+\infty}(-1)^{n}\frac{2}{n^2}$ **17.** $\displaystyle\sum_{n=1}^{+\infty}(-1)^{n+1}\frac{1}{(2n-1)^2}$

18. $\displaystyle\sum_{n=1}^{+\infty}(-1)^{n+1}\frac{n}{(n+1)^2}$ **19.** $\displaystyle\sum_{n=1}^{+\infty}(-1)^{n}\frac{1}{n^2}$ **20.** $\displaystyle\sum_{n=1}^{+\infty}(-1)^{n+1}\frac{1}{n^n}$

21. $\displaystyle\sum_{n=1}^{+\infty}(-1)^{n+1}\frac{1}{(n+1)\ln(n+1)}$ **22.** $\displaystyle\sum_{n=1}^{+\infty}(-1)^{n}\frac{1}{n!}$

In Exercises 23 through 30, find the sum of the given infinite series, accurate to three decimal places.

23. $\displaystyle\sum_{n=1}^{+\infty}(-1)^{n+1}\frac{1}{2^n}$ **24.** $\displaystyle\sum_{n=1}^{+\infty}(-1)^{n+1}\frac{1}{n^4}$ **25.** $\displaystyle\sum_{n=1}^{+\infty}(-1)^{n+1}\frac{1}{n!}$ **26.** $\displaystyle\sum_{n=1}^{+\infty}(-1)^{n+1}\frac{2}{3^n}$

27. $\displaystyle\sum_{n=1}^{+\infty}(-1)^{n+1}\frac{1}{(2n)^3}$ **28.** $\displaystyle\sum_{n=1}^{+\infty}(-1)^{n}\frac{1}{(2n+1)^3}$ **29.** $\displaystyle\sum_{n=1}^{+\infty}(-1)^{n+1}\frac{1}{n2^n}$ **30.** $\displaystyle\sum_{n=1}^{+\infty}(-1)^{n+1}\frac{1}{(2n)!}$

11.8
ABSOLUTE AND CONDITIONAL CONVERGENCE, THE RATIO TEST, AND THE ROOT TEST

If all the terms of a given infinite series are replaced by their absolute values and the resulting series is convergent, then the given series is said to be *absolutely convergent*. Following is the formal definition.

11.8.1 DEFINITION The infinite series $\displaystyle\sum_{n=1}^{+\infty}u_n$ is said to be **absolutely convergent** if the series $\displaystyle\sum_{n=1}^{+\infty}|u_n|$ is convergent.

▶ ILLUSTRATION 1   Consider the series

$$\sum_{n=1}^{+\infty}(-1)^{n+1}\frac{2}{3^n}=\frac{2}{3}-\frac{2}{3^2}+\frac{2}{3^3}-\frac{2}{3^4}+\ldots+(-1)^{n+1}\frac{2}{3^n}+\ldots \qquad (1)$$

This series will be absolutely convergent if the series

$$\sum_{n=1}^{+\infty} \frac{2}{3^n} = \frac{2}{3} + \frac{2}{3^2} + \frac{2}{3^3} + \frac{2}{3^4} + \ldots + \frac{2}{3^n} + \ldots$$

is convergent. Because this is the geometric series with $r = \frac{1}{3} < 1$, it is convergent. Therefore series (1) is absolutely convergent.   ◀

▶ ILLUSTRATION 2   A convergent series that is not absolutely convergent is the series

$$\sum_{n=1}^{+\infty} \frac{(-1)^{n+1}}{n}$$

In Example 1 of Section 11.7 this series was proved to be convergent. The series is not absolutely convergent because the series of absolute values is the harmonic series, which is divergent.   ◀

The series of Illustration 2 is an example of a *conditionally convergent* series.

11.8.2 DEFINITION   A series that is convergent, but not absolutely convergent, is said to be **conditionally convergent**.

It is possible, then, for a series to be convergent but not absolutely convergent. If a series is absolutely convergent, it must be convergent, however, and this is given by the next theorem.

11.8.3 THEOREM   If the infinite series $\sum_{n=1}^{+\infty} u_n$ is absolutely convergent, it is convergent and

$$\left| \sum_{n=1}^{+\infty} u_n \right| \le \sum_{n=1}^{+\infty} |u_n|$$

PROOF   Consider the three infinite series $\sum_{n=1}^{+\infty} u_n$, $\sum_{n=1}^{+\infty} |u_n|$, and $\sum_{n=1}^{+\infty} (u_n + |u_n|)$, and let their sequences of partial sums be $\{s_n\}$, $\{t_n\}$, and $\{r_n\}$, respectively. For every positive integer n, $u_n + |u_n|$ is either 0 or $2|u_n|$; so we have the inequality

$$0 \le u_n + |u_n| \le 2|u_n| \tag{2}$$

Because $\sum_{n=1}^{+\infty} |u_n|$ is convergent, it has a sum, which we denote by T. $\{t_n\}$ is an increasing sequence of positive numbers; so $t_n \le T$ for all positive integers n. From (2) it follows that

$$0 \le r_n \le 2t_n \le 2T$$

Therefore the sequence $\{r_n\}$ has an upper bound of $2T$. Thus, by Theorem 11.5.1 the series $\sum_{n=1}^{+\infty} (u_n + |u_n|)$ is convergent. Let its sum be R. Because from

(2), $\{r_n\}$ is an increasing sequence, it may be concluded from Theorem 11.2.7 that $R \leq 2T$.

Each of the series $\sum\limits_{n=1}^{+\infty} (u_n + |u_n|)$ and $\sum\limits_{n=1}^{+\infty} |u_n|$ is convergent; hence, from Theorem 11.4.3 the series

$$\sum_{n=1}^{+\infty} [(u_n + |u_n|) - |u_n|] = \sum_{n=1}^{+\infty} u_n$$

is also convergent.

Let the sum of the series $\sum\limits_{n=1}^{+\infty} u_n$ be S. Then, also from Theorem 11.4.3, $S = R - T$. And because $R \leq 2T$, $S \leq 2T - T = T$.

Because $\sum\limits_{n=1}^{+\infty} u_n$ is convergent and has the sum S, it follows from Theorem 11.4.2 that $\sum\limits_{n=1}^{+\infty} (-u_n)$ is convergent and has the sum $-S$. Because $\sum\limits_{n=1}^{+\infty} |-u_n|$ and $\sum\limits_{n=1}^{+\infty} |u_n|$ both equal T, we can replace $\sum\limits_{n=1}^{+\infty} u_n$ by $\sum\limits_{n=1}^{+\infty} (-u_n)$ in the above discussion and show that $-S \leq T$. Because $S \leq T$ and $-S \leq T$, we have $|S| \leq T$; therefore $\left| \sum\limits_{n=1}^{+\infty} u_n \right| \leq \sum\limits_{n=1}^{+\infty} |u_n|$, and the theorem is proved.   ■

EXAMPLE 1

Determine if the series

$$\sum_{n=1}^{+\infty} \frac{\cos \frac{1}{3}n\pi}{n^2}$$

is convergent or divergent.

SOLUTION

Denote the given series by $\sum\limits_{n=1}^{+\infty} u_n$. Therefore

$$\sum_{n=1}^{+\infty} u_n = \frac{\frac{1}{2}}{1^2} - \frac{\frac{1}{2}}{2^2} - \frac{1}{3^2} - \frac{\frac{1}{2}}{4^2} + \frac{\frac{1}{2}}{5^2} + \frac{1}{6^2} + \frac{\frac{1}{2}}{7^2} - \ldots + \frac{\cos \frac{1}{3}n\pi}{n^2} + \ldots$$

$$= \frac{1}{2} - \frac{1}{8} - \frac{1}{9} - \frac{1}{32} + \frac{1}{50} + \frac{1}{36} + \frac{1}{98} - \ldots$$

This is a series of positive and negative terms. We can prove this series is convergent if we can show that it is absolutely convergent.

$$\sum_{n=1}^{+\infty} |u_n| = \sum_{n=1}^{+\infty} \frac{\left| \cos \frac{1}{3}n\pi \right|}{n^2}$$

Because

$$\left| \cos \tfrac{1}{3}n\pi \right| \leq 1 \qquad \text{for all } n$$

$$\frac{\left| \cos \tfrac{1}{3}n\pi \right|}{n^2} \leq \frac{1}{n^2} \qquad \text{for all positive integers } n$$

The series $\displaystyle\sum_{n=1}^{+\infty}\frac{1}{n^2}$ is the p series, with $p = 2$, and is therefore convergent. So by the comparison test $\displaystyle\sum_{n=1}^{+\infty}|u_n|$ is convergent. The given series is therefore absolutely convergent; hence, by Theorem 11.8.3 it is convergent.

Observe that the terms of the series $\displaystyle\sum_{n=1}^{+\infty}|u_n|$ neither increase monotonically nor decrease monotonically. For example, $|u_4| = \frac{1}{32}$, $|u_5| = \frac{1}{50}$, $|u_6| = \frac{1}{36}$; and so $|u_5| < |u_4|$, but $|u_6| > |u_5|$.

The *ratio test*, given in the next theorem, is used frequently to determine whether a given series is absolutely convergent.

11.8.4 THEOREM
Ratio Test

Let $\displaystyle\sum_{n=1}^{+\infty} u_n$ be a given infinite series for which every u_n is nonzero. Then

(i) if $\displaystyle\lim_{n \to +\infty} \left|\frac{u_{n+1}}{u_n}\right| = L < 1$, the given series is absolutely convergent;

(ii) if $\displaystyle\lim_{n \to +\infty} \left|\frac{u_{n+1}}{u_n}\right| = L > 1$ or if $\displaystyle\lim_{n \to +\infty} \left|\frac{u_{n+1}}{u_n}\right| = +\infty$, the series is divergent;

(iii) if $\displaystyle\lim_{n \to +\infty} \left|\frac{u_{n+1}}{u_n}\right| = 1$, no conclusion regarding convergence may be made from this test.

PROOF OF (i)   It is given that $L < 1$. Let R be a number such that $L < R < 1$. Let $R - L = \epsilon < 1$. Because $\displaystyle\lim_{n \to +\infty} \left|\frac{u_{n+1}}{u_n}\right| = L$, there exists an integer $N > 0$ such that

if $n \geq N$ then $\left|\left|\dfrac{u_{n+1}}{u_n}\right| - L\right| < \epsilon$

Therefore

if $n \geq N$ then $0 < \left|\dfrac{u_{n+1}}{u_n}\right| < L + \epsilon = R$ $\qquad$ (3)

Let n take on the successive values $N, N + 1, N + 2, \ldots$, and so forth. We obtain, from (3),

$|u_{N+1}| < R|u_N|$

$|u_{N+2}| < R|u_{N+1}| < R^2|u_N|$

$|u_{N+3}| < R|u_{N+2}| < R^3|u_N|$

$\cdots$

In general,

$|u_{N+k}| < R^k|u_N|$ for every positive integer k $\qquad$ (4)

The series

$$\sum_{k=1}^{+\infty} |u_N|R^k = |u_N|R + |u_N|R^2 + \ldots + |u_N|R^n + \ldots$$

is convergent because it is a geometric series whose ratio is less than 1. So from (4) and the comparison test it follows that the series $\sum\limits_{k=1}^{+\infty} |u_{N+k}|$ is convergent. The series $\sum\limits_{k=1}^{+\infty} |u_{N+k}|$ differs from the series $\sum\limits_{n=1}^{+\infty} |u_n|$ in only the first N terms. Therefore $\sum\limits_{n=1}^{+\infty} |u_n|$ is convergent; so the given series is absolutely convergent.

PROOF OF (ii)   If $\lim\limits_{n \to +\infty} \left| \dfrac{u_{n+1}}{u_n} \right| = L > 1$ or $\lim\limits_{n \to +\infty} \left| \dfrac{u_{n+1}}{u_n} \right| = +\infty$, then in either case there is an integer $N > 0$ such that if $n \geq N$, then $\left| \dfrac{u_{n+1}}{u_n} \right| > 1$. Let n take on the successive values $N, N + 1, N + 2, \ldots$, and so on. We obtain

$$|u_{N+1}| > |u_N|$$
$$|u_{N+2}| > |u_{N+1}| > |u_N|$$
$$|u_{N+3}| > |u_{N+2}| > |u_N|$$
$$\cdots$$

Thus if $n > N$, then $|u_n| > |u_N|$. Hence $\lim\limits_{n \to +\infty} u_n \neq 0$; so the given series is divergent.

PROOF OF (iii)   If the ratio test is applied to the p series, we have

$$\lim\limits_{n \to +\infty} \left| \frac{u_{n+1}}{u_n} \right| = \lim\limits_{n \to +\infty} \left| \frac{\dfrac{1}{(n+1)^p}}{\dfrac{1}{n^p}} \right|$$

$$= \lim\limits_{n \to +\infty} \left| \left(\frac{n}{n+1} \right)^p \right|$$

$$= 1$$

Because the p series diverges if $p \leq 1$ and converges if $p > 1$, we have shown that it is possible to have both convergent and divergent series for which $\lim\limits_{n \to +\infty} \left| \dfrac{u_{n+1}}{u_n} \right| = 1$. This proves part (iii).   ∎

EXAMPLE 2
Determine if the series

$$\sum\limits_{n=1}^{+\infty} (-1)^{n+1} \frac{n}{2^n}$$

is convergent or divergent.

SOLUTION

$u_n = (-1)^{n+1} \dfrac{n}{2^n}$ and $u_{n+1} = (-1)^{n+2} \dfrac{n+1}{2^{n+1}}$. Therefore

$$\left| \frac{u_{n+1}}{u_n} \right| = \frac{n+1}{2^{n+1}} \cdot \frac{2^n}{n}$$

$$= \frac{n+1}{2n}$$

So

$$\lim_{n \to +\infty} \left| \frac{u_{n+1}}{u_n} \right| = \lim_{n \to +\infty} \frac{1 + \dfrac{1}{n}}{2}$$

$$= \frac{1}{2}$$

$$< 1$$

Therefore, by the ratio test, the given series is absolutely convergent and hence, by Theorem 11.8.3, it is convergent.

EXAMPLE 3

In Example 2 of Section 11.7 the series

$$\sum_{n=1}^{+\infty} (-1)^n \frac{n+2}{n(n+1)}$$

was shown to be convergent. Is this series absolutely convergent or conditionally convergent?

SOLUTION

To test for absolute convergence we apply the ratio test. In the solution of Example 2 of Section 11.7 it was shown that the ratio $\dfrac{|u_{n+1}|}{|u_n|} = \dfrac{n^2 + 3n}{n^2 + 4n + 4}$.

Hence

$$\lim_{n \to +\infty} \left| \frac{u_{n+1}}{u_n} \right| = \lim_{n \to +\infty} \frac{1 + \dfrac{3}{n}}{1 + \dfrac{4}{n} + \dfrac{4}{n^2}}$$

$$= 1$$

So the ratio test fails. Because

$$|u_n| = \frac{n+2}{n(n+1)}$$

$$= \frac{n+2}{n+1} \cdot \frac{1}{n}$$

$$> \frac{1}{n}$$

the comparison test can be applied. And because the series $\sum\limits_{n=1}^{+\infty} \dfrac{1}{n}$ is the harmonic series, which diverges, we conclude that the series $\sum\limits_{n=1}^{+\infty} |u_n|$ is divergent and hence $\sum\limits_{n=1}^{+\infty} u_n$ is not absolutely convergent. Therefore the series is conditionally convergent.

It should be noted that the ratio test does not include all possibilities for $\lim\limits_{n\to+\infty} \left| \dfrac{u_{n+1}}{u_n} \right|$ because it is possible that the limit does not exist and is not $+\infty$. The discussion of such cases is beyond the scope of this book.

The proof of the ratio test was based on using the comparison test with the geometric series. Another test whose proof is similar is the *root test*.

11.8.5 THEOREM
Root Test

Let $\sum\limits_{n=1}^{+\infty} u_n$ be a given infinite series for which every u_n is nonzero. Then

(i) if $\lim\limits_{n\to+\infty} \sqrt[n]{|u_n|} = L < 1$, the given series is absolutely convergent;

(ii) if $\lim\limits_{n\to+\infty} \sqrt[n]{|u_n|} = L > 1$, or if $\lim\limits_{n\to+\infty} \sqrt[n]{|u_n|} = +\infty$, the series is divergent;

(iii) if $\lim\limits_{n\to+\infty} \sqrt[n]{|u_n|} = 1$, no conclusion regarding convergence may be made from this test.

Because of the similarity of the proof of the root test with that of the ratio test it is left for you to do as an exercise (see Exercises 26 through 28).

EXAMPLE 4
Use the root test to determine if the series

$$\sum_{n=1}^{+\infty} (-1)^n \frac{3^{2n+1}}{n^{2n}}$$

is convergent or divergent.

SOLUTION
By applying the root test we have

$$\lim_{n\to+\infty} \sqrt[n]{|u_n|} = \lim_{n\to+\infty} \left(\frac{3^{2n+1}}{n^{2n}} \right)^{1/n}$$

$$= \lim_{n\to+\infty} \frac{3^{2+(1/n)}}{n^2}$$

$$= 0$$

$$< 1$$

Therefore, by the root test, the given series is absolutely convergent. Hence by Theorem 11.8.3 it is convergent.

The ratio test and the root test are closely related; however, the ratio test is usually easier to apply. If the terms of the series contain factorials, this is certainly the case. If the terms contain powers, as in Example 4, it may be advantageous to use the root test. The next example provides a series for which the root test is the better one to apply.

EXAMPLE 5
Determine if the series

$$\sum_{n=1}^{+\infty} \frac{1}{[\ln(n+1)]^n}$$

is convergent or divergent.

SOLUTION

$$\lim_{n \to +\infty} \sqrt[n]{|u_n|} = \lim_{n \to +\infty} \sqrt[n]{\left| \frac{1}{[\ln(n+1)]^n} \right|}$$

$$= \lim_{n \to +\infty} \left| \frac{1}{\ln(n+1)} \right|$$

$$= 0$$

$$< 1$$

From the root test, the given series is absolutely convergent. Therefore, by Theorem 11.8.3 it is convergent.

There are series for which the root test can be used to show convergence but for which the ratio test fails. Such a series appears in Exercise 25.

EXERCISES 11.8

In Exercises 1 through 20, determine if the given series is absolutely convergent, conditionally convergent, or divergent. Prove your answer.

1. $\sum_{n=1}^{+\infty} \left(-\frac{2}{3}\right)^n$

2. $\sum_{n=1}^{+\infty} (-1)^n \frac{2^n}{n^3}$

3. $\sum_{n=1}^{+\infty} (-1)^{n+1} \frac{2^n}{n!}$

4. $\sum_{n=1}^{+\infty} n \left(\frac{2}{3}\right)^n$

5. $\sum_{n=1}^{+\infty} \frac{n^2}{n!}$

6. $\sum_{n=1}^{+\infty} (-1)^{n+1} \frac{1}{(2n-1)!}$

7. $\sum_{n=1}^{+\infty} (-1)^n \frac{n!}{2^{n+1}}$

8. $\sum_{n=1}^{+\infty} (-1)^{n+1} \frac{1}{n(n+2)}$

9. $\sum_{n=1}^{+\infty} \frac{1 - 2 \sin n}{n^3}$

10. $\sum_{n=1}^{+\infty} (-1)^n \frac{1}{(n+1)^3}$

11. $\sum_{n=1}^{+\infty} (-1)^{n+1} \frac{3^n}{n!}$

12. $\sum_{n=1}^{+\infty} (-1)^n \frac{n^2+1}{n^3}$

13. $\sum_{n=2}^{+\infty} (-1)^{n+1} \frac{1}{n(\ln n)^2}$

14. $\sum_{n=1}^{+\infty} \frac{\cos n}{n^2}$

15. $\sum_{n=1}^{+\infty} \frac{\sin \pi n}{n}$

16. $\sum_{n=2}^{+\infty} (-1)^{n+1} \frac{n}{\ln n}$

17. $\sum_{n=2}^{+\infty} \frac{1}{(\ln n)^n}$

18. $\sum_{n=1}^{+\infty} \frac{\left(1 + \frac{1}{n}\right)^{2n}}{e^n}$

19. $\sum_{n=1}^{+\infty} \frac{n^n}{n!}$

20. $\sum_{n=1}^{+\infty} \frac{1 \cdot 3 \cdot 5 \cdot \ldots \cdot (2n-1)}{1 \cdot 4 \cdot 7 \cdot \ldots \cdot (3n-2)}$

21. If $|r| < 1$, prove that the series $\sum\limits_{n=1}^{+\infty} r^n \sin nt$ is absolutely convergent for all values of t.

22. Prove that if $\sum\limits_{n=1}^{+\infty} u_n$ is absolutely convergent and $u_n \neq 0$ for all n, then $\sum\limits_{n=1}^{+\infty} \dfrac{1}{|u_n|}$ is divergent.

23. Prove that if $\sum\limits_{n=1}^{+\infty} u_n$ is absolutely convergent, then $\sum\limits_{n=1}^{+\infty} u_n^2$ is convergent.

24. Show by means of an example that the converse of Exercise 23 is not true.

25. Given the series $\sum\limits_{n=1}^{+\infty} \dfrac{1}{2^{n+1+(-1)^n}}$ (a) Show that the ratio test fails for this series. (b) Use the root test to determine if the series is convergent or divergent.

26. Prove part (i) of the root test (Theorem 11.8.5). (*Hint:* Because $L < 1$, let R be a number such that $L < R < 1$, and let $R - L = \epsilon < 1$. Show that there is an integer N such that if $n > N$, then $|u_n| < R^n$. Then use the comparison test.)

27. Prove part (ii) of the root test. See the hint for Exercise 26.

28. Prove part (iii) of the root test by applying it to the two series $\sum\limits_{n=1}^{+\infty} \dfrac{1}{n}$ and $\sum\limits_{n=1}^{+\infty} \dfrac{1}{n^2}$. $\left(\textit{Hint:} \text{ Determine } \lim\limits_{n \to +\infty} \sqrt[n]{n} \text{ by letting } \sqrt[n]{n} = e^{(\ln n)/n} \text{ and using L'Hôpital's rule to find } \lim\limits_{n \to +\infty} \dfrac{\ln n}{n}.\right)$

11.9
A SUMMARY OF TESTS FOR CONVERGENCE OR DIVERGENCE OF AN INFINITE SERIES

To conclude the discussion of infinite series of constant terms we summarize the various tests that can be used to determine the convergence or divergence of a given series. In Sections 11.3 through 11.8 we had a number of these tests, and to gain proficiency in recognizing and applying the appropriate one requires considerable practice. You will gain this practice by doing Exercises 11 through 48 in the Review Exercises that follow. Here is a list of these tests. You might attempt them in the indicated order. If a particular step is not applicable or no conclusion can be made, continue on to the next one. Of course, sometimes more than one test can be used, but you should select the most efficient one.

1. Compute $\lim\limits_{n \to +\infty} u_n$. If $\lim\limits_{n \to +\infty} u_n \neq 0$, then the series diverges. If $\lim\limits_{n \to +\infty} u_n = 0$, no conclusion can be made.

2. Examine the series to determine if it is one of the special types:

   (i) A geometric series: $\sum\limits_{n=1}^{+\infty} ar^{n-1}$. It converges to the sum $\dfrac{a}{1-r}$ if $|r| < 1$; it diverges if $|r| \geq 1$.

   (ii) A p series: $\sum\limits_{n=1}^{+\infty} \dfrac{1}{n^p}$ (where p is a constant). It converges if $p > 1$; it diverges if $p \leq 1$.

   (iii) An alternating series: $\sum\limits_{n=1}^{+\infty} (-1)^{n+1} a_n$ or $\sum\limits_{n=1}^{+\infty} (-1)^n a_n$. Apply the alternating-series test (Theorem 11.7.2): If $a_n > 0$ and $a_{n+1} < a_n$ for all positive integers n, and $\lim\limits_{n \to +\infty} a_n = 0$, then the alternating series is convergent.

3. Try the ratio test (Theorem 11.8.4): Let $\sum\limits_{n=1}^{+\infty} u_n$ be a given infinite series for which every u_n is nonzero. Then

   (i) if $\lim\limits_{n \to +\infty} \left| \dfrac{u_{n+1}}{u_n} \right| = L < 1$, the given series is absolutely convergent;

   (ii) if $\lim\limits_{n \to +\infty} \left| \dfrac{u_{n+1}}{u_n} \right| = L > 1$ or if $\lim\limits_{n \to +\infty} \left| \dfrac{u_{n+1}}{u_n} \right| = +\infty$, the series is divergent;

   (iii) if $\lim\limits_{n \to +\infty} \left| \dfrac{u_{n+1}}{u_n} \right| = 1$, no conclusion regarding convergence may be made from this test.

4. Try the root test (Theorem 11.8.5): Let $\sum\limits_{n=1}^{+\infty} u_n$ be a given infinite series for which every u_n is nonzero. Then

   (i) if $\lim\limits_{n \to +\infty} \sqrt[n]{|u_n|} = L < 1$, the given series is absolutely convergent;

   (ii) if $\lim\limits_{n \to +\infty} \sqrt[n]{|u_n|} = L > 1$, or if $\lim\limits_{n \to +\infty} \sqrt[n]{|u_n|} = +\infty$, the series is divergent;

   (iii) if $\lim\limits_{n \to +\infty} \sqrt[n]{|u_n|} = 1$, no conclusion regarding convergence may be made from this test.

5. Try the integral test (Theorem 11.6.1): Let f be a function that is continuous, decreasing, and positive valued for all $x \geq 1$. Then the infinite series

$$\sum_{n=1}^{+\infty} f(n) = f(1) + f(2) + f(3) + \ldots + f(n) + \ldots$$

is convergent if the improper integral

$$\int_1^{+\infty} f(x)\,dx$$

exists, and it is divergent if $\lim\limits_{b \to +\infty} \int_1^b f(x)\,dx = +\infty$.

6. Try the comparison test (Theorem 11.5.2): Let the series $\sum\limits_{n=1}^{+\infty} u_n$ be a series of positive terms.

   (i) If $\sum\limits_{n=1}^{+\infty} v_n$ is a series of positive terms that is known to be convergent, and $u_n \leq v_n$ for all positive integers n, then $\sum\limits_{n=1}^{+\infty} u_n$ is convergent.

(ii) If $\displaystyle\sum_{n=1}^{+\infty} w_n$ is a series of positive terms that is known to be divergent, and $u_n \geq w_n$ for all positive integers n, then $\displaystyle\sum_{n=1}^{+\infty} u_n$ is divergent.

or the limit comparison test (Theorem 11.5.3): Let $\displaystyle\sum_{n=1}^{+\infty} u_n$ and $\displaystyle\sum_{n=1}^{+\infty} v_n$ be two series of positive terms.

(i) If $\displaystyle\lim_{n \to +\infty} \frac{u_n}{v_n} = c > 0$, then the two series either both converge or both diverge.

(ii) If $\displaystyle\lim_{n \to +\infty} \frac{u_n}{v_n} = 0$, and if $\displaystyle\sum_{n=1}^{+\infty} v_n$ converges, then $\displaystyle\sum_{n=1}^{+\infty} u_n$ converges.

(iii) If $\displaystyle\lim_{n \to +\infty} \frac{u_n}{v_n} = +\infty$, and if $\displaystyle\sum_{n=1}^{+\infty} v_n$ diverges, then $\displaystyle\sum_{n=1}^{+\infty} u_n$ diverges.

REVIEW EXERCISES FOR CHAPTER 11

In Exercises 1 through 8, write the first four numbers of the sequence and find the limit of the sequence, if it exists.

1. $\left\{\dfrac{3n}{n+2}\right\}$

2. $\left\{\dfrac{(-1)^{n-1}}{(n+1)^2}\right\}$

3. $\left\{\dfrac{n^2-1}{n^2+1}\right\}$

4. $\left\{\dfrac{n+3n^2}{4+2n^3}\right\}$

5. $\{2 + (-1)^n\}$

6. $\left\{\dfrac{n^2}{\ln(n+1)}\right\}$

7. $\left\{\left(1+\dfrac{1}{n}\right)^{2n}\right\}$

8. $\left\{\dfrac{(n+2)^2}{n+4} - \dfrac{(n+2)^2}{n}\right\}$

In Exercises 9 and 10, find the first four elements of the sequence of partial sums $\{s_n\}$, and find a formula for s_n in terms of n. Also determine if the infinite series is convergent or divergent; if it is convergent, find its sum.

9. $\displaystyle\sum_{n=1}^{+\infty} \frac{3}{4^{n+1}}$

10. $\displaystyle\sum_{n=1}^{+\infty} \ln\left(\frac{2n-1}{2n+1}\right)$

In Exercises 11 through 20, determine if the series is convergent or divergent. If the series is convergent, find its sum.

11. $\displaystyle\sum_{n=1}^{+\infty} \left(\frac{3}{4}\right)^n$

12. $\displaystyle\sum_{n=1}^{+\infty} e^{-2n}$

13. $\displaystyle\sum_{n=1}^{+\infty} \frac{n-1}{n+1}$

14. $\displaystyle\sum_{n=0}^{+\infty} [(-1)^n + (-1)^{n+1}]$

15. $\displaystyle\sum_{n=0}^{+\infty} \sin^n \frac{1}{3}\pi$

16. $\displaystyle\sum_{n=0}^{+\infty} \cos^n \frac{1}{3}\pi$

17. $\displaystyle\sum_{n=1}^{+\infty} \frac{1}{(3n-1)(3n+2)}$ (*Hint:* To find the sum, first find the sequence of partial sums.)

18. $\displaystyle\sum_{n=1}^{+\infty} \frac{3}{2}\left(\frac{1}{5}\right)^n$

19. $\displaystyle\sum_{n=1}^{+\infty} \frac{\left[\sin\dfrac{3}{n}\pi + 2\right]}{3^n}$

20. $\displaystyle\sum_{n=1}^{+\infty} \left(\frac{1}{4^n} + \frac{1}{3^n}\right)$

In Exercises 21 through 38, determine if the series is convergent or divergent.

21. $\displaystyle\sum_{n=1}^{+\infty} \frac{2}{n^2 + 6n}$

22. $\displaystyle\sum_{n=1}^{+\infty} \frac{1}{(2n+1)^3}$

23. $\displaystyle\sum_{n=1}^{+\infty} \cos\left(\frac{\pi}{2n^2 - 1}\right)$

24. $\displaystyle\sum_{n=1}^{+\infty} \frac{3 + \sin n}{n^2}$

25. $\displaystyle\sum_{n=1}^{+\infty} \frac{(n!)^2}{(2n)!}$

26. $\displaystyle\sum_{n=1}^{+\infty} \frac{n}{\sqrt{3n+2}}$

27. $\displaystyle\sum_{n=1}^{+\infty} (-1)^n \ln \frac{1}{n}$

28. $\displaystyle\sum_{n=1}^{+\infty} \frac{(-1)^{n+1}}{1+\sqrt{n}}$

29. $\displaystyle\sum_{n=2}^{+\infty} \frac{1}{n(\ln n)^2}$

30. $\displaystyle\sum_{n=1}^{+\infty} \frac{\ln n}{n^2}$

31. $\displaystyle\sum_{n=1}^{+\infty} \left(\frac{2}{5n} - \frac{3}{2n}\right)$

32. $\displaystyle\sum_{n=0}^{+\infty} \frac{n!}{10^n}$

33. $\displaystyle\sum_{n=1}^{+\infty} \frac{1}{1+2\ln n}$

34. $\displaystyle\sum_{n=1}^{+\infty} \frac{|\sec n|}{n^{3/4}}$

35. $\displaystyle\sum_{n=1}^{+\infty} \frac{\cos n}{n^3}$

36. $\displaystyle\sum_{n=1}^{+\infty} n3^{-n^2}$

37. $\displaystyle\sum_{n=1}^{+\infty} \frac{1}{2^n + \sin n}$

38. $\displaystyle\sum_{n=1}^{+\infty} \frac{(n+2)^2}{(n+3)!}$

In Exercises 39 through 48, determine if the given series is absolutely convergent, conditionally convergent, or divergent. Prove your answer.

39. $\displaystyle\sum_{n=0}^{+\infty} (-1)^n \frac{n^2}{3^n}$

40. $\displaystyle\sum_{n=0}^{+\infty} (-1)^n \frac{5^{2n+1}}{(2n+1)!}$

41. $\displaystyle\sum_{n=1}^{+\infty} (-1)^{n-1} \frac{1}{(n+1)^{3/4}}$

42. $\displaystyle\sum_{n=1}^{+\infty} (-1)^{n-1} \frac{6^n}{5^{n+1}}$

43. $\displaystyle\sum_{n=1}^{+\infty} (-1)^n \frac{n!}{10n}$

44. $\displaystyle\sum_{n=1}^{+\infty} (-1)^n \frac{\sqrt{2n-1}}{n}$

45. $\displaystyle\sum_{n=1}^{+\infty} (-1)^{n-1} \frac{2^{3n}}{n^n}$

46. $\displaystyle\sum_{n=1}^{+\infty} (-1)^{n-1} \frac{1}{[\ln(n+2)]^n}$

47. $\displaystyle\sum_{n=1}^{+\infty} c_n$, where $c_n = \begin{cases} -\dfrac{1}{n} & \text{if } n \text{ is a perfect square} \\ \dfrac{1}{n^2} & \text{if } n \text{ is not a perfect square} \end{cases}$

48. $\displaystyle\sum_{n=1}^{+\infty} c_n$, where $c_n = \begin{cases} -\dfrac{1}{n} & \text{if } \frac{1}{4}n \text{ is an integer} \\ \dfrac{1}{n^2} & \text{if } \frac{1}{4}n \text{ is not an integer} \end{cases}$

49. Express the nonterminating repeating decimal 1.324 24 24 . . . as a common fraction.

50. A ball is dropped from a height of 18 m. Each time it strikes the ground, it bounces back to a height of two-thirds the distance from which it fell. Find the total distance traveled by the ball before it comes to rest.

51. The path of each swing, after the first, of a pendulum bob is 80 percent as long as the path of the previous swing from one side to the other side. If the path of the first swing is 18 in. long, and air resistance eventually brings the pendulum to rest, how far does the bob travel before it comes to rest?

T W E L V E

POWER SERIES

$$\int_0^{1/2} e^{-t^2}\, dt = \frac{1}{2} - \frac{1}{24} + \frac{1}{320} - \frac{1}{3576} + \cdots$$

$$\cos x = \sum_{n=0}^{+\infty} \frac{(-1)^n x^{2n}}{(2n)!} = 1 - \frac{x^2}{2!} + \frac{x^4}{4!} - \frac{x^6}{6!} + \cdots$$

12.1
INTRODUCTION TO
POWER SERIES

The infinite series in Chapter 11 involved constant terms. We now discuss an important type of series of variable terms called *power series*, which can be considered as a generalization of a polynomial function. You will learn in this chapter how power series can be used to calculate values of such functions as $\sin x$, e^x, $\ln x$, and $\sqrt{x}$, which cannot be evaluated by the familiar operations of arithmetic used for determining rational function values.

12.1.1 DEFINITION

A **power series** in $x - a$ is a series of the form

$$c_0 + c_1(x - a) + c_2(x - a)^2 + \ldots + c_n(x - a)^n + \ldots \tag{1}$$

We use the notation $\sum\limits_{n=0}^{+\infty} c_n(x - a)^n$ to represent series (1). (Note that we take $(x - a)^0 = 1$, even when $x = a$, for convenience in writing the general term.) If x is a particular number, the power series (1) becomes an infinite series of constant terms. A special case of (1) is obtained when $a = 0$, and the series becomes a power series in x, which is

$$\sum_{n=0}^{+\infty} c_n x^n = c_0 + c_1 x + c_2 x^2 + \ldots + c_n x^n + \ldots \tag{2}$$

In addition to power series in $x - a$ and x, there are power series of the form

$$\sum_{n=0}^{+\infty} c_n[\phi(x)]^n = c_0 + c_1\phi(x) + c_2[\phi(x)]^2 + \ldots + c_n[\phi(x)]^n + \ldots$$

where ϕ is a function of x. Such a series is called a power series in $\phi(x)$. In this book we are concerned exclusively with power series of the forms (1) and (2), and when the term "power series" is used, we mean either of these forms. The discussion of the theory of power series is confined to series (2). The more general power series (1) can be obtained from (2) by the translation $x = \bar{x} - a$; therefore our results can be applied to series (1) as well.

In dealing with an infinite series of constant terms we were concerned with the question of convergence or divergence of the series. In considering a power series we ask, For what values of x does the power series converge? For each value of x for which the power series converges, the series represents the number that is the sum of the series. Therefore, a power series defines a function. The function f, with function values

$$f(x) = \sum_{n=0}^{+\infty} c_n x^n \tag{3}$$

has as its domain all values of x for which the power series in (3) converges. It is apparent that every power series (2) is convergent for $x = 0$. There are some series (see Example 3) that are convergent for no other value of x, and there are also series that converge for every value of x (see Example 2).

The following three examples illustrate how the ratio test can be used to determine the values of x for which a power series is convergent. When $n!$ is used in representing the nth term of a power series (as in Example 2), we take $0! = 1$ so that the expression for the nth term will hold when $n = 0$.

EXAMPLE 1

Find the values of x for which the power series

$$\sum_{n=1}^{+\infty} (-1)^{n+1} \frac{2^n x^n}{n3^n}$$

is convergent.

SOLUTION

For the given series,

$$u_n = (-1)^{n+1} \frac{2^n x^n}{n3^n} \quad \text{and} \quad u_{n+1} = (-1)^{n+2} \frac{2^{n+1} x^{n+1}}{(n+1)3^{n+1}}$$

So

$$\lim_{n \to +\infty} \left| \frac{u_{n+1}}{u_n} \right| = \lim_{n \to +\infty} \left| \frac{2^{n+1} x^{n+1}}{(n+1)3^{n+1}} \cdot \frac{n3^n}{2^n x^n} \right|$$

$$= \lim_{n \to +\infty} \frac{2}{3} |x| \frac{n}{n+1}$$

$$= \tfrac{2}{3} |x|$$

Therefore the power series is absolutely convergent when $\frac{2}{3}|x| < 1$ or, equivalently, when $|x| < \frac{3}{2}$. The series is divergent when $\frac{2}{3}|x| > 1$ or, equivalently, when $|x| > \frac{3}{2}$. When $\frac{2}{3}|x| = 1$ (i.e., when $x = \pm\frac{3}{2}$), the ratio test fails. When $x = \frac{3}{2}$, the given power series becomes

$$\frac{1}{1} - \frac{1}{2} + \frac{1}{3} - \frac{1}{4} + \ldots + (-1)^{n+1} \frac{1}{n} + \ldots$$

which is convergent, as was shown in Example 1 of Section 11.7. When $x = -\frac{3}{2}$, we have

$$-\frac{1}{1} - \frac{1}{2} - \frac{1}{3} - \frac{1}{4} - \ldots - \frac{1}{n} - \ldots$$

which by Theorem 11.4.2 is divergent. We conclude, then, that the given power series is convergent when $-\frac{3}{2} < x \le \frac{3}{2}$. The series is absolutely convergent when $-\frac{3}{2} < x < \frac{3}{2}$ and is conditionally convergent when $x = \frac{3}{2}$. If $x \le -\frac{3}{2}$ or $x > \frac{3}{2}$, the series is divergent.

EXAMPLE 2

Find the values of x for which the power series $\sum_{n=0}^{+\infty} \frac{x^n}{n!}$ is convergent.

SOLUTION

For the given series, $u_n = \dfrac{x^n}{n!}$ and $u_{n+1} = \dfrac{x^{n+1}}{(n+1)!}$. So by applying the ratio test,

$$\lim_{n \to +\infty} \left| \frac{u_{n+1}}{u_n} \right| = \lim_{n \to +\infty} \left| \frac{x^{n+1}}{(n+1)!} \cdot \frac{n!}{x^n} \right|$$

$$= |x| \lim_{n \to +\infty} \frac{1}{n+1}$$

$$= 0$$

$$< 1$$

Therefore the given power series is absolutely convergent for all values of x.

EXAMPLE 3

Find the values of x for which the power series $\displaystyle\sum_{n=0}^{+\infty} n! x^n$ is convergent.

SOLUTION

For the given series, $u_n = n! x^n$ and $u_{n+1} = (n+1)! x^{n+1}$. Applying the ratio test we have

$$\lim_{n \to +\infty} \left| \frac{u_{n+1}}{u_n} \right| = \lim_{n \to +\infty} \left| \frac{(n+1)! x^{n+1}}{n! x^n} \right|$$

$$= \lim_{n \to +\infty} |(n+1)x|$$

$$= \begin{cases} 0 & \text{if } x = 0 \\ +\infty & \text{if } x \neq 0 \end{cases}$$

It follows that the series is divergent for all values of x except 0.

In the next example the root test is used to determine when a power series is convergent.

EXAMPLE 4

Find the values of x for which the power series $\displaystyle\sum_{n=1}^{+\infty} n^3 x^n$ is convergent.

SOLUTION

We use the root test and compute $\displaystyle\lim_{n \to +\infty} \sqrt[n]{|u_n|}$.

$$\lim_{n \to +\infty} \sqrt[n]{|n^3 x^n|} = \lim_{n \to +\infty} n^{3/n} |x| \tag{4}$$

To determine $\displaystyle\lim_{n \to +\infty} n^{3/n}$, let $y = n^{3/n}$. Then $\ln y = \dfrac{3}{n} \ln n$. Thus

$$\lim_{n \to +\infty} \ln y = \lim_{n \to +\infty} \frac{3 \ln n}{n} \tag{5}$$

To compute the limit on the right-hand side of (5) we first find $\lim\limits_{z \to +\infty} \dfrac{3 \ln z}{z}$, where z is a real number. Because $\lim\limits_{z \to +\infty} \ln z = +\infty$ and $\lim\limits_{z \to +\infty} z = +\infty$, we apply L'Hôpital's rule and get

$$\lim_{z \to +\infty} \frac{3 \ln z}{z} = \lim_{z \to +\infty} \frac{3}{z}$$
$$= 0$$

Thus from Theorem 11.1.3, $\lim\limits_{n \to +\infty} \dfrac{3 \ln n}{n} = 0$; hence, from (5),

$$\lim_{n \to +\infty} \ln y = 0$$

$$\lim_{n \to +\infty} y = 1$$

Substituting this result in (4) we have

$$\lim_{n \to +\infty} \sqrt[n]{|n^3 x^n|} = |x|$$

Therefore the power series is absolutely convergent when $|x| < 1$. The series is divergent when $|x| > 1$. When $x = 1$, the given power series becomes $\sum\limits_{n=1}^{+\infty} n^3$, which is divergent because $\lim\limits_{n \to +\infty} n^3 \neq 0$. Similarly, the power series is divergent when $x = -1$.

12.1.2 THEOREM   If the power series $\sum\limits_{n=0}^{+\infty} c_n x^n$ is convergent for $x = x_1$ $(x_1 \neq 0)$, then it is absolutely convergent for all values of x for which $|x| < |x_1|$.

PROOF   If $\sum\limits_{n=0}^{+\infty} c_n x_1{}^n$ is convergent, then $\lim\limits_{n \to +\infty} c_n x_1{}^n = 0$. Therefore, if we take $\epsilon = 1$ in Definition 3.7.1, there exists an integer $N > 0$ such that

$$\text{if } n \geq N \quad \text{then} \quad |c_n x_1{}^n| < 1$$

Now if x is any number such that $|x| < |x_1|$, then if $n \geq N$

$$\left| c_n x^n \right| = \left| c_n x_1{}^n \frac{x^n}{x_1{}^n} \right|$$

$$\left| c_n x^n \right| = \left| c_n x_1{}^n \right| \left| \frac{x}{x_1} \right|^n$$

$$\left| c_n x^n \right| < \left| \frac{x}{x_1} \right|^n \tag{6}$$

The series

$$\sum_{n=N}^{+\infty} \left| \frac{x}{x_1} \right|^n \tag{7}$$

is convergent because it is a geometric series with $r = |x/x_1| < 1$ (because $|x| < |x_1|$). Compare the series $\sum_{n=N}^{+\infty} |c_n x^n|$, where $|x| < |x_1|$, with series (7). From (6) and the comparison test, $\sum_{n=N}^{+\infty} |c_n x^n|$ is convergent for $|x| < |x_1|$. So the given power series is absolutely convergent for all values of x for which $|x| < |x_1|$.   ∎

▶ ILLUSTRATION 1   An illustration of Theorem 12.1.2 is given in Example 1. The power series is convergent for $x = \frac{3}{2}$ and is absolutely convergent for all values of x for which $|x| < \frac{3}{2}$.   ◀

The following theorem is a corollary of Theorem 12.1.2.

12.1.3 THEOREM   If the power series $\sum_{n=0}^{+\infty} c_n x^n$ is divergent for $x = x_2$, it is divergent for all values of x for which $|x| > |x_2|$.

PROOF   Suppose that the given power series is convergent for some number x for which $|x| > |x_2|$. Then by Theorem 12.1.2 the series must converge when $x = x_2$. However, this contradicts the hypothesis. Therefore the given power series is divergent for all values of x for which $|x| > |x_2|$.   ∎

▶ ILLUSTRATION 2   To illustrate Theorem 12.1.3, consider again the power series of Example 1. It is divergent for $x = -\frac{3}{2}$ and is also divergent for all values of x for which $|x| > \left|-\frac{3}{2}\right|$.   ◀

From Theorems 12.1.2 and 12.1.3 the following important theorem can be proved.

12.1.4 THEOREM   Let $\sum_{n=0}^{+\infty} c_n x^n$ be a given power series. Then exactly one of the following conditions holds:

(i) the series converges only when $x = 0$;
(ii) the series is absolutely convergent for all values of x;
(iii) there exists a number $R > 0$ such that the series is absolutely convergent for all values of x for which $|x| < R$ and is divergent for all values of x for which $|x| > R$.

PROOF   If x is replaced by zero in the given power series, we have $c_0 + 0 + 0 + \ldots$, which is obviously convergent. Therefore every power series of the form $\sum_{n=0}^{+\infty} c_n x^n$ is convergent when $x = 0$. If this is the only value of x for which the series converges, then condition (i) holds.

Suppose that the given series is convergent for $x = x_1$ where $x_1 \neq 0$. Then it follows from Theorem 12.1.2 that the series is absolutely convergent for all values of x for which $|x| < |x_1|$. Now if in addition there is no value of x for which the given series is divergent, then the series is absolutely convergent for all values of x. This is condition (ii).

If the given series is convergent for $x = x_1$ where $x_1 \neq 0$ and is divergent for $x = x_2$ where $|x_2| > |x_1|$, it follows from Theorem 12.1.3 that the series is divergent for all values of x for which $|x| > |x_2|$. Hence $|x_2|$ is an upper bound of the set of values of $|x|$ for which the series is absolutely convergent. Therefore, by the axiom of completeness (11.2.5) this set of numbers has a least upper bound, which is the number R of condition (iii). This proves that exactly one of the three conditions holds. ∎

series convergent
for $|x| < R$

series divergent for $|x| > R$

FIGURE 12.1.1

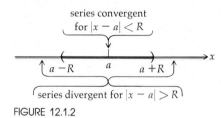

series convergent
for $|x - a| < R$

series divergent for $|x - a| > R$

FIGURE 12.1.2

Theorem 12.1.4(iii) can be illustrated on the number line. See Fig. 12.1.1. If instead of the power series $\sum_{n=0}^{+\infty} c_n x^n$ we have the series $\sum_{n=0}^{+\infty} c_n(x - a)^n$, then in conditions (i) and (iii) of Theorem 12.1.4 x is replaced by $x - a$. The conditions become

(i) the series converges only when $x = a$;

(iii) there exists a number $R > 0$ such that the series is absolutely convergent for all values of x for which $|x - a| < R$ and is divergent for all values of x for which $|x - a| > R$. (See Fig. 12.1.2 for an illustration of this on the number line.)

The set of all values of x for which a given power series is convergent is called the **interval of convergence** of the power series. The number R of condition (iii) of Theorem 12.1.4 is called the **radius of convergence** of the power series. If condition (i) holds, $R = 0$; if condition (ii) holds, we write $R = +\infty$.

▶ ILLUSTRATION 3   For the power series of Example 1, $R = \frac{3}{2}$ and the interval of convergence is $(-\frac{3}{2}, \frac{3}{2}]$. In Example 2, $R = +\infty$, and the interval of convergence is written as $(-\infty, +\infty)$. ◀

If R is the radius of convergence of the power series $\sum_{n=0}^{+\infty} c_n x^n$, the interval of convergence is one of the following intervals: $(-R, R)$, $[-R, R]$, $(-R, R]$, or $[-R, R)$. For the more general power series $\sum_{n=0}^{+\infty} c_n(x - a)^n$, the interval of convergence is one of the following:

$$(a - R, a + R) \quad [a - R, a + R] \quad (a - R, a + R] \quad [a - R, a + R)$$

A given power series defines a function having the interval of convergence as its domain. The most useful method at our disposal for determining the interval of convergence of a power series is the ratio test. However, the ratio test will not reveal anything about the convergence or divergence of the power series at the endpoints of the interval of convergence. At an endpoint, a power series may be either absolutely convergent, conditionally convergent, or divergent. If a power series converges absolutely at one endpoint, it follows from the definition of absolute convergence that the series is absolutely convergent at each endpoint (see Exercise 33). If a power series converges at one endpoint and diverges at the other, the series is conditionally convergent at the endpoint at which it converges (see Exercise 34). There are cases for which the convergence or divergence of a power series at the endpoints cannot be determined by the methods of elementary calculus.

EXAMPLE 5

Determine the interval of convergence of the power series $\sum\limits_{n=1}^{+\infty} n(x-2)^n$.

SOLUTION
The given series is

$$(x-2) + 2(x-2)^2 + \ldots + n(x-2)^n + (n+1)(x-2)^{n+1} + \ldots$$

Applying the ratio test we have

$$\lim_{n \to +\infty} \left| \frac{u_{n+1}}{u_n} \right| = \lim_{n \to +\infty} \left| \frac{(n+1)(x-2)^{n+1}}{n(x-2)^n} \right|$$

$$= |x-2| \lim_{n \to +\infty} \frac{n+1}{n}$$

$$= |x-2|$$

The given series then will be absolutely convergent if $|x-2| < 1$ or, equivalently, $-1 < x - 2 < 1$ or, equivalently, $1 < x < 3$.

When $x = 1$, the series is $\sum\limits_{n=1}^{+\infty} (-1)^n n$, which is divergent because $\lim\limits_{n \to +\infty} u_n \neq 0$. When $x = 3$, the series is $\sum\limits_{n=1}^{+\infty} n$, which is also divergent because $\lim\limits_{n \to +\infty} u_n \neq 0$. Therefore the interval of convergence is $(1, 3)$. So the given power series defines a function having the interval $(1, 3)$ as its domain.

EXAMPLE 6
Determine the interval of convergence of the power series

$$\sum_{n=1}^{+\infty} \frac{x^n}{2 + n^2}$$

SOLUTION
The given series is

$$\frac{x}{2 + 1^2} + \frac{x^2}{2 + 2^2} + \frac{x^3}{2 + 3^2} + \ldots + \frac{x^n}{2 + n^2} + \frac{x^{n+1}}{2 + (n+1)^2} + \ldots$$

Applying the ratio test we have

$$\lim_{n \to +\infty} \left| \frac{u_{n+1}}{u_n} \right| = \lim_{n \to +\infty} \left| \frac{x^{n+1}}{2 + (n+1)^2} \cdot \frac{2 + n^2}{x^n} \right|$$

$$= |x| \lim_{n \to +\infty} \frac{2 + n^2}{2 + n^2 + 2n + 1}$$

$$= |x|$$

So the given series will be absolutely convergent if $|x| < 1$ or, equivalently, $-1 < x < 1$. When $x = 1$, the series is

$$\frac{1}{2 + 1^2} + \frac{1}{2 + 2^2} + \frac{1}{2 + 3^2} + \ldots + \frac{1}{2 + n^2} + \ldots$$

Because $\dfrac{1}{2 + n^2} < \dfrac{1}{n^2}$ for all positive integers n, and because $\displaystyle\sum_{n=1}^{+\infty} \frac{1}{n^2}$ is a convergent p series, it follows from the comparison test that the given power series is convergent when $x = 1$. When $x = -1$, the series is $\displaystyle\sum_{n=1}^{+\infty} \frac{(-1)^n}{2 + n^2}$, which is convergent because we have just seen that it is absolutely convergent. Hence the interval of convergence of the given power series is $[-1, 1]$.

EXERCISES 12.1

In Exercises 1 through 28, find the interval of convergence of the given power series.

1. $\displaystyle\sum_{n=0}^{+\infty} \frac{x^n}{n+1}$

2. $\displaystyle\sum_{n=0}^{+\infty} \frac{x^n}{n^2 + 1}$

3. $\displaystyle\sum_{n=0}^{+\infty} \frac{x^n}{n^2 - 3}$

4. $\displaystyle\sum_{n=0}^{+\infty} \frac{n^2 x^n}{2^n}$

5. $\displaystyle\sum_{n=1}^{+\infty} \frac{2^n x^n}{n^2}$

6. $\displaystyle\sum_{n=1}^{+\infty} \frac{x^n}{2^n \sqrt{n}}$

7. $\displaystyle\sum_{n=1}^{+\infty} \frac{nx^n}{3^n}$

8. $\displaystyle\sum_{n=1}^{+\infty} (-1)^n \frac{x^{2n}}{(2n)!}$

9. $\displaystyle\sum_{n=1}^{+\infty} (-1)^{n+1} \frac{x^{2n-1}}{(2n-1)!}$

10. $\displaystyle\sum_{n=1}^{+\infty} \frac{n+1}{n^{2n}} x^n$

11. $\displaystyle\sum_{n=0}^{+\infty} \frac{(x+3)^n}{2^n}$

12. $\displaystyle\sum_{n=0}^{+\infty} \frac{x^n}{(n+1)5^n}$

13. $\displaystyle\sum_{n=1}^{+\infty} (-1)^n \frac{x^n}{(2n-1)3^{2n-1}}$

14. $\displaystyle\sum_{n=1}^{+\infty} (-1)^{n+1} \frac{(n+1)x^n}{n!}$

15. $\displaystyle\sum_{n=1}^{+\infty} (-1)^{n+1} \frac{(x-1)^n}{n}$

16. $\displaystyle\sum_{n=1}^{+\infty} \frac{(x+2)^n}{(n+1)2^n}$

17. $\displaystyle\sum_{n=0}^{+\infty} (\sinh 2n)x^n$

18. $\displaystyle\sum_{n=1}^{+\infty} \frac{x^n}{\ln(n+1)}$

19. $\displaystyle\sum_{n=2}^{+\infty} (-1)^{n+1} \frac{x^n}{n(\ln n)^2}$

20. $\displaystyle\sum_{n=1}^{+\infty} \frac{(x+5)^{n-1}}{n^2}$

21. $\displaystyle\sum_{n=1}^{+\infty} \frac{n^2}{5^n} (x-1)^n$

22. $\displaystyle\sum_{n=0}^{+\infty} \frac{4^{n+1} x^{2n}}{n+3}$

23. $\displaystyle\sum_{n=1}^{+\infty} \frac{\ln n(x-5)^n}{n+1}$

24. $\displaystyle\sum_{n=1}^{+\infty} \frac{x^n}{n^n}$

25. $\displaystyle\sum_{n=1}^{+\infty} (-1)^n \frac{1 \cdot 3 \cdot 5 \cdot \ldots \cdot (2n-1)}{2 \cdot 4 \cdot 6 \cdot \ldots \cdot 2n} x^{2n+1}$

26. $\displaystyle\sum_{n=1}^{+\infty} n^n(x-3)^n$

27. $\displaystyle\sum_{n=1}^{+\infty} \frac{n!x^n}{n^n}$

28. $\displaystyle\sum_{n=1}^{+\infty} \frac{(-1)^{n+1} 1 \cdot 3 \cdot 5 \cdot \ldots \cdot (2n-1)}{2 \cdot 4 \cdot 6 \cdot \ldots \cdot 2n} x^n$

29. If a and b are positive integers, find the radius of convergence of the power series $\sum\limits_{n=1}^{+\infty} \dfrac{(n+a)!}{n!(n+b)!} x^n$.

30. If $\sum\limits_{n=1}^{+\infty} a_n$ is an absolutely convergent series, prove that the power series $\sum\limits_{n=1}^{+\infty} a_n x^n$ is absolutely convergent when $|x| \le 1$.

31. Prove that if the radius of convergence of the power series $\sum\limits_{n=1}^{+\infty} u_n x^n$ is r, then the radius of convergence of the series $\sum\limits_{n=1}^{+\infty} u_n x^{2n}$ is $\sqrt{r}$.

32. Prove that if $\lim\limits_{n \to +\infty} \sqrt[n]{|u_n|} = L \; (L \ne 0)$, then the radius of convergence of the power series $\sum\limits_{n=1}^{+\infty} u_n x^n$ is $\dfrac{1}{L}$.

33. Prove that if a power series converges absolutely at one endpoint of its interval of convergence, then the power series is absolutely convergent at each endpoint.

34. Prove that if a power series converges at one endpoint of its interval of convergence and diverges at the other endpoint, then the power series is conditionally convergent at the endpoint at which it converges.

12.2 DIFFERENTIATION OF POWER SERIES

You learned in Section 12.1 that a power series $\sum\limits_{n=0}^{+\infty} c_n x^n$ defines a function whose domain is the interval of convergence of the series.

▶ ILLUSTRATION 1   Consider the geometric series with $a = 1$ and $r = x$, which is $\sum\limits_{n=0}^{+\infty} x^n$. By Theorem 11.3.5 this series converges to the sum $1/(1-x)$ if $|x| < 1$. Therefore the power series $\sum\limits_{n=0}^{+\infty} x^n$ defines the function f for which $f(x) = 1/(1-x)$ and $|x| < 1$. Hence

$$1 + x + x^2 + x^3 + \ldots + x^n + \ldots = \frac{1}{1-x} \qquad \text{if } |x| < 1 \qquad (1)$$

◀

The series in (1) can be used to form other power series whose sums can be determined.

▶ ILLUSTRATION 2   If in (1) x is replaced by $-x$, we have

$$1 - x + x^2 - x^3 + \ldots + (-1)^n x^n + \ldots = \frac{1}{1+x} \qquad \text{if } |x| < 1 \qquad (2)$$

Let $x = x^2$ in (1) and get

$$1 + x^2 + x^4 + x^6 + \ldots + x^{2n} + \ldots = \frac{1}{1-x^2} \qquad \text{if } |x| < 1 \qquad (3)$$

If x is replaced by $-x^2$ in (1), we obtain

$$1 - x^2 + x^4 - x^6 + \ldots + (-1)^n x^{2n} + \ldots = \frac{1}{1 + x^2} \qquad \text{if } |x| < 1 \quad (4)$$

◀

In this section and the next, other interesting series are obtained from those like the above by differentiation and integration. We prove that if R (where $R \neq 0$) is the radius of convergence of a power series that defines a function f, then f is differentiable on the open interval $(-R, R)$ and the derivative of f can be obtained by differentiating the power series term by term. Furthermore, we show that f is integrable on every closed subinterval of $(-R, R)$, and the integral of f is evaluated by integrating the power series term by term. First some preliminary theorems are needed.

12.2.1 THEOREM   If $\displaystyle\sum_{n=0}^{+\infty} c_n x^n$ is a power series having a radius of convergence of $R > 0$, then the series $\displaystyle\sum_{n=1}^{+\infty} n c_n x^{n-1}$ also has R as its radius of convergence.

This theorem states that the series, obtained by differentiating term by term each term of a given power series, will have the same radius of convergence as the given series.

PROOF   Let x be any number in the open interval $(-R, R)$. Then $|x| < R$. Choose a number x_1 such that $|x| < |x_1| < R$. Because $|x_1| < R$, $\displaystyle\sum_{n=0}^{+\infty} c_n x_1^n$ is convergent. Hence $\displaystyle\lim_{n \to +\infty} c_n x_1^n = 0$. So if we take $\epsilon = 1$ in Definition 3.7.1, there exists a number $N > 0$ such that

if $n > N$ then $|c_n x_1^n| < 1$

Let M be the largest of the numbers $|c_1 x_1|, |c_2 x_1^2|, |c_3 x_1^3|, \ldots, |c_N x_1^N|, 1$. Then

$$|c_n x_1^n| \leq M \qquad \text{for all positive integers } n \quad (5)$$

Now

$$|n c_n x^{n-1}| = \left| n c_n \cdot \frac{x^{n-1}}{x_1^n} \cdot x_1^n \right|$$

$$= n \frac{|c_n x_1^n|}{|x_1|} \left| \frac{x}{x_1} \right|^{n-1}$$

From (5) and the above equation,

$$|n c_n x^{n-1}| \leq n \frac{M}{|x_1|} \left| \frac{x}{x_1} \right|^{n-1} \quad (6)$$

If the ratio test is applied to the series

$$\frac{M}{|x_1|} \sum_{n=1}^{+\infty} n \left| \frac{x}{x_1} \right|^{n-1} \tag{7}$$

then

$$\lim_{n \to +\infty} \left| \frac{u_{n+1}}{u_n} \right| = \lim_{n \to +\infty} \left| \frac{(n+1)|x|^n}{|x_1|^n} \cdot \frac{|x_1|^{n-1}}{n|x|^{n-1}} \right|$$

$$= \left| \frac{x}{x_1} \right| \lim_{n \to +\infty} \frac{n+1}{n}$$

$$= \left| \frac{x}{x_1} \right| < 1$$

Therefore series (7) is absolutely convergent; so from (6) and the comparison test, the series $\sum_{n=1}^{+\infty} nc_n x^{n-1}$ is also absolutely convergent. Because x is any number in $(-R, R)$, it follows that if the radius of convergence of $\sum_{n=1}^{+\infty} nc_n x^{n-1}$ is R', then $R' \geq R$.

To complete the proof we must show that R' cannot be greater than R. Assume that $R' > R$, and let x_2 be a number such that $R < |x_2| < R'$. Because $|x_2| > R$, it follows that

$$\sum_{n=0}^{+\infty} c_n x_2{}^n \text{ is divergent} \tag{8}$$

Because $|x_2| < R'$, it follows that $\sum_{n=1}^{+\infty} nc_n x_2{}^{n-1}$ is absolutely convergent. Furthermore,

$$|x_2| \sum_{n=1}^{+\infty} |nc_n x_2{}^{n-1}| = \sum_{n=1}^{+\infty} |nc_n x_2{}^n|$$

and so from Theorem 11.4.2,

$$\sum_{n=1}^{+\infty} |nc_n x_2{}^n| \text{ is convergent} \tag{9}$$

If n is any positive integer,

$$|c_n x_2{}^n| \leq n|c_n x_2{}^n| = |nc_n x_2{}^n| \tag{10}$$

From statement (9), inequality (10), and the comparison test it follows that $\sum_{n=1}^{+\infty} |c_n x_2{}^n|$ is convergent. Therefore the series $\sum_{n=0}^{+\infty} c_n x_2{}^n$ is convergent, which contradicts statement (8). Hence the assumption that $R' > R$ is false. Therefore R' cannot be greater than R; and because it was shown that $R' \geq R$, it follows that $R' = R$, which proves the theorem. ∎

▶ ILLUSTRATION 3   We verify Theorem 12.2.1 for the power series

$$\sum_{n=0}^{+\infty} \frac{x^{n+1}}{(n+1)^2} = x + \frac{x^2}{4} + \frac{x^3}{9} + \ldots + \frac{x^{n+1}}{(n+1)^2} + \frac{x^{n+2}}{(n+2)^2} + \ldots$$

The radius of convergence is found by applying the ratio test.

$$\lim_{n \to +\infty} \left| \frac{u_{n+1}}{u_n} \right| = \lim_{n \to +\infty} \left| \frac{(n+1)^2 x^{n+2}}{(n+2)^2 x^{n+1}} \right|$$

$$= |x| \lim_{n \to +\infty} \left| \frac{n^2 + 2n + 1}{n^2 + 4n + 4} \right|$$

$$= |x|$$

Hence the power series is convergent when $|x| < 1$; so its radius of convergence $R = 1$.

The power series obtained from the given series by differentiating term by term is

$$\sum_{n=0}^{+\infty} \frac{(n+1)x^n}{(n+1)^2} = \sum_{n=0}^{+\infty} \frac{x^n}{n+1}$$

$$= 1 + \frac{x}{2} + \frac{x^2}{3} + \frac{x^3}{4} + \ldots + \frac{x^n}{n+1} + \frac{x^{n+1}}{n+2} + \ldots$$

Applying the ratio test for this power series we have

$$\lim_{n \to +\infty} \left| \frac{u_{n+1}}{u_n} \right| = \lim_{n \to +\infty} \left| \frac{(n+1)x^{n+1}}{(n+2)x^n} \right|$$

$$= |x| \lim_{n \to +\infty} \left| \frac{n+1}{n+2} \right|$$

$$= |x|$$

This power series is convergent when $|x| < 1$; thus its radius of convergence $R' = 1$. Because $R = R'$, Theorem 12.2.1 is verified.   ◀

12.2.2 THEOREM   If the radius of convergence of the power series $\sum_{n=0}^{+\infty} c_n x^n$ is $R > 0$, then R is also the radius of convergence of the series $\sum_{n=2}^{+\infty} n(n-1)c_n x^{n-2}$.

PROOF   If Theorem 12.2.1 is applied to the series $\sum_{n=1}^{+\infty} nc_n x^{n-1}$, the desired result follows.   ■

We are now in a position to prove the theorem regarding term-by-term differentiation of a power series.

12.2.3 THEOREM   Let $\displaystyle\sum_{n=0}^{+\infty} c_n x^n$ be a power series whose radius of convergence is $R > 0$. Then if f is the function defined by

$$f(x) = \sum_{n=0}^{+\infty} c_n x^n \tag{11}$$

$f'(x)$ exists for every x in the open interval $(-R, R)$ and it is given by

$$f'(x) = \sum_{n=1}^{+\infty} n c_n x^{n-1}$$

PROOF   Let x and a be two distinct numbers in the open interval $(-R, R)$. Taylor's formula (formula (9) in Section 9.5), with $n = 1$, is

$$f(x) = f(a) + \frac{f'(a)}{1!}(x - a) + \frac{f''(\xi)}{2!}(x - a)^2$$

From this formula with $f(x) = x^n$ it follows that for every positive integer n,

$$x^n = a^n + na^{n-1}(x - a) + \tfrac{1}{2}n(n - 1)(\xi_n)^{n-2}(x - a)^2 \tag{12}$$

where ξ_n is between a and x for every positive integer n. From (11),

$$f(x) - f(a) = \sum_{n=0}^{+\infty} c_n x^n - \sum_{n=0}^{+\infty} c_n a^n$$

$$= c_0 + \sum_{n=1}^{+\infty} c_n x^n - c_0 - \sum_{n=1}^{+\infty} c_n a^n$$

$$= \sum_{n=1}^{+\infty} c_n(x^n - a^n)$$

Dividing by $x - a$ (because $x \neq a$) and using (12) we have from the above equation

$$\frac{f(x) - f(a)}{x - a} = \frac{1}{x - a}\sum_{n=1}^{+\infty} c_n\big[na^{n-1}(x - a) + \tfrac{1}{2}n(n - 1)(\xi_n)^{n-2}(x - a)^2\big]$$

So

$$\frac{f(x) - f(a)}{x - a} = \sum_{n=1}^{+\infty} n c_n a^{n-1} + \tfrac{1}{2}(x - a)\sum_{n=2}^{+\infty} n(n - 1)c_n(\xi_n)^{n-2} \tag{13}$$

Because a is in $(-R, R)$, we conclude from Theorem 12.2.1 that $\displaystyle\sum_{n=1}^{+\infty} n c_n a^{n-1}$ is absolutely convergent.

Because both a and x are in $(-R, R)$, there is some number $K > 0$ such that $|a| < K < R$ and $|x| < K < R$. It follows from Theorem 12.2.2 that

$$\sum_{n=2}^{+\infty} n(n - 1)c_n K^{n-2}$$

is absolutely convergent. Then because

$$|n(n - 1)c_n(\xi_n)^{n-2}| < |n(n - 1)c_n K^{n-2}| \tag{14}$$

for each ξ_n, we can conclude from the comparison test that

$$\sum_{n=2}^{+\infty} n(n-1)c_n(\xi_n)^{n-2}$$

is absolutely convergent.

From (13),

$$\left| \frac{f(x) - f(a)}{x - a} - \sum_{n=1}^{+\infty} nc_n a^{n-1} \right| = \left| \tfrac{1}{2}(x - a) \sum_{n=2}^{+\infty} n(n-1)c_n(\xi_n)^{n-2} \right| \qquad (15)$$

However, from Theorem 11.8.3, if $\displaystyle\sum_{n=1}^{+\infty} u_n$ is absolutely convergent, then

$$\left| \sum_{n=1}^{+\infty} u_n \right| \leq \sum_{n=1}^{+\infty} |u_n|$$

Applying this to the right side of (15) we obtain

$$\left| \frac{f(x) - f(a)}{x - a} - \sum_{n=1}^{+\infty} nc_n a^{n-1} \right| \leq \tfrac{1}{2}|x - a| \sum_{n=2}^{+\infty} n(n-1)|c_n|\,|\xi_n|^{n-2} \qquad (16)$$

From (14) and (16),

$$\left| \frac{f(x) - f(a)}{x - a} - \sum_{n=1}^{+\infty} nc_n a^{n-1} \right| \leq \tfrac{1}{2}|x - a| \sum_{n=2}^{+\infty} n(n-1)|c_n|K^{n-2} \qquad (17)$$

where $0 < K < R$. Because the series on the right side of (17) is absolutely convergent, the limit of the right side, as x approaches a, is zero. Therefore, from (17) and the squeeze theorem (1.8.1),

$$\lim_{x \to a} \frac{f(x) - f(a)}{x - a} = \sum_{n=1}^{+\infty} nc_n a^{n-1}$$

$$\Leftrightarrow \quad f'(a) = \sum_{n=1}^{+\infty} nc_n a^{n-1}$$

and because a may be any number in the open interval $(-R, R)$, the theorem is proved. ∎

EXAMPLE 1
Let f be the function defined by the power series of Illustration 3. (a) Find the domain of f; (b) write the power series that defines the function f' and find the domain of f'.

SOLUTION

(a) $\displaystyle f(x) = \sum_{n=0}^{+\infty} \frac{x^{n+1}}{(n+1)^2}$

The domain of f is the interval of convergence of the power series. In Illustration 3 we showed that the radius of convergence of the power series is 1; that is, the series converges when $|x| < 1$. Consider now the power series when $|x| = 1$. When $x = 1$, the series is

$$1 + \frac{1}{4} + \frac{1}{9} + \ldots + \frac{1}{(n+1)^2} + \ldots$$

which is convergent because it is the p series with $p = 2$. When $x = -1$, we have the series $\sum_{n=0}^{+\infty} \frac{(-1)^{n+1}}{(n+1)^2}$, which is convergent because it is absolutely convergent. Hence the domain of f is the interval $[-1, 1]$.

(b) From Theorem 12.2.3 it follows that f' is defined by

$$f'(x) = \sum_{n=0}^{+\infty} \frac{x^n}{n+1} \qquad (18)$$

and that $f'(x)$ exists for every x in the open interval $(-1, 1)$. In Illustration 3 the radius of convergence of the power series in (18) was shown to be 1. We now consider the power series in (18) when $x = \pm 1$. When $x = 1$, the series is

$$1 + \frac{1}{2} + \frac{1}{3} + \frac{1}{4} + \ldots + \frac{1}{n+1} + \ldots$$

which is the harmonic series and hence is divergent. When $x = -1$, the series is

$$1 - \frac{1}{2} + \frac{1}{3} - \frac{1}{4} + \ldots + (-1)^n \frac{1}{n+1} + \ldots$$

which is a convergent alternating series. Therefore the domain of f' is the interval $[-1, 1)$.

Example 1 illustrates the fact that if a function f is defined by a power series and this power series is differentiated term by term, the resulting power series, which defines f', has the same radius of convergence but not necessarily the same interval of convergence.

EXAMPLE 2
Obtain a power-series representation of

$$\frac{1}{(1-x)^2}$$

SOLUTION
From (1),

$$\frac{1}{1-x} = 1 + x + x^2 + x^3 + \ldots + x^n + \ldots \qquad \text{if } |x| < 1$$

Using Theorem 12.2.3 and differentiating on both sides of the above we get

$$\frac{1}{(1-x)^2} = 1 + 2x + 3x^2 + \ldots + nx^{n-1} + \ldots \qquad \text{if } |x| < 1$$

EXAMPLE 3
Show that

$$e^x = \sum_{n=0}^{+\infty} \frac{x^n}{n!}$$

$$= 1 + x + \frac{x^2}{2!} + \frac{x^3}{3!} + \ldots + \frac{x^n}{n!} + \ldots$$

for all real values of x.

SOLUTION

In Example 2 of Section 12.1 we showed that the power series $\sum_{n=0}^{+\infty} \frac{x^n}{n!}$ is absolutely convergent for all real values of x. Therefore, if f is the function defined by

$$f(x) = \sum_{n=0}^{+\infty} \frac{x^n}{n!} \tag{19}$$

the domain of f is the set of all real numbers; that is, the interval of convergence is $(-\infty, +\infty)$. It follows from Theorem 12.2.3 that for all real values of x,

$$f'(x) = \sum_{n=1}^{+\infty} \frac{nx^{n-1}}{n!} \tag{20}$$

Because $\dfrac{n}{n!} = \dfrac{1}{(n-1)!}$, (20) can be written as

$$f'(x) = \sum_{n=1}^{+\infty} \frac{x^{n-1}}{(n-1)!}$$

$\Leftrightarrow$ $$f'(x) = \sum_{n=0}^{+\infty} \frac{x^n}{n!} \tag{21}$$

From (19) and (21), $f'(x) = f(x)$ for all real values of x. Therefore the function f satisfies the differential equation

$$\frac{dy}{dx} = y$$

for which, from Theorem 6.6.1, the general solution is $y = Ce^x$. Hence for some constant C, $f(x) = Ce^x$. From (19), $f(0) = 1$. (Remember that we take $x^0 = 1$ even when $x = 0$ for convenience in writing the general term.) Therefore $C = 1$; so $f(x) = e^x$, and we have the desired result.

EXAMPLE 4
Use the result of Example 3 to find a power-series representation of e^{-x}.

SOLUTION
If x is replaced by $-x$ in the series for e^x, it follows that

$$e^{-x} = 1 - x + \frac{x^2}{2!} - \frac{x^3}{3!} + \ldots + (-1)^n \frac{x^n}{n!} + \ldots$$

for all real values of x.

EXAMPLE 5
Use the series of Example 4 to find the value of e^{-1} correct to five decimal places.

SOLUTION
If $x = 1$ in the series for e^{-x},

$$e^{-1} = 1 - 1 + \frac{1}{2!} - \frac{1}{3!} + \frac{1}{4!} - \frac{1}{5!} + \frac{1}{6!} - \frac{1}{7!} + \frac{1}{8!} - \frac{1}{9!} + \frac{1}{10!} - \cdots$$

$$= 1 - 1 + \frac{1}{2} - \frac{1}{6} + \frac{1}{24} - \frac{1}{120} + \frac{1}{720} - \frac{1}{5040} + \frac{1}{40{,}320}$$

$$- \frac{1}{362{,}880} + \frac{1}{3{,}628{,}800} - \cdots$$

$$\approx 1 - 1 + 0.5 - 0.166667 + 0.041667 - 0.008333 + 0.001389$$
$$- 0.000198 + 0.000025 - 0.000003 + 0.0000003 - \cdots$$

This is a convergent alternating series for which $|u_{n+1}| < |u_n|$. So if the first ten terms are used to approximate the sum, by Theorem 11.7.4 the error is less than the absolute value of the eleventh term. Adding the first ten terms we obtain 0.367880. Rounding off to five decimal places gives

$$e^{-1} \approx 0.36788$$

In computation with infinite series two kinds of errors occur. One is the error given by the remainder after the first n terms. The other is the round-off error that occurs when each term of the series is approximated by a decimal with a finite number of places. In particular, in Example 5 we wanted the result accurate to five decimal places; so each term was rounded off to six decimal places. After computing the sum, we rounded off this result to five decimal places. Of course, the error given by the remainder can be reduced by considering additional terms of the series, while the round-off error can be reduced by using more decimal places.

If you take a course in differential equations, you will learn that it is possible to express solutions of many differential equations as power series. In the following example we have this situation.

EXAMPLE 6
Show that

$$y = x + \sum_{n=0}^{+\infty} \frac{x^n}{n!} \tag{22}$$

is a solution of the differential equation $\dfrac{d^2y}{dx^2} - y + x = 0$.

SOLUTION

The power series in (22) is convergent for all values of x. Therefore, from Theorem 12.2.3, for all x,

$$\frac{dy}{dx} = 1 + \sum_{n=1}^{+\infty} \frac{nx^{n-1}}{n!} \qquad \frac{d^2y}{dx^2} = \sum_{n=2}^{+\infty} \frac{(n-1)x^{n-2}}{(n-1)!}$$

$$= 1 + \sum_{n=1}^{+\infty} \frac{x^{n-1}}{(n-1)!} \qquad = \sum_{n=2}^{+\infty} \frac{x^{n-2}}{(n-2)!}$$

$$= \sum_{n=0}^{+\infty} \frac{x^n}{n!}$$

Thus

$$\frac{d^2y}{dx^2} - y + x = \sum_{n=0}^{+\infty} \frac{x^n}{n!} - \left(x + \sum_{n=0}^{+\infty} \frac{x^n}{n!} \right) + x$$

$$= 0$$

Hence the differential equation is satisfied; so (22) is a solution.

EXERCISES 12.2

In Exercises 1 through 10, a function f is defined by a power series. In each exercise do the following: (a) Find the radius of convergence of the given power series and the domain of f; (b) write the power series that defines the function f' and find its radius of convergence by using methods of Section 12.1 (thus verifying Theorem 12.2.1); (c) find the domain of f'.

1. $f(x) = \sum_{n=1}^{+\infty} \frac{x^n}{n^2}$

2. $f(x) = \sum_{n=1}^{+\infty} (-1)^{n-1} \frac{x^n}{n}$

3. $f(x) = \sum_{n=1}^{+\infty} \frac{x^n}{\sqrt{n}}$

4. $f(x) = \sum_{n=2}^{+\infty} \frac{(x-2)^n}{\sqrt{n-1}}$

5. $f(x) = \sum_{n=1}^{+\infty} (-1)^{n-1} \frac{x^{2n-1}}{(2n-1)!}$

6. $f(x) = \sum_{n=0}^{+\infty} \frac{x^{2n}}{(n!)^2}$

7. $f(x) = \sum_{n=1}^{+\infty} (n+1)(3x-1)^n$

8. $f(x) = \sum_{n=1}^{+\infty} \frac{x^{2n-2}}{(2n-2)!}$

9. $f(x) = \sum_{n=1}^{+\infty} \frac{(x-1)^n}{n3^n}$

10. $f(x) = \sum_{n=2}^{+\infty} (-1)^n \frac{(x-3)^n}{n(n-1)}$

11. Use the result of Example 2 to find a power-series representation of $\frac{1}{(1-x)^3}$.

12. Use the result of Example 3 to find a power-series representation of $e^{\sqrt{x}}$.

13. Obtain a power-series representation of $\frac{1}{(1+x)^2}$ if $|x| < 1$ by differentiating series (2) term by term.

14. Obtain a power-series representation of $\frac{x}{(1+x^2)^2}$ if $|x| < 1$ by differentiating series (4) term by term.

15. (a) Use series (1) to find a power-series representation for $\frac{1}{1-2x}$. (b) Differentiate term by term the series found in part (a) to find a power-series representation for $\frac{2}{(1-2x)^2}$.

16. (a) Use series (2) to find a power-series representation for $\dfrac{1}{1 + x^3}$. (b) Differentiate term by term the series found in part (a) to find a power-series representation for $\dfrac{-3x^2}{(1 + x^3)^2}$.

17. (a) Use the result of Example 3 to find a power-series representation for e^{x^2}. (b) Differentiate term by term the series found in part (a) to find a power-series representation for xe^{x^2}.

18. Let f be the function defined by $f(x) = \displaystyle\sum_{n=0}^{+\infty} (-1)^n \dfrac{x^n}{3^n(n + 2)}$. (a) Find the domain of f. (b) Find $f'(x)$ and find the domain of f'.

19. Use the result of Example 4 to find the value of $\dfrac{1}{\sqrt{e}}$ correct to five decimal places.

20. If $f(x) = \displaystyle\sum_{n=0}^{+\infty} (-1)^n \dfrac{x^{2n}}{3^n}$, find $f'(\tfrac{1}{2})$ correct to four decimal places.

21. Use the results of Examples 3 and 4 to find a power-series representation of (a) $\sinh x$ and (b) $\cosh x$.

22. Show that each of the power series in parts (a) and (b) of Exercise 21 can be obtained from the other by term-by-term differentiation.

23. Use the result of Example 2 to find the sum of the series $\displaystyle\sum_{n=1}^{+\infty} \dfrac{n}{2^n}$.

24. (a) Find a power-series representation for $\dfrac{e^x - 1}{x}$. (b) By differentiating term by term the power series in part (a), show that $\displaystyle\sum_{n=1}^{+\infty} \dfrac{n}{(n + 1)!} = 1$.

25. (a) Find a power-series representation for $x^2 e^{-x}$. (b) By differentiating term by term the power series in part (a), show that $\displaystyle\sum_{n=1}^{+\infty} (-2)^{n+1} \dfrac{n + 2}{n!} = 4$.

26. (a) Find a power-series representation for e^{-x^2}. (b) By differentiating the power series in part (a) twice term by term, show that $\displaystyle\sum_{n=1}^{+\infty} (-1)^{n+1} \dfrac{2n + 1}{2^n n!} = 1$.

27. Suppose a function f has the power-series representation $\displaystyle\sum_{n=0}^{+\infty} c_n x^n$, where if R is the radius of convergence, $R > 0$. If $f'(x) = f(x)$ and $f(0) = 1$, find the power series by using only properties of power series and nothing about the exponential function.

28. (a) Use only properties of power series to find a power-series representation of the function f if $f(x) > 0$ and $f'(x) = 2xf(x)$ for all x, and $f(0) = 1$. (b) Verify your result in part (a) by solving the differential equation $\dfrac{dy}{dx} = 2xy$ with the initial condition $y = 1$ when $x = 0$.

29. Suppose a function f has the power-series representation $\displaystyle\sum_{n=0}^{+\infty} c_n x^n$. If f is an even function, show that $c_n = 0$ when n is odd.

30. Assume that the constant 0 has a power-series representation $\displaystyle\sum_{n=0}^{+\infty} c_n x^n$, where the radius of convergence $R > 0$. Prove that $c_n = 0$ for all n.

In Exercises 31 through 35, show that the power series is a solution of the differential equation.

31. $y = \displaystyle\sum_{n=0}^{+\infty} \dfrac{2^n}{n!} x^n$; $\dfrac{dy}{dx} - 2y = 0$

32. $y = \displaystyle\sum_{n=0}^{+\infty} \dfrac{1}{2^n n!} x^{2n}$; $\dfrac{dy}{dx} - xy = 0$

33. $y = \sum\limits_{n=1}^{+\infty} \dfrac{(-1)^{n+1}}{(2n-1)!} x^{2n-1}; \dfrac{d^2y}{dx^2} + y = 0$

34. $y = x + \sum\limits_{n=0}^{+\infty} (-1)^n \dfrac{x^{2n}}{(2n)!}; \dfrac{d^2y}{dx^2} + y - x = 0$

35. $y = \sum\limits_{n=0}^{+\infty} (-1)^n \dfrac{2^n n!}{(2n+1)!} x^{2n+1}; \dfrac{d^2y}{dx^2} + x\dfrac{dy}{dx} + y = 0$

**12.3
INTEGRATION OF POWER
SERIES**

The theorem regarding the term-by-term integration of a power series is a consequence of Theorem 12.2.3.

12.3.1 THEOREM   Let $\sum\limits_{n=0}^{+\infty} c_n x^n$ be a power series whose radius of convergence is $R > 0$. Then if f is the function defined by

$$f(x) = \sum_{n=0}^{+\infty} c_n x^n$$

f is integrable on every closed subinterval of $(-R, R)$, and the integral of f is evaluated by integrating the given power series term by term; that is, if x is in $(-R, R)$, then

$$\int_0^x f(t)\,dt = \sum_{n=0}^{+\infty} \frac{c_n}{n+1} x^{n+1}$$

Furthermore, R is the radius of convergence of the resulting series.

PROOF   Let g be the function defined by

$$g(x) = \sum_{n=0}^{+\infty} \frac{c_n}{n+1} x^{n+1}$$

Because the terms of the power-series representation of $f(x)$ are the derivatives of the terms of the power-series representation of $g(x)$, the two series have, by Theorem 12.2.1, the same radius of convergence. By Theorem 12.2.3

$$g'(x) = f(x) \qquad \text{for every } x \text{ in } (-R, R)$$

By Theorem 12.2.2 it follows that $f'(x) = g''(x)$ for every x in $(-R, R)$. Because f is differentiable on $(-R, R)$, f is continuous there; consequently f is continuous on every closed subinterval of $(-R, R)$. From Theorem 4.8.2 we conclude that if x is in $(-R, R)$, then

$$\int_0^x f(t)\,dt = g(x) - g(0)$$
$$= g(x)$$

$$\Leftrightarrow \quad \int_0^x f(t)\,dt = \sum_{n=0}^{+\infty} \frac{c_n}{n+1} x^{n+1} \qquad\qquad \blacksquare$$

Theorem 12.3.1 often is used to compute a definite integral that cannot be evaluated directly by finding an antiderivative of the integrand. Examples 1 and 2 illustrate the technique. The definite integral $\int_0^x e^{-t^2}\,dt$ appearing in these two examples is similar to the one that represents the measure of the area of a region under the "normal probability curve."

918   Chapter 12   POWER SERIES

EXAMPLE 1

Find a power-series representation of $\int_0^x e^{-t^2}\, dt$.

SOLUTION

From Example 4 of Section 12.2,

$$e^{-x} = \sum_{n=0}^{+\infty} \frac{(-1)^n x^n}{n!}$$

for all values of x. If x is replaced by t^2,

$$e^{-t^2} = 1 - t^2 + \frac{t^4}{2!} - \frac{t^6}{3!} + \ldots + (-1)^n \frac{t^{2n}}{n!} + \ldots \qquad \text{for all values of } t$$

Applying Theorem 12.3.1 we integrate term by term and obtain

$$\int_0^x e^{-t^2}\, dt = \sum_{n=0}^{+\infty} \int_0^x (-1)^n \frac{t^{2n}}{n!}\, dt$$

$$= x - \frac{x^3}{3} + \frac{x^5}{2! \cdot 5} - \frac{x^7}{3! \cdot 7} + \ldots + (-1)^n \frac{x^{2n+1}}{n!(2n+1)} + \ldots$$

The power series represents the integral for all values of x.

EXAMPLE 2

Use the result of Example 1 to compute accurate to three decimal places the value of $\int_0^{1/2} e^{-t^2}\, dt$.

SOLUTION

Replace x by $\frac{1}{2}$ in the power series obtained in Example 1 to obtain

$$\int_0^{1/2} e^{-t^2}\, dt = \frac{1}{2} - \frac{1}{24} + \frac{1}{320} - \frac{1}{5376} + \ldots$$

$$\approx 0.5 - 0.0417 + 0.0031 - 0.0002 + \ldots$$

This is a convergent alternating series with $|u_{n+1}| < |u_n|$. Thus if we use the first three terms to approximate the sum, by Theorem 11.7.4 the error is less than the absolute value of the fourth term. From the first three terms,

$$\int_0^{1/2} e^{-t^2}\, dt \approx 0.461$$

EXAMPLE 3

Obtain a power-series representation of $\ln(1 + x)$.

SOLUTION

Consider the function f defined by $f(t) = \dfrac{1}{1 + t}$. A power-series representation of this function is given by series (2) in Section 12.2, which is

$$\frac{1}{1 + t} = 1 - t + t^2 - t^3 + \ldots + (-1)^n t^n + \ldots \qquad \text{if } |t| < 1$$

Apply Theorem 12.3.1 and integrate term by term to obtain

$$\int_0^x \frac{dt}{1+t} = \sum_{n=0}^{+\infty} \int_0^x (-1)^n t^n \, dt \qquad \text{if } |x| < 1$$

Therefore

$$\ln(1+x) = x - \frac{x^2}{2} + \frac{x^3}{3} - \frac{x^4}{4} + \ldots + (-1)^n \frac{x^{n+1}}{n+1} + \ldots \qquad \text{if } |x| < 1$$

$$\Leftrightarrow \quad \ln(1+x) = \sum_{n=1}^{+\infty} (-1)^{n-1} \frac{x^n}{n} \qquad \text{if } |x| < 1 \tag{1}$$

Because $|x| < 1$, $|1 + x| = 1 + x$. Thus the absolute-value bars are not needed when writing $\ln(1 + x)$.

In Example 3, Theorem 12.3.1 allows us to conclude that the power series in (1) represents the function only for values of x in the open interval $(-1, 1)$. However, the power series is convergent at the right endpoint 1, as was shown in Example 1 of Section 11.7. When $x = -1$, the power series becomes the negative of the harmonic series and is divergent. Hence the interval of convergence of the power series in (1) is $(-1, 1]$.

In the following illustration we show that the power series in (1) represents $\ln(1 + x)$ at $x = 1$ by proving that the sum of the series $\sum\limits_{n=1}^{+\infty} \dfrac{(-1)^{n-1}}{n}$ is $\ln 2$.

▶ ILLUSTRATION 1   For the infinite series $\sum\limits_{n=1}^{+\infty} \dfrac{(-1)^{n-1}}{n}$, the nth partial sum is

$$s_n = 1 - \frac{1}{2} + \frac{1}{3} - \frac{1}{4} + \ldots + (-1)^{n-1} \frac{1}{n} \tag{2}$$

It follows from Definition 11.3.2 that if we show $\lim\limits_{n \to +\infty} s_n = \ln 2$, we will have proved that the sum of the series is $\ln 2$.

From algebra comes the following formula for the sum of a finite geometric series:

$$a + ar + ar^2 + ar^3 + \ldots + ar^{n-1} = \frac{a - ar^n}{1 - r}$$

From this formula with $a = 1$ and $r = -t$,

$$1 - t + t^2 - t^3 + \ldots + (-t)^{n-1} = \frac{1 - (-t)^n}{1 + t}$$

which may be written as

$$1 - t + t^2 - t^3 + \ldots + (-1)^{n-1} t^{n-1} = \frac{1}{1+t} + (-1)^{n+1} \frac{t^n}{1+t}$$

Integrating from 0 to 1 we get

$$\int_0^1 \left[1 - t + t^2 - t^3 + \ldots + (-1)^{n-1}t^{n-1}\right] dt$$

$$= \int_0^1 \frac{dt}{1+t} + (-1)^{n+1} \int_0^1 \frac{t^n}{1+t} dt$$

which gives

$$1 - \frac{1}{2} + \frac{1}{3} - \frac{1}{4} + \ldots + (-1)^{n-1}\frac{1}{n} = \ln 2 + (-1)^{n+1} \int_0^1 \frac{t^n}{1+t} dt \qquad (3)$$

Referring to (2), we see that the left side of (3) is s_n. Letting

$$R_n = (-1)^{n+1} \int_0^1 \frac{t^n}{1+t} dt$$

(3) may be written as

$$s_n = \ln 2 + R_n \qquad\qquad\qquad (4)$$

Because $\dfrac{t^n}{1+t} \le t^n$ for all t in $[0, 1]$ it follows from Theorem 4.6.8 that

$$\int_0^1 \frac{t^n}{1+t} dt \le \int_0^1 t^n \, dt$$

Hence

$$0 \le |R_n| = \int_0^1 \frac{t^n}{1+t} dt \le \int_0^1 t^n \, dt = \frac{1}{n+1}$$

Because $\lim\limits_{n \to +\infty} \dfrac{1}{n+1} = 0$, it follows from the above inequality and the squeeze theorem (1.8.1) that $\lim\limits_{n \to +\infty} R_n = 0$. Therefore, from (4),

$$\lim_{n \to +\infty} s_n = \ln 2 + \lim_{n \to +\infty} R_n$$

$$= \ln 2$$

Thus

$$\sum_{n=1}^{+\infty} (-1)^{n-1}\frac{1}{n} = 1 - \frac{1}{2} + \frac{1}{3} - \frac{1}{4} + \ldots \qquad (5)$$

$$= \ln 2 \qquad\qquad\qquad \blacktriangleleft$$

The solution of Example 3 shows that the power series in (1) represents $\ln(x + 1)$ if $|x| < 1$. Hence, with the result of Illustration 1 we can conclude that the power series in (1) represents $\ln(x + 1)$ for all x in its interval of convergence $(-1, 1]$.

Although it is interesting that the sum of the series in (5) is $\ln 2$, this series converges too slowly to use it to calculate $\ln 2$. We now proceed to obtain a power series for computation of natural logarithms.

From (1),

$$\ln(1 + x) = x - \frac{x^2}{2} + \frac{x^3}{3} - \ldots + (-1)^{n-1}\frac{x^n}{n} + \ldots \qquad \text{for } x \text{ in } (-1, 1] \qquad (6)$$

If x is replaced by $-x$ in this series,

$$\ln(1 - x) = -x - \frac{x^2}{2} - \frac{x^3}{3} - \frac{x^4}{4} - \ldots - \frac{x^n}{n} - \ldots \qquad \text{for } x \text{ in } [-1, 1) \qquad (7)$$

Subtracting term by term (7) from (6) we obtain

$$\ln\frac{1+x}{1-x} = 2\left(x + \frac{x^3}{3} + \frac{x^5}{5} + \ldots + \frac{x^{2n-1}}{2n-1} + \ldots\right) \qquad \text{if } |x| < 1 \qquad (8)$$

The series in (8) can be used to compute the value of the natural logarithm of any positive number.

▶ ILLUSTRATION 2   If y is any positive number, let

$$y = \frac{1+x}{1-x} \quad \text{and then} \quad x = \frac{y-1}{y+1} \quad \text{and } |x| < 1$$

For instance, if $y = 2$, then $x = \frac{1}{3}$. From (8),

$$\ln 2 = 2\left(\frac{1}{3} + \frac{1}{3^4} + \frac{1}{5 \cdot 3^5} + \frac{1}{7 \cdot 3^7} + \frac{1}{9 \cdot 3^9} + \frac{1}{11 \cdot 3^{11}} + \ldots\right)$$

$$= 2\left(\frac{1}{3} + \frac{1}{81} + \frac{1}{1215} + \frac{1}{15,309} + \frac{1}{177,147} + \frac{1}{1,948,617} + \ldots\right)$$

$$\approx 2(0.333333 + 0.012346 + 0.000823 + 0.000065 + 0.000006$$
$$+ 0.000001 + \ldots)$$

Using the first six terms in parentheses, multiplying by 2, and rounding off to five decimal places we get

$$\ln 2 \approx 0.69315 \qquad \qquad ◀$$

EXAMPLE 4
Obtain a power-series representation of $\tan^{-1} x$.

SOLUTION
From series (4) in Section 12.2,

$$\frac{1}{1+x^2} = 1 - x^2 + x^4 - x^6 + \ldots + (-1)^n x^{2n} + \ldots \qquad \text{if } |x| < 1$$

Apply Theorem 12.3.1 and integrate term by term to obtain

$$\int_0^x \frac{1}{1+t^2}\, dt = x - \frac{x^3}{3} + \frac{x^5}{5} - \ldots + (-1)^n \frac{x^{2n+1}}{2n+1} + \ldots$$

Therefore

$$\tan^{-1} x = \sum_{n=0}^{+\infty} (-1)^n \frac{x^{2n+1}}{2n+1} \qquad \text{if } |x| < 1 \tag{9}$$

Although Theorem 12.3.1 allows us to conclude that the power series in (9) represents $\tan^{-1} x$ only for values of x such that $|x| < 1$, it can be shown that the interval of convergence of the power series is $[-1, 1]$ and that the power series is a representation of $\tan^{-1} x$ for all x in its interval of convergence. (You are asked to do this in Exercise 36.) Therefore

$$\tan^{-1} x = \sum_{n=0}^{+\infty} (-1)^n \frac{x^{2n+1}}{2n+1}$$

$$= x - \frac{x^3}{3} + \frac{x^5}{5} - \dots \qquad \text{if } |x| \le 1 \tag{10}$$

▶ ILLUSTRATION 3   If $x = 1$ in (10),

$$\frac{\pi}{4} = 1 - \frac{1}{3} + \frac{1}{5} - \frac{1}{7} + \dots + (-1)^n \frac{1}{2n+1} + \dots \qquad ◀$$

The series in Illustration 3 is not suitable for computing π because it converges too slowly. The following example gives a better method.

EXAMPLE 5
Prove that $\frac{1}{4}\pi = \tan^{-1} \frac{1}{2} + \tan^{-1} \frac{1}{3}$. Use this formula and the power series for $\tan^{-1} x$ of Example 4 to compute accurate to five significant figures the value of π.

SOLUTION
Let $\alpha = \tan^{-1} \frac{1}{2}$ and $\beta = \tan^{-1} \frac{1}{3}$. Then

$$\tan(\alpha + \beta) = \frac{\tan \alpha + \tan \beta}{1 - \tan \alpha \tan \beta}$$

$$= \frac{\frac{1}{2} + \frac{1}{3}}{1 - \frac{1}{2} \cdot \frac{1}{3}}$$

$$= \frac{3 + 2}{6 - 1}$$

$$= 1$$

$$= \tan \tfrac{1}{4}\pi$$

Therefore

$$\tfrac{1}{4}\pi = \alpha + \beta$$

$$\tfrac{1}{4}\pi = \tan^{-1} \tfrac{1}{2} + \tan^{-1} \tfrac{1}{3} \tag{11}$$

From formula (10) with $x = \frac{1}{2}$,

$$\tan^{-1}\frac{1}{2} = \frac{1}{2} - \frac{1}{3}\left(\frac{1}{2}\right)^3 + \frac{1}{5}\left(\frac{1}{2}\right)^5 - \frac{1}{7}\left(\frac{1}{2}\right)^7 + \frac{1}{9}\left(\frac{1}{2}\right)^9$$

$$- \frac{1}{11}\left(\frac{1}{2}\right)^{11} + \frac{1}{13}\left(\frac{1}{2}\right)^{13} - \frac{1}{15}\left(\frac{1}{2}\right)^{15} + \cdots$$

$$= \frac{1}{2} - \frac{1}{24} + \frac{1}{160} - \frac{1}{896} + \frac{1}{4608} - \frac{1}{22,528}$$

$$+ \frac{1}{106,492} - \frac{1}{491,520} + \cdots$$

$$\approx 0.500000 - 0.041667 + 0.006250 - 0.001116 + 0.000217$$

$$- 0.000044 + 0.000009 - 0.000002 + \cdots$$

Because the series is alternating and $|u_{n+1}| < |u_n|$, it follows from Theorem 11.7.4 that if the first seven terms are used to approximate the sum of the series, the error is less than the absolute value of the eighth term. Therefore

$$\tan^{-1}\tfrac{1}{2} \approx 0.463648 \tag{12}$$

From formula (10) with $x = \frac{1}{3}$,

$$\tan^{-1}\frac{1}{3} = \frac{1}{3} - \frac{1}{3}\left(\frac{1}{3}\right)^3 + \frac{1}{5}\left(\frac{1}{3}\right)^5 - \frac{1}{7}\left(\frac{1}{3}\right)^7 + \frac{1}{9}\left(\frac{1}{3}\right)^9 - \frac{1}{11}\left(\frac{1}{3}\right)^{11} + \cdots$$

$$= \frac{1}{3} - \frac{1}{81} + \frac{1}{1215} - \frac{1}{15,309} + \frac{1}{177,147} - \frac{1}{1,948,617} + \cdots$$

$$\approx 0.333333 - 0.012346 + 0.000823 - 0.000065$$

$$+ 0.000006 - 0.0000005 + \cdots$$

If the first five terms are used to approximate the sum,

$$\tan^{-1}\tfrac{1}{3} \approx 0.321751 \tag{13}$$

By substituting from (12) and (13) into (11),

$$\tfrac{1}{4}\pi \approx 0.463648 + 0.321751$$

$$\approx 0.78540$$

We multiply by 4 and the result to five significant figures is $\pi \approx 3.1416$.

EXERCISES 12.3

In Exercises 1 through 4, find a power-series representation of the given integral. Also find the radius of convergence of the resulting power series.

1. $\int_0^x e^t \, dt$ **2.** $\int_0^x \dfrac{dt}{t^2 + 4}$ **3.** $\int_2^x \dfrac{dt}{4 - t}$ **4.** $\int_0^x \ln(1 + t) \, dt$

In Exercises 5 through 8, compute accurate to three decimal places the value of the given integral by two methods: (a) Use the second fundamental theorem of the calculus; (b) use the result of the indicated exercise.

5. $\int_0^1 e^t \, dt$; Exercise 1 **6.** $\int_0^1 \frac{dt}{t^2 + 4}$; Exercise 2 **7.** $\int_2^3 \frac{dt}{4 - t}$; Exercise 3 **8.** $\int_0^{1/3} \ln(1 + t) \, dt$; Exercise 4

In Exercises 9 through 12, find a power-series representation of the given integral. Also find the radius of convergence of the resulting power series.

9. $\int_0^x f(t) \, dt$, where $f(t) = \begin{cases} \dfrac{e^t - 1}{t} & \text{if } t \neq 0 \\ 1 & \text{if } t = 0 \end{cases}$

10. $\int_0^x f(t) \, dt$, where $f(t) = \begin{cases} \dfrac{\ln(1 + t)}{t} & \text{if } t \neq 0 \\ 1 & \text{if } t = 0 \end{cases}$

11. $\int_0^x h(t) \, dt$, where $h(t) = \begin{cases} \dfrac{\sinh t}{t} & \text{if } t \neq 0 \\ 1 & \text{if } t = 0 \end{cases}$

12. $\int_0^x g(t) \, dt$, where $g(t) = \begin{cases} \dfrac{\tan^{-1} t}{t} & \text{if } t \neq 0 \\ 1 & \text{if } t = 0 \end{cases}$

In Exercises 13 through 16, compute accurate to three decimal places the value of the definite integral obtained by replacing x by the given number in the indicated exercise.

13. $x = 1$; Exercise 9 **14.** $x = \frac{1}{2}$; Exercise 10 **15.** $x = 1$; Exercise 11 **16.** $x = \frac{1}{4}$; Exercise 12

In Exercises 17 through 24, compute accurate to three decimal places the value of the given integral by using series.

17. $\int_0^{1/2} \frac{dx}{1 + x^3}$ **18.** $\int_0^{1/2} e^{-x^3} \, dx$ **19.** $\int_0^1 e^{-x^2} \, dx$ **20.** $\int_0^{1/3} \frac{dx}{1 + x^4}$ **21.** $\int_0^{1/2} \tan^{-1} x^2 \, dx$

22. $\int_0^{1/2} \cosh x^2 \, dx$ **23.** $\int_0^1 x \sinh \sqrt{x} \, dx$ **24.** $\int_0^1 g(x) \, dx$, where $g(x) = \begin{cases} \dfrac{\cosh x - 1}{x} & \text{if } x \neq 0 \\ 0 & \text{if } x = 0 \end{cases}$

25. Given $\cosh x = \sum_{n=0}^{+\infty} \frac{x^{2n}}{(2n)!}$ for all x. Obtain a power-series representation for $\sinh x$ by integrating term by term from 0 to x the given series.

26. Find a power-series representation for $\ln(1 + ax)$ by integrating term by term from 0 to x a power-series representation for $\frac{1}{1 + at}$.

27. Use the power series in Eq. (9) to compute $\tan^{-1} \frac{1}{4}$ accurate to four decimal places.

28. Use the power series in Eq. (8) to compute $\ln 3$ accurate to four decimal places.

29. If $f'(x) = \sum_{n=0}^{+\infty} (-1)^n \frac{(x - 1)^n}{n!}$, find $f(\frac{3}{4})$ accurate to three decimal places.

30. If $g'(x) = \sum_{n=0}^{+\infty} (-1)^n \frac{x^n}{n^2 + 3}$, find $g(1)$ accurate to two decimal places.

31. Find a power-series representation for $\tanh^{-1} x$ by integrating term by term from 0 to x a power-series representation for $(1 - t^2)^{-1}$.

32. Find a power series for xe^x by multiplying the series for e^x by x, and then integrate the resulting series term by term from 0 to 1 and show that $\sum_{n=1}^{+\infty} \frac{1}{n!(n + 2)} = \frac{1}{2}$.

33. By integrating term by term from 0 to x a power-series representation for $\ln(1 - t)$, show that

$$\sum_{n=2}^{+\infty} \frac{x^n}{(n - 1)n} = x + (1 - x) \ln(1 - x)$$

34. By integrating term by term from 0 to x a power-series representation for $\ln(1 + t)$, show that

$$\sum_{n=0}^{+\infty} \frac{(-1)^n}{(n + 1)(n + 2)} = 2 \ln 2 - 1$$

35. By integrating term by term from 0 to x a power-series representation for $t \tan^{-1} t$, show that

$$\sum_{n=1}^{+\infty} (-1)^{n+1} \frac{x^{2n+1}}{(2n - 1)(2n + 1)} = \tfrac{1}{2}[(x^2 + 1) \tan^{-1} x - x]$$

36. Show that the interval of convergence of the power series in Eq. (9) is $[-1, 1]$ and that the power series is a representation of $\tan^{-1} x$ for all x in its interval of convergence.

37. Find the power series in x of $f(x)$ if $f''(x) = -f(x)$, $f(0) = 0$, and $f'(0) = 1$. Also, find the radius of convergence of the resulting series.

38. Integrate term by term from 0 to x a power-series representation for $(1 - t^2)^{-1}$ to obtain the power series in Eq. (8) for $\ln \dfrac{1 + x}{1 - x}$.

12.4
TAYLOR SERIES

If f is the function defined by

$$f(x) = \sum_{n=0}^{+\infty} c_n x^n$$

$$= c_0 + c_1 x + c_2 x^2 + c_3 x^3 + \ldots + c_n x^n + \ldots \tag{1}$$

whose radius of convergence is $R > 0$, it follows from successive applications of Theorem 12.2.3 that f has derivatives of all orders on $(-R, R)$. Such a function is said to be *infinitely differentiable* on $(-R, R)$. Successive differentiations of the function in (1) give

$$f'(x) = c_1 + 2c_2 x + 3c_3 x^2 + 4c_4 x^3 + \ldots + nc_n x^{n-1} + \ldots \tag{2}$$

$$f''(x) = 2c_2 + 2 \cdot 3c_3 x + 3 \cdot 4c_4 x^2 + \ldots + (n - 1)nc_n x^{n-2} + \ldots \tag{3}$$

$$f'''(x) = 2 \cdot 3c_3 + 2 \cdot 3 \cdot 4c_4 x + \ldots + (n - 2)(n - 1)nc_n x^{n-3} + \ldots \tag{4}$$

$$f^{(iv)} = 2 \cdot 3 \cdot 4c_4 + \ldots + (n - 3)(n - 2)(n - 1)nc_n x^{n-4} + \ldots \tag{5}$$

etc. If $x = 0$ in (1),

$$f(0) = c_0$$

If $x = 0$ in (2),

$$f'(0) = c_1$$

If in (3) $x = 0$,

$$f''(0) = 2c_2 \quad \Leftrightarrow \quad c_2 = \frac{f''(0)}{2!}$$

From (4), if $x = 0$,

$$f'''(0) = 2 \cdot 3c_3 \quad \Leftrightarrow \quad c_3 = \frac{f'''(0)}{3!}$$

In a similar manner, from (5), if $x = 0$,

$$f^{(iv)}(0) = 2 \cdot 3 \cdot 4c_4 \quad \Leftrightarrow \quad c_4 = \frac{f^{(iv)}(0)}{4!}$$

In general,

$$c_n = \frac{f^{(n)}(0)}{n!} \qquad \text{for every positive integer } n \qquad (6)$$

Formula (6) also holds when $n = 0$ if we take $f^{(0)}(0)$ to be $f(0)$ and $0! = 1$. So from (1) and (6) the power series of f in x can be written as

$$\sum_{n=0}^{+\infty} \frac{f^{(n)}(0)}{n!} x^n = f(0) + f'(0)x + \frac{f''(0)}{2!} x^2 + \ldots + \frac{f^{(n)}(0)}{n!} x^n + \ldots \qquad (7)$$

In a more general sense, consider the function f as a power series in $x - a$; that is,

$$f(x) = \sum_{n=0}^{+\infty} c_n(x - a)^n$$

$$= c_0 + c_1(x - a) + c_2(x - a)^2 + \ldots + c_n(x - a)^n + \ldots \qquad (8)$$

If the radius of convergence of this series is R, then f is infinitely differentiable on $(a - R, a + R)$. Successive differentiations of the function in (8) give

$$f'(x) = c_1 + 2c_2(x - a) + 3c_3(x - a)^2 + 4c_4(x - a)^3 + \ldots$$
$$+ nc_n(x - a)^{n-1} + \ldots$$

$$f''(x) = 2c_2 + 2 \cdot 3c_3(x - a) + 3 \cdot 4c_4(x - a)^2 + \ldots$$
$$+ (n - 1)nc_n(x - a)^{n-2} + \ldots$$

$$f'''(x) = 2 \cdot 3c_3 + 2 \cdot 3 \cdot 4c_4(x - a) + \ldots$$
$$+ (n - 2)(n - 1)nc_n(x - a)^{n-3} + \ldots$$

etc. Letting $x = a$ in the power-series representations of f and its derivatives we get

$$c_0 = f(a) \qquad c_1 = f'(a) \qquad c_2 = \frac{f''(a)}{2!} \qquad c_3 = \frac{f'''(a)}{3!}$$

and in general

$$c_n = \frac{f^{(n)}(a)}{n!} \qquad (9)$$

From (8) and (9) the power series of f in $x - a$ can be written as

$$\sum_{n=0}^{+\infty} \frac{f^{(n)}(a)}{n!}(x-a)^n = f(a) + f'(a)(x-a) + \frac{f''(a)}{2!}(x-a)^2$$

$$+ \ldots + \frac{f^{(n)}(a)}{n!}(x-a)^n + \ldots \qquad (10)$$

The series in (10) is called the **Taylor series** of f at a. The special case of (10), when $a = 0$, is (7), which is called the **Maclaurin series**.

Observe that the nth partial sum of infinite series (10) is the nth degree Taylor polynomial of the function f at the number a, discussed in Section 9.5.

EXAMPLE 1
Find the Maclaurin series for e^x.

SOLUTION
If $f(x) = e^x$, $f^{(n)}(x) = e^x$ for all x; therefore $f^{(n)}(0) = 1$ for all n. So from (7) we have the Maclaurin series:

$$e^x = 1 + x + \frac{x^2}{2!} + \frac{x^3}{3!} + \ldots + \frac{x^n}{n!} + \ldots \qquad (11)$$

Series (11) is the same as the series obtained in Example 3 of Section 12.2.

EXAMPLE 2
Find the Taylor series for $\sin x$ at a.

SOLUTION
If $f(x) = \sin x$, then we have $f'(x) = \cos x$, $f''(x) = -\sin x$, $f'''(x) = -\cos x$, $f^{(iv)}(x) = \sin x$, and so forth. Thus, from formula (9), $c_0 = \sin a$, $c_1 = \cos a$, $c_2 = (-\sin a)/2!$, $c_3 = (-\cos a)/3!$, $c_4 = (\sin a)/4!$, and so on. The required Taylor series is obtained from (10), and it is

$$\sin x = \sin a + (\cos a)(x-a) - (\sin a)\frac{(x-a)^2}{2!}$$

$$- (\cos a)\frac{(x-a)^3}{3!} + (\sin a)\frac{(x-a)^4}{4!} + \ldots$$

We can deduce that a power-series representation of a function is unique. That is, if two functions have the same function values in some interval containing the number a, and if both functions have a power-series representation in $x - a$, then these series must be the same because the coefficients in the series are obtained from the values of the functions and their derivatives at a. Therefore, if a function has a power-series representation in $x - a$, this series must be its Taylor series at a. Hence the Taylor series for a given function does not have to be obtained by using formula (10). Any method that gives a power series in $x - a$ representing the function will be the Taylor series of the function at a.

▶ ILLUSTRATION 1   To find the Taylor series for e^x at a, write $e^x = e^a e^{x-a}$ and then use series (11), where x is replaced by $x - a$. Then

$$e^x = e^a \left[1 + (x - a) + \frac{(x - a)^2}{2!} + \frac{(x - a)^3}{3!} + \ldots + \frac{(x - a)^n}{n!} + \ldots \right] \quad ◀$$

▶ ILLUSTRATION 2   The series for $\ln(1 + x)$ found in Example 3 of Section 12.3 can be used to find the Taylor series for $\ln x$ at a ($a > 0$) by writing

$$\ln x = \ln[a + (x - a)]$$

$$\ln x = \ln a + \ln\left(1 + \frac{x - a}{a} \right) \tag{12}$$

Because

$$\ln(1 + t) = \sum_{n=1}^{+\infty} (-1)^{n-1} \frac{t^n}{n} \qquad \text{if } -1 < t \le 1$$

then

$$\ln\left(1 + \frac{x - a}{a} \right) = \frac{x - a}{a} - \frac{(x - a)^2}{2a^2} + \frac{(x - a)^3}{3a^3} - \ldots$$

Therefore, from (12),

$$\ln x = \ln a + \frac{x - a}{a} - \frac{(x - a)^2}{2a^2} + \frac{(x - a)^3}{3a^3} - \ldots$$

and the series represents $\ln x$ if $-1 < \dfrac{x - a}{a} \le 1$ or, equivalently, $0 < x \le 2a$.

◀

A natural question that arises is: If a function has a Taylor series in $x - a$ having radius of convergence $R > 0$, does this series represent the function for all values of x in the interval $(a - R, a + R)$? For most elementary functions the answer is yes. However, there are functions for which the answer is no. The following example shows this.

EXAMPLE 3
Let f be the function defined by

$$f(x) = \begin{cases} e^{-1/x^2} & \text{if } x \ne 0 \\ 0 & \text{if } x = 0 \end{cases}$$

Find the Maclaurin series for f, and show that it converges for all values of x but that it represents $f(x)$ only when $x = 0$.

SOLUTION
To find $f'(0)$, use the definition of a derivative.

$$f'(0) = \lim_{x \to 0} \frac{e^{-1/x^2} - 0}{x - 0}$$

$$= \lim_{x \to 0} \frac{\dfrac{1}{x}}{e^{1/x^2}}$$

Because $\lim_{x \to 0} (1/x) = +\infty$ and $\lim_{x \to 0} e^{1/x^2} = +\infty$, L'Hôpital's rule can be used. Therefore

$$f'(0) = \lim_{x \to 0} \frac{-\dfrac{1}{x^2}}{e^{1/x^2}\left(-\dfrac{2}{x^3}\right)}$$

$$= \lim_{x \to 0} \frac{x}{2e^{1/x^2}}$$

$$= 0$$

By a similar method, using the definition of a derivative and L'Hôpital's rule, we get 0 for every derivative. So $f^{(n)}(0) = 0$ for all n. Therefore the Maclaurin series for the given function is $0 + 0 + 0 + \ldots + 0 + \ldots$. This series converges to 0 for all x; however, if $x \neq 0$, $f(x) = e^{-1/x^2}$ and $e^{-1/x^2} \neq 0$.

A theorem that gives a test for determining whether a function is represented by its Taylor series is the following one.

12.4.1 THEOREM   Let f be a function such that f and all of its derivatives exist in some interval $(a - r, a + r)$. Then the function is represented by its Taylor series

$$\sum_{n=0}^{+\infty} \frac{f^{(n)}(a)}{n!} (x - a)^n$$

for all x such that $|x - a| < r$ if and only if

$$\lim_{n \to +\infty} R_n(x) = \lim_{n \to +\infty} \frac{f^{(n+1)}(\xi_n)}{(n + 1)!} (x - a)^{n+1}$$

$$= 0$$

where each ξ_n is between x and a.

PROOF   In the interval $(a - r, a + r)$, the function f satisfies the hypothesis of Theorem 9.5.1 for which

$$f(x) = P_n(x) + R_n(x) \tag{13}$$

where $P_n(x)$ is the nth-degree Taylor polynomial of f at a and $R_n(x)$ is the remainder, given by

$$R_n(x) = \frac{f^{(n+1)}(\xi_n)}{(n + 1)!} (x - a)^{n+1} \tag{14}$$

where each ξ_n is between x and a.

Now $P_n(x)$ is the nth partial sum of the Taylor series of f at a. So if we show that $\lim\limits_{n \to +\infty} P_n(x)$ exists and equals $f(x)$ if and only if $\lim\limits_{n \to +\infty} R_n(x) = 0$, the theorem will be proved. From (13),

$$P_n(x) = f(x) - R_n(x) \tag{15}$$

If $\lim\limits_{n \to +\infty} R_n(x) = 0$, it follows from (15) that

$$\lim_{n \to +\infty} P_n(x) = f(x) - \lim_{n \to +\infty} R_n(x)$$
$$= f(x) - 0$$
$$= f(x)$$

Now under the hypothesis that $\lim\limits_{n \to +\infty} P_n(x) = f(x)$ we wish to show that $\lim\limits_{n \to +\infty} R_n(x) = 0$. From (13),

$$R_n(x) = f(x) - P_n(x)$$

Thus

$$\lim_{n \to +\infty} R_n(x) = f(x) - \lim_{n \to +\infty} P_n(x)$$
$$= f(x) - f(x)$$
$$= 0$$

This proves the theorem. ∎

Theorem 12.4.1 also holds for other forms of the remainder $R_n(x)$ besides the Lagrange form.

It is often difficult to apply Theorem 12.4.1 in practice because the values of ξ_n are arbitrary. However, sometimes an upper bound for $R_n(x)$ can be found, and it may be possible to prove that the limit of the upper bound is zero as $n \to +\infty$. The following limit is helpful in some cases:

$$\lim_{n \to +\infty} \frac{x^n}{n!} = 0 \qquad \text{for all } x \tag{16}$$

This follows from Example 2 of Section 12.1 where it was shown that the power series $\sum\limits_{n=0}^{+\infty} \dfrac{x^n}{n!}$ is convergent for all values of x and hence the limit of its nth term must be zero. In a similar manner, because $\sum\limits_{n=0}^{+\infty} \dfrac{(x-a)^n}{n!}$ is convergent for all values of x,

$$\lim_{n \to +\infty} \frac{(x-a)^n}{n!} = 0 \qquad \text{for all } x \tag{17}$$

EXAMPLE 4

Use Theorem 12.4.1 to show that the Maclaurin series for e^x, found in Example 1, represents the function for all values of x.

SOLUTION

The Maclaurin series for e^x is series (11) and

$$R_n(x) = \frac{e^{\xi_n}}{(n+1)!} x^{n+1}$$

where each ξ_n is between 0 and x.

We must show that $\lim\limits_{n \to +\infty} R_n = 0$ for all x. There are three cases: $x > 0$, $x < 0$, and $x = 0$.

If $x > 0$, then $0 < \xi_n < x$; hence $e^{\xi_n} < e^x$. So

$$0 < \frac{e^{\xi_n}}{(n+1)!} x^{n+1} < e^x \frac{x^{n+1}}{(n+1)!} \qquad (18)$$

From (16) it follows that $\lim\limits_{n \to +\infty} \dfrac{x^{n+1}}{(n+1)!} = 0$, and so

$$\lim\limits_{n \to +\infty} e^x \frac{x^{n+1}}{(n+1)!} = 0$$

Therefore, from (18) and the squeeze theorem (1.8.1) it follows that $\lim\limits_{n \to +\infty} R_n(x) = 0$.

If $x < 0$, then $x < \xi_n < 0$ and $0 < e^{\xi_n} < 1$. Therefore, if $x^{n+1} > 0$,

$$0 < \frac{e^{\xi_n}}{(n+1)!} x^{n+1} < \frac{x^{n+1}}{(n+1)!}$$

and if $x^{n+1} < 0$,

$$\frac{x^{n+1}}{(n+1)!} < \frac{e^{\xi_n}}{(n+1)!} x^{n+1} < 0$$

In either case, because $\lim\limits_{n \to +\infty} \dfrac{x^{n+1}}{(n+1)!} = 0$, we conclude that $\lim\limits_{n \to +\infty} R_n = 0$.

Finally, if $x = 0$, the series has the sum of 1, which is e^0. Hence series (11) represents e^x for all values of x.

From the results of the above example we can write

$$e^x = \sum_{n=0}^{+\infty} \frac{x^n}{n!}$$

$$= 1 + x + \frac{x^2}{2!} + \frac{x^3}{3!} + \ldots \qquad \text{for all } x \qquad (19)$$

and this agrees with Example 3 of Section 12.2.

EXAMPLE 5

Show that the Taylor series for $\sin x$ at a, found in Example 2, represents the function for all values of x.

SOLUTION
We use Theorem 12.4.1; so we must show that

$$\lim_{n \to +\infty} R_n(x) = \lim_{n \to +\infty} \frac{f^{(n+1)}(\xi_n)}{(n+1)!}(x-a)^{n+1}$$
$$= 0$$

Because $f(x) = \sin x$, $f^{(n+1)}(\xi_n)$ will be one of the following numbers: $\cos \xi_n$, $\sin \xi_n$, $-\cos \xi_n$, or $-\sin \xi_n$. In any case, $|f^{(n+1)}(\xi_n)| \le 1$. Hence

$$0 < |R_n(x)| \le \frac{|x-a|^{n+1}}{(n+1)!} \tag{20}$$

From (17), $\displaystyle\lim_{n \to +\infty} \frac{|x-a|^{n+1}}{(n+1)!} = 0$. Thus, by the squeeze theorem (1.8.1) and (20) it follows that $\displaystyle\lim_{n \to +\infty} R_n(x) = 0$.

EXAMPLE 6
Compute the value of $\sin 47°$ accurate to four decimal places.

SOLUTION
In Examples 2 and 5 the Taylor series for $\sin x$ at a was obtained. It represents the function for all values of x. The series is

$$\sin x = (\sin a) + (\cos a)(x-a) - (\sin a)\frac{(x-a)^2}{2!} - (\cos a)\frac{(x-a)^3}{3!} + \ldots$$

To make $x - a$ small we must choose a value of a near the value of x for which the function value is being computed. The sine and cosine of a must also be known. We therefore choose $a = \frac{1}{4}\pi$ and have

$$\sin x = \sin \tfrac{1}{4}\pi + (\cos \tfrac{1}{4}\pi)(x - \tfrac{1}{4}\pi) - (\sin \tfrac{1}{4}\pi)\frac{(x - \tfrac{1}{4}\pi)^2}{2!}$$
$$- (\cos \tfrac{1}{4}\pi)\frac{(x - \tfrac{1}{4}\pi)^3}{3!} + \ldots \tag{21}$$

Because $47°$ is equivalent to $\frac{47}{180}\pi$ radians $= (\frac{1}{4}\pi + \frac{1}{90}\pi)$ radians, from Eq. (21),

$$\sin \tfrac{47}{180}\pi = \tfrac{1}{2}\sqrt{2} + \tfrac{1}{2}\sqrt{2} \cdot \tfrac{1}{90}\pi - \tfrac{1}{2}\sqrt{2} \cdot \tfrac{1}{2}(\tfrac{1}{90}\pi)^2 - \tfrac{1}{2}\sqrt{2} \cdot \tfrac{1}{6}(\tfrac{1}{90}\pi)^3 + \ldots$$
$$\approx \tfrac{1}{2}\sqrt{2}(1 + 0.03490 - 0.00061 - 0.000002 + \ldots)$$

Taking $\sqrt{2} \approx 1.41421$ and using the first three terms of the series we get

$$\sin \tfrac{47}{180}\pi \approx (0.70711)(1.03429)$$
$$\approx 0.73136$$

Rounding off to four decimal places gives $\sin 47° \approx 0.7314$. The error introduced by using the first three terms is $R_2(\frac{47}{180}\pi)$, and from (20),

$$\left| R_2\left(\frac{47}{180}\pi\right) \right| \leq \frac{(\frac{1}{90}\pi)^3}{3!} \approx 0.00001$$

The result, then, is between $0.73136 - 0.00001$ and $0.73136 + 0.00001$; that is, the result is between 0.73135 and 0.73137. Thus to four decimal place accuracy, we have

$$\sin \tfrac{47}{180}\pi \approx 0.7314$$

The following Maclaurin series represent the given function for all values of x:

$$\sin x = \sum_{n=0}^{+\infty} \frac{(-1)^n x^{2n+1}}{(2n+1)!}$$

$$= x - \frac{x^3}{3!} + \frac{x^5}{5!} - \frac{x^7}{7!} + \ldots \tag{22}$$

$$\cos x = \sum_{n=0}^{+\infty} \frac{(-1)^n x^{2n}}{(2n)!}$$

$$= 1 - \frac{x^2}{2!} + \frac{x^4}{4!} - \frac{x^6}{6!} + \ldots \tag{23}$$

$$\sinh x = \sum_{n=0}^{+\infty} \frac{x^{2n+1}}{(2n+1)!}$$

$$= x + \frac{x^3}{3!} + \frac{x^5}{5!} + \frac{x^7}{7!} + \ldots \tag{24}$$

$$\cosh x = \sum_{n=0}^{+\infty} \frac{x^{2n}}{(2n)!}$$

$$= 1 + \frac{x^2}{2!} + \frac{x^4}{4!} + \frac{x^6}{6!} + \ldots \tag{25}$$

Series (22) is a direct result of Examples 2 and 5 with $a = 0$. You are asked to verify series (23), (24), and (25) in Exercises 1, 2, and 3.

EXAMPLE 7
Evaluate

$$\int_{1/2}^{1} \frac{\sin x}{x}\, dx$$

accurate to five decimal places.

SOLUTION

An antiderivative of the integrand in terms of elementary functions cannot be found. However, from series (22),

$$\frac{\sin x}{x} = \frac{1}{x} \cdot \sin x$$

$$= \frac{1}{x}\left(x - \frac{x^3}{3!} + \frac{x^5}{5!} - \frac{x^7}{7!} + \frac{x^9}{9!} - \cdots\right)$$

$$= 1 - \frac{x^2}{3!} + \frac{x^4}{5!} - \frac{x^6}{7!} + \frac{x^8}{9!} - \cdots$$

which is true for all $x \neq 0$. Using term-by-term integration we get

$$\int_{1/2}^{1} \frac{\sin x}{x}\, dx = x - \frac{x^3}{3 \cdot 3!} + \frac{x^5}{5 \cdot 5!} - \frac{x^7}{7 \cdot 7!} + \frac{x^9}{9 \cdot 9!} - \cdots \Big]_{1/2}^{1}$$

$$\approx (1 - 0.0555555 + 0.0016667 - 0.0000283 + 0.0000003 - \ldots)$$
$$-(0.5 - 0.0069444 + 0.0000521 - 0.0000002 + \ldots)$$

In each set of parentheses there is a convergent alternating series with $|u_{n+1}| < |u_n|$. In the first set of parentheses we use the first four terms because the error obtained is less than 0.0000003. In the second set of parentheses we use the first three terms where the error obtained is less than 0.0000002. Doing the arithmetic and rounding off to five decimal places we get

$$\int_{1/2}^{1} \frac{\sin x}{x}\, dx \approx 0.45298$$

EXERCISES 12.4

1. Prove that the series $\displaystyle\sum_{n=0}^{+\infty} \frac{(-1)^n x^{2n}}{(2n)!}$ represents $\cos x$ for all values of x.

2. Prove that the series $\displaystyle\sum_{n=0}^{+\infty} \frac{x^{2n+1}}{(2n+1)!}$ represents $\sinh x$ for all values of x.

3. Prove that the series $\displaystyle\sum_{n=0}^{+\infty} \frac{x^{2n}}{(2n)!}$ represents $\cosh x$ for all values of x.

4. Obtain the Maclaurin series for the cosine function by differentiating the Maclaurin series for the sine function. Also obtain the Maclaurin series for the sine function by differentiating the one for the cosine function.

5. Obtain the Maclaurin series for the hyperbolic sine function by differentiating the Maclaurin series for the hyperbolic cosine function. Also differentiate the Maclaurin series for the hyperbolic sine function to obtain the one for the hyperbolic cosine function.

6. Find the Taylor series for e^x at 3 by using the Maclaurin series for e^x.

7. Use the Maclaurin series for $\ln(1 + x)$ to find the Taylor series for $\ln x$ at 2.

8. Given $\ln 2 = 0.6931$, use the series obtained in Exercise 7 to find $\ln 3$ accurate to four decimal places.

In Exercises 9 through 14, find a power-series representation for the given function at the number a, and determine its radius of convergence.

9. $f(x) = \ln(x + 1);\ a = 1$ **10.** $f(x) = \sqrt[3]{x};\ a = 1$ **11.** $f(x) = \sqrt{x};\ a = 4$

12. $f(x) = \dfrac{1}{x};\ a = 1$ **13.** $f(x) = \cos x;\ a = \frac{1}{3}\pi$ **14.** $f(x) = 2^x;\ a = 0$

15. Find the Maclaurin series for $\sin^2 x$. (*Hint:* Use $\sin^2 x = \frac{1}{2}(1 - \cos 2x)$.)

16. Find the Maclaurin series for $\cos^2 x$. (*Hint:* Use $\cos^2 x = \frac{1}{2}(1 + \cos 2x)$.)

17. (a) Find the first three nonzero terms of the Maclaurin series for $\tan x$. (b) Use the result of part (a) and term-by-term differentiation to find the first three nonzero terms of the Maclaurin series for $\sec^2 x$. (c) Use the result of part (a) and term-by-term integration to find the first three nonzero terms of the Maclaurin series for $\ln|\sec x|$.

18. (a) Find the first three nonzero terms of the Taylor series for $\cot x$ at $\frac{1}{2}\pi$. (b) Use the result of part (a) and term-by-term integration to find the first three nonzero terms of the Taylor series for $\ln \sin x$ at $\frac{1}{2}\pi$.

In Exercises 19 through 24, use a power series to compute to the indicated accuracy the value of the given quantity.

19. $\cos 58°$; four decimal places   **20.** $\sqrt[5]{e}$; four decimal places   **21.** $\sqrt[5]{30}$; five decimal places

22. $\sinh \frac{1}{2}$; five decimal places   **23.** $\ln(0.8)$; four decimal places   **24.** $\sqrt[3]{29}$; three decimal places.

25. Compute the value of e correct to seven decimal places, and prove that your answer has the required accuracy.

In Exercises 26 through 31, compute accurate to three decimal places the value of the definite integral.

26. $\displaystyle\int_0^1 \sqrt{x}e^{-x^2}\,dx$ **27.** $\displaystyle\int_0^{1/2} \sin x^2\,dx$ **28.** $\displaystyle\int_0^1 \cos\sqrt{x}\,dx$ **29.** $\displaystyle\int_0^{0.1} \ln(1 + \sin x)\,dx$

30. $\displaystyle\int_0^{1/3} f(x)\,dx$, where $f(x) = \begin{cases} \dfrac{\sin x}{x} & \text{if } x \neq 0 \\ 1 & \text{if } x = 0 \end{cases}$ **31.** $\displaystyle\int_0^1 g(x)\,dx$, where $g(x) = \begin{cases} \dfrac{1 - \cos x}{x} & \text{if } x \neq 0 \\ 0 & \text{if } x = 0 \end{cases}$

32. The function E defined by

$$E(x) = \frac{2}{\sqrt{\pi}} \int_0^x e^{-t^2}\,dt$$

is called the *error function*, and it is important in mathematical statistics. Find the Maclaurin series for the error function.

33. Determine a_n ($n = 0, 1, 2, 3, 4$) so that the polynomial $f(x) = 3x^4 - 17x^3 + 35x^2 - 32x + 17$ is written in the form

$$f(x) = a_4(x - 1)^4 + a_3(x - 1)^3 + a_2(x - 1)^2 + a_1(x - 1) + a_0$$

34. Determine a_n ($n = 0, 1, 2, 3$) so that the polynomial $f(x) = 4x^3 - 5x^2 + 2x - 3$ is written in the form

$$f(x) = a_3(x + 2)^3 + a_2(x + 2)^2 + a_1(x + 2) + a_0$$

12.5
THE BINOMIAL SERIES

In algebra you learned that the binomial theorem expresses $(a + b)^m$ as a sum of powers of a and b, where m is a positive integer, as follows:

$$(a + b)^m = a^m + ma^{m-1}b + \frac{m(m - 1)}{2!} a^{m-2}b^2 + \cdots$$

$$+ \frac{m(m - 1) \cdot \ldots \cdot (m - k + 1)}{k!} a^{m-k}b^k + \cdots + b^m$$

We now take $a = 1$ and $b = x$ and apply the binomial theorem to the expression $(1 + x)^m$, where m is not a positive integer. We obtain the power series

$$1 + mx + \frac{m(m - 1)}{2!} x^2 + \frac{m(m - 1)(m - 2)}{3!} x^3 + \cdots$$

$$+ \frac{m(m - 1)(m - 2) \cdot \ldots \cdot (m - n + 1)}{n!} x^n + \cdots \qquad (1)$$

Series (1) is the Maclaurin series for $(1 + x)^m$. It is called a **binomial series**. To find the radius of convergence of series (1), apply the ratio test and get

$$\lim_{n \to +\infty} \left| \frac{u_{n+1}}{u_n} \right| = \lim_{n \to +\infty} \left| \frac{\dfrac{m(m - 1) \cdot \ldots \cdot (m - n + 1)(m - n)}{(n + 1)!} x^{n+1}}{\dfrac{m(m - 1) \cdot \ldots \cdot (m - n + 1)}{n!} x^n} \right|$$

$$= \lim_{n \to +\infty} \left| \frac{m - n}{n + 1} \right| |x|$$

$$= \lim_{n \to +\infty} \left| \frac{\dfrac{m}{n} - 1}{1 + \dfrac{1}{n}} \right| |x|$$

$$= |x|$$

So the series is convergent if $|x| < 1$. We now prove that series (1) represents $(1 + x)^m$ for all real numbers m if x is in the open interval $(-1, 1)$. This is not done by calculating $R_n(x)$ and showing that its limit is zero because this is quite difficult, as you will soon see if you attempt to do so. Instead the following method is used. Let

$$f(x) = 1 + \sum_{n=1}^{+\infty} \frac{m(m - 1) \cdot \ldots \cdot (m - n + 1)}{n!} x^n \qquad |x| < 1 \qquad (2)$$

We wish to show that $f(x) = (1 + x)^m$, where $|x| < 1$. By Theorem 12.2.3,

$$f'(x) = \sum_{n=1}^{+\infty} \frac{m(m - 1) \cdot \ldots \cdot (m - n + 1)}{(n - 1)!} x^{n-1} \qquad |x| < 1 \qquad (3)$$

Multiplying on both sides of (3) by x we get, from Theorem 11.4.2,

$$xf'(x) = \sum_{n=1}^{+\infty} \frac{m(m - 1) \cdot \ldots \cdot (m - n + 1)}{(n - 1)!} x^n \qquad (4)$$

By rewriting the right side of (3),

$$f'(x) = m + \sum_{n=2}^{+\infty} \frac{m(m - 1) \cdot \ldots \cdot (m - n + 1)}{(n - 1)!} x^{n-1} \qquad (5)$$

The summation in (5) is rewritten with the lower limit decreased by 1 and n replaced by $n + 1$. Then

$$f'(x) = m + \sum_{n=1}^{+\infty} (m - n) \frac{m(m - 1) \cdot \ldots \cdot (m - n + 1)}{n!} x^n \qquad (6)$$

In (4) the numerator and the denominator are multiplied by n to obtain

$$xf'(x) = \sum_{n=1}^{+\infty} n \frac{m(m - 1) \cdot \ldots \cdot (m - n + 1)}{n!} x^n \qquad (7)$$

Because the series in (6) and (7) are absolutely convergent for $|x| < 1$, then by Theorem 11.4.3 they can be added term by term, and the resulting series will be absolutely convergent for $|x| < 1$. Therefore, from the addition,

$$(1 + x)f'(x) = m \left[1 + \sum_{n=1}^{+\infty} \frac{m(m - 1) \cdot \ldots \cdot (m - n + 1)}{n!} x^n \right]$$

Because by (2) the expression in brackets is $f(x)$, then

$$(1 + x)f'(x) = mf(x)$$

$$\frac{f'(x)}{f(x)} = \frac{m}{1 + x}$$

The left side of the above equation is $D_x[\ln f(x)]$; thus

$$\frac{d}{dx} [\ln f(x)] = \frac{m}{1 + x}$$

However, we also know that

$$\frac{d}{dx} [\ln(1 + x)^m] = \frac{m}{1 + x}$$

Because $\ln f(x)$ and $\ln(1 + x)^m$ have the same derivative, they differ by a constant. Hence

$$\ln f(x) = \ln(1 + x)^m + C$$

From (2), $f(0) = 1$. Therefore $C = 0$; so

$$f(x) = (1 + x)^m$$

We have proved the general binomial theorem, which is now stated.

12.5.1 THEOREM
Binomial Theorem

If m is any real number, then

$$(1 + x)^m = 1 + \sum_{n=1}^{+\infty} \frac{m(m - 1)(m - 2) \cdot \ldots \cdot (m - n + 1)}{n!} x^n \qquad (8)$$

for all values of x such that $|x| < 1$.

If m is a positive integer, the binomial series will terminate after a finite number of terms.

EXAMPLE 1

Express

$$\frac{1}{\sqrt{1+x}}$$

as a power series in x.

SOLUTION

From Theorem 12.5.1, when $|x| < 1$,

$$(1+x)^{-1/2} = 1 - \frac{1}{2}x + \frac{(-\frac{1}{2})(-\frac{1}{2}-1)}{2!}x^2 + \frac{(-\frac{1}{2})(-\frac{1}{2}-1)(-\frac{1}{2}-2)}{3!}x^3$$

$$+ \ldots + \frac{(-\frac{1}{2})(-\frac{3}{2})(-\frac{5}{2})\cdot \ldots \cdot(-\frac{1}{2}-n+1)}{n!}x^n + \ldots$$

$$= 1 - \frac{1}{2}x + \frac{1 \cdot 3}{2^2 \cdot 2!}x^2 - \frac{1 \cdot 3 \cdot 5}{2^3 \cdot 3!}x^3 + \ldots$$

$$+ (-1)^n \frac{1 \cdot 3 \cdot 5 \cdot \ldots \cdot(2n-1)}{2^n n!}x^n + \ldots$$

EXAMPLE 2

From the result of Example 1 obtain a binomial series for $(1-x^2)^{-1/2}$, and use it to find a power series for $\sin^{-1} x$.

SOLUTION

Replace x by $-x^2$ in the series for $(1+x)^{-1/2}$ and get for $|x| < 1$

$$(1-x^2)^{-1/2} = 1 + \frac{1}{2}x^2 + \frac{1 \cdot 3}{2^2 \cdot 2!}x^4 + \frac{1 \cdot 3 \cdot 5}{2^3 \cdot 3!}x^6 + \ldots$$

$$+ \frac{1 \cdot 3 \cdot 5 \cdot \ldots \cdot(2n-1)}{2^n n!}x^{2n} + \ldots$$

We apply Theorem 12.3.1 and integrate term by term to obtain

$$\int_0^x \frac{dt}{\sqrt{1-t^2}} = x + \frac{1}{2} \cdot \frac{x^3}{3} + \frac{1 \cdot 3}{2^2 \cdot 2!} \cdot \frac{x^5}{5} + \frac{1 \cdot 3 \cdot 5}{2^3 \cdot 3!} \cdot \frac{x^7}{7} + \ldots$$

$$+ \frac{1 \cdot 3 \cdot 5 \cdot \ldots \cdot(2n-1)}{2^n n!} \cdot \frac{x^{2n+1}}{2n+1} + \ldots$$

Therefore

$$\sin^{-1} x = x + \sum_{n=1}^{+\infty} \frac{1 \cdot 3 \cdot 5 \cdot \ldots \cdot(2n-1)}{2^n n!} \cdot \frac{x^{2n+1}}{2n+1} \qquad \text{for } |x| < 1$$

EXAMPLE 3

Compute the value of $\sqrt[3]{25}$ accurate to three decimal places by using the binomial series for $(1+x)^{1/3}$.

SOLUTION
From Eq. (8),

$$(1 + x)^{1/3} = 1 + \frac{1}{3}x + \left(\frac{1}{3}\right)\left(-\frac{2}{3}\right)\frac{x^2}{2!} + \left(\frac{1}{3}\right)\left(-\frac{2}{3}\right)\left(-\frac{5}{3}\right)\frac{x^3}{3!} + \dots \quad \text{if } |x| < 1 \tag{9}$$

Because $\sqrt[3]{25} = \sqrt[3]{27}\sqrt[3]{\frac{25}{27}}$, we can write

$$\sqrt[3]{25} = 3(1 - \tfrac{2}{27})^{1/3} \tag{10}$$

From Eq. (9) with $x = -\frac{2}{27}$,

$$\left(1 - \frac{2}{27}\right)^{1/3} = 1 + \frac{1}{3}\left(-\frac{2}{27}\right) - \frac{2}{3^2 \cdot 2!}\left(-\frac{2}{27}\right)^2 + \frac{2 \cdot 5}{3^3 \cdot 3!}\left(-\frac{2}{27}\right)^3 + \dots$$

$$\approx 1 - 0.0247 - 0.0006 - 0.00003 - \dots \tag{11}$$

If the first three terms of the above series are used, it follows from (14) of Section 12.4 that the remainder is

$$R_2\left(-\frac{2}{27}\right) = \frac{f'''(\xi_2)}{3!}\left(-\frac{2}{27}\right)^3$$

$$= \left(\frac{1}{3}\right)\left(-\frac{2}{3}\right)\left(-\frac{5}{3}\right)\left(\frac{1}{3!}\right)(1 + \xi_2)^{(1/3)-3}\left(-\frac{2}{27}\right)^3$$

where $-\frac{2}{27} < \xi_2 < 0$. Therefore

$$\left|R_2\left(-\frac{2}{27}\right)\right| = \left(\frac{2 \cdot 5}{3^3 \cdot 3!}\right)\frac{(1 + \xi_2)^{1/3}}{(1 + \xi_2)^3}\left(\frac{2}{27}\right)^3 \tag{12}$$

Because $-\frac{2}{27} < \xi_2 < 0$, it follows that

$$(1 + \xi_2)^{1/3} < 1^{1/3} = 1 \quad \text{and} \quad \frac{1}{(1 + \xi_2)^3} < \frac{1}{(\frac{25}{27})^3} \tag{13}$$

Furthermore,

$$\frac{2 \cdot 5}{3^3 \cdot 3!} = \frac{2 \cdot 5}{3 \cdot 3 \cdot 3 \cdot 1 \cdot 2 \cdot 3}$$

$$= \frac{2}{3} \cdot \frac{5}{6} \cdot \frac{1}{9} < \frac{1}{9} \tag{14}$$

and so from inequalities (13) and (14) in (12),

$$\left|R_2\left(-\frac{2}{27}\right)\right| < \frac{1}{9} \cdot \frac{1}{(\frac{25}{27})^3} \cdot \left(\frac{2}{27}\right)^3 = \frac{8}{140,625} < 0.00006$$

Using, then, the first three terms of series (11) gives

$$\left(1 - \frac{2}{27}\right)^{1/3} \approx 0.9747$$

with an error less than 0.00006. From (10) we obtain

$$\sqrt[3]{25} \approx 3(0.9747)$$
$$= 2.9241$$

with an error less than $3(0.00006) = 0.00018$. Rounding off to three decimal places gives $\sqrt[3]{25} \approx 2.924$.

EXERCISES 12.5

In Exercises 1 through 10, use a binomial series to find the Maclaurin series for the given function. Determine the radius of convergence of the resulting series.

1. $f(x) = \sqrt{1 + x}$ **2.** $f(x) = (3 - x)^{-2}$ **3.** $f(x) = (4 + x)^{-1/2}$ **4.** $f(x) = \sqrt[3]{8 + x}$

5. $f(x) = \sqrt[3]{1 - x^3}$ **6.** $f(x) = (4 + x^2)^{-1}$ **7.** $f(x) = (9 + x^4)^{-1/2}$ **8.** $f(x) = \dfrac{x}{\sqrt{1 - x}}$

9. $f(x) = \dfrac{x^2}{\sqrt{1 + x}}$ **10.** $f(x) = \dfrac{x}{\sqrt[3]{1 + x^2}}$

In Exercises 11 through 16, compute the value of the given quantity accurate to three decimal places by using a binomial series.

11. $\sqrt{24}$ **12.** $\sqrt{51}$ **13.** $\sqrt[4]{630}$ **14.** $\sqrt[3]{66}$ **15.** $\dfrac{1}{\sqrt[3]{128}}$ **16.** $\dfrac{1}{\sqrt[5]{31}}$

17. Integrate term by term from 0 to x the binomial series for $(1 + t^2)^{-1/2}$ to obtain the Maclaurin series for $\sinh^{-1} x$. Determine the radius of convergence.

18. Use a method similar to that of Exercise 17 to find the Maclaurin series for $\tanh^{-1} x$. Determine the radius of convergence.

19. (a) Express $\sqrt[4]{1 + x}$ as a power series in x. (b) Use the result of part (a) to express $\sqrt[4]{1 + x^2}$ as a power series in x. (c) Use the result of part (b) to compute accurate to three decimal places the value of $\int_0^{1/2} \sqrt[4]{1 + x^2}\, dx$.

20. Use the procedure indicated in Exercise 19 to compute accurate to three decimal places the value of $\int_0^{1/4} (1 - \sqrt{x})^{2/3}\, dx$.

In Exercises 21 through 28, compute accurate to three decimal places the value of the definite integral.

21. $\int_0^{1/3} \sqrt{1 + x^3}\, dx$ **22.** $\int_0^{2/5} \sqrt[3]{1 + x^4}\, dx$ **23.** $\int_0^1 \sqrt[3]{8 + x^2}\, dx$

24. $\int_0^{1/2} \sqrt{1 - x^3}\, dx$ **25.** $\int_0^{1/2} \dfrac{dx}{\sqrt{1 + x^4}}$ **26.** $\int_0^{1/3} \dfrac{dx}{\sqrt[3]{x^2 + 1}}$

27. $\int_0^{1/2} \dfrac{dx}{\sqrt{1 - x^3}}$ **28.** $\int_0^{1/2} f(x)\, dx$, where $f(x) = \begin{cases} \dfrac{\sin^{-1} x}{x} & \text{if } x \neq 0 \\ 1 & \text{if } x = 0 \end{cases}$

29. Find the Maclaurin series for $\int_0^x \dfrac{t^p}{\sqrt{1 - t^2}}\, dt$ if $p \geq 0$. Determine the radius of convergence of the series.

REVIEW EXERCISES FOR CHAPTER 12

In Exercises 1 through 12, find the interval of convergence of the given power series.

1. $\displaystyle\sum_{n=1}^{+\infty} \frac{x^n}{\sqrt{n}}$

2. $\displaystyle\sum_{n=1}^{+\infty} \frac{(x-2)^n}{n}$

3. $\displaystyle\sum_{n=1}^{+\infty} \frac{x^n}{3^n(n^2+n)}$

4. $\displaystyle\sum_{n=0}^{+\infty} \frac{x^n}{2^n}$

5. $\displaystyle\sum_{n=0}^{+\infty} \frac{n!}{2^n}(x-3)^n$

6. $\displaystyle\sum_{n=1}^{+\infty} \frac{(-1)^{n-1}x^{2n-1}}{(2n-1)!}$

7. $\displaystyle\sum_{n=1}^{+\infty} \frac{n^2}{6^n}(x+1)^n$

8. $\displaystyle\sum_{n=1}^{+\infty} n(2x-1)^n$

9. $\displaystyle\sum_{n=1}^{+\infty} \frac{(-1)^{n-1}(x-1)^n}{n2^n}$

10. $\displaystyle\sum_{n=1}^{+\infty} n^n x^n$

11. $\displaystyle\sum_{n=0}^{+\infty} (\sin 2n)x^n$

12. $\displaystyle\sum_{n=1}^{+\infty} (-1)^{n+1} \frac{x^n}{(n+1)\ln(n+1)}$

In Exercises 13 through 16, a function f is defined by a power series. In each exercise do the following: (a) Find the radius of convergence of the given power series and the domain of f; (b) write the power series that defines the function f' and state its radius of convergence; (c) find the domain of f'.

13. $\displaystyle f(x) = \sum_{n=1}^{+\infty} (-1)^n \frac{x^{2n}}{2n}$

14. $\displaystyle f(x) = \sum_{n=1}^{+\infty} \frac{x^n}{n^3}$

15. $\displaystyle f(x) = \sum_{n=0}^{+\infty} \frac{x^n}{(n!)^2}$

16. $\displaystyle f(x) = \sum_{n=1}^{+\infty} \frac{(x+1)^n}{n2^n}$

In Exercises 17 through 20, find a power-series representation of the given integral. Also find the radius of convergence of the resulting power series.

17. $\displaystyle\int_0^x \frac{dt}{t^2+16}$

18. $\displaystyle\int_3^x \frac{dt}{t-1}$

19. $\displaystyle\int_0^x f(t)\,dt$, where $f(t) = \begin{cases} \dfrac{\sin t}{t} & \text{if } t \neq 0 \\ 1 & \text{if } t = 0 \end{cases}$

20. $\displaystyle\int_0^x f(t)\,dt$, where $f(t) = \begin{cases} \dfrac{1-\cos t}{t} & \text{if } t \neq 0 \\ 0 & \text{if } t = 0 \end{cases}$

In Exercises 21 through 24, compute accurate to three decimal places the value of the definite integral obtained by replacing x by the given number in the indicated exercise.

21. $x = 3$; Exercise 17

22. $x = 4$; Exercise 18

23. $x = \frac{1}{2}$; Exercise 19

24. $x = \frac{1}{4}$; Exercise 20

In Exercises 25 through 34, use a power series to compute accurate to four decimal places the value of the given quantity.

25. $\tan^{-1}\frac{1}{5}$

26. $\sin 0.3$

27. $\sqrt[3]{130}$

28. $\sin^{-1} 1$

29. $\cos 3°$

30. $\sqrt[4]{e}$

31. $\ln 5$

32. $\displaystyle\int_0^1 \cos x^3\,dx$

33. $\displaystyle\int_0^{1/4} \sqrt{x}\,\sin x\,dx$

34. $\displaystyle\int_0^{1/2} \frac{dx}{1+x^5}$

In Exercises 35 through 38, find the Maclaurin series for the given function, and find its interval of convergence.

35. $f(x) = a^x \ (a > 0)$

36. $f(x) = \dfrac{1}{2-x}$

37. $f(x) = \sin^3 x$

38. $f(x) = \sqrt{x+1}$

In Exercises 39 through 42, find the Taylor series for the given function at the indicated number.

39. $f(x) = \sin 3x$; at $-\frac{1}{3}\pi$

40. $f(x) = \dfrac{1}{x}$; at 2

41. $f(x) = \ln|x|$; at -1

42. $f(x) = e^{x-2}$; at 2

Exercises 43 through 47 pertain to the functions J_0 and J_1 defined by power series as follows:

$$J_0(x) = \sum_{n=0}^{+\infty} (-1)^n \frac{x^{2n}}{n!n!2^{2n}} \qquad J_1(x) = \sum_{n=0}^{+\infty} (-1)^n \frac{x^{2n+1}}{n!(n+1)!2^{2n+1}}$$

The functions J_0 and J_1 are called *Bessel functions of the first kind* of orders zero and one, respectively.

43. Show that both J_0 and J_1 converge for all real values of x.

44. Show that $y = J_0(x)$ is a solution of the differential equation

$$x \frac{d^2y}{dx^2} + \frac{dy}{dx} + xy = 0$$

45. Show that $J_0'(x) = -J_1(x)$.   **46.** Show that $D_x(xJ_1(x)) = xJ_0(x)$.

47. Show that $y = J_1(x)$ is a solution of the differential equation

$$x^2 \frac{d^2y}{dx^2} + x \frac{dy}{dx} + (x^2 - 1)y = 0$$

P A R T 3

VECTORS AND FUNCTIONS OF MORE THAN ONE VARIABLE

The approach to *vectors* in Chapters 13 and 14 is modern, and it serves both as an introduction to the viewpoint of linear algebra and to that of classical vector analysis. The first four sections of Chapter 13 on vectors in the plane can be taken up after Chapter 4 if you wish to study vectors earlier in the course. Chapter 14 treats vectors in three-dimensional space, and if desired, topics in the first two sections of this chapter may be studied concurrently with the corresponding ones in Chapter 13. Applications of vectors to geometry and physics are given. The geometrical applications include length of arc, tangent and normal vectors to curves, and curvature. For applications in physics, we use vectors to calculate work and to discuss motion along a curve. Chapter 14 also contains subject matter from *solid analytic geometry*. These topics include planes, lines, cylinders, surfaces of revolution, and quadric surfaces.

The functions you studied in Parts 1 and 2 were of a single variable. In Chapter 15, we extend the idea of a function to more than one variable. Limits, continuity, and *partial differentiation* of these functions are considered. The discussion and examples are applied mainly to functions of two and three variables; however, statements of most of the definitions and theorems are extended to functions of *n* variables. The applications in Chapter 15 are to find rates of change and to compute approximations by using differentials. In Chapter 16, a section on *directional derivatives* and *gradients* is followed by a section that shows the application of the gradient to finding an equation of the tangent plane to a surface. Additional applications of partial derivatives in Chapter 16 include the solution of extrema problems, Lagrange multipliers, and the method of least squares.

The concept of the definite integral is extended to functions of more than one variable in Chapter 17. The *double integral* of a function of two variables and the *triple integral* of a function of three variables are discussed. These integrals are used to calculate volumes of solids, areas of surfaces, centers of mass, and moments of inertia. The treatment here is less formal and more intuitive than in previous chapters. This is necessary because the proofs of most of the theorems of this chapter belong to a course in advanced calculus. This situation also prevails with the material of Chapter 18 where we study *vector fields*, which are functions that associate vectors with points in space. The divergence and curl of a vector field are introduced. *Line integrals* are defined and applied to find the work done by a force field in moving a particle along a curve. Line integrals independent of the path are then discussed and the results are used to prove the Law of Conservation of Energy, a major concept in physics. The final two sections in this book are devoted to three important theorems in vector calculus: Green's Theorem, Gauss's Divergence Theorem, and Stokes's Theorem. Applications involving *surface integrals* include the mass of a surface and the flux of a velocity field across a surface.

THIRTEEN

VECTORS IN THE PLANE AND PARAMETRIC EQUATIONS

13.1
VECTORS IN THE PLANE

The applications of mathematics are often concerned with quantities that possess both magnitude and direction. An example of such a quantity is *velocity*. For instance, an airplane's velocity has magnitude (the speed of the airplane) and direction (which determines the course of the airplane). Other examples of such quantities are *force*, *displacement*, and *acceleration*. Physicists and engineers refer to a directed line segment as a *vector*, and the quantities that have both magnitude and direction are called **vector quantities**. In contrast, a quantity that has magnitude but not direction is called a **scalar quantity**. Examples of scalar quantities are length, area, volume, and speed. The study of vectors is called **vector analysis**.

The approach to vector analysis can be on either a geometric or an analytic basis. If the geometric approach is taken, we first define a directed line segment as a line segment from a point P to a point Q and denote this directed line segment by $\overrightarrow{PQ}$. The point P is called the **initial point**, and the point Q is called the **terminal point**. Then two directed line segments $\overrightarrow{PQ}$ and $\overrightarrow{RS}$ are said to be equal if they have the same *length* and *direction*, and we write $\overrightarrow{PQ} = \overrightarrow{RS}$ (see Fig. 13.1.1). The directed line segment PQ is called the **vector** from P to Q. A vector is denoted by a single letter, set in boldface type, such as **A**. In some books, a letter in lightface type, with an arrow above it, is used to indicate a vector, for example $\vec{A}$. When doing your work, you may use that notation or $\underline{A}$ to distinguish the symbol for a vector from the symbol for a real number.

Continuing with the geometric approach to vector analysis, note that if the directed line segment $\overrightarrow{PQ}$ is the vector **A**, and $\overrightarrow{PQ} = \overrightarrow{RS}$, the directed line segment $\overrightarrow{RS}$ is also the vector **A**. Then a vector is considered to remain unchanged if it is moved parallel to itself. With this interpretation of a vector, we can assume for convenience that every vector has its initial point at some fixed reference point. By taking this point as the origin of a rectangular cartesian coordinate system, a vector can be defined analytically in terms of real numbers. Such a definition permits the study of vector analysis from a purely mathematical viewpoint.

In this book we use the analytic approach; however, the geometric interpretation is used for illustrative purposes. A vector in the plane is denoted by an ordered pair of real numbers and the notation $\langle x, y \rangle$ is used instead of (x, y) to avoid confusing the notation for a vector with the notation for a point. V_2 is the set of all such ordered pairs. Following is the formal definition.

$\overrightarrow{PQ} = \overrightarrow{RS}$

FIGURE 13.1.1

13.1.1 DEFINITION   A **vector in the plane** is an ordered pair of real numbers $\langle x, y \rangle$. The numbers x and y are called the **components** of the vector $\langle x, y \rangle$.

There is a one-to-one correspondence between the vectors $\langle x, y \rangle$ in the plane and the points (x, y) in the plane. Let the vector **A** be the ordered pair of real numbers $\langle a_1, a_2 \rangle$. If A is the point (a_1, a_2), then the vector **A** may be represented geometrically by the directed line segment $\overrightarrow{OA}$. Such a directed line segment is called a **representation** of vector **A**. Any directed line segment that is equal to $\overrightarrow{OA}$ is also a representation of vector **A**. The particular representation of a vector that has its initial point at the origin is called the **position representation** of the vector.

▶ ILLUSTRATION 1   The vector $\langle 2, 3\rangle$ has as its position representation the directed line segment from the origin to the point $(2, 3)$. The representation of the vector $\langle 2, 3\rangle$ whose initial point is (h, k) has as its terminal point $(h + 2, k + 3)$; refer to Fig. 13.1.2.   ◀

The vector $\langle 0, 0\rangle$ is called the **zero vector**, and it is denoted by **0**; that is,

$$\mathbf{0} = \langle 0, 0\rangle$$

Any point is a representation of the zero vector.

13.1.2 DEFINITION   The **magnitude** of a vector is the length of any of its representations, and the **direction** of a nonzero vector is the direction of any of its representations.

The magnitude of the vector **A** is denoted by $\|\mathbf{A}\|$.

13.1.3 THEOREM   If **A** is the vector $\langle a_1, a_2\rangle$, then $\|\mathbf{A}\| = \sqrt{a_1{}^2 + a_2{}^2}$.

PROOF   Because by Definition 13.1.2 $\|\mathbf{A}\|$ is the length of any of the representations of **A**, then $\|\mathbf{A}\|$ will be the length of the position representation of **A**, which is the distance from the origin to the point (a_1, a_2). So from the formula for the distance between two points,

$$\|\mathbf{A}\| = \sqrt{(a_1 - 0)^2 + (a_2 - 0)^2}$$
$$= \sqrt{a_1{}^2 + a_2{}^2}$$ ■

Observe that $\|\mathbf{A}\|$ is a nonnegative number and not a vector. From Theorem 13.1.3 it follows that $\|\mathbf{0}\| = 0$.

▶ ILLUSTRATION 2   If $\mathbf{A} = \langle -3, 5\rangle$, then

$$\|\mathbf{A}\| = \sqrt{(-3)^2 + 5^2}$$
$$= \sqrt{34}$$ ◀

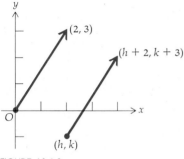

FIGURE 13.1.2

EXAMPLE 1
Let the vector **A** be $\langle -4, 5\rangle$ and the point P be $(6, -2)$. (a) Draw the position representation of **A** and also the particular representation of **A** having P as its initial point. (b) Find the magnitude of **A**.

SOLUTION
(a) Let A be the point $(-4, 5)$. Figure 13.1.3 shows $\overrightarrow{OA}$, which is the position representation of vector **A**. Let $\overrightarrow{PQ}$ be the particular representation of vector **A** having P as its initial point. If $Q = (x, y)$, then

$$x - 6 = -4 \qquad y + 2 = 5$$
$$x = 2 \qquad\qquad y = 3$$

Therefore $Q = (2, 3)$ and $\overrightarrow{PQ}$ appears in Fig. 13.1.3.
(b) From Theorem 13.1.3,

$$\|\mathbf{A}\| = \sqrt{(-4)^2 + 5^2}$$
$$= \sqrt{41}$$

FIGURE 13.1.3

FIGURE 13.1.4

FIGURE 13.1.5   FIGURE 13.1.6

FIGURE 13.1.7

The **direction angle** of any nonzero vector is the angle θ measured from the positive side of the x axis counterclockwise to the position representation of the vector. If θ is measured in radians, $0 \leq \theta < 2\pi$. If $\mathbf{A} = \langle a_1, a_2 \rangle$, then

$$\tan \theta = \frac{a_2}{a_1} \qquad \text{if } a_1 \neq 0 \tag{1}$$

If $a_1 = 0$ and $a_2 > 0$, then $\theta = \frac{1}{2}\pi$; if $a_1 = 0$ and $a_2 < 0$, then $\theta = \frac{3}{2}\pi$. Figures 13.1.4 through 13.1.6 show the direction angle θ for specific vectors whose position representations are drawn.

EXAMPLE 2
Find the radian measure of the direction angle of each of the following vectors: (a) $\langle -1, 1 \rangle$; (b) $\langle 0, -5 \rangle$; (c) $\langle 1, -2 \rangle$.

SOLUTION
The position representation of each of the vectors in (a), (b), and (c) is shown in Figs. 13.1.7, 13.1.8, and 13.1.9, respectively. (a) $\tan \theta = -1$, and $\frac{1}{2}\pi < \theta < \pi$; so $\theta = \frac{3}{4}\pi$. (b) $\tan \theta$ does not exist, and $a_2 < 0$; thus $\theta = \frac{3}{2}\pi$. (c) $\tan \theta = -2$, and $\frac{3}{2}\pi < \theta < 2\pi$; therefore $\theta = \tan^{-1}(-2) + 2\pi$; that is, $\theta \approx 5.176$.

FIGURE 13.1.8

Observe that if $\mathbf{A} = \langle a_1, a_2 \rangle$ and θ is the direction angle of $\mathbf{A}$, then

$$a_1 = \|\mathbf{A}\| \cos \theta \qquad a_2 = \|\mathbf{A}\| \sin \theta \tag{2}$$

See Fig. 13.1.10, where the point (a_1, a_2) is in the first quadrant.

FIGURE 13.1.9

FIGURE 13.1.10

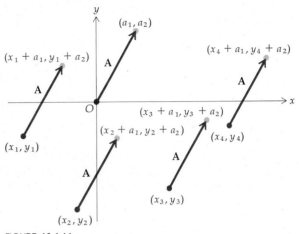

FIGURE 13.1.11

If the vector $\mathbf{A} = \langle a_1, a_2 \rangle$, then the representation of $\mathbf{A}$ whose initial point is (x, y) has as its endpoint $(x + a_1, y + a_2)$. In this way a vector may be thought of as a translation of the plane into itself. Figure 13.1.11 illustrates five representations of the vector $\mathbf{A} = \langle a_1, a_2 \rangle$. In each case $\mathbf{A}$ translates the point (x_i, y_i) into the point $(x_i + a_1, y_i + a_2)$.

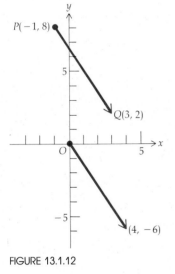

FIGURE 13.1.12

EXAMPLE 3

Suppose P is the point $(-1, 8)$ and Q is the point $(3, 2)$. Find the vector $\mathbf{A}$ having $\overrightarrow{PQ}$ as a representation. Draw $\overrightarrow{PQ}$ and the position representation of $\mathbf{A}$.

SOLUTION

Figure 13.1.12 shows the directed line segment $\overrightarrow{PQ}$. Let $\mathbf{A} = \langle a_1, a_2 \rangle$. Because $\overrightarrow{PQ}$ is a representation of vector $\mathbf{A}$, the vector $\mathbf{A}$ translates the point $P(-1, 8)$ into the point $Q(3, 2)$. But vector $\langle a_1, a_2 \rangle$ translates the point $(-1, 8)$ into the point $\langle -1 + a_1, 8 + a_2 \rangle$. Thus

$$-1 + a_1 = 3 \qquad 8 + a_2 = 2$$
$$a_1 = 4 \qquad a_2 = -6$$

Therefore $\mathbf{A} = \langle 4, -6 \rangle$. Figure 13.1.12 also shows the position representation of $\mathbf{A}$.

The following definition gives the method for adding two vectors.

13.1.4 DEFINITION   The **sum** of two vectors $\mathbf{A} = \langle a_1, a_2 \rangle$ and $\mathbf{B} = \langle b_1, b_2 \rangle$ is the vector $\mathbf{A} + \mathbf{B}$ defined by

$$\mathbf{A} + \mathbf{B} = \langle a_1 + b_1, a_2 + b_2 \rangle$$

▶ ILLUSTRATION 3   If $\mathbf{A} = \langle 3, -1 \rangle$ and $\mathbf{B} = \langle -4, 5 \rangle$, then

$$\mathbf{A} + \mathbf{B} = \langle 3 + (-4), -1 + 5 \rangle$$
$$= \langle -1, 4 \rangle \qquad \qquad ◀$$

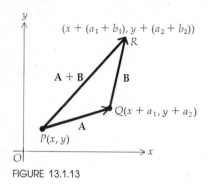

FIGURE 13.1.13

The geometric interpretation of the sum of two vectors is shown in Fig. 13.1.13. Let $\mathbf{A} = \langle a_1, a_2 \rangle$ and $\mathbf{B} = \langle b_1, b_2 \rangle$, and let P be the point (x, y). Then $\mathbf{A}$ translates the point P into the point $(x + a_1, y + a_2) = Q$. The vector $\mathbf{B}$ translates the point Q into the point $((x + a_1) + b_1, (y + a_2) + b_2)$ or, equivalently, $(x + (a_1 + b_1), y + (a_2 + b_2)) = R$. Furthermore,

$$\mathbf{A} + \mathbf{B} = \langle a_1 + b_1, a_2 + b_2 \rangle$$

Therefore the vector $\mathbf{A} + \mathbf{B}$ translates the point P into the point $(x + (a_1 + b_1), y + (a_2 + b_2)) = R$. Thus, in Fig. 13.1.13 $\overrightarrow{PQ}$ is a representation of the vector $\mathbf{A}$, $\overrightarrow{QR}$ is a representation of the vector $\mathbf{B}$, and $\overrightarrow{PR}$ is a representation of the vector $\mathbf{A} + \mathbf{B}$. The representations of the vectors $\mathbf{A}$ and $\mathbf{B}$ are adjacent sides of a parallelogram, and the representation of the vector $\mathbf{A} + \mathbf{B}$ is a diagonal of the parallelogram. This diagonal is called the **resultant** of the vectors $\mathbf{A}$ and $\mathbf{B}$. The rule for the addition of vectors is sometimes referred to as the **parallelogram law**.

Force is a vector quantity where the magnitude is expressed in force units and the direction angle is determined by the direction of the force. It is shown in physics that two forces applied to an object at a particular point can be replaced by an equivalent force that is their resultant.

EXAMPLE 4

Two forces of magnitudes 200 lb and 250 lb make an angle of $\frac{1}{3}\pi$ with each other and are applied to an object at the same point. Find (a) the magnitude of the resultant force and (b) the angle it makes with the force of 200 lb.

SOLUTION

Refer to Fig. 13.1.14, where the axes are chosen so that the position representation of the force of 200 lb is along the positive side of the x axis. The vector $\mathbf{A}$ represents this force, and $\mathbf{A} = \langle 200, 0 \rangle$. The vector $\mathbf{B}$ represents the force of 250 lb. From formulas (2), if $\mathbf{B} = \langle b_1, b_2 \rangle$, then

$$b_1 = 250 \cos \tfrac{1}{3}\pi \qquad b_2 = 250 \sin \tfrac{1}{3}\pi$$
$$= 125 \qquad\qquad \approx 216.5$$

Thus $\mathbf{B} = \langle 125, 216.5 \rangle$. The resultant force is $\mathbf{A} + \mathbf{B}$, and

$$\mathbf{A} + \mathbf{B} = \langle 200, 0 \rangle + \langle 125, 216.5 \rangle$$
$$= \langle 325, 216.5 \rangle$$

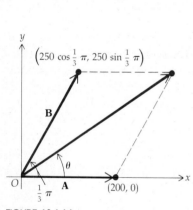

FIGURE 13.1.14

(a) $\|\mathbf{A} + \mathbf{B}\| = \sqrt{(325)^2 + (216.5)^2}$
$$\approx 390.5$$

(b) If θ is the angle the vector $\mathbf{A} + \mathbf{B}$ makes with $\mathbf{A}$, then

$$\tan \theta = \frac{216.5}{325}$$

$$\tan \theta \approx 0.6662$$

$$\theta \approx 0.5877$$

| 13.1.5 DEFINITION | If $\mathbf{A} = \langle a_1, a_2 \rangle$, then the vector $\langle -a_1, -a_2 \rangle$ is defined to be the **negative** of $\mathbf{A}$, denoted by $-\mathbf{A}$. |

FIGURE 13.1.15

If the directed line segment $\overrightarrow{PQ}$ is a representation of the vector $\mathbf{A}$, then the directed line segment $\overrightarrow{QP}$ is a representation of $-\mathbf{A}$. Any directed line segment that is parallel to $\overrightarrow{PQ}$, has the same length as $\overrightarrow{PQ}$, and has a direction opposite to that of $\overrightarrow{PQ}$ is also a representation of $-\mathbf{A}$. See Fig. 13.1.15. We now define subtraction of two vectors.

| 13.1.6 DEFINITION | The **difference** of the two vectors $\mathbf{A}$ and $\mathbf{B}$, denoted by $\mathbf{A} - \mathbf{B}$, is the vector obtained by adding $\mathbf{A}$ to the negative of $\mathbf{B}$; that is, $$\mathbf{A} - \mathbf{B} = \mathbf{A} + (-\mathbf{B})$$ |

Thus if $\mathbf{A} = \langle a_1, a_2 \rangle$ and $\mathbf{B} = \langle b_1, b_2 \rangle$, then $-\mathbf{B} = \langle -b_1, -b_2 \rangle$, and

$$\mathbf{A} - \mathbf{B} = \langle a_1 - b_1, a_2 - b_2 \rangle$$

▶ ILLUSTRATION 4   If $\mathbf{A} = \langle 4, -2 \rangle$ and $\mathbf{B} = \langle 6, -3 \rangle$, then

$$\begin{aligned}
\mathbf{A} - \mathbf{B} &= \langle 4, -2 \rangle - \langle 6, -3 \rangle \\
&= \langle 4, -2 \rangle + \langle -6, 3 \rangle \\
&= \langle -2, 1 \rangle
\end{aligned}$$
◀

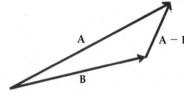

FIGURE 13.1.16

To interpret the difference of two vectors geometrically, let the representations of the vectors $\mathbf{A}$ and $\mathbf{B}$ have the same initial point. Then the directed line segment from the endpoint of the representation of $\mathbf{B}$ to the endpoint of the representation of $\mathbf{A}$ is a representation of the vector $\mathbf{A} - \mathbf{B}$. This obeys the parallelogram law $\mathbf{B} + (\mathbf{A} - \mathbf{B}) = \mathbf{A}$. See Fig. 13.1.16.

Suppose that P is the point (a_1, a_2) and Q is the point (b_1, b_2). We shall use the notation $\mathbf{V}(\overrightarrow{PQ})$ to denote the vector having the directed line segment $\overrightarrow{PQ}$ as a representation. See Fig. 13.1.17, showing representations of the vectors $\mathbf{V}(\overrightarrow{PQ})$, $\mathbf{V}(\overrightarrow{OP})$, and $\mathbf{V}(\overrightarrow{OQ})$. Observe that

$$\mathbf{V}(\overrightarrow{PQ}) = \mathbf{V}(\overrightarrow{OQ}) - \mathbf{V}(\overrightarrow{OP})$$
$$\mathbf{V}(\overrightarrow{PQ}) = \langle b_1, b_2 \rangle - \langle a_1, a_2 \rangle$$
$$\mathbf{V}(\overrightarrow{PQ}) = \langle b_1 - a_1, b_2 - a_2 \rangle$$

FIGURE 13.1.17

▶ ILLUSTRATION 5   If P is the point $(-6, 7)$ and Q is the point $(2, 9)$, then

$$\begin{aligned}
\mathbf{V}(\overrightarrow{PQ}) &= \langle 2 - (-6), 9 - 7 \rangle \\
&= \langle 8, 2 \rangle
\end{aligned}$$
◀

Another operation with vectors is *scalar multiplication*. Following is the definition of the multiplication of a vector by a scalar (a real number).

| 13.1.7 DEFINITION | If c is a scalar and $\mathbf{A}$ is the vector $\langle a_1, a_2 \rangle$, then the product of c and $\mathbf{A}$, denoted by $c\mathbf{A}$, is a vector and is given by $$\begin{aligned} c\mathbf{A} &= c\langle a_1, a_2 \rangle \\ &= \langle ca_1, ca_2 \rangle \end{aligned}$$ |

▶ ILLUSTRATION 6   If $\mathbf{A} = \langle 4, -5 \rangle$, then

$$3\mathbf{A} = 3\langle 4, -5 \rangle$$
$$= \langle 12, -15 \rangle$$ ◀

EXAMPLE 5
If $\mathbf{A}$ is any vector and c is any scalar, show that (a) $0(\mathbf{A}) = \mathbf{0}$, and (b) $c(\mathbf{0}) = \mathbf{0}$.

SOLUTION
We apply Definition 13.1.7.

(a) $0(\mathbf{A}) = 0\langle a_1, a_2 \rangle$ (b) $c(\mathbf{0}) = c\langle 0, 0 \rangle$
         $= \langle 0 \cdot a_1, 0 \cdot a_2 \rangle$ $= \langle c \cdot 0, c \cdot 0 \rangle$
         $= \langle 0, 0 \rangle$ $= \langle 0, 0 \rangle$
         $= \mathbf{0}$ $= \mathbf{0}$

The magnitude of the vector $c\mathbf{A}$ is computed as follows:

$$\|c\mathbf{A}\| = \sqrt{(ca_1)^2 + (ca_2)^2}$$
$$= \sqrt{c^2(a_1{}^2 + a_2{}^2)}$$
$$= \sqrt{c^2}\sqrt{a_1{}^2 + a_2{}^2}$$
$$= |c|\,\|\mathbf{A}\|$$

3A

A

FIGURE 13.1.18

Therefore the magnitude of $c\mathbf{A}$ is the absolute value of c times the magnitude of $\mathbf{A}$.

The geometric interpretation of the vector $c\mathbf{A}$ is given in Figs. 13.1.18 and 13.1.19. If $c > 0$, then $c\mathbf{A}$ is a vector whose representation has a length c times the magnitude of $\mathbf{A}$ and the same direction as $\mathbf{A}$; an example of this is shown in Fig. 13.1.18, where $c = 3$. If $c < 0$, then $c\mathbf{A}$ is a vector whose representation has a length that is $|c|$ times the magnitude of $\mathbf{A}$ and a direction opposite to that of $\mathbf{A}$. This is shown in Fig. 13.1.19, where $c = -\frac{1}{2}$.

A

$-\frac{1}{2}\mathbf{A}$

FIGURE 13.1.19

The following theorem gives laws satisfied by the operations of vector addition and scalar multiplication of any vectors in V_2.

13.1.8 THEOREM   If $\mathbf{A}$, $\mathbf{B}$, and $\mathbf{C}$ are any vectors in V_2, and c and d are any scalars, then vector addition and scalar multiplication satisfy the following properties:

(i) $\mathbf{A} + \mathbf{B} = \mathbf{B} + \mathbf{A}$ (commutative law)
(ii) $\mathbf{A} + (\mathbf{B} + \mathbf{C}) = (\mathbf{A} + \mathbf{B}) + \mathbf{C}$ (associative law)
(iii) There is a vector $\mathbf{0}$ in V_2 for which
       $\mathbf{A} + \mathbf{0} = \mathbf{A}$ (existence of additive identity)
(iv) There is a vector $-\mathbf{A}$ in V_2 such that
       $\mathbf{A} + (-\mathbf{A}) = \mathbf{0}$ (existence of negative)
(v) $(cd)\mathbf{A} = c(d\mathbf{A})$ (associative law)
(vi) $c(\mathbf{A} + \mathbf{B}) = c\mathbf{A} + c\mathbf{B}$ (distributive law)
(vii) $(c + d)\mathbf{A} = c\mathbf{A} + d\mathbf{A}$ (distributive law)
(viii) $1(\mathbf{A}) = \mathbf{A}$ (existence of scalar multiplicative identity)

PROOF   We give the proofs of (i), (iv), and (vi) and leave the others as exercises (see Exercises 40 through 43). Let $\mathbf{A} = \langle a_1, a_2 \rangle$ and $\mathbf{B} = \langle b_1, b_2 \rangle$.

PROOF OF (i)   By the commutative law for real numbers,

$$a_1 + b_1 = b_1 + a_1 \qquad a_2 + b_2 = b_2 + a_2$$

Therefore

$$
\begin{aligned}
\mathbf{A} + \mathbf{B} &= \langle a_1, a_2 \rangle + \langle b_1, b_2 \rangle \\
&= \langle a_1 + b_1, a_2 + b_2 \rangle \\
&= \langle b_1 + a_1, b_2 + a_2 \rangle \\
&= \langle b_1, b_2 \rangle + \langle a_1, a_2 \rangle \\
&= \mathbf{B} + \mathbf{A}
\end{aligned}
$$

PROOF OF (iv)   The vector $-\mathbf{A}$ is given by Definition 13.1.5 and

$$
\begin{aligned}
\mathbf{A} + (-\mathbf{A}) &= \langle a_1, a_2 \rangle + \langle -a_1, -a_2 \rangle \\
&= \langle a_1 + (-a_1), a_2 + (-a_2) \rangle \\
&= \langle 0, 0 \rangle \\
&= \mathbf{0}
\end{aligned}
$$

PROOF OF (vi)

$$
\begin{aligned}
c(\mathbf{A} + \mathbf{B}) &= c(\langle a_1, a_2 \rangle + \langle b_1, b_2 \rangle) \\
&= c(\langle a_1 + b_1, a_2 + b_2 \rangle) \\
&= \langle c(a_1 + b_1), c(a_2 + b_2) \rangle \\
&= \langle ca_1 + cb_1, ca_2 + cb_2 \rangle \\
&= \langle ca_1, ca_2 \rangle + \langle cb_1, cb_2 \rangle \\
&= c\langle a_1, a_2 \rangle + c\langle b_1, b_2 \rangle \\
&= c\mathbf{A} + c\mathbf{B} \qquad\qquad \blacksquare
\end{aligned}
$$

EXAMPLE 6
Verify (ii), (iii), (v), (vii), and (viii) of Theorem 13.1.8 if $\mathbf{A} = \langle 3, 4 \rangle$, $\mathbf{B} = \langle -2, 1 \rangle$, $\mathbf{C} = \langle 5, -3 \rangle$, $c = 2$, and $d = -6$.

SOLUTION

$$
\begin{aligned}
\mathbf{A} + (\mathbf{B} + \mathbf{C}) &= \langle 3, 4 \rangle + (\langle -2, 1 \rangle + \langle 5, -3 \rangle) \\
&= \langle 3, 4 \rangle + \langle 3, -2 \rangle \\
&= \langle 6, 2 \rangle
\end{aligned}
$$

$$
\begin{aligned}
(\mathbf{A} + \mathbf{B}) + \mathbf{C} &= (\langle 3, 4 \rangle + \langle -2, 1 \rangle) + \langle 5, -3 \rangle \\
&= \langle 1, 5 \rangle + \langle 5, -3 \rangle \\
&= \langle 6, 2 \rangle
\end{aligned}
$$

Therefore $\mathbf{A} + (\mathbf{B} + \mathbf{C}) = (\mathbf{A} + \mathbf{B}) + \mathbf{C}$.

$$\begin{aligned}
\mathbf{A} + \mathbf{0} &= \langle 3,4 \rangle + \langle 0,0 \rangle \\
&= \langle 3,4 \rangle \\
&= \mathbf{A}
\end{aligned}$$

$$\begin{aligned}
(cd)\mathbf{A} &= [(2)(-6)]\langle 3,4 \rangle & c(d\mathbf{A}) &= 2(-6\langle 3,4 \rangle) \\
&= (-12)\langle 3,4 \rangle & &= 2\langle -18, -24 \rangle \\
&= \langle -36, -48 \rangle & &= \langle -36, -48 \rangle
\end{aligned}$$

Thus $(cd)\mathbf{A} = c(d\mathbf{A})$.

$$\begin{aligned}
(c + d)\mathbf{A} &= [2 + (-6)]\langle 3,4 \rangle & c\mathbf{A} + d\mathbf{A} &= 2\langle 3,4 \rangle + (-6)\langle 3,4 \rangle \\
&= (-4)\langle 3,4 \rangle & &= \langle 6,8 \rangle + \langle -18, -24 \rangle \\
&= \langle -12, -16 \rangle & &= \langle -12, -16 \rangle
\end{aligned}$$

Therefore $(c + d)\mathbf{A} = c\mathbf{A} + d\mathbf{A}$.

$$\begin{aligned}
1\mathbf{A} &= 1\langle 3,4 \rangle \\
&= \langle (1)(3), (1)(4) \rangle \\
&= \langle 3,4 \rangle \\
&= \mathbf{A}
\end{aligned}$$

Theorem 13.1.8 is important because every algebraic law for the operations of vector addition and scalar multiplication of vectors in V_2 can be derived from the eight properties stated in the theorem. These laws are similar to the laws of arithmetic of real numbers. Furthermore, in linear algebra, a *real vector space* is defined as a set of vectors together with a set of real numbers (*scalars*) and the two operations of vector addition and scalar multiplication that satisfy the eight properties given in Theorem 13.1.8. Following is the formal definition.

13.1.9 DEFINITION   A **real vector space** V is a set of elements, called *vectors*, together with a set of real numbers, called *scalars*, with two operations called *vector addition* and *scalar multiplication* such that for every pair of vectors $\mathbf{A}$ and $\mathbf{B}$ in V and for every scalar c, a vector $\mathbf{A} + \mathbf{B}$ and a vector $c\mathbf{A}$ are defined so that properties (i)–(viii) of Theorem 13.1.8 are satisfied.

From Definition 13.1.9 and Theorem 13.1.8 it follows that V_2 is a real vector space.

We now take an arbitrary vector in V_2 and write it in a special form.

$$\langle a_1, a_2 \rangle = \langle a_1, 0 \rangle + \langle 0, a_2 \rangle$$

$$\langle a_1, a_2 \rangle = a_1\langle 1,0 \rangle + a_2\langle 0,1 \rangle \tag{3}$$

Because the magnitude of each of the two vectors $\langle 1,0 \rangle$ and $\langle 0,1 \rangle$ is one unit, they are called **unit vectors**. We introduce the following notations for these two unit vectors:

$$\mathbf{i} = \langle 1, 0 \rangle \qquad \mathbf{j} = \langle 0, 1 \rangle \qquad (4)$$

The position representation of each of these unit vectors is shown in Fig. 13.1.20. From (3) and (4),

$$\langle a_1, a_2 \rangle = a_1 \mathbf{i} + a_2 \mathbf{j} \qquad (5)$$

Equation (5) states that any vector in V_2 can be written as a linear combination of the two vectors $\mathbf{i}$ and $\mathbf{j}$. Because of this statement and the fact that $\mathbf{i}$ and $\mathbf{j}$ are independent (their position representations are not collinear), the vectors $\mathbf{i}$ and $\mathbf{j}$ are said to form a **basis** for the vector space V_2. See Exercise 44 for an example of a basis consisting of nonunit vectors. The number of elements in a basis of a vector space is called the **dimension** of the vector space. Therefore V_2 is a two-dimensional vector space.

FIGURE 13.1.20

▶ ILLUSTRATION 7   From (5),

$$\langle 3, -4 \rangle = 3\mathbf{i} - 4\mathbf{j} \qquad\qquad ◀$$

Let $\mathbf{A}$ be the vector $\langle a_1, a_2 \rangle$ and θ the direction angle of $\mathbf{A}$. See Fig. 13.1.21, where the point (a_1, a_2) is in the second quadrant and the position representation of $\mathbf{A}$ is shown. Because $\mathbf{A} = a_1 \mathbf{i} + a_2 \mathbf{j}$, $a_1 = \|\mathbf{A}\| \cos \theta$, and $a_2 = \|\mathbf{A}\| \sin \theta$, we can write

$$\mathbf{A} = \|\mathbf{A}\| \cos \theta \mathbf{i} + \|\mathbf{A}\| \sin \theta \mathbf{j}$$
$$\mathbf{A} = \|\mathbf{A}\|(\cos \theta \mathbf{i} + \sin \theta \mathbf{j}) \qquad (6)$$

Equation (6) expresses the vector $\mathbf{A}$ in terms of its magnitude, the cosine and sine of its direction angle, and the unit vectors $\mathbf{i}$ and $\mathbf{j}$.

FIGURE 13.1.21

EXAMPLE 7
Express the vector $\langle -5, -2 \rangle$ in the form of Eq. (6).

SOLUTION
Refer to Fig. 13.1.22, which shows the position representation of the vector $\langle -5, -2 \rangle$.

$$\|\langle -5, -2 \rangle\| = \sqrt{(-5)^2 + (-2)^2} \qquad \cos \theta = -\frac{5}{\sqrt{29}} \qquad \sin \theta = -\frac{2}{\sqrt{29}}$$
$$= \sqrt{29}$$

Therefore from (6) we have

$$\langle -5, -2 \rangle = \sqrt{29}\left(-\frac{5}{\sqrt{29}} \mathbf{i} - \frac{2}{\sqrt{29}} \mathbf{j} \right)$$

FIGURE 13.1.22

13.1.10 THEOREM   If the nonzero vector $\mathbf{A} = a_1\mathbf{i} + a_2\mathbf{j}$, then the unit vector $\mathbf{U}$ having the same direction as $\mathbf{A}$ is given by

$$\mathbf{U} = \frac{a_1}{\|\mathbf{A}\|}\,\mathbf{i} + \frac{a_2}{\|\mathbf{A}\|}\,\mathbf{j} \qquad (7)$$

PROOF   The vector $\mathbf{U}$ defined by (7) can be written as

$$\mathbf{U} = \frac{1}{\|\mathbf{A}\|}\,(\mathbf{a_1}\mathbf{i} + \mathbf{a_2}\mathbf{j})$$

$$= \frac{1}{\|\mathbf{A}\|}\,(\mathbf{A})$$

Therefore $\mathbf{U}$ is a positive scalar times the vector $\mathbf{A}$; so the direction of $\mathbf{U}$ is the same as the direction of $\mathbf{A}$. Furthermore

$$\|\mathbf{U}\| = \sqrt{\left(\frac{a_1}{\|\mathbf{A}\|}\right)^2 + \left(\frac{a_2}{\|\mathbf{A}\|}\right)^2}$$

$$= \frac{\sqrt{a_1{}^2 + a_2{}^2}}{\|\mathbf{A}\|}$$

$$= \frac{\|\mathbf{A}\|}{\|\mathbf{A}\|}$$

$$= 1$$

Therefore $\mathbf{U}$ is the unit vector having the same direction as $\mathbf{A}$, and the theorem is proved.   ∎

EXAMPLE 8
Given $\mathbf{A} = \langle 3, 1 \rangle$ and $\mathbf{B} = \langle -2, 4 \rangle$, find the unit vector having the same direction as $\mathbf{A} - \mathbf{B}$.

SOLUTION

$$\mathbf{A} - \mathbf{B} = \langle 3, 1 \rangle - \langle -2, 4 \rangle$$

$$= \langle 5, -3 \rangle$$

$$= 5\mathbf{i} - 3\mathbf{j}$$

Thus

$$\|\mathbf{A} - \mathbf{B}\| = \sqrt{5^2 + (-3)^2}$$

$$= \sqrt{34}$$

By Theorem 13.1.10, the desired unit vector is

$$\mathbf{U} = \frac{5}{\sqrt{34}}\,\mathbf{i} - \frac{3}{\sqrt{34}}\,\mathbf{j}$$

EXAMPLE 9

If $\mathbf{A} = 3\mathbf{i} - 2\mathbf{j}$, $\mathbf{B} = 2\mathbf{i} - 3\mathbf{j}$, and $\mathbf{C} = 2\mathbf{i} + 7\mathbf{j}$, find scalars h and k such that $\mathbf{C} = h\mathbf{A} + k\mathbf{B}$.

SOLUTION

We wish to find h and k such that

$$2\mathbf{i} + 7\mathbf{j} = h(3\mathbf{i} - 2\mathbf{j}) + k(2\mathbf{i} - 3\mathbf{j})$$

$$2\mathbf{i} + 7\mathbf{j} = (3h\mathbf{i} - 2h\mathbf{j}) + (2k\mathbf{i} - 3k\mathbf{j})$$

$$2\mathbf{i} + 7\mathbf{j} = (3h + 2k)\mathbf{i} + (-2h - 3k)\mathbf{j}$$

The vectors on the two sides of this equation are equal if and only if

$$3h + 2k = 2$$

$$-2h - 3k = 7$$

We solve these two equations simultaneously and obtain $h = 4$ and $k = -5$.

EXERCISES 13.1

In Exercises 1 through 4, (a) draw the position representation of the given vector $\mathbf{A}$ and also the particular representation through the given point P. (b) Find the magnitude of $\mathbf{A}$.

1. $\mathbf{A} = \langle 3, 4 \rangle$; $P = (2, 1)$
2. $\mathbf{A} = \langle -2, 5 \rangle$; $P = (-3, 4)$
3. $\mathbf{A} = \langle e, -\frac{1}{2} \rangle$; $P = (-2, -e)$
4. $\mathbf{A} = \langle 4, 0 \rangle$; $P = (2, 6)$

In Exercises 5 and 6, find the exact radian measure of the direction angle of the given vector. In part (c) also approximate the radian measure to the nearest hundredth.

5. (a) $\langle 1, -1 \rangle$; (b) $\langle -3, 0 \rangle$; (c) $\langle 5, 2 \rangle$
6. (a) $\langle \sqrt{3}, 1 \rangle$; (b) $\langle 0, 4 \rangle$; (c) $\langle -3, 2 \rangle$

In Exercises 7 through 10, find the vector $\mathbf{A}$ having $\overrightarrow{PQ}$ as a representation. Draw $\overrightarrow{PQ}$ and the position representation of $\mathbf{A}$.

7. $P = (3, 7)$; $Q = (5, 4)$
8. $P = (5, 4)$; $Q = (3, 7)$
9. $P = (-5, -3)$; $Q = (0, 3)$
10. $P = (-\sqrt{2}, 0)$; $Q = (0, 0)$

In Exercises 11 through 14, find the point S so that $\overrightarrow{PQ}$ and $\overrightarrow{RS}$ are each representations of the same vector.

11. $P = (2, 5)$; $Q = (1, 6)$; $R = (-3, 2)$
12. $P = (-2, 0)$; $Q = (-3, -4)$; $R = (4, 2)$
13. $P = (0, 3)$; $Q = (5, -2)$; $R = (7, 0)$
14. $P = (-1, 4)$; $Q = (2, -3)$; $R = (-5, -2)$

In Exercises 15 and 16, find the sum of the given pairs of vectors and illustrate geometrically.

15. (a) $\langle 2, 4 \rangle$, $\langle -3, 5 \rangle$; (b) $\langle -3, 0 \rangle$, $\langle 4, -5 \rangle$
16. (a) $\langle 0, 3 \rangle$, $\langle -2, 3 \rangle$; (b) $\langle 2, 3 \rangle$, $\langle -\sqrt{2}, -1 \rangle$

In Exercises 17 and 18, subtract the second vector from the first and illustrate geometrically.

17. (a) $\langle -3, -4 \rangle$, $\langle 6, 0 \rangle$; (b) $\langle 1, e \rangle$, $\langle -3, 2e \rangle$
18. (a) $\langle 0, 5 \rangle$, $\langle 2, 8 \rangle$; (b) $\langle 3, 7 \rangle$, $\langle 3, 7 \rangle$

In Exercises 19 and 20, find the given vector or scalar if $\mathbf{A} = \langle 2, 4 \rangle$, $\mathbf{B} = \langle 4, -3 \rangle$, and $\mathbf{C} = \langle -3, 2 \rangle$.

19. (a) $\mathbf{A} + \mathbf{B}$; (b) $\|\mathbf{C} - \mathbf{B}\|$; (c) $\|7\mathbf{A} - \mathbf{B}\|$
20. (a) $\mathbf{A} - \mathbf{B}$; (b) $\|\mathbf{C}\|$; (c) $\|2\mathbf{A} + 3\mathbf{B}\|$

In Exercises 21 through 24, find the given vector or scalar if $\mathbf{A} = 2\mathbf{i} + 3\mathbf{j}$ and $\mathbf{B} = 4\mathbf{i} - \mathbf{j}$.

21. (a) $5\mathbf{A}$; (b) $-6\mathbf{B}$; (c) $\mathbf{A} + \mathbf{B}$; (d) $\|\mathbf{A} + \mathbf{B}\|$ **22.** (a) $-2\mathbf{A}$; (b) $3\mathbf{B}$; (c) $\mathbf{A} - \mathbf{B}$; (d) $\|\mathbf{A} - \mathbf{B}\|$

23. (a) $\|\mathbf{A}\| + \|\mathbf{B}\|$; (b) $5\mathbf{A} - 6\mathbf{B}$; (c) $\|5\mathbf{A} - 6\mathbf{B}\|$; (d) $\|5\mathbf{A}\| - \|6\mathbf{B}\|$

24. (a) $\|\mathbf{A}\| - \|\mathbf{B}\|$; (b) $3\mathbf{B} - 2\mathbf{A}$; (c) $\|3\mathbf{B} - 2\mathbf{A}\|$; (d) $\|3\mathbf{B}\| - \|2\mathbf{A}\|$

In Exercises 25 and 26, $\mathbf{A} = -4\mathbf{i} + 2\mathbf{j}$, $\mathbf{B} = -\mathbf{i} + 3\mathbf{j}$, and $\mathbf{C} = 5\mathbf{i} - \mathbf{j}$.

25. Find: (a) $5\mathbf{A} - 2\mathbf{B} - 2\mathbf{C}$ and (b) $\|5\mathbf{A} - 2\mathbf{B} - 2\mathbf{C}\|$.   **26.** Find: (a) $3\mathbf{B} - 2\mathbf{A} - \mathbf{C}$ and (b) $\|3\mathbf{B} - 2\mathbf{A} - \mathbf{C}\|$.

In Exercises 27 and 28, $\mathbf{A} = 8\mathbf{i} + 5\mathbf{j}$ and $\mathbf{B} = 3\mathbf{i} - \mathbf{j}$.

27. Find a unit vector having the same direction as $\mathbf{A} + \mathbf{B}$.   **28.** Find a unit vector having the same direction as $\mathbf{A} - \mathbf{B}$.

In Exercises 29 through 32, write the given vector in the form $r(\cos\theta\mathbf{i} + \sin\theta\mathbf{j})$, where r is the magnitude and θ is the direction angle. Also find a unit vector having the same direction.

29. (a) $3\mathbf{i} - 4\mathbf{j}$; (b) $2\mathbf{i} + 2\mathbf{j}$ **30.** (a) $8\mathbf{i} + 6\mathbf{j}$; (b) $2\sqrt{5}\mathbf{i} + 4\mathbf{j}$ **31.** (a) $-4\mathbf{i} + 4\sqrt{3}\mathbf{j}$; (b) $-16\mathbf{i}$ **32.** (a) $3\mathbf{i} - 3\mathbf{j}$; (b) $2\mathbf{j}$

33. If $\mathbf{A} = -2\mathbf{i} + \mathbf{j}$, $\mathbf{B} = 3\mathbf{i} - 2\mathbf{j}$, and $\mathbf{C} = 5\mathbf{i} - 4\mathbf{j}$, find scalars h and k such that $\mathbf{C} = h\mathbf{A} + k\mathbf{B}$.

34. If $\mathbf{A} = 5\mathbf{i} - 2\mathbf{j}$, $\mathbf{B} = -4\mathbf{i} + 3\mathbf{j}$, and $\mathbf{C} = -6\mathbf{i} + 8\mathbf{j}$, find scalars h and k such that $\mathbf{B} = h\mathbf{C} - k\mathbf{A}$.

35. If $\mathbf{A} = \mathbf{i} - 2\mathbf{j}$, $\mathbf{B} = -2\mathbf{i} + 4\mathbf{j}$, and $\mathbf{C} = 7\mathbf{i} - 5\mathbf{j}$, show that $\mathbf{C}$ cannot be written in the form $h\mathbf{A} + k\mathbf{B}$, where h and k are scalars.

36. Two forces of magnitudes 340 lb and 475 lb make an angle of $34.6°$ with each other and are applied to an object at the same point. Find (a) the magnitude of the resultant force and (b) to the nearest tenth of a degree the angle it makes with the force of 475 lb.

37. Two forces of magnitudes 60 lb and 80 lb make an angle of $30°$ with each other and are applied to an object at the same point. Find (a) the magnitude of the resultant force and (b) to the nearest degree the angle it makes with the force of 60 lb.

38. A force of magnitude 22 lb and one of magnitude 34 lb are applied to an object at the same point and make an angle of θ with each other. If the resultant force has a magnitude of 46 lb, find θ to the nearest degree.

39. A force of magnitude 112 lb and one of 84 lb are applied to an object at the same point, and the resultant force has a magnitude of 162 lb. Find to the nearest tenth of a degree the angle made by the resultant force with the force of 112 lb.

40. Prove Theorem 13.1.8(ii).   **41.** Prove Theorem 13.1.8(v).

42. Prove Theorem 13.1.8(vii).   **43.** Prove Theorem 13.1.8(iii) and (viii).

44. Two vectors are said to be *independent* if and only if their position representations are not collinear. Furthermore, two vectors $\mathbf{A}$ and $\mathbf{B}$ are said to form a *basis* for the vector space V_2 if and only if any vector in V_2 can be written as a linear combination of $\mathbf{A}$ and $\mathbf{B}$. A theorem can be proved which states that two vectors form a basis for the vector space V_2 if they are independent. Show that this theorem holds for the two vectors $\langle 2, 5\rangle$ and $\langle 3, -1\rangle$ by doing the following: (a) Verify that the vectors are independent by showing that their position representations are not collinear; (b) verify that the vectors form a basis by showing that any vector $a_1\mathbf{i} + a_2\mathbf{j}$ can be written as $c(2\mathbf{i} + 5\mathbf{j}) + d(3\mathbf{i} - \mathbf{j})$, where c and d are scalars. (*Hint:* Find c and d in terms of a_1 and a_2.)

45. Refer to the first two sentences of Exercise 44. A theorem can be proved which states that two vectors form a basis for the vector space V_2 only if they are independent. Show that this theorem holds for the two vectors $\langle 3, -2\rangle$ and $\langle -6, 4\rangle$ by doing the following: (a) Verify that the vectors are dependent (not independent) by showing that their position representations are collinear; (b) verify that the vectors do not form a basis by taking a particular vector and showing that it cannot be written as $c(3\mathbf{i} - 2\mathbf{j}) + d(-6\mathbf{i} + 4\mathbf{j})$, where c and d are scalars.

46. A set of vectors $\mathbf{V}_1, \mathbf{V}_2, \mathbf{V}_3, \ldots, \mathbf{V}_n$ is said to be *linearly dependent* if and only if there are scalars $k_1, k_2, k_3, \ldots, k_n$, not all zero, such that

$$k_1\mathbf{V}_1 + k_2\mathbf{V}_2 + k_3\mathbf{V}_3 + \ldots + k_n\mathbf{V}_n = \mathbf{0}$$

Show that if $\mathbf{V}_1 = 3\mathbf{i} - 2\mathbf{j}$, $\mathbf{V}_2 = \mathbf{i} + 4\mathbf{j}$, and $\mathbf{V}_3 = 2\mathbf{i} + 5\mathbf{j}$, then $\mathbf{V}_1, \mathbf{V}_2$, and $\mathbf{V}_3$ are linearly dependent.

47. Given $\mathbf{A} = \langle 2, -5\rangle$; $\mathbf{B} = \langle 3, 1\rangle$; $\mathbf{C} = \langle -4, 2\rangle$. (a) Find $\mathbf{A} + (\mathbf{B} + \mathbf{C})$ and illustrate geometrically. (b) Find $(\mathbf{A} + \mathbf{B}) + \mathbf{C}$ and illustrate geometrically.

48. Prove analytically the triangle inequality for vectors $\|\mathbf{A} + \mathbf{B}\| \le \|\mathbf{A}\| + \|\mathbf{B}\|$.

49. Let $\overrightarrow{PQ}$ be a representation of vector $\mathbf{A}$, $\overrightarrow{QR}$ be a representation of vector $\mathbf{B}$, and $\overrightarrow{RS}$ be a representation of vector $\mathbf{C}$. Prove that if $\overrightarrow{PQ}, \overrightarrow{QR}$, and $\overrightarrow{RS}$ are sides of a triangle, then $\mathbf{A} + \mathbf{B} + \mathbf{C} = \mathbf{0}$.

13.2
DOT PRODUCT

In Section 13.1 addition and subtraction of vectors and multiplication of a vector by a scalar were defined. However, the multiplication of two vectors was not considered. We now define a multiplication operation on two vectors that gives what is called the *dot product*.

13.2.1 DEFINITION   If $\mathbf{A} = \langle a_1, a_2 \rangle$ and $\mathbf{B} = \langle b_1, b_2 \rangle$ are two vectors in V_2, then the **dot product** of $\mathbf{A}$ and $\mathbf{B}$, denoted by $\mathbf{A} \cdot \mathbf{B}$, is given by

$$\mathbf{A} \cdot \mathbf{B} = \langle a_1, a_2 \rangle \cdot \langle b_1, b_2 \rangle$$
$$= a_1 b_1 + a_2 b_2$$

The dot product of two vectors is a real number (or scalar) and not a vector. It is sometimes called the **scalar product** or **inner product**.

▶ ILLUSTRATION 1   If $\mathbf{A} = \langle 2, -3 \rangle$ and $\mathbf{B} = \langle -\frac{1}{2}, 4 \rangle$, then

$$\mathbf{A} \cdot \mathbf{B} = \langle 2, -3 \rangle \cdot \langle -\tfrac{1}{2}, 4 \rangle$$
$$= (2)(-\tfrac{1}{2}) + (-3)(4)$$
$$= -13 \qquad\qquad\qquad ◀$$

The following dot products are useful and are easily verified (see Exercise 5).

$$\mathbf{i} \cdot \mathbf{i} = 1 \qquad\qquad\qquad\qquad\qquad\qquad (1)$$
$$\mathbf{j} \cdot \mathbf{j} = 1 \qquad\qquad\qquad\qquad\qquad\qquad (2)$$
$$\mathbf{i} \cdot \mathbf{j} = 0 \qquad\qquad\qquad\qquad\qquad\qquad (3)$$

The next theorem states that dot multiplication is commutative and distributive with respect to vector addition.

13.2.2 THEOREM   If $\mathbf{A}$, $\mathbf{B}$, and $\mathbf{C}$ are any vectors in V_2, then

(i) $\mathbf{A} \cdot \mathbf{B} = \mathbf{B} \cdot \mathbf{A}$ (commutative law)
(ii) $\mathbf{A} \cdot (\mathbf{B} + \mathbf{C}) = \mathbf{A} \cdot \mathbf{B} + \mathbf{A} \cdot \mathbf{C}$ (distributive law)

The proofs of (i) and (ii) are left as exercises (see Exercises 6 and 7).
Because $\mathbf{A} \cdot \mathbf{B}$ is a scalar, the expression $(\mathbf{A} \cdot \mathbf{B}) \cdot \mathbf{C}$ is meaningless. Hence we do not consider associativity of dot multiplication.
Some other laws of dot multiplication are given in the following theorem.

13.2.3 THEOREM   If $\mathbf{A}$ and $\mathbf{B}$ are any vectors in V_2 and c is any scalar, then

(i) $c(\mathbf{A} \cdot \mathbf{B}) = (c\mathbf{A}) \cdot \mathbf{B}$
(ii) $\mathbf{0} \cdot \mathbf{A} = 0$
(iii) $\mathbf{A} \cdot \mathbf{A} = \|\mathbf{A}\|^2$

The proofs are left as exercises (see Exercises 8 through 10).

We now consider what is meant by the angle between two vectors, and this leads to another expression for the dot product of two vectors.

13.2.4 DEFINITION   Let **A** and **B** be two nonzero vectors such that **A** is not a scalar multiple of **B**. If $\overrightarrow{OP}$ is the position representation of **A** and $\overrightarrow{OQ}$ is the position representation of **B**, then the **angle between the vectors A and B** is defined to be the angle of positive measure between $\overrightarrow{OP}$ and $\overrightarrow{OQ}$ interior to the triangle determined by the points O, P, and Q. If $\mathbf{A} = c\mathbf{B}$, where c is a scalar, then if $c > 0$ the angle between the vectors has radian measure 0; if $c < 0$, the angle between the vectors has radian measure π.

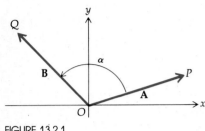

FIGURE 13.2.1

The symbol used to denote the angle between two vectors is also used to denote the measure of that angle. From Definition 13.2.4, if α is the radian measure of the angle between two vectors, then $0 \leq \alpha \leq \pi$. Figure 13.2.1 shows the angle between two vectors if **A** is not a scalar multiple of **B**.

The following theorem is perhaps the most important fact about the dot product of two vectors.

13.2.5 THEOREM   If α is the angle between the two nonzero vectors **A** and **B**, then

$$\mathbf{A} \cdot \mathbf{B} = \|\mathbf{A}\| \|\mathbf{B}\| \cos \alpha \tag{4}$$

PROOF   Let $\mathbf{A} = a_1\mathbf{i} + a_2\mathbf{j}$ and $\mathbf{B} = b_1\mathbf{i} + b_2\mathbf{j}$. Let $\overrightarrow{OP}$ be the position representation of **A** and OQ be the position representation of **B**. Then the angle between the vectors **A** and **B** is the angle at the origin in triangle POQ (see Fig. 13.2.2); P is the point (a_1, a_2) and Q is the point (b_1, b_2). In triangle POQ, $\|\mathbf{A}\|$ is the length of the side OP and $\|\mathbf{B}\|$ is the length of the side OQ. So from the law of cosines,

FIGURE 13.2.2

$$\cos \alpha = \frac{\|\mathbf{A}\|^2 + \|\mathbf{B}\|^2 - |\overline{PQ}|^2}{2\|\mathbf{A}\| \|\mathbf{B}\|}$$

$$= \frac{(a_1{}^2 + a_2{}^2) + (b_1{}^2 + b_2{}^2) - [(a_1 - b_1)^2 + (a_2 - b_2)^2]}{2\|\mathbf{A}\| \|\mathbf{B}\|}$$

$$= \frac{2a_1b_1 + 2a_2b_2}{2\|\mathbf{A}\| \|\mathbf{B}\|}$$

$$= \frac{a_1b_1 + a_2b_2}{\|\mathbf{A}\| \|\mathbf{B}\|}$$

Hence

$$\cos \alpha = \frac{\mathbf{A} \cdot \mathbf{B}}{\|\mathbf{A}\| \|\mathbf{B}\|}$$

$$\mathbf{A} \cdot \mathbf{B} = \|\mathbf{A}\| \|\mathbf{B}\| \cos \alpha \qquad \blacksquare$$

Theorem 13.2.5 states that the dot product of two vectors is the product of the magnitudes of the vectors and the cosine of the angle between them.

▶ ILLUSTRATION 2   If $\mathbf{A} = 3\mathbf{i} - 2\mathbf{j}$, $\mathbf{B} = 2\mathbf{i} + \mathbf{j}$, and α is the angle between $\mathbf{A}$ and $\mathbf{B}$, then from Theorem 13.2.5,

$$
\begin{aligned}
\cos \alpha &= \frac{\mathbf{A} \cdot \mathbf{B}}{\|\mathbf{A}\| \, \|\mathbf{B}\|} \\[2mm]
&= \frac{(3)(2) + (-2)(1)}{\sqrt{9 + 4}\sqrt{4 + 1}} \\[2mm]
&= \frac{6 - 2}{\sqrt{13}\sqrt{5}} \\[2mm]
&= \frac{4}{\sqrt{65}}
\end{aligned}
$$
◀

We learned in Section 13.1 that if two nonzero vectors are scalar multiples of each other, they have either the same or opposite directions. We have, then, the following definition.

13.2.6 DEFINITION   Two vectors are said to be **parallel** if and only if one of the vectors is a scalar multiple of the other.

▶ ILLUSTRATION 3   The vectors $\langle 3, -4 \rangle$ and $\langle \frac{3}{4}, -1 \rangle$ are parallel because $\langle 3, -4 \rangle = 4 \langle \frac{3}{4}, -1 \rangle$. ◀

If $\mathbf{A}$ is any vector, $\mathbf{0} = 0\mathbf{A}$; thus, from Definition 13.2.6 it follows that the zero vector is parallel to any vector.

It is left as an exercise for you to show that two nonzero vectors are parallel if and only if the radian measure of the angle between them is 0 or π (see Exercise 37).

If $\mathbf{A}$ and $\mathbf{B}$ are nonzero vectors, then from (4) it follows that

$$\cos \alpha = 0 \quad \text{if and only if} \quad \mathbf{A} \cdot \mathbf{B} = 0$$

Because $0 \le \alpha \le \pi$, it follows from this statement that

$$\alpha = \tfrac{1}{2}\pi \quad \text{if and only if} \quad \mathbf{A} \cdot \mathbf{B} = 0$$

We have, then, the following definition.

13.2.7 DEFINITION   Two vectors $\mathbf{A}$ and $\mathbf{B}$ are said to be **orthogonal (perpendicular)** if and only if $\mathbf{A} \cdot \mathbf{B} = 0$.

▶ ILLUSTRATION 4   The vectors $\langle -4, 5 \rangle$ and $\langle 10, 8 \rangle$ are orthogonal because

$$
\begin{aligned}
\langle -4, 5 \rangle \cdot \langle 10, 8 \rangle &= (-4)(10) + (5)(8) \\
&= 0
\end{aligned}
$$
◀

If $\mathbf{A}$ is any vector, $\mathbf{0} \cdot \mathbf{A} = 0$, and therefore it follows from Definition 13.2.7 that the zero vector is orthogonal to any vector.

EXAMPLE 1

Given $\mathbf{A} = 3\mathbf{i} + 2\mathbf{j}$ and $\mathbf{B} = 2\mathbf{i} + k\mathbf{j}$, where k is a scalar, find (a) k such that $\mathbf{A}$ and $\mathbf{B}$ are orthogonal; (b) k such that $\mathbf{A}$ and $\mathbf{B}$ are parallel.

SOLUTION

(a) By Definition 13.2.7, $\mathbf{A}$ and $\mathbf{B}$ are orthogonal if and only if $\mathbf{A} \cdot \mathbf{B} = 0$; that is,

$$(3)(2) + 2(k) = 0$$
$$k = -3$$

(b) From Definition 13.2.6, $\mathbf{A}$ and $\mathbf{B}$ are parallel if and only if there is some scalar c such that $\langle 3, 2 \rangle = c \langle 2, k \rangle$; that is,

$$3 = 2c \quad \text{and} \quad 2 = ck$$

Solving these two equations simultaneously we obtain $k = \frac{4}{3}$.

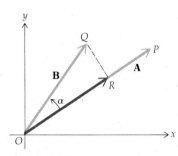

FIGURE 13.2.3

A geometric interpretation of the dot product is obtained by considering the *scalar projection* of a vector onto another vector. See Fig. 13.2.3, where $\overrightarrow{OP}$ and $\overrightarrow{OQ}$ are the position representations of vectors $\mathbf{A}$ and $\mathbf{B}$, respectively. Point R is the foot of the perpendicular from Q to the line containing $\overrightarrow{OP}$. The scalar projection of $\mathbf{B}$ onto $\mathbf{A}$ is the magnitude of the vector having $\overrightarrow{OR}$ as its position representation. Following is the formal definition.

13.2.8 DEFINITION

If $\mathbf{A}$ and $\mathbf{B}$ are nonzero vectors, the scalar projection of $\mathbf{B}$ onto $\mathbf{A}$ is defined to be $\|\mathbf{B}\| \cos \alpha$, where α is the angle between $\mathbf{A}$ and $\mathbf{B}$.

Observe that the scalar projection may be either positive or negative, depending on the sign of $\cos \alpha$.

From (4),

$$\mathbf{A} \cdot \mathbf{B} = \|\mathbf{A}\|(\|\mathbf{B}\| \cos \alpha) \tag{5}$$

Thus the dot product of $\mathbf{A}$ and $\mathbf{B}$ is the magnitude of $\mathbf{A}$ multiplied by the scalar projection of $\mathbf{B}$ onto $\mathbf{A}$. See Fig. 13.2.4(a) and (b). Because dot multiplication is commutative, $\mathbf{A} \cdot \mathbf{B}$ is also the magnitude of $\mathbf{B}$ multiplied by the scalar projection of $\mathbf{A}$ onto $\mathbf{B}$.

If $\mathbf{B} = b_1\mathbf{i} + b_2\mathbf{j}$, then

$$\mathbf{i} \cdot \mathbf{B} = b_1 \quad \text{and} \quad \mathbf{j} \cdot \mathbf{B} = b_2$$

$\|\mathbf{B}\| \cos \alpha > 0$ $\|\mathbf{B}\| \cos \alpha < 0$

(a)   (b)

FIGURE 13.2.4

Hence the dot product of $\mathbf{i}$ and $\mathbf{B}$ gives the component of $\mathbf{B}$ in the direction of $\mathbf{i}$ and the dot product of $\mathbf{j}$ and $\mathbf{B}$ gives the component of $\mathbf{B}$ in the direction of $\mathbf{j}$. To generalize this result, let $\mathbf{U}$ be any unit vector. Then from (5), if α is the angle between $\mathbf{U}$ and $\mathbf{B}$,

$$\mathbf{U} \cdot \mathbf{B} = \|\mathbf{U}\| \|\mathbf{B}\| \cos \alpha$$
$$= \|\mathbf{B}\| \cos \alpha$$

Therefore $\mathbf{U} \cdot \mathbf{B}$ is the scalar projection of $\mathbf{B}$ onto $\mathbf{U}$, which is called the *component* of the vector $\mathbf{B}$ in the direction of $\mathbf{U}$. More generally, the **component** of a vector $\mathbf{B}$ in the direction of a vector $\mathbf{A}$ is the scalar projection of $\mathbf{B}$ onto a unit vector in the direction of $\mathbf{A}$.

The following theorem can be used to compute the scalar projection of one vector onto another.

13.2.9 THEOREM   The scalar projection of the vector $\mathbf{B}$ onto the vector $\mathbf{A}$ is

$$\frac{\mathbf{A} \cdot \mathbf{B}}{\|\mathbf{A}\|}$$

PROOF   From Definition 13.2.8, the scalar projection of $\mathbf{B}$ onto $\mathbf{A}$ is $\|\mathbf{B}\| \cos \alpha$, where α is the angle between $\mathbf{A}$ and $\mathbf{B}$. From Theorem 13.2.5,

$$\|\mathbf{A}\| \|\mathbf{B}\| \cos \alpha = \mathbf{A} \cdot \mathbf{B}$$

$$\|\mathbf{B}\| \cos \alpha = \frac{\mathbf{A} \cdot \mathbf{B}}{\|\mathbf{A}\|}$$

■

Refer again to Fig. 13.2.3. If $\mathbf{C}$ is the vector having $\overrightarrow{OR}$ as its position representation, then $\mathbf{C}$ is called the **vector projection** of $\mathbf{B}$ onto $\mathbf{A}$. To determine $\mathbf{C}$, we multiply $\|\mathbf{B}\| \cos \alpha$ by the unit vector having the same direction as $\mathbf{A}$. Thus

$$\mathbf{C} = (\|\mathbf{B}\| \cos \alpha) \frac{\mathbf{A}}{\|\mathbf{A}\|}$$

$$= \frac{\|\mathbf{A}\|(\|\mathbf{B}\| \cos \alpha)}{\|\mathbf{A}\|^2} \mathbf{A}$$

$$= \left(\frac{\mathbf{A} \cdot \mathbf{B}}{\|\mathbf{A}\|^2}\right) \mathbf{A} \quad \text{(from Theorem 13.2.5)}$$

We state this result as a theorem.

13.2.10 THEOREM   The vector projection of the vector $\mathbf{B}$ onto the vector $\mathbf{A}$ is

$$\left(\frac{\mathbf{A} \cdot \mathbf{B}}{\|\mathbf{A}\|^2}\right) \mathbf{A}$$

EXAMPLE 2
Given the vectors

$$\mathbf{A} = -5\mathbf{i} + \mathbf{j} \qquad \mathbf{B} = 4\mathbf{i} + 2\mathbf{j}$$

Find: (a) the scalar projection of $\mathbf{B}$ onto $\mathbf{A}$; (b) the vector projection of $\mathbf{B}$ onto $\mathbf{A}$. (c) Show on a figure the position representations of $\mathbf{A}$, $\mathbf{B}$, and the vector projection of $\mathbf{B}$ onto $\mathbf{A}$.

SOLUTION

We first compute $\mathbf{A} \cdot \mathbf{B}$ and $\|\mathbf{A}\|$.

$$\mathbf{A} \cdot \mathbf{B} = \langle -5, 1 \rangle \cdot \langle 4, 2 \rangle \qquad \|\mathbf{A}\| = \sqrt{(-5)^2 + 1^2}$$
$$= -20 + 2 \qquad\qquad\qquad = \sqrt{26}$$
$$= -18$$

(a) From Theorem 13.2.9, the scalar projection of $\mathbf{B}$ onto $\mathbf{A}$ is

$$\frac{\mathbf{A} \cdot \mathbf{B}}{\|\mathbf{A}\|} = -\frac{18}{\sqrt{26}}$$

(b) From Theorem 13.2.10, the vector projection of $\mathbf{B}$ onto $\mathbf{A}$ is

$$\left(\frac{\mathbf{A} \cdot \mathbf{B}}{\|\mathbf{A}\|^2} \right) \mathbf{A} = -\tfrac{18}{26}(-5\mathbf{i} + \mathbf{j})$$
$$= -\tfrac{9}{13}(-5\mathbf{i} + \mathbf{j})$$
$$= \tfrac{45}{13}\mathbf{i} - \tfrac{9}{13}\mathbf{j}$$

(c) Figure 13.2.5 shows the position representations of $\mathbf{A}$, $\mathbf{B}$, and $\mathbf{C}$, where $\mathbf{C}$ is the vector projection of $\mathbf{B}$ onto $\mathbf{A}$.

FIGURE 13.2.5

In Section 5.4 we stated that if a constant force of F pounds moves an object a distance d feet along a straight line and the force is acting in the direction of motion, then if W is the number of foot-pounds in the work done by the force, $W = Fd$. Suppose, however, that the constant force is not directed along the line of motion. In this case the physicist defines the **work** done as the *product of the component of the force along the line of motion times the displacement*. If the object moves from the point A to the point B, we call the vector, having $\overrightarrow{AB}$ as a representation, the **displacement vector** and denote it by $\mathbf{V}(\overrightarrow{AB})$. So if the magnitude of a constant force vector $\mathbf{F}$ is expressed in pounds and the distance from A to B is expressed in feet, and α is the radian measure of the angle between the vectors $\mathbf{F}$ and $\mathbf{V}(\overrightarrow{AB})$, then if W is the number of foot-pounds in the work done by the force $\mathbf{F}$ in moving an object from A to B,

$$W = (\|\mathbf{F}\| \cos \alpha)\|\mathbf{V}(\overrightarrow{AB})\|$$
$$= \|\mathbf{F}\| \|\mathbf{V}(\overrightarrow{AB})\| \cos \alpha$$
$$= \mathbf{F} \cdot \mathbf{V}(\overrightarrow{AB})$$

EXAMPLE 3

Suppose that a force $\mathbf{F}$ has a magnitude of 6 lb and $\frac{1}{6}\pi$ is the radian measure of the angle giving its direction. Find the work done by $\mathbf{F}$ in moving an object along a straight line from the origin to the point $P(7, 1)$, where distance is measured in feet.

FIGURE 13.2.6

Figure 13.2.6 shows the position representations of **F** and $\mathbf{V}(\overrightarrow{OP})$. Because $\mathbf{F} = \langle 6\cos\frac{1}{6}\pi, 6\sin\frac{1}{6}\pi \rangle$, and $\mathbf{V}(\overrightarrow{OP}) = \langle 7, 1 \rangle$, then if W ft-lb is the work done, we have

$$\mathbf{W} = \mathbf{F} \cdot \mathbf{V}(\overrightarrow{OP})$$
$$= \langle 6\cos\tfrac{1}{6}\pi, 6\sin\tfrac{1}{6}\pi \rangle \cdot \langle 7, 1 \rangle$$
$$= \langle 3\sqrt{3}, 3 \rangle \cdot \langle 7, 1 \rangle$$
$$= 21\sqrt{3} + 3$$
$$\approx 39.37$$

Therefore the work done is approximately 39.37 ft-lb.

Vectors have geometric representations that are independent of the coordinate system used. Because of this, vector analysis can be used to prove certain theorems of plane geometry. This is illustrated in the following example.

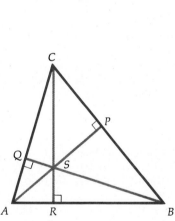

FIGURE 13.2.7

EXAMPLE 4
Prove by vector analysis that the altitudes of a triangle meet in a point.

SOLUTION
Let ABC be a triangle having altitudes AP and BQ intersecting at point S. Draw a line through C and S intersecting AB at point R. We wish to prove that RC is perpendicular to AB (see Fig. 13.2.7).

Let $\overrightarrow{AB}$, $\overrightarrow{BC}$, $\overrightarrow{AC}$, $\overrightarrow{AS}$, $\overrightarrow{BS}$, $\overrightarrow{CS}$ be representations of vectors. Let $\mathbf{V}(\overrightarrow{AB})$ be the vector having directed line segment $\overrightarrow{AB}$ as a representation. In a similar manner let $\mathbf{V}(\overrightarrow{BC})$, $\mathbf{V}(\overrightarrow{AC})$, $\mathbf{V}(\overrightarrow{AS})$, and $\mathbf{V}(\overrightarrow{BS})$, and $\mathbf{V}(\overrightarrow{CS})$ be the vectors having the directed line segment in parentheses as a representation.

Because AP is an altitude of the triangle,

$$\mathbf{V}(\overrightarrow{AS}) \cdot \mathbf{V}(\overrightarrow{BC}) = 0 \tag{6}$$

Also, because BQ is an altitude of the triangle,

$$\mathbf{V}(\overrightarrow{BS}) \cdot \mathbf{V}(\overrightarrow{AC}) = 0 \tag{7}$$

To prove that RC is perpendicular to AB we shall show that

$$\mathbf{V}(\overrightarrow{CS}) \cdot \mathbf{V}(\overrightarrow{AB}) = 0$$
$$\mathbf{V}(\overrightarrow{CS}) \cdot \mathbf{V}(\overrightarrow{AB}) = \mathbf{V}(\overrightarrow{CS}) \cdot [\mathbf{V}(\overrightarrow{AC}) + \mathbf{V}(\overrightarrow{CB})]$$
$$= \mathbf{V}(\overrightarrow{CS}) \cdot \mathbf{V}(\overrightarrow{AC}) + \mathbf{V}(\overrightarrow{CS}) \cdot \mathbf{V}(\overrightarrow{CB})$$
$$= [\mathbf{V}(\overrightarrow{CB}) + \mathbf{V}(\overrightarrow{BS})] \cdot \mathbf{V}(\overrightarrow{AC}) + [\mathbf{V}(\overrightarrow{CA}) + \mathbf{V}(\overrightarrow{AS})] \cdot \mathbf{V}(\overrightarrow{CB})$$
$$= \mathbf{V}(\overrightarrow{CB}) \cdot \mathbf{V}(\overrightarrow{AC}) + \mathbf{V}(\overrightarrow{BS}) \cdot \mathbf{V}(\overrightarrow{AC})$$
$$+ \mathbf{V}(\overrightarrow{CA}) \cdot \mathbf{V}(\overrightarrow{CB}) + \mathbf{V}(\overrightarrow{AS}) \cdot \mathbf{V}(\overrightarrow{CB})$$

Replacing $\mathbf{V}(\overrightarrow{CA})$ by $-\mathbf{V}(\overrightarrow{AC})$ and using (6) and (7) we obtain

$$\mathbf{V}(\overrightarrow{CS}) \cdot \mathbf{V}(\overrightarrow{AB}) = \mathbf{V}(\overrightarrow{CB}) \cdot \mathbf{V}(\overrightarrow{AC}) + 0 + [-\mathbf{V}(\overrightarrow{AC})] \cdot \mathbf{V}(\overrightarrow{CB}) + 0$$
$$= 0$$

Therefore altitudes AP, BQ, and RC meet in a point.

EXERCISES 13.2

In Exercises 1 through 4, find $\mathbf{A} \cdot \mathbf{B}$.

1. $\mathbf{A} = \langle -1, 2 \rangle$; $\mathbf{B} = \langle -4, 3 \rangle$

2. $\mathbf{A} = \langle \frac{1}{3}, -\frac{1}{2} \rangle$; $\mathbf{B} = \langle \frac{5}{2}, \frac{4}{3} \rangle$

3. $\mathbf{A} = 2\mathbf{i} - \mathbf{j}$; $\mathbf{B} = \mathbf{i} + 3\mathbf{j}$

4. $\mathbf{A} = -2\mathbf{i}$; $\mathbf{B} = -\mathbf{i} + \mathbf{j}$

5. Show that $\mathbf{i} \cdot \mathbf{i} = 1$; $\mathbf{j} \cdot \mathbf{j} = 1$; $\mathbf{i} \cdot \mathbf{j} = 0$.

6. Prove Theorem 13.2.2(i).

7. Prove Theorem 13.2.2(ii).

8. Prove Theorem 13.2.3(i).

9. Prove Theorem 13.2.3(ii).

10. Prove Theorem 13.2.3(iii).

In Exercises 11 through 14, if α is the angle between $\mathbf{A}$ and $\mathbf{B}$, find $\cos \alpha$.

11. $\mathbf{A} = \langle 4, 3 \rangle$; $\mathbf{B} = \langle 1, -1 \rangle$

12. $\mathbf{A} = \langle -2, -3 \rangle$; $\mathbf{B} = \langle 3, 2 \rangle$

13. $\mathbf{A} = 5\mathbf{i} - 12\mathbf{j}$; $\mathbf{B} = 4\mathbf{i} + 3\mathbf{j}$

14. $\mathbf{A} = 2\mathbf{i} + 4\mathbf{j}$; $\mathbf{B} = -5\mathbf{j}$

15. Find k such that the radian measure of the angle between the vectors in Example 1 of this section is $\frac{1}{4}\pi$.

16. Given $\mathbf{A} = k\mathbf{i} - 2\mathbf{j}$ and $\mathbf{B} = k\mathbf{i} + 6\mathbf{j}$, where k is a scalar. Find k such that $\mathbf{A}$ and $\mathbf{B}$ are orthogonal.

17. Given $\mathbf{A} = 5\mathbf{i} - k\mathbf{j}$ and $\mathbf{B} = k\mathbf{i} + 6\mathbf{j}$, where k is a scalar. Find (a) k such that $\mathbf{A}$ and $\mathbf{B}$ are orthogonal; (b) k such that $\mathbf{A}$ and $\mathbf{B}$ are parallel.

18. Find k such that the vectors given in Exercise 16 have opposite directions.

19. Given $\mathbf{A} = 5\mathbf{i} + 12\mathbf{j}$; $\mathbf{B} = \mathbf{i} + k\mathbf{j}$, where k is a scalar. Find k such that the radian measure of the angle between $\mathbf{A}$ and $\mathbf{B}$ is $\frac{1}{3}\pi$.

20. Find two unit vectors each having a representation whose initial point is $(2, 4)$ and which is tangent to the parabola $y = x^2$ there.

21. Find two unit vectors each having a representation whose initial point is $(2, 4)$ and which is normal to the parabola $y = x^2$ there.

22. If $\mathbf{A} = 2\mathbf{i} - 7\mathbf{j}$, find the unit vectors that are orthogonal to $\mathbf{A}$.

23. If $\mathbf{A}$ is the vector $a_1\mathbf{i} + a_2\mathbf{j}$, find the unit vectors that are orthogonal to $\mathbf{A}$.

24. If $\mathbf{A} = 5\mathbf{i} - 9c\mathbf{j}$ and $\mathbf{B} = 7\mathbf{i} - 4c\mathbf{j}$, show that there is no real value for c such that $\mathbf{A}$ and $\mathbf{B}$ are orthogonal.

25. If $\mathbf{A} = -8\mathbf{i} + 4\mathbf{j}$ and $\mathbf{B} = 7\mathbf{i} - 6\mathbf{j}$, find (a) the scalar projection of $\mathbf{A}$ onto $\mathbf{B}$ and (b) the vector projection of $\mathbf{A}$ onto $\mathbf{B}$.

26. For the vectors of Exercise 25, find (a) the scalar projection of $\mathbf{B}$ onto $\mathbf{A}$ and (b) the vector projection of $\mathbf{B}$ onto $\mathbf{A}$.

27. Find the component of the vector $\mathbf{A} = 5\mathbf{i} - 6\mathbf{j}$ in the direction of the vector $\mathbf{B} = 7\mathbf{i} + \mathbf{j}$.

28. For the vectors $\mathbf{A}$ and $\mathbf{B}$ of Exercise 27, find the component of the vector $\mathbf{B}$ in the direction of vector $\mathbf{A}$.

29. A vector $\mathbf{F}$ represents a force that has a magnitude of 8 lb and $\frac{1}{3}\pi$ is the radian measure of its direction angle. Find the work done by the force in moving an object (a) along the x axis from the origin to the point $(6, 0)$ and (b) along the y axis from the origin to the point $(0, 6)$. Distance is measured in feet.

30. A vector $\mathbf{F}$ represents a force that has a magnitude of 10 lb and $\frac{1}{4}\pi$ is the radian measure of its direction angle. Find the work done by the force in moving an object along the y axis from the point $(0, -2)$ to the point $(0, 5)$. Distance is measured in feet.

31. A vector $\mathbf{F}$ represents a force that has a magnitude of 9 lb and $\frac{2}{3}\pi$ is the radian measure of its direction angle. Find the work done by the force in moving an object from the origin to the point $(-4, -2)$. Distance is measured in feet.

32. Two forces represented by the vectors $\mathbf{F}_1$ and $\mathbf{F}_2$ act on a particle and cause it to move along a straight line from the point $(2, 5)$ to the point $(7, 3)$. If $\mathbf{F}_1 = 3\mathbf{i} - \mathbf{j}$ and $\mathbf{F}_2 = -4\mathbf{i} + 5\mathbf{j}$, the magnitudes of the forces are measured in pounds, and distance is measured in feet, find the work done by the two forces acting together.

33. If $\mathbf{A}$ and $\mathbf{B}$ are vectors, prove that

$$(\mathbf{A} + \mathbf{B}) \cdot (\mathbf{A} + \mathbf{B}) = \mathbf{A} \cdot \mathbf{A} + 2\mathbf{A} \cdot \mathbf{B} + \mathbf{B} \cdot \mathbf{B}$$

34. Prove by vector analysis that the medians of a triangle meet in a point.

35. Prove by vector analysis that the line segment joining the midpoints of two sides of a triangle is parallel to the third side and its length is one-half the length of the third side.

36. Prove by vector analysis that the line segment joining the midpoints of the nonparallel sides of a trapezoid is parallel to the parallel sides and its length is one-half the sum of the lengths of the parallel sides.

37. Prove that two nonzero vectors are parallel if and only if the radian measure of the angle between them is 0 or π.

38. If **A** and **B** are vectors, prove that $\|\mathbf{A} \cdot \mathbf{B}\| \leq \|\mathbf{A}\| \, \|\mathbf{B}\|$, where equality pertains if and only if there exists a scalar c such that $\mathbf{A} = c\mathbf{B}$.

13.3
VECTOR-VALUED FUNCTIONS AND PARAMETRIC EQUATIONS

Suppose a particle moves so that the coordinates (x, y) of its position at any time t are given by the equations

$$x = f(t) \quad \text{and} \quad y = g(t) \tag{1}$$

Then for every number t in the domain common to f and g there is a vector $f(t)\mathbf{i} + g(t)\mathbf{j}$, and the endpoints of the position representations of these vectors trace a curve C traveled by the particle. This leads us to consider a function whose domain is a set of real numbers and whose range is a set of vectors. Such a function is called a *vector-valued function*.

13.3.1 DEFINITION

Let f and g be two real-valued functions of a real variable t. Then for every number t in the domain common to f and g there is a vector **R** defined by

$$\mathbf{R}(t) = f(t)\mathbf{i} + g(t)\mathbf{j} \tag{2}$$

and **R** is called a **vector-valued function**.

▶ ILLUSTRATION 1   Let **R** be the vector-valued function defined by

$$\mathbf{R}(t) = \sqrt{t - 2}\,\mathbf{i} + (t - 3)^{-1}\mathbf{j}$$

If $f(t) = \sqrt{t - 2}$ and $g(t) = (t - 3)^{-1}$, the domain of **R** is the set of values of t for which both $f(t)$ and $g(t)$ are defined. Because $f(t)$ is defined for $t \geq 2$ and $g(t)$ is defined for all real numbers except 3, the domain of **R** is $\{t \mid t \geq 2, t \neq 3\}$.   ◀

Equation (2) defines a curve C. This same curve C is also defined by Eqs. (1). Equation (2) is called a **vector equation** of C, and Eqs. (1) are called **parametric equations** of C. The variable t is a parameter. The curve C is also called a **graph**; that is, the set of all points (x, y) satisfying (1) is the graph of the vector-valued function **R**.

A vector equation of a curve, as well as parametric equations of a curve, gives the curve a direction at each point. That is, if we think of the curve as being traced by a particle, we can consider the positive direction along a curve as the direction in which the particle moves as the parameter t increases. In such a case as this, t may be taken to be the measure of the time, and the vector $\mathbf{R}(t)$ is called the **position vector**. Sometimes $\mathbf{R}(t)$ is referred to as the **radius vector**.

If the parameter t is eliminated from the pair of Eqs. (1), we obtain one equation in x and y, which is called a **cartesian equation** of C. It may happen that elimination of the parameter leads to a cartesian equation whose graph contains more points than the graph defined by either the vector equation or the parametric equations. This situation occurs in Example 2.

EXAMPLE 1

Given the vector equation

$$\mathbf{R}(t) = 2 \cos t\mathbf{i} + 2 \sin t\mathbf{j}$$

(a) Draw a sketch of the graph of this equation, and (b) find a cartesian equation of the graph.

SOLUTION

(a) The domain of $\mathbf{R}$ is the set of all real numbers. The values of x and y for particular values of t can be tabulated. We find the magnitude of the position vector. For every t,

$$\|\mathbf{R}(t)\| = \sqrt{4 \cos^2 t + 4 \sin^2 t}$$
$$= 2\sqrt{\cos^2 t + \sin^2 t}$$
$$= 2$$

Therefore the endpoint of the position representation of each vector $\mathbf{R}(t)$ is two units from the origin. By letting t take on all numbers in the closed interval $[0, 2\pi]$, we obtain a circle having its center at the origin and radius 2. This is the entire graph because any value of t will give a point on this circle. A sketch of the circle is shown in Fig. 13.3.1. Parametric equations of the graph are

$$x = 2 \cos t \quad \text{and} \quad y = 2 \sin t$$

(b) A cartesian equation of the graph can be found by eliminating t from the two parametric equations, which when squaring on both sides of each equation and adding gives

$$x^2 + y^2 = 4$$

FIGURE 13.3.1

EXAMPLE 2

Given the parametric equations

$$x = \cosh t \quad \text{and} \quad y = \sinh t$$

(a) Draw a sketch of the graph defined by these equations, and (b) find a cartesian equation of the graph.

SOLUTION

(a) Squaring on both sides of the given equations and subtracting we have

$$x^2 - y^2 = \cosh^2 t - \sinh^2 t$$

From the identity $\cosh^2 t - \sinh^2 t = 1$, this equation becomes

$$x^2 - y^2 = 1$$

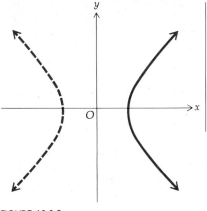

y

O

x

FIGURE 13.3.2

This is an equation of an equilateral hyperbola. However, observe that for t any real number, $\cosh t$ is never less than 1. Thus the curve defined by the given parametric equations consists of only the points on the right branch of the hyperbola. A sketch of this curve is shown in Fig. 13.3.2. The "dashed" curve in the figure is the left branch of the equilateral hyperbola.

(b) A cartesian equation is

$$x^2 - y^2 = 1 \qquad x \geq 1$$

We have seen that by eliminating t from parametric equations (1) we obtain a cartesian equation. The cartesian equation either implicitly or explicitly defines y as one or more functions of x. That is, if $x = f(t)$ and $y = g(t)$, then $y = h(x)$. If h is a differentiable function of x and f is a differentiable function of t, then from the chain rule,

$$\frac{dy}{dt} = \frac{dy}{dx} \cdot \frac{dx}{dt}$$

If $\dfrac{dx}{dt} \neq 0$, we can divide on both sides of the above equation by $\dfrac{dx}{dt}$ and obtain

$$\frac{dy}{dx} = \frac{\dfrac{dy}{dt}}{\dfrac{dx}{dt}} \tag{3}$$

Equation (3) enables us to find the derivative of y with respect to x directly from the parametric equations.

Because $\dfrac{d^2y}{dx^2} = \dfrac{d}{dx}\left(\dfrac{dy}{dx}\right)$, then $\dfrac{d^2y}{dx^2} = \dfrac{d(y')}{dx}$. Thus

$$\frac{d^2y}{dx^2} = \frac{\dfrac{d(y')}{dt}}{\dfrac{dx}{dt}} \tag{4}$$

EXAMPLE 3
Given the parametric equations

$$x = 3t^2 \quad \text{and} \quad y = 4t^3$$

find $\dfrac{dy}{dx}$ and $\dfrac{d^2y}{dx^2}$ without eliminating t.

SOLUTION

Because $\dfrac{dy}{dt} = 12t^2$ and $\dfrac{dx}{dt} = 6t$, we have, from (3),

$$\frac{dy}{dx} = \frac{12t^2}{6t}$$

$$= 2t$$

Since $y' = 2t$, $\dfrac{d(y')}{dt} = 2$. Then from (4),

$$\frac{d^2y}{dx^2} = \frac{\dfrac{d(y')}{dt}}{\dfrac{dx}{dt}}$$

$$= \frac{2}{6t}$$

$$= \frac{1}{3t}$$

EXAMPLE 4

(a) Draw a sketch of the graph defined by the parametric equations of Example 3, and (b) find a cartesian equation of the graph.

SOLUTION

The parametric equations are

$$x = 3t^2 \quad \text{and} \quad y = 4t^3$$

(a) We observe that x is nonnegative. Thus the graph is restricted to the first and fourth quadrants. Table 13.3.1 gives values of x and y for particular values of t. Because $\dfrac{dy}{dx} = 2t$, then when $t = 0$, $\dfrac{dy}{dx} = 0$. Hence at the point $(0, 0)$, the tangent line is horizontal. With this fact and points obtained from Table 13.3.1, we obtain the sketch of the graph appearing in Fig. 13.3.3.

(b) From the two parametric equations, we get $x^3 = 27t^6$ and $y^2 = 16t^6$. Solving each of these equations for t^6 and eliminating t^6 we have

$$\frac{x^3}{27} = \frac{y^2}{16}$$

$$16x^3 = 27y^2 \tag{5}$$

which is the cartesian equation desired.

▶ ILLUSTRATION 2   If in Eq. (5) we differentiate implicitly,

$$48x^2 = 54y\frac{dy}{dx}$$

$$\frac{dy}{dx} = \frac{8x^2}{9y}$$

Substituting for x and y in terms of t from the given parametric equations

Table 13.3.1

t	x	y
0	0	0
$\frac{1}{2}$	$\frac{3}{4}$	$\frac{1}{2}$
1	3	4
2	12	32
$-\frac{1}{2}$	$\frac{3}{4}$	$-\frac{1}{2}$
-1	3	-4
-2	12	-32

FIGURE 13.3.3

we obtain

$$\frac{dy}{dx} = \frac{8(3t^2)^2}{9(4t^3)}$$

$$= 2t$$

which agrees with the value of $\dfrac{dy}{dx}$ found in Example 3.   ◄

It follows from (3) that if at a particular point, $\dfrac{dy}{dt} = 0$ and $\dfrac{dx}{dt} \neq 0$, then $\dfrac{dy}{dx} = 0$, and the graph of the pair of parametric equations has a horizontal tangent line at the point. Furthermore, if at a particular point, $\dfrac{dx}{dt} = 0$ and $\dfrac{dy}{dt} \neq 0$, then $\dfrac{dy}{dx}$ does not exist at the point, and the graph may have a vertical tangent line there. These tangent lines can be helpful when drawing a sketch of the graph of the parametric equations.

EXAMPLE 5

Given the parametric equations

$$x = 4 - t^2 \quad \text{and} \quad y = t^2 + 4t$$

Find the horizontal and vertical tangent lines of the graph of this pair of equations and draw a sketch of the graph.

SOLUTION

$$\frac{dx}{dt} = -2t \qquad \frac{dy}{dt} = 2t + 4$$

Therefore $\dfrac{dy}{dx} = \dfrac{2t + 4}{-2t}$; that is,

$$\frac{dy}{dx} = \frac{t + 2}{-t}$$

When $t = -2$, $\dfrac{dy}{dx} = 0$, $x = 0$, and $y = -4$. Thus the graph has a horizontal tangent line at $(0, -4)$. When $t = 0$, $\dfrac{dy}{dx}$ does not exist, $x = 4$, and $y = 0$; the graph has a vertical tangent line at $(4, 0)$. Table 13.3.2 gives values of x and y for particular values of t. With points obtained from these values and knowing the horizontal and vertical tangent lines, we have the sketch of the graph shown in Fig. 13.3.4.

In Section 9.1, where Cauchy's mean-value theorem (9.1.3) was stated and proved, we indicated that a geometric interpretation would be given in

Table 13.3.2

t	x	y
-4	-12	0
-3	-5	-3
-2	0	-4
-1	3	-3
0	4	0
1	3	5
2	0	12
3	-5	21

FIGURE 13.3.4

FIGURE 13.3.5

this section because parametric equations are needed. Recall that the theorem states that if f and g are two functions such that (i) f and g are continuous on $[a, b]$, (ii) f and g are differentiable on (a, b), and (iii) for all x in (a, b) $g'(x) \neq 0$, then there exists a number z in the open interval (a, b) such that

$$\frac{f(b) - f(a)}{g(b) - g(a)} = \frac{f'(z)}{g'(z)}$$

Figure 13.3.5 shows a curve having the parametric equations $x = g(t)$ and $y = f(t)$, where $a \leq t \leq b$. The slope of the curve in the figure at a particular point is given by

$$\frac{dy}{dx} = \frac{f'(t)}{g'(t)}$$

and the slope of the line segment through the points $A(g(a), f(a))$ and $B(g(b), f(b))$ is given by

$$\frac{f(b) - f(a)}{g(b) - g(a)}$$

Cauchy's mean-value theorem states that the slopes are equal for at least one value of t between a and b. For the curve shown in Fig. 13.3.5 there are four values of t satisfying the conclusion of the theorem: $t = z_1$, $t = z_2$, $t = z_3$, and $t = z_4$.

We now show how parametric equations can be used to define a curve that is described by a physical motion. The curve we consider is a **cycloid**, which is the curve traced by a point on the circumference of a circle as the circle rolls along a straight line. Suppose the circle has radius a. Let the fixed straight line on which the circle rolls be the x axis, and let the origin be one of the points at which the given point P comes in contact with the x axis. See Fig. 13.3.6, showing the circle after it has rolled through an angle of t radians.

FIGURE 13.3.6

From Fig. 13.3.6,

$$\mathbf{V}(\overrightarrow{OT}) + \mathbf{V}(\overrightarrow{TA}) + \mathbf{V}(\overrightarrow{AP}) = \mathbf{V}(\overrightarrow{OP}) \tag{6}$$

The length of the arc PT is $\|\mathbf{V}(\overrightarrow{OT})\| = at$. Because the direction of $\mathbf{V}(\overrightarrow{OT})$ is along the positive x axis,

$$\mathbf{V}(\overrightarrow{OT}) = at\mathbf{i} \tag{7}$$

Also, $\|\mathbf{V}(\overrightarrow{TA})\| = a - a\cos t$. And because the direction of $\mathbf{V}(\overrightarrow{TA})$ is the same as the direction of $\mathbf{j}$,

$$\mathbf{V}(\overrightarrow{TA}) = a(1 - \cos t)\mathbf{j} \tag{8}$$

$\|\mathbf{V}(\overrightarrow{AP})\| = a\sin t$, and the direction of $\mathbf{V}(\overrightarrow{AP})$ is the same as the direction of $-\mathbf{i}$; thus

$$\mathbf{V}(\overrightarrow{AP}) = -a\sin t\mathbf{i} \tag{9}$$

Substituting from (7), (8), and (9) into (6) we obtain

$$at\mathbf{i} + a(1 - \cos t)\mathbf{j} - a\sin t\mathbf{i} = \mathbf{V}(\overrightarrow{OP})$$

$$\Leftrightarrow \quad \mathbf{V}(\overrightarrow{OP}) = a(t - \sin t)\mathbf{i} + a(1 - \cos t)\mathbf{j} \tag{10}$$

Equation (10) is a vector equation of the cycloid. So parametric equations of the cycloid are

$$x = a(t - \sin t) \quad \text{and} \quad y = a(1 - \cos t) \tag{11}$$

where t is any real number. A sketch of a portion of the cycloid is shown in Fig. 13.3.7.

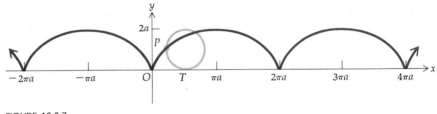

FIGURE 13.3.7

EXERCISES 13.3

In Exercises 1 through 6, find the domain of the vector-valued function $\mathbf{R}$.

1. $\mathbf{R}(t) = (1/t)\mathbf{i} + \sqrt{4 - t}\,\mathbf{j}$
2. $\mathbf{R}(t) = (t^2 + 3)\mathbf{i} + (t - 1)\mathbf{j}$
3. $\mathbf{R}(t) = (\sin^{-1} t)\mathbf{i} + (\cos^{-1} t)\mathbf{j}$
4. $\mathbf{R}(t) = \ln(t + 1)\mathbf{i} + (\tan^{-1} t)\mathbf{j}$
5. $\mathbf{R}(t) = \sqrt{t^2 - 9}\,\mathbf{i} + \sqrt{t^2 + 2t - 8}\,\mathbf{j}$
6. $\mathbf{R}(t) = \sqrt{t - 4}\,\mathbf{i} + \sqrt{4 - t}\,\mathbf{j}$

In Exercises 7 through 12, find $\dfrac{dy}{dx}$ and $\dfrac{d^2y}{dx^2}$ without eliminating the parameter.

7. $x = 3t, y = 2t^2$
8. $x = 1 - t^2, y = 1 + t$
9. $x = t^2 e^t, y = t\ln t$
10. $x = e^{2t}, y = 1 + \cos t$
11. $x = a\cos t, y = b\sin t$
12. $x = a\cosh t, y = b\sinh t$

In Exercises 13 through 19, draw a sketch of the graph of the given vector equation, and find a cartesian equation of the graph.

13. $R(t) = t^2 i + (t + 1) j$ **14.** $R(t) = (t - 2) i + (t^2 + 4) j$ **15.** $R(t) = 3 \cosh t i + 5 \sinh t j$

16. $R(t) = \dfrac{4}{t^2} i + \dfrac{4}{t} j$ **17.** $R(t) = \sec t i + \tan t j;\ t \text{ in } (-\tfrac{1}{2}\pi, \tfrac{1}{2}\pi)$ **18.** $R(t) = \cos t i + \cos t j;\ t \text{ in } [0, \tfrac{1}{2}\pi]$

19. $R(t) = 4 \cos t i + 3 \sin t j;\ t \text{ in } [0, 2\pi]$

20. Find an equation of the tangent line to the curve $x = 1 + 3 \sin t$, $y = 2 - 5 \cos t$, at the point where $t = \tfrac{1}{6}\pi$.

21. Find an equation of the tangent line to the curve $x = 2 \sin t$, $y = 5 \cos t$, at the point where $t = \tfrac{1}{3}\pi$.

In Exercises 22 through 24, find the horizontal and vertical tangent lines of the graph of the given pair of parametric equations, and draw a sketch of the graph.

22. $x = t^2 + t,\ y = t^2 - t$ **23.** $x = 4t^2 - 4t,\ y = 1 - 4t^2$ **24.** $x = \dfrac{3at}{1 + t^3},\ y = \dfrac{3at^2}{1 + t^3}$

25. Parametric equations for the *trochoid* are

$$x = at - b \sin t \quad \text{and} \quad y = a - b \cos t$$

Show that the trochoid has no vertical tangent line if $a > b > 0$.

26. A projectile moves so that the coordinates of its position at any time t are given by the parametric equations $x = 60t$ and $y = 80t - 16t^2$. Draw a sketch of the path of the projectile.

27. Find $\dfrac{dy}{dx}, \dfrac{d^2 y}{dx^2}$, and $\dfrac{d^3 y}{dx^3}$ at the point on the cycloid having Eqs. (11) for which y has its largest value when x is in the closed interval $[0, 2\pi a]$.

28. Show that the slope of the tangent line at $t = t_1$ to the cycloid having Eqs. (11) is $\cot \tfrac{1}{2} t_1$. Deduce, then, that the tangent line is vertical when $t = 2n\pi$, where n is any integer.

29. A *hypocycloid* is the curve traced by a point P on a circle of radius b that is rolling inside a fixed circle of radius a, $a > b$. If the origin is at the center of the fixed circle, $A (a, 0)$ is one of the points at which the point P comes in contact with the fixed circle, B is the moving point of tangency of the two circles, and the parameter t is the number of radians in the angle AOB, prove that parametric equations of the hypocycloid are

$$x = (a - b) \cos t + b \cos \frac{a - b}{b} t \quad \text{and} \quad y = (a - b) \sin t - b \sin \frac{a - b}{b} t$$

30. If $a = 4b$ in Exercise 29, we have a *hypocycloid of four cusps*. Show that parametric equations of this curve are $x = a \cos^3 t$ and $y = a \sin^3 t$.

31. Use the parametric equations of Exercise 30 to find a cartesian equation of the hypocycloid of four cusps, and draw a sketch of the graph of the resulting equation.

32. Parametric equations for the *tractrix* are

$$x = t - a \tanh \frac{t}{a} \quad \text{and} \quad y = a \operatorname{sech} \frac{t}{a}$$

Draw a sketch of the curve for $a = 4$.

33. Prove that the parameter t in the parametric equations of a tractrix (see Exercise 32) is the x intercept of the tangent line.

34. Show that the tractrix of Exercise 32 is a curve such that the length of the segment of every tangent line from the point of tangency to the point of intersection with the x axis is constant and equal to a.

35. Find the area of the region bounded by the x axis and one arch of the cycloid having Eqs. (11).

36. Find the centroid of the region of Exercise 35.

13.4
CALCULUS OF VECTOR-VALUED FUNCTIONS

The definitions of limits, continuity, derivatives, and indefinite integrals of vector-valued functions involve the corresponding definitions for real-valued functions.

13.4.1 DEFINITION

Let $\mathbf{R}$ be a vector-valued function whose function values are given by

$$\mathbf{R}(t) = f(t)\mathbf{i} + g(t)\mathbf{j}$$

Then the **limit of $\mathbf{R}(t)$ as t approaches t_1** is defined by

$$\lim_{t \to t_1} \mathbf{R}(t) = \left[\lim_{t \to t_1} f(t) \right]\mathbf{i} + \left[\lim_{t \to t_1} g(t) \right]\mathbf{j}$$

if $\lim_{t \to t_1} f(t)$ and $\lim_{t \to t_1} g(t)$ both exist.

▶ ILLUSTRATION 1   If $\mathbf{R}(t) = \cos t\mathbf{i} + 2e^t\mathbf{j}$, then

$$\lim_{t \to 0} \mathbf{R}(t) = (\lim_{t \to 0} \cos t)\mathbf{i} + (\lim_{t \to 0} 2e^t)\mathbf{j}$$
$$= \mathbf{i} + 2\mathbf{j}$$

◀

13.4.2 DEFINITION

The vector-valued function $\mathbf{R}$ is **continuous** at t_1 if and only if the following three conditions are satisfied:

(i) $\mathbf{R}(t_1)$ exists;
(ii) $\lim_{t \to t_1} \mathbf{R}(t)$ exists;
(iii) $\lim_{t \to t_1} \mathbf{R}(t) = \mathbf{R}(t_1)$.

From Definitions 13.4.1 and 13.4.2 it follows that the vector-valued function $\mathbf{R}$, defined by $\mathbf{R}(t) = f(t)\mathbf{i} + g(t)\mathbf{j}$, is continuous at t_1 if and only if f and g are continuous there.

In the following definition the expression

$$\frac{\mathbf{R}(t + \Delta t) - \mathbf{R}(t)}{\Delta t}$$

is used to indicate the division of a vector by a scalar. This expression means

$$\frac{1}{\Delta t}[\mathbf{R}(t + \Delta t) - \mathbf{R}(t)]$$

13.4.3 DEFINITION

If $\mathbf{R}$ is a vector-valued function, then the **derivative** of $\mathbf{R}$ is a vector-valued function, denoted by $\mathbf{R}'$ and defined by

$$\mathbf{R}'(t) = \lim_{\Delta t \to 0} \frac{\mathbf{R}(t + \Delta t) - \mathbf{R}(t)}{\Delta t}$$

if this limit exists.

The notation $D_t\mathbf{R}(t)$ is sometimes used in place of $\mathbf{R}'(t)$.

The following theorem follows from Definition 13.4.3 and the definition of the derivative of a real-valued function.

13.4.4 THEOREM   If **R** is a vector-valued function defined by

$$\mathbf{R}(t) = f(t)\mathbf{i} + g(t)\mathbf{j} \tag{1}$$

then

$$\mathbf{R}'(t) = f'(t)\mathbf{i} + g'(t)\mathbf{j}$$

if $f'(t)$ and $g'(t)$ exist.

PROOF   From Definition 13.4.3,

$$\mathbf{R}'(t) = \lim_{\Delta t \to 0} \frac{\mathbf{R}(t + \Delta t) - \mathbf{R}(t)}{\Delta t}$$

$$= \lim_{\Delta t \to 0} \frac{[f(t + \Delta t)\mathbf{i} + g(t + \Delta t)\mathbf{j}] - [f(t)\mathbf{i} + g(t)\mathbf{j}]}{\Delta t}$$

$$= \lim_{\Delta t \to 0} \frac{[f(t + \Delta t) - f(t)]}{\Delta t}\mathbf{i} + \lim_{\Delta t \to 0} \frac{[g(t + \Delta t) - g(t)]}{\Delta t}\mathbf{j}$$

$$= f'(t)\mathbf{i} + g'(t)\mathbf{j} \qquad\blacksquare$$

The direction of $\mathbf{R}'(t)$ is along the tangent line at the point $(f(t), g(t))$ to the graph of $\mathbf{R}(t)$ having vector equation (1). That is, the direction of $\mathbf{R}'(t)$ is given by θ $(0 \le \theta < 2\pi)$, where $\tan \theta = g'(t)/f'(t)$; that is,

$$\tan \theta = \frac{\dfrac{dy}{dt}}{\dfrac{dx}{dt}}$$

$$= \frac{dy}{dx}$$

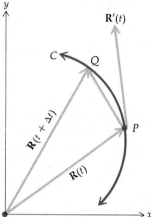

FIGURE 13.4.1

A geometric interpretation of Definition 13.4.3 is obtained by considering representations of the vectors $\mathbf{R}(t)$, $\mathbf{R}(t + \Delta t)$, and $\mathbf{R}'(t)$. Refer to Fig. 13.4.1. The curve C is traced by the endpoint of the position representation of $\mathbf{R}(t)$ as t assumes all values in the domain of **R**. Let $\overrightarrow{OP}$ be the position representation of $\mathbf{R}(t)$ and $\overrightarrow{OQ}$ be the position representation of $\mathbf{R}(t + \Delta T)$. Then $\mathbf{R}(t + \Delta t) - \mathbf{R}(t)$ is a vector for which $\overrightarrow{PQ}$ is a representation. If the vector $\mathbf{R}(t + \Delta t) - \mathbf{R}(t)$ is multiplied by the scalar $1/\Delta t$, we obtain a vector having the same direction and whose magnitude is $1/|\Delta t|$ times the magnitude of $\mathbf{R}(t + \Delta t) - \mathbf{R}(t)$. As Δt approaches zero, the vector $[\mathbf{R}(t + \Delta t) - \mathbf{R}(t)]/\Delta t$ approaches a vector having one of its representations tangent to the curve C at the point P.

▶ ILLUSTRATION 2   If $\mathbf{R}(t) = (2 + \sin t)\mathbf{i} + \cos t\mathbf{j}$, then

$$\mathbf{R}'(t) = \cos t\mathbf{i} - \sin t\mathbf{j}$$

◀

Higher order derivatives of vector-valued functions are defined as for higher order derivatives of real-valued functions. So if $\mathbf{R}$ is a vector-valued function defined by $\mathbf{R}(t) = f(t)\mathbf{i} + g(t)\mathbf{j}$, the second derivative of $\mathbf{R}$, denoted by $\mathbf{R}''(t)$, is given by

$$\mathbf{R}''(t) = D_t[\mathbf{R}'(t)]$$

The notation $D_t{}^2\mathbf{R}(t)$ can be used in place of $\mathbf{R}''(t)$. By applying Theorem 13.4.4 to $\mathbf{R}'(t)$,

$$\mathbf{R}''(t) = f''(t)\mathbf{i} + g''(t)\mathbf{j}$$

if $f''(t)$ and $g''(t)$ exist.

▶ ILLUSTRATION 3   If $\mathbf{R}(t) = (\ln t)\mathbf{i} + \left(\dfrac{1}{t}\right)\mathbf{j}$, then

$$\mathbf{R}'(t) = \frac{1}{t}\mathbf{i} - \frac{1}{t^2}\mathbf{j} \qquad \mathbf{R}''(t) = -\frac{1}{t^2}\mathbf{i} + \frac{2}{t^3}\mathbf{j} \qquad ◀$$

13.4.5 DEFINITION   A vector-valued function $\mathbf{R}$ is said to be **differentiable** on an interval if $\mathbf{R}'(t)$ exists for all values of t in the interval.

The following theorems give differentiation formulas for vector-valued functions. The proofs are based on Theorem 13.4.4 and theorems on differentiation of real-valued functions.

13.4.6 THEOREM   If $\mathbf{R}$ and $\mathbf{Q}$ are differentiable vector-valued functions on an interval, then $\mathbf{R} + \mathbf{Q}$ is differentiable on the interval, and

$$D_t[\mathbf{R}(t) + \mathbf{Q}(t)] = D_t\mathbf{R}(t) + D_t\mathbf{Q}(t)$$

The proof of this theorem is left as an exercise (see Exercise 25).

EXAMPLE 1
If $\mathbf{R}(t) = t^2\mathbf{i} + (t - 1)\mathbf{j}$ and $\mathbf{Q}(t) = \sin t\mathbf{i} + \cos t\mathbf{j}$, verify Theorem 13.4.6.

SOLUTION

$$\begin{aligned}
D_t[\mathbf{R}(t) + \mathbf{Q}(t)] &= D_t([t^2\mathbf{i} + (t - 1)\mathbf{j}] + [\sin t\mathbf{i} + \cos t\mathbf{j}]) \\
&= D_t[(t^2 + \sin t)\mathbf{i} + (t - 1 + \cos t)\mathbf{j}] \\
&= (2t + \cos t)\mathbf{i} + (1 - \sin t)\mathbf{j} \\
D_t\mathbf{R}(t) + D_t\mathbf{Q}(t) &= D_t[t^2\mathbf{i} + (t - 1)\mathbf{j}] + D_t(\sin t\mathbf{i} + \cos t\mathbf{j}) \\
&= (2t\mathbf{i} + \mathbf{j}) + (\cos t\mathbf{i} - \sin t\mathbf{j}) \\
&= (2t + \cos t)\mathbf{i} + (1 - \sin t)\mathbf{j}
\end{aligned}$$

Hence $D_t[\mathbf{R}(t) + \mathbf{Q}(t)] = D_t\mathbf{R}(t) + D_t\mathbf{Q}(t)$.

13.4.7 THEOREM   If $\mathbf{R}$ and $\mathbf{Q}$ are differentiable vector-valued functions on an interval, then $\mathbf{R} \cdot \mathbf{Q}$ is differentiable on the interval, and

$$D_t[\mathbf{R}(t) \cdot \mathbf{Q}(t)] = [D_t\mathbf{R}(t)] \cdot \mathbf{Q}(t) + \mathbf{R}(t) \cdot [D_t\mathbf{Q}(t)]$$

PROOF   Let $\mathbf{R}(t) = f_1(t)\mathbf{i} + g_1(t)\mathbf{j}$ and $\mathbf{Q}(t) = f_2(t)\mathbf{i} + g_2(t)\mathbf{j}$. Then by Theorem 13.4.4,

$$D_t\mathbf{R}(t) = f_1'(t)\mathbf{i} + g_1'(t)\mathbf{j} \qquad D_t\mathbf{Q}(t) = f_2'(t)\mathbf{i} + g_2'(t)\mathbf{j}$$

$$\mathbf{R}(t) \cdot \mathbf{Q}(t) = [f_1(t)][f_2(t)] + [g_1(t)][g_2(t)]$$

Therefore

$$
\begin{aligned}
D_t[\mathbf{R}(t) \cdot \mathbf{Q}(t)] \\
&= [f_1'(t)][f_2(t)] + [f_1(t)][f_2'(t)] + [g_1'(t)][g_2(t)] + [g_1(t)][g_2'(t)] \\
&= \{[f_1'(t)][f_2(t)] + [g_1'(t)][g_2(t)]\} + \{[f_1(t)][f_2'(t)] + [g_1(t)][g_2'(t)]\} \\
&= [D_t\mathbf{R}(t)] \cdot \mathbf{Q}(t) + \mathbf{R}(t) \cdot [D_t\mathbf{Q}(t)] \qquad\blacksquare
\end{aligned}
$$

EXAMPLE 2
Verify Theorem 13.4.7 for the vectors of Example 1.

SOLUTION
The vectors are

$$\mathbf{R}(t) = t^2\mathbf{i} + (t-1)\mathbf{j} \qquad \mathbf{Q}(t) = \sin t\,\mathbf{i} + \cos t\,\mathbf{j}$$

Thus $\mathbf{R}(t) \cdot \mathbf{Q}(t) = t^2 \sin t + (t-1) \cos t$. Therefore

$$
\begin{aligned}
D_t[\mathbf{R}(t) \cdot \mathbf{Q}(t)] &= 2t \sin t + t^2 \cos t + \cos t + (t-1)(-\sin t) \\
&= (t+1) \sin t + (t^2 + 1) \cos t \qquad\qquad (2)
\end{aligned}
$$

Because $D_t\mathbf{R}(t) = 2t\mathbf{i} + \mathbf{j}$ and $D_t\mathbf{Q}(t) = \cos t\,\mathbf{i} - \sin t\,\mathbf{j}$, we have

$$
\begin{aligned}
[D_t\mathbf{R}(t)] \cdot \mathbf{Q}(t) &+ \mathbf{R}(t) \cdot [D_t\mathbf{Q}(t)] \\
&= (2t\mathbf{i} + \mathbf{j}) \cdot (\sin t\,\mathbf{i} + \cos t\,\mathbf{j}) + [t^2\mathbf{i} + (t-1)\mathbf{j}] \cdot (\cos t\,\mathbf{i} - \sin t\,\mathbf{j}) \\
&= (2t \sin t + \cos t) + [t^2 \cos t - (t-1) \sin t] \\
&= (t+1) \sin t + (t^2 + 1) \cos t \qquad\qquad (3)
\end{aligned}
$$

Comparing (2) and (3) we see that Theorem 13.4.7 holds for these vectors.

13.4.8 THEOREM   If $\mathbf{R}$ is a differentiable vector-valued function on an interval and f is a differentiable real-valued function on the interval, then

$$D_t\{[f(t)][\mathbf{R}(t)]\} = [D_t f(t)]\mathbf{R}(t) + f(t)\,D_t\mathbf{R}(t)$$

The proof is left as an exercise (see Exercise 26).

We need to apply the following theorem in some future discussions. It is the chain-rule for vector-valued functions. The proof, which is left as an exercise (see Exercise 27), is based on Theorem 13.4.4 and the chain rule for real-valued functions.

13.4.9 THEOREM   Suppose that $\mathbf{F}$ is a vector-valued function, h is a real-valued function, and $\mathbf{G}$ is the vector-valued function defined by $\mathbf{G}(t) = \mathbf{F}(h(t))$. If $\phi = h(t)$ and both $\dfrac{d\phi}{dt}$ and $D_\phi\mathbf{G}(t)$ exist, then $D_t\mathbf{G}(t)$ exists and is given by

$$D_t\mathbf{G}(t) = [D_\phi\mathbf{G}(t)]\frac{d\phi}{dt} \qquad\qquad (4)$$

▶ ILLUSTRATION 4   Let the functions $\mathbf{F}$ and h of Theorem 13.4.9 be defined by

$$\mathbf{F}(\phi) = \phi^2\mathbf{i} + e^{\phi}\mathbf{j} \quad \text{and} \quad h(t) = \sin t$$

If $\phi = h(t)$ and $\mathbf{G}(t) = \mathbf{F}(h(t))$, we have

$$\phi = \sin t \quad \text{and} \quad \mathbf{G}(t) = \sin^2 t\mathbf{i} + e^{\sin t}\mathbf{j}$$

Computing $D_t\mathbf{G}(t)$ by Theorem 13.4.4 we have

$$D_t\mathbf{G}(t) = 2\sin t\cos t\mathbf{i} + e^{\sin t}\cos t\mathbf{j} \tag{5}$$

We show that we get the same result if we apply Theorem 13.4.9. Because $\mathbf{G}(t)$ can also be written as $\phi^2\mathbf{i} + e^{\phi}\mathbf{j}$, we have

$$\begin{aligned} D_{\phi}\mathbf{G}(t) &= D_{\phi}[\phi^2\mathbf{i} + e^{\phi}\mathbf{j}] \\ &= 2\phi\mathbf{i} + e^{\phi}\mathbf{j} \end{aligned}$$

But $\phi = \sin t$; thus

$$D_{\phi}\mathbf{G}(t) = 2\sin t\mathbf{i} + e^{\sin t}\mathbf{j} \tag{6}$$

and

$$\frac{d\phi}{dt} = \cos t \tag{7}$$

Substituting from (6) and (7) into the right side of (4) we have

$$\begin{aligned} D_t\mathbf{G}(t) &= [2\sin t\mathbf{i} + e^{\sin t}\mathbf{j}]\cos t \\ &= 2\sin t\cos t\mathbf{i} + e^{\sin t}\cos t\mathbf{j} \end{aligned}$$

which agrees with (5).   ◀

We now define an indefinite integral (or antiderivative) of a vector-valued function.

13.4.10 DEFINITION   If $\mathbf{Q}$ is the vector-valued function given by

$$\mathbf{Q}(t) = f(t)\mathbf{i} + g(t)\mathbf{j}$$

then the **indefinite integral of $\mathbf{Q}(t)$** is defined by

$$\int \mathbf{Q}(t)\,dt = \mathbf{i}\int f(t)\,dt + \mathbf{j}\int g(t)\,dt \tag{8}$$

This definition is consistent with the definition of an indefinite integral of a real-valued function because if we take the derivative on both sides of (8) with respect to t,

$$D_t\int \mathbf{Q}(t)\,dt = \mathbf{i}D_t\int f(t)\,dt + \mathbf{j}D_t\int g(t)\,dt$$

$$D_t\int \mathbf{Q}(t)\,dt = \mathbf{i}f(t) + \mathbf{j}g(t)$$

For each of the indefinite integrals on the right side of (8) there occurs an arbitrary scalar constant. When each of these scalars is multiplied by either $\mathbf{i}$ or $\mathbf{j}$, there occurs an arbitrary constant vector in the sum. So

$$\int \mathbf{Q}(t)\,dt = \mathbf{R}(t) + \mathbf{C}$$

where $D_t\mathbf{R}(t) = \mathbf{Q}(t)$ and $\mathbf{C}$ is an arbitrary constant vector.

EXAMPLE 3
Find the most general vector-valued function whose derivative is

$$\mathbf{Q}(t) = \sin t\mathbf{i} - 3\cos t\mathbf{j}$$

SOLUTION
If $D_t\mathbf{R}(t) = \mathbf{Q}(t)$, then $\mathbf{R}(t) = \int \mathbf{Q}(t)\,dt$; that is,

$$\mathbf{R}(t) = \mathbf{i}\int \sin t\,dt - 3\mathbf{j}\int \cos t\,dt$$

$$= \mathbf{i}(-\cos t + C_1) - 3\mathbf{j}(\sin t + C_2)$$

$$= -\cos t\mathbf{i} - 3\sin t\mathbf{j} + (C_1\mathbf{i} - 3C_2\mathbf{j})$$

$$= -\cos t\mathbf{i} - 3\sin t\mathbf{j} + \mathbf{C}$$

EXAMPLE 4
Find the vector $\mathbf{R}(t)$ for which

$$D_t\mathbf{R}(t) = e^{-t}\mathbf{i} + e^t\mathbf{j} \quad \text{and} \quad \mathbf{R}(0) = \mathbf{i} + \mathbf{j}$$

SOLUTION

$$\mathbf{R}(t) = \mathbf{i}\int e^{-t}\,dt + \mathbf{j}\int e^t\,dt$$

$$= \mathbf{i}(-e^{-t} + C_1) + \mathbf{j}(e^t + C_2)$$

Because $\mathbf{R}(0) = \mathbf{i} + \mathbf{j}$,

$$\mathbf{i} + \mathbf{j} = \mathbf{i}(-1 + C_1) + \mathbf{j}(1 + C_2)$$

Therefore

$$C_1 - 1 = 1 \qquad C_2 + 1 = 1$$

$$C_1 = 2 \qquad\qquad C_2 = 0$$

Hence

$$\mathbf{R}(t) = (-e^{-t} + 2)\mathbf{i} + e^t\mathbf{j}$$

The following theorem will be useful later.

13.4.11 THEOREM   If $\mathbf{R}$ is a differentiable vector-valued function on an interval and $\|\mathbf{R}(t)\|$ is constant for all t in the interval, then the vectors $\mathbf{R}(t)$ and $D_t\mathbf{R}(t)$ are orthogonal.

PROOF   Let $\|\mathbf{R}(t)\| = k$. Then by Theorem 13.2.3(iii),

$$\mathbf{R}(t) \cdot \mathbf{R}(t) = k^2$$

FIGURE 13.4.2

Differentiating on both sides with respect to t and using Theorem 13.4.7 we obtain

$$[D_t\mathbf{R}(t)] \cdot \mathbf{R}(t) + \mathbf{R}(t) \cdot [D_t\mathbf{R}(t)] = 0$$

$$2\mathbf{R}(t) \cdot D_t\mathbf{R}(t) = 0$$

Because the dot product of $\mathbf{R}(t)$ and $D_t\mathbf{R}(t)$ is zero, it follows from Definition 13.2.7 that $\mathbf{R}(t)$ and $D_t\mathbf{R}(t)$ are orthogonal.   ∎

The geometric interpretation of Theorem 13.4.11 is evident. If the vector $\mathbf{R}(t)$ has constant magnitude, then the position representation $\overrightarrow{OP}$ of $\mathbf{R}(t)$ has its terminal point P on the circle with its center at the origin and radius k. So the graph of $\mathbf{R}$ is this circle. Because $D_t\mathbf{R}(t)$ and $\mathbf{R}(t)$ are orthogonal, $\overrightarrow{OP}$ is perpendicular to a representation of $D_t\mathbf{R}(t)$. Figure 13.4.2 shows a sketch of a quarter circle, the position representation $\overrightarrow{OP}$ of $\mathbf{R}(t)$, and the representation $\overrightarrow{PB}$ of $D_t\mathbf{R}(t)$.

EXERCISES 13.4

In Exercises 1 through 5, find the indicated limit, if it exists.

1. $\mathbf{R}(t) = (3t - 2)\mathbf{i} + t^2\mathbf{j}$; $\lim\limits_{t \to 2} \mathbf{R}(t)$ **2.** $\mathbf{R}(t) = (t - 2)\mathbf{i} + \dfrac{t^2 - 4}{t - 2}\mathbf{j}$; $\lim\limits_{t \to 2} \mathbf{R}(t)$ **3.** $\mathbf{R}(t) = 2\sin t\mathbf{i} + \cos t\mathbf{j}$; $\lim\limits_{t \to \pi/2} \mathbf{R}(t)$

4. $\mathbf{R}(t) = \dfrac{t^2 - 2t - 3}{t - 3}\mathbf{i} + \dfrac{t^2 - 5t + 6}{t - 3}\mathbf{j}$; $\lim\limits_{t \to 3} \mathbf{R}(t)$ **5.** $\mathbf{R}(t) = e^{t+1}\mathbf{i} + |t + 1|\mathbf{j}$; $\lim\limits_{t \to -1} \mathbf{R}(t)$

In Exercises 6 through 14, find $\mathbf{R}'(t)$ and $\mathbf{R}''(t)$.

6. $\mathbf{R}(t) = (t^2 - 3)\mathbf{i} + (2t + 1)\mathbf{j}$ **7.** $\mathbf{R}(t) = e^{2t}\mathbf{i} + \ln t\mathbf{j}$ **8.** $\mathbf{R}(t) = \cos 2t\mathbf{i} + \tan t\mathbf{j}$

9. $\mathbf{R}(t) = \tan^{-1} t\mathbf{i} + 2^t\mathbf{j}$ **10.** $\mathbf{R}(t) = \dfrac{t - 1}{t + 1}\mathbf{i} + \dfrac{t - 2}{t}\mathbf{j}$ **11.** $\mathbf{R}(t) = (t^2 + 4)^{-1}\mathbf{i} + \sqrt{1 - 5t}\mathbf{j}$

12. $\mathbf{R}(t) = \sqrt{2t + 1}\mathbf{i} + (t - 1)^2\mathbf{j}$ **13.** $\mathbf{R}(t) = 5\sin 2t\mathbf{i} - \sec 4t\mathbf{j}$ **14.** $\mathbf{R}(t) = (e^{3t} + 2)\mathbf{i} + 2e^{3t}\mathbf{j}$

In Exercises 15 and 16, find $D_t\|\mathbf{R}(t)\|$.
15. $\mathbf{R}(t) = (t - 1)\mathbf{i} + (2 - t)\mathbf{j}$ **16.** $\mathbf{R}(t) = (e^t + 1)\mathbf{i} + (e^t - 1)\mathbf{j}$

In Exercises 17 and 18, verify Theorem 13.4.6 for the given vectors.
17. $\mathbf{R}(t) = (t^2 + e^t)\mathbf{i} + (t - e^{2t})\mathbf{j}$; $\mathbf{Q}(t) = (t^3 + 2e^t)\mathbf{i} - (3t + e^{2t})\mathbf{j}$ **18.** $\mathbf{R}(t) = \cos 2t\mathbf{i} - \sin 2t\mathbf{j}$; $\mathbf{Q}(t) = \sin^2 t\mathbf{i} + \cos 2t\mathbf{j}$

In Exercises 19 and 20, verify Theorem 13.4.7 for the vectors of the indicated exercise.
19. Exercise 17   **20.** Exercise 18

In Exercises 21 through 24, find $\mathbf{R}'(t) \cdot \mathbf{R}''(t)$.
21. $\mathbf{R}(t) = (2t^2 - 1)\mathbf{i} + (t^2 + 3)\mathbf{j}$ **22.** $\mathbf{R}(t) = \ln(t - 1)\mathbf{i} - 3t^{-1}\mathbf{j}$
23. $\mathbf{R}(t) = e^{2t}\mathbf{i} + e^{-2t}\mathbf{j}$ **24.** $\mathbf{R}(t) = -\cos 2t\mathbf{i} + \sin 2t\mathbf{j}$
25. Prove Theorem 13.4.6.   **26.** Prove Theorem 13.4.8.   **27.** Prove Theorem 13.4.9.

In Exercises 28 through 33, find the most general vector whose derivative has the given function value.

28. $(t^2 - 9)\mathbf{i} + (2t - 5)\mathbf{j}$

29. $\tan t\mathbf{i} - \dfrac{1}{t}\mathbf{j}$

30. $3^t\mathbf{i} - 2^t\mathbf{j}$

31. $e^{3t}\mathbf{i} + \dfrac{1}{t - 1}\mathbf{j}$

32. $\dfrac{1}{4 + t^2}\mathbf{i} - \dfrac{4}{1 - t^2}\mathbf{j}$

33. $\ln t\mathbf{i} + t^2\mathbf{j}$

34. If $\mathbf{R}'(t) = t^2\mathbf{i} + \dfrac{1}{t - 2}\mathbf{j}$, and $\mathbf{R}(3) = 2\mathbf{i} - 5\mathbf{j}$, find $\mathbf{R}(t)$.

35. If $\mathbf{R}'(t) = \sin^2 t\mathbf{i} + 2\cos^2 t\mathbf{j}$, and $\mathbf{R}(\pi) = \mathbf{0}$, find $\mathbf{R}(t)$.

36. If $\mathbf{R}'(t) = e^t \sin t\mathbf{i} + e^t \cos t\mathbf{j}$, and $\mathbf{R}(0) = \mathbf{i} - \mathbf{j}$, find $\mathbf{R}(t)$.

In Exercises 37 and 38, for the given vector equation find a cartesian equation of the curve that is traced by the endpoint of the position representation of $\mathbf{R}'(t)$. Find $\mathbf{R}(t) \cdot \mathbf{R}'(t)$. Interpret the result geometrically.

37. $\mathbf{R}(t) = \cos t\mathbf{i} + \sin t\mathbf{j}$

38. $\mathbf{R}(t) = \cosh t\mathbf{i} - \sinh t\mathbf{j}$

In Exercises 39 and 40, if $\alpha(t)$ is the radian measure of the angle between $\mathbf{R}(t)$ and $\mathbf{Q}(t)$, find $D_t\alpha(t)$.

39. $\mathbf{R}(t) = 3e^{2t}\mathbf{i} - 4e^{2t}\mathbf{j}$ and $\mathbf{Q}(t) = 6e^{3t}\mathbf{j}$

40. $\mathbf{R}(t) = 2t\mathbf{i} + (t^2 - 1)\mathbf{j}$ and $\mathbf{Q}(t) = 3t\mathbf{i}$

41. Suppose that $\mathbf{R}$ and $\mathbf{R}'$ are vector-valued functions defined on an interval and $\mathbf{R}'$ is differentiable on the interval. Prove

$$D_t[\mathbf{R}'(t) \cdot \mathbf{R}(t)] = \|\mathbf{R}'(t)\|^2 + \mathbf{R}(t) \cdot \mathbf{R}''(t)$$

42. If $\|\mathbf{R}(t)\| = h(t)$, prove that $\mathbf{R}(t) \cdot \mathbf{R}'(t) = [h(t)][h'(t)]$.

43. If the vector-valued function $\mathbf{R}$ and the real-valued function f are both differentiable on an interval and $f(t) \neq 0$ on the interval, prove that $\mathbf{R}/f$ is also differentiable on the interval and

$$D_t\left[\frac{\mathbf{R}(t)}{f(t)}\right] = \frac{f(t)\mathbf{R}'(t) - f'(t)\mathbf{R}(t)}{[f(t)]^2}$$

44. Prove that if $\mathbf{A}$ and $\mathbf{B}$ are constant vectors and f and g are integrable functions, then

$$\int [\mathbf{A}f(t) + \mathbf{B}g(t)]\,dt = \mathbf{A}\int f(t)\,dt + \mathbf{B}\int g(t)\,dt$$

(*Hint:* Express $\mathbf{A}$ and $\mathbf{B}$ in terms of $\mathbf{i}$ and $\mathbf{j}$.)

13.5
LENGTH OF ARC

In Section 5.5 we obtained a formula for finding the length of arc of the graph of a function. Such a graph is a special kind of curve because the graph of a function cannot be intersected by a vertical line in more than one point.

We now develop a method for finding the length of arc of some other kinds of curves. Let C be the curve having parametric equations

$$x = f(t) \quad \text{and} \quad y = g(t) \tag{1}$$

and suppose that f and g are continuous on the closed interval $[a, b]$. We wish to assign a number L to represent the number of units in the length of arc of C from $t = a$ to $t = b$. We proceed as in Section 5.5.

Let Δ be a partition of the closed interval $[a, b]$ formed by dividing the interval into n subintervals by choosing $n - 1$ numbers between a and b. Let $t_0 = a$ and $t_n = b$, and let $t_1, t_2, \ldots, t_{n-1}$ be intermediate numbers:

$$t_0 < t_1 < \ldots < t_{n-1} < t_n$$

FIGURE 13.5.1

The ith subinterval is $[t_{i-1}, t_i]$ and the number of units in its length, denoted by $\Delta_i t$, is $t_i - t_{i-1}$, where $i = 1, 2, \ldots, n$. Let $\|\Delta\|$ be the norm of the partition; so each $\Delta_i t \le \|\Delta\|$.

Associated with each number t_i is a point $P_i(f(t_i), g(t_i))$ on C. From each point P_{i-1} draw a line segment to the next point P_i. See Fig. 13.5.1. The number of units in the length of the line segment from P_{i-1} to P_i is denoted by $|\overline{P_{i-1}P_i}|$. From the distance formula we have

$$|\overline{P_{i-1}P_i}| = \sqrt{[f(t_i) - f(t_{i-1})]^2 + [g(t_i) - g(t_{i-1})]^2} \tag{2}$$

The sum of the numbers of units of lengths of the n line segments is

$$\sum_{i=1}^{n} |\overline{P_{i-1}P_i}| \tag{3}$$

Our intuitive notion of the length of the arc from $t = a$ to $t = b$ leads us to define the number of units of the length of arc as the limit of the sum in (3) as $\|\Delta\|$ approaches zero.

13.5.1 DEFINITION   Let the curve C have parametric equations $x = f(t)$ and $y = g(t)$. Suppose there exists a number L having the following property: For any $\epsilon > 0$ there is a $\delta > 0$ such that for every partition Δ of the interval $[a, b]$ for which $\|\Delta\| < \delta$, then

$$\left| \sum_{i=1}^{n} |\overline{P_{i-1}P_i}| - L \right| < \epsilon$$

Then we write

$$L = \lim_{\|\Delta\| \to 0} \sum_{i=1}^{n} |\overline{P_{i-1}P_i}| \tag{4}$$

and L units is called the **length of arc** of the curve C from the point $(f(a), g(a))$ to the point $(f(b), g(b))$.

The arc of the curve is rectifiable if the limit in (4) exists.

If f' and g' are continuous on $[a, b]$, we proceed as follows to find a formula for evaluating the limit in (4).

Because f' and g' are continuous on $[a, b]$, they are continuous on each subinterval of the partition Δ. So the hypothesis of the mean-value theorem (Theorem 3.3.2) is satisfied by f and g on each $[t_{i-1}, t_i]$; therefore there are numbers z_i and w_i in the open interval (t_{i-1}, t_i) such that

$$f(t_i) - f(t_{i-1}) = f'(z_i)\Delta_i t \quad \text{and} \quad g(t_i) - g(t_{i-1}) = g'(w_i)\Delta_i t \tag{5}$$

Substituting from (5) into (2) we obtain

$$|\overline{P_{i-1}P_i}| = \sqrt{[f'(z_i)\Delta_i t]^2 + [g'(w_i)\Delta_i t]^2}$$
$$|\overline{P_{i-1}P_i}| = \sqrt{[f'(z_i)]^2 + [g'(w_i)]^2}\,\Delta_i t \tag{6}$$

where z_i and w_i are in the open interval (t_{i-1}, t_i). Then from (4) and (6), if the limit exists,

$$L = \lim_{\|\Delta\| \to 0} \sum_{i=1}^{n} \sqrt{[f'(z_i)]^2 + [g'(w_i)]^2}\,\Delta_i t \tag{7}$$

The sum in (7) is not a Riemann sum because z_i and w_i are not necessarily the same numbers. So we cannot apply the definition of a definite integral to evaluate the limit in (7). However, there is a theorem that can be applied to evaluate this limit. We state the thoerem, but a proof is not given because it is beyond the scope of this book. You can find a proof in an advanced calculus text.

13.5.2 THEOREM   If the functions F and G are continuous on the closed interval $[a, b]$, then the function $\sqrt{F^2 + G^2}$ is also continuous on $[a, b]$, and if Δ is a partition of the interval $[a, b]$ ($\Delta: a = t_0 < t_1 < \ldots < t_{i-1} < t_i < \ldots < t_n = b$), and z_i and w_i are any numbers in (t_{i-1}, t_i), then

$$\lim_{\|\Delta\| \to 0} \sum_{i=1}^{n} \sqrt{[F(z_i)]^2 + [G(w_i)]^2}\,\Delta_i t = \int_a^b \sqrt{[F(t)]^2 + [G(t)]^2}\,dt \tag{8}$$

Applying (8) to (7), where F is f' and G is g', we have

$$L = \int_a^b \sqrt{[f'(t)]^2 + [g'(t)]^2}\,dt$$

This result is stated as a theorem.

13.5.3 THEOREM   Let the curve C have parametric equations $x = f(t)$ and $y = g(t)$, and suppose that f' and g' are continuous on the closed interval $[a, b]$. Then if L units is the length of arc of the curve C from the point $(f(a), g(a))$ to the point $(f(b), g(b))$,

$$L = \int_a^b \sqrt{[f'(t)]^2 + [g'(t)]^2}\,dt \tag{9}$$

EXAMPLE 1
Find the length of the arc of the curve having parametric equations

$$x = t^3 \quad \text{and} \quad y = 2t^2$$

in each of the following cases: (a) from $t = 0$ to $t = 1$; (b) from $t = -2$ to $t = 0$.

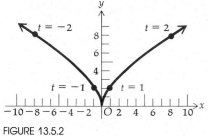

FIGURE 13.5.2

SOLUTION

A sketch of the curve appears in Fig. 13.5.2. Let

$$f(t) = t^3 \qquad g(t) = 2t^2$$
$$f'(t) = 3t^2 \qquad g'(t) = 4t$$

The curve has parametric equations $x = f(t)$ and $y = g(t)$. We apply Theorem 13.5.3 in parts (a) and (b) where L_a units is the length of the arc from $t = 0$ to $t = 1$ and L_b units is the length of the arc from $t = -2$ to $t = 0$.

(a) $L_a = \int_0^1 \sqrt{9t^4 + 16t^2}\, dt$ (b) $L_b = \int_{-2}^0 \sqrt{9t^4 + 16t^2}\, dt$

$\qquad = \int_0^1 \sqrt{t^2}\sqrt{9t^2 + 16}\, dt$ $\qquad = \int_{-2}^0 \sqrt{t^2}\sqrt{9t^2 + 16}\, dt$

$\qquad = \int_0^1 t\sqrt{9t^2 + 16}\, dt$ $\qquad = \int_{-2}^0 -t\sqrt{9t^2 + 16}\, dt$

$\qquad = \frac{1}{18} \cdot \frac{2}{3}(9t^2 + 16)^{3/2}\Big]_0^1$ $\qquad = -\frac{1}{18} \cdot \frac{2}{3}(9t^2 + 16)^{3/2}\Big]_{-2}^0$

$\qquad = \frac{1}{27}[(25)^{3/2} - (16)^{3/2}]$ $\qquad = -\frac{1}{27}[(16)^{3/2} - (52)^{3/2}]$

$\qquad = \frac{1}{27}(125 - 64)$ $\qquad = \frac{1}{27}(104\sqrt{13} - 64)$

$\qquad = \frac{61}{27}$ $\qquad \approx 11.5$

Observe in the third integral in part (a) we replaced $\sqrt{t^2}$ by t because $0 \le t \le 1$. However, in the third integral in part (b) we replaced $\sqrt{t^2}$ by $-t$ because $-2 \le t \le 0$.

For the curve C having parametric equations (1), let s units be the length of arc of C from the point $(f(t_0), g(t_0))$ to the point $(f(t), g(t))$, and let s increase as t increases. Then s is a function of t and is given by

$$s = \int_{t_0}^t \sqrt{[f'(u)]^2 + [g'(u)]^2}\, du \tag{10}$$

From the first fundamental theorem of the calculus (Theorem 4.8.1),

$$\frac{ds}{dt} = \sqrt{[f'(t)]^2 + [g'(t)]^2} \tag{11}$$

A vector equation of C is

$$\mathbf{R}(t) = f(t)\mathbf{i} + g(t)\mathbf{j} \tag{12}$$

Because

$$\mathbf{R}'(t) = f'(t)\mathbf{i} + g'(t)\mathbf{j}$$

then

$$\|\mathbf{R}'(t)\| = \sqrt{[f'(t)]^2 + [g'(t)]^2} \tag{13}$$

Substituting from (13) into (11) we obtain

$$\|\mathbf{R}'(t)\| = \frac{ds}{dt} \tag{14}$$

From (14) it follows that if s units is the length of arc of curve C having vector equation (12) measured from some fixed point to the point $(f(t), g(t))$ where s increases as t increases, then the derivative of s with respect to t is the magnitude of the derivative of the position vector at the point $(f(t), g(t))$.

We substitute from (13) into (9) and obtain $L = \int_a^b \|\mathbf{R}'(t)\| \, dt$. Thus Theorem 13.5.3 can be stated in terms of vectors in the following way.

13.5.4 THEOREM   Let the curve C have the vector equation $\mathbf{R}(t) = f(t)\mathbf{i} + g(t)\mathbf{j}$, and suppose that f' and g' are continuous on the closed interval $[a, b]$. Then the length of arc of C, traced by the terminal point of the position representation of $\mathbf{R}(t)$ as t increases from a to b, is determined by

$$L = \int_a^b \|\mathbf{R}'(t)\| \, dt \tag{15}$$

EXAMPLE 2

Find the length of the arc traced by the terminal point of the position representation of $\mathbf{R}(t)$ as t increases from 1 to 4 if

$$\mathbf{R}(t) = e^t \sin t\mathbf{i} + e^t \cos t\mathbf{j}$$

SOLUTION

$$\mathbf{R}'(t) = (e^t \sin t + e^t \cos t)\mathbf{i} + (e^t \cos t - e^t \sin t)\mathbf{j}$$

$$\begin{aligned}
\|\mathbf{R}'(t)\| &= \sqrt{(e^t \sin t + e^t \cos t)^2 + (e^t \cos t - e^t \sin t)^2} \\
&= \sqrt{e^{2t}}\sqrt{\sin^2 t + 2 \sin t \cos t + \cos^2 t + \cos^2 t - 2 \sin t \cos t + \sin^2 t} \\
&= e^t \sqrt{2}
\end{aligned}$$

From (15),

$$\begin{aligned}
L &= \int_1^4 \sqrt{2} e^t \, dt \\
&= \sqrt{2} e^t \Big]_1^4 \\
&= \sqrt{2}(e^4 - e)
\end{aligned}$$

An alternate form of formula (9) for the length of arc of a curve C, having parametric equations $x = f(t)$ and $y = g(t)$, is obtained by replacing $f'(t)$ by $\dfrac{dx}{dt}$ and $g'(t)$ by $\dfrac{dy}{dt}$. This form is

$$L = \int_a^b \sqrt{\left(\frac{dx}{dt}\right)^2 + \left(\frac{dy}{dt}\right)^2} \, dt \tag{16}$$

Now suppose that we wish to find the length of arc of a curve C whose polar equation is $r = F(\theta)$. If (x, y) is the cartesian representation of a point P on C and (r, θ) is a polar representation of P, then

$$x = r \cos \theta \quad \text{and} \quad y = r \sin \theta \tag{17}$$

Replacing r by $F(\theta)$ in Eqs. (17) we have

$$x = F(\theta) \cos \theta \quad \text{and} \quad y = F(\theta) \sin \theta \tag{18}$$

Equations (18) can be considered as parametric equations of C where θ is the parameter instead of t. Therefore, if F' is continuous on the closed interval $[\alpha, \beta]$, the formula for the length of arc of the curve C whose polar equation is $r = F(\theta)$ is obtained from (16) by taking $t = \theta$. So

$$L = \int_\alpha^\beta \sqrt{\left(\frac{dx}{d\theta}\right)^2 + \left(\frac{dy}{d\theta}\right)^2}\, d\theta \tag{19}$$

From (17),

$$\frac{dx}{d\theta} = \cos \theta \frac{dr}{d\theta} - r \sin \theta \quad \text{and} \quad \frac{dy}{d\theta} = \sin \theta \frac{dr}{d\theta} + r \cos \theta$$

Therefore

$$\sqrt{\left(\frac{dx}{d\theta}\right)^2 + \left(\frac{dy}{d\theta}\right)^2} = \sqrt{\left(\cos \theta \frac{dr}{d\theta} - r \sin \theta\right)^2 + \left(\sin \theta \frac{dr}{d\theta} + r \cos \theta\right)^2}$$

$$= \sqrt{\cos^2 \theta \left(\frac{dr}{d\theta}\right)^2 - 2r \sin \theta \cos \theta \frac{dr}{d\theta} + r^2 \sin^2 \theta + \sin^2 \theta \left(\frac{dr}{d\theta}\right)^2 + 2r \sin \theta \cos \theta \frac{dr}{d\theta} + r^2 \cos^2 \theta}$$

$$= \sqrt{(\cos^2 \theta + \sin^2 \theta)\left(\frac{dr}{d\theta}\right)^2 + (\sin^2 \theta + \cos^2 \theta)r^2}$$

$$= \sqrt{\left(\frac{dr}{d\theta}\right)^2 + r^2}$$

Substituting this into (19) we obtain

$$L = \int_\alpha^\beta \sqrt{\left(\frac{dr}{d\theta}\right)^2 + r^2}\, d\theta \tag{20}$$

EXAMPLE 3
Find the length of the cardioid $r = 2(1 + \cos \theta)$.

SOLUTION
A sketch of the curve is shown in Fig. 13.5.3. To obtain the length of the entire curve we can let θ take on values from 0 to 2π or we can make use of the symmetry of the curve and find half the length by letting θ take on values from 0 to π.

Because $r = 2(1 + \cos \theta)$, $\dfrac{dr}{d\theta} = -2 \sin \theta$. Substituting into (20), integrating from 0 to π, and multiplying by 2 we have

$$L = 2 \int_0^\pi \sqrt{(-2 \sin \theta)^2 + 4(1 + \cos \theta)^2}\, d\theta$$

$$= 4 \int_0^\pi \sqrt{\sin^2 \theta + 1 + 2 \cos \theta + \cos^2 \theta}\, d\theta$$

$$= 4\sqrt{2} \int_0^\pi \sqrt{1 + \cos \theta}\, d\theta$$

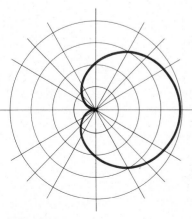

FIGURE 13.5.3

To evaluate this integral, use the identity $\cos^2 \frac{1}{2}\theta = \frac{1}{2}(1 + \cos\theta)$, which gives $\sqrt{1 + \cos\theta} = \sqrt{2} |\cos \frac{1}{2}\theta|$. Because $0 \leq \theta \leq \pi$, $0 \leq \frac{1}{2}\theta \leq \frac{1}{2}\pi$; thus $\cos \frac{1}{2}\theta \geq 0$. Therefore $\sqrt{1 + \cos\theta} = \sqrt{2} \cos \frac{1}{2}\theta$. So

$$L = 4\sqrt{2} \int_0^\pi \sqrt{2} \cos \frac{1}{2}\theta \, d\theta$$

$$= 16 \sin \frac{1}{2}\theta \Big]_0^\pi$$

$$= 16$$

EXERCISES 13.5

In Exercises 1 through 28, find the length of the given arc. When a appears, $a > 0$.

1. $x = \frac{1}{2}t^2 + t$, $y = \frac{1}{2}t^2 - t$; from $t = 0$ to $t = 1$.

2. $x = 3t^2$, $y = 2t^3$; from $t = 0$ to $t = 3$.

3. $x = t^2 + 2t$, $y = t^2 - 2t$; from $t = 0$ to $t = 2$.

4. $x = t^3$, $y = 3t^2$; from $t = -2$ to $t = 0$.

5. $\mathbf{R}(t) = 2t^2\mathbf{i} + 2t^3\mathbf{j}$; from $t = 1$ to $t = 2$.

6. $\mathbf{R}(t) = t\mathbf{i} + \cosh t\mathbf{j}$; from $t = 0$ to $t = 3$.

7. $\mathbf{R}(t) = 3e^{2t}\mathbf{i} - 4e^{2t}\mathbf{j}$; from $t = 0$ to $t = \ln 5$.

8. $\mathbf{R}(t) = (t^2 + 3)\mathbf{i} + 3t^2\mathbf{j}$; from $t = 1$ to $t = 4$.

9. $\mathbf{R}(t) = e^t \cos t\mathbf{i} + e^t \sin t\mathbf{j}$; from $t = 0$ to $t = 1$.

10. $\mathbf{R}(t) = \ln \sin t\mathbf{i} + (t + 1)\mathbf{j}$; from $t = \frac{1}{6}\pi$ to $t = \frac{1}{2}\pi$.

11. $\mathbf{R}(t) = \tan^{-1} t\mathbf{i} + \frac{1}{2}\ln(t^2 + 1)\mathbf{j}$; from $t = 0$ to $t = 1$.

12. $\mathbf{R}(t) = a(\cos t + t \sin t)\mathbf{i} + a(\sin t - t \cos t)\mathbf{j}$; from $t = 0$ to $t = \frac{1}{3}\pi$.

13. The entire hypocycloid of four cusps: $x = a \cos^3 t$, $y = a \sin^3 t$.

14. $x = e^{-t} \cos t$, $y = e^{-t} \sin t$; from $t = 0$ to $t = \pi$.

15. $x = 4 \sin 2t$, $y = 4 \cos 2t$; from $t = 0$ to $t = \pi$.

16. One arch of the cycloid: $x = a(t - \sin t)$, $y = a(1 - \cos t)$.

17. The tractrix $x = t - a \tanh \dfrac{t}{a}$, $y = a \operatorname{sech} \dfrac{t}{a}$ from $t = -a$ to $t = 2a$.

18. The circumference of the circle: $\mathbf{R}(t) = a \cos t\mathbf{i} + a \sin t\mathbf{j}$.

19. The circumference of the circle: $r = 5 \cos\theta$.

20. The circumference of the circle: $r = a \sin\theta$.

21. The circumference of the circle: $r = a$.

22. The entire curve: $r = 1 - \sin\theta$.

23. The entire curve: $r = 3 \cos^2 \frac{1}{2}\theta$.

24. $r = a\theta$; from $\theta = 0$ to $\theta = 2\pi$.

25. $r = e^{2\theta}$; from $\theta = 0$ to $\theta = 4$.

26. $r = a\theta^2$; from $\theta = 0$ to $\theta = \pi$.

27. $r = a \sin^3 \frac{1}{3}\theta$; from $\theta = 0$ to $\theta = \theta_1$.

28. $r = \sin^2 \frac{1}{2}\theta$; from $\theta = 0$ to $\theta = \frac{1}{2}\pi$.

29. Find the distance traveled by a thumbtack in the tread of a bicycle tire if the radius of the tire is 40 cm and the bicycle goes a distance of 50π m. (*Hint:* The path of the thumbtack is a cycloid.)

13.6

THE UNIT TANGENT AND UNIT NORMAL VECTORS AND ARC LENGTH AS A PARAMETER

A unit vector is a vector having a magnitude of 1, examples of which are the two unit vectors $\mathbf{i}$ and $\mathbf{j}$. With each point on a curve in the plane we now associate two other unit vectors, the *unit tangent vector* and the *unit normal vector*. These vectors occur in many applications of vector-valued functions.

13.6.1 DEFINITION   If $\mathbf{R}(t)$ is the position vector of curve C at a point P on C, then the **unit tangent vector** of C at P, denoted by $\mathbf{T}(t)$, is the unit vector in the direction of $D_t\mathbf{R}(t)$ if $D_t\mathbf{R}(t) \neq 0$.

The unit vector in the direction of $D_t\mathbf{R}(t)$ is given by $D_t\mathbf{R}(t)/\|D_t\mathbf{R}(t)\|$; thus

$$\mathbf{T}(t) = \frac{D_t\mathbf{R}(t)}{\|D_t\mathbf{R}(t)\|} \tag{1}$$

Because $\mathbf{T}(t)$ is a unit vector, it follows from Theorem 13.4.11 that $D_t\mathbf{T}(t)$ must be orthogonal to $\mathbf{T}(t)$. $D_t\mathbf{T}(t)$ is not necessarily a unit vector. However, the vector $D_t\mathbf{T}(t)/\|D_t\mathbf{T}(t)\|$ is of unit magnitude and has the same direction as $D_t\mathbf{T}(t)$. Therefore $D_t\mathbf{T}(t)/\|D_t\mathbf{T}(t)\|$ is a unit vector that is orthogonal to $\mathbf{T}(t)$, and it is called the *unit normal vector*.

13.6.2 DEFINITION   If $\mathbf{T}(t)$ is the unit tangent vector of curve C at a point P on C, then the **unit normal vector**, denoted by $\mathbf{N}(t)$, is the unit vector in the direction of $D_t\mathbf{T}(t)$.

From Definition 13.6.2 and the previous discussion,

$$\mathbf{N}(t) = \frac{D_t\mathbf{T}(t)}{\|D_t\mathbf{T}(t)\|} \tag{2}$$

EXAMPLE 1
Given the curve having parametric equations

$$x = t^3 - 3t \quad \text{and} \quad y = 3t^2$$

find $\mathbf{T}(t)$ and $\mathbf{N}(t)$. Draw a sketch of a portion of the curve at $t = 2$ and draw the representations of $\mathbf{T}(2)$ and $\mathbf{N}(2)$ having their initial point at $t = 2$.

SOLUTION
A vector equation of the curve is

$$\mathbf{R}(t) = (t^3 - 3t)\mathbf{i} + 3t^2\mathbf{j}$$

Thus

$$D_t\mathbf{R}(t) = (3t^2 - 3)\mathbf{i} + 6t\mathbf{j} \qquad \|D_t\mathbf{R}(t)\| = \sqrt{(3t^2 - 3)^2 + 36t^2}$$
$$= \sqrt{9(t^4 + 2t^2 + 1)}$$
$$= 3(t^2 + 1)$$

From (1),

$$\mathbf{T}(t) = \frac{D_t\mathbf{R}(t)}{\|D_t\mathbf{R}(t)\|}$$

$$= \frac{t^2 - 1}{t^2 + 1}\mathbf{i} + \frac{2t}{t^2 + 1}\mathbf{j}$$

Differentiating $\mathbf{T}(t)$ with respect to t we obtain

$$D_t\mathbf{T}(t) = \frac{4t}{(t^2 + 1)^2}\mathbf{i} + \frac{2 - 2t^2}{(t^2 + 1)^2}\mathbf{j}$$

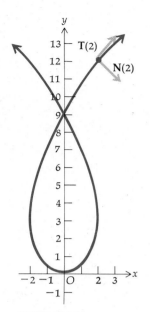

FIGURE 13.6.1

Therefore

$$\|D_t\mathbf{T}(t)\| = \sqrt{\frac{16t^2}{(t^2+1)^4} + \frac{4 - 8t^2 + 4t^4}{(t^2+1)^4}}$$

$$= \sqrt{\frac{4 + 8t^2 + 4t^4}{(t^2+1)^4}}$$

$$= \sqrt{\frac{4(t^2+1)^2}{(t^2+1)^4}}$$

$$= \frac{2}{t^2+1}$$

From (2),

$$\mathbf{N}(t) = \frac{D_t\mathbf{T}(t)}{\|D_t\mathbf{T}(t)\|}$$

$$= \frac{2t}{t^2+1}\mathbf{i} + \frac{1-t^2}{t^2+1}\mathbf{j}$$

We find $\mathbf{R}(t)$, $\mathbf{T}(t)$, and $\mathbf{N}(t)$ when $t = 2$.

$$\mathbf{R}(2) = 2\mathbf{i} + 12\mathbf{j} \qquad \mathbf{T}(2) = \tfrac{3}{5}\mathbf{i} + \tfrac{4}{5}\mathbf{j} \qquad \mathbf{N}(2) = \tfrac{4}{5}\mathbf{i} - \tfrac{3}{5}\mathbf{j}$$

The required sketch is shown in Fig. 13.6.1.

From (1),

$$D_t\mathbf{R}(t) = \|D_t\mathbf{R}(t)\|\mathbf{T}(t) \tag{3}$$

Equation (3) expresses the vector $D_t\mathbf{R}(t)$ as a scalar times the unit tangent vector. Figure 13.6.2 shows a portion of a curve C with the position representation of $\mathbf{R}(t)$ and the representations of $\mathbf{T}(t)$ and $D_t\mathbf{R}(t)$ whose initial points are at the point P on C.

We now use (3) to compute $D_t{}^2\mathbf{R}(t)$ by applying Theorem 13.4.8.

$$D_t{}^2\mathbf{R}(t) = (D_t\|D_t\mathbf{R}(t)\|)\mathbf{T}(t) + \|D_t\mathbf{R}(t)\|(D_t\mathbf{T}(t)) \tag{4}$$

From (2),

$$D_t\mathbf{T}(t) = \|D_t\mathbf{T}(t)\|\mathbf{N}(t) \tag{5}$$

Substituting from (5) into (4) we get

$$D_t{}^2\mathbf{R}(t) = (D_t\|D_t\mathbf{R}(t)\|)\mathbf{T}(t) + (\|D_t\mathbf{R}(t)\| \, \|D_t\mathbf{T}(t)\|)\mathbf{N}(t) \tag{6}$$

Equation (6) expresses the vector $D_t{}^2\mathbf{R}(t)$ as a scalar times the unit tangent vector plus a scalar times the unit normal vector. The coefficient of $\mathbf{T}(t)$ on the right side of (6) is the component of the vector $D_t{}^2\mathbf{R}(t)$ in the direction of the unit tangent vector. The coefficient of $\mathbf{N}(t)$ on the right side of (6) is the component of $D_t{}^2\mathbf{R}(t)$ in the direction of the unit normal vector.

Figure 13.6.3 shows the position representation of $\mathbf{R}(t)$ and the same portion of the curve C as shown in Fig. 13.6.2. Also shown in Fig. 13.6.3 are the representations of the following vectors, all of whose initial points are at the point P on C:

$$D_t{}^2\mathbf{R}(t) \qquad \mathbf{T}(t) \qquad (D_t\|D_t\mathbf{R}(t)\|)\mathbf{T}(t) \qquad \mathbf{N}(t) \qquad (\|D_t\mathbf{R}(t)\| \, \|D_t\mathbf{T}(t)\|)\mathbf{N}(t)$$

FIGURE 13.6.2

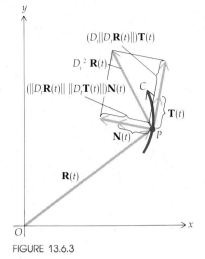

FIGURE 13.6.3

Observe that the representation of the unit normal vector $\mathbf{N}(t)$ is on the concave side of the curve. This fact is proved in general in Section 13.7.

Sometimes instead of a parameter t, we wish to use as a parameter the number of units of arc length s from an arbitrarily chosen point $P_0(x_0, y_0)$ on curve C to the point $P(x, y)$ on C. Let s increase as t increases so that s is positive if the length of arc is measured in the direction of increasing t and s is negative if the length of arc is measured in the opposite direction. Therefore s units is a directed distance. Also, $\dfrac{ds}{dt} > 0$. To each value of s there corresponds a unique point P on the curve C. Consequently, the coordinates of P are functions of s, and s is a function of t. From Eq. (14) in Section 13.5 we have

$$\|D_t\mathbf{R}(t)\| = \frac{ds}{dt} \tag{7}$$

Substituting from (7) into (3) we get

$$D_t\mathbf{R}(t) = \frac{ds}{dt}\,\mathbf{T}(t) \tag{8}$$

If the parameter is s instead of t, we have from (8), by taking $t = s$ and noting that $\dfrac{ds}{ds} = 1$,

$$D_s\mathbf{R}(s) = \mathbf{T}(s)$$

This result is stated as a theorem.

13.6.3 THEOREM   If the vector equation of a curve C is $\mathbf{R}(s) = f(s)\mathbf{i} + g(s)\mathbf{j}$, where s units is the length of arc measured from a particular point P_0 on C to the point P, then the unit tangent vector of C at P is given by

$$\mathbf{T}(s) = D_s\mathbf{R}(s)$$

if it exists.

Now suppose that the parametric equations of a curve C involve a parameter t, and we wish to find parametric equations of C, with s, the number of units of arc length measured from some fixed point, as the parameter. Often the operations involved are quite complicated. However, the method used is illustrated in the following example.

EXAMPLE 2
Suppose that parametric equations of the curve C are

$$x = t^3 \quad \text{and} \quad y = t^2 \qquad t \geq 0$$

Find parametric equations of C having s as a parameter, where s units is the arc length measured from the point where $t = 0$.

SOLUTION
The point where $t = 0$ is the origin. A vector equation of C is

$$\mathbf{R}(t) = t^3\mathbf{i} + t^2\mathbf{j}$$

Because $\dfrac{ds}{dt} = \|D_t R(t)\|$, we differentiate the above vector and get

$$D_t R(t) = 3t^2 \mathbf{i} + 2t\mathbf{j}$$

$$\|D_t R(t)\| = \sqrt{9t^4 + 4t^2}$$
$$= \sqrt{t^2}\sqrt{9t^2 + 4}$$
$$= t\sqrt{9t^2 + 4} \quad \text{(because } t \geq 0\text{)}$$

Therefore

$$\frac{ds}{dt} = t\sqrt{9t^2 + 4}$$

$$s = \int t\sqrt{9t^2 + 4}\, dt$$

$$= \tfrac{1}{18}\int \sqrt{9t^2 + 4}(18t\, dt)$$

$$= \tfrac{1}{27}(9t^2 + 4)^{3/2} + C$$

Because $s = 0$ when $t = 0$, we obtain $C = -\tfrac{8}{27}$. Therefore

$$s = \tfrac{1}{27}(9t^2 + 4)^{3/2} - \tfrac{8}{27} \tag{9}$$

Solving (9) for t in terms of s we have

$$(9t^2 + 4)^{3/2} = 27s + 8$$

$$9t^2 + 4 = (27s + 8)^{2/3}$$

Because $t \geq 0$,

$$t = \tfrac{1}{3}\sqrt{(27s + 8)^{2/3} - 4}$$

Substituting this value of t into the given parametric equations for C we obtain

$$x = \tfrac{1}{27}\left[(27s + 8)^{2/3} - 4\right]^{3/2} \quad \text{and} \quad y = \tfrac{1}{9}\left[(27s + 8)^{2/3} - 4\right] \tag{10}$$

Since $D_s R(s) = T(s)$, then if $R(s) = x(s)\mathbf{i} + y(s)\mathbf{j}$, $T(s) = \left(\dfrac{dx}{ds}\right)\mathbf{i} + \left(\dfrac{dy}{ds}\right)\mathbf{j}$.

Thus, because $T(s)$ is a unit vector,

$$\left(\frac{dx}{ds}\right)^2 + \left(\frac{dy}{ds}\right)^2 = 1 \tag{11}$$

Equation (11) can be used to check Eqs. (10). This check is left as an exercise (see Exercise 17).

EXERCISES 13.6

In Exercises 1 through 10, for the given curve find $T(t)$ and $N(t)$, and at $t = t_1$ draw a sketch of a portion of the curve and draw the representations of $T(t_1)$ and $N(t_1)$ having initial point at $t = t_1$.

1. $x = \tfrac{1}{3}t^3 - t,\ y = t^2;\ t_1 = 2$

2. $x = \tfrac{1}{2}t^2,\ y = \tfrac{1}{3}t^3;\ t_1 = 1$

3. $R(t) = e^t\mathbf{i} + e^{-t}\mathbf{j};\ t_1 = 0$

4. $R(t) = e^{-2t}\mathbf{i} + e^{2t}\mathbf{j};\ t_1 = 0$

5. $R(t) = 3\cos t\mathbf{i} + 3\sin t\mathbf{j};\ t_1 = \tfrac{1}{2}\pi$

6. $x = e^t\sin t,\ y = e^t\cos t;\ t_1 = 0$

7. $x = \cos kt$, $y = \sin kt$, $k > 0$; $t_1 = \pi/k$
8. $x = t - \sin t$, $y = 1 - \cos t$; $t_1 = \pi$
9. $\mathbf{R}(t) = \ln \cos t\mathbf{i} + \ln \sin t\mathbf{j}$, $0 < t < \frac{1}{2}\pi$; $t_1 = \frac{1}{4}\pi$
10. $\mathbf{R}(t) = t \cos t\mathbf{i} + t \sin t\mathbf{j}$; $t_1 = 0$

In Exercises 11 through 16, find $\mathbf{T}(t)$ and $\mathbf{N}(t)$ for the curve having the given vector equation.

11. $\mathbf{R}(t) = t^2\mathbf{i} + e^t\mathbf{j}$
12. $\mathbf{R}(t) = t\mathbf{i} + \cosh t\mathbf{j}$; $t \geq 0$
13. $\mathbf{R}(t) = \sin^3 t\mathbf{i} + \cos^3 t\mathbf{j}$; $0 \leq t \leq \frac{1}{2}\pi$
14. $\mathbf{R}(t) = \sqrt{1 + t^2}\mathbf{i} + t\mathbf{j}$
15. $\mathbf{R}(t) = 4t\mathbf{i} + 2t^2\mathbf{j}$
16. $\mathbf{R}(t) = (1 + t)\mathbf{i} + \ln \cos t\mathbf{j}$
17. Check Eqs. (10) of the solution of Example 2 by using Eq. (11).

In Exercises 18 through 23, find parametric equations of the curve having arc length s as a parameter, where s is measured from the point where $t = 0$. Check your result by using Eq. (11).

18. $x = a \cos t$, $y = a \sin t$
19. $x = 2 + \cos t$, $y = 3 + \sin t$
20. $x = 2(\cos t + t \sin t)$, $y = 2(\sin t - t \cos t)$
21. $y = x^{3/2}$
22. $x = t^2 - 1$, $y = \frac{1}{3}t^3$
23. One cusp of the hypocycloid of four cusps: $\mathbf{R}(t) = a \cos^3 t\mathbf{i} + a \sin^3 t\mathbf{j}$, $0 \leq t \leq \frac{1}{2}\pi$.
24. Given the cycloid $x = 2(t - \sin t)$, $y = 2(1 - \cos t)$, express the arc length s as a function of t, where s is measured from the point where $t = 0$.
25. For the curve having parametric equations $x = e^t \cos t$ and $y = e^t \sin t$, express the arc length s as a function of t, where s is measured from the point where $t = 0$.
26. If the vector equation of curve C is $\mathbf{R}(t) = 2 \sin t\mathbf{i} + \sin 2t\mathbf{j}$, find the cosine of the angle between the vectors $\mathbf{R}(\frac{1}{6}\pi)$ and $\mathbf{T}(\frac{1}{6}\pi)$.
27. If the vector equation of curve C is $\mathbf{R}(t) = 3t^2\mathbf{i} + (t^3 - 3t)\mathbf{j}$, find the cosine of the angle between the vectors $\mathbf{R}(2)$ and $\mathbf{T}(2)$.
28. If the vector equation of curve C is $\mathbf{R}(t) = (4 - 3t^2)\mathbf{i} + (t^3 - 3t)\mathbf{j}$, find the radian measure of the angle between the vectors $\mathbf{N}(1)$ and $D_t^2\mathbf{R}(1)$.

**13.7
CURVATURE**

FIGURE 13.7.1

Differential geometry and *curvilinear motion* (motion along a curved path) involve a study of curves by means of the calculus of vector-valued functions. One of the important concepts in this subject is *curvature*, which gives the rate of change of the direction of a curve with respect to a change in its length. Related to this idea is the angle giving the direction of the unit tangent vector associated with a curve C. Thus we let ϕ be the radian measure of the angle from the direction of the positive x axis counterclockwise to the direction of the unit tangent vector $\mathbf{T}(t)$. See Fig. 13.7.1. We compute $D_\phi\mathbf{T}(t)$. Because $\|\mathbf{T}(t)\| = 1$, it follows from Eq. (6) in Section 13.1 that

$$\mathbf{T}(t) = \cos \phi\mathbf{i} + \sin \phi\mathbf{j}$$

Differentiating with respect to ϕ we obtain

$$D_\phi\mathbf{T}(t) = -\sin \phi\mathbf{i} + \cos \phi\mathbf{j}$$

Since $\|D_\phi\mathbf{T}(t)\| = \sqrt{(-\sin \phi)^2 + (\cos \phi)^2}$, which is 1, $D_\phi\mathbf{T}(t)$ is a unit vector. Later in this section we show the relationship of the vector $D_\phi\mathbf{T}(t)$ to the unit normal vector $\mathbf{N}(t)$.

To lead up to the definition of curvature, consider the vector $D_s\mathbf{T}(t)$, where s units is the arc length measured from an arbitrarily chosen point on C to point P and s increases as t increases. By the chain rule (Theorem 13.4.9),

$$D_s\mathbf{T}(t) = D_\phi\mathbf{T}(t)\frac{d\phi}{ds}$$

Hence

$$\left\|D_s\mathbf{T}(t)\right\| = \left\|D_\phi\mathbf{T}(t)\frac{d\phi}{ds}\right\|$$

$$= \left\|D_\phi\mathbf{T}(t)\right\|\left|\frac{d\phi}{ds}\right|$$

But because $D_\phi\mathbf{T}(t)$ is a unit vector, $\left\|D_\phi\mathbf{T}(t)\right\| = 1$; thus

$$\left\|D_s\mathbf{T}(t)\right\| = \left|\frac{d\phi}{ds}\right| \tag{1}$$

The number $\left|\dfrac{d\phi}{ds}\right|$ is the absolute value of the rate of change of the measure of the angle giving the direction of the unit tangent vector $\mathbf{T}(t)$ at a point on a curve with respect to the measure of arc length along the curve. This number is what we shall define to be the curvature of the curve at the point; however, before giving the formal definition, we show that taking it as this number is consistent with our intuitive notion of curvature. For example, at point P on C, ϕ is the radian measure of the angle giving the direction of the vector $\mathbf{T}(t)$, and s units is the arc length from a point P_0 on C to P. Let Q be a point on C for which the radian measure of the angle giving the direction of $\mathbf{T}(t + \Delta t)$ at Q is $\phi + \Delta\phi$, and $s + \Delta s$ units is the arc length from P_0 to Q. Then the arc length from P to Q is Δs units, and the ratio $\Delta\phi/\Delta s$ seems like a good measure of what we would intuitively think of as the *average curvature* along arc PQ.

▶ ILLUSTRATION 1   See Fig. 13.7.2(a), (b), (c) and (d): In (a), $\Delta\phi > 0$ and $\Delta s > 0$; in (b), $\Delta\phi > 0$ and $\Delta s < 0$; in (c), $\Delta\phi < 0$ and $\Delta s > 0$; and in (d), $\Delta\phi < 0$ and $\Delta s < 0$.   ◀

FIGURE 13.7.2

We now define the *curvature vector* of a curve at a point, and its magnitude, the curvature of the curve at the point.

13.7.1 DEFINITION   If $\mathbf{T}(t)$ is the unit tangent vector to a curve C at a point P, s is the arc length measured from an arbitrarily chosen point on C to P, and s increases as t increases, then the **curvature vector** of C at P, denoted by $\mathbf{K}(t)$, is given by

$$\mathbf{K}(t) = D_s\mathbf{T}(t) \tag{2}$$

The **curvature** of C at P, denoted by $K(t)$, is the magnitude of the curvature vector; that is,

$$K(t) = \|D_s\mathbf{T}(t)\| \tag{3}$$

Observe that by substituting from (1) in (3) the curvature $K(t)$ can also be defined by

$$K(t) = \left|\frac{d\phi}{ds}\right| \tag{4}$$

To find the curvature vector and the curvature for a particular curve it is convenient to have a formula expressing the curvature vector in terms of derivatives with respect to t. By the chain rule,

$$D_t\mathbf{T}(t) = D_s\mathbf{T}(t)\,\frac{ds}{dt}$$

From Eq. (14) in Section 13.5, $\dfrac{ds}{dt} = \|D_t\mathbf{R}(t)\|$. Thus

$$D_t\mathbf{T}(t) = [D_s\mathbf{T}(t)]\|D_t\mathbf{R}(t)\|$$

$$D_s\mathbf{T}(t) = \frac{D_t\mathbf{T}(t)}{\|D_t\mathbf{R}(t)\|} \tag{5}$$

Substituting from (5) in (2) we obtain

$$\mathbf{K}(t) = \frac{D_t\mathbf{T}(t)}{\|D_t\mathbf{R}(t)\|} \tag{6}$$

Because $K(t) = \|\mathbf{K}(t)\|$, the curvature is given by

$$K(t) = \left\|\frac{D_t\mathbf{T}(t)}{\|D_t\mathbf{R}(t)\|}\right\| \tag{7}$$

EXAMPLE 1
Given the circle with radius a:

$$x = a \cos t \qquad y = a \sin t \qquad a > 0$$

find the curvature vector and the curvature at any t.

SOLUTION
The vector equation of the circle is

$$\mathbf{R}(t) = a \cos t\,\mathbf{i} + a \sin t\,\mathbf{j}$$

Thus

$$D_t\mathbf{R}(t) = -a\sin t\mathbf{i} + a\cos t\mathbf{j} \qquad \|D_t\mathbf{R}(t)\| = \sqrt{(-a\sin t)^2 + (a\cos t)^2}$$
$$= a$$

Therefore

$$\mathbf{T}(t) = \frac{D_t\mathbf{R}(t)}{\|D_t\mathbf{R}(t)\|} \qquad\qquad D_t\mathbf{T}(t) = -\cos t\mathbf{i} - \sin t\mathbf{j}$$

$$= -\sin t\mathbf{i} + \cos t\mathbf{j}$$

$$\frac{D_t\mathbf{T}(t)}{\|D_t\mathbf{R}(t)\|} = -\frac{\cos t}{a}\mathbf{i} - \frac{\sin t}{a}\mathbf{j}$$

Hence the curvature vector and the curvature are given by

$$\mathbf{K}(t) = -\frac{1}{a}\cos t\mathbf{i} - \frac{1}{a}\sin t\mathbf{j} \qquad K(t) = \|\mathbf{K}(t)\|$$
$$= \frac{1}{a}$$

The result of Example 1 states that the curvature of a circle is constant, which is what you would expect. Furthermore, it is the reciprocal of the radius.

Let us return now to the unit vector $D_\phi\mathbf{T}(t)$ defined by

$$D_\phi\mathbf{T}(t) = -\sin \phi\mathbf{i} + \cos \phi\mathbf{j} \tag{8}$$

Because $\mathbf{T}(t)$ has constant magnitude, it follows from Theorem 13.4.11 that $D_\phi\mathbf{T}(t)$ is orthogonal to $\mathbf{T}(t)$. Replacing $-\sin \phi$ by $\cos(\frac{1}{2}\pi + \phi)$ and $\cos \phi$ by $\sin(\frac{1}{2}\pi + \phi)$, we write (8) as

$$D_\phi\mathbf{T}(t) = \cos(\tfrac{1}{2}\pi + \phi)\mathbf{i} + \sin(\tfrac{1}{2}\pi + \phi)\mathbf{j}$$

Thus the vector $D_\phi\mathbf{T}(t)$ is a unit vector orthogonal to $\mathbf{T}(t)$ in the direction $\frac{1}{2}\pi$ counterclockwise from the direction of $\mathbf{T}(t)$. The unit normal vector $\mathbf{N}(t)$ is also orthogonal to $\mathbf{T}(t)$. By the chain rule,

$$D_t\mathbf{T}(t) = D_\phi\mathbf{T}(t)\frac{d\phi}{dt} \tag{9}$$

Because the direction of $\mathbf{N}(t)$ is the same as the direction of $D_t\mathbf{T}(t)$, it follows from (9) that the direction of $\mathbf{N}(t)$ is the same as the direction of $D_\phi\mathbf{T}(t)$ if $\frac{d\phi}{dt} > 0$ (i.e., if $\mathbf{T}(t)$ turns counterclockwise as t increases), and the direction of $\mathbf{N}(t)$ is opposite that of $D_\phi\mathbf{T}(t)$ if $\frac{d\phi}{dt} < 0$ (i.e., if $\mathbf{T}(t)$ turns clockwise as t increases). Because both $D_\phi\mathbf{T}(t)$ and $\mathbf{N}(t)$ are unit vectors, we conclude that

$$D_\phi\mathbf{T}(t) = \begin{cases} \mathbf{N}(t) & \text{if } \dfrac{d\phi}{dt} > 0 \\[2mm] -\mathbf{N}(t) & \text{if } \dfrac{d\phi}{dt} < 0 \end{cases} \tag{10}$$

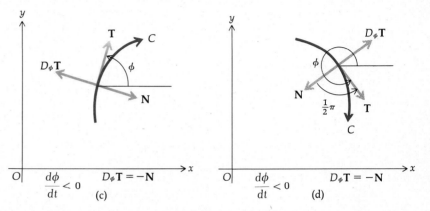

FIGURE 13.7.3

▶ ILLUSTRATION 2   In Fig. 13.7.3(a), (b), (c), and (d) various situations are shown; in (a) and (b), $\dfrac{d\phi}{dt} > 0$, and in (c) and (d), $\dfrac{d\phi}{dt} < 0$. The positive direction along the curve C is indicated by the tip of the arrow on C. In each figure are shown the angle of radian measure ϕ and representations of the vectors $\mathbf{T}(t)$, $D_\phi\mathbf{T}(t)$, and $\mathbf{N}(t)$.   ◀

Observe in Fig. 13.7.3 that *the representation of the unit normal vector* $\mathbf{N}(t)$ *is always on the concave side of the curve.*

Suppose we are given a curve C and at a particular point P the curvature exists and is $K(t)$, where $K(t) \neq 0$. Consider the circle that is tangent to curve C at P and has curvature $K(t)$ at P. From Example 1, the radius of this circle is $1/K(t)$ and its center is on a line perpendicular to the tangent line in the direction of $\mathbf{N}(t)$. This circle is called the **circle of curvature**, and its radius is the *radius of curvature* of C at P. The circle of curvature is sometimes referred to as the **osculating circle**.

13.7.2 DEFINITION   If $K(t)$ is the curvature of a curve C at point P and $K(t) \neq 0$, then the **radius of curvature** of C at P, denoted by $\rho(t)$, is defined by

$$\rho(t) = \frac{1}{K(t)}$$

EXAMPLE 2
Given that a vector equation of a curve C is

$$\mathbf{R}(t) = 2t\mathbf{i} + (t^2 - 1)\mathbf{j}$$

(a) Find the unit tangent vector, the curvature, and the radius of curvature at $t = 1$. (b) Draw a sketch of a portion of the curve, the unit tangent vector, and the circle of curvature at $t = 1$.

FIGURE 13.7.4

Table 13.7.1

t	x	y
-2	-4	3
-1	-2	0
0	0	-1
1	2	0
2	4	3

SOLUTION

$$D_t\mathbf{R}(t) = 2\mathbf{i} + 2t\mathbf{j} \qquad \|D_t\mathbf{R}(t)\| = 2\sqrt{1 + t^2}$$

$$\mathbf{T}(t) = \frac{D_t\mathbf{R}(t)}{\|D_t\mathbf{R}(t)\|}$$

$$= \frac{1}{\sqrt{1 + t^2}}\mathbf{i} + \frac{t}{\sqrt{1 + t^2}}\mathbf{j}$$

$$D_t\mathbf{T}(t) = -\frac{t}{(1 + t^2)^{3/2}}\mathbf{i} + \frac{1}{(1 + t^2)^{3/2}}\mathbf{j}$$

$$\mathbf{K}(t) = \frac{D_t\mathbf{T}(t)}{\|D_t\mathbf{R}(t)\|}$$

$$= -\frac{t}{2(1 + t^2)^2}\mathbf{i} + \frac{1}{2(1 + t^2)^2}\mathbf{j}$$

$$K(t) = \|\mathbf{K}(t)\|$$

$$= \sqrt{\frac{t^2}{4(1 + t^2)^4} + \frac{1}{4(1 + t^2)^4}}$$

$$= \frac{1}{2(1 + t^2)^{3/2}}$$

(a) $\mathbf{T}(1) = \dfrac{1}{\sqrt{2}}\mathbf{i} + \dfrac{1}{\sqrt{2}}\mathbf{j} \qquad K(1) = \dfrac{1}{4\sqrt{2}} \qquad \rho(1) = 4\sqrt{2}$

(b) Figure 13.7.4 shows the required sketch. The accompanying Table 13.7.1 gives the corresponding values of x and y when t is -2, -1, 0, 1, and 2.

We now find a formula for computing the curvature directly from parametric equations of the curve, $x = f(t)$ and $y = g(t)$. From (4), we have $K(t) = \left|\dfrac{d\phi}{ds}\right|$. We compute $\dfrac{d\phi}{ds}$.

$$\frac{d\phi}{ds} = \frac{\dfrac{d\phi}{dt}}{\dfrac{ds}{dt}}$$

With the assumption that s and t increase together, $\dfrac{ds}{dt} > 0$. Thus

$$\frac{d\phi}{ds} = \frac{\dfrac{d\phi}{dt}}{\sqrt{\left(\dfrac{dx}{dt}\right)^2 + \left(\dfrac{dy}{dt}\right)^2}} \qquad (11)$$

To find $\dfrac{d\phi}{dt}$ we observe that because ϕ is the radian measure of the angle giving the direction of the unit tangent vector, $\tan \phi = \dfrac{dy}{dx}$. Therefore

$$\tan \phi = \frac{\dfrac{dy}{dt}}{\dfrac{dx}{dt}}$$

Differentiating implicitly with respect to t the left and right members of this equation, we obtain

$$\sec^2 \phi \, \frac{d\phi}{dt} = \frac{\left(\dfrac{dx}{dt}\right)\left(\dfrac{d^2 y}{dt^2}\right) - \left(\dfrac{dy}{dt}\right)\left(\dfrac{d^2 x}{dt^2}\right)}{\left(\dfrac{dx}{dt}\right)^2}$$

$$\frac{d\phi}{dt} = \frac{\left(\dfrac{dx}{dt}\right)\left(\dfrac{d^2 y}{dt^2}\right) - \left(\dfrac{dy}{dt}\right)\left(\dfrac{d^2 x}{dt^2}\right)}{\sec^2 \phi \left(\dfrac{dx}{dt}\right)^2} \qquad (12)$$

Because $\sec^2 \phi = 1 + \tan^2 \phi$, we have

$$\sec^2 \phi = 1 + \frac{\left(\dfrac{dy}{dt}\right)^2}{\left(\dfrac{dx}{dt}\right)^2}$$

Substituting this expression for $\sec^2 \phi$ in (12) we get

$$\frac{d\phi}{dt} = \frac{\left(\dfrac{dx}{dt}\right)\left(\dfrac{d^2 y}{dt^2}\right) - \left(\dfrac{dy}{dt}\right)\left(\dfrac{d^2 x}{dt^2}\right)}{\left(\dfrac{dx}{dt}\right)^2 + \left(\dfrac{dy}{dt}\right)^2} \qquad (13)$$

Substituting from (13) into (11), and because $K(t) = \left|\dfrac{d\phi}{ds}\right|$, we have

$$K(t) = \frac{\left|\left(\dfrac{dx}{dt}\right)\left(\dfrac{d^2 y}{dt^2}\right) - \left(\dfrac{dy}{dt}\right)\left(\dfrac{d^2 x}{dt^2}\right)\right|}{\left[\left(\dfrac{dx}{dt}\right)^2 + \left(\dfrac{dy}{dt}\right)^2\right]^{3/2}} \qquad (14)$$

EXAMPLE 3

Find the curvature of the curve in Example 2 by using formula (14).

SOLUTION

Parametric equations of C are $x = 2t$ and $y = t^2 - 1$. Hence

$$\frac{dx}{dt} = 2 \qquad \frac{d^2x}{dt^2} = 0 \qquad \frac{dy}{dt} = 2t \qquad \frac{d^2y}{dt^2} = 2$$

Therefore, from (14),

$$K(t) = \frac{|2(2) - 2t(0)|}{[(2)^2 + (2t)^2]^{3/2}}$$

$$= \frac{4}{(4 + 4t^2)^{3/2}}$$

$$= \frac{1}{2(1 + t^2)^{3/2}}$$

Suppose a cartesian equation of a curve is given in either of the forms $y = F(x)$ or $x = G(y)$. Special cases of formula (14) can be used to find the curvature of a curve in such situations.

If $y = F(x)$ is an equation of a curve C, a set of parametric equations of C is $x = t$ and $y = F(t)$. Then

$$\frac{dx}{dt} = 1 \qquad \frac{d^2x}{dt^2} = 0 \qquad \frac{dy}{dt} = \frac{dy}{dx} \qquad \frac{d^2y}{dt^2} = \frac{d^2y}{dx^2}$$

Substituting into (14) we obtain

$$K = \frac{\left|\dfrac{d^2y}{dx^2}\right|}{\left[1 + \left(\dfrac{dy}{dx}\right)^2\right]^{3/2}} \qquad (15)$$

Similarly, if an equation of a curve C is $x = G(y)$,

$$K = \frac{\left|\dfrac{d^2x}{dy^2}\right|}{\left[1 + \left(\dfrac{dx}{dy}\right)^2\right]^{3/2}} \qquad (16)$$

EXAMPLE 4

If an equation of curve C is

$$y = \frac{1}{x}$$

find the radius of curvature of C at the point $(1, 1)$, and draw a sketch of the curve and the circle of curvature at $(1, 1)$.

FIGURE 13.7.5

SOLUTION

$$\frac{dy}{dx} = -\frac{1}{x^2} \qquad \frac{d^2y}{dx^2} = \frac{2}{x^3}$$

We compute K from (15) and then $\rho = 1/K$.

$$K = \frac{\left|\dfrac{2}{x^3}\right|}{\left[1 + \dfrac{1}{x^4}\right]^{3/2}} \qquad \rho = \frac{(x^4 + 1)^{3/2}}{2|x^3|}$$

$$= \frac{2|x^3|}{(x^4 + 1)^{3/2}}$$

Therefore at $(1, 1)$, $\rho = \sqrt{2}$. The required sketch appears in Fig. 13.7.5.

EXERCISES 13.7

In Exercises 1 through 4, find the curvature K and the radius of curvature ρ at the point where $t = t_1$. Use formula (7) to find K. Draw a sketch showing a portion of the curve, the unit tangent vector, and the circle of curvature at $t = t_1$.

1. $\mathbf{R}(t) = t^2\mathbf{i} + (2t + 1)\mathbf{j}$; $t_1 = 1$
2. $\mathbf{R}(t) = (t^2 - 2t)\mathbf{i} + (t^3 - t)\mathbf{j}$; $t_1 = 1$
3. $\mathbf{R}(t) = 2e^t\mathbf{i} + 2e^{-t}\mathbf{j}$; $t_1 = 0$
4. $\mathbf{R}(t) = \sin t\mathbf{i} + \sin 2t\mathbf{j}$; $t_1 = \frac{1}{2}\pi$

In Exercises 5 and 6, find the curvature K by using formula (14). Then find K and ρ at the point where $t = t_1$, and draw a sketch showing a portion of the curve, the unit tangent vector, and the circle of curvature at $t = t_1$.

5. $x = \dfrac{1}{1 + t}$, $y = \dfrac{1}{1 - t}$; $t_1 = 0$
6. $x = e^t + e^{-t}$, $y = e^t - e^{-t}$; $t_1 = 0$

In Exercises 7 through 14, find the curvature K and the radius of curvature ρ at the given point. Draw a sketch showing a portion of the curve, a piece of the tangent line, and the circle of curvature at the given point.

7. $y = 2\sqrt{x}$; $(0, 0)$
8. $y^2 = x^3$; $(\frac{1}{4}, \frac{1}{8})$
9. $y = e^x$; $(0, 1)$
10. $y = \ln x$; $(e, 1)$
11. $x = \sin y$; $(\frac{1}{2}, \frac{1}{6}\pi)$
12. $4x^2 + 9y^2 = 36$; $(0, 2)$
13. $x = \sqrt{y - 1}$; $(2, 5)$
14. $x = \tan y$; $(1, \frac{1}{4}\pi)$

In Exercises 15 through 22, find the radius of curvature at any point on the given curve.

15. $y = \sin^{-1} x$
16. $y = \ln \sec x$
17. $4x^2 - 9y^2 = 16$
18. $x = \tan^{-1} y$
19. $x^{1/2} + y^{1/2} = a^{1/2}$
20. $\mathbf{R}(t) = e^t \sin t\mathbf{i} + e^t \cos t\mathbf{j}$

21. The cycloid $x = a(t - \sin t)$, $y = a(1 - \cos t)$
22. The tractrix $x = t - a \tanh \dfrac{t}{a}$, $y = a \operatorname{sech} \dfrac{t}{a}$

23. Show that the curvature of the catenary $y = a \cosh(x/a)$ at any point (x, y) on the curve is a/y^2. Draw the circle of curvature at $(0, a)$. Show that the curvature K is an absolute maximum at the point $(0, a)$ without referring to $K'(x)$.

In Exercises 24 through 28, find a point on the given curve at which the curvature is an absolute maximum.

24. $y = e^x$
25. $y = 6x - x^2$
26. $y = \sin x$
27. $\mathbf{R}(t) = (2t - 3)\mathbf{i} + (t^2 - 1)\mathbf{j}$
28. $y = x^2 - 2x + 3$
29. Find an equation of the circle of curvature for $y = e^x$ at the point $(0, 1)$.

30. If a polar equation of a curve is $r = F(\theta)$, prove that the curvature K is given by the formula

$$K = \frac{\left| r^2 + 2\left(\dfrac{dr}{d\theta}\right)^2 - r\left(\dfrac{d^2r}{d\theta^2}\right) \right|}{\left[r^2 + \left(\dfrac{dr}{d\theta}\right)^2 \right]^{3/2}}$$

In Exercises 31 through 34, find the curvature K and the radius of curvature ρ at the indicated point. Use the formula of Exercise 30 to find K.

31. $r = 4 \cos 2\theta$; $\theta = \frac{1}{12}\pi$ **32.** $r = 1 - \sin\theta$; $\theta = 0$ **33.** $r = a \sec^2 \frac{1}{2}\theta$; $\theta = \frac{2}{3}\pi$ **34.** $r = a\theta$; $\theta = 1$

35. The center of the circle of curvature of a curve C at a point P is called the *center of curvature at P*. Prove that the coordinates of the center of curvature of a curve at $P(x, y)$ are given by

$$x_c = x - \frac{\left(\dfrac{dy}{dx}\right)\left[1 + \left(\dfrac{dy}{dx}\right)^2\right]}{\dfrac{d^2y}{dx^2}} \qquad y_c = y + \frac{\left(\dfrac{dy}{dx}\right)^2 + 1}{\dfrac{d^2y}{dx^2}}$$

In Exercises 36 through 38, find the curvature K, the radius of curvature ρ, and the center of curvature at the given point. Draw a sketch of the curve and the circle of curvature.

36. $y = \ln x$; $(1, 0)$ **37.** $y = x^4 - x^2$; $(0, 0)$ **38.** $y = \cos x$; $(\frac{1}{3}\pi, \frac{1}{2})$

In Exercises 39 through 42, find the coordinates of the center of curvature at any point.

39. $y^2 = 4px$ **40.** $y^3 = a^2x$
41. $\mathbf{R}(t) = a \cos t\mathbf{i} + b \sin t\mathbf{j}$ **42.** $\mathbf{R}(t) = a \cos^3 t\mathbf{i} + a \sin^3 t\mathbf{j}$
43. Show that the curvature of a straight line is zero at every point.

13.8
PLANE MOTION

Our previous discussions of the motion of a particle were confined to rectilinear motion. In this connection we defined the velocity and acceleration of a particle moving along a straight line. Consider now the motion of a particle along a curve in the plane. This is called **curvilinear motion**.

Suppose that C is the plane curve having parametric equations $x = f(t)$ and $y = g(t)$, where t units denote time. Then

$$\mathbf{R}(t) = f(t)\mathbf{i} + g(t)\mathbf{j}$$

is a vector equation of C. As t varies, the endpoint $P(f(t), g(t))$ of $\overrightarrow{OP}$ moves along the curve C. The position at time t units of a particle moving along C is the point $P(f(t), g(t))$. The **velocity vector** of the particle at time t units is defined to be $\mathbf{R}'(t)$ and is denoted by the symbol $\mathbf{V}(t)$.

13.8.1 DEFINITION   Let C be the curve having parametric equations $x = f(t)$ and $y = g(t)$. If a particle is moving along C so that its position at any time t units is the point (x, y), then the **instantaneous velocity** of the particle at time t units is determined by the velocity vector

$$\mathbf{V}(t) = f'(t)\mathbf{i} + g'(t)\mathbf{j}$$

if $f'(t)$ and $g'(t)$ exist.

Because the direction of $\mathbf{R}'(t)$ at the point $P(f(t), g(t))$ is along the tangent line to the curve C at P, the velocity vector $\mathbf{V}(t)$ has this direction at P.

The magnitude of the velocity vector is a measure of the **speed** of the particle at time t and is given by

$$\|\mathbf{V}(t)\| = \sqrt{[f'(t)]^2 + [g'(t)]^2} \tag{1}$$

Note that the velocity is a vector and the speed is a scalar. As shown in Section 13.5, the expression on the right side of (1) is ds/dt. So the speed is the rate of change of s with respect to t, and we write

$$\|\mathbf{V}(t)\| = \frac{ds}{dt} \tag{2}$$

The **acceleration vector** of the particle at time t units is denoted by $\mathbf{A}(t)$ and is defined to be the derivative of the velocity vector or, equivalently, the second derivative of the position vector.

13.8.2 DEFINITION   The **instantaneous acceleration** at time t units of a particle moving along a curve C, having parametric equations $x = f(t)$ and $y = g(t)$, is determined by the acceleration vector

$$\mathbf{A}(t) = \mathbf{V}'(t) \quad \Leftrightarrow \quad \mathbf{A}(t) = \mathbf{R}''(t)$$

where $\mathbf{R}(t) = f(t)\mathbf{i} + g(t)\mathbf{j}$ and $\mathbf{R}''(t)$ exists.

Figure 13.8.1 shows the representations of the velocity vector and the acceleration vector whose initial point is the point P on C.

FIGURE 13.8.1

EXAMPLE 1

A particle is moving along the curve having parametric equations

$$x = 4 \cos \tfrac{1}{2}t \quad \text{and} \quad y = 4 \sin \tfrac{1}{2}t$$

If x and y are centimeter measures, find the speed and the magnitude of the particle's acceleration vector at t sec. Draw a sketch of the particle's path, and also draw the representations of the velocity and acceleration vectors having initial point where $t = \tfrac{1}{3}\pi$.

SOLUTION

A vector equation of C is

$$\mathbf{R}(t) = 4 \cos \tfrac{1}{2}t\mathbf{i} + 4 \sin \tfrac{1}{2}t\mathbf{j}$$

$$
\begin{aligned}
\mathbf{V}(t) &= \mathbf{R}'(t) & \mathbf{A}(t) &= \mathbf{V}'(t) \\
&= -2 \sin \tfrac{1}{2}t\,\mathbf{i} + 2 \cos \tfrac{1}{2}t\mathbf{j} & &= -\cos \tfrac{1}{2}t\mathbf{i} - \sin \tfrac{1}{2}t\mathbf{j} \\
\|\mathbf{V}(t)\| &= \sqrt{(-2 \sin \tfrac{1}{2}t)^2 + (2 \cos \tfrac{1}{2}t)^2} & \|\mathbf{A}(t)\| &= \sqrt{(-\cos \tfrac{1}{2}t)^2 + (-\sin \tfrac{1}{2}t)^2} \\
&= \sqrt{4 \sin^2 \tfrac{1}{2}t + 4 \cos^2 \tfrac{1}{2}t} & &= 1 \\
&= 2
\end{aligned}
$$

Therefore, the speed of the particle is constant and is 2 cm/sec. The magnitude of the acceleration vector is also constant and is 1 cm/sec².

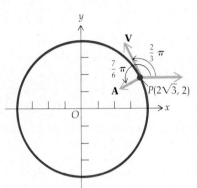

FIGURE 13.8.2

Eliminating t between the parametric equations of C, we obtain the cartesian equation

$$x^2 + y^2 = 16$$

which is a circle with its center at the origin and radius 4. We now find the velocity and acceleration vectors at $t = \frac{1}{3}\pi$.

$$\mathbf{V}(\tfrac{1}{3}\pi) = -2\sin\tfrac{1}{6}\pi\mathbf{i} + 2\cos\tfrac{1}{6}\pi\mathbf{j} \qquad \mathbf{A}(\tfrac{1}{3}\pi) = -\cos\tfrac{1}{6}\pi\mathbf{i} - \sin\tfrac{1}{6}\pi\mathbf{j}$$
$$= -\mathbf{i} + \sqrt{3}\mathbf{j} \qquad\qquad\qquad = -\tfrac{1}{2}\sqrt{3}\mathbf{i} - \tfrac{1}{2}\mathbf{j}$$

The direction of $\mathbf{V}(\tfrac{1}{3}\pi)$ is given by

$$\tan\theta_1 = -\sqrt{3} \qquad \tfrac{1}{2}\pi < \theta_1 < \pi$$

and the direction of $\mathbf{A}(\tfrac{1}{3}\pi)$ is given by

$$\tan\theta_2 = \frac{1}{\sqrt{3}} \qquad \pi < \theta_2 < \tfrac{3}{2}\pi$$

Thus $\theta_1 = \frac{2}{3}\pi$ and $\theta_2 = \frac{7}{6}\pi$. Figure 13.8.2 shows the particle's path and representations of the velocity and acceleration vectors having initial point where $t = \frac{1}{3}\pi$.

EXAMPLE 2

The position of a moving particle at time t units is given by the vector equation

$$\mathbf{R}(t) = e^{-2t}\mathbf{i} + 3e^t\mathbf{j}$$

Find $\mathbf{V}(t)$, $\mathbf{A}(t)$, $\|\mathbf{V}(t)\|$, $\|\mathbf{A}(t)\|$. Draw a sketch of the path of the particle and the representations of the velocity and acceleration vectors having initial point where $t = \frac{1}{2}$.

SOLUTION

$$\mathbf{V}(t) = \mathbf{R}'(t) \qquad\qquad \mathbf{A}(t) = \mathbf{V}'(t)$$
$$= -2e^{-2t}\mathbf{i} + 3e^t\mathbf{j} \qquad\quad = 4e^{-2t}\mathbf{i} + 3e^t\mathbf{j}$$
$$\|\mathbf{V}(t)\| = \sqrt{4e^{-4t} + 9e^{2t}} \qquad \|\mathbf{A}(t)\| = \sqrt{16e^{-4t} + 9e^{2t}}$$
$$\|\mathbf{V}(\tfrac{1}{2})\| = \sqrt{4e^{-2} + 9e} \qquad \|\mathbf{A}(\tfrac{1}{2})\| = \sqrt{16e^{-2} + 9e}$$
$$\approx 5.00 \qquad\qquad\qquad \approx 5.16$$

Parametric equations of the path of the particle are

$$x = e^{-2t} \quad \text{and} \quad y = 3e^t$$

We eliminate t between these two equations and obtain

$$e^{2t} = \frac{1}{x} \quad \text{and} \quad e^{2t} = \frac{y^2}{9}$$

$$\frac{1}{x} = \frac{y^2}{9}$$

$$xy^2 = 9$$

FIGURE 13.8.3

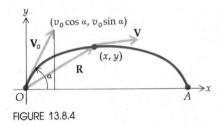

FIGURE 13.8.4

Because $x > 0$ and $y > 0$, the path of the particle is the portion of the curve $xy^2 = 9$ in the first quadrant. Figure 13.8.3 shows the path of the particle and representations of the velocity and acceleration vectors when $t = \frac{1}{2}$. The slope of the representation of $\mathbf{V}(\frac{1}{2})$ is $-\frac{3}{2}e^{3/2} \approx -6.7$, and the slope of the representation of $\mathbf{A}(\frac{1}{2})$ is $\frac{3}{4}e^{3/2} \approx 3.4$.

We now derive the equations of motion of a projectile by assuming that the projectile is moving in a vertical plane. We also assume that the only force acting on the projectile is its weight, which has a downward direction and a magnitude of mg lb, where m slugs is its mass and g ft/sec^2 is the constant of acceleration caused by gravity. We are neglecting the force attributed to air resistance (which for heavy bodies traveling at small speeds has no noticeable effect). The positive direction is taken as vertically upward and horizontally to the right.

Suppose, then, that a projectile is shot from a gun having an angle of elevation of radian measure α. Let the number of feet per second in the initial speed, or *muzzle speed*, be denoted by v_0. The coordinate axes are set up so that the gun is located at the origin. Refer to Fig. 13.8.4. The initial velocity vector, $\mathbf{V}_0$, of the projectile is given by

$$\mathbf{V}_0 = v_0 \cos \alpha \mathbf{i} + v_0 \sin \alpha \mathbf{j} \qquad (3)$$

Let t sec be the time that has elapsed since the gun was fired, x ft be the horizontal distance of the projectile from the starting point at t sec, and y ft be the vertical distance of the projectile from the starting point at t sec. $\mathbf{R}(t)$ is the position vector of the projectile at t sec, $\mathbf{V}(t)$ is the velocity vector of the projectile at t sec, and $\mathbf{A}(t)$ is the acceleration vector of the projectile at t sec.

Because x is a function of t, we write $x(t)$. Similarly, y is a function of t and we write $y(t)$. Then

$$\mathbf{R}(t) = x(t)\mathbf{i} + y(t)\mathbf{j} \qquad (4)$$

$$\mathbf{V}(t) = \mathbf{R}'(t) \qquad (5)$$

$$\mathbf{A}(t) = \mathbf{V}'(t) \qquad (6)$$

Because the only force acting on the projectile has a magnitude of mg lb and is in the downward direction, then if $\mathbf{F}$ denotes this force,

$$\mathbf{F} = -mg\mathbf{j} \qquad (7)$$

Newton's second law of motion states that the net force acting on a body is its "mass times acceleration." So

$$\mathbf{F} = m\mathbf{A} \qquad (8)$$

From (7) and (8),

$$m\mathbf{A} = -mg\mathbf{j}$$

$$\mathbf{A} = -g\mathbf{j} \qquad (9)$$

Because $\mathbf{A}(t) = \mathbf{V}'(t)$, we have, from (9),

$$\mathbf{V}'(t) = -g\mathbf{j} \tag{10}$$

Integrating on both sides of (10) with respect to t we obtain

$$\mathbf{V}(t) = -gt\mathbf{j} + \mathbf{C}_1 \tag{11}$$

where $\mathbf{C}_1$ is a vector constant of integration.

When $t = 0$, $\mathbf{V} = \mathbf{V}_0$. So $\mathbf{C}_1 = \mathbf{V}_0$. Therefore, from (11),

$$\mathbf{V}(t) = -gt\mathbf{j} + \mathbf{V}_0$$

or, because $\mathbf{V}(t) = \mathbf{R}'(t)$,

$$\mathbf{R}'(t) = -gt\mathbf{j} + \mathbf{V}_0 \tag{12}$$

Integrating on both sides of the vector equation (12) with respect to t we obtain

$$\mathbf{R}(t) = -\tfrac{1}{2}gt^2\mathbf{j} + \mathbf{V}_0 t + \mathbf{C}_2$$

where $\mathbf{C}_2$ is a vector constant of integration.

When $t = 0$, $\mathbf{R} = \mathbf{0}$ because the projectile is at the origin at the start. So $\mathbf{C}_2 = \mathbf{0}$. Therefore

$$\mathbf{R}(t) = -\tfrac{1}{2}gt^2\mathbf{j} + \mathbf{V}_0 t \tag{13}$$

Substituting the value of $\mathbf{V}_0$ from (3) into (13) we obtain

$$\mathbf{R}(t) = -\tfrac{1}{2}gt^2\mathbf{j} + (v_0 \cos \alpha \mathbf{i} + v_0 \sin \alpha \mathbf{j})t$$

$$\mathbf{R}(t) = tv_0 \cos \alpha \mathbf{i} + (tv_0 \sin \alpha - \tfrac{1}{2}gt^2)\mathbf{j} \tag{14}$$

Equation (14) gives the position vector of the projectile at time t sec. From this equation we can discuss the motion of the projectile. We are usually concerned with the following questions:

1. What is the range of the projectile? The range is the distance $|\overline{OA}|$ along the x axis (see Fig. 13.8.4).
2. What is the total time of flight, that is, the time it takes the projectile to go from O to A?
3. What is the maximum height of the projectile?
4. What is a cartesian equation of the curve traveled by the projectile?
5. What is the velocity vector of the projectile at impact?

These questions are answered in the following example.

EXAMPLE 3

A projectile is shot from a gun at an angle of elevation of radian measure $\tfrac{1}{6}\pi$. Its muzzle speed is 480 ft/sec. Find (a) the position vector of the projectile at any time; (b) the time of flight; (c) the range; (d) the maximum height; (e) the velocity vector of the projectile at impact; (f) the position vector and the velocity vector at 2 sec; (g) the speed at 2 sec; (h) a cartesian equation of the curve traveled by the projectile.

SOLUTION

From (3) with $v_0 = 480$ and $\alpha = \frac{1}{6}\pi$, the initial velocity vector is

$$\mathbf{V}_0 = 480 \cos \tfrac{1}{6}\pi \mathbf{i} + 480 \sin \tfrac{1}{6}\pi \mathbf{j}$$
$$= 240\sqrt{3}\mathbf{i} + 240\mathbf{j}$$

(a) We can obtain the position vector at t sec by applying (14); we get

$$\mathbf{R}(t) = 240\sqrt{3}t\mathbf{i} + (240t - \tfrac{1}{2}gt^2)\mathbf{j}$$

By letting $g = 32$ we have

$$\mathbf{R}(t) = 240\sqrt{3}t\mathbf{i} + (240t - 16t^2)\mathbf{j} \tag{15}$$

Thus if (x, y) is the position of the projectile at t sec,

$$x = 240\sqrt{3}t \quad \text{and} \quad y = 240t - 16t^2 \tag{16}$$

(b) To determine the time of flight, we must find t when $y = 0$. We set $y = 0$ in the second equation of (16), and we have

$$240t - 16t^2 = 0$$

$$t(240 - 16t) = 0$$

$$t = 0 \qquad t = 15$$

The value $t = 0$ occurs when the projectile is fired. The value $t = 15$ gives the time of flight. Thus the time of flight is 15 sec.

(c) To find the range, we determine x when $t = 15$. From the first equation of (16) with $t = 15$, we obtain $x = 3600\sqrt{3}$. Hence the range is $3600\sqrt{3}$ ft ≈ 6235 ft.

(d) The maximum height is attained when the vertical component of the velocity vector is 0, that is, when $\dfrac{dy}{dt} = 0$. We compute $\dfrac{dy}{dt}$ from the second equation of (16) and get

$$\frac{dy}{dt} = 240 - 32t$$

Setting $\dfrac{dy}{dt} = 0$ we obtain $t = \frac{15}{2}$, which is half the total time of flight. When $t = \frac{15}{2}$, $y = 900$. So the maximum height attained is 900 ft.

(e) Because the time of flight is 15 sec, the velocity vector at impact is $\mathbf{V}(15)$. Because $\mathbf{V}(t) = \mathbf{R}'(t)$, we get, from (15),

$$\mathbf{V}(t) = 240\sqrt{3}\mathbf{i} + (240 - 32t)\mathbf{j} \tag{17}$$

$$\mathbf{V}(15) = 240\sqrt{3}\mathbf{i} - 240\mathbf{j}$$

(f) If $t = 2$ in (15) and (17), we have

$$\mathbf{R}(2) = 480\sqrt{3}\mathbf{i} + 416\,\mathbf{j} \qquad \mathbf{V}(2) = 240\sqrt{3}\mathbf{i} + 176\mathbf{j}$$

(g) $\|\mathbf{V}(2)\| = \sqrt{(240\sqrt{3})^2 + (176)^2}$
$$= 32\sqrt{199}$$

Therefore at 2 sec the speed is $32\sqrt{199}$ ft/sec ≈ 451.4 ft/sec.

(h) To find a cartesian equation of the curve traveled by the projectile, we eliminate t between the parametric equations (16). Substituting the value of t from the first equation into the second, we have

$$y = 240 \left(\frac{x}{240\sqrt{3}} \right) - 16 \left(\frac{x}{240\sqrt{3}} \right)^2$$

$$y = \frac{1}{\sqrt{3}} x - \frac{1}{10,800} x^2$$

which is an equation of a parabola.

EXERCISES 13.8

In Exercises 1 through 8, a particle is moving along the curve having the given parametric equations, where t sec is the time. Find: (a) the velocity vector $\mathbf{V}(t)$; (b) the acceleration vector $\mathbf{A}(t)$; (c) the speed at $t = t_1$; (d) the magnitude of the acceleration vector at $t = t_1$. Draw a sketch of the path of the particle and the representations of the velocity vector and the acceleration vector at $t = t_1$.

1. $x = t^2 + 4$, $y = t - 2$; $t_1 = 3$
2. $x = \ln(t - 2)$, $y = t^3 - 1$; $t_1 = 3$
3. $x = 5 \cos 2t$, $y = 3 \sin 2t$; $t_1 = \frac{1}{4}\pi$
4. $x = 2/t$, $y = -\frac{1}{4}t$; $t_1 = 4$
5. $x = t$, $y = \ln \sec t$; $t_1 = \frac{1}{4}\pi$
6. $x = 2 \cos t$, $y = 3 \sin t$; $t_1 = \frac{1}{3}\pi$
7. $x = \sin t$, $y = \tan t$; $t_1 = \frac{1}{6}\pi$
8. $x = e^{2t}$, $y = e^{3t}$; $t_1 = 0$

In Exercises 9 through 16, the position of a moving particle at t sec is determined from a vector equation. Find: (a) $\mathbf{V}(t_1)$; (b) $\mathbf{A}(t_1)$; (c) $\|\mathbf{V}(t_1)\|$; (d) $\|\mathbf{A}(t_1)\|$. Draw a sketch of a portion of the path of the particle containing the position of the particle at $t = t_1$, and draw the representations of $\mathbf{V}(t_1)$ and $\mathbf{A}(t_1)$ having initial point where $t = t_1$.

9. $\mathbf{R}(t) = (2t - 1)\mathbf{i} + (t^2 + 1)\mathbf{j}$; $t_1 = 3$
10. $\mathbf{R}(t) = (1 - t)\mathbf{i} + (t^2 - 1)\mathbf{j}$; $t_1 = -1$
11. $\mathbf{R}(t) = e^t\mathbf{i} + e^{2t}\mathbf{j}$; $t_1 = \ln 2$
12. $\mathbf{R}(t) = (t^2 + 3t)\mathbf{i} + (1 - 3t^2)\mathbf{j}$; $t_1 = \frac{1}{2}$
13. $\mathbf{R}(t) = \cos 2t\mathbf{i} - 3 \sin t\mathbf{j}$; $t_1 = \pi$
14. $\mathbf{R}(t) = e^{-t}\mathbf{i} + e^{2t}\mathbf{j}$; $t_1 = \ln 2$
15. $\mathbf{R}(t) = 2(1 - \cos t)\mathbf{i} + 2(1 - \sin t)\mathbf{j}$; $t_1 = \frac{5}{6}\pi$
16. $\mathbf{R}(t) = \ln(t + 2)\mathbf{i} + \frac{1}{3}t^2\mathbf{j}$; $t_1 = 1$

In Exercises 17 through 20, find the position vector $\mathbf{R}(t)$.

17. $\mathbf{V}(t) = \dfrac{1}{(t - 1)^2}\mathbf{i} - (t + 1)\mathbf{j}$, and $\mathbf{R}(0) = 3\mathbf{i} + 2\mathbf{j}$
18. $\mathbf{V}(t) = (2t - 1)\mathbf{i} + 3t^{-2}\mathbf{j}$, and $\mathbf{R}(1) = 4\mathbf{i} - 3\mathbf{j}$

19. $\mathbf{A}(t) = e^{-t}\mathbf{i} + 2e^{2t}\mathbf{j}$, $\mathbf{V}(0) = 2\mathbf{i} + \mathbf{j}$, and $\mathbf{R}(0) = 3\mathbf{j}$
20. $\mathbf{A}(t) = 2 \cos 2t\mathbf{i} + 2 \sin 2t\mathbf{j}$, $\mathbf{V}(0) = \mathbf{i} + \mathbf{j}$, and $\mathbf{R}(0) = \frac{1}{2}\mathbf{i} - \frac{1}{2}\mathbf{j}$

21. A projectile is shot from a gun at an angle of elevation of $45°$ with a muzzle speed of 2500 ft/sec. Find (a) the range of the projectile; (b) the maximum height reached; (c) the velocity at impact.

22. A projectile is shot from a gun at an angle of elevation of $60°$. The muzzle speed is 160 ft/sec. Find (a) the position vector of the projectile at t sec; (b) the time of flight; (c) the range; (d) the maximum height reached; (e) the velocity at impact; (f) the speed at 4 sec.

23. A projectile is shot from the top of a building 96 ft high from a gun at an angle of $30°$ with the horizontal. If the muzzle speed is 1600 ft/sec, find the time of flight and the distance from the base of the building to the point where the projectile lands.

24. The muzzle speed of a gun is 160 ft/sec. At what angle of elevation should the gun be fired so that a projectile will hit an object on the same level as the gun and a distance of 400 ft from it?

25. What is the muzzle speed of a gun if a projectile fired from it has a range of 2000 ft and reaches a maximum height of 1000 ft?

26. A ball is thrown horizontally from the top of a cliff 256 ft high with an initial speed of 50 ft/sec. Find the time of flight of the ball and the distance from the base of the cliff to the point where the ball lands.

27. A person throws a ball with an initial speed of 60 ft/sec at an angle of elevation of 60° toward a tall building that is 25 ft from the person. If the person's hand is 5 ft from the ground, show that the ball hits the building, and find the direction of the ball when it hits the building.

28. From the top of a building 60 ft high, a girl tosses a rock toward the ground at an angle of 45° with the horizontal at an initial speed of 15 ft/sec. Determine the distance on the ground from the base of the building to the landing position of the rock.

29. Solve Exercise 28 if the girl tosses the rock horizontally with an initial speed of 15 ft/sec.

30. At what angle of elevation should a gun be fired to obtain the maximum range for a given muzzle speed?

13.9 (*Supplementary*)
TANGENTIAL AND NORMAL COMPONENTS OF ACCELERATION

If a particle is moving along a curve C having the vector equation

$$\mathbf{R}(t) = f(t)\mathbf{i} + g(t)\mathbf{j}$$

the velocity vector at a point P is given by

$$\mathbf{V}(t) = D_t\mathbf{R}(t) \tag{1}$$

From Eq. (8) in Section 13.6, if $\mathbf{T}(t)$ is the unit tangent vector at P, s is the length of arc of C from a fixed point P_0 to P, and s increases as t increases,

$$D_t\mathbf{R}(t) = \frac{ds}{dt}\mathbf{T}(t) \tag{2}$$

Substituting from (2) into (1) we have

$$\mathbf{V}(t) = \frac{ds}{dt}\mathbf{T}(t) \tag{3}$$

Equation (3) expresses the velocity vector at a point as a scalar times the unit tangent vector at the point. The coefficient of $\mathbf{T}(t)$ is called the **tangential component of the velocity vector** and it is $\dfrac{ds}{dt}$. We now proceed to express the acceleration vector at a point in terms of a vector tangent to the direction of motion and a vector normal to the direction of motion.

The acceleration vector at P is given by

$$\mathbf{A}(t) = D_t{}^2\mathbf{R}(t) \tag{4}$$

From (6) in Section 13.6,

$$D_t{}^2\mathbf{R}(t) = \left(D_t\|D_t\mathbf{R}(t)\|\right)\mathbf{T}(t) + \left(\|D_t\mathbf{R}(t)\|\,\|D_t\mathbf{T}(t)\|\right)\mathbf{N}(t) \tag{5}$$

From (14) in Section 13.5,

$$\frac{ds}{dt} = \|D_t\mathbf{R}(t)\| \tag{6}$$

Differentiating with respect to t on both sides of (6) we obtain

$$\frac{d^2s}{dt^2} = D_t \|D_t \mathbf{R}(t)\| \tag{7}$$

Furthermore,

$$\|D_t\mathbf{R}(t)\| \, \|D_t\mathbf{T}(t)\| = \|D_t\mathbf{R}(t)\|^2 \left\| \frac{D_t\mathbf{T}(t)}{\|D_t\mathbf{R}(t)\|} \right\| \tag{8}$$

Applying (6) above and Eq. (7) of Section 13.7 to the right side of (8) we have

$$\|D_t\mathbf{R}(t)\| \, \|D_t\mathbf{T}(t)\| = \left(\frac{ds}{dt}\right)^2 K(t) \tag{9}$$

Substituting from (4), (7), and (9) into (5) we obtain

$$\mathbf{A}(t) = \frac{d^2s}{dt^2}\mathbf{T}(t) + \left(\frac{ds}{dt}\right)^2 K(t)\mathbf{N}(t) \tag{10}$$

Equation (10) expresses the acceleration vector as the sum of a scalar times the unit tangent vector and a scalar times the unit normal vector; that is, it converts $\mathbf{A}(t)$ into the sum of a vector tangent to the direction of motion and a vector normal to the direction of motion. The coefficient of $\mathbf{T}(t)$ is called the **tangential component of the acceleration vector** and is denoted by $A_T(t)$, whereas the coefficient of $\mathbf{N}(t)$ is called the **normal component of the acceleration vector** and is denoted by $A_N(t)$. Thus

$$A_T(t) = \frac{d^2s}{dt^2} \tag{11}$$

and

$$A_N(t) = \left(\frac{ds}{dt}\right)^2 K(t) \quad \Leftrightarrow \quad A_N(t) = \frac{\left(\frac{ds}{dt}\right)^2}{\rho(t)} \tag{12}$$

Because $\mathbf{A}(t) = D_t\mathbf{V}(t)$, $\mathbf{A}(t)$ is the rate of change of $\mathbf{V}(t)$. A change in $\mathbf{V}(t)$ can be caused by either a change in its magnitude or a change in its direction. Since $\|\mathbf{V}(t)\|$ is the measure of the speed of the particle at time t units and $\dfrac{ds}{dt} = \|\mathbf{V}(t)\|$, then $A_T(t)$ is the rate of change of the measure of the speed of the particle; that is, $A_T(t)$ is related to the change in the magnitude of $\mathbf{V}(t)$. Because $A_N(t)$ involves the curvature $K(t)$, $A_N(t)$ is related to the change in the direction of $\mathbf{V}(t)$. These results are important in mechanics. From Newton's second law of motion,

$$\mathbf{F} = m\mathbf{A} \tag{13}$$

where $\mathbf{F}$ is the force vector applied to a moving object, m is the constant measure of the mass of the object and $\mathbf{A}$ is the acceleration vector of the object. Substituting from (10) in (13) and letting $v = \dfrac{ds}{dt}$ we have

$$\mathbf{F}(t) = m\frac{dv}{dt}\mathbf{T}(t) + mv^2 K(t)\mathbf{N}(t) \tag{14}$$

Thus in curvilinear motion the normal component of $\mathbf{F}$ is

$$mv^2 K(t) \quad \Leftrightarrow \quad \frac{mv^2}{\rho(t)}$$

which is the magnitude of the force normal to the curve necessary to keep the object on the curve. For example, if an automobile is going around a curve at a high speed, then the normal force must have a large magnitude to keep the car on the road. Also, if the curve is sharp, the radius of curvature is a small number; so the magnitude of the normal force must be a large number.

Substituting from (11) and (12) into (10) we have

$$\mathbf{A}(t) = A_T(t)\mathbf{T}(t) + A_N(t)\mathbf{N}(t) \tag{15}$$

from which it follows that

$$\|\mathbf{A}(t)\| = \sqrt{[A_T(t)]^2 + [A_N(t)]^2}$$

Solving this equation for $A_N(t)$, and noting from (12) that $A_N(t)$ is nonnegative, we have

$$A_N(t) = \sqrt{\|\mathbf{A}(t)\|^2 - [A_T(t)]^2} \tag{16}$$

EXAMPLE 1

A particle is moving along the curve having the vector equation

$$\mathbf{R}(t) = t\mathbf{i} + e^t\mathbf{j}$$

Find the tangential and normal components of the acceleration vector.

SOLUTION

$$\begin{aligned}
\mathbf{V}(t) &= D_t\mathbf{R}(t) & \mathbf{A}(t) &= D_t\mathbf{V}(t) \\
&= \mathbf{i} + e^t\mathbf{j} & &= e^t\mathbf{j} \\
\|\mathbf{V}(t)\| &= \sqrt{1 + e^{2t}} & \|\mathbf{A}(t)\| &= e^t
\end{aligned}$$

Because $\dfrac{ds}{dt} = \|\mathbf{V}(t)\|$, $\dfrac{ds}{dt} = \sqrt{1 + e^{2t}}$, and $\dfrac{d^2s}{dt^2} = \dfrac{e^{2t}}{\sqrt{1 + e^{2t}}}$. Hence

$$A_T(t) = \frac{e^{2t}}{\sqrt{1 + e^{2t}}} \qquad A_N(t) = \sqrt{\|\mathbf{A}(t)\|^2 - [A_T(t)]^2}$$

$$= \sqrt{e^{2t} - \frac{e^{4t}}{1 + e^{2t}}}$$

$$= \frac{e^t}{\sqrt{1 + e^{2t}}}$$

EXAMPLE 2

A particle is moving along the curve having the vector equation

$$\mathbf{R}(t) = (t^2 - 1)\mathbf{i} + (\tfrac{1}{3}t^3 - t)\mathbf{j}$$

Find each of the following vectors: $\mathbf{V}(t)$, $\mathbf{A}(t)$, $\mathbf{T}(t)$, and $\mathbf{N}(t)$. Also find the following scalars: $\|\mathbf{V}(t)\|$, $A_T(t)$, $A_N(t)$, and $K(t)$. Find the particular values when $t = 2$. Draw a sketch showing a portion of the curve at the point where $t = 2$, and representations of $\mathbf{V}(2)$, $\mathbf{A}(2)$, $A_T(2)\mathbf{T}(2)$, and $A_N(2)\mathbf{N}(2)$, having their initial point at $t = 2$.

SOLUTION

Because $\mathbf{V}(t) = D_t\mathbf{R}(t)$ and $\mathbf{A}(t) = D_t\mathbf{V}(t)$, we have

$$\mathbf{V}(t) = 2t\mathbf{i} + (t^2 - 1)\mathbf{j} \qquad \mathbf{A}(t) = 2\mathbf{i} + 2t\mathbf{j}$$

$$\|\mathbf{V}(t)\| = \sqrt{4t^2 + (t^2 - 1)^2} \qquad \|\mathbf{A}(t)\| = \sqrt{4 + 4t^2}$$
$$= \sqrt{t^4 + 2t^2 + 1} \qquad\qquad = 2\sqrt{1 + t^2}$$
$$= t^2 + 1$$

Therefore, $\dfrac{ds}{dt} = t^2 + 1$. Hence

$$A_T(t) = \frac{d^2s}{dt^2} \qquad A_N(t) = \sqrt{\|\mathbf{A}(t)\|^2 - [A_T(t)]^2}$$
$$= \sqrt{4 + 4t^2 - 4t^2}$$
$$= 2t \qquad\qquad = 2$$

$$\mathbf{T}(t) = \frac{\mathbf{V}(t)}{\|\mathbf{V}(t)\|}$$

$$= \frac{2t}{t^2 + 1}\mathbf{i} + \frac{t^2 - 1}{t^2 + 1}\mathbf{j}$$

To compute $\mathbf{N}(t)$ we use the following formula that comes from (10):

$$\mathbf{N}(t) = \frac{1}{(D_t s)^2 K(t)}\left[\mathbf{A}(t) - (D_t^2 s)\mathbf{T}(t)\right] \tag{17}$$

$$\mathbf{A}(t) - (D_t^2 s)\mathbf{T}(t) = 2\mathbf{i} + 2t\mathbf{j} - 2t\left(\frac{2t}{t^2 + 1}\mathbf{i} + \frac{t^2 - 1}{t^2 + 1}\mathbf{j}\right)$$

$$\mathbf{A}(t) - (D_t^2 s)\mathbf{T}(t) = \frac{2}{t^2 + 1}\left[(1 - t^2)\mathbf{i} + 2t\mathbf{j}\right] \tag{18}$$

From (17), $\mathbf{N}(t)$ is a scalar times the vector in (18). Because $\mathbf{N}(t)$ is a unit vector, $\mathbf{N}(t)$ can be obtained by dividing the vector in (18) by its magnitude. Thus

$$\mathbf{N}(t) = \frac{(1 - t^2)\mathbf{i} + 2t\mathbf{j}}{\sqrt{(1 - t^2)^2 + (2t)^2}}$$

$$= \frac{1 - t^2}{1 + t^2}\mathbf{i} + \frac{2t}{1 + t^2}\mathbf{j}$$

FIGURE 13.9.1

The curvature $K(t)$ is found from the first equation in (12). Because $\dfrac{ds}{dt} = t^2 + 1$, we get

$$K(t) = \frac{2}{(t^2 + 1)^2}$$

The required vectors and scalars at $t = 2$ are as follows:

$$\mathbf{V}(2) = 4\mathbf{i} + 3\mathbf{j} \qquad \mathbf{A}(2) = 2\mathbf{i} + 4\mathbf{j}$$

$$\|\mathbf{V}(2)\| = 5 \qquad A_T(2) = 4$$

$$\mathbf{T}(2) = \tfrac{4}{5}\mathbf{i} + \tfrac{3}{5}\mathbf{j} \qquad \mathbf{N}(2) = -\tfrac{3}{5}\mathbf{i} + \tfrac{4}{5}\mathbf{j}$$

$$A_N(2) = 2 \qquad K(2) = \tfrac{2}{25}$$

The sketch appears in Fig. 13.9.1.

EXERCISES 13.9

In Exercises 1 through 4, a particle is moving along the curve having the given vector equation. Find the vectors $\mathbf{V}(t)$ and $\mathbf{A}(t)$ and the scalars $A_T(t)$ and $A_N(t)$.

1. $\mathbf{R}(t) = t\mathbf{i} + t^2\mathbf{j}$
2. $\mathbf{R}(t) = 2\sin 4t\mathbf{i} + 2\cos 4t\mathbf{j}$
3. $\mathbf{R}(t) = (\cos t + t \sin t)\mathbf{i} + (\sin t - t \cos t)\mathbf{j}; \ t \geq 0$
4. $\mathbf{R}(t) = (t^3 - 3t)\mathbf{i} + 3t^2\mathbf{j}$

In Exercises 5 through 8, a particle is moving along the curve having the given vector equation. Find $\mathbf{V}(t_1)$, $\mathbf{A}(t_1)$, $A_T(t_1)$, and $A_N(t_1)$ for the given value of t_1.

5. $\mathbf{R}(t) = e^{-t}\mathbf{i} + e^t\mathbf{j}; \ t_1 = 0$
6. $\mathbf{R}(t) = \cos^2 t\mathbf{i} + \sin^2 t\mathbf{j}; \ t_1 = \tfrac{1}{6}\pi$
7. $\mathbf{R}(t) = \sin^3 t\mathbf{i} + \cos^3 t\mathbf{j}; \ t_1 = \tfrac{1}{4}\pi$
8. $\mathbf{R}(t) = e^{-2t}\mathbf{i} + e^{2t}\mathbf{j}; \ t_1 = \ln 2$

In Exercises 9 and 10, find the following scalars: $\|\mathbf{V}(t)\|$, $A_T(t)$, $A_N(t)$, and $K(t)$.

9. $\mathbf{R}(t) = t^2\mathbf{i} + t^3\mathbf{j}; \ t \geq 0$
10. $\mathbf{R}(t) = (t^2 + 4)\mathbf{i} + (2t - 5)\mathbf{j}$

In Exercises 11 through 16, a particle is moving along the curve having the given vector equation. In each exercise, find the vectors $\mathbf{V}(t)$, $\mathbf{A}(t)$, $\mathbf{T}(t)$, and $\mathbf{N}(t)$, and the following scalars for an arbitrary value of t: $\|\mathbf{V}(t)\|$, $A_T(t)$, $A_N(t)$, and $K(t)$. Also find the particular values when $t = t_1$. At $t = t_1$, draw a sketch of a portion of the curve and representations of the vectors $\mathbf{V}(t_1)$, $\mathbf{A}(t_1)$, $A_T(t_1)\mathbf{T}(t_1)$, and $A_N(t_1)\mathbf{N}(t_1)$.

11. $\mathbf{R}(t) = (2t + 3)\mathbf{i} + (t^2 - 1)\mathbf{j}; \ t_1 = 2$
12. $\mathbf{R}(t) = (t - 1)\mathbf{i} + t^2\mathbf{j}; \ t_1 = 1$
13. $\mathbf{R}(t) = 5\cos 3t\mathbf{i} + 5\sin 3t\mathbf{j}; \ t_1 = \tfrac{1}{6}\pi$
14. $\mathbf{R}(t) = 3t^2\mathbf{i} + 2t^3\mathbf{j}; \ t_1 = 1$
15. $\mathbf{R}(t) = e^t\mathbf{i} + e^{-t}\mathbf{j}; \ t_1 = 0$
16. $\mathbf{R}(t) = \cos t^2\mathbf{i} + \sin t^2\mathbf{j}; \ t_1 = \tfrac{1}{2}\sqrt{\pi}$

In Exercises 17 and 18, a particle is moving along the curve having the given cartesian equation. At the given point find (a) the position vector, (b) the velocity vector, (c) the acceleration vector, (d) A_T, and (e) A_N.

17. $y = 4x^2; (1, 4)$
18. $y^2 = x^3; (4, 8)$
19. A particle is moving along the parabola $y^2 = 8x$ and its speed is constant. Find each of the following when the particle is at $(2, 4)$: the position vector, the velocity vector, the acceleration vector, the unit tangent vector, the unit normal vector, A_T, and A_N.

20. A particle is moving along the top branch of the hyperbola $y^2 - x^2 = 9$, such that $\dfrac{dx}{dt}$ is a positive constant. Find each of the following when the particle is at $(4, 5)$: the position vector, the velocity vector, the acceleration vector, the unit tangent vector, the unit normal vector, A_T, and A_N.

REVIEW EXERCISES FOR CHAPTER 13

In Exercises 1 through 18, $\mathbf{A} = 4\mathbf{i} - 6\mathbf{j}$, $\mathbf{B} = \mathbf{i} + 7\mathbf{j}$, and $\mathbf{C} = 9\mathbf{i} - 5\mathbf{j}$.

1. Find $3\mathbf{B} - 7\mathbf{A}$.
2. Find $5\mathbf{B} - 3\mathbf{C}$.
3. Find $\|3\mathbf{B} - 7\mathbf{A}\|$.
4. Find $\|5\mathbf{B} - 3\mathbf{C}\|$.
5. Find $\|3\mathbf{B}\| - \|7\mathbf{A}\|$.
6. Find $\|5\mathbf{B}\| - \|3\mathbf{C}\|$.
7. Find $(\mathbf{A} - \mathbf{B}) \cdot \mathbf{C}$.
8. Find $(\mathbf{A} \cdot \mathbf{B})\mathbf{C}$.
9. Find a vector having the same direction as $2\mathbf{A} + \mathbf{B}$.
10. Find the unit vectors that are orthogonal to $\mathbf{B}$.
11. Find scalars h and k such that $\mathbf{A} = h\mathbf{B} + k\mathbf{C}$.
12. Find scalars h and k such that $h\mathbf{A} + k\mathbf{B} = -\mathbf{C}$.
13. Find the scalar projection of $\mathbf{A}$ onto $\mathbf{B}$.
14. Find the scalar projection of $\mathbf{C}$ onto $\mathbf{A}$.
15. Find the vector projection of $\mathbf{A}$ onto $\mathbf{B}$.
16. Find the vector projection of $\mathbf{C}$ onto $\mathbf{A}$.
17. Find the component of $\mathbf{B}$ in the direction of $\mathbf{A}$.
18. Find $\cos \alpha$ if α is the angle between $\mathbf{A}$ and $\mathbf{C}$.

In Exercises 19 and 20, for the vector-valued function, find (a) the domain of $\mathbf{R}$; (b) $\lim\limits_{t \to 1} \mathbf{R}(t)$; (c) $D_t\mathbf{R}(t)$.

19. $\mathbf{R}(t) = \dfrac{1}{t+1}\mathbf{i} + \dfrac{\sqrt{t-1}}{t-1}\mathbf{j}$

20. $\mathbf{R}(t) = |t - 1|\mathbf{i} + \ln t\mathbf{j}$

In Exercises 21 and 22, find $\dfrac{dy}{dx}$ and $\dfrac{d^2y}{dx^2}$ without eliminating the parameter.

21. $x = 9t^2 - 1$, $y = 3t + 1$

22. $x = e^{2t}$, $y = e^{-3t}$

In Exercises 23 and 24, find equations of the horizontal and vertical tangent lines, and then draw a sketch of the graph of the given pair of parametric equations.

23. $x = 12 - t^2$, $y = 12t - t^3$

24. $x = \dfrac{2at^2}{1+t^2}$, $y = \dfrac{2at^3}{1+t^2}$, $a > 0$ (the cissoid of Diocles)

25. If $\mathbf{R}(t) = \ln(t^2 - 1)\mathbf{i} - 2t^{-3}\mathbf{j}$, find $\mathbf{R}'(t) \cdot \mathbf{R}''(t)$.
26. Find the length of the arc of the curve having parametric equations $x = t^2$, $y = t^3$, from $t = 1$ to $t = 2$.
27. Find the length of the arc of the curve $\mathbf{R}(t) = (2 - t)\mathbf{i} + t^2\mathbf{j}$ from $t = 0$ to $t = 3$.
28. Find the length of the arc of the curve $r = 3 \sec \theta$ from $\theta = 0$ to $\theta = \frac{1}{4}\pi$.
29. (a) Show that the curve defined by the parametric equations $x = a \sin t$ and $y = b \cos t$ is an ellipse. (b) If s is the measure of the length of arc of the ellipse of part (a), show that

$$s = 4 \int_0^{\pi/2} a\sqrt{1 - k^2 \sin^2 t}\, dt$$

where $k^2 = (a^2 - b^2)/a^2 < 1$. This integral is called an *elliptic integral* and cannot be evaluated exactly in terms of elementary functions. Tables are available that give the value of the integral in terms of k.

30. Draw a sketch of the graph of the vector equation $\mathbf{R}(t) = e^t\mathbf{i} + e^{-t}\mathbf{j}$, and find a cartesian equation of the graph.
31. Show that the curvature of the curve $y = \ln x$ at any point (x, y) is $x/(x^2 + 1)^{3/2}$. Also show that the absolute maximum curvature is $\frac{2}{9}\sqrt{3}$, which occurs at the point $(\frac{1}{2}\sqrt{2}, -\frac{1}{2} \ln 2)$.
32. Find the curvature at any point of the branch of the hyperbola defined by $x = a \cosh t$, $y = b \sinh t$. Also show that the curvature is an absolute maximum at the vertex.
33. Find the radius of curvature at any point on the curve $x = a(\cos t + t \sin t)$, $y = a(\sin t - t \cos t)$.
34. Find the curvature, the radius of curvature, and the center of curvature of the curve $y = e^{-x}$ at the point $(0, 1)$.
35. Find the curvature and radius of curvature of the curve $\mathbf{R}(t) = 3t^2\mathbf{i} + (t^3 - 3t)\mathbf{j}$ at the point where $t = 2$.
36. If $\mathbf{R}(t) = e^{\lambda t}\mathbf{i} + e^{-\lambda t}\mathbf{j}$, where λ is a constant, show that $\mathbf{R}(t)$ satisfies the equation $\mathbf{R}''(t) - \lambda^2\mathbf{R}(t) = \mathbf{0}$.
37. A particle is moving along a curve having the vector equation $\mathbf{R}(t) = 3t\mathbf{i} + (4t - t^2)\mathbf{j}$. (a) Find a cartesian equation of the path of the particle. (b) Find the velocity vector and the acceleration vector. (c) Find $\mathbf{V}(1)$ and $\mathbf{A}(1)$.
38. Follow the instructions of Exercise 37 if $\mathbf{R}(t) = 2e^t\mathbf{i} + 3e^{-t}\mathbf{j}$.

39. For the hypocycloid of four cusps, $x = a \cos^3 t$ and $y = a \sin^3 t$, find $\dfrac{dy}{dx}$ and $\dfrac{d^2y}{dx^2}$ without eliminating the parameter.

40. If a particle is moving along a curve, under what conditions will the acceleration vector and the unit tangent vector have the same or opposite directions?

In Exercises 41 and 42, for the given curve find $\mathbf{T}(t)$ and $\mathbf{N}(t)$, and at $t = t_1$ draw a sketch of a portion of the curve and draw the representations of $\mathbf{T}(t_1)$ and $\mathbf{N}(t_1)$ having initial point at $t = t_1$.

41. $\mathbf{R}(t) = (e^t + e^{-t})\mathbf{i} + 2t\mathbf{j}$; $t_1 = 2$ **42.** $\mathbf{R}(t) = 3(\cos t + t \sin t)\mathbf{i} + 3(\sin t - t \cos t)\mathbf{j}$, $t > 0$; $t_1 = \frac{1}{2}\pi$

43. Given the curve having parametric equations $x = 4t$, $y = \frac{1}{3}(2t + 1)^{3/2}$, $t \geq 0$, find parametric equations having the measure of arc length s as a parameter, where arc length is measured from the point where $t = 0$. Check your result by using Eq. (11) of Section 13.6.

44. Find the radian measure of the angle of elevation at which a gun should be fired in order to obtain the maximum range for a given muzzle speed.

45. Find a formula for obtaining the maximum height reached by a projectile fired from a gun having a given muzzle speed of v_0 ft/sec and an angle of elevation of radian measure α.

46. A girl throws a ball horizontally from the top of a cliff 288 ft high at an initial speed of 32 ft/sec. Find (a) the time of flight, and (b) the distance from the base of the cliff to the point where the ball hits the ground.

47. Prove by vector analysis that the diagonals of a parallelogram bisect each other.

48. Find the position vector $\mathbf{R}(t)$ if the acceleration vector $\mathbf{A}(t) = t^2\mathbf{i} - \dfrac{1}{t^2}\mathbf{j}$ and $\mathbf{V}(1) = \mathbf{j}$, and $\mathbf{R}(1) = -\frac{1}{4}\mathbf{i} + \frac{1}{2}\mathbf{j}$.

49. An *epicycloid* is the curve traced by a point P on the circumference of a circle of radius b which is rolling externally on a fixed circle of radius a. If the origin is at the center of the fixed circle, $A(a, 0)$ is one of the points at which the given point P comes in contact with the fixed circle, B is the moving point of tangency of the two circles, and the parameter t is the radian measure of the angle AOB, prove that parametric equations of the epicycloid are

$$x = (a + b)\cos t - b\cos\frac{a + b}{b}t \quad \text{and} \quad y = (a + b)\sin t - b\sin\frac{a + b}{b}t$$

50. Given triangle ABC, points D, E, and F are on the sides AB, BC, and AC, respectively, and $\mathbf{V}(\overrightarrow{AD}) = \frac{1}{3}\mathbf{V}(\overrightarrow{AB})$; $\mathbf{V}(\overrightarrow{BE}) = \frac{1}{3}\mathbf{V}(\overrightarrow{BC})$; $\mathbf{V}(\overrightarrow{CF}) = \frac{1}{3}\mathbf{V}(\overrightarrow{CA})$. Prove $\mathbf{V}(\overrightarrow{AE}) + \mathbf{V}(\overrightarrow{BF}) + \mathbf{V}(\overrightarrow{CD}) = \mathbf{0}$.

Exercises 51 through 54 pertain to Supplementary Section 13.9. In Exercises 51 and 52, find (a) the velocity and acceleration vectors, (b) the speed, and (c) the tangential and normal components of acceleration.

51. $\mathbf{R}(t) = \cosh 2t\mathbf{i} + \sinh 2t\mathbf{j}$ **52.** $\mathbf{R}(t) = (2\tan^{-1} t - t)\mathbf{i} + \ln(1 + t^2)\mathbf{j}$

53. Find the tangential and normal components of the acceleration vector for the particle of Exercise 37.

54. Find the tangential and normal components of the acceleration vector for the particle of Exercise 38.

F O U R T E E N

VECTORS IN THREE-DIMENSIONAL SPACE
AND SOLID ANALYTIC GEOMETRY

14.1
THE THREE-DIMENSIONAL
NUMBER SPACE

Until now we have been concerned with the number line R (the one-dimensional number space) and the number plane R^2 (the two-dimensional number space). We identified the real numbers in R with points on a horizontal axis and the real number pairs in R^2 with points in a geometric plane. In an analogous fashion we now introduce the set of all ordered triples of real numbers.

14.1.1 DEFINITION   The set of all ordered triples of real numbers is called the **three-dimensional number space** and is denoted by R^3. Each ordered triple (x, y, z) is called a **point** in the three-dimensional number space.

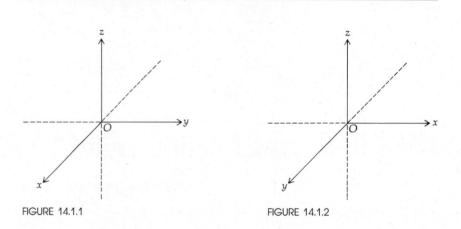

FIGURE 14.1.1   FIGURE 14.1.2

To represent R^3 in a geometric three-dimensional space, consider the directed distances of a point from three mutually perpendicular planes. The planes are formed by first considering three mutually perpendicular lines that intersect at a point that we call the *origin* and denote by the letter O. These lines, called the coordinate axes, are designated as the x axis, the y axis, and the z axis. Usually the x axis and the y axis are taken in a horizontal plane, and the z axis is vertical. A positive direction is selected on each axis. If the positive directions are chosen as in Fig. 14.1.1, the coordinate system is called a *right-handed system*. This terminology follows from the fact that if the right hand is placed so the thumb is pointed in the positive direction of the x axis and the index finger is pointed in the positive direction of the y axis, then the middle finger is pointed in the positive direction of the z axis. If the middle finger is pointed in the negative direction of the z axis, then the coordinate system is called *left handed*. A left-handed system is shown in Fig. 14.1.2. In general, we use a right-handed system. The three axes determine three coordinate planes: the xy plane containing the x and y axes, the xz plane containing the x and z axes, and the yz plane containing the y and z axes.

An ordered triple of real numbers (x, y, z) is associated with each point P in a geometric three-dimensional space. The directed distance of P from the yz plane is called the x *coordinate*, the directed distance of P from the xz plane is called the y *coordinate*, and the z *coordinate* is the directed distance of P

FIGURE 14.1.3

FIGURE 14.1.4

from the xy plane. These three coordinates are called the *rectangular cartesian coordinates* of the point, and there is a one-to-one correspondence (called a *rectangular cartesian coordinate system*) between all such ordered triples of real numbers and the points in a geometric three-dimensional space. Hence we identify R^3 with the geometric three-dimensional space, and we call an ordered triple (x, y, z) a point. The point $(3, 2, 4)$ is shown in Fig. 14.1.3, and the point $(4, -2, -5)$ is shown in Fig. 14.1.4. The three coordinate planes divide the space into eight parts, called *octants*. The first octant is the one in which all three coordinates are positive.

A line is parallel to a plane if and only if the distance from any point on the line to the plane is the same.

▶ ILLUSTRATION 1   A line parallel to the yz plane, one parallel to the xz plane, and one parallel to the xy plane are shown in Fig. 14.1.5(a), (b), and (c), respectively.   ◀

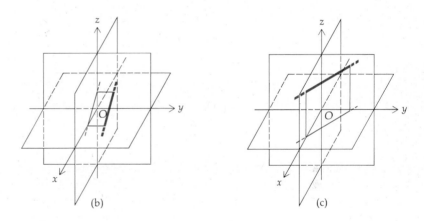

FIGURE 14.1.5

We consider all lines lying in a given plane as being parallel to the plane, in which case the distance from any point on the line to the plane is zero. The following theorem follows immediately.

14.1.2 THEOREM

(i) A line is parallel to the yz plane if and only if all points on the line have equal x coordinates.

(ii) A line is parallel to the xz plane if and only if all points on the line have equal y coordinates.

(iii) A line is parallel to the xy plane if and only if all points on the line have equal z coordinates.

In three-dimensional space, if a line is parallel to each of two intersecting planes, it is parallel to the line of intersection of the two planes. Also, if a given line is parallel to a second line, then the given line is parallel to any plane containing the second line. Theorem 14.1.3 follows from these two facts from solid geometry and from Theorem 14.1.2.

14.1.3 THEOREM

(i) A line is parallel to the x axis if and only if all points on the line have equal y coordinates and equal z coordinates.

(ii) A line is parallel to the y axis if and only if all points on the line have equal x coordinates and equal z coordinates.

(iii) A line is parallel to the z axis if and only if all points on the line have equal x coordinates and equal y coordinates.

▶ ILLUSTRATION 2   A line parallel to the x axis, a line parallel to the y axis, and a line parallel to the z axis are shown in Fig. 14.1.6(a), (b) and (c), respectively.   ◀

(a)   (b)   (c)

FIGURE 14.1.6

The formulas for finding the directed distance from one point to another on a line parallel to a coordinate axis follow from the definition of directed distance given in Section 0.3 and are stated in the following theorem.

14.1.4 THEOREM

(i) If $A(x_1, y, z)$ and $B(x_2, y, z)$ are two points on a line parallel to the x axis, then the directed distance from A to B, denoted by $\overline{AB}$, is given by

$$\overline{AB} = x_2 - x_1$$

(ii) If $C(x, y_1, z)$ and $D(x, y_2, z)$ are two points on a line parallel to the y axis, then the directed distance from C to D, denoted by $\overline{CD}$, is given by

$$\overline{CD} = y_2 - y_1$$

(iii) If $E(x, y, z_1)$ and $F(x, y, z_2)$ are two points on a line parallel to the z axis, then the directed distance from E to F, denoted by $\overline{EF}$, is given by

$$\overline{EF} = z_2 - z_1$$

▶ ILLUSTRATION 3   The directed distance $\overline{PQ}$ from the point $P(2, -5, -4)$ to the point $Q(2, -3, -4)$ is given by Theorem 14.1.4(ii).

$$\overline{PQ} = (-3) - (-5)$$
$$= 2$$

◀

The following theorem gives a formula for finding the undirected distance between any two points in three-dimensional space.

14.1.5 THEOREM   The undirected distance between the two points $P_1(x_1, y_1, z_1)$ and $P_2(x_2, y_2, z_2)$ is given by

$$|\overline{P_1P_2}| = \sqrt{(x_2 - x_1)^2 + (y_2 - y_1)^2 + (z_2 - z_1)^2}$$

PROOF   Construct a rectangular parallelepiped having P_1 and P_2 as opposite vertices and faces parallel to the coordinate planes (see Fig. 14.1.7).

By the Pythagorean theorem,

$$|\overline{P_1P_2}|^2 = |\overline{P_1A}|^2 + |\overline{AP_2}|^2 \tag{1}$$

Because

$$|\overline{P_1A}|^2 = |\overline{P_1B}|^2 + |\overline{BA}|^2 \tag{2}$$

we obtain, by substituting from (2) into (1),

$$|\overline{P_1P_2}|^2 = |\overline{P_1B}|^2 + |\overline{BA}|^2 + |\overline{AP_2}|^2 \tag{3}$$

Applying Theorem 14.1.4(i), (ii), and (iii) to the right side of (3) we obtain

$$|\overline{P_1P_2}|^2 = (x_2 - x_1)^2 + (y_2 - y_1)^2 + (z_2 - z_1)^2$$
$$|\overline{P_1P_2}| = \sqrt{(x_2 - x_1)^2 + (y_2 - y_1)^2 + (z_2 - z_1)^2}$$

and the theorem is proved.   ■

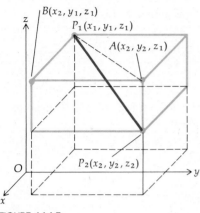

FIGURE 14.1.7

EXAMPLE 1
Find the undirected distance between the points $P(-3, 4, -1)$ and $Q(2, 5, -4)$.

SOLUTION
From Theorem 14.1.5,

$$|\overline{PQ}| = \sqrt{(2 + 3)^2 + (5 - 4)^2 + (-4 + 1)^2}$$
$$= \sqrt{35}$$

The formula for the distance between two points in R^3 is merely an extension of the corresponding formula for the distance between two points in R^2 given in Theorem 0.3.1. It is noteworthy that the undirected distance between two points x_2 and x_1 in R is given by

$$|x_2 - x_1| = \sqrt{(x_2 - x_1)^2}$$

The formulas for the coordinates of the midpoint of a line segment are derived by forming congruent triangles and proceeding in a manner analogous to the two-dimensional case. These formulas are given in Theorem 14.1.6, and the proof is left as an exercise (see Exercise 18).

14.1.6 THEOREM   The coordinates of the midpoint of the line segment having endpoints $P_1(x_1, y_1, z_1)$ and $P_2(x_2, y_2, z_2)$ are given by

$$\bar{x} = \frac{x_1 + x_2}{2} \qquad \bar{y} = \frac{y_1 + y_2}{2} \qquad \bar{z} = \frac{z_1 + z_2}{2}$$

14.1.7 DEFINITION

The **graph of an equation** in R^3 is the set of all points (x, y, z) whose coordinates are numbers satisfying the equation.

The graph of an equation in R^3 is called a **surface**. One particular surface is the sphere, which is now defined.

14.1.8 DEFINITION

A **sphere** is the set of all points in three-dimensional space equidistant from a fixed point. The fixed point is called the **center** of the sphere and the measure of the constant distance is called the **radius** of the sphere.

14.1.9 THEOREM

An equation of the sphere of radius r and center at (h, k, l) is

$$(x - h)^2 + (y - k)^2 + (z - l)^2 = r^2 \qquad (4)$$

PROOF   Let the point (h, k, l) be denoted by C (see Fig. 14.1.8). The point $P(x, y, z)$ is a point on the sphere if and only if

$$|\overline{CP}| = r$$

$$\Leftrightarrow \quad \sqrt{(x - h)^2 + (y - k)^2 + (z - l)^2} = r$$

Squaring on both sides of the above equation we obtain the desired result.   ∎

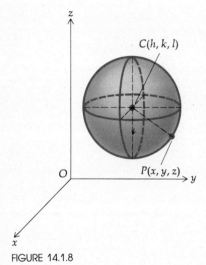

FIGURE 14.1.8

If the center of the sphere is at the origin, then $h = 0$, $k = 0$, $l = 0$; so an equation of this sphere is

$$x^2 + y^2 + z^2 = r^2$$

If we expand the terms of (4) and regroup the terms, we have

$$x^2 + y^2 + z^2 - 2hx - 2ky - 2lz + (h^2 + k^2 + l^2 - r^2) = 0$$

This equation is of the form

$$x^2 + y^2 + z^2 + Gx + Hy + Iz + J = 0 \qquad (5)$$

where G, H, I, and J are constants. Equation (5) is called the *general form* of an equation of a sphere, whereas (4) is called the *center-radius* form. Because every sphere has a center and a radius, its equation can be put in the center-radius form and hence the general form.

It can be shown that any equation of the form (5) can be put in the form

$$(x - h)^2 + (y - k)^2 + (z - l)^2 = K \qquad (6)$$

where

$$h = -\tfrac{1}{2}G \qquad k = -\tfrac{1}{2}H \qquad l = -\tfrac{1}{2}I \qquad K = \tfrac{1}{4}(G^2 + H^2 + I^2 - 4J)$$

It is left as an exercise to show this (see Exercise 19).

If $K > 0$, then (6) is of the form of Eq. (4); so the graph of the equation is a sphere having its center at (h, k, l) and radius $\sqrt{K}$. If $K = 0$, the graph of the equation is the point (h, k, l). This is called a **point-sphere**. If $K < 0$, the graph is the empty set because the sum of the squares of three real numbers is nonnegative. We state this result as a theorem.

14.1.10 THEOREM   The graph of any second-degree equation in x, y, and z, of the form

$$x^2 + y^2 + z^2 + Gx + Hy + Iz + J = 0$$

is either a sphere, a point-sphere, or the empty set.

EXAMPLE 2

Draw a sketch of the graph of the equation

$$x^2 + y^2 + z^2 - 6x - 4y + 2z = 2$$

SOLUTION

Regrouping terms and completing the squares we have

$$x^2 - 6x + 9 + y^2 - 4y + 4 + z^2 + 2z + 1 = 2 + 9 + 4 + 1$$

$$(x - 3)^2 + (y - 2)^2 + (z + 1)^2 = 16$$

So the graph is a sphere having its center at $(3, 2, -1)$ and radius 4. A sketch of the graph is shown in Fig. 14.1.9.

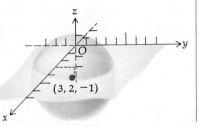

(3, 2, −1)

FIGURE 14.1.9

EXAMPLE 3

Find an equation of the sphere having the points $A(-5, 6, -2)$ and $B(9, -4, 0)$ as endpoints of a diameter.

SOLUTION

The center of the sphere is the midpoint of the line segment AB. Let this point be $C(\bar{x}, \bar{y}, \bar{z})$. By Theorem 14.1.6 we get

$$\bar{x} = \frac{9 - 5}{2} \qquad \bar{y} = \frac{-4 + 6}{2} \qquad \bar{z} = \frac{0 - 2}{2}$$

$$= 2 \qquad\qquad = 1 \qquad\qquad = -1$$

Thus C is the point $(2, 1, -1)$. The radius of the sphere is $|\overline{CB}|$. Hence

$$r = \sqrt{(9 - 2)^2 + (-4 - 1)^2 + (0 + 1)^2}$$

$$= \sqrt{75}$$

Therefore, from Theorem 14.1.9 an equation of the sphere is

$$(x - 2)^2 + (y - 1)^2 + (z + 1)^2 = 75$$

$$x^2 + y^2 + z^2 - 4x - 2y + 2z - 69 = 0$$

EXERCISES 14.1

In Exercises 1 through 5, the given points A and B are opposite vertices of a rectangular parallelepiped having its faces parallel to the coordinate planes. In each exercise, (a) draw a sketch of the figure, (b) find the coordinates of the other six vertices, (c) find the length of the diagonal AB.

1. $A(0, 0, 0)$; $B(7, 2, 3)$ **2.** $A(1, 1, 1)$; $B(3, 4, 2)$ **3.** $A(-1, 1, 2)$; $B(2, 3, 5)$

4. $A(2, -1, -3)$; $B(4, 0, -1)$ **5.** $A(1, -1, 0)$; $B(3, 3, 5)$

6. The vertex opposite one corner of a room is 18 ft east, 15 ft south, and 12 ft up from the first corner. (a) Draw a sketch of the figure; (b) determine the length of the diagonal joining two opposite vertices; (c) find the coordinates of all eight vertices of the room.

In Exercises 7 through 11, find (a) the undirected distance between the points A and B and (b) the midpoint of the line segment joining A and B.

7. $A(3, 4, 2)$; $B(1, 6, 3)$ **8.** $A(4, -3, 2)$; $B(-2, 3, -5)$ **9.** $A(2, -4, 1)$; $B(\frac{1}{2}, 2, 3)$

10. $A(-2, -\frac{1}{2}, 5)$; $B(5, 1, -4)$ **11.** $A(-5, 2, 1)$; $B(3, 7, -2)$

12. Prove that the three points $(1, -1, 3)$, $(2, 1, 7)$, and $(4, 2, 6)$ are the vertices of a right triangle, and find its area.

13. A line is drawn through the point $(6, 4, 2)$ perpendicular to the yz plane. Find the coordinates of the points on this line at a distance of 10 units from the point $(0, 4, 0)$.

14. Solve Exercise 13 if the line is drawn perpendicular to the xy plane.

15. Prove that the three points $(-3, 2, 4)$, $(6, 1, 2)$, and $(-12, 3, 6)$ are collinear by using the distance formula.

16. Find the vertices of the triangle whose sides have midpoints at $(3, 2, 3)$, $(-1, 1, 5)$, and $(0, 3, 4)$.

17. For the triangle having vertices at $A(2, -5, 3)$, $B(-1, 7, 0)$, and $C(-4, 9, 7)$ find (a) the length of each side and (b) the midpoint of each side.

18. Prove Theorem 14.1.6.

19. Show that any equation of the form $x^2 + y^2 + z^2 + Gx + Hy + Iz + J = 0$ can be put in the form

$$(x - h)^2 + (y - k)^2 + (z - l)^2 = K$$

In Exercises 20 through 25, determine the graph of the given equation.

20. $x^2 + y^2 + z^2 - 8y + 6z - 25 = 0$ **21.** $x^2 + y^2 + z^2 - 8x + 4y + 2z - 4 = 0$

22. $x^2 + y^2 + z^2 - x - y - 3z + 2 = 0$ **23.** $x^2 + y^2 + z^2 - 6z + 9 = 0$

24. $x^2 + y^2 + z^2 - 8x + 10y - 4z + 13 = 0$ **25.** $x^2 + y^2 + z^2 - 6x + 2y - 4z + 19 = 0$

In Exercises 26 through 28, find an equation of the sphere satisfying the given conditions.

26. A diameter is the line segment having endpoints at $(6, 2, -5)$ and $(-4, 0, 7)$.

27. It is concentric with the sphere having equation $x^2 + y^2 + z^2 - 2y + 8z - 9 = 0$ and has radius 3.

28. It contains the points $(0, 0, 4)$, $(2, 1, 3)$, and $(0, 2, 6)$ and has its center in the yz plane.

29. Prove analytically that the four diagonals joining opposite vertices of a rectangular parallelepiped bisect each other.

30. If P, Q, R, and S are four points in three-dimensional space and A, B, C, and D are the midpoints of PQ, QR, RS, and SP, respectively, prove analytically that $ABCD$ is a parallelogram.

31. Prove analytically that the four diagonals of a cube have the same length.

14.2
VECTORS IN THREE-DIMENSIONAL SPACE

The presentation of topics in solid analytic geometry is simplified by the use of vectors in three-dimensional space. The definitions and theorems given in Section 13.1 for vectors in the plane are easily extended.

14.2.1 DEFINITION

A **vector in three-dimensional space** is an ordered triple of real numbers $\langle x, y, z \rangle$. The numbers x, y, and z are called the **components** of the vector $\langle x, y, z \rangle$.

We let V_3 be the set of all ordered triples $\langle x, y, z \rangle$ for which x, y, and z are real numbers. In this chapter a vector is always in V_3 unless otherwise stated.

Just as for vectors in V_2, a vector in V_3 can be represented by a directed line segment. If $\mathbf{A} = \langle a_1, a_2, a_3 \rangle$, then the directed line segment having its initial point at the origin and its terminal point at the point (a_1, a_2, a_3) is called the **position representation** of $\mathbf{A}$. A directed line segment having its initial point at (x, y, z) and its terminal point at $(x + a_1, y + a_2, z + a_3)$ is also a representation of the vector $\mathbf{A}$. See Fig. 14.2.1.

FIGURE 14.2.1

The **zero vector** is the vector $\langle 0,0,0 \rangle$ and is denoted by **0**. Any point is a representation of the zero vector.

The **magnitude** of a vector is the length of any of its representations. If the vector $\mathbf{A} = \langle a_1, a_2, a_3 \rangle$, the magnitude of $\mathbf{A}$ is denoted by $\|\mathbf{A}\|$, and it follows that

$$\|\mathbf{A}\| = \sqrt{a_1{}^2 + a_2{}^2 + a_3{}^2}$$

The **direction** of a nonzero vector in V_3 is given by three angles, called the *direction angles* of the vector.

14.2.2 DEFINITION   The **direction angles** of a nonzero vector are the three angles that have the smallest nonnegative radian measures α, β, and γ measured from the positive x, y, and z axes, respectively, to the position representation of the vector.

The radian measure of each direction angle of a vector is greater than or equal to 0 and less than or equal to π. The direction angles having radian measures α, β, and γ of the vector $\mathbf{A} = \langle a_1, a_2, a_3 \rangle$ are shown in Fig. 14.2.2. In this figure the components of $\mathbf{A}$ are all positive numbers, and the direction angles of this vector all have positive radian measure less than $\frac{1}{2}\pi$. From the figure we see that triangle POR is a right triangle and

$$\cos \alpha = \frac{a_1}{\|\mathbf{A}\|}$$

It can be shown that the same formula holds if $\frac{1}{2}\pi \leq \alpha \leq \pi$. Similar formulas can be found for $\cos \beta$ and $\cos \gamma$, and we have

$$\cos \alpha = \frac{a_1}{\|\mathbf{A}\|} \qquad \cos \beta = \frac{a_2}{\|\mathbf{A}\|} \qquad \cos \gamma = \frac{a_3}{\|\mathbf{A}\|} \tag{1}$$

The three numbers $\cos \alpha$, $\cos \beta$, and $\cos \gamma$ are called the **direction cosines** of vector $\mathbf{A}$. The zero vector has no direction angles and hence no direction cosines.

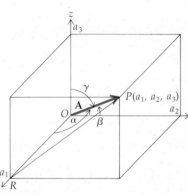

FIGURE 14.2.2

▶ ILLUSTRATION 1   We find the magnitude and direction cosines of the vector $\mathbf{A} = \langle 3, 2, -6 \rangle$.

$$\|\mathbf{A}\| = \sqrt{(3)^2 + (2)^2 + (-6)^2}$$
$$= 7$$

From Eqs. (1),

$$\cos \alpha = \tfrac{3}{7} \qquad \cos \beta = \tfrac{2}{7} \qquad \cos \gamma = -\tfrac{6}{7} \qquad\qquad ◀$$

If the magnitude of a vector and its direction cosines are known, the vector is uniquely determined because from (1) it follows that

$$a_1 = \|\mathbf{A}\| \cos \alpha \qquad a_2 = \|\mathbf{A}\| \cos \beta \qquad a_3 = \|\mathbf{A}\| \cos \gamma \tag{2}$$

The three direction cosines of a vector are not independent of each other, as we see by the following theorem.

14.2.3 THEOREM   If $\cos \alpha$, $\cos \beta$, and $\cos \gamma$ are the direction cosines of a vector, then

$$\cos^2 \alpha + \cos^2 \beta + \cos^2 \gamma = 1$$

PROOF   If $\mathbf{A} = \langle a_1, a_2, a_3 \rangle$, then the direction cosines of $\mathbf{A}$ are given by (1) and

$$\cos^2 \alpha + \cos^2 \beta + \cos^2 \gamma = \frac{a_1{}^2}{\|\mathbf{A}\|^2} + \frac{a_2{}^2}{\|\mathbf{A}\|^2} + \frac{a_3{}^2}{\|\mathbf{A}\|^2}$$

$$= \frac{a_1{}^2 + a_2{}^2 + a_3{}^2}{\|\mathbf{A}\|^2}$$

$$= \frac{\|\mathbf{A}\|^2}{\|\mathbf{A}\|^2}$$

$$= 1 \qquad \blacksquare$$

▶ ILLUSTRATION 2   We verify Theorem 14.2.3 for the vector of Illustration 1.

$$\cos^2 \alpha + \cos^2 \beta + \cos^2 \gamma = (\tfrac{3}{7})^2 + (\tfrac{2}{7})^2 + (-\tfrac{6}{7})^2$$

$$= \tfrac{9}{49} + \tfrac{4}{49} + \tfrac{36}{49}$$

$$= \tfrac{49}{49}$$

$$= 1 \qquad \blacktriangleleft$$

The vector $\mathbf{A} = \langle a_1, a_2, a_3 \rangle$ is a unit vector if $\|\mathbf{A}\| = 1$, and from Eqs. (1) the components of a unit vector are its direction cosines.

The operations of addition, subtraction, and scalar multiplication of vectors in V_3 are given definitions analogous to the corresponding definitions for vectors in V_2.

14.2.4 DEFINITION   If $\mathbf{A} = \langle a_1, a_2, a_3 \rangle$ and $\mathbf{B} = \langle b_1, b_2, b_3 \rangle$, then the sum of these vectors is given by

$$\mathbf{A} + \mathbf{B} = \langle a_1 + b_1, a_2 + b_2, a_3 + b_3 \rangle$$

EXAMPLE 1

Given $\mathbf{A} = \langle 5, -2, 6 \rangle$ and $\mathbf{B} = \langle 8, -5, -4 \rangle$, find $\mathbf{A} + \mathbf{B}$.

SOLUTION

$$\mathbf{A} + \mathbf{B} = \langle 5 + 8, (-2) + (-5), 6 + (-4) \rangle$$

$$= \langle 13, -7, 2 \rangle$$

The geometric interpretation of the sum of two vectors in V_3 is similar to that for vectors in V_2. See Fig. 14.2.3. If P is the point (x, y, z), and $\mathbf{A} = \langle a_1, a_2, a_3 \rangle$, and $\overrightarrow{PQ}$ is a representation of $\mathbf{A}$, then Q is the point $(x + a_1, y + a_2, z + a_3)$. Let $\mathbf{B} = \langle b_1, b_2, b_3 \rangle$ and let $\overrightarrow{QR}$ be a representation of $\mathbf{B}$. Then R is the point $(x + (a_1 + b_1), y + (a_2 + b_2), z + (a_3 + b_3))$. There-

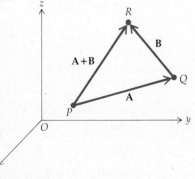

FIGURE 14.2.3

fore $\overrightarrow{PR}$ is a representation of the vector $\mathbf{A} + \mathbf{B}$, and the parallelogram law holds

14.2.5 DEFINITION   If $\mathbf{A} = \langle a_1, a_2, a_3 \rangle$, then the vector $\langle -a_1, -a_2, -a_3 \rangle$ is defined to be the **negative** of $\mathbf{A}$, denoted by $-\mathbf{A}$.

14.2.6 DEFINITION   The **difference** of the two vectors $\mathbf{A}$ and $\mathbf{B}$, denoted by $\mathbf{A} - \mathbf{B}$, is defined by

$$\mathbf{A} - \mathbf{B} = \mathbf{A} + (-\mathbf{B})$$

From Definitions 14.2.5 and 14.2.6 it follows that if $\mathbf{A} = \langle a_1, a_2, a_3 \rangle$ and $\mathbf{B} = \langle b_1, b_2, b_3 \rangle$, then $-\mathbf{B} = \langle -b_1, -b_2, -b_3 \rangle$ and

$$\mathbf{A} - \mathbf{B} = \langle a_1 - b_1, a_2 - b_2, a_3 - b_3 \rangle$$

EXAMPLE 2

For the vectors $\mathbf{A}$ and $\mathbf{B}$ of Example 1 find $\mathbf{A} - \mathbf{B}$.

SOLUTION

$$\begin{aligned} \mathbf{A} - \mathbf{B} &= \langle 5, -2, 6 \rangle - \langle 8, -5, -4 \rangle \\ &= \langle 5, -2, 6 \rangle + \langle -8, 5, 4 \rangle \\ &= \langle -3, 3, 10 \rangle \end{aligned}$$

The difference of two vectors in V_3 is also interpreted geometrically as it is in V_2. See Fig. 14.2.4. A representation of the vector $\mathbf{A} - \mathbf{B}$ is obtained by choosing representations of $\mathbf{A}$ and $\mathbf{B}$ having the same initial point. Then a representation of the vector $\mathbf{A} - \mathbf{B}$ is the directed line segment from the terminal point of the representation of $\mathbf{B}$ to the terminal point of the representation of $\mathbf{A}$.

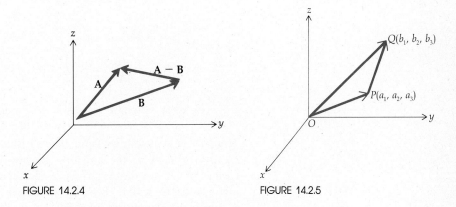

FIGURE 14.2.4   FIGURE 14.2.5

Figure 14.2.5 shows the points $P(a_1, a_2, a_3)$ and $Q(b_1, b_2, b_3)$, and the directed line segments $\overrightarrow{PQ}$, $\overrightarrow{OP}$, and $\overrightarrow{OQ}$. Observe that

$$\begin{aligned} \mathbf{V}(\overrightarrow{PQ}) &= \mathbf{V}(\overrightarrow{OQ}) - \mathbf{V}(\overrightarrow{OP}) \\ &= \langle b_1, b_2, b_3 \rangle - \langle a_1, a_2, a_3 \rangle \end{aligned}$$

Therefore

$$\mathbf{V}(\overrightarrow{PQ}) = \langle b_1 - a_1, b_2 - a_2, b_3 - a_3 \rangle$$

▶ ILLUSTRATION 3   Figure 14.2.6 shows the directed line segment $\overrightarrow{PQ}$, where P is the point $(1, 3, 5)$ and Q is the point $(2, -1, 4)$.

$$\mathbf{V}(\overrightarrow{PQ}) = \langle 2 - 1, -1 - 3, 4 - 5 \rangle$$
$$= \langle 1, -4, -1 \rangle \qquad \blacktriangleleft$$

14.2.7 DEFINITION   If c is a scalar and $\mathbf{A}$ is the vector $\langle a_1, a_2, a_3 \rangle$, then the product of c and $\mathbf{A}$, denoted by $c\mathbf{A}$, is a vector and is given by

$$c\mathbf{A} = c\langle a_1, a_2, a_3 \rangle$$
$$= \langle ca_1, ca_2, ca_3 \rangle$$

FIGURE 14.2.6

EXAMPLE 3
Given $\mathbf{A} = \langle -4, 7, -2 \rangle$, find $3\mathbf{A}$ and $-5\mathbf{A}$.

SOLUTION

$$3\mathbf{A} = 3\langle -4, 7, -2 \rangle \qquad -5\mathbf{A} = (-5)\langle -4, 7, -2 \rangle$$
$$= \langle -12, 21, -6 \rangle \qquad \qquad = \langle 20, -35, 10 \rangle$$

Suppose that $\mathbf{A} = \langle a_1, a_2, a_3 \rangle$ is a nonzero vector having direction cosines $\cos \alpha$, $\cos \beta$, and $\cos \gamma$, and let c be any nonzero scalar. Then $c\mathbf{A} = \langle ca_1, ca_2, ca_3 \rangle$; and if $\cos \alpha_1$, $\cos \beta_1$, and $\cos \gamma_1$ are the direction cosines of $c\mathbf{A}$, we have, from Eqs. (1),

$$\cos \alpha_1 = \frac{ca_1}{\|c\mathbf{A}\|} \qquad \cos \beta_1 = \frac{ca_2}{\|c\mathbf{A}\|} \qquad \cos \gamma_1 = \frac{ca_3}{\|c\mathbf{A}\|}$$

$$\cos \alpha_1 = \frac{c}{|c|} \frac{a_1}{\|\mathbf{A}\|} \qquad \cos \beta_1 = \frac{c}{|c|} \frac{a_2}{\|\mathbf{A}\|} \qquad \cos \gamma_1 = \frac{c}{|c|} \frac{a_3}{\|\mathbf{A}\|}$$

$$\cos \alpha_1 = \frac{c}{|c|} \cos \alpha \qquad \cos \beta_1 = \frac{c}{|c|} \cos \beta \qquad \cos \gamma_1 = \frac{c}{|c|} \cos \gamma \qquad (3)$$

Thus if $c > 0$, it follows from Eqs. (3) that the direction cosines of vector $c\mathbf{A}$ are the same as the direction cosines of $\mathbf{A}$. And if $c < 0$, the direction cosines of $c\mathbf{A}$ are the negatives of the direction cosines of $\mathbf{A}$. Therefore, if c is a nonzero scalar, then the vector $c\mathbf{A}$ is a vector whose magnitude is $|c|$ times the magnitude of $\mathbf{A}$. If $c > 0$, $c\mathbf{A}$ has the same direction as $\mathbf{A}$, whereas if $c < 0$, the direction of $c\mathbf{A}$ is opposite that of $\mathbf{A}$.

The operations of vector addition and scalar multiplication of any vectors in V_3 satisfy properties identical with those given in Theorem 13.1.8. These are given in the following theorem, and the proofs are left as exercises (see Exercises 19 and 20).

14.2.8 THEOREM   If **A**, **B**, and **C** are any vectors in V_3 and c and d are any scalars, then vector addition and scalar multiplication satisfy the following properties:

(i) $\mathbf{A} + \mathbf{B} = \mathbf{B} + \mathbf{A}$ (commutative law)

(ii) $\mathbf{A} + (\mathbf{B} + \mathbf{C}) = (\mathbf{A} + \mathbf{B}) + \mathbf{C}$ (associative law)

(iii) There is a vector **0** in V_3 for which

$$\mathbf{A} + \mathbf{0} = \mathbf{A}   \text{(existence of additive identity)}$$

(iv) There is a vector $-\mathbf{A}$ in V_3 such that

$$\mathbf{A} + (-\mathbf{A}) = \mathbf{0}   \text{(existence of negative)}$$

(v) $(cd)\mathbf{A} = c(d\mathbf{A})$ (associative law)

(vi) $c(\mathbf{A} + \mathbf{B}) = c\mathbf{A} + c\mathbf{B}$ (distributive law)

(vii) $(c + d)\mathbf{A} = c\mathbf{A} + d\mathbf{A}$ (distributive law)

(viii) $1(\mathbf{A}) = \mathbf{A}$ (existence of scalar multiplicative identity)

From Definition 13.1.9 and Theorem 14.2.8 it follows that V_3 is a real vector space. The three unit vectors

$$\mathbf{i} = \langle 1,0,0 \rangle   \mathbf{j} = \langle 0,1,0 \rangle   \mathbf{k} = \langle 0,0,1 \rangle$$

form a basis for the vector space V_3 because any vector $\langle a_1, a_2, a_3 \rangle$ can be written in terms of them as follows:

$$\langle a_1, a_2, a_3 \rangle = a_1 \langle 1,0,0 \rangle + a_2 \langle 0,1,0 \rangle + a_3 \langle 0,0,1 \rangle$$

Hence, if $\mathbf{A} = \langle a_1, a_2, a_3 \rangle$, we also can write

$$\mathbf{A} = a_1 \mathbf{i} + a_2 \mathbf{j} + a_3 \mathbf{k} \tag{4}$$

Because there are three elements in a basis, V_3 is a three-dimensional vector space.

Substituting from (2) into (4) we have

$$\mathbf{A} = \|\mathbf{A}\| \cos \alpha \mathbf{i} + \|\mathbf{A}\| \cos \beta \mathbf{j} + \|\mathbf{A}\| \cos \gamma \mathbf{k}$$

$$\mathbf{A} = \|\mathbf{A}\|(\cos \alpha \mathbf{i} + \cos \beta \mathbf{j} + \cos \gamma \mathbf{k}) \tag{5}$$

Equation (5) enables us to express any nonzero vector in terms of its magnitude and direction cosines.

EXAMPLE 4

Express the vector of Illustration 1 in terms of its magnitude and direction cosines.

SOLUTION

In Illustration 1, we have $\mathbf{A} = \langle 3, 2, -6 \rangle$, $\|\mathbf{A}\| = 7$, $\cos \alpha = \frac{3}{7}$, $\cos \beta = \frac{2}{7}$, and $\cos \gamma = -\frac{6}{7}$. Hence, from (5),

$$\mathbf{A} = 7(\tfrac{3}{7}\mathbf{i} + \tfrac{2}{7}\mathbf{j} - \tfrac{6}{7}\mathbf{k})$$

14.2.9 THEOREM   If the nonzero vector $\mathbf{A} = a_1\mathbf{i} + a_2\mathbf{j} + a_3\mathbf{k}$, then the unit vector $\mathbf{U}$ having the same direction as $\mathbf{A}$ is given by

$$\mathbf{U} = \frac{a_1}{\|\mathbf{A}\|}\mathbf{i} + \frac{a_2}{\|\mathbf{A}\|}\mathbf{j} + \frac{a_3}{\|\mathbf{A}\|}\mathbf{k}$$

The proof of Theorem 14.2.9 is analogous to the proof of Theorem 13.1.10 for a vector in V_2 and is left as an exercise (see Exercise 46).

EXAMPLE 5

Given the points $R(2, -1, 3)$ and $S(3, 4, 6)$, find the unit vector having the same direction as $\mathbf{V}(\vec{RS})$.

SOLUTION

$$\mathbf{V}(\vec{RS}) = \langle 3, 4, 6 \rangle - \langle 2, -1, 3 \rangle \qquad \|\mathbf{V}(\vec{RS})\| = \sqrt{1^2 + 5^2 + 3^2}$$
$$= \mathbf{i} + 5\mathbf{j} + 3\mathbf{k} \qquad\qquad\qquad = \sqrt{35}$$

Therefore, by Theorem 14.2.9 the desired unit vector is

$$\mathbf{U} = \frac{1}{\sqrt{35}}\mathbf{i} + \frac{5}{\sqrt{35}}\mathbf{j} + \frac{3}{\sqrt{35}}\mathbf{k}$$

The definition of the dot product of two vectors in V_3 is an extension of the definition for vectors in V_2.

14.2.10 DEFINITION   If $\mathbf{A} = \langle a_1, a_2, a_3 \rangle$ and $\mathbf{B} = \langle b_1, b_2, b_3 \rangle$, then the **dot product** of $\mathbf{A}$ and $\mathbf{B}$, denoted by $\mathbf{A} \cdot \mathbf{B}$, is given by

$$\mathbf{A} \cdot \mathbf{B} = \langle a_1, a_2, a_3 \rangle \cdot \langle b_1, b_2, b_3 \rangle$$
$$= a_1 b_1 + a_2 b_2 + a_3 b_3$$

▶ **ILLUSTRATION 4**   If $\mathbf{A} = \langle 4, 2, -6 \rangle$ and $\mathbf{B} = \langle -5, 3, -2 \rangle$, then

$$\mathbf{A} \cdot \mathbf{B} = \langle 4, 2, -6 \rangle \cdot \langle -5, 3, -2 \rangle$$
$$= 4(-5) + 2(3) + (-6)(-2)$$
$$= -20 + 6 + 12$$
$$= -2 \qquad\qquad\qquad ◀$$

For the unit vectors $\mathbf{i}$, $\mathbf{j}$, and $\mathbf{k}$,

$$\mathbf{i} \cdot \mathbf{i} = 1 \qquad \mathbf{j} \cdot \mathbf{j} = 1 \qquad \mathbf{k} \cdot \mathbf{k} = 1$$
$$\mathbf{i} \cdot \mathbf{j} = 0 \qquad \mathbf{i} \cdot \mathbf{k} = 0 \qquad \mathbf{j} \cdot \mathbf{k} = 0$$

Laws of dot multiplication that are given in the following Theorem 14.2.11 are the same as those in Theorems 13.2.2 and 13.2.3 for vectors in V_2. The proofs are left as exercises (see Exercises 27 and 28).

14.2.11 THEOREM   If **A**, **B**, and **C** are any vectors in V_3 and c is a scalar, then

(i) $\mathbf{A} \cdot \mathbf{B} = \mathbf{B} \cdot \mathbf{A}$ (commutative law)
(ii) $\mathbf{A} \cdot (\mathbf{B} + \mathbf{C}) = \mathbf{A} \cdot \mathbf{B} + \mathbf{A} \cdot \mathbf{C}$ (distributive law)
(iii) $c(\mathbf{A} \cdot \mathbf{B}) = (c\mathbf{A}) \cdot \mathbf{B}$
(iv) $\mathbf{0} \cdot \mathbf{A} = 0$
(v) $\mathbf{A} \cdot \mathbf{A} = \|\mathbf{A}\|^2$

We now define the angle between two vectors and then express the dot product in terms of the cosine of the radian measure of this angle.

14.2.12 DEFINITION   Let **A** and **B** be two nonzero vectors in V_3 such that **A** is not a scalar multiple of **B**. If $\overrightarrow{OP}$ is the position representation of **A** and $\overrightarrow{OQ}$ is the position representation of **B**, then the **angle between the vectors A** and **B** is defined to be the angle of positive measure between $\overrightarrow{OP}$ and $\overrightarrow{OQ}$ interior to the triangle POQ. If $\mathbf{A} = c\mathbf{B}$, where c is a scalar, then if $c > 0$, the angle between the vectors has radian measure 0, and if $c < 0$, the angle between the vectors has radian measure π.

FIGURE 14.2.7

Figure 14.2.7 shows the angle θ between the two vectors if **A** is not a scalar multiple of **B**.

14.2.13 THEOREM   If θ is the angle between the two nonzero vectors **A** and **B** in V_3, then

$$\mathbf{A} \cdot \mathbf{B} = \|\mathbf{A}\| \, \|\mathbf{B}\| \cos \theta \qquad\qquad (6)$$

The proof of Theorem 14.2.13 is comparable to the proof of Theorem 13.2.5 for vectors in V_2 and is left as an exercise (see Exercise 47).

The definition of *parallel* vectors in V_3 is analogous to Definition 13.2.6 for vectors in V_2; that is, two vectors in V_3 are **parallel** if and only if one of the vectors is a scalar multiple of the other. As with vectors in V_2, we can prove from this definition and Theorem 14.2.13 that two nonzero vectors in V_3 are parallel if and only if the radian measure of the angle between them is 0 or π.

The definition of *orthogonal* vectors in V_3 corresponds to Definition 13.2.7 for vectors in V_2; that is, if **A** and **B** are two vectors in V_3, **A** and **B** are said to be **orthogonal** if and only if $\mathbf{A} \cdot \mathbf{B} = 0$.

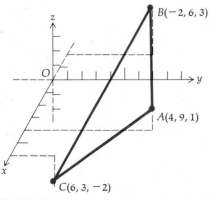

FIGURE 14.2.8

EXAMPLE 6
Prove by using vectors that the points $A(4, 9, 1)$, $B(-2, 6, 3)$, and $C(6, 3, -2)$ are the vertices of a right triangle.

SOLUTION
Triangle CAB appears in Fig. 14.2.8. From the figure it looks as if the angle at A is the one that may be a right angle. We shall find $\mathbf{V}(\overrightarrow{AB})$ and $\mathbf{V}(\overrightarrow{AC})$ and if the dot product of these two vectors is zero, the angle is a right angle.

$$\mathbf{V}(\overrightarrow{AB}) = \langle -2 - 4, 6 - 9, 3 - 1 \rangle \qquad \mathbf{V}(\overrightarrow{AC}) = \langle 6 - 4, 3 - 9, -2 - 1 \rangle$$

$$= \langle -6, -3, 2 \rangle \qquad\qquad = \langle 2, -6, -3 \rangle$$

$$\mathbf{V}(\overrightarrow{AB}) \cdot \mathbf{V}(\overrightarrow{AC}) = \langle -6, -3, 2 \rangle \cdot \langle 2, -6, -3 \rangle$$

$$= -12 + 18 - 6$$

$$= 0$$

Therefore $\mathbf{V}(\overrightarrow{AB})$ and $\mathbf{V}(\overrightarrow{AC})$ are orthogonal; thus the angle at A in triangle CAB is a right angle, and CAB is a right triangle.

If $\mathbf{U}$ is a unit vector in the direction of $\mathbf{A}$, and θ is the angle between $\mathbf{A}$ (or $\mathbf{U}$) and a vector $\mathbf{B}$, then, from (6),

$$\mathbf{U} \cdot \mathbf{B} = \|\mathbf{U}\| \|\mathbf{B}\| \cos \theta$$

$$= \|\mathbf{B}\| \cos \theta$$

As with vectors in V_2, $\|\mathbf{B}\| \cos \theta$ is the **scalar projection** of $\mathbf{B}$ onto $\mathbf{A}$ and the **component** of $\mathbf{B}$ in the direction of $\mathbf{A}$.

The formula of Theorem 13.2.9 giving the scalar projection of $\mathbf{B}$ onto $\mathbf{A}$ holds for vectors in V_3, and the proof is identical to that for vectors in V_2. Thus the scalar projection of $\mathbf{B}$ onto $\mathbf{A}$ is

$$\frac{\mathbf{A} \cdot \mathbf{B}}{\|\mathbf{A}\|} \tag{7}$$

Theorem 13.2.10, for vectors in V_2, gives a formula for computing the **vector projection** of the vector $\mathbf{B}$ onto the vector $\mathbf{A}$. This theorem and its proof are identical for vectors in V_3, and the vector is

$$\left(\frac{\mathbf{A} \cdot \mathbf{B}}{\|\mathbf{A}\|^2} \right) \mathbf{A} \tag{8}$$

EXAMPLE 7

Given the vectors

$$\mathbf{A} = 6\mathbf{i} - 3\mathbf{j} + 2\mathbf{k} \qquad \mathbf{B} = 2\mathbf{i} + \mathbf{j} - 3\mathbf{k}$$

Find: (a) $\cos \theta$ if θ is the angle between $\mathbf{A}$ and $\mathbf{B}$; (b) the component of $\mathbf{B}$ in the direction of $\mathbf{A}$; (c) the vector projection of $\mathbf{B}$ onto $\mathbf{A}$.

SOLUTION

We first compute $\mathbf{A} \cdot \mathbf{B}$, $\|\mathbf{A}\|$, and $\|\mathbf{B}\|$.

$$\mathbf{A} \cdot \mathbf{B} = \langle 6, -3, 2 \rangle \cdot \langle 2, 1, -3 \rangle \qquad \|\mathbf{A}\| = \sqrt{36 + 9 + 4} \qquad \|\mathbf{B}\| = \sqrt{4 + 1 + 9}$$

$$= 12 - 3 - 6 \qquad\qquad\qquad = \sqrt{49} \qquad\qquad = \sqrt{14}$$

$$= 3 \qquad\qquad\qquad\qquad = 7$$

(a) From Theorem 14.2.13,

$$\cos \theta = \frac{\mathbf{A} \cdot \mathbf{B}}{\|\mathbf{A}\| \|\mathbf{B}\|}$$

$$= \frac{3}{7\sqrt{14}}$$

(b) The component of **B** in the direction of **A** is the scalar projection of **B** onto **A**, which is

$$\|\mathbf{B}\| \cos \theta = \sqrt{14}\left(\frac{3}{7\sqrt{14}}\right)$$

$$= \tfrac{3}{7}$$

(c) From (8), the vector projection of **B** onto **A** is

$$\left(\frac{\mathbf{A} \cdot \mathbf{B}}{\|\mathbf{A}\|^2}\right)\mathbf{A} = \tfrac{3}{49}(6\mathbf{i} - 3\mathbf{j} + 2\mathbf{k})$$

$$= \tfrac{18}{49}\mathbf{i} - \tfrac{9}{49}\mathbf{j} + \tfrac{6}{49}\mathbf{k}$$

EXAMPLE 8

Find the distance from the point $P(4, 1, 6)$ to the line through the points $A(8, 3, 2)$ and $B(2, -3, 5)$.

SOLUTION

FIGURE 14.2.9

Figure 14.2.9 shows the point P and a sketch of the line through A and B. The point M is the foot of the perpendicular line from P to the line through A and B. Let d units be the distance $|\overline{PM}|$. Thus from the Pythagorean theorem,

$$d = \sqrt{|\overline{AP}|^2 - |\overline{AM}|^2} \tag{9}$$

To apply (9) we need to compute $|\overline{AP}|$, which is the magnitude of $\mathbf{V}(\overrightarrow{AP})$, and $|\overline{AM}|$, which is the scalar projection of $\mathbf{V}(\overrightarrow{AP})$ onto $\mathbf{V}(\overrightarrow{AB})$. We first find $\mathbf{V}(\overrightarrow{AP})$ and $\mathbf{V}(\overrightarrow{AB})$.

$$\mathbf{V}(\overrightarrow{AP}) = \langle 4 - 8, 1 - 3, 6 - 2 \rangle \qquad \mathbf{V}(\overrightarrow{AB}) = \langle 2 - 8, -3 - 3, 5 - 2 \rangle$$

$$= \langle -4, -2, 4 \rangle \qquad\qquad\qquad = \langle -6, -6, 3 \rangle$$

We compute $|\overline{AP}|$ by finding $\|\mathbf{V}(\overrightarrow{AP})\|$, and we compute $|\overline{AM}|$ by using formula (7) with $\mathbf{A} = \mathbf{V}(\overrightarrow{AB})$ and $\mathbf{B} = \mathbf{V}(\overrightarrow{AP})$.

$$|\overline{AP}| = \|\mathbf{V}(\overrightarrow{AP})\| \qquad\qquad |\overline{AM}| = \frac{\mathbf{V}(\overrightarrow{AB}) \cdot \mathbf{V}(\overrightarrow{AP})}{\|\mathbf{V}(\overrightarrow{AB})\|}$$

$$= \sqrt{(-4)^2 + (-2)^2 + 4^2}$$

$$= \sqrt{36} \qquad\qquad\qquad\qquad = \frac{\langle -6, -6, 3 \rangle \cdot \langle -4, -2, 4 \rangle}{\sqrt{(-6)^2 + (-6)^2 + 3^2}}$$

$$= 6 \qquad\qquad\qquad\qquad\qquad = \frac{24 + 12 + 12}{\sqrt{81}}$$

$$= \tfrac{48}{9}$$

Substituting these values of $|\overline{AP}|$ and $|\overline{AM}|$ in (9) we get

$$d = \sqrt{6^2 - \left(\tfrac{48}{9}\right)^2}$$

$$= 6\sqrt{1 - \tfrac{64}{81}}$$

$$= \tfrac{2}{3}\sqrt{17}$$

EXERCISES 14.2

In Exercises 1 through 6, $\mathbf{A} = \langle 1, 2, 3 \rangle$, $\mathbf{B} = \langle 4, -3, -1 \rangle$, $\mathbf{C} = \langle -5, -3, 5 \rangle$, and $\mathbf{D} = \langle -2, 1, 6 \rangle$.

1. Find: (a) $\mathbf{A} + 5\mathbf{B}$; (b) $7\mathbf{C} - 5\mathbf{D}$; (c) $\|7\mathbf{C}\| - \|5\mathbf{D}\|$; (d) $\|7\mathbf{C} - 5\mathbf{D}\|$.

2. Find: (a) $2\mathbf{A} - \mathbf{C}$; (b) $\|2\mathbf{A}\| - \|\mathbf{C}\|$; (c) $4\mathbf{B} + 6\mathbf{C} - 2\mathbf{D}$; (d) $\|4\mathbf{B}\| + \|6\mathbf{C}\| - \|2\mathbf{D}\|$.

3. Find: (a) $\mathbf{C} + 3\mathbf{D} - 8\mathbf{A}$; (b) $\|\mathbf{A}\| \|\mathbf{B}\| (\mathbf{C} - \mathbf{D})$.

4. Find: (a) $3\mathbf{A} - 2\mathbf{B} + \mathbf{C} - 12\mathbf{D}$; (b) $\|\mathbf{A}\|\mathbf{C} - \|\mathbf{B}\|\mathbf{D}$.

5. Find scalars a and b such that $a(\mathbf{A} + \mathbf{B}) + b(\mathbf{C} + \mathbf{D}) = \mathbf{0}$.

6. Find scalars a, b, and c such that $a\mathbf{A} + b\mathbf{B} + c\mathbf{C} = \mathbf{D}$.

In Exercises 7 through 10, find the direction cosines of the vector $\mathbf{V}(\overrightarrow{P_1 P_2})$ and check the answers by verifying that the sum of their squares is 1.

7. $P_1(3, -1, -4)$; $P_2(7, 2, 4)$ **8.** $P_1(-2, 6, 5)$; $P_2(2, 4, 1)$

9. $P_1(4, -3, -1)$; $P_2(-2, -4, -8)$ **10.** $P_1(1, 3, 5)$; $P_2(2, -1, 4)$

11. Use the points P_1 and P_2 of Exercise 7 and find the point Q such that $\mathbf{V}(\overrightarrow{P_1 P_2}) = 3\mathbf{V}(\overrightarrow{P_1 Q})$.

12. Use the points P_1 and P_2 of Exercise 10 and find the point R such that $\mathbf{V}(\overrightarrow{P_1 R}) = -2\mathbf{V}(\overrightarrow{P_2 R})$.

13. Given $P_1(3, 2, -4)$ and $P_2(-5, 4, 2)$, find the point P_3 such that $4\mathbf{V}(\overrightarrow{P_1 P_2}) = -3\mathbf{V}(\overrightarrow{P_2 P_3})$.

14. Given $P_1(7, 0, -2)$ and $P_2(2, -3, 5)$, find the point P_3 such that $\mathbf{V}(\overrightarrow{P_1 P_3}) = 5\mathbf{V}(\overrightarrow{P_2 P_3})$.

In Exercises 15 and 16, express the given vector in terms of its magnitude and direction cosines.

15. (a) $-6\mathbf{i} + 2\mathbf{j} + 3\mathbf{k}$; (b) $-2\mathbf{i} + \mathbf{j} - 3\mathbf{k}$ **16.** (a) $2\mathbf{i} - 2\mathbf{j} + \mathbf{k}$; (b) $3\mathbf{i} + 4\mathbf{j} - 5\mathbf{k}$

In Exercises 17 and 18, find the unit vector having the same direction as $\mathbf{V}(\overrightarrow{P_1 P_2})$.

17. (a) $P_1(4, -1, -6)$ and $P_2(5, 7, -2)$; (b) $P_1(-2, 5, 3)$ and $P_2(-4, 7, 5)$

18. (a) $P_1(3, 0, -1)$ and $P_2(-3, 8, -1)$; (b) $P_1(-8, -5, 2)$ and $P_2(-3, -9, 4)$

19. Prove Theorem 14.2.8(i), (iii), (iv), (vi), and (viii).   **20.** Prove Theorem 14.2.8(ii), (v), and (vii).

In Exercises 21 through 26, $\mathbf{A} = \langle -4, -2, 4 \rangle$; $\mathbf{B} = \langle 2, 7, -1 \rangle$; $\mathbf{C} = \langle 6, -3, 0 \rangle$, and $\mathbf{D} = \langle 5, 4, -3 \rangle$.

21. Find: (a) $\mathbf{A} \cdot (\mathbf{B} + \mathbf{C})$; (b) $(\mathbf{A} \cdot \mathbf{B})(\mathbf{C} \cdot \mathbf{D})$; (c) $\mathbf{A} \cdot \mathbf{D} - \mathbf{B} \cdot \mathbf{C}$; (d) $(\mathbf{B} \cdot \mathbf{D})\mathbf{A} - (\mathbf{D} \cdot \mathbf{A})\mathbf{B}$.

22. Find: (a) $\mathbf{A} \cdot \mathbf{B} + \mathbf{A} \cdot \mathbf{C}$; (b) $(\mathbf{A} \cdot \mathbf{B})(\mathbf{B} \cdot \mathbf{C})$; (c) $(\mathbf{A} \cdot \mathbf{B})\mathbf{C} + (\mathbf{B} \cdot \mathbf{C})\mathbf{D}$; (d) $(2\mathbf{A} + 3\mathbf{B}) \cdot (4\mathbf{C} - \mathbf{D})$.

23. Find: (a) $\cos \theta$ if θ is the angle between $\mathbf{A}$ and $\mathbf{C}$; (b) the component of $\mathbf{C}$ in the direction of $\mathbf{A}$; (c) the vector projection of $\mathbf{C}$ onto $\mathbf{A}$.

24. Find: (a) $\cos \theta$ if θ is the angle between $\mathbf{B}$ and $\mathbf{D}$; (b) the component of $\mathbf{B}$ in the direction of $\mathbf{D}$; (c) the vector projection of $\mathbf{B}$ onto $\mathbf{D}$.

25. Find: (a) the scalar projection of $\mathbf{A}$ onto $\mathbf{B}$; (b) the vector projection of $\mathbf{A}$ onto $\mathbf{B}$.

26. Find: (a) the scalar projection of $\mathbf{D}$ onto $\mathbf{C}$; (b) the vector projection of $\mathbf{D}$ onto $\mathbf{C}$.

27. Prove Theorem 14.2.11(i) and (iii).   **28.** Prove Theorem 14.2.11(ii), (iv), and (v).

29. Find the distance from the point $(2, -1, -4)$ to the line through the points $(3, -2, 2)$ and $(-9, -6, 6)$.

30. Find the distance from the point $(3, 2, 1)$ to the line through the points $(1, 2, 9)$ and $(-3, -6, -3)$.

31. Prove by using vectors that the points $(2, 2, 2)$, $(2, 0, 1)$, $(4, 1, -1)$, and $(4, 3, 0)$ are the vertices of a rectangle.

32. Prove by using vectors that the points $(2, 2, 2)$, $(0, 1, 2)$, $(-1, 3, 3)$, and $(3, 0, 1)$ are the vertices of a parallelogram.

33. Find the area of the triangle having vertices at $(-2, 3, 1)$, $(1, 2, 3)$, and $(3, -1, 2)$.

34. Prove by using vectors that the points $(-2, 1, 6)$, $(2, 4, 5)$, and $(-1, -2, 1)$ are the vertices of a right triangle, and find the area of the triangle.

35. If $\mathbf{A} = 3\mathbf{i} + 5\mathbf{j} - 3\mathbf{k}$, $\mathbf{B} = -\mathbf{i} - 2\mathbf{j} + 3\mathbf{k}$, and $\mathbf{C} = 2\mathbf{i} - \mathbf{j} + 4\mathbf{k}$, find the component of $\mathbf{B}$ in the direction of $\mathbf{A} - 2\mathbf{C}$.

36. Find the cosines of the angles of the triangle having vertices at $A(0, 0, 0)$, $B(4, -1, 3)$, and $C(1, 2, 3)$.

37. If a force has the vector representation $\mathbf{F} = 3\mathbf{i} - 2\mathbf{j} + \mathbf{k}$, find the work done by the force in moving an object from the point $P_1(-2, 4, 3)$ along a straight line to the point $P_2(1, -3, 5)$. The magnitude of the force is measured in pounds and distance is measured in feet. (*Hint:* Review Section 13.2.)

38. If a force has the vector representation $\mathbf{F} = 5\mathbf{i} - 3\mathbf{k}$, find the work done by the force in moving an object from the point $P_1(4, 1, 3)$ along a straight line to the point $P_2(-5, 6, 2)$. The magnitude of the force is measured in pounds and distance is measured in feet. (See hint for Exercise 37.)

39. A force is represented by the vector $\mathbf{F}$, it has a magnitude of 10 lb, and direction cosines of $\mathbf{F}$ are $\cos \alpha = \frac{1}{6}\sqrt{6}$ and $\cos \beta = \frac{1}{3}\sqrt{6}$. If the force moves an object from the origin along a straight line to the point $(7, -4, 2)$, find the work done. Distance is measured in feet. (See hint for Exercise 37.)

40. If $\mathbf{A}$ and $\mathbf{B}$ are nonzero vectors, prove that the vector $\mathbf{A} - c\mathbf{B}$ is orthogonal to $\mathbf{B}$ if $c = \mathbf{A} \cdot \mathbf{B}/\|\mathbf{B}\|^2$.

41. If $\mathbf{A} = 12\mathbf{i} + 9\mathbf{j} - 5\mathbf{k}$ and $\mathbf{B} = 4\mathbf{i} + 3\mathbf{j} - 5\mathbf{k}$, use the result of Exercise 40 to find the value of the scalar c so that the vector $\mathbf{B} - c\mathbf{A}$ is orthogonal to $\mathbf{A}$.

42. For the vectors of Exercise 41, use the result of Exercise 40, to find the value of the scalar d so that the vector $\mathbf{A} - d\mathbf{B}$ is orthogonal to $\mathbf{B}$.

43. Prove that if $\mathbf{A}$ and $\mathbf{B}$ are any vectors, then the vectors $\|\mathbf{B}\|\mathbf{A} + \|\mathbf{A}\|\mathbf{B}$ and $\|\mathbf{B}\|\mathbf{A} - \|\mathbf{A}\|\mathbf{B}$ are orthogonal.

44. Prove that if $\mathbf{A}$ and $\mathbf{B}$ are any nonzero vectors and $\mathbf{C} = \|\mathbf{B}\|\mathbf{A} + \|\mathbf{A}\|\mathbf{B}$, then the angle between $\mathbf{A}$ and $\mathbf{C}$ has the same measure as the angle between $\mathbf{B}$ and $\mathbf{C}$.

45. If the radian measure of each direction angle of a vector is the same, what is it? Prove your answer.

46. Prove Theorem 14.2.9.   **47.** Prove Theorem 14.2.13.

48. Three vectors in V_3 are said to be *independent* if and only if their position representations do not lie in a plane, and three vectors $\mathbf{E}_1$, $\mathbf{E}_2$, and $\mathbf{E}_3$ are said to form a *basis* for the vector space V_3 if and only if any vector in V_3 can be written as a linear combination of $\mathbf{E}_1$, $\mathbf{E}_2$, and $\mathbf{E}_3$. A theorem can be proved which states that three vectors form a basis for the vector space V_3 if they are independent. Show that this theorem holds for the three vectors $\langle 1, 0, 0 \rangle$, $\langle 1, 1, 0 \rangle$, and $\langle 1, 1, 1 \rangle$ by doing the following: (a) Verify that the vectors are independent by showing that their position representations are not coplanar; (b) verify that the vectors form a basis by showing that any vector $\mathbf{A}$ can be written

$$\mathbf{A} = r\langle 1, 0, 0 \rangle + s\langle 1, 1, 0 \rangle + t\langle 1, 1, 1 \rangle \tag{10}$$

where r, s, and t are scalars. (c) If $\mathbf{A} = \langle 6, -2, 5 \rangle$, find the particular values of r, s, and t so that (10) holds.

49. See Exercise 48. (a) Verify that the vectors $\langle 2, 0, 1 \rangle$, $\langle 0, -1, 0 \rangle$, and $\langle 1, -1, 0 \rangle$ form a basis for V_3 by showing that any vector $\mathbf{A}$ can be written

$$\mathbf{A} = r\langle 2, 0, 1 \rangle + s\langle 0, -1, 0 \rangle + t\langle 1, -1, 0 \rangle \tag{11}$$

where r, s, and t are scalars. (b) If $\mathbf{A} = \langle -2, 3, 5 \rangle$, find the particular values of r, s, and t such that (11) holds.

50. Refer to the first sentence of Exercise 48. A theorem can be proved which states that three vectors form a basis for the vector space V_3 only if they are independent. Show that this theorem is valid for the three vectors $\mathbf{F}_1 = \langle 1, 0, 1 \rangle$, $\mathbf{F}_2 = \langle 1, 1, 1 \rangle$, and $\mathbf{F}_3 = \langle 2, 1, 2 \rangle$ by doing the following: (a) Verify that $\mathbf{F}_1$, $\mathbf{F}_2$, and $\mathbf{F}_3$ are not independent by showing that their position representations are coplanar; (b) verify that the vectors do not form a basis by showing that every vector in V_3 cannot be written as a linear combination of $\mathbf{F}_1$, $\mathbf{F}_2$, and $\mathbf{F}_3$.

14.3
PLANES

The graph of an equation in two variables, x and y, is a curve in the xy plane. The simplest kind of curve in two-dimensional space is a straight line, and the general equation of a straight line is of the form $Ax + By + C = 0$, which is an equation of the first degree. In three-dimensional space the graph of an equation in three variables, x, y, and z, is a surface. The simplest kind of surface is a *plane*, and we shall see that an equation of a plane is an equation of the first degree in three variables.

14.3.1 DEFINITION   If $\mathbf{N}$ is a given nonzero vector and P_0 is a given point, then the set of all points P for which $V(\overrightarrow{P_0P})$ and $\mathbf{N}$ are orthogonal is defined to be a **plane** through P_0 having $\mathbf{N}$ as a **normal vector**.

Figure 14.3.1 shows a portion of a plane through the point $P_0(x_0, y_0, z_0)$ and the representation of the normal vector **N** having its initial point at P_0.

In plane analytic geometry we can obtain an equation of a line if a point on the line and its direction (slope) are given. In an analogous manner, in solid analytic geometry an equation of a plane can be determined by knowing a point in the plane and the direction of a normal vector.

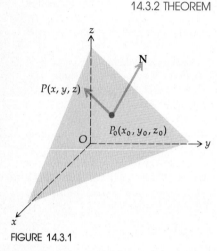

14.3.2 THEOREM   If $P_0(x_0, y_0, z_0)$ is a point in a plane and $\langle a, b, c \rangle$ is a normal vector to the plane then an equation of the plane is

$$a(x - x_0) + b(y - y_0) + c(z - z_0) = 0 \tag{1}$$

PROOF   Refer to Fig. 14.3.1 where $\mathbf{N} = \langle a, b, c \rangle$. Let $P(x, y, z)$ be any point in the plane. $\mathbf{V}(\overrightarrow{P_0P})$ is the vector having $\overrightarrow{P_0P}$ as a representation; so

$$\mathbf{V}(\overrightarrow{P_0P}) = \langle x - x_0, y - y_0, z - z_0 \rangle \tag{2}$$

From Definition 14.3.1 and the fact that the dot product of two orthogonal vectors is zero we have

$$\mathbf{V}(\overrightarrow{P_0P}) \cdot \langle a, b, c \rangle = 0$$

From (2) and the above equation,

$$a(x - x_0) + b(y - y_0) + c(z - z_0) = 0 \tag{3}$$

which is the desired equation.   ∎

FIGURE 14.3.1

EXAMPLE 1

Find an equation of the plane containing the point $(2, 1, 3)$ and having $3\mathbf{i} - 4\mathbf{j} + \mathbf{k}$ as a normal vector.

SOLUTION

Using (1) where the point (x_0, y_0, z_0) is $(2, 1, 3)$ and the vector $\langle a, b, c \rangle$ is $\langle 3, -4, 1 \rangle$, we have as an equation of the required plane

$$3(x - 2) - 4(y - 1) + (z - 3) = 0$$

$$3x - 4y + z - 5 = 0$$

14.3.3 THEOREM   If a, b, and c are not all zero, the graph of an equation of the form

$$ax + by + cz + d = 0 \tag{4}$$

is a plane and $\langle a, b, c \rangle$ is a normal vector to the plane.

PROOF   Suppose that $b \neq 0$. Then the point $(0, -d/b, 0)$ is on the graph of the equation because its coordinates satisfy the equation. The given equation can be written as

$$a(x - 0) + b\left(y + \frac{d}{b}\right) + c(z - 0) = 0$$

which from Theorem 14.3.2 is an equation of a plane through the point $(0, -d/b, 0)$ and for which $\langle a, b, c \rangle$ is a normal vector. This proves the theorem if $b \neq 0$. A similar argument holds if $b = 0$ and either $a \neq 0$ or $c \neq 0$.   ∎

Equations (1) and (4) are called *cartesian equations* of a plane. Equation (1) is analogous to the point-slope form of an equation of a line in two dimensions. Equation (4) is the general first-degree equation in three variables and is called a *linear equation*.

A plane is determined by three noncollinear points, by a line and a point not on the line, by two intersecting lines, or by two parallel lines.

EXAMPLE 2

Find an equation of the plane through the points $P(1, 3, 2)$, $Q(3, -2, 2)$, and $R(2, 1, 3)$.

SOLUTION

From Theorem 14.3.3 the graph of an equation of the form (4) is a plane. If this equation is satisfied by the coordinates of points P, Q, and R, the plane will contain the points. Replacing x, y, and z in (4) by the coordinates of the three points we obtain the three equations

$$a + 3b + 2c + d = 0$$

$$3a - 2b + 2c + d = 0$$

$$2a + b + 3c + d = 0$$

We solve this system of equations for a, b, and c in terms of d, and we obtain

$$a = -\tfrac{5}{9}d \qquad b = -\tfrac{2}{9}d \qquad c = \tfrac{1}{9}d$$

Replacing a, b, and c in (4) by these values we have

$$-\tfrac{5}{9}dx - \tfrac{2}{9}dy + \tfrac{1}{9}dz + d = 0$$

Multiplying on both sides of this equation by $-9/d$ we get

$$5x + 2y - z - 9 = 0$$

which is the required equation.

To draw a sketch of a plane from its equation it is convenient to find the points at which the plane intersects each of the coordinate axes. The x coordinate of the point at which the plane intersects the x axis is called the x *intercept* of the plane; the y coordinate of the point at which the plane intersects the y axis is called the y *intercept* of the plane; and the z *intercept* of the plane is the z coordinate of the point at which the plane intersects the z axis.

▶ ILLUSTRATION 1   We wish to draw a sketch of the plane having the equation

$$2x + 4y + 3z = 8$$

By substituting zero for y and z we obtain $x = 4$; so the x intercept of the plane is 4. The y intercept and the z intercept are obtained in a similar manner; they are 2 and $\tfrac{8}{3}$, respectively. Plotting the points corresponding to these intercepts and connecting them with lines we have the sketch of the plane shown in Fig. 14.3.2. Note that only a portion of the plane is shown in the figure. ◀

FIGURE 14.3.2

$x - 2z = 0$

$y - 3z = 0$

P_2

P_1

FIGURE 14.3.3

▶ ILLUSTRATION 2   To draw a sketch of the plane having the equation

$$3x + 2y - 6z = 0$$

first notice that because the equation is satisfied when x, y, and z are all zero, the plane intersects each of the axes at the origin. If $x = 0$ in the given equation, we obtain $y - 3z = 0$, which is a line in the yz plane; this is the line of intersection of the yz plane with the given plane. Similarly, the line of intersection of the xz plane with the given plane is obtained by setting $y = 0$, and we get $x - 2z = 0$. Drawing a sketch of each of these two lines and drawing a line segment from a point on one of the lines to a point on the other line we obtain Fig. 14.3.3.   ◀

In Illustration 2 the line in the yz plane and the line in the xz plane used to draw the sketch of the plane are called the **traces** of the given plane in the yz plane and the xz plane, respectively. The equation $x = 0$ is an equation of the yz plane because the point (x, y, z) is in the yz plane if and only if $x = 0$. Similarly, the equations $y = 0$ and $z = 0$ are equations of the xz plane and the xy plane, respectively.

A plane parallel to the yz plane has an equation of the form $x = k$, where k is a constant. Figure 14.3.4 shows a sketch of the plane having the equation $x = 3$. A plane parallel to the xz plane has an equation of the form $y = k$, and a plane parallel to the xy plane has an equation of the form $z = k$. Figures 14.3.5 and 14.3.6 show sketches of the planes having the equations $y = -5$ and $z = 6$, respectively.

FIGURE 14.3.4

FIGURE 14.3.5

FIGURE 14.3.6

14.3.4 DEFINITION   The **angle between two planes** is defined to be the angle between the normal vectors of the two planes.

EXAMPLE 3

Find the radian measure of the angle between the planes

$$5x - 2y + 5z - 12 = 0 \quad \text{and} \quad 2x + y - 7z + 11 = 0$$

SOLUTION

Let $\mathbf{N}_1$ be a normal vector to the first plane and $\mathbf{N}_2$ be a normal vector to the second plane. Then

$$\mathbf{N}_1 = 5\mathbf{i} - 2\mathbf{j} + 5\mathbf{k} \qquad \mathbf{N}_2 = 2\mathbf{i} + \mathbf{j} - 7\mathbf{k}$$

From Definition 14.3.4, the angle between the two planes is the angle between $\mathbf{N}_1$ and $\mathbf{N}_2$. Thus from Theorem 14.2.13, if θ is the radian measure of this angle,

$$\cos \theta = \frac{\mathbf{N}_1 \cdot \mathbf{N}_2}{\|\mathbf{N}_1\| \, \|\mathbf{N}_2\|}$$

$$= \frac{\langle 5, -2, 5 \rangle \cdot \langle 2, 1, -7 \rangle}{\sqrt{25 + 4 + 25}\sqrt{4 + 1 + 49}}$$

$$= \frac{10 - 2 - 35}{\sqrt{54}\sqrt{54}}$$

$$= -\tfrac{27}{54}$$

$$= -\tfrac{1}{2}$$

Therefore $\theta = \tfrac{2}{3}\pi$.

FIGURE 14.3.7

14.3.5 DEFINITION   Two planes are **parallel** if and only if their normal vectors are parallel.

Recall that two vectors are parallel if and only if one of the vectors is a scalar multiple of the other. Thus from Definition 14.3.5 it follows that if we have one plane with a normal vector $\mathbf{N}_1$ and another plane with a normal vector $\mathbf{N}_2$, then the two planes are parallel if and only if

$$\mathbf{N}_1 = k\mathbf{N}_2$$

where k is a constant. Figure 14.3.7 shows sketches of two parallel planes and representations of some of their normal vectors.

14.3.6 DEFINITION   Two planes are **perpendicular** if and only if their normal vectors are orthogonal.

From Definition 14.3.6 and the fact that two vectors are orthogonal if and only if their dot product is zero it follows that two planes having normal vectors $\mathbf{N}_1$ and $\mathbf{N}_2$ are perpendicular if and only if

$$\mathbf{N}_1 \cdot \mathbf{N}_2 = 0 \tag{5}$$

EXAMPLE 4

Find an equation of the plane containing the point $(4, 0, -2)$ and perpendicular to each of the planes

$$x - y + z = 0 \quad \text{and} \quad 2x + y - 4z - 5 = 0$$

FIGURE 14.3.8

FIGURE 14.3.9

FIGURE 14.3.10

SOLUTION

Let M be the required plane and $\langle a, b, c \rangle$, $a \neq 0$, be a normal vector of M. Let M_1 be the plane having the equation $x - y + z = 0$. By Theorem 14.3.3, a normal vector of M_1 is $\langle 1, -1, 1 \rangle$. Because M and M_1 are perpendicular, it follows from (5) that

$$\langle a, b, c \rangle \cdot \langle 1, -1, 1 \rangle = 0$$

$$a - b + c = 0 \tag{6}$$

Let M_2 be the plane having the equation $2x + y - 4z - 5 = 0$. A normal vector of M_2 is $\langle 2, 1, -4 \rangle$. Because M and M_2 are perpendicular,

$$\langle a, b, c \rangle \cdot \langle 2, 1, -4 \rangle = 0$$

$$2a + b - 4c = 0 \tag{7}$$

Solving (6) and (7) simultaneously for b and c in terms of a we get $b = 2a$ and $c = a$. Therefore a normal vector of M is $\langle a, 2a, a \rangle$. Because $(4, 0, -2)$ is a point in M, it follows from Theorem 14.3.2 that an equation of M is

$$a(x - 4) + 2a(y - 0) + a(z + 2) = 0$$

Because $a \neq 0$, we divide by a and combine terms to obtain

$$x + 2y + z - 2 = 0$$

Consider now the plane having the equation $ax + by + d = 0$ and the xy plane whose equation is $z = 0$. Normal vectors to these planes are $\langle a, b, 0 \rangle$ and $\langle 0, 0, 1 \rangle$, respectively. Because $\langle a, b, 0 \rangle \cdot \langle 0, 0, 1 \rangle = 0$, the two planes are perpendicular. This means that a plane having an equation with no z term is perpendicular to the xy plane. Figure 14.3.8 illustrates this. In a similar manner we can conclude that a plane having an equation with no x term is perpendicular to the yz plane (see Fig. 14.3.9), and a plane having an equation with no y term is perpendicular to the xz plane (see Fig. 14.3.10).

Vectors can be used to find the distance from a point to a plane. The following example illustrates the procedure.

EXAMPLE 5

Find the distance from the point $(1, 4, 6)$ to the plane

$$2x - y + 2z + 10 = 0$$

SOLUTION

Let P be the point $(1, 4, 6)$ and choose any point Q in the plane. For simplicity choose the point Q as the point where the plane intersects the x axis, that is, the point $(-5, 0, 0)$. The vector having $\overrightarrow{PQ}$ as a representation is given by

$$\mathbf{V}(\overrightarrow{PQ}) = -6\mathbf{i} - 4\mathbf{j} - 6\mathbf{k}$$

A normal vector to the given plane is

$$\mathbf{N} = 2\mathbf{i} - \mathbf{j} + 2\mathbf{k}$$

The negative of **N** is also a normal vector to the given plane and

$$-\mathbf{N} = -2\mathbf{i} + \mathbf{j} - 2\mathbf{k}$$

We are not certain which of the two vectors, **N** or $-\mathbf{N}$, makes the smaller angle with vector $\mathbf{V}(\overrightarrow{PQ})$. Let **N'** be the one of the two vectors **N** or $-\mathbf{N}$ that makes an angle of radian measure $\theta < \frac{1}{2}\pi$ with $\mathbf{V}(\overrightarrow{PQ})$. In Fig. 14.3.11 there is a portion of the given plane containing the point $Q(-5, 0, 0)$, the representation of the vector **N'** having its initial point at Q, the point $P(1, 4, 6)$, the directed line segment $\overrightarrow{PQ}$, and the point R, which is the foot of the perpendicular from P to the plane. For simplicity the coordinate axes are not included in this figure. The distance $|\overline{RP}|$ is the required distance, which we call d. Because d is an undirected distance it is nonnegative. We see from Fig. 14.3.11 that d is the absolute value of the scalar projection of $\mathbf{V}(\overrightarrow{PQ})$ onto **N'**. Thus from (7) in Section 14.2 we get

$$d = \frac{|\mathbf{N'} \cdot \mathbf{V}(\overrightarrow{PQ})|}{\|\mathbf{N'}\|}$$

FIGURE 14.3.11

Because we have the absolute value of the dot product in the numerator and the magnitude of **N'** in the denominator we can replace **N'** by **N**, and we have

$$d = \frac{|\mathbf{N} \cdot \mathbf{V}(\overrightarrow{PQ})|}{\|\mathbf{N}\|}$$

$$= \frac{|\langle 2, -1, 2 \rangle \cdot \langle -6, -4, -6 \rangle|}{\sqrt{4 + 1 + 4}}$$

$$= \frac{|-12 + 4 - 12|}{\sqrt{9}}$$

$$= \tfrac{20}{3}$$

EXERCISES 14.3

In Exercises 1 through 6, find an equation of the plane containing the given point P and having the given vector **N** as a normal vector.

1. $P(3, 1, 2)$; $\mathbf{N} = \langle 1, 2, -3 \rangle$ **2.** $P(-3, 2, 5)$; $\mathbf{N} = \langle 6, -3, -2 \rangle$ **3.** $P(0, -1, 2)$; $\mathbf{N} = \langle 0, 1, -1 \rangle$

4. $P(-1, 8, 3)$; $\mathbf{N} = \langle -7, -1, 1 \rangle$ **5.** $P(2, 1, -1)$; $\mathbf{N} = -\mathbf{i} + 3\mathbf{j} + 4\mathbf{k}$ **6.** $P(1, 0, 0)$; $\mathbf{N} = \mathbf{i} + \mathbf{k}$

In Exercises 7 and 8, find an equation of the plane containing the given three points.

7. $(3, 4, 1)$, $(1, 7, 1)$, $(-1, -2, 5)$ **8.** $(0, 0, 2)$, $(2, 4, 1)$, $(-2, 3, 3)$

In Exercises 9 through 14, draw a sketch of the given plane and find two unit vectors that are normal to the plane.

9. $2x - y + 2z - 6 = 0$ **10.** $4x - 4y + 2z - 9 = 0$ **11.** $4x + 3y - 12z = 0$

12. $y + 2z - 4 = 0$ **13.** $3x + 2z - 6 = 0$ **14.** $z = 5$

In Exercises 15 through 20, find an equation of the plane satisfying the given conditions.

15. Perpendicular to the line through the points $(2, 2, -4)$ and $(7, -1, 3)$ and containing the point $(-5, 1, 2)$.

16. Parallel to the plane $4x - 2y + z - 1 = 0$ and containing the point $(2, 6, -1)$.

17. Perpendicular to the plane $x + 3y - z - 7 = 0$ and containing the points $(2, 0, 5)$ and $(0, 2, -1)$.
18. Perpendicular to each of the planes $x - y + z = 0$ and $2x + y - 4z - 5 = 0$ and containing the point $(4, 0, -2)$.
19. Perpendicular to the yz plane, containing the point $(2, 1, 1)$, and making an angle of radian measure $\cos^{-1} \frac{2}{3}$ with the plane $2x - y + 2z - 3 = 0$.
20. Containing the point $P(-3, 5, -2)$ and perpendicular to the representations of the vector $\mathbf{V}(\overrightarrow{OP})$.

In Exercises 21 through 23, find the cosine of the angle between the two given planes.

21. $2x - y - 2z - 5 = 0$ and $6x - 2y + 3z + 8 = 0$　　　　22. $2x - 5y + 3z - 1 = 0$ and $y - 5z + 3 = 0$
23. $3x + 4y = 0$ and $4x - 7y + 4z - 6 = 0$
24. Find the distance from the plane $2x + 2y - z - 6 = 0$ to the point $(2, 2, -4)$.
25. Find the distance from the plane $5x + 11y + 2z - 30 = 0$ to the point $(-2, 6, 3)$.
26. Find the perpendicular distance between the parallel planes $4x - 8y - z + 9 = 0$ and $4x - 8y - z - 6 = 0$.
27. Find the perpendicular distance between the parallel planes $4y - 3z - 6 = 0$ and $8y - 6z - 27 = 0$.
28. Prove that the undirected distance from the plane $ax + by + cz + d = 0$ to the point (x_0, y_0, z_0) is given by

$$\frac{|ax_0 + by_0 + cz_0 + d|}{\sqrt{a^2 + b^2 + c^2}}$$

29. Prove that the perpendicular distance between the two parallel planes $ax + by + cz + d_1 = 0$ and $ax + by + cz + d_2 = 0$ is given by

$$\frac{|d_1 - d_2|}{\sqrt{a^2 + b^2 + c^2}}$$

30. If a, b, and c are nonzero and are the x intercept, y intercept, and z intercept, respectively, of a plane, prove that an equation of the plane is

$$\frac{x}{a} + \frac{y}{b} + \frac{z}{c} = 1$$

This is called the *intercept form* of an equation of a plane.

14.4
LINES IN R^3

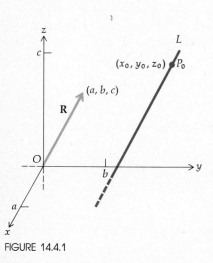

FIGURE 14.4.1

Let L be a line in R^3 such that it contains a given point $P_0(x_0, y_0, z_0)$ and is parallel to the representations of a given vector $\mathbf{R} = \langle a, b, c \rangle$. Figure 14.4.1 shows a sketch of line L and the position representation of vector $\mathbf{R}$. Line L is the set of points $P(x, y, z)$ such that $\mathbf{V}(\overrightarrow{P_0P})$ is parallel to the vector $\mathbf{R}$. So P is on the line L if and only if there is a nonzero scalar t such that

$$\mathbf{V}(\overrightarrow{P_0P}) = t\mathbf{R} \qquad (1)$$

Because $\mathbf{V}(\overrightarrow{P_0P}) = \langle x - x_0, y - y_0, z - z_0 \rangle$, we obtain, from (1),

$$\langle x - x_0, y - y_0, z - z_0 \rangle = t\langle a, b, c \rangle$$

from which it follows that

$$x - x_0 = ta \qquad y - y_0 = tb \qquad z - z_0 = tc$$
$$\Leftrightarrow \qquad x = x_0 + ta \qquad y = y_0 + tb \qquad z = z_0 + tc \qquad (2)$$

Letting the parameter t be any real number (i.e., t takes on all values in the interval $(-\infty, +\infty)$), the point P may be any point on the line L. Therefore Eqs. (2) represent the line L; these equations are called **parametric equations** of the line.

▶ ILLUSTRATION 1   From Eqs. (2), parametric equations of the line L that is parallel to the representations of the vector $\mathbf{R} = \langle 11, 8, 10 \rangle$ and that contains the point $(8, 12, 6)$ are

$$x = 8 + 11t \qquad y = 12 + 8t \qquad z = 6 + 10t$$

Figure 14.4.2 shows a sketch of the line and the position representation of $\mathbf{R}$.   ◀

If none of the numbers a, b, or c is zero, we can eliminate t from Eqs. (2) and obtain

$$\frac{x - x_0}{a} = \frac{y - y_0}{b} = \frac{z - z_0}{c} \tag{3}$$

These equations are called **symmetric equations** of the line. Equations (3) are equivalent to the system of three equations

$$b(x - x_0) = a(y - y_0)$$
$$c(x - x_0) = a(z - z_0) \tag{4}$$
$$c(y - y_0) = b(z - z_0)$$

FIGURE 14.4.2

Actually, the three equations in (4) are not independent because any one of them can be derived from the other two. Each of the equations in (4) is an equation of a plane containing the line L represented by Eqs. (3). Any two of these planes have as their intersection the line L; hence any two of the Eqs. (4) define the line. However, there is an unlimited number of planes containing a given line, and because any two of them will determine the line, there is an unlimited number of pairs of equations that represent a line.

The vector $\mathbf{R} = \langle a, b, c \rangle$ determines the direction of the line having symmetric equations (3), and the numbers a, b, and c are called **direction numbers** of the line. Any vector parallel to $\mathbf{R}$ has either the same or the opposite direction as $\mathbf{R}$; hence such a vector can be used in place of $\mathbf{R}$ in the above discussion. Because the components of any vector parallel to $\mathbf{R}$ are proportional to the components of $\mathbf{R}$, any set of three numbers proportional to a, b, and c also can serve as a set of direction numbers of the line. So a line has an unlimited number of sets of direction numbers. A set of direction numbers of a line is written in brackets as $[a, b, c]$.

▶ ILLUSTRATION 2   If $[2, 3, -4]$ represents a set of direction numbers of a line, other sets of direction numbers of the same line can be represented as $[4, 6, -8]$, $[1, \frac{3}{2}, -2]$, and $[2/\sqrt{29}, 3/\sqrt{29}, -4/\sqrt{29}]$.   ◀

▶ ILLUSTRATION 3   A set of direction numbers of the line of Illustration 1 is $[11, 8, 10]$, and the line contains the point $(8, 12, 6)$. Thus, from (3), symmetric equations of this line are

$$\frac{x - 8}{11} = \frac{y - 12}{8} = \frac{z - 6}{10}$$
                                                                                 ◀

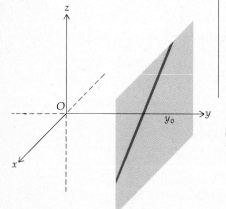

FIGURE 14.4.3

EXAMPLE 1

Find two sets of symmetric equations of the line through the two points $(-3, 2, 4)$ and $(6, 1, 2)$.

SOLUTION

Let P_1 be the point $(-3, 2, 4)$ and P_2 be the point $(6, 1, 2)$. Then the required line is parallel to the representations of the vector $\mathbf{V}(\overrightarrow{P_1P_2})$, and so the components of this vector constitute a set of direction numbers of the line. $\mathbf{V}(\overrightarrow{P_1P_2}) = \langle 9, -1, -2 \rangle$. Taking P_0 as the point $(-3, 2, 4)$ we have, from (3), the equations

$$\frac{x+3}{9} = \frac{y-2}{-1} = \frac{z-4}{-2}$$

Another set of symmetric equations of this line, obtained by taking P_0 as the point $(6, 1, 2)$, is

$$\frac{x-6}{9} = \frac{y-1}{-1} = \frac{z-2}{-2}$$

If one of the numbers a, b, or c is zero, we do not use symmetric equations (3). However, suppose, for example, that $b = 0$ and neither a nor c is zero. Then equations of the line are

$$\frac{x-x_0}{a} = \frac{z-z_0}{c} \quad \text{and} \quad y = y_0 \tag{5}$$

A line having symmetric equations (5) lies in the plane $y = y_0$ and hence is parallel to the xz plane. Figure 14.4.3 shows such a line.

EXAMPLE 2

Given the two planes

$$x + 3y - z - 9 = 0 \quad \text{and} \quad 2x - 3y + 4z + 3 = 0$$

For the line of intersection of these two planes, find (a) a set of symmetric equations and (b) a set of parametric equations.

SOLUTION

(a) A set of symmetric equations is of the form (3). To obtain this form we solve the pair of given equations for x and y in terms of z. The computations is as follows:

$$
\begin{array}{ll}
\begin{aligned}
x + 3y - \ z - 9 &= 0 \\
2x - 3y + 4z + 3 &= 0 \ (+) \\
\hline
3x \qquad + 3z - 6 &= 0 \\
x &= -z + 2
\end{aligned}
&
\begin{aligned}
2x + 6y - 2z - 18 &= 0 \\
2x - 3y + 4z + \ 3 &= 0 \ (-) \\
\hline
9y - 6z - 21 &= 0 \\
y &= \tfrac{2}{3}z + \tfrac{7}{3}
\end{aligned}
\end{array}
$$

We now solve each equation for z and obtain

$$\frac{x-2}{-1} = z \qquad \frac{y - \tfrac{7}{3}}{\tfrac{2}{3}} = z$$

Thus a set of symmetric equations is

$$\frac{x-2}{-1} = \frac{y-\frac{7}{3}}{\frac{2}{3}} = \frac{z-0}{1}$$

$$\Leftrightarrow \quad \frac{x-2}{-3} = \frac{y-\frac{7}{3}}{2} = \frac{z-0}{3}$$

(b) A set of parametric equations is obtained by setting each of the ratios in part (a) equal to t, and we have

$$\frac{x-2}{-3} = t \qquad \frac{y-\frac{7}{3}}{2} = t \qquad \frac{z-0}{3} = t$$

$$x = 2 - 3t \qquad y = \tfrac{7}{3} + 2t \qquad z = 3t$$

EXAMPLE 3

Find the direction cosines of a vector whose representations are parallel to the line of Example 2.

SOLUTION

From the symmetric equations of the line in Example 2 a set of direction numbers of the line is $[-3, 2, 3]$. Therefore the vector $\langle -3, 2, 3 \rangle$ has its representations parallel to the line. Because $\sqrt{(-3)^2 + 2^2 + 3^2} = \sqrt{22}$, the direction cosines of this vector are

$$\cos \alpha = -\frac{3}{\sqrt{22}} \qquad \cos \beta = \frac{2}{\sqrt{22}} \qquad \cos \gamma = \frac{3}{\sqrt{22}}$$

EXAMPLE 4

Find equations of the line through the point $(1, -1, 1)$ perpendicular to the line

$$3x = 2y = z \tag{6}$$

and parallel to the plane

$$x + y - z = 0 \tag{7}$$

SOLUTION

Let $[a, b, c]$ be a set of direction numbers of the required line. Equations (6) can be written as

$$\frac{x-0}{\frac{1}{3}} = \frac{y-0}{\frac{1}{2}} = \frac{z-0}{1}$$

which are symmetric equations of a line. A set of direction numbers of this line is $[\frac{1}{3}, \frac{1}{2}, 1]$. Because the required line is perpendicular to this line, it follows that the vectors $\langle a, b, c \rangle$ and $\langle \frac{1}{3}, \frac{1}{2}, 1 \rangle$ are orthogonal. So

$$\langle a, b, c \rangle \cdot \langle \tfrac{1}{3}, \tfrac{1}{2}, 1 \rangle = 0$$

$$\tfrac{1}{3}a + \tfrac{1}{2}b + c = 0 \tag{8}$$

A normal vector to the plane (7) is $\langle 1, 1, -1 \rangle$. Because the required line is parallel to this plane, it is perpendicular to representations of the normal vector. Hence the vectors $\langle a, b, c \rangle$ and $\langle 1, 1, -1 \rangle$ are orthogonal; so

$$\langle a, b, c \rangle \cdot \langle 1, 1, -1 \rangle = 0$$

$$a + b - c = 0 \tag{9}$$

Assuming $c \neq 0$, we solve (8) and (9) simultaneously for a and b in terms of c and get $a = 9c$ and $b = -8c$. The required line then has the set of direction numbers $[9c, -8c, c]$ and contains the point $(1, -1, 1)$. Therefore symmetric equations of the line are

$$\frac{x-1}{9c} = \frac{y+1}{-8c} = \frac{z-1}{c}$$

$$\Leftrightarrow \quad \frac{x-1}{9} = \frac{y+1}{-8} = \frac{z-1}{1}$$

EXAMPLE 5

If l_1 is the line through $A(1, 2, 7)$ and $B(-2, 3, -4)$ and l_2 is the line through $C(2, -1, 4)$ and $D(5, 7, -3)$, prove that l_1 and l_2 are skew lines (i.e., they do not lie in one plane).

SOLUTION

To show that two lines do not lie in one plane we demonstrate that they do not intersect and are not parallel. Parametric equations of a line are

$$x = x_0 + ta \qquad y = y_0 + tb \qquad z = z_0 + tc$$

where $[a, b, c]$ is a set of direction numbers of the line and (x_0, y_0, z_0) is any point on the line. Because $\mathbf{V}(\overrightarrow{AB}) = \langle -3, 1, -11 \rangle$, a set of direction numbers of l_1 is $[-3, 1, -11]$. Taking A as the point P_0 we have as parametric equations of l_1

$$x = 1 - 3t \qquad y = 2 + t \qquad z = 7 - 11t \tag{10}$$

Because $\mathbf{V}(\overrightarrow{CD}) = \langle 3, 8, -7 \rangle$ and l_2 contains the point C, parametric equations of l_2 are

$$x = 2 + 3s \qquad y = -1 + 8s \qquad z = 4 - 7s \tag{11}$$

Because the sets of direction numbers are not proportional, l_1 and l_2 are not parallel. For the lines to intersect, there have to be a value of t and a value of s that give the same point (x_1, y_1, z_1) in both sets of Eqs. (10) and (11). Therefore we equate the right sides of the respective equations and obtain

$$1 - 3t = 2 + 3s$$

$$2 + t = -1 + 8s$$

$$7 - 11t = 4 - 7s$$

Solving the first two equations simultaneously we obtain $s = \frac{8}{27}$ and $t = -\frac{17}{27}$. This set of values does not satisfy the third equation; hence the two lines do not intersect. Thus l_1 and l_2 are skew lines.

EXERCISES 14.4

In Exercises 1 through 8, find parametric and symmetric equations for the line satisfying the given conditions.

1. Through the two points $(1, 2, 1)$ and $(5, -1, 1)$.
2. Through the point $(5, 3, 2)$ with direction numbers $[4, 1, -1]$.
3. Through the origin and perpendicular to the line $\frac{1}{4}(x - 10) = \frac{1}{3}y = \frac{1}{2}z$ at their intersection.
4. Through the origin and perpendicular to the lines having direction numbers $[4, 2, 1]$ and $[-3, -2, 1]$.
5. Perpendicular to the lines having direction numbers $[-5, 1, 2]$ and $[2, -3, -4]$ at the point $(-2, 0, 3)$.
6. Through the point $(-3, 1, -5)$ and perpendicular to the plane $4x - 2y + z - 7 = 0$.
7. Through the point $(4, -5, 20)$ and perpendicular to the plane $x + 3y - 6z - 8 = 0$.
8. Through the point $(2, 0, -4)$ and parallel to each of the planes $2x + y - z = 0$ and $x + 3y + 5z = 0$.
9. Find a set of symmetric equations for the line

$$\begin{cases} 4x - 3y + z - 2 = 0 \\ 2x + 5y - 3z + 4 = 0 \end{cases}$$

10. Show that the lines

$$\frac{x + 1}{2} = \frac{y + 4}{-5} = \frac{z - 2}{3} \quad \text{and} \quad \frac{x - 3}{-2} = \frac{y + 14}{5} = \frac{z - 8}{-3}$$

are coincident.

11. Prove that the line $\frac{1}{2}(x - 3) = \frac{1}{3}(y + 2) = \frac{1}{4}(z + 1)$ lies in the plane $x - 2y + z = 6$.
12. Prove that the line $x + 1 = -\frac{1}{2}(y - 6) = z$ lies in the plane $3x + y - z = 3$.

The planes through a line that are perpendicular to the coordinate planes are called the **projecting planes** of the line. In Exercises 13 through 16, find equations of the projecting planes of the given line and draw a sketch of the line.

13. $\begin{cases} 3x - 2y + 5z - 30 = 0 \\ 2x + 3y - 10z - 6 = 0 \end{cases}$ 14. $\begin{cases} x + y - 3z + 1 = 0 \\ 2x - y - 3z + 14 = 0 \end{cases}$ 15. $\begin{cases} x - 2y - 3z + 6 = 0 \\ x + y + z - 1 = 0 \end{cases}$ 16. $\begin{cases} 2x - y + z - 7 = 0 \\ 4x - y + 3z - 13 = 0 \end{cases}$

17. Find the cosine of the smallest angle between the vector whose representations are parallel to the line $x = 2y + 4$, $z = -y + 4$, and the vector whose representations are parallel to the line $x = y + 7$, $2z = y + 2$.
18. Find an equation of the plane containing the point $(6, 2, 4)$ and the line $\frac{1}{5}(x - 1) = \frac{1}{6}(y + 2) = \frac{1}{7}(z - 3)$.

In Exercises 19 and 20, find an equation of the plane containing the given intersecting lines.

19. $\frac{x - 2}{4} = \frac{y + 3}{-1} = \frac{z + 2}{3}$ and $\begin{cases} 3x + 2y + z + 2 = 0 \\ x - y + 2z - 1 = 0 \end{cases}$ 20. $\frac{x}{2} = \frac{y - 2}{3} = \frac{z - 1}{1}$ and $\frac{x}{1} = \frac{y - 2}{-1} = \frac{z - 1}{1}$

21. Show that the lines

$$\begin{cases} 3x - y - z = 0 \\ 8x - 2y - 3z + 1 = 0 \end{cases} \quad \text{and} \quad \begin{cases} x - 3y + z + 3 = 0 \\ 3x - y - z + 5 = 0 \end{cases}$$

are parallel, and find an equation of the plane determined by these lines.

22. Show that the lines

$$\frac{x + 2}{5} = \frac{y - 1}{-2} = z + 4 \quad \text{and} \quad \frac{x - 3}{-5} = \frac{y + 4}{2} = \frac{z - 3}{-1}$$

are parallel, and find an equation of the plane determined by these lines.

23. Find the coordinates of the point of intersection of the line $\frac{1}{4}(x - 2) = -\frac{1}{2}(y + 3) = \frac{1}{7}(z - 1)$ and the plane $5x - y + 2z - 12 = 0$.
24. Find equations of the line through the point $(1, -1, 1)$, perpendicular to the line $3x = 2y = z$, and parallel to the plane $x + y - z = 0$.
25. Find equations of the line through the point $(3, 6, 4)$, intersecting the z axis, and parallel to the plane $x - 3y + 5z - 6 = 0$.
26. Find the perpendicular distance from the origin to the line $x = -2 + \frac{6}{7}t$, $y = 7 - \frac{2}{7}t$, $z = 4 + \frac{3}{7}t$.

27. Find the perpendicular distance from the point $(-1, 3, -1)$ to the line $x - 2z = 7$, $y = 1$.

28. Find equations of the line through the origin, perpendicular to the line $x = y - 5$, $z = 2y - 3$, and intersecting the line $y = 2x + 1$, $z = x + 2$.

29. Prove that the lines

$$\frac{x-1}{5} = \frac{y-2}{-2} = \frac{z+1}{-3} \quad \text{and} \quad \frac{x-2}{1} = \frac{y+1}{-3} = \frac{z+3}{2}$$

are skew lines.

30. Find equations of the line through the point $(3, -4, -5)$ that intersects each of the skew lines of Exercise 29.

31. What are the symmetric equations of a line if the two direction numbers a and b are zero?

14.5
CROSS PRODUCT

If **A** and **B** are two nonparallel vectors, representations of the two vectors with the same initial point determine a plane as indicated in Fig. 14.5.1. We show that a vector whose representations are perpendicular to this plane is given by the vector operation called the *cross product* of **A** and **B**. The cross product is a vector operation for vectors in V_3 that we did not have for vectors in V_2. We first define this operation and then consider its algebraic and geometric properties.

14.5.1 DEFINITION   If $\mathbf{A} = \langle a_1, a_2, a_3 \rangle$ and $\mathbf{B} = \langle b_1, b_2, b_3 \rangle$, then the **cross product** of **A** and **B**, denoted by $\mathbf{A} \times \mathbf{B}$, is given by

$$\mathbf{A} \times \mathbf{B} = \langle a_2 b_3 - a_3 b_2, \, a_3 b_1 - a_1 b_3, \, a_1 b_2 - a_2 b_1 \rangle \tag{1}$$

Because the cross product of two vectors is a vector, the cross product also is called the **vector product**. The operation of obtaining the cross product is called **vector multiplication**.

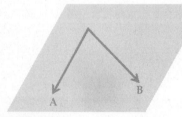

FIGURE 14.5.1

▶ **ILLUSTRATION 1**   If $\mathbf{A} = \langle 2, 1, -3 \rangle$ and $\mathbf{B} = \langle 3, -1, 4 \rangle$, then from Definition 14.5.1,

$$\begin{aligned}
\mathbf{A} \times \mathbf{B} &= \langle 2, 1, -3 \rangle \times \langle 3, -1, 4 \rangle \\
&= \langle (1)(4) - (-3)(-1), (-3)(3) - (2)(4), (2)(-1) - (1)(3) \rangle \\
&= \langle 4 - 3, -9 - 8, -2 - 3 \rangle \\
&= \langle 1, -17, -5 \rangle \\
&= \mathbf{i} - 17\mathbf{j} - 5\mathbf{k} \qquad \blacktriangleleft
\end{aligned}$$

There is a mnemonic device for remembering formula (1) that makes use of determinant notation. A second-order determinant is defined by the equation

$$\begin{vmatrix} a & b \\ c & d \end{vmatrix} = ad - bc$$

where a, b, and c are real numbers. For example,

$$\begin{vmatrix} 3 & 6 \\ -2 & 5 \end{vmatrix} = 3(5) - (6)(-2)$$

$$= 27$$

Therefore formula (1) can be written as

$$\mathbf{A} \times \mathbf{B} = \begin{vmatrix} a_2 & a_3 \\ b_2 & b_3 \end{vmatrix} \mathbf{i} - \begin{vmatrix} a_1 & a_3 \\ b_2 & b_3 \end{vmatrix} \mathbf{j} + \begin{vmatrix} a_1 & a_2 \\ b_1 & b_2 \end{vmatrix} \mathbf{k}$$

The right side of the above expression can be written symbolically as

$$\begin{vmatrix} \mathbf{i} & \mathbf{j} & \mathbf{k} \\ a_1 & a_2 & a_3 \\ b_1 & b_2 & b_3 \end{vmatrix}$$

which is the notation for a third-order determinant. However, observe that the first row contains vectors and not real numbers as is customary with determinant notation.

▶ ILLUSTRATION 2   We use the mnemonic device employing determinant notation to find the cross product of the vectors of Illustration 1.

$$\mathbf{A} \times \mathbf{B} = \begin{vmatrix} \mathbf{i} & \mathbf{j} & \mathbf{k} \\ 2 & 1 & -3 \\ 3 & -1 & 4 \end{vmatrix}$$

$$= \begin{vmatrix} 1 & -3 \\ -1 & 4 \end{vmatrix} \mathbf{i} - \begin{vmatrix} 2 & -3 \\ 3 & 4 \end{vmatrix} \mathbf{j} + \begin{vmatrix} 2 & 1 \\ 3 & -1 \end{vmatrix} \mathbf{k}$$

$$= [(1)(4) - (-3)(-1)]\mathbf{i} - [(2)(4) - (-3)(3)]\mathbf{j} + [(2)(-1) - (1)(3)]\mathbf{k}$$

$$= \mathbf{i} - 17\mathbf{j} - 5\mathbf{k} \qquad ◀$$

14.5.2 THEOREM   If **A** is any vector in V_3, then

(i) $\mathbf{A} \times \mathbf{A} = \mathbf{0}$
(ii) $\mathbf{0} \times \mathbf{A} = \mathbf{0}$
(iii) $\mathbf{A} \times \mathbf{0} = \mathbf{0}$

PROOF OF (i)   If $\mathbf{A} = \langle a_1, a_2, a_3 \rangle$, then by Definition 14.5.1,

$$\mathbf{A} \times \mathbf{A} = \langle a_2 a_3 - a_3 a_2, \; a_3 a_1 - a_1 a_3, \; a_1 a_2 - a_2 a_1 \rangle$$
$$= \langle 0, 0, 0 \rangle$$
$$= \mathbf{0}$$

The proofs of (ii) and (iii) are left as exercises (see Exercise 13).   ■

By applying Definition 14.5.1 to pairs of unit vectors **i**, **j**, and **k** we obtain the following:

$$\mathbf{i} \times \mathbf{i} = \mathbf{0} \qquad \mathbf{j} \times \mathbf{j} = \mathbf{0} \qquad \mathbf{k} \times \mathbf{k} = \mathbf{0}$$

$$\mathbf{i} \times \mathbf{j} = \mathbf{k} \qquad \mathbf{j} \times \mathbf{k} = \mathbf{i} \qquad \mathbf{k} \times \mathbf{i} = \mathbf{j}$$

$$\mathbf{j} \times \mathbf{i} = -\mathbf{k} \qquad \mathbf{k} \times \mathbf{j} = -\mathbf{i} \qquad \mathbf{i} \times \mathbf{k} = -\mathbf{j}$$

As an aid in remembering the above cross products, first notice that the cross product of any one of the unit vectors **i**, **j**, or **k** with itself is the zero vector. The other six cross products can be obtained from Fig. 14.5.2 by

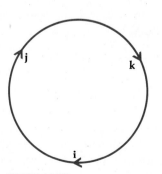

FIGURE 14.5.2

applying the following rule: The cross product of two consecutive vectors in the clockwise direction is the next vector; and the cross product of two consecutive vectors in the counterclockwise direction is the negative of the next vector.

It can be seen easily that cross multiplication of two vectors is not commutative because in particular $\mathbf{i} \times \mathbf{j} \neq \mathbf{j} \times \mathbf{i}$. However, we have $\mathbf{i} \times \mathbf{j} = \mathbf{k}$ and $\mathbf{j} \times \mathbf{i} = -\mathbf{k}$; so $\mathbf{i} \times \mathbf{j} = -(\mathbf{j} \times \mathbf{i})$. In general, if $\mathbf{A}$ and $\mathbf{B}$ are any vectors in V_3, $\mathbf{A} \times \mathbf{B} = -(\mathbf{B} \times \mathbf{A})$, which we state and prove as a theorem.

14.5.3 THEOREM   If $\mathbf{A}$ and $\mathbf{B}$ are any vectors in V_3,
$$\mathbf{A} \times \mathbf{B} = -(\mathbf{B} \times \mathbf{A})$$

PROOF   If $\mathbf{A} = \langle a_1, a_2, a_3 \rangle$ and $\mathbf{B} = \langle b_1, b_2, b_3 \rangle$, then by Definition 14.5.1,
$$\begin{aligned} \mathbf{A} \times \mathbf{B} &= \langle a_2 b_3 - a_3 b_2, a_3 b_1 - a_1 b_3, a_1 b_2 - a_2 b_1 \rangle \\ &= -1 \langle a_3 b_2 - a_2 b_3, a_1 b_3 - a_3 b_1, a_2 b_1 - a_1 b_2 \rangle \\ &= -(\mathbf{B} \times \mathbf{A}) \end{aligned}$$ ∎

Cross multiplication of vectors is not associative. This is shown by the following example:
$$\mathbf{i} \times (\mathbf{i} \times \mathbf{j}) = \mathbf{i} \times \mathbf{k} \qquad (\mathbf{i} \times \mathbf{i}) \times \mathbf{j} = \mathbf{0} \times \mathbf{j}$$
$$= -\mathbf{j} \qquad\qquad\qquad = \mathbf{0}$$

Thus
$$\mathbf{i} \times (\mathbf{i} \times \mathbf{j}) \neq (\mathbf{i} \times \mathbf{i}) \times \mathbf{j}$$

Cross multiplication of vectors is distributive with respect to vector addition, as given by the following theorem.

14.5.4 THEOREM   If $\mathbf{A}$, $\mathbf{B}$, and $\mathbf{C}$ are any vectors in V_3, then
$$\mathbf{A} \times (\mathbf{B} + \mathbf{C}) = \mathbf{A} \times \mathbf{B} + \mathbf{A} \times \mathbf{C} \tag{2}$$

To prove Theorem 14.5.4 let $\mathbf{A} = \langle a_1, a_2, a_3 \rangle$, $\mathbf{B} = \langle b_1, b_2, b_3 \rangle$, and $\mathbf{C} = \langle c_1, c_2, c_3 \rangle$, and then show that the components of the vector on the left side of (2) are the same as the components of the vector on the right side of (2). The details are left as an exercise (see Exercise 35).

14.5.5 THEOREM   If $\mathbf{A}$ and $\mathbf{B}$ are any two vectors in V_3 and c is a scalar, then

(i) $(c\mathbf{A}) \times \mathbf{B} = \mathbf{A} \times (c\mathbf{B})$;
(ii) $(c\mathbf{A}) \times \mathbf{B} = c(\mathbf{A} \times \mathbf{B})$.

The proof of Theorem 14.5.5 is left as an exercise (see Exercise 36).

Theorems 14.5.4 and 14.5.5 can be applied to compute the cross product of two vectors by using laws of algebra, provided the order of the vectors in cross multiplication is not changed, because it is prohibited by Theorem 14.5.3. The following illustration demonstrates this procedure.

▶ ILLUSTRATION 3   We find the cross product of the vectors in Illustration 1 by applying Theorems 14.5.4 and 14.5.5.

$$\mathbf{A} \times \mathbf{B} = (2\mathbf{i} + \mathbf{j} - 3\mathbf{k}) \times (3\mathbf{i} - \mathbf{j} + 4\mathbf{k})$$

$$= 6(\mathbf{i} \times \mathbf{i}) - 2(\mathbf{i} \times \mathbf{j}) + 8(\mathbf{i} \times \mathbf{k}) + 3(\mathbf{j} \times \mathbf{i}) - 1(\mathbf{j} \times \mathbf{j})$$
$$+ 4(\mathbf{j} \times \mathbf{k}) - 9(\mathbf{k} \times \mathbf{i}) + 3(\mathbf{k} \times \mathbf{j}) - 12(\mathbf{k} \times \mathbf{k})$$

$$= 6(\mathbf{0}) - 2(\mathbf{k}) + 8(-\mathbf{j}) + 3(-\mathbf{k}) - 1(\mathbf{0}) + 4(\mathbf{i}) - 9(\mathbf{j}) + 3(-\mathbf{i}) - 12(\mathbf{0})$$

$$= -2\mathbf{k} - 8\mathbf{j} - 3\mathbf{k} + 4\mathbf{i} - 9\mathbf{j} - 3\mathbf{i}$$

$$= \mathbf{i} - 17\mathbf{j} - 5\mathbf{k} \qquad \blacktriangleleft$$

The method used in Illustration 3 gives a way of finding the cross product without having to remember formula (1) or to use determinant notation. Actually all the steps shown in the solution need not be included because the various cross products of the unit vectors can be obtained immediately by using Fig. 14.5.2 and the corresponding rule.

There are two *triple products* that we shall consider. One is the product $\mathbf{A} \cdot (\mathbf{B} \times \mathbf{C})$, called the **triple scalar product** of the vectors $\mathbf{A}$, $\mathbf{B}$, and $\mathbf{C}$. Actually, the parentheses are not needed because $\mathbf{A} \cdot \mathbf{B}$ is a scalar, and therefore $\mathbf{A} \cdot \mathbf{B} \times \mathbf{C}$ can be interpreted in only one way.

14.5.6 THEOREM   If $\mathbf{A}$, $\mathbf{B}$, and $\mathbf{C}$ are vectors in V_3, then

$$\mathbf{A} \cdot \mathbf{B} \times \mathbf{C} = \mathbf{A} \times \mathbf{B} \cdot \mathbf{C} \qquad (3)$$

Theorem 14.5.6 can be proved by letting

$$\mathbf{A} = \langle a_1, a_2, a_3 \rangle \qquad \mathbf{B} = \langle b_1, b_2, b_3 \rangle \qquad \mathbf{C} = \langle c_1, c_2, c_3 \rangle$$

and then by showing that the scalar on the left side of (3) is equal to the scalar on the right. The details are left as an exercise (see Exercise 37).

▶ ILLUSTRATION 4   We verify Theorem 14.5.6 if

$$\mathbf{A} = \langle 1, -1, 2 \rangle \qquad \mathbf{B} = \langle 3, 4, -2 \rangle \qquad \mathbf{C} = \langle -5, 1, -4 \rangle$$

$$\mathbf{B} \times \mathbf{C} = (3\mathbf{i} + 4\mathbf{j} - 2\mathbf{k}) \times (-5\mathbf{i} + \mathbf{j} - 4\mathbf{k})$$

$$= 3\mathbf{k} - 12(-\mathbf{j}) - 20(-\mathbf{k}) - 16\mathbf{i} + 10\mathbf{j} - 2(-\mathbf{i})$$

$$= -14\mathbf{i} + 22\mathbf{j} + 23\mathbf{k}$$

$$\mathbf{A} \cdot (\mathbf{B} \times \mathbf{C}) = \langle 1, -1, 2 \rangle \cdot \langle -14, 22, 23 \rangle$$

$$= -14 - 22 + 46$$

$$= 10$$

$$\mathbf{A} \times \mathbf{B} = (\mathbf{i} - \mathbf{j} + 2\mathbf{k}) \times (3\mathbf{i} + 4\mathbf{j} - 2\mathbf{k})$$

$$= 4\mathbf{k} - 2(-\mathbf{j}) - 3(-\mathbf{k}) + 2\mathbf{i} + 6\mathbf{j} + 8(-\mathbf{i})$$

$$= -6\mathbf{i} + 8\mathbf{j} + 7\mathbf{k}$$

$$(\mathbf{A} \times \mathbf{B}) \cdot \mathbf{C} = \langle -6, 8, 7 \rangle \cdot \langle -5, 1, -4 \rangle$$

$$= 30 + 8 - 28$$

$$= 10$$

This verifies the theorem for these three vectors.   ◀

The other triple product is $\mathbf{A} \times (\mathbf{B} \times \mathbf{C})$, called the **triple vector product**.

14.5.7 THEOREM   If $\mathbf{A}$, $\mathbf{B}$, and $\mathbf{C}$ are vectors in V_3, then

$$\mathbf{A} \times (\mathbf{B} \times \mathbf{C}) = (\mathbf{A} \cdot \mathbf{C})\mathbf{B} - (\mathbf{A} \cdot \mathbf{B})\mathbf{C} \tag{4}$$

The proof of Theorem 14.5.7 is similar to the proof of Theorem 14.5.6. By using components of the vectors $\mathbf{A}$, $\mathbf{B}$, and $\mathbf{C}$ we can show that the vector on the left side of (4) is the same as the vector on the right. The computation is left as an exercise (see Exercise 38).

▶ ILLUSTRATION 5   We verify Theorem 14.5.7 for the vectors $\mathbf{A}$, $\mathbf{B}$, and $\mathbf{C}$ of Illustration 4. Because $\mathbf{B} \times \mathbf{C} = -14\mathbf{i} + 22\mathbf{j} + 23\mathbf{k}$

$$\mathbf{A} \times (\mathbf{B} \times \mathbf{C}) = \begin{vmatrix} \mathbf{i} & \mathbf{j} & \mathbf{k} \\ 1 & -1 & 2 \\ -14 & 22 & 23 \end{vmatrix}$$

$$= -23\mathbf{i} - 28\mathbf{j} + 22\mathbf{k} - 14\mathbf{k} - 44\mathbf{i} - 23\mathbf{j}$$

$$= -67\mathbf{i} - 51\mathbf{j} + 8\mathbf{k} \tag{5}$$

$$\begin{aligned} \mathbf{A} \cdot \mathbf{C} &= \langle 1, -1, 2 \rangle \cdot \langle -5, 1, -4 \rangle & \mathbf{A} \cdot \mathbf{B} &= \langle 1, -1, 2 \rangle \cdot \langle 3, 4, -2 \rangle \\ &= -5 - 1 - 8 & &= 3 - 4 - 4 \\ &= -14 & &= -5 \end{aligned}$$

Thus

$$\begin{aligned} (\mathbf{A} \cdot \mathbf{C})\mathbf{B} - (\mathbf{A} \cdot \mathbf{B})\mathbf{C} &= -14\langle 3, 4, -2 \rangle - (-5)\langle -5, 1, -4 \rangle \\ &= \langle -42, -56, 28 \rangle - \langle 25, -5, 20 \rangle \\ &= \langle -67, -51, 8 \rangle \\ &= -67\mathbf{i} - 51\mathbf{j} + 8\mathbf{k} \end{aligned} \tag{6}$$

By comparing (5) with (6), Theorem 14.5.7 is verified for these three vectors.   ◀

The following theorem is used for a geometric interpretation of the cross product.

14.5.8 THEOREM   If $\mathbf{A}$ and $\mathbf{B}$ are two vectors in V_3 and θ is the radian measure of the angle between $\mathbf{A}$ and $\mathbf{B}$, then

$$\|\mathbf{A} \times \mathbf{B}\| = \|\mathbf{A}\| \, \|\mathbf{B}\| \sin \theta$$

PROOF   From Theorem 14.2.11(v),

$$\|\mathbf{A} \times \mathbf{B}\|^2 = (\mathbf{A} \times \mathbf{B}) \cdot (\mathbf{A} \times \mathbf{B}) \tag{7}$$

With the notation $\mathbf{U}$, $\mathbf{V}$, and $\mathbf{W}$ as the vectors, we have from Theorem 14.5.6

$$(\mathbf{U} \times \mathbf{V}) \cdot \mathbf{W} = \mathbf{U} \cdot (\mathbf{V} \times \mathbf{W}) \tag{8}$$

If in (8) we let $\mathbf{U} = \mathbf{A}$, $\mathbf{V} = \mathbf{B}$, and $\mathbf{W} = \mathbf{A} \times \mathbf{B}$, we have

$$(\mathbf{A} \times \mathbf{B}) \cdot (\mathbf{A} \times \mathbf{B}) = \mathbf{A} \cdot [\mathbf{B} \times (\mathbf{A} \times \mathbf{B})] \tag{9}$$

Applying Theorem 14.5.7 to the vector in brackets on the right side of (9) we get

$$\begin{aligned}
(\mathbf{A} \times \mathbf{B}) \cdot (\mathbf{A} \times \mathbf{B}) &= \mathbf{A} \cdot [(\mathbf{B} \cdot \mathbf{B})\mathbf{A} - (\mathbf{B} \cdot \mathbf{A})\mathbf{B}] \\
&= (\mathbf{A} \cdot \mathbf{A})(\mathbf{B} \cdot \mathbf{B}) - (\mathbf{A} \cdot \mathbf{B})(\mathbf{A} \cdot \mathbf{B}) \\
&= \|\mathbf{A}\|^2 \|\mathbf{B}\|^2 - (\mathbf{A} \cdot \mathbf{B})^2 \tag{10}
\end{aligned}$$

From Theorem 14.2.13, if θ is the angle between $\mathbf{A}$ and $\mathbf{B}$,

$$\mathbf{A} \cdot \mathbf{B} = \|\mathbf{A}\| \|\mathbf{B}\| \cos \theta \tag{11}$$

Substituting from (11) into (10) we have

$$\begin{aligned}
(\mathbf{A} \times \mathbf{B}) \cdot (\mathbf{A} \times \mathbf{B}) &= \|\mathbf{A}\|^2 \|\mathbf{B}\|^2 - \|\mathbf{A}\|^2 \|\mathbf{B}\|^2 \cos^2 \theta \\
&= \|\mathbf{A}\|^2 \|\mathbf{B}\|^2 (1 - \cos^2 \theta) \tag{12}
\end{aligned}$$

Substituting from (7) into (12), and because $1 - \cos^2 \theta = \sin^2 \theta$, we have

$$\|\mathbf{A} \times \mathbf{B}\|^2 = \|\mathbf{A}\|^2 \|\mathbf{B}\|^2 \sin^2 \theta \tag{13}$$

Because $0 \le \theta \le \pi$, $\sin \theta \ge 0$. Therefore, from (13),

$$\|\mathbf{A} \times \mathbf{B}\| = \|\mathbf{A}\| \|\mathbf{B}\| \sin \theta \qquad\blacksquare$$

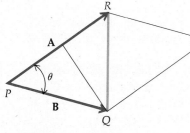

FIGURE 14.5.3

We consider now a geometric interpretation of $\|\mathbf{A} \times \mathbf{B}\|$. Let $\overrightarrow{PR}$ be a representation of $\mathbf{A}$ and let $\overrightarrow{PQ}$ be a representation of $\mathbf{B}$. Then the angle between the vectors $\mathbf{A}$ and $\mathbf{B}$ is the angle at P in triangle RPQ (see Fig. 14.5.3). Let the radian measure of this angle be θ. Therefore the number of square units in the area of the parallelogram having $\overrightarrow{PR}$ and $\overrightarrow{PQ}$ as adjacent sides is $\|\mathbf{A}\| \|\mathbf{B}\| \sin \theta$ because the altitude of the parallelogram has length $\|\mathbf{B}\| \sin \theta$ units and the length of the base is $\|\mathbf{A}\|$ units. So from (7) it follows that $\|\mathbf{A} \times \mathbf{B}\|$ square units is the area of this parallelogram.

EXAMPLE 1
Show that the quadrilateral having vertices at $P(1, -2, 3)$, $Q(4, 3, -1)$, $R(2, 2, 1)$, and $S(5, 7, -3)$ is a parallelogram, and find its area.

SOLUTION
Figure 14.5.4 shows the quadrilateral $PQSR$.

$$\begin{aligned}
\mathbf{V}(\overrightarrow{PQ}) &= \langle 4 - 1, 3 + 2, -1 - 3 \rangle & \mathbf{V}(\overrightarrow{PR}) &= \langle 2 - 1, 2 + 2, 1 - 3 \rangle \\
&= \langle 3, 5, -4 \rangle & &= \langle 1, 4, -2 \rangle \\
\mathbf{V}(\overrightarrow{RS}) &= \langle 5 - 2, 7 - 2, -3 - 1 \rangle & \mathbf{V}(\overrightarrow{QS}) &= \langle 5 - 4, 7 - 3, -3 + 1 \rangle \\
&= \langle 3, 5, -4 \rangle & &= \langle 1, 4, -2 \rangle
\end{aligned}$$

Because $\mathbf{V}(\overrightarrow{PQ}) = \mathbf{V}(\overrightarrow{RS})$ and $\mathbf{V}(\overrightarrow{PR}) = \mathbf{V}(\overrightarrow{QS})$, it follows that $\overrightarrow{PQ}$ is parallel to $\overrightarrow{RS}$ and $\overrightarrow{PR}$ is parallel to $\overrightarrow{QS}$. Therefore $PQSR$ is a parallelogram.

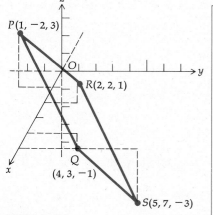

FIGURE 14.5.4

Let $\mathbf{A} = \mathbf{V}(\overrightarrow{PR})$ and $\mathbf{B} = \mathbf{V}(\overrightarrow{PQ})$; then

$$\begin{aligned}
\mathbf{A} \times \mathbf{B} &= (\mathbf{i} + 4\mathbf{j} - 2\mathbf{k}) \times (3\mathbf{i} + 5\mathbf{j} - 4\mathbf{k}) \\
&= 3(\mathbf{i} \times \mathbf{i}) + 5(\mathbf{i} \times \mathbf{j}) - 4(\mathbf{i} \times \mathbf{k}) + 12(\mathbf{j} \times \mathbf{i}) + 20(\mathbf{j} \times \mathbf{j}) - 16(\mathbf{j} \times \mathbf{k}) \\
&\qquad\qquad - 6(\mathbf{k} \times \mathbf{i}) - 10(\mathbf{k} \times \mathbf{j}) + 8(\mathbf{k} \times \mathbf{k}) \\
&= 3(\mathbf{0}) + 5(\mathbf{k}) - 4(-\mathbf{j}) + 12(-\mathbf{k}) + 20(\mathbf{0}) - 16(\mathbf{i}) - 6(\mathbf{j}) - 10(-\mathbf{i}) + 8(\mathbf{0}) \\
&= -6\mathbf{i} - 2\mathbf{j} - 7\mathbf{k}
\end{aligned}$$

Thus

$$\begin{aligned}
\|\mathbf{A} \times \mathbf{B}\| &= \sqrt{36 + 4 + 49} \\
&= \sqrt{89}
\end{aligned}$$

Therefore the area of the parallelogram is $\sqrt{89}$ square units.

The following theorem, which gives a method for determining if two vectors in V_3 are parallel, follows from Theorem 14.5.8.

14.5.9 THEOREM   If $\mathbf{A}$ and $\mathbf{B}$ are two vectors in V_3, $\mathbf{A}$ and $\mathbf{B}$ are parallel if and only if $\mathbf{A} \times \mathbf{B} = \mathbf{0}$.

PROOF   If either $\mathbf{A}$ or $\mathbf{B}$ is the zero vector, then from Theorem 14.5.2, $\mathbf{A} \times \mathbf{B} = \mathbf{0}$. Because the zero vector is parallel to any vector, the theorem holds.

If neither $\mathbf{A}$ nor $\mathbf{B}$ is the zero vector, $\|\mathbf{A}\| \neq 0$ and $\|\mathbf{B}\| \neq 0$. Therefore by Theorem 14.5.8, $\|\mathbf{A} \times \mathbf{B}\| = 0$ if and only if $\sin \theta = 0$. Because $\|\mathbf{A} \times \mathbf{B}\| = 0$ if and only if $\mathbf{A} \times \mathbf{B} = \mathbf{0}$ and $\sin \theta = 0$ $(0 \leq \theta \leq \pi)$ if and only if $\theta = 0$ or $\theta = \pi$, we can conclude that

$$\mathbf{A} \times \mathbf{B} = \mathbf{0} \quad \text{if and only if} \quad \theta = 0 \text{ or } \theta = \pi$$

However, two nonzero vectors are parallel if and only if the radian measure of the angle between the two vectors is 0 or π. Thus the theorem follows. ∎

14.5.10 THEOREM   If $\mathbf{A}$ and $\mathbf{B}$ are two vectors in V_3, then the vector $\mathbf{A} \times \mathbf{B}$ is orthogonal to both $\mathbf{A}$ and $\mathbf{B}$.

PROOF   From Theorem 14.5.6,

$$\mathbf{A} \cdot \mathbf{A} \times \mathbf{B} = \mathbf{A} \times \mathbf{A} \cdot \mathbf{B}$$

From Theorem 14.5.2(i), $\mathbf{A} \times \mathbf{A} = \mathbf{0}$. Therefore, from the above equation,

$$\begin{aligned}
\mathbf{A} \cdot \mathbf{A} \times \mathbf{B} &= \mathbf{0} \cdot \mathbf{B} \\
&= 0
\end{aligned}$$

Because the dot product of $\mathbf{A}$ and $\mathbf{A} \times \mathbf{B}$ is zero, it follows that $\mathbf{A}$ and $\mathbf{A} \times \mathbf{B}$ are orthogonal.

Also from Theorem 14.5.6,

$$\mathbf{A} \times \mathbf{B} \cdot \mathbf{B} = \mathbf{A} \cdot \mathbf{B} \times \mathbf{B}$$

Again applying Theorem 14.5.2(i) we get $\mathbf{B} \times \mathbf{B} = \mathbf{0}$; thus from the above equation,

$$\mathbf{A} \times \mathbf{B} \cdot \mathbf{B} = \mathbf{A} \cdot \mathbf{0}$$
$$= 0$$

Therefore, because the dot product of $\mathbf{A} \times \mathbf{B}$ and $\mathbf{B}$ is zero, $\mathbf{A} \times \mathbf{B}$ and $\mathbf{B}$ are orthogonal and the theorem is proved.   ■

From Theorem 14.5.10 we can conclude that if representations of the vectors $\mathbf{A}$, $\mathbf{B}$, and $\mathbf{A} \times \mathbf{B}$ have the same initial point, then the representation of $\mathbf{A} \times \mathbf{B}$ is perpendicular to the plane formed by the representations of $\mathbf{A}$ and $\mathbf{B}$.

EXAMPLE 2
Given the points $P(-1, -2, -3)$, $Q(-2, 1, 0)$, and $R(0, 5, 1)$, find a unit vector whose representations are perpendicular to the plane through the points P, Q, and R.

SOLUTION
Let $\mathbf{A} = \mathbf{V}(\overrightarrow{PQ})$ and $\mathbf{B} = \mathbf{V}(\overrightarrow{PR})$. Then

$$\mathbf{A} = \langle -2 + 1, 1 + 2, 0 + 3 \rangle \qquad \mathbf{B} = \langle 0 + 1, 5 + 2, 1 + 3 \rangle$$
$$= \langle -1, 3, 3 \rangle \qquad\qquad\qquad = \langle 1, 7, 4 \rangle$$

The plane through P, Q, and R is the plane formed by $\overrightarrow{PQ}$ and $\overrightarrow{PR}$, which are, respectively, representations of vectors $\mathbf{A}$ and $\mathbf{B}$. Therefore any representation of the vector $\mathbf{A} \times \mathbf{B}$ is perpendicular to this plane.

$$\mathbf{A} \times \mathbf{B} = (-\mathbf{i} + 3\mathbf{j} + 3\mathbf{k}) \times (\mathbf{i} + 7\mathbf{j} + 4\mathbf{k})$$
$$= -9\mathbf{i} + 7\mathbf{j} - 10\mathbf{k}$$

The desired vector is a unit vector parallel to $\mathbf{A} \times \mathbf{B}$. To find this unit vector we apply Theorem 14.2.9 and divide $\mathbf{A} \times \mathbf{B}$ by $\|\mathbf{A} \times \mathbf{B}\|$ and obtain

$$\frac{\mathbf{A} \times \mathbf{B}}{\|\mathbf{A} \times \mathbf{B}\|} = -\frac{9}{\sqrt{230}} \mathbf{i} + \frac{7}{\sqrt{230}} \mathbf{j} - \frac{10}{\sqrt{230}} \mathbf{k}$$

The following two examples show the use of the cross product to find an equation of a plane. These examples involve the use of the same information as in Examples 2 and 4 of Section 14.3, respectively.

EXAMPLE 3
Find an equation of the plane through the points $P(1, 3, 2)$, $Q(3, -2, 2)$, and $R(2, 1, 3)$.

SOLUTION
$\mathbf{V}(\overrightarrow{QR}) = -\mathbf{i} + 3\mathbf{j} + \mathbf{k}$ and $\mathbf{V}(\overrightarrow{PR}) = \mathbf{i} - 2\mathbf{j} + \mathbf{k}$. A normal vector to the required plane is the cross product $\mathbf{V}(\overrightarrow{QR}) \times \mathbf{V}(\overrightarrow{PR})$, which is

$$(-\mathbf{i} + 3\mathbf{j} + \mathbf{k}) \times (\mathbf{i} - 2\mathbf{j} + \mathbf{k}) = 5\mathbf{i} + 2\mathbf{j} - \mathbf{k}$$

So if $P_0 = (1, 3, 2)$ and $\mathbf{N} = \langle 5, 2, -1 \rangle$, from Theorem 14.3.2 an equation of the required plane is

$$5(x - 1) + 2(y - 3) - (z - 2) = 0$$

$$5x + 2y - z - 9 = 0$$

This result agrees with that of Example 2 in Section 14.3.

EXAMPLE 4

Find an equation of the plane containing the point $(4, 0, -2)$ and perpendicular to each of the planes

$$x - y + z = 0 \quad \text{and} \quad 2x + y - 4z - 5 = 0$$

SOLUTION

By Theorem 14.3.3, a normal vector to the plane $x - y + z = 0$ is $\langle 1, -1, 1 \rangle$ and a normal vector to the plane $2x + y - 4z - 5 = 0$ is $\langle 2, 1, -4 \rangle$. Thus a normal vector to the required plane is orthogonal to both $\langle 1, -1, 1 \rangle$ and $\langle 2, 1, -4 \rangle$. By Theorem 14.5.10, such a vector is

$$\langle 1, -1, 1 \rangle \times \langle 2, 1, -4 \rangle = \begin{vmatrix} \mathbf{i} & \mathbf{j} & \mathbf{k} \\ 1 & -1 & 1 \\ 2 & 1 & -4 \end{vmatrix}$$

$$= 3\mathbf{i} + 6\mathbf{j} + 3\mathbf{k}$$

The required plane contains the point $(4, 0, -2)$ and has $\langle 3, 6, 3 \rangle$ as a normal vector. From Theorem 14.3.2, an equation of this plane is

$$3(x - 4) + 6(y - 0) + 3(z + 2) = 0$$

$$x + 2y + z - 2 = 0$$

This equation is the same as the one obtained in Example 4 of Section 14.3.

A geometric interpretation of the triple scalar product is obtained by considering a parallelepiped having the edges $\overrightarrow{PQ}$, $\overrightarrow{PR}$, and $\overrightarrow{PS}$ and letting $\mathbf{A} = \mathbf{V}(\overrightarrow{PQ})$, $\mathbf{B} = \mathbf{V}(\overrightarrow{PR})$, and $\mathbf{C} = \mathbf{V}(\overrightarrow{PS})$. See Fig. 14.5.5. The vector $\mathbf{A} \times \mathbf{B}$ is a normal vector to the plane of $\overrightarrow{PQ}$ and $\overrightarrow{PR}$. The vector $-(\mathbf{A} \times \mathbf{B})$ is also a normal vector to this plane. We are not certain which of the two vectors, $\mathbf{A} \times \mathbf{B}$ or $-(\mathbf{A} \times \mathbf{B})$, makes the smaller angle with $\mathbf{C}$. Let $\mathbf{N}$ be the one of the two vectors $\mathbf{A} \times \mathbf{B}$ or $-(\mathbf{A} \times \mathbf{B})$ that makes an angle of radian measure $\theta < \frac{1}{2}\pi$ with $\mathbf{C}$. Then the representations of $\mathbf{N}$ and $\mathbf{C}$ having their initial points at P are on the same side of the plane of $\overrightarrow{PQ}$ and $\overrightarrow{PR}$ as shown in Fig. 14.5.5. The area of the base of the parallelepiped is $\|\mathbf{A} \times \mathbf{B}\|$ square units. If h units is the length of the altitude of the parallelepiped, and if V cubic units is the volume of the parallelepiped,

$$V = \|\mathbf{A} \times \mathbf{B}\| h \tag{14}$$

FIGURE 14.5.5

Consider now the dot product $\mathbf{N} \cdot \mathbf{C}$. By Theorem 14.2.13,

$$\mathbf{N} \cdot \mathbf{C} = \|\mathbf{N}\| \, \|\mathbf{C}\| \cos \theta$$

But $h = \|\mathbf{C}\| \cos \theta$; thus

$$\mathbf{N} \cdot \mathbf{C} = \|\mathbf{N}\| h \tag{15}$$

Because $\mathbf{N}$ is either $\mathbf{A} \times \mathbf{B}$ or $-(\mathbf{A} \times \mathbf{B})$, it follows that $\|\mathbf{N}\| = \|\mathbf{A} \times \mathbf{B}\|$. Thus, from (15),

$$\mathbf{N} \cdot \mathbf{C} = \|\mathbf{A} \times \mathbf{B}\| h \tag{16}$$

Comparing (14) and (16) we have

$$V = \mathbf{N} \cdot \mathbf{C}$$

It follows that the measure of the volume of the parallelepiped is either $(\mathbf{A} \times \mathbf{B}) \cdot \mathbf{C}$ or $-(\mathbf{A} \times \mathbf{B}) \cdot \mathbf{C}$; that is, the measure of the volume of the parallelepiped is the absolute value of the triple scalar product $\mathbf{A} \times \mathbf{B} \cdot \mathbf{C}$.

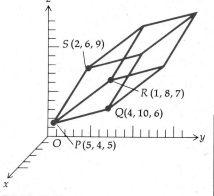

FIGURE 14.5.6

EXAMPLE 5

Find the volume of the parallelepiped having vertices $P(5, 4, 5)$, $Q(4, 10, 6)$, $R(1, 8, 7)$, and $S(2, 6, 9)$ and edges $\overrightarrow{PQ}$, $\overrightarrow{PR}$, and $\overrightarrow{PS}$.

SOLUTION

Figure 14.5.6 shows the parallelepiped. Let $\mathbf{A} = \mathbf{V}(\overrightarrow{PQ})$; then $\mathbf{A} = \langle -1, 6, 1 \rangle$. Let $\mathbf{B} = \mathbf{V}(\overrightarrow{PR})$; then $\mathbf{B} = \langle -4, 4, 2 \rangle$. Let $\mathbf{C} = \mathbf{V}(\overrightarrow{PS})$; then $\mathbf{C} = \langle -3, 2, 4 \rangle$. Thus

$$\mathbf{A} \times \mathbf{B} = (-\mathbf{i} + 6\mathbf{j} + \mathbf{k}) \times (-4\mathbf{i} + 4\mathbf{j} + 2\mathbf{k})$$
$$= 8\mathbf{i} - 2\mathbf{j} + 20\mathbf{k}$$

Therefore

$$\mathbf{A} \times \mathbf{B} \cdot \mathbf{C} = \langle 8, -2, 20 \rangle \cdot \langle -3, 2, 4 \rangle$$
$$= -24 - 4 + 80$$
$$= 52$$

Hence the volume is 52 cubic units.

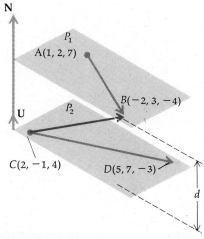

FIGURE 14.5.7

EXAMPLE 6

Find the distance between the two skew lines l_1 and l_2 of Example 5 in Section 14.4.

SOLUTION

The line l_1 contains the points $A(1, 2, 7)$ and $B(-2, 3, -4)$. The line l_2 contains the points $C(2, -1, 4)$ and $D(5, 7, -3)$. Because l_1 and l_2 are skew lines, there are parallel planes P_1 and P_2 containing the lines l_1 and l_2, respectively. See Fig. 14.5.7. Let d units be the distance between planes P_1 and P_2. The distance between l_1 and l_2 is also d units. A normal vector to the two planes is

$$\mathbf{N} = \mathbf{V}(\overrightarrow{AB}) \times \mathbf{V}(\overrightarrow{CD})$$

Let $\mathbf{U}$ be a normal vector in the direction of $\mathbf{N}$. Then

$$\mathbf{U} = \frac{\mathbf{V}(\overrightarrow{AB}) \times \mathbf{V}(\overrightarrow{CD})}{\|\mathbf{V}(\overrightarrow{AB}) \times \mathbf{V}(\overrightarrow{CD})\|} \tag{17}$$

Now we take two points, one in each plane (e.g., B and C). Then the scalar projection of $\mathbf{V}(\overrightarrow{CB})$ onto $\mathbf{N}$ is $\mathbf{V}(\overrightarrow{CB}) \cdot \mathbf{U}$, and

$$d = \left| \mathbf{V}(\overrightarrow{CB}) \cdot \mathbf{U} \right| \tag{18}$$

We now perform the computations.

$$\mathbf{V}(\overrightarrow{AB}) = \langle -2 - 1, 3 - 2, -4 - 7 \rangle \qquad \mathbf{V}(\overrightarrow{CD}) = \langle 5 - 2, 7 + 1, -3 - 4 \rangle$$
$$= \langle -3, 1, -11 \rangle \qquad\qquad\qquad = \langle 3, 8, -7 \rangle$$

Thus

$$\mathbf{V}(\overrightarrow{AB}) \times \mathbf{V}(\overrightarrow{CD}) = \begin{vmatrix} \mathbf{i} & \mathbf{j} & \mathbf{k} \\ -3 & 1 & -11 \\ 3 & 8 & -7 \end{vmatrix}$$
$$= 27(3\mathbf{i} - 2\mathbf{j} - \mathbf{k})$$

Therefore, from (17),

$$\mathbf{U} = \frac{27(3\mathbf{i} - 2\mathbf{j} - \mathbf{k})}{\sqrt{27^2(3^2 + 2^2 + 1^2)}}$$

$$\mathbf{U} = \frac{1}{\sqrt{14}}(3\mathbf{i} - 2\mathbf{j} - \mathbf{k}) \tag{19}$$

Furthermore

$$\mathbf{V}(\overrightarrow{CB}) = \langle -2 - 2, 3 + 1, -4 - 4 \rangle$$
$$\mathbf{V}(\overrightarrow{CB}) = \langle -4, 4, -8 \rangle \tag{20}$$

Substituting from (20) and (19) into (18) we get

$$d = \left| \langle -4, 4, -8 \rangle \cdot \frac{1}{\sqrt{14}} \langle 3, -2, -1 \rangle \right|$$

$$= \frac{1}{\sqrt{14}} \left| -12 - 8 + 8 \right|$$

$$= \frac{12}{\sqrt{14}}$$

$$\approx 3.21$$

EXERCISES 14.5

In Exercises 1 through 12, $\mathbf{A} = \langle 1, 2, 3 \rangle$, $\mathbf{B} = \langle 4, -3, -1 \rangle$, $\mathbf{C} = \langle -5, -3, 5 \rangle$, $\mathbf{D} = \langle -2, 1, 6 \rangle$, $\mathbf{E} = \langle 4, 0, -7 \rangle$, and $\mathbf{F} = \langle 0, 2, 1 \rangle$.

1. Find $\mathbf{A} \times \mathbf{B}$.   2. Find $\mathbf{D} \times \mathbf{E}$.   3. Find $(\mathbf{C} \times \mathbf{D}) \cdot (\mathbf{E} \times \mathbf{F})$.   4. Find $(\mathbf{C} \times \mathbf{E}) \cdot (\mathbf{D} \times \mathbf{F})$.

5. Verify Theorem 14.5.3 for vectors $\mathbf{A}$ and $\mathbf{B}$.   6. Verify Theorem 14.5.4 for vectors $\mathbf{A}$, $\mathbf{B}$, and $\mathbf{C}$.

7. Verify Theorem 14.5.5(i) for vectors $\mathbf{A}$ and $\mathbf{B}$ and $c = 3$.   8. Verify Theorem 14.5.5(ii) for vectors $\mathbf{A}$ and $\mathbf{B}$ and $c = 3$.

9. Verify Theorem 14.5.6 for vectors $\mathbf{A}$, $\mathbf{B}$, and $\mathbf{C}$.   10. Verify Theorem 14.5.7 for vectors $\mathbf{A}$, $\mathbf{B}$, and $\mathbf{C}$.

11. Find $(\mathbf{A} + \mathbf{B}) \times (\mathbf{C} - \mathbf{D})$ and $(\mathbf{D} - \mathbf{C}) \times (\mathbf{A} + \mathbf{B})$, and verify that they are equal.

12. Find $\|\mathbf{A} \times \mathbf{B}\| \|\mathbf{C} \times \mathbf{D}\|$.

13. Prove Theorem 14.5.2(ii) and (iii).

14. Given the two unit vectors $\mathbf{A} = \frac{4}{9}\mathbf{i} + \frac{7}{9}\mathbf{j} - \frac{4}{9}\mathbf{k}$ and $\mathbf{B} = -\frac{2}{3}\mathbf{i} + \frac{2}{3}\mathbf{j} + \frac{1}{3}\mathbf{k}$. If θ is the angle between $\mathbf{A}$ and $\mathbf{B}$, find $\sin \theta$ in two ways: (a) by using the cross product (Theorem 14.5.8); (b) by using the dot product and a trigonometric identity.

15. Follow the instructions of Exercise 14 for the two unit vectors

$$\mathbf{A} = \frac{1}{\sqrt{3}}\mathbf{i} - \frac{1}{\sqrt{3}}\mathbf{j} + \frac{1}{\sqrt{3}}\mathbf{k} \quad \text{and} \quad \mathbf{B} = \frac{1}{3\sqrt{3}}\mathbf{i} + \frac{5}{3\sqrt{3}}\mathbf{j} + \frac{1}{3\sqrt{3}}\mathbf{k}$$

16. Show that the quadrilateral having vertices at $(1, 1, 3)$, $(-2, 1, -1)$, $(-5, 4, 0)$, and $(8, 4, -4)$ is a parallelogram, and find its area.

17. Show that the quadrilateral having vertices at $(1, -2, 3)$, $(4, 3, -1)$, $(2, 2, 1)$, and $(5, 7, -3)$ is a parallelogram, and find its area.

18. Find the area of the parallelogram $PQRS$ if $\mathbf{V}(\overrightarrow{PQ}) = 3\mathbf{i} - 2\mathbf{j}$ and $\mathbf{V}(\overrightarrow{PS}) = 3\mathbf{j} + 4\mathbf{k}$.

19. Find the area of the triangle having vertices at $(0, 2, 2)$, $(8, 8, -2)$, and $(9, 12, 6)$.

20. Find the area of the triangle having vertices at $(4, 5, 6)$, $(4, 4, 5)$, and $(3, 5, 5)$.

In Exercises 21 and 22, use the cross product to find an equation of the plane containing the given three points.

21. $(-2, 2, 2)$, $(-8, 1, 6)$, $(3, 4, -1)$ 22. $(2, 3, 0)$, $(2, 0, 4)$, $(0, 3, 4)$

23. Do Exercise 18 in Exercises 14.3 by using the cross product.

24. Find a unit vector whose representations are perpendicular to the plane containing $\overrightarrow{PQ}$ and $\overrightarrow{PR}$ if $\overrightarrow{PQ}$ is a representation of the vector $\mathbf{i} + 3\mathbf{j} - 2\mathbf{k}$ and $\overrightarrow{PR}$ is a representation of the vector $2\mathbf{i} - \mathbf{j} - \mathbf{k}$.

In Exercises 25 through 27, find a unit vector whose representations are perpendicular to the plane through the given points P, Q, and R.

25. $P(5, 2, -1)$, $Q(2, 4, -2)$, $R(11, 1, 4)$ 26. $P(-2, 1, 0)$, $Q(2, -2, -1)$ $R(-5, 0, 2)$ 27. $P(1, 4, 2)$, $Q(3, 2, 4)$, $R(4, 3, 1)$

28. Find the volume of the parallelepiped having edges $\overrightarrow{PQ}$, $\overrightarrow{PR}$, and $\overrightarrow{PS}$ if the points P, Q, R, and S are, respectively, $(1, 3, 4)$, $(3, 5, 3)$, $(2, 1, 6)$, and $(2, 2, 5)$.

29. Find the volume of the parallelepiped $PQRS$ if the vectors $\mathbf{V}(\overrightarrow{PQ})$, $\mathbf{V}(\overrightarrow{PR})$, and $\mathbf{V}(\overrightarrow{PS})$ are, respectively, $\mathbf{i} + 3\mathbf{j} + 2\mathbf{k}$, $2\mathbf{i} + \mathbf{j} - \mathbf{k}$, and $\mathbf{i} - 2\mathbf{j} + \mathbf{k}$.

30. If $\mathbf{A}$ and $\mathbf{B}$ are any two vectors in V_3, prove that $(\mathbf{A} - \mathbf{B}) \times (\mathbf{A} + \mathbf{B}) = 2(\mathbf{A} \times \mathbf{B})$.

In Exercises 31 and 32, find the perpendicular distance between the two given skew lines.

31. $\dfrac{x - 1}{5} = \dfrac{y - 2}{3} = \dfrac{z + 1}{2}$ and $\dfrac{x + 2}{4} = \dfrac{y + 1}{2} = \dfrac{z - 3}{-3}$ 32. $\dfrac{x + 1}{2} = \dfrac{y + 2}{-4} = \dfrac{z - 1}{-3}$ and $\dfrac{x - 1}{5} = \dfrac{y - 1}{3} = \dfrac{z + 1}{2}$

33. Let P, Q, and R be three noncollinear points in R^3 and let $\overrightarrow{OP}$, $\overrightarrow{OQ}$, and $\overrightarrow{OR}$ be the position representations of vectors $\mathbf{A}$, $\mathbf{B}$, and $\mathbf{C}$, respectively. Prove that the representations of the vector $\mathbf{A} \times \mathbf{B} + \mathbf{B} \times \mathbf{C} + \mathbf{C} \times \mathbf{A}$ are perpendicular to the plane containing the points P, Q, and R.

34. Find an equation of the plane containing the endpoints of the position representations of the vectors $2\mathbf{i} - \mathbf{j} + 3\mathbf{k}$, $-\mathbf{i} + \mathbf{j} + 2\mathbf{k}$, and $5\mathbf{i} + \mathbf{j} - \mathbf{k}$.

35. Prove Theorem 14.5.4.   36. Prove Theorem 14.5.5.   37. Prove Theorem 14.5.6.   38. Prove Theorem 14.5.7.

39. Let $\overrightarrow{OP}$ be the position representation of vector $\mathbf{A}$, $\overrightarrow{OQ}$ be the position representation of vector $\mathbf{B}$, and $\overrightarrow{OR}$ be the position representation of vector $\mathbf{C}$. Prove that the area of triangle PQR is $\frac{1}{2}\|(\mathbf{B} - \mathbf{A}) \times (\mathbf{C} - \mathbf{A})\|$.

14.6
CYLINDERS AND SURFACES OF REVOLUTION

As mentioned previously, the graph of an equation in three variables is a **surface**. A surface is represented by an equation if the coordinates of every point on the surface satisfy the equation and if every point whose coordinates satisfy the equation lies on the surface. We have already discussed two kinds of surfaces, a plane and a sphere. Another kind of surface that is fairly simple is a cylinder. You are probably familiar with right-circular cylinders from previous experience. We now consider a more general cylindrical surface.

14.6.1 DEFINITION

A **cylinder** is a surface that is generated by a line moving along a given plane curve in such a way that it always remains parallel to a fixed line not lying in the plane of the given curve. The moving line is called a **generator** of the cylinder and the given plane curve is called a **directrix** of the cylinder. Any position of a generator is called a **ruling** of the cylinder.

This discussion is confined to cylinders having a directrix in a coordinate plane and rulings perpendicular to that plane. If the rulings of a cylinder are perpendicular to the plane of a directrix, the cylinder is said to be perpendicular to the plane.

The familiar right-circular cylinder is one for which a directrix is a circle in a plane perpendicular to the cylinder.

▶ ILLUSTRATION 1 In Fig. 14.6.1 there is a cylinder whose directrix is the parabola $y^2 = 8x$ in the xy plane and whose rulings are parallel to the z axis. The cylinder is called a **parabolic cylinder**. An **elliptic cylinder** is shown in Fig. 14.6.2; its directrix is the ellipse $9x^2 + 16y^2 = 144$ in the xy plane, and its rulings are parallel to the z axis. Figure 14.6.3 shows a **hyperbolic cylinder** having as a directrix the hyperbola $25x^2 - 4y^2 = 100$ in the xy plane and rulings parallel to the z axis.   ◀

Consider the problem of finding an equation of a cylinder having a directrix in a coordinate plane and rulings parallel to the coordinate axis not in that plane. To be specific, take the directrix in the xy plane and the rulings parallel to the z axis. Refer to Fig. 14.6.4. Suppose that an equation of the directrix in the xy plane is $y = f(x)$. If the point $(x_0, y_0, 0)$ in the xy plane satisfies this equation, any point (x_0, y_0, z) in three-dimensional space, where z is any real number, will satisfy the same equation because z does not appear in the equation. The points having representations (x_0, y_0, z) all lie on the line parallel to the z axis through the point $(x_0, y_0, 0)$. This line is a ruling of the cylinder. Hence any point whose x and y coordinates satisfy the equation $y = f(x)$ lies on the cylinder. Conversely, if the point $P(x, y, z)$ lies on the cylinder (see Fig. 14.6.5), then the point $(x, y, 0)$ lies on the directrix of the cylinder in the xy plane, and hence the x and y coordinates of P satisfy the equation $y = f(x)$. Therefore, if $y = f(x)$ is considered as an equation of a graph in three-dimensional space, the graph is a cylinder whose rulings are parallel to the z axis and which has as a directrix the curve $y = f(x)$ in the plane $z = 0$. A similar discussion pertains when the directrix is in either of the other coordinate planes. The results are summarized in the following theorem.

FIGURE 14.6.1

FIGURE 14.6.2

FIGURE 14.6.3

FIGURE 14.6.4   FIGURE 14.6.5   FIGURE 14.6.6

14.6.2 THEOREM   In three-dimensional space, the graph of an equation in two of the three variables x, y, and z is a cylinder whose rulings are parallel to the axis associated with the missing variable and whose directrix is a curve in the plane associated with the two variables appearing in the equation.

▶ ILLUSTRATION 2   It follows from Theorem 14.6.2 that an equation of the parabolic cylinder of Fig. 14.6.1 is $y^2 = 8x$, considered as an equation in R^3. Similarly, equations of the elliptic cylinder of Fig. 14.6.2 and the hyperbolic cylinder of Fig. 14.6.3 are, respectively,

$$9x^2 + 16y^2 = 144 \quad \text{and} \quad 25x^2 - 4y^2 = 100$$

both considered as equations in R^3.   ◀

A **cross section** of a surface in a plane is the set of all points of the surface that lie in the given plane. If a plane is parallel to the plane of the directrix of a cylinder, the cross section of the cylinder is the same as the directrix. For example, the cross section of the elliptic cylinder of Fig. 14.6.2 in any plane parallel to the xy plane is an ellipse.

FIGURE 14.6.7

EXAMPLE 1
Draw a sketch of the graph of each of the following equations: (a) $y = \ln z$; (b) $z^2 = x^3$.

SOLUTION
(a) The graph is a cylinder whose directrix lying in the yz plane is the curve $y = \ln z$ and whose rulings are parallel to the x axis. A sketch of the graph is shown in Fig. 14.6.6.
(b) The graph is a cylinder whose directrix is in the xz plane and whose rulings are parallel to the y axis. An equation of the directrix is the curve $z^2 = x^3$ in the xz plane. A sketch of the graph is shown in Fig. 14.6.7.

14.6.3 DEFINITION If a plane curve is revolved about a fixed line lying in the plane of the curve, the surface generated is called a **surface of revolution**. The fixed line is called the **axis** of the surface of revolution, and the plane curve is called the **generating curve**.

Figure 14.6.8 shows a surface of revolution whose generating curve is the curve C in the yz plane and whose axis is the z axis. A sphere is a particular example of a surface of revolution because a sphere can be generated by revolving a semicircle about a diameter.

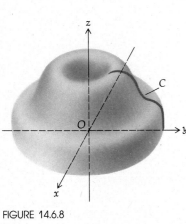

▶ ILLUSTRATION 3   Figure 14.6.9 shows a sphere that can be generated by revolving the semicircle $y^2 + z^2 = r^2$, $z \geq 0$, about the y axis. Another example of a surface of revolution is a right-circular cylinder for which the generating curve and the axis are parallel straight lines. If the generating curve is the line $z = k$ in the xz plane and the axis is the x axis, we obtain the right-circular cylinder shown in Fig. 14.6.10.   ◀

We now find an equation of the surface generated by revolving about the y axis the curve in the yz plane having the two-dimensional equation

$$z = f(y) \tag{1}$$

Refer to Fig. 14.6.11. Let $P(x, y, z)$ be any point on the surface of revolution. Through P, pass a plane perpendicular to the y axis. Denote the point of intersection of this plane with the y axis by $Q(0, y, 0)$, and let $P_0(0, y, z_0)$ be the point of intersection of the plane with the generating curve. Because the cross section of the surface with the plane through P is a circle, P is on the surface if and only if

$$|\overline{QP}|^2 = |\overline{QP_0}|^2 \tag{2}$$

Because $|\overline{QP}| = \sqrt{x^2 + z^2}$ and $|\overline{QP_0}| = z_0$, we obtain from (2)

$$x^2 + z^2 = z_0{}^2 \tag{3}$$

The point P_0 is on the generating curve; so its coordinates must satisfy (1). Therefore

$$z_0 = f(y) \tag{4}$$

From (3) and (4), the point P is on the surface of revolution if and only if

$$x^2 + z^2 = [f(y)]^2 \tag{5}$$

Equation (5) is the desired equation of the surface of revolution. Because (5) is equivalent to

$$\pm\sqrt{x^2 + z^2} = f(y)$$

we can obtain (5) by replacing z in (1) by $\pm\sqrt{x^2 + z^2}$.

FIGURE 14.6.8

FIGURE 14.6.9

FIGURE 14.6.10

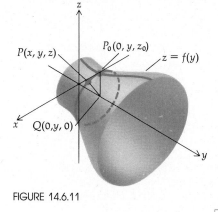

FIGURE 14.6.11

In a similar manner we can show that if the curve in the yz plane having the two-dimensional equation

$$y = g(z) \tag{6}$$

is revolved about the z axis, an equation of the surface of revolution generated is obtained by replacing y in (6) by $\pm\sqrt{x^2 + y^2}$. Analogous remarks hold when a curve in any coordinate plane is revolved about either one of the coordinate axes in that plane. In summary, the graphs of the following equations are surfaces of revolution having the indicated axis: $x^2 + y^2 = [F(z)]^2$ —z axis; $x^2 + z^2 = [F(y)]^2$—y axis; $y^2 + z^2 = [F(x)]^2$—x axis. In each case, cross sections of the surface in planes perpendicular to the axis are circles having centers on the axis.

EXAMPLE 2

Find an equation of the surface of revolution generated by revolving the parabola $y^2 = 4x$ in the xy plane about the x axis. Draw a sketch of the graph of the surface.

SOLUTION

In the equation of the parabola we replace y by $\pm\sqrt{y^2 + z^2}$ and obtain

$$y^2 + z^2 = 4x$$

A sketch of the graph is shown in Fig. 14.6.12. The same surface is generated if the parabola $z^2 = 4x$ in the xz plane is revolved about the x axis.

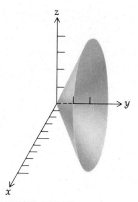

FIGURE 14.6.12

The surface obtained in Example 2 is called a **paraboloid of revolution**. If an ellipse is revolved about one of its axes, the surface obtained is called an **ellipsoid of revolution**. **A hyperboloid of revolution** is obtained when a hyperbola is revolved about an axis.

EXAMPLE 3

Draw a sketch of the surface $x^2 + z^2 - 4y^2 = 0$, if $y \geq 0$.

SOLUTION

The given equation is of the form $x^2 + z^2 = [F(y)]^2$; so its graph is a surface of revolution having the y axis as axis. We solve the given equation for y and obtain

$$2y = \pm\sqrt{x^2 + z^2}$$

Hence the generating curve can be either the straight line $2y = x$ in the xy plane or the straight line $2y = z$ in the yz plane. By drawing sketches of the two possible generating curves and using the fact that cross sections of the surface in planes perpendicular to the y axis are circles having centers on the y axis, we obtain the surface shown in Fig. 14.6.13 (note that because $y \geq 0$ there is only one nappe of the cone).

FIGURE 14.6.13

The surface obtained in Example 3 is called a **right-circular cone**.

EXERCISES 14.6

In Exercises 1 through 4, draw a sketch of the cross section of the given cylinder in the given plane.

1. $4x^2 + y^2 = 16$; xy plane   **2.** $4z^2 - y^2 = 4$; yz plane   **3.** $z = e^x$; xz plane   **4.** $x = |y|$; xy plane

In Exercises 5 through 12, draw a sketch of the cylinder having the given equation.

5. $4x^2 + 9y^2 = 36$ **6.** $z = \sin y$ **7.** $y = |z|$ **8.** $x^2 - z^2 = 4$
9. $z = 2x^2$ **10.** $z^2 = 4y^2$ **11.** $y = \cosh x$ **12.** $x^2 = y^3$

In Exercises 13 through 20, find an equation of the surface of revolution generated by revolving the given plane curve about the indicated axis. Draw a sketch of the surface.

13. $x^2 = 4y$ in the xy plane, about the y axis.   **14.** $x^2 + 4z^2 = 16$ in the xz plane, about the z axis.
15. $x^2 + 4z^2 = 16$ in the xz plane, about the x axis.   **16.** $x^2 = 4y$ in the xy plane, about the x axis.
17. $y = 3z$ in the yz plane, about the y axis.   **18.** $9y^2 - 4z^2 = 144$ in the yz plane, about the z axis.
19. $y = \sin x$ in the xy plane, about the x axis.   **20.** $y^2 = z^3$ in the yz plane, about the z axis.

In Exercises 21 through 28, find a generating curve and the axis for the given surface of revolution. Draw a sketch of the surface.

21. $x^2 + y^2 + z^2 = 16$ **22.** $x^2 + z^2 = y$ **23.** $x^2 + y^2 - z^2 = 4$ **24.** $y^2 + z^2 = e^{2x}$
25. $x^2 + z^2 = |y|$ **26.** $4x^2 + 9y^2 + 4z^2 = 36$ **27.** $9x^2 - y^2 + 9z^2 = 0$ **28.** $4x^2 + 4y^2 - z = 9$
29. The tractrix

$$x = t - a \tanh \frac{t}{a} \qquad y = a \operatorname{sech} \frac{t}{a}$$

from $x = -a$ to $x = 2a$ is revolved about the x axis. Draw a sketch of the surface of revolution.

14.7
QUADRIC SURFACES

We learned in Chapter 10 that the graph of a second-degree equation in two variables x and y,

$$Ax^2 + Bxy + Cy^2 + Dx + Ey + F = 0$$

is a conic section. The graph of a second-degree equation in three variables x, y, and z,

$$Ax^2 + By^2 + Cz^2 + Dxy + Exz + Fyz + Gx + Hy + Iz + J = 0 \quad (1)$$

is called a **quadric surface**. The simplest types of quadric surfaces are the parabolic, elliptic, and hyperbolic cylinders that were discussed in the preceding section. There are six other types of quadric surfaces that we now consider. We choose the coordinate axes so the equations are in their simplest form. In the discussion of each of these surfaces we refer to the cross sections of the surfaces in planes parallel to the coordinate planes. These cross sections help to visualize the surface.

The ellipsoid

$$\frac{x^2}{a^2} + \frac{y^2}{b^2} + \frac{z^2}{c^2} = 1 \quad (2)$$

where a, b, and c are positive (see Fig. 14.7.1).

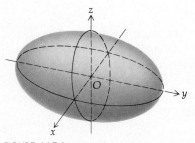

FIGURE 14.7.1

If in (2) z is replaced by zero, we obtain the cross section of the ellipsoid in the xy plane, which is the ellipse

$$\frac{x^2}{a^2} + \frac{y^2}{b^2} = 1$$

To obtain the cross sections of the surface with the planes $z = k$, we replace z by k in the equation of the ellipsoid and get

$$\frac{x^2}{a^2} + \frac{y^2}{b^2} = 1 - \frac{k^2}{c^2}$$

If $|k| < c$, the cross section is an ellipse and the lengths of the semiaxes decrease to zero as $|k|$ increases to the value c. If $|k| = c$, the intersection of a plane $z = k$ with the ellipsoid is the single point $(0, 0, k)$. If $|k| > c$, there is no intersection. The discussion is similar if we consider cross sections formed by planes parallel to either of the other coordinate planes.

The numbers a, b, and c are the lengths of the semiaxes of the ellipsoid. If any two of these three numbers are equal, we have an ellipsoid of revolution, which is also called a **spheroid**. A spheroid for which the third number is greater than the two equal numbers is said to be **prolate**. A prolate spheroid is shaped like a football. An **oblate** spheroid is obtained if the third number is less than the two equal numbers. If all three numbers, a, b, and c in the equation of an ellipsoid are equal, the ellipsoid is a **sphere**.

The elliptic hyperboloid of one sheet

$$\frac{x^2}{a^2} + \frac{y^2}{b^2} - \frac{z^2}{c^2} = 1 \tag{3}$$

where a, b, and c are positive (see Fig. 14.7.2).

The cross sections in the planes $z = k$ are the ellipses

$$\frac{x^2}{a^2} + \frac{y^2}{b^2} = 1 + \frac{k^2}{c^2}$$

When $k = 0$, the lengths of the semiaxes of the ellipse are smallest, and these lengths increase as $|k|$ increases. The cross sections in the planes $x = k$ are the hyperbolas

$$\frac{y^2}{b^2} - \frac{z^2}{c^2} = 1 - \frac{k^2}{a^2}$$

If $|k| < a$, the transverse axis of the hyperbola is parallel to the y axis, and if $|k| > a$, the transverse axis is parallel to the z axis. If $k = a$, the hyperbola degenerates into two straight lines:

$$\frac{y}{b} - \frac{z}{c} = 0 \quad \text{and} \quad \frac{y}{b} + \frac{z}{c} = 0$$

FIGURE 14.7.2

In an analogous manner, the cross sections in the planes $y = k$ are also hyperbolas. The axis of this hyperboloid is the z axis.

If $a = b$, the surface is a hyperboloid of revolution for which the axis is the line containing the conjugate axis.

The elliptic hyperboloid of two sheets

$$-\frac{x^2}{a^2} - \frac{y^2}{b^2} + \frac{z^2}{c^2} = 1 \qquad\qquad (4)$$

where a, b, and c are positive (see Fig. 14.7.3).

Replacing z by k in (4) we obtain

$$\frac{x^2}{a^2} + \frac{y^2}{b^2} = \frac{k^2}{c^2} - 1$$

If $|k| < c$, there is no intersection of the plane $z = k$ with the surface; hence there are no points of the surface between the planes $z = -c$ and $z = c$. If $|k| = c$, the intersection of the plane $z = k$ with the surface is the single point $(0, 0, k)$. When $|k| > c$, the cross section of the surface in the plane $z = k$ is an ellipse, and the lengths of the semiaxes of the ellipse increase as $|k|$ increases.

The cross sections of the surface in the planes $x = k$ are the hyperbolas

$$\frac{z^2}{c^2} - \frac{y^2}{b^2} = 1 + \frac{k^2}{a^2}$$

whose transverse axes are parallel to the z axis. In a similar fashion, the cross sections in the planes $y = k$ are the hyperbolas

$$\frac{z^2}{c^2} - \frac{x^2}{a^2} = 1 + \frac{k^2}{b^2}$$

for which the transverse axes are also parallel to the z axis.

If $a = b$, the surface is a hyperboloid of revolution in which the axis is the line containing the transverse axis of the hyperbola.

Each of the above three quadric surfaces is symmetric with respect to each of the coordinate planes and symmetric with respect to the origin. Their graphs are called **central quadrics** and their center is at the origin. The graph of any equation of the form

$$\pm\frac{x^2}{a^2} \pm \frac{y^2}{b^2} \pm \frac{z^2}{c^2} = 1$$

where a, b, and c are positive, is a central quadric.

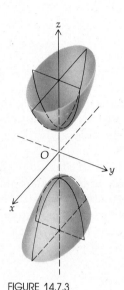

FIGURE 14.7.3

EXAMPLE 1

Draw a sketch of the graph of the equation

$$4x^2 - y^2 + 25z^2 = 100$$

and name the surface.

SOLUTION

The given equation can be written as

$$\frac{x^2}{25} - \frac{y^2}{100} + \frac{z^2}{4} = 1$$

which is of the form of (3) with y and z interchanged. Hence the surface is an elliptic hyperboloid of one sheet whose axis is the y axis. The cross sections in the planes $y = k$ are the ellipses

$$\frac{x^2}{25} + \frac{z^2}{4} = 1 + \frac{k^2}{100}$$

The cross sections in the planes $x = k$ are the hyperbolas

$$\frac{z^2}{4} - \frac{y^2}{100} = 1 - \frac{k^2}{25}$$

and the cross sections in the planes $z = k$ are the hyperbolas

$$\frac{x^2}{25} - \frac{y^2}{100} = 1 - \frac{k^2}{4}$$

A sketch of the surface appears in Fig. 14.7.4.

FIGURE 14.7.4

EXAMPLE 2
Draw a sketch of the graph of the equation

$$4x^2 - 25y^2 - z^2 = 100$$

and name the surface.

SOLUTION

By dividing on both sides of the equation by 100 we can write the given equation as

$$\frac{x^2}{25} - \frac{y^2}{4} - \frac{z^2}{100} = 1$$

which is of the form of (4) with x and z interchanged; thus the surface is an elliptic hyperboloid of two sheets whose axis is the x axis. The cross sections

FIGURE 14.7.5

FIGURE 14.7.6

in the planes $x = k$, where $|k| > 5$, are the ellipses

$$\frac{y^2}{4} + \frac{z^2}{100} = \frac{k^2}{25} - 1$$

The planes $x = k$, where $|k| < 5$, do not intersect the surface. The cross sections in the planes $y = k$ are the hyperbolas

$$\frac{x^2}{25} - \frac{z^2}{100} = 1 + \frac{k^2}{4}$$

and the cross sections in the planes $z = k$ are the hyperbolas

$$\frac{x^2}{25} - \frac{y^2}{4} = 1 + \frac{k^2}{100}$$

The required sketch is shown in Fig. 14.7.5.

The following two quadrics are called noncentral quadrics.

The elliptic paraboloid

$$\frac{x^2}{a^2} + \frac{y^2}{b^2} = \frac{z}{c} \tag{5}$$

where a and b are positive and $c \neq 0$. Figure 14.7.6 shows the surface if $c > 0$.

Substituting k for z in (5) we obtain

$$\frac{x^2}{a^2} + \frac{y^2}{b^2} = \frac{k}{c}$$

When $k = 0$, this equation becomes $b^2x^2 + a^2y^2 = 0$, which represents a single point, the origin. If $k \neq 0$ and k and c have the same sign, the equation is that of an ellipse. So we conclude that cross sections of the surface in the planes $z = k$, where k and c have the same sign, are ellipses and the lengths of the semiaxes increase as $|k|$ increases. If k and c have opposite signs, the planes $z = k$ do not intersect the surface. The cross sections of the surface with the planes $x = k$ and $y = k$ are parabolas. When $c > 0$, the parabolas open upward, as shown in Fig. 14.7.6; when $c < 0$, the parabolas open downward.

If $a = b$, the surface is a paraboloid of revolution.

The hyperbolic paraboloid

$$\frac{y^2}{b^2} - \frac{x^2}{a^2} = \frac{z}{c} \tag{6}$$

where a and b are positive and $c \neq 0$. The surface appears in Fig. 14.7.7 for $c > 0$.

The cross sections of the surface in the planes $z = k$, where $k \neq 0$, are hyperbolas having their transverse axes parallel to the y axis if k and c have the same sign and parallel to the x axis if k and c have opposite signs. The cross section of the surface in the plane $z = 0$ consists of two straight lines through the origin. The cross sections in the planes $x = k$ are parabolas

FIGURE 14.7.7

opening upward if $c > 0$ and opening downward if $c < 0$. The cross sections in the planes $y = k$ are parabolas opening downward if $c > 0$ and opening upward if $c < 0$.

EXAMPLE 3
Draw a sketch of the graph of the equation

$$3y^2 + 12z^2 = 16x$$

and name the surface.

SOLUTION
The given equation can be written as

$$\frac{y^2}{16} + \frac{z^2}{4} = \frac{x}{3}$$

which is of the form of (5) with x and z interchanged. Hence the graph of the equation is an elliptic paraboloid whose axis is the x axis. The cross sections in the planes $x = k$, where $k > 0$, are the ellipses

$$\frac{y^2}{16} + \frac{z^2}{4} = \frac{k}{3}$$

The planes $x = k$, where $k < 0$, do not intersect the surface. The cross sections in the planes $y = k$ are the parabolas $12z^2 = 16x - 3k^2$, and the cross sections in the planes $z = k$ are the parabolas $3y^2 = 16x - 12k^2$. Figure 14.7.8 shows a sketch of the elliptic paraboloid.

FIGURE 14.7.8

EXAMPLE 4
Draw a sketch of the graph of the equation

$$3y^2 - 12z^2 = 16x$$

and name the surface.

SOLUTION
If the given equation is written as

$$\frac{y^2}{16} - \frac{z^2}{4} = \frac{x}{3}$$

it is of the form of (6) with x and z interchanged. The surface is therefore a hyperbolic paraboloid. The cross sections in the planes $x = k$, where $k \neq 0$, are the hyperbolas

$$\frac{y^2}{16} - \frac{z^2}{4} = \frac{k}{3}$$

The cross section in the yz plane ($x = 0$) consists of the two lines $y = 2z$ and $y = -2z$. In the planes $z = k$ the cross sections are the parabolas $3y^2 = 16x + 12k^2$; in the planes $y = k$ the cross sections are the parabolas $12z^2 = 3k^2 - 16x$. A sketch of the hyperbolic paraboloid appears in Fig. 14.7.9.

FIGURE 14.7.9

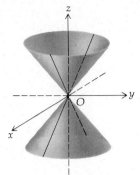

FIGURE 14.7.10

The elliptic cone

$$\frac{x^2}{a^2} + \frac{y^2}{b^2} - \frac{z^2}{c^2} = 0 \tag{7}$$

where a, b, and c are positive (see Fig. 14.7.10).

The intersection of the plane $z = 0$ with the surface is a single point, the origin. The cross sections of the surface in the planes $z = k$, where $k \neq 0$, are ellipses, and the lengths of the semiaxes increase as k increases. Cross sections in the planes $x = 0$ and $y = 0$ are pairs of intersecting lines. In the planes $x = k$ and $y = k$, where $k \neq 0$, the cross sections are hyperbolas.

EXAMPLE 5
Draw a sketch of the graph of the equation

$$4x^2 - y^2 + 25z^2 = 0$$

and name the surface.

SOLUTION
The given equation can be written as

$$\frac{x^2}{25} - \frac{y^2}{100} + \frac{z^2}{4} = 0$$

which is of the form of (7) with y and z interchanged. Therefore the surface is an elliptic cone having the y axis as its axis. The surface intersects the plane $y = 0$ at the origin only. The intersection of the surface with the plane $x = 0$ is the pair of intersecting lines $y = \pm 5z$, and the intersection with the plane $z = 0$ is the pair of intersecting lines $y = \pm 2x$. The cross sections in the planes $y = k$, where $k \neq 0$, are the ellipses

$$\frac{x^2}{25} + \frac{z^2}{4} = \frac{k^2}{100}$$

In the planes $x = k$ and $z = k$, where $k \neq 0$, the cross sections are, respectively, the hyperbolas

$$\frac{y^2}{100} - \frac{z^2}{4} = \frac{k^2}{25} \quad \text{and} \quad \frac{y^2}{100} - \frac{x^2}{25} = \frac{k^2}{4}$$

Figure 14.7.11 shows a sketch of the surface.

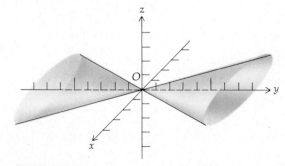

FIGURE 14.7.11

Equation (1) is the general equation of the second degree in x, y, and z. It can be shown that by translation and rotation of the three-dimensional coordinate axes (the study of which is beyond the scope of this book) this equation can be reduced to one of the following two forms:

$$Ax^2 + By^2 + Cz^2 + J = 0 \qquad (8)$$

$$Ax^2 + By^2 + Iz = 0 \qquad (9)$$

Graphs of the equations of the second degree will either be one of the above six types of quadrics or else will degenerate into a cylinder, plane, line, point, or the empty set.

The nondegenerate curves associated with equations of the form (8) are the central quadrics and the elliptic cone, whereas those associated with equations of the form (9) are the noncentral quadrics. Following are examples of some degenerate cases:

$x^2 - y^2 = 0$; two planes, $x - y = 0$ and $x + y = 0$

$z^2 = 0$; one plane, the xy plane

$x^2 + y^2 = 0$; one line, the z axis

$x^2 + y^2 + z^2 = 0$; a single point, the origin

$x^2 + y^2 + z^2 + 1 = 0$; the empty set

EXERCISES 14.7

In Exercises 1 through 6, name the surface having the given equation.

1. $9x^2 - 4y^2 + 36z^2 = 36$ **2.** $4x^2 - 16y^2 + 9z^2 = 0$ **3.** $5x^2 - 2z^2 - 3y = 0$

4. $25x^2 = 4y^2 + z^2 + 100$ **5.** $4y^2 - 25x^2 = 100$ **6.** $3y^2 + 7z^2 = 6x$

In Exercises 7 through 18, draw a sketch of the graph of the given equation and name the surface.

7. $4x^2 + 9y^2 + z^2 = 36$ **8.** $4x^2 - 9y^2 - z^2 = 36$ **9.** $4x^2 + 9y^2 - z^2 = 36$ **10.** $4x^2 - 9y^2 + z^2 = 36$

11. $x^2 = y^2 - z^2$ **12.** $x^2 = y^2 + z^2$ **13.** $\dfrac{x^2}{36} + \dfrac{z^2}{25} = 4y$ **14.** $\dfrac{y^2}{25} + \dfrac{x^2}{36} = 4$

15. $\dfrac{x^2}{36} - \dfrac{z^2}{25} = 9y$ **16.** $x^2 = 2y + 4z$ **17.** $x^2 + 16z^2 = 4y^2 - 16$ **18.** $9y^2 - 4z^2 + 18x = 0$

19. Find the values of k for which the intersection of the plane $x + ky = 1$ and the elliptic hyperboloid of two sheets $y^2 - x^2 - z^2 = 1$ is (a) an ellipse and (b) a hyperbola.

20. Find the vertex and focus of the parabola that is the intersection of the plane $y = 2$ with the hyperbolic paraboloid $\dfrac{y^2}{16} - \dfrac{x^2}{4} = \dfrac{z}{9}$.

21. Find the vertex and focus of the parabola that is the intersection of the plane $x = 1$ with the hyperbolic paraboloid $\dfrac{z^2}{4} - \dfrac{x^2}{9} = \dfrac{y}{3}$.

22. Find the area of the plane section formed by the intersection of the plane $y = 3$ with the solid bounded by the ellipsoid $\dfrac{x^2}{9} + \dfrac{y^2}{25} + \dfrac{z^2}{4} = 1$.

23. Show that the intersection of the surface $x^2 - 4y^2 - 9z^2 = 36$ and the plane $x + z = 9$ is an ellipse.

24. Show that the intersection of the hyperbolic paraboloid $\dfrac{y^2}{b^2} - \dfrac{x^2}{a^2} = \dfrac{z}{c}$ and the plane $z = bx + ay$ consists of two intersecting straight lines.

In Exercises 25 through 27, use the method of parallel plane sections (Section 5.3) to find the volume of the given solid. The measure of the area of the region enclosed by the ellipse having semiaxes a and b is πab.

25. The solid bounded by the ellipsoid $36x^2 + 9y^2 + 4z^2 = 36$.

26. The solid bounded by the ellipsoid $\dfrac{x^2}{a^2} + \dfrac{y^2}{b^2} + \dfrac{z^2}{c^2} = 1$.

27. The solid bounded by the plane $z = h$, where $h > 0$, and the elliptic paraboloid $\dfrac{x^2}{a^2} + \dfrac{y^2}{b^2} = \dfrac{z}{c}$, where $c > 0$.

14.8
CURVES IN R^3

We now consider vector-valued functions in three-dimensional space.

14.8.1 DEFINITION   Let f_1, f_2, and f_3 be three real-valued functions of a real variable t. Then for every number t in the domain common to f_1, f_2, and f_3 there is a vector **R** defined by

$$\mathbf{R}(t) = f_1(t)\mathbf{i} + f_2(t)\mathbf{j} + f_3(t)\mathbf{k} \tag{1}$$

and **R** is called a **vector-valued function**.

The graph of a vector-valued function in three-dimensional space is obtained analogously to the way the graph of a vector-valued function in two dimensions was obtained in Section 13.3. As t assumes all values in the domain of **R**, the terminal point of the position representation of the vector $\mathbf{R}(t)$ traces a curve C, and this curve is called the graph of (1). A point on the curve C has the cartesian representation (x, y, z), where

$$x = f_1(t) \qquad y = f_2(t) \qquad z = f_3(t) \tag{2}$$

Equations (2) are called **parametric equations** of C, whereas Eq. (1) is called a **vector equation** of C. By eliminating t from Eqs. (2) we obtain two equations in x, y, and z. These equations are called **cartesian equations** of C. Each cartesian equation is an equation of a surface, and curve C is the intersection of the two surfaces. The equations of any two surfaces containing C may be taken as the cartesian equations defining C.

▶ ILLUSTRATION 1   We draw a sketch of the curve having the vector equation

$$\mathbf{R}(t) = a \cos t\,\mathbf{i} + b \sin t\,\mathbf{j} + t\mathbf{k}$$

Parametric equations of the given curve are

$$x = a \cos t \qquad y = b \sin t \qquad z = t$$

Table 14.8.1

t	x	y	z
0	a	0	0
$\dfrac{\pi}{4}$	$\dfrac{a}{\sqrt{2}}$	$\dfrac{b}{\sqrt{2}}$	$\dfrac{\pi}{4}$
$\dfrac{\pi}{2}$	0	b	$\dfrac{\pi}{2}$
$\dfrac{3\pi}{4}$	$-\dfrac{a}{\sqrt{2}}$	$\dfrac{b}{\sqrt{2}}$	$\dfrac{3\pi}{4}$
π	$-a$	0	π
$\dfrac{3\pi}{2}$	0	$-b$	$\dfrac{3\pi}{2}$

To eliminate t from the first two equations we write them as

$$\frac{x^2}{a^2} = \cos^2 t \quad \text{and} \quad \frac{y^2}{b^2} = \sin^2 t$$

Adding corresponding members of these two equations we obtain

$$\frac{x^2}{a^2} + \frac{y^2}{b^2} = 1$$

Therefore the curve lies entirely on the elliptical cylinder whose directrix is an ellipse in the xy plane and whose rulings are parallel to the z axis. Table 14.8.1 gives sets of values of x, y, and z for specific values of t. A sketch of the curve is shown in Fig. 14.8.1. ◀

The curve of Illustration 1 is called a **helix**. If $a = b$, the helix is a **circular helix** and it lies on the right-circular cylinder $x^2 + y^2 = a^2$.

▶ ILLUSTRATION 2   The curve having the vector equation

$$\mathbf{R}(t) = t\mathbf{i} + t^2\mathbf{j} + t^3\mathbf{k}$$

is called a **twisted cubic**. Parametric equations of the twisted cubic are

$$x = t \qquad y = t^2 \qquad z = t^3$$

Eliminating t from the first two of these equations yields $y = x^2$, which is a cylinder whose directrix in the xy plane is a parabola. The twisted cubic lies on this cylinder. Figure 14.8.2 shows a sketch of the cylinder and the portion of the twisted cubic from $t = 0$ to $t = 2$. ◀

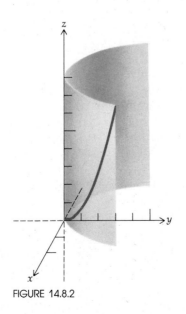

FIGURE 14.8.1

FIGURE 14.8.2

Many of the definitions and theorems pertaining to vector-valued functions in two dimensions can be extended to vector-valued functions in three dimensions.

14.8.2 DEFINITION   If $\mathbf{R}(t) = f_1(t)\mathbf{i} + f_2(t)\mathbf{j} + f_3(t)\mathbf{k}$, then

$$\lim_{t \to t_1} \mathbf{R}(t) = \lim_{t \to t_1} f_1(t)\mathbf{i} + \lim_{t \to t_1} f_2(t)\mathbf{j} + \lim_{t \to t_1} f_3(t)\mathbf{k}$$

if $\lim_{t \to t_1} f_1(t)$, $\lim_{t \to t_1} f_2(t)$, and $\lim_{t \to t_1} f_3(t)$ all exist.

14.8.3 DEFINITION   The vector-valued function $\mathbf{R}$ is continuous at t_1 if and only if

(i) $\mathbf{R}(t_1)$ exists;

(ii) $\lim_{t \to t_1} \mathbf{R}(t)$ exists;

(iii) $\lim_{t \to t_1} \mathbf{R}(t) = \mathbf{R}(t_1)$.

14.8.4 DEFINITION   The derivative of the vector-valued function $\mathbf{R}$ is a vector-valued function, denoted by $\mathbf{R}'$ and defined by

$$\mathbf{R}'(t) = \lim_{\Delta t \to 0} \frac{\mathbf{R}(t + \Delta t) - \mathbf{R}(t)}{\Delta t}$$

if this limit exists.

14.8.5 THEOREM   If $\mathbf{R}$ is the vector-valued function defined by

$$\mathbf{R}(t) = f_1(t)\mathbf{i} + f_2(t)\mathbf{j} + f_3(t)\mathbf{k}$$

and $\mathbf{R}'(t)$ exists, then

$$\mathbf{R}'(t) = f_1'(t)\mathbf{i} + f_2'(t)\mathbf{j} + f_3'(t)\mathbf{k}$$

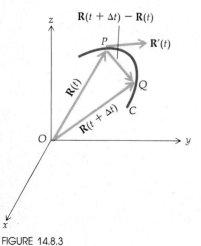

FIGURE 14.8.3

The proof of Theorem 14.8.5 is left as an exercise (see Exercise 13).

The geometric interpretation for the derivative of $\mathbf{R}$ is the same as that for the derivative of a vector-valued function in R^2. Figure 14.8.3 shows a portion of the curve C, which is the graph of $\mathbf{R}$. In the figure, $\overrightarrow{OP}$ is the position representation of $\mathbf{R}(t)$, $\overrightarrow{OQ}$ is the position representation of $\mathbf{R}(t + \Delta t)$, and so $\overrightarrow{PQ}$ is a representation of the vector $\mathbf{R}(t + \Delta t) - \mathbf{R}(t)$. As Δt approaches zero, the vector $[\mathbf{R}(t + \Delta t) - \mathbf{R}(t)]/\Delta t$ has a representation approaching a directed line segment tangent to the curve C at P.

The definition of the **unit tangent vector** is analogous to Definition 13.6.1 for vectors in the plane. So if $\mathbf{T}(t)$ denotes the unit tangent vector to curve C having vector equation (1), then

$$\mathbf{T}(t) = \frac{D_t \mathbf{R}(t)}{\|D_t \mathbf{R}(t)\|} \tag{3}$$

FIGURE 14.8.4

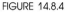

▶ ILLUSTRATION 3   We find the unit tangent vector for the twisted cubic of Illustration 2.

Because $\mathbf{R}(t) = t\mathbf{i} + t^2\mathbf{j} + t^3\mathbf{k}$,

$$D_t\mathbf{R}(t) = \mathbf{i} + 2t\mathbf{j} + 3t^2\mathbf{k} \qquad \|D_t\mathbf{R}(t)\| = \sqrt{1 + 4t^2 + 9t^4}$$

From (3), then,

$$\mathbf{T}(t) = \frac{1}{\sqrt{1 + 4t^2 + 9t^4}}(\mathbf{i} + 2t\mathbf{j} + 3t^2\mathbf{k})$$

Therefore, in particular,

$$\mathbf{T}(1) = \frac{1}{\sqrt{14}}\mathbf{i} + \frac{2}{\sqrt{14}}\mathbf{j} + \frac{3}{\sqrt{14}}\mathbf{k}$$

Figure 14.8.4 shows the representation of $\mathbf{T}(1)$ at the point $(1, 1, 1)$.   ◀

Theorems 13.4.6, 13.4.7, and 13.4.8 regarding derivatives of sums and products of two-dimensional vector-valued functions also hold for vectors in three dimensions. The following theorem regarding the derivative of the cross product of two vector-valued functions is similar to the corresponding formula for the derivative of the product of real-valued functions; however, it is important to maintain the correct order of the vector-valued functions because the cross product is not commutative.

14.8.6 THEOREM   If $\mathbf{R}$ and $\mathbf{Q}$ are vector-valued functions, then

$$D_t[\mathbf{R}(t) \times \mathbf{Q}(t)] = \mathbf{R}(t) \times \mathbf{Q}'(t) + \mathbf{R}'(t) \times \mathbf{Q}(t)$$

for all values of t for which $\mathbf{R}'(t)$ and $\mathbf{Q}'(t)$ exist.

The proof of Theorem 14.8.6 is left as an exercise (see Exercise 14).

The length of an arc of a curve C in three-dimensional space can be defined in exactly the same way as we defined the length of an arc of a curve in the plane (see Definition 13.5.1). If C is the curve having parametric equations (2), f_1', f_2', f_3' are continuous on the closed interval $[a, b]$, and no two values of t give the same point (x, y, z) on C, then we can prove (as we did for the plane) a theorem similar to Theorem 13.5.3, which states that the length of arc, L units, of the curve C from the point $(f_1(a), f_2(a), f_3(a))$ to the point $(f_1(b), f_2(b), f_3(b))$ is determined by

$$L = \int_a^b \sqrt{[f_1'(t)]^2 + [f_2'(t)]^2 + [f_3'(t)]^2}\, dt \qquad (4)$$

If s is the measure of the length of arc of C from the fixed point $(f_1(t_0), f_2(t_0), f_3(t_0))$ to the variable point $(f_1(t), f_2(t), f_3(t))$ and s increases as t increases, then s is a function of t and is given by

$$s = \int_{t_0}^t \sqrt{[f_1'(u)]^2 + [f_2'(u)]^2 + [f_3'(u)]^2}\, du \qquad (5)$$

As shown in Section 13.5 for plane curves, it follows that if (1) is a vector equation of C, then

$$\frac{ds}{dt} = \|D_t\mathbf{R}(t)\| \tag{6}$$

and the length of arc, L units, given by (4), also can be determined by

$$L = \int_a^b \|D_t\mathbf{R}(t)\| \, dt \tag{7}$$

EXAMPLE 1
Given the circular helix

$$\mathbf{R}(t) = a \cos t\mathbf{i} + a \sin t\mathbf{j} + t\mathbf{k}$$

where $a > 0$, find the length of arc from $t = 0$ to $t = 2\pi$.

SOLUTION

$$D_t\mathbf{R}(t) = -a \sin t\mathbf{i} + a \cos t\mathbf{j} + \mathbf{k}$$

So from (7),

$$L = \int_0^{2\pi} \sqrt{(-a \sin t)^2 + (a \cos t)^2 + 1} \, dt$$

$$= \int_0^{2\pi} \sqrt{a^2 + 1} \, dt$$

$$= 2\pi\sqrt{a^2 + 1}$$

Thus the length of arc is $2\pi\sqrt{a^2 + 1}$ units.

The definitions of the **curvature vector** $\mathbf{K}(t)$ and the **curvature** $K(t)$ at a point P on a curve C in R^3 are the same as for plane curves given in Definition 13.7.1. Hence, if $\mathbf{T}(t)$ is the unit tangent vector to C at P and s is the measure of the arc length from an arbitrarily chosen point on C to P, where s increases as t increases, then

$$\mathbf{K}(t) = D_s\mathbf{T}(t)$$

$$\Leftrightarrow \quad \mathbf{K}(t) = \frac{D_t\mathbf{T}(t)}{\|D_t\mathbf{R}(t)\|} \tag{8}$$

and

$$K(t) = \|D_s\mathbf{T}(t)\|$$

$$\Leftrightarrow \quad K(t) = \left\|\frac{D_t\mathbf{T}(t)}{\|D_t\mathbf{R}(t)\|}\right\| \tag{9}$$

Taking the dot product of $\mathbf{K}(t)$ and $\mathbf{T}(t)$ and using (8) we get

$$\mathbf{K}(t) \cdot \mathbf{T}(t) = \frac{D_t\mathbf{T}(t)}{\|D_t\mathbf{R}(t)\|} \cdot \mathbf{T}(t)$$

$$= \frac{1}{\|D_t\mathbf{R}(t)\|} D_t\mathbf{T}(t) \cdot \mathbf{T}(t) \qquad (10)$$

Theorem 13.4.11 states that if a vector-valued function in a plane has a constant magnitude, it is orthogonal to its derivative. This theorem and its proof also hold for vectors in three dimensions. Therefore, because $\|\mathbf{T}(t)\| = 1$, we can conclude from (10) that $\mathbf{K}(t) \cdot \mathbf{T}(t) = 0$. And so the curvature vector and the unit tangent vector of a curve at a point are orthogonal.

The **unit normal vector** is defined as the unit vector having the same direction as the curvature vector, provided that the curvature vector is not the zero vector. So if $\mathbf{N}(t)$ denotes the unit normal vector to a curve C at a point P, then if $\mathbf{K}(t) \neq \mathbf{0}$,

$$\mathbf{N}(t) = \frac{\mathbf{K}(t)}{\|\mathbf{K}(t)\|} \qquad (11)$$

From (11) and the previous discussion it follows that the unit normal vector and the unit tangent vector are orthogonal. Thus the angle between these two vectors has a radian measure of $\frac{1}{2}\pi$, and from Theorem 14.5.8,

$$\|\mathbf{T}(t) \times \mathbf{N}(t)\| = \|\mathbf{T}(t)\| \, \|\mathbf{N}(t)\| \sin \tfrac{1}{2}\pi$$
$$= 1$$

Therefore the cross product of $\mathbf{T}(t)$ and $\mathbf{N}(t)$ is a unit vector. By Theorem 14.5.10, $\mathbf{T}(t) \times \mathbf{N}(t)$ is orthogonal to both $\mathbf{T}(t)$ and $\mathbf{N}(t)$; hence the vector $\mathbf{B}(t)$, defined by

$$\mathbf{B}(t) = \mathbf{T}(t) \times \mathbf{N}(t) \qquad (12)$$

is a unit vector orthogonal to $\mathbf{T}(t)$ and $\mathbf{N}(t)$ and is called the **unit binormal vector** to the curve C at P.

The three mutually orthogonal unit vectors, $\mathbf{T}(t)$, $\mathbf{N}(t)$, and $\mathbf{B}(t)$, of a curve C are called the **moving trihedral** of C (see Fig. 14.8.5).

FIGURE 14.8.5

EXAMPLE 2

Find the moving trihedral and the curvature at any point of the circular helix of Example 1.

SOLUTION

A vector equation of the circular helix is

$$\mathbf{R}(t) = a \cos t\mathbf{i} + a \sin t\mathbf{j} + t\mathbf{k}$$

So $D_t\mathbf{R}(t) = -a \sin t\mathbf{i} + a \cos t\mathbf{j} + \mathbf{k}$ and $\|D_t\mathbf{R}(t)\| = \sqrt{a^2 + 1}$. From (3),

$$\mathbf{T}(t) = \frac{1}{\sqrt{a^2 + 1}}(-a \sin t\mathbf{i} + a \cos t\mathbf{j} + \mathbf{k})$$

Thus

$$D_t\mathbf{T}(t) = \frac{1}{\sqrt{a^2 + 1}}(-a\cos t\mathbf{i} - a\sin t\mathbf{j})$$

Applying (8) we obtain

$$\mathbf{K}(t) = \frac{1}{a^2 + 1}(-a\cos t\mathbf{i} - a\sin t\mathbf{j})$$

Because the curvature $K(t) = \|\mathbf{K}(t)\|$, we have

$$K(t) = \frac{a}{a^2 + 1}$$

Thus the curvature of the circular helix is constant. From (11),

$$\mathbf{N}(t) = -\cos t\mathbf{i} - \sin t\mathbf{j}$$

Applying (12) we have

$$\mathbf{B}(t) = \frac{1}{\sqrt{a^2 + 1}}(-a\sin t\mathbf{i} + a\cos t\mathbf{j} + \mathbf{k}) \times (-\cos t\mathbf{i} - \sin t\mathbf{j})$$

$$= \frac{1}{\sqrt{a^2 + 1}}(\sin t\mathbf{i} - \cos t\mathbf{j} + a\mathbf{k})$$

A thorough treatment of curves and surfaces by means of calculus is presented in a course in differential geometry. The use of calculus of vectors further enhances this subject. The previous discussion has been but a short introduction.

We now consider briefly the motion of a particle along a curve in three-dimensional space. If the parameter t in the vector equation (1) measures time, then the position at t of a particle moving along the curve C, having vector equation (1), is the point $P(f_1(t), f_2(t), f_3(t))$. The **velocity vector**, $\mathbf{V}(t)$, and the **acceleration vector**, $\mathbf{A}(t)$, are defined as in the plane. The vector $\mathbf{R}(t)$ is called the **position vector**, and

$$\mathbf{V}(t) = D_t\mathbf{R}(t) \qquad \mathbf{A}(t) = D_t\mathbf{V}(t)$$

The speed of the particle at t is the magnitude of the velocity vector. By applying (6) we can write

$$\|\mathbf{V}(t)\| = \frac{ds}{dt}$$

EXAMPLE 3

A particle is moving along the curve having parametric equations

$$x = 3t \qquad y = t^2 \qquad z = \tfrac{2}{3}t^3$$

Find the velocity and acceleration vectors and the speed of the particle at $t = 1$. Draw a sketch of a portion of the curve at $t = 1$, and draw representations of the velocity and acceleration vectors there.

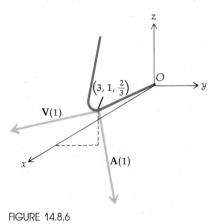

FIGURE 14.8.6

SOLUTION

A vector equation of the curve is

$$R(t) = 3t\mathbf{i} + t^2\mathbf{j} + \tfrac{2}{3}t^3\mathbf{k}$$

Therefore

$$V(t) = D_t R(t) \qquad\qquad A(t) = D_t V(t)$$
$$= 3\mathbf{i} + 2t\mathbf{j} + 2t^2\mathbf{k} \qquad\qquad = 2\mathbf{j} + 4t\mathbf{k}$$
$$\|V(t)\| = \sqrt{9 + 4t^2 + 4t^4}$$

Thus

$$V(1) = 3\mathbf{i} + 2\mathbf{j} + 2\mathbf{k} \qquad A(1) = 2\mathbf{j} + 4\mathbf{k} \qquad \|V(1)\| = \sqrt{17}$$

The required sketch is shown in Fig. 14.8.6.

EXERCISES 14.8

In Exercises 1 through 5, find the unit tangent vector for the curve having the given vector equation.

1. $R(t) = (t + 1)\mathbf{i} - t^2\mathbf{j} + (1 - 2t)\mathbf{k}$ 2. $R(t) = \sin 2t\mathbf{i} + \cos 2t\mathbf{j} + 2t^{3/2}\mathbf{k}$ 3. $R(t) = e^t \cos t\mathbf{i} + e^t \sin t\mathbf{j} + e^t\mathbf{k}$
4. $R(t) = t^2\mathbf{i} + (t + \tfrac{1}{3}t^3)\mathbf{j} + (t - \tfrac{1}{3}t^3)\mathbf{k}$ 5. $R(t) = 2t \cos t\mathbf{i} + 5t\mathbf{j} + 2t \sin t\mathbf{k}$
6. Find the unit tangent vector for the curve having the vector equation $R(t) = 4 \cosh 2t\mathbf{i} + 4 \sinh 2t\mathbf{j} + 6t\mathbf{k}$ at the point where $t = 0$.

In Exercises 7 through 11, find the length of arc of the curve from t_1 to t_2.

7. The curve of Exercise 1; $t_1 = -1$; $t_2 = 2$.   8. The curve of Exercise 2; $t_1 = 0$; $t_2 = 1$.
9. The curve of Exercise 3; $t_1 = 0$; $t_2 = 3$.   10. The curve of Exercise 4; $t_1 = 0$; $t_2 = 1$.
11. $R(t) = 4t^{3/2}\mathbf{i} - 3 \sin t\mathbf{j} + 3 \cos t\mathbf{k}$; $t_1 = 0$; $t_2 = 2$.
12. Prove that the unit tangent vector of the circular helix of Example 1 makes an angle of constant radian measure with the unit vector $\mathbf{k}$.
13. Prove Theorem 14.8.5.   14. Prove Theorem 14.8.6.
15. Write a vector equation of the curve of intersection of the surfaces $y = e^x$ and $z = xy$.
16. Write a vector equation of the curve of intersection of the surfaces $x = \ln(z^2 + 2)$ and $y = xz^3$.
17. Find the cosine of the angle between the vector $\mathbf{j}$ and the unit tangent vector to the curve $R(t) = \cos 2t\mathbf{i} - 3t\mathbf{j} + 2 \sin 2t\mathbf{k}$ at the point where $t = \pi$.
18. Find the curvature at the point where $t = 1$ of the curve $R(t) = t^2\mathbf{i} + (4 + t)\mathbf{j} + (3 - 2t)\mathbf{k}$.
19. Find the moving trihedral and the curvature at the point where $t = 1$ of the twisted cubic of Illustration 2.
20. Find the moving trihedral and the curvature at any point of the curve $R(t) = \cosh t\mathbf{i} + \sinh t\mathbf{j} + t\mathbf{k}$.

In Exercises 21 through 24, find the moving trihedral and the curvature of the given curve at $t = t_1$, if they exist.

21. The curve of Exercise 1; $t_1 = -1$.   22. The curve of Exercise 2; $t_1 = 0$.
23. The curve of Exercise 3; $t_1 = 0$.   24. The curve of Exercise 4; $t_1 = 1$.

In Exercises 25 through 28, a particle is moving along the given curve. Find the velocity vector, the acceleration vector, and the speed at $t = t_1$. Draw a sketch of a portion of the curve at $t = t_1$, and draw the velocity and acceleration vectors there.

25. The circular helix of Example 1; $t_1 = \tfrac{1}{2}\pi$.   26. $x = t$, $y = \tfrac{1}{2}t^2$, $z = \tfrac{1}{3}t^3$; $t_1 = 2$.
27. $x = e^{2t}$, $y = e^{-2t}$, $z = te^{2t}$; $t_1 = 1$.   28. $x = \tfrac{1}{2}(t^2 + 1)^{-1}$, $y = \ln(1 + t^2)$, $z = \tan^{-1} t$; $t_1 = 1$.

29. Prove that if $\mathbf{R}(t) = f_1(t)\mathbf{i} + f_2(t)\mathbf{j} + f_3(t)\mathbf{k}$ is a vector equation of curve C, and $K(t)$ is the curvature of C, then

$$K(t) = \frac{\|D_t\mathbf{R}(t) \times D_t^2\mathbf{R}(t)\|}{\|D_t\mathbf{R}(t)\|^3}$$

30. Use the formula of Exercise 29 to show that the curvature of the circular helix of Example 1 is $a/(a^2 + 1)$.

In Exercises 31 and 32, find the curvature of the given curve at the indicated point.

31. $x = t$, $y = t^2$, $z = t^3$; the origin.   32. $x = e^t$, $y = e^{-t}$, $z = t$; $t = 0$.

33. Prove that if $\mathbf{R}(t) = f_1(t)\mathbf{i} + f_2(t)\mathbf{j} + f_3(t)\mathbf{k}$ is a vector equation of curve C, $K(t)$ is the curvature of C at a point P, and s units is the arc length measured from an arbitrarily chosen point on C to P, then $D_s\mathbf{R}(t) \cdot D_s^3\mathbf{R}(t) = -[K(t)]^2$.

34. A particle is moving along a curve with vector equation $\mathbf{R}(t) = \tan t\mathbf{i} + \sinh 2t\mathbf{j} + \text{sech } t\mathbf{k}$. Prove that the velocity vector and the acceleration vector are orthogonal at $t = 0$.

35. Prove that if the speed of a moving particle is constant, its acceleration vector is always orthogonal to its velocity vector.

36. Prove that for the twisted cubic of Illustration 2, if $t \neq 0$, no two of the vectors $\mathbf{R}(t)$, $\mathbf{V}(t)$, and $\mathbf{A}(t)$ are orthogonal.

14.9
CYLINDRICAL AND
SPHERICAL COORDINATES

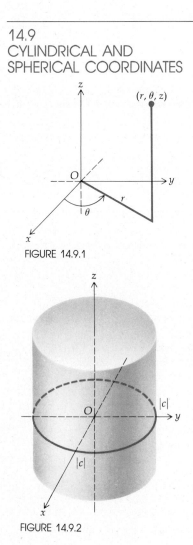

FIGURE 14.9.1

FIGURE 14.9.2

Cylindrical and *spherical coordinates* are generalizations of polar coordinates to three-dimensional space. The **cylindrical coordinate** representation of a point P is (r, θ, z), where r and θ are the polar coordinates of the projection of P on a polar plane and z is the directed distance from this polar plane to P. See Fig. 14.9.1.

EXAMPLE 1
Draw a sketch of the graph of each of the following equations where c is a constant: (a) $r = c$; (b) $\theta = c$; (c) $z = c$.

SOLUTION

(a) For a point $P(r, \theta, z)$ on the graph of $r = c$, θ and z can have any values and r is a constant. The graph is a right-circular cylinder having radius $|c|$ and the z axis as its axis. A sketch of the graph is shown in Fig. 14.9.2.

(b) For all points $P(r, \theta, z)$ on the graph of $\theta = c$, r and z can assume any value while θ remains constant. The graph is a plane through the z axis. See Fig. 14.9.3 for a sketch of the graph.

(c) The graph of $z = c$ is a plane parallel to the polar plane at a directed distance of c units from it. Figure 14.9.4 shows a sketch of the graph.

The name "cylindrical coordinates" comes from the fact that the graph of $r = c$ is a right-circular cylinder as in Example 1(a). Cylindrical coordinates are often used in a physical problem when there is an axis of symmetry.

Suppose that a cartesian coordinate system and a cylindrical coordinate system are placed so the xy plane is the polar plane of the cylindrical coordinate system and the positive side of the x axis is the polar axis as shown in Fig. 14.9.5. Then the point P has (x, y, z) and (r, θ, z) as two sets of coordinates that are related by the equations

$$x = r \cos \theta \qquad y = r \sin \theta \qquad z = z \qquad (1)$$

$$r^2 = x^2 + y^2 \qquad \tan \theta = \frac{y}{x} \quad \text{if } x \neq 0 \qquad z = z \qquad (2)$$

FIGURE 14.9.3

FIGURE 14.9.4   FIGURE 14.9.5

EXAMPLE 2

Find an equation in cartesian coordinates of the following surfaces whose equations are expressed in cylindrical coordinates, and identify the surface: (a) $r = 6 \sin \theta$; (b) $r(3 \cos \theta + 2 \sin \theta) + 6z = 0$.

SOLUTION

(a) Multiplying on both sides of the equation by r we get $r^2 = 6r \sin \theta$. Because $r^2 = x^2 + y^2$ and $r \sin \theta = y$, then $x^2 + y^2 = 6y$. This equation can be written in the form $x^2 + (y - 3)^2 = 9$, which shows that its graph is a right-circular cylinder whose cross section in the xy plane is the circle with its center at $(0, 3)$ and radius 3.

(b) Replacing $r \cos \theta$ with x and $r \sin \theta$ with y we obtain the equation $3x + 2y + 6z = 0$. Hence the graph is a plane through the origin and has $\langle 3, 2, 6 \rangle$ as a normal vector.

EXAMPLE 3

Find an equation in cylindrical coordinates for each of the following surfaces whose equations are given in cartesian coordinates, and identify the surface: (a) $x^2 + y^2 = z$; (b) $x^2 - y^2 = z$.

SOLUTION

(a) The equation is similar to Eq. (5) of Section 14.7; so the graph is an elliptic paraboloid. If $x^2 + y^2$ is replaced by r^2, the equation becomes $r^2 = z$.

(b) The equation is similar to Eq. (6) of Section 14.7 with x and y interchanged. The graph is therefore a hyperbolic paraboloid having the z axis as its axis. When x is replaced by $r \cos \theta$ and y is replaced by $r \sin \theta$, we get the equation $r^2 \cos^2 \theta - r^2 \sin^2 \theta = z$; and because $\cos^2 \theta - \sin^2 \theta = \cos 2\theta$, we can write this as $z = r^2 \cos 2\theta$.

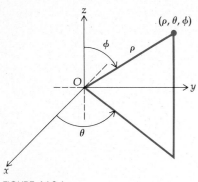

FIGURE 14.9.6

In a spherical coordinate system there is a polar plane and an axis perpendicular to the polar plane, with the origin of the z axis at the pole of the polar plane. A point is located by three numbers, and the **spherical coordinate** representation of a point P is (ρ, θ, ϕ), where $\rho = |\overline{OP}|$, θ is the radian measure of the polar angle of the projection of P on the polar plane, and ϕ is the nonnegative radian measure of the smallest angle measured from the positive side of the z axis to the line OP. See Fig. 14.9.6. The origin has the spherical coordinate representation $(0, \theta, \phi)$, where θ and ϕ may have any values. If the point $P(\rho, \theta, \phi)$ is not the origin, then $\rho > 0$ and $0 \le \phi \le \pi$, where $\phi = 0$ if P is on the positive side of the z axis and $\phi = \pi$ if P is on the negative side of the z axis.

FIGURE 14.9.7

FIGURE 14.9.8

EXAMPLE 4

Draw a sketch of the graph of each of the following equations where c is a constant: (a) $\rho = c$, and $c > 0$; (b) $\theta = c$; (c) $\phi = c$, and $0 < c < \pi$.

SOLUTION

(a) Every point $P(\rho, \theta, \phi)$ on the graph of $\rho = c$ has the same value of ρ, θ may be any number, and $0 \le \phi \le \pi$. It follows that the graph is a sphere of radius c and has its center at the pole. Figure 14.9.7 shows a sketch of the sphere.

(b) For any point $P(\rho, \theta, \phi)$ on the graph of $\theta = c$, ρ may be any nonnegative number, ϕ may be any number in the closed interval $[0, \pi]$, and θ is constant. The graph is a half plane containing the z axis and is obtained by rotating about the z axis through an angle of c radians that half of the xz plane for which $x \ge 0$. Figure 14.9.8 shows sketches of the half planes for $\theta = \frac{1}{4}\pi$, $\theta = \frac{2}{3}\pi$, $\theta = \frac{4}{3}\pi$, and $\theta = -\frac{1}{6}\pi$.

(c) The graph of $\phi = c$ contains all the points $P(\rho, \theta, \phi)$ for which ρ is any nonnegative number, θ is any number, and ϕ is the constant c. The graph is half of a cone having its vertex at the origin and the z axis as its axis. Figure 14.9.9(a) and (b) each show a sketch of the half cone for $0 < c < \frac{1}{2}\pi$ and $\frac{1}{2}\pi < c < \pi$, respectively.

(a)   (b)

FIGURE 14.9.9

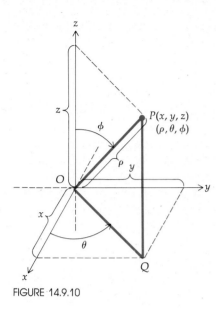

FIGURE 14.9.10

Because the graph of $\rho = c$ is a sphere as seen in Example 4(a), we have the name "spherical coordinates." In a physical problem when there is a point that is a center of symmetry, spherical coordinates are often used.

By placing a spherical coordinate system and a cartesian coordinate system together as shown in Fig. 14.9.10, we obtain relationships between the spherical coordinates and the cartesian coordinates of a point P from

$$x = |\overline{OQ}| \cos \theta \qquad y = |\overline{OQ}| \sin \theta \qquad z = |\overline{QP}|$$

Because $|\overline{OQ}| = \rho \sin \phi$ and $|\overline{QP}| = \rho \cos \phi$, these equations become

$$x = \rho \sin \phi \cos \theta \qquad y = \rho \sin \phi \sin \theta \qquad z = \rho \cos \phi \qquad (3)$$

By squaring each of the equations in (3) and adding,

$$x^2 + y^2 + z^2 = \rho^2 \sin^2 \phi \cos^2 \theta + \rho^2 \sin^2 \phi \sin^2 \theta + \rho^2 \cos^2 \phi$$

$$x^2 + y^2 + z^2 = \rho^2 \sin^2 \phi (\cos^2 \theta + \sin^2 \theta) + \rho^2 \cos^2 \phi$$

$$x^2 + y^2 + z^2 = \rho^2 (\sin^2 \phi + \cos^2 \phi)$$

$$x^2 + y^2 + z^2 = \rho^2$$

EXAMPLE 5
Find an equation in cartesian coordinates of the following surfaces whose equations are expressed in spherical coordinates, and identify the surface: (a) $\rho \cos \phi = 4$; (b) $\rho \sin \phi = 4$.

SOLUTION
(a) Because $z = \rho \cos \phi$, the equation becomes $z = 4$. Hence the graph is a plane parallel to the xy plane and 4 units above it.
(b) For spherical coordinates $\rho \geq 0$ and $\sin \phi \geq 0$ (because $0 \leq \phi \leq \pi$); therefore, by squaring on both sides of the given equation we obtain the equivalent equation $\rho^2 \sin^2 \phi = 16$, which in turn is equivalent to

$$\rho^2 (1 - \cos^2 \phi) = 16$$

$$\rho^2 - \rho^2 \cos^2 \phi = 16$$

Replacing ρ^2 by $x^2 + y^2 + z^2$ and $\rho \cos \phi$ by z we get

$$x^2 + y^2 + z^2 - z^2 = 16$$

$$x^2 + y^2 = 16$$

Therefore the graph is the right-circular cylinder having the z axis as its axis and radius 4.

EXAMPLE 6
Find an equation in spherical coordinates for (a) the elliptic paraboloid of Example 3(a); (b) the plane of Example 2(b).

SOLUTION
(a) A cartesian equation of the elliptic paraboloid of Example 3(a) is $x^2 + y^2 = z$. Replacing x by $\rho \sin \phi \cos \theta$, y by $\rho \sin \phi \sin \theta$, and z by

$\rho \cos \phi$ we get

$$\rho^2 \sin^2 \phi \cos^2 \theta + \rho^2 \sin^2 \phi \sin^2 \theta = \rho \cos \phi$$

$$\rho^2 \sin^2 \phi(\cos^2 \theta + \sin^2 \theta) = \rho \cos \phi$$

which is equivalent to the two equations

$$\rho = 0 \quad \text{and} \quad \rho \sin^2 \phi = \cos \phi$$

The origin is the only point whose coordinates satisfy $\rho = 0$. Because the origin $(0, 0, \frac{1}{2}\pi)$ lies on $\rho \sin^2 \phi = \cos \phi$, we can disregard the equation $\rho = 0$. Furthermore, $\sin \phi \neq 0$ because there is no value of ϕ for which both $\sin \phi$ and $\cos \phi$ are 0. Therefore the equation $\rho \sin^2 \phi = \cos \phi$ can be written as $\rho = \csc^2 \phi \cos \phi$, or, equivalently, $\rho = \csc \phi \cot \phi$.

(b) A cartesian equation of the plane of Example 2(b) is $3x + 2y + 6z = 0$. By using Eqs. (3) this equation becomes

$$3\rho \sin \phi \cos \theta + 2\rho \sin \phi \sin \theta + 6\rho \cos \phi = 0$$

Applications of cylindrical and spherical coordinates in engineering appear in Section 17.7.

EXERCISES 14.9

1. Find the cartesian coordinates of the point having the given cylindrical coordinates: (a) $(3, \frac{1}{2}\pi, 5)$; (b) $(7, \frac{2}{3}\pi, -4)$; (c) $(1, 1, 1)$.

2. Find a set of cylindrical coordinates of the point having the given cartesian coordinates: (a) $(4, 4, -2)$; (b) $(-3\sqrt{3}, 3, 6)$; (c) $(1, 1, 1)$.

3. Find the cartesian coordinates of the point having the given spherical coordinates: (a) $(4, \frac{1}{6}\pi, \frac{1}{4}\pi)$; (b) $(4, \frac{1}{2}\pi, \frac{1}{3}\pi)$; (c) $(\sqrt{6}, \frac{1}{3}\pi, \frac{3}{4}\pi)$.

4. Find a set of spherical coordinates of the point having the given cartesian coordinates: (a) $(1, -1, -\sqrt{2})$; (b) $(-1, \sqrt{3}, 2)$; (a) $(2, 2, 2)$.

5. Find a set of cylindrical coordinates of the point having the given spherical coordinates: (a) $(4, \frac{2}{3}\pi, \frac{5}{6}\pi)$; (b) $(\sqrt{2}, \frac{3}{4}\pi, \pi)$; (c) $(2\sqrt{3}, \frac{1}{3}\pi, \frac{1}{4}\pi)$.

6. Find a set of spherical coordinates of the point having the given cylindrical coordinates: (a) $(3, \frac{1}{6}\pi, 3)$; (b) $(3, \frac{1}{2}\pi, 2)$; (c) $(2, \frac{5}{6}\pi, -4)$.

In Exercises 7 through 12 find an equation in cylindrical coordinates of the given surface, and identify the surface.

7. $x^2 + y^2 + 4z^2 = 16$ 8. $x^2 - y^2 = 9$ 9. $x^2 + y^2 = 3z$
10. $9x^2 + 4y^2 = 36$ 11. $x^2 - y^2 = 3z^2$ 12. $x^2 + y^2 = z^2$

In Exercises 13 through 17, find an equation in spherical coordinates of the given surface, and identify the surface.

13. $x^2 + y^2 + z^2 - 9z = 0$ 14. $x^2 + y^2 = z^2$ 15. $x^2 + y^2 = 9$
16. $x^2 + y^2 = 2z$ 17. $x^2 + y^2 + z^2 - 8x = 0$

In Exercises 18 through 22, find an equation in cartesian coordinates for the surface whose equation is given in cylindrical coordinates. In Exercises 18 and 19, identify the surface.

18. $r = 3 \cos \theta$ 19. (a) $r = 4$; (b) $\theta = \frac{1}{4}\pi$ 20. $r = 3 + 2 \cos \theta$ 21. $r^2 \cos 2\theta = z^3$ 22. $z^2 \sin^3 \theta = r^3$

In Exercises 23 through 28, find an equation in cartesian coordinates for the surface whose equation is given in spherical coordinates. In Exercise 23 through 25, identify the surface.

23. (a) $\rho = 9$; (b) $\theta = \frac{1}{4}\pi$; (c) $\phi = \frac{1}{4}\pi$ **24.** $\rho = 9 \sec \phi$ **25.** $\rho = 6 \csc \phi$

26. $\rho = 3 \cos \phi$ **27.** $\rho = 2 \tan \theta$ **28.** $\rho = 6 \sin \phi \sin \theta + 3 \cos \phi$

29. A curve C in R^3 has the following parametric equations in cylindrical coordinates: $r = F_1(t)$, $\theta = F_2(t)$, $z = F_3(t)$. Use formula (4) of Section 14.8 and formulas (1) of this section to prove that if L units is the length of arc of C from the point where $t = a$ to the point where $t = b$, then

$$L = \int_a^b \sqrt{\left(\frac{dr}{dt}\right)^2 + r^2 \left(\frac{d\theta}{dt}\right)^2 + \left(\frac{dz}{dt}\right)^2} \, dt$$

30. A curve C in R^3 has the following parametric equations in spherical coordinates: $\rho = G_1(t)$, $\theta = G_2(t)$, $\phi = G_3(t)$. Use formula (4) of Section 14.8 and formulas (3) of this section to prove that if L units is the length of arc of C from the point where $t = a$ to the point where $t = b$, then

$$L = \int_a^b \sqrt{\left(\frac{d\rho}{dt}\right)^2 + \rho^2 \sin^2 \phi \left(\frac{d\theta}{dt}\right)^2 + \rho^2 \left(\frac{d\phi}{dt}\right)^2} \, dt$$

31. (a) Show that parametric equations for the circular helix of Example 1, Section 14.8, are $r = a$, $\theta = t$, $z = t$.
   (b) Use the formula of Exercise 29 to find the length of arc of the circular helix of part (a) from $t = 0$ to $t = 2\pi$. Check your result with that of Example 1, Section 14.8.

32. A *conical helix* winds around a cone in a way similar to that in which a circular helix winds around a cylinder. Use the formula of Exercise 30 to find the length of arc from $t = 0$ to $t = 2\pi$ of the conical helix having parametric equations $\rho = t$, $\theta = t$, $\phi = \frac{1}{4}\pi$.

REVIEW EXERCISES FOR CHAPTER 14

1. Draw a sketch of the graph of $x = 3$ in R, R^2, and R^3.

2. Draw a sketch of the set of points satisfying the simultaneous equations $x = 6$ and $y = 3$ in R^2 and R^3.

In Exercises 3 through 12, describe in words the set of points in R^3 satisfying the given equation or the given pair of equations. Draw a sketch of the graph.

3. $\begin{cases} y = 0 \\ z = 0 \end{cases}$ **4.** $\begin{cases} x = z \\ y = z \end{cases}$ **5.** $\begin{cases} x^2 + z^2 = 4 \\ y = 0 \end{cases}$ **6.** $y^2 - z^2 = 0$

7. $x = y$ **8.** $x^2 + y^2 + z^2 = 25$ **9.** $x^2 + y^2 = 9z$ **10.** $x^2 + y^2 = z^2$

11. $x^2 - y^2 = z^2$ **12.** $x^2 + z^2 = 4$

In Exercises 13 through 22, $\mathbf{A} = -\mathbf{i} + 3\mathbf{j} + 2\mathbf{k}$, $\mathbf{B} = 2\mathbf{i} + \mathbf{j} - 4\mathbf{k}$, $\mathbf{C} = \mathbf{i} + 2\mathbf{j} - 2\mathbf{k}$, $\mathbf{D} = 3\mathbf{j} - \mathbf{k}$, and $\mathbf{E} = 5\mathbf{i} - 2\mathbf{j}$.

13. Find $6\mathbf{C} + 4\mathbf{D} - \mathbf{E}$.   **14.** Find $3\mathbf{A} - 2\mathbf{B} + \mathbf{C}$.   **15.** Find $\mathbf{D} \cdot \mathbf{B} \times \mathbf{C}$.

16. Find $(\mathbf{A} \times \mathbf{C}) - (\mathbf{D} \times \mathbf{E})$.   **17.** Find $\|\mathbf{A} \times \mathbf{B}\| \|\mathbf{D} \times \mathbf{E}\|$.   **18.** Find $2\mathbf{B} \cdot \mathbf{C} + 3\mathbf{D} \cdot \mathbf{E}$.

19. Find the scalar projection of $\mathbf{A}$ onto $\mathbf{B}$.   **20.** Find the scalar projection of $\mathbf{C}$ onto $\mathbf{D}$.

21. Find the vector projection of $\mathbf{E}$ onto $\mathbf{C}$.   **22.** Find the vector projection of $\mathbf{D}$ onto $\mathbf{E}$.

In Exercises 23 through 28, there is only one way that a meaningful expression can be obtained by inserting parentheses. Insert the parentheses and find the indicated vector or scalar if $\mathbf{A} = \langle 3, -2, 4 \rangle$, $\mathbf{B} = \langle -5, 7, 2 \rangle$, and $\mathbf{C} = \langle 4, 6, -1 \rangle$.

23. $\mathbf{A}\mathbf{B} \cdot \mathbf{C}$ **24.** $\mathbf{A} \cdot \mathbf{B}\mathbf{C}$ **25.** $\mathbf{A} + \mathbf{B} \cdot \mathbf{C}$

26. $\mathbf{B} \cdot \mathbf{A} - \mathbf{C}$ **27.** $\mathbf{A} \times \mathbf{B} \cdot \mathbf{A} + \mathbf{B} - \mathbf{C}$ **28.** $\mathbf{A} \times \mathbf{B} \cdot \mathbf{C} \times \mathbf{A}$

29. A line is drawn through the point $(-3, 5, 1)$ perpendicular to the xz plane. Find the coordinates of the points on this line at a distance of 13 units from $(-2, 0, 0)$.

30. Find an equation of the sphere having as a diameter the line segment with endpoints $(3, 5, -4)$ and $(-1, 7, 4)$.

31. Find an equation of the sphere concentric with the sphere $x^2 + y^2 + z^2 + 4x + 2y - 6z + 10 = 0$ and containing the point $(-4, 2, 5)$.

32. Prove that the points $(4, 1, -1)$, $(2, 0, 1)$, and $(4, 3, 0)$ are the vertices of a right triangle, and find the area of the triangle.

33. Find a generating curve and the axis for the surface of revolution having the equation $x^2 + z^2 = e^{4y}$.

34. Find an equation of the surface of revolution generated by revolving the ellipse $9x^2 + 4z^2 = 36$ in the xz plane about the x axis. Draw a sketch of the surface.

35. Determine the value of c such that the vectors $3\mathbf{i} + c\mathbf{j} - 3\mathbf{k}$ and $5\mathbf{i} - 4\mathbf{j} + \mathbf{k}$ are orthogonal.

36. Show that there are representations of the three vectors $\mathbf{A} = 5\mathbf{i} + \mathbf{j} - 3\mathbf{k}$, $\mathbf{B} = \mathbf{i} + 3\mathbf{j} - 2\mathbf{k}$, and $\mathbf{C} = -4\mathbf{i} + 2\mathbf{j} + \mathbf{k}$ which form a triangle.

37. Given the points $A(5, 9, -3)$ and $B(-2, 4, -5)$, find (a) the direction cosines of $\mathbf{V}(\overrightarrow{AB})$ and (b) the unit vector having the same direction as $\mathbf{V}(\overrightarrow{AB})$.

38. If $\mathbf{A} = \mathbf{i} + \mathbf{j} - \mathbf{k}$, $\mathbf{B} = 2\mathbf{i} - \mathbf{j} + \mathbf{k}$, $\mathbf{C} = 3\mathbf{i} - 2\mathbf{j} + 4\mathbf{k}$, and $\mathbf{D} = 5\mathbf{i} + 6\mathbf{j} - 8\mathbf{k}$, find scalars a, b, and c such that

$$a\mathbf{A} + b\mathbf{B} + c\mathbf{C} = \mathbf{D}$$

39. If $\mathbf{A} = \langle 7, -1, 5 \rangle$ and $\mathbf{B} = \langle -2, 3, 1 \rangle$, find (a) the scalar projection of $\mathbf{B}$ onto $\mathbf{A}$ and (b) the vector projection of $\mathbf{B}$ onto $\mathbf{A}$.

40. Find an equation of the plane containing the points $(1, 7, -3)$ and $(3, 1, 2)$ and which does not intersect the x axis.

41. Find an equation of the plane through the three points $(-1, 2, 1)$, $(1, 4, 0)$, and $(1, -1, 3)$ by two methods: (a) using the cross product; (b) without using the cross product.

42. Find an equation of the plane that contains the point $(7, -2, -5)$ and is perpendicular to the line through the points $(-3, 0, 4)$ and $(3, 2, 1)$.

43. Find the distance from the origin to the plane through the point $(-6, 3, -2)$ and having $5\mathbf{i} - 3\mathbf{j} + 4\mathbf{k}$ as a normal vector.

44. Find two unit vectors orthogonal to $\mathbf{i} - 3\mathbf{j} + 4\mathbf{k}$ and whose representations are parallel to the yz plane.

45. Find the distance from the point $P(4, 6, -4)$ to the line through the two points $A(2, 2, 1)$ and $B(4, 3, -1)$.

46. Find the distance from the plane $9x - 2y + 6z + 44 = 0$ to the point $(-3, 2, 0)$.

47. If θ is the angle between the vectors $\mathbf{A} = 2\mathbf{i} + \mathbf{j} + \mathbf{k}$ and $\mathbf{B} = 4\mathbf{i} - 3\mathbf{j} + 5\mathbf{k}$, find $\cos \theta$ in two ways: (a) by using the dot product; (b) by using the cross product and a trigonometric identity.

48. Prove that the lines

$$\frac{x - 1}{1} = \frac{y + 2}{2} = \frac{z - 2}{2} \quad \text{and} \quad \frac{x - 2}{2} = \frac{y - 5}{3} = \frac{z - 5}{1}$$

are skew lines, and find the distance between them.

49. Find symmetric and parametric equations of the line through the origin and perpendicular to each of the lines of Exercise 48.

50. Find symmetric and parametric equations of the line through the two points $(-3, 5, 2)$ and $(-1, -3, 4)$.

51. Show that the lines

$$\frac{x - 2}{3} = \frac{y + 3}{-1} = \frac{z - 5}{4} \quad \text{and} \quad \frac{x + 1}{-6} = \frac{y + 2}{2} = \frac{z - 1}{-8}$$

are coincident.

52. Find an equation of the plane containing the line $\frac{1}{2}(x - 3) = -(y + 5) = -\frac{1}{3}(z + 2)$ and the point $(5, 0, -4)$.

53. Find the area of the cross section of the ellipsoid $\dfrac{x^2}{4} + \dfrac{y^2}{9} + \dfrac{z^2}{25} = 1$ in the plane $z = 4$.

54. Find the area of the parallelogram two of whose sides are the position representations of the vectors $2\mathbf{j} - 3\mathbf{k}$ and $5\mathbf{i} + 4\mathbf{k}$.

55. Find the volume of the parallelepiped having vertices at $(1, 3, 0)$, $(2, -1, 3)$, $(-2, 2, -1)$, and $(-1, 1, 2)$.

56. Find the length of the arc of the curve

$$x = t \cos t \qquad y = t \sin t \qquad z = t$$

from $t = 0$ to $t = \frac{1}{2}\pi$.

57. Find the length of the arc of the curve $\mathbf{R}(t) = (2 - 3t)\mathbf{i} + (4t - 1)\mathbf{j} + t^2\mathbf{k}$ from $t = 0$ to $t = \frac{5}{2}$.

58. A particle is moving along the curve with parametric equations

$$x = \ln(t^2 + 1) \qquad y = t^2 + 1 \qquad z = \tan^{-1} t$$

Find the velocity vector, the acceleration vector, and the speed at $t = 0$.

59. A particle is moving along the curve of Exercise 56. Find the velocity vector, the acceleration vector, and the speed at $t = \frac{1}{2}\pi$. Draw a sketch of a portion of the curve at $t = \frac{1}{2}\pi$, and draw the representations of the velocity and acceleration vectors there.

60. Find the unit tangent vector and the curvature at any point on the curve having the vector equation $\mathbf{R}(t) = e^t\mathbf{i} + e^{-t}\mathbf{j} + 2t\mathbf{k}$.

61. Find the moving trihedral of the curve $\mathbf{R}(t) = 3 \sin 2t\mathbf{i} - 4t\mathbf{j} - 3 \cos 2t\mathbf{k}$ at the point where $t = \frac{1}{2}\pi$.

62. Find a set of spherical coordinates for the point having cartesian coordinates $(-3, \sqrt{3}, 2)$.

63. Find a set of cylindrical coordinates for the point having spherical coordinates $(3, \pi, \frac{1}{3}\pi)$.

64. Find an equation in spherical coordinates of the graph of each of the equations: (a) $x^2 + y^2 + 4z^2 = 4$; (b) $4x^2 - 4y^2 + 9z^2 = 36$.

65. Find an equation in cylindrical coordinates of the graph of each of the equations: (a) $(x + y)^2 + 1 = z$; (b) $25x^2 + 4y^2 = 100$.

66. If $\mathbf{R}$, $\mathbf{Q}$, and $\mathbf{W}$ are three vector-valued functions whose derivatives with respect to t exist, prove that

$$D_t[\mathbf{R}(t) \cdot \mathbf{Q}(t) \times \mathbf{W}(t)] = D_t\mathbf{R}(t) \cdot \mathbf{Q}(t) \times \mathbf{W}(t) + \mathbf{R}(t) \cdot D_t\mathbf{Q}(t) \times \mathbf{W}(t) + \mathbf{R}(t) \cdot \mathbf{Q}(t) \times D_t\mathbf{W}(t)$$

67. If $\mathbf{A}$ is any vector, prove that $\mathbf{A} = (\mathbf{A} \cdot \mathbf{i})\mathbf{i} + (\mathbf{A} \cdot \mathbf{j})\mathbf{j} + (\mathbf{A} \cdot \mathbf{k})\mathbf{k}$.

F I F T E E N

DIFFERENTIAL CALCULUS OF FUNCTIONS OF
MORE THAN ONE VARIABLE

15.1
FUNCTIONS OF MORE
THAN ONE VARIABLE

We now extend the concept of a function to functions of n variables, and in succeeding sections we extend to functions of n variables the concepts of the *limit* of a function, *continuity* of a function, and *derivative* of a function. A thorough treatment of these topics belongs to a course in advanced calculus. In this book most of our discussion of functions of more than one variable is confined to those of two and three variables; however, we make the definitions for functions of n variables and then show the applications of these definitions to functions of two and three variables. We also show that when each of these definitions is applied to a function of one variable, we have the definition previously given.

To extend the concept of a function to functions of any number of variables we must first consider points in n-dimensional number space. Just as we denoted a point in R by a real number x, a point in R^2 by an ordered pair of real numbers (x, y), and a point in R^3 by an ordered triple of real numbers (x, y, z), a point in n-dimensional number space, R^n, is represented by an ordered n-tuple of real numbers customarily denoted by $P = (x_1, x_2, \ldots, x_n)$. In particular, if $n = 1$, let $P = x$; if $n = 2$, $P = (x, y)$; if $n = 3$, $P = (x, y, z)$; if $n = 6$, $P = (x_1, x_2, x_3, x_4, x_5, x_6)$.

15.1.1 DEFINITION   The set of all ordered n-tuples of real numbers is called the **n-dimensional number space** and is denoted by R^n. Each ordered n-tuple $(x_1, x_2, \ldots, x_n)$ is called a **point** in the n-dimensional number space.

15.1.2 DEFINITION   A **function of n variables** is a set of ordered pairs of the form (P, w) in which no two distinct ordered pairs have the same first element. P is a point in n-dimensional number space and w is a real number. The set of all possible values of P is called the **domain** of the function, and the set of all possible values of w is called the **range** of the function.

From this definition, the domain of a function of n variables is a set of points in R^n and the range is a set of real numbers or, equivalently, a set of points in R. When $n = 1$, we have a function of one variable; thus the domain is a set of points in R or, equivalently, a set of real numbers, and the range is a set of real numbers. Hence Definition 0.5.1 is a special case of Definition 15.1.2. If $n = 2$, we have a function of two variables, and the domain is a set of points in R^2 or, equivalently, a set of ordered pairs of real numbers (x, y). The range is a set of real numbers.

▶ ILLUSTRATION 1   Let the function f of two variables x and y be the set of all ordered pairs of the form (P, z) such that

$$z = \sqrt{25 - x^2 - y^2}$$

The domain of f is the set $\{(x, y) \mid x^2 + y^2 \le 25\}$. This is the set of points in the xy plane on the circle $x^2 + y^2 = 25$ and in the interior region bounded by the circle.

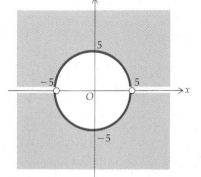

FIGURE 15.1.1

Because $z = \sqrt{25 - (x^2 + y^2)}$, then $0 \leq z \leq 5$; therefore the range of f is the set of all real numbers in the closed interval $[0, 5]$. In Fig. 15.1.1 there is a sketch showing as a shaded region in R^2 the set of points in the domain of f. ◀

▶ ILLUSTRATION 2 The function g of two variables x and y is the set of all ordered pairs of the form (P, z) such that

$$z = \frac{\sqrt{x^2 + y^2 - 25}}{y}$$

The domain of g is the set $\{(x, y) \mid x^2 + y^2 \geq 25 \text{ and } y \neq 0\}$. This is the set of points, not on the x axis, that are either on the circle $x^2 + y^2 = 25$ or in the exterior region bounded by the circle. A sketch showing as a shaded region in R^2 the set of points in the domain of g appears in Fig. 15.1.2. ◀

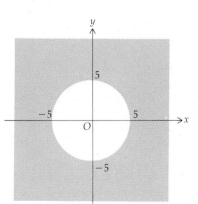

FIGURE 15.1.2

▶ ILLUSTRATION 3 The function h of two variables x and y is the set of all ordered pairs of the form (P, z) such that

$$z = \frac{1}{\sqrt{x^2 + y^2 - 25}}$$

The domain of h is the set $\{(x, y) \mid x^2 + y^2 > 25\}$. This is the set of points in the exterior region bounded by the circle $x^2 + y^2 = 25$. In Fig. 15.1.3 there is a sketch showing as a shaded region in R^2 the set of points in the domain of h. ◀

If f is a function of n variables, then according to Definition 15.1.2, f is a set of ordered pairs of the form (P, w), where $P = (x_1, x_2, \ldots, x_n)$ is a point in R^n and w is a real number. The particular value of w that corresponds to a point P is denoted by the symbol $f(P)$ or $f(x_1, x_2, \ldots, x_n)$. In particular, if $n = 2$ and $P = (x, y)$, we can represent the function value by either $f(P)$ or $f(x, y)$. Similarly, if $n = 3$ and $P = (x, y, z)$, we denote the function value by either $f(P)$ or $f(x, y, z)$. Note that if $n = 1$, $P = x$; hence if f is a function of one variable, $f(P) = f(x)$. Therefore this notation is consistent with the notation for function values of one variable.

A function f of n variables can be defined by the equation

$$w = f(x_1, x_2, \ldots, x_n)$$

The variables $x_1, x_2, \ldots, x_n$ are called the *independent variables*, and w is called the *dependent variable*.

▶ ILLUSTRATION 4 Let f be the function of Illustration 1; that is,

$$f(x, y) = \sqrt{25 - x^2 - y^2}$$

FIGURE 15.1.3

Then

$$f(3, -4) = \sqrt{25 - 3^2 - (-4)^2} \qquad f(-2, 1) = \sqrt{25 - (-2)^2 - 1^2}$$
$$= \sqrt{25 - 9 - 16} \qquad\qquad\qquad = \sqrt{25 - 4 - 1}$$
$$= 0 \qquad\qquad\qquad\qquad\qquad = 2\sqrt{5}$$

$$f(u, 3v) = \sqrt{25 - u^2 - (3v)^2}$$
$$= \sqrt{25 - u^2 - 9v^2}$$

◀

EXAMPLE 1

The function g is defined by

$$g(x, y, z) = x^3 - 4yz^2$$

Find: (a) $g(1, 3, -2)$; (b) $g(2a, -4b, 3c)$; (c) $g(x^2, y^2, z^2)$; (d) $g(y, z, -x)$.

SOLUTION

(a) $g(1, 3, -2) = 1^3 - 4(3)(-2)^2$ (b) $g(2a, -4b, 3c) = (2a)^3 - 4(-4b)(3c)^2$
$\qquad\qquad = 1 - 48 \qquad\qquad\qquad\qquad = 8a^3 + 144bc^2$
$\qquad\qquad = -47$

(c) $g(x^2, y^2, z^2) = (x^2)^3 - 4y^2(z^2)^2$ (d) $g(y, z, -x) = y^3 - 4z(-x)^2$
$\qquad\qquad = x^6 - 4y^2z^4 \qquad\qquad\qquad\qquad = y^3 - 4x^2z$

15.1.3 DEFINITION   If f is a function of a single variable and g is a function of two variables, then the **composite function** $f \circ g$ is the function of two variables defined by

$$(f \circ g)(x, y) = f(g(x, y))$$

and the domain of $f \circ g$ is the set of all points (x, y) in the domain of g such that $g(x, y)$ is in the domain of f.

EXAMPLE 2

Given $f(t) = \ln t$ and $g(x, y) = x^2 + y$, find $h(x, y)$ if $h = f \circ g$, and determine the domain of h.

SOLUTION

$$h(x, y) = (f \circ g)(x, y)$$
$$= f(g(x, y))$$
$$= f(x^2 + y)$$
$$= \ln(x^2 + y)$$

The domain of g is the set of all points in R^2, and the domain of f is $(0, +\infty)$. Therefore the domain of h is the set $\{(x, y) \mid x^2 + y > 0\}$.

Definition 15.1.3 can be extended to a composite function of n variables as follows.

15.1.4 DEFINITION   If f is a function of a single variable and g is a function of n variables, then the **composite function** $f \circ g$ is the function of n variables defined by

$$(f \circ g)(x_1, x_2, \ldots, x_n) = f(g(x_1, x_2, \ldots, x_n))$$

and the domain of $f \circ g$ is the set of all points $(x_1, x_2, \ldots, x_n)$ in the domain of g such that $g(x_1, x_2, \ldots, x_n)$ is in the domain of f.

EXAMPLE 3

Given $F(x) = \sin^{-1} x$ and

$$G(x, y, z) = \sqrt{x^2 + y^2 + z^2 - 4}$$

find the function $F \circ G$ and its domain.

SOLUTION

$$
\begin{aligned}
(F \circ G)(x, y, z) &= F(G(x, y, z)) \\
&= F(\sqrt{x^2 + y^2 + z^2 - 4}) \\
&= \sin^{-1}\sqrt{x^2 + y^2 + z^2 - 4}
\end{aligned}
$$

The domain of G is the set $\{(x, y, z) \mid x^2 + y^2 + z^2 - 4 \geq 0\}$, and the domain of F is $[-1, 1]$. So the domain of $F \circ G$ is the set of all points (x, y, z) in R^3 such that $0 \leq x^2 + y^2 + z^2 - 4 \leq 1$ or, equivalently, $4 \leq x^2 + y^2 + z^2 \leq 5$.

A **polynomial function** of two variables x and y is a function f such that $f(x, y)$ is the sum of terms of the form $cx^n y^m$, where c is a real number and n and m are nonnegative integers. The **degree** of the polynomial function is determined by the largest sum of the exponents of x and y appearing in any one term.

▶ ILLUSTRATION 5   (a) The function f defined by

$$f(x, y) = x^3 + 2x^2 y^2 - y^3$$

is a polynomial function of degree 4 because the term of highest degree is $2x^2 y^2$.

(b) If

$$g(x, y) = 6x^3 y^2 - 5xy^3 + 7x^2 y - 2x^2 + y + 4$$

g is a polynomial function of degree 5.   ◀

The graph of a function f of a single variable consists of the set of points (x, y) in R^2 for which $y = f(x)$. Similarly, the graph of a function of two variables is a set of points in R^3.

15.1.5 DEFINITION   If f is a function of two variables, then the **graph** of f is the set of all points (x, y, z) in R^3 for which (x, y) is a point in the domain of f and $z = f(x, y)$.

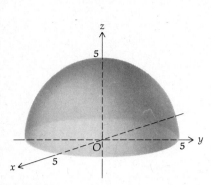

FIGURE 15.1.4

Hence the graph of a function f of two variables is a surface that is the set of all points in three-dimensional space whose cartesian coordinates are given by the ordered triples of real numbers (x, y, z). Because the domain of f is a set of points in the xy plane, and because for each ordered pair (x, y) in the domain of f there corresponds a unique value of z, no line perpendicular to the xy plane can intersect the graph of f in more than one point.

▶ ILLUSTRATION 6   The function of Illustration 1 is the function f that is the set of all ordered pairs of the form (P, z) such that

$$z = \sqrt{25 - x^2 - y^2}$$

So the graph of f is the hemisphere on and above the xy plane having a radius of 5 and its center at the origin. A sketch of the graph of this hemisphere is shown in Fig. 15.1.4.   ◀

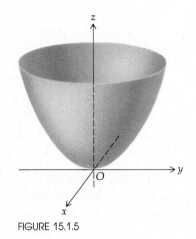

FIGURE 15.1.5

EXAMPLE 4
Draw a sketch of the graph of the function f having function values $f(x, y) = x^2 + y^2$.

SOLUTION
The graph of f is the surface having the equation $z = x^2 + y^2$. The trace of the surface in the xy plane is found by using the equation $z = 0$ simultaneously with the equation of the surface. We obtain $x^2 + y^2 = 0$, which is the origin. The traces in the xz and yz planes are found by using the equations $y = 0$ and $x = 0$, respectively, with the equation $z = x^2 + y^2$. These traces are the parabolas $z = x^2$ and $z = y^2$. The cross section of the surface in a plane $z = k$, parallel to the xy plane, is a circle with its center on the z axis and radius $\sqrt{k}$. With this information we have the required sketch shown in Fig. 15.1.5.

FIGURE 15.1.6

Another useful method of representing a function of two variables geometrically is similar to that of representing a three-dimensional landscape by a two-dimensional topographical map. Suppose that the surface $z = f(x, y)$ is intersected by the plane $z = k$, and the curve of intersection is projected onto the xy plane. This projected curve has $f(x, y) = k$ as an equation, and the curve is called the **level curve** (or **contour curve**) of the function f at k. Each point on the level curve corresponds to the unique point on the surface that is k units above it if k is positive, or k units below it if k is negative. By considering different values for the constant k we obtain a set of level curves called a **contour map**. The set of all possible values of k is the range of the function f, and each level curve, $f(x, y) = k$, in the contour map consists of the points (x, y) in the domain of f having equal function values of k. For example, for the function f of Example 4 the level curves are circles with the center at the origin. The particular level curves for $z = 1, 2, 3, 4, 5,$ and 6 are shown in Fig. 15.1.6.

FIGURE 15.1.7

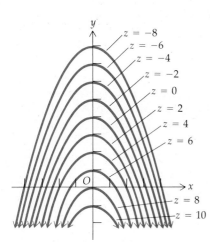

FIGURE 15.1.8

A contour map shows the variation of z with x and y. The level curves are usually shown for values of z at constant intervals, and the values of z are changing more rapidly when the level curves are close together than when they are far apart; that is, when the level curves are close together the surface is steep, and when the level curves are far apart the elevation of the surface is changing slowly. On a two-dimensional topographical map of a landscape, a general notion of its steepness is obtained by considering the spacing of its level curves. Also on a topographical map if the path of a level curve is followed, the elevation remains constant.

EXAMPLE 5

Let f be the function for which $f(x, y) = 8 - x^2 - 2y$. Draw a sketch of the graph of f and a contour map of f showing the level curves of f at 10, 8, 6, 4, 2, 0, -2, -4, -6, and -8.

SOLUTION

A sketch of the graph of f is shown in Fig. 15.1.7. This is the surface $z = 8 - x^2 - 2y$. The trace in the xy plane is obtained by setting $z = 0$, which gives the parabola $x^2 = -2(y - 4)$. Setting $y = 0$ and $x = 0$, we obtain the traces in the xz and yz planes, which are, respectively, the parabola $x^2 = -(z - 8)$ and the line $2y + z = 8$. The cross section of the surface made by the plane $z = k$ is a parabola having its vertex on the line $2y + z = 8$ in the yz plane and opening to the left. The cross sections for $z = 8, 6, 4, 2, -2, -4, -6$, and -8 are shown in the figure.

The level curves of f are the parabolas $x^2 = -2(y - 4 + \frac{1}{2}k)$. The contour map of f with sketches of the required level curves is shown in Fig. 15.1.8.

To illustrate a use of level curves, suppose that the temperature at any point of a flat metal plate is given by the function f; that is, if t degrees is the temperature, then at the point (x, y), $t = f(x, y)$. Then the curves having equations of the form $f(x, y) = k$, where k is a constant, are curves on which the temperature is constant. These are the level curves of f and are called **isothermals**. Furthermore, if V volts gives the amount of electric potential at any point (x, y) of the xy plane, and $V = f(x, y)$, then the level curves of f are called **equipotential curves** because the electric potential at each point of such a curve is the same.

For an application of level curves in economics, consider the productivity (or output) of an industry that is dependent on several inputs. Among the inputs may be the number of machines used in production, the number of person-hours available, the amount of working capital to be had, the quantity of material used, and the amount of land available. Suppose the amounts of the inputs are given by x and y, the amount of the output is given by z, and $z = f(x, y)$. Such a function is called a **production function**, and the level curves of f, having equations of the form $f(x, y) = k$, where k is a constant, are called **constant product curves**.

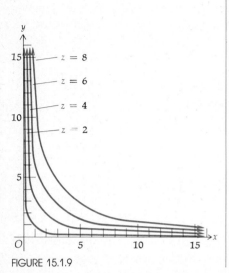

FIGURE 15.1.9

EXAMPLE 6

Let f be the production function for which

$$f(x, y) = 2x^{1/2}y^{1/2}$$

Draw a contour map of f showing the constant product curves at 8, 6, 4, and 2.

SOLUTION

The contour map consists of the curves that are the intersection of the surface

$$z = 2x^{1/2}y^{1/2} \tag{1}$$

with the planes $z = k$, where $k = 8, 6, 4$, and 2. Substituting $z = 8$ in Eq. (1) we obtain $4 = x^{1/2}y^{1/2}$ or, equivalently,

$$xy = 16 \quad x > 0 \quad \text{and} \quad y > 0 \tag{2}$$

The curve in the xy plane represented by (2) is a branch of a hyperbola lying in the first quadrant. With each of the numbers 6, 4, and 2 we also obtain a branch of a hyperbola in the first quadrant. These are the constant product curves, and they are shown in Fig. 15.1.9.

The following definition extends the notion of the graph of a function to a function of n variables.

15.1.6 DEFINITION   If f is a function of n variables, then the **graph** of f is the set of all points $(x_1, x_2, \ldots, x_n, w)$ in R^{n+1} for which $(x_1, x_2, \ldots, x_n)$ is a point in the domain of f and $w = f(x_1, x_2, \ldots, x_n)$.

Analogous to level curves for a function of two variables is a similar situation for functions of three variables. If f is a function whose domain is a set of points in R^3, then if k is a number in the range of f, the graph of the equation $f(x, y, z) = k$ is a surface. This surface is called the **level surface** of f at k. Every surface in three-dimensional space can be considered as a level surface of some function of three variables. For example, if the function g is defined by the equation $g(x, y, z) = x^2 + y^2 - z$, then the surface shown in Fig. 15.1.5 is the level surface of g at 0. Similarly, the surface having equation $z - x^2 - y^2 + 5 = 0$ is the level surface of g at 5.

EXERCISES 15.1

1. Let the function f of two variables x and y be the set of all ordered pairs of the form (P, z) such that

$$z = \frac{x + y}{x - y} \quad \Leftrightarrow \quad f(x, y) = \frac{x + y}{x - y}$$

Find: (a) $f(-3, 4)$; (b) $f(\frac{1}{2}, \frac{1}{3})$; (c) $f(x + 1, y - 1)$; (d) $f(-x, y) - f(x, -y)$. (e) Draw a sketch showing as a region in R^2 the set of points in the domain of f.

2. Let the function g of two variables x and y be the set of all ordered pairs of the form (P, z) such that

$$z = \sqrt{x^2 - y} \quad \Leftrightarrow \quad g(x, y) = \sqrt{x^2 - y}$$

Find: (a) $g(3, 5)$; (b) $g(-4, -9)$; (c) $g(x + 2, 4x + 4)$; (d) $g\left(\dfrac{1}{x}, \dfrac{-3}{x^2}\right)$. (e) Draw a sketch showing as a region in R^2 the set of points in the domain of g.

3. Let the function g of three variables x, y, and z be the set of all ordered pairs of the form (P, w) such that

$$w = \sqrt{4 - x^2 - y^2 - z^2} \quad \Leftrightarrow \quad g(x, y, z) = \sqrt{4 - x^2 - y^2 - z^2}$$

Find: (a) $g(1, -1, -1)$; (b) $g(-1, \frac{1}{2}, \frac{3}{2})$; (c) $g(\frac{1}{2}x; \frac{1}{2}y; \frac{1}{2}z)$; (d) $[g(x, y, z)]^2 - [g(x + 2, y + 2, z)]^2$. (e) Draw a sketch showing as a region in R^3 the set of points in the domain of g.

4. Let the function f of three variables x, y, and z be the set of all ordered pairs of the form (P, w) such that

$$w = \frac{4}{x^2 + y^2 + z^2 - 9} \quad \Leftrightarrow \quad f(x, y, z) = \frac{4}{x^2 + y^2 + z^2 - 9}$$

Find: (a) $f(1, 2, 3)$; (b) $f(2, -\frac{1}{2}, \frac{3}{2})$; (c) $f\left(-\dfrac{2}{x}, \dfrac{2}{x}, -\dfrac{1}{x}\right)$; (d) $f(x + 2, 1, x - 2)$. (e) Draw a sketch showing as a surface in R^3 the set of points *not* in the domain of f.

In Exercises 5 through 20, determine the domain of the function f, and draw a sketch showing as a region in R^2 the set of points in the domain.

5. $f(x, y) = \dfrac{1}{x^2 + y^2 - 1}$

6. $f(x, y) = \dfrac{4}{4 - x^2 - y^2}$

7. $f(x, y) = \sqrt{1 - x^2 - y^2}$

8. $f(x, y) = \sqrt{16 - x^2 - 4y^2}$

9. $f(x, y) = \dfrac{\sqrt{25 - x^2 - y^2}}{x}$

10. $f(x, y) = \dfrac{\sqrt{16 - x^2 - 4y^2}}{y}$

11. $f(x, y) = \sqrt{x^2 - y^2 - 25}$

12. $f(x, y) = \sqrt{x^2 + 4y^2 - 16}$

13. $f(x, y) = \dfrac{1}{\sqrt{25 - x^2 - y^2}}$

14. $f(x, y) = \dfrac{1}{\sqrt{16 - x^2 - 4y^2}}$

15. $f(x, y) = \dfrac{x^2 - y^2}{x - y}$

16. $f(x, y) = \sqrt{\dfrac{x - y}{x + y}}$

17. $f(x, y) = \dfrac{x}{|y|}$

18. $f(x, y) = \dfrac{x + y}{xy}$

19. $f(x, y) = \ln(xy - 1)$

20. $f(x, y) = \sin^{-1}(x + y)$

In Exercises 21 through 23, find the domain and range of the function f.

21. $f(x, y, z) = (x + y)\sqrt{z - 2}$

22. $f(x, y, z) = |x|e^{y/z}$

23. $f(x, y, z) = \sin^{-1} x + \cos^{-1} y + \tan^{-1} z$

In Exercises 24 through 30, find the domain and range of the function f, and draw a sketch of the graph.

24. $f(x, y) = \sqrt{x + y}$

25. $f(x, y) = 4x^2 + 9y^2$

26. $f(x, y) = 144 - 9x^2 - 16y^2$

27. $f(x, y) = 16 - x^2 - y^2$

28. $f(x, y) = \sqrt{100 - 25x^2 - 4y^2}$

29. $f(x, y) = \sqrt{10 - x - y^2}$

30. $f(x, y) = \begin{cases} 2 & \text{if } x \neq y \\ 0 & \text{if } x = y \end{cases}$

In Exercises 31 through 36, draw a sketch of a contour map of the function f showing the level curves of f at the given numbers.

31. The function of Exercise 25 at 36, 25, 16, 9, 4, 1, and 0.   **32.** The function of Exercise 24 at 5, 4, 3, 2, 1, and 0.

33. The function of Exercise 27 at 8, 4, 0, -4, and -8.   **34.** The function of Exercise 28 at 10, 8, 6, 5, and 0.

35. The function f for which $f(x, y) = \frac{1}{2}(x^2 + y^2)$ at 8, 6, 4, 2, and 0.

36. The function f for which $f(x, y) = (x - 3)/(y + 2)$ at 4, 2, 1, $\frac{1}{2}$, $\frac{1}{4}$, 0, $-\frac{1}{4}$, $-\frac{1}{2}$, -1, -2, and -4.

In Exercises 37 and 38, find $h(x, y)$ if $h = f \circ g$; also find the domain of h.

37. $f(t) = \sin^{-1} t$; $g(x, y) = \sqrt{1 - x^2 - y^2}$ **38.** $f(t) = e^t$; $g(x, y) = y \ln x$

39. Given $f(x, y) = x - y$, $g(t) = \sqrt{t}$, $h(s) = s^2$. Find (a) $(g \circ f)(5, 1)$; (b) $f(h(3), g(9))$; (c) $f(g(x), h(y))$; (d) $g((h \circ f)(x, y))$; (e) $(g \circ h)(f(x, y))$.

40. Given $f(x, y) = x/y^2$, $g(x) = x^2$, $h(x) = \sqrt{x}$. Find (a) $(h \circ f)(2, 1)$; (b) $f(g(2), h(4))$; (c) $f(g(\sqrt{x}), h(x^2))$; (d) $h((g \circ f)(x, y))$; (e) $(h \circ g)(f(x, y))$.

41. The electric potential at a point (x, y) of the xy plane is V volts and $V = 4/\sqrt{9 - x^2 - y^2}$. Draw the equipotential curves for V at 16, 12, 8, 4, 1, $\frac{1}{2}$, and $\frac{1}{4}$.

42. The production function f for a certain commodity has function values $f(x, y) = 4x^{1/3}y^{2/3}$, where x and y give the amounts of two inputs. Draw a contour map of f showing the constant product curves at 16, 12, 8, 4, and 2.

43. Suppose the number of units of a commodity produced is z, and $z = 6xy$, where x is the number of machines used in production and y is the number of person-hours available. Then the function f defined by $f(x, y) = 6xy$ is a production function. Draw a contour map of f showing the constant product curves for z at 30, 24, 18, 12, and 6.

44. The temperature at a point (x, y) of a flat metal plate is t degrees and $t = 4x^2 + 2y^2$. Draw the isothermals for t at 12, 8, 4, 1, and 0.

In Exercises 45 and 46, draw sketches of the level surfaces of the function f at the given numbers.

45. $f(x, y, z) = x^2 + y^2 - 4z$ at 8, 4, 0, -4, and -8.   **46.** $f(x, y, z) = x^2 + y^2 + z^2$ at 9, 4, 1, and 0.

15.2
LIMITS OF FUNCTIONS OF MORE THAN ONE VARIABLE

In R the distance between two points is the absolute value of the difference of two real numbers. That is, $|x - a|$ is the distance between the points x and a. In R^2 the distance between the two points $P(x, y)$ and $P_0(x_0, y_0)$ is given by $\sqrt{(x - x_0)^2 + (y - y_0)^2}$. In R^3 the distance between the two points $P(x, y, z)$ and $P_0(x_0, y_0, z_0)$ is given by $\sqrt{(x - x_0)^2 + (y - y_0)^2 + (z - z_0)^2}$. In R^n the distance between two points is defined analogously.

15.2.1 DEFINITION   If $P(x_1, x_2, \ldots, x_n)$ and $A(a_1, a_2, \ldots, a_n)$ are two points in R^n, then the distance between P and A, denoted by $\|P - A\|$, is given by

$$\|P - A\| = \sqrt{(x_1 - a_1)^2 + (x_2 - a_2)^2 + \ldots + (x_n - a_n)^2} \tag{1}$$

If in R, $P = x$ and $A = a$, (1) becomes

$$\|x - a\| = \sqrt{(x - a)^2}$$
$$= |x - a| \tag{2}$$

If in R^2, $P = (x, y)$ and $A = (x_0, y_0)$, (1) becomes

$$\|(x, y) - (x_0, y_0)\| = \sqrt{(x - x_0)^2 + (y - y_0)^2} \tag{3}$$

And if in R^3, $P = (x, y, z)$ and $A = (x_0, y_0, z_0)$, (1) becomes

$$\|(x, y, z) - (x_0, y_0, z_0)\| = \sqrt{(x - x_0)^2 + (y - y_0)^2 + (z - z_0)^2} \tag{4}$$

$\|P - A\|$ is read as "the distance between P and A." It is a nonnegative number.

15.2.2 DEFINITION   If A is a point in R^n and r is a positive number, then the **open ball** $B(A;r)$ is defined to be the set of all points P in R^n such that $\|P - A\| < r$.

15.2.3 DEFINITION   If A is a point in R^n and r is a positive number, then the **closed ball** $B[A;r]$ is defined to be the set of all points P in R^n such that $\|P - A\| \leq r$.

To illustrate these definitions, we show what they mean in R, R^2, and R^3. First of all, if a is a point in R, then the open ball $B(a;r)$ is the set of all points x in R such that

$$|x - a| < r \tag{5}$$

The set of all points x satisfying (5) is the set of all points in the open interval $(a - r, a + r)$; so the open ball $B(a;r)$ in R (see Fig. 15.2.1) is simply an open interval having its midpoint at a and endpoints at $a - r$ and $a + r$. The closed ball $B[a;r]$ in R (Fig. 15.2.2) is the closed interval $[a - r, a + r]$.

If (x_0, y_0) is a point in R^2, then the open ball $B((x_0, y_0);r)$ is the set of all points (x, y) in R^2 such that

$$\|(x, y) - (x_0, y_0)\| < r \tag{6}$$

From (3) it follows that (6) is equivalent to

$$\sqrt{(x - x_0)^2 + (y - y_0)^2} < r$$

So the open ball $B((x_0, y_0);r)$ in R^2 (Fig. 15.2.3) consists of all points in the interior region bounded by the circle having its center at (x_0, y_0) and radius r. An open ball in R^2 is sometimes called an *open disk*. The closed ball, or closed disk, $B[(x_0, y_0);r]$ in R^2 (Fig. 15.2.4) is the set of all points in the open ball $B((x_0, y_0);r)$ and on the circle having its center at (x_0, y_0) and radius r.

If (x_0, y_0, z_0) is a point in R^3, then the open ball $B((x_0, y_0, z_0);r)$ is the set of all points (x, y, z) in R^3 such that

$$\|(x, y, z) - (x_0, y_0, z_0)\| < r \tag{7}$$

From (4) and (7) it follows that the open ball $B((x_0, y_0, z_0);r)$ in R^3 (Fig. 15.2.5) consists of all points in the interior region bounded by the sphere having its center at P_0 and radius r. Similarly, the closed ball $B[(x_0, y_0, z_0);r]$ in R^3 (Fig. 15.2.6) consists of all points in the open ball $B((x_0, y_0, z_0);r)$ and

open ball $B(a; r)$ in R

FIGURE 15.2.1

closed ball $B[a; r]$ in R

FIGURE 15.2.2

open ball $B((x_0, y_0); r)$ in R^2

FIGURE 15.2.3

closed ball $B[(x_0, y_0); r]$ in R^2

FIGURE 15.2.4

open ball $B((x_0, y_0, z_0); r)$ in R^3

FIGURE 15.2.5

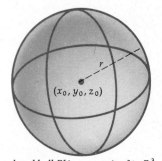

closed ball $B[(x_0, y_0, z_0); r]$ in R^3

FIGURE 15.2.6

on the sphere having its center at (x_0, y_0, z_0) and radius r. We are now in a position to define what is meant by the limit of a function of n variables.

15.2.4 DEFINITION   Let f be a function of n variables that is defined on some open ball $B(A; r)$, except possibly at the point A itself. Then the **limit of $f(P)$ as P approaches A is L**, written as

$$\lim_{P \to A} f(P) = L \tag{8}$$

if for any $\epsilon > 0$, however small, there exists a $\delta > 0$ such that

$$\text{if } 0 < \|P - A\| < \delta \quad \text{then} \quad |f(P) - L| < \epsilon \tag{9}$$

If f is a function of one variable and if in the above definition $A = a$ in R and $P = x$, then (8) becomes

$$\lim_{x \to a} f(x) = L$$

and (9) becomes

$$\text{if } 0 < |x - a| < \delta \quad \text{then} \quad |f(x) - L| < \epsilon$$

So the definition (1.2.1) of the limit of a function of one variable is a special case of Definition 15.2.4.

We now state the definition of the limit of a function of two variables. It is the special case of Definition 15.2.4, where A is the point (x_0, y_0) and P is the point (x, y).

15.2.5 DEFINITION   Let f be a function of two variables that is defined on some open disk $B((x_0, y_0); r)$, except possibly at the point (x_0, y_0) itself. Then the **limit of $f(x, y)$ as (x, y) approaches (x_0, y_0) is L**, written as

$$\lim_{(x,y) \to (x_0, y_0)} f(x, y) = L$$

if for any $\epsilon > 0$, however small, there exists a $\delta > 0$ such that

$$\text{if } 0 < \sqrt{(x - x_0)^2 + (y - y_0)^2} < \delta \quad \text{then} \quad |f(x, y) - L| < \epsilon \tag{10}$$

In words, Definition 15.2.5 states that the function values $f(x, y)$ approach a limit L as the point (x, y) approaches the point (x_0, y_0) if the absolute value of the difference between $f(x, y)$ and L can be made arbitrarily small by taking the point (x, y) sufficiently close to (x_0, y_0) but not equal to (x_0, y_0). In Definition 15.2.5 nothing is said about the function value at the point (x_0, y_0); that is, it is not necessary that the function be defined at (x_0, y_0) for $\lim_{(x,y) \to (x_0, y_0)} f(x, y)$ to exist.

A geometric interpretation of Definition 15.2.5 is illustrated in Fig. 15.2.7. The portion, above the open disk $B((x_0, y_0); \delta)$, of the surface having equation $z = f(x, y)$ is shown. We see that $f(x, y)$ on the z axis will lie between $L - \epsilon$ and $L + \epsilon$ whenever the point (x, y) in the xy plane is in the open disk $B((x_0, y_0); \delta)$. Another way of stating this is that $f(x, y)$ on the z axis can be restricted to lie between $L - \epsilon$ and $L + \epsilon$ by restricting the point (x, y) in the xy plane to be in the open disk $B((x_0, y_0); \delta)$.

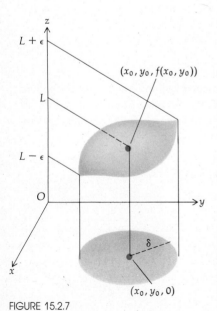

FIGURE 15.2.7

▶ ILLUSTRATION 1   We apply Definition 15.2.5 to prove that

$$\lim_{(x,y)\to(1,3)} (2x + 3y) = 11$$

The first requirement of the definition is that $2x + 3y$ must be defined on some open disk having its center at the point $(1, 3)$, except possibly at $(1, 3)$. Because $2x + 3y$ is defined at every point (x, y), any open disk having its center at $(1, 3)$ will satisfy this requirement. Now, we must show that for any $\epsilon > 0$ there exists a $\delta > 0$ such that

$$\text{if } 0 < \sqrt{(x - 1)^2 + (y - 3)^2} < \delta \quad \text{then} \quad |(2x + 3y) - 11| < \epsilon \qquad (11)$$

From the triangle inequality,

$$|2x + 3y - 11| = |2x - 2 + 3y - 9|$$
$$\leq 2|x - 1| + 3|y - 3|$$

Because

$$|x - 1| \leq \sqrt{(x - 1)^2 + (y - 3)^2} \quad \text{and} \quad |y - 3| \leq \sqrt{(x - 1)^2 + (y - 3)^2}$$

it follows that

$$\text{if } 0 < \sqrt{(x - 1)^2 + (y - 3)^2} < \delta \quad \text{then} \quad 2|x - 1| + 3|y - 3| < 2\delta + 3\delta$$

This statement indicates that a suitable choice for δ is $5\delta = \epsilon$, that is, $\delta = \frac{1}{5}\epsilon$. With this δ we have the following argument:

$$0 < \sqrt{(x - 1)^2 + (y - 3)^2} < \delta$$
$$\Rightarrow \quad |x - 1| < \delta \quad \text{and} \quad |y - 3| < \delta$$
$$\Rightarrow \quad 2|x - 1| + 3|y - 3| < 5\delta$$
$$\Rightarrow \quad |2(x - 1) + 3(y - 3)| < 5(\tfrac{1}{5}\epsilon)$$
$$\Rightarrow \quad |2x + 3y - 11| < \epsilon$$

We have demonstrated that for any $\epsilon > 0$ choose $\delta = \frac{1}{5}\epsilon$ and statement (11) is true. This proves that $\lim_{(x,y)\to(1,3)} (2x + 3y) = 11$.   ◀

EXAMPLE 1
Prove that

$$\lim_{(x,y)\to(1,2)} (3x^2 + y) = 5$$

by applying Definition 15.2.5.

SOLUTION
Because $3x^2 + y$ is defined at every point (x, y), any open disk having its center at $(1, 2)$ will satisfy the first requirement of Definition 15.2.5. We must show that for any $\epsilon > 0$ there exists a $\delta > 0$ such that

$$\text{if } 0 < \sqrt{(x - 1)^2 + (y - 2)^2} < \delta \quad \text{then} \quad |(3x^2 + y) - 5| < \epsilon \qquad (12)$$

From the triangle inequality,

$$|3x^2 + y - 5| = |3x^2 - 3 + y - 2|$$
$$\leq 3|x^2 - 1| + |y - 2|$$

Thus

$$|3x^2 + y - 5| \leq 3|x - 1||x + 1| + |y - 2| \qquad (13)$$

Because

$$|x - 1| \leq \sqrt{(x - 1)^2 + (y - 2)^2} \quad \text{and} \quad |y - 2| \leq \sqrt{(x - 1)^2 + (y - 2)^2}$$

it follows that

$$\text{if} \quad 0 < \sqrt{(x - 1)^2 + (y - 2)^2} < \delta \quad \text{then} \quad |x - 1| < \delta \quad \text{and} \quad |y - 2| < \delta$$

Observe that on the right end of (13), in addition to the expressions $|x - 1|$ and $|y - 2|$, we have the expression $|x + 1|$. Thus to prove (12) we wish to place a restriction on δ that will give us an inequality involving $|x + 1|$. One such restriction is to choose the radius of the open disk required by Definition 15.2.5 to be less than or equal to 1. Then

$$0 < \sqrt{(x - 1)^2 + (y - 2)^2} < \delta \quad \text{and} \quad \delta \leq 1$$
$$\Rightarrow \qquad |x - 1| < 1$$
$$\Rightarrow \qquad -1 < x - 1 < 1$$
$$\Rightarrow \qquad 1 < x + 1 < 3$$
$$\Rightarrow \qquad |x + 1| < 3$$

Now

$$|x - 1| < \delta \quad \text{and} \quad |x + 1| < 3 \quad \text{and} \quad |y - 2| < \delta$$
$$\Rightarrow \quad 3|x - 1||x + 1| + |y - 2| < 3 \cdot \delta \cdot 3 + \delta$$

Because our goal is to have $3|x - 1||x + 1| + |y - 2| < \epsilon$, we should require $10\delta \leq \epsilon$, that is $\delta \leq \frac{1}{10}\epsilon$. This means that we have put two restrictions on δ: $\delta \leq 1$ and $\delta \leq \frac{1}{10}\epsilon$. For both restrictions to hold, we take $\delta = \min(1, \frac{1}{10}\epsilon)$. With this δ we have the following argument:

$$0 < \sqrt{(x - 1)^2 + (y - 2)^2} < \delta \quad \text{and} \quad \delta = \min(1, \tfrac{1}{10}\epsilon)$$
$$\Rightarrow \quad |x - 1| < \tfrac{1}{10}\epsilon \quad \text{and} \quad |x - 1| < 1 \quad \text{and} \quad |y - 2| < \tfrac{1}{10}\epsilon$$
$$\Rightarrow \quad |x - 1| < \tfrac{1}{10}\epsilon \quad \text{and} \quad |x + 1| < 3 \quad \text{and} \quad |y - 2| < \tfrac{1}{10}\epsilon$$
$$\Rightarrow \quad 3|x - 1||x + 1| + |y - 2| < 3 \cdot \tfrac{1}{10}\epsilon \cdot 3 + \tfrac{1}{10}\epsilon$$
$$\Rightarrow \quad |3(x - 1)(x + 1) + y - 2| < \epsilon$$
$$\Rightarrow \qquad |(3x^2 + y) - 5| < \epsilon$$

We have demonstrated that for any $\epsilon > 0$, choose $\delta = \min(1, \frac{1}{10}\epsilon)$ and statement (12) is true. This proves that $\lim\limits_{(x,y) \to (1,2)} (3x^2 + y) = 5$.

The limit theorems of Section 1.3 and their proofs, with minor modifications, apply to functions of more than one variable. For example, corresponding to Limit Theorem 1 in Section 1.3 we have

$$\lim_{(x,y)\to(a,b)} (mx + ny + d) = ma + nb + d$$

and the proof is a generalization of the proof in Illustration 1. We use the limit theorems without restating them and their proofs.

▶ ILLUSTRATION 2   By applying the limit theorems on sums and products,

$$\lim_{(x,y)\to(-2,1)} (x^3 + 2x^2y - y^2 + 2) = (-2)^3 + 2(-2)^2(1) - (1)^2 + 2$$
$$= 1 \qquad \blacktriangleleft$$

EXAMPLE 2
Find

$$\lim_{(x,y)\to(0,0)} \frac{x^4 - y^4}{x^2 + y^2}$$

SOLUTION

$$\lim_{(x,y)\to(0,0)} \frac{x^4 - y^4}{x^2 + y^2} = \lim_{(x,y)\to(0,0)} \frac{(x^2 - y^2)(x^2 + y^2)}{x^2 + y^2}$$
$$= \lim_{(x,y)\to(0,0)} (x^2 - y^2)$$
$$= 0 - 0$$
$$= 0$$

Analogous to Theorem 1.7.1 for functions of a single variable is the following theorem regarding the limit of a composite function of two variables.

15.2.6 THEOREM   If g is a function of two variables and $\displaystyle\lim_{(x,y)\to(x_0,y_0)} g(x, y) = b$, and f is a function of a single variable continuous at b, then

$$\lim_{(x,y)\to(x_0,y_0)} (f \circ g)(x, y) = f(b)$$

$$\Leftrightarrow \qquad \lim_{(x,y)\to(x_0,y_0)} f(g(x, y)) = f\left(\lim_{(x,y)\to(x_0,y_0)} g(x, y) \right)$$

The proof of this theorem is similar to the proof of Theorem 1.7.1 and is left as an exercise (see Exercise 49).

EXAMPLE 3
Use Theorem 15.2.6 to find $\displaystyle\lim_{(x,y)\to(2,1)} \ln(xy - 1)$.

SOLUTION
Let g be the function such that $g(x, y) = xy - 1$, and let f be the function such that $f(t) = \ln t$.

$$\lim_{(x,y)\to(2,1)} (xy - 1) = 1$$

and because f is continuous at 1, from Theorem 15.2.6,

$$\lim_{(x,y)\to(2,1)} \ln(xy - 1) = \ln\left(\lim_{(x,y)\to(2,1)} (xy - 1) \right)$$

$$= \ln 1$$

$$= 0$$

We now introduce the concept of an *accumulation point*, which is needed to continue the discussion of limits of functions of two variables.

15.2.7 DEFINITION   A point P_0 is said to be an **accumulation point** of a set S of points in R^n if every open ball $B(P_0; r)$ contains infinitely many points of S.

▶ ILLUSTRATION 3   If S is the set of all points in R^2 on the positive side of the x axis, the origin will be an accumulation point of S because no matter how small we take the value of r, every open disk having its center at the origin and radius r will contain infinitely many points of S. This is an example of a set having an accumulation point for which the accumulation point is not a point of the set. Any point of this set S also will be an accumulation point of S.   ◀

▶ ILLUSTRATION 4   If S is the set of all points in R^2 for which the cartesian coordinates are positive integers, then this set has no accumulation point. This can be seen by considering the point (m, n), where m and n are positive integers. Then an open disk having its center at (m, n) and radius less than 1 will contain no points of S other than (m, n); therefore Definition 15.2.7 will not be satisfied (see Fig. 15.2.8).   ◀

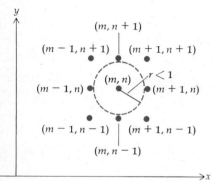

FIGURE 15.2.8

We now consider the limit of a function of two variables as a point (x, y) approaches a point (x_0, y_0), where (x, y) is restricted to a specific set of points.

15.2.8 DEFINITION   Let f be a function defined on a set of points S in R^2, and let (x_0, y_0) be an accumulation point of S. Then the **limit of $f(x, y)$ as (x, y) approaches (x_0, y_0) in S is L**, written as

$$\lim_{\substack{(x,y)\to(x_0,y_0) \\ (P \text{ in } S)}} f(x, y) = L \tag{14}$$

if for any $\epsilon > 0$, however small, there exists a $\delta > 0$ such that

$$\text{if}\quad 0 < \|(x, y) - (x_0, y_0)\| < \delta \quad \text{then}\quad |f(x, y) - L| < \epsilon$$

and (x, y) is in S.

In some cases the limit in (14) becomes the limit of a function of one variable. For example, consider $\lim\limits_{(x,y)\to(0,0)} f(x, y)$. Then if S_1 is the set of all points on the positive side of the x axis,

$$\lim_{\substack{(x,y)\to(0,0) \\ (P \text{ in } S_1)}} f(x, y) = \lim_{x\to 0^+} f(x, 0)$$

If S_2 is the set of all points on the negative side of the y axis,

$$\lim_{\substack{(x,y)\to(0,0) \\ (P \text{ in } S_2)}} f(x, y) = \lim_{y\to 0^-} f(0, y)$$

If S_3 is the set of all points on the x axis,

$$\lim_{\substack{(x,y)\to(0,0) \\ (P \text{ in } S_3)}} f(x, y) = \lim_{x\to 0} f(x, 0)$$

If S_4 is the set of all points on the parabola $y = x^2$,

$$\lim_{\substack{(x,y)\to(0,0) \\ (P \text{ in } S_4)}} f(x, y) = \lim_{x\to 0} f(x, x^2)$$

15.2.9 THEOREM   Suppose that the function f is defined for all points on an open disk having its center at (x_0, y_0), except possibly at (x_0, y_0) itself, and

$$\lim_{(x,y)\to(x_0,y_0)} f(x, y) = L$$

Then if S is any set of points in R^2 having (x_0, y_0) as an accumulation point,

$$\lim_{\substack{(x,y)\to(x_0,y_0) \\ (P \text{ in } S)}} f(x, y)$$

exists and always has the value L.

PROOF   Because $\lim\limits_{(x,y)\to(x_0,y_0)} f(x, y) = L$, then by Definition 15.2.5, for any $\epsilon > 0$ there exists a $\delta > 0$ such that

$$\text{if } 0 < \|(x, y) - (x_0, y_0)\| < \delta \quad \text{then} \quad |f(x, y) - L| < \epsilon$$

The above will be true if we further restrict (x, y) by the requirement that (x, y) be in a set S, where S is any set of points having (x_0, y_0) as an accumulation point. Therefore, by Definition 15.2.8,

$$\lim_{\substack{(x,y)\to(x_0,y_0) \\ (P \text{ in } S)}} f(x, y) = L$$

and L does not depend on the set S through which (x, y) is approaching (x_0, y_0). This proves the theorem.   ∎

The following is an immediate consequence of Theorem 15.2.9.

15.2.10 THEOREM   If the function f has different limits as (x, y) approaches (x_0, y_0) through two distinct sets of points having (x_0, y_0) as an accumulation point, then $\lim\limits_{(x,y)\to(x_0,y_0)} f(x, y)$ does not exist.

PROOF   Let S_1 and S_2 be two distinct sets of points in R^2 having (x_0, y_0) as an accumulation point, and let

$$\lim_{\substack{(x,y)\to(x_0,y_0) \\ (P \text{ in } S_1)}} f(x, y) = L_1 \quad \text{and} \quad \lim_{\substack{(x,y)\to(x_0,y_0) \\ (P \text{ in } S_2)}} f(x, y) = L_2$$

Now assume that $\lim\limits_{(x,y)\to(x_0,y_0)} f(x, y)$ exists. Then by Theorem 15.2.9 L_1 must equal L_2, but by hypothesis $L_1 \neq L_2$, and so we have a contradiction. Therefore $\lim\limits_{(x,y)\to(x_0,y_0)} f(x, y)$ does not exist.   ∎

EXAMPLE 4
Given

$$f(x, y) = \frac{xy}{x^2 + y^2}$$

find $\lim\limits_{(x,y)\to(0,0)} f(x, y)$ if it exists.

SOLUTION
The function f is defined at all points in R^2 except at $(0, 0)$. Let S_1 be the set of all points on the x axis, and S_2 be the set of all points on the line $y = x$. Then

$$\lim_{\substack{(x,y)\to(0,0) \\ (P\ in\ S_1)}} f(x, y) = \lim_{x\to 0} f(x, 0) \qquad \lim_{\substack{(x,y)\to(0,0) \\ (P\ in\ S_2)}} f(x, y) = \lim_{x\to 0} f(x, x)$$

$$= \lim_{x\to 0} \frac{0}{x^2 + 0} \qquad\qquad = \lim_{x\to 0} \frac{x^2}{x^2 + x^2}$$

$$= \lim_{x\to 0} 0 \qquad\qquad\qquad = \lim_{x\to 0} \tfrac{1}{2}$$

$$= 0 \qquad\qquad\qquad\qquad = \tfrac{1}{2}$$

Because

$$\lim_{\substack{(x,y)\to(0,0) \\ (P\ in\ S_1)}} f(x, y) \neq \lim_{\substack{(x,y)\to(0,0) \\ (P\ in\ S_2)}} f(x, y)$$

it follows from Theorem 15.2.10 that $\lim\limits_{(x,y)\to(0,0)} f(x, y)$ does not exist.

EXAMPLE 5
Given

$$f(x, y) = \frac{3x^2 y}{x^2 + y^2}$$

find $\lim\limits_{(x,y)\to(0,0)} f(x, y)$ if it exists.

SOLUTION
The function f is defined everywhere in R^2 except at the origin. Let S_1 be the set of all points on the x axis. Then

$$\lim_{\substack{(x,y)\to(0,0) \\ (P\ in\ S_1)}} f(x, y) = \lim_{x\to 0} \frac{0}{x^2 + 0}$$

$$= 0$$

Let S_2 be the set of all points on any line through the origin; that is, for any point (x, y) in S_2, $y = mx$.

$$\lim_{\substack{(x,y)\to(0,0)\\(P \text{ in } S_2)}} f(x, y) = \lim_{x\to 0} \frac{3x^2(mx)}{x^2 + m^2x^2}$$

$$= \lim_{x\to 0} \frac{3mx}{1 + m^2}$$

$$= 0$$

Even though the same limit of 0 is obtained if (x, y) approaches $(0, 0)$ through a set of points on any line through the origin, we cannot conclude that $\lim_{(x,y)\to(0,0)} f(x, y)$ exists and is zero (see Example 6). However, let us attempt to prove that $\lim_{(x,y)\to(0,0)} f(x, y) = 0$. Any open disk having its center at the origin will satisfy the first requirement of Definition 15.2.5. If we can show that for any $\epsilon > 0$ there exists a $\delta > 0$ such that

$$\text{if } 0 < \sqrt{x^2 + y^2} < \delta \quad \text{then} \quad \left| \frac{3x^2y}{x^2 + y^2} \right| < \epsilon \tag{15}$$

then we have proved that $\lim_{(x,y)\to(0,0)} f(x, y) = 0$

Because $x^2 \leq x^2 + y^2$ and $|y| \leq \sqrt{x^2 + y^2}$

$$\left| \frac{3x^2y}{x^2 + y^2} \right| = \frac{3x^2|y|}{x^2 + y^2}$$

$$\leq \frac{3(x^2 + y^2)\sqrt{x^2 + y^2}}{x^2 + y^2}$$

$$= 3\sqrt{x^2 + y^2}$$

Thus a suitable choice for δ is $3\delta = \epsilon$, that is, $\delta = \frac{1}{3}\epsilon$. With this δ we have the following argument:

$$0 < \sqrt{x^2 + y^2} < \delta$$

$$\Rightarrow \quad \frac{3(x^2 + y^2)\sqrt{x^2 + y^2}}{x^2 + y^2} < 3\delta$$

$$\Rightarrow \quad \frac{3x^2|y|}{x^2 + y^2} < 3(\tfrac{1}{3}\epsilon)$$

$$\Rightarrow \quad \left| \frac{3x^2y}{x^2 + y^2} \right| < \epsilon$$

Thus if $\delta = \frac{1}{3}\epsilon$, statement (15) holds. Hence we have proved that $\lim_{(x,y)\to(0,0)} f(x, y) = 0$.

EXAMPLE 6
Given

$$f(x, y) = \frac{x^2 y}{x^4 + y^2}$$

find $\lim\limits_{(x,y)\to(0,0)} f(x, y)$ if it exists.

SOLUTION
The function f is defined everywhere in R^2 except at $(0,0)$. Let S_1 be the set of all points on either the x axis or the y axis. So if (x, y) is in S_1, $xy = 0$. Therefore

$$\lim_{\substack{(x,y)\to(0,0)\\(P \text{ in } S_1)}} f(x, y) = 0$$

Let S_2 be the set of all points on any line through the origin; so if (x, y) is a point in S_2, $y = mx$. Let S_3 be the set of all points on the parabola $y = x^2$. Then

$$\lim_{\substack{(x,y)\to(0,0)\\(P \text{ in } S_2)}} f(x, y) = \lim_{x\to 0} f(x, mx) \qquad \lim_{\substack{(x,y)\to(0,0)\\(P \text{ in } S_3)}} f(x, y) = \lim_{x\to 0} f(x, x^2)$$

$$= \lim_{x\to 0} \frac{mx^3}{x^4 + m^2 x^2} \qquad\qquad = \lim_{x\to 0} \frac{x^4}{x^4 + x^4}$$

$$= \lim_{x\to 0} \frac{mx}{x^2 + m^2} \qquad\qquad = \lim_{x\to 0} \tfrac{1}{2}$$

$$= 0 \qquad\qquad\qquad = \tfrac{1}{2}$$

Because

$$\lim_{\substack{(x,y)\to(0,0)\\(P \text{ in } S_3)}} f(x, y) \neq \lim_{\substack{(x,y)\to(0,0)\\(P \text{ in } S_1)}} f(x, y)$$

it follows that $\lim\limits_{(x,y)\to(0,0)} f(x, y)$ does not exist.

EXAMPLE 7
Given

$$f(x, y) = \begin{cases} (x + y) \sin \dfrac{1}{x} & \text{if } x \neq 0 \\[2mm] 0 & \text{if } x = 0 \end{cases}$$

find $\lim\limits_{(x,y)\to(0,0)} f(x, y)$ if it exists.

SOLUTION
The function f is defined at all points in R^2. Let S_1 be the set of all points on the y axis. Then

$$\lim_{\substack{(x,y)\to(0,0)\\(P \text{ in } S_1)}} f(x, y) = \lim_{x\to 0} 0$$

$$= 0$$

Let S_2 be the set of all points on any line through the origin except points on the y axis; that is, if (x, y) is a point in S_2, $y = kx$, where $x \neq 0$. Then

$$\lim_{\substack{(x,y) \to (0,0) \\ (P \text{ in } S_2)}} f(x, y) = \lim_{x \to 0} (x + kx) \sin \frac{1}{x}$$

To find the above limit we make use of the fact that $\lim\limits_{x \to 0} (x + kx) = 0$. Because $\lim\limits_{x \to 0} \sin(1/x)$ does not exist, the theorem about the limit of a product cannot be applied. However, because $\lim\limits_{x \to 0} (x + kx) = 0$, it follows that for any $\epsilon > 0$ there exists a $\delta > 0$ such that

$$\text{if } 0 < |x| < \delta \quad \text{then} \quad |x + kx| < \epsilon$$

A δ is, in fact, $\epsilon/|1 + k|$.

Because

$$\left| (x + kx) \sin \frac{1}{x} \right| = |x| |1 + k| \left| \sin \frac{1}{x} \right|$$

$$\leq |x| |1 + k| |1|$$

then if $\delta = \epsilon/|1 + k|$

$$0 < |x| < \delta$$

$$\Rightarrow \quad \left| (x + kx) \sin \frac{1}{x} \right| < \delta |1 + k|$$

$$\Rightarrow \quad \left| (x + kx) \sin \frac{1}{x} \right| < \frac{\epsilon}{|1 + k|} \cdot |1 + k|$$

$$\Rightarrow \quad \left| (x + kx) \sin \frac{1}{x} \right| < \epsilon$$

Thus

$$\lim_{\substack{(x,y) \to (0,0) \\ (P \text{ in } S_2)}} f(x, y) = 0 \tag{16}$$

Let S_3 be the set of all points (x, y) for which $y = kx^n$, where n is any positive integer and $x \neq 0$. Then by an argument similar to that used for proving (16) it follows that

$$\lim_{\substack{(x,y) \to (0,0) \\ (P \text{ in } S_3)}} f(x, y) = \lim_{x \to 0} (x + kx^n) \sin \frac{1}{x}$$

$$= 0$$

We now attempt to find a $\delta > 0$ for any $\epsilon > 0$ such that

$$\text{if } 0 < \|(x, y) - (0,0)\| < \delta \quad \text{then} \quad |f(x, y) - 0| < \epsilon \tag{17}$$

which will prove that $\lim\limits_{(x,y) \to (0,0)} f(x, y) = 0$. We distinguish two cases: $x = 0$ and $x \neq 0$.

Case 1: If $x = 0$, $|f(x, y) - 0| = |0 - 0|$, which is less than ϵ for any $\delta > 0$.

Case 2: If $x \neq 0$, $|f(x, y) - 0| = |(x + y) \sin(1/x)|$.

$$\left|(x + y) \sin \frac{1}{x}\right| = |x + y|\left|\sin \frac{1}{x}\right|$$

$$\leq |x + y|(1)$$
$$\leq |x| + |y|$$
$$\leq \sqrt{x^2 + y^2} + \sqrt{x^2 + y^2}$$
$$= 2\sqrt{x^2 + y^2}$$

Then if $\delta = \frac{1}{2}\epsilon$

$$0 < \sqrt{x^2 + y^2} < \delta$$

$$\Rightarrow \quad \left|(x + y) \sin \frac{1}{x}\right| < 2(\tfrac{1}{2}\epsilon)$$

$$\Rightarrow \quad \left|(x + y) \sin \frac{1}{x}\right| < \epsilon$$

Therefore, in both cases we have found a $\delta > 0$ for any $\epsilon > 0$ such that (17) holds, and this proves that $\lim\limits_{(x,y)\to(0,0)} f(x, y) = 0$.

EXERCISES 15.2

In Exercises 1 through 8, evaluate the given limit by use of limit theorems.

1. $\lim\limits_{(x,y)\to(2,3)} (3x^2 + xy - 2y^2)$

2. $\lim\limits_{(x,y)\to(-1,4)} (5x^2 - 2xy + y^2)$

3. $\lim\limits_{(x,y)\to(2,-1)} \dfrac{3x - 2y}{x + 4y}$

4. $\lim\limits_{(x,y)\to(-2,4)} y\sqrt[3]{x^3 + 2y}$

5. $\lim\limits_{(x,y)\to(0,0)} \dfrac{e^x + e^y}{\cos x + \sin y}$

6. $\lim\limits_{(x,y)\to(0,0)} \dfrac{\sin^2 x + \cos^2 y}{e^{2x} + e^{2y}}$

7. $\lim\limits_{(x,y)\to(0,1)} \dfrac{x^4 - (y - 1)^4}{x^2 + (y - 1)^2}$

8. $\lim\limits_{(x,y)\to(1,1)} \dfrac{(x - 1)^{4/3} - (y - 1)^{4/3}}{(x - 1)^{2/3} + (y - 1)^{2/3}}$

In Exercises 9 through 16, establish the limit by finding a $\delta > 0$ for any $\epsilon > 0$ such that Definition 15.2.5 holds.

9. $\lim\limits_{(x,y)\to(3,2)} (3x - 4y) = 1$

10. $\lim\limits_{(x,y)\to(2,4)} (5x - 3y) = -2$

11. $\lim\limits_{(x,y)\to(-1,3)} (3x - 2y) = -9$

12. $\lim\limits_{(x,y)\to(-2,1)} (5x + 4y) = -6$

13. $\lim\limits_{(x,y)\to(1,1)} (x^2 + y^2) = 2$

14. $\lim\limits_{(x,y)\to(2,3)} (2x^2 - y^2) = -1$

15. $\lim\limits_{(x,y)\to(2,4)} (x^2 + 2x - y) = 4$

16. $\lim\limits_{(x,y)\to(3,-1)} (x^2 + y^2 - 4x + 2y) = -4$

In Exercises 17 through 22, prove that for the given function f, $\lim\limits_{(x,y)\to(0,0)} f(x, y)$ does not exist.

17. $f(x, y) = \dfrac{x^2 - y^2}{x^2 + y^2}$

18. $f(x, y) = \dfrac{x^2}{x^2 + y^2}$

19. $f(x, y) = \dfrac{x^4 y^4}{(x^2 + y^4)^3}$

20. $f(x, y) = \dfrac{x^4 + 3x^2y^2 + 2xy^3}{(x^2 + y^2)^2}$ **21.** $f(x, y) = \dfrac{x^9 y}{(x^6 + y^2)^2}$ **22.** $f(x, y) = \dfrac{x^2 y^2}{x^4 + y^4}$

In Exercises 23 through 26, prove that $\lim\limits_{(x,y) \to (0,0)} f(x, y)$ exists.

23. $f(x, y) = \dfrac{xy}{\sqrt{x^2 + y^2}}$ **24.** $f(x, y) = \dfrac{x^3 + y^3}{x^2 + y^2}$

25. $f(x, y) = \begin{cases} (x + y) \sin \dfrac{1}{x} \sin \dfrac{1}{y} & \text{if } x \neq 0 \text{ and } y \neq 0 \\ 0 & \text{if either } x = 0 \text{ or } y = 0 \end{cases}$ **26.** $f(x, y) = \begin{cases} \dfrac{1}{x} \sin(xy) & \text{if } x \neq 0 \\ y & \text{if } x = 0 \end{cases}$

In Exercises 27 through 29, show the application of Theorem 15.2.6 to find the indicated limit.

27. $\lim\limits_{(x,y) \to (2,2)} \tan^{-1} \dfrac{y}{x}$ **28.** $\lim\limits_{(x,y) \to (-2,3)} \left[\!\left[5x + \tfrac{1}{2}y^2 \right]\!\right]$ **29.** $\lim\limits_{(x,y) \to (4,2)} \sqrt{\dfrac{1}{3x - 4y}}$

In Exercises 30 through 35, determine if the indicated limit exists.

30. $\lim\limits_{(x,y) \to (0,0)} \dfrac{x^2 y^2}{x^4 + y^4}$ **31.** $\lim\limits_{(x,y) \to (0,0)} \dfrac{x^2 y^2}{x^2 + y^2}$ **32.** $\lim\limits_{(x,y) \to (0,0)} \dfrac{x^2 y^4}{x^4 + y^4}$ **33.** $\lim\limits_{(x,y) \to (0,0)} \dfrac{x^2 + y}{x^2 + y^2}$

34. $f(x, y) = \begin{cases} \dfrac{xy}{x^2 + y^2} + y \sin \dfrac{1}{x} & \text{if } x \neq 0 \\ 0 & \text{if } x = 0 \end{cases}$; $\lim\limits_{(x,y) \to (0,0)} f(x, y)$

35. $f(x, y) = \begin{cases} x \sin \dfrac{1}{y} + y \sin \dfrac{1}{x} & \text{if } x \neq 0 \text{ and } y \neq 0 \\ 0 & \text{if either } x = 0 \text{ or } y = 0 \end{cases}$; $\lim\limits_{(x,y) \to (0,0)} f(x, y)$

36. (a) Give a definition, similar to Definition 15.2.5, of the limit of a function of three variables as a point (x, y, z) approaches a point (x_0, y_0, z_0). (b) Give a definition, similar to Definition 15.2.8, of the limit of a function of three variables as a point (x, y, z) approaches a point (x_0, y_0, z_0) in a specific set of points S in R^3.

37. (a) State and prove a theorem similar to Theorem 15.2.9 for a function f of three variables. (b) State and prove a theorem similar to Theorem 15.2.10 for a function f of three variables.

In Exercises 38 through 41, evaluate the given limit by the use of limit theorems.

38. $\lim\limits_{(x,y,z) \to (-2,1,4)} (4x^2 y - 3xyz^2 + 7y^2 z^3)$ **39.** $\lim\limits_{(x,y,z) \to (\pi/3, 1, \pi)} \dfrac{\sec xy + \sec yz}{y - \sec z}$

40. $\lim\limits_{(x,y,z) \to (0,2,0)} \dfrac{x^2 y^2 + y^2 z^2}{x^2 + z^2}$ **41.** $\lim\limits_{(x,y,z) \to (0,0,0)} \dfrac{(e^x + e^y + e^z)^2}{e^{2x} + e^{2y} + e^{2z}}$

In Exercises 42 through 45, use the definitions and theorems of Exercises 36 and 37 to prove that $\lim\limits_{(x,y,z) \to (0,0,0)} f(x, y, z)$ does not exist.

42. $f(x, y, z) = \dfrac{x^3 + yz^2}{x^4 + y^2 + z^4}$ **43.** $f(x, y, z) = \dfrac{x^2 + y^2 - z^2}{x^2 + y^2 + z^2}$

44. $f(x, y, z) = \dfrac{x^4 + yx^3 + z^2 x^2}{x^4 + y^4 + z^4}$ **45.** $f(x, y, z) = \dfrac{x^2 y^2 z^2}{x^6 + y^6 + z^6}$

In Exercises 46 and 47, use the definition in Exercise 36(a) to prove that $\lim\limits_{(x,y,z)\to(0,0,0)} f(x, y, z)$ exists.

46. $f(x, y, z) = \dfrac{y^3 + xz^2}{x^2 + y^2 + z^2}$

47. $f(x, y, z) = \begin{cases} (x + y + z) \sin \dfrac{1}{x} \sin \dfrac{1}{y} & \text{if } x \neq 0 \text{ and } y \neq 0 \\ 0 & \text{if either } x = 0 \text{ or } y = 0 \end{cases}$

48. Suppose that f and g are functions of two variables satisfying the following conditions:

    (i) $f(tx, ty) = t^n f(x, y); \; g(tx, ty) = t^n g(x, y)$ for some n and for all t;
    (ii) $g(1, 1) \neq 0$ and $g(1, 0) \neq 0$;
    (iii) $g(1, 1) \cdot f(1, 0) \neq g(1, 0) \cdot f(1, 1)$.

    Show that $\lim\limits_{(x,y)\to(0,0)} \dfrac{f(x, y)}{g(x, y)}$ does not exist.

49. Prove Theorem 15.2.6.

15.3
CONTINUITY OF FUNCTIONS OF MORE THAN ONE VARIABLE

We now define continuity of a function of n variables at a point in R^n. Observe that Definition 1.6.1 of continuity of a function of one variable at a number a is a special case of this definition.

15.3.1 DEFINITION   Suppose that f is a function of n variables and A is a point in R^n. Then f is said to be **continuous** at the point A if and only if the following three conditions are satisfied:

    (i) $f(A)$ exists;

    (ii) $\lim\limits_{P \to A} f(P)$ exists;

    (iii) $\lim\limits_{P \to A} f(P) = f(A)$.

If one or more of these three conditions fails to hold at the point A, then f is said to be **discontinuous** at A.

If f is a function of two variables, A is the point (x_0, y_0), and P is a point (x, y), then Definition 15.3.1 becomes the following.

15.3.2 DEFINITION   The function f of two variables x and y is said to be **continuous** at the point (x_0, y_0) if and only if the following three conditions are satisfied:

    (i) $f(x_0, y_0)$ exists;

    (ii) $\lim\limits_{(x,y)\to(x_0,y_0)} f(x, y)$ exists;

    (iii) $\lim\limits_{(x,y)\to(x_0,y_0)} f(x, y) = f(x_0, y_0)$.

EXAMPLE 1
Given

$$f(x, y) = \begin{cases} \dfrac{3x^2y}{x^2 + y^2} & \text{if } (x, y) \neq (0,0) \\ 0 & \text{if } (x, y) = (0,0) \end{cases}$$

determine if f is continuous at $(0,0)$.

SOLUTION
We check the three conditions of Definition 15.3.2 at the point $(0,0)$.

(i) $f(0,0) = 0$. Therefore condition (i) holds.

(ii) $\displaystyle \lim_{(x,y)\to(0,0)} f(x, y) = \lim_{(x,y)\to(0,0)} \frac{3x^2y}{x^2 + y^2}$

$$= 0$$

This fact was proved in Example 5, Section 15.2.

(iii) $\displaystyle \lim_{(x,y)\to(0,0)} f(x, y) = f(0,0)$

Therefore f is continuous at $(0,0)$.

EXAMPLE 2
Let the function f be defined by

$$f(x, y) = \begin{cases} \dfrac{xy}{x^2 + y^2} & \text{if } (x, y) \neq (0,0) \\ 0 & \text{if } (x, y) = (0,0) \end{cases}$$

Determine if f is continuous at $(0,0)$.

SOLUTION
Checking the conditions of Definition 15.3.2, we have the following:

(i) $f(0,0) = 0$; so condition (i) holds.
(ii) When $(x, y) \neq (0,0)$, $f(x, y) = xy/(x^2 + y^2)$. In Example 4, Section 15.2, we showed that $\displaystyle \lim_{(x,y)\to(0,0)} xy/(x^2 + y^2)$ does not exist; so $\displaystyle \lim_{(x,y)\to(0,0)} f(x, y)$ does not exist. Therefore condition (ii) fails to hold.

Thus f is discontinuous at $(0,0)$.

If a function f of two variables is discontinuous at the point (x_0, y_0) but $\displaystyle \lim_{(x,y)\to(x_0,y_0)} f(x, y)$ exists, then f is said to have a **removable discontinuity** at (x_0, y_0) because if f is redefined at (x_0, y_0) so that

$$f(x_0, y_0) = \lim_{(x,y)\to(x_0,y_0)} f(x, y)$$

then the new function is continuous at (x_0, y_0). If the discontinuity is not removable, it is called an **essential discontinuity**.

▶ ILLUSTRATION 1   (a) If $g(x, y) = 3x^2y/(x^2 + y^2)$, then g is discontinuous at the origin because $g(0, 0)$ is not defined. However, in Example 5, Section 15.2, we showed that $\displaystyle\lim_{(x,y)\to(0,0)} 3x^2y/(x^2 + y^2) = 0$. Therefore the discontinuity is removable if $g(0, 0)$ is redefined to be 0. (Refer to Example 1.)

(b) Let $h(x, y) = xy/(x^2 + y^2)$. Then h is discontinuous at the origin because $h(0, 0)$ is not defined. In Example 4, Section 15.2, we showed that $\displaystyle\lim_{(x,y)\to(0,0)} xy/(x^2 + y^2)$ does not exist. Therefore the discontinuity is essential. (Refer to Example 2.)   ◀

The theorems about continuity for functions of a single variable can be extended to functions of two variables.

15.3.3 THEOREM   If f and g are two functions that are continuous at the point (x_0, y_0), then

(i) $f + g$ is continuous at (x_0, y_0);
(ii) $f - g$ is continuous at (x_0, y_0);
(iii) fg is continuous at (x_0, y_0);
(iv) f/g is continuous at (x_0, y_0), provided that $g(x_0, y_0) \neq 0$.

The proof of this theorem is analogous to the proof of the corresponding theorem (1.6.2) for functions of one variable, and hence it is omitted.

15.3.4 THEOREM   A polynomial function of two variables is continuous at every point in R^2.

PROOF   Every polynomial function is the sum of products of the functions defined by $f(x, y) = x, g(x, y) = y$, and $h(x, y) = c$, where c is a real number. Because f, g, and h are continuous at every point in R^2, the theorem follows by repeated applications of Theorem 15.3.3, parts (i) and (iii).   ■

15.3.5 THEOREM   A rational function of two variables is continuous at every point in its domain.

PROOF   A rational function is the quotient of two polynomial functions f and g that are continuous at every point in R^2, by Theorem 15.3.4. If (x_0, y_0) is any point in the domain of f/g, then $g(x_0, y_0) \neq 0$; so by Theorem 15.3.3(iv) f/g is continuous there.   ■

EXAMPLE 3
Let the function f be defined by

$$f(x, y) = \begin{cases} x^2 + y^2 & \text{if } x^2 + y^2 \leq 1 \\ 0 & \text{if } x^2 + y^2 > 1 \end{cases}$$

Discuss the continuity of f. What is the region of continuity of f?

SOLUTION

The function f is defined at all points in R^2. Therefore condition (i) of Definition 15.3.2 holds for every point (x_0, y_0).

Consider the points (x_0, y_0) if $x_0^2 + y_0^2 \neq 1$.

If $x_0^2 + y_0^2 < 1$,

$$\lim_{(x,y)\to(x_0,y_0)} f(x, y) = \lim_{(x,y)\to(x_0,y_0)} (x^2 + y^2)$$

$$= x_0^2 + y_0^2$$

$$= f(x_0, y_0)$$

If $x_0^2 + y_0^2 > 1$,

$$\lim_{(x,y)\to(x_0,y_0)} f(x, y) = \lim_{(x,y)\to(x_0,y_0)} 0$$

$$= 0$$

$$= f(x_0, y_0)$$

Thus f is continuous at all points (x_0, y_0) for which $x_0^2 + y_0^2 \neq 1$.

To determine the continuity of f at all points (x_0, y_0) for which $x_0^2 + y_0^2 = 1$, we determine if $\lim\limits_{(x,y)\to(x_0,y_0)} f(x, y)$ exists and equals 1.

Let S_1 be the set of all points (x, y) such that $x^2 + y^2 \leq 1$, and S_2 be the set of all points (x, y) such that $x^2 + y^2 > 1$. Then

$$\lim_{\substack{(x,y)\to(x_0,y_0)\\ (P \text{ in } S_1)}} f(x, y) = \lim_{\substack{(x,y)\to(x_0,y_0)\\ (P \text{ in } S_1)}} (x^2 + y^2) \qquad \lim_{\substack{(x,y)\to(x_0,y_0)\\ (P \text{ in } S_2)}} f(x, y) = \lim_{\substack{(x,y)\to(x_0,y_0)\\ (P \text{ in } S_2)}} 0$$

$$= x_0^2 + y_0^2 \qquad\qquad = 0$$

$$= 1$$

Because

$$\lim_{\substack{(x,y)\to(x_0,y_0)\\ (P \text{ in } S_1)}} f(x, y) \neq \lim_{\substack{(x,y)\to(x_0,y_0)\\ (P \text{ in } S_2)}} f(x, y)$$

we conclude that $\lim\limits_{(x,y)\to(x_0,y_0)} f(x, y)$ does not exist. Hence f is discontinuous at all points (x_0, y_0) for which $x_0^2 + y_0^2 = 1$. The region of continuity of f consists of all points in the xy plane except those on the circle $x^2 + y^2 = 1$.

15.3.6 DEFINITION   The function f of n variables is said to be **continuous on an open ball** if it is continuous at every point of the open ball.

As an illustration of the above definition, the function of Example 3 is continuous on every open disk that does not contain a point of the circle $x^2 + y^2 = 1$.

The following theorem states that a continuous function of a continuous function is continuous. It is analogous to Theorem 1.7.2.

15.3.7 THEOREM   Suppose that f is a function of a single variable and g is a function of two variables. Suppose further that g is continuous at (x_0, y_0) and f is continuous at $g(x_0, y_0)$. Then the composite function $f \circ g$ is continuous at (x_0, y_0).

The proof of this theorem, which makes use of Theorem 15.2.6, is similar to the proof of Theorem 1.7.2 and is left as an exercise (see Exercise 15).

▶ ILLUSTRATION 2   Let

$$h(x, y) = \ln(xy - 1)$$

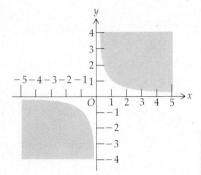

FIGURE 15.3.1

We discuss the continuity of h. If g is the function defined by $g(x, y) = xy - 1$, g is continuous at all points in R^2. The natural logarithmic function is continuous on its entire domain, which is the set of all positive numbers. So if f is the function defined by $f(t) = \ln t$, f is continuous for all $t > 0$. Then the function h is the composite function $f \circ g$ and, by Theorem 15.3.7, is continuous at all points (x, y) in R^2 for which $xy - 1 > 0$ or, equivalently, $xy > 1$. The shaded region of Fig. 15.3.1 is the region of continuity of h. ◀

EXAMPLE 4
Given

$$f(x, y) = \frac{1}{\sqrt{x^2 + y^2 - 25}}$$

discuss the continuity of f, and draw a sketch showing as a shaded region in R^2 the region of continuity of f.

SOLUTION
The domain of f is the set of all points in the exterior region bounded by the circle $x^2 + y^2 = 25$. Function f is the quotient of functions g and h for which

$$g(x, y) = 1 \qquad h(x, y) = \sqrt{x^2 + y^2 - 25}$$

The function g is a constant function and is therefore continuous everywhere. It follows from Theorem 15.3.7 that h is continuous at all points in R^2 for which $x^2 + y^2 > 25$. Therefore by Theorem 15.3.3(iv), f is continuous at all points in its domain. The shaded region of Fig. 15.3.2 is the region of continuity of f.

FIGURE 15.3.2

EXERCISES 15.3
In Exercises 1 through 14, discuss the continuity of the given function.

1. $f(x, y) = \dfrac{x^2}{y - 1}$ **2.** $F(x, y) = \dfrac{1}{x - y}$ **3.** $h(x, y) = \sin \dfrac{y}{x}$ **4.** $f(x, y) = \ln xy^2$

5. $f(x, y) = \dfrac{4x^2y + 3y^2}{2x - y}$ **6.** $g(x, y) = \dfrac{5xy^2 + 2y}{16 - x^2 - 4y^2}$ **7.** $g(x, y) = \ln(25 - x^2 - y^2)$ **8.** $f(x, y) = \cos^{-1}(x + y)$

9. $f(x, y) = \begin{cases} \dfrac{xy}{\sqrt{x^2 + y^2}} & \text{if } (x, y) \neq (0, 0) \\ 0 & \text{if } (x, y) = (0, 0) \end{cases}$ (Hint: See Exercise 23, Exercises 15.2.)

10. $h(x, y) = \begin{cases} \dfrac{x^2y}{x^4 + y^2} & \text{if } (x, y) \neq (0, 0) \\ 0 & \text{if } (x, y) = (0, 0) \end{cases}$ (Hint: See Example 6, Section 15.2.)

11. $f(x, y) = \begin{cases} \dfrac{x + y}{x^2 + y^2} & \text{if } (x, y) \neq (0, 0) \\ 0 & \text{if } (x, y) = (0, 0) \end{cases}$ **12.** $f(x, y) = \begin{cases} \dfrac{x^3 + y^3}{x^2 + y^2} & \text{if } (x, y) \neq (0, 0) \\ 0 & \text{if } (x, y) = (0, 0) \end{cases}$

13. $G(x, y) = \begin{cases} \dfrac{xy}{|x| + |y|} & \text{if } (x, y) \neq (0, 0) \\ 0 & \text{if } (x, y) = (0, 0) \end{cases}$

14. $F(x, y) = \begin{cases} \dfrac{x^2 y^2}{|x^3| + |y^3|} & \text{if } (x, y) \neq (0, 0) \\ 0 & \text{if } (x, y) = (0, 0) \end{cases}$

15. Prove Theorem 15.3.7.

In Exercises 16 through 25, determine the region of continuity of f, and draw a sketch showing as a shaded region in R^2 the region of continuity of f.

16. $f(x, y) = \dfrac{y}{\sqrt{x^2 - y^2 - 4}}$

17. $f(x, y) = \dfrac{xy}{\sqrt{16 - x^2 - y^2}}$

18. $f(x, y) = \dfrac{x^2 + y^2}{\sqrt{9 - x^2 - y^2}}$

19. $f(x, y) = \dfrac{x}{\sqrt{4x^2 + 9y^2 - 36}}$

20. $f(x, y) = \ln(x^2 + y^2 - 9) - \ln(1 - x^2 - y^2)$
22. $f(x, y) = \sin^{-1}(xy)$

21. $f(x, y) = \sec^{-1}(xy)$
23. $f(x, y) = \sin^{-1}(x + y) + \ln(xy)$

24. $f(x, y) = \begin{cases} \dfrac{x^2 - y^2}{x - y} & \text{if } x \neq y \\ x - y & \text{if } x = y \end{cases}$

25. $f(x, y) = \begin{cases} \dfrac{\sin(x + y)}{x + y} & \text{if } x + y \neq 0 \\ 1 & \text{if } x + y = 0 \end{cases}$

In Exercises 26 through 31, the function is discontinuous at the origin because $f(0, 0)$ does not exist. Determine if the discontinuity is removable or essential. If the discontinuity is removable, redefine $f(0, 0)$ so that the new function is continuous at $(0, 0)$.

26. $f(x, y) = \dfrac{x}{x^2 + y^2}$

27. $f(x, y) = (x + y) \sin \dfrac{x}{x^2 + y^2}$

28. $f(x, y) = \dfrac{x^2 y^2}{x^2 + y^2}$

29. $f(x, y) = \dfrac{x^3 y^2}{x^6 + y^4}$

30. $f(x, y) = \dfrac{2y^2 - 3xy}{\sqrt{x^2 + y^2}}$

31. $f(x, y) = \dfrac{x^3 - 4xy^2}{x^2 + y^2}$

32. The function F is defined by

$$F(x, y) = \begin{cases} x^2 - 3y^2 & \text{if } x^2 - 3y^2 \leq 1 \\ 2 & \text{if } x^2 - 3y^2 > 1 \end{cases}$$

Show that the region of continuity of F consists of all points in R^2 except those on the hyperbola $x^2 - 3y^2 = 1$.

33. The function G is defined by

$$G(x, y) = \begin{cases} x^2 + 4y^2 & \text{if } x^2 + 4y^2 \leq 5 \\ 3 & \text{if } x^2 + 4y^2 > 5 \end{cases}$$

Show that the region of continuity of G consists of all points in R^2 except those on the ellipse $x^2 + 4y^2 = 5$.

34. (a) Give a definition of continuity at a point for a function of three variables, similar to Definition 15.3.2. (b) State theorems for functions of three variables similar to Theorems 15.3.3 and 15.3.7. (c) Define a polynomial function of three variables and a rational function of three variables.

In Exercises 35 through 38, use the definitions and theorems of Exercise 34 to discuss the continuity of the given function.

35. $f(x, y, z) = \dfrac{xz}{\sqrt{x^2 + y^2 + z^2 - 1}}$

36. $f(x, y, z) = \ln(36 - 4x^2 - y^2 - 9z^2)$

37. $f(x, y, z) = \begin{cases} \dfrac{3xyz}{x^2 + y^2 + z^2} & \text{if } (x, y, z) \neq (0, 0, 0) \\ 0 & \text{if } (x, y, z) = (0, 0, 0) \end{cases}$

38. $f(x, y, z) = \begin{cases} \dfrac{xz - y^2}{x^2 + y^2 + z^2} & \text{if } (x, y, z) \neq (0, 0, 0) \\ 0 & \text{if } (x, y, z) = (0, 0, 0) \end{cases}$

15.4
PARTIAL DERIVATIVES

The discussion of the differentiation of real-valued functions of n variables is reduced to the one-dimensional case by treating a function of n variables as a function of one variable at a time and holding the others fixed. This leads to the concept of a *partial derivative*. We first define the partial derivative of a function of two variables.

15.4.1 DEFINITION

Let f be a function of two variables, x and y. The **partial derivative of f with respect to x** is that function, denoted by $D_1 f$, such that its function value at any point (x, y) in the domain of f is given by

$$D_1 f(x, y) = \lim_{\Delta x \to 0} \frac{f(x + \Delta x, y) - f(x, y)}{\Delta x} \tag{1}$$

if this limit exists. Similarly, the **partial derivative of f with respect to y** is that function, denoted by $D_2 f$, such that its function value at any point (x, y) in the domain of f is given by

$$D_2 f(x, y) = \lim_{\Delta y \to 0} \frac{f(x, y + \Delta y) - f(x, y)}{\Delta y} \tag{2}$$

if this limit exists.

The process of finding a partial derivative is called **partial differentiation**.

$D_1 f$ is read as "D sub 1 of f," and this denotes the function that is the partial derivative of f with respect to the first variable. $D_1 f(x, y)$ is read as "D sub 1 of f of x and y," and this denotes the function value of $D_1 f$ at the point (x, y). Other notations for $D_1 f$ are f_1, f_x, and $\dfrac{\partial f}{\partial x}$. Other notations for $D_1 f(x, y)$ are $f_1(x, y), f_x(x, y)$, and $\dfrac{\partial f(x, y)}{\partial x}$. Similarly, other notations for $D_2 f$ are f_2, f_y, and $\dfrac{\partial f}{\partial y}$; other notations for $D_2 f(x, y)$ are $f_2(x, y), f_y(x, y)$, and $\dfrac{\partial f(x, y)}{\partial y}$.

If $z = f(x, y)$, we can write $\dfrac{\partial z}{\partial x}$ for $D_1 f(x, y)$. A partial derivative cannot be thought of as a ratio of ∂z and ∂x because neither of these symbols has a separate meaning. The notation $\dfrac{dy}{dx}$ can be regarded as the quotient of two differentials when y is a function of the single variable x, but there is not a similar interpretation for $\dfrac{\partial z}{\partial x}$.

EXAMPLE 1
Given

$$f(x, y) = 3x^2 - 2xy + y^2$$

find $D_1 f(x, y)$ and $D_2 f(x, y)$ by applying Definition 15.4.1.

SOLUTION

$$D_1 f(x, y) = \lim_{\Delta x \to 0} \frac{f(x + \Delta x, y) - f(x, y)}{\Delta x}$$

$$= \lim_{\Delta x \to 0} \frac{3(x + \Delta x)^2 - 2(x + \Delta x)y + y^2 - (3x^2 - 2xy + y^2)}{\Delta x}$$

$$= \lim_{\Delta x \to 0} \frac{3x^2 + 6x\,\Delta x + 3(\Delta x)^2 - 2xy - 2y\,\Delta x + y^2 - 3x^2 + 2xy - y^2}{\Delta x}$$

$$= \lim_{\Delta x \to 0} \frac{6x\,\Delta x + 3(\Delta x)^2 - 2y\,\Delta x}{\Delta x}$$

$$= \lim_{\Delta x \to 0} (6x + 3\,\Delta x - 2y)$$

$$= 6x - 2y$$

$$D_2 f(x, y) = \lim_{\Delta y \to 0} \frac{f(x, y + \Delta y) - f(x, y)}{\Delta y}$$

$$= \lim_{\Delta y \to 0} \frac{3x^2 - 2x(y + \Delta y) + (y + \Delta y)^2 - (3x^2 - 2xy + y^2)}{\Delta y}$$

$$= \lim_{\Delta y \to 0} \frac{3x^2 - 2xy - 2x\,\Delta y + y^2 + 2y\,\Delta y + (\Delta y)^2 - 3x^2 + 2xy - y^2}{\Delta y}$$

$$= \lim_{\Delta y \to 0} \frac{-2x\,\Delta y + 2y\,\Delta y + (\Delta y)^2}{\Delta y}$$

$$= \lim_{\Delta y \to 0} (-2x + 2y + \Delta y)$$

$$= -2x + 2y$$

If (x_0, y_0) is a particular point in the domain of f, then

$$D_1 f(x_0, y_0) = \lim_{\Delta x \to 0} \frac{f(x_0 + \Delta x, y_0) - f(x_0, y_0)}{\Delta x} \qquad (3)$$

if this limit exists, and

$$D_2 f(x_0, y_0) = \lim_{\Delta y \to 0} \frac{f(x_0, y_0 + \Delta y) - f(x_0, y_0)}{\Delta y} \qquad (4)$$

if this limit exists.

▶ ILLUSTRATION 1   We apply formula (3) to find $D_1 f(3, -2)$ for the function f of Example 1.

$$D_1 f(3, -2) = \lim_{\Delta x \to 0} \frac{f(3 + \Delta x, -2) - f(3, -2)}{\Delta x}$$

$$= \lim_{\Delta x \to 0} \frac{3(3 + \Delta x)^2 - 2(3 + \Delta x)(-2) + (-2)^2 - (27 + 12 + 4)}{\Delta x}$$

$$= \lim_{\Delta x \to 0} \frac{27 + 18\,\Delta x + 3(\Delta x)^2 + 12 + 4\,\Delta x + 4 - 43}{\Delta x}$$

$$= \lim_{\Delta x \to 0} (18 + 3\,\Delta x + 4)$$

$$= 22 \qquad \blacktriangleleft$$

Alternate formulas to (3) and (4) for $D_1 f(x_0, y_0)$ and $D_2 f(x_0, y_0)$ are given by

$$D_1 f(x_0, y_0) = \lim_{x \to x_0} \frac{f(x, y_0) - f(x_0, y_0)}{x - x_0} \tag{5}$$

if this limit exists, and

$$D_2 f(x_0, y_0) = \lim_{y \to y_0} \frac{f(x_0, y) - f(x_0, y_0)}{y - y_0} \tag{6}$$

if this limit exists.

▶ ILLUSTRATION 2   We apply formula (5) to find $D_1 f(3, -2)$ for the function f of Example 1.

$$D_1 f(3, -2) = \lim_{x \to 3} \frac{f(x, -2) - f(3, -2)}{x - 3}$$

$$= \lim_{x \to 3} \frac{3x^2 + 4x + 4 - 43}{x - 3}$$

$$= \lim_{x \to 3} \frac{3x^2 + 4x - 39}{x - 3}$$

$$= \lim_{x \to 3} \frac{(3x + 13)(x - 3)}{x - 3}$$

$$= \lim_{x \to 3} (3x + 13)$$

$$= 22 \qquad \blacktriangleleft$$

▶ ILLUSTRATION 3   In Example 1 we showed that

$$D_1 f(x, y) = 6x - 2y$$

Therefore

$$D_1 f(3, -2) = 18 + 4$$
$$= 22$$

This result agrees with those of Illustrations 1 and 2.   ◀

Comparing Definition 15.4.1 with the definition of an ordinary derivative (2.2.1), we see that $D_1 f(x, y)$ is the ordinary derivative of f if f is considered as a function of one variable x (i.e., y is held constant), and $D_2 f(x, y)$ is the ordinary derivative of f if f is considered as a function of one variable y (and x is held constant). So the results in Example 1 can be obtained more easily by applying the theorems for ordinary differentiation if y is considered constant when finding $D_1 f(x, y)$ and if x is considered constant when finding $D_2 f(x, y)$. The following example illustrates this.

EXAMPLE 2
Given

$$f(x, y) = 3x^3 - 4x^2y + 3xy^2 + 7x - 8y$$

find $f_x(x, y)$ and $f_y(x, y)$.

SOLUTION
Treating f as a function of x and holding y constant we have

$$f_x(x, y) = 9x^2 - 8xy + 3y^2 + 7$$

Considering f as a function of y and holding x constant we have

$$f_y(x, y) = -4x^2 + 6xy - 8$$

EXAMPLE 3
Given

$$f(x, y) = \begin{cases} \dfrac{xy(x^2 - y^2)}{x^2 + y^2} & \text{if } (x, y) \neq (0, 0) \\ 0 & \text{if } (x, y) = (0, 0) \end{cases}$$

Show that (a) $f_1(0, y) = -y$ for all y and (b) $f_2(x, 0) = x$ for all x.

SOLUTION
(a) If $y \neq 0$, from (5),

$$f_1(0, y) = \lim_{x \to 0} \frac{f(x, y) - f(0, y)}{x - 0}$$

$$= \lim_{x \to 0} \frac{\dfrac{xy(x^2 - y^2)}{x^2 + y^2} - 0}{x}$$

$$= \lim_{x \to 0} \frac{y(x^2 - y^2)}{x^2 + y^2}$$

$$= -\frac{y^3}{y^2}$$

$$= -y$$

If $y = 0$, from (5),

$$f_1(0, 0) = \lim_{x \to 0} \frac{f(x, 0) - f(0, 0)}{x - 0}$$

$$= \lim_{x \to 0} \frac{0 - 0}{x}$$

$$= \lim_{x \to 0} 0$$

$$= 0$$

Because $f_1(0, y) = -y$ if $y \neq 0$ and $f_1(0,0) = 0$, we can conclude that $f_1(0, y) = -y$ for all y.

(b) If $x \neq 0$, from (6),

$$f_2(x, 0) = \lim_{y \to 0} \frac{f(x, y) - f(x, 0)}{y - 0}$$

$$= \lim_{y \to 0} \frac{\dfrac{xy(x^2 - y^2)}{x^2 + y^2} - 0}{y}$$

$$= \lim_{y \to 0} \frac{x(x^2 - y^2)}{x^2 + y^2}$$

$$= \frac{x^3}{x^2}$$

$$= x$$

If $x = 0$, from (6),

$$f_2(0,0) = \lim_{y \to 0} \frac{f(0, y) - f(0,0)}{y - 0}$$

$$= \lim_{y \to 0} \frac{0 - 0}{y}$$

$$= \lim_{y \to 0} 0$$

$$= 0$$

Because $f_2(x, 0) = x$ if $x \neq 0$ and $f_2(0,0) = 0$, then $f_2(x, 0) = x$ for all x.

Geometric interpretations of the partial derivatives of a function of two variables are similar to those of a function of one variable. The graph of a function f of two variables is a surface having equation $z = f(x, y)$. If y is held constant (say, $y = y_0$), then $z = f(x, y_0)$ is an equation of the trace of this surface in the plane $y = y_0$. The curve can be represented by the two equations

$$y = y_0 \quad \text{and} \quad z = f(x, y) \tag{7}$$

because the curve is the intersection of these two surfaces.

Then $D_1 f(x_0, y_0)$ is the slope of the tangent line to the curve given by Eqs. (7) at the point $P_0(x_0, y_0, f(x_0, y_0))$ in the plane $y = y_0$. In an analogous fashion, $D_2 f(x_0, y_0)$ represents the slope of the tangent line to the curve having equations

$$x = x_0 \quad \text{and} \quad z = f(x, y)$$

at the point P_0 in the plane $x = x_0$. Figure 15.4.1(a) and (b) shows the portions of the curves and the tangent lines.

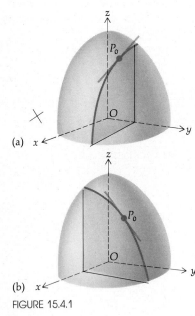

(a)

(b)

FIGURE 15.4.1

EXAMPLE 4

Find the slope of the tangent line to the curve of intersection of the surface

$$z = \tfrac{1}{2}\sqrt{24 - x^2 - 2y^2}$$

with the plane $y = 2$ at the point $(2, 2, \sqrt{3})$.

SOLUTION

The required slope is the value of $\dfrac{\partial z}{\partial x}$ at the point $(2, 2, \sqrt{3})$.

$$\frac{\partial z}{\partial x} = \frac{-x}{2\sqrt{24 - x^2 - 2y^2}}$$

So at $(2, 2, \sqrt{3})$,

$$\frac{\partial z}{\partial x} = \frac{-2}{2\sqrt{12}}$$

$$= -\frac{1}{2\sqrt{3}}$$

Because every derivative is a measure of a rate of change, a partial derivative can be so interpreted. If f is a function of the two variables x and y, the partial derivative of f with respect to x at the point $P_0(x_0, y_0)$ gives the instantaneous rate of change, at P_0, of $f(x, y)$ per unit change in x (x alone varies and y is held fixed at y_0). Similarly, the partial derivative of f with respect to y at P_0 gives the instantaneous rate of change, at P_0, of $f(x, y)$ per unit change in y.

EXAMPLE 5

According to the *ideal gas law* for a confined gas, if P newtons per square centimeter is the pressure, V cubic centimeters is the volume, and T degrees is the temperature, we have the formula

$$PV = kT \tag{8}$$

where k is a constant of proportionality. Suppose that the volume of a gas in a certain container is 100 cm^3 and the temperature is $90°$ and $k = 8$. (a) Find the instantaneous rate of change of P per unit change in T if V remains fixed at 100. (b) Use the result of part (a) to approximate the change in the pressure if the temperature is increased to $92°$. (c) Find the instantaneous rate of change of V per unit change in P if T remains fixed at 90. (d) Suppose that the temperature is held constant. Use the result of part (c) to find the approximate change in the volume necessary to produce the same change in the pressure as obtained in part (b).

SOLUTION
Substituting $V = 100$, $T = 90$, and $k = 8$ in Eq. (8) we obtain $P = 7.2$.
(a) Solving (8) for P when $k = 8$ we get

$$P = \frac{8T}{V}$$

The instantaneous rate of change of P per unit change in T if V remains fixed is $\dfrac{\partial P}{\partial T}$, and

$$\frac{\partial P}{\partial T} = \frac{8}{V}$$

When $T = 90$ and $V = 100$, $\dfrac{\partial P}{\partial T} = 0.08$, which is the answer required.

(b) From the result of part (a), when T is increased by 2 (and V remains fixed), an approximate increase in P is $2(0.08) = 0.16$. We conclude, then, that if the temperature is increased from $90°$ to $92°$, the increase in the pressure is approximately 0.16 nt/cm^2.

(c) Solving (8) for V when $k = 8$ we obtain

$$V = \frac{8T}{P}$$

The instantaneous rate of change of V per unit change in P if T remains fixed is $\dfrac{\partial V}{\partial P}$, and

$$\frac{\partial V}{\partial P} = -\frac{8T}{P^2}$$

When $T = 90$ and $P = 7.2$,

$$\frac{\partial V}{\partial P} = -\frac{8(90)}{(7.2)^2}$$

$$= -\frac{125}{9}$$

which is the instantaneous rate of change of V per unit change in P when $T = 90$ and $P = 7.2$ if T remains fixed at 90.

(d) If P is to be increased by 0.16 and T is held fixed, then from the result of part (c) the change in V should be approximately $(0.16)(-\frac{125}{9}) = -\frac{20}{9}$. Hence the volume should be decreased by approximately $\frac{20}{9}$ cm^3 if the pressure is to be increased from 7.2 nt/cm^2 to 7.36 nt/cm^2.

We now extend the concept of partial derivative to functions of n variables.

15.4.2 DEFINITION   Let $P(x_1, x_2, \ldots, x_n)$ be a point in R^n, and let f be a function of the n variables $x_1, x_2, \ldots, x_n$. Then the partial derivative of f with respect to x_k is that function, denoted by $D_k f$, such that its function value at any point P in the domain of f is given by

$$D_k f(x_1, x_2, \ldots, x_n)$$

$$= \lim_{\Delta x_k \to 0} \frac{f(x_1, x_2, \ldots, x_{k-1}, x_k + \Delta x_k, x_{k+1}, \ldots, x_n) - f(x_1, x_2, \ldots, x_n)}{\Delta x_k}$$

if this limit exists.

In particular, if f is a function of the three variables x, y, and z, then the partial derivatives of f are given by

$$D_1 f(x, y, z) = \lim_{\Delta x \to 0} \frac{f(x + \Delta x, y, z) - f(x, y, z)}{\Delta x}$$

$$D_2 f(x, y, z) = \lim_{\Delta y \to 0} \frac{f(x, y + \Delta y, z) - f(x, y, z)}{\Delta y}$$

$$D_3 f(x, y, z) = \lim_{\Delta z \to 0} \frac{f(x, y, z + \Delta z) - f(x, y, z)}{\Delta z}$$

if these limits exist.

EXAMPLE 6

Given $f(x, y, z) = x^2y + yz^2 + z^3$, verify that

$$xf_1(x, y, z) + yf_2(x, y, z) + zf_3(x, y, z) = 3f(x, y, z)$$

SOLUTION

Holding y and z constant we get

$$f_1(x, y, z) = 2xy$$

Holding x and z constant we obtain

$$f_2(x, y, z) = x^2 + z^2$$

Holding x and y constant we get

$$f_3(x, y, z) = 2yz + 3z^2$$

Therefore

$$
\begin{aligned}
xf_1(x, y, z) + yf_2(x, y, z) + zf_3(x, y, z) &= x(2xy) + y(x^2 + z^2) + z(2yz + 3z^2) \\
&= 2x^2y + x^2y + yz^2 + 2yz^2 + 3z^3 \\
&= 3(x^2y + yz^2 + z^3) \\
&= 3f(x, y, z)
\end{aligned}
$$

EXERCISES 15.4

In Exercises 1 through 6, apply Definition 15.4.1 to find each of the partial derivatives.

1. $f(x, y) = 6x + 3y - 7$; $D_1 f(x, y)$ **2.** $f(x, y) = 4x^2 - 3xy$; $D_1 f(x, y)$ **3.** $f(x, y) = 3xy + 6x - y^2$; $D_2 f(x, y)$

4. $f(x, y) = xy^2 - 5y + 6$; $D_2 f(x, y)$ **5.** $f(x, y) = \sqrt{x^2 + y^2}$; $f_x(x, y)$ **6.** $f(x, y) = \dfrac{x + 2y}{x^2 - y}$; $f_y(x, y)$

In Exercises 7 through 10, apply Definition 15.4.2 to find each of the partial derivatives.

7. $f(x, y, z) = x^2y - 3xy^2 + 2yz$; $D_2 f(x, y, z)$ **8.** $f(x, y, z) = x^2 + 4y^2 + 9z^2$; $D_1 f(x, y, z)$

9. $f(x, y, z, r, t) = xyr + yzt + yrt + zrt$; $f_r(x, y, z, r, t)$

10. $f(r, s, t, u, v, w) = 3r^2st + st^2v - 2tuv^2 - tvw + 3uw^2$; $f_v(r, s, t, u, v, w)$

11. Given $f(x, y) = x^2 - 9y^2$. Find $D_1 f(2, 1)$ by (a) applying formula (3); (b) applying formula (5); (c) applying formula (1) and then replacing x and y by 2 and 1, respectively.

12. For the function in Exercise 11, find $D_2 f(2, 1)$ by (a) applying formula (4); (b) applying formula (6); (c) applying formula (2) and then replacing x and y by 2 and 1, respectively.

In Exercises 13 through 24, find the indicated partial derivative by holding all but one of the variables constant and applying theorems for ordinary differentiation.

13. $f(x, y) = 4y^3 + \sqrt{x^2 + y^2}$; $D_1 f(x, y)$ **14.** $f(x, y) = \dfrac{x + y}{\sqrt{y^2 - x^2}}$; $D_2 f(x, y)$ **15.** $f(\theta, \phi) = \sin 3\theta \cos 2\phi$; $f_\phi(\theta, \phi)$

16. $f(r, \theta) = r^2 \cos \theta - 2r \tan \theta$; $f_\theta(r, \theta)$ **17.** $z = e^{y/x} \ln \dfrac{x^2}{y}$; $\dfrac{\partial z}{\partial y}$ **18.** $r = e^{-\theta} \cos(\theta + \phi)$; $\dfrac{\partial r}{\partial \theta}$

19. $u = (x^2 + y^2 + z^2)^{-1/2}$; $\dfrac{\partial u}{\partial z}$ **20.** $u = \tan^{-1}(xyzw)$; $\dfrac{\partial u}{\partial w}$

21. $f(x, y, z) = 4xyz + \ln(2xyz); f_3(x, y, z)$

22. $f(x, y, z) = e^{xy} \sinh 2z - e^{xy} \cosh 2z; f_z(x, y, z)$

23. $f(x, y, z) = e^{xyz} + \tan^{-1} \dfrac{3xy}{z^2}; f_y(x, y, z)$

24. $f(r, \theta, \phi) = 4r^2 \sin \theta + 5e^r \cos \theta \sin \phi - 2 \cos \phi; f_2(r, \theta, \phi)$

25. If $f(r, \theta) = r \tan \theta - r^2 \sin \theta$, find (a) $f_1(\sqrt{2}, \frac{1}{4}\pi)$; (b) $f_2(3, \pi)$.

26. If $f(x, y, z) = e^{xy^2} + \ln(y + z)$, find (a) $f_1(3, 0, 17)$; (b) $f_2(1, 0, 2)$; (c) $f_3(0, 0, 1)$.

In Exercises 27 and 28, find $f_x(x, y)$ and $f_y(x, y)$.

27. $f(x, y) = \displaystyle\int_x^y \ln \sin t \, dt$

28. $f(x, y) = \displaystyle\int_x^y e^{\cos t} \, dt$

29. Given $u = \sin \dfrac{r}{t} + \ln \dfrac{t}{r}$. Verify $t \dfrac{\partial u}{\partial t} + r \dfrac{\partial u}{\partial r} = 0$.

30. Given $w = x^2 y + y^2 z + z^2 x$. Verify $\dfrac{\partial w}{\partial x} + \dfrac{\partial w}{\partial y} + \dfrac{\partial w}{\partial z} = (x + y + z)^2$.

31. Given $f(x, y) = \begin{cases} \dfrac{x^3 + y^3}{x^2 + y^2} & \text{if } (x, y) \neq (0, 0) \\ 0 & \text{if } (x, y) = (0, 0) \end{cases}$

32. Given $f(x, y) = \begin{cases} \dfrac{x^2 - xy}{x + y} & \text{if } (x, y) \neq (0, 0) \\ 0 & \text{if } (x, y) = (0, 0) \end{cases}$

Find (a) $f_1(0, 0)$; (b) $f_2(0, 0)$.

Find (a) $f_1(0, y)$ if $y \neq 0$; (b) $f_1(0, 0)$.

33. For the function of Exercise 32 find (a) $f_2(x, 0)$ if $x \neq 0$; (b) $f_2(0, 0)$.

34. Find the slope of the tangent line to the curve of intersection of the surface $36x^2 - 9y^2 + 4z^2 + 36 = 0$ with the plane $x = 1$ at the point $(1, \sqrt{12}, -3)$. Interpret this slope as a partial derivative.

35. Find the slope of the tangent line to the curve of intersection of the surface $z = x^2 + y^2$ with the plane $y = 1$ at the point $(2, 1, 5)$. Draw a sketch. Interpret this slope as a partial derivative.

36. Find equations of the tangent line to the curve of intersection of the surface $x^2 + y^2 + z^2 = 9$ with the plane $y = 2$ at the point $(1, 2, 2)$.

37. The temperature at any point (x, y) of a flat plate is T degrees and $T = 54 - \frac{2}{3}x^2 - 4y^2$. If distance is measured in centimeters, find the rate of change of the temperature with respect to the distance moved along the plate in the directions of the positive x and y axes, respectively, at the point $(3, 1)$.

38. Use the ideal gas law for a confined gas (see Example 5) to show that

$$\frac{\partial V}{\partial T} \cdot \frac{\partial T}{\partial P} \cdot \frac{\partial P}{\partial V} = -1$$

39. If V dollars is the present value of an ordinary annuity of equal payments of \$100 per year for t years at an interest rate of $100i$ percent per year, then

$$V = 100 \left[\frac{1 - (1 + i)^{-t}}{i} \right]$$

(a) Find the instantaneous rate of change of V per unit change in i if t remains fixed at 8. (b) Use the result of part (a) to find the approximate change in the present value if the interest rate changes from 6% to 7% and the time remains fixed at 8 years. (c) Find the instantaneous rate of change of V per unit change in t if i remains fixed at 0.06. (d) Use the result of part (c) to find the approximate change in the present value if the time is decreased from 8 to 7 years and the interest rate remains fixed at 6%.

40. Suppose that x is the number of tens of thousands of dollars in the inventory carried in a store, y is the number of clerks in the store, P is the number of dollars in the weekly profit of the store, and

$$P = 3000 + 240y + 20y(x - 2y) - 10(x - 12)^2$$

where $15 \leq x \leq 25$ and $5 \leq y \leq 12$. At present the inventory is \$180,000 and there are 8 clerks. (a) Find the instantaneous rate of change of P per unit change in x if y remains fixed at 8. (b) Use the result of part (a) to find the approximate change in the weekly profit if the inventory changes from \$180,000 to \$200,000 and the number of clerks remains fixed at 8. (c) Find the instantaneous rate of change of P per unit change in y if x remains fixed at 18. (d) Use the result of part (c) to find the approximate change in the weekly profit if the number of clerks is increased from 8 to 10 and the inventory remains fixed at \$180,000.

41. If S m^2 is a person's body surface area, then a formula giving an approximate value of S is

$$S = 2W^{0.4}H^{0.7}$$

where W kg is the person's weight and H meters is the person's height. When $W = 70$ and $H = 1.8$, find $\dfrac{\partial S}{\partial W}$ and $\dfrac{\partial S}{\partial H}$, and interpret the results.

15.5
DIFFERENTIABILITY AND
THE TOTAL DIFFERENTIAL

We shall define *differentiability* of functions of more than one variable by means of an equation involving the increment of a function. To motivate this definition we first obtain a representation for the increment of a function of a single variable that is similar to what will appear in our Definition 15.5.2 of differentiability. We discussed the increment of a function of a single variable in Section 2.2, and recall from that section that if f is a differentiable function of x, and $y = f(x)$, then

$$f'(x) = \lim_{\Delta x \to 0} \frac{\Delta y}{\Delta x}$$

where Δx and Δy are increments of x and y and

$$\Delta y = f(x + \Delta x) - f(x)$$

When $|\Delta x|$ is small and $\Delta x \neq 0$, $\Delta y / \Delta x$ differs from $f'(x)$ by a small number that depends on Δx, which we shall denote by ϵ. Thus

$$\epsilon = \frac{\Delta y}{\Delta x} - f'(x) \qquad \text{if } \Delta x \neq 0 \tag{1}$$

where ϵ is a function of Δx. From (1) we obtain

$$\Delta y = f'(x)\,\Delta x + \epsilon\,\Delta x$$

where ϵ is a function of Δx and $\epsilon \to 0$ as $\Delta x \to 0$.

From the above it follows that if the function f is differentiable at x_0, the increment of f at x_0, denoted by $\Delta f(x_0)$, is given by

$$\Delta f(x_0) = f'(x_0)\,\Delta x + \epsilon\,\Delta x \tag{2}$$

where $\lim_{\Delta x \to 0} \epsilon = 0$.

For functions of two or more variables an equation corresponding to Eq. (2) is used to define differentiability of a function. And from the definition we determine criteria for a function to be differentiable at a point. We give the details for a function of two variables and begin by defining the *increment* of such a function.

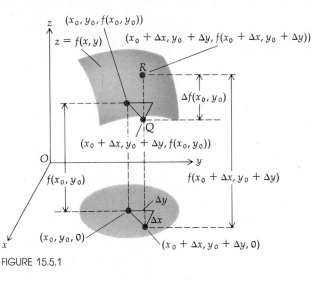

FIGURE 15.5.1

15.5.1 DEFINITION   If f is a function of two variables x and y, then the **increment of f** at the point (x_0, y_0), denoted by $\Delta f(x_0, y_0)$, is given by

$$\Delta f(x_0, y_0) = f(x_0 + \Delta x, y_0 + \Delta y) - f(x_0, y_0) \tag{3}$$

Figure 15.5.1 illustrates Eq. (3) for a function that is continuous on an open disk containing the points (x_0, y_0) and $(x_0 + \Delta x, y_0 + \Delta y)$. The figure shows a portion of the surface $z = f(x, y)$. $\Delta f(x_0, y_0) = \overline{QR}$, where Q is the point $(x_0 + \Delta x, y_0 + \Delta y, f(x_0, y_0))$ and R is the point $(x_0 + \Delta x, y_0 + \Delta y, f(x_0 + \Delta x, y_0 + \Delta y))$.

▶ ILLUSTRATION 1   For the function f defined by

$$f(x, y) = 3x - xy^2$$

we find the increment of f at any point (x_0, y_0).

$$\begin{aligned}
\Delta f(x_0, y_0) &= f(x_0 + \Delta x, y_0 + \Delta y) - f(x_0, y_0) \\
&= 3(x_0 + \Delta x) - (x_0 + \Delta x)(y_0 + \Delta y)^2 - (3x_0 - x_0 y_0{}^2) \\
&= 3x_0 + 3\,\Delta x - x_0 y_0{}^2 - y_0{}^2\,\Delta x - 2x_0 y_0\,\Delta y - 2y_0\,\Delta x\,\Delta y \\
&\qquad\qquad - x_0(\Delta y)^2 - \Delta x(\Delta y)^2 - 3x_0 + x_0 y_0{}^2 \\
&= 3\,\Delta x - y_0{}^2\,\Delta x - 2x_0 y_0\,\Delta y - 2y_0\,\Delta x\,\Delta y - x_0(\Delta y)^2 - \Delta x(\Delta y)^2
\end{aligned}$$

◀

15.5.2 DEFINITION   If f is a function of two variables x and y and the increment of f at (x_0, y_0) can be written as

$$\Delta f(x_0, y_0) = D_1 f(x_0, y_0)\,\Delta x + D_2 f(x_0, y_0)\,\Delta y + \epsilon_1\,\Delta x + \epsilon_2\,\Delta y \tag{4}$$

where ϵ_1 and ϵ_2 are functions of Δx and Δy such that $\epsilon_1 \to 0$ and $\epsilon_2 \to 0$ as $(\Delta x, \Delta y) \to (0, 0)$, then f is said to be **differentiable** at (x_0, y_0).

▶ ILLUSTRATION 2   We use Definition 15.5.2 to prove that the function of Illustration 1 is differentiable at all points in R^2. We must show that for all points (x_0, y_0) in R^2 we can find an ϵ_1 and an ϵ_2 such that

$$\Delta f(x_0, y_0) - D_1 f(x_0, y_0)\Delta x - D_2 f(x_0, y_0)\Delta y = \epsilon_1 \Delta x + \epsilon_2 \Delta y$$

and $\epsilon_1 \to 0$ and $\epsilon_2 \to 0$ as $(\Delta x, \Delta y) \to (0, 0)$.
    Because $f(x, y) = 3x - xy^2$,

$$D_1 f(x_0, y_0) = 3 - y_0{}^2 \quad \text{and} \quad D_2 f(x_0, y_0) = -2x_0 y_0$$

With these values and the value of $\Delta f(x_0, y_0)$ from Illustration 1,

$$\Delta f(x_0, y_0) - D_1 f(x_0, y_0)\Delta x - D_2 f(x_0, y_0)\Delta y = -x_0(\Delta y)^2 - 2y_0\Delta x\Delta y - \Delta x(\Delta y)^2$$

The right side of the above equation can be written in the following ways:

$$\left[-2y_0\Delta y - (\Delta y)^2\right]\Delta x + (-x_0\Delta y)\Delta y$$

$$(-2y_0\Delta y)\Delta x + (-\Delta x\Delta y - x_0\Delta y)\Delta y$$

$$\left[-(\Delta y)^2\right]\Delta x + (-2y_0\Delta x - x_0\Delta y)\Delta y$$

$$0 \cdot \Delta x + \left[-2y_0\Delta x - \Delta x\Delta y - x_0\Delta y\right]\Delta y$$

So there are four possible pairs of values for ϵ_1 and ϵ_2:

$$\epsilon_1 = -2y_0\Delta y - (\Delta y)^2 \quad \text{and} \quad \epsilon_2 = -x_0\Delta y$$

$$\epsilon_1 = -2y_0\Delta y \qquad\qquad \text{and} \quad \epsilon_2 = -\Delta x\Delta y - x_0\Delta y$$

$$\epsilon_1 = -(\Delta y)^2 \qquad\qquad \text{and} \quad \epsilon_2 = -2y_0\Delta x - x_0\Delta y$$

$$\epsilon_1 = 0 \qquad\qquad\qquad \text{and} \quad \epsilon_2 = -2y_0\Delta x - \Delta x\Delta y - x_0\Delta y$$

For each pair,

$$\lim_{(\Delta x, \Delta y)\to(0,0)} \epsilon_1 = 0 \quad \text{and} \quad \lim_{(\Delta x, \Delta y)\to(0,0)} \epsilon_2 = 0$$

It should be noted that it is only necessary to find one pair of values for ϵ_1 and ϵ_2.   ◀

15.5.3 THEOREM   If a function f of two variables is differentiable at a point, it is continuous at that point.

PROOF   If f is differentiable at the point (x_0, y_0), it follows from Definition 15.5.2 that

$$f(x_0 + \Delta x, y_0 + \Delta y) - f(x_0, y_0) = D_1 f(x_0, y_0)\Delta x + D_2 f(x_0, y_0)\Delta y$$
$$+ \epsilon_1 \Delta x + \epsilon_2 \Delta y$$

where $\epsilon_1 \to 0$ and $\epsilon_2 \to 0$ as $(\Delta x, \Delta y) \to (0, 0)$. Therefore

$$f(x_0 + \Delta x, y_0 + \Delta y) = f(x_0, y_0) + D_1 f(x_0, y_0)\Delta x + D_2 f(x_0, y_0)\Delta y$$
$$+ \epsilon_1 \Delta x + \epsilon_2 \Delta y$$

Taking the limit on both sides of the above as $(\Delta x, \Delta y) \rightarrow (0, 0)$ we obtain

$$\lim_{(\Delta x, \Delta y) \rightarrow (0,0)} f(x_0 + \Delta x, y_0 + \Delta y) = f(x_0, y_0) \tag{5}$$

If we let $x_0 + \Delta x = x$ and $y_0 + \Delta y = y$, then "$(\Delta x, \Delta y) \rightarrow (0, 0)$" is equivalent to "$(x, y) \rightarrow (x_0, y_0)$." Thus, from (5),

$$\lim_{(x,y) \rightarrow (x_0, y_0)} f(x, y) = f(x_0, y_0)$$

which proves that f is continuous at (x_0, y_0),   ■

Theorem 15.5.3 states that for a function of two variables *differentiability implies continuity*. However, the mere existence of the partial derivatives $D_1 f$ and $D_2 f$ at a point does not imply differentiability at that point. The following example illustrates this.

EXAMPLE 1
Given

$$f(x, y) = \begin{cases} \dfrac{xy}{x^2 + y^2} & \text{if } (x, y) \neq (0, 0) \\ 0 & \text{if } (x, y) = (0, 0) \end{cases}$$

prove that $D_1 f(0, 0)$ and $D_2 f(0, 0)$ exist but that f is not differentiable at $(0, 0)$.

SOLUTION

$$D_1 f(0,0) = \lim_{x \rightarrow 0} \frac{f(x, 0) - f(0, 0)}{x - 0} \qquad D_2 f(0,0) = \lim_{y \rightarrow 0} \frac{f(0, y) - f(0, 0)}{y - 0}$$

$$= \lim_{x \rightarrow 0} \frac{0 - 0}{x} \qquad\qquad = \lim_{y \rightarrow 0} \frac{0 - 0}{y}$$

$$= \lim_{x \rightarrow 0} 0 \qquad\qquad\qquad = \lim_{y \rightarrow 0} 0$$

$$= 0 \qquad\qquad\qquad\qquad = 0$$

Therefore both $D_1 f(0, 0)$ and $D_2 f(0, 0)$ exist.
    In Example 4 of Section 15.2 we demonstrated that for this function $\lim_{(x,y) \rightarrow (0,0)} f(x, y)$ does not exist; hence f is not continuous at $(0, 0)$. Because f is not continuous at $(0, 0)$, it follows from Theorem 15.5.3 that f is not differentiable there.

The following theorem gives conditions guaranteeing that a function is differentiable at a point. It is easier to apply than Definition 15.5.2.

15.5.4 THEOREM   Let f be a function of two variables x and y. Suppose that $D_1 f$ and $D_2 f$ exist on an open disk $B(P_0; r)$, where P_0 is the point (x_0, y_0). Then if $D_1 f$ and $D_2 f$ are continuous at P_0, f is differentiable at P_0.

To prove Theorem 15.5.4 we need a preliminary theorem. This theorem, its proof, and the proof of Theorem 15.5.4 appear in Supplementary Section 15.8.

EXAMPLE 2
Use Theorem 15.5.4 to prove that the function defined by

$$f(x, y) = x^3 + 3xy - 5y^3$$

is differentiable everywhere.

SOLUTION
We compute the partial derivatives:

$$D_1 f(x, y) = 3x^2 + 3y \qquad D_2 f(x, y) = 3x - 15y^2$$

Because $D_1 f$ and $D_2 f$ are continuous everywhere, it follows from Theorem 15.5.4 that f is differentiable everywhere.

The argument given for the polynomial function f in Example 2 can be applied to any polynomial function. Thus all polynomial functions are differentiable everywhere.

Observe that the conditions given in Theorem 15.5.4 are sufficient to prove that a function is differentiable at a point. However, they are not necessary conditions. That is, it is possible for a function to be differentiable at a point even if its partial derivatives are not continuous there. An example of such a function appears in Exercises 42 through 45. A function satisfying the hypothesis of Theorem 15.5.4 is said to be **continuously differentiable** at the point P_0. Thus continuous differentiability at a point is a sufficient condition, but not a necessary one, for differentiability at the point.

EXAMPLE 3
Given

$$f(x, y) = \begin{cases} \dfrac{x^2 y^2}{x^2 + y^2} & \text{if } (x, y) \neq (0, 0) \\ 0 & \text{if } (x, y) = (0, 0) \end{cases}$$

use Theorem 15.5.4 to prove that f is differentiable at $(0, 0)$.

SOLUTION
To find $D_1 f$ consider two cases: $(x, y) = (0, 0)$ and $(x, y) \neq (0, 0)$. If $(x, y) = (0, 0)$, we have

$$D_1 f(0, 0) = \lim_{x \to 0} \frac{f(x, 0) - f(0, 0)}{x - 0}$$

$$= \lim_{x \to 0} \frac{0 - 0}{x}$$

$$= 0$$

If $(x, y) \neq (0, 0)$, $f(x, y) = x^2 y^2 / (x^2 + y^2)$. To find $D_1 f(x, y)$ we use the theorem for the ordinary derivative of a quotient and consider y as a constant.

$$D_1 f(x, y) = \frac{2xy^2(x^2 + y^2) - 2x(x^2 y^2)}{(x^2 + y^2)^2}$$

$$= \frac{2xy^4}{(x^2 + y^2)^2}$$

The function $D_1 f$ is therefore defined by

$$D_1 f(x, y) = \begin{cases} \dfrac{2xy^4}{(x^2 + y^2)^2} & \text{if } (x, y) \neq (0, 0) \\ 0 & \text{if } (x, y) = (0, 0) \end{cases}$$

In the same manner we obtain the function $D_2 f$ defined by

$$D_2 f(x, y) = \begin{cases} \dfrac{2x^4 y}{(x^2 + y^2)^2} & \text{if } (x, y) \neq (0, 0) \\ 0 & \text{if } (x, y) = (0, 0) \end{cases}$$

Both $D_1 f$ and $D_2 f$ exist on every open disk having its center at the origin. It remains to show that $D_1 f$ and $D_2 f$ are continuous at $(0, 0)$.

Because $D_1 f(0, 0) = 0$, $D_1 f$ will be continuous at $(0, 0)$ if

$$\lim_{(x,y) \to (0,0)} D_1 f(x, y) = 0$$

Therefore we must show that for any $\epsilon > 0$ there exists a $\delta > 0$ such that

$$\text{if } 0 < \sqrt{x^2 + y^2} < \delta \quad \text{then} \quad \left| \frac{2xy^4}{(x^2 + y^2)^2} \right| < \epsilon \tag{6}$$

$$\left| \frac{2xy^4}{(x^2 + y^2)^2} \right| = \frac{2|x|y^4}{(x^2 + y^2)^2}$$

$$\leq \frac{2\sqrt{x^2 + y^2}(\sqrt{x^2 + y^2})^4}{(x^2 + y^2)^2}$$

$$= 2\sqrt{x^2 + y^2}$$

Thus a suitable choice for δ is $2\delta = \epsilon$, that is, $\delta = \frac{1}{2}\epsilon$. With this δ we have the following argument:

$$0 < \sqrt{x^2 + y^2} < \delta \quad \text{and} \quad \delta = \tfrac{1}{2}\epsilon$$

$$\Rightarrow \qquad 2\sqrt{x^2 + y^2} < 2(\tfrac{1}{2}\epsilon)$$

$$\Rightarrow \qquad \frac{2\sqrt{x^2 + y^2}(\sqrt{x^2 + y^2})^4}{(x^2 + y^2)^2} < \epsilon$$

$$\Rightarrow \qquad \frac{2|x|y^4}{(x^2 + y^2)^2} < \epsilon$$

$$\Rightarrow \qquad \left| \frac{2xy^4}{(x^2 + y^2)^2} \right| < \epsilon$$

We have therefore shown that (6) holds. Hence $D_1 f$ is continuous at $(0, 0)$. In the same way we can show that $D_2 f$ is continuous at $(0, 0)$. It follows from Theorem 15.5.4 that f is differentiable at $(0, 0)$.

Refer back to Eq. (4). The expression involving the first two terms on the right side, $D_1 f(x_0, y_0) \Delta x + D_2 f(x_0, y_0) \Delta y$, is called the *principal part* of $\Delta f(x_0, y_0)$ or the *total differential* of the function f at (x_0, y_0). We make this as a formal definition.

15.5.5 DEFINITION   If f is a function of two variables x and y, and f is differentiable at (x, y), then the **total differential** of f is the function df having function values given by

$$df(x, y, \Delta x, \Delta y) = D_1 f(x, y) \Delta x + D_2 f(x, y) \Delta y \tag{7}$$

Note that df is a function of the four variables $x, y, \Delta x$, and Δy. If $z = f(x, y)$, we sometimes write dz in place of $df(x, y, \Delta x, \Delta y)$, and then Eq. (7) is written as

$$dz = D_1 f(x, y) \Delta x + D_2 f(x, y) \Delta y \tag{8}$$

If in particular $f(x, y) = x$, then $z = x$, $D_1 f(x, y) = 1$, and $D_2 f(x, y) = 0$; so Eq. (8) gives $dz = \Delta x$. Because $z = x$, for this function $dx = \Delta x$. In a similar fashion, if we take $f(x, y) = y$, then $z = y$, $D_1 f(x, y) = 0$, and $D_2 f(x, y) = 1$; thus (8) gives $dz = \Delta y$. Because $z = y$, then for this function $dy = \Delta y$. Hence we define the differentials of the independent variables as $dx = \Delta x$ and $dy = \Delta y$. Then (8) can be written as

$$dz = D_1 f(x, y) \, dx + D_2 f(x, y) \, dy \tag{9}$$

and at the point (x_0, y_0),

$$dz = D_1 f(x_0, y_0) \, dx + D_2 f(x_0, y_0) \, dy \tag{10}$$

In Eq. (4) let $\Delta z = \Delta f(x_0, y_0)$, $dx = \Delta x$, and $dy = \Delta y$. Then

$$\Delta z = D_1 f(x_0, y_0) \, dx + D_2 f(x_0, y_0) \, dy + \epsilon_1 \, dx + \epsilon_2 \, dy \tag{11}$$

By comparing (10) and (11), observe that when dx (i.e., Δx) and dy (i.e., Δy) are close to zero, and because then ϵ_1 and ϵ_2 also will be close to zero, dz is an approximation to Δz. Before giving an example, we write (9) with the notation $\dfrac{\partial z}{\partial x}$ and $\dfrac{\partial z}{\partial y}$ instead of $D_1 f(x, y)$ and $D_2 f(x, y)$, respectively:

$$dz = \frac{\partial z}{\partial x} \, dx + \frac{\partial z}{\partial y} \, dy \tag{12}$$

EXAMPLE 4

A closed metal container in the shape of a right-circular cylinder is to have an inside height of 6 in., an inside radius of 2 in., and a thickness of 0.1 in. If the cost of the metal to be used is 10 cents per cubic inch, find by differentials the approximate cost of the metal to be used in the manufacture of the container.

SOLUTION

The formula for the volume of a right-circular cylinder, where the volume is V in.^{3}, the radius is r in., and the height is h in., is

$$V = \pi r^2 h \tag{13}$$

The exact volume of metal in the container is the difference between the volumes of two right-circular cylinders for which $r = 2.1$, $h = 6.2$, and $r = 2$, $h = 2$, respectively.

ΔV would give the exact volume of metal, but because only an approximate value is wanted, we find dV instead. From (12),

$$dV = \frac{\partial V}{\partial r}\, dr + \frac{\partial V}{\partial h}\, dh \tag{14}$$

From (13),

$$\frac{\partial V}{\partial r} = 2\pi r h \quad \text{and} \quad \frac{\partial V}{\partial h} = \pi r^2$$

Substituting these values into (14) gives

$$dV = 2\pi r h\, dr + \pi r^2\, dh$$

Because $r = 2$, $h = 6$, $dr = 0.1$, and $dh = 0.2$,

$$dV = 2\pi(2)(6)(0.1) + \pi(2)^2(0.2)$$
$$= 3.2\pi$$

Hence $\Delta V \approx 3.2\pi$, and so there is approximately 3.2π in.^{3} of metal in the container. Because the cost of the metal is 10 cents per cubic inch, and $10 \cdot 3.2\pi = 32\pi$ and $32\pi \approx 100.53$, the approximate cost of the metal to be used in the manufacture of the container is $1.

The concepts of differentiability and the total differential are now extended to a function of n variables.

15.5.6 DEFINITION   If f is a function of the n variables $x_1, x_2, \ldots, x_n$, and $\bar{P}$ is the point $(\bar{x}_1, \bar{x}_2, \ldots, \bar{x}_n)$, then the **increment of f** at $\bar{P}$ is given by

$$\Delta f(\bar{P}) = f(\bar{x}_1 + \Delta x_1, \bar{x}_2 + \Delta x_2, \ldots, \bar{x}_n + \Delta x_n) - f(\bar{P}) \tag{15}$$

15.5.7 DEFINITION   If f is a function of the n variables $x_1, x_2, \ldots, x_n$, and the increment of f at the point $\bar{P}$ can be written as

$$\Delta f(\bar{P}) = D_1 f(\bar{P})\, \Delta x_1 + D_2 f(\bar{P})\, \Delta x_2 + \ldots + D_n f(\bar{P})\, \Delta x_n$$
$$+ \epsilon_1\, \Delta x_1 + \epsilon_2\, \Delta x_2 + \ldots + \epsilon_n\, \Delta x_n \tag{16}$$

where $\epsilon_1 \to 0$, $\epsilon_2 \to 0$, $\ldots$, $\epsilon_n \to 0$, as

$$(\Delta x_1, \Delta x_2, \ldots, \Delta x_n) \to (0, 0, \ldots, 0)$$

then f is said to be **differentiable** at $\bar{P}$.

Analogously to Theorem 15.5.4 it can be proved that sufficient conditions for a function f of n variables to be differentiable at a point $\bar{P}$ are that $D_1 f, D_2 f, \ldots, D_n f$ all exist on an open ball $B(\bar{P}; r)$ and that $D_1 f, D_2 f, \ldots, D_n f$ are all continuous at $\bar{P}$. As was the case for functions of two variables, it follows that for functions of n variables differentiability implies continuity. However, the existence of the partial derivatives $D_1 f, D_2 f, \ldots, D_n f$ at a point does not imply differentiability of the function at the point.

15.5.8 DEFINITION   If f is a function of the n variables $x_1, x_2, \ldots, x_n$ and f is differentiable at P, then the **total differential** of f is the function df having function values given by

$$df(P, \Delta x_1, \Delta x_2, \ldots, \Delta x_n) = D_1 f(P) \Delta x_1 + D_2 f(P) \Delta x_2 + \ldots + D_n f(P) \Delta x_n$$

Letting $w = f(x_1, x_2, \ldots, x_n)$, defining $dx_1 = \Delta x_1$, $dx_2 = \Delta x_2, \ldots, dx_n = \Delta x_n$, and using the notation $\dfrac{\partial w}{\partial x_i}$ instead of $D_i f(P)$, we can write the equation of Definition 15.5.8 as

$$dw = \frac{\partial w}{\partial x_1} dx_1 + \frac{\partial w}{\partial x_2} dx_2 + \ldots + \frac{\partial w}{\partial x_n} dx_n \tag{17}$$

EXAMPLE 5
The dimensions of a box are measured to be 10 cm, 12 cm, and 15 cm, and the measurements are correct to 0.02 cm. Find approximately the greatest error if the volume of the box is calculated from the given measurements. Also find the approximate percent error.

SOLUTION
Let V cm^3 be the volume of a box whose dimensions are x cm, y cm, and z cm. Then

$$V = xyz$$

The exact value of the error is found from ΔV; however, we use dV as an approximation to ΔV. From Eq. (17) for three independent variables,

$$dV = \frac{\partial V}{\partial x} dx + \frac{\partial V}{\partial y} dy + \frac{\partial V}{\partial z} dz$$

$$dV = yz \, dx + xz \, dy + xy \, dz \tag{18}$$

From the given information $|\Delta x| \leq 0.02$, $|\Delta y| \leq 0.02$, and $|\Delta z| \leq 0.02$. To find the greatest error in the volume we take the greatest error in the measurements of the three dimensions. So taking $dx = 0.02$, $dy = 0.02$, $dz = 0.02$, and $x = 10$, $y = 12$, $z = 15$, we have, from Eq. (18),

$$dV = (12)(15)(0.02) + (10)(15)(0.02) + (10)(12)(0.02)$$

$$= 9$$

Thus $\Delta V \approx 9$, and therefore the greatest possible error in the calculation of the volume from the given measurements is approximately 9 cm^3.

The relative error is found by dividing the error by the actual value. Hence the relative error in computing the volume from the given measurements is $\dfrac{\Delta V}{V} \approx \dfrac{dV}{V}$. Because $\dfrac{dV}{V} = \dfrac{9}{1800}$, $\dfrac{\Delta V}{V} \approx 0.005$. So the approximate percent error is 0.5%.

EXERCISES 15.5

1. If $f(x, y) = 3x^2 + 2xy - y^2$, find: (a) $\Delta f(1, 4)$, the increment of f at $(1, 4)$; (b) $\Delta f(1, 4)$ when $\Delta x = 0.03$ and $\Delta y = -0.02$; (c) $df(1, 4, \Delta x, \Delta y)$, the total differential of f at $(1, 4)$; (d) $df(1, 4, 0.03, -0.02)$.

2. If $f(x, y) = 2x^2 + 5xy + 4y^2$, find: (a) $\Delta f(2, -1)$, the increment of f at $(2, -1)$; (b) $\Delta f(2, -1)$ when $\Delta x = -0.01$ and $\Delta y = 0.02$; (c) $df(2, -1, \Delta x, \Delta y)$, the total differential of f at $(2, -1)$; (d) $df(2, -1, -0.01, 0.02)$.

3. If $g(x, y) = xye^{xy}$, find: (a) $\Delta g(2, -4)$, the increment of g at $(2, -4)$; (b) $\Delta g(2, -4)$ when $\Delta x = -0.1$ and $\Delta y = 0.2$; (c) $dg(2, -4, \Delta x, \Delta y)$, the total differential of g at $(2, -4)$; (d) $dg(2, -4, -0.1, 0.2)$.

4. If $h(x, y) = (x + y)/(x - y)$, find: (a) $\Delta h(3, 0)$, the increment of h at $(3, 0)$; (b) $\Delta h(3, 0)$ when $\Delta x = 0.04$ and $\Delta y = 0.03$; (c) $dh(3, 0, \Delta x, \Delta y)$, the total differential of h at $(3, 0)$; (d) $dh(3, 0, 0.04, 0.03)$.

5. If $F(x, y, z) = xy + \ln(yz)$, find: (a) $\Delta F(4, 1, 5)$, the increment of F at $(4, 1, 5)$; (b) $\Delta F(4, 1, 5)$ when $\Delta x = 0.02$, $\Delta y = 0.04$, and $\Delta z = -0.03$; (c) $dF(4, 1, 5, \Delta x, \Delta y, \Delta z)$, the total differential of F at $(4, 1, 5)$; (d) $dF(4, 1, 5, 0.02, 0.04, -0.03)$.

6. If $G(x, y, z) = x^2y + 2xyz - z^3$, find: (a) $\Delta G(-3, 0, 2)$, the increment of G at $(-3, 0, 2)$; (b) $\Delta G(-3, 0, 2)$ when $\Delta x = 0.01$, $\Delta y = 0.03$, $\Delta z = -0.01$; (c) $dG(-3, 0, 2, \Delta x, \Delta y, \Delta z)$, the total differential of G at $(-3, 0, 2)$; (d) $dG(-3, 0, 2, 0.01, 0.03, -0.01)$.

In Exercises 7 through 14, find the total differential dw.

7. $w = 4x^3 - xy^2 + 3y - 7$ 8. $w = y \tan x^2 - 2xy$ 9. $w = x \cos y - y \sin x$ 10. $w = xe^{2y} + e^{-y}$

11. $w = \ln(x^2 + y^2 + z^2)$ 12. $w = \dfrac{xyz}{x + y + z}$ 13. $w = x \tan^{-1} z - \dfrac{y^2}{z}$ 14. $w = e^{yz} - \cos xz$

In Exercises 15 through 18, prove that f is differentiable at all points in its domain by doing each of the following: (a) Find $\Delta f(x_0, y_0)$ for the given function; (b) find an ϵ_1 and an ϵ_2 so that Eq. (4) holds; (c) show that the ϵ_1 and the ϵ_2 found in part (b) both approach zero as $(\Delta x, \Delta y) \to (0, 0)$.

15. $f(x, y) = x^2y - 2xy$ 16. $f(x, y) = 2x^2 + 3y^2$ 17. $f(x, y) = \dfrac{x^2}{y}$ 18. $f(x, y) = \dfrac{y}{x}$

19. Given $f(x, y) = \begin{cases} x + y - 2 & \text{if } x = 1 \text{ or } y = 1 \\ 2 & \text{if } x \neq 1 \text{ and } y \neq 1 \end{cases}$

   Prove that $D_1 f(1, 1)$ and $D_2 f(1, 1)$ exist but f is not differentiable at $(1, 1)$.

20. Given $f(x, y) = \begin{cases} \dfrac{3x^2y^2}{x^4 + y^4} & \text{if } (x, y) \neq (0, 0) \\ 0 & \text{if } (x, y) = (0, 0) \end{cases}$

   Prove that $D_1 f(0, 0)$ and $D_2 f(0, 0)$ exist but f is not differentiable at $(0, 0)$.

In Exercises 21 through 27, use Theorem 15.5.4 to prove that the given function is differentiable at all points in its domain.

21. $g(x, y) = 2x^4 - 3x^2y^2 + x^{-2}y^{-2}$ 22. $f(x, y) = \dfrac{3x - 4y}{x^2 + 8y}$ 23. $f(x, y) = 3 \ln xy + 5 \sin x$

24. $g(x, y) = y \ln x - \dfrac{x}{y}$

25. $h(x, y) = \tan^{-1}(x + y) + \dfrac{1}{x - y}$

26. $f(x, y) = e^{2x} \sin y + e^{-2x} \cos y$

27. $f(x, y) = ye^{3x} - xe^{-3y}$

28. Given $f(x, y) = \begin{cases} \dfrac{3x^2 y}{x^2 + y^2} & \text{if } (x, y) \neq (0, 0) \\ 0 & \text{if } (x, y) = (0, 0) \end{cases}$

This function is continuous at $(0, 0)$ (see Example 5, Section 15.2, and Illustration 1, Section 15.3). Prove that $D_1 f(0, 0)$ and $D_2 f(0, 0)$ exist but $D_1 f$ and $D_2 f$ are not continuous at $(0, 0)$.

29. Given $f(x, y) = \begin{cases} \dfrac{xy(x^2 - y^2)}{x^2 + y^2} & \text{if } (x, y) \neq (0, 0) \\ 0 & \text{if } (x, y) = (0, 0) \end{cases}$

Prove that f is differentiable at $(0, 0)$ by using Theorem 15.5.4.

In Exercises 30 through 32, prove that f is differentiable at all points in R^3 by doing each of the following: (a) Find $\Delta f(x_0, y_0, z_0)$; (b) find an $\epsilon_1, \epsilon_2,$ and ϵ_3 such that Eq. (16) holds; (c) show that the $\epsilon_1, \epsilon_2,$ and ϵ_3 found in (b) all approach zero as $(\Delta x, \Delta y, \Delta z)$ approaches $(0, 0, 0)$.

30. $f(x, y, z) = 3x + 2y - 4z$

31. $f(x, y, z) = xy - xz + z^2$

32. $f(x, y, z) = 2x^2 z - 3yz^2$

33. Given $f(x, y, z) = \begin{cases} \dfrac{xy^2 z}{x^4 + y^4 + z^4} & \text{if } (x, y, z) \neq (0, 0, 0) \\ 0 & \text{if } (x, y, z) = (0, 0, 0) \end{cases}$

(a) Show that $D_1 f(0, 0, 0)$, $D_2 f(0, 0, 0)$, and $D_3 f(0, 0, 0)$ exist; (b) make use of the fact that differentiability implies continuity to prove that f is not differentiable at $(0, 0, 0)$.

34. Given $f(x, y, z) = \begin{cases} \dfrac{xyz^2}{x^2 + y^2 + z^2} & \text{if } (x, y, z) \neq (0, 0, 0) \\ 0 & \text{if } (x, y, z) = (0, 0, 0) \end{cases}$

Prove that f is differentiable at $(0, 0, 0)$.

35. A closed container in the shape of a rectangular solid is to have an inside length of 8 m, an inside width of 5 m, an inside height of 4 m, and a thickness of 4 cm. Use differentials to approximate the amount of material needed to construct the container.

36. Use the total differential to find approximately the greatest error in calculating the area of a right triangle from the lengths of the legs if they are measured to be 6 cm and 8 cm, respectively, with a possible error of 0.1 cm for each measurement. Also find the approximate percent error.

37. Find approximately, by using the total differential, the greatest error in calculating the length of the hypotenuse of the right triangle from the measurements of Exercise 36. Also find the approximate percent error.

38. If the ideal gas law (see Example 5, Section 15.4) is used to find P when T and V are given, but there is an error of 0.3% in measuring T and an error of 0.8% in measuring V, find approximately the greatest percent error in P.

39. The specific gravity s of an object is given by the formula

$$s = \frac{A}{A - W}$$

where A is the number of pounds in the weight of the object in air and W is the number of pounds in the weight of the object in water. If the weight of an object in air is read as 20 lb with a possible error of 0.01 lb and its weight in water is read as 12 lb with a possible error of 0.02 lb, find approximately the largest possible error in calculating s from these measurements. Also find the largest possible relative error.

40. A wooden box is to be made of lumber that is $\frac{2}{3}$ in. thick. The inside length is to be 6 ft, the inside width is to be 3 ft, the inside depth is to be 4 ft, and the box is to have no top. Use the total differential to find the approximate amount of lumber to be used in the box.

41. A company has contracted to manufacture 10,000 closed wooden crates having dimensions 3 m, 4 m, and 5 m. The cost of the wood to be used is $1 per square meter. If the machines that are used to cut the pieces of wood have a possible error of 0.5 cm in each dimension, find approximately, by using the total differential, the greatest possible error in the estimate of the cost of the wood.

In Exercises 42 through 45, we show that a function may be differentiable at a point even though it is not continuously differentiable there. Hence the conditions of Theorem 15.5.4 are sufficient but not necessary for differentiability. The function f in these exercises is defined by

$$f(x, y) = \begin{cases} (x^2 + y^2) \sin \dfrac{1}{\sqrt{x^2 + y^2}} & \text{if } (x, y) \neq (0, 0) \\ 0 & \text{if } (x, y) = (0, 0) \end{cases}$$

42. Find $\Delta f(0, 0)$.   **43.** Find $D_1 f(x, y)$ and $D_2 f(x, y)$.

44. Prove that f is differentiable at $(0, 0)$ by using Definition 15.5.2 and the results of Exercises 42 and 43.

45. Prove that $D_1 f$ and $D_2 f$ are not continuous at $(0, 0)$.

15.6
THE CHAIN RULE

In Section 2.7 we had the chain rule (Theorem 2.7.1) for functions of a single variable. With the Leibniz notation the theorem is as follows: If y is a function of u and $\dfrac{dy}{du}$ exists, and u is a function of x and $\dfrac{du}{dx}$ exists, then y is a function of x and $\dfrac{dy}{dx}$ exists and is given by

$$\frac{dy}{dx} = \frac{dy}{du} \cdot \frac{du}{dx} \tag{1}$$

We now consider the chain rule for a function of two variables, where each of these variables is also a function of two variables.

15.6.1 THEOREM
The Chain Rule

If u is a differentiable function of x and y, defined by $u = f(x, y)$, and $x = F(r, s)$, $y = G(r, s)$, and $\dfrac{\partial x}{\partial r}, \dfrac{\partial x}{\partial s}, \dfrac{\partial y}{\partial r}$, and $\dfrac{\partial y}{\partial s}$ all exist, then u is a function of r and s and

$$\frac{\partial u}{\partial r} = \left(\frac{\partial u}{\partial x}\right)\left(\frac{\partial x}{\partial r}\right) + \left(\frac{\partial u}{\partial y}\right)\left(\frac{\partial y}{\partial r}\right) \tag{2}$$

$$\frac{\partial u}{\partial s} = \left(\frac{\partial u}{\partial x}\right)\left(\frac{\partial x}{\partial s}\right) + \left(\frac{\partial u}{\partial y}\right)\left(\frac{\partial y}{\partial s}\right) \tag{3}$$

PROOF   We prove (2). The proof of (3) is similar.

If s is held fixed and r is changed by an amount Δr, then x is changed by an amount Δx and y is changed by an amount Δy. Thus

$$\Delta x = F(r + \Delta r, s) - F(r, s) \tag{4}$$

and

$$\Delta y = G(r + \Delta r, s) - G(r, s) \tag{5}$$

Because f is differentiable,

$$\Delta f(x, y) = D_1 f(x, y) \Delta x + D_2 f(x, y) \Delta y + \epsilon_1 \Delta x + \epsilon_2 \Delta y \tag{6}$$

where ϵ_1 and ϵ_2 both approach zero as $(\Delta x, \Delta y)$ approaches $(0, 0)$. Furthermore, we require that $\epsilon_1 = 0$ and $\epsilon_2 = 0$ when $(\Delta x, \Delta y) = (0, 0)$. We make this requirement so that ϵ_1 and ϵ_2, which are functions of Δx and Δy, will be continuous at $(\Delta x, \Delta y) = (0, 0)$.

If in (6) we replace $\Delta f(x, y)$ by Δu, $D_1 f(x, y)$ by $\dfrac{\partial u}{\partial x}$, and $D_2 f(x, y)$ by $\dfrac{\partial u}{\partial y}$ and divide on both sides by Δr ($\Delta r \neq 0$), we obtain

$$\frac{\Delta u}{\Delta r} = \frac{\partial u}{\partial x} \frac{\Delta x}{\Delta r} + \frac{\partial u}{\partial y} \frac{\Delta y}{\Delta r} + \epsilon_1 \frac{\Delta x}{\Delta r} + \epsilon_2 \frac{\Delta y}{\Delta r}$$

Taking the limit on both sides of the above as Δr approaches zero we get

$$\lim_{\Delta r \to 0} \frac{\Delta u}{\Delta r} = \frac{\partial u}{\partial x} \lim_{\Delta r \to 0} \frac{\Delta x}{\Delta r} + \frac{\partial u}{\partial y} \lim_{\Delta r \to 0} \frac{\Delta y}{\Delta r} + \left(\lim_{\Delta r \to 0} \epsilon_1 \right) \lim_{\Delta r \to 0} \frac{\Delta x}{\Delta r} + \left(\lim_{\Delta r \to 0} \epsilon_2 \right) \lim_{\Delta r \to 0} \frac{\Delta y}{\Delta r} \tag{7}$$

Because u is a function of x and y and both x and y are functions of r and s, u is a function of r and s. Because s is held fixed and r is changed by an amount Δr,

$$\lim_{\Delta r \to 0} \frac{\Delta u}{\Delta r} = \lim_{\Delta r \to 0} \frac{u(r + \Delta r, s) - u(r, s)}{\Delta r}$$

$$= \frac{\partial u}{\partial r} \tag{8}$$

Also

$$\lim_{\Delta r \to 0} \frac{\Delta x}{\Delta r} = \frac{\partial x}{\partial r} \quad \text{and} \quad \lim_{\Delta r \to 0} \frac{\Delta y}{\Delta r} = \frac{\partial y}{\partial r} \tag{9}$$

Because $\dfrac{\partial x}{\partial r}$ and $\dfrac{\partial y}{\partial r}$ exist, F and G are each continuous with respect to the variable r. (*Note:* The existence of the partial derivatives of a function does not imply continuity with respect to all of the variables simultaneously, as we saw in the preceding section, but as with functions of a single variable it does imply continuity of the function with respect to each variable separately.) Hence, from (4),

$$\lim_{\Delta r \to 0} \Delta x = \lim_{\Delta r \to 0} \left[F(r + \Delta r, s) - F(r, s) \right]$$

$$= F(r, s) - F(r, s)$$

$$= 0$$

and from (5),

$$\lim_{\Delta r \to 0} \Delta y = \lim_{\Delta r \to 0} \left[G(r + \Delta r, s) - G(r, s) \right]$$

$$= G(r, s) - G(r, s)$$

$$= 0$$

Therefore, as Δr approaches zero, both Δx and Δy approach zero. And because both ϵ_1 and ϵ_2 approach zero as $(\Delta x, \Delta y)$ approaches $(0, 0)$, we can conclude that

$$\lim_{\Delta r \to 0} \epsilon_1 = 0 \quad \text{and} \quad \lim_{\Delta r \to 0} \epsilon_2 = 0 \qquad (10)$$

Now, it is possible that for certain values of Δr, $\Delta x = 0$ and $\Delta y = 0$. Because we required in such a case that $\epsilon_1 = 0$ and $\epsilon_2 = 0$, the limits in (10) are still zero. Substituting from (8), (9), and (10) into (7) we obtain

$$\frac{\partial u}{\partial r} = \left(\frac{\partial u}{\partial x} \right) \left(\frac{\partial x}{\partial r} \right) + \left(\frac{\partial u}{\partial y} \right) \left(\frac{\partial y}{\partial r} \right)$$

which proves (2).   ∎

EXAMPLE 1
Given

$$u = \ln \sqrt{x^2 + y^2} \qquad x = re^s \qquad y = re^{-s}$$

find $\dfrac{\partial u}{\partial r}$ and $\dfrac{\partial u}{\partial s}$.

SOLUTION

$$\frac{\partial u}{\partial x} = \frac{x}{x^2 + y^2} \qquad \frac{\partial u}{\partial y} = \frac{y}{x^2 + y^2} \qquad \frac{\partial x}{\partial r} = e^s$$

$$\frac{\partial x}{\partial s} = re^s \qquad \frac{\partial y}{\partial r} = e^{-s} \qquad \frac{\partial y}{\partial s} = -re^{-s}$$

From (2) we get

$$\frac{\partial u}{\partial r} = \frac{x}{x^2 + y^2}(e^s) + \frac{y}{x^2 + y^2}(e^{-s})$$

$$= \frac{xe^s + ye^{-s}}{x^2 + y^2}$$

From (3) we get

$$\frac{\partial u}{\partial s} = \frac{x}{x^2 + y^2}(re^s) + \frac{y}{x^2 + y^2}(-re^{-s})$$

$$= \frac{r(xe^s - ye^{-s})}{x^2 + y^2}$$

As mentioned earlier, the symbols $\dfrac{\partial u}{\partial r}$, $\dfrac{\partial u}{\partial s}$, $\dfrac{\partial u}{\partial x}$, $\dfrac{\partial u}{\partial y}$, and so forth, must not be considered as fractions. The symbols ∂u, ∂x, and so on, have no meaning by themselves. For functions of one variable, the chain rule, given by Eq. (1), is easily remembered by thinking of an ordinary derivative as the quotient of two differentials, but there is no similar interpretation for partial derivatives.

Another troublesome notational problem arises when considering u as a function of x and y and then as a function of r and s. If $u = f(x, y)$, $x = F(r, s)$, and $y = G(r, s)$, then $u = f(F(r, s), G(r, s))$. (It is incorrect to write $u = f(r, s)$.)

▶ ILLUSTRATION 1   In Example 1,

$$u = f(x, y) \qquad x = F(r, s) \qquad y = G(r, s)$$
$$= \ln \sqrt{x^2 + y^2} \qquad = re^s \qquad = re^{-s}$$

Thus

$$u = f(F(r, s), G(r, s))$$
$$= \ln \sqrt{r^2 e^{2s} + r^2 e^{-2s}}$$

Observe that

$$f(r, s) = \ln \sqrt{r^2 + s^2}$$

That is, $f(r, s) \neq u$.   ◀

If we let $f(F(r, s), G(r, s)) = h(r, s)$, then (2) and (3) can be written respectively as

$$h_1(r, s) = f_1(x, y)F_1(r, s) + f_2(x, y)G_1(r, s)$$

$$h_2(r, s) = f_1(x, y)F_2(r, s) + f_2(x, y)G_2(r, s)$$

In the statement of Theorem 15.6.1 the independent variables are r and s, and u is the dependent variable. The variables x and y can be called the intermediate variables. We now extend the chain rule to n intermediate variables and m independent variables.

15.6.2 THEOREM
The General Chain Rule

Suppose that u is a differentiable function of the n variables $x_1, x_2, \ldots, x_n$, and each of these variables is in turn a function of the m variables $y_1, y_2, \ldots, y_m$. Suppose further that each of the partial derivatives $\dfrac{\partial x_i}{\partial y_j}$ $(i = 1, 2, \ldots, n;$ $j = 1, 2, \ldots, m)$ exists. Then u is a function of $y_1, y_2, \ldots, y_m$, and

$$\frac{\partial u}{\partial y_1} = \left(\frac{\partial u}{\partial x_1}\right)\left(\frac{\partial x_1}{\partial y_1}\right) + \left(\frac{\partial u}{\partial x_2}\right)\left(\frac{\partial x_2}{\partial y_1}\right) + \cdots + \left(\frac{\partial u}{\partial x_n}\right)\left(\frac{\partial x_n}{\partial y_1}\right)$$

$$\frac{\partial u}{\partial y_2} = \left(\frac{\partial u}{\partial x_1}\right)\left(\frac{\partial x_1}{\partial y_2}\right) + \left(\frac{\partial u}{\partial x_2}\right)\left(\frac{\partial x_2}{\partial y_2}\right) + \cdots + \left(\frac{\partial u}{\partial x_n}\right)\left(\frac{\partial x_n}{\partial y_2}\right)$$

$$\vdots$$

$$\frac{\partial u}{\partial y_m} = \left(\frac{\partial u}{\partial x_1}\right)\left(\frac{\partial x_1}{\partial y_m}\right) + \left(\frac{\partial u}{\partial x_2}\right)\left(\frac{\partial x_2}{\partial y_m}\right) + \cdots + \left(\frac{\partial u}{\partial x_n}\right)\left(\frac{\partial x_n}{\partial y_m}\right)$$

The proof is an extension of the proof of Theorem 15.6.1.

Observe that in the general chain rule there are as many terms on the right side of each equation as there are intermediate variables.

EXAMPLE 2
Given

$$u = xy + xz + yz \qquad x = r \qquad y = r \cos t \qquad z = r \sin t$$

find $\dfrac{\partial u}{\partial r}$ and $\dfrac{\partial u}{\partial t}$.

SOLUTION
From the chain rule,

$$\frac{\partial u}{\partial r} = \left(\frac{\partial u}{\partial x}\right)\left(\frac{\partial x}{\partial r}\right) + \left(\frac{\partial u}{\partial y}\right)\left(\frac{\partial y}{\partial r}\right) + \left(\frac{\partial u}{\partial z}\right)\left(\frac{\partial z}{\partial r}\right)$$

$$= (y + z)(1) + (x + z)(\cos t) + (x + y)(\sin t)$$
$$= y + z + x \cos t + z \cos t + x \sin t + y \sin t$$
$$= r \cos t + r \sin t + r \cos t + (r \sin t)(\cos t) + r \sin t + (r \cos t)(\sin t)$$
$$= 2r(\cos t + \sin t) + r(2 \sin t \cos t)$$
$$= 2r(\cos t + \sin t) + r \sin 2t$$

$$\frac{\partial u}{\partial t} = \left(\frac{\partial u}{\partial x}\right)\left(\frac{\partial x}{\partial t}\right) + \left(\frac{\partial u}{\partial y}\right)\left(\frac{\partial y}{\partial t}\right) + \left(\frac{\partial u}{\partial z}\right)\left(\frac{\partial z}{\partial t}\right)$$

$$= (y + z)(0) + (x + z)(-r \sin t) + (x + y)(r \cos t)$$
$$= (r + r \sin t)(-r \sin t) + (r + r \cos t)(r \cos t)$$
$$= -r^2 \sin t - r^2 \sin^2 t + r^2 \cos t + r^2 \cos^2 t$$
$$= r^2(\cos t - \sin t) + r^2(\cos^2 t - \sin^2 t)$$
$$= r^2(\cos t - \sin t) + r^2 \cos 2t$$

Now suppose that u is a differentiable function of the two variables x and y, and both x and y are differentiable functions of the single variable t. Then u is a function of the single variable t, and so instead of the partial derivative of u with respect to t we have the ordinary derivative of u with respect to t, which is given by

$$\frac{du}{dt} = \left(\frac{\partial u}{\partial x}\right)\left(\frac{dx}{dt}\right) + \left(\frac{\partial u}{\partial y}\right)\left(\frac{dy}{dt}\right) \tag{11}$$

We call $\dfrac{du}{dt}$ given by Eq. (11) the **total derivative** of u with respect to t.

If u is a differentiable function of the n variables $x_1, x_2, \ldots, x_n$ and each x_i is a differentiable function of the single variable t, then u is a function of t

and the total derivative of u with respect to t is given by

$$\frac{du}{dt} = \left(\frac{\partial u}{\partial x_1}\right)\left(\frac{dx_1}{dt}\right) + \left(\frac{\partial u}{\partial x_2}\right)\left(\frac{dx_2}{dt}\right) + \cdots + \left(\frac{\partial u}{\partial x_n}\right)\left(\frac{dx_n}{dt}\right)$$

EXAMPLE 3
Given

$$u = x^2 + 2xy + y^2 \qquad x = t \cos t \qquad y = t \sin t$$

find $\dfrac{du}{dt}$ by two methods: (a) Use the chain rule; (b) express u in terms of t before differentiating.

SOLUTION
(a) We compute the derivatives:

$$\frac{\partial u}{\partial x} = 2x + 2y \qquad\qquad \frac{\partial u}{\partial y} = 2x + 2y$$

$$\frac{dx}{dt} = \cos t - t \sin t \qquad \frac{dy}{dt} = \sin t + t \cos t$$

Thus from (11),

$$\frac{du}{dt} = (2x + 2y)(\cos t - t \sin t) + (2x + 2y)(\sin t + t \cos t)$$

$$= 2(x + y)(\cos t - t \sin t + \sin t + t \cos t)$$
$$= 2(t \cos t + t \sin t)(\cos t - t \sin t + \sin t + t \cos t)$$
$$= 2t(\cos^2 t - t \sin t \cos t + \sin t \cos t + t \cos^2 t + \sin t \cos t$$
$$\qquad\qquad -t \sin^2 t + \sin^2 t + t \sin t \cos t)$$
$$= 2t[1 + 2 \sin t \cos t + t(\cos^2 t - \sin^2 t)]$$
$$= 2t(1 + \sin 2t + t \cos 2t)$$

(b) $$u = (t \cos t)^2 + 2(t \cos t)(t \sin t) + (t \sin t)^2$$
$$= t^2 \cos^2 t + t^2(2 \sin t \cos t) + t^2 \sin^2 t$$
$$= t^2 + t^2 \sin 2t$$

Therefore

$$\frac{du}{dt} = 2t + 2t \sin 2t + 2t^2 \cos 2t$$

EXAMPLE 4
If f is a differentiable function and a and b are constants, prove that $z = f(\tfrac{1}{2}bx^2 - \tfrac{1}{3}ay^3)$ satisfies the partial differential equation

$$ay^2 \frac{\partial z}{\partial x} + bx \frac{\partial z}{\partial y} = 0$$

SOLUTION

Let $u = \frac{1}{2}bx^2 - \frac{1}{3}ay^3$. We wish to show that $z = f(u)$ satisfies the given equation. By the chain rule,

$$\frac{\partial z}{\partial x} = \left(\frac{dz}{du}\right)\left(\frac{\partial u}{\partial x}\right) \qquad \frac{\partial z}{\partial y} = \left(\frac{dz}{du}\right)\left(\frac{\partial u}{\partial y}\right)$$

$$= f'(u)(bx) \qquad\qquad = f'(u)(-ay^2)$$

Therefore

$$ay^2 \frac{\partial z}{\partial x} + bx \frac{\partial z}{\partial y} = ay^2[f'(u)(bx)] + bx[f'(u)(-ay^2)]$$

$$= 0$$

which is what we wished to prove.

EXAMPLE 5

Use the ideal gas law (see Example 5, Section 15.4) with $k = 10$ to find the rate at which the temperature is changing at the instant when the volume of the gas is 120 cm^3 and the gas is under a pressure of 8 nt/cm^2 if the volume is increasing at the rate of 2 cm^3/min and the pressure is decreasing at the rate of 0.1 nt/cm^2 per min.

SOLUTION

Let t min be the time that has elapsed since the volume of the gas started to increase, T degrees be the temperature at t min, P nt/cm^2 be the pressure at t min, and V cm^3 be the volume of the gas at t min.

From the ideal gas law,

$$T = \frac{PV}{10}$$

At the given instant, $P = 8$, $V = 120$, $\dfrac{dP}{dt} = -0.1$, and $\dfrac{dV}{dt} = 2$. From the chain rule,

$$\frac{dT}{dt} = \frac{\partial T}{\partial P}\frac{dP}{dt} + \frac{\partial T}{\partial V}\frac{dV}{dt}$$

$$= \frac{V}{10}\frac{dP}{dt} + \frac{P}{10}\frac{dV}{dt}$$

$$= \frac{120}{10}(-0.1) + \frac{8}{10}(2)$$

$$= -1.2 + 1.6$$

$$= 0.4$$

Therefore the temperature is increasing at the rate of 0.4 degree per minute at the given instant.

EXERCISES 15.6

In Exercises 1 through 6, find the indicated partial derivative by two methods: (a) Use the chain rule; (b) make the substitutions for x and y before differentiating.

1. $u = x^2 - y^2$; $x = 3r - s$; $y = r + 2s$; $\dfrac{\partial u}{\partial r}$; $\dfrac{\partial u}{\partial s}$

2. $u = 3x - 4y^2$; $x = 5pq$; $y = 3p^2 - 2q$; $\dfrac{\partial u}{\partial p}$; $\dfrac{\partial u}{\partial q}$

3. $u = 3x^2 + xy - 2y^2 + 3x - y$; $x = 2r - 3s$; $y = r + s$; $\dfrac{\partial u}{\partial r}$; $\dfrac{\partial u}{\partial s}$

4. $u = x^2 + y^2$; $x = \cosh r \cos t$; $y = \sinh r \sin t$; $\dfrac{\partial u}{\partial r}$; $\dfrac{\partial u}{\partial t}$

5. $u = e^{y/x}$; $x = 2r \cos t$; $y = 4r \sin t$; $\dfrac{\partial u}{\partial r}$; $\dfrac{\partial u}{\partial t}$

6. $V = \pi x^2 y$; $x = \cos z \sin t$; $y = z^2 e^t$; $\dfrac{\partial V}{\partial z}$; $\dfrac{\partial V}{\partial t}$

In Exercises 7 through 14, find the indicated partial derivative by using the chain rule.

7. $u = x^2 + xy$; $x = r^2 + s^2$; $y = 3r - 2s$; $\dfrac{\partial u}{\partial r}$; $\dfrac{\partial u}{\partial s}$

8. $u = xy + xz + yz$; $x = rs$; $y = r^2 - s^2$; $z = (r - s)^2$; $\dfrac{\partial u}{\partial r}$; $\dfrac{\partial u}{\partial s}$

9. $u = \sin^{-1}(3x + y)$; $x = r^2 e^s$; $y = \sin rs$; $\dfrac{\partial u}{\partial r}$; $\dfrac{\partial u}{\partial s}$

10. $u = \sin(xy)$; $x = 2ze^t$; $y = t^2 e^{-z}$; $\dfrac{\partial u}{\partial t}$; $\dfrac{\partial u}{\partial z}$

11. $u = \cosh \dfrac{y}{x}$; $x = 3r^2 s$; $y = 6se^r$; $\dfrac{\partial u}{\partial r}$; $\dfrac{\partial u}{\partial s}$

12. $u = xe^{-y}$; $x = \tan^{-1}(rst)$; $y = \ln(3rs + 5st)$; $\dfrac{\partial u}{\partial r}$; $\dfrac{\partial u}{\partial s}$; $\dfrac{\partial u}{\partial t}$

13. $u = x^2 + y^2 + z^2$; $x = r \sin \phi \cos \theta$; $y = r \sin \phi \sin \theta$; $z = r \cos \phi$; $\dfrac{\partial u}{\partial r}$; $\dfrac{\partial u}{\partial \phi}$; $\dfrac{\partial u}{\partial \theta}$

14. $u = x^2 yz$; $x = \dfrac{r}{s}$; $y = re^s$; $z = re^{-s}$; $\dfrac{\partial u}{\partial r}$; $\dfrac{\partial u}{\partial s}$

In Exercises 15 through 18, find the total derivative $\dfrac{du}{dt}$ by two methods: (a) Use the chain rule; (b) make the substitutions for x and y or for x, y, and z before differentiating.

15. $u = ye^x + xe^y$; $x = \cos t$; $y = \sin t$

16. $u = \ln xy + y^2$; $x = e^t$; $y = e^{-t}$

17. $u = \sqrt{x^2 + y^2 + z^2}$; $x = \tan t$; $y = \cos t$; $z = \sin t$; $0 < t < \tfrac{1}{2}\pi$

18. $u = \dfrac{t + e^x}{y - e^t}$; $x = 3 \sin t$; $y = \ln t$

In Exercises 19 through 22, find the total derivative $\dfrac{du}{dt}$ by using the chain rule; do not express u as a function of t before differentiating.

19. $u = \tan^{-1}\left(\dfrac{y}{x}\right)$; $x = \ln t$; $y = e^t$

20. $u = xy + xz + yz$; $x = t \cos t$; $y = t \sin t$; $z = t$

21. $u = \dfrac{x + t}{y + t}$; $x = \ln t$; $y = \ln \dfrac{1}{t}$
    22. $u = \ln(x^2 + y^2 + t^2)$; $x = t \sin t$; $y = \cos t$

In Exercises 23 through 26, assume that the given equation defines z as a function of x and y. Differentiate implicitly to find $\dfrac{\partial z}{\partial x}$ and $\dfrac{\partial z}{\partial y}$.

23. $3x^2 + y^2 + z^2 - 3xy + 4xz - 15 = 0$
    24. $z = (x^2 + y^2) \sin xz$

25. $ye^{xyz} \cos 3xz = 5$
    26. $ze^{yz} + 2xe^{xz} - 4e^{xy} = 3$

27. If f is a differentiable function of the variable u, let $u = bx - ay$ and prove that $z = f(bx - ay)$ satisfies the equation $a\left(\dfrac{\partial z}{\partial x}\right) + b\left(\dfrac{\partial z}{\partial y}\right) = 0$, where a and b are constants.

28. If f is a differentiable function of two variables u and v, let $u = x - y$ and $v = y - x$; prove that $z = f(x - y, y - x)$ satisfies the equation $\dfrac{\partial z}{\partial x} + \dfrac{\partial z}{\partial y} = 0$.

29. At a given instant, the length of one leg of a right triangle is 10 cm and it is increasing at the rate of 1 cm/min, and the length of the other leg of the right triangle is 12 cm and it is decreasing at the rate of 2 cm/min. Find the rate of change of the measure of the acute angle opposite the leg of length 12 cm at the given instant.

30. The height of a right-circular cone is increasing at the rate of 40 cm/min and the radius is decreasing at the rate of 15 cm/min. Find the rate of change of the volume at the instant when the height is 200 cm and the radius is 60 cm.

31. The height of a right-circular cylinder is decreasing at the rate of 10 cm/min and the radius is increasing at the rate of 4 cm/min. Find the rate of change of the volume at the instant when the height is 50 cm and the radius is 16 cm.

32. Water is flowing into a tank in the form of a right-circular cylinder at the rate of $\frac{4}{3}\pi$ m³/min. The tank is stretching in such a way that even though it remains cylindrical, its radius is increasing at the rate of 0.2 cm/min. How fast is the surface of the water rising when the radius is 2 m and the volume of water in the tank is 20π m³?

33. A quantity of gas obeys the ideal gas law (see Example 5, Section 15.4) with $k = 12$, and the gas is in a container that is being heated at a rate of 3° per minute. If at the instant when the temperature is 300°, the pressure is 6 nt/cm² and is decreasing at the rate of 0.1 nt/cm² per minute, find the rate of change of the volume at that instant.

34. A vertical wall makes an angle of radian measure $\frac{2}{3}\pi$ with the ground. A ladder of length 20 ft is leaning against the wall and its top is sliding down the wall at the rate of 3 ft/sec. How fast is the area of the triangle formed by the ladder, the wall, and the ground changing when the ladder makes an angle of $\frac{1}{6}\pi$ radians with the ground?

35. Suppose that f is a differentiable function of x and y, and $u = f(x, y)$. Then if $x = \cosh v \cos w$ and $y = \sinh v \sin w$, express $\dfrac{\partial u}{\partial v}$ and $\dfrac{\partial u}{\partial w}$ in terms of $\dfrac{\partial u}{\partial x}$ and $\dfrac{\partial u}{\partial y}$.

36. Suppose that f is a differentiable function of x, y, and z, and $u = f(x, y, z)$. Then if $x = r \sin \phi \cos \theta$, $y = r \sin \phi \sin \theta$, and $z = r \cos \phi$, express $\dfrac{\partial u}{\partial r}, \dfrac{\partial u}{\partial \phi}$, and $\dfrac{\partial u}{\partial \theta}$ in terms of $\dfrac{\partial u}{\partial x}, \dfrac{\partial u}{\partial y}$, and $\dfrac{\partial u}{\partial z}$.

37. If $u = f(x, y)$ and $v = g(x, y)$, then the equations
$$\dfrac{\partial u}{\partial x} = \dfrac{\partial v}{\partial y} \quad \text{and} \quad \dfrac{\partial v}{\partial x} = -\dfrac{\partial u}{\partial y}$$
are called the *Cauchy-Riemann* equations. Show that the Cauchy-Riemann equations are satisfied if $u = \ln(x^2 + y^2)$ and $v = \tan^{-1}\dfrac{y}{x}$.

38. Suppose that f and g are differentiable functions of x and y, and $u = f(x, y)$ and $v = g(x, y)$. Show that if the Cauchy-Riemann equations (see Exercise 37) hold and if $x = r \cos \theta$ and $y = r \sin \theta$, then
$$\dfrac{\partial u}{\partial r} = \dfrac{1}{r}\dfrac{\partial v}{\partial \theta} \quad \text{and} \quad \dfrac{\partial v}{\partial r} = -\dfrac{1}{r}\dfrac{\partial u}{\partial \theta}.$$

39. If f is a differentiable function of x and y, and $u = f(x, y)$, $x = r \cos \theta$, and $y = r \sin \theta$, show that

$$\frac{\partial u}{\partial x} = \frac{\partial u}{\partial r} \cos \theta - \frac{\partial u}{\partial \theta} \frac{\sin \theta}{r}$$

$$\frac{\partial u}{\partial y} = \frac{\partial u}{\partial r} \sin \theta + \frac{\partial u}{\partial \theta} \frac{\cos \theta}{r}$$

15.7
HIGHER ORDER PARTIAL
DERIVATIVES

If f is a function of two variables, then in general $D_1 f$ and $D_2 f$ are also functions of two variables. And if the partial derivatives of these functions exist, they are called second partial derivatives of f. In contrast, $D_1 f$ and $D_2 f$ are called first partial derivatives of f. There are four second partial derivatives of a function of two variables. If f is a function of the two variables x and y, the notations

$$D_2(D_1 f) \qquad D_{12}f \qquad f_{12} \qquad f_{xy} \qquad \frac{\partial^2 f}{\partial y \, \partial x}$$

all denote the second partial derivative of f, which is obtained by first partial-differentiating f with respect to x and then partial-differentiating the result with respect to y. This second partial derivative is defined by

$$f_{12}(x, y) = \lim_{\Delta y \to 0} \frac{f_1(x, y + \Delta y) - f_1(x, y)}{\Delta y} \tag{1}$$

if this limit exists. The notations

$$D_1(D_1 f) \qquad D_{11}f \qquad f_{11} \qquad f_{xx} \qquad \frac{\partial^2 f}{\partial x^2}$$

all denote the second partial derivative of f, which is obtained by partial-differentiating twice with respect to x. We have the definition

$$f_{11}(x, y) = \lim_{\Delta x \to 0} \frac{f_1(x + \Delta x, y) - f_1(x, y)}{\Delta x} \tag{2}$$

if this limit exists. The other two second partial derivatives are defined in an analogous way.

$$f_{21}(x, y) = \lim_{\Delta x \to 0} \frac{f_2(x + \Delta x, y) - f_2(x, y)}{\Delta x} \tag{3}$$

$$f_{22}(x, y) = \lim_{\Delta y \to 0} \frac{f_2(x, y + \Delta y) - f_2(x, y)}{\Delta y} \tag{4}$$

if these limits exist.

The definitions of higher order partial derivatives are similar. Again there are various notations for a specific derivative. For example,

$$D_{112}f \qquad f_{112} \qquad f_{xxy} \qquad \frac{\partial^3 f}{\partial y \, \partial x \, \partial x} \qquad \frac{\partial^3 f}{\partial y \, \partial x^2}$$

all stand for the third partial derivative of f, which is obtained by partial-differentiating twice with respect to x and then once with respect to y. In the subscript notation, the order of partial differentiation is from left to right; in the notation $\dfrac{\partial^3 f}{\partial y\, \partial x\, \partial x}$, the order is from right to left.

EXAMPLE 1
Given

$$f(x, y) = e^x \sin y + \ln xy$$

Find: (a) $D_{11}f(x, y)$; (b) $D_{12}f(x, y)$; (c) $\dfrac{\partial^3 f}{\partial x\, \partial y^2}$.

SOLUTION

$$D_1 f(x, y) = e^x \sin y + \frac{1}{xy}(y)$$

$$= e^x \sin y + \frac{1}{x}$$

(a) $D_{11}f(x, y) = e^x \sin y - \dfrac{1}{x^2}$ (b) $D_{12}f(x, y) = e^x \cos y$

(c) To find $\dfrac{\partial^3 f}{\partial x\, \partial y^2}$ we partial-differentiate twice with respect to y and then once with respect to x. We have, then,

$$\frac{\partial f}{\partial y} = e^x \cos y + \frac{1}{y} \qquad \frac{\partial^2 f}{\partial y^2} = -e^x \sin y - \frac{1}{y^2} \qquad \frac{\partial^3 f}{\partial x\, \partial y^2} = -e^x \sin y$$

Higher order partial derivatives of a function of n variables have definitions that are analogous to the definitions of higher order partial derivatives of a function of two variables. If f is a function of n variables, there may be n^2 second partial derivatives of f at a particular point. That is, for a function of three variables, if all the second-order partial derivatives exist, there are nine of them: $f_{11}, f_{12}, f_{13}, f_{21}, f_{22}, f_{23}, f_{31}, f_{32},$ and f_{33}.

EXAMPLE 2
Given

$$f(x, y, z) = \sin(xy + 2z)$$

Find $D_{132}f(x, y, z)$.

SOLUTION

$$D_1 f(x, y, z) = y \cos(xy + 2z)$$
$$D_{13}f(x, y, z) = -2y \sin(xy + 2z)$$
$$D_{132}f(x, y, z) = -2 \sin(xy + 2z) - 2xy \cos(xy + 2z)$$

EXAMPLE 3
Given

$$f(x, y) = x^3 y - y \cosh xy$$

Find: (a) $f_{xy}(x, y)$; (b) $f_{yx}(x, y)$.

SOLUTION

(a) $f_x(x, y) = 3x^2 y - y^2 \sinh xy$

   $f_{xy}(x, y) = 3x^2 - 2y \sinh xy - xy^2 \cosh xy$

(b) $f_y(x, y) = x^3 - \cosh xy - xy \sinh xy$

   $f_{yx}(x, y) = 3x^2 - y \sinh xy - y \sinh xy - xy^2 \cosh xy$

   $= 3x^2 - 2y \sinh xy - xy^2 \cosh xy$

Observe from the above results that for the function of Example 3 the "mixed" partial derivatives $f_{xy}(x, y)$ and $f_{yx}(x, y)$ are equal. So for this particular function, when finding the second partial derivative with respect to x and then y, the order of differentiation is immaterial. This condition holds for many functions. However, the following example shows that it is not always true.

EXAMPLE 4
Let f be the function defined by

$$f(x, y) = \begin{cases} (xy) \dfrac{x^2 - y^2}{x^2 + y^2} & \text{if } (x, y) \neq (0, 0) \\ 0 & \text{if } (x, y) = (0, 0) \end{cases}$$

Find $f_{12}(0, 0)$ and $f_{21}(0, 0)$.

SOLUTION
In Example 3, Section 15.4, we showed that for this function

$$f_1(0, y) = -y \qquad \text{for all } y \tag{5}$$

and

$$f_2(x, 0) = x \qquad \text{for all } x \tag{6}$$

From formula (1),

$$f_{12}(0, 0) = \lim_{\Delta y \to 0} \frac{f_1(0, 0 + \Delta y) - f_1(0, 0)}{\Delta y}$$

But from (5), $f_1(0, \Delta y) = -\Delta y$ and $f_1(0, 0) = 0$, and so

$$f_{12}(0, 0) = \lim_{\Delta y \to 0} \frac{-\Delta y - 0}{\Delta y}$$

$$= \lim_{\Delta y \to 0} (-1)$$

$$= -1$$

From formula (3),

$$f_{21}(0,0) = \lim_{\Delta x \to 0} \frac{f_2(0 + \Delta x, 0) - f_2(0,0)}{\Delta x}$$

From (6), $f_2(\Delta x, 0) = \Delta x$ and $f_2(0,0) = 0$. Therefore

$$f_{21}(0,0) = \lim_{\Delta x \to 0} \frac{\Delta x - 0}{\Delta x}$$

$$= \lim_{\Delta x \to 0} 1$$

$$= 1$$

For the function in Example 4 the mixed partial derivatives $f_{12}(x, y)$ and $f_{21}(x, y)$ are not equal at $(0,0)$. A set of conditions for which $f_{12}(x_0, y_0)$ and $f_{21}(x_0, y_0)$ are equal is given by Theorem 15.7.1, which follows. The function of Example 4 does not satisfy the hypothesis of this theorem because both f_{12} and f_{21} are discontinuous at $(0,0)$. It is left as an exercise to show this (see Exercise 24).

15.7.1 THEOREM   Suppose that f is a function of two variables x and y defined on an open disk $B((x_0, y_0); r)$ and f_x, f_y, f_{xy}, and f_{yx} also are defined on B. Furthermore, suppose that f_{xy} and f_{yx} are continuous on B. Then

$$f_{xy}(x_0, y_0) = f_{yx}(x_0, y_0)$$

PROOF   Consider a square having its center at (x_0, y_0) and the length of its side $2|h|$ such that $0 < \sqrt{2}|h| < r$. Then all the points in the interior of the square and on the sides of the square are in the open disk B (see Fig. 15.7.1). So the points $(x_0 + h, y_0 + h), (x_0 + h, y_0)$, and $(x_0, y_0 + h)$ are in B. Let Δ be defined by

$$\Delta = f(x_0 + h, y_0 + h) - f(x_0 + h, y_0) - f(x_0, y_0 + h) + f(x_0, y_0) \quad (7)$$

Consider the function G defined by

$$G(x) = f(x, y_0 + h) - f(x, y_0) \quad (8)$$

Then

$$G(x + h) = f(x + h, y_0 + h) - f(x + h, y_0)$$

So (7) can be written as

$$\Delta = G(x_0 + h) - G(x_0) \quad (9)$$

From (8),

$$G'(x) = f_x(x, y_0 + h) - f_x(x, y_0) \quad (10)$$

Now, because $f_x(x, y_0 + h)$ and $f_x(x, y_0)$ are defined on B, $G'(x)$ exists if x is in the closed interval having endpoints at x_0 and $x_0 + h$. Hence G is continuous if x is in this closed interval. By the mean-value theorem (3.3.2)

FIGURE 15.7.1

there is a number c_1 between x_0 and $x_0 + h$ such that

$$G(x_0 + h) - G(x_0) = hG'(c_1) \tag{11}$$

Substituting from (11) into (9) we get

$$\Delta = hG'(c_1) \tag{12}$$

From (12) and (10),

$$\Delta = h[f_x(c_1, y_0 + h) - f_x(c_1, y_0)] \tag{13}$$

Now if g is the function defined by

$$g(y) = f_x(c_1, y) \tag{14}$$

we can write (13) as

$$\Delta = h[g(y_0 + h) - g(y_0)] \tag{15}$$

From (14),

$$g'(y) = f_{xy}(c_1, y) \tag{16}$$

Because $f_{xy}(c_1, y)$ is defined on B, $g'(y)$ exists if y is in the closed interval having endpoints at y_0 and $y_0 + h$; hence g is continuous if y is in this closed interval. Therefore, by the mean-value theorem there is a number d_1 between y_0 and $y_0 + h$ such that

$$g(y_0 + h) - g(y_0) = hg'(d_1) \tag{17}$$

Substituting from (17) into (15) we get $\Delta = h^2 g'(d_1)$; so from (16) it follows that

$$\Delta = h^2 f_{xy}(c_1, d_1) \tag{18}$$

for some point (c_1, d_1) in the open disk B. We define a function ϕ by

$$\phi(y) = f(x_0 + h, y) - f(x_0, y) \tag{19}$$

and so $\phi(y + h) = f(x_0 + h, y + h) - f(x_0, y + h)$. Therefore (7) can be written as

$$\Delta = \phi(y_0 + h) - \phi(y_0) \tag{20}$$

From (19),

$$\phi'(y) = f_y(x_0 + h, y) - f_y(x_0, y) \tag{21}$$

ϕ' exists if y is in the closed interval having y_0 and $y_0 + h$ as endpoints because, by hypothesis, each term on the right side of (21) exists on B. Therefore ϕ is continuous on this closed interval. So by the mean-value theorem there is a number d_2 between y_0 and $y_0 + h$ such that

$$\phi(y_0 + h) - \phi(y_0) = h\phi'(d_2) \tag{22}$$

From (20), (21), and (22) it follows that

$$\Delta = h[f_y(x_0 + h, d_2) - f_y(x_0, d_2)] \tag{23}$$

Define the function χ by

$$\chi(x) = f_y(x, d_2) \tag{24}$$

and write (23) as

$$\Delta = h[\chi(x_0 + h) - \chi(x_0)] \tag{25}$$

From (24),

$$\chi'(x) = f_{yx}(x, d_2) \tag{26}$$

and by the mean-value theorem there is a number c_2 between x_0 and $x_0 + h$ such that

$$\chi(x_0 + h) - \chi(x_0) = h\chi'(c_2) \tag{27}$$

From (25), (26), and (27),

$$\Delta = h^2 f_{yx}(c_2, d_2) \tag{28}$$

Equating the right sides of (18) and (28) we get

$$h^2 f_{xy}(c_1, d_1) = h^2 f_{yx}(c_2, d_2)$$

and because $h \neq 0$, we can divide by h^2, which gives

$$f_{xy}(c_1, d_1) = f_{yx}(c_2, d_2) \tag{29}$$

where (c_1, d_1) and (c_2, d_2) are in B.

Because c_1 and c_2 are each between x_0 and $x_0 + h$, we can write $c_1 = x_0 + \epsilon_1 h$, where $0 < \epsilon_1 < 1$, and $c_2 = x_0 + \epsilon_2 h$, where $0 < \epsilon_2 < 1$. Similarly, because both d_1 and d_2 are between y_0 and $y_0 + h$, we can write $d_1 = y_0 + \epsilon_3 h$, where $0 < \epsilon_3 < 1$, and $d_2 = y_0 + \epsilon_4 h$, where $0 < \epsilon_4 < 1$. Making these substitutions in (29) gives

$$f_{xy}(x_0 + \epsilon_1 h, y_0 + \epsilon_3 h) = f_{yx}(x_0 + \epsilon_2 h, y_0 + \epsilon_4 h) \tag{30}$$

Because f_{xy} and f_{yx} are continuous on B, upon taking the limit on both sides of (30) as h approaches zero we obtain

$$f_{xy}(x_0, y_0) = f_{yx}(x_0, y_0) \qquad \blacksquare$$

EXAMPLE 5

Suppose $u = f(x, y)$, $x = F(r, s)$, and $y = G(r, s)$, and assume $f_{xy} = f_{yx}$. Prove by using the chain rule that

$$\frac{\partial^2 u}{\partial r^2} = f_{xx}(x, y)[F_r(r, s)]^2 + 2f_{xy}(x, y)F_r(r, s)G_r(r, s) + f_{yy}(x, y)[G_r(r, s)]^2$$
$$+ f_x(x, y)F_{rr}(r, s) + f_y(x, y)G_{rr}(r, s)$$

SOLUTION

From the chain rule (Theorem 15.6.1),

$$\frac{\partial u}{\partial r} = f_x(x, y)F_r(r, s) + f_y(x, y)G_r(r, s)$$

Taking the partial derivative again with respect to r, and using the formula for the derivative of a product and the chain rule, we obtain

$$\frac{\partial^2 u}{\partial r^2} = [f_{xx}(x, y)F_r(r, s) + f_{xy}(x, y)G_r(r, s)]F_r(r, s) + f_x(x, y)F_{rr}(r, s)$$
$$+ [f_{yx}(x, y)F_r(r, s) + f_{yy}(x, y)G_r(r, s)]G_r(r, s) + f_y(x, y)G_{rr}(r, s)$$

Multiplying and combining terms, and using the fact that $f_{xy}(x, y)$ and $f_{yx}(x, y)$ are equal, we get

$$\frac{\partial^2 u}{\partial r^2} = f_{xx}(x, y)[F_r(r, s)]^2 + 2f_{xy}(x, y)F_r(r, s)G_r(r, s) + f_{yy}(x, y)[G_r(r, s)]^2$$
$$+ f_x(x, y)F_{rr}(r, s) + f_y(x, y)G_{rr}(r, s)$$

which is what we wished to prove.

As a result of Theorem 15.7.1, if the function f of two variables has continuous partial derivatives on some open disk, then the order of partial differentiation can be changed without affecting the result; that is,

$$D_{112}f = D_{121}f = D_{211}f$$

$$D_{1122}f = D_{1212}f = D_{1221}f = D_{2112}f = D_{2121}f = D_{2211}f$$

and so forth. In particular, assuming that all of the partial derivatives are continuous on some open disk, we can prove that $D_{211}f = D_{112}f$ by applying Theorem 15.7.1 repeatedly. Doing this we have

$$D_{211}f = D_1(D_{21}f) = D_1(D_{12}f) = D_1[D_2(D_1f)] = D_2[D_1(D_1f)]$$
$$= D_2(D_{11}f) = D_{112}f$$

EXERCISES 15.7

In Exercises 1 through 10, do each of the following: (a) Find $D_{11}f(x, y)$; (b) find $D_{22}f(x, y)$; (c) show that $D_{12}f(x, y) = D_{21}f(x, y)$.

1. $f(x, y) = \dfrac{x^2}{y} - \dfrac{y}{x^2}$

2. $f(x, y) = 2x^3 - 3x^2y + xy^2$

3. $f(x, y) = e^{2x} \sin y$

4. $f(x, y) = e^{-x/y} + \ln \dfrac{y}{x}$

5. $f(x, y) = (x^2 + y^2) \tan^{-1} \dfrac{y}{x}$

6. $f(x, y) = \sin^{-1} \dfrac{3y}{x^2}$

7. $f(x, y) = 4x \sinh y + 3y \cosh x$

8. $f(x, y) = x \cos y - ye^x$

9. $f(x, y) = e^x \cos y + \tan^{-1} x \cdot \ln y$

10. $f(x, y) = 3x \cosh y - y \sin^{-1} e^x$

In Exercises 11 through 18, find the indicated partial derivatives.

11. $f(x, y) = 2x^3y + 5x^2y^2 - 3xy^2$; (a) $f_{121}(x, y)$; (b) $f_{211}(x, y)$

12. $G(x, y) = 3x^3y^2 + 5x^2y^3 + 2x$; (a) $G_{yyx}(x, y)$; (b) $G_{yxy}(x, y)$

13. $f(x, y, z) = ye^x + ze^y + e^z$; (a) $f_{xz}(x, y, z)$; (b) $f_{yz}(x, y, z)$

14. $g(x, y, z) = \sin(xyz)$; (a) $g_{23}(x, y, z)$; (b) $g_{12}(x, y, z)$

15. $f(w, z) = w^2 \cos e^z$; (a) $f_{121}(w, z)$; (b) $f_{212}(w, z)$

16. $f(u, v) = \ln \cos(u - v)$; (a) $f_{uuv}(u, v)$; (b) $f_{vuv}(u, v)$

17. $g(r, s, t) = \ln(r^2 + 4s^2 - 5t^2)$; (a) $g_{132}(r, s, t)$; (b) $g_{122}(r, s, t)$

18. $f(x, y, z) = \tan^{-1}(3xyz)$; (a) $f_{113}(x, y, z)$; (b) $f_{123}(x, y, z)$

In Exercises 19 through 22, show that $u(x, y)$ satisfies the equation

$$\frac{\partial^2 u}{\partial x^2} + \frac{\partial^2 u}{\partial y^2} = 0$$

which is known as *Laplace's equation* in R^2.

19. $u(x, y) = \ln(x^2 + y^2)$

20. $u(x, y) = e^x \sin y + e^y \cos x$

21. $u(x, y) = \tan^{-1} \dfrac{y}{x} + \dfrac{x}{x^2 + y^2}$

22. $u(x, y) = \tan^{-1} \dfrac{2xy}{x^2 - y^2}$

23. Laplace's equation in R^3 is

$$\frac{\partial^2 u}{\partial x^2} + \frac{\partial^2 u}{\partial y^2} + \frac{\partial^2 u}{\partial z^2} = 0$$

Show that $u(x, y, z) = (x^2 + y^2 + z^2)^{-1/2}$ satisfies this equation.

24. For the function of Example 4, show that f_{12} is discontinuous at $(0,0)$ and hence that the hypothesis of Theorem 15.7.1 is not satisfied if $(x_0, y_0) = (0,0)$.

In Exercises 25 through 27, find $f_{12}(0,0)$ and $f_{21}(0,0)$, if they exist.

25. $f(x, y) = \begin{cases} \dfrac{2xy}{x^2 + y^2} & \text{if } (x, y) \neq (0,0) \\ 0 & \text{if } (x, y) = (0,0) \end{cases}$

26. $f(x, y) = \begin{cases} \dfrac{x^2 y^2}{x^4 + y^4} & \text{if } (x, y) \neq (0,0) \\ 0 & \text{if } (x, y) = (0,0) \end{cases}$

27. $f(x, y) = \begin{cases} x^2 \tan^{-1} \dfrac{y}{x} - y^2 \tan^{-1} \dfrac{x}{y} & \text{if } x \neq 0 \text{ and } y \neq 0 \\ 0 & \text{if either } x = 0 \text{ or } y = 0 \end{cases}$

28. Given that $u = f(x, y)$, $x = F(t)$, and $y = G(t)$, and assuming that $f_{xy} = f_{yx}$, prove by using the chain rule that

$$\frac{d^2 u}{dt^2} = f_{xx}(x, y)[F'(t)]^2 + 2f_{xy}(x, y)F'(t)G'(t) + f_{yy}(x, y)[G'(t)]^2 + f_x(x, y)F''(t) + f_y(x, y)G''(t)$$

29. Given that $u = f(x, y)$, $x = F(r, s)$, and $y = G(r, s)$, and assuming that $f_{xy} = f_{yx}$, prove by using the chain rule that

$$\frac{\partial^2 u}{\partial r \partial s} = f_{xx}(x, y)F_r(r, s)F_s(r, s) + f_{xy}(x, y)[F_s(r, s)G_r(r, s) + F_r(r, s)G_s(r, s)]$$
$$+ f_{yy}(x, y)G_r(r, s)G_s(r, s) + f_x(x, y)F_{sr}(r, s) + f_y(x, y)G_{sr}(r, s)$$

30. Given $u = e^y \cos x$, $x = 2t$, $y = t^2$. Find $\dfrac{d^2 u}{dt^2}$ in three ways: (a) by first expressing u in terms of t; (b) by using the formula of Exercise 28; (c) by using the chain rule.

31. Given $u = 3xy - 4y^2$, $x = 2se^r$, $y = re^{-s}$. Find $\dfrac{\partial^2 u}{\partial r^2}$ in three ways: (a) by first expressing u in terms of r and s; (b) by using the formula of Example 5; (c) by using the chain rule.

32. For u, x, and y as given in Exercise 31, find $\dfrac{\partial^2 u}{\partial s \partial r}$ in three ways: (a) by first expressing u in terms of r and s; (b) by using the formula of Exercise 29; (c) by using the chain rule.

33. Given $u = 9x^2 + 4y^2$, $x = r \cos \theta$, $y = r \sin \theta$. Find $\dfrac{\partial^2 u}{\partial r^2}$ in three ways: (a) by first expressing u in terms of r and θ; (b) by using the formula of Example 5; (c) by using the chain rule.

34. For u, x, and y as given in Exercise 33, find $\dfrac{\partial^2 u}{\partial \theta^2}$ in three ways: (a) by first expressing u in terms of r and θ; (b) by using the formula of Example 5; (c) by using the chain rule.

35. For u, x, and y as given in Exercise 33, find $\dfrac{\partial^2 u}{\partial r\, \partial \theta}$ in three ways: (a) by first expressing u in terms of r and θ; (b) by using the formula of Exercise 29; (c) by using the chain rule.

36. Suppose that $u = f(x, y)$ and $v = g(x, y)$, and that f and g and their first and second partial derivatives are continuous. Prove that if u and v satisfy the Cauchy-Riemann equations (see Exercise 37 in Exercises 15.6), they also satisfy Laplace's equation (see Exercises 19 through 22).

37. The one-dimensional heat-conduction partial differential equation is

$$\frac{\partial u}{\partial t} = k^2 \frac{\partial^2 u}{\partial x^2}$$

Show that if f is a function of x satisfying the equation

$$\frac{d^2 f}{dx^2} + \lambda^2 f(x) = 0$$

and g is a function of t satisfying the equation $\dfrac{dg}{dt} + k^2 \lambda^2 g(t) = 0$, then if $u = f(x)g(t)$, the partial differential equation is satisfied. k and λ are constants.

38. The partial differential equation for a vibrating string is

$$\frac{\partial^2 u}{\partial t^2} = a^2 \frac{\partial^2 u}{\partial x^2}$$

Show that if f is a function of x satisfying the equation $\dfrac{d^2 f}{dx^2} + \lambda^2 f(x) = 0$ and g is a function of t satisfying the equation

$\dfrac{d^2 g}{dt^2} + a^2 \lambda^2 g(t) = 0$, then if $u = f(x)g(t)$, the partial differential equation is satisfied. a and λ are constants.

39. Prove that if f and g are two arbitrary functions of a real variable having continuous second derivatives and $u = f(x + at) + g(x - at)$, then u satisfies the partial differential equation of the vibrating string given in Exercise 38. (*Hint:* Let $v = x + at$ and $w = x - at$; then u is a function of v and w, and v and w are in turn functions of x and t.)

40. Prove that if f is a function of two variables and all the partial derivatives of f up to the fourth order are continuous on some open disk, then

$$D_{1122}f = D_{2121}f$$

15.8 (*Supplementary*) SUFFICIENT CONDITIONS FOR DIFFERENTIABILITY

The proof of Theorem 15.5.4, which gives sufficient conditions for a function of two variables to be differentiable at a point, was delayed until this section. We first give a theorem that is needed for the proof. It is the mean-value theorem for a function of a single variable applied to a function of two variables.

15.8.1 THEOREM   Let f be a function of two variables defined for all x in the closed interval $[a,b]$ and all y in the closed interval $[c,d]$.

(i) If $D_1 f(x, y_0)$ exists for some y_0 in $[c,d]$ and for all x in $[a,b]$, then there is a number ξ_1 in the open interval (a,b) such that

$$f(b, y_0) - f(a, y_0) = (b - a)D_1 f(\xi_1, y_0) \tag{1}$$

(ii) If $D_2 f(x_0, y)$ exists for some x_0 in $[a,b]$ and for all y in $[c,d]$, then there is a number ξ_2 in the open interval (c,d) such that

$$f(x_0, d) - f(x_0, c) = (d - c)D_2 f(x_0, \xi_2) \tag{2}$$

Before proving this theorem, we interpret it geometrically. For part (i) refer to Fig. 15.8.1, which shows the portion of the surface $z = f(x, y)$ above the rectangular region in the xy plane bounded by the lines $x = a, x = b, y = c$, and $y = d$. The plane $y = y_0$ intersects the surface in the curve represented by the two equations $y = y_0$ and $z = f(x, y)$. The slope of the line through the points $A(a, y_0, f(a, y_0))$ and $B(b, y_0, f(b, y_0))$ is

$$\frac{f(b, y_0) - f(a, y_0)}{b - a}$$

Theorem 15.8.1(i) states that there is some point $(\xi_1, y_0, f(\xi_1, y_0))$ on the curve between the points A and B where the tangent line is parallel to the secant line through A and B; that is, there is some number ξ_1 in (a,b) such that $D_1 f(\xi_1, y_0) = [f(b, y_0) - f(a, y_0)]/(b - a)$, and this is illustrated in the figure, for which $D_1 f(\xi_1, y_0) < 0$.

Figure 15.8.2 illustrates part (ii) of Theorem 15.8.1. The plane $x = x_0$ intersects the surface $z = f(x, y)$ in the curve represented by the two equations $x = x_0$ and $z = f(x, y)$. The slope of the line through the points

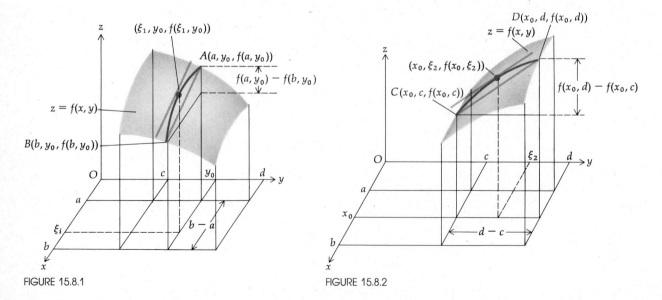

FIGURE 15.8.1   FIGURE 15.8.2

$C(x_0, c, f(x_0, c))$ and $D(x_0, d, f(x_0, d))$ is $[f(x_0, d) - f(x_0, c)]/(d - c)$, and Theorem 15.8.1(ii) states that there is some point $(x_0, \xi_2, f(x_0, \xi_2))$ on the curve between the points C and D where the tangent line is parallel to the secant line through C and D; that is, there is some number ξ_2 in (c, d) such that $D_2 f(x_0, \xi_2) = [f(x_0, d) - f(x_0, c)]/(d - c)$.

PROOF OF THEOREM 15.8.1(i)   Let g be the function of one variable x defined by

$$g(x) = f(x, y_0)$$

Then

$$g'(x) = D_1 f(x, y_0)$$

Because $D_1 f(x, y_0)$ exists for all x in $[a, b]$, it follows that $g'(x)$ exists for all x in $[a, b]$, and therefore g is continuous on $[a, b]$. So by the mean-value theorem (3.3.2) for ordinary derivatives there exists a number ξ_1 in (a, b) such that

$$g'(\xi_1) = \frac{g(b) - g(a)}{b - a}$$

$$\Leftrightarrow \quad D_1 f(\xi_1, y_0) = \frac{f(b, y_0) - f(a, y_0)}{b - a}$$

from which we obtain

$$f(b, y_0) - f(a, y_0) = (b - a)D_1 f(\xi_1, y_0)$$

The proof of part (ii) is similar to the proof of part (i) and is left as an exercise (see Exercise 17). ∎

Equation (1) can be written in the form

$$f(x_0 + h, y_0) - f(x_0, y_0) = hD_1 f(\xi_1, y_0) \tag{3}$$

where ξ_1 is between x_0 and $x_0 + h$ and h is either positive or negative (see Exercise 1).

Equation (2) can be written in the form

$$f(x_0, y_0 + k) - f(x_0, y_0) = kD_2 f(x_0, \xi_2) \tag{4}$$

where ξ_2 is between y_0 and $y_0 + k$ and k is either positive or negative (see Exercise 2).

EXAMPLE 1
Given

$$f(x, y) = \frac{2xy}{3 + x}$$

find a ξ_1 required by Theorem 15.8.1 if x is in $[2, 5]$ and $y = 4$.

SOLUTION

$$D_1 f(x, y) = \frac{6y}{(3 + x)^2}$$

By Theorem 15.8.1(i) there is a number ξ_1 in the open interval $(2, 5)$ such that

$$f(5, 4) - f(2, 4) = (5 - 2)D_1 f(\xi_1, 4)$$

$$5 - \frac{16}{5} = 3 \cdot \frac{24}{(3 + \xi_1)^2}$$

$$\frac{9}{5} = \frac{72}{(3 + \xi_1)^2}$$

$$(3 + \xi_1)^2 = 40$$

$$3 + \xi_1 = \pm 2\sqrt{10}$$

But because $2 < \xi_1 < 5$, we take only the $+$ sign and obtain

$$\xi_1 = 2\sqrt{10} - 3$$

EXAMPLE 2
Given

$$f(x, y) = 3xe^y - 2ye^x$$

find a ξ_2 required by Theorem 15.8.1 if y is in $[0, 4]$ and $x = 3$.

SOLUTION

$$D_2 f(x, y) = 3xe^y - 2e^x$$

By Theorem 15.8.1(ii) there is a number ξ_2 in the open interval $(0, 4)$ such that

$$f(3, 4) - f(3, 0) = (4 - 0)D_2 f(3, \xi_2)$$

$$(9e^4 - 8e^3) - 9 = 4(9e^{\xi_2} - 2e^3)$$

$$9e^4 - 8e^3 - 9 = 36e^{\xi_2} - 8e^3$$

$$36e^{\xi_2} = 9e^4 - 9$$

$$e^{\xi_2} = \tfrac{1}{4}(e^4 - 1)$$

$$e^{\xi_2} \approx 13.40$$

$$\xi_2 \approx \ln 13.40$$

$$\xi_2 \approx 2.60$$

Following is a restatement of Theorem 15.5.4 along with its proof.

15.5.4 THEOREM   Let f be a function of two variables x and y. Suppose that $D_1 f$ and $D_2 f$ exist on an open disk $B(P_0; r)$, where P_0 is the point (x_0, y_0). Then if $D_1 f$ and $D_2 f$ are continuous at P_0, f is differentiable at P_0.

PROOF   Choose the point $(x_0 + \Delta x, y_0 + \Delta y)$ so that it is in $B(P_0; r)$. Then

$$\Delta f(x_0, y_0) = f(x_0 + \Delta x, y_0 + \Delta y) - f(x_0, y_0)$$

Subtracting and adding $f(x_0 + \Delta x, y_0)$ to the right side of the above equation we get

$$\Delta f(x_0, y_0) = [f(x_0 + \Delta x, y_0 + \Delta y) - f(x_0 + \Delta x, y_0)]$$
$$+ [f(x_0 + \Delta x, y_0) - f(x_0, y_0)] \tag{5}$$

Because $D_1 f$ and $D_2 f$ exist on $B(P_0; r)$ and $(x_0 + \Delta x, y_0 + \Delta y)$ is in $B(P_0; r)$, it follows from (4) that

$$f(x_0 + \Delta x, y_0 + \Delta y) - f(x_0 + \Delta x, y_0) = (\Delta y)D_2 f(x_0 + \Delta x, \xi_2) \tag{6}$$

where ξ_2 is between y_0 and $y_0 + \Delta y$.

From (3) it follows that

$$f(x_0 + \Delta x, y_0) - f(x_0, y_0) = (\Delta x)D_1 f(\xi_1, y_0) \tag{7}$$

where ξ_1 is between x_0 and $x_0 + \Delta x$.

Substituting from (6) and (7) in (5) we obtain

$$\Delta f(x_0, y_0) = (\Delta y)D_2 f(x_0 + \Delta x, \xi_2) + (\Delta x)D_1 f(\xi_1, y_0) \tag{8}$$

Because $(x_0 + \Delta x, y_0 + \Delta y)$ is in $B(P_0; r)$, ξ_2 is between y_0 and $y_0 + \Delta y$, and $D_2 f$ is continuous at P_0, it follows that

$$\lim_{(\Delta x, \Delta y) \to (0,0)} D_2 f(x_0 + \Delta x, \xi_2) = D_2 f(x_0, y_0) \tag{9}$$

and because ξ_1 is between x_0 and $x_0 + \Delta x$ and $D_1 f$ is continuous at P_0,

$$\lim_{(\Delta x, \Delta y) \to (0,0)} D_1 f(\xi_1, y_0) = D_1 f(x_0, y_0) \tag{10}$$

If

$$\epsilon_1 = D_1 f(\xi_1, y_0) - D_1 f(x_0, y_0) \tag{11}$$

it follows from (10) that

$$\lim_{(\Delta x, \Delta y) \to (0,0)} \epsilon_1 = 0 \tag{12}$$

and if

$$\epsilon_2 = D_2 f(x_0 + \Delta x, \xi_2) - D_2 f(x_0, y_0) \tag{13}$$

it follows from (9) that

$$\lim_{(\Delta x, \Delta y) \to (0,0)} \epsilon_2 = 0 \tag{14}$$

Substituting from (11) and (13) into (8) we get

$$\Delta f(x_0, y_0) = \Delta y[D_2 f(x_0, y_0) + \epsilon_2] + \Delta x[D_1 f(x_0, y_0) + \epsilon_1]$$

from which we obtain

$$\Delta f(x_0, y_0) = D_1 f(x_0, y_0)\Delta x + D_2 f(x_0, y_0)\Delta y + \epsilon_1 \Delta x + \epsilon_2 \Delta y \tag{15}$$

From (12), (14), and (15), Definition 15.5.2 holds; so f is differentiable at (x_0, y_0).   ∎

EXERCISES 15.8

1. Show that Eq. (1) may be written in the form (3), where ξ_1 is between x_0 and $x_0 + h$.
2. Show that Eq. (2) may be written in the form (4), where ξ_2 is between y_0 and $y_0 + k$.

In Exercises 3 through 8, apply Theorem 15.8.2(i) to find a ξ_1.

3. $f(x, y) = x^2 + 3xy - y^2$; x is in $[1, 3]$; $y = 4$

4. $f(x, y) = x^3 - y^2$; x is in $[2, 6]$; $y = 3$

5. $f(x, y) = \dfrac{4x}{x + y}$; x is in $[0, 4]$; $y = -6$

6. $f(x, y) = \dfrac{2x - y}{2y + x}$; x is in $[-3, 3]$; $y = 5$

7. $f(x, y) = \cos x + y$; x is in $[-\pi, \pi]$; $y = 4$

8. $f(x, y) = \ln(x + y)$; x is in $[0, 2]$; $y = 1$

In Exercises 9 through 14, apply Theorem 15.8.1(ii) to find a ξ_2.

9. The function of Exercise 3; y is in $[-2, 2]$; $x = 0$

10. The function of Exercise 4; y is in $[-3, -1]$; $x = 7$

11. The function of Exercise 5; y is in $[-2, 2]$; $x = 4$

12. The function of Exercise 6; y is in $[0, 4]$; $x = 2$

13. $f(x, y) = e^y \tan x$; y is in $[3, 5]$; $x = \frac{1}{4}\pi$

14. $f(x, y) = \cos x + \sin y$; y is in $[-\frac{1}{6}\pi, \frac{1}{6}\pi]$; $x = \frac{1}{3}\pi$

In Exercises 15 and 16, use Theorem 15.5.4 to prove that f is differentiable at $(0, 0)$.

15. $f(x, y) = \begin{cases} \dfrac{x^3 y^3}{x^3 + y^3} & \text{if } (x, y) \neq (0, 0) \\ 0 & \text{if } (x, y) = (0, 0) \end{cases}$

16. $f(x, y) = \begin{cases} \dfrac{x^4 + y^4}{x^2 + y^2} & \text{if } (x, y) \neq (0, 0) \\ 0 & \text{if } (x, y) = (0, 0) \end{cases}$

17. Prove Theorem 15.5.4(ii).

REVIEW EXERCISES FOR CHAPTER 15

In Exercises 1 through 4, determine the domain of the function f, and draw a sketch showing as a region in R^2 the set of points in the domain.

1. $f(x, y) = \sqrt{x^2 + y^2 - 1}$

2. $f(x, y) = \dfrac{\sqrt{4 - x^2 - y^2}}{x}$

3. $f(x, y) = \ln(xy - 4)$

4. $f(x, y) = \sin^{-1} \sqrt{1 - x^2 - y^2}$

In Exercises 5 through 8, determine the domain of the function f.

5. $f(x, y, z) = \dfrac{xy}{\sqrt{z}}$

6. $f(x, y, z) = \dfrac{xy + xz + yz}{xyz}$

7. $f(x, y, z) = \dfrac{x}{|y| - |z|}$

8. $f(x, y, z) = \ln(x^2 + y^2 + z^2 - 4)$

9. The production function for a certain commodity is f, where $f(x, y) = 4x^{1/2}y$, and x and y give the amounts of two inputs. Draw a contour map of f showing the constant product curves at 16, 8, 4, and 2.

10. The temperature at a point (x, y) of a flat metal plate is t degrees, and $t = x^2 + 2y$. Draw the isothermals when t is 0, 2, 4, 6, and 8.

In Exercises 11 through 24, find the indicated partial derivatives.

11. $f(x, y) = 2x^2y - 3xy^2 + 4x - 2y$; (a) $D_1 f(x, y)$; (b) $D_2 f(x, y)$; (c) $D_{11} f(x, y)$; (d) $D_{22} f(x, y)$; (e) $D_{12} f(x, y)$; (f) $D_{21} f(x, y)$.

12. $f(x, y) = (4x^2 - 2y)^3$; (a) $f_1(x, y)$; (b) $f_2(x, y)$; (c) $f_{11}(x, y)$; (d) $f_{22}(x, y)$; (e) $f_{12}(x, y)$; (f) $f_{21}(x, y)$.

13. $f(x, y) = \dfrac{x^2 - y}{3y^2}$; (a) $f_x(x, y)$; (b) $f_y(x, y)$; (c) $f_{xy}(x, y)$; (d) $f_{yx}(x, y)$.

14. $f(r, s) = re^{2rs}$; (a) $D_r f(r, s)$; (b) $D_s f(r, s)$; (c) $D_{rs} f(r, s)$; (d) $D_{sr} f(r, s)$.
15. $g(s, t) = \sin(st^2) + te^s$; (a) $D_s g(s, t)$; (b) $D_t g(s, t)$; (c) $D_{st} g(s, t)$; (d) $D_{ts} g(s, t)$.

16. $h(x, y) = \tan^{-1} \dfrac{x^3}{y^2}$; (a) $D_1 h(x, y)$; (b) $D_2 h(x, y)$; (c) $D_{11} h(x, y)$; (d) $D_{22} h(x, y)$.

17. $f(x, y) = e^{x/y} + \ln \dfrac{x}{y}$; (a) $f_x(x, y)$; (b) $f_y(x, y)$; (c) $f_{xx} f(x, y)$; (d) $f_{yy}(x, y)$.

18. $f(x, y) = \ln \sqrt{x^2 + y^2}$; (a) $f_1(x, y)$; (b) $f_{11}(x, y)$; (c) $f_{12}(x, y)$; (d) $f_{121}(x, y)$.

19. $f(x, y, z) = \dfrac{x}{x^2 + y^2 + z^2}$; (a) $D_1 f(x, y, z)$; (b) $D_2 f(x, y, z)$; (c) $D_3 f(x, y, z)$.

20. $f(x, y, z) = \sqrt{x^2 + 3yz - z^2}$; (a) $f_x(x, y, z)$; (b) $f_y(x, y, z)$; (c) $f_z(x, y, z)$.
21. $f(u, v, w) = \ln(u^2 + 4v^2 - 5w^2)$; (a) $f_{uwv}(u, v, w)$; (b) $f_{uvv}(u, v, w)$.
22. $f(r, s, t) = t^2 e^{4rst}$; (a) $f_r(r, s, t)$; (b) $f_{rt}(r, s, t)$; (c) $f_{rts}(r, s, t)$.

23. $f(r, s, t) = \dfrac{\ln 4rs}{t^2}$; (a) $D_1 f(r, s, t)$; (b) $D_{13} f(r, s, t)$; (c) $D_{131} f(r, s, t)$.

24. $f(u, v, w) = w \cos 2v + 3v \sin u - 2uv \tan w$; (a) $D_2 f(u, v, w)$; (b) $D_1 f(u, v, w)$; (c) $D_{131} f(u, v, w)$.
25. If $w = x^2 y - y^2 x + y^2 z - z^2 y + z^2 x - x^2 z$, show that
26. If $u = (x^2 + y^2 + z^2)^{-1/2}$, show that

$$\frac{\partial w}{\partial x} + \frac{\partial w}{\partial y} + \frac{\partial w}{\partial z} = 0$$

$$\frac{\partial^2 u}{\partial x^2} + \frac{\partial^2 u}{\partial y^2} + \frac{\partial^2 u}{\partial z^2} = 0$$

In Exercises 27 and 28, find $\dfrac{\partial u}{\partial t}$ and $\dfrac{\partial u}{\partial s}$ by two methods.

27. $u = y \ln(x^2 + y^2)$, $x = 2s + 3t$, $y = 3t - 2s$

28. $u = e^{2x+y} \cos(2y - x)$, $x = 2s^2 - t^2$, $y = s^2 + 2t^2$

29. If $u = 3x^2 y + 2xy - 3yz - 2z^2$, $x = e^{3rs}$, $y = r^3 s^2$, and $z = \ln 4$, find $\dfrac{\partial u}{\partial r}$ by two methods: (a) Use the chain rule; (b) make the substitutions for x, y, and z before differentiating.

30. If $u = e^{x^2 + y^2} - \dfrac{3x}{y} + 3z$, $x = \sin \theta$, $y = \cos \theta$, and $z = \tan \theta$, find the total derivative $\dfrac{du}{d\theta}$ by two methods: (a) Do not express u in terms of θ before differentiating; (b) express u in terms of θ before differentiating.

31. If $u = xy + x^2$, $x = 4 \cos t$, and $y = 3 \sin t$, find the value of the total derivative $\dfrac{du}{dt}$ at $t = \tfrac{1}{4}\pi$ by two methods: (a) Do not express u in terms of t before differentiating; (b) express u in terms of t before differentiating.
32. If $f(x, y) = x^2 + ye^x$, find: (a) $\Delta f(0, 2)$, the increment of f at $(0, 2)$; (b) $\Delta f(0, 2)$ when $\Delta x = -0.1$ and $\Delta y = 0.2$; (c) $df(0, 2, \Delta x, \Delta y)$, the total differential of f at $(0, 2)$; (d) $df(0, 2, -0.1, 0.2)$.
33. If $f(x, y, z) = 3xy^2 - 5xz^2 - 2xyz$, find: (a) $\Delta f(-1, 3, 2)$, the increment of f at $(-1, 3, 2)$; (b) $\Delta f(-1, 3, 2)$ when $\Delta x = 0.02$, $\Delta y = -0.01$, and $\Delta z = -0.02$; (c) $df(-1, 3, 2, \Delta x, \Delta y, \Delta z)$, the total differential of f at $(-1, 3, 2)$; (d) $df(-1, 3, 2, 0.02, -0.01, -0.02)$.

34. Given $f(x) = x^2 + 1$, $g(x, y) = \dfrac{2x}{3y}$, and $h(x) = \dfrac{1}{x}$, find: (a) $(h \circ g)(-3, 4)$; (b) $g(f(3), h(\tfrac{1}{4}))$; (c) $g(f(x), h(y))$; (d) $f((h \circ g)(x, y))$.

In Exercises 35 through 37, evaluate the given limit by the use of limit theorems.

35. $\displaystyle \lim_{(x,y) \to (e,0)} \ln \left(\frac{x^2}{y + 1} \right)$

36. $\displaystyle \lim_{(x,y) \to (0, \pi/2)} \frac{xy^2 + e^x}{\cos x + \sin y}$

37. $\displaystyle \lim_{(x,y) \to (1,3)} \sin^{-1} \left(\frac{3x}{2y} \right)$

In Exercises 38 through 40, establish the limit by finding a $\delta > 0$ for any $\epsilon > 0$ such that Definition 15.2.5 holds.

38. $\displaystyle\lim_{(x,y)\to(4,-1)} (4x - 5y) = 21$

39. $\displaystyle\lim_{(x,y)\to(2,-2)} (3x^2 - 4y^2) = -4$

40. $\displaystyle\lim_{(x,y)\to(3,1)} (x^2 - y^2 + 2x - 4y) = 10$

In Exercises 41 through 44, determine if the indicated limit exists.

41. $\displaystyle\lim_{(x,y)\to(0,0)} \frac{x^3 y^3}{x^2 + y^2}$

42. $\displaystyle\lim_{(x,y)\to(0,0)} \frac{x^4 - y^4}{x^4 + y^4}$

43. $\displaystyle\lim_{(x,y)\to(0,0)} \frac{x^9 y}{(x^6 + y^2)^2}$

44. $\displaystyle\lim_{(x,y)\to(0,0)} \frac{2x^3 + 4x^2 y}{x^2 + y^2}$

In Exercises 45 through 49, discuss the continuity of f.

45. $f(x, y) = \dfrac{x^2 + 4y^2}{x^2 - 4y^2}$

46. $f(x, y) = \dfrac{1}{\cos \frac{1}{2}\pi x} + \dfrac{1}{\cos \frac{1}{2}\pi y}$

47. $f(x, y) = \dfrac{1}{\cos^2 \frac{1}{2}\pi x + \cos^2 \frac{1}{2}\pi y}$

48. $f(x, y) = \begin{cases} \dfrac{x^4 - y^4}{x^4 + y^4} & \text{if } (x, y) \neq (0, 0) \\ 0 & \text{if } (x, y) = (0, 0) \end{cases}$ (*Hint:* See Exercise 42.)

49. $f(x, y) = \begin{cases} \dfrac{x^3 y^3}{x^2 + y^2} & \text{if } (x, y) \neq (0, 0) \\ 0 & \text{if } (x, y) = (0, 0) \end{cases}$ (*Hint:* See Exercise 41.)

50. Suppose α is the radian measure of an acute angle of a right triangle and $\sin \alpha$ is determined by a/c, where a cm is the length of the side opposite the angle and c cm is the length of the hypotenuse. If by measurement a is found to be 3.52 and c is found to be 7.14, and there is a possible error of 0.01 in each, find the possible error in the computation of $\sin \alpha$ from these measurements.

51. A painting contractor charges \$2 per square meter for painting the four walls and ceiling of a room. If the dimensions of the ceiling are measured to be 4 m and 5 m, the height of the room is measured to be 3 m, and these measurements are correct to 0.5 cm, find approximately, by using the total differential, the greatest error in estimating the cost of the job from these measurements.

52. At a given instant, the length of one side of a rectangle is 6 cm and it is increasing at the rate of 1 cm/sec, and the length of another side of the rectangle is 10 cm and it is decreasing at the rate of 2 cm/sec. Find the rate of change of the area of the rectangle at the given instant.

53. The radius of a right-circular cylinder is decreasing at the rate of 5 cm/min and the height is increasing at the rate of 12 cm/min. Find the rate of change of the volume at the instant when the radius is 20 cm and the height is 40 cm.

54. Find the slope of the tangent line to the curve of intersection of the surface $25x^2 - 16y^2 + 9z^2 - 4 = 0$ with the plane $x = 4$ at the point $(4, 9, 10)$.

55. Use the ideal gas law (see Example 5, Section 15.4) with $k = 5$ to find the rate of change of the pressure at the instant when the volume of the gas is 80 cm^3 and the temperature is $75°$ if the volume is increasing at the rate of 3 cm^3/min and the temperature is increasing at the rate of $0.3°$ per minute.

In Exercises 56 and 57, prove that the function f is differentiable at all points in its domain by showing that Definition 15.5.2 holds.

56. $f(x, y) = 3xy^2 - 4x^2 + y^2$

57. $f(x, y) = \dfrac{2x + y}{y^2}$

58. If $f(x, y)$ units are produced by x workers and y machines, then $D_x f(x, y)$ is called the *marginal productivity of labor* and $D_y f(x, y)$ is called the *marginal productivity of machines*. Suppose that

$$f(x, y) = x^2 + 6xy + 3y^2$$

where $5 \leq x \leq 30$ and $4 \leq y \leq 12$. (a) Find the number of units produced in a day when the labor force for that day consists of 15 workers, and 8 machines are used. (b) Use the marginal productivity of labor to determine the approximate number of additional units that can be produced in 1 day if the labor force is increased from 15 to 16 and the number of machines remains fixed at 8. (c) Use the marginal productivity of machines to determine the approximate number of additional units that can be produced in 1 day if the number of machines is increased from 8 to 9 and the number of workers remains fixed at 15.

59. Find (a) $f_2(x, 0)$ if $x \neq 0$ and (b) $f_2(0, 0)$ if

$$f(x, y) = \begin{cases} \dfrac{12x^2y - 3y^2}{x^2 + y} & \text{if } (x, y) \neq (0, 0) \\ 0 & \text{if } (x, y) = (0, 0) \end{cases}$$

60. Verify that $u(x, y) = (\sinh x)(\sin y)$ satisfies Laplace's equation in R^2:

$$\frac{\partial^2 u}{\partial x^2} + \frac{\partial^2 u}{\partial y^2} = 0$$

61. If f is a differentiable function of the variable u, let $u = x^2 + y^2$ and prove that $z = xy + f(x^2 + y^2)$ satisfies the equation

$$y \frac{\partial z}{\partial x} - x \frac{\partial z}{\partial y} = y^2 - x^2$$

62. Laplace's equation in polar coordinates is

$$r^2 \frac{\partial^2 u}{\partial r^2} + r \frac{\partial u}{\partial r} + \frac{\partial^2 u}{\partial \theta^2} = 0$$

Verify that $u(r, \theta) = r^n \sin n\theta$, where n is a constant, satisfies this equation.

63. Verify that $u(x, y, z) = e^{3x + 4y} \sin 5z$ satisfies Laplace's equation in R^3:

$$\frac{\partial^2 u}{\partial x^2} + \frac{\partial^2 u}{\partial y^2} + \frac{\partial^2 u}{\partial z^2} = 0$$

64. Verify that $u(x, t) = A \cos(kat) \sin(kx)$, where A and k are arbitrary constants, satisfies the partial differential equation for a vibrating string:

$$\frac{\partial^2 u}{\partial t^2} = a^2 \frac{\partial^2 u}{\partial x^2}$$

65. Verify that

$$u(x, t) = \sin \frac{n\pi x}{L} e^{(-n^2\pi^2k^2/L^2)t}$$

satisfies the one-dimensional heat-conduction partial differential equation:

$$\frac{\partial u}{\partial t} = k^2 \frac{\partial^2 u}{\partial x^2}$$

66. Given

$$f(x, y) = \begin{cases} \dfrac{x^2y}{x^4 + y^2} & \text{if } (x, y) \neq (0, 0) \\ 0 & \text{if } (x, y) = (0, 0) \end{cases}$$

Prove that $D_1 f(0, 0)$ and $D_2 f(0, 0)$ exist but that f is not differentiable at $(0, 0)$. (*Hint:* See Example 6, Section 15.2, and Exercise 10 in Exercises 15.3).

67. Given

$$f(x, y, z) = \begin{cases} \dfrac{x^2 y^2 z^2}{(x^2 + y^2 + z^2)^2} & \text{if } (x, y, z) \neq (0, 0, 0) \\ 0 & \text{if } (x, y, z) = (0, 0, 0) \end{cases}$$

Prove that f is differentiable at $(0, 0, 0)$.

68. Let f be the function defined by

$$f(x, y) = \begin{cases} \dfrac{e^{-1/x^2} y}{e^{-2/x^2} + y^2} & \text{if } x \neq 0 \\ 0 & \text{if } x = 0 \end{cases}$$

Prove that f is discontinuous at the origin.

69. For the function of Exercise 68, prove that $D_1 f(0, 0)$ and $D_2 f(0, 0)$ both exist.

70. If f is a differentiable function of x and y and $u = f(x, y)$, $x = r \cos \theta$, and $y = r \sin \theta$, show that

$$\left(\frac{\partial u}{\partial r} \right)^2 + \frac{1}{r^2} \left(\frac{\partial u}{\partial \theta} \right)^2 = \left(\frac{\partial u}{\partial x} \right)^2 + \left(\frac{\partial u}{\partial y} \right)^2$$

S I X T E E N

DIRECTIONAL DERIVATIVES, GRADIENTS, AND APPLICATIONS OF PARTIAL DERIVATIVES

$$\nabla f(x, y) = f_x(x, y)\mathbf{i} + f_y(x, y)\mathbf{j}$$

16.1
DIRECTIONAL DERIVATIVES AND GRADIENTS

We now generalize the definition of a partial derivative to obtain the rate of change of a function with respect to any direction. This leads to the concept of a *directional derivative*.

Let f be a function of the two variables x and y and let $P(x, y)$ be a point in the xy plane. Suppose that $\mathbf{U}$ is the unit vector making an angle of radian measure θ with the positive side of the x axis. Then

$$\mathbf{U} = \cos \theta \mathbf{i} + \sin \theta \mathbf{j}$$

Figure 16.1.1 shows the representation of $\mathbf{U}$ having its initial point at the point $P(x, y)$.

16.1.1 DEFINITION   Let f be a function of two variables x and y. If $\mathbf{U}$ is the unit vector $\cos \theta \mathbf{i} + \sin \theta \mathbf{j}$, then the **directional derivative** of f in the direction of $\mathbf{U}$, denoted by $D_{\mathbf{U}}f$, is given by

$$D_{\mathbf{U}}f(x, y) = \lim_{h \to 0} \frac{f(x + h \cos \theta, y + h \sin \theta) - f(x, y)}{h}$$

if this limit exists.

FIGURE 16.1.1

FIGURE 16.1.2

The directional derivative gives the rate of change of the function values $f(x, y)$ with respect to the direction of the unit vector $\mathbf{U}$. This is illustrated in Fig. 16.1.2. An equation of the surface S in the figure is $z = f(x, y)$. $P_0(x_0, y_0, z_0)$ is a point on the surface, and the points $R(x_0, y_0, 0)$ and $Q(x_0 + h \cos \theta, y_0 + h \sin \theta, 0)$ are points in the xy plane. The plane through R and Q, parallel to the z axis, makes an angle of θ radians with the positive direction on the x axis. This plane intersects the surface S in the curve C. The directional derivative $D_{\mathbf{U}}f$, evaluated at P_0, is the slope of the tangent line to the curve C at P_0 in the plane of R, Q, and P_0.

If $\mathbf{U} = \mathbf{i}$, then $\cos \theta = 1$ and $\sin \theta = 0$, and from Definition 16.1.1,

$$D_{\mathbf{i}}f(x, y) = \lim_{h \to 0} \frac{f(x + h, y) - f(x, y)}{h}$$

which is the partial derivative of f with respect to x.

If $\mathbf{U} = \mathbf{j}$, then $\cos \theta = 0$ and $\sin \theta = 1$, and

$$D_{\mathbf{j}}f(x, y) = \lim_{h \to 0} \frac{f(x, y + h) - f(x, y)}{h}$$

which is the partial derivative of f with respect to y.

So f_x and f_y are special cases of the directional derivative in the directions of the unit vectors $\mathbf{i}$ and $\mathbf{j}$, respectively.

▶ ILLUSTRATION 1   We apply Definition 16.1.1 to find $D_{\mathbf{U}}f$ if

$$f(x, y) = 3x^2 - y^2 + 4x$$

and **U** is the unit vector in the direction $\frac{1}{6}\pi$. Then $\mathbf{U} = \cos\frac{1}{6}\pi\mathbf{i} + \sin\frac{1}{6}\pi\mathbf{j}$; that is, $\mathbf{U} = \frac{1}{2}\sqrt{3}\mathbf{i} + \frac{1}{2}\mathbf{j}$. Thus from Definition 16.1.1,

$$D_{\mathbf{U}}f(x, y) = \lim_{h \to 0} \frac{f(x + \frac{1}{2}\sqrt{3}h, y + \frac{1}{2}h) - f(x, y)}{h}$$

$$= \lim_{h \to 0} \frac{3(x + \frac{1}{2}\sqrt{3}h)^2 - (y + \frac{1}{2}h)^2 + 4(x + \frac{1}{2}\sqrt{3}h) - (3x^2 - y^2 + 4x)}{h}$$

$$= \lim_{h \to 0} \frac{3x^2 + 3\sqrt{3}hx + \frac{9}{4}h^2 - y^2 - hy - \frac{1}{4}h^2 + 4x + 2\sqrt{3}h - 3x^2 + y^2 - 4x}{h}$$

$$= \lim_{h \to 0} \frac{3\sqrt{3}hx + \frac{9}{4}h^2 - hy - \frac{1}{4}h^2 + 2\sqrt{3}h}{h}$$

$$= \lim_{h \to 0} (3\sqrt{3}x + \frac{9}{4}h - y - \frac{1}{4}h + 2\sqrt{3})$$

$$= 3\sqrt{3}x - y + 2\sqrt{3} \qquad \blacktriangleleft$$

We now proceed to obtain a formula that will enable us to calculate a directional derivative in a way that is shorter than if we used the definition. Let g be the function of the single variable t, with x, y, and θ fixed, such that

$$g(t) = f(x + t \cos \theta, y + t \sin \theta) \qquad (1)$$

and let $\mathbf{U} = \cos \theta \mathbf{i} + \sin \theta \mathbf{j}$. Then by the definition of an ordinary derivative,

$$g'(0) = \lim_{h \to 0} \frac{f(x + (0 + h)\cos \theta, y + (0 + h)\sin \theta) - f(x + 0\cos \theta, y + 0\sin \theta)}{h}$$

$$g'(0) = \lim_{h \to 0} \frac{f(x + h \cos \theta, y + h \sin \theta) - f(x, y)}{h}$$

Because the right side of the above is $D_{\mathbf{U}}f(x, y)$,

$$g'(0) = D_{\mathbf{U}}f(x, y) \qquad (2)$$

We now find $g'(t)$ by applying the chain rule to the right side of (1), which gives

$$g'(t) = f_1(x + t \cos \theta, y + t \sin \theta)\frac{\partial(x + t \cos \theta)}{\partial t}$$

$$+ f_2(x + t \cos \theta, y + t \sin \theta)\frac{\partial(y + t \sin \theta)}{\partial t}$$

$$= f_1(x + t \cos \theta, y + t \sin \theta)\cos \theta + f_2(x + t \cos \theta, y + t \sin \theta)\sin \theta$$

Therefore

$$g'(0) = f_x(x, y)\cos \theta + f_y(x, y)\sin \theta \qquad (3)$$

From (2) and (3) the following theorem is obtained.

16.1.2 THEOREM   If f is a differentiable function of x and y, and $\mathbf{U} = \cos\theta\mathbf{i} + \sin\theta\mathbf{j}$, then

$$D_{\mathbf{U}}f(x, y) = f_x(x, y)\cos\theta + f_y(x, y)\sin\theta$$

▶ ILLUSTRATION 2   For the function f and the unit vector $\mathbf{U}$ of Illustration 1, we find $D_{\mathbf{U}}f$ by Theorem 16.1.2.

Because $f(x, y) = 3x^2 - y^2 + 4x$, $f_x(x, y) = 6x + 4$ and $f_y(x, y) = -2y$. Because $\mathbf{U} = \cos\frac{1}{6}\pi\mathbf{i} + \sin\frac{1}{6}\pi\mathbf{j}$, then from Theorem 16.1.2,

$$D_{\mathbf{U}}f(x, y) = (6x + 4)\tfrac{1}{2}\sqrt{3} + (-2y)\tfrac{1}{2}$$
$$= 3\sqrt{3}x + 2\sqrt{3} - y$$

which agrees with the result in Illustration 1.   ◀

The directional derivative can be written as the dot product of two vectors. Because

$$f_x(x, y)\cos\theta + f_y(x, y)\sin\theta = (\cos\theta\mathbf{i} + \sin\theta\mathbf{j}) \cdot [f_x(x, y)\mathbf{i} + f_y(x, y)\mathbf{j}]$$

it follows from Theorem 16.1.2 that

$$D_{\mathbf{U}}f(x, y) = (\cos\theta\mathbf{i} + \sin\theta\mathbf{j}) \cdot [f_x(x, y)\mathbf{i} + f_y(x, y)\mathbf{j}] \qquad (4)$$

The second vector on the right side of (4) is a very important one, and it is called the *gradient* of the function f. The symbol used for the gradient of f is ∇f, where ∇ is an inverted capital delta and is read "del." Sometimes the abbreviation *grad f* is used.

16.1.3 DEFINITION   If f is a function of two variables x and y, and f_x and f_y exist, then the **gradient** of f, denoted by ∇f (read "del f"), is defined by

$$\nabla f(x, y) = f_x(x, y)\mathbf{i} + f_y(x, y)\mathbf{j}$$

From Definition 16.1.3, Eq. (4) can be written as

$$D_{\mathbf{U}}f(x, y) = \mathbf{U} \cdot \nabla f(x, y) \qquad (5)$$

Therefore any directional derivative of a differentiable function can be obtained by dot-multiplying the gradient by a unit vector in the desired direction.

EXAMPLE 1

If

$$f(x, y) = \tfrac{1}{16}x^2 + \tfrac{1}{9}y^2$$

find the gradient of f at the point $(4, 3)$. Also find the rate of change of $f(x, y)$ in the direction $\frac{1}{4}\pi$ at $(4, 3)$.

SOLUTION
Because $f_x(x, y) = \frac{1}{8}x$ and $f_y(x, y) = \frac{2}{9}y$,

$$\nabla f(x, y) = \tfrac{1}{8}x\mathbf{i} + \tfrac{2}{9}y\mathbf{j} \qquad \nabla f(4, 3) = \tfrac{1}{2}\mathbf{i} + \tfrac{2}{3}\mathbf{j}$$

The rate of change of $f(x, y)$ in the direction $\frac{1}{4}\pi$ at $(4, 3)$ is $D_\mathbf{U} f(4, 3)$, where $\mathbf{U}$ is the unit vector

$$\frac{1}{\sqrt{2}}\mathbf{i} + \frac{1}{\sqrt{2}}\mathbf{j}$$

We find $D_\mathbf{U} f(4, 3)$ by dot-multiplying $\nabla f(4, 3)$ by $\mathbf{U}$.

$$D_\mathbf{U} f(4, 3) = \left(\frac{1}{\sqrt{2}}\mathbf{i} + \frac{1}{\sqrt{2}}\mathbf{j}\right) \cdot \left(\frac{1}{2}\mathbf{i} + \frac{2}{3}\mathbf{j}\right)$$

$$= \frac{7}{6\sqrt{2}}$$

If α is the radian measure of the angle between the two vectors $\mathbf{U}$ and ∇f, then

$$\mathbf{U} \cdot \nabla f(x, y) = \|\mathbf{U}\| \, \|\nabla f(x, y)\| \cos \alpha \tag{6}$$

From (5) and (6) it follows that

$$D_\mathbf{U} f(x, y) = \|\mathbf{U}\| \, \|\nabla f(x, y)\| \cos \alpha \tag{7}$$

We see from (7) that $D_\mathbf{U} f$ will be a maximum when $\cos \alpha = 1$, that is, when $\mathbf{U}$ is in the direction of ∇f; and in this case, $D_\mathbf{U} f = \|\nabla f\|$. Hence the gradient of a function is in the direction in which the function has its maximum rate of change. In particular, on a two-dimensional topographical map of a landscape where z units is the elevation at a point (x, y) and $z = f(x, y)$, the direction in which the rate of change of z is the greatest is given by $\nabla f(x, y)$; that is, $\nabla f(x, y)$ points in the direction of steepest ascent. This accounts for the name *gradient* (the grade is steepest in the direction of the gradient).

▶ ILLUSTRATION 3   In Fig. 16.1.3 there is a contour map showing the level curves of the function of Example 1 at 1, 2, and 3. The level curves are ellipses. The figure also shows the representation of $\nabla f(4, 3)$ having its initial point at $(4, 3)$.   ◀

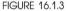

FIGURE 16.1.3

EXAMPLE 2
Given

$$f(x, y) = 2x^2 - y^2 + 3x - y$$

find the maximum value of $D_\mathbf{U} f$ at the point where $x = 1$ and $y = -2$.

SOLUTION
Because $f_x(x, y) = 4x + 3$ and $f_y(x, y) = -2y - 1$,

$$\nabla f(x, y) = (4x + 3)\mathbf{i} + (-2y - 1)\mathbf{j} \qquad \nabla f(1, -2) = 7\mathbf{i} + 3\mathbf{j}$$

Thus the maximum value of $D_\mathbf{U} f$ at the point $(1, -2)$ is

$$\|\nabla f(1, -2)\| = \sqrt{49 + 9}$$
$$= \sqrt{58}$$

EXAMPLE 3

The temperature at any point (x, y) of a rectangular plate lying in the xy plane is determined by

$$T(x, y) = x^2 + y^2$$

(a) Find the rate of change of the temperature at the point $(3, 4)$ in the direction making an angle of radian measure $\frac{1}{3}\pi$ with the positive x direction; (b) find the direction for which the rate of change of the temperature at the point $(-3, 1)$ is a maximum.

SOLUTION
(a) We wish to find $D_\mathbf{U} T(x, y)$, where

$$\mathbf{U} = \cos \tfrac{1}{3}\pi \mathbf{i} + \sin \tfrac{1}{3}\pi \mathbf{j} \qquad \nabla T(x, y) = T_x(x, y)\mathbf{i} + T_y(x, y)\mathbf{j}$$
$$= \tfrac{1}{2}\mathbf{i} + \tfrac{1}{2}\sqrt{3}\mathbf{j} \qquad\qquad = 2x\mathbf{i} + 2y\mathbf{j}$$

Therefore

$$D_\mathbf{U} T(x, y) = \mathbf{U} \cdot \nabla T(x, y)$$
$$= (\tfrac{1}{2}\mathbf{i} + \tfrac{1}{2}\sqrt{3}\mathbf{j}) \cdot (2x\mathbf{i} + 2y\mathbf{j})$$
$$= x + \sqrt{3}y$$

Thus

$$D_\mathbf{U} T(3, 4) = 3 + 4\sqrt{3}$$
$$\approx 9.93$$

So at $(3, 4)$ the temperature is increasing at the rate of approximately 9.93 units per unit change in the distance measured in the direction of $\mathbf{U}$.
(b) $D_\mathbf{U} T(-3, 1)$ is a maximum when $\mathbf{U}$ is in the direction of $\nabla T(-3, 1)$. Because $\nabla T(-3, 1) = -6\mathbf{i} + 2\mathbf{j}$, the radian measure of the angle giving the direction of $\nabla T(-3, 1)$ is θ, where $\tan \theta = -\frac{1}{3}$. Thus $\theta = \pi - \tan^{-1} \frac{1}{3}$. Therefore the rate of change of the temperature at the point $(-3, 1)$ is a maximum in the direction making an angle of radian measure $\pi - \tan^{-1} \frac{1}{3}$ with the positive side of the x axis.

We extend the definition of a directional derivative to a function of three variables. In three-dimensional space the direction of a vector is determined by its direction cosines. If $\cos \alpha$, $\cos \beta$, and $\cos \gamma$ are the direction cosines of the unit vector $\mathbf{U}$, then $\mathbf{U} = \cos \alpha \mathbf{i} + \cos \beta \mathbf{j} + \cos \gamma \mathbf{k}$.

16.1.4 DEFINITION Suppose that f is a function of three variables x, y, and z. If $\mathbf{U}$ is the unit vector $\cos \alpha \mathbf{i} + \cos \beta \mathbf{j} + \cos \gamma \mathbf{k}$, then the directional derivative of f in the direction of $\mathbf{U}$, denoted by $D_\mathbf{U} f$, is given by

$$D_\mathbf{U} f(x, y, z) = \lim_{h \to 0} \frac{f(x + h \cos \alpha, y + h \cos \beta, z + h \cos \gamma) - f(x, y, z)}{h}$$

if this limit exists.

The directional derivative of a function of three variables gives the rate of change of the function values $f(x, y, z)$ with respect to distance in three-dimensional space measured in the direction of the unit vector $\mathbf{U}$.

The following theorem, which provides a method for calculating a directional derivative for a function of three variables, is proved in a manner similar to the proof of Theorem 16.1.2.

16.1.5 THEOREM   If f is a differentiable function of x, y, and z and

$$\mathbf{U} = \cos \alpha \mathbf{i} + \cos \beta \mathbf{j} + \cos \gamma \mathbf{k}$$

then

$$D_{\mathbf{U}} f(x, y, z) = f_x(x, y, z) \cos \alpha + f_y(x, y, z) \cos \beta + f_z(x, y, z) \cos \gamma \qquad (8)$$

EXAMPLE 4
Given

$$f(x, y, z) = 3x^2 + xy - 2y^2 - yz + z^2$$

find the rate of change of $f(x, y, z)$ at $(1, -2, -1)$ in the direction of the vector $2\mathbf{i} - 2\mathbf{j} - \mathbf{k}$.

SOLUTION
The unit vector in the direction of $2\mathbf{i} - 2\mathbf{j} - \mathbf{k}$ is

$$\mathbf{U} = \tfrac{2}{3}\mathbf{i} - \tfrac{2}{3}\mathbf{j} - \tfrac{1}{3}\mathbf{k}$$

So from (8),

$$D_{\mathbf{U}} f(x, y, z) = \tfrac{2}{3}(6x + y) - \tfrac{2}{3}(x - 4y - z) - \tfrac{1}{3}(-y + 2z)$$

Therefore the rate of change of $f(x, y, z)$ at $(1, -2, -1)$ in the direction of $\mathbf{U}$ is given by

$$D_{\mathbf{U}} f(1, -2, -1) = \tfrac{2}{3}(4) - \tfrac{2}{3}(10) - \tfrac{1}{3}(0)$$
$$= -4$$

16.1.6 DEFINITION   If f is a function of three variables x, y, and z and the first partial derivatives f_x, f_y, and f_z exist, then the **gradient** of f, denoted by ∇f, is defined by

$$\nabla f(x, y, z) = f_x(x, y, z)\mathbf{i} + f_y(x, y, z)\mathbf{j} + f_z(x, y, z)\mathbf{k}$$

Just as for functions of two variables, it follows from Theorem 16.1.5 and Definition 16.1.6 that if $\mathbf{U} = \cos \alpha \mathbf{i} + \cos \beta \mathbf{j} + \cos \gamma \mathbf{k}$, then

$$D_{\mathbf{U}} f(x, y, z) = \mathbf{U} \cdot \nabla f(x, y, z) \qquad (9)$$

Also, the directional derivative is a maximum when $\mathbf{U}$ is in the direction of the gradient, and the maximum directional derivative is the magnitude of the gradient.

Applications of the gradient occur in physics in problems in heat conduction and electricity. Suppose that $w = f(x, y, z)$. A level surface of this function f at the constant k is given by

$$f(x, y, z) = k \tag{10}$$

If w is the number of degrees in the temperature at point (x, y, z), then all points on the surface of Eq. (10) have the same temperature of k degrees, and the surface is called an **isothermal surface**. If w is the number of volts in the electric potential at point (x, y, z), then all points on the surface are at the same potential, and the surface is called an **equipotential surface**. The gradient vector at a point gives the direction of greatest rate of change of w. So if the level surface of Eq. (10) is an isothermal surface, $\nabla f(x, y, z)$ gives the direction of the greatest rate of change of temperature at (x, y, z). If (10) is an equation of an equipotential surface, then $\nabla f(x, y, z)$ gives the direction of the greatest rate of change of potential at (x, y, z).

EXAMPLE 5

If $V(x, y, z)$ volts is the electric potential at any point (x, y, z) in three-dimensional space and

$$V(x, y, z) = \frac{1}{\sqrt{x^2 + y^2 + z^2}}$$

find: (a) the rate of change of V at the point $(2, 2, -1)$ in the direction of the vector $2\mathbf{i} - 3\mathbf{j} + 6\mathbf{k}$; (b) the direction of the greatest rate of change of V at $(2, 2, -1)$.

SOLUTION

(a) A unit vector in the direction of $2\mathbf{i} - 3\mathbf{j} + 6\mathbf{k}$ is

$$\mathbf{U} = \tfrac{2}{7}\mathbf{i} - \tfrac{3}{7}\mathbf{j} + \tfrac{6}{7}\mathbf{k}$$

We wish to find $D_\mathbf{U}V(2, 2, -1)$.

$$\nabla V(x, y, z) = V_x(x, y, z)\mathbf{i} + V_y(x, y, z)\mathbf{j} + V_z(x, y, z)\mathbf{k}$$

$$= \frac{-x}{(x^2 + y^2 + z^2)^{3/2}}\mathbf{i} + \frac{-y}{(x^2 + y^2 + z^2)^{3/2}}\mathbf{j} + \frac{-z}{(x^2 + y^2 + z^2)^{3/2}}\mathbf{k}$$

Then

$$D_\mathbf{U}V(2, 2, -1) = \mathbf{U} \cdot \nabla V(2, 2, -1)$$

$$= (\tfrac{2}{7}\mathbf{i} - \tfrac{3}{7}\mathbf{j} + \tfrac{6}{7}\mathbf{k}) \cdot (-\tfrac{2}{27}\mathbf{i} - \tfrac{2}{27}\mathbf{j} + \tfrac{1}{27}\mathbf{k})$$

$$= -\tfrac{4}{189} + \tfrac{6}{189} + \tfrac{6}{189}$$

$$= \tfrac{8}{189}$$

$$\approx 0.042$$

Therefore, at $(2, 2, -1)$ the potential is increasing at the rate of approximately 0.042 volt per unit change in the distance measured in the direction of $\mathbf{U}$.

(b) $\nabla V(2, 2, -1) = -\frac{2}{27}\mathbf{i} - \frac{2}{27}\mathbf{j} + \frac{1}{27}\mathbf{k}$. A unit vector in the direction of $\nabla V(2, 2, -1)$ is

$$\frac{\nabla V(2, 2, -1)}{\|\nabla V(2, 2, -1)\|} = \frac{-\frac{2}{27}\mathbf{i} - \frac{2}{27}\mathbf{j} + \frac{1}{27}\mathbf{k}}{\frac{3}{27}}$$

$$= -\tfrac{2}{3}\mathbf{i} - \tfrac{2}{3}\mathbf{j} + \tfrac{1}{3}\mathbf{k}$$

The direction cosines of this vector are $-\frac{2}{3}$, $-\frac{2}{3}$, and $\frac{1}{3}$, which give the direction of the greatest rate of change of V at $(2, 2, -1)$.

EXERCISES 16.1

In Exercises 1 through 6, find the directional derivative of the given function in the direction of the given unit vector $\mathbf{U}$ by using either Definition 16.1.1 or Definition 16.1.4, and then verify your result by applying either Theorem 16.1.2 or Theorem 16.1.5, whichever one applies.

1. $f(x, y) = 2x^2 + 5y^2$; $\mathbf{U} = \cos \frac{1}{4}\pi \mathbf{i} + \sin \frac{1}{4}\pi \mathbf{j}$

2. $g(x, y) = 3x^2 - 4y^2$; $\mathbf{U} = \cos \frac{1}{3}\pi \mathbf{i} + \sin \frac{1}{3}\pi \mathbf{j}$

3. $h(x, y, z) = 3x^2 + y^2 - 4z^2$; $\mathbf{U} = \cos \frac{1}{3}\pi \mathbf{i} + \cos \frac{1}{4}\pi \mathbf{j} + \cos \frac{2}{3}\pi \mathbf{k}$

4. $f(x, y, z) = 6x^2 - 2xy + yz$; $\mathbf{U} = \frac{3}{7}\mathbf{i} + \frac{2}{7}\mathbf{j} + \frac{6}{7}\mathbf{k}$

5. $g(x, y) = \dfrac{1}{x - y}$; $\mathbf{U} = -\frac{12}{13}\mathbf{i} + \frac{5}{13}\mathbf{j}$

6. $f(x, y) = \dfrac{1}{x^2 + y^2}$; $\mathbf{U} = \frac{3}{5}\mathbf{i} - \frac{4}{5}\mathbf{j}$

In Exercises 7 through 14, find the gradient of the given function.

7. $f(x, y) = 4x^2 - 3xy + y^2$

8. $g(x, y) = \dfrac{xy}{x^2 + y^2}$

9. $g(x, y) = \ln \sqrt{x^2 + y^2}$

10. $f(x, y) = e^y \tan 2x$

11. $f(x, y, z) = \dfrac{x - y}{x + z}$

12. $f(x, y, z) = 3z \ln(x + y)$

13. $g(x, y, z) = xe^{-2y} \sec z$

14. $g(x, y, z) = e^{2z}(\sin x - \cos y)$

In Exercises 15 through 22, find the value of the directional derivative at the particular point P_0 for the given function in the direction of $\mathbf{U}$.

15. $f(x, y) = x^2 - 2xy^2$; $\mathbf{U} = \cos \pi \mathbf{i} + \sin \pi \mathbf{j}$; $P_0 = (1, -2)$

16. $g(x, y) = 3x^3y + 4y^2 - xy$; $\mathbf{U} = \cos \frac{1}{4}\pi \mathbf{i} + \sin \frac{1}{4}\pi \mathbf{j}$; $P_0 = (0, 3)$

17. $g(x, y) = y^2 \tan^2 x$; $\mathbf{U} = -\frac{1}{2}\sqrt{3}\mathbf{i} + \frac{1}{2}\mathbf{j}$; $P_0 = (\frac{1}{3}\pi, 2)$

18. $f(x, y) = xe^{2y}$; $\mathbf{U} = \frac{1}{2}\mathbf{i} + \frac{1}{2}\sqrt{3}\mathbf{j}$; $P_0 = (2, 0)$

19. $h(x, y, z) = \cos(xy) + \sin(yz)$; $\mathbf{U} = -\frac{1}{3}\mathbf{i} + \frac{2}{3}\mathbf{j} + \frac{2}{3}\mathbf{k}$; $P_0 = (2, 0, -3)$

20. $f(x, y, z) = \ln(x^2 + y^2 + z^2)$; $\mathbf{U} = \dfrac{1}{\sqrt{3}}\mathbf{i} - \dfrac{1}{\sqrt{3}}\mathbf{j} - \dfrac{1}{\sqrt{3}}\mathbf{k}$; $P_0 = (1, 3, 2)$

21. $f(x, y) = e^{-3x}\cos 3y$; $\mathbf{U} = \cos(-\frac{1}{12}\pi)\mathbf{i} + \sin(-\frac{1}{12}\pi)\mathbf{j}$; $P_0 = (-\frac{1}{12}\pi, 0)$

22. $g(x, y, z) = \cos 2x \cos 3y \sinh 4z$; $\mathbf{U} = \dfrac{1}{\sqrt{3}}\mathbf{i} - \dfrac{1}{\sqrt{3}}\mathbf{j} + \dfrac{1}{\sqrt{3}}\mathbf{k}$; $P_0 = (\frac{1}{2}\pi, 0, 0)$

In Exercises 23 through 26, a function f, a point P, and a unit vector $\mathbf{U}$ are given. Find (a) the gradient of f at P and (b) the rate of change of the function value in the direction of $\mathbf{U}$ at P.

23. $f(x, y) = x^2 - 4y$; $P = (-2, 2)$; $\mathbf{U} = \cos \frac{1}{3}\pi \mathbf{i} + \sin \frac{1}{3}\pi \mathbf{j}$

24. $f(x, y) = e^{2xy}$; $P = (2, 1)$; $\mathbf{U} = \frac{4}{5}\mathbf{i} - \frac{3}{5}\mathbf{j}$

25. $f(x, y, z) = y^2 + z^2 - 4xz;\ P = (-2, 1, 3);\ \mathbf{U} = \frac{2}{7}\mathbf{i} - \frac{6}{7}\mathbf{j} + \frac{3}{7}\mathbf{k}$

26. $f(x, y, z) = 2x^3 + xy^2 + xz^2;\ P = (1, 1, 1);\ \mathbf{U} = \frac{1}{7}\sqrt{21}\mathbf{j} - \frac{2}{7}\sqrt{7}\mathbf{k}$

27. Draw a contour map showing the level curves of the function of Exercise 23 at 8, 4, 0, -4, and -8. Also show the representation of $\nabla f(-2, 2)$ having its initial point at $(-2, 2)$.

28. Draw a contour map showing the level curves of the function of Exercise 24, at e^8, e^4, 1, e^{-4}, and e^{-8}. Also show the representation of $\nabla f(2, 1)$ having its initial point at $(2, 1)$.

In Exercises 29 through 32, find $D_{\mathbf{U}}f$ at the given point P for which $\mathbf{U}$ is a unit vector in the direction of $\overline{PQ}$. Also at P find $D_{\mathbf{U}}f$, if $\mathbf{U}$ is a unit vector for which $D_{\mathbf{U}}f$ is a maximum.

29. $f(x, y) = e^x \tan^{-1} y;\ P(0, 1),\ Q(3, 5)$

30. $f(x, y) = e^x \cos y + e^y \sin x;\ P(1, 0),\ Q(-3, 3)$

31. $f(x, y, z) = x - 2y + z^2;\ P(3, 1, -2),\ Q(10, 7, 4)$

32. $f(x, y, z) = x^2 + y^2 - 4xz;\ P(3, 1, -2),\ Q(-6, 3, 4)$

33. Find the direction from the point $(1, 3)$ for which the value of f does not change if $f(x, y) = e^{2y} \tan^{-1} \dfrac{y}{3x}$.

34. The density is $\rho(x, y)$ slugs/ft^2 at any point of a rectangular plate in the xy plane and $\rho(x, y) = 1/\sqrt{x^2 + y^2 + 3}$. (a) Find the rate of change of the density at the point $(3, 2)$ in the direction of the unit vector $\cos \frac{2}{3}\pi \mathbf{i} + \sin \frac{2}{3}\pi \mathbf{j}$. (b) Find the direction and magnitude of the greatest rate of change of ρ at $(3, 2)$.

35. The temperature is $T(x, y)$ degrees at any point of a rectangular plate lying in the xy plane, and $T(x, y) = 3x^2 + 2xy$. Distance is measured in meters. (a) Find the maximum rate of change of the temperature at the point $(3, -6)$ on the plate. (b) Find the direction for which this maximum rate of change at $(3, -6)$ occurs.

36. The temperature is $T(x, y, z)$ degrees at any point of a solid in three-dimensional space, and $T(x, y, z) = 60/(x^2 + y^2 + z^2 + 3)$. Distance is measured in inches. (a) Find the rate of change of the temperature at the point $(3, -2, 2)$ in the direction of the vector $-2\mathbf{i} + 3\mathbf{j} - 6\mathbf{k}$. (b) Find the direction and magnitude of the greatest rate of change of T at $(3, -2, 2)$.

37. The electric potential is $V(x, y)$ volts at any point in the xy plane, and $V(x, y) = e^{-2x} \cos 2y$. Distance is measured in feet. (a) Find the rate of change of the potential at the point $(0, \frac{1}{4}\pi)$ in the direction of the unit vector $\cos \frac{1}{6}\pi \mathbf{i} + \sin \frac{1}{6}\pi \mathbf{j}$. (b) Find the direction and magnitude of the greatest rate of change of V at $(0, \frac{1}{4}\pi)$.

38. An equation of the surface of a mountain is $z = 1200 - 3x^2 - 2y^2$, where distance is measured in meters, the x axis points to the east, and the y axis points to the north. A mountain climber is at the point corresponding to $(-10, 5, 850)$. (a) What is the direction of steepest ascent? (b) If the climber moves in the east direction, is she ascending or descending, and what is her rate? (c) If the climber moves in the southwest direction, is she ascending or descending, and what is her rate? (d) In what direction is she traveling a level path?

16.2
TANGENT PLANES AND NORMALS TO SURFACES

Let S be the surface having the equation

$$F(x, y, z) = 0 \qquad (1)$$

and suppose that $P_0(x_0, y_0, z_0)$ is a point on S. Then $F(x_0, y_0, z_0) = 0$. Suppose further that C is a curve on S through P_0 and that a set of parametric equations of C is

$$x = f(t) \qquad y = g(t) \qquad z = h(t) \qquad (2)$$

Let the value of the parameter t at the point P_0 be t_0. A vector equation of C is

$$\mathbf{R}(t) = f(t)\mathbf{i} + g(t)\mathbf{j} + h(t)\mathbf{k} \qquad (3)$$

Because curve C is on surface S, we have, upon substituting from (2) in (1),

$$F(f(t), g(t), h(t)) = 0 \qquad (4)$$

Let $G(t) = F(f(t), g(t), h(t))$. If F_x, F_y, and F_z are continuous and not all zero at P_0, and if $f'(t_0)$, $g'(t_0)$, and $h'(t_0)$ exist, then the total derivative of F with respect to t at P_0 is given by

$$G'(t_0) = F_x(x_0, y_0, z_0)f'(t_0) + F_y(x_0, y_0, z_0)g'(t_0) + F_z(x_0, y_0, z_0)h'(t_0)$$

The right side of this equation can be written as

$$[F_x(x_0, y_0, z_0)\mathbf{i} + F_y(x_0, y_0, z_0)\mathbf{j} + F_z(x_0, y_0, z_0)\mathbf{k}] \cdot [f'(t_0)\mathbf{i} + g'(t_0)\mathbf{j} + h'(t_0)\mathbf{k}]$$

Thus

$$G'(t_0) = \nabla F(x_0, y_0, z_0) \cdot D_t \mathbf{R}(t_0)$$

Since $G'(t) = 0$ for all t under consideration (because of (4)), $G'(t_0) = 0$; so it follows from the above that

$$\nabla F(x_0, y_0, z_0) \cdot D_t \mathbf{R}(t_0) = 0 \qquad (5)$$

From Section 14.8 we know that $D_t \mathbf{R}(t_0)$ has the same direction as a tangent vector to curve C at P_0. Therefore, from (5) we can conclude that the gradient vector of F at P_0 is orthogonal to a tangent vector of every curve C on S through the point P_0. We are led, then, to the following definition.

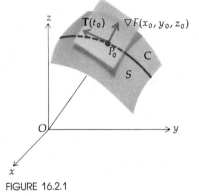

FIGURE 16.2.1

16.2.1 DEFINITION   A vector that is orthogonal to a tangent vector of every curve C through a point P_0 on a surface S is called a **normal vector** to S at P_0.

From this definition and the preceding discussion we have the following theorem.

16.2.2 THEOREM   If an equation of a surface S is $F(x, y, z) = 0$, and F_x, F_y, and F_z are continuous and not all zero at the point $P_0(x_0, y_0, z_0)$ on S, then a normal vector to S at P_0 is $\nabla F(x_0, y_0, z_0)$.

The concept of a normal vector is used to define the *tangent plane* to a surface at a point.

16.2.3 DEFINITION   If an equation of a surface S is $F(x, y, z) = 0$, then the **tangent plane** of S at a point $P_0(x_0, y_0, z_0)$ is the plane through P_0 having as a normal vector $\nabla F(x_0, y_0, z_0)$.

An equation of the tangent plane of the above definition is

$$F_x(x_0, y_0, z_0)(x - x_0) + F_y(x_0, y_0, z_0)(y - y_0) + F_z(x_0, y_0, z_0)(z - z_0) = 0 \qquad (6)$$

Refer to Fig. 16.2.1, which shows the tangent plane to the surface S at P_0 and the representation of the gradient vector having its initial point at P_0.

A vector equation of the tangent plane given by (6) is

$$\nabla F(x_0, y_0, z_0) \cdot [(x - x_0)\mathbf{i} + (y - y_0)\mathbf{j} + (z - z_0)\mathbf{k}] = 0 \tag{7}$$

EXAMPLE 1

Find an equation of the tangent plane to the elliptic paraboloid

$$4x^2 + y^2 - 16z = 0$$

at the point $(2, 4, 2)$.

SOLUTION
Let $F(x, y, z) = 4x^2 + y^2 - 16z$. Then

$$\nabla F(x, y, z) = 8x\mathbf{i} + 2y\mathbf{j} - 16\mathbf{k} \qquad \nabla F(2, 4, 2) = 16\mathbf{i} + 8\mathbf{j} - 16\mathbf{k}$$

From (7) it follows that an equation of the tangent plane is

$$16(x - 2) + 8(y - 4) - 16(z - 2) = 0$$
$$2x + y - 2z - 4 = 0$$

16.2.4 DEFINITION

The **normal line** to a surface S at a point P_0 on S is the line through P_0 having as a set of direction numbers the components of any normal vector to S at P_0.

If an equation of a surface S is $F(x, y, z) = 0$, it follows from Definition 16.2.4 that symmetric equations of the normal line to S at $P_0(x_0, y_0, z_0)$ are

$$\frac{x - x_0}{F_x(x_0, y_0, z_0)} = \frac{y - y_0}{F_y(x_0, y_0, z_0)} = \frac{z - z_0}{F_z(x_0, y_0, z_0)} \tag{8}$$

The denominators in (8) are components of $\nabla F(x_0, y_0, z_0)$, which is a normal vector to S at P_0; thus (8) follows from Definition 16.2.4. The normal line at a point on a surface is perpendicular to the tangent plane there.

EXAMPLE 2

Find symmetric equations of the normal line to the surface of Example 1 at $(2, 4, 2)$.

SOLUTION
Because $\nabla F(2, 4, 2) = 16\mathbf{i} + 8\mathbf{j} - 16\mathbf{k}$, it follows that symmetric equations of the required normal line are

$$\frac{x - 2}{2} = \frac{y - 4}{1} = \frac{z - 2}{-2}$$

16.2.5 DEFINITION

The **tangent line** to a curve C at a point P_0 is the line through P_0 having as a set of direction numbers the components of the unit tangent vector to C at P_0.

FIGURE 16.2.2

From Definitions 16.2.3 and 16.2.5, all the tangent lines at the point P_0 to the curves lying on a given surface lie in the tangent plane to the surface at P_0. Refer to Fig. 16.2.2, showing sketches of a surface and the tangent plane at P_0. Some of the curves through P_0 and their tangent lines are also sketched in the figure.

Consider a curve C that is the intersection of two surfaces having equations

$$F(x, y, z) = 0 \quad \text{and} \quad G(x, y, z) = 0$$

respectively. We shall show how to obtain equations of the tangent line to C at a point $P_0(x_0, y_0, z_0)$. Because this tangent line lies in each of the tangent planes to the given surfaces at P_0, it is the line of intersection of the two tangent planes. Let N_1 be a normal vector at P_0 to the surface having the equation $F(x, y, z) = 0$, and let N_2 be a normal vector at P_0 to the surface having the equation $G(x, y, z) = 0$. Then

$$N_1 = \nabla F(x_0, y_0, z_0) \quad \text{and} \quad N_2 = \nabla G(x_0, y_0, z_0)$$

Both N_1 and N_2 are orthogonal to the unit tangent vector to C at P_0; so if N_1 and N_2 are not parallel, it follows from Theorem 14.5.10 that the unit tangent vector has the direction that is the same as, or opposite to, the direction of $N_1 \times N_2$. Therefore the components of $N_1 \times N_2$ serve as a set of direction numbers of the tangent line. From this set of direction numbers and the coordinates of P_0 we can obtain symmetric equations of the required tangent line. This is illustrated in the following example.

EXAMPLE 3
Find symmetric equations of the tangent line to the curve of intersection of the surfaces

$$3x^2 + 2y^2 + z^2 = 49 \quad \text{and} \quad x^2 + y^2 - 2z^2 = 10$$

at the point $(3, -3, 2)$.

SOLUTION
Let

$$F(x, y, z) = 3x^2 + 2y^2 + z^2 - 49 \quad \text{and} \quad G(x, y, z) = x^2 + y^2 - 2z^2 - 10$$

Then $\nabla F(x, y, z) = 6x\mathbf{i} + 4y\mathbf{j} + 2z\mathbf{k}$ and $\nabla G(x, y, z) = 2x\mathbf{i} + 2y\mathbf{j} - 4z\mathbf{k}$. Therefore

$$
\begin{aligned}
N_1 &= \nabla F(3, -3, 2) & N_2 &= \nabla G(3, -3, 2) \\
&= 18\mathbf{i} - 12\mathbf{j} + 4\mathbf{k} & &= 6\mathbf{i} - 6\mathbf{j} - 8\mathbf{k} \\
&= 2(9\mathbf{i} - 6\mathbf{j} + 2\mathbf{k}) & &= 2(3\mathbf{i} - 3\mathbf{j} - 4\mathbf{k})
\end{aligned}
$$

$$
\begin{aligned}
N_1 \times N_2 &= 4(9\mathbf{i} - 6\mathbf{j} + 2\mathbf{k}) \times (3\mathbf{i} - 3\mathbf{j} - 4\mathbf{k}) \\
&= 4(30\mathbf{i} + 42\mathbf{j} - 9\mathbf{k}) \\
&= 12(10\mathbf{i} + 14\mathbf{j} - 3\mathbf{k})
\end{aligned}
$$

Therefore a set of direction numbers of the required tangent line is $[10, 14, -3]$. Symmetric equations of the line are, then,

$$\frac{x-3}{10} = \frac{y+3}{14} = \frac{z-2}{-3}$$

If two surfaces have a common tangent plane at a point, the two surfaces are said to be **tangent** at that point. It follows from Definition 16.2.3 that two surfaces S_1 and S_2, whose equations are $F(x, y, z) = 0$ and $G(x, y, z) = 0$, respectively, are tangent at the point $P_0(x_0, y_0, z_0)$ if for some constant k
$$\nabla F(x_0, y_0, z_0) = k\,\nabla G(x_0, y_0, z_0).$$

EXERCISES 16.2

In Exercises 1 through 12, find an equation of the tangent plane and equations of the normal line to the given surface at the indicated point.

1. $x^2 + y^2 + z^2 = 17$; $(2, -2, 3)$
2. $4x^2 + y^2 + 2z^2 = 26$; $(1, -2, 3)$
3. $x^2 + y^2 - 3z = 2$; $(-2, -4, 6)$
4. $x^2 + y^2 - z^2 = 6$; $(3, -1, 2)$
5. $y = e^x \cos z$; $(1, e, 0)$
6. $z = e^{3x} \sin 3y$; $(0, \frac{1}{6}\pi, 1)$
7. $x^2 = 12y$; $(6, 3, 3)$
8. $z = x^{1/2} + y^{1/2}$; $(1, 1, 2)$
9. $x^{1/2} + y^{1/2} + z^{1/2} = 4$; $(4, 1, 1)$
10. $zx^2 - xy^2 - yz^2 = 18$; $(0, -2, 3)$
11. $x^{2/3} + y^{2/3} + z^{2/3} = 14$; $(-8, 27, 1)$
12. $x^{1/2} + z^{1/2} = 8$; $(25, 2, 9)$

In Exercises 13 through 20, if the two given surfaces intersect in a curve, find equations of the tangent line to the curve of intersection at the given point; if the two given surfaces are tangent at the given point, prove it.

13. $x^2 + y^2 - z = 8$, $x - y^2 + z^2 = -2$; $(2, -2, 0)$
14. $x^2 + y^2 - 2z + 1 = 0$, $x^2 + y^2 - z^2 = 0$; $(0, 1, 1)$
15. $y = x^2$, $y = 16 - z^2$; $(4, 16, 0)$
16. $x = 2 + \cos \pi yz$, $y = 1 + \sin \pi xz$; $(3, 1, 2)$
17. $y = e^x \sin 2\pi z + 2$, $z = y^2 - \ln(x + 1) - 3$; $(0, 2, 1)$
18. $x^2 - 3xy + y^2 = z$, $2x^2 + y^2 - 3z + 27 = 0$; $(1, -2, 11)$
19. $x^2 + z^2 + 4y = 0$, $x^2 + y^2 + z^2 - 6z + 7 = 0$; $(0, -1, 2)$
20. $x^2 + y^2 + z^2 = 8$, $yz = 4$; $(0, 2, 2)$
21. Show that the spheres $x^2 + y^2 + z^2 = a^2$ and $(x - b)^2 + y^2 + z^2 = (b - a)^2$ are tangent at the point $(a, 0, 0)$.
22. Show that the surfaces $xyz = 36$ and $4x^2 + y^2 + 9z^2 = 108$ are tangent at the point $(3, 6, 2)$.
23. Two surfaces are said to be *perpendicular* at a point P_0 of intersection if the normal vectors to the surfaces at P_0 are orthogonal. Show that the surface $x^2 - 2yz + y^3 = 4$ is perpendicular to every member of the family of surfaces $x^2 + (4c - 2)y^2 - cz^2 + 1 = 0$ at the point $(1, -1, 2)$.
24. Prove that every normal line to the sphere $x^2 + y^2 + z^2 = a^2$ passes through the center of the sphere.

16.3
EXTREMA OF FUNCTIONS OF TWO VARIABLES

An important application of the derivative of a function of a single variable is in the study of extreme values of a function, which leads to a variety of problems involving maximum and minimum. This was discussed in Chapter 3, where we proved theorems involving the first and second derivatives, from which relative maximum and minimum values of a function of one variable were determined. In extending the theory to functions of two variables, you will see that it is similar to the one-variable case; however, more complications arise.

16.3.1 DEFINITION   The function f of two variables is said to have a **relative maximum value** at the point (x_0, y_0) if there exists an open disk $B((x_0, y_0); r)$ such that $f(x_0, y_0) \geq f(x, y)$ for all (x, y) in B.

16.3.2 DEFINITION   The function f of two variables is said to have a **relative minimum value** at the point (x_0, y_0) if there exists an open disk $B((x_0, y_0); r)$ such that $f(x_0, y_0) \leq f(x, y)$ for all (x, y) in B.

▶ ILLUSTRATION 1   In Fig. 16.3.1 there is the graph of the function f defined by

$$f(x, y) = \sqrt{25 - x^2 - y^2}$$

Let B be any open disk $((0, 0); r)$ for which $r \leq 5$. From Definition 16.3.1 it follows that f has a relative maximum value of 5 at the point where $x = 0$ and $y = 0$.

In Fig. 16.3.2 there appears a sketch of the graph of the function g for which

$$g(x, y) = x^2 + y^2$$

Let B be any open disk $((0, 0); r)$. Then from Definition 16.3.2, g has a relative minimum value of 0 at the origin.   ◀

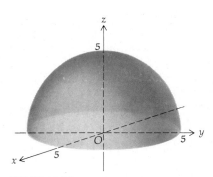

FIGURE 16.3.1

Analogous to Theorem 3.1.3 for functions of a single variable is the following one for functions of two variables.

16.3.3 THEOREM   If $f(x, y)$ exists at all points in some open disk $B((x_0, y_0); r)$ and if f has a relative extremum at (x_0, y_0), then if $f_x(x_0, y_0)$ and $f_y(x_0, y_0)$ exist,

$$f_x(x_0, y_0) = 0 \quad \text{and} \quad f_y(x_0, y_0) = 0$$

Before proving Theorem 16.3.3 we give an informal geometric argument. Let f be a function satisfying the hypothesis of Theorem 16.3.3, and let f have a relative maximum value at (x_0, y_0). Consider the curve of intersection of the plane $y = y_0$ with the surface $z = f(x, y)$ (refer to Fig. 16.3.3). This curve is represented by the equations

$$y = y_0 \quad \text{and} \quad z = f(x, y) \tag{1}$$

Because f has a relative maximum value at the point where $x = x_0$, $y = y_0$, it follows that the curve having Eqs. (1) has a horizontal tangent line in the plane $y = y_0$ at $(x_0, y_0, f(x_0, y_0))$. The slope of this tangent line is $f_x(x_0, y_0)$; so $f_x(x_0, y_0) = 0$. In a similar way we can consider the curve of intersection of the plane $x = x_0$ with the surface $z = f(x, y)$ and obtain $f_y(x_0, y_0) = 0$. A similar discussion can be given if f has a relative minimum value at (x_0, y_0). Following is the formal proof.

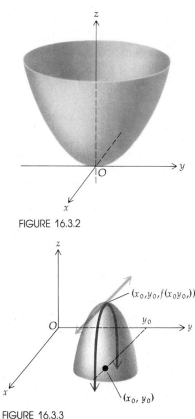

FIGURE 16.3.2

FIGURE 16.3.3

PROOF OF THEOREM 16.3.3   We prove that if f has a relative maximum value at (x_0, y_0) and if $f_x(x_0, y_0)$ exists, then $f_x(x_0, y_0) = 0$. By the definition of a partial derivative,

$$f_x(x_0, y_0) = \lim_{\Delta x \to 0} \frac{f(x_0 + \Delta x, y_0) - f(x_0, y_0)}{\Delta x}$$

Because f has a relative maximum value at (x_0, y_0), by Definition 16.3.1,

$$f(x_0 + \Delta x, y_0) - f(x_0, y_0) \leq 0$$

whenever Δx is sufficiently small so that $(x_0 + \Delta x, y_0)$ is in B. If Δx approaches zero from the right, $\Delta x > 0$; therefore

$$\frac{f(x_0 + \Delta x, y_0) - f(x_0, y_0)}{\Delta x} \leq 0$$

Hence, by Theorem 1.10.3, if $f_x(x_0, y_0)$ exists, $f_x(x_0, y_0) \leq 0$.

Similarly, if Δx approaches zero from the left, $\Delta x < 0$; so

$$\frac{f(x_0 + \Delta x, y_0) - f(x_0, y_0)}{\Delta x} \geq 0$$

Therefore, by Theorem 1.10.4, if $f_x(x_0, y_0)$ exists, $f_x(x_0, y_0) \geq 0$. We conclude, then, that because $f_x(x_0, y_0)$ exists, both inequalities, $f_x(x_0, y_0) \leq 0$ and $f_x(x_0, y_0) \geq 0$, must hold. Consequently $f_x(x_0, y_0) = 0$.

The proof that $f_y(x_0, y_0) = 0$, if $f_y(x_0, y_0)$ exists and f has a relative maximum value at (x_0, y_0), is analogous and is left as an exercise (see Exercise 33). The proof of the theorem when $f(x_0, y_0)$ is a relative minimum value is also left as an exercise (see Exercise 34).   ∎

16.3.4 DEFINITION   A point (x_0, y_0) for which both $f_x(x_0, y_0) = 0$ and $f_y(x_0, y_0) = 0$ is called a **critical point**.

Theorem 16.3.3 states that a necessary condition for a function of two variables to have a relative extremum at a point, where its first partial derivatives exist, is that this point be a critical point. It is possible for a function of two variables to have a relative extremum at a point at which the partial derivatives do not exist, but we do not consider this situation in this book. Furthermore, the vanishing of the first partial derivatives of a function of two variables is not a sufficient condition for the function to have a relative extremum at the point. Such a situation occurs in the following illustration.

▶ ILLUSTRATION 2   Let the function f be defined by

$$f(x, y) = y^2 - x^2$$

Then

$$f_x(x, y) = -2x \qquad f_y(x, y) = 2y$$

Both $f_x(0, 0)$ and $f_y(0, 0)$ equal zero. A sketch of the graph of f appears in Fig. 16.3.4; it is saddle-shaped at points close to the origin. It is apparent that f does not satisfy either Definition 16.3.1 or 16.3.2 when $(x_0, y_0) = (0, 0)$.   ◀

FIGURE 16.3.4

In Illustration 2, the point $(0, 0)$ is called a *saddle point* of the function f.

There is a second-derivative test that gives conditions that guarantee a function to have a relative extremum at a point where the first partial derivatives equal zero. However, sometimes it is possible to determine relative extrema of a function by Definitions 16.3.1 and 16.3.2, as shown in the following illustration.

$(3, -1, 11)$

FIGURE 16.3.5

▶ ILLUSTRATION 3   Let f be the function defined by

$$f(x, y) = 6x - 4y - x^2 - 2y^2$$

We determine if f has any relative extrema.

Because f and its first partial derivatives exist at all (x, y) in R^2, Theorem 16.3.3 is applicable. Differentiating we obtain

$$f_x(x, y) = 6 - 2x \quad \text{and} \quad f_y(x, y) = -4 - 4y$$

Setting $f_x(x, y)$ and $f_y(x, y)$ equal to zero we get $x = 3$ and $y = -1$. See Fig. 16.3.5 for a sketch of the graph of the equation

$$z = 6x - 4y - x^2 - 2y^2$$

It is a paraboloid having a vertical axis, with vertex at $(3, -1, 11)$ and opening downward. We can conclude that $f(x, y) \leq f(3, -1)$ for all (x, y); hence, by Definition 16.3.1, $f(3, -1) = 11$ is a relative maximum function value.   ◀

The basic test for determining relative maxima and minima for functions of two variables is the second-derivative test, which is given in the next theorem.

16.3.5 THEOREM
Second-Derivative Test

Let f be a function of two variables such that f and its first- and second-order partial derivatives are continuous on some open disk $B((a, b); r)$. Suppose further that $f_x(a, b) = 0$ and $f_y(a, b) = 0$. Then

(i) f has a relative minimum value at (a, b) if

$$f_{xx}(a, b)f_{yy}(a, b) - f_{xy}^2(a, b) > 0 \quad \text{and} \quad f_{xx}(a, b) > 0 \text{ (or } f_{yy}(a, b) > 0)$$

(ii) f has a relative maximum value at (a, b) if

$$f_{xx}(a, b)f_{yy}(a, b) - f_{xy}^2(a, b) > 0 \quad \text{and} \quad f_{xx}(a, b) < 0 \text{ (or } f_{yy}(a, b) < 0)$$

(iii) $f(a, b)$ is not a relative extremum if

$$f_{xx}(a, b)f_{yy}(a, b) - f_{xy}^2(a, b) < 0$$

(iv) We can make no conclusion if

$$f_{xx}(a, b)f_{yy}(a, b) - f_{xy}^2(a, b) = 0$$

PROOF OF (i)   For simplicity of notation let us define

$$\phi(x, y) = f_{xx}(x, y)f_{yy}(x, y) - f_{xy}^2(x, y)$$

We are given $\phi(a, b) > 0$ and $f_{xx}(a, b) > 0$, and we wish to prove that $f(a, b)$ is a relative minimum function value. Because f_{xx}, f_{xy}, and f_{yy} are continuous on $B((a, b); r)$, it follows that ϕ is also continuous on B. Hence there exists an open disk $B'((a, b); r')$, where $r' \leq r$, such that $\phi(x, y) > 0$ and $f_{xx}(x, y) > 0$ for every point (x, y) in B'. Let h and k be constants, not both zero, such that the point $(a + h, b + k)$ is in B'. Then the two equations

$$x = a + ht \quad \text{and} \quad y = b + kt \qquad 0 \leq t \leq 1$$

define all the points on the line segment from (a, b) to $(a + h, b + k)$, and all these points are in B'. Let F be the function of one variable defined by

$$F(t) = f(a + ht, b + kt) \tag{2}$$

By Taylor's formula (formula (9), Section 9.5),

$$F(t) = F(0) + F'(0)t + \frac{F''(\xi)}{2!} t^2 \tag{3}$$

where ξ is between 0 and t. If $t = 1$ in Eq. (3), we get

$$F(1) = F(0) + F'(0) + \tfrac{1}{2}F''(\xi) \tag{4}$$

where $0 < \xi < 1$. Because $F(0) = f(a, b)$ and $F(1) = f(a + h, b + k)$, it follows from (4) that

$$f(a + h, b + k) = f(a, b) + F'(0) + \tfrac{1}{2}F''(\xi) \tag{5}$$

where $0 < \xi < 1$.

To find $F'(t)$ and $F''(t)$ from (2) we use the chain rule and obtain

$$F'(t) = hf_x(a + ht, b + kt) + kf_y(a + ht, b + kt) \tag{6}$$

and

$$F''(t) = h^2 f_{xx} + hk f_{yx} + hk f_{xy} + k^2 f_{yy}$$

where each second partial derivative is evaluated at $(a + ht, b + kt)$. From Theorem 15.7.1 it follows that $f_{xy}(x, y) = f_{yx}(x, y)$ for all (x, y) in B'. So

$$F''(t) = h^2 f_{xx} + 2hk f_{xy} + k^2 f_{yy} \tag{7}$$

where each second partial derivative is evaluated at $(a + ht, b + kt)$. Substituting 0 for t in Eq. (6) and ξ for t in Eq. (7) we get

$$F'(0) = hf_x(a, b) + kf_y(a, b)$$
$$= 0 \tag{8}$$

and

$$F''(\xi) = h^2 f_{xx} + 2hk f_{xy} + k^2 f_{yy} \tag{9}$$

where each second partial derivative is evaluated at $(a + h\xi, b + k\xi)$, where $0 < \xi < 1$. Substituting from (8) and (9) into (5) we obtain

$$f(a + h, b + k) - f(a, b) = \tfrac{1}{2}(h^2 f_{xx} + 2hk f_{xy} + k^2 f_{yy}) \tag{10}$$

The terms in parentheses on the right side of (10) can be written as

$$h^2 f_{xx} + 2hk f_{xy} + k^2 f_{yy} = f_{xx}\left[h^2 + 2hk \frac{f_{xy}}{f_{xx}} + \left(k \frac{f_{xy}}{f_{xx}} \right)^2 - \left(k \frac{f_{xy}}{f_{xx}} \right)^2 + k^2 \frac{f_{yy}}{f_{xx}} \right]$$

So from (10),

$$f(a + h, b + k) - f(a, b) = \frac{f_{xx}}{2}\left[\left(h + \frac{f_{xy}}{f_{xx}} k \right)^2 + \frac{f_{xx} f_{yy} - f_{xy}{}^2}{f_{xx}{}^2} k^2 \right] \tag{11}$$

Because $f_{xx}f_{yy} - f_{xy}^2$ evaluated at $(a + h\xi, b + k\xi)$ equals

$$\phi(a + h\xi, b + k\xi) > 0$$

it follows that the expression in brackets on the right side of Eq. (11) is positive. Furthermore, because $f_{xx}(a + h\xi, b + k\xi) > 0$, it follows from (11) that $f(a + h, b + k) - f(a, b) > 0$. Hence we have proved that

$$f(a + h, b + k) > f(a, b)$$

for every point $(a + h, b + k) \neq (a, b)$ in B'. Therefore, by Definition 16.3.2, $f(a, b)$ is a relative minimum value of f.

The proof of part (ii) is similar and is left as an exercise (see Exercise 35). The proof of part (iii) is also left as an exercise (see Exercise 36). Part (iv) is included to cover all possible cases. ∎

EXAMPLE 1
If

$$f(x, y) = 2x^4 + y^2 - x^2 - 2y$$

determine the relative extrema of f if there are any.

SOLUTION
To apply the second-derivative test, we find the first and second partial derivatives of f.

$$f_x(x, y) = 8x^3 - 2x \qquad f_y(x, y) = 2y - 2$$

$$f_{xx}(x, y) = 24x^2 - 2 \qquad f_{yy}(x, y) = 2 \qquad f_{xy}(x, y) = 0$$

Setting $f_x(x, y) = 0$ we get $x = -\frac{1}{2}$, $x = 0$, and $x = \frac{1}{2}$. Setting $f_y(x, y) = 0$ we obtain $y = 1$. Therefore f_x and f_y are both 0 at the points $(-\frac{1}{2}, 1)$, $(0, 1)$, and $(\frac{1}{2}, 1)$, and these are the critical points of f. The results of applying the second-derivative test at these points are summarized in Table 16.3.1.

Table 16.3.1

Critical point	f_{xx}	f_{yy}	f_{xy}	$f_{xx}f_{yy} - f_{xy}^2$	Conclusion
$(-\frac{1}{2}, 1)$	4	2	0	8	f has a relative minimum value
$(0, 1)$	-2	2	0	-4	f does not have a relative extremum
$(\frac{1}{2}, 1)$	4	2	0	8	f has a relative minimum value

At the point $(-\frac{1}{2}, 1)$, $f_{xx} > 0$ and $f_{xx}f_{yy} - f_{xy}^2 > 0$; thus from Theorem 16.3.5(i), f has a relative minimum value at $(-\frac{1}{2}, 1)$. At $(0, 1)$, $f_{xx}f_{yy} - f_{xy}^2 < 0$; so from Theorem 16.3.5(iii), f does not have a relative extremum at $(0, 1)$. Because $f_{xx} > 0$ and $f_{xx}f_{yy} - f_{xy}^2 > 0$ at $(\frac{1}{2}, 1)$, f has a relative minimum value there by Theorem 16.3.5(i).

Because $f(-\frac{1}{2}, 1) = -\frac{9}{8}$ and $f(\frac{1}{2}, 1) = -\frac{9}{8}$, we conclude that f has a relative minimum value of $-\frac{9}{8}$ at each of the points $(-\frac{1}{2}, 1)$ and $(\frac{1}{2}, 1)$.

We now discuss absolute extrema of functions of two variables.

16.3.6 DEFINITION

The function f of two variables is said to have an **absolute maximum value** on its domain D in the xy plane if there is some point (x_0, y_0) in D such that $f(x_0, y_0) \geq f(x, y)$ for all points (x, y) in D. In such a case, $f(x_0, y_0)$ is the absolute maximum value of f on D.

16.3.7 DEFINITION

The function f of two variables is said to have an **absolute minimum value** on its domain D in the xy plane if there is some point (x_0, y_0) in D such that $f(x_0, y_0) \leq f(x, y)$ for all (x, y) in D. In such a case, $f(x_0, y_0)$ is the absolute minimum value of f on D.

For functions of a single variable we had the following theorem (3.1.9), called the extreme-value theorem: If the function f is continuous on the closed interval $[a, b]$, then f has an absolute maximum value and an absolute minimum value on $[a, b]$. We learned that an absolute extremum of a function continuous on a closed interval must be either a relative extremum or a function value at an endpoint of the interval. We have a corresponding situation for functions of two variables. In the statement of the extreme-value theorem for functions of two variables we refer to a *closed region* in the xy plane. By a **closed region** we mean that the region includes its *boundary*. In the following illustration we give some closed regions and identify the boundary of each region.

FIGURE 16.3.6

▶ ILLUSTRATION 4   (a) A closed disk is a closed region. The boundary is the circumference of the disk. See Fig. 16.3.6.

     (b) The sides of a triangle together with the region enclosed by the triangle is a closed region. The boundary consists of the sides of the triangle. See Fig. 16.3.7.

     (c) The edges of a rectangle together with the region enclosed by the rectangle is a closed region. The boundary consists of the edges of the rectangle. See Fig. 16.3.8. ◀

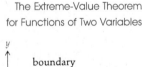

FIGURE 16.3.7

16.3.8 THEOREM

The Extreme-Value Theorem for Functions of Two Variables

Let R be a closed region in the xy plane, and let f be a function of two variables that is continuous on R. Then there is at least one point in R where f has an absolute maximum value and at least one point in R where f has an absolute minimum value.

The proof of Theorem 16.3.8 is omitted because it is beyond the scope of this book.

     If f is a function satisfying Theorem 16.3.8 and if both $f_x(x, y)$ and $f_y(x, y)$ exist at all points of R, then the absolute extrema of f occur at either a point (x_0, y_0), where $f_x(x_0, y_0) = 0$ and $f_y(x_0, y_0) = 0$, or at a point on the boundary of R.

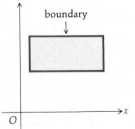

FIGURE 16.3.8

EXAMPLE 2

A manufacturer who is a monopolist makes two types of lamps. From experience the manufacturer has determined that if x lamps of the first type and y lamps of the second type are made, they can be sold for $(100 - 2x)$ dollars each and $(125 - 3y)$ dollars each, respectively. The cost of manufacturing x lamps of the first type and y lamps of the second type is $(12x + 11y + 4xy)$ dollars. How many lamps of each type should be produced to realize the greatest profit, and what is the greatest profit?

SOLUTION

The number of dollars in the revenue received from the lamps of the first type is $x(100 - 2x)$, and the number of dollars in the revenue received from the lamps of the second type is $y(125 - 3y)$. Hence, if $f(x, y)$ dollars is the manufacturer's profit,

$$f(x, y) = x(100 - 2x) + y(125 - 3y) - (12x + 11y + 4xy)$$

$$f(x, y) = 88x + 114y - 2x^2 - 3y^2 - 4xy \qquad (12)$$

Because x and y both represent the number of lamps, we require that $x \geq 0$ and $y \geq 0$ and allow x and y to be any nonnegative real numbers. Furthermore, $(100 - 2x)$ dollars is the selling price of lamps of the first type. Thus we require that $100 - 2x \geq 0$ or, equivalently, $x \leq 50$. Similarly, because $(125 - 3y)$ dollars is the selling price of lamps of the second type, we require that $y \leq \frac{125}{3}$. Therefore the domain of f is the closed region defined by the set

$$\{(x, y) | 0 \leq x \leq 50 \text{ and } 0 \leq y \leq \frac{125}{3}\} \qquad (13)$$

This region is rectangular and is shown in Fig. 16.3.9. The boundary of the region consists of the edges of the rectangle. Because f is a polynomial function, it is continuous everywhere. Hence f is continuous on the closed region defined by set (13); so the extreme-value theorem can be applied. The critical points of f are found by determining where $f_x(x, y) = 0$ and $f_y(x, y) = 0$.

$$f_x(x, y) = 88 - 4x - 4y$$

$$f_y(x, y) = 114 - 6y - 4x$$

FIGURE 16.3.9

Setting $f_x(x, y) = 0$ and $f_y(x, y) = 0$ we have

$$x + y = 22$$

$$2x + 3y = 57$$

Solving these two equations simultaneously we obtain $x = 9$ and $y = 13$. To apply the second-derivative test we find the second partial derivatives.

$$f_{xx}(x, y) = -4 \qquad f_{yy}(x, y) = -6 \qquad f_{xy}(x, y) = -4$$

At the point $(9, 13)$,

$$f_{xx}(9, 13) = -4 < 0$$

$$f_{xx}(9, 13)f_{yy}(9, 13) - f_{xy}{}^2(9, 13) = (-4)(-6) - (-4)^2$$
$$= 8 > 0$$

It follows, then, by Theorem 16.3.5(ii) that f will have a relative maximum value at $(9, 13)$.

From Eq. (12),

$$f(x, y) = x(88 - 2x) + y(114 - 3y) - 4xy \tag{14}$$

Thus

$$f(9, 13) = 9(70) + 13(75) - 468$$
$$= 1137$$

The absolute maximum value of f must occur at either $(9, 13)$ or on the boundary of the domain of f. Let us compare $f(9, 13)$ with the function values on the boundary.

For that part of the boundary on the x axis with $x \in [0, 50]$ we have the function values computed from (14) as follows:

$$f(x, 0) = 88x - 2x^2$$

Let

$$g(x) = 88x - 2x^2 \qquad x \in [0, 50]$$

Then

$$g'(x) = 88 - 4x \quad \text{and} \quad g''(x) = -4$$

Because $g'(22) = 0$ and $g''(22) < 0$, g has a relative maximum value of 968 at $x = 22$. Furthermore, $g(0) = 0$ and $g(50) < 0$. Because $f(9, 13) = 1137 > 968$, the absolute maximum value of f does not occur on the x axis.

For that part of the boundary on the y axis with $y \in [0, \frac{125}{3}]$, from (14),

$$f(0, y) = 114y - 3y^2$$

Let

$$h(y) = 114y - 3y^2 \qquad y \in [0, \tfrac{125}{3}]$$

Then

$$h'(y) = 114 - 6y \quad \text{and} \quad h''(y) = -6$$

Because $h'(19) = 0$ and $h''(19) < 0$, h has a relative maximum value of 1083 at $y = 19$. Moreover, $h(0) = 0$ and $h(\frac{125}{3}) < 0$. Because $f(9, 13) = 1137 > 1083$, the absolute maximum value of f does not occur on the y axis.

We now consider the part of the boundary on the line $x = 50$ with $y \in [0, \frac{125}{3}]$. From (14),

$$f(50, y) = y(114 - 3y) - 600 - 200y \tag{15}$$

$$f(0, y) = y(114 - 3y) \tag{16}$$

By comparing (15) and (16),

$$f(50, y) < f(0, y) \tag{17}$$

Because $f(9, 13) > f(0, y)$ for all y in $[0, \frac{125}{3}]$, then, from inequality (17),

$$f(9, 13) > f(50, y) \qquad \text{for } y \in [0, \tfrac{125}{3}]$$

Hence the absolute maximum value of f does not occur on the line $x = 50$.

Finally, we have that part of the boundary on the line $y = \frac{125}{3}$ with $x \in [0, 50]$. From (14),

$$f(x, \tfrac{125}{3}) = x(88 - 2x) - \tfrac{1375}{3} - \tfrac{500}{3}x \qquad (18)$$

$$f(x, 0) = x(88 - 2x) \qquad (19)$$

From (18) and (19) it follows that $f(x, \frac{125}{3}) < f(x, 0)$. Therefore, because $f(9, 13) > f(x, 0)$ for all x in $[0, 50]$, we can conclude that it is also greater than $f(x, \frac{125}{3})$ for all x in $[0, 50]$. Thus the absolute maximum value cannot occur on the line $y = \frac{125}{3}$.

Therefore the absolute maximum value of f is not on the boundary, and thus it is at the point $(9, 13)$. We conclude then that 9 lamps of the first type and 13 lamps of the second type should be produced for the greatest profit of $1137.

EXAMPLE 3

Determine the relative dimensions of a rectangular box, without a top and having a specific volume, if the least amount of material is to be used in its manufacture.

SOLUTION

Let x units be the length of the base of the box, y units be the width of the base of the box, z units be the depth of the box, and S square units be the surface area of the box. The box has a specific volume; therefore, if V cubic units is the volume of the box, V is a constant. Figure 16.3.10 shows the box.

Each of the variables x, y, and z is in the interval $(0, +\infty)$. We have the equations

$$S = xy + 2xz + 2yz \quad \text{and} \quad V = xyz$$

Solving the second equation for z in terms of x, y, and the constant V, we get $z = \dfrac{V}{xy}$, and substituting this into the first equation gives

$$S = xy + \frac{2V}{y} + \frac{2V}{x} \qquad (20)$$

Differentiating we get

$$\frac{\partial S}{\partial x} = y - \frac{2V}{x^2} \qquad \frac{\partial S}{\partial y} = x - \frac{2V}{y^2}$$

$$\frac{\partial^2 S}{\partial x^2} = \frac{4V}{x^3} \qquad \frac{\partial^2 S}{\partial y\, \partial x} = 1 \qquad \frac{\partial^2 S}{\partial y^2} = \frac{4V}{y^3}$$

Setting $\dfrac{\partial S}{\partial x} = 0$ and $\dfrac{\partial S}{\partial y} = 0$ and solving simultaneously we get

$$x^2 y - 2V = 0$$

$$xy^2 - 2V = 0$$

FIGURE 16.3.10

from which it follows that $x = \sqrt[3]{2V}$ and $y = \sqrt[3]{2V}$. For these values of x and y,

$$\frac{\partial^2 S}{\partial x^2} = \frac{4V}{(\sqrt[3]{2V})^3}$$

$$= 2 > 0$$

and

$$\frac{\partial^2 S}{\partial x^2} \cdot \frac{\partial^2 S}{\partial y^2} - \left(\frac{\partial^2 S}{\partial y \, \partial x}\right)^2 = \frac{4V}{(\sqrt[3]{2V})^3} \cdot \frac{4V}{(\sqrt[3]{2V})^3} - 1$$

$$= 3 > 0$$

From Theorem 16.3.5(i) it follows that S has a relative minimum value when $x = \sqrt[3]{2V}$ and $y = \sqrt[3]{2V}$. Recall that x and y are both in the interval $(0, +\infty)$, and notice from Eq. (20) that S is very large when x and y are either close to zero or very large. We therefore conclude that the relative minimum value of S is an absolute minimum value of S.

Because $z = V/(xy)$, then when $x = \sqrt[3]{2V}$ and $y = \sqrt[3]{2V}$,

$$z = \frac{V}{\sqrt[3]{4V^2}}$$

$$= \frac{\sqrt[3]{2V}}{2}$$

Hence the box should have a square base and a depth that is one-half the length of a side of the base.

In Section 15.1 we referred to a production function where the independent variables can be any of several inputs, such as the number of machines used in production, the number of person-hours available, the amount of working capital to be had, the quantity of materials used, and the amount of land available. Consider now a production function f of two variables, where the amounts of the inputs are given by x and y, and z gives the amount of the output; then $z = f(x, y)$. Suppose that the prices of the two inputs are a dollars and b dollars per unit, respectively, and that the price of the output is c dollars (a, b, and c are constants). This situation could occur if there were so many producers in the market that a change in the output of any particular producer would not affect the price of the commodity. Such a market is called a **perfectly competitive market**. If P dollars is the total profit, and because the total profit is obtained by subtracting the total cost from the total revenue, then

$$P = cz - (ax + by)$$

and because $z = f(x, y)$,

$$P = cf(x, y) - ax - by$$

It is, of course, desired to maximize P. This is illustrated by an example.

EXAMPLE 4

Suppose that the production of a certain commodity depends upon two inputs. The amounts of these are given by $100x$ and $100y$, whose prices are, respectively, \$7 and \$4. The amount of the output is given by $100z$, the price per unit of which is \$9. Furthermore, the production function f has the functions values

$$f(x, y) = \frac{x}{3} + \frac{y}{3} + 5 - \frac{1}{x} - \frac{1}{y}$$

Determine the greatest profit.

SOLUTION
If P dollars is the profit,

$$P = 9(100z) - 7(100x) - 4(100y) \tag{21}$$

Letting $z = f(x, y)$ we get

$$P = 900\left(\frac{x}{3} + \frac{y}{3} + 5 - \frac{1}{x} - \frac{1}{y}\right) - 700x - 400y$$

$$= 4500 - \frac{900}{x} - \frac{900}{y} - 400x - 100y$$

x and y are both in the interval $(0, +\infty)$. Hence

$$\frac{\partial P}{\partial x} = \frac{900}{x^2} - 400 \quad \text{and} \quad \frac{\partial P}{\partial y} = \frac{900}{y^2} - 100$$

Also,

$$\frac{\partial^2 P}{\partial x^2} = -\frac{1800}{x^3} \qquad \frac{\partial^2 P}{\partial y^2} = -\frac{1800}{y^3} \qquad \frac{\partial^2 P}{\partial y\,\partial x} = 0$$

Setting $\dfrac{\partial P}{\partial x} = 0$ and $\dfrac{\partial P}{\partial y} = 0$ we have

$$\frac{900}{x^2} - 400 = 0 \quad \text{and} \quad \frac{900}{y^2} - 100 = 0$$

from which we obtain $x = \frac{3}{2}$ and $y = 3$ (the negative result is rejected because x and y must be positive). At $(\frac{3}{2}, 3)$,

$$\frac{\partial^2 P}{\partial x^2} \cdot \frac{\partial^2 P}{\partial y^2} - \left(\frac{\partial^2 P}{\partial y\,\partial x}\right)^2 = \left(-\frac{1800}{\frac{27}{8}}\right)\left(-\frac{1800}{27}\right) - (0)^2 > 0$$

From the above and the fact that at $(\frac{3}{2}, 3)$, $\dfrac{\partial^2 P}{\partial x^2} < 0$, it follows from Theorem 16.3.5(ii) that P has a relative maximum value at $(\frac{3}{2}, 3)$. Because x and y are both in the interval $(0, +\infty)$, and P is a negative number when x and y are either close to zero or very large, we conclude that the relative maximum value of P is an absolute maximum value. Because $z = f(x, y)$, the value of z at $(\frac{3}{2}, 3)$ is $f(\frac{3}{2}, 3) = \frac{11}{2}$. Hence, from Eq. (21),

$$P_{max} = 900 \cdot \tfrac{11}{2} - 700 \cdot \tfrac{3}{2} - 400 \cdot 3$$
$$= 2700$$

The greatest profit then is $2700.

The discussion of the extrema of functions of two variables can be extended to functions of three or more variables. The definitions of relative extrema and critical point are easily made. For example, if f is a function of the three variables x, y, and z, and

$$f_x(x_0, y_0, z_0) = 0 \qquad f_y(x_0, y_0, z_0) = 0 \qquad f_z(x_0, y_0, z_0) = 0$$

then (x_0, y_0, z_0) is a critical point of f. Such a point is obtained by solving simultaneously three equations in three unknowns. For a function of n variables, the critical points are found by setting all the n first partial derivatives equal to zero and solving simultaneously the n equations in n unknowns. The extension of Theorem 16.3.5 to functions of three or more variables is given in advanced calculus texts.

EXERCISES 16.3

In Exercises 1 through 12, determine the relative extrema of f, if there are any.

1. $f(x, y) = x^3 + y^2 - 6x^2 + y - 1$

2. $f(x, y) = x^2 - 4xy + y^3 + 4y$

3. $f(x, y) = \dfrac{1}{x} - \dfrac{64}{y} + xy$

4. $f(x, y) = 18x^2 - 32y^2 - 36x - 128y - 110$

5. $f(x, y) = e^{xy}$

6. $f(x, y) = x^3 + y^3 - 18xy$

7. $f(x, y) = 4xy^2 - 2x^2y - x$

8. $f(x, y) = \dfrac{2x + 2y + 1}{x^2 + y^2 + 1}$

9. $f(x, y) = x^3 + y^3 + 3y^2 - 3x - 9y + 2$

10. $f(x, y) = \sin x + \sin y; \ 0 \le x \le \pi; \ 0 \le y \le \pi$

11. $f(x, y) = \sin(x + y) + \sin x + \sin y; \ 0 \le x \le 2\pi; \ 0 \le y \le 2\pi$

12. $f(x, y) = e^x \sin y$

13. Find the three positive numbers whose sum is 24 such that their product is as great as possible.

14. Find the three positive numbers whose product is 24 such that their sum is as small as possible.

15. Find the point in the plane $3x + 2y - z = 5$ that is closest to the point $(1, -2, 3)$, and find the minimum distance.

16. Find the points on the surface $y^2 - xz = 4$ that are closest to the origin, and find the minimum distance.

17. Find the points on the curve of intersection of the ellipsoid $x^2 + 4y^2 + 4z^2 = 4$ and the plane $x - 4y - z = 0$ that are closest to the origin, and find the minimum distance.

18. A manufacturing plant has two classifications for its workers, A and B. Class A workers earn $14 per run, and class B workers earn $13 per run. For a certain production run it is determined that in addition to the salaries of the workers, if x class A workers and y class B workers are used, the number of dollars in the cost of the run is $y^3 + x^2 - 8xy + 600$. How many workers of each class should be used so that the cost of the run is a minimum if at least three workers of each class are required for a run?

19. An injection of x mg of drug A and y mg of drug B causes a response of R units, and $R = x^2y^3(c - x - y)$, where c is a positive constant. What quantity of each drug will cause the maximum response?

20. Suppose that t hours after the injection of x mg of adrenalin the response is R units, and $R = te^{-t}(c - x)x$, where c is a positive constant. What values of x and t will cause the maximum response?

21. Find the volume of the largest rectangular parallelepiped that can be inscribed in the ellipsoid $36x^2 + 9y^2 + 4z^2 = 36$ if the edges are parallel to the coordinate axes.

22. A rectangular box without a top is to be made at a cost of $10 for the material. If the material for the bottom of the box costs $0.15 per square foot and the material for the sides costs $0.30 per square foot, find the dimensions of the box of greatest volume that can be made.

23. A closed rectangular box to contain 16 ft^3 is to be made of three kinds of material. The cost of the material for the top and the bottom is $0.18 per ft^2, the cost of the material for the front and the back is $0.16 per ft^2, and the cost of the material for the other two sides is $0.12 per ft^2. Find the dimensions of the box such that the cost of the materials is a minimum.

24. Suppose that T degrees is the temperature at any point (x, y, z) on the sphere $x^2 + y^2 + z^2 = 4$, and $T = 100xy^2z$. Find the points on the sphere where the temperature is the greatest and also the points where the temperature is the least. Also find the temperature at these points.

25. Suppose that when the production of a particular commodity requires x machine-hours and y person-hours, the cost of production is given by $f(x, y)$, where $f(x, y) = 2x^3 - 6xy + y^2 + 500$. Determine the number of machine-hours and the number of person-hours needed to produce the commodity at the least cost.

26. A clothing store sells two kinds of shirts that are similar but are made by different manufacturers. The cost to the store of the first kind is $40 and the cost of the second kind is $50. It has been determined by experience that if the selling price of the first kind is x dollars and the selling price of the second kind is y dollars, then the number sold monthly of the first kind is $3200 - 50x + 25y$, and the number sold monthly of the second kind is $25x - 25y$. What should be the selling price of each kind of shirt for the greatest gross profit?

27. The production function f for a certain commodity has function values $f(x, y) = 4 - \dfrac{8}{xy}$. The amounts of the two inputs are given by $100x$ and $100y$, whose prices per unit are, respectively, $10 and $5, and the amount of the output is given by $100z$, whose price per unit is $20. Determine the greatest profit.

28. Solve Exercise 27 if $f(x, y) = x + \frac{5}{2}y - \frac{1}{8}x^2 - \frac{1}{4}y^2 - \frac{9}{8}$; the prices per unit of the two inputs are $4 and $8 (instead of $10 and $5, respectively), and the price per unit of the output is $16.

29. A monopolist produces staplers and staples having demand equations $x = 10/(pq)$ and $y = 20/(pq)$, where $1000x$ staplers are demanded if the price is p dollars per stapler and $1000y$ boxes of staples are demanded if the price per box of staples is q dollars. It costs $2 to produce each stapler and $1 to produce each box of staples. Determine the price of each commodity in order to have the greatest total profit.

30. If the demand equations in Exercise 29 are $x = 11 - 2p - 2q$ and $y = 19 - 2p - 3q$, show that to have the greatest total profit the staplers should be free and the staples should be expensive.

31. Determine the relative dimensions of a rectangular box, without a top, to be made from a given amount of material for the box to have the greatest possible volume.

32. Prove that the box having the largest volume that can be placed inside a sphere is in the shape of a cube.

33. Prove that $f_y(x_0, y_0) = 0$ if $f_y(x_0, y_0)$ exists and f has a relative maximum value at (x_0, y_0).

34. Prove Theorem 16.3.3 when $f(x_0, y_0)$ is a relative minimum value.

35. Prove Theorem 16.3.5(ii).   36.   Prove Theorem 16.3.5(iii).

16.4
LAGRANGE MULTIPLIERS

In the solution of Example 3 of Section 16.3 we minimized the function having function values $xy + 2xz + 2yz$, subject to the condition that x, y, and z satisfy the equation $xyz = V$. Compare this with Example 1 of Section 16.3, in which we found the relative extrema of the function f for which $f(x, y) = 2x^4 + y^2 - x^2 - 2y$. These are essentially two different kinds of problems because in the first case we had an additional condition, called a *constraint* (or *side condition*). Such a problem is called one in *constrained extrema*, whereas that of the second type is called a problem in *free extrema*.

The solution of Example 3 of Section 16.3 involved obtaining a function of the two variables x and y by replacing z in the first equation by its value from the second equation. Because it is not always feasible to solve the constraint for one of the variables in terms of the others, there is another procedure that can be used to find the critical points to solve a problem in constrained extrema. It is due to Joseph L. Lagrange (1736–1813) and is known as the method of Lagrange multipliers. Before discussing the theory of this method, we outline the procedure and illustrate it by an example.

Suppose that we wish to find the relative extrema of a function f of the three variables x, y, and z, subject to the constraint $g(x, y, z) = 0$. We introduce a new variable λ, called a **Lagrange multiplier**, and form the auxiliary function F for which

$$F(x, y, z, \lambda) = f(x, y, z) + \lambda g(x, y, z)$$

The problem, then, becomes one of finding the critical points of the function F of the four variables x, y, z, and λ. The values of x, y, and z that give the relative extrema of f are among these critical points. The critical points of F are the values of x, y, z, and λ for which the four first partial derivatives of F are zero:

$$F_x = 0 \qquad F_y = 0 \qquad F_z = 0 \qquad F_\lambda = 0$$

EXAMPLE 1

Solve Example 3 of Section 16.3 by the method of Lagrange multipliers.

SOLUTION

The variables x, y, and z and the constant V are as defined in the solution of Example 3 of Section 16.3. Let

$$S = f(x, y, z) \qquad \text{and} \quad g(x, y, z) = xyz - V$$
$$= xy + 2xz + 2yz$$

We wish to minimize the function f subject to the constraint that

$$g(x, y, z) = 0$$

Form the function F for which

$$F(x, y, z, \lambda) = f(x, y, z) + \lambda g(x, y, z)$$
$$= xy + 2xz + 2yz + \lambda(xyz - V)$$

To find the critical points of F we compute the four partial derivatives F_x, F_y, F_z, and F_λ and set the function values equal to zero.

$$F_x(x, y, z, \lambda): \ y + 2z + \lambda yz = 0 \tag{1}$$

$$F_y(x, y, z, \lambda): \ x + 2z + \lambda xz = 0 \tag{2}$$

$$F_z(x, y, z, \lambda): \ 2x + 2y + \lambda xy = 0 \tag{3}$$

$$F_\lambda(x, y, z, \lambda): \ xyz - V = 0 \tag{4}$$

Subtracting corresponding members of Eq. (2) from those of Eq. (1) we obtain

$$y - x + \lambda z(y - x) = 0$$

$$(y - x)(1 + \lambda z) = 0$$

giving the two equations

$$y = x \tag{5}$$

and because $z \neq 0$,

$$\lambda = -\frac{1}{z} \tag{6}$$

Substituting from (6) into (2) we get $x + 2z - x = 0$, giving $z = 0$, which is impossible because z is in the interval $(0, +\infty)$. Substituting from (5) into (3) gives

$$2x + 2x + \lambda x^2 = 0$$

$$x(4 + \lambda x) = 0$$

$$\lambda = -\frac{4}{x} \quad \text{(because } x \neq 0\text{)}$$

If in Eq. (2), $\lambda = -4/x$, then

$$x + 2z - \frac{4}{x}(xz) = 0$$

$$x + 2z - 4z = 0$$

$$z = \frac{x}{2} \tag{7}$$

Substituting from (5) and (7) into (4) we obtain $\frac{1}{2}x^3 - V = 0$, from which $x = \sqrt[3]{2V}$. From (5) and (7) it follows that $y = \sqrt[3]{2V}$ and $z = \frac{1}{2}\sqrt[3]{2V}$. Therefore $(\sqrt[3]{2V}, \sqrt[3]{2V}, \frac{1}{2}\sqrt[3]{2V})$ is a critical point of the function F, and, as shown in Example 3 of Section 16.3, f has an absolute minimum value at this point.

Note in the solution that the equation $F_\lambda(x, y, z, \lambda) = 0$ is equivalent to the constraint given by the equation $V = xyz$.

The validity of the method of Lagrange multipliers can be shown by considering the general problem of constrained extrema. Suppose we wish to find the relative extrema of a function f of three variables x, y, and z, subject to the constraint

$$g(x, y, z) = 0 \tag{8}$$

Assume that (8) can be solved for z to obtain

$$z = h(x, y) \tag{9}$$

where h is defined on an open disk $B((x_0, y_0); r)$ and $f(x, y, h(x, y))$ has a relative extremum at $(x_0, y_0, h(x_0, y_0))$. Also assume that the first partial derivatives of f, g, and h exist on B and $g_3(x, y, h(x, y)) \neq 0$ on B. Because f has a relative extremum at $(x_0, y_0, h(x_0, y_0))$, the first partial derivatives of f are zero there. We compute these first partial derivatives by the chain rule.

$$\text{at } (x_0, y_0, h(x_0, y_0)) \qquad f_1 + f_3 \frac{\partial h}{\partial x} = 0 \quad \text{and} \quad f_2 + f_3 \frac{\partial h}{\partial y} = 0 \qquad (10)$$

If in (8) we differentiate implicitly with respect to x and then y, and consider z as the differentiable function h of x and y, then at a point (x, y) in the open disk B

$$g_1 + g_3 \frac{\partial h}{\partial x} = 0 \quad \text{and} \quad g_2 + g_3 \frac{\partial h}{\partial y} = 0$$

or, equivalently, because $g_3 \neq 0$ on the open disk B,

$$\text{at } (x, y) \text{ in } B \qquad \frac{\partial h}{\partial x} = -\frac{g_1}{g_3} \quad \text{and} \quad \frac{\partial h}{\partial y} = -\frac{g_2}{g_3} \qquad (11)$$

where the function values of g_1, g_2, and g_3 are at $(x, y, h(x, y))$. If the values of $\frac{\partial h}{\partial x}$ and $\frac{\partial h}{\partial y}$ from Eqs. (11) are substituted in Eqs. (10), then at the point $(x_0, y_0, h(x_0, y_0))$,

$$f_1 + f_3 \left(-\frac{g_1}{g_3} \right) = 0 \quad \text{and} \quad f_2 + f_3 \left(-\frac{g_2}{g_3} \right) = 0$$

Furthermore,

$$f_3 - g_3 \left(\frac{f_3}{g_3} \right) = 0$$

everywhere $g_3 \neq 0$. Thus at $(x_0, y_0, h(x_0, y_0))$,

$$f_1 + g_1 \left(-\frac{f_3}{g_3} \right) = 0 \qquad f_2 + g_2 \left(-\frac{f_3}{g_3} \right) = 0 \qquad f_3 + g_3 \left(-\frac{f_3}{g_3} \right) = 0 \qquad (12)$$

If $\lambda = -f_3/g_3$ then Eqs. (12) can be written as

$$f_1 + \lambda g_1 = 0 \qquad f_2 + \lambda g_2 = 0 \qquad f_3 + \lambda g_3 = 0 \qquad (13)$$

Furthermore, because f has a relative extremum at $(x_0, y_0, h(x_0, y_0))$ and this extremum is subject to the constraint $g(x, y, z) = 0$, then

$$g(x_0, y_0, h(x_0, y_0)) = 0 \qquad (14)$$

If

$$F(x, y, z, \lambda) = f(x, y, z) + \lambda g(x, y, z) \qquad (15)$$

and if $z_0 = h(x_0, y_0)$, then (13) and (14) are equivalent to the equations

$$F_x = 0 \qquad F_y = 0 \qquad F_z = 0 \qquad F_\lambda = 0 \qquad \text{at } (x_0, y_0, z_0) \qquad (16)$$

Therefore we can conclude that a point (x_0, y_0, z_0) at which the function f has a relative extremum is among the critical points of the function F defined by (15).

Observe that Eqs. (13) can be written as the vector equation

$$\mathbf{\nabla} f + \lambda \mathbf{\nabla} g = \mathbf{0} \qquad \text{at } (x_0, y_0, z_0) \qquad \text{where } \mathbf{\nabla} g \neq \mathbf{0}$$

This vector equation together with the equation $g(x_0, y_0, z_0) = 0$ give another form of Eqs. (16).

EXAMPLE 2

Use the method of Lagrange multipliers to find the shortest distance from the origin to the plane $Ax + By + Cz = D$.

SOLUTION

Let w units be the distance from the origin to a point (x, y, z) in the plane. Then

$$w = \sqrt{x^2 + y^2 + z^2}$$

Because w will be a minimum when w^2 is a minimum, we form the function f for which

$$f(x, y, z) = x^2 + y^2 + z^2$$

We wish to find the minimum value of f subject to the constraint

$$Ax + By + Cz - D = 0$$

With the assumption that there is such a minimum value it will occur at a critical point of the function F such that

$$F(x, y, z, \lambda) = x^2 + y^2 + z^2 + \lambda(Ax + By + Cz - D)$$

To find the critical points of F we compute the partial derivatives of F and set them equal to zero.

$$F_x(x, y, z, \lambda): \quad 2x + \lambda A = 0 \tag{17}$$

$$F_y(x, y, z, \lambda): \quad 2y + \lambda B = 0 \tag{18}$$

$$F_z(x, y, z, \lambda): \quad 2z + \lambda C = 0 \tag{19}$$

$$F_\lambda(x, y, z, \lambda): \quad Ax + By + Cz - D = 0 \tag{20}$$

From (17), (18), and (19),

$$x = -\tfrac{1}{2}\lambda A \qquad y = -\tfrac{1}{2}\lambda B \qquad z = -\tfrac{1}{2}\lambda C \tag{21}$$

Substituting these values of x, y, and z into (20) we get

$$-\tfrac{1}{2}\lambda(A^2 + B^2 + C^2) = D$$

$$-\tfrac{1}{2}\lambda = \frac{D}{A^2 + B^2 + C^2} \tag{22}$$

In Eqs. (21) we replace $-\frac{1}{2}\lambda$ by its value from (22) and obtain

$$x = \frac{AD}{A^2 + B^2 + C^2} \qquad y = \frac{BD}{A^2 + B^2 + C^2} \qquad z = \frac{CD}{A^2 + B^2 + C^2} \qquad (23)$$

The point having these coordinates is the one and only critical point of F. Therefore the minimum distance from the origin to the plane is the distance from the origin to the point (x_0, y_0, z_0), where x_0, y_0, and z_0 are the values of x, y, and z in Eqs. (23). The minimum distance then is

$$\sqrt{x_0{}^2 + y_0{}^2 + z_0{}^2} = \sqrt{\frac{A^2 D^2}{A^2 + B^2 + C^2} + \frac{B^2 D^2}{A^2 + B^2 + C^2} + \frac{C^2 D^2}{A^2 + B^2 + C^2}}$$

$$= \frac{|D|}{\sqrt{A^2 + B^2 + C^2}}$$

EXAMPLE 3

Solve Example 4 of Section 16.3 by the method of Lagrange multipliers.

SOLUTION

We wish to maximize the function P defined by

$$P(x, y, z) = 900z - 700x - 400y$$

subject to the constraint given by the equation

$$z = \frac{x}{3} + \frac{y}{3} + 5 - \frac{1}{x} - \frac{1}{y}$$

which we write as

$$\frac{1}{x} + \frac{1}{y} + z - \frac{x}{3} - \frac{y}{3} - 5 = 0$$

and let $g(x, y, z)$ be the left side of this equation. Then let

$$F(x, y, z, \lambda) = P(x, y, z) + \lambda g(x, y, z)$$

$$= 900z - 700x - 400y + \lambda \left(\frac{1}{x} + \frac{1}{y} + z - \frac{x}{3} - \frac{y}{3} - 5 \right)$$

We find the four partial derivatives F_x, F_y, F_z, and F_λ and set them equal to zero.

$$F_x(x, y, z, \lambda): \quad -700 - \frac{\lambda}{x^2} - \frac{\lambda}{3} = 0 \qquad F_y(x, y, z, \lambda): \quad -400 - \frac{\lambda}{y^2} - \frac{\lambda}{3} = 0$$

$$F_z(x, y, z, \lambda): \quad 900 + \lambda = 0 \qquad F_\lambda(x, y, z, \lambda): \quad \frac{1}{x} + \frac{1}{y} + z - \frac{x}{3} - \frac{y}{3} - 5 = 0$$

Solving these equations simultaneously we obtain

$$\lambda = -900 \qquad x = \tfrac{3}{2} \qquad y = 3 \qquad z = \tfrac{11}{2}$$

The values of x, y, and z agree with those found previously; $P(\frac{3}{2}, 3, \frac{11}{2}) = 2700$. P is shown to have an absolute maximum value of 2700 in the same way as before.

If several constraints are imposed, the method of Lagrange multipliers can be extended by using several multipliers. In particular, if we wish to find critical points of the function having function values $f(x, y, z)$ subject to the two side conditions $g(x, y, z) = 0$ and $h(x, y, z) = 0$, we find the critical points of the function F of the five variables x, y, z, λ, and μ for which

$$F(x, y, z, \lambda, \mu) = f(x, y, z) + \lambda g(x, y, z) + \mu h(x, y, z)$$

The following example illustrates the method.

EXAMPLE 4

Find the relative extrema of the function f if

$$f(x, y, z) = xz + yz$$

and the point (x, y, z) lies on the intersection of the surfaces $x^2 + z^2 = 2$ and $yz = 2$.

SOLUTION

We form the function F for which

$$F(x, y, z, \lambda, \mu) = xz + yz + \lambda(x^2 + z^2 - 2) + \mu(yz - 2)$$

Finding the five partial derivatives and setting them equal to zero we have

$$F_x(x, y, z, \lambda, \mu): \quad z + 2\lambda x = 0 \tag{24}$$

$$F_y(x, y, z, \lambda, \mu): \quad z + \mu z = 0 \tag{25}$$

$$F_z(x, y, z, \lambda, \mu): \quad x + y + 2\lambda z + \mu y = 0 \tag{26}$$

$$F_\lambda(x, y, z, \lambda, \mu): \quad x^2 + z^2 - 2 = 0 \tag{27}$$

$$F_\mu(x, y, z, \lambda, \mu): \quad yz - 2 = 0 \tag{28}$$

From (25) we obtain $\mu = -1$ and $z = 0$. We reject $z = 0$ because this contradicts (28). From (24) we obtain, if $x \neq 0$,

$$\lambda = -\frac{z}{2x} \tag{29}$$

Substituting from (29) and $\mu = -1$ into (26) we get

$$x + y - \frac{z^2}{x} - y = 0$$

$$x^2 = z^2 \tag{30}$$

Substituting from (30) into (27) we have $2x^2 - 2 = 0$, or $x^2 = 1$. This gives two values for x, namely 1 and -1; and for each of these values of x we get, from (30), the two values 1 and -1 for z. Obtaining the corresponding values for y from (28) we have four sets of solutions for the five Eqs. (24) through (28). These solutions are

$$
\begin{array}{llllll}
x = & 1 & y = & 2 & z = & 1 & \lambda = -\tfrac{1}{2} & \mu = -1 \\
x = & 1 & y = & -2 & z = & -1 & \lambda = \tfrac{1}{2} & \mu = -1 \\
x = & -1 & y = & 2 & z = & 1 & \lambda = \tfrac{1}{2} & \mu = -1 \\
x = & -1 & y = & -2 & z = & -1 & \lambda = -\tfrac{1}{2} & \mu = -1
\end{array}
$$

The first and fourth sets of solutions give $f(x, y, z) = 3$, and the second and third sets of solutions give $f(x, y, z) = 1$. Hence f has a relative maximum function value of 3 and a relative minimum function value of 1.

EXERCISES 16.4

In Exercises 1 through 4, use the method of Lagrange multipliers to find the critical points of the given function subject to the given constraint.

1. $f(x, y) = 25 - x^2 - y^2$ with constraint $x^2 + y^2 - 4y = 0$
2. $f(x, y) = 4x^2 + 2y^2 + 5$ with constraint $x^2 + y^2 - 2y = 0$
3. $f(x, y, z) = x^2 + y^2 + z^2$ with constraint $3x - 2y + z - 4 = 0$
4. $f(x, y, z) = x^2 + y^2 + z^2$ with constraint $y^2 - x^2 = 1$

In Exercises 5 through 8, use the method of Lagrange multipliers to find the relative extrema of f subject to the indicated constraint. Also find the points at which the extrema occur. Assume that the relative extrema exist.

5. $f(x, y) = x^2 + y$ with constraint $x^2 + y^2 = 9$
6. $f(x, y) = x^2 y$ with constraint $x^2 + 8y^2 = 24$
7. $f(x, y, z) = xyz$ with constraint $x^2 + 2y^2 + 4z^2 = 4$
8. $f(x, y, z) = y^3 + xz^2$ with constraint $x^2 + y^2 + z^2 = 1$

In Exercises 9 and 10, find a relative minimum value of f subject to the indicated constraint. Asssume that a relative minimum value exists.

9. $f(x, y, z) = x^2 + y^2 + z^2$ with constraint $xyz = 1$
10. $f(x, y, z) = xyz$ with constraint $x^2 + y^2 + z^2 = 1$

In Exercises 11 and 12, find a relative maximum value of f subject to the indicated constraint. Assume that a relative maximum value exists.

11. $f(x, y, z) = x + y + z$ with constraint $x^2 + y^2 + z^2 = 9$
12. $f(x, y, z) = xyz$ with constraint $2xy + 3xz + yz = 72$
13. Find a relative minimum value of the function f for which $f(x, y, z) = x^2 + 4y^2 + 16z^2$ with the constraint (a) $xyz = 1$; (b) $xy = 1$; (c) $x = 1$.
14. Use the method of Lagrange multipliers to find the shortest distance from the point $(1, 3, 0)$ to the plane $4x + 2y - z = 5$.
15. Use Lagrange multipliers to find the shortest distance from the point $(1, -1, -1)$ to the plane $x + 4y + 3z = 2$.
16. Find the least and greatest distances from the origin to a point on the ellipse $x^2 + 4y^2 = 16$.
17. Find the least and greatest distances from the origin to a point on the ellipsoid $9x^2 + 4y^2 + z^2 = 36$.
18. If $f(x, y, z) = 2x^2 + 3y^2 + z^2$, use the method of Lagrange multipliers to find the point in the plane $x + y + z = 5$ at which $f(x, y, z)$ is least.
19. Use the method of Lagrange multipliers to find a relative minimum function value of f if $f(x, y, z) = x^2 + y^2 + z^2$ with the two constraints $x + 2y + 3z = 6$ and $x - y - z = -1$.
20. Use the method of Lagrange multipliers to find a relative minimum function value of f if $f(x, y, z) = x^2 + y^2 + z^2$ with the two constraints $x + y + 2z = 1$ and $3x - 2y + z = -4$.
21. Use the method of Lagrange multipliers to find a relative maximum function value of f if $f(x, y, z) = xyz$ with the two constraints $x + y + z = 4$ and $x - y - z = 3$.

22. Use the method of Lagrange multipliers to find a relative maximum function value of f if $f(x, y, z) = x^3 + y^3 + z^3$ with the two constraints $x + y + z = 1$ and $x + y - z = 0$.

In Exercises 23 through 32, use the method of Lagrange multipliers to solve the indicated exercise of Exercises 16.3.

23. Exercise 13   **24.** Exercise 14   **25.** Exercise 15   **26.** Exercise 16   **27.** Exercise 17
28. Exercise 22   **29.** Exercise 23   **30.** Exercise 24   **31.** Exercise 31   **32.** Exercise 32

33. A circular disk is in the shape of the region bounded by the circle $x^2 + y^2 = 1$. If T degrees is the temperature at any point (x, y) of the disk and $T = 2x^2 + y^2 - y$, find the hottest and coldest points on the disk.

34. A company has three factories, each manufacturing the same product. If factory A produces x units, factory B produces y units, and factory C produces z units, their respective manufacturing costs are $(3x^2 + 200)$ dollars, $(y^2 + 400)$ dollars, and $(2z^2 + 300)$ dollars. If an order for 1100 units is to be filled, use the method of Lagrange multipliers to determine how the production should be distributed among the three factories to minimize the total manufacturing cost.

16.5
OBTAINING A FUNCTION FROM ITS GRADIENT AND EXACT DIFFERENTIALS

In Chapter 18, when studying vector fields, we shall wish to determine if a given vector-valued function is the gradient of some real-valued function f, and if it is, we will want to find such a function f. First let us consider the problem of how to obtain f if its gradient is known. That is, we are given

$$\nabla f(x, y) = f_x(x, y)\mathbf{i} + f_y(x, y)\mathbf{j} \tag{1}$$

and we wish to find $f(x, y)$.

▶ ILLUSTRATION 1   Suppose

$$\nabla f(x, y) = (y^2 + 2x + 4)\mathbf{i} + (2xy + 4y - 5)\mathbf{j} \tag{2}$$

Then because Eq. (1) must be satisfied, it follows that

$$f_x(x, y) = y^2 + 2x + 4 \tag{3}$$

$$f_y(x, y) = 2xy + 4y - 5 \tag{4}$$

By integrating both members of (3) with respect to x,

$$f(x, y) = y^2 x + x^2 + 4x + g(y) \tag{5}$$

Observe that the "constant" of integration is a function of y and independent of x because we are integrating with respect to x. If we now differentiate both members of (5) partially with respect to y, we obtain

$$f_y(x, y) = 2xy + g'(y) \tag{6}$$

Equations (4) and (6) give two expressions for $f_y(x, y)$. Hence

$$2xy + 4y - 5 = 2xy + g'(y)$$

Therefore

$$g'(y) = 4y - 5$$

$$g(y) = 2y^2 - 5y + C$$

Substituting this value of $g(y)$ into (5) we have

$$f(x, y) = y^2 x + x^2 + 4x + 2y^2 - 5y + C$$

◀

EXAMPLE 1
Find $f(x, y)$ if

$$\nabla f(x, y) = e^{y^2} \cos x \mathbf{i} + 2y e^{y^2} \sin x \mathbf{j}$$

SOLUTION
Because Eq. (1) must hold,

$$f_x(x, y) = e^{y^2} \cos x \tag{7}$$

$$f_y(x, y) = 2y e^{y^2} \sin x \tag{8}$$

Integrating both members of (8) with respect to y we obtain

$$f(x, y) = e^{y^2} \sin x + g(x) \tag{9}$$

where $g(x)$ is independent of y. We now partially differentiate both members of (9) with respect to x and get

$$f_x(x, y) = e^{y^2} \cos x + g'(x) \tag{10}$$

By equating the right members of (7) and (10),

$$e^{y^2} \cos x = e^{y^2} \cos x + g'(x)$$

$$g'(x) = 0$$

$$g(x) = C$$

We substitute this value of $g(x)$ into (9) and obtain

$$f(x, y) = e^{y^2} \sin x + C$$

All vectors of the form $M(x, y)\mathbf{i} + N(x, y)\mathbf{j}$ are not necessarily gradients, as shown in the next illustration.

▶ ILLUSTRATION 2   We show that there is no function f such that

$$\nabla f(x, y) = 3y \mathbf{i} - 2x \mathbf{j} \tag{11}$$

Assume that there is such a function. Then it follows that

$$f_x(x, y) = 3y \tag{12}$$

$$f_y(x, y) = -2x \tag{13}$$

We integrate both members of (12) with respect to x and obtain

$$f(x, y) = 3xy + g(y)$$

We partially differentiate both members of this equation with respect to y and have

$$f_y(x, y) = 3x + g'(y) \tag{14}$$

Equating the right members of (13) and (14) we obtain

$$-2x = 3x + g'(y)$$

$$-5x = g'(y)$$

If both members of this equation are differentiated with respect to x, it must follow that

$$-5 = 0$$

which, of course, is not true. Thus the assumption that $3y\mathbf{i} - 2x\mathbf{j}$ is a gradient leads to a contradiction.   ◀

We now investigate a condition that must be satisfied for a vector to be a gradient.

Suppose that M_y and N_x are continuous on an open disk B in R^2. If

$$M(x, y)\mathbf{i} + N(x, y)\mathbf{j} \tag{15}$$

is a gradient on B, then there is a function f such that

$$f_x(x, y) = M(x, y) \tag{16}$$

$$f_y(x, y) = N(x, y) \tag{17}$$

for all (x, y) in B. Because $M_y(x, y)$ exists on B, then, from (16),

$$M_y(x, y) = f_{xy}(x, y) \tag{18}$$

Furthermore, because $N_x(x, y)$ exists on B, it follows from (17) that

$$N_x(x, y) = f_{yx}(x, y) \tag{19}$$

Because M_y and N_x are continuous on B, their equivalents f_{xy} and f_{yx} are also continuous on B. Thus, from Theorem 15.7.1, $f_{xy}(x, y) = f_{yx}(x, y)$ at all points in B. Therefore the left members of (18) and (19) are equal at all points in B. We have proved that if M_y and N_x are continuous on an open disk B in R^2, a necessary condition for vector (15) to be a gradient on B is that

$$M_y(x, y) = N_x(x, y) \tag{20}$$

Equation (20) is also a sufficient condition for vector (15) to be a gradient on B. A rigorous proof of the sufficiency of the condition involves concepts beyond the scope of this book. If (20) holds, we can show how to find a function f such that vector (15) is a gradient. However, the proof that, whenever (20) holds, such a function exists belongs to a course in advanced calculus. The method for finding f is a generalization of that used in Illustration 1 and Example 1 and in Example 2 below. We have the following theorem.

16.5.1 THEOREM   Suppose that M and N are functions of two variables x and y defined on an open disk $B((x_0, y_0); r)$ in R^2, and M_y and N_x are continuous on B. Then the vector

$$M(x, y)\mathbf{i} + N(x, y)\mathbf{j}$$

is a gradient on B if and only if

$$M_y(x, y) = N_x(x, y)$$

at all points in B.

▶ ILLUSTRATION 3   (a) We apply Theorem 16.5.1 to the vector in the right member of Eq. (2) in Illustration 1. Let

$$M(x, y) = y^2 + 2x + 4 \qquad N(x, y) = 2xy + 4y - 5$$

$$M_y(x, y) = 2y \qquad\qquad N_x(x, y) = 2y$$

Thus $M_y(x, y) = N_x(x, y)$, and therefore the vector is a gradient.

(b) If we apply Theorem 16.5.1 to the vector in the right member of Eq. (11) in Illustration 2, with $M(x, y) = 3y$ and $N(x, y) = -2x$, we obtain

$$M_y(x, y) = 3 \qquad N_x(x, y) = -2$$

Hence $M_y(x, y) \neq N_x(x, y)$; thus the vector is not a gradient.   ◀

EXAMPLE 2

Determine if the vector

$$(e^{-y} - 2x)\mathbf{i} - (xe^{-y} + \sin y)\mathbf{j}$$

is a gradient $\mathbf{V}f(x, y)$, and if it is, then find $f(x, y)$.

SOLUTION

We apply Theorem 16.5.1. Let

$$M(x, y) = e^{-y} - 2x \qquad N(x, y) = -xe^{-y} - \sin y$$

$$M_y(x, y) = -e^{-y} \qquad N_x(x, y) = -e^{-y}$$

Therefore $M_y(x, y) = N_x(x, y)$; so the given vector is a gradient $\mathbf{V}f(x, y)$. Furthermore,

$$f_x(x, y) = e^{-y} - 2x \tag{21}$$

$$f_y(x, y) = -xe^{-y} - \sin y \tag{22}$$

Integrating both members of (21) with respect to x we obtain

$$f(x, y) = xe^{-y} - x^2 + g(y) \tag{23}$$

where $g(y)$ is independent of x. We now partially differentiate both members of (23) with respect to y and have

$$f_y(x, y) = -xe^{-y} + g'(y) \tag{24}$$

We equate the right members of (24) and (22) and get

$$-xe^{-y} + g'(y) = -xe^{-y} - \sin y$$

$$g'(y) = -\sin y$$

$$g(y) = \cos y + C$$

We substitute this expression for $g(y)$ into (23) and have

$$f(x, y) = xe^{-y} - x^2 + \cos y + C$$

Related to the concept of determining if a given vector-valued function is a gradient of a real-valued function f is that of determining if an expression of the form $M(x, y)\,dx + N(x, y)\,dy$ is the total differential of a function f. Such an expression is said to be an *exact differential*.

16.5.2 DEFINITION   The differential expression

$$M(x, y)\,dx + N(x, y)\,dy \qquad (25)$$

is called **exact** on an open disk B in R^2 if there exists a function f such that

$$f_x(x, y) = M(x, y) \quad \text{and} \quad f_y(x, y) = N(x, y)$$

at all points (x, y) in B.

▶ ILLUSTRATION 4   The expression

$$xy^4\,dx + 2x^2y^3\,dy \qquad (26)$$

is an exact differential because if

$$f(x, y) = \tfrac{1}{2}x^2y^4$$

then $f_x(x, y) = xy^4$ and $f_y(x, y) = 2x^2y^3$. Observe that (26) is the total differential of f because if $z = \tfrac{1}{2}x^2y^4$, then

$$dz = xy^4\,dx + 2x^2y^3\,dy \qquad ◀$$

The following theorem, which provides a test for determining if an expression of the form (25) is exact, follows immediately from Definition 16.5.2 and Theorem 16.5.1.

16.5.3 THEOREM   Suppose that M and N are functions of the two variables x and y defined on an open disk $B((x_0, y_0); r)$ in R^2, and M_y and N_x are continuous on B. Then the differential expression

$$M(x, y)\,dx + N(x, y)\,dy$$

is exact on B if and only if

$$M_y(x, y) = N_x(x, y)$$

at all points in B.

EXAMPLE 3
Determine if the differential expression is exact:

(a) $(y \sin x - 3 \cos y)\,dx + (3x \sin y - \cos x)\,dy$ (b) $(3 + r \cos \theta)\,dr + r^2 \sin \theta\,d\theta$

SOLUTION
(a) Let

$$M(x, y) = y \sin x - 3 \cos y \qquad N(x, y) = 3x \sin y - \cos x$$

$$M_y(x, y) = \sin x + 3 \sin y \qquad N_x(x, y) = 3 \sin y + \sin x$$

Because $M_y(x, y) = N_x(x, y)$, we have an exact differential.

(b) Let

$$M(r, \theta) = 3 + r \cos \theta \qquad N(r, \theta) = r^2 \sin \theta$$

$$M_\theta(r, \theta) = -r \sin \theta \qquad N_r(r, \theta) = 2r \sin \theta$$

Because $M_\theta(r, \theta) \neq N_r(r, \theta)$, the differential expression is not exact.

If $M(x, y) \, dx + N(x, y) \, dx$ is an exact differential then the differential equation

$$M(x, y) \, dx + N(x, y) \, dy = 0 \tag{27}$$

$$\Leftrightarrow \qquad M(x, y) + N(x, y) \frac{dy}{dx} = 0$$

is said to be an *exact differential equation*. If a differential equation of the form (27) is exact then there is a function f such that $f_x(x, y) = M(x, y)$ and $f_y(x, y) = N(x, y)$, and the general solution of the equation is given by $f(x, y) = C$, where C is an arbitrary constant.

▶ ILLUSTRATION 5   Equating the exact differential expression (26) to zero, we have the exact differential equation

$$xy^4 \, dx + 2x^2 y^3 \, dy = 0$$

The general solution of this equation is

$$\tfrac{1}{2}x^2 y^4 = \bar{C}$$

$$\Leftrightarrow \qquad x^2 y^4 = C$$

where $C = 2\bar{C}$.   ◀

EXAMPLE 4
Determine if the following differential equation is exact. If it is exact, find the general solution.

$$(3x^2 y + xy^2 + e^x) \, dx + (x^3 + x^2 y + \cos y) \, dy = 0 \tag{28}$$

SOLUTION
Let,

$$M(x, y) = 3x^2 y + xy^2 + e^x \qquad N(x, y) = x^3 + x^2 y + \cos y$$

$$M_y(x, y) = 3x^2 + 2xy \qquad N_x(x, y) = 3x^2 + 2xy$$

Therefore $M_y(x, y) = N_x(x, y)$; thus the differential equation is exact. Hence the general solution is $f(x, y) = C$, where

$$f_x(x, y) = 3x^2 y + xy^2 + e^x \tag{29}$$

$$f_y(x, y) = x^3 + x^2 y + \cos y \tag{30}$$

Integrating both members of (29) with respect to x we obtain

$$f(x, y) = x^3 y + \tfrac{1}{2}x^2 y^2 + e^x + g(y) \tag{31}$$

where $g(y)$ is independent of x. Partial differentiating both members of (31) with respect to y gives

$$f_y(x, y) = x^3 + x^2y + g'(y) \tag{32}$$

We equate the right members of (32) and (30) and get

$$x^3 + x^2y + g'(y) = x^3 + x^2y + \cos y$$

$$g'(y) = \cos y$$

$$g(y) = \sin y + \bar{C}$$

Substituting this expression in (31) we have

$$f(x, y) = x^3y + \tfrac{1}{2}x^2y^2 + e^x + \sin y + \bar{C}$$

Therefore the general solution is

$$x^3y + \tfrac{1}{2}x^2y^2 + e^x + \sin y + \bar{C} = 0$$

$$\Leftrightarrow \quad 2x^3y + x^2y^2 + 2e^x + 2 \sin y = C$$

where $C = -2\bar{C}$.

Theorem 16.5.1 can be extended to functions of three variables.

16.5.4 THEOREM Let M, N, and R be functions of three variables x, y, and z defined on an open ball $B((x_0, y_0, z_0); r)$ in R^3, and M_y, M_z, N_x, N_z, R_x, and R_y are continuous on B. Then the vector $M(x, y, z)\mathbf{i} + N(x, y, z)\mathbf{j} + R(x, y, z)\mathbf{k}$ is a gradient on B if and only if $M_y(x, y, z) = N_x(x, y, z)$, $M_z(x, y, z) = R_x(x, y, z)$, and $N_z(x, y, z) = R_y(x, y, z)$.

The proof of the "only if" part of Theorem 16.5.3 is similar to the proof of the "only if" part of Theorem 16.5.1 and is left as an exercise (see Exercise 35). The proof of the "if" part is beyond the scope of this book.

EXAMPLE 5
Determine if the following vector is a gradient $\nabla f(x, y, z)$, and if it is, then find $f(x, y, z)$:

$$(e^x \sin z + 2yz)\mathbf{i} + (2xz + 2y)\mathbf{j} + (e^x \cos z + 2xy + 3z^2)\mathbf{k}$$

SOLUTION
We apply Theorem 16.5.4. Let

$M(x, y, z) = e^x \sin z + 2yz$	$N(x, y, z) = 2xz + 2y$	$R(x, y, z) = e^x \cos z + 2xy + 3z^2$
$M_y(x, y, z) = 2z$	$N_x(x, y, z) = 2z$	$R_x(x, y, z) = e^x \cos z + 2y$
$M_z(x, y, z) = e^x \cos z + 2y$	$N_z(x, y, z) = 2x$	$R_y(x, y, z) = 2x$

Therefore

$$M_y(x, y, z) = N_x(x, y, z) \qquad M_z(x, y, z) = R_x(x, y, z) \qquad N_z(x, y, z) = R_y(x, y, z)$$

Thus the given vector is a gradient $\nabla f(x, y, z)$. Furthermore,

$$f_x(x, y, z) = e^x \sin z + 2yz \tag{33}$$

$$f_y(x, y, z) = 2xz + 2y \tag{34}$$

$$f_z(x, y, z) = e^x \cos z + 2xy + 3z^2 \tag{35}$$

Integrating both members of (33) with respect to x we have

$$f(x, y, z) = e^x \sin z + 2xyz + g(y, z) \tag{36}$$

where $g(y, z)$ is independent of x. We partially differentiate both members of (36) with respect to y and obtain

$$f_y(x, y, z) = 2xz + g_y(y, z) \tag{37}$$

Equating the right members of (37) and (34) we have

$$2xz + g_y(y, z) = 2xz + 2y$$

$$g_y(y, z) = 2y$$

We now integrate both members of this equation with respect to y and get

$$g(y, z) = y^2 + h(z) \tag{38}$$

where h is independent of x and y. Substituting from (38) into (36) we obtain

$$f(x, y, z) = e^x \sin z + 2xyz + y^2 + h(z) \tag{39}$$

We now partially differentiate with respect to z both members of (39):

$$f_z(x, y, z) = e^x \cos z + 2xy + h'(z) \tag{40}$$

Equating the right members of (40) and (35) we have

$$e^x \cos z + 2xy + h'(z) = e^x \cos z + 2xy + 3z^2$$

$$h'(z) = 3z^2$$

$$h(z) = z^3 + C \tag{41}$$

We substitute from (41) into (39) and obtain

$$f(x, y, z) = e^x \sin z + 2xyz + y^2 + z^3 + C$$

The following definition is the extension of Definition 16.5.2 to functions of three variables.

16.5.5 DEFINITION   The differential expression

$$M(x, y, z)\, dx + N(x, y, z)\, dy + R(x, y, z)\, dz$$

is called **exact** on an open ball B in R^3 if there exists a function f such that

$$f_x(x, y, z) = M(x, y, z) \qquad f_y(x, y, z) = N(x, y, z) \qquad f_z(x, y, z) = R(x, y, z)$$

at all points (x, y, z) in B.

▶ ILLUSTRATION 6   The expression

$$(e^x \sin z + 2yz)\, dx + (2xz + 2y)\, dy + (e^x \cos z + 2xy + 3z^2)\, dz$$

is an exact differential because it is the total differential of the function f found in Example 5.   ◀

EXERCISES 16.5

In Exercises 1 through 12, determine if the vector is a gradient. If it is, find a function having the given gradient.

1. $4x\mathbf{i} - 3y\mathbf{j}$

2. $y^2\mathbf{i} + 3x^2\mathbf{j}$

3. $(6x - 5y)\mathbf{i} - (5x - 6y^2)\mathbf{j}$

4. $(4y^2 + 6xy - 2)\mathbf{i} + (3x^2 + 8xy + 1)\mathbf{j}$

5. $(6x^2y^2 - 14xy + 3)\mathbf{i} + (4x^3y - 7x^2 - 8)\mathbf{j}$

6. $(2x + \ln y)\mathbf{i} + \left(y^2 + \dfrac{x}{y}\right)\mathbf{j}$

7. $\left(\dfrac{1}{x^2} + \dfrac{1}{y^2}\right)\mathbf{i} + \left(\dfrac{1 - 2x}{y^3}\right)\mathbf{j}$

8. $\left(\dfrac{2x - 1}{y}\right)\mathbf{i} + \left(\dfrac{x - x^2}{y^2}\right)\mathbf{j}$

9. $2x \sec 2y\mathbf{i} + 2x^2 \sec 2y \tan 2y\mathbf{j}$

10. $(2xy - y \sin x)\mathbf{i} + (x^2 + \cos x)\mathbf{j}$

11. $(2x \cos y - 1)\mathbf{i} - x^2 \sin y\mathbf{j}$

12. $(ye^x + x)\mathbf{i} + (xe^y - y)\mathbf{j}$

In Exercises 13 through 16, determine if the expression is an exact differential. If it is, find a function for which the given expression is the total differential.

13. $(2xy + y^2 + 1)\, dx + (x^2 + 2xy + x)\, dy$

14. $3(2x^2 + 6xy)\, dx + 3(3x^2 + 8y)\, dy$

15. $(e^y - 2xy)\, dx + (xe^y - x^2)\, dy$

16. $(\sin 2x - \tan y)\, dx - x \sec^2 y\, dy$

In Exercises 17 through 24, determine if the differential equation is exact. If it is exact, find the general solution.

17. $(x + y)\, dx + (2y + x)\, dy = 0$

18. $3x - y + (x + 3y)\dfrac{dy}{dx} = 0$

19. $ye^x - 2x + e^x\dfrac{dy}{dx} = 0$

20. $\dfrac{y}{x - 1}\, dx + \left(\ln(2x - 2) + \dfrac{1}{y}\right)dy = 0$

21. $(xy^2 + 2y)\, dx + (2y^3 - x^2y + 2x)\, dy = 0$

22. $e^{2x}\left(\dfrac{dy}{dx} + 2y\right) = x^2$

23. $\left(\dfrac{x^2}{y^2}\cos\dfrac{x}{y}\right)\dfrac{dy}{dx} = \dfrac{x}{y}\cos\dfrac{x}{y} + \sin\dfrac{x}{y} + \cos x$

24. $(2y \cot x - 3x^2)\dfrac{dy}{dx} = y^2 \csc^2 x + 6xy - 2$

In Exercises 25 through 30, determine if the vector is a gradient. If it is, find a function having the given gradient.

25. $(2y - 5z)\mathbf{i} + (2x + 8z)\mathbf{j} - (5x - 8y)\mathbf{k}$

26. $4x^3\mathbf{i} + 9y^2\mathbf{j} - 2z\mathbf{k}$

27. $e^y \sin z\mathbf{i} + xe^y \sin z\mathbf{j} + xe^y \cos z\mathbf{k}$

28. $(2xy + 7z^3)\mathbf{i} + (x^2 + 2y^2 - 3z)\mathbf{j} + (21xz^2 - 4y)\mathbf{k}$

29. $e^x(e^z - \ln y)\mathbf{i} + (e^y \ln z - e^xy^{-1})\mathbf{j} + (e^{x+z} + e^yz^{-1})\mathbf{k}$

30. $\dfrac{1}{y + z}\mathbf{i} - \dfrac{x - z}{(y + z)^2}\mathbf{j} - \dfrac{x + y}{(y + z)^2}\mathbf{k}$

In Exercises 31 through 34, determine if the expression is an exact differential. If it is, find a function for which the given expression is the total differential.

31. $(4xy + 3yz - 2)\, dx + (2x^2 + 3xz - 5z^2)\, dy + (3xy - 10yz + 1)\, dz$

32. $e^z \cos x\, dx + z \sin y\, dy + (e^z \sin x - \cos y)\, dz$

33. $z \tan y \, dx + xz \sec^2 y \, dy + x \tan y \, dz$

34. $(2y + z) \, dx + (2x - 3z + 4yz) \, dy + (x - 3y + 2y^2) \, dz$

35.   Prove the "only if" part of Theorem 16.5.4.

36.   Prove that $\nabla f(x, y) = a\mathbf{i} + b\mathbf{j}$ if and only if $f(x, y) = ax + by + c$, where a, b, and c are constants.

37.   Prove that a first-order differential equation whose variables are separable is exact.

16.6 (Supplementary) THE METHOD OF LEAST SQUARES

Suppose that we wish to find a mathematical model for some data given by a set of points $(x_1, y_1), (x_2, y_2), \ldots, (x_n, y_n)$. In particular, y_i may be the number of dollars in a manufacturer's weekly profit when x_i is the number of units sold in a week, or y_i could be a company's total annual sales when x_i years have elapsed since the start of the company. The number of new cases of a certain disease could be y_i when x_i is the number of days since the outbreak of an epidemic of the disease. The desired model is a relationship involving x and y that can be used to make future predictions. Such a relationship is afforded by a line that "fits" the data.

Table 16.6.1

x	1	2	3	4
y	5	8	7	12

▶ ILLUSTRATION 1   Table 16.6.1 gives a company's total annual receipts from sales during its first 4 years of operation, where x is the number of years of operation and y is the number of millions of dollars in annual sales.

Suppose that one wanted to make a prediction of the fifth year's receipts from this information. The four points (x_i, y_i) are plotted in Fig. 16.6.1. There are various lines that one could consider "fit" the four points, and three such lines are shown in the figure. Each line gives a different estimate of the receipts for the fifth year.   ◀

As observed in Illustration 1, there are often several lines that "fit" a set of data points. We would like to obtain the "line of best fit." To arrive at a suitable definition for such a line we first indicate how well a particular line fits a set of data points by measuring the vertical distances from the points to the line. For instance, in Fig. 16.6.2 there are n data points, and the line $y = mx + b$. The point (x_i, y_i) is the ith data point, and corresponding to it on the line is the point $(x_i, mx_i + b)$. The **deviation** (or **error**) between

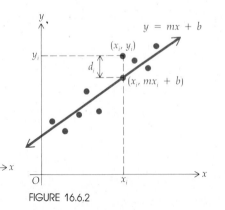

FIGURE 16.6.1

FIGURE 16.6.2

the ith data point and the line is defined to be d_i, where

$$d_i = y_i - (mx_i + b) \tag{1}$$

Note that d_i is positive if the point (x_i, y_i) is above the line and negative if it is below the line. If the line is to fit the data points well, it may seem reasonable to require that

$$d_1 + d_2 + \ldots + d_n$$

be small. However, such a requirement would allow for the situation in which the sum of some positive deviations and negative deviations would be small while the points were not close to the line. To avoid this possibility we consider the sum of the squares of the deviations

$$d_1{}^2 + d_2{}^2 + \ldots + d_n{}^2 = \sum_{i=1}^{n} d_i{}^2$$

Observe that $\sum_{i=1}^{n} d_i{}^2$ is never negative and is zero only if each d_i is zero, in which case all the data points lie on the line. We shall take as the line of best fit the one for which $\sum_{i=1}^{n} d_i{}^2$ is an absolute minimum. This line is called the **regression line** of y on x, and the process for finding it is called the **method of least squares.**

▶ ILLUSTRATION 2   We apply the method of least squares to find the regression line for the data points of Illustration 1. These points are $(1, 5)$, $(2, 8)$, $(3, 7)$, and $(4, 12)$. Let the regression line have the equation

$$y = mx + b$$

We wish to find m and b. From (1), the deviations between the data points and the line are

$$
\begin{aligned}
d_1 &= 5 - \big[m(1) + b\big] & d_2 &= 8 - \big[m(2) + b\big] \\
&= 5 - m - b & &= 8 - 2m - b
\end{aligned}
$$

$$
\begin{aligned}
d_3 &= 7 - \big[m(3) + b\big] & d_4 &= 12 - \big[m(4) + b\big] \\
&= 7 - 3m - b & &= 12 - 4m - b
\end{aligned}
$$

Thus

$$\sum_{i=1}^{4} d_i{}^2 = (5 - m - b)^2 + (8 - 2m - b)^2 + (7 - 3m - b)^2 + (12 - 4m - b)^2$$

Observe that $\sum_{i=1}^{4} d_i{}^2$ is a function of m and b. Denote this function by f so that

$$f(m, b) = (5 - m - b)^2 + (8 - 2m - b)^2 + (7 - 3m - b)^2 + (12 - 4m - b)^2 \tag{2}$$

We apply the second-derivative test to find the values of m and b that give the absolute minimum value of f.

$$f_m(m, b) = 2(5 - m - b)(-1) + 2(8 - 2m - b)(-2)$$
$$+ 2(7 - 3m - b)(-3) + 2(12 - 4m - b)(-4)$$
$$= 2(-5 + m + b - 16 + 4m + 2b - 21 + 9m + 3b - 48 + 16m + 4b)$$
$$= 2(-90 + 30m + 10b)$$

$$f_b(m, b) = 2(5 - m - b)(-1) + 2(8 - 2m - b)(-1)$$
$$+ 2(7 - 3m - b)(-1) + 2(12 - 4m - b)(-1)$$
$$= 2(-5 + m + b - 8 + 2m + b - 7 + 3m + b - 12 + 4m + b)$$
$$= 2(-32 + 10m + 4b)$$

$$f_{mm}(m, b) = 60 \qquad f_{mb}(m, b) = 20 \qquad f_{bb}(m, b) = 8 \qquad f_{bm}(m, b) = 20$$

Setting $f_m(m, b) = 0$ and $f_b(m, b) = 0$ we have

$$-9 + 3m + b = 0$$
$$-16 + 5m + 2b = 0$$

Solving these two equations simultaneously we obtain $m = 2$ and $b = 3$. For all values of m and b,

$$f_{mm}(m, b) = 60 > 0$$

and

$$f_{mm}(m, b) \cdot f_{bb}(m, b) - f_{mb}{}^2(m, b) = (60)(8) - (20)^2$$
$$= 80 > 0$$

Therefore, from Theorem 16.3.5(i), f has a relative minimum value when $m = 2$ and $b = 3$. Furthermore, there is only one relative extremum for f. Also, m and b are both in the interval $(-\infty, +\infty)$, and from (2) we observe that $f(m, b)$ is large when either the absolute value of m or the absolute value of b is large. Thus we conclude that the relative minimum value of f is an absolute minimum value. Hence the regression line is

$$y = 2x + 3 \tag{3}$$

Figure 16.6.3 shows the four points and this line. From (3), when $x = 5$, $y = 13$. Therefore the fifth year's receipts are estimated to be \$13 million.

◀

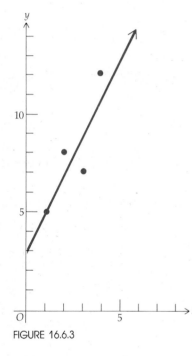

FIGURE 16.6.3

We now give the general procedure for using the method of least squares to find the regression line $y = mx + b$ for a set of n data points (x_1, y_1), $(x_2, y_2), \ldots, (x_n, y_n)$. The sum of the squares of the deviations between the points and the line is

$$\sum_{i=1}^{n} d_i{}^2 = \sum_{i=1}^{n} [y_i - (mx_i + b)]^2$$

Remember that x_i and y_i are constants and m and b are variables, so that

$$f(m, b) = \sum_{i=1}^{n} (y_i - mx_i - b)^2$$

We wish to find the values of m and b that make $f(m, b)$ an absolute minimum. We first find the partial derivatives $f_m(m, b)$ and $f_b(m, b)$.

$$f_m(m, b) = \sum_{i=1}^{n} \frac{\partial}{\partial m} [(y_i - mx_i - b)^2]$$

$$= \sum_{i=1}^{n} 2(y_i - mx_i - b)(-x_i)$$

$$= 2 \sum_{i=1}^{n} (-x_i y_i + mx_i^2 + bx_i)$$

$$= 2 \left[-\sum_{i=1}^{n} x_i y_i + m \sum_{i=1}^{n} x_i^2 + b \sum_{i=1}^{n} x_i \right]$$

$$f_b(m, b) = \sum_{i=1}^{n} \frac{\partial}{\partial b} [(y_i - mx_i - b)^2]$$

$$= \sum_{i=1}^{n} 2(y_i - mx_i - b)(-1)$$

$$= 2 \sum_{i=1}^{n} (-y_i + mx_i + b)$$

$$= 2 \left[-\sum_{i=1}^{n} y_i + m \sum_{i=1}^{n} x_i + nb \right]$$

Setting $f_m(m, b) = 0$ and $f_b(m, b) = 0$ we obtain

$$\left(\sum_{i=1}^{n} x_i^2 \right) m + \left(\sum_{i=1}^{n} x_i \right) b = \sum_{i=1}^{n} x_i y_i \tag{4}$$

$$\left(\sum_{i=1}^{n} x_i \right) m + nb = \sum_{i=1}^{n} y_i \tag{5}$$

These are two simultaneous equations in m and b. Solving Eq. (5) for b we have

$$b = \frac{1}{n} \left[\sum_{i=1}^{n} y_i - m \sum_{i=1}^{n} x_i \right] \tag{6}$$

Substituting this value of b in (4) we get

$$m = \frac{n \sum_{i=1}^{n} x_i y_i - \sum_{i=1}^{n} x_i \sum_{i=1}^{n} y_i}{n \sum_{i=1}^{n} x_i^2 - \left(\sum_{i=1}^{n} x_i \right)^2} \tag{7}$$

In Exercise 18 you are asked to supply the details involved to obtain Eqs. (6) and (7) from (4) and (5). In Exercise 19 you are asked to use the second-derivative test to show that f has a relative minimum value for the values of m and b in (6) and (7). As in Illustration 2 the relative minimum value of f is an absolute minimum value.

Observe that there are four distinct summations appearing in Eqs. (6) and (7). A convenient way of computing these summations is shown in the following example.

Table 16.6.2

x	1	3	5	7	9	11
y	3	5	6	5	7	8

Table 16.6.3

x_i	y_i	x_i^2	$x_i y_i$
1	3	1	3
3	5	9	15
5	6	25	30
7	5	49	35
9	7	81	63
11	8	121	88
$\sum$ 36	34	286	234

EXAMPLE 1

Find the regression line for the data in Table 16.6.2.

SOLUTION

The computation of the four summations appearing in Eqs. (6) and (7) is shown in Table 16.6.3. From the table,

$$\sum_{i=1}^{6} x_i = 36 \qquad \sum_{i=1}^{6} y_i = 34 \qquad \sum_{i=1}^{6} x_i^2 = 286 \qquad \sum_{i=1}^{6} x_i y_i = 234$$

Substituting these values first in (7) and then in (6) we get

$$m = \frac{6(234) - 36(34)}{6(286) - 36(36)} \qquad b = \tfrac{1}{6}[34 - \tfrac{3}{7}(36)]$$

$$= \tfrac{3}{7} \qquad\qquad = \tfrac{65}{21}$$

Therefore the regression line has the equation

$$y = \tfrac{3}{7}x + \tfrac{65}{21}$$

$$9x - 21y + 65 = 0$$

EXAMPLE 2

A rare antique was purchased in 1965 for $1200. Its value was $1800 in 1970, $2500 in 1975, and $3500 in 1980. If the value of the antique were to appreciate according to the same pattern through 1990, estimate the value of the antique in 1990 by the method of least squares.

Table 16.6.4

x	0	1	2	3
y	1200	1800	2500	3500

Table 16.6.5

x_i	y_i	x_i^2	$x_i y_i$
0	1200	0	0
1	1800	1	1800
2	2500	4	5000
3	3500	9	10,500
$\sum$ 6	9000	14	17,300

SOLUTION

To find a regression line $y = mx + b$, we let x be the number of 5-year periods since 1965 and let y dollars be the value of the antique $5x$ years since 1965. Thus we have the data points given in Table 16.6.4.

Table 16.6.5 shows the computation of the four summations appearing in Eqs. (6) and (7). From the table,

$$\sum_{i=1}^{4} x_i = 6 \qquad \sum_{i=1}^{4} y_i = 9000 \qquad \sum_{i=1}^{4} x_i^2 = 14 \qquad \sum_{i=1}^{4} x_i y_i = 17,300$$

With these values and $n = 4$ we obtain, from (7) and (6),

$$m = \frac{4(17,300) - 6(9000)}{4(14) - 14(14)} \qquad b = \tfrac{1}{4}[9000 - 760(6)]$$

$$= 760 \qquad\qquad = 1110$$

Hence the regression line has the equation

$$y = 760x + 1110$$

For the year 1990, $x = 5$. For this value of x we have

$$y = 760(5) + 1110$$
$$= 4910$$

Thus in 1990, the value of the antique is estimated to be $4910.

Table 16.6.6

x	1	2	3	4	5
y	20	24	30	35	42

Table 16.6.7

x_i	y_i	x_i^2	$x_i y_i$
1	20	1	20
2	24	4	48
3	30	9	90
4	35	16	140
5	42	25	210
$\sum$ 15	151	55	508

EXAMPLE 3

In Table 16.6.6, x days have elapsed since the outbreak of a particular disease, and y is the number of new cases of the disease on the xth day. (a) Find the regression line for the data points (x_i, y_i). (b) Use the regression line to estimate the number of new cases of the disease on the sixth day.

SOLUTION

(a) The required line has the equation $y = mx + b$. To determine m and b we first find the summations in Eqs. (6) and (7) from the computation in Table 16.6.7. From the table,

$$\sum_{i=1}^{5} x_i = 15 \qquad \sum_{i=1}^{5} y_i = 151 \qquad \sum_{i=1}^{5} x_i^2 = 55 \qquad \sum_{i=1}^{5} x_i y_i = 508$$

From (7) and (6) with these values and $n = 5$ we obtain

$$m = \frac{5(508) - (15)(151)}{5(55) - (15)(15)} \qquad b = \tfrac{1}{5}[151 - 5.5(15)]$$
$$= 5.5 \qquad\qquad\qquad\qquad = 13.7$$

Therefore the regression line has the equation

$$y = 5.5x + 13.7$$

(b) From the equation of the regression line, when $x = 6$, then $y = 46.7$. Therefore, on the sixth day of the epidemic 47 new cases are estimated.

EXAMPLE 4

A monopolist wants to determine the selling price of a particular commodity in order to maximize the daily profits. A linear demand equation was obtained by testing the commodity in five similar regions. The test results are shown in Table 16.6.8, where u units are demanded per day when p dollars is the price per unit. If the monopolist's daily overhead is $1200 and in addition it costs $1.50 to manufacture each unit, what should be the selling price to yield the greatest daily profit?

Table 16.6.8

	Region A	Region B	Region C	Region D	Region E
p	5.00	5.25	5.50	6.00	6.25
u	920	860	820	730	700

Table 16.6.9

p_i	u_i	$p_i{}^2$	$p_i u_i$
5.00	920	25.00	4600
5.25	860	27.56	4515
5.50	820	30.25	4510
6.00	730	36.00	4380
6.25	700	39.06	4375
$\sum$ 28.00	4030	157.87	22,380

SOLUTION

As the demand curve we take the regression line for the data in Table 16.6.8. To obtain an equation of this line we first find m and b from Eqs. (7) and (6) from the computation in Table 16.6.9. From the table,

$$\sum_{i=1}^{5} p_i = 28.00 \qquad \sum_{i=1}^{5} u_i = 4030 \qquad \sum_{i=1}^{5} p_i{}^2 = 157.87 \qquad \sum_{i=1}^{5} p_i u_i = 22,380$$

Therefore

$$m = \frac{5(22,380) - (28.00)(4030)}{5(157.87) - (28.00)(28.00)} \qquad b = \tfrac{1}{5}[4030 - (-175.7)(28.00)]$$
$$= -175.7 \qquad\qquad\qquad = 1790$$

Thus the regression line has the equation

$$u = -176p + 1790$$

which we take as the monopolist's demand equation for the commodity. Because the total revenue is the product of the price per unit and the number of units demanded, then if $R(p)$ dollars is the total daily revenue,

$$R(p) = p(-176p + 1790)$$
$$R(p) = -176p^2 + 1790p \tag{8}$$

The total cost is the sum of the overhead and the cost of manufacturing u units. The daily overhead is \$1200, and it costs \$1.50 to manufacture each unit. Therefore, if $C(p)$ is the total daily cost,

$$C(p) = 1200 + 1.50(-176p + 1790)$$
$$C(p) = 3885 - 264p \tag{9}$$

If $S(p)$ dollars is the total daily profit, $S(p) = R(p) - C(p)$. Thus, from (8) and (9) we have

$$S(p) = -176p^2 + 1790p - (3885 - 264p)$$
$$S(p) = -176p^2 + 2054p - 3885$$

We wish to find the value of p that makes $S(p)$ an absolute maximum.

$$S'(p) = -352p + 2054$$
$$S''(p) = -352$$

Setting $S'(p) = 0$ we get

$$-352p + 2054 = 0$$
$$p = 5.84$$

Because $S''(5.84) < 0$, it follows that S has a relative maximum value when $p = 5.84$. This is an absolute maximum value because S is continuous everywhere and S has only one relative extremum. Therefore the selling price should be \$5.84 per unit to yield the greatest daily profit.

EXERCISES 16.6

In Exercises 1 through 6, find the regression line for the given set of data points. Make a figure showing the data points and the regression line.

1. $(2, 0), (4, -1), (6, 2)$

2. $(1, 1), (2, 3), (3, 4)$

3. $(0, 2), (1, 1), (2, 3), (3, 2)$

4. $(1, -2), (3, 0), (5, 1), (7, 4)$

5. $(-2, 6), (-1, 4), (0, 1), (1, -1), (2, -2)$

6. $(0, 1), (2, 3), (4, 3), (6, 5), (8, 6)$

7. An early abstract painting was sold by the artist in 1915 for $100. Because of its historical importance its value has increased over the years. Its value was $4600 in 1935, $11,000 in 1955, and $20,000 in 1975. With the assumption that the value of the painting will appreciate according to the same pattern through 1995, use the method of least squares to estimate its value in 1995.

8. A 1980 model car was sold as a used car in 1981 for $6800. Its value was $6200 in 1982, $5700 in 1983, and $4800 in 1985. Use the method of least squares to estimate what its value was in 1984.

9. A motion picture has been playing at Cinema One for 5 weeks, and the weekly attendance (to the nearest 100) for each week is given in the following table.

Week Number	1	2	3	4	5
Attendance	5000	4500	4100	3900	3500

Assume that the weekly attendance will continue to decline according to the same pattern until it reaches 1500. (a) Use the regression line for the data in the table to determine the expected attendance for the sixth week. (b) The film will move over to the smaller Cinema Two when the weekly attendance drops below 2250. How many weeks is the film expected to play in Cinema One?

10. The following table gives a manufacturer's monthly production and profit for the first 5 months of the year, where x thousands of units were produced and y thousands of dollars was the profit.

	Jan.	Feb.	Mar.	April	May
x	65	72	82	90	100
y	30	35	42	48	60

If the production for June is to be 105,000 units, use the regression line for the data in the table to estimate that month's profit.

11. The following table gives data for five patients having a particular surgical operation at a certain hospital, where x years is the patient's age and y days is the length of time the patient remained in the hospital while recuperating after the surgery.

	Patient A	Patient B	Patient C	Patient D	Patient E
x	54	46	40	36	30
y	15	12	9	10	8

(a) Find an equation of the regression line for the data in the table. (b) Use the regression line to estimate the length of stay in the hospital for a 42-year-old person having the surgery.

12. The score on a student's entrance examination was used to predict the student's grade-point average at the end of the freshman year. The following table gives the data for six students, where x is the test score and y is the grade-point average.

	Student A	Student B	Student C	Student D	Student E	Student F
x	92	81	73	98	79	85
y	3.4	2.7	3.1	3.8	2.2	3.0

(a) Find an equation of the regression line for the data in the table. (b) Use the regression line to estimate a student's grade-point average at the end of the freshman year if the student had a score of 88 on the college entrance examination.

13. Five joggers were given examinations to determine their maximum oxygen uptake, a measure used to denote a person's cardiovascular fitness. The results are given in the following table, where x is the number of seconds in the jogger's best time for a mile run and y is the number of milliliters per minute per kilogram of body weight in the jogger's maximum oxygen uptake.

	Jogger A	Jogger B	Jogger C	Jogger D	Jogger E
x	300.5	350.6	407.3	326.2	512.8
y	350.2	325.8	375.6	418.5	400.2

(a) Find an equation of the regression line for the data in the table. (b) Use the regression line to estimate a jogger's maximum oxygen uptake if the jogger's best time for a mile run is 340.4 sec.

14. Five trees had their sap analyzed for the amount of a plant hormone that causes the detachment of leaves. For the trees in the following table, when x micrograms (μg) of plant hormone were released, y leaves were detached.

	Oak tree	Maple tree	Birch tree	Pine tree	Locust tree
x	28	57	38	75	82
y	208	350	300	620	719

(a) Find an equation of the regression line for the data in the table. (b) Use the regression line to estimate the number of leaves detached from another kind of tree when 100 μg of plant hormone are released.

15. In deserts, water is a factor that sharply limits plant activity. In the following table x is the number of millimeters of precipitation per year for six different regions, and y is the number of kilograms per hectare in the net photosynthate produced.

	Region A	Region B	Region C	Region D	Region E	Region F
x	100	200	400	500	600	650
y	1000	1900	3200	4400	5800	6400

(a) Find an equation of the regression line for the data in the table. (b) Use the regression line to estimate the net photosynthate produced in a region having an annual precipitation of 300 mm.

16. In the following table, for five healthy children, w kg is the body weight and y mm of mercury is the mean arterial pressure (average of systolic and diastolic blood pressures).

	Child A	Child B	Child C	Child D	Child E
w	20	30	35	40	50
y	70	85	90	96	100

(a) An equation that "fits" the data in this table is

$$y = m(\ln w) + b$$

To find such an equation let $x = \ln w$, and use the method of least squares for the points (x_i, y_i). (b) Use the result of part (a) to estimate the mean arterial pressure of a healthy child of weight 45 kg.

17. In order to set a selling price for a certain cosmetic, a monopolist test-marketed the cosmetic in four cities of the same size and at different prices. The following table gives the results, where u units were demanded per day when p dollars was the price per unit.

	City A	City B	City C	City D
p	8.50	9.00	9.50	10.00
u	1160	1080	980	740

From the data in the table a linear demand equation was obtained by the method of least squares. If the monopolist must pay \$5 to produce each unit and the daily overhead is \$900, what should be the selling price to maximize the daily profit?

18. Obtain Eqs. (6) and (7) by solving Eqs. (4) and (5) for m and b.

19. If $f(m, b) = \sum_{i=1}^{n} (y_i - mx_i - b)^2$, use the second-derivative test to prove that the values of m and b in (6) and (7) give a relative minimum value of f. (*Hint:* First show that $f_{mm}(m, b) > 0$. To show that $f_{mm}(m, b) \cdot f_{bb}(m, b) - f_{mb}^2(m, b) > 0$, you must prove that $\sum_{i=1}^{n} x_i^2 > \left(\sum_{i=1}^{n} x_i \right)^2$. To prove this, let $\bar{x} = \dfrac{1}{n} \sum_{i=1}^{n} x_i$ and apply properties of the sigma notation to the inequality $\sum_{i=1}^{n} (\bar{x} - x_i)^2 > 0$.)

REVIEW EXERCISES FOR CHAPTER 16

In Exercises 1 through 5, find the value of the directional derivative at the particular point P_0 for the given function in the direction of $\mathbf{U}$.

1. $f(x, y) = 3x^2 - 2xy + 1$; $\mathbf{U} = \frac{3}{5}\mathbf{i} - \frac{4}{5}\mathbf{j}$; $P_0 = (5, 10)$

2. $g(x, y) = \tan^{-1}\dfrac{y}{x}$; $\mathbf{U} = \dfrac{2}{\sqrt{13}}\mathbf{i} - \dfrac{3}{\sqrt{13}}\mathbf{j}$; $P_0 = (4, -4)$

3. $h(x, y) = e^x + y^2 \cos x$; $\mathbf{U} = \frac{1}{2}\sqrt{2}\mathbf{i} - \frac{1}{2}\sqrt{2}\mathbf{j}$; $P_0 = (0, 3)$

4. $f(x, y) = x^2 - 2x^2y + \ln x$; $\mathbf{U} = \cos \pi \mathbf{i} + \sin \pi \mathbf{j}$; $P_0 = (1, -2)$

5. $f(x, y, z) = xy^2z - 3xyz + 2xz^2$; $\mathbf{U} = -\frac{2}{3}\mathbf{i} + \frac{2}{3}\mathbf{j} - \frac{1}{3}\mathbf{k}$; $P_0 = (2, 1, 1)$

In Exercises 6 through 9, find (a) the gradient of f at P_0; (b) the rate of change of the function value in the direction of $\mathbf{U}$ at P_0.

6. $f(x, y) = 3x^2 - 2xy^3$; $\mathbf{U} = \cos \frac{1}{6}\pi \mathbf{i} + \sin \frac{1}{6}\pi \mathbf{j}$; $P_0 = (-3, 1)$

7. $f(x, y) = \frac{1}{2}\ln(x^2 + y^2)$; $\mathbf{U} = \frac{1}{2}\mathbf{i} + \frac{1}{2}\sqrt{3}\mathbf{j}$; $P_0 = (1, 1)$

8. $f(x, y, z) = yz - y^2 - xz$; $\mathbf{U} = \frac{6}{7}\mathbf{i} + \frac{3}{7}\mathbf{j} + \frac{2}{7}\mathbf{k}$; $P_0 = (1, 2, 3)$

9. $f(x, y, z) = x^3 + y^3 + 2xyz$; $\mathbf{U} = \dfrac{3}{\sqrt{14}}\mathbf{i} - \dfrac{2}{\sqrt{14}}\mathbf{j} + \dfrac{1}{\sqrt{14}}\mathbf{k}$; $P_0 = (2, -1, 0)$

In Exercises 10 through 13, determine if the vector is a gradient. If it is, find a function having the given gradient.

10. $(e^x \tan y - \sec y)\mathbf{i} - \sec y(x \tan y - e^x \sec y)\mathbf{j}$

11. $2xe^{x^2} \ln y \mathbf{i} + \dfrac{e^{x^2}}{y}\mathbf{j}$

12. $y(\cos x - z \sin x)\mathbf{i} + z(\cos x + \sin y)\mathbf{j} - (\cos y - y \cos x)\mathbf{k}$

13. $\left(\dfrac{-y}{(x + z)^2} + \dfrac{1}{x^2} \right)\mathbf{i} + \dfrac{1}{x + z}\mathbf{j} + \left(\dfrac{-y}{(x + z)^2} + \dfrac{2}{z^2} \right)\mathbf{k}$

In Exercises 14 through 16, find an equation of the tangent plane and equations of the normal line to the given surface at the indicated point.

14. $3x^2 + 2xy - y^2 = 15$; $(2, 3, 4)$ **15.** $x^2 + 2y + z = 8$; $(2, 1, 2)$ **16.** $z = x^2 + 2xy$; $(1, 3, 7)$

17. Find symmetric equations of the tangent line to the curve of intersection of the surfaces $x^2 - 3xy + y^2 = z$ and $2x^2 + y^2 - 3z + 27 = 0$ at the point $(1, -2, 11)$.

18. Find equations of the tangent line to the curve of intersection of the surface $z = 3x^2 + y^2 + 1$ with the plane $x = 2$ at the point $(2, -1, 14)$.

19. An equation of the surface of a mountain is $z = 900 - 3xy$, where distance is measured in meters, the x axis points to the west, and the y axis points to the south. A mountain climber is at the point corresponding to $(50, 4, 300)$. (a) What is the direction of steepest ascent? (b) Is the climber ascending or descending when moving in the north direction? (c) In what direction is the climber traveling a level path?

20. If $f(x, y, z) = \sinh(x + z) \cosh y$, find the rate of change of $f(x, y, z)$ with respect to distance in R^3 at the point $P(1, 1, 0)$ in the direction of $\overrightarrow{PQ}$ if Q is the point $(-1, 0, 2)$.

In Exercises 21 and 22, determine the relative extrema of f, if there are any.

21. $f(x, y) = x^3 + y^3 + 3xy$ **22.** $f(x, y) = 2x^2 - 3xy + 2y^2 + 10x - 11y$

In Exercises 23 through 26, use the method of Lagrange multipliers to find the critical point(s) of the given function subject to the indicated constraint. Determine if the function has a relative maximum or a relative minimum value at any critical point.

23. $f(x, y) = 5 + x^2 - y^2$ with constraints $x^2 - 2y^2 = 5$ **24.** $f(x, y, z) = x^2 + y^2 + z^2$ with constraint $x^2 - y^2 = 1$

25. $f(x, y, z) = y + xz - 2x^2 - y^2 - z^2$ with constraint $z = 35 - x - y$

26. $f(x, y, z) = xz^2 + y^3$ with constraint $x^2 + y^2 + z^2 = 1$

27. Use the method of Lagrange multipliers to find the shortest distance from the point $(4, 1, 2)$ to the plane $x - y + 2z = 0$.

28. Use the method of Lagrange multipliers to find the point on the surface $z = x^2 + y^2 + 2$ that is closest to the origin.

In Exercises 29 through 32, determine if the expression is an exact differential. If it is, find a function for which the given expression is the total differential.

29. $(6x^2y + 5x) dx + (2x^3 + 3y) dy$ **30.** $\sin x \sin y \, dx - \cos x \cos y \, dy$

31. $e^x \tan y \, dx + (e^x \sec^2 y - z \csc^2 y) dy + \cot y \, dz$ **32.** $\left(\dfrac{1}{x} - ze^{yz}\right) dx - xz^2 e^{yz} \, dy - xe^{yz}(yz + 1) dz$

In Exercises 33 through 36, determine if the differential equation is exact. If it is exact, find the general solution.

33. $(3x^2y^2 + 2y^3) dx + (2x^3y + 6xy^2) dy = 0$ **34.** $(2xy + 3 - 2 \sin x) dx + (x^2 + 3x) dy = 0$

35. $\dfrac{dy}{dx} = \dfrac{ye^{-x} - \sin x}{e^{-x} + 2y}$ **36.** $2y \sin xy + (2x \sin xy + y^3) \dfrac{dy}{dx} = 0$

37. Find three numbers whose sum is 100 and the sum of whose squares is least.

38. A manufacturer produces daily x units of commodity A and y units of commodity B. If $P(x, y)$ dollars is the daily profit from their sale, and $P(x, y) = 33x + 66y + xy - x^2 - 3y^2$, how many units of each commodity should be produced daily for the manufacturer to receive the greatest profit?

39. Find the dimensions of the rectangular parallelepiped of greatest volume that can be inscribed in the ellipsoid $x^2 + 9y^2 + z^2 = 9$. Assume that the edges are parallel to the coordinate axes.

40. The temperature is T degrees at any point (x, y) of the curve $4x^2 + 12y^2 = 1$, and $T = 4x^2 + 24y^2 - 2x$. Find the points on the curve where the temperature is the greatest and where it is the least. Also find the temperature at these points.

41. The temperature is T degrees at any point (x, y) on a heated circular plate and

$$T = \frac{44}{x^2 + y^2 + 9}$$

where distance is measured in centimeters from the origin at the center of the plate. (a) Find the rate of change of the temperature at the point $(3, 2)$ in the direction of the vector $\cos \frac{1}{6}\pi \mathbf{i} + \sin \frac{1}{6}\pi \mathbf{j}$. (b) Find the direction and magnitude of the greatest rate of change of T at the point $(3, 2)$.

42. A rectangular crate without a top is to have a surface area of 216 ft². What are the dimensions for a crate of greatest volume?

43. For the crate of Exercise 42, suppose instead of a surface area of 216 ft², the sum of the lengths of the edges is 216 ft. What then are the dimensions for a crate of greatest volume?

44. A piece of wire L ft long is cut into three pieces. One piece is bent into the shape of a circle, a second piece is bent into the shape of a square, and the third piece is bent into the shape of an equilateral triangle. How should the wire be cut so that (a) the combined area of the three figures is as small as possible and (b) the combined area of the three figures is as large as possible?

45. Determine the relative dimensions of a rectangular box, without a top and having a specific surface area, if the volume is to be a maximum.

46. Find the greatest and least distances from the origin to the curve of intersection of the surfaces $x^2 = 2yz$ and $x^2 + 3y^2 + 2z^2 = 30$.

Exercises 47 through 50 pertain to Supplementary Section 16.6. In Exercises 47 and 48, find the regression line for the given data points. Make a figure showing the data points and the regression line.

47. $(2, 0), (2, 1), (3, 2), (5, 3), (6, -1)$ **48.** $(0, 4), (1, 3), (2, 4), (3, -2), (4, -1)$

49. In the following table, a patient's systolic blood pressure and corresponding heart rate are given, where x mm of mercury is the systolic blood pressure and y beats per minute is the heart rate.

	Patient A	Patient B	Patient C	Patient D	Patient E	Patient F
x	110	117	133	146	115	127
y	70	74	80	65	60	77

(a) Find an equation of the regression line for the data in the table. (b) Use the regression line to estimate a patient's heart rate if the systolic blood pressure is 85 mm of mercury.

50. A breakfast cereal was test-sold in four cities of the same size at different prices; the results are shown in the table, where x cents was the price per box and y thousands of cases were sold per week.

	City A	City B	City C	City D
x	130	140	150	160
y	100	85	75	63

(a) Find an equation of the regression line for the data in the table. Use the regression line in part (a) as the demand curve to estimate the weekly sales if the price per box is (b) $1.20 and (c) $1.70.

S E V E N T E E N

MULTIPLE INTEGRATION

**17.1
THE DOUBLE INTEGRAL**

The definite integral of a function of a single variable can be extended to a function of several variables. We call an integral of a function of a single variable a **single integral** to distinguish it from a *multiple integral*, which involves a function of several variables. There are physical and geometric applications of multiple integrals similar to those for single integrals.

In the discussion of a single integral we required that the function be defined on a closed interval in the set of real numbers R. For the *double integral* of a function of two variables we require that the function be defined on a closed region in R^2. A closed region is one that includes its boundary. In this chapter, when we refer to a region, it is assumed to be closed. The simplest kind of closed region in R^2 is a closed rectangular region, which we proceed to define. Consider two distinct points $A(a_1, a_2)$ and $B(b_1, b_2)$ such that $a_1 \leq b_1$ and $a_2 \leq b_2$. These two points determine a rectangle having sides parallel to the coordinate axes. Refer to Fig. 17.1.1. The two points, together with the points (b_1, a_2) and (a_1, b_2), are called the *vertices* of the rectangle. The line segments joining consecutive vertices are called the *edges* of the rectangle. The set of all points interior to the rectangle is called an *open rectangular region*, and the set of all points in the open rectangle, together with the points on the edges of the rectangle, is called a *closed rectangular region*.

Let the closed rectangular region of Fig. 17.1.1 be denoted by R, and let f be a function defined on R. The region R can be considered as a **region of integration**. The first step is to define a **partition**, Δ, of R. We draw lines parallel to the coordinate axes and obtain a network of rectangular subregions that cover R. The **norm** of this partition, denoted by $\|\Delta\|$, is determined by the length of the longest diagonal of a rectangular subregion of the partition. The length of the diagonal is chosen because it represents the greatest distance between any two points in a rectangular subregion. Number the subregions in some arbitrary way and let the total be n. Denote the width of the ith subregion by $\Delta_i x$ units and its height by $\Delta_i y$ units. Then if $\Delta_i A$ square units is the area of the ith rectangular subregion,

$$\Delta_i A = \Delta_i x \, \Delta_i y$$

Let (ξ_i, γ_i) be an arbitrary point in the ith subregion and let $f(\xi_i, \gamma_i)$ be the function value there. Consider the product $f(\xi_i, \gamma_i) \Delta_i A$. Associated with each of the n subregions is such a product, and their sum is

$$\sum_{i=1}^{n} f(\xi_i, \gamma_i) \Delta_i A \tag{1}$$

There are many sums of the form (1) because the norm of the partition can be any positive number and each point (ξ_i, γ_i) can be any point in the ith subregion. If all such sums can be made arbitrarily close to one number L by taking partitions with sufficiently small norms, then L is defined to be the limit of these sums as the norm of the partition of R approaches zero. We have the following definition.

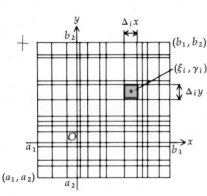

FIGURE 17.1.1

17.1.1 DEFINITION   Let f be a function defined on a closed rectangular region R. The number L is said to be the **limit** of sums of the form $\sum_{i=1}^{n} f(\xi_i, \gamma_i)\Delta_i A$ if L satisfies the property that for any $\epsilon > 0$ there exists a $\delta > 0$ such that for every partition Δ for which $\|\Delta\| < \delta$ and for all possible selections of the point (ξ_i, γ_i) in the ith rectangle, $i = 1, 2, \ldots, n$,

$$\left| \sum_{i=1}^{n} f(\xi_i, \gamma_i)\Delta_i A - L \right| < \epsilon$$

If such a number L exists, we write

$$\lim_{\|\Delta\| \to 0} \sum_{i=1}^{n} f(\xi_i, \gamma_i)\Delta_i A = L$$

If there is a number L satisfying Definition 17.1.1, it can be shown that it is unique. The proof is similar to the proof of the theorem (1.2.2) regarding the uniqueness of the limit of a function.

17.1.2 DEFINITION   A function f of two variables is said to be **integrable** on a closed rectangular region R if f is defined on R and the number L of Definition 17.1.1 exists. This number L is called the **double integral** of f on R, and we write

$$\lim_{\|\Delta\| \to 0} \sum_{i=1}^{n} f(\xi_i, \gamma_i)\Delta_i A = \iint_R f(x, y)\, dA \tag{2}$$

Other symbols for the double integral in (2) are

$$\iint_R f(x, y)\, dx\, dy \quad \text{and} \quad \iint_R f(x, y)\, dy\, dx$$

The following theorem, stated without proof, gives a sufficient condition for which a function of two variables is integrable.

17.1.3 THEOREM   If a function of two variables is continuous on a closed rectangular region R, then it is integrable on R.

The approximation of the value of a double integral is shown in the following example.

EXAMPLE 1
Find an approximate value of the double integral

$$\iint_R (2x^2 - 3y)\, dA$$

where R is the rectangular region having vertices $(-1, 1)$ and $(2, 3)$. Take a partition of R formed by the lines $x = 0$, $x = 1$, and $y = 2$, and take (ξ_i, γ_i) at the center of the ith subregion.

FIGURE 17.1.2

FIGURE 17.1.3

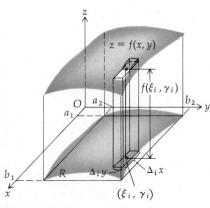

FIGURE 17.1.4

SOLUTION

Refer to Fig. 17.1.2, which shows the region R partitioned into six subregions that are squares having sides one unit in length. So for each i, $\Delta_i A = 1$. In each of the subregions the point (ξ_i, γ_i) is at the center of the square. With $f(x, y) = 2x^2 - 3y$, an approximation to the given double integral is given by

$$\iint\limits_{R} (2x^2 - 3y)\, dA \approx f(-\tfrac{1}{2}, \tfrac{3}{2}) \cdot 1 + f(\tfrac{1}{2}, \tfrac{3}{2}) \cdot 1 + f(\tfrac{3}{2}, \tfrac{3}{2}) \cdot 1$$
$$+ f(\tfrac{3}{2}, \tfrac{5}{2}) \cdot 1 + f(\tfrac{1}{2}, \tfrac{5}{2}) \cdot 1 + f(-\tfrac{1}{2}, \tfrac{5}{2}) \cdot 1$$
$$= -4 \cdot 1 - 4 \cdot 1 + 0 \cdot 1 - 3 \cdot 1 - 7 \cdot 1 - 7 \cdot 1$$
$$= -25$$

The exact value of the double integral in Example 1 is -24, as will be shown in Example 1 of Section 17.2.

We now consider the double integral of a function over a more general region. In Definition 5.5.2 a smooth function was defined as one that has a continuous derivative, and a smooth curve is the graph of a smooth function. Let R be a closed region whose boundary consists of a finite number of arcs of smooth curves that are joined together to form a closed curve. As we did with a rectangular region, draw lines parallel to the coordinate axes, which gives a rectangular partition of the region R. Discard the subregions that contain points not in R and consider only those that lie entirely in R (these are shaded in Fig. 17.1.3). Let the number of these shaded subregions be n and proceed in a manner analogous to the procedure used for a rectangular region. Definitions 17.1.1 and 17.1.2 apply when the region R is the more general one described above. You should intuitively realize that as the norm of the partition approaches zero, n increases without bound, and the area of the region omitted (i.e., the discarded rectangles) approaches zero. Actually, it can be proved that if a function is integrable on a region R, the limit of the approximating sums of the form (1) is the same no matter how we subdivide R, as long as each subregion has a shape to which an area can be assigned.

Just as the integral of a function of a single variable is interpreted geometrically in terms of the area of a plane region, the double integral can be interpreted geometrically in terms of the volume of a three-dimensional solid. Suppose that the function f is continuous on a closed region R in R^2. Furthermore, for simplicity in this discussion, assume that $f(x, y)$ is nonnegative on R. The graph of the equation $z = f(x, y)$ is a surface lying above the xy plane, as shown in Fig. 17.1.4. The figure shows a particular rectangular subregion of R, having dimensions of measures $\Delta_i x$ and $\Delta_i y$. The figure also shows a rectangular solid having this subregion as a base and $f(\xi_i, \gamma_i)$ as the measure of the altitude, where (ξ_i, γ_i) is a point in the ith subregion. The volume of the rectangular solid is determined by

$$\Delta_i V = f(\xi_i, \gamma_i)\, \Delta_i A$$
$$= f(\xi_i, \gamma_i)\, \Delta_i x\, \Delta_i y \qquad (3)$$

The number given in (3) is the measure of the volume of the thin rectangular solid shown in Fig. 17.1.4; thus the sum given in (1) is the sum of the measures of the volumes of n such solids. This sum approximates the measure of the volume of the three-dimensional solid shown in Fig. 17.1.4. The solid is bounded above by the graph of f and below by the region R in the xy plane. The sum in (1) also approximates the number given by the double integral

$$\iint_R f(x, y)\, dA$$

It can be proved that the volume of the three-dimensional solid of Fig. 17.1.4 is the value of the double integral. This fact is stated in the following theorem, for which a formal proof is not given.

17.1.4 THEOREM   Let f be a function of two variables that is continuous on a closed region R in the xy plane and $f(x, y) \geq 0$ for all (x, y) in R. If V is the measure of the volume of the solid S having the region R as its base and having an altitude of measure $f(x, y)$ at the point (x, y) in R, then

$$V = \lim_{||\Delta|| \to 0} \sum_{i=1}^{n} f(\xi_i, \gamma_i)\, \Delta_i A$$

$$= \iint_R f(x, y)\, dA$$

EXAMPLE 2
Approximate the volume of the solid bounded by the surface

$$f(x, y) = 4 - \tfrac{1}{9}x^2 - \tfrac{1}{16}y^2$$

the planes $x = 3$ and $y = 2$, and the three coordinate planes. To find an approximate value of the double integral take a partition of the region in the xy plane by drawing the lines $x = 1$, $x = 2$, and $y = 1$, and take (ξ_i, γ_i) at the center of the ith subregion.

SOLUTION
The solid is shown in Fig. 17.1.5. The rectangular region R is the rectangle in the xy plane bounded by the coordinate axes and the lines $x = 3$ and $y = 2$. From Theorem 17.1.4, if V cubic units is the volume of the solid,

$$V = \iint_R (4 - \tfrac{1}{9}x^2 - \tfrac{1}{16}y^2)\, dA$$

Figure 17.1.5 shows R partitioned into six subregions that are squares having sides of length one unit. Therefore, for each i, $\Delta_i A = 1$. The point (ξ_i, γ_i) in each subregion is at the center of the square. Then an approximation of V

FIGURE 17.1.5

is given by an approximation of the double integral. Therefore

$$V \approx f(\tfrac{1}{2}, \tfrac{1}{2}) \cdot 1 + f(\tfrac{3}{2}, \tfrac{1}{2}) \cdot 1 + f(\tfrac{5}{2}, \tfrac{1}{2}) \cdot 1 + f(\tfrac{1}{2}, \tfrac{3}{2}) \cdot 1 + f(\tfrac{3}{2}, \tfrac{3}{2}) \cdot 1 + f(\tfrac{5}{2}, \tfrac{3}{2}) \cdot 1$$

$$= (4 - \tfrac{25}{576}) + (4 - \tfrac{17}{64}) + (4 - \tfrac{409}{576}) + (4 - \tfrac{97}{576}) + (4 - \tfrac{25}{64}) + (4 - \tfrac{481}{576})$$

$$= 24 - \tfrac{695}{288}$$

$$\approx 21.59$$

Thus the volume is approximately 21.59 cubic units.

The exact value of the volume in Example 2 is shown to be 21.5 cubic units in Example 2 of Section 17.2.

Analogous to properties of the definite integral of a function of a single variable are several properties of the double integral, and the most important ones are given in the following theorems.

17.1.5 THEOREM   If c is a constant and the function f is integrable on a closed region R, then cf is integrable on R and

$$\iint\limits_{R} cf(x, y)\, dA = c \iint\limits_{R} f(x, y)\, dA$$

17.1.6 THEOREM   If the functions f and g are integrable on a closed region R, then the function $f + g$ is integrable on R and

$$\iint\limits_{R} [f(x, y) + g(x, y)]\, dA = \iint\limits_{R} f(x, y)\, dA + \iint\limits_{R} g(x, y)\, dA$$

The result of Theorem 17.1.6 can be extended to any finite number of functions that are integrable. The proofs of Theorems 17.1.5 and 17.1.6 follow directly from the definition of a double integral. These proofs are left as exercises (see Exercises 23 and 24).

17.1.7 THEOREM   If the functions f and g are integrable on the closed region R and furthermore $f(x, y) \geq g(x, y)$ for all (x, y) in R, then

$$\iint\limits_{R} f(x, y)\, dA \geq \iint\limits_{R} g(x, y)\, dA$$

Theorem 17.1.7 is analogous to Theorem 4.6.8 for the definite integral of a function of a single variable. The proof is similar and is left as an exercise (see Exercise 25).

17.1.8 THEOREM   Let the function f be integrable on a closed region R, and suppose that m and M are two numbers such that $m \leq f(x, y) \leq M$ for all (x, y) in R. Then if A is the measure of the area of region R,

$$mA \leq \iint\limits_{R} f(x, y)\, dA \leq MA$$

The proof of Theorem 17.1.8 is left as an exercise (see Exercise 26). The proof is similar to that of Theorem 4.6.9 and is based on Theorem 17.1.7.

17.1.9 THEOREM   Suppose that the function f is continuous on the closed region R and that region R is composed of the two subregions R_1 and R_2 that have no points in common except for points on parts of their boundaries. Then

$$\iint\limits_{R} f(x, y)\,dA = \iint\limits_{R_1} f(x, y)\,dA + \iint\limits_{R_2} f(x, y)\,dA$$

The proof of Theorem 17.1.9 is also left as an exercise and depends on the definition of a double integral and limit theorems (see Exercise 27).

EXERCISES 17.1

1. Find an approximate value of the double integral $\iint\limits_{R}(3x - 2y + 1)\,dA$, where R is the rectangular region having vertices $(0, -2)$ and $(3, 0)$. Take a partition of R formed by the lines $x = 1$, $x = 2$, and $y = -1$, and take (ξ_i, γ_i) at the center of the ith subregion.

2. Find an approximate value of the double integral $\iint\limits_{R}(y^2 - 4x)\,dA$, where R is the rectangular region having vertices $(-1, 0)$ and $(1, 3)$. Take a partition of R formed by the lines $x = 0$, $y = 1$, and $y = 2$, and take (ξ_i, γ_i) at the center of the ith subregion.

In Exercises 3 through 8, find an approximate value of the given double integral, where R is the rectangular region having the vertices P and Q, Δ is a partition of R, and (ξ_i, γ_i) is at the center of each subregion.

3. $\iint\limits_{R}(x^2 + y)\,dA$; $P(0, 0)$; $Q(4, 2)$; Δ: $x_1 = 0$, $x_2 = 1$, $x_3 = 2$, $x_4 = 3$, $y_1 = 0$, $y_2 = 1$.

4. $\iint\limits_{R}(2 - x - y)\,dA$; $P(0, 0)$; $Q(6, 4)$; Δ: $x_1 = 0$, $x_2 = 2$, $x_3 = 4$, $y_1 = 0$, $y_2 = 2$.

5. $\iint\limits_{R}(xy + 3y^2)\,dA$; $P(-2, 0)$; $Q(4, 6)$; Δ: $x_1 = -2$, $x_2 = 0$, $x_3 = 2$, $y_1 = 0$, $y_2 = 2$, $y_3 = 4$.

6. $\iint\limits_{R}(xy + 3y^2)\,dA$; $P(0, -2)$; $Q(6, 4)$; Δ: $x_1 = 0$, $x_2 = 2$, $x_3 = 4$, $y_1 = -2$, $y_2 = 0$, $y_3 = 2$.

7. $\iint\limits_{R}(x^2y - 2xy^2)\,dA$; $P(-3, -2)$; $Q(1, 6)$; Δ: $x_1 = -3$, $x_2 = -1$, $y_1 = -2$, $y_2 = 0$, $y_3 = 2$, $y_4 = 4$.

8. $\iint\limits_{R}(x^2y - 2xy^2)\,dA$; $P(-3, -2)$; $Q(1, 6)$; Δ: $x_1 = -3$, $x_2 = -2$, $x_3 = -1$, $x_4 = 0$, $y_1 = -2$, $y_2 = -1$, $y_3 = 0$, $y_4 = 1$, $y_5 = 2$, $y_6 = 3$, $y_7 = 4$, $y_8 = 5$.

In Exercises 9 through 12, find an approximate value of the given double integral, where R is the rectangular region having the vertices P and Q, Δ is a partition of R, and (ξ_i, γ_i) is an arbitrary point in each subregion.

9. The double integral, P, Q, and Δ are the same as in Exercise 3; $(\xi_1, \gamma_1) = (\frac{1}{4}, \frac{1}{2})$; $(\xi_2, \gamma_2) = (\frac{7}{4}, 0)$; $(\xi_3, \gamma_3) = (\frac{5}{2}, \frac{1}{4})$; $(\xi_4, \gamma_4) = (4, 1)$; $(\xi_5, \gamma_5) = (\frac{3}{4}, \frac{7}{4})$; $(\xi_6, \gamma_6) = (\frac{5}{4}, \frac{3}{2})$; $(\xi_7, \gamma_7) = (\frac{5}{2}, 2)$; $(\xi_8, \gamma_8) = (3, 1)$.

10. The double integral, P, Q, and Δ are the same as in Exercise 4; $(\xi_1, \gamma_1) = (\frac{1}{2}, \frac{3}{2})$; $(\xi_2, \gamma_2) = (3, 1)$; $(\xi_3, \gamma_3) = (\frac{11}{2}, \frac{1}{2})$; $(\xi_4, \gamma_4) = (2, 2)$; $(\xi_5, \gamma_5) = (2, 2)$; $(\xi_6, \gamma_6) = (5, 3)$.

11. The double integral, P, Q, and Δ are the same as in Exercise 5; $(\xi_1, \gamma_1) = (-\frac{1}{2}, \frac{1}{2})$; $(\xi_2, \gamma_2) = (1, \frac{3}{2})$; $(\xi_3, \gamma_3) = (\frac{5}{2}, 2)$; $(\xi_4, \gamma_4) = (-\frac{3}{2}, \frac{7}{2})$; $(\xi_5, \gamma_5) = (0, 3)$; $(\xi_6, \gamma_6) = (4, 4)$; $(\xi_7, \gamma_7) = (-1, \frac{9}{2})$; $(\xi_8, \gamma_8) = (1, \frac{9}{2})$; $(\xi_9, \gamma_9) = (3, \frac{9}{2})$.

12. The double integral, P, Q, and Δ are the same as in Exercise 5; $(\xi_1, \gamma_1) = (-2, 0)$; $(\xi_2, \gamma_2) = (0, 0)$; $(\xi_3, \gamma_3) = (2, 0)$; $(\xi_4, \gamma_4) = (-2, 2)$; $(\xi_5, \gamma_5) = (0, 2)$; $(\xi_6, \gamma_6) = (2, 2)$; $(\xi_7, \gamma_7) = (-2, 4)$; $(\xi_8, \gamma_8) = (0, 4)$; $(\xi_9, \gamma_9) = (2, 4)$.

13. Approximate the volume of the solid in the first octant bounded by the sphere $x^2 + y^2 + z^2 = 64$, the planes $x = 3$, $y = 3$, and the three coordinate planes. To find an approximate value of the double integral take a partition of the region in the xy plane formed by the lines $x = 1$, $x = 2$, $y = 1$, and $y = 2$, and take (ξ_i, γ_i) at the center of the ith subregion.

14. Approximate the volume of the solid bounded by the planes $z = 2x + y + 4$, $x = 2$, $y = 3$, and the three coordinate planes. To find an approximate value of the double integral take a partition of the region in the xy plane formed by the lines $x = 1$, $y = 1$, and $y = 2$, and take (ξ_i, γ_i) at the center of the ith subregion.

15. Approximate the volume of the solid bounded by the surface $z = 10 - \frac{1}{4}x^2 - \frac{1}{9}y^2$, the planes $x = 2$, $y = 2$, and the three coordinate planes. To find an approximate value of the double integral take a partition of the region in the xy plane formed by the lines $x = 1$ and $y = 1$, and take (ξ_i, γ_i) at the center of the ith subregion.

16. Approximate the volume of the solid bounded by the surface $100z = 300 - 25x^2 - 4y^2$, the planes $x = -1$, $x = 3$, $y = -3$, $y = 5$ and the xy plane. To find an approximate value of the double integral take a partition of the region in the xy plane formed by the lines $x = 1$, $y = -1$, $y = 1$, and $y = 3$, and take (ξ_i, γ_i) at the center of the ith subregion.

In Exercises 17 through 22, apply Theorem 17.1.8 to find a closed interval containing the value of the given double integral.

17. $\iint\limits_R (2x + 5y)\, dA$, where R is the rectangular region having vertices $(0, 0)$, $(1, 0)$, $(1, 2)$, and $(0, 2)$.

18. $\iint\limits_R (x^2 + y^2)\, dA$, where R is the rectangular region having vertices $(0, 0)$, $(1, 0)$, $(1, 1)$, and $(0, 1)$.

19. $\iint\limits_R e^{xy}\, dA$, where R is the rectangular region having vertices $(0, 0)$, $(1, 0)$, $(1, 1)$, and $(0, 1)$.

20. $\iint\limits_R (\sin x + \sin y)\, dA$, where R is the rectangular region having vertices $(0, 0)$, $(\pi, 0)$, (π, π), and $(0, \pi)$. (*Hint:* Use the result of Exercise 10 in Exercises 16.3.)

21. $\iint\limits_R [\sin(x + y) + \sin x + \sin y]\, dA$, where R is the rectangular region having vertices $(0, 0)$, $(\pi, 0)$, (π, π), and $(0, \pi)$. (*Hint:* Use the result of Exercise 11 in Exercises 16.3.)

22. $\iint\limits_R \dfrac{2x + 2y + 1}{x^2 + y^2 + 1}\, dA$, where R is the rectangular region having vertices $(-1, -1)$, $(1, -1)$, $(1, 1)$, and $(-1, 1)$. (*Hint:* Use the result of Exercise 8 in Exercises 16.3.)

23. Prove Theorem 17.1.5.   24. Prove Theorem 17.1.6.   25. Prove Theorem 17.1.7.
26. Prove Theorem 17.1.8.   27. Prove Theorem 17.1.9.

17.2
EVALUATION OF
DOUBLE INTEGRALS
AND ITERATED INTEGRALS

For functions of a single variable, the second fundamental theorem of the calculus provides a method for evaluating a definite integral by finding an antiderivative (or indefinite integral) of the integrand. There is a corresponding method for evaluating a double integral that involves performing successive single integrations. A rigorous development of this method belongs to a course in advanced calculus. Our discussion is an intuitive one, and we use the geometric interpretation of the double integral as the measure of a volume. We first develop the method for the double integral on a rectangular region.

FIGURE 17.2.1

Let f be a given function that is integrable on a closed rectangular region R in the xy plane bounded by the lines $x = a_1$, $x = b_1$, $y = a_2$, and $y = b_2$. Assume that $f(x, y) \geq 0$ for all (x, y) in R. Refer to Fig. 17.2.1, which shows a sketch of the graph of the equation $z = f(x, y)$ when (x, y) is in R. The number that represents the value of the double integral

$$\iint_R f(x, y) \, dA$$

is the measure of the volume of the solid between the surface and the region R. This number can be found by the method of parallel plane sections, discussed in Section 5.3, as we now do.

Let y be a number in $[a_2, b_2]$. Consider the plane parallel to the xz plane through the point $(0, y, 0)$. Let $A(y)$ be the measure of the area of the plane region of intersection of this plane with the solid. By the method of parallel plane sections, the measure of the volume of the solid is expressed by

$$\int_{a_2}^{b_2} A(y) \, dy$$

Because the volume of the solid also is determined by the double integral,

$$\iint_R f(x, y) \, dA = \int_{a_2}^{b_2} A(y) \, dy \tag{1}$$

By using (1) we can find the value of the double integral of the function f on R by evaluating a single integral of $A(y)$. We now must find $A(y)$ when y is given. Because $A(y)$ is the measure of the area of a plane region, we can find it by integration. In Fig. 17.2.1, notice that the upper boundary of the plane region is the graph of the equation $z = f(x, y)$ when x is in $[a_1, b_1]$. Therefore $A(y) = \int_{a_1}^{b_1} f(x, y) \, dx$. Substituting from this equation into (1) we obtain

$$\iint_R f(x, y) \, dA = \int_{a_2}^{b_2} \left[\int_{a_1}^{b_1} f(x, y) \, dx \right] dy \tag{2}$$

The integral on the right side of (2) is called an **iterated integral**. Usually the brackets are omitted when writing an iterated integral. So (2) can be written as

$$\iint_R f(x, y) \, dA = \int_{a_2}^{b_2} \int_{a_1}^{b_1} f(x, y) \, dx \, dy \tag{3}$$

When evaluating the "inner integral" in (3), remember that x is the variable of integration and y is considered a constant. This is comparable to considering y as a constant when finding the partial derivative of $f(x, y)$ with respect to x.

By considering plane sections parallel to the yz plane we obtain an iterated integral that interchanges the order of integration; we have

$$\iint_R f(x, y) \, dA = \int_{a_1}^{b_1} \int_{a_2}^{b_2} f(x, y) \, dy \, dx \tag{4}$$

A sufficient condition for (3) and (4) to be valid is that the function be continuous on the rectangular region R.

FIGURE 17.2.2

FIGURE 17.2.3

FIGURE 17.2.4

EXAMPLE 1

Evaluate the double integral

$$\iint_R (2x^2 - 3y)\, dA$$

if R is the region consisting of all points (x, y) for which $-1 \le x \le 2$ and $1 \le y \le 3$.

SOLUTION

$a_1 = -1$, $b_1 = 2$, $a_2 = 1$, and $b_2 = 3$. So from (3),

$$\iint_R (2x^2 - 3y)\, dA = \int_1^3 \int_{-1}^2 (2x^2 - 3y)\, dx\, dy$$

$$= \int_1^3 \left[\int_{-1}^2 (2x^2 - 3y)\, dx \right] dy$$

$$= \int_1^3 \left[\tfrac{2}{3}x^3 - 3xy \right]_{-1}^2 dy$$

$$= \int_1^3 (6 - 9y)\, dy$$

$$= -24$$

In Example 1 of Section 17.1 we found an approximate value of the double integral in the above example to be -25.

EXAMPLE 2

Find the volume of the solid bounded by the surface

$$f(x, y) = 4 - \tfrac{1}{9}x^2 - \tfrac{1}{16}y^2$$

the planes $x = 3$ and $y = 2$, and the three coordinate planes.

SOLUTION

Figure 17.2.2 shows the graph of the equation $z = f(x, y)$ in the first octant and the given solid. If V cubic units is the volume of the solid, then from Theorem 17.1.4,

$$V = \lim_{\|\Delta\| \to 0} \sum_{i=1}^n f(\xi_i, \gamma_i)\, \Delta_i A$$

$$= \iint_R f(x, y)\, dA$$

$$= \int_0^3 \int_0^2 (4 - \tfrac{1}{9}x^2 - \tfrac{1}{16}y^2)\, dy\, dx$$

$$= \int_0^3 \left[4y - \tfrac{1}{9}x^2y - \tfrac{1}{48}y^3 \right]_0^2 dx$$

$$= \int_0^3 (\tfrac{47}{6} - \tfrac{2}{9}x^2)\, dx$$

$$= \tfrac{47}{6}x - \tfrac{2}{27}x^3 \Big]_0^3$$

$$= 21.5$$

The volume is therefore 21.5 cubic units.

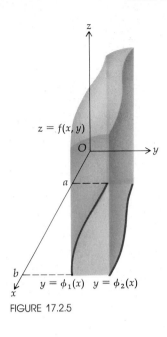

$z = f(x, y)$

$y = \phi_1(x)$ $y = \phi_2(x)$

FIGURE 17.2.5

FIGURE 17.2.6

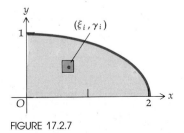

(ξ_i, γ_i)

FIGURE 17.2.7

In Example 2 of Section 17.1 an approximate value of the volume in Example 2 was found to be 21.59 cubic units.

Suppose now that R is a region in the xy plane that is bounded by the lines $x = a$ and $x = b$, where $a < b$, and by the curves $y = \phi_1(x)$ and $y = \phi_2(x)$, where ϕ_1 and ϕ_2 are two functions continuous on the closed interval $[a, b]$; furthermore, $\phi_1(x) \leq \phi_2(x)$ whenever $a \leq x \leq b$ (see Fig. 17.2.3). Let Δ be a partition of the interval $[a, b]$ defined by Δ: $a = x_0 < x_1 < \ldots < x_n = b$. Consider the region R of Fig. 17.2.3 to be divided into vertical strips with widths of measure $\Delta_i x$. A particular strip is shown in the figure. The intersection of the surface $z = f(x, y)$ and a plane $x = \xi_i$, where $x_{i-1} \leq \xi_i \leq x_i$, is a curve. A segment of this curve is over the ith vertical strip. The region under this curve segment and above the xy plane is shown in Fig. 17.2.4, and the measure of the area of this region is given by

$$\int_{\phi_1(\xi_i)}^{\phi_2(\xi_i)} f(\xi_i, y) \, dy$$

The measure of the volume of the solid bounded above by the surface $z = f(x, y)$ and below by the ith vertical strip is approximately equal to

$$\left[\int_{\phi_1(\xi_i)}^{\phi_2(\xi_i)} f(\xi_i, y) \, dy \right] \Delta_i x$$

If we take the limit, as the norm of Δ approaches zero, of the sum of these measures of volume for n vertical strips of R from $x = a$ to $x = b$, we obtain the measure of the volume of the solid bounded above by the surface $z = f(x, y)$ and below by the region R in the xy plane. (See Fig. 17.2.5.) This is the double integral of f on R; that is,

$$\lim_{\|\Delta\| \to 0} \sum_{i=1}^{n} \left[\int_{\phi_1(\xi_i)}^{\phi_2(\xi_i)} f(\xi_i, y) \, dy \right] \Delta_i x = \int_a^b \int_{\phi_1(x)}^{\phi_2(x)} f(x, y) \, dy \, dx$$

$$= \iint_R f(x, y) \, dy \, dx \qquad (5)$$

Sufficient conditions for (5) to be valid are that f be continuous on the closed region R and that ϕ_1 and ϕ_2 be smooth functions.

EXAMPLE 3
Express as both a double integral and an iterated integral the measure of the volume of the solid above the xy plane bounded by the elliptic paraboloid $z = x^2 + 4y^2$ and the cylinder $x^2 + 4y^2 = 4$. Evaluate the iterated integral to find the volume of the solid.

SOLUTION
The solid is shown in Fig. 17.2.6. We find the volume of the portion of the solid in the first octant, which from properties of symmetry is one-fourth of the required volume. The region R in the xy plane is that bounded by the x and y axes and the ellipse $x^2 + 4y^2 = 4$. This region is shown in Fig. 17.2.7, which also shows the ith subregion of a rectangular partition of R, where (ξ_i, γ_i) is any point in this ith subregion. If V cubic units is the volume of the

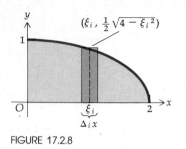

FIGURE 17.2.8

given solid, then by Theorem 17.1.4,

$$V = 4 \lim_{\|\Delta\| \to 0} \sum_{i=1}^{n} (\xi_i^2 + 4\gamma_i^2) \Delta_i A$$

$$= 4 \iint_R (x^2 + 4y^2) \, dA$$

To express the measure of the volume as an iterated integral we divide the region R into n vertical strips. Figure 17.2.8 shows the region R and the ith vertical strip having width of $\Delta_i x$ units and length of $\frac{1}{2}\sqrt{4 - \xi_i^2}$ units, where $x_{i-1} \leq \xi_i \leq x_i$. From (5),

$$V = 4 \lim_{\|\Delta\| \to 0} \sum_{i=1}^{n} \left[\int_0^{\sqrt{4 - \xi_i^2}/2} (\xi_i^2 + 4y^2) \, dy \right] \Delta_i x$$

$$= 4 \int_0^2 \int_0^{\sqrt{4 - x^2}/2} (x^2 + 4y^2) \, dy \, dx$$

$$= 4 \int_0^2 \left[x^2 y + \frac{4}{3} y^3 \right]_0^{\sqrt{4 - x^2}/2} dx$$

$$= 4 \int_0^2 \left[\frac{1}{2} x^2 \sqrt{4 - x^2} + \frac{1}{6} (4 - x^2)^{3/2} \right] dx$$

$$= \frac{4}{3} \int_0^2 (x^2 + 2) \sqrt{4 - x^2} \, dx$$

$$= -\frac{1}{3} x (4 - x^2)^{3/2} + 2x \sqrt{4 - x^2} + 8 \sin^{-1} \tfrac{1}{2} x \Big]_0^2$$

$$= 4\pi$$

Therefore the volume is 4π cubic units.

Suppose the region R is bounded by the curves $x = \lambda_1(y)$ and $x = \lambda_2(y)$ and the lines $y = c$ and $y = d$, where $c < d$, and λ_1 and λ_2 are two functions continuous on the closed interval $[c, d]$ for which $\lambda_1(y) \leq \lambda_2(y)$ whenever $c \leq y \leq d$. Consider a partition Δ of the interval $[c, d]$ and divide the region into horizontal strips, the measures of whose widths are $\Delta_i y$. See Fig. 17.2.9, showing the ith horizontal strip. The intersection of the surface $z = f(x, y)$ and a plane $y = \gamma_i$, where $y_{i-1} \leq \gamma_i \leq y_i$, is a curve, and a segment of this curve is over the ith horizontal strip. Then, as in the derivation of (5), the measure of the volume of the solid bounded above by the surface $z = f(x, y)$ and below by the ith vertical strip is approximately equal to

$$\left[\int_{\lambda_1(\gamma_i)}^{\lambda_2(\gamma_i)} f(x, \gamma_i) \, dx \right] \Delta_i y$$

Taking the limit, as $\|\Delta\|$ approaches zero, of the sum of these measures of volume for n horizontal strips of R from $y = c$ to $y = d$, we obtain the measure of the volume of the solid bounded above by the surface $z = f(x, y)$ and below by the region R in the xy plane. This measure of volume is the double integral

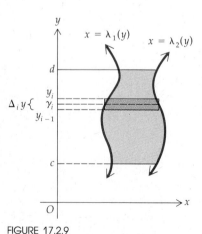

FIGURE 17.2.9

of f on R. Hence

$$\lim_{||\Delta|| \to 0} \sum_{i=1}^{n} \left[\int_{\lambda_1(\gamma_i)}^{\lambda_2(\gamma_i)} f(x, \gamma_i) \, dx \right] \Delta_i y = \int_c^d \int_{\lambda_1(y)}^{\lambda_2(y)} f(x, y) \, dx \, dy$$

$$= \iint_R f(x, y) \, dx \, dy \tag{6}$$

Sufficient conditions for (6) to be valid are that λ_1 and λ_2 be smooth functions and f be continuous on R. In applying both (5) and (6), sometimes it may be necessary to subdivide a region R into subregions on which these sufficient conditions hold.

EXAMPLE 4
Express the volume of the solid of Example 3 by an iterated integral in which the order of integration is the reverse of that of Example 3. Compute the volume.

SOLUTION
Again we find the volume of the solid in the first octant and multiply the result by 4. Figure 17.2.10 shows the region R in the xy plane and the ith horizontal strip whose width has a measure of $\Delta_i y$ and whose length has a measure of $2\sqrt{1 - y_i^2}$. Then by (6),

$$V = 4 \lim_{||\Delta|| \to 0} \sum_{i=1}^{n} \left[\int_0^{2\sqrt{1-y_i^2}} (x^2 + 4y_i^2) \, dx \right] \Delta_i y$$

$$= 4 \int_0^1 \int_0^{2\sqrt{1-y^2}} (x^2 + 4y^2) \, dx \, dy$$

$$= 4 \int_0^1 \left[\tfrac{1}{3}x^3 + 4y^2 x \right]_0^{2\sqrt{1-y^2}} dy$$

$$= 4 \int_0^1 \left[\tfrac{8}{3}(1 - y^2)^{3/2} + 8y^2\sqrt{1 - y^2} \right] dy$$

$$= \tfrac{32}{3} \int_0^1 (2y^2 + 1)\sqrt{1 - y^2} \, dy$$

$$= -\tfrac{16}{3}y(1 - y^2)^{3/2} + 8y\sqrt{1 - y^2} + 8 \sin^{-1} y \Big]_0^1$$

$$= 4\pi$$

Hence the volume is 4π cubic units, which agrees with the answer of Example 3.

From the solutions of Examples 3 and 4 we see that the double integral $\iint_R (x^2 + 4y^2) \, dA$ can be evaluated by either of the iterated integrals

$$\int_0^2 \int_0^{\sqrt{4-x^2}/2} (x^2 + 4y^2) \, dy \, dx \quad \text{or} \quad \int_0^1 \int_0^{2\sqrt{1-y^2}} (x^2 + 4y^2) \, dx \, dy$$

$(2\sqrt{1 - y_i^2}, y_i)$

$\Delta_i y \left\{ y_i \right.$

FIGURE 17.2.10

If in either (5) or (6), $f(x, y) = 1$ for all x and y, then the measure A of the area of a region R is expressed as a double integral. We have

$$A = \iint_R dy\,dx \quad \Leftrightarrow \quad A = \iint_R dx\,dy \qquad (7)$$

$y = x^2$

$(2, 4)$

$y = -x^2 + 4x$

$\Delta_i y$ {

$\Delta_i x$

FIGURE 17.2.11

EXAMPLE 5

Find by double integration the area of the region in the xy plane bounded by the curves $y = x^2$ and $y = 4x - x^2$.

SOLUTION

The region is shown in Fig. 17.2.11. From (7),

$$A = \iint_R dy\,dx$$

$$= \int_0^2 \int_{x^2}^{4x-x^2} dy\,dx$$

$$= \int_0^2 (4x - x^2 - x^2)\,dx$$

$$= 2x^2 - \tfrac{2}{3}x^3 \Big]_0^2$$

$$= \tfrac{8}{3}$$

Hence the area of the region is $\tfrac{8}{3}$ square units.

EXERCISES 17.2

In Exercises 1 through 10, evaluate the iterated integral.

1. $\displaystyle\int_1^2 \int_0^{2x} xy^3\,dy\,dx$

2. $\displaystyle\int_0^4 \int_0^y dx\,dy$

3. $\displaystyle\int_0^4 \int_0^y \sqrt{9 + y^2}\,dx\,dy$

4. $\displaystyle\int_{-1}^1 \int_1^{e^x} \frac{x}{y}\,dy\,dx$

5. $\displaystyle\int_1^4 \int_{y^2}^y \sqrt{\frac{y}{x}}\,dx\,dy$

6. $\displaystyle\int_1^4 \int_{x^2}^x \sqrt{\frac{y}{x}}\,dy\,dx$

7. $\displaystyle\int_0^1 \int_0^1 |x - y|\,dy\,dx$

8. $\displaystyle\int_0^3 \int_0^x x^2 e^{xy}\,dy\,dx$

9. $\displaystyle\int_{\pi/2}^\pi \int_0^x \sin(4x - y)\,dy\,dx$

10. $\displaystyle\int_{\pi/2}^\pi \int_0^{y^2} \sin\frac{x}{y}\,dx\,dy$

In Exercises 11 through 18, find the exact value of the double integral.

11. The double integral is the same as in Exercise 1 in Exercises 17.1.

12. The double integral is the same as in Exercise 2 in Exercises 17.1.

13. The double integral is the same as in Exercise 3 in Exercises 17.1.

14. The double integral is the same as in Exercise 6 in Exercises 17.1.

15. $\displaystyle\iint_R \sin x\,dA$; R is the region bounded by the lines $y = 2x$, $y = \tfrac{1}{2}x$, and $x = \pi$.

16. $\displaystyle\iint_R \cos(x + y)\,dA$; R is the region bounded by the lines $y = x$ and $x = \pi$, and the x axis.

17. $\displaystyle\iint_R x^2 \sqrt{9 - y^2}\,dA$; R is the region bounded by the circle $x^2 + y^2 = 9$.

18. $\displaystyle\iint\limits_{R} \frac{y^2}{x^2}\, dA$; R is the region bounded by the lines $y = x$ and $y = 2$, and the hyperbola $xy = 1$.

19. Find the volume of the solid under the plane $z = 4x$ and above the circle $x^2 + y^2 = 16$ in the xy plane. Draw a sketch of the solid.

20. Find the volume of the solid bounded by the planes $x = y + 2z + 1$, $x = 0$, $y = 0$, $z = 0$, and $3y + z - 3 = 0$. Draw a sketch of the solid.

21. Find the volume of the solid in the first octant bounded by the two cylinders $x^2 + y^2 = 4$ and $x^2 + z^2 = 4$. Draw a sketch of the solid.

22. Find the volume of the solid in the first octant bounded by the paraboloid $z = 9 - x^2 - 3y^2$. Draw a sketch of the solid.

23. Find the volume of the solid in the first octant bounded by the surfaces $x + z^2 = 1$, $x = y$, and $x = y^2$. Draw a sketch of the solid.

24. Find by double integration the volume of the portion of the solid bounded by the sphere $x^2 + y^2 + z^2 = 16$ that lies in the first octant. Draw a sketch of the solid.

In Exercises 25 through 28, use double integrals to find the area of the region bounded by the given curves in the xy plane. Draw a sketch of the region.

25. $y = x^3$ and $y = x^2$ **26.** $y^2 = 4x$ and $x^2 = 4y$ **27.** $y = x^2 - 9$ and $y = 9 - x^2$ **28.** $x^2 + y^2 = 16$ and $y^2 = 6x$

29. Express as an iterated integral the measure of the volume of the solid bounded by the ellipsoid

$$\frac{x^2}{a^2} + \frac{y^2}{b^2} + \frac{z^2}{c^2} = 1$$

30. Use double integration to find the area of the region in the first quadrant bounded by the parabola $y^2 = 4x$, the circle $x^2 + y^2 = 5$, and the x axis by two methods: (a) Integrate first with respect to x; (b) integrate first with respect to y. Compare the two methods of solution.

31. Find, by two methods, the volume of the solid below the plane $3x + 8y + 6z = 24$ and above the region in the first quadrant of the xy plane bounded by the parabola $y^2 = 2x$, the line $2x + 3y = 10$, and the x axis: (a) Integrate first with respect to x; (b) integrate first with respect to y. Compare the two methods of solution.

32. Given the iterated integral $\int_0^a \int_0^x \sqrt{a^2 - x^2}\, dy\, dx$. (a) Draw a sketch of the solid the measure of whose volume is represented by the given iterated integral; (b) evaluate the iterated integral; (c) write the iterated integral that gives the measure of the volume of the same solid with the order of integration reversed.

33. Given the iterated integral $\frac{2}{3} \int_0^a \int_0^{\sqrt{a^2 - x^2}} (2x + y)\, dy\, dx$. The instructions are the same as for Exercise 32.

34. Use double integration to find the volume of the solid common to two right-circular cylinders of radius r units, whose axes intersect at right angles. (See Exercise 16 in Exercises 5.3.)

In Exercises 35 and 36, the iterated integral cannot be evaluated exactly in terms of elementary functions by the given order of integration. Reverse the order of integration and perform the computation.

35. $\displaystyle\int_0^4 \int_{\sqrt{x}}^2 \sin \pi y^3 \, dy\, dx$

36. $\displaystyle\int_0^1 \int_y^1 e^{x^2} \, dx\, dy$

**17.3
CENTER OF MASS AND
MOMENTS OF INERTIA**

In Chapter 5 we used single integrals to find the center of mass of a homogeneous lamina. In using single integrals we can consider only laminae of constant area density (except in special cases); however, with double integrals we can find the center of mass of either a homogeneous or a nonhomogeneous lamina.

Suppose we are given a lamina having the shape of a closed region R in the xy plane. Let $\rho(x, y)$ be the measure of the area density of the lamina at any point (x, y) of R where ρ is continuous on R. To find the total mass of the lamina we proceed as follows. Let Δ be a partition of R into n rectangles. If (ξ_i, γ_i) is any point in the ith rectangle having an area of measure $\Delta_i A$, then an approximation to the measure of the mass of the ith rectangle is given by $\rho(\xi_i, \gamma_i)\,\Delta_i A$, and the measure of the total mass of the lamina is approximated by

$$\sum_{i=1}^{n} \rho(\xi_i, \gamma_i)\,\Delta_i A$$

Taking the limit of the above sum as the norm of Δ approaches zero, we express the measure M of the mass of the lamina by

$$M = \lim_{||\Delta|| \to 0} \sum_{i=1}^{n} \rho(\xi_i, \gamma_i)\,\Delta_i A$$
$$= \iint_{R} \rho(x, y)\,dA \tag{1}$$

The measure of the moment of mass of the ith rectangle with respect to the x axis is approximated by $\gamma_i \rho(\xi_i, \gamma_i)\,\Delta_i A$. The sum of the measures of the moments of mass of the n rectangles with respect to the x axis is then approximated by the sum of n such terms. The measure M_x of the moment of mass with respect to the x axis of the entire lamina is given by

$$M_x = \lim_{||\Delta|| \to 0} \sum_{i=1}^{n} \gamma_i \rho(\xi_i, \gamma_i)\,\Delta_i A$$
$$= \iint_{R} y\rho(x, y)\,dA \tag{2}$$

Analogously, the measure M_y of its moment of mass with respect to the y axis is given by

$$M_y = \lim_{||\Delta|| \to 0} \sum_{i=1}^{n} \xi_i \rho(\xi_i, \gamma_i)\,\Delta_i A$$
$$= \iint_{R} x\rho(x, y)\,dA \tag{3}$$

The center of mass of the lamina is denoted by the point $(\bar{x}, \bar{y})$ and

$$\bar{x} = \frac{M_y}{M} \qquad \bar{y} = \frac{M_x}{M}$$

EXAMPLE 1

A lamina in the shape of an isosceles right triangle has an area density that varies as the square of the distance from the vertex of the right angle. If mass is measured in kilograms and distance is measured in meters, find the mass and the center of mass of the lamina.

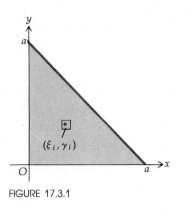

FIGURE 17.3.1

SOLUTION
Choose the coordinate axes so that the vertex of the right triangle is at the origin and the sides of length a meters of the triangle are along the coordinate axes (see Fig. 17.3.1). Let $\rho(x, y)$ be the number of kilograms per square meter in the area density of the lamina at the point (x, y). Then $\rho(x, y) = k(x^2 + y^2)$, where k is a constant. Therefore, if M kg is the mass of the lamina, we have, from (1),

$$
\begin{aligned}
M &= \lim_{\|\Delta\| \to 0} \sum_{i=1}^{n} k(\xi_i^2 + \gamma_i^2)\,\Delta_i A \\
&= k \iint_R (x^2 + y^2)\, dA \\
&= \int_0^a \int_0^{a-x} (x^2 + y^2)\, dy\, dx \\
&= k \int_0^a \left[yx^2 + \tfrac{1}{3}y^3 \right]_0^{a-x} dx \\
&= k \int_0^a (\tfrac{1}{3}a^3 - a^2 x + 2ax^2 - \tfrac{4}{3}x^3)\, dx \\
&= k(\tfrac{1}{3}a^4 - \tfrac{1}{2}a^4 + \tfrac{2}{3}a^4 - \tfrac{1}{3}a^4) \\
&= \tfrac{1}{6}ka^4
\end{aligned}
$$

To find the center of mass, observe that because of symmetry it must lie on the line $y = x$. Therefore, if we find $\bar{x}$, we also have $\bar{y}$. From (3),

$$
\begin{aligned}
M_y &= \lim_{\|\Delta\| \to 0} \sum_{i=1}^{n} k\xi_i(\xi_i^2 + \gamma_i^2)\,\Delta_i A \\
&= k \iint_R x(x^2 + y^2)\, dA \\
&= k \int_0^a \int_0^{a-x} x(x^2 + y^2)\, dy\, dx \\
&= k \int_0^a \left[x^3 y + \tfrac{1}{3}xy^3 \right]_0^{a-x} dx \\
&= k \int_0^a (\tfrac{1}{3}a^3 x - a^2 x^2 + 2ax^3 - \tfrac{4}{3}x^4)\, dx \\
&= k(\tfrac{1}{6}a^5 - \tfrac{1}{3}a^5 + \tfrac{1}{2}a^5 - \tfrac{4}{15}a^5) \\
&= \tfrac{1}{15}ka^5
\end{aligned}
$$

Because $M\bar{x} = M_y$, then $M\bar{x} = \tfrac{1}{15}ka^5$; and because $M = \tfrac{1}{6}ka^4$, we get $\bar{x} = \tfrac{2}{5}a$. Therefore the center of mass is at the point $(\tfrac{2}{5}a, \tfrac{2}{5}a)$.

17.3.1 DEFINITION   The **moment of inertia** of a particle, whose mass is m kg, about an axis is defined to be mr^2 kg-m^2, where r meters is the perpendicular distance from the particle to the axis.

If we have a system of n particles, the moment of inertia of the system is defined as the sum of the moments of inertia of all the particles. That is, if the ith particle has a mass of m_i kg and is at a distance of r_i meters from the axis, then I kg-m^2 is the moment of inertia of the system, where

$$I = \sum_{i=1}^{n} m_i r_i^2 \tag{4}$$

Extending this concept of moment of inertia to a continuous distribution of mass in a plane such as rods or laminae by processes similar to those previously used, we have the following definition.

17.3.2 DEFINITION   Suppose that we are given a continuous distribution of mass occupying a region R in the xy plane, and suppose that the measure of area density of this distribution at the point (x, y) is $\rho(x, y)$ kg/m^2, where ρ is continuous on R. Then the **moment of inertia I_x kg-m^2 about the x axis** of this distribution of mass is determined by

$$I_x = \lim_{\|\Delta\| \to 0} \sum_{i=1}^{n} \gamma_i^2 \rho(\xi_i, \gamma_i) \Delta_i A$$

$$= \iint_R y^2 \rho(x, y) \, dA \tag{5}$$

Similarly, the measure I_y of the **moment of inertia about the y axis** is given by

$$I_y = \lim_{\|\Delta\| \to 0} \sum_{i=1}^{n} \xi_i^2 \rho(\xi_i, \gamma_i) \Delta_i A$$

$$= \iint_R x^2 \rho(x, y) \, dA \tag{6}$$

and the measure I_0 of the **moment of inertia about the origin**, or the z axis, is given by

$$I_0 = \lim_{\|\Delta\| \to 0} \sum_{i=1}^{n} (\xi_i^2 + \gamma_i^2) \rho(\xi_i, \gamma_i) \Delta_i A$$

$$= \iint_R (x^2 + y^2) \rho(x, y) \, dA \tag{7}$$

The number I_0 is the measure of what is called the **polar moment of inertia**.

EXAMPLE 2
A homogeneous straight wire has a constant linear density of ρ kg/m. Find the moment of inertia of the wire about an axis perpendicular to the wire and passing through one end.

SOLUTION
Let the wire be of length a meters, and suppose that it extends along the x axis from the origin. We find its moment of inertia about the y axis. Divide the wire into n segments; the length of the ith segment is $\Delta_i x$ meters. The mass

of the ith segment is then $\rho\,\Delta_i x$ kg. Assume that the mass of the ith segment is concentrated at a single point ξ_i, where $x_{i-1} \le \xi_i \le x_i$. The moment of inertia of the ith segment about the y axis lies between $\rho x_{i-1}{}^2\,\Delta_i x$ kg-m^2 and $\rho x_i{}^2\,\Delta_i x$ kg-m^2 and is approximated by $\rho\xi_i{}^2\,\Delta_i x$ kg-m^2, where $x_{i-1} \le \xi_i \le x_i$. If the moment of inertia of the wire about the y axis is I_y kg-m^2, then

$$
\begin{aligned}
I_y &= \lim_{\|\Delta\|\to 0} \sum_{i=1}^{n} \rho\xi_i{}^2\,\Delta_i x \\
&= \int_0^a \rho x^2\,dx \\
&= \tfrac{1}{3}\rho a^3
\end{aligned}
$$

Therefore the moment of inertia is $\tfrac{1}{3}\rho a^3$ kg-m^2.

EXAMPLE 3

A homogeneous rectangular lamina has constant area density of ρ slugs/ft^2. Find the moment of inertia of the lamina about one corner.

SOLUTION

Suppose that the lamina is bounded by the lines $x = a$, $y = b$, the x axis, and the y axis. See Fig. 17.3.2. If I_0 slug-ft^2 is the moment of inertia about the origin, then

$$
\begin{aligned}
I_0 &= \lim_{\|\Delta\|\to 0} \sum_{i=1}^{n} \rho(\xi_i{}^2 + \gamma_i{}^2)\,\Delta_i A \\
&= \iint_R \rho(x^2 + y^2)\,dA \\
&= \rho \int_0^b \int_0^a (x^2 + y^2)\,dx\,dy \\
&= \rho \int_0^b \left[\tfrac{1}{3}x^3 + xy^2\right]_0^a dy \\
&= \rho \int_0^b (\tfrac{1}{3}a^3 + ay^2)\,dy \\
&= \tfrac{1}{3}\rho ab(a^2 + b^2)
\end{aligned}
$$

The moment of inertia is therefore $\tfrac{1}{3}\rho ab(a^2 + b^2)$ slug-ft^2.

FIGURE 17.3.2

It is possible to find the distance from any axis L at which the mass of a lamina can be concentrated without affecting the moment of inertia of the lamina about L. The measure of this distance, denoted by r, is called the *radius of gyration* of the lamina about L. That is, if the mass M kg of a lamina is concentrated at a point r meters from L, the moment of inertia of the lamina about L is the same as that of a particle of mass M kg at a distance of r meters from L; this moment of inertia is Mr^2 kg-m^2. Thus we have the following definition.

17.3.3 DEFINITION

If I is the measure of the moment of inertia about an axis L of a distribution of mass in a plane and M is the measure of the total mass of the distribution, then the **radius of gyration** of the distribution about L has measure r, where

$$r^2 = \frac{I}{M}$$

EXAMPLE 4

Suppose that a lamina is in the shape of a semicircle and the measure of the area density of the lamina at any point is proportional to the measure of the distance of the point from the diameter. If mass is measured in kilograms and distance is measured in meters, find the radius of gyration of the lamina about the x axis.

SOLUTION

Choose the x and y axes such that the semicircle is the top half of the circle $x^2 + y^2 = a^2$. See Fig. 17.3.3. The area density of the lamina at the point (x, y) is then ky kg/m^2. So if M kg is the mass of the lamina, we have

$$M = \lim_{\|\Delta\| \to 0} \sum_{i=1}^{n} k\gamma_i \Delta_i A$$

$$= \iint_R ky \, dA$$

$$= \int_0^a \int_{-\sqrt{a^2 - y^2}}^{\sqrt{a^2 - y^2}} ky \, dx \, dy$$

$$= k \int_0^a \left[yx \right]_{-\sqrt{a^2 - y^2}}^{\sqrt{a^2 - y^2}} dy$$

$$= 2k \int_0^a y\sqrt{a^2 - y^2} \, dy$$

$$= -\tfrac{2}{3}k(a^2 - y^2)^{3/2} \Big]_0^a$$

$$= \tfrac{2}{3}ka^3$$

If I_x kg-m^2 is the moment of inertia of the lamina about the x axis, then

$$I_x = \lim_{\|\Delta\| \to 0} \sum_{i=1}^{n} \gamma_i^2 (k\gamma_i) \Delta_i A$$

$$= \iint_R ky^3 \, dy \, dx$$

$$= \int_{-a}^a \int_0^{\sqrt{a^2 - x^2}} ky^3 \, dy \, dx$$

$$= k \int_{-a}^a \left[\tfrac{1}{4}y^4 \right]_0^{\sqrt{a^2 - x^2}} dx$$

$$= \tfrac{1}{4}k \int_{-a}^a (a^4 - 2a^2x^2 + x^4) \, dx$$

$$= \tfrac{1}{4}k(2a^5 - \tfrac{4}{3}a^5 + \tfrac{2}{5}a^5)$$

$$= \tfrac{4}{15}ka^5$$

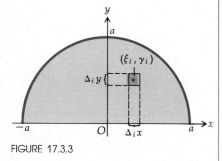

FIGURE 17.3.3

Therefore, if r meters is the radius of gyration,

$$r^2 = \frac{\frac{4}{15}ka^5}{\frac{2}{3}ka^3}$$

$$= \frac{2}{5}a^2$$

Thus $r = \frac{1}{5}\sqrt{10}a$. The radius of gyration is therefore $\frac{1}{5}\sqrt{10}a$ meters.

EXERCISES 17.3

In Exercises 1 through 12, find the mass and center of mass of the given lamina if the area density is as indicated. Mass is measured in kilograms and distance is measured in meters.

1. A lamina in the shape of the rectangular region bounded by the lines $x = 3$ and $y = 2$ and the coordinate axes. The area density at any point is xy^2 kg/m².
2. A lamina in the shape of the rectangular region bounded by the lines $x = 4$ and $y = 5$ and the coordinates axes. The area density at any point is $(x^2 + y)$ kg/m².
3. A lamina in the shape of the triangular region whose sides are segments of the coordinate axes and the line $x + 2y = 6$. The area density at any point is y^2 kg/m².
4. A lamina in the shape of the region in the first quadrant bounded by the parabola $y = x^2$, the line $y = 1$, and the y axis. The area density at any point is $(x + y)$ kg/m².
5. A lamina in the shape of the region in the first quadrant bounded by the parabola $x^2 = 8y$, the line $y = 2$, and the y axis. The area density varies as the distance from the line $y = -1$.
6. A lamina in the shape of the region bounded by the curve $y = e^x$, the line $x = 1$, and the coordinate axes. The area density varies as the distance from the x axis.
7. A lamina in the shape of the region in the first quadrant bounded by the circle $x^2 + y^2 = a^2$ and the coordinate axes. The area density varies as the sum of the distances from the two straight edges.
8. A lamina in the shape of the region bounded by the triangle whose sides are segments of the coordinate axes and the line $3x + 2y = 18$. The area density varies as the product of the distances from the coordinate axes.
9. A lamina in the shape of the region bounded by the curve $y = \sin x$ and the x axis from $x = 0$ to $x = \pi$. The area density varies as the distance from the x axis.
10. A lamina in the shape of the region bounded by the curve $y = \sqrt{x}$ and the line $y = x$. The area density varies as the distance from the y axis.
11. A lamina in the shape of the region in the first quadrant bounded by the circle $x^2 + y^2 = 4$ and the line $x + y = 2$. The area density at any point is xy kg/m².
12. A lamina in the shape of the region bounded by the circle $x^2 + y^2 = 1$ and the lines $x = 1$ and $y = 1$. The area density at any point is xy kg/m².

In Exercises 13 through 18, find the moment of inertia of the given homogeneous lamina about the indicated axis if the area density is ρ kg/m² and distance is measured in meters.
13. A lamina in the shape of the region bounded by $4y = 3x$, $x = 4$, and the x axis; about the x axis.
14. The lamina of Exercise 13; about the line $x = 4$.
15. A lamina in the shape of the region bounded by a circle of radius a units; about its center.
16. A lamina in the shape of the region bounded by the parabola $x^2 = 4 - 4y$ and the x axis; about the x axis.
17. The lamina of Exercise 16; about the origin.
18. A lamina in the shape of the region bounded by a triangle of sides of lengths a meters, b meters, and c meters; about the side of length a meters.

In Exercises 19 through 22, find for the given lamina each of the following: (a) the moment of inertia about the x axis; (b) the moment of inertia about the y axis, (c) the radius of gyration about the x axis, (d) the polar moment of inertia.
19. The lamina of Exercise 1.   20. The lamina of Exercise 4.
21. The lamina of Exercise 9.   22. The lamina of Exercise 10.

23. A homogeneous lamina of area density ρ slugs/ft^2 is in the shape of the region bounded by an isosceles triangle having a base of length b ft and an altitude of length h ft. Find the radius of gyration of the lamina about its line of symmetry.

24. A homogeneous lamina of area density ρ slugs/ft^2 is in the shape of the region bounded by the curve $x = \sqrt{y}$, the x axis, and the line $x = a$, where $a > 0$. Find the moment of inertia of the lamina about the line $x = a$.

25. A lamina is in the shape of the region enclosed by the parabola $y = 2x - x^2$ and the x axis. Find the moment of inertia of the lamina about the line $y = 4$ if the area density varies as its distance from the line $y = 4$. Mass is measured in kilograms and distance is measured in meters.

17.4
THE DOUBLE INTEGRAL
IN POLAR COORDINATES

FIGURE 17.4.1

We now show how the double integral of a function on a closed region in the polar coordinate plane can be defined. We begin by considering the simplest kind of region. Let R be the region bounded by the rays $\theta = \alpha$ and $\theta = \beta$ and by the circles $r = a$ and $r = b$. Then let Δ be a *partition* of this region obtained by drawing rays through the pole and circles having centers at the pole. This is shown in Fig. 17.4.1. We obtain a network of subregions that we call "curved" rectangles. The norm $\|\Delta\|$ of the partition is the length of the longest of the diagonals of the "curved" rectangles. Let the number of subregions be n, and let $\Delta_i A$ be the measure of area of the ith "curved" rectangle. Because the area of the ith subregion is the difference of the areas of two circular sectors,

$$\Delta_i A = \tfrac{1}{2}r_i{}^2(\theta_i - \theta_{i-1}) - \tfrac{1}{2}r_{i-1}{}^2(\theta_i - \theta_{i-1})$$
$$= \tfrac{1}{2}(r_i - r_{i-1})(r_i + r_{i-1})(\theta_i - \theta_{i-1})$$

Let $\bar{r}_i = \tfrac{1}{2}(r_i + r_{i-1})$, $\Delta_i r = r_i - r_{i-1}$, and $\Delta_i\theta = \theta_i - \theta_{i-1}$. Then

$$\Delta_i A = \bar{r}_i \Delta_i r \Delta_i \theta$$

Take the point $(\bar{r}_i, \bar{\theta}_i)$ in the ith subregion, where $\theta_{i-1} \le \bar{\theta}_i \le \theta_i$, and form the sum

$$\sum_{i=1}^{n} f(\bar{r}_i, \bar{\theta}_i)\Delta_i A = \sum_{i=1}^{n} f(\bar{r}_i, \bar{\theta}_i)\bar{r}_i \Delta_i r \Delta_i\theta \tag{1}$$

It can be shown that if f is continuous on the region R, then the limit of the sum in (1), as $\|\Delta\|$ approaches zero, exists and that this limit will be the double integral of f on R. We write either

$$\lim_{\|\Delta\| \to 0} \sum_{i=1}^{n} f(\bar{r}_i, \bar{\theta}_i)\Delta_i A = \iint\limits_{R} f(r, \theta)\, dA \tag{2}$$

$$\Leftrightarrow \quad \lim_{\|\Delta\| \to 0} \sum_{i=1}^{n} f(\bar{r}_i, \bar{\theta}_i)\bar{r}_i \Delta_i r \Delta_i\theta = \iint\limits_{R} f(r, \theta)r\, dr\, d\theta \tag{3}$$

Observe that in polar coordinates, $dA = r\, dr\, d\theta$.

The double integral can be shown to be equal to an iterated integral having one of two possible forms:

$$\iint\limits_{R} f(r, \theta)\, dA = \int_{\alpha}^{\beta} \int_{a}^{b} f(r, \theta)r\, dr\, d\theta$$
$$= \int_{a}^{b} \int_{\alpha}^{\beta} f(r, \theta)r\, d\theta\, dr \tag{4}$$

FIGURE 17.4.2

FIGURE 17.4.3

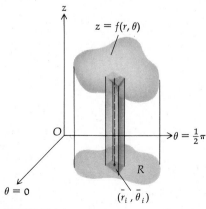

FIGURE 17.4.4

We can define the double integral of a continuous function f of two variables on closed regions of the polar coordinate plane other than the one previously considered. For example, consider the region R bounded by the curves $r = \phi_1(\theta)$ and $r = \phi_2(\theta)$, where ϕ_1 and ϕ_2 are smooth functions, and by the lines $\theta = \alpha$ and $\theta = \beta$. See Fig. 17.4.2. In the figure, $\phi_1(\theta) \leq \phi_2(\theta)$ for all θ in the closed interval $[\alpha, \beta]$. Then it can be shown that the double integral of f on R exists and equals an iterated integral, and we have

$$\iint\limits_R f(r, \theta)\, dA = \int_\alpha^\beta \int_{\phi_1(\theta)}^{\phi_2(\theta)} f(r, \theta) r\, dr\, d\theta \qquad (5)$$

If the region R is bounded by the curves $\theta = \chi_1(r)$ and $\theta = \chi_2(r)$, where χ_1 and χ_2 are smooth functions, and by the circles $r = a$ and $r = b$, as shown in Fig. 17.4.3, where $\chi_1(r) \leq \chi_2(r)$ for all r in the closed interval $[a, b]$, then

$$\iint\limits_R f(r, \theta)\, dA = \int_a^b \int_{\chi_1(r)}^{\chi_2(r)} f(r, \theta) r\, d\theta\, dr \qquad (6)$$

We can interpret the double integral of a function on a closed region in the polar coordinate plane as the measure of the volume of a solid by using cylindrical coordinates. Figure 17.4.4 shows a solid having as its base a region R in the polar coordinate plane and bounded above by the surface $z = f(r, \theta)$, where f is continuous on R and $f(r, \theta) \geq 0$ on R. Take a partition of R giving a network of n "curved" rectangles. Construct the n solids for which the ith solid has as its base the ith "curved" rectangle and as the measure of its altitude $f(\bar{r}_i, \bar{\theta}_i)$, where $(\bar{r}_i, \bar{\theta}_i)$ is in the ith subregion. Figure 17.4.4 shows the ith solid. The measure of the volume of the ith solid is

$$f(\bar{r}_i, \bar{\theta}_i)\, \Delta_i A = f(\bar{r}_i, \bar{\theta}_i)\bar{r}_i\, \Delta_i r \Delta_i \theta$$

The sum of the measures of the volumes of the n solids is

$$\sum_{i=1}^n f(\bar{r}_i, \bar{\theta}_i)\bar{r}_i\, \Delta_i r\, \Delta_i \theta$$

If V is the measure of the volume of the given solid, then

$$V = \lim_{\|\Delta\| \to 0} \sum_{i=1}^n f(\bar{r}_i, \bar{\theta}_i)\bar{r}_i\, \Delta_i r\, \Delta_i \theta$$

$$= \iint\limits_R f(r, \theta) r\, dr\, d\theta \qquad (7)$$

EXAMPLE 1
Find the volume of the solid in the first octant bounded by the cone $z = r$ and the cylinder $r = 3 \sin \theta$.

$\theta = \frac{1}{2}\pi$

$\theta = 0$

FIGURE 17.4.5

SOLUTION

The solid and the ith element are shown in Fig. 17.4.5. Using (7) with $f(r, \theta) = r$, we have, where V cubic units is the volume of the given solid,

$$V = \lim_{\|\Delta\| \to 0} \sum_{i=1}^{n} \bar{r}_i \cdot \bar{r}_i \Delta_i r \Delta_i \theta$$

$$= \iint_R r^2 \, dr \, d\theta$$

$$= \int_0^{\pi/2} \int_0^{3 \sin \theta} r^2 \, dr \, d\theta$$

$$= \int_0^{\pi/2} \left[\tfrac{1}{3} r^3 \right]_0^{3 \sin \theta} d\theta$$

$$= 9 \int_0^{\pi/2} \sin^3 \theta \, d\theta$$

$$= -9 \cos \theta + 3 \cos^3 \theta \Big]_0^{\pi/2}$$

$$= 6$$

The volume is therefore 6 cubic units.

EXAMPLE 2

Find the mass of the lamina in the shape of the region inside the semicircle $r = a \cos \theta$, $0 \le \theta \le \frac{1}{2}\pi$, and whose measure of area density at any point is proportional to the measure of its distance from the pole. The mass is measured in kilograms and distance is measured in meters.

SOLUTION

Figure 17.4.6 shows a sketch of the lamina and the ith "curved" rectangle. The area density at the point (r, θ) is kr kg/m², where k is a constant. If M kg is the mass of the lamina, then

$$M = \lim_{\|\Delta\| \to 0} \sum_{i=1}^{n} (k\bar{r}_i)\bar{r}_i \Delta_i r \Delta_i \theta$$

$$= \iint_R kr^2 \, dr \, d\theta$$

$$= k \int_0^{\pi/2} \int_0^{a \cos \theta} r^2 \, dr \, d\theta$$

$$= \tfrac{1}{3} ka^3 \int_0^{\pi/2} \cos^3 \theta \, d\theta$$

$$= \tfrac{1}{3} ka^3 \left[\sin \theta - \tfrac{1}{3} \sin^3 \theta \right]_0^{\pi/2}$$

$$= \tfrac{2}{9} ka^3$$

Therefore the mass is $\tfrac{2}{9} ka^3$ kg.

$\theta = \frac{1}{2}\pi$

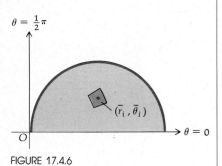

$(\bar{r}_i, \bar{\theta}_i)$

O $\theta = 0$

FIGURE 17.4.6

EXAMPLE 3
Find the center of mass of the lamina in Example 2.

SOLUTION
Let the cartesian coordinates of the center of mass of the lamina be $\bar{x}$ and $\bar{y}$, where, as is customary, the x axis is along the polar axis and the y axis is along the $\frac{1}{2}\pi$ axis. Let the cartesian coordinate representation of the point $(\bar{r}_i, \bar{\theta}_i)$ be $(\bar{x}_i, \bar{y}_i)$. Then if M_x kg-m is the moment of mass of the lamina with respect to the x axis,

$$M_x = \lim_{||\Delta|| \to 0} \sum_{i=1}^{n} \bar{y}_i (k\bar{r}_i)\bar{r}_i \, \Delta_i r \, \Delta_i \theta$$

Replacing $\bar{y}_i$ by $\bar{r}_i \sin \bar{\theta}_i$ we get

$$M_x = \lim_{||\Delta|| \to 0} \sum_{i=1}^{n} k\bar{r}_i^3 \sin \bar{\theta}_i \, \Delta_i r \, \Delta_i \theta$$

$$= \iint_R kr^3 \sin \theta \, dr \, d\theta$$

$$= k \int_0^{\pi/2} \int_0^{a \cos \theta} r^3 \sin \theta \, dr \, d\theta$$

$$= \tfrac{1}{4}ka^4 \int_0^{\pi/2} \cos^4 \theta \sin \theta \, d\theta$$

$$= -\tfrac{1}{20}ka^4 \cos^5 \theta \Big]_0^{\pi/2}$$

$$= \tfrac{1}{20}ka^4$$

If M_y kg-m is the moment of mass of the lamina with respect to the y axis, then

$$M_y = \lim_{||\Delta|| \to 0} \sum_{i=1}^{n} \bar{x}_i (k\bar{r}_i)\bar{r}_i \, \Delta_i r \, \Delta_i \theta$$

Replacing $\bar{x}_i$ by $\bar{r}_i \cos \bar{\theta}_i$ we have

$$M_y = \lim_{||\Delta|| \to 0} \sum_{i=1}^{n} k\bar{r}_i^3 \cos \bar{\theta}_i \, \Delta_i r \, \Delta_i \theta$$

$$= \iint_R kr^3 \cos \theta \, dr \, d\theta$$

$$= k \int_0^{\pi/2} \int_0^{a \cos \theta} r^3 \cos \theta \, dr \, d\theta$$

$$= \tfrac{1}{4}ka^4 \int_0^{\pi/2} \cos^5 \theta \, d\theta$$

$$= \tfrac{1}{4}ka^4 \left[\sin \theta - \tfrac{2}{3} \sin^3 \theta + \tfrac{1}{5} \sin^5 \theta \right]_0^{\pi/2}$$

$$= \tfrac{2}{15}ka^4$$

Therefore

$$\bar{x} = \frac{M_y}{M} \qquad \bar{y} = \frac{M_x}{M}$$

$$= \frac{\frac{2}{15}ka^4}{\frac{2}{9}ka^3} \qquad = \frac{\frac{1}{20}ka^4}{\frac{2}{9}ka^3}$$

$$= \frac{3}{5}a \qquad = \frac{9}{40}a$$

Hence the center of mass is at the point $(\frac{3}{5}a, \frac{9}{40}a)$.

The area of a region in the polar plane can be found by double integration. The following example illustrates the method.

EXAMPLE 4
Find by double integration the area of the region enclosed by one leaf of the rose $r = \sin 3\theta$.

SOLUTION
Figure 17.4.7 shows the region and the ith "curved" rectangle. If A square units is the area of the region, then

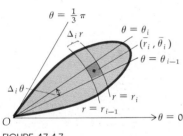

FIGURE 17.4.7

$$A = \lim_{\|\Delta\| \to 0} \sum_{i=1}^{n} \Delta_i A$$

$$= \lim_{\|\Delta\| \to 0} \sum_{i=1}^{n} \bar{r}_i \Delta_i r \, \Delta_i \theta$$

$$= \iint_R r \, dr \, d\theta$$

$$= \int_0^{\pi/3} \int_0^{\sin 3\theta} r \, dr \, d\theta$$

$$= \frac{1}{2} \int_0^{\pi/3} \sin^2 3\theta \, d\theta$$

$$= \frac{1}{4}\theta - \frac{1}{24}\sin 6\theta \Big]_0^{\pi/3}$$

$$= \frac{1}{12}\pi$$

Hence the area is $\frac{1}{12}\pi$ square units.

Sometimes it is easier to evaluate a double integral by using polar coordinates instead of cartesian coordinates. Such a situation is shown in the following example.

EXAMPLE 5

Evaluate the double integral $\iint\limits_{R} e^{-(x^2+y^2)} \, dA$, where the region R is in the first quadrant and bounded by the circle $x^2 + y^2 = a^2$ and the coordinate axes.

SOLUTION

Because $x^2 + y^2 = r^2$, and $dA = r \, dr \, d\theta$,

$$\iint\limits_{R} e^{-(x^2+y^2)} \, dA = \iint\limits_{R} e^{-r^2} r \, dr \, d\theta$$

$$= \int_0^{\pi/2} \int_0^a e^{-r^2} r \, dr \, d\theta$$

$$= -\tfrac{1}{2} \int_0^{\pi/2} \left[e^{-r^2} \right]_0^a d\theta$$

$$= -\tfrac{1}{2} \int_0^{\pi/2} (e^{-a^2} - 1) \, d\theta$$

$$= \tfrac{1}{4}\pi(1 - e^{-a^2})$$

EXERCISES 17.4

In Exercises 1 through 6, use double integrals to find the area of the given region.
1. The region inside the cardioid $r = 2(1 + \sin \theta)$.   2. One leaf of the rose $r = a \cos 2\theta$.
3. The region inside the cardioid $r = a(1 + \cos \theta)$ and outside the circle $r = a$.
4. The region inside the circle $r = 1$ and outside the lemniscate $r^2 = \cos 2\theta$.
5. The region inside the large loop of the limaçon $r = 2 - 4 \sin \theta$ and outside the small loop.
6. The region inside the limaçon $r = 3 - \cos \theta$ and outside the circle $r = 5 \cos \theta$.

In Exercises 7 through 12, find the volume of the given solid.
7. The solid bounded by the ellipsoid $z^2 + 9r^2 = 9$.
8. The solid cut out of the sphere $z^2 + r^2 = 4$ by the cylinder $r = 1$.
9. The solid cut out of the sphere $z^2 + r^2 = 16$ by the cylinder $r = 4 \cos \theta$.
10. The solid above the polar plane bounded by the cone $z = 2r$ and the cylinder $r = 1 - \cos \theta$.
11. The solid bounded by the paraboloid $z = 4 - r^2$, the cylinder $r = 1$, and the polar plane.
12. The solid above the paraboloid $z = r^2$ and below the plane $z = 2r \sin \theta$.

In Exercises 13 through 19, find the mass and center of mass of the given lamina if the area density is as indicated. Mass is measured in kilograms and distance is measured in meters.
13. A lamina in the shape of the region of Exercise 1. The area density varies as the distance from the pole.
14. A lamina in the shape of the region of Exercise 2. The area density varies as the distance from the pole.
15. A lamina in the shape of the region inside the limaçon $r = 2 - \cos \theta$. The area density varies as the distance from the pole.
16. A lamina in the shape of the region bounded by the limaçon $r = 2 + \cos \theta, 0 \le \theta \le \pi$, and the polar axis. The area density at any point is $k \sin \theta$ kg/m^2.
17. The lamina of Exercise 16. The area density at any point is $kr \sin \theta$ kg/m^2.
18. A lamina in the shape of the region of Exercise 6. The area density varies as the distance from the pole.
19. A lamina in the shape of the region of Exercise 5. The area density varies as the distance from the pole.

In Exercises 20 through 24, find the moment of inertia of the given lamina about the indicated axis or point if the area density is as indicated. Mass is measured in kilograms and distance is measured in meters.

20. A lamina in the shape of the region enclosed by the circle $r = \sin \theta$; about the $\frac{1}{2}\pi$ axis. The area density at any point is k kg/m^2.

21. The lamina of Exercise 20; about the polar axis. The area density at any point is k kg/m^2.

22. A lamina in the shape of the region bounded by the cardioid $r = a(1 - \cos \theta)$; about the pole. The area density at any point is k kg/m^2.

23. A lamina in the shape of the region bounded by the cardioid $r = a(1 + \cos \theta)$ and the circle $r = 2a \cos \theta$; about the pole. The area density at any point is k kg/m^2.

24. A lamina in the shape of the region enclosed by the lemniscate $r^2 = a^2 \cos 2\theta$; about the polar axis. The area density at any point is k kg/m^2.

25. A homogeneous lamina is in the shape of the region enclosed by one loop of the lemniscate $r^2 = \cos 2\theta$. Find the radius of gyration of the lamina about an axis perpendicular to the polar plane at the pole.

26. A lamina is in the shape of the region enclosed by the circle $r = 4$, and the area density varies as the distance from the pole. Find the radius of gyration of the lamina about an axis perpendicular to the polar plane at the pole.

27. Evaluate by polar coordinates the double integral $\iint\limits_{R} e^{x^2 + y^2} \, dA$, where R is the region bounded by the circles $x^2 + y^2 = 1$ and $x^2 + y^2 = 9$.

28. Evaluate by polar coordinates the double integral $\iint\limits_{R} \frac{x}{\sqrt{x^2 + y^2}} \, dA$, where R is the region in the first quadrant bounded by the circle $x^2 + y^2 = 1$ and the coordinate axes.

17.5
AREA OF A SURFACE

The double integral can be used to determine the area of the portion of the surface $z = f(x, y)$ that lies over a closed region R in the xy plane. To show this we must first define what we mean by the measure of this area and then obtain a formula for computing it. Assume that f and its first partial derivatives are continuous on R, and suppose that $f(x, y) > 0$ on R. Let Δ be a partition of R into n rectangular subregions. The ith rectangle has dimensions of measures $\Delta_i x$ and $\Delta_i y$ and an area of measure $\Delta_i A$. Let (ξ_i, γ_i) be any point in the ith rectangle, and at the point $Q(\xi_i, \gamma_i, f(\xi_i, \gamma_i))$ on the surface consider the tangent plane to the surface. Project vertically upward the ith rectangle onto the tangent plane and let $\Delta_i \sigma$ be the measure of the area of this projection. Figure 17.5.1 shows the region R, the portion of the surface above R, the ith rectangular subregion of R, and the projection of the ith rectangle onto the tangent plane to the surface at Q. The number $\Delta_i \sigma$ is an approximation to the measure of the area of the piece of the surface that lies above the ith rectangle. Because there are n such pieces, the summation

$$\sum_{i=1}^{n} \Delta_i \sigma$$

is an approximation to the measure σ of the area of the portion of the surface that lies above R. This leads to defining σ as follows:

$$\sigma = \lim_{||\Delta|| \to 0} \sum_{i=1}^{n} \Delta_i \sigma \qquad (1)$$

We now need to obtain a formula for computing the limit in Eq. (1). To do this we find a formula for computing $\Delta_i \sigma$ as the measure of the area

FIGURE 17.5.1

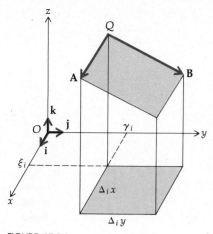

FIGURE 17.5.2

of a parallelogram. For simplicity in computation take the point (ξ_i, γ_i) in the ith rectangle at the corner (x_{i-1}, y_{i-1}). Let **A** and **B** be vectors having as representations the directed line segments having initial points at Q and forming the two adjacent sides of the parallelogram whose area has measure $\Delta_i\sigma$. See Fig. 17.5.2. Then $\Delta_i\sigma = \|\mathbf{A} \times \mathbf{B}\|$. Because

$$\mathbf{A} = \Delta_i x \mathbf{i} + f_x(\xi_i, \gamma_i)\,\Delta_i x \mathbf{k}$$

and

$$\mathbf{B} = \Delta_i y \mathbf{j} + f_y(\xi_i, \gamma_i)\,\Delta_i y \mathbf{k}$$

it follows that

$$\mathbf{A} \times \mathbf{B} = \begin{vmatrix} \mathbf{i} & \mathbf{j} & \mathbf{k} \\ \Delta_i x & 0 & f_x(\xi_i, \gamma_i)\,\Delta_i x \\ 0 & \Delta_i y & f_y(\xi_i, \gamma_i)\,\Delta_i y \end{vmatrix}$$
$$= -\Delta_i x\,\Delta_i y f_x(\xi_i, \gamma_i)\mathbf{i} - \Delta_i x\,\Delta_i y f_y(\xi_i, \gamma_i)\mathbf{j} + \Delta_i x\,\Delta_i y\mathbf{k}$$

Therefore

$$\Delta_i\sigma = \|\mathbf{A} \times \mathbf{B}\|$$
$$= \sqrt{f_x{}^2(\xi_i, \gamma_i) + f_y{}^2(\xi_i, \gamma_i) + 1}\,\Delta_i x\,\Delta_i y \qquad (2)$$

Substituting from (2) into (1) we get

$$\sigma = \lim_{\|\Delta\| \to 0} \sum_{i=1}^{n} \sqrt{f_x{}^2(\xi_i, \gamma_i) + f_y{}^2(\xi_i, \gamma_i) + 1}\,\Delta_i x\,\Delta_i y$$

This limit is a double integral that exists on R because of the continuity of f_x and f_y on R. We have, then, the following theorem.

17.5.1 THEOREM   Suppose that f and its first partial derivatives are continuous on the closed region R in the xy plane. Then if σ is the measure of the area of the surface $z = f(x, y)$ that lies over R,

$$\sigma = \iint_R \sqrt{f_x{}^2(x, y) + f_y{}^2(x, y) + 1}\,dx\,dy \qquad (3)$$

FIGURE 17.5.3

EXAMPLE 1
Find the area of the surface that is cut from the cylinder $x^2 + z^2 = 16$ by the planes $x = 0$, $x = 2$, $y = 0$, and $y = 3$.

SOLUTION
The given surface is shown in Fig. 17.5.3. The region R is the rectangle in the first quadrant of the xy plane bounded by the lines $x = 2$ and $y = 3$. The surface has the equation $x^2 + z^2 = 16$. Solving for z we get $z = \sqrt{16 - x^2}$. Hence $f(x, y) = \sqrt{16 - x^2}$. So if σ is the measure of the area of the surface, then

from Theorem 17.5.1,

$$\sigma = \iint\limits_{R} \sqrt{f_x{}^2(x, y) + f_y{}^2(x, y) + 1}\; dx\, dy$$

$$= \int_0^3 \int_0^2 \sqrt{\left(\frac{-x}{\sqrt{16 - x^2}}\right)^2 + 0 + 1}\; dx\, dy$$

$$= \int_0^3 \int_0^2 \frac{4}{\sqrt{16 - x^2}}\; dx\, dy$$

$$= 4 \int_0^3 \left[\sin^{-1} \tfrac{1}{4}x\right]_0^2 dy$$

$$= 4 \int_0^3 \tfrac{1}{6}\pi\, dy$$

$$= 2\pi$$

The surface area is therefore 2π square units.

EXAMPLE 2

Find the area of the paraboloid $z = x^2 + y^2$ below the plane $z = 4$.

SOLUTION

FIGURE 17.5.4

Figure 17.5.4 shows the given surface. From the equation of the paraboloid we see that $f(x, y) = x^2 + y^2$. The closed region in the xy plane bounded by the circle $x^2 + y^2 = 4$ is the region R. If σ square units is the required surface area, then from Theorem 17.5.1,

$$\sigma = \iint\limits_{R} \sqrt{f_x{}^2(x, y) + f_y{}^2(x, y) + 1}\; dx\, dy$$

$$= \iint\limits_{R} \sqrt{4(x^2 + y^2) + 1}\; dx\, dy$$

Because the integrand contains the terms $4(x^2 + y^2)$, the evaluation of the double integral is simplified by using polar coordinates. Then $x^2 + y^2 = r^2$. Because $dx\, dy = dA$, then $dx\, dy = r\, dr\, d\theta$. Furthermore, the limits for r are from 0 to 2 and the limits for θ are from 0 to 2π. Therefore

$$\sigma = \iint\limits_{R} \sqrt{4r^2 + 1}\, r\, dr\, d\theta$$

$$= \int_0^{2\pi} \int_0^2 \sqrt{4r^2 + 1}\, r\, dr\, d\theta$$

$$= \int_0^{2\pi} \left[\tfrac{1}{12}(4r^2 + 1)^{3/2}\right]_0^2 d\theta$$

$$= \tfrac{1}{6}\pi(17\sqrt{17} - 1)$$

Hence the area of the paraboloid below the given plane is $\tfrac{1}{6}\pi(17\sqrt{17} - 1)$ square units.

FIGURE 17.5.5

EXAMPLE 3

Find the area of the top half of the sphere

$$x^2 + y^2 + z^2 = a^2$$

SOLUTION

The hemisphere is shown in Fig. 17.5.5. Solving the equation of the sphere for z and setting this equal to $f(x, y)$ we get

$$f(x, y) = \sqrt{a^2 - x^2 - y^2}$$

Because $f_x(x, y) = -x/\sqrt{a^2 - x^2 - y^2}$ and $f_y(x, y) = -y/\sqrt{a^2 - x^2 - y^2}$, f_x and f_y are not defined on the circle $x^2 + y^2 = a^2$, which is the boundary of the region R in the xy plane. Furthermore, the double integral obtained from (3) is

$$\iint_R \frac{a}{\sqrt{a^2 - x^2 - y^2}}\, dx\, dy$$

which is improper because the integrand has an infinite discontinuity at each point of the boundary of R. We can take care of this situation by considering the region R' as that bounded by the circle $x^2 + y^2 = b^2$, where $b < a$, and then take the limit as $b \to a^-$. Furthermore, the computation is simplified if the double integral is evaluated by an iterated integral using polar coordinates. Then

$$\sigma = \lim_{b \to a^-} \int_0^b \int_0^{2\pi} \frac{a}{\sqrt{a^2 - r^2}}\, r\, d\theta\, dr$$

$$= 2\pi a \lim_{b \to a^-} \int_0^b \frac{r}{\sqrt{a^2 - r^2}}\, dr$$

$$= 2\pi a \lim_{b \to a^-} \left[-\sqrt{a^2 - r^2} \right]_0^b$$

$$= 2\pi a \lim_{b \to a^-} \left[-\sqrt{a^2 - b^2} + a \right]$$

$$= 2\pi a^2$$

The area of the hemisphere is therefore $2\pi a^2$ square units.

Consider now the curve $y = F(x)$ with $a \le x \le b$, $F(x) > 0$ on $[a, b]$ and F' continuous on $[a, b]$. If this curve is rotated about the x axis, we obtain a surface of revolution. From Section 14.6 an equation of this surface is

$$y^2 + z^2 = [F(x)]^2 \tag{4}$$

Figure 17.5.6 shows the surface of revolution. In the figure the xy plane is in the plane of the paper; however, we still have a right-handed system. We wish to obtain a formula for finding the measure of the area of this surface of revolution by using Theorem 17.5.1. From properties of symmetry, the measure of the area of the surface above the xz plane and in front of the xy

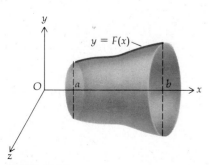

FIGURE 17.5.6

plane is one-fourth of the measure of the area of the entire surface. Solving (4) for z and neglecting the negative square root because $z \geq 0$, we get $f(x, y) = \sqrt{[F(x)]^2 - y^2}$. The region R in the xy plane is that bounded by the x axis, the curve $y = F(x)$, and the lines $x = a$ and $x = b$. Computing the partial derivatives of f we obtain

$$f_x(x, y) = \frac{F(x)F'(x)}{\sqrt{[F(x)]^2 - y^2}} \qquad f_y(x, y) = \frac{-y}{\sqrt{[F(x)]^2 - y^2}}$$

We see that $f_x(x, y)$ and $f_y(x, y)$ do not exist on part of the boundary of R (when $y = -F(x)$ and when $y = F(x)$). The double integral obtained from (3) is

$$\iint_R \sqrt{\frac{[F(x)]^2[F'(x)]^2}{[F(x)]^2 - y^2} + \frac{y^2}{[F(x)]^2 - y^2} + 1}\, dy\, dx$$

$$= \iint_R \frac{F(x)\sqrt{[F'(x)]^2 + 1}}{\sqrt{[F(x)]^2 - y^2}}\, dy\, dx$$

This double integral is improper because the integrand has an infinite discontinuity at each point of the boundary of R where $y = -F(x)$ and $y = F(x)$. Hence we evaluate the double integral by an iterated integral for which the inner integral is improper.

$$\sigma = 4 \int_a^b \left[F(x)\sqrt{[F'(x)]^2 + 1} \int_0^{F(x)} \frac{dy}{\sqrt{[F(x)]^2 - y^2}} \right] dx \qquad (5)$$

$$\int_0^{F(x)} \frac{dy}{\sqrt{[F(x)]^2 - y^2}} = \lim_{b \to F(x)^-} \int_0^b \frac{dy}{\sqrt{[F(x)]^2 - y^2}}$$

$$= \lim_{b \to F(x)^-} \left[\sin^{-1} \frac{y}{F(x)} \right]_0^b$$

$$= \lim_{b \to F(x)^-} \sin^{-1} \frac{b}{F(x)}$$

$$= \tfrac{1}{2}\pi$$

Therefore, from (5),

$$\sigma = 2\pi \int_a^b F(x)\sqrt{[F'(x)]^2 + 1}\, dx$$

We state this result as a theorem, where F is replaced by f.

17.5.2 THEOREM   Suppose that the function f is positive on $[a, b]$ and f' is continuous on $[a, b]$. If σ is the measure of the area of the surface of revolution obtained by revolving the curve $y = f(x)$, with $a \leq x \leq b$, about the x axis, then

$$\sigma = 2\pi \int_a^b f(x)\sqrt{[f'(x)]^2 + 1}\, dx$$

FIGURE 17.5.7

EXAMPLE 4

Find the area of the paraboloid of revolution generated by revolving the top half of the parabola $y^2 = 4px$, with $0 \leq x \leq h$, about the x axis.

SOLUTION

The paraboloid of revolution is shown in Fig. 17.5.7. Solving the equation of the parabola for y ($y \geq 0$) we obtain $y = 2p^{1/2}x^{1/2}$. So if σ square units is the area of the surface, from Theorem 17.5.2, with $f(x) = 2p^{1/2}x^{1/2}$,

$$\sigma = 2\pi \int_0^h 2p^{1/2}x^{1/2} \sqrt{\frac{p}{x} + 1}\, dx$$

$$= 4\pi p^{1/2} \int_0^h \sqrt{p + x}\, dx$$

$$= \tfrac{8}{3}\pi p^{1/2}(p + x)^{3/2}\Big]_0^h$$

$$= \tfrac{8}{3}\pi(\sqrt{p(p + h)^3} - p^2)$$

The area of the paraboloid of revolution is therefore $\tfrac{8}{3}\pi(\sqrt{p(p + h)^3} - p^2)$ square units.

EXERCISES 17.5

1. Find the area of the surface cut from the plane $2x + y + z = 4$ by the planes $x = 0$, $x = 1$, $y = 0$, and $y = 1$.
2. Find the area of the surface cut from the plane $z - 2x - y = 5$ by the planes $x = 0$, $x = 2$, $y = 0$, and $y = 4$.
3. Find the area of the portion of the surface of the plane $36x + 16y + 9z = 144$ that is cut by the coordinate planes.
4. Find the area of the surface that is cut from the plane $z = ax + by$ by the planes $x = 0$, $x = a$, $y = 0$, and $y = b$, where $a > 0$ and $b > 0$.
5. Find the area of the surface in the first octant that is cut from the cylinder $x^2 + y^2 = 9$ by the plane $x = z$.
6. Find the area of the surface that is cut from the cylinder $x^2 + y^2 = 25$ by the planes $x = 0$, $x = 1$, $z = 1$, and $z = 3$.
7. Let R be the triangular region in the xy plane with vertices at $(0, 0, 0)$, $(0, 4, 0)$, and $(2, 4, 0)$. Find the surface area of the part of the graph of $z - 5x - y^2 = 2$ that lies above R.
8. Find the area of the surface in the first octant cut from the cone $x^2 + y^2 = z^2$ by the plane $x + y = 4$.
9. Find the area of the portion of the surface of the sphere $x^2 + y^2 + z^2 = 4x$ that is cut out by one nappe of the cone $y^2 + z^2 = x^2$.
10. Find the area of the portion of the surface of the sphere $x^2 + y^2 + z^2 = 36$ that lies within the cylinder $x^2 + y^2 = 9$.
11. Find the area of the portion of the surface of the sphere $x^2 + y^2 + z^2 = 4z$ that lies within the paraboloid $x^2 + y^2 = 3z$.
12. For the sphere and paraboloid of Exercise 11, find the area of the portion of the surface of the paraboloid that lies within the sphere.
13. Find the area of the surface of the portion of the cylinder $x^2 + z^2 = 4$ that is inside the cylinder $x^2 + y^2 = 4$.
14. Find the area of the surface of the portion of the cone $x^2 + y^2 = z^2$ that is inside the cylinder $x^2 + y^2 = 2x$.
15. Find the area of the surface of the portion of the cone $x^2 + y^2 = z^2$ between the cylinder $y^2 = x$ and the plane $x - y = 2$.
16. Find the area of the surface of the portion of the solid $xy = az$ in the first octant that lies inside the cylinder $x^2 + y^2 = a^2$.
17. The line segment from the origin to the point (a, b) is revolved about the x axis. Find the area of the surface of the cone generated.
18. Derive the formula for the area of the surface of a sphere by revolving a semicircle about its diameter.

19. Find the area of the surface of revolution obtained by revolving the arc of the catenary $y = a \cosh(x/a)$ from $x = 0$ to $x = a$ about the y axis.
20. Find the area of the surface of revolution obtained by revolving the catenary of Exercise 19 about the x axis.
21. The loop of the curve $18y^2 = x(6 - x)^2$ is revolved about the x axis. Find the area of the surface of revolution generated.
22. Find the area of the surface of revolution generated by revolving the arc of the curve $y = \ln x$ from $x = 1$ to $x = 2$ about the y axis.
23. Find the area of the portion of the plane $x = z$ that lies between the planes $y = 0$ and $y = 6$ and within the hyperboloid $9x^2 - 4y^2 + 16z^2 = 144$.
24. Find the area of the surface cut from the hyperbolic paraboloid $y^2 - x^2 = 6z$ by the cylinder $x^2 + y^2 = 36$.
25. Suppose that f and its first partial derivatives are continuous on the closed region R in the xy plane. Show that if σ is the measure of the area of the portion of the surface $z = f(x, y)$ that lies over R, then

$$\sigma = \iint\limits_{R} \|\nabla g(x, y, z)\| \, dx \, dy$$

where $g(x, y, z) = z - f(x, y)$.

17.6
THE TRIPLE INTEGRAL

The extension of the double integral to the triple integral is analogous to the extension of the single integral to the double integral. The simplest type of region in R^3 is a rectangular parallelepiped that is bounded by six planes: $x = a_1$, $x = a_2$, $y = b_1$, $y = b_2$, $z = c_1$, and $z = c_2$, with $a_1 < a_2$, $b_1 < b_2$, and $c_1 < c_2$. Let f be a function of three variables and suppose that f is continuous on such a region S. A partition of this region is formed by dividing S into rectangular boxes by drawing planes parallel to the coordinate planes. Denote such a partition by Δ and suppose that n is the number of boxes. Let $\Delta_i V$ be the measure of the volume of the ith box. Choose an arbitrary point (ξ_i, γ_i, μ_i) in the ith box. Form the sum

$$\sum_{i=1}^{n} f(\xi_i, \gamma_i, \mu_i) \Delta_i V \tag{1}$$

Refer to Fig. 17.6.1, which shows the rectangular parallelepiped together with the ith box. The *norm* $\|\Delta\|$ of the partition is the length of the longest diagonal of the boxes. If the sums of form (1) approach a limit as $\|\Delta\|$ approaches zero for any choices of the points (ξ_i, γ_i, μ_i), then we call this limit the **triple integral** of f on S and write

$$\lim_{\|\Delta\| \to 0} \sum_{i=1}^{n} f(\xi_i, \gamma_i, \mu_i) \Delta_i V = \iiint\limits_{S} f(x, y, z) \, dV$$

A sufficient condition for the triple integral of f on S to exist is that f be continuous on S.

Analogous to a double integral being equal to a twice-iterated integral, the triple integral is equal to a thrice-iterated integral. When S is the rectangular parallelepiped described above, and f is continuous on S, then

$$\iiint\limits_{S} f(x, y, z) \, dV = \int_{a_1}^{a_2} \int_{b_1}^{b_2} \int_{c_1}^{c_2} f(x, y, z) \, dz \, dy \, dx$$

FIGURE 17.6.1

EXAMPLE 1

Evaluate the triple integral $\iiint\limits_{S} xy \sin yz \, dV$ if S is the rectangular paral-

lelepiped bounded by the planes $x = \pi$, $y = \frac{1}{2}\pi$, $z = \frac{1}{3}\pi$, and the coordinate planes.

SOLUTION

$$
\begin{aligned}
\iiint\limits_{S} xy \sin yz \, dV &= \int_0^\pi \int_0^{\pi/2} \int_0^{\pi/3} xy \sin yz \, dz \, dy \, dx \\
&= \int_0^\pi \int_0^{\pi/2} \Big[-x \cos yz \Big]_0^{\pi/3} dy \, dx \\
&= \int_0^\pi \int_0^{\pi/2} x(1 - \cos \tfrac{1}{3}\pi y) \, dy \, dx \\
&= \int_0^\pi x \left(y - \frac{3}{\pi} \sin \frac{1}{3}\pi y \right) \Big]_0^{\pi/2} dx \\
&= \int_0^\pi x \left(\frac{\pi}{2} - \frac{3}{\pi} \sin \frac{\pi^2}{6} \right) dx \\
&= \frac{x^2}{2} \left(\frac{\pi}{2} - \frac{3}{\pi} \sin \frac{\pi^2}{6} \right) \Big]_0^\pi \\
&= \frac{\pi}{4} \left(\pi^2 - 6 \sin \frac{\pi^2}{6} \right)
\end{aligned}
$$

We now discuss how to define the triple integral of a continuous function of three variables on a region in R^3 other than a rectangular parallelepiped. Let S be the closed three-dimensional region that is bounded by the planes $x = a$ and $x = b$, the cylinders $y = \phi_1(x)$ and $y = \phi_2(x)$, and the surfaces $z = F_1(x, y)$ and $z = F_2(x, y)$, where the functions ϕ_1, ϕ_2, F_1, and F_2 are smooth (i.e., they have continuous derivatives or partial derivatives). See Fig. 17.6.2. Construct planes parallel to the coordinate planes, thereby forming a set of rectangular parallelepipeds that completely cover S. The parallelepipeds that are entirely inside S or on the boundary of S form a **partition** Δ of S. Choose some system of numbering so that they are numbered from 1 to n. The norm $\|\Delta\|$ of this partition of S is the length of the longest diagonal of any parallelepiped belonging to the partition. Let the measure of the volume of the ith parallelepiped be $\Delta_i V$. Let f be a function of three variables that is continuous on S, and let (ξ_i, γ_i, μ_i) be an arbitrary point in the ith parallelepiped. Form the sum

$$
\sum_{i=1}^{n} f(\xi_i, \gamma_i, \mu_i) \Delta_i V \tag{2}
$$

If the sums of form (2) have a limit as $\|\Delta\|$ approaches zero, and if this limit is independent of the choice of the partitioning planes and the choices of the arbitrary points (ξ_i, γ_i, μ_i) in each parallelepiped, then this limit is called the

FIGURE 17.6.2

triple integral of f on S, and we write

$$\lim_{||\Delta|| \to 0} \sum_{i=1}^{n} f(\xi_i, \gamma_i, \mu_i) \Delta_i V = \iiint_S f(x, y, z) \, dV \tag{3}$$

It can be proved in advanced calculus that a sufficient condition for the limit in (3) to exist is that f be continuous on S. Furthermore, under the condition imposed on the functions ϕ_1, ϕ_2, F_1, and F_2 that they be smooth, it can also be proved that the triple integral can be evaluated by the iterated integral.

$$\int_a^b \int_{\phi_1(x)}^{\phi_2(x)} \int_{F_1(x,y)}^{F_2(x,y)} f(x, y, z) \, dz \, dy \, dx$$

Just as the double integral can be interpreted as the measure of the area of a plane region when $f(x, y) = 1$ on R, the triple integral can be interpreted as the measure of the volume of a three-dimensional region. If $f(x, y, z) = 1$ on S, then Eq. (3) becomes

$$\lim_{||\Delta|| \to 0} \sum_{i=1}^{n} \Delta_i V = \iiint_S dV$$

and the triple integral is the measure of the volume of the region S.

EXAMPLE 2
Find by triple integration the volume of the solid of Example 3 in Section 17.2.

SOLUTION
The solid lies above the xy plane bounded by the elliptic paraboloid $z = x^2 + 4y^2$ and the cylinder $x^2 + 4y^2 = 4$. It is shown in Fig. 17.6.3. If V cubic units is the volume of the solid, then

$$V = \lim_{||\Delta|| \to 0} \sum_{i=1}^{n} \Delta_i V \tag{4}$$

$$= \iiint_S dV$$

where S is the region bounded by the solid. The z limits are from 0 (the value of z on the xy plane) to $x^2 + 4y^2$ (the value of z on the elliptic paraboloid). The y limits for one-fourth of the volume are from 0 (the value of y on the xz plane) to $\frac{1}{2}\sqrt{4 - x^2}$ (the value of y on the cylinder). The x limits for the first octant are from 0 to 2. We evaluate the triple integral in (4) by an iterated integral and obtain

$$V = 4 \int_0^2 \int_0^{\sqrt{4-x^2}/2} \int_0^{x^2+4y^2} dz \, dy \, dx$$

$$= 4 \int_0^2 \int_0^{\sqrt{4-x^2}/2} (x^2 + 4y^2) \, dy \, dx$$

This is the same twice-iterated integral that we obtained in Example 3 in Section 17.2, and so the remainder of the solution is the same.

FIGURE 17.6.3

EXAMPLE 3
Find the volume of the solid bounded by the cylinder $x^2 + y^2 = 25$, the plane $x + y + z = 8$ and the xy plane.

SOLUTION
The solid is shown in Fig. 17.6.4. The z limits for the iterated integral are from 0 to $8 - x - y$ (the value of z on the plane). The y limits are obtained from the boundary region in the xy plane, which is the circle $x^2 + y^2 = 25$. Hence the y limits are from $-\sqrt{25 - x^2}$ to $\sqrt{25 - x^2}$. The x limits are from -5 to 5. If V cubic units is the required volume,

$$
\begin{aligned}
V &= \lim_{\|\Delta\| \to 0} \sum_{i=1}^{n} \Delta_i V \\
&= \iiint_{S} dV \\
&= \int_{-5}^{5} \int_{-\sqrt{25-x^2}}^{\sqrt{25-x^2}} \int_{0}^{8-x-y} dz \, dy \, dx \\
&= \int_{-5}^{5} \int_{-\sqrt{25-x^2}}^{\sqrt{25-x^2}} (8 - x - y) \, dy \, dx \\
&= \int_{-5}^{5} \left[(8 - x)y - \tfrac{1}{2}y^2 \right]_{-\sqrt{25-x^2}}^{\sqrt{25-x^2}} dx \\
&= 2 \int_{-5}^{5} (8 - x)\sqrt{25 - x^2} \, dx \\
&= 16 \int_{-5}^{5} \sqrt{25 - x^2} \, dx + \int_{-5}^{5} \sqrt{25 - x^2}(-2x) \, dx \\
&= 16(\tfrac{1}{2}x\sqrt{25 - x^2} + \tfrac{25}{2} \sin^{-1} \tfrac{1}{5}x) + \tfrac{2}{3}(25 - x^2)^{3/2} \Big]_{-5}^{5} \\
&= 200\pi
\end{aligned}
$$

The volume is therefore 200π cubic units.

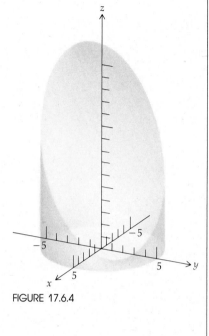

FIGURE 17.6.4

EXAMPLE 4
Find the mass of the solid above the xy plane bounded by the cone $9x^2 + z^2 = y^2$ and the plane $y = 9$ if the measure of the volume density at any point (x, y, z) in the solid is proportional to the measure of the distance of the point from the xy plane.

SOLUTION
Figure 17.6.5 shows the solid. Let M kg be the mass of the solid, and let the distance be measured in meters. Then the volume density at any point (x, y, z) in the solid is kz kg/m^3, where k is a constant. Then if (ξ_i, γ_i, μ_i) is any point

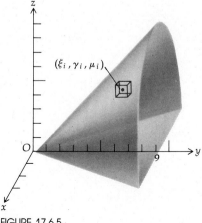

FIGURE 17.6.5

in the ith rectangular parallelepiped of the partition,

$$M = \lim_{||\Delta|| \to 0} \sum_{i=1}^{n} k\mu_i \Delta_i V$$

$$= \iiint_S kz \, dV$$

$$= 2k \int_0^9 \int_0^{y/3} \int_0^{\sqrt{y^2 - 9x^2}} z \, dz \, dx \, dy$$

$$= 2k \int_0^9 \int_0^{y/3} \left[\tfrac{1}{2}z^2 \right]_0^{\sqrt{y^2 - 9x^2}} dx \, dy$$

$$= k \int_0^9 \int_0^{y/3} (y^2 - 9x^2) \, dx \, dy$$

$$= \tfrac{2}{9}k \int_0^9 y^3 \, dy$$

$$= \tfrac{729}{2}k$$

The mass is therefore $\tfrac{729}{2}k$ kg.

EXERCISES 17.6

In Exercises 1 through 8, evaluate the iterated integral.

1. $\int_0^1 \int_0^{1-x} \int_{2y}^{1+y^2} x \, dz \, dy \, dx$

2. $\int_1^2 \int_0^x \int_1^{x+xy} xy \, dz \, dy \, dx$

3. $\int_0^1 \int_0^x \int_0^{x+y} (x + y + z) \, dz \, dy \, dx$

4. $\int_0^2 \int_0^{\sqrt{4-y^2}} \int_0^{2-y} z \, dx \, dz \, dy$

5. $\int_{-1}^0 \int_e^{2e} \int_0^{\pi/3} y \ln z \tan x \, dx \, dz \, dy$

6. $\int_1^2 \int_y^{y^2} \int_0^{\ln x} ye^z \, dz \, dx \, dy$

7. $\int_0^{\pi/2} \int_z^{\pi/2} \int_0^{xz} \cos \frac{y}{z} \, dy \, dx \, dz$

8. $\int_0^2 \int_0^y \int_0^{\sqrt{3}z} \frac{z}{x^2 + z^2} \, dx \, dz \, dy$

In Exercises 9 through 18, evaluate the triple integral.

9. $\iiint_S y \, dV$ if S is the region bounded by the tetrahedron formed by the plane $12x + 20y + 15z = 60$ and the coordinate planes.

10. $\iiint_S (x^2 + z^2) \, dV$ if S is the same region as in Exercise 9.

11. $\iiint_S z \, dV$ if S is the region bounded by the tetrahedron having vertices $(0, 0, 0)$, $(1, 1, 0)$, $(1, 0, 0)$, and $(1, 0, 1)$.

12. $\iiint_S yz \, dV$ if S is the same region as in Exercise 11.

13. $\iiint_S xy \, dV$ if S is the rectangular parallelepiped in the first octant bounded by the coordinate planes and the planes $x = 2$, $y = 3$, and $z = 4$.

14. $\iiint_S x \, dV$ if S is the tetrahedron bounded by the planes $x + 2y + 3z = 6$, $x = 0$, $y = 0$, and $z = 0$.

15. $\iiint_S dV$ if S is the region bounded by the surfaces $z = x^2 + y^2$ and $z = 27 - 2x^2 - 2y^2$.

16. $\iiint\limits_{S} y^2 \, dV$ if S is the region bounded by the cylinders $x^2 + y = 1$ and $z^2 + y = 1$ and the plane $y = 0$.

17. $\iiint\limits_{S} (xz + 3z) \, dV$ if S is the region bounded by the cylinder $x^2 + z^2 = 9$ and the planes $x + y = 3$, $z = 0$, and $y = 0$, above the xy plane.

18. $\iiint\limits_{S} xyz \, dV$ if S is the region bounded by the cylinders $x^2 + y^2 = 4$ and $x^2 + z^2 = 4$.

In Exercises 19 through 32, use triple integration.

19. Find the volume of the solid in the first octant bounded below by the xy plane, above by the plane $z = y$, and laterally by the cylinder $y^2 = x$ and the plane $x = 1$.

20. Find the volume of the solid in the first octant bounded by the cylinder $x^2 + z^2 = 16$, the plane $x + y = 2$, and the three coordinate planes.

21. Find the volume of the solid in the first octant bounded by the cylinders $x^2 + y^2 = 4$ and $x^2 + 2z = 4$ and the three coordinate planes.

22. Find the volume of the solid bounded by the elliptic cone $4x^2 + 9y^2 - 36z^2 = 0$ and the plane $z = 1$.

23. Find the volume of the solid above the elliptic paraboloid $3x^2 + y^2 = z$ and below the cylinder $x^2 + z = 4$.

24. Find the volume of the solid enclosed by the sphere $x^2 + y^2 + z^2 = a^2$.

25. Find the volume of the solid enclosed by the ellipsoid

$$\frac{x^2}{a^2} + \frac{y^2}{b^2} + \frac{z^2}{c^2} = 1$$

26. Find the volume of the solid bounded by the cylinders $z = 5x^2$ and $z = 3 - x^2$, the plane $y + z = 4$, and the xz plane.

27. Find the mass of the homogeneous solid bounded by the cylinder $z = 4 - x^2$, the plane $y = 5$, and the coordinate planes if the volume density at any point is ρ kg/m^3.

28. Find the mass of the solid enclosed by the tetrahedron formed by the plane $100x + 25y + 16z = 400$ and the coordinate planes if the volume density varies as the distance from the yz plane. The volume density is measured in kilograms per cubic meter.

29. Find the mass of the solid bounded by the cylinders $x = z^2$ and $y = x^2$, and the planes $x = 1$, $y = 0$, and $z = 0$. The volume density varies as the product of the distances from the three coordinate planes, and it is measured in kilograms per cubic meter.

30. Find the mass of the solid bounded by the surface $z = 4 - 4x^2 - y^2$ and the xy plane. The volume density at any point of the solid is $3z|x|$ kg/m^3.

31. Find the mass of the solid bounded by the surface $z = xy$ and the planes $x = 1$, $y = 1$ and $z = 0$. The volume density at any point of the solid is $3\sqrt{x^2 + y^2}$ kg/m^3.

32. A solid has the shape of a right-circular cylinder of base radius r meters and height h meters. Find the mass of the solid if the volume density varies as the distance from one of the bases. The volume density is measured in kilograms per cubic meter.

17.7
THE TRIPLE INTEGRAL
IN CYLINDRICAL AND
SPHERICAL COORDINATES

If a region S in R^3 has an axis of symmetry, triple integrals on S are easier to evaluate if cylindrical coordinates are used. If there is symmetry with respect to a point, it is convenient to choose that point as the origin and to use spherical coordinates. In this section we discuss the triple integral in these coordinates and apply them to physical problems.

To define the triple integral in cylindrical coordinates we construct a partition of the region S by drawing planes through the z axis, planes perpendicular to the z axis, and right-circular cylinders having the z axis as

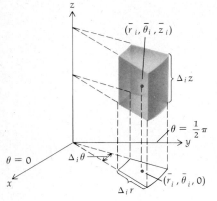

FIGURE 17.7.1

axis. A typical subregion is shown in Fig. 17.7.1. The elements of the constructed partition lie entirely in S. We call this partition a **cylindrical partition**. The measure of the length of the longest "diagonal" of any of the subregions is the **norm** of the partition. Let n be the number of subregions of the partition and $\Delta_i V$ be the measure of the volume of the ith subregion. The measure of the area of the base is $\bar{r}_i \Delta_i r \Delta_i \theta$, where $\bar{r}_i = \frac{1}{2}(r_i + r_{i-1})$. Hence if $\Delta_i z$ is the measure of the altitude of the ith subregion,

$$\Delta_i V = \bar{r}_i \Delta_i r \Delta_i \theta \Delta_i z$$

Let f be a function of r, θ, and z, and suppose that f is continuous on S. Choose a point $(\bar{r}_i, \bar{\theta}_i, \bar{z}_i)$ in the ith subregion such that $\theta_{i-1} \le \bar{\theta}_i \le \theta_i$, and $z_{i-1} \le \bar{z}_i \le z_i$. Form the sum

$$\sum_{i=1}^{n} f(\bar{r}_i, \bar{\theta}_i, \bar{z}_i) \Delta_i V = \sum_{i=1}^{n} f(\bar{r}_i, \bar{\theta}_i, \bar{z}_i) \bar{r}_i \Delta_i r \Delta_i \theta \Delta_i z \tag{1}$$

As the norm of Δ approaches zero, it can be shown, under suitable conditions on S, that the limit of the sums of form (1) exist. This limit is called the **triple integral in cylindrical coordinates** of the function f on S, and we write

$$\lim_{\|\Delta\| \to 0} \sum_{i=1}^{n} f(\bar{r}_i, \bar{\theta}_i, \bar{z}_i) \Delta_i V = \iiint_S f(r, \theta, z)\, dV \tag{2}$$

$$\Leftrightarrow \quad \lim_{\|\Delta\| \to 0} \sum_{i=1}^{n} f(\bar{r}_i, \bar{\theta}_i, \bar{z}_i) \bar{r}_i \Delta_i r \Delta_i \theta \Delta_i z = \iiint_S f(r, \theta, z) r\, dr\, d\theta\, dz \tag{3}$$

Note that in cylindrical coordinates, $dV = r\, dr\, d\theta\, dz$. We can evaluate the triple integral in (2) and (3) by an iterated integral. For instance, suppose that the region S in R^3 is bounded by the planes $\theta = \alpha$ and $\theta = \beta$, with $\alpha < \beta$, by the cylinders $r = \lambda_1(\theta)$ and $r = \lambda_2(\theta)$, where λ_1 and λ_2 are smooth on $[\alpha, \beta]$ and $\lambda_1(\theta) \le \lambda_2(\theta)$ for $\alpha \le \theta \le \beta$, and by the surfaces $z = F_1(r, \theta)$ and $z = F_2(r, \theta)$, where F_1 and F_2 are functions of two variables that are smooth on some region R in the polar plane bounded by the curves $r = \lambda_1(\theta)$, $r = \lambda_2(\theta)$, $\theta = \alpha$, and $\theta = \beta$. Also, suppose that $F_1(r, \theta) \le F_2(r, \theta)$ for every point (r, θ) in R. Then the triple integral can be evaluated by an iterated integral by the formula

$$\iiint_S f(r, \theta, z) r\, dr\, d\theta\, dz = \int_\alpha^\beta \int_{\lambda_1(\theta)}^{\lambda_2(\theta)} \int_{F_1(r,\theta)}^{F_2(r,\theta)} f(r, \theta, z) r\, dz\, dr\, d\theta \tag{4}$$

There are five other iterated integrals that can be used to evaluate the triple integral in (4) because there are six possible permutations of the three variables r, θ, and z.

Triple integrals and cylindrical coordinates are especially useful in finding the moment of inertia of a solid with respect to the z axis because the distance from the z axis to a point in the solid is determined by the coordinate r.

FIGURE 17.7.2

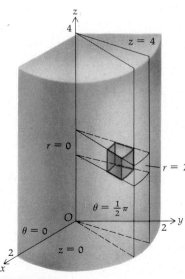

FIGURE 17.7.3

EXAMPLE 1

A homogeneous solid in the shape of a right-circular cylinder has a radius of 2 m and an altitude of 4 m. Find the moment of inertia of the solid with respect to its axis.

SOLUTION

Choose the coordinate planes so that the xy plane is the plane of the base of the solid and the z axis is the axis of the solid. Figure 17.7.2 shows the portion of the solid in the first octant together with the ith subregion of a cylindrical partition. Using cylindrical coordinates and taking the point $(\bar{r}_i, \bar{\theta}_i, \bar{z}_i)$ in the ith subregion with k kg/m^3 as the volume density at any point, then if I_z kg-m^2 is the moment of inertia of the solid with respect to the z axis,

$$I_z = \lim_{||\Delta|| \to 0} \sum_{i=1}^{n} \bar{r}_i^2 k\, \Delta_i V$$

$$= \iiint_S kr^2\, dV$$

There are six different possible orders of integration. Figure 17.7.2 shows the order $dz\, dr\, d\theta$. Using this order, we have

$$I_z = \iiint_S kr^2\, dz\, r\, dr\, d\theta$$

$$= 4k \int_0^{\pi/2} \int_0^2 \int_0^4 r^3\, dz\, dr\, d\theta$$

In the first integration the blocks are summed from $z = 0$ to $z = 4$; the blocks become a column. In the second integration the columns are summed from $r = 0$ to $r = 2$; the columns become a wedge-shaped slice of the cylinder. In the third integration the wedge-shaped slice is rotated from $\theta = 0$ to $\theta = \frac{1}{2}\pi$; this sweeps the wedge about the entire three-dimensional region in the first octant. We multiply by 4 to obtain the entire volume. Performing the integration we obtain

$$I_z = 16k \int_0^{\pi/2} \int_0^2 r^3\, dr\, d\theta$$

$$= 64k \int_0^{\pi/2} d\theta$$

$$= 32k\pi$$

Hence the moment of inertia is $32k\pi$ kg-m^2.

EXAMPLE 2

Solve Example 1 by taking the order of integration as (a) $dr\, dz\, d\theta$; (b) $d\theta\, dr\, dz$.

SOLUTION

(a) Figure 17.7.3 represents the order $dr\, dz\, d\theta$. It shows the block summed from $r = 0$ to $r = 2$ to give a wedge-shaped sector. We then sum from

$z = 0$ to $z = 4$ to give a wedge-shaped slice. The slice is rotated from $\theta = 0$ to $\theta = \frac{1}{2}\pi$ to cover the first octant. Then

$$I_z = 4k \int_0^{\pi/2} \int_0^4 \int_0^2 r^3 \, dr \, dz \, d\theta$$

$$= 32k\pi$$

(b) Figure 17.7.4 represents the order $d\theta \, dr \, dz$. It shows the blocks summed from $\theta = 0$ to $\theta = \frac{1}{2}\pi$ to give a hollow ring inside the cylinder. These hollow rings are summed from $r = 0$ to $r = 2$ to give a horizontal slice of the cylinder. The horizontal slices are summed from $z = 0$ to $z = 4$. Therefore

$$I_z = 4k \int_0^4 \int_0^2 \int_0^{\pi/2} r^3 \, d\theta \, dr \, dz$$

$$= 32k\pi$$

FIGURE 17.7.4

FIGURE 17.7.5

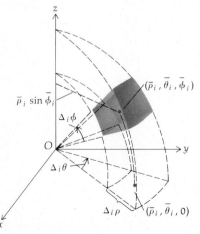

FIGURE 17.7.6

EXAMPLE 3

Find the mass of a solid hemisphere of radius a meters if the volume density at any point is proportional to the distance of the point from the axis of the solid and is measured in kilograms per cubic meter.

SOLUTION

If we choose the coordinate planes so that the origin is at the center of the sphere and the z axis is the axis of the solid, then an equation of the hemispherical surface above the xy plane is $z = \sqrt{a^2 - x^2 - y^2}$. Figure 17.7.5 shows this surface and the solid together with the ith subregion of a cylindrical partition. An equation of the hemisphere in cylindrical coordinates is $z = \sqrt{a^2 - r^2}$. If $(\bar{r}_i, \bar{\theta}_i, \bar{z}_i)$ is a point in the ith subregion, the volume density at this point is $k\bar{r}_i$ kg/m³, where k is a constant; and if M kg is the mass of the solid, then

$$M = \lim_{\|\Delta\| \to 0} \sum_{i=1}^{n} k\bar{r}_i \, \Delta_i V$$

$$= \iiint_S kr \, dV$$

$$= k \int_0^{2\pi} \int_0^a \int_0^{\sqrt{a^2 - r^2}} r^2 \, dz \, dr \, d\theta$$

$$= k \int_0^{2\pi} \int_0^a r^2 \sqrt{a^2 - r^2} \, dr \, d\theta$$

$$= k \int_0^{2\pi} \left[-\tfrac{1}{4}r(a^2 - r^2)^{3/2} + \tfrac{1}{8}a^2 r \sqrt{a^2 - r^2} + \tfrac{1}{8}a^4 \sin^{-1} \frac{r}{a} \right]_0^a d\theta$$

$$= \tfrac{1}{16}ka^4\pi \int_0^{2\pi} d\theta$$

$$= \tfrac{1}{8}ka^4\pi^2$$

The mass of the solid hemisphere is therefore $\tfrac{1}{8}ka^4\pi^2$ kg.

EXAMPLE 4
Find the center of mass of the solid of Example 3.

SOLUTION
Let the cartesian coordinate representation of the center of mass be $(\bar{x}, \bar{y}, \bar{z})$. Because of symmetry, $\bar{x} = 0$ and $\bar{y} = 0$. We need to calculate $\bar{z}$. If M_{xy} kg-m is the moment of mass of the solid with respect to the xy plane, then

$$
\begin{aligned}
M_{xy} &= \lim_{\|\Delta\| \to 0} \sum_{i=1}^{n} \bar{z}_i (k\bar{r}_i) \,\Delta_i V \\
&= \iiint_S kzr \, dV \\
&= k \int_0^{2\pi} \int_0^a \int_0^{\sqrt{a^2 - r^2}} zr^2 \, dz \, dr \, d\theta \\
&= \tfrac{1}{2} k \int_0^{2\pi} \int_0^a (a^2 - r^2) r^2 \, dr \, d\theta \\
&= \tfrac{1}{15} k a^5 \int_0^{2\pi} d\theta \\
&= \tfrac{2}{15} k a^5 \pi
\end{aligned}
$$

Because $M\bar{z} = M_{xy}$, we get $\bar{z} = M_{xy}/M$; thus

$$
\begin{aligned}
\bar{z} &= \frac{\tfrac{2}{15} k a^5 \pi}{\tfrac{1}{8} k a^4 \pi^2} \\
&= \frac{16}{15\pi} a
\end{aligned}
$$

The center of mass is therefore on the axis of the solid at a distance of $\dfrac{16a}{15\pi}$ meters from the plane of the base.

We now proceed to define the triple integral in spherical coordinates. A spherical partition of the three-dimensional region S is formed by planes containing the z axis, spheres with centers at the origin, and circular cones having vertices at the origin and the z axis as the axis. A typical subregion of the partition is shown in Fig. 17.7.6. If $\Delta_i V$ is the measure of the volume of the ith subregion, and $(\bar{\rho}_i, \bar{\theta}_i, \bar{\phi}_i)$ is a point in it, we can get an approximation to $\Delta_i V$ by considering the region as if it were a rectangular parallelepiped and taking the product of the measures of the three dimensions. These measures are $\bar{\rho}_i \sin \bar{\phi}_i \Delta_i \theta$, $\bar{\rho}_i \Delta_i \phi$, and $\Delta_i \rho$. Figures 17.7.7 and 17.7.8 show how the first two measures are obtained, and Fig. 17.7.6 shows the dimension of measure $\Delta_i \rho$. Hence

$$
\Delta_i V = \bar{\rho}_i^{\,2} \sin \bar{\phi}_i \, \Delta_i \rho \, \Delta_i \theta \, \Delta_i \phi
$$

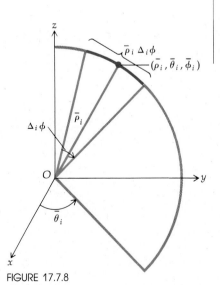

FIGURE 17.7.7

FIGURE 17.7.8

The **triple integral in spherical coordinates** of a function f on S is given by

$$\lim_{||\Delta||\to 0} \sum_{i=1}^{n} f(\bar{\rho}_i, \bar{\theta}_i, \bar{\phi}_i)\Delta_i V = \iiint_S f(\rho, \theta, \phi)\, dv \qquad (5)$$

$$\Leftrightarrow \quad \lim_{||\Delta||\to 0} \sum_{i=1}^{n} f(\bar{\rho}_i, \bar{\theta}_i, \bar{\phi}_i)\bar{\rho}_i{}^2 \sin \bar{\phi}_i\, \Delta_i\rho\, \Delta_i\theta\, \Delta_i\phi$$

$$= \iiint_S f(\rho, \theta, \phi)\rho^2 \sin \phi\, d\rho\, d\theta\, d\phi \qquad (6)$$

Observe that in spherical coordinates, $dV = \rho^2 \sin \phi\, d\rho\, d\theta\, d\phi$. The triple integrals in (5) or (6) can be evaluated by an iterated integral. Spherical coordinates are especially useful in some problems involving spheres, as shown in the following example.

EXAMPLE 5

Find the mass of the solid hemisphere of Example 3 if the volume density at any point is proportional to the distance of the point from the center of the base.

SOLUTION

If $(\bar{\rho}_i, \bar{\theta}_i, \bar{\phi}_i)$ is a point in the ith subregion of a spherical partition, the volume density at this point is $k\bar{\rho}_i$ kg/m^3, where k is a constant. If M kg is the mass of the solid, then

$$M = \lim_{||\Delta||\to 0} \sum_{i=1}^{n} k\bar{\rho}_i \Delta_i V$$

$$= \iiint_S k\rho\, dV$$

$$= 4k \int_0^{\pi/2} \int_0^{\pi/2} \int_0^{a} \rho^3 \sin \phi\, d\rho\, d\phi\, d\theta$$

$$= a^4 k \int_0^{\pi/2} \int_0^{\pi/2} \sin \phi\, d\phi\, d\theta$$

$$= a^4 k \int_0^{\pi/2} \left[-\cos \phi \right]_0^{\pi/2} d\theta$$

$$= a^4 k \int_0^{\pi/2} d\theta$$

$$= \tfrac{1}{2} a^4 k\pi$$

Hence the mass of the solid hemisphere is $\frac{1}{2}a^4 k\pi$ kg.

It is interesting to compare the solution of Example 5, which uses spherical coordinates, with what is entailed when using cartesian coordinates. By the latter method, a partition of S is formed by dividing S into rectangular boxes by drawing planes parallel to the coordinate planes. If (ξ_i, γ_i, μ_i) is any point in the ith subregion, and because $\rho = \sqrt{x^2 + y^2 + z^2}$, then

$$M = \lim_{||\Delta|| \to 0} \sum_{i=1}^{n} k\sqrt{\xi_i^2 + \gamma_i^2 + \mu_i^2}\, \Delta_i V$$

$$= \iiint_S k\sqrt{x^2 + y^2 + z^2}\, dV$$

$$= 4k \int_0^a \int_0^{\sqrt{a^2-z^2}} \int_0^{\sqrt{a^2-y^2-z^2}} \sqrt{x^2 + y^2 + z^2}\, dx\, dy\, dz$$

The computation involved in evaluating this integral is obviously much more complicated than that using spherical coordinates.

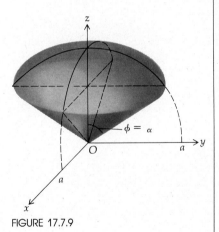

FIGURE 17.7.9

EXAMPLE 6
A homogeneous solid is bounded above by the sphere $\rho = a$ and below by the cone $\phi = \alpha$, where $0 < \alpha < \frac{1}{2}\pi$. Find the moment of inertia of the solid about the z axis.

SOLUTION
The solid is shown in Fig. 17.7.9. Let k kg/m^3 be the constant volume density at any point of the solid. Form a spherical partition of the solid and let $(\bar{\rho}_i, \bar{\theta}_i, \bar{\phi}_i)$ be a point in the ith subregion. The measure of the distance of the point $(\bar{\rho}_i, \bar{\theta}_i, \bar{\phi}_i)$ from the z axis is $\bar{\rho}_i \sin \bar{\phi}_i$. Hence if I_z kg-m^2 is the moment of inertia of the given solid about the z axis, then

$$I_z = \lim_{||\Delta|| \to 0} \sum_{i=1}^{n} (\bar{\rho}_i \sin \bar{\phi}_i)^2 k\, \Delta_i V$$

$$= \iiint_S k\rho^2 \sin^2 \phi\, dV$$

$$= k \int_0^\alpha \int_0^{2\pi} \int_0^a (\rho^2 \sin^2 \phi)\rho^2 \sin \phi\, d\rho\, d\theta\, d\phi$$

$$= \tfrac{1}{5}ka^5 \int_0^\alpha \int_0^{2\pi} \sin^3 \phi\, d\theta\, d\phi$$

$$= \tfrac{2}{5}ka^5\pi \int_0^\alpha \sin^3 \phi\, d\phi$$

$$= \tfrac{2}{5}ka^5\pi \left[-\cos \phi + \tfrac{1}{3} \cos^3 \phi \right]_0^\alpha$$

$$= \tfrac{2}{15}ka^5\pi(\cos^3 \alpha - 3 \cos \alpha + 2)$$

The moment of inertia of the solid about the z axis is therefore

$$\tfrac{2}{15}ka^5\pi(\cos^3 \alpha - 3 \cos \alpha + 2) \text{ kg-m}^2$$

EXERCISES 17.7

In Exercises 1 through 6, evaluate the iterated integral.

1. $\int_0^{\pi/4} \int_0^a \int_0^{r \cos \theta} r \sec^3 \theta \, dz \, dr \, d\theta$ **2.** $\int_0^{\pi/4} \int_{2 \sin \theta}^{2 \cos \theta} \int_0^{r \sin \theta} r^2 \cos \theta \, dz \, dr \, d\theta$ **3.** $\int_0^{\pi} \int_2^4 \int_0^1 re^z \, dz \, dr \, d\theta$

4. $\int_0^{2\pi} \int_0^{\pi} \int_0^2 \rho^3 \sin \phi \, d\rho \, d\phi \, d\theta$ **5.** $\int_0^{\pi/4} \int_0^{2a \cos \phi} \int_0^{2\pi} \rho^2 \sin \phi \, d\theta \, d\rho \, d\phi$ **6.** $\int_{\pi/4}^{\pi/2} \int_{\pi/4}^{\phi} \int_0^{a \csc \theta} \rho^3 \sin^2 \theta \sin \phi \, d\rho \, d\theta \, d\phi$

7. Find the volume of the solid enclosed by the sphere $x^2 + y^2 + z^2 = a^2$ by using (a) cylindrical coordinates and (b) spherical coordinates.

8. If S is the solid in the first octant bounded by the sphere $x^2 + y^2 + z^2 = 16$ and the coordinate planes, evaluate the triple integral $\iiint\limits_S xyz \, dV$ by three methods: (a) using spherical coordinates; (b) using rectangular coordinates; (c) using cylindrical coordinates.

In Exercises 9 through 16, use cylindrical coordinates.

9. Find the volume of the solid in the first octant bounded by the cylinder $x^2 + y^2 = 1$ and the plane $z = x$.

10. Find the volume of the solid bounded by the paraboloid $x^2 + y^2 + z = 1$ and the xy plane.

11. Find the volume of the solid bounded by the paraboloid $x^2 + y^2 + z = 12$ and the plane $z = 8$.

12. Find the volume of the solid bounded by the cylinder $x^2 + y^2 = 2y$, the paraboloid $x^2 + y^2 = 2z$ and the xy plane.

13. Find the mass of the solid bounded by a sphere of radius a meters if the volume density varies as the square of the distance from the center. The volume density is measured in kilograms per cubic meter.

14. Find the mass of the solid in the first octant inside the cylinder $x^2 + y^2 = 4x$ and under the sphere $x^2 + y^2 + z^2 = 16$. The volume density varies as the distance from the xy plane, and it is measured in kilograms per cubic meter.

15. Find the moment of inertia with respect to the z axis of the homogeneous solid bounded by the cylinder $r = 5$, the cone $z = r$, and the xy plane. The volume density at any point is k slugs/ft^3.

16. Find the moment of inertia of the solid bounded by a right-circular cylinder of altitude h meters and radius a meters, with respect to the axis of the cylinder. The volume density varies as the distance from the axis of the cylinder, and it is measured in kilograms per cubic meter.

17. Find the mass of the solid in Exercise 13 by using spherical coordinates.

18. Use spherical coordinates to find the center of mass of the solid bounded by the hemisphere of Example 5. The volume density is the same as that in Example 5.

In Exercises 19 through 22, use spherical coordinates.

19. Find the volume of the solid inside the sphere $x^2 + y^2 + z^2 = 4z$ and above the cone $x^2 + y^2 = z^2$.

20. Find the volume of the solid inside the sphere $x^2 + y^2 + z^2 = 2z$ and above the paraboloid $x^2 + y^2 = z$.

21. Find the moment of inertia with respect to the z axis of the homogeneous solid inside the cylinder $x^2 + y^2 - 2x = 0$, below the cone $x^2 + y^2 = z^2$, and above the xy plane. The volume density at any point is k kg/m^3.

22. Find the moment of inertia with respect to the z axis of the homogeneous solid bounded by the sphere $x^2 + y^2 + z^2 = 4$. The volume density at any point is k slugs/ft^3.

In Exercises 23 through 28, use the coordinate system that you decide is best for the problem.

23. Find the mass of a solid hemisphere of radius 2 m if the volume density varies as the distance from the center of the base and is measured in kilograms per cubic meter.

24. Find the mass of the homogeneous solid inside the paraboloid $3x^2 + 3y^2 = z$ and outside the cone $x^2 + y^2 = z^2$ if the constant volume density is k kg/m^3.

25. Find the moment of inertia about a diameter of the solid between two concentric spheres having radii a ft and $2a$ ft. The volume density varies inversely as the square of the distance from the center, and it is measured in slugs per cubic foot.

26. Find the mass of the solid of Exercise 25. The volume density is the same as that in Exercise 25.

27. Find the center of mass of the solid inside the paraboloid $x^2 + y^2 = z$ and outside the cone $x^2 + y^2 = z^2$. The constant volume density is k kg/m^3.

28. Find the moment of inertia with respect to the z axis of the homogeneous solid of Exercise 27.

In Exercises 29 through 32, evaluate the iterated integral by using either cylindrical or spherical coordinates.

29. $\int_0^4 \int_0^3 \int_0^{\sqrt{9-x^2}} \sqrt{x^2 + y^2} \, dy \, dx \, dz$

30. $\int_0^1 \int_0^{\sqrt{1-x^2}} \int_0^{\sqrt{1-x^2-y^2}} \dfrac{z}{\sqrt{x^2 + y^2}} \, dz \, dy \, dx$

31. $\int_0^1 \int_0^{\sqrt{1-y^2}} \int_{\sqrt{x^2+y^2}}^{\sqrt{2-x^2-y^2}} z^2 \, dz \, dx \, dy$

32. $\int_0^2 \int_0^{\sqrt{4-y^2}} \int_0^{\sqrt{4-x^2-y^2}} \dfrac{1}{x^2 + y^2 + z^2} \, dz \, dx \, dy$

REVIEW EXERCISES FOR CHAPTER 17

In Exercises 1 through 8, evaluate the given iterated integral.

1. $\int_0^1 \int_x^{\sqrt{x}} x^2 y \, dy \, dx$

2. $\int_{-2}^2 \int_{-\sqrt{4-y^2}}^{\sqrt{4-y^2}} xy \, dx \, dy$

3. $\int_0^{\pi/2} \int_0^{2 \sin \theta} r \cos^2 \theta \, dr \, d\theta$

4. $\int_0^{\pi} \int_0^{3(1 + \cos \theta)} r^2 \sin \theta \, dr \, d\theta$

5. $\int_0^1 \int_0^z \int_0^{y+z} e^x e^y e^z \, dx \, dy \, dz$

6. $\int_1^2 \int_3^x \int_0^{\sqrt{3}y} \dfrac{y}{y^2 + z^2} \, dz \, dy \, dx$

7. $\int_0^{\pi/2} \int_{\pi/6}^{\pi/2} \int_0^2 \rho^3 \sin \phi \cos \phi \, d\rho \, d\phi \, d\theta$

8. $\int_0^a \int_0^{\pi/2} \int_0^{\sqrt{a^2-z^2}} zre^{-r^2} \, dr \, d\theta \, dz$

In Exercises 9 through 12, evaluate the multiple integral.

9. $\iint\limits_R xy \, dA$; R is the region in the first quadrant bounded by the circle $x^2 + y^2 = 1$ and the coordinate axes.

10. $\iint\limits_R (x + y) \, dA$; R is the region bounded by the curve $y = \cos x$ and the x axis from $x = -\frac{1}{2}\pi$ to $x = \frac{1}{2}\pi$.

11. $\iiint\limits_S z^2 \, dV$; S is the region bounded by the cylinders $x^2 + z = 1$ and $y^2 + z = 1$ and the xy plane.

12. $\iiint\limits_S y \cos(x + z) \, dV$; S is the region bounded by the cylinder $x = y^2$ and the planes $x + z = \frac{1}{2}\pi$, $y = 0$, and $z = 0$.

13. Evaluate by polar coordinates the double integral

$$\iint\limits_R \frac{1}{x^2 + y^2} \, dA$$

where R is the region in the first quadrant bounded by the two circles $x^2 + y^2 = 1$ and $x^2 + y^2 = 4$.

14. Evaluate by polar coordinates the iterated integral $\int_0^1 \int_{\sqrt{3}y}^{\sqrt{4-y^2}} \ln(x^2 + y^2) \, dx \, dy$.

In Exercises 15 and 16, evaluate the iterated integral by reversing the order of integration.

15. $\int_0^1 \int_x^1 \sin y^2 \, dy \, dx$

16. $\int_0^1 \int_0^{\cos^{-1} y} e^{\sin x} \, dx \, dy$

In Exercises 17 and 18, use double integrals to find the area of the region bounded by the given curves in the xy plane. Draw a sketch of the region.

17. $y = x^2$ and $y = x^4$

18. $y = \sqrt{x}$ and $y = x^3$

In Exercises 19 and 20, evaluate the iterated integral by changing to either cylindrical or spherical coordinates.

19. $\int_0^3 \int_0^{\sqrt{9-x^2}} \int_0^2 \sqrt{x^2 + y^2} \, dz \, dy \, dx$

20. $\int_0^2 \int_0^{\sqrt{4-x^2}} \int_0^{\sqrt{4-x^2-y^2}} z \sqrt{4 - x^2 - y^2} \, dz \, dy \, dx$

21. Use double integration to find the area of the region in the first quadrant bounded by the parabolas $x^2 = 4y$ and $x^2 = 8 - 4y$ by two methods: (a) Integrate first with respect to x; (b) integrate first with respect to y.

22. Use double integration to find the area of the region in the xy plane bounded by the parabolas $y = 9 - x^2$ and $y = x^2 + 1$.

23. Use double integration to find the volume of the solid bounded by the planes $x = y$, $x = 0$, $y = 0$, $z = 0$, $x = 1$, and $z = 1$ by two methods: (a) Integrate first with respect to x; (b) integrate first with respect to y.

24. Use double integration to find the volume of the solid above the xy plane bounded by the cylinder $x^2 + y^2 = 16$ and the plane $z = 2y$ by two methods: (a) Integrate first with respect to x; (b) integrate first with respect to y.

25. Find the volume of the solid above the xy plane bounded by the surfaces $x^2 = 4y$, $y^2 = 4x$, and $x^2 = z - y$.

26. Find the mass of the lamina in the shape of the region bounded by the parabola $y = x^2$ and the line $x - y + 2 = 0$ if the area density at any point is x^2y^2 kg/m².

27. Find the area of the surface of the cylinder $x^2 + y^2 = 9$ lying in the first octant and between the planes $x = z$ and $3x = z$.

28. Find the area of the surface of the part of the cylinder $x^2 + y^2 = a^2$ that lies inside the cylinder $y^2 + z^2 = a^2$.

29. Use double integration to find the area of the region inside the circle $r = 1$ and to the right of the parabola $r(1 + \cos \theta) = 1$.

30. Find the mass of the lamina in the shape of the region exterior to the limaçon $r = 3 - \cos \theta$ and interior to the circle $r = 5 \cos \theta$ if the area density at any point is $2|\sin \theta|$ kg/m².

31. Find the center of mass of the rectangular lamina bounded by the lines $x = 3$ and $y = 2$ and the coordinate axes if the area density at any point is xy^2 kg/m².

32. Find the center of mass of the lamina in the shape of the region bounded by the parabolas $x^2 = 4 + 4y$ and $x^2 = 4 - 8y$ if the area density at any point is kx^2 kg/m².

33. Find the mass of the lamina in the shape of the region bounded by the polar axis and the curve $r = \cos 2\theta$, where $0 \leq \theta \leq \frac{1}{4}\pi$. The area density at any point is $r\theta$ kg/m².

34. Find the moment of inertia about the x axis of the lamina in the shape of the region bounded by the circle $x^2 + y^2 = a^2$ if the area density at any point is $k\sqrt{x^2 + y^2}$ kg/m².

35. Use cylindrical coordinates to find the volume of the solid bounded by the paraboloid $x^2 + y^2 = 4z$, the cylinder $x^2 + y^2 = 4ay$, and the plane $z = 0$.

36. Use spherical coordinates to find the mass of a spherical solid of radius a meters if the volume density at each point is proportional to the distance of the point from the center of the sphere. The volume density is measured in kilograms per cubic meter.

37. Use triple integration to find the volume of the solid bounded by the plane $z = 1$ and the smaller segment of the sphere $x^2 + y^2 + z^2 = 4$ cut off by this plane.

38. Use triple integration to find the volume of the solid in the first octant bounded by the plane $y + z = 8$, the cylinder $y = 2x^2$, the xy plane, and the yz plane.

39. Find the moment of inertia about the x axis of the lamina in the shape of the region bounded by the curve $y = e^x$, the line $x = 2$, and the coordinate axes if the area density at any point is xy kg/m².

40. Find the moment of inertia of the lamina of Exercise 39 about the y axis.

41. Find the moment of inertia with respect to the $\frac{1}{2}\pi$ axis of the homogeneous lamina in the shape of the region bounded by the curve $r^2 = 4 \cos 2\theta$ if the area density at any point is k kg/m².

42. Find the mass of the lamina of Exercise 41.

43. Find the polar moment of inertia and the corresponding radius of gyration of the lamina of Exercise 41.

44. Find the moment of inertia about the y axis of the lamina in the shape of the region bounded by the parabola $y = x - x^2$ and the line $x + y = 0$ if the area density at any point is $(x + y)$ kg/m².

45. Find the mass of the solid bounded by the spheres $x^2 + y^2 + z^2 = 4$ and $x^2 + y^2 + z^2 = 9$ if the volume density at any point is $k\sqrt{x^2 + y^2 + z^2}$ kg/m³.

46. Find the moment of inertia about the z axis of the solid of Exercise 45.

47. The homogeneous solid bounded by the cone $z^2 = 4x^2 + 4y^2$ between the planes $z = 0$ and $z = 4$ has a volume density at any point of k kg/m³. Find the moment of inertia about the z axis for this solid.

48. Find the center of mass of the solid bounded by the sphere $x^2 + y^2 + z^2 - 6z = 0$ and the cone $x^2 + y^2 = z^2$, and above the cone, if the volume density at any point is kz kg/m³.

E I G H T E E N

INTRODUCTION TO THE CALCULUS OF VECTOR FIELDS

18.1
VECTOR FIELDS

A *vector field* associates a vector with a point in space. For instance, if **F** is a vector-valued function defined on some open ball B in R^3 such that

$$\mathbf{F}(x, y, z) = M(x, y, z)\mathbf{i} + N(x, y, z)\mathbf{j} + R(x, y, z)\mathbf{k} \tag{1}$$

then **F** associates with each point (x, y, z) in B a vector, and **F** is called a **vector field**. A vector field defined by (1) has as its domain a subset of R^3 and as its range a subset of V_3. If the domain of a vector field is a set of points in a plane and its range is a set of vectors in V_2, then the vector field has an equation of the form

$$\mathbf{F}(x, y) = M(x, y)\mathbf{i} + N(x, y)\mathbf{j} \tag{2}$$

If, instead of a vector, a scalar is associated with each point in space, we have a **scalar field**; thus a scalar field is a real-valued function. An example of a scalar field is obtained by expressing the temperature at a point as a function of the coordinates of the point.

For an example of a vector field, consider the flow of a fluid, such as water through a pipe or blood through an artery. Assume that the fluid consists of infinitely many particles and that the velocity of a particle depends only on its position; thus the velocity is independent of time, and because of this fact the fluid flow is designated as a *steady-state* fluid flow. At a point (x, y, z) the velocity of the fluid is given by $\mathbf{F}(x, y, z)$ defined by an equation of the form (1). Thus **F** is a vector field called the **velocity field** of the fluid. Velocity fields can describe other motions, such as that of a wind or the rotation of a wheel. The vector fields that occur in this book will all be independent of time; they are termed **steady-state vector fields**.

We cannot show in a figure representations of all the vectors of a particular vector field. However, by drawing representations of some of the vectors we may get a visual description of the vector field, as indicated in the following example.

Table 18.1.1

(x, y)	$\mathbf{F}(x, y)$
$(1, 1)$	$-\mathbf{i} + \mathbf{j}$
$(1, -1)$	$\mathbf{i} + \mathbf{j}$
$(-1, 1)$	$-\mathbf{i} - \mathbf{j}$
$(-1, -1)$	$\mathbf{i} - \mathbf{j}$
$(1, 2)$	$-2\mathbf{i} + \mathbf{j}$
$(1, -2)$	$2\mathbf{i} + \mathbf{j}$
$(-1, 2)$	$-2\mathbf{i} - \mathbf{j}$
$(-1, -2)$	$2\mathbf{i} - \mathbf{j}$
$(2, 1)$	$-\mathbf{i} + 2\mathbf{j}$
$(2, -1)$	$\mathbf{i} + 2\mathbf{j}$
$(-2, 1)$	$-\mathbf{i} - 2\mathbf{j}$
$(-2, -1)$	$\mathbf{i} - 2\mathbf{j}$
$(2, 2)$	$-2\mathbf{i} + 2\mathbf{j}$
$(2, -2)$	$2\mathbf{i} + 2\mathbf{j}$
$(-2, 2)$	$-2\mathbf{i} - 2\mathbf{j}$
$(-2, -2)$	$2\mathbf{i} - 2\mathbf{j}$

FIGURE 18.1.1

EXAMPLE 1
(a) Show on a figure the representations, having initial point at (x, y), of the vectors in the vector field

$$\mathbf{F}(x, y) = -y\mathbf{i} + x\mathbf{j}$$

where x is ± 1 or ± 2 and y is ± 1 or ± 2. (b) Prove that each representation is tangent to a circle having its center at the origin and has a length equal to the radius of the circle.

SOLUTION
(a) Table 18.1.1 gives the vectors $\mathbf{F}(x, y)$ associated with the sixteen points (x, y). Representations of these vectors appear in Fig. 18.1.1.
(b) Let

$$\mathbf{R}(x, y) = x\mathbf{i} + y\mathbf{j}$$

be the position vector whose terminal point is at (x, y). Then

$$\mathbf{R}(x, y) \cdot \mathbf{F}(x, y) = (x\mathbf{i} + y\mathbf{i}) \cdot (-y\mathbf{i} + x\mathbf{j})$$
$$= -xy + xy$$
$$= 0$$

Therefore $\mathbf{R}$ and $\mathbf{F}$ are orthogonal. Thus the representation of $\mathbf{F}$ whose initial point is at (x, y) is tangent to the circle having its center at the origin and radius $\|\mathbf{R}(x, y)\|$. Because

$$\|\mathbf{F}(x, y)\| = \sqrt{(-y)^2 + x^2}$$
$$= \|\mathbf{R}(x, y)\|$$

the length of each representation is equal to the radius of the circle.

The vector field in Example 1 is similar to a velocity field determined by a wheel rotating at the origin.

An example of a vector field in V_3 arises from Newton's inverse square law of gravitational attraction. This law states that the measure of the magnitude of the gravitational force between two particles of mass M units and m units, respectively, is

$$\frac{GMm}{d^2}$$

where d units is the distance between the particles and G is a gravitational constant. Thus if a particle of mass M units is at the origin and a particle of mass 1 unit $(m = 1)$ is at the point $P(x, y, z)$, then if $\mathbf{F}(x, y, z)$ is the gravitational force exerted by the particle at the origin on the particle at P,

$$\|\mathbf{F}(x, y, z)\| = \frac{GM(1)}{\|\mathbf{R}(x, y, z)\|^2} \tag{3}$$

where $\mathbf{R}(x, y, z) = x\mathbf{i} + y\mathbf{j} + z\mathbf{k}$. To obtain the force vector $\mathbf{F}(x, y, z)$, we also need the direction of $\mathbf{F}$. Because this direction is toward the origin, it is the same as the direction of the unit vector $-\dfrac{1}{\|\mathbf{R}\|}\mathbf{R}$. With this direction and the magnitude given by (3), we have

$$\mathbf{F}(x, y, z) = \frac{GM}{\|\mathbf{R}(x, y, z)\|^2}\left(-\frac{\mathbf{R}(x, y, z)}{\|\mathbf{R}(x, y, z)\|}\right)$$

Because $\|\mathbf{R}(x, y, z)\| = \sqrt{x^2 + y^2 + z^2}$, we obtain

$$\mathbf{F}(x, y, z) = \frac{-GM}{(x^2 + y^2 + z^2)^{3/2}}(x\mathbf{i} + y\mathbf{j} + z\mathbf{k}) \tag{4}$$

The vector field defined by (4) is called a **force field**. Figure 18.1.2 shows some of the representations of the vectors in this force field where the object at the origin is a sphere (for instance, the earth) and $\|\mathbf{R}\|$ is greater than the radius of the sphere. Each representation points toward the origin. Representations of vectors at points near the origin are longer than those at points

FIGURE 18.1.2

farther away from the origin, and the lengths are the same at points having the same distance from the origin. Because of these properties, the force field defined by (4) is designated as a *central force field*.

The gradient of a scalar field is a vector field. If ϕ is a scalar field and **F** is the vector field defined by $\mathbf{F} = \nabla\phi$, then **F** is called a **gradient vector field**, and ϕ is termed a **potential function** for **F**. A gradient vector field is also called a **conservative vector field**. The terminology *conservative* will be apparent after you read Section 18.3.

▶ ILLUSTRATION 1   Consider the vector field defined by

$$\mathbf{F}(x, y) = (y^2 + 2x + 4)\mathbf{i} + (2xy + 4y - 5)\mathbf{j}$$

From Illustration 1 of Section 16.5 it follows that if

$$\phi(x, y) = y^2x + x^2 + 4x + 2y^2 - 5y$$

then

$$\mathbf{F}(x, y) = \nabla\phi(x, y)$$

Thus **F** is a conservative vector field and ϕ is a potential function for **F**.
◀

The following illustration shows that the gravitational force field defined by (4) is conservative.

▶ ILLUSTRATION 2   In Example 5 of Section 16.1 we showed that if

$$V(x, y, z) = \frac{1}{\sqrt{x^2 + y^2 + z^2}}$$

then

$$\nabla V(x, y, z) = \frac{-1}{(x^2 + y^2 + z^2)^{3/2}} (x\mathbf{i} + y\mathbf{j} + z\mathbf{k})$$

Thus if

$$\phi(x, y, z) = \frac{GM}{\sqrt{x^2 + y^2 + z^2}}$$

$$\nabla\phi(x, y, z) = \frac{-GM}{(x^2 + y^2 + z^2)^{3/2}} (x\mathbf{i} + y\mathbf{j} + z\mathbf{k}) \tag{5}$$

Comparing (5) and (4) we observe that

$$\mathbf{F}(x, y, z) = \nabla\phi(x, y, z)$$

Therefore **F** is conservative and ϕ is a potential function for **F**.   ◀

In the above two illustrations it is a simple matter to prove that the vector field is conservative because we know a function ϕ for which **F** is its gradient. A more difficult problem is to decide if a given vector field is conservative, and if it is, to find a potential function. We apply Theorems 16.5.1 and 16.5.4 for this purpose, as shown in the following two examples.

EXAMPLE 2

If **F** is the vector field defined by

$$F(x, y) = \frac{1}{y}\mathbf{i} - \frac{x}{y^2}\mathbf{j}$$

prove that **F** is conservative, and find a potential function for **F**.

SOLUTION

Let

$$M(x, y) = \frac{1}{y} \qquad N(x, y) = -\frac{x}{y^2}$$

$$M_y(x, y) = -\frac{1}{y^2} \qquad N_x(x, y) = -\frac{1}{y^2}$$

Because $M_y(x, y) = N_x(x, y)$, it follows from Theorem 16.5.1 that **F** is a gradient and hence a conservative vector field. A potential function ϕ satisfies the equation

$$\mathbf{F}(x, y) = \nabla\phi(x, y)$$

Therefore

$$\phi_x(x, y) = \frac{1}{y} \qquad \phi_y(x, y) = -\frac{x}{y^2} \tag{6}$$

Integrating with respect to x both members of the first of Eqs. (6) we have

$$\phi(x, y) = \frac{x}{y} + g(y) \tag{7}$$

where $g(y)$ is independent of x. Partial differentiating both members of (7) with respect to y gives

$$\phi_y(x, y) = -\frac{x}{y^2} + g'(y) \tag{8}$$

We equate the right members of (8) and the second of Eqs. (6) and get

$$-\frac{x}{y^2} + g'(y) = -\frac{x}{y^2}$$

$$g'(y) = 0$$

$$g(y) = C$$

With this value of $g(y)$ in (7) we have

$$\phi(x, y) = \frac{x}{y} + C$$

which is the required potential function.

EXAMPLE 3

If $\mathbf{F}$ is the vector field defined by

$$\mathbf{F}(x, y, z) = (z^2 + 1)\mathbf{i} + 2yz\mathbf{j} + (2xz + y^2)\mathbf{k}$$

prove that $\mathbf{F}$ is conservative, and find a potential function for $\mathbf{F}$.

SOLUTION

To determine if $\mathbf{F}$ is conservative we apply Theorem 16.5.4 to ascertain if $\mathbf{F}$ is a gradient. Let

$$M(x, y, z) = z^2 + 1 \qquad N(x, y, z) = 2yz \qquad R(x, y, z) = 2xz + y^2$$
$$M_y(x, y, z) = 0 \qquad N_x(x, y, z) = 0 \qquad R_x(x, y, z) = 2z$$
$$M_z(x, y, z) = 2z \qquad N_z(x, y, z) = 2y \qquad R_y(x, y, z) = 2y$$

Therefore

$$M_y(x, y, z) = N_x(x, y, z) \quad M_z(x, y, z) = R_x(x, y, z) \quad N_z(x, y, z) = R_y(x, y, z)$$

Thus by Theorem 16.5.4, $\mathbf{F}$ is a gradient and hence a conservative vector field. We now find a potential function ϕ such that $\mathbf{F}(x, y, z) = \nabla\phi(x, y, z)$; therefore

$$\phi_x(x, y, z) = z^2 + 1 \qquad \phi_y(x, y, z) = 2yz \qquad \phi_z(x, y, z) = 2xz + y^2 \quad (9)$$

Integrating with respect to x both members of the first of Eqs. (9) we have

$$\phi(x, y, z) = xz^2 + x + g(y, z) \tag{10}$$

We partial differentiate with respect to y both members of Eq. (10) and obtain

$$\phi_y(x, y, z) = g_y(y, z) \tag{11}$$

Equating the right members of (11) and the second of Eqs. (9) we have

$$g_y(y, z) = 2yz$$

We now integrate with respect to y both members of this equation and get

$$g(y, z) = y^2 z + h(z) \tag{12}$$

Substituting from (12) into (10) we obtain

$$\phi(x, y, z) = xz^2 + x + y^2 z + h(z) \tag{13}$$

We partial differentiate with respect to z both members of (13) and get

$$\phi_z(x, y, z) = 2xz + y^2 + h'(z) \tag{14}$$

We equate the right members of (14) and the third of Eqs. (9) and have

$$2xz + y^2 + h'(z) = 2xz + y^2$$
$$h'(z) = 0$$
$$h(z) = C \tag{15}$$

Substituting from (15) into (13) we obtain the required potential function ϕ such that

$$\phi(x, y, z) = xz^2 + x + y^2z + C$$

There are two fields involving derivatives that are associated with a vector field $\mathbf{F}$. One is called the *curl* of $\mathbf{F}$, which is a vector field, and the other is called the *divergence* of $\mathbf{F}$, which is a scalar field. Before giving their definitions, we show how the symbol $\mathbf{V}$ is used as an operator. Recall that if f is a scalar function of three variables x, y, and z, the gradient of f is

$$\mathbf{V}f(x, y, z) = f_x(x, y, z)\mathbf{i} + f_y(x, y, z)\mathbf{j} + f_z(x, y, z)\mathbf{k} \qquad (16)$$

We shall now let the del operator in three dimensions be used to denote

$$\mathbf{i}\frac{\partial}{\partial x} + \mathbf{j}\frac{\partial}{\partial y} + \mathbf{k}\frac{\partial}{\partial z}$$

Therefore, $\mathbf{V}$ operating on the scalar function f means

$$\mathbf{V}f = \frac{\partial f}{\partial x}\mathbf{i} + \frac{\partial f}{\partial y}\mathbf{j} + \frac{\partial f}{\partial z}\mathbf{k}$$

which agrees with (16).

18.1.1 DEFINITION   Let $\mathbf{F}$ be a vector field on some open ball B in R^3 such that

$$\mathbf{F}(x, y, z) = M(x, y, z)\mathbf{i} + N(x, y, z)\mathbf{j} + R(x, y, z)\mathbf{k}$$

Then the **curl** of $\mathbf{F}$ is defined by

$$\text{curl } \mathbf{F}(x, y, z) = \left(\frac{\partial R}{\partial y} - \frac{\partial N}{\partial z}\right)\mathbf{i} + \left(\frac{\partial M}{\partial z} - \frac{\partial R}{\partial x}\right)\mathbf{j} + \left(\frac{\partial N}{\partial x} - \frac{\partial M}{\partial y}\right)\mathbf{k}$$

if these partial derivatives exist.

A mnemonic device for computing curl $\mathbf{F}$ is to extend the notation for the cross product of two vectors to the "cross product" of the operator $\mathbf{V}$ and the vector field $\mathbf{F}$ and write

$$\text{curl } \mathbf{F} = \mathbf{V} \times \mathbf{F}$$

$$= \begin{vmatrix} \mathbf{i} & \mathbf{j} & \mathbf{k} \\ \dfrac{\partial}{\partial x} & \dfrac{\partial}{\partial y} & \dfrac{\partial}{\partial z} \\ M & N & R \end{vmatrix}$$

As we indicated when determinant notation was first used for the cross product of two vectors, the elements of the determinant were not all real numbers as is customary. In the above "determinant" the first row contains vectors, the second row contains partial derivative operators, and the third row consists of scalar functions.

EXAMPLE 4

Find curl $\mathbf{F}$ if $\mathbf{F}$ is the vector field defined by

$$\mathbf{F}(x, y, z) = e^{2x}\mathbf{i} + 3x^2yz\mathbf{j} + (2y^2z + x)\mathbf{k}$$

SOLUTION

$$\text{curl } \mathbf{F}(x, y, z) = \begin{vmatrix} \mathbf{i} & \mathbf{j} & \mathbf{k} \\ \dfrac{\partial}{\partial x} & \dfrac{\partial}{\partial y} & \dfrac{\partial}{\partial z} \\ e^{2x} & 3x^2yz & 2y^2z + x \end{vmatrix}$$

$$= (4yz - 3x^2y)\mathbf{i} + (0 - 1)\mathbf{j} + (6xyz - 0)\mathbf{k}$$
$$= (4yz - 3x^2y)\mathbf{i} - \mathbf{j} + 6xyz\mathbf{k}$$

18.1.2 DEFINITION   Let $\mathbf{F}$ be a vector field on some open ball B in R^3 such that

$$\mathbf{F}(x, y, z) = M(x, y, z)\mathbf{i} + N(x, y, z)\mathbf{j} + R(x, y, z)\mathbf{k}$$

Then the **divergence** of $\mathbf{F}$, denoted by div $\mathbf{F}$, is defined by

$$\text{div } \mathbf{F}(x, y, z) = \frac{\partial M}{\partial x} + \frac{\partial N}{\partial y} + \frac{\partial R}{\partial z}$$

if these partial derivatives exist.

We extend the notation for the dot product of two vectors to the "dot product" of the operator $\mathbf{V}$ and the vector field $\mathbf{F}$ to compute the divergence of $\mathbf{F}$, and we write

$$\text{div } \mathbf{F} = \mathbf{V} \cdot \mathbf{F}$$

$$= \left(\mathbf{i}\frac{\partial}{\partial x} + \mathbf{j}\frac{\partial}{\partial y} + \mathbf{k}\frac{\partial}{\partial z} \right) \cdot (M\mathbf{i} + N\mathbf{j} + R\mathbf{k})$$

$$= \frac{\partial M}{\partial x} + \frac{\partial N}{\partial y} + \frac{\partial R}{\partial z}$$

EXAMPLE 5

Find div $\mathbf{F}$ if $\mathbf{F}$ is the vector field of Example 4.

SOLUTION

$$\text{div } \mathbf{F}(x, y, z) = \mathbf{V} \cdot \mathbf{F}(x, y, z)$$

$$= \frac{\partial}{\partial x}(e^{2x}) + \frac{\partial}{\partial y}(3x^2yz) + \frac{\partial}{\partial z}(2y^2z + x)$$

$$= 2e^{2x} + 3x^2z + 2y^2$$

A physical significance of curl **F** and div **F** in the study of fluid motion will be discussed in Sections 18.4 and 18.5. In this section we are concerned with learning how to compute them and to prove some properties involving them. Two such properties are given in the following two theorems. You are asked to prove these theorems in Exercises 47 and 48.

18.1.3 THEOREM   Suppose **F** is a vector field on an open ball B in R^3 such that

$$\mathbf{F}(x, y, z) = M(x, y, z)\mathbf{i} + N(x, y, z)\mathbf{j} + R(x, y, z)\mathbf{k}$$

If the second partial derivatives of M, N, and R are continuous on B, then

$$\text{div}(\text{curl } \mathbf{F}) = 0$$

18.1.4 THEOREM   If f is a scalar field on an open ball B in R^3 and the second partial derivatives of f are continuous on B, then

$$\text{curl}(\nabla f) = \mathbf{0} \tag{17}$$

Equation (17) states that the curl of the gradient of f equals the zero vector. Consider now the divergence of the gradient of f, that is, $\nabla \cdot (\nabla f)$, which can also be written as $\nabla \cdot \nabla f$ or $\nabla^2 f$. By definition,

$$\nabla^2 f(x, y, z) = \left(\mathbf{i}\frac{\partial}{\partial x} + \mathbf{j}\frac{\partial}{\partial y} + \mathbf{k}\frac{\partial}{\partial z} \right) \cdot \left(\frac{\partial f}{\partial x}\mathbf{i} + \frac{\partial f}{\partial y}\mathbf{j} + \frac{\partial f}{\partial z}\mathbf{k} \right)$$

$$\nabla^2 f(x, y, z) = \frac{\partial^2 f}{\partial x^2} + \frac{\partial^2 f}{\partial y^2} + \frac{\partial^2 f}{\partial z^2} \tag{18}$$

The expression on the right side of Eq. (18) is called the **Laplacian** of f. The following equation obtained by setting the Laplacian equal to zero is called **Laplace's equation**:

$$\frac{\partial^2 f}{\partial x^2} + \frac{\partial^2 f}{\partial y^2} + \frac{\partial^2 f}{\partial z^2} = 0$$

A scalar function satisfying Laplace's equation is said to be **harmonic**. These functions have important applications in physics in the study of heat transfer, electromagnetic radiation, acoustics, and so on.

If **F** is a vector field on some open disk B in R^2 such that $\mathbf{F}(x, y) = M(x, y)\mathbf{i} + N(x, y)\mathbf{j}$, then the curl of **F** and divergence of **F** in two dimensions are defined by

$$\text{curl } \mathbf{F}(x, y) = \left(\frac{\partial N}{\partial x} - \frac{\partial M}{\partial y} \right)\mathbf{k} \qquad \text{div } \mathbf{F}(x, y) = \frac{\partial M}{\partial x} + \frac{\partial N}{\partial y}$$

if these partial derivatives exist. The Laplacian in two dimensions is defined by

$$\nabla^2 f(x, y) = \frac{\partial^2 f}{\partial x^2} + \frac{\partial^2 f}{\partial y^2}$$

EXAMPLE 6

If $\mathbf{F}(x, y) = 3x^2 y \mathbf{i} - 2xy^3 \mathbf{j}$, find (a) curl $\mathbf{F}(x, y)$ and (b) div $\mathbf{F}(x, y)$.

SOLUTION

Because $\mathbf{F}(x, y) = 3x^2 y \mathbf{i} - 2xy^3 \mathbf{j}$, $M(x, y) = 3x^2 y$ and $N(x, y) = -2xy^3$. Therefore,

(a) curl $\mathbf{F}(x, y) = \left(\dfrac{\partial N}{\partial x} - \dfrac{\partial M}{\partial y} \right) \mathbf{k}$ (b) div $\mathbf{F}(x, y) = \dfrac{\partial M}{\partial x} + \dfrac{\partial N}{\partial y}$

$\qquad\qquad\quad = (-2y^3 - 3x^2) \mathbf{k}$ $\qquad\qquad\qquad\qquad = 6xy - 6xy^2$

EXERCISES 18.1

In Exercises 1 through 6, show on a figure the representations having initial point at (x, y) of the vectors in the given vector field, where x is ± 1 or ± 2 and y is ± 1 or ± 2.

1. $\mathbf{F}(x, y) = x\mathbf{i} - y\mathbf{j}$ **2.** $\mathbf{F}(x, y) = -x\mathbf{i} + y\mathbf{j}$ **3.** $\mathbf{F}(x, y) = 4y\mathbf{i} + 3x\mathbf{j}$

4. $\mathbf{F}(x, y) = -3y\mathbf{i} + 4x\mathbf{j}$ **5.** $\mathbf{F}(x, y) = \dfrac{1}{\sqrt{x^2 + y^2}} (x\mathbf{i} + y\mathbf{j})$ **6.** $\mathbf{F}(x, y) = y\mathbf{i} + 2\mathbf{j}$

In Exercises 7 through 14, find a conservative vector field having the given potential function.

7. $f(x, y) = 3x^2 + 2y^3$ **8.** $f(x, y) = 2x^4 - 5x^2 y^2 + 4y^4$ **9.** $f(x, y) = \tan^{-1} x^2 y$

10. $f(x, y) = ye^x - xe^y$ **11.** $f(x, y, z) = 2x^3 - 3x^2 y + xy^2 - 4y^3$ **12.** $f(x, y, z) = \sqrt{x^2 + y^2 + z^2}$

13. $f(x, y, z) = x^2 y e^{-4z}$ **14.** $f(x, y, z) = z \sin(x^2 - y)$

In Exercises 15 through 20, determine if the given vector field is conservative.

15. $\mathbf{F}(x, y) = (3x^2 - 2y^2)\mathbf{i} + (3 - 4xy)\mathbf{j}$ **16.** $\mathbf{F}(x, y) = (e^x e^y + 6e^{2x})\mathbf{i} + (e^x e^y - 2e^y)\mathbf{j}$

17. $\mathbf{F}(x, y) = y \cos(x + y)\mathbf{i} - x \sin(x + y)\mathbf{j}$

18. $\mathbf{F}(x, y, z) = (3x^2 + 2yz)\mathbf{i} + (2xz + 6yz)\mathbf{j} + (2xy + 3y^2 - 2z)\mathbf{k}$

19. $\mathbf{F}(x, y, z) = (2ye^{2x} + e^z)\mathbf{i} + (3ze^{3y} + e^{2x})\mathbf{j} + (xe^z + e^{3y})\mathbf{k}$

20. $\mathbf{F}(x, y, z) = y \sec^2 x\mathbf{i} + (\tan x - z \sec^2 y)\mathbf{j} + x \sec z \tan z\mathbf{k}$

In Exercises 21 through 32, prove that the given vector field is conservative, and find a potential function.

21. $\mathbf{F}(x, y) = y\mathbf{i} + x\mathbf{j}$ **22.** $\mathbf{F}(x, y) = x\mathbf{i} + y\mathbf{j}$ **23.** $\mathbf{F}(x, y) = e^x \sin y\mathbf{i} + e^x \cos y\mathbf{j}$

24. $\mathbf{F}(x, y) = (\sin y \sinh x + \cos y \cosh x)\mathbf{i} + (\cos y \cosh x - \sin y \sinh x)\mathbf{j}$

25. $\mathbf{F}(x, y) = (2xy^2 - y^3)\mathbf{i} + (2x^2 y - 3xy^2 + 2)\mathbf{j}$ **26.** $\mathbf{F}(x, y) = (3x^2 + 2y - y^2 e^x)\mathbf{i} + (2x - 2ye^x)\mathbf{j}$

27. $\mathbf{F}(x, y, z) = (x^2 - y)\mathbf{i} - (x - 3z)\mathbf{j} + (z + 3y)\mathbf{k}$ **28.** $\mathbf{F}(x, y, z) = yz\mathbf{i} + xz\mathbf{j} + xy\mathbf{k}$

29. $\mathbf{F}(x, y, z) = (ze^x + e^y)\mathbf{i} + (xe^y - e^z)\mathbf{j} + (-ye^z + e^x)\mathbf{k}$

30. $\mathbf{F}(x, y, z) = (\tan y + 2xy \sec z)\mathbf{i} + (x \sec^2 y + x^2 \sec z)\mathbf{j} + \sec z(x^2 y \tan z - \sec z)\mathbf{k}$

31. $\mathbf{F}(x, y, z) = (2x \cos y - 3)\mathbf{i} - (x^2 \sin y + z^2)\mathbf{j} - (2yz - 2)\mathbf{k}$

32. $\mathbf{F}(x, y, z) = (2y^3 - 8xz^2)\mathbf{i} + (6xy^2 + 1)\mathbf{j} - (8x^2 z + 3z^2)\mathbf{k}$

In Exercises 33 through 42, find curl $\mathbf{F}$ and div $\mathbf{F}$ for the given vector field.

33. $\mathbf{F}(x, y) = 2x\mathbf{i} + 3y\mathbf{j}$ **34.** $\mathbf{F}(x, y) = \cos x\mathbf{i} - \sin y\mathbf{j}$ **35.** $\mathbf{F}(x, y) = e^x \cos y\mathbf{i} + e^x \sin y\mathbf{j}$

36. $\mathbf{F}(x, y) = -\dfrac{y}{x}\mathbf{i} + \dfrac{1}{x}\mathbf{j}$ **37.** $\mathbf{F}(x, y, z) = x^2\mathbf{i} + y^2\mathbf{j} + z^2\mathbf{k}$ **38.** $\mathbf{F}(x, y, z) = xz^2\mathbf{i} + y^2\mathbf{j} + x^2 z\mathbf{k}$

39. $\mathbf{F}(x, y, z) = \cos y\mathbf{i} + \cos z\mathbf{j} + \cos x\mathbf{k}$ **40.** $\mathbf{F}(x, y, z) = (y^2 + z^2)\mathbf{i} + xe^y \cos z\mathbf{j} - xe^y \cos z\mathbf{k}$

41. $\mathbf{F}(x, y, z) = \sqrt{x^2 + y^2 + 1}\,\mathbf{i} + \sqrt{x^2 + y^2 + 1}\,\mathbf{j} + z^2\mathbf{k}$

42. $\mathbf{F}(x, y, z) = \dfrac{x}{(x^2 + y^2)^{3/2}}\,\mathbf{i} + \dfrac{y}{(x^2 + y^2)^{3/2}}\,\mathbf{j} + \mathbf{k}$

In Exercises 43 through 46, prove that the given scalar function is harmonic by showing that its Laplacian is zero.

43. $f(x, y) = e^y \sin x + e^x \cos y$

44. $f(x, y) = \ln(\sqrt{x^2 + y^2})$

45. $f(x, y, z) = 2x^2 + 3y^2 - 5z^2$

46. $f(x, y, z) = (x^2 + y^2 + z^2)^{-1/2}$

47. Prove Theorem 18.1.3.

48. Prove Theorem 18.1.4.

18.2
LINE INTEGRALS

In Chapter 4 the concept of area was used to motivate the definition of the definite integral. To motivate the definition of an integral of a vector field, we use the physical concept of work.

We showed in Section 13.2 that if a constant force of vector measure $\mathbf{F}$ moves a particle along a straight line from a point A to a point B, then if $\mathbf{W}$ is the measure of the work done,

$$\mathbf{W} = \mathbf{F} \cdot \mathbf{V}(\overrightarrow{AB}) \tag{1}$$

Suppose now that the force vector is not constant, and instead of the motion being along a straight line, it is along a curve. Let the force that is exerted on the particle at the point (x, y) in some open disk B in R^2 be given by the force field

$$\mathbf{F}(x, y) = M(x, y)\mathbf{i} + N(x, y)\mathbf{j} \tag{2}$$

where M and N are continuous on B. Let C be a curve lying in B and having the vector equation

$$\mathbf{R}(t) = f(t)\mathbf{i} + g(t)\mathbf{j} \qquad a \le t \le b$$

We require the functions f and g be such that f' and g' are continuous on $[a, b]$. We wish to define the work done by the variable force of vector measure $\mathbf{F}$ in moving the particle along C from the point $(f(a), g(a))$ to $(f(b), g(b))$. At a point $(f(t), g(t))$ on C the force vector is

$$\mathbf{F}(f(t), g(t)) = M(f(t), g(t))\mathbf{i} + N(f(t), g(t))\mathbf{j} \tag{3}$$

Let Δ be a partition of the interval $[a, b]$:

$$a = t_0 < t_1 < t_2 < \ldots < t_{n-1} < t_n = b$$

On C let P_i be the point $(x_i, y_i) = (f(t_i), g(t_i))$. Refer to Fig. 18.2.1. The vector $\mathbf{V}(\overrightarrow{P_{i-1}P_i}) = \mathbf{R}(t_i) - \mathbf{R}(t_{i-1})$; therefore

$$\mathbf{V}(\overrightarrow{P_{i-1}P_i}) = f(t_i)\mathbf{i} + g(t_i)\mathbf{j} - [f(t_{i-1})\mathbf{i} + g(t_{i-1})\mathbf{j}]$$

$$\mathbf{V}(\overrightarrow{P_{i-1}P_i}) = [f(t_i) - f(t_{i-1})]\mathbf{i} + [g(t_i) - g(t_{i-1})]\mathbf{j} \tag{4}$$

Because f' and g' are continuous on $[a, b]$, it follows from the mean-value theorem (3.3.2) that there are numbers c_i and d_i in the open interval (t_{i-1}, t_i) such that

$$f(t_i) - f(t_{i-1}) = f'(c_i)(t_i - t_{i-1}) \tag{5}$$

$$g(t_i) - g(t_{i-1}) = g'(d_i)(t_i - t_{i-1}) \tag{6}$$

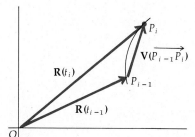

FIGURE 18.2.1

Letting $\Delta_i t = t_i - t_{i-1}$, and substituting from (5) and (6) into (4), we obtain

$$\mathbf{V}(\overrightarrow{P_{i-1}P_i}) = [f'(c_i)\mathbf{i} + g'(d_i)\mathbf{j}]\,\Delta_i t \tag{7}$$

For each i consider the vector

$$\mathbf{F}_i = M(f(c_i), g(c_i))\mathbf{i} + N(f(d_i), g(d_i))\mathbf{j} \tag{8}$$

Each of the vectors $\mathbf{F}_i$ $(i = 1, 2, \ldots, n)$ is an approximation to the force vector $\mathbf{F}(f(t), g(t))$, given by (3), along the arc of C from P_{i-1} to P_i. Observe that even though c_i and d_i are in general different numbers in the open interval (t_{i-1}, t_i), the values of the vectors $\mathbf{F}(f(t), g(t))$ are close to the vector $\mathbf{F}_i$. Furthermore, we approximate the arc of C from P_{i-1} to P_i by the line segment $\overrightarrow{P_{i-1}P_i}$. Thus we apply formula (1) and obtain an approximation for the work done by the vector $\mathbf{F}(f(t), g(t))$ in moving a particle along the arc of C from P_{i-1} to P_i. Denoting this approximation by $\Delta_i W$, we have, from formula (1) and Eqs. (8) and (7),

$$\Delta_i W = [M(f(c_i), g(c_i))\mathbf{i} + N(f(d_i), g(d_i))\mathbf{j}] \cdot [f'(c_i)\mathbf{i} + g'(d_i)\mathbf{j}]\,\Delta_i t$$

$$\Leftrightarrow \quad \Delta_i W = [M(f(c_i), g(c_i))f'(c_i)]\,\Delta_i t + [N(f(d_i), g(d_i))g'(d_i)]\,\Delta_i t$$

An approximation of the measure of the work done by $F(f(t), g(t))$ along C is $\sum_{i=1}^{n} \Delta_i W$ or, equivalently,

$$\sum_{i=1}^{n} [M(f(c_i), g(c_i))f'(c_i)]\,\Delta_i t + \sum_{i=1}^{n} [N(f(d_i), g(d_i))g'(d_i)]\,\Delta_i t$$

Each of these sums is a Riemann sum. The first summation is a Riemann sum for the function having function values $M(f(t), g(t))f'(t)$, and the second summation is a Riemann sum for the function having function values $N(f(t), g(t))g'(t)$. If n increases without bound, these two sums approach the definite integral:

$$\int_a^b [M(f(t), g(t))f'(t) + N(f(t), g(t))g'(t)]\,dt$$

We therefore have the following definition. In the definition the notation $\mathbf{F}(\mathbf{R}(t))$ is used in place of $\mathbf{F}(f(t), g(t))$.

18.2.1 DEFINITION Let C be a curve lying in an open disk B in R^2 for which a vector equation of C is $\mathbf{R}(t) = f(t)\mathbf{i} + g(t)\mathbf{j}$, where f' and g' are continuous on $[a, b]$. Furthermore, let a force field on B be defined by $\mathbf{F}(x, y) = M(x, y)\mathbf{i} + N(x, y)\mathbf{j}$, where M and N are continuous on B. Then if W is the measure of the work done by a force of vector measure $\mathbf{F}$ in moving a particle along C from $(f(a), g(a))$ to $(f(b), g(b))$,

$$W = \int_a^b [M(f(t), g(t))f'(t) + N(f(t), g(t))g'(t)]\,dt \tag{9}$$

or, equivalently, by using vector notation,

$$W = \int_a^b \langle M(f(t), g(t)), N(f(t), g(t))\rangle \cdot \langle f'(t), g'(t)\rangle\,dt \tag{10}$$

$$\Leftrightarrow \quad W = \int_a^b \mathbf{F}(\mathbf{R}(t)) \cdot \mathbf{R}'(t)\,dt \tag{11}$$

EXAMPLE 1

Suppose a particle moves along the parabola $y = x^2$ from the point $(-1, 1)$ to the point $(2, 4)$. Find the total work done if the motion is caused by the force field $\mathbf{F}(x, y) = (x^2 + y^2)\mathbf{i} + 3x^2 y\mathbf{j}$. Assume the arc is measured in meters and the force is measured in newtons.

SOLUTION

Parametric equations of the parabola are

$$x = t \quad \text{and} \quad y = t^2 \qquad -1 \le t \le 2$$

Thus a vector equation of the parabola is

$$\mathbf{R}(t) = t\mathbf{i} + t^2\mathbf{j} \quad \text{and} \quad R'(t) = \mathbf{i} + 2t\mathbf{j}$$

Because $\mathbf{F}(x, y) = \langle x^2 + y^2, 3x^2y \rangle$, then

$$\mathbf{F}(\mathbf{R}(t)) = \mathbf{F}(t, t^2)$$
$$= \langle t^2 + t^4, 3t^4 \rangle$$

If W joules is the work done, then, from (11),

$$W = \int_{-1}^{2} \mathbf{F}(\mathbf{R}(t)) \cdot \mathbf{R}'(t)\, dt \tag{12}$$

$$= \int_{-1}^{2} \langle t^2 + t^4, 3t^4 \rangle \cdot \langle 1, 2t \rangle\, dt$$

$$= \int_{-1}^{2} (t^2 + t^4 + 6t^5)\, dt$$

$$= \frac{t^3}{3} + \frac{t^5}{5} + t^6 \Big]_{-1}^{2}$$

$$= \tfrac{8}{3} + \tfrac{32}{5} + 64 - (-\tfrac{1}{3} - \tfrac{1}{5} + 1)$$

$$= \tfrac{363}{5}$$

Therefore the work done is $\frac{363}{5}$ joules.

The integrals in Eqs. (9) and (11) are called *line integrals*. For the line integral of Eq. (9), a common notation involving the differential form $M(x, y)\, dx + N(x, y)\, dy$ is

$$\int_C M(x, y)\, dx + N(x, y)\, dy$$

This notation is suggested by the fact that because parametric equations of C are $x = f(t)$ and $y = g(t)$, then $dx = f'(t)\, dt$ and $dy = g'(t)\, dt$. A vector notation for the line integral of Eq. (11) is

$$\int_C \mathbf{F} \cdot d\mathbf{R}$$

This notation is suggested by considering the vector equation of C, which is $\mathbf{R}(t) = f(t)\mathbf{i} + g(t)\mathbf{j}$, and letting $d\mathbf{R} = \mathbf{R}'(t)\, dt$. Then

$$\mathbf{F}(\mathbf{R}(t)) \cdot d\mathbf{R} = \mathbf{F}(\mathbf{R}(t)) \cdot \mathbf{R}'(t)\, dt$$

We have, then, the following formal definition.

18.2.2 DEFINITION Let C be a curve lying in an open disk B in R^2 and having the vector equation

$$\mathbf{R}(t) = f(t)\mathbf{i} + g(t)\mathbf{j} \qquad a \le t \le b$$

such that f' and g' are continuous on $[a, b]$. Let $\mathbf{F}$ be a vector field on B defined by

$$\mathbf{F}(x, y) = M(x, y)\mathbf{i} + N(x, y)\mathbf{j}$$

where M and N are continuous on B. Then the **line integral** of $M(x, y)\,dx + N(x, y)\,dy$ over C is given by

$$\int_C M(x, y)\,dx + N(x, y)\,dy$$

$$= \int_a^b \left[M(f(t), g(t))f'(t) + N(f(t), g(t))g'(t) \right] dt \qquad (13)$$

or, equivalently, by using vector notation, the **line integral** of $\mathbf{F}$ over C is given by

$$\int_C \mathbf{F} \cdot d\mathbf{R} = \int_C \mathbf{F}(\mathbf{R}(t)) \cdot \mathbf{R}'(t)\,dt \qquad (14)$$

Both the differential form notation of Eq. (13) and the vector notation of Eq. (14) are used for line integrals. Consquently, we shall apply the two notations.

▶ ILLUSTRATION 1   In Example 1, the integral in Eq. (12) that defines W is a line integral. With vector notation, this line integral can be denoted by

$$\int_C \mathbf{F} \cdot d\mathbf{R}$$

where $\mathbf{F}(x, y) = (x^2 + y^2)\mathbf{i} + 3x^2 y\mathbf{j}$ and $\mathbf{R}(t) = t\mathbf{i} + t^2\mathbf{j}$. With differential form notation this line integral is written as

$$\int_C (x^2 + y^2)\,dx + 3x^2 y\,dy \qquad (15)$$

◀

If an equation of C is of the form $y = F(x)$, then x may be used as a parameter in place of t. In a similar manner, if an equation of C is of the form $x = G(y)$, then y may be used as a parameter in place of t.

▶ ILLUSTRATION 2   In Example 1 and Illustration 1, the equation of C is $y = x^2$, which is of the form $y = F(x)$. Therefore we can use x as a parameter instead of t. Thus, in integral (15) of Illustration 1 we can replace y by x^2 and dy by $2x\,dx$, and we have

$$W = \int_{-1}^2 (x^2 + x^4)\,dx + 3x^2 x^2 (2x\,dx)$$

$$= \int_{-1}^2 (x^2 + x^4 + 6x^5)\,dx$$

This integral is the same as the third one appearing in the solution of Example 1, except that the variable is x instead of t. ◀

If the curve C in the definition of the line integral is the closed interval $[a, b]$ on the x axis, then $y = 0$ and $dy = 0$. Thus

$$\int_C M(x, y)\,dx + N(x, y)\,dy = \int_a^b M(x, 0)\,dx$$

Therefore, in such a case, the line integral reduces to a definite integral.

In the definition of a line integral, if C has the vector equation

$$\mathbf{R}(t) = f(t)\mathbf{i} + g(t)\mathbf{j} \qquad a \le t \le b$$

we required that f' and g' be continuous on $[a, b]$. If, in addition to this continuity requirement, $f'(t)$ and $g'(t)$ are not both zero at every point in the open interval (a, b), then C is said to be **smooth** on $[a, b]$. If an interval I can be partitioned into a finite number of subintervals on which C is smooth, then C is said to be **sectionally smooth** on I.

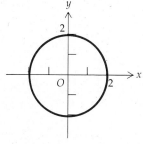

FIGURE 18.2.2

▶ ILLUSTRATION 3   (a) The curve defined by the vector equation

$$\mathbf{R}(t) = 2\cos t\mathbf{i} + 2\sin t\mathbf{j} \qquad 0 \le t \le 2\pi$$

is a circle with center at the origin and radius 2. See Fig. 18.2.2. For this curve,

$$f(t) = 2\cos t \qquad g(t) = 2\sin t$$
$$f'(t) = -2\sin t \qquad g'(t) = 2\cos t$$

Because f' and g' are continuous for all t and $f'(t)$ and $g'(t)$ are not both zero anywhere, the circle is a smooth curve.

(b) The curve defined by the vector equation

$$\mathbf{R}(t) = a(t - \sin t)\mathbf{i} + a(1 - \cos t)\mathbf{j} \qquad t \in (-\infty, +\infty)$$

is a cycloid. This curve was discussed in Sec. 13.3, and a sketch of its graph on the interval $[-2\pi a, 4\pi a]$ is shown in Fig. 18.2.3. For the cycloid,

$$f(t) = a(t - \sin t) \qquad g(t) = a(1 - \cos t)$$
$$f'(t) = a(1 - \cos t) \qquad g'(t) = a\sin t$$

The functions f' and g' are continuous for all t, but $f'(t)$ and $g'(t)$ are both zero if $t = 2\pi n$, where n is any integer. Therefore the cycloid is not smooth. However, the cycloid is sectionally smooth, because it is smooth in each subinterval $[2\pi n, 2\pi(n + 1)]$, where n is any integer.   ◀

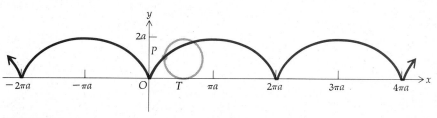

FIGURE 18.2.3

Figures 18.2.4 and 18.2.5 show two other sectionally smooth curves. We can extend the concept of a line integral to include curves that are sectionally smooth.

18.2.3 DEFINITION

Let the curve C consist of the smooth arcs $C_1, C_2, \ldots, C_n$. Then the **line integral** of $M(x, y)\,dx + N(x, y)\,dy$ over C is defined by

$$\int_C M(x, y)\,dx + N(x, y)\,dy = \sum_{i=1}^{n} \left(\int_{C_i} M(x, y)\,dx + N(x, y)\,dy \right)$$

or, equivalently, by using vector notation, the line integral of $\mathbf{F}$ over C is defined by

$$\int_C \mathbf{F} \cdot d\mathbf{R} = \sum_{i=1}^{n} \left(\int_{C_i} \mathbf{F}(\mathbf{R}(t)) \cdot \mathbf{R}'(t)\,dt \right)$$

FIGURE 18.2.4

FIGURE 18.2.5

FIGURE 18.2.6

EXAMPLE 2

Evaluate the line integral

$$\int_C 4xy\,dx + (2x^2 - 3xy)\,dy$$

if the curve C consists of the line segment from $(-3, -2)$ to $(1, 0)$ and the first quadrant arc of the circle $x^2 + y^2 = 1$ from $(1, 0)$ to $(0, 1)$, traversed in the counterclockwise direction.

SOLUTION

Figure 18.2.6 shows the curve C composed of arcs C_1 and C_2. The arc C_1 is the line segment. An equation of the line through $(-3, -2)$ and $(1, 0)$ is $x - 2y = 1$. Therefore C_1 can be represented parametrically by

$$x = 1 + 2t \qquad y = t \qquad -2 \le t \le 0$$

The arc C_2, which is the first quadrant arc of the circle $x^2 + y^2 = 1$, can be represented parametrically by

$$x = \cos t \qquad y = \sin t \qquad 0 \le t \le \tfrac{1}{2}\pi$$

Applying (13) of Definition 18.2.2 for each of the arcs C_1 and C_2 we have

$$\int_{C_1} 4xy\,dx + (2x^2 - 3xy)\,dy$$

$$= \int_{-2}^{0} 4(1 + 2t)t(2\,dt) + [2(1 + 2t)^2 - 3(1 + 2t)t]\,dt$$

$$= \int_{-2}^{0} (8t + 16t^2 + 2 + 8t + 8t^2 - 3t - 6t^2)\,dt$$

$$= \int_{-2}^{0} (18t^2 + 13t + 2)\,dt$$

$$= 6t^3 + \tfrac{13}{2}t^2 + 2t \Big]_{-2}^{0}$$

$$= -(-48 + 26 - 4)$$

$$= 26$$

and

$$\int_{C_2} 4xy \, dx + (2x^2 - 3xy) \, dy$$

$$= \int_0^{\pi/2} 4 \cos t \sin t(-\sin t \, dt) + [2 \cos^2 t - 3 \cos t \sin t](\cos t \, dt)$$

$$= \int_0^{\pi/2} (-4 \cos t \sin^2 t + 2 \cos^3 t - 3 \cos^2 t \sin t) \, dt$$

$$= \int_0^{\pi/2} \left[-4 \cos t \sin^2 t + 2 \cos t(1 - \sin^2 t) - 3 \cos^2 t \sin t \right] dt$$

$$= \int_0^{\pi/2} (2 \cos t - 6 \cos t \sin^2 t - 3 \cos^2 t \sin t) \, dt$$

$$= 2 \sin t - 2 \sin^3 t + \cos^3 t \Big]_0^{\pi/2}$$

$$= 2 - 2 - 1$$

$$= -1$$

Therefore, from Definition 18.2.3,

$$\int_C 4xy \, dx + (2x^2 - 3xy) \, dy = 26 + (-1)$$
$$= 25$$

We now extend the concept of a line integral to three dimensions.

18.2.4 DEFINITION   Let C be a curve lying in an open ball B in R^3 and having the vector equation

$$\mathbf{R}(t) = f(t)\mathbf{i} + g(t)\mathbf{j} + h(t)\mathbf{k} \qquad a \le t \le b$$

such that f', g', and h' are continuous on $[a, b]$. Let $\mathbf{F}$ be a vector field on B defined by

$$\mathbf{F}(x, y, z) = M(x, y, z)\mathbf{i} + N(x, y, z)\mathbf{j} + R(x, y, z)\mathbf{k}$$

where M, N, and R are continuous on B. Then the line integral of $M(x, y, z) \, dx + N(x, y, z) \, dy + R(x, y, z) \, dz$ over C is given by

$$\int_C M(x, y, z) \, dx + N(x, y, z) \, dy + R(x, y, z) \, dz =$$

$$\int_a^b \left[M(f(t), g(t), h(t))f'(t) + N(f(t), g(t), h(t))g'(t) + R(f(t), g(t), h(t))h'(t) \right] dt \quad (16)$$

or, equivalently, by using vector notation, the line integral of $\mathbf{F}$ over C is given by

$$\int_C \mathbf{F} \cdot d\mathbf{R} = \int_C \mathbf{F}(\mathbf{R}(t)) \cdot \mathbf{R}'(t) \, dt \qquad (17)$$

We can define the work done by a force field in moving a particle along a curve in R^3 as a line integral just as we did in Definition 18.2.1 for a curve in R^2. Such a definition is applied in the following example.

EXAMPLE 3

A particle traverses the twisted cubic

$$\mathbf{R}(t) = t\mathbf{i} + t^2\mathbf{j} + t^3\mathbf{k} \qquad 0 \le t \le 1$$

Find the total work done if the motion is caused by the force field

$$\mathbf{F}(x, y, z) = e^x\mathbf{i} + xe^z\mathbf{j} + x \sin \pi y^2 \mathbf{k}$$

Assume that the arc is measured in meters and the force is measured in newtons.

SOLUTION

$$\mathbf{R}(t) = t\mathbf{i} + t^2\mathbf{j} + t^3\mathbf{k} \qquad \mathbf{R}'(t) = \mathbf{i} + 2t\mathbf{j} + 3t^2\mathbf{k}$$

Because $\mathbf{F}(x, y, z) = \langle e^x, xe^z, x \sin \pi y^2 \rangle$, then

$$\mathbf{F}(\mathbf{R}(t)) = \mathbf{F}(t, t^2, t^3)$$
$$= \langle e^t, te^{t^3}, t \sin \pi t^4 \rangle$$

If W joules is the work done, then from the line integral (17),

$$W = \int_C \mathbf{F} \cdot d\mathbf{R}$$

$$= \int_0^1 \mathbf{F}(\mathbf{R}(t)) \cdot \mathbf{R}'(t) \, dt$$

$$= \int_0^1 \langle e^t, te^{t^3}, t \sin \pi t^4 \rangle \cdot \langle 1, 2t, 3t^2 \rangle \, dt$$

$$= \int_0^1 (e^t + 2t^2 e^{t^3} + 3t^3 \sin \pi t^4) \, dt$$

$$= e^t + \frac{2}{3}e^{t^3} - \frac{3}{4\pi} \cos \pi t^4 \Big]_0^1$$

$$= e + \frac{2}{3}e - \frac{3}{4\pi} \cos \pi - 1 - \frac{2}{3} + \frac{3}{4\pi} \cos 0$$

$$= \frac{5}{3}e + \frac{3}{2\pi} - \frac{5}{3}$$

Therefore the work done is $\left[\dfrac{5}{3}(e - 1) + \dfrac{3}{2\pi} \right]$ joules.

EXAMPLE 4

Evaluate the line integral

$$\int_C 3x \, dx + 2xy \, dy + z \, dz$$

if the curve C is the circular helix defined by the parametric equations

$$x = \cos t \qquad y = \sin t \qquad z = t \qquad 0 \le t \le 2\pi$$

SOLUTION
From (16),

$$\int_C 3x\,dx + 2xy\,dy + z\,dz$$

$$= \int_0^{2\pi} 3\cos t(-\sin t\,dt) + 2(\cos t)(\sin t)(\cos t\,dt) + t\,dt$$

$$= \int_0^{2\pi} (-3\sin t\cos t + 2\cos^2 t\sin t + t)\,dt$$

$$= -\tfrac{3}{2}\sin^2 t - \tfrac{2}{3}\cos^3 t + \tfrac{1}{2}t^2\Big]_0^{2\pi}$$

$$= -\tfrac{3}{2}(0) - \tfrac{2}{3}(1) + \tfrac{1}{2}(4\pi^2) + \tfrac{3}{2}(0) + \tfrac{2}{3}(1) + \tfrac{1}{2}(0)$$

$$= 2\pi^2$$

EXERCISES 18.2

In Exercises 1 through 22, evaluate the line integral over the given curve.

1. $\int_C \mathbf{F}\cdot d\mathbf{R}$; $\mathbf{F}(x,y) = y\mathbf{i} + x\mathbf{j}$; C: $\mathbf{R}(t) = t\mathbf{i} + t^2\mathbf{j}$, $0 \le t \le 1$.

2. $\int_C \mathbf{F}\cdot d\mathbf{R}$; $\mathbf{F}(x,y) = 2xy\mathbf{i} - 3x\mathbf{j}$; C: $\mathbf{R}(t) = 3t^2\mathbf{i} - t\mathbf{j}$, $0 \le t \le 1$.

3. $\int_C \mathbf{F}\cdot d\mathbf{R}$; $\mathbf{F}(x,y) = 2xy\mathbf{i} + (x - 2y)\mathbf{j}$; C: $\mathbf{R}(t) = \sin t\mathbf{i} - 2\cos t\mathbf{j}$, $0 \le t \le \pi$.

4. $\int_C \mathbf{F}\cdot d\mathbf{R}$; $\mathbf{F}(x,y) = xy\mathbf{i} - y^2\mathbf{j}$; C: $\mathbf{R}(t) = t^2\mathbf{i} + t^3\mathbf{j}$, from the point $(1,1)$ to the point $(4,-8)$.

5. $\int_C \mathbf{F}\cdot d\mathbf{R}$; $\mathbf{F}(x,y) = (x - y)\mathbf{i} + (y + x)\mathbf{j}$; C: the circle $x^2 + y^2 = 4$ from the point $(2,0)$ in the counterclockwise direction.

6. $\int_C \mathbf{F}\cdot d\mathbf{R}$; $\mathbf{F}(x,y) = (x - 2y)\mathbf{i} + xy\mathbf{j}$; C: $\mathbf{R}(t) = 3\cos t\mathbf{i} + 2\sin t\mathbf{j}$, $0 \le t \le \tfrac{1}{2}\pi$.

7. $\int_C \mathbf{F}\cdot d\mathbf{R}$; $\mathbf{F}(x,y) = y\sin x\mathbf{i} - \cos x\mathbf{j}$; C: the line segment from $(\tfrac{1}{2}\pi, 0)$ to $(\pi, 1)$.

8. $\int_C \mathbf{F}\cdot d\mathbf{R}$; $\mathbf{F}(x,y) = 9x^2y\mathbf{i} + (5x^2 - y)\mathbf{j}$; C: the curve $y = x^3 + 1$ from $(1,2)$ to $(3,28)$.

9. $\int_C (x^2 + xy)\,dx + (y^2 - xy)\,dy$; C: the line $y = x$ from the origin to the point $(2,2)$.

10. The line integral of Exercise 9; C: the parabola $x^2 = 2y$ from the origin to the point $(2,2)$.

11. The line integral of Exercise 9; C: the x axis from the origin to $(2,0)$ and then the line $x = 2$ from $(2,0)$ to $(2,2)$.

12. $\int_C yx^2\,dx + (x + y)\,dy$; C: the line $y = -x$ from the origin to the point $(1,-1)$.

13. The line integral of Exercise 12; C: the curve $y = -x^3$ from the origin to the point $(1,-1)$.

14. The line integral of Exercise 12; C: the y axis from the origin to $(0,-1)$ and then the line $y = -1$ from $(0,-1)$ to $(1,-1)$.

15. $\int_C 3xy\,dx + (4x^2 - 3y)\,dy$; C: the line $y = 2x + 3$ from $(0,3)$ to $(3,9)$ and then the parabola $y = x^2$ from $(3,9)$ to $(5,25)$.

16. $\int_C (xy - z)\,dx + e^x\,dy + y\,dz$; C: the line segment from $(1,0,0)$ to $(3,4,8)$.

17. $\int_C (x + y)\,dx + (y + z)\,dy + (x + z)\,dz$; C: the line segment from the origin to the point $(1, 2, 4)$.

18. The line integral of Exercise 16; C: $\mathbf{R}(t) = (t + 1)\mathbf{i} + t^2\mathbf{j} + t^3\mathbf{k}$, $0 \le t \le 2$.

19. $\int_C \mathbf{F} \cdot d\mathbf{R}$; $\mathbf{F}(x, y, z) = z\mathbf{i} + x\mathbf{j} + y\mathbf{k}$; C: the circular helix $\mathbf{R}(t) = a \cos t\mathbf{i} + a \sin t\mathbf{j} + t\mathbf{k}$; $0 \le t \le 2\pi$.

20. $\int_C \mathbf{F} \cdot d\mathbf{R}$; $\mathbf{F}(x, y, z) = 2xy\mathbf{i} + (6y^2 - xz)\mathbf{j} + 10z\mathbf{k}$; C: the twisted cubic $\mathbf{R}(t) = t\mathbf{i} + t^2\mathbf{j} + t^3\mathbf{k}$, $0 \le t \le 1$.

21. The line integral of Exercise 20; C: the line segment from the origin to the point $(0, 0, 1)$; then the line segment from $(0, 0, 1)$ to $(0, 1, 1)$; then the line segment from $(0, 1, 1)$ to $(1, 1, 1)$.

22. The line integral of Exercise 20; C: the line segment from the origin to the point $(1, 1, 1)$.

In Exercises 23 through 36, find the total work done in moving a particle along the given arc C if the motion is caused by the given force field. Assume the arc is measured in meters and the force is measured in newtons.

23. $\mathbf{F}(x, y) = 2xy\mathbf{i} + (x^2 + y^2)\mathbf{j}$; C: the line segment from the origin to the point $(1, 1)$.

24. The force field of Exercise 23; C: the arc of the parabola $y^2 = x$ from the origin to the point $(1, 1)$.

25. $\mathbf{F}(x, y) = (y - x)\mathbf{i} + x^2 y\mathbf{j}$; C: the line segment from the point $(1, 1)$ to $(2, 4)$.

26. The force field of Exercise 25; C: the arc of the parabola $y = x^2$ from the point $(1, 1)$ to $(2, 4)$.

27. The force field of Exercise 25; C: the line segment from $(1, 1)$ to $(2, 2)$ and then the line segment from $(2, 2)$ to $(2, 4)$.

28. $\mathbf{F}(x, y) = -x^2 y\mathbf{i} + 2y\mathbf{j}$; C: the line segment from $(a, 0)$ to $(0, a)$.

29. The force field of Exercise 28; C: $\mathbf{R}(t) = a \cos t\mathbf{i} + a \sin t\mathbf{j}$, $0 \le t \le \frac{1}{2}\pi$.

30. The force field of Exercise 28; C: the line segment from $(a, 0)$ to (a, a) and then the line segment from (a, a) to $(0, a)$.

31. $\mathbf{F}(x, y, z) = (y + z)\mathbf{i} + (x + z)\mathbf{j} + (x + y)\mathbf{k}$; C: the line segment from the origin to the point $(1, 1, 1)$.

32. $\mathbf{F}(x, y, z) = z^2\mathbf{i} + y^2\mathbf{j} + xz\mathbf{k}$; C: the line segment from the origin to the point $(4, 0, 3)$.

33. $\mathbf{F}(x, y, z) = e^x\mathbf{i} + e^y\mathbf{j} + e^z\mathbf{k}$; C: $\mathbf{R}(t) = t\mathbf{i} + t^2\mathbf{j} + t^3\mathbf{k}$, $0 \le t \le 2$.

34. $\mathbf{F}(x, y, z) = (xyz + x)\mathbf{i} + (x^2 z + y)\mathbf{j} + (x^2 y + z)\mathbf{k}$; C: the arc of Exercise 33.

35. The force field of Exercise 34; C: the line segment from the origin to the point $(1, 0, 0)$; then the line segment from $(1, 0, 0)$ to $(1, 1, 0)$; then the line segment from $(1, 1, 0)$ to $(1, 1, 1)$.

36. $\mathbf{F}(x, y, z) = x\mathbf{i} + y\mathbf{j} + (yz - x)\mathbf{k}$; C: $\mathbf{R}(t) = 2t\mathbf{i} + t^2\mathbf{j} + 4t^3\mathbf{k}$, $0 \le t \le 1$.

18.3
LINE INTEGRALS
INDEPENDENT OF
THE PATH

We learned in Section 18.2 that the value of a line integral is determined by the integrand and a curve C between two points P_1 and P_2. However, under certain conditions the value of a line integral depends only on the integrand and the points P_1 and P_2 and not on the path from P_1 to P_2. Such a line integral is said to be *independent of the path*.

▶ ILLUSTRATION 1   Suppose a force field

$$\mathbf{F}(x, y) = (y^2 + 2x + 4)\mathbf{i} + (2xy + 4y - 5)\mathbf{j}$$

moves a particle from the origin to the point $(1, 1)$. We show that the total work done is the same if the path is along (a) the line segment from the origin to $(1, 1)$; (b) the segment of the parabola $y = x^2$ from the origin to $(1, 1)$; and (c) the segment of the curve $x = y^3$ from the origin to $(1, 1)$.

If W is the measure of the work done, then

$$W = \int_C (y^2 + 2x + 4)\,dx + (2xy + 4y - 5)\,dy \tag{1}$$

FIGURE 18.3.1

(a) See Fig. 18.3.1. An equation of C is $y = x$. We use x as the parameter and let $y = x$ and $dy = dx$ in Eq. (1). Then

$$W = \int_0^1 (x^2 + 2x + 4)\,dx + (2x^2 + 4x - 5)\,dx$$

$$= \int_0^1 (3x^2 + 6x - 1)\,dx$$

$$= x^3 + 3x^2 - x\Big]_0^1$$

$$= 3$$

FIGURE 18.3.2

(b) See Fig. 18.3.2. An equation of C is $y = x^2$. Again taking x as the parameter and in Eq. (1) letting $y = x^2$ and $dy = 2x\,dx$, we have

$$W = \int_0^1 (x^4 + 2x + 4)\,dx + (2x^3 + 4x^2 - 5)2x\,dx$$

$$= \int_0^1 (5x^4 + 8x^3 - 8x + 4)\,dx$$

$$= x^5 + 2x^4 - 4x^2 + 4x\Big]_0^1$$

$$= 3$$

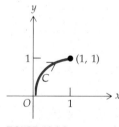

FIGURE 18.3.3

(c) See Fig. 18.3.3. An equation of C is $x = y^3$. We take y as the parameter and in Eq. (1) let $x = y^3$ and $dx = 3y^2\,dy$. Then

$$W = \int_0^1 (y^2 + 2y^3 + 4)3y^2\,dy + (2y^4 + 4y - 5)\,dy$$

$$= \int_0^1 (6y^5 + 5y^4 + 12y^2 + 4y - 5)\,dy$$

$$= y^6 + y^5 + 4y^3 + 2y^2 - 5y\Big]_0^1$$

$$= 3 \qquad\qquad\blacktriangleleft$$

In Illustration 1 we see that the value of the line integral is the same over three different paths from $(0, 0)$ to $(1, 1)$. Actually the value of the line integral is the same over any sectionally smooth curve from the origin to $(1, 1)$; so this line integral is independent of the path. (This fact is proved in Illustration 2, which follows Theorem 18.3.1.)

We now state and prove Theorem 18.3.1, which not only gives conditions for which the value of a line integral is independent of the path but also gives a formula for finding the value of such a line integral. The theorem is stated and proved with vector notation.

18.3.1 THEOREM   Let C be any sectionally smooth curve lying in an open disk B in R^2 from the point (x_1, y_1) to the point (x_2, y_2). If $\mathbf{F}$ is a conservative vector field continuous on B and ϕ is a potential function for $\mathbf{F}$, then the line integral

$$\int_C \mathbf{F} \cdot d\mathbf{R}$$

is independent of the path C, and

$$\int_C \mathbf{F} \cdot d\mathbf{R} = \phi(x_2, y_2) - \phi(x_1, y_1)$$

PROOF   We give the proof if C is smooth. If C is only sectionally smooth, then consider each piece separately; the following proof applies to each smooth piece.

Let parametric equations of C be

$$x = f(t) \qquad y = g(t) \qquad t_1 \leq t \leq t_2$$

Thus a vector equation of C is

$$\mathbf{R}(t) = f(t)\mathbf{i} + g(t)\mathbf{j} \qquad t_1 \leq t \leq t_2$$

Furthermore, the point (x_1, y_1) is $(f(t_1), g(t_1))$ and the point (x_2, y_2) is $(f(t_2), g(t_2))$. Because ϕ is a potential function for $\mathbf{F}$, $\nabla\phi(x, y) = \mathbf{F}(x, y)$ where $\mathbf{F}(x, y) = M(x, y)\mathbf{i} + N(x, y)\mathbf{j}$. Hence

$$\int_C \mathbf{F} \cdot d\mathbf{R} = \int_C \nabla\phi \cdot d\mathbf{R}$$

$$= \int_{t_1}^{t_2} \nabla\phi(\mathbf{R}(t)) \cdot \mathbf{R}'(t)\, dt$$

$$= \int_{t_1}^{t_2} \nabla\phi(f(t), g(t)) \cdot \mathbf{R}'(t)\, dt$$

$$= \int_{t_1}^{t_2} \langle M(f(t), g(t)), N(f(t), g(t)) \rangle \cdot \langle f'(t), g'(t) \rangle\, dt$$

$$= \int_{t_1}^{t_2} \big[M(f(t), g(t))f'(t)\, dt + N(f(t), g(t))g'(t)\, dt \big] \qquad (2)$$

Observe that because $M(x, y)\, dx + N(x, y)\, dy = d\phi(x, y)$, then

$$M(f(t), g(t))f'(t)\, dt + N(f(t), g(t))g'(t)\, dt = d\phi(f(t), g(t)) \qquad (3)$$

Substituting from (3) in (2) and then applying the second fundamental theorem of the calculus (Theorem 4.8.2) we get

$$\int_C \mathbf{F} \cdot d\mathbf{R} = \int_{t_1}^{t_2} d\phi(f(t), g(t))$$

$$= \phi(f(t), g(t)) \Big]_{t_1}^{t_2}$$

$$= \phi(f(t_2), g(t_2)) - \phi(f(t_1), g(t_1))$$

$$= \phi(x_2, y_2) - \phi(x_1, y_1)$$

which is what we wished to prove. ■

Recall from Section 16.5 that stating that a vector field $M(x, y)\mathbf{i} + N(x, y)\mathbf{j}$ is conservative is equivalent to stating that the differential form $M(x, y)\, dx + N(x, y)\, dy$ is exact. Thus we conclude from Theorem 18.3.1 that the line integral $\int_C M(x, y)\, dx + N(x, y)\, dy$ is independent of the path C if the integrand is an exact differential.

Because of the resemblance of Theorem 18.3.1 to the second fundamental theorem of the calculus, it is sometimes called the *fundamental theorem for line integrals*.

▶ ILLUSTRATION 2   We use Theorem 18.3.1 to evaluate the line integral in Illustration 1. The line integral is

$$\int_C (y^2 + 2x + 4)\, dx + (2xy + 4y - 5)\, dy \qquad (4)$$

With vector notation this line integral is

$$\int_C \mathbf{F} \cdot d\mathbf{R}$$

where

$$\mathbf{F}(x, y) = (y^2 + 2x + 4)\mathbf{i} + (2xy + 4y - 5)\mathbf{j}$$

In Illustration 1 of Section 18.1 we showed that $\mathbf{F}$ is a conservative vector field having the potential function

$$\phi(x, y) = y^2x + x^2 + 4x + 2y^2 - 5y$$

Therefore, from Theorem 18.3.1, the line integral is independent of the path, and C can be any sectionally smooth curve from $(0, 0)$ to $(1, 1)$. Furthermore, from Theorem 18.3.1,

$$\int_C (y^2 + 2x + 4)\,dx + (2xy + 4y - 5)\,dy = \phi(1, 1) - \phi(0, 0)$$
$$= 3 - 0$$
$$= 3$$

This result agrees with that of Illustration 1.

Observe that the integrand of line integral (4) is an exact differential because

$$d\phi(x, y) = (y^2 + 2x + 4)\,dx + (2xy + 4y - 5)\,dy \qquad \blacktriangleleft$$

EXAMPLE 1
Evaluate the line integral

$$\int_C \mathbf{F} \cdot d\mathbf{R}$$

if $\mathbf{F}(x, y) = (e^{-y} - 2x)\mathbf{i} - (xe^{-y} + \sin y)\mathbf{j}$ and C is the first quadrant arc of the circle

$$\mathbf{R}(t) = \pi \cos t\,\mathbf{i} + \pi \sin t\,\mathbf{j} \qquad 0 \le t \le \tfrac{1}{2}\pi$$

SOLUTION
In Example 2 of Section 16.5 we showed that

$$\mathbf{V}(xe^{-y} - x^2 + \cos y) = (e^{-y} - 2x)\mathbf{i} - (xe^{-y} + \sin y)\mathbf{j}$$

Therefore $\mathbf{F}$ is a conservative vector field, and we apply Theorem 18.3.1 with $\phi(x, y) = xe^{-y} - x^2 + \cos y$. The point where $t = 0$ is $(\pi, 0)$ and the point where $t = \tfrac{1}{2}\pi$ is $(0, \pi)$.

$$\int_C \mathbf{F} \cdot d\mathbf{R} = \phi(0, \pi) - \phi(\pi, 0)$$
$$= \cos \pi - (\pi - \pi^2 + 1)$$
$$= \pi^2 - \pi - 2$$

If the value of a line integral is independent of the path, it is not necessary to find a potential function ϕ. We show the procedure in the next example.

EXAMPLE 2

If $\mathbf{F}(x, y) = \dfrac{1}{y}\mathbf{i} - \dfrac{x}{y^2}\mathbf{j}$ and C is any sectionally smooth curve from the point $(5, -1)$ to the point $(9, -3)$, show that the value of the line integral

$$\int_C \mathbf{F} \cdot d\mathbf{R}$$

is independent of the path, and evaluate it.

SOLUTION

We showed in Example 2 of Section 18.1 that $\mathbf{F}$ is a conservative vector field. Therefore the line integral is independent of the path. We take for the path the line segment from $(5, -1)$ to $(9, -3)$. An equation of the line is $x + 2y = 3$. By letting $y = -t$ and $x = 3 + 2t$, a vector equation of the line is

$$\mathbf{R}(t) = (3 + 2t)\mathbf{i} - t\mathbf{j} \qquad 1 \leq t \leq 3$$

We compute the value of the line integral by applying Definition 18.2.2.

$$\int_C \mathbf{F} \cdot d\mathbf{R} = \int_C \mathbf{F}(\mathbf{R}(t)) \cdot \mathbf{R}'(t)\, dt$$

$$= \int_1^3 \mathbf{F}(3 + 2t, -t) \cdot \langle 2, -1 \rangle\, dt$$

$$= \int_1^3 \left\langle -\frac{1}{t}, -\frac{3 + 2t}{t^2} \right\rangle \cdot \langle 2, -1 \rangle\, dt$$

$$= \int_1^3 \left(-\frac{2}{t} + \frac{3 + 2t}{t^2} \right) dt$$

$$= \int_1^3 \frac{3}{t^2}\, dt$$

$$= -\frac{3}{t} \Big]_1^3$$

$$= 2$$

▶ ILLUSTRATION 3   In Example 2 of Section 18.1 we obtained the following potential function for the conservative vector field $\mathbf{F}(x, y) = \dfrac{1}{y}\mathbf{i} - \dfrac{x}{y^2}\mathbf{j}$:

$$\phi(x, y) = \frac{x}{y} + K$$

where K is an arbitrary constant. Thus if Theorem 18.3.1 is used to evaluate the line integral of Example 2 above, we have

$$\int_C \mathbf{F} \cdot d\mathbf{R} = \phi(9, -3) - \phi(5, -1)$$

$$= \left(\frac{9}{-3} + K \right) - \left(\frac{5}{-1} + K \right)$$

$$= 2$$

which agrees with our previous result.   ◀

FIGURE 18.3.4

Observe in the above illustration that the arbitrary constant K does not appear because $K - K = 0$. Hereafter when applying Theorem 18.3.1 we shall omit the arbitrary constant for the potential function; what we are doing then is choosing as our potential function the one for which $K = 0$.

If for a curve C defined by the vector equation

$$\mathbf{R}(t) = f(t)\mathbf{i} + g(t)\mathbf{j} \qquad a \le t \le b$$

the initial point $A(f(a), g(a))$ and the point $B(f(b), g(b))$ coincide, then curve C is said to be **closed**. Figure 18.3.4 shows a closed curve where the points A and B coincide. The following theorem regarding the line integral of a conservative vector field over a sectionally smooth closed curve follows immediately from Theorem 18.3.1.

18.3.2 THEOREM If C is any sectionally smooth closed curve lying in some open disk B in R^2 and $\mathbf{F}$ is a conservative vector field on B, then

$$\int_C \mathbf{F} \cdot d\mathbf{R} = 0$$

PROOF We apply Theorem 18.3.1, and because C is closed, the point (x_1, y_1) coincides with the point (x_2, y_2). Therefore

$$\int_C \mathbf{F} \cdot d\mathbf{R} = \phi(x_2, y_2) - \phi(x_1, y_1)$$

$$= 0 \qquad \blacksquare$$

EXAMPLE 3

A particle moves on the circle

$$\mathbf{R}(t) = 2 \cos t\mathbf{i} + 2 \sin t\mathbf{j} \qquad 0 \le t \le 2\pi$$

Find the total work done if the motion is caused by the force field

$$\mathbf{F}(x, y) = \left(4 \ln 3y + \frac{1}{x} \right)\mathbf{i} + \frac{4x}{y}\mathbf{j}$$

SOLUTION
Let

$$M(x, y) = 4 \ln 3y + \frac{1}{x} \qquad N(x, y) = \frac{4x}{y}$$

$$M_y(x, y) = \frac{4}{y} \qquad N_x(x, y) = \frac{4}{y}$$

Because $M_y(x, y) = N_x(x, y)$, $\mathbf{F}$ is conservative. Furthermore, C is a closed curve. Therefore, if W is the measure of the work done, we have, from Theorem 18.3.2,

$$W = \int_C \mathbf{F} \cdot d\mathbf{R}$$

$$= 0$$

We now extend our discussion to functions of three variables. The statement of the following theorem and its proof are analogous to Theorem 18.3.1. The proof is left as an exercise (see Exercise 32).

18.3.3 THEOREM   Let C be any sectionally smooth curve lying in an open ball B in R^3 from the point (x_1, y_1, z_1) to (x_2, y_2, z_2). If $\mathbf{F}$ is a conservative vector field continuous on B and ϕ is a potential function for $\mathbf{F}$, then the line integral

$$\int_C \mathbf{F} \cdot d\mathbf{R}$$

is independent of the path C, and

$$\int_C \mathbf{F} \cdot d\mathbf{R} = \phi(x_2, y_2, z_2) - \phi(x_1, y_1, z_1)$$

▶ ILLUSTRATION 4   In Example 5 of Sec. 16.5 we showed that the vector field defined by

$$\mathbf{F}(x, y, z) = (e^x \sin z + 2yz)\mathbf{i} + (2xz + 2y)\mathbf{j} + (e^x \cos z + 2xy + 3z^2)\mathbf{k}$$

is a gradient $\nabla f(x, y, z)$, and

$$f(x, y, z) = e^x \sin z + 2xyz + y^2 + z^3$$

Thus $\mathbf{F}$ is a conservative vector field. Therefore, if C is any sectionally smooth curve from $(0, 0, 0)$ to $(1, -2, \pi)$, it follows from Theorem 18.3.3 that the line integral

$$\int_C \mathbf{F} \cdot d\mathbf{R}$$

is independent of the path and its value is

$$f(1, -2, \pi) - f(0, 0, 0) = (e \sin \pi - 4\pi + 4 + \pi^3) - 0$$
$$= \pi^3 - 4\pi + 4 \qquad ◀$$

EXAMPLE 4
Show that the line integral

$$\int_C (4x + 2y - z)\,dx + (2x - 2y + z)\,dy + (-x + y + 2z)\,dz$$

is independent of the path, and evaluate the integral if C is any sectionally smooth curve from $(4, -2, 1)$ to $(-1, 2, 0)$.

SOLUTION
The line integral is independent of the path if the integrand is an exact differential. Let

$M(x, y, z) = 4x + 2y - z$	$N(x, y, z) = 2x - 2y + z$	$R(x, y, z) = -x + y + 2z$
$M_y(x, y, z) = 2$	$N_x(x, y, z) = 2$	$R_x(x, y, z) = -1$
$M_z(x, y, z) = -1$	$N_z(x, y, z) = 1$	$R_y(x, y, z) = 1$

Because

$$M_y(x, y, z) = N_x(x, y, z) \qquad M_z(x, y, z) = R_x(x, y, z) \qquad N_z(x, y, z) = R_y(x, y, z)$$

the integrand is an exact differential and the line integral is independent of the path. We take for the path the line segment from $(4, -2, 1)$ to $(-1, 2, 0)$.

A set of direction numbers of this line is $[5, -4, 1]$. Therefore equations of the line are

$$\frac{x+1}{5} = \frac{y-2}{-4} = \frac{z}{1}$$

Parametric equations of the line segment are

$$x = -5t - 1 \qquad y = 4t + 2 \qquad z = -t \qquad -1 \le t \le 0$$

Therefore

$$\int_C (4x + 2y - z)\,dx + (2x - 2y + z)\,dy + (-x + y + 2z)\,dz$$

$$= \int_{-1}^0 [4(-5t-1) + 2(4t+2) - (-t)](-5\,dt)$$

$$+ \int_{-1}^0 [2(-5t-1) - 2(4t+2) + (-t)](4\,dt)$$

$$+ \int_{-1}^0 [-(-5t-1) + (4t+2) + 2(-t)](-dt)$$

$$= \int_{-1}^0 (-28t - 27)\,dt$$

$$= -14t^2 - 27t \Big]_{-1}^0$$

$$= -13$$

The line integral in Example 4 can be computed by finding a potential function for the conservative vector field $(4x + 2y - z)\mathbf{i} + (2x - 2y + z)\mathbf{j} + (-x + y + 2z)\mathbf{k}$ and applying Theorem 18.3.3. You are asked to do this in Exercise 31.

EXAMPLE 5
Suppose $\mathbf{F}$ is the gravitational force field exerted by a particle of mass M units at the origin on a particle of mass 1 unit at the point $P(x, y, z)$. Then from Section 18.1,

$$\mathbf{F}(x, y, z) = \frac{-GM}{(x^2 + y^2 + z^2)^{3/2}} (x\mathbf{i} + y\mathbf{j} + z\mathbf{k})$$

Find the work done by the force $\mathbf{F}$ in moving a particle of mass 1 unit along a smooth curve C from $(0, 3, 4)$ to $(2, 2, 1)$.

SOLUTION
In Illustration 2 of Section 18.1 we showed that $\mathbf{F}$ is conservative and that a potential function for $\mathbf{F}$ is given by

$$\phi(x, y, z) = \frac{GM}{\sqrt{x^2 + y^2 + z^2}}$$

If W is the measure of the work done in moving a particle of mass 1 unit along C,

$$W = \int_C \mathbf{F} \cdot d\mathbf{R}$$

From Theorem 18.3.3, the line integral is independent of the path, and

$$W = \phi(2, 2, 1) - \phi(0, 3, 4)$$

$$= \frac{GM}{\sqrt{2^2 + 2^2 + 1^2}} - \frac{GM}{\sqrt{0^2 + 3^2 + 4^2}}$$

$$= \frac{GM}{3} - \frac{GM}{5}$$

$$= \tfrac{2}{15}GM$$

We now show how the results of this section lead to an important conclusion in physics. If the motion of a particle is caused by a conservative force field $\mathbf{F}$, the **potential energy** of the particle at the point (x, y, z) is defined to be a scalar field E such that

$$\mathbf{F}(x, y, z) = -\nabla E(x, y, z)$$

That is, $-E$ is a potential function of $\mathbf{F}$. We shall use the notation $E(P)$ to denote the potential energy of the particle at the point P. If W is the measure of the work done by $\mathbf{F}$ in moving a particle along a sectionally smooth curve C from point A to point B, then from Theorem 18.3.3,

$$W = \int_C \mathbf{F} \cdot d\mathbf{R}$$

$$W = -E(x, y, z)\Big]_A^B$$

$$W = -[E(B) - E(A)]$$

$$W = E(A) - E(B) \tag{5}$$

Thus W is the difference in the potential energies of the particle at A and B.

Now suppose that the particle is at point A at time t_1 and at point B at time t_2 and the curve C has the vector equation

$$\mathbf{R}(t) = f(t)\mathbf{i} + g(t)\mathbf{j} + h(t)\mathbf{k} \qquad t_1 \leq t \leq t_2$$

Then the velocity and acceleration vectors at t are $\mathbf{V}(t)$ and $\mathbf{A}(t)$ defined by

$$\mathbf{V}(t) = \mathbf{R}'(t) \qquad \mathbf{A}(t) = \mathbf{V}'(t)$$

The speed of the particle at t is denoted by $v(t)$, where $v(t) = \|\mathbf{V}(t)\|$. Then another formula for computing W is given by

$$W = \int_C \mathbf{F} \cdot d\mathbf{R}$$

$$W = \int_{t_1}^{t_2} \mathbf{F}(\mathbf{R}(t)) \cdot \mathbf{R}'(t)\, dt$$

$$W = \int_{t_1}^{t_2} \mathbf{F}(\mathbf{R}(t)) \cdot \mathbf{V}(t)\, dt \tag{6}$$

Newton's second law of motion states that if a force $\mathbf{F}$ is acting on a particle of mass m units, then

$$\mathbf{F}(\mathbf{R}(t)) = m\mathbf{A}(t)$$

$$\Leftrightarrow \quad \mathbf{F}(\mathbf{R}(t)) = m\mathbf{V}'(t) \tag{7}$$

Substituting from (7) in (6) we have

$$W = \int_{t_1}^{t_2} m[\mathbf{V}'(t) \cdot \mathbf{V}(t)] \, dt \tag{8}$$

Because $D_t[\mathbf{V}(t) \cdot \mathbf{V}(t)] = 2\mathbf{V}'(t) \cdot \mathbf{V}(t)$ and $\mathbf{V}(t) \cdot \mathbf{V}(t) = [v(t)]^2$, we get, from (8),

$$W = \tfrac{1}{2}m \int_{t_1}^{t_2} D_t[\mathbf{V}(t) \cdot \mathbf{V}(t)] \, dt$$

$$W = \tfrac{1}{2}m \int_{t_1}^{t_2} D_t[v(t)]^2 \, dt$$

$$W = \tfrac{1}{2}m[v(t)]^2 \Big|_{t_1}^{t_2}$$

$$W = \tfrac{1}{2}m[v(t_2)]^2 - \tfrac{1}{2}m[v(t_1)]^2 \tag{9}$$

In physics the **kinetic energy** of a particle is defined to be $\tfrac{1}{2}mv^2$. Therefore Eq. (9) states that the work done in moving a particle along C from point A to point B is the change in kinetic energy of the particle. If we use the notation $K(P)$ to indicate the kinetic energy of a particle at point P, Eq. (9) can be written as

$$W = K(B) - K(A) \tag{10}$$

From Eqs. (5) and (10) we have

$$E(A) - E(B) = K(B) - K(A)$$

$$E(A) + K(A) = E(B) + K(B) \tag{11}$$

This equation states that the sums of the potential and kinetic energies are equal at the initial point A and the terminal point B. Because A and B can be any points on C, the sum of the two energies is constant along C; that is, the total energy of the particle remains unchanged during the motion. This fact is a major concept in physics called the *Law of Conservation of Energy*. It is for this reason that the terminology *conservative* is used for a force field that is a gradient.

EXERCISES 18.3

In Exercises 1 through 12, use the results of the indicated exercises in Exercises 18.1 to prove that the value of the given line integral is independent of the path. Then evaluate the line integral by applying either Theorem 18.3.1 or 18.3.3 and using the potential function found in the indicated exercise. In each exercise C is any sectionally smooth curve from the point A to the point B.

1. $\int_C y \, dx + x \, dy$; A is $(1, 4)$ and B is $(3, 2)$; Exercise 21.

2. $\int_C x \, dx + y \, dy$; A is $(-5, 2)$ and B is $(1, 3)$; Exercise 22.

3. $\int_C e^x \sin y \, dx + e^x \cos y \, dy$; A is $(0,0)$ and B is $(2, \frac{1}{2}\pi)$; Exercise 23.

4. $\int_C (\sin y \sinh x + \cos y \cosh x) \, dx + (\cos y \cosh x - \sin y \sinh x) \, dy$; A is $(1,0)$ and B is $(2,\pi)$; Exercise 24.

5. $\int_C (2xy^2 - y^3) \, dx + (2x^2y - 3xy^2 + 2) \, dy$; A is $(-3, -1)$ and B is $(1,2)$; Exercise 25.

6. $\int_C (3x^2 + 2y - y^2e^x) \, dx + (2x - 2ye^x) \, dy$; A is $(0,2)$ and B is $(1, -3)$; Exercise 26.

7. $\int_C (x^2 - y) \, dx - (x - 3z) \, dy + (z + 3y) \, dz$; A is $(-3,1,2)$ and B is $(3,0,4)$; Exercise 27.

8. $\int_C yz \, dx + xz \, dy + xy \, dz$; A is $(0, -2, 5)$ and B is $(4, 1, -3)$; Exercise 28.

9. $\int_C (ze^x + e^y) \, dx + (xe^y - e^z) \, dy + (-ye^z + e^x) \, dz$; A is $(1,0,2)$ and B is $(0,2,1)$; Exercise 29.

10. $\int_C (\tan y + 2xy \sec z) \, dx + (x \sec^2 y + x^2 \sec z) \, dy + \sec z(x^2y \tan z - \sec z) \, dz$; A is $(2, \frac{1}{4}\pi, 0)$ and B is $(3, \pi, \pi)$; Exercise 30.

11. $\int_C (2x \cos y - 3) \, dx - (x^2 \sin y + z^2) \, dy - (2yz - 2) \, dz$; A is $(-1,0,3)$ and B is $(1, \pi, 0)$; Exercise 31.

12. $\int_C (2y^3 - 8xz^2) \, dx + (6xy^2 + 1) \, dy - (8x^2z + 3z^2) \, dz$; A is $(2,0,0)$ and B is $(3,2,1)$; Exercise 32.

In Exercises 13 through 20, show that the value of the line integral $\int_C \mathbf{F} \cdot d\mathbf{R}$ for the given $\mathbf{F}$ and C is independent of the path, and evaluate the line integral.

13. $\mathbf{F}(x, y) = 2(x - y)\mathbf{i} + 2(3y - x)\mathbf{j}$; C is the first quadrant arc of the circle $x^2 + y^2 = 9$ from the point on the x axis to the point on the y axis.

14. $\mathbf{F}(x, y) = (3x^2 + 6xy - 2y^2)\mathbf{i} + (3x^2 - 4xy + 3y^2)\mathbf{j}$; C is the first quadrant arc of the ellipse $4x^2 + 9y^2 = 36$ from the point on the x axis to the point on the y axis.

15. $\mathbf{F}(x, y) = (4e^{2x} - 3e^xe^y)\mathbf{i} + (2e^{2y} - 3e^xe^y)\mathbf{j}$; C is the arc of the parabola $y^2 = 4x$ from the vertex to the endpoint of the latus rectum in the first quadrant.

16. $\mathbf{F}(x, y) = e^x \cos y\mathbf{i} - e^x \sin y\mathbf{j}$; C is the segment of the line $3x + 4y = 12$ from the point where it intersects the y axis to the point where it intersects the x axis.

17. $\mathbf{F}(x, y, z) = 2x\mathbf{i} + 3y^2\mathbf{j} + \mathbf{k}$; C is the trace of the ellipsoid $4x^2 + 4y^2 + z^2 = 9$ in the xz plane from the positive x axis to the positive z axis.

18. $\mathbf{F}(x, y, z) = (2xy + z^2)\mathbf{i} + (x^2 - 2yz)\mathbf{j} + (2xz - y^2)\mathbf{k}$; C is the trace of the sphere $x^2 + y^2 + z^2 = 1$ in the yz plane from the positive y axis to the positive z axis.

19. $\mathbf{F}(x, y, z) = 2ye^{2x}\mathbf{i} + e^{2x}\mathbf{j} + 3z^2\mathbf{k}$; C is any sectionally smooth curve from the point $(\ln 2, 1, 1)$ to the point $(\ln 2, 2, 2)$.

20. $\mathbf{F}(x, y, z) = \left(\frac{1}{z} - \frac{y}{x^2}\right)\mathbf{i} + \left(\frac{1}{x} + \frac{z}{y^2}\right)\mathbf{j} - \left(\frac{1}{y} + \frac{x}{z^2}\right)\mathbf{k}$; C is any sectionally smooth curve from the point $(1, 2, -1)$ to the point $(2, -4, -2)$.

In Exercises 21 through 30, show that the value of the line integral is independent of the path, and compute the value in any convenient manner. In each exercise C is any sectionally smooth curve from the point A to the point B.

21. $\int_C (2y - x) \, dx + (y^2 + 2x) \, dy$; A is $(0, -1)$ and B is $(1,2)$.

22. $\int_C (\ln x + 2y) \, dx + (e^y + 2x) \, dy$; A is $(3, 1)$ and B is $(1, 3)$.

23. $\int_C \tan y \, dx + x \sec^2 y \, dy$; A is $(-2, 0)$ and B is $(4, \frac{1}{4}\pi)$.

24. $\int_C \sin y \, dx + (\sin y + x \cos y) \, dy$; A is $(-2, 0)$ and B is $(2, \frac{1}{6}\pi)$.

25. $\displaystyle\int_C \frac{2y}{(xy+1)^2}\,dx + \frac{2x}{(xy+1)^2}\,dy;$ A is $(0,2)$ and B is $(1,0)$.

26. $\displaystyle\int_C \frac{x}{x^2+y^2+z^2}\,dx + \frac{y}{x^2+y^2+z^2}\,dy + \frac{z}{x^2+y^2+z^2}\,dz;$ A is $(1,0,0)$ and B is $(1,2,3)$.

27. $\displaystyle\int_C (y+z)\,dx + (x+z)\,dy + (x+y)\,dz;$ A is $(0,0,0)$ and B is $(1,1,1)$.

28. $\displaystyle\int_C (yz+x)\,dx + (xz+y)\,dy + (xy+z)\,dz;$ A is $(0,0,0)$ and B is $(1,1,1)$.

29. $\displaystyle\int_C (e^x \sin y + yz)\,dx + (e^x \cos y + z \sin y + xz)\,dy + (xy - \cos y)\,dz;$ A is $(2,0,1)$ and B is $(0,\pi,3)$.

30. $\displaystyle\int_C (2x \ln yz - 5ye^x)\,dx - (5e^x - x^2y^{-1})\,dy + (x^2z^{-1} + 2z)\,dz;$ A is $(2,1,1)$ and B is $(3,1,e)$.

31. Evaluate the line integral in Example 4 by finding a potential function for the conservative vector field $(4x + 2y - z)\mathbf{i} + (2x - 2y + z)\mathbf{j} + (-x + y + 2z)\mathbf{k}$ and applying Theorem 18.3.3.

32. Prove Theorem 18.3.3.

In Exercises 33 through 36, find the total work done in moving a particle along the given arc C if the motion is caused by the given force field. Assume the arc is measured in meters and the force is measured in newtons. (*Hint:* First show that $\mathbf{F}$ is conservative.)

33. $\mathbf{F}(x,y) = 3(x+y)^2\mathbf{i} + 3(x+y)^2\mathbf{j};$ C: the arc of the parabola $y = x^2$ from the vertex to the point $(2,4)$.

34. $\mathbf{F}(x,y) = (2xy - 5y + 2y^2)\mathbf{i} + (x^2 - 5x + 4xy)\mathbf{j};$ C: the quarter circle $\mathbf{R}(t) = 2\cos t\mathbf{i} + 2\sin t\mathbf{j},$ $0 \le t \le \frac{1}{2}\pi.$

35. $\mathbf{F}(x,y,z) = 2y^2z^3\mathbf{i} + 4xyz^3\mathbf{j} + 6xy^2z^2\mathbf{k};$ C: the arc of the twisted cubic $\mathbf{R}(t) = t\mathbf{i} + t^2\mathbf{j} + t^3\mathbf{k}$ from $t = 1$ to $t = 2$.

36. $\mathbf{F}(x,y,z) = 4y^2z\mathbf{i} + 8xyz\mathbf{j} + 4(3z^3 + xy^2)\mathbf{k};$ C: the arc of the circular helix $\mathbf{R}(t) = 3\cos t\mathbf{i} + 3\sin t\mathbf{j} + \mathbf{k}$ from $t = 0$ to $t = \frac{1}{3}\pi.$

37. If $\mathbf{F}$ is the inverse-square force field defined by

$$\mathbf{F}(x,y,z) = \frac{k(x\mathbf{i} + y\mathbf{j} + z\mathbf{k})}{(x^2+y^2+z^2)^{3/2}}$$

find the work done by $\mathbf{F}$ in moving a particle along the segment of a straight line from the point $(3,0,0)$ to $(3,0,4)$. Evaluate the line integral by two methods: (a) Use a potential function for $\mathbf{F}$; (b) do not use a potential function for $\mathbf{F}$.

18.4
GREEN'S THEOREM

There is a theorem that expresses a double integral over a plane region R in terms of a line integral around a curve that is a boundary of R. This theorem is named after the English mathematician and physicist George Green (1793–1841), who introduced the theorem in a paper on the applications of mathematics to electricity and magnetism. Before stating the theorem, it is necessary to review and introduce some terminology pertaining to plane curves. A *smooth* curve was defined in Section 18.2 and a *closed* curve was defined in Section 18.3. A curve C is said to be *simple* if it does not intersect itself. That is, if a vector equation of C is

$$\mathbf{R}(t) = f(t)\mathbf{i} + g(t)\mathbf{j}$$

and if A is the point $(f(a), g(a))$ and B is the point $(f(b), g(b))$, then C is **simple** between A and B if $(f(t_1), g(t_1))$ is not the same point as $(f(t_2), g(t_2))$ for all t_1 and t_2 in the open interval (a,b).

The circle and the ellipse are examples of smooth simple closed curves. In Fig. 18.4.1(a)–(d) there are further examples of smooth curves that may or may not be simple and closed. In (a) the curve is both simple and closed; in (b) the curve is simple but not closed; in (c) the curve is closed but not simple; and in (d) the curve is neither simple nor closed.

In the statement of Green's theorem we refer to a line integral around a sectionally smooth simple closed curve C that forms the boundary of a region R in the plane, and the direction along C is counterclockwise. In Fig. 18.4.2 such a region R is shown with the required boundary curve C. The line integral around C in the counterclockwise direction is denoted by $\oint_C$.

18.4.1 THEOREM
Green's Theorem

Let M and N be functions of two variables x and y such that they have continuous first partial derivatives on an open disk B in R^2. If C is a sectionally smooth simple closed curve lying entirely in B, and if R is the region bounded by C, then

$$\oint_C M(x, y)\, dx + N(x, y)\, dy = \iint\limits_R \left(\frac{\partial N}{\partial x} - \frac{\partial M}{\partial y} \right) dA \tag{1}$$

The proof of Green's theorem for all regions bounded by curves that are sectionally smooth, simple, and closed belongs to a course in advanced calculus. However, we shall prove the theorem for a particular kind of region, one for which each horizontal line and each vertical line intersect it in at most two points. The proof follows.

PROOF Let R be a region in the xy plane that can be defined by either

$$R = \{(x, y) \mid a \le x \le b, f_1(x) \le y \le f_2(x)\} \tag{2}$$

or

$$R = \{(x, y) \mid c \le y \le d, g_1(y) \le x \le g_2(y)\} \tag{3}$$

where the functions f_1, f_2, g_1, and g_2 are smooth. Figure 18.4.3 shows such

$A = B$

(a) Simple and closed

B

A

(b) Simple but not closed

$A = B$

(c) Closed but not simple

B

A

(d) Neither simple nor closed

FIGURE 18.4.1

R

C

FIGURE 18.4.2

FIGURE 18.4.3

FIGURE 18.4.4

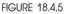

FIGURE 18.4.5

a region R. Region R is regarded as being defined by (2) in Fig. 18.4.4 and by (3) in Fig. 18.4.5.

The proof of (1) will consist of showing that

$$\oint_C M(x, y)\, dx = -\iint_R \frac{\partial M}{\partial y}\, dA \tag{4}$$

and

$$\oint_C N(x, y)\, dy = \iint_R \frac{\partial N}{\partial x}\, dA \tag{5}$$

To prove (4) we treat R as a region defined by (2). Refer to Fig. 18.4.4. Let C_1 be the graph of $y = f_1(x)$ from $x = a$ to $x = b$; that is, C_1 is the lower part of the boundary curve C going from left to right. Let C_2 be the graph of $y = f_2(x)$ from $x = b$ to $x = a$; that is, C_2 is the upper part of the boundary curve C going from right to left. Consider the line integral $\oint_C M(x, y)\, dx$.

$$\oint_C M(x, y)\, dx = \int_{C_1} M(x, y)\, dx + \int_{C_2} M(x, y)\, dx$$

$$= \int_a^b M(x, f_1(x))\, dx + \int_b^a M(x, f_2(x))\, dx$$

$$= \int_a^b M(x, f_1(x))\, dx - \int_a^b M(x, f_2(x))\, dx$$

$$= \int_a^b \left[M(x, f_1(x)) - M(x, f_2(x)) \right]\, dx \tag{6}$$

We now deal with the double integral $\iint_R \dfrac{\partial M}{\partial y}\, dA$, where R is still considered to be defined by (2). Then

$$\iint_R \frac{\partial M}{\partial y}\, dA = \int_a^b \int_{f_1(x)}^{f_2(x)} \frac{\partial M}{\partial y}\, dy\, dx$$

$$= \int_a^b \left(\int_{f_1(x)}^{f_2(x)} \frac{\partial M}{\partial y}\, dy \right) dx$$

$$= \int_a^b M(x, y) \Big]_{f_1(x)}^{f_2(x)}\, dx$$

$$= \int_a^b \left[M(x, f_2(x)) - M(x, f_1(x)) \right]\, dx \tag{7}$$

By comparing (6) and (7) it follows that (4) holds.

To prove (5), R is regarded as a region defined by (3), as in Fig. 18.4.5. The details of the proof are left as an exercise (see Exercise 43).

By adding corresponding members of Eqs. (4) and (5) we obtain Green's theorem for this region R. ∎

▶ ILLUSTRATION 1   We use Green's theorem to evaluate the line integral $\oint_C y^2\, dx + 4xy\, dy$, where C is the closed curve consisting of the arc of

FIGURE 18.4.6

the parabola $y = x^2$ from the origin to the point $(2, 4)$ and the line segment from $(2, 4)$ to the origin. The region R with the boundary C is shown in Fig. 18.4.6. From Green's theorem,

$$\oint_C y^2\, dx + 4xy\, dy = \iint_R \left[\frac{\partial}{\partial x}(4xy) - \frac{\partial}{\partial y}(y^2) \right] dA$$

$$= \int_0^2 \int_{x^2}^{2x} (4y - 2y)\, dy\, dx$$

$$= \int_0^2 y^2 \Big]_{x^2}^{2x}\, dx$$

$$= \int_0^2 (4x^2 - x^4)\, dx$$

$$= \tfrac{4}{3}x^3 - \tfrac{1}{5}x^5 \Big]_0^2$$

$$= \tfrac{64}{15}$$

To show the advantage of using Green's theorem we evaluate the same line integral by the method of Section 18.2. If C_1 is the arc of the parabola $y = x^2$ from $(0, 0)$ to $(2, 4)$ and C_2 is the line segment from $(2, 4)$ to $(0, 0)$, then

$$\oint_C y^2\, dx + 4xy\, dy = \oint_{C_1} y^2\, dx + 4xy\, dy + \oint_{C_2} y^2\, dx + 4xy\, dy$$

Parametric equations for C_1 are

$$x = t \qquad y = t^2 \qquad 0 \le t \le 2$$

Therefore

$$\oint_{C_1} y^2\, dx + 4xy\, dy = \int_0^2 (t^2)^2\, dt + 4(t)(t^2)(2t\, dt)$$

$$= \int_0^2 9t^4\, dt$$

$$= \tfrac{9}{5}t^5 \Big]_0^2$$

$$= \tfrac{288}{5}$$

Arc C_2 can be represented parametrically by

$$x = t \qquad y = 2t \qquad \text{from } t = 2 \text{ to } t = 0$$

Thus

$$\oint_{C_2} y^2\, dx + 4xy\, dy = \int_2^0 (2t)^2\, dt + 4(t)(2t)(2\, dt)$$

$$= \int_2^0 20t^2\, dt$$

$$= \tfrac{20}{3}t^3 \Big]_2^0$$

$$= -\tfrac{160}{3}$$

Hence

$$\oint_C y^2\,dx + 4xy\,dy = \tfrac{288}{5} - \tfrac{160}{3}$$

$$= \tfrac{64}{15}$$

which agrees with the result obtained by using Green's theorem.   ◄

EXAMPLE 1

Use Green's theorem to find the total work done in moving an object in the counterclockwise direction once around the circle $x^2 + y^2 = a^2$ if the motion is caused by the force field $\mathbf{F}(x, y) = (\sin x - y)\mathbf{i} + (e^y - x^2)\mathbf{j}$. Assume the arc is measured in meters and the force is measured in newtons.

SOLUTION

If W joules is the work done, then

$$W = \oint_C (\sin x - y)\,dx + (e^y - x^2)\,dy$$

where C is the circle $x^2 + y^2 = a^2$. From Green's theorem,

$$W = \iint_R \left[\frac{\partial}{\partial x}(e^y - x^2) - \frac{\partial}{\partial y}(\sin x - y) \right] dA$$

$$= \iint_R (-2x + 1)\,dA$$

We use polar coordinates to evaluate the double integral, with $x = r\cos\theta$ and $dA = r\,dr\,d\theta$. Then

$$W = \int_0^{2\pi}\int_0^a (-2r\cos\theta + 1)r\,dr\,d\theta$$

$$= \int_0^{2\pi}\int_0^a (-2r^2\cos\theta + r)\,dr\,d\theta$$

$$= \int_0^{2\pi} \left[-\frac{2}{3}r^3\cos\theta + \frac{r^2}{2} \right]_0^a d\theta$$

$$= \int_0^{2\pi} \left(-\frac{2}{3}a^3\cos\theta + \frac{a^2}{2} \right) d\theta$$

$$= \left[-\frac{2}{3}a^3\sin\theta + \frac{a^2}{2}\theta \right]_0^{2\pi}$$

$$= \pi a^2$$

Therefore the work done is πa^2 joules.

The following theorem, which is a consequence of Green's theorem, gives a useful method for computing the area of a region bounded by a sectionally smooth simple closed curve.

18.4.2 THEOREM   If R is a region having as its boundary a sectionally smooth simple closed curve C, and A square units is the area of R, then

$$A = \frac{1}{2} \oint_C x\,dy - y\,dx$$

PROOF   In the statement of Green's theorem, let $M(x, y) = -\frac{1}{2}y$ and $N(x, y) = \frac{1}{2}x$. Then

$$\oint_C -\frac{1}{2}y\,dx + \frac{1}{2}x\,dy = \iint_R \left[\frac{\partial}{\partial x}\left(\frac{1}{2}x\right) - \frac{\partial}{\partial y}\left(-\frac{1}{2}y\right) \right] dA$$

$$= \iint_R (\tfrac{1}{2} + \tfrac{1}{2})\,dA$$

$$= \iint_R dA$$

Because $\iint_R dA$ is the measure of the area of R,

$$\frac{1}{2} \oint_C x\,dy - y\,dx = A \qquad \blacksquare$$

EXAMPLE 2

Use Theorem 18.4.2 to find the area of the region enclosed by the ellipse

$$\frac{x^2}{a^2} + \frac{y^2}{b^2} = 1$$

SOLUTION

Parametric equations for the ellipse are

$$x = a\cos t \qquad y = b\sin t \qquad 0 \le t \le 2\pi$$

Then $dx = -a\sin t\,dt$ and $dy = b\cos t\,dt$. If C is the ellipse and A square units is the area of the region enclosed by C, then, from Theorem 18.4.2,

$$A = \frac{1}{2} \oint_C x\,dy - y\,dx$$

$$= \frac{1}{2} \int_0^{2\pi} [(a\cos t)(b\cos t\,dt) - (b\sin t)(-a\sin t\,dt)]$$

$$= \frac{1}{2} \int_0^{2\pi} ab(\cos^2 t + \sin^2 t)\,dt$$

$$= \tfrac{1}{2}ab \int_0^{2\pi} dt$$

$$= \pi ab$$

Therefore the area is πab square units.

EXAMPLE 3

Use Green's theorem to evaluate the line integral

$$\oint_C (x^4 - 3y)\,dx + (2y^3 + 4x)\,dy$$

if C is the ellipse $\dfrac{x^2}{9} + \dfrac{y^2}{4} = 1$.

SOLUTION
From Green's theorem,

$$\oint_C (x^4 - 3y)\,dx + (2y^3 + 4x)\,dy = \iint_R \left[\frac{\partial}{\partial x}(2y^3 + 4x) - \frac{\partial}{\partial y}(x^4 - 3y) \right] dA$$

$$= \iint_R (4 + 3)\,dA$$

$$= 7 \iint_R dA$$

The double integral $\displaystyle\iint_R dA$ is the measure of the area of the region enclosed by the ellipse. From Example 2 with $a = 3$ and $b = 2$, the measure of the area of the region enclosed by the ellipse is 6π. Therefore

$$\oint_C (x^4 - 3y)\,dx + (2y^3 + 4x)\,dy = 42\pi$$

There are two vector forms of Green's theorem that we shall proceed to obtain. Let C be a sectionally smooth simple closed curve in the xy plane. Suppose a vector equation of C is

$$\mathbf{R}(s) = x\mathbf{i} + y\mathbf{j}$$

and $x = f(s)$ and $y = g(s)$, where s units is the length of arc measured in the counterclockwise direction from a particular point P_0 on C to the point P on C. Then if $\mathbf{T}(s)$ is the unit tangent vector of C at P, from Theorem 13.6.3 we have $\mathbf{T}(s) = D_s\mathbf{R}(s)$. Thus

$$\mathbf{T}(s) = \frac{dx}{ds}\mathbf{i} + \frac{dy}{ds}\mathbf{j} \tag{8}$$

The vector $\mathbf{N}(s)$ defined by

$$\mathbf{N}(s) = \frac{dy}{ds}\mathbf{i} - \frac{dx}{ds}\mathbf{j} \tag{9}$$

is a unit normal vector of C at P. To verify this fact observe that $\mathbf{T}(s) \cdot \mathbf{N}(s) = 0$ and the magnitudes of $\mathbf{T}(s)$ and $\mathbf{N}(s)$ are equal. The unit normal vector defined by (9) is selected rather than its negative so that when the direction along C is counterclockwise, $\mathbf{N}(s)$ will point outward from the region R bounded

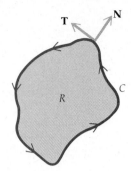

FIGURE 18.4.7

by C, and it is called the **unit outward normal**. See Fig. 18.4.7. Let $\mathbf{F}(x, y) = M(x, y)\mathbf{i} + N(x, y)\mathbf{j}$ where M and N satisfy the hypothesis of Green's theorem. Because

$$\mathbf{F}(x, y) \cdot \mathbf{N}(s)\,ds = [M(x, y)\mathbf{i} + N(x, y)\mathbf{j}] \cdot \left(\frac{dy}{ds}\mathbf{i} - \frac{dx}{ds}\mathbf{j}\right)ds$$

$$= M(x, y)\,dy - N(x, y)\,dx$$

then

$$\oint_C \mathbf{F}(x, y) \cdot \mathbf{N}(s)\,ds = \oint_C -N(x, y)\,dx + M(x, y)\,dy \tag{10}$$

We apply Green's theorem to the line integral on the right side of (10) and we have

$$\oint_C \mathbf{F}(x, y) \cdot \mathbf{N}(s)\,ds = \iint_R \left[\frac{\partial M}{\partial x} - \frac{\partial}{\partial y}(-N)\right]dA$$

$$= \iint_R \left(\frac{\partial M}{\partial x} + \frac{\partial N}{\partial y}\right)dA$$

$$= \iint_R \operatorname{div} \mathbf{F}\,dA$$

This vector form of Green's theorem is stated formally as the following theorem, named after the German mathematician and scientist Karl F. Gauss (1777–1855).

18.4.3 THEOREM

Gauss's Divergence Theorem in the Plane

Let the functions M and N, the curve C, and the region R be as defined in Green's theorem. If $\mathbf{F}(x, y) = M(x, y)\mathbf{i} + N(x, y)\mathbf{j}$ and $\mathbf{N}(s)$ is the unit outward normal vector of C at P, where s units is the length of arc measured in the counterclockwise direction from a particular point P_0 on C to P, then

$$\oint_C \mathbf{F} \cdot \mathbf{N}\,ds = \iint_R \operatorname{div} \mathbf{F}\,dA$$

EXAMPLE 4

Verify Gauss's divergence theorem in the plane if $\mathbf{F}(x, y) = 2y\mathbf{i} + 5x\mathbf{j}$ and R is the region bounded by the circle $x^2 + y^2 = 1$.

SOLUTION

The boundary of R is the unit circle that can be represented parametrically by

$$x = \cos s \qquad y = \sin s \qquad 0 \le s \le 2\pi$$

where s units is the length of arc from the point where $s = 0$ to the point P on C. Then a vector equation of C is

$$\mathbf{R}(s) = \cos s\,\mathbf{i} + \sin s\,\mathbf{j} \qquad 0 \le s \le 2\pi$$

From (9) the unit outward normal is

$$\mathbf{N}(s) = \cos s\mathbf{i} + \sin s\mathbf{j}$$

At a point $P(\cos s, \sin s)$ on C, $\mathbf{F}$ is $2 \sin s\mathbf{i} + 5 \cos s\mathbf{j}$. Therefore

$$\oint_C \mathbf{F} \cdot \mathbf{N}\, ds = \int_0^{2\pi} (2 \sin s\mathbf{i} + 5 \cos s\mathbf{j}) \cdot (\cos s\mathbf{i} + \sin s\mathbf{j})\, ds$$

$$= \int_0^{2\pi} (2 \sin s \cos s + 5 \sin s \cos s)\, ds$$

$$= 7 \int_0^{2\pi} \sin s \cos s\, ds$$

$$= \tfrac{7}{2} \sin^2 s \Big]_0^{2\pi}$$

$$= 0$$

Because $M = 2y$, $\dfrac{\partial M}{\partial x} = 0$, and because $N = 5x$, $\dfrac{\partial N}{\partial y} = 0$. Thus

$$\iint_R \operatorname{div} \mathbf{F}\, dA = \iint_R \left(\frac{\partial M}{\partial x} + \frac{\partial N}{\partial y} \right) dA$$

$$= 0$$

We have therefore verified Gauss's divergence theorem in the plane for this $\mathbf{F}$ and R.

Observe in Example 4 that $\displaystyle\iint_R \operatorname{div} \mathbf{F}\, dA$ is easier to compute than

$\displaystyle\oint_C \mathbf{F} \cdot \mathbf{N}\, ds$.

If $\mathbf{F}$ is a vector field and div $\mathbf{F} = 0$, then $\mathbf{F}$ is said to be **divergence free**. The vector field in Example 4 is divergence free. In the study of hydrodynamics (fluid motion), if the velocity field of a fluid is divergence free, the fluid is called **incompressible**. In the theory of electricity and magnetism, a vector field that is divergence free is said to be **solenoidal**.

We now use Gauss's divergence theorem in the plane to give a physical interpretation of the divergence of a vector field. Let the functions M and N, the region R, and the curve C be as defined in Green's theorem. Suppose $\mathbf{F}$ is the velocity field of a two-dimensional fluid (constant depth) and $\mathbf{F}$ is defined by $\mathbf{F}(x, y) = M(x, y)\mathbf{i} + N(x, y)\mathbf{j}$. Suppose that the fluid flows through region R having curve C as its boundary, for which the direction along C is counterclockwise. We assume that the fluid has a constant density in R, and for convenience let the density have unit measure. The flow of the velocity field $\mathbf{F}$ across C is the rate at which the fluid crosses C in a direction perpendicular to C. We shall show how this flow can be expressed as a line integral.

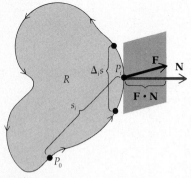

FIGURE 18.4.8

Let s units denote the length of arc of curve C measured from a particular point P_0 to a point P. Divide the curve C into n arcs and let $\Delta_i s$ be the length of the ith arc containing the point $P_i(x_i, y_i)$, where s_i is the length of arc of C from P_0 to P_i. Because $\mathbf{F}$ is continuous, an approximation of the velocity of the fluid at each point of the ith arc is $\mathbf{F}(x_i, y_i)$. The amount of fluid that crosses the arc per unit of time is given approximately by the area of a parallelogram having one pair of opposite sides of length $\Delta_i s$ and an altitude of length $\mathbf{F}(x_i, y_i) \cdot \mathbf{N}(s_i)$, where $\mathbf{N}(s_i)$ is the unit outward normal vector of C at $P_i(x_i, y_i)$. See Fig. 18.4.8. The measure of the area of the parallelogram is $\mathbf{F}(x_i, y_i) \cdot \mathbf{N}(s_i) \Delta_i s$. The total amount of fluid that crosses C per unit of time is approximately

$$\sum_{i=1}^{n} \mathbf{F}(x_i, y_i) \cdot \mathbf{N}(s_i) \Delta_i s$$

Taking the limit of this summation as n increases without bound and as each $\Delta_i s$ approaches zero we obtain the line integral

$$\oint_C \mathbf{F}(x, y) \cdot \mathbf{N}(s) \, ds$$

which is called the **flux** of $\mathbf{F}$ across C.

Now let $\bar{P}(\bar{x}, \bar{y})$ be a particular point in region R. Consider a circle having its center at $\bar{P}$ and having a small radius δ, and denote this circle by C_δ. Let R_δ be the region enclosed by C_δ. Then

$$\text{flux of } \mathbf{F} \text{ across } C_\delta = \oint_{C_\delta} \mathbf{F}(x, y) \cdot \mathbf{N} \, ds$$

Applying Theorem 18.4.3 we have

$$\text{flux of } \mathbf{F} \text{ across } C_\delta = \iint_{R_\delta} \text{div } \mathbf{F} \, dA$$

If M_x and N_y are continuous on R_δ, then div $\mathbf{F}$ is continuous there, and for small δ, div $\mathbf{F}$ on R_δ is approximately div $\mathbf{F}(\bar{x}, \bar{y})$. Thus

$$\text{flux of } \mathbf{F} \text{ across } C_\delta \approx \iint_{R} \text{div } \mathbf{F}(\bar{x}, \bar{y}) \, dA$$

Because div $\mathbf{F}(\bar{x}, \bar{y})$ is constant and $\iint_{R_\delta} dA$ is the area of a circle of radius δ, we have

$$\text{flux of } \mathbf{F} \text{ across } C_\delta \approx \text{div } \mathbf{F}(\bar{x}, \bar{y})(\pi\delta^2) \tag{11}$$

Remember that the flux of $\mathbf{F}$ across C_δ is the total amount of fluid that crosses C_δ per unit of time. Therefore, from (11), div $\mathbf{F}(x, y)$ can be interpreted as the approximate measure of the rate of flow of the fluid per unit area away from the point $(\bar{x}, \bar{y})$. If div $\mathbf{F}(\bar{x}, \bar{y}) > 0$, the fluid is said to have a **source** at $(\bar{x}, \bar{y})$. If div $\mathbf{F}(\bar{x}, \bar{y}) < 0$, the fluid has a **sink** at $(\bar{x}, \bar{y})$. If $\mathbf{F}$ is divergence free at all points in a region, then there are no sources or sinks in the region. As mentioned above, in such a case the fluid is incompressible.

The word *flux* normally means flow; however, the terminology *flux* is applied to vector fields in general, not just to those associated with the velocity of a fluid. Thus if **F** is a vector field

$$\text{flux of } \mathbf{F} \text{ across } C = \oint_C \mathbf{F} \cdot \mathbf{N} \, ds \tag{12}$$

EXAMPLE 5

The velocity field of a fluid is defined by

$$\mathbf{F}(x, y) = (5x - y)\mathbf{i} + (x^2 - 3y)\mathbf{j}$$

Find the rate of flow of the fluid out of a region R bounded by a smooth closed curve C and whose area is 150 cm^2.

SOLUTION

The rate of flow of the fluid is given by the flux of **F** across C. From (12) and Gauss's divergence theorem in the plane,

$$\begin{aligned}
\text{flux} &= \oint_C \mathbf{F} \cdot \mathbf{N} \, ds \\
&= \iint_R \text{div } \mathbf{F} \, dA \\
&= \iint_R \left[\frac{\partial}{\partial x} (5x - y) + \frac{\partial}{\partial y} (x^2 - 3y) \right] dA \\
&= \iint_R (5 - 3) \, dA \\
&= 2 \iint_R dA
\end{aligned}$$

Because the area of R is 150 cm^2, $\iint_R dA = 150$. Thus

$$\text{flux} = 300$$

Hence the rate of flow of the fluid out of the region is 300 cm^2 per unit of time.

To obtain the second vector form of Green's theorem we consider the dot product of $\mathbf{F}(x, y)$ and the unit tangent vector $\mathbf{T}(s)$ defined by Eq. (8). We have

$$\mathbf{F}(x, y) \cdot \mathbf{T}(s) \, ds = [M(x, y)\mathbf{i} + N(x, y)\mathbf{j}] \cdot \left(\frac{dx}{ds} \mathbf{i} + \frac{dy}{ds} \mathbf{j} \right) ds$$

$$= M(x, y) \, dx + N(x, y) \, dy$$

Thus

$$\oint_C \mathbf{F}(x, y) \cdot \mathbf{T}(s)\, ds = \oint_C M(x, y)\, dx + N(x, y)\, dy \tag{13}$$

The curl of $\mathbf{F}$ in two dimensions was defined in Section 18.1 as

$$\text{curl } \mathbf{F}(x, y) = \left(\frac{\partial N}{\partial x} - \frac{\partial M}{\partial y}\right)\mathbf{k}$$

Therefore

$$\text{curl } \mathbf{F}(x, y) \cdot \mathbf{k} = \left(\frac{\partial N}{\partial x} - \frac{\partial M}{\partial y}\right)\mathbf{k} \cdot \mathbf{k}$$

$$\text{curl } \mathbf{F}(x, y) \cdot \mathbf{k} = \frac{\partial N}{\partial x} - \frac{\partial M}{\partial y} \tag{14}$$

Hence from (13) and (14), Eq. (1) of Green's theorem can be written

$$\oint_C \mathbf{F}(x, y) \cdot \mathbf{T}(s)\, ds = \iint_R \text{curl } \mathbf{F}(x, y) \cdot \mathbf{k}\, dA$$

This vector form of Green's theorem is stated formally as the following theorem named after the Irish mathematician and physicist George Stokes (1819–1903).

18.4.4 THEOREM
Stokes's Theorem in the Plane

Let the functions M and N, the curve C, and the region R be as defined in Green's theorem. If $\mathbf{F}(x, y) = M(x, y)\mathbf{i} + N(x, y)\mathbf{j}$ and $\mathbf{T}(s)$ is the unit tangent vector of C at P, where s units is the length of arc measured from a particular point P_0 on C to P, then

$$\oint_C \mathbf{F} \cdot \mathbf{T}\, ds = \iint_R \text{curl } \mathbf{F} \cdot \mathbf{k}\, dA$$

EXAMPLE 6
Verify Stokes's theorem in the plane for $\mathbf{F}$ and R of Example 4.

SOLUTION
As in Example 4, the vector field $\mathbf{F}$ is defined by

$$\mathbf{F}(x, y) = 2y\mathbf{i} + 5x\mathbf{j}$$

and a vector equation of C is

$$\mathbf{R}(s) = \cos s\, \mathbf{i} + \sin s\, \mathbf{j} \qquad 0 \leq s \leq 2\pi$$

Because $\mathbf{T}(s) = D_s\mathbf{R}(s)$,

$$\mathbf{T}(s) = -\sin s\, \mathbf{i} + \cos s\, \mathbf{j}$$

At a point $P(\cos s, \sin s)$ on C, $\mathbf{F}$ is $2 \sin s\mathbf{i} + 5 \cos s\mathbf{j}$. Therefore

$$\oint_C \mathbf{F} \cdot \mathbf{T} \, ds = \int_0^{2\pi} (2 \sin s\mathbf{i} + 5 \cos s\mathbf{j}) \cdot (-\sin s\mathbf{i} + \cos s\mathbf{j}) \, ds$$

$$= \int_0^{2\pi} (-2 \sin^2 s + 5 \cos^2 s) \, ds$$

$$= -2 \int_0^{2\pi} \frac{1 - \cos 2s}{2} \, ds + 5 \int_0^{2\pi} \frac{1 + \cos 2s}{2} \, ds$$

$$= -s + \tfrac{1}{2} \sin 2s + \tfrac{5}{2}s + \tfrac{5}{4} \sin 2s \Big]_0^{2\pi}$$

$$= \tfrac{3}{2}s + \tfrac{7}{4} \sin 2s \Big]_0^{2\pi}$$

$$= 3\pi$$

Because $N = 5x$, $\dfrac{\partial N}{\partial x} = 5$, and because $M = 2y$, $\dfrac{\partial M}{\partial y} = 2$. Thus

$$\iint_R \text{curl } \mathbf{F} \cdot \mathbf{k} \, dA = \iint_R \left(\frac{\partial N}{\partial x} - \frac{\partial M}{\partial y} \right) dA$$

$$= \iint_R (5 - 2) \, dA$$

$$= 3 \iint_R dA$$

$$= 3\pi$$

Therefore Stokes's theorem in the plane is verified for this $\mathbf{F}$ and R.

$$\oint_C \mathbf{F} \cdot \mathbf{T} \, ds > 0$$

FIGURE 18.4.9

If $\mathbf{F}$ is the velocity field of a fluid, the dot product $\mathbf{F} \cdot \mathbf{T}$ is the tangential component of $\mathbf{F}$ and the line integral $\oint_C \mathbf{F} \cdot \mathbf{T} \, ds$ is called the **circulation** of $\mathbf{F}$ around the closed curve C. In an intuitive sense, we can think of the circulation as being the sum of the tangential components of $\mathbf{F}$ around C. If motion around C is in the counterclockwise direction and $\oint_C \mathbf{F} \cdot \mathbf{T} \, ds > 0$, then the fluid is circulating counterclockwise; see Fig. 18.4.9.

If $\oint_C \mathbf{F} \cdot \mathbf{T} \, ds < 0$, the circulation of the fluid is clockwise; see Fig. 18.4.10.

Let $\bar{P}(\bar{x}, \bar{y})$ be a particular point in region R and let C_δ be the circle having its center at $\bar{P}$ and having a small radius δ. If R_δ is the region enclosed by C_δ,

$$\oint_{C_\delta} \mathbf{F} \cdot \mathbf{T} \, ds = \iint_{R_\delta} \text{curl } \mathbf{F} \cdot \mathbf{k} \, dA$$

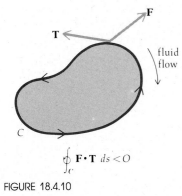

$$\oint_C \mathbf{F} \cdot \mathbf{T} \, ds < 0$$

FIGURE 18.4.10

If M_y and N_x are continuous on R_δ, then curl $\mathbf{F} \cdot \mathbf{k}$ is continuous there and for small δ, curl $\mathbf{F} \cdot \mathbf{k}$ on R_δ is approximately curl $\mathbf{F}(\bar{x}, \bar{y}) \cdot \mathbf{k}$. Therefore

$$\oint_{C_\delta} \mathbf{F} \cdot \mathbf{T} \, ds \approx \text{curl } \mathbf{F}(\bar{x}, \bar{y}) \cdot \mathbf{k} \iint_{R_\delta} dA$$

$$\oint_{C_\delta} \mathbf{F} \cdot \mathbf{T} \, ds \approx \text{curl } \mathbf{F}(\bar{x}, \bar{y}) \cdot \mathbf{k}(\pi \delta^2) \tag{15}$$

From (15), we interpret curl $\mathbf{F}(\bar{x}, \bar{y}) \cdot \mathbf{k}$ as the approximate measure of the rate of circulation per unit area in the counterclockwise direction at the point $\bar{P}$. When $\mathbf{F}$ and $\mathbf{T}$ are orthogonal vectors, $\mathbf{F} \cdot \mathbf{T} = 0$ and then curl $\mathbf{F} = \mathbf{0}$. In such a case $\mathbf{F}$ is said to be **irrotational**. This terminology is used even if $\mathbf{F}$ is not the velocity field of a fluid.

EXERCISES 18.4

In Exercises 1 through 8, evaluate the line integral by Green's theorem. Then verify the result by the method of Section 18.2.

1. $\oint_C 4y \, dx + 3x \, dy$, where C is the square with vertices at $(0, 0)$, $(1, 0)$, $(1, 1)$, and $(0, 1)$.

2. $\oint_C y^2 \, dx + x^2 \, dy$, where C is the square of Exercise 1.

3. $\oint_C 2xy \, dx - x^2 y \, dy$, where C is the triangle with vertices at $(0, 0)$, $(1, 0)$, and $(0, 1)$.

4. The line integral of Exercise 3, where C is the triangle with vertices at $(0, 0)$, $(1, 0)$, and $(1, 1)$.

5. $\oint_C x^2 y \, dx - y^2 x \, dy$, where C is the circle $x^2 + y^2 = 1$.

6. $\oint_C (x^2 - y^2) \, dx + 2xy \, dy$, where C is the circle $x^2 + y^2 = 1$.

7. The line integral of Exercise 5, where C is the closed curve consisting of the arc of $4y = x^3$ from $(0, 0)$ to $(2, 2)$ and the line segment from $(2, 2)$ to $(0, 0)$.

8. The line integral of Exercise 6, where C is the closed curve of Exercise 7.

In Exercises 9 through 20, use Green's theorem to evaluate the line integral.

9. $\oint_C (x + y) \, dx + xy \, dy$, where C is the closed curve determined by the x axis, the line $x = 2$, and the curve $4y = x^3$.

10. $\oint_C y^2 \, dx + x^2 \, dy$, where C is the closed curve determined by the x axis, the line $x = 1$, and the curve $y = x^2$.

11. $\oint_C (-x^2 + x) \, dy$, where C is the closed curve determined by the line $x - 2y = 0$ and the parabola $x = 2y^2$.

12. $\oint_C (x^2 + y) \, dx$, where C is the closed curve determined by the x axis and the parabola $y = 4 - x^2$.

13. $\oint_C \cos y \, dx + \cos x \, dy$, where C is the rectangle with vertices at $(0, 0)$, $(\frac{1}{3}\pi, 0)$, $(\frac{1}{3}\pi, \frac{1}{4}\pi)$, and $(0, \frac{1}{4}\pi)$.

14. $\oint_C e^{x+y} \, dx + e^{x+y} \, dy$, where C is the circle $x^2 + y^2 = 4$.

15. $\oint_C (\sin^4 x + e^{2x}) \, dx + (\cos^3 y - e^y) \, dy$, where C is the curve $x^4 + y^4 = 16$.

16. $\oint_C x \sin y \, dx - y \cos x \, dy$, where C is the rectangle with vertices at $(0, 0)$, $(\frac{1}{2}\pi, 0)$, $(\frac{1}{4}\pi, \frac{1}{2}\pi)$, and $(0, \frac{1}{4}\pi)$.

17. $\displaystyle\oint_C \frac{x^2 y}{x^2 + 1}\, dx - \tan^{-1} x\, dy$, where C is the ellipse $4x^2 + 25y^2 = 100$.

18. $\displaystyle\oint_C e^y \cos x\, dx + e^y \sin x\, dy$, where C is the curve $x^6 + y^4 = 10$.

19. $\displaystyle\oint_C (e^x - x^2 y)\, dx + 3x^2 y\, dy$, where C is the closed curve determined by $y = x^2$ and $x = y^2$.

20. $\displaystyle\oint_C \tan y\, dx - x \tan^2 y\, dy$, where C is the ellipse $x^2 + 4y^2 = 1$.

In Exercises 21 through 26, use Theorem 18.4.2 to find the area of the given region.
21. The region having as its boundary the quadrilateral with vertices at $(0,0)$, $(4,0)$, $(3,2)$, and $(1,1)$.
22. The region having as its boundary the circle $x^2 + y^2 = a^2$.
23. The region bounded by the graphs of $y = x^2$ and $y = \sqrt{x}$.
24. The region bounded by the parabola $y = 2x^2$ and the line $y = 8x$.
25. The region bounded by the hypocycloid having parametric equations

$$x = a \cos^3 t \qquad y = a \sin^3 t \qquad a > 0 \qquad 0 \le t \le 2\pi$$

26. The region bounded below by the x axis and above by one arch of the cycloid having parametric equations

$$x = t - \sin t \qquad y = 1 - \cos t \qquad 0 \le t \le 2\pi$$

In Exercises 27 through 30, verify Gauss's divergence theorem in the plane and Stokes's theorem in the plane for the given $\mathbf{F}$ and R.
27. $\mathbf{F}(x, y) = 3x\mathbf{i} + 2y\mathbf{j}$ and R is the region bounded by the circle $x^2 + y^2 = 1$.
28. $\mathbf{F}(x, y) = 3y\mathbf{i} - 2x\mathbf{j}$ and R is the region bounded by $x^{2/3} + y^{2/3} = 1$.
29. $\mathbf{F}(x, y) = x^2\mathbf{i} + y^2\mathbf{j}$ and R is the region bounded by the ellipse $4x^2 + 25y^2 = 100$.
30. $\mathbf{F}(x, y) = y^2\mathbf{i} + x^2\mathbf{j}$ and R is the region bounded by the circle $x^2 + y^2 = 4$.

In Exercises 31 through 34, use Green's theorem to find the total work done in moving an object in the counterclockwise direction once around the given curve C if the motion is caused by the given force field $\mathbf{F}(x, y)$. Assume the arc is measured in meters and the force is measured in newtons.
31. C is the ellipse $x^2 + 4y^2 = 16$; $\mathbf{F}(x, y) = (3x + y)\mathbf{i} + (4x - 5y)\mathbf{j}$.
32. C is the circle $x^2 + y^2 = 25$; $\mathbf{F}(x, y) = (e^x + y^2)\mathbf{i} + (x^2 y + \cos y)\mathbf{j}$.
33. C is the triangle with vertices at $(0,0)$, $(2,0)$, and $(0,2)$; $\mathbf{F}(x, y) = (e^{x^2} + y^2)\mathbf{i} + (e^{y^2} + x^2)\mathbf{j}$.
34. C consists of the top half of the ellipse $9x^2 + 4y^2 = 36$ and the interval $[-2, 2]$ on the x axis; $\mathbf{F}(x, y) = (xy + y^2)\mathbf{i} + xy\mathbf{j}$.

In Exercises 35 through 38, find the rate of flow of the fluid out of a region R bounded by the given curve C if $\mathbf{F}$ is the velocity field of the fluid. Assume the velocity is measured in centimeters per second and the area of R is measured in square centimeters.
35. $\mathbf{F}(x, y) = (y^2 + 6x)\mathbf{i} + (2y - x^2)\mathbf{j}$; C is the ellipse $x^2 + 4y^2 = 4$.
36. $\mathbf{F}(x, y) = (5x - y^2)\mathbf{i} + (3x - 2y)\mathbf{j}$; C is the right triangle having vertices at $(1,2)$, $(4,2)$, and $(4,6)$.
37. $\mathbf{F}(x, y) = x^3\mathbf{i} + y^3\mathbf{j}$; C is the circle $x^2 + y^2 = 1$.
38. $\mathbf{F}(x, y) = xy^2\mathbf{i} + yx^2\mathbf{j}$; C is the circle $x^2 + y^2 = 9$.

In Exercises 39 through 42, $\mathbf{F}$ is the velocity field of a fluid around the closed curve C, where motion around C is in the counterclockwise direction. Use Stokes's theorem in the plane to compute $\displaystyle\oint_C \mathbf{F} \cdot \mathbf{T}\, ds$, and from the result determine which of the following applies: (i) the circulation of the fluid is counterclockwise; (ii) the circulation of the fluid is clockwise; (iii) $\mathbf{F}$ is irrotational.
39. $\mathbf{F}(x, y) = 4y\mathbf{i} + 6x\mathbf{j}$; C is the triangle having vertices at $(0,0)$, $(3,0)$, and $(3,5)$.
40. $\mathbf{F}(x, y) = 8y\mathbf{i} + 3x\mathbf{j}$; C is the ellipse $4x^2 + 9y^2 = 1$.

41. $\mathbf{F}(x, y) = \sin^2 x\mathbf{i} + \cos^2 y\mathbf{j}$; C is the ellipse $9x^2 + y^2 = 9$.

42. $\mathbf{F}(x, y) = y^3\mathbf{i} + x^3\mathbf{j}$; C is the circle $x^2 + y^2 = 25$.

43. Prove $\oint_C N(x, y)\, dy = \iint\limits_R \dfrac{\partial N}{\partial x}\, dA$ if R is the region defined by $R = \{(x, y) | c \le y \le d,\ g_1(y) \le x \le g_2(y)\}$, where g_1 and g_2 are smooth.

18.5 (*Supplementary*) SURFACE INTEGRALS, GAUSS'S DIVERGENCE THEOREM, AND STOKES'S THEOREM

The discussion in Section 18.4 was concerned with a treatment of Green's theorem and its two vector forms, Gauss's divergence theorem in the plane and Stokes's theorem in the plane. These ideas can be extended to three-dimensional space. The concept of a line integral can be generalized to that of an integral defined on a surface, and the two vector forms of Green's theorem can be generalized to Gauss's divergence theorem and Stokes's theorem in three dimensions. A rigorous presentation of these topics belongs to a course in advanced calculus; however, in this section we give a brief intuitive introduction.

Let D be a closed region in the xy plane. In this section we are using the symbol D instead of R to denote a region in the xy plane to avoid confusion with the function defined by $R(x, y, z)$ used later in the discussion. Suppose S is a surface lying over D and having the equation $z = f(x, y)$, where f and its first partial derivatives are continuous on D. Then if σ is the measure of the area of the surface S, we have, from Theorem 17.5.1,

$$\sigma = \iint\limits_D \sqrt{f_x{}^2(x, y) + f_y{}^2(x, y) + 1}\ dx\, dy \tag{1}$$

We can generalize the integral in (1) by considering a function G of the three variables x, y, and z, where G is continuous on S. We proceed in a manner similar to the discussion in Section 17.5 that precedes the statement of Theorem 17.5.1. Let Δ be a partition of region D into n rectangular subregions, where the ith rectangle has dimensions of measures $\Delta_i x$ and $\Delta_i y$ and an area of measure $\Delta_i A$. Let (ξ_i, γ_i) be any point in the ith rectangle, and at the point $Q(\xi_i, \gamma_i, f(\xi_i, \gamma_i))$ on the surface S consider the tangent plane to the surface. Project vertically upward the ith rectangle onto the tangent plane, and let $\Delta_i \sigma$ be the measure of the area of this projection. See Fig. 18.5.1. The number $\Delta_i \sigma$ is an approximation to the measure of the area of the piece of the surface that lies above the ith rectangle. We showed in Section 17.5 that

$$\Delta_i \sigma = \sqrt{f_x{}^2(\xi_i, \gamma_i) + f_y{}^2(\xi_i, \gamma_i) + 1}\ \Delta_i A \tag{2}$$

If we form the sum

$$\sum_{i=1}^{n} G(\xi_i, \gamma_i, f(\xi_i, \gamma_i))\, \Delta_i \sigma$$

and take the limit of this sum as the norm of the partition approaches zero, we have

$$\lim_{\|\Delta\| \to 0} \sum_{i=1}^{n} G(\xi_i, \gamma_i, f(\xi_i, \gamma_i))\, \Delta_i \sigma \tag{3}$$

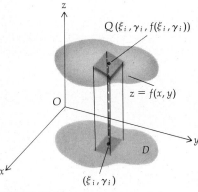

FIGURE 18.5.1

This limit is called the **surface integral** of G over S and is denoted by

$$\iint_S G(x, y, z)\, d\sigma$$

To obtain a formula for evaluating this surface integral, we substitute from (2) into (3) and we have

$$\lim_{\|\Delta\| \to 0} \sum_{i=1}^{n} G(\xi_i, \gamma_i, f(\xi_i, \gamma_i)) \sqrt{f_x^{\,2}(\xi_i, \gamma_i) + f_y^{\,2}(\xi_i, \gamma_i) + 1}\, \Delta_i A$$

This limit is a double integral over the region R in the xy plane. Thus

$$\iint_S G(x, y, z)\, d\sigma = \iint_D G(x, y, f(x, y)) \sqrt{f_x^{\,2}(x, y) + f_y^{\,2}(x, y) + 1}\, dA \qquad (4)$$

If $G(x, y, z) = 1$, then (4) becomes

$$\iint_S d\sigma = \iint_D \sqrt{f_x^{\,2}(x, y) + f_y^{\,2}(x, y) + 1}\, dA$$

Comparing this equation with (1), we observe that for this G the surface integral of G over S gives the measure of the area of surface S.

For the surface integral in (4), $z = f(x, y)$ is an equation of the surface S that is projected onto the region D in the xy plane. If an equation of surface S is of the form $y = g(x, z)$ and S is projected onto a region D in the xz plane, and g and its first partial derivatives are continuous on R, then

$$\iint_S G(x, y, z)\, d\sigma = \iint_D G(x, g(x, z), z) \sqrt{g_x^{\,2}(x, z) + g_z^{\,2}(x, z) + 1}\, dA \qquad (5)$$

Furthermore, if an equation of surface S is of the form $x = h(y, z)$, and S is projected onto a region D in the yz plane, and h and its first partial derivatives are continuous on D, then

$$\iint_S G(x, y, z)\, d\sigma = \iint_D G(h(y, z), y, z) \sqrt{h_y^{\,2}(y, z) + h_z^{\,2}(y, z) + 1}\, dA \qquad (6)$$

EXAMPLE 1
Evaluate the surface integral

$$\iint_S x^2 z^2\, d\sigma$$

where S is the portion of the cone $x^2 + y^2 = z^2$ between the planes $z = 1$ and $z = 2$.

SOLUTION
Figure 18.5.2 shows the surface S and the projection of S onto the region D in the xy plane. Region D is bounded by the two circles of radii 1 and 2 whose centers are at the origin. We solve the equation of S for z, where $z \geq 0$,

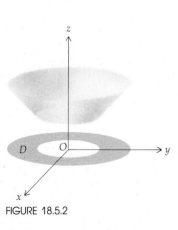

FIGURE 18.5.2

and we obtain $z = \sqrt{x^2 + y^2}$. Therefore

$$f(x, y) = \sqrt{x^2 + y^2} \qquad f_x(x, y) = \frac{x}{\sqrt{x^2 + y^2}} \qquad f_y(x, y) = \frac{y}{\sqrt{x^2 + y^2}}$$

From (4), with $G(x, y, z) = x^2 z^2$, we obtain

$$\iint_S x^2 z^2 \, d\sigma = \iint_D x^2 (x^2 + y^2) \sqrt{\frac{x^2}{x^2 + y^2} + \frac{y^2}{x^2 + y^2} + 1} \, dA$$

$$= \iint_D x^2 (x^2 + y^2) \sqrt{2} \, dA$$

We evaluate the double integral by using polar coordinates, where $x = r \cos \theta$, $x^2 + y^2 = r^2$, and $dA = r \, dr \, d\theta$. Therefore

$$\iint_S x^2 z^2 \, d\sigma = \sqrt{2} \int_0^{2\pi} \int_1^2 (r^2 \cos^2 \theta) r^2 (r \, dr \, d\theta)$$

$$= \sqrt{2} \int_0^{2\pi} \int_1^2 \cos^2 \theta r^5 \, dr \, d\theta$$

$$= \sqrt{2} \int_0^{2\pi} \left[\cos^2 \theta \frac{r^6}{6} \right]_1^2 \, d\theta$$

$$= \frac{21\sqrt{2}}{2} \int_0^{2\pi} \frac{1 + \cos 2\theta}{2} \, d\theta$$

$$= \frac{21\sqrt{2}}{4} \left[\theta + \frac{\sin 2\theta}{2} \right]_0^{2\pi}$$

$$= \frac{21\pi}{\sqrt{2}}$$

If the measure of the area density at the point (x, y, z) on a surface S is $\rho(x, y, z)$, and if M is the measure of the mass of S, then

$$M = \iint_S \rho(x, y, z) \, d\sigma \qquad\qquad (7)$$

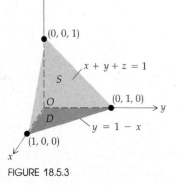

$x + y + z = 1$

S

$(0, 0, 1)$

$(0, 1, 0)$

$y = 1 - x$

$(1, 0, 0)$

FIGURE 18.5.3

EXAMPLE 2

Find the mass of the portion of the plane $x + y + z = 1$ in the first octant if the area density at any point (x, y, z) on the surface is kx^2 kg/m^2, where k is a constant.

SOLUTION

Figure 18.5.3 shows S, which is the surface of the given plane in the first octant, and the region D, which is the projection of S onto the xy plane. We solve the equation of the plane for z and obtain $z = 1 - x - y$. Therefore

$$f(x, y) = 1 - x - y \qquad f_x(x, y) = -1 \qquad f_y(x, y) = -1$$

From (7), with $\rho(x, y, z) = kx^2$, if M kg is the mass of the surface,

$$
M = \iint\limits_{S} kx^2 \, d\sigma
$$

$$
= \iint\limits_{D} kx^2 \sqrt{f_x{}^2(x, y) + f_y{}^2(x, y) + 1} \, dA
$$

$$
= \iint\limits_{D} kx^2 \sqrt{(-1)^2 + (-1)^2 + 1} \, dA
$$

$$
= \sqrt{3}k \int_0^1 \int_0^{1-x} x^2 \, dy \, dx
$$

$$
= \sqrt{3}k \int_0^1 \left[x^2 y \right]_0^{1-x} dx
$$

$$
= \sqrt{3}k \int_0^1 (x^2 - x^3) \, dx
$$

$$
= \sqrt{3}k \left[\frac{x^3}{3} - \frac{x^4}{4} \right]_0^1
$$

$$
= \tfrac{1}{12}\sqrt{3}k
$$

Thus the mass is $\tfrac{1}{12}\sqrt{3}k$ kg.

We now give an application of surface integrals to fluid flow. Let **F** be the velocity field of a fluid defined by

$$
\mathbf{F}(x, y, z) = M(x, y, z)\mathbf{i} + N(x, y, z)\mathbf{j} + R(x, y, z)\mathbf{k}
$$

Furthermore, suppose that the fluid flows through a surface S having the equation $z = f(x, y)$, which lies over a closed region D in the xy plane. Assume that f and its first partial derivatives are continuous on D. At each point of S there are two unit normal vectors to S. The unit normal having a positive **k** component is called the **unit upper normal**.

As in our discussion preceding Eq. (2), take a partition of D into n rectangular subregions. Choose a point (ξ_i, γ_i) in the ith rectangle. Project vertically upward the ith rectangle onto the tangent plane at the point $Q(\xi_i, \gamma_i, f(\xi_i, \gamma_i))$ on S and let $\Delta_i\sigma$, given by (2), be an approximation to the measure of the area of this projection. Again refer to Fig. 18.5.1. Now let $\mathbf{N}_i$ be the unit upper normal to S at point Q and let $\mathbf{F}_i$ be the velocity vector of the fluid at Q. The amount of fluid that crosses the projection per unit of time is given approximately by the volume of the parallelepiped having a base of area $\Delta_i\sigma$ square units and an altitude of length $\mathbf{F}_i \cdot \mathbf{N}_i$ units. See Fig. 18.5.4. The measure of the volume of the parallelepiped is $\mathbf{F}_i \cdot \mathbf{N}_i \, \Delta_i\sigma$. The total amount of fluid that crosses S per unit of time is given approximately by

$$
\sum_{i=1}^{n} \mathbf{F}_i \cdot \mathbf{N}_i \, \Delta_i\sigma
$$

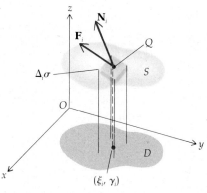

FIGURE 18.5.4

Taking the limit of this summation as n increases without bound and each $\Delta_i \sigma$ approaches zero, we obtain the surface integral

$$\iint\limits_{S} \mathbf{F} \cdot \mathbf{N}\, d\sigma \tag{8}$$

which is called the **flux** of $\mathbf{F}$ across S.

To evaluate surface integral (8), write the equation of S in the form $g(x, y, z) = 0$, where

$$g(x, y, z) = z - f(x, y)$$

From Theorem 16.2.2, a unit normal vector of the surface defined by $g(x, y, z) = 0$ is

$$\mathbf{N} = \frac{\nabla g}{\|\nabla g\|}$$

$$= \frac{-f_x(x, y)\mathbf{i} - f_y(x, y)\mathbf{j} + \mathbf{k}}{\sqrt{f_x^2(x, y) + f_y^2(x, y) + 1}}$$

Thus

$$\iint\limits_{S} \mathbf{F} \cdot \mathbf{N}\, d\sigma = \iint\limits_{S} (M\mathbf{i} + N\mathbf{j} + R\mathbf{k}) \cdot \left(\frac{-f_x\mathbf{i} - f_y\mathbf{j} + \mathbf{k}}{\sqrt{f_x^2 + f_y^2 + 1}} \right) d\sigma$$

$$= \iint\limits_{D} \frac{-Mf_x - Nf_y + R}{\sqrt{f_x^2 + f_y^2 + 1}} \left(\sqrt{f_x^2 + f_y^2 + 1} \right) dA$$

Therefore, we conclude that

$$\iint\limits_{S} \mathbf{F} \cdot \mathbf{N}\, d\sigma = \iint\limits_{D} (-Mf_x - Nf_y + R)\, dA \tag{9}$$

EXAMPLE 3

The velocity field of a fluid is given by

$$\mathbf{F}(x, y, z) = y\mathbf{i} - x\mathbf{j} + 8\mathbf{k} \tag{10}$$

and surface S is that part of the sphere $x^2 + y^2 + z^2 = 9$ that is above the region D in the xy plane enclosed by the circle $x^2 + y^2 = 4$. Find the flux of $\mathbf{F}$ across S.

SOLUTION

Figure 18.5.5 shows the surface S and the region D in the xy plane. We solve the equation of the sphere for z, with $z > 0$, and obtain $z = \sqrt{9 - x^2 - y^2}$. Therefore

$$f(x, y) = \sqrt{9 - x^2 - y^2} \qquad f_x = \frac{-x}{\sqrt{9 - x^2 - y^2}} \qquad f_y = \frac{-y}{\sqrt{9 - x^2 - y^2}}$$

$$f_x = -\frac{x}{z} \qquad\qquad f_y = -\frac{y}{z}$$

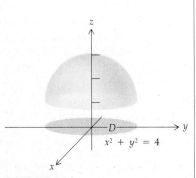

$x^2 + y^2 = 4$

FIGURE 18.5.5

From the definition of flux we have

$$\text{flux of } \mathbf{F} \text{ across } S = \iint_S \mathbf{F} \cdot \mathbf{N} \, d\sigma$$

From (10), $M = y$, $N = -x$, and $R = 8$. Therefore, from (9),

$$\text{flux of } \mathbf{F} \text{ across } S = \iint_D (-Mf_x - Nf_y + R) \, dA$$

$$= \iint_D \left[-y \left(-\frac{x}{z} \right) - (-x) \left(-\frac{y}{z} \right) + 8 \right] dA$$

$$= 8 \iint_D dA$$

Because D is the region enclosed by the circle $x^2 + y^2 = 4$, $A = 4\pi$. Thus

$$\text{flux of } \mathbf{F} \text{ across } S = 8(4\pi)$$

$$= 32\pi$$

We conclude, then, that the rate of flow of the fluid across S is 32π cubic units of length per unit of time.

Suppose S is a closed surface, examples of which are rectangular parallelepipeds, spheres, and ellipsoids. When using (8) to compute the flux of $\mathbf{F}$ across a closed surface, we select $\mathbf{N}$ as a **unit outward normal**, which is a normal whose direction is away from the solid bounded by the surface. In particular, if S is an ellipsoid, as shown in Fig. 18.5.6, we consider S as consisting of an upper surface S_1 and a lower surface S_2, as indicated in the figure. In such a case the flux of $\mathbf{F}$ across S is

$$\iint_S \mathbf{F} \cdot \mathbf{N} \, d\sigma = \iint_{S_1} \mathbf{F} \cdot \mathbf{N}_1 \, d\sigma + \iint_{S_2} \mathbf{F} \cdot \mathbf{N}_2 \, d\sigma$$

For the surface integral across S_1, $\mathbf{N}_1$ is a unit upper normal, and for the surface integral across S_2, $\mathbf{N}_2$ is a unit lower normal (where the component of $\mathbf{k}$ is negative).

If $\mathbf{N}$ is a unit lower normal, then instead of using (9) to compute the surface integral we apply the following formula:

$$\iint_S \mathbf{F} \cdot \mathbf{N} \, d\sigma = \iint_D (Mf_x + Nf_y - R) \, dA \qquad (11)$$

This formula is proved in a manner similar to that used to prove (9).

FIGURE 18.5.6

EXAMPLE 4

Suppose the velocity field of a fluid is given by $\mathbf{F}(x, y, z) = 5z\mathbf{k}$, and S is the sphere $x^2 + y^2 + z^2 = 16$. Find the flux of $\mathbf{F}$ across S if length is measured in centimeters and time is measured in hours.

SOLUTION

Figure 18.5.7 shows the sphere and the region D in the xy plane, which is the circle $x^2 + y^2 = 16$. Because $\mathbf{F}(x, y, z) = 5z\mathbf{k}$, $M = 0$, $N = 0$, and $R = 5z$.

FIGURE 18.5.7

The flux of $\mathbf{F}$ across S is

$$\iint_S \mathbf{F} \cdot \mathbf{N}\, d\sigma = \iint_{S_1} \mathbf{F} \cdot \mathbf{N}_1\, d\sigma + \iint_{S_2} \mathbf{F} \cdot \mathbf{N}_2\, d\sigma \tag{12}$$

where S_1 is the top half of the sphere and S_2 is the bottom half. For S_1, $\mathbf{N}_1$ is a unit upper normal and an equation of S_1 is $z = \sqrt{16 - x^2 - y^2}$. Thus $f(x, y) = \sqrt{16 - x^2 - y^2}$. From (9),

$$\iint_{S_1} \mathbf{F} \cdot \mathbf{N}_1\, d\sigma = \iint_D [-Mf_x - Nf_y + R]\, dA$$

$$= \iint_D 5z\, dA$$

$$= 5 \iint_D \sqrt{16 - x^2 - y^2}\, dA$$

$$= 5 \int_0^{2\pi} \int_0^4 \sqrt{16 - r^2}\, r\, dr\, d\theta$$

$$= 5 \int_0^{2\pi} -\frac{1}{3}(16 - r^2)^{3/2} \Big]_0^4 d\theta$$

$$= \frac{320}{3} \int_0^{2\pi} d\theta$$

$$= \frac{640}{3}\pi$$

For S_2, $\mathbf{N}_2$ is a unit lower normal and as an equation of S_2 we have $z = -\sqrt{16 - x^2 - y^2}$. Therefore $f(x, y) = -\sqrt{16 - x^2 - y^2}$. From (11),

$$\iint_{S_2} \mathbf{F} \cdot \mathbf{N}_2\, d\sigma = \iint_D [Mf_x + Nf_y - R]\, dA$$

$$= \iint_D -5z\, dA$$

$$= 5 \iint_D \sqrt{16 - x^2 - y^2}\, dA$$

As in the computation for the flux of $\mathbf{F}$ across S_1 we get

$$\iint_{S_2} \mathbf{F} \cdot \mathbf{N}\, d\sigma = \frac{640}{3}\pi$$

Hence from (12),

$$\iint_S \mathbf{F} \cdot \mathbf{N}\, d\sigma = \frac{640}{3}\pi + \frac{640}{3}\pi$$

$$= \frac{1280}{3}\pi$$

Therefore the rate of flow of the fluid across the sphere is $\frac{1280}{3}\pi$ cm^3/hr.

The concept of flux is not limited to velocity fields of fluids. For instance, if $\mathbf{F}$ is an electric field, then surface integral (8) is an electric flux, and if $\mathbf{F}$ is a magnetic field, the surface integral is a magnetic flux. Surface integral (8) could also represent a flux of heat.

In Section 18.4 we had Gauss's divergence theorem in the plane (Theorem 18.4.3), which is a special case of Gauss's divergence theorem in three-dimensional space, which we now state.

18.5.1 THEOREM
Gauss's Divergence Theorem

Let M, N, and R be functions of three variables x, y, and z such that they have continuous first partial derivatives on an open ball B in R^3. Let S be a sectionally smooth closed surface lying in B, and let E be the region bounded by S. If $\mathbf{F}(x, y, z) = M(x, y, z)\mathbf{i} + N(x, y, z)\mathbf{j} + R(x, y, z)\mathbf{k}$, and $\mathbf{N}$ is a unit outward normal vector of S, then

$$\iint_S \mathbf{F} \cdot \mathbf{N}\, d\sigma = \iiint_E \operatorname{div} \mathbf{F}\, dV \tag{13}$$

Gauss's divergence theorem states that the flux of $\mathbf{F}$ across the boundary S of a region E in R^3 is the triple integral of the divergence of $\mathbf{F}$ over E. The proof of the theorem is beyond the scope of this book; it can be found in an advanced calculus text. The following example verifies it for the $\mathbf{F}$ and S of Example 4.

EXAMPLE 5
Use Gauss's divergence theorem to solve Example 4.

SOLUTION
From Gauss's divergence theorem, the flux of $\mathbf{F}$ across S is

$$\iint_S \mathbf{F} \cdot \mathbf{N}\, d\sigma = \iiint_E \operatorname{div} \mathbf{F}\, dV$$

Because $\mathbf{F}(x, y, z) = 5z\mathbf{k}$, $\operatorname{div} \mathbf{F} = \dfrac{\partial}{\partial z}(5z)$; that is, $\operatorname{div} \mathbf{F} = 5$. Thus

$$\iint_S \mathbf{F} \cdot \mathbf{N}\, d\sigma = 5 \iiint_E dV$$

Because the volume of E is the volume of a sphere of radius 4, we have

$$\iint_S \mathbf{F} \cdot \mathbf{N}\, d\sigma = 5\left[\tfrac{4}{3}\pi(4)^3\right]$$

$$= \tfrac{1280}{3}\pi$$

This result agrees with that of Example 4.

By comparing the solutions in Examples 4 and 5 observe how Gauss's divergence theorem can simplify the computation of a surface integral.

EXAMPLE 6
If $\mathbf{F}(x, y, z) = x^2 y\mathbf{i} + y^2\mathbf{j} + xz\mathbf{k}$, and S is the cube in the first octant bounded by the planes $x = 1$, $y = 1$, $z = 1$, and the coordinate planes, find the flux of $\mathbf{F}$ across S.

FIGURE 18.5.8

SOLUTION
The cube appears in Fig. 18.5.8. The flux of $\mathbf{F}$ across S is

$$\iint_S \mathbf{F} \cdot \mathbf{N} \, d\sigma$$

To compute this surface integral directly we would have to evaluate six surface integrals, one for each face of the cube. By applying Gauss's divergence theorem with

$$\text{div } \mathbf{F} = \frac{\partial}{\partial x} (x^2 y) + \frac{\partial}{\partial y} (y^2) + \frac{\partial}{\partial z} (xz)$$

$$= 2xy + 2y + x$$

we have

$$\iint_S \mathbf{F} \cdot \mathbf{N} \, d\sigma = \iiint_E \text{div } \mathbf{F} \, dV$$

$$= \int_0^1 \int_0^1 \int_0^1 (2xy + 2y + x) \, dz \, dy \, dx$$

$$= \int_0^1 \int_0^1 (2xy + 2y + x) \, dy \, dx$$

$$= \int_0^1 \left[xy^2 + y^2 + xy \right]_0^1 dx$$

$$= \int_0^1 (2x + 1) \, dx$$

$$= x^2 + x \Big]_0^1$$

$$= 2$$

In Section 18.4 we had a second vector form of Green's theorem, known as Stokes's theorem in the plane (Theorem 18.4.4):

$$\oint_C \mathbf{F} \cdot \mathbf{T} \, ds = \iint_D \text{curl } \mathbf{F} \cdot \mathbf{k} \, dA$$

where C is a sectionally smooth simple closed curve in R^2 and D is the region bounded by C. We now extend this theorem to three-dimensional space.

18.5.2 THEOREM
Stokes's Theorem

Let M, N, and R be functions of three variables x, y, and z such that they have continuous first partial derivatives on an open ball B in R^3. Let S be a sectionally smooth surface lying in B and let C be a sectionally smooth simple closed curve that is the boundary of S. If $\mathbf{F}(x, y, z)$ is $M(x, y, z)\mathbf{i} + N(x, y, z)\mathbf{j} + R(x, y, z)\mathbf{k}$, $\mathbf{N}$ is a unit upward normal vector of S and $\mathbf{T}$ is a unit tangent vector to C, then

$$\oint_C \mathbf{F} \cdot \mathbf{T} \, ds = \iint_S \text{curl } \mathbf{F} \cdot \mathbf{N} \, d\sigma \tag{14}$$

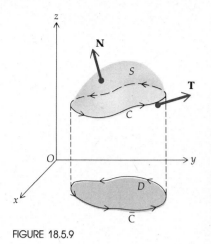

FIGURE 18.5.9

Stokes's theorem states that the line integral of the tangential component of a vector field **F** around the boundary C of a surface S can be computed by evaluating the surface integral of the normal component of the curl of **F** over S.

Theorem 18.5.2 is restricted to surfaces for which **N** is an upward normal of S. A complete statement of Stokes's theorem, involving surfaces having an orientation and for which a unit normal **N** can be adequately defined, can be found in an advanced calculus text. The proof of the theorem can also be found there.

Figure 18.5.9 shows a surface S with boundary curve C for which Theorem 18.5.2 applies. An equation of S is of the form $z = f(x, y)$, where f has continuous first partial derivatives on region D that is the projection of S onto the xy plane. Curve $\bar{C}$ is the projection of C onto the xy plane, and D and $\bar{C}$ are defined as in Green's theorem (Theorem 18.4.1). The positive direction along C is the same as the positive direction along $\bar{C}$, which is counterclockwise. Figure 18.5.9 also shows representations of vectors **N** and **T**.

EXAMPLE 7
Verify Stokes's theorem if $\mathbf{F}(x, y, z) = -4y\mathbf{i} + 2z\mathbf{j} + 3x\mathbf{k}$ and S is the portion of the paraboloid $z = 10 - x^2 - y^2$ above the plane $z = 1$.

SOLUTION
Figure 18.5.10 shows the surface S and the region D, which is the projection of S onto the xy plane. Region D is bounded by the circle $x^2 + y^2 = 9$. The curve C, which is the boundary of S, is the circle having its center at $(0, 0, 1)$ and radius 3 in the plane $z = 1$. Parametric equations of C are

$$x = 3 \cos s \qquad y = 3 \sin s \qquad z = 1 \qquad 0 \leq s \leq 2\pi$$

Therefore a vector equation of C is

$$\mathbf{R}(s) = 3 \cos s\mathbf{i} + 3 \sin s\mathbf{j} + \mathbf{k} \qquad 0 \leq s \leq 2\pi$$

Because $\mathbf{T}(s) = D_s\mathbf{R}(s)$,

$$\mathbf{T}(s) = -3 \sin s\mathbf{i} + 3 \cos s\mathbf{j}$$

At a point $P(3 \cos s, 3 \sin s, 1)$ on C, **F** is $-12 \sin s\mathbf{i} + 2\mathbf{j} + 9 \cos s\mathbf{k}$. Therefore,

$$\oint_C \mathbf{F} \cdot \mathbf{T} \, ds = \int_0^{2\pi} (-12 \sin s\mathbf{i} + 2\mathbf{j} + 9 \cos s\mathbf{k}) \cdot (-3 \sin s\mathbf{i} + 3 \cos s\mathbf{j}) \, ds$$

$$= \int_0^{2\pi} (36 \sin^2 s + 6 \cos s) \, ds$$

$$= 36 \int_0^{2\pi} \frac{1 - \cos 2s}{2} \, ds + 6 \int_0^{2\pi} \cos s \, ds$$

$$= 18s - 9 \sin 2s + 6 \sin s \Big]_0^{2\pi}$$

$$= 36\pi$$

FIGURE 18.5.10

We now compute the surface integral on the right side of Eq. (14).

$$\text{curl } \mathbf{F} = \begin{vmatrix} \mathbf{i} & \mathbf{j} & \mathbf{k} \\ \dfrac{\partial}{\partial x} & \dfrac{\partial}{\partial y} & \dfrac{\partial}{\partial z} \\ -4y & 2z & 3x \end{vmatrix}$$

$$= -2\mathbf{i} - 3\mathbf{j} + 4\mathbf{k}$$

Thus

$$\iint_S \text{curl } \mathbf{F} \cdot \mathbf{N} \, d\sigma = \iint_S (-2\mathbf{i} - 3\mathbf{j} + 4\mathbf{k}) \cdot \mathbf{N} \, d\sigma$$

To evaluate this surface integral we apply (9) because $\mathbf{N}$ is a unit upper normal. The vector field is $-2\mathbf{i} - 3\mathbf{j} + 4\mathbf{k}$; thus $M = -2$, $N = -3$, and $R = 4$. Because an equation of the surface is $z = 10 - x^2 - y^2$,

$$f(x, y) = 10 - x^2 - y^2 \qquad f_x(x, y) = -2x \qquad f_y(x, y) = -2y$$

Therefore, from (9),

$$\iint_S \text{curl } \mathbf{F} \cdot \mathbf{N} \, d\sigma = \iint_D [-(-2)(-2x) - (-3)(-2y) + 4] \, dA$$

$$= \iint_D (-4x - 6y + 4) \, dA$$

$$= \int_0^{2\pi} \int_0^3 (-4r \cos \theta - 6r \sin \theta + 4) r \, dr \, d\theta$$

$$= \int_0^{2\pi} \left[-\frac{4}{3} r^3 \cos \theta - 2r^3 \sin \theta + 2r^2 \right]_0^3 \, d\theta$$

$$= \int_0^{2\pi} (-36 \cos \theta - 54 \sin \theta + 18) \, d\theta$$

$$= -36 \sin \theta + 54 \cos \theta + 18\theta \Big]_0^{2\pi}$$

$$= 36\pi$$

which agrees with the result obtained from the line integral. We have therefore verified Stokes's theorem for this particular case.

EXAMPLE 8

Use Stokes's theorem to evaluate the line integral

$$\oint_C \mathbf{F} \cdot \mathbf{T} \, ds$$

if $\mathbf{F}(x, y, z) = xz\mathbf{i} + xy\mathbf{j} + y^2\mathbf{k}$ and C is the boundary of the surface consisting

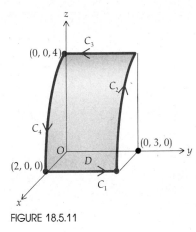

$(0, 0, 4)$

C_3

C_2

C_4

$(0, 3, 0)$

O

$(2, 0, 0)$

D

C_1

x

FIGURE 18.5.11

of the portion of the cylinder $z = 4 - x^2$ in the first octant that is cut off by the coordinate planes and the plane $y = 3$.

SOLUTION

Figure 18.5.11 shows the surface S and the boundary curve C that is composed of the four arcs C_1, C_2, C_3, and C_4. From Stokes's theorem,

$$\oint_C \mathbf{F} \cdot \mathbf{T}\, ds = \iint_S \text{curl } \mathbf{F} \cdot \mathbf{N}\, d\sigma$$

$$\text{curl } \mathbf{F} = \begin{vmatrix} \mathbf{i} & \mathbf{j} & \mathbf{k} \\ \dfrac{\partial}{\partial x} & \dfrac{\partial}{\partial y} & \dfrac{\partial}{\partial z} \\ xz & xy & y^2 \end{vmatrix}$$

$$= 2y\mathbf{i} + x\mathbf{j} + y\mathbf{k}$$

Thus

$$\oint_C \mathbf{F} \cdot \mathbf{T}\, ds = \iint_S (2y\mathbf{i} + x\mathbf{j} + y\mathbf{k}) \cdot \mathbf{N}\, d\sigma$$

Because $\mathbf{N}$ is a unit upper normal vector, we compute the value of the surface integral by applying (9). Because the vector field is $2y\mathbf{i} + x\mathbf{j} + y\mathbf{k}$, then $M = 2y$, $N = x$, and $R = y$. An equation of S is $z = 4 - x^2$. Therefore

$$f(x, y) = 4 - x^2 \qquad f_x(x, y) = -2x \qquad f_y(x, y) = 0$$

Hence, from (9), we have

$$\oint_C \mathbf{F} \cdot \mathbf{T}\, ds = \iint_D \left[-(2y)(-2x) - x(0) + y \right] dA$$

$$= \iint_D (4xy + y)\, dA$$

Region D is enclosed by the rectangle in the xy plane bounded by the x and y axes and the lines $x = 2$ and $y = 3$. Therefore

$$\oint_C \mathbf{F} \cdot \mathbf{T}\, ds = \int_0^2 \int_0^3 (4xy + y)\, dy\, dx$$

$$= \int_0^2 \left[2xy^2 + \frac{1}{2} y^2 \right]_0^3 dx$$

$$= \int_0^2 (18x + \tfrac{9}{2})\, dx$$

$$= 9x^2 + \tfrac{9}{2}x \Big]_0^2$$

$$= 45$$

EXERCISES 18.5

In Exercises 1 through 10, evaluate the surface integral $\iint\limits_{S} G(x, y, z)\, d\sigma$ for the given G and S.

1. $G(x, y, z) = z$; S is the hemisphere $x^2 + y^2 + z^2 = 4$ above the xy plane.
2. $G(x, y, z) = x$; S is the portion of the plane $x + y + z = 1$ in the first octant.
3. $G(x, y, z) = x$; S is the portion of the cylinder $z = x^2$ in the first octant bounded by the coordinate planes and the planes $x = 1$ and $y = 2$.
4. $G(x, y, z) = y$; S is the portion of the cylinder $z = 4 - y^2$ in the first octant bounded by the coordinate planes and the plane $x = 3$.
5. $G(x, y, z) = z^2$; S is the portion of the cone $x^2 + y^2 = z^2$ between the planes $z = 1$ and $z = 2$.
6. $G(x, y, z) = xyz$; S is the portion of the cone $x^2 + y^2 = z^2$ between the planes $z = 1$ and $z = 2$.
7. $G(x, y, z) = x + y$; S is the portion of the plane $4x + 3y + 6z = 12$ in the first octant.
8. $G(x, y, z) = \sqrt{x^2 + y^2 + z^2}$; S is the portion of the cone $x^2 + y^2 = z^2$ between the xy plane and the plane $z = 2$.
9. $G(x, y, z) = xyz$; S is the portion of the cylinder $x^2 + z^2 = 4$ between the planes $y = 1$ and $y = 3$.
10. $G(x, y, z) = x^2$; S is the hemisphere $x^2 + y^2 + z^2 = 9$ above the xy plane.

In Exercises 11 through 14, find the mass of the given surface S if the area density at any point (x, y, z) on the surface is $\rho(x, y, z)$ kg/m^2.

11. S is the hemisphere $x^2 + y^2 + z^2 = 4$ above the xy plane; $\rho(x, y, z) = k\sqrt{x^2 + y^2 + z^2}$, where k is a constant.
12. S is the portion of the plane $3x + 2y + z = 6$ in the first octant; $\rho(x, y, z) = y + 2z$.
13. S is the portion of the cone $x^2 + y^2 = z^2$ between the planes $z = 2$ and $z = 3$; $\rho(x, y, z) = y^2 z^2$.
14. S is the portion of the sphere $x^2 + y^2 + z^2 = 16$ in the first octant; $\rho(x, y, z) = kz^2$, where k is a constant.
15. The velocity field of a fluid is $\mathbf{F}(x, y, z) = -2y\mathbf{i} + 2x\mathbf{j} + 5\mathbf{k}$, and surface S is that part of the sphere $x^2 + y^2 + z^2 = 16$ that is above the region D in the xy plane enclosed by the circle $x^2 + y^2 = 9$. Find the flux of $\mathbf{F}$ across S.
16. The velocity field of a fluid is given by $\mathbf{F}(x, y, z) = 3x\mathbf{i} + 3y\mathbf{j} + 6z\mathbf{k}$, and surface S is that portion of the paraboloid $z = 4 - x^2 - y^2$ above the xy plane. Find the flux of $\mathbf{F}$ across S.
17. Suppose $\mathbf{F}(x, y, z) = x^2\mathbf{i} + xy\mathbf{j} + 2z\mathbf{k}$, and S is the cube in the first octant bounded by the coordinate planes and the planes $x = 1$, $y = 1$, and $z = 1$. Find the flux of $\mathbf{F}$ across S by evaluating six surface integrals, one for each face of the cube.
18. If $\mathbf{F}(x, y, z) = 3x\mathbf{i} + y^2\mathbf{j} + yz\mathbf{k}$ and S is the cube of Exercise 17, find the flux of $\mathbf{F}$ across S by evaluating six surface integrals, one for each face of the cube.
19. Solve Exercise 17 by using Gauss's divergence theorem.
20. Solve Exercise 18 by using Gauss's divergence theorem.

In Exercises 21 and 22, verify Gauss's divergence theorem for the given $\mathbf{F}$ and S.
21. $\mathbf{F}(x, y, z) = 2z\mathbf{k}$; S is the sphere $x^2 + y^2 + z^2 = 1$.   22. $\mathbf{F}(x, y, z) = 3z\mathbf{k}$; S is the sphere $x^2 + y^2 + z^2 = 9$.

In Exercises 23 through 28, use Gauss's divergence theorem to evaluate the surface integral $\iint\limits_{S} \mathbf{F} \cdot \mathbf{N}\, d\sigma$ for the given $\mathbf{F}$ and S.

23. $\mathbf{F}(x, y, z) = x\mathbf{i} + y\mathbf{j} + z\mathbf{k}$; S is the sphere $x^2 + y^2 + z^2 = 4$.
24. $\mathbf{F}(x, y, z) = x\mathbf{i} + y\mathbf{j} + z\mathbf{k}$; S is the boundary of the region enclosed on the side by the cylinder $x^2 + y^2 = 9$, below by the xy plane, and above by the plane $z = 4$.
25. $\mathbf{F}(x, y, z) = x^2\mathbf{i} + y^2\mathbf{j} + z^2\mathbf{k}$; S is the surface of Exercise 24.
26. $\mathbf{F}(x, y, z) = x^2\mathbf{i} + y^2\mathbf{j} + z^2\mathbf{k}$; S is the surface of Exercise 23.
27. $\mathbf{F}(x, y, z) = 2x\mathbf{i} + 2yz\mathbf{j} + 3z\mathbf{k}$; S is the boundary of the region enclosed by the coordinate planes and the plane $x + y + z = 1$.
28. $\mathbf{F}(x, y, z) = \dfrac{x\mathbf{i} + y\mathbf{j} + z\mathbf{k}}{x^2 + y^2 + z^2}$; S is the boundary of the region outside the sphere $x^2 + y^2 + z^2 = 1$ and inside the sphere $x^2 + y^2 + z^2 = 4$.

In Exercises 29 through 32, verify Stokes's theorem for the given $\mathbf{F}$ and S.

29. $\mathbf{F}(x, y, z) = -3y\mathbf{i} + 3x\mathbf{j} + 2\mathbf{k}$; S is the portion of the plane $z = 1$ inside the cylinder $x^2 + y^2 = 9$.

30. $\mathbf{F}(x, y, z) = 2z\mathbf{i} + 3x\mathbf{j} + 4z\mathbf{k}$; S is the portion of the paraboloid $z = 4 - x^2 - y^2$ above the xy plane.

31. $\mathbf{F}(x, y, z) = y^2\mathbf{i} + x\mathbf{j} + z^2\mathbf{k}$; S is the portion of the paraboloid $z = x^2 + y^2$ below the plane $z = 1$.

32. $\mathbf{F}(x, y, z) = xy\mathbf{i} + y^2\mathbf{j} + 2\mathbf{k}$; S is the surface of Exercise 31.

In Exercises 33 through 36, use Stokes's theorem to evaluate the line integral $\oint_C \mathbf{F} \cdot \mathbf{T}\, ds$ for the given $\mathbf{F}$ and C.

33. $\mathbf{F}(x, y, z) = -y\mathbf{i} + x\mathbf{j} + z\mathbf{k}$; C is the circle $x^2 + y^2 = 4$ in the xy plane.

34. $\mathbf{F}(x, y, z) = yz\mathbf{i} + xy\mathbf{j} + xz\mathbf{k}$; C is the square having vertices at $(0, 0, 0)$, $(2, 0, 0)$, $(0, 2, 0)$, and $(2, 2, 0)$.

35. $\mathbf{F}(x, y, z) = 4y\mathbf{i} - 3z\mathbf{j} + x\mathbf{k}$; C is the triangle having vertices at $(1, 0, 0)$, $(0, 1, 0)$, and $(0, 0, 1)$.

36. $\mathbf{F}(x, y, z) = (y - x)\mathbf{i} + (x - z)\mathbf{j} + (x - y)\mathbf{k}$; C is the triangle having vertices at $(2, 0, 0)$, $(0, 2, 0)$, and $(0, 0, 1)$.

37. Let C be a sectionally smooth simple closed curve that is the boundary of a surface S having the equation $z = f(x, y)$. If $\mathbf{F}(x, y, z) = ze^y\mathbf{i} + xze^y\mathbf{j} + xe^y\mathbf{k}$, use Stokes's theorem to show that $\oint_C \text{curl } \mathbf{F} \cdot \mathbf{N}\, d\sigma = 0$.

REVIEW EXERCISES FOR CHAPTER 18

In Exercises 1 and 2, find a conservative vector field having the given potential function.

1. (a) $f(x, y) = 2x^2y + 3xy^3$; (b) $f(x, y, z) = xe^y - yze^y$

2. (a) $f(x, y) = e^x \cos y + x \sin y$; (b) $f(x, y, z) = \dfrac{1}{x^2 + y^2 + z^2}$

In Exercises 3 through 6, prove that the given vector field is conservative, and find a potential function.

3. $\mathbf{F}(x, y) = \dfrac{2y^2}{1 + 4x^2y^4}\mathbf{i} + \dfrac{4xy}{1 + 4x^2y^4}\mathbf{j}$

4. $\mathbf{F}(x, y, z) = (6x - 4y)\mathbf{i} + (z - 4x)\mathbf{j} + (y - 8z)\mathbf{k}$

5. $\mathbf{F}(x, y, z) = z^2 \sec^2 x\mathbf{i} + 2ye^{3z}\mathbf{j} + (3y^2e^{3z} + 2z \tan x)\mathbf{k}$

6. $\mathbf{F}(x, y) = (y \sin x - \sin y)\mathbf{i} - (x \cos y + \cos x)\mathbf{j}$

In Exercises 7 through 10, find curl $\mathbf{F}$ and div $\mathbf{F}$ for the given vector field.

7. $\mathbf{F}(x, y, z) = e^{yz}\mathbf{i} + e^{xz}\mathbf{j} + e^{xy}\mathbf{k}$

8. $\mathbf{F}(x, y) = \sin y\mathbf{i} + \sin x\mathbf{k}$

9. $\mathbf{F}(x, y) = \dfrac{1}{y}\mathbf{i} - \dfrac{2x}{y}\mathbf{j}$

10. $\mathbf{F}(x, y, z) = \dfrac{x}{y}\mathbf{i} + \dfrac{y}{z}\mathbf{j} + \dfrac{z}{x}\mathbf{k}$

In Exercises 11 through 18, evaluate the line integral over the given curve.

11. $\displaystyle\int_C \mathbf{F} \cdot d\mathbf{R}$; $\mathbf{F}(x, y) = 3y\mathbf{i} - 4x\mathbf{j}$; C: $\mathbf{R}(t) = 2t^2\mathbf{i} - t\mathbf{j}$, $0 \le t \le 1$.

12. $\displaystyle\int_C \mathbf{F} \cdot d\mathbf{R}$; $\mathbf{F}(x, y) = (x + y)\mathbf{i} + (y - x)\mathbf{j}$; C: $\mathbf{R}(t) = t^3\mathbf{i} + t^2\mathbf{j}$ from the point $(8, 4)$ to the point $(1, 1)$.

13. $\displaystyle\int_C (2x + 3y)\, dx + xy\, dy$; C: $\mathbf{R}(t) = 4 \sin t\mathbf{i} - \cos t\mathbf{j}$; $0 \le t \le \frac{1}{2}\pi$.

14. $\displaystyle\int_C (2x + y)\, dx + (x - 2y)\, dy$; C: $x^2 + y^2 = 9$.

15. $\displaystyle\int_C y^2\, dx + z^2\, dy + x^2\, dz$; C: $\mathbf{R}(t) = (t - 1)\mathbf{i} + (t + 1)\mathbf{j} + t^2\mathbf{k}$; $0 \le t \le 1$.

16. $\displaystyle\int_C xe^y\, dx - xe^z\, dy + e^z\, dz$; C: $\mathbf{R}(t) = t\mathbf{i} + t^2\mathbf{j} + t^3\mathbf{k}$; $0 \le t \le 1$.

17. $\int_C \mathbf{F} \cdot d\mathbf{R}$; $\mathbf{F}(x, y, z) = 3xy\mathbf{i} + (4y^2 - xz)\mathbf{j} + 6z\mathbf{k}$; C: the twisted cubic $\mathbf{R}(t) = t\mathbf{i} + t^2\mathbf{j} + t^3\mathbf{k}$, $0 \le t \le 1$.

18. $\int_C \mathbf{F} \cdot d\mathbf{R}$; $\mathbf{F}(x, y, z) = 2x\mathbf{i} + 3y\mathbf{j} + z\mathbf{k}$; C: the circular helix $\mathbf{R}(t) = 2\cos t\mathbf{i} + 2\sin t\mathbf{j} + t\mathbf{k}$; $0 \le t \le 2\pi$.

In Exercises 19 through 26, prove that the value of the given line integral is independent of the path, and compute the value in any convenient manner. In each exercise C is any sectionally smooth curve from the point A to the point B.

19. $\int_C 2xe^y \, dx + x^2 e^y \, dy$; A is $(1, 0)$ and B is $(3, 2)$.

20. $\int_C \left(\frac{1}{y} - y \right) dx + \left(-\frac{x}{y^2} - x \right) dy$; A is $(0, 1)$ and B is $(6, 3)$.

21. $\int_C \mathbf{F} \cdot d\mathbf{R}$; $\mathbf{F}(x, y) = (\cos y - y \cos x)\mathbf{i} - (\sin x + x \sin y)\mathbf{j}$; A is $(0, \frac{1}{2}\pi)$ and B is $(\pi, 0)$.

22. $\int_C \mathbf{F} \cdot d\mathbf{R}$; $\mathbf{F}(x, y) = (2xy - 2y)\mathbf{i} + (x^2 - 2x + 3y^2)\mathbf{j}$; A is $(2, -1)$ and B is $(3, 2)$.

23. $\int_C 3y \, dx + (3x + 4y) \, dy - 2z \, dz$; A is $(0, 1, -1)$ and B is $(1, 2, 0)$.

24. $\int_C z \sin y \, dx + xz \cos y \, dy + x \sin y \, dz$; A is $(0, 0, 0)$ and B is $(2, 3, \frac{1}{2}\pi)$.

25. $\int_C \mathbf{F} \cdot d\mathbf{R}$; $\mathbf{F}(x, y, z) = \left(\frac{1}{y} - \frac{2z}{x^2} \right)\mathbf{i} - \left(\frac{1}{z} + \frac{x}{y^2} \right)\mathbf{j} + \left(\frac{2}{x} + \frac{y}{z^2} \right)\mathbf{k}$; A is $(2, -1, 1)$ and B is $(4, 2, -2)$.

26. $\int_C \mathbf{F} \cdot d\mathbf{R}$; $\mathbf{F}(x, y, z) = (2xy + 3yz)\mathbf{i} + (x^2 - 4yz + 3xz)\mathbf{j} + (3xy - 2y^2)\mathbf{k}$; A is $(0, 2, 1)$ and B is $(1, -1, 4)$.

In Exercises 27 through 30, use Green's theorem to evaluate the line integral.

27. $\oint_C (3x + 2y) \, dx + (3x + y^2) \, dy$, where C is the ellipse $16x^2 + 9y^2 = 144$.

28. $\oint_C \ln(y + 1) \, dx - \frac{xy}{y + 1} \, dy$, where C is the closed curve determined by the curve $\sqrt{x} + \sqrt{y} = 2$ and the intervals $[0, 4]$ on the x and y axes.

29. $\oint_C e^x \sin y \, dx + e^x \cos y \, dy$, where C is any smooth closed curve.

30. $\oint_C (x^2 - y^3) \, dx + (y^2 + x^3) \, dy$; where C is the circle $x^2 + y^2 = 1$.

In Exercises 31 and 32, use Theorem 18.4.2 to find the area of the given region.

31. The region enclosed by the parabola $y = x^2$ and the line $y = x + 2$.

32. The region enclosed by the two parabolas $y = x^2$ and $x^2 = 18 - y$.

In Exercises 33 through 36, find the total work done in moving an object along the given arc C if the motion is caused by the given force field. Assume the arc is measured in meters and the force is measured in newtons.

33. $F(x, y) = 2x^2 y\mathbf{i} + (x^2 + 3y)\mathbf{j}$; C: the arc of the parabola $y = 3x^2 + 2x + 4$ from $(0, 4)$ to $(1, 9)$.

34. $F(x, y) = xy^2\mathbf{i} - x^2 y\mathbf{j}$; C: the arc of the circle $x^2 + y^2 = 4$ from $(2, 0)$ to $(0, 2)$.

35. $F(x, y, z) = (xy - z)\mathbf{i} + y\mathbf{j} + z\mathbf{k}$; C: the line segment from the origin to the point $(4, 1, 2)$.

36. $F(x, y, z) = xyz\mathbf{i} + e^y\mathbf{j} + (x + z)\mathbf{k}$; C: $\mathbf{R}(t) = 3t\mathbf{i} + t^2\mathbf{j} + 2t\mathbf{k}$; $0 \le t \le 3$.

In Exercises 37 and 38, verify Gauss's divergence theorem in the plane and Stokes's theorem in the plane for the given F and R.

37. $\mathbf{F}(x, y) = 4y\mathbf{i} + 3x\mathbf{j}$, and R is the region bounded by $x^{2/3} + y^{2/3} = 1$.

38. $\mathbf{F}(x, y) = 3x^2\mathbf{i} + 4y^2\mathbf{j}$ and R is the region bounded by the ellipse $9x^2 + 16y^2 = 144$.

In Exercises 39 and 40, use Green's theorem to find the total work done in moving an object in the counterclockwise direction once around the given curve C if the motion is caused by the given force field $F(x, y)$. Assume the arc is measured in meters and the force is measured in newtons.

39. C is the circle $x^2 + y^2 = 4$; $F(x, y) = (xy^2 + \cos x)\mathbf{i} + (x^2 + e^y)\mathbf{j}$.

40. C is the ellipse $9x^2 + y^2 = 9$; $F(x, y) = (2x - 3y)\mathbf{i} + (x + 2y)\mathbf{j}$.

In Exercises 41 and 42, find the rate of flow of the fluid out of a region R bounded by the given curve C if F is the velocity field of the fluid. Assume the velocity is measured in centimeters per second and the area of R is measured in square centimeters.

41. $F(x, y) = (4x - 3y)\mathbf{i} + (5y - 4x^2)\mathbf{j}$; C is the right triangle with vertices at $(0, 1)$, $(0, 4)$, and $(4, 4)$.

42. $F(x, y) = (y^2 + 12x)\mathbf{i} + (4y - x^2)\mathbf{j}$; C is the ellipse $x^2 + 4y^2 = 16$.

43. Find the value of the line integral

$$\int_C \frac{-y}{x^2 + y^2} \, dx + \frac{x}{x^2 + y^2} \, dy$$

if C is the arc of the circle $x^2 + y^2 = 4$ from $(\sqrt{2}, \sqrt{2})$ to $(-\sqrt{2}, \sqrt{2})$.

44. Apply Green's theorem to compute the area of the quadrilateral having vertices at the points $(0, 0)$, $(3, 2)$, $(1, 5)$, and $(-2, 1)$.

Exercises 45 through 52 pertain to Supplementary Section 18.5.

45. Evaluate the surface integral $\iint_S x \, d\sigma$, where S is the portion of the cylinder $z = 9 - x^2$ in the first octant bounded by the coordinate planes and the plane $y = 2$.

46. Evaluate the surface integral $\iint_S xyz \, d\sigma$, where S is the portion of the cylinder $y^2 + z^2 = 9$ between the planes $x = 1$ and $x = 4$.

47. Find the mass of the portion of the sphere $x^2 + y^2 + z^2 = 4$ in the first octant if the area density at any point (x, y, z) on the surface is kz^2 kg/m², where k is a constant.

48. Suppose surface S is that part of the sphere $x^2 + y^2 + z^2 = 9$ that is above the region D in the xy plane enclosed by the circle $x^2 + y^2 = 1$. If the velocity field of a fluid is given by $F(x, y, z) = -y\mathbf{i} + x\mathbf{j} + 3\mathbf{k}$, find the flux of F across S.

49. Use Gauss's divergence theorem to evaluate the surface integral $\iint_S F \cdot N \, d\sigma$ if $F(x, y, z) = 2x\mathbf{i} + y\mathbf{j} + 2z\mathbf{k}$ and S is the boundary of the region enclosed on the side by the cylinder $x^2 + y^2 = 16$, below by the xy plane and above by the plane $z = 2$.

50. Verify Gauss's divergence theorem if $F(x, y, z) = \frac{1}{2}z\mathbf{i}$ and S is the sphere $x^2 + y^2 + z^2 = 4$.

51. Verify Stokes's theorem if $F(x, y, z) = z\mathbf{i} + 4x\mathbf{j} + 2z\mathbf{k}$ and S is the portion of the paraboloid $z = 9 - x^2 - y^2$ above the xy plane.

52. Use Stokes's theorem to evaluate the line integral $\oint_C F \cdot T \, ds$ if $F(x, y, z) = -2y\mathbf{i} + 3x\mathbf{j} + z\mathbf{k}$ and C is the circle $x^2 + y^2 = 1$ in the xy plane.

APPENDIX

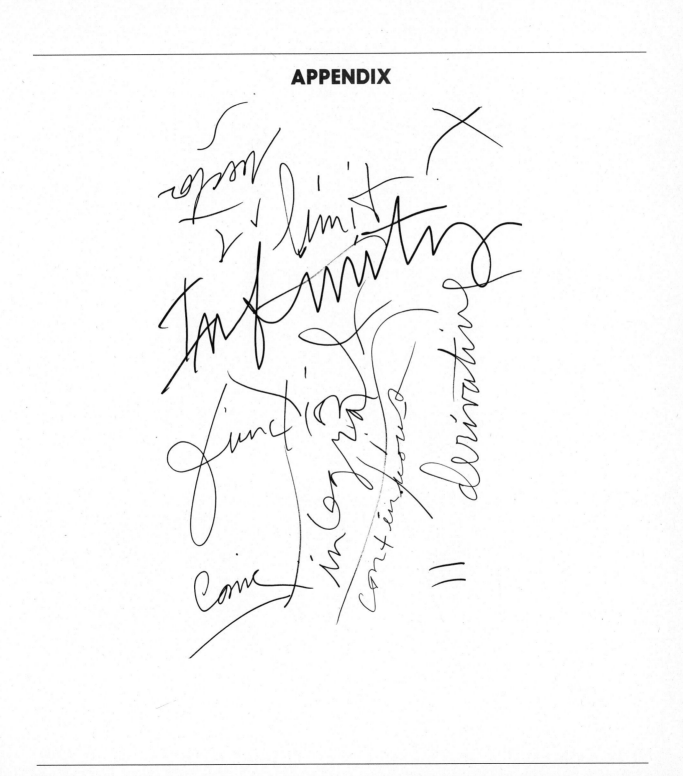

TABLES

Table 1 Powers and Roots

n	n^2	$\sqrt{n}$	n^3	$\sqrt[3]{n}$	n	n^2	$\sqrt{n}$	n^3	$\sqrt[3]{n}$
1	1	1.000	1	1.000	51	2,601	7.141	132,651	3.708
2	4	1.414	8	1.260	52	2,704	7.211	140,608	3.732
3	9	1.732	27	1.442	53	2,809	7.280	148,877	3.756
4	16	2.000	64	1.587	54	2,916	7.348	157,464	3.780
5	25	2.236	125	1.710	55	3,025	7.416	166,375	3.803
6	36	2.449	216	1.817	56	3,136	7.483	175,616	3.826
7	49	2.646	343	1.913	57	3,249	7.550	185,193	3.848
8	64	2.828	512	2.000	58	3,364	7.616	195,112	3.871
9	81	3.000	729	2.080	59	3,481	7.681	205,379	3.893
10	100	3.162	1,000	2.154	60	3,600	7.746	216,000	3.915
11	121	3.317	1,331	2.224	61	3,721	7.810	226,981	3.936
12	144	3.464	1,728	2.289	62	3,844	7.874	238,328	3.958
13	169	3.606	2,197	2.351	63	3,969	7.937	250,047	3.979
14	196	3.742	2,744	2.410	64	4,096	8.000	262,144	4.000
15	225	3.873	3,375	2.466	65	4,225	8.062	274,625	4.021
16	256	4.000	4,096	2.520	66	4,356	8.124	287,496	4.041
17	289	4.123	4,913	2.571	67	4,489	8.185	300,763	4.062
18	324	4.243	5,832	2.621	68	4,624	8.246	314,432	4.082
19	361	4.359	6,859	2.668	69	4,761	8.307	328,509	4.102
20	400	4.472	8,000	2.714	70	4,900	8.367	343,000	4.121
21	441	4.583	9,261	2.759	71	5,041	8.426	357,911	4.141
22	484	4.690	10,648	2.802	72	5,184	8.485	373,248	4.160
23	529	4.796	12,167	2.844	73	5,329	8.544	389,017	4.179
24	576	4.899	13,824	2.884	74	5,476	8.602	405,224	4.198
25	625	5.000	15,625	2.924	75	5,625	8.660	421,875	4.217
26	676	5.099	17,576	2.962	76	5,776	8.718	438,976	4.236
27	729	5.196	19,683	3.000	77	5,929	8.775	456,533	4.254
28	784	5.291	21,952	3.037	78	6,084	8.832	474,552	4.273
29	841	5.385	24,389	3.072	79	6,241	8.888	493,039	4.291
30	900	5.477	27,000	3.107	80	6,400	8.944	512,000	4.309
31	961	5.568	29,791	3.141	81	6,561	9.000	531,441	4.327
32	1,024	5.657	32,768	3.175	82	6,724	9.055	551,368	4.344
33	1,089	5.745	35,937	3.208	83	6,889	9.110	571,787	4.362
34	1,156	5.831	39,304	3.240	84	7,056	9.165	592,704	4.380
35	1,225	5.916	42,875	3.271	85	7,225	9.220	614,125	4.397
36	1,296	6.000	46,656	3.302	86	7,396	9.274	636,056	4.414
37	1,369	6.083	50,653	3.332	87	7,569	9.327	658,503	4.431
38	1,444	6.164	54,872	3.362	88	7,744	9.381	681,472	4.448
39	1,521	6.245	59,319	3.391	89	7,921	9.434	704,969	4.465
40	1,600	6.325	64,000	3.420	90	8,100	9.487	729,000	4.481
41	1,681	6.403	68,921	3.448	91	8,281	9.539	753,571	4.498
42	1,764	6.481	74,088	3.476	92	8,464	9.592	778,688	4.514
43	1,849	6.557	79,507	3.503	93	8,649	9.643	804,357	4.531
44	1,936	6.633	85,184	3.530	94	8,836	9.695	830,584	4.547
45	2,025	6.708	91,125	3.557	95	9,025	9.747	857,375	4.563
46	2,116	6.782	97,336	3.583	96	9,216	9.798	884,736	4.579
47	2,209	6.856	103,823	3.609	97	9,409	9.849	912,673	4.595
48	2,304	6.928	110,592	3.634	98	9,604	9.899	941,192	4.610
49	2,401	7.000	117,649	3.659	99	9,801	9.950	970,299	4.626
50	2,500	7.071	125,000	3.684	100	10,000	10.000	1,000,000	4.642

Table 2   Natural Logarithms

N	0	1	2	3	4	5	6	7	8	9
1.0	0000	0100	0198	0296	0392	0488	0583	0677	0770	0862
1.1	0953	1044	1133	1222	1310	1398	1484	1570	1655	1740
1.2	1823	1906	1989	2070	2151	2231	2311	2390	2469	2546
1.3	2624	2700	2776	2852	2927	3001	3075	3148	3221	3293
1.4	3365	3436	3507	3577	3646	3716	3784	3853	3920	3988
1.5	4055	4121	4187	4253	4318	4383	4447	4511	4574	4637
1.6	4700	4762	4824	4886	4947	5008	5068	5128	5188	5247
1.7	5306	5365	5423	5481	5539	5596	5653	5710	5766	5822
1.8	5878	5933	5988	6043	6098	6152	6206	6259	6313	6366
1.9	6419	6471	6523	6575	6627	6678	6729	6780	6831	6881
2.0	6931	6981	7031	7080	7129	7178	7227	7275	7324	7372
2.1	7419	7467	7514	7561	7608	7655	7701	7747	7793	7839
2.2	7885	7930	7975	8020	8065	8109	8154	8198	8242	8286
2.3	8329	8372	8416	8459	8502	8544	8587	8629	8671	8713
2.4	8755	8796	8838	8879	8920	8961	9002	9042	9083	9123
2.5	9163	9203	9243	9282	9322	9361	9400	9439	9478	9517
2.6	9555	9594	9632	9670	9708	9746	9783	9821	9858	9895
2.7	9933	9969	*0006	*0043	*0080	*0116	*0152	*0188	*0225	*0260
2.8	1.0296	0332	0367	0403	0438	0473	0508	0543	0578	0613
2.9	0647	0682	0716	0750	0784	0818	0852	0886	0919	0953
3.0	1.0986	1019	1053	1086	1119	1151	1184	1217	1249	1282
3.1	1314	1346	1378	1410	1442	1474	1506	1537	1569	1600
3.2	1632	1663	1694	1725	1756	1787	1817	1848	1878	1909
3.3	1939	1969	2000	2030	2060	2090	2119	2149	2179	2208
3.4	2238	2267	2296	2326	2355	2384	2413	2442	2470	2499
3.5	1.2528	2556	2585	2613	2641	2669	2698	2726	2754	2782
3.6	2809	2837	2865	2892	2920	2947	2975	3002	3029	3056
3.7	3083	3110	3137	3164	3191	3218	3244	3271	3297	3324
3.8	3350	3376	3403	3429	3455	3481	3507	3533	3558	3584
3.9	3610	3635	3661	3686	3712	3737	3762	3788	3813	3838
4.0	1.3863	3888	3913	3938	3962	3987	4012	4036	4061	4085
4.1	4110	4134	4159	4183	4207	4231	4255	4279	4303	4327
4.2	4351	4375	4398	4422	4446	4469	4493	4516	4540	4563
4.3	4586	4609	4633	4656	4679	4702	4725	4748	4770	4793
4.4	4816	4839	4861	4884	4907	4929	4951	4974	4996	5019
4.5	1.5041	5063	5085	5107	5129	5151	5173	5195	5217	5239
4.6	5261	5282	5304	5326	5347	5369	5390	5412	5433	5454
4.7	5476	5497	5518	5539	5560	5581	5602	5623	5644	5665
4.8	5686	5707	5728	5748	5769	5790	5810	5831	5851	5872
4.9	5892	5913	5933	5953	5974	5994	6014	6034	6054	6074
5.0	1.6094	6114	6134	6154	6174	6194	6214	6233	6253	6273
5.1	6292	6312	6332	6351	6371	6390	6409	6429	6448	6467
5.2	6487	6506	6525	6544	6563	6582	6601	6620	6639	6658
5.3	6677	6696	6715	6734	6752	6771	6790	6808	6827	6845
5.4	6864	6882	6901	6919	6938	6956	6974	6993	7011	7029

Table 2   (Continued)

N	0	1	2	3	4	5	6	7	8	9
5.5	1.7047	7066	7084	7102	7120	7138	7156	7174	7192	7210
5.6	7228	7246	7263	7281	7299	7317	7334	7352	7370	7387
5.7	7405	7422	7440	7457	7475	7492	7509	7527	7544	7561
5.8	7579	7596	7613	7630	7647	7664	7681	7699	7716	7733
5.9	7750	7766	7783	7800	7817	7834	7851	7867	7884	7901
6.0	1.7918	7934	7951	7967	7984	8001	8017	8034	8050	8066
6.1	8083	8099	8116	8132	8148	8165	8181	8197	8213	8229
6.2	8245	8262	8278	8294	8310	8326	8342	8358	8374	8390
6.3	8405	8421	8437	8453	8469	8485	8500	8516	8532	8547
6.4	8563	8579	8594	8610	8625	8641	8656	8672	8687	8703
6.5	1.8718	8733	8749	8764	8779	8795	8810	8825	8840	8856
6.6	8871	8886	8901	8916	8931	8946	8961	8976	8991	9006
6.7	9021	9036	9051	9066	9081	9095	9110	9125	9140	9155
6.8	9169	9184	9199	9213	9228	9242	9257	9272	9286	9301
6.9	9315	9330	9344	9359	9373	9387	9402	9416	9430	9445
7.0	1.9459	9473	9488	9502	9516	9530	9544	9559	9573	9587
7.1	9601	9615	9629	9643	9657	9671	9685	9699	9713	9727
7.2	9741	9755	9769	9782	9796	9810	9824	9838	9851	9865
7.3	9879	9892	9906	9920	9933	9947	9961	9974	9988	*0001
7.4	2.0015	0028	0042	0055	0069	0082	0096	0109	0122	0136
7.5	2.0149	0162	0176	0189	0202	0215	0229	0242	0255	0268
7.6	0281	0295	0308	0321	0334	0347	0360	0373	0386	0399
7.7	0412	0425	0438	0451	0464	0477	0490	0503	0516	0528
7.8	0541	0554	0567	0580	0592	0605	0618	0630	0643	0656
7.9	0669	0681	0694	0707	0719	0732	0744	0757	0769	0782
8.0	2.0794	0807	0819	0832	0844	0857	0869	0882	0894	0906
8.1	0919	0931	0943	0956	0968	0980	0992	1005	1017	1029
8.2	1041	1054	1066	1078	1090	1102	1114	1126	1138	1150
8.3	1163	1175	1187	1199	1211	1223	1235	1247	1258	1270
8.4	1282	1294	1306	1318	1330	1342	1353	1365	1377	1389
8.5	2.1401	1412	1424	1436	1448	1459	1471	1483	1494	1506
8.6	1518	1529	1541	1552	1564	1576	1587	1599	1610	1622
8.7	1633	1645	1656	1668	1679	1691	1702	1713	1725	1736
8.8	1748	1759	1770	1782	1793	1804	1815	1827	1838	1849
8.9	1861	1872	1883	1894	1905	1917	1928	1939	1950	1961
9.0	2.1972	1983	1994	2006	2017	2028	2039	2050	2061	2072
9.1	2083	2094	2105	2116	2127	2138	2148	2159	2170	2181
9.2	2192	2203	2214	2225	2235	2246	2257	2268	2279	2289
9.3	2300	2311	2322	2332	2343	2354	2364	2375	2386	2396
9.4	2407	2418	2428	2439	2450	2460	2471	2481	2492	2502
9.5	2.2513	2523	2534	2544	2555	2565	2576	2586	2597	2607
9.6	2618	2628	2638	2649	2659	2670	2680	2690	2701	2711
9.7	2721	2732	2742	2752	2762	2773	2783	2793	2803	2814
9.8	2824	2834	2844	2854	2865	2875	2885	2895	2905	2915
9.9	2925	2935	2946	2956	2966	2976	2986	2996	3006	3016

Use ln 10 = 2.30259 to find logarithms of numbers greater than 10 or less than 1. *Example:* ln 220 = ln 2.2 + 2 ln 10 = 0.7885 + 2(2.30259) = 5.3937.

Table 3 Exponential Functions

x	e^x	$\log_{10}(e^x)$	e^{-x}	x	e^x	$\log_{10}(e^x)$	e^{-x}
0.00	1.0000	0.00000	1.000000	**0.50**	1.6487	0.21715	0.606531
0.01	1.0101	.00434	0.990050	0.51	1.6653	.22149	.600496
0.02	1.0202	.00869	.980199	0.52	1.6820	.22583	.594521
0.03	1.0305	.01303	.970446	0.53	1.6989	.23018	.588605
0.04	1.0408	.01737	.960789	0.54	1.7160	.23452	.582748
0.05	1.0513	0.02171	0.951229	**0.55**	1.7333	0.23886	0.576950
0.06	1.0618	.02606	.941765	0.56	1.7507	.24320	.571209
0.07	1.0725	.03040	.932394	0.57	1.7683	.24755	.565525
0.08	1.0833	.03474	.923116	0.58	1.7860	.25189	.559898
0.09	1.0942	.03909	.913931	0.59	1.8040	.35623	.554327
0.10	1.1052	0.04343	0.904837	**0.60**	1.8221	0.26058	0.548812
0.11	1.1163	.04777	.895834	0.61	1.8404	.26492	.543351
0.12	1.1275	.05212	.886920	0.62	1.8589	.26926	.537944
0.13	1.1388	.05646	.878095	0.63	1.8776	.27361	.532592
0.14	1.1503	.06080	.869358	0.64	1.8965	.27795	.527292
0.15	1.1618	0.06514	0.860708	**0.65**	1.9155	0.28229	0.522046
0.16	1.1735	.06949	.852144	0.66	1.9348	.28663	.516851
0.17	1.1853	.07383	.843665	0.67	1.9542	.29098	.511709
0.18	1.1972	.07817	.835270	0.68	1.9739	.29532	.506617
0.19	1.2092	.08252	.826959	0.69	1.9937	.29966	.501576
0.20	1.2214	0.08686	0.818731	**0.70**	2.0138	0.30401	0.496585
0.21	1.2337	.09120	.810584	0.71	2.0340	.30835	.491644
0.22	1.2461	.09554	.802519	0.72	2.0544	.31269	.486752
0.23	1.2586	.09989	.794534	0.73	2.0751	.31703	.481909
0.24	1.2712	.10423	.786628	0.74	2.0959	.32138	.477114
0.25	1.2840	0.10857	0.778801	**0.75**	2.1170	0.32572	0.472367
0.26	1.2969	.11292	.771052	0.76	2.1383	.33006	.467666
0.27	1.3100	.11726	.763379	0.77	2.1598	.33441	.463013
0.28	1.3231	.12160	.755784	0.78	2.1815	.33875	.458406
0.29	1.3364	.12595	.748264	0.79	2.2034	.34309	.453845
0.30	1.3499	0.13029	0.740818	**0.80**	2.2255	0.34744	0.449329
0.31	1.3634	.13463	.733447	0.81	2.2479	.35178	.444858
0.32	1.3771	.13897	.726149	0.82	2.2705	.35612	.440432
0.33	1.3910	.14332	.718924	0.83	2.2933	.36046	.436049
0.34	1.4049	.14766	.711770	0.84	2.3164	.36481	.431711
0.35	1.4191	0.15200	0.704688	**0.85**	2.3396	0.36915	0.427415
0.36	1.4333	.15635	.697676	0.86	2.3632	.37349	.423162
0.37	1.4477	.16069	.690734	0.87	2.3869	.37784	.418952
0.38	1.4623	.16503	.683861	0.88	2.4109	.38218	.414783
0.39	1.4770	.16937	.677057	0.89	2.4351	.38652	.410656
0.40	1.4918	0.17372	0.670320	**0.90**	2.4596	0.39087	0.406570
0.41	1.5068	.17806	.663650	0.91	2.4843	.39521	.402524
0.42	1.5220	.18240	.657047	0.92	2.5093	.39955	.398519
0.43	1.5373	.18675	.650509	0.93	2.5345	.40389	.394554
0.44	1.5527	.19109	.644036	0.94	2.5600	.40824	.390628
0.45	1.5683	0.19543	0.637628	**0.95**	2.5857	0.41258	0.386741
0.46	1.5841	.19978	.631284	0.96	2.6117	.41692	.382893
0.47	1.6000	.20412	.625002	0.97	2.6379	.42127	.379083
0.48	1.6161	.20846	.618783	0.98	2.6645	.42561	.375311
0.49	1.6323	.21280	.612626	0.99	2.6912	.42995	.371577
0.50	1.6487	0.21715	0.606531	**1.00**	2.7183	0.43429	0.367879

Table 3   (Continued)

x	e^x	$\log_{10}(e^x)$	e^{-x}	x	e^x	$\log_{10}(e^x)$	e^{-x}
1.00	2.7183	0.43429	0.367879	**1.50**	4.4817	0.65144	0.223130
1.01	2.7456	.43864	.364219	1.51	4.5267	.65578	.220910
1.02	2.7732	.44298	.360595	1.52	4.5722	.66013	.218712
1.03	2.8011	.44732	.357007	1.53	4.6182	.66447	.216536
1.04	2.8292	.45167	.353455	1.54	4.6646	.66881	.214381
1.05	2.8577	0.45601	0.349938	**1.55**	4.7115	0.67316	0.212248
1.06	2.8864	.46035	.346456	1.56	4.7588	.67750	.210136
1.07	2.9154	.46470	.343009	1.57	4.8066	.68184	.208045
1.08	2.9447	.46904	.339596	1.58	4.8550	.68619	.205975
1.09	2.9743	.47338	.336216	1.59	4.9037	.69053	.203926
1.10	3.0042	0.47772	0.332871	**1.60**	4.9530	0.69487	0.201897
1.11	3.0344	.48207	.329559	1.61	5.0028	.69921	.199888
1.12	3.0649	.48641	.326280	1.62	5.0531	.70356	.197899
1.13	3.0957	.49075	.323033	1.63	5.1039	.70790	.195930
1.14	3.1268	.49510	.319819	1.64	5.1552	.71224	.193980
1.15	3.1582	0.49944	0.316637	**1.65**	5.2070	0.71659	0.192050
1.16	3.1899	.50378	.313486	1.66	5.2593	.72093	.190139
1.17	3.2220	.50812	.310367	1.67	5.3122	.72527	.188247
1.18	3.2544	.51247	.307279	1.68	5.3656	.72961	.186374
1.19	3.2871	.51681	.304221	1.69	5.4195	.73396	.184520
1.20	3.3201	0.52115	0.301194	**1.70**	5.4739	0.73830	0.182684
1.21	3.3535	.52550	.298197	1.71	5.5290	.74264	.180866
1.22	3.3872	.52984	.295230	1.72	5.5845	.74699	.179066
1.23	3.4212	.53418	.292293	1.73	5.6407	.75133	.177284
1.24	3.4556	.53853	.289384	1.74	5.6973	.75567	.175520
1.25	3.4903	0.54287	0.286505	**1.75**	5.7546	0.76002	0.173774
1.26	3.5254	.54721	.283654	1.76	5.8124	.76436	.172045
1.27	3.5609	.55155	.280832	1.77	5.8709	.76870	.170333
1.28	3.5966	.55590	.278037	1.78	5.9299	.77304	.168638
1.29	3.6328	.56024	.275271	1.79	5.9895	.77739	.166960
1.30	3.6693	0.56458	0.272532	**1.80**	6.0496	0.78173	0.165299
1.31	3.7062	.56893	.269820	1.81	6.1104	.78607	.163654
1.32	3.7434	.57327	.267135	1.82	6.1719	.79042	.162026
1.33	3.7810	.57761	.264477	1.83	6.2339	.79476	.160414
1.34	3.8190	.58195	.261846	1.84	6.2965	.79910	.158817
1.35	3.8574	0.58630	0.259240	**1.85**	6.3598	0.80344	0.157237
1.36	3.8962	.59064	.256661	1.86	6.4237	.80779	.155673
1.37	3.9354	.59498	.254107	1.87	6.4483	.81213	.154124
1.38	3.9749	.59933	.251579	1.88	6.5535	.81647	.152590
1.39	4.0149	.60367	.249075	1.89	6.6194	.82082	.151072
1.40	4.0552	0.60801	0.246597	**1.90**	6.6859	0.82516	0.149569
1.41	4.0960	.61236	.244143	1.91	6.7531	.82950	.148080
1.42	4.1371	.61670	.241714	1.92	6.8210	.83385	.146607
1.43	4.1787	.62104	.239309	1.93	6.8895	.83819	.145148
1.44	4.2207	.62538	.236928	1.94	6.9588	.84253	.143704
1.45	4.2631	0.62973	0.234570	**1.95**	7.0287	0.84687	0.142274
1.46	4.3060	.63407	.232236	1.96	7.0993	.85122	.140858
1.47	4.3492	.63841	.229925	1.97	7.1707	.85556	.139457
1.48	4.3929	.64276	.227638	1.98	7.2427	.85990	.138069
1.49	4.4371	.64710	.225373	1.99	7.3155	.86425	.136695
1.50	4.4817	0.65144	0.223130	**2.00**	7.3891	0.86859	0.135335

Table 3 *(Continued)*

x	e^x	$\log_{10}(e^x)$	e^{-x}	x	e^x	$\log_{10}(e^x)$	e^{-x}
2.00	7.3891	0.86859	0.135335	**2.50**	12.182	1.08574	0.082085
2.01	7.4633	.87293	.133989	2.51	12.305	1.09008	.081268
2.02	7.5383	.87727	.132655	2.52	12.429	1.09442	.080460
2.03	7.6141	.88162	.131336	2.53	12.554	1.09877	.079659
2.04	7.6906	.88596	.130029	2.54	12.680	1.10311	.078866
2.05	7.7679	0.89030	0.128735	**2.55**	12.807	1.10745	0.078082
2.06	7.8460	.89465	.127454	2.56	12.936	1.11179	.077305
2.07	7.9248	.89899	.126186	2.57	13.066	1.11614	.076536
2.08	8.0045	.90333	.124930	2.58	13.197	1.12048	.075774
2.09	8.0849	.90756	.123687	2.59	13.330	1.12482	.075020
2.10	8.1662	0.91202	0.122456	**2.60**	13.464	1.12917	0.074274
2.11	8.2482	.91636	.121238	2.61	13.599	1.13351	.073535
2.12	8.3311	.92070	.120032	2.62	13.736	1.13785	.072803
2.13	8.4149	.92505	.118837	2.63	13.874	1.14219	.072078
2.14	8.4994	.92939	.117655	2.64	14.013	1.14654	.071361
2.15	8.5849	0.93373	0.116484	**2.65**	14.154	1.15088	0.070651
2.16	8.6711	.93808	.115325	2.66	14.296	1.15522	.069948
2.17	8.7583	.94242	.114178	2.67	14.440	1.15957	.069252
2.18	8.8463	.94676	.113042	2.68	14.585	1.16391	.068563
2.19	8.9352	.95110	.111917	2.69	14.732	1.16825	.067881
2.20	9.0250	0.95545	0.110803	**2.70**	14.880	1.17260	0.067206
2.21	9.1157	.95979	.109701	2.71	15.029	1.17694	.066537
2.22	9.2073	.96413	.108609	2.72	15.180	1.18128	.065875
2.23	9.2999	.96848	.107528	2.73	15.333	1.18562	.065219
2.24	9.3933	.97282	.106459	2.74	15.487	1.18997	.064570
2.25	9.4877	0.97716	0.105399	**2.75**	15.643	1.19431	0.063928
2.26	9.5831	.98151	.104350	2.76	15.800	1.19865	.063292
2.27	9.6794	.98585	.103312	2.77	15.959	1.20300	.062662
2.28	9.7767	.99019	.102284	2.78	16.119	1.20734	.062039
2.29	9.8749	.99453	.101266	2.79	16.281	1.21168	.061421
2.30	9.9742	0.99888	0.100259	**2.80**	16.445	1.21602	0.060810
2.31	10.074	1.00322	.099261	2.81	16.610	1.22037	.060205
2.32	10.176	1.00756	.098274	2.82	16.777	1.22471	.059606
2.33	10.278	1.01191	.097296	2.83	16.945	1.22905	.059013
2.34	10.381	1.01625	.096328	2.84	17.116	1.23340	.058426
2.35	10.486	1.02059	0.095369	**2.85**	17.288	1.23774	0.057844
2.36	10.591	1.02493	.094420	2.86	17.462	1.24208	.057269
2.37	10.697	1.02928	.093481	2.87	17.637	1.24643	.056699
2.38	10.805	1.03362	.092551	2.88	17.814	1.25077	.056135
2.39	10.913	1.03796	.091630	2.89	17.993	1.25511	.055576
2.40	11.023	1.04231	0.090718	**2.90**	18.174	1.25945	0.055023
2.41	11.134	1.04665	.089815	2.91	18.357	1.26380	.054476
2.42	11.246	1.05099	.088922	2.92	18.541	1.26814	.053934
2.43	11.359	1.05534	.088037	2.93	18.728	1.27248	.053397
2.44	11.473	1.05968	.087161	2.94	18.916	1.27683	.052866
2.45	11.588	1.06402	0.086294	**2.95**	19.106	1.28117	0.052340
2.46	11.705	1.06836	.085435	2.96	19.298	1.28551	.051819
2.47	11.822	1.07271	.084585	2.97	19.492	1.28985	.051303
2.48	11.941	1.07705	.083743	2.98	19.688	1.29420	.050793
2.49	12.061	1.08139	.082910	2.99	19.886	1.29854	.050287
2.50	12.182	1.08574	0.082085	**3.00**	20.086	1.30288	0.049787

Table 3 *(Continued)*

x	e^x	$\log_{10}(e^x)$	e^{-x}	x	e^x	$\log_{10}(e^x)$	e^{-x}
3.00	20.086	1.30288	0.049787	**3.50**	33.115	1.52003	0.030197
3.01	20.287	1.30723	.049292	3.51	33.448	1.52437	.029897
3.02	20.491	1.31157	.048801	3.52	33.784	1.52872	.029599
3.03	20.697	1.31591	.048316	3.53	34.124	1.53306	.029305
3.04	20.905	1.32026	.047835	3.54	34.467	1.53740	.029013
3.05	21.115	1.32460	0.047359	**3.55**	34.813	1.54175	0.028725
3.06	21.328	1.32894	.046888	3.56	35.163	1.54609	.028439
3.07	21.542	1.33328	.046421	3.57	35.517	1.55043	.028156
3.08	21.758	1.33763	.045959	3.58	35.874	1.55477	.027876
3.09	21.977	1.34197	.045502	3.59	36.234	1.55912	.027598
3.10	22.198	1.34631	0.045049	**3.60**	36.598	1.56346	0.027324
3.11	22.421	1.35066	.044601	3.61	36.966	1.56780	.027052
3.12	22.646	1.35500	.044157	3.62	37.338	1.57215	.026783
3.13	22.874	1.35934	.043718	3.63	37.713	1.57649	.026516
3.14	23.104	1.36368	.043283	3.64	38.092	1.58083	.026252
3.15	23.336	1.36803	0.042852	**3.65**	38.475	1.58517	0.025991
3.16	23.571	1.37237	.042426	3.66	38.861	1.58952	.025733
3.17	23.807	1.36671	.042004	3.67	39.252	1.59386	.025476
3.18	24.047	1.38106	.041586	3.68	39.646	1.59820	.025223
3.19	24.288	1.38540	.041172	3.69	40.045	1.60255	.024972
3.20	24.533	1.38974	0.040764	**3.70**	40.447	1.60689	0.024724
3.21	24.779	1.39409	.040357	3.71	40.854	1.61123	.024478
3.22	25.028	1.39843	.039955	3.72	41.264	1.61558	.024234
3.23	25.280	1.40277	.039557	3.73	41.679	1.61992	.023993
3.24	25.534	1.40711	.039164	3.74	42.098	1.62426	.023754
3.25	25.790	1.41146	0.038774	**3.75**	42.521	1.62860	0.023518
3.26	26.050	1.41580	.038388	3.76	42.948	1.63295	.023284
3.27	26.311	1.42014	.038006	3.77	43.380	1.63729	.023052
3.28	26.576	1.42449	.037628	3.78	43.816	1.64163	.022823
3.29	26.843	1.42883	.037254	3.79	44.256	1.64598	.022596
3.30	27.113	1.44317	0.036883	**3.80**	44.701	1.65032	0.022371
3.31	27.385	1.43751	.036516	3.81	45.150	1.65466	.022148
3.32	27.660	1.44186	.036153	3.82	45.604	1.65900	.021928
3.33	27.938	1.44620	.035793	3.83	46.063	1.66335	.021710
3.34	28.219	1.45054	.035437	3.84	46.525	1.66769	.021494
3.35	28.503	1.45489	0.035084	**3.85**	46.993	1.67203	0.021280
3.36	28.789	1.45923	.034735	3.86	47.465	1.67638	.021068
3.37	29.079	1.46357	.034390	3.87	47.942	1.68072	.020858
3.38	29.371	1.46792	.034047	3.88	48.424	1.68506	.020651
3.39	29.666	1.47226	.033709	3.89	48.911	1.68941	.020445
3.40	29.964	1.47660	0.033373	**3.90**	49.402	1.69375	0.020242
3.41	30.265	1.48094	.033041	3.91	49.899	1.69809	.020041
3.42	30.569	1.48529	.032712	3.92	50.400	1.70243	.019840
3.43	30.877	1.48963	.032387	3.93	50.907	1.70678	.019644
3.44	31.187	1.49397	.032065	3.94	51.419	1.71112	.019448
3.45	31.500	1.49832	0.031746	**3.95**	51.935	1.71546	0.019255
3.46	31.817	1.50266	.031430	3.96	52.457	1.71981	.019063
3.47	32.137	1.50700	.031117	3.97	52.985	1.72415	.018873
3.48	32.460	1.51134	.030807	3.98	53.517	1.72849	.018686
3.49	32.786	1.51569	.030501	3.99	54.055	1.73283	.018500
3.50	33.115	1.52003	0.030197	**4.00**	54.598	1.73718	0.018316

Table 3   *(Continued)*

x	e^x	$\log_{10}(e^x)$	e^{-x}	x	e^x	$\log_{10}(e^x)$	e^{-x}
4.00	54.598	1.73718	0.018316	**4.50**	90.017	1.95433	0.011109
4.01	55.147	1.74152	.018133	4.51	90.922	1.95867	.010998
4.02	55.701	1.74586	.017953	4.52	91.836	1.96301	.010889
4.03	56.261	1.75021	.017774	4.53	92.759	1.96735	.010781
4.04	56.826	1.75455	.017597	4.54	93.691	1.97170	.010673
4.05	57.397	1.75889	0.017422	**4.55**	94.632	1.97604	0.010567
4.06	57.974	1.76324	.017249	4.56	95.583	1.98038	.010462
4.07	58.577	1.76758	.017077	4.57	96.544	1.98473	.010358
4.08	59.145	1.77192	.016907	4.58	97.514	1.98907	.010255
4.09	59.740	1.77626	.016739	4.59	98.494	1.99341	.010153
4.10	60.340	1.78061	0.016573	**4.60**	99.484	1.99775	0.010052
4.11	60.947	1.78495	.016408	4.61	100.48	2.00210	.009952
4.12	61.559	1.78929	.016245	4.62	101.49	2.00644	.009853
4.13	62.178	1.79364	.016083	4.63	102.51	2.01078	.009755
4.14	62.803	1.79798	.015923	4.64	103.54	2.01513	.009658
4.15	63.434	1.80232	0.015764	**4.65**	104.58	2.01947	0.009562
4.16	64.072	1.80667	.015608	4.66	105.64	2.02381	.009466
4.17	64.715	1.81101	.015452	4.67	106.70	2.02816	.009372
4.18	65.366	1.81535	.015299	4.68	107.77	2.03250	.009279
4.19	66.023	1.81969	.015146	4.69	108.85	2.03684	.009187
4.20	66.686	1.82404	0.014996	**4.70**	109.95	2.04118	0.009095
4.21	67.357	1.82838	.014846	4.71	111.05	2.04553	.009005
4.22	68.033	1.83272	.014699	4.72	112.17	2.04987	.008915
4.23	68.717	1.83707	.014552	4.73	113.30	2.05421	.008826
4.24	69.408	1.84141	.014408	4.74	114.43	2.05856	.008739
4.25	70.105	1.84575	0.014264	**4.75**	115.58	2.06290	0.008652
4.26	70.810	1.85009	.014122	4.76	116.75	2.06724	.008566
4.27	71.522	1.85444	.013982	4.77	117.92	2.07158	.008480
4.28	72.240	1.85878	.013843	4.78	119.10	2.07593	.008396
4.29	72.966	1.86312	.013705	4.79	120.30	2.08027	.008312
4.30	73.700	1.86747	0.013569	**4.80**	121.51	2.08461	0.008230
4.31	74.440	1.87181	.013434	4.81	122.73	2.08896	.008148
4.32	75.189	1.87615	.013300	4.82	123.97	2.09330	.008067
4.33	75.944	1.88050	.013168	4.83	125.21	2.09764	.007987
4.34	76.708	1.88484	.013037	4.84	126.47	2.10199	.007907
4.35	77.478	1.88918	0.012907	**4.85**	127.74	2.10633	0.007828
4.36	78.257	1.89352	.012778	4.86	129.02	2.11067	.007750
4.37	79.044	1.89787	.012651	4.87	130.32	2.11501	.007673
4.38	79.838	1.90221	.012525	4.88	131.63	2.11936	.007597
4.39	80.640	1.90655	.012401	4.89	132.95	2.12370	.007521
4.40	81.451	1.91090	0.012277	**4.90**	134.29	2.12804	0.007477
4.41	82.269	1.91524	.012155	4.91	135.64	2.13239	.007372
4.42	83.096	1.91958	.012034	4.92	137.00	2.13673	.007299
4.43	83.931	1.92392	.011914	4.93	138.38	2.14107	.007227
4.44	84.775	1.92827	.011796	4.94	139.77	2.14541	.007155
4.45	85.627	1.93261	0.011679	**4.95**	141.17	2.14976	0.007083
4.46	86.488	1.93695	.011562	4.96	142.59	2.15410	.007013
4.47	87.357	1.94130	.011447	4.97	144.03	2.15844	.006943
4.48	88.235	1.94564	.011333	4.98	145.47	2.16279	.006874
4.49	89.121	1.94998	.011221	4.99	146.94	2.16713	.006806
4.50	90.017	1.95433	0.011109	**5.00**	148.41	2.17147	0.006738

Table 3   *(Continued)*

x	e^x	$\log_{10}(e^x)$	e^{-x}	x	e^x	$\log_{10}(e^x)$	e^{-x}
5.00	148.41	2.17147	0.006738	**5.50**	244.69	2.38862	0.0040868
5.01	149.90	2.17582	.006671	5.55	257.24	2.41033	.0038875
5.02	151.41	2.18016	.006605	5.60	270.43	2.43205	.0036979
5.03	152.93	2.18450	.006539	5.65	284.29	2.45376	.0035175
5.04	154.47	2.18884	.006474	5.70	298.87	2.47548	.0033460
5.05	156.02	2.19319	0.006409	**5.75**	314.19	2.49719	0.0031828
5.06	157.59	2.19753	.006346	5.80	330.30	2.51891	.0030276
5.07	159.17	2.20187	.006282	5.85	347.23	2.54062	.0028799
5.08	160.77	2.20622	.006220	5.90	365.04	2.56234	.0027394
5.09	162.39	2.21056	.006158	5.95	383.75	2.58405	.0026058
5.10	164.02	2.21490	0.006097	**6.00**	403.43	2.60577	0.0024788
5.11	165.67	2.21924	.006036	6.05	424.11	2.62748	.0023579
5.12	167.34	2.22359	.005976	6.10	445.86	2.64920	.0022429
5.13	169.02	2.22793	.005917	6.15	468.72	2.67091	.0021335
5.14	170.72	2.23227	.005858	6.20	492.75	2.69263	.0020294
5.15	172.43	2.23662	0.005799	**6.25**	518.01	2.71434	0.0019305
5.16	174.16	2.24096	.005742	6.30	544.57	2.73606	.0018363
5.17	175.91	2.24530	.005685	6.35	572.49	2.75777	.0017467
5.18	177.68	2.24965	.005628	6.40	601.85	2.77948	.0016616
5.19	179.47	2.25399	.005572	6.45	632.70	2.80120	.0015805
5.20	181.27	2.25833	0.005517	**6.50**	665.14	2.82291	0.0015034
5.21	183.09	2.26267	.005462	6.55	699.24	2.84463	.0014301
5.22	184.93	2.26702	.005407	6.60	735.10	2.86634	.0013604
5.23	186.79	2.27136	.005354	6.65	772.78	2.88806	.0012940
5.24	188.67	2.27570	.005300	6.70	812.41	2.90977	.0012309
5.25	190.57	2.28005	0.005248	**6.75**	854.06	2.93149	0.0011709
5.26	192.48	2.28439	.005195	6.80	897.85	2.95320	.0011138
5.27	194.42	2.28873	.005144	6.85	943.88	2.97492	.0010595
5.28	196.37	2.29307	.005092	6.90	992.27	2.99663	.0010078
5.29	198.34	2.29742	.005042	6.95	1043.1	3.01835	.0009586
5.30	200.34	2.30176	0.004992	**7.00**	1096.6	3.04006	0.0009119
5.31	202.35	2.30610	.004942	7.05	1152.9	3.06178	.0008674
5.32	204.38	2.31045	.004893	7.10	1212.0	3.08349	.0008251
5.33	206.44	2.31479	.004844	7.15	1274.1	3.10521	.0007849
5.34	208.51	2.31913	.004796	7.20	1339.4	3.12692	.0007466
5.35	210.61	2.32348	0.004748	**7.25**	1408.1	3.14863	0.0007102
5.36	212.72	2.32782	.004701	7.30	1480.3	3.17035	.0006755
5.37	214.86	2.33216	.004654	7.35	1556.2	3.19206	.0006426
5.38	217.02	2.33650	.004608	7.40	1636.0	3.21378	.0006113
5.39	219.20	2.34085	.004562	7.45	1719.9	3.23549	.0005814
5.40	221.41	2.34519	0.004517	**7.50**	1808.0	3.25721	0.0005531
5.41	223.63	2.34953	.004472	7.55	1900.7	3.27892	.0005261
5.42	225.88	2.35388	.004427	7.60	1998.2	3.30064	.0005005
5.43	228.15	2.35822	.004383	7.65	2100.6	3.32235	.0004760
5.44	230.44	2.36256	.004339	7.70	2208.3	3.34407	.0004528
5.45	232.76	2.36690	0.004296	**7.75**	2321.6	3.36578	0.0004307
5.46	235.10	2.37125	.004254	7.80	2440.6	3.38750	.0004097
5.47	237.46	2.37559	.004211	7.85	2565.7	3.40921	.0003898
5.48	239.85	2.37993	.004169	7.90	2697.3	3.43093	.0003707
5.49	242.26	2.38428	.004128	7.95	2835.6	3.45264	.0003527
5.50	244.69	2.38862	0.004087	**8.00**	2981.0	3.47436	0.0003355

Table 3 *(Continued)*

x	e^x	$\log_{10}(e^x)$	e^{-x}	x	e^x	$\log_{10}(e^x)$	e^{-x}
8.00	2981.0	3.47436	0.0003355	**9.00**	8103.1	3.90865	0.0001234
8.05	3133.8	3.49607	.0003191	9.05	8518.5	3.93037	.0001174
8.10	3294.5	3.51779	.0003035	9.10	8955.3	3.95208	.0001117
8.15	3463.4	3.53950	.0002887	9.15	9414.4	3.97379	.0001062
8.20	3641.0	3.56121	.0002747	9.20	9897.1	3.99551	.0001010
8.25	3827.6	3.58293	0.0002613	**9.25**	10405	4.01722	0.0000961
8.30	4023.9	3.60464	.0002485	9.30	10938	4.03894	.0000914
8.35	4230.2	3.62636	.0002364	9.35	11499	4.06065	.0000870
8.40	4447.1	3.64807	.0002249	9.40	12088	4.08237	.0000827
8.45	4675.1	3.66979	.0002139	9.45	12708	4.10408	.0000787
8.50	4914.8	3.69150	0.0002036	**9.50**	13360	4.12580	0.0000749
8.55	5166.8	3.71322	.0001935	9.55	14045	4.14751	.0000712
8.60	5431.7	3.73493	.0001841	9.60	14765	4.16923	.0000677
8.65	5710.0	3.75665	.0001751	9.65	15522	4.19094	.0000644
8.70	6002.9	3.77836	.0001666	9.70	16318	4.21266	.0000613
8.75	6310.7	3.80008	0.0001585	**9.75**	17154	4.23437	0.0000583
8.80	6634.2	3.82179	.0001507	9.80	18034	4.25609	.0000555
8.85	6974.4	3.84351	.0001434	9.85	18958	4.27780	.0000527
8.90	7332.0	3.86522	.0001364	9.90	19930	4.29952	.0000502
8.95	7707.9	3.88694	.0001297	9.95	20952	4.32123	0.0000477
9.00	8103.1	3.90865	0.0001234	**10.00**	22026	4.34294	0.0000454

Table 4 Hyperbolic Functions

x	$\sinh x$	$\cosh x$	$\tanh x$
0	.00000	1.0000	.00000
0.1	.10017	1.0050	.09967
0.2	.20134	1.0201	.19738
0.3	.30452	1.0453	.29131
0.4	.41075	1.0811	.37995
0.5	.52110	1.1276	.46212
0.6	.63665	1.1855	.53705
0.7	.75858	1.2552	.60437
0.8	.88811	1.3374	.66404
0.9	1.0265	1.4331	.71630
1.0	1.1752	1.5431	.76159
1.1	1.3356	1.6685	.80050
1.2	1.5095	1.8107	.83365
1.3	1.6984	1.9709	.86172
1.4	1.9043	2.1509	.88535
1.5	2.1293	2.3524	.90515
1.6	2.3756	2.5775	.92167
1.7	2.6456	2.8283	.93541
1.8	2.9422	3.1075	.94681
1.9	3.2682	3.4177	.95624
2.0	3.6269	3.7622	.96403
2.1	4.0219	4.1443	.97045
2.2	4.4571	4.5679	.97574
2.3	4.9370	5.0372	.98010
2.4	5.4662	5.5569	.98367
2.5	6.0502	6.1323	.98661
2.6	6.6947	6.7690	.98903
2.7	7.4063	7.4735	.99101
2.8	8.1919	8.2527	.99263
2.9	9.0596	9.1146	.99396
3.0	10.018	10.068	.99505
3.1	11.076	11.122	.99595
3.2	12.246	12.287	.99668
3.3	13.538	13.575	.99728
3.4	14.965	14.999	.99777
3.5	16.543	16.573	.99818
3.6	18.285	18.313	.99851
3.7	20.211	20.236	.99878
3.8	22.339	22.362	.99900
3.9	24.691	24.711	.99918
4.0	27.290	27.308	.99933
4.1	30.162	30.178	.99945
4.2	33.336	33.351	.99955
4.3	36.843	36.857	.99963
4.4	40.719	40.732	.99970
4.5	45.003	45.014	.99975
4.6	49.737	49.747	.99980
4.7	54.969	54.978	.99983
4.8	60.751	60.759	.99986
4.9	67.141	67.149	.99989
5.0	74.203	74.210	.99991

Table 5 Trigonometric Functions

Degrees	Radians	Sin	Cos	Tan	Cot		
0	0.0000	0.0000	1.0000	0.0000		1.5708	90
1	0.0175	0.0175	0.9998	0.0175	57.290	1.5533	89
2	0.0349	0.0349	0.9994	0.0349	28.636	1.5359	88
3	0.0524	0.0523	0.9986	0.0524	19.081	1.5184	87
4	0.0698	0.0698	0.9976	0.0699	14.301	1.5010	86
5	0.0873	0.0872	0.9962	0.0875	11.430	1.4835	85
6	0.1047	0.1045	0.9945	0.1051	9.5144	1.4661	84
7	0.1222	0.1219	0.9925	0.1228	8.1443	1.4486	83
8	0.1396	0.1392	0.9903	0.1405	7.1154	1.4312	82
9	0.1571	0.1564	0.9877	0.1584	6.3138	1.4137	81
10	0.1745	0.1736	0.9848	0.1763	5.6713	1.3963	80
11	0.1920	0.1908	0.9816	0.1944	5.1446	1.3788	79
12	0.2094	0.2079	0.9781	0.2126	4.7046	1.3614	78
13	0.2269	0.2250	0.9744	0.2309	4.3315	1.3439	77
14	0.2443	0.2419	0.9703	0.2493	4.0108	1.3265	76
15	0.2618	0.2588	0.9659	0.2679	3.7321	1.3090	75
16	0.2793	0.2756	0.9613	0.2867	3.4874	1.2915	74
17	0.2967	0.2924	0.9563	0.3057	3.2709	1.2741	73
18	0.3142	0.3090	0.9511	0.3249	3.0777	1.2566	72
19	0.3316	0.3256	0.9455	0.3443	2.9042	1.2392	71
20	0.3491	0.3420	0.9397	0.3640	2.7475	1.2217	70
21	0.3665	0.3584	0.9336	0.3839	2.6051	1.2043	69
22	0.3840	0.3746	0.9272	0.4040	2.4751	1.1868	68
23	0.4014	0.3907	0.9205	0.4245	2.3559	1.1694	67
24	0.4189	0.4067	0.9135	0.4452	2.2460	1.1519	66
25	0.4363	0.4226	0.9063	0.4663	2.1445	1.1345	65
26	0.4538	0.4384	0.8988	0.4877	2.0503	1.1170	64
27	0.4712	0.4540	0.8910	0.5095	1.9626	1.0996	63
28	0.4887	0.4695	0.8829	0.5317	1.8807	1.0821	62
29	0.5061	0.4848	0.8746	0.5543	1.8040	1.0647	61
30	0.5236	0.5000	0.8660	0.5774	1.7321	1.0472	60
31	0.5411	0.5150	0.8572	0.6009	1.6643	1.0297	59
32	0.5585	0.5299	0.8480	0.6249	1.6003	1.0123	58
33	0.5760	0.5446	0.8387	0.6494	1.5399	0.9948	57
34	0.5934	0.5592	0.8290	0.6745	1.4826	0.9774	56
35	0.6109	0.5736	0.8192	0.7002	1.4281	0.9599	55
36	0.6283	0.5878	0.8090	0.7265	1.3764	0.9425	54
37	0.6458	0.6018	0.7986	0.7536	1.3270	0.9250	53
38	0.6632	0.6157	0.7880	0.7813	1.2799	0.9076	52
39	0.6807	0.6293	0.7771	0.8098	1.2349	0.8901	51
40	0.6981	0.6428	0.7660	0.8391	1.1918	0.8727	50
41	0.7156	0.6561	0.7547	0.8693	1.1504	0.8552	49
42	0.7330	0.6691	0.7431	0.9004	1.1106	0.8378	48
43	0.7505	0.6820	0.7314	0.9325	1.0724	0.8203	47
44	0.7679	0.6947	0.7193	0.9657	1.0355	0.8029	46
45	0.7854	0.7071	0.7071	1.0000	1.0000	0.7854	45
		Cos	Sin	Cot	Tan	Radians	Degrees

Table 6   The Greek Alphabet

α	alpha	ι	iota	ρ	rho
β	beta	κ	kappa	σ	sigma
γ	gamma	λ	lambda	τ	tau
δ	delta	μ	mu	υ	upsilon
ϵ	epsilon	ν	nu	ϕ	phi
ζ	zeta	ξ	xi	χ	chi
η	eta	o	omicron	ψ	psi
θ	theta	π	pi	ω	omega

FORMULAS FROM GEOMETRY

The following symbols are used for the measure:
r: radius h: altitude b: base a: base C: circumference A: area S: surface area
B: area of base V: volume

Circle: $A = \pi r^2$; $C = 2\pi r$

Triangle: $A = \frac{1}{2}bh$

Rectangle and parallelogram: $A = bh$

Trapezoid: $A = \frac{1}{2}(a + b)h$

Right circular cylinder: $V = \pi r^2 h$; $S = 2\pi rh$

Right circular cone: $V = \frac{1}{3}\pi r^2 h$; $S = \pi r \sqrt{r^2 + h^2}$

Sphere: $V = \frac{4}{3}\pi r^3$; $S = 4\pi r^2$

Prism (with parallel bases): $V = Bh$

Pyramid: $V = \frac{1}{3}Bh$

FORMULAS FROM TRIGONOMETRY

THE EIGHT FUNDAMENTAL TRIGONOMETRIC IDENTITIES

$$\sin x \csc x = 1 \qquad \cos x \sec x = 1 \qquad \tan x \cot x = 1 \qquad \tan x = \frac{\sin x}{\cos x} \qquad \cot x = \frac{\cos x}{\sin x}$$

$$\sin^2 x + \cos^2 x = 1 \qquad 1 + \tan^2 x = \sec^2 x \qquad 1 + \cot^2 x = \csc^2 x$$

SUM AND DIFFERENCE IDENTITIES

$$\sin(u + v) = \sin u \cos v + \cos u \sin v \qquad \sin(u - v) = \sin u \cos v - \cos u \sin v$$
$$\cos(u + v) = \cos u \cos v - \sin u \sin v \qquad \cos(u - v) = \cos u \cos v + \sin u \sin v$$

$$\tan(u + v) = \frac{\tan u + \tan v}{1 - \tan u \tan v} \qquad \tan(u - v) = \frac{\tan u - \tan v}{1 + \tan u \tan v}$$

MULTIPLE-MEASURE IDENTITIES

$$\sin 2u = 2 \sin u \cos u$$
$$\cos 2u = \cos^2 u - \sin^2 u \qquad \cos 2u = 1 - 2\sin^2 u \qquad \cos 2u = 2\cos^2 u - 1$$

$$\tan 2u = \frac{2\tan u}{1 - \tan^2 u}$$

$$\sin^2 u = \frac{1 - \cos 2u}{2} \qquad \cos^2 u = \frac{1 + \cos 2u}{2} \qquad \tan^2 u = \frac{1 - \cos 2u}{1 + \cos 2u}$$

$$\sin^2 \tfrac{1}{2} t = \frac{1 - \cos t}{2} \qquad \cos^2 \tfrac{1}{2} t = \frac{1 + \cos t}{2}$$

$$\tan \tfrac{1}{2} t = \frac{1 - \cos t}{\sin t} \qquad \tan \tfrac{1}{2} t = \frac{\sin t}{1 + \cos t}$$

IDENTITIES FOR THE PRODUCT, SUM, AND DIFFERENCE OF SINE AND COSINE

$$\sin u \cos v = \tfrac{1}{2}[\sin(u + v) + \sin(u - v)] \qquad \cos u \sin v = \tfrac{1}{2}[\sin(u + v) - \sin(u - v)]$$
$$\cos u \cos v = \tfrac{1}{2}[\cos(u + v) + \cos(u - v)] \qquad \sin u \sin v = \tfrac{1}{2}[\cos(u - v) - \cos(u + v)]$$

$$\sin s + \sin t = 2 \sin\left(\frac{s + t}{2}\right) \cos\left(\frac{s - t}{2}\right) \qquad \sin s - \sin t = 2 \cos\left(\frac{s + t}{2}\right) \sin\left(\frac{s - t}{2}\right)$$

$$\cos s + \cos t = 2 \cos\left(\frac{s + t}{2}\right) \cos\left(\frac{s - t}{2}\right) \qquad \cos s - \cos t = -2 \sin\left(\frac{s + t}{2}\right) \sin\left(\frac{s - t}{2}\right)$$

SOME REDUCTION FORMULAS

$$\sin(-x) = -\sin x \qquad \cos(-x) = \cos x \qquad \tan(-x) = -\tan x$$
$$\sin(\tfrac{1}{2}\pi - x) = \cos x \qquad \cos(\tfrac{1}{2}\pi - x) = \sin x \qquad \tan(\tfrac{1}{2}\pi - x) = \cot x$$
$$\sin(\tfrac{1}{2}\pi + x) = \cos x \qquad \cos(\tfrac{1}{2}\pi + x) = -\sin x \qquad \tan(\tfrac{1}{2}\pi + x) = -\cot x$$
$$\sin(\pi - x) = \sin x \qquad \cos(\pi - x) = -\cos x \qquad \tan(\pi - x) = -\tan x$$

LAW OF SINES and LAW OF COSINES

a, b, and c represent the measures of the sides of a triangle; α, β, and γ represent the measures of the angles opposite the sides of measures a, b, and c, respectively.

$$\frac{a}{\sin \alpha} = \frac{b}{\sin \beta} = \frac{c}{\sin \gamma} \qquad c^2 = a^2 + b^2 - 2ab \cos \gamma$$

ANSWERS TO ODD-NUMBERED EXERCISES

EXERCISES 0.1 (PAGE 16)

1. $(-2, +\infty)$ **3.** $(-\infty, \frac{3}{4}]$ **5.** $[4, 8]$ **7.** $(-\frac{5}{3}, \frac{4}{3}]$ **9.** $(-\infty, -\frac{1}{2}) \cup (0, +\infty)$ **11.** $(-\infty, -1) \cup (\frac{1}{3}, 3)$ **13.** $(-\infty, -2) \cup (2, +\infty)$
15. $(-\infty, -5) \cup (3, +\infty)$ **17.** $[-1, \frac{1}{2}]$ **19.** $(-3, \frac{3}{4})$ **21.** $(\frac{3}{2}, \frac{31}{14}] \cup (\frac{7}{3}, +\infty)$ **23.** $\{-\frac{5}{2}, 1\}$ **25.** $\{-\frac{1}{4}, 4\}$ **27.** $\{-\frac{2}{3}, \frac{1}{3}\}$ **29.** $\{\frac{4}{3}, 3\}$
31. $[\frac{3}{8}, +\infty)$ **33.** $(-\infty, -2] \cup [5, +\infty)$ **35.** $(-\infty, 1] \cup [4, +\infty)$ **37.** $(-11, 3)$ **39.** $[\frac{2}{3}, 2]$ **41.** $(-\infty, -2) \cup (12, +\infty)$
43. $[-\frac{1}{2}, 4]$ **45.** $(-\infty, 1) \cup (4, +\infty)$ **47.** $(1, +\infty)$ **49.** $[-\frac{9}{2}, \frac{3}{2}]$ **51.** $(-\infty, \frac{10}{9}) \cup (2, +\infty)$ **53.** $|x| > |9|$ **55.** $|x - 2| > 2$

EXERCISES 0.2 (PAGE 22)

1. (a)(1, 2); (b) $(-1, -2)$; (c) $(-1, 2)$; (d) $(-2, 1)$ **3.** (a) $(2, -2)$; (b) $(-2, 2)$; (c) $(-2, -2)$; (d) does not apply
5. (a) $(-1, 3)$; (b) $(1, -3)$; (c) $(1, 3)$; (d) $(-3, -1)$ **43.** (a) $y = 0$; (b) $x = 0$; (c) $xy = 0$ (Sketches of the graphs for Exercises 7 through 41 appear in Figs. EX 0.2-7 through EX 0.2-41.)

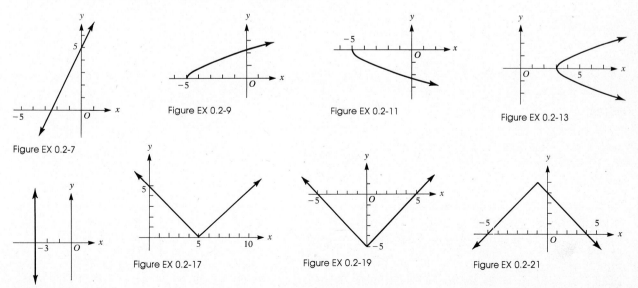

Figure EX 0.2-7

Figure EX 0.2-9

Figure EX 0.2-11

Figure EX 0.2-13

Figure EX 0.2-15

Figure EX 0.2-17

Figure EX 0.2-19

Figure EX 0.2-21

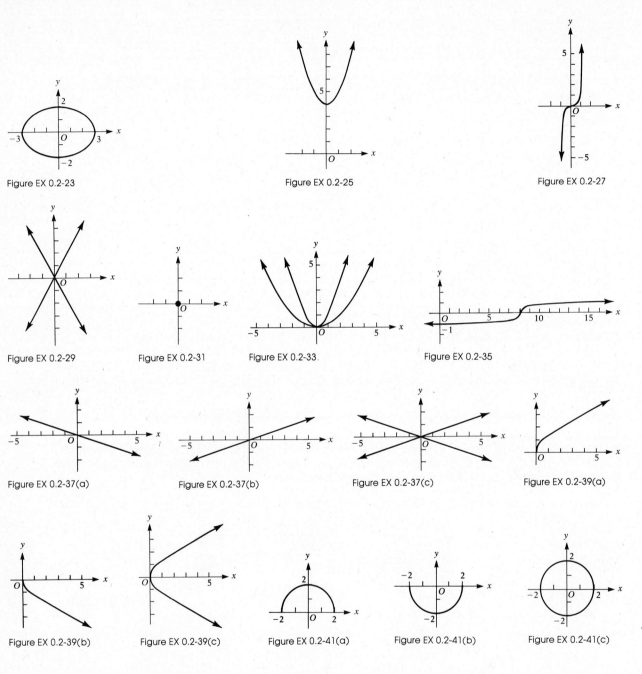

Figure EX 0.2-23

Figure EX 0.2-25

Figure EX 0.2-27

Figure EX 0.2-29

Figure EX 0.2-31

Figure EX 0.2-33

Figure EX 0.2-35

Figure EX 0.2-37(a)

Figure EX 0.2-37(b)

Figure EX 0.2-37(c)

Figure EX 0.2-39(a)

Figure EX 0.2-39(b)

Figure EX 0.2-39(c)

Figure EX 0.2-41(a)

Figure EX 0.2-41(b)

Figure EX 0.2-41(c)

EXERCISES 0.3 (PAGE 31)

1. $\sqrt{26}; \frac{1}{2}\sqrt{89}; \frac{1}{2}\sqrt{53}$ **3.** $|\overline{AB}| = \sqrt{41}, |\overline{AC}| = \sqrt{41}, |\overline{BC}| = \sqrt{82}$, and $|\overline{AB}|^2 + |\overline{AC}|^2 = |\overline{BC}|^2; \frac{41}{2}$ **9.** $\sqrt{578}$ **11.** $(-8, 12)$
13. $(-2 + \frac{3}{2}\sqrt{3}, \frac{3}{2} + 2\sqrt{3})$ and $(-2 - \frac{3}{2}\sqrt{3}, \frac{3}{2} - 2\sqrt{3})$ **15.** $77x^2 + 90xy + 21y^2 - 122x - 66y - 55 = 0$
17. $x^2 + y^2 - 8x + 6y = 0$ **19.** $x^2 + y^2 + 10x + 24y + 160 = 0$ **21.** $x^2 + y^2 - 2x - 4y - 8 = 0$ **23.** $(3, 4); 4$
25. $(0, -\frac{2}{3}); \frac{5}{3}$ **27.** circle   **29.** the empty set   **31.** circle   **37.** $(-\frac{9}{4}, \frac{17}{4}), (\frac{1}{2}, \frac{11}{2}), (\frac{13}{4}, \frac{27}{4})$ **39.** $(\frac{9}{4}, \frac{11}{4})$ **41.** $(-\frac{3}{2}, \frac{7}{2})$ **43.** $(5, -7)$
45. $2x + y = 5$

EXERCISES 0.4 (PAGE 43)

1. -1 **3.** $-\frac{1}{7}$ **5.** $4x - y = 11$ **7.** $2x + y + 1 = 0$ **9.** $13x - 4y - 7 = 0$ **11.** $4x - 3y + 12 = 0$ **13.** $x + 2y - 4 = 0$

15. $y = -7$ **17.** $\sqrt{3}x - y + (2\sqrt{3} - 5) = 0$ **19.** $x + y = 0$ **21.** $\frac{2}{3}$ **23.** 0 **25.** $\frac{x}{\frac{8}{3}} + \frac{y}{8} = 1$ **29.** (a) collinear;

(b) not collinear **31.** (a) not collinear; (b) collinear **33.** slopes of two sides are $-\frac{4}{3}$ **35.** (a) $2x + 3y + 7 = 0$; (b) $\sqrt{13}$

37. (a) $-\frac{A}{B}$; (b) $-\frac{C}{B}$; (c) $-\frac{C}{A}$; (d) $Bx - Ay = 0$ **39.** $9x - 4y - 11 = 0$; $y = 1$; $9x + 4y - 19 = 0$. They intersect at $(\frac{5}{3}, 1)$.

41. $2x + 3y - 12 = 0$; $(2 + 2\sqrt{2})x + (3 - 3\sqrt{2})y = 12$; $(2 - 2\sqrt{2})x + (3 + 3\sqrt{2}y) = 12$ **43.** (a) $x = 1$; (b) $y = 1$; (c) $x = 1$;
(d) $2x + y - 3 = 0$; (e) $x - 2y + 1 = 0$; (f) $x + y - 2 = 0$ **45.** $x^2 + y^2 + 6x + 10y + 9 = 0$

47. $x^2 + y^2 - 4x - 4y - 2 = 0$ **49.** $3x + 4y - 19 = 0$ **55.** $r^2(m^2 + 1) = b^2$

EXERCISES 0.5 (PAGE 52)

1. (a) yes, $x \geq 4$; (b) yes, $|x| \geq 2$; (c) yes, $|x| \leq 2$; (d) no **3.** (a) yes, $(-\infty, +\infty)$; (b) no; (c) yes; $(-\infty, +\infty)$; (d) yes; $(-\infty, +\infty)$

5. (a) 5; (b) -5; (c) -1; (d) $2a + 1$; (e) $2x + 1$; (f) $4x - 1$; (g) $4x - 2$; (h) $2x + 2h - 1$; (i) $2x + 2h - 2$; (j) 2 **7.** (a) -5;
(b) -6; (c) -3; (d) 30; (e) $2h^2 + 9h + 4$; (f) $8x^4 + 10x^2 - 3$; (g) $2x^4 - 7x^2$; (h) $2x^2 + (4h + 5)x + (2h^2 + 5h - 3)$;

(i) $2x^2 + 5x + (2h^2 + 5h - 6)$; (j) $4x + 2h + 5$ **9.** (a) 1; (b) $\sqrt{11}$; (c) 2; (d) 5; (e) $\sqrt{4x + 9}$; (f) $\dfrac{2}{\sqrt{2x + 2h + 3} + \sqrt{2x + 3}}$

11. (a) $x^2 + x - 6$; domain: $(-\infty, +\infty)$; (b) $-x^2 + x - 4$, domain: $(-\infty, +\infty)$; (c) $x^3 - 5x^2 - x + 5$, domain: $(-\infty, +\infty)$;

(d) $\dfrac{x - 5}{x^2 - 1}$, domain: $\{x | x \neq -1, x \neq 1\}$; (e) $\dfrac{x^2 - 1}{x - 5}$, domain: $\{x | x \neq 5\}$ **13.** (a) $\dfrac{x^2 + 2x - 1}{x^2 - x}$, domain: $\{x | x \neq 0, x \neq 1\}$;

(b) $\dfrac{x^2 + 1}{x^2 - x}$, domain: $\{x | x \neq 0, x \neq 1\}$; (c) $\dfrac{x + 1}{x^2 - x}$, domain: $\{x | x \neq 0, x \neq 1\}$; (d) $\dfrac{x^2 + x}{x - 1}$, domain: $\{x | x \neq 0, x \neq 1\}$;

(e) $\dfrac{x - 1}{x^2 + x}$, domain: $\{x | x \neq -1, x \neq 0, x \neq 1\}$ **15.** (a) $\sqrt{x} + x^2 - 1$, domain: $[0, +\infty)$; (b) $\sqrt{x} - x^2 + 1$, domain: $[0, +\infty)$;

(c) $\sqrt{x}(x^2 - 1)$, domain: $[0, +\infty)$; (d) $\dfrac{\sqrt{x}}{x^2 - 1}$, domain: $[0, 1) \cup (1, +\infty)$; (e) $\dfrac{x^2 - 1}{\sqrt{x}}$, domain: $(0, +\infty)$

17. (a) $x^2 + 3x - 1$, domain: $(-\infty, +\infty)$; (b) $x^2 - 3x + 3$, domain: $(-\infty, +\infty)$;

(c) $3x^3 - 2x^2 + 3x - 2$, domain: $(-\infty, +\infty)$; (d) $\dfrac{x^2 + 1}{3x - 2}$, domain: $\{x | x \neq \frac{2}{3}\}$; (e) $\dfrac{3x - 2}{x^2 + 1}$, domain: $(-\infty, +\infty)$

19. (a) $\dfrac{x^2 + 2x - 2}{x^2 - x - 2}$, domain: $\{x | x \neq -1, x \neq 2\}$; (b) $\dfrac{-x^2 - 2}{x^2 - x - 2}$, domain: $\{x | x \neq -1, x \neq 2\}$;

(c) $\dfrac{x}{x^2 - x - 2}$, domain: $\{x | x \neq -1, x \neq 2\}$; (d) $\dfrac{x - 2}{x^2 + x}$, domain: $\{x | x \neq -1, x \neq 0, x \neq 2\}$;

(e) $\dfrac{x^2 + x}{x - 2}$, domain: $\{x | x \neq -1, x \neq 2\}$ **21.** (a) $x + 5$, domain: $(-\infty, +\infty)$; (b) $x + 5$, domain: $(-\infty, +\infty)$;

(c) $x - 4$, domain: $(-\infty, +\infty)$; (d) $x + 4$, domain: $(-\infty, +\infty)$ **23.** (a) $x^2 - 6$, domain: $(-\infty, +\infty)$;
(b) $x^2 - 10x + 24$, domain: $(-\infty, +\infty)$; (c) $x - 10$, domain: $(-\infty, +\infty)$; (d) $x^4 - 2x^2$, domain: $(-\infty, +\infty)$

25. (a) $\sqrt{x^2 - 4}$, domain: $(-\infty, -2] \cup [2, +\infty)$; (b) $x - 4$, domain: $[2, +\infty)$; (c) $\sqrt{\sqrt{x - 2} - 2}$, domain: $[6, +\infty)$;

(d) $x^4 - 4x^2 + 2$, domain: $(-\infty, +\infty)$ **27.** (a) $\dfrac{1}{\sqrt{x}}$, domain: $(0, +\infty)$; (b) $\dfrac{1}{\sqrt{x}}$, domain: $(0, +\infty)$; (c) x, domain: $\{x | x \neq 0\}$;

(d) $\sqrt[4]{x}$, domain: $[0, +\infty)$ **29.** (a) $|x + 2|$, domain: $(-\infty, +\infty)$; (b) $||x| + 2|$, domain: $(-\infty, +\infty)$; (c) $|x|$, domain: $(-\infty, +\infty)$;
(d) $||x + 2| + 2|$, domain: $(-\infty, +\infty)$ **31.** (a) $2x^2 - 3$, domain: $(-\infty, +\infty)$; (b) $4x^2 - 12x + 9$, domain: $(-\infty, +\infty)$;
(c) $4x - 9$, domain: $(-\infty, +\infty)$ **33.** (a) 0; (b) 4; (c) $4 - 2x$ **35.** (a) 1; (b) -1; (c) 1; (d) -1; (e) 1 if $x \leq 0$, -1 if $x > 0$;
(f) 1 if $x \geq -1$, -1 if $x < -1$; (g) 1; (h) -1 if $x \neq 0$, 1 if $x = 0$ **37.** (a) even; (b) neither; (c) odd; (d) even; (e) odd;
(f) odd; (g) neither; (h) even; (i) even; (j) odd **39.** (a) even; (b) odd; (c) even; (d) even

EXERCISES 0.6 (PAGE 63)

1. (a) $\frac{1}{3}\pi$; (b) $\frac{3}{4}\pi$; (c) $\frac{7}{6}\pi$; (d) $-\frac{5}{3}\pi$; (e) $\frac{1}{9}\pi$; (f) $\frac{5}{3}\pi$; (g) $-\frac{5}{12}\pi$; (h) $\frac{5}{8}\pi$ **3.** (a) $45°$; (b) $120°$; (c) $330°$; (d) $-90°$; (e) $28°39'$;
(f) $540°$; (g) $-114°36'$; (h) $15°$ **5.** (a) $\frac{1}{2}$; (b) $\frac{1}{2}\sqrt{2}$; (c) 1; (d) $-\frac{1}{2}$ **7.** (a) $-\frac{1}{2}\sqrt{3}$; (b) $\frac{1}{2}\sqrt{2}$; (c) -1; (d) 0 **9.** (a) $\sqrt{3}$; (b) 1; (c) -1;
(d) 1 **11.** (a) $\frac{2}{3}\sqrt{3}$; (b) $\sqrt{2}$; (c) $-\frac{1}{3}\sqrt{3}$; (d) 1 **13.** (a) $\frac{1}{4}(\sqrt{6}-\sqrt{2})$; (b) $\frac{1}{4}(\sqrt{2}-\sqrt{6})$ **15.** (a) $\frac{1}{2}\sqrt{2+\sqrt{3}}$; (b) $\frac{1}{2}\sqrt{2-\sqrt{2}}$

23. $3\sin t - 4\sin^3 t$ **27.** (a) $\frac{1}{2}\sqrt{2-\sqrt{3}}$; (b) $\frac{1}{2}\sqrt{2-\sqrt{2}}$; (c) $-\sqrt{\dfrac{2+\sqrt{2}}{2-\sqrt{2}}}$; (d) $-\sqrt{\dfrac{2+\sqrt{3}}{2-\sqrt{3}}}$ **29.** (a) $\frac{1}{2}\sqrt{2-\sqrt{3}}$; (b) $-\frac{1}{2}\sqrt{2+\sqrt{2}}$;

(c) $-\sqrt{\dfrac{2+\sqrt{3}}{2-\sqrt{3}}}$; (d) $\sqrt{\dfrac{2-\sqrt{2}}{2+\sqrt{2}}}$ **31.** (a) $\frac{1}{2}\pi$; (b) π; (c) $\frac{1}{4}\pi$, $\frac{5}{4}\pi$; (d) 0 **33.** (a) 0, π; (b) $\frac{1}{2}\pi$, $\frac{3}{2}\pi$; (c) 0, π; (d) $\frac{1}{2}\pi$, $\frac{3}{2}\pi$

35. (a) $\frac{7}{6}\pi$, $\frac{11}{6}\pi$; (b) $\frac{1}{3}\pi$, $\frac{5}{3}\pi$; (c) $\frac{3}{4}\pi$, $\frac{7}{4}\pi$; (d) $\frac{1}{6}\pi$, $\frac{5}{6}\pi$ **37.** (a) $\frac{1}{2}\pi$, $\frac{3}{2}\pi$; (b) 0, π

EXERCISES 0.7 (PAGE 67)

1. (a) $\frac{3}{5}$; (b) 1 **3.** (a) $-\frac{7}{4}$; (b) $\frac{3}{55}$ **5.** (a) $63°$; (b) $19°$ **7.** $27°$, $45°$, $108°$ **9.** $3x - y + 7 = 0$; $x + 3y - 11 = 0$
13. $-\frac{19}{3} - \frac{1}{3}\sqrt{370}$ **17.** $\frac{23}{19}$, $\frac{69}{121}$, $-\frac{23}{24}$

REVIEW EXERCISES FOR CHAPTER 0 (PAGE 68)

1. $[5, +\infty)$ **3.** $(-\infty, -\frac{1}{3}) \cup (1, +\infty)$ **5.** $(-\frac{3}{2}, 1)$ **7.** $[-\frac{12}{5}, \frac{6}{5}]$ **9.** $(-\frac{4}{3}, 4)$ **11.** $(-\infty, -6) \cup (-1, +\infty)$ **13.** $\{-3, \frac{9}{2}\}$
15. $\{-2, 2\}$ **17.** see Fig. REV 0-17 **19.** see Fig. REV. 0-19 **21.** see Fig. REV 0-21

Figure REV 0-17

Figure REV 0-19

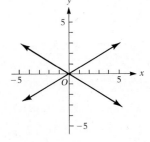

Figure REV 0-21

23. -4, 14 **25.** $x^2 + y^2 + 6x - 8y + 15 = 0$ **29.** 15 **31.** $(-2, -5)$; $(3, -6)$ **35.** $x^2 + y^2 - 2x - 8y - 3 = 0$
37. $(\frac{3}{2}, -1)$; 1 **39.** $3x - y + 9 = 0$ **41.** $5x + 2y - 19 = 0$ **43.** $7x^2 + 7y^2 + 11x - 19y - 6 = 0$ **45.** $12x + 3y - 2 = 0$
47. $0 \le k \le 2$ **49.** (a) 35; (b) $3x^4 + x^2 + 5$; (c) $6x + 3h - 1$ **51.** (a) odd; (b) even; (c) neither; (d) odd
53. (a) $x^2 + 4x - 7$, domain: $(-\infty, +\infty)$; (b) $x^2 - 4x - 1$, domain: $(-\infty, +\infty)$;

(c) $4x^3 - 3x^2 - 16x + 12$, domain: $(-\infty, +\infty)$; (d) $\dfrac{x^2 - 4}{4x - 3}$, domain: $\{x | x \ne \frac{3}{4}\}$; (e) $\dfrac{4x - 3}{x^2 - 4}$, domain: $\{x | x \ne \pm 2\}$;

(f) $16x^2 - 24x + 5$, domain: $(-\infty, +\infty)$; (g) $4x^2 - 19$, domain: $(-\infty, +\infty)$

55. (a) $\dfrac{x^2 - 2x + 1}{x^2 - 2x - 3}$, domain: $\{x | x \ne -1, x \ne 3\}$; (b) $\dfrac{-x^2 + 4x + 1}{x^2 - 2x - 3}$, domain: $\{x | x \ne -1, x \ne 3\}$;

(c) $\dfrac{x}{x^2 - 2x - 3}$, domain: $\{x | x \ne -1, x \ne 3\}$; (d) $\dfrac{x + 1}{x^2 - 3x}$, domain: $\{x | x \ne -1, x \ne 0, x \ne 3\}$;

(e) $\dfrac{x^2 - 3x}{x + 1}$, domain: $\{x | x \ne -1, x \ne 3\}$; (f) $-\dfrac{x + 1}{2x + 3}$, domain: $\{x | x \ne -\frac{3}{2}, x \ne -1\}$; (g) $\dfrac{1}{x - 2}$, domain: $\{x | x \ne 2, x \ne 3\}$

57. (a) $\frac{1}{2}\sqrt{3}$; (b) $-\frac{1}{2}\sqrt{2}$; (c) $\frac{1}{3}\sqrt{3}$; (d) $\sqrt{\dfrac{2+\sqrt{3}}{2-\sqrt{3}}}$; (e) -1; (f) $-\dfrac{2}{\sqrt{2-\sqrt{2}}}$ **59.** (a) $\frac{1}{6}\pi$, $\frac{5}{6}\pi$; (b) 0; (c) $\frac{3}{4}\pi$, $\frac{7}{4}\pi$; (d) $\frac{1}{6}\pi$, $\frac{7}{6}\pi$;

(e) $\frac{2}{3}\pi$, $\frac{4}{3}\pi$; (f) $\frac{1}{4}\pi$, $\frac{3}{4}\pi$ **61.** $52°$, $90°$, $106°$, $112°$ **63.** $x = 0$; $3x + 4y = 0$ **67.** medians intersect at $(\frac{13}{3}, \frac{13}{3})$; altitudes intersect at (9, 7); center of circumscribed circle is at (2, 3)

EXERCISES 1.1 (PAGE 78)

(Sketches of the graphs appear in Figs. EX 1.1-1 through EX 1.1-39.)

1. domain: $(-\infty, +\infty)$; range: $(-\infty, +\infty)$ **3.** domain: $(-\infty, +\infty)$; range: $[-6, +\infty)$ **5.** domain: $[-1, +\infty)$; range: $[0, +\infty)$
7. domain: $[\frac{4}{3}, +\infty)$; range: $[0, +\infty)$ **9.** domain: $(-\infty, -2] \cup [2, +\infty)$; range: $[0, +\infty)$ **11.** domain: $(-\infty, +\infty)$;
range: $[0, +\infty)$ **13.** domain: $\{x | x \neq 1\}$; range: $\{y | y \neq -2\}$ **15.** domain: $(-\infty, +\infty)$; range: $\{-2, 2\}$
17. domain: $(-\infty, +\infty)$, range: $\{y | y \neq 3\}$ **19.** domain: $(-\infty, +\infty)$; range: $[-4, +\infty)$ **21.** domain: $(-\infty, +\infty)$;
range: $(-\infty, 4]$ **23.** domain: $(-\infty, +\infty)$; range: $(-\infty, -2) \cup \{0\} \cup (1, +\infty)$ **25.** domain: $\{x | x \neq -5 \text{ and } x \neq -1\}$;
range: $\{y | y \neq -7 \text{ and } y \neq -3\}$ **27.** domain: $(-\infty, -1] \cup [4, +\infty)$; range: $[0, +\infty)$ **29.** domain: $\{x | x \neq 2\}$; range: $[0, +\infty)$
31. domain: $\{x | x \neq -5\}$; range: $[-6, +\infty)$ **33.** domain: $(-\infty, +\infty)$; range: $[1, +\infty)$ **35.** domain: $(-\infty, +\infty)$;
range: $\{\text{integers}\}$ **37.** domain: $(-\infty, +\infty)$; range: $[0, 1)$ **39.** domain: $\{x | x \neq 0\}$; range: $(-\infty, -1] \cup \{0\} \cup (\frac{1}{2}, 1]$

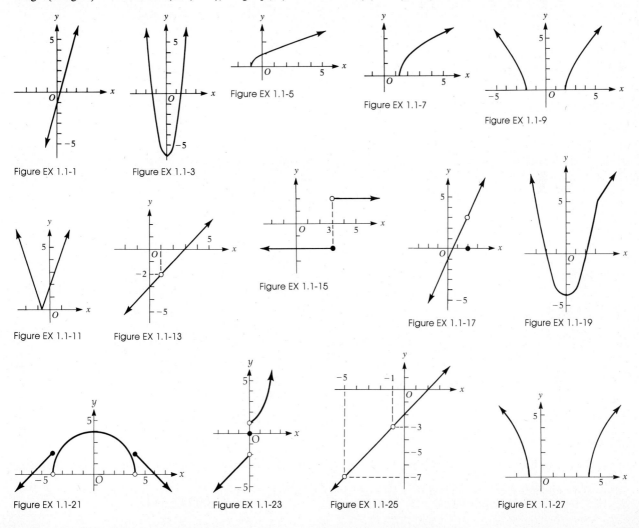

Figure EX 1.1-1

Figure EX 1.1-3

Figure EX 1.1-5

Figure EX 1.1-7

Figure EX 1.1-9

Figure EX 1.1-11

Figure EX 1.1-13

Figure EX 1.1-15

Figure EX 1.1-17

Figure EX 1.1-19

Figure EX 1.1-21

Figure EX 1.1-23

Figure EX 1.1-25

Figure EX 1.1-27

Figure EX 1.1-29

Figure EX 1.1-31

Figure EX 1.1-33

Figure EX 1.1-35

Figure EX 1.1-37

Figure EX 1.1-39

EXERCISES 1.2 (PAGE 89)

1. (b) 0.005 **3.** (b) 0.005 **5.** (b) $\frac{1}{1400}$ **7.** (b) $\frac{1}{3000}$ **9.** (b) 0.002 **11.** (b) 0.01 **13.** (b) 0.006 **15.** $\delta = \epsilon$ **17.** $\delta = \frac{1}{2}\epsilon$
19. $\delta = \frac{1}{5}\epsilon$ **21.** $\delta = \frac{1}{3}\epsilon$ **23.** $\delta = \frac{1}{3}\epsilon$ **25.** $\delta = \epsilon$ **27.** $\delta = \min(1, \frac{1}{3}\epsilon)$ **29.** $\delta = \min(1, \frac{1}{8}\epsilon)$ **31.** $\delta = \min(1, \frac{1}{6}\epsilon)$
33. $\delta = \min(1, 6\epsilon)$ **35.** $\delta = \min(1, \frac{1}{2}\epsilon)$ **37.** $\delta = \min(\frac{1}{2}, \frac{1}{4}\epsilon)$ **39.** $\delta = \min(1, \frac{2}{3}\epsilon)$ **41.** $\delta = \min(1, (2\sqrt{2} + 3)\epsilon)$
43. $\delta = \min(1, 2\epsilon)$ **45.** $\delta = \min(1, \frac{1}{2}(9\sqrt{7} + 21)\epsilon)$ **47.** $\delta = \min(a, \epsilon/3a)$ **49.** $\delta = \min(a, \sqrt{a}\epsilon)$ **51.** $\delta = \min(-a, \epsilon/7a^2)$

EXERCISES 1.3 (PAGE 99)

1. 8 **3.** 7 **5.** 0 **7.** $\frac{1}{2}$ **9.** $-\frac{1}{22}$ **11.** $\frac{1}{4}$ **13.** -6 **15.** $\frac{1}{7}$ **17.** $\frac{16}{7}$ **19.** 12 **21.** $\frac{3}{2}$ **23.** $\frac{2}{3}$ **25.** $\sqrt{\frac{6}{5}}$ **27.** $\frac{1}{6}$ **29.** $\frac{1}{4}\sqrt{2}$ **31.** $\frac{1}{3}$ **33.** -1
35. $\frac{11}{17}$ **41.** (a) 0

EXERCISES 1.4 (PAGE 104)

1. (a) -3; (b) 2; (c) does not exist because $\lim\limits_{x \to 1^+} f(x) \neq \lim\limits_{x \to 1^-} f(x)$ **3.** (a) 8; (b) 0;
(c) does not exist because $\lim\limits_{t \to -4^+} f(t) \neq \lim\limits_{t \to -4^-} f(t)$ **5.** (a) 4; (b) 4; (c) 4 **7.** (a) 5; (b) 5; (c) 5 **9.** (a) 0; (b) 0; (c) 0 **11.** (a) 0;
(b) 0; (c) 0 **13.** (a) -4; (b) -4; (c) -4 **15.** (a) 1; (b) -1; (c) does not exist because $\lim\limits_{x \to 0^+} f(x) \neq \lim\limits_{x \to 0^-} f(x)$ **17.** (a) 2; (b) 0;
(c) does not exist because $\lim\limits_{x \to -2^-} f(x) \neq \lim\limits_{x \to -2^+} f(x)$; (d) 0; (e) -2; (f) does not exist because $\lim\limits_{x \to 2^-} f(x) \neq \lim\limits_{x \to 2^+} f(x)$ **19.** (a) 0;
(b) 0; (c) 0 **21.** (a) 0; (b) 0; (c) 0; (d) 0; (e) 0; (f) 0 **23.** (a) -2; (b) 2; (c) does not exist **25.** (a) 2; (b) 1; (c) does not exist
27. (a) -1; (b) 1; (c) does not exist **29.** -6 **31.** $a = -\frac{3}{2}$, $b = 1$ **35.** (b) 40; (c) 35; (d) 140; (e) 130

EXERCISES 1.5 (PAGE 117)

1. $+\infty$ **3.** $+\infty$ **5.** $-\infty$ **7.** $-\infty$ **9.** $+\infty$ **11.** $+\infty$ **13.** $+\infty$ **15.** $-\infty$ **17.** $-\infty$ **19.** $+\infty$ **21.** $-\infty$ **23.** $+\infty$ **25.** $+\infty$ **27.** $-\infty$
29. $-\infty$ **31.** $-\infty$ **33.** (a) $x = 0$; (b) $x = 0$; (c) $x = 0$; (d) $x = 0$ **35.** $x = 4$ **37.** $x = -3$ **39.** $x = -3$ **41.** $x = -5, x = -3$
43. $x = \frac{7}{2}, x = -\frac{5}{2}$ **45.** $x = 2$ **47.** take $\delta = \sqrt{3}/N$

EXERCISES 1.6 (PAGE 126)

1. -3; $f(-3)$ does not exist **3.** -3; $\lim\limits_{x \to -3} g(x) \neq g(-3)$ **5.** $-3, 2$; $h(-3)$ and $h(2)$ do not exist **7.** 4;

$\lim\limits_{x \to 4} f(x)$ does not exist **9.** $-2, 2$; $F(-2)$ and $F(2)$ do not exist **11.** $-2, 2$; $G(-2)$ and $G(2)$ do not exist **13.** 0;

$\lim\limits_{x \to 0} f(x)$ does not exist **15.** $2, \lim\limits_{t \to 2} g(t) \neq g(2)$ **17.** 0; $\lim\limits_{x \to 0} g(x)$ does not exist **19.** 3; $f(3)$ does not exist **21.** all integers;

$\lim\limits_{x \to k} [\![x]\!]$ does not exist if k is any integer **23.** removable; 4 **25.** essential **27.** removable; 0 **29.** essential **31.** removable;

$-\frac{1}{6}$ **33.** all real numbers **35.** all real numbers except 3 **37.** all real numbers except -2 and 2

39. all real numbers except 2 **41.** all real numbers except 1 **43.** (b) 50 and 200 **45.** f is continuous everywhere

47. let $y(0) = \dfrac{1}{3a^2}$ **51.** $f(x) = \begin{cases} 0 & \text{if } x < a \\ 1 & \text{if } a \le x \end{cases}$; $g(x) = \begin{cases} 1 & \text{if } x < a \\ 0 & \text{if } a \le x \end{cases}$

EXERCISES 1.7 (PAGE 135)

1. $(f \circ g)(x) = \sqrt{x^2 - 16}$; continuous at all numbers in $(-\infty, -4) \cup (4, +\infty)$ **3.** $(f \circ g)(x) = \sqrt[4]{x^2 + 1}$;

continuous at all real numbers **5.** $(f \circ g)(x) = x^{3/2}$; continuous at all positive numbers **7.** $(f \circ g)(x) = \dfrac{1}{\sqrt{x} - 2}$;

continuous at all numbers in $(2, +\infty)$ **9.** $(f \circ g)(x) = \dfrac{1}{\sqrt{x} - 2}$; continuous at all positive numbers except 4

11. continuous; discontinuous; discontinuous; continuous; discontinuous; continuous **13.** continuous; continuous;
discontinuous; continuous; discontinuous; continuous **15.** continuous; continuous; continuous; continuous;
discontinuous **17.** continuous; discontinuous; discontinuous; discontinuous; continuous **19.** continuous; continuous;
continuous; continuous; discontinuous; discontinuous **21.** (a) $[-3, 3]$; (b) $(-\infty, -4] \cup (4, +\infty)$ **23.** (a) $(-5, 2]$;
(b) $(-\infty, +\infty)$ **25.** (a) $(-\infty, 0) \cup (0, +\infty)$; (b) $(0, +\infty)$ **27.** See Fig. EX 1.7-27 **29.** See Fig. EX 1.7-29

Figure EX 1.7-27

Figure EX 1.7-29

33. (a) $V(x) = x(8 - 2x)(15 - 2x)$; (b) $[0, 4]$ **35.** (a) $A(x) = x(120 - x)$; (b) $[0, 120]$ **37.** $k = 5$ **39.** $c = -3, k = 4$
41. $c = \frac{1}{2}(1 + \sqrt{5})$ **43.** $c = -4$ **45.** f is discontinuous at -2 **47.** f is discontinuous at 1 **49.** no; f will be continuous on
$[a, c]$ if $\lim\limits_{x \to b} g(x)$ exists and is equal to $h(b)$ **51.** $\sqrt{10} - 3$

EXERCISES 1.8 (PAGE 147)

1. 4 **3.** $\frac{9}{7}$ **5.** $\frac{3}{5}$ **7.** $\frac{1}{9}$ **9.** 0 **11.** 0 **13.** 12 **15.** $\frac{1}{2}$ **17.** $+\infty$ **19.** 0 **21.** 0 **23.** -1 **25.** 3 **27.** 0 **29.** -4 **31.** 1 **33.** 0

EXERCISES 1.9 (PAGE 156)

1. 1 **3.** -1 **5.** 1 **7.** 2 **9.** -2 **11.** $\sqrt{3}$ **13.** no

EXERCISES 1.10 (PAGE 160)

1. any interval (a, b) where $-2 < a < 3$ and $b > 3$ **3.** any interval (a, b) where $0 \leq a < 1$ and $b > 1$ **5.** any interval (a, b) where $a < -3$ and $-3 < b < 3$ **7.** any interval (a, b) where $-4 \leq a < 0$ and $b > 0$ **9.** (a) $(-\infty, -1) \cup (0, +\infty)$; (b) (a, b) where $a < k < b \leq -1$ or $0 \leq a < k < b$ **11.** (a) all numbers in the interval $(-\frac{7}{2}, \frac{3}{2})$; (b) (a, b) where $-\frac{7}{2} \leq a < b \leq -2$ or $-2 \leq a < b \leq \frac{3}{2}$

REVIEW EXERCISES FOR CHAPTER 1 (PAGE 161)

1. 9 **3.** -6 **5.** $\sqrt[3]{2}$ **7.** $-\frac{1}{6}$ **9.** $\delta = \frac{1}{4}\epsilon$ **11.** $\delta = \min(1, \frac{1}{9}\epsilon)$ **13.** $\delta = \min(1, \epsilon)$ **15.** $\delta = \min(1, \epsilon)$ **17.** $\delta = \min(1, 5\epsilon)$ **19.** -2, 1; $f(-2)$ and $f(1)$ do not exist **21.** -2; $\lim\limits_{x \to -2} g(x)$ does not exist **23.** (a) domain: $(-\infty, -2] \cup [2, +\infty)$, range: $[0, +\infty)$; (c) f is continuous at all numbers in $(-\infty, -2) \cup (2 +\infty)$ **25.** (a) domain: $\{x | x \neq -\frac{5}{2}, x \neq 2\}$, range: $\{y | y \neq \frac{1}{2}, y \neq 5\}$; (c) f is continuous at all numbers in its domain **27.** $-\frac{5}{2}$ **29.** 8 **31.** $\frac{1}{3}$ **33.** $-\infty$ **35.** $+\infty$ **37.** $-\infty$ **39.** $\frac{2}{3}$ **41.** 0 **43.** $\frac{1}{3}$ **45.** $\dfrac{2}{3a^{1/3}}$ **47.** $-\frac{1}{2}$ **49.** (a) yes; (b) no **51.** removable; $\frac{6}{5}$ **53.** essential **55.** $x = -2, x = 2$ **57.** $(f \circ g)(x) = \sqrt{x - 1}$; continuous at all numbers in $(1, +\infty)$ **59.** $(f \circ g)(x) = \begin{cases} 1 & \text{if } x < -1 \\ 0 & \text{if } x = -1 \\ -1 & \text{if } -1 < x < 1 \\ 0 & \text{if } x = 1 \\ 1 & \text{if } x > 1 \end{cases}$; continuous at all real numbers except -1 and 1 **61.** $c = \sqrt{7}$ **63.** 0 **65.** $a = 10, b = -3$ **67.** take $\delta = \min(1, \frac{1}{3}\epsilon^2)$ **69.** the signum function **71.** (b) $f(x) = \begin{cases} 1 & \text{if } x \neq 0 \\ 0 & \text{if } x = 0 \end{cases}$ **75.** (b) all values of a; (c) all nonintegers **77.** (b) any interval (a, b) where $-2 \leq a < -\frac{3}{2} < b \leq \frac{3}{2}$

EXERCISES 2.1 (PAGE 171)

1. $-2x_1$ **3.** $-4x_1 + 4$ **5.** $3x_1^2$ **7.** $6x_1 - 12$ **9.** $-\dfrac{1}{2\sqrt{4 - x_1}}$ **11.** $3x_1^2 - 12x_1 + 9$ **13.** $8x + y + 9 = 0$ **15.** $6x - y - 16 = 0$ **17.** $2x + 5y - 17 = 0$ **19.** $2x + 3y - 12 = 0$ **21.** $4x + y = 0$ **23.** $x - 12y + 16 = 0$ **25.** $8x - y - 5 = 0$ **27.** $4x + 4y - 11 = 0$ **29.** (a) $V(x) = 4x^3 - 46x^2 + 120x$; (b) $\frac{5}{3}$ in. **31.** (a) $A(x) = 120x - x^2$; (b) 60 m **33.** (a) $P(x) = -2x^2 + 380x - 12,000$; (b) \$95 **37.** $x = 1$

EXERCISES 2.2 (PAGE 179)

1. 7 **3.** 0 **5.** $-4x$ **7.** $8x + 5$ **9.** $-3x^2$ **11.** $\dfrac{1}{2\sqrt{x}}$ **13.** $-\dfrac{1}{(x+1)^2}$ **15.** $-\dfrac{2}{x^3} - 1$ **17.** $-\dfrac{1}{2(x+1)^{3/2}}$ **19.** -6 **21.** $-\frac{3}{128}$

23. $-\frac{1}{4}$ **25.** 3 **27.** -12 **29.** $-\frac{1}{27}$ **31.** $-\dfrac{8}{x^3} + 3$ **33.** $-\dfrac{7}{2\sqrt{2-7x}}$ **35.** $-\dfrac{2}{3(1-2x)^{2/3}}$ **37.** $g(a)$ **39.** $\dfrac{1}{3(x-1)^{2/3}}$; no

41. $\dfrac{-1}{(1-3x)^{2/3}}; \dfrac{1}{3}$ **43.** $\dfrac{2a}{x^3}$

EXERCISES 2.3 (PAGE 186)

1. (b) yes; (c) $1, -1$; (d) no **3.** (b) yes; (c) $-1, 1$; (d) no **5.** (b) yes; (c) 0, 1; (d) no **7.** (b) yes; (c) 0, 0; (d) yes **9.** (b) yes; (c) does not exist, 0; (d) no **11.** (b) yes; (c) 8, 8; (d) yes **13.** (b) yes; (c) neither exists; (d) no **15.** (b) yes; (c) $-6, -6$; (d) yes **17.** (b) no; (c) 1, 0; (d) no **19.** (b) no; (c) 12, 12; (d) no **29.** (a) 3; (b) no **31.** $a = 8, b = -9$ **33.** (a) 0; (b) 1; (c) does not exist **39.** (a) $I(x) = \begin{cases} 15x & \text{if } 0 \le x \le 150 \\ 22.5x - 0.05x^2 & \text{if } 150 < x \le 250 \end{cases}$ **41.** (a) $n > 1$; (b) $n > 1$

EXERCISES 2.4 (PAGE 195)

1. 7 **3.** $-2 - 2x$ **5.** $3x^2 - 6x + 5$ **7.** $x^7 - 4x^3$ **9.** $t^3 - t$ **11.** $4\pi r^2$ **13.** $2x + 3 - \dfrac{2}{x^3}$ **15.** $16x^3 + \dfrac{1}{x^5}$ **17.** $-\dfrac{6}{x^3} - \dfrac{20}{x^5}$

19. $3\sqrt{3}s^2 - 2\sqrt{3}s$ **21.** $70x^6 + 60x^4 - 15x^2 - 6$ **23.** $-18y^2(7 - 3y^3)$ **25.** $10x^4 - 24x^3 + 12x^2 + 2x - 3$

27. $-\dfrac{1}{(x-1)^2}$ **29.** $-\dfrac{4(x+1)}{(x-1)^3}$ **31.** $\dfrac{5(1-2t^2)}{(1+2t^2)^2}$ **33.** $\dfrac{48y^2}{(y^3+8)^2}$ **35.** $\dfrac{6(x^2+10x+1)}{(x+5)^2}$ **37.** $12x - y = 20$

39. $x + 20y + 96 = 0$ **41.** $2x - y = 3$ **43.** $x + 8y + 2 = 0; x + 8y - 2 = 0$ **45.** $28x - y = 99; 4x - y = 3$

49. $2(3x + 2)(6x^2 + 2x - 3)$ **51.** $3(2x^2 + x + 1)^2(4x + 1)$

EXERCISES 2.5 (PAGE 204)

1. $v(t) = 6t$; 18 **3.** $v(t) = -\dfrac{1}{4t^2}; -1$ **5.** $v(t) = 6t^2 - 2t$; 8 **7.** $v(t) = \dfrac{8}{(4+t)^2}; \dfrac{1}{2}$ **9.** $t < -3$, moving to right;

$-3 < t < 1$, moving to left; $t > 1$, moving to right; changes direction when $t = -3$ and $t = 1$

11. $t < -2$, moving to right; $-2 < t < \frac{1}{2}$, moving to left; $t > \frac{1}{2}$, moving to right; changes direction when $t = -2$ and $t = \frac{1}{2}$

13. $t < -3$, moving to left; $-3 < t < 3$, moving to right; $t > 3$, moving to left; changes direction when $t = -3$ and $t = 3$

15. (a) -32 ft/sec; (b) -64 ft/sec; (c) 4 sec; (d) -128 ft/sec **17.** 160 cm/sec **19.** (a) $(20t_1 + 24)$ ft/sec; (b) $\frac{6}{5}$ sec

21. (a) 8.6; (b) 8.3; (c) 8.1; (d) 8 **23.** (a) $20,400 per year; (b) 23.1 percent **25.** (a) 18,750 liters per minute;

(b) 17,500 liters per minute **27.** $\dfrac{dR}{dt} = -\dfrac{E}{I^2}$ **29.** (a) -2.9 degrees per hour; (b) -3 degrees per hour

31. (a) 920 people per year; (b) 1400 people per year; (c) 6.1 percent; (d) 6.4 percent **33.** (a) profitable; (b) not profitable; (c) 90

EXERCISES 2.6 (PAGE 214)

3. $3 \cos x$ **5.** $\sec^2 x - \csc^2 x$ **7.** $2(\cos t - t \sin t)$ **9.** $x \cos x$ **11.** $4 \cos 2x$ **13.** $-x^2 \sin x$ **15.** $3 \sec x(2 \tan^2 x + 1)$

17. $-\csc y(2 \cot^2 y + 1)$ **19.** $-\dfrac{2(z+1)\sin z + 2 \cos z}{(z+1)^2}$ **21.** $\dfrac{1}{\cos x - 1}$ **23.** $\dfrac{1 - 4 \sec t + \sin^2 t}{\cos t(\cos t - 4)^2}$ **25.** $\dfrac{2 \cos y}{(1 - \sin y)^2}$

27. $(1 - \cos x)(x + \cos x) + (1 - \sin x)(x - \sin x)$ **29.** $-\dfrac{5 \csc t \cot t}{(\csc t + 2)^2}$ **31.** 1 **33.** $-\dfrac{2}{\pi}$ **35.** π^2 **37.** 2 **39.** $\sqrt{2}$ **41.** $-\dfrac{10}{3}$

43. (a) $x - y = 0$; (b) $x - 2y + \sqrt{3} - \frac{1}{3}\pi = 0$; (c) $x + y - \pi = 0$ **45.** (a) $x - y = 0$; (b) $4x - 2y + 2 - \pi = 0$;
(c) $4x - 2y - 2 + \pi = 0$ **47.** (a) $4 \cos t$; (b) $v(0) = 4$, $v(\frac{1}{3}\pi) = 2$, $v(\frac{1}{2}\pi) = 0$, $v(\frac{2}{3}\pi) = -2$, $v(\pi) = -4$ **49.** (a) $3 \sin t$;
(b) $v(0) = 0$, $v(\frac{1}{6}\pi) = \frac{3}{2}$, $v(\frac{1}{3}\pi) = \frac{3}{2}\sqrt{3}$, $v(\frac{1}{2}\pi) = 3$, $v(\frac{2}{3}\pi) = \frac{3}{2}\sqrt{3}$, $v(\frac{5}{6}\pi) = \frac{3}{2}$, $v(\pi) = 0$ **51.** (a) $\dfrac{\sqrt{2}}{9} W$; (b) $2W$

EXERCISES 2.7 (PAGE 225)

1. $6(2x + 1)^2$ **3.** $8(x + 2)(x^2 + 4x - 5)^3$ **5.** $2(2t^4 - 7t^3 + 2t - 1)(8t^3 - 21t^2 + 2)$ **7.** $\dfrac{-4x}{(x^2 + 4)^3}$ **9.** $-12(\sin 3x + \cos 4x)$

11. $2 \sec 2x \tan^3 2x$ **13.** $2 \sec^2 x \tan x(2 \tan^2 x + 1)$ **15.** $4 \cot t \csc^2 t$ **17.** $6(3u^2 + 5)^2(3u - 1)(12u^2 - 3u + 5)$

19. $-2(2x - 5)^{-2}(4x + 3)^{-3}(12x - 17)$ **21.** $2(r^2 + 1)^2(2r^2 + 5r - 3)(10r^3 + 20r^2 - 5r + 5)$ **23.** $\dfrac{18(y - 7)}{(y + 2)^3}$

25. $-\dfrac{9(2x - 1)^2(2x^2 - 2x + 1)}{(3x^2 + x - 2)^4}$ **27.** $\dfrac{2z(z^2 - 5)^2(z^2 + 22)}{(z^2 + 4)^3}$ **29.** $6t \sin(6t^2 - 2)$ **31.** $6(\tan^2 x - x^2)(\tan x \sec^2 x - x)$

33. $\dfrac{6 \cos 2y(\sin^2 2y + 2)}{(\cos^2 2y + 1)^2}$ **35.** $-12 \cos 3x \sin(\sin 3x)$ **37.** $24x + y + 39 = 0$; $y = 0$; $y = 1$; $y = 0$; $24x - y - 39 = 0$

39. (a) $\dfrac{8t(t^2 - 1)}{(t^2 + 1)^3}$ cm/sec; (b) 0 cm/sec; $\frac{48}{125}$ cm/sec **41.** (a) $(5\pi \cos \pi t - 3\pi \sin \pi t)$ cm/sec; (b) $\sqrt{2}\pi$ cm/sec, 3π cm/sec
43. (a) $6000 \sin \frac{12}{5}\pi \approx 1854$; (b) 6000 **45.** (a) $5\sqrt{3}$; (b) 10 **47.** 2.7 km/min **53.** (a) $3x^4$; (b) $6x^5$

EXERCISES 2.8 (PAGE 231)

1. $x^{-1/2}(2 - \frac{5}{2}x^{-1})$ **3.** $\dfrac{4x}{\sqrt{1 + 4x^2}}$ **5.** $\dfrac{-2}{(5 - 3x)^{1/3}}$ **7.** $\dfrac{y}{(25 - y^2)^{3/2}}$ **9.** $-\dfrac{\sin \sqrt{t}}{\sqrt{t}}$ **11.** $-\dfrac{\sqrt{3}}{2\sqrt{r}} \csc^2 \sqrt{3r}$

13. $-\dfrac{3 \cos 3x}{2(\sin 3x)^{3/2}}$ **15.** $\dfrac{x}{\sqrt{x^2 + 1}} \sec^2 \sqrt{x^2 + 1}$ **17.** $\dfrac{17}{2(3x + 1)^{3/2}(2x - 5)^{1/2}}$ **19.** $\dfrac{6x^2 - 10x + 1}{3\sqrt[3]{2x^3 - 5x^2 + x)^2}}$ **21.** $\dfrac{1}{\sqrt{2t}}\left(1 - \dfrac{1}{t}\right)$

23. $\frac{1}{4}x(5 - x^2)^{-1/2}(x^3 + 1)^{-3/4}(-7x^3 + 15x - 4)$ **25.** $\dfrac{1}{x^2\sqrt{x^2 - 1}}$ **27.** $\dfrac{\cos t}{(1 - \sin t)^{3/2}(1 + \sin t)^{1/2}}$

29. $\dfrac{\sec \sqrt{y}}{2\sqrt{y}}(\tan^2 \sqrt{y} + \sec^2 \sqrt{y})$ **31.** $\dfrac{x + 5}{6\sqrt{x - 1}\sqrt[3]{x + 1)^4}}$ **33.** $\dfrac{-1}{4\sqrt{9 + \sqrt{9 - x}}\sqrt{9 - x}}$ **35.** $\dfrac{\sin 4z}{(1 + \cos^2 2z)^{3/2}}$ **37.** $2x + 3y = 6$;
$2\sqrt[3]{4}x + 3y = 3\sqrt[3]{4}$; $x = 1$; $2\sqrt[3]{4}x - 3y = \sqrt[3]{4}$; $2x - 3y + 2 = 0$ **39.** $4x - 5y + 9 = 0$ **41.** $x + 4y = 0$ **43.** (a) 0; (b) $\frac{1}{2}$;
(c) no value of t **45.** 100 **47.** $\dfrac{2x(x^2 - 4)}{|x^2 - 4|}$ **49.** $3x|x|$ **51.** $\dfrac{4x^3 - 1}{x^2}$

EXERCISES 2.9 (PAGE 236)

1. $-\dfrac{x}{y}$ **3.** $\dfrac{8y-3x^2}{3y^2-8x}$ **5.** $-\dfrac{y^2}{x^2}$ **7.** $-\dfrac{\sqrt{y}}{\sqrt{x}}$ **9.** $\dfrac{x-xy^2}{x^2y-y}$ **11.** $\dfrac{3x^2-4xy-1}{2x^2+2}$ **13.** $\dfrac{y^{2/3}+y}{24x^{2/3}y^{5/3}-x}$

15. $\dfrac{y+4\sqrt{xy}}{\sqrt{x}-x}$ **17.** $\dfrac{2+5x^2-4x^{3/2}y}{2x^{1/2}(x^2+3)}$ **19.** $\dfrac{\sin(x-y)}{\sin(x-y)-1}$ **21.** $\dfrac{\tan x\sec^2 x}{\cot y\csc^2 y}$ **23.** $\dfrac{y\sin x-\sin y}{x\cos y+\cos x}$

25. $\dfrac{2\tan x\sec^2 x+\csc^2(x-y)}{2\tan y\sec^2 y+\csc^2(x-y)}$ **27.** $\dfrac{3x^2-4y}{4x-3y^2}$ **29.** $\dfrac{3x^2-y^3}{x^3-6xy}$ **31.** $\dfrac{x^3+8y^3}{4x^3-3x^2y}$ **33.** $2x+y=4$ **35.** $5x-7y+11=0$

37. -1 **39.** $(1,0)$ **41.** (a) $f_1(x)=2\sqrt{x-2}$, domain: $x\ge 2$; $f_2(x)=-2\sqrt{x-2}$, domain: $x\ge 2$; (d) $f_1'(x)=(x-2)^{-1/2}$, domain: $x>2$; $f_2'(x)=-(x-2)^{-1/2}$, domain: $x>2$; (e) $\dfrac{2}{y}$; (f) $x-y-1=0$; $x+y-1=0$ **43.** (a) $f_1(x)=\sqrt{x^2-9}$, domain: $|x|\ge 3$; $f_2(x)=-\sqrt{x^2-9}$, domain: $|x|\ge 3$; (d) $f_1'(x)=x(x^2-9)^{-1/2}$, domain: $|x|>3$; $f_2'(x)=-x(x^2-9)^{-1/2}$, domain: $|x|>3$; (e) $\dfrac{x}{y}$; (f) $5x+4y+9=0$; $5x-4y+9=0$ **45.** (a) $f_1(x)=\sqrt{8-x^2+2x}+2$, domain: $-2\le x\le 4$; $f_2(x)=-\sqrt{8-x^2+2x}+2$, domain: $-2\le x\le 4$; (d) $f_1'(x)=(1-x)(8-x^2+2x)^{-1/2}$, domain: $-2<x<4$; $f_2'(x)=(x-1)(8-x^2+2x)^{-1/2}$, domain: $-2<x<4$; (e) $\dfrac{1-x}{y-2}$; (f) $y-5=0$, $y+1=0$

EXERCISES 2.10 (PAGE 243)

1. -3 **3.** -2 **5.** $-\dfrac{\sqrt{3}}{2}$ **7.** $-\tfrac{3}{4}$ **9.** $\tfrac{9}{5}$ ft/sec **11.** $\dfrac{1}{2\pi}$ ft min **13.** $\dfrac{5}{8\pi}$ m/min **15.** $\tfrac{25}{3}$ ft/sec **17.** $\dfrac{6}{25\pi}$ m/min

19. 1800 lb/ft² per min **21.** 128π cm²/sec **23.** 14 ft/sec **25.** \$1020 per week **27.** 875 units per month **29.** decreasing at the rate of 55 shirts per week **33.** 22 m³/min **35.** $\tfrac{1}{194}(3\sqrt{97}+97)$ ft/sec ≈ 0.65 ft/sec **37.** $\tfrac{2000}{9}$ ft/sec **39.** decreasing at the rate of $\tfrac{1}{48}$ rad/sec

EXERCISES 2.11 (PAGE 251)

1. $f'(x)=5x^4-6x^2+1$; $f''(x)=20x^3-12x$ **3.** $g'(s)=8s^3-12s^2+7$; $g''(s)=24s^2-24s$ **5.** $f'(x)=\tfrac{3}{2}x^{3/2}-5$; $f''(x)=\tfrac{3}{4}x^{1/2}$ **7.** $f'(x)=x(x^2+1)^{-1/2}$; $f''(x)=(x^2+1)^{-3/2}$ **9.** $f'(t)=-8t\sin t^2$; $f''(t)=-16t^2\cos t^2-8\sin t^2$

11. $G'(x)=-2\cot x\csc^2 x$; $G''(x)=2\csc^2 x(3\cot^2 x+1)$ **13.** $g'(x)=\dfrac{8x}{(x^2+4)^2}$; $g''(x)=\dfrac{32-24x^2}{(x^2+4)^3}$

15. $f'(x)=\dfrac{\cos x}{2\sqrt{\sin x+1}}$; $f''(x)=-\tfrac{1}{4}\sqrt{\sin x+1}$ **17.** $24x$ **19.** $1152(2x-1)^{-5}$ **21.** $108\sec^2 3x(3\tan^2 3x+1)$

23. $-32(\sin 2x+\cos 2x)$ **29.** $-\dfrac{3a^4x^2}{y^7}$ **31.** $-\tfrac{7}{4}$; 5 **33.** $v=3t^2-18t+15$; $a=6t-18$; when $0<t<1$, the particle is at the right of the origin, it is moving to the right, the velocity is decreasing, and the speed is decreasing; when $1<t<\dfrac{9-\sqrt{21}}{2}$, the particle is at the right of the origin, it is moving to the left, the velocity is decreasing, and the speed is increasing; when $\dfrac{9-\sqrt{21}}{2}<t<3$, the particle is at the left of the origin, it is moving to the left, the velocity is decreasing, and the speed is increasing; when $3<t<5$, the particle is at the left of the origin, it is moving to the left, the velocity is increasing, and the speed is decreasing; when $5<t<\dfrac{9+\sqrt{21}}{2}$, the particle is at the left of the origin, it is moving to the right, the velocity is increasing, and the speed is increasing; when $\dfrac{9+\sqrt{21}}{2}<t<+\infty$, the particle is at the right of the origin, it is moving to the right, the velocity is increasing, and the speed is increasing

35. $\frac{3}{2}$ sec; $\frac{7}{4}$ ft; $-\frac{1}{4}$ ft/sec   **37.** $\frac{1}{2}$ sec; $\frac{249}{80}$ ft; $-\frac{11}{8}$ ft/sec   **39.** $\frac{3}{2}$ sec; $\frac{4}{3}\sqrt{6}$ ft; $\frac{2}{3}\sqrt{6}$ ft/sec   **47.** $f'(x) = 2|x|$, domain: $(-\infty, +\infty)$;

$f''(x) = 2\dfrac{|x|}{x}$, domain: $(-\infty, 0) \cup (0, +\infty)$ **49.** $f'(x) = 4|x^3|$, domain: $(-\infty, +\infty)$; $f''(x) = 12x|x|$, domain: $(-\infty, +\infty)$

51. $f'''(x) = 24|x|$ **53.** $h''(x) = (f'' \circ g)(x)(g'(x))^2 + (f' \circ g)(x)g''(x)$

EXERCISES 2.12 (PAGE 262)

1. 4; 5   **3.** $\frac{1}{6} \approx 0.167$; $\sqrt[3]{10} - 2 \approx 0.154$ **5.** (a) $(-3 - 4x)\Delta x - 2(\Delta x)^2$; (b) $(-3 - 4x)\Delta x$; (c) $-2(\Delta x)^2$

7. (a) $(3x^2 - 2x)\Delta x + (3x - 1)(\Delta x)^2 + (\Delta x)^3$; (b) $(3x^2 - 2x)\Delta x$; (c) $(3x - 1)(\Delta x)^2 + (\Delta x)^3$ **9.** (a) $\dfrac{-2\,\Delta x}{(x - 1)(x + \Delta x - 1)}$;

(b) $\dfrac{-2\,\Delta x}{(x - 1)^2}$; (c) $\dfrac{2(\Delta x)^2}{(x - 1)^2(x + \Delta x - 1)}$ **11.** (a) 0.0309; (b) 0.03; (c) 0.0009   **13.** (a) $\frac{1}{42} \approx 0.0238$; (b) $\frac{1}{40} = 0.025$;

(c) $-\frac{1}{840} \approx -0.0012$ **15.** (a) -0.875; (b) -1.5; (c) 0.625   **17.** $3(3x^2 - 2x + 1)^2(6x - 2)\,dx$ **19.** $\dfrac{(14x^2 + 18x)\,dx}{3(2x + 3)^{2/3}}$

21. $\dfrac{(1 - 2\sin x + 2\cos x)\,dx}{(2 - \sin x)^2}$ **23.** $2\tan x \sec^2 x(2\tan^2 x + 1)\,dx$ **25.** $\dfrac{3x}{4y}$ **27.** $-\dfrac{\sqrt{y}}{\sqrt{x}}$ **29.** 1   **31.** (a) $dx = 4t\,dt$, $dy = 3\,dt$;

(b) $\dfrac{3}{4t}$; (c) $\dfrac{9}{4y}$ **33.** (a) $dx = -2\sin t\,dt$, $dy = 3\cos t\,dt$; (b) $-\dfrac{3\cos t}{2\sin t}$; (c) $-\dfrac{9x}{4y}$ **35.** 0.719   **37.** -1.881 **39.** (a) 6.75 cm³;

(b) 0.3 cm²   **41.** $\frac{12}{5}\pi$ m³   **43.** 0.4π cm²   **45.** 0.9π cm³   **47.** 4 percent   **49.** 10 ft³

REVIEW EXERCISES FOR CHAPTER 2 (PAGE 264)

1. $-\dfrac{8}{x^3} + \dfrac{12}{x^5}$ **3.** $60t^4 - 39t^2 - 6t - 4$ **5.** $\dfrac{-6x^2}{(x^3 - 1)^2}$ **7.** $4(2s^3 - 3s + 7)^3(6s^2 - 3)$ **9.** $\dfrac{1 - 2x^3}{3x^{2/3}(x^3 + 1)^{4/3}}$

11. $x(4x^2 - 13)(x^2 - 1)^{1/2}(x^2 - 4)^{-1/2}$ **13.** $(x + 1)\sin x + x\cos x$ **15.** $\tan\dfrac{1}{x} - \dfrac{1}{x}\sec^2\dfrac{1}{x}$

17. $\dfrac{2(1 + t^2)\sec^2 t \tan t - 2t\sec^2 t}{(1 + t^2)^2}$ **19.** $-3\sin 3w \cos(\cos 3w) + 6\sin 4w$ **21.** $\dfrac{8x}{3y^2 - 8y}$ **23.** $\dfrac{1}{\sqrt{x^2 - 1}(x - \sqrt{x^2 - 1})}$

25. $\dfrac{\cos x \cos y}{\sin x \sin y}$ **27.** $2x + 3[x^3 + (x^4 + x)^2]^2[3x^2 + 2(x^4 + x)(4x^3 + 1)]$ **29.** $\dfrac{y - \sec^2 x}{\sec^2 y - x}$ **31.** $x - 2y + 9 = 0$;

$27x - 54y - 7 = 0$ **33.** $5x - 4y - 6 = 0$; $4x + 5y - 13 = 0$ **35.** $(-1, 0)$ **37.** $-3(3 - 2x)^{-5/2}$ **39.** $x < -3$ or $x > -1$

41. (a) $t = 3$, $t = 8$; (b) when $t = 0$, $v = 24$, and the particle is moving to the right; when $t = 3$, $v = -15$, and the particle is moving to the left; when $t = 8$, $v = 40$, and the particle is moving to the right   **43.** $v^2 = 4(5 - s)(s - 3)$; $a = 16 - 4s$

45. (a) 32 ft/sec; (b) 256 ft; (c) 7 sec; (d) -128 ft/sec   **47.** (a) 8.005; (b) 8   **49.** (a) 3312.2; (b) 3212.5   **51.** $-\frac{1}{3}$

53. $\dfrac{2}{\sqrt{4x - 3}}$ **55.** $\frac{1}{2}$ **57.** $2(|x + 1| - |x|)\left(\dfrac{x + 1}{|x + 1|} - \dfrac{x}{|x|}\right)$ **59.** (b) continuous at 3; (c) not differentiable at 3

61. (b) 0; (c) 0   **63.** $a = -\frac{1}{2}$, $b = \frac{3}{2}$ **69.** 10.6 knots   **71.** $\dfrac{512}{625\pi}$ in./sec ≈ 0.26 in./sec   **73.** 9.6 ft/sec   **75.** (a) -0.16;

(b) -0.64 **77.** $\dfrac{dy}{dt} = \left(\dfrac{4y - 24x^2}{81y^2 - 4x}\right)\left(\dfrac{3t^2 + 2tx}{3x^2 - t^2}\right)$ **79.** 2π in.³   **81.** an error of approximately $\dfrac{\pi^2}{1610t}$ sec

85. $f(x) = |x|$, $g(x) = x^2$ **89.** no; if $f(x) = \dfrac{1}{x - 1}$ and $g(x) = x + 1$, then f and g are differentiable at 0; however $(f \circ g)(x) = \dfrac{1}{x}$

and $f \circ g$ is not differentiable at 0

EXERCISES 3.1 (PAGE 277)

1. $-5, \frac{1}{3}$ **3.** $-3, -1, 1$ **5.** $0, 2$ **7.** $-2, 0, 2$ **9.** no critical numbers **11.** no critical numbers
13. $\frac{1}{6}k\pi$, where k is any integer **15.** $\frac{1}{8}(2k+1)\pi$, where k is any integer **17.** $\frac{1}{4}k\pi$, where k is any integer **19.** $-1, \frac{1}{3}, 2$
21. abs min: $f(2) = -2$ **23.** no absolute extrema **25.** abs min: $f(-\frac{2}{3}\pi) = -1$; abs max: $f(0) = 2$
27. abs min: $f(-3) = 0$ **29.** abs min: $h(5) = 1$ **31.** abs min: $F(4) = 1$ **33.** abs min: $g(0) = 2$ **35.** abs max: $f(5) = 2$
37. abs min: $f(2) = 0$ **39.** abs min: $g(0) = 1$ **41.** abs min: $g(-3) = -46$; abs max: $g(-1) = -10$
43. abs min: $f(-2) = 0$; abs max: $f(-4) = 144$ **45.** abs min: $f(2) = 0$; abs max: $f(3) = 25$ **47.** abs min: $f(-\frac{1}{2}\pi) = -2$;
abs max: $f(\frac{1}{2}\pi) = 2$ **49.** no absolute extrema **51.** abs min: $f(-1) = -1$; abs max: $f(2) = \frac{1}{2}$ **53.** abs min: $f(1) = -2$;
abs max: $f(0) = -\frac{1}{3}$ **55.** abs min: $F(-3) = -13$; abs max: $F(3) = 7$ **57.** abs min: $f(-1) = 0$; abs max: $f(1) = \sqrt[3]{4}$

EXERCISES 3.2 (PAGE 283)

1. $\frac{2}{3}$ in. **3.** 60 m by 60 m **5.** 60 m by 120 m **7.** $\frac{1}{2}$ **9.** $12\sqrt{3}$ sq units **11.** from A to P to C, where P is 8 km down the
beach from B **13.** from A to P to C, where P is 4 km down the river from B **15.** $\frac{1}{4}\pi$ **17.** 225 **19.** 400 **21.** 30
23. radius is $3\sqrt{2}$ in., height is $6\sqrt{2}$ in. **25.** $\frac{50}{3}$ in. by $\frac{50}{3}$ in. by $\frac{100}{3}$ in. **27.** (a) $\sqrt{50 - 6\sqrt{41}}$ units ≈ 3.4 units;

(b) $\sqrt{50 + 6\sqrt{41}}$ units ≈ 9.4 units **29.** $\frac{2}{3}k$ **33.** breadth is $48\sqrt{3}$ cm; depth is $48\sqrt{6}$ cm **35.** (a) radius of circle is $\dfrac{5}{\pi+4}$ ft

and length of side of square is $\dfrac{10}{\pi+4}$ ft; (b) radius of circle is $\dfrac{5}{\pi}$ ft and there is no square

EXERCISES 3.3 (PAGE 290)

1. 2 **3.** $\frac{1}{4}\pi$ **5.** $\frac{1}{2}$ **7.** $\frac{8}{27}$ **9.** 0 **11.** (b) (i), (ii), (iii) satisfied; (c) $(\frac{3}{4}, -\frac{9}{8}\sqrt[3]{6})$ **13.** (b) (i) not satisfied **15.** (b) (ii) not satisfied
17. 4 **19.** $\cos c = \dfrac{2}{\pi}$; $c \approx 0.8807$ **21.** (i) not satisfied **23.** (ii) not satisfied **29.** $\frac{3}{2}$

EXERCISES 3.4 (PAGE 296)

1. (a) and (b) $f(2) = -5$, rel min; (c) $[2, +\infty)$; (d) $(-\infty, 2]$ **3.** (a) and (b) $f(-\frac{1}{3}) = \frac{5}{27}$, rel max; $f(1) = -1$, rel min;
(c) $(-\infty, -\frac{1}{3}], [1, +\infty)$; (d) $[-\frac{1}{3}, 1]$ **5.** (a) and (b) $f(0) = 2$, rel max; $f(3) = -25$, rel min; (c) $(-\infty, 0], [3, +\infty)$; (d) $[0, 3]$
7. (a) and (b) f has a rel max value of 4 if $x = (4k+1)\pi$, where k is any integer; f has a rel min value of -4 if
$x = (4k+3)\pi$, where k is any integer; (c) $[(4k-1)\pi, (4k+1)\pi]$, where k is any integer; (d) $[(4k+1)\pi, (4k+3)\pi]$, where
k is any integer **9.** (a) and (b) $f(0) = 0$, rel min; $f(1) = \frac{1}{4}$, rel max; $f(2) = 0$, rel min; (c) $[0, 1], [2, +\infty)$;
(d) $(-\infty, 0], [1, 2]$ **11.** (a) and (b) $f(-2) = -\frac{1}{15}$, rel max; $f(-1) = -\frac{23}{15}$, rel min; $f(1) = \frac{53}{15}$, rel max; $f(2) = \frac{31}{15}$, rel min;
(c) $(-\infty, -2], [-1, 1], [2, +\infty)$; $[-2, -1], [1, 2]$ **13.** (a) and (b) no relative extrema; (c) $(0, +\infty)$; (d) nowhere
15. (a) and (b) $f(\sqrt[3]{2}) = \frac{3}{2}\sqrt[3]{2}$, rel min; (c) $(-\infty, 0), [\sqrt[3]{2}, +\infty)$; (d) $(0, \sqrt[3]{2}]$ **17.** (a) and (b) $f(\frac{1}{5}) = \frac{3456}{3125}$, rel max;
$f(1) = 0$, rel min; (c) $(-\infty, \frac{1}{5}], [1, +\infty)$; (d) $[\frac{1}{5}, 1]$ **19.** (a) and (b) $f(2) = 4$, rel max; (c) $(-\infty, 2]$; (d) $[2, 3]$
21. (a) and (b) $f(-1) = 2$, rel max; $f(1) = -2$, rel min; (c) $(-\infty, -1], [1, +\infty)$; (d) $[-1, 1]$
23. (a) and (b) $f(4) = 2$, rel max; (c) $(-\infty, -2]$; (d) $[4, +\infty)$ **25.** (a) and (b) $f(\frac{1}{8}) = -\frac{1}{4}$, rel min; (c) $[\frac{1}{8}, +\infty)$; (d) $(-\infty, \frac{1}{8}]$
27. (a) and (b) f has a rel max value of $-\frac{1}{2}$ if $x = \frac{1}{4}\pi + \frac{1}{2}k\pi$, where k is any integer; f has a rel min value of $\frac{1}{2}$ if $x = \frac{1}{2}k\pi$,
where k is any integer; (c) $[\frac{1}{2}k\pi, \frac{1}{4}\pi + \frac{1}{2}k\pi), (\frac{1}{4}\pi + \frac{1}{2}k\pi, \frac{1}{2}\pi + \frac{1}{2}k\pi]$, where k is any integer;
(d) $[\frac{1}{4}\pi + \frac{1}{2}k\pi, \frac{3}{8}\pi + \frac{1}{2}k\pi), (\frac{3}{8}\pi + \frac{1}{2}k\pi, \frac{1}{2}\pi + \frac{1}{2}k\pi]$, where k is any integer **29.** (a) and (b) $f(-1) = 0$, rel max;
$f(1) = -\sqrt[3]{4}$, rel min; (c) $(-\infty, -1], [1, +\infty)$; (d) $[-1, 1]$ **31.** (a) and (b) $f(-2) = 5$, rel max; $f(0) = 1$, rel min;
(c) $(-\infty, -2], [0, +\infty)$; (d) $[-2, 0]$ **33.** (a) and (b) $f(-1) = 2$, rel max; $f(0) = 1$, rel min; $f(2) = 5$, rel max;

(c) $(-\infty, -1]$, $[0, 2]$; (d) $[-1, 0]$, $[2, +\infty)$ **35.** (a) and (b) $f(-9) = -8$, rel min; $f(-4) = -5$, rel min; $f(2) = -7$, rel min; $f(-7) = -4$, rel max; $f(0) = -3$, rel max; (c) $[-9, -7]$, $[-4, 0]$, $[2, +\infty)$; (d) $(-\infty, -9]$, $[-7, -4]$, $[0, 2]$ **37.** (a) and (b) no relative extrema; (c) $[0, +\infty)$; (d) nowhere **39.** (a) and (b) $f(4) = \frac{1}{4}\sqrt[3]{4}$, rel max; (c) $(-4, 4]$; (d) $(-\infty, -4)$, $[4, +\infty)$ **43.** $a = -3$, $b = 7$ **45.** $a = -2$, $b = 9$, $c = -12$, $d = 7$

EXERCISES 3.5 (PAGE 304)

1. $(0, 0)$; concave downward for $x < 0$; concave upward for $x > 0$ **3.** $(-\frac{1}{2}, \frac{15}{2})$; concave downward for $x < -\frac{1}{2}$; concave upward for $x > -\frac{1}{2}$ **5.** $(0, 0)$, $[4, -256)$; concave upward for $x < 0$ and $x > 4$; concave downward for $0 < x < 4$ **7.** $(1, 0)$; concave downward for $x < 1$; concave upward for $x > 1$ **9.** $(-2, 0)$; concave upward for $x < -2$; concave downward for $x > -2$ **11.** $(1, \frac{1}{2})$, $(-1, \frac{1}{2})$; concave upward for $x < -1$ and $x > 1$; concave downward for $-1 < x < 1$ **13.** $(-\frac{2}{3}\pi, 0)$, $(-\frac{1}{3}\pi, 0)$, $(0, 0)$, $(\frac{1}{3}\pi, 0)$, $(\frac{2}{3}\pi, 0)$; concave upward for $-\pi < x < -\frac{2}{3}\pi$, $-\frac{1}{3}\pi < x < 0$, and $\frac{1}{3}\pi < x < \frac{2}{3}\pi$; concave downward for $-\frac{2}{3}\pi < x < -\frac{1}{3}\pi$, $0 < x < \frac{1}{3}\pi$, and $\frac{2}{3}\pi < x < \pi$ **15.** $(0, 0)$; concave downward for $-\pi < x < 0$; concave upward for $0 < x < \pi$ **17.** no pt of infl; concave upward for $x < 2$; concave downward for $x > 2$ **19.** $(0, 0)$; concave upward for $x < 0$; concave downward for $x > 0$ **21.** $(0, 0)$; concave downward for $x < 0$; concave upward for $x > 0$ **23.** $(2, 3)$; concave upward for $x < 2$; concave downward for $x > 2$ **25.** $a = -1$, $b = 3$ **27.** $a = 2$, $b = -6$, $c = 0$, $d = 3$ **29.** (a) $(\frac{1}{2}\pi + k\pi, 0)$, where k is any integer; 1 at $(\frac{1}{2}\pi + k\pi, 0)$ if k is an even integer, -1 at $(\frac{1}{2}\pi + k\pi, 0)$ if k is an odd integer **31.** $(\frac{1}{2}\pi + k\pi, 0)$, where k is any integer; -1 at all pts of infl. (Sketches of the graphs for Exercises 33–47 appear in Figs. EX 3.5-33 through 3.5-47.) **49.** yes **51.** 10:40 A.M.

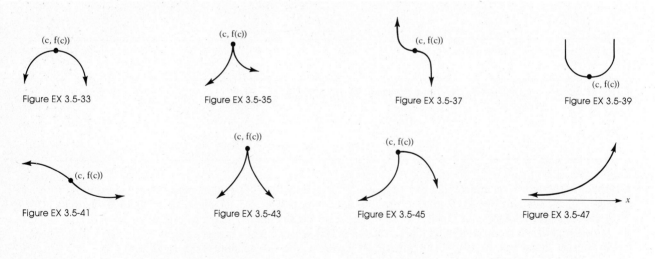

Figure EX 3.5-33   Figure EX 3.5-35   Figure EX 3.5-37   Figure EX 3.5-39

Figure EX 3.5-41   Figure EX 3.5-43   Figure EX 3.5-45   Figure EX 3.5-47

EXERCISES 3.6 (PAGE 310)

1. $f(\frac{1}{3}) = \frac{2}{3}$, rel min; concave upward everywhere **3.** $f(\frac{3}{2}) = \frac{81}{4}$, rel max; $f(-1) = -11$, rel min; $(\frac{1}{4}, \frac{37}{8})$, pt. of infl.; concave upward for $x < \frac{1}{4}$; concave downward for $x > \frac{1}{4}$ **5.** $g(0) = 3$, rel max; $g(2) = \frac{5}{3}$, rel min; $(1, \frac{7}{3})$, pt. of infl.; concave downward for $x < 1$; concave upward for $x > 1$ **7.** $f(4) = 0$, rel min; concave upward everywhere

9. $h(-\frac{3}{4}) = -\frac{99}{256}$, rel min; $h(0) = 0$, rel max; $h(1) = -\frac{5}{6}$, rel min; pts. of infl. at $x = \dfrac{1 \pm \sqrt{37}}{2}$; concave upward for $x < \dfrac{1 - \sqrt{37}}{12}$ and $x > \dfrac{1 + \sqrt{37}}{12}$; concave downward for $\dfrac{1 - \sqrt{37}}{12} < x < \dfrac{1 + \sqrt{37}}{12}$

11. $F(\frac{1}{3}\pi) = -1$, rel min; $F(0) = 1$, rel max; $(\frac{1}{6}\pi, 0)$, pt. of infl.; concave downward for $-\frac{1}{6}\pi < x < \frac{1}{6}\pi$; concave upward for $\frac{1}{6}\pi < x < \frac{1}{2}\pi$ **13.** $f(-\frac{1}{2}) = -\frac{27}{16}$, rel min; $(-2, 0)$, $(-1, -1)$, pts. of infl.; concave upward for $x < -2$ and $x > -1$; concave downward for $-2 < x < -1$ **15.** $f(1) = 8$, rel min; $(3, \frac{16}{3}\sqrt{3})$, pt. of infl.; concave upward for $0 < x < 3$; concave downward for $x > 3$ **17.** $h(-2) = -2$, rel min; concave upward for $x > -3$ **19.** $F(27) = 9$, rel max; $(0, 0)$, $(216, 0)$, pts. of infl.; concave upward for $x < 0$ and $x > 216$; concave downward for $0 < x < 216$ **21.** $f(-1) = -2$, rel min; $f(1) = 2$, rel max; $(0, 0)$, $(-\frac{1}{2}\sqrt{2}, -\frac{7}{8}\sqrt{2})$, $(\frac{1}{2}\sqrt{2}, \frac{7}{8}\sqrt{2})$, pts. of infl.; concave upward for $x < -\frac{1}{2}\sqrt{2}$ and $0 < x < \frac{1}{2}\sqrt{2}$; concave downward for $-\frac{1}{2}\sqrt{2} < x < 0$ and $x > \frac{1}{2}\sqrt{2}$ **23.** $f(0) = 0$, rel min; concave upward everywhere **25.** no relative extrema; $(-\frac{3}{2}\pi, -\frac{3}{2}\pi)$, $(-\frac{1}{2}\pi, -\frac{1}{2}\pi)$, $(\frac{1}{2}\pi, \frac{1}{2}\pi)$, $(\frac{3}{2}\pi, \frac{3}{2}\pi)$, pts. of infl.; concave downward for $-2\pi < x < -\frac{3}{2}\pi$, $-\frac{1}{2}\pi < x < \frac{1}{2}\pi$ and $\frac{3}{2}\pi < x < 2\pi$; concave upward for $-\frac{3}{2}\pi < x < -\frac{1}{2}\pi$ and $\frac{1}{2}\pi < x < \frac{3}{2}\pi$ **27.** $\cos k\pi = 1$, where k is any even integer, rel max; $\cos k\pi = -1$, where k is any odd integer, rel min **29.** $\csc(\frac{1}{2}\pi + 2k\pi) = 1$, where k is any integer, rel min; $\csc(\frac{3}{2}\pi + 2k\pi) = -1$, where k is any integer, rel max **43.** f has a rel min at $x = \frac{1}{2}\sqrt{2}$ and a rel max at $x = -\frac{1}{2}\sqrt{3}$

EXERCISES 3.7 (PAGE 319)

1. 0 **3.** $\frac{2}{5}$ **5.** $-\frac{2}{3}$ **7.** $\frac{7}{3}$ **9.** 0 **11.** $+\infty$ **13.** $\frac{1}{2}$ **15.** $+\infty$ **17.** $-\infty$ **19.** 1 **21.** -1 **23.** (a) $C(x) = 48x + 97{,}200x^{-1}$, $(0, +\infty)$; (b) $+\infty$, $+\infty$ **25.** (a) $A(x) = 3x + 30 + 48x^{-1}$, $(0, +\infty)$; (b) $+\infty$, $+\infty$ **27.** 0 **29.** $-\infty$ **31.** 0 **33.** take $N = 1 + \dfrac{1}{\epsilon}$

35. take $N = \sqrt{\dfrac{2}{\epsilon} - 1}$ **37.** take $N = \dfrac{\epsilon - 7}{2\epsilon}$

EXERCISES 3.8 (PAGE 327)

1. $y = 2$, $x = 3$ **3.** $y = 1$, $x = 0$ **5.** $y = 0$, $x = -2$, $x = 2$ **7.** $y = 4$, $x = -3$, $x = 3$ **9.** $y = 0$, $x = \frac{2}{3}$, $x = -\frac{5}{2}$ **11.** $x = -\sqrt{2}$, $x = \sqrt{2}$ **13.** $y = \frac{2}{3}$, $x = \frac{4}{3}$ **15.** $y = 1$, $y = -1$ **17.** $y = -1$, $y = 1$, $x = 3$ **19.** $y = 2$, $x = -1$, $x = 1$ **21.** $x = 1$, $y = x + 1$ **23.** $x = 3$, $y = x + 3$ **25.** $x = 0$, $y = 2x$ **27.** $x = 0$, $y = x$ **29.** $x = -2$, $y = x - 6$ **31.** $x = 0$, $y = x + 2$

EXERCISES 3.9 (PAGE 331)

(Sketches of the graphs appear in Figs. EX 3.9-1 through EX 3.9-49)

1. $f(-1) = 5$, rel max; $f(1) = -3$, rel min; $(0, 1)$, pt. of infl.; f increasing on $(-\infty, -1]$ and $[1, +\infty)$; f decreasing on $[-1, 1]$; graph concave downward for $x < 0$; graph concave upward for $x > 0$ **3.** $f(\frac{3}{2}) = -\frac{27}{16}$, rel min; $(0, 0)$, $(1, -1)$, pts. of infl.; f increasing on $[\frac{3}{2}, +\infty)$; f decreasing on $(-\infty, \frac{3}{2}]$; graph concave upward for $x < 0$ and $x > 1$; graph concave downward for $0 < x < 1$ **5.** $f(-3) = 5$, rel max; $f(-\frac{1}{3}) = -\frac{121}{27}$, rel min; $(-\frac{5}{3}, \frac{7}{27})$, pt. of infl.; f increasing on $(-\infty, -3]$ and $[-\frac{1}{3}, +\infty)$; f decreasing on $[-3, -\frac{1}{3}]$; graph concave downward for $x < -\frac{5}{3}$; graph concave upward for $x > -\frac{5}{3}$ **7.** $f(0) = 1$, rel min; $(\frac{1}{2}, \frac{23}{16})$, $(1, 2)$, pts. of infl.; f decreasing on $(-\infty, 0]$; f increasing on $[0, +\infty)$; graph concave upward for $x < \frac{1}{2}$ and $x > 1$; graph concave downward for $\frac{1}{2} < x < 1$ **9.** $f(-1) = \frac{7}{12}$, rel min; $f(0) = 1$, rel max; $f(2) = -\frac{5}{3}$, rel min; pts. of infl. at $x = \frac{1}{3}(1 \pm \sqrt{7})$; f decreasing on $(-\infty, -1]$ and $[0, 2]$; f increasing on $[-1, 0]$ and $[2, +\infty)$; graph concave upward for $x < \frac{1}{3}(1 - \sqrt{7})$ and $x > \frac{1}{3}(1 + \sqrt{7})$; graph concave downward for $\frac{1}{3}(1 - \sqrt{7}) < x < \frac{1}{3}(1 + \sqrt{7})$ **11.** $f(0) = 0$, rel min; no pts. of infl.; f decreasing on $(-\infty, 0]$; f increasing on $[0, +\infty)$; graph concave upward everywhere **13.** $f(0) = 0$, rel min; no pts. of infl.; f decreasing on $(-\infty, 0]$; f increasing on $[0, +\infty)$; graph concave upward everywhere **15.** no relative extrema; $(0, 0)$ pt. of infl. with horizontal tangent; f increasing on $(-\infty, +\infty)$; graph concave downward on $(-\infty, 0)$; graph concave upward on $(0, +\infty)$ **17.** no relative extrema; $(2, 0)$ pt. of infl.; f decreasing on $(-\infty, +\infty)$; graph

Figure EX 3.9-1

Figure EX 3.9-3

Figure EX 3.9-5

Figure EX 3.9-7

Figure EX 3.9-9

Figure EX 3.9-11

Figure EX 3.9-13

Figure EX 3.9-15

Figure EX 3.9-17

Figure EX 3.9-19

concave upward for $x < 2$; graph concave downward for $x > 2$ **19.** $f\left(\frac{4}{5}\right) = \frac{26,244}{3,125}$, rel max; $f(2) = 0$, rel min; pts. of infl. at $(-1, 0)$ and $x = \frac{1}{10}(8 \pm 3\sqrt{6})$; f increasing on $(-\infty, \frac{4}{5}]$ and $[2, +\infty)$; f decreasing on $[\frac{4}{5}, 2]$; graph concave downward for $x < -1$ and $\frac{1}{10}(8 - 3\sqrt{6}) < x < \frac{1}{10}(8 + 3\sqrt{6})$; graph concave upward for $-1 < x < \frac{1}{10}(8 - 3\sqrt{6})$ and $x > \frac{1}{10}(8 + 3\sqrt{6})$
21. $f(-\frac{4}{3}) = \frac{256}{81}$, rel max; $f(0) = 0$, rel min; $(-1, 2)$, pt. of infl.; f increasing on $(-\infty, -\frac{4}{3}]$ and $[0, +\infty)$; f decreasing on $[-\frac{4}{3}, 0]$; graph concave downward for $x < -1$; graph concave upward for $x > -1$
23. $f(-\frac{1}{2}\pi) = -3$, rel min; $f(0) = 3$, rel max; $f(\frac{1}{2}\pi) = 0$, rel min; $(-\frac{3}{4}\pi, 0)$, $(-\frac{1}{4}\pi, 0)$, $(\frac{1}{4}\pi, 0)$, $(\frac{3}{4}\pi, 0)$, pts. of infl.; f decreasing on $[-\pi, -\frac{1}{2}\pi]$ and $[0, \frac{1}{2}\pi]$; f increasing on $[-\frac{1}{2}\pi, 0]$ and $[\frac{1}{2}\pi, \pi]$; graph concave downward for $-\pi < x < -\frac{3}{4}\pi$, $-\frac{1}{4}\pi < x < \frac{1}{4}\pi$, and $\frac{3}{4}\pi < x < \pi$; graph concave upward for $-\frac{3}{4}\pi < x < -\frac{1}{4}\pi$ and $\frac{1}{4}\pi < x < \frac{3}{4}\pi$ **25.** no relative extrema; no pts. of infl.; f increasing on $[0, \frac{1}{2}\pi]$ and $[\frac{1}{2}\pi, \pi]$; graph concave downward for $0 < x < \frac{1}{2}\pi$ and $\frac{1}{2}\pi < x < \pi$
27. no relative extrema; $(0, 0)$, pt. of infl.; f increasing on $(-\pi, \pi)$; graph concave downward for $-\pi < x < 0$; graph concave upward for $0 < x < \pi$; $x = -\pi$ and $x = \pi$ are asymptotes **29.** $f(-\frac{7}{4}\pi) = \sqrt{2}$, rel max; $f(-\frac{3}{4}\pi) = -\sqrt{2}$, rel min; $f(\frac{1}{4}\pi) = \sqrt{2}$, rel max; $f(\frac{5}{4}\pi) = -\sqrt{2}$, rel min; $(-\frac{5}{4}\pi, 0)$, $(-\frac{1}{4}\pi, 0)$, $(\frac{3}{4}\pi, 0)$, $(\frac{7}{4}\pi, 0)$, pts. of infl.; f increasing on $[-2\pi, -\frac{7}{4}\pi]$, $[-\frac{3}{4}\pi, \frac{1}{4}\pi]$, and $[\frac{5}{4}\pi, 2\pi]$; f decreasing on $[-\frac{7}{4}\pi, -\frac{3}{4}\pi]$, and $[\frac{1}{4}\pi, \frac{5}{4}\pi]$; graph concave downward for $-2\pi < x < -\frac{5}{4}\pi$, $-\frac{1}{4}\pi < x < \frac{3}{4}\pi$, and $\frac{7}{4}\pi < x < 2\pi$; graph concave upward for $-\frac{5}{4}\pi < x < -\frac{1}{4}\pi$ and $\frac{3}{4}\pi < x < \frac{7}{4}\pi$ **31.** $f(0) = 0$, rel max; $f(2) = 4$, rel min.; no pts. of infl.; f increasing on $(-\infty, 0]$ and $[2, +\infty)$; f decreasing on $[0, 1)$ and $(1, 2]$; graph concave downward for $x < 1$; graph concave upward for $x > 1$; $x = 1$ and $y = x + 1$ are asymptotes **33.** $f(0) = -1$, rel max; no pts. of infl.; f increasing on $(-\infty, -1)$ and $(-1, 0]$; f decreasing on $[0, 1)$ and $(1, +\infty)$; graph concave upward for $x < -1$ and $x > 1$; graph concave downward for $-1 < x < 1$; $y = 1$, $x = -1$, and $x = 1$ are asymptotes

35. $f(-1) = -1$, rel min; $f(1) = 1$, rel max; $(-\sqrt{3}, -\frac{1}{2}\sqrt{3})$ and $(\sqrt{3}, \frac{1}{2}\sqrt{3})$, pts. of infl.; f decreasing in $(-\infty, -1]$ and $[1, +\infty)$; f increasing on $[-1, 1]$; graph concave downward for $x < -\sqrt{3}$ and $0 < x < \sqrt{3}$; graph concave upward for $-\sqrt{3} < x < 0$ and $x > \sqrt{3}$ **37.** $f(0) = 0$, rel min; $f(1) = 1$, rel max; f decreasing on $(-\infty, 0)$ and $[1, +\infty)$; f increasing on $(0, 1]$; graph concave downward for $x < 0$ and $x > 0$ **39.** $f(1) = -1$, rel min; f decreasing on $(-\infty, 1]$; f increasing on $[1, +\infty)$; graph concave upward for all x **41.** no relative extrema; $(3, 2)$, pt. of infl.; f increasing on $(-\infty, +\infty)$; graph concave downward for $x > 3$; graph concave upward for $x < 3$ **43.** no relative extrema; $(3, 2)$, pt. of infl. with horizontal tangent; f increasing on $(-\infty, +\infty)$; graph concave downward for $x < 3$; graph concave upward for $x > 3$ **45.** $f(0) = 0$, rel min; $f(\frac{16}{5}) = \frac{512}{125}\sqrt{5}$, rel max; pt. of infl. at $x = \frac{1}{15}(48 - 8\sqrt{6})$; f decreasing on $(-\infty, 0]$ and $[\frac{16}{5}, 4]$; f increasing on $[0, \frac{16}{5}]$; graph concave upward for $x < \frac{1}{15}(48 - 8\sqrt{6})$; graph concave downward for $\frac{1}{15}(48 - 8\sqrt{6}) < x < 4$ **47.** $f(-\frac{2}{3}) = \frac{4}{9}\sqrt{6}$, rel max; f increasing on $(-\infty, -\frac{2}{3}]$; f decreasing on $[-\frac{2}{3}, 0]$; graph concave downward for $x < 0$ **49.** $f(-1) = 0$, rel max; $f(1) = -\sqrt[3]{4}$, rel min; $(2, 0)$, pt. of infl.; f increasing on $(-\infty, -1]$ and $[1, +\infty)$; f decreasing on $[-1, 1]$; graph concave upward for $x < -1$ and $-1 < x < 2$; graph concave downward for $x > 2$

Figure EX 3.9-21

Figure EX 3.9-23

Figure EX 3.9-25

Figure EX 3.9-27

Figure EX 3.9-29

Figure EX 3.9-31

Figure EX 3.9-33

Figure EX 3.9-35

Figure EX 3.9-37

Figure EX 3.9-39

Figure EX 3.9-41

Figure EX 3.9-43

Figure EX 3.9-45

Figure EX 3.9-47

Figure EX 3.9-49

EXERCISES 3.10 (PAGE 341)

1. $f(0) = 0$, abs min   **3.** no absolute extrema   **5.** $g(\frac{1}{4}) = \frac{3}{4}$, abs min   **7.** no absolute extrema   **9.** $f(-2) = -13$, abs min   **11.** $g(-1) = -2$, abs max   **13.** $f(0) = 0$, abs min; $f(\sqrt{2}) = \frac{1}{18}\sqrt{3}$, abs max   **15.** $f(-\frac{1}{6}\pi) = \sqrt{3}$, abs min   **17.** 45 m by 60 m   **19.** 6 inches by 9 inches   **21.** 12 inches by 4 inches by 6 inches   **23.** $2x - y + 1 = 0$ **25.** $(\frac{9}{5}, \frac{3}{5})$ **27.** height, $4\sqrt[3]{4}$ inches; radius, $\sqrt[3]{4}$ inches   **29.** 1 month; 7.5 percent   **31.** radius of semicircle, $\dfrac{32}{4 + \pi}$ ft; height of rectangle, $\dfrac{32}{4 + \pi}$ ft   **33.** 90 km/hr   **35.** $\frac{36}{23}$ sec   **37.** $5\sqrt{5}$ ft   **39.** $\sqrt{2}$ **41.** $2\sqrt{2}$ **43.** $\frac{4}{3}a$

EXERCISES 3.11 (PAGE 355)

1. (a) $C'(x) = 3 + 2x$; (b) \$83; (c) \$84   **3.** (a) 50 cents per liter; (b) 25   **5.** (a) $Q(x) = x + 4 + \dfrac{8}{x}$; (b) $C'(x) = 2x + 4$; (c) \$9.66   **7.** (a) $Q(x) = 3x - 6 + \dfrac{4}{x}$; (b) $C'(x) = 6x - 6$; (c) $[1, +\infty)$; (d) \$0.93   **9.** (a) $[2, +\infty)$; (b) $C'(x) = x^2 - 4x + 5$; (c) decreasing on $[0, 2]$ and increasing on $[2, +\infty)$; (d) graph concave downward for $0 < x < 2$, graph concave upward for $x > 2$, pt. of infl. at $(2, \frac{20}{3})$, infl. tangent: $3x - 3y + 14 = 0$ **11.** (a) $C(x) = 3x + 400$; (b) $Q(x) = 3 + \dfrac{400}{x}$; (c) $C'(x) = 3$; (e) 953   **13.** (a) $R'(x) = 600 - \frac{3}{20}x^2$; (b) \$540; (c) \$536.95

15. (a) $R(x) = 100x\sqrt{36 - x^2}$; (b) $R'(x) = \dfrac{3600 - 200x^2}{\sqrt{36 - x^2}}$; (c) \$1800   **17.** 80, \$3400   **19.** (a) $P(x) = -2x^2 + 120x - 300$; (c) $R'(x) = 140 - 2x$, $C'(x) = 2x + 20$; (d) \$1500   **21.** (a) $R'(x) = 6 - \frac{1}{3}\sqrt{x - 100} - \dfrac{x}{10\sqrt{x - 100}}$; $C'(x) = 2$; (b) 200 or 100   **23.** (a) 1875; (b) 62.5 cents; (c) \$503.12   **25.** (a) 1812; (b) 63.8 cents; (c) \$397.30   **27.** (a) \$8.67; (b) \$9.18   **31.** 120   **33.** 12 orders of 1000 cases each   **35.** $\sqrt{\dfrac{RS}{2K}}$ orders of $\sqrt{\dfrac{2KR}{S}}$ cases each

EXERCISES 3.12 (PAGE 362)

1. 4.1179   **3.** -1.1673 **5.** 2.649   **7.** 0.507   **9.** -1.128 **11.** 1.73205   **13.** 1.81712   **15.** 0.7391   **17.** 0.8767   **19.** 2.0288, 4.9132   **21.** 3.14159

REVIEW EXERCISES FOR CHAPTER 3 (PAGE 363)

1. abs min: $f(-5) = 0$ **3.** abs min: $f(1) = -\frac{1}{2}$; abs max: $f(2) = 64$ **5.** abs min: $f(3) = 0$; abs max: $f(0) = 9$
7. abs min: $f(\sqrt{6}) = 0$; abs max: $f(5) = 361$ **9.** abs min: $f(-\frac{1}{6}\pi) = -2$; abs max: $f(\frac{1}{6}\pi) = 2$ **11.** abs max: $f(\frac{1}{12}\pi) = \sqrt{3}$
13. $f(-2) = 0$, rel max; $f(0) = -4$, rel min; $(-1, -2)$, pt. of infl.; f increasing on $(-\infty, -2]$ and $[0, +\infty)$;
f decreasing on $[-2, 0]$; graph concave downward for $x < -1$; graph concave upward for $x > -1$

15. $f\left(\frac{8}{5}\right) = \frac{839,808}{3125}$, rel max; $f(4) = 0$, rel min; pts. of infl. at $x = -2$ and $x = \frac{1}{3}(8 \pm 3\sqrt{6})$;
f increasing on $(-\infty, \frac{8}{5}]$ and $[4, +\infty)$; f decreasing on $[\frac{8}{5}, 4]$; graph concave upward for $-2 < x < \frac{1}{3}(8 - 3\sqrt{6})$ and
$\frac{1}{3}(8 + 3\sqrt{6}) < x < +\infty$; graph concave downward for $-\infty < x < -2$ and $\frac{1}{3}(8 - 3\sqrt{6}) < x < \frac{1}{3}(8 + 3\sqrt{6})$
17. no relative extrema; $(4, -3)$, pt. of infl.; f increasing on $(-\infty, +\infty)$; graph concave upward for $x < 4$;
graph concave downward for $x > 4$ **19.** $f(0) = 0$, rel min; $(\frac{1}{3}, \frac{1}{3})$, $(-\frac{1}{3}, \frac{1}{3})$, pts. of infl.; f decreasing on $(-\infty, 0]$;
f increasing on $[0, +\infty)$; graph concave downward for $x < -\frac{1}{3}$ and $x > \frac{1}{3}$; graph concave upward for $-\frac{1}{3} < x < \frac{1}{3}$;
$y = \frac{4}{3}$ is an asymptote **21.** $f(0) = 0$, rel max; f increasing on $(-\infty, -2)$ and $(-2, 0]$; f decreasing on $[0, 2)$ and $(2, +\infty)$;
graph concave upward for $x < -2$ and $x > 2$; graph concave downward for $-2 < x < 2$;
$y = 5$, $x = -2$, and $x = 2$ are asymptotes **23.** $f(0) = 0$, rel max; $f(6) = 12$, rel min; no pts. of infl.;
f increasing on $(-\infty, 0]$ and $[6, +\infty)$; f decreasing on $[0, 3)$ and $(3, 6]$; graph concave downward for $x < 3$;
graph concave upward for $x > 3$; $x = 3$ and $y = x + 3$ are asymptotes **25.** $f(1) = 0$, rel min; f decreasing on $(-\infty, 1]$;
f increasing on $[1, +\infty)$; graph concave downward for $x < 1$ and $x > 1$ **27.** no relative extrema; $(3, 1)$, pt. of infl.;
f increasing on $(-\infty, +\infty)$; graph concave downward for $x < 3$; graph concave upward for $x > 3$ **29.** $f(-1) = 0$, rel min;
$f(0) = 9$, rel max; $f(3) = 0$, rel min; pts. of infl. at $x = \pm\frac{3}{5}\sqrt{5}$; f decreasing on $(-\infty, -1]$ and $[0, 3]$;
f increasing on $[-1, 0]$ and $[3, +\infty)$; graph concave upward for $x < -\frac{3}{5}\sqrt{5}$ and $x > \frac{3}{5}\sqrt{5}$;
graph concave downward for $-\frac{3}{5}\sqrt{5} < x < \frac{3}{5}\sqrt{5}$ **31.** $f(-\frac{1}{4}\pi) = -\sqrt{2}$, rel min; $f(\frac{3}{4}\pi) = \sqrt{2}$, rel max; $(\frac{1}{4}\pi, 0)$, pt. of infl.;
f decreasing on $[-\frac{3}{4}\pi, -\frac{1}{4}\pi]$ and $[\frac{3}{4}\pi, \frac{7}{4}\pi]$; f increasing on $[-\frac{1}{4}\pi, \frac{3}{4}\pi]$; graph concave upward for $-\frac{3}{4}\pi < x < \frac{1}{4}\pi$;
graph concave downward for $\frac{1}{4}\pi < x < \frac{5}{4}\pi$ **33.** $-\infty$ **35.** 0 **37.** $\frac{1}{4}(1 - \sqrt{13})$ **39.** $-\frac{13}{4}$ **45.** $\sqrt{A^2 + B^2}$
47. pt. of infl. at $(0, 2)$; graph concave downward for $x < 0$; graph concave upward for $x > 0$ **49.** $a = 2$, $b = -6$, $c = 3$
55. $y = \frac{3}{4}$ and $x = -\frac{7}{4}$ **57.** 6, 6 **61.** $\frac{1}{2}\sqrt{17}$ **63.** 20 inches by 20 inches by 10 inches
65. $\sqrt[3]{2k}$ inches by $\sqrt[3]{2k}$ inches by $\frac{1}{2}\sqrt[3]{2k}$ inches **67.** 1500 **69.** 12 km from point on bank nearest A **73.** $\frac{3}{2}$ sec;
velocity of particle moving horizontally is 1 cm/sec; velocity of particle moving vertically is 3 cm/sec
75. $(h^{2/3} + w^{2/3})^{3/2}$ meters **77.** 9 m by 18 m **79.** radius is $\frac{2}{3} r$ cm and altitude is $3k$ cm

81. the wire should be cut in half **87.** (a) $P(x) = \dfrac{36 - 4x}{x + 3}$; (b) $R(x) = \dfrac{3600x - 400x^2}{x + 3}$;

(c) $R'(x) = -\dfrac{400(x + 9)(x - 3)}{(x + 3)^2}$; (d) \$1200 **89.** 1000, \$11 **91.** 120 **93.** 1.168 **95.** 1.9622

EXERCISES 4.1 (PAGE 376)

1. $\frac{3}{5}x^5 + C$ **3.** $-\dfrac{1}{2x^2} + C$ **5.** $2u^{5/2} + C$ **7.** $3x^{2/3} + C$ **9.** $\frac{9}{10}t^{10/3} + C$ **11.** $x^4 + \frac{1}{3}x^3 + C$ **13.** $\frac{1}{9}y^6 - \frac{3}{4}y^4 + C$
15. $3t - t^2 + \frac{1}{3}t^3 + C$ **17.** $\frac{8}{5}x^5 + x^4 - 2x^3 - 2x^2 + 5x + C$ **19.** $\frac{2}{5}x^{5/2} + \frac{2}{3}x^{3/2} + C$ **21.** $\frac{2}{5}x^{5/2} - \frac{1}{2}x^2 + C$
23. $-\dfrac{1}{x^2} - \dfrac{3}{x} + 5x + C$ **25.** $\frac{2}{5}x^{5/2} + \frac{8}{3}x^{3/2} - 8x^{1/2} + C$ **27.** $\frac{3}{4}x^{4/3} + \frac{3}{2}x^{2/3} + C$ **29.** $-3\cos t - 2\sin t + C$ **31.** $\sec x + C$
33. $-4\csc x + 2\tan x + C$ **35.** $-2\cot\theta - 3\tan\theta + \theta + C$ **37.** $y = x^2 - 3x + 2$ **39.** $3y = -2x^3 + 3x^2 + 2x + 6$
41. $12y = -x^4 + 6x^2 - 20x + 27$ **43.** 117π m³ **45.** g is not differentiable on $(-1, 1)$

EXERCISES 4.2 (PAGE 385)

1. $-\frac{1}{6}(1 - 4y)^{3/2} + C$ **3.** $-\frac{3}{8}(6 - 2x)^{4/3} + C$ **5.** $\frac{1}{3}(x^2 - 9)^{3/2} + C$ **7.** $\frac{1}{33}(x^3 - 1)^{11} + C$ **9.** $-\frac{3}{8}(9 - 4x^2)^{5/3} + C$

11. $\dfrac{1}{32(1 - 2y^4)^4} + C$ **13.** $\frac{3}{11}(x - 2)^{11/3} + C$ **15.** $\frac{2}{5}(x + 2)^{5/2} - \frac{4}{3}(x + 2)^{3/2} + C$ **17.** $-\frac{2}{5}(1 - r)^{-5} + \frac{1}{3}(1 - r)^{-6} + C$

19. $-\frac{3}{4}(3-2x)^{3/2} + \frac{3}{10}(3-2x)^{5/2} - \frac{1}{28}(3-2x)^{7/2} + C$ **21.** $\frac{1}{4}\sin 4\theta + C$ **23.** $-2\cos x^3 + C$ **25.** $\frac{1}{5}\tan 5x + C$
27. $-\frac{1}{6}\csc 3y^2 + C$ **29.** $\frac{1}{6}(2 + \sin x)^6 + C$ **31.** $-2\left(1 + \dfrac{1}{3x}\right)^{3/2} + C$ **41.** $\frac{2}{3}\sqrt{x^3 + 3x^2 + 1} + C$

43. $-\dfrac{1}{4(3x^4 + 2x^2 + 1)} + C$ **45.** $\frac{3}{4}(3 - y)^{4/3} - 18(3 - y)^{1/3} + C$ **47.** $\frac{2}{5}(r^{1/3} + 2)^5 + C$ **49.** $\sqrt{x^2 + 4} + \dfrac{4}{\sqrt{x^2 + 4}} + C$
51. $\cos(\cos x) + C$ **53.** (a) $2x^4 + 4x^3 + 3x^2 + x + C$; (b) $\frac{1}{8}(2x + 1)^4 + C$ **55.** (a) $\frac{2}{3}x^{3/2} - 2x + 2x^{1/2} + C$;
(b) $\frac{2}{3}(\sqrt{x} - 1)^3 + C$ **57.** (a) $\sin^2 x + C$; (b) $-\cos^2 x + C$; (c) $-\frac{1}{2}\cos 2x + C$ **59.** $\frac{1}{6}$ coulombs   **61.** \$325   **63.** 3.1 μm^3

EXERCISES 4.3 (PAGE 394)
1. $y = 2x^2 - 5x + C$ **3.** $y = x^3 + x^2 - 7x + C$ **5.** $3x^2y + Cy + 2 = 0$ **7.** $2\sqrt{1 + u^2} = 3v^2 + C$
9. $\tan x - \tan y + y = C$ **11.** $12y = 5x^4 + 6x^2 + C_1x + C_2$ **13.** $9s = -\sin 3t - \cos 3t + C_1t + C_2$
15. $3y = x^3 - 3x^2 - 12x + 18$ **17.** $4\sin 3x + 6\cos 2y + 7 = 0$ **19.** $u = 3v^4 + 4v^3 + 2v^2 + 2v$ **21.** $s = \frac{1}{3}(2t + 4)^{3/2} - \frac{8}{3}$
23. $v = 2 + 5t - t^2$, $s = 2t + \frac{5}{2}t^2 - \frac{1}{3}t^3$ **25.** $v = \frac{1}{3}t^3 + t^2 - 4$, $s = \frac{1}{12}t^4 + \frac{1}{3}t^3 - 4t + 1$
27. $v = -2\sqrt{2}\sin(2t - \frac{1}{4}\pi)$, $s = \sqrt{2}\cos(2t - \frac{1}{4}\pi)$ **29.** $1600s = v^2 + 1200$ **31.** $5s^2 + 4s = v^2 + 12$ **33.** (a) $\frac{3}{4}$ sec;
(b) 20 ft/sec; (c) $\frac{5}{8}$ sec; (d) $\frac{25}{4}$ ft   **35.** (a) $\frac{1}{4}\sqrt{555}$ sec; (b) $8\sqrt{555}$ ft/sec   **37.** (a) 3.4 sec; (b) 99 ft/sec

39. (a) $v^2 = -64s + 1600$; (b) 24 ft/sec   **41.** $\dfrac{3}{2\pi}$ cm to the right of the origin   **43.** 1.62 m/sec^2

45. (a) 3.47 sec; (b) 48.22 m   **47.** 20 m/sec or 72 km/hr   **49.** $x^2 + 2y^2 = C$

EXERCISES 4.4 (PAGE 408)
1. 51   **3.** 147   **5.** 2025   **7.** $\frac{73}{12}$ **9.** $\frac{63}{4}$ **11.** $\frac{7}{12}$ **13.** 10,400   **15.** $2^n - 1$ **17.** $\frac{100}{101}$ **19.** $n^4 - \frac{2}{3}n^3 - 3n^2 - \frac{4}{3}n$ **21.** $\frac{8}{3}$ sq units
23. 15 sq units   **25.** $\frac{5}{3}$ sq units   **27.** 9 sq units   **29.** $\frac{3}{3}$ sq units   **31.** $\frac{17}{4}$ sq units   **33.** $\frac{27}{4}$ sq units   **35.** $\frac{1}{2}m(b^2 - a^2)$ sq units
37. $\frac{1}{2}h(b_1 + b_2)$ sq units

EXERCISES 4.5 (PAGE 416)
1. $\frac{247}{32}$ **3.** $\frac{1469}{1320}$ **5.** 0.835   **7.** $\frac{1}{24}(10 + \sqrt{2} + 3\sqrt{3})\pi$ **9.** 9   **11.** $\frac{8}{3}$ **13.** $\frac{15}{4}$ **15.** 66   **17.** 4   **19.** 20 sq units   **21.** 44 sq units
23. $\frac{5}{3}$ sq units   **25.** $\frac{305}{6}$ sq units   **27.** $\frac{31}{4}$ sq units   **29.** $\displaystyle\int_0^2 x^2\,dx$ **31.** $\displaystyle\int_0^1 \frac{1}{x^2}\,dx$

EXERCISES 4.6 (PAGE 427)
1. 12   **3.** $4\sqrt{5}$ **5.** -5 **7.** 15   **9.** 0   **11.** -21 **13.** $-\frac{3}{2}$ **15.** $4 + \pi$ **17.** $\frac{33}{2}\pi$ **19.** $[24, 56]$ **21.** $[0, 64]$ **23.** $[0, 27]$
25. $[0, 576]$ **27.** $[\frac{1}{12}\pi, \frac{1}{12}\sqrt{3}\pi]$ **29.** $[0, 6]$ **31.** $[-3, \frac{3}{2}]$ **33.** $[-5\pi, 0]$ **35.** $\geq$ **37.** $\leq$

EXERCISES 4.7 (PAGE 431)
1. $\frac{2}{3}\sqrt{3}$ **3.** $\frac{1}{2}\sqrt[3]{30}$ **5.** $-2 + \sqrt{21}$ **7.** 0   **17.** average value is $\frac{1}{2}$ at $x = \frac{1}{2}$ **19.** $\dfrac{2}{\pi}$; 0.69   **21.** $v = 32t$; 32   **23.** π

EXERCISES 4.8 (PAGE 440)

1. 12   **3.** 36   **5.** $\frac{3}{2}$ **7.** $\frac{3}{16}$ **9.** $\frac{134}{3}$ **11.** -8 **13.** 1   **15.** $\frac{2}{3}(27 - 2\sqrt{2})$ **17.** $2 - \sqrt[3]{2}$ **19.** $\frac{104}{5}$ **21.** $\frac{29}{2}$ **23.** $\frac{2}{3}\sqrt{2}$ **25.** $\frac{256}{15}$ **27.** $\frac{5}{6}$

29. $\frac{6215}{12}$ **31.** $\frac{11}{6}$ **33.** 0   **35.** $\frac{3}{2}$ **37.** $\sqrt{4 + x^6}$ **39.** $-\sqrt{\sin x}$ **41.** $\dfrac{2}{3 + x^2}$ **43.** average value is 6 at $x = \sqrt{3}$ **45.** 27   **47.** $\dfrac{4}{\pi}$ **51.** $\frac{188}{3}$

EXERCISES 4.9 (PAGE 448)

1. $\frac{32}{3}$ sq units   **3.** $\frac{22}{3}$ sq units   **5.** $\frac{32}{3}$ sq units   **7.** $\frac{343}{6}$ sq units   **9.** 1 sq unit   **11.** 1 sq unit   **13.** $\frac{32}{3}$ sq units   **15.** $\frac{32}{3}$ sq units
17. $\frac{1}{6}$ sq units   **19.** $\frac{12}{5}$ sq units   **21.** $\frac{9}{2}$ sq units   **23.** $\frac{8}{3}\sqrt{2}$ sq units   **25.** $\frac{5}{12}$ sq units   **27.** $\frac{27}{10}$ sq units   **29.** $\frac{64}{3}$ sq units
31. $\frac{253}{12}$ sq units   **33.** $\frac{37}{12}$ sq units   **35.** $(\sqrt{2} - 1)$ sq units   **37.** $\frac{7}{3}$ sq units   **39.** 12 sq units   **41.** $\frac{128}{5}$ sq units   **43.** 64 sq units
45. $(\frac{1}{2}\pi - 1)$ sq units   **47.** $(1 - \frac{1}{4}\pi)$ sq units   **49.** $\frac{16}{3}p^2$ sq units   **51.** 32   **53.** $\frac{3}{2}K$

EXERCISES 4.10 (PAGE 456)

1. (a) $R(x) = 12x - \frac{3}{2}x^2$; (b) $p = 12 - \frac{3}{2}x$ **3.** $C(x) = \frac{6}{5}\sqrt{5x + 4} + \frac{38}{5}$ **5.** $p(x + 5) = 4x + 22$ **7.** $C(x) = 3x^2 - 17x + 47$
9. (a) $R(x) = 16x - x^3$; (b) $p = 16 - x^2$ **11.** $C(x) = x^3 + 4x^2 + 4x + 6$ **13.** \$3375   **15.** 450 sq units   **17.** 375
19. (a) \$96,000; (b) 5 years   **21.** \$50,000   **23.** 10th   **25.** \$57.60   **27.** (a) \$5.40   **29.** 4800   **31.** 79 words per minute
33. \$2.82   **35.** \$5

REVIEW EXERCISES FOR CHAPTER 4 (PAGE 458)

1. $\frac{1}{2}x^4 - \frac{1}{3}x^3 + 3x + C$ **3.** $2y^2 + 4y^{3/2} + C$ **5.** $-\frac{1}{3}\cos 3t + C$ **7.** $-\frac{1}{3}\cos^3 x + C$ **9.** $-\dfrac{2}{3x^3} + \dfrac{5}{x} + C$

11. $\frac{5}{54}(2 + 3x^2)^9 + C$ **13.** $\frac{2}{3}x\sqrt{3x} + \frac{2}{3}\sqrt{5x} + C$ **15.** $-\dfrac{1}{24(x^4 + 2x^2)^6} + C$ **17.** $\frac{1}{6}(2s + 3)^{3/2} - \frac{3}{2}(2s + 3)^{1/2} + C$

19. $\frac{1}{3}\tan 3\theta - \theta + C$ **21.** $5 \sin x - 3 \sec x + C$ **23.** $\frac{1}{420}(4x + 3)^{3/2}(30x^2 - 18x + 79) + C$ **25.** 6   **27.** $-\frac{27}{4}$ **29.** 0   **31.** $\frac{5}{4}$
33. $\frac{1}{2}$ **35.** $\frac{652}{15}$ **37.** $4 - \frac{1}{2}\pi$ **39.** $\frac{1}{2}(y^2 - 1)^{-1} = x^{-1} + C$ **41.** $y = \frac{1}{15}(2x - 1)^{5/2} + C_1 x + C_2$ **43.** (a) $\frac{1}{9}(x^3 + 1)^3 + C$;
(b) $\frac{1}{9}x^9 + \frac{1}{3}x^6 + \frac{1}{3}x^3 + C$ **45.** $y = 10x - 2x^2 - 9$ **47.** 1.46 cm^3 **49.** (a) $V = \frac{2}{3}(t + 1)^{3/2} + \frac{1}{2}t^2 + \frac{74}{3}$; (b) 64 cm^3

51. $v = 3 \sin 2t + 3$, $s = -\frac{3}{2}\cos 2t + 3t + \frac{1}{2}(5 - 3\pi)$ **53.** $\dfrac{25\sqrt{817} - 625}{8}$ sec ≈ 11.2 sec   **55.** (a) 1 sec; (b) -80 ft/sec

57. 4, 100, 656, 560   **63.** $[0, \pi]$ **65.** $\frac{313}{2}$ **67.** $-(3x^2 - 4)^{3/2}$ **69.** 12; $\sqrt{3}$ **71.** 0   **73.** $\frac{42,304}{175}$ **77.** 18 sq units   **79.** $\frac{224}{3}$ sq units
81. $\frac{1}{3}(40\sqrt{5} - 20)$ sq units   **83.** 36 sq units   **85.** $\frac{1}{12}$ sq unit   **87.** $2\sqrt{2}$ sq units   **89.** $(\frac{1}{2}\pi - 1)$ sq units   **91.** (b) 47°; (c) 60°;
(d) 73°; (e) 75°; (f) 67.5°; (g) $\dfrac{3}{\pi}(20\pi - 3\sqrt{3} + 6)$ degrees $\approx 60.8°$ **93.** $\sqrt{3} + \frac{1}{6}\pi$ **99.** \$850   **101.** 1190 units

103. (b) \$20,000; (c) \$40,000; (d) \$20,000; (e) 0; (f) \$720,000   **105.** \$18

EXERCISES 5.1 (PAGE 469)

1. 64π cu units   **3.** $\frac{704}{3}\pi$ cu units   **5.** $\frac{384}{5}\pi$ cu units   **7.** $\frac{3456}{35}\pi$ cu units   **9.** $\frac{256}{15}\pi$ cu units   **11.** $\frac{128}{5}\pi$ cu units   **13.** $\frac{128}{5}\pi$ cu units
15. $\frac{4}{3}\pi r^3$ cu units   **17.** π cu units   **19.** $\frac{1}{2}\pi^2$ cu units   **21.** $(4\pi - \frac{1}{2}\pi^2)$ cu units   **23.** $(\sqrt{3}\pi - \frac{1}{3}\pi^2)$ cu units   **25.** $\frac{1250}{3}\pi$ cu units
27. $\frac{64}{5}\pi$ cu units   **29.** $\frac{261}{32}\pi$ cu units   **31.** 16π cu units   **33.** 180π cm^3 **35.** $\frac{16}{3}\pi$ cu units   **37.** 2   **39.** $(\frac{8}{3}\pi^2 - 2\sqrt{3}\pi)$ cu units

EXERCISES 5.2 (PAGE 474)

13. $\frac{1}{2}\pi$ cu units 15. $\frac{3}{10}\pi$ cu units 17. $\frac{2}{5}\pi$ cu units 19. $\frac{49}{30}\pi$ cu units 21. 16π cu units 23. $\frac{512}{15}\pi$ cu units
25. $\frac{32}{15}\pi a^3$ cu units 27. $\frac{32}{15}\pi$ cu units 29. $\frac{1}{2}(2+\sqrt{2})\pi$ cu units 31. π cu units 33. $\frac{224}{3}\pi$ cu units 35. 24π cu units
37. $\sqrt[7]{2744}$

EXERCISES 5.3 (PAGE 477)

1. $32\sqrt{2}$ cu units 3. $\frac{1372}{3}\sqrt{3}$ cm³ 5. $\frac{686}{3}$ cm³ 7. $\frac{8}{3}r^3$ cu units 9. 14 in.³ 11. $\frac{128}{3}$ cu units 13. 1944 cm³ 15. $\frac{3969}{10}$ cu units
17. $\frac{2}{3}r^3$ cm³

EXERCISES 5.4 (PAGE 483)

1. $\frac{158}{3}$ ft-lb 3. $\frac{1076}{15}$ joules 5. $\frac{1}{2}(3+\sqrt{393})$ 7. 180 in.-lb 9. 8 joules 11. 1350 ergs 13. 6562.5w ft-lb 15. $256\pi w$ ft-lb
17. 100,000 ft-lb 19. 5500 ft-lb 21. $1{,}017{,}938\pi$ joules 23. $9{,}196{,}875\pi$ joules 25. $\frac{144}{550}w$ sec 27. $2\sqrt{3}$ ft

EXERCISES 5.5 (PAGE 489)

1. $\sqrt{10}$ 3. $\sqrt{97}$ 5. $\frac{14}{3}$ 7. $\frac{33}{16}$ 9. $\frac{1}{27}(97^{3/2}-125)$ 11. 12 13. $\frac{22}{3}$ 15. $\frac{9}{8}$ 17. $\dfrac{8a^3-(a^2+3b^2)^{3/2}}{8(a^2-b^2)}$ 19. $2\sqrt{3}-\frac{4}{3}$ 21. 2

EXERCISES 5.6 (PAGE 496)

1. 250 lb 3. 4000 dynes 5. $\frac{3}{2}$ m/sec² 7. $\frac{8}{3}$ slugs 9. 4 11. 6 13. 54 kg; $\frac{11}{3}$ m from one end 15. 171 slugs;
5.92 in. from one end 17. 42 g; $\frac{44}{7}$ cm from one end 19. 31.5 kg; $\frac{18}{7}$ m from the end having the greater density
21. 16 slugs; $\frac{16}{5}$ ft from one end 23. 1.2 m from the end having the greater density 25. $\dfrac{20}{L^2}x^2$ slugs/ft 27. 12 kg/m;

EXERCISES 5.7 (PAGE 504)

1. $(2, \frac{1}{3})$ 3. $\frac{29}{7}$ 5. $(\frac{2}{3}, 1)$ 7. $(0, \frac{8}{3})$ 9. $(0, \frac{12}{5})$ 11. $(\frac{16}{13}, \frac{64}{21})$ 13. $(\frac{1}{2}, -\frac{3}{2})$ 15. $(0, 2)$ 17. $\frac{3}{5}p$ 21. the point on the bisecting
radial line whose distance from the center of the circle is $\dfrac{4}{3\pi}$ times the radius 23. $\frac{1}{3}\pi r^2 h$ cu units 25. $\dfrac{(4+3\pi)\pi r^3}{3\sqrt{2}}$ cu units

EXERCISES 5.8 (PAGE 512)

1. $(0, \frac{8}{5}, 0)$ 3. $(\frac{1}{2}, 0, 0)$ 5. $(\frac{16}{5}, 0, 0)$ 7. $(2, \frac{10}{9}, 0)$ 9. $(0, \frac{7}{3}, 0)$ 11. $(0, \frac{13}{5}, 0)$ 13. $(\frac{5}{16}, 4, 0)$ 15. $(2, \frac{5}{8}p, 0)$
17. on axis and $\frac{3}{4}h$ units from vertex 19. $(\frac{4}{3}, 0, 0)$ 21. $(\frac{8}{9}, 0, 0)$ 23. 2

EXERCISES 5.9 (PAGE 519)

1. 320ρ lb **3.** 64ρ lb **5.** 2.25ρ lb **7.** 941,760 nt **9.** 4,087,500 nt **11.** $\sqrt[3]{16.31}$ m ≈ 2.54 m **15.** $14,000\rho$ lb
17. $100,000\rho$ ft-lb **19.** 756 lb **21.** $250\sqrt{409}\rho$ lb **23.** $11,250\sqrt{3}\rho$ lb

REVIEW EXERCISES FOR CHAPTER 5 (PAGE 520)

1. $\frac{1}{3}\pi$ cu units **3.** π cu units **5.** π cu units **7.** 250π cu units **9.** 1024 cu units **11.** 3π cu units **13.** $\frac{25}{6}\pi$ cu units
15. $\frac{16}{3}\sqrt{3}$ **17.** 6000 ergs **19.** 400 ft-lb **21.** 44,145,000 joules **23.** $\frac{2752}{3}w\pi$ ft-lb **25.** $\frac{13,600}{3}w\pi$ ft-lb **27.** 90 ft³ **29.** 558π cm³
31. $\frac{1}{5320}[(2251)^{3/2} - \frac{1}{8}(10999)^{3/2}]$ **33.** $\frac{2}{13}$ **35.** $(\frac{3}{2}, 2)$ **37.** $\frac{104}{3}$ slugs; $\frac{298}{65}$ in. from the left end **39.** $(\frac{9}{8}, \frac{18}{5})$ **41.** $(\frac{9}{20}, \frac{9}{20})$
43. $\frac{256}{3}\pi$ m³ **45.** $\frac{1}{4}\pi^2$ cu units **47.** $2 - \sqrt{2}$ **49.** $(\frac{7}{8}, 0, 0)$ **51.** $(\frac{5}{6}, 0, 0)$ **53.** $\frac{18}{5}\rho$ lb **55.** $\frac{5120}{3}$ lb

EXERCISES 6.1 (PAGE 534)

1. one-to-one **3.** not one-to-one **5.** one-to-one **7.** one-to-one **9.** one-to-one **11.** one-to-one **13.** one-to-one
15. one-to-one **17.** not one-to-one **19.** $f^{-1}(x) = \frac{1}{5}(x + 7)$; domain: $(-\infty, +\infty)$, range: $(-\infty, +\infty)$ **21.** no inverse
23. $f^{-1}(x) = 4 - \sqrt[3]{x}$; domain: $(-\infty, +\infty)$, range: $(-\infty, +\infty)$ **25.** $h^{-1}(x) = \frac{1}{2}x^2 + 3$; domain: $[0, +\infty)$, range: $[3, +\infty)$
27. $F^{-1}(x) = x^3 - 1$; domain: $(-\infty, +\infty)$, range: $(-\infty, +\infty)$ **29.** no inverse **31.** $f^{-1}(x) = \frac{1}{32}x^5$; domain: $(-\infty, +\infty)$, range:
$(-\infty, +\infty)$ **33.** $f^{-1}(x) = \frac{x+3}{1-x}$; domain: $\{x|x \neq 1\}$; range: $\{y|y \neq 1\}$ **35.** $g^{-1}(x) = \sqrt{x-5}$; domain: $[5, +\infty)$, range:
$[0, +\infty)$ **37.** $f^{-1}(x) = \frac{1}{2}(\sqrt[3]{x} - 1)$; domain: $[0, 8]$, range: $[-\frac{1}{2}, \frac{1}{2}]$ **39.** $F^{-1}(x) = \sqrt{9 - x^2}$; domain: $[0, 3]$, range: $[0, 3]$
41. (b) $f^{-1}(x) = \frac{1}{4}(x + 3)$ **43.** (b) $f^{-1}(x) = \sqrt[3]{x - 2}$ **45.** (b) $f^{-1}(x) = \frac{4x - 1}{3 - 2x}$ **47.** (b) $f_1(x) = x^2 + 4, x \geq 0; f_2(x) = x^2 + 4,$
$x \leq 0$; (c) $f_1^{-1}(x) = \sqrt{x - 4}$, domain: $[4, +\infty); f_2^{-1}(x) = -\sqrt{x^2 - 4}$; domain: $[4, +\infty)$ **49.** (b) $f_1(x) = \sqrt{9 - x^2}, 0 \leq x \leq 3;$
$f_2(x) = \sqrt{9 - x^2}, -3 \leq x \leq 0$; (c) $f_1^{-1}(x) = \sqrt{9 - x^2}$, domain: $[0, 3]; f_2^{-1}(x) = -\sqrt{9 - x^2}$, domain: $[0, 3]$
51. $f^{-1}(x) = \begin{cases} x & \text{if } x < 1 \\ \sqrt{x} & \text{if } 1 \leq x \leq 81 \\ \frac{1}{729}x^2 & \text{if } x > 81 \end{cases}$

EXERCISES 6.2 (PAGE 542)

11. $\frac{2}{3}$ **13.** $\frac{1}{10}$ **15.** $\frac{1}{12}$ **17.** $\frac{1}{21}$ **19.** -2 **21.** $-\frac{1}{4}$ **23.** $-\frac{1}{\sqrt{3}}$ **25.** $-\frac{1}{4}$ **27.** $\frac{1}{3}$ **29.** $\frac{1}{\sqrt{15}}$ **31.** (b) 6; (c) $\frac{1}{6}$ **33.** (a) $f_1(x) = \sqrt{9 - x^2}$;
$f_2(x) = -\sqrt{9 - x^2}$; (b) neither has an inverse; (c) $\frac{dy}{dx} = -\frac{x}{y}, \frac{dx}{dy} = -\frac{y}{x}$ **35.** (a) $f(x) = \frac{4}{x}$; (b) $f^{-1}(x) = \frac{4}{x}$, domain: $\{x|x \neq 0\}$;
(c) $\frac{dy}{dx} = -\frac{y}{x}, \frac{dx}{dy} = -\frac{x}{y}$ **37.** (a) $f(x) = \frac{2x^2 + 1}{3x}$; (b) no inverse; (c) $\frac{dy}{dx} = \frac{4x - 3y}{3x}, \frac{dx}{dy} = \frac{3x}{4x - 3y}$

EXERCISES 6.3 (PAGE 553)

1. $\frac{5}{4 + 5x}$ **3.** $\frac{5}{8 + 10x}$ **5.** $\frac{6}{3t + 1}$ **7.** $\frac{6\ln(3t + 1)}{3t + 1}$ **9.** $-\frac{2x}{12 - 3x^2}$ **11.** $\frac{5\cos 5y}{\sin 5y}$ **13.** $-\frac{\sin(\ln x)}{x}$ **15.** $2\sec 2x$
17. $\frac{\sec^2 x}{2\tan x} = \csc 2x$ **19.** $-\frac{17}{2(2w - 5)(3w + 1)}$ **21.** $\frac{\ln x - 1}{(\ln x)^2}$ **23.** $\frac{1 - 2x - x^2}{3(x + 1)(x^2 + 1)}$ **25.** $\frac{1}{2(1 + \sqrt{x + 1})}$ **27.** $-\frac{xy + y}{xy + x}$

29. $x + y$ **31.** $\dfrac{4x^2y - xy - 2y}{6xy^2 + x}$ **43.** $2x + 4y + 4\ln 2 - 1 = 0$ **45.** $40x - y = 40$ **47.** $-\tfrac{1}{2}$
49. (a) \$5 per \$1 change in budget; (b) \$688

EXERCISES 6.4 (PAGE 560)

1. $\dfrac{3x^2}{x^3 + 1}$ **3.** $-\dfrac{3\sin 3x}{\cos 3x}$ **5.** $4\sec 4x$ **7.** $\dfrac{4 - x^2}{x(x^2 + 4)}$ **9.** $2x(x + 1)^6(x - 1)^2(6x^2 - 2x - 1)$ **11.** $\dfrac{-13x^2 - 4x + 8}{(x - 4)^4}$
13. $\dfrac{8x^9 - 4x^7 + 15x^2 + 10}{5(x^7 + 1)^{6/5}}$ **15.** $-\tfrac{1}{2}\ln|3 - 2x| + C$ **17.** $\tfrac{3}{2}\ln(x^2 + 4) + C$ **19.** $\tfrac{1}{5}\ln|5x^3 - 1| + C$
21. $\tfrac{1}{2}\ln|1 + 2\sin t| + C$ **23.** $\tfrac{1}{5}\ln|1 - \cos 5x| + C$ **25.** $\ln|\sec 2x + \tan 2x| + \tfrac{3}{2}\ln|\cos 2x| + C$
27. $x^2 + 4\ln|x^2 - 4| + C$ **29.** $\ln|\ln x| + C$ **31.** $\tfrac{1}{3}\ln^3 3x + C$ **33.** $\ln|(\ln x)^2 + \ln x| + C$ **35.** $\ln|\sec(\ln x)| + C$ **37.** $\ln 5$
39. $4 + \ln 2$ **41.** $\tfrac{1}{2}\ln(4 + 2\sqrt{3})$ **43.** $\dfrac{1}{\ln 4}$ **47.** $\tfrac{1}{4}\ln 5 \approx 0.402$ **49.** $2000\ln 2$ lb/ft² ≈ 1386 lb/ft²
51. $\ln 4$ sq units ≈ 1.386 sq units   **53.** $\pi(11 + 8\ln 2)$ cu units ≈ 51.978 cu units   **55.** $\ln(2 + \sqrt{3})$

EXERCISES 6.5 (PAGE 569)

1. $5e^{5x}$ **3.** $-6xe^{-3x^2}$ **5.** $-e^{\cos x}\sin x$ **7.** $e^{2x}\cos e^x + e^x\sin e^x$ **9.** $\dfrac{e^{\sqrt{x}}\sec^2 e^{\sqrt{x}}}{2\sqrt{x}}$ **11.** $\dfrac{4}{(e^x + e^{-x})^2}$ **13.** $2x$
15. $2e^{2x}\sec e^{2x}\tan e^{2x} + 2e^{2\sec x}\sec x\tan x$ **17.** $\dfrac{e^x(1 - e^y)}{e^y(e^x - 1)}$ **19.** $-\dfrac{y^2 + 2ye^{2x}}{2e^{2x} + 3xy}$ **21.** $-\tfrac{1}{5}e^{2-5x} + C$ **23.** $e^x - e^{-x} + C$
25. $\dfrac{1}{6(1 - 2e^{3x})} + C$ **27.** $e^x - 3\ln(e^x + 3) + C$ **29.** e^2 **31.** 2 **33.** $\tfrac{1}{2}$ **35.** $\tfrac{1}{2}(e^4 - 1)$ **39.** $f(1) = \dfrac{1}{e}$, rel max;
$(2, 2e^{-2})$, pt. of infl.; f increasing on $(-\infty, 1]$; f decreasing on $[1, +\infty)$; graph concave downward for $x < 2$; graph concave upward for $x > 2$ **41.** 2.66 **43.** 2.56 **45.** $x + 2y = 1 + \ln 2$ **47.** $(e^3 + 1)$ ft ≈ 20.586 ft
49. -9.17 lb/ft² per sec   **51.** 0.006 **53.** $\tfrac{1}{2}\pi w(e^{-2} - e^{-8})$ ft-lb

EXERCISES 6.6 (PAGE 582)
1. 68.4 years   **3.** 38,720   **5.** \$2734   **7.** 16,000   **9.** (a) 96 percent; (b) 66 percent   **11.** 1389
13. (a) \$52.59; (b) 10.52 percent   **15.** 15.9 years   **17.** 8.7 years   **19.** (a) 34.7; (b) 55.5   **21.** 40 min   **23.** 43.9 g
25. 11.6 kg   **27.** 15,327 years ago   **29.** $\ln\left|\ln\dfrac{a}{y}\right| + kt = C$

EXERCISES 6.7 (PAGE 590)

1. $(5\ln 3)3^{5x}$ **3.** $4^{3t^2}(\ln 4)6t$ **5.** $4^{\sin 2x}(2\ln 4)\cos 2x$ **7.** $2^{5x}3^{4x^2}(5\ln 2 + 8x\ln 3)$ **9.** $\dfrac{1}{x^2}\log_{10}\dfrac{e}{x}$ **11.** $\dfrac{\log_a e}{2x\sqrt{\log_a x}}$
13. $\dfrac{(\log_{10} e)^2}{(x + 1)\log_{10}(x + 1)}$ **15.** $3^{t^2}\sec 3^{t^2}\tan 3^{t^2}(2t\ln 3)$ **17.** $x^{\sqrt{x}-(1/2)}(1 + \tfrac{1}{2}\ln x)$ **19.** $z^{\cos z - 1}[\cos z - z(\ln z)\sin z]$

21. $(\sin x)^{\tan x}[1 + (\ln \sin x) \sec^2 x]$ **23.** $x^{e^x - 1}e^x(x \ln x + 1)$ **25.** $\dfrac{3^{2x}}{2 \ln 3} + C$ **27.** $\dfrac{a^t e^t}{1 + \ln a} + C$ **29.** $\dfrac{10^{x^3}}{3 \ln 10} + C$

31. $\dfrac{6^{e^y}}{\ln 6} + C$ **33.** 0.621 **35.** 2.999 **45.** (a) 61 sales per day; (b) 2.26 sales per day

49. (a) $y = 200 \cdot 2^{t/10}$; (b) \$12,800; (c) \$877 per year **51.** $\left(e - 1 - \dfrac{1}{\ln 2}\right)$ sq units ≈ 0.276 sq units

53. $\left(\dfrac{4}{\ln 3} - 1\right) \pi$ cu units ≈ 8.297 cu units

REVIEW EXERCISES FOR CHAPTER 6 (PAGE 591)

1. (a) $f^{-1}(x) = \sqrt[3]{x + 4}$, domain: $(-\infty, +\infty)$, range: $(-\infty, +\infty)$ **3.** no inverse **5.** $f^{-1}(x) = \dfrac{4}{3 - x}$; domain: $\{x | x \neq 3\}$;

range: $\{y | y \neq 0\}$ **7.** (b) $f^{-1}(x) = x^3 - 1$ **9.** $\frac{1}{4}$ **11.** $\frac{1}{12}$ **13.** $\dfrac{8 \ln x}{x}$ **15.** $(-4 \ln 2)2^{\cos 4x} \sin 4x$

17. $\dfrac{(4 - x^2)e^{x/(4+x^2)}}{(4 + x^2)^2}$ **19.** $\dfrac{\log_{10} e}{(1 - x^2) \sqrt{\log_{10} \dfrac{1 + x}{1 - x}}}$ **21.** 0 **23.** $x^{xe^x + ex - 1}(xe^x \ln^2 x + e^x \ln x + 1)$ **25.** $\frac{3}{2} \ln(1 + e^{2x}) + C$

27. $\dfrac{1}{3}\left(e^{3x} + \dfrac{2^{3x}}{\ln 2} + C\right)$ **29.** $\frac{1}{6}\sqrt{1 + e^{6x^2}} + C$ **31.** $\dfrac{2}{3 \ln 2}\sqrt{3 \cdot 2^x + 4} + C$ **33.** $\frac{1}{3}(e^8 - 1)$

35. $\frac{3}{2} \ln 2$ **37.** $1 + 5 \ln \frac{3}{4}$ **39.** $-\dfrac{ye^x + e^y + 1}{e^x + xe^y + 1}$ **41.** 5.004 **43.** $v = e^t - e^{-t} + 1$; $s = e^t + e^{-t} + t$ **45.** $\frac{1}{2}\pi(1 - e^{-2b})$; $\frac{1}{2}\pi$

47. 8212 years **49.** $g(x) = -e^x$; domain: $(-\infty, +\infty)$ **53.** $3000 \ln \frac{3}{2}$ in.-lb **55.** 8.66 years **57.** $187,500$

59. (a) 32; (b) 43.5; (c) 3.83 **61.** 8.63 min **63.** $(1, e)$ **71.** $\operatorname{sgn} t(1 - e^{-|t|})$

EXERCISES 7.1 (PAGE 603)

1. (a) $\frac{1}{6}\pi$; (b) $-\frac{1}{6}\pi$; (c) $\frac{1}{3}\pi$; (d) $\frac{2}{3}\pi$ **3.** (a) $\frac{1}{6}\pi$; (b) $-\frac{1}{3}\pi$; (c) $\frac{1}{6}\pi$; (d) $\frac{7}{6}\pi$ **5.** (a) $\frac{1}{2}\pi$; (b) $-\frac{1}{2}\pi$;
(c) $\frac{1}{2}\pi$; (d) $-\frac{1}{2}\pi$; (e) 0 **7.** (a) $\frac{2}{3}\sqrt{2}$; (b) $\frac{1}{4}\sqrt{2}$; (c) $2\sqrt{2}$; (d) $\frac{3}{4}\sqrt{2}$; (e) 3 **9.** (a) $\frac{2}{3}\sqrt{2}$; (b) $-\frac{1}{4}\sqrt{2}$; (c) $-2\sqrt{2}$; (d) $\frac{3}{4}\sqrt{2}$; (e) -3
11. (a) $-\frac{2}{5}\sqrt{5}$; (b) $\frac{1}{5}\sqrt{5}$; (c) $-\frac{1}{2}$; (d) $\sqrt{5}$; (e) $-\frac{1}{5}\sqrt{5}$ **13.** (a) $-\frac{2}{3}$; (b) $-\frac{1}{5}\sqrt{5}$; (c) $\frac{2}{5}\sqrt{5}$; (d) $\frac{1}{2}\sqrt{5}$; (e) $-\frac{3}{5}\sqrt{5}$ **15.** (a) $\sqrt{3}$; (b) $\frac{1}{7}\sqrt{21}$
17. (a) $\frac{1}{2}\sqrt{3}$; (b) $\frac{1}{2}\sqrt{3}$ **19.** $\frac{119}{169}$ **21.** $\frac{2}{9}(1 + \sqrt{10})$ **23.** $\frac{1}{27}(7\sqrt{5} + 8\sqrt{2})$ **25.** $\frac{1}{39}(48 - 25\sqrt{3})$ **27.** $\frac{1}{15}(4\sqrt{10} + \sqrt{5})$

39. (a) $\left\{t \,\middle|\, t = \dfrac{1}{4\pi} \sin^{-1} \dfrac{y}{2} - \dfrac{1}{8} + k \cdot \dfrac{1}{2}\right\} \cup \left\{t \,\middle|\, t = \dfrac{1}{8} - \dfrac{1}{4\pi} \sin^{-1} \dfrac{y}{2} + k \cdot \dfrac{1}{2}\right\}$, where k is any integer; (b) $\frac{1}{12}, \frac{5}{12}, \frac{7}{12}$

EXERCISES 7.2 (PAGE 610)

1. $\dfrac{1}{\sqrt{4 - x^2}}$ **3.** $\dfrac{2}{1 + 4x^2}$ **5.** $-\dfrac{1}{\sqrt{x - x^2}}$ **7.** 0 **9.** $-\dfrac{x}{|x|\sqrt{1 - x^2}}$ **11.** $\dfrac{4}{4 + x^2}$ **13.** $\sin^{-1} 2y + \dfrac{2y}{\sqrt{1 - 4y^2}}$

15. $2x \sec^{-1} \dfrac{1}{x} - \dfrac{x|x|}{\sqrt{1 - x^2}}$ **17.** $-\dfrac{\cos x}{|\cos x|}$ **19.** $\dfrac{2x}{(1 + x^4)\tan^{-1} x^2}$ **21.** $2\sqrt{4 - x^2}$ **23.** $-\dfrac{3}{\sqrt{4e^{6x} - 1}}$ **25.** $\cot^{-1} x$

27. $-e^x - \dfrac{1}{\sqrt{1 - x^2}}$ **29.** $\dfrac{(1 + y^2)(3x^2 + \sin y)}{1 - x \cos y(1 + y^2)}$ **31.** $2\sqrt{3}x - 6y + 2\pi - \sqrt{3} = 0$; $6\sqrt{3}x + 6y - 2\pi - 3\sqrt{3} = 0$ **33.** $\sqrt{10}$ ft

35. 0.078 rad/sec **37.** $\frac{32}{3}\pi$ km/min **39.** 8 ft/sec **41.** $\dfrac{6}{x\sqrt{x^2 - 64}}$

EXERCISES 7.3 (PAGE 614)

1. $\frac{1}{2}\sin^{-1}2x + C$ **3.** $\frac{1}{12}\tan^{-1}\frac{3x}{4} + C$ **5.** $\frac{1}{2}\tan^{-1}\frac{x-1}{2} + C$ **7.** $\frac{1}{16}\sec^{-1}\frac{x}{4} + C$ **9.** $\frac{\sqrt{5}}{5}\sin^{-1}\frac{\sqrt{10}}{2}x + C$

11. $\frac{1}{6}\sin^{-1}\frac{3r^2}{4} + C$ **13.** $\frac{1}{\sqrt{7}}\tan^{-1}\frac{e^x}{\sqrt{7}} + C$ **15.** $2\tan^{-1}\sqrt{x} + C$ **17.** $\frac{2}{\sqrt{7}}\tan^{-1}\frac{2x-1}{\sqrt{7}} + C$ **19.** $\cos^{-1}\frac{1-x}{4} + C$

21. $\cos^{-1}\frac{1+x}{2} - \sqrt{3-2x-x^2} + C$ **23.** $\sin^{-1}\frac{1+x}{\sqrt{5}} - \sqrt{4-2x-x^2} + C$

25. $\frac{1}{2}x^2 + 2x + \frac{5}{4}\ln(2x^2 - 4x + 3) - \frac{1}{4}\sqrt{2}\tan^{-1}\sqrt{2}(x-1) + C$ **27.** $\frac{1}{4}\pi + \frac{1}{2}\ln 2$ **29.** $\frac{1}{3}\pi$ **31.** $\tan^{-1}e - \frac{1}{4}\pi$ **33.** $\frac{1}{4}\pi$
35. $\frac{1}{3}\pi$ sq units **37.** $2\pi r$

EXERCISES 7.4 (PAGE 624)

17. $\frac{4}{5}\operatorname{sech}^2\frac{4x+1}{5}$ **19.** $2e^{2y}\cosh e^{2y}$ **21.** $-8\operatorname{sech}^2 4w \tanh 4w$ **23.** e^{2x} **25.** $2\operatorname{csch} 2t$ **27.** $2x\operatorname{sech} x^2$

29. $x^{\sinh x - 1}(x\cosh x \ln x + \sinh x)$ **35.** $\cosh^2(e^t) + C$ **37.** $2\cosh\sqrt{x} + C$ **39.** $x - \frac{1}{3}\coth 3x + C$ **41.** $\frac{1}{6}\tanh^6 x + C$

43. $\frac{1}{2}\ln^2\cosh x + C$ **45.** $\frac{4}{3}$ **47.** $\frac{1}{4}\sinh^4 2$ **49.** $\operatorname{sech}(0) = 1$, rel max; pts of infl at $x = \pm\ln(1 + \sqrt{2})$; sech increasing on $(-\infty, 0]$; sech decreasing on $[0, +\infty)$; graph concave upward for $x < -\ln(1 + \sqrt{2})$ and $x > \ln(1 + \sqrt{2})$; graph concave

downward for $-\ln(1 + \sqrt{2}) < x < \ln(1 + \sqrt{2})$ **51.** $a^2\sinh\frac{x_1}{a}$ sq units **53.** $v = e^{-ct/2}[(B - \frac{1}{2}cA)\sinh t + (A - \frac{1}{2}cB)\cosh t]$;

$a = e^{-ct/2}[(A - cB + \frac{1}{4}c^2 A)\sinh t + (B - cA + \frac{1}{4}c^2 B)\cosh t]$; $a = K_1 s + K_2 v$, where $K_1 = 1 - \frac{1}{4}c^2$ and $K_2 = -c$

EXERCISES 7.5 (PAGE 630)

11. $\ln\frac{1}{4}(1 + \sqrt{17})$ **13.** $\frac{1}{2}\ln 3$ **15.** $\frac{2x}{\sqrt{x^4 + 1}}$ **17.** $\frac{4}{1 - 16x^2}$ **19.** $-\frac{1}{2x + 3x^2}$ **21.** $2x\left(\cosh^{-1}x^2 + \frac{x^2}{\sqrt{x^4 - 1}}\right)$ **23.** $|\sec x|$

25. $-\frac{\csc x \cot x}{|\cot x|}$ **27.** $\frac{6z(\coth^{-1}z^2)^2}{1 - z^4}$ **29.** $-e^x\csc e^x$ **31.** $\sinh^{-1} x$

REVIEW EXERCISES FOR CHAPTER 7 (PAGE 631)

1. (a) $\frac{4}{3}$; (b) $\frac{3}{5}$; (c) $\frac{3}{4}$; (d) $-\frac{3}{4}$ **3.** (a) $-\frac{120}{169}$; (b) $\frac{3}{4}$ **5.** $\frac{2^x \ln 2}{1 + 2^{2x}}$ **7.** $6\sinh^2 2x \cosh 2x$ **9.** $\frac{x}{|x|(x^2 + 1)}$ **11.** $\frac{\operatorname{sech}\sqrt{x}}{2\sqrt{x}}$

13. $-4e^{2x}\operatorname{csch}^2 e^{2x}\coth e^{2x}$ **15.** $(\cosh x)^{1/x}\left(\frac{x\tanh x - \ln\cosh x}{x^2}\right)$ **19.** $\frac{6xy - 9y^4 - x^4 y^2}{3x^2 + 18xy^3 + 2x^5 y}$ **21.** $\frac{3}{2}\sin^{-1}x^2 + C$

23. $x - \frac{1}{3}\tanh 3x + C$ **25.** $\frac{2}{\sqrt{31}}\tan^{-1}\frac{4t + 3}{\sqrt{31}} + C$ **27.** $\frac{\sqrt{2}}{4}\sec^{-1}\frac{\sqrt{2}e^x}{4} + C$ **29.** $\frac{2}{3}\pi + \sqrt{3} - 2$ **31.** $2\sqrt{2}\sinh\frac{1}{2}$

33. (a) $\frac{1}{120\pi}\cos^{-1}\frac{x}{20}$; (b) $\frac{1}{360}$; (c) 0.0035; (d) $\frac{1}{180}$; (e) 0.0048 **37.** $9\sin^{-1}\frac{2}{3}\sqrt{2}$ sq units **39.** $\frac{1}{4}\pi$ **41.** (a) 120 rad/hr;

(b) 60 rad/hr **43.** decreasing at the rate of 60 rad/hr **45.** $\frac{1}{10}\pi$ hr; he walks all the way **47.** $\ln(2 + \sqrt{3})$

49. $2w\sinh^{-1}2w + \frac{2w^2}{\sqrt{4w^2 + 1}}$ **51.** $\frac{2e^{2x}}{\sqrt{e^{4x} - 1}}$

EXERCISES 8.2 (PAGE 640)

1. $\frac{1}{3}xe^{3x} - \frac{1}{9}e^{3x} + C$ **3.** $x \sec x - \ln|\sec x + \tan x| + C$ **5.** $x \ln 2x - x + C$ **7.** $x \ln^2 x - 2x \ln x + 2x + C$

9. $\frac{1}{2}\tan^{-1} x(x^2 + 1) - \frac{1}{2}x + C$ **11.** $\dfrac{e^x}{x+1} + C$ **13.** $-\cos x \ln(\cos x) + \cos x + C$ **15.** $\frac{1}{2}e^x(\cos x + \sin x) + C$

17. $-x^2\sqrt{1-x^2} - \frac{2}{3}(1-x^2)^{3/2} + C$ **19.** $x^2 \cosh x - 2x \sinh x + 2 \cosh x + C$ **21.** $2\sqrt{z} \cot^{-1}\sqrt{z} + \ln(1+z) + C$

23. $2\sqrt{x} \sin\sqrt{x} + 2 \cos\sqrt{x} + C$ **25.** $\dfrac{36}{\ln 3} - \dfrac{36}{(\ln 3)^2} + \dfrac{16}{(\ln 3)^3}$ **27.** $\frac{9}{16}$ **29.** $\frac{1}{4}(3e^4 + 1)$ **31.** $\frac{4}{25}(e^{3\pi/4} + 1)$ **33.** $\frac{4}{3}\pi - \sqrt{3}$

35. $(e^2 + 1)$ sq units **37.** $\frac{1}{2}\pi(3e^4 + 1)$ cu units **39.** $(8 - 24e^{-2})$ sq units **41.** $2(1 - e^{-6})$kg; $\dfrac{e^6 - 7}{e^6 - 1}$ m from one end

43. $(0.267, 0.604)$ **45.** $\frac{1}{4}(1 - 9e^{-8})w\pi$ ft-lb **47.** (b) $\frac{1}{5}\sec^4 x \tan x + \frac{4}{15}\sec^2 x \tan x + \frac{8}{15}\tan x + C$

49. (b) $\frac{1}{4}x^4 \ln x - \frac{1}{16}x^4 + C$ **51.** $C(x) = x \ln x - x + 6$ **53.** $(0.69, 0, 0)$ **55.** $(4 - \frac{1}{4}\pi)\rho$ lb

EXERCISES 8.3 (PAGE 646)

1. $\frac{1}{5}\sin^5 x + C$ **3.** $-\frac{1}{16}\cos^4 x + C$ **5.** $\frac{1}{3}\cos^3 x - \cos x + C$ **7.** $\frac{3}{8}z - \frac{1}{4}\sin 2z + \frac{1}{32}\sin 4z + C$ **9.** $\frac{1}{2}x + \frac{1}{2}\sin x + C$

11. $\frac{1}{3}\sin^3 x - \frac{1}{5}\sin^5 x + C$ **13.** $-\frac{1}{3}\cos^3 x + \frac{2}{5}\cos^5 x - \frac{1}{7}\cos^7 x + C$ **15.** $\frac{1}{8}t - \frac{1}{96}\sin 12t + C$

17. $\frac{3}{2}\sin^{2/3} 3x - \frac{1}{8}\sin^{8/3} 3x + C$ **19.** $\frac{1}{14}\sin 7x + \frac{1}{2}\sin x + C$ **21.** $-\frac{1}{16}\cos 8y + \frac{1}{4}\cos 2y + C$

23. $t - \sin t - \frac{1}{8}\sin 4t + \frac{1}{5}\sin 5t - \frac{1}{12}\sin 6t + C$ **25.** $\frac{2}{3}$ **27.** $\frac{3}{8}$ **29.** $\frac{1}{8}$ **31.** $\frac{1}{8}(\sqrt{2} - 1)$ **33.** (a) $\sin^2 x + C$; (b) $-\cos^2 x + C$;

(c) $-\frac{1}{2}\cos 2x + C$ **37.** (a) $\sin^2 x + C$; (b) $-\cos^2 x + C$; (c) $-\frac{1}{2}\cos 2x + C$ **39.** $\frac{3}{2}\pi$ sq units **41.** $\frac{3}{8}\pi^2$ cu units

43. $\frac{5}{8}\pi^2$ cu units **45.** $(\frac{1}{2}\pi - 1, \frac{1}{8}\pi)$ **47.** $\left(\dfrac{\pi^2 + 4}{4\pi}, 0, 0\right)$

EXERCISES 8.4 (PAGE 651)

1. $\frac{1}{5}\tan 5x - x + C$ **3.** $-\frac{1}{4}\cot 2x^2 - \frac{1}{2}x^2 + C$ **5.** $-\frac{1}{2}\csc^2 t - \ln|\sin t| + C$ **7.** $\frac{1}{15}\tan^5 3x - \frac{1}{9}\tan^3 3x + \frac{1}{3}\tan 3x - x + C$

9. $\frac{1}{3}\tan^3 x + \tan x + C$ **11.** $-\frac{1}{2}\csc x \cot x + \frac{1}{2}\ln|\csc x - \cot x| + C$ **13.** $\frac{1}{3}\tan^3 e^x - \tan e^x + e^x + C$

15. $\frac{1}{7}\tan^7 x + \frac{1}{9}\tan^9 x + C$ **17.** $-\frac{1}{9}\cot^3 3x - \frac{1}{15}\cot^5 3x + C$ **19.** $\frac{1}{2}(\tan 2x - \cot 2x) + C$ **21.** $2 \sec w - \tan w + C$

23. $\frac{1}{12}\tan^4 3x - \frac{1}{6}\tan^2 3x + \frac{1}{3}\ln|\sec 3x| + C$ **25.** $u - 2 \tan \frac{1}{4}u + C$ **27.** $-\frac{1}{3}\csc^3 x + C$

29. $\frac{1}{4}\tan^4(\ln x) + \frac{1}{3}\tan^6(\ln x) + \frac{1}{8}\tan^8(\ln x) + C$ **31.** $\frac{1}{4} - \frac{1}{8}\ln 2$ **33.** $\frac{56}{15}$ **35.** $\frac{1}{3}$ **37.** $(1 - \frac{1}{4}\pi)$ sq units **39.** $\frac{4}{3}\pi$ cu units

EXERCISES 8.5 (PAGE 656)

1. $-\dfrac{(4-x^2)^{1/2}}{4x} + C$ **3.** $\dfrac{1}{2}\ln\left|\dfrac{x}{2 + \sqrt{x^2+4}}\right| + C$ **5.** $\dfrac{1}{5}\ln\left|\dfrac{5 - \sqrt{25 - x^2}}{x}\right| + C$ **7.** $\ln|x + \sqrt{x^2 - a^2}| + C$

9. $\dfrac{1}{4}\tan^{-1}\dfrac{1}{2}x - \dfrac{x}{2(x^2 + 4)} + C$ **11.** $-\frac{1}{9}x(4x^2 - 9)^{-1/2} + C$ **13.** $\frac{1}{3}\ln(\sqrt{t^4 + 25} - 5) - \frac{2}{3}\ln t + C$

15. $\ln|x + 2 + \sqrt{4x + x^2}| + C$ **17.** $\dfrac{x+2}{9\sqrt{5 - 4x - x^2}} + C$ **19.** $\dfrac{\tan x}{4\sqrt{4 - \tan^2 x}} + C$ **21.** $\frac{1}{3}\sqrt{\ln^2 w - 4}(8 + \ln^2 w) + C$

23. $-\dfrac{e^t + 4}{9\sqrt{e^{2t} + 8e^t + 7}} + C$ **25.** $\frac{128}{3} - 24\sqrt{3}$ **27.** $\frac{1}{27}(6 - 2\sqrt{3})$ **29.** $\frac{1}{2}\cos^{-1}\frac{1}{3} - \frac{1}{6}\pi$ **31.** $\frac{625}{16}\pi$ **33.** (a) $\sec^{-1}\frac{2}{3}x + C$;

(b) $-\frac{2}{3}\sqrt{3 - 2x^2} + C$ **35.** $\ln \frac{1}{3}(2\sqrt{5} + \sqrt{10} - \sqrt{2} - 1) + \sqrt{10} - \sqrt{2}$ **37.** $\frac{81}{16}\pi^2$ cu units

39. $\dfrac{392}{60 + 27 \ln 3}$ cm from the left end **41.** $\left(\dfrac{20 - 15 \cos^{-1}\frac{3}{5}}{5 \ln 3 - 4}, \dfrac{26}{225(5 \ln 3 - 4)}\right)$ **43.** $(\frac{8}{3}\pi + 3\sqrt{3})\rho$ lb

45. $(\frac{512}{3}\pi + 192\sqrt{3})\rho$ oz

EXERCISES 8.6 (PAGE 668)

1. $\frac{1}{4} \ln \left| \frac{x-2}{x+2} \right| + C$ **3.** $\ln|C(x-2)^2(x+2)^3|$ **5.** $\ln \left| \frac{C(w+4)^3}{2w-1} \right|$ **7.** $\frac{1}{4} \ln \left| \frac{Cx^4(2x+1)^3}{2x-1} \right|$

9. $\frac{1}{9} \ln \left| \frac{x+3}{x} \right| - \frac{1}{3x} + C$ **11.** $2 \ln \left| \frac{x+1}{x} \right| - \frac{1}{x} - \frac{1}{x+1} + C$ **13.** $\frac{3}{x+1} + \ln|x+1| - \frac{1}{2} \ln|2x+3| + C$

15. $\frac{5}{16(z+2)} - \frac{7}{16(z-2)} + \frac{1}{32} \ln \left| \frac{z+2}{z-2} \right| + C$ **17.** $\frac{1}{2}x^2 + 2x - \frac{3}{x-1} - \ln|x^2+2x-3| + C$

19. $-\ln(3x+2)^{2/3}(x-1)^2 - \frac{1}{3(3x+2)} - \frac{3}{x-1} + C$ **21.** $4 \ln \frac{4}{3} - \frac{3}{2}$ **23.** $\ln \frac{27}{4} - 2$ **25.** $13 \ln 2 - 4 \ln 5$ **27.** $\ln \frac{7}{2} - \frac{5}{3}$

29. $\ln 4.5$ sq units   **31.** $2\pi(2 + 6 \ln 3 - 2 \ln 2)$ cu units   **33.** $\left(\frac{6 \ln 3 - 2 \ln 2 + 2}{2 \ln 3 - \ln 2}, \frac{2 \ln 2 - 2 \ln 3 + \frac{35}{24}}{2 \ln 3 - \ln 2} \right)$

35. (a) $f(t) = \frac{5000}{1 + 4999e^{-0.5t}}$; (b) 1328; (c) 4075; (d) 5000   **37.** 50 percent   **39.** \$11,201,300   **41.** $\frac{31}{19}$ **43.** 7.4 lb

EXERCISES 8.7 (PAGE 674)

1. $\frac{1}{2} \ln \left| \frac{Cx^2}{2x^2+1} \right|$ **3.** $\frac{1}{8} \ln \left| \frac{2x-1}{2x+1} \right| - \frac{1}{4} \tan^{-1} 2x + C$ **5.** $\frac{1}{10} \ln|(t^2+1)(2t+1)^3| + \frac{2}{3} \tan^{-1} t + C$

7. $\ln|x-1| + \tan^{-1} x + C$ **9.** $\frac{1}{2} \ln \left(\frac{Cx^2}{x^2+x+1} \right) - \frac{1}{\sqrt{3}} \tan^{-1} \left(\frac{2x+1}{\sqrt{3}} \right)$ **11.** $\ln \left(\frac{Cx^2}{x^2+1} \right) - \frac{1}{2} \tan^{-1} x - \frac{x}{2(x^2+1)}$

13. $\frac{5}{2} \ln(z^2 - 2z + 5) - \frac{65}{16} \tan^{-1} \left(\frac{z-1}{2} \right) + \frac{-47z+15}{8(z^2-2z+5)} + C$

15. $\frac{5}{162} \ln|9x^2 + 3x + 1| - \frac{2}{81} \ln|3x-1| + \frac{5}{9\sqrt{3}} \tan^{-1} \left(\frac{6x+1}{\sqrt{3}} \right) + C$ **17.** $\frac{1}{6} \tan^{-1} \frac{2}{3}x + \frac{x}{4x^2+9} + C$

19. $\ln|1 + \tan x| + \frac{2}{\sqrt{3}} \tan^{-1} \left(\frac{2 \tan x - 1}{\sqrt{3}} \right) + C$ **21.** $6 \ln 2$ **23.** $\ln \frac{12}{5} + \frac{3}{2} \ln \frac{20}{13}$ **25.** $\frac{1}{4}\pi$ **27.** $\frac{3}{4} \ln 2 + \frac{5}{8}\pi$ **29.** $\frac{3}{8} \ln \frac{125}{41}$

31. $\frac{1}{16}$ sq units   **33.** $(\frac{2}{3}\sqrt{3}\pi^2 - \frac{2}{3}\pi \ln 3)$ cu units   **35.** $\frac{3}{50} \ln \frac{(t_1+2)^2}{4(t_1^2+1)} - \frac{7}{5(t_1+2)} - \frac{4}{25} \tan^{-1} t_1 + \frac{7}{10}$

EXERCISES 8.8 (PAGE 678)

1. $\frac{6}{\sqrt{15}} \tan^{-1} \left(\frac{1}{\sqrt{15}} \tan \frac{x}{2} \right) + C$ **3.** $\frac{\sqrt{15}}{5} \ln \left| \frac{\sqrt{15} + \tan \frac{1}{2}x}{\sqrt{15} - \tan \frac{1}{2}x} \right| + C$ **5.** $\frac{1}{2} \ln|\tan \frac{1}{2}x| - \frac{1}{4} \tan^2 \frac{1}{2}x + C$

7. $\frac{1}{4} \ln \left| \frac{3 \tan \frac{1}{2}x - 9}{3 \tan \frac{1}{2}x - 1} \right| + C$ **9.** $\frac{1}{5} \ln \left| \frac{3 \tan \frac{1}{2}x - 1}{\tan \frac{1}{2}x + 3} \right| + C$ **11.** $\sqrt{2} \ln \left| \frac{\tan x + \sqrt{2}}{\tan x - \sqrt{2}} \right| + C$

13. $\frac{1}{2}x + \frac{\sqrt{3}}{6} \ln \left| \frac{\sqrt{3} - \tan \frac{1}{2}x}{\sqrt{3} + \tan \frac{1}{2}x} \right| + C$ **15.** $-\frac{1}{2} \ln|13 - \tan^2 x| + C$

17. $2 \tan^{-1}(2 + \tan \frac{1}{2}x) + C$ **19.** $\frac{1}{4} \ln 3$ **21.** $2\sqrt{3} \ln(1 + \sqrt{3})$ **23.** $\frac{1}{2}\sqrt{3} \ln(1 + \frac{1}{2}\sqrt{3})$

EXERCISES 8.9 (PAGE 681)

1. $\frac{2}{3}x^{3/2} - 3x + 18\sqrt{x} - 54 \ln(3 + \sqrt{x}) + C$ **3.** $\ln \left| \frac{\sqrt{1+4x} - 1}{\sqrt{1+4x} + 1} \right| + C$ **5.** $-2\sqrt{1+x} + \sqrt{2} \ln \left| \frac{\sqrt{1+x} + \sqrt{2}}{\sqrt{1+x} - \sqrt{2}} \right| + C$

7. $\frac{3}{2}(x-2)^{2/3} - 3(x-2)^{1/3} + 3 \ln|1 + (x-2)^{1/3}| + C$ **9.** $2\sqrt{2x} + 2\sqrt{x+4} + 4\sqrt{2} \ln \left| \frac{(2\sqrt{2} - \sqrt{x+4})(\sqrt{x} - 2)}{x-4} \right| + C$

11. $\frac{1}{9}\sqrt{1+2x^3}(2x^3+7)+C$ **13.** $3\tan^{-1}\sqrt[6]{x}+\dfrac{3\sqrt[6]{x}}{1+\sqrt[3]{x}}+C$

15. $\begin{cases} \ln\left|\dfrac{Cx}{x+2+2\sqrt{1+x+x^2}}\right| & \text{if } x>0 \\[3mm] -\ln\left|\dfrac{Cx}{x+2-2\sqrt{1+x+x^2}}\right| & \text{if } x<0 \end{cases}$

17. $\begin{cases} -\sin^{-1}\left(\dfrac{1-2x}{\sqrt{13}x}\right)+C & \text{if } x>0 \\[3mm] \sin^{-1}\left(\dfrac{1-2x}{\sqrt{13}x}\right)+C & \text{if } x<0 \end{cases}$

19. $\begin{cases} \dfrac{\sqrt{5x^2+8x-1}}{x}-4\sin^{-1}\left(\dfrac{1-4x}{\sqrt{21}x}\right)+C & \text{if } x>0 \\[3mm] \dfrac{\sqrt{5x^2+8x-1}}{x}+4\sin^{-1}\left(\dfrac{1-4x}{\sqrt{21}x}\right)+C & \text{if } x<0 \end{cases}$

21. $4-2\ln 3$ **23.** $\ln\frac{11}{10}$ **25.** $\frac{1}{9}(54-2\sqrt[3]{3})$

27. $\frac{3}{2}\pi-\frac{152}{35}$ **29.** $2\ln|\sqrt{x}-1|+C$

EXERCISES 8.10 (PAGE 692)

1. approx: 4.250; exact: 4   **3.** 0.696; exact: ln 2   **5.** approx: 0; exact: 0   **7.** 0.880   **9.** 0.248   **11.** 3.689   **13.** $-0.5\le\epsilon_T\le 0$
15. $-0.007\le\epsilon_T\le-0.001$ **17.** $0\le\epsilon_T\le 0.072$ **19.** 4.000   **21.** 0.693   **23.** 0.6045   **25.** 0   **27.** $-0.0005\le\epsilon_s\le 0$

29. 0.237   **31.** 1.568   **33.** 1.402   **35.** 3.090   **37.** (a) 0.3401; (b) 0.3414

39. $\displaystyle\int_0^\pi\sqrt{1+\cos^2 x}\,dx;\ \frac{1}{6}\pi(2\sqrt{6}+\sqrt{2}+1)\approx 3.829$ **43.** 0.8427

EXERCISES 8.11 (PAGE 698)

1. $\sinh^{-1}\frac{1}{2}x+C=\ln\frac{1}{2}(x+\sqrt{x^2+4})+C$ **3.** $\frac{1}{2}\cosh^{-1}x^2+C=\frac{1}{2}\ln(x^2+\sqrt{x^4-1})+C$

5. $\begin{cases} -\frac{1}{12}\tanh^{-1}(\frac{3}{4}x)+C & \text{if }|x|<\frac{4}{3} \\ -\frac{1}{12}\coth^{-1}(\frac{3}{4}x)+C & \text{if }|x|>\frac{4}{3} \end{cases}=\frac{1}{24}\ln\left|\dfrac{4-3x}{4+3x}\right|+C$ **7.** $\sinh^{-1}\dfrac{\sin x}{\sqrt{3}}+C=\ln(\sin x+\sqrt{4-\cos^2 x})+C$

9. $\frac{1}{\sqrt{5}}\cosh^{-1}(\sqrt{5}e^t)+C=\frac{1}{\sqrt{5}}\ln(\sqrt{5}e^t+\sqrt{5e^{2t}-1})+C$

11. $\begin{cases} \frac{1}{\sqrt{6}}\tanh^{-1}\left(\dfrac{x+2}{\sqrt{6}}\right)+C & \text{if }|x+2|<\sqrt{6} \\[3mm] \frac{1}{\sqrt{6}}\coth^{-1}\left(\dfrac{x+2}{\sqrt{6}}\right)+C & \text{if }|x+2|>\sqrt{6} \end{cases}=\frac{1}{2\sqrt{6}}\ln\left|\dfrac{\sqrt{6}+2+x}{\sqrt{6}-2-x}\right|+C$

13. $\begin{cases} -\tanh^{-1}(x+5)+C & \text{if }|x+5|<1 \\ -\coth^{-1}(x+5)+C & \text{if }|x+5|>1 \end{cases}=-\frac{1}{2}\ln\left|\dfrac{x+6}{x+4}\right|+C$

15. $\frac{3}{2}\sinh^{-1}\left(\dfrac{2\ln w}{3}\right)+C=\frac{3}{2}\ln(2\ln w+\sqrt{4\ln^2 w+9})+C$ **17.** $\ln\left(\dfrac{5+\sqrt{21}}{3+\sqrt{5}}\right)$ **19.** $\ln 3$ **21.** $\frac{1}{3}\ln\dfrac{7+\sqrt{40}}{4+\sqrt{7}}$

EXERCISES 8.12 (PAGE 703)

1. $\frac{1}{3}x-\frac{2}{9}\ln|2+3x|+C$ **3.** $x+\dfrac{36}{6-x}+12\ln|6-x|+C$ **5.** $\frac{1}{30}(3x-1)(1+2x)^{3/2}+C$

7. $2\sqrt{1+2x}+\ln\left|\dfrac{\sqrt{1+2x}-1}{\sqrt{1+2x}+1}\right|+C$ **9.** $\frac{1}{4}\ln\left|\dfrac{x+2}{x-2}\right|+C$ **11.** $\ln|x+3+\sqrt{x^2+6x}|+C$

13. $\sqrt{9x^2 + 4} - 2\ln\left|\dfrac{2 + \sqrt{9x^2 + 4}}{3x}\right| + C$ **15.** $\sqrt{9 - 4x^2} - 3\ln\left|\dfrac{3 + \sqrt{9 - 4x^2}}{2x}\right| + C$

17. $\dfrac{1}{3}(x^2 - x - 6)\sqrt{4x - x^2} + 4\cos^{-1}\left(\dfrac{2 - x}{2}\right) + C$ **19.** $-\frac{1}{5}\sin^4 x\cos x - \frac{4}{15}\sin^2 x\cos x - \frac{8}{15}\cos x + C$

21. $-\frac{1}{6}\csc^5 x\cot x - \frac{5}{24}\csc^3 x\cot x - \frac{5}{16}\csc x\cot x + \frac{5}{16}\ln|\csc x - \cot x| + C$

23. $t^4\sin t + 4t^3\cos t - 12t^2\sin t - 24t\cos t + 24\sin t + C$ **25.** $x\sec^{-1} 3x - \frac{1}{3}\ln|3x + \sqrt{9x^2 - 1}| + C$

27. $e^x(x^4 - 4x^3 + 12x^2 - 24x + 24) + C$ **29.** $\dfrac{e^{4x}}{32}(8x^2 - 4x + 1) + C$ **31.** $\dfrac{x^4}{16}(4\ln 3x - 1) + C$

33. $\frac{2}{29}e^{2x}\sin 5x - \frac{5}{29}e^{2x}\cos 5x + C$ **35.** $\frac{3}{5}y\cosh 5y - \frac{3}{25}\sinh 5y + C$ **37.** $\frac{2}{3}\ln 2$ **39.** $\frac{15}{2} - 8\ln 2$ **41.** $\frac{32}{5}\ln 2 - \frac{31}{25}$

43. $\frac{15}{2} - 8\ln 2$ **45.** $\frac{1}{3}\pi + \frac{1}{2}\sqrt{3}$ **47.** $\frac{1}{4} - \frac{1}{8}\sqrt{2}$

REVIEW EXERCISES FOR CHAPTER 8 (PAGE 704)

1. $\frac{1}{8}x - \frac{1}{128}\sin 16x + C$ **3.** $-2\sqrt{4 - e^x} + C$ **5.** $(x + 1)\tan^{-1}\sqrt{x} - \sqrt{x} + C$ **7.** $\frac{1}{2}x + \frac{3}{4}\sin\frac{2}{3}x + C$

9. $\ln|x - 1| - 2(x - 1)^{-1} - (x - 1)^{-2} + C$ **11.** $\frac{1}{4}\sin 2x - \frac{1}{8}\sin 4x + C$ **13.** $3\ln\left|\dfrac{x^{1/3}}{1 + x^{1/3}}\right| + C$

15. $\frac{1}{3}\tan 3x - \frac{1}{3}\cot 3x + \frac{2}{3}\ln|\tan 3x| + C$ **17.** $2t + \ln\dfrac{t^2}{(t + 2)^{10}} - \dfrac{15}{t + 2} + C$ **19.** $x - \tan^{-1} x + \dfrac{1}{2}\ln\left|\dfrac{x - 1}{x + 1}\right| + C$

21. $\frac{1}{16}x - \frac{1}{192}\sin 12x - \frac{1}{144}\sin^3 6x + C$ **23.** $\sin^{-1}\left(\dfrac{r + 2}{\sqrt{7}}\right) + C$ **25.** $\frac{1}{2}x^2\sin x^2 + \frac{1}{2}\cos x^2 + C$

27. $\frac{2}{17}e^{t/2}(4\sin 2t + \cos 2t) + C$ **29.** $\frac{1}{4}\tan^{-1}(\frac{1}{2}\sin^2 x) + C$ **31.** $\begin{cases}\dfrac{1}{n}\left(-\cos nx + \dfrac{2}{3}\cos^3 nx - \dfrac{1}{5}\cos^5 nx\right) + C & \text{if } n \neq 0 \\ C & \text{if } n = 0\end{cases}$

33. $-\frac{1}{4}\csc^3 x\cot x - \frac{3}{8}\csc x\cot x + \frac{3}{8}\ln|\csc x - \cot x| + C$ **35.** $2\ln|y - 2| - 8(y - 2)^{-1} - \frac{9}{2}(y - 2)^{-2} + C$

37. $-\tan^{-1}(\cos x) + C$ **39.** $2\sin^{-1}\left(\dfrac{t - 2}{2}\right) + \dfrac{1}{2}(t - 2)(4t - t^2)^{1/2} + C$ **41.** $\dfrac{1}{3}\ln\left|1 - \dfrac{1}{x^3}\right| + C$ **43.** $\frac{1}{3}\sin^{-1}(\frac{3}{2}e^x) + C$

45. $-\frac{1}{15}\cot^5 3x - \frac{1}{9}\cot^3 3x + C$ **47.** $\frac{1}{3}x^3\sin^{-1} x + \frac{1}{9}(x^2 + 2)\sqrt{1 - x^2} + C$ **49.** $\tan^{-1}(\cos x) + C$

51. $\frac{2}{3}\sec^{-1}|2\sin 3t| + C$ **53.** $-\dfrac{(x^2 + a^2)^{3/2}}{3a^2 x^3} + C$ **55.** $\sqrt{2t} - \sqrt{1 - 2t}\sin^{-1}\sqrt{2t} + C$

57. $\frac{4}{3}\sqrt{2 + \sqrt{x - 1}}(\sqrt{x - 1} - 4) + C$

59. $\dfrac{\sqrt{2}}{4}\ln\left|\dfrac{\tan x - \sqrt{2\tan x} + 1}{\tan x + \sqrt{2\tan x} + 1}\right| + \dfrac{\sqrt{2}}{2}\tan^{-1}(\sqrt{2\tan x} - 1) + \dfrac{\sqrt{2}}{2}\tan^{-1}(\sqrt{2\tan x} + 1) + C$

61. $\begin{cases}\dfrac{x^{n+1}\ln x}{n + 1} - \dfrac{x^{n+1}}{(n + 1)^2} + C & \text{if } n \neq -1 \\ \frac{1}{2}\ln^2 x + C & \text{if } n = -1\end{cases}$ **63.** 4 **65.** $\frac{1}{2} + 2\ln\frac{6}{5}$ **67.** $\frac{16}{3} - \frac{8}{3}\sqrt{2}$ **69.** $\frac{4}{3}\sqrt{3} - \frac{1}{3}\pi$ **71.** $\frac{4}{3}$

73. $\frac{1}{2} - \frac{1}{4}\ln 2$ **75.** $a^2(\frac{1}{3}\pi - \frac{2}{8}\sqrt{3})$ **77.** $\frac{1}{2}\ln\frac{9}{2} - \frac{1}{6}\pi$ **79.** 5 **81.** $\frac{1}{6} + \ln\frac{3}{2}$ **83.** $\frac{4}{3}$ **85.** $1 - \frac{1}{2}\ln 3$ **87.** $\frac{1}{24}\pi$

89. $\sqrt{3} - \frac{1}{2}\ln(2 + \sqrt{3})$ **91.** $\frac{1}{3}\ln\frac{3}{2}$ **93.** $\frac{256}{15}$ **95.** $\frac{1}{3}k(1 - e^{-9})$ kg; $\dfrac{e^9 - 10}{3(e^9 - 1)}$ m from one end

97. $9\sqrt{2} - 3\sqrt{5} + \frac{3}{2}\ln(2\sqrt{10} + 3\sqrt{5} - 4\sqrt{2} - 6)$ **99.** $\frac{1}{8}\pi$ sq units **101.** $\pi(e^2\ln 2 + \frac{1}{2}e^2 + \frac{1}{8})$ cu units

103. (a) $x = 300\left(\dfrac{18^t - 17^t}{3\cdot 18^t - 2\cdot 17^t}\right)$; (b) 35.94 lb **105.** $\left(0, \dfrac{32}{15\pi}\right)$ **107.** $(\frac{1}{4}\pi - \frac{1}{2}, \frac{1}{2})$ **109.** (a) 4824; (b) 8.18 weeks

111. (a) 2.977; (b) 2.958 **113.** (a) 1.624; (b) 1.563 **117.** $\dfrac{x}{2}\sqrt{9x^2 + 1} + \dfrac{1}{6}\ln|3x + \sqrt{9x^2 + 1}| + C$

119. $-\frac{1}{96}\sin^2 4x\cos 4x - \frac{1}{48}\cos 4x + C$ **121.** $\frac{1}{8}(e^2 + 3)$ **123.** 3ρ lb **125.** $\left(\dfrac{\pi^2 - 4}{4\pi}, 0, 0\right)$

EXERCISES 9.1 (PAGE 716)

1. $\frac{4}{3}$ **3.** $\frac{1}{2}\pi$ **5.** $-\frac{11}{4}$ **7.** 1 **9.** 2 **11.** $+\infty$ **13.** $-\frac{1}{8}$ **15.** $\ln\frac{2}{3}$ **17.** $\frac{1}{6}$ **19.** $\frac{1}{3}$ **21.** $\frac{1}{2}$ **23.** 1 **25.** $\frac{3}{5}$ **27.** 2 **29.** -1 **31.** $\frac{Et}{L}$

33. $a = -3$, $b = \frac{9}{2}$ **37.** (b) $f'(0) = -\frac{1}{2}$

EXERCISES 9.2 (PAGE 723)

1. 0 **3.** 0 **5.** 1 **7.** 0 **9.** 0 **11.** $\frac{1}{2}$ **13.** 1 **15.** $\frac{1}{2}$ **17.** 1 **19.** e^2 **21.** e^a **23.** 1 **25.** e^2 **27.** $e^{-1/3}$ **29.** $\frac{1}{2}$ **31.** 0 **33.** 1 **35.** 1

37. $\dfrac{1}{\ln 3}$ **41.** $f(e) = e^{1/e}$, rel max; $y = 1$ is an asymptote **43.** (a) 0; (b) 0; no

EXERCISES 9.3 (PAGE 731)

1. 1 **3.** $-\dfrac{1}{2\ln 5}$ **5.** $\dfrac{1}{(\ln 2)^2}$ **7.** divergent **9.** divergent **11.** $\frac{1}{3}\pi$ **13.** 2 **15.** 1 **17.** divergent **19.** (a) divergent; (b) 0

21. (a) 0 **23.** $\frac{1}{2}\pi$ **25.** $\frac{1}{2}\pi$ **27.** (a) 0.565; (b) 0.287 **29.** (a) 0.203; (b) 0.188 **31.** $\dfrac{1}{k}$ **33.** $\dfrac{1000}{0.08 + \ln 2}$ dollars $\approx \$1293.41$

37. $n > 1$ **39.** $\frac{1}{3}$; $\frac{1}{9}\ln\frac{32}{27}$

EXERCISES 9.4 (PAGE 737)

1. 2 **3.** -4 **5.** $\frac{1}{3}\pi$ **7.** divergent **9.** divergent **11.** divergent **13.** divergent **15.** divergent **17.** 0 **19.** divergent **21.** 0

23. $\frac{1}{3}\pi$ **25.** 0 **27.** $n > -1$; $\dfrac{1}{n+1}$ **29.** $n > -1$; $\dfrac{2}{(n+1)^3}$ **31.** yes; 6π

EXERCISES 9.5 (PAGE 745)

1. $P_3(x) = -1 - (x - 1) - (x - 1)^2 - (x - 1)^3$; $R_3(x) = \dfrac{(x-1)^4}{(\xi-2)^5}$, ξ between 1 and x

3. $P_3(x) = 8 + 3(x - 4) + \frac{3}{16}(x - 4)^2 - \frac{1}{128}(x - 4)^3$; $R_3(x) = \dfrac{3(x-4)^4}{128\xi^{5/2}}$, ξ between 4 and x

5. $P_3(x) = \frac{1}{2} + \frac{1}{2}\sqrt{3}(x - \frac{1}{6}\pi) - \frac{1}{4}(x - \frac{1}{6}\pi)^2 - \frac{1}{12}\sqrt{3}(x - \frac{1}{6}\pi)^3$; $R_3(x) = \frac{1}{24}\sin\xi(x - \frac{1}{6}\pi)^4$, ξ between $\frac{1}{6}\pi$ and x

7. $P_4(x) = x + \frac{1}{6}x^3$; $R_4(x) = \frac{1}{120}\cosh\xi x^5$, ξ between 0 and x **9.** $P_3(x) = x - 1 - \frac{1}{2}(x - 1)^2 + \frac{1}{3}(x - 1)^3$;
$R_3(x) = -\frac{1}{4}\xi^{-4}(x - 1)^4$; ξ between 1 and x **11.** $P_3(x) = -\ln 2 - \sqrt{3}(x - \frac{1}{3}\pi) - 2(x - \frac{1}{3}\pi)^2 - \frac{4}{3}\sqrt{3}(x - \frac{1}{3}\pi)^3$;
$R_3(x) = -\frac{1}{12}(2\sec^2\xi\tan^2\xi + \sec^4\xi)(x - \frac{1}{3}\pi)^4$, ξ between $\frac{1}{3}\pi$ and x **13.** $P_3(x) = 1 + \frac{3}{2}x + \frac{3}{8}x^2 - \frac{1}{16}x^3$;

$R_3(x) = \frac{3}{128}(1 + \xi)^{-5/2}x^4$, ξ between 0 and x **15.** 2.71828 **17.** $|error| < \dfrac{(0.1)^4}{24} < 0.000005$ **19.** 0.515 **21.** 0.1823

23. $\dfrac{55\sqrt{2}}{672}$, $|error| < \frac{1}{7680}\sqrt{2}$ **27.** $x \approx \dfrac{\pi}{2(1 + m)}$ **29.** $2(x - 1) + 5(x - 1)^2 + 3(x - 1)^3 + (x - 1)^4$

REVIEW EXERCISES FOR CHAPTER 9 (PAGE 746)

1. $+\infty$ **3.** $\frac{1}{3}$ **5.** $+\infty$ **7.** 1   **9.** $-\frac{1}{2}$ **11.** e^{12} **13.** 0   **15.** 1   **17.** e **19.** divergent   **21.** $\frac{1}{2}$ **23.** divergent   **25.** $\dfrac{32}{\ln 2}$ **27.** divergent

29. $\frac{1}{4}\pi$ **31.** divergent   **33.** (b) $f'(0) = -1$ **35.** (a) no; (b) 0   **37.** (a) divergent; (b) 0   **39.** $P_6(x) = 1 - \frac{1}{2}x^2 + \frac{1}{24}x^4 - \frac{1}{720}x^6$;
$R_6(x) = \frac{1}{5040}\sin \xi x^7$, ξ between 0 and x **41.** $P_4(x) = \frac{1}{3} - \frac{1}{54}(x-9) + \frac{1}{648}(x-9)^2 - \frac{5}{34,992}(x-9)^3 + \frac{35}{2,519,424}(x-9)^4$;
$R_4(x) = -\frac{63}{256}\xi^{-11/2}(x-9)^5$, ξ between 9 and x **43.** $+\infty$ **45.** $\frac{64}{3}$ **47.** 1   **49.** (a) 0.148; (b) 0.699; (c) 0.018   **51.** \$152,500

EXERCISES 10.1 (PAGE 754)

1. (a) $(-4, \frac{1}{4}\pi)$; (b) $(4, -\frac{7}{4}\pi)$; (c) $(-4, -\frac{3}{4}\pi)$ **3.** (a) $(-2, \frac{3}{2}\pi)$; (b) $(2, -\frac{3}{2}\pi)$; (c) $(-2, -\frac{1}{2}\pi)$ **5.** (a) $(-\sqrt{2}, \frac{3}{4}\pi)$; (b) $(\sqrt{2}, -\frac{1}{4}\pi)$;
(c) $(-\sqrt{2}, -\frac{5}{4}\pi)$ **7.** (a) $(-2, \frac{3}{4}\pi)$; (b) $(-2, -\frac{5}{4}\pi)$; (c) $(2, \frac{15}{4}\pi)$ **9.** $(3, \frac{4}{3}\pi)$; $(-3, \frac{1}{3}\pi)$ **11.** $(-4, -\frac{7}{6}\pi)$; $(4, -\frac{1}{6}\pi)$
13. $(-2, \frac{3}{4}\pi)$; $(2, \frac{7}{4}\pi)$ **15.** $(2, 2\pi + 6)$; $(-2, 6 - \pi)$ **17.** (a) $(-3, 0)$; (b) $(-1, -1)$; (c) $(2, -2\sqrt{3})$; (d) $(\frac{1}{2}\sqrt{3}, -\frac{1}{2})$

19. (a) $(\sqrt{2}, \frac{7}{4}\pi)$; (b) $(2, \frac{2}{3}\pi)$; (c) $(2\sqrt{2}, \frac{1}{4}\pi)$; (d) $(5, \pi)$ **21.** $r = |a|$ **23.** $r = \dfrac{2}{1 - \cos\theta}$ **25.** $r = 6\sin\theta$ **27.** $r^2 = 4\cos 2\theta$

29. $r = \dfrac{3a\sin 2\theta}{2(\sin^3\theta + \cos^3\theta)}$ **31.** $(x^2 + y^2)^2 = 4xy$ **33.** $(x^2 + y^2)^3 = x^2$ **35.** $y = x\tan(x^2 + y^2)$ **37.** $x = -1$
39. $4x^2 - 5y^2 - 36y - 36 = 0$

EXERCISES 10.2 (PAGE 767)

1. (a) line through the pole with slope $\sqrt{3}$; (b) circle with center at the pole and radius $\frac{1}{3}\pi$ **3.** (a) line through the pole
with slope $\tan^{-1} 2$; circle with center at the pole and radius 2   **5.** (a) line parallel to the $\frac{1}{2}\pi$ axis and 4 units to the right of
it; (b) circle tangent to the $\frac{1}{2}\pi$ axis with center on the polar axis and radius 2   **7.** (a) line parallel to the polar axis and 4
units below it; (b) circle tangent to the polar axis with center on the extension of the $\frac{1}{2}\pi$ axis and radius 2   **9.** cardioid
11. cardioid   **13.** limaçon   **15.** limaçon   **17.** limaçon   **19.** three-leafed rose   **21.** eight-leafed rose   **23.** four-leafed rose
37. $(\frac{3}{2}, \frac{1}{6}\pi)$; $(\frac{3}{2}, \frac{5}{6}\pi)$ **39.** pole; $(\sqrt{2}, \frac{1}{4}\pi)$ **41.** $(\frac{1}{2}\pi, \frac{1}{8}\pi)$ **43.** pole; $(\sqrt{15}, \cos^{-1}\frac{1}{4})$; $(\sqrt{15}, \pi - \cos^{-1}\frac{1}{4})$ **45.** $(3, \frac{1}{3}\pi)$; $(3, -\frac{1}{3}\pi)$
47. $(2\sqrt{2}, \frac{1}{4}\pi)$ **49.** $(1, \frac{3}{2}\pi)$; $(1, \frac{1}{2}\pi)$; $(\frac{1}{2}, \frac{4}{3}\pi)$; $(\frac{1}{2}, \frac{2}{3}\pi)$; pole; $(0.22, 2.47)$; $(0.22, 3.82)$ **51.** pole;
$(\frac{1}{2}\sqrt{2}, \frac{1}{8}(2n+1)\pi)$, where n is 0, 1, . . . , 7   **55.** pole; $(\frac{1}{2}\sqrt{2}, \frac{1}{6}(2n+1)\pi)$, where n is 0, 1, 2, 3, 4, 5

EXERCISES 10.3 (PAGE 771)

1. $\frac{2}{3}\pi$ sq units   **3.** 4π sq units   **5.** 4 sq units   **7.** $\frac{9}{16}\pi^3$ sq units   **9.** $\frac{9}{8}\pi$ sq units   **11.** $(\frac{11}{4}\pi - \frac{11}{4}\sin^{-1}\frac{1}{3} - 3\sqrt{2})$ sq units
13. $(\frac{19}{3}\pi - \frac{11}{2}\sqrt{3})$ sq units   **15.** $(\frac{8}{3}\pi - 9)$ sq units   **17.** $a^2(2 - \frac{1}{4}\pi)$ sq units   **19.** $\frac{1}{2}(\pi + 1)$ sq units   **21.** $\frac{1}{2}\pi a^2$ sq units
23. 4 sq units   **25.** $16a^2\pi^3$ sq units

EXERCISES 10.4 (PAGE 780)

1. $(0, 1)$; $y = -1$; 4   **3.** $(-2, 0)$; $x = 2$; 8   **5.** $(0, -\frac{1}{4})$; $y = \frac{1}{4}$; 1   **7.** $(\frac{9}{8}, 0)$; $x = -\frac{9}{8}$; $\frac{9}{2}$ **9.** $y^2 = 20x$ **11.** $x^2 = -8y$
13. $y^2 = 2x$ **15.** $y^2 = -6x$ **17.** $y^2 = 10x$ **19.** $x^2 = -y$ **21.** $x'^2 + y'^2 = 13$ **23.** $y'^2 = 6x'$ **25.** $y' = 2x'^3$ **27.** $(-3, \frac{1}{4})$;
$(-3, -\frac{3}{4})$; $x = -3$; $y = \frac{5}{4}$ **29.** $(1, -5)$; $(-\frac{1}{2}, -5)$; $y = -5$; $x = \frac{1}{2}$ **31.** $(\frac{2}{3}, 1)$; $(\frac{7}{24}, 1)$; $y = 1$; $x = \frac{25}{24}$
33. $y^2 + 20x - 8y - 24 = 0$ **35.** $x^2 + 2x - 8y + 41 = 0$ **37.** $x^2 - 6x - 6y - 3 = 0$; $x^2 - 6x + 6y + 21 = 0$
39. $y^2 - 4x - 4y - 12 = 0$ **41.** (a) $-\frac{5}{2}$; (b) $x^2 = -10y$ **43.** 16.6 m   **47.** $\frac{32}{43}$ in.   **49.** (a) $y = x$; (b) $(\frac{1}{2}, \frac{1}{2})$; (c) $2\sqrt{2}$
53. translate the axes so that the new origin is $(\frac{1}{4}\pi, -3)$

EXERCISES 10.5 (PAGE 790)

1. center: $(0, 0)$; vertices: $(\pm 3, 0)$; foci: $(\pm \sqrt{5}, 0)$; extremities of minor axis: $(0, \pm 2)$ **3.** center: $(0, 0)$; vertices: $(0, \pm 5)$; foci: $(0, \pm \sqrt{21})$; extremities of minor axis: $(\pm 2, 0)$ **5.** center: $(0, 0)$; vertices: $(\pm 3, 0)$; foci: $(\pm \sqrt{3}, 0)$; extremities of minor axis: $(0, \pm \sqrt{6})$ **7.** center: $(0, 0)$; vertices: $(0, \pm \frac{1}{2})$; foci: $(0, \pm \frac{1}{4}\sqrt{3})$; extremities of minor axis: $(\pm \frac{1}{4}, 0)$ **9.** center: $(2, 3)$; vertices: $(-1, 3)$, $(5, 3)$; foci: $(2 \pm \sqrt{3}, 3)$; extremities of minor axis: $(2, 3 \pm \sqrt{6})$ **11.** center: $(0, \frac{1}{2})$; vertices: $(0, \frac{1}{2} \pm \frac{1}{2}\sqrt{17})$; foci: $(0, \frac{1}{2} \pm \frac{1}{10}\sqrt{170})$; extremities of minor axis: $(\pm \frac{1}{10}\sqrt{255}, \frac{1}{2})$ **13.** point-circle $(-\frac{2}{3}, 4)$ **15.** center: $(1, 0)$; vertices: $(1 \pm \sqrt{5}, 0)$; foci: $(1 \pm \sqrt{2}, 0)$; extremities of minor axis: $(1, \pm \sqrt{3})$ **17.** point-ellipse $(1, -1)$

19. $16x^2 + 25y^2 = 100$ **21.** $3x^2 + 2y^2 = 54$ **23.** $x^2 + 4y^2 = 4$ **25.** $\dfrac{(x-4)^2}{25} + \dfrac{(y+2)^2}{9} = 1$ **27.** $\dfrac{(y-3)^2}{64} + \dfrac{(x+1)^2}{48} = 1$

29. $2x + 3y - 12 = 0$ **33.** $\frac{42}{5}$ m **35.** $9x^2 + 25y^2 = 56{,}250{,}000$ **37.** 98π in.^{3} **39.** $(h \pm \sqrt{a^2 - b^2}, k)$ **41.** $18\pi\rho$ lb

EXERCISES 10.6 (PAGE 801)

1. center: $(0, 0)$; vertices: $(\pm 2, 0)$; foci: $(\pm \sqrt{13}, 0)$; asymptotes: $y = \pm \frac{3}{2}x$ **3.** center: $(0, 0)$; vertices: $(\pm 5, 0)$; foci: $(\pm \sqrt{29}, 0)$; asymptotes: $y = \pm \frac{2}{5}x$ **5.** center: $(0, 0)$; vertices: $(0, \pm 2)$; foci: $(0, \pm 2\sqrt{5})$; asymptotes: $y = \pm \frac{1}{2}x$ **7.** center: $(0, 0)$; vertices $(0, \pm \frac{1}{3})$; foci: $(0, \pm \frac{5}{12})$; asymptotes: $y = \pm \frac{4}{3}x$ **9.** center: $(-4, -1)$; vertices: $(-10, -1)$, $(2, -1)$; foci: $(-4 \pm 6\sqrt{2}, -1)$; asymptotes: $x - y + 3 = 0$, $x + y + 5 = 0$ **11.** center: $(-3, -1)$; vertices: $(-3, -1 \pm \frac{2}{3}\sqrt{2})$; foci: $(-3, -1 \pm \frac{2}{3}\sqrt{6})$; asymptotes: $\pm x + \sqrt{2}y + \sqrt{2} \pm 3 = 0$ **13.** center: $(-1, 4)$; vertices: $(-1, 4 \pm 2\sqrt{7})$; foci: $(-1, -3)$, $(-1, 11)$; asymptotes: $\pm 2x + \sqrt{3}y \pm 2 - 4\sqrt{3} = 0$ **15.** center: $(1, -2)$; vertices: $(1, -5)$, $(1, 1)$; foci: $(1, -2 \pm \sqrt{13})$; asymptotes: $3x + 2y + 1 = 0$, $3x - 2y - 7 = 0$ **17.** $9x^2 - 4y^2 = 36$ **19.** $32y^2 - 33x^2 = 380$

21. $25x^2 - 144y^2 = 14{,}400$ **23.** $\dfrac{(y+1)^2}{144} - \dfrac{(x+2)^2}{81} = 1$ **25.** $72(x-3)^2 - 91(y-4)^2 = 128$

27. (a) $\dfrac{(x+2)^2}{1} - \dfrac{(y+1)^2}{4} = 1$; (b) $2x - y + 3 = 0$, $2x + y + 5 = 0$ **29.** $2x - \sqrt{2}y + 1 = 0$ **31.** $7x^2 - 4y^2 = 28$

33. the right branch of the hyperbola $16x^2 - 9y^2 = 14{,}400$ **35.** $\frac{4}{3}\pi ab^2$ cu units **37.** $(3, 4)$

EXERCISES 10.7 (PAGE 807)

1. (a) $\bar{x}^2 - \bar{y}^2 = 16$ **3.** (a) $\bar{x}\bar{y} = -4$ **5.** $16\bar{y}^2 - 9\bar{x}^2 = 36$ **7.** $\bar{x}^2 + 4\sqrt{2}\bar{y} = 0$ **9.** $\bar{y}^2 - \bar{x}^2 = 32$ **11.** $9\bar{x}^2 + 4\bar{y}^2 = 36$ **13.** $3\bar{x}'^2 + \bar{y}'^2 = 18$ **15.** $\bar{x}'^2 + 4\bar{y}'^2 = 16$ **17.** $\sqrt{2}\bar{x}'^2 = \bar{y}'$ **19.** $\bar{x}'^2 - 4\bar{y}'^2 = 16$ **21.** $5\bar{x}^2 + 6\sqrt{5}\bar{y} = 0$

EXERCISES 10.8 (PAGE 817)

1. (a) $e = \frac{1}{3}\sqrt{5}$; foci: $(\pm \sqrt{5}, 0)$; directrices: $x = \pm \frac{9}{5}\sqrt{5}$ **3.** (a) $e = \frac{1}{5}\sqrt{21}$; foci: $(0, \pm \sqrt{21})$; directrices: $y = \pm \frac{25}{21}\sqrt{21}$ **5.** (a) $e = \frac{1}{5}\sqrt{29}$; foci: $(\pm \sqrt{29}, 0)$; directrices: $x = \pm \frac{25}{29}\sqrt{29}$ **7.** (a) $e = \frac{5}{3}$; foci: $(\pm 5, 0)$; directrices: $x = \pm \frac{9}{5}$ **9.** (a) parabola; (b) hyperbola; (c) ellipse; (d) circle **11.** (a) 1; (b) parabola; (c) $r \cos \theta = -2$ **13.** (a) $\frac{1}{2}$; (b) ellipse; (c) $r \sin \theta = 5$ **15.** (a) $\frac{2}{3}$; (b) ellipse; (c) $r \cos \theta = -3$ **17.** (a) $\frac{6}{5}$; (b) hyperbola; (c) $2r \sin \theta = -3$ **19.** (a) $\frac{2}{7}$; (b) ellipse; (c) $r \sin \theta = -5$ **21.** (a) $\frac{5}{4}$; (b) hyperbola; (c) $r \cos \theta = 2$ **23.** $r = \dfrac{8}{1 - \sin \theta}$ **25.** $r = \dfrac{36}{3 + 4 \cos \theta}$ **27.** $r = \dfrac{3}{2 - \cos \theta}$ **29.** (a) $r = \dfrac{5}{1 - 3 \cos \theta}$; (b) $3r \cos \theta = -5$ **31.** $\frac{16}{3}\sqrt{3}\pi$ sq units **37.** (a) $r = \dfrac{40{,}000{,}000}{1 - \cos \theta}$; (b) 20,000,000 miles **39.** $r^2 = \dfrac{a^2(1 - e^2)}{1 - e^2 \cos^2 \theta}$

EXERCISES 10.9 (PAGE 827)

1. $\sqrt{3}$ **3.** $\sqrt{2} + 1$ **5.** $2\sqrt{3}$ **7.** $-\frac{7}{3}\sqrt{3}$ **9.** horizontal tangent lines at $(7, \frac{1}{2}\pi)$, $(1, \frac{3}{2}\pi)$, $(2, 3.87)$, $(2, 5.55)$; vertical tangent lines at $(5.34, 0.46)$, $(5.34, 2.68)$ **11.** horizontal tangent lines at $(4.73, 1.95)$, $(4.73, 4.34)$;

vertical tangent lines at $(2, 0)$, $(6, \pi)$ **13.** horizontal tangent lines at $(2, \frac{3}{2}\pi)$, $(\frac{1}{2}, \frac{1}{6}\pi)$, $(\frac{1}{2}, \frac{5}{6}\pi)$;
vertical tangent lines at $(0, \frac{1}{2}\pi)$, $(\frac{3}{2}, \frac{7}{6}\pi)$, $(\frac{3}{2}, \frac{11}{6}\pi)$ **15.** horizontal tangent lines at $(3.68, 0.98)$, $(3.68, 5.31)$, $(-0.68, 2.68)$,
$(-0.68, 3.61)$; vertical tangent lines at $(5, 0)$, $(-1, \pi)$, $(1, 1.91)$, $(1, 4.37)$ **17.** horizontal tangent lines at $(-1, \frac{1}{2}\pi)$,
$(-1, \frac{3}{2}\pi)$, $(\frac{2}{3}, 0.42)$, $(\frac{2}{3}, 2.72)$, $(\frac{2}{3}, 3.56)$, $(\frac{2}{3}, 5.86)$; vertical tangent lines at $(1, 0)$, $(1, \pi)$, $(-\frac{2}{3}, 1.15)$, $(-\frac{2}{3}, 1.99)$, $(-\frac{2}{3}, 4.29)$,
$(-\frac{2}{3}, 5.13)$ **19.** horizontal tangent lines at $(0, 0)$, $(\sqrt[4]{12}, \frac{1}{3}\pi)$, $(-\sqrt[4]{12}, \frac{1}{3}\pi)$; vertical tangent lines at $(0, \frac{1}{2}\pi)$, $(\sqrt{2}, \frac{1}{6}\pi)$,
$(-\sqrt{2}, \frac{1}{6}\pi)$ **21.** $\chi = \frac{3}{4}\pi$, $\alpha = 1 + \frac{3}{4}\pi$ **23.** $\chi = 153°26'$; $\alpha = 130°57'$ **25.** $\chi = 38°9'$; $\alpha = 128°9'$ **27.** $\chi = \frac{5}{6}\pi$; $\alpha = \pi$
29. $90°$ **31.** $60°$ **33.** $0°$ at pole; $90°$ at $(1, 0)$; $90°$ at $(1, \pi)$ **35.** $0°$ at $(0, \frac{1}{2}\pi)$; $79°6'$ at $(\frac{1}{2}\sqrt{3}, \frac{1}{6}\pi)$; $79°6'$ at $(-\frac{1}{2}\sqrt{3}, \frac{5}{6}\pi)$

REVIEW EXERCISES FOR CHAPTER 10 (PAGE 828)

1. $r^2(4\cos^2\theta - 9\sin^2\theta) = 36$ **3.** $r = 9\cos\theta - 8\sin\theta$ **5.** $4x^4 + 8x^2y^2 + 4y^4 + 36x^3 + 36xy^2 - 81y^2 = 0$
7. $(x^2 + y^2)^2 = 4x^3 - 12xy^2$ **15.** the pole **17.** (a) $(16\pi - 24\sqrt{3})$ sq units; (b) $(32\pi + 24\sqrt{3})$ sq units

19. $a^2(\frac{1}{3}\pi + \frac{1}{2}\sqrt{3})$ sq units **21.** $\left(\dfrac{e^{4k\pi} - 1}{4k}\right)$ sq units **25.** $r^2 - 2r_0 r\cos(\theta - \theta_0) + r_0^2 = a^2$ **27.** (a) $x^2 = -12y$; (b) 12

29. (a) $(0, \frac{3}{2})$; (b) $y = -3$ **31.** (a) $(2, 3)$; (b) $y = 3$; (c) $(0, 3)$; (d) $x = 4$ **33.** $(x - 5)^2 = 8(y - 1)$ **35.** 9π cm
37. center: $(0, 0)$; vertices: $(\pm 5, 0)$; foci: $(\pm\sqrt{21}, 0)$; extremities of minor axis: $(0, \pm 2)$ **39.** center: $(0, 0)$; vertices: $(\pm 3, 0)$;
foci: $(\pm 5, 0)$; asymptotes: $y = \pm\frac{4}{3}x$ **41.** center: $(-3, 8)$; vertices: $(-3, 4)$, $(-3, 12)$; foci: $(-3, 8 \pm 2\sqrt{3})$;
extremities of minor axis: $(-5, 8)$, $(-1, 8)$ **43.** center: $(-1, 3)$; vertices: $(-2, 3)$, $(0, 3)$; foci: $(-1 \pm\sqrt{26}, 3)$;
asymptotes: $5x + y + 2 = 0$, $5x - y + 8 = 0$ **45.** $\dfrac{x^2}{20} + \dfrac{(y - 2)^2}{36} = 1$ **47.** $\dfrac{(x + 2)^2}{4} - \dfrac{(y - 1)^2}{5} = 1$ **49.** (a) $e = \frac{1}{3}\sqrt{34}$;
foci: $(\pm\sqrt{34}, 0)$; directrices: $x = \pm\frac{9}{34}\sqrt{34}$ **51.** (a) $e = \frac{1}{2}$; (b) ellipse; (c) $r\sin\theta = -2$ **53.** (a) $e = \frac{3}{2}$; (b) hyperbola;
(c) $3r\cos\theta = 4$ **55.** $r = \dfrac{8}{1 - 3\cos\theta}$ **57.** $r = \dfrac{6}{1 - \sin\theta}$ **59.** $21\bar{x}'^2 - 49\bar{y}'^2$ **61.** $r = \dfrac{1}{1 - \cos\theta}$ **63.** $4\sqrt{3}\pi a^2 b$ cu units
65. 600 mi. **67.** $(3 \pm\frac{3}{4}\sqrt{6})$ million miles **69.** the left branch of the hyperbola $16x^2 - 9y^2 = 1,440,000$
75. first quadrant part of a parabola **77.** $3\sqrt{3}$ **79.** horizontal tangent lines at $(6, \frac{3}{2}\pi)$, $(\frac{3}{2}, \frac{1}{6}\pi)$, $(\frac{3}{2}, \frac{5}{6}\pi)$;
vertical tangent lines at $(0, \frac{1}{2}\pi)$, $(\frac{9}{2}, \frac{7}{6}\pi)$, $(\frac{9}{2}, \frac{11}{6}\pi)$ **81.** $\chi = 1.98$, $\alpha = 0.93$ **83.** $\frac{1}{2}\pi$ at pole; $\frac{1}{6}\pi$ at $(3, \frac{2}{3}\pi)$, $\frac{1}{6}\pi$ at $(3, \frac{4}{3}\pi)$

EXERCISES 11.1 (PAGE 841)

1. $\frac{1}{2}$ **3.** divergent **5.** -2 **7.** 0 **9.** 1 **11.** divergent **13.** divergent **15.** $e^{1/3}$ **17.** 1 **19.** 0 **21.** choose $N = 1 + \dfrac{3}{\epsilon}$

23. choose $N = \dfrac{6}{\epsilon} - \dfrac{3}{2}$ **25.** choose $N = \dfrac{1}{5}\sqrt{\dfrac{2 - 5\epsilon}{\epsilon}}$

EXERCISES 11.2 (PAGE 848)

1. increasing **3.** decreasing **5.** not monotonic **7.** not monotonic **9.** decreasing **11.** increasing after the first two terms
13. increasing **15.** decreasing **19.** not bounded **31.** $\left\{\dfrac{(-1)^{n+1}}{n}\right\}$

EXERCISES 11.3 (PAGE 857)

1. $s_n = \dfrac{n}{2n + 1}$; $\dfrac{1}{2}$ **3.** $s_n = \dfrac{5n}{3n + 1}$; $\dfrac{5}{3}$ **5.** $s_n = \ln\left(\dfrac{1}{n + 1}\right)$; divergent **7.** $s_n = \dfrac{5}{2}\left(1 - \dfrac{1}{5^n}\right)$; $\dfrac{5}{2}$

9. $\displaystyle\sum_{n=1}^{+\infty} \frac{2}{(3n-2)(3n+1)}; \frac{2}{3}$ **11.** $1 - \displaystyle\sum_{n=1}^{+\infty} \frac{2}{3^n}; 0$ **13.** $\displaystyle\sum_{n=1}^{+\infty} \ln\left(\frac{2n+1}{2n-1}\right);$ divergent **15.** divergent **17.** 2 **19.** divergent

21. 1 **23.** $\dfrac{1}{e-1}$ **25.** divergent **27.** $\frac{3}{11}$ **29.** $\frac{137}{111}$ **31.** 8 m **33.** $\frac{3100}{9}$ m **35.** 1500

EXERCISES 11.4 (PAGE 863)

1. divergent **3.** divergent **5.** 3 **7.** $\frac{10}{3}$ **9.** $\dfrac{63 \cdot 2^{10} + 1}{2^{16}}$ **11.** divergent **13.** $\frac{3}{2}$ **15.** divergent **17.** divergent **19.** 2

21. divergent **23.** $\displaystyle\sum_{n=1}^{+\infty} \frac{1}{n}$ and $\displaystyle\sum_{n=1}^{+\infty} \frac{1}{n+1}$ are both divergent; $\displaystyle\sum_{n=1}^{+\infty} \frac{1}{n^2+n}$ is convergent

---

EXERCISES 11.5 (PAGE 874)

1. convergent **3.** convergent **5.** divergent **7.** convergent **9.** divergent **11.** convergent **13.** divergent **15.** convergent
17. divergent **19.** convergent **21.** divergent **23.** convergent **25.** convergent

---

EXERCISES 11.6 (PAGE 878)

1. divergent **3.** convergent **5.** convergent **7.** convergent **9.** divergent **11.** convergent **13.** convergent

15. convergent **17.** convergent **19.** convergent **21.** divergent **27.** $0.7032 < \displaystyle\sum_{m=50}^{100} \frac{1}{m} < 0.7134$

EXERCISES 11.7 (PAGE 884)

1. convergent **3.** convergent **5.** convergent **7.** convergent **9.** convergent **11.** divergent **13.** convergent **15.** $|R_4| < \frac{1}{3}$

17. $|R_4| < \frac{1}{81}$ **19.** $|R_4| < \frac{1}{25}$ **21.** $|R_4| < \dfrac{1}{6 \ln 6}$ **23.** 0.332 **25.** 0.632 **27.** 0.113 **29.** 0.406

EXERCISES 11.8 (PAGE 891)

1. absolutely convergent **3.** absolutely convergent **5.** absolutely convergent **7.** divergent **9.** absolutely convergent
11. absolutely convergent **13.** absolutely convergent **15.** absolutely convergent **17.** absolutely convergent
19. divergent **25.** (b) convergent

REVIEW EXERCISES FOR CHAPTER 11 (PAGE 894)

1. $1, \frac{3}{2}, \frac{9}{5}, 2; 3$ **3.** $0, \frac{3}{5}, \frac{4}{5}, \frac{15}{17}; 1$ **5.** $1, 3, 1, 3;$ no limit **7.** $4, \dfrac{81}{16}, \dfrac{4096}{729}, \dfrac{390,625}{65,536}; e^2$ **9.** $\frac{3}{16}, \frac{15}{64}, \frac{63}{256}, \frac{255}{1024}; s_n = \dfrac{4^n - 1}{4^{n+1}}; \frac{1}{4}$

11. convergent; 3 **13.** divergent **15.** convergent; $4 + 2\sqrt{3}$ **17.** convergent; $\frac{1}{6}$ **19.** $\frac{649}{729}$ **21.** convergent **23.** divergent

25. convergent   **27.** divergent   **29.** convergent   **31.** divergent   **33.** divergent   **35.** convergent   **37.** convergent
39. absolutely convergent   **41.** conditionally convergent   **43.** divergent   **45.** absolutely convergent
47. absolutely convergent   **49.** $\frac{437}{330}$ **51.** 90 in.

EXERCISES 12.1 (PAGE 905)

1. $[-1, 1)$ **3.** $[-1, 1]$ **5.** $[-\frac{1}{2}, \frac{1}{2}]$ **7.** $(-3, 3)$ **9.** $(-\infty, +\infty)$ **11.** $(-5, -1)$ **13.** $(-9, 9]$ **15.** $(0, 2]$ **17.** $\left(-\dfrac{1}{e^2}, \dfrac{1}{e^2}\right)$
19. $[-1, 1]$ **21.** $(-4, 6)$ **23.** $[4, 6)$ **25.** $[-1, 1]$ **27.** $(-e, e)$ **29.** $+\infty$

EXERCISES 12.2 (PAGE 915)

1. (a) $R = 1$; $[-1, 1]$; (b) $\displaystyle\sum_{n=1}^{+\infty} \frac{x^{n-1}}{n}$; $R = 1$; (c) $[-1, 1)$ **3.** (a) $R = 1$; $[-1, 1)$; (b) $\displaystyle\sum_{n=1}^{+\infty} \sqrt{n}\, x^{n-1}$; $R = 1$; (c) $(-1, 1)$

5. (a) $R = +\infty$; $(-\infty, +\infty)$; (b) $\displaystyle\sum_{n=1}^{+\infty} (-1)^{n-1} \frac{x^{2n-2}}{(2n-2)!}$; $R = +\infty$; (c) $(-\infty, +\infty)$ **7.** (a) $R = \frac{1}{3}$; $(0, \frac{2}{3})$;

(b) $\displaystyle\sum_{n=1}^{+\infty} 3n(n+1)(3x-1)^n$; $R = \frac{1}{3}$; $(0, \frac{2}{3})$ **9.** (a) $R = 3$; $[-2, 4)$; (b) $\displaystyle\sum_{n=1}^{+\infty} \frac{(x-1)^{n-1}}{3^n}$; $R = 3$; (c) $(-2, 4)$

11. $\dfrac{1}{2} \displaystyle\sum_{n=2}^{+\infty} n(n-1)x^{n-2}$ **13.** $\displaystyle\sum_{n=0}^{+\infty} (-1)^n(n+1)x^n$ **15.** (a) $\displaystyle\sum_{n=0}^{+\infty} 2^n x^n$ if $|x| < \frac{1}{2}$; (b) $\displaystyle\sum_{n=1}^{+\infty} n2^n x^{n-1}$ if $|x| < \frac{1}{2}$

17. (a) $\displaystyle\sum_{n=0}^{+\infty} \frac{x^{2n}}{n!}$; (b) $\displaystyle\sum_{n=0}^{+\infty} \frac{x^{2n+1}}{n!}$ **19.** 0.60653 **21.** (a) $\displaystyle\sum_{n=0}^{+\infty} \frac{x^{2n+1}}{(2n+1)!}$; (b) $\displaystyle\sum_{n=0}^{+\infty} \frac{x^{2n}}{(2n)!}$

23. 2 **25.** (a) $\displaystyle\sum_{n=0}^{+\infty} (-1)^n \frac{x^{n+2}}{n!}$ **27.** $\displaystyle\sum_{n=0}^{+\infty} \frac{x^n}{n!}$

EXERCISES 12.3 (PAGE 923)

1. $\displaystyle\sum_{n=0}^{+\infty} \frac{x^{n+1}}{(n+1)!}$; $R = +\infty$ **3.** $\displaystyle\sum_{n=0}^{+\infty} \frac{x^{n+1} - 2^{n+1}}{4^{n+1}(n+1)}$; $R = 4$ **5.** 1.718 **7.** 0.693 **9.** $\displaystyle\sum_{n=1}^{+\infty} \frac{x^n}{n(n!)}$; $R = +\infty$

11. $\displaystyle\sum_{n=1}^{+\infty} \frac{x^{2n+1}}{(2n+1)(2n+1)!}$; $+\infty$ **13.** 1.318 **15.** 1.057 **17.** 0.485 **19.** 0.747 **21.** 0.041 **23.** 0.450 **25.** $\displaystyle\sum_{n=0}^{+\infty} \frac{x^{2n+1}}{(2n+1)!}$

27. 0.2450 **29.** 0.221 **31.** $\displaystyle\sum_{n=0}^{+\infty} \frac{x^{2n+1}}{2n+1}$ **37.** $\displaystyle\sum_{n=0}^{+\infty} \frac{(-1)^n x^{2n+1}}{(2n+1)!}$; $R = +\infty$

EXERCISES 12.4 (PAGE 934)

7. $\ln 2 + \displaystyle\sum_{n=1}^{+\infty} (-1)^{n-1} \frac{(x-2)^n}{n2^n}$ **9.** $\ln 2 + \displaystyle\sum_{n=1}^{+\infty} (-1)^{n-1} \frac{(x-1)^n}{n2^n}$; $R = 2$

11. $2 + \frac{1}{4}(x-4) + 2 \displaystyle\sum_{n=2}^{+\infty} (-1)^{n-1} \frac{1 \cdot 3 \cdot 5 \cdot \ldots \cdot (2n-3)(x-4)^n}{2 \cdot 4 \cdot 6 \cdot \ldots \cdot (2n) \cdot 4^n}$; $R = 4$

13. $\frac{1}{2} - \frac{1}{2}\sqrt{3}(x - \frac{1}{3}\pi) - \frac{1}{4}(x - \frac{1}{3}\pi)^2 + \frac{1}{12}\sqrt{3}(x - \frac{1}{3}\pi)^3 + \frac{1}{48}(x - \frac{1}{3}\pi)^4 - \cdots$; $R = +\infty$

15. $\dfrac{1}{2} \displaystyle\sum_{n=1}^{+\infty} \frac{(-1)^{n-1}(2x)^{2n}}{(2n)!}$ **17.** (a) $x + \frac{1}{3}x^3 + \frac{2}{15}x^5$; (b) $1 + x^2 + \frac{2}{3}x^4$; (c) $\frac{1}{2}x^2 + \frac{1}{12}x^4 + \frac{1}{45}x^6$ **19.** 0.5299 **21.** 1.97435

23. -0.2231 **25.** 2.7182818 **27.** 0.0415 **29.** 0.0048 **31.** 0.2398 **33.** $a_4 = 3$; $a_3 = -5$; $a_2 = 2$; $a_1 = -1$; $a_0 = 6$

EXERCISES 12.5 (PAGE 940)

1. $1 + \dfrac{x}{2} + \displaystyle\sum_{n=2}^{+\infty} \dfrac{(-1)^{n+1} \cdot 1 \cdot 3 \cdot \ldots \cdot (2n-3)}{2^n n!} x^n; R = 1$ **3.** $\dfrac{1}{2} + \dfrac{1}{2}\displaystyle\sum_{n=1}^{+\infty} \dfrac{(-1)^n \cdot 1 \cdot 3 \cdot 5 \cdot \ldots \cdot (2n-1)x^n}{8^n n!}; R = 4$

5. $1 + \displaystyle\sum_{n=1}^{+\infty} \dfrac{(-1)2 \cdot 5 \cdot 8 \cdot \ldots \cdot (3n-4)}{3^n n!} x^{3n}; R = 1$ **7.** $\dfrac{1}{3} + \dfrac{1}{3}\displaystyle\sum_{n=1}^{+\infty} \dfrac{(-1)^n \cdot 1 \cdot 3 \cdot 5 \cdot \ldots \cdot (2n-1)x^{4n}}{18^n n!}; R = \sqrt{3}$

9. $x^2 + \displaystyle\sum_{n=1}^{+\infty} \dfrac{(-1)^n \cdot 1 \cdot 3 \cdot 5 \cdot \ldots \cdot (2n-1)x^{n+2}}{2^n n!}; R = 1$ **11.** 4.899   **13.** 5.010   **15.** 0.198

17. $x + \displaystyle\sum_{n=1}^{+\infty} \dfrac{(-1)^n \cdot 1 \cdot 3 \cdot 5 \cdot \ldots \cdot (2n-1)x^{2n+1}}{2^n n!(2n+1)}; R = 1$ **19.** (a) $1 + \dfrac{1}{4}x + \displaystyle\sum_{n=2}^{+\infty} \dfrac{(-1)^{n+1} \cdot 3 \cdot 7 \cdot 11 \cdot \ldots \cdot (4n-5)}{4^n n!} x^n;$

(b) $1 + \dfrac{1}{4}x^2 + \displaystyle\sum_{n=2}^{+\infty} \dfrac{(-1)^{n+1} \cdot 3 \cdot 7 \cdot 11 \cdot \ldots \cdot (4n-5)}{4^n n!} x^{2n};$ (c) 0.510   **21.** 0.335   **23.** 2.0271   **25.** 0.4970   **27.** 0.5082

29. $\dfrac{1}{p+1} x^{p+1} + \displaystyle\sum_{n=1}^{+\infty} \dfrac{1 \cdot 3 \cdot 5 \cdot \ldots \cdot (2n-1)x^{2n+p+1}}{2^n n!(2n+p+1)}; R = 1$

REVIEW EXERCISES FOR CHAPTER 12 (PAGE 941)

1. $[-1, 1)$ **3.** $[-3, 3]$ **5.** $x = 3$ **7.** $(-7, 5)$ **9.** $(-1, 3]$ **11.** $x = 0$ **13.** (a) $R = 1, [-1, 1]$;

(b) $\displaystyle\sum_{n=1}^{+\infty} (-1)^n x^{2n-1}, R = 1$; (c) $(-1, 1)$ **15.** (a) $R = +\infty, (-\infty, +\infty)$; (b) $\displaystyle\sum_{n=1}^{+\infty} \dfrac{nx^{n-1}}{(n!)^2}, R = +\infty$; (c) $(-\infty, +\infty)$

17. $\displaystyle\sum_{n=0}^{+\infty} \dfrac{(-1)^n x^{2n+1}}{2^{4n+4}(2n+1)}; R = 4$ **19.** $\displaystyle\sum_{n=0}^{+\infty} \dfrac{(-1)^n x^{2n+1}}{(2n+1)(2n+1)!}; R = +\infty$ **21.** 0.161   **23.** 0.493   **25.** 0.1974   **27.** 5.0658

29. 0.9986   **31.** 1.6094   **33.** 0.0124   **35.** $\displaystyle\sum_{n=0}^{+\infty} \dfrac{(\ln a)^n x^n}{n!}; (-\infty, +\infty)$ **37.** $\dfrac{3}{4}\displaystyle\sum_{n=1}^{+\infty} \dfrac{(-1)^{n-1}(3^{2n}-1)x^{2n+1}}{(2n+1)!}; (-\infty, +\infty)$

39. $\displaystyle\sum_{n=1}^{+\infty} \dfrac{(-1)^n(3x+\pi)^{2n-1}}{(2n-1)!}$ **41.** $\displaystyle\sum_{n=1}^{+\infty} -\dfrac{(x+1)^n}{n}$

EXERCISES 13.1 (PAGE 957)

1. (b) 5   **3.** (b) $\frac{1}{2}\sqrt{1+4e^2}$ **5.** (a) $\frac{3}{4}\pi$; (b) π; (c) $\tan^{-1} 0.4 \approx 0.38$ **7.** $\langle 2, -3 \rangle$ **9.** $\langle 5, 6 \rangle$ **11.** $(-4, 3)$ **13.** $(12, -5)$

15. (a) $\langle -1, 9 \rangle$; (b) $\langle 1, -5 \rangle$ **17.** (a) $\langle -9, -4 \rangle$; (b) $\langle 4, -e \rangle$ **19.** (a) $\langle 6, 1 \rangle$; (b) $\sqrt{74}$; (c) $\sqrt{1061}$ **21.** (a) $10\mathbf{i} + 15\mathbf{j}$;

(b) $-24\mathbf{i} + 6\mathbf{j}$; (c) $6\mathbf{i} + 2\mathbf{j}$; (d) $2\sqrt{10}$ **23.** (a) $\sqrt{13} + \sqrt{17}$; (b) $-14\mathbf{i} + 21\mathbf{j}$; (c) $7\sqrt{13}$; (d) $5\sqrt{13} - 6\sqrt{17}$ **25.** (a) $-28\mathbf{i} + 6\mathbf{j}$;

(b) $2\sqrt{205}$ **27.** $\dfrac{11}{\sqrt{137}}\mathbf{i} + \dfrac{4}{\sqrt{137}}\mathbf{j}$ **29.** (a) $5(\frac{3}{5}\mathbf{i} - \frac{4}{5}\mathbf{j}), \frac{3}{5}\mathbf{i} - \frac{4}{5}\mathbf{j}$; (b) $2\sqrt{2}(\cos \frac{1}{4}\pi\mathbf{i} + \sin \frac{1}{4}\pi\mathbf{j}), \frac{1}{2}\sqrt{2}\mathbf{i} + \frac{1}{2}\sqrt{2}\mathbf{j}$

31. (a) $8(\cos \frac{2}{3}\pi\mathbf{i} + \sin \frac{2}{3}\pi\mathbf{j}), -\frac{1}{2}\mathbf{i} + \frac{1}{2}\sqrt{3}\mathbf{j}$; (b) $16(\cos \pi\mathbf{i} + \sin \pi\mathbf{j}), -\mathbf{i}$ **33.** $h = 2, k = 3$ **37.** (a) 135 lb; (b) $17°$ **39.** $29.0°$

47. (a) $\langle 1, -2 \rangle$; (b) $\langle 1, -2 \rangle$

EXERCISES 13.2 (PAGE 966)

1. 10   **3.** -1 **11.** $\frac{1}{10}\sqrt{2}$ **13.** $-\frac{16}{65}$ **15.** $10, -\frac{2}{3}$ **17.** (a) 0; (b) no k **19.** $\dfrac{-240 + \sqrt{85,683}}{407}$ **21.** $-\frac{4}{17}\sqrt{17}\mathbf{i} + \frac{1}{17}\sqrt{17}\mathbf{j}$;

$\frac{4}{17}\sqrt{17}\mathbf{i} - \frac{1}{17}\sqrt{17}\mathbf{j}$ **23.** $\pm\dfrac{1}{\|A\|}(a_2\mathbf{i} - a_1\mathbf{j})$ **25.** (a) $-\dfrac{80}{\sqrt{85}}$; (b) $\dfrac{-112}{17}\mathbf{i} + \dfrac{96}{17}\mathbf{j}$ **27.** $\frac{29}{50}\sqrt{50}$ **29.** (a) 24 ft-lb; (b) $24\sqrt{3}$ ft-lb

31. $(18 - 9\sqrt{3})$ ft-lb ≈ 2.41 ft-lb

EXERCISES 13.3 (PAGE 973)

1. $(-\infty, 0) \cup (0, 4]$ **3.** $[-1, 1]$ **5.** all real numbers not in $(-4, 3)$ **7.** $\frac{4}{3}t, \frac{4}{9}$ **9.** $\frac{1 + \ln t}{te^t(2 + t)}; \frac{(2 + t) - (1 + \ln t)(2 + 4t + t^2)}{t^3 e^{2t}(2 + t)^3}$

11. $-\frac{b}{a}\cot t; -\frac{b}{a^2}\csc^3 t$ **13.** $(y - 1)^2 = x$ **15.** $25x^2 - 9y^2 = 225$ **17.** $x^2 - y^2 = 1, x \geq 1$ **19.** $9x^2 + 16y^2 = 144$

21. $5\sqrt{3}x + 2y = 20$ **23.** $y = 1; x = -1$ **27.** $\frac{dy}{dx} = 0; \frac{d^2y}{dx^2} = -\frac{1}{4a}; \frac{d^3y}{dx^3} = 0$ **31.** $x^{2/3} + y^{2/3} = a^{2/3}$ **35.** $3\pi a^2$ sq units

EXERCISES 13.4 (PAGE 981)

1. $4\mathbf{i} + 4\mathbf{j}$ **3.** $2\mathbf{i}$ **5.** $\mathbf{i}$ **7.** $\mathbf{R}'(t) = 2e^{2t}\mathbf{i} + \frac{1}{t}\mathbf{j}; \mathbf{R}''(t) = 4e^{2t}\mathbf{i} - \frac{1}{t^2}\mathbf{j}$ **9.** $\mathbf{R}'(t) = (1 + t^2)^{-1}\mathbf{i} + 2^t \ln 2\mathbf{j};$

$\mathbf{R}''(t) = -2t(1 + t^2)^{-2}\mathbf{i} + 2^t(\ln 2)^2\mathbf{j}$ **11.** $\mathbf{R}'(t) = -2t(t^2 + 4)^{-2}\mathbf{i} - \frac{5}{2}(1 - 5t)^{-1/2}\mathbf{j};$
$\mathbf{R}''(t) = (6t^2 - 8)(t^2 + 4)^{-3}\mathbf{i} - \frac{25}{4}(1 - 5t)^{-3/2}\mathbf{j}$ **13.** $\mathbf{R}'(t) = 10\cos 2t\mathbf{i} - 4\sec 4t\tan 4t\mathbf{j};$
$\mathbf{R}''(t) = -20\sin 2t\mathbf{i} + (16\sec 4t - 32\sec^3 4t)\mathbf{j}$ **15.** $(2t - 3)(2t^2 - 6t + 5)^{-1/2}$ **21.** $20t$ **23.** $8e^{4t} - 8e^{-4t}$
29. $\ln|\sec t|\mathbf{i} - \ln|t|\mathbf{j} + \mathbf{C}$ **31.** $\frac{1}{3}e^{3t}\mathbf{i} + \ln|t - 1|\mathbf{j} + \mathbf{C}$ **33.** $(t \ln t - t)\mathbf{i} + \frac{1}{3}t^3\mathbf{j} + \mathbf{C}$
35. $\frac{1}{2}(t - \frac{1}{2}\sin 2t - \pi)\mathbf{i} + (t + \frac{1}{2}\sin 2t - \pi)\mathbf{j}$ **37.** $x^2 + y^2 = 1; 0$ **39.** 0

EXERCISES 13.5 (PAGE 988)

1. $1 + \frac{1}{2}\sqrt{2}\ln(1 + \sqrt{2})$ **3.** $2\sqrt{10} + \sqrt{2}\ln(2 + \sqrt{5})$ **5.** $\frac{2}{27}[(40)^{3/2} - (13)^{3/2}]$ **7.** 120 **9.** $\sqrt{2}(e - 1)$ **11.** $\ln(1 + \sqrt{2})$ **13.** $6a$
15. 8π **17.** $a[\ln\cosh 2 + \ln\cosh 1]$ **19.** 5π **21.** $2\pi a$ **23.** 12 **25.** $\frac{1}{2}\sqrt{5}(e^8 - 1)$ **27.** $\frac{1}{2}a(\theta_1 - \frac{3}{2}\sin\frac{2}{3}\theta_1)$ **29.** 200 m

EXERCISES 13.6 (PAGE 992)

1. $\mathbf{T}(t) = \frac{t^2 - 1}{t^2 + 1}\mathbf{i} + \frac{2t}{t^2 + 1}\mathbf{j}; \mathbf{N}(t) = \frac{2t}{t^2 + 1}\mathbf{i} + \frac{1 - t^2}{t^2 + 1}\mathbf{j}$ **3.** $\mathbf{T}(t) = \frac{e^{2t}}{\sqrt{e^{4t} + 1}}\mathbf{i} - \frac{1}{\sqrt{e^{4t} + 1}}\mathbf{j}; \mathbf{N}(t) = \frac{1}{\sqrt{e^{4t} + 1}}\mathbf{i} + \frac{e^{2t}}{\sqrt{e^{4t} + 1}}\mathbf{j}$
5. $\mathbf{T}(t) = -\sin t\mathbf{i} + \cos t\mathbf{j}; \mathbf{N}(t) = -\cos t\mathbf{i} - \sin t\mathbf{j}$ **7.** $\mathbf{T}(t) = -\sin kt\mathbf{i} + \cos kt\mathbf{j}; \mathbf{N}(t) = -\cos kt\mathbf{i} - \sin kt\mathbf{j}$
9. $\mathbf{T}(t) = -(1 + \cot^4 t)^{-1/2}\mathbf{i} + (1 + \tan^4 t)^{-1/2}\mathbf{j}; \mathbf{N}(t) = \frac{-\cos^2 t}{\sqrt{\sin^4 t + \cos^4 t}}\mathbf{i} - \frac{\sin^2 t}{\sqrt{\sin^4 t + \cos^4 t}}\mathbf{j}$

11. $\mathbf{T}(t) = \frac{2t}{\sqrt{4t^2 + e^{2t}}}\mathbf{i} + \frac{e^t}{\sqrt{4t^2 + e^{2t}}}\mathbf{j}; \mathbf{N}(t) = \frac{e^t(1 - t)}{|t - 1|\sqrt{4t^2 + e^{2t}}}\mathbf{i} + \frac{2t(t - 1)}{|t - 1|\sqrt{4t^2 + e^{2t}}}\mathbf{j}$ **13.** $\mathbf{T}(t) = \sin t\mathbf{i} - \cos t\mathbf{j};$

$\mathbf{N}(t) = \cos t\mathbf{i} + \sin t\mathbf{j}$ **15.** $\mathbf{T}(t) = \frac{1}{\sqrt{1 + t^2}}\mathbf{i} + \frac{t}{\sqrt{1 + t^2}}\mathbf{j}; \mathbf{N}(t) = -\frac{t}{\sqrt{1 + t^2}}\mathbf{i} + \frac{1}{\sqrt{1 + t^2}}\mathbf{j}$ **19.** $x = 2 + \cos s, y = 3 + \sin s$

21. $x = \frac{1}{9}[(27s + 8)^{2/3} - 4], y = \frac{1}{27}[(27s + 8)^{2/3} - 4]^{3/2}$ **23.** $x = a\left(\frac{3a - 2s}{3a}\right)^{3/2}, y = a\left(\frac{2s}{3a}\right)^{3/2}$ **25.** $s = \sqrt{2}(e^t - 1)$
27. $\frac{27}{185}\sqrt{37}$

EXERCISES 13.7 (PAGE 1001)

1. $\frac{1}{8}\sqrt{2}; 4\sqrt{2}$ **3.** $\frac{1}{4}\sqrt{2}; 2\sqrt{2}$ **5.** $\frac{\sqrt{2}|1 - t^2|^3}{(1 + 6t^2 + t^4)^{3/2}}; \sqrt{2}; \frac{1}{2}\sqrt{2}$ **7.** $\frac{1}{2}; 2$ **9.** $\frac{1}{4}\sqrt{2}; 2\sqrt{2}$ **11.** $\frac{4}{49}\sqrt{7}; \frac{7}{4}\sqrt{7}$

3. $\frac{2}{289}\sqrt{17}; \frac{17}{2}\sqrt{17}$ **15.** $\frac{(2 - x^2)^{3/2}}{|x|}$ **17.** $\frac{1}{576}(16x^2 + 81y^2)^{3/2}$ **19.** $\frac{2(x + y)^{3/2}}{a^{1/2}}$ **21.** $4|a\sin\frac{1}{2}t|$ **25.** $(3, 9)$ **27.** $(-3, -1)$

29. $x^2 + y^2 + 4x - 6y + 5 = 0$ **31.** $\frac{23}{58}\sqrt{7}$; $\frac{14}{23}\sqrt{7}$ **33.** $\frac{1}{16|a|}$; $16|a|$ **37.** 2; $\frac{1}{2}$; $(0, -\frac{1}{2})$ **39.** $\left(3x + 2p, -\dfrac{y^3}{4p^2}\right)$

41. $\left(\dfrac{a^2 - b^2}{a}\cos^3 t, \dfrac{b^2 - a^2}{b}\sin^3 t\right)$

EXERCISES 13.8 (PAGE 1008)

1. (a) $2ti + j$; (b) $2i$; (c) $\sqrt{37}$; (d) 2 **3.** (a) $-10\sin 2ti + 6\cos 2tj$; (b) $-20\cos 2ti - 12\sin 2tj$; (c) 10; (d) 12
5. (a) $i + \tan tj$; (b) $\sec^2 tj$; (c) $\sqrt{2}$; (d) 2 **7.** (a) $\cos ti + \sec^2 tj$; (b) $-\sin ti + 2\sec^2 t \tan tj$; (c) $\frac{1}{6}\sqrt{91}$; (d) $\frac{1}{18}\sqrt{849}$
9. (a) $2i + 6j$; (b) $2j$; (c) $2\sqrt{10}$; (d) 2 **11.** (a) $2i + 8j$; (b) $2i + 16j$; (c) $2\sqrt{17}$; (d) $2\sqrt{65}$ **13.** (a) $3j$; (b) $-4i$; (c) 3; (d) 4
15. (a) $i + \sqrt{3}j$; (b) $-\sqrt{3}i + j$; (c) 2; (d) 2 **17.** $\dfrac{2t-3}{t-1}i + \dfrac{4 - 2t - t^2}{2}j$ **19.** $(e^{-t} + 3t - 1)i + (\frac{1}{2}e^{2t} + \frac{5}{2})j$ **21.** (a) $\dfrac{390{,}625}{2}$ ft;

(b) $\dfrac{390{,}625}{8}$ ft; (c) $1250\sqrt{2}i - 1250\sqrt{2}j$ **23.** $(25 + \sqrt{631})$ sec; $(20{,}000\sqrt{3} + 800\sqrt{1893})$ ft **25.** 283 m/sec **27.** 40° 8′

29. 52.49 m **31.** $\dfrac{1}{2g}v_0^2 \sin^2 \alpha$

EXERCISES 13.9 (PAGE 1013)

1. $V(t) = i + 2tj$; $A(t) = 2j$; $A_T(t) = \dfrac{4t}{\sqrt{1 + 4t^2}}$; $A_N(t) = \dfrac{2}{\sqrt{1 + 4t^2}}$ **3.** $V(t) = t\cos ti + t\sin tj$;

$A(t) = (\cos t - t\sin t)i + (\sin t + t\cos t)j$; $A_T(t) = 1$; $A_N(t) = t$ **5.** $V(0) = -i + j$; $A(0) = i + j$; $A_T(0) = 0$; $A_N(0) = \sqrt{2}$

7. $V(\frac{1}{4}\pi) = \frac{3}{4}\sqrt{2}i - \frac{3}{4}\sqrt{2}j$; $A(\frac{1}{4}\pi) = \frac{3}{4}\sqrt{2}i + \frac{3}{4}\sqrt{2}j$; $A_T(\frac{1}{4}\pi) = 0$; $A_N(\frac{1}{4}\pi) = \frac{3}{2}$ **9.** $\|V(t)\| = t\sqrt{4 + 9t^2}$; $A_T(t) = \dfrac{4 + 18t^2}{\sqrt{4 + 9t^2}}$;

$A_N(t) = \dfrac{6t}{\sqrt{4 + 9t^2}}$; $K(t) = \dfrac{6}{t(4 + 9t^2)^{3/2}}$ **11.** $V(t) = 2i + 2tj$; $A(t) = 2j$; $T(t) = \dfrac{1}{\sqrt{1 + t^2}}i + \dfrac{t}{\sqrt{1 + t^2}}j$;

$N(t) = \dfrac{-t}{\sqrt{1 + t^2}}i + \dfrac{1}{\sqrt{1 + t^2}}j$; $\|V(t)\| = 2\sqrt{1 + t^2}$; $A_T(t) = \dfrac{2t}{\sqrt{1 + t^2}}$; $A_N(t) = \dfrac{2}{\sqrt{1 + t^2}}$; $K(t) = \dfrac{1}{2(1 + t^2)^{3/2}}$;

$V(2) = 2i + 4j$; $T(2) = \dfrac{1}{\sqrt{5}}i + \dfrac{2}{\sqrt{5}}j$; $A(2) = 2j$; $N(2) = \dfrac{-2}{\sqrt{5}}i + \dfrac{1}{\sqrt{5}}j$; $\|V(2)\| = 2\sqrt{5}$; $A_T(2) = \dfrac{4}{\sqrt{5}}$; $A_N(2) = \dfrac{2}{\sqrt{5}}$;

$K(2) = \dfrac{1}{10\sqrt{5}}$ **13.** $V(t) = -15\sin 3ti + 15\cos 3tj$; $A(t) = -45\cos 3ti - 45\sin 3tj$; $T(t) = -\sin 3ti + \cos 3tj$;

$N(t) = -\cos 3ti - \sin 3tj$; $\|V(t)\| = 15$; $A_T(t) = 0$; $A_N(t) = 45$; $K(t) = \frac{1}{5}$; $V(\frac{1}{3}\pi) = -15j$; $A(\frac{1}{3}\pi) = 45i$; $T(\frac{1}{3}\pi) = -j$;

$N(\frac{1}{3}\pi) = i$; $\|V(\frac{1}{3}\pi)\| = 15$ **15.** $V(t) = e^ti - e^{-t}j$; $A(t) = e^ti + e^{-t}j$; $T(t) = \dfrac{e^{2t}}{\sqrt{e^{4t} + 1}}i - \dfrac{1}{\sqrt{e^{4t} + 1}}j$;

$N(t) = \dfrac{1}{\sqrt{e^{4t} + 1}}i + \dfrac{e^{2t}}{\sqrt{e^{4t} + 1}}j$; $\|V(t)\| = \dfrac{\sqrt{e^{4t} + 1}}{e^t}$; $A_T(t) = \dfrac{e^{4t} - 1}{e^t\sqrt{e^{4t} + 1}}$; $A_N(t) = \dfrac{2e^t}{\sqrt{e^{4t} + 1}}$; $K(t) = \dfrac{2e^{3t}}{(e^{4t} + 1)^{3/2}}$;

$V(0) = i - j$; $A(0) = i + j$; $T(0) = \dfrac{1}{\sqrt{2}}i - \dfrac{1}{\sqrt{2}}j$; $N(0) = \dfrac{1}{\sqrt{2}}i + \dfrac{1}{\sqrt{2}}j$; $\|V(0)\| = \sqrt{2}$; $A_T(0) = 0$; $A_N(0) = \sqrt{2}$; $K(0) = \dfrac{1}{\sqrt{2}}$

17. (a) $i + 4j$; (b) $i + 8j$; (c) $8j$; (d) $\frac{64}{65}\sqrt{65}$; (e) $\frac{8}{65}\sqrt{65}$ **19.** $R = 2i + 4j$; if k is constant speed, $V = \dfrac{k}{\sqrt{2}}i + \dfrac{k}{\sqrt{2}}j$;

$A = \dfrac{k^2}{16}i - \dfrac{k^2}{16}j$; $T = \dfrac{1}{\sqrt{2}}i + \dfrac{1}{\sqrt{2}}j$; $N = \dfrac{1}{\sqrt{2}}i - \dfrac{1}{\sqrt{2}}j$; $A_T = 0$; $A_N = \dfrac{k^2\sqrt{2}}{16}$

REVIEW EXERCISES FOR CHAPTER 13 (PAGE 1014)

1. $-25\mathbf{i} + 63\mathbf{j}$ **3.** $\sqrt{4594}$ **5.** $15\sqrt{2} - 14\sqrt{13}$ **7.** 92 **9.** $\dfrac{9}{\sqrt{106}}\mathbf{i} - \dfrac{5}{\sqrt{106}}\mathbf{j}$ **11.** $h = -\frac{1}{2};\ k = \frac{1}{2}$ **13.** $-\frac{19}{5}\sqrt{2}$ **15.** $-\frac{19}{25}\mathbf{i} - \frac{133}{25}\mathbf{j}$

17. $-\frac{19}{26}\sqrt{13}$ **19.** (a) all real numbers in $[0, +\infty)$ except 1; (b) $\frac{1}{2}\mathbf{i} + \frac{1}{2}\mathbf{j}$; (c) $\dfrac{-1}{(t+1)^2}\mathbf{i} + \dfrac{2t^{1/2} - t - 1}{2t^{1/2}(t-1)^2}\mathbf{j}$ **21.** $\dfrac{dy}{dx} = \dfrac{1}{6t}$;

$\dfrac{d^2y}{dx^2} = -\dfrac{1}{108t^3}$ **23.** $x = 12;\ y = 16;\ y = -16$ **25.** $-\dfrac{4(t^{12} + t^{10}) + 144(t^2 - 1)^3}{t^9(t^2 - 1)^3}$ **27.** $\frac{3}{2}\sqrt{37} + \frac{1}{4}\ln(6 + \sqrt{37})$ **33.** $|at|$

35. $\frac{2}{75};\ \frac{75}{2}$ **37.** (a) $9y = 12x - x^2$; (b) $\mathbf{V}(t) = 3\mathbf{i} + (4 - 2t)\mathbf{j},\ \mathbf{A}(t) = -2\mathbf{j}$; (c) $\mathbf{V}(1) = 3\mathbf{i} + 2\mathbf{j},\ \mathbf{A}(1) = -2\mathbf{j}$

39. $\dfrac{dy}{dx} = -\tan t;\ \dfrac{d^2y}{dx^2} = \dfrac{1}{3a}\sec^4 t \csc t$ **41.** $\mathbf{T}(t) = \dfrac{e^t - e^{-t}}{e^t + e^{-t}}\mathbf{i} + \dfrac{2}{e^t + e^{-t}}\mathbf{j};\ \mathbf{N}(t) = \dfrac{2}{e^t + e^{-t}}\mathbf{i} - \dfrac{e^t - e^{-t}}{e^t + e^{-t}}\mathbf{j}$

43. $x = 2(3s + 17\sqrt{17})^{2/3} - 34;\ y = \frac{1}{3}[(3s + 17\sqrt{17})^{2/3} - 16]^{3/2}$ **45.** $h = \dfrac{1}{2g}(v_0 \sin \alpha)^2$ **51.** $\mathbf{V}(t) = 2\sinh 2t\,\mathbf{i} + 2\cosh 2t\,\mathbf{j}$;

$\mathbf{A}(t) = 4\cosh 2t\,\mathbf{i} + 4\sinh 2t\,\mathbf{j};\ \|\mathbf{V}(t)\| = 2\sqrt{\cosh 4t};\ A_T(t) = \dfrac{4\sinh 4t}{\sqrt{\cosh 4t}};\ A_N(t) = \dfrac{4}{\sqrt{\cosh 4t}}$ **53.** $A_T = \dfrac{4t - 8}{\sqrt{4t^2 - 16t + 25}}$;

$A_N = \dfrac{6}{\sqrt{4t^2 - 16t + 25}}$

EXERCISES 14.1 (PAGE 1023)

1. (b) $(7, 2, 0), (0, 0, 3), (0, 2, 0), (0, 2, 3), (7, 0, 3), (7, 0, 0)$; (c) $\sqrt{62}$
3. (b) $(2, 1, 2), (-1, 3, 2), (-1, 1, 5), (2, 3, 2), (-1, 3, 5), (2, 1, 5)$; (c) $\sqrt{22}$ **5.** (b) $(3, -1, 0)$,
$(3, 3, 0), (1, 3, 0), (1, 3, 5), (1, -1, 5), (3, -1, 5)$; (c) $3\sqrt{5}$ **7.** (a) 3; (b) $(2, 5, \frac{5}{2})$ **9.** (a) $\frac{13}{2}$; (b) $(\frac{5}{4}, -1, 2)$ **11.** (a) $7\sqrt{2}$;
(b) $(-1, \frac{9}{2}, -\frac{1}{2})$ **13.** $(\pm 4\sqrt{6}, 4, 2)$ **17.** (a) $|\overline{AB}| = 9\sqrt{2};\ |\overline{AC}| = 2\sqrt{62};\ |\overline{BC}| = \sqrt{62}$; (b) midpoint of AB: $(\frac{1}{2}, 1, \frac{3}{2})$;
midpoint of AC: $(-1, 2, 5)$; midpoint of BC: $(-\frac{5}{2}, 8, \frac{7}{2})$ **21.** sphere with center at $(4, -2, -1)$ and $r = 5$
23. the point $(0, 0, 3)$ **25.** the empty set **27.** $x^2 + (y - 1)^2 + (z + 4)^2 = 9$

EXERCISES 14.2 (PAGE 1034)

1. (a) $\langle 21, -13, -2 \rangle$; (b) $\langle -25, -26, 5 \rangle$; (c) $7\sqrt{59} - 5\sqrt{41}$; (d) $\sqrt{1326}$ **3.** (a) $\langle -19, -16, -1 \rangle$;
(b) $\langle -6\sqrt{91}, -8\sqrt{91}, -2\sqrt{91} \rangle$ **5.** $a = 0,\ b = 0$ **7.** $\dfrac{4}{\sqrt{89}};\ \dfrac{3}{\sqrt{89}};\ \dfrac{8}{\sqrt{89}}$ **9.** $-\dfrac{6}{\sqrt{86}};\ -\dfrac{1}{\sqrt{86}};\ -\dfrac{7}{\sqrt{86}}$ **11.** $(\frac{13}{3}, 0, -\frac{4}{3})$

13. $(\frac{17}{3}, \frac{4}{3}, -6)$ **15.** (a) $7(-\frac{6}{7}\mathbf{i} + \frac{2}{7}\mathbf{j} + \frac{3}{7}\mathbf{k})$; (b) $\sqrt{14}\left(-\dfrac{2}{\sqrt{14}}\mathbf{i} + \dfrac{1}{\sqrt{14}}\mathbf{j} - \dfrac{3}{\sqrt{14}}\mathbf{k}\right)$ **17.** (a) $\langle \frac{1}{9}, \frac{8}{9}, \frac{4}{9} \rangle$; (b) $\left\langle -\dfrac{1}{\sqrt{3}}, \dfrac{1}{\sqrt{3}}, \dfrac{1}{\sqrt{3}} \right\rangle$

21. (a) -44; (b) -468; (c) -31; (d) $\langle -84, 198, 124 \rangle$ **23.** (a) $-\frac{1}{3}\sqrt{5}$; (b) -3; (c) $\langle 2, 1, -2 \rangle$ **25.** (a) $-\frac{13}{9}\sqrt{6}$;
(b) $\langle -\frac{26}{27}, -\frac{91}{27}, \frac{13}{27} \rangle$ **29.** $\frac{1}{11}\sqrt{4422}$ **33.** $\frac{7}{2}\sqrt{3}$ sq units **35.** $-\frac{43}{30}\sqrt{6}$ **37.** 25 ft-lb **39.** $\frac{5}{3}\sqrt{6}$ ft-lb **41.** $\frac{2}{5}$
45. $\cos^{-1}\dfrac{1}{\sqrt{3}}$ or $\cos^{-1}\left(-\dfrac{1}{\sqrt{3}}\right)$ **49.** $r = 5,\ s = 9,\ t = -12$

EXERCISES 14.3 (PAGE 1041)

1. $x + 2y - 3z + 1 = 0$ **3.** $y - z + 3 = 0$ **5.** $x - 3y - 4z - 3 = 0$ **7.** $3x + 2y + 6z = 23$ **9.** $\langle \frac{2}{3}, -\frac{1}{3}, \frac{2}{3} \rangle$; $\langle -\frac{2}{3}, \frac{1}{3}, -\frac{2}{3} \rangle$
11. $\langle \frac{4}{13}, \frac{3}{13}, -\frac{12}{13} \rangle$; $\langle -\frac{4}{13}, -\frac{3}{13}, \frac{12}{13} \rangle$ **13.** $\left\langle \dfrac{3}{\sqrt{13}}, 0, \dfrac{2}{\sqrt{13}} \right\rangle$; $\left\langle -\dfrac{3}{\sqrt{13}}, 0, -\dfrac{2}{\sqrt{13}} \right\rangle$ **15.** $5x - 3y + 7z + 14 = 0$
17. $2x - y - z + 1 = 0$ **19.** $4y - 3z - 1 = 0$ and $z = 1$ **21.** $\frac{8}{21}$ **23.** $-\frac{16}{45}$ **25.** $\frac{16}{15}\sqrt{6}$ **27.** $\frac{3}{2}$

EXERCISES 14.4 (PAGE 1047)

1. $x = 1 + 4t,\ y = 2 - 3t,\ z = 1;\ \dfrac{x-1}{4} = \dfrac{y-2}{-3},\ z = 1$ **3.** $x = 13t,\ y = -12t,\ z = -8t;\ \dfrac{x}{13} = \dfrac{y}{-12} = \dfrac{z}{-8}$

5. $x = 2t - 2,\ y = -16t;\ z = 13t + 3;\ \dfrac{x+2}{2} = \dfrac{y}{-16} = \dfrac{z-3}{13}$ **7.** $x = 4 + t,\ y = -5 + 3t,\ z = 20 - 6t;$

$\dfrac{x-4}{1} = \dfrac{y+5}{3} = \dfrac{z-20}{-6}$ **9.** $\dfrac{x-\frac{1}{2}}{2} = \dfrac{y}{7} = \dfrac{z-\frac{10}{7}}{13}$ **13.** $8x - y - 66 = 0;\ 13x - 5z - 102 = 0;\ 13y - 40z + 42 = 0$

15. $4x + y + 3 = 0;\ 3x - z + 4 = 0;\ 3y + 4z - 7 = 0$ **17.** $\frac{5}{18}\sqrt{6}$ **19.** $4x + 7y - 3z + 7 = 0$ **21.** $4x + 2y - 3z + 5 = 0$

23. $(\frac{5}{3}, -\frac{17}{6}, \frac{5}{12})$ **25.** $\dfrac{x-3}{1} = \dfrac{y-6}{2} = \dfrac{z-4}{1}$ **27.** $\frac{2}{3}\sqrt{70}$ **31.** $x = x_0,\ y = y_0$

EXERCISES 14.5 (PAGE 1059)

1. $\langle 7, 13, -11 \rangle$ **3.** -490 **11.** $\langle 9, -1, -23 \rangle$ **15.** $\frac{2}{3}\sqrt{2}$ **17.** $\sqrt{89}$ sq units **19.** $9\sqrt{29}$ sq units **21.** $5x - 2y + 7z = 0$

23. $x + 2y + z - 2 = 0$ **25.** $\dfrac{1}{\sqrt{3}}(\mathbf{i} + \mathbf{j} - \mathbf{k})$ **27.** $\dfrac{1}{\sqrt{6}}\mathbf{i} + \dfrac{2}{\sqrt{6}}\mathbf{j} + \dfrac{1}{\sqrt{6}}\mathbf{k}$ **29.** 20 cu units **31.** $\dfrac{38}{3\sqrt{78}}$

EXERCISES 14.6 (PAGE 1064)

13. $x^2 + z^2 = 4y$ **15.** $x^2 + 4y^2 + 4z^2 = 16$ **17.** $y^2 = 9x^2 + 9z^2$ **19.** $y^2 + z^2 = \sin^2 x$ **21.** $x^2 + z^2 = 16;\ x$ axis
23. $x^2 - z^2 = 4;\ z$ axis **25.** $z = \sqrt{|y|};\ y$ axis **27.** $y^2 = 9x^2;\ y$ axis

EXERCISES 14.7 (PAGE 1071)

1. elliptic hyperboloid of one sheet **3.** hyperbolic paraboloid **5.** hyperbolic cylinder **7.** ellipsoid
9. elliptic hyperboloid of one sheet **11.** elliptic cone **13.** elliptic paraboloid **15.** hyperbolic paraboloid
17. elliptic hyperboloid of two sheets **19.** (a) $1 < |k| < \sqrt{2}$; (b) $|k| < 1$ **21.** vertex: $(1, -\frac{1}{3}, 0)$; focus: $(1, 0, 0)$

25. 8π cu units **27.** $\dfrac{abh^2}{2c}\,\pi$ cu units

EXERCISES 14.8 (PAGE 1079)

1. $\dfrac{1}{\sqrt{4t^2 + 5}}(\mathbf{i} - 2t\mathbf{j} - 2\mathbf{k})$ **3.** $\mathbf{T}(t) = \frac{1}{3}\sqrt{3}[(\cos t - \sin t)\mathbf{i} + (\cos t + \sin t)\mathbf{j} + \mathbf{k}]$

5. $\dfrac{1}{\sqrt{4t^2 + 29}}[2(\cos t - t \sin t)\mathbf{i} + 5\mathbf{j} + 2(\sin t + t \cos t)\mathbf{k}]$ **7.** $\sqrt{21} + \frac{3}{2} + \frac{5}{4}\ln(4 + \sqrt{21})$ **9.** $\sqrt{3}(e^3 - 1)$ **11.** 13

15. $\mathbf{R}(t) = t\mathbf{i} + e^t\mathbf{j} + te^t\mathbf{k}$ **17.** $-\frac{2}{3}$ **19.** $\mathbf{T}(1) = \frac{1}{14}\sqrt{14}\mathbf{i} + \frac{1}{7}\sqrt{14}\mathbf{j} + \frac{3}{14}\sqrt{14}\mathbf{k}$; $\mathbf{N}(1) = -\frac{11}{266}\sqrt{266}\mathbf{i} - \frac{4}{133}\sqrt{266}\mathbf{j} + \frac{9}{266}\sqrt{266}\mathbf{k}$;

$\mathbf{B}(1) = \dfrac{\sqrt{19}}{19}(3\mathbf{i} - 3\mathbf{j} + \mathbf{k})$; $K(1) = \frac{1}{98}\sqrt{266}$ **21.** $\mathbf{T}(-1) = \frac{1}{3}(\mathbf{i} + 2\mathbf{j} - 2\mathbf{k})$; $\mathbf{N}(-1) = \frac{2}{15}\sqrt{5}\mathbf{i} - \frac{1}{3}\sqrt{5}\mathbf{j} - \frac{4}{15}\sqrt{5}\mathbf{k}$;

$\mathbf{B}(-1) = -\frac{2}{3}\sqrt{5}\mathbf{i} - \frac{1}{3}\sqrt{5}\mathbf{k}$; $K(-1) = \frac{2}{15}\sqrt{5}$ **23.** $\mathbf{T}(0) = \frac{1}{3}\sqrt{3}(\mathbf{i} + \mathbf{j} + \mathbf{k})$; $\mathbf{N}(0) = -\frac{1}{2}\sqrt{2}(\mathbf{i} - \mathbf{j})$; $\mathbf{B}(0) = -\frac{1}{6}\sqrt{6}(\mathbf{i} + \mathbf{j} - 2\mathbf{k})$;
$K(0) = \frac{1}{3}\sqrt{2}$ **25.** $\mathbf{V}(\frac{1}{2}\pi) = -a\mathbf{i} + \mathbf{k}$; $\mathbf{A}(\frac{1}{2}\pi) = -a\mathbf{j}$; $\|\mathbf{V}(\frac{1}{2}\pi)\| = \sqrt{a^2 + 1}$ **27.** $\mathbf{V}(1) = 2e^2\mathbf{i} - 2e^{-2}\mathbf{j} + 3e^2\mathbf{k}$;
$\mathbf{A}(1) = 4e^2\mathbf{i} + 4e^{-2}\mathbf{j} + 8e^2\mathbf{k}$; $\|\mathbf{V}(1)\| = \sqrt{13e^4 + 4e^{-4}}$ **31.** 2

EXERCISES 14.9 (PAGE 1084)

1. (a) $(0, 3, 5)$; (b) $(-\frac{7}{2}, \frac{7}{2}\sqrt{3}, -4)$; (c) $(\cos 1, \sin 1, 1)$ **3.** (a) $(\sqrt{6}, \sqrt{2}, 2\sqrt{2})$; (b) $(0, 2\sqrt{3}, 2)$; (c) $(\frac{1}{2}\sqrt{3}, \frac{3}{2}, -\sqrt{3})$
5. (a) $(2, \frac{2}{3}\pi, -2\sqrt{3})$; (b) $(0, \frac{3}{4}\pi, -\sqrt{2})$; (c) $(\sqrt{6}, \frac{1}{3}\pi, \sqrt{6})$ **7.** ellipsoid; $r^2 + 4z^2 = 16$ **9.** elliptic paraboloid; $r^2 = 3z$
11. elliptic cone; $r^2 \cos 2\theta = 3z^2$ **13.** sphere; $\rho^2 - 9\rho \cos \phi = 0$ **15.** right circular cylinder; $\rho^2 \sin^2 \phi = 9$ **17.** sphere;
$\rho = 8 \sin \phi \cos \theta$ **19.** (a) right circular cylinder; $x^2 + y^2 = 16$; (b) plane through z axis; $y = x$ **21.** $x^2 - y^2 = z^3$
23. (a) sphere; $x^2 + y^2 + z^2 = 81$; (b) plane through z axis; $x = y$; (c) cone with vertex at origin; $z = \sqrt{x^2 + y^2}$
25. right circular cylinder; $x^2 + y^2 = 36$ **27.** $x\sqrt{x^2 + y^2 + z^2} = 2y$ **31.** (b) $2\pi\sqrt{a^2 + 1}$

REVIEW EXERCISES FOR CHAPTER 14 (PAGE 1085)

1. a point on the x axis in R, a line parallel to the y axis in R^2, a plane parallel to the yz plane in R^3 **3.** the x axis
5. the circle in the xz plane with center at the origin and radius 2 **7.** the plane perpendicular to the xy plane and
intersecting the xy plane in the line $y = x$ **9.** the solid of revolution generated by revolving $y^2 = 9z$ about the z axis
11. the solid of revolution generated by revolving $y = x$ about the x axis **13.** $\mathbf{i} + 26\mathbf{j} - 16\mathbf{k}$ **15.** -3 **17.** $7\sqrt{1270}$
19. $-\frac{1}{3}\sqrt{21}$ **21.** $\frac{1}{3}(\mathbf{i} + 2\mathbf{j} - 2\mathbf{k})$ **23.** $\langle 60, -40, 80 \rangle$ **25.** 16 **27.** 295 **29.** $(-3, \sqrt{167}, 1), (-3, -\sqrt{167}, 1)$
31. $(x + 2)^2 + (y + 1)^2 + (z - 3)^2 = 17$ **33.** $z^2 = e^{4y}$ or $x^2 = e^{4y}$; the y axis **35.** 3

37. (a) $\cos \alpha = -\dfrac{7}{\sqrt{78}}$, $\cos \beta = -\dfrac{5}{\sqrt{78}}$, $\cos \gamma = -\dfrac{2}{\sqrt{78}}$; (b) $-\dfrac{7}{\sqrt{78}}\mathbf{i} - \dfrac{5}{\sqrt{78}}\mathbf{j} - \dfrac{2}{\sqrt{78}}\mathbf{k}$ **39.** (a) $-\frac{4}{5}\sqrt{3}$; (b) $-\frac{28}{25}\mathbf{i} + \frac{4}{25}\mathbf{j} - \frac{4}{5}\mathbf{k}$

41. $x - 6y - 10z + 23 = 0$ **43.** $\frac{47}{10}\sqrt{2}$ **45.** 3 **47.** $\frac{1}{3}\sqrt{3}$ **49.** $\dfrac{x}{4} = \dfrac{y}{-3} = \dfrac{z}{1}$, $x = 4t, y = -3t, z = t$ **53.** $\frac{54}{23}\pi$ sq units

55. 24 cu units **57.** $\frac{25}{4}[\sqrt{2} + \ln(\sqrt{2} + 1)]$ **59.** $\mathbf{V}(\frac{1}{2}\pi) = -\frac{1}{2}\pi\mathbf{i} + \mathbf{j} + \mathbf{k}$; $\mathbf{A}(\frac{1}{2}\pi) = -2\mathbf{i} - \frac{1}{2}\pi\mathbf{j}$; $\|\mathbf{V}(\frac{1}{2}\pi)\| = \frac{1}{2}\sqrt{8 + \pi^2}$

61. $\mathbf{T}(\frac{1}{2}\pi) = -\dfrac{3}{\sqrt{13}}\mathbf{i} - \dfrac{2}{\sqrt{13}}\mathbf{j}$; $\mathbf{N}(\frac{1}{2}\pi) = -\mathbf{k}$; $\mathbf{B}(\frac{1}{2}\pi) = \dfrac{2}{\sqrt{13}}\mathbf{i} - \dfrac{3}{\sqrt{13}}\mathbf{j}$ **63.** $(\frac{3}{2}\sqrt{3}, \pi, \frac{3}{2})$
65. (a) $z = r^2(1 + \sin 2\theta) + 1$; (b) $r^2(25 \cos^2 \theta + 4 \sin^2\theta) = 100$

EXERCISES 15.1 (PAGE 1096)

1. (a) $-\frac{1}{4}$; (b) 5; (c) $\dfrac{x + y}{x - y + 2}$; (d) 0; (e) $\{(x, y)|x \neq y\}$ **3.** (a) 1; (b) $\frac{1}{2}\sqrt{2}$; (c) $\frac{1}{2}\sqrt{16 - x^2 - y^2 - z^2}$; (d) $4x + 4y + 8$;
(e) $\{(x, y, z)|x^2 + y^2 + z^2 \leq 4\}$ **5.** $\{(x, y)|x^2 + y^2 \neq 1\}$ **7.** $\{(x, y)|x^2 + y^2 \leq 1\}$ **9.** $\{(x, y)|x^2 + y^2 \leq 5$ and $x \neq 0\}$
11. $\{(x, y)|x^2 - y^2 \geq 25\}$ **13.** $\{(x, y)|x^2 + y^2 < 25\}$ **15.** $\{(x, y)|x \neq y\}$ **17.** $\{(x, y)|y \neq 0\}$ **19.** $\{(x, y)|xy > 1\}$
21. domain: $\{(x, y, z)|z \geq 2\}$; range: $(-\infty, +\infty)$ **23.** domain: $\{(x, y, z)||x| \leq 1$ and $|y| \leq 1\}$; range: $(-\pi, 2\pi)$
25. domain: R^2; range: $[0, +\infty)$ **27.** domain: R^2; range: $(-\infty, 16]$ **29.** domain: $\{(x, y)|x + y^2 \leq 10\}$; range: $[0, +\infty)$
37. $h(x, y) = \sin^{-1}\sqrt{1 - x^2 - y^2}$; domain: $\{(x, y)|x^2 + y^2 \leq 1\}$ **39.** (a) 2; (b) 6; (c) $\sqrt{x - y^2}$; (d) $|x - y|$; (e) $|x - y|$

EXERCISES 15.2 (PAGE 1110)

1. 0 **3.** -4 **5.** 2 **7.** 0 **9.** $\delta = \frac{1}{4}\epsilon$ **11.** $\delta = \frac{1}{4}\epsilon$ **13.** $\delta = \min(1, \frac{1}{6}\epsilon)$ **15.** $\delta = \min(1, \frac{1}{8}\epsilon)$ **23.** 0 **25.** 0 **27.** $\frac{1}{4}\pi$ **29.** $\frac{1}{2}$
31. limit exists and is 0; take $\delta = \sqrt{\epsilon}$ **33.** limit does not exist **35.** limit exists and is 0; take $\delta = \frac{1}{2}\epsilon$ **39.** $\frac{1}{2}$ **41.** 3
47. 0; take $\delta = \frac{1}{3}\epsilon$

EXERCISES 15.3 (PAGE 1116)

1. continuous at every point (x, y) in R^2 that is not on the line $y = 1$ **3.** continuous at every point (x, y) in R^2 that is not
on the y axis **5.** continuous at every point (x, y) in R^2 that is not on the line $y = 2x$ **7.** continuous at every point (x, y)

in R^2 that is interior to the circle $x^2 + y^2 = 25$ **9.** continuous at every point in R^2 **11.** continuous at every point $(x, y) \neq (0, 0)$ in R^2 **13.** continuous at every point in R^2 **17.** all points (x, y) in R^2 that are interior to the circle $x^2 + y^2 = 16$ **19.** all points (x, y) in R^2 that are exterior to the ellipse $4x^2 + 9y^2 = 36$ **21.** all points (x, y) in R^2 for which $|xy| > 1$ **23.** all points (x, y) in R^2 in the third quadrant for which $|x + y| < 1$ **25.** all points in R^2 **27.** removable; $f(0, 0) = 0$ **29.** essential **31.** removable; $f(0, 0) = 0$ **35.** continuous at every point (x, y, z) in R^3 that is exterior to the sphere $x^2 + y^2 + z^2 = 1$ **37.** continuous at all points in R^3

EXERCISES 15.4 (PAGE 1125)

1. 6 **3.** $3x - 2y$ **5.** $\dfrac{x}{\sqrt{x^2 + y^2}}$ **7.** $x^2 - 6xy + 2z$ **9.** $xy + yt + zt$ **11.** 4 **13.** $\dfrac{x}{\sqrt{x^2 + y^2}}$ **15.** $-2 \sin 3\theta \sin 2\phi$

17. $\dfrac{e^{y/x}}{xy}\left(y \ln \dfrac{x^2}{y} - x\right)$ **19.** $\dfrac{-z}{(x^2 + y^2 + z^2)^{3/2}}$ **21.** $4xy + \dfrac{1}{z}$ **23.** $xze^{xyz} + \dfrac{3xz^2}{z^4 + 9x^2y^2}$ **25.** (a) -1; (b) 12

27. $-\ln \sin x$; $\ln \sin y$ **31.** (a) 1; (b) 1 **33.** (a) -2; (b) 0 **35.** 4 **37.** -4 deg/cm; -8 deg/cm **39.** (a) $\dfrac{100}{i^2}\left[\dfrac{9i + 1}{(1 + i)^9} - 1\right]$;

(b) $\dfrac{1}{0.0036}\left[\dfrac{1.54}{(1.06)^9} - 1\right] \approx -24.4$; (c) $\dfrac{5000 \ln 1.06}{3(1.06)^t}$; (d) $-\dfrac{5000 \ln 1.06}{3(1.06)^8} \approx -61$ **41.** 0.0943 m²/kg; 6.42 m²/m

EXERCISES 15.5 (PAGE 1136)

1. (a) $3(\Delta x)^2 + 2(\Delta x)(\Delta y) - (\Delta y)^2 + 14 \Delta x - 6 \Delta y$; (b) 0.5411; (c) $14 \Delta x - 6 \Delta y$; (d) 0.54
3. (a) $(2 + \Delta x)(-4 + \Delta y)e^{(2+\Delta x)(-4+\Delta y)} + 8e^{-8}$; (b) -0.0026; (c) $28e^{-8} \Delta x - 14e^{-8} \Delta y$; (d) -0.0019
5. (a) $\Delta x + 4 \Delta y + (\Delta x)(\Delta y) + \ln(1 + \Delta y) + \ln(5 + \Delta z) - \ln 5$; (b) 0.2141; (c) $\Delta x + 5 \Delta y + \frac{1}{5} \Delta z$; (d) 0.214
7. $(12x^2 - y^2) \, dx + (3 - 2xy) \, dy$ **9.** $(\cos y - y \cos x) \, dx + (-x \sin y - \sin x) \, dy$

11. $\dfrac{2x \, dx + 2y \, dy + 2z \, dz}{x^2 + y^2 + z^2}$ **13.** $\tan^{-1} z \, dx - \dfrac{2y}{z} \, dy + \left(\dfrac{x}{1 + z^2} + \dfrac{y^2}{z^2}\right) dz$

15. (a) $2(x_0 y_0 - y_0) \Delta x + (x_0^2 - 2x_0) \Delta y + (y_0 \Delta x + \Delta x \Delta y) \Delta x + 2(x_0 \Delta x - \Delta x) \Delta y$; (b) $\epsilon_1 = y_0 \Delta x + \Delta x \Delta y$;

$\epsilon_2 = 2(x_0 \Delta x - \Delta x)$ **17.** (a) $\dfrac{2x_0 y_0 \Delta x + y_0(\Delta x)^2 - x_0^2 \Delta y}{y_0^2 + y_0 \Delta y}$; (b) $\epsilon_1 = \dfrac{y_0^2 \Delta x}{y_0^3 + y_0^2 \Delta y}$; $\epsilon_2 = \dfrac{x_0^2 \Delta y - 2x_0 y_0 \Delta x}{y_0^3 + y_0^2 \Delta y}$

31. (a) $(y_0 - z_0) \Delta x + x_0 \Delta y + (2z_0 - x_0) \Delta z - (\Delta z)(\Delta x) + (\Delta x)(\Delta y) + (\Delta z)(\Delta z)$; (b) $\epsilon_1 = -\Delta z, \epsilon_2 = \Delta x, \epsilon_3 = \Delta z$
35. 7.36 m³ **37.** 0.14 cm; 1.4 percent **39.** $\frac{13}{1600}$; 0.325 percent **41.** $2400

43. $D_1 f(x, y) = \begin{cases} 2x \sin \dfrac{1}{\sqrt{x^2 + y^2}} - \dfrac{x}{\sqrt{x^2 + y^2}} \cos \dfrac{1}{\sqrt{x^2 + y^2}} & \text{if } (x, y) \neq (0, 0) \\ 0 & \text{if } xy = (0, 0) \end{cases}$;

$D_2 f(x, y) = \begin{cases} 2y \sin \dfrac{1}{\sqrt{x^2 + y^2}} - \dfrac{y}{\sqrt{x^2 + y^2}} \cos \dfrac{1}{\sqrt{x^2 + y^2}} & \text{if } (x, y) \neq (0, 0) \\ 0 & \text{if } (x, y) = (0, 0) \end{cases}$

EXERCISES 15.6 (PAGE 1145)

1. $\dfrac{\partial u}{\partial r}$: (a) $6x - 2y$; (b) $16r - 10s$; $\dfrac{\partial u}{\partial s}$: (a) $-2x - 4y$; (b) $-10r - 6s$ **3.** $\dfrac{\partial u}{\partial r}$: (a) $13x - 2y + 5$; (b) $24r - 41s + 5$;

$\dfrac{\partial u}{\partial s}$: (a) $-17x - 7y - 10$; (b) $-41r + 44s - 16$ **5.** $\dfrac{\partial u}{\partial r}$: (a) $\dfrac{2e^{y/x}}{x^2}(2x \sin t - y \cos t)$; (b) 0;

$\dfrac{\partial u}{\partial t}$: (a) $\dfrac{2re^{y/x}}{x^2}(y \sin t + 2x \cos t)$; (b) $2e^{2 \tan t} \sec^2 t$ **7.** $\dfrac{\partial u}{\partial r} = 2r(2x + y) + 3x$; $\dfrac{\partial u}{\partial s} = 2s(2x + y) - 2x$

9. $\dfrac{\partial u}{\partial r} = \dfrac{6re^s + s \cos rs}{\sqrt{1 - (3x+y)^2}}$; $\dfrac{\partial u}{\partial s} = \dfrac{3r^2e^s + r \cos rs}{\sqrt{1 - (3x+y)^2}}$ **11.** $\dfrac{\partial u}{\partial r} = \dfrac{6s}{x^2} \sinh \dfrac{y}{x}(xe^r - ry)$; $\dfrac{\partial u}{\partial s} = \dfrac{3}{x^2} \sinh \dfrac{y}{x}(2xe^r - yr^2) = 0$

13. $\dfrac{\partial u}{\partial r} = 2x \sin \phi \cos \theta + 2y \sin \phi \sin \theta + 2z \cos \phi$; $\dfrac{\partial u}{\partial \phi} = 2xr \cos \phi \cos \theta + 2yr \cos \phi \sin \theta - 2zr \sin \phi$;

$\dfrac{\partial u}{\partial \theta} = -2xr \sin \phi \sin \theta + 2yr \sin \phi \cos \theta$ **15.** (a) $e^x(\cos t - y \sin t) + e^y(x \cos t - \sin t)$;

(b) $e^{\cos t}(\cos t - \sin^2 t) + e^{\sin t}(\cos^2 t - \sin t)$ **17.** (a) $\dfrac{x \sec^2 t - y \sin t + z \cos t}{\sqrt{x^2 + y^2 + z^2}}$; (b) $\tan t \sec t$ **19.** $\dfrac{txe^t - y}{t(x^2 + y^2)}$

21. $\dfrac{x + y + 2t + ty - tx}{t(y + t)^2}$ **23.** $\dfrac{\partial z}{\partial x} = \dfrac{3y - 6x - 4z}{2z + 4x}$; $\dfrac{\partial z}{\partial y} = \dfrac{3x - 2y}{2z + 4x}$ **25.** $\dfrac{\partial z}{\partial x} = -\dfrac{z}{x}$; $\dfrac{\partial z}{\partial y} = \dfrac{xyz + 1}{3xy \tan 3xz - xy^2}$

29. decreasing at a rate of $\frac{8}{61}$ rad/min   **31.** increasing at a rate of 3840π cm³/min   **33.** increasing at a rate of 16 cm³/min

35. $\dfrac{\partial u}{\partial v} = \cos w \sinh v \dfrac{\partial u}{\partial x} + \sin w \cosh v \dfrac{\partial u}{\partial y}$; $\dfrac{\partial u}{\partial w} = -\sin w \cosh v \dfrac{\partial u}{\partial x} + \cos w \sinh v \dfrac{\partial u}{\partial y}$

EXERCISES 15.7 (PAGE 1153)

1. (a) $\dfrac{2}{y} - \dfrac{6y}{x^4}$; (b) $\dfrac{2x^2}{y^3}$ **3.** (a) $4e^{2x} \sin y$; (b) $-e^{2x} \sin y$ **5.** (a) $2 \tan^{-1} \dfrac{y}{x} - \dfrac{2xy}{x^2 + y^2}$; (b) $2 \tan^{-1} \dfrac{y}{x} + \dfrac{2xy}{x^2 + y^2}$

7. (a) $3y \cosh x$; (b) $4x \sinh y$ **9.** (a) $e^x \cos y - \dfrac{2x \ln y}{(1 + x^2)^2}$; (b) $-e^x \cos y - \dfrac{\tan^{-1} x}{y^2}$ **11.** (a) $12x + 20y$; (b) $12x + 20y$

13. (a) 0; (b) e^y **15.** (a) $-2e^z \sin e^z$; (b) $-2we^z(\sin e^z + e^z \cos e^z)$ **17.** (a) $\dfrac{-320rst}{(r^2 + 4s^2 - 5t^2)^3}$; (b) $\dfrac{16r(5t^2 + 12s^2 - r^2)}{(r^2 + 4s^2 - 5t^2)^3}$

25. neither exist   **27.** $f_{12}(0, 0) = -1$; $f_{21}(0, 0) = 1$ **31.** $6se^{r-s}(2 + r) - 8e^{-2s}$ **33.** $10 \cos^2 \theta + 8$ **35.** $-10r \sin 2\theta$

EXERCISES 15.8 (PAGE 1160)

3. $\xi_1 = 2$ **5.** $\xi_1 = 6 - 2\sqrt{3}$ **7.** 0   **9.** 0   **11.** $\xi_2 = 2\sqrt{3} - 4$ **13.** $\xi_2 = 3 + \ln\left(\dfrac{e^2 - 1}{2}\right)$

REVIEW EXERCISES FOR CHAPTER 15 (PAGE 1160)

1. $\{(x, y)|x^2 + y^2 \geq 1\}$ **3.** $\{(x, y)|xy > 4\}$ **5.** $\{(x, y, z)|z > 0\}$ **7.** $\{(x, y, z)|y \neq \pm z\}$ **11.** (a) $4xy - 3y^2 + 4$;

(b) $2x^2 - 6xy - 2$; (c) $4y$; (d) $-6x$; (e) $4x - 6y$; (f) $4x - 6y$ **13.** (a) $\dfrac{2x}{3y^2}$; (b) $\dfrac{y - 2x^2}{3y^3}$; (c) $-\dfrac{4x}{3y^3}$; (d) $\dfrac{4x}{3y^3}$

15. (a) $t^2 \cos st^2 + te^s$; (b) $2st \cos st^2 + e^s$; (c) $2t(\cos st^2 - st^2 \sin st^2) + e^s$; (d) $2t(\cos st^2 - st^2 \sin st^2) + e^s$

17. (a) $\dfrac{1}{y} e^{x/y} + \dfrac{1}{x}$; (b) $-\dfrac{x}{y^2} e^{x/y} - \dfrac{1}{y}$; (c) $\dfrac{1}{y^2} e^{x/y} - \dfrac{1}{x^2}$; (d) $\dfrac{2x}{y^3} e^{x/y} + \dfrac{x^2}{y^4} e^{x/y} + \dfrac{1}{y^2}$ **19.** (a) $\dfrac{y^2 + z^2 - x^2}{(x^2 + y^2 + z^2)^2}$;

(b) $\dfrac{-2xy}{(x^2 + y^2 + z^2)^2}$; (c) $\dfrac{-2xz}{(x^2 + y^2 + z^2)^2}$ **21.** (a) $\dfrac{-320uvw}{(u^2 + 4v^2 - 5w^2)^3}$; (b) $\dfrac{16u(12v^2 + 5w^2 - u^2)}{(u^2 + 4v^2 - 5w^2)^3}$

23. (a) $\dfrac{1}{rt^2}$; (b) $\dfrac{-2}{rt^3}$; (c) $\dfrac{2}{r^2t^3}$ **27.** (a) $\dfrac{\partial u}{\partial t} = \dfrac{6y(x + y)}{x^2 + y^2} + 3 \ln(x^2 + y^2)$; $\dfrac{\partial u}{\partial s} = \dfrac{4y(x - y)}{x^2 + y^2} - 2 \ln(x^2 + y^2)$;

(b) $\dfrac{\partial u}{\partial t} = (3t - 2s) \dfrac{18t}{4s^2 + 9t^2} + 3 \ln(8s^2 + 18t^2)$; $\dfrac{\partial u}{\partial s} = (3t - 2s) \dfrac{8s}{4s^2 + 9t^2} - 2 \ln(8s^2 + 18t^2)$

29. (a) $18xyse^{3rs} + 6yse^{3rs} + 9x^2r^2s^2 + 6xr^2s^2 - 9zr^2s^2$; (b) $6r^3s^3e^{3rs}(3e^{3rs} + 1) + 3r^2s^2(3e^{6rs} + 2e^{3rs} - 3 \ln 4)$

31. (a) $3x \cos t - 4(y + 2x) \sin t$; (b) $12 \cos 2t - 16 \sin 2t$; $\dfrac{du}{dt}\bigg|_{t=\pi/4} = -16$

33. (a) $3(-1 + \Delta x)(3 + \Delta y)^2 - 5(-1 + \Delta x)(2 + \Delta z)^2 - 2(-1 + \Delta x)(3 + \Delta y)(2 + \Delta z) - 5$; (b) -0.48;
(c) $-5\,\Delta x - 14\,\Delta y + 26\,\Delta z$; (d) -0.48 **35.** 2 **37.** $\frac{1}{6}\pi$ **39.** $\delta = \min(1, \frac{1}{35}\epsilon)$ **41.** limit exists and is 0; take $\delta = \sqrt[4]{\epsilon}$
43. limit does not exist **45.** continuous at all points (x, y) in R^2 not on the lines $x = \pm 2y$ **47.** continuous at all points
(x, y) in R^2 except $(x, y) = (2n + 1, 2m + 1)$, where n and m are any integers **49.** continuous at all points in R^2 **51.** 39¢
53. -3200π cm^3/min **55.** $-\frac{201}{1280}$ nt/cm^2 per min **59.** (a) 12; (b) -3

EXERCISES 16.1 (PAGE 1173)

1. $2\sqrt{2}x + 5\sqrt{2}y$ **3.** $3x + \sqrt{2}y + 4z$ **5.** $\dfrac{17}{13(x - y)^2}$ **7.** $(8x - 3y)\mathbf{i} + (2y - 3x)\mathbf{j}$ **9.** $\dfrac{x}{x^2 + y^2}\mathbf{i} + \dfrac{y}{x^2 + y^2}\mathbf{j}$

11. $\dfrac{y + z}{(x + z)^2}\mathbf{i} - \dfrac{1}{x + z}\mathbf{j} - \dfrac{x - y}{(x + z)^2}\mathbf{k}$ **13.** $e^{-2y}\sec z(\mathbf{i} - 2x\mathbf{j} + x\tan z\mathbf{k})$ **15.** 6 **17.** -42 **19.** -2 **21.** $-3e^{\pi/4}\cos\frac{1}{12}\pi$

23. (a) $\langle -4, -4 \rangle$; (b) $-2 - 2\sqrt{3}$ **25.** $\langle -12, 2, 14 \rangle$; (b) $\frac{4}{9}$ **29.** $\frac{3}{20}\pi + \frac{2}{3}$; $\frac{1}{4}\sqrt{\pi^2 + 4}$ **31.** $-\frac{29}{11}$; $\sqrt{21}$ **33.** $\theta = \tan^{-1}\dfrac{3}{3\pi + 1}$

35. (a) $6\sqrt{2}$ degrees per meter; (b) direction of the vector $\dfrac{1}{\sqrt{2}}\mathbf{i} + \dfrac{1}{\sqrt{2}}\mathbf{j}$ **37.** (a) -1;
(b) direction of the vector $-\mathbf{j}$ and magnitude 2

EXERCISES 16.2 (PAGE 1178)

1. $2x - 2y + 3z = 17$; $\dfrac{x - 2}{2} = \dfrac{y + 2}{-2} = \dfrac{z - 3}{3}$ **3.** $4x + 8y + 3z + 22 = 0$; $\dfrac{x + 2}{4} = \dfrac{y + 4}{8} = \dfrac{z - 6}{3}$ **5.** $ex - y = 0$;
$\dfrac{x - 1}{-e} = \dfrac{y - e}{1}$, $z = 0$ **7.** $x - y - 3 = 0$; $\dfrac{x - 6}{1} = \dfrac{y - 3}{-1}$, $z = 3$ **9.** $x + 2y + 2z - 8 = 0$; $\dfrac{x - 4}{1} = \dfrac{y - 1}{2} = \dfrac{z - 1}{2}$
11. $3x - 2y - 6z + 84 = 0$; $\dfrac{x + 8}{-3} = \dfrac{y - 27}{2} = \dfrac{z - 1}{6}$ **13.** $\dfrac{x - 2}{4} = \dfrac{y + 2}{-1} = \dfrac{z}{20}$ **15.** $x = 4$, $y = 16$

17. $\dfrac{x}{1 - 8\pi} = \dfrac{y - 2}{-2\pi} = \dfrac{z - 1}{-1}$ **19.** surfaces are tangent

EXERCISES 16.3 (PAGE 1190)

1. $f(4, -\frac{1}{2}) = -\frac{133}{4}$, rel min **3.** $f(-\frac{1}{4}, 16) = -12$, rel max **5.** no relative extrema **7.** no relative extrema
9. $f(1, 1) = -5$, rel min; $f(-1, -3) = 31$, rel max **11.** $f(\frac{1}{3}\pi, \frac{1}{3}\pi) = \frac{3}{2}\sqrt{3}$, rel max; $f(\frac{5}{3}\pi, \frac{5}{3}\pi) = -\frac{3}{2}\sqrt{3}$, rel min **13.** 8, 8, 8
15. $(\frac{41}{14}, -\frac{5}{7}, \frac{33}{14})$; $\frac{9}{14}\sqrt{14}$ **17.** $\left(0, \dfrac{1}{\sqrt{17}}, -\dfrac{4}{\sqrt{17}}\right)$ and $\left(0, -\dfrac{1}{\sqrt{17}}, \dfrac{4}{\sqrt{17}}\right)$; 1 **19.** $\frac{1}{3}c$ mg of drug A and $\frac{1}{2}c$ mg of drug B
21. $\frac{16}{3}\sqrt{3}$ cu units **23.** length of the base is $2\frac{2}{3}$ ft; width of the base is 2 ft; depth is 3 ft
25. 3 machine-hours and 9 person-hours **27.** \$2000 **29.** there is no greatest total profit **31.** $l:w:h = 1:1:\frac{1}{2}$

EXERCISES 16.4 (PAGE 1198)

1. $(0, 0)$ and $(0, 4)$ **3.** $(\frac{5}{3}, -\frac{4}{3}, \frac{2}{3})$ **5.** $f(\pm\frac{1}{2}\sqrt{35}, \frac{1}{2}) = \frac{27}{4}$, rel max; $f(0, 3) = 3$, rel min; $f(0, -3) = -3$, rel min
7. $f(-\frac{2}{3}\sqrt{3}, -\frac{1}{3}\sqrt{6}, -\frac{1}{3}\sqrt{3}) = f(-\frac{2}{3}\sqrt{3}, \frac{1}{3}\sqrt{6}, \frac{1}{3}\sqrt{3}) = f(\frac{2}{3}\sqrt{3}, -\frac{1}{3}\sqrt{6}, \frac{1}{3}\sqrt{3}) = f(\frac{2}{3}\sqrt{3}, \frac{1}{3}\sqrt{6}, -\frac{1}{3}\sqrt{3}) = -\frac{2}{9}\sqrt{6}$, rel min;
$f(\frac{2}{3}\sqrt{3}, \frac{1}{3}\sqrt{6}, \frac{1}{3}\sqrt{3}) = f(-\frac{2}{3}\sqrt{3}, -\frac{1}{3}\sqrt{6}, \frac{1}{3}\sqrt{3}) = f(\frac{2}{3}\sqrt{3}, -\frac{1}{3}\sqrt{6}, -\frac{1}{3}\sqrt{3}) = f(-\frac{2}{3}\sqrt{3}, \frac{1}{3}\sqrt{6}, -\frac{1}{3}\sqrt{3}) = \frac{2}{9}\sqrt{6}$, rel max **9.** 3
11. $3\sqrt{3}$ **13.** (a) 12; (b) 4; (c) 1 **15.** $\frac{4}{13}\sqrt{26}$ **17.** 2, 6 **19.** $\frac{37}{13}$ **21.** $\frac{7}{32}$ **33.** hottest at $(\pm\frac{1}{2}\sqrt{3}, -\frac{1}{2})$; coldest at $(0, \frac{1}{2})$

EXERCISES 16.5 (PAGE 1207)

1. $f(x, y) = 2x^2 - \frac{3}{2}y^2 + C$ **3.** $f(x, y) = 3x^2 - 5xy + 2y^3 + C$ **5.** $f(x, y) = 2x^3y^2 - 7x^2y + 3x - 8y + C$

7. $f(x, y) = \dfrac{2x^2 - 2y^2 - x}{2xy^2} + C$ **9.** $f(x, y) = x^2 \sec^2 y + C$ **11.** $f(x, y) = x^2 \cos y - x + C$ **13.** not an exact differential

15. $f(x, y) = xe^y - x^2y + C$ **17.** $\frac{1}{2}x^2 + y^2 + xy = C$ **19.** $ye^x - x^2 = C$ **21.** not exact **23.** $x \sin \dfrac{x}{y} + \sin x = C$

25. $f(x, y, z) = 2xy - 5xz + 8yz + C$ **27.** $f(x, y, z) = xe^y \sin z + C$ **29.** $f(x, y, z) = e^{x+z} + e^y \ln z - e^x \ln y + C$
31. $f(x, y, z) = 2x^2y + 3xyz - 5yz^2 - 2x + z + C$ **33.** $f(x, y, z) = xz \tan y + C$

EXERCISES 16.6 (PAGE 1215)

1. $3x - 6y - 10 = 0$ **3.** $2x - 10y + 17 = 0$ **5.** $21x + 10y - 16 = 0$ **7.** $25,400 **9.** (a) 3120; (b) 9
11. (a) $81x - 284y - 270 = 0$; (b) 11 days **13.** (a) $y = 0.1560x + 314.9$; (b) 368.0 milliliters per minute per kilogram
15. (a) $17,070x - 1,743y - 375,900 = 0$; (b) 2722 kilograms per hectare **17.** $8.94

REVIEW EXERCISES FOR CHAPTER 16 (PAGE 1217)

1. 14 **3.** $-\frac{5}{2}\sqrt{2}$ **5.** $-\frac{8}{3}$ **7.** (a) $\frac{1}{2}i + \frac{1}{2}j$; (b) $\frac{1}{4}(1 + \sqrt{3})$ **9.** (a) $12i + 3j - 4k$; (b) $\frac{13}{7}\sqrt{14}$ **11.** $f(x, y) = e^{x^2} \ln y + C$

13. $f(x, y, z) = \dfrac{y}{x+z} - \dfrac{1}{x} - \dfrac{2}{z} + C$ **15.** $4x + 2y + z - 12 = 0$; $\dfrac{x-2}{4} = \dfrac{y-1}{2} = \dfrac{z-2}{1}$ **17.** $\dfrac{x-1}{17} = \dfrac{y+2}{20} = \dfrac{z-11}{-4}$

19. (a) the direction of the vector $-12i - 150j$; (b) ascending; (c) the direction of either of the vectors: $\dfrac{25}{\sqrt{629}}i - \dfrac{2}{\sqrt{629}}j$ or

$-\dfrac{25}{\sqrt{629}}i + \dfrac{2}{\sqrt{629}}j$ **21.** $f(-1, -1) = 1$, rel max **23.** $f(\pm\sqrt{5}, 0) = 10$, rel min **25.** $f(9, 11, 15) = -362$, rel max **27.** $\frac{7}{6}\sqrt{6}$

29. $f(x, y) = 4x^3y + 5x^2 + 3y^2 + C$ **31.** $f(x, y, z) = e^x \tan y + z \cot y + C$ **33.** $x^3y^2 + 2xy^3 = C$

35. $ye^{-x} + y^2 - \cos x = C$ **37.** $\frac{100}{3}, \frac{100}{3}, \frac{100}{3}$ **39.** $2\sqrt{3}$ by $\frac{2}{3}\sqrt{3}$ by $2\sqrt{3}$ **41.** (a) $-\dfrac{3\sqrt{3}+2}{11}$ degrees per cm; (b) $\frac{2}{11}\sqrt{13}$ degrees

per cm in the direction of the vector $-\dfrac{3}{\sqrt{13}}i - \dfrac{2}{\sqrt{13}}j$ **43.** 18 ft by 18 ft by 18 ft **45.** square base and a depth that
is one-half the length of a side of the base **47.** $5x + 66y - 84 = 0$ **49.** (a) $177x - 2692y + 169,066 = 0$;
(b) 68 beats per minute

EXERCISES 17.1 (PAGE 1227)

1. 45 **3.** 50 **5.** 1368 **7.** 704 **9.** $\frac{203}{4}$ **11.** 1376 **13.** 68.6 **15.** 38.2 **17.** $[0, 24]$ **19.** $[1, e]$ **21.** $[0, \frac{3}{2}\sqrt{3\pi^2}]$

EXERCISES 17.2 (PAGE 1234)

1. 42 **3.** $\frac{98}{3}$ **5.** $-\frac{49}{3}$ **7.** $\frac{1}{3}$ **9.** $\frac{1}{3}$ **11.** 45 **13.** $\frac{152}{3}$ **15.** $\frac{3}{2}\pi$ **17.** $\frac{864}{5}$ **19.** $\frac{512}{3}$ cu units **21.** $\frac{16}{3}$ cu units **23.** $\dfrac{15\pi - 32}{120}$ cu units

25. $\frac{1}{12}$ sq units **27.** 72 sq units **29.** $\dfrac{c}{ab} \displaystyle\int_{-a}^{a} \int_{-(b/a)\sqrt{a^2-x^2}}^{(b/a)\sqrt{a^2-x^2}} \sqrt{a^2b^2 - b^2x^2 - a^2y^2} \, dy \, dx$

31. $\frac{337}{30}$ cu units **33.** (b) $\frac{2}{3}a^3$; (c) $\dfrac{2}{3} \displaystyle\int_{0}^{a} \int_{0}^{\sqrt{a^2-x^2}} (2x + y) \, dx \, dy$ **35.** 0

EXERCISES 17.3 (PAGE 1241)

1. 12 kg; $(2, \frac{3}{2})$ **3.** $\frac{27}{2}$ kg; $(\frac{6}{5}, \frac{9}{5})$ **5.** $\frac{176}{15}k$ kg; $(\frac{35}{22}, \frac{102}{77})$ **7.** $\frac{2}{3}ka^3$ kg; $(\frac{3}{32}a(2 + \pi), \frac{3}{32}a(2 + \pi))$ **9.** $\frac{1}{4}k\pi$ kg; $\left(\dfrac{\pi}{2}, \dfrac{16}{9\pi}\right)$ **11.** $\frac{4}{3}$ kg; $(\frac{6}{5}, \frac{6}{5})$ **13.** 9ρ kg-m^2 **15.** $\frac{1}{2}\pi\rho a^4$ kg-m^2 **17.** $\frac{96}{35}\rho$ kg-m^2 **19.** (a) $\frac{144}{5}$ kg-m^2; (b) 54 kg-m^2; (c) $\frac{2}{3}\sqrt{15}$ m; (d) $\frac{414}{5}$ kg-m^2 **21.** (a) $\frac{3}{32}\pi k$ kg-m^2; (b) $\frac{1}{24}\pi(2\pi^2 - 3)k$ kg-m^2; (c) $\frac{1}{4}\sqrt{6}$ m; $(\frac{1}{12}\pi^3 - \frac{1}{32}\pi)k$ kg-m^2 **23.** $\frac{1}{12}b\sqrt{6}$ ft **25.** $\dfrac{19{,}904}{315}k$ kg-m^2

EXERCISES 17.4 (PAGE 1247)

1. 6π sq units **3.** $\frac{1}{4}a^2(8 + \pi)$ sq units **5.** $(4\pi + 12\sqrt{3})$ sq units **7.** 4π cu units **9.** $\frac{128}{9}(3\pi - 4)$ cu units **11.** $\frac{7}{2}\pi$ cu units **13.** $\frac{40}{3}\pi k$ kg; $(0, \frac{21}{10})$ **15.** $\frac{22}{3}k\pi$ kg; $(-\frac{57}{44}, 0)$ **17.** $\frac{20}{3}k$ kg; $(\frac{23}{25}, \frac{531}{1280}\pi)$ **19.** $\frac{112}{3}\pi k$ kg; $(0, -\frac{24}{7})$ **21.** $\frac{5}{64}k\pi$ kg-m^2 **23.** $\frac{11}{16}k\pi a^4$ kg-m^2 **25.** $\frac{1}{4}\sqrt{2\pi}$ m **27.** $\pi e(e^8 - 1)$

EXERCISES 17.5 (PAGE 1253)

1. $\sqrt{6}$ sq units **3.** $2\sqrt{1633}$ sq units **5.** 9 sq units **7.** $\frac{1}{12}(135\sqrt{10} - 13\sqrt{26})$ sq units **9.** 8π sq units **11.** 12π sq units **13.** 32 sq units **15.** $9\sqrt{2}$ sq units **17.** $\pi b\sqrt{a^2 + b^2}$ sq units **19.** $2\pi a^2(1 - e^{-1})$ sq units **21.** 12π sq units **23.** $\frac{7}{3}[2 + \sqrt{2}\ln(1 + \sqrt{2})]$ sq units

EXERCISES 17.6 (PAGE 1258)

1. $\frac{1}{10}$ **3.** $\frac{7}{8}$ **5.** $-e(\ln 2)^2$ **7.** $\frac{1}{2}\pi - 1$ **9.** $\frac{15}{2}$ **11.** $\frac{1}{24}$ **13.** 36 **15.** $\frac{243}{2}\pi$ **17.** $\frac{648}{5}$ **19.** $\frac{1}{4}$ cu units **21.** $\frac{3}{2}\pi$ cu units **23.** 4π cu units **25.** $\frac{4}{3}\pi abc$ cu units **27.** $\frac{80}{3}\rho$ kg **29.** $\frac{1}{28}k$ kg **31.** $\frac{2}{3}(2\sqrt{2} - 1)$ kg

EXERCISES 17.7 (PAGE 1266)

1. $\frac{1}{3}a^3$ **3.** $6\pi(e - 1)$ **5.** πa^3 **7.** $\frac{4}{3}\pi a^3$ **9.** $\frac{1}{3}$ **11.** 8π **13.** $\frac{4}{3}a^5\pi k$ kg **15.** $1250\pi k$ slug-ft^2 **17.** $\frac{4}{3}a^5\pi k$ kg **19.** 8π **21.** $\frac{512}{45}k$ kg-m^2 **23.** $8k\pi$ kg **25.** $\frac{56}{9}\pi a^3k$ slug-ft^2 **27.** $(0, 0, \frac{1}{2})$ **29.** 18π **31.** $\frac{1}{15}\pi(2\sqrt{2} - 1)$

REVIEW EXERCISES FOR CHAPTER 17 (PAGE 1267)

1. $\frac{1}{40}$ **3.** $\frac{1}{8}\pi$ **5.** $\frac{1}{8}e^4 - \frac{3}{4}e^2 + e - \frac{3}{8}$ **7.** $\frac{3}{4}\pi$ **9.** $\frac{1}{8}$ **11.** $\frac{1}{3}$ **13.** $\frac{1}{2}\pi \ln 2$ **15.** $\frac{1}{2}(1 - \cos 1)$ **17.** $\frac{4}{15}$ sq units **19.** 9π **21.** $\frac{8}{3}$ sq units **23.** $\frac{1}{2}$ cu units **25.** $\frac{1104}{35}$ cu units **27.** 18 sq units **29.** $(\frac{1}{2}\pi - \frac{2}{3})$ sq units **31.** $(2, \frac{3}{2})$ **33.** $\frac{1}{108}(3\pi - 7)$ kg **35.** $6\pi a^4$ cu units **37.** $\frac{4}{3}\pi$ cu units **39.** $\frac{1}{64}(7e^8 + 1)$ kg-m^2 **41.** $k(\pi + \frac{8}{3})$ kg-m^2 **43.** $2k\pi$ kg-m^2; $\frac{1}{2}\sqrt{2\pi}$ m **45.** $65k\pi$ kg **47.** $\frac{32}{5}k\pi$ kg-m^2

EXERCISES 18.1 (PAGE 1278)

7. $6x\mathbf{i} + 6y^2\mathbf{j}$ **9.** $\dfrac{2xy}{1 + x^4y^2}\mathbf{i} + \dfrac{x^2}{1 + x^4y^2}\mathbf{j}$ **11.** $(6x^2 - 6xy + y^2)\mathbf{i} + (-3x^2 + 2xy - 12y^2)\mathbf{j}$ **13.** $2xye^{-4z}\mathbf{i} + x^2e^{-4z}\mathbf{j} - 4x^2ye^{-4z}\mathbf{k}$ **15.** conservative **17.** not conservative **19.** conservative **21.** $\phi(x, y) = xy + C$

23. $\phi(x, y) = e^x \sin y + C$ **25.** $\phi(x, y) = x^2y^2 - xy^3 + 2y + C$ **27.** $\phi(x, y, z) = \frac{1}{3}x^3 + \frac{1}{2}z^2 - xy + 3yz + C$
29. $\phi(x, y, z) = ze^x + xe^y - ye^z + C$ **31.** $\phi(x, y, z) = x^2 \cos y - yz^2 - 3x + 2z + C$ **33.** 0; 5 **35.** $2e^x \sin y\mathbf{k}$; $2e^x \cos y$
37. 0; $2x + 2y + 2z$ **39.** $\sin z\mathbf{i} + \sin x\mathbf{j} + \sin y\mathbf{k}$; 0 **41.** $\dfrac{x - y}{\sqrt{x^2 + y^2 + 1}}\mathbf{k}$; $\dfrac{x + y}{\sqrt{x^2 + y^2 + 1}} + 2z$

EXERCISES 18.2 (PAGE 1287)

1. 1 **3.** $\pi - \frac{8}{3}$ **5.** 8π **7.** 1 **9.** $\frac{16}{3}$ **11.** $\frac{4}{3}$ **13.** $-\frac{5}{12}$ **15.** $\frac{1477}{2}$ **17.** $\frac{35}{2}$ **19.** $\pi(a^2 + 2a)$ **21.** 8 **23.** $\frac{4}{3}$ joules **25.** $20\frac{3}{4}$ joules
27. $27\frac{3}{4}$ joules **29.** $(\frac{1}{16}\pi a^4 + a^2)$ joules **31.** 3 joules **33.** $(e^2 + e^4 + e^8 - 3)$ joules **35.** $2\frac{1}{2}$ joules

EXERCISES 18.3 (PAGE 1297)

1. 2 **3.** e^2 **5.** -4 **7.** 15 **9.** $-4e$ **11.** -14 **13.** 18 **15.** $2e^2 - 3e^3 + e^4$ **17.** $\frac{3}{4}$ **19.** 11 **21.** $\frac{13}{2}$ **23.** 4 **25.** 0 **27.** 3 **29.** 4
31. -13 **33.** 216 joules **35.** 32,766 joules **37.** $\frac{2}{15}k$ joules

EXERCISES 18.4 (PAGE 1312)

1. -1 **3.** $-\frac{5}{12}$ **5.** $-\frac{1}{2}\pi$ **7.** $-\frac{32}{15}$ **9.** $-\frac{3}{7}$ **11.** $\frac{1}{3}$ **13.** $\frac{1}{24}(5 - 4\sqrt{2})\pi$ **15.** 0 **17.** -10π **19.** $\frac{41}{70}$ **21.** $\frac{2}{3}$ sq units **23.** $\frac{1}{3}$ sq unit
25. $\frac{3}{8}\pi a^2$ sq units **31.** 24π joules **33.** 0 **35.** 16π cm²/sec **37.** $\frac{3}{2}\pi$ cm²/sec **39.** 15; (i) **41.** 0; (iii)

EXERCISES 18.5 (PAGE 1326)

1. 8π **3.** $\frac{1}{6}(5\sqrt{5} - 1)$ **5.** $\frac{15}{2}\sqrt{2}\pi$ **7.** $\frac{7}{3}\sqrt{61}$ **9.** 0 **11.** $16k\pi$ kg **13.** $\frac{665}{6}\sqrt{2}\pi$ kg **15.** 0 **17.** $\frac{7}{2}$ **19.** $\frac{7}{2}$ **21.** $\frac{8}{3}\pi$ **23.** 32π
25. 72π **27.** $\frac{11}{12}$ **29.** 54π **31.** π **33.** 8π **35.** -1

REVIEW EXERCISES FOR CHAPTER 18 (PAGE 1327)

1. (a) $(4xy + 3y^3)\mathbf{i} + (2x^2 + 9xy^2)\mathbf{j}$; (b) $e^y\mathbf{i} + (xe^y - ze^y - yze^y)\mathbf{j} - ye^y\mathbf{k}$ **3.** $\phi(x, y) = \tan^{-1} 2xy^2 + C$
5. $\phi(x, y, z) = z^2 \tan x + y^2e^{3z} + C$ **7.** $x(e^{xy} - e^{xz})\mathbf{i} + y(e^{yz} - e^{xy})\mathbf{j} + z(e^{xz} - e^{yz})\mathbf{k}$; 0 **9.** $\dfrac{1 - 2y}{y^2}\mathbf{k}$; $\dfrac{2x}{y^2}$ **11.** $-\frac{4}{3}$
13. $\frac{44}{3} - 3\pi$ **15.** $\frac{27}{10}$ **17.** $\frac{87}{20}$ **19.** $9e^2 - 1$ **21.** π **23.** 13 **25.** 4 **27.** 12π **29.** 0 **31.** $\frac{9}{2}$ sq units **33.** $\frac{1568}{15}$ joules **35.** $\frac{23}{6}$ joules
39. 0 **41.** 54 cm²/sec **43.** $\frac{1}{2}\pi$ **45.** $\frac{1}{6}(37\sqrt{37} - 1)$ **47.** $\frac{8}{3}k\pi$ kg **49.** 160π

INDEX

44. $\int \dfrac{u^2\,du}{\sqrt{a^2-u^2}} = -\dfrac{u}{2}\sqrt{a^2-u^2} + \dfrac{a^2}{2}\sin^{-1}\dfrac{u}{a} + C$

45. $\int \dfrac{du}{u\sqrt{a^2-u^2}} = -\dfrac{1}{a}\ln\left|\dfrac{a+\sqrt{a^2-u^2}}{u}\right| + C$

$= -\dfrac{1}{a}\cosh^{-1}\dfrac{a}{u} + C$

46. $\int \dfrac{du}{u^2\sqrt{a^2-u^2}} = -\dfrac{\sqrt{a^2-u^2}}{a^2u} + C$

47. $\int (a^2-u^2)^{3/2}\,du = -\dfrac{u}{8}(2u^2-5a^2)\sqrt{a^2-u^2} + \dfrac{3a^4}{8}\sin^{-1}\dfrac{u}{a} + C$

48. $\int \dfrac{du}{(a^2-u^2)^{3/2}} = \dfrac{u}{a^2\sqrt{a^2-u^2}} + C$

Forms Containing $2au - u^2$

49. $\int \sqrt{2au-u^2}\,du = \dfrac{u-a}{2}\sqrt{2au-u^2} + \dfrac{a^2}{2}\cos^{-1}\left(1-\dfrac{u}{a}\right) + C$

50. $\int u\sqrt{2au-u^2}\,du = \dfrac{2u^2-au-3a^2}{6}\sqrt{2au-u^2}$

$+ \dfrac{a^3}{2}\cos^{-1}\left(1-\dfrac{u}{a}\right) + C$

51. $\int \dfrac{\sqrt{2au-u^2}\,du}{u} = \sqrt{2au-u^2} + a\cos^{-1}\left(1-\dfrac{u}{a}\right) + C$

52. $\int \dfrac{\sqrt{2au-u^2}\,du}{u^2} = -\dfrac{2\sqrt{2au-u^2}}{u} - \cos^{-1}\left(1-\dfrac{u}{a}\right) + C$

53. $\int \dfrac{du}{\sqrt{2au-u^2}} = \cos^{-1}\left(1-\dfrac{u}{a}\right) + C$

54. $\int \dfrac{u\,du}{\sqrt{2au-u^2}} = -\sqrt{2au-u^2} + a\cos^{-1}\left(1-\dfrac{u}{a}\right) + C$

55. $\int \dfrac{u^2\,du}{\sqrt{2au-u^2}} = -\dfrac{(u+3a)}{2}\sqrt{2au-u^2} + \dfrac{3a^2}{2}\cos^{-1}\left(1-\dfrac{u}{a}\right) + C$

56. $\int \dfrac{du}{u\sqrt{2au-u^2}} = -\dfrac{\sqrt{2au-u^2}}{au} + C$

57. $\int \dfrac{du}{(2au-u^2)^{3/2}} = \dfrac{u-a}{a^2\sqrt{2au-u^2}} + C$

58. $\int \dfrac{u\,du}{(2au-u^2)^{3/2}} = \dfrac{u}{a\sqrt{2au-u^2}} + C$

Forms Containing Trigonometric Functions

59. $\int \sin u\,du = -\cos u + C$

60. $\int \cos u\,du = \sin u + C$

61. $\int \tan u\,du = \ln|\sec u| + C$

62. $\int \cot u\,du = \ln|\sin u| + C$

63. $\int \sec u\,du = \ln|\sec u + \tan u| + C = \ln|\tan(\tfrac{1}{4}\pi + \tfrac{1}{2}u)| + C$

64. $\int \csc u\,du = \ln|\csc u - \cot u| + C = \ln|\tan \tfrac{1}{2}u| + C$

65. $\int \sec^2 u\,du = \tan u + C$

66. $\int \csc^2 u\,du = -\cot u + C$

67. $\int \sec u \tan u\,du = \sec u + C$

68. $\int \csc u \cot u\,du = -\csc u + C$

69. $\int \sin^2 u\,du = \tfrac{1}{2}u - \tfrac{1}{4}\sin 2u + C$

70. $\int \cos^2 u\,du = \tfrac{1}{2}u + \tfrac{1}{4}\sin 2u + C$

71. $\int \tan^2 u\,du = \tan u - u + C$

72. $\int \cot^2 u\,du = -\cot u - u + C$

73. $\int \sin^n u\,du = -\dfrac{1}{n}\sin^{n-1} u \cos u + \dfrac{n-1}{n}\int \sin^{n-2} u\,du$

74. $\int \cos^n u\,du = \dfrac{1}{n}\cos^{n-1} u \sin u + \dfrac{n-1}{n}\int \cos^{n-2} u\,du$

75. $\int \tan^n u\,du = \dfrac{1}{n-1}\tan^{n-1} u - \int \tan^{n-2} u\,du$

76. $\int \cot^n u\,du = -\dfrac{1}{n-1}\cot^{n-1} u - \int \cot^{n-2} u\,du$

77. $\int \sec^n u\,du = \dfrac{1}{n-1}\sec^{n-2} u \tan u + \dfrac{n-2}{n-1}\int \sec^{n-2} u\,du$

78. $\int \csc^n u\,du = -\dfrac{1}{n-1}\csc^{n-2} u \cot u + \dfrac{n-2}{n-1}\int \csc^{n-2} u\,du$

79. $\int \sin mu \sin nu\,du = -\dfrac{\sin(m+n)u}{2(m+n)} + \dfrac{\sin(m-n)u}{2(m-n)} + C$

80. $\int \cos mu \cos nu\,du = \dfrac{\sin(m+n)u}{2(m+n)} + \dfrac{\sin(m-n)u}{2(m-n)} + C$

81. $\int \sin mu \cos nu\,du = -\dfrac{\cos(m+n)u}{2(m+n)} - \dfrac{\cos(m-n)u}{2(m-n)} + C$

82. $\int u \sin u\,du = \sin u - u \cos u + C$

83. $\int u \cos u\,du = \cos u + u \sin u + C$

84. $\int u^2 \sin u\,du = 2u \sin u + (2-u^2)\cos u + C$